Atomic, Molecular, & Optical Physics
HANDBOOK

Atomic, Molecular, & Optical Physics Handbook

Editor:

Gordon W. F. Drake
Department of Physics, University of Windsor, Windsor, Ontario, Canada
A36@server.uwindsor.ca

Associate Editor:

Nigel E. Hedgecock
Department of Physics, University of Windsor, Windsor, Ontario, Canada
A43@server.uwindsor.ca

Advisory Board:

Brian R. Judd – Mathematical Methods
Department of Physics, The Johns Hopkins University, Baltimore, Maryland

W. E. Baylis – Atoms
Department of Physics, University of Windsor, Windsor, Ontario, Canada
baylis@uwindsor.ca

Kate P. Kirby – Molecules, Theory
Harvard-Smithsonian Center for Astrophysics, Cambridge, Massachusetts
kirby@cfa.harvard.edu

Peter L. Smith – Molecules, Experiment
Harvard-Smithsonian Center for Astrophysics, Cambridge, Massachusetts
plsmith@cfa.harvard.edu

M. R. Flannery – Scattering, Theory
School of Physics, Georgia Institute of Technology, Atlanta, Georgia
flannery@eikonal.physics.gatech.edu

Robert N. Compton – Scattering, Experiment
Oak Ridge National Laboratory, Oak Ridge, Tennessee
ahd@ornl.gov

Pierre Meystre – Optical Physics
Optical Sciences Center, The University of Arizona, Tucson, Arizona
pierre@rhea.opt-sci.arizona.edu

Alexander Dalgarno – Applications
Harvard-Smithsonian Center for Astrophysics, Cambridge, Massachusetts
adalgarno@cfa.harvard.edu

Atomic, Molecular, & Optical Physics
HANDBOOK

Editor

GORDON W. F. DRAKE

Department of Physics
University of Windsor
Windsor, Ontario

American Institute of Physics   **Woodbury, New York**

AIP Press
American Institute of Physics
500 Sunnyside Boulevard
Woodbury, NY 11797-2999

Library of Congress Cataloging-in-Publication Data
Atomic, molecular, and optical physics handbook / editor Gordon W. F. Drake.
     p. cm.
   Includes bibliographical references and index.
   ISBN 1-56396-242-X (book)
   ISBN 1-56396-596-8 (CD-ROM)
   ISBN 1-56396-597-6 (set)
   1. Atoms—Handbooks, manuals, etc. 2. Physics—Handbooks, manuals, etc. 3. Molecules—Handbooks, manuals, etc. 4. Optics—Handbooks, manuals, etc. I. Drake, Gordon W. F.
QC173.A827 1996
539.7--dc20   95-50875
                                                                CIP

10 9 8 7 6 5 4 3 2

Contents

SCATTERING THEORY

SCATTERING EXPERIMENT

QUANTUM OPTICS

APPLICATIONS

List of Contributors

Nigel G. Adams, *Department of Chemistry, University of Georgia, Athens, Georgia 30602*

M. Ya. Amusia, *Argonne National Laboratory, Argonne, Illinois 60439, and A. F. Ioffe Physical-Technical Institute, St. Petersburg 194021, Russia*

Nils Andersen, *Niels Bohr Institute, Ørsted Laboratory, Universitetsparken 5, DK-2100 Copenhagen, Denmark*

N. R. Badnell, *Department of Physics, University of Strathclyde, Glasgow G40NG, United Kingdom*

K. Bartschat, *Department of Physics and Astronomy, Drake University, Des Moines, Iowa 50311*

W. E. Baylis, *Department of Physics, University of Windsor, Windsor, Ontario N9B 3P4, Canada*

Hans Bichsel, *1211 22nd Avenue East, Seattle, Washington 98112-3534*

Robert W. Boyd, *Institute of Optics, University of Rochester, Rochester, New York 14627*

Henry Buijs, *BOMEM, Hartmann and Braun, 450 avenue St.-Jean-Baptiste, Québec, Québec G2E 5S5, Canada*

P. G. Burke, *Department of Applied Mathematics and Theoretical Physics, The Queen's University of Belfast, Belfast BT7 1NN Northern Ireland*

C. D. Caldwell, *Department of Physics, University of Central Florida, Orlando, Florida 32816-2385*

R. Y. Chiao, *Department of Physics, University of California, Berkeley, California 94720*

C. L. Cocke, *Department of Physics, Kansas State University, Manhattan, Kansas 66506*

James S. Cohen, *Theoretical Division, Los Alamos National Laboratory, Los Alamos, New Mexico 87545*

F. B. M. Copeland, *The Queen's University of Belfast, Belfast BT7 1NN, Northern Ireland*

Bernd Crasemann, *Department of Physics, University of Oregon, Eugene, Oregon 97403*

David R. Crosley, *SRI International, Menlo Park, California 94025*

D. S. F. Crothers, *The Queen's University of Belfast, Belfast BT7 1NN, Northern Ireland*

L. J. Curtis, *Department of Physics and Astronomy, University of Toledo, Toledo, Ohio 43606*

A. Dalgarno, *Harvard-Smithsonian Center for Astrophysics, 60 Garden Street, Cambridge, Massachusetts 02138*

Abigail J. Dobbyn, *Max-Planck-Institut für Strömungsforschung, 37073 Göttingen, Germany*

G. W. F. Drake, *Department of Physics, University of Windsor, Windsor, Ontario N9B 3P4, Canada*

J. H. Eberly, *Department of Physics and Astronomy, and Institute of Optics, University of Rochester, Rochester, New York 14627*

Guy T. Emery, *Bowdoin College, Brunswick, Maine 04011*

V. Engel, *Institut für Physikalische Chemie, Universität Würzburg, Marcusstraße 9-11, 97070 Würzburg, Germany*

Paul Engelking, *Department of Chemistry and Chemical Physics Institute, University of Oregon, Eugene, Oregon 97403-1253*

K. M. Evenson, *Time and Frequency Division, National Institute of Standards and Technology, Boulder, Colorado 80303*

James M. Farrar, *Department of Chemistry, University of Rochester, Rochester, New York 14627*

Gordon Feldman, *The Henry A. Rowland Department of Physics and Astronomy, The Johns Hopkins University, Baltimore, Maryland 21218*

Paul D. Feldman, *Department of Physics and Astronomy, The Johns Hopkins University, Baltimore, Maryland 21218*

Charlotte F. Fischer, *Department of Computer Science, Box 1679B, Vanderbilt University, Nashville, Tennessee 37235*

M. R. Flannery, *School of Physics, Georgia Institute of Technology, Atlanta, Georgia 30332-0430*

D. R. Flower, *Department of Physics, University of Durham, Durham DH1 3LE, United Kingdom*

A. L. Ford, *Department of Physics, Texas A&M University, College Station, Texas 77843*

J. L. Fox, *Institute for Terrestrial and Planetary Atmospheres and Marine Sciences Research Center, The University at Stony Brook, Stony Brook, New York 11794-5000*

M. Freyberger, *Universität Ulm, Abteilung für Quantenphysik, 89069 Ulm, Federal Republic of Germany*

Thomas Fulton, *The Henry A. Rowland Department of Physics and Astronomy, The Johns Hopkins University, Baltimore, Maryland 21218*

Alexander L. Gaeta, *School of Applied and Engineering Physics, Cornell University, Ithaca, New York 14853*

Alan Gallagher, *JILA, University of Colorado and National Institute of Standards and Technology, Boulder, Colorado 80309-0440*

T. F. Gallagher, *Department of Physics, University of Virginia, Charlottesville, Virginia 22901*

M. Gargaud, *Observatoire de Bordeaux (INSU/CNRS:URA 352), B.P. 89, 33270-Floirac, France*

Alan Garscadden, *Wright Laboratory, Wright Patterson Air Force Base, Ohio 45433*

J. T. Glass, *The Queen's University of Belfast, Belfast BT7 1NN, Northern Ireland*

S. P. Goldman, *Department of Physics, The University of Western Ontario London, Ontario N6A 3K7, Canada*

I. P. Grant, *University of Oxford, Mathematical Institute, 24/29 St. Giles', Oxford OX1 3LB, England*

D. C. Griffin, *Department of Physics, Rollins College, Winter Park, Florida 32789*

William G. Harter, *Department of Physics, University of Arkansas, Fayetteville, Arkansas 72701*

Eric Herbst, *Department of Physics and Department of Astronomy, The Ohio State University, Columbus, Ohio 43210*

Robert Nyden Hill, *Department of Physics and Astronomy, University of Delaware, Newark, Delaware 19716*

David L. Huestis, *Molecular Physics Laboratory, SRI International, Menlo Park, California 94025*

M. A. Iken, *School of Physics, Georgia Institute of Technology, Atlanta, Georgia 30332-0430*

Mitio Inokuti, *Argonne National Laboratory, Argonne, Illinois 60439*

T. Ishihara, *Institute of Applied Physics, University of Tsukuba, Tsukuba, Ibaraki 305, Japan*

Juha Javanainen, *Department of Physics, University of Connecticut, Storrs, Connecticut 06269-3046*

E. T. Jensen, *Department of Physics, University of Windsor, Windsor, Ontario N9B 3P4, Canada*

B. R. Judd, *Department of Physics and Astronomy, The Johns Hopkins University, Baltimore, Maryland 21218*

I. Kanik, *Jet Propulsion Laboratory, California Institute of Technology, Pasadena, California 91109*

Kate P. Kirby, *Harvard-Smithsonian Center for Astrophysics, 60 Garden Street, Cambridge, Massachusetts 02138*

M. O. Krause, *Oak Ridge National Laboratory, Oak Ridge, Tennessee 37831-6201*

K. C. Kulander, *Physics Division, Lawrence Livermore National Laboratory, Livermore, California 94551, and JILA, University of Colorado, Boulder, Colorado 80309*

P. G. Kwiat, *Department of Physics, University of California, Berkeley, California 94720*

Yuan T. Lee, *Lawrence Berkeley Laboratory, University of California, Berkeley, California 94720*

S. Lepp, *Department of Physics, University of Nevada, Las Vegas, 4505 Maryland Pkwy., Las Vegas, Nevada 89154*

M. Lewenstein, *Service des Photons, Atomes et Molécules, Centre d'Etudes de Saclay, Gif-sur-Yvette 91191, France, and Centre for Theoretical Physics, Polish Academy of Sciences, 02-668 Warsaw, Poland*

James D. Louck, *Los Alamos National Laboratory, Los Alamos, New Mexico 87545*

J. H. Macek, *Department of Physics and Astronomy, University of Tennessee, Knoxville, Tennessee 37996-1501 and Oak Ridge National Laboratory, Post Office Box 2009, Oak Ridge, Tennessee 37831*

M. L. Mandich, *AT&T Bell Laboratories, Murray Hill, New Jersey 07974*

E. J. Mansky, *Controlled Fusion Atomic Data Center, Oak Ridge National Laboratory, Oak Ridge, Tennessee 37831-6372*

S. T. Manson, *Department of Physics and Astronomy, Georgia State University, Atlanta, Georgia 30303*

W. C. Martin, *National Institute of Standards and Technology, Gaithersburg, Maryland 20899*

R. McCarroll, *Laboratoire de Dynamique Moléculaire et Atomique (CNRS:URA 774) Université P. et M. Curie, 4 place Jussieu, 75252-Paris Cedex 05, France*

J. W. McConkey, *Department of Physics, University of Windsor, Windsor, Ontario N9B 3P4, Canada*

R. P. McEachran, *Department of Physics and Astronomy, York University, Toronto, Ontario M3J 1P3, Canada*

J. H. McGuire, *Department of Physics, Tulane University, New Orleans, Louisiana 70118*

Dieter Meschede, *Institut für Angewandte Physik der Rheinische Friedrich-Wilhelms-Universität Bonn, Wegelerstraße 8, D-53125 Bonn, Germany*

Pierre Meystre, *Optical Sciences Center, University of Arizona, Tucson, Arizona 85721*

P. W. Milonni, *Theoretical Division (T-4), Los Alamos National Laboratory, Los Alamos, New Mexico 87545*

Peter J. Mohr, *National Institute of Standards and Technology, Gaithersburg, Maryland 20899*

David H. Mordaunt, *Max-Planck-Institut für Strömungsforschung, 37073 Göttingen, Germany*

John D. Morgan III, *Department of Physics and Astronomy, University of Delaware, Newark, Delaware 19716*

E. E. Nikitin, *Department of Chemistry, Technion, Israel Institute of Technology, Haifa 32000, Israel*

R. F. O'Connell, *Department of Physics and Astronomy, Louisiana State University, Baton Rouge, Louisiana 70803*

R. E. Olson, *Physics Department, University of Missouri, Rolla, Missouri 65401*

J. Paldus, *Department of Applied Mathematics, University of Waterloo, Waterloo, Ontario N2L 3G1, Canada*

G. Peach, *Department of Physics and Astronomy, University College London, London WC1E 6BT, United Kingdom*

David J. Pegg, *Department of Physics, University of Tennessee, Knoxville, Tennessee 37996 and Physics Division, Oak Ridge National Laboratory, Oak Ridge, Tennessee 37831*

Ronald Phaneuf, *Department of Physics, University of Nevada, Reno, Nevada 89557-0058*

M. S. Pindzola, *Department of Physics, Auburn University, Auburn, Alabama 36849*

Eric H. Pinnington, *Department of Physics, University of Alberta, Edmonton, Alberta T6G 2J1, Canada*

Richard C. Powell, *Optical Sciences Center, University of Arizona, Tuscon, Arizona 85715*

J. F. Reading, *Department of Physics, Texas A&M University, College Station, Texas 77843*

J. Sapirstein, *Department of Physics, University of Notre Dame, Notre Dame, Indiana 46556*

Axel Schenzle, *Sektion Physik der Ludwig-Maximilians-Universität München, Theresienstraße 37, D-80333 München, Germany*

Reinhard Schinke, *Max-Planck-Institut für Strömungsforschung, 37073 Göttingen, Germany*

W. Schleich, *Universität Ulm, Abteilung für Quantenphysik, 89069 Ulm, Federal Republic of Germany*

D. R. Schultz, *Physics Division, Oak Ridge National Laboratory, Oak Ridge, Tennessee 37831-6373.*

Peter L. Smith, *Harvard-Smithsonian Center for Astrophysics, Cambridge, Massachusetts 02138.*

Anthony F. Starace, *Department of Physics and Astronomy, The University of Nebraska, Lincoln, Nebraska 68588-0111.*

G. Stark, *Department of Physics, Wellesley College, Wellesley, Massachusetts 02181.*

A. D. Stauffer, *Department of Physics and Astronomy, York University, Toronto, Ontario M3J 1P3, Canada.*

A. M. Steinberg, *Department of Physics, University of California, Berkeley, California 94720.*

Stig Stenholm, *Research Institute for Theoretical Physics, P. O. Box 9, FIN-00014 University of Helsinki, Finland.*

Jack C. Straton, *Department of Physics, Portland State University, Portland, Oregon 97207-0751.*

M. R. Strayer, *Physics Division, Oak Ridge National Laboratory, Oak Ridge, Tennessee 37831-6373.*

C. R. Stroud, Jr., *Department of Physics and Astronomy, and Institute of Optics, University of Rochester, Rochester, New York 14627.*

Arthur G. Suits, *Lawrence Berkeley Laboratory, University of California, Berkeley, California 94720.*

A. Temkin, *Laboratory for Astronomy and Solar Physics, Goddard Space Flight Center, National Aeronautics and Space Administration, Greenbelt, Maryland 20771.*

S. Trajmar, *Jet Propulsion Laboratory, California Institute of Technology, Pasadena, California 91109.*

T. Uzer, *School of Physics, Georgia Institute of Technology, Atlanta, Georgia 30332-0430.*

K. Vogel, *Universität Ulm, Abteilung für Quantenphysik, 89069 Ulm, Federal Republic of Germany.*

Jon Weisheit, *Department of Space Physics and Astronomy, Rice University, Houston, Texas 77251-1892.*

W. L. Wiese, *National Institute of Standards and Technology, Gaithersburg, Maryland 20899.*

Martin Wilkens, *Fakultät für Physik, Universität Konstanz, 78434 Konstanz, Germany.*

David R. Yarkony, *Department of Chemistry, The Johns Hopkins University, Baltimore, Maryland 21218.*

Preface

The principal aim of this handbook is to provide a ready reference that covers the key ideas, techniques, formulas, and results of atomic, molecular, and optical physics. It is intended to be useful to graduate students and established researchers in the field, as well as workers in related fields such as plasma physics, astrophysics, space science, chemistry, engineering, gaseous electronics and microelectronics, surface science, plasma processing, and atmospheric and environmental studies. It should also be useful to senior undergraduate students as a resource in the preparation of essays or special projects. Each chapter is written by an expert at an authoritative and reasonably comprehensive level. The contents therefore go well beyond what would typically be found in an encyclopedia article on the same subject.

In the course of the past century, the field of atomic, molecular and optical physics has grown tremendously to cover a vast wealth of phenomena ranging from the interface with nuclear and elementary particle physics to the properties of macromolecules, clusters, and surfaces. The ever-increasing power and versatility of lasers continues to open up new areas of study. The parallel advance of theory and experiment has always been a prominent feature of the field in the best scientific tradition. The result today is a closely-woven interplay between the two at all levels of complexity. The cover drawing of the Kapitza-Dirac effect symbolizes diversity flowing from the essential unity of physics.

The ambitious project of trying to capture all of this in a single volume began with a survey of a broad sample of members of the Division of Atomic, Molecular, and Optical Physics (DAMOP) of The American Physical Society, together with other internationally known researchers in the field. The respondents provided a list of key words describing their work, and suggestions for topics that should be covered. The enthusiastic response to the initial survey provided the encouragement to proceed with the formation of an Editorial Board, and the organization of the survey responses into an initial index for the book. The index entries were then arranged into broad subject areas and, with the help of the Editorial Board, broken down into individual chapters, together with suggestions for experts in the area to write them. The result is the present 88 chapters divided into the seven broad areas of Mathematical Methods, Atomic Physics, Molecular Physics, Scattering Theory, Scattering Experiment, Quantum Optics, and Applications. Although the boundaries are by no means well defined, the organization of the chapters into these broad areas is indicated in the Table of Contents. The chapters are extensively cross-referenced and collaboratively written so that the book forms a coherent whole. The comprehensive index of about 4300 entries makes it easy to find information on a given topic.

An entire book could be (and in many cases has been) written on the contents of any one chapter. In order to keep the subject matter within the bounds of a single volume, authors were asked to focus on the most important definitions, techniques, and results, and to provide references to more extensive discussions or tabulations of data. They were asked to write down for all to read the things that a graduate student should know in order to begin research in their area. In the interest of saving space, references are intended to be a guide to the literature, and *not* a comprehensive listing. In many cases, only a recent work is cited, from which the reader can gain access to the older literature.

Other handbooks exist that cover certain aspects of the present work in more detail. Of particular relevance are the two volume *Handbook of Optics*, sponsored by the Optical Society of America, and *Basic Plasma Physics II*, edited by A. A. Galeev and R. N. Sudan. Other excellent textbooks and treatises are available on individual topics. These are referenced as appropriate. The classic work *Quantum Mechanics of One- and Two-electron Atoms* by H. A. Bethe and E. E. Salpeter, although now somewhat outdated, remains a basic reference and source of inspiration. However, none of these provides both the depth and breadth of coverage of the field of atomic, molecular, and optical physics that is the goal of the present volume. Readers are encouraged to communicate to me or members of the Editorial Board their criticisms and suggestions for future editions.

This work would not have been possible without the very able and diligent work of the Associate Editor, Nigel Hedgecock, and the other members of the Editorial Board. Their advice and expertise have been invaluable in assembling this material. I would also like to acknowledge the skills and patience of my assistants Chandra Boon, Mark Cassar, and Peter Drake in updating and maintaining the LaTeX files for each chapter.

Gordon W. F. Drake
Windsor, Ontario, Canada
November 1995

1

Units and Constants

W. E. Baylis and G. W. F. Drake

Department of Physics, University of Windsor, Windsor, Ontario N9B 3P4, Canada

1.1 PHYSICAL CONSTANTS

The currently accepted values for the physical constants are listed in Table 1.1, based on the 1986 CODATA(Committee on Data for Science and Technology) recommendations [1]. Although improved measurements are now available for several of the entries [2], they cannot readily be incorporated with the 1986 values until the next complete least-squares adjustment of the constants. The recommended values therefore remain as tabulated. The one exception is the value of the Rydberg constant R_∞, which has been updated to reflect the fifty-fold improvement in its measurement in units of m^{-1} (or equivalently in units of Hz, since the conversion factor c is a defined physical constant) [3]. Because the uncertainties are correlated, the full variance matrix must be used in calculating uncertainties for further quantities in terms of those tabulated.

1.2 ELECTROMAGNETIC UNITS

The standard electromagnetic units adopted by most scientific journals and elementary texts belong to the *système international* (SI) or rationalized MKSA (meters, kilograms, seconds, and amperes) units. However, many authors working with microscopic phenomena prefer *Gaussian* units, and theoretical physicists often use *Heaviside-Lorentz* (H-L) units. In this Handbook, SI units are used together with *atomic units*. The current

section is meant as a reference relating these different systems.

The relations among different sets of units are not simple conversions since the same symbol in different systems can have different physical dimensions. To clarify the meanings of the units, we summarize basic electromagnetic relations for SI, Gaussian, and H-L systems below. The Coulomb law for the magnitude F of the force acting on each of two static charges q and Q separated by a distance r in a homogeneous medium of permittivity ϵ can be written

$$F = \frac{1}{4\pi\epsilon}\frac{qQ}{r^2} \, , \tag{1.1}$$

where in a vacuum, ϵ is[1]

$$\epsilon_0 = \begin{cases} \left(\mu_0 c^2\right)^{-1} , & \text{SI} \\ (4\pi)^{-1} , & \text{Gaussian} \\ 1, & \text{H-L} , \end{cases} \tag{1.2}$$

with the closely related permeability of vacuum given by

$$\mu_0 = \begin{cases} 4\pi \times 10^{-7}\,\text{N/A}^2 , & \text{SI} \\ 4\pi , & \text{Gaussian} \\ 1, & \text{H-L} . \end{cases} \tag{1.3}$$

---

[1]We deviate here from Jackson [4] who takes $\epsilon_0 = \mu_0 = 1$ in Gaussian units and must introduce additional constants to relate the units. The physically important quantities are the *relative values* $\epsilon_r \equiv \epsilon/\epsilon_0$ and $\mu_r \equiv \mu/\mu_0$, which in traditional Gaussian-unit notation are written without the r subscript.

2 / Atomic, Molecular, & Optical Physics Handbook

Table 1.1. Table of physical constants. Uncertainties are given in parentheses.

Quantity	Symbol	Value	Units
speed of light in vacuum	c	2.997 924 58	10^8 m s^{-1}
gravitational constant	G	6.672 59(85)	10^{-11} m^{3}kg^{-1}m^{-2}
Planck constant	h	6.626 075 5(40)	10^{-34} J s
	$\hbar = h/2\pi$	1.054 572 66(63)	10^{-34} J s
elementary charge	e	1.602 177 33(49)	10^{-19} C
		4.803 206 8(15)	10^{-10} esu
inverse fine structure constant, $4\pi\epsilon_0\hbar c/e^2$	α^{-1}	137.035 989 5(61)	
magnetic flux quantum, $h/2e$	Φ_0	2.067 834 61(61)	10^{-15} Wb
atomic mass unit, $\frac{1}{12}m(^{12}\mathrm{C})$	$m_\mathrm{u} = u$	1.660 540 2(10)	10^{-27} kg
	$m_\mathrm{u}c^2$	931.494 32(28)	MeV
electron mass	m_e	9.109 389 7(54)	10^{-31} kg
		5.485 799 03(13)	10^{-4} u
muon mass	m_μ	0.113 428 913(17)	u
proton mass	m_p	1.007 276 470(12)	u
neutron mass	m_n	1.008 664 904(14)	u
deuteron mass	m_d	2.013 553 214(24)	u
α-particle mass	m_α	4.001 506 178(84)	u
Rydberg constant, $m_\mathrm{e}c\alpha^2/2h$	R_∞	1.097 373 156 834(24)	10^7 m^{-1}
	$R_\infty c$	3.289 841 960 305(72)	10^{15} Hz
	$R_\infty hc$	13.605 698 1(40)	eV
		2.179 874 1(13)	10^{-18} J
Bohr radius, $\alpha/4\pi R_\infty$	a_0	0.529 177 249(24)	10^{-10} m
Hartree energy, $e^2/[4\pi\epsilon_0]a_0 = 2R_\infty hc$	E_h	27.211 396 1(81)	eV
	E_h/h	6.579 683 920 61(14)	10^{15}Hz
	E_h/hc	2.194 746 313 668(48)	10^7 m^{-1}
Compton wavelength, αa_0	$\lambda_\mathrm{C} = \bar{\lambda}_\mathrm{C}/2\pi$	3.861 593 23(35)	10^{-13} m
classical electron radius, $\alpha^2 a_0$	r_e	2.817 940 92(38)	10^{-15} m
Thomson cross section, $8\pi r_\mathrm{e}^2/3$	σ_e	0.665 246 16(18)	10^{-28} m^2
Bohr magneton, $e\hbar/2m_\mathrm{e}$	μ_B	9.274 015 4(31)	10^{-24} J T^{-1}
		5.788 382 63(52)	10^{-5} eV T^{-1}
electron magnetic moment	$\mu_\mathrm{e}/\mu_\mathrm{B}$	1.001 159 652 193(10)	
muon magnetic moment	μ_μ/μ_B	4.841 970 97(71)	10^{-3}
proton magnetic moment	$\mu_\mathrm{p}/\mu_\mathrm{B}$	1.521 032 202(15)	10^{-3}
neutron magnetic moment	$\mu_\mathrm{n}/\mu_\mathrm{B}$	1.041 875 63(25)	10^{-3}
deuteron magnetic moment	$\mu_\mathrm{d}/\mu_\mathrm{B}$	0.466 975 447(91)	10^{-3}
electron g factor $2(1 + a_\mathrm{e})$	g_μ	2.002 319 304 386(20)	
muon g factor $2(1 + a_\mu)$	g_μ	2.002 331 846(17)	
proton gyromagnetic ratio	γ_p	2.675 221 28(81)	10^8 s^{-1}T^{-1}
Avogadro constant	N_A	6.022 136 7(36)	10^{23} mol^{-1}
Faraday constant, $N_\mathrm{A}e$	F	9.648 530 9(29)	10^4 C mol^{-1}
Boltzmann constant	k_B	1.380 658(12)	10^{-23} J K^{-1}
		8.617 385(73)	10^{-5} eV K^{-1}
	$k_\mathrm{B}/E_\mathrm{h}$	3.166 830(27)	10^{-6} K^{-1}
molar gas constant	R	8.314 510(70)	J mol^{-1} K^{-1}
molar volume (ideal gas), RT/P			
$\quad T = 273.15$ K, $P = 101.325$ kPa	V_m	0.022 414 10(19)	m^3 mol^{-1}
$\quad T = 273.15$ K, $P = 100$ kPa	V_m	0.022 711 08(19)	m^3 mol^{-1}
Stefan-Boltzmann constant $\pi^2 k_\mathrm{B}^4/(60\hbar^3 c^2)$	σ	5.670 51(19)	10^{-8} W m^{-2} K^{-4}
first radiation constant, $2\pi hc^2$	c_1	3.741 774 9(22)	10^{-16} W m^2
second radiation constant, hc/k_B	c_2	0.014 387 69(12)	m K
Wien displacement constant, $\lambda_\mathrm{max}T = c_2/4.965\,114\,23$	b	2.897 756(24)	10^{-3} m K

Note that ϵ_0 and μ_0 are dimensionless in H-L and Gaussian units, but not in the SI units. Current or electric charge is an independent quantity in the MKSA system but can be expressed in purely mechanical dimensions in the H-L and Gaussian systems. Thus, in Gaussian units, 1 statcoulomb = 1 dyne$^{1/2}$cm, but in SI, even though the ampere is *defined* in terms of the attractive force between thin parallel wires carrying equal currents, there is no mechanical equivalent for the ampere or the coulomb. To establish such an equivalence, one can supplement the SI units by assigning a dimensionless number to ϵ_0 or to μ_0. Gaussian and H-L units arise from two different assignments. The result of assigning a number to ϵ_0 is analogous to the relation $1 \text{ s} = \dot{3} \times 10^8$ m established between time and distance units if one sets the speed of light $c = 1$, a convention often used in conjunction with H-L units. (Note that for simplicity, the pure number 2.99792458, equal numerically to the defined speed of light in vacuum in units of 10^8 m/s, is represented by $\dot{3}$.) Thus, although within the Gaussian system, where the assignment $4\pi\epsilon_0 = 1$ is made, it is justified to assert that 1 coulomb *equals* $\dot{3} \times 10^9$ statcoulombs, this is not true in pure SI, where there is no equivalent mechanical unit for charge.

Maxwell's macroscopic equations can be written

$$\lambda \nabla \cdot \mathbf{D} = \rho,$$
$$\lambda c' \nabla \times \mathbf{H} - \lambda \frac{\partial \mathbf{D}}{\partial t} = \mathbf{j},$$
$$c' \nabla \times \mathbf{E} + \frac{\partial \mathbf{B}}{\partial t} = 0, \qquad (1.4)$$
$$\nabla \cdot \mathbf{B} = 0,$$

with the macroscopic field variables related to the polarizations $\mathbf{P}$ and $\mathbf{M}$ by

$$\lambda \mathbf{D} = \epsilon_0 \mathbf{E} + \mathbf{P} = \epsilon \mathbf{E}$$
$$\lambda \mathbf{H} = \mu_0^{-1} \mathbf{B} - \mathbf{M} = \mathbf{B}/\mu \qquad (1.5)$$

(the last equalities for $\mathbf{D}$ and $\mathbf{H}$ hold only for homogeneous media) and

$$\lambda = \begin{cases} 1, & \text{SI} \\ \epsilon_0 = \mu_0^{-1}, & \text{Gaussian or H-L}, \end{cases} \qquad (1.6)$$

$$c' = \begin{cases} 1, & \text{SI} \\ c, & \text{Gaussian or H-L}. \end{cases} \qquad (1.7)$$

In Gaussian or H-L units, fields $\mathbf{E}, \mathbf{B}, \mathbf{D}, \mathbf{H}$, and polarizations (dipole moments per unit volume) $\mathbf{P}, \mathbf{M}$ all have the same dimensions, whereas in SI units the microscopic fields $\mathbf{E}$ and $\mathbf{B}$ have dimensions that are generally distinct from each other as well as from $\mathbf{P}$ (or $\mathbf{D}$) and $\mathbf{M}$ (or $\mathbf{H}$) respectively. In all three unit systems, the dimensionless ratio ϵ/ϵ_0 is called the dielectric constant (or relative permittivity) of the medium, and the (dimensionless) fine-structure constant is

$$\alpha = \frac{1}{4\pi\epsilon_0} \frac{e^2}{\hbar c} \qquad (1.8)$$

with a numerical value $\alpha^{-1} = 137.0359895\,(61)$.

In atomic units (see Sect. 1.3), the factor $e^2/(4\pi\epsilon_0)$, the electron mass m_{e}, and $\hbar$, Planck's constant divided by 2π, are all equal to 1. In Gaussian and H-L systems, these conditions determine a numerical value for all electro-mechanical units. Thus in Gaussian units, the electronic charge is $e = 1$ whereas in H-L units $e = \sqrt{4\pi}$. In the SI system, on the other hand, the three conditions $e^2/(4\pi\epsilon_0) = m_{\mathrm{e}} = \hbar = 1$ determine numerical values for mechanical units but not for electromagnetic ones. A complete determination of values requires that ϵ_0 also be assigned a value. The choice most consistent with previous work is to take $e = 1 = 4\pi\epsilon_0$. This choice is made here.

Since a volt is a joule/coulomb and a statvolt is an erg/statcoulomb, 1 volt corresponds to (but is not generally *equal* to, since the physical dimensions may differ)

$$\frac{10^7 \text{ erg}}{2\,997\,924\,580 \text{ statcoulomb}} = \frac{1}{300} \text{ statvolt} . \qquad (1.9)$$

In Gaussian units, the unit of magnetic field, namely Gauss ($\mathbf{B}$) or Oersted ($\mathbf{H}$) has the same physical size and dimension as the unit of electric field, namely statvolt/cm, which in turn corresponds to an SI field of $\dot{3} \times 10^4$ V/m. However, the tesla (1 T = 1 weber/m^2), the SI unit of magnetic field $\mathbf{B}$ (older texts refer to $\mathbf{B}$ as the *magnetic induction*), has the physical dimensions of V s/m^2. To find the correspondence to Gaussian units, one must multiply by the speed of light c:

$$1 \text{T} \, c = \dot{3} \times 10^8 \text{ V/m} \qquad (1.10)$$

which corresponds to 10^4 statvolt/cm and hence to 10^4 gauss.

Below are two tables relating basic mechanical and electromagnetic quantities. Caution is required both because the same symbol often stands for quantities of different physical dimensions in different systems of units, and because factors of 2π sometimes enter frequencies, depending on whether the units are cycles/s (Hz) or radians/s. The double-headed arrows ($\leftrightarrow$) indicate a correspondence between quantities whose dimensions are not necessarily equal. Thus for example, the force on an electron due to a Gaussian electric field of 1 statvolt/cm is the same as due to an SI electric field of $\dot{3} \times 10^4$ V/m. The correspondences between Gaussian and SI electrostatic quantities become equalities if and only if $4\pi\epsilon_0 = 1$. Thus they are equalities within the Gaussian system but not within the less constrained SI scheme. The SI and Gaussian units of magnetic field have different dimensions unless both ϵ_0 and c are set equal to dimensionless numbers. *Natural H-L*

Table 1.2. Conversion factors for various physical quantities.

Quantity	SI units	Gaussian units	Natural H-L units: $\hbar = c = \epsilon_0 = 1$
length	1 m	$= 10^2$ cm	$= 1$ m
mass	1 kg	$= 10^3$ g	$\leftrightarrow 3.519\,341\,3\,(21) \times 10^8$ m^{-1}
time	1 s	$= 1$ s	$\leftrightarrow 2.997\,924\,58 \times 10^8$ m
velocity	1 m/s	$= 10^2$ cm/s	$\leftrightarrow 3.335\,640\,951\,98 \cdots \times 10^{-9}$
energy	1 J$=1$ kg m^2/s^2	$= 10^7$ erg	$\leftrightarrow 3.163\,026\,2\,(19) \times 10^{25}$ m^{-1}
action	1 J s	$= 10^7$ erg s	$\leftrightarrow 0.948\,251\,40\,(57) \times 10^{34}$
force	1 N $= 1$ J/m	$= 10^5$ dyne	$\leftrightarrow 3.163\,026\,2\,(19) \times 10^{25}$ m^{-2}
power	1 W $= 1$ J/s	$= 10^7$ erg/s	$\leftrightarrow 1.055\,071\,97\,(63) \times 10^{33}$ m^{-2}
intensity	1 W/m^2	$= 10^3$ erg/cm^2	$\leftrightarrow 1.055\,071\,97\,(63) \times 10^{33}$ m^{-2}
charge	1 C $= 1$ A s	$\leftrightarrow \dot{3} \times 10^9$ statcoul	$\leftrightarrow 1.993\,684\,01\,(61) \times 10^{13}$
potential	1 V $= 1$ J/C	$\leftrightarrow 1/\dot{3}00$ statvolt	$\leftrightarrow 1.586\,523\,3\,(11) \times 10^{12}$ m^{-1}
electric field	1 V/m $= 1$ N/C	$\leftrightarrow \left(\dot{3} \times 10^4\right)^{-1}$ statvolt/cm	$\leftrightarrow 1.586\,523\,3\,(11) \times 10^{12}$ m^{-2}
magnetic field	1 T $= 1$ N/(A m)	$\leftrightarrow 10^4$ gauss	$\leftrightarrow 4.756\,277\,3\,(32) \times 10^{20}$ m^{-2}

Table 1.3. Physical quantities in atomic units with $\hbar = e = m_e = 4\pi\epsilon_0 = 1$, and $\alpha^{-1} = 137.035\,989\,5\,(61)$.

Quantity	Unit	Value
length	a_0	$0.529\,177\,249\,(24) \times 10^{-10}$ m
mass	m_e	$0.910\,938\,97\,(54) \times 10^{-30}$ kg
time	$\hbar/E_h$	$2.418\,884\,326\,555\,(53) \times 10^{-17}$ s
velocity	$v_B \equiv \alpha c$	$2.187\,691\,42(10) \times 10^6$ m/s
energy	E_h	$4.359\,748\,2(26) \times 10^{-18}$ J
action	$\hbar$	$1.054\,572\,66\,(63) \times 10^{-34}$ J s
force	E_h/a_0	$0.823\,872\,95\,(49) \times 10^{-7}$ N
power	$E_h^2/\hbar$	$0.180\,237\,98\,(11)$ W
intensity	$\dfrac{E_h^2}{\hbar a_0^2}$	$64.364\,142\,(39) \times 10^{18}$ W/m^2
charge	e	$1.602\,177\,33\,(49) \times 10^{-19}$ C
potential	E_h/e	$27.211\,396\,1\,(81)$ V
electric field	$\dfrac{E_h}{ea_0}$	$0.514\,220\,82\,(15) \times 10^{12}$ V/m
magnetic field	$\dfrac{E_h}{ea_0\alpha c}$	$2.350\,518\,09\,(71) \times 10^5$ tesla

units can be considered SI units supplemented by the conditions $\epsilon_0 = c = \hbar = 1$. They are listed here in units of meters, although eV are also often used: 1 eV $= 5.067\,728\,9\,(15) \times 10^6$ m$^{-1} \times \hbar c$. The correspondences may be considered equalities within the natural H-L system but not within SI. More electromagnetic conversions can be found in Jackson [4]. The data here are based mainly on the 1986 adjustment by Cohen and Taylor [1] with the 1993 correction to the Rydberg constant[3].

A few additional energy conversion factors are

$$1 \text{ eV} = 1.602\,177\,33\,(49) \times 10^{-19} \text{ J}$$
$$= 2.417\,988\,36\,(72) \times 10^{14} \text{ Hz} \times h$$

$$= 8065.541\,0(24) \text{ cm}^{-1} \times hc$$
$$= 3.674\,930\,9\,(11) \times 10^{-2}\ E_h$$
$$= 1.106\,044\,5\,(10) \times 10^4 \text{ K} \times k_B$$
$$= 23.049\,524\,(20) \text{ kcal/mol}$$
$$= 96.485\,309\,(82) \text{ kJ/mol}$$

The basic unit of temperature, the kelvin, is equivalent to about 0.7 cm^{-1}, i.e., the value of the Boltzmann constant k_B expressed in wavenumber units per kelvin is 0.695 039(6) cm^{-1}/K. Since K is the internationally accepted symbol for the kelvin [5], this suggests that the use of the letter K as a symbol for 1 cm^{-1} (1 kayser) should be discontinued.

1.3 ATOMIC UNITS

Atomic and molecular calculations based on the Schrödinger equation are most conveniently done in atomic units (a.u.), and then the final result converted to the correct SI units as listed in Table 1.3. In atomic units, $\hbar = m_e = e = 4\pi\epsilon_0 = 1$. The atomic units of length, velocity, time, and energy are then

$$\begin{aligned}
\text{length}: \quad & a_0 = \frac{\hbar^2}{m_e e^2} = \frac{\hbar}{\alpha m_e c}, \\
\text{velocity}: \quad & v_B = e^2/\hbar = \alpha c, \\
\text{time}: \quad & \tau_0 = \frac{\hbar^3}{m_e e^4} = \frac{\hbar}{\alpha^2 m_e c^2}, \\
\text{energy}: \quad & E_h = e^2/a_0 = \alpha^2 m_e c^2,
\end{aligned}$$

where, from the definition (1.8), the numerical value of c is $\alpha^{-1} = 137.035\,9895(61)$ a.u. For the lowest $1s$ state of hydrogen (with infinite nuclear mass), a_0 is the Bohr radius, v_B is the Bohr velocity, $2\pi\tau_0$ is the time to complete a Bohr orbit, and E_h (the Hartree energy) is twice the ionization energy. To include finite nuclear mass effects, replace m_e by the electron reduced mass $\mu = M/(M + m_e)$.

Atomic energies are often expressed in units of the Rydberg (Ry). The Rydberg for an atom having nuclear mass M is

$$1 \text{ Ry} = R_M = M(M + m_e)^{-1} R_\infty, \qquad (1.11)$$

with

$$R_\infty = m_e c \alpha^2/(2h) = 10\,973\,731.568\,3(3) \text{ m}^{-1} \quad (1.12)$$

which is equivalent to $13.605\,698(4)$ eV. The Rydberg constant R_∞ is thus the limit value for infinite nuclear mass.

The energy equivalent of the electron mass, $m_e c^2$, is $510.999\,06(15)$ keV. This energy is a natural unit for relativistic atomic theory. For example, for inner-shell energies in the heaviest elements, the binding energy of the $1s$ electron in hydrogenic Lr ($Z{=}103$) is $0.338\,42\,m_e c^2$.

1.4 MATHEMATICAL CONSTANTS

A selection of the most important mathematical constants is listed in Table 1.4. More extensive tabulations and formulas can be found in the standard mathematical works [6,7] (see also Chap. 3).

1.4.1 Series Summation Formula

The Riemann zeta function defined by $\zeta(n) = \sum_{i=1}^\infty i^{-n}$ (see Table 1.4) is particularly useful in summing slowly convergent series of the form

$$S = \sum_{i=1}^\infty T_i. \qquad (1.13)$$

For example, suppose that the series

$$T_i = t_2 i^{-2} + t_3 i^{-3} + \cdots \qquad (1.14)$$

for the individual terms in S is rapidly convergent for $i > N$, where N is some suitably large integer. Then

$$S = \sum_{i=1}^N T_i + t_2 \zeta^N(2) + t_3 \zeta^N(3) + \cdots, \qquad (1.15)$$

where $\zeta^N(n) = \zeta(n) - \sum_{i=1}^N i^{-n}$ is the zeta function with the first N terms subtracted. For N sufficiently large, only the first few t_j coefficients need be known, and they can be adequately estimated by solving the system of equations

Table 1.4. Values of e, π, Euler's constant γ, and the Riemann zeta function $\zeta(n)$.

Constant	Value
e	$2.718\,281\,828\,459\,045\,235\,360\,287\,471\,352\,66$
π	$3.141\,592\,653\,589\,793\,238\,462\,643\,383\,279\,50$
$\pi^{1/2}$	$1.772\,453\,850\,905\,516\,027\,298\,167\,483\,341\,14$
γ	$0.577\,215\,664\,901\,532\,860\,606\,512\,090\,082\,40$
$\zeta(2)$	$1.644\,934\,066\,848\,226\,436\,472\,415\,166\,646\,02$
$\zeta(3)$	$1.202\,056\,903\,159\,594\,285\,399\,738\,161\,511\,45$
$\zeta(4)$	$1.082\,323\,233\,711\,138\,191\,516\,003\,696\,541\,16$
$\zeta(5)$	$1.036\,927\,755\,143\,369\,926\,331\,365\,486\,457\,03$
$\zeta(6)$	$1.017\,343\,061\,984\,449\,139\,714\,517\,929\,790\,92$
$\zeta(7)$	$1.008\,349\,277\,381\,922\,826\,839\,797\,549\,849\,80$
$\zeta(8)$	$1.004\,077\,356\,197\,944\,339\,378\,685\,238\,508\,65$
$\zeta(9)$	$1.002\,008\,392\,826\,082\,214\,417\,852\,769\,232\,41$
$\zeta(10)$	$1.000\,994\,575\,127\,818\,085\,337\,145\,958\,900\,31$

$$T_N = t_2 N^{-2} + t_3 N^{-3} + \cdots + t_{k+2} N^{-k-2}, \qquad (1.16a)$$

$$T_{N-1} = t_2 (N-1)^{-2} + t_3 (N-1)^{-3} + \cdots$$
$$+ t_{k+2}(N-1)^{-k-2}, \qquad (1.16b)$$

$$\vdots$$

$$T_{N-k} = t_2 (N-k)^{-2} + t_3 (N-k)^{-3} + \cdots$$
$$+ t_{k+2}(N-k)^{-k-2}, \qquad (1.16c)$$

where $k + 1 \le N$ is the number of terms retained in Eq. (1.14).

REFERENCES

1. E. R. Cohen and B. N. Taylor, Rev. Mod. Phys. **59**, 1121 (1987).
2. B. N. Taylor and E. R. Cohen, J. Res. Natl. Inst. Stand. Technol. **95**, 497 (1990); E. R. Cohen and B. N. Taylor, Physics Today, **44** (8), BG 9 (1995).
3. F. Nez, D. Plimmer, S. Bourzeix, L. Julien, F. Biraben, R. Felder, Y. Millerieux, and P. De Natale, Europhys. Lett. **24**, 635 (1993); F. Nez, D. Plimmer, S. Bourzeix, L. Julien, F. Biraben, R. Felder, O. Acef, J. J. Zondy, P. Laurent, A. Clairon, M. Abed, Y. Millerieux, and P. Juncar, Phys. Rev. Lett. **69**, 2326 (1992).
4. J. D. Jackson, *Classical Electrodynamics*, 2nd edition (Wiley, New York, 1975), Appendix.
5. *The International System of Units (SI)*, NIST Spec. Publ. 330, edited by B. N. Taylor (U. S. Government Printing Office, Washington, 1991), p. 29.
6. M. Abramowitz and I. A. Stegun, *Handbook of Mathematical Functions* (Dover, New York, 1965).
7. I. S. Gradshteyn and I. M. Ryzhik, Table of Integrals, Series, and Products (Academic Press, New York, 1965).

2

Angular Momentum Theory

James D. Louck

Los Alamos National Laboratory, Los Alamos, New Mexico 87545

2.1 INTRODUCTION

Angular momentum theory in its quantum mechanical applications, which is the subject of this section, is the study of the group of 2×2 unitary unimodular matrices and its irreducible representations. It is the mathematics of implementing into physical theory the basic tenet that isolated physical systems are invariant to rotations of the system in physical 3-space, denoted $\mathbf{R}^3$, or, equivalently, to the orientation of a Cartesian reference system used to describe the system. That it is the group of 2×2 unimodular matrices that is basic in quantum theory in place of the more obvious group of 3×3 real, orthogonal matrices representing transformations of the coordinates of the constituent particles of the system, or of the reference frame, is a consequence of the Hilbert space structure of the state space of quantum systems and the impossibility of assigning overall phase factors to such states because measurements depend only on the absolute value of transition amplitudes.

The exact relationship between the group $SU(2)$ of 2×2 unimodular unitary matrices and the group $SO(3,\mathbf{R})$ of 3×3 real, proper, orthogonal matrices is an important one for keeping the quantum theory of angular momentum, with its numerous conventions and widespread applications across all fields of quantum

physics, free of ambiguities. These notations and relations are fixed at the outset.

Presentation of a point in $\mathbf{R}^3$:

$$\mathbf{x} = \text{col}\,(x_1, x_2, x_3) \qquad \text{column matrix,}$$
$$\mathbf{x}^T = (x_1, x_2, x_3) \qquad \text{row matrix,}$$
$$X = \begin{pmatrix} x_3 & x_1 - ix_2 \\ x_1 + ix_2 & -x_3 \end{pmatrix} \quad \begin{smallmatrix} \text{2×2 traceless Hermitian ma-} \\ \text{trix; Cartan's representation.} \end{smallmatrix}$$

A one-to-one correspondence between the set $\mathbf{R}^3$ of points in 3-space and the set $\mathbf{H}^2$ of 2×2 traceless Hermitian matrices is obtained from $x_i = \frac{1}{2}\,\text{Tr}\,(\sigma_i X)$, where the σ_i denote the matrices (Pauli matrices)

$$\sigma_1 = \begin{pmatrix} 0 & 1 \\ 1 & 0 \end{pmatrix},\ \sigma_2 = \begin{pmatrix} 0 & -i \\ i & 0 \end{pmatrix},\ \sigma_3 = \begin{pmatrix} 1 & 0 \\ 0 & -1 \end{pmatrix}.$$

$$(2.1)$$

Mappings of $\mathbf{R}^3$ onto itself:

$$\mathbf{x} \to \mathbf{x}' = R\mathbf{x},$$
$$X \to X' = UXU^\dagger,$$

where $\dagger$ denotes Hermitian conjugation of a matrix or an operator.

Two-to-one homomorphism of $SU(2)$ onto $SO(3, \mathbf{R})$:

$$R_{ij} = R_{ij}(U) = \tfrac{1}{2}\mathrm{Tr}(\sigma_i U \sigma_j U^\dagger), \qquad (2.2)$$

$$\begin{pmatrix} \xi \\ \mathbf{x}' \end{pmatrix} = \begin{pmatrix} 1 & 0 & 0 & 0 \\ 0 & & & \\ 0 & & R(U) & \\ 0 & & & \end{pmatrix} = \begin{pmatrix} \xi \\ \mathbf{x} \end{pmatrix}$$

$$= A^\dagger (U \times U^*) A \begin{pmatrix} \xi \\ \mathbf{x} \end{pmatrix}, \qquad (2.3)$$

where ξ is an indeterminate, A is the unitary matrix given by

$$A = \frac{1}{\sqrt{2}} \begin{pmatrix} 1 & 0 & 0 & 1 \\ 0 & 1 & -i & 0 \\ 0 & 1 & i & 0 \\ 1 & 0 & 0 & -1 \end{pmatrix},$$

$U \times U^*$ denotes the matrix direct product, and $*$ denotes complex conjugation. There is a simple unifying theme in almost all the applications. The basic mathematical notions that are implemented over and over again in various contexts are: group action on the underlying coordinates and momenta of the physical system and the corresponding group action in the associated Hilbert space of states; the determination of those subspaces that are mapped irreducibly onto themselves by the group action; the Lie algebra and its actions as derived from the group actions, and conversely; the construction of composite objects from elementary constituents, using the notion of tensor product space and Kronecker products of representations, which are the basic precepts in quantum theory for building complex systems from simpler ones; the reduction of the Kronecker product of irreducible representations into irreducibles with the associated Wigner-Clebsch-Gordan and Racah coefficients determining not only this reduction, but also having a dual role in the construction of the irreducible state spaces themselves; and, finally, the repetition of this process for many-particle systems with the attendant theory of $3n\text{-}j$ coefficients. The universality of this methodology may be attributed to being able, in favorable situations, to separate the particular consequences of physical law (e.g., the Coulomb force) from the implications of symmetry imposed on the system by our underlying conceptions of space and time. Empirical models based on symmetry that attempt to identify the more important ingredients underlying observed physical phenomena are also of great importance.

The group actions in complex systems are often modeled after the following examples for the actions of the groups $SO(3, \mathbf{R})$ and $SU(2)$ on functions defined over the 2-sphere $\mathbf{S}^2 \subset \mathbf{R}^3$:

Hilbert space:

$$\mathbf{V} = \{f \,|\, f \text{ is a polynomial satisfying } \nabla^2 f(\mathbf{x}) = 0\}.$$

Inner or scalar product:

$$(f, f') = \int_{\text{unit sphere}} f^*(\mathbf{x}) f'(\mathbf{x})\, dS,$$

where $f(\mathbf{x}) = f(X)$ for $\mathbf{x}$ presented in the Cartan matrix form X.

Group actions:

$$(O_R f)(\mathbf{x}) = f(R^{-1}\mathbf{x}), \quad \text{each } f \in \mathbf{V}, \text{ each } \mathbf{x} \in \mathbf{R}^3,$$
$$(T_U f)(X) = f(U^\dagger X U), \quad \text{each } f \in \mathbf{V}, \text{ each } X \in \mathbf{H}^2.$$

Operator properties:

(a) O_R is a unitary operator on $\mathbf{V}$; that is, $(O_R f, O_R f') = (f, f')$.

(b) T_U is a unitary operator on $\mathbf{V}$; that is, $(T_U f, T_U f') = (f, f')$.

(c) $R \to O_R$ is a unitary representation of $SO(3, \mathbf{R})$; that is, $O_{R_1} O_{R_2} = O_{R_1 R_2}$.

(d) $U \to T_U$ is a unitary representation of $SU(2)$; that is, $T_{U_1} T_{U_2} = T_{U_1 U_2}$.

(e) $O_{R(U)} = T_U = T_{-U}$ is an operator identity on the space $\mathbf{V}$.

One parameter subgroups:

$$\begin{aligned} U_j(t) &= \exp(-it\sigma_j/2), \quad t \in \mathbf{R}, \ j = 1, 2, 3 \,; \\ R_j(t) &= R(U_j(t)) = \exp(-itM_j), \\ & \qquad t \in \mathbf{R}, \ j = 1, 2, 3 \,; \end{aligned}$$

where

$$M_1 = i \begin{pmatrix} 0 & 0 & 0 \\ 0 & 0 & -1 \\ 0 & 1 & 0 \end{pmatrix}, M_2 = i \begin{pmatrix} 0 & 0 & 1 \\ 0 & 0 & 0 \\ -1 & 0 & 0 \end{pmatrix}, M_3 = i \begin{pmatrix} 0 & -1 & 0 \\ 1 & 0 & 0 \\ 0 & 0 & 0 \end{pmatrix}. \qquad (2.4)$$

Infinitesimal generators:

$$\begin{aligned} L_j &= i(dO_{R_j(t)}/dt)_{t=0}\,, \\ L_j &= i(dT_{U_j(t)}/dt)_{t=0}\,, \\ (L_j f)(\mathbf{x}) &= -i\left(x_k \frac{\partial}{\partial x_l} - x_l \frac{\partial}{\partial x_k} \right) f(\mathbf{x}), \qquad (2.5) \end{aligned}$$

j, k, l cyclic in 1, 2, 3.

Historically, the algebra of angular momentum came about through the quantum rule of replacing the linear momentum $\mathbf{p}$ of a classical point particle, which is located at position $\mathbf{r}$, by $\mathbf{p} \to -i\hbar\nabla$, thus replacing the classical angular momentum $\mathbf{r} \times \mathbf{p}$ about the origin of a chosen Cartesian inertial system by the angular momentum operator:

$$\mathbf{L} = -i\mathbf{r} \times \nabla \quad (\text{in units of } \hbar). \qquad (2.6)$$

The quantal angular momentum properties of this simple one-particle system are then to be inferred from the properties of these operators and their actions in the associated Hilbert space. This remains the method of introducing angular momentum theory in most textbooks because of its simplicity and historical roots. It also leads to focusing the developments of the theory on the algebra of operators in contrast to emphasizing the associated group transformations of the Hilbert space, although the two viewpoints are intimately linked, as illustrated above. Both perspectives will be presented here.

2.2 ORBITAL ANGULAR MOMENTUM

The model provided by orbital angular momentum operators is the paradigm for standardizing many of the conventions and relations used in more abstract and general treatments. These basic results for the orbital angular momentum operator $\mathbf{L} = -i\mathbf{r} \times \nabla$ acting in the vector space $\mathbf{V}$ are given in this section both in Cartesian coordinates $\mathbf{x} = \mathrm{col}\,(x_1, x_2, x_3)$ and spherical polar coordinates:

$$\mathbf{x} = (r\sin\theta\cos\phi, r\sin\theta\sin\phi, r\cos\theta),$$
$$0 \le r < \infty, \ 0 \le \phi < 2\pi, \ 0 \le \theta \le \pi.$$

2.2.1 Cartesian Representation

Commutation relations:
Cartesian form:

$$[L_1, L_2] = iL_3, \quad [L_2, L_3] = iL_1, \quad [L_3, L_1] = iL_2.$$

Cartan form:

$$[L_3, L_+] = L_+, \quad [L_3, L_-] = -L_-, \quad [L_+, L_-] = 2L_3.$$

Squared orbital angular momentum:

$$\mathbf{L}^2 = L_1^2 + L_2^2 + L_3^2 = L_-L_+ + L_3(L_3 + 1)$$
$$= L_+L_- + L_3(L_3 - 1)$$
$$= -r^2\nabla^2 + (\mathbf{x}\cdot\nabla)^2 + (\mathbf{x}\cdot\nabla).$$

$\mathbf{L}^2, L_3$ form a complete set of commuting Hermitian operators in $\mathbf{V}$ with eigenfunctions

$$\mathcal{Y}_{lm}(\mathbf{x}) = \left[\frac{2l+1}{4\pi}(l+m)!(l-m)!\right]^{1/2}$$
$$\times \sum_k \frac{(-x_1 - ix_2)^{k+m}(x_1 - ix_2)^k x_3^{l-m-2k}}{2^{2k+m}(k+m)!k!(l-m-2k)!},$$

where $l = 0, 1, 2, \ldots,; \quad m = l, l-1, \ldots, -l$.
Homogeneous polynomial solutions of Laplace's equation:

$$\mathcal{Y}_{lm}(\lambda\mathbf{x}) = \lambda^l \mathcal{Y}_{lm}(\mathbf{x}),$$
$$(\mathbf{x}\cdot\nabla)\mathcal{Y}_{lm}(\mathbf{x}) = l\mathcal{Y}_{lm}(\mathbf{x}),$$
$$\nabla^2 \mathcal{Y}_{lm}(\mathbf{x}) = 0.$$

Complex conjugate:

$$\mathcal{Y}_{lm}^*(\mathbf{x}) = (-1)^m \mathcal{Y}_{l,-m}(\mathbf{x}).$$

Action of angular momentum operators:

$$L_\pm \mathcal{Y}_{lm}(\mathbf{x}) = [(l \mp m)(l \pm m + 1)]^{1/2}\mathcal{Y}_{l,m\pm 1}(\mathbf{x}),$$
$$L_3 \mathcal{Y}_{lm}(\mathbf{x}) = m\mathcal{Y}_{lm}(\mathbf{x}),$$
$$\mathbf{L}^2 \mathcal{Y}_{lm}(\mathbf{x}) = l(l+1)\mathcal{Y}_{lm}.$$

Highest weight eigenfunction:

$$L_+\mathcal{Y}_{ll}(\mathbf{x}) = 0, \quad L_3\mathcal{Y}_{ll}(\mathbf{x}) = l\mathcal{Y}_{ll}(\mathbf{x}),$$
$$\mathcal{Y}_{ll}(\mathbf{x}) = \frac{1}{2^l l!}\left[\frac{(2l+1)!}{4\pi}\right]^{1/2}(-x_1 - ix_2)^l.$$

Generation from highest weight:

$$\mathcal{Y}_{lm}(\mathbf{x}) = \left[\frac{(l+m)}{(2l)!(l-m)!}\right]^{1/2} L_-^{l-m}Y_{ll}(\mathbf{x}).$$

Relation to Gegenbauer and Jacobi polynomials:

$$\mathcal{Y}_{lm}(\mathbf{x}) = r^{l-|m|}\mathcal{Y}_m(x_1, x_2)$$
$$\times [(2l+1)(l+m)!(l-m)!/2]^{1/2}H_{l,|m|}(x_3/r),$$
$$H_{l\lambda}(z) = \frac{(2\lambda)!}{2^\lambda \lambda!}C_{l-\lambda}^{\lambda+\frac{1}{2}}(z) = \frac{(l+\lambda)!}{2^\lambda l!}P_{l-\lambda}^{(\lambda,\lambda)}(z),$$
$$0 \le \lambda \le l = 0, 1, 2, \ldots,$$

where the $\mathcal{Y}_m(x_1, x_2)$ are homogeneous polynomial solutions of degree $|m|$ of Laplace's equation in 2-space, $\mathbf{R}^2$:

$$\mathcal{Y}_m(x_1, x_2) = \begin{cases} (-x_1 - ix_2)^m/\sqrt{2\pi}, & m \ge 0, \\ (x_1 - ix_2)^{-m}/\sqrt{2\pi}, & m \le 0. \end{cases}$$

(see Sect. 2.2.2 for the definition of Gegenbauer and Jacobi polynomials.)
Orthogonal group action:

$$(O_R\mathcal{Y}_{lm})(\mathbf{x}) = \mathcal{Y}_{lm}(R^{-1}\mathbf{x}) = \sum_{m'}\mathcal{D}_{m'm}^l(R)\mathcal{Y}_{lm'}(\mathbf{x}),$$

where the functions $\mathcal{D}_{m'm}^l(R) = D_{m'm}^l(U(R))$ are defined in Sect. 2.4 for various parametrizations of R.
Unitary group action:

$$(T_U\mathcal{Y}_{lm})(X) = \mathcal{Y}_{lm}(U^\dagger XU) = \sum_{m'}D_{m'm}^l(U)\mathcal{Y}_{lm'}(X),$$

where the functions $D_{m'm}^l(U)$ are defined in Sects. 2.3 and 2.4.

Orthogonality on the unit sphere:

$$\int_{\text{unit sphere}} \mathcal{Y}^*_{l'm'}(\mathbf{x})\mathcal{Y}_{lm}(\mathbf{x})\, dS = \delta_{l'l}\delta_{m'm}\,.$$

Product of solid harmonics:

$$\mathcal{Y}_{k\mu}(\mathbf{x})\mathcal{Y}_{lm}(\mathbf{x}) = \sum_{l'}\langle l'||\mathcal{Y}_k||l\rangle C^{lkl'}_{m,\mu,m+\mu}\mathcal{Y}_{l',m+\mu}(\mathbf{x})$$

$$= \sum_{l'}(l'||\mathcal{Y}_k||l)\begin{pmatrix} l & k & l' \\ m & \mu & -m-\mu \end{pmatrix}$$

$$\times (-1)^{m+\mu}\mathcal{Y}_{l',m+\mu}(\mathbf{x})\,,$$

$$\langle l'||\mathcal{Y}_k||l\rangle = r^{l+k-l'}\left[\frac{(2l+1)(2k+1)}{4\pi(2l'+1)}\right]^{1/2}C^{lkl'}_{000}\,,$$

$$(l'||\mathcal{Y}_k||l) = (-1)^{l'}\left[\frac{(2l+1)(2k+1)(2l'+1)}{4\pi}\right]^{1/2}$$

$$\times \begin{pmatrix} l' & k & l \\ 0 & 0 & 0 \end{pmatrix}\,,$$

where $C^{l_1l_2l}_{m_1m_2m}$ and

$$\begin{pmatrix} l_1 & l_2 & l \\ m_1 & m_2 & -m \end{pmatrix} = \frac{(-1)^{l_1-l_2+m}}{\sqrt{2l+1}}C^{l_1l_2l}_{m_1m_2m}$$

denote Wigner-Clebsch-Gordan coefficients and 3-*j* coefficients, respectively (see Sect. 2.8).

Vector addition theorem for solid harmonics:

$$\mathcal{Y}_{lm}(\mathbf{z}+\mathbf{z}') = \sum_{k\mu}\left[\frac{4\pi(2l+1)!}{(2l-2k+1)!(2k+1)!}\right]^{1/2}C^{l-k,k,l}_{m-\mu,\mu,m}$$

$$\times \mathcal{Y}_{l-k,m-\mu}(\mathbf{z})\mathcal{Y}_{k\mu}(\mathbf{z}')\,,\quad \mathbf{z},\mathbf{z}'\in\mathbf{C}^3\,,$$

$$C^{l-kkl}_{m-\mu,\mu,m} = \left[\binom{l+m}{k+\mu}\binom{l-m}{k-\mu}\Big/\binom{2l}{2k}\right]^{1/2}\,.$$

Rotational invariant in two vectors:

$$I_l(\mathbf{x},\mathbf{y}) = \frac{1}{2^l}\sum_k(-1)^k\binom{l}{k}\binom{2l-2k}{l}$$

$$\times (\mathbf{x}\cdot\mathbf{y})^{l-2k}(\mathbf{x}\cdot\mathbf{x})^k(\mathbf{y}\cdot\mathbf{y})^k$$

$$= (\mathbf{x}\cdot\mathbf{x})^{l/2}(\mathbf{y}\cdot\mathbf{y})^{l/2}C^{(1/2)}_l(\hat{\mathbf{x}}\cdot\hat{\mathbf{y}})$$

$$= \frac{4\pi}{2l+1}\sum_m(-1)^m\mathcal{Y}_{lm}(\mathbf{x})\mathcal{Y}_{l,-m}(\mathbf{y})\,,$$

where $C^{(1/2)}_l(z)$ is a Gegenbauer polynomial (see Sect. 2.2.2) and

$$\hat{\mathbf{x}} = \mathbf{x}/|\mathbf{x}|,\quad \hat{\mathbf{y}} = \mathbf{y}/|\mathbf{y}|,\quad \cos\theta = \hat{\mathbf{x}}\cdot\hat{\mathbf{y}}\,.$$

Legendre polynomials:

$$P_l(\cos\theta) = \frac{4\pi}{2l+1}\sum_m(-1)^m\mathcal{Y}_{l,-m}(\hat{\mathbf{y}})\mathcal{Y}_{lm}(\hat{\mathbf{x}})\,,$$

$$\left[\frac{4\pi}{2l+1}\right]^{1/2}\mathcal{Y}_{l0}(\mathbf{x})$$

$$= \frac{1}{2^l}\sum_k(-1)^k\binom{l}{k}\binom{2l-2k}{l}x_3^{l-2k}(\mathbf{x}\cdot\mathbf{x})^k$$

$$= r^l P_l(x_3/r)\,.$$

Rayleigh plane wave expansion:

$$e^{i\mathbf{k}\cdot\mathbf{x}} = 4\pi\sum_{l=0}^{\infty}\sum_{m=-l}^{l}i^l j_l(kr)\mathcal{Y}^*_{lm}(\hat{\mathbf{k}})\mathcal{Y}_{lm}(\hat{\mathbf{x}})\,,$$

$$j_l(kr) = \left(\frac{\pi}{2kr}\right)^{1/2}J_{l+1/2}(kr)\,.$$

Relations in potential theory:

$$\mathcal{Y}_{lm}(\nabla)\left(\frac{1}{r}\right) = \frac{(-1)^l(2l)!}{2^l l!}\frac{\mathcal{Y}_{lm}(\mathbf{x})}{r^{2l+1}}\,,$$

$$1/R = \sum_{l=0}^{\infty}I_l(\mathbf{x},\mathbf{y})/r^{2l+1}\,,$$

$$I_l(\mathbf{x},\mathbf{y})/r^{2l+1} = \frac{(-1)^l}{l!}(\mathbf{y}\cdot\nabla)^l\left(\frac{1}{r}\right)$$

for $\mathbf{R} = \mathbf{x}-\mathbf{y},\quad r = (\mathbf{x}\cdot\mathbf{x})^{1/2},\quad s = (\mathbf{y}\cdot\mathbf{y})^{1/2}$,

$$1/R = \sum_l P_l(\cos\theta)\frac{s^l}{r^{l+1}}\,,\quad s\le r, \cos\theta = \hat{\mathbf{x}}\cdot\hat{\mathbf{y}}\,.$$

Rotational invariants in three vectors:

$$I_{(l_1l_2l_3)}(\mathbf{x}^1,\mathbf{x}^2,\mathbf{x}^3) = \frac{(4\pi)^{3/2}}{[(2l_1+1)(2l_2+1)(2l_3+1)]^{1/2}}$$

$$\times \sum_{m_1m_2m_3}\begin{pmatrix} l_1 & l_2 & l_3 \\ m_1 & m_2 & m_3 \end{pmatrix}$$

$$\times \mathcal{Y}_{l_1m_1}(\mathbf{x}^1)\mathcal{Y}_{l_2m_2}(\mathbf{x}^2)\mathcal{Y}_{l_3m_3}(\mathbf{x}^3)\,,$$

where $\begin{pmatrix} l_1 & l_2 & l_3 \\ m_1 & m_2 & m_3 \end{pmatrix}$ is a 3-*j* coefficient (see Sect. 2.8).

$$I_{(l_1l_2l_3)}(\mathbf{x}^1,\mathbf{x}^2,0) = \delta_{l_1l_2}\delta_{l_30}(-1)^{l_1}I_{l_1}(\mathbf{x}^1,\mathbf{x}^2)/(2l_1+1)^{1/2}\,.$$

Product law:

$$I_{(l)}(\mathbf{x})I_{(k)}(\mathbf{x})$$

$$= \sum_{(j)}\left[\prod_{\alpha=1}^{3}(-1)^{j_\alpha}(2j_\alpha+1)\begin{pmatrix} l_\alpha & k_\alpha & j_\alpha \\ 0 & 0 & 0 \end{pmatrix}\right.$$

$$\left.\times (\mathbf{x}^\alpha\cdot\mathbf{x}^\alpha)^{(l_\alpha+k_\alpha-j_\alpha)/2}\right]\begin{Bmatrix} l_1 & l_2 & l_3 \\ k_1 & k_2 & k_3 \\ j_1 & j_2 & j_3 \end{Bmatrix}I_{(j)}(\mathbf{x})\,,$$

where $l = (l_1,l_2,l_3)$, etc., $\mathbf{x} = (\mathbf{x}^1,\mathbf{x}^2,\mathbf{x}^3)$.

Coplanar vectors:

$$I_{(l)}(\mathbf{x}^1,\mathbf{x}^2,\alpha\mathbf{x}^1+\beta\mathbf{x}^2)$$

$$= \sum_{kl}\left[\frac{(2l_3+1)!}{(2l_3-2k)!(2k)!}\right]^{1/2}$$

$$\times \alpha^{l_3-k}\beta^k(-1)^{l_1+l_3+k}(2l+1)$$

$$\times \begin{pmatrix} l_3-k & l_1 & l \\ 0 & 0 & 0 \end{pmatrix}\begin{pmatrix} k & l_2 & l \\ 0 & 0 & 0 \end{pmatrix}\begin{Bmatrix} l_3-k & l_3 & k \\ l_2 & l & l_1 \end{Bmatrix}$$

$$\times (\mathbf{x}^1\cdot\mathbf{x}^1)^{(l_1+l_3-l-k)/2}(\mathbf{x}^2\cdot\mathbf{x}^2)^{(l_2+k-l)/2}I_l(\mathbf{x}^1,\mathbf{x}^2)\,.$$

The bracket symbols in these relations are 6-j and 9-j coefficients [Sects. 2.10 and 2.11].

Cartan's vectors of zero length:

$$\begin{aligned}
\boldsymbol{\alpha} &= \left(-z_1^2 + z_2^2, -i(z_1^2 + z_2^2), 2z_1 z_2\right), \\
\boldsymbol{\alpha}\cdot\boldsymbol{\alpha} &= \alpha_1^2 + \alpha_2^2 + \alpha_3^2 = 0, \\
\mathbf{z} &= (z_1, z_2) \in \mathbf{C}^2.
\end{aligned}$$

Solutions of Laplace's equation using vectors of zero length:

$$\nabla^2 (\boldsymbol{\alpha}\cdot\mathbf{x})^l = 0, \quad l = 0, 1, \ldots.$$

Solid harmonics for vectors of zero length:

$$(-1)^m Y_{l,-m}(\boldsymbol{\alpha}) = \frac{(2l)!}{l!} \left[\frac{2l+1}{4\pi}\right]^{1/2} P_{lm}(z_1, z_2),$$

$$P_{lm}(z_1, z_2) = \frac{z_1^{l+m} z_2^{l-m}}{\sqrt{(l+m)!(l-m)!}}.$$

Orbital angular momentum operators for vectors of zero length:

$$\mathbf{J} = -i(\boldsymbol{\alpha} \times \nabla_\alpha),$$

$$J_+ = z_1 \frac{\partial}{\partial z_2}, \quad J_- = z_2 \frac{\partial}{\partial z_1},$$

$$J_3 = \frac{1}{2}\left(z_1 \frac{\partial}{\partial z_1} - z_2 \frac{\partial}{\partial z_2}\right).$$

Rotational invariant for vectors of zero length:

$$(\boldsymbol{\alpha}\cdot\mathbf{x})^l = \left[\frac{4\pi}{2l+1}\right]^{1/2} 2^l l! \sum_m P_{lm}(\mathbf{z}) Y_{lm}(\mathbf{x}).$$

Spinorial invariant under $\mathbf{z}^i \to U\mathbf{z}^i (i = 1, 2, 3)$:

$$\sum_{m_1 m_2 m_3} \begin{pmatrix} j_1 & j_2 & j_3 \\ m_1 & m_2 & m_3 \end{pmatrix} P_{j_1 m_1}(\mathbf{z}^1) P_{j_2 m_2}(\mathbf{z}^2) P_{j_3 m_3}(\mathbf{z}^3)$$

$$= [(j_1 + j_2 + j_3 + 1)!]^{-1/2}$$

$$\times \frac{(z_{12}^{12})^{j_1 + j_2 - j_3}(z_{12}^{31})^{j_3 + j_1 - j_2}(z_{12}^{23})^{j_2 + j_3 - j_1}}{[(j_1 + j_2 - j_3)!(j_3 + j_1 - j_2)!(j_2 + j_3 - j_1)!]^{1/2}},$$

$$z_{12}^{ij} = z_1^i z_2^j - z_1^j z_2^i.$$

This relation is invariant under the transformation

$$\mathbf{z} \to U\mathbf{z} = \left((U\mathbf{z})_1, (U\mathbf{z})_2\right)$$

$$= (u_{11} z_1 + u_{12} z_2, u_{21} z_1 + u_{22} z_2),$$

where $U \in SU(2)$. Transformation properties of vectors of zero length:

$$\boldsymbol{\alpha} \to R\boldsymbol{\alpha}, \quad \boldsymbol{\alpha} = \mathrm{col}(\alpha_1, \alpha_2, \alpha_3),$$

where $\mathbf{z} \to U\mathbf{z}$ and R is given in terms of U in Sect. 2.1. Simultaneous eigenvectors of $\mathbf{L}^2$ and $\mathbf{J}^2$:

$$\mathbf{L}^2(\boldsymbol{\alpha}\cdot\mathbf{x})^l = l(l+1)(\boldsymbol{\alpha}\cdot\mathbf{x})^l, \quad l = 0, 1, \ldots,$$

$$\mathbf{J}^2(\boldsymbol{\alpha}\cdot\mathbf{x})^l = l(l+1)(\boldsymbol{\alpha}\cdot\mathbf{x})^l, \quad l = 0, 1, \ldots.$$

2.2.2 Spherical Polar Coordinate Representation

The results given in Sect. 2.2.1 may be presented in any system of coordinates well-defined in terms of Cartesian coordinates. The principal relations for spherical polar coordinates are given in this section, where a vector in $\mathbf{R}^3$ is now given in the form

$$\mathbf{x} = r\hat{\mathbf{x}} = r(\sin\theta\cos\phi, \sin\theta\sin\phi, \cos\theta),$$

$$0 \leq \theta \leq \pi, \quad 0 \leq \phi < 2\pi.$$

Orbital angular momentum operators:

$$\begin{aligned}
L_1 &= i\cos\phi\cot\theta \frac{\partial}{\partial\phi} + i\sin\phi\frac{\partial}{\partial\theta}, \\
L_2 &= i\sin\phi\cot\theta \frac{\partial}{\partial\phi} - i\cos\phi\frac{\partial}{\partial\theta}, \\
L_3 &= -i\frac{\partial}{\partial\phi}, \\
L_\pm &= e^{\pm i\phi}\left(\pm\frac{\partial}{\partial\theta} + i\cot\theta\frac{\partial}{\partial\phi}\right), \\
\mathbf{L}^2 &= -\frac{1}{\sin\theta}\frac{\partial}{\partial\theta}\left(\sin\theta\frac{\partial}{\partial\theta}\right) - \frac{1}{\sin^2\theta}\frac{\partial^2}{\partial\phi^2}.
\end{aligned}$$

Laplacian:

$$\begin{aligned}
\mathbf{x}\cdot\nabla &= r\frac{\partial}{\partial r}, \\
\nabla^2 &= \frac{1}{r^2}\left[\left(r\frac{\partial}{\partial r}\right)^2 + r\frac{\partial}{\partial r} - \mathbf{L}^2\right].
\end{aligned}$$

Spherical harmonics (solid harmonics on the unit sphere $\mathbf{S}^2$):

$$Y_{lm}(\theta, \phi) = (-1)^m \left[\frac{2l+1}{4\pi}(l+m)!(l-m)!\right]^{1/2} e^{im\phi}$$

$$\times \sum_k \frac{(-1)^k (\sin\theta)^{2k+m}(\cos\theta)^{l-2k-m}}{2^{2k+m}(k+m)!k!(l-2k-m)!}.$$

Orthogonality on the unit sphere:

$$\int_0^{2\pi} d\phi \int_0^\pi d\theta \sin\theta\, Y_{l'm'}^*(\theta, \phi) Y_{lm}(\theta, \phi) = \delta_{l'l}\delta_{m'm}.$$

Relation to Legendre, Jacobi, and Gegenbauer polynomials:

$$Y_{lm}(\theta, \phi) = (-1)^m \left[\frac{(2l+1)(l-m)!}{4\pi(l+m)!}\right]^{1/2} P_l^m(\cos\theta)e^{im\phi},$$

$$P_l^m(\cos\theta) = \frac{(l+m)!}{l!}\left(\frac{\sin\theta}{2}\right)^m P_{l-m}^{(m,m)}(\cos\theta).$$

Jacobi polynomials:

$$P_n^{(\alpha,\beta)}(x) = \sum_s \binom{n+\alpha}{s}\binom{n+\beta}{n-s}$$

$$\times \left(\frac{x-1}{2}\right)^{n-s}\left(\frac{x+1}{2}\right)^s, \quad n = 0, 1, \ldots,$$

where α, β are arbitrary parameters and

$$\begin{pmatrix} z \\ k \end{pmatrix} = \begin{cases} z(z-1)\cdots(z-k+1)/k! & \text{for } k = 1, 2, \ldots \\ 1 & \text{for } k = 0 \\ 0 & \text{for } k = -1, -2, \ldots \end{cases}$$

Relations between Jacobi polynomials for $n + \alpha$, $n + \beta$, $n + \alpha + \beta$ nonnegative integers:

$$P_n^{\alpha, \beta}(x) = \frac{(n+\alpha)!(n+\beta)!}{n!(n+\alpha+\beta)!} \left(\frac{x+1}{2}\right)^{-\beta} P_{n+\beta}^{(\alpha, -\beta)}(x),$$

$$P_n^{\alpha, \beta}(x) = \frac{(n+\alpha)!(n+\beta)!}{n!(n+\alpha+\beta)!} \left(\frac{x-1}{2}\right)^{-\alpha} P_{n+\alpha}^{(-\alpha, \beta)}(x),$$

$$P_n^{\alpha, \beta}(x) = \left(\frac{x-1}{2}\right)^{-\alpha} \left(\frac{x+1}{2}\right)^{-\beta} P_{n+\alpha+\beta}^{(-\alpha, -\beta)}(x).$$

Nonstandard form (α arbitrary):

$$P_n^{(\alpha, \alpha)}(x) = \sum_s \frac{(-1)^s (\alpha+s+1)_{n-s}(1-x^2)^s x^{n-2s}}{2^{2s} s!(n-2s)!},$$

$$(z)_k = z(z+1)\cdots(z+k-1), \ k = 1, 2, \ldots; \ (z)_0 = 1.$$

Gegenbauer polynomials ($\alpha > -1/2$):

$$C_n^{(\alpha)}(x) = \frac{(2\alpha)_n}{(\alpha+1/2)_n} P_n^{(\alpha-\frac{1}{2}, \alpha-\frac{1}{2})}(x)$$

$$= \sum_s \frac{(-1)^s (\alpha)_{n-s}(2x)^{n-2s}}{s!(n-2s)!}.$$

2.3 ABSTRACT ANGULAR MOMENTUM

Abstract angular momentum theory addresses the problem of constructing all finite Hermitian matrices, up to equivalence, that satisfy the same commutation relations

$$[J_1, J_2] = iJ_3, [J_2, J_3] = iJ_1, [J_3, J_1] = iJ_2 \quad (2.7)$$

as some set of Hermitian operators J_1, J_2, J_3 appropriately defined in some Hilbert space; that is, of constructing all finite Hermitian matrices M_i such that under the correspondence $J_i \rightarrow M_i (i = 1, 2, 3)$ the commutation relations are still obeyed. If M_1, M_2, M_3 is such a set of Hermitian matrices, then AM_1A^{-1}, AM_2A^{-1}, AM_3A^{-1}, is another such set, where A is an arbitrary unitary matrix. This defines what is meant by equivalence. The commutation relations (2.7) may also be formulated as:

$$[J_3, J_\pm] = \pm J_\pm, \quad [J_+, J_-] = 2J_3,$$
$$J_\pm = J_1 \pm iJ_2, \quad J_+^\dagger = J_-. \quad (2.8)$$

The squared angular momentum

$$\mathbf{J}^2 = J_1^2 + J_2^2 + J_3^2 = J_-J_+ + J_3(J_3+1)$$
$$= J_+J_- + J_3(J_3-1) \quad (2.9)$$

commutes with each J_i, and J_3 is, by convention, taken with $\mathbf{J}^2$ as a pair of commuting Hermitian operators to be diagonalized.

Examples of matrices satisfying relations (2.7) are provided by $J_i \rightarrow \sigma_i/2$ [the 2×2 Hermitian Pauli matrices defined in Eq. (2.1)] and $J_i \rightarrow M_i$ [the 3×3 matrices defined in Eq. (2.4)], these latter matrices being equivalent to those obtained from the matrices of the orbital angular momentum operators for $l = 1$.

One could determine all Hermitian matrices solving Eqs. (2.7) and (2.8) by using only matrix theory, but it is customary in quantum mechanics to formulate the problem using Hilbert space concepts appropriate to that theory. Thus, one takes the viewpoint that the J_i are linear Hermitian operators with an action defined in a separable Hilbert space $\mathcal{H}$ such that $J_i: \mathcal{H} \rightarrow \mathcal{H}$.

One then seeks to decompose the Hilbert space into a direct sum of subspaces that are irreducible with respect to this action; that is, subspaces that cannot be further decomposed as a direct sum of subspaces that all the J_i leave invariant (map vectors in the space into vectors in the space). In this section, the solution of this fundamental problem for angular momentum theory is given. These results set the notation and phase conventions for all of angular momentum theory, in all of its varied realizations, and the relations are therefore sometimes referred to as standard. The method most often used to solve the posed problem is called the method of highest weights.

The solution of this problem is among the most important in quantum theory because of its generality and applicability to a wide range of problems. The space $\mathcal{H}$ can be written as a direct sum

$$\mathcal{H} = \sum_{j=0,\frac{1}{2},1,\ldots} \oplus n_j \mathcal{H}_j, \text{ each } \mathcal{H}_j \perp \mathcal{H}_{j'}, \quad j \neq j', \quad (2.10)$$

in which $\mathcal{H}_j$ denotes a vector space of dimension $2j + 1$ that is invariant and irreducible under the action of the set of operators $J_i, i = 1, 2, 3$, and where the direct sum is over all half intergers $j = 0, \frac{1}{2}, 1, \ldots$. There may be multiple occurrences, n_j in number, of the same space $\mathcal{H}_j$ for given j, or no such space, $n_j = 0$, in the direct sum. Abstractly, in so far as angular momentum properties are concerned, each repeated space $\mathcal{H}_j$ is *identical*. Such spaces may, however, be distinguished by their properties with respect to other physical observables, but not by the angular momentum operators themselves. The result, Eq. (2.10), applies to any physical system, no matter how complex, in which rotational symmetry, hence $SU(2)$ symmetry, is present, even in situations of higher symmetry where $SU(2)$ is a subgroup. Indeed, the resolution of the terms in Eq. (2.10) for various physical systems constitutes "spectroscopy" in the broadest sense.

The characterization of the space $\mathcal{H}_j$ with respect to angular momentum properties is given by the following results, where basis vectors are denoted in the Dirac bra-ket notation.

Orthonormal basis:

$$\{|jm\rangle \mid m = -j, -j+1, \ldots, j\}. \qquad (2.11)$$

$$\langle jm' | jm\rangle = \delta_{m',m}. \qquad (2.12)$$

Simultaneous eigenvectors:

$$\mathbf{J}^2|jm\rangle = j(j+1)|jm\rangle, \quad J_3|jm\rangle = m|jm\rangle. \qquad (2.13)$$

Action of angular momentum operators:

$$
\begin{aligned}
J_+|jm\rangle &= [(j-m)(j+m+1)]^{1/2}|jm+1\rangle, \\
J_-|jm\rangle &= [(j+m)(j-m+1)]^{1/2}|jm-1\rangle.
\end{aligned} \qquad (2.14)
$$

Defining properties of highest weight vector:

$$J_+|jj\rangle = 0, \quad J_3|jj\rangle = j|jj\rangle.$$

Generation of general vector from highest weight:

$$|jm\rangle = \left[\frac{(j+m)!}{(2j)!(j-m)!}\right]^{1/2} J_-^{j-m}|jj\rangle.$$

Necessary property of lowest weight vector:

$$J_-|j,-j\rangle = 0, \quad J_3|j,-j\rangle = -j|j,-j\rangle.$$

Operator in $\mathcal{H}$ corresponding to a rotation by angle ψ about direction $\hat{\mathbf{n}}$ in $\mathbf{R}^3$:

$$
\begin{aligned}
T_{U(\psi,\hat{\mathbf{n}})} &= \exp(-i\psi\hat{\mathbf{n}}\cdot\mathbf{J}), \\
\hat{\mathbf{n}}\cdot\hat{\mathbf{n}} &= n_1^2 + n_2^2 + n_3^2 = 1, \\
\hat{\mathbf{n}}\cdot\mathbf{J} &= n_1 J_1 + n_2 J_2 + n_3 J_3, \\
U(\psi,\hat{\mathbf{n}}) &= \exp(-i\psi\hat{\mathbf{n}}\cdot\boldsymbol{\sigma}/2) \\
&= \sigma_0 \cos(\tfrac{1}{2}\psi) - i(\hat{\mathbf{n}}\cdot\boldsymbol{\sigma})\sin(\tfrac{1}{2}\psi) \\
&= \begin{pmatrix} \cos(\tfrac{1}{2}\psi) - in_3\sin(\tfrac{1}{2}\psi) & (-in_1 - n_2)\sin(\tfrac{1}{2}\psi) \\ (-in_1 + n_2)\sin(\tfrac{1}{2}\psi) & \cos(\tfrac{1}{2}\psi) + in_3\sin(\tfrac{1}{2}\psi) \end{pmatrix}, \\
& \qquad\qquad 0 \le \psi \le 2\pi,
\end{aligned} \qquad (2.15)
$$

where σ_0 denotes the 2×2 unit matrix.

Action of $T_{U(\psi,\hat{\mathbf{n}})}$ on $\mathcal{H}_j$:

$$T_U|jm\rangle = \sum_{m'} D^j_{m',m}(U)|jm'\rangle, \qquad (2.16)$$

in which $U = U(\psi,\hat{\mathbf{n}})$ and $D^j_{m',m}(U)$ denotes a homogeneous polynomial of degree $2j$ defined on the elements $u_{ij} = U_{ij}(\psi,\hat{\mathbf{n}})$ in row i and column j of the matrix $U(\psi,\hat{\mathbf{n}})$ given by Eq. (2.15). The explicit form of this polynomial is

$$
\begin{aligned}
D^j_{m'm}(U) &= [(j+m)!(j-m)!(j+m')!(j-m')!]^{1/2} \\
&\times \sum_{\boxed{\alpha}} \frac{(u_{11})^{\alpha_{11}}(u_{12})^{\alpha_{12}}(u_{21})^{\alpha_{21}}(u_{22})^{\alpha_{22}}}{\alpha_{11}!\alpha_{12}!\alpha_{21}!\alpha_{22}!}.
\end{aligned}
$$
$$\qquad (2.17)$$

The notation $\boxed{\alpha}$ symbolizes a 2×2 array of nonnegative integers with certain constraints:

$$
\begin{array}{cc|c}
& \boxed{\begin{array}{cc} \alpha_{11} & \alpha_{12} \\ \alpha_{21} & \alpha_{22} \end{array}} & \begin{array}{c} j+m' \\ j-m' \end{array} \\
\hline
& j+m \quad j-m &
\end{array}
$$

In this array the α_{ij} are nonnegative integers subject to the row and column constraints (sums) indicated by the (nonnegative) integers $j \pm m, j \pm m'$. Explicitly,

$$\alpha_{11} + \alpha_{12} = j + m', \quad \alpha_{21} + \alpha_{22} = j - m',$$
$$\alpha_{11} + \alpha_{21} = j + m, \quad \alpha_{12} + \alpha_{22} = j - m.$$

The summation is over all such arrays. Any one of the α_{ij} may serve as a single summation index if one wishes to eliminate the redundancy inherent in the square-array notation. The form (2.17) is very useful for obtaining symmetry relations for these polynomials (see Sect. 2.4.6).

Unitary property on $\mathcal{H}$:

$$\langle T_U\Psi | T_U\Psi\rangle, \text{ each } \Psi \in \mathcal{H}.$$

Irreducible unitary matrix representation of $SU(2)$:

$$
\begin{aligned}
(D^j(U))_{j-m'+1,j-m+1} &= D^j_{m',m}(U), \\
m' = j, j-1, \ldots, -j \quad & m = j, j-1, \ldots, -j,
\end{aligned} \qquad (2.18)
$$

denotes the element in row $j-m'+1$ and column $j-m+1$. Then, dimension $D^j(U) = 2j+1$ and

$$D^j(U)D^j(U') = D^j(UU'), \quad U \in SU(2), \quad U' \in SU(2),$$
$$(D^j(U))^\dagger = (D^j(U))^{-1} = D^j(U^\dagger).$$

Kronecker (direct) product representation:

$$D^{j_1}(U) \times D^{j_2}(U)$$

is a $(2j_1+1)(2j_2+1)$ dimensional reducible representation of $SU(2)$. One can also effect the reduction of this representation into irreducible ones by abstract methods. The results are given in Sect. 2.8.

2.4 REPRESENTATION FUNCTIONS

2.4.1 Parametrizations of the Groups $SU(2)$ and $SO(3,R)$

The irreducible representations of the quantal rotation group, $SU(2)$, are among the most important quantities in all of angular momentum theory: These are the unitary matrices of dimension $2j + 1$, denoted by $D^j(U)$, where this notation is used to signify that the elements of this matrix, denoted $D^j_{m'm}(U)$, are functions of the elements u_{ij} of the 2×2 unitary umimodular matrix $U \in SU(2)$. It has become standard to enumerate the rows and columns of these matrices in the order $j, j - 1, \ldots, -j$ as read from top to bottom down the rows and from left to right across the columns [see also Eq. (2.18)]. These matrices may be presented in a variety of parametrizations, all of which are useful. In order to make comparisons between the group $SO(3, \mathbf{R})$ and the group $SU(2)$, it is most useful to parametrize these groups so that they are related according to the two-to-one homomorphism given by Eq. (2.2).

The general parametrization of the group $SU(2)$ is given in terms of the Euler-Rodrigues parameters corresponding to points belonging to the surface of the unit sphere $\mathbf{S}^3$ in $\mathbf{R}^4$,

$$\alpha_0^2 + \alpha_1^2 + \alpha_2^2 + \alpha_3^2 = 1. \tag{2.19}$$

Each $U \in SU(2)$ can be written in the form:

$$U(\alpha_0, \boldsymbol{\alpha}) = \begin{pmatrix} \alpha_0 - i\alpha_3 & -i\alpha_1 - \alpha_2 \\ -i\alpha_1 + \alpha_2 & \alpha_0 + i\alpha_3 \end{pmatrix}$$
$$= \alpha_0\sigma_0 - i\boldsymbol{\alpha}\cdot\boldsymbol{\sigma}. \tag{2.20}$$

The $R \in SO(3, \mathbf{R})$ corresponding to this U in the two-to-one homomorphism given by Eq. (2.2) is:

$$R(\alpha_0, \boldsymbol{\alpha}) =$$
$$\begin{pmatrix} \alpha_0^2 + \alpha_1^2 - \alpha_2^2 - \alpha_3^2 & 2\alpha_1\alpha_2 - 2\alpha_0\alpha_3 & 2\alpha_1\alpha_3 + 2\alpha_0\alpha_2 \\ 2\alpha_1\alpha_2 + 2\alpha_0\alpha_3 & \alpha_0^2 + \alpha_2^2 - \alpha_3^2 - \alpha_1^2 & 2\alpha_2\alpha_3 - 2\alpha_0\alpha_1 \\ 2\alpha_1\alpha_3 - 2\alpha_0\alpha_2 & 2\alpha_2\alpha_3 + 2\alpha_0\alpha_1 & \alpha_0^2 + \alpha_3^2 - \alpha_1^2 - \alpha_2^2 \end{pmatrix}. \tag{2.21}$$

The procedure of parametrization is implemented uniformly by first parametrizing the points on the unit sphere $\mathbf{S}^3$ so as to cover the points in $\mathbf{S}^3$ exactly once, thus obtaining a parametrization of each $U \in SU(2)$. Equation (2.21) is then used to obtain the corresponding parametrization of each $R \in SO(3, \mathbf{R})$, where one notes that $R(-\alpha_0, -\boldsymbol{\alpha}) = R(\alpha_0, \boldsymbol{\alpha})$. Because of this two-to-one correspondence $\pm U \to R$, the domain of the parameters that cover the unit sphere $\mathbf{S}^3$ exactly once will cover the group $SO(3, \mathbf{R})$ exactly twice. This is taken into account uniformly by redefining the domain for $SO(3, \mathbf{R})$ so as to cover only the upper hemisphere ($\alpha_0 \geq 0$) of $\mathbf{S}^3$.

In the active viewpoint (reference frame fixed with points being transformed into new points), an arbitrary vector $\mathbf{x} = \text{col}(x_1, x_2, x_3) \in \mathbf{R}^3$ is transformed to the new vector $\mathbf{x}' = \text{col}(x_1', x_2', x_3')$ by the rule $\mathbf{x}' = R\mathbf{x}$, or, equivalently, in terms of the Cartan matrix: $X' = UXU^\dagger$. In the passive viewpoint, the basic inertial reference system, which is taken to be a right-handed triad of unit vectors $(\hat{\mathbf{e}}_1, \hat{\mathbf{e}}_2, \hat{\mathbf{e}}_3)$, is transformed by R to a new right-handed triad $(\hat{\mathbf{f}}_1, \hat{\mathbf{f}}_2, \hat{\mathbf{f}}_3)$ by the rule

$$\hat{\mathbf{f}}_j = \sum_i R_{ij}\hat{\mathbf{e}}_i, \quad i = 1, 2, 3,$$

so that $\hat{\mathbf{e}}_i \cdot \hat{\mathbf{f}}_j = R_{ij}$. In this viewpoint, the coordinates of one and the same point P undergo a *redescription* under the change of frame. If the coordinates of P are (x_1, x_2, x_3) relative to the frame $(\hat{\mathbf{e}}_1, \hat{\mathbf{e}}_2, \hat{\mathbf{e}}_3)$ and (x_1', x_2', x_3') relative to the frame $(\hat{\mathbf{f}}_1, \hat{\mathbf{f}}_2, \hat{\mathbf{f}}_3)$, then

$$x_1\hat{\mathbf{e}}_1 + x_2\hat{\mathbf{e}}_2 + x_3\hat{\mathbf{e}}_3 = x_1'\hat{\mathbf{f}}_1 + x_2'\hat{\mathbf{f}}_2 + x_3'\hat{\mathbf{f}}_3,$$

so that $\mathbf{x}' = R^T\mathbf{x}$.

Rotation about direction $\hat{\mathbf{n}} \in \mathbf{S}^2$ by positive angle ψ (right-hand rule):

$$(\alpha_0, \boldsymbol{\alpha}) = \left(\cos\tfrac{1}{2}\psi, \, \hat{\mathbf{n}}\sin\tfrac{1}{2}\psi\right), \quad 0 \leq \psi \leq 2\pi,$$
$$U(\psi, \hat{\mathbf{n}}) = \exp(-i\tfrac{1}{2}\psi\hat{\mathbf{n}}\cdot\boldsymbol{\sigma})$$
$$= \begin{pmatrix} \cos\tfrac{1}{2}\psi - in_3\sin\tfrac{1}{2}\psi & (-in_1 - n_2)\sin\tfrac{1}{2}\psi \\ (-in_1 + n_2)\sin\tfrac{1}{2}\psi & \cos\tfrac{1}{2}\psi + in_3\sin\tfrac{1}{2}\psi \end{pmatrix},$$

$$R(\psi, \hat{\mathbf{n}}) = \exp(-i\psi\hat{\mathbf{n}}\cdot\mathbf{M}), \quad 0 \leq \psi \leq \pi$$
$$= I_3 - i\sin\psi(\hat{\mathbf{n}}\cdot\mathbf{M}) - (\hat{\mathbf{n}}\cdot\mathbf{M})^2(1 - \cos\psi)$$
$$= \begin{pmatrix} R_{11} & R_{12} & R_{13} \\ R_{21} & R_{22} & R_{23} \\ R_{31} & R_{32} & R_{33} \end{pmatrix},$$

$R_{11} = n_1^2 + (1 - n_1^2)\cos\psi,$
$R_{21} = n_1n_2(1 - \cos\psi) + n_3\sin\psi,$
$R_{31} = n_1n_3(1 - \cos\psi) - n_2\sin\psi,$
$R_{12} = n_1n_2(1 - \cos\psi) - n_3\sin\psi,$
$R_{22} = n_2^2 + (1 - n_2^2)\cos\psi,$
$R_{32} = n_2n_3(1 - \cos\psi) + n_1\sin\psi,$
$R_{13} = n_1n_3(1 - \cos\psi) + n_2\sin\psi,$
$R_{23} = n_2n_3(1 - \cos\psi) - n_1\sin\psi,$
$R_{33} = n_3^2 + (1 - n_3^2)\cos\psi.$

The unit vector $\hat{\mathbf{n}} \in \mathbf{S}^2$ can be further parametrized in terms of the usual spherical polar coordinates:

$$\hat{\mathbf{n}} = (\sin\theta\cos\phi, \sin\theta\sin\phi, \cos\theta),$$
$$0 \leq \theta \leq \pi, \quad 0 \leq \phi < 2\pi.$$

Euler angle parametrization:

$$U(\alpha\beta\gamma) = e^{-i\alpha\sigma_3/2}e^{-i\beta\sigma_2/2}e^{-i\gamma\sigma_3/2}$$
$$= \begin{pmatrix} e^{-i\alpha/2}\cos(\tfrac{1}{2}\beta)e^{-i\gamma/2} & -e^{-i\alpha/2}\sin(\tfrac{1}{2}\beta)e^{i\gamma/2} \\ e^{i\alpha/2}\sin(\tfrac{1}{2}\beta)e^{-i\gamma/2} & e^{i\alpha/2}\cos(\tfrac{1}{2}\beta)e^{i\gamma/2} \end{pmatrix},$$
$$0 \leq \alpha < 2\pi, \, 0 \leq \beta \leq \pi \text{ or } 2\pi \leq \beta \leq 3\pi, \, 0 \leq \gamma < 2\pi,$$
$$U(\alpha, \beta + 2\pi, \gamma) = -U(\alpha\beta\gamma);$$
$$R(\alpha\beta\gamma) = e^{-i\alpha M_3}e^{-i\beta M_2}e^{-i\gamma M_3}$$

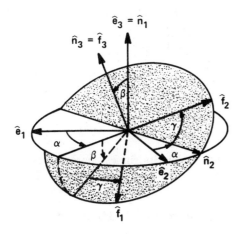

Figure 2.1. Euler Angles. The Three Euler Angles $(\alpha\beta\gamma)$ are Defined by a Sequence of Three Rotations. Reprinted with the permission of Cambridge University Press.

$$= \begin{pmatrix} \cos\alpha & -\sin\alpha & 0 \\ \sin\alpha & \cos\alpha & 0 \\ 0 & 0 & 1 \end{pmatrix} \begin{pmatrix} \cos\beta & 0 & \sin\beta \\ 0 & 1 & 0 \\ -\sin\beta & 0 & \cos\beta \end{pmatrix}$$

$$\times \begin{pmatrix} \cos\gamma & -\sin\gamma & 0 \\ \sin\gamma & \cos\gamma & 0 \\ 0 & 0 & 1 \end{pmatrix}$$

$$= \left(\begin{array}{c|c|c} \cos\alpha\cos\beta\cos\gamma & -\cos\alpha\cos\beta\sin\gamma & \cos\alpha\sin\beta \\ -\sin\alpha\sin\gamma & -\sin\alpha\cos\gamma & \\ \hline \sin\alpha\cos\beta\cos\gamma & -\sin\alpha\cos\beta\sin\gamma & \sin\alpha\sin\beta \\ +\cos\alpha\sin\gamma & +\cos\alpha\cos\gamma & \\ \hline -\sin\beta\cos\gamma & \sin\beta\sin\gamma & \cos\beta \end{array} \right)$$

$$0 \le \alpha < 2\pi, \quad 0 \le \beta \le \pi, \quad 0 \le \gamma < 2\pi.$$

This matrix corresponds to the sequence of frame rotations given by

$$\begin{aligned} \text{rotate by } \gamma \text{ about } \hat{\mathbf{e}}_3 &= (0,0,1), \\ \text{rotate by } \beta \text{ about } \hat{\mathbf{e}}_2 &= (0,1,0), \\ \text{rotate by } \alpha \text{ about } \hat{\mathbf{e}}_3 &= (0,0,1). \end{aligned}$$

Equivalently, it corresponds to the sequence of frame rotations given by

$$\begin{aligned} \text{rotate by } \alpha \text{ about } \hat{\mathbf{n}}_1 &= (0,0,1), \\ \text{rotate by } \beta \text{ about } \hat{\mathbf{n}}_2 &= (-\sin\alpha, \cos\alpha, 0), \\ \text{rotate by } \gamma \text{ about } \hat{\mathbf{n}}_3 &= (\cos\alpha\sin\beta, \sin\alpha\sin\beta, \cos\beta). \end{aligned}$$

This latter sequence of rotations is depicted in Fig. 2.1 in obtaining the frame $(\hat{\mathbf{f}}_1, \hat{\mathbf{f}}_2, \hat{\mathbf{f}}_3)$ from $(\hat{\mathbf{e}}_1, \hat{\mathbf{e}}_2, \hat{\mathbf{e}}_3)$.

The four *complex numbers*

$$(a,b,c,d) = (\alpha_0 + i\alpha_3, i\alpha_1 - \alpha_2, i\alpha_1 + \alpha_2, \alpha_0 - i\alpha_3)$$

are called the *Cayley-Klein parameters*, whereas the four *real numbers* $(\alpha_0, \boldsymbol{\alpha})$ defining a point on the surface of the unit sphere in four-space, $\mathbf{S}^3$, are known as the *Euler-Rodrigues parameters*. The three ratios α_i/α_0 form the *homogeneous* or *symmetric* Euler parameters.

2.4.2 Explicit Forms of Representation Functions

The general form of the representation functions is given in its most basic and symmetric form in Eq. (2.17). This form applies to every parametrization, it being necessary only to introduce the explicit parametrizations of $U \in SU(2)$ or $R \in SO(3, \mathbf{R})$ given in Sect. 2.4.1 to obtain the explicit results given in this section. A choice is also made for the single independent summation parameter in the α-array. The notation for functions is abused by writing

$$D^j(\omega) = D^j(U(\omega)),$$
$$\omega = \text{set of parameters of } U \in \widetilde{SU}(2).$$

Euler-Rodrigues representation $((\alpha_0, \boldsymbol{\alpha}) \in \mathbf{S}^3)$:

$$D^j_{m'm}(\alpha_0, \boldsymbol{\alpha})$$
$$= [(j+m')!(j-m')!(j+m)!(j-m)!]^{1/2}$$
$$\times \sum_s \frac{(\alpha_0 - i\alpha_3)^{j+m-s}(-i\alpha_1 - \alpha_2)^{m'-m+s}}{(j+m-s)!(m'-m+s)!}$$
$$\times \frac{(-i\alpha_1 + \alpha_2)^s(\alpha_0 + i\alpha_3)^{j-m'-s}}{s!(j-m'-s)!}. \tag{2.22}$$

Quaternionic multiplication rule for points on the sphere $\mathbf{S}^3$:

$$(\alpha_0', \boldsymbol{\alpha}')(\alpha_0, \boldsymbol{\alpha}) = (\alpha_0'', \boldsymbol{\alpha}''),$$
$$\alpha_0'' = \alpha_0'\alpha_0 - \boldsymbol{\alpha}'\cdot\boldsymbol{\alpha},$$
$$\boldsymbol{\alpha}'' = \alpha_0'\boldsymbol{\alpha} + \alpha_0\boldsymbol{\alpha}' + \boldsymbol{\alpha}'\times\boldsymbol{\alpha};$$
$$D^j(\alpha_0', \boldsymbol{\alpha}')D^j(\alpha_0, \boldsymbol{\alpha}) = D^j(\alpha_0'', \boldsymbol{\alpha}'').$$

The $(\psi, \hat{\mathbf{n}})$ parameters:

$$\alpha_0 = \cos\tfrac{1}{2}\psi, \quad \boldsymbol{\alpha} = \hat{\mathbf{n}}\sin\tfrac{1}{2}\psi.$$

Euler angle parametrization:

$$D^j_{m'm}(\alpha\beta\gamma) = e^{-im'\alpha}d^j_{m'm}(\beta)e^{-im\gamma},$$
$$d^j_{m'm}(\beta) = \langle jm'|e^{-i\beta J_2}|jm\rangle$$
$$= [(j+m')!(j-m')!(j+m)!(j-m)!]^{1/2}$$
$$\times \sum_s \frac{(-1)^{m'-m+s}\left(\cos\tfrac{1}{2}\beta\right)^{2j+m-m'-2s}}{(j+m-s)!s!(m'-m+s)!}$$
$$\times \frac{\left(\sin\tfrac{1}{2}\beta\right)^{m'-m+2s}}{(j-m'-s)!}. \tag{2.23}$$

Explicit matrices:

$$d^{1/2}(\beta) = \begin{pmatrix} \cos\tfrac{1}{2}\beta & -\sin\tfrac{1}{2}\beta \\ \sin\tfrac{1}{2}\beta & \cos\tfrac{1}{2}\beta \end{pmatrix},$$

$$d^1(\beta) = \begin{pmatrix} \frac{1+\cos\beta}{2} & \frac{-\sin\beta}{\sqrt{2}} & \frac{1-\cos\beta}{2} \\ \frac{\sin\beta}{\sqrt{2}} & \cos\beta & \frac{-\sin\beta}{\sqrt{2}} \\ \frac{1-\cos\beta}{2} & \frac{\sin\beta}{\sqrt{2}} & \frac{1+\cos\beta}{2} \end{pmatrix}.$$

Formal polynomial form (z_{ij} are indeterminates):

$$D^j_{m'm}(Z) = [(j+m')!(j-m')!(j+m)!(j-m)!]^{1/2}$$
$$\times \sum_{\boxed{\alpha}} \prod_{i,j=1}^{2} (z_{ij})^{\alpha_{ij}}/(\alpha_{ij})!, \qquad (2.24)$$

$$D^j(Z')D^j(Z) = D^j(Z'Z).$$

Boson operator form:

Put $a^j_i = z_{ij}(i,j=1,2)$ in Eq. (2.24). Let $\bar{a}^j_i$ denote the Hermitian conjugate boson so that

$$\left[a^k_l, a^j_i\right] = 0, \quad \left[\bar{a}^k_l, \bar{a}^j_i\right] = 0, \quad \left[\bar{a}^k_l, a^j_i\right] = \delta^{kj}\delta_{li}.$$

Then the boson polynomials are orthogonal in the boson inner product:

$$\langle 0 \mid D^{j'}_{\mu'\mu}(\bar{A})D^j_{m'm}(A) \mid 0\rangle = (2j)!\delta_{j'j}\delta_{\mu'm'}\delta_{\mu m}.$$

2.4.3 Relations to Special Functions

Jacobi polynomials (see Sect. 2.2.2):

$$d^j_{m'm}(\beta) = \left[\frac{(j+m)!(j-m)!}{(j+m')!(j-m')!}\right]^{1/2} \left(\sin\tfrac{1}{2}\beta\right)^{m-m'}$$
$$\times \left(\cos\tfrac{1}{2}\beta\right)^{m'+m} P^{(m-m',m+m')}_{j-m}(\cos\beta),$$
$$d^j_{m'm}(\beta) = (-1)^{m'-m}d^j_{-m',-m}(\beta) = (-1)^{m'-m}d^j_{mm'}(\beta)$$
$$= d^j_{mm'}(-\beta).$$

Legendre polynomials:

$$D^l_{m0}(\beta) = (-1)^m \left[\frac{(l-m)!}{(l+m)!}\right]^{1/2} P^m_l(\cos\beta)$$
$$= \left[\frac{(l+m)!}{(l-m)!}\right]^{1/2} P^{-m}_l(\cos\beta).$$

Spherical harmonics:

$$Y_{lm}(\beta\alpha) = \left(\frac{2l+1}{4\pi}\right)^{1/2} e^{im\alpha}d^l_{m0}(\beta)$$
$$= \left(\frac{2l+1}{4\pi}\right)^{1/2} D^{l*}_{m0}(\alpha\beta\gamma),$$
$$Y^*_{lm}(\beta\alpha) = (-1)^m Y_{l,-m}(\beta\alpha).$$

Gegenbauer polynomials:

$$d^l_{m0}(\beta) = (-1)^m[(l+m)!(l-m)!]^{1/2}$$
$$\times \frac{(2m)!}{m!} \left(\frac{\sin\beta}{2}\right)^m C^{(m+1/2)}_{l-m}(\cos\beta), \quad m \geq 0.$$

Solutions of Laplace's equation in $\mathbf{R}^4$ (see Sect. 2.6):

$$\nabla^2_4 D^j_{m'm}(x_0,\mathbf{x}) = 0, \quad (x_0,\mathbf{x}) \in \mathbf{R}^4, \quad \nabla^2_4 = \sum_{\mu=0}^{3} \frac{\partial^2}{\partial x^2_\mu}.$$

Replace the Euler-Rodrigues parameters $(\alpha_0,\boldsymbol{\alpha})$ in Eq. (2.22) by an arbitrary point $(x_0,\mathbf{x}) \in \mathbf{R}^4$.

2.4.4 Orthogonality Properties

Inner (scalar) product:

$$(\Psi,\Phi) = \int d\Omega\, \Psi^*(\mathbf{x})\Phi(\mathbf{x}),$$
$$d\Omega = \text{invariant surface measure for } \mathbf{S}^3,$$
$$\int_{\mathbf{S}^3} d\Omega = 2\pi^2.$$

Spherical polar coordinate for $\mathbf{S}^3$:

$$(\alpha_0,\boldsymbol{\alpha})$$
$$= (\cos\phi\sin\theta\sin\chi, \sin\phi\sin\theta\sin\chi, \cos\theta\sin\chi, \cos\chi),$$
$$0 \leq \theta \leq \pi, \quad 0 \leq \phi < 2\pi, \quad 0 \leq \chi \leq \pi,$$
$$d\Omega = d\omega\sin^2\chi\, d\chi,$$
$$d\omega = d\phi\sin\theta\, d\theta = \text{invariant surface measure for } \mathbf{S}^2;$$

$$\int_0^{2\pi} d\phi \int_0^\pi d\theta\sin\theta \int_0^\pi d\chi\sin^2\chi\, D^{j*}_{m'm}(\alpha_0,\boldsymbol{\alpha})D^{j'}_{\mu'\mu}(\alpha_0,\boldsymbol{\alpha})$$
$$= \frac{2\pi^2}{2j+1}\delta_{jj'}\delta_{m'\mu'}\delta_{m\mu}.$$

Coordinates $(\psi,\hat{\mathbf{n}})$ for $\mathbf{S}^3$:

$$(\alpha_0,\boldsymbol{\alpha}) = (\cos\tfrac{\psi}{2}, \hat{\mathbf{n}}\sin\tfrac{\psi}{2}), \quad 0 \leq \psi \leq 2\pi, \quad \hat{\mathbf{n}}\cdot\hat{\mathbf{n}} = 1,$$
$$d\Omega = dS(\hat{\mathbf{n}})\sin^2\tfrac{\psi}{2}\frac{d\psi}{2},$$
$$dS(\hat{\mathbf{n}}) = d\omega \text{ for } \hat{\mathbf{n}} = (\sin\theta\cos\phi, \sin\theta\sin\phi, \cos\theta),$$
$$\int dS(\hat{\mathbf{n}})\int_0^{2\pi} \frac{d\psi}{2}\left(\sin\tfrac{\psi}{2}\right)^2 D^{j*}_{m'm}(\psi,\hat{\mathbf{n}})\, D^{j'}_{\mu'\mu}(\psi,\hat{\mathbf{n}})$$
$$= \frac{2\pi^2}{2j+1}\delta_{jj'}\delta_{m'\mu'}\delta_{m\mu},$$

Euler angles for $\mathbf{S}^3$ ($SU(2)$):

$$(\alpha_0,\boldsymbol{\alpha}) = (\cos\tfrac{\beta}{2}\cos\tfrac{1}{2}(\gamma+\alpha), \sin\tfrac{\beta}{2}\sin\tfrac{1}{2}(\gamma-\alpha),$$
$$\sin\tfrac{\beta}{2}\cos\tfrac{1}{2}(\gamma-\alpha), \cos\tfrac{\beta}{2}\sin\tfrac{1}{2}(\gamma+\alpha)),$$
$$d\Omega = \tfrac{1}{8}d\alpha\, d\gamma\sin\beta\, d\beta, \qquad (2.25)$$

$$\frac{1}{8}\int_0^{2\pi} d\alpha \int_0^{2\pi} d\gamma \int_0^{\pi} d\beta \, \sin\beta D_{m'm}^{j*}(\alpha\beta\gamma) D_{\mu'\mu}^{j'}(\alpha\beta\gamma)$$

$$+\frac{1}{8}\int_0^{2\pi} d\alpha \int_0^{2\pi} d\gamma \int_{2\pi}^{3\pi} d\beta \, \sin\beta D_{m'm}^{j*}(\alpha\beta\gamma) D_{\mu'\mu}^{j'}(\alpha\beta\gamma)$$

$$=\frac{2\pi^2}{2j+1}\delta_{jj'}\delta_{m'\mu'}\delta_{m\mu}. \tag{2.26}$$

Euler angles for hemisphere of $\mathbf{S}^3$ ($SO(3,\mathbf{R})$; j' and j both integral):

$$\int_0^{2\pi} d\alpha \int_0^{2\pi} d\gamma \int_0^{\pi} d\beta \, \sin\beta D_{m'm}^{j*}(\alpha\beta\gamma) D_{\mu'\mu}^{j'}(\alpha\beta\gamma)$$

$$=\frac{8\pi^2}{2j+1}\delta_{jj'}\delta_{m'\mu'}\delta_{m\mu}. \tag{2.27}$$

Formal polynomials [see Eq. (2.24)]:

$$(D_{m'm}^j, D_{\mu'\mu}^{j'}) = (2j)!\delta_{jj'}\delta_{m'\mu'}\delta_{m\mu},$$

with inner product

$$(P,P') = P^*(\frac{\partial}{\partial Z})P'(Z)|_{Z=0},$$

where $P^*(\frac{\partial}{\partial Z})$ is the complex conjugate polynomial P^* of P in which each z_{ij} is replaced by $\frac{\partial}{\partial z_{ij}}$.

Boson polynomials:

$$\langle D_{m'm}^j | D_{\mu'\mu}^{j'}\rangle = (2j)!\delta_{jj'}\delta_{m'\mu'}\delta_{m\mu},$$

with inner product $\langle P|P'\rangle = \langle 0|P^*(\bar{A})P'(A)|0\rangle$.

2.4.5 Recurrence Relations

Many useful relations between the representation functions may be derived as special cases of general relations between these functions and the WCG-coefficients given in Sect. 2.8.1. The simplest of these are obtained from the reduction

$$D^j \times D^{1/2} = D^{j+1/2} \oplus D^{j-1/2}.$$

Such relations are usually presented in terms of the Euler angle realization of U, leading to the following relations between the functions $d_{m',m}^j(\beta)$:

$$(j-m+1)^{1/2}\cos(\tfrac{1}{2}\beta)d_{m'-1/2,m-1/2}^{j+1/2}(\beta)$$

$$+ (j+m+1)^{1/2}\sin(\tfrac{1}{2}\beta)d_{m'-1/2,m+1/2}^{j+1/2}(\beta)$$

$$= (j-m'+1)^{1/2}d_{m'm}^j(\beta),$$

$$-(j-m+1)^{1/2}\sin(\tfrac{1}{2}\beta)d_{m'+1/2,m-1/2}^{j+1/2}(\beta)$$

$$+ (j+m+1)^{1/2}\cos(\tfrac{1}{2}\beta)d_{m'+1/2,m+1/2}^{j+1/2}(\beta)$$

$$= (j+m'+1)^{1/2}d_{m'm}^j(\beta),$$

$$(j+m)^{1/2}\cos(\tfrac{1}{2}\beta)d_{m'-1/2,m-1/2}^{j-1/2}(\beta)$$

$$- (j-m)^{1/2}\sin(\tfrac{1}{2}\beta)d_{m'-1/2,m+1/2}^{j-1/2}(\beta)$$

$$= (j+m')^{1/2}d_{m'm}^j(\beta),$$

$$(j+m)^{1/2}\sin(\tfrac{1}{2}\beta)d_{m'+1/2,m-1/2}^{j-1/2}(\beta)$$

$$+ (j-m)^{1/2}\cos(\tfrac{1}{2}\beta)d_{m'+1/2,m+1/2}^{j-1/2}(\beta)$$

$$= (j-m')^{1/2}d_{m'm}^j(\beta).$$

Two useful relations implied by the above are:

$$[(j-m)(j+m+1)]^{1/2}\sin\beta \, d_{m',m+1}^j(\beta)$$

$$+ [(j+m)(j-m+1)]^{1/2}\sin\beta \, d_{m',m-1}^j(\beta)$$

$$= 2(m\cos\beta - m')d_{m'm}^j(\beta),$$

$$[(j+m)(j-m+1)]^{1/2} d_{m',m-1}^j(\beta)$$

$$+ [(j+m')(j-m'+1)]^{1/2} d_{m'-1,m}^j(\beta)$$

$$= (m-m')\cot(\tfrac{1}{2}\beta)d_{m'm}^j(\beta).$$

By considering

$$D^j \times D^1 = D^{j+1} \oplus D^j \oplus D^{j-1},$$

one can also readily derive the matrix elements of the direction cosines specifying the orientation of the body-fixed frame $(\hat{\mathbf{f}}_1, \hat{\mathbf{f}}_2, \hat{\mathbf{f}}_3)$ of a symmetric rotor relative to the inertial frame $(\hat{\mathbf{e}}_1, \hat{\mathbf{e}}_2, \hat{\mathbf{e}}_3)$:

$$\lambda_{\mu,\nu}\Psi_{m,m'}^j = \sum_{j'}\left[\frac{2j+1}{2j'+1}\right]^{1/2}$$

$$\times C_{m\mu m+\mu}^{j1j'} C_{m'\nu m'+\nu}^{j1j'} \Psi_{m+\mu,m'+\nu}^{j'},$$

where the wave functions are those defined for integral j by Eq. (2.37), for half-integral j by Eq. (2.36), and

$$\lambda_{\mu,\nu} = \hat{\mathbf{e}}_\mu \cdot \hat{\mathbf{f}}_\nu^* = (D_{\mu,\nu}^1)^*, \quad \mu,\nu = -1,0,+1;$$

$$\hat{\mathbf{e}}_{+1} = -(\hat{\mathbf{e}}_1 + i\hat{\mathbf{e}}_2)/\sqrt{2}, \quad \hat{\mathbf{e}}_0 = \hat{\mathbf{e}}_3,$$

$$\hat{\mathbf{e}}_{-1} = (\hat{\mathbf{e}}_1 - i\hat{\mathbf{e}}_2)/\sqrt{2},$$

$$\hat{\mathbf{f}}_{+1} = -(\hat{\mathbf{f}}_1 + i\hat{\mathbf{f}}_2)/\sqrt{2}, \quad \hat{\mathbf{f}}_0 = \hat{\mathbf{f}}_3,$$

$$\hat{\mathbf{f}}_{-1} = (\hat{\mathbf{f}}_1 - i\hat{\mathbf{f}}_2)/\sqrt{2}.$$

2.4.6 Symmetry Relations

Symmetry relations for the representation functions $D_{m',m}^j(Z)$ defined by Eq. (2.24) are associated with the action of a finite group G on the set $M(2,2)$ of complex 2×2 matrices: $g : M(2,2) \to M(2,2)$, $g \in G$. Equivalently, if $Z \in M(2,2)$ is parametrized by a set Ω of parameters $\omega \in \Omega$ (parameter space), then g may be taken to act directly in the parameter space $g : \Omega \to \Omega$.

The action, denoted $\Box$, of a group $G = \{e, g, g', \ldots\}$ ($e =$ identity) on a set $\mathbf{X} = \{x, x', \ldots\}$ must satisfy the rules

$$g : \mathbf{X} \to \mathbf{X}, \quad e \Box x = x, \text{ all } x \in \mathbf{X},$$
$$g' \Box (g \Box x) = (g'g) \Box x, \text{ all } g', g \in G, \text{ all } x \in \mathbf{X}. \tag{2.28}$$

Using $\cdot$ to denote the action of G on $M(2,2)$ and $\Box$ to denote the action of G on Ω, one has the relation:

$$(g \cdot Z)(\omega) = Z(g^{-1} \Box \omega).$$

Only finite subgroups G of the unitary group $U(2)$ (group of 2×2 unitary matrices) are considered here: $G \subset U(2)$.

Generally, when G acts on $M(2,2)$, it effects a unitary linear transformation of the set of functions $\{D^j_{m',m}\}$ (j fixed) defined over $Z \in M(2,2)$. For certain groups G, or for some elements of G, a *single function* $D^j_{\mu',\mu} \in \{D^j_{m',m}\}$ occurs in the transformation, so that

$$(g D^j_{m',m})(Z) = D^j_{m',m}(g^{-1} \cdot Z)$$
$$= g_{m',m} D^j_{\mu',\mu}(Z), \tag{2.29}$$

$$(\mu', \mu) \in \{(\lambda'm', \lambda m), (\lambda m, \lambda'm') \mid \lambda' = \pm 1, \lambda = \pm 1\},$$

where $g_{m',m}$ is a complex number of unit modulus. (Juxtaposition gF denotes that g acts in the space of functions over Z.) Relation (2.29) is called a *symmetry relation of $D^j_{m',m}$ with respect to g.* Usually not all elements in G correspond to symmetry relations. In a symmetry relation, the action of the group is effectively transferred to the discrete quantum labels themselves:

$$g : m' \to \mu' = m'(g),$$
$$m \to \mu = m(g). \tag{2.30}$$

In terms of a parametrization Ω of $M(2,2)$, relation (2.29) is written

$$(g D^j_{m',m})(\omega) = D^j_{m',m}(g^{-1} \Box \omega)$$
$$= g_{m'm} D^j_{\mu',\mu}(\omega). \tag{2.31}$$

In practice, action symbols such as $\cdot$ and $\Box$ are often dropped in favor of juxtaposition, when the context is clear. Moreover the set of complex matrices $M(2,2)$ may be replaced by $U(2)$ or $SU(2)$ whenever the action conditions (2.28) are satisfied. Relations (2.29)–(2.31) are illustrated below by examples.

There are several finite subgroups of interest with various group-subgroup relations between them:

a. Pauli group:

$$P = \{\sigma_\mu, -\sigma_\mu, i\sigma_\mu, -i\sigma_\mu \mid \mu = 0, 1, 2, 3\}, \quad |P| = 16.$$

Each element of this group is an element of $U(2)$. The action of the group P may therefore be defined on the group $U(2)$ by left and right actions as discussed in Sect. 2.5.1.

b. Symmetric group S_4:

$S_4 = \{p \mid p$ is a permutation of the four EulerRodrigues parameters $(\alpha_0, \alpha_1, \alpha_2, \sigma_3)\}$, $|S_4| = 24$.
Points in $\mathbf{S}^3$ are mapped to distinct points in $\mathbf{S}^3$; hence, one can take $Z \in SU(2)$, and define the group action directly from $U(\alpha_0, \boldsymbol{\alpha})$ in Eq. (2.20). It is simpler, however, to define the action of the group directly on the representation functions (2.22). Not all elements of this group define a symmetry in the sense defined by Eq. (2.29) (see below).

c. Abelian group T:

$$T = \{(\tau_0, \tau_1, \tau_2, \tau_3) \mid \text{ each } \tau_\mu = \pm 1\}, \quad |T| = 16.$$

Group multiplication is component-wise multiplication and the identity is $(1,1,1,1)$. The action of an element of T is defined directly on the Euler-Rodrigues parameters by component-wise multiplication, thus mapping points in $\mathbf{S}^3$ to points in $\mathbf{S}^3$; hence, one can take $Z \in SU(2)$. This group is isomorphic to the direct product group $S_2 \times S_2 \times S_2 \times S_2$, $S_2 =$ symmetric group on two distinct objects.

d. Group G:

$$G = \langle \mathcal{R}, \mathcal{C}, \mathcal{T}, \mathcal{K} \rangle, \quad |G| = 32,$$

where $\mathcal{R}, \mathcal{C}, \mathcal{T}, \mathcal{K}$ denote the operations of row interchange, column interchange, transposition, and conjugation (see below) of an arbitrary matrix.

$$Z = \begin{pmatrix} a & b \\ c & d \end{pmatrix}$$

The notation $\langle \ \rangle$ designates that the enclosed elements generate G.

It is impossible to give here all the interrelationships among the groups defined in (a)–(d). Instead, some relations are listed as obtained directly from either $D^j_{m',m}(Z)$ defined by Eq. (2.24) or $D^j_{m',m}(\alpha_0, \boldsymbol{\alpha})$ defined by Eq. (2.22). The actions of the groups T and G defined in (c) and (d) are fully given.
Abelian group T of order 16:

Generators:

$$T = \langle t_0, t_1, t_2, t_3 \rangle, \quad t_0 = (-1, 1, 1, 1), \quad t_1 = (1, -1, 1, 1)$$
$$t_2 = (1, 1, -1, 1), \quad t_3 = (1, 1, 1, -1).$$

Group action:

$$t \cdot a = (\tau_0 \alpha_0, \tau_1 \alpha_1, \tau_2 \alpha_2, \tau_3 \alpha_3),$$
$$\text{each } t = (\tau_0, \tau_1, \tau_2, \tau_3) \in T,$$
$$\text{each } a = (\alpha_0, \alpha_1, \alpha_2, \alpha_3) \in \mathbf{S}^3,$$

$$(tF)(a) = F(t \cdot a),$$

$$t_0 D^j_{m',m} = (-1)^{m'-m} D^j_{-m,-m'},$$

$$t_1 D^j_{m',m} = (-1)^{m'-m} D^j_{m,m'},$$

$$t_2 D^j_{m',m} = D^j_{m,m'},$$

$$t_3 D^j_{m',m} = D^j_{-m,-m'}.$$

Group G of order 32:
   Generators:

$$G = \langle \mathcal{R}, \mathcal{C}, \mathcal{T}, \mathcal{K} \rangle,$$

Generator actions: Generator actions:

$$(\mathcal{R}F)\begin{pmatrix} a & b \\ c & d \end{pmatrix} = F\begin{pmatrix} c & d \\ a & b \end{pmatrix}, \qquad \text{row interchange}$$

$$(\mathcal{C}F)\begin{pmatrix} a & b \\ c & d \end{pmatrix} = F\begin{pmatrix} b & a \\ d & c \end{pmatrix}, \qquad \text{column interchange}$$

$$(\mathcal{T}F)\begin{pmatrix} a & b \\ c & d \end{pmatrix} = F\begin{pmatrix} a & c \\ b & d \end{pmatrix}, \qquad \text{transposition}$$

$$(\mathcal{K}F)\begin{pmatrix} a & b \\ c & d \end{pmatrix} = F\begin{pmatrix} d & -c \\ -b & a \end{pmatrix}, \qquad \text{conjugation}$$

Subgroup H :

$$H = \langle \mathcal{R}, \mathcal{C}, \mathcal{T} \rangle$$
$$= \{1, \mathcal{R}, \mathcal{C}, \mathcal{T}, \mathcal{R}\mathcal{C} = \mathcal{C}\mathcal{R}, \mathcal{T}\mathcal{R} = \mathcal{C}\mathcal{T}, \mathcal{T}\mathcal{C} = \mathcal{R}\mathcal{T}, \mathcal{R}\mathcal{C}\mathcal{T}\}$$

with relations in H given by

$$\mathcal{R}^2 = \mathcal{C}^2 = \mathcal{T}^2 = 1,$$

$$\mathcal{T}\mathcal{R}\mathcal{C} = \mathcal{T}\mathcal{C}\mathcal{R} = \mathcal{R}\mathcal{C}\mathcal{T} = \mathcal{C}\mathcal{R}\mathcal{T},$$

$$\mathcal{R}\mathcal{T}\mathcal{C} = \mathcal{C}\mathcal{T}\mathcal{R} = \mathcal{T}.$$

Adjoining the idempotent element $\mathcal{K}$ to H gives the group G of order 32:

$$G = \{H, H\mathcal{K}, H\mathcal{K}\mathcal{R}, H\mathcal{K}\mathcal{R}\mathcal{K}\}.$$

Symmetry relations:

$$\mathcal{R}D^j_{m',m} = D^j_{-m',m},$$

$$\mathcal{C}D^j_{m',m} = D^j_{m',-m}, \qquad (2.32)$$

$$\mathcal{T}D^j_{m',m} = D^j_{m,m'},$$

$$\mathcal{K}D^j_{m',m} = (-1)^{m'-m} D^j_{-m',-m}.$$

These function relations are valid for $D^j_{m',m}$ defined over the arbitrary matrix Z defined by Eq. (2.24). They are also true for $Z = U \in SU(2)$, but now the operations $\mathcal{R}$ and $\mathcal{C}$ change the sign of the determinant of the matrix Z so that the transformed matrix no longer belongs to $SU(2)$. It does, however, belong to $U(2)$, the group of all 2×2 unitary matrices. The *special* irreducible representation functions of $U(2)$ defined by Eq. (2.24),

$$D^j_{m',m}(U), \quad U \in U(2),$$

possess each of the 32 symmetries corresponding to the operations in the group G. [There exist other irreducible representations of $U(2)$, involving $\det U$.] The operation $\mathcal{K}$ is closely related to complex conjugation, since for each $U \in U(2), U = (u_{ij})$, one can write

$$U^* = (\det U)^{-1} \begin{pmatrix} u_{22} & -u_{21} \\ -u_{12} & u_{11} \end{pmatrix},$$

(2.33)

$$(\mathcal{K}D^j_{m',m})(U) = (\det U)^{2j} D^j_{m',m}(U^*)$$
$$= (\det U)^{2j} D^{j*}_{m',m}(U)$$
$$= (-1)^{m'-m} D^j_{-m',-m}(U).$$

Relations (2.32) and (2.34) are valid in an arbitrary parametrization of $U \in U(2)$. In terms of the parametrization

$$U(\chi, \alpha_0, \boldsymbol{\alpha}) = e^{i\chi/2} U(\alpha_0, \boldsymbol{\alpha}), \quad 0 \le \chi \le 2\pi,$$

where $U(\alpha_0, \boldsymbol{\alpha}) \in SU(2)$ is the Euler-Rodrigues parametrization, the actions of $\mathcal{R}, \mathcal{C}, \mathcal{T}$, and $\mathcal{K}$ correspond to the following transformations in parameter space:

$$\mathcal{R}: \quad \chi \to \chi + \pi, (\alpha_0, \alpha_1, \alpha_2, \alpha_3) \to (-\alpha_1, \alpha_0, -\alpha_3, \alpha_2),$$
$$\mathcal{C}: \quad \chi \to \chi + \pi, (\alpha_0, \alpha_1, \alpha_2, \alpha_3) \to (-\alpha_1, \alpha_0, \alpha_3, -\alpha_2),$$
$$\mathcal{T}: \quad \chi \to \chi, (\alpha_0, \alpha_1, \alpha_2, \alpha_3) \to (\alpha_0, \alpha_1, -\alpha_2, \alpha_3),$$
$$\mathcal{C}: \quad \chi \to \chi, (\alpha_0, \alpha_1, \alpha_2, \alpha_3) \to (\alpha_0, -\alpha_1, \alpha_2, -\alpha_3).$$

The new angle $\chi' = \chi + \pi$ is to be identified with the corresponding point on the unit circle so that these mappings are always in the parameter space, which is the sphere $\mathbf{S}^3$ together with the unit circle for χ. Observe that the following identities hold for functions over $SU(2)$; hence over $U(2)$:

$$\mathcal{C} = T_{t_1} T_{t_3}, \quad \mathcal{T} = T_{t_2}.$$

Abelian subgroup of S_4:
   Generators:
   $K = \langle (0,3), (1,2) \rangle$, where $(0,3)$ and $(1,2)$ denote transpositions in S_4, $|K| = 4$.

Group action in parameter space:

$$(0,3)(\alpha_0, \alpha_1, \alpha_2, \alpha_3) = (\alpha_3, \alpha_1, \alpha_2, \alpha_0),$$
$$(1,2)(\alpha_0, \alpha_1, \alpha_2, \alpha_3) = (\alpha_0, \alpha_2, \alpha_1, \alpha_3).$$

Symmetry relations:

$$(0,3)D^j_{m',m} = (-i)^{m'+m} D^j_{-m,-m'},$$
$$(1,2)D^j_{m',m} = (-i)^{m'-m} D^j_{m,m'}.$$

Diagonal subgroup Σ of the direct product group $P \times P$ (P=Pauli group):

Group elements:

$$\Sigma = \{(\sigma,\sigma)|\sigma \in P\}, \quad |\Sigma| = 16.$$

Group action:

$$(\sigma,\sigma) : U \to \sigma U \sigma^T, \quad \text{each } \sigma \in P,$$
$$[(\sigma,\sigma)F](U) = F(\sigma^T U \sigma).$$

Example: $\sigma = i\sigma_2$:

$$(\sigma,\sigma) : (\alpha_0, \alpha_1, \alpha_2, \alpha_3) \to (\alpha_0, -\alpha_1, \alpha_2, -\alpha_3),$$
$$[(\sigma,\sigma)F](\alpha_0, \alpha_1, \alpha_2, \alpha_3) = F(\alpha_0, -\alpha_1, \alpha_2, -\alpha_3),$$
$$(\sigma,\sigma) = t_1 t_2 \text{ on functions over } U(2).$$

The relations presented above barely touch on the interrelations among the finite groups introduced in (a)–(d). Symmetry relations (2.32) and (2.34), however, give the symmetries of the $d_{m'm}^j(\beta)$ given in Sect. 2.4.3 in the Euler angle parametrization. In general, it is quite tedious to present the above symmetries in terms of Euler angles, with χ adjoined when necessary, because *the Euler angles are not uniquely determined by the points of* $\mathbf{S}^3$.

2.5 GROUP AND LIE ALGEBRA ACTIONS

The concept of a group acting on a set is fundamental to applications of group theory to physical problems. Because of the unity that this notion brings to angular momentum theory, it is well worth a brief review in a setting in which a matrix group acts on the set of complex matrices. Thus, let $G \subseteq GL(n, \mathbf{C})$ and $H \subseteq GL(m, \mathbf{C})$ denote arbitrary subgroups, respectively, of the general linear groups of $n \times n$ and $m \times m$ nonsingular complex matrices, and let $M(n, m)$ denote the set of $n \times m$ complex matrices. A matrix $Z \in M(n, m)$ has row and column entries $(z_i^\alpha), i = 1, 2, \ldots, n$; $\alpha = 1, 2, \ldots, m$.

2.5.1 Matrix Group Actions

Left and right translations of $Z \in M(n, m)$:

$$L_g Z = gZ, \quad \text{each } g \in G, \text{ each } Z \in M(n, m),$$
$$R_h Z = Z h^T, \quad \text{each } h \in H, \text{ each } Z \in M(n, m).$$

(T denotes matrix transposition.)
Left and right translations commute:

$$Z' = L_g(R_h Z) = R_h(L_g Z), \quad \text{each } g \in G, h \in H,$$
$$Z \in M(n, m).$$

Equivalent form as a transformation on $\mathbf{z} \in \mathbf{C}^{nm}$:

$$\mathbf{z}' = (g \times h)\mathbf{z},$$

where $\times$ denotes the direct product of g and h; the column matrix $\mathbf{z}$ (resp., $\mathbf{z}'$) is obtained from the columns of Z (resp., Z'), $\mathbf{z}^\alpha, \alpha = 1, 2, \ldots, m$, of the $n \times m$ matrix Z as successive entries in a single column vector $\mathbf{z} \in \mathbf{C}^{nm}$. Left and right translations in function space:

$$(\mathcal{L}_g f)(Z) = f(g^T Z), \quad \text{each } g \in G,$$
$$(\mathcal{R}_h f)(Z) = f(Zh), \quad \text{each } h \in H,$$

where $f(Z) = f(z_i^\alpha)$, and the commuting property holds for all well-defined functions f:

$$\mathcal{L}_g(\mathcal{R}_h f) = \mathcal{R}_h(\mathcal{L}_g f).$$

2.5.2 Lie Algebra Actions

Lie algebra of left and right translations:

$$(D_X f)(Z) = i\frac{d}{dt} f(e^{-itX^T} Z)|_{t=0},$$
$$(D^Y f)(Z) = i\frac{d}{dt} f(Ze^{-itY})|_{t=0};$$

$$
\begin{aligned}
D_X &= \text{tr}(Z^T X \partial/\partial Z), \quad \text{each } X \in L(G),\\
D^Y &= \text{tr}(Y^T Z^T \partial/\partial Z), \quad \text{each } Y \in L(H),\\
L(G) &= \text{Lie algebra of } G,\\
L(H) &= \text{Lie algebra of } H.
\end{aligned}
$$

Linear derivations:

$$D_{\alpha X + \beta X'} = \alpha D_X + \beta D_{X'}, \quad \alpha, \beta \in \mathbf{C},$$
$$[D_X, D_{X'}] = D_{[X,X']},$$

D^Y obeys these same rules.
Commuting property of left and right derivations:

$$[D_X, D^Y] = 0, \quad X \in L(G), \quad Y \in L(H).$$

Basis set:

$$
\begin{aligned}
D_X &= \sum_{i,j=1}^n x_{ij} D_{ij}, \quad X = (x_{ij}),\\
D^Y &= \sum_{\alpha,\beta=1}^m y^{\alpha\beta} D^{\alpha\beta}, \quad Y = (y^{\alpha\beta}),\\
D_{ij} &= \sum_{\alpha=1}^m z_i^\alpha \frac{\partial}{\partial z_j^\alpha},\\
D^{\alpha\beta} &= \sum_{i=1}^n z_i^\alpha \frac{\partial}{\partial z_i^\beta}.
\end{aligned}
$$

Commutation rules:

$$[D_{ij}, D_{kl}] = \delta_{jk} D_{il} - \delta_{il} D_{kj},$$
$$[D^{\alpha\beta}, D^{\gamma\epsilon}] = \delta^{\beta\gamma} D^{\alpha\epsilon} - \delta^{\alpha\epsilon} D^{\gamma\beta},$$
$$[D_{ij}, D^{\alpha\beta}] = 0,$$

where $i, j, k, l = 1, 2, \ldots, n$ and $\alpha, \beta, \gamma, \epsilon = 1, 2, \ldots, m$. The operator sets $\{D_{ij}\}$ and $\{D^{\alpha\beta}\}$ are realizations of the Weyl generators of $GL(n, \mathbf{C})$ and $GL(m, \mathbf{C})$, respectively.

2.5.3 Hilbert Spaces

Space of polynomials with inner product:

$$(P, P') = P^*(\partial/\partial Z)P'(Z)|_{Z=0}.$$

Bargmann space of entire functions with inner product:

$$\langle F, F'\rangle = \int F^*(Z)F'(Z)\,d\mu(Z),$$

$$d\mu(Z) = \pi^{-nm}\exp\left(-\sum_{i,\alpha} z_i^{\alpha*}z_i^\alpha\right)\prod_{i,\alpha} dx_i^\alpha dy_i^\alpha,$$

$$z_i^\alpha = x_i^\alpha + iy_i^\alpha, i=1,2,\dots,n; \quad \alpha = 1,2,\dots,m.$$

Numerical equality of inner products:

$$(P, P') = \langle P, P'\rangle.$$

2.5.4 Relation to Angular Momentum Theory

Spinorial Realization of Sects. 2.5.2 and 2.5.3:

$G = SU(2)$, $H = 1$, $Z \in M(2,1)$, $\mathbf{z} = \mathrm{col}(z_1, z_2)$,

$X = $ set of 2×2 traceless, Hermitian matrices,

$(\mathcal{R}_U f)(\mathbf{z}) = f(U^T\mathbf{z})$,

$D_{\sigma_i/2} = (\mathbf{z}^T\sigma_i\partial/\partial\mathbf{z})/2$,

$$J_\pm = D_{(\sigma_1\pm\sigma_2)/2} = (D_{\sigma_1}\pm iD_{\sigma_2})/2, \quad J_3 = D_{\sigma_3/2},$$

$$J_+ = z_1\partial/\partial z_2, \quad J_- = z_2\partial/\partial z_1,$$
$$J_3 = (1/2)(z_1\partial/\partial z_1 - z_2\partial/\partial z_2),$$
$$(P, P') = P^*(\partial/\partial z_1, \partial/\partial z_2)P(z_1, z_2)|_{z_1=z_2=0}.$$

Orthonormal basis:

$$P_{jm}(z_1, z_2) = z_1^{j+m}z_2^{j-m}/[(j+m)!(j-m)!]^{1/2},$$
$$j = 0, 1/2, 1, 3/2, \dots; \quad m = j, j-1, \dots, -j.$$

Standard action:

$$\mathbf{J}^2 P_{jm}(\mathbf{z}) = j(j+1)P_{jm}(\mathbf{z}),$$
$$J_3 P_{jm}(\mathbf{z}) = mP_{jm}(\mathbf{z}),$$
$$J_\pm P_{jm}(\mathbf{z}) = [(j\mp m)(j\pm m+1)]^{\frac12}P_{j,m\pm1}(\mathbf{z}).$$

Group transformation:

$$(\mathcal{R}_U P_{jm})(\mathbf{z}) = \sum_{m'} D^j_{m'm}(U)P_{jm'}(\mathbf{z}),$$

where the representation functions are given by Eq. (2.17).

The 2-Spinorial Realization of Sects. 2.5.2 and 2.5.3:

$$G = H = SU(2), \quad Z \in M(2,2),$$
$$Z = [\mathbf{z}^1, \mathbf{z}^2], \quad \mathbf{z}^\alpha = \mathrm{col}(z_1^\alpha, z_2^\alpha),$$

$X = Y = $ set of 2×2 traceless, Hermitian matrices,

$$(\mathcal{R}_U f)(Z) = f(U^T Z), \quad (\mathcal{L}_V f)(Z) = f(ZV),$$
$$U, V \in SU(2),$$
$$D_{\sigma_i/2} = \mathrm{tr}(Z^T\sigma_i\partial/\partial Z)/2,$$
$$D^{\sigma_i/2} = \mathrm{tr}(\sigma_i Z^T\partial/\partial Z)/2.$$

$$M_\pm = D_{(\sigma_1\pm i\sigma_2)/2} = (D_{\sigma_1}\pm iD_{\sigma_2})/2, M_3 = D_{\sigma_3/2},$$
$$K_\pm = D^{(\sigma_1\pm i\sigma_2)/2} = (D^{\sigma_1}\pm iD^{\sigma_2})/2, K_3 = D^{\sigma_3/2},$$

$$M_+ = \sum_{\alpha=1}^2 z_1^\alpha\partial/\partial z_2^\alpha, \quad M_- = \sum_{\alpha=1}^2 z_2^\alpha\partial/\partial z_1^\alpha,$$

$$M_3 = \frac12\sum_{\alpha=1}^2(z_1^\alpha\partial/\partial z_1^\alpha - z_2^\alpha\partial/\partial z_2^\alpha),$$

$$K_+ = \sum_{i=1}^2 z_i^1\partial/\partial z_i^2, \quad K_- = \sum_{i=1}^2 z_i^2\partial/\partial z_i^1,$$

$$K_3 = \frac12\sum_{i=1}^2(z_i^1\partial/\partial z_i^1 - z_i^2\partial/\partial z_i^2).$$

Mutual commutivity of Lie algebras:

$$[M_i, K_j] = 0, \quad i, j = 1, 2, 3.$$

Inner product:

$$(P, P') = P^*(Z)P'(\partial/\partial Z)|_{Z=0},$$

Orthogonal basis [see Eq. (2.24)]:

$$D^j_{mm'}(Z), j = 0, 1/2, 1, 3/2, \dots,$$
$$m = j, j-1, \dots, -j; m' = j, j-1, \dots, -j;$$
$$(D^j_{mm'}, D^{j'}_{\mu\mu'}) = (2j)!\delta_{jj'}\delta_{m\mu}\delta_{m'\mu'}.$$

Equality of Casimir operators:

$$\mathbf{M}^2 = \mathbf{K}^2 = M_1^2 + M_2^2 + M_3^2 = K_1^2 + K_2^2 + K_3^2.$$

Standard actions:

$$\mathbf{M}^2 D^j_{mm'}(Z) = \mathbf{K}^2 D^j_{mm'}(Z) = j(j+1)D^j_{mm'}(Z),$$
$$M_3 D^j_{mm'}(Z) = mD^j_{mm'}(Z),$$
$$K_3 D^j_{mm'}(Z) = m'D^j_{mm'}(Z),$$
$$M_\pm D^j_{mm'}(Z) = [(j\mp m)(j\pm m+1)]^{1/2}D^j_{m\pm1,m'}(Z),$$
$$K_\pm D^j_{mm'}(Z) = [(j\mp m')(j\pm m'+1)]^{1/2}D^j_{m,m'\pm1}(Z).$$

Special values:

$$D^j \begin{pmatrix} 1 & 0 \\ 0 & 1 \end{pmatrix} = I_{2j+1} = \text{ unit matrix},$$

$$D^j_{mm'} \begin{pmatrix} z_1 & 0 \\ z_2 & 0 \end{pmatrix} = \delta_{jm'} P_{jm}(z_1, z_2),$$

$$D^j_{mm'} \begin{pmatrix} 0 & z_1 \\ 0 & z_2 \end{pmatrix} = \delta_{jm} P_{jm'}(z_1, z_2),$$

$$D^j_{mm'} \begin{pmatrix} z_1 & 0 \\ 0 & z_2 \end{pmatrix} = \delta_{mm'} z_1^{j+m} z_2^{j-m},$$

$$D^j_{jj}(Z) = (z_1^1)^{2j}.$$

Symmetry relation:

$$(D^j(Z))^T = D^j(Z^T).$$

Generation from highest weight:

$$D^j_{mm'}(Z) = \left[\frac{(j+m)!}{(2j)!(j-m)!} \times \frac{(j+m')!}{(2j)!(j-m')!} \right]^{1/2}$$
$$\times M_-^{j-m} K_-^{j-m'} D^j_{jj}(Z).$$

Generating functions:

$$(\mathbf{x}^T Z \mathbf{y})^{2j} / (2j)! = \sum_{m,m'} P_{jm}(\mathbf{x}) D^j_{mm'}(Z) P_{jm'}(\mathbf{y}),$$

$$\exp(t\mathbf{x}^T Z \mathbf{y}) = \sum_j t^{2j} \sum_{m,m'} P_{jm}(\mathbf{x}) D^j_{mm'}(Z) P_{jm'}(\mathbf{y}),$$

$$\mathbf{x} = \text{col}(x_1, x_2), \quad \mathbf{y} = \text{col}(y_1, y_2),$$
$$Z = (z_i^\alpha), \ i, \alpha = 1, 2; \text{ all indeterminates}.$$

2.6 DIFFERENTIAL OPERATOR REALIZATIONS OF ANGULAR MOMENTUM

Differential operators realizing the standard commutation relations (2.7) and (2.8) can be obtained from the 2-spinorial realizations given in Sect. 2.5.4 by specializing the matrix Z to the appropriate unitary unimodular matrix $U \in SU(2)$ and using the chain rule of elementary calculus. Similarly, one obtains the explicit functions $D^j_{mm'}$ simply by substituting for Z the parametrized U in Eq. (2.24). This procedure is used in this section to obtain all the realizations listed. The notations $\mathbf{M} = (M_1, M_2, M_3)$ and $\mathbf{K} = (K_1, K_2, K_3)$ and the associated $M_\pm$ and $K_\pm$ refer to the differential operators given by the 2-spinorial realization now transformed to the parameters in question.

Euler angles with $Z = U(\alpha\beta\gamma)$ (Sect. 2.4.1):

$$M_3 = i\partial/\partial\alpha, \quad K_3 = i\partial/\partial\gamma,$$

$$\tfrac{1}{2}(e^{i\alpha} M_+ - e^{-i\alpha} M_-) = \tfrac{1}{2}(e^{-i\gamma} K_- - e^{i\gamma} K_+) = \frac{\partial}{\partial\beta},$$

$$(e^{i\alpha} M_+ + e^{-i\alpha} M_-)/2 = -(\cot\beta) M_3 + (\sin\beta)^{-1} K_3,$$
$$(e^{-i\gamma} K_- + e^{i\gamma} K_+)/2 = (\cot\beta) K_3 - (\sin\beta)^{-1} M_3,$$

$$M_+ = e^{-i\alpha}[\partial/\partial\beta - (\cot\beta) M_3 + (\sin\beta)^{-1} K_3],$$
$$M_- = e^{i\alpha}[-\partial/\partial\beta - (\cot\beta) M_3 + (\sin\beta)^{-1} K_3],$$
$$K_+ = e^{-i\gamma}[-\partial/\partial\beta + (\cot\beta) K_3 - (\sin\beta)^{-1} M_3],$$
$$K_- = e^{i\gamma}[\partial/\partial\beta + (\cot\beta) K_3 - (\sin\beta)^{-1} M_3].$$

Euler angles with $U = U^*(\alpha\beta\gamma)$ [replace i by $-i$ in the above relations]:

$$M_3 = -i\partial/\partial\alpha, \quad K_3 = -i\partial/\partial\gamma, \tag{2.34}$$
$$M_\pm = e^{\pm i\alpha}[\pm\partial/\partial\beta - (\cot\beta) M_3 + (\sin\beta)^{-1} K_3],$$
$$K_\pm = e^{\pm i\gamma}[\mp\partial/\partial\beta + (\cot\beta) K_3 - (\sin\beta)^{-1} M_3].$$

Since $D^j(U^*) = (D^j(U))^*$, which is denoted $D^{j*}(U)$, these operators have the standard action on the complex conjugate functions $D^{j*}_{mm'}(U)$.

Quaternionic variables:

$$Z = \begin{pmatrix} x_0 - ix_3 & -ix_1 - x_2 \\ -ix_1 + x_2 & x_0 + ix_3 \end{pmatrix}, \quad (x_0, \mathbf{x}) \in \mathbf{R}^4.$$

$$(x_0', \mathbf{x}')(x_0, \mathbf{x}) = (x_0' x_0 - \mathbf{x}' \cdot \mathbf{x}, x_0' \mathbf{x} + x_0 \mathbf{x}' + \mathbf{x}' \times \mathbf{x}),$$

$$\partial/\partial Z = \frac{1}{2} \begin{pmatrix} \partial/\partial x_0 + i\partial/\partial x_3 & i\partial/\partial x_1 - \partial/\partial x_2 \\ i\partial/\partial x_1 + \partial/\partial x_2 & \partial/\partial x_0 - i\partial/\partial x_3 \end{pmatrix},$$

$$L_{jk} = -i(x_j \partial/\partial x_k - x_k \partial/\partial x_j), \quad j,k = 0,1,2,3,$$
$$\mathbf{L} = -i\mathbf{x} \times \nabla, \ L_1 = L_{23}, \ L_2 = L_{31}, \ L_3 = L_{12},$$
$$\mathbf{A} = (A_1, A_2, A_3) = (L_{01}, L_{02}, L_{03}),$$
$$K_1 = (K_+ + K_-)/2 = (L_1 - A_1)/2,$$
$$K_2 = -i(K_+ - K_-)/2 = (L_2 - A_2)/2,$$
$$K_3 = (L_3 - A_3)/2,$$
$$M_1 = (M_+ + M_-)/2 = -(L_1 + A_1)/2,$$
$$M_2 = -i(M_+ - M_-)/2 = (L_2 + A_2)/2,$$
$$M_3 = -(L_3 + A_3)/2,$$

Commutation rules:

$$[M_j, K_k] = 0, \quad j,k = 1,2,3,$$
$$\mathbf{M} \times \mathbf{M} = i\mathbf{M}, \quad \mathbf{K} \times \mathbf{K} = i\mathbf{K},$$
$$\mathbf{L} \times \mathbf{L} = i\mathbf{L}, \quad \mathbf{A} \times \mathbf{A} = i\mathbf{L},$$
$$[L_j, A_k] = ie_{jkl} A_l,$$

where $e_{jkl} = 1$ for j,k,l an even permutation of $1, 2, 3$; $e_{jkl} = -1$ for an odd permutation of $1, 2, 3$; $e_{jkl} = 0$, otherwise.

The $\mathbf{M} = (M_1, M_2, M_3)$ and $\mathbf{K} = (K_1, K_2, K_3)$ operators have the standard action given in Sect. 2.5.4 on the functions $D^j_{mm'}(x_0, \mathbf{x})$ defined by Eq. (2.22) (replace α_0 by x_0 and $\boldsymbol{\alpha}$ by $\mathbf{x}$).

Additional relations:

$$K_0 = (1/2) \sum_{\mu=0}^{3} x_\mu \frac{\partial}{\partial x_\mu} \,,$$

$$\mathbf{K}^2 = \mathbf{M}^2 = -\frac{1}{4} R^2 \nabla_4^2 + K_0^2 + K_0 \,,$$

$$R^2 = x_0^2 + \mathbf{x} \cdot \mathbf{x} \,,$$

$$\nabla_4^2 = \frac{\partial^2}{\partial x_0^2} + \nabla^2 = \sum_\mu \frac{\partial^2}{\partial x_\mu^2} \,,$$

$$K_0 D_{m,m'}^j(x_0, \mathbf{x}) = j D_{mm'}^j(x_0, \mathbf{x}) \,,$$

$$\nabla_4^2 D_{m'm}^j(x_0, \mathbf{x}) = 0 \,,$$

$$(M_1, -M_2, M_3) = \left(\sum_i R_{i1} K_i, \sum_i R_{i2} K_i, \sum_i R_{i3} K_i \right) \,,$$

$$R_{ij} = \frac{(x_0^2 - \mathbf{x} \cdot \mathbf{x}) \delta_{ij} - 2 e_{ijk} x_0 x_k + 2 x_i x_j}{x_0^2 + \mathbf{x} \cdot \mathbf{x}} \,,$$

$$\text{each } (x_0, \mathbf{x}) \in \mathbf{R}^4 \,.$$

The relation $R_{ij} = R_{ij}(x_0, \mathbf{x})$ is a mapping of all points of four-space (except the origin) onto the group of proper, orthogonal matrices; for $x_0^2 + \mathbf{x} \cdot \mathbf{x} = 1$, it is just the Euler-Rodrigues parametrization, Eq. (2.21).

The operators $\mathbf{R} = (-M_1, M_2, -M_3)$ and $\mathbf{K} = (K_1, K_2, K_3)$ have the standard action on $(-1)^{j+m} \times D_{-m,m'}^j(x_0, \mathbf{x})$, so that the orbital angular momentum in $\mathbf{R}^3$ is given by the addition $\mathbf{L} = \mathbf{R} + \mathbf{K}$. Thus, one finds:

$$\sum_{m,m'} C_{mm'M}^{jjL} (-1)^{j+m} D_{-m,m'}^j(x_0, \mathbf{x})$$

$$= A_{2j,L} R^{2j-L} \mathcal{Y}_{LM}(\mathbf{x}) C_{2j-L}^{(L+1)}(x_0/R) \,,$$

$$A_{2j,L} = (2i)^L (-1)^{2j} L! \left[\frac{4\pi (2j-L)!}{(2j+L+1)!} \right]^{1/2} \,.$$

2.7 THE SYMMETRIC ROTOR AND REPRESENTATION FUNCTIONS

The rigid rotor is an important physical object and its quantum description enters into many physical theories. This description is an application of angular momentum theory with subtleties that need to be made explicit. It is customary to describe the classical rotor in terms of a right-handed triad of unit vectors $(\hat{\mathbf{f}}_1, \hat{\mathbf{f}}_2, \hat{\mathbf{f}}_3)$ fixed in the rotor and constituting a principal axes system located at the center of mass. The instantaneous orientation of this *body-fixed frame* relative to a right-handed triad of unit vector $(\hat{\mathbf{e}}_1, \hat{\mathbf{e}}_2, \hat{\mathbf{e}}_3)$ specifying an inertial frame, also located at the center of mass, is then given, say, in terms of Euler angles (one could use for this purpose any parametrization of a proper, orthogonal matrix). For Euler angles, the relationship is

$$\hat{\mathbf{f}}_j = \sum_i R_{ij}(\alpha\beta\gamma) \hat{\mathbf{e}}_i \,. \qquad (2.35)$$

The Hamiltonian for the rigid rotor is then of the form

$$H = A \mathcal{P}_1^2 + B \mathcal{P}_2^2 + C \mathcal{P}_3^2 \,,$$

where $A, B,$ and C are physical constants related to the reciprocals of the principal moments of inertia, and the angular momenta $\mathcal{P}_j$ $(j = 1, 2, 3)$ are the components of the total angular momentum $\mathcal{J}$ referred to the body-fixed frame:

$$\mathcal{P}_j = \hat{\mathbf{f}}_j \cdot \mathcal{J} = \sum_i R_{ij}(\alpha\beta\gamma) \mathcal{J}_i \,,$$

$$\mathcal{J} = \hat{\mathbf{e}}_1 \mathcal{J}_1 + \hat{\mathbf{e}}_2 \mathcal{J}_2 + \hat{\mathbf{e}}_3 \mathcal{J}_3 \,.$$

For the symmetric rotor (taking $A = B$), the Hamiltonian can be written in the form

$$H = a \mathcal{P}^2 + b \mathcal{P}_3^2 \,.$$

It is in the interpretation of this Hamiltonian for quantum mechanics that the subtleties already enter, since the nature of angular momentum components referred to a moving reference system must be treated correctly. Relation (2.35) shows that the body-fixed axes cannot commute with the components of the total angular momentum $\mathcal{J}$ referred to the frame $(\hat{\mathbf{e}}_1, \hat{\mathbf{e}}_2, \hat{\mathbf{e}}_3)$. A position vector $\mathbf{x}$ and the orbital angular momentum $\mathbf{L}$, with components both referred to an inertial frame, satisfy the commutation relations $[L_j, x_k] = i e_{jkl} x_l$, and for a rigid body thought of as a collection of point particles rotating together, the same conditions are to be enforced. Relative to the body-fixed frame, the vector $\mathbf{x}$ is expressed as

$$\sum_k x_k \hat{\mathbf{e}}_k = \sum_h a_h \hat{\mathbf{f}}_h, \quad \text{each } a_h = \text{constant},$$

$$x_k = \sum_h a_h R_{kh}(\alpha\beta\gamma) \,.$$

The direction cosines $R_{kh} = R_{kh}(\alpha\beta\gamma) = \hat{\mathbf{e}}_k \cdot \hat{\mathbf{f}}_h$ and the physical total angular momentum components referred to an inertial frame must satisfy

$$[\mathcal{J}_j, R_{kh}] = i e_{jkl} R_{lh}, \quad \text{each } h = 1, 2, 3 \,,$$

in complete analogy to $[L_j, x_k] = i e_{jkl} x_l$.

The description of the angular momentum associated with a symmetric rigid rotor and the angular momentum states is summarized as follows [compare Eq. (2.35)]: Physical total angular momentum $\mathcal{J}$ with components referred to $(\hat{\mathbf{e}}_1, \hat{\mathbf{e}}_2, \hat{\mathbf{e}}_3)$:

$$\mathcal{J}_1 = i \cos\alpha \cot\beta \frac{\partial}{\partial\alpha} + i \sin\alpha \frac{\partial}{\partial\beta} - i \frac{\cos\alpha}{\sin\beta} \frac{\partial}{\partial\gamma} \,,$$

$$\mathcal{J}_2 = i \sin\alpha \cot\beta \frac{\partial}{\partial\alpha} - i \cos\alpha \frac{\partial}{\partial\beta} - i \frac{\sin\alpha}{\sin\beta} \frac{\partial}{\partial\gamma} \,,$$

$$\mathcal{J}_3 = -i \frac{\partial}{\partial\alpha} \,.$$

Physical angular momentum $\boldsymbol{\mathcal{J}}$ with components referred to $(\hat{\mathbf{f}}_1, \hat{\mathbf{f}}_2, \hat{\mathbf{f}}_3)$:

$$\mathcal{P}_1 = -i \cos\gamma \cot\beta \frac{\partial}{\partial\gamma} - i\sin\gamma \frac{\partial}{\partial\beta} + i\frac{\cos\gamma}{\sin\beta}\frac{\partial}{\partial\alpha},$$

$$\mathcal{P}_2 = i \sin\gamma \cot\beta \frac{\partial}{\partial\gamma} - i\cos\gamma \frac{\partial}{\partial\beta} - i\frac{\sin\gamma}{\sin\beta}\frac{\partial}{\partial\alpha},$$

$$\mathcal{P}_3 = -i\frac{\partial}{\partial\gamma}.$$

Standard commutation of the $\mathcal{J}_i$:

$$[\mathcal{J}_i, \mathcal{J}_j] = i\mathcal{J}_k, \quad i,j,k \text{ cyclic}.$$

$\mathcal{J}_i$ can stand to either side:

$$\mathcal{P}_j = \sum_i R_{ij}(\alpha\beta\gamma)\mathcal{J}_i, \quad \mathcal{J}_i = \sum_i \mathcal{J}_i R_{ij}(\alpha\beta\gamma).$$

The famous Van Vleck factor of $-i$ in the commutation of the $\mathcal{P}_i$:

$$[\mathcal{P}_i, \mathcal{P}_j] = -i\mathcal{P}_k, \quad i,j,k \text{ cyclic}.$$

Mutual commutivity of the $\mathcal{J}_j$ and $\mathcal{P}_i$:

$$[\mathcal{P}_i, \mathcal{J}_j] = 0, \quad i,j = 1,2,3.$$

Same invariant (squared) total angular momentum:

$$\mathcal{P}_1^2 + \mathcal{P}_2^2 + \mathcal{P}_3^2 = \mathcal{J}_1^2 + \mathcal{J}_2^2 + \mathcal{J}_3^2 = \boldsymbol{\mathcal{J}}^2$$

$$= -\csc^2\beta \left(\frac{\partial^2}{\partial\alpha^2} + \frac{\partial^2}{\partial\gamma^2} - 2\cos\beta \frac{\partial^2}{\partial\alpha\partial\gamma} \right)$$

$$- \frac{\partial^2}{\partial\beta^2} - \cot\beta \frac{\partial}{\partial\beta}.$$

Standard actions:

$$\boldsymbol{\mathcal{J}}^2 D^{j*}_{mm'}(\alpha\beta\gamma) = j(j+1)D^{j*}_{mm'}(\alpha\beta\gamma),$$

$$\mathcal{J}_3 D^{j*}_{mm'}(\alpha\beta\gamma) = m D^{j*}_{mm'}(\alpha\beta\gamma),$$

$$\mathcal{P}_3 D^{j*}_{mm'}(\alpha\beta\gamma) = m' D^{j*}_{mm'}(\alpha\beta\gamma);$$

$$\mathcal{J}_\pm D^{j*}_{mm'}(\alpha\beta\gamma) = [(j \mp m)(j \pm m + 1)]^{1/2}$$
$$\times D^{j*}_{m\pm1,m'}(\alpha\beta\gamma),$$

$$(\mathcal{P}_1 - i\mathcal{P}_2)D^{j*}_{mm'}(\alpha\beta\gamma) = [(j - m')(j + m' + 1)]^{1/2}$$
$$\times D^{j*}_{m,m'+1}(\alpha\beta\gamma),$$

$$(\mathcal{P}_1 + i\mathcal{P}_2)D^{j*}_{mm'}(\alpha\beta\gamma) = [(j + m')(j - m' + 1)]^{1/2}$$
$$\times D^{j*}_{m,m'-1}(\alpha\beta\gamma).$$

Normalized wave functions:
Integral or half-integral j ($SU(2)$ solid body):

$$\left\langle \alpha\beta\gamma \left| \begin{matrix} j \\ mm' \end{matrix} \right. \right\rangle = \sqrt{\frac{2j+1}{2\pi^2}} D^{j*}_{mm'}(\alpha\beta\gamma), \qquad (2.36)$$

with inner product

$$\langle F|F'\rangle = \int d\Omega \, \langle \alpha\beta\gamma|F\rangle^* \langle \alpha\beta\gamma|F'\rangle,$$

where $d\Omega$ is defined by Eq. (2.25) and the integration extends over all α, β, γ given by Eq. (2.26).

Integral j (collection of "rigid" point particles):

$$\Psi^j_{mm'}(\alpha\beta\gamma) = \sqrt{\frac{2j+1}{8\pi^2}} D^{j*}_{mm'}(\alpha\beta\gamma), \qquad (2.37)$$

with inner product

$$(\Psi, \Psi') = \int_0^{2\pi} d\alpha \int_0^\pi d\beta \sin\beta \int_0^{2\pi} d\gamma \, F^*(\alpha\beta\gamma)F'(\alpha\beta\gamma).$$

The concept of a solid (inpenetrable) body is conceptually distinct from that of a collection of point particles moving collectively together in translation and rotation.

2.8 WIGNER-CLEBSCH-GORDAN AND 3-*J* COEFFICIENTS

Wigner-Clebsch-Gordan (WCG) coefficients (also called vector coupling coefficients) enter the theory of angular momentum in several ways: (1) as the coefficients in the real, orthogonal matrix that reduces the Kronecker product of two irreducible representations of the quantal rotation group into a direct sum of irreducibles; (2) as the coupling coefficients for constructing basis states of sharp angular momentum in the tensor product space from basis states of sharp angular momentum spanning the two constituent spaces; (3) as purely combinatoric objects in the expansion of a power of a 3×3 determinant; and (4) as coupling coefficients in the algebra of tensor operators. These perspectives are intimately connected, but have a different focus: the first considers the group itself to be primary and views the Lie algebra as the secondary or derived concept; the second considers the Lie algebra and the construction of the vector spaces that carry irreducible representations as primary, and views the representations carried by these spaces as derived quantities; the third is a mathematical construction, at first seeming almost empty of angular momentum concepts, yet the most revealing in showing the symmetry and other properties of the WCG-coefficients; and the fourth is the natural extension of (2) to operators, recognizing that the set of mappings of a vector space into itself is also a vector space. The subject of tensor operator algebra is considered in the next section because of its special importance for physical applications. This section summarizes formulas relating to the first three viewpoints, giving also the explicit mathematical expression of the coefficients in their several forms.

Either viewpoint, (1) or (2), may be taken as an interpretation of the *Clebsch-Gordan series*, which expresses abstractly the reduction of a Kronecker product of matrices (denoted $\times$) into a direct sum (denoted $\oplus$) of matrices:

$$[j_1] \times [j_2] = \sum_{j=|j_1-j_2|}^{j_1+j_2} \oplus [j] \qquad (2.38)$$

$$= [|j_1 - j_2|] \oplus [|j_1 - j_2| + 1] \oplus \cdots \oplus [j_1 + j_2].$$

Given two angular momenta $j_1 \in \{0, \frac{1}{2}, 1, \ldots\}$ and $j_2 \in \{0, \frac{1}{2}, 1, \ldots\}$, the Clebsch-Gordan (CG) series also expresses the rule of *addition of two angular momenta*:

$$j = j_1 + j_2, \; j_1 + j_2 - 1, \; \ldots, \; |j_1 - j_2|.$$

The integers $\epsilon_{j_1 j_2 j}$ defined by

$$\epsilon_{j_1 j_2 j} = \begin{cases} 1, & \text{for } j_1, j_2, j \text{ satisfying the CG-series rule,} \\ 0, & \text{otherwise} \end{cases} \qquad (2.39)$$

are useful in many relations between angular momentum quantities. The notation $(j_1 j_2 j)$ is used to symbolize the CG-series relation between three angular momentum quantum numbers.

The representation function and Lie algebra interpretations of the CG-series (2.38) are, respectively:

$$C(D^{j_1} \times D^{j_2})C^T = \sum_j \oplus \epsilon_{j_1 j_2 j} D^j \,,$$

$$C(J_i^{(j_1)} \times J_i^{(j_2)})C^T = \sum_j \oplus \epsilon_{j_1 j_2 j} J_i^{(j)}, \quad i = 1, 2, 3 \,.$$

The notation $J_i^{(j)}$ denotes the $(2j+1) \times (2j+1)$ matrix with elements

$$J_{m',m}^{(j)} = \langle jm' \mid J_i \mid jm \rangle, \quad m', m = j, j-1, \ldots, -j \,.$$

The elements of the real, orthogonal matrix C of dimension $(2j_1+1)(2j_2+1)$ that effects these reductions are the WCG-coefficients:

$$(C)_{jm;m_1m_2} = C_{m_1 m_2 m}^{j_1 j_2 j} \,.$$

The pairs, (jm) and $(m_1 m_2)$, index rows and columns, respectively, of the matrix C:

$$(jm) : j = j_1 + j_2, \ldots, |j_1 - j_2|, \; m = j, \ldots, -j \,;$$
$$(m_1 m_2) : m_1 = j_1, \ldots, -j_1; \; m_2 = j_2, \ldots, -j_2 \,.$$

Sum rule on projection quantum numbers:

$$C_{m_1 m_2 m}^{j_1 j_2 j} = 0, \quad \text{for } m_1 + m_2 \neq m \,. \qquad (2.40)$$

Clebsch-Gordan series rule on angular momentum quantum numbers:

$$C_{m_1 m_2 m}^{j_1 j_2 j} = 0, \text{ for } \epsilon_{j_1 j_2 j} = 0 \,. \qquad (2.41)$$

In presenting formulas that express relations relating to the conceptual framework described above, it is best to use a notation for a WCG-coefficient giving it as an element of an orthogonal matrix. For the expression of symmetry relations, the 3-j coefficient or 3-j symbol notation is most convenient. The following notations are used here:

WCG-coefficient notation:

$$\mid (j_1 j_2)jm \rangle = \sum_{m_1, m_2} C_{m_1 m_2 m}^{j_1 j_2 j} \mid j_1 m_1 \rangle \otimes \mid j_2 m_2 \rangle \,,$$

$$\mid j_1 m_1; j_2 m_2 \rangle = \mid j_1 m_1 \rangle \otimes \mid j_2 m_2 \rangle \,,$$

$$C_{m_1 m_2 m}^{j_1 j_2 j} = \langle j_1 m_1; j_2 m_2 \mid (j_1 j_2)jm \rangle \,,$$

$$\langle (j_1 j_2)j'm' \mid (j_1 j_2)jm \rangle = \delta_{j'j}\delta_{m'm} \,.$$

The 3-j coefficient notation:

$$\begin{pmatrix} j_1 & j_2 & j \\ m_1 & m_2 & -m \end{pmatrix}$$
$$= (-1)^{j_1 - j_2 + m}(2j+1)^{-1/2} C_{m_1 m_2 m}^{j_1 j_2 j} \,. \qquad (2.42)$$

Orthogonality of WCG-coefficients:
  Orthogonality of rows:

$$\sum_{m_1 m_2} C_{m_1 m_2 m}^{j_1 j_2 j} C_{m_1 m_2 m'}^{j_1 j_2 j'} = \delta_{jj'}\delta_{mm'} \,. \qquad (2.43)$$

Orthogonality of columns (three forms):

$$\sum_{jm} C_{m_1 m_2 m}^{j_1 j_2 j} C_{m_1' m_2' m}^{j_1 j_2 j} = \delta_{m_1 m_1'}\delta_{m_2 m_2'} \,,$$

$$\sum_j C_{m_1 m-m_1, m}^{j_1 j_2 j} C_{m_1', m-m_1', m}^{j_1 j_2 j} = \delta_{m_1 m_1'} \,, \qquad (2.44)$$

$$\sum_j C_{m-m_2, m_2, m}^{j_1 j_2 j} C_{m-m_2', m_2', m}^{j_1 j_2 j} = \delta_{m_2 m_2'} \,.$$

Orthogonality of 3-j coefficients (symbols):

$$\sum_{m_1 m_2} \begin{pmatrix} j_1 & j_2 & j_3 \\ m_1 & m_2 & m_3 \end{pmatrix} \begin{pmatrix} j_1 & j_2 & j_3' \\ m_1 & m_2 & m_3' \end{pmatrix}$$
$$= \delta_{j_3 j_3'}\delta_{m_3 m_3'}/(2j_3+1) \,, \qquad (2.45)$$

$$\sum_{j_3 m_3} (2j_3+1) \begin{pmatrix} j_1 & j_2 & j_3 \\ m_1 & m_2 & m_3 \end{pmatrix} \begin{pmatrix} j_1 & j_2 & j_3 \\ m_1' & m_2' & m_3 \end{pmatrix}$$
$$= \delta_{m_1 m_1'}\delta_{m_2 m_2'} \,. \qquad (2.46)$$

The integers $\epsilon_{j_1 j_2 j}$ $(j_3 = j)$ are sometimes included in the orthogonality relations (2.43) and (2.45) to incorporate the extended definition (2.41) of the WCG-coefficients.

2.8.1 Kronecker Product Reduction

Product form:

$$D_{m_1' m_1}^{j_1}(U) D_{m_2' m_2}^{j_2}(U)$$
$$= \sum_j C_{m_1', m_2', m_1'+m_2'}^{j_1 j_2 j} C_{m_1, m_2, m_1+m_2}^{j_1 j_2 j} D_{m_1'+m_2', m_1+m_2}^{j}(U) \,.$$

Singly coupled form:

$$\sum_{m_1 m_2} C^{j_1 j_2 j}_{m_1 m_2 m} D^{j_1}_{m'_1 m_1}(U) D^{j_2}_{m'_2 m_2}(U)$$
$$= C^{j_1 j_2 j}_{m'_1, m'_2, m'_1 + m'_2} D^{j}_{m'_1 + m'_2, m}(U).$$

Doubly coupled (reduction) form:

$$\sum_{\substack{m'_1 m'_2 \\ m_1 m_2}} C^{j_1 j_2 j}_{m'_1 m'_2 m'} C^{j_1 j_2 j}_{m_1 m_2 m} D^{j_1}_{m'_1 m_1}(U) D^{j_2}_{m'_2 m_2}(U)$$
$$= \delta_{j' j} D^{j}_{m' m}(U).$$

Integral relation:

$$\int d\Omega \, D^{j_1}_{m'_1 m_1}(U) D^{j_2}_{m'_2 m_2}(U) D^{j*}_{m' m_1}(U)$$
$$= \frac{2\pi^2}{2j+1} C^{j_1 j_2 j}_{m'_1 m'_2 m'} C^{j_1 j_2 j}_{m_1 m_2 m},$$

in any parametrization of $U \in SU(2)$ that covers $\mathbf{S}^3$ exactly once.

Gaunt's integral:

$$\int_0^{2\pi} d\alpha \int_0^{\pi} \sin\beta \, d\beta \, Y^*_{lm}(\beta\alpha) Y_{l_1 m_1}(\beta\alpha) Y_{l_2 m_2}(\beta\alpha)$$
$$= \left[\frac{(2l_1+1)(2l_2+1)}{4\pi(2l+1)} \right]^{1/2} C^{l_1 l_2 l}_{000} C^{l_1 l_2 l}_{m_1 m_2 m'}.$$

Integral over three Legendre functions:

$$\int_0^{\pi} \sin\beta \, d\beta \, P_l(\cos\beta) P_{l_1}(\cos\beta) P_{l_2}(\cos\beta) = \frac{2(C^{l_1 l_2 l}_{000})^2}{2l+1}.$$

$C^{l_1 l_2 l}_{000}$

$$= \left[\frac{(2l+1)(l_1+l_2-l)!(l_1-l_2+l)!(-l_1+l_2+l)!}{(l_1+l_2+l+1)!} \right]^{\frac{1}{2}}$$
$$\times \frac{(-1)^{L-1}L!}{(L-l_1)!(L-l_2)!(L-l)!}, \quad L = \tfrac{1}{2}(l_1+l_2+l),$$
$$\text{for } l_1+l_2+l \text{ even,}$$

$C^{l_1 l_2 l}_{000} = 0, \quad \text{for } l_1+l_2+l \text{ odd.}$

2.8.2 Tensor Product Space Construction

Orthonormal basis of H_{j_1}:

$$\{ |\, j_1 m_1 \rangle \mid m_1 = j_1, j_1-1, \ldots, -j_1 \}.$$

Orthonormal basis of H_{j_2}:

$$\{ |\, j_2 m_2 \rangle \mid m_2 = j_2, j_2-1, \ldots, -j_2 \}.$$

Uncoupled basis of $H_{j_1} \otimes H_{j_2}$:

$$\{ |\, j_1 m_1 \rangle \otimes |\, j_2 m_2 \rangle \mid m_1 = j_1, j_1-1, \ldots, -j_1;$$
$$m_2 = j_2, j_2-1, \ldots, -j_2 \}.$$

Coupled basis of $H_{j_1} \otimes H_{j_2}$:

$$\{ |\, (j_1 j_2) j m \rangle \mid j = j_1+j_2, j_1+j_2-1, \ldots, |j_1-j_2|;$$
$$m = j, j-1, \ldots, -j \},$$
$$|\, (j_1 j_2) j m \rangle = \sum_{m_1, m_2} C^{j_1 j_2 j}_{m_1 m_2 m} |\, j_1 m_1 \rangle \otimes |\, j_2 m_2 \rangle.$$

Unitary transformations of spaces:

$$T_U |\, j_1 m_1 \rangle = \sum_{m'_1} D^{j_1}_{m'_1 m_1}(U) |\, j_1 m'_1 \rangle,$$
$$m_1 = j_1, j_1-1, \ldots, -j_1, \text{ each } U \in SU(2);$$
$$T_V |\, j_2 m_2 \rangle = \sum_{m'_2} D^{j}_{m'_2 m_2}(V) |\, j_2 m'_2 \rangle,$$
$$m_2 = j_2, j_2-1, \ldots, -j_2, \text{ each } V \in SU(2);$$
$$T_{(U,V)} |\, j_1 m_1 \rangle \otimes |\, j_2 m_2 \rangle = T_U |\, j_1 m_1 \rangle \otimes T_V |\, j_2 m_2 \rangle$$
$$= \sum_{m'_1, m'_2} D^{j_1}_{m'_1 m_1}(U) D^{j_2}_{m'_2 m_2}(V) |\, j_1 m'_1 \rangle \otimes |\, j_2 m'_2 \rangle,$$
$$\text{each } U \in SU(2), \text{ each } V \in SU(2);$$
$$T_{(U,U)} |\, (j_1 j_2) j m \rangle = \sum_{m'} D^{j}_{m' m}(U) |\, (j_1 j_2) j m' \rangle,$$
$$m = j, j-1, \ldots, -j; \text{ each } U \in SU(2).$$

Representation of direct product group $SU(2) \times SU(2)$:

$$T_{(U,V)} T_{(U',V')} = T_{(UU', VV')}.$$

Representation of $SU(2)$ as diagonal subgroup of $SU(2) \times SU(2)$:

$$T_{(U,U)} T_{(U',U')} = T_{(UU', UU')},$$
$$T_U = T_{(U,U)}.$$

2.8.3 Explicit Forms of WCG-Coefficients

Wigner's form:

$$C^{j_1 j_2 j}_{m_1 m_2 m} = \delta(m_1+m_2, m)(2j+1)^{1/2}$$
$$\times \left[\frac{(j+j_1-j_2)!(j-j_1+j_2)!(j_1+j_2-j)!}{(j+j_1+j_2+1)!} \right]^{1/2}$$
$$\times \left[\frac{(j+m)!(j-m)!}{(j_1+m_1)!(j_1-m_1)!(j_2+m_2)!(j_2-m_2)!} \right]^{1/2}$$
$$\times \sum_s \frac{(-1)^{j_2+m_2+s}(j_2+j+m_1-s)!(j_1-m_1+s)!}{s!(j-j_1+j_2-s)!(j+m-s)!(j_1-j_2-m+s)!}.$$

Racah's form:

$$C^{j_1 j_2 j}_{m_1 m_2 m} = \delta(m_1 + m_2, m)$$

$$\times \left[\frac{(2j+1)(j_1 + j_2 - j)!}{(j_1 + j_2 + j + 1)!(j + j_1 - j_2)!(j + j_2 - j_1)!} \right]^{1/2}$$

$$\times \left[\frac{(j_1 - m_1)!(j_2 - m_2)!(j - m)!(j + m)!}{(j_1 + m_1)!(j_2 + m_2)!} \right]^{1/2}$$

$$\times \sum_t \frac{(-1)^{j_1 - m_1 + t}(j_1 + m_1 + t)!(j + j_2 - m_1 - t)!}{t!(j - m - t)!(j_1 - m_1 - t)!(j_2 - j + m_1 + t)!}.$$

Van der Waerden's form:

$$C^{j_1 j_2 j}_{m_1 m_2 m} = \delta(m_1 + m_2, m)$$

$$\times \left[\frac{(2j+1)(j_1 + j_2 - j)!(j + j_1 - j_2)!(j + j_2 - j_1)!}{(j_1 + j_2 + j + 1)!} \right]^{1/2}$$

$$\times [(j_1 + m_1)!(j_1 - m_1)!(j_2 + m_2)!(j_2 - m_2)!]^{1/2}$$

$$\times [(j + m)!(j - m)!]^{1/2}$$

$$\times \sum_k (-1)^k [k!(j_1 + j_2 - j - k)!(j_1 - m_1 - k)!$$

$$\times (j_2 + m_2 - k)!(j - j_2 + m_1 + k)!(j - j_1 - m_2 + k)!]^{-1}$$

Regge's formula and its combinatoric structure:

$$(\det A)^k = \sum_{\boxed{\alpha}} A(\alpha) \prod_{i,j=1}^3 (a_{ij})^{\alpha_{ij}}, \quad A = (a_{ij}), \quad (2.47)$$

where the summation is over all nonnegative integers α_{ij} that satisfy the row and column sum constraints [see Eq. (2.17)] given by

$$\alpha = \begin{array}{|ccc|c} \alpha_{11} & \alpha_{12} & \alpha_{13} & k \\ \alpha_{21} & \alpha_{22} & \alpha_{23} & k \\ \alpha_{31} & \alpha_{32} & \alpha_{33} & k \\ \hline k & k & k \end{array} \qquad (2.48)$$

The coefficients $A(\alpha)$ are constrained sums over multinomial coefficients:

$$A(\alpha) = \sum (-1)^{\phi(K)} \binom{k}{k_{123}, k_{132}, k_{231}, k_{213}, k_{312}, k_{321}},$$

where the summation is carried out over all nonnegative integers $k_{i_1 i_2 i_3}$ such that

$$\alpha_{11} = k_{123} + k_{132}, \quad \alpha_{12} = k_{231} + k_{213}, \quad \alpha_{13} = k_{312} + k_{321},$$
$$\alpha_{21} = k_{312} + k_{213}, \quad \alpha_{22} = k_{123} + k_{321}, \quad \alpha_{23} = k_{231} + k_{132},$$
$$\alpha_{31} = k_{231} + k_{321}, \quad \alpha_{32} = k_{312} + k_{132}, \quad \alpha_{33} = k_{123} + k_{213},$$
$$\phi(K) = \sum_{\pi \in A_3} k_\pi = k_{132} + k_{213} + k_{321}.$$

The general multinomial coefficient is the integer defined by

$$\binom{k}{k_1, k_2, \ldots, k_s} = k!/k_1!k_2!\cdots k_s!, \quad k = \sum_i k_i.$$

Relation (2.47) generalizes in the obvious way to an $n \times n$ determinant, using the symmetric group S_n and its S_{n-1} subgroups $S_{n-1}^{(j)}$, where j denotes that this is the permutation group on the integers $1, 2, \ldots, n$ with j deleted.

Regge's formula for the 3-j coefficient is:

$$\begin{pmatrix} j_1 & j_2 & j_3 \\ m_1 & m_2 & m_3 \end{pmatrix} = \delta(m_1 + m_2 + m_3, 0) \left[\prod_{ij} (\alpha_{ij})! \right]^{1/2}$$

$$\times \frac{A(\alpha)}{k! [(k+1)!]^{1/2}}, \qquad (2.49)$$

$$k = j_1 + j_2 + j_3,$$
$$\alpha_{11} = j_1 + m_1 \quad \alpha_{12} = j_2 + m_2 \quad \alpha_{13} = j_3 + m_3$$
$$\alpha_{21} = j_1 - m_1 \quad \alpha_{22} = j_2 - m_2 \quad \alpha_{23} = j_3 - m_3$$
$$\alpha_{31} = j_2 + j_3 - j_1 \quad \alpha_{32} = j_3 + j_1 - j_2 \quad \alpha_{33} = j_1 + j_2 - j_3$$

Equation (2.49) shows that WCG-coefficients and 3-j coefficients are *sums over integers*, except for a multiplicative normalization factor.

Schwinger's generating function:

$$\exp(t \det A) = \sum_k \frac{t^k}{k!} \sum_{\boxed{\alpha}} A(\alpha) \prod_{i,j} (a_{ij})^{\alpha_{ij}}.$$

The general definition of the $_pF_q$ hypergeometric function depending on p numerator parameters, q denominator parameters, and a single variable z is:

$$_pF_q \begin{pmatrix} a_1 \ldots, a_p \\ b_1 \ldots, b_q \end{pmatrix} = \sum_{n=0}^{\infty} \frac{(a_1)_n \cdots (a_p)_n}{(b_1)_n \cdots (b_q)_n} \frac{z^n}{n!},$$

$$(a)_n = a(a+1) \cdots (a+n-1), \quad (a)_0 = 1.$$

Such a series is terminating if at least one of the numerator parameters is a negative integer (and all other factors are well-defined). Both WCG-coefficients and Racah 6-j coefficients relate to special series of this type, evaluated at $z = 1$:

$$C^{abc}_{\alpha\beta\gamma} = [(2c+1)(a+\alpha)!(a-\alpha)!(b+\beta)!(b-\beta)!(c+\gamma)!(c-\gamma)!]^{1/2}$$

$$\times (-1)^{a+b+\gamma+\delta_1} \Delta(abc)$$

$$\times \frac{_3F_2 \begin{pmatrix} \epsilon_1 - \delta_1, \epsilon_2 - \delta_1, \epsilon_3 - \delta_1 \\ \delta_2 - \delta_1 + 1, \delta_3 - \delta_1 + 1 \end{pmatrix}}{(\delta_2 - \delta_1)!(\delta_3 - \delta_1)!(\delta_1 - \epsilon_1)!(\delta_1 - \epsilon_2)!(\delta_1 - \epsilon_3)!},$$

$\delta_1 = \min(a + \alpha + b + \beta, b - \beta + c + \gamma, a + \alpha + c + \gamma)$, $(\delta_1, \delta_2, \delta_3) =$ any permutation of $(a + \alpha + b + \beta, b - \beta + c + \gamma, a + \alpha + c + \gamma)$, after δ_1 is fixed, $(\epsilon_1, \epsilon_2, \epsilon_3) =$ any permutation of $(a + \alpha, b + \alpha + \gamma, c + \gamma)$;

$$\Delta(abc) = \left[\frac{(a+b-c)!(a-b+c)!(-a+b+c)!}{(a+b+c+1)!} \right]^{1/2}.$$

The quantity $\Delta(abc)$ is called a *triangle coefficient*.

All 72 Regge symmetries are consequences of known symmetries of the $_3F_2$ hypergeometric series.

2.8.4 Symmetries of WCG-Coefficients in 3-j Symbol Form

There are 72 known symmetries (up to sign changes) of the 3-j coefficient. There are at least four ways of verifying these symmetries: (1) directly from the van der Waerden form of the coefficients; (2) directly from Regge's generating function; (3) from the known symmetries of the $_3F_2$ hypergeometric series; and (4) directly from the symmetries of the representation functions $D^j_{mm'}(U)$. The set of 72 symmetries is succinctly expressed in terms of the coefficient $A(\alpha)$ defined in Sect. 2.8.3 with α_{ij} entries given by Eqs. (2.48) and (2.49):

$$
A \begin{bmatrix} j_1+m_1 & j_2+m_2 & j_3+m_3 \\ j_1-m_1 & j_2-m_2 & j_3-m_3 \\ j_2+j_3-j_1 & j_3+j_1-j_2 & j_1+j_2-j_3 \end{bmatrix} .
$$

This coefficient has determinantal symmetry; that is, it is invariant under even permutations of its rows or columns and under transposition, and is multiplied by the factor $(-1)^{j_1+j_2+j_3}$ under odd permutations of its rows or columns. These 72 determinantal operations may be generated from the three operations C_{12}, C_{13}, T consisting of interchange of columns 1 and 2, interchange of columns 1 and 3, and transposition, since the first two operations generate the symmetric group S_3 of permutations of columns, and the symmetric group S_3' of permutations of rows is then given by TS_3T. The transposition T itself generates a group $\{e, T\}$ isomorphic to the symmetric group S_2. Thus, the 72 element determinantal group is the direct product group $S_3 \times S_3' \times \{e, T\}$. The three relations between 3-j coefficients corresponding to the generators C_{12}, C_{13}, T are

$$
\begin{pmatrix} j_1 & j_2 & j_3 \\ m_1 & m_2 & m_3 \end{pmatrix} = (-1)^{j_1+j_2+j_3} \begin{pmatrix} j_2 & j_1 & j_3 \\ m_2 & m_1 & m_3 \end{pmatrix} ,
$$

$$
\begin{pmatrix} j_1 & j_2 & j_3 \\ m_1 & m_2 & m_3 \end{pmatrix} = (-1)^{j_1+j_2+j_3} \begin{pmatrix} j_3 & j_2 & j_1 \\ m_3 & m_2 & m_1 \end{pmatrix} ,
$$

$$
\begin{pmatrix} j_1 & j_2 & j_3 \\ m_1 & m_2 & m_3 \end{pmatrix}
$$
$$
= \begin{pmatrix} \frac{j_1+j_2+m_1+m_2}{2} & \frac{j_1+j_2-m_1-m_2}{2} & j_3 \\ \frac{j_1-j_2+m_1-m_2}{2} & \frac{j_1-j_2-m_1+m_2}{2} & j_2-j_1 \end{pmatrix} .
$$

All 72 relations among 3-j coefficients can be obtained from these 3. The 12 "classical" symmetries of the 3-j symbol

$$
\begin{pmatrix} a & b & c \\ \alpha & \beta & \gamma \end{pmatrix}
$$

are expressed by:

1. even permutations of the columns leave the coefficient invariant;

2. odd permutations of the columns change the sign by the factor $(-1)^{a+b+c}$;

3. simultaneous sign reversal of the projection quantum numbers changes the sign by $(-1)^{a+b+c}$.

The 72 corresponding symmetries of the WCG-coefficients (up to sign changes and dimensional factors) are best obtained from those of the 3j-coefficients by using Eq. (2.42).

2.8.5 Recurrence Relations

Three-term:

$$
[(J+1)(J-2j_1)]^{1/2} \begin{pmatrix} j_1 & j_2 & j_3 \\ m_1 & m_2 & m_3 \end{pmatrix}
$$
$$
= [(j_2+m_2)(j_3-m_3)]^{1/2} \begin{pmatrix} j_1 & j_2-\frac{1}{2} & j_3-\frac{1}{2} \\ m_1 & m_2-\frac{1}{2} & m_3+\frac{1}{2} \end{pmatrix}
$$
$$
- [(j_2-m_2)(j_3+m_3)]^{1/2} \begin{pmatrix} j_1 & j_2-\frac{1}{2} & j_3-\frac{1}{2} \\ m_1 & m_2+\frac{1}{2} & m_3-\frac{1}{2} \end{pmatrix} ,
$$

$$
[(J-2j_2)(J+1-2j_3)]^{1/2} \begin{pmatrix} j_1 & j_2 & j_3 \\ m_1 & m_2 & m_3 \end{pmatrix}
$$
$$
+ [(j_2+m_2+1)(j_3+m_3)]^{1/2} \begin{pmatrix} j_1 & j_2-\frac{1}{2} & j_3+\frac{1}{2} \\ m_1 & m_2-\frac{1}{2} & m_3+\frac{1}{2} \end{pmatrix}
$$
$$
+ [(j_2-m_2+1)(j_3-m_3)]^{1/2} \begin{pmatrix} j_1 & j_2-\frac{1}{2} & j_3+\frac{1}{2} \\ m_1 & m_2+\frac{1}{2} & m_3-\frac{1}{2} \end{pmatrix} = 0 ;
$$

$$
(j_2+m_2)^{1/2} \begin{pmatrix} j_1 & j_2 & j_3 \\ m_1 & m & m_3 \end{pmatrix}
$$
$$
= [(j_3-j_1+j_2)(J+1)(j_3-m_3)]^{1/2}
$$
$$
\times \begin{pmatrix} j_1 & j_2-\frac{1}{2} & j_3-\frac{1}{2} \\ m_1 & m_2-\frac{1}{2} & m_3-\frac{1}{2} \end{pmatrix}
$$
$$
- [(j_1-j_3+j_2)(J-2j_2+1)(j_3+m_3+1)]^{1/2}
$$
$$
\times \begin{pmatrix} j_1 & j_2-\frac{1}{2} & j_3+\frac{1}{2} \\ m_1 & m_2-\frac{1}{2} & m_3+\frac{1}{2} \end{pmatrix} ;
$$

$$
\begin{pmatrix} j_1 & j_2 & j_3 \\ j_2-m_3 & -j_2 & m_3 \end{pmatrix} = - \left[\frac{2j_2(j_3+m_3)}{(j_3-j_1+j_2)(J+1)} \right]^{1/2}
$$
$$
\times \begin{pmatrix} j_1 & j_2-\frac{1}{2} & j_3-\frac{1}{2} \\ j_2-m_3 & -j_2+\frac{1}{2} & m_3-\frac{1}{2} \end{pmatrix} ,
$$

$j_3 = j_1+j_2, j_1+j_2 \quad -1, \ldots, j_1-j_2+1$ for $j_1 \geq j_2$;

$$
\begin{pmatrix} j_1 & j_2 & j_3 \\ j_2-m_3 & -j_2 & m_3 \end{pmatrix} = - \left[\frac{2j_2(j_3-m_3+1)}{(j_1-j_3+j_2)(J-2j_2+1)} \right]^{\frac{1}{2}}
$$
$$
\times \begin{pmatrix} j_1 & j_2-\frac{1}{2} & j_3+\frac{1}{2} \\ j_2-m_3 & -j_2+\frac{1}{2} & m_3-\frac{1}{2} \end{pmatrix} ,
$$

$j_3 = j_1+j_2-1, j_1 \quad +j_2-2, \ldots, j_1-j_2$ for $j_1 \geq j_2$.

Four-term:

$$[(J+1)(J-2j_1)(J-2j_2)(J-2j_3+1)]^{1/2} \begin{pmatrix} j_1 & j_2 & j_3 \\ m_1 & m_2 & m_3 \end{pmatrix}$$

$$= [(j_2-m_2)(j_2+m_2+1)(j_3+m_3)(j_3+m_3-1)]^{1/2}$$

$$\times \begin{pmatrix} j_1 & j_2 & j_3-1 \\ m_1 & m_2+1 & m_3-1 \end{pmatrix}$$

$$- 2m_2[(j_3+m_3)(j_3-m_3)]^{1/2} \begin{pmatrix} j_1 & j_2 & j_3-1 \\ m_1 & m_2 & m_3 \end{pmatrix}$$

$$- [(j_2+m_2)(j_2-m_2+1)(j_3-m_3)(j_3-m_3-1)]^{1/2}$$

$$\times \begin{pmatrix} j_1 & j_2 & j_3-1 \\ m_1 & m_2-1 & m_3+1 \end{pmatrix} .$$

Five-term:

$$C^{bdf}_{\beta,\delta,\beta+\delta}$$

$$= \left[\frac{(b+d-f)(b+f-d+1)(d-\delta)(f+\beta+\delta+1)}{(2d)(2f+1)(2d)(2f+2)}\right]^{1/2}$$

$$\times C^{bd-1/2\,f+1/2}_{\beta,\delta+1/2,\beta+\delta+1/2}$$

$$- \left[\frac{(b+d-f)(b+f-d+1)(d+\delta)(f-\beta-\delta+1)}{(2d)(2f+1)(2d)(2f+2)}\right]^{1/2}$$

$$\times C^{bd-1/2\,f+1/2}_{\beta,\delta-1/2,\beta+\delta-1/2}$$

$$+ \left[\frac{(d+f-b)(b+d+f+1)(d+\delta)(f+\beta+\delta)}{(2d)(2f+1)(2d)(2f)}\right]^{1/2}$$

$$\times C^{bd-1/2\,f-1/2}_{\beta,\delta-1/2,\beta+\delta-1/2}$$

$$+ \left[\frac{(d+f-b)(b+d+f+1)(d-\delta)(f-\beta-\delta)}{(2d)(2f+1)(2d)(2f)}\right]^{1/2}$$

$$\times C^{bd-1/2\,f-1/2}_{\beta,\delta+1/2,\beta+\delta+1/2} .$$

This relation may be used to prove the limit relation (2.50) from the similar recurrence relation (2.84c) for the Racah coefficients.

2.8.6 Limiting Properties and Asymptotic Forms

$$\lim_{a\to\infty} C^{a\ b\ a+\rho}_{a-\alpha,\beta,a-\alpha+\beta} = \delta_{\rho\beta},$$

$$\lim_{j\to\infty} (-1)^{a+b+2j-\tau}[(2c+1)(2j-2\sigma+1)]^{1/2}$$

$$\times \left\{ \begin{matrix} j-\tau & a & j-\sigma \\ b & j & c \end{matrix} \right\} = C^{abc}_{\rho\sigma\tau}, \qquad (2.50)$$

where the brace symbol is a 6-j coefficient (see Sect. 2.10).

$$C^{jkj+\Delta}_{m,\mu,m+\mu} \sim (-1)^{\Delta-\mu} D^k_{\mu\Delta} \begin{pmatrix} \cos\frac{1}{2}\beta & \sin\frac{1}{2}\beta \\ -\sin\frac{1}{2}\beta & \cos\frac{1}{2}\beta \end{pmatrix}$$

$$= d^k_{\mu\Delta}(\beta), \text{ for large } j;$$

$$\cos\tfrac{1}{2}\beta = \sqrt{\frac{j+m}{2j}}, \qquad \sin\tfrac{1}{2}\beta = \sqrt{\frac{j-m}{2j}},$$

$$C^{jkj}_{m0m} \sim P_k(\cos\beta), \quad \text{for large } j;$$

$$(-1)^k[(2j+1)(2J+1)]^{1/2}W(j,k,J+m,J;j,J)$$
$$\sim P_k(\cos\beta), \text{ first for large } J, \text{ then large } j.$$

2.8.7 WCG-Coefficients as Discretized Representation Functions

$$C^{j_1j_2j}_{m_1m_2m} = \delta_{m_1+m_2,m}(-1)^{j_1-m_1}$$

$$\times \left[\frac{(2j+1)(j_1+j_2-j)!}{(j_1+j_2+j+1)!}\right]^{1/2} \qquad (2.51)$$

$$\times D^j_{m,j_1-j_2}\begin{pmatrix} \sqrt{j_1+m_1} & \sqrt{j_2+m_2} \\ -\sqrt{j_1-m_1} & \sqrt{j_2-m_2} \end{pmatrix}_{\text{symbolic powers}},$$

where in evaluating this result one first substitutes

$$u_{11} = \sqrt{j_1+m_1}, \qquad u_{12} = \sqrt{j_2+m_2},$$
$$u_{21} = -\sqrt{j_1-m_1}, \qquad u_{22} = \sqrt{j_2-m_2},$$

into the form (2.17), followed by the replacement of ordinary powers by generalized powers:

$$(\pm\sqrt{k})^s \to (\pm1)^s \left[\frac{k!}{(k-s)}\right]^{1/2}.$$

2.9 TENSOR OPERATOR ALGEBRA

2.9.1 Conceptual Framework

A tensor operator can be characterized in terms of its algebraic properties with respect to the angular momentum $\mathbf{J}$ or in terms of its transformation properties under unitary transformations generated by $\mathbf{J}$. Both viewpoints are essential.

A tensor operator $\mathbf{T}$ with respect to the group $SU(2)$ is a set of linear operators

$$\mathbf{T} = \{T_1, T_2, \ldots, T_n\},$$

where each operator in the set acts in the space $\mathcal{H}$ defined by Eq. (2.10) and maps this space into itself $T_i : \mathcal{H} \to \mathcal{H}, i = 1,2,\ldots,n$, and where this set of operators has the following properties with respect to the angular momentum $\mathbf{J}$, which acts in the same space $\mathcal{H}$ in the standard way:

1. Commutation relations with respect to the angular momentum **J**:

$$[J_i, T_j] = \sum_{k=1}^{n} t_{kj}^{(i)} T_k \,,$$

where the $t_{kj}^{(i)}$ are scalars (invariants) with respect to **J**.

2. Unitary transformation with respect to $SU(2)$ rotations:

$$e^{-i\psi \hat{\mathbf{n}} \cdot \mathbf{J}} T_i e^{i\psi \hat{\mathbf{n}} \cdot \mathbf{J}} = \sum_{j=1}^{n} D_{ji}(U) T_j \,, \quad U = U(\psi, \hat{\mathbf{n}}) \,,$$

where the matrix $D(U)$ is an $n \times n$ unitary matrix representation of $SU(2)$. Reduction of this representation into its irreducible constituents gives the notion of an irreducible tensor operator $\mathbf{T}^J$ of rank J. An irreducible tensor operator $\mathbf{T}^J$ of rank J is a set of $2J+1$ operators

$$\mathbf{T}^J = \{ T_M^J \mid M = J, J-1, \ldots, -J \}$$

with the following properties with respect to $SU(2)$:

1. Commutation relations with respect to the angular momentum **J**:

$$[J_+, T_M^J] = [(J-M)(J+M+1)]^{1/2} T_{M+1}^J \,,$$
$$[J_-, T_M^J] = [(J+M)(J-M+1)]^{1/2} T_{M-1}^J \,,$$
$$[J_3, T_M^J] = M T_M^J \,, \qquad (2.52)$$
$$\sum_i [J_i, [J_i, T_M^J]] = J(J+1) T_M^J \,.$$

2. Generation from highest "weight":

$$T_M^J = \left[\frac{(J+M)!}{(2J)!(J-M)!} \right]^{1/2} [J_-, T_J^J]_{(J-M)} \,,$$

where $[A, B]_{(k)} = [A, [A,B]_{(k-1)}], k = 1, 2, \ldots$, with $[A,B]_{(0)} = B$, denotes the k-fold commutator of A with B.

3. Unitary transformation with respect to $SU(2)$ rotations:

$$e^{-i\psi \hat{\mathbf{n}} \cdot \mathbf{J}} T_M^J e^{i\psi \hat{\mathbf{n}} \cdot \mathbf{J}} = \sum_{M'} D_{M'M}^J(U) T_M^J \,, \quad U = U(\psi, \hat{\mathbf{n}}) \,.$$

$$(2.53)$$

Angular momentum operators act in Hilbert spaces by acting linearly on the vectors in such spaces. The concept of a tensor operator generalizes this by replacing the irreducible space H_J by the irreducible tensor $\mathbf{T}^J$,

and angular momentum operator action on H_J by commutator action on $\mathbf{T}^J$, as symbolized, respectively, by

$$\mathbf{J} : \{\text{states}\} \rightarrow \{\text{states}\},$$
$$\{\text{commutator action of } \mathbf{J}\} : \{\text{tensor operators}\}$$
$$\rightarrow \{\text{tensor operators}\}.$$

Just as exponentiation of the standard generator action (2.13) and (2.14) gives relation (2.16), so does the exponentiation of the commutator action (2.52) give relation (2.53), when one uses the Baker-Campell-Hausdorff identity:

$$e^{tA} B e^{-tA} = \sum_k \frac{t^k}{k!} [A, B]_{(k)} \,.$$

Thus, the linear vector space of states is replaced by the linear vector space of operators. Abstractly, relations (2.13) and (2.52) are identical: only the rule of action and the object of that action has changed.

An example of an irreducible tensor of rank 1 is the angular momentum **J** itself, which has the special property $\mathbf{J} : \mathcal{H}_j \rightarrow \mathcal{H}_j$. Thus, relations (2.52) and (2.53) are realized as:

$$T_1^1 = J_{+1} = -(J_1 + iJ_2)/\sqrt{2} \,,$$
$$T_0^1 = J_0 = J_3 \,,$$
$$T_{-1}^1 = J_{-1} = (J_1 - iJ_2)/\sqrt{2} \,;$$

$$[J_+, T_\mu^1] = [(1-\mu)(2+\mu)]^{1/2} T_{\mu+1}^1 \,,$$
$$[J_-, T_\mu^1] = [(1+\mu)(2-\mu)]^{1/2} T_{\mu-1}^1 \,,$$
$$[J_3, T_\mu^1] = \mu T_\mu^1 \,, \quad \mu = 1, 0, -1 \,;$$

$$e^{-i\psi \hat{\mathbf{n}} \cdot \mathbf{J}} \mathbf{J} e^{i\psi \hat{\mathbf{n}} \cdot \mathbf{J}} = \mathbf{J} \cos\psi + \hat{\mathbf{n}}(\hat{\mathbf{n}} \cdot \mathbf{J})(1 - \cos\psi)$$
$$- (\hat{\mathbf{n}} \times \mathbf{J}) \sin\psi \,,$$
$$e^{-i\psi \hat{\mathbf{n}} \cdot \mathbf{J}} T_\mu^1 e^{i\psi \hat{\mathbf{n}} \cdot \mathbf{J}} = \sum_\nu D_{\nu\mu}^1(\psi, \hat{\mathbf{n}}) T_\nu^1 \,.$$

2.9.2 Universal Enveloping Algebra of J

The universal enveloping algebra $A(\mathbf{J})$ of **J** is the set of all complex polynomial operators in the components $\mathbf{J}_i$ of **J**, or equivalently in (J_+, J_3, J_-). The irreducible tensor operators spanning this algebra are the analogues of the solid harmonics $\mathcal{Y}_{lm}(\mathbf{x})$ and are characterized by the following properties: Basis set:

$$T_k^k = a_k J_+^k, \quad a_k \text{ arbitrary constant} \,,$$
$$T_\mu^k = a_k \left[\frac{(k+\mu)!}{(2k)!(k-\mu)!} \right] [J_-, J_+^k]_{(k-\mu)} \,,$$
$$\mu = k, k-1, \ldots, -k; \qquad k = 0, 1, 2, \ldots .$$

Standard action with respect to **J**:

$$[J_\pm, \mathcal{T}_\mu^k] = [(k \mp \mu)(k \pm \mu + 1)]^{1/2} \mathcal{T}_{\mu\pm1}^k \,,$$

$$[J_3, \mathcal{T}_\mu^k] = \mu \mathcal{T}_\mu^k \,,$$

$$\sum_{i=1}^{3} [J_i, [J_i, \mathcal{T}_\mu^k]] = k(k+1)\mathcal{T}_\mu^k \,.$$

Unitary transformation:

$$e^{-i\psi \hat{\mathbf{n}} \cdot \mathbf{J}} \mathcal{T}_\mu^k e^{i\psi \hat{\mathbf{n}} \cdot \mathbf{J}} = \sum_\nu D_{\nu\mu}^k(\psi, \hat{\mathbf{n}}) \mathcal{T}_\nu^k \,.$$

2.9.3 Algebra of Irreducible Tensor Operators

Irreducible tensor operators possess, as linear operators acting in the same space, properties (a), (b), and (c) below, and an additional multiplication property (d), which constructs new irreducible tensor operators out of two given ones and is called *coupling of irreducible tensor operators*. Property (d) extends also to tensor operators acting in the tensor product space associated with kinematically independent systems. It is important that associativity extend to the product (2.54), as well as to the product (2.55). Commutivity in these products is generally invalid. The coupling properties given in (d) and (e) are analogous to the coupling of basis state vectors. The operation of Hermitian conjugation of operators, which is the analogue of complex conjugation of states, is also important, and has the properties presented under (f).

(a) Multiplication of an irreducible tensor operator by a complex number or an invariant with respect to angular momentum $\mathbf{J}$ gives an irreducible tensor operator of the same rank.

(b) Addition of two irreducible tensor operator of the same rank gives an irreducible tensor of that rank.

(c) Ordinary multiplication (juxtaposition) of three irreducible tensor operators is associative, but the multiplication of two is noncommutative, in general.

(d) Two irreducible tensor operators $\mathbf{S}^{k_1}$ and $\mathbf{T}^{k_2}$ of different or the same ranks acting in the same space may be multiplied to obtain new irreducible tensor operators of ranks given by the angular momentum addition rule (Clebsch-Gordan series):

$$[\mathbf{S}^{k_1} \times \mathbf{T}^{k_2}]_\mu^k = \sum_{\mu_1, \mu_2} C_{\mu_1 \mu_2 \mu}^{k_1 k_2 k} S_{\mu_1}^{k_1} T_{\mu_2}^{k_2} \,, \qquad (2.54)$$

$$\mu = k, k-1, \ldots, -k \,;$$

$$\text{rank} = k \in \{k_1 + k_2, k_1 + k_2 - 1, \ldots, |k_1 - k_2|\} \,.$$

The following symbol denotes the irreducible tensor operator with the components (2.54):

$$[\mathbf{S}^{k_1} \times \mathbf{T}^{k_2}]^k \,.$$

(e) Two irreducible tensor operators $\mathbf{S}^{k_1}$ and $\mathbf{T}^{k_2}$ of different or the same ranks acting in different Hilbert spaces, say $\mathcal{H}$ and $\mathcal{K}$, may first be multiplied by the tensor product rule so as to act in the tensor product space $\mathcal{H} \otimes \mathcal{K}$, that is,

$$S_{\mu_1}^{k_1} \otimes T_{\mu_2}^{k_2} : \mathcal{H} \otimes \mathcal{K} \to \mathcal{H} \otimes \mathcal{K} \,,$$

and then coupled to obtain new irreducible tensor operators, acting in the same tensor product space $\mathcal{H} \otimes \mathcal{K}$:

$$[\mathbf{S}^{k_1} \otimes \mathbf{T}^{k_2}]_\mu^k = \sum_{\mu_1, \mu_2} C_{\mu_1 \mu_2 \mu}^{k_1 k_2 k} S_{\mu_1}^{k_1} \otimes T_{\mu_2}^{k_2} \,, \qquad (2.55)$$

$$\mu = k, k-1, \ldots, -k \,.$$

The following symbol denotes the tensor operator with the components (2.55):

$$[\mathbf{S}^{k_1} \otimes \mathbf{T}^{k_2}]^k, k \in \{k_1 + k_2, k_1 + k_2 - 1, \ldots, |k_1 - k_2|\} \,.$$

(f) The *conjugate tensor operator* to $\mathbf{T}^J$, denoted by $\mathbf{T}^{J\dagger}$, is the set of operators with components $T_M^{J\dagger}$ defined by

$$\left\langle j'm' \mid T_M^{J\dagger} \mid jm \right\rangle = \langle jm | T_M^J | j'm' \rangle^* \,.$$

These components satisfy the following relations:

$$[J_\pm, T_M^{J\dagger}] = -[(J \pm M)(J \mp M + 1)]^{1/2} T_{M\mp1}^{J\dagger}$$

$$[J_3, T_M^{J\dagger}] = -M T_M^{J\dagger} \,,$$

$$\sum_i [J_i, [J_i, T_M^{J\dagger}]] = J(J+1) T_M^{J\dagger} \,;$$

$$e^{-i\psi \hat{\mathbf{n}} \cdot \mathbf{J}} T_M^{J\dagger} e^{i\psi \hat{\mathbf{n}} \cdot \mathbf{J}} = \sum_{M'} D_{M'M}^{J*}(\psi, \hat{\mathbf{n}}) T_M^{J\dagger} \,;$$

$$I^J = \sum_M T_M^J T_M^{J\dagger} = \text{(invariant operator to } SU(2) \text{ rotations)},$$

$$e^{-i\psi \hat{\mathbf{n}} \cdot \mathbf{J}} I^J e^{i\psi \hat{\mathbf{n}} \cdot \mathbf{J}} = I^J \,.$$

An important invariant operator is

$$I^{k_1 k_2 k} = \sum_{\mu_1 \mu_2 \mu} C_{\mu_1 \mu_2 \mu}^{k_1 k_2 k} T_{\mu_1}^{k_1} T_{\mu_2}^{k_2} T_\mu^{k\dagger} \,.$$

(g) Other definitions of conjugation:

$$T_M^T \to (-1)^{J-M} T_{-M}^J \,, \quad T_M^J \to (-1)^{J+M} T_{-M}^J \,.$$

2.9.4 Wigner-Eckart Theorem

The Wigner-Eckart theorem establishes the form of the matrix elements of an arbitrary irreducible tensor operator:

$$\langle j'm' | T_M^J | jm \rangle = \langle j' \| \mathbf{T}^J \| j \rangle C_{mMm'}^{jJj'}$$

$$= (j' \| \mathbf{T}^J \| j)(-1)^{j+J+m'} \begin{pmatrix} j & J & j' \\ m & M & -m' \end{pmatrix} \,.$$

Reduced matrix elements with respect to WCG-coefficients:

$$\langle j' \| \mathbf{T}^J \| j \rangle = \sum_{\mu M} C^{jJj'}_{\mu M \mu'} \langle j' \mu' | T^J_M | j \mu \rangle \,,$$

each $\mu' = j', j' - 1, \ldots, -j'$ (the reduced matrix element is independent of μ).

Reduced matrix elements with respect to 3-j coefficients:

$$(j' \| \mathbf{T}^J \| j) = (-1)^{2J} \sqrt{2j'+1} \, \langle j' \| \mathbf{T}^J \| j \rangle \,.$$

Examples of irreducible tensor operators include:

1. The solid harmonics with respect to the orbital angular momentum $\mathbf{L}$:

$$\boldsymbol{\mathcal{Y}}^k(\mathbf{x}) = \{ \mathcal{Y}_{k\mu}(\mathbf{x}) : \mu = k, \ldots, -k \},$$

$$\mathcal{Y}_{k\mu} \mid lm \rangle = \sum_{l'} \langle l' \| \boldsymbol{\mathcal{Y}}^k \| l \rangle C^{lkl'}_{m,\mu,m+\mu} | l', m + \mu \rangle \,,$$

where

$$\langle \mathbf{x} \mid lm \rangle = \mathcal{Y}_{lm}(\mathbf{x}) \,,$$

$$\langle l' \| \boldsymbol{\mathcal{Y}}^k \| l \rangle = r^{l+k-l'} \left[\frac{(2l+1)(2k+1)}{4\pi(2l'+1)} \right]^{1/2} C^{lkl'}_{000} \,,$$

$$\mathcal{Y}_{k\mu}(\mathbf{x}) \mathcal{Y}_{lm}(\mathbf{x}) = \sum_{l'} \langle l' \| \boldsymbol{\mathcal{Y}}^k \| l \rangle C^{lkl'}_{m,\mu,m+\mu} \mathcal{Y}_{l',m+\mu}(\mathbf{x}) \,,$$

$$[\boldsymbol{\mathcal{Y}}^{k_1}(\mathbf{x}) \otimes \boldsymbol{\mathcal{Y}}^{k_2}(\mathbf{x})]^k_\mu = \sum_{\mu_1 \mu_2} C^{k_1 k_2 k}_{\mu_1 \mu_2 \mu} \mathcal{Y}_{k_1 \mu_1}(\mathbf{x}) \mathcal{Y}_{k_2 \mu_2}(\mathbf{x}) \,,$$

$$[\boldsymbol{\mathcal{Y}}^{k_1}(\mathbf{x}) \otimes \boldsymbol{\mathcal{Y}}^{k_2}(\mathbf{x})]^k_\mu = \langle k \| \boldsymbol{\mathcal{Y}}^{k_1} \| k_2 \rangle \mathcal{Y}_{k\mu}(\mathbf{x}) \,.$$

2. The polynomial operator $\boldsymbol{\mathcal{T}}^k$ in the components of $\mathbf{J}$ (Sect. 2.9.2):

$$\langle j'm' | \mathcal{T}^k_\mu | jm \rangle = \delta_{j'j} \langle j \| \boldsymbol{\mathcal{T}}^k \| j \rangle C^{jkj}_{m\mu m'} \,,$$

$$\langle j \| \boldsymbol{\mathcal{T}}^k \| j \rangle = a_k (-1)^k \left[\frac{(2j+k+1)!k!k!}{(2j+1)(2j-k)!(2k)!} \right]^{\frac{1}{2}} \,.$$

3. Polynomials in the components of an arbitrary vector operator $\mathbf{V}$, which has the defining relations:

$$[J_i, V_j] = i e_{ijk} V_k \,,$$

$$[J_\pm, V_\mu] = [(1 \mp \mu)(2 \pm \mu)]^{1/2} V_{\mu \pm 1} \,,$$

$$[J_3, V_\mu] = \mu V_\mu \,,$$

$$V_{+1} = -(V_1 + iV_2)/\sqrt{2}, V_0 = V_3 \,,$$

$$V_{-1} = (V_1 - iV_2)/\sqrt{2} \,.$$

This construction parallels exactly that given in Sect. 2.9.2 upon replacing $\mathbf{J}$ by $\mathbf{V}$. The explicit form of the resulting polynomials may be quite different since no assumptions are made concerning commutation relations between the components V_i of $\mathbf{V}$. The solid harmonics in the gradient operator ∇ constitute an irreducible tensor operator with respect to the orbital angular momentum $\mathbf{L}$.

2.9.5 Unit Tensor Operators or Wigner Operators

A unit tensor operator is an irreducible tensor operator $\hat{\mathbf{T}}^{J,\Delta}$, indexed not only by the angular momentum quantum number J, but also by an additional label Δ, which specifies that this irreducible tensor operator has reduced matrix elements given by

$$\langle j' \| \hat{\mathbf{T}}^{J,\Delta} \| j \rangle = \delta_{j',j+\Delta} \,.$$

This condition is to be true for all $j = 0, 1/2, 1, \ldots$. There is a unit tensor operator defined for each

$$\Delta = J, J - 1, \ldots, -J \,.$$

The special symbol

$$\left\langle \begin{array}{cc} & J + \Delta \\ 2J & \\ & 0 \\ & \bullet \end{array} \right\rangle$$

denotes a unit tensor operator, replacing the boldface symbol $\hat{\mathbf{T}}^{J,\Delta}$, while the symbol

$$\left\langle \begin{array}{cc} & J + \Delta \\ 2J & 0 \\ & J + M \end{array} \right\rangle, \quad M = J, J - 1, \ldots, -J$$

denotes the components. In the same way that abstract angular momentum $\mathbf{J}$ and state vectors $\{|jm\rangle\}$ extract the intrinsic structure of all realizations of angular momentum theory, as given in Sect. 2.3, so does the notion of a unit tensor operator extract the intrinsic structure of the concept of irreducible tensor operator by disregarding the physical content of the theory, which is carried in the structure of the reduced matrix elements. Physical theory is regained from the fact that the unit tensor operators are the basis for arbitrary tensor operators, which is the structural content of the Wigner-Eckart theorem. The concept of a unit tensor operator was introduced by Racah, but it was Biedenharn who recognized the full significance of this concept not only for $SU(2)$, but for all the unitary groups.

All of the content of physical tensor operator theory can be regained from the properties of unit tensor operators or Wigner operators as summarized below:
Notation (double Gel'fand patterns):

$$\left\langle \begin{array}{cc} & J + \Delta \\ 2J & 0 \\ & J + M \end{array} \right\rangle, \quad \begin{array}{c} M, \Delta = J, J-1, \ldots, -J \\ 2J = 0, 1, 2, \ldots \end{array}$$

Definition (shift action):

$$\left\langle \begin{array}{cc} & J + \Delta \\ 2J & 0 \\ & J + M \end{array} \right\rangle | jm \rangle = C^{jJj+\Delta}_{m,M,m+M} | j + \Delta, m + M \rangle \quad (2.56)$$

for all $j = 0, \frac{1}{2}, \ldots$; $m = j, j-1, \ldots, -j$.

Conjugation:

$$\left\langle \begin{matrix} & J+\Delta & \\ 2J & & 0 \\ & J+M & \end{matrix} \right\rangle^{\dagger} | jm \rangle = C^{j-\Delta\, Jj}_{m-M,M,m} | j-\Delta, m-M \rangle. \tag{2.57}$$

Orthogonality:

$$\sum_{M} \left\langle \begin{matrix} & J+\Delta' & \\ 2J & & 0 \\ & J+M & \end{matrix} \right\rangle \left\langle \begin{matrix} & J+\Delta & \\ 2J & & 0 \\ & J+M & \end{matrix} \right\rangle^{\dagger} = \delta_{\Delta'\Delta} \mathbf{I}^{J}_{\Delta}, \tag{2.58}$$

$$\sum_{\Delta} \left\langle \begin{matrix} & J+\Delta & \\ 2J & & 0 \\ & J+M' & \end{matrix} \right\rangle^{\dagger} \left\langle \begin{matrix} & J+\Delta & \\ 2J & & 0 \\ & J+M & \end{matrix} \right\rangle = \delta_{M'M}, \tag{2.59}$$

$$\sum_{m} \langle jm | \left\langle \begin{matrix} & J'+\Delta' & \\ 2J' & & 0 \\ & J'+M' & \end{matrix} \right\rangle \left\langle \begin{matrix} & J+\Delta & \\ 2J & & 0 \\ & J+M & \end{matrix} \right\rangle^{\dagger} | jm \rangle$$
$$= \frac{2j+1}{2J+1} \delta_{J'J} \delta_{M'M} \delta_{\Delta'\Delta}. \tag{2.60}$$

The invariant operator $\mathbf{I}^{J}_{\Delta}$ is defined by its action on an arbitrary vector $\psi_{j} \in \mathcal{H}_{j}$:

$$\mathbf{I}^{J}_{\Delta} \psi_{j} = \epsilon_{j-\Delta,J,j} \psi_{j}.$$

Tensor operator property:

$$e^{-i\psi\hat{\mathbf{n}}\cdot\mathbf{J}} \left\langle \begin{matrix} & J+\Delta & \\ 2J & & 0 \\ & J+M & \end{matrix} \right\rangle e^{i\psi\hat{\mathbf{n}}\cdot\mathbf{J}}$$
$$= \sum_{M'} D^{J}_{M'M}(\psi,\hat{\mathbf{n}}) \left\langle \begin{matrix} & J+\Delta & \\ 2J & & 0 \\ & J+M' & \end{matrix} \right\rangle. \tag{2.61}$$

Basis property (Wigner-Eckart theorem):

$$T^{J}_{M} | jm \rangle = \left(\sum_{\Delta} \langle j+\Delta \| \mathbf{T}^{J} \| j \rangle \left\langle \begin{matrix} & J+\Delta & \\ 2J & & 0 \\ & J+M & \end{matrix} \right\rangle \right) | jm \rangle. \tag{2.62}$$

Characteristic null space:

The characteristic null space of the Wigner operator defined by Eq. (2.56) is the *set of irreducible subspaces* $\mathcal{H}_{j} \subset \mathcal{H}$ given be

$$\{ \mathcal{H}_{j} : 2j = 0, 1, \ldots, J-\Delta-1 \}.$$

Coupling law:

$$\sum_{\alpha\beta} C^{abc}_{\alpha\beta\gamma} \left\langle \begin{matrix} & b+\sigma & \\ 2b & & 0 \\ & b+\beta & \end{matrix} \right\rangle \left\langle \begin{matrix} & a+\rho & \\ 2a & & 0 \\ & a+\alpha & \end{matrix} \right\rangle$$
$$= \mathbf{W}^{abc}_{\rho,\sigma,\rho+\sigma} \left\langle \begin{matrix} & c+\rho+\sigma & \\ 2c & & 0 \\ & c+\gamma & \end{matrix} \right\rangle, \tag{2.63}$$

where $\mathbf{W}^{abc}_{\rho\sigma\tau}$ is an invariant operator (commutes with $\mathbf{J}$) and is called a *Racah invariant*. Its relationship to Racah coefficients and 6-j coefficients is given in Sect. 2.10.

Product law:

$$\left\langle \begin{matrix} & b+\sigma & \\ 2b & & 0 \\ & b+\beta & \end{matrix} \right\rangle \left\langle \begin{matrix} & a+\rho & \\ 2a & & 0 \\ & a+\alpha & \end{matrix} \right\rangle$$
$$= \sum_{c} \mathbf{W}^{abc}_{\rho,\sigma,\rho+\sigma} C^{abc}_{\alpha,\beta,\alpha+\beta} \left\langle \begin{matrix} & c+\rho+\sigma & \\ 2c & & 0 \\ & c+\alpha+\beta & \end{matrix} \right\rangle. \tag{2.64}$$

Racah invariant:

$$\mathbf{W}^{abc}_{\rho\sigma\tau} = \sum_{\alpha\beta\gamma} C^{abc}_{\alpha\beta\gamma} \left\langle \begin{matrix} & b+\sigma & \\ 2b & & 0 \\ & b+\beta & \end{matrix} \right\rangle \left\langle \begin{matrix} & a+\rho & \\ 2a & & 0 \\ & a+\alpha & \end{matrix} \right\rangle \left\langle \begin{matrix} & c+\tau & \\ 2c & & 0 \\ & c+\gamma & \end{matrix} \right\rangle^{\dagger}. \tag{2.65}$$

The notation $\mathbf{W}^{abc}_{\rho\sigma\tau}$ for a Racah invariant is designed to "match" that of the WCG-coefficient on the left, the latter being associated with the lower group theoretical labels, for example,

$$\begin{pmatrix} & 2a & & 0 \\ & & a+\alpha & \end{pmatrix} \rightarrow | a\alpha \rangle,$$

the state vector having a group transformation law under the action of $SU(2)$, and the former with the shift labels of a unit tensor operator,

$$\begin{pmatrix} & \alpha+\rho & \\ 2\alpha & & 0 \end{pmatrix},$$

and having no associated group transformation law. The invariant operator defined by Eq. (2.65) has real eigenvalues, hence, is a Hermitian operator,

$$\mathbf{W}^{abc\dagger}_{\rho\sigma\tau} = \mathbf{W}^{abc}_{\rho\sigma\tau}, \tag{2.66}$$

which is diagonal on an arbitrary state vector in $\mathcal{H}_{j}$ (see Sect. 2.10).

The Racah invariant operator does not commute with a unit tensor operator, and it makes a difference whether it is written to the left or to the right of such a unit tensor operator. The convention here writes it to the left.

Relation (2.65) is taken as the *definition* of $\mathbf{W}^{abc}_{\rho\sigma\tau}$ and the following properties all follow from this expression:

Domain of definition:

$$\mathbf{W}^{abc}_{\rho\sigma\tau} : a, b, c \in \{0, 1/2, 1, 3/2, \ldots\};$$
$$\rho = a, a-1, \ldots, -a$$
$$\sigma = b, b-1, \ldots, -b$$
$$\tau = c, c-1, \ldots, -c;$$

$$\mathbf{W}^{abc}_{\rho\sigma\tau} = \mathbf{0}, \text{ unless } \rho+\sigma = \tau; \text{ if } \epsilon_{abc} = 0.$$

Orthogonality relations:

$$\sum_{\rho\sigma} \mathbf{W}^{abc}_{\rho\sigma\tau} \mathbf{W}^{abd}_{\rho\sigma\tau'} = \delta_{cd} \delta_{\tau\tau'} \epsilon_{abc} \mathbf{I}^{c}_{\tau}, \tag{2.67}$$

$$\sum_{c\tau} \mathbf{W}^{abc}_{\rho\sigma\tau} \mathbf{W}^{abc}_{\rho'\sigma'\tau} = \delta_{\rho\rho'} \delta_{\sigma\sigma'} \mathbf{I}^{ab}_{\rho\sigma}, \tag{2.68}$$

where the $\mathbf{I}$ invariant operators in these expressions have the following eigenvalues on an arbitrary vector $\psi_j \in \mathcal{H}_j$:

$$\mathbf{I}_\tau^c \psi_j = \epsilon_{j-\tau,c,j}\psi_j,$$
$$\mathbf{I}_{\rho\sigma}^{ab}\psi_j = \epsilon_{j-\sigma-\rho,a,j-\sigma}\epsilon_{j-\sigma,b,j}\psi_j.$$

The orthogonality relations for Racah invariants parallel exactly those of WCG-coefficients.

Using the orthogonality relations (2.67) for Racah invariants, the following two relations now follow from Eqs. (2.63) and (2.64), respectively:
WCG and Racah operator coupling:

$$\sum_{\rho\sigma}\sum_{\alpha\beta}\mathbf{W}_{\rho\sigma\rho+\sigma}^{abd}C_{\alpha,\beta,\alpha+\beta}^{abc}\left\langle \begin{matrix} b+\sigma \\ 2b \quad\quad 0 \\ b+\beta \end{matrix} \right\rangle\left\langle \begin{matrix} a+\rho \\ 2a \quad\quad 0 \\ a+\alpha \end{matrix} \right\rangle$$

$$= \delta_{cd}\epsilon_{abc}\mathbf{I}_\tau^d\left\langle \begin{matrix} c+\tau \\ 2c \quad\quad 0 \\ c+\gamma \end{matrix} \right\rangle. \quad (2.69)$$

Racah operator coupling of shift patterns:

$$\sum_{\rho\sigma}\mathbf{W}_{\rho\sigma\tau}^{abc}\left\langle \begin{matrix} b+\sigma \\ 2b \quad\quad 0 \\ b+\beta \end{matrix} \right\rangle\left\langle \begin{matrix} a+\rho \\ 2a \quad\quad 0 \\ a+\alpha \end{matrix} \right\rangle$$

$$= C_{\alpha,\beta,\alpha+\beta}^{abc}\left\langle \begin{matrix} c+\tau \\ 2c \quad\quad 0 \\ c+\alpha+\beta \end{matrix} \right\rangle. \quad (2.70)$$

Relations (2.56)–(2.70) capture the *full content* of irreducible tensor operator algebra through the concept of unit tensor operators that have only 0 or 1 for their reduced matrix elements. Using the Wigner-Eckart theorem (2.62), the relations between general tensor operators can be reconstructed. Unit tensor operators were invented to exhibit in the most elementary way possible the abstract and intrinsic structure of the irreducible tensor operator algebra, stripping away the details of particular physical applications, thus giving the theory universal application. It accomplishes the same goal for tensor operator theory that the abstract multiplet theory in Sect. 2.3 accomplishes for representation theory.

Physical theory is regained through the concept of reduced matrix element. The coupling rule (2.54) is now transformed to a rule empty of WCG-coefficient content and becomes a rule for coupling of reduced matrix elements using the invariant Racah operators:

$$\langle(\alpha')j'\|[\mathbf{S}^{k_1}\times\mathbf{T}^{k_2}]^k\|(\alpha)j\rangle$$
$$= (-1)^{k_1+k_2-k}\sum_{(\alpha'')j''}W_{j''-j,j'-j'',j'-j}^{k_2k_1k}(j')$$
$$\times \langle(\alpha')j'\|\mathbf{S}^{k_1}\|(\alpha'')j''\rangle\langle(\alpha'')j''\|\mathbf{T}^{k_2}\|(\alpha)j\rangle. \quad (2.71)$$

This coupling rule is invariant to all $SU(2)$ rotations, and reveals the true role of the Racah coefficients and

reduced matrix elements in physical theory as invariant objects under $SU(2)$ rotations. It now becomes imperative to understand Racah coefficients as objects free of their original definition in terms of WCG-coefficients.

2.10 RACAH COEFFICIENTS

Relation (2.65) is taken, initially, as the definition of the Racah coefficient with appropriate adjustments of notations to conform to Racah's W-notation and to Wigner's 6-j notation. Corresponding to each of Eqs. (2.63)–(2.65), (2.69), and (2.70), there is a corresponding numerical relationship between WCG-coefficients and Racah coefficients. Despite the present day popularity of expressing all such relations in terms of the 3-j and 6-j notation, this temptation is resisted here for this particular set of relations because of their fundamental origins. The relation between the Racah invariant notation and Racah's original W-notation is

$$\mathbf{W}_{\rho\sigma\tau}^{abc}\mid jm\rangle = \mathbf{W}_{\rho\sigma\tau}^{abc}(j)\mid jm\rangle,$$
$$\mathbf{W}_{\rho\sigma\tau}^{abc}(j) = 0 \text{ if } \tau \neq \rho+\sigma, \text{ or } \epsilon_{abc} = 0,$$
$$\mathbf{W}_{\rho\sigma\tau}^{abc}(j) = [(2c+1)(2j-2\sigma+1)]^{1/2}$$
$$\times W(j-\tau,a,j,b;j-\sigma,c),$$
$$[(2e+1)(2f+1)]^{1/2}W(abcd;ef) = W_{e-a,c-e,c-a}^{bdf}(c),$$

$W(abcd;ef) = 0$ unless the triples of nonnegative integers and half-integers (abe), (cde), (acf), (bdf) satisfy the triangle conditions.

2.10.1 Basic Relations Between WCG and Racah Coefficients

$$\sum_{\beta\delta}C_{\beta\delta\gamma}^{bdf}C_{\alpha+\beta,\delta,\alpha+\gamma}^{edc}C_{\alpha,\beta,\alpha+\beta}^{abe}$$

$$= [(2e+1)(2f+1)]^{1/2}W(abcd;ef)C_{\alpha,\gamma,\alpha+\gamma}^{afc}$$
$$\times C_{\alpha+\beta,\delta,\alpha+\beta+\delta}^{edc}C_{\alpha,\beta,\alpha+\beta}^{abe}$$
$$= \sum_f[(2e+1)(2f+1)]^{1/2}W(abcd;ef)$$
$$\times C_{\beta,\delta,\beta+\delta}^{bdf}C_{\alpha,\beta+\delta,\alpha+\beta+\delta}^{afc},$$
$$\delta_{cc'}[(2e+1)(2f+1)]^{1/2}W(abcd;ef)$$
$$= \sum_{\beta\delta}C_{\beta,\delta,\beta+\delta}^{bdf}C_{\gamma-\delta,\delta,\gamma}^{edc}C_{\gamma-\beta-\delta,\beta,\gamma-\delta}^{abe}C_{\gamma-\beta-\delta,\beta+\delta,\gamma}^{abc'},$$
$$\sum_{\beta\delta e}[(2e+1)(2f+1)]^{1/2}W(abcd;ef)$$
$$\times C_{\beta\delta\gamma}^{bdf'}C_{\alpha+\beta,\delta,\alpha+\gamma}^{edc}C_{\alpha,\beta,\alpha+\beta}^{abe}$$
$$= \delta_{ff'}C_{\alpha,\gamma,\alpha+\gamma}^{afc},$$
$$\sum_e[(2e+1)(2f+1)]^{1/2}W(abcd;ef)$$

$$\times C^{edc}_{\alpha+\beta,\delta,\alpha+\beta+\delta} C^{abe}_{\alpha,\beta,\alpha+\beta}$$
$$= C^{bdf}_{\beta,\delta,\beta+\delta} C^{afc}_{\alpha,\beta+\delta,\alpha+\beta+\delta} \, .$$

2.10.2 Orthogonality and Explicit Form

Orthogonality relations for Racah coefficients:

$$\sum_e (2e+1)(2f+1)W(abcd; ef)W(abcd; ef')$$
$$= \delta_{ff'}\epsilon_{afc}\epsilon_{bdf} \, , \tag{2.72}$$

$$\sum_f (2e+1)(2f+1)W(abcd; ef)W(abcd; e'f)$$
$$= \delta_{ee'}\epsilon_{abe}\epsilon_{edc} \, . \tag{2.73}$$

Definition of 6-j coefficients:

$$\left\{ \begin{matrix} a\,b\,e \\ d\,c\,f \end{matrix} \right\} = (-1)^{a+b+c+d}W(abcd; ef) \, . \tag{2.74}$$

Orthogonality of 6-j coefficients:

$$\sum_e (2e+1)(2f+1) \left\{ \begin{matrix} a\,b\,e \\ d\,c\,f \end{matrix} \right\} \left\{ \begin{matrix} a\,b\,e \\ d\,c\,f' \end{matrix} \right\} = \delta_{ff'}\epsilon_{afc}\epsilon_{bdf} \, ,$$
$$\tag{2.75}$$

$$\sum_f [(2e+1)(2f+1)] \left\{ \begin{matrix} a\,b\,e \\ d\,c\,f \end{matrix} \right\} \left\{ \begin{matrix} a\,b\,e' \\ d\,c\,f \end{matrix} \right\} = \delta_{ee'}\epsilon_{abe}\epsilon_{edc} \, .$$
$$\tag{2.76}$$

Explicit form of Racah coefficients:

$$W(abcd; ef) = \Delta(abe)\Delta(cde)\Delta(acf)\Delta(bdf)$$
$$\times \sum_k \frac{(-1)^{a+b+c+d+k}(k+1)!}{(k-a-b-e)!(k-c-d-e)!}$$
$$\times \frac{1}{(k-a-c-f)!(k-b-d-f)!}$$
$$\times \frac{1}{(a+b+c+d-k)!}$$
$$\times \frac{1}{(a+d+e+f-k)!(b+c+e+f-k)!} \, , \tag{2.77}$$

where $\Delta(abc)$ denotes the *triangle coefficient*, defined for every triple a, b, c of integers and half-odd integers satisfying the triangle conditions by:

$$\Delta(abc) = \left[\frac{(a+b-c)!(a-b+c)!(-a+b+c)!}{(a+b+c+1)!} \right]^{1/2} \, . \tag{2.78}$$

2.10.3 The Fundamental Identities Between Racah Coefficients

Each of the three relations given in this section is between Racah coefficients alone. Each expresses a fundamental mathematical property. The Biedenharn-Elliott identity is a consequence of the associativity rule for the open product of three irreducible tensor operators; the Racah sum rule is a consequence of the commutivity of a mapping diagram associated with the coupling of three angular momenta; and the triangle coupling rule is a consequence of the associativity of the open product of three symplection polynomials (see Ref. [1]). As such, these three relations between Racah coefficients, together with the orthogonality relations, are the building blocks on which is constructed a theory of these coefficients that stands on its own, independent of the WCG-coefficient origins. Indeed, the latter is recovered through the limit relation (2.50).

Biedenharn-Elliott identity:

$$W(a'ab'b; c'e)W(a'ed'd; b'c)$$
$$= \sum_f (2f+1)W(abcd; ef)W(c'bd'd; b'f)W(a'ad'f; c'c), \tag{2.79a}$$

$$\left\{ \begin{matrix} a'\,a\,c' \\ b\,b'e \end{matrix} \right\} \left\{ \begin{matrix} a'e\,b' \\ d\,d'c \end{matrix} \right\}$$
$$= \sum_f (-1)^{\phi}(2f+1) \left\{ \begin{matrix} a\,b\,e \\ d\,c\,f \end{matrix} \right\} \left\{ \begin{matrix} c'\,b\,b' \\ d\,d'\,f \end{matrix} \right\} \left\{ \begin{matrix} a'\,a\,c' \\ f\,d'\,c \end{matrix} \right\},$$
$$\phi = f + e + (a' + a + b' + b + c' + c + d' + d) \, . \tag{2.79b}$$

Racah sum rule:

$$\sum_f (-1)^{b+d-f}(2f+1)W(abcd; ef)W(adcb; gf)$$
$$= (-1)^{e+g-a-c}W(bacd; eg) \, , \tag{2.80a}$$

$$\sum_f (-1)^{e+g+f}(2f+1) \left\{ \begin{matrix} a\,b\,e \\ d\,c\,f \end{matrix} \right\} \left\{ \begin{matrix} a\,d\,g \\ b\,c\,f \end{matrix} \right\} = \left\{ \begin{matrix} a\,b\,e \\ c\,d\,g \end{matrix} \right\} \, . \tag{2.80b}$$

Triangle sum rule:

$$[\Delta(acf)\Delta(bdf)]^{-1}$$
$$= (2f+1)\sum_e [\Delta(abe)\Delta(cde)]^{-1}W(abcd; ef) \, , \tag{2.81a}$$

$$(-1)^{a+b+c+d}[\Delta(acf)\Delta(bdf)]^{-1}$$
$$= (2f+1)\sum_e [\Delta(abe)\Delta(cde)]^{-1} \left\{ \begin{matrix} a\,b\,e \\ d\,c\,f \end{matrix} \right\} \, . \tag{2.81b}$$

2.10.4 Schwinger-Bargmann Generating Function and Its Combinatorics

Triangles associated with the 6-j symbol $\begin{Bmatrix} j_1 j_2 j_3 \\ j_4 j_5 j_6 \end{Bmatrix}$:

$$(j_1 j_2 j_3), \ (j_3 j_4 j_5), \ (j_1 j_5 j_6), \ (j_2 j_4 j_6).$$

Points in $\mathbf{R}^3$ associated with the triangles:

$$(j_1 j_2 j_3) \rightarrow (x_1, x_2, x_3), \quad (j_3 j_4 j_5) \rightarrow (y_3, x_4, x_5),$$
$$(j_1 j_5 j_6) \rightarrow (y_1, y_5, x_6), \quad (j_2 j_4 j_6) \rightarrow (y_2, y_4, y_6).$$

Tetrahedron associated with the points:

The points define the vertices of a general tetrahedron with lines joining each pair of points that share a common subscript, and the lines are labeled by the product of the common coordinates (see Fig. 2.2).

Monomial term:

Define the *triangle monomial* associated with a triangle $(j_a j_b j_c)$ and its associated point (z_a, z_b, z_c) in $\mathbf{R}^3$ by

$$(z_a, z_b, z_c)^{(j_a j_b j_c)} = z_a^{j_b + j_c - j_a} z_b^{j_c + j_a - j_b} z_c^{j_a + j_b - j_c},$$
$$(j_a j_b j_c) = (j_b + j_c - j_a, j_a + j_c - j_b, j_a + j_b - j_c). \tag{2.82}$$

Cubic graph (tetrahedral $\mathbf{T}_4$) functions:

Interchange the symbols x and y in the coordinates of the vertices of the tetrahedron and define the following polynomials on the vertices and edges of the tetrahedron with this modified labeling.

Vertex function: multiply together the coordinates of each vertex and sum over all such vertices to obtain

$$V_3 = y_1 y_2 y_3 + x_3 x_4 x_5 + x_1 x_5 x_6 + x_2 x_4 x_6 \, ;$$

Edge function: multiply together the coordinates of a given edge and the opposite edge and sum over all such pairs to obtain

$$E_4 = x_1 y_1 x_4 y_4 + x_2 y_2 x_5 y_5 + x_3 y_3 x_6 y_6 \, .$$

Generating function:

$$(1 + V_3 + E_4)^{-2} = \sum_{\Delta} T(\Delta) Z^{\Delta}, \tag{2.83a}$$

$$Z^{\Delta} = (x_1, x_2, x_3)^{(j_1 j_2 j_3)} (y_3, x_4, x_5)^{(j_3 j_4 j_5)}$$
$$\times (y_1, y_5, x_6)^{(j_1 j_5 j_6)} (y_2, y_4, y_6)^{(j_2 j_4 j_6)}, \tag{2.83b}$$

$$\Delta = \begin{bmatrix} (j_1 j_2 j_3) \\ (j_3 j_4 j_5) \\ (j_1 j_5 j_6) \\ (j_2 j_4 j_6) \end{bmatrix} ;$$

$$T(\Delta) = \sum_k (-1)^k (k+1) \binom{k}{k_1, k_2, k_3, k_4, k_5, k_6, k_7}, \tag{2.83c}$$

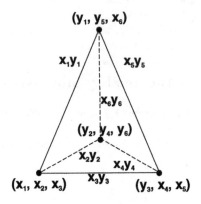

(y_1, y_5, x_6)

$x_1 y_1$ $x_5 y_5$

$x_6 y_6$

(y_2, y_4, y_6)

$x_2 y_2$ $x_4 y_4$

(x_1, x_2, x_3) $x_3 y_3$ (y_3, x_4, x_5)

Figure 2.2. Labeled Cubic Graph (Tetrahedron) Associated with 6-j Coefficient.

$$k_i = k - t_i, \quad i = 1, 2, 3, 4,$$
$$k_j = e_{j-4} - k, \quad j = 5, 6, 7 \, ;$$
$$t_i = \text{triangle sum} = \text{vertex sum},$$
$$e_j = \text{opposite edge sum, in pairs},$$
$$t_1 = j_1 + j_2 + j_3, \quad t_2 = j_3 + j_4 + j_5,$$
$$t_3 = j_1 + j_5 + j_6, \quad t_4 = j_2 + j_4 + j_6,$$
$$e_1 = (j_2 + j_5) + (j_3 + j_6),$$
$$e_2 = (j_1 + j_4) + (j_3 + j_6),$$
$$e_3 = (j_1 + j_4) + (j_2 + j_5).$$

The summation in Eq. (2.83b) is over the infinite set of all tetrahedra; that is, over the infinite set of arrays Δ having nonnegative integral entries.

2.10.5 Symmetries of 6-j Coefficients

There are 144 symmetry relations among the Racah (6-j) coefficients. The 24 classical ones, given already by Racah, and corresponding to the tetrahedral point group T_d of rotations-inversions (isomorphic to the symmetric group S_4) mapping the regular tetrahedron onto itself, are realized in the 6-j symbol

$$\begin{Bmatrix} a\,b\,e \\ d\,c\,f \end{Bmatrix}$$

as permutations of its columns and the exchange of any pair of letters in the top row with the corresponding pair in the bottom row. Regge discovered the 6-fold increase in symmetry by noting that each term in the summation in Eq. (2.77) is invariant not only to the classical 24 symmetries, but also under certain linear transformations of the quantum labels. These symmetries are also implicit in Schwinger's generating function.

The full set, including the original 24 substitutions, of linear transformations of the letters a, b, c, d, e, f thus yields a group of linear transformation isomorphic to $S_4 \times S_3$. The column permutations and row-pair interchanges described above applied to each of the six symbols in the equalities below yield the set of 144 relationships:

$$\begin{Bmatrix} abe \\ dcf \end{Bmatrix} = \begin{Bmatrix} a & \dfrac{b+c+e-f}{2} & \dfrac{b+e+f-c}{2} \\ d & \dfrac{b+c+f-e}{2} & \dfrac{c+e+f-b}{2} \end{Bmatrix}$$

$$= \begin{Bmatrix} \dfrac{a+d+e-f}{2} & b & \dfrac{a+e+f-d}{2} \\ \dfrac{a+d+f-e}{2} & c & \dfrac{d+e+f-a}{2} \end{Bmatrix}$$

$$= \begin{Bmatrix} \dfrac{a+b+d-c}{2} & \dfrac{a+b+c-d}{2} & e \\ \dfrac{a+c+d-b}{2} & \dfrac{b+c+d-a}{2} & f \end{Bmatrix}$$

$$= \begin{Bmatrix} \dfrac{a+b+d-c}{2} & \dfrac{b+c+e-f}{2} & \dfrac{a+e+f-d}{2} \\ \dfrac{a+c+d-b}{2} & \dfrac{b+c+f-e}{2} & \dfrac{d+e+f-a}{2} \end{Bmatrix}$$

$$= \begin{Bmatrix} \dfrac{a+d+e-f}{2} & \dfrac{a+b+c-d}{2} & \dfrac{b+e+f-c}{2} \\ \dfrac{a+d+f-e}{2} & \dfrac{b+c+d-a}{2} & \dfrac{c+e+f-b}{2} \end{Bmatrix}.$$

2.10.6   Further Properties

Recurrence relations:
Three-term:

$$[(a+b+e+1)(b+e-a)$$
$$\times (c+d+e+1)(d+e-c)]^{1/2} \begin{Bmatrix} a & b & e \\ d & c & f \end{Bmatrix}$$
$$= -2e[(b+d+f+1)(b+d-f)]^{1/2}$$
$$\times \begin{Bmatrix} a & b-\frac{1}{2} & e-\frac{1}{2} \\ d-\frac{1}{2} & c & f \end{Bmatrix}$$
$$+ [(a+b-e+1)(a+e-b)(c+d-e+1)$$
$$\times (c+e-d)]^{1/2} \begin{Bmatrix} a & b & e-1 \\ d & c & f \end{Bmatrix}, \tag{2.84a}$$

$$[(a+c+f+1)(c+e-d)$$
$$\times (d+e-c+1)(b+d-f+1)]^{1/2} \begin{Bmatrix} a & b & e \\ d & c & f \end{Bmatrix}$$
$$= [(a+c-f)(a+e-b)$$
$$\times (b+f+d+2)(b+e-a+1)]^{1/2}$$
$$\times \begin{Bmatrix} a+\frac{1}{2} & b+\frac{1}{2} & e \\ d+\frac{1}{2} & c-\frac{1}{2} & f \end{Bmatrix}.$$
$$+ [(c+f-a)(c+e-d)(b-a-c+d+1)]^{1/2}$$
$$\times \begin{Bmatrix} a & b & e \\ d+\frac{1}{2} & c-\frac{1}{2} & f-\frac{1}{2} \end{Bmatrix}. \tag{2.84b}$$

Five-term:

$$(2c+1)(2d)(2f+1) \begin{Bmatrix} a & b & e \\ d & c & f \end{Bmatrix}$$
$$= [(b+d-f)(b+f-d+1)(d+e-c)(c+e-d+1)$$
$$\times (c+f-a+1)(a+c+f+2)]^{1/2} \begin{Bmatrix} a & b & e \\ d-\frac{1}{2} & c+\frac{1}{2} & f+\frac{1}{2} \end{Bmatrix}$$
$$+ [(b+d-f)(b+f-d+1)(c+d-e)(c+d+e+1)$$
$$\times (a+c-f)(a+f-c+1)]^{1/2} \begin{Bmatrix} a & b & e \\ d-\frac{1}{2} & c-\frac{1}{2} & f+\frac{1}{2} \end{Bmatrix}$$
$$- [(d+f-b)(b+d+f+1)(c+d-e)(c+d+e+1)$$
$$\times (c+f-a)(a+c+f+1)]^{1/2} \begin{Bmatrix} a & b & e \\ d-\frac{1}{2} & c-\frac{1}{2} & f-\frac{1}{2} \end{Bmatrix}$$
$$+ [(d+f-b)(b+d+f+1)(d+e-c)(c+e-d+1)$$
$$\times (a+f-c)(a+c-f+1)]^{1/2} \begin{Bmatrix} a & b & e \\ d-\frac{1}{2} & c+\frac{1}{2} & f-\frac{1}{2} \end{Bmatrix}. \tag{2.84c}$$

Relation to hypergeometric series:

$$\begin{Bmatrix} abe \\ dcf \end{Bmatrix} = (-1)^{a+b+c+d} W(abcd; ef)$$

$$= \Delta(abe)\Delta(cde)\Delta(acf)\Delta(bdf) \times \frac{(-1)^{\beta_1}(\beta_1+1)!}{(\beta_2-\beta_1)!(\beta_3-\beta_1)!}$$

$$\times \frac{{}_4F_3 \begin{pmatrix} \alpha_1-\beta_1, & \alpha_2-\beta_1, & \alpha_3-\beta_1, & \alpha_4-\beta_1, \\ -\beta_1-1, & \beta_2-\beta_1+1, & \beta_3-\beta_1+1, & \end{pmatrix}}{(\beta_1-\alpha_1)!(\beta_1-\alpha_2)!(\beta_1-\alpha_3)!(\beta_1-\alpha_4)!},$$

$$\beta_1 = \min(a+b+c+d, a+d+e+f, b+c+e+f),$$

The parameters β_2 and β_3 are identified in either way with the pair remaining in the 3-tuple

$$(a+b+c+d, a+d+e+f, b+c+e+f)$$

after deleting β_1. The $(\alpha_1, \alpha_2, \alpha_3, \alpha_4)$ may be identified with any permutation of the 4-tuple

$$(a+b+e, c+d+e, a+c+f, b+d+f).$$

The ${}_4F_3$ series is Saalschützian:

$$1 + \sum(\text{numerator parameters})$$
$$= \sum(\text{denominator parameters}).$$

2.11   THE 9-J COEFFICIENTS

2.11.1   Hilbert Space and Tensor Operator Actions

Let $\mathbf{T}^a(1)$ and $\mathbf{T}^b(2)$ denote irreducible tensor operators of ranks a and b with respect to kinematically independent angular momentum operators $\mathbf{J}(1)$ and $\mathbf{J}(2)$ that act, respectively, in separable Hilbert spaces $\mathcal{H}(1)$ and $\mathcal{H}(2)$. Let $\mathcal{H}(1)$ and $\mathcal{H}(2)$ be reduced, respectively, into a direct sum of spaces $\mathcal{H}_{j_1}(1)$ and $\mathcal{H}_{j_2}(2)$. The angular momentum $\mathbf{J}(1)$ has the standard action on the orthonormal basis $\{|\, j_1 m_1\rangle \mid m_1 = j_1, j_1 - 1, \ldots, -j_1\}$ of $\mathcal{H}_{j_1}(1)$, and $\mathbf{J}(2)$ has the standard action on the orthonormal basis $\{|\, j_2 m_2\rangle \mid m_2 = j_2, j_2 - 1, \ldots, -j_2\}$ of $\mathcal{H}_{j_2}(2)$. The irreducible tensor operators $\mathbf{T}^a(1)$ and $\mathbf{T}^b(2)$ also have the standard actions in their respective Hilbert spaces $\mathcal{H}(1)$ and $\mathcal{H}(2)$, as given by Wigner-Eckart theorem. The total angular momentum $\mathbf{J}$ has the standard action on the coupled orthonormal basis of the tensor product space $\mathcal{H}_{j_1} \otimes \mathcal{H}_{j_2}$:

$$| \,(j_1 j_2)jm\rangle = \sum_{m_1 m_2} C^{j_1 j_2 j}_{m_1 m_2 m} \, | \, j_1 m_1\rangle \otimes |\, j_2 m_2\rangle \,.$$

$$(2.85)$$

The tensor product operator $\mathbf{T}^a(1) \otimes \mathbf{T}^b(2)$ acts in the tensor product space $\mathcal{H}(1) \otimes \mathcal{H}(2)$ according to the rule:

$$\big(\mathbf{T}^a(1) \otimes \mathbf{T}^b(2)\big)\big(|\, j_1 m_1\rangle \otimes |\, j_2 m_2\rangle\big)$$
$$= \mathbf{T}^a(1)\,|\, j_1 m_1\rangle \otimes \mathbf{T}^b(2)\,|\, j_2 m_2\rangle \,,$$

so that

$$\big(\mathbf{T}^a(1) \otimes \mathbf{T}^b(2)\big) \, | \,(j_1 j_2)\, jm\rangle$$
$$= \sum_{m_1 m_2} C^{j_1 j_2 j}_{m_1 m_2 m} \, \mathbf{T}^a(1)\,|\, j_1 m_1\rangle \otimes \mathbf{T}^b(2)\,|\, j_2 m_2\rangle \,.$$

$$(2.86a)$$

The angular momentum quantities called 9-j coefficients arise when the coupled tensor operators $\mathbf{T}^{(ab)c}$ with components γ defined by

$$T^{(ab)c}_{\quad \gamma} = \sum_{\alpha\beta} C^{abc}_{\alpha\beta\gamma} T^a_\alpha(1) \otimes T^b_\beta(2) \,, \qquad (2.86b)$$

$$\gamma = c, c-1, \ldots, -c \,,$$

are considered. The quantity $\mathbf{T}^{(ab)c}$ is an irreducible tensor operator of rank c with respect to the total angular momentum $\mathbf{J}$ for all a, b that yield c under the rule of addition of angular momentum.

2.11.2 9-j Invariant Operators

The entire angular momentum content of relation (2.86c) is captured by taking the irreducible tensor operators $\mathbf{T}^a(1)$ and $\mathbf{T}^b(2)$ to be unit tensor operators acting in the respective spaces $\mathcal{H}(1)$ and $\mathcal{H}(2)$:

$$\mathbf{T}^{(ab)c}_{(\rho\sigma)\gamma} = \sum_{\alpha\beta} C^{abc}_{\alpha\beta\gamma} \left\langle \begin{matrix} & \alpha+\rho & \\ 2a & & 0 \\ & a+\alpha & \end{matrix} \right\rangle_1 \otimes \left\langle \begin{matrix} & b+\sigma & \\ 2b & & 0 \\ & \alpha+\beta & \end{matrix} \right\rangle_2 \,. \quad (2.87)$$

The placement of the unit tensor operators shows in which space they act, so that the additional identification

by indices 1 and 2 could be eliminated. For each given $c \in \{0, 1/2, 1, 3/2, 2, \ldots\}$ and all a, b such that the triangle relation (abc) is satisfied, and, for each such pair a, b, all ρ, σ with $\rho \in \{a, a-1, \ldots, -a\}$, $\sigma \in \{b, b-1, \ldots, -b\}$, an irreducible tensor operator of rank c with respect to the total angular momentum $\mathbf{J}$ with components γ is defined by Eq. (2.87). By the Wigner-Eckart theorem, it must be possible to write

$$\sum_{\alpha\beta} C^{abc}_{\alpha\beta\gamma} \left\langle \begin{matrix} & \alpha+\rho & \\ 2a & & 0 \\ & a+\alpha & \end{matrix} \right\rangle_1 \otimes \left\langle \begin{matrix} & b+\sigma & \\ 2b & & 0 \\ & b+\beta & \end{matrix} \right\rangle_2$$

$$= \sum_\tau \left[\begin{matrix} abc \\ \rho\sigma\tau \end{matrix} \right] \left\langle \begin{matrix} & c+\tau & \\ 2c & & 0 \\ & c+\gamma & \end{matrix} \right\rangle \,. \qquad (2.88)$$

where: (i) The unit tensor operator on the right-hand side is a irreducible tensor operator with respect to $\mathbf{J}$; that is, has the action on the coupled states given by

$$\left\langle \begin{matrix} & c+\tau & \\ 2c & & 0 \\ & c+\gamma & \end{matrix} \right\rangle | \,(j_1 j_2)jm\rangle = C^{jc\ j+\tau}_{m,\gamma,m+\gamma} | \,(j_1 j_2)j+\tau, m+\gamma\rangle \,;$$

$$(2.89)$$

and (ii) the symbol $\left[\begin{matrix} abc \\ \rho\sigma\tau \end{matrix} \right]$ denotes an invariant operator with respect to the total angular momentum $\mathbf{J}$. Using the orthogonality of unit tensor operators, we can also write relation (2.88) in the form:

$$\left[\begin{matrix} abc \\ \rho\sigma\tau \end{matrix} \right] = \sum_{\alpha\beta\gamma} C^{abc}_{\alpha\beta\gamma} \left\langle \begin{matrix} & a+\rho & \\ 2a & & 0 \\ & a+\alpha & \end{matrix} \right\rangle_1 \otimes \left\langle \begin{matrix} & b+\sigma & \\ 2b & & 0 \\ & b+\beta & \end{matrix} \right\rangle_2 \left\langle \begin{matrix} & c+\tau & \\ 2c & & 0 \\ & c+\gamma & \end{matrix} \right\rangle^\dagger .$$

$$(2.90)$$

This form is taken as the definition of the 9-j invariant operator. Its eigenvalues in the coupled basis define the 9-j coefficient:

$$\left[\begin{matrix} abc \\ \rho\sigma\tau \end{matrix} \right] | \,(j_1 j_2)jm\rangle$$

$$= \left\langle (j_1+\rho, j_2+\sigma)j+\tau \, \right\| \left[\begin{matrix} abc \\ \rho\sigma\tau \end{matrix} \right] \left\| (j_1 j_2)j \right\rangle | \,(j_1 j_2)jm\rangle$$

$$= [(2j+1)(2c+1)(2j_1+2\rho+1)(2j_2+2\sigma+1)]^{1/2}$$

$$\times \left\{ \begin{matrix} j_1 & j_2 & j \\ a & b & c \\ j_1+\rho & j_2+\sigma & j+\tau \end{matrix} \right\} | \,(j_1 j_2)jm\rangle \,. \qquad (2.91)$$

The 9-j invariant operators play exactly the same role in the tensor product space of two irreducible angular momentum spaces as do the Racah invariants in one such irreducible angular momentum space.

The full content of the coupling law (2.86c) for physical irreducible tensor operators is regained in the coupling law for reduced matrix elements:

$$\langle (\alpha'_1 \alpha'_2 j'_1 j'_2)j' \| [\mathbf{T}^a(1) \times \mathbf{T}^b(2)]^c \| (\alpha_1 \alpha_2 j_1 j_2)j \rangle$$

$$= \left[\begin{matrix} j_1 & j_2 & j \\ a & b & c \\ j'_1 & j'_2 & j' \end{matrix} \right] \langle \,(\alpha'_1)j'_1 \| \mathbf{T}^a(1) \| (\alpha_1)j_1 \rangle$$

$$\times \langle (\alpha'_2)j'_2 \| \mathbf{T}^b(2) \| (\alpha_2)j_2 \rangle \,; \qquad (2.92a)$$

$$\begin{bmatrix} j_1 j_2 \, j \\ a \ b \ c \\ j_1' j_2' j' \end{bmatrix} = [(2j_1' + 1)(2j_2' + 1)(2j + 1)(2c + 1)]^{1/2}$$

$$\times \begin{Bmatrix} j_1 & j_2 & j \\ a & b & c \\ j_1' & j_2' & j' \end{Bmatrix}. \tag{2.92b}$$

2.11.3 Basic Relations Between 9-j Coefficients and 6-j Coefficients

Orthogonality of 9-j coefficients:

$$\sum_{hi} (2c + 1)(2f + 1)(2h + 1)(2i + 1)$$

$$\times \begin{Bmatrix} a & b & c \\ d & e & f \\ h & i & j \end{Bmatrix} \begin{Bmatrix} a & b & c' \\ d & e & f' \\ h & i & j \end{Bmatrix} = \delta_{cc'} \delta_{ff'},$$

where this relation is to be applied only to triples (abc), (def), (cfj), (abc'), (def'), $(c'f'j)$ for which the triangle conditions hold.

Definition of 9-j coefficients in terms of 3-j coefficients:

$$\delta_{j_{33} j_{33}'} (2j_{33} + 1)^{-1} \begin{Bmatrix} j_{11} & j_{12} & j_{13} \\ j_{21} & j_{22} & j_{23} \\ j_{31} & j_{32} & j_{33} \end{Bmatrix}$$

$$= \sum_{\substack{\text{all } m_{ij} \\ \text{except } m_{33}}} \begin{pmatrix} j_{11} & j_{12} & j_{13} \\ m_{11} & m_{12} & m_{13} \end{pmatrix} \begin{pmatrix} j_{21} & j_{22} & j_{23} \\ m_{21} & m_{22} & m_{23} \end{pmatrix} \begin{pmatrix} j_{31} & j_{32} & j_{33} \\ m_{31} & m_{32} & m_{33} \end{pmatrix}$$

$$\times \begin{pmatrix} j_{11} & j_{21} & j_{31} \\ m_{11} & m_{21} & m_{31} \end{pmatrix} \begin{pmatrix} j_{12} & j_{22} & j_{32} \\ m_{12} & m_{22} & m_{32} \end{pmatrix} \begin{pmatrix} j_{13} & j_{23} & j_{33}' \\ m_{13} & m_{23} & m_{33} \end{pmatrix}. \tag{2.93}$$

Definition of 9-j coefficients in terms of 6-j coefficients:

$$\begin{Bmatrix} j_{11} \, j_{12} \, j_{13} \\ j_{21} \, j_{22} \, j_{23} \\ j_{31} \, j_{32} \, j_{33} \end{Bmatrix} = \sum_k (-1)^{2k} (2k + 1)$$

$$\times \begin{Bmatrix} j_{11} & j_{21} & j_{31} \\ j_{32} & j_{33} & k \end{Bmatrix} \begin{Bmatrix} j_{12} & j_{22} & j_{32} \\ j_{21} & k & j_{23} \end{Bmatrix} \begin{Bmatrix} j_{13} & j_{23} & j_{33} \\ k & j_{11} & j_{12} \end{Bmatrix}. \tag{2.94}$$

Basic relation:

$$(-1)^\phi \begin{Bmatrix} j_{12} \, j_{22} \, j_{32} \\ j_{11} \, j_{21} \, j_{31} \\ j_{13} \, j_{23} \, j_{33} \end{Bmatrix} \begin{pmatrix} j_{31} & j_{32} & j_{33} \\ m_{31} & m_{32} & m_{33} \end{pmatrix}$$

$$= \sum_{\text{all } m_{(1i)} m_{(2i)}} \begin{pmatrix} j_{11} & j_{21} & j_{31} \\ m_{11} & m_{21} & m_{31} \end{pmatrix} \begin{pmatrix} j_{12} & j_{22} & j_{32} \\ m_{12} & m_{22} & m_{32} \end{pmatrix}$$

$$\times \begin{pmatrix} j_{13} & j_{23} & j_{33} \\ m_{13} & m_{23} & m_{33} \end{pmatrix} \begin{pmatrix} j_{11} & j_{12} & j_{13} \\ m_{11} & m_{12} & m_{13} \end{pmatrix} \begin{pmatrix} j_{21} & j_{22} & j_{23} \\ m_{21} & m_{22} & m_{23} \end{pmatrix},$$

$$\phi = \sum_{kl} j_{kl}. \tag{2.95}$$

Additional relations:

$$\sum_{kl} (-1)^{2b+l+h-f} (2k + 1)(2l + 1)$$

$$\times \begin{Bmatrix} a & b & c \\ e & d & f \\ k & l & i \end{Bmatrix} \begin{Bmatrix} a & e & k \\ d & b & l \\ g & h & i \end{Bmatrix} = \begin{Bmatrix} a & b & c \\ d & e & f \\ g & h & i \end{Bmatrix},$$

$$\sum_c (2c + 1) \begin{Bmatrix} a & b & c \\ d & e & f \\ g & h & i \end{Bmatrix} \begin{Bmatrix} a & b & c \\ f & i & j \end{Bmatrix}$$

$$= (-1)^{2j} \begin{Bmatrix} d & e & f \\ b & j & h \end{Bmatrix} \begin{Bmatrix} g & h & i \\ j & a & d \end{Bmatrix},$$

$$\sum_{klm} (2k + 1)(2l + 1)(2m + 1)$$

$$\times \begin{Bmatrix} a & b & c \\ d & e & f \\ k & l & m \end{Bmatrix} \begin{Bmatrix} k & l & m \\ a' & b' & c' \\ d' & e' & f' \end{Bmatrix}$$

$$\times \begin{Bmatrix} a & d & k \\ a' & d' & k' \end{Bmatrix} \begin{Bmatrix} b & e & l \\ b' & e' & l' \end{Bmatrix} \begin{Bmatrix} c & f & m \\ c' & f' & m' \end{Bmatrix}$$

$$= \begin{Bmatrix} a & b & c \\ d & e & f \\ k' & l' & m' \end{Bmatrix} \begin{Bmatrix} k' & l' & m' \\ a' & b' & c' \\ d & e & f \end{Bmatrix}.$$

2.11.4 Symmetry Relations for 9-j Coefficients and Reduction to 6-j Coefficients

The 9-j coefficient

$$\begin{Bmatrix} j_{11} & j_{12} & j_{13} \\ j_{21} & j_{22} & j_{23} \\ j_{31} & j_{32} & j_{33} \end{Bmatrix}$$

is invariant under even permutation of its rows, even permutation of its columns, and under the interchange of rows and columns (matrix transposition). It is multiplied by the factor $(-1)^\phi$ [see Eq. (2.95)] under odd permutations of its rows or columns. These 72 symmetries are all consequences of the 72 symmetries of the 3-j coefficient in relation (2.93).

Reduction to 6-j coefficients:

$$\begin{Bmatrix} a & b & e \\ c & d & e \\ f & f & 0 \end{Bmatrix} = \begin{Bmatrix} 0 & e & e \\ f & d & b \\ f & c & a \end{Bmatrix} = \begin{Bmatrix} e & 0 & e \\ c & f & a \\ d & f & b \end{Bmatrix}$$

$$= \begin{Bmatrix} f & f & 0 \\ d & c & e \\ b & a & e \end{Bmatrix} = \begin{Bmatrix} f & b & d \\ 0 & e & e \\ f & a & c \end{Bmatrix} = \begin{Bmatrix} a & f & c \\ e & 0 & e \\ b & f & d \end{Bmatrix}$$

$$= \begin{Bmatrix} b & a & e \\ f & f & 0 \\ d & c & e \end{Bmatrix} = \begin{Bmatrix} e & d & c \\ e & b & a \\ 0 & f & f \end{Bmatrix} = \begin{Bmatrix} c & e & d \\ a & e & b \\ f & 0 & f \end{Bmatrix}$$

$$= \frac{(-1)^{b+c+e+f}}{[(2e+1)(2f+1)]^{1/2}} \left\{ \begin{matrix} abe \\ dcf \end{matrix} \right\}.$$

2.11.5 Explicit Algebraic Form of 9-j Coefficients

$$\left\{ \begin{matrix} a & b & c \\ d & e & f \\ h & i & j \end{matrix} \right\} = (1)^{c+f-j} \frac{(dah)(bei)(jhi)}{(def)(bac)(jcf)}$$

$$\times \sum_{xyz} \frac{(-1)^{x+y+z}}{x!y!z!} \times \frac{(2f-x)!(2a-z)!}{(2i+1+y)!(a+d+h+1-z)!}$$

$$\times \frac{(d+e-f+x)!(c+j-f+x)!}{(e+f-d-x)!(c+f-j-x)!}$$

$$\times \frac{(e+i-b+y)!(h+i-j+y)!}{(b+e-i-y)!(h+j-i-y)!}$$

$$\times \frac{(b+c-a+z)!}{(a+d-h-z)!(a+c-b-z)!}$$

$$\times \frac{(a+d+j-i-y-z)!}{(d+i-b-f+x+y)!(b+j-a-f+x+z)!},$$

$$(abc) = \left[\frac{(a-b+c)!(a+b-c)!(a+b+c+1)!}{(b+c-a)!} \right]^{1/2}.$$

2.11.6 Racah Operators

A Racah operator is denoted

$$\left\{ \begin{matrix} & a+\rho & \\ 2a & & 0 \\ & a+\sigma & \end{matrix} \right\} \qquad \begin{matrix} \rho,\sigma = a, a-1, \ldots, -a, \\ 2a = 0,1,2,\ldots, \end{matrix}$$

and is a special case of the operator defined by Eq. (2.87):

$$\left\{ \begin{matrix} & a+\rho & \\ 2a & & 0 \\ & a+\sigma & \end{matrix} \right\} | (j_1 j_2)jm \rangle = \left[\frac{(2a+1)(2j_2+1)}{(2j_2+2\sigma+1)} \right]^{1/2}$$

$$\times \mathbf{T}_{\rho\sigma 0}^{aa0} | (j_1 j_2)jm \rangle. \qquad (2.96)$$

Thus, a Racah operator is an invariant operator with respect to the total angular momentum **J**. Alternative definitions are:

$$\left\{ \begin{matrix} & a+\rho & \\ 2a & & 0 \\ & a+\sigma & \end{matrix} \right\} = (-1)^{a+\sigma} \sum_a \left\langle \begin{matrix} & a+\rho & \\ 2a & & 0 \\ & a+\alpha & \end{matrix} \right\rangle \otimes \left\langle \begin{matrix} & a-\sigma & \\ 2a & & 0 \\ & a+\alpha & \end{matrix} \right\rangle^{\dagger},$$

$$\left\{ \begin{matrix} & a+\rho & \\ 2a & & 0 \\ & a+\sigma & \end{matrix} \right\} | (j_1 j_2)jm \rangle = [(2j_1+2\rho+1)(2j_2+1)]^{1/2}$$

$$\times W(j, j_1, j_2+\sigma, a; j_2, j_1+\rho) | (j_1+\rho, j_2+\sigma)jm \rangle$$

with conjugate

$$\left\{ \begin{matrix} & a+\rho & \\ 2a & & 0 \\ & a+\sigma & \end{matrix} \right\}^{\dagger} | (j_1 j_2)jm \rangle = [(2j_1+1)(2j_2-2\sigma+1)]^{1/2}$$

$$\times W(j, j_1-\rho, j_2, a; j_2-\sigma, j_1) | (j_1-\rho, j_2-\sigma)jm \rangle.$$

Racah operators satisfy orthogonality relations similar in form to Wigner operators. The open product rule is:

$$\left\{ \begin{matrix} & b+\sigma & \\ 2b & & 0 \\ & b+\beta & \end{matrix} \right\} \left\{ \begin{matrix} & a+\rho & \\ 2a & & 0 \\ & a+\alpha & \end{matrix} \right\}$$

$$= \sum_c \overline{\mathbf{W}}_{\rho,\sigma,\rho+\sigma}^{abc} \underline{\mathbf{W}}_{\alpha,\beta,\alpha+\beta}^{abc} \left\{ \begin{matrix} & c+\rho+\sigma & \\ 2c & & 0 \\ & c+\alpha+\beta & \end{matrix} \right\}. \qquad (2.97)$$

In this result $\overline{\mathbf{W}}_{\rho\sigma\tau}^{abc}$ and $\underline{\mathbf{W}}_{\alpha,\beta,\alpha+\beta}^{abc}$ denote Racah invariants with respect to the angular momenta $\mathbf{J}(1)$ and $\mathbf{J}(2)$, respectively, so that

$$\overline{\mathbf{W}}_{\rho\sigma\tau}^{abc} | (j_1 j_2)jm \rangle = \mathbf{W}_{\rho\sigma\tau}^{abc}(j_1) | (j_1 j_2)jm \rangle,$$

$$\underline{\mathbf{W}}_{\alpha\beta\gamma}^{abc} | (j_1 j_2)jm \rangle = \mathbf{W}_{\alpha\beta\gamma}^{abc}(j_2) | (j_1 j_2)jm \rangle.$$

The matrix elements of relation (2.97) lead to the Biedenharn-Elliott identity. There are five versions of this relationship in complete analogy to relations (2.63)–(2.65) and (2.69)–(2.70) for Wigner operators.

Racah operators are a basis for all invariant operators acting in the tensor product space spanned by the coupled basis vectors (2.85) and are the natural way of formulating interactions in that space. Their algebra is a fascinating study, initiated already in a different guise in the work of Schwinger (see Ref. [2]). Little use has been made of this concept in physical applications.

Additional relations between Racah coefficients or 6-j coefficients may be derived from the various versions of the rule (2.97) or directly from relation (2.79b) by using the orthogonality relations (2.75). Two of these are:

$$\sum_e (-1)^{a+b+e}(2e+1)$$

$$\times \left\{ \begin{matrix} a & b & e \\ d & c & g \end{matrix} \right\} \left\{ \begin{matrix} a' & a & c' \\ b & b' & e \end{matrix} \right\} \left\{ \begin{matrix} a' & e & b' \\ d & d' & c \end{matrix} \right\}$$

$$= (-1)^{\phi_1} \left\{ \begin{matrix} c' & b & b' \\ d & d' & g \end{matrix} \right\} \left\{ \begin{matrix} a' & a & c' \\ g & d' & c \end{matrix} \right\},$$

$$\phi_1 = g + a' + b' + c' + c + d' + d;$$

$$\sum_{e,e'} (-1)^{a-c'+e-e'}(2e+1)(2e'+1)(2f+1)$$

$$\times \left\{ \begin{matrix} c' & b & e' \\ d & d' & f \end{matrix} \right\} \left\{ \begin{matrix} a & b & e \\ d & c & g \end{matrix} \right\} \left\{ \begin{matrix} a' & a & c' \\ b & e' & e \end{matrix} \right\} \left\{ \begin{matrix} a' & e & e' \\ d & d' & c \end{matrix} \right\}$$

$$= \delta_{fg}(-1)^{\phi_2} \left\{ \begin{matrix} a' & a & c' \\ g & d' & c \end{matrix} \right\},$$

$$\phi_2 = g + a' - b + c + d' + d.$$

The W-coefficient form of these relations is obtained by deleting all phase factors and making the substitution (2.74), ignoring the phase factor. There are no phase factors in the corresponding W-coefficient relations.

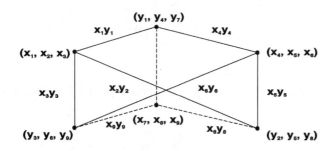

Figure 2.3. Labeled Cubic Graph Associated with the 9-j Coefficient.

2.11.7 Schwinger-Wu Generating Function and Its Combinatorics

Triangles associated with the 9-j coefficient $\left\{ \begin{matrix} j_1 j_2 j_3 \\ j_4 j_5 j_6 \\ j_7 j_8 j_9 \end{matrix} \right\}$:

$(j_1 j_2 j_3), (j_4 j_5 j_6), (j_7 j_8 j_9), (j_1 j_4 j_7), (j_2 j_5 j_8), (j_3 j_6 j_9)$.

Points in $\mathbf{R}^3$ associated with the triangles:

$$(j_1 j_2 j_3) \rightarrow (x_1, x_2, x_3), \quad (j_4 j_5 j_6) \rightarrow (x_4, x_5, x_6),$$
$$(j_7 j_8 j_9) \rightarrow (x_7, x_8, x_9), \quad (j_1 j_4 j_7) \rightarrow (y_1, y_4, y_7),$$
$$(j_2 j_5 j_8) \rightarrow (y_2, y_5, y_8), \quad (j_3 j_6 j_9) \rightarrow (y_3, y_6, y_9).$$

Cubic graph $\mathbf{C}_6$ in $\mathbf{R}^3$ associated with the points:

The points define the vertices of a cubic graph $\mathbf{C}_6$ on six points with lines joining each pair of points that share a common subscript, and the lines are labeled by the product of the common coordinates (see Fig. 2.3). Cubic graph $\mathbf{C}_6$ functions:

Interchange the symbols x and y in the coordinates of the vertices of the cubic graph $\mathbf{C}_6$, and define the following polynomials on the vertices and edges of the $\mathbf{C}_6$ with this modified labeling:

Vertex function: multiply together the coordinates of each pair of adjacent vertices, divide out the coordinates with a common subscript, and sum over all pairs of vertices to obtain

$$V_4 = y_1 y_2 x_6 x_9 + y_1 y_3 x_5 x_8 + y_2 y_3 x_4 x_7$$
$$+ y_4 y_5 x_3 x_9 + y_4 y_6 x_2 x_8 + y_5 y_6 x_1 x_7$$
$$+ y_7 y_8 x_3 x_6 + y_7 y_9 x_2 x_5 + y_8 y_9 x_1 x_4.$$

Edge function:

$$E_6 = \det \begin{bmatrix} x_1 y_1 & x_2 y_2 & x_3 y_3 \\ x_4 y_4 & x_5 y_5 & x_6 y_6 \\ x_7 y_7 & x_8 y_8 & x_9 y_9 \end{bmatrix}.$$

Generating function (see Ref. [3]):

$$(1 - V_4 + E_6)^{-2} = \sum_\Delta C(\Delta) Z^\Delta,$$

$$Z^\Delta = \prod_{\text{all vertices}} (z_a, z_b, z_c)^{(j_a j_b j_c)} \quad \text{(see Eq. (2.82))},$$

$$\Delta = \begin{bmatrix} (j_1 j_2 j_3) \\ (j_4 j_5 j_6) \\ (j_7 j_8 j_9) \\ (j_1 j_4 j_7) \\ (j_2 j_5 j_8) \\ (j_3 j_6 j_9) \end{bmatrix},$$

$$C(\Delta) = \sum_k \sum_\square \sum_a (-1)^{k_{10}+k_{11}+k_{12}} (k+1)$$
$$\times \binom{k}{k_1, \ldots, k_9, k_{10}, \ldots, k_{15}},$$

where summation $\sum_\square$ is over all 3×3 square arrays of nonnegative integers $k_j (j = 1, 2, \ldots, 9)$ with fixed row and column sums given by

$$\begin{bmatrix} k_1 & k_2 & k_3 \\ k_4 & k_5 & k_6 \\ k_7 & k_8 & k_9 \end{bmatrix} \begin{matrix} k - t_1 \\ k - t_2 \\ k - t_3 \end{matrix}$$
$$k - t_4 \quad k - t_5 \quad k - t_6$$

and for each such array the summation $\sum_a$ is over all nonnegative integers a such that the following quantities are nonnegative integers:

$$k_{10} = -a + k_1 - k + j_2 + j_3 + j_4 + j_7,$$
$$k_{11} = -a + k_6 - k + j_3 + j_4 + j_5 + j_9,$$
$$k_{12} = -a + k_8 - k + j_2 + j_5 + j_7 + j_9,$$
$$k_{13} = a + k_5 - k_1 - j_3 + j_6 - j_7 + j_8,$$
$$k_{14} = a + k_2 - k_6 + j_1 - j_4 + j_8 - j_9,$$
$$k_{15} = a.$$

Note that

$$\sum_{i=10}^{15} k_i = -2k + \sum_{i=1}^{9} j_i.$$

The t_i are the following triangle sums:

$$t_1 = j_1 + j_2 + j_3, t_2 = j_4 + j_5 + j_6,$$
$$t_3 = j_7 + j_8 + j_9,$$
$$t_4 = j_1 + j_4 + j_7, t_5 = j_2 + j_5 + j_8,$$
$$t_6 = j_3 + j_6 + j_9.$$

The 9-j coefficient is given by

$$\left\{ \begin{matrix} j_1 j_2 j_3 \\ j_4 j_5 j_6 \\ j_7 j_8 j_9 \end{matrix} \right\} = \Delta(j_1 j_2 j_3) \Delta(j_4 j_5 j_6) \Delta(j_7 j_8 j_9)$$

$$\times \Delta(j_1 j_4 j_7) \Delta(j_2 j_5 j_8) \Delta(j_3 j_6 j_9) C(\Delta).$$

The coefficient $C(\Delta)$ is an integer associated with each cubic graph $\mathbf{C}_6$ that counts the number of occurrences of the monomial term Z^Δ in the expansion of $(1 - V_4 + E_6)^{-2}$.

2.12 TENSOR SPHERICAL HARMONICS

Tensor spherical or solid harmonics are special cases of the coupling of two irreducible tensor operators in the tensor product space given in Sect. 2.8.2. They are defined by

$$\boldsymbol{\mathcal{Y}}^{(ls)jm} = \sum_\nu C^{l\ s\ j}_{m-\nu,\nu,m} \mathcal{Y}_{l,m-\nu} \otimes \xi_\nu$$

and belong to the tensor product space $\mathcal{H}_l \otimes \mathcal{H}'_s$, where the orthonormal bases of the spaces $\mathcal{H}_l$ and $\mathcal{H}'_s$ are:

$$\{\mathcal{Y}_{l\mu} : \mu = l, l-1, \ldots, -l\},$$
$$\{\xi_\nu : \nu = s, s-1, \ldots, -s\}.$$

The orbital angular momentum $\mathbf{L}$ has the standard action on the solid harmonics, and a second set of kinematically independent angular momentum operators $\mathbf{S}$ has the standard action on the basis set of $\mathcal{H}'_s$.
The total angular momentum is:

$$\mathbf{J} = \mathbf{L} \otimes \mathbf{1}' + \mathbf{1} \otimes \mathbf{S},$$

The set of vectors

$$\{\boldsymbol{\mathcal{Y}}^{(ls)jm} : m = j, j-1, \ldots, -j; (lsj) \text{ obey the}$$
$$\text{triangle conditions}\}$$

has the following properties:
Orthogonality:

$$\left\langle \boldsymbol{\mathcal{Y}}^{(l's)j'm'}, \boldsymbol{\mathcal{Y}}^{(ls)jm} \right\rangle$$
$$= \sum_{\nu\nu'} C^{l'\ s\ j'}_{m'-\nu',\nu',m'} C^{l\ s\ j}_{m-\nu,\nu,m} (\mathcal{Y}_{l',m'-\nu'}, \mathcal{Y}_{l,m-\nu})(\xi_{\nu'}, \xi_\nu)'$$
$$= \delta_{j'j}\delta_{l'l}\delta_{m'm},$$

where $\langle\,,\,\rangle$ denotes the inner product in the space $\mathcal{H}_l \otimes \mathcal{H}'_s$, $(\,,\,)$ the inner product in $\mathcal{H}_l$, and $(\,,\,)'$ the inner product in $\mathcal{H}'_s$.
Operator actions:

$$\mathbf{J}^2 \boldsymbol{\mathcal{Y}}^{(ls)jm} = j(j+1)\boldsymbol{\mathcal{Y}}^{(ls)jm},$$
$$J_3 \boldsymbol{\mathcal{Y}}^{(ls)jm} = m\boldsymbol{\mathcal{Y}}^{(ls)jm},$$
$$(\mathbf{L}^2 \otimes \mathbf{1}')\boldsymbol{\mathcal{Y}}^{(ls)jm} = l(l+1)\boldsymbol{\mathcal{Y}}^{(ls)jm},$$
$$(\mathbf{1} \otimes \mathbf{S}^2)\boldsymbol{\mathcal{Y}}^{(ls)jm} = s(s+1)\boldsymbol{\mathcal{Y}}^{(ls)jm},$$
$$\mathbf{J}^2 = \mathbf{L}^2 \otimes \mathbf{1}' + \mathbf{1} \otimes \mathbf{S}^2 + 2\sum_i L_i \otimes S_i,$$

$$J_\pm \boldsymbol{\mathcal{Y}}^{(ls)jm} = [(j \mp m)(j \pm m + 1)]^{1/2}\boldsymbol{\mathcal{Y}}^{(ls)j,m\pm1}.$$

Transformation property under unitary rotations:

$$\exp(-i\psi\hat{\mathbf{n}}{\cdot}\mathbf{J})\boldsymbol{\mathcal{Y}}^{(ls)jm} = \sum_{m'} D^j_{m'm}(\psi,\hat{\mathbf{n}})\boldsymbol{\mathcal{Y}}^{(ls)jm'}.$$

Special realization:
The eigenvectors ξ_ν are often replaced by column matrices:

$$\xi_\nu = \mathrm{col}(0\cdots010\cdots0),$$
$$1 \text{ in position } s - \nu + 1, \quad \nu = s, s-1, \ldots, -s.$$

The operators $\mathbf{S} = (S_1, S_2, S_3)$ are correspondingly replaced by their standard $(2s+1) \times (2s+1)$ matrix representations $S_i^{(s)}$. The tensor product of operators becomes a $(2s+1) \times (2s+1)$ matrix containing both operators and numerical matrix elements, e.g.,

$$J_i = L_i I_{2s+1} + S_i^{(s)},$$

in which L_i is a differential operator multiplying the unit matrix, that is, L_i is repeated $2s+1$ times along the diagonal.

2.12.1 Spinor Spherical Harmonics as Matrix Functions

Choose $\xi_{+1/2} = \mathrm{col}(1,0), \xi_{-1/2} = \mathrm{col}(0,1)$, and $\mathbf{S} = \boldsymbol{\sigma}/2$. The spinor spherical harmonics or Pauli central field spinors are the following, where $j \in \{1/2, 3/2, \ldots\}$:

$$\boldsymbol{\mathcal{Y}}^{(j-\frac{1}{2},\frac{1}{2})jm} = \begin{pmatrix} \sqrt{\dfrac{j+m}{2j}}\ \mathcal{Y}_{j-\frac{1}{2},m-\frac{1}{2}} \\[2ex] \sqrt{\dfrac{j-m}{2j}}\ \mathcal{Y}_{j-\frac{1}{2},m+\frac{1}{2}} \end{pmatrix},$$

$$\boldsymbol{\mathcal{Y}}^{(j+\frac{1}{2},\frac{1}{2})jm} = \begin{pmatrix} -\sqrt{\dfrac{j-m+1}{2j+2}}\ \mathcal{Y}_{j+\frac{1}{2},m-\frac{1}{2}} \\[2ex] \sqrt{\dfrac{j+m+1}{2j+2}}\ \mathcal{Y}_{j+\frac{1}{2},m+\frac{1}{2}} \end{pmatrix}.$$

2.12.2 Vector Spherical Harmonics as Matrix Functions

Choose $\xi_{+1} = \mathrm{col}(1,0,0), \xi_0 = \mathrm{col}(0,1,0), \xi_{-1} = \mathrm{col}(0,0,1)$, and $\mathbf{S}$ the 3×3 angular momentum matrices given by

$$S_+ = \begin{pmatrix} 0 & \sqrt{2} & 0 \\ 0 & 0 & \sqrt{2} \\ 0 & 0 & 0 \end{pmatrix}, \quad S_- = \begin{pmatrix} 0 & 0 & 0 \\ \sqrt{2} & 0 & 0 \\ 0 & \sqrt{2} & 0 \end{pmatrix},$$

$$S_3 = \begin{pmatrix} 1 & 0 & 0 \\ 0 & 0 & 0 \\ 0 & 0 & -1 \end{pmatrix}.$$

The vector spherical harmonics are the following, where $j \in \{0, 1, 2, \ldots\}$:

$$\boldsymbol{\mathcal{Y}}^{(j-1,1)jm} = \begin{pmatrix} \sqrt{\dfrac{(j+m-1)(j+m)}{2j(2j-1)}}\, \mathcal{Y}_{j-1,m-1} \\[3ex] \sqrt{\dfrac{(j-m)(j+m)}{j(2j-1)}}\, \mathcal{Y}_{j-1,m} \\[3ex] \sqrt{\dfrac{(j-m-1)(j-m)}{2j(2j-1)}}\, \mathcal{Y}_{j-1,m+1} \end{pmatrix},$$

$$\boldsymbol{\mathcal{Y}}^{(j1)jm} = \begin{pmatrix} -\sqrt{\dfrac{(j+m)(j-m+1)}{2j(j+1)}}\, \mathcal{Y}_{j,m-1} \\[3ex] \dfrac{m}{\sqrt{j(j+1)}}\, \mathcal{Y}_{j,m} \\[3ex] \sqrt{\dfrac{(j-m)(j+m+1)}{2j(j+1)}}\, \mathcal{Y}_{j,m+1} \end{pmatrix},$$

$$\boldsymbol{\mathcal{Y}}^{(j+1,1)jm} = \begin{pmatrix} \sqrt{\dfrac{(j-m+1)(j-m+2)}{2(j+1)(2j+3)}}\, \mathcal{Y}_{j+1,m-1} \\[3ex] -\sqrt{\dfrac{(j-m+1)(j+m+1)}{(j+1)(2j+3)}}\, \mathcal{Y}_{j+1,m} \\[3ex] \sqrt{\dfrac{(j+m+2)(j+m+1)}{2(j+1)(2j+3)}}\, \mathcal{Y}_{j+1,m+1} \end{pmatrix}.$$

Explicit operator actions:

$$\nabla^2 \boldsymbol{\mathcal{Y}}^{(l1)jm} = 0,$$
$$\mathbf{J}^2 \boldsymbol{\mathcal{Y}}^{(l1)jm} = j(j+1)\boldsymbol{\mathcal{Y}}^{(l1)jm},$$
$$J_3 \boldsymbol{\mathcal{Y}}^{(l1)jm} = m\boldsymbol{\mathcal{Y}}^{(l1)jm},$$
$$\mathbf{L}^2 \boldsymbol{\mathcal{Y}}^{(l1)jm} = l(l+1)\boldsymbol{\mathcal{Y}}^{(l1)jm},$$
$$\mathbf{S}^2 \boldsymbol{\mathcal{Y}}^{(l1)jm} = 2\boldsymbol{\mathcal{Y}}^{(l1)jm}.$$

2.12.3 Vector Solid Harmonics as Vector Functions

Vector spherical and solid harmonics can also be defined and their properties presented in terms of the ordinary solid harmonics, using the vectors $\mathbf{x}$, ∇, and $\mathbf{L}$, and the operations of divergence and curl:

Defining equations:

$$\boldsymbol{\mathcal{Y}}^{(l+1,1)lm} = -[(l+1)(2l+1)]^{-1/2}[(l+1)\mathbf{x} + i\mathbf{x}\times\mathbf{L}]\mathcal{Y}_{lm},$$
$$\boldsymbol{\mathcal{Y}}^{(l1)lm} = [l(l+1)]^{-1/2}\mathbf{L}\mathcal{Y}_{lm},$$
$$r^2\boldsymbol{\mathcal{Y}}^{(l-1,1)lm} = -[l(2l+1)]^{-1/2} \times [-l\mathbf{x} + i\mathbf{x}\times\mathbf{L}]\mathcal{Y}_{lm}.$$

Eigenvalue properties:

$$\mathbf{J}^2 \boldsymbol{\mathcal{Y}}^{(l1)jm} = j(j+1)\boldsymbol{\mathcal{Y}}^{(l1)jm},$$
$$\mathbf{L}^2 \boldsymbol{\mathcal{Y}}^{(l1)jm} = l(l+1)\boldsymbol{\mathcal{Y}}^{(l1)jm},$$
$$\mathbf{S}^2 \boldsymbol{\mathcal{Y}}^{(l1)jm} = 2\boldsymbol{\mathcal{Y}}^{(l1)jm},$$
$$J_3 \boldsymbol{\mathcal{Y}}^{(l1)jm} = m\boldsymbol{\mathcal{Y}}^{(l1)jm},$$
$$\nabla^2 \boldsymbol{\mathcal{Y}}^{(l1)jm} = \mathbf{0},$$
$$2i\mathbf{L}\times\boldsymbol{\mathcal{Y}}^{(l1)jm} = [j(j+1) - l(l+1) - 2]\boldsymbol{\mathcal{Y}}^{(l1)jm}.$$

Orthogonality:

$$\int dS_{\hat{\mathbf{x}}}\, \boldsymbol{\mathcal{Y}}^{(l'1)j'm'}(\mathbf{x}) \cdot \boldsymbol{\mathcal{Y}}^{(l1)jm}(\mathbf{x}) = \delta_{l'l}\delta_{j'j}\delta_{m'm}\, r^{l'+l},$$

where the integration is over the unit sphere in $\mathbf{R}^3$.
Complex conjugation:

$$\boldsymbol{\mathcal{Y}}^{(l1)jm*} = (-1)^{l+1-j}(-1)^m \boldsymbol{\mathcal{Y}}^{(l1)j,-m}.$$

Vector and gradient formulas:

$$\mathbf{x}\mathcal{Y}_{lm} = -\left[\frac{l+1}{2l+1}\right]^{1/2} \boldsymbol{\mathcal{Y}}^{(l+1,1)lm} + \left[\frac{l}{2l+1}\right]^{1/2} r^2 \boldsymbol{\mathcal{Y}}^{(l-1,1)lm},$$

$$[(l+1)\nabla + i\nabla\times\mathbf{L}](F\mathcal{Y}_{lm}) = -[(l+1)(2l+1)]^{1/2}\left(\frac{1}{r}\frac{dF}{dr}\right) \boldsymbol{\mathcal{Y}}^{(l+1,1)lm},$$

$$[-l\nabla + i\nabla\times\mathbf{L}](F\mathcal{Y}_{lm}) = -[l(2l+1)]^{1/2}\left[r\frac{dF}{dr} + (2l+1)F\right] \boldsymbol{\mathcal{Y}}^{(l-1,1)lm},$$

$$\nabla(F\mathcal{Y}_{lm}) = -\left[\frac{l+1}{2l+1}\right]^{1/2}\left(\frac{1}{r}\frac{dF}{dr}\right) \boldsymbol{\mathcal{Y}}^{(l+1,1)lm} + \left[\frac{l}{2l+1}\right]^{1/2}\left(r\frac{dF}{dr} + (2l+1)F\right) \boldsymbol{\mathcal{Y}}^{(l-1,1)lm},$$

$$i\nabla\times\mathbf{L}(F\mathcal{Y}_{lm}) = -l\left(\frac{l+1}{2l+1}\right)^{1/2}\left(\frac{1}{r}\frac{dF}{dr}\right) \boldsymbol{\mathcal{Y}}^{(l+1,1)lm} - (l+1)\left(\frac{l}{2l+1}\right)^{1/2}\left(r\frac{dF}{dr} + (2l+1)F\right) \times \boldsymbol{\mathcal{Y}}^{(l-1,1)lm}.$$

Curl equations:

$$i\nabla\times(F\boldsymbol{\mathcal{Y}}^{(l+1,1)lm}) = -\left[\frac{l}{2l+1}\right]^{1/2}\left(r\frac{dF}{dr} + (2l+3)F\right)\boldsymbol{\mathcal{Y}}^{(l1)lm},$$

$$i\nabla\times(F\boldsymbol{\mathcal{Y}}^{(l1)lm}) = -\left[\frac{l}{2l+1}\right]^{1/2}\left(\frac{1}{r}\frac{dF}{dr}\right)\boldsymbol{\mathcal{Y}}^{(l+1,1)lm} - \left[\frac{l+1}{2l+1}\right]^{1/2}\left(r\frac{dF}{dr} + (2l+1)F\right)\boldsymbol{\mathcal{Y}}^{(l-1,1)lm},$$

$$i\nabla\times(F\boldsymbol{\mathcal{Y}}^{(l-1,1)lm}) = -\left[\frac{l+1}{2l+1}\right]^{1/2}\left(\frac{1}{r}\frac{dF}{dr}\right)\boldsymbol{\mathcal{Y}}^{(l1)lm}.$$

Divergence equations:

$$\nabla \cdot (F \boldsymbol{\mathcal{Y}}^{(l+1,1)lm})$$
$$= -\left[\frac{l+1}{2l+1}\right]^{1/2}\left(r\frac{dF}{dr} + (2l+3)F\right)\mathcal{Y}_{lm},$$
$$\nabla \cdot (F\boldsymbol{\mathcal{Y}}^{(l1)lm}) = 0,$$
$$\nabla \cdot (F\boldsymbol{\mathcal{Y}}^{(l-1,1)lm}) = \left[\frac{l}{2l+1}\right]^{1/2}\left(\frac{1}{r}\frac{dF}{dr}\right)\mathcal{Y}_{lm}.$$

Parity property:

$$\boldsymbol{\mathcal{Y}}^{(l+\delta,1)lm}(-\mathbf{x}) = (-1)^{l+\delta}\boldsymbol{\mathcal{Y}}^{(l+\delta,1)lm}(\mathbf{x}).$$

Scalar product:

$$\boldsymbol{\mathcal{Y}}^{(l'1)j'm'}\cdot\boldsymbol{\mathcal{Y}}^{(l1)jm}$$
$$= \sum_{l''} r^{l+l'-l''}\left[\frac{(2j+1)(2j'+1)(2l+1)(2l'+1)}{4\pi(2l''+1)}\right]^{1/2}$$
$$\times (-1)^{l+j'+l''}C_{000}^{ll'l''}C_{m,m',m+m'}^{jj'l''}\left\{\begin{matrix} l' & j' & 1 \\ j & l & l'' \end{matrix}\right\}\mathcal{Y}_{l'',m+m'}.$$

Cross product:

$$\boldsymbol{\mathcal{Y}}^{(l'1)j'm'}\times\boldsymbol{\mathcal{Y}}^{(l1)jm} = (-i\sqrt{2})\sum_{l''j''} r^{l+l'-l''}$$
$$\times\left[\frac{(2j+1)(2j'+1)(3)(2l+1)(2l'+1)}{4\pi}\right]^{1/2}$$
$$\times C_{000}^{ll'l''}C_{m,m',m+m'}^{jj'j''}\left\{\begin{matrix} l & 1 & j \\ l' & 1 & j' \\ l'' & 1 & j'' \end{matrix}\right\}\boldsymbol{\mathcal{Y}}^{(l'1)j'',m+m'}.$$

Conversion to spherical harmonic form:

$$\boldsymbol{\mathcal{Y}}^{(l+\delta,1)lm}(\mathbf{x}) = r^{l+\delta}\boldsymbol{\mathcal{Y}}^{(l+\delta,1)lm}(\hat{\mathbf{x}}),$$

with appropriate modification of F to account for the factor $r^{l+\delta}$.

2.13 RECOUPLING THEORY AND 3N-J COEFFICIENTS

2.13.1 Composite Angular Momentum Systems

An "elementary" angular momentum system is one whose state space can be written as a direct sum of vector spaces $\mathcal{H}_j$ with orthonormal basis

$$\{|jm\rangle \mid m = j, j-1, \ldots, -j\}$$

on which the angular momentum $\mathbf{J}$ has the standard action, and which under unitary transformation by $\exp(-i\psi\hat{\mathbf{n}}\cdot\mathbf{J})$ undergoes the standard unitary transformation. A composite angular momentum system is one whose state space is a direct sum of the tensor product spaces $\mathcal{H}_{j_1 j_2 \cdots j_n}$ of dimension $\prod_{\alpha=1}^{n}(2j_\alpha + 1)$ with orthonormal basis in the tensor product space of the elementary systems given by

$$|j_1 m_1\rangle \otimes |j_2 m_2\rangle \otimes \cdots \otimes |j_n m_n\rangle, \quad (2.98)$$

each $m_\alpha = j_\alpha, j_\alpha - 1, \ldots, -j_\alpha$.

The following properties then hold for the composite system:

Independent rotations of the elementary parts:

$$\left(\exp\left(-i\psi_1\hat{\mathbf{n}}_1\cdot\mathbf{J}(1)\right)\otimes\cdots\otimes\exp\left(-i\psi_n\hat{\mathbf{n}}_n\cdot\mathbf{J}(n)\right)\right)$$
$$\times |j_1 m_1\rangle\otimes\cdots\otimes j_n m_n\rangle$$
$$= \exp\left(-i\psi_1\hat{\mathbf{n}}_1\cdot\mathbf{J}(1)\right)|j_1 m_1\rangle\otimes\cdots\otimes$$
$$\otimes\exp\left(-i\psi_n\hat{\mathbf{n}}_n\cdot\mathbf{J}(n)\right)|j_n m_n\rangle$$
$$= \sum_{m_1'\cdots m_n'}\left[D^{j_1}(U_1)\times\cdots\times D^{j_n}(U_n)\right]_{m_1'\cdots m_n';m_1\cdots m_n}$$
$$\times |j_1 m_1'\rangle\otimes\cdots\otimes|j_n m_n'\rangle, \quad (2.99)$$

$$\left[D^{j_1}(U_1)\times\cdots\times D^{j_n}(U_n)\right]_{m_1'\cdots m_n';m_1\cdots m_n}$$
$$= D^{j_1}_{m_1'm_1}(U_1)\cdots D^{j_n}_{m_n'm_n}(U_n),$$
$$U_\alpha = U(\psi_\alpha, \hat{\mathbf{n}}_\alpha) \in SU(2), \quad \alpha = 1, 2, \ldots, n.$$

Multiple Kronecker (direct) product group $SU(2) \times \cdots \times SU(2)$:

Group elements:

$$(U_1, \ldots, U_n), \quad \text{each } U_a \in SU(2).$$

Multiplication rule:

$$(U_1', \ldots, U_n')(U_1, \ldots, U_n) = (U_1'U_1, \ldots, U_n'U_n).$$

Irreducible representations:

$$D^{j_1}(U_1)\times\cdots\times D^{j_n}(U_n). \quad (2.100)$$

Rotation of the composite system as a unit:

Common rotation:

$$U_1 = U_2 = \cdots = U_n = U \in SU(2).$$

Diagonal subgroup $SU(2) \subset SU(2) \times \cdots \times SU(2)$:

$$(U, U, \ldots, U), \quad \text{each } U \in SU(2).$$

Reducibile representation of $SU(2)$:

$$D^{j_1}(U)\otimes\cdots\otimes D^{j_n}(U). \quad (2.101)$$

Total angular momentum of the composite system:

$$\mathbf{J} = \mathbf{J}(1) + \mathbf{J}(2) + \cdots + \mathbf{J}(n),$$

in which the k-th term in the sum is to be interpreted as the tensor product operator:

$\mathbf{I}_1 \otimes \cdots \otimes \mathbf{J}(k) \otimes \cdots \otimes \mathbf{I}_n, \mathbf{I}_\alpha = $ unit operator in $\mathcal{H}_{j_\alpha}$.

The basic problem for composite systems:

The basic problem is to reduce the n-fold direct product representation (2.101) of $SU(2)$ into a direct sum of irreducible representations, or equivalently, to find all subspaces $\mathcal{H}_j \subset \mathcal{H}_{j_1 j_2 \cdots j_n}, j \in \{0, 1/2, 1, \ldots\}$, with orthonormal bases sets $\{|jm\rangle| \ m = j, j-1, \ldots, -j\}$ on which the total angular momentum $\mathbf{J}$ has the standard action.

Form of the solution:

$$| (j_1 j_2 \cdots j_n)(k)jm \rangle \qquad (2.102)$$
$$= \sum_{\substack{\text{all } m_\alpha \\ \sum m_\alpha = m}} C^{j_1 j_2 \cdots j_n j}_{m_1 m_2 \cdots m_n m}(k)| \ j_1 m_1 \rangle \otimes | \ j_2 m_2 \rangle \otimes \cdots \otimes |j_n m_n\rangle \,,$$

$m = j, j-1, \ldots, -j$; index set (k) unspecified.

Diagonal operators:

$$\mathbf{J}^2(\alpha) = J_1^2(\alpha) + J_2^2(\alpha) + J_3^2(\alpha),$$
$$\qquad (2.103)$$
$$\mathbf{J}^2(\alpha) | (j_1 j_2 \cdots j_n)(k)jm \rangle$$
$$= j_\alpha(j_\alpha + 1) | (j_1 j_2 \cdots j_n)(k)jm \rangle \,,$$

$$\alpha = 1, 2, \ldots, n \,.$$

Total angular momentum properties imposed:

$$\mathbf{J}^2 | (j_1 j_2 \cdots j_n)(k)jm \rangle = j(j+1) | (j_1 j_2 \cdots j_n)(k)jm \rangle \,,$$
$$J_3 | (j_1 j_2 \cdots j_n)(k)jm \rangle = m | (j_1 j_2 \cdots j_n)(k)jm \rangle \,,(2.104)$$
$$J_\pm | (j_1 j_2 \cdots j_n)(k)jm \rangle$$
$$= [(j \mp m)(j \pm m + 1)]^{1/2} | (j_1 j_2 \cdots j_n)(k)jm \pm 1 \rangle \,.$$

Properties of the index set (k):
Reduction of Kronecker product:

$$D^{j_1} \times D^{j_2} \times \cdots \times D^{j_n} = \sum_j \oplus n_j D^j \,,$$
$$\qquad (2.105)$$
$$\prod_\alpha (2j_\alpha + 1) = \sum_j n_j(2j + 1) \,.$$

For fixed $j_1, j_2, \ldots, j_n$, and j, the index set (k) must enumerate exactly n_j perpendicular spaces $\mathcal{H}_j$.
Incompleteness of set of operators:
There are $2n$ commuting Hermitian operators diagonal on the basis (2.98):

$$\{\mathbf{J}^2(\alpha), J_3(\alpha) \ | \ \alpha = 1, 2, \ldots, n\}. \qquad (2.106a)$$

There are $n+2$ commuting Hermitian operators diagonal on the basis (2.102):

$$\{\mathbf{J}^2, J_3; \mathbf{J}^2(\alpha) \ | \ \alpha = 1, 2, \ldots, n\}. \qquad (2.106b)$$

There are $n - 2$ additional commuting Hermitian operators, or other rules, required to complete set (2.106b) and determine the indexing set (k).
Basic content of recoupling theory:

Recoupling theory is the study of completing the operator set (2.106b), or the specification of other rules, that uniquely determine the irreducible representation spaces $\mathcal{H}_j$ occuring in the Kronecker product reduction (2.105), and of the inter-relations between different methods of effecting this reduction. In effect, it is a study of the properties of the multiplicity space

$$\mathcal{H}_j \oplus \mathcal{H}_j \oplus \cdots \oplus \mathcal{H}_j \ (n_j \text{ terms}) \,.$$

2.13.2 Binary Coupling Theory: Combinatorics

Binary coupling of angular momenta refers to the selecting any pair of angular momentum operators from the set of individual system angular momenta

$$\{\mathbf{J}(1), \mathbf{J}(2), \ldots, \mathbf{J}(n)\},$$

and carrying out the "addition of angular momenta" for that pair by coupling the corresponding states in the tensor product space by the standard use of $SU(2)$ WCG-coefficients; this is followed by addition of a new pair, which may be a pair distinct from the first pair, or the addition of one new angular momentum to the sum of the first pair, etc. If the order $1, 2, \ldots, n$ of the angular momenta is kept fixed in

$$\mathbf{J}_1 + \mathbf{J}_2 + \cdots + \mathbf{J}_n \,, \qquad (2.107)$$

one is led to the *problem of parentheses*. (To avoid misleading parentheses, the notation $\mathbf{J}_\alpha = \mathbf{J}(\alpha)$ is used in this section.) This is the problem of introducing pairs of parentheses into expression (2.107) that specify the coupling procedure that is to be implemented. The procedure is clear from the following cases for $n = 2, 3$, and 4:

$$n = 2: \quad \mathbf{J}_1 + \mathbf{J}_2 \,;$$
$$n = 3: \quad (\mathbf{J}_1 + \mathbf{J}_2) + \mathbf{J}_3 \,,$$
$$\qquad \quad \mathbf{J}_1 + (\mathbf{J}_2 + \mathbf{J}_3) \,;$$
$$n = 4: \quad (\mathbf{J}_1 + \mathbf{J}_2) + (\mathbf{J}_3 + \mathbf{J}_4) \,,$$
$$\qquad \quad [(\mathbf{J}_1 + \mathbf{J}_2) + \mathbf{J}_3] + \mathbf{J}_4 \,,$$
$$\qquad \quad [\mathbf{J}_1 + (\mathbf{J}_2 + \mathbf{J}_3)] + \mathbf{J}_4 \,,$$
$$\qquad \quad \mathbf{J}_1 + [(\mathbf{J}_2 + \mathbf{J}_3) + \mathbf{J}_4] \,,$$
$$\qquad \quad \mathbf{J}_1 + [\mathbf{J}_2 + (\mathbf{J}_3 + \mathbf{J}_4)] \,.$$

It is customary to use the ordered sequence

$$j_1 j_2 \cdots j_n \qquad (2.108)$$

of angular momentum quantum numbers in place of the angular momentum operators in (2.107). Thus, the five placement of parentheses for $n = 4$ becomes:

$$(j_1 j_2)(j_3 j_4), [(j_1 j_2) j_3] j_4, [j_1(j_2 j_3)] j_4,$$
$$j_1[(j_2 j_3) j_4], j_1[j_2(j_3 j_4)].$$

(It is also customary to omit the last parentheses pair.) A sequence (2.108) into which pairwise insertions of parentheses has been completed is called a *binary bracketing* of the sequence, and denoted by $(j_1 j_2 \cdots j_n)^B$. This symbol may also be called a *coupling symbol*. The total number of coupling symbols, that is, the total number of elements a_n in the set

$$\{(j_1 j_2 \cdots j_n)^B \mid B \text{ is a binary bracketing}\}$$

is given by the Catalan numbers:

$$a_n = \frac{1}{n}\binom{2n-2}{n-1}, \qquad n = 2, 3, \ldots.$$

Effect of permuting the angular momenta:

Since the position of an individual vector space in the tensor product $\mathcal{H}_{j_1} \otimes \cdots \otimes \mathcal{H}_{j_n}$ is kept fixed, the meaning of a permutation of the j_α in the sequence (2.108) corresponding to a given binary bracketing is to permute the positions of the terms in the summation for the total angular momentum, e.g., $(j_1 j_2) j_3 \rightarrow (j_3 j_1) j_2$ corresponds to

$$(\mathbf{J}_1 \otimes \mathbf{I}_2 \otimes \mathbf{I}_3 + \mathbf{I}_1 \otimes \mathbf{J}_2 \otimes \mathbf{I}_3) + \mathbf{I}_1 \otimes \mathbf{I}_2 \otimes \mathbf{J}_3$$
$$= (\mathbf{I}_1 \otimes \mathbf{I}_2 \otimes \mathbf{J}_3 + \mathbf{J}_1 \otimes \mathbf{I}_2 \otimes \mathbf{I}_3) + \mathbf{I}_1 \otimes \mathbf{J}_2 \otimes \mathbf{I}_3.$$

Total number of binary bracketing schemes including permutations:

The number of symbols in the set

$$\left\{ (j_{\alpha_1} j_{\alpha_2} \cdots j_{\alpha_n})^B \middle| \begin{array}{l} B \text{ is a binary bracketing and} \\ \alpha_1 \alpha_2 \cdots \alpha_n \text{ is a permutation of} \\ 1, 2, \ldots, n \end{array} \right\}$$

$$(2.109)$$

is $c_n = n! a_n = (n)_{n-1} = n(n+1) \cdots (2n-2)$.
Caution: One should not assign numbers to the symbols j_α, since these symbols serve as *noncommuting, nonassociative distinct objects* in a counting process.
Binary subproducts:

A *binary subproduct* in the coupling symbol $(j_{\alpha_1} j_{\alpha_2} \cdots j_{\alpha_n})^B$ is the subset of symbols between a given parentheses pair, say, $\{xy\}$. The symbols x and y may themselves contain binary subproducts. *Commutation of a binary subproduct* is the operation $\{xy\} \rightarrow \{yx\}$. For example, the coupling symbol $\{[(j_1 j_2) j_3] j_4\}$ contains three binary subproducts, $\{xy\}, [xy], (xy)$.
Equivalence relation:

Two coupling symbols are defined to be *equivalent*

$$(j_{\alpha_1} j_{\alpha_2} \cdots j_{\alpha_n})^B \sim (j_{\alpha_1'} j_{\alpha_2'} \cdots j_{\alpha_n'})^B$$

if one can be obtained from the other by commutation of the symbols in the binary subproducts. Such commutations change the overall phase of the state vector (2.102) corresponding to a particular coupling symbol, and such states are counted as being the same (equivalent).
Number of inequivalent coupling schemes:

The equivalence relation under commutation of binary subproducts partitions the set (2.109) into equivalence classes, each containing 2^{n-1} elements. There are $d_n = c_n / 2^{n-1} = (2n-3)!!$ inequivalent coupling schemes in binary coupling theory. Thus, for $n = 4$, there are $5!! = 5 \times 3 \times 1 = 15$ inequivalent binary coupling schemes.
Type of a coupling symbol:

The *type* of the coupling symbol $(j_{\alpha_1} j_{\alpha_2} \cdots j_{\alpha_n})^B$ is defined to be the symbol obtained by setting all the j_α equal to a common symbol, say, x. Thus, the type of the coupling symbol $\{[(j_1 j_2) j_3] j_4\}$ is $\{[(x^2) x] x\}$.

The Wedderburn-Etherington number b_n gives the number of coupling symbols of distinct types, counting two symbols as equivalent if they are related by commutation of binary subproducts. A closed form of these numbers is not known, although generating functions exist. The first few numbers are:

n	1	2	3	4	5	6	7	8	9	10
b_n	1	1	1	2	3	6	11	23	46	98

There are 15 nontrivial coupling schemes for 4 angular momenta, and they are classified into 2 types, allowing commutation of binary subproducts:

$$\text{Type } [(x^2)x]x$$
$$[(j_1 j_2) j_3] j_4, [(j_2 j_3) j_1] j_4, [(j_3 j_1) j_2] j_4$$
$$[(j_1 j_2) j_4] j_3, [(j_2 j_4) j_1] j_3, [(j_4 j_1) j_2] j_3$$
$$[(j_1 j_3) j_4] j_2, [(j_3 j_4) j_1] j_2, [(j_4 j_1) j_3] j_2$$
$$[(j_2 j_3) j_4] j_1, [(j_3 j_4) j_2] j_1, [(j_4 j_2) j_3] j_1$$
$$\text{Type } (x^2)(x^2)$$
$$(j_1 j_2)(j_3 j_4), (j_1 j_3)(j_2 j_4), (j_2 j_3)(j_1 j_4)$$

2.13.3 Implementation of Binary Couplings

Each binary coupling scheme specifies uniquely a set of *intermediate angular momentum operators*. For example, the intermediate angular momenta associated with the coupling symbol $[(j_1 j_2) j_3] j_4$ are

$$\mathbf{J}(1) + \mathbf{J}(2) = \mathbf{J}(12), \quad \mathbf{J}(12) + \mathbf{J}(3) = \mathbf{J}(123),$$
$$\mathbf{J}(123) + \mathbf{J}(4) = \mathbf{J},$$

where $\mathbf{J}$ is the total angular momentum. Each coupling symbol $(j_{\alpha_1} j_{\alpha_2} \cdots j_{\alpha_n})^B$, defines exactly $n - 2$ intermediate angular momentum operators $\mathbf{K}(\lambda), \lambda = 1, 2, \ldots, n - 2$. The squares of these operators complete the set of operators (2.106b) for each coupling symbol; that is, the states vectors satisfying Eqs. (2.104)–(2.105) and the following equations are unique, up to an overall choice of phase factor:

$$
\begin{aligned}
\mathbf{K}^2(\lambda) & \left| (j_{\alpha_1} j_{\alpha_2} \cdots j_{\alpha_n})^B (k_1 k_2 \cdots k_{n-2}) jm \right\rangle \\
& = k_\lambda(k_\lambda+1) \left| (j_{\alpha_1} j_{\alpha_2} \cdots j_{\alpha_n})^B (k_1 k_2 \cdots k_{n-2}) jm \right\rangle, \\
& \quad \lambda = 1, 2, \ldots, n - 2, \quad n > 2.
\end{aligned}
\tag{2.110}
$$

The intermediate angular momentum operators $\mathbf{K}^2(\lambda)$ depend, of course, on the choice of binary couplings implicit in the symbol $(j_{\alpha_1} j_{\alpha_2} \cdots j_{\alpha_n})^B$. The vectors have the following properties:

Orthonormal basis of $\mathcal{H}_{j_1}(1) \otimes \cdots \otimes \mathcal{H}_{j_n}(n)$:

$$
\left\langle (j_\alpha)^B (k') jm | (j_\alpha)^B (k) jm \right\rangle = \prod_\lambda \delta_{k'_\lambda k_\lambda},
$$

$$
\begin{aligned}
(j_\alpha) &= (j_{\alpha_1}, j_{\alpha_2}, \ldots, j_{\alpha_n}), \\
(k) &= (k_1, k_2, \ldots, k_{n-2}), \\
(k') &= (k'_1, k'_2, \ldots, k'_{n-2}).
\end{aligned}
$$

The range of each k_λ is uniquely determined by the Clebsch-Gordan series and the binary couplings in the coupling symbol. Together these ranges enumerate exactly the multiplicity n_j of $\mathcal{H}_j$ occuring in the reduction of the multiple Kronecker product.

Uniqueness of state vectors:

$$
\begin{aligned}
& \left| (j_{\alpha_1} j_{\alpha_2} \cdots j_{\alpha_n})^B (k_1 k_2 \cdots k_{n-2}) jm \right\rangle \\
& = \sum_{\sum m_\alpha = m} C \left[\begin{pmatrix} j_{\alpha_1} \cdots j_{\alpha_n} \\ m_{\alpha_1} \cdots m_{\alpha_n} \end{pmatrix}^B \begin{matrix} j \\ m \end{matrix} \right] (k_1, \ldots, k_{n-2}) \\
& \quad \times |j_1 m_1\rangle \otimes \cdots \otimes |j_n m_n\rangle.
\end{aligned}
$$

In the C-coefficient, the $\begin{pmatrix} j_\alpha \\ m_\alpha \end{pmatrix}$ are paired in the binary bracketing. Each such C-coefficient is a summation over a unique product of $n - 1$ $SU(2)$ WCG-coefficients.

Equivalent basis vectors:

$$
\begin{aligned}
& \left| (j_{\alpha_1} j_{\alpha_2} \cdots j_{\alpha_n})^B (k_1 k_2 \cdots k_{n-2}) jm \right\rangle \\
& = \pm \left| (j_{\alpha'_1} j_{\alpha'_2} \cdots j_{\alpha'_n})^B (k_1 k_2 \cdots k_{n-2}) jm \right\rangle,
\end{aligned}
$$

if and only if $(j_{\alpha_1} j_{\alpha_2} \cdots j_{\alpha_n})^B \sim (j_{\alpha'_1} j_{\alpha'_2} \cdots j_{\alpha'_n})^B$. Inequivalent basis vector are orthonormal in all quantum numbers labeling the state vector.

Recoupling coefficients:

A recoupling coefficient is a transformation coefficient

$$
\left\langle (j_\beta)^B (l) jm \mid (j_\alpha)^B (k) jm \right\rangle
$$

relating any two orthonormal bases of the space $\mathcal{H}_{j_1} \otimes \cdots \otimes \mathcal{H}_{j_n}$, say, the one defined by Eqs. (2.104), (2.105),

and (2.110) for a prescribed coupling symbol and a second one, again defined by these relations but for a different coupling symbol. For example, for $n = 3$, there are 3 inequivalent coupling symbols and 3 recoupling coefficients; for $n = 4$, there are 15 inequivalent coupling symbols and 105 recoupling coefficients. Each coefficient is, of course, expressible as a sum over products of $2(n-1)$ WCG-coefficients, obtained simply by taking the inner product:

$$
\begin{aligned}
& \left\langle (j_\beta)^B (l) jm \mid (j_\alpha)^B (k) jm \right\rangle \\
& = \sum_{\sum m_\alpha = m} C \left[\begin{pmatrix} j_{\beta_1} \cdots j_{\beta_n} \\ m_{\beta_1} \cdots m_{\beta_n} \end{pmatrix}^B \begin{matrix} j \\ m \end{matrix} \right] (l) \\
& \quad \times C \left[\begin{pmatrix} j_{\alpha_1} \cdots j_{\alpha_n} \\ m_{\alpha_1} \cdots m_{\alpha_n} \end{pmatrix}^B \begin{matrix} j \\ m \end{matrix} \right] (k).
\end{aligned}
\tag{2.111}
$$

The fundamental theorem of binary coupling theory states:

> *Each recoupling coefficient is expressible as a sum over products of Racah coefficients, the only other quantities occurring in the summation being phase and dimension factors.*

In every instance, the summation over projection quantum numbers in the right-hand side of Eq. (2.111) is re-expressible as a sum over Racah coefficients.

2.13.4 Construction of All Transformation Coefficients in Binary Coupling Theory

Augmented notation:

The coupling symbol $(j_{\alpha_1} j_{\alpha_2} \cdots j_{\alpha_n})^B$ contains all information as to how n angular momenta are to be coupled, but is not specific in how the intermediate angular momentum quantum numbers $(k_1 k_2 \cdots k_{n-2})$, are to be matched with the binary couplings implicit in the coupling symbol. For explicit calculations, it is necessary to remedy this deficiency in notation. This may be done by attaching the $n - 2$ intermediate angular momentum quantum numbers and the total angular momentum j as subscripts to the $n - 1$ parentheses pairs in the coupling symbol. For example, for $(j_1 j_2 j_3 j_4 j_5)^B = \{[(j_1 j_2)(j_3 j_4)] j_5\}$, this results in the replacement

$$
\{[(j_1 j_2)(j_3 j_4)] j_5\}(k_1 k_2 k_3) \rightarrow \{[(j_1 j_2)_{k_1} (j_3 j_4)_{k_2}]_{k_3} j_5\}_j.
$$

The basic coupling symbol structure is regained simply by ignoring all inferior letters.

Basic rules for commutation and association:

Let x, y, z denote arbitrary disjoint contiguous sub-coupling symbols $\{[(x)(y)](z)\}$ contained in the coupling symbol $(j_{\alpha_1} j_{\alpha_2} \cdots j_{\alpha_n})^B$. Let a, b, c denote the intermediate angular momenta associated with addition of the angular momenta represented in x, y, z, respectively, d

the angular mometum representing the sum of a and b, and k the sum of d and c. Symbolically, this subcoupling may be presented as

$$\sum \mathbf{J}(x) = \mathbf{J}(a), \quad \sum \mathbf{J}(y) = \mathbf{J}(b), \quad \sum \mathbf{J}(z) = \mathbf{J}(c),$$
$$\mathbf{J} = \cdots \{[\mathbf{J}(a) + \mathbf{J}(b)] + \mathbf{J}(c)\} \cdots;$$
$$\mathbf{J}(a) + \mathbf{J}(b) = \mathbf{J}(d); \quad \mathbf{J}(d) + \mathbf{J}(c) = \mathbf{J}(k)$$

with augmented coupling symbol

$$(j_{\alpha_1} j_{\alpha_2} \cdots j_{\alpha_n})^B = \cdots \{[(x)_a (y)_b]_d (z)_c\}_k \cdots.$$

There are only two basic operations in constructing the recoupling coefficient between any two coupling schemes:
with the transformation of state vector given by

$$| \cdots [(x)_a (y)_b]_d \cdots \rangle \to (-1)^{a+b-d} | \cdots [(y)_b (x)_a]_d \cdots \rangle$$
$$= | \cdots [(x)_a (y)_b]_d \cdots \rangle.$$

Association of symbols:

$$[(x)_a (y)_b](z)_c \to (x)_a [(y)_b (z)_c]$$

with the transformation of state vector given by

$$| \cdots \{[(x)_a (y)_b]_d (z)_c\}_k \cdots \rangle$$
$$\to \sum_e [(2d+1)(2e+1)]^{1/2} W(abkc; de)$$
$$\times | \cdots \{(x)_a [(y)_b (z)_c]_e\}_k \cdots \rangle$$
$$= | \cdots \{[(x)_a (y)_b]_d (z)_c\}_k \cdots \rangle.$$

The basic result for the calculation of all recoupling coefficients is:
Each pair of coupling symbols for n angular momenta may be brought into coincidence by a series of commutations and associations performed on either of the symbols.

This result solves completely the construction of all transformation coefficients and sets the stage for the formulation of still deeper questions arising in recoupling theory, as summarized in Sect. 2.13.5. The following examples illustrate the content of the preceding abstract constructions.
Examples:
WCG-coefficient form:

$$\langle \{[(ab)_e c]_f d\}_g \mid [(ac)_h (bd)_k]_g \rangle$$
$$= \sum_{\alpha+\beta+\gamma+\delta=m} C^{a\,b\,e}_{\alpha,\beta,\alpha+\beta} C^{e\,c\,f}_{\alpha+\beta,\gamma,\alpha+\beta+\gamma} C^{f\,d\,g}_{\alpha+\beta+\gamma,\delta,m}$$
$$\times C^{a\,c\,h}_{\alpha,\gamma,\alpha+\gamma} C^{b\,d\,k}_{\beta,\delta,\beta+\delta} C^{h\,k\,g}_{\alpha+\gamma,\beta+\delta,m}.$$

Racah (6-j) coefficient as recoupling coefficient:

$$(ac)(bd) \xrightarrow{R} [(ac)b]d \xrightarrow{\phi} [b(ac)]d \xrightarrow{R} [(ba)c]d \xrightarrow{\phi} [(ab)c]d,$$

where ϕ denotes that the transformation corresponds to a transposition (phase factor change) and R denotes that the transformation is a Racah coefficient transformation:

$$\langle \{[(ab)_e c]_f d\}_g \mid [(ac)_h (bd)_k]_g \rangle$$
$$= (-1)^{e+h-a-f} [(2f+1)(2k+1)]^{1/2} W(hbgd; fk)$$
$$\times [(2e+1)(2h+1)]^{1/2} W(bafc; eh).$$

9-j coefficient as recoupling coefficient:

$$(ac)(bd) \xrightarrow{R} [(ac)b]d \xrightarrow{\phi} [b(ac)]d \xrightarrow{R} [(ba)c]d$$
$$\xrightarrow{\phi} [(ab)c]d \xrightarrow{R} (ab)(cd),$$
$$\langle [(ab)_e (cd)_f]_g \mid [(ac)_h (bd)_k]_g \rangle$$
$$= [(2e+1)(2f+1)(2h+1)(2k+1)]^{1/2}$$
$$\times \sum_l (-1)^{2l} (2l+1) \left\{ \begin{matrix} ach \\ lbe \end{matrix} \right\} \left\{ \begin{matrix} bdk \\ ghl \end{matrix} \right\} \left\{ \begin{matrix} efg \\ dlc \end{matrix} \right\}$$
$$= [(2e+1)(2f+1)(2h+1)(2k+1)]^{1/2} \left\{ \begin{matrix} abe \\ cdf \\ hkg \end{matrix} \right\}.$$

2.13.5 Unsolved Problems in Recoupling Theory

(1) Define a *route* between two coupling symbols for n angular momenta to be any sequence of transpositions and associations that carries one symbol into the other. Each such route then gives rise to a unique expression for the corresponding recoupling coefficient in terms of 6-j coefficients. In general, there are several routes between the same pair of coupling symbols, leading therefore to identities between 6-j coefficients. How many nontrivial routes are there between two given coupling symbols, leading to nontrivial relations between 6-j coefficients (trivial means related by a phase factor)?

(2) Only 6-j coefficients arise in all possible couplings of three angular momenta; only 6-j and 9-j coefficients arise in all possible couplings of four angular momenta; in addition to 6-j and 9-j coefficients, two new "classes" of coefficients, called 12-j coefficients of the first and second kind, arise in the coupling of five angular momenta; in addition to 6-j, 9-j, and the two classes of 12-j coefficients, five new classes of 15-j coefficients arise in the coupling of six angular momenta, $\cdots$. What are the classes of $3n$-j coefficients? The nonconstructive answer is that a summation over 6-j coefficients arising in the coupling of n angular momenta is of a new class if it cannot be expressed as a simple product (no sums) of previously defined coefficients occurring in the coupling of $n-1$ or less angular momenta.

(3) Toward answering the question of classes of $3n$-j coefficients, one is lead into the classification problem

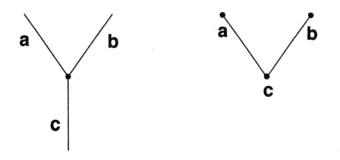

Figure 2.4. The Fundamental Triangle $[(ab)c]$ can be Realized by Lines or Points.

of planar cubic graphs. It is known that every $3n\text{-}j$ coefficient corresponds to a planar cubic graph, but the converse is not true. For small n, the relation between the coupling of n angular momenta, the number of new classes of $3(n-1)\text{-}j$ coefficients, and the number of nonisomorphic planar cubic graphs on $2(n-1)$ points is:

n	classes of $3(n-1)-j$ coefficients	cubic graphs on $2(n-1)$ points
3	1	1
4	1	2
5	2	5
6	5	19
7	18	87
8	84	?
9	576	?

The geometrical object for $n = 3$ is a planar graph isomorphic to the tetrahedron in 3-space. The classification of all nonisomorphic cubic graphs on $2(n-1)$ points is an unsolved problem in mathematics, as is the classification of classes for $3(n-1)\text{-}j$ coefficients.

(4) There are (at least) two methods of realizing the basic triangles of angular momentum theory in terms of graphs. The fundamental structural element $[(ab)c]$ is represented either in terms of its points or in terms of its lines (see Fig. 2.4):

The first representation leads to the interpretation of recoupling coefficients as functions defined on pairs of labeled binary trees (Ref. [1]); the second to the diagrams of the Jucys school [4, 5]. Either method leads to the relationship of recoupling coefficients to cubic graphs.

(5) The approach of classifying $3n-j$ coefficients through the use of unit tensor operator couplings, Racah operators, 9-j invariant operators, and general invariant operators is undeveloped.

Acknowlegments

This contribution on angular momentum theory is dedicated to Lawrence C. Biedenharn, whose tireless and continuing efforts in bringing understanding and structure to this complex subject is everywhere imprinted.

The author expresses his gratitude to Debi Erpenbeck, whose artful mastery of TeX and scrupulous attention to detail allowed the numerous complex relations to be displayed in two-column format.

Thanks are also given to Professors Brian Judd and Gordon Drake for the opportunity to make this contribution.

Author's note. It is quite impossible to attribute credits fairly in this subject because of its diverse origins across all areas of physics, chemistry, and mathematics. Any attempt to do so would likely be as misleading as it is informative. Most of the material is rooted in the very foundations of quantum theory itself, and the physical problems it addresses, making it still more difficult to assess unambiguous credit of ideas. Pragmatically, there is also the problem of confidence in the detailed correctness of complicated relationships, which prejudices one to cite those relationships personally checked. This accounts for the heavy use of formulas from Ref. cite46-2, which is, by far, the most often used source. But most of that material itself is derived from other primary sources, and an inadequate attempt was made there to indicate the broad base of origins. While one might expect to find in a reference book a comprehensive list of credits for most of the formulas, it has been necessary to weigh the relative merits of presenting a mature subject from a viewpoint of conceptual unity versus credits for individual contributions. The first position was adopted. Nonetheless, there is an obligation to indicate the origins of a subject, noting those works that have been most influential in its developments. The list of textbooks and seminal articles given in the references is intended to serve this purpose, however inadequately.

Exerpts and Fig. 2.1 are reprinted from Biedenharn and Louck [1] with permission of Cambridge University Press. Tables 2.2–2.4 have been adapted from Edmonds [6] by permission of Princeton University Press. Thanks are given for this cooperation.

2.14 TABLES

Table 2.1. The solid and spherical harmonics $\mathcal{Y}_{lm}$, and the tensor harmonics $\mathcal{T}_\mu^k$ (labeled by $k = l$ and $\mu = m$) for $l = 0$, 1, 2, 3, and 4.

l	m	$\sqrt{4\pi}\,\mathcal{Y}_{lm}(\mathbf{r})$	$\sqrt{4\pi}\,\mathcal{Y}_{lm}(\theta,\varphi)$	$\mathcal{T}_m^l$
1	± 1	$\mp\sqrt{\frac{3}{2}}\,(x \pm iy)$	$\mp\sqrt{\frac{3}{2}}\,e^{\pm i\varphi}\sin\theta$	$\mp\sqrt{2}\,J_\pm$
	0	$\sqrt{3}\,z$	$\sqrt{3}\cos\theta$	$2J_3$
2	± 2	$\frac{1}{2}\sqrt{\frac{15}{2}}\,(x \pm iy)^2$	$\frac{1}{2}\sqrt{\frac{15}{2}}\,e^{\pm 2i\varphi}\sin^2\theta$	$\sqrt{6}\,J_\pm^2$
	± 1	$\mp\sqrt{\frac{15}{2}}\,(x \pm iy)z$	$\mp\sqrt{\frac{15}{2}}\,e^{\pm i\varphi}\sin\theta\cos\theta$	$\mp\sqrt{6}\,J_\pm(2J_3 \pm 1)$
	0	$\frac{1}{2}\sqrt{5}\,(3z^2 - r^2)$	$\frac{1}{2}\sqrt{5}\,(3\cos^2\theta - 1)$	$2(3J_3^2 - J^2)$
3	± 3	$\mp\frac{1}{4}\sqrt{35}\,(x \pm iy)^3$	$\mp\frac{1}{4}\sqrt{35}\,e^{\pm 3i\varphi}\sin^3\theta$	$\mp 2\sqrt{5}\,J_\pm^3$
	± 2	$\frac{1}{2}\sqrt{\frac{105}{2}}\,(x \pm iy)^2 z$	$\frac{1}{2}\sqrt{\frac{105}{2}}\,e^{\pm 2i\varphi}\sin^2\theta\cos\theta$	$2\sqrt{30}\,J_\pm^2(J_3 \pm 1)$
	± 1	$\mp\frac{1}{4}21(x \pm iy)(5z^2 - r^2)$	$\mp\frac{1}{4}\sqrt{21}\,e^{\pm i\varphi}\sin\theta(5\cos^2\theta - 1)$	$\mp 2\sqrt{3}\,J_\pm(5J_3^2 - J^2 \pm 5J_3 + 2)$
	0	$\frac{1}{2}\sqrt{7}\,(5z^2 - 3r^2)z$	$\frac{1}{2}\sqrt{7}\,(5\cos^2\theta - 3)\cos\theta$	$4(5J_3^2 - 3J^2 + 1)J_3$
4	± 4	$\frac{3}{16}\sqrt{70}\,(x \pm iy)^4$	$\frac{3}{16}\sqrt{70}\,e^{\pm 4i\varphi}\sin^4\theta$	$\sqrt{70}\,J_\pm^4$
	± 3	$\mp\frac{3}{4}\sqrt{35}\,(x \pm iy)^3 z$	$\mp\frac{3}{4}\sqrt{35}\,e^{\pm 3i\varphi}\sin^3\theta\cos\theta$	$\mp 2\sqrt{35}\,J_\pm^3(2J_3 \pm 3)$
	± 2	$\frac{3}{8}\sqrt{10}\,(x \pm iy)^2(7z^2 - r^2)$	$\frac{3}{8}\sqrt{10}\,e^{\pm 2i\varphi}\sin^2\theta(7\cos^2\theta - 1)$	$2\sqrt{10}\,J_\pm^2(7J_3^2 - J^2 \pm 14J_3 + 9)$
	± 1	$\mp\frac{3}{4}\sqrt{5}\,(x \pm iy)(7z^2 - 3r^2)z$	$\mp\frac{3}{4}\sqrt{5}\,e^{\pm i\varphi}\sin\theta(7\cos^2\theta - 3)\cos\theta$	$\mp\sqrt{5}\,J_\pm(28J_3^3 - 12J^2 J_3 \pm 42J_3^2$ $-6J^2 + 38J_3 \pm 12)$
	0	$\frac{15}{8}(7z^4 - 6z^2 r^2 + \frac{3}{5}r^4)$	$\frac{15}{8}(7\cos^4\theta - 6\cos^2\theta + \frac{3}{5})$	$70J_3^4 - 60J^2 J_3^2 + 6(J^2)^2 + 50J_3^2$ $-12J^2$

Table 2.2. The 3-j coefficients for all M's $= 0$, or $J_3 = 0, \frac{1}{2}$.

$$
\begin{pmatrix} J_1 & J_2 & J_3 \\ 0 & 0 & 0 \end{pmatrix} = (-1)^{\frac{1}{2}J}\left[\frac{(J_1 + J_3 - J_3)!(J_1 + J_3 - J_2)!(J_2 + J_3 - J_1)!}{(J - 1 + J_2 + J_3 + 1)!}\right]^{1/2}
$$

$$
\times \frac{(\frac{1}{2}J)!}{(\frac{1}{2}J - J_1)!(\frac{1}{2}J - J_2)!(\frac{1}{2}J - J_3)!}, \quad J \text{ even}
$$

$$
\begin{pmatrix} J_1 & J_2 & J_3 \\ 0 & 0 & 0 \end{pmatrix} = 0, \quad J \text{ odd, where } J = J_1 + J_2 + J_3
$$

$$
\begin{pmatrix} J & J & 0 \\ M & -M & 0 \end{pmatrix} = (-1)^{J-M}\frac{1}{(2J+1)^{1/2}}
$$

$$
\begin{pmatrix} J + \frac{1}{2} & J & \frac{1}{2} \\ M & -M - \frac{1}{2} & \frac{1}{2} \end{pmatrix} = (-1)^{J-M-\frac{1}{2}}\left[\frac{J - M + \frac{1}{2}}{(2J+2)(2J+1)}\right]^{1/2}
$$

Table 2.3. The 3-j coefficients for $J_3 = 1, \frac{3}{2}, 2$.

$$\begin{pmatrix} J+1 & J & 1 \\ M & -M-1 & 1 \end{pmatrix} = (-1)^{J-M-1}\left[\frac{(J-M)(J-M+1)}{(2J+3)(2J+2)(2J+1)}\right]^{1/2}$$

$$\begin{pmatrix} J+1 & J & 1 \\ M & -M & 0 \end{pmatrix} = (-1)^{J-M-1}\left[\frac{2(J+M+1)(J-M+1)}{(2J+3)(2J+2)(2J+1)}\right]^{1/2}$$

$$\begin{pmatrix} J & J & 1 \\ M & -M-1 & 1 \end{pmatrix} = (-1)^{J-M}\left[\frac{2(J-M)(J+M+1)}{(2J+2)(2J+1)(2J)}\right]^{1/2}$$

$$\begin{pmatrix} J & J & 1 \\ M & -M & 0 \end{pmatrix} = (-1)^{J-M}\frac{M}{[(2J+1)(J+1)J]^{1/2}}$$

$$\begin{pmatrix} J+\frac{3}{2} & J & \frac{3}{2} \\ M & -M-\frac{3}{2} & \frac{3}{2} \end{pmatrix} = (-1)^{J-M+\frac{1}{2}}\left[\frac{(J-M-\frac{1}{2})(J-M+\frac{1}{2})(J-M+\frac{3}{2})}{(2J+4)(2J+3)(2J+2)(2J+1)}\right]^{1/2}$$

$$\begin{pmatrix} J+\frac{3}{2} & J & \frac{3}{2} \\ M & -M-\frac{1}{2} & \frac{1}{2} \end{pmatrix} = (-1)^{J-M+\frac{1}{2}}\left[\frac{3(J-M+\frac{1}{2})(J-M+\frac{3}{2})(J+M+\frac{3}{2})}{(2J+4)(2J+3)(2J+2)(2J+1)}\right]^{1/2}$$

$$\begin{pmatrix} J+\frac{1}{2} & J & \frac{3}{2} \\ M & -M-\frac{3}{2} & \frac{3}{2} \end{pmatrix} = (-1)^{J-M-\frac{1}{2}}\left[\frac{3(J-M-\frac{1}{2})(J-M+\frac{1}{2})(J+M+\frac{3}{2})}{(2J+3)(2J+2)(2J+1)2J}\right]^{1/2}$$

$$\begin{pmatrix} J+\frac{1}{2} & J & \frac{3}{2} \\ M & -M-\frac{1}{2} & \frac{1}{2} \end{pmatrix} = (-1)^{J-M-\frac{1}{2}}\left[\frac{J-M+\frac{1}{2}}{(2J+3)(2J+2)(2J+1)2J}\right]^{1/2}(J+3M+\frac{3}{2})$$

$$\begin{pmatrix} J+2 & J & 2 \\ M & -M-2 & 2 \end{pmatrix} = (-1)^{J-M}\left[\frac{(J-M-1)(J-M)(J-M+1)(J-M+2)}{(2J+5)(2J+4)(2J+3)(2J+2)(2J+1)}\right]^{1/2}$$

$$\begin{pmatrix} J+2 & J & 2 \\ M & -M-1 & 1 \end{pmatrix} = (-1)^{J-M}\left[\frac{(J+M+2)(J-M+2)(J-M+1)(J-M)}{(2J+5)(2J+4)(2J+3)(2J+2)(2J+1)}\right]^{1/2}$$

$$\begin{pmatrix} J+2 & J & 2 \\ M & -M & 0 \end{pmatrix} = (-1)^{J-M}\left[\frac{6(J+M+2)(J+M+1)(J-M+2)(J-M+1)}{(2J+5)(2J+4)(2J+3)(2J+2)(2J+1)}\right]^{1/2}$$

$$\begin{pmatrix} J+1 & J & 2 \\ M & -M-2 & 2 \end{pmatrix} = 2(-1)^{J-M+1}\left[\frac{(J-M-1)(J-M)(J-M+1)(J+M+2)}{(2J+4)(2J+3)(2J+2)(2J+1)2J}\right]^{1/2}$$

$$\begin{pmatrix} J+1 & J & 2 \\ M & -M-1 & 1 \end{pmatrix} = (-1)^{J-M+1}2(J+2M+2)\left[\frac{(J-M+1)(J-M)}{(2J+4)(2J+3)(2J+2)(2J+1)2J}\right]^{1/2}$$

$$\begin{pmatrix} J+1 & J & 2 \\ M & -M & 0 \end{pmatrix} = (-1)^{J-M+1}2M\left[\frac{6(J+M+1)(J-M+1)}{(2J+4)(2J+3)(2J+2)(2J+1)2J}\right]^{1/2}$$

$$\begin{pmatrix} J & J & 2 \\ M & -M-2 & 2 \end{pmatrix} = (-1)^{J-M}\left[\frac{6(J-M-1)(J-M)(J+M+1)(J+M+2)}{(2J+3)(2J+2)(2J+1)(2J)(2J-1)}\right]^{1/2}$$

$$\begin{pmatrix} J & J & 2 \\ M & -M-1 & 1 \end{pmatrix} = (-1)^{J-M}(1+2M)\left[\frac{6(J+M+1)(J-M)}{(2J+3)(2J+2)(2J+1)(2J)(2J-1)}\right]^{1/2}$$

$$\begin{pmatrix} J & J & 2 \\ M & -M & 0 \end{pmatrix} = (-1)^{J-M}\frac{2\left[3M^2 - J(J+1)\right]}{[(2J+3)(2J+2)(2J+1)(2J)(2J-1)]^{1/2}}$$

Table 2.4. The 6-j coefficients for $d = 0, \frac{1}{2}, 1, \frac{3}{2}, 2$, with $s = a + b + c$.

$$\begin{Bmatrix} a & b & c \\ 0 & e & f \end{Bmatrix} = (-1)^s \left[(2b+1)(2c+1)\right]^{-1/2} \delta_{bf}\delta_{ce}$$

$$\begin{Bmatrix} a & b & c \\ \frac{1}{2} & c-\frac{1}{2} & b+\frac{1}{2} \end{Bmatrix} = (-1)^s \left[\frac{(s-2b)(s-2c+1)}{(2b+1)(2b+2)2c(2c+1)}\right]^{1/2}$$

$$\begin{Bmatrix} a & b & c \\ \frac{1}{2} & c-\frac{1}{2} & b-\frac{1}{2} \end{Bmatrix} = (-1)^s \left[\frac{(s+1)(s-2a)}{2b(2b+1)2c(2c+1)}\right]^{1/2}$$

$$\begin{Bmatrix} a & b & c \\ 1 & c-1 & b-1 \end{Bmatrix} = (-1)^s \left[\frac{s(s+1)(s-2a-1)(s-2a)}{(2b-1)2b(2b+1)(2c-1)2c(2c+1)}\right]^{1/2}$$

$$\begin{Bmatrix} a & b & c \\ 1 & c-1 & b \end{Bmatrix} = (-1)^s \left[\frac{2(s+1)(s-2a)(s-2b)(s-2c+1)}{2b(2b+1)(2b+2)(2c-1)2c(2c+1)}\right]^{1/2}$$

$$\begin{Bmatrix} a & b & c \\ 1 & c-1 & b+1 \end{Bmatrix} = (-1)^s \left[\frac{(s-2b-1)(s-2b)(s-2c+1)(s-2c+2)}{(2b+1)(2b+2)(2b+3)(2c-1)2c(2c+1)}\right]^{1/2}$$

$$\begin{Bmatrix} a & b & c \\ 1 & c & b \end{Bmatrix} = (-1)^{s+1} \frac{2\left[b(b+1) + c(c+1) - a(a+1)\right]}{\left[2b(2b+1)(2b+2)2c(2c+1)(2c+2)\right]^{1/2}} ,$$

$$\begin{Bmatrix} a & b & c \\ \frac{3}{2} & c-\frac{3}{2} & b-\frac{3}{2} \end{Bmatrix} = (-1)^s \left[\frac{(s-1)s(s+1)(s-2a-2)(s-2a-1)(s-2a)}{(2b-2)(2b-1)2b(2b+1)(2c-2)(2c-1)2c(2c+1)}\right]^{1/2}$$

$$\begin{Bmatrix} a & b & c \\ \frac{3}{2} & c-\frac{3}{2} & b-\frac{1}{2} \end{Bmatrix} = (-1)^s \left[\frac{3s(s+1)(s-2a-1)(s-2a)(s-2b)(s-2b+1)}{(2b-1)2b(2b+1)(2b+2)(2c-2)(2c-1)2c(2c+1)}\right]^{1/2}$$

$$\begin{Bmatrix} a & b & c \\ \frac{3}{2} & c-\frac{3}{2} & b+\frac{1}{2} \end{Bmatrix} = (-1)^s \left[\frac{3(s+1)(s-2a)(s-2b-1)(s-2b)(s-2c+1)(s-2c+2)}{2b(2b+1)(2b+2)(2b+3)(2c-2)(2c-1)2c(2c+1)}\right]^{1/2}$$

$$\begin{Bmatrix} a & b & c \\ \frac{3}{2} & c-\frac{3}{2} & b+\frac{3}{2} \end{Bmatrix} = (-1)^s \left[\frac{(s-2b-2)(s-2b-1)(s-2b)(s-2c+1)(s-2c+2)(s-2c+3)}{(2b+1)(2b+2)(2b+3)(2b+4)(2c-2)(2c-1)2c(2c+1)}\right]^{1/2}$$

$$\begin{Bmatrix} a & b & c \\ \frac{3}{2} & c-\frac{1}{2} & b-\frac{1}{2} \end{Bmatrix} = (-1)^s \frac{\left[2(s-2b)(s-2c) - (s+2)(s-2a-1)\right]\left[(s+1)(s-2a)\right]^{1/2}}{\left[(2b-1)2b(2b+1)(2b+2)(2c-1)2c(2c+1)(2c+2)\right]^{1/2}}$$

$$\begin{Bmatrix} a & b & c \\ \frac{3}{2} & c-\frac{1}{2} & b+\frac{1}{2} \end{Bmatrix} = (-1)^s \frac{\left[(s-2b-1)(s-2c) - 2(s+2)(s-2a)\right]\left[(s-2b)(s-2c+1)\right]^{1/2}}{\left[2b(2b+1)(2b+2)(2b+3)2c(2c+1)(2c+2)(2c+3)\right]^{1/2}} ,$$

Table 2.4. *Continued.*

$$\begin{Bmatrix} a & b & c \\ 2 & c-2 & b-2 \end{Bmatrix} = (-1)^s \left[\frac{(s-2)(s-1)s(s+1)}{(2b-3)(2b-2)(2b-1)2b(2b+1)} \right]^{1/2}$$
$$\times \left[\frac{(s-2a-3)(s-2a-2)(s-2a-1)(s-2a)}{(2c-3)(2c-2)(2c-1)2c(2c+1)} \right]^{1/2}$$

$$\begin{Bmatrix} a & b & c \\ 2 & c-2 & b-1 \end{Bmatrix} = (-1)^s 2 \left[\frac{(s-1)s(s+1)}{(2b-2)(2b-1)2b(2b+1)(2b+2)} \right]^{1/2}$$
$$\times \left[\frac{(s-2a-2)(s-2a-1)(s-2a)(s-2b)(s-2c+1)}{(2c-3)(2c-2)(2c-1)2c(2c+1)} \right]^{1/2}$$

$$\begin{Bmatrix} a & b & c \\ 2 & c-2 & b \end{Bmatrix} = (-1)^s \left[\frac{6s(s+1)(s-2a-1)(s-2a)}{(2b-1)2b(2b+1)(2b+2)(2b+3)} \right]^{1/2}$$
$$\times \left[\frac{(s-2b)(s-2c+1)(s-2c+2)}{(2c-3)(2c-2)(2c-1)2c(2c+1)} \right]^{1/2}$$

$$\begin{Bmatrix} a & b & c \\ 2 & c-2 & b+1 \end{Bmatrix} = (-1)^s 2 \left[\frac{(s+1)(s-2a)(s-2b-2)(s-2b-1)(s-2b)}{2b(2b+1)(2b+2)(2b+3)(2b+4)} \right]^{1/2}$$
$$\times \left[\frac{(s-2c+1)(s-2c+2)(s-2c+3)}{(2c-3)(2c-2)(2c-1)2c(2c+1)} \right]^{1/2}$$

$$\begin{Bmatrix} a & b & c \\ 2 & c-2 & b+2 \end{Bmatrix} = (-1)^s \left[\frac{(s-2b-3)(s-2b-2)(s-2b-1)(s-2b)}{(2b+1)(2b+2)(2b+3)(2b+4)(2b+5)} \right]^{1/2}$$
$$\times \left[\frac{(s-2c+1)(s-2c+2)(s-2c+3)(s-2c+4)}{(2c-3)(2c-2)(2c-1)2c(2c+1)} \right]^{1/2}$$

$$\begin{Bmatrix} a & b & c \\ 2 & c-1 & b-1 \end{Bmatrix} = (-1)^s \frac{4\left[(a+b)(a-b+1)-(c-1)(c-b+1)\right]}{[(2b-2)(2b-1)2b(2b+1)(2b+2)]^{1/2}}$$
$$\times \left[\frac{s(s+1)(s-2a-1)(s-2a)}{(2c-2)(2c-1)2c(2c+1)(2c+2)} \right]^{1/2}$$

$$\begin{Bmatrix} a & b & c \\ 2 & c-1 & b \end{Bmatrix} = (-1)^s 2 \frac{\left[(a-b+1)(a-b)-c^2+1\right]}{[(2b-1)2b(2b+1)(2b+2)(2b+3)]^{1/2}}$$
$$\times \left[\frac{6(s+1)(s-2a)(s-2b)(s-2c+1)}{(2c-2)(2c-1)2c(2c+1)(2c+2)} \right]^{1/2}$$

$$\begin{Bmatrix} a & b & c \\ 2 & c-1 & b+1 \end{Bmatrix} = (-1)^s \frac{4\left[(a+b+2)(a-b-1)-(c-1)(b+c+2)\right]}{[2b(2b+1)(2b+2)(2b+3)(2b+4)]^{1/2}}$$
$$\times \left[\frac{(s-2b-1)(s-2b)(s-2c+1)(s-2c+2)}{(2c-2)(2c-1)2c(2c+1)(2c+2)} \right]^{1/2},$$

$$\begin{Bmatrix} a & b & c \\ 2 & c & b \end{Bmatrix} = (-1)^s \frac{2\left[3X(X-1)-4b(b+1)c(c+1)\right]}{[(2b-1)2b(2b+1)(2b+2)(2b+3)]^{1/2}}$$
$$\times \left[\frac{1}{(2c-1)2c(2c+1)(2c+2)(2c+3)} \right]^{1/2},$$

where $X = b(b+1) + c(c+1) - a(a+1)$

REFERENCES

Books and Monoghraphs

1. L. C. Biedenharn and J. D. Louck in, *Encyclopidia of Mathematics and Its Applications*, edited by G.-C. Rota (Addison- Wesley, Reading, 1981); presently by (Cambridge University Press, Cambridge), Vols. 8 & 9.

2. J. Schwinger, *On Angular Momentum*. U. S. Atomic Energy Commission Report NYO-3071, 1952 (unpublished). Published in *Quantum Theory of Angular Momentum* edited by L. C. Biedenharn and H. van Dam (Academic Press, New York, 1965), pp. 229–279.

3. B. R. Judd and G. M. S. Lister, J. Phys. A **20**, 3159 (1987).

4. A. P. Jucys, I. B. Levinson, and V. V. Vanagas, *The Theory of Angular Momentum* (Mathematicheskii apparat teorii momenta kolichestva dvizheniya.) (Vilnius, USSR., 1960). Translated from the Russian by A. Sen and A. R. Sen (Jerusalem, Israel, 1962).

5. A. P. Jucys and A. A. Bandzaitis, *Angular Momentum Theory in Quantum Physics* (Moksias, Vilnius, 1977).

6. A. R. Edmonds, *Angular Momentum in Quantum Mechanics* (Princeton University, Princeton, 1957).

7. E. Cartan, Thesis, Paris, Nony, 1894. [*Ouevres Complète*, Part 1, pp. 137–287. Gauthier-Villars, Paris, 1952.]

8. H. Weyl, *Gruppentheorie und Quantenmechanik* (Hirzel, Leipzig, 1928), 1st ed. Translated by H. P. Robertson as *The Theory of Groups and Quantum Mechanics* (Methuen, London, 1931).

9. P. A. M. Dirac, *The Principles of Quantum Mechanics* (Oxford University Press, London, 1958), 4th ed.

10. M. Born and P. Jordan, *Elementare Quantenmechanik* (Springer, Berlin, 1930).

11. H. B. G. Casimir, Thesis, University of Leyden, Wolters, Groningen, 1931. [*Koninkl. Ned. Akad. Wetenschap, Proc.* **34**, 844 (1931)].

12. B. L. van der Waerden, *Die gruppentheoretische Methode in der Quantenmechanik* (Springer, Berlin, 1932).

13. W. Pauli, in *Handbuch der Physik*, edited by H. Geiger and K. Scheel (Springer, Berlin, 1933), Vol. 24, Part 1, pp. 83–272. Later published in *Encyclopedia of Physics*, edited by S. Flügge (Springer, Berlin, 1958), Vol. 5, Part 1, pp. 45, 46.

14. H. Weyl, *The Structure and Representations of Continuous Groups* Lectures at the Institute for Advanced Study. Princeton, 1934–1935 (unpublished). Notes by R. Brauer.

15. E. U. Condon and G. H. Shortley, *The Theory of Atomic Spectra*. (Cambridge University, London, 1935).

16. H. A. Kramers, *Quantum Mechanics* (North-Holland, Amsterdam, 1957). Translation by D. ter Haar of Kramer's monograph published in the *Hand- und Jahrbuch der chemischen Physik* (1937).

17. G. Szegö, *Orthogonal Polynomials* (Edwards, Ann Arbor, 1948).

18. I. M. Gel'fand and Z. Ya. Shapiro, Amer. Math. Soc. Transl. **2**, 207 (1956).

19. A. Erdelyi, W. Magnus, F. Oberhettinger, and G. F. Tricomi, *Higher Transcendental Functions* , (McGraw-Hill, New York, 1953), Vol. 1.

20. E. P. Wigner, *Application of Group Theory to the Special Functions of Mathematical Physics*. Lecture notes, 1955 (unpublished).

21. H. C. Brinkman, *Applications of Spinor Invariants in Atomic Physics* (North-Holland, Amsterdam, 1956).

22. M. E. Rose, *Elementary Theory of Angular Momentum* (Wiley, New York, 1957).

23. U. Fano and G. Racah, *Irreducible Tensorial Sets* (Academic Press, New York, 1959).

24. E. P. Wigner, *Group Theory and Its Application to the Quantum Mechanics of Atomic Spectra* (Academic Press, New York, 1959). Translation from the 1931 German edition by J. J. Griffin.

25. J. C. Slater, *Quantum Theory of Atomic Structure* (McGraw-Hill, New York, 1960), Vol. 2.

26. V. Heine, *Group Theory and Quantum Mechanics; An Introduction to Its Present Usage* (Pergamon, New York, 1960).

27. W. T. Sharp, thesis, Princeton University, 1960; issued as Report AECL-1098, Atomic Energy of Canada, Chalk River, Ontario, 1960.

28. D. M. Brink and G. R. Satchler, *Angular Momentum* (Oxford University, London, 1962).

29. M. Hamermesh, *Group Theory and Its Applications to Physical Problems* (Addison-Wesley, Reading, 1962).

30. G. W. Mackey, *The Mathematical Foundations of Quantum Mechanics* (Benjamin, New York, 1963).

31. A. de-Shalit and I. Talmi, *Nuclear Shell Theory (Pure and Applied Physics Series)* (Academic Press, New York, 1963), Vol. 14.

32. R. P. Feynman, *Feynman Lectures on Physics* (Addison-Wesley, Reading, 1963), Chap. 34.

33. R. Judd, *Operator Techniques in Atomic Spectroscopy* (McGraw-Hill, New York, 1963).

34. A. S. Davydov, *Quantum Mechanics* (Pergamon, London, Addison-Wesley, Reading, 1965). Translation from the Russian of *Kvantovaya Mekhanika* (Moscow, 1963), with revisions and additions by D. ter Haar.

35. I. M. Gel'fand, R. A. Minlos, and Z. Ya. Shapiro, *Representations of the Rotation and Lorentz Groups and Their Applications* (Macmillan, New York, 1963). Translated from the Russian by G. Cummins and T. Boddington.

36. R. Hagedorn, *Selected Topics on Scattering Theory: Part IV, Angular Momentum* Lectures given at the Max-Planck-Institut für Physik, Munich, 1963.

37. M. A. Naimark, *Linear Representations of the Lorentz Group* (Pergamon Press, New York, 1964).

38. L. C. Biedenharn and H. van Dam, *Quantum Theory of Angular Momentum* (Academic Press, New York, 1965).

39. B. L. van der Waerden, *Sources of Quantum Mechanics* (North-Holland, Amsterdam, 1967).

40. B. R. Judd, *Second Quantization and Atomic Spectroscopy* (Johns Hopkins Press, Baltimore, 1967).

41. N. Vilenkin, *Special Functions and the Theory of Group Representations* Translated from the Russian Amer. Math. Soc. Transl.   (Amer. Math. Soc., Providence, 1968), Vol. 22.

42. J. D. Talman, *Special Functions: A Group Theoretic Approach* (Benjamin, New York, 1968). Based on E. P. Wigner's lectures; see Ref. [2].

43. B. G. Wybourne, *Symmetry Principles and Atomic Spec-*

troscopy (Wiley-Interscience, New York, 1970).

44. E. A. El Baz and B. Castel, *Graphical Methods of Spin Algebras in Atomic, Nuclear, and Particle Physics* (Dekker, New York, 1972).

45. R. Gilmore, *Lie Groups, Lie Algebras, and Some of Their Applications* (Wiley, New York, 1974).

46. D. A. Varshalovich, A. N. Moskalev, and V. K. Khersonskiĭ, *Quantum Theory of Angular Momentum* (Nauka, Leningrad, 1975) (in Russian).

47. R. D. Cowan, *The Theory of Atomic Structure and Spectra* (Univiversity of California Press, Berkeley, 1981).

48. R. N. Zare, *Angular Momentum* (Wiley-Interscience, New York, 1988).

Articles

49. A. Clebsch, "Theorie der binären algebraischen Formen," Teubner, Leipzig (1872).

50. P. Gordan, "Über das Formensystem binärer Formen," Teubner, Leipzig (1875).

51. W. Heisenberg, Z. Phys. **33**, 879 (1925).

52. M. Born and P. Jordan, Z. Phys. **34**, 858 (1925).

53. P. A. M. Dirac, Proc. Roy. Soc. A **109**, 642 (1925).

54. M. Born, W. Heisenberg, and P. Jordan, Z. Phys. **35**, 557 (1926).

55. W. Pauli, Z. Phys. **36**, 336 (1926).

56. E. P. Wigner, Z. Phys. **43**, 624 (1927).

57. E. P. Wigner, Z. Phys. **45**, 601 (1927).

58. C. Eckart, Rev. Mod. Phys. **2**, 305 (1930).

59. E. P. Wigner, Göttinger Nachrichten, Math.-Phys. 546 (1932).

60. J. H. Van Vleck, Phys. Rev. **47**, 487 (1935).

61. E. P. Wigner, "On the Matrices which Reduce the Kronecker Products of Representations of S. R. Groups," 1940 (unpublished). Published in *Quantum Theory of Angular Momentum* edited by L. C. Biedenharn and H. van Dam (Academic Press, New York, 1965), pp. 87–133.

62. G. Racah, Phys. Rev. **62**, 438 (1942); *ibid.* **63**, 367 (1943).

63. I. M. Gel'fand and M. L. Tseitlin, Dokl. Akad. Nauk SSSR, 825 (1950).

64. J. H. Van Vleck, Rev. Mod. Phys. **23**, 213 (1951).

65. H. A. Jahn, Proc. Roy. Soc. A **205**, 192 (1951).

66. L. C. Biedenharn, J. M. Blatt, and M. E. Rose, Rev. Mod. Phys. **24**, 249 (1952).

67. L. C. Biedenharn, *Notes on Multipole Fields.* Lecture notes at Yale University, New Haven, 1952 (unpublished).

68. J. P. Elliott, Proc. Roy. Soc. A **218**, 345 (1953).

69. L. C. Biedenharn, J. Math. Phys. **31**, 287 (1953).

70. H. A. Jahn and J. Hope, Phys. Rev. **93**, 318 (1954).

71. T. Regge, Nuovo Cimento **10**, 544 (1958).

72. T. Regge, Nuovo Cimento **11**, 116 (1959).

73. V. Bargmann, Commun. Pure Appl. Math. **14**, 187 (1961).

74. A. Giovannini and D. A. Smith, in *Spectroscopic and Group Theoretic Methods in Physics* (Racah Memorial Volume) edited by F. Block, S. G. Cohen, A. de-Shalit, S. Sambursky, and I. Talmi (Wiley Interscience, New York, 1968), pp. 89–97.

75. V. Bargmann, Rev. Mod. Phys. **34**, 829 (1962).

76. L. Michel, in *Lecture Notes in Physics: Group Representations in Mathematics and Physics, Battelle Recontres,* edited by V. Bargmann (Springer, Berlin, 1970), pp. 36–143.

77. A. C. T. Wu, J. Math. Phys. **13**, 84 (1972).

78. Ya. A. Smorodinskiĭ and L. A. Shelepin, Sov. Phys. Usp. **15**, 1 (1972). [*Usp. Fiz. Nauk* **106**, 3 (1972)].

79. L. A. Shelepin, in *Group-Theoretical Methods in Physics* edited by D. V. Skobel'tsyn (1975), pp. 1–109. A special research report translated from the Russian by Consultants Bureau, New York, and London.

80. Ya. A. Smorodinskiĭ, Sov. Phys. JETP **48**, 403 (1978).

81. B. R. Judd, in *Symmetries in Science* edited by B. Gruber and R. S. Millman (Plenum, New York, 1980), pp. 151–160.

3

Group Theory for Atomic Shells

B. R. Judd

Department of Physics and Astronomy, The Johns Hopkins University, Baltimore, Maryland 21218

3.1 GENERATORS

3.1.1 Group Elements

An element S_a of a Lie group $\mathcal{G}$ corresponding to an infinitesimal transformation can be written in the form

$$S_a = 1 + \delta a^\sigma X_\sigma , \qquad (3.1)$$

where the δa^σ are the infinitesimal parameters and the X_σ are the generators [1]. Summation over the repeated Greek index is implied. Transformations corresponding to finite parameters can be found by exponentiation:

$$S_a \to \exp[a^1 X_1]\exp[a^2 X_2]\cdots\exp[a^r X_r] . \qquad (3.2)$$

The generators necessarily form a Lie algebra, that is, they close under commutation:

$$[X_\rho, X_\sigma] = c_{\rho\sigma}^\tau X_\tau . \qquad (3.3)$$

In terms of the structure constants $c_{\rho\sigma}^\tau$, the metric tensor is defined as

$$g_{\rho\sigma} = c_{\rho\lambda}^\mu \, c_{\sigma\mu}^\lambda . \qquad (3.4)$$

3.1.2 Conditions on the Structure Constants

For an Abelian group, all the generators commute with one another:

$$c_{\rho\sigma}^\tau = 0 . \qquad (3.5)$$

The operators X_σ, $(\sigma = 1, 2, \ldots, p < r)$ form the generators of a subgroup if [2]

$$c_{\rho\sigma}^\tau = 0 , \quad (\rho, \sigma \le p, \tau > p) . \qquad (3.6)$$

The subgroup is invariant if the stronger condition

$$c_{\rho\sigma}^\tau = 0 , \quad (\rho \le p, \tau > p) \qquad (3.7)$$

is satisfied. A group is *simple* if it contains no invariant subgroup (besides the unit element). A group is *semisimple* if it contains no Abelian invariant subgroup (besides the unit element). A necessary and sufficient condition that a group be semisimple is that

$$\det | \, g_{\rho\tau} \, | \ne 0 . \qquad (3.8)$$

All simple groups are semisimple. For semisimple groups, the inverse tensor $g^{\mu\nu}$ can be formed, thus permitting suffixes to be raised. The quadratic operator

$$C = g^{\rho\sigma} X_\rho X_\sigma \qquad (3.9)$$

commutes with all generators of the group and is called Casimir's operator [1]. If the generators of a group $\mathcal{G}$ can be broken up into two sets such that each member of one set commutes with all the members of the other, that is, if

$$c_{\rho\sigma}^\tau = 0 , \quad (\rho \le p, \sigma > p) , \qquad (3.10)$$

then the two sets form the generators of two invariant subgroups, $\mathcal{H}$ and $\mathcal{K}$. The group $\mathcal{G}$ is the direct product of $\mathcal{H}$ and $\mathcal{K}$ and is written as $\mathcal{H} \times \mathcal{K}$.

3.1.3 Cartan-Weyl Form

By taking suitable linear combinations H_i and E_α of the generators X_σ, the basic commutation relations (3.3) can be thrown into the so-called Cartan-Weyl form [1]

$$[H_i, H_j] = 0 , \qquad (3.11)$$
$$[H_i, E_\alpha] = \alpha_i E_\alpha , \qquad (3.12)$$
$$[E_\alpha, E_{-\alpha}] = \alpha^i H_i , \qquad (3.13)$$
$$[E_\alpha, E_\beta] = N_{\alpha\beta} E_{\alpha+\beta} . \qquad (3.14)$$

The Roman symbols $i, j, \ldots$ run over an l-dimensional space (the *weight* space of *rank* l) in which the numbers α_i can be visualized as the components of the vectors (called *roots*). The E_α are shift operators, the displacements being specified by the components of $\boldsymbol{\alpha}$. The operator $E_{\alpha+\beta}$ in Eq. (3.14) is to be interpreted as 0 if $\boldsymbol{\alpha} + \boldsymbol{\beta}$ is not a root. The coefficient $N_{\alpha\beta}$ depends on the choice of normalization.

3.1.4 Atomic Operators as Generators

The pairs $a_\xi^\dagger a_\eta$ of creation and annihilation operators for either bosons or fermions, as defined in Sect. 6.1.1 close under commutation and form a Lie algebra. The coupled forms $\mathbf{W}^{(\kappa k)}$, defined in Eq. (6.18), are often used to play the role of the generators for electrons in an atomic shell.

3.2 CLASSIFICATION OF LIE ALGEBRAS

3.2.1 Introduction

The semisimple Lie algebras have been classified by Cartan [3]. They consist of four main classes A_l, B_l, C_l, D_l, and five exceptions G_2, F_4, E_6, E_7, E_8. Each algebra is charaterized by an array of roots in the l-dimensional weight space; they are conveniently specified by a set of mutually orthogonal unit vectors $\mathbf{e}_i$. The total number of generators (those of type E_α plus the l generators of type H_i) gives the *order* of the algebra.

Table 3.1. Generators of the Lie groups for the atomic l shell. The subscripts i and j run over all $4l + 2$ states of a single electron.

Group	Generators
$SO(8l + 5)^{\text{a}}$	$a_i^\dagger a_j^\dagger , a_i^\dagger a_j , a_i a_j , a_i^\dagger , a_j$
$SO(8l + 4)^{\text{a}}$	$a_i^\dagger a_j^\dagger , a_i^\dagger a_j , a_i a_j$
$U(4l + 2)^{\text{b}}$	$\mathbf{W}^{(\kappa k)}$ ($\kappa = 0, 1$; $k = 0, 1, \ldots, 2l$)
$SU(4l + 2)^{\text{b}}$	$\mathbf{W}^{(\kappa k)}$ (As above, with $\kappa = k = 0$ excluded)
$Sp(4l + 2)^{\text{c}}$	$\mathbf{W}^{(\kappa k)}$ (As above, with $\kappa + k$ odd)
$U(2l + 1)^{\text{d}}$	$\mathbf{W}^{(0k)}$ ($k = 0, 1, \ldots, 2l$)
$SU(2l + 1)^{\text{d}}$	$\mathbf{W}^{(0k)}$ ($k = 1, 2, \ldots, 2l$)
$SO(2l + 1)^{\text{d}}$	$\mathbf{W}^{(0k)}$ ($k = 1, 3, 5, \ldots, 2l - 1$)
G_2^{d}	$\mathbf{W}^{(01)} , \mathbf{W}^{(05)}$ (for $l = 3$)
$SO_L(3)^{\text{e}}$	$\mathbf{W}^{(01)}$ (or $\mathbf{L}$)
$SO_S(3)^{\text{e}}$	$\mathbf{W}^{(10)}$ (or $\mathbf{S}$)
$U_A(2l + 1) \times U_B(2l + 1)^{\text{f}}$	$W_{0q}^{(0k)} + W_{0q}^{(1k)} , W_{0q}^{(0k)} - W_{0q}^{(1k)}$ ($k = 0, 1, \ldots, 2l$)
$SO_\lambda(2l + 1) \times SO_\mu(2l + 1) \times SO_\nu(2l + 1) \times SO_\xi(2l + 1)^{\text{g}}$	$(\boldsymbol{\theta}^\dagger \boldsymbol{\theta})^{(k)}$ (k odd, $\boldsymbol{\theta} \equiv \boldsymbol{\lambda}, \boldsymbol{\mu}, \boldsymbol{\nu}, \boldsymbol{\xi}$)
$U_\lambda(2^l) \times U_\mu(2^l) \times U_\nu(2^l) \times U_\xi(2^l)^{\text{h}}$	$\mathbf{q}_\theta^\dagger \mathbf{q}_\theta$ (all components, $\theta \equiv \lambda, \mu, \nu, \xi$)

a Refs. [4,5]	c Refs. [6,7]	e Ref. [8]	g Ref. [9] and Eqs. (6.69)–(6.72)
b Refs. [1,6]	d Ref. [6]	f Ref. [10]	h Ref. [11] .

3.2.2 The Semisimple Lie Algebras

A_l. The roots are conveniently represented by the vectors $\mathbf{e}_i - \mathbf{e}_j$ ($i, j = 1, 2, \ldots, l+1$). They are all perpendicular to $\Sigma \mathbf{e}_k$ and do not extend beyond the l-dimensional weight space. The order of the algebra is $l(l + 2)$. The group for which this algebra can serve as a basis is the special unitary group $SU(l + 1)$.

B_l. The roots are $\pm \mathbf{e}_i$ and $\pm \mathbf{e}_i \pm \mathbf{e}_j$ ($i, j = 1, 2, \ldots, l; i \neq j$). The order of the algebra is $l(2l + 1)$. A corresponding group is the special orthogonal (or rotation) group in $2l + 1$ dimensions, $SO(2l + 1)$.

C_l. The roots are $\pm 2\mathbf{e}_i$ and $\pm \mathbf{e}_i \pm \mathbf{e}_j$ ($i, j = 1, 2, \ldots, l; i \neq j$). The order of the algebra is $l(2l + 1)$. A corresponding group is the symplectic group in $2l$ dimensions, $Sp(2l)$. A rotation of the roots yields $C_2 = B_2$.

D_l. The roots are $\pm \mathbf{e}_i \pm \mathbf{e}_j$ ($i, j = 1, 2, \ldots, l; i \neq j$). The order of the algebra is $l(2l - 1)$. A corresponding group is the special orthogonal (or rotation) group $SO(2l)$. A rotation of the roots yields $D_3 = A_3$. Also, $D_2 = A_1 \times A_1$.

E_6, E_7, E_8. The roots are given by Racah [1]. The respective orders are 78, 133, and 248.

F_4. The roots consist of the roots of B_4 together with the 16 vectors $\frac{1}{2}(\pm \mathbf{e}_1 \pm \mathbf{e}_2 \pm \mathbf{e}_3 \pm \mathbf{e}_4)$. The order of the algebra is 52.

G_2. The roots consist of the roots of A_2 together with the six vectors $\pm (2\mathbf{e}_i - \mathbf{e}_j - \mathbf{e}_k)$ ($i \neq j \neq k = 1, 2, 3$).

The order of the algebra is 14.

Examples of Lie groups used in atomic shell theory, together with their generators, are given in Table 3.1.

3.3 IRREDUCIBLE REPRESENTATIONS

3.3.1 Labels

If n atomic states of a collection transform among themselves under an arbitrary action of the generators of a group $\mathcal{G}$, then the states form a *representation* of $\mathcal{G}$. The representation is *irreducible* if n' linear combinations of the states cannot be found that also exhibit that property, where $n' < n$. The commuting generators H_i of $\mathcal{G}$ can be simultaneously diagonalized within the n states: their eigenvalues $(m_1, m_2, \ldots, m_l)$ for an eigenstate ψ specify the *weight* of the eigenstate. The weight above is said to be *higher* than $(m'_1, m'_2, \ldots, m'_l)$ if the first non-vanishing term in the sequence $m_1 - m'_1, m_2 - m'_2, \ldots$ is positive. An irreducible representation (IR) of a semisimple group is uniquely specified (to within an equivalence) by its highest weight [1], which can therefore be used as a defining label.

3.3.2 Dimensions

The dimensions of the IR's of various groups are expressed in terms of the highest weights and set out in Table 3.2. General algebraic expressions have been

Table 3.2. Dimensions D of the irreducible representations (IR's) of various Lie groups.

Group	IR	D
$SO(2)$	M	1
$SO(3)$	D_J	$2J+1$
$SO(4) = SO_A(3) \times SO_B(3)$	$D_J \times D_K$	$(2J+1)(2K+1)$
$SO(5)$	$(w_1 w_2)$	$(w_1 + w_2 + 2)(w_1 - w_2 + 1)(2w_1 + 3)(2w_2 + 1)/6$
$SO(6)$	$(w_1 w_2 w_3)$	$(w_1 - w_2 + 1)(w_1 - w_3 + 2)(w_2 - w_3 + 1)$ $\times (w_1 + w_2 + 3)(w_1 + w_3 + 2)(w_2 + w_3 + 1)/12$
$SO(7)$	$(w_1 w_2 w_3)$	$(w_1 + w_2 + 4)(w_1 + w_3 + 3)(w_2 + w_3 + 2)$ $\times (w_1 - w_2 + 1)(w_1 - w_3 + 2)(w_2 - w_3 + 1)$ $\times (2w_1 + 5)(2w_2 + 3)(2w_3 + 1)/720$
G_2	$(u_1 u_2)$	$(u_1 + u_2 + 3)(u_1 + 2)(2u_1 + u_2 + 5)(u_1 + 2u_2 + 4)$ $\times (u_1 - u_2 + 1)(u_2 + 1)/120$
$SU(3)$ or $U(3)$	$[\lambda_1 \lambda_2 \lambda_3]$	$(\lambda_1 - \lambda_2 + 1)(\lambda_1 - \lambda_3 + 2)(\lambda_2 - \lambda_3 + 1)/2$
$SU(4)$ or $U(4)$	$[\lambda_1 \lambda_2 \lambda_3 \lambda_4]$	As for $(w_1 w_2 w_3)$ of $SO(6)^a$
$Sp(4)$	$\langle \sigma_1 \sigma_2 \rangle$	As for $(w_1 w_2)$ of $SO(5)^b$
$Sp(6)$	$\langle \sigma_1 \sigma_2 \sigma_3 \rangle$	$(\sigma_1 - \sigma_2 + 1)(\sigma_1 - \sigma_3 + 2)(\sigma_1 + \sigma_2 + 5)$ $\times (\sigma_1 + \sigma_3 + 4)(\sigma_2 + \sigma_3 + 3)(\sigma_2 - \sigma_3 + 1)$ $\times (\sigma_1 + 3)(\sigma_2 + 2)(\sigma_3 + 1)/720$

^{a} Subject to the conditions $w_1 = (\lambda_1 + \lambda_2 - \lambda_3 - \lambda_4)/2$, $\;w_2 = (\lambda_1 - \lambda_2 + \lambda_3 - \lambda_4)/2$, $\;w_3 = (\lambda_1 - \lambda_2 - \lambda_3 + \lambda_4)/2$.
^{b} Subject to the conditions $w_1 = (\sigma_1 + \sigma_2)/2$, $\;w_2 = (\sigma_1 - \sigma_2)/2$.

given by Wybourne (Ref. [12], pp. 137–8). Numerical tabulations have been made by Butler in the appendix to another book by Wybourne [13], and also by McKay and Patera [14]. The latter defines the IR's by specifying the coordinates of the weights with respect to the simple roots of Dynkin [15].

3.3.3 Casimir's Operator

The eigenvalues of Casimir's operator C, defined in Eq. (3.9), can be expressed in terms of the highest weights of an IR [1]. A complete algebraic listing for all the semisimple Lie groups has been given by Wybourne (Ref. [12], p. 140). Sometimes Casimir's operator is given in terms of the spherical tensors $\mathbf{W}^{(\kappa k)}$, or of their special cases $\mathbf{V}^{(k)}$ ($= 2^{\frac{1}{2}} \mathbf{W}^{(0k)}$) for which the single-electron reduced matrix element satisfies

$$(nl \parallel v^{(k)} \parallel nl) = (2k+1)^{\frac{1}{2}}. \qquad (3.15)$$

The eigenvalues of several operators of that form are given in Table 3.3.

3.4 BRANCHING RULES

3.4.1 Introduction

If a group $\mathcal{H}$ shares some of its generators with a group $\mathcal{G}$, the first can be considered a subgroup of the second. That is, $\mathcal{G} \supset \mathcal{H}$. Many of the groups in Table 3.1 can be put in extended group–subgroup sequences. The

IR's of a subgroup that together span an IR of the group constitute a *branching* rule.

3.4.2 $U(n) \supset SU(n)$

The group $U(n)$ differs from $SU(n)$ in that the former contains among its generators a scalar (such as $W^{(00)}$) that, by itself, forms an invariant subgroup. Thus $U(n)$ is not semisimple. The scalar in question commutes with all the generators of the group and so is of type H_i. Its presence enlarges the dimension, l, of the weight space by 1, an extension that can be accommodated by the unit vectors $\mathbf{e}_i$ of A_l given in Sect. 3.2.2. The reduction $U(n) \supset SU(n)$ leads to the branching rule

$$[\lambda_1 \lambda_2 \cdots \lambda_n] \to [\lambda_1 - a, \; \lambda_2 - a, \cdots, \lambda_n - a], \quad (3.16)$$

where, in the IR of $SU(n)$ on the right,

$$a = (\lambda_1 + \lambda_2 + \cdots + \lambda_n)/n. \qquad (3.17)$$

To avoid fractional weights, the IR's of $SU(n)$ are frequently replaced by those of $U(n)$ for which the λ_i are integers. The weights λ_1, λ_2, ... can be interpreted as the number of cells in successive rows of a *Young Tableau*. When the n states of a single particle are taken as a basis for the IR $[10\ldots0]$ of $U(n)$, thus corresponding to a tableau comprising a single cell, the tableaux comprising N cells can be interpreted in two ways, namely, (1) as an IR of $U(n)$ for a system of N particles, and (2) as an IR of S_N, the finite group of permutations on N objects. A given tableau corresponds to as many permutations as

Table 3.3. Eigenvalues of Casimir's operator C for groups used in the atomic l shell.

Group	IR	Operator	Eigenvalue
$SU(2l+1)$	$[\lambda]$[a]	$\sum_{k>0}(\mathbf{V}^{(k)})^2$	$3N + 2Nl - \frac{1}{2}N^2 - 2S(S+1) - N^2/(2l+1)$
$SO(2l+1)$	W[b]	$\sum_{k_{\text{odd}}}(\mathbf{V}^{(k)})^2$	$\frac{1}{2}\sum_{i=1}^{l} w_i(w_i + 1 + 2l - 2i)$
G_2	$(u_1 u_2)$	$\frac{1}{4}[(\mathbf{V}^{(1)})^2 + (\mathbf{V}^{(5)})^2]$	$(u_1^2 + u_2^2 + u_1 u_2 + 5u_1 + 4u_2)/12$

[a] Appropriate for terms of l^N with total spin S. See Ref. [7], p. 125.
[b] Defined by the l weights $(w_1 w_2 \cdots w_l)$.

there are ways of entering the numbers 1, 2, ..., N in the cells such that the numbers increase going from left to right along the rows, and from top to bottom down the columns. A tableau possessing cells numbered in this way is called *standard*; it defines a permutation corresponding to a symmetrization with respect to the numbers in the rows, followed by an antisymmetrization with respect to the numbers in the columns [16].

3.4.3 Canonical Reductions

A group–subgroup sequence of the type

$$U(n) \supset U(n-1) \supset U(n-2) \supset \cdots \supset U(1) \quad (3.18)$$

is called *canonical* [17]. The branching rules for those IR's $[\lambda_1' \lambda_2' \cdots \lambda_{n-1}']$ of $U(n-1)$ contained in $[\lambda_1 \lambda_2 \cdots \lambda_n]$ of $U(n)$ have been given by Weyl [18] in terms of the "betweenness" conditions

$$\lambda_1 \geq \lambda_1' \geq \lambda_2 \geq \lambda_2' \cdots \geq \lambda_{n-1}' \geq \lambda_n. \quad (3.19)$$

The possibility of using the scheme of Eq. (3.18) in the theory of complex atomic spectra has been explored by Harter and Patterson [19, 20], and by Drake and Schlesinger [21] (see also Sect. 4.4.1).

3.4.4 Other Reductions

The algebraic formulae for $U(n) \supset SO(n)$ and $U(n) \supset Sp(n)$ have been given by Littlewood [22] and in a rather more accessible form by Wybourne [13]. Special cases have been tabulated by Butler (in Tables C-1 through C-15 in Ref. [13]). Another set of tables, in which Dynkin's labeling scheme is used, has been given by McKay and Patera [14]. Descriptions of how to apply the mechanics of the mathematics to the Young tableaux that describe the IR's of $U(n)$ can be found in the articles of Jahn [23] (with particular reference to $SO(5)$) and Flowers [24] (for $Sp(2j+1)$). For the atomic l shell, the reductions $SO(2l+1) \supset SO(3)$ and (for f electrons) $SO(7) \supset G_2$ and $G_2 \supset SO(3)$ are important. The sources cited in the previous Section are useful here.

It is important to recognize that the embedding of one group in another can often be performed in inequivalent ways, depending on which generators are discarded in the reduction process. Thus the use of $SO(5) \supset SO(3)$ in the atomic d shell involves a different $SO(3)$ group from that derived from the canonical sequence $SO(5) \supset SO(4) \supset SO(3)$.

3.5 KRONECKER PRODUCTS

3.5.1 Outer Products of Tableaux

Consider the tableau $[\lambda_1 \lambda_2 \cdots \lambda_n]$, where the total number of cells is N. A preliminary definition is required. If among the first r terms of any permutation of the N factors of the product, $x_1^{\lambda^1} x_2^{\lambda^2} \cdots x_n^{\lambda^n}$, the number of times x_1 occurs is $\geq$ the number of times x_2 occurs $\geq$ the number of times x_3 occurs, etc. for all values of r, this permutation is called a *lattice* permutation. The prescription of Littlewood [22] for finding the tableaux appearing in the Kronecker product of $[\lambda_1 \lambda_2 \cdots \lambda_n]$ with $[\mu_1 \mu_2 \cdots \mu_m]$ is as follows. The acceptable tableaux are those that can be built by adding to the tableau $[\lambda_1 \lambda_2 \cdots \lambda_n]$, μ_1 cells containing the same symbol α, then μ_2 cells containing the same symbol β, etc., subject to two conditions:

1. After the addition of each set of cells labeled by a common symbol we must have a permissible tableau with no two identical symbols in the same column;

2. If the total set of added symbols is read from right to left in the consecutive rows of the final tableau, we obtain a lattice permutation of $\alpha^{\mu_1} \beta^{\mu_2} \gamma^{\mu_3} \cdots$.

Examples of this procedure have been given (see Refs. [22] p. 96, [7] p. 136, [13] p. 24). An extensive tabulation involving tableaux with $N < 8$ has been calculated by Butler and given by Wybourne (Ref. [13], Table B-1).

3.5.2 Other Outer Products

The rules for constructing the Kronecker products for $U(n)$ follow by interpreting the Young tableaux of the previous section as IR's of $U(n)$. The known branching rules for reductions to subgroups enable the Kronecker products for the subgroups to be found. Many examples for $SO(n)$, $Sp(n)$, and G_2 can be found in the book by Wybourne (Ref. [13], Tables D-1 through D-15, and E-4).

3.5.3 Plethysms

Sometimes a particle can be thought of as being composite (as when the six orbital states $s + d$ of a single electron are taken to span the IR [200] of $SU(3)$). When the n' component states of a particle form a basis for an IR $[\lambda']$ of $U(n)$ other than $[10 \ldots 0]$, the process of finding which IR's of $U(n)$ occur for N-particle states whose permutation symmetries are determined by a given Young tableau $[\lambda]$ with N cells is called a *plethysm* (Ref. [22], p. 289) and written as $[\lambda'] \otimes [\lambda]$. The special techniques for doing this have been described by Wybourne [13]. An elementary method, which is often adequate in many cases, runs as follows:

1. Expand $[\lambda']^N$ by repeated use of Table B-1 from Ref. [13]. The resulting tableaux $[\lambda'']$ are independent of n.

2. Choose a small value of n, and strike out all tableaux from the set $[\lambda'']$ that possess more than n rows (since they are unacceptable as IR's of $U(n)$).

3. Interpret the remaining tableaux $[\lambda'']$ as IR's of $U(n)$ and find their dimensions from Tables A-2 through A-17 of Ref. [13]. Check that the sum of the dimensions is $(\dim[\lambda'])^N$.

4. Interpret the various tableaux $[\lambda]$ possessing the same number N of cells as IR's of $U(\dim[\lambda'])$, and find their dimensions from Ref. [13].

5. Match the dimensions of parts (3) and (4), remembering that each tableau $[\lambda]$ occurs as often as the number of its standard forms. This determines the possible ways of assigning the IR's $[\lambda'']$ of $U(n)$ to each $[\lambda]$.

6. Proceed to higher n to remove ambiguities and to include the tableaux struck out in step 2.

This procedure can be extended to calculate the plethysms for other groups. Examples of the type $W \otimes [\lambda]$ and $U \otimes [\lambda]$, where W and U are IR's of $SO(7)$ and G_2, have been given for $[\lambda] \equiv [2]$ and [11] corresponding to the separation of W^2 and U^2 into their symmetric and anti-symmetric parts [25]. The technique of plethysm is also useful for mixed atomic configurations (see Sect. 3.6.1).

3.6 ATOMIC STATES

3.6.1 Shell Structure

The 2^{4l+2} states of the l shell span the elementary spinor IR $(\frac{1}{2}\frac{1}{2}\cdots\frac{1}{2})$ of $SO(8l+5)$, which decomposes into the two IR's $(\frac{1}{2}\frac{1}{2}\cdots\pm\frac{1}{2})$ of $SO(8l+4)$, corresponding to an even and an odd number N of electrons [4]. The states of l^N span the IR $[1^N 0^{4l+2-N}]$ of $U(4l+2)$, corresponding to the antisymmetric Young tableau comprising a single column of N cells. The separation of spin and orbit through the subgroup $U(2) \times U(2l+1)$ yields the tableau products $[\lambda] \times [\tilde{\lambda}]$, where $[\tilde{\lambda}]$ is the tableau obtained by reflecting $[\lambda]$ in a diagonal line [1]. The IR's of the subgroup $U(2) \times SO(2l+1)$ are denoted by S and W [6]. An alternative way of reaching this subgroup from $U(4l+2)$ involves the intermediary $Sp(4l+2)$, whose IR's $\langle 1^v 0^{2l+1-v} \rangle$ possess as a basis the states with seniority v [7]. A subgroup of $SO(2l+1)$ is the $SO(3)$ group whose IR's specify L, the total orbital angular momentum.

Alternatives to this classic sequence are provided by the last three groups listed in Table 3.1, together with their respective subgroups. For $U_A(2l+1) \times U_B(2l+1)$, the shell is factored by considering spin-up and spin-down electrons as distinct (and statistically independent) particles [26]. A further factorization by means of the quasiparticles, θ, leads to four independent spaces. The 2^l states in each space span the elementary spinor $(\frac{1}{2}^l)$ of $SO_\theta(2l+1)$, which can be regarded as a fictitious particle (or quark), q_θ [27].

The standard classification of the states of the d shell is given in Table 3.4. The component M_Q of the quasispin Q (defined in Eqs. (6.33)–(6.35)) is listed, as well as the seniority, $v = 2l + 1 - 2Q$, the IR's W of $SO(5)$, and the value of L (as a spectroscopic symbol). Only states in the first half of the shell appear; the classification for the second half is the same as the first except that the signs of M_Q are reversed. A general rule for arbitrary l is exemplified by noting that every W (the IR of $SO(2l+1)$) occurs with two spins (S_1 and S_2) and two quasispins (Q_1 and Q_2) such that $S_1 = Q_2$ and $S_2 = Q_1$. No duplicated spectroscopic terms appear in Table 3.4. The generators of $SO(5)$ do not commute with the inter-electronic Coulomb interaction; thus the separations effected by $SO(5)$ merely define (to within a phase) a basis. The analog of Table 3.4 has been given by Wybourne [28] for the f shell. As Racah [6] showed, the group G_2 can be used to help distinguish repeated terms, but a few duplications remain. They are distinguished by Nielson and Koster [29] in their tables of spectroscopic coefficients by the letters A and B. The scope for applications of group theory becomes enlarged when the states of a single electron embrace more than one l value. Extensions of the standard model have been made by Feneuille [30] with particular reference to the

Table 3.4. The states of the d shell.

d^N	M_Q	$^{2S+1}[\lambda]$	v	W	L
d^0	$-\frac{5}{2}$	$^1[0]$	0	(00)	S
d^1	-2	$^2[1]$	1	(10)	D
d^2	$-\frac{3}{2}$	$^1[2]$	0	(00)	S
			2	(20)	DG
		$^3[11]$	2	(11)	PF
d^3	-1	$^2[21]$	1	(10)	D
			3	(21)	$PDFGH$
		$^4[111]$	3	(11)	PF
d^4	$-\frac{1}{2}$	$^1[22]$	0	(00)	S
			2	(20)	DG
			4	(22)	$SDFGI$
		$^3[211]$	2	(11)	PF
			4	(21)	$PDFGH$
		$^5[1111]$	4	(10)	D
d^5	0	$^2[221]$	1	(10)	D
			3	(21)	$PDFGH$
			5	(22)	$SDFGI$
		$^4[2111]$	3	(11)	PF
			5	(20)	DG
		$^6[11111]$	5	(00)	S

configurations $(d+s)^N$, for which quasiparticles have also been considered [31]. The group $SU(3)$ has been used for $(d+s)^N p^M$ [32]. The mixed configurations $(s+f)^4$ have found a use in the quark model of the atomic f shell [27].

3.6.2 Hydrogen and Hydrogen-Like Atoms

The nonrelativistic hydrogen atom possesses an $SO(4)$ symmetry associated with the invariance of the Runge-Lenz vector, which indicates the direction of the major axis of the classical elliptic orbit [5]. The quantum-mechanical form of this vector can be written in dimensionless units as

$$\mathbf{a} = [(\mathbf{l} \times \mathbf{p}) - (\mathbf{p} \times \mathbf{l}) + 2Z\mathbf{r}/ra_0]/2p_0, \qquad (3.20)$$

where $a_0 = \hbar^2/me^2$ is the Bohr radius, Ze is the nuclear charge, p_0 is related to the principal quantum number n by $p_0 = Z/na_0$, and where the momentum $\mathbf{p}$ and angular momentum $\mathbf{l}$ of the electron in its orbit are measured in units of $\hbar$. The analysis is best carried out in momentum space [33]. The four coordinates to which $SO(4)$ refers can be taken from Eqs. (9.43)–(9.46) or directly as kp_x, kp_y, kp_z, and $kp_0(1 - p^2/p_0^2)/2$, where $k = 2p_0/(p^2 + p_0^2)$. The generators of $SO(4)$ are provided by the 6 components of the two mutually commuting vectors $(\mathbf{l} + \mathbf{a})/2$ and $(\mathbf{l} - \mathbf{a})/2$, each of which behaves as an angular momentum vector. The

equivalence $SO(4) = SO(3) \times SO(3)$ corresponds to the isomorphism $D_2 = A_1 \times A_1$ of Sect. 3.2.2.

Hydrogenic eigenfunctions belonging to various energies can be selected to form bases for a number of groups. The inclusion of all the levels up to a given n yields the IR $(n - 1, 0)$ of $SO(5)$. Levels of a given l and all n form an infinite basis for an IR of the noncompact group $SO(2,1)$ [34]. All the bound hydrogenic states span an IR of $SO(4,2)$, as do the states in the continuum [35]. Subgroups of $SO(4,2)$ and their generators have been listed by Wybourne [12] in his Table 21.2.

To the extent that the central potential of a complex atom resembles the r^{-1} dependence for a bare nucleus, the group $SO(4)$ can be used to label the states [36].

3.7 THE GENERALIZED WIGNER-ECKART THEOREM

3.7.1 Operators

All atomic operators involving only the electrons can be built from their creation and annihilation operators. The appropriate group labels for an atomic operator acting on N electrons, each with n relevant component states, reduces to working out the various parts of the Kronecker products $[10\ldots0]^N \times [0\ldots0 - 1]^N$ of $U(n)$. Subgroups of $U(n)$ can further define these parts, which may be limited by Hermiticity constraints. The group labels for the Coulomb interaction for f electrons were first given by Racah [6]. Interactions involving electron spin were classified later [37–39]. Operators that represent the effects of configuration interaction on the d and f shells have also been studied [25, 40–44].

3.7.2 The Theorem

Let the ket, operator T, and bra of a matrix element be labeled by an IR (R_a, R_c, R_b) of a group $\mathcal{G}$, each with a component (i_a, i_c, i_b). Suppose the supplementary labels γ_k are also required to complete the definitions. The generalized Wigner-Eckart theorem is

$$\langle \gamma_a R_a i_a \mid T(\gamma_c R_c i_c) \mid \gamma_b R_b i_b \rangle$$
$$= \sum_\beta A_\beta (\beta R_a i_a \mid R_b i_b, R_c i_c), \qquad (3.21)$$

where β distinguishes the IR's R_a should they appear more than once in the reduction of the Kronecker product $R_b \times R_c$. The *reduced* matrix element A_β is independent of the i_k [6,8]. The second factor on the right-hand side of Eq. (3.21) is a Clebsch-Gordan (CG) coefficient for the group $\mathcal{G}$.

If the specification Ri can be replaced by $R\tau ri$, where r denotes an IR of a subgroup $\mathcal{H}$ of $\mathcal{G}$, and τ is an additional symbol that may be necessary to make the classification unambiguous, the CG coefficient for $\mathcal{G}$ factorizes according to the Racah lemma [6]

$$
\begin{aligned}
&(\beta R_a \tau_a r_a i_a \mid R_b \tau_b r_b i_b , R_c \tau_c r_c i_c) \\
&= \sum_\alpha (\alpha r_a i_a \mid r_b i_b , r_c i_c)(\beta R_a \tau_a r_a \mid R_b \tau_b r_b + R_c \tau_c r_c)_\alpha .
\end{aligned}
$$

$$(3.22)$$

The first factor on the right is a CG coefficient for the group $\mathcal{H}$; the second factor is an *isoscalar* factor [45].

3.7.3 Calculation of the Isoscalar Factors

The group $\mathcal{H}$ above is often $SO(3)$, whose Clebsch-Gordan coefficients (and their related 3-j symbols) are well-known (See Chap. 1). The principal difficulty in establishing comparable formulae for the isoscalar factors lies in giving algebraic meaning to β. Several methods are available for obtaining numerical results as follows.

Extraction from Tabulated Quantities. If R_b or R_c correspond to the IR's labeling a single electron, the factorization of the known [29] coefficients of fractional parentage (cfp) according to formulae of the type [6]

$$
\begin{aligned}
&(d^N SLv \{\mid d^{N-1} S'L'v') \\
&= (d^N Sv \{\mid d^{N-1} S'v')(W'L' + (10)d \mid WL)
\end{aligned}
$$

$$(3.23)$$

yields some isoscalar factors. In this example, W and W' are the IR's of $SO(5)$ defined by the triples NSv and $N'S'v'$ (with $N' = N-1$) as in Table 3.4. This approach can be applied to the f shell to give isoscalar factors for $SO(7)$ and G_2. The two-electron cfp of Donlan [46] further extend the range to IR's R_b and R_c describing two-electron systems. Isoscalar factors found in this way have the advantage that their relative phases as well as the significance of the indices β and τ coincide with current usage.

Evaluation Using Casimir's Operator. Two commuting copies (b and c) are made of the generators of the group $\mathcal{G}$ to form the generators of the direct product $\mathcal{G}_b \times \mathcal{G}_c$ [47]. Corresponding generators of $\mathcal{G}_b$ and $\mathcal{G}_c$ are added to give the generators of $\mathcal{G}_a$. Each quadratic operator $(T_a)^2$ appearing in the expression for Casimir's operator C_a for $\mathcal{G}_a$ (as listed in Table 3.3) is written as $(T_b + T_c)^2$. On expanding the expressions of this type, the terms $(T_b)^2$ and $(T_c)^2$ yield Casimir's operators C_b and C_c for $\mathcal{G}_b$ and $\mathcal{G}_c$. Their eigenvalues can be written down in terms of the highest weights of the IR's appearing in the isoscalar factor of Eq. (3.22). If the cross products of the type $(T_b \cdot T_c)$ can be evaluated within the uncoupled states $\mid R_b \tau_b r_b , R_c \tau_c r_c)$, then our knowledge of the eigenvalues of C_a for the coupled states $\mid \beta R_a \tau_a r_a)$ provides the equations for determining (to within the freedom implied by β) the isoscalar factors relating the uncoupled to the coupled states. The evaluation of the cross products is straightforward when $\mathcal{H} = SO(3)$, since the relevant 6-j symbols are readily available [48]. Examples of this method can be found in the literature [40].

3.7.4 Generalizations of Angular Momentum Theory

CG coefficients, n-j symbols, reduced matrix elements, and the entire apparatus of angular momentum theory all have their generalizations to groups other than $SO(3)$. An interchange of two columns of a 3-j symbol has its analog in the interchange of two parts of an isoscalar factor. For IR's W and L of $SO(2l+1)$ and $SO(3)$, there are two possibilities:
(1) The interchange of the two parts separated by the plus sign, namely,

$$
\begin{aligned}
&(W_a \tau_a L_a \mid W_b \tau_b L_b + W_c \tau_c L_c) \\
&= (-1)^t (W_a \tau_a L_a \mid W_c \tau_c L_c + W_b \tau_b L_b), \quad (3.24)
\end{aligned}
$$

where $t = L_a - L_b - L_c + x$, with x dependent on the IR's W only; or,
(2) The *reciprocity* relation of Racah [6]:

$$
\begin{aligned}
&(W_a \tau_a L_a \mid W_b \tau_b L_b + W_c \tau_c L_c) \\
&= (-1)^{t'} [(2L_b + 1) \dim W_a / (2L_a + 1) \dim W_b]^{\frac{1}{2}} \\
&\quad \times (W_b \tau_b L_b \mid W_a \tau_a L_a + W_c \tau_c L_c), \quad (3.25)
\end{aligned}
$$

where $t' = L_a - L_b - L_c + x'$, with x' dependent on the IR's W, but taken to be l by Racah for $W_c = (10\ldots0)$.

Reduced matrix elements in $SO(3)$ can be further reduced by the extraction of isoscalar factors. When W_a occurs once in the decomposition of $W \times W_b$ we have

$$
\begin{aligned}
&(\gamma_a W_a \tau_a L_a \| T^{(WL)} \| \gamma_b W_b \tau_b L_b) \\
&= [(2L_a + 1)/ \dim W_a]^{\frac{1}{2}} (\gamma_a W_a \|\| T^{(W)} \|\| \gamma_b W_b) \\
&\quad \times (W_b \tau_b L_b + WL \mid W_a \tau_a L_a). \quad (3.26)
\end{aligned}
$$

Analogs of the n-j symbols are discussed by Butler [49].

3.8 CHECKS

The existence of numerical checks is useful when using group theory in atomic physics. The CG coefficients, isoscalar factors, and the various generalizations of the n-j symbols are often calculated in ways that conceal the simplicity and structure of the answer. Practitioners are familiar with several empirical rules:

1. Numbers with different irrationalities, such as $\sqrt{2}$ and $\sqrt{3}$, are never added to one another.

2. The denominators of fractions seldom involve high primes.

3. High primes are uncommon, but when they appear, it is usually in diagonal matrix elements rather than off-diagonal ones.

4. A sum of a number of terms frequently factors in what appears to be an unexpected way, and similar sums often exhibit similar factors.

Guided by these rules, one will find that such errors as do arise occur with phases rather than with magnitudes.

REFERENCES

1. G. Racah, *Group Theory and Spectroscopy* (Springer-Verlag, New York, 1965), Vol. 37.
2. L. P. Eisenhart, *Continuous Groups of Transformations* (Dover, New York, 1961).
3. E. Cartan, Ph. D. thesis, University of Paris, 1894.
4. B. R. Judd, in *Group Theory and Its Applications*, edited by E. M. Loebl (Academic, New York, 1968).
5. E. U. Condon and H. Odabasi, *Atomic Structure* (Cambridge University Press, New York, 1980).
6. G. Racah, Phys. Rev. **76**, 1352 (1949).
7. B. R. Judd, *Operator Techniques in Atomic Spectroscopy* (McGraw-Hill, New York, 1963).
8. E. P. Wigner, *Group Theory* (Academic Press, New York, 1959).
9. L. Armstrong and B. R. Judd, Proc. R. Soc. London Ser. A **315**, 27, and 39 (1970).
10. B. R. Judd, Phys. Rev. **162**, 28 (1967).
11. B. R. Judd and G. M. S. Lister, J. Phys. A **25**, 2615 (1992).
12. B. G. Wybourne, *Classical Groups for Physicists* (Wiley-Interscience, New York, 1974).
13. B. G. Wybourne, *Symmetry Principles and Atomic Spectroscopy* (Wiley-Interscience, New York, 1970).
14. W. G. McKay and J. Patera, *Tables of Dimensions, Indices, and Branching Rules for Representations of Simple Lie Algebras* (Dekker, New York, 1981).
15. E. B. Dynkin, *Maximal Subgroups of the Classical Groups* American Mathematical Society Translations, Ser. 2, **6**, 245 (1965).
16. D. E. Rutherford, *Substitutional Analysis* (Edinburgh University Press, Edinburgh, 1948).
17. M. Moshinsky, *Group Theory and the Many-Body Problem* (Gordon and Breach, New York, 1968).
18. H. Weyl, *The Theory of Groups and Quantum Mechanics* (Dover, New York, undated).
19. W. G. Harter, Phys. Rev. A **8**, 2819 (1973); *Principles of Symmetry, Dynamics and Spectroscopy* (Wiley-Interscience, New York, 1993).
20. W. G. Harter and C. W. Patterson, *A Unitary Calculus for Electronic Orbitals* (Springer, Berlin, 1976), Vol. 49.
21. G. W. F. Drake and M. Schlesinger, Phys. Rev. A **15**, 1990 (1977); R. D. Kent and M. Schlesinger, Phys. Rev. A **50**, 186 (1994).
22. D. E. Littlewood, *The Theory of Group Characters* (Clarendon, Oxford, 1950).
23. H. A. Jahn, Proc. R. Soc. London Ser. A **201**, 516 (1950).
24. B. H. Flowers, Proc. R. Soc. London Ser. A **212**, 248 (1952).
25. B. R. Judd and H. T. Wadzinski, J. Math. Phys. **8**, 2125 (1967).
26. C. L. B. Shudeman, J. Franklin Inst. **224**, 501 (1937).
27. B. R. Judd and G. M. S. Lister, Phys. Rev. Lett. **67**, 1720 (1991).
28. B. G. Wybourne, *Spectroscopic Properties of Rare Earths* (Wiley-Interscience, New York, 1965), p. 15.
29. C. W. Nielson and G. F. Koster, *Spectroscopic Coefficients for the p^n, d^n, and f^n Configurations* (MIT Press, Cambridge, 1963).
30. S. Feneuille, J. Phys. (Paris), **28**, 61, 315, 701, and 497 (1967).
31. S. Feneuille, J. Phys. (Paris), **30**, 923 (1969).
32. S. Feneuille, A. Crubellier, and T. Haskell, J. Phys. (Paris), **31**, 25 (1970).
33. M. J. Englefield, *Group Theory and the Coulomb Problem* (Wiley-Interscience, New York, 1972).
34. L. Armstrong, J. Phys. (Paris), **31**, 17 (1970).
35. C. E. Wulfman, in *Group Theory and Its Applications*, edited by E. M. Loebl (Academic Press, New York, 1971), Vol. 2.
36. D. R. Herrick, Adv. Chem. Phys. **52**, 1 (1982).
37. A. G. McLellan, Proc. Phys. Soc. London **76**, 419 (1960).
38. B. R. Judd, Physica **33**, 174 (1967).
39. B. R. Judd, H. M. Crosswhite, and H. Crosswhite, Phys. Rev. **169**, 130 (1968).
40. B. R. Judd, Phys. Rev. **141**, 4 (1966).
41. B. R. Judd and M. A. Suskin, J. Opt. Soc. Am. B **1**, 261 (1984).
42. B. R. Judd and R. C. Leavitt, J. Phys. B **19**, 485 (1986).
43. R. C. Leavitt, J. Phys. A **20**, 3171 (1987).
44. R. C. Leavitt, J. Phys. B **21**, 2363 (1988).
45. A. R. Edmonds, Proc. R. Soc. London Ser. A **268**, 567 (1962).
46. V. L. Donlan, *Air Force Material Laboratory Report No. AFML-TR-70-249*, Wright-Patterson Air Force Base, Ohio, 1970.
47. P. Nutter and C. Nielsen, *Fractional Parentage Coefficients of Terms of f^n*. II. Direct Evaluation of Racah's Factored Forms by a Group Theoretical Approach, Technical Memorandum T-133 (Raytheon, Waltham, 1963).
48. M. Rotenberg, R. Bivins, N. Metropolis, and J. K. Wooten, *The 3-j and 6-j Symbols* (MIT Press, Cambridge).
49. P. H. Butler, Phil. Trans. R. Soc. London Ser. A **277**, 545 (1975).

4

Dynamical Groups

J. Paldus

Department of Applied Mathematics, University of Waterloo, Waterloo, Ontario, N2L 3G1 Canada

4.1 INTRODUCTION

The well known *symmetry (invariance, degeneracy) groups* or *algebras* of quantum mechanical Hamiltonians provide quantum numbers (conservation laws, integrals of motion) for state labeling and the associated selection rules (*cf.* Chap. 3). In addition, it is often advantageous to employ much larger groups, referred to as the *dynamical groups (noninvariance groups, dynamical algebras, spectrum generating algebras)*, which may or may not be the invariance groups of the studied system [1–5]. In all known cases, they are Lie groups (LG), or rather corresponding Lie algebras (LA), and one usually requires that all states of interest of a system be contained in a single irreducible representation (irrep). In a weaker sense, one regards any group (or corresponding algebra)

as a dynamical group if the Hamiltonian can be expressed in terms of its generators [6]. In nuclear physics, one sometimes distinguishes *exact* (baryon number preserving), *almost exact* (e.g., total isospin), *approximate* (e.g., SU(3) of the "eightfold way") and *model* (e.g., nuclear shell model) *dynamical symmetries* [7]. The dynamical groups of interest in atomic and molecular physics can be conveniently classified by their topological characteristic of compactness. Noncompact LG's (LA's) generally arise in simple problems involving an infinite number of bound states, while those involving a finite number of bound states (e.g., molecular vibrations or *ab initio* models of electronic structure) exploit compact LG's.

We follow the convention of designating Lie groups by capital letters and Lie algebras by lower case letters, e.g., the Lie algebra of the rotation group SO(3) is designated as so(3).

4.2 NONCOMPACT DYNAMICAL GROUPS

As an illustration we present basic facts concerning LAs that are useful for centrosymmetric Kepler-type problems, their realizations and typical applications. Recall that *a realization* of a LA is a homomorphism associating a concrete set of physically relevant operators with each abstract basis of the given LA. The physical operators we will use are general (intrinsic) position vectors $\mathbf{R} = (X_1, X_2, \ldots, X_N)$ in $\mathbf{R}^N$ and their corresponding momenta $\mathbf{P} = (P_1, P_2, \ldots, P_N)$, satisfying the basic commutation relations ($\hbar = 1$)

$$[X_j, X_k] = [P_j, P_k] = 0, \quad [X_j, P_k] = i\delta_{jk}I. \quad (4.1)$$

4.2.1 Realizations of so(2,1)

This important LA is a simple noncompact analogue of the well known rotation group LA so(3), [*cf.* Sects. 2.1, 2.2 and 3.2]. Designating its three generators by T_j ($j = 1, 2, 3$), its structure constants (Sects. 2.2.1 and 3.1.1) are defined by

$$\begin{aligned}
[T_1, T_2] &= i\gamma T_3, \\
[T_2, T_3] &= iT_1, \\
[T_3, T_1] &= iT_2,
\end{aligned} \quad (4.2)$$

with $\gamma = -1$, while $\gamma = 1$ gives so(3). Defining the so-called ladder (raising and lowering) operators

$$T_{\pm} = T_1 \pm iT_2, \quad (4.3)$$

we also have that

$$[T_+, T_-] = 2\gamma T_3, \quad [T_3, T_{\pm}] = \pm T_3. \quad (4.4)$$

The Casimir operator then has the form

$$T^2 = \gamma(T_1^2 + T_2^2) + T_3^2 = \gamma T_+ T_- + T_3^2 - T_3. \quad (4.5)$$

With a Hermitian scalar product satisfying $T_j^{\dagger} = T_j$ ($j = 1, 2, 3$), so that $T_{\pm}^{\dagger} = T_{\mp}$, the unitary irreps (unirreps) carried by the simultaneous eigenstates of T^2 and T_3 have the form (*cf.* Sects. 2.2.1 and 2.3)

$$\begin{aligned}
T^2|kq\rangle &= k(k+1)|kq\rangle, \\
T_3|kq\rangle &= q|kq\rangle, \\
T_{\pm}|kq\rangle &= \sqrt{\gamma(k \mp q)(k \pm q + 1)}\,|k, q \pm 1\rangle.
\end{aligned} \quad (4.6)$$

For so(3), ($\gamma = 1$), only finite dimensional irreps $\mathcal{D}^{(k)}$, $k = 0, 1, 2, \ldots$ with $|q| \leq k$ are possible (Sects. 2.3 and 2.4). In contrast, there are no nontrivial finite dimensional unirreps of so(2,1); (for classification, see e.g., Ref. [8]). The relevant class $\mathcal{D}^+(k)$ of so(2,1) unirreps for bound state problems has a T_3 eigenspectrum

bounded from below and is given by $q = -k + \mu$; $\mu = 0, 1, 2, \ldots$, and $k < 0$ or, equivalently, $\mathcal{D}^+(-k - 1)$ with $q = k + 1 + \mu$; $\mu = 0, 1, 2, \ldots$; $k > -1$, since $k_1 = -k - 1$ defines an equivalent unirrep and $k_1(k_1 + 1) = k(k + 1)$. There exists a similar class of irreps with the T_3 spectrum bounded from above and two classes (principal and supplementary) with unbounded T_3 spectra (which may be exploited in scattering problems).

For problems involving only central potentials, a useful realization is given in terms of the radial distance $R = |\mathbf{R}|$ and the radial momentum

$$P_R = -\frac{i}{R}\frac{\partial}{\partial R}R = -i\left(\frac{\partial}{\partial R} + \frac{1}{R}\right) = \frac{1}{R}(\mathbf{R}\cdot\mathbf{P} - iI), \quad (4.7)$$

so that $[R, P_R] = iI$. Recall that

$$P^2 = P_R^2 + \frac{L^2}{R^2}, \quad \mathbf{L} = \mathbf{R} \wedge \mathbf{P}. \quad (4.8)$$

The general form of the desired so(2,1) realization is (*cf.* [8–10])

$$\begin{aligned}
\left.\begin{array}{c}T_1 \\ T_3\end{array}\right\} &= \tfrac{1}{2}R^{-\nu}(\nu^{-2}R^2 P_R^2 + \xi \mp R^{2\nu}), \\
T_2 &= \tfrac{1}{2}[2\nu^{-1}RP_R - i(1 - \nu^{-1})I],
\end{aligned} \quad (4.9)$$

where ξ is either a *c*-number (scalar operator) or an operator which commutes with both R and P_R, and ν is an arbitrary real number.

To interrelate this realization with so(2,1) unirreps $\mathcal{D}^+(k)$ or $\mathcal{D}^+(-k-1)$, we have to establish the connection between the quantum numbers k, q and the parameters ξ and ν. Considering the Casimir operator T^2 in Eq. (4.5), we find that in our realization (4.9)

$$T^2 = \xi + (1 - \nu^2)/4\nu^2, \quad (4.10)$$

so that

$$k = \tfrac{1}{2}\left[-1 \pm \sqrt{4\xi + \nu^{-2}}\right] \quad (4.11)$$

and

$$q = q_0 + \mu, \quad \mu = 0, 1, 2, \ldots \quad (4.12)$$

where

$$q_0 = k + 1 = \tfrac{1}{2}\left[1 \pm \sqrt{4\xi + \nu^{-2}}\right], \quad k > -1. \quad (4.13)$$

4.2.2 Hydrogenic Realization of so(4,2)

To obtain suitable hydrogenic realizations of so(4,2) it is best to proceed from so(4) (the dynamical symmetry group for the bound states of the nonrelativistic Kepler problem), and merge it with so(2,1) [8].

The so(4) LA can be realized either as a direct sum so(4) = so(3) $\oplus$ so(3), or by supplementing so(3) with an appropriately scaled quantum mechanical analogue of the Laplace-Runge-Lenz (LRL) vector (*cf.* Sect. 3.6.2). In the first case, we use two commuting angular momentum vectors **M** and **N** (*cf.* Sect. 2.6),

$$[M_j, M_k] = i\varepsilon_{jk\ell}M_\ell, \quad [N_j, N_k] = i\varepsilon_{jk\ell}N_\ell,$$
$$[M_j, N_k] = 0, \quad (j, k, \ell = 1, 2, 3) \tag{4.14}$$

while in the second case we use the components of the total angular momentum vector **J** and LRL-like vector **V** with commutation relations

$$[J_j, J_k] = i\varepsilon_{jk\ell}J_\ell,$$
$$[V_j, V_k] = i\sigma\varepsilon_{jk\ell}J_\ell, \tag{4.15}$$
$$[J_j, V_k] = i\varepsilon_{jk\ell}V_\ell, \quad (j, k, \ell = 1, 2, 3),$$

with $\sigma = 1$. For $\sigma = -1$ we obtain so(3,1) (the LA of the homogeneous Lorentz group), which is relevant to the scattering problem of a particle in the Coulomb (or Kepler) potential (see below). For $\sigma = 0$ we get e(3) (the LA of the three-dimensional Euclidean group), [11–13]. Note that Eqs. (4.14) and (4.15) are interrelated by

$$\mathbf{M} = \tfrac{1}{2}(\mathbf{J} + \mathbf{V}), \quad \mathbf{N} = \tfrac{1}{2}(\mathbf{J} - \mathbf{V}), \tag{4.16}$$

so that $\mathbf{J} = \mathbf{M} \oplus \mathbf{N}$ and $\mathbf{V} = 2\mathbf{M} - \mathbf{J}$. The two Casimir operators C_1 and C_2 are

$$C_1 = \sigma J^2 + V^2$$
$$= \sigma J_+ J_- + V_+ V_- + V_3^2 + \sigma J_3(J_3 - 2),$$
$$C_2 = (\mathbf{V} \cdot \mathbf{J}) = (\mathbf{J} \cdot \mathbf{V}) \tag{4.17}$$
$$= \tfrac{1}{2}(V_+ J_- + V_- J_+) + V_3 J_3,$$

where again

$$X_\pm = X_1 \pm iX_2, \quad X = J \text{ or } V. \tag{4.18}$$

For so(3,1) and e(3), only infinite dimensional nontrivial irreps are possible, while for so(4), only finite dimensional ones are. To get unirreps, we require **J** and **V** to be Hermitian. Using $\{J^2, J_3, C_1, C_2\}$ as a complete set of commuting operators for so(4), we label the basis vectors by the four quantum numbers as $|\gamma jm\rangle \equiv |(j_0, \eta)jm\rangle$, so that

$$J^2|\gamma jm\rangle = j(j+1)|\gamma jm\rangle,$$
$$J_3|\gamma jm\rangle = m|\gamma jm\rangle,$$
$$C_1|\gamma jm\rangle = (j_0^2 - \eta^2 - 1)|\gamma jm\rangle, \tag{4.19}$$
$$C_2|\gamma jm\rangle = j_0\eta|\gamma jm\rangle,$$

with $2|j_0|$ being a nonnegative integer and

$$j = |j_0|, |j_0| + 1, \dots, \eta - 1; \eta = |j_0| + k, \quad k = 1, 2, \dots \tag{4.20}$$

(see, e.g., Ref. [10] for the action of $J_\pm, V_3$ and $V_\pm$).

To obtain the hydrogenic (or Kepler) realization of so(4), we consider the quantum mechanical analog of the classical LRL vector

$$\tilde{\mathbf{V}} = \tfrac{1}{2}(\mathbf{p} \wedge \mathbf{L} - \mathbf{L} \wedge \mathbf{p}) - Zr^{-1}\mathbf{r}$$
$$= \tfrac{1}{2}\mathbf{r}p^2 - \mathbf{p}(\mathbf{r} \cdot \mathbf{p}) + \mathbf{r}H, \tag{4.21}$$
$$\mathbf{L} = \mathbf{r} \wedge \mathbf{p},$$

which commutes with the hydrogenic Hamiltonian

$$H = \tfrac{1}{2}p^2 - \mathcal{Z}r^{-1}. \tag{4.22}$$

Note that

$$[\mathbf{L}, H] = [\tilde{\mathbf{V}}, H] = 0,$$
$$(\mathbf{L} \cdot \tilde{\mathbf{V}}) = (\tilde{\mathbf{V}} \cdot \mathbf{L}) = 0, \tag{4.23}$$
$$\tilde{V}^2 = 2H(L^2 + 1) + \mathcal{Z}^2,$$

while the components of **L** and $\tilde{\mathbf{V}}$ satisfy the commutation relations

$$[L_j, L_k] = i\epsilon_{jk\ell}L_\ell,$$
$$\left[L_j, \tilde{V}_k\right] = i\epsilon_{jk\ell}\tilde{V}_\ell, \tag{4.24}$$
$$\left[\tilde{V}_j, \tilde{V}_k\right] = (-2H)i\epsilon_{jk\ell}L_\ell.$$

Thus, restricting ourselves to a specific bound state energy level E_n, we can replace H by E_n and define

$$V_j = (-2E_n)^{-1/2}\tilde{V}_j \quad (j = 1, 2, 3), \tag{4.25}$$

obtaining the so(4) commutation relations (4.15) (with **J** replaced by **L**). This is Pauli's hydrogenic realization of so(4) [14,15]. [In a similar way we can consider continuum states $E > 0$ and define $\mathbf{V} = (2E)^{-1/2}\tilde{\mathbf{V}}$, obtaining an so(3,1) realization.] The last identity of (4.23) now becomes

$$V^2 = -(L^2 + 1) - \mathcal{Z}^2/2E_n, \tag{4.26}$$

which immediately implies Bohr's formula, since $V^2 + L^2 = 4M^2 = -1 - \mathcal{Z}^2/2E_n$, so that

$$E_n = -\mathcal{Z}^2/2(2j_1 + 1)^2 = -\mathcal{Z}^2/2n^2, \tag{4.27}$$

where $n = 2j_1 + 1$ and j_1 is the angular momentum quantum number for **M**, Eq. (4.16). In terms of the irrep labels (4.20), we have that $j_0 = 0, \eta = n$, so that $|\gamma\ell m\rangle = |(0, n)\ell m\rangle \equiv |n\ell m\rangle, \ell = 0, 1, \dots, n - 1$.

Using the stepwise merging of so(4) and so(2,1) [adding first T_2 which leads to so(4,1) and subsequently T_1 and T_3], we arrive at the hydrogenic realization of so(4,2) having fifteen generators **L, A, B, Γ**, T_1, T_2, T_3, namely (*cf.* [8, 10])

$$\mathbf{L} = \mathbf{R} \wedge \mathbf{P},$$

$$\left.\begin{array}{c} \mathbf{A} \\ \mathbf{B} \end{array}\right\} = \tfrac{1}{2}\mathbf{R}P^2 - \mathbf{P}(\mathbf{R}\cdot\mathbf{P}) \mp \mathbf{R},$$

$$\mathbf{\Gamma} = \mathbf{R}\mathbf{P},$$

$$\left.\begin{array}{c} T_1 \\ T_3 \end{array}\right\} = \tfrac{1}{2}(RP^2 \mp R) = \tfrac{1}{2}(RP_R^2 + L^2 R^{-1} \mp R),$$

$$T_2 = \mathbf{R}\cdot\mathbf{P} - iI = RP_R. \qquad (4.28)$$

Relabeling these generators by the elements of an anti-symmetric 6×6 matrix according to the scheme

$$L_{jk} \leftrightarrow \begin{bmatrix} 0 & L_3 & -L_2 & A_1 & B_1 & \Gamma_1 \\ & 0 & L_1 & A_2 & B_2 & \Gamma_2 \\ & & 0 & A_3 & B_3 & \Gamma_3 \\ & & & 0 & T_2 & T_1 \\ & & & & 0 & T_3 \\ & & & & & 0 \end{bmatrix} \qquad (4.29)$$

we can write the commutation relations in standard form

$$[L_{jk}, L_{\ell m}] = i(g_{j\ell}L_{km} + g_{km}L_{j\ell} - g_{k\ell}L_{jm} - g_{jm}L_{k\ell}), \qquad (4.30)$$

with the diagonal metric tensor g_{jk} defined by the matrix $G = \text{diag}[1, 1, 1, 1, -1, -1]$. The matrix form (4.29) also implies the subalgebra structure

$$\text{so}(4,2) \supset \text{so}(4,1) \supset \text{so}(4) \supset \text{so}(3), \qquad (4.31)$$

with so(4,1) generated by $\mathbf{L}, \mathbf{A}, \mathbf{B}$ and T_2, and so(4) by $\mathbf{L}, \mathbf{A}$. [$\mathbf{L}, \mathbf{B}$ also generate so(3,1).]

The three independent Casimir operators (quadratic, cubic and quartic) are (*cf.* [16]), (summation over all indices is implied)

$$\begin{aligned} Q_2 &= \tfrac{1}{2}L_{jk}L^{jk} \\ &= L^2 + A^2 - B^2 - \Gamma^2 + T_3^2 - T_1^2 - T_2^2, \\ Q_3 &= \tfrac{1}{48}\varepsilon_{ijk\ell mn}L^{ij}L^{k\ell}L^{mn} \\ &= T_1(\mathbf{B}\cdot\mathbf{L}) + T_2(\mathbf{\Gamma}\cdot\mathbf{L}) + T_3(\mathbf{A}\cdot\mathbf{L}) + \mathbf{A}\cdot(\mathbf{B}\wedge\mathbf{\Gamma}), \\ Q_4 &= L_{jk}L^{k\ell}L_{\ell m}L^{mj}. \end{aligned} \qquad (4.32)$$

For our hydrogenic realization $Q_2 = -3$, $Q_3 = Q_4 = 0$. Thus, our hydrogenic realization implies a single unirrep of so(4,2) adapted to the chain (4.31).

4.3 HAMILTONIAN TRANSFORMATION AND SIMPLE APPLICATIONS

The basic idea is to transform the relevant Schrö-dinger equation into an eigenvalue problem for one of the operators from the complete set of commuting operators in our realizations, e.g., T_3 for so(2,1). Instead of using a rather involved "tilting" transformation (*cf.* [1], p. 20

and [8]), we can rely on a simple scaling transformation [9, 17]

$$\mathbf{r} = \lambda\mathbf{R}, \quad \mathbf{p} = \lambda^{-1}\mathbf{P}, \quad r = \lambda R, \quad p_r = \lambda^{-1}P_R, \quad (4.33)$$

where

$$r = \left(\sum_{j=1}^{N} x_j^2\right)^{1/2}, \qquad (4.34)$$

and

$$p^2 = p_r^2 + r^{-2}[\tfrac{1}{4}(N-1)(N-3) + L^2], \qquad (4.35)$$

with p_r defined analogously to P_R in Eq. (4.7); r and p are the physical operators in terms of which the Hamiltonian of the studied system is expressed. Recall that L^2 has eigenvalues (*cf.* [18])

$$\ell(\ell + N - 2), \quad \ell = 0, 1, 2, \ldots \quad \text{for} \quad N \geq 2, \qquad (4.36)$$

and we can set $\ell = 0$ for $N = 1$ (angular momentum term vanishes in one-dimensional case). The units in which $m = e = \hbar = 1$, $c \approx 137$ are used throughout.

4.3.1 *N*-Dimensional Isotropic Harmonic Oscillator

Considering the Hamiltonian

$$H = \tfrac{1}{2}p^2 + \tfrac{1}{2}\omega^2 r^2, \qquad (4.37)$$

with p^2 in the form (4.35), transforming the corresponding Schrödinger equation using the scaling transformation (4.33) and multiplying by $\tfrac{1}{4}\lambda^2$, we get for the radial component

$$\tfrac{1}{2}(\tfrac{1}{4}P_R^2 + R^{-2}\xi + \tfrac{1}{4}\omega^2\lambda^4 R^2 - \tfrac{1}{2}\lambda^2 E)\psi_R(\lambda R) = 0, \qquad (4.38)$$

with

$$\xi = \tfrac{1}{16}(N-1)(N-3) + \tfrac{1}{4}\ell(\ell + N - 2). \qquad (4.39)$$

Choosing λ such that $(\omega/2)^2 \lambda^4 = 1$, we can rewrite (4.38) using the so(2,1) realization (4.9) with $\nu = 2$ as

$$(T_3 - \tfrac{1}{4}\lambda^2 E)\psi_R(\lambda R) = 0. \qquad (4.40)$$

Thus, using the second equation of (4.6) we can interrelate $\psi_R(\lambda R)$ with $|kq\rangle$ and set $\tfrac{1}{4}\lambda^2 E = q$, so that

$$E = 4q/\lambda^2 = 2q\omega, \qquad (4.41)$$

with q given by Eqs. (4.12) and (4.13), i.e.,

$$q = q_0 + \mu, \quad \mu = 0, 1, 2, \ldots \qquad (4.42)$$

and

$$q_0 = k + 1 = \tfrac{1}{2}[1 \pm (\ell + \tfrac{1}{2}N - 1)]. \qquad (4.43)$$

Now, for $N = 1$ we set $\ell = 0$ so that $q_0 = \tfrac{1}{4}$ and $q_0 = \tfrac{3}{4}$, yielding for $E = 2q\omega$ the values $(\tfrac{1}{2} + 2\mu)\omega$ and $(\tfrac{1}{2} + 2\mu + 1)\omega, \mu = 0, 1, 2, \ldots$. Combining both sets we thus get for $N = 1$ the well known result

$$E \equiv E_n = (n + \tfrac{1}{2})\omega, \quad n = 0, 1, 2, \ldots. \qquad (4.44)$$

Similarly, for the general case $N \geq 2$ we choose the upper sign in (4.43) [so that $k > -1$] and get

$$E \equiv E_n = (n + \tfrac{1}{2}N)\omega, \quad n = 0, 1, 2, \ldots \qquad (4.45)$$

where we identified $(\ell + 2\mu)$ with the principal quantum number n.

4.3.2   *N*-Dimensional Hydrogenic Atom

Applying the scaling transformation (4.33) to the hydrogenic Hamiltonian (4.22) in N-dimensions, we get for the radial component (after multiplying from the left by $\lambda^2 R$)

$$\left[\tfrac{1}{2}(RP_R^2 + R^{-1}\xi - 2\lambda^2 ER) - \lambda \mathcal{Z}\right]\psi_R(\lambda R) = 0, \quad (4.46)$$

where now

$$\xi = \tfrac{1}{4}(N-1)(N-3) + \ell(\ell + N - 2). \qquad (4.47)$$

In this case we must set $2\lambda^2 E = -1$ and use realization (4.9) with $\nu = 1$ to obtain

$$(T_3 - \lambda \mathcal{Z})\psi_R(\lambda R) = 0. \qquad (4.48)$$

This immediately implies that

$$\lambda \mathcal{Z} = q \qquad (4.49)$$

and

$$q_0 = k + 1 = \tfrac{1}{2}[1 \pm (2\ell + N - 2)]. \qquad (4.50)$$

Choosing the upper sign [since $\ell \geq 0$ and $k > -1$], so that $q_0 = \ell + \tfrac{1}{2}(N-1)$, and identifying q with the principal quantum number n, we have finally that

$$E \equiv E_n = -\frac{1}{2\lambda^2} = -\frac{\mathcal{Z}^2}{2n^2}. \qquad (4.51)$$

The N-dimensional relativistic hydrogenic atom can be treated in the same way, using either the Klein-Gordon or Dirac-Coulomb equations (*cf.* [8–10]).

4.3.3   Perturbed Hydrogenic Systems

The so(4,2) based Lie algebraic formalism can be conveniently exploited to carry out large order perturbation theory (*cf.* Ref. [19] and Chap. 5) for hydrogenic systems described by the Schrödinger equation

$$[H_0 + \varepsilon V(\mathbf{r})]\psi(\mathbf{r}) = (E_0 + \Delta E)\psi(\mathbf{r}), \qquad (4.52)$$

with H_0 given by Eq. (4.22) and E_0 by Eq. (4.51). Applying transformation (4.33), using Eqs. (4.49), (4.51) and multiplying on the left by $\lambda^2 R$, we get

$$\left[\tfrac{1}{2}RP^2 - q + \tfrac{1}{2}R + \varepsilon\lambda^2 RV(\lambda \mathbf{R}) - \lambda^2 R\Delta E\right]\Psi(\mathbf{R}) = 0, \qquad (4.53)$$

where we set $\psi(\lambda \mathbf{R}) \equiv \Psi(\mathbf{R})$. For the important case of a 3-dimensional hydrogenic atom $[N = 3, \xi = \ell(\ell + 1), q \equiv n]$ we get using the so(4,2) realization (4.28) [or so(2,1) realization (4.9) with $\nu = 1$]

$$(K + \varepsilon W - S\Delta E)\Psi(\mathbf{R}) = 0, \qquad (4.54)$$

with

$$\begin{aligned} K &= T_3 - n, \\ W &= \lambda^2 RV(\lambda \mathbf{R}), \\ S &= \lambda^2 R. \end{aligned} \qquad (4.55)$$

We also have that $\lambda = n/\mathcal{Z}$ and for the ground state case $n = q = 1$. Although Eq. (4.54) has the form of a generalized eigenvalue problem requiring perturbation theory formalism with a nonorthogonal basis (where S represents an overlap), T_3 is Hermitian with respect to a $(1/R)$ scalar product, and the required matrix elements can therefore be easily evaluated [8, 10, 19].

For central field perturbations, $V(\mathbf{r}) = V(r)$, the problem reduces to one dimension and since $R = T_3 - T_1$, the so(2,1) hydrogenic realization ($\nu = 1$) can be employed. For problems of a hydrogenic atom in a magnetic field (Zeeman effect) [19, 20] or a one-electron diatomic ion [21], the so(4,2) formalism is required (note, however, that the LoSurdo-Stark effect can also be treated as a one-dimensional problem using parabolic coordinates [22]).

The main advantage of the LA approach stems from the fact that the spectrum of T_3 is discrete, so that no integration over continuum states is required. Moreover, the relevant perturbations are closely packed around the diagonal in this representation, so that infinite sums are replaced by small finite sums.

For example, for the LoSurdo-Stark problem when $V(\mathbf{r}) = \mathcal{F}z$, where $\mathcal{F}$ designates electric field strength in the z-direction, we get Eq. (4.54) with $\varepsilon = \mathcal{F}$ and $W = (n/\mathcal{Z})^3 RZ, S = (n/\mathcal{Z})^2 R$. Since both R and Z are easily expressed in terms of so(4,2) generators,

$$Z = B_3 - A_3, \quad R = T_3 - T_1, \qquad (4.56)$$

we can easily compute all the required matrix elements (*cf.* [8, 10]).

Similarly, considering the Zeeman effect with

$$V(\mathbf{r}) = \tfrac{1}{2}\mathcal{B}L_3 + \tfrac{1}{8}\mathcal{B}^2(r^2 - z^2), \qquad (4.57)$$

where $\mathcal{B}$ designates magnetic field strength in the z-direction, we have for the ground state when $n = 1$, $\ell = m = 0$ that $\varepsilon = \tfrac{1}{8}\mathcal{B}^2$, $K = T_3 - 1$, $W = \mathcal{Z}^{-4}R(R^2 - Z^2)$ and $S = \mathcal{Z}^{-2}R$. Again, the matrix elements of W and S are obtained from those of Z and R, [*cf.* Eq. (4.56)] by matrix multiplication (for tables and programs, see [10]).

One can treat one-electron diatomic ions [8, 21] and screened Coulomb potentials, including charmonium and harmonium [6, 10, 23], in a similar way.

Note, finally, that we can also formulate the perturbed problem (4.54) in a standard form not involving the "overlap" by defining the scaling factor as $\lambda = (-2E)^{-1/2}$, where E is now the exact energy $E = E_0 + \Delta E$. Eq. (4.54) then becomes

$$(T_3 + \varepsilon W - \lambda \mathcal{Z})\Psi(\mathbf{R}) = 0, \qquad (4.58)$$

with the eigenvalue $\lambda \mathcal{Z}$. In this case any conventional perturbation formalism applies, but the desired energy has to be found from $\lambda \mathcal{Z}$ [24].

4.4 COMPACT DYNAMICAL GROUPS

Unitary groups U(n) and their LAs often play the role of (compact) dynamical groups since

(1) quantum mechanical observables are Hermitian and the LA of U(n) is comprised of Hermitian operators [under the $\exp(iA)$ mapping],

(2) any compact Lie group is isomorphic to a subgroup of some U(n),

(3) "nothing of algebraic import is lost by the unitary restriction" [25].

All U(n) irreps have finite dimension and are thus relevant to problems involving a finite number of bound states [3–6, 25–32].

4.4.1 Unitary Group and Its Representations

The unitary group U(n) has n^2 generators E_{ij} spanning its LA and satisfying the commutation relations

$$[E_{ij}, E_{k\ell}] = \delta_{jk}E_{i\ell} - \delta_{i\ell}E_{kj} \qquad (4.59)$$

and the Hermitian property

$$E_{ij}^{\dagger} = E_{ji}. \qquad (4.60)$$

They are classified as *raising* $(i < j)$, *lowering* $(i > j)$ and *weight* $(i = j)$ generators according to whether they raise, lower and preserve the weight, respectively. The *weight vector* is a vector of the carrier space of an irrep which is a simultaneous eigenvector of all weight generators E_{ii} of U(n) (comprising its Cartan subalgebra), and the vector $\mathbf{m} = (m_1, m_2, \ldots, m_n)$ with integer components, consisting of corresponding eigenvalues, is called a *weight*. The *highest weight* $\mathbf{m}_n$ (in lexical ordering),

$$\mathbf{m}_n = (m_{1n}, m_{2n}, \ldots, m_{nn}), \qquad (4.61)$$

with

$$m_{1n} \geq m_{2n} \geq \cdots \geq m_{nn}, \qquad (4.62)$$

uniquely labels U(n) irreps, $\Gamma(\mathbf{m}_n)$, and may be represented by a Young pattern. Subducing $\Gamma(\mathbf{m}_r)$ of U(r) to U$(r-1)$, embedded as U$(r-1) \oplus 1$ in U(r), gives [30]

$$\Gamma(\mathbf{m}_r) \downarrow \mathrm{U}(r-1) = \bigoplus \Gamma(\mathbf{m}_{r-1}), \qquad (4.63)$$

where the sum extends over all U$(r - 1)$ weights $\mathbf{m}_{r-1} = (m_{1,r-1}, m_{2,r-1}, \ldots m_{r-1,r-1})$ satisfying the so-called "betweenness conditions" [27]

$$m_{ir} \geq m_{i,r-1} \geq m_{i+1,r} \qquad (i = 1, \ldots, r-1). \qquad (4.64)$$

Two irreps $\Gamma(\mathbf{m}_n)$ and $\Gamma(\mathbf{m}'_n)$ of U(n) yield the same irrep when restricted to SU(n) if $m_i = m'_i + h$, $i = 1, \ldots, n$. The SU(n) irreps are thus labeled with highest weights with $m_{nn} = 0$. The dimension of $\Gamma(\mathbf{m}_n)$ of U(n) is given by the Weyl dimension formula [25]

$$\dim \Gamma(\mathbf{m}_n) = \prod_{i<j}(m_{in} - m_{jn} + j - i) \Big/ 1!2! \cdots (n-1)!. \qquad (4.65)$$

The U(n) Casimir operators have the form

$$C_k^{\mathrm{U}(n)} = \sum_{i_1, i_2, \ldots, i_k = 1}^{n} E_{i_1 i_2} E_{i_2 i_3} \cdots E_{i_{k-1} i_k} E_{i_k i_1}. \qquad (4.66)$$

The first order Casimir operator is given by the sum of weight generators and equals the sum of the highest weight components.

Since U(1) is Abelian, the Gel'fand-Tsetlin [31] canonical chain (*cf.* Sect. 3.4.3)

$$\mathrm{U}(n) \supset \mathrm{U}(n-1) \supset \cdots \supset \mathrm{U}(1) \qquad (4.67)$$

can be used to label uniquely the basis vectors of the carrier space of $\Gamma(\mathbf{m}_n)$ by triangular Gel'fand tableaux $[m]$ defined by

$$[m] = \begin{bmatrix} \mathbf{m}_n \\ \mathbf{m}_{n-1} \\ \cdots \\ \cdots \\ \mathbf{m}_2 \\ \mathbf{m}_1 \end{bmatrix} = \begin{bmatrix} m_{1n} & m_{2n} & \cdots\cdots & m_{nn} \\ & m_{1,n-1} & \cdots\cdots m_{n-1,n-1} \\ & & \cdots \\ & & \cdots \\ & m_{12}\, m_{22} \\ & m_{11} \end{bmatrix},$$

(4.68)

with entries satisfying betweenness conditions (4.64). Matrix representatives of weight generators are diagonal

$$\langle [m']|E_{ii}|[m]\rangle = \delta_{[m],[m']}\left(\sum_{j=1}^{i} m_{ji} - \sum_{j=1}^{i-1} m_{j,i-1}\right),$$

(4.69)

while those for other generators are rather involved [31, 32]. Note that only elementary ($E_{i,i+1}$) raising generators are required since

$$\langle [m']|E_{ij}|[m]\rangle = \langle [m]|E_{ji}|[m']\rangle$$

(4.70)

and

$$E_{ij} = [E_{i,i+1}, E_{i+1,j}].$$

(4.71)

In special cases required in applications (*cf.* [28, 29] and Sect. 4.4.4) efficient algorithms exist for the computation of explicit representations.

4.4.2 Orthogonal Group O(n) and Its Representations

Since O(n) is a proper subgroup of U(n), its representation theory has a similar structure. The suitable generators are

$$F_{ij} = E_{ij} - E_{ji}, \quad F_{ji} = -F_{ij},$$
$$F_{ii} = 0, \quad F_{ij}^\dagger = -F_{ji}$$

(4.72)

and satisfy the commutation relations

$$[F_{ij}, F_{k\ell}] = \delta_{jk}F_{i\ell} + \delta_{i\ell}F_{jk} - \delta_{ik}F_{j\ell} - \delta_{j\ell}F_{ik}.$$

(4.73)

The canonical chain has the form

$$\mathrm{O}(n) \supset \mathrm{O}(n-1) \supset \cdots \supset \mathrm{O}(2).$$

(4.74)

The components of the highest weight $\mathbf{m}_n$,

$$\mathbf{m}_n = (m_{1n}, m_{2n}, \ldots, m_{kn}),$$

(4.75)

satisfy the conditions

$$m_{1n} \geq m_{2n} \geq \cdots \geq m_{kn} \geq 0 \quad \text{for} \quad n = 2k+1,$$

(4.76)

and

$$m_{1n} \geq m_{2n} \geq \cdots \geq |m_{kn}| \quad \text{for} \quad n = 2k,$$

(4.77)

where m_{in} are simultaneously integers or half-odd integers. The former are referred to as *tensor representations* (since they arise as tensor products of fundamental irreps), while those with half-odd integer components are called *spinor representations*. Note that for $n = 2k$, we have two lowest (mirror-conjugated) spinor representations, namely $\mathbf{m}^{(+)} = (\frac{1}{2}, \frac{1}{2}, \ldots, \frac{1}{2})$ and $\mathbf{m}^{(-)} = (\frac{1}{2}, \ldots, \frac{1}{2}, -\frac{1}{2})$. Only tensor representations can be labeled by Young tableaux.

Subducing O(n) to O($n-1$), the betweenness conditions (branching rules) have the form

$$m_{in} \geq m_{i,n-1} \geq m_{i+1,n} \quad (i = 1, \ldots, k-1)$$

(4.78)

together with

$$m_{k,2k+1} \geq |m_{k,2k}|$$

(4.79)

when $n = 2k+1$. The $m_{i,n-1}$ components are integral (half-odd integral) if the m_{in} are integral (half-odd integral). The U(n) $\supset$ O(n) [or SU(n) $\supset$ SO(n)] subduction rules are more involved (*cf.* [33]).

4.4.3 Clifford Algebras and Spinor Representations

While all reps of U(n) or SL(n) arise as tensor powers of the standard rep, only half of the reps of SO(m) or O(m) arise this way, since SO(m) is not simply connected when $m > 2$. A double covering of SO(m) leads to *spin groups* Spin(m). The best way to proceed is, however, to construct the so-called *Clifford algebras* C_m, whose multiplicative group (consisting of invertible elements) contains a subgroup which provides a double cover of SO(m). The key fact is that C_{2k} is isomorphic with gl(2^k) and C_{2k+1} with gl(2^k)$\oplus$ gl(2^k). The reps of C_m thus provide the required *spinor reps*.

A Clifford algebra C_m is an associative algebra generated by *Clifford numbers* α_i satisfying the anticommutation relations

$$\{\alpha_i, \alpha_j\} = 2\delta_{ij} \quad (i, j = 1, \ldots, m).$$

(4.80)

Since $\alpha_i^2 = 1$, $\dim C_m = 2^m$ and a general element of C_m is a product of Clifford numbers $\alpha_1^{\nu_1} \alpha_2^{\nu_2} \cdots \alpha_m^{\nu_m}$ with $\nu_i = 0$ or 1.

To see the relation with so($m+1$), note that

$$F_{0k} = -\tfrac{1}{2}i\alpha_k, \quad F_{jk} = \tfrac{1}{4}[\alpha_j, \alpha_k] = \tfrac{1}{2}\alpha_j\alpha_k, \quad (j \neq k)$$

(4.81)

satisfy the commutation relations (4.73).

As an example, C_2 can be realized by Pauli matrices by setting

$$\alpha_1 = \boldsymbol{\sigma}_1 = \begin{pmatrix} 0 & 1 \\ 1 & 0 \end{pmatrix}, \quad \alpha_2 = \boldsymbol{\sigma}_2 = \begin{pmatrix} 0 & i \\ -i & 0 \end{pmatrix}.$$

(4.82)

Clearly, the four matrices $\mathbf{1}_2, \alpha_1, \alpha_2$ and $\alpha_1 \alpha_2$ are linearly independent (note that $\boldsymbol{\sigma}_3 = i\boldsymbol{\sigma}_1\boldsymbol{\sigma}_2$), so that C_2 is isomorphic to $\text{gl}(2, \mathbb{C})$.

Similarly, considering Dirac-Pauli matrices

$$\gamma_0 = \begin{pmatrix} -i\mathbf{1}_2 & 0 \\ 0 & -i\mathbf{1}_2 \end{pmatrix} = i\gamma_4,$$

$$\gamma_k = \begin{pmatrix} 0 & i\boldsymbol{\sigma}_k \\ -i\boldsymbol{\sigma}_k & 0 \end{pmatrix}, \quad (k = 1, 2, 3)$$

(4.83)

we have that

$$\{\gamma_i, \gamma_j\} = 2\delta_{ij}, \quad (i, j = 1, \ldots, 4)$$

(4.84)

so that $\gamma_i (i = 1, \ldots, 4)$ or $(i = 0, \ldots, 3)$ represent Clifford numbers for C_4 and $\mathbf{1}_4$, γ_i, $\gamma_i\gamma_j$ $(i < j)$, $\gamma_i\gamma_j\gamma_k$ $(i < j < k)$ and $\gamma_5 \equiv \gamma_1\gamma_2\gamma_3\gamma_4 = i\gamma_0\gamma_1\gamma_2\gamma_3$ form an additive basis for $\text{gl}(4, \mathbb{C})$ (the γ_i themselves are said to form a multiplicative basis). For general construction of C_m Clifford numbers in terms of direct products of Pauli matrices see Ref. [34].

4.4.4 Bosonic and Fermionic Realizations of U(n)

Designating by $b_i^\dagger$ (b_i) the boson creation (annihilation) operators (*cf.* Sect. 6.1.1) satisfying the commutation relations $[b_i, b_j] = [b_i^\dagger, b_j^\dagger] = 0$, $[b_i, b_j^\dagger] = \delta_{ij}$, we obtain a possible U(n) realization by defining its n^2 generators as follows

$$G_{ij} = b_i^\dagger b_j.$$

(4.85)

The first order Casimir operator, Eq. (4.66) with $k = 1$, then represents the total number operator

$$\hat{N} \equiv C_1^{\text{U}(n)} = \sum_{i=1}^{n} G_{ii} = \sum_{i=1}^{n} b_i^\dagger b_i,$$

(4.86)

and the physically relevant states, being totally symmetric, carry single row irreps $\Gamma(N\dot{0}) \equiv \Gamma(N0\cdots0)$.

Similarly for fermion creation (annihilation) operators $X_I^\dagger (X_I)$ that are associated with some orthonormal spin orbital set $\{|I\rangle\}$, $I = 1, 2, \ldots, 2n$, and satisfy the anticommutation relations $\{X_I, X_J\} = \{X_I^\dagger, X_J^\dagger\} = 0$, $\{X_I, X_J^\dagger\} = \delta_{IJ}$, the operators

$$e_{IJ} = X_I^\dagger X_J$$

(4.87)

again represent the U($2n$) generators satisfying (4.59) and (4.60). The first-order Casimir then represents the total number operator $\hat{N} = \sum_I X_I^\dagger X_I$, while the possible physical states are characterized by totally antisymmetric single column irreps $\Gamma(1^N\dot{0}) \equiv \Gamma(11\cdots10\cdots0)$.

4.4.5 Vibron Model

Similar to the unified description of nuclear collective rovibrational states using the interacting boson model [35], one can build an analogous model for molecular rotation-vibration spectra [36]. For diatomics, an appropriate dynamical group is U(4) [36, 37] and, generally, for rotation-vibration spectra in r-dimensions one requires U($r + 1$). For triatomics, the U(4) generating algebra is generalized to U(4) $\otimes$ U(4), and for the $(k + 1)$ atomic molecule to U$^{(1)}$(4) $\otimes \cdots \otimes$ U$^{(k)}$(4) [36, 37].

For the bosonic realization of U(4), we need four creation $(b_i^\dagger, i = 1, \ldots, 4)$ and four annihilation (b_i) operators (*cf.* Sect. 4.4.4). The Hamiltonian may be generally expressed as a multilinear form in terms of boson number preserving products $(b_i^\dagger b_j)$, so that using (4.85) we can write

$$H = h^{(0)} + \sum_{ij} h_{ij}^{(1)} G_{ij} + \frac{1}{2} \sum_{ijk\ell} h_{ijk\ell}^{(2)} G_{ij} G_{k\ell} + \cdots.$$

(4.88)

The energy levels (as a function of $0, 1, 2, \ldots$-body matrix elements $h^0, h_{ij}^{(1)}, h_{ijk\ell}^{(2)}$, etc.) are then determined by diagonalizing H in an appropriate space, which is conveniently provided by the carrier space of the totally symmetric irrep $\Gamma(N\,0\,0\,0) \equiv \Gamma(N\dot{0})$ of U(4).

The requirement that the resulting states be characterized by angular momentum J and parity P quantum numbers necessitates that the boson operators involved have definite transformation properties under rotations and reflections (*cf.* [36]). The boson operators are thus subdivided into the scalar operators $(\sigma^\dagger, \sigma)$, $J = 0$, and vector operators $(\pi_\mu^\dagger, \pi_\mu; \mu = 0, \pm 1)$, $J = 1$ with parity $P = (-)^J$. All commutators vanish except for

$$[\sigma, \sigma^\dagger] = 1, \quad [\pi_\mu, \pi_{\mu'}^\dagger] = \delta_{\mu\mu'}.$$

(4.89)

Since H preserves the total number of vibrons $N = n_\sigma + n_\pi$, the second order Hamiltonian (4.88) within the irrep $\Gamma(N\dot{0})$ can be expressed in terms of four independent parameters (apart from an overall constant) as

$$\begin{aligned}
H = {}& e^{(0)} + e^{(1)}[\pi^\dagger \times \tilde{\pi}]_0^{(0)} \\
& + e_1^{(2)} \left[[\pi^\dagger \times \pi^\dagger]^{(0)} \times [\tilde{\pi} \times \tilde{\pi}]^{(0)} \right]_0^{(0)} \\
& + e_2^{(2)} \left[[\pi^\dagger \times \pi^\dagger]^{(2)} \times [\tilde{\pi} \times \tilde{\pi}]^{(2)} \right]_0^{(0)} \\
& + e_3^{(2)} \left[[\pi^\dagger \times \pi^\dagger]^{(0)} \times [\tilde{\sigma} \times \tilde{\sigma}]^{(0)} \right]
\end{aligned}$$

(4.90)

$$+ \left[[\sigma^\dagger \times \sigma^\dagger]^{(0)} \times [\tilde{\pi} \times \tilde{\pi}]^{(0)} \right]_0^{(0)} + \cdots,$$

where $\tilde{\sigma} = \sigma, \tilde{\pi}_\mu = (-)^{1-\mu} \pi_{-\mu}$ and square brackets indicate the SU(2) couplings.

In special cases the eigenvalue problem for H can be solved analytically, assuming that H can be expressed in terms of Casimir operators of a complete chain of subgroups of U(4) [referred to as dynamical symmetries]. Requiring that the chain contain the physical rotation group $O(3)$, one has two possibilities

$$\begin{array}{ll}
(I) & \text{U}(4) \supset \text{O}(4) \supset \text{O}(3) \supset \text{O}(2), \\
(II) & \text{U}(4) \supset \text{U}(3) \supset \text{O}(3) \supset \text{O}(2).
\end{array} \quad (4.91)$$

These imply labels (quantum numbers): N [total vibron number defining a totally symmetric irrep of U(4)], $\omega = N, N-2, N-4, \ldots, 1$ or 0 [defining a totally symmetric irrep of O(4)] and $n_\pi = N, N-1, \ldots, 0$ [defining the U(3) irrep], in addition to the $O(3) \supset O(2)$ labels $J, M; |M| \leq J$. In terms of these labels one finds for the respective Hamiltonians

$$\begin{array}{l}
H^{(I)} = F + A C_2^{\text{O}(4)} + B C_2^{\text{O}(3)}, \\
H^{(II)} = F + \varepsilon C_1^{\text{U}(3)} + \alpha C_2^{\text{U}(3)} + \beta C_2^{\text{O}(3)},
\end{array} \quad (4.92)$$

where $F, A, B, \varepsilon, \alpha, \beta$ are free parameters and $C_i^{\text{U}(k)}$, $C_i^{\text{O}(k)}$ are relevant Casimir operators, the following expressions [36, 37] for their eigenvalues

$$\begin{array}{l}
E^{(I)}(N, \omega, J, M) = F + A\omega(\omega + 2) + BJ(J+1), \\
E^{(II)}(N, n_\pi, J, M) = F + \epsilon n_\pi + \alpha n_\pi(n_\pi + 3) \\
\qquad\qquad\qquad\qquad + \beta J(J+1).
\end{array} \quad (4.93)$$

The limit (I) is appropriate for rigid diatomics and limit (II) for nonrigid ones [36, 37].

4.4.6 Many-Electron Correlation Problem

In atomic and molecular electronic structure calculations one employs a spin-independent model Hamiltonian

$$H = \sum_{i,j} h_{ij} \sum_{\sigma=1}^{2} X_{i\sigma}^\dagger X_{j\sigma}$$

$$+ \frac{1}{2} \sum_{i,j,k,\ell} v_{ij,k\ell} \sum_{\sigma,\tau=1}^{2} X_{i\sigma}^\dagger X_{j\tau}^\dagger X_{\ell\tau} X_{k\sigma}, \quad (4.94)$$

where $X_I^\dagger \equiv X_{i\sigma}^\dagger$ (X_I) designate the creation (annihilation) operators associated with the orthonormal spin orbitals $|I\rangle \equiv |i\sigma\rangle = |i\rangle \otimes |\sigma\rangle; i = 1, \ldots, n; \sigma = 1, 2$ [$\sigma = 1, 2$ labeling the spin-up and spin-down eigenstates of S_z], and $h_{ij} = \langle i|\hat{h}|j\rangle, v_{ij,k\ell} = \langle i(1)j(2)|\hat{v}|k(1)\ell(2)\rangle$ are the one- and two-electron integrals in the orbital basis $\{|i\rangle\}$.

As stated in Sect. 4.4.4, $e_{IJ} \equiv e_{i\sigma,j\tau} = X_{i\sigma}^\dagger X_{j\tau}$ may then be regarded as U(2n) generators, and the appropriate U(2n) irrep for N-electron states is $\Gamma(1^N \dot{0})$.

Similar to the nuclear many-body problem [38], one defines mutually commuting partial traces of spin orbital generators e_{IJ}, Eq. (4.87),

$$\begin{array}{l}
E_{ij} = \sum_{\sigma=1}^{2} e_{i\sigma,j\sigma} = \sum_{\sigma=1}^{2} X_{i\sigma}^\dagger X_{j\sigma}, \\[2mm]
\mathcal{E}_{\sigma\tau} = \sum_{i=1}^{n} e_{i\sigma,i\tau} = \sum_{i=1}^{n} X_{i\sigma}^\dagger X_{i\tau},
\end{array} \quad (4.95)$$

which again satisfy the unitary group commutation relations (4.59) and property (4.60), and may thus be considered as the generators of the orbital group U(n) and the spin group U(2). The Hamiltonian (4.94) is thus expressible in terms of orbital U(n) generators

$$H = \sum_{i,j} h_{ij} E_{ij} + \frac{1}{2} \sum_{i,j,k,\ell} v_{ij,k\ell}(E_{ik}E_{j\ell} - \delta_{jk}E_{i\ell}). \quad (4.96)$$

We can thus achieve an automatic spin adaptation by exploiting the chain

$$\text{U}(2n) \supset \text{U}(n) \otimes \text{U}(2) \quad (4.97)$$

and diagonalize H within the carrier space of two-column U(n) irreps $\Gamma(2^a 1^b 0^c) \equiv \Gamma(a, b, c)$ with [28, 39]

$$a = \tfrac{1}{2}N - S, \quad b = 2S, \quad c = n - a - b = n - \tfrac{1}{2}N - S, \quad (4.98)$$

considering the states of multiplicity $(2S + 1)$ involving n orbitals and N electrons. The dimension of each spin-adapted subproblem equals [28, 39]

$$\dim \Gamma(2^a 1^b 0^c) = \frac{b+1}{n+1} \binom{n+1}{a} \binom{n+1}{c}, \quad (4.99)$$

where $\binom{m}{n}$ designate binomial coefficients.

Exploiting simplified irrep labeling by triples of integers (a, b, c), Eq. (4.98), at each level of the canonical chain (4.67), one achieves more efficient state labeling by replacing Gel'fand tableaux (4.68) by $n \times 3$ *ABC* [39] or *Paldus tableaux* (see e.g., Refs. [29, 40–45])

$$[P] = [a_i b_i c_i], \quad (4.100)$$

where $a_i + b_i + c_i = i$. Another convenient labeling uses the ternary *step numbers* $d_i, 0 \leq d_i \leq 3$ (e.g., Refs. [39, 40, 46, 47])

$$d_i = 1 + 2(a_i - a_{i-1}) - (c_i - c_{i-1}). \quad (4.101)$$

An efficient and transparent representation of this basis can be achieved in terms of *Shavitt graphs* and *distinct row tables* ([40], *cf.* also [6, 28, 41]). An efficient evaluation of generator matrix representatives, as well as of their products, is formulated in terms of products of *segment values*, whose explicit form has been derived in several different ways (*cf.* [6, 39–41, 45, 47, 48]). Since the dimension (4.99) rapidly increases with n and N, various truncated schemes (limited CI) are often employed. The unitary group formalism, especially that of the universal enveloping algebra of $U(n)$, proved to be of great usefulness both in coupled cluster (*cf.* Chap. 5) and other approaches (see Refs. [6, 49] and references therein).

4.4.7 Clifford Algebra Unitary Group Approach

The Clifford algebra unitary group approach (CAUGA) exploits a realization of the spinor algebra of the rotation group $SO(2n + 1)$ in the covering algebra of $U(2^n)$ to obtain explicit representation matrices for the $U(n)$ [or $SO(2n + 1)$ or $SO(2n)$] generators in the basis adapted to the chain [50–52]

$$U(2^n) \supset \text{Spin}(m) \supset SO(m) \supset U(n),$$
$$(m = 2n + 1 \text{ or } m = 2n) \tag{4.102}$$

supplemented, if desired, by the canonical chain (4.67) for $U(n)$.

To realize the connection with the fermionic Grassmann algebra generated by the creation ($X_I^\dagger$) and annihilation (X_I) operators, $I = 1, \ldots, 2n$, note that it is isomorphic with the Clifford algebra C_{4n} when we define [49]

$$\alpha_I = X_I + X_I^\dagger, \quad \alpha_{I+2n} = i(X_I - X_I^\dagger),$$
$$(I = 1, \ldots, 2n). \tag{4.103}$$

For practical applications, the most important is the final imbedding $U(2^n) \supset U(n)$, (for the role of intermediate groups, see [50]). All states of an n-orbital model, regardless the electron number N and the total spin S, are contained in a single two-box totally symmetric irrep $\langle 2\dot{0}\rangle$ of $U(2^n)$ [51, 52]. To simplify the notation, one employs the one-to-one correspondence between the Clifford algebra monomials, labeled by the occupation numbers $m_i = 0$ or $m_i = 1$ ($i = 1, \ldots, n$), and "multiparticle" single-column $U(n)$ states labeled by

$$p \equiv p\{m_i\} = 2^n - (m_1 m_2 \cdots m_n)_2, \tag{4.104}$$

where the occupation number array $(m_1 \cdots m_n)$ is interpreted as a binary integer, which we then regard as one-box states $|p\rangle$ of $U(2^n)$. The orbital $U(n)$ generators Λ_{ij} may then be expressed as simple linear combinations of $U(2^n)$ generators $E_{pq} = |p\rangle\langle q|$ with coefficients equal to ± 1 [51, 52].

Generally, any p-column $U(n)$ irrep is contained at least once in the totally symmetric p-box irrep of $U(2^n)$. For many-electron problems, one thus requires a two-box irrep $\langle 2\dot{0}\rangle$. Any state arising in the $U(n)$ irrep $\Gamma(a, b, c)$ can then be represented as a linear combination of two-box states, labeled by the Weyl tableaux $[i|j] \equiv \boxed{i \; j}$. In particular, the highest state of $\Gamma(a, b, c)$ is represented by $[2^c|2^{b+c}]$. Once this representation is available, it is straightforward to compute explicit representations of $U(n)$ generators, since E_{pq} act trivially on $[i|j]$ (see [52]). Defining unnormalized states $(i|j)$ as

$$(i|j) = \sqrt{1 + \delta_{ij}} \; [i|j], \tag{4.105}$$

we have

$$E_{pq}(i|j) = \delta_{qi}(p|j) + \delta_{qj}(i|p). \tag{4.106}$$

The main features of CAUGA may thus be summarized as follows: CAUGA

(1) effectively reduces an N-electron problem to a number of two-boson problems;

(2) enables an exploitation of an arbitrary coupling scheme (being particularly suited for the valence bond method);

(3) can be applied to particle-number nonconserving operators;

(4) easily extends to fermions with an arbitrary spin;

(5) drastically simplifies evaluation of explicit representations of $U(n)$ generators and of their products;

(6) can be exploited in other than shell-model approaches (see, e.g., Refs. [53–55]).

REFERENCES

1. A. O. Barut, *Dynamical Groups and Generalized Symmetries in Quantum Theory* (University of Canterbury, Christchurch, 1971).
2. A. Bohm, Y. Ne'eman, A. O. Barut *et al.*, *Dynamical Groups and Spectrum Generating Algebras* (World Scientific, Singapore, 1988), Vols. 1 & 2.
3. R. Gilmore, *Lie Groups, Lie Algebras, and Some of Their Applications* (Wiley, New York, 1974).
4. B. G. Wybourne, *Classical Groups for Physicists* (Wiley, New York, 1974).
5. J.-Q. Chen, *Group Representation Theory for Physicists* (World Scientific, Singapore, 1989).
6. J. Paldus, in *Mathematical Frontiers in Computational Chemical Physics*, edited by D. G. Truhlar (Springer-Verlag, New York, 1988), pp. 262–299; I. Shavitt, *ibid.* pp. 300–349; J. Paldus, in *Contemporary Mathematics* (AMS, Providence, 1994), Vol. 160, pp. 209–236.
7. J. C. Parikh, *Group Symmetries in Nuclear Structure* (Plenum, New York, 1978).
8. B. G. Adams, J. Čížek, and J. Paldus, Int. J. Quantum Chem. **21**, 153 (1982); *idem*, Adv. Quantum Chem. **19**, 1 (1988); also reprinted in Ref. [2], Vol. 1, pp. 103–207.

9. J. Čížek and J. Paldus, Int. J. Quantum Chem. **12**, 875 (1977).

10. B. G. Adams, *Algebraic Approach to Simple Quantum Systems* (Springer, Berlin, 1994).

11. A. Bohm, Nuovo Cimento A**43**, 665 (1966).

12. A. Bohm, *Quantum Mechanics* (Springer, New York, 1979).

13. M. A. Naimark, *Linear Representations of the Lorentz Group* (Pergamon, New York, 1964).

14. W. Pauli, Z. Phys. **36**, 336 (1926); V. Fock, Z. Phys. **98**, 145 (1935).

15. A. O. Barut, Phys. Rev. **135**, B839 (1964).

16. A. O. Barut and G. L. Bornzin, J. Math. Phys. **12**, 841 (1971).

17. G. A. Zaicev, *Algebraic Problems of Mathematical and Theoretical Physics* (Science Publ. House, Moscow, 1974), (in Russian).

18. A. Joseph, Rev. Mod. Phys. **39**, 829 (1967).

19. M. Bednar, Ann. Phys. (N. Y.) **75**, 305 (1973); J. Čížek and E. R. Vrscay, in *Group Theoretical Methods in Physics* (Academic, New York, 1977), pp. 155–160; *idem*, Int. J. Quantum Chem. **21**, 27 (1982).

20. J. E. Avron, B. G. Adams, J. Čížek, M. Clay, L. Glasser, P. Otto, J. Paldus, and E. R. Vrscay, Phys. Rev. Lett. **43**, 691 (1979).

21. J. Čížek, M. Clay, and J. Paldus, Phys. Rev. A**22**, 793 (1980).

22. H. J. Silverstone, B. G. Adams, J. Čížek, and P. Otto, Phys. Rev. Lett. **43**, 1498 (1979).

23. E. R. Vrscay, Phys. Rev. A**31**, 2054 (1985); A**33**, 1433 (1986).

24. H. J. Silverstone and R. K. Moats, Phys. Rev. A**23**, 1645 (1981).

25. H. Weyl, *Classical Groups* (Princeton Univ. Press, Princeton, 1939).

26. A. O. Barut and R. Raczka, *Theory of Group Representations and Applications* (Polish Sci. Publishers, Warszawa, 1977).

27. J. D. Louck, Am. J. Phys. **38**, 3 (1970).

28. J. Paldus, in *Theoretical Chemistry: Advances and Perspectives*, edited by H. Eyring and D. J. Henderson (Academic, New York, 1976), Vol. 2, pp. 131–290.

29. F. A. Matsen and R. Pauncz, *The Unitary Group in Quantum Chemistry* (Elsevier, Amsterdam, 1986).

30. H. Weyl, *The Theory of Groups and Quantum Mechanics* (Dover, New York, 1964).

31. I. M. Gel'fand and M. L. Tsetlin, Dokl. Akad. Nauk SSSR **71**, 825, 1070 (1950).

32. G. E. Baird and L. C. Biedenharn, J. Math. Phys. **4**, 1449 (1963).

33. J. Deneen and C. Quesne, J. Phys. A**16**, 2995 (1983).

34. H. Boerner, *Representations of Groups* (North-Holland, Amsterdam, 1970), 2nd ed.; A. Ramakrishnan, *L-Matrix Theory or the Grammar of Dirac Matrices* (Tata McGraw-Hill, India, 1972).

35. A. Arima and F. Iachello, Phys. Rev. Lett. **35**, 1069 (1975); *Interacting Bosons in Nuclear Physics*, edited by F. Iachello (Plenum, New York, 1979); *idem, Interacting Bose-Fermi Systems in Nuclei* (Plenum, New York, 1981).

36. F. Iachello and R. D. Levine, *Algebraic Theory of Molecules* (Oxford University Press, Oxford, 1995).

37. F. Iachello, Chem. Phys. Lett. **78**, 581 (1981); F. Iachello and R. D. Levine, J. Chem. Phys. **77**, 3046 (1982); O. S. van Roosmalen, F. Iachello, R. D. Levine, and A. E. L. Dieperink, J. Chem. Phys. **79**, 2515 (1983).

38. M. Moshinsky, *Group Theory and the Many-Body Problem* (Gordon and Breach, New York, 1968).

39. J. Paldus, J. Chem. Phys. **61**, 5321 (1974).

40. I. Shavitt, Int. J. Quantum Chem. Symp. **11**, 131 (1977); *ibid.* **12**, 5 (1978).

41. *The Unitary Group for the Evaluation of Electronic Energy Matrix Elements*, Lecture Notes in Chemistry, edited by J. Hinze (Springer-Verlag, Berlin, 1981), Vol. 22.

42. R. Pauncz, *Spin Eigenfunctions: Construction and Use* (Plenum, New York, 1979), Chap. 9.

43. S. Wilson, *Electron Correlation in Molecules* (Clarendon, Oxford, 1984), Chap. 5.

44. R. McWeeny, *Methods of Molecular Quantum Mechanics* (Academic, New York, 1989), 2nd. ed., Chap. 10.

45. M. D. Gould and G. S. Chandler, Int. J. Quantum Chem. **25**, 553, 603, 1089 (1984); **26**, 441 (1984); E**27**, 878 (1985).

46. J. Paldus, Phys. Rev. A**14**, 1620 (1976).

47. J. Paldus and M. J. Boyle, Phys. Scr. **21**, 295 (1980).

48. G. W. F. Drake and M. Schlesinger, Phys. Rev. A**15**, 1990 (1977).

49. J. Paldus and B. Jeziorski, Theor. Chim. Acta **86**, 83 (1993).

50. R. S. Nikam and C. R. Sarma, J. Math. Phys. **25**, 1199 (1984); C. R. Sarma and J. Paldus, *ibid.* **26**, 1140 (1985); M. D. Gould and J. Paldus, *ibid.* **28**, 2304 (1987).

51. J. Paldus and C. R. Sarma, J. Chem. Phys. **83**, 5135 (1985).

52. J. Paldus, M.-J. Gao, and J.-Q. Chen, Phys. Rev. A**35**, 3197 (1987).

53. J. Paldus, in *Relativistic and Electron Correlation Effects in Molecules and Solids*, edited by G. L. Malli, NATO ASI Series B (Plenum, New York, 1994), Vol. 318, pp. 207–282; J. Paldus and X. Li, in *Symmetries in Science VI: From the Rotation Group to Quantum Algebras*, edited by B. Gruber (Plenum, New York, 1993), pp. 573–592.

54. J. Paldus and J. Planelles, Theor. Chim. Acta **89**, 13 (1994); J. Planelles, J. Paldus, and X. Li, *ibid.* **89**, 33, 59 (1994).

55. X. Li and J. Paldus, Int. J. Quantum Chem. **41**, 117 (1992); *idem*, J. Chem. Phys. **101**, 8812 (1994); **102**, 8059 (1995); P. Piecuch, X. Li, and J. Paldus, Chem. Phys. Lett. **230**, 377 (1994) and references therein.

5

Perturbation Theory

J. Paldus

Department of Applied Mathematics, University of Waterloo, Waterloo, Ontario N2L 3G1, Canada

5.1 INTRODUCTION

Perturbation theory (PT) represents one of the bridges that takes us from a simpler, exactly solvable (unperturbed) problem to a corresponding real (perturbed) problem by expressing its solutions as a series expansion in a suitably chosen "small" parameter ε in such a way that the problem reduces to the unperturbed problem when $\varepsilon = 0$. It originated in classical mechanics and eventually developed into an important branch of applied mathematics enabling physicists and engineers to obtain approximate solutions of various systems of differential equations (see, e.g., Refs. [1–4]). For the problems of atomic and molecular structure and dynamics, the perturbed problem is usually given by the time-independent or time-dependent Schrödinger equation [5–7].

5.2 MATRIX PERTURBATION THEORY (PT)

A prototype of a time-independent PT considers an eigenvalue problem for the Hamiltonian H of the form

$$H = H_0 + V, \quad V = \sum_{i=1}^{\infty} \varepsilon^i V_i, \qquad (5.1)$$

acting in a (finite-dimensional) Hilbert space $\mathcal{V}_n$, assuming that the spectral resolution of the unperturbed operator H_0 is known; i.e.,

$$H_0 = \sum_i \omega_i P_i, \quad P_i P_j = \delta_{ij} P_j, \quad \sum_i P_i = I, \qquad (5.2)$$

where ω_i are distinct eigenvalues of H_0, the P_i form a complete orthonormal set of Hermitian idempotents and I is the identity operator on $\mathcal{V}_n$. The PT problem for H can then be formulated within the Lie algebra $\mathcal{A}$ (see Sect. 3.2) generated by H_0 and V [8].

5.2.1 Basic Concepts

Define the diagonal part $\langle X \rangle$ of a general operator $X \in \mathcal{A}$ by

$$\langle X \rangle = \sum_i P_i X P_i, \qquad (5.3)$$

and recall that the adjoint action of $X \in \mathcal{A}$, ad $X : \mathcal{A} \to \mathcal{A}$, is defined by

$$\text{ad}\, X(Y) = [X, Y], \quad (\forall\, Y \in \mathcal{A}), \qquad (5.4)$$

where the square bracket denotes the commutator. The key problem of PT is the 'inversion' of this operation, i.e., the solution of the equation [8, 9]

$$\text{ad} H_0(X) \equiv [H_0, X] = Y. \qquad (5.5)$$

Assuming that $\langle Y \rangle = 0$, then

$$X = R(Y) + \langle A \rangle, \qquad (5.6)$$

where $A \in \mathcal{A}$ is arbitrary and

$$R(Y) = \sum_{i \neq j} \Delta_{ij}^{-1} P_i Y P_j, \qquad (5.7)$$

with $\Delta_{ij} = \omega_i - \omega_j$, represents the solution of (5.5) with the vanishing diagonal part $\langle R(Y) \rangle = 0$.

5.2.2 Level-Shift Operators

To solve the PT problem for H, Eq. (5.1), we search for a unitary level-shift transformation U [8],

$$UHU^\dagger = U(H_0 + V)U^\dagger = H_0 + E, \quad (U^\dagger U = UU^\dagger = I), \qquad (5.8)$$

where the level-shift operator E satisfies the condition

$$E = \langle E \rangle. \qquad (5.9)$$

To guarantee the unitarity of U, we express it in the form

$$U = e^G, \quad G^\dagger = -G, \quad \langle G \rangle = 0. \qquad (5.10)$$

Using the Haussdorff formula

$$e^A B e^{-A} = \sum_{k=0}^{\infty} (k!)^{-1} (\text{ad} A)^k B, \qquad (5.11)$$

and defining the operator

$$F = [H_0, G], \qquad (5.12)$$

we find from Eq. (5.8) that

$$E + F = V + \tfrac{1}{2}[H_0, V + E] \\ + \sum_{k=2}^{\infty} (k!)^{-1} B_k (\text{ad} G)^k (V - E), \qquad (5.13)$$

where we used the identity [10]

$$\left(\sum_{k=0}^{\infty} \frac{B_k}{k!} X^k \right) \left(\sum_{k=0}^{\infty} \frac{1}{(k+1)!} X^k \right) = I, \qquad (5.14)$$

and B_k designates the Bernoulli numbers [11]

$$B_0 = 1, \quad B_1 = -\frac{1}{2}, \quad B_2 = \frac{1}{6}, \\ B_{2k+1} = 0 \quad (k \geq 1), \qquad (5.15) \\ B_4 = \frac{1}{30}, \quad B_6 = \frac{1}{42}, \quad \text{etc.}$$

5.2.3 General Formalism

Introducing the PT expansion for relevant operators,

$$X = \sum_{i=1}^{\infty} \varepsilon^i X_i, \quad X = E, F, G; \quad F_i = [H_0, G_i], \qquad (5.16)$$

Eq. (5.13) leads to the following system of equations

$$E_1 + F_1 = V_1, \\ E_2 + F_2 = V_2 + \tfrac{1}{2}[G_1, V_1 + E_1], \\ E_3 + F_3 = V_3 + \tfrac{1}{2}[G_1, V_2 + E_2] \qquad (5.17) \\ \quad + \tfrac{1}{2}[G_2, V_1 + E_1] \\ \quad + \tfrac{1}{12}[G_1, [G_1, V_1 - E_1]], \quad \text{etc.,}$$

which can be solved recursively for E_i and G_i by taking their diagonal part and applying operator R, Eq. (5.7), since

$$\langle E_i \rangle = E_i, \qquad \langle G_i \rangle = \langle F_i \rangle = 0,$$

$$RF_i = G_i, \qquad RE_i = 0. \qquad (5.18)$$

We thus get

$$E_1 = \langle V_1 \rangle,$$
$$E_2 = \langle V_2 \rangle + \tfrac{1}{2}\langle [RV_1, V_1] \rangle, \qquad (5.19)$$
$$E_3 = \langle V_3 \rangle + \langle [RV_1, V_2] \rangle + \tfrac{1}{6}\langle [RV_1, [RV_1, 2V_1 + E_1]] \rangle,$$

etc.,

and

$$G_1 = RV_1,$$
$$G_2 = RV_2 + \tfrac{1}{2}R[RV_1, V_1 + E_1], \text{ etc.} \qquad (5.20)$$

Since

$$\langle R(X) \rangle = R\langle X \rangle = 0, \quad \langle R(X)Y \rangle = -\langle XR(Y) \rangle,$$
$$R(X\langle Y \rangle) = R(X)\langle Y \rangle, \quad R(\langle X \rangle Y) = \langle X \rangle R(Y), \qquad (5.21)$$

these relationships can be transformed to a more conventional form

$$E_2 = \langle V_2 \rangle - \langle V_1 R V_1 \rangle,$$
$$E_3 = \langle V_3 \rangle - \langle V_1 R V_2 \rangle - \langle V_2 R V_1 \rangle$$
$$\quad + \tfrac{1}{6}\langle R(V_1)R(V_1)[2V_1 + \langle V_1 \rangle] \rangle \qquad (5.22)$$
$$\quad - \tfrac{1}{3}\langle R(V_1)[2V_1 + \langle V_1 \rangle]R(V_1) \rangle$$
$$\quad + \tfrac{1}{6}\langle [2V_1 + \langle V_1 \rangle]R(V_1)R(V_1) \rangle, \text{ etc.}$$

However, in this way certain nonphysical terms arise that exactly cancel when the commutator form is employed (see Sect. 5.4.7).

5.2.4   Nondegenerate Case

In the nondegenerate case, when $P_i = |i\rangle\langle i|$, with $|i\rangle$ representing the eigenvector of H_0 associated with the eigenvalue ω_i, the level-shift operator is diagonal and its explicit PT expansion (as well as that for the corresponding eigenvectors) is easily obtained from Eqs. (5.19) and (5.20). Writing x_{ij} for the matrix element $\langle i|X|j \rangle$, we get

$$(e_1)_{ii} = (v_1)_{ii},$$

$$(e_2)_{ii} = (v_2)_{ii} - \sum_j{}' \frac{(v_1)_{ij}(v_1)_{ji}}{\Delta_{ji}},$$

$$(e_3)_{ii} = (v_3)_{ii} - \sum_j{}' \frac{(v_1)_{ij}(v_2)_{ji} + (v_2)_{ij}(v_1)_{ji}}{\Delta_{ji}} \qquad (5.23)$$

$$\quad + \sum_{j,k}{}' \frac{(v_1)_{ij}(v_1)_{jk}(v_1)_{ki}}{\Delta_{ji}\Delta_{ki}}$$

$$\quad - (v_1)_{ii}\sum_j{}' \frac{(v_1)_{ij}(v_1)_{ji}}{\Delta_{ji}^2},$$

etc.,

the prime on the summation symbols indicating that the terms with the vanishing denominator are to be deleted.

Note that in contrast to PT expansions which directly expand the level-shift transformation U, $U = 1 + \varepsilon U_1 + \varepsilon^2 U_2 + \cdots$, the above Lie algebraic formulation has the advantage that U stays unitary in every order of PT. This is particularly useful in spectroscopic applications, such as line broadening.

5.3   TIME-INDEPENDENT PERTURBATION THEORY

For stationary problems, particularly those arising in atomic and molecular electronic structure studies relying on *ab initio* model Hamiltonians, the PT of Sect. 5.2 can be given a more explicit form which avoids *a priori* the nonphysical, size inextensive terms [6, 7, 12].

5.3.1   General Formulation

We wish to find the eigenvalues and eigenvectors of the full (perturbed) problem

$$K|\Psi_i \rangle \equiv (K_0 + W)|\Psi_i \rangle = k_i|\Psi_i \rangle, \qquad (5.24)$$

assuming we know those of the unperturbed problem

$$K_0|\Phi_i \rangle = \kappa_i|\Phi_i \rangle, \quad \langle \Phi_i|\Phi_j \rangle = \delta_{ij}. \qquad (5.25)$$

For simplicity, we restrict ourselves to the nondegenerate case ($\kappa_i \neq \kappa_j$ if $i \neq j$) and consider only the first order perturbation [see Eq. (5.1), $\varepsilon V_1 \equiv W$, $V_i = 0$ for $i \geq 2$]. Using the intermediate normalization for $|\Psi_i \rangle$,

$$\langle \Psi_i|\Phi_i \rangle = 1, \qquad (5.26)$$

the asymmetric energy formula gives

$$k_i = \kappa_i + \langle \Phi_i|W|\Psi_i \rangle. \qquad (5.27)$$

The idempotent Hermitian projectors

$$P_i = |\Phi_i \rangle\langle \Phi_i|, \quad Q_i = P_i^\perp = 1 - P_i = \sum_{j(\neq i)} |\Phi_j \rangle\langle \Phi_j|, \qquad (5.28)$$

commute with K_0, so that

$$(\lambda - K_0)Q_i|\Psi_i \rangle = Q_i(\lambda - k_i + W)|\Psi_i \rangle, \qquad (5.29)$$

λ being an arbitrary scalar (note that we write λI simply as λ). Since the resolvent $(\lambda - K_0)^{-1}$ of K_0 is nonsingular on the orthogonal complement of the ith eigenspace, we get

$$Q_i|\Psi_i \rangle = |\Psi_i \rangle - |\Phi_i \rangle = R_i(\lambda)(\lambda - k_i + W)|\Psi_i \rangle, \qquad (5.30)$$

where

$$R_i \equiv R_i(\lambda) = (\lambda - K_0)^{-1} Q_i$$
$$= Q_i(\lambda - K_0)^{-1} = \sum_{j(\neq i)} \frac{|\Phi_j\rangle\langle\Phi_j|}{\lambda - \kappa_j}, \quad (5.31)$$

assuming $(\lambda \neq \kappa_j)$. Iterating this relationship, we get prototypes of the desired PT expansion for $|\Psi_i\rangle$,

$$|\Psi_i\rangle = \sum_{n=0}^{\infty} [R_i(\lambda - k_i + W)]^n |\Phi_i\rangle, \quad (5.32)$$

and, from Eq. (5.27), for k_i,

$$k_i = \kappa_i + \sum_{n=0}^{\infty} \langle\Phi_i|W[R_i(\lambda - k_i + W)]^n|\Phi_i\rangle. \quad (5.33)$$

5.3.2 Brillouin-Wigner and Rayleigh-Schrödinger PT (RSPT)

So far, the parameter λ was arbitrary, as long as $\lambda \neq \kappa_j$ ($j \neq i$). The following two choices lead to the two basic types of many-body perturbation theory (MBPT):

Brillouin-Wigner (BW) PT. Setting $\lambda = k_i$ gives

$$k_i = \kappa_i + \sum_{n=0}^{\infty} \langle\Phi_i|W(R_i^{(BW)}W)^n|\Phi_i\rangle, \quad (5.34)$$

$$|\Psi_i\rangle = \sum_{n=0}^{\infty} (R_i^{(BW)}W)^n|\Phi_i\rangle, \quad (5.35)$$

where

$$R_i^{(BW)} = \sum_{j(\neq i)} \frac{|\Phi_j\rangle\langle\Phi_j|}{k_i - \kappa_j}. \quad (5.36)$$

Rayleigh-Schrödinger (RS) PT. Setting $\lambda = \kappa_i$ gives

$$k_i = \kappa_i + \sum_{n=0}^{\infty} \langle\Phi_i|W[R_i^{(RS)}(\kappa_i - k_i + W)]^n|\Phi_i\rangle, \quad (5.37)$$

$$|\Psi_i\rangle = \sum_{n=0}^{\infty} [R_i^{(RS)}(\kappa_i - k_i + W)]^n|\Phi_i\rangle, \quad (5.38)$$

where now

$$R_i \equiv R_i^{(RS)} = \sum_{j(\neq i)} \frac{|\Phi_j\rangle\langle\Phi_j|}{\kappa_i - \kappa_j}. \quad (5.39)$$

The main distinction between these two PTs lies in the fact that the BW form has the exact eigenvalues appearing in the denominators, and thus leads to polynomial expressions for k_i. Although these are not difficult to solve numerically, since the eigenvalues are separated, the resulting energies are never size extensive and thus unusable for extended systems. They are also unsuitable for finite systems when the particle number changes, as in various dissociation processes. From now on, we thus investigate only the RSPT.

5.3.3 Bracketing Theorem and RSPT

Expressions (5.37) and (5.38) are not explicit, since they involve the exact eigenvalues k_i on the right-hand side. To achieve an order by order separation, set

$$k_i \equiv k = \sum_{j=0}^{\infty} k^{(j)}, \quad |\Psi_i\rangle \equiv |\Psi\rangle = \sum_{j=0}^{\infty} |\Psi^{(j)}\rangle, \quad (5.40)$$

where the superscript (j) indicates the jth-order in the perturbation W. We only consider the eigenvalue expressions, since the corresponding eigenvectors are easily recovered from them by removing the bra state and the first interaction W [see Eqs. (5.37) and (5.38)]. We also simplify the mean value notation writing for a general operator X,

$$\langle X \rangle \equiv \langle\Phi_i|X|\Phi_i\rangle. \quad (5.41)$$

Substituting the first expansion (5.40) into Eq. (5.37) and collecting the terms of the same order in W, we get

$$k^{(0)} = \kappa_i,$$
$$k^{(1)} = \langle W \rangle,$$
$$k^{(2)} = \langle WRW \rangle,$$
$$k^{(3)} = \langle W(RW)^2 \rangle - \langle W \rangle\langle WR^2W \rangle, \quad (5.42)$$
$$k^{(4)} = \langle W(RW)^3 \rangle - \langle W \rangle(\langle WR(RW)^2 \rangle + \langle (WR)^2 RW \rangle)$$
$$+ \langle W \rangle^2 \langle WR^3W \rangle - \langle WRW \rangle\langle WR^2W \rangle, \text{ etc.}$$

The general expression has the form

$$k^{(n)} = \langle W(RW)^{n-1} \rangle + \mathcal{R}^{(n)}, \quad (5.43)$$

the first term on the right-hand side being referred to as the *principal* nth-order term, while $\mathcal{R}^{(n)}$ designates the so-called *renormalization* terms that are obtained by the *bracketing theorem* [12, 13] as follows:

1. Insert the bracketings $\langle\cdots\rangle$ around the W, WRW, ..., $WR\cdots RW$ operator strings of the principal term in all possible ways.

2. Bracketings involving the rightmost and/or the leftmost interaction vanish.

3. The sign of each bracketed term is given by $(-1)^{n_B}$, where n_B is the number of bracketings.

4. Bracketings within bracketings are allowed, e.g., $\langle WR\langle WR\langle W\rangle RW\rangle RW\rangle = \langle W\rangle\langle WR^2W\rangle^2$.

5. The total number of bracketings (including the principal term) is $(2n-2)!/[n!(n-1)!]$.

5.4 FERMIONIC MANY-BODY PERTURBATION THEORY

5.4.1 Normal Product Form of PT

Consider the eigenvalue problem for a general *ab initio* or semi-empirical electronic Hamiltonian H with one- and two-body components Z and V, namely,

$$H|\Psi_i\rangle = E_i|\Psi_i\rangle, \quad H = Z + V = \sum_i z(i) + \sum_{i<j} v(i,j), \tag{5.44}$$

and a corresponding unperturbed problem

$$H_0|\Phi_i\rangle = \varepsilon_i|\Phi_i\rangle, \quad H_0 = Z + U, \quad \langle\Phi_i|\Phi_j\rangle = \delta_{ij}, \tag{5.45}$$

with U representing some approximation to V. In the case that U is also a one-electron operator, $U = \Sigma_i u(i)$, the unperturbed problem (5.45) is separable and reduces to a one-electron problem,

$$(z + u)|\mu\rangle = \omega_\mu|\mu\rangle, \tag{5.46}$$

which is assumed to be solved. Choosing the orthonormal spin orbitals $\{|\mu\rangle\}$ as a basis of the second quantization representation [see Chap. 6, Eq. (6.8)], the N-electron solutions of (5.45) can be represented as

$$|\Phi_i\rangle = a_{\mu_1}^\dagger a_{\mu_2}^\dagger \cdots a_{\mu_N}^\dagger|0\rangle, \tag{5.47}$$

$$\varepsilon_i = \sum_{j=1}^N \omega_{\mu_j}, \tag{5.48}$$

the state label i representing the occupied spin orbital set $\{\mu_1, \mu_2, \ldots, \mu_N\}$, while the one- and two-body operators take the form

$$X = \sum_{\mu,\nu} \langle\mu|x|\nu\rangle a_\mu^\dagger a_\nu, \quad X = Z, U; \ x = z, u, \tag{5.49}$$

$$V = \frac{1}{2} \sum_{\mu,\nu,\sigma,\tau} \langle\mu\nu|v|\sigma\tau\rangle a_\mu^\dagger a_\nu^\dagger a_\tau a_\sigma. \tag{5.50}$$

Considering, for simplicity, a nondegenerate ground state $|\Phi\rangle \equiv |\Phi_0\rangle = a_1^\dagger a_2^\dagger \cdots a_N^\dagger|0\rangle$, we define the *normal product form* of these operators relative to $|\Phi\rangle$

$$X_N \equiv X - \langle X\rangle = \sum_{\mu,\nu} \langle\mu|x|\nu\rangle N[a_\mu^\dagger a_\nu], \tag{5.51a}$$

$$(X = Z, U, G; \quad x = z, u, g)$$

$$V_N \equiv V - \langle V\rangle - G_N$$
$$= \frac{1}{2} \sum_{\mu,\nu,\sigma,\tau} \langle\mu\nu|v|\sigma\tau\rangle N[a_\mu^\dagger a_\nu^\dagger a_\tau a_\sigma]$$
$$= \frac{1}{4} \sum_{\mu,\nu,\sigma,\tau} \langle\mu\nu|v|\sigma\tau\rangle_A N[a_\mu^\dagger a_\nu^\dagger a_\tau a_\sigma], \tag{5.51b}$$

where

$$\langle\mu|g|\nu\rangle = \sum_{\sigma=1}^N \langle\mu\sigma|v|\nu\sigma\rangle_A, \tag{5.52}$$

$$\langle\mu\nu|v|\sigma\tau\rangle_A = \langle\mu\nu|v|\sigma\tau\rangle - \langle\mu\nu|v|\tau\sigma\rangle, \tag{5.53}$$

$\langle X\rangle = \langle\Phi|X|\Phi\rangle$, and $N[\cdots]$ designates the normal product relative to $|\Phi\rangle$ [12]. (Recall that $N[x_1 x_2 \cdots x_k] = \pm b_{\mu_1}^\dagger \cdots b_{\mu_i}^\dagger b_{\mu_{i+1}} \cdots b_{\mu_k}$, where $x_i = b_{\mu_i}$ or $b_{\mu_i}^\dagger$ are the annihilation and creation operators of the particle-hole formalism relative to $|\Phi\rangle$, i.e., $b_\mu = a_\mu^\dagger$ for $\mu \leq N$ and $b_\mu = a_\mu$ for $\mu > N$, the sign being determined by the parity of the permutation $p : j \mapsto \mu_j$.)

Defining

$$K = H - \langle H\rangle, \quad K_0 = H_0 - \langle H_0\rangle = H_0 - \varepsilon_0, \tag{5.54}$$

we can return to Eqs. (5.24) and (5.25), where now

$$k_i = E_i - \langle H\rangle, \quad \kappa_i = \varepsilon_i - \varepsilon_0, \quad \varepsilon_0 = \sum_{\mu=1}^N \omega_\mu, \tag{5.55}$$

and

$$W = K - K_0 = V - U - \langle V - U\rangle. \tag{5.56}$$

With this choice, $\langle W\rangle = 0$, so that for the reference state $|\Phi\rangle$, Eqs. (5.42) simplify to (we drop the subscript 0 for simplicity)

$$\begin{aligned} k^{(0)} &= 0, \quad k^{(1)} = 0, \\ k^{(2)} &= \langle WRW\rangle, \\ k^{(3)} &= \langle WRWRW\rangle, \\ k^{(4)} &= \langle W(RW)^3\rangle - \langle WRW\rangle\langle WR^2W\rangle, \text{ etc.} \end{aligned} \tag{5.57}$$

Note that W is also in the N-product form,

$$W = W_1 + W_2, \quad W_1 = G_N - U_N, \quad W_2 = V_N. \tag{5.58}$$

5.4.2 Møller-Plesset and Epstein-Nesbet PT

Choosing $U = G$ we have

$$H_0 = Z + G \equiv F, \tag{5.59}$$

so that Eqs. (5.46) represent Hartree-Fock (HF) equations, and ω_μ and $|\mu\rangle$ the canonical HF orbital energies and spin orbitals, respectively. Since $\langle H\rangle = \sum_{\mu=1}^N (\langle\mu|z|\mu\rangle + \frac{1}{2}\langle\mu|g|\mu\rangle)$ is the HF energy, $k = k_0$ gives directly the ground state correlation energy. (Note, however, that the N-product form of PT eliminates the first-order contribution $k^{(1)} = \langle W\rangle$ in any basis, even when F is not diagonal.) With this choice, $W_1 = 0$,

$W = V_N$, and the denominators in (5.39) are given by the differences of HF orbital energies

$$\kappa_0 - \kappa_j = \sum_{i=1}^{\lambda} (\omega_{\mu_i} - \omega_{\nu_i}) \equiv \Delta\langle\{\mu_i\}; \{\nu_j\}\rangle, \qquad (5.60)$$

assuming that $|\Phi_j\rangle$ is a λ-times excited configuration relative to $|\Phi\rangle$ obtained through excitations $\mu_i \to \nu_i$, $i = 1, \ldots, \lambda$. Using the Slater rules (or the second quantization algebra), we can express the second-order contribution in terms of the two-electron integrals and HF orbital energies as

$$k^{(2)} = \frac{1}{2} \sum_{a,b,r,s} \frac{\langle ab|v|rs\rangle(\langle rs|v|ab\rangle - \langle rs|v|ba\rangle)}{\omega_a + \omega_b - \omega_r - \omega_s}, \quad (5.61)$$

where the summations over $a, b\,(r, s)$ extend over all occupied (unoccupied) spin orbitals in $|\Phi\rangle$. Obtaining the corresponding higher-order corrections becomes more and more laborious and, beginning with the fourth-order, important cancellations arise between the principal and renormalization terms, even when the N-product form is employed. These will be addressed in Sect. 5.4.7.

The above outlined PT with H_0 given by the HF operator is often referred to as the *Møller-Plesset* PT [14]. In this version, the two-electron integrals enter the denominators only through the HF orbital energies. In an alternative, less often employed variant, referred to as the *Epstein-Nesbet* PT [15], the whole diagonal part of H is used as the unperturbed Hamiltonian, i.e.,

$$H_0 = \sum_i \langle \Phi_i|H|\Phi_i\rangle P_i. \qquad (5.62)$$

With this choice, the denominators are given as differences of the diagonal elements of the configuration interaction matrix.

5.4.3 Diagrammatic PT and Time Independent Wick's Theorem

To facilitate the evaluation of higher order terms, and especially to derive the general properties and characteristics of the MBPT, it is useful to employ a diagrammatic representation (see, e.g., [6, 7, 12, 16]). Representing all the operators in Eqs. (5.42) and (5.43) or (5.57) in the second quantized form, we have to deal with the reference state (called the Fermi vacuum) mean values of the strings of annihilation and creation operators (or with these strings acting on the reference in the case of a wave function). This is efficiently done using Wick's theorem and its diagrammatic representation via a special form of Feynman diagrams.

Wick's theorem expresses an arbitrary product of creation and annihilation operators as a normal product

(relative to $|\Phi_0\rangle$) and normal products with all possible contractions of these operators (see, e.g., [12]),

$$x_1 x_2 \cdots x_k = N[x_1 x_2 \cdots x_k] + \Sigma N[x_1 \overbrace{x_2 \cdots} x_k], (5.63)$$
$$(x_i = a^\dagger_{\mu_i} \text{ or } x_i = a_{\mu_i})$$

where

$$\overline{a^\dagger_\mu a^\dagger_\nu} = \overline{a_\mu a_\nu} = 0,$$
$$\overline{a^\dagger_\mu a_\nu} = h(\mu)\delta_{\mu\nu}, \quad \overline{a_\mu a^\dagger_\nu} = p(\mu)\delta_{\mu\nu}, \qquad (5.64)$$

and

$$h(\mu) = 1, \ p(\mu) = 0 \text{ if } |\mu\rangle \text{ is occupied in } |\Phi_0\rangle$$
$$\text{(hole states)},$$
$$h(\mu) = 0, \ p(\mu) = 1 \text{ if } |\mu\rangle \text{ is unoccupied in } |\Phi_0\rangle$$
$$\text{(particle states)}. \qquad (5.65)$$

The N-product with contractions is defined as a product of individual contractions times the N-product of uncontracted operators (defining $N[\emptyset] \equiv 1$ for an empty set) with the sign given by the parity of the permutation reordering the operators into their final order.

Note that the Fermi vacuum mean value of an N-product vanishes unless all operators are contracted. Thus, $\langle x_1 x_2 \cdots x_k\rangle$ is given by the sum over all possible fully contracted terms (vacuum terms). Similar rules easily follow for the expressions of the type $(x_1 x_2 \cdots x_k)|\Phi\rangle$. Moreover, if some operators on the left-hand side of (5.63) are already in the N-product form, all the terms involving contractions between these operators vanish.

5.4.4 Diagrammatic Representation

We associate with various operators suitable vertices with incident oriented lines representing the creation (outgoing lines) and annihilation (ingoing lines) operators that are involved in their second quantization representation. A few typical diagrams representing operators $(-U)$, W_1 and V are shown in Figs. 5.1(a), 5.1(b) and 5.1(c), 5.1(d), respectively. Using the N-product form of PT with HF orbitals (Sect. 5.4.2), we only need the two-electron operator V or V_N, which can be represented using either non-antisymmetrized vertices [Fig. 5.1(c)], leading to the *Goldstone diagrams* [17], or antisymmetrized vertices [Fig. 5.1(d)], yielding the *Hugenholtz diagrams* [18], [see Eq. (5.51b)].

5.4.5 Vacuum and Wave Function Diagrams

Applying Wick's theorem to the strings of operators involved, we represent the individual contractions

Figure 5.1. Diagrammatic representation of one- and two-electron operators.

by joining corresponding oriented lines. To obtain a non-vanishing contribution, only contractions preserving the orientation need be considered [see Eqs. (5.64)]. The resulting *internal* lines have either the left-right orientation (*hole* lines) or the right-left one (*particle* lines). Only fully contracted terms, represented by the so-called *vacuum diagrams* (having only internal lines), can contribute to the energy, while those representing wave function contributions have uncontracted or *free* lines extending to the left. When the operators involved are in the N-product form, no contractions of oriented lines issuing from the same vertex are allowed. The projection-like operators R, Eq. (5.39), or their powers, lead to the denominators, Eq. (5.60), given by the difference of hole and particle orbital energies associated with, respectively, hole and particle lines passing through the interval separating the corresponding two neighboring vertices. Clearly, there must always be at least one pair of such lines lest the denominator vanish. Thus, for example, the second-order contribution $\langle WRW \rangle$ is represented either by the two Goldstone diagrams [17] [Fig. 5.2(a) and (b)] or by the single Hugenholtz diagram [18] [Fig. 5.2(c)]. The rules for the energy (vacuum) diagram evaluation are as follows:

1. Associate appropriate matrix elements with all vertices and form their product. The outgoing (ingoing) lines on each vertex define the bra (ket) states of a given matrix element, and for the Goldstone diagrams, the oriented lines attached to the same node are associated with the same electron number, (e.g.,

Figure 5.2. The second-order Goldstone (a), (b) and Hugenholtz (c) diagrams.

Figure 5.3. Hugenholtz diagrams for the third-order energy contribution.

for the leftmost vertex in diagram (a) of Fig. 5.2 we have $\langle ab|\hat{v}|rs \rangle \equiv \langle a(1)b(2)|v|r(1)s(2) \rangle$).

2. Associate a denominator, Eq. (5.60), or its appropriate power, with every neighboring pair of vertices (and, for the wave function diagrams, also with the free lines extending to the left of the leftmost vertex; with each pair of such free lines associate also the corresponding pair of particle creation and hole annihilation operators).

3. Sum over all hole and particle labels.

4. Multiply each diagram contribution by the weight factor given by the reciprocal value of the order of the group of automorphisms of the diagram (stripped of summation labels) and by the sign $(-1)^{h+\ell}$, where h designates the number of internal hole lines and ℓ gives the number of closed loops of oriented lines (for Hugenholtz diagrams, use any of its Goldstone representatives to determine the correct phase).

Applying these rules to diagrams (a) and (b) of Fig. 5.2 we clearly recover Eq. (5.61) or, using the Hugenholtz diagram of Fig. 5.2, the equivalent expression

$$k^{(2)} = \frac{1}{4} \sum_{a,b,r,s} \langle ab|v|rs \rangle_A \langle rs|v|ab \rangle_A \, \Delta^{-1}(a,b;r,s) \,.$$

(5.66)

The possible third-order diagrams are shown in Fig. 5.3 with the central vertex involving particle-particle, hole-hole and particle-hole interaction [12, 16].

5.4.6 Hartree-Fock Diagrams

In the general case (non-HF orbitals and/or not normal product form of PT), the one-electron terms, as well as the contractions between operators associated with the same vertex, can occur (the latter are always the hole lines). Representing the W_1 and $(-U)$ operators as shown in Fig. 5.1, the one-body perturbation W_1 represents in fact the three diagrams as shown in Fig. 5.4. The second-order contribution of this type is then represented by the diagrams in Fig. 5.5, which in fact represents nine diagrams which result when each W_1 vertex is replaced by three vertices as shown in Fig. 5.4.

Figure 5.4. Schematic representation of $W_1 = G_N - U_N$.

Figure 5.6. The fourth-order unlinked diagrams.

Using HF orbitals, all these terms cancel, as seen above. For this reason, the diagrams involving contractions of lines issuing from the same vertex are referred to as the *Hartree-Fock diagrams*. Note, however, that even when not employing the canonical HF orbitals, it is convenient to introduce W_1 vertices of the normal product form PT and replace all nine HF-type diagrams by a single diagram of Fig. 5.5 (clearly, this feature provides even greater efficiency in higher orders of PT).

5.4.7 Linked and Connected Cluster Theorems

Using the N-product form of PT, the first nonvanishing renormalization term occurs in the fourth-order [see Eqs. (5.57)]. For a system consisting of N noninteracting species, the energy given by this nonphysical term is proportional to N^2, and thus violates the size extensivity of the theory. It was first shown by Brueckner [19] that in the fourth-order these terms are in fact exactly canceled by the corresponding contributions originating in the principal term. A general proof of this cancellation in an arbitrary order was given first by Goldstone [17] using the time-dependent PT formalism (Sect. 5.5).

To comprehend this cancellation, consider the fourth-order energy contribution arising from the so-called unlinked diagrams (no such contribution can arise in the second- or the third-order) shown in Fig. 5.6. An *unlinked diagram* is defined as a diagram containing a disconnected vacuum diagram (for the energy diagrams, the terms unlinked and disconnected are synonymous). The numerators associated with both diagrams being identical, we only consider the denominators. Designating

the denominator associated with the top and the bottom part by A and B, respectively, we find for the overall contribution

$$
\frac{1}{B} \cdot \frac{1}{A+B} \cdot \frac{1}{B} + \frac{1}{A} \cdot \frac{1}{A+B} \cdot \frac{1}{B}
$$
$$
= \left(\frac{1}{B} + \frac{1}{A}\right) \frac{1}{(A+B)B} = \frac{1}{AB^2}. \quad (5.67)
$$

Thus, the contribution from these terms exactly cancels that from the renormalization term $\langle WRW\rangle\langle WR^2W\rangle$.

Generalizing Eq. (5.67), we obtain the *factorization lemma* of Frantz and Mills [20], which implies the cancellation of renormalization terms by the unlinked terms originating from the principal term. This result holds for the energy as well as for the wave function contributions in every order of PT, as ascertained by the *linked cluster theorem*, which states that

$$
\Delta E = k = \sum_{n=0}^{\infty} \langle \Phi|W|\Psi^{(n)}\rangle = \sum_{n=0}^{\infty} \langle W(RW)^n\rangle_{\mathrm{L}}, \quad (5.68)
$$

$$
|\Psi\rangle = \sum_{n=0}^{\infty} |\Psi^{(n)}\rangle = \sum_{n=0}^{\infty} \{(RW)^n|\Phi\rangle\}_{\mathrm{L}}, \quad (5.69)
$$

where the subscript L indicates that only linked diagrams (or terms) are to be considered. This enables us to obtain general, explicit expressions for the nth-order PT contributions by first constructing all possible linked diagrams involving n vertices and by converting them into the explicit algebraic expressions using the rules of Sect. 5.4.5. Note that linked energy diagrams are always connected, but the linked wave function diagrams are either connected or disconnected, each disconnected component possessing at least one pair of particle-hole free lines extending to the left.

To reveal a deeper structure of the result (5.69), define the *cluster operator* T that generates all connected wave function diagrams,

$$
T|\Phi\rangle = \sum_{n=1}^{\infty} \{(RW)^n|\Phi\rangle\}_{\mathrm{C}}, \quad (5.70)
$$

the subscript C indicating that only contributions from *connected* diagrams are to be included. Since the general

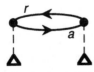

Figure 5.5. The second-order one-particle contribution.

component with r disconnected parts can be shown to be represented by the term $(r!)^{-1}T^r|\Phi\rangle$, the general structure of the exact wave function $|\Psi\rangle$ is given by the *connected cluster theorem*, which states that

$$|\Psi\rangle = e^T |\Phi\rangle. \qquad (5.71)$$

In other words, the wave operator $\mathcal{W}$ which transforms the unperturbed independent particle model wave function $|\Phi\rangle$ into the exact one according to

$$|\Psi\rangle = \mathcal{W}|\Phi\rangle, \qquad (5.72)$$

is given by the exponential of the cluster operator T,

$$\mathcal{W} = e^T, \qquad (5.73)$$

which in turn is given by the connected wave function diagrams. This is in fact the basis of the coupled cluster methods [16, 21, 22] (see Sect. 5.4.8).

The contributions to T may be further classified by their excitation rank i,

$$T = \sum_{i=1}^{N} T_i, \qquad (5.74)$$

where T_i designates connected diagrams with i pairs of free particle-hole lines, producing i-times excited components of $|\Psi\rangle$ when acting on $|\Phi\rangle$.

5.4.8 Coupled Cluster Theory

Summing all HF diagrams (see Sect. 5.4.6) is equivalent to solving the HF equations. Depending on the average electron density of the system, it may be essential to sum certain types of PT diagrams to infinite order at the post-HF level. A frequently used approach that is capable of recovering a large part of the electronic correlation energy is based on the connected cluster theorem (Sect. 5.4.7), referred to in this context as the exponential cluster Ansatz for the wave operator. Using this Ansatz, one derives a system of energy-independent nonlinear coupled cluster (CC) equations [16, 22] determining the cluster amplitudes of T. These CC equations can be regarded as recurrence relations generating the MBPT series [16], so that by solving these equations one in fact implicitly generates all the MBPT diagrams and sums them to infinite order. Since the solution of CC equations is equivalent to the exact solution of the Schrödinger equation, we must clearly introduce a suitable truncation scheme, which implies that only diagrams of certain types are summed.

Generally, using the cluster expansion (5.71) in the N-product form of the Schrödinger equation,

$$H_N|\Psi\rangle \equiv (H - \langle H \rangle)|\Psi\rangle = \Delta E|\Psi\rangle, \quad \Delta E = E - E_0, \qquad (5.75)$$

premultiplying with the inverse of the wave operator, and using the Hausdorff formula (5.11) yields

$$e^{-T}H_N e^T|\Phi\rangle = \sum_{n=0}^{\infty} \frac{[\mathrm{ad}(-T)]^n H_N}{n!} = \Delta E|\Phi\rangle. \qquad (5.76)$$

In fact, this expansion terminates, so that using (5.74) and projecting onto $|\Phi\rangle$ we obtain the energy expression

$$\Delta E = \langle H_N T_2 \rangle + \tfrac{1}{2}\langle H_N T_1^2 \rangle, \qquad (5.77)$$

while the projection onto the manifold of excited states $\{|\Phi_i\rangle\}$ relative to $|\Phi\rangle \equiv |\Phi_0\rangle$ gives the system of CC equations

$$\langle \Phi_i | H_N + [H_N, T] + \tfrac{1}{2}[[H_N, T], T] + \cdots |\Phi\rangle = 0. \qquad (5.78)$$

Approximating, e.g., T by the most important pair cluster component $T \approx T_2$ gives the so-called CCD (coupled clusters with doubles) approximation

$$\langle \Phi_i^{(2)} | H_N + [H_N, T_2] + \tfrac{1}{2}[[H_N, T_2], T_2]|\Phi\rangle = 0, \qquad (5.79)$$

the superscript (2) indicating pair excitations relative to $|\Phi\rangle$.

Equivalently, (5.77) and (5.78) can be written in the form

$$\Delta E = \langle H_N e^T \rangle_C, \qquad (5.80)$$

$$\langle \Phi_i | (H_N e^T)_C |\Phi\rangle = 0, \qquad (5.81)$$

the subscript C again indicating that only connected diagrams are to be considered. The general form of CC equations is

$$a_i + \sum_j b_{ij} t_j + \sum_{j \le k} c_{ijk} t_j t_k + \cdots = 0, \qquad (5.82)$$

where $a_i = \langle \Phi_i | H_N |\Phi_0\rangle$, $b_{ij} = \langle \Phi_i | H_N |\Phi_j\rangle_C$, $c_{ijk} = \langle \Phi_i | H_N |\Phi_j \otimes \Phi_k\rangle_C$, etc. Writing the diagonal linear term b_{ii} in the form

$$b_{ii} = \Delta_i + b'_{ii}, \qquad (5.83)$$

this system can be solved iteratively by rewriting it in the form

$$t_i^{(n+1)} = \Delta_i^{-1}\Big[a_i + b'_{ii}t_i^{(n)} + {\sum_j}' b_{ij}t_j^{(n)} + \sum_{j \le k} c_{ijk}t_j^{(n)}t_k^{(n)} + \cdots\Big]. \qquad (5.84)$$

Starting with the zeroth approximation $t_i^{(0)} = 0$, the first iteration is

$$t_i^{(1)} = \Delta_i^{-1}a_i, \qquad (5.85)$$

which yields the second order PT energy when used in (5.77). Clearly, the successive iterations generate higher and higher orders of the PT. At any truncation level, a size extensive result is obtained.

5.5 TIME-DEPENDENT PERTURBATION THEORY

5.5.1 Evolution Operator PT Expansion

By introducing the *evolution operator* $U(t, t_0)$

$$|\Psi(t)\rangle = U(t, t_0)|\Psi(t_0)\rangle, \qquad (5.86)$$

time-dependent Schrödinger equation

$$i\hbar \frac{\partial}{\partial t}|\Psi(t)\rangle = H|\Psi(t)\rangle \qquad (5.87)$$

becomes

$$i\hbar \frac{\partial}{\partial t} U(t, t_0) = H U(t, t_0). \qquad (5.88)$$

Clearly,

$$
\begin{aligned}
U(t_0, t_0) &= 1, \\
U(t, t_0) &= U(t, t')U(t', t_0), \\
U(t, t_0)^{-1} &= U(t_0, t) = U^\dagger(t, t_0).
\end{aligned}
\qquad (5.89)
$$

If the Hamiltonian is time independent then

$$U(t, t_0) = \exp\left[-\frac{i}{\hbar}H(t - t_0)\right]. \qquad (5.90)$$

In the *interaction picture* (subscript I)

$$|\Psi(t)\rangle_I = \exp(\frac{i}{\hbar}H_0 t)|\Psi(t)\rangle, \qquad (5.91)$$

where now

$$H = H_0 + V, \qquad (5.92)$$

the Schrödinger equation becomes

$$i\hbar \frac{\partial}{\partial t}|\Psi(t)\rangle_I = V(t)_I |\Psi(t)\rangle_I, \qquad (5.93)$$

known as *Tomonaga-Schwinger equation* [23]. Analogously, the evolution operator in this picture (we drop the subscript I from now on) satisfies

$$i\hbar \frac{\partial}{\partial t} U(t, t_0) = V(t)U(t, t_0), \qquad (5.94)$$

with the initial condition $U(t_0, t_0) = 1$. This differential equation is equivalent to an integral equation

$$U(t, t_0) = 1 - \frac{i}{\hbar}\int_{t_0}^{t} V(t_1)U(t_1, t_0)\, dt_1. \qquad (5.95)$$

Iterating we get [24]

$$U(t, t_0)$$

$$
= \sum_{n=0}^{\infty}\left(-\frac{i}{\hbar}\right)^n \int_{t_0}^{t} dt_1 \int_{t_0}^{t_1} dt_2 \cdots \int_{t_0}^{t_{n-1}} dt_n\, V(t_1)V(t_2)\cdots V(t_n)
$$

$$
= \sum_{n=0}^{\infty}\frac{(-i/\hbar)^n}{n!}\int_{t_0}^{t} dt_1 \cdots \int_{t_0}^{t} dt_n T[V(t_1)\cdots V(t_n)],
$$

$$(5.96)$$

where $T[\cdots]$ designates the time-ordering or chronological operator.

5.5.2 Gell-Mann and Low Formula

For a time-independent perturbation, one introduces the so-called *adiabatic switching* by writing

$$H_\alpha(t) = H_0 + \lambda e^{-\alpha|t|}V, \quad \alpha > 0 \qquad (5.97)$$

so that $H_\alpha(t \to \pm\infty) = H_0$ and $H_\alpha(t \to 0) = H = H_0 + \lambda V$. Then

$$|\Psi(t)\rangle_I = U_\alpha(t, -\infty|\lambda)|\Phi_0\rangle, \qquad (5.98)$$

with $U_\alpha(t, -\infty|\lambda)$ obtained with $V_\alpha(t) = \lambda e^{-\alpha|t|}V$ (all in the interaction picture). The desired energy is then given by the Gell-Mann and Low formula [25]

$$\Delta E = \lim_{\alpha \to 0+} i\hbar\alpha\lambda \frac{\partial}{\partial \lambda}\ln\langle\Phi_0|U_\alpha(0, -\infty|\lambda)|\Phi_0\rangle, \qquad (5.99a)$$

or

$$\Delta E = \frac{1}{2}\lim_{\alpha \to 0+} i\hbar\alpha\lambda \frac{\partial}{\partial \lambda}\ln\langle\Phi_0|U_\alpha(\infty, -\infty|\lambda)|\Phi_0\rangle, \qquad (5.99b)$$

which result from the asymmetric energy formula (5.27). One can similarly obtain the perturbation expansion for the one- or two-particle Green functions, e.g.,

$$G_{\mu\nu}(t, t') = \lim_{\alpha \to 0+} \frac{\langle T\{a_\mu(t)a_\nu^\dagger(t')U_\alpha(\infty, -\infty|\lambda)\}\rangle}{\langle U_\alpha(\infty, -\infty|\lambda)\rangle}, \qquad (5.100)$$

with the operators in the interaction representation and the expectation values in the noninteracting ground state $|\Phi_0\rangle$. Analogous expressions result for $G(\mathbf{r}t, \mathbf{r}'t')$, etc., when the creation and annihilation operators are replaced by the corresponding field operators.

5.5.3 Potential Scattering and Quantum Dynamics

The Schrödinger equation for a free particle of energy $E = \hbar^2 k^2/2m$, moving in the potential $V(\mathbf{r})$,

$$(\nabla^2 + k^2)\psi(k, \mathbf{r}) = v(\mathbf{r})\psi(k, \mathbf{r}), \quad v(\mathbf{r}) = (2m/\hbar^2)V(\mathbf{r}),$$
(5.101)

has the formal solution

$$\psi(k, \mathbf{r}) = \Phi(k, \mathbf{r}) + \int G_0(k, \mathbf{r}, \mathbf{r}')v(\mathbf{r}')\psi(k, \mathbf{r}') \, d\mathbf{r}',$$
(5.102)

where $\Phi(k, \mathbf{r})$ is a solution of the homogeneous equation $[v(\mathbf{r}) \equiv 0]$ and $G_0(k, \mathbf{r}, \mathbf{r}')$ is a classical Green function

$$(\nabla^2 + k^2)G_0(k, \mathbf{r}, \mathbf{r}') = \delta(\mathbf{r} - \mathbf{r}').$$
(5.103)

For an in-going plane wave $\Phi(k, \mathbf{r}) \equiv \Phi_{\mathbf{k}_i}(\mathbf{r}) = (2\pi)^{3/2}\exp(i\mathbf{k}_i \cdot \mathbf{r})$ with the initial wave vector $\mathbf{k}_i$ and appropriate asymptotic boundary conditions (outgoing spherical wave with positive phase velocity), when $G_0(k, \mathbf{r}, \mathbf{r}') \equiv G_0^{(+)}(|\mathbf{r} - \mathbf{r}'|) = -(4\pi|\mathbf{r} - \mathbf{r}'|)^{-1}e^{ik|\mathbf{r}-\mathbf{r}'|}$, Eq. (5.102) is referred to as the *Lippmann-Schwinger equation* [26]. It can be equivalently transformed into the integral equation for the Green function

$$G^{(+)}(\mathbf{r}, \mathbf{r}') = G_0^{(+)}(\mathbf{r}, \mathbf{r}') + \int G_0^{(+)}(\mathbf{r}, \mathbf{r}'')v(\mathbf{r}'')$$
$$\times G^{(+)}(\mathbf{r}'', \mathbf{r}') \, d\mathbf{r}'',$$
(5.104)

representing a special case of Dyson equation.

In the time-dependent case, considering the scattering of a spinless massive particle by a time-dependent potential $V(\mathbf{r}, t)$, we get similarly

$$\psi(\mathbf{r}, t) = \Phi(\mathbf{r}, t) + \int G_0(\mathbf{r}, \mathbf{r}'; t, t')V(\mathbf{r}', t')\psi(\mathbf{r}', t') \, d\mathbf{r}'dt',$$
(5.105)

where the zero-order time-dependent Green function now satisfies the equation

$$(i\hbar\frac{\partial}{\partial t} - H_0)G_0(\mathbf{r}, \mathbf{r}'; t, t') = \delta(\mathbf{r} - \mathbf{r}')\delta(t - t').$$
(5.106)

Again, for causal propagation one chooses the *time-retarded* or *causal* Green function or propagator $G_0^{(+)}(\mathbf{r}, \mathbf{r}'; t, t')$.

5.5.4 Born Series

Iteration of Eq. (5.105) gives the *Born sequence*

$$\psi_0(\mathbf{r}, t) = \Phi(\mathbf{r}, t),$$
$$\psi_1(\mathbf{r}, t) = \Phi(\mathbf{r}, t) + \int G_0^{(+)}(\mathbf{r}, \mathbf{r}'; t, t')$$
$$\times V(\mathbf{r}', t')\Phi(\mathbf{r}', t') \, d\mathbf{r}'dt',$$
(5.107a)

$$\psi_2(\mathbf{r}, t) = \Phi(\mathbf{r}, t) + \int G_0^{(+)}(\mathbf{r}, \mathbf{r}'; t, t')$$
$$\times V(\mathbf{r}', t')\psi_1(\mathbf{r}', t') \, d\mathbf{r}'dt',$$
(5.107b)

and, generally

$$\psi_n(\mathbf{r}, t) = \Phi(\mathbf{r}, t) + \int G_0^{(+)}(\mathbf{r}, \mathbf{r}'; t, t')$$
$$\times V(\mathbf{r}', t')\psi_{n-1}(\mathbf{r}', t') \, d\mathbf{r}'dt'.$$
(5.108)

Summing individual contributions gives the *Born series* for $\psi(\mathbf{r}, t) \equiv \psi^{(+)}(\mathbf{r}, t)$,

$$\psi(\mathbf{r}, t) = \sum_{n=0}^{\infty} \chi_n(\mathbf{r}, t),$$
(5.109)

where

$$\chi_0(\mathbf{r}, t) = \Phi(\mathbf{r}, t),$$
$$\chi_n(\mathbf{r}, t) = \int \mathcal{G}_n(\mathbf{r}, \mathbf{r}'; t, t')\Phi(\mathbf{r}', t') \, d\mathbf{r}'dt',$$
(5.110)

with

$$\mathcal{G}_n(\mathbf{r}, \mathbf{r}'; t, t') = \int \mathcal{G}_1(\mathbf{r}, \mathbf{r}''; t, t'')$$
$$\times \mathcal{G}_{n-1}(\mathbf{r}'', \mathbf{r}'; t'', t') \, d\mathbf{r}''dt'', \quad (n > 1)$$
$$\mathcal{G}_1(\mathbf{r}, \mathbf{r}'; t, t') = G_0^{(+)}(\mathbf{r}, \mathbf{r}'; t, t')V(\mathbf{r}', t').$$
(5.111)

In a similar way we obtain the Born series for the scattering amplitudes or transition matrix elements.

5.5.5 Variation of Constants Method

An alternative way of formulating the time-dependent PT is the method of variation of the constants [27]. Start again with the time-dependent Schrödinger equation (5.87) with $H = H_0 + V$, and assume that H_0 is time-independent, while V is a time-dependent perturbation. Designating the eigenvalues and eigenstates of H_0 by ε_i and $|\Phi_i\rangle$, respectively [see Eq. (5.45)], the general solution of the unperturbed time-dependent Schrödinger equation

$$i\hbar\frac{\partial}{\partial t}|\Psi_0\rangle = H_0|\Psi_0\rangle$$
(5.112)

has the form

$$|\Psi_0\rangle = \sum_j c_j|\Phi_j\rangle \exp\left(-\frac{i}{\hbar}\varepsilon_j t\right),$$
(5.113)

with c_j representing arbitrary constants, and the sum indicating both the summation over the discrete part and the integration over the continuum part of the spectrum of H_0.

In the spirit of the general variation of the constants procedure, write the unknown perturbed wave function $|\Psi(t)\rangle$, Eq. (5.87), in the form

$$|\Psi(t)\rangle = \sum_j C_j(t)|\Phi_j\rangle \exp\left(-\frac{i}{\hbar}\varepsilon_j t\right), \qquad (5.114)$$

where the $C_j(t)$ are now functions of time. Substituting this Ansatz into the time-dependent Schrödinger equation [Eq. (5.87)] gives

$$\dot{C}_j(t) = (i\hbar)^{-1} \sum_k C_k(t) V_{jk} \exp[(i/\hbar)\Delta_{jk} t], \qquad (5.115)$$

where

$$\Delta_{jk} = \varepsilon_j - \varepsilon_k, \qquad V_{jk} = \langle\Phi_j|V|\Phi_k\rangle. \qquad (5.116)$$

Introducing again the 'small' parameter λ by writing the Hamiltonian H in the form

$$H = H_0 + \lambda V(t), \qquad (5.117)$$

and expanding the 'coefficients' $C_j(t)$ in powers of λ,

$$C_j \equiv C_j(t) = \sum_{k=0}^{\infty} C_j^{(k)}(t)\lambda^k, \qquad (5.118)$$

gives the system of first order differential equations

$$\dot{C}_j^{(n+1)} = (i\hbar)^{-1} \sum_k C_k^{(n)} V_{jk} \exp[(i/\hbar)\Delta_{jk} t],$$
$$n = 0, 1, 2, \ldots, \qquad (5.119)$$

with the initial condition $\dot{C}_j^{(0)} = 0$, which implies that $C_j^{(0)}$ are time independent, so that $C_j^{(0)} = c_j$, obtaining (5.113) in the zeroth order. The system (5.119) can be integrated to any prescribed order. For example, if the system is initially in a stationary state $|\Phi_i\rangle$, then set

$$C_j^{(0)} = \begin{cases} \delta_{ji} & \text{for discrete states,} \\ \delta(j-i) & \text{for continuous states,} \end{cases} \qquad (5.120)$$

so that

$$C_{j(i)}^{(1)}(t) = (i\hbar)^{-1} \int_{-\infty}^{t} V_{ji} \exp[(i/\hbar)\Delta_{ji} t'] \, dt', \qquad (5.121)$$

assuming $C_{j(i)}^{(1)}(-\infty) = 0$. Clearly, $|C_{j(i)}^{(1)}(t)|^2$ gives the first order transition probability for the transition from the initial state $|\Phi_i\rangle$ to a particular state $|\Phi_j\rangle$. These in turn will yield the first order differential cross sections (see, e.g., [28]).

REFERENCES

1. T. Kato, *Perturbation Theory for Linear Operators* (Springer, Berlin, 1966).
2. H. Baumgärtel, *Analytic Perturbation Theory for Matrices and Operators* (Akademie, Berlin, 1984).
3. E. J. Hinch, *Perturbation Methods* (Cambridge University Press, Cambridge, 1991).
4. V. N. Bogaevski and A. Povzner, *Algebraic Methods in Nonlinear Perturbation Theory* (Springer, Berlin, 1991).
5. E. M. Corson, *Perturbation Methods in the Quantum Mechanics of n-Electron Systems* (Blackie & Son, London, 1951).
6. I. Lindgren and J. Morrison, *Atomic Many-Body Theory* (Springer, Berlin, 1982).
7. E. K. U. Gross, E. Runge, and O. Heinonen, *Many-Particle Theory* (Hilger, New York, 1991).
8. H. Primas, Helv. Phys. Acta **34**, 331 (1961); Rev. Mod. Phys. **35**, 710 (1963).
9. M. Rosenblum, Duke Math. J. **23**, 263 (1956).
10. G. Arfken, *Mathematical Methods for Physicists* (Academic, New York, 1985), p. 327.
11. *Encyclopedic Dictionary of Mathematics*, edited by S. Iyanaga and Y. Kawada (MIT Press, Cambridge, 1980), Appendix B, Table 3, p. 1494.
12. J. Paldus and J. Čížek, Adv. Quantum Chem. **9**, 105 (1975).
13. H. J. Silverstone and T. T. Holloway, J. Chem. Phys. **52**, 1472 (1970), and references therein.
14. C. Møller and M. S. Plesset, Phys. Rev. **46**, 618 (1934).
15. P. S. Epstein, Phys. Rev. **28**, 695 (1926); R. K. Nesbet, Proc. R. Soc. London A**250**, 312 (1955).
16. J. Paldus, in *Methods in Computational Molecular Physics*, edited by S. Wilson and G. H. F. Diercksen, NATO ASI Series B (Plenum, New York, 1992), Vol. 293, pp. 99–194.
17. J. Goldstone, Proc. R. Soc. London A**239**, 267 (1957).
18. H. M. Hugenholtz, Physica (Utrecht) **23**, 481 (1957).
19. K. A. Brueckner, Phys. Rev. **100**, 36 (1955).
20. L. M. Frantz and R. L. Mills, Nucl. Phys. **15**, 16 (1960).
21. F. Coester, Nucl. Phys. **7**, 421 (1958); F. Coester and H. Kümmel, Nucl. Phys. **17**, 477 (1960).
22. J. Čížek, J. Chem. Phys. **45**, 4256 (1966); Adv. Chem. Phys. **14**, 35 (1969).
23. F. J. Dyson, Phys. Rev. **75**, 486 (1949).
24. S. Tomonaga, Prog. Theor. Phys. (Kyoto) **1**, 27 (1946); J. Schwinger, Phys. Rev. **74**, 1439 (1948).
25. M. Gell-Mann and F. Low, Phys. Rev. **84**, 350 (1951).
26. B. A. Lippmann and J. Schwinger, Phys. Rev. **79**, 469 (1950).
27. P. A. M. Dirac, Proc. R. Soc. London A**112**, 661 (1926); **114**, 243 (1926).
28. C. J. Joachain, *Quantum Collision Theory* (Elsevier, New York, 1975).

6

Second Quantization

B. R. Judd

Department of Physics and Astronomy, The Johns Hopkins University, Baltimore, Maryland 21218

6.1 BASIC PROPERTIES

6.1.1 Definitions

The creation operator $a_\xi^\dagger$ creates the quantum state ξ. The annihilation (or destruction) operator a_η annihilates the quantum state η. The vacuum (or reference) state $|0\rangle$ satisfies the equation

$$a_\eta|0\rangle = 0. \tag{6.1}$$

Bosons satisfy the commutation relations

$$[a_\xi^\dagger, a_\eta^\dagger] = 0, \tag{6.2}$$
$$[a_\xi, a_\eta] = 0, \tag{6.3}$$
$$[a_\xi, a_\eta^\dagger] = \delta(\xi, \eta), \tag{6.4}$$

where $[A, B] \equiv AB - BA$. Fermions satisfy the anticommutation relations

$$[a_\xi^\dagger, a_\eta^\dagger]_+ = 0, \tag{6.5}$$
$$[a_\xi, a_\eta]_+ = 0, \tag{6.6}$$
$$[a_\xi, a_\eta^\dagger]_+ = \delta(\xi, \eta), \tag{6.7}$$

where $[A, B]_+ \equiv AB + BA$.

6.1.2 Representation of States

For an electron in an atom, characterized by the quantum number quartet $(n\,\ell\,m_s\,m_\ell)$, the identification $\xi \equiv (n\,\ell\,m_s\,m_\ell)$ for fermions is made. For normalized

Slater determinants $\{\alpha\beta\ldots\nu\}$ characterized by the electron states $\alpha,\beta,\ldots,\nu$, the equivalences

$$a_\alpha^\dagger a_\beta^\dagger \ldots a_\nu^\dagger |0\rangle \equiv \{\alpha\beta\ldots\nu\}, \qquad (6.8)$$

$$\langle 0|a_\nu \ldots a_\beta a_\alpha \equiv \{\alpha\beta\ldots\nu\}^* \qquad (6.9)$$

are valid, where the asterisk denotes the complex conjugate.

For a normalized boson state $\{\cdots\}$ in which the label ξ appears N_ξ times, the additional factor

$$[N_\alpha! N_\beta! \ldots N_\nu!]^{-\frac{1}{2}} \qquad (6.10)$$

must be included on the left-hand sides of the equivalences (6.8) and (6.9).

6.1.3   Representation of Operators

For an operator F, consisting of the sum of operators f_i acting on the single electron i,

$$F \equiv \sum_{\xi,\eta} a_\xi^\dagger \langle \xi|f|\eta\rangle a_\eta. \qquad (6.11)$$

For an operator G, consisting of the sum of operators g_{ij} acting on the pair of electrons i and j,

$$G \equiv \frac{1}{2} \sum_{\xi,\eta,\zeta,\lambda} a_\xi^\dagger a_\eta^\dagger \langle \xi_1\eta_2|g_{12}|\zeta_1\lambda_2\rangle a_\lambda a_\zeta. \qquad (6.12)$$

For an N-particle system $|\Psi\rangle$,

$$\sum_\xi a_\xi^\dagger a_\xi |\Psi\rangle = N|\Psi\rangle. \qquad (6.13)$$

The representations of single-particle and two-particle operators for bosons are identical to those given above for fermions [1].

6.2   TENSORS

6.2.1   Construction

If the description ξ for a single fermion or boson state includes an angular momentum quantum number t and the corresponding magnetic quantum number m_t, then the $2t+1$ components of a creation operator $a_\sigma^\dagger$, where $\sigma \equiv (t, m_t)$ and $-t \le m_t \le t$, satisfy the commutation relations of Racah [2] for an irreducible spherical tensor of rank t with respect to the total angular momentum $\mathbf{T}$, given by

$$\mathbf{T} = \sum a_\xi^\dagger \langle \xi|\mathbf{t}|\eta\rangle a_\eta. \qquad (6.14)$$

That is, with the phase conventions of Condon and Shortley [3],

$$[T_z, a_\sigma^\dagger] = m_t a_\sigma^\dagger, \qquad (6.15)$$

$$[T_x \pm iT_y, a_\sigma^\dagger] = [t(t+1) - m_t(m_t \pm 1)]^{\frac{1}{2}} a_\tau^\dagger, \qquad (6.16)$$

where $\tau \equiv (t, m_t \pm 1)$.

A spherical tensor $\mathbf{a}$ constructed from annihilation operators possesses the components $\tilde{a}_\sigma$, which satisfy

$$\tilde{a}_\sigma = (-1)^p a_\zeta, \qquad (6.17)$$

with $p = t - m_t$ and $\zeta \equiv (t, -m_t)$.

The $4\ell + 2$ components of the creation operator for an electron in the atomic ℓ shell form a double tensor of rank $\frac{1}{2}$ with respect to the total spin $\mathbf{S}$, and rank ℓ with respect to the total angular momentum $\mathbf{L}$.

6.2.2   Coupled Forms

Tensors formed from annihilation and creation operators can be coupled by means of the usual rules of angular momentum theory [4]. The double tensor defined for electrons in the ℓ shell by

$$\mathbf{W}^{(\kappa k)} = -(\mathbf{a}^\dagger \mathbf{a})^{(\kappa k)}, \qquad (6.18)$$

possesses a rank κ with respect to $\mathbf{S}$, and rank k with respect to $\mathbf{L}$. Its reduced matrix element, defined here as in Eq. (5.4.1) of Edmonds [4], for a single electron in both the spin and orbital spaces, is given by

$$(s\,\ell\,||\,W^{(\kappa k)}\,||\,s\,\ell\,) = [(2\kappa+1)(2k+1)]^{\frac{1}{2}}. \qquad (6.19)$$

The connections to tensors whose matrix elements have been tabulated [5,6] are

$$\mathbf{W}^{(0k)} = [(2k+1)/2]^{\frac{1}{2}} \mathbf{U}^{(k)}, \qquad (6.20)$$

$$\mathbf{W}^{(1k)} = [2(2k+1)]^{\frac{1}{2}} \mathbf{V}^{(k1)}. \qquad (6.21)$$

For terms with common spin S, say ψ and ψ',

$$(\psi||W^{(0k)}||\psi') = [(2S+1)(2k+1)/2]^{\frac{1}{2}} (\psi||U^{(k)}||\psi'). \qquad (6.22)$$

This result is obtained because the ranks assigned to the tensors imply that $\mathbf{W}^{(0k)}$ is to be reduced with respect to both the spin $\mathbf{S}$ and the orbit $\mathbf{L}$, while $\mathbf{U}^{(k)}$ is to be reduced only with respect to $\mathbf{L}$.

The following relations hold for electrons with azimuthal quantum numbers ℓ [7]:

$$\mathbf{S} = [(2\ell+1)/2]^{\frac{1}{2}} \mathbf{W}^{(10)}, \qquad (6.23)$$

$$\mathbf{L} = [2\ell(\ell+1)(2\ell+1)/3]^{\frac{1}{2}} \mathbf{W}^{(01)}, \qquad (6.24)$$

$$\sum_i (\mathbf{s}_i \mathbf{C}_i^{(2)})^{(1)} = -\left[\frac{\ell(\ell+1)(2\ell+1)}{10(2\ell-1)(2\ell+3)}\right]^{\frac{1}{2}} \mathbf{W}^{(12)1}, \qquad (6.25)$$

$$\sum_i (\mathbf{s}_i \cdot \boldsymbol{\ell}_i) = -[\ell(\ell+1)(2\ell+1)/2]\mathbf{W}^{(11)0}, \qquad (6.26)$$

where the tensor $\mathbf{C}^k$ of Racah [2] is related to the spherical harmonics by

$$C_q^{(k)} = [4\pi/(2k+1)]^{\frac{1}{2}} Y_{kq} , \qquad (6.27)$$

and where the tensors of the type $\mathbf{W}^{(\kappa k)K}$ indicate that the spin and orbital ranks are coupled to a resultant K.

6.2.3 Coefficients of Fractional Parentage

Let ψ and $\bar{\psi}$ denote terms of ℓ^N and ℓ^{N-1} characterized by (S, L) and $(\bar{S}, \bar{L})$. The coefficients of fractional parentage (cfp) $(\psi\{|\bar{\psi})$ of Racah [8] allow one to calculate an antisymmetrized function ψ by vector-coupling $\bar{\psi}$ to the spin and orbit of the Nth electron:

$$|\psi\rangle = \sum_{\bar{\psi}} |\bar{\psi}, {}^2\ell, SL\rangle \, (\bar{\psi}|\}\psi) , \qquad (6.28)$$

where the sum over $\bar{\psi}$ includes $\bar{S}, \bar{L}$, and any other quantum numbers necessary to define the spectroscopic terms of ℓ^{N-1}. The cfp's are given by

$$(\psi||a^\dagger||\bar{\psi}) = (-1)^N [N(2S+1)(2L+1)]^{\frac{1}{2}} (\psi\{|\bar{\psi}), \quad (6.29)$$

$$(\bar{\psi}||a||\psi) = (-1)^g [N(2S+1)(2L+1)]^{\frac{1}{2}} (\bar{\psi}|\}\psi), \quad (6.30)$$

where $g = N + \bar{S} + \bar{L} - s - S - \ell - L$. A tabulation for the p, d, and f shells has been given by Nielson and Koster [5].

Two-electron cfp are given by

$$(\psi||(\mathbf{a}^\dagger \mathbf{a}^\dagger)^{(\kappa k)}||\tilde{\psi}) = [N(N-1)(2S+1)(2L+1)]^{\frac{1}{2}}$$
$$\times (\psi\{|\tilde{\psi}, \ell^2(\kappa k)), \qquad (6.31)$$

where $\tilde{\psi}$ denotes a term of ℓ^{N-2}, and the symbols κ and k stand for the S and the L of a term of ℓ^2. A tabulation for the p, d, and f shells has been given by Donlan [9].

If, through successive applications of the two-particle operators $(\mathbf{aa})^{(00)}$, a state of ℓ^N can be reduced to ℓ^v, but no further, then v is the seniority number of Racah [8].

If the ranks s and ℓ of $\mathbf{a}^\dagger$ are coupled to $\bar{S}$ and $\bar{L}$ of $\bar{\psi}$, the term

$$(\mathbf{a}^\dagger|\bar{\psi}))^{(SL)} \qquad (6.32)$$

either vanishes, or is a term of ℓ^N characterized by S and L. Such a term is said to possess the *godparent* $\bar{\psi}$. Redmond [10] has used the notion of godparents to generate an explicit formula for the single particle cfp [7].

6.3 QUASISPIN

6.3.1 Fermions

For electrons, the components $Q_\pm (\equiv Q_x \pm i Q_y)$ and Q_z of the quasispin $\mathbf{Q}$ are defined by [7, 11]

$$Q_+ = [(2\ell+1)/2]^{\frac{1}{2}} (\mathbf{a}^\dagger \, \mathbf{a}^\dagger)^{(00)} , \qquad (6.33)$$

$$Q_- = -[(2\ell+1)/2]^{\frac{1}{2}} (\mathbf{aa})^{(00)} , \qquad (6.34)$$

$$Q_z = -[(2\ell+1)/8]^{\frac{1}{2}} [(\mathbf{a}^\dagger \, \mathbf{a})^{(00)} + (\mathbf{aa}^\dagger)^{(00)}] . \qquad (6.35)$$

The term quasispin comes from the fact that the components of $\mathbf{Q}$ satisfy the commutation relations of an angular momentum vector. The eigenvalues M_Q of Q_z, for a state of ℓ^N, are given by

$$M_Q = -(2\ell+1-N)/2 . \qquad (6.36)$$

The shift operators Q_+ and Q_- connect states of the ℓ shell possessing the same value of the seniority v of Racah [8]. A string of such connected states defines the extrema of M_Q, from which it follows that

$$Q = (2\ell+1-v)/2 . \qquad (6.37)$$

6.3.2 Bosons

For real vibrational modes created by $a_\nu^\dagger$ ($\nu = 1, 2, \dots, d$), the analogs of (6.33)–(6.35) are

$$P_+ = -\frac{1}{2} \sum_\nu a_\nu^\dagger a_\nu^\dagger , \qquad (6.38)$$

$$P_- = \frac{1}{2} \sum_\nu a_\nu a_\nu , \qquad (6.39)$$

$$P_z = \frac{1}{4} \sum_\nu (a_\nu^\dagger a_\nu + a_\nu a_\nu^\dagger) , \qquad (6.40)$$

and $\mathbf{P}$ is an angular momentum vector [12]. The eigenvalues M_P for an n-boson state are given by

$$M_P = (2n+d)/4 , \qquad (6.41)$$

and can therefore be quarter-integral. Successive application of the operator P_+ to a state $|n_0\rangle$, for which $P_-|n_0\rangle = 0$, generates an infinite ladder of states characterized by

$$P = (2n+d-4)/4 . \qquad (6.42)$$

6.3.3 Triple Tensors

The creation and annihilation operators $a_\xi^\dagger$ and a_ξ for a given state ξ can be regarded as the two components of a tensor of rank $\frac{1}{2}$ with respect to quasispin (either $\mathbf{Q}$ or $\mathbf{P}$). For electrons, this leads to triple tensors $\mathbf{a}^{(qs\ell)}$ (for which $q = s = \frac{1}{2}$) satisfying

$$a_\lambda^{(qs\ell)} a_\mu^{(qs\ell)} + a_\mu^{(qs\ell)} a_\lambda^{(qs\ell)}$$
$$= (-1)^{x+1} \delta(m_q, -m_q') \delta(m_s, -m_s') \delta(m_\ell, -m_\ell') , \quad (6.43)$$

where $\lambda \equiv (m_q m_s m_\ell)$, $\mu \equiv (m'_q m'_s m'_\ell)$, and $x = q+s+\ell+m_q+m_s+m_\ell$. In terms of the coupled tensor

$$\mathbf{X}^{(K\kappa k)} = (\mathbf{a}^{(qs\ell)}\mathbf{a}^{(qs\ell)})^{(K\kappa k)}, \qquad (6.44)$$

the angular momenta $\mathbf{Q}$, $\mathbf{S}$, and $\mathbf{L}$ are given by

$$\mathbf{Q} = -[(2\ell+1)/4]^{\frac{1}{2}}\mathbf{X}^{(100)}, \qquad (6.45)$$
$$\mathbf{S} = -[(2\ell+1)/4]^{\frac{1}{2}}\mathbf{X}^{(010)}, \qquad (6.46)$$
$$\mathbf{L} = -[\ell(\ell+1)(2\ell+1)/3]^{\frac{1}{2}}\mathbf{X}^{(001)}. \qquad (6.47)$$

Furthermore, the components of $\mathbf{X}^{(K\kappa k)}$ for which $M_K = 0$ are identical to the corresponding components of $2^{\frac{1}{2}}(\mathbf{a}^\dagger \mathbf{a})^{(\kappa k)}$ when $K+\kappa+k$ is odd; and

$$\mathbf{X}^{(K\kappa k)} = -(2\ell+1)^{\frac{1}{2}}\delta(K,0)\delta(\kappa,0)\delta(k,0) \qquad (6.48)$$

when $K+\kappa+k$ is even.

6.3.4 Conjugation

Creation and annihilation operators can be interchanged by the operation of the conjugation operator C [7,13]. For electrons in the atomic ℓ shell,

$$Ca_\xi^{(qs\ell)}C^{-1} = (-1)^{q-m_q}a_\eta^{(qs\ell)}, \qquad (6.49)$$

where $\xi \equiv (m_q m_s m_\ell)$ and $\eta \equiv ((-m_q)m_s m_\ell)$. In terms of the tensors $\mathbf{a}^\dagger$ and $\mathbf{a}$,

$$C\mathbf{a}^\dagger C^{-1} = \mathbf{a}, \quad C\mathbf{a}C^{-1} = -\mathbf{a}^\dagger. \qquad (6.50)$$

Furthermore,

$$CX_\lambda^{(K\kappa k)}C^{-1} = (-1)^{K-M_K}X_\mu^{(K\kappa k)}, \qquad (6.51)$$

where $\lambda \equiv (M_K M_\kappa M_k)$ and $\mu \equiv ((-M_K)M_\kappa M_k)$, and

$$C|QM_Q\rangle = (-1)^{Q-M_Q}|Q-M_Q\rangle. \qquad (6.52)$$

Thus, from (6.36), the action of C takes N into $4\ell+2-N$; that is, C interchanges electrons and holes. When the case $\kappa = k = 0$ is excluded, application of (6.51) and (6.52) yields

$$(\ell^N\psi\|W^{(\kappa k)}\|\ell^N\psi') = (-1)^y(\ell^{4\ell+2-N}\psi\|W^{(\kappa k)}\|\ell^{4\ell+2-N}\psi'), \qquad (6.53)$$

where $y = \kappa+k+\frac{1}{2}(v'-v)+1$, and where the seniorities v and v' are implied by ψ and ψ'. A similar application to reduced matrix elements of $\mathbf{a}^\dagger$ and $\mathbf{a}$ gives the following relation between cfp:

$$(\ell^{N+1}\psi\{|\ell^N\psi') = (-1)^z(\ell^{4\ell+1-N}\psi|\}\ell^{4\ell+2-N}\psi')$$
$$\times \left[\frac{(4\ell+2-N)(2S'+1)(2L'+1)}{(N+1)(2S+1)(2L+1)}\right]^{\frac{1}{2}} \qquad (6.54)$$

where $z = S+S'-s+L+L'-\ell+\frac{1}{2}(v+v'-1)$. The phases y and z stem from the conventions of angular momentum theory, which enter via quasispin. Racah [2,8] did not use this concept, and his phase choices are slightly different from the ones above.

For a Cartesian component Q_u of the quasispin $\mathbf{Q}$,

$$CQ_u C^{-1} = -Q_u. \qquad (6.55)$$

Thus, C is the analog of the time-reversal operator T, for which

$$TL_u T^{-1} = -L_u, \qquad (6.56)$$
$$TS_u T^{-1} = -S_u. \qquad (6.57)$$

Both C and T are antiunitary; thus,

$$CiC^{-1} = -i. \qquad (6.58)$$

6.3.5 Dependence on Electron Number

Application of the Wigner-Eckart theorem to matrix elements whose component parts have well-defined quasispin ranks yields the dependence of the matrix elements on the electron number N [14,15]. For $\kappa + k$ even and nonzero, the quasispin rank of $\mathbf{W}^{(\kappa k)}$ is 1, and

$$(\ell^N\psi\|W^{(\kappa k)}\|\ell^N\psi')$$
$$= \frac{(2\ell+1-N)}{(2\ell+1-v)}(\ell^v\psi\|W^{(\kappa k)}\|\ell^v\psi'). \qquad (6.59)$$

For $\kappa + k$ odd, $\mathbf{W}^{(\kappa k)}$ is necessarily a quasispin scalar, and the matrix elements are diagonal with respect to the seniority and independent of N. These properties were first stated in Eqs. (69) and (70) of Ref. [8].

Application of these ideas to single-electron cfp yields

$$(\ell^N\psi\{|\ell^{N-1}\bar\psi)$$
$$= [(N-v)(v+2)/2N]^{\frac{1}{2}}(\ell^{v+2}\psi\{|\ell^{v+1}\bar\psi). \qquad (6.60)$$

6.3.6 The Half-Filled Shell

Selection rules for operators of good quasispin rank K, taken between states of the half-filled shell (for which $M_Q = 0$), can be found by inspecting the 3-j symbol

$$\begin{pmatrix} Q & K & Q' \\ 0 & 0 & 0 \end{pmatrix},$$

which appears when the Wigner-Eckart theorem is applied in quasispin space. This 3-j symbol vanishes unless $Q+K+Q'$ is even. An equivalent result can be obtained for $\mathbf{W}^{(\kappa k)}$ by referring to (6.53) and insisting that y be even.

6.4 COMPLEMENTARITY

6.4.1 Spin-Quasispin Interchange

The operator R formally interchanges spin and quasispin. The result for the creation and annihilation operators for electrons can be expressed in terms of triple tensors:

$$Ra_\xi^{(qs\ell)}R^{-1} = a_\eta^{(qs\ell)}, \qquad (6.61)$$

where $\xi \equiv (m_q m_s m_\ell)$ and $\eta \equiv (m_s m_q m_\ell)$. For the tensors $\mathbf{X}^{(K\kappa k)}$ defined in (6.44), we get

$$RX_\lambda^{(K\kappa k)}R^{-1} = X_\mu^{(\kappa K k)}, \qquad (6.62)$$

where $\lambda \equiv (M_K M_\kappa M_k)$ and $\mu \equiv (M_\kappa M_K M_k)$. For states of the ℓ shell,

$$R|\gamma Q M_Q S M_S\rangle = (-1)^t|\gamma S M_S Q M_Q\rangle, \qquad (6.63)$$

where the quasispin of the ket on the right is S and the spin is Q. The phase factor t depends on S and Q and on phase choices made for the coefficients of fractional parentage. The symbol γ denotes the additional labels necessary to completely define the state in question, including L and M_L.

For every γ, Racah [16] observed that there are two possible pairs (v_1, S_1) and (v_2, S_2) satisfying

$$v_1 + 2S_2 = v_2 + 2S_1 = 2\ell + 1. \qquad (6.64)$$

From (6.37) it follows that

$$S_1 = Q_2, \qquad S_2 = Q_1. \qquad (6.65)$$

6.4.2 Matrix Elements

Application of the complementarity operator R to the component parts of a matrix element leads to the equation

$$\langle\gamma Q M_Q S M_S|X_\lambda^{(K\kappa k)}|\gamma' Q' M_Q' S' M_S'\rangle$$
$$= (-1)^y\langle\gamma S M_S Q M_Q|X_\mu^{(\kappa K k)}|\gamma' S' M_S' Q' M_Q'\rangle, \quad (6.66)$$

where λ and μ have the same significance as in (6.62), and where y, like t of (6.63), depends on the spins and quasispins but not on the associated magnetic quantum numbers. Equation (6.66) leads to a useful special case when $M_K = M_\kappa = 0$ and the tensors $\mathbf{X}$ are converted to those of type $\mathbf{W}$, defined in (6.18). The sum $K + \kappa + k$ is taken to be odd, with the scalars $\kappa = k = 0$ and $K = k = 0$ excluded. Application of the Wigner-Eckart theorem to the spin and orbital spaces yields

$$\frac{(\gamma Q M_Q S||W^{(\kappa k)}||\gamma' Q' M_Q' S')}{(\gamma S M_S Q||W^{(Kk)}||\gamma' S' M_S' Q')}$$
$$= (-1)^z \frac{\begin{pmatrix} Q & K & Q' \\ -M_Q & 0 & M_Q \end{pmatrix}}{\begin{pmatrix} S & \kappa & S' \\ -M_S & 0 & M_S \end{pmatrix}}, \qquad (6.67)$$

where $z = y + Q - M_Q - S + M_S$. An equivalent form is

$$\frac{(\ell^N \gamma v_1 S_1||W^{(\kappa k)}||\ell^N \gamma' v_1' S_1')}{(\ell^{N'} \gamma v_2 S_2||W^{(Kk)}||\ell^{N'} \gamma' v_2' S_2')}$$
$$= (-1)^z \frac{\begin{pmatrix} \frac{1}{2}(2\ell+1-v_1) & K & \frac{1}{2}(2\ell+1-v_1') \\ \frac{1}{2}(2\ell+1-N) & 0 & \frac{1}{2}(N-2\ell-1) \end{pmatrix}}{\begin{pmatrix} \frac{1}{2}(2\ell+1-v_2) & \kappa & \frac{1}{2}(2\ell+1-v_2') \\ \frac{1}{2}(2\ell+1-N') & 0 & \frac{1}{2}(N'-2\ell-1) \end{pmatrix}},$$
$$(6.68)$$

where (6.64) is satisfied both for the unprimed and primed quantities.

6.5 QUASIPARTICLES

Sets of linear combinations of the creation and annihilation operators for electrons in the ℓ shell can be constructed such that every member of one set anticommutes with a member of a different set. To preserve the tensorial character of these quasiparticle operators with respect to $\mathbf{L}$, it is convenient to define [17]

$$\lambda_q^\dagger = 2^{-\frac{1}{2}}[a_{\frac{1}{2},q}^\dagger + (-1)^{\ell-q}a_{\frac{1}{2},-q}], \qquad (6.69)$$

$$\mu_q^\dagger = 2^{-\frac{1}{2}}[a_{\frac{1}{2},q}^\dagger - (-1)^{\ell-q}a_{\frac{1}{2},-q}], \qquad (6.70)$$

$$\nu_q^\dagger = 2^{-\frac{1}{2}}[a_{-\frac{1}{2},q}^\dagger + (-1)^{\ell-q}a_{-\frac{1}{2},-q}], \qquad (6.71)$$

$$\xi_q^\dagger = 2^{-\frac{1}{2}}[a_{-\frac{1}{2},q}^\dagger - (-1)^{\ell-q}a_{-\frac{1}{2},-q}]. \qquad (6.72)$$

The four tensors $\boldsymbol{\theta}^\dagger(\equiv \boldsymbol{\lambda}^\dagger, \boldsymbol{\mu}^\dagger, \boldsymbol{\nu}^\dagger, \text{ or } \boldsymbol{\xi}^\dagger)$ anticommute with each other; the first two act in the spin-up space, the second two in the spin-down space. The tensors $\boldsymbol{\theta}$, whose components $\tilde{\theta}_q$ are defined as in (6.17) with $t = \ell$ and $m_t = q$, are related to their adjoints by the equations

$$\boldsymbol{\lambda}^\dagger = \boldsymbol{\lambda}, \qquad \boldsymbol{\mu}^\dagger = -\boldsymbol{\mu}, \qquad (6.73)$$

$$\boldsymbol{\nu}^\dagger = \boldsymbol{\nu}, \qquad \boldsymbol{\xi}^\dagger = -\boldsymbol{\xi}. \qquad (6.74)$$

Under the action of the complementarity operator R (see Eq. (6.61)) [18],

$$R\boldsymbol{\lambda}R^{-1} = \boldsymbol{\lambda}, \qquad R\boldsymbol{\mu}R^{-1} = \boldsymbol{\mu}, \qquad (6.75)$$

$$R\boldsymbol{\nu}R^{-1} = \boldsymbol{\nu}, \qquad R\boldsymbol{\xi}R^{-1} = -\boldsymbol{\xi}. \qquad (6.76)$$

The tensors $\boldsymbol{\lambda}, \boldsymbol{\mu}$, and $\boldsymbol{\nu}$, for a given component q, form a vector with respect to $\mathbf{S} + \mathbf{Q}$. Every component of $\boldsymbol{\xi}$ is scalar with respect to $\mathbf{S} + \mathbf{Q}$ [19].

The compound quasiparticle operators defined by [17]

$$\Theta_q^\dagger = 2^{-\frac{1}{2}} [\theta_q^\dagger, \theta_0^\dagger],\qquad (6.77)$$

where $q > 0$ and $\theta \equiv \lambda,\ \mu,\ \nu,$ or ξ satisfy the anticommutation relations

$$[\Theta_q^\dagger, \Theta_{q'}^\dagger]_+ = 0,\qquad (6.78)$$

$$[\Theta_q, \Theta_{q'}]_+ = 0,\qquad (6.79)$$

$$[\Theta_q^\dagger, \Theta_{q'}]_+ = \delta(q, q'),\qquad (6.80)$$

for $q, q' > 0$. The $\Theta_q^\dagger$ with $q > 0$ can thus be regarded as the creation operators for a fermion quasiparticle with ℓ components.

REFERENCES

1. E. K. U. Gross, E. Runge, and O. Heinonen, *Many-Particle Theory* (Hilger, New York, 1991).
2. G. Racah, Phys. Rev. **62**, 438 (1942).
3. E. U. Condon and G. H. Shortley, *The Theory of Atomic Spectra* (Cambridge University Press, New York, 1935).
4. A. R. Edmonds, *Angular Momentum in Quantum Mechanics* (Princeton University Press, Princeton, 1957).
5. C. W. Nielson and G. F. Koster, *Spectroscopic Coefficients for the p^n, d^n, and f^n Configurations* (MIT Press, Cambridge, 1963).
6. R. Karazija, J. Vizbaraitė, Z. Rudzikas, and A. P. Jucys, *Tables for the Calculation of Matrix Elements of Atomic Operators* (Academy of Sciences Computing Center, Moscow, 1967).
7. B. R. Judd, *Second Quantization and Atomic Spectroscopy* (Johns Hopkins, Baltimore, 1967).
8. G. Racah, Phys. Rev. **63**, 367 (1943).
9. V. L. Donlan, *Air Force Materials Laboratory Report No. AFML-TR-70-249*, Wright-Patterson Air Force Base, Ohio (1970).
10. P. J. Redmond, Proc. R. Soc. London A222, 84 (1954).
11. B. H. Flowers and S. Szpikowski, Proc. Phys. Soc. London **84**, 673 (1964).
12. B. R. Judd, J. Phys. C **14**, 375 (1981).
13. J. S. Bell, Nucl. Phys. **12**, 117 (1959).
14. H. Watanabe, Prog. Theor. Phys. **32**, 106 (1964).
15. R. D. Lawson and M. H. Macfarlane, Nucl. Phys. **66**, 80 (1965).
16. G. Racah, Phys. Rev. **76**, 1352 (1949), Table I.
17. L. Armstrong and B. R. Judd, Proc. R. Soc. London A**315**, 27, and 39 (1970).
18. B. R. Judd and S. Li, J. Phys. B **22**, 2851 (1989).
19. B. R. Judd, G. M. S. Lister, and M. A. Suskin, J. Phys. B **19**, 1107 (1986).

7

Density Matrices

K. Bartschat

Department of Physics and Astronomy, Drake University, Des Moines, Iowa 50311

7.1 INTRODUCTION

The density operator was first introduced by J. von Neumann [1] in 1927 and has since been widely used in quantum statistics. Over the past decades, however, the application of density matrices has spread to many other fields of physics. Density matrices have been used to describe, for example, coherence and correlation phenomena, alignment and orientation and their effect on the polarization of emitted radiation, quantum beat spectroscopy, optical pumping, and scattering processes, particularly when spin-polarized projectiles and/or targets are involved. A thorough introduction to the theory of density matrices and their applications with emphasis on atomic physics can be found in the book by Blum [2] from which many equations have been extracted for use in this article.

The main advantage of the density matrix formalism is its ability to deal with pure and mixed states in the same consistent manner. The preparation of the initial state as well as the details regarding the observation of the final state can be treated in a systematic way. In particular, averages over quantum numbers of unpolarized beams in the initial state and incoherent sums over non-observed quantum numbers in the final state can be accounted for via the reduced density matrix. Furthermore, expansion of the density matrix in terms of irreducible tensor operators and the corresponding state multipoles allows for the use of advanced angular momentum techniques, as outlined in Chaps. 2, 3, and 12.

7.2   BASIC FORMULAE

7.2.1   Pure States

Consider a system in a quantum state that is represented by a single wave function $|\Psi\rangle$. The density operator for this situation is defined as

$$\rho = |\Psi\rangle\langle\Psi|. \qquad (7.1)$$

If $|\Psi\rangle$ is normalized to unity, i.e., if

$$\langle\Psi \mid \Psi\rangle = 1, \qquad (7.2)$$

then

$$\rho^2 = \rho. \qquad (7.3)$$

Equation (7.3) is the basic equation for identifying *pure* quantum mechanical states represented by a density operator.

Next, consider the expansion of $|\Psi\rangle$ in terms of a complete orthonormal set of basis functions $\{|\Phi_n\rangle\}$, i.e.,

$$|\Psi\rangle = \sum_n c_n |\Phi_n\rangle. \qquad (7.4)$$

The density operator then becomes

$$\rho = \sum_{n,m} c_n c_m^* |\Phi_n\rangle\langle\Phi_m| = \rho_{nm}|\Phi_n\rangle\langle\Phi_m|, \qquad (7.5)$$

where the star denotes the complex conjugate quantity. Note that the density matrix elements $\rho_{nm} = \langle\Phi_n|\,\rho\,|\Phi_m\rangle$ depend on the choice of the basis and that the density matrix is Hermitian, i.e.,

$$\rho_{mn}^* = \rho_{nm}. \qquad (7.6)$$

Finally, if $|\Psi\rangle = |\Phi_i\rangle$ is one of the basis functions, then

$$\rho_{mn} = \delta_{ni}\delta_{mi}, \qquad (7.7)$$

where δ_{ni} is the Kronecker δ. Hence, the density matrix is diagonal in this representation with only one nonvanishing element.

7.2.2   Mixed States

The above concepts can be extended to treat statistical ensembles of pure quantum states. In the simplest case, such *mixed* states can be represented by a diagonal density matrix of the form

$$\rho = \sum_n w_n|\Psi_n\rangle\langle\Psi_n|, \qquad (7.8)$$

where the weight w_n is the fraction of systems in the pure quantum state $|\Psi_n\rangle$. The standard normalization for the trace of ρ is

$$\mathrm{Tr}\{\rho\} = \sum_n w_n = 1. \qquad (7.9)$$

Since the trace is invariant under unitary transformations of the basis functions, Eq. (7.9) also holds if the $|\Psi_n\rangle$ states themselves are expanded in terms of basis functions as in Eq. (7.4). For a pure state and the normalization (7.9), one finds in an arbitrary basis

$$\mathrm{Tr}\{\rho\} = \mathrm{Tr}\{\rho^2\} = 1. \qquad (7.10)$$

7.2.3   Expectation Values

The density operator contains the maximum available information about a physical system. Consequently, it can be used to calculate expectation values for any operator $\mathcal{A}$ that represents a physical observable. In general,

$$\langle\mathcal{A}\rangle = \mathrm{Tr}\{\mathcal{A}\rho\}/\mathrm{Tr}\{\rho\}, \qquad (7.11)$$

where $\mathrm{Tr}\{\rho\}$ in the denominator of Eq. (7.11) ensures the correct result even for a normalization that is different from Eq. (7.9). The invariance of the trace operation ensures the same result – independent of the particular choice of the basis representation.

7.2.4   The Liouville Equation

Suppose Eq. (7.8) is valid for a time $t = 0$. If the functions $|\Psi_n(\mathbf{r}, t)\rangle$ obey the Schrödinger equation, i.e.

$$i\frac{\partial}{\partial t}|\Psi_n(\mathbf{r}, t)\rangle = H(t)\,|\Psi_n(\mathbf{r}, t)\rangle, \qquad (7.12)$$

the density operator at the time t can be written as

$$\rho(t) = U(t)\,\rho(0)\,U^{\dagger}(t)\,. \tag{7.13}$$

In (7.13), $U(t)$ is the time evolution operator which relates the wave functions at times $t = 0$ and t according to

$$|\Psi_n(\mathbf{r}, t)\rangle = U(t)\,|\Psi_n(\mathbf{r}, 0)\rangle\,, \tag{7.14}$$

and $U^{\dagger}(t)$ denotes its adjoint. Note that

$$U(t) = e^{-iHt}\,, \tag{7.15}$$

if the Hamiltonian H is time-independent.

Differentiation of Eq. (7.13) with respect to time and inserting (7.14) into the Schrödinger equation (7.12) yields the equation of motion

$$i\frac{\partial}{\partial t}\rho(t) = [H(t), \rho(t)]\,, \tag{7.16}$$

where $[\mathcal{A}, \mathcal{B}]$ denotes a commutator.

The Liouville equation (7.16) can be used to determine the density matrix and to treat transitions from nonequilibrium to equilibrium states in quantum mechanical systems. Especially for approximate solutions in the presence of small time-dependent perturbation terms in an otherwise time-independent Hamiltonian, i.e., for

$$H(t) = H_0 + V(t)\,, \tag{7.17}$$

the interaction picture is preferably used. The Liouville equation then becomes

$$i\frac{\partial}{\partial t}\rho_I(t) = [V_I(t), \rho_I(t)]\,, \tag{7.18}$$

where the subscript I denotes the operator in the interaction picture. In first-order perturbation theory, Eq. (7.18) can be integrated to yield

$$\rho_I(t) = \rho_I(0) - i\int_0^t [V_I(\tau), \rho_I(0)]\,d\tau\,, \tag{7.19}$$

and higher-order terms can be obtained through subsequent iterations.

7.2.5 Systems in Thermal Equilibrium

According to quantum statistics, the density operator for a system which is in thermal equilibrium with a surrounding reservoir $\mathcal{R}$ at a temperature T (canonical ensemble), can be expressed as

$$\rho = \frac{\exp[-\beta H]}{Z}\,, \tag{7.20}$$

where H is the Hamiltonian, and $\beta = 1/k_B T$ with k_B being the Boltzmann constant. The partition sum

$$Z = \text{Tr}\{\exp[-\beta H]\}\,, \tag{7.21}$$

ensures the normalization condition (7.9). Expectation values are calculated according to (7.11), and extensions to other types of ensembles are straightforward.

7.2.6 Relaxation Processes

Transitions from nonequilibrium to equilibrium states can also be described within the density matrix formalism. One of the basic problems is to account for irreversibility in the energy (and sometimes particle) exchange between the system of interest, $\mathcal{S}$, and the reservoir, $\mathcal{R}$. This is usually achieved by assuming that the interaction of the system with the reservoir is negligible and, therefore, the density matrix representation for the reservoir at any time t is the same as the representation for $t = 0$.

Another important assumption that is frequently made is the Markov approximation. In this approximation, one assumes that the system "forgets" all knowledge of the past, so that the density matrix elements at the time $t + \Delta t$ depend only on the values of these elements, and their first derivatives, at the time t. When Eq. (7.19) is put back into (7.18), the result in the Markov approximation can be rewritten as

$$\frac{\partial}{\partial t}\rho_{SI}(t) = -i\,\text{Tr}_{\mathcal{R}}[V_I(t), \rho_{SI}(0)\rho_{\mathcal{R}}(0)]$$
$$- \int_0^t d\tau\,\text{Tr}_{\mathcal{R}}[V_I(t), [V_I(\tau), \rho_{SI}(t)\rho_{\mathcal{R}}(0)]]\,, \tag{7.22}$$

where $\text{Tr}_{\mathcal{R}}$ denotes the trace with regard to all variables of the reservoir. Note that the integral over $d\tau$ contains the system density matrix in the interaction picture, ρ_{SI}, at the time t, rather than at all times τ which are integrated over (the Markov approximation), and that the density matrix for the reservoir is taken as $\rho_{\mathcal{R}}(0)$ at all times. For more details, see Chapter 7 of Blum [2] and references therein.

Equations such as (7.22) are the basis for the master or rate equation approach used, for example, in quantum optics for the theory of lasers and the coupling of atoms to cavity modes. For more details, see Chaps. 66, 67, 68, and 75.

7.3 SPIN AND LIGHT POLARIZATIONS

Density matrices are frequently used to describe the polarization state of spin-polarized particle beams as well as light. The latter can either be emitted from excited atomic or molecular ensembles or can be used, for example, for laser pumping purposes.

7.3.1 Spin-Polarized Electrons

The spin polarization of an electron beam with respect to a given quantization axis $\hat{\mathbf{n}}$ is defined as [3]

$$P_{\hat{n}} = \frac{N_\uparrow - N_\downarrow}{N_\uparrow + N_\downarrow}, \qquad (7.23)$$

where $N_\uparrow (N_\downarrow)$ is the number of electrons with spin up (down) with regard to this axis. An arbitrary polarization state is described by the density matrix

$$\rho = \frac{1}{2} \begin{pmatrix} 1 + P_z & P_x - iP_y \\ P_x + iP_y & 1 - P_z \end{pmatrix}, \qquad (7.24)$$

where $P_{x,y,z}$ are the cartesian components of the spin polarization vector. The individual components can be obtained from the density matrix as

$$P_i = \mathrm{Tr}\{\sigma_i \rho\}, \qquad (7.25)$$

where the σ_i $(i = x, y, z)$ are the standard Pauli spin matrices.

7.3.2   Light Polarization

Another important use of the density matrix formalism is the description of light polarization in terms of the so-called Stokes parameters [4]. For a given direction of observation, the general polarization state of light can be fully determined by the measurement of one circular and two independent linear polarizations. Using the notation of Born and Wolf [5], the density matrix is given by

$$\rho = \frac{I_{\mathrm{tot}}}{2} \begin{pmatrix} 1 - P_3 & P_1 - iP_2 \\ P_1 + iP_2 & 1 + P_3 \end{pmatrix}, \qquad (7.26)$$

where P_1 and P_2 are linear light polarizations while P_3 is the circular polarization (see also Sect. 7.7). In (7.26), the density matrix is normalized in such as way that

$$\mathrm{Tr}\{\rho\} = I_{\mathrm{tot}}, \qquad (7.27)$$

where I_{tot} is the total light intensity. Other frequently used names for the various Stokes parameters are

$$P_1 = \eta_3 = M, \qquad (7.28)$$
$$P_2 = \eta_1 = C, \qquad (7.29)$$
$$P_3 = -\eta_2 = S. \qquad (7.30)$$

The Stokes parameters of electric dipole radiation can be related directly to the charge distribution of the emitting atomic ensemble. As discussed in detail in Chap. 44, one finds, for example,

$$L_\perp = -P_3 \qquad (7.31)$$

for the angular momentum transfer perpendicular to the scattering plane in collisional (de-)excitation, and

$$\gamma = \tfrac{1}{2}\arg\{P_1 + iP_2\} \qquad (7.32)$$

for the alignment angle.

7.4   ATOMIC COLLISIONS

7.4.1   Scattering Amplitudes

Transitions from an initial state $|J_0 M_0; \mathbf{k}_0 m_0\rangle$ to a final state $|J_1 M_1; \mathbf{k}_1 m_1\rangle$ are described by scattering amplitudes

$$f(M_1 m_1; M_0 m_0) = \langle J_1 M_1; \mathbf{k}_1 m_1 \mid \mathcal{T} \mid J_0 M_0; \mathbf{k}_0 m_0 \rangle, \qquad (7.33)$$

where $\mathcal{T}$ is the transition operator. Furthermore, J_0 (J_1) is the total electronic angular momentum in the initial (final) state of the target and M_0 (M_1) its corresponding z-component, while $\mathbf{k}_0$ $(\mathbf{k}_1)$ is the initial (final) momentum of the projectile and m_0 (m_1) its spin component.

7.4.2   Reduced Density Matrices

While the scattering amplitudes are the central elements in a theoretical description, some restrictions usually need to be taken into account in a practical experiment. The most important ones are: (i) there is no "pure" initial state, and (ii) not all possible quantum numbers are simultaneously determined in the final state. The solution to this problem can be found by using the density matrix formalism. First, the complete density operator after the collision process is given by [2]

$$\rho_{\mathrm{out}} = \mathcal{T} \rho_{\mathrm{in}} \mathcal{T}^\dagger, \qquad (7.34)$$

where ρ_{in} is the density operator before the collision. The corresponding matrix elements are given by

$$(\rho_{\mathrm{out}})_{m_1' m_1}^{\mathbf{k}_1, M_1' M_1} = \sum_{m_0' m_0 M_0' M_0} \rho_{m_0' m_0} \, \rho_{M_0' M_0} \, f(M_1' m_1'; M_0' m_0') \\ \times f^*(M_1 m_1; M_0 m_0), \qquad (7.35)$$

where the term $\rho_{m_0' m_0} \rho_{M_0' M_0}$ describes the preparation of the initial state (i). Secondly, "reduced" density matrices account for (ii). For example, if only the scattered projectiles are observed, the corresponding elements of the reduced density matrix are obtained by summing over the atomic quantum numbers as follows:

$$(\rho_{\mathrm{out}})_{m_1' m_1}^{\mathbf{k}_1} = \sum_{M_1} (\rho_{\mathrm{out}})_{m_1' m_1}^{\mathbf{k}_1, M_1 M_1}. \qquad (7.36)$$

The differential cross section for unpolarized projectile and target beams is given by

$$\frac{d\sigma}{d\Omega} = C \sum_{m_1} (\rho_{\mathrm{out}})_{m_1 m_1}^{\mathbf{k}_1}, \qquad (7.37)$$

where C is a constant that depends on the normalization of the continuum waves in a numerical calculation.

On the other hand, if only the atoms are observed (for example, by analyzing the light emitted in optical transitions), the elements

$$(\rho_{\text{out}})_{M_1' M_1} = \int d^3 \mathbf{k}_1 \sum_{m_1} (\rho_{\text{out}})_{m_1 m_1}^{\mathbf{k}_1, M_1' M_1} \qquad (7.38)$$

determine the integrated Stokes parameters [6,7], i.e., the polarization of the emitted light. They contain information about the angular momentum distribution in the excited target ensemble.

Finally, for electron–photon coincidence experiments without spin analysis in the final state, the elements

$$(\rho_{\text{out}})_{M_1' M_1}^{\mathbf{k}_1} = \sum_{m_1} (\rho_{\text{out}})_{m_1 m_1}^{\mathbf{k}_1, M_1' M_1} \qquad (7.39)$$

simultaneously contain information about the projectiles and the target. This information can be extracted by measuring the angle-differential Stokes parameters. In particular, for unpolarized electrons and atoms, the "natural coordinate system", where the quantization axis coincides with the normal to the scattering plane, allows for a simple physical interpretation of the various parameters [8] (see Chap. 44).

The density matrix formalism outlined above is very useful for obtaining a qualitative description of the geometrical and sometimes also of the dynamical symmetries of the collision process [9]. Two explicit examples are discussed in Sect. 7.7.

7.5 IRREDUCIBLE TENSOR OPERATORS

The general density matrix theory can be formulated in a very elegant fashion by decomposing the density operator in terms of irreducible components whose matrix elements then become the state multipoles. In such a formulation, full advantage can be taken of the most sophisticated techniques developed in angular momentum algebra (see Chap. 2).

7.5.1 Definition

The density operator for an ensemble of particles in quantum states labeled as $|JM\rangle$ where J and M are the total angular momentum and its magnetic component, respectively, can be written as

$$\rho = \sum_{J' J M' M} \rho_{M' M}^{J' J} |J'M'\rangle\langle JM|, \qquad (7.40)$$

where

$$\rho_{M' M}^{J' J} = \langle J'M' \mid \rho \mid JM \rangle \qquad (7.41)$$

are the matrix elements. (For simplicity, interactions outside the single manifold of momentum states $|JM\rangle$ are neglected). Alternatively, one may write

$$\rho = \sum_{J' J K Q} \langle T(J'J)_{KQ}^{\dagger} \rangle \, T(J'J)_{KQ}, \qquad (7.42)$$

where the irreducible tensor operators are defined in terms of 3-j symbols as

$$T(J'J)_{KQ} = \sum_{M'M} (-1)^{J'-M'} \sqrt{2K+1}$$
$$\times \begin{pmatrix} J' & J & K \\ M' & -M & -Q \end{pmatrix} |J'M'\rangle\langle JM|, \quad (7.43)$$

and the state multipoles or statistical tensors are given by

$$\langle T(J'J)_{KQ}^{\dagger} \rangle = \sum_{M'M} (-1)^{J'-M'} \sqrt{2K+1}$$
$$\times \begin{pmatrix} J' & J & K \\ M' & -M & -Q \end{pmatrix} \langle J'M' \mid \rho \mid JM \rangle. \quad (7.44)$$

Hence, the selection rules for the 3-j symbols imply that

$$|J - J'| \le K \le J + J', \qquad (7.45)$$
$$M' - M = Q. \qquad (7.46)$$

Equation (7.44) can be inverted through the orthogonality condition of the 3-j symbols to give

$$\langle J'M' \mid \rho \mid JM \rangle = \sum_{KQ} (-1)^{J'-M'} \sqrt{2K+1}$$
$$\times \begin{pmatrix} J' & J & K \\ M' & -M & -Q \end{pmatrix} \langle T(J'J)_{KQ}^{\dagger} \rangle. \quad (7.47)$$

7.5.2 Transformation Properties

Suppose a coordinate system (X_2, Y_2, Z_2) is obtained from another coordinate system (X_1, Y_1, Z_1) through a rotation by a set of three Euler angles (γ, β, α) as defined in Edmonds [10]. The irreducible tensor operators (7.43) defined in the (X_1, Y_1, Z_1) system are then related to the operators $\langle T(J'J)_{KQ}^{\dagger} \rangle$ in the (X_2, Y_2, Z_2) system by

$$T(J'J)_{KQ} = \sum_{q} T(J'J)_{Kq} \, D(\gamma, \beta, \alpha)_{qQ}^{K}, \qquad (7.48)$$

where

$$D(\gamma, \beta, \alpha)_{M'M}^{J} = e^{iM'\gamma} d(\beta)_{M'M}^{J} e^{iM\alpha} \qquad (7.49)$$

is a rotation matrix (see Chap. 2). Note that the rank K of the tensor operator is invariant under such rotations. Similarly,

$$\langle T(J'J)^{\dagger}_{KQ}\rangle = \sum_q \langle T(J'J)^{\dagger}_{Kq}\rangle\, D(\gamma,\beta,\alpha)^{K\,*}_{qQ} \qquad (7.50)$$

holds for the state multipoles.

The irreducible tensor operators fulfill the orthogonality condition

$$\text{Tr}\{T(J'J)_{KQ}\,T(J'J)^{\dagger}_{K'Q'}\} = \delta_{K'K}\,\delta_{Q'Q}, \qquad (7.51)$$

with

$$T(J'J)_{00} = \frac{1}{\sqrt{2J+1}}\,\delta_{J'J}\,\mathbf{1} \qquad (7.52)$$

being proportional to the unit operator $\mathbf{1}$, it follows that all tensor operators have vanishing trace, except for the monopole $T(J'J)_{00}$.

Reduced tensor operators fulfill the Wigner-Eckart theorem (see Sect. 2.9.4)

$$\langle J'M' \mid T(J'J)_{KQ} \mid JM\rangle$$
$$= (-1)^{J'-M'}\begin{pmatrix} J' & K & J \\ -M' & Q & M \end{pmatrix}$$
$$\times \langle J'\|T_K\|J\rangle, \qquad (7.53)$$

where the reduced matrix element is simply given by

$$\langle J'\|T_K\|J\rangle = \frac{1}{\sqrt{2K+1}}. \qquad (7.54)$$

7.5.3 Symmetry Properties of State Multipoles

The Hermiticity condition for the density matrix implies

$$\langle T(J'J)^{\dagger}_{KQ}\rangle^{*} = (-1)^{J'-J+Q}\langle T(JJ')^{\dagger}_{K-Q}\rangle, \qquad (7.55)$$

which, for sharp angular momentum $J'=J$, yields

$$\langle T(J)^{\dagger}_{KQ}\rangle^{*} = (-1)^{Q}\langle T(J)^{\dagger}_{K-Q}\rangle. \qquad (7.56)$$

Hence, the state multipoles $\langle T(J)^{\dagger}_{K0}\rangle$ are real numbers.

Furthermore, the transformation property (7.50) of the state multipoles imposes restrictions on nonvanishing state multipoles to describe systems with given symmetry properties. In detail, one finds:

1. For spherically symmetric systems,

$$\langle T(J'J)^{\dagger}_{KQ}\rangle = \langle T(J'J)^{\dagger}_{KQ}\rangle_{\text{rot}} \qquad (7.57)$$

for *all* sets of Euler angles. This implies that only the monopole term $\langle T(J)^{\dagger}_{00}\rangle$ can be different from zero.

2. For axially symmetric systems,

$$\langle T(J'J)^{\dagger}_{KQ}\rangle = \langle T(J'J)^{\dagger}_{KQ}\rangle_{\text{rot}} \qquad (7.58)$$

for *all* Euler angles ϕ that describe a rotation around the z-axis. Since this angle enters via a factor $\exp\{-iQ\phi\}$ into the general transformation formula (7.50), it follows that only state multipoles with $Q=0$, i.e., $\langle T(J'J)^{\dagger}_{K0}\rangle$, can be different from zero in such a situation.

3. For planar symmetric systems with fixed $J'=J$,

$$\langle T(J)^{\dagger}_{KQ}\rangle = (-1)^{K}\langle T(J)^{\dagger}_{KQ}\rangle^{*} \qquad (7.59)$$

if the system properties are invariant under reflection in the xz-plane. Hence, state multipoles with even rank K are real numbers, while those with odd rank are purely imaginary in this case.

The above results can be applied immediately to the description of atomic collisions where the incident beam axis is the quantization axis (the so-called "collision system"). For example, impact excitation of unpolarized targets by unpolarized projectiles without observation of the scattered projectiles is symmetric both with regard to rotation around the incident beam axis and with regard to reflection in any plane containing this axis. Consequently, the state multipoles $\langle T(J)^{\dagger}_{00}\rangle$, $\langle T(J)^{\dagger}_{20}\rangle$, $\langle T(J)^{\dagger}_{40}\rangle, \ldots$ fully characterize the atomic ensemble of interest. Using Eq. (7.50), similar relationships can be derived for state multipoles defined with regard to other coordinate systems, such as the "natural system" where the quantization axis coincides with the normal vector to the scattering plane (see Chap. 44).

7.5.4 Orientation and Alignment

From the above discussion, it is apparent that the description of systems that do not exhibit spherical symmetry requires the knowledge of state multipoles with rank $K \neq 0$. Frequently, the multipoles with $K=1$ and $K=2$ are determined via the angular correlation and the polarization of radiation emitted from an ensemble of collisionally excited targets. The state multipoles with $K=1$ are proportional to the spherical components of the angular momentum expectation value and, therefore, give rise to a nonvanishing circular light polarization (see also Sect. 7.7). This corresponds to a sense of rotation or an orientation in the ensemble which is therefore called oriented (see Sect. 44.1, 44.1).

On the other hand, nonvanishing multipoles with rank $K=2$ describe the alignment of the system. Some authors, however, use the terms "alignment" or "orientation" synonymously for all nonvanishing state multipoles with ranks $K \neq 0$, thereby describing any system with anisotropic occupation of magnetic sublevels as "aligned" or "oriented". For details on alignment and orientation, see Chap. 44.

7.5.5 Coupled Systems

Tensor operators and state multipoles for coupled systems are constructed as direct products ($\otimes$) of the operators for the individual systems. For example, the density operator for two subsystems in basis states $|L, M_L\rangle$ and $|S, M_S\rangle$ is constructed as [2]

$$\rho = \sum_{KQkq} \langle T(L)_{KQ}^{\dagger} \otimes T(S)_{kq}^{\dagger}\rangle [T(L)_{KQ} \otimes T(S)_{kq}]. \tag{7.60}$$

If the two systems are uncorrelated, the state multipoles factor as

$$\langle T(L)_{KQ}^{\dagger} \otimes T(S)_{kq}^{\dagger}\rangle = \langle T(L)_{KQ}^{\dagger}\rangle\langle T(S)_{kq}^{\dagger}\rangle; \tag{7.61}$$

More generally, irreducible representations of coupled operators can be defined in terms of a 9-j symbol as

$$T(J', J)_{K'Q'} = \sum_{KQkq} \hat{K}\hat{k}\hat{J}\hat{J}'\, (KQ, kq|K'Q')$$
$$\times \begin{Bmatrix} K & k & K' \\ L & S & J' \\ L & S & J \end{Bmatrix} T(L)_{KQ} \otimes T(S)_{kq}, \tag{7.62}$$

where $\hat{x} \equiv \sqrt{2x+1}$, and $(j_1 m_1, j_2 m_2|j_3 m_3)$ is a standard Clebsch-Gordan coefficient.

7.6 TIME EVOLUTION OF STATE MULTIPOLES

7.6.1 Perturbation Coefficients

From the general expansions

$$\rho(t) = \sum_{j'jkq} \langle T(j'j;t)_{kq}^{\dagger}\rangle T(j'j)_{kq} \tag{7.63}$$

in terms of irreducible components, together with equations (7.42) for time $t = 0$ and (7.13) for the time development of the density operator, it follows that

$$\langle T(j'j;t)_{kq}^{\dagger}\rangle = \sum_{J'JKQ} \langle T(J'J;0)_{KQ}^{\dagger}\rangle\, G(J'J, j'j;t)_{Kk}^{Qq}, \tag{7.64}$$

where the perturbation coefficients are defined as

$$G(J'J, j'j;t)_{Kk}^{Qq} = \mathrm{Tr}\left\{U(t)T(J'J)_{KQ}U(t)^{\dagger}T(j'j)_{kq}^{\dagger}\right\}. \tag{7.65}$$

Hence, these coefficients relate the state multipoles at time t to those at $t = 0$.

7.6.2 Quantum Beats

An important application of the perturbation coefficients is the coherent excitation of several quantum states which subsequently decay by optical transitions. Such an excitation may be performed, for example, in beam-foil experiments or electron-atom collisions where the energy width of the electron beam is too large to resolve the fine structure (or hyperfine structure) of the target states.

Suppose, for instance, that explicitly relativistic effects, such as the spin-orbit interaction between the projectile and the target, can be neglected *during* a collision process between an incident electron and a target atom. In that case, the orbital angular momentum (L) system of the collisionally excited target states may be oriented, depending on the scattering angle of the projectile. On the other hand, the spin (S) system remains unaffected (unpolarized), provided that both the target and the projectile beams are unpolarized. During the lifetime of the excited target states, however, the spin-orbit interaction *within* the target produces an exchange of orientation between the L and the S systems, which results in a net loss of orientation in the L system.

This effect can be observed directly through the intensity and the polarization of the light emitted from the excited target ensemble. The perturbation coefficients for the fine structure interaction are found to be [11, 2]

$$G(L;t)_K = \frac{\exp\{-\gamma t\}}{2S+1} \sum_{J'J} (2J'+1)(2J+1)$$
$$\times \begin{Bmatrix} L & J' & S \\ J & L & K \end{Bmatrix}^2 \cos(\omega_{J'} - \omega_J)t, \tag{7.66}$$

where $\omega_{J'} - \omega_J$ corresponds to the (angular) frequency difference between the various multiplet states with total electronic angular momenta J' and J, respectively. Also, γ is the natural width of the spectral line; for simplicity, the same lifetime has been assumed in Eq. (7.66) for all states of the multiplet.

Note that the perturbation coefficients are independent of the multipole component Q in this case, and that there is no mixing between different multipole ranks K. Similar results can be derived [2, 11] for the hyperfine interaction and also to account for the combined effect of fine and hyperfine structure. The cosine terms represent correlation between the signal from different fine structure states, and they lead to oscillations in the intensity as well as the measured Stokes parameters in a time-resolved experiment.

Finally, generalized perturbation coefficients have been derived for the case where both the L and the S systems may be oriented and/or aligned during the collision process [12]. This can happen when spin polarized projectiles and/or target beams are prepared.

7.6.3 Time Integration Over Quantum Beats

If the excitation and decay times cannot be resolved in a given experimental setup, the perturbation coefficients need to be integrated over time. As a result, the quantum beats disappear, but a net effect may still be visible through a depolarization of the emitted radiation. For the case of atomic fine structure interaction discussed above, one finds [2, 11]

$$\bar{G}(L)_K = \int_0^\infty G(L;t)_K\, dt$$

$$= \frac{1}{2S+1} \sum_{J'J} (2J'+1)(2J+1)$$

$$\times \left\{ \begin{matrix} L & J' & S \\ J & L & K \end{matrix} \right\}^2 \frac{\gamma}{\gamma^2 + \omega_{J'J}^2}, \qquad (7.67)$$

where $\omega_{J'J} = \omega_{J'} - \omega_J$. Note that the amount of depolarization depends on the relationship between the fine structure splitting and the natural line width. For $|\omega_{J'J}| \gg \gamma$ (if $J' \neq J$), the terms with $J' = J$ dominate and cause the maximum depolarization; for the opposite case $|\omega_{J'J}| \ll \gamma$, the sum rule for the 6-j symbols can be applied and no depolarization is observed.

Similar depolarizations can be caused through hyperfine structure effects, as well as through external fields. An important example of the latter case is the Hanle effect (see Sect. 17.3.1).

7.7 EXAMPLES

In this section, two examples of the reduced density formalism are discussed explicitly. These are: (i) the change of the spin polarization of initially polarized spin-$\frac{1}{2}$ projectiles after scattering from unpolarized targets, and (ii) the Stokes parameters describing the angular distribution and the polarization of light as detected in projectile-photon coincidence experiments after collisional excitation.

7.7.1 Generalized *STU*-Parameters

For spin-polarized projectile scattering from unpolarized targets, the generalized *STU*-parameters [9] contain information about the projectile spin polarization after the collision. These parameters can be expressed in terms of the elements (7.36).

To analyze this problem explicitly, one defines the quantities

$$\langle m_1' m_0'; m_1 m_0 \rangle = \frac{1}{2J_0+1} \sum_{M_1 M_0} f(M_1 m_1'; M_0 m_0')$$

$$\times f^*(M_1 m_1; M_0 m_0) \qquad (7.68)$$

which contain the *maximum information* that can be obtained from the scattering process, if only the polarization of the projectiles is prepared before the collision and measured thereafter.

Next, the number of independent parameters that can be determined in such an experiment needs to be examined. For spin-$\frac{1}{2}$ particles, there are $2 \times 2 \times 2 \times 2 = 16$ possible combinations of $\{m_1' m_0'; m_1 m_0\}$ and, therefore, 16 complex or 32 real parameters (in the most general case of spin-S particles, there would be $[2S+1]^4$ combinations). However, from the definition (7.68) and the Hermiticity of the reduced density matrix contained therein, it follows that

$$\langle m_1' m_0'; m_1 m_0 \rangle = \langle m_1 m_0; m_1' m_0' \rangle^*. \qquad (7.69)$$

Furthermore, parity conservation of the interaction or the equivalent reflection invariance with regard to the scattering plane yields the additional relationship [9]

$$f(M_1 m_1; M_0 m_0) = (-1)^{J_1 - M_1 + \frac{1}{2} - m_1 + J_0 - M_0 + \frac{1}{2} - m_0}$$

$$\times \Pi_1 \Pi_0 f(-M_1 - m_1; -M_0 - m_0), \qquad (7.70)$$

where Π_1 and Π_0 are ± 1, depending on the parities of the atomic states involved. Hence,

$$\langle m_1' m_0'; m_1 m_0 \rangle = (-1)^{m_1' - m_1 + m_0' - m_0}$$

$$\times \langle -m_1' - m_0'; -m_1 - m_0 \rangle. \qquad (7.71)$$

Note that Eqs. (7.70), (7.71) hold for the collision frame where the quantization axis ($\hat{\mathbf{z}}$) is taken as the incident beam axis and the scattering plane is the xz-plane. Similar formulas can be derived for the natural frame (see Sect. 7.4.2)

Consequently, *eight* independent parameters are sufficient to characterize the reduced spin density matrix of the scattered projectiles. These can be chosen as the *absolute* differential cross section

$$\sigma_u = \frac{1}{2} \sum_{m_1, m_0} \langle m_1 m_0; m_1 m_0 \rangle \qquad (7.72)$$

for the scattering of unpolarized projectiles from unpolarized targets and the seven *relative* parameters

$$S_A = -\frac{2}{\sigma_u} \Im\{\langle \tfrac{1}{2} -\tfrac{1}{2}; \tfrac{1}{2}\tfrac{1}{2}\rangle\}, \qquad (7.73)$$

$$S_P = -\frac{2}{\sigma_u} \Im\{\langle \tfrac{1}{2}\tfrac{1}{2}; -\tfrac{1}{2}\tfrac{1}{2}\rangle\}, \qquad (7.74)$$

$$T_y = \frac{1}{\sigma_u} \left\{ \langle -\tfrac{1}{2} -\tfrac{1}{2}; \tfrac{1}{2}\tfrac{1}{2}\rangle - \langle -\tfrac{1}{2}\tfrac{1}{2}; \tfrac{1}{2} -\tfrac{1}{2}\rangle \right\}, \qquad (7.75)$$

$$T_x = \frac{1}{\sigma_u} \left[\langle -\tfrac{1}{2} -\tfrac{1}{2}; \tfrac{1}{2}\tfrac{1}{2}\rangle + \langle -\tfrac{1}{2}\tfrac{1}{2}; \tfrac{1}{2} -\tfrac{1}{2}\rangle \right], \qquad (7.76)$$

$$T_z = \frac{1}{\sigma_u} \left[\langle \tfrac{1}{2}\tfrac{1}{2}; \tfrac{1}{2}\tfrac{1}{2}\rangle - \langle \tfrac{1}{2}\tfrac{1}{2}; -\tfrac{1}{2}\tfrac{1}{2}\rangle \right], \qquad (7.77)$$

$$U_{xz} = \frac{2}{\sigma_u} \Re\{\langle \tfrac{1}{2}\tfrac{1}{2}; -\tfrac{1}{2}\tfrac{1}{2}\rangle\}, \qquad (7.78)$$

$$U_{zx} = -\frac{2}{\sigma_u} \Re\{\langle \tfrac{1}{2} -\tfrac{1}{2}; \tfrac{1}{2}\tfrac{1}{2}\rangle\}, \qquad (7.79)$$

Figure 7.1. Physical meaning of the generalized STU-parameters: the polarization function S_P gives the polarization of an initially unpolarized projectile beam after the collision while the asymmetry function S_A determines a left-right asymmetry in the differential cross section for scattering of a spin-polarized beam. Furthermore, the contraction parameters (T_x, T_y, T_z) describe the change of an initial polarization component along the three cartesian axes while the parameters U_{xz} and U_{zx} determine the rotation of a polarization component in the scattering plane.

Figure 7.2. Geometry of electron-photon coincidence experiments.

where $\Re\{x\}$ and $\Im\{x\}$ denote the real and imaginary parts of the complex quantity x, respectively. Note that normalization constants have been omitted in Eq. (7.72) to simplify the notation.

Therefore, the most general form for the polarization vector after scattering, $\mathbf{P}'$, for an initial polarization vector $\mathbf{P} = (P_x, P_y, P_z)$ is given by

$$\frac{(S_P + T_y P_y)\hat{\mathbf{y}} + (T_x P_x + U_{xz} P_z)\hat{\mathbf{x}} + (T_z P_z - U_{zx} P_x)\hat{\mathbf{z}}}{1 + S_A P_y} .$$

$$(7.80)$$

The physical meaning of the above relation is illustrated in Fig. 7.1.

The following geometries are particularly suitable for the experimental determination of the individual parameters. σ_u and S_P can be measured with unpolarized incident projectiles. A transverse polarization component perpendicular to the scattering plane ($\mathbf{P} = P_y \hat{\mathbf{y}}$) is needed to obtain S_A and T_y. Finally, the measurement of T_x, U_{zx}, T_z, and U_{xz} requires both transverse ($P_x \hat{\mathbf{x}}$) and longitudinal ($P_z \hat{\mathbf{z}}$) projectile polarization components in the scattering plane.

7.7.2 Radiation from Excited States: Stokes Parameters

The state multipole description is also widely used for the parametrization of the Stokes parameters that de-

scribe the polarization of light emitted in optical decays of excited atomic ensembles. The general case of excitation by spin-polarized projectiles has been treated by Bartschat and collaborators [6]. The basic experimental setup for electron-photon coincidence experiments and the definition of the Stokes parameters are illustrated in Figs. 7.2 and 7.3.

For impact excitation of an atomic state with total electronic angular momentum J and an electric dipole transition to a state with J_f, the photon intensity in a direction $\hat{\mathbf{n}} = (\Theta_\gamma, \Phi_\gamma)$ is given by

$$
\begin{aligned}
I(\Theta_\gamma, \Phi_\gamma) = C \Bigg(& \frac{2\,(-1)^{J-J_f}}{3\sqrt{2J+1}} \langle T(J)_{00}^\dagger \rangle \\
& - \begin{Bmatrix} 1 & 1 & 2 \\ J & J & J_f \end{Bmatrix} \big[\Re\{\langle T(J)_{22}^\dagger \rangle\} \sin^2\Theta_\gamma \cos 2\Phi_\gamma \\
& - \Re\{\langle T(J)_{21}^\dagger \rangle\} \sin 2\Theta_\gamma \cos \Phi_\gamma \\
& + \sqrt{\tfrac{1}{6}} \langle T(J)_{20}^\dagger \rangle \, (3\cos^2\Theta_\gamma - 1) \\
& - \Im\{\langle T(J)_{22}^\dagger \rangle\} \sin^2\Theta_\gamma \sin 2\Phi_\gamma \\
& + \Im\{\langle T(J)_{21}^\dagger \rangle\} \sin 2\Theta_\gamma \sin \Phi_\gamma \big] \Bigg),
\end{aligned}
$$

$$(7.81)$$

where

$$C = \frac{e^2 \omega^4}{2\pi c^3} \, |\langle J_f \|\mathbf{r}\| J \rangle|^2 \, (-1)^{J-J_f} \qquad (7.82)$$

is a constant containing the frequency ω of the transition as well as the reduced radial dipole matrix element.

Similarly, the product of the intensity I and the circular light polarization P_3 can be written in terms of state multipoles as

$$
(I \cdot P_3)(\Theta_\gamma, \Phi_\gamma) = -C \begin{Bmatrix} 1 & 1 & 1 \\ J & J & J_f \end{Bmatrix}
$$
$$
\times \Big(\Im\{\langle T(J)_{11}^\dagger \rangle\} 2\sin\Theta_\gamma \sin\Phi_\gamma
$$

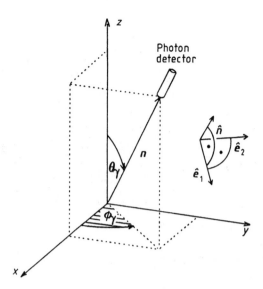

Figure 7.3. Definition of the Stokes parameters: Photons are observed in a direction $\hat{n}$ with polar angles $(\Theta_\gamma, \Phi_\gamma)$ in the collision system. The three unit vectors $(\hat{n}, \hat{e}_1, \hat{e}_2)$ define the helicity system of the photons, $\hat{e}_1 = (\Theta_\gamma + 90°, \Phi_\gamma)$ lies in the plane spanned by $\hat{n}$ and $\hat{z}$ and is perpendicular to $\hat{n}$ while $\hat{e}_2 = (\Theta_\gamma, \Phi_\gamma + 90°)$ is perpendicular to both $\hat{n}$ and $\hat{e}_1$. In addition to the circular polarization P_3, the linear polarizations P_1 and P_2 are defined with respect to axes in the plane spanned by $\hat{e}_1$ and $\hat{e}_2$. Counting from the direction of $\hat{e}_1$, the axes are located at $(0°, 90°)$ for P_1 and at $(45°, 135°)$ for P_2, respectively.

$$
\begin{aligned}
&- \Re\{\langle T(J)_{11}^\dagger\rangle\}2\sin\Theta_\gamma\cos\Phi_\gamma \\
&+ \sqrt{2}\langle T(J)_{10}^\dagger\rangle\cos\Theta_\gamma\Bigg),
\end{aligned}
\tag{7.83}
$$

so that P_3 can be calculated as

$$
P_3(\Theta_\gamma, \Phi_\gamma) = (I \cdot P_3)(\Theta_\gamma, \Phi_\gamma)/I(\Theta_\gamma, \Phi_\gamma).
\tag{7.84}
$$

Note that each state multipole gives rise to a characteristic angular dependence in the formulas for the Stokes parameters, and that perturbation coefficients may need to be applied to deal, for example, with depolarization effects due to internal or external fields. General formulas for $P_1 = \eta_3$ and $P_2 = \eta_1$ can be found in Ref. [6].

As pointed out before, some of the state multipoles may vanish, depending on the experimental arrangement. A detailed analysis of the information contained in the state multipoles and the generalized Stokes parameters (which are defined for specific values of the projectile spin polarization) has been given by Andersen and Bartschat [13]. They re-analyzed the experiment performed by Sohn and Hanne [14] and showed how the density matrix of the excited atomic ensemble can be determined by a measurement of the generalized Stokes parameters. In some cases, this will allow for the extraction of a complete set of scattering amplitudes for the collision process. Such a "perfect scattering experiment" has been called for by Bederson many years ago [15] and is now within reach even for fairly complex excitation processes.

7.8 SUMMARY

The basic formulas dealing with density matrices in quantum mechanics, with particular emphasis on reduced matrix theory and its applications in atomic physics, have been summarized. More details are given in the introductory monograph by Blum [2] and the references listed below.

REFERENCES

1. J. von Neumann, Göttinger Nachrichten **245** (1927).
2. K. Blum, *Density Matrix Theory and Applications* (Plenum, New York, 1981).
3. J. Kessler, *Polarized Electrons* (Springer, New York, 1985).
4. W. E. Baylis, J. Bonenfant, J. Derbyshire. and J. Huschilt, Am. J. Phys. **61**, 534 (1993).
5. M. Born and E. Wolf, *Principles of Optics* (Pergamon, New York, 1970).
6. K. Bartschat, K. Blum, G. F. Hanne, and J. Kessler, J. Phys. B **14**, 3761 (1981).
7. K. Bartschat and K. Blum, Z. Phys. A **304**, 85 (1982).
8. N. Andersen, J. W. Gallagher, and I. V. Hertel, Phys. Rep. **165**, 1 (1988).
9. K. Bartschat, Phys. Rep. **180**, 1 (1989).
10. A. R. Edmonds, *Angular Momentum in Quantum Mechanics* (Princeton University Press, Princeton, 1957).
11. U. Fano and J. H. Macek, Rev. Mod. Phys. **45**, 553 (1973).
12. K. Bartschat, H. J. Andrä, and K. Blum, Z. Phys. A **314**, 257 (1983).
13. N. Andersen and K. Bartschat, J. Phys. B **27**, 3189 (1994).
14. M. Sohn and G. F. Hanne, J. Phys. B **25**, 4627 (1992).
15. B. Bederson, Comments At. Mol. Phys. **1**, 41 and 65 (1969).

8

Computational Techniques

D.R. Schultz and M.R. Strayer

Physics Division, Oak Ridge National Laboratory, Oak Ridge, Tennessee 37831-6373

8.1 INTRODUCTION

Essential to all fields of physics is the ability to perform numerical computations accurately and efficiently. Whether the specific approach involves perturbation theory, close coupling expansion, solution of classical equations of motion, or fitting and smoothing of data, basic computational techniques such as integration, differentiation, interpolation, matrix and eigenvalue manipulation, Monte Carlo sampling, and solution of differential equations must be among the standard tool kit.

This chapter outlines a portion of this tool kit with the aim of giving guidance and organization to a wide array of computational techniques. Having digested the present overview, the reader is then referred to detailed treatments given in many of the large number of texts existing on numerical analysis and computational techniques (see Refs. [1–5]), and mathematical physics (see Refs. [6–10]). We also summarize, especially in the sections on differential equations and computational linear algebra, the role of software packages readily available to aid in implementing practical solutions.

8.2 REPRESENTATION OF FUNCTIONS

The ability to represent functions in terms of polynomials or other basic functions is the key to interpolating or fitting data, and to approximating numerically the operations of integration and differentiation. In addition, using methods such as Fourier analysis, knowledge of the properties of functions beyond even their intermediate values, derivatives, and antiderivatives may be determined (e.g., the "spectral" properties).

8.2.1 Interpolation

Given the value of a function $f(x)$ at a set of points $x_1, x_2, \ldots, x_n$, the function is often required at some other values between these abscissae. The process known as *interpolation* seeks to estimate these unknown values by adjusting the parameters of a known function to approximate the local or global behavior of $f(x)$. One of the most useful representations of a function for these purposes utilizes the *algebraic polynomials*, $P_n(x) = a_0 + a_1 x + \cdots + a_n x^n$, where the coefficients are real constants and the exponents are nonnegative integers. The utility stems from the fact that given any continuous function defined on a closed interval, there exists an algebraic polynomial which is as close to that function as desired (Weierstrass Theorem).

One simple application of these polynomials is the power series expansion of the function $f(x)$ about some point, x_0, i.e.,

$$f(x) = \sum_{k=0}^{\infty} a_k (x - x_0)^k . \tag{8.1}$$

A familiar example is the *Taylor expansion* in which the coefficients are given by

$$a_k = \frac{f^{(k)}(x_0)}{k!}, \tag{8.2}$$

where $f^{(k)}$ indicates the kth derivative of the function. This form, though quite useful in the derivation of formal techniques, is not very useful for interpolation since it assumes the function and its derivatives are known, and since it is guaranteed to be a good approximation only very near the point x_0 about which the expansion has been made.

Lagrange Interpolation. The polynomial of degree $n - 1$ which passes through all n points $(x_1, f(x_1)), (x_2, f(x_2)), \ldots, (x_n, f(x_n))$ is given by

$$P(x) = \sum_{k=1}^{n} f(x_k) \prod_{i=1, i \neq k}^{n} \frac{x - x_i}{x_k - x_i} \tag{8.3}$$

$$= \sum_{k=1}^{n} f(x_k) L_{nk}(x), \tag{8.4}$$

where $L_{nk}(x)$ are the *Lagrange interpolating polynomials*. Perhaps the most familiar example is that of linear interpolation between the points $(x_1, y_1 \equiv f(x_1))$ and $(x_2, y_2 \equiv f(x_2))$, namely,

$$P(x) = \frac{x - x_2}{x_1 - x_2} y_1 + \frac{x - x_1}{x_2 - x_1} y_2 . \tag{8.5}$$

In practice, it is difficult to estimate the formal error bound for this method, since it depends on knowledge of the $(n + 1)^{th}$ derivative. Alternatively, one uses *iterated interpolation* in which successively higher order approximations are tried until appropriate agreement is obtained. *Neville's algorithm* defines a recursive procedure to yield an arbitrary order interpolant from polynomials of lower order. This method, and subtle refinements of it, form the basis for most "recommended" polynomial interpolation schemes (see Ref. [3]).

One important caution to bear in mind is that the more points that are used in constructing the interpolant, and therefore the higher the polynomial order, the greater will be the oscillation in the interpolating function. This highly oscillating polynomial most likely will not correspond more closely to the desired function than polynomials of lower order, and, as a general rule of thumb, fewer than six points should be used.

Cubic Splines. By dividing the interval of interest into a number of subintervals and in each using a polynomial of only modest order, one may avoid the oscillatory nature of high-order (many-point) interpolants. This approach utilizes *piecewise polynomial functions*, the simplest of which is just a linear segment. However, such a straight line approximation has a discontinuous derivative at the data points — a property that one may wish to avoid especially if the derivative of the function is also desired — and which clearly does not provide a smooth interpolant. The solution is therefore to choose the polynomial of lowest order that has enough free parameters (the constants $a_0, a_1, \ldots$) to satisfy the constraints that the function and its derivative are continuous across the subintervals, as well as specifying the derivative at the endpoints x_0 and x_n.

Piecewise cubic polynomials satisfy these constraints, and have a continuous second derivative as well. *Cubic spline interpolation* does not, however, guarantee that the derivatives of the interpolant agree with those of the function at the data points, much less globally. The cubic polynomial in each interval has four undetermined coefficients,

$$P_i(x) = a_i + b_i(x - x_i) + c_i(x - x_i)^2 + d_i(x - x_i)^3 \tag{8.6}$$

for $i = 0, 1, \ldots, n - 1$. Applying the constraints, a system of equations is found which may be solved once the endpoint derivatives are specified. If the second derivatives at the endpoints are set to zero, then the result is termed a *natural spline* and its shape is like that which a long flexible rod would take if forced to pass through all the data points. A *clamped spline* results if the first derivatives are specified at the endpoints, and is usually a better approximation since it incorporates more information about the function (if one has a reasonable way to determine or approximate these first derivatives).

The set of equations in the unknowns, along with the boundary conditions, constitute a *tridiagonal system* or *matrix*, and is therefore amenable to solution by algorithms designed for speed and efficiency for such systems (see Sect. 8.4; also [1–3]). Other alternatives of potentially significant utility are schemes based on the use of rational functions and orthogonal polynomials.

Rational Function Interpolation. If the function which one seeks to interpolate has one or more poles for real x, then polynomial approximations are not good, and a better method is to use quotients of polynomials, so-called *rational functions*. This occurs since the inverse powers of the dependent variable will fit the region near the pole better if the order is large enough. In fact, if the function is free of poles on the real axis but its analytic continuation in the complex plane has poles, the polynomial approximation may also be poor. It is this property that slows or prevents the convergence of power series. Numerical algorithms very similar to those used to generate iterated polynomial interpolants exist [1, 3] and can be useful for functions which are not amenable to polynomial interpolation. Rational function interpolation is related to the method of *Padé approximation* used to improve convergence of power series, and which is a rational function analog of Taylor expansion.

Orthogonal Function Interpolation. Interpolation using functions other than the algebraic polynomials can be defined and are often useful. Particularly worthy of mention are schemes based on *orthogonal polynomials* since they play a central role in numerical quadrature. A set of functions $\phi_1(x), \phi_2(x), \ldots, \phi_n(x)$ defined on the interval [a,b] is said to be orthogonal with respect to a weight function $\mathcal{W}(x)$ if the inner product defined by

$$\langle \phi_i | \phi_j \rangle = \int_a^b \phi_i(x) \phi_j(x) \mathcal{W}(x) dx \qquad (8.7)$$

is zero for $i \neq j$ and positive for $i = j$. In this case, for any polynomial $P(x)$ of degree at most n, there exists unique constants α_k such that

$$P(x) = \sum_{k=0}^{n} \alpha_k \phi_k(x). \qquad (8.8)$$

Among the more commonly used orthogonal polynomials are *Legendre*, *Laguerre*, and *Chebyshev* polynomials.

Chebyshev Interpolation. The significant advantages of employing a representation of a function in terms of Chebyshev polynomials, $T_k(x)$ (see Refs. [4,6] for tabulations, recurrence formulas, orthogonality properties, etc. of these polynomials), i.e.,

$$f(x) = \sum_{k=0}^{\infty} a_k T_k(x), \qquad (8.9)$$

stems from the fact that (*i*) the expansion rapidly converges, (*ii*) the polynomials have a simple form, and (*iii*) the polynomial approximates very closely the solution of

the *minimax* problem. This latter property refers to the requirement that the expansion *mini*mizes the *maxi*mum magnitude of the error of the approximation. In particular, the Chebyshev series expansion can be truncated so that for a given n it yields the most accurate approximation to the function. Thus, Chebyshev polynomial interpolation is essentially as "good" as one can hope to do. Since these polynomials are defined on the interval [-1,1], if the endpoints of the interval in question are a and b, the change of variable

$$y = \frac{x - \frac{1}{2}(b+a)}{\frac{1}{2}(b-a)} \qquad (8.10)$$

will effect the proper transformation. Press *et al.* [3], for example, give convenient and efficient routines for computing the Chebyshev expansion of a function.

8.2.2 Fitting

Fitting of data stands in distinction from interpolation in that the data may have some uncertainty, and therefore, simply determining a polynomial which passes through the points may not yield the best approximation of the underlying function. In fitting, one is concerned with minimizing the deviations of some model function from the data points in an optimal or best fit manner. For example, given a set of data points, even a low-order interpolating polynomial might have significant oscillation, when, in fact, if one accounts for the statistical uncertainties in the data, the best fit may be obtained simply by considering the points to lie on a line.

In addition, most of the traditional methods of assigning this quality of best fit to a particular set of parameters of the model function rely on the assumption that the random deviations are described by a Gaussian (normal) distribution. Results of physical measurements, for example the counting of events, is often closer to a Poisson distribution which tends (not necessarily uniformly) to a Gaussian in the limit of a large number of events, or may even contain "outliers" which lie far outside a Gaussian distribution. In these cases, fitting methods might significantly distort the parameters of the model function in trying to force these different distributions to the Gaussian form. Thus, the *least squares* and *chi-square* fitting procedures discussed below should be used with this caveat in mind. Other techniques, often termed "robust" [3, 11], should be used when the distribution is not Gaussian, or replete with outliers.

Least Squares. In this common approach to fitting, we wish to determine the m parameters a_l of some function $f(x; a_1, a_2, \ldots, a_m)$ depending in this example on one variable, x. In particular, we seek to minimize the sum of the squares of the deviations

$$\sum_{k=1}^{n} [y(x_k) - f(x_k; a_1, a_2, \ldots, a_m)]^2 \qquad (8.11)$$

by adjusting the parameters, where the $y(x_k)$ are the n data points. In the simplest case, the model function is just a straight line, $f(x; a_1, a_2) = a_1 x + a_2$. Elementary multivariate calculus implies that a minimum occurs if

$$a_1 \sum_{k=1}^{n} x_i^2 + a_2 \sum_{k=1}^{n} x_i = \sum_{k=1}^{n} x_i y_i, \qquad (8.12)$$

$$a_1 \sum_{k=1}^{n} x_i + a_2 n = \sum_{k=1}^{n} y_i, \qquad (8.13)$$

which are called the *normal equations*. Solution of these equations is straightforward, and an error estimate of the fit can be found [3]. In particular, variances may be computed for each parameter, as well as measures of the correlation between uncertainties and an overall estimate of the "goodness of fit" of the data.

Chi-square Fitting. If the data points each have associated with them a different standard deviation, σ_k, the least square principle is modified by minimizing the *chi-square*, defined as

$$\chi^2 \equiv \sum_{k=1}^{n} \left[\frac{y_k - f(x_k; a_1, a_2, \ldots, a_m)}{\sigma_k} \right]^2. \qquad (8.14)$$

Assuming that the uncertainties in the data points are normally distributed, the chi-square value gives a measure of the goodness of fit. If there are n data points and m adjustable parameters, then the probability that χ^2 should exceed a particular value purely by chance is

$$Q = Q\left(\frac{n-m}{2}, \frac{\chi^2}{2} \right), \qquad (8.15)$$

where $Q(a, x) = \Gamma(a, x)/\Gamma(a)$ is the incomplete gamma function. For small values of Q, the deviations of the fit from the data are unlikely to be by chance, and values close to one are indications of better fits. In terms of the chi-square, reasonable fits often have $\chi^2 \approx n - m$.

Other important applications of the chi-square method include simulation and estimating standard deviations. For example, if one has some idea of the actual (i.e., non-Gaussian) distribution of uncertainties of the data points, Monte Carlo simulation can be used to generate a set of test data points subject to this presumed distribution, and the fitting procedure performed on the simulated data set. This allows one to test the accuracy or applicability of the model function chosen. In other situations, if the uncertainties of the data points are unknown, one can assume that they are all equal to some value, say σ, fit using the chi-square procedure, and solve for the value of σ. Thus, some measure of the uncertainty from this statistical point of view can be provided.

General Least Squares. The least squares procedure can be generalized usually by allowing any linear combination of basis functions to determine the model function

$$f(x; a_1, a_2, \ldots, a_m) = \sum_{l=1}^{m} a_l \psi_l(x). \qquad (8.16)$$

The basis functions need not be polynomials. Similarly, the formula for chi-square can be generalized, and normal equations determined through minimization. The equations may be written in compact form by defining a matrix **A** with elements

$$A_{i,j} = \frac{\psi_j(x_i)}{\sigma_i}, \qquad (8.17)$$

and a column vector **B** with elements

$$B_i = \frac{y_i}{\sigma_i}. \qquad (8.18)$$

Then the normal equations are [3]

$$\sum_{j=1}^{m} \alpha_{kj} a_j = \beta_k, \qquad (8.19)$$

where

$$[\alpha] = \mathbf{A}^T \mathbf{A}, \quad [\beta] = \mathbf{A}^T \mathbf{B}, \qquad (8.20)$$

and a_j are the adjustable parameters. These equations may be solved using standard methods of computational linear algebra such as Gauss-Jordan elimination. Difficulties involving sensitivity to round-off errors can be avoided by using carefully developed codes to perform this solution [3]. We note that elements of the inverse of the matrix α are related to the variances associated with the free parameters and to the covariances relating them.

Statistical Analysis of Data. Data generated by an experiment, or perhaps from a Monte Carlo simulation, have uncertainties due to the statistical, or random, character of the processes by which they are acquired. Therefore, one must be able to describe statistically certain features of the data such as their mean, variance and skewness, and the degree to which correlations exist, either between one portion of the data and another, or between the data and some other standard or model distribution. A very readable introduction to this type analysis has been given by Young [12], while more comprehensive treatments are also available [13].

8.2.3 Fourier Analysis

The *Fourier transform* takes, for example, a function of time, into a function of frequency, or vice versa, namely

$$\tilde{\varphi}(\omega) = \frac{1}{\sqrt{2\pi}} \int_{-\infty}^{\infty} \varphi(t)e^{i\omega t} dt, \qquad (8.21)$$

$$\varphi(t) = \frac{1}{\sqrt{2\pi}} \int_{-\infty}^{\infty} \tilde{\varphi}(\omega)e^{-i\omega t} d\omega. \qquad (8.22)$$

In this case, the time history of the function $\varphi(t)$ may be termed the "signal" and $\tilde{\varphi}(\omega)$ the "frequency spectrum". Also, if the frequency is related to the energy by $E = \hbar\omega$, one obtains an "energy spectrum" from a signal, and thus the name *spectral methods* for techniques based on the Fourier analysis of signals.

The Fourier transform also defines the relationship between the spatial and momentum representations of wave functions, i.e.,

$$\psi(x) = \frac{1}{\sqrt{2\pi}} \int_{-\infty}^{\infty} \tilde{\psi}(p)e^{ipx} dp, \qquad (8.23)$$

$$\tilde{\psi}(p) = \frac{1}{\sqrt{2\pi}} \int_{-\infty}^{\infty} \psi(x)e^{-ipx} dx. \qquad (8.24)$$

Along with the closely related *sine, cosine,* and *Laplace transforms,* the Fourier transform is an extraordinarily powerful tool in the representation of functions, spectral analysis, convolution of functions, filtering, and analysis of correlation. Good introductions to these techniques with particular attention to applications in physics can be found in [6, 7, 14]. To implement the Fourier transform numerically, the integral tranform pair can be converted to sums

$$\tilde{\varphi}(\omega_j) = \frac{1}{\sqrt{2N2\pi}} \sum_{k=0}^{2N-1} \varphi(t_k)e^{i\omega_j t_k}, \qquad (8.25)$$

$$\varphi(t_k) = \frac{1}{\sqrt{2N2\pi}} \sum_{j=0}^{2N-1} \tilde{\varphi}(\omega_j)e^{-i\omega_j t_k}, \qquad (8.26)$$

where the functions are "sampled" at $2N$ points. These equations define the *discrete Fourier transform* (DFT). Two cautions in using the DFT are as follows.

First, if a continuous function of time that is sampled at, for simplicity, uniformly spaced intervals, (i.e., $t_{i+1} = t_i + \Delta$), then there is a critical frequency $\omega_c = \pi/\Delta$, known as the *Nyquist frequency,* which limits the fidelity of the DFT of this function in that it is *aliased.* That is, components outside the frequency range $-\omega_c$ to ω_c are falsely transformed into this range due to the finite sampling. This effect can be remediated by filtering or windowing techniques. If, however, the function is *bandwidth limited* to frequencies smaller than ω_c, then the DFT does not suffer from this effect, and the signal is completely determined by its samples. Second, implementing the DFT directly from the above equations would require approximately N^2 multiplications to perform the Fourier transform of a function sampled at N points. A variety of *fast Fourier transform* (FFT) algorithms have been developed (e.g., the Danielson-Lanczos

and Cooley-Tukey methods) which require only on the order of $(N/2)\log_2 N$ multiplications. Thus, for even moderately large sets of points, the FFT methods are indeed much faster than the direct implementation of the DFT. Issues involved in sampling, aliasing, and selection of algorithms for the FFT are discussed in great detail, for example, in [3, 15, 16].

8.2.4 Approximating Integrals

Polynomial Quadrature. Definite integrals may be approximated through a procedure known as numerical quadrature by replacing the integral by an appropriate sum, i.e.,

$$\int_a^b f(x)dx \approx \sum_{k=0}^{n} a_k f(x_k). \qquad (8.27)$$

Most formulas for such approximation are based on the interpolating polynomials described in Sect. 8.2.1, especially the Lagrange polynomials, in which case the coefficients a_k are given by

$$a_k = \int_a^b L_{nk}(x_k)dx. \qquad (8.28)$$

If first or second degree Lagrange polynomials are used with a uniform spacing between the data points, one obtains the *trapezoidal* and *Simpson's rules,* i.e.,

$$\int_a^b f(x)dx \approx \frac{\delta}{2}[f(a) + f(b)] + \mathcal{O}\left(\delta^3 f^{(2)}(\zeta)\right), \qquad (8.29)$$

$$\int_a^b f(x)dx \approx \frac{\delta}{3}[f(a) + 4f(\frac{\delta}{2}) + f(b)] + \mathcal{O}\left(\delta^5 f^{(4)}(\zeta)\right), \qquad (8.30)$$

respectively, with $\delta = b - a$, and for some ζ in $[a, b]$.

Other commonly used formulas based on low-order polynomials, and generally referred to as *Newton-Cotes* formulas, are described and discussed in detail in numerical analysis texts (see Refs. [1, 2]). Since potentially unwanted rapid oscillations in interpolants may arise, it is generally the case that increasing the order of the quadrature scheme too greatly does not generally improve the accuracy of the approximation. Dividing the interval $[a, b]$ into a number of subintervals and summing the result of application of a low-order formula in each subinterval is usually a much better approach. This procedure, referred to as *composite quadrature,* may be combined with choosing the data points at a nonuniform spacing, decreasing the spacing where the function varies rapidly, and increasing the spacing for economy where the function is smooth to construct an *adaptive quadrature.*

Gaussian Quadrature. If the function whose definite integral is to be approximated can be evaluated explicitly, then the data points (abscissas) can be chosen in a manner in which significantly greater accuracy may be obtained than using Newton-Cotes formulas of equal order. *Gaussian quadrature* is a procedure in which the error in the approximation is minimized owing to this freedom to choose both data points (abscissas) and coefficients. By utilizing orthogonal polynomials and choosing the abscissas at the roots of the polynomials in the interval under consideration, it can be shown that the coefficients may be optimally chosen by solving a simple set of linear equations. Thus, a Gaussian quadrature scheme approximates the definite integral of a function multiplied by the weight function appropriate to the orthogonal polynomial being used as

$$\int_a^b \mathcal{W}(x)f(x)dx \approx \sum_{k=1}^{n} a_k f(x_k), \qquad (8.31)$$

where the function is to be evaluated at the abscissas given by the roots of the orthogonal polynomial, x_k. In this case, the coefficients a_k are often referred to as "weights," but should not be confused with the weight function $\mathcal{W}(x)$ (see Sect. 8.2.1). Since the Legendre polynomials are orthogonal over the interval [-1,1] with respect to the weight function $\mathcal{W}(x) \equiv 1$, this equation has a particularly simple form, leading immediately to the *Gauss-Legendre quadrature*. If $f(x)$ contains as a factor the weight function of another of the orthogonal polynomials, the corresponding *Gauss-Laguerre* or *Gauss-Chebyshev* quadrature should be used.

The roots and coefficients have been tabulated [4] for many common choices of the orthogonal polynomials (e.g., Legendre, Laguerre, Chebyshev) and for various orders. Simple computer subroutines are also available which conveniently compute them [3]. Since the various orthogonal polynomials are defined over different intervals, use of the change of variables such as that given in equation 8.10 may be required. So, for Gauss-Legendre quadrature we make use of the transformation

$$\int_a^b f(x)\,dx \approx \frac{(b-a)}{2} \int_{-1}^{1} f\left(\frac{(b-a)y + b + a}{2}\right) dy.$$
$$(8.32)$$

Other Methods. Especially for multidimensional integrals which can not be reduced analytically to seperable or iterated integrals of lower dimension, *Monte Carlo* integration may provide the only means of finding a good approximation. This method is described in Sect. 8.5.3. Also, a convenient quadrature scheme can easily be devised based on the cubic spline interpolation described in Sect. 8.2.1 since in each subinterval, the definite integral of a cubic polynomial of known coefficients is evident.

8.2.5 Approximating Derivatives

Numerical Differentiation. The calculation of derivatives from a numerical representaion of a function is generally less stable than the calculation of integrals because differentation tends to enhance fluctuations and worsen the convergence properties of power series. For example, if $f(x)$ is twice continuously differentiable on [a,b], then differentiation of the linear Lagrange interpolation formula 8.5 yields

$$f^{(1)}(x_0) = \frac{f(x_0 + \delta) - f(x_0)}{\delta} + \mathcal{O}\left(\delta f^{(2)}(\zeta)\right) \quad (8.33)$$

for some x_0 and ζ in [a,b], where $\delta = b - a$. In the limit $\delta \to 0$, Eq. 8.33 coincides with the definition of the derivative. However, in practical calculations with finite precision arithmetic, δ cannot be taken too small because of numerical cancellation in the calculation of $f(a + \delta) - f(a)$.

In practice, increasing the order of the polynomial used decreases the truncation error, but at the expense of increasing round-off error, the upshot being that three- and five-point approximations are usually most useful. Various three- and five-point formulas are given in standard texts [2, 4, 17]. Two common five-point formulas (centered and forward/backward) are

$$f^{(1)}(x_0) = \frac{1}{12\delta}[f(x_0 - 2\delta) - 8f(x_0 - \delta) + 8f(x_0 + \delta)$$
$$- f(x_0 + 2\delta)] + \mathcal{O}\left(\delta^4 f^{(5)}(\zeta)\right) \quad (8.34)$$

$$f^{(1)}(x_0) = \frac{1}{12\delta}[-25f(x_0) + 48f(x_0 + \delta) - 36f(x_0 + 2\delta)$$
$$+ 16f(x_0 + 3\delta) - 3f(x_0 + 4\delta)]$$
$$+ \mathcal{O}\left(\delta^4 f^{(5)}(\zeta)\right) \quad (8.35)$$

The second formula is useful for evaluating the derivative at the left or right endpoint of the interval, depending on whether δ is positive or negative, respectively.

Derivatives of Interpolated Functions. An interpolating function can be directly differentiated to obtain the derivative at any desired point. For example, if $f(x) \approx a_0 + a_1 x + a_2 x^2$, then $f^{(1)}(x) = a_1 + 2a_2 x$. However, this approach may fail to give the best approximation to $f^{(1)}(x)$ if the original interpolation was optimized to give the best possible representaion of $f(x)$.

8.3 DIFFERENTIAL AND INTEGRAL EQUATIONS

The subject of differential and integral equations is immense in both richness and scope. The discussion here focuses on techniques and algorithms, rather than the formal aspect of the theory. Further information can be found elsewhere under the broad catagories of finite element and finite difference methods. The *Numerov method*, which is particularly useful in integrating the Schrödinger equation, is described in great detail in Ref. [8].

8.3.1 Ordinary Differential Equations

An *ordinary differential equation* is an equation involving an unknown function and one or more of its derivatives that depends on only one independent variable [18]. The *order* of a differential equation is the order of the highest derivative appearing in the equation. A *solution* of a general differential equation of order n,

$$f(t, y, \dot{y}, \ldots, y^{(n)}) = 0, \qquad (8.36)$$

is a real-valued function $y(t)$ having the following properties: (1) $y(t)$ and its first n derivatives exist, so $y(t)$ and its first $n-1$ derivatives must be continuous, and (2) $y(t)$ satisfies the differential equation for all t. A unique solution requires the specification of n conditions on $y(t)$ and its derivatives. The conditions may be specified as n initial conditions at a single t to give an *initial value problem*, or at the end points of an interval to give a *boundary value problem*.

Consider first solutions to the simple equation

$$\dot{y} = f(t, y), \qquad y(a) = A. \qquad (8.37)$$

The methods discussed below can easily be extended to systems of first-order differential equations and to higher-order differential equations. The methods are referred to as *discrete variable methods* and generate a sequence of approximate values for $y(t), y_1, y_2, y_3, \ldots$ at points $t_1, t_2, t_3, \ldots$. For simplicity, the discussion assumes a constant spacing h between t points. We shall first describe a class of methods known as *one-step* methods [19]. They have no memory of the solutions at past times; given y_i, there is a recipe for y_{i+1} that depends only on information at t_i. Errors enter into numerical solutions from two sources. The first is *discretization error* and depends on the method being used. The second is *computational error* which includes such things as round off error.

For a solution on the interval [a,b], let the t points be equally spaced; so for some positive integer n and $h = (b-a)/n$, $t_i = a + ih$, $i = 0, 1, \ldots, n$. If $a < b$, h is positive and the integration is forward; if $a > b$, h is negative and the integration is backward. The latter case could occur in solving for the initial point of a solution curve given the terminal point. A general one-step method can then be written in the form

$$y_{i+1} = y_i + h\Delta(t_i, y_i), \qquad y_0 = y(t_0), \quad (8.38)$$

where Δ is a function that characterizes the method. Different Δ functions are displayed, giving rise to the Taylor series methods and the Runge-Kutta methods.

Taylor Series Algorithm. To obtain an approximate solution of order p on $[a, b]$, generate the sequence

$$y_{i+1} = y_i + h\left[f(t_i, y_i) + \cdots + f^{(p-1)}(t_i, y_i)\frac{h^{p-1}}{p!}\right],$$
$$t_{i+1} = t_i + h, \quad i = 0, 1, \ldots, n-1 \qquad (8.39)$$

where $t_0 = a$ and $y_0 = A$. The Taylor method of order $p = 1$ is known as *Euler's method*:

$$y_{i+1} = y_i + hf(t_i, y_i),$$
$$t_{i+1} = t_i + h. \qquad (8.40)$$

Taylor series methods can be quite effective if the total derivatives of f are not too difficult to evaluate. Software packages are available that perform exact differentiation, (ADIFOR, MAPLE, MATHEMATICA, etc.) facilitating the use of this approach.

Runge-Kutta Methods. Runge-Kutta methods are designed to approximate Taylor series methods [20], but have the advantage of not requiring explicit evaluations of the derivatives of $f(t, y)$. The basic idea is to use a linear combination of values of $f(t, y)$ to approximate $y(t)$. This linear combination is matched up as closely as possible with a Taylor series for $y(t)$ to obtain methods of the highest possible order p. Euler's method is an example using one function evaluation.

To obtain an approximate solution of order $p = 2$, let $h = (b-a)/n$ and generate the sequences

$$y_{i+1} = y_i + h\left[(1-\gamma)f(t_i, y_i)\right.$$
$$\left. + \gamma f(t_i + \frac{h}{2\gamma}, y_i + \frac{h}{2\gamma}f(t_i, y_i))\right],$$
$$t_{i+1} = t_i + h, \quad i = 0, 1, \ldots, n-1, \qquad (8.41)$$

where $\gamma \neq 0$, $t_0 = a$, $y_0 = A$.

Euler's method is the special case, $\gamma = 0$, and has order 1; the improved Euler method has $\gamma = 1/2$ and the Euler-Cauchy method has $\gamma = 1$.

The Adams-Bashforth and Adams-Moulton Formulas. These formulas furnish important and widely-used examples of multistep methods [21]. On reaching a mesh point t_i with approximate solution $y_i \cong y(t_i)$, there are (usually) available approximate solutions $y_{i+1-j} \cong y(t_{i+1-j})$ for $j = 2, 3, \ldots, p$. From the differential equation itself, approximations to the derivatives $\dot{y}(t_{i+1-j})$ can be obtained.

An attractive feature of the approach is the form of the underlying polynomial approximation, $P(t)$, to $\dot{y}(t)$ because it can be used to approximate $y(t)$ between mesh points:

$$y(t) \cong y_i + \int_{t_i}^{t} P(t)\, dt. \qquad (8.42)$$

The lowest-order Adams-Bashforth formula arises from interpolating the single value $f_i = f(t_i, y_i)$ by

$P(t)$. The interpolating polynomial is constant so its integration from t_i to t_{i+1} results in $hf(t_i, y_i)$ and the first order Adams-Bashforth formula:

$$y_{i+1} = y_i + hf(t_i, y_i). \tag{8.43}$$

This is just the forward Euler formula. For constant step size h, the second-order Adams-Bashforth formula is

$$y_{i+1} = y_i + h[(3/2)f(t_i, y_i) - (1/2)f(t_{i-1}, y_{i-1})]. \tag{8.44}$$

The lowest-order Adams-Moulton formula involves interpolating the single value $f_{i+1} = f(x_{i+1}, y_{i+1})$ and leads to the backward Euler formula

$$y_{i+1} = y_i + hf(t_{i+1}, y_{i+1}), \tag{8.45}$$

which defines y_{i+1} implicitly. From its definition it is clear that it has the same accuracy as the forward Euler method; its advantage is vastly superior stability. The second-order Adams-Moulton method also does not use previously computed solution values; it is called the trapezoidal rule because it generalizes the trapezoidal rule for integrals to differential equations:

$$y_{i+1} = y_i + \frac{h}{2}[f(t_{i+1}, y_{i+1}) + f(t_i, y_i)]. \tag{8.46}$$

The Adams-Moulton formula of order p is more accurate than the Adams-Bashforth formula of the same order, so that it can use a larger step size; the Adams-Moulton formula is also more stable. A code based on such methods is more complex than a Runge-Kutta code because it must cope with the difficulties of starting the integration and changing the step size. Modern Adams codes attempt to select the most efficient formula at each step, as well as to choose an optimal step size h to achieve a specified accuracy.

8.3.2 Differencing Algorithms for Partial Differential Equations

The modern approach to evolve differencing schemes for most physical problems is based on flux conservation methods [22]. One begins by writing the balance equations for a single cell, and subsequently applying quadratures and interpolation formulas. Such approaches have been successful for the full spectrum of hyperbolic, elliptic, and parabolic equations. For simplicity, we begin by discussing systems involving only one space variable.

As a prototype, consider the parabolic equation

$$c\frac{\partial}{\partial t}u(x, t) = \sigma\frac{\partial^2}{\partial x^2}u(x, t), \tag{8.47}$$

where c and σ are constants and $u(x, t)$ is the solution. We begin by establishing a grid of points on the xt-plane

with step size h in the x direction and step size k in the t-direction. Let spatial grid points be denoted by $x_n = x_0 + nh$ and time grid points by $t_j = t_0 + jk$, where n and j are integers and (x_0, t_0) is the origin of the space–time grid. The points ξ_{n-1} and ξ_n are introduced to establish a "control interval". We begin with a conservation statement

$$\int_{\xi_{n-1}}^{\xi_n} dx\,[r(x, t_{j+1}) - r(x, t_j)]$$
$$= \int_{t_j}^{t_{j+1}} dt\,[q(\xi_{n-1}, t) - q(\xi_n, t)]. \tag{8.48}$$

This equation states that the change in the field density on the interval (ξ_{n-1}, ξ_n) from time $t = t_j$ to time $t = t_{j+1}$ is given by the flux into this interval at ξ_{n-1} minus the flux out of the interval at ξ_n from time t_j to time t_{j+1}. This expresses the conservation of material in the case that no sources or sinks are present. We relate the field variable u to the physical variables (the density r and the flux q). We consider the case in which density is assumed to have the form

$$r(x, t) = cu(x, t) + b \tag{8.49}$$

with c and b constants, thus

$$c\int_{\xi_{n-1}}^{\xi_n} dx\,[u(x, t_{j+1}) - u(x, t_j)]$$
$$\approx c[u(x_n, t_{j+1}) - u(x_n, t_j)]h. \tag{8.50}$$

When developing conservation law equations, there are two commonly used strategies for approximating the right-hand-side of 8.48: (i) left end-point quadrature

$$\int_{t_j}^{t_{j+1}} dt\,[q(\xi_{n-1}, t) - q(\xi_n, t)]$$
$$\approx [q(\xi_{n-1}, t_j) - q(\xi_n, t_j)]k, \tag{8.51}$$

and (ii) right end-point quadrature

$$\int_{t_j}^{t_{j+1}} dt\,[q(\xi_{n-1}, t) - q(\xi_n, t)]$$
$$\approx [q(\xi_{n-1}, t_{j+1}) - q(\xi_n, t_{j+1})]k. \tag{8.52}$$

Combining (8.48) with the respective approximations yields: from (i) an explicit method

$$c\,[u(x_n, t_{j+1}) - u(x_n, t_j)]\,h$$
$$\approx [q(\xi_{n-1}, t_j) - q(\xi_n, t_j)]\,k, \tag{8.53}$$

and from (ii) an implicit method

$$c\,[u(x_n, t_{j+1}) - u(x_n, t_j)]\,h$$
$$\approx [q(\xi_{n-1}, t_{j+1}) - q(\xi_n, t_{j+1})]\,k. \tag{8.54}$$

Using centered finite difference formulas to approximate the fluxes at the control points ξ_{n-1} and ξ_n yields

$$q(\xi_{n-1}, t_j) = -\sigma \frac{u(x_n, t_j) - u(x_{n-1}, t_j)}{h}, \qquad (8.55)$$

and

$$q(\xi_n, t_j) = -\sigma \frac{u(x_{n+1}, t_j) - u(x_n, t_j)}{h} \qquad (8.56)$$

where σ is a constant. We also obtain similar formulas for the fluxes at time t_{j+1}.

We have used a lower case u to denote the continuous field variable, $u = u(x, t)$. Note that all of the quadrature and difference formulas involving u are stated as approximate equalities. In each of these approximate equality statements, the amount by which the right side differs from the left side is called the truncation error. If u is a well-behaved function (has enough smooth derivatives), then it can be shown that these truncation errors approach zero as the grid spacings, h and k, approach zero.

If U_n^j denotes the exact solution on the grid, we have from (i) the result

$$c\left(U_n^{j+1} - U_n^j\right) h^2 = \sigma k \left(U_{n-1}^j + U_{n+1}^j - 2U_n^j\right). \quad (8.57)$$

This is an explicit method since it provides the solution to the difference equation at time t_{j+1}, knowing the values at time t_j.

If we use the numerical approximations (ii) we obtain the result

$$c\left(U_n^{j+1} - U_n^j\right) h^2 = \sigma k \left(U_{n-1}^{j+1} + U_{n+1}^{j+1} - 2U_n^{j+1}\right). \quad (8.58)$$

Note that this equation defines the solution at time t_{j+1} implicitly, since a system of algebraic equations is required to be satisfied.

8.3.3 Variational Methods

Perhaps the most widely used approximation procedures in AMO physics are the variational methods. We shall outline in detail the Rayleigh-Ritz method [23]. This method is limited to boundary value problems which can be formulated in terms of the minimization of a functional $J[u]$. For definiteness we consider the case of a differential operator defined by

$$L u(\mathbf{x}) = f(\mathbf{x}) \qquad (8.59)$$

with $\mathbf{x} = x_i, i = 1, 2, 3$ in R, for example, and with $u = 0$ on the boundary of R. The function $f(\mathbf{x})$ is the source. It is assumed that L is always nonsingular and in addition, for the Ritz method L is Hermitian. The real-valued functions u are in the Hilbert space Ω of the operator L. We construct the functional $J[u]$ defined as

$$J[u] = \int_\Omega dx \, [u(x) L u(x) - 2u(x) f(x)]. \qquad (8.60)$$

The variational ansatz considers a subspace of Ω, Ω_n, spanned by a class of functions $\phi_n(x)$, and we construct the function $u^n \approx u$

$$u^n(x) = \sum_{i=1}^n c_i \phi_i(x). \qquad (8.61)$$

We solve for the coefficients c_i by minimizing $J[u^n]$

$$\partial_{c_i} J[u^n] = 0, \quad i = 1, \dots, n. \qquad (8.62)$$

These equations are simply cast into a set of well-behaved algebraic equations

$$\sum_{j=1}^n A_{i,j} c_j = g_i, \quad i = 1, \dots, n, \qquad (8.63)$$

with $A_{i,j} = \int_\Omega dx \phi_i(x) L \phi_j(x)$, and $g_i = \int_\Omega dx \phi_i(x) f(x)$. Under very general conditions, the functions u^n converge uniformly to u. The main drawback of the Ritz method is in the assumption of Hermiticity of the operator L. For the *Galerkin Method* we relax this assumption with no other changes. Thus we obtain an identical set of equations, as above with the exception that the function g is no longer symmetric. The convergence of the sequence of solutions u^n to u is no longer guaranteed, unless the operator can be separated into a symmetric part L_0, $L = L_0 + K$ so that $L_0^{-1} K$ is bounded.

8.3.4 Finite Elements

As discussed in Sect. 8.3.2, in the finite difference method for classical partial differential equations, the solution domain is approximated by a grid of uniformly spaced nodes. At each node, the governing differential equation is approximated by an algebraic expression which references adjacent grid points. A system of equations is obtained by evaluating the previous algebraic approximations for each node in the domain. Finally, the system is solved for each value of the dependent variable at each node.

In the finite element method [24], the solution domain can be discretized into a number of uniform or nonuniform finite elements that are connected via nodes. The change of the dependent variable with regard to location is approximated within each element by an interpolation function. The interpolation function is defined relative to the values of the variable at the nodes associated with each element. The original boundary value problem is then replaced with an equivalent integral formulation. The interpolation functions are then substituted into the integral equation, integrated, and combined with the results from all other elements in the solution domain.

The results of this procedure can be reformulated into a matrix equation of the form

$$\sum_{j=1}^{n} A_{i,j} c_j = g_i\,, \quad i = 1, \ldots, n\,, \tag{8.64}$$

with $A_{i,j} = \int_{\Omega} dx \phi_i(x) L \phi_j(x)$, and $g_i = \int_{\Omega} dx \phi_i(x) f(x)$ exactly as obtained in Sect 8.3.3. The only difference arises in the definitions of the support functions $\phi_i(x)$. In general, if these functions are piecewise polynomials on some finite domain, they are called finite elements or splines. Finite elements make it possible to deal in a systematic fashion with regions having curved boundaries of an arbitrary shape. Also, one can systematically estimate the accuracy of the solution in terms of the parameters that label the finite element family, and the solutions are no more difficult to generate than more complex variational methods.

In one space dimension, the simplest finite element family begins with the set of step functions defined by

$$\phi_i(x) = \begin{cases} 1 & x_{i-1} \le x \le x_i \\ 0 & \text{otherwise.} \end{cases} \tag{8.65}$$

The use of these simple "hat" functions as a basis provides no advantage over the usual finite difference schemes. However, for certain problems in two or more dimensions, finite element methods have distinct advantages over other methods. Generally, the use of finite elements requires complex, sophisticated computer programs for implementation. The use of higher-order polynomials, commonly called splines, as a basis has been extensively used in atomic and molecular physics. An extensive literature is available [25, 26].

We illustrate the use of the finite element method by applying it to the Schrödinger equation. In this case, the linear operator L is $H - E$ where, as usual, E is the energy and the Hamiltonian H is the sum of the kinetic and potential energies, that is, $L = H - E = T + V - E$ and $Lu(x) = 0$. We define the finite elements through support points, or knots, given by the sequence $\{x_1, x_2, x_3, \ldots\}$ which are not necessarily spaced uniformly. Since the "hat" functions have vanishing derivatives, we employ the next more complex basis, i.e., "tent" functions, which are piecewise linear functions given by

$$\phi_i(x) = \begin{cases} \dfrac{x - x_{i-1}}{x_i - x_{i-1}} & x_{i-1} \le x \le x_i \\[2mm] \dfrac{x_{i+1} - x}{x_{i+1} - x_i} & x_i \le x \le x_{i+1} \\[2mm] 0 & \text{otherwise,} \end{cases} \tag{8.66}$$

and for which the derivative is given by

$$\frac{d}{dx} \phi_i(x) = \begin{cases} \dfrac{1}{x_i - x_{i-1}} & x_{i-1} \le x \le x_i \\[2mm] \dfrac{-1}{x_{i+1} - x_i} & x_i \le x \le x_{i+1} \\[2mm] 0 & \text{otherwise.} \end{cases} \tag{8.67}$$

The functions have a maximum value of one at the midpoint of the interval $[x_{i-1}, x_{i+1}]$, with partially overlapping adjacent elements. In fact, the overlaps may be represented by a matrix $\mathbf{O}$ with elements

$$O_{ij} = \int_{-\infty}^{\infty} dx\, \phi_i(x) \phi_j(x)\,. \tag{8.68}$$

Thus, if $i = j$

$$O_{ii} = \int_{x_{i-1}}^{x_i} dx\, \frac{(x - x_{i-1})^2}{(x_i - x_{i-1})^2} + \int_{x_i}^{x_{i+1}} dx\, \frac{(x - x_i)^2}{(x_{i+1} - x_i)^2}$$
$$= \tfrac{1}{3}(x_{i+1} - x_{i-1})\,, \tag{8.69}$$

if $i = j - 1$

$$O_{ij} = \int_{x_i}^{x_{i+1}} dx\, \frac{(x - x_i)(x_{i+1} - x)}{(x_{i+1} - x_i)^2}$$
$$= \tfrac{1}{6}(x_{i+1} - x_i)\,, \tag{8.70}$$

if $i = j + 1$

$$O_{ij} = \int_{x_{i-1}}^{x_i} dx\, \frac{(x - x_{i-1})(x_i - x)}{(x_i - x_{i-1})^2}$$
$$= \tfrac{1}{6}(x_i - x_{i-1})\,, \tag{8.71}$$

and $O_{ij} = 0$ otherwise.

The potential energy is represented by the matrix

$$V_{ij} = \int_{-\infty}^{\infty} dx\, \phi_i(x) V(x) \phi_j(x)\,, \tag{8.72}$$

which may be well approximated by

$$V_{ij} \approx V(x_i) \int_{-\infty}^{\infty} dx\, \phi_i(x) \phi_j(x) \tag{8.73}$$
$$= V(x_i) O_{ij}$$

if $x_j - x_i$ is small. The kinetic energy, $T = -\tfrac{1}{2} d^2/dx^2$, is similarly given by

$$T_{ij} = -\frac{1}{2} \int_{-\infty}^{\infty} dx \phi_i(x) \frac{d^2}{dx^2} \phi_j(x)\,, \tag{8.74}$$

which we compute by integrating by parts since the tent functions have a singular second derivative

$$T_{ij} = \frac{1}{2} \int_{-\infty}^{\infty} dx (\frac{d}{dx} \phi_i(x))(\frac{d}{dx} \phi_j(x))\,, \tag{8.75}$$

which in turn is evalutated to yield

$$T_{ij} = \begin{cases} \dfrac{x_{i+1} - x_{i-1}}{2(x_i - x_{i-1})(x_{i+1} - x_i)} & i = j \\[3mm] \dfrac{1}{2(x_i - x_{i+1})} & i = j - 1 \\[3mm] \dfrac{1}{2(x_{i-1} - x_i)} & i = j + 1 \\[3mm] 0 & \text{otherwise.} \end{cases} \tag{8.76}$$

Finally, since the Hamiltonian matrix is $H_{ij} = T_{ij} + V_{ij}$, the solution vector $u_i(x)$ may be found by solving the eigenvalue equation

$$[H_{ij} - EO_{ij}]u_i(x) = 0. \qquad (8.77)$$

8.3.5 Integral Equations

Central to much of practical and formal scattering theory is the integral equation and techniques of its solution. For example, in atomic collision theory, the Schrödinger differential equation

$$(E - H_0(\mathbf{r}))\psi(\mathbf{r}) = V(\mathbf{r})\psi(\mathbf{r}) \qquad (8.78)$$

where the Hamiltonian $H_0 \equiv -(\hbar^2/2m)\nabla^2 + V_0$ may be solved by exploiting the solution for a delta function source, i.e.,

$$(E - H_0)G(\mathbf{r}, \mathbf{r}') = \delta(\mathbf{r} - \mathbf{r}'). \qquad (8.79)$$

In terms of this *Green's function* $G(\mathbf{r}, \mathbf{r}')$, and any solution $\chi(\mathbf{r})$ of the homogeneous equation [i.e. with $V(\mathbf{r}) = 0$], the general solution is

$$\psi(\mathbf{r}) = \chi(\mathbf{r}) + \int \mathbf{dr'}\, \mathbf{G}(\mathbf{r}, \mathbf{r}')\mathbf{V}(\mathbf{r}')\psi(\mathbf{r}') \qquad (8.80)$$

for which, given a choice of the functions $G(\mathbf{r}, \mathbf{r}')$ and $\chi(\mathbf{r})$, particular boundary conditions are determined. This integral equation is the Lippmann-Schwinger equation of potential scattering. Further topics on scattering theory are covered in other chapters (see especially Chaps. 45 to 56) and in standard texts such as those by Joachain [27], Rodberg and Thaler [28], and Goldberger and Watson [29]. Owing especially to the wide variety of specialized techniques for solving integral equations, we survey briefly only a few of the most widely applied methods.

Integral Transforms. Certain classes of integral equations may be solved using integral transforms such as the Fourier or Laplace transforms. These integral transforms typically have the form

$$f(x) = \int dx'\, K(x, x')g(x'), \qquad (8.81)$$

where $f(x)$ is the integral transform of $g(x')$ by the kernel $K(x, x')$. Such a pair of functions is the solution of the Schrödinger equation (spatial wave function) and its Fourier transform (momentum representation wave function). Arfken [6], Morse and Feshbach [9], and Courant and Hilbert [10] give other examples, as well as being excellent references for the application of integral equations and Green's functions in mathematical physics. In their analytic form, these transform methods provide a powerful method of solving integral equations for special cases, and, in addition, they may be implemented by performing the transform numerically.

Power Series Solution. For an equation of the form (in one dimension for simplicity)

$$\psi(r) = \chi(r) + \lambda \int dr'\, K(r, r')\psi(r'), \qquad (8.82)$$

a solution may be found by iteration. That is, as a first approximation, set $\psi_0(r) = \chi(r)$ so that

$$\psi_1(r) = \chi(r) + \lambda \int dr'\, K(r, r')\chi(r'). \qquad (8.83)$$

This may be repeated to form a power series solution, i.e.,

$$\psi_n(r) = \sum_{k=0}^{n} \lambda^k I_k(r), \qquad (8.84)$$

where

$$I_0(r) = \chi(r), \qquad (8.85)$$

$$I_1(r) = \int dr'\, K(r, r')\chi(r'), \qquad (8.86)$$

$$I_2(r) = \int dr'' \int dr'\, K(r, r')K(r', r'')\chi(r''), \qquad (8.87)$$

$$I_n(r) = \int dr' \cdots \int dr^{(n)}\, K(r, r')K(r, r'') \qquad (8.88)$$
$$\cdots K(r^{(n-1)}, r^{(n)}).$$

If the series converges, then the solution $\psi(r)$ is approached by the expansion. When the Schrödinger equation is cast as an integral equation for scattering in a potential, this iteration scheme leads to the *Born series*, the first term of which is the incident, unperturbed wave, and the second term is usually referred to simply as the *Born approximation*.

Separable Kernels. If the kernel is separable, i.e.,

$$K(r, r') = \sum_{k=1}^{n} f_k(r)g_k(r'), \qquad (8.89)$$

where n is finite, then substitution into the prototype integral equation (8.82) yields

$$\psi(r) = \chi(r) + \lambda \sum_{k=1}^{n} f_k(r) \int dr'\, g(r')\psi(r'). \qquad (8.90)$$

Multiplying by $f_k(r)$, integrating over r, and rearranging, yields the set of algebraic equations

$$c_j = b_j + \lambda \sum_{k=1}^{n} a_{jk}c_k, \qquad (8.91)$$

where

$$c_k = \int dr'\, g_k(r')\psi(r'), \qquad (8.92)$$

$$b_k = \int dr\, f_k(r)\chi(r), \qquad (8.93)$$

$$a_{jk} = \int dr\, g_j(r)f_k(r), \qquad (8.94)$$

or, if **c** and **b** denote vectors, and **A** denotes the matrix of constants a_{jk},

$$\mathbf{c} = (\mathbf{1} - \lambda\mathbf{A})^{-1}\mathbf{b}. \qquad (8.95)$$

The eigenvalues are the roots of the determinantal equation. Substituting these into $(\mathbf{1} - \lambda\mathbf{A})\mathbf{c} = 0$ yields the constants c_k which determine the solution of the original equation. This derivation may be found in Arfken [6], along with an explicit example. Even if the kernel is not exactly separable, if it is approximately so, then this procedure can yield a result which can be substituted into the original equation as a first step in an iterative solution.

Numerical Integration. Perhaps the most straightforward method of solving an integral equation is to apply a numerical integration formula such as Gaussian quadrature. An equation of the form

$$\psi(r) = \int dr'\, K(r,r')\chi(r') \qquad (8.96)$$

can be approximated as

$$\psi(r_j) = \sum_{k=1}^{n} w_k K(r_j, r'_k)\chi(r_k), \qquad (8.97)$$

where w_k are quadrature weights, if the kernel is well behaved. However, such an approach is not without pitfalls. In light of the previous subsection, this approach is equivalent to replacing the integral equation by a set of algebraic equations. In this example we have

$$\psi_j = \sum_{k=1}^{n} M_{jk}\chi_k, \qquad (8.98)$$

so that the solution of the equation is found by inverting the matrix **M**. Since there is no guarantee that this matrix is not ill-conditioned, the numerical procedure may not produce meaningful results. In particular, only certain classes of integral equations and kernels will lead to stable solutions.

Having only scratched the surface regarding the very rich field of integral equations, the interested reader is encouraged to explore the references given here.

8.4 COMPUTATIONAL LINEAR ALGEBRA

Previous sections of this chapter have dealt with interpolation, differential equations, and related topics. Generally, discretization methodologies lead to classes of algebraic equations. Over the past 15 years enormous progress has been made in developing algorithms for solving linear algebraic equations, and many very good books have been written on this topic [30]. A large body of numerical software is freely available 24 hours a day via an electronic service called *Netlib*. In addition to LAPACK and LINPACK, there are dozens of other libraries, technical reports on various parallel computers and software, test data, facilities to automatically translate Fortran programs to C, bibliographies, names and addresses of scientists and mathematicians, and so on. One can communicate with Netlib in one of two ways, by e-mail or (much more easily) via an X-window interface called *Xnetlib*. Using e-mail, one sends messages of the form "send subroutine-name from library-name" or "send index for library-name" to the address `netlib@ornl.gov` or `netlib@research.att.com`. The message will be automatically read and the corresponding subroutine mailed back. Xnetlib (which can be obtained and installed by sending the message "send xnetlib.shar from xnetlib" to `netlib@ornl.gov`) is an X-window interface which lets one point at and click on subroutines, which are then automatically transferred back into the user's directory. There are also index search features to help find the appropriate subroutine. To get started using Netlib, send the one-line message "send index" to `netlib@ornl.gov`.

Here we discuss methods for solving systems of equations such as

$$a_{11}x_1 + a_{12}x_2 + \cdots + a_{1n}x_n = b_1,$$
$$a_{21}x_1 + a_{22}x_2 + \cdots + a_{2n}x_n = b_2,$$
$$\vdots$$
$$a_{m1}x_1 + a_{m2}x_2 + \cdots + a_{mn}x_n = b_m. \qquad (8.99)$$

In these equations a_{ij} and b_i form the set of known quantities, and the x_i must be determined. The solution to these equations can found if they are linearly independent. Numerically, problems can arise due to truncation and roundoff errors that lead to an approximate linear dependence [31]. In this case the set of equations are approximately singular and special methods must be invoked. Much of the complexity of modern algorithms comes from minimizing the effects of such errors. For relatively small sets of nonsingular equations, direct methods in which the solution is obtained after a definite number of operations can work well. However, for very large systems iterative techniques are preferable [32].

A great many algorithms are available for solving (8.99), depending on the structure of the coefficients. For example, if the matrix of coefficients **A** is dense, using Gaussian elimination takes $2n^3/3$ operations; if **A** is also symmetric and positive definite, using the Cholesky algorithm takes a factor of two fewer operations. If **A** is triangular, i.e., either zero above the diagonal or zero below the diagonal, we can solve the above by simple substitution in only n^2 operations. For example, if **A** arises from solving certain elliptic partial differential equations, such as Poisson's equation, then **Ax=b** can be solved using multigrid methods in only n operations.

We shall outline below how to solve (8.99) using elementary Gaussian elimination. More advanced methods, such as *conjugate gradient, generalized minimum residuals*, and the *Lanczos method* are treated elsewhere [33].

To solve $\mathbf{Ax=b}$, we first use Gaussian elimination to factor the matrix $\mathbf{A}$ as $\mathbf{PA=LU}$, where $\mathbf{L}$ is lower triangular, $\mathbf{U}$ is upper triangular, and $\mathbf{P}$ is a matrix which permutes the rows of $\mathbf{A}$. Then we solve the triangular system $\mathbf{Ly=Pb}$ and $\mathbf{Ux=y}$. These last two operations are easily performed using standard linear algebra libraries. The factorization $\mathbf{PA=LU}$ takes most of the time. Reordering the rows of $\mathbf{A}$ with $\mathbf{P}$ is called pivoting and is necessary for numerical stability. In the standard partial pivoting scheme, $\mathbf{L}$ has ones on its diagonal and other entries bounded in absolute value by one. The simplest version of Gaussian elimination involves adding multiples of one row of $\mathbf{A}$ to others to zero out subdiagonal entries, and overwriting $\mathbf{A}$ with $\mathbf{L}$ and $\mathbf{U}$.

We first describe the decomposition of $\mathbf{PA}$ into a product of upper and lower triangular matrices,

$$\mathbf{A}' = \mathbf{LU}, \qquad (8.100)$$

where the matrix $\mathbf{A}'$ is defined by $\mathbf{A}'=\mathbf{PA}$. A very nice algorithm for pivoting is given in Ref. [3] and will not be discussed further. Writing out the indices,

$$A'_{ij} = \sum_{k=1}^{\min(i,j)} L_{ik}U_{kj}. \qquad (8.101)$$

We shall make the choice

$$L_{ii} = 1. \qquad (8.102)$$

These equations have the remarkable property that the elements A'_{ij} of each row can be scanned in turn, writing L_{ij} and U_{ij} into the locations A'_{ij} as we go. At each position (i,j), only the current A'_{ij} and already-calculated values of $L_{i'j'}$ and $U_{i'j'}$ are required. To see how this works, consider the first few rows. If $i=1$,

$$A'_{1j} = U_{1j}, \qquad (8.103)$$

defining the first row of $\mathbf{L}$ and $\mathbf{U}$. The U_{1j} are written over the A'_{1j}, which are no longer needed. If $i=2$,

$$\begin{aligned} A'_{21} &= L_{21}U_{11}, && j=1 \\ A'_{2j} &= L_{21}U_{1j} + U_{2j}, && j \geq 2. \end{aligned} \qquad (8.104)$$

The first line gives L_{21}, and the second U_{2j}, in terms of existing elements of $\mathbf{L}$ and $\mathbf{U}$. The U_{2j} and L_{21} are written over the A'_{2j}. (Remember that $L_{ii} = 1$ by definition.) If $i=3$,

$$\begin{aligned} A'_{31} &= L_{31}U_{11}, && j=1 \\ A'_{32} &= L_{31}U_{12} + L_{32}U_{22}, && j=2 \\ A'_{3j} &= L_{31}U_{1j} + L_{32}U_{2j} + U_{3j}, && j \geq 3, \end{aligned} \qquad (8.105)$$

yielding in turn L_{31}, L_{32}, and U_{3j}, which are written over A'_{3j}.

The algorithm should now be clear. At the *ith* row

$$\begin{aligned} L_{ij} &= U_{jj}^{-1}\left(A'_{ij} - \sum_{k=1}^{j-1} L_{ik}U_{kj}\right), && j \leq i-1 \\[2mm] U_{ij} &= A'_{ij} - \sum_{k=1}^{i-1} L_{ik}U_{kj}, && j \geq i \end{aligned} \qquad (8.106)$$

We observe from the first line of these equations that the algorithm may run into numerical inaccuracies if any U_{jj} becomes very small. Now $U_{11} = A'_{11}$, while in general $U_{ii} = A'_{ii} - \cdots$. Thus the absolute values of the U_{ii} are maximized if the rows are rearranged so that the absolutely largest elements of $\mathbf{A}'$ in each column lie on the diagonal. A little thought shows that the solutions are unchanged by permuting the rows (same equations, different order).

The *LU decomposition* can now be used to solve the system. This relies on the fact that the inversion of a triangular matrix is a simple process of back substitution. We replace (8.99) by two systems of equations. Written out in full, the equations for a typical column of $\mathbf{y}$ look like

$$\begin{aligned} L_{11}y_1 &= b'_1, \\ L_{21}y_1 + L_{22}y_2 &= b'_2, \\ L_{31}y_1 + L_{32}y_2 + L_{33}y_3 &= b'_3, \\ \vdots\ \ \vdots\, , \end{aligned} \qquad (8.107)$$

where the vector $\mathbf{b}'$ is $\mathbf{p}'=\mathbf{Pb}$. Thus from successive rows we obtain $y_1, y_2, y_3, \ldots$ in turn

$$\begin{aligned} U_{11}x_1 &= y_1, \\ U_{12}x_1 + U_{22}x_2 &= y_2, \\ U_{13}x_1 + U_{23}x_2 + U_{33}x_3 &= y_3, \\ \vdots\ \ \vdots\, , \end{aligned} \qquad (8.108)$$

and from successive rows of the latter we obtain $x_1, x_2, x_3, \ldots$ in turn.

Library software also exists for evaluating all the error bounds for dense and band matrices (see discussion of *Netlib* in Sect. 8.4). Gaussian elimination with pivoting is almost always numerically stable, so the error bound one expects from solving these equations is of the order of $n\epsilon$, where ϵ is related to the condition number of the matrix $\mathbf{A}$. A good discussion of errors and conditioning is given in Ref. [3].

8.5 MONTE CARLO METHODS

Owing to the continuing rapid development of computational facilities and the ever-increasing desire to perform *ab initio* calcalutions, the use of Monte Carlo methods is becoming widespread as a means to evaluate previously intractable multidimensional integrals and to enable complex modeling and simulation. For example, a wide range of applications broadly classified as *Quantum Monte Carlo* have been used to compute, for example, the ground state eigenfunctions of simple molecules. Also, guided random walks have found application in the computation of Green functions, and variables chosen randomly, subject to particular constaints, have been used to mimic the electronic distribution of atoms. The latter application, used in the *classical trajectory Monte Carlo technique* described in Chap. 56, allows the statistical quasiquantal representation of ion-atom collisions.

Here we summarize the basic tools needed in these methods, and how they may be used to produce specific distributions and make tractable the evaluation of multidimensional integrals with complicated boundaries. Detailed descriptions of these methods can be found in References [3, 8, 34].

8.5.1 Random Numbers

An essential ingredient of any Monte Carlo procedure is the availability of a computer-generated sequence of random numbers which is not periodic and is free of other significant statistical correlations. Often such numbers are termed *pseudorandom* or *quasirandom*, in distinction to truly random physical processes. While the quality of random number generators supplied with computers has greatly improved over time, it is important to be aware of the potential dangers which can be present. For example, many systems are supplied with a random number generator based on the *linear congruential method*. Typically a sequence of integers $n_1, n_2, n_3, \ldots$ is first produced between 0 and $N - 1$ by using the recurrence relation

$$n_{i+1} = (an_i + b) \bmod N, \qquad 0 \le i < N - 1 \quad (8.109)$$

where a, b, N and the seed value n_0 are positive integers. Real numbers between 0 and (strictly) 1 are then obtained by dividing by N. The period of this sequence is at most N, and depends on the judicious choice of the constants, with N being limited by the wordsize of the computer. A user who is unsure that the character of the random numbers generated on a particular computer platform is proper can perform additional randomizing shuffles or use a portable random number generator, both procedures being described in detail by Knuth [5] and Press *et al.* [3], for example.

8.5.2 Distributions of Random Numbers

Most distributions of random numbers begin with sequences generated uniformly between a lower and an upper limit, and are therefore called *uniform deviates*. However, it is often useful to draw the random numbers from other distributions, such as the Gaussian, Poisson, exponential, gamma, or binomial distributions. These are particularly useful in modeling data or supplying input for an event generator or simulator. In addition, as described below, choosing the random numbers according to some weighting function can signficantly improve the efficiency of integration schemes based on Monte Carlo sampling.

Perhaps the most direct way to produce the required distribution is the *transformation method*. If we have a sequence of uniform deviates x on $(0, 1)$ and wish to find a new sequence y which is distributed with probability given by some function $f(y)$, it can be shown that the required transformation is given by

$$y(x) = \left[\int_0^y f(y) dy \right]^{-1}. \quad (8.110)$$

Evidently, the indefinite integral must be both known and invertible, either analytically or numerically. Since this is seldom the case for distributions of interest, other less direct methods are most often applied. However, even these other methods often rely on the transformation method as one "stage" of the procedure. The transformation method may also be generalized to more than one dimension [3].

A more widely applicable approach is the *rejection method*, also known as *von Neumann rejection*. In this case, if one wishes to find a sequence y distributed according to $f(y)$, first choose another function $\tilde{f}(y)$, called the comparison function, which is everywhere greater than $f(y)$ on the desired interval. In addition, a way must exist to generate y according to the comparison function, such as use of the transformation method. Thus, the comparison function must be simpler or better known than the distribution to be found. One simple choice is a constant function which is larger than the maximum value of $f(y)$, but choices which are "closer" to $f(y)$ will be much more efficient.

To proceed, y is generated uniformly according to $\tilde{f}(y)$ and another deviate x is chosen uniformly on $(0, 1)$. One then simply rejects or accepts y depending on whether x is greater than or less than the ratio $f(y)/\tilde{f}(y)$, respectively. The fraction of trial numbers accepted simply depends on the ratio of the area under the desired function to that under the comparison function. Clearly, the efficiency of this scheme depends on how few of the numbers initially generated must be rejected, and therefore on how closely the comparison function approximates the desired distribution. The Lorentzian distribution, for which the inverse definite integral is known (the tangent function), is a good comparison function for a variety of "bell-shaped" distributions such as the Gaussian (normal), Poisson, and gamma distributions.

Especially for distributions which are functions of more than one variable and possess complicated boundaries, the rejection method is impractical and the transformation method simply inapplicable. In the 1950's, a method to generate distributions for such situations was developed and applied in the study of statistical mechanics where multidimensional integrals (e.g., the partition function) must often be solved numerically, and is known as the *Metropolis algorithm*. This procedure, or its variants, has more recently been adopted to aid in the computation of eigenfunctions of complicated Hamiltonians and scattering operators. In essence, the Metropolis method generates a random walk through the space of the dependent variables, and in the limit of a large number of steps in the walk, the points visited approximate the desired distribution.

In its simplest form, the Metropolis method generates this distribution of points by stepping through this space, most frequently taking a step "downhill" but sometimes taking a step "uphill". That is, given a set of coordinates $\mathbf{q}$ and a desired distribution function $f(\mathbf{q})$, a trial step is taken from the *ith* configuration $\mathbf{q}_i$ to the next, depending on whether the ratio $f(\mathbf{q}_i+1)/f(\mathbf{q}_i)$ is greater or less than one. If the ratio is greater than one, the step is accepted, but if it is less than one, the step is accepted with a probability given by the ratio.

8.5.3 Monte Carlo Integration

The basic idea of Monte Carlo integration is that if a large number of points is generated uniformly randomly in some n-dimensional space, the number falling inside a given region is proportional to the volume, or definite integral, of the function defining that region. Though this idea is as true in one dimension as it is in n, unless there is a large number ("large" could be as little as three) of dimensions or the boundaries are quite complicated, the numerical quadrature schemes described previously are more accurate and efficient. However, since the Monte Carlo approach is based on just sampling the function at representative points rather than evaluating the function at a large number of finely spaced quadrature points, its advantage for very large problems is apparent.

For simplicity, consider the Monte Carlo method for integrating a function of only one variable; the generalization to n dimensions being straightforward. If we generate N random points uniformly on (a, b), then in the limit of large N the integral is

$$\int_a^b f(x)dx \approx \frac{1}{N}\langle f(x)\rangle \pm \sqrt{\frac{\langle f^2(x)\rangle - \langle f(x)\rangle^2}{N}}, \quad (8.111)$$

where

$$\langle f(x)\rangle \equiv \frac{1}{N}\sum_{i=1}^{N} f(x_i) \quad (8.112)$$

is the arithmetic mean. The probable error given is appropriately a statistical one rather than a rigorous error bound and is the one standard error limit. From this one can see that the error decreases as only $N^{1/2}$, more slowly than the rate of decrease for the quadrature schemes based on interpolation. Also, the accuracy is greater for relatively smooth functions, since the Monte Carlo generation of points is unlikely to sample narrowly peaked features of the integrand well. To estimate the integral of a multidimensional function with complicated boundaries, simply find an enclosing volume and generate points uniformly randomly within it. Keeping the enclosing volume as close as possible to the volume of interest miminizes the number of points which fall outside, and therefore increases the efficiency of the procedure.

The Monte Carlo integral is related to techniques for generating random numbers according to prescribed distributions described in Sect. 8.5.2 If we consider a normalized distribution $w(x)$, known as the *weight function*, then with the change of variables defined by

$$y(x) = \int_a^x w(x')dx', \quad (8.113)$$

the Monte Carlo estimate of the integral becomes

$$\int_a^b f(x)dx \approx \frac{1}{N}\left\langle \frac{f(x(y))}{w(x(y))}\right\rangle, \quad (8.114)$$

assuming that the transformation is invertible. Choosing $w(x)$ to behave approximately as $f(x)$ allows a more efficient generation of points within the boundaries of the integrand. This occurs since the uniform distribution of points y results in values of x distributed according to w and therefore "close" to f. This procedure, generally termed the *reduction of variance* of the Monte Carlo integration, improves the efficiency of the procedure to the extent that the transformed function f/w can be made smooth, and that the sampled region is as small as possible but still contains the volume to be estimated.

REFERENCES

1. J. Stoer and R. Bulirsch, *Introduction to Numerical Analysis* (Springer-Verlag, New York, 1980).
2. R. L. Burden, J. D. Faires, and A. C. Reynolds, *Numerical Analysis* (Prindle, Weber, and Schmidt, Boston, 1981).

3. W. H. Press, B. P. Flannery, S. A. Teukolsky, and W. T. Vetterling, *Numerical Recipes, The Art of Scientific Computing* (Cambridge University Press, Cambridge, 1992).

4. *Handbook of Mathematical Functions, Applied Mathematics Series*, edited by M. Abramowitz and I. A. Stegun (National Bureau of Standards, Washington, 1664; reprinted by Dover, New York 1968), Vol. 55.

5. D. E. Knuth, *The Art of Computer Programming* (Addison-Wesley, Reading, 1981), Vol. 2.

6. G. Arfken, *Mathematical Methods for Physicists* (Academic Press, Orlando, 1985).

7. H. Jeffreys and B. S. Jeffreys, *Methods of Mathematical Physics* (Cambridge University Press, Cambridge, 1966).

8. S. E. Koonin and D. C. Meredith, *Computational Physics* (Addison-Wesley, Redwood City, 1990).

9. P. M. Morse and H. Feshbach, *Methods of Theoretical Physics* (McGraw-Hill, New York, 1953).

10. R. Courant and D. Hilbert, *Methods of Mathematical Physics* (Interscience, New York, 1953).

11. P. J. Huber, *Robust Statistics* (Wiley, New York, 1981).

12. H. D. Young, *Statistical Treatment of Experimental Data* (McGraw-Hill, New York, 1962).

13. P. R. Bevington, *Data Reduction and Error Analysis for the Physical Sciences* (McGraw-Hill, New York, 1969).

14. D. C. Champeney, *Fourier Transforms and Their Physical Applications* (Academic Press, New York, 1973).

15. R. W. Hamming, *Numerical Methods for Scientists and Engineers* (McGraw-Hill, New York, 1973).

16. D. F. Elliott and K. R. Rao, *Fast Transforms: Algorithms, Analyses, Applications* (Academic Press, New York, 1982).

17. *The Chemical Rubber Company Mathematical Handbook* (CRC Press, Boca Raton, various editions).

18. J. Lambert, *Numerical Methods for Ordinary Differential Equations* (Wiley, New York, 1991).

19. L. F. Shampine, *Numerical Solution of Ordinary Differential Equations* (Chapman and Hall, New York, 1994).

20. J. Butcher, *The Numerical Analysis of Ordinary Differential Equations: Runge-Kutta and General Linear Methods* (Wiley, New York, 1987).

21. G. Hall and J. Watt, *Modern Numerical Methods for Ordinary Differential Equations* (Clarendon Press, Oxford, 1976).

22. *A Survey of Numerical Methods for Partial Differential Equations*, edited by I. Gladwell and R. Wait (Clarendon, Oxford, 1979).

23. K. Rektorys, *Variational Methods in Mathematics, Science, and Engineering* (D. Reidel Pubishing Company, Boston, 1980), 2nd ed.

24. D. Cook, D. S. Malkus, M. E. Plesha, *Concepts and Applications of Finite Element Analysis* (Wiley, New York, 1989), 3rd ed.

25. G. Nurnberger, *Approximation by Spline Functions* (Springer-Verlag, Berlin, 1989).

26. C. DeBoor, *Practical Guide to Splines* (Springer-Verlag, New York, 1978), and references therein.

27. C. J. Joachain, *Quantum Collision Theory* (Elsevier Science Publishers, New York, 1983).

28. L. S. Rodberg and R. M. Thaler, *Introduction to the Quantum Theory of Scattering* (Academic Press, New York, 1967).

29. M. L. Goldberger and K. M. Watson, *Collision Theory* (Wiley, New York, 1964).

30. P. G. Ciarlet, *Introduction to Numerical Linear Algebra and Optimisation* (Cambridge University Press, Cambridge, 1989).

31. G. Golub and C. Van Loan, *Matrix Computations* (Johns Hopkins University Press, Baltimore, 1989), 2nd ed.

32. W. Hackbusch, *Iterative Solution of Large Sparse Systems of Equations* (Springer-Verlag, New York, 1994).

33. A. George and J. Liu, *Computer Solution of Large Sparse Positive Definite Systems* (Prentice-Hall, Englewood Cliffs, 1981).

34. M. H. Kalos and P. A. Whitlock, *The Basics of Monte Carlo Methods* (Wiley, New York, 1986).

9

Hydrogenic Wave Functions

Robert Nyden Hill

Department of Physics and Astronomy, University of Delaware, Newark, Delaware 19716

9.1 INTRODUCTION

This chapter summarizes the solutions of the one–electron nonrelativistic Schrödinger equation, and the one–electron relativistic Dirac equation, for the Coulomb potential. The standard notations and conventions used in the mathematics literature for special functions have been chosen in preference to the notations customarily used in the physics literature whenever there is a conflict. This has been done to facilitate the use of standard reference works such as Abramowitz and Stegun [1], the Bateman project [2, 3], Gradshteyn and Ryzhik [4], Jahnke and Emde [5], Luke [6, 7], Magnus, Oberhettinger, and Soni [8], Olver [9], and Szego [10]. The section on special functions contains many of the formulas which are needed to check the results quoted in the previous sections, together with a number of other useful formulas. It includes a brief introduction to asymptotic methods. A reference to the numerical evaluation of special functions is given.

9.2 SCHRÖDINGER EQUATION

The nonrelativistic Schrödinger equation for a hydrogenic ion of nuclear charge Z in atomic units is

$$\left(-\tfrac{1}{2}\nabla^2 - \frac{Z}{r}\right)\psi\left(\mathbf{r}\right) = E\psi\left(\mathbf{r}\right) . \tag{9.1}$$

9.2.1 Spherical Coordinates

The separable solutions of Eq. (9.1) in spherical coordinates are

$$\psi\left(\mathbf{r}\right) = Y_{\ell m}\left(\theta, \phi\right) R_\ell\left(r\right) , \tag{9.2}$$

where $Y_{\ell m}\left(\theta, \phi\right)$ is a spherical harmonic as defined by Edmonds [11] and $R_\ell\left(r\right)$ is a solution of the radial equation

$$\left\{-\tfrac{1}{2}\left[\frac{d^2}{dr^2}+\frac{2}{r}\frac{d}{dr}-\frac{\ell(\ell+1)}{r^2}\right]-\frac{Z}{r}\right\}R_\ell(r)=ER_\ell(r).$$
(9.3)

The general solution to Eq. (9.3) is

$$R_\ell(r)=r^\ell\exp(ikr)\left[A\,_1F_1(a;c;z)+BU(a,c,z)\right],$$
(9.4)

where $_1F_1$ and U are the regular and irregular solutions of the confluent hypergeometric equation defined in Eqs. (9.130), and (9.131) below, and

$$k=\sqrt{2E},$$
(9.5)
$$a=\ell+1-ik^{-1}Z,$$
(9.6)
$$c=2\ell+2,$$
(9.7)
$$z=-2ikr.$$
(9.8)

A and B are arbitrary constants. The solution for Eq. (9.4) has an $r^{-\ell-1}$ singularity at $r=0$ unless $B=0$ or a is a non-positive integer. The leading term for small r is proportional to r^ℓ when $B=0$ and/or a is a non-positive integer. The large r behavior of the solution for Eq. (9.4) follows from Eqs. (9.134), (9.135), and (9.164) below. Bound state solutions, with energy

$$E=-\tfrac{1}{2}Z^2n^{-2}$$
(9.9)

are obtained when $a=-n+\ell+1$ where $n>\ell$ is the principal quantum number. The properly normalized bound state solutions are

$$R_{n,\ell}(r)=\frac{2Z}{n^2}\sqrt{\frac{Z(n-\ell-1)!}{(n+\ell)!}}\left(\frac{2Zr}{n}\right)^\ell$$
$$\times\exp(-Zr/n)L_{n-\ell-1}^{(2\ell+1)}(2Zr/n),$$
(9.10)

where $L_{n-\ell-1}^{(2\ell+1)}$ is the generalized Laguerre polynomial defined in Eq. (9.187). The relation in Eq. (9.188) shows that $_1F_1$ and U are linearly dependent in this case, so that Eq. (9.4) is no longer the general solution of Eq. (9.3). A linearly independent solution for this case can be obtained by replacing the $L_{n-\ell-1}^{(2\ell+1)}(2Zr/n)$ in Eq. (9.10) by the second (irregular) solution $M_{n-\ell-1}^{(2\ell+1)}(2Zr/n)$ of the Laguerre equation (see Eqs. (9.194), (9.196), and (9.197) below). The first three $R_{n,\ell}$ are

$$R_{1,0}(r)=2Z^{3/2}\exp(-Zr),$$
(9.11)
$$R_{2,0}(r)=(\tfrac{1}{2}Z)^{3/2}(2-Zr)\exp(-\tfrac{1}{2}Zr),$$
(9.12)
$$R_{2,1}(r)=(\tfrac{1}{2}Z)^{3/2}(\tfrac{1}{3})^{1/2}Zr\exp(-\tfrac{1}{2}Zr).$$
(9.13)

Additional explicit expressions, together with graphs of some of them, can be found in Pauling and Wilson [12]. The $R_{n,\ell}$ can be expanded in powers of $1/n$ [13]:

$$R_{n,\ell}(r)=-\left[\frac{2Z^2(n+\ell)!}{(n-\ell-1)!\,n^{2\ell+4}}\right]^{1/2}$$
$$\times r^{-1/2}\sum_{k=0}^{\infty}g_k^{(\ell)}\left[(8Zr)^{1/2}\right]n^{-2k},$$
(9.14)

where the functions $g_k^{(\ell)}(z)$ are finite linear combinations of Bessel functions:

$$g_k^{(\ell)}(z)=z^{3k}\sum_{m=0}^{k}a_{k,m}^{(\ell)}J_{2\ell+2m+k+1}(z).$$
(9.15)

The coefficients $a_{k,m}^{(\ell)}$ in Eq. (9.15) are calculated recursively from

$$a_{k,m}^{(\ell)}=\frac{(2\ell+2m+k+1)}{32(2k+m)(2\ell+m+2k+1)(2\ell+2m+k-1)}$$
$$\times\left[(2\ell+2m+k-1)a_{k-1,m}^{(\ell)}\right.$$
$$\left.+32(k-m+1)(2\ell+m-k)a_{k,m-1}^{(\ell)}\right],$$
(9.16)

starting with the initial condition

$$a_{0,0}^{(\ell)}=1.$$
(9.17)

The expansion (9.14) converges uniformly in r for r in any bounded region of the complex r plane. However, it converges fast enough so that a few terms give a good description of $R_{n,\ell}$ only if r is small. The square root in Eq. (9.14) has not been expanded in inverse powers of n because it has a branch point at $1/n=1/\ell$ which would reduce the radius of convergence of the expansion to $1/\ell$. In some cases, large n expansions of matrix elements can be obtained by inserting Eq. (9.14) for $R_{n,\ell}$ and integrating term by term; examples can be found in Drake and Hill [13]. An asymptotic expansion in powers of $1/n$, which is valid from r equal to an arbitrary fixed positive number through the turning point at $r=2n^2/Z$ out to $r=\infty$, can be assembled from Eqs. (9.133), (9.166) – (9.181), and (9.188) below.

The $R_{n,\ell}$ are not a complete set because the continuum has been left out. The *Sturmian* functions $\rho_{k,\ell}$, given by

$$\rho_{k,\ell}(\beta;r)=\sqrt{\frac{\beta^3k!}{\Gamma(k+2\ell+3)}}(\beta r)^\ell e^{-\beta r/2}$$
$$\times L_k^{(2\ell+2)}(\beta r),$$
(9.18)

do form a complete orthonormal set. The positive constant β, which is independent of k and ℓ, sets the length scale for the basis set (9.18).

9.2.2 Parabolic Coordinates

The Schrödinger equation (9.1) is separable in parabolic coordinates ξ,η,ϕ, which are related to spherical coordinates r,θ,ϕ via

$$\xi = r + z = r\left[1 + \cos\left(\theta\right)\right], \qquad (9.19)$$

$$\eta = r - z = r\left[1 - \cos\left(\theta\right)\right], \qquad (9.20)$$

$$\phi = \phi. \qquad (9.21)$$

This separability in a second coordinate system is related to the existence of a "hidden" $O(4)$ symmetry, which is also responsible for the degeneracy of the bound states [14]. The solutions in parabolic coordinates are particularly convenient for derivations of the Stark effect and the Rutherford scattering cross section. The separable solutions of Eq. (9.1) in parabolic coordinates are

$$\psi\left(\mathbf{r}\right) = \exp\left(im\phi\right) N\left(\eta\right) \Xi\left(\xi\right), \qquad (9.22)$$

where

$$N\left(\eta\right) = \eta^{|m|/2} \exp\left(\tfrac{1}{2}ik_1\eta\right)\left[A\,_1F_1\left(a_1; c; -ik_1\eta\right) \\ + BU\left(a_1, c, -ik_1\eta\right)\right], \qquad (9.23)$$

$$\Xi\left(\xi\right) = \xi^{|m|/2} \exp\left(\tfrac{1}{2}ik_2\xi\right)\left[C\,_1F_1\left(a_2; c; -ik_2\xi\right) \\ + DU\left(a_2, c, -ik_2\xi\right)\right], \qquad (9.24)$$

with $_1F_1$ and U defined in Eqs. (9.130), (9.131) below, and

$$k_1 = \pm k_2 = \pm\sqrt{2E}, \qquad (9.25)$$

$$a_1 = \tfrac{1}{2}\left(|m| + 1\right) - ik_1^{-1}\mu, \qquad (9.26)$$

$$a_2 = \tfrac{1}{2}\left(|m| + 1\right) - ik_2^{-1}\left(Z - \mu\right), \qquad (9.27)$$

$$c = |m| + 1. \qquad (9.28)$$

A, B, C, and D are arbitrary constants. μ is the separation constant. An important special case is the well-known *Coulomb function*

$$\psi_C\left(\mathbf{r}\right) = \Gamma\left(1 - ik^{-1}Z\right)\exp\left(\tfrac{1}{2}\pi k^{-1}Z + i\mathbf{k}\cdot\mathbf{r}\right) \\ \times\,_1F_1\left[ik^{-1}Z; 1; i\left(kr - \mathbf{k}\cdot\mathbf{r}\right)\right], \qquad (9.29)$$

which is obtained by orienting the z-axis in the $\mathbf{k}$ direction and taking $m = 0$, $-k_1 = k_2 = |\mathbf{k}|$, $\mu = Z + \tfrac{1}{2}i|\mathbf{k}|$. ψ_C is normalized to unit incoming flux [see (9.34) below]. In applications, Z is often replaced by $-Z_1 Z_2$, so that the Coulomb potential in Eq. (9.1) becomes $+Z_1 Z_2/r$. Equation (9.232), the addition theorem for the spherical harmonics [11] (p. 63 Eq. 4.6.6), and the $\lambda = c = 1$ special case of Eq. (9.163) below can be used to expand ψ_C into an infinite sum of solutions of the form (9.2):

$$\psi_C\left(\mathbf{r}\right) = 4\pi\sum_{\ell=0}^{\infty}\sum_{m=-\ell}^{\ell}\frac{\Gamma\left(\ell + 1 - ik^{-1}Z\right)}{\left(2\ell + 1\right)!} \\ \times\left(-2ik\right)^{\ell} e^{\pi k^{-1}Z/2} Y_{\ell m}^{*}\left(\theta_k, \phi_k\right) Y_{\ell m}\left(\theta, \phi\right) \\ \times r^{\ell} e^{ikr}\,_1F_1\left(\ell + 1 - ik^{-1}Z; 2\ell + 2; -2ikr\right), \qquad (9.30)$$

where k, θ_k, and ϕ_k are the spherical coordinates of $\mathbf{k}$. ψ_C can be split into an incoming plane wave and an outgoing spherical wave with the aid of Eq. (9.134) below:

$$\psi_C\left(\mathbf{r}\right) = \psi_{\text{in}}\left(\mathbf{r}\right) + \psi_{\text{out}}\left(\mathbf{r}\right), \qquad (9.31)$$

where

$$\psi_{\text{in}}\left(\mathbf{r}\right) = \exp\left(i\mathbf{k}\cdot\mathbf{r} - \tfrac{1}{2}\pi k^{-1}Z\right) \\ \times U\left[ik^{-1}Z; 1; i\left(kr - \mathbf{k}\cdot\mathbf{r}\right)\right], \qquad (9.32)$$

$$\psi_{\text{out}}\left(\mathbf{r}\right) = -\frac{\Gamma\left(1 - ik^{-1}Z\right)}{\Gamma\left(ik^{-1}Z\right)}\exp\left(ikr - \tfrac{1}{2}\pi k^{-1}Z\right) \\ \times U\left[1 - ik^{-1}Z; 1; -i\left(kr - \mathbf{k}\cdot\mathbf{r}\right)\right]. \qquad (9.33)$$

The functions ψ_{in} and ψ_{out} can be expanded for $kr - \mathbf{k}\cdot\mathbf{r}$ large with the aid of Eq. (9.164). The result is

$$\psi_{\text{in}}\left(\mathbf{r}\right) \sim \exp\left[i\mathbf{k}\cdot\mathbf{r} - ik^{-1}Z\ln\left(kr - \mathbf{k}\cdot\mathbf{r}\right)\right] \\ \times\sum_{n=0}^{\infty}\frac{\left(-i\right)^n}{n!}\left[\frac{\Gamma\left(ik^{-1}Z + n\right)}{\Gamma\left(ik^{-1}Z\right)}\right]^2\left(kr - \mathbf{k}\cdot\mathbf{r}\right)^{-n}, \qquad (9.34)$$

$$\psi_{\text{out}}\left(\mathbf{r}\right) \sim -\frac{i\Gamma\left(1 - ik^{-1}Z\right)}{\Gamma\left(ik^{-1}Z\right)\left(kr - \mathbf{k}\cdot\mathbf{r}\right)} \\ \times\exp\left[ikr - ik^{-1}Z\ln\left(kr - \mathbf{k}\cdot\mathbf{r}\right)\right] \\ \times\sum_{n=0}^{\infty}\frac{i^n}{n!}\left[\frac{\Gamma\left(1 - ik^{-1}Z + n\right)}{\Gamma\left(1 - ik^{-1}Z\right)}\right]^2\left(kr - \mathbf{k}\cdot\mathbf{r}\right)^{-n}. \qquad (9.35)$$

Because Eq. (9.1) is an elliptic partial differential equation, its solutions must be analytic functions of the Cartesian coordinates (except at $r = 0$, where the solutions have cusps). The $n = 0$ special case of Eq. (9.138) shows that ψ_{in} and ψ_{out} are logarithmically singular at $\mathbf{k}\cdot\mathbf{r} = kr$. Thus ψ_{in} and ψ_{out} are not solutions to Eq. (9.1) at $\mathbf{k}\cdot\mathbf{r} = kr$. The logarithmic singularity cancels when ψ_{in} and ψ_{out} are added to form ψ_C, which *is* a solution to Eq. (9.1).

Bound state solutions, with energy

$$E = -\tfrac{1}{2}Z^2\left(n_1 + n_2 + |m| + 1\right)^{-2}, \qquad (9.36)$$

are obtained when $a_1 = -n_1$ and $a_2 = -n_2$ where n_1 and n_2 are non-negative integers. The properly normalized bound state solutions, which can be put into one–one correspondence with the bound state solutions in spherical coordinates, are

$$\psi_{n_1, n_2, m}\left(\eta, \xi, \phi\right) \\ = \sqrt{\frac{\beta^{2|m|+4}n_1! n_2!}{2\pi Z\left(n_1 + |m|\right)!\left(n_2 + |m|\right)!}}\exp\left[im\phi - \tfrac{1}{2}\beta\left(\eta + \xi\right)\right] \\ \times\left(\eta\xi\right)^{|m|/2} L_{n_1}^{(|m|)}\left(\beta\eta\right) L_{n_2}^{(|m|)}\left(\beta\xi\right), \qquad (9.37)$$

where

$$\beta = Z\left(n_1 + n_2 + |m| + 1\right)^{-1}. \qquad (9.38)$$

9.2.3 Momentum Space

The nonrelativistic Schrödinger equation (9.1) becomes the integral equation

$$\tfrac{1}{2}p^2\phi\left(\mathbf{p}\right) - \frac{Z}{2\pi^2}\int\frac{\phi\left(\mathbf{p}'\right)}{\left(\mathbf{p}-\mathbf{p}'\right)^2}\,d^3\mathbf{p}' = E\phi\left(\mathbf{p}\right) \qquad (9.39)$$

in momentum space. Its solutions are related to the solutions in coordinate space via the Fourier transforms

$$\psi\left(\mathbf{r}\right) = \left(2\pi\right)^{-3/2}\int\exp\left(i\mathbf{p}\cdot\mathbf{r}\right)\phi\left(\mathbf{p}\right)\,d^3\mathbf{p}\,, \qquad (9.40)$$

$$\phi\left(\mathbf{p}\right) = \left(2\pi\right)^{-3/2}\int\exp\left(-i\mathbf{p}\cdot\mathbf{r}\right)\psi\left(\mathbf{r}\right)\,d^3\mathbf{r}\,. \qquad (9.41)$$

A trick of Fock's [15] can be used to expose the "hidden" $O(4)$ symmetry of hydrogen and construct the bound state solutions to (9.39). Let p, θ_p, ϕ_p and p', θ'_p, ϕ'_p be the spherical coordinates of $\mathbf{p}$ and $\mathbf{p}'$. Change variables from p, p' to χ, χ' via $p = \sqrt{-2E}\tan\left(\chi/2\right)$ and $p' = \sqrt{-2E}\tan\left(\chi'/2\right)$. This brings (9.39) to the form

$$2\pi^2 Z^{-1}\sqrt{-2E}\left[\sec^4\left(\chi/2\right)\phi\left(\mathbf{p}\right)\right]$$
$$= \int\frac{\left[\sec^4\left(\chi'/2\right)\phi\left(\mathbf{p}'\right)\right]\sin^2\left(\chi'\right)\,d\chi'\sin\left(\theta_p\right)\,d\theta_p\,d\phi_p}{2 - 2\left[\cos\left(\chi\right)\cos\left(\chi'\right) + \sin\left(\chi\right)\sin\left(\chi'\right)\cos\left(\gamma'\right)\right]}\,, \qquad (9.42)$$

where γ' is the angle between $\mathbf{p}$ and $\mathbf{p}'$. Equation (9.42) is solved by introducing spherical coordinates and spherical harmonics in four dimensions via a natural extension of the procedure used in three dimensions. Going to polar coordinates on x and y yields the cylindrical coordinates r_2, ϕ, z; the further step of going to polar coordinates on r_2 and z yields spherical coordinates r_3, θ, ϕ. If there is a fourth coordinate w, spherical coordinates in four dimensions are obtained via the additional step of going to polar coordinates on r_3 and w. The result is

$$x = r_4\sin\left(\chi\right)\sin\left(\theta\right)\cos\left(\phi\right)\,, \qquad (9.43)$$
$$y = r_4\sin\left(\chi\right)\sin\left(\theta\right)\sin\left(\phi\right)\,, \qquad (9.44)$$
$$z = r_4\sin\left(\chi\right)\cos\left(\theta\right)\,, \qquad (9.45)$$
$$w = r_4\cos\left(\chi\right)\,. \qquad (9.46)$$

The volume element, which is easily obtained via the same series of transformations, is

$$dV = r_4^3\,dr_4\,d\Omega_4\,, \qquad (9.47)$$
$$d\Omega_4 = \sin^2\left(\chi\right)\,d\chi\sin\left(\theta\right)\,d\theta\,d\phi\,. \qquad (9.48)$$

The four-dimensional spherical harmonics [2, Vol. 2, Chap. XI] are

$$Y_{n,\ell,m}\left(\chi,\theta,\phi\right) = 2^{\ell+1}\ell!\sqrt{\frac{n\left(n-\ell-1\right)!}{2\pi\left(n+\ell\right)!}}\sin^\ell\left(\chi\right)$$
$$\times C_{n-\ell-1}^{\ell+1}\left[\cos\left(\chi\right)\right]Y_{\ell m}\left(\theta,\phi\right)\,, \qquad (9.49)$$

where $C_{n-\ell-1}^{\ell+1}$ is a Gegenbauer polynomial and $n \ge \ell+1$ is an integer. They have the orthonormality property

$$\int Y_{n,\ell,m}^*\left(\chi,\theta,\phi\right)Y_{n',\ell',m'}\left(\chi,\theta,\phi\right)\,d\Omega_4$$
$$= \delta_{n,n'}\delta_{\ell,\ell'}\delta_{m,m'}\,. \qquad (9.50)$$

Equations (9.229) and (9.230) with $\lambda = 1$, equation (9.231), and the addition theorem for the three dimensional spherical harmonics $Y_{\ell m}$ can be used to show that

$$\{1 - 2\left[\cos\left(\chi\right)\cos\left(\chi'\right)\right.$$
$$\left.+\sin\left(\chi\right)\sin\left(\chi'\right)\cos\left(\gamma'\right)\right]t + t^2\}^{-1}$$
$$= \sum_{n=1}^{\infty}\sum_{\ell=0}^{n-1}\sum_{m=-\ell}^{\ell}\frac{2\pi^2}{n}t^{n-1}Y_{n,\ell,m}\left(\chi,\theta,\phi\right)Y_{n,\ell,m}^*\left(\chi',\theta',\phi'\right)$$

$$(9.51)$$

holds for $|t| < 1$, where γ' is the angle between $\mathbf{p}$ and $\mathbf{p}'$. Multiply both sides of (9.51) by $Y_{n,\ell,m}\left(\chi',\theta',\phi'\right)\,d\Omega'_4$ (where $d\Omega'_4$ is $d\Omega_4$ with χ, θ, ϕ replaced by χ', θ', ϕ') and use the orthogonality relation (9.50). The result can be rearranged to the form

$$2\pi^2 n^{-1}t^{n-1}Y_{n,\ell,m}\left(\chi,\theta,\phi\right) =$$
$$\int\frac{Y_{n,\ell,m}\left(\chi',\theta',\phi'\right)\sin^2\left(\chi'\right)\,d\chi'\sin\left(\theta\right)\,d\theta\,d\phi}{1 - 2\left[\cos\left(\chi\right)\cos\left(\chi'\right) + \sin\left(\chi\right)\sin\left(\chi'\right)\cos\left(\gamma'\right)\right]t + t^2}\,. $$
$$(9.52)$$

Analytic continuation can be used to show that (9.52) is valid for all complex t despite the fact that (9.51) is restricted to $|t| < 1$. Comparing the $t = 1$ case of (9.52) with (9.42) shows that $E = -\tfrac{1}{2}Z^2 n^{-2}$ in agreement with (9.9), and that

$$\phi\left(\mathbf{p}\right) = \left\{\begin{array}{c}\text{normalizing}\\\text{factor}\end{array}\right\}\cos^4\left(\chi/2\right)Y_{n,\ell,m}\left(\chi,\theta,\phi\right)\,. $$
$$(9.53)$$

Transforming from χ back to p brings these to the form

$$\phi\left(\mathbf{p}\right) = Y_{\ell m}\left(\theta_p,\phi_p\right)F_{n,\ell}\left(p\right)\,, \qquad (9.54)$$

where the properly normalized radial functions are

$$F_{n,\ell}\left(p\right) = 2^{2\ell+2}n^2\ell!\sqrt{\frac{2\left(n-\ell-1\right)!}{\pi Z^3\left(n+\ell\right)!}}\left(\frac{np}{Z}\right)^\ell$$
$$\times\frac{Z^{2\ell+4}}{\left(n^2p^2+Z^2\right)^{\ell+2}}C_{n-\ell-1}^{\ell+1}\left(\frac{n^2p^2-Z^2}{n^2p^2+Z^2}\right)\,. $$
$$(9.55)$$

The first three $F_{n,\ell}$ are

$$F_{1,0}\left(p\right) = 4\sqrt{\frac{2}{\pi Z^3}}\frac{Z^4}{\left(p^2+Z^2\right)^2}\,, \qquad (9.56)$$

$$F_{2,0}\left(p\right) = \frac{32}{\sqrt{\pi Z^3}}\frac{Z^4\left(4p^2-Z^2\right)}{\left(4p^2+Z^2\right)^3}\,, \qquad (9.57)$$

$$F_{2,1}\left(p\right) = \frac{128}{\sqrt{3\pi Z^3}}\frac{Z^5 p}{\left(4p^2+Z^2\right)^3}\,. \qquad (9.58)$$

The $F_{n,\ell}$ satisfy the integral equation

$$\tfrac{1}{2}p^2 F_{n,\ell}(p) - \frac{Z}{\pi p}\int_0^\infty Q_\ell\left(\frac{p^2+p'^2}{2pp'}\right) F_{n,\ell}(p')\, p'\, dp'$$
$$= E F_{n,\ell}(p)\,, \tag{9.59}$$

which can be obtained by inserting (9.54) in (9.39). Here Q_ℓ is the Legendre function of the second kind, which is defined in (9.233) below.

9.3 DIRAC EQUATION

The relativistic Dirac equation for a hydrogenic ion of nuclear charge Z can be reduced to dimensionless form by using the Compton wavelength $\hbar/(mc)$ for the length scale and the rest mass energy mc^2 for the energy scale. The result is

$$\left(-i\boldsymbol{\alpha}\cdot\nabla + \beta - \frac{Z\alpha}{r}\right)\psi(\mathbf{r}) = E\psi(\mathbf{r})\,, \tag{9.60}$$

where $\alpha = e^2/(\hbar c)$ is the fine structure constant, and $\boldsymbol{\alpha}$, β are the usual Dirac matrices:

$$\boldsymbol{\alpha} = \begin{pmatrix} 0 & \boldsymbol{\sigma} \\ \boldsymbol{\sigma} & 0 \end{pmatrix}, \quad \beta = \begin{pmatrix} I & 0 \\ 0 & -I \end{pmatrix}. \tag{9.61}$$

Here $\boldsymbol{\sigma}$ is a vector whose components are the two by two Pauli matrices, and I is the two by two identity matrix given by

$$\sigma_x = \begin{pmatrix} 0 & 1 \\ 1 & 0 \end{pmatrix}, \quad \sigma_y = \begin{pmatrix} 0 & -i \\ i & 0 \end{pmatrix},$$
$$\sigma_z = \begin{pmatrix} 1 & 0 \\ 0 & -1 \end{pmatrix}, \quad I = \begin{pmatrix} 1 & 0 \\ 0 & 1 \end{pmatrix}. \tag{9.62}$$

The solutions to Eq. (9.60) in spherical coordinates have the form

$$\psi(\mathbf{r}) = \begin{pmatrix} G(r)\,\chi_\kappa^m(\theta,\phi) \\ iF(r)\,\chi_{-\kappa}^m(\theta,\phi) \end{pmatrix}, \tag{9.63}$$

where, for positive energy states, $G(r)$ is the radial part of the large component and $iF(r)$ is the radial part of the small component. For negative energy states, $G(r)$ is the radial part of the small component and $iF(r)$ is the radial part of the large component. χ is the two component spinor

$$\chi_\kappa^m = \begin{pmatrix} -\dfrac{\kappa}{|\kappa|}\left(\dfrac{\kappa+\frac{1}{2}-m}{2\kappa+1}\right)^{1/2} Y_{|\kappa+\frac{1}{2}|-\frac{1}{2},m-\frac{1}{2}} \\[2ex] \left(\dfrac{\kappa+\frac{1}{2}+m}{2\kappa+1}\right)^{1/2} Y_{|\kappa+\frac{1}{2}|-\frac{1}{2},m+\frac{1}{2}} \end{pmatrix}. \tag{9.64}$$

The relativistic quantum number κ is related to the total angular momentum quantum number j by

$$\kappa = \pm\left(j+\tfrac{1}{2}\right)\,. \tag{9.65}$$

Because j takes on the values $\frac{1}{2}, \frac{3}{2}, \frac{5}{2}, \ldots$, κ is restricted to the values $\pm 1, \pm 2, \pm 3, \ldots$. The spinor χ_κ^m obeys the useful relations

$$\boldsymbol{\sigma}\cdot\hat{\mathbf{r}}\,\chi_\kappa^m = -\chi_{-\kappa}^m\,, \tag{9.66}$$
$$\boldsymbol{\sigma}\cdot\mathbf{L}\,\chi_\kappa^m = -(\kappa+1)\chi_\kappa^m\,, \tag{9.67}$$

where $\hat{\mathbf{r}} = \mathbf{r}/r$ and $\mathbf{L} = \mathbf{r}\times\mathbf{p}$ with $\mathbf{p} = -i\nabla$. Equations (9.66), (9.67), and the identity

$$\boldsymbol{\sigma}\cdot\mathbf{p} = (\boldsymbol{\sigma}\cdot\hat{\mathbf{r}})\left(\hat{\mathbf{r}}\cdot\mathbf{p} + \frac{i\boldsymbol{\sigma}\cdot\mathbf{L}}{r}\right) \tag{9.68}$$

can be used to derive the radial equations, which are

$$\left(\frac{d}{dr} + \frac{1+\kappa}{r}\right)G(r) - \left(1+E+\frac{Z\alpha}{r}\right)F(r) = 0\,, \tag{9.69}$$

$$\left(\frac{d}{dr} + \frac{1-\kappa}{r}\right)F(r) - \left(1-E-\frac{Z\alpha}{r}\right)G(r) = 0\,. \tag{9.70}$$

Equations (9.158), (9.159), (9.161), and (9.162) below can be used to show that the general solution to Eqs. (9.69) and (9.70) is

$$G(r) = r^\gamma \exp(-\lambda r)(1+E)^{1/2}\{A[f_2(r)+f_1(r)] + B[f_4(r)+f_3(r)]\}\,, \tag{9.71}$$

$$F(r) = r^\gamma \exp(-\lambda r)(1-E)^{1/2}\{A[f_2(r)-f_1(r)] + B[f_4(r)-f_3(r)]\}\,, \tag{9.72}$$

where

$$f_1(r) = (Z\alpha\lambda^{-1}-\kappa)\,{}_1F_1(a;c;2\lambda r)\,, \tag{9.73}$$
$$f_2(r) = a\,{}_1F_1(a+1;c;2\lambda r)\,, \tag{9.74}$$
$$f_3(r) = U(a,c,2\lambda r)\,, \tag{9.75}$$
$$f_4(r) = (Z\alpha\lambda^{-1}+\kappa)\,U(a+1,c,2\lambda r)\,, \tag{9.76}$$
$$\lambda = (1+E)^{1/2}(1-E)^{1/2}\,, \tag{9.77}$$
$$\gamma = -1 + (\kappa^2 - Z^2\alpha^2)^{1/2}\,, \tag{9.78}$$
$$a = 1 + \gamma - \lambda^{-1}EZ\alpha\,, \tag{9.79}$$
$$c = 3 + 2\gamma\,. \tag{9.80}$$

A and B are arbitrary constants. Because γ is in general not an integer, the solutions have a branch point at $r = 0$, and become infinite at $r = 0$ when $\kappa = \pm 1$, which makes γ negative. The solutions for $E < -1$ and $E > +1$ are in the continuum, which implies that one of the factors $(1+E)^{1/2}$, $(1-E)^{1/2}$ is real with the other imaginary. Square integrable solutions, with energy

$$E_{n,\kappa} = \frac{Z}{|Z|}\left[1 + \frac{Z^2\alpha^2}{(n+1+\gamma)^2}\right]^{-1/2}, \quad (9.81)$$

are obtained when $a = -n$ where n is a non-negative integer. The properly normalized square integrable solutions are

$$G_{n,\kappa}(r) = C_{n,\kappa}(2\lambda r)^\gamma \exp(-\lambda r)(1+E_{n,\kappa})^{1/2}$$
$$\times \left[g_{n,\kappa}^{(2)}(r) + g_{n,\kappa}^{(1)}(r)\right], \quad (9.82)$$

$$F_{n,\kappa}(r) = C_{n,\kappa}(2\lambda r)^\gamma \exp(-\lambda r)(1-E_{n,\kappa})^{1/2}$$
$$\times \left[g_{n,\kappa}^{(2)}(r) - g_{n,\kappa}^{(1)}(r)\right], \quad (9.83)$$

$$g_{n,\kappa}^{(1)}(r) = \left(Z\alpha\lambda^{-1} - \kappa\right)^{1/2} L_n^{(2+2\gamma)}(2\lambda r), \quad (9.84)$$

$$g_{n,\kappa}^{(2)}(r) = -(n+2+2\gamma)\left(Z\alpha\lambda^{-1} - \kappa\right)^{-1/2}$$
$$\times L_{n-1}^{(2+2\gamma)}(2\lambda r), \quad (9.85)$$

$$C_{n,\kappa} = \sqrt{\frac{2\lambda^4 n!}{Z\alpha\Gamma(n+3+2\gamma)}}. \quad (9.86)$$

When $n = 0$, $|Z\alpha\lambda^{-1}| = |\kappa|$, and the value of κ whose sign is the same as the sign of $Z\alpha\lambda^{-1}$ is not permitted. Also, $L_{-1}^{(2+2\gamma)}(2\lambda r)$ is counted as zero, so that $g_{0,\kappa}^{(2)}(r) = 0$.

The eigenvalues and eigenfunctions for the first four states for $Z > 0$ will now be written out explicitly in terms of the variable $\rho = Z\alpha r$. For the $1S_{1/2}$ ground state, with $n = 0$, $j = \frac{1}{2}$, $\kappa = -1$, the formulae are

$$E_{0,-1} = \sqrt{1 - Z^2\alpha^2}, \quad (9.87)$$

$$G_{0,-1}(r) = \sqrt{\frac{4Z^3\alpha^3(1+E_{0,-1})}{\Gamma(1+2E_{0,-1})}}(2\rho)^{E_{0,-1}-1}e^{-\rho}, \quad (9.88)$$

$$F_{0,-1}(r) = -\sqrt{\frac{4Z^3\alpha^3(1-E_{0,-1})}{\Gamma(1+2E_{0,-1})}}(2\rho)^{E_{0,-1}-1}e^{-\rho}. \quad (9.89)$$

The formulae for the $2S_{1/2}$ excited state, with $n = 1$, $j = \frac{1}{2}$, $\kappa = -1$, and for the $2P_{1/2}$ excited state, with $n = 1$, $j = \frac{1}{2}$, $\kappa = 1$, can be written together. They are

$$E_{1,\kappa} = \left[\frac{1}{2} + \frac{1}{2}\sqrt{1-Z^2\alpha^2}\right]^{1/2}, \quad (9.90)$$

$$G_{1,\kappa}(r) = \sqrt{\frac{Z^3\alpha^3(2E_{1,\kappa}-\kappa)(1+E_{1,\kappa})}{2E_{1,\kappa}^2\Gamma(4E_{1,\kappa}+1)}}\rho_1^{2E_{1,\kappa}^2-2}e^{-\rho_1/2}$$
$$\times [(2E_{1,\kappa}-\kappa-1)(2E_{1,\kappa}+\kappa)-\rho_1], \quad (9.91)$$

$$F_{1,\kappa}(r) = -\sqrt{\frac{Z^3\alpha^3(2E_{1,\kappa}-\kappa)(1-E_{1,\kappa})}{2E_{1,\kappa}^2\Gamma(4E_{1,\kappa}+1)}}\rho_1^{2E_{1,\kappa}^2-2}$$
$$\times e^{-\rho_1/2}[(2E_{1,\kappa}-\kappa+1)(2E_{1,\kappa}+\kappa)-\rho_1], \quad (9.92)$$

where $\rho_1 = \rho/E_{1,\kappa}$. For the $2P_{3/2}$ excited state, with $n = 0$, $j = \frac{1}{2}$, $\kappa = -2$, the formulae are

$$E_{0,-2} = \sqrt{1 - \frac{1}{4}Z^2\alpha^2}, \quad (9.93)$$

$$G_{0,-2}(r) = \sqrt{\frac{Z^3\alpha^3(1+E_{0,-2})}{2\Gamma(1+4E_{0,-2})}}\rho^{2E_{0,-2}-1}e^{-\rho/2}, \quad (9.94)$$

$$F_{0,-2}(r) = -\sqrt{\frac{Z^3\alpha^3(1-E_{0,-2})}{2\Gamma(1+4E_{0,-2})}}\rho^{2E_{0,-2}-1}e^{-\rho/2}. \quad (9.95)$$

9.4 THE COULOMB GREEN'S FUNCTION

The abstract Green's operator for a Hamiltonian H is the inverse $G(E) = (H-E)^{-1}$. It is used to write the solution to $(H-E)|\xi\rangle = |\eta\rangle$ in the form $|\xi\rangle = G|\eta\rangle$. It has the spectral representation

$$G(E) = \sum_j \frac{1}{E_j - E}|e_j\rangle\langle e_j|. \quad (9.96)$$

The sum over j in (9.96) runs over *all* of the spectrum of H, including the continuum. For the bound state part of the spectrum, the numbers E_j and vectors $|e_j\rangle$ are the eigenvalues and eigenvectors of H. For the continuous spectrum, $|e_j\rangle\langle e_j|$ is a projection valued measure [16]. The representation (9.96) shows that $G(E)$ has first order poles at the eigenvalues. The reduced Green's operator (also known as the generalized Green's operator), which is the ordinary Green's operator with the singular terms subtracted out, remains finite when E is at an eigenvalue. It can be calulated from

$$G^{(\text{red})}(E_k) = \lim_{E\to E_k}\left\{\frac{\partial}{\partial E}[(E-E_k)G(E)]\right\}. \quad (9.97)$$

The coordinate and momentum space representatives of the abstract Green's operator are the Green's functions. The nonrelativistic Coulomb Green's function has been discussed by Hostler and Schwinger [17,18]. A unified treatment of the Coulomb Green's functions for the Schrödinger and Dirac equations has been given by Swainson and Drake [19]. Reduced Green's functions are discussed in the third of the Swainson–Drake papers, and in the paper of Hill and Huxtable [20].

9.4.1 The Green's Function for the Schrödinger Equation

The Green's function $G^{(S)}$ for the Schrödinger equation (9.98) is a solution of

$$\left(-\tfrac{1}{2}\nabla^2 - \frac{Z}{r} - E\right) G^{(S)}(\mathbf{r}, \mathbf{r}'; E) = \delta(\mathbf{r} - \mathbf{r}') \ . \quad (9.98)$$

An explicit closed form expression for $G^{(S)}$ is

$$G^{(S)}(\mathbf{r}, \mathbf{r}'; E) = \frac{\Gamma(1-\nu)}{2\pi|\mathbf{r} - \mathbf{r}'|} \left[W_{\nu,\frac{1}{2}}(z_2) \frac{\partial}{\partial z_1} M_{\nu,\frac{1}{2}}(z_1) \right.$$
$$\left. - M_{\nu,\frac{1}{2}}(z_1) \frac{\partial}{\partial z_2} W_{\nu,\frac{1}{2}}(z_2) \right] , \quad (9.99)$$

where $M_{\nu,\frac{1}{2}}$ and $W_{\nu,\frac{1}{2}}$ are the Whittaker functions defined in (9.132) and (9.133) below, and

$$\nu = Z(-2E)^{-1/2} , \quad (9.100)$$
$$z_1 = (-2E)^{1/2}(r + r' - |\mathbf{r} - \mathbf{r}'|) , \quad (9.101)$$
$$z_2 = (-2E)^{1/2}(r + r' + |\mathbf{r} - \mathbf{r}'|) . \quad (9.102)$$

The branch on which $(-2E)^{1/2}$ is positive should be taken when $E < 0$. When $E > 0$, the branch which corresponds to incoming (or outgoing) waves at infinity can be selected with the aid of the asymptotic approximation

$$G^{(S)}(\mathbf{r}, \mathbf{r}'; E) \approx \frac{\Gamma(1-\nu)}{2\pi|\mathbf{r} - \mathbf{r}'|} z_2^\nu \exp\left(-\tfrac{1}{2} z_2\right) , \quad (9.103)$$

which holds when $z_2 \gg z_1$. This approximation is obtained by using (9.130), (9.132), (9.133), and (9.164) in (9.99). A number of useful expansions for $G^{(S)}$ can be obtained from the integral representation

$$G^{(S)}(\mathbf{r}, \mathbf{r}'; E)$$
$$= \frac{2Z}{\nu} \int_0^\infty \left[\coth\left(\tfrac{1}{2}\rho\right)\right]^{2\nu} \sinh(\rho)$$
$$\times I_0\left\{\nu^{-1}Z \sinh(\rho)(2rr')^{1/2}\left[1 + \cos(\Theta)\right]\right\}$$
$$\times \exp\left[-\nu^{-1}Z(r + r')\cosh(\rho)\right] d\rho , \quad (9.104)$$

where Θ is the angle between $\mathbf{r}$ and $\mathbf{r}'$. These expansions, and other integral representations, can be found in [17–19]. The partial wave expansion of $G^{(S)}$ is

$$G^{(S)}(\mathbf{r}, \mathbf{r}'; E) = \sum_{\ell,m} g_\ell^{(S)}(r, r'; \nu) Y_{\ell m}(\theta, \phi) Y_{\ell m}^*(\theta', \phi') .$$
$$(9.105)$$

The radial Green's function $g_\ell^{(S)}$ is a solution of the radial equation

$$\left\{-\tfrac{1}{2}\left[\frac{d^2}{dr^2} + \frac{2}{r}\frac{d}{dr} - \frac{\ell(\ell+1)}{r^2}\right] \right.$$
$$\left. - \frac{Z}{r} - E\right\} g_\ell^{(S)}(r, r'; \nu) = \frac{\delta(r - r')}{rr'} . \quad (9.106)$$

The standard method for calculating the Green's function of a second order ordinary differential equation [21] (pp. 354–355) yields

$$g_\ell^{(S)}(r, r'; \nu)$$
$$= \frac{(2Z)^{2\ell+2}\Gamma(\ell+1-\nu)}{(2\ell+1)! \, \nu^{2\ell+1}} \exp\left[-\nu^{-1}Z(r + r')\right]$$
$$\times (rr')^\ell \, {}_1F_1\left(\ell+1-\nu; 2\ell+2; 2\nu^{-1}Zr_<\right)$$
$$\times U\left(\ell+1-\nu, 2\ell+2, 2\nu^{-1}Zr_>\right) , \quad (9.107)$$

where $r_<$ is the smaller of the pair r, r' and $r_>$ is the larger of the pair r, r'. Matrix elements of $g_\ell^{(S)}$ can be calculated with the aid of the formula for the double Laplace transform, which is

$$\int_0^\infty dr \int_0^\infty dr' \, (rr')^{\ell+1} \exp(-\lambda r - \lambda' r') g_\ell^{(S)}(r, r'; \nu)$$
$$= \frac{2(2\ell+1)!}{\ell-\nu+1}\left(\frac{\nu}{2Z}\right)^{2\ell+3}\left[\frac{4Z^2}{(\nu\lambda + Z)(\nu\lambda' + Z)}\right]^{2\ell+2}$$
$$\times {}_2F_1\left(2\ell+2, \ell-\nu+1; \ell-\nu+2; 1-\zeta\right) \quad (9.108)$$

where

$$\zeta = \frac{2\nu Z(\lambda + \lambda')}{(\nu\lambda + Z)(\nu\lambda' + Z)} . \quad (9.109)$$

Matrix elements with respect to Slater orbitals can be calculated from (9.109) by taking derivatives with respect to λ and/or λ' to bring down powers of r and r'. Matrix elements with respect to Laguerre polynomials can be calculated by using (9.109) to evaluate integrals over the generating function (9.199) for the Laguerre polynomial [20]. Other methods of calculating matrix elements are discussed in Swainson and Drake [19]. The Green's function $\tilde{G}^{(S)}$ in momentum space is related to the coordinate space Green's function $G^{(S)}$ via the Fourier transforms

$$G^{(S)}(\mathbf{r}, \mathbf{r}'; E) = (2\pi)^{-3}\int \exp\left[i(\mathbf{p}\cdot\mathbf{r} - \mathbf{p}'\cdot\mathbf{r}')\right]$$
$$\times \tilde{G}^{(S)}(\mathbf{p}, \mathbf{p}'; E) \, d^3\mathbf{p} \, d^3\mathbf{p}' , \quad (9.110)$$
$$\tilde{G}^{(S)}(\mathbf{p}, \mathbf{p}'; E) = (2\pi)^{-3}\int \exp\left[-i(\mathbf{p}\cdot\mathbf{r} - \mathbf{p}'\cdot\mathbf{r}')\right]$$
$$\times G^{(S)}(\mathbf{r}, \mathbf{r}'; E) \, d^3\mathbf{r} \, d^3\mathbf{r}' . \quad (9.111)$$

The Green's function $\tilde{G}^{(S)}$ is a solution of

$$\left(\tfrac{1}{2}p^2 - E\right)\tilde{G}^{(S)}(\mathbf{p}, \mathbf{p}'; E) - \frac{Z}{2\pi^2}\int \frac{1}{(\mathbf{p} - \mathbf{p}'')^2}$$
$$\times \tilde{G}^{(S)}(\mathbf{p}'', \mathbf{p}'; E) \, d^3\mathbf{p}'' = \delta(\mathbf{p} - \mathbf{p}') . \quad (9.112)$$

An explicit closed form expression for $\tilde{G}^{(S)}$ is

$$\tilde{G}^{(S)}(\mathbf{p}, \mathbf{p}'; E)$$
$$= \frac{\delta(\mathbf{p} - \mathbf{p}')}{\tfrac{1}{2}p^2 - E} + \frac{Z}{2\pi^2|\mathbf{p} - \mathbf{p}'|^2\left(\tfrac{1}{2}p^2 - E\right)\left[\tfrac{1}{2}(p')^2 - E\right]}$$
$$\times \left\{1 + \frac{\nu q}{1-\nu}\left[\left(\frac{1-q}{1+q}\right){}_2F_1\left(1, 1-\nu; 2-\nu; \frac{1-q}{1+q}\right)\right.\right.$$
$$\left.\left. - \left(\frac{1+q}{1-q}\right){}_2F_1\left(1, 1-\nu; 2-\nu; \frac{1+q}{1-q}\right)\right]\right\} , \quad (9.113)$$

where

$$q = \sqrt{\frac{2E\,|\mathbf{p}-\mathbf{p}'|^2}{4E^2 - 4E\,\mathbf{p}\cdot\mathbf{p}' + (pp')^2}}\,. \qquad (9.114)$$

9.4.2 The Green's Function for the Dirac Equation

The Green's function G_{D} for the Dirac equation (9.60) is a 4×4 matrix valued solution of

$$\left(-i\boldsymbol{\alpha}\cdot\nabla + \beta - \frac{Z\alpha}{r} - E\right) G_{\mathrm{D}}\,(\mathbf{r},\mathbf{r}';E)$$
$$= \delta\,(\mathbf{r}-\mathbf{r}')\,I_4\,, \qquad (9.115)$$

where I_4 is the 4×4 identity matrix. The partial wave expansion of G_{D} is

$$G_{\mathrm{D}}\,(\mathbf{r},\mathbf{r}';E) = \sum_{\kappa,m} \begin{pmatrix} G_{11}^{\kappa,m} & -iG_{12}^{\kappa,m} \\ iG_{21}^{\kappa,m} & G_{22}^{\kappa,m} \end{pmatrix}, \qquad (9.116)$$

where

$$G_{11}^{\kappa,m} = \chi_{\kappa}^{m}\,(\theta,\phi)\,\chi_{\kappa}^{m\dagger}\,(\theta',\phi')\,g_{11}\,(r,r';E)\,, \quad (9.117)$$
$$G_{12}^{\kappa,m} = \chi_{\kappa}^{m}\,(\theta,\phi)\,\chi_{-\kappa}^{m\dagger}\,(\theta',\phi')\,g_{12}\,(r,r';E)\,, \quad (9.118)$$
$$G_{21}^{\kappa,m} = \chi_{-\kappa}^{m}\,(\theta,\phi)\,\chi_{\kappa}^{m\dagger}\,(\theta',\phi')\,g_{21}\,(r,r';E)\,, \quad (9.119)$$
$$G_{22}^{\kappa,m} = \chi_{-\kappa}^{m}\,(\theta,\phi)\,\chi_{-\kappa}^{m\dagger}\,(\theta',\phi')\,g_{22}\,(r,r';E)\,. \quad (9.120)$$

The identity

$$\delta\,(\mathbf{r}-\mathbf{r}')\,I_4 = \frac{\delta\,(r-r')}{rr'}$$
$$\times \sum_{\kappa,m} \begin{pmatrix} \chi_{\kappa}^{m}\,(\theta,\phi)\,\chi_{\kappa}^{m\dagger}\,(\theta',\phi') & 0 \\ 0 & \chi_{-\kappa}^{m}\,(\theta,\phi)\,\chi_{-\kappa}^{m\dagger}\,(\theta',\phi') \end{pmatrix}$$
$$(9.121)$$

can be used to show that the radial functions $g_{jk}\,(r,r';E)$ satisfy the equation

$$\begin{pmatrix} \left(1-E-\dfrac{Z\alpha}{r}\right) & -\left(\dfrac{d}{dr}+\dfrac{1-\kappa}{r}\right) \\[2mm] \left(\dfrac{d}{dr}+\dfrac{1+\kappa}{r}\right) & -\left(1+E+\dfrac{Z\alpha}{r}\right) \end{pmatrix}$$
$$\times \begin{pmatrix} g_{11}\,(r,r';E) & g_{12}\,(r,r';E) \\ g_{21}\,(r,r';E) & g_{22}\,(r,r';E) \end{pmatrix}$$
$$= \frac{\delta\,(r-r')}{rr'}\begin{pmatrix} 1 & 0 \\ 0 & 1 \end{pmatrix}. \qquad (9.122)$$

The solution to Eq. (9.122) is

$$\begin{pmatrix} g_{11}\,(r,r';E) & g_{12}\,(r,r';E) \\ g_{21}\,(r,r';E) & g_{22}\,(r,r';E) \end{pmatrix} = \frac{(2\lambda)^{1+2\gamma}\,\Gamma\,(a)}{\Gamma\,(3+2\gamma)}$$
$$\times \left[\Theta\,(r'-r)\begin{pmatrix} G_{<}\,(r) \\ F_{<}\,(r) \end{pmatrix}\big(\,G_{>}\,(r') \quad F_{>}\,(r')\,\big) \right.$$
$$\left. + \Theta\,(r-r')\begin{pmatrix} G_{>}\,(r) \\ F_{>}\,(r) \end{pmatrix}\big(\,G_{<}\,(r') \quad F_{<}\,(r')\,\big) \right],$$
$$(9.123)$$

where a is defined by (9.79), Θ is the Heaviside unit function, defined by

$$\Theta\,(x) = \begin{cases} 1, & x>0, \\ \frac{1}{2}, & x=0, \\ 0, & x<0, \end{cases} \qquad (9.124)$$

and the functions $G_{<}$, $F_{<}$, $G_{>}$, and $F_{>}$ are special cases of the homogeneous solutions (9.71) — (9.80):

$$G_{<}\,(r) = r^{\gamma}\exp\,(-\lambda r)\,(1+E)^{1/2}\,[f_2\,(r)+f_1\,(r)]\,, \qquad (9.125)$$

$$F_{<}\,(r) = r^{\gamma}\exp\,(-\lambda r)\,(1-E)^{1/2}\,[f_2\,(r)-f_1\,(r)]\,, \qquad (9.126)$$

$$G_{>}\,(r) = r^{\gamma}\exp\,(-\lambda r)\,(1+E)^{1/2}\,[f_4\,(r)+f_3\,(r)]\,, \qquad (9.127)$$

$$F_{>}\,(r) = r^{\gamma}\exp\,(-\lambda r)\,(1-E)^{1/2}\,[f_4\,(r)-f_3\,(r)]\,. \qquad (9.128)$$

The functions $G_{<}\,(r)$ and $F_{<}\,(r)$ obey the boundary conditions at $r=0$. The functions $G_{>}\,(r)$ and $F_{>}\,(r)$ obey the boundary conditions at $r=\infty$. Integral representations and expansions for the Dirac Green's function can be found in [19] and [22]. Matrix element evaluation is discussed in [19].

9.5 SPECIAL FUNCTIONS

This section contains a brief list of formulae for the special functions which appear in the solutions discussed above. Derivations, and many additional formulae, can be found in the standard reference works listed in the bibliography. For numerically useful approximations and available software packages, see Lozier and Olver [23].

9.5.1 Confluent Hypergeometric Functions

The confluent hypergeometric differential equation is

$$\left[z\frac{d^2}{dz^2} + (c-z)\frac{d}{dz} - a\right]w\,(z) = 0\,. \qquad (9.129)$$

Equation (9.129) has a regular singular point at $r = 0$ with indices 0 and $1 - c$ and an irregular singular point at ∞. The regular solution to Eq. (9.129) is the confluent hypergeometric function, denoted by $_1F_1$ in generalized hypergeometric series notation. It can be defined by the series

$$_1F_1\left(a;c;z\right) = \frac{\Gamma\left(c\right)}{\Gamma\left(a\right)} \sum_{n=0}^{\infty} \frac{\Gamma\left(a+n\right)}{\Gamma\left(c+n\right)} \frac{z^n}{n!}. \qquad (9.130)$$

The series (9.130) for $_1F_1$ converges for all finite z if c is not a negative integer or zero. It reduces to a polynomial of degree n in z if $a = -n$ where n is a positive integer and c is not a negative integer or zero. The function $_1F_1\left(a;c;z\right)$ is denoted by the symbol $M\left(a,c,z\right)$ in Abramowitz and Stegun [1], in Jahnke and Emde [5], and in Olver [9], by $_1F_1\left(a;c;z\right)$ in both of Luke's books [6,7] and in Magnus, Oberhettinger, and Soni [8], and by $\Phi\left(a,c;z\right)$ in the Bateman project [2,3] and Gradshteyn and Ryzhik [4]. The irregular solution to Eq. (9.129) is

$$U\left(a,c,z\right) = \frac{\Gamma\left(1-c\right)}{\Gamma\left(1+a-c\right)} {}_1F_1\left(a;c;z\right) + \frac{\Gamma\left(c-1\right)}{\Gamma\left(a\right)}$$
$$\times z^{1-c} {}_1F_1\left(1+a-c;2-c;z\right). \qquad (9.131)$$

The function $U\left(a,c,z\right)$ is multiple-valued, with principal branch $-\pi < \arg z \le \pi$. It is denoted by the symbol $U\left(a,c,z\right)$ in Abramowitz and Stegun [1], in Magnus, Oberhettinger, and Soni [8], and in Olver [9], by $\psi\left(a;c;z\right)$ in the first of Luke's books [6], by $U\left(a;c;z\right)$ in the second of Luke's books [7], and by $\Psi\left(a,c;z\right)$ in the Bateman project [2,3] and Gradshteyn and Ryzhik [4].

The Whittaker functions $M_{\kappa,\mu}$ and $W_{\kappa,\mu}$, which are related to $_1F_1$ and U via

$$M_{\kappa,\mu}\left(z\right) = \exp\left(-\tfrac{1}{2}z\right) z^{\mu+\frac{1}{2}} {}_1F_1\left(\mu + \tfrac{1}{2} - \kappa; 2\mu+1; z\right), \qquad (9.132)$$

$$W_{\kappa,\mu}\left(z\right) = \exp\left(-\tfrac{1}{2}z\right) z^{\mu+\frac{1}{2}} U\left(\mu + \tfrac{1}{2} - \kappa, 2\mu+1, z\right), \qquad (9.133)$$

are sometimes used instead of $_1F_1$ and U. For numerical evaluation and a program, see [24].

The regular solution can be written as a linear combination of irregular solutions via

$$_1F_1\left(a;c;z\right) = \frac{\Gamma\left(c\right)}{\Gamma\left(c-a\right)} e^{i\pi\epsilon a} U\left(a,c,z\right) + \frac{\Gamma\left(c\right)}{\Gamma\left(a\right)}$$
$$\times e^{z+i\pi\epsilon\left(a-c\right)} U\left(c-a,c,-z\right), \qquad (9.134)$$

$$\epsilon = \begin{cases} +1, & \operatorname{Im} z > 0, \\ -1, & \operatorname{Im} z < 0. \end{cases} \qquad (9.135)$$

$_1F_1$ can also be obtained from U as the discontinuity across a branch cut:

$$z^{c-1} \exp\left(-z\right) {}_1F_1\left(a;c;z\right)$$
$$= \frac{\Gamma\left(1-a\right)\Gamma\left(c\right)}{2\pi i} \left[\left(ze^{-i\pi}\right)^{c-1} U\left(c-a,c,ze^{-i\pi}\right)\right.$$
$$\left. - \left(ze^{i\pi}\right)^{c-1} U\left(c-a,c,ze^{i\pi}\right)\right]. \qquad (9.136)$$

The Wronskian of the two solutions is

$$_1F_1\left(a;c;z\right) \frac{d}{dz} U\left(a,c,z\right) - U\left(a,c,z\right) \frac{d}{dz} {}_1F_1\left(a;c;z\right)$$
$$= -\Gamma\left(c\right) z^{-c} \exp\left(z\right)/\Gamma\left(a\right). \qquad (9.137)$$

A formula for $U\left(a,c,z\right)$ when c is the integer $n+1$ can be obtained by taking the $c \to n+1$ limit of the right-hand side of Eq. (9.131) to obtain

$$U\left(a,n+1,z\right) = \frac{\left(-1\right)^{n+1}}{\Gamma\left(a-n\right)} \left[\frac{1}{n!} \ln\left(z\right) {}_1F_1\left(a;n+1;z\right)\right.$$
$$\left. + \sum_{k=-n}^{\infty} \frac{\Gamma\left(a+k\right) a_k z^k}{\Gamma\left(a\right)\left(k+n\right)!}\right], \qquad (9.138)$$

where

$$a_k = \begin{cases} \left(-1\right)^{k+1}\left(-k-1\right)!, & -n \le k \le -1, \\[2mm] \left[\Psi\left(k+a\right) - \Psi\left(k+1\right) \right. & \\ \left. - \Psi\left(k+n+1\right)\right]/k!, & k \ge 0. \end{cases} \qquad (9.139)$$

Here Ψ is the logarithmic derivative of the gamma function:

$$\Psi\left(z\right) = \Gamma'\left(z\right)/\Gamma\left(z\right). \qquad (9.140)$$

n is a non-negative integer. When $n = 0$, the sum from $k = -n$ to -1 is omitted.

The basic integral representations for $_1F_1$ and U are

$$_1F_1\left(a;c;z\right) = \frac{\Gamma\left(c\right)}{\Gamma\left(a\right)\Gamma\left(c-a\right)} \int_0^1 e^{zt} t^{a-1}\left(1-t\right)^{c-a-1} dt, \qquad (9.141)$$

$$U\left(a,c,z\right) = \frac{1}{\Gamma\left(a\right)} \int_0^{\infty} e^{-zt} t^{a-1}\left(1+t\right)^{c-a-1} dt. \qquad (9.142)$$

The basic transformation formulae for $_1F_1$ and U are

$$_1F_1\left(a;c;z\right) = e^z {}_1F_1\left(c-a;c;-z\right), \qquad (9.143)$$
$$U\left(a,c,z\right) = z^{1-c} U\left(a-c+1, 2-c, z\right). \qquad (9.144)$$

The recurrence relations among contiguous functions are

$$\left(z+2a-c\right) {}_1F_1\left(a;c;z\right) = \left(a-c\right) {}_1F_1\left(a-1;c;z\right)$$
$$+ a {}_1F_1\left(a+1;c;z\right), \qquad (9.145)$$
$$\left(z+a-1\right) {}_1F_1\left(a;c;z\right) = \left(a-c\right) {}_1F_1\left(a-1;c;z\right)$$
$$+ \left(c-1\right) {}_1F_1\left(a;c-1;z\right), \qquad (9.146)$$
$$c {}_1F_1\left(a;c;z\right) = c {}_1F_1\left(a-1;c;z\right)$$
$$+ z {}_1F_1\left(a;c+1;z\right), \qquad (9.147)$$

$$(a + 1 - c) \, {}_1F_1 (a; c; z) = a \, {}_1F_1 (a + 1; c; z)$$
$$+ (1 - c) \, {}_1F_1 (a; c - 1; z) , \qquad (9.148)$$

$$c (z + a) \, {}_1F_1 (a; c; z) = ac \, {}_1F_1 (a + 1; c; z)$$
$$+ (c - a) z \, {}_1F_1 (a; c + 1; z) , \qquad (9.149)$$

$$c (z + c - 1) \, {}_1F_1 (a; c; z) = c (c - 1) \, {}_1F_1 (a; c - 1; z)$$
$$+ (c - a) \, {}_1F_1 (a; c + 1; z) , \qquad (9.150)$$

$$(z + 2a - c) U (a; c; z) = U (a - 1; c; z)$$
$$+ a (a - c + 1) U (a + 1; c; z) , \qquad (9.151)$$

$$(z + a - 1) U (a; c; z) = U (a - 1; c; z)$$
$$+ (c - a - 1) U (a; c - 1; z) , \qquad (9.152)$$

$$(c - a) U (a; c; z) = -U (a - 1; c; z)$$
$$+ zU (a; c + 1; z) , \qquad (9.153)$$

$$(a + 1 - c) U (a; c; z) = aU (a + 1; c; z)$$
$$+ U (a; c - 1; z) , \qquad (9.154)$$

$$(z + a) U (a; c; z) = a (a - c + 1) U (a + 1; c; z)$$
$$+ zU (a; c + 1; z) , \qquad (9.155)$$

$$(z + c - 1) U (a; c; z) = (c - a - 1) (c - 1)$$
$$\times U (a; c - 1; z) + zU (a; c + 1; z) . \qquad (9.156)$$

Useful differentiation formulae include

$$\frac{d}{dz} \, {}_1F_1 (a; c; z) = ac^{-1} \, {}_1F_1 (a + 1; c + 1; z) , \qquad (9.157)$$

$$\frac{d}{dz} \left[z^a \, {}_1F_1 (a; c; z) \right] = az^{a-1} \, {}_1F_1 (a + 1; c; z) , \qquad (9.158)$$

$$\frac{d}{dz} \left[e^{-z} z^{c-a-1} \, {}_1F_1 (a + 1; c; z) \right]$$
$$= (c - a - 1) e^{-z} z^{c-a-2} \, {}_1F_1 (a; c; z) , \qquad (9.159)$$

$$\frac{d}{dz} U (a, c, z) = -aU (a + 1, c + 1, z) , \qquad (9.160)$$

$$\frac{d}{dz} \left[z^a U (a, c, z) \right] = a (a - c + 1) z^{a-1} U (a + 1, c, z) , \qquad (9.161)$$

$$\frac{d}{dz} \left[e^{-z} z^{c-a-1} U (a + 1, c, z) \right] = -e^{-z} z^{c-a-2} U (a, c, z) . \qquad (9.162)$$

An important multiplication theorem is

$${}_1F_1 (a; c; z_1 z_2)$$
$$= \sum_{k=0}^{\infty} \frac{\Gamma (a + k) \Gamma (\lambda + 2k)}{k! \, \Gamma (a) \, \Gamma (\lambda + k)} (-z_1)^k \, {}_2F_1 (-k, \lambda + k; c; z_1)$$
$$\times \, {}_1F_1 (a + k; \lambda + 2k + 1; z_2) . \qquad (9.163)$$

The fundamental asymptotic expansion for large z is

$$U (a, c, z)$$
$$\sim z^{-a} \sum_{n=0}^{\infty} \frac{\Gamma (a + n) \Gamma (1 + a - c + n)}{n! \, \Gamma (a) \, \Gamma (1 + a - c)} (-z)^{-n} ,$$
$$-\tfrac{3}{2}\pi < \arg z < \tfrac{3}{2}\pi . \qquad (9.164)$$

The asymptotic expansion of ${}_1F_1$ for large z is obtained by using Eq. (9.164) and

$$\exp \left[z + i\pi\epsilon (a - c) \right] U (c - a, c, -z)$$
$$\sim e^z z^{a-c} \sum_{n=0}^{\infty} \frac{\Gamma (c - a + n) \Gamma (1 - a + n)}{n! \Gamma (c - a) \Gamma (1 - a)} z^{-n} ,$$
$$-\tfrac{5}{2}\pi < \arg z < \tfrac{5}{2}\pi , \qquad (9.165)$$

which is a consequence of Eq. (9.164), on the right-hand side of Eq. (9.134). In the asymptotic expansion Eq. (9.165), and in the asymptotic expansion for ${}_1F_1$, the change in the factor $\exp [i\pi\epsilon (a - c)]$ as $\arg z$ passes through zero is compensated by the phase change which comes from a factor $(-z)^{a-c}$ in the asymptotic expansion of $U (c - a, c, -z)$. The change in the factor $\exp (i\pi\epsilon a)$ in the first term of Eq. (9.134) as $\arg z$ passes through zero is not compensated by any other phase change. However, this discontinuity occurs in a region in which this first term is negligible compared to the second term. This is an example of the Stokes phenomenon [25], which occurs because the single-valued function ${}_1F_1$ is being approximated by multiple-valued functions. The large z asymptotic expansion of ${}_1F_1$ is valid for $-\tfrac{3}{2}\pi < \arg z < \tfrac{3}{2}\pi$, which is the overlap of the domain of validity of the expansions (9.164) and (9.165).

Uniform asymptotic expansions for the Whittaker functions $M_{\kappa,\mu}$ and $W_{\kappa,\mu}$ introduced in (9.132), (9.133) have been constructed via Olver's method. The following result [9], (p. 412, Ex. 7.3), which holds for x positive, κ large and positive, and μ unrestricted, gives the flavor of these approximations:

$$W_{\kappa,\mu} (4\kappa x) = \frac{2^{4/3} \pi^{1/2} \kappa^{\kappa + (1/6)}}{\phi_n (\kappa, \mu) \exp (\kappa)} \left(\frac{x\zeta}{x - 1} \right)^{1/4}$$
$$\times \left\{ \text{Ai} \left[(4\kappa)^{2/3} \zeta \right] \sum_{s=0}^{n} \frac{A_s (\zeta)}{(4\kappa)^{2s}} + \frac{\text{Ai}' \left[(4\kappa)^{2/3} \zeta \right]}{(4\kappa)^{2/3}} \right.$$
$$\left. \times \sum_{s=0}^{n} \frac{B_s (\zeta)}{(4\kappa)^{2s}} + \epsilon_{2n+1,2} (4\kappa, \zeta) \right\} . \qquad (9.166)$$

Here Ai is the Airy function, and $\epsilon_{2n+1,2}$ is an error term which tends to zero faster than the last term kept when $\kappa \to \infty$ with n fixed. ζ is related to x by

$$\tfrac{4}{3} \zeta^{3/2} = (x^2 - x)^{1/2} - \ln \left[x^{1/2} + (x - 1)^{1/2} \right] ,$$
$$x \geq 1, \qquad (9.167)$$

$$\tfrac{4}{3} (-\zeta)^{3/2} = \cos^{-1} \left(x^{1/2} \right) - (x - x^2)^{1/2} ,$$
$$0 < x \leq 1. \qquad (9.168)$$

ζ is an analytic function of x in the neighborhood of $x = 1$; conversely, x is an analytic function of ζ in the neighborhood of $\zeta = 0$. The differential form of the relations (9.167), (9.168) is

$$2\zeta^{1/2}\,d\zeta = x^{-1/2}\,(x-1)^{1/2}\,dx \quad x \geq 1, \qquad (9.169)$$

$$2\left(-\zeta\right)^{1/2}\,d\zeta = x^{-1/2}\,(1-x)^{1/2}\,dx \quad 0 < x \leq 1. \qquad (9.170)$$

A Taylor series expansion of ζ about $x = 1$ is most easily constructed by expanding the $x^{-1/2}$ in Eq. (9.169) or Eq. (9.170) about $x = 1$ and integrating term by term. The opening terms of such an expansion are

$$\zeta = 2^{-2/3}\left[(x-1) - \tfrac{1}{5}(x-1)^2 \right.$$
$$\left. + \tfrac{17}{175}(x-1)^3\right] + O\left[(x-1)^4\right]. \qquad (9.171)$$

The coefficient functions A_s and B_s are calculated recursively from

$$B_s(\zeta) = \frac{1}{2\zeta^{1/2}}\int_0^\zeta \left[\psi(\eta)A_s(\eta) - A_s''(\eta)\right]\frac{d\eta}{\eta^{1/2}},$$
$$x \geq 1, \qquad (9.172)$$

$$B_s(\zeta) = \frac{1}{2\left(-\zeta\right)^{1/2}}\int_\zeta^0 \left[\psi(\eta)A_s(\eta) \right.$$
$$\left. -A_s''(\eta)\right]\frac{d\eta}{(-\eta)^{1/2}}, \quad 0 < x \leq 1, \qquad (9.173)$$

$$A_{s+1}(\zeta) = -\tfrac{1}{2}B_s'(\zeta) + \tfrac{1}{2}\int \psi(\zeta)B_s(\zeta)\,d\zeta, \qquad (9.174)$$

$$\psi(\zeta) = \frac{(4\mu^2-1)\zeta}{x(x-1)} + \frac{(3-8x)\zeta}{4x(x-1)^3} + \frac{5}{16\zeta^2}. \qquad (9.175)$$

The functions $A_s(\zeta)$, $B_s(\zeta)$, and $\psi(\zeta)$, which appear to be singular at $\zeta = 0$, are actually analytic functions of ζ at $\zeta = 0$. The first few coefficient functions are:

$$A_0(\zeta) = 1, \qquad (9.176)$$

$$B_0(\zeta) = \tfrac{1}{4}\zeta^{-1/2}\left(\frac{x}{x-1}\right)^{3/2}\left[\left(8\mu^2 - \tfrac{1}{2}\right)\left(\frac{x}{x-1}\right)^2\right.$$
$$\left. -\left(\frac{x}{x-1}\right) + \tfrac{5}{6}\right] - \tfrac{5}{48}\zeta^{-2}, \qquad (9.177)$$

$$A_1(\zeta) = \tfrac{1}{4}\zeta^{-1}\left[B_0(\zeta) - \psi(\zeta)\right] + \zeta\left[B_0(\zeta)\right]^2. \qquad (9.178)$$

The function ϕ_n is calculated from

$$\phi_n(\kappa,\mu) = \sum_{s=0}^n \frac{A_s(\infty)}{(4\kappa)^{2s}} - \sum_{s=0}^{n-1} \lim_{\zeta\to\infty} \frac{\zeta^{1/2}B_s(\zeta)}{(4\kappa)^{2s+1}}. \qquad (9.179)$$

The first two ϕ_n are

$$\phi_0(\kappa,\mu) = 1, \qquad (9.180)$$

$$\phi_1(\kappa,\mu) = 1 - \left(\frac{12\mu^2-1}{24\kappa}\right) + \left(\frac{12\mu^2-1}{24\kappa}\right)^2. \qquad (9.181)$$

A bound for the error term $\epsilon_{2n+1,2}$ has been given by Olver [9] (p. 410, Eq. 7.13). The extension to the complex case can be found in Skovgarrd [26]. Skovgarrd's expansions are in powers of $(4\kappa)^{-1}$ instead of $(4\kappa)^{-2}$, because the factor $1/\phi_n(\kappa,\mu)$ has been expanded out in inverse powers of 4κ in his results; as a consequence, his coefficient functions A_s and B_s differ from Olver's. The corresponding asymptotic expansions for $U\left(\mu + \tfrac{1}{2} - \kappa, 2\mu+1, 4\kappa z\right)$ and for the Laguerre polynomial $L^{(2\mu)}_{\kappa-\mu-1/2}(4\kappa z)$ can be constructed with the aid of Eqs. (9.133) and (9.188).

Formulae for matrix element integrals can be obtained by inserting the integral representation (9.141) and interchanging the orders of integration. An example is

$$\int_0^\infty z^\mu\,{}_1F_1(a_1; c_1; \lambda_1 z)\,{}_1F_1(a_2; c_2; \lambda_2 z)\,dz$$
$$= F_2(\mu+1, a_1, a_2, c_1, c_2; -\lambda_1, -\lambda_2). \qquad (9.182)$$

Here F_2 is one of the hypergeometric functions of two variables introduced by Appell, which can be defined either by the integral representation [2], (Vol. 1, p. 230, Eq. 2)

$$F_2(\alpha,\beta,\beta',\gamma,\gamma';x,y)$$
$$= \frac{\Gamma(\gamma)\Gamma(\gamma')}{\Gamma(\beta)\Gamma(\beta')\Gamma(\gamma-\beta)\Gamma(\gamma'-\beta')}$$
$$\times \int_0^1 du \int_0^1 dv\; u^{\beta-1}v^{\beta'-1}(1-u)^{\gamma-\beta-1}$$
$$\times (1-v)^{\gamma'-\beta'-1}(1-ux-vy)^{-\alpha}, \qquad (9.183)$$

or by the series expansion [2], (Vol. 1, p. 224, Eq. 7)

$$F_2(\alpha,\beta,\beta',\gamma,\gamma';x,y)$$
$$= \sum_{m=0}^\infty \sum_{n=0}^\infty \frac{\Gamma(\alpha+m+n)}{m!n!\Gamma(\alpha)}$$
$$\times \frac{\Gamma(\beta+m)\Gamma(\beta'+n)\Gamma(\gamma)\Gamma(\gamma')}{\Gamma(\beta)\Gamma(\beta')\Gamma(\gamma+m)\Gamma(\gamma'+n)}x^m y^n, \qquad (9.184)$$

which converges for $|x| + |y| < 1$. Numerical evaluation of F_2 is particularly convenient in the special cases where it can be expressed in terms of ordinary hypergeometric functions ${}_2F_1$ [2] (Vol. 1, Chap. 2), which are easy to calculate, or in terms of elementary functions. The key is the formula [2] (Vol. 1, p. 238, Eq. 3)

$$F_2(\alpha,\beta,\beta',\alpha,\alpha;x,y)$$
$$= (1-x)^{-\beta}(1-y)^{-\beta'}\,{}_2F_1\left[\beta,\beta';\alpha;\frac{xy}{(1-x)(1-y)}\right], \qquad (9.185)$$

which shows that it is necessary to get Appell functions F_2 in which the first, fourth, and fifth parameters are equal. This can be achieved by exploiting whatever freedom in the choice of a_1, a_2, c_1, and c_2 is available, and by using identities such as

$$F_2(\alpha+2,\beta,\beta',\alpha+1,\alpha+1;x,y)$$
$$= (\alpha+1)^{-1}\beta'yF_2(\alpha+1,\beta,\beta'+1,\alpha+1,\alpha+1;x,y)$$
$$+ \big[F_2(\alpha+1,\beta,\beta',\alpha+1,\alpha+1;x,y) + 2(\alpha+1)^{-2}$$
$$\times \beta\beta'xyF_2(\alpha+2,\beta+1,\beta'+1,\alpha+2,\alpha+2;x,y)\big]$$
$$+ (\alpha+1)^{-1}\beta xF_2(\alpha+1,\beta+1,\beta',\alpha+1,\alpha+1;x,y)$$
$$(9.186)$$

to obtain Appell functions F_2 for which the reduction (9.185) can be used.

9.5.2 Laguerre Polynomials

The Laguerre polynomials $L_n^{(\alpha)}$ are the polynomial solutions of the differential equation

$$\left[z\frac{d^2}{dz^2} + (\alpha+1-z)\frac{d}{dz} + n\right]L_n^{(\alpha)}(z) = 0. \quad (9.187)$$

They are a special case of the confluent hypergeometric function:

$$L_n^{(\alpha)}(z) = \frac{\Gamma(n+\alpha+1)}{n!\Gamma(\alpha+1)}{}_1F_1(-n;\alpha+1;z)$$
$$= \frac{(-1)^n}{n!}U(-n,\alpha+1,z), \quad (9.188)$$

and are given explicitly by

$$L_n^{(\alpha)}(z) = \sum_{k=0}^{n}\frac{\Gamma(\alpha+n+1)}{k!(n-k)!\Gamma(\alpha+k+1)}(-z)^k. \quad (9.189)$$

The $L_n^{(\alpha)}$ are sometimes called generalized Laguerre polynomials, because $L_n^{(0)}$, which is often denoted by L_n, is the polynomial introduced by Laguerre. This Laguerre polynomial differs from the "associated Laguerre function" for which the symbol L_q^p (with p and q both integers) is often used in the physics literature. The relation between the two is

$$\left[L_q^p(z)\right]_{\text{physics}} = (-1)^{p+q}q!L_{q-p}^{(p)}(z). \quad (9.190)$$

The first three $L_n^{(\alpha)}$ are

$$L_0^{(\alpha)}(z) = 1, \quad (9.191)$$
$$L_1^{(\alpha)}(z) = \alpha+1-z, \quad (9.192)$$
$$L_2^{(\alpha)}(z) = \tfrac{1}{2}(\alpha+1)(\alpha+2) - (\alpha+2)z + \tfrac{1}{2}z^2. \quad (9.193)$$

Equation (9.188) shows that the irregular solution U does not supply a linearly independent second solution. A second solution which remains linearly independent and finite when α is a positive integer is

$$M_n^{(\alpha)}(z) = -\Gamma(\alpha)\big[z^{-\alpha}{}_1F_1(-n-\alpha;1-\alpha;z)$$
$$- \cos(\pi\alpha)\Gamma(1-\alpha)L_n^{(\alpha)}(z)\big]. \quad (9.194)$$

The Wronskian of the two solutions is

$$L_n^{(\alpha)}(z)\frac{d}{dz}M_n^{(\alpha)}(z) - M_n^{(\alpha)}(z)\frac{d}{dz}L_n^{(\alpha)}(z)$$
$$= \Gamma(n+\alpha+1)z^{-\alpha-1}\exp(z)/n!. \quad (9.195)$$

A formula for $M_n^{(\alpha)}(z)$ when α is a positive integer m can be obtained by taking the $\alpha\to m$ limit of the right-hand side of (9.194) to obtain

$$M_n^{(m)}(z) = \ln(z)L_n^{(m)}(z) + \sum_{k=-m}^{\infty}\frac{(n+m)!}{(k+m)!}b_kz^k, \quad (9.196)$$

where

$$b_k = \begin{cases} -(-k-1)!/(n-k)!, & -m \le k \le -1, \\ (-1)^k[\Psi(n+1-k)-\Psi(k+1) \\ \quad -\Psi(k+m+1)]/[k!(n-k)!], \\ \hspace{3cm} 0 \le k \le n, \\ (-1)^n(k-n-1)!/k!, & k \ge n+1. \end{cases} \quad (9.197)$$

The Rodrigues formula is

$$L_n^{(\alpha)}(z) = \frac{1}{n!}x^{-\alpha}\exp(x)\frac{d^n}{dx^n}\big[x^{n+\alpha}\exp(-x)\big]. \quad (9.198)$$

The generating function is

$$(1-w)^{-\alpha-1}\exp\left(-\frac{wz}{1-w}\right) = \sum_{n=0}^{\infty}L_n^{(\alpha)}(z)w^n. \quad (9.199)$$

The Christoffel-Darboux formula is

$$\sum_{k=0}^{n}\frac{k!}{\Gamma(k+\alpha+1)}L_k^{(\alpha)}(w)L_k^{(\alpha)}(z)$$
$$= \frac{(n+1)!\big[L_n^{(\alpha)}(w)L_{n+1}^{(\alpha)}(z) - L_{n+1}^{(\alpha)}(w)L_n^{(\alpha)}(z)\big]}{\Gamma(n+\alpha+1)(w-z)}. \quad (9.200)$$

The orthonormality relation is

$$\int_0^{\infty}x^{\alpha}\exp(-x)L_n^{(\alpha)}(x)L_{n'}^{(\alpha)}(x)\,dx$$
$$= (n!)^{-1}\Gamma(n+\alpha+1)\delta_{n,n'}. \quad (9.201)$$

Other useful integration formulae include

$$\int_0^\infty x^{\alpha+1} \exp(-x) L_n^{(\alpha)}(x) L_{n'}^{(\alpha)}(x) \, dx$$

$$= \frac{\Gamma(n+\alpha+1)}{n!} [-n\delta_{n,n'+1} + (2n+\alpha+1)\delta_{n,n'}$$
$$- (n+\alpha+1)\delta_{n,n'-1}], \qquad (9.202)$$

$$\int_0^\infty x^{\alpha-1} \exp(-x) L_n^{(\alpha)}(x) L_{n'}^{(\alpha)}(x) \, dx$$

$$= \frac{\Gamma(n_< + \alpha + 1)}{n_<! \, \alpha}, \quad n_< = \min(n, n'), \qquad (9.203)$$

$$\int_0^\infty x^{\alpha-2} \exp(-x) L_n^{(\alpha)}(x) L_{n'}^{(\alpha)}(x) \, dx$$

$$= \frac{\Gamma(n_< + \alpha + 1)}{n_<! \, \alpha(\alpha+2)} \{\alpha[(\alpha+1)(n+n') + 4\alpha + 5]$$
$$\times (1 - n_<) + 2(n+1)(n'+1) + 2\alpha(n+n')\},$$
$$n_< = \min(n, n'). \qquad (9.204)$$

The differentiation and recursion relations are

$$z\frac{d}{dz}L_n^{(\alpha)}(z) = nL_n^{(\alpha)}(z) - (n+\alpha)L_{n-1}^{(\alpha)}(z), \qquad (9.205)$$

$$L_{n+1}^{(\alpha)}(z) = (n+1)^{-1}\left[(2n+\alpha+1-z)L_n^{(\alpha)}(z)\right.$$
$$\left. - (n+\alpha)L_{n-1}^{(\alpha)}(z)\right]. \qquad (9.206)$$

Additional relations can be obtained as special cases of the relations listed above for the confluent hypergeometric function by using the relation (9.189).

The coefficients c_k in a Laguerre polynomial expansion such as

$$F(x) = \sum_{k=0}^\infty \frac{k!}{\Gamma(k+\alpha+1)} c_k$$
$$\times (\beta x)^{\alpha/2} \exp\left(-\tfrac{1}{2}\beta x\right) L_k^{(\alpha)}(\beta x) \qquad (9.207)$$

are given by the integral

$$c_k = \beta \int_0^\infty F(x)(\beta x)^{\alpha/2} \exp\left(-\tfrac{1}{2}\beta x\right) L_k^{(\alpha)}(\beta x) \, dx. \qquad (9.208)$$

The rate of convergence of the expansion (9.207) is determined by the asymptotic behavior of the integral (9.208) for large k. A convenient way of extracting this asymptotic behavior will be described with the special case

$$\alpha = 0, \qquad (9.209)$$
$$F(x) = (x+c)^\nu \exp(-\gamma x), \qquad (9.210)$$

as an example. Since the discussion is intended to be illustrative rather than exhaustive, only the case $\gamma < \tfrac{1}{2}\beta$ will be considered. The method is based on the generating function (9.199). The generating function $g(z)$ for the coefficients c_k is given in general by

$$g(z) = \sum_{k=0}^\infty c_k z^n$$

$$= \beta^{(2+\alpha)/2}(1-z)^{-\alpha-1} G\left[\frac{\beta(1+z)}{2(1-z)}\right], \qquad (9.211)$$

where G is the Laplace transform

$$G(\lambda) = \int_0^\infty x^{\alpha/2} F(x) \exp(-\lambda x) \, dx. \qquad (9.212)$$

The asymptotic analysis of the expansion coefficient c_k begins with the Cauchy integral for c_k, which is

$$c_k = \frac{1}{2\pi i} \int_C z^{-k-1} g(z) \, dz, \qquad (9.213)$$

where the contour C is a small circle which runs counterclockwise around the origin. The contour C is deformed to give integrals which can be evaluated via standard methods for the asymptotic evaluation of integrals [27, 28]. For the example (9.209), (9.210),

$$g(z) = \beta c^{\nu+1}(1-z)^{-1} U[1, \nu+2, t(z)], \qquad (9.214)$$

$$t(z) = \frac{(\beta+2\gamma)c + (\beta-2\gamma)cz}{2(1-z)}. \qquad (9.215)$$

This $g(z)$ has a branch point on the negative real axis at $z = -(\beta+2\gamma)/(\beta-2\gamma)$, and a combination of a branch point and an essential singularity on the positive real axis at $z = 1$. It is convenient to take the associated branch cuts to run from $-\infty$ to $-(\beta+2\gamma)/(\beta-2\gamma)$ on the negative real axis, and from $+1$ to $+\infty$ on the positive real axis. The contour C can be deformed into the sum of two contours which run clockwise around the branch cuts. Then

$$c_k = c_k^{(1)} + c_k^{(2)}, \qquad (9.216)$$

where $c_k^{(1)}$ is the contribution from a contour which runs clockwise around the branch cut from $-\infty$ to $-(\beta+2\gamma)/(\beta-2\gamma)$ on the negative real axis, and $c_k^{(2)}$ is the contribution from a contour which runs clockwise around the branch cut from $+1$ to $+\infty$ on the positive real axis. The asymptotic behavior of $c_k^{(1)}$ can be extracted in straightforward fashion via the method of Darboux [9] (pp. 309–315, 321) [28] (pp. 116–122). The result is

$$c_k^{(1)} = \left[\frac{2\beta}{\beta+2\gamma}\right]\left[\frac{4\beta}{(\beta+2\gamma)(\beta-2\gamma)}\right]^\nu (-1)^k$$
$$\times k^\nu \left(\frac{\beta-2\gamma}{\beta+2\gamma}\right)^k [1 + O(k^{-1})]. \qquad (9.217)$$

The contribution $c_k^{(2)}$ requires a somewhat different strategy. The first step writes it as an integral along the real axis from $+1$ to $+\infty$ of the jump across the branch cut. Evaluating the jump yields

$$c_k^{(2)} = \frac{\beta}{\Gamma(-\nu)} \int_0^\infty dx\, x^\nu \left[\beta + \left(\tfrac{1}{2}\beta - \gamma\right)x\right]^{-\nu-1}$$
$$\times \exp\left[-\left(\tfrac{1}{2}\beta - \gamma\right)c - (k+1)\ln(1+x) - \beta c x^{-1}\right]. \tag{9.218}$$

The integrand in (9.218) has a saddle point at $x \approx [\beta c/(k+1)]^{1/2}$. The asymptotics for $(k+1)c$ large can be extracted via the method of steepest descent (also known as the saddle point method) [9] (pp. 136–138), [28] (pp. 85–103), [27], (see Chap. 8). However, if $k+1$ is large with $(k+1)c$ small or moderate, the steepest descent approximation breaks down because the saddle point is too close to the branch point of the integrand at $x = 0$. The breakdown can be cured by using the small x approximations $\left[\beta + \left(\tfrac{1}{2}\beta - \gamma\right)x\right]^{-\nu-1} \approx \beta^{-\nu-1}$ and $\ln(1+x) \approx x - \tfrac{1}{2}(k+1)^{-1}\beta c$ to obtain

$$c_k^{(2)} = \frac{2\beta}{\Gamma(-\nu)} \left[\frac{c}{(k+1)\beta}\right]^{(\nu+1)/2} \exp(\beta c)$$
$$\times K_{\nu+1}\left[2\sqrt{(k+1)\beta c}\right]\left[1 + O\left(k^{-1/2}\right)\right]. \tag{9.219}$$

The function $K_{\nu+1}$ which appears in (9.219) is a modified Bessel function of the third kind in standard notation. The large z approximation $K_{\nu+1}(z) \approx \pi^{1/2}(2z)^{-1/2}\exp(-z)$ can be used to recover the result of a steepest descent approximation to (9.218), which is

$$c_k^{(2)} = \frac{\pi^{1/2}\beta c^{\nu+1}}{\Gamma(-\nu)}\left[\beta c\left(k+\tfrac{1}{2}\right)\right]^{-(2\nu+3)/4}$$
$$\times \exp\left\{\gamma c - 2\left[\beta c\left(k+\tfrac{1}{2}\right)\right]^{1/2}\right\}\left[1 + O\left(k^{-1}\right)\right]. \tag{9.220}$$

The approximation (9.219) remains valid as $c \to 0$. The approximation (9.220) does not.

The generating function method outlined above has one very nice feature: if the singularity in the complex z-plane which dominates the asymptotics is known, the analysis can be inverted to obtain a "convergence acceleration function" which builds in this singularity, and has no other singularities in the finite complex z plane. The difference between the original $F(x)$ and this convergence acceleration function will have an expansion of the form (9.207) which converges faster than the expansion of $F(x)$. Examples can be found in [29], [30], and [31].

The contributions $c_k^{(1)}$ and $c_k^{(2)}$ exhibit two typical features. The most rapidly varying part, which is

$$\left[(\beta - 2\gamma)/(\beta + 2\gamma)\right]^k \tag{9.221}$$

for $c_k^{(1)}$, and

$$\exp\left\{-2\left[\beta c\left(k+\tfrac{1}{2}\right)\right]^{1/2}\right\} \tag{9.222}$$

for the steepest descent approximation to $c_k^{(2)}$, is determined by the location of the associated singularity. The next most important part, which is k^ν for $c_k^{(1)}$, and $\left(k+\tfrac{1}{2}\right)^{-(2\nu+3)/4}$ for the steepest descent approximation to $c_k^{(2)}$, is determined by the nature of the singularity (i.e., by the value of ν).

9.5.3 Gegenbauer Polynomials

The Gegenbauer polynomials are a special case of the Jacobi polynomial, and of the hypergeometric function:

$$C_n^\lambda(z) = \frac{\Gamma\left(\lambda+\tfrac{1}{2}\right)\Gamma(2\lambda+n)}{\Gamma(2\lambda)\Gamma\left(\lambda+n+\tfrac{1}{2}\right)}P_n^{(\lambda-\tfrac{1}{2},\lambda-\tfrac{1}{2})}(z)$$
$$= \frac{\Gamma(2\lambda+n)}{n!\Gamma(2\lambda)}\,{}_2F_1\left(-n, n+2\lambda; \lambda+\tfrac{1}{2}; \tfrac{1}{2}-\tfrac{1}{2}z\right), \tag{9.223}$$

and are given explicitly by

$$C_n^\lambda(z) = \sum_{k=0}^{[k/2]} \frac{(-1)^k \Gamma(\lambda+n-k)}{k!\,(n-2k)!\,\Gamma(\lambda)}(2z)^{n-2k}. \tag{9.224}$$

The first three C_n^λ are

$$C_0^\lambda(z) = 1, \tag{9.225}$$
$$C_1^\lambda(z) = 2\lambda z, \tag{9.226}$$
$$C_2^\lambda(z) = 2\lambda(\lambda+1)z^2 - \lambda. \tag{9.227}$$

Additional C_n^λ can be obtained with the aid of the recursion relation

$$C_{n+1}^\lambda(z) = (n+1)^{-1}\left[2(n+\lambda)zC_n^\lambda(z) - (n+2\lambda-1)C_{n-1}^\lambda(z)\right], \tag{9.228}$$

which is valid for $n \geq 1$. A generating function is

$$\left(1 - 2zt + t^2\right)^{-\lambda} = \sum_{n=0}^\infty C_n^\lambda(z)t^n. \tag{9.229}$$

An important addition theorem, which can be used to derive addition theorems for spherical harmonics in any number of dimensions, is

$$C_n^\lambda[\cos(\gamma)] = \frac{\Gamma(2\lambda-1)}{[\Gamma(\lambda)]^2}\sum_{\ell=0}^n \frac{2^{2\ell}(2\lambda+2\ell-1)}{\Gamma(n+2\lambda+\ell)}$$
$$\times \Gamma(n-\ell+1)[\Gamma(\lambda+\ell)]^2 \sin^\ell(\chi)\sin^\ell(\chi')$$
$$\times C_{n-\ell}^{\lambda+\ell}[\cos(\chi)]C_{n-\ell}^{\lambda+\ell}[\cos(\chi')]$$
$$\times C_\ell^{\lambda-\tfrac{1}{2}}[\cos(\gamma')], \tag{9.230}$$
$$\cos(\gamma) = \cos(\chi)\cos(\chi') + \sin(\chi)\sin(\chi')\cos(\gamma'). \tag{9.231}$$

9.5.4 Legendre Functions

The Legendre polynomials $P_\ell(z)$ are a special case of the Gegenbauer polynomial, and of the hypergeometric function:

$$P_\ell(z) = C_n^{1/2}(z)$$
$$= {}_2F_1\left(-n, n+1; 1; \tfrac{1}{2} - \tfrac{1}{2}z\right). \qquad (9.232)$$

The second (irregular) solution to the Legendre equation, which appears in (9.43) above, can be defined by the integral representation

$$Q_\ell(z) = 2^{-\ell-1} \int_{-1}^{1} \left(1 - t^2\right)^\ell (z - t)^{-\ell-1} \, dt. \qquad (9.233)$$

The first two Q_ℓ are

$$Q_0(z) = \tfrac{1}{2} \ln\left(\frac{z+1}{z-1}\right), \qquad (9.234)$$

$$Q_1(z) = \tfrac{1}{2} z \ln\left(\frac{z+1}{z-1}\right) - 1, \qquad (9.235)$$

Additional Q_ℓ can be obtained with the aid of the recursion relation

$$Q_{\ell+1}(z) = \left[(2\ell+1) z Q_\ell(z) - \ell Q_{\ell-1}(z)\right] / (\ell+1), \qquad (9.236)$$

which is valid for $\ell \geq 1$.

REFERENCES

1. M. Abramowitz and I. A. Stegun, *Handbook of Mathematical Functions* (Dover, New York, 1965).
2. A. Erdelyi, W. Magnus, F. Oberhettinger, and F. G. Tricomi, *Higher Transcendental Functions* (McGraw-Hill, New York, 1953, 1955), Vols. 1, 2, and 3.
3. A. Erdelyi, W. Magnus, F. Oberhettinger, and F. G. Tricomi, *Tables of Integral Transforms* (McGraw–Hill, New York, 1954), Vols. 1 and 2.
4. I. S. Gradshteyn and I. W. Ryzhik, *Tables of Integrals, Series, and Products* (Academic, 1965), 4th ed.
5. E. Jahnke and F. Emde, *Tables of Functions with Formulae and Curves* (Dover, 1945), 4th ed.
6. Y. L. Luke, *The Special Functions and Their Approximations* (Academic, New York, 1969), Vols. 1 and 2.
7. Y. L. Luke, *Mathematical Functions and Their Approximations* (Academic, New York, 1975).
8. W. Magnus, F. Oberhettinger, and R. P. Soni, *Formulas and Theorems for the Special Functions of Mathematical Physics* (Springer, New York, 1966), 3rd ed.
9. F. W. J. Olver, *Asymptotics and Special Functions* (Academic, New York, 1974).
10. G. Szego, *Orthogonal Polynomials* (American Mathematical Society, Providence, 1975), 4th ed. Vol. 23.
11. A. R. Edmonds, *Angular Momentum in Quantum Mechanics* (Princeton University Press, Princeton, 1960), Sec. 2.5.
12. L. Pauling and E. Bright Wilson, *Introduction to Quantum Mechanics With Applications to Chemistry* (McGraw-Hill, 1935), Sec. 21.
13. G. W. F. Drake and R. N. Hill, J. Phys. B **26**, 3159 (1993).
14. L. I. Schiff, *Quantum Mechanics* (McGraw-Hill, New York, 1968), 3rd ed. pp. 234–239; J. D. Talman, *Special Functions, A Group Theoretic Approach* (W. A. Benjamin, New York, 1968), pp. 186–188.
15. V. Fock, Z. Physik **98**, 145 (1935). The relation of the Runge–Lenz vector to the "hidden" $O(4)$ symmetry is discussed in Section 3.6.3 and in Leonard I. Schiff, *Quantum Mechanics*, 3rd ed. (McGraw-Hill, New York, 1968), pp. 234–239.
16. M. Reed and B. Simon, *Methods of Modern Mathematical Physics. I. Functional Analysis* (Academic Press, New York and London 1972) Chapters VII The Spectral Theorem and VIII Unbounded Operators. See Theorem VIII.6 on pp. 263–264.
17. L. Hostler, J. Math. Phys. **5**, 591 (1964); **5**, 1235 (1964).
18. J. Schwinger, J. Math. Phys. **5**, 1606 (1964).
19. R. A. Swainson and G. W. F. Drake, J. Phys. A **24**, 79 (1991); **24**, 95 (1991); **24**, 1801 (1991).
20. R. N. Hill and B. D. Huxtable, J. Math. Phys. **23**, 2365 (1982).
21. R. Courant and D. Hilbert, *Methods of Mathematical Physics* (Interscience, New York, 1953), Vol. 1.
22. D. J. Hylton, J. Math. Phys. **25**, 1125 (1984).
23. D. W. Lozier and F. W. J. Olver, in *Proceedings of Symposia in Applied Mathematics Vol. 48, Mathematics of Computation 1943–1993: A Half–Century of Computational Mathematics*, edited by Walter Gautschi (Am. Math. Soc., Providence, 1994).
24. I. J. Thompson and A. R. Barnett, J. Computational Phys. **64**, 490 (1986); Computer Phys. Commun. **36**, 363 (1985).
25. R. E. Meyer, SIAM Review **31**, 435 (1989).
26. H. Skovgarrd, *Uniform Asymptotic Expansions of Confluent Hypergeometric Functions and Whittaker Functions* (Gjellerups, 1966).
27. N. Bleistein and R. A. Handelsman, *Asymptotic Expansions of Integrals* (Holt, Rinehart, & Winston, New York, 1986).
28. R. Wong, *Asymptotic Approximations of Integrals* (Academic, San Diego, 1989).
29. R. C. Forrey and R. N. Hill, Ann. Phys. **226**, 88–157 (1993).
30. R. N. Hill, Phys. Rev. A **51**, 4433 (1995).
31. R. C. Forrey, R. N. Hill, and R. D. Sharma, Phys. Rev. A **52**, 2948 (1995).

10

Atomic Spectroscopy

W. C. Martin and W. L. Wiese

National Institute of Standards and Technology, Gaithersburg, Maryland 20899

10.1 INTRODUCTION

This chapter outlines the main concepts of atomic structure, with some emphasis on terminology and notation. Atomic radiation is discussed, in particular the wavelengths, intensities, and shapes of spectral lines, and a few remarks are made regarding continuous spectra. We include updated tabulations of ionization energies for the neutral atoms and transition probabilities for persistent lines of selected neutral atoms. Some sources of additional atomic spectroscopic data are mentioned.

Experimental techniques and the details of atomic theoretical methods are not covered in this chapter; these and a number of other subjects pertinent to atomic spectroscopy are treated in one or more of at least fifteen other chapters in this book.

10.2 FREQUENCY, WAVENUMBER, WAVELENGTH

The photon energy due to an electron transition between an upper atomic level k (of energy E_k) and a lower level i is

$$\Delta E = E_k - E_i = h\nu = hc\sigma = hc/\lambda_{\text{vac}}, \quad (10.1)$$

where ν is the frequency, σ the wavenumber in vacuum, and λ_{vac} the wavelength in vacuum. The most accurate spectroscopic measurements are determinations of transition frequencies, the unit being the Hertz (1 Hz = 1 s^{-1}) or one of its multiples. A measurement of any one of the entities frequency, wavenumber, or wavelength (in vacuum) is an equally accurate determination of the others since the speed of light is exactly defined [1]. The most common wavelength units are the nanometer (nm), the Ångström (1 Å = 10^{-1} nm) and the micrometer (μm). The SI wavenumber unit is the inverse meter, but in practice wavenumbers are usually expressed in inverse centimeters: 1 cm^{-1} = 10^2 m^{-1}, equivalent to 2.99792458×10^4 MHz. Energy units and conversion factors are further discussed in Chap. 1.

10.3 ATOMIC STATES, SHELLS, AND CONFIGURATIONS

A one-electron atomic *state* is defined by the quantum numbers $nlm_l\,m_s$ or $nljm_j$, with n and l representing the principal quantum number and the orbital angular momentum quantum number, respectively. The allowed values of n are the positive integers, and $l = 0, 1, \ldots, n-1$. The quantum number j represents the angular momentum obtained by coupling the orbital and spin angular momenta of an electron, i.e., $\mathbf{j} = \mathbf{l} + \mathbf{s}$, so that $j = l \pm {}^1\!/_2$. The magnetic quantum numbers m_l, m_s, and m_j represent the projections of the corresponding angular momenta along a particular direction; thus, for example, $m_l = -l, -l+1 \cdots l$ and $m_s = \pm\, {}^1\!/_2$.

The central field approximation for a many-electron atom leads to wave functions expressed in terms of products of such one-electron states [2,3]. Those electrons having the same principal quantum number n belong to the *shell* for that number. Electrons having both the same n value and l value belong to a *subshell*, all electrons in a particular subshell being *equivalent*. The notation for a *configuration* of N equivalent electrons is nl^N, the superscript usually being omitted for $N = 1$. A configuration of several subshells is written as $nl^N n'l'^M \cdots$. The numerical values of l are replaced by letters in writing a configuration, according to the code s, p, d for $l = 0, 1, 2$ and f, g, $h \ldots$ for $l = 3, 4, 5 \ldots$, the letter j being omitted.

The Pauli exclusion principle prohibits atomic states having two electrons with all four quantum numbers the same. Thus the maximum number of equivalent electrons is $2(2l+1)$. A subshell having this number of electrons is *full, complete*, or *closed*, and a subshell having a smaller number of electrons is *unfilled, incomplete*, or *open*. The $3p^6$ configuration thus represents a full subshell and $3s^2\,3p^6\,3d^{10}$ represents a full shell for $n = 3$.

The *parity* of a configuration is *even* or *odd* according to whether $\Sigma_i l_i$ is even or odd, the sum being taken over all electrons (in practice only those in open subshells need be considered).

10.4 HYDROGEN AND HYDROGEN-LIKE IONS

The quantum numbers n, l, and j are appropriate [4]. A particular *level* is denoted either by nl_j or by $nl\ ^2L_J$ with $L = l$ and $J = j$. The latter notation is somewhat redundant for one-electron spectra, but is useful for consistency with more complex structures. The L values are written with the same letter code used for l values, but with capital letters. The *multiplicity* of the L *term* is equal to $2S + 1 = 2s + 1 = 2$. Written as a superscript, this number expresses the *doublet* character of the structure: each term for $L \geq 1$ has two levels, with $J = L \pm {}^1\!/_2$, respectively.

The Coulomb interaction between the nucleus and the single electron is dominant, so that the largest energy separations are associated with levels having different n. The hyperfine splitting of the ^{1}H $1s$ ground level [1420.405 751 766 7(10) MHz] results from the interaction of the proton and electron magnetic moments and gives rise to the famous 21 cm line. The separations of the $2n - 1$ excited levels having the same n are largely determined by relativistic contributions, including the spin-orbit interaction, with the result that each of the $n - 1$ pairs of levels having the same j value is almost degenerate; the separation of the two levels in each pair is mainly due to relatively small Lamb shifts (see Sects. 28.3 and 27.11).

10.5 ALKALIS AND ALKALI-LIKE SPECTRA

In the central field approximation there exists no angular-momentum coupling between a closed subshell and an electron outside the subshell, since the net spin and orbital angular momenta of the subshell are both zero. The nlj quantum numbers are, then, again appropriate for a single electron outside closed subshells. However, the electrostatic interactions of this electron with the core electrons and with the nucleus yield a strong l-dependence of the energy levels [5]. The differing extent of "core penetration" for ns and np electrons can in some cases, for example, give an energy difference comparable to or exceeding the difference between the np and $(n + 1)p$ levels. The spin-orbit fine-structure

separation between the nl $(l>0)$ levels having $j = l - {}^1\!/_2$ and $l + {}^1\!/_2$, respectively, is relatively small.

10.6 HELIUM AND HELIUM-LIKE IONS; *LS* COUPLING

The energy structure of the normal $1snl$ configurations is dominated by the electron-nucleus and electron-electron Coulomb contributions [4]. In helium and in helium-like ions of the lighter elements, the separations of levels having the same n and having $l = s$, p, or d are mainly determined by direct and exchange electrostatic interactions between the electrons — the spin-orbit, spin-other orbit, and other relativistic contributions are much smaller. This is the condition for LS coupling, in which:

(a) The orbital angular momenta of the electrons are coupled to give a total orbital angular momentum $\mathbf{L} = \Sigma_i\,\mathbf{l}_i$.

(b) The spins of the electrons are coupled to give a total spin $\mathbf{S} = \Sigma_i\,\mathbf{s}_i$.

The combination of a particular S value with a particular L value comprises a spectroscopic *term*, the notation for which is ^{2S+1}L. The quantum number $2S+1$ is the *multiplicity* of the term. The $\mathbf{S}$ and $\mathbf{L}$ vectors are coupled to obtain the total angular momentum, $\mathbf{J} = \mathbf{S} + \mathbf{L}$, for a *level* of the term; the level is denoted as $^{2S+1}L_J$.

For $1snl$ configurations, $L = l$ and $S = 0$ or 1, i.e., the terms are *singlets* ($S = 0$) or *triplets* ($S = 1$). As examples of the He I structure, the ionization energy (energy required to remove one of the $1s$ electrons in the $1s^2$ ground configuration) is 24.5874 eV, the $1s2s\ ^3S - {}^1S$ separation is 0.7962 eV, the $1s2p\ ^3P^\circ - {}^1P^\circ$ separation is 0.2539 eV, and the $1s2p\ ^3P_2^\circ - {}^3P_0^\circ$ fine-structure spread is only 1.32×10^{-4} eV.

10.7 HIERARCHY OF ATOMIC STRUCTURE IN *LS* COUPLING

The centrality of LS coupling in the analysis and theoretical interpretation of atomic spectra has led to the acceptance of notations and nomenclature well adapted to discussions of particular structures and spectra [2]. The main elements of the nomenclature are shown in Table 10.1, most of the structural entities having already been defined in the above discussions of simple spectra. The quantum numbers in the table represent a full description for complex configurations, and the accepted names for transitions between the structural elements are also given.

As an example, the Ca I $3d4p\ ^3D_2^\circ$ *level* belongs to the $^3D^\circ$ *term* which, in turn, belongs to the $3d4p\ ^3(P^\circ\ D^\circ\ F^\circ)$ triplet *triad*. The $3d4p$ *configuration* also has a $^1(P^\circ\ D^\circ\ F^\circ)$ singlet *triad*. The $3d4s$ configuration has only *monads*, one ^1D and one ^3D. The $3d4s\ ^3D_2 -$

Table 10.1. Atomic structural hierarchy in LS coupling and names for the groups of all transitions between structural entities.

Structural entity	Quantum numbers[a]	Group of all transitions
Configuration	$(n_i l_i)^{N_i}$	Transition array
Polyad	$(n_i l_i)^{N_i} \gamma S_1 L_1 nl$ $SL, SL' \cdots$	Supermultiplet
Term	$(n_i l_i)^{N_i} \gamma SL$	Multiplet
Level	$(n_i l_i)^{N_i} \gamma SLJ$	Line
State	$(n_i l_i)^{N_i} \gamma SLJM$	Line component

[a] The configuration may include several open subshells, as indicated by the i subscripts. The letter γ represents any additional quantum numbers, such as ancestral terms, necessary to specify a particular term.

$3d4p$ $^3D_3^\circ$ *line* belongs to the corresponding $^3D - {}^3D^\circ$ triplet *multiplet*, and this multiplet belongs to the great CaI $3d4s$ $^3D - 3d4p$ $^3(P^\circ\ D^\circ\ F^\circ)$ *supermultiplet* of three triplet multiplets discussed by Russell and Saunders in their classic paper on the alkaline-earth spectra [6]. The $3d4s - 3d4p$ *transition array* includes both the singlet and triplet supermultiplets, as well as any (LS-forbidden) *intercombination* or *intersystem* lines arising from transitions between levels of the singlet *system* and those of the triplet *system*. The order of the two terms in the transitions as written above, with the lower-energy term on the left, is standard in atomic spectroscopy. Examples of notations for complex configurations are given in Sect. 10.9.

10.8 ALLOWED TERMS OR LEVELS FOR EQUIVALENT ELECTRONS

10.8.1 *LS* Coupling

The allowed LS terms of a configuration consisting of two nonequivalent groups of electrons are obtained by coupling the **S** and **L** vectors of the groups in all possible ways, and the procedure may be extended to any number of such groups. Thus the allowed terms for any configuration can be obtained from a table of the allowed terms for groups of equivalent electrons.

The configuration l^N has more than one allowed term of certain LS types if $l > 1$ and $2 < N < 4l$ ($d^3 - d^7$, $f^3 - f^{11}$, etc.). The recurring terms of a particular LS term type from d^N and f^N configurations are assigned sequential index numbers in the tables of Nielson and Koster [7]; the index numbers stand for additional numbers having

group-theoretical significance that serve to differentiate the recurring terms, except for a few terms of f^5 and f^9, f^6 and f^8, and f^7. These remaining terms, which occur only in pairs, are further labeled A or B to indicate Racah's separation of the two terms.

The index numbers of Nielson and Koster are in practice the most frequently used labels for the recurring terms of f^N configurations. Use of their index numbers for the recurring terms of d^N configurations has perhaps the disadvantage of substituting an arbitrary number for a quantum number (the seniority) that itself distinguishes the recurring terms in all cases. The actual value of the seniority number is rarely needed, however, and a consistent notation for the d^N and f^N configurations is desirable. A table of the allowed LS terms of the l^N electrons for $l \leq 3$ is given in Ref. [8], with all recurring terms having the index numbers of Nielson and Koster as a following on-line integer. The theoretical group labels are also listed. Thus the d^3 ^2D term having seniority 3 is designated $^2D2$, instead of ^2_3D, in this scheme; and the level having $J = {}^3\!/_2$ is designated $^2D_{3/2}2$.

10.8.2 *jj* Coupling

The allowed J values for a group of N equivalent electrons having the same j value, l_j^N, are given in Table 10.2 for $j = {}^1\!/_2, {}^3\!/_2, {}^5\!/_2$, and ${}^7\!/_2$ (sufficient for $l \leq 3$). The $l_{7/2}^4$ group has two allowed levels for each of the J values 2 and 4. The subscripts distinguishing the two levels in each case are the seniority numbers [9].

The allowed levels of the configuration nl^N may be obtained by dividing the electrons into sets of two groups $nl_{l+1/2}^Q$ $nl_{l-1/2}^R$, $Q + R = N$. The possible sets run from $Q = N - 2l$ (or zero if $N < 2l$) up to $Q = N$ or $Q = 2l + 2$, whichever is smaller. The (degenerate) levels for a set with both Q and R nonzero have wave functions defined by the quantum numbers $(\alpha J_1, \beta J_2)J$, with J_1 and J_2 deriving from the Q and R groups, respectively. The symbols α and β represent any additional quantum numbers required to identify levels. The J values of the allowed levels for each $(\alpha J_1, \beta J_2)$ subset are obtained by combining J_1 and J_2 in the usual way.

10.9 NOTATIONS FOR DIFFERENT COUPLING SCHEMES

In this section we give enough examples to make clear the meaning of the different coupling-scheme notations. Not all the configurations in the examples have been identified experimentally, and some of the examples of a particular coupling scheme given for heuristic purposes may be physically inappropriate. Cowan [3] describes the physical conditions for the different coupling schemes and

Table 10.2. Allowed J values for l_j^N equivalent electrons (jj coupling).

l_j^N	Allowed J values
$l_{1/2}$	$1/2$
$l_{1/2}^2$	0
$l_{3/2}$ and $l_{3/2}^3$	$3/2$
$l_{3/2}^2$	$0, 2$
$l_{3/2}^4$	0
$l_{5/2}$ and $l_{5/2}^5$	$5/2$
$l_{5/2}^2$ and $l_{5/2}^4$	$0, 2, 4$
$l_{5/2}^3$	$3/2, 5/2, 9/2$
$l_{5/2}^6$	0
$l_{7/2}$ and $l_{7/2}^7$	$7/2$
$l_{7/2}^2$ and $l_{7/2}^6$	$0, 2, 4, 6$
$l_{7/2}^3$ and $l_{7/2}^5$	$3/2, 5/2, 7/2, 9/2, 11/2, 15/2$
$l_{7/2}^4$	$0, {}_2 2, {}_4 2, {}_2 4, {}_4 4, 5, 6, 8$
$l_{7/2}^8$	0

gives experimental examples.

10.9.1 *LS* Coupling (Russell-Saunders Coupling)

Some of the examples given below indicate notations bearing on the order of coupling of the electrons.

1. $3d^7 \, {}^4F_{7/2}$

2. $3d^7({}^4F)4s4p({}^3P^\circ) \; {}^6F_{9/2}^\circ$

3. $4f^7({}^8S^\circ)6s6p^2({}^4P) \; {}^{11}P_5^\circ$

4. $3p^5({}^2P^\circ)3d^2({}^1G) \; {}^2F_{7/2}^\circ$

5. $4f^{10}({}^3K2)6s6p({}^1P^\circ) \; {}^3L_6^\circ$

6. $4f^7({}^8S^\circ)5d \, ({}^7D^\circ)6p \; {}^8F_{13/2}$

7. $4f^7({}^8S^\circ)5d \, ({}^9D^\circ)6s \, ({}^8D^\circ)7s \; {}^9D_5^\circ$

8. $4f^7({}^8S^\circ)5d \, ({}^9D^\circ)6s6p({}^3P^\circ) \; {}^{11}F_8$

9. $4f^7({}^8S^\circ)5d^2({}^1G) \, ({}^8G^\circ)6p \; {}^7F_0$

10. $4f({}^2F^\circ) \; 5d^2({}^1G)6s \, ({}^2G) \; {}^1P_1^\circ$

In the second example, the seven $3d$ electrons couple to give a ${}^4F$ term, and the $4s$ and $4p$ electrons couple to form the ${}^3P^\circ$ term; the final ${}^6F^\circ$ term is one of nine possible terms obtained by coupling the ${}^4F$ grandparent

and ${}^3P^\circ$ parent terms. The next three examples are similar to the second. The meaning of the index number 2 following the ${}^3K$ symbol in the fifth example is explained in Sect. 10.8.1.

The coupling in example 6 is appropriate if the interaction of the $5d$ and $4f$ electrons is sufficiently stronger than the $5d$-$6p$ interaction. The ${}^7D^\circ$ parent term results from coupling the $5d$ electron to the ${}^8S^\circ$ grandparent, and the $6p$ electron is then coupled to the ${}^7D^\circ$ parent to form the final ${}^8F$ term. A space is inserted between the $5d$ electron and the ${}^7D^\circ$ parent to emphasize that the latter is formed by coupling a term (${}^8S^\circ$) listed to the left of the space. Example 7 illustrates a similar coupling order carried to a further stage; the ${}^8D^\circ$ parent term results from the coupling of the $6s$ electron to the ${}^9D^\circ$ grandparent.

Example 8 is similar to examples 2–5, but in 8 the first of the two terms that couple to form the final ${}^{11}F$ term, i.e., the ${}^9D^\circ$ term, is itself formed by the coupling of the $5d$ electron to the ${}^8S^\circ$ core term. Example 9 shows an ${}^8G^\circ$ parent term formed by coupling the ${}^8S^\circ$ and ${}^1G$ grandparent terms. A space is again used to emphasize that the following (${}^8G^\circ$) term is formed by the coupling of terms listed before the space.

A different order of coupling is indicated in the final example, the $5d^2 \, {}^1G$ term being coupled first to the external $6s$ electron instead of directly to the $4f$ core electron. The $4f({}^2F^\circ)$ core term is isolated by a space to denote that it is coupled (to the $5d^2({}^1G)6s \, {}^2G$ term) only after the other electrons have been coupled. The notation in this particular case (with a single $4f$ electron) could be simplified by writing the $4f$ electron after the ${}^2G$ term to which it is coupled. It appears more important, however, to retain the convention of giving the core portion of the configuration first.

The notations in examples 1–5 are in the form recommended by Russell, Shenstone, and Turner [10], and used in both the *Atomic Energy States* [11] and *Atomic Energy Levels* [8, 12] compilations. The spacings used in the remaining examples allow different orders of coupling of the electrons to be indicated without the use of additional parentheses, brackets, etc.

Some authors assign a short name to each (final) term, so that the configuration can be omitted in tables of classified lines, etc. The most common scheme distinguishes the low terms of a particular *SL* type by the prefixes *a*, *b*, *c*, ..., and the high terms by *z*, *y*, *x*, ... [12].

10.9.2 *jj* Coupling of Equivalent Electrons

This scheme is used, for example, in relativistic calculations. The lower-case j indicates the angular momentum of one electron ($j = l \pm 1/2$) or of each electron in an l_j^N group. Various ways of indicating which of the two possible j values applies to such a group without writing the j-value subscript have been used by different authors; we give the j values explicitly in the examples below. We use the symbols J_i and j to represent total angular momenta.

1. $(6p_{1/2}^2)_0$

2. $(6p_{1/2}^2\, 6p_{3/2})_{3/2}^\circ$

3. $(6p_{1/2}^2\, 6p_{3/2}^2)_2$

4. $4d_{5/2}^3\, 4d_{3/2}^2\, (^9\!/_2, 2)_{11/2}$

The relatively large spin-orbit interaction of the $6p$ electrons produces jj-coupling structures for the $6p^2$, $6p^3$, and $6p^4$ ground configurations of neutral Pb, Bi, and Po, respectively; the notations for the ground levels of these atoms are given as the first three examples above. The configuration in the first example shows the notation for equivalent electrons having the same j value l_j^N, in this case two $6p$ electrons each having $j = 1/2$. A convenient notation for a particular level ($J = 0$) of such a group is also indicated. The second example extends this notation to the case of a $6p^3$ configuration divided into two groups according to the two possible j values. A similar notation is shown for the $6p^4$ level in the third example; this level might also be designated $(6p_{3/2}^{-2})_2$, the negative superscript indicating the two $6p$ holes. The $(J_1, J_2)_J$ term and level notation shown on the right in the fourth example is convenient because each of the two electron groups $4d_{5/2}^3$ and $4d_{3/2}^2$ has more than one allowed total J_i value. The assumed convention is that J_1 applies to the group on the left ($J_1 = 9/2$ for the $4d_{5/2}^3$ group) and J_2 to that on the right.

10.9.3 J_1j or J_1J_2 Coupling

1. $3d^9(^2D_{5/2})4p_{3/2}\ (^5\!/_2, ^3\!/_2)_3^\circ$

2. $4f^{11}(^2H_{9/2}^\circ\, 2)6s6p(^3\mathrm{P}_1^\circ)\ (^9\!/_2, 1)_{7/2}$

3. $4f^9(^6H^\circ)5d\ (^7H_8^\circ)6s6p(^3P_0^\circ)\ (8,0)_8$

4. $4f^{12}(^3H_6)\, 5d(^2D)6s6p(^3P^\circ)\ (^4F_{3/2}^\circ)\ (6, ^3\!/_2)_{13/2}^\circ$

5. $5f^4(^5I_4)6d_{3/2}\ (4, ^3\!/_2)_{11/2}\, 7s7p(^1P_1^\circ)\ (^{11}\!/_2, 1)_{9/2}^\circ$

6. $5f_{7/2}^4 5f_{5/2}^5\ (8, ^5\!/_2)_{21/2}\, 7p_{3/2}\ (^{21}\!/_2, ^3\!/_2)_{10}$

7. $5f_{7/2}^3 5f_{5/2}^3\ (^9\!/_2, ^9\!/_2)_9\, 7s7p(^3P_2^\circ)\ (9,2)_7^\circ$

The first five examples all have core electrons in LS coupling, whereas jj coupling is indicated for the $5f$ core electrons in the last two examples. Since the J_1 and J_2 values in the final (J_1, J_2) term have already been given as subscripts in the configuration, the (J_1, J_2) term notations are redundant in all these examples.

Unless separation of the configuration and final term designations is desired, as in some data tables, one may obtain a more concise notation by simply enclosing the entire configuration in brackets and adding the final J value as a subscript. Thus, the level in the first example can be designated as $[3d^9(^2D_{5/2})4p_{3/2}]_3^\circ$. If the configuration and coupling order are assumed to be known, still shorter designations may be used; for example, the fourth level above might then be given as $[(^3H_6)\,(^3P^\circ)\,(^4F_{3/2}^\circ)]_{13/2}$ or $(^3H_6,\, ^3P^\circ,\, ^4F_{3/2}^\circ)_{13/2}$. Similar economies of notation are of course possible, and often useful, in all coupling schemes.

10.9.4 J_1l or J_1L_2 Coupling (J_1K Coupling)

1. $3p^5(^2P_{1/2}^\circ)5g\ ^2[^9\!/_2]_5^\circ$

2. $4f^2(^3H_4)5g\ ^2[3]_{5/2}$

3. $4f^{13}(^2F_{7/2}^\circ)5d^2(^1D)\ ^1[^7\!/_2]_{7/2}^\circ$

4. $4f^{13}(^2F_{5/2}^\circ)5d6s(^3D)\ ^3[^9\!/_2]_{11/2}^\circ$

The final terms in the first two examples result from coupling a parent-level $\mathbf{J}_1$ to the orbital angular momentum of a $5g$ electron to obtain a resultant $\mathbf{K}$, the K value being enclosed in brackets. The spin of the external electron is then coupled with the $\mathbf{K}$ angular momentum to obtain a pair of J values, $J = K \pm 1/2$ (for $K \neq 0$). The multiplicity (2) of such pair terms is usually omitted from the term symbol, but other multiplicities occur in the more general $J_1 L_2$ coupling (examples 3 and 4). The last two examples are straightforward extensions of $J_1 l$ coupling, with the $\mathbf{L}_2$ and $\mathbf{S}_2$ momenta of the "external" term (^1D and ^3D in examples 3 and 4, respectively) replacing the $\mathbf{l}$ and $\mathbf{s}$ momenta of a single external electron.

10.9.5 LS_1 Coupling (LK Coupling)

1. $3s^2 3p(^2P^\circ)4f\, G\ ^2[^7\!/_2]_3$

2. $3d^7(^4P)4s4p(^3P^\circ)\, D^\circ\ ^3[^5\!/_2]_{7/2}^\circ$

The orbital angular momentum of the core is coupled with the orbital angular momentum of the external electron(s) to give the total orbital angular momentum $\mathbf{L}$. The letter symbol for the final L value is listed with the configuration because this angular momentum is then coupled with the spin of the core ($\mathbf{S}_1$) to obtain the resultant $\mathbf{K}$ angular momentum of the final term (in brackets). The multiplicity of the $[K]$ term arises from the spin of the external electron(s).

10.9.6 Coupling Schemes and Term Symbols

The coupling schemes outlined above include those now most frequently used in calculations of atomic structure [3]. Any term symbol gives the values of two angular momenta that may be coupled to give the total electronic angular momentum of a level (indicated by the J value). For configurations of more than one unfilled subshell, the angular momenta involved in the final coupling derive from two groups of electrons (either group may consist of only one electron). These are often an inner group of coupled electrons and an outer group of coupled electrons, respectively. In any case the quantum numbers for the two groups can be distinguished by subscripts 1 and 2, so that quantum numbers represented by capital letters without subscripts are total quantum numbers for both groups. Thus, the quantum numbers for the two vectors that couple to give the final J are related to the term symbol as follows:

Coupling scheme	Quantum numbers for vectors that couple to give J	Term symbol
LS	L, S	^{2S+1}L
$J_1 J_2$	J_1, J_2	(J_1, J_2)
$J_1 L_2 (\to K)$	K, S_2	$^{2S_2+1}[K]$
$LS_1 (\to K)$	K, S_2	$^{2S_2+1}[K]$

The parity is indicated by appended degree symbols on odd parity terms.

10.10 EIGENVECTOR COMPOSITION OF LEVELS

The wave functions of levels are often expressed as eigenvectors that are linear combinations of basis states in one of the standard coupling schemes. Thus, the wave function $\Psi(\alpha J)$ for a level labeled αJ might be expressed in terms of LS coupling basis states $\Phi(\gamma SLJ)$:

$$\Psi(\alpha J) = \sum_{\gamma SL} c(\gamma SLJ) \, \Phi(\gamma SLJ) \,. \quad (10.2)$$

The $c(\gamma SLJ)$ are expansion coefficients, and

$$\sum_{\gamma SL} |c(\gamma SLJ)|^2 = 1 \,. \quad (10.3)$$

The squared expansion coefficients for the various γSL terms in the composition of the αJ level are conveniently expressed as percentages, whose sum is 100 %. Thus the percentage contributed by the pure Russell-Saunders state γSLJ is equal to $100 \times |c(\gamma SLJ)|^2$. The notation for Russell-Saunders basis states has been used

only for concreteness; the eigenvectors may be expressed in any coupling scheme, and the coupling schemes may be different for different configurations included in a single calculation (with configuration interaction). "Intermediate coupling" conditions for a configuration are such that calculations in both LS and jj coupling yield some eigenvectors representing significant mixtures of basis states.

The largest percentage in the composition of a level is called the *purity* of the level in that coupling scheme. The coupling scheme (or combination of coupling schemes if more than one configuration is involved) that results in the largest average purity for all the levels in a calculation is usually best for naming the levels. With regard to any particular calculation, one does well to remember that, as with other calculated quantities, the resulting eigenvectors depend on a specific theoretical model and are subject to the inaccuracies of whatever approximations the model involves.

Theoretical calculations of experimental energy level structures have yielded many eigenvectors having significantly less than 50% purity in any coupling scheme. Since many of the corresponding levels have nevertheless been assigned names by spectroscopists, some caution is advisable in the acceptance of level designations found in the literature.

10.11 GROUND LEVELS AND IONIZATION ENERGIES FOR THE NEUTRAL ATOMS

Fortunately, the ground levels of the neutral atoms have reasonably meaningful LS-coupling names, the corresponding eigenvector percentages lying in the range from $\sim 55\%$ to 100%. These names are listed in Table 10.3, except for Pa, U, and Np; the lowest few ground-configuration levels of these atoms comprise better $5f^N (L_1 S_1 J_1) 6d_j 7s^2 \, (J_1 j)$ terms than LS-coupling terms. As noted in Sect. 10.9.2, the jj-coupling names given there for the ground levels of Pb, Bi, and Po are more appropriate than the alternative LS-coupling designations in Table 10.3.

The ionization energies in the table are based on a recent survey of the literature, several of the values having been published in 1994 or scheduled for publication in 1995 [13]. The uncertainties are mainly in the range from less than one to several units in the last decimal place, but a few of the values may be in error by 20 or more units in the final place; i.e., the error of some of the two place values could be greater than 0.2 eV. Although no more than four decimal places are given, the accuracies of some of the better known values would, in eV units, be limited only by the uncertainty in the conversion factor, $1.239\,842\,44(37) \times 10^{-4}$ eV/cm^{-1}.

Table 10.3. Ground levels and ionization energies for the neutral atoms.

Z	Element	Ground configuration[a]	Ground level	Ionization energy (eV)	Z	Element	Ground configuration[a]	Ground level	Ionization energy (eV)
1	H	$1s$	$^2S_{1/2}$	13.5984	53	I	[Kr] $4d^{10}\ 5s^2\ 5p^5$	$^2P^\circ_{3/2}$	10.4513
2	He	$1s^2$	$^1S_0$	24.5874	54	Xe	[Kr] $4d^{10}\ 5s^2\ 5p^6$	$^1S_0$	12.1298
3	Li	$1s^2\ 2s$	$^2S_{1/2}$	5.3917	55	Cs	[Xe] $\quad\quad 6s$	$^2S_{1/2}$	3.8939
4	Be	$1s^2\ 2s^2$	$^1S_0$	9.3227	56	Ba	[Xe] $\quad\quad 6s^2$	$^1S_0$	5.2117
5	B	$1s^2\ 2s^2\ 2p$	$^2P^\circ_{1/2}$	8.2980	57	La	[Xe] $\quad 5d\ 6s^2$	$^2D_{3/2}$	5.5770
6	C	$1s^2\ 2s^2\ 2p^2$	$^3P_0$	11.2603	58	Ce	[Xe] $4f\quad 5d\ 6s^2$	$^1G^\circ_4$	5.5387
7	N	$1s^2\ 2s^2\ 2p^3$	$^4S^\circ_{3/2}$	14.5341	59	Pr	[Xe] $4f^3\quad 6s^2$	$^4I^\circ_{9/2}$	5.464
8	O	$1s^2\ 2s^2\ 2p^4$	$^3P_2$	13.6181	60	Nd	[Xe] $4f^4\quad 6s^2$	$^5I_4$	5.5250
9	F	$1s^2\ 2s^2\ 2p^5$	$^2P^\circ_{3/2}$	17.4228	61	Pm	[Xe] $4f^5\quad 6s^2$	$^6H^\circ_{5/2}$	5.58
10	Ne	$1s^2\ 2s^2\ 2p^6$	$^1S_0$	21.5646	62	Sm	[Xe] $4f^6\quad 6s^2$	$^7F_0$	5.6436
11	Na	[Ne] $3s$	$^2S_{1/2}$	5.1391	63	Eu	[Xe] $4f^7\quad 6s^2$	$^8S^\circ_{7/2}$	5.6704
12	Mg	[Ne] $3s^2$	$^1S_0$	7.6462	64	Gd	[Xe] $4f^7\ 5d\ 6s^2$	$^9D^\circ_2$	6.1501
13	Al	[Ne] $3s^2\ 3p$	$^2P^\circ_{1/2}$	5.9858	65	Tb	[Xe] $4f^9\quad 6s^2$	$^6H^\circ_{15/2}$	5.8638
14	Si	[Ne] $3s^2\ 3p^2$	$^3P_0$	8.1517	66	Dy	[Xe] $4f^{10}\quad 6s^2$	$^5I_8$	5.9389
15	P	[Ne] $3s^2\ 3p^3$	$^4S^\circ_{3/2}$	10.4867	67	Ho	[Xe] $4f^{11}\quad 6s^2$	$^4I^\circ_{15/2}$	6.0215
16	S	[Ne] $3s^2\ 3p^4$	$^3P_2$	10.3600	68	Er	[Xe] $4f^{12}\quad 6s^2$	$^3H_6$	6.1077
17	Cl	[Ne] $3s^2\ 3p^5$	$^2P^\circ_{3/2}$	12.9676	69	Tm	[Xe] $4f^{13}\quad 6s^2$	$^2F^\circ_{7/2}$	6.1843
18	Ar	[Ne] $3s^2\ 3p^6$	$^1S_0$	15.7596	70	Yb	[Xe] $4f^{14}\quad 6s^2$	$^1S_0$	6.2542
19	K	[Ar] $\quad 4s$	$^2S_{1/2}$	4.3407	71	Lu	[Xe] $4f^{14}\ 5d\ 6s^2$	$^2D_{3/2}$	5.4259
20	Ca	[Ar] $\quad 4s^2$	$^1S_0$	6.1132	72	Hf	[Xe] $4f^{14}\ 5d^2\ 6s^2$	$^3F_2$	6.8251
21	Sc	[Ar] $3d\ 4s^2$	$^2D_{3/2}$	6.5615	73	Ta	[Xe] $4f^{14}\ 5d^3\ 6s^2$	$^4F_{3/2}$	7.5496
22	Ti	[Ar] $3d^2\ 4s^2$	$^3F_2$	6.8281	74	W	[Xe] $4f^{14}\ 5d^4\ 6s^2$	$^5D_0$	7.8640
23	V	[Ar] $3d^3\ 4s^2$	$^4F_{3/2}$	6.7463	75	Re	[Xe] $4f^{14}\ 5d^5\ 6s^2$	$^6S_{5/2}$	7.8335
24	Cr	[Ar] $3d^5\ 4s$	$^7S_3$	6.7665	76	Os	[Xe] $4f^{14}\ 5d^6\ 6s^2$	$^5D_4$	8.4382
25	Mn	[Ar] $3d^5\ 4s^2$	$^6S_{5/2}$	7.4340	77	Ir	[Xe] $4f^{14}\ 5d^7\ 6s^2$	$^4F_{9/2}$	8.9670
26	Fe	[Ar] $3d^6\ 4s^2$	$^5D_4$	7.9024	78	Pt	[Xe] $4f^{14}\ 5d^9\ 6s$	$^3D_3$	8.9587
27	Co	[Ar] $3d^7\ 4s^2$	$^4F_{9/2}$	7.8810	79	Au	[Xe] $4f^{14}\ 5d^{10}\ 6s$	$^2S_{1/2}$	9.2255
28	Ni	[Ar] $3d^8\ 4s^2$	$^3F_4$	7.6398	80	Hg	[Xe] $4f^{14}\ 5d^{10}\ 6s^2$	$^1S_0$	10.4375
29	Cu	[Ar] $3d^{10}\ 4s$	$^2S_{1/2}$	7.7264	81	Tl	[Xe] $4f^{14}\ 5d^{10}\ 6s^2\ 6p$	$^2P^\circ_{1/2}$	6.1082
30	Zn	[Ar] $3d^{10}\ 4s^2$	$^1S_0$	9.3942	82	Pb	[Xe] $4f^{14}\ 5d^{10}\ 6s^2\ 6p^2$	$^3P_0$	7.4167
31	Ga	[Ar] $3d^{10}\ 4s^2\ 4p$	$^2P^\circ_{1/2}$	5.9993	83	Bi	[Xe] $4f^{14}\ 5d^{10}\ 6s^2\ 6p^3$	$^4S^\circ_{3/2}$	7.2856
32	Ge	[Ar] $3d^{10}\ 4s^2\ 4p^2$	$^3P_0$	7.8994	84	Po	[Xe] $4f^{14}\ 5d^{10}\ 6s^2\ 6p^4$	$^3P_2$	8.4167
33	As	[Ar] $3d^{10}\ 4s^2\ 4p^3$	$^4S^\circ_{3/2}$	9.7886	85	At	[Xe] $4f^{14}\ 5d^{10}\ 6s^2\ 6p^5$	$^2P^\circ_{3/2}$	
34	Se	[Ar] $3d^{10}\ 4s^2\ 4p^4$	$^3P_2$	9.7524	86	Rn	[Xe] $4f^{14}\ 5d^{10}\ 6s^2\ 6p^6$	$^1S_0$	10.7485
35	Br	[Ar] $3d^{10}\ 4s^2\ 4p^5$	$^2P^\circ_{3/2}$	11.8138	87	Fr	[Rn] $\quad 7s$	$^2S_{1/2}$	4.0727
36	Kr	[Ar] $3d^{10}\ 4s^2\ 4p^6$	$^1S_0$	13.9996	88	Ra	[Rn] $\quad 7s^2$	$^1S_0$	5.2784
37	Rb	[Kr] $\quad 5s$	$^2S_{1/2}$	4.1771	89	Ac	[Rn] $\quad 6d\ 7s^2$	$^2D_{3/2}$	5.17
38	Sr	[Kr] $\quad 5s^2$	$^1S_0$	5.6949	90	Th	[Rn] $\quad 6d^2\ 7s^2$	$^3F_2$	6.3067
39	Y	[Kr] $4d\ 5s^2$	$^2D_{3/2}$	6.2171	91	Pa	[Rn] $5f^2\ (^3H_4)\ 6d\ 7s^2$	$(4,\tfrac{3}{2})_{11/2}$	5.89
40	Zr	[Kr] $4d^2\ 5s^2$	$^3F_2$	6.6339	92	U	[Rn] $5f^3\ (^4I^\circ_{9/2})\ 6d\ 7s^2$	$(^9/_2,\tfrac{3}{2})^\circ_6$	6.1941
41	Nb	[Kr] $4d^4\ 5s$	$^6D_{1/2}$	6.7589	93	Np	[Rn] $5f^4\ (^5I_4)\ 6d\ 7s^2$	$(4,\tfrac{3}{2})_{11/2}$	6.2657
42	Mo	[Kr] $4d^5\ 5s$	$^7S_3$	7.0924	94	Pu	[Rn] $5f^6\quad 7s^2$	$^7F_0$	6.0262
43	Tc	[Kr] $4d^5\ 5s^2$	$^6S_{5/2}$	7.28	95	Am	[Rn] $5f^7\quad 7s^2$	$^8S^\circ_{7/2}$	5.9738
44	Ru	[Kr] $4d^7\ 5s$	$^5F_5$	7.3605	96	Cm	[Rn] $5f^7\ 6d\ 7s^2$	$^9D^\circ_2$	6.02
45	Rh	[Kr] $4d^8\ 5s$	$^4F_{9/2}$	7.4589	97	Bk	[Rn] $5f^9\quad 7s^2$	$^6H^\circ_{15/2}$	6.23
46	Pd	[Kr] $4d^{10}$	$^1S_0$	8.3369	98	Cf	[Rn] $5f^{10}\quad 7s^2$	$^5I_8$	6.30
47	Ag	[Kr] $4d^{10}\ 5s$	$^2S_{1/2}$	7.5763	99	Es	[Rn] $5f^{11}\quad 7s^2$	$^4I^\circ_{15/2}$	6.42
48	Cd	[Kr] $4d^{10}\ 5s^2$	$^1S_0$	8.9938	100	Fm	[Rn] $5f^{12}\quad 7s^2$	$^3H_6$	6.50
49	In	[Kr] $4d^{10}\ 5s^2\ 5p$	$^2P^\circ_{1/2}$	5.7864	101	Md	[Rn] $5f^{13}\quad 7s^2$	$^2F^\circ_{7/2}$	6.58
50	Sn	[Kr] $4d^{10}\ 5s^2\ 5p^2$	$^3P_0$	7.3439	102	No	[Rn] $5f^{14}\quad 7s^2$	$^1S_0$	6.65
51	Sb	[Kr] $4d^{10}\ 5s^2\ 5p^3$	$^4S^\circ_{3/2}$	8.6084	103	Lr	[Rn] $5f^{14}\quad 7s^2\ 7p$?	$^2P^\circ_{1/2}$?	4.9 ?
52	Te	[Kr] $4d^{10}\ 5s^2\ 5p^4$	$^3P_2$	9.0096	104	Rf	[Rn] $5f^{14}\ 6d^2\ 7s^2$?	$^3F_2$?	6.0 ?

[a] An element symbol in brackets represents the electrons in the ground configuration of that element.

10.12 ZEEMAN EFFECT

The Zeeman effect for "weak" magnetic fields (the anomalous Zeeman effect) is of special interest because of the importance of Zeeman data in the analysis and theoretical interpretation of complex spectra. In a weak field, the J value remains a good quantum number although in general a level is split into magnetic sublevels [3]. The g value of such a level may be defined by the expression for the energy shift of its magnetic sublevel having magnetic quantum number M, which has one of the $2J + 1$ values, $-J, -J+1, \ldots, J$:

$$\Delta E = gM\mu_{\mathrm{B}}B . \qquad (10.4)$$

The magnetic flux density is B and μ_{B} is the Bohr magneton ($\mu_{\mathrm{B}} = e\hbar/2m_{\mathrm{e}}$).

The wavenumber shift $\Delta\sigma$ corresponding to this energy shift is

$$\Delta\sigma = gM(0.466\,86B \text{ cm}^{-1}), \qquad (10.5)$$

with B representing the numerical value of the magnetic flux density in teslas. The quantity in parentheses, the Lorentz unit, is of the order of 1 or 2 cm^{-1} for typical flux densities used to obtain Zeeman-effect data with classical spectroscopic methods. Accurate data can be obtained with much smaller fields, of course, by using higher-resolution techniques such as laser spectroscopy. Most of the g values now available for atomic energy levels were derived by application of the above formula (for each of the two combining levels) to measurements of optical Zeeman patterns. A single transverse-Zeeman-effect pattern (two polarizations, resolved components, and sufficiently complete) can yield the J value and the g value for each of the two levels involved.

Neglecting a number of higher-order effects, we can evaluate the g value of a level βJ belonging to a pure LS-coupling term using the formula

$$g_{\beta SLJ} = 1 + (g_{\mathrm{e}} - 1) \frac{J(J+1) - L(L+1) + S(S+1)}{2J(J+1)} . \qquad (10.6)$$

The independence of this expression from any other quantum numbers (represented by β) such as the configuration, etc., is important. The expression is derived from vector coupling formulas by assuming a g value of unity for a pure orbital angular momentum and writing the g value for a pure electron spin as g_{e} [14]. A value of 2 for g_{e} yields the Landé formula. If the anomalous magnetic moment of the electron is taken into account, the value of g_{e} is 2.002 319 3. "Schwinger" g values obtained with this more accurate value for g_{e} are given for levels of SL terms in Ref. [8].

The usefulness of g_{SLJ} values is enhanced by their relation to the g values in intermediate coupling. In the notation used in Eq. (10.2) for the wave function of a level βJ in intermediate coupling, the corresponding g value is given by

$$g_{\beta J} = \sum_{\gamma SL} g_{SLJ} \, |c(\gamma SLJ)|^2 , \qquad (10.7)$$

where the summation is over the same set of quantum numbers as for the wave function. The $g_{\beta J}$ value is thus a weighted average of the Landé g_{SLJ} values, the weighting factors being just the corresponding component percentages.

Formulas for magnetic splitting factors in the J_1J_2 and J_1L_2 coupling schemes are given in Refs. [8] and [14]. Some higher-order effects that must be included in more accurate Zeeman-effect calculations are treated by Bethe and Salpeter [4] and by Wybourne [14], for example. High precision calculations for helium are given in Ref. [15]. See also Chaps. 13 and 15.

10.13 TERM SERIES, QUANTUM DEFECTS, AND SPECTRAL-LINE SERIES

The Bohr energy levels for hydrogen or for a hydrogenic (one-electron) ion are given by

$$E_n = -\frac{Z^2}{n^2} , \qquad (10.8)$$

in units of the Rydberg for the appropriate nuclear mass. For a multielectron atom, the deviations of a series of (core)nl levels from hydrogenic E_n values may be due mainly to core penetration by the nl electron (low l-value series), or core polarization by the nl electron (high l-value series), or a combination of the two effects. In either case it can be shown that these deviations can be approximately represented by a constant quantum defect δ_l in the Rydberg formula,

$$E_{nl} = -\frac{Z_c^2}{(n - \delta_l)^2} = -\frac{Z_c^2}{(n^*)^2} , \qquad (10.9)$$

where Z_c is the charge of the core and $n^* = n - \delta$ is the effective principal quantum number. If the core includes only closed subshells, the E_{nl} values are with respect to a value of zero for the (core)$^1S_0$ level, i.e., the $^1S_0$ level is the limit of the (core)nl series. If the quantities in Eq. (10.9) are taken as positive, they represent term values or ionization energies; the term value of the ground level of an atom or ion with respect to the ground level of the next higher ion is thus the principal ionization energy.

If the core has one or more open subshells, the series limit may be the baricenter of the entire core configuration, or any appropriate sub-structure of the core, down to and including a single level. The E_{nl} values refer to the series of corresponding (core)nl structures built on the particular limit structure. The value of the quantum defect depends to some extent on which (core)nl structures are represented by the series formula.

The quantum defect in general also has an energy dependence that must be taken into account if lower members of a series are to be accurately represented by Eq. (10.9). For an unperturbed series, this dependence can be expressed by the extended Ritz formula

$$\delta = n - n^*$$
$$= \delta_0 + \frac{a}{(n - \delta_0)^2} + \frac{b}{(n - \delta_0)^4} + \cdots, \qquad (10.10)$$

with $\delta_0, a, b \ldots$ constants for the series (δ_0 being the limit value of the quantum defect for high series members) [16]. The value of a is usually positive for core-penetration series and negative for core-polarization series. A recent discussion of the foundations of the Ritz expansion and application to high precision calculations in helium is given in Ref. [17].

A spectral-line series results from either emission or absorption transitions involving a common lower level and a series of successive (core)nl upper levels differing only in their n values. The principal series of Na I, $3s\,^2S_{1/2} - np\,^2P^\circ_{1/2,3/2}$ ($n \geq 3$), is an example. The regularity of successive upper term values with increasing n [Eqs. (10.9), (10.10)] is of course observed in line series; the intervals between successive lines decrease in a regular manner towards higher wavenumbers, and the series of increasing wavenumbers converges towards the term value of the lower level as a limit.

10.14   SEQUENCES

Several types of sequences of elements and/or ionization stages are useful because of regularities in the progressive values of parameters relating to structure and other properties along the sequences. All sequence names may refer either to the atoms and/or ions of the sequence or to their spectra.

10.14.1   Isoelectronic Sequence

A neutral atom and those ions of other elements having the same number of electrons as the atom comprise an isoelectronic sequence. (Note that a negative ion having this number of electrons is a member of the sequence.) An isoelectronic sequence is named according to its neutral member; for example, the Na I isolectronic sequence.

10.14.2   Isoionic, Isonuclear, and Homologous Sequences

An isoionic sequence comprises atoms or ions of different elements having the same charge. Such sequences

have probably been most useful along the d- and f-shell rows of the periodic table. Isoionic analyses have also been carried out along p-shell rows, however, and a fine-structure regularity covering spectra of the p-shell atoms throughout the periodic table is known [18].

The atom and successive ions of a particular element comprise the isonuclear sequence for that element.

The elements of a particular column and subgroup in the periodic table are homologous. Thus the C, Si, Ge, Sn, and Pb atoms belong to a homologous sequence having np^2 ground configurations (Table 10.3). The singly ionized atoms of these elements comprise another example of a homologous sequence.

10.15   SPECTRAL WAVELENGTH RANGES, DISPERSION OF AIR

The ranges of most interest for optical atomic spectroscopy are:

$\sim 2 - 20\,\mu$m	mid-infrared (ir)
$700 - 2000\,$nm	near ir
$400 - 700\,$nm	visible
$200 - 400\,$nm	near ultraviolet (uv)
$100 - 200\,$nm	vacuum uv or far uv
$10 - 100\,$nm	extreme uv (euv or xuv)
$< 10\,$nm	soft x-ray, x-ray

The above correspondence of names to ranges should not be taken as exact; the variation as to the extent of some of the named ranges found in the literature is considerable.

Wavelengths in standard air are often tabulated for the region longer than 200 nm. These wavelengths can be related to energy-level differences by conversion to the corresponding (vacuum) wavenumbers or frequencies [19].

10.16   WAVELENGTH (FREQUENCY) STANDARDS

In 1992 the Comité International des Poids et Mesures recommended values for six frequencies for the absorbing molecule $^{127}I_2$ (515 to 640 nm), one frequency for the ^{40}Ca atom (657 nm), and one for methane (3.39 μm) as standards for suitably stabilized lasers [20]. These frequencies range from 88 376 181.600 18 MHz (3.392 231 397 327 μm, fractional uncertainty 3×10^{-12}) for a particular hyperfine-structure (hfs) component in methane to 582 490 603.37 MHz (514.673 466 4 nm, fractional uncertainty 2.5×10^{-10}) for a particular hfs component in $^{127}I_2$. Reference [20] also gives accurate values for hfs frequency intervals in $^{127}I_2$ that connect each of the six standard frequencies to other hfs components of the same or nearby transitions.

Extensive tables of wavenumbers for molecular transitions in the mid-ir range 2.3 to 20.5 μm are included in a calibration atlas published in 1991 [21]. Measurements of U and Th lines (575 to 692 nm) suitable for wavenumber calibration at uncertainty levels of 0.0003 or 0.0004 cm^{-1} were reported in Ref. [22]. Comprehensive tables of lines for U [23], Th [24], and I$_2$ [25] are useful for calibration at uncertainty levels of 0.002 to 0.003 cm^{-1}, the atlas of the Th spectrum extending down to 278 nm.

A 1974 compilation gives reference wavelengths for some 5400 lines of 38 elements covering the range 1.5 nm to 2.5 μm, with most uncertainties between 10^{-5} and 2×10^{-4} nm [26]. Norlén's tables of Ar I and Ar II wavelengths range from 156 nm to 2.4 μm, with uncertainties of 1 to 3$\times10^{-5}$ nm for the more accurate wavelengths shorter than $\sim$ 850 nm [27]. The wavelengths for some 1100 Fe lines selected from the Fe/Ne hollow-cathode spectrum have been recommended for reference standards over the range 183 nm to 4.2 μm, with wavenumber uncertainties 0.001 to 0.002 cm^{-1} [28]. Wavelengths for about 3000 vuv and uv lines (110 to 400 nm) from a Pt/Ne hollow-cathode lamp have been determined with uncertainties of 0.0002 nm or less [29].

10.17 SPECTRAL LINES: SELECTION RULES, INTENSITIES, TRANSITION PROBABILITIES, f VALUES, AND LINE STRENGTHS

The selection rules for discrete transitions are given in Table 10.4.

10.17.1 Emission Intensities (Transition Probabilities)

The total power ϵ radiated in a spectral line of frequency ν per unit source volume and per unit solid angle is

$$\epsilon_{\text{line}} = (4\pi)^{-1} \, h\nu \, A_{ki} \, N_k \,, \quad (10.11)$$

where A_{ki} is the atomic transition probability and N_k the number per unit volume (number density) of excited atoms in the upper (initial) level k. For a homogeneous light source of length l and for the optically thin case, where all radiation escapes, the total emitted line intensity (SI quantity: radiance) is

$$I_{\text{line}} = \epsilon_{\text{line}} l = \int_0^{+\infty} I(\lambda) \, d\lambda$$
$$= (4\pi)^{-1} (hc/\lambda_0) \, A_{ki} \, N_k \, l, \quad (10.12)$$

where $I(\lambda)$ is the specific intensity at wavelength λ, and λ_0 the wavelength at line center.

10.17.2 Absorption f Values

In absorption, the reduced absorption

$$W(\lambda) = [I(\lambda) - I'(\lambda)]/I(\lambda) \quad (10.13)$$

is used, where $I(\lambda)$ is the incident intensity at wavelength λ, e.g., from a source providing a continuous background, and $I'(\lambda)$ the intensity after passage through the absorbing medium. The reduced line intensity from a homogeneous and optically thin absorbing medium of length l follows as

$$W_{ik} = \int_0^{+\infty} W(\lambda) \, d\lambda = \frac{e^2}{4\epsilon_0 \, m_e \, c^2} \lambda_0^2 \, N_i \, f_{ik} \, l \,,$$
$$(10.14)$$

where f_{ik} is the atomic (absorption) oscillator strength (dimensionless).

10.17.3 Line Strengths

A_{ki} and f_{ik} are the principal atomic quantities related to line intensities. In theoretical work, the *line strength* S is also widely used (see Chap. 21):

$$S = S(i,k) = S(k,i) = |R_{ik}|^2 \,, \quad (10.15)$$

$$R_{ik} = \langle \psi_k | P | \psi_i \rangle \,, \quad (10.16)$$

where ψ_i and ψ_k are the initial- and final-state wave functions and R_{ik} is the *transition matrix element* of the appropriate multipole operator P (R_{ik} involves an integration over spatial and spin coordinates of all N electrons of the atom or ion).

10.17.4 Relationships Between A, f, and S

The relationships between A, f, and S for electric dipole (E1, or allowed) transitions in SI units (A in s^{-1}, λ in m, S in m^2 C^2) are

$$A_{ki} = \frac{2\pi e^2}{m_e c \epsilon_0 \lambda^2} \frac{g_i}{g_k} f_{ik} = \frac{16\pi^3}{3h \epsilon_0 \lambda^3 g_k} S \,. \quad (10.17)$$

Numerically, in customary units (A in s^{-1}, λ in Å, S in atomic units),

$$A_{ki} = \frac{6.6702 \times 10^{15}}{\lambda^2} \frac{g_i}{g_k} f_{ik} = \frac{2.0261 \times 10^{18}}{\lambda^3 g_k} S \,,$$
$$(10.18)$$

Table 10.4. Selection rules for discrete transitions.

	Electric dipole (E1) ("allowed")	Magnetic dipole (M1) ("forbidden")	Electric quadrupole (E2) ("forbidden")
Rigorous rules	1. $\Delta J = 0, \pm 1$ (except $0 \not\to 0$)	$\Delta J = 0, \pm 1$ (except $0 \not\to 0$)	$\Delta J = 0, \pm 1, \pm 2$ (except $0 \not\to 0$, $1/2 \not\to 1/2, 0 \not\to 1$)
	2. $\Delta M = 0, \pm 1$ (except $0 \not\to 0$ when $\Delta J = 0$)	$\Delta M = 0, \pm 1$ (except $0 \not\to 0$ when $\Delta J = 0$)	$\Delta M = 0, \pm 1, \pm 2$
With negligible configuration interaction	3. Parity change 4. One electron jumping, with $\Delta l = \pm 1$, Δn arbitrary	No parity change No change in electron configuration; i.e., for all electrons, $\Delta l = 0, \Delta n = 0$	No parity change *No* change in electron configuration; *or* one electron jumping with $\Delta l = 0, \pm 2, \Delta n$ arbitrary
For *LS* coupling only	5. $\Delta S = 0$	$\Delta S = 0$	$\Delta S = 0$
	6. $\Delta L = 0, \pm 1$ (except $0 \not\to 0$)	$\Delta L = 0$ $\Delta J = \pm 1$	$\Delta L = 0, \pm 1, \pm 2$ (except $0 \not\to 0, 0 \not\to 1$)

and for S and ΔE in atomic units,

$$f_{ik} = \tfrac{2}{3} (\Delta E / g_i) S . \qquad (10.19)$$

g_i and g_k are the statistical weights, which are obtained from the appropriate angular momentum quantum numbers. Thus for the lower (upper) level of a spectral line, $g_{i(k)} = 2J_{i(k)} + 1$, and for the lower (upper) term of a multiplet,

$$\bar{g}_{i(k)} = \sum_{J_{i(k)}} (2J_{i(k)} + 1)$$
$$= (2L_{i(k)} + 1)(2S_{i(k)} + 1) . \qquad (10.20)$$

The A_{ki} values for strong lines of selected elements are given in Table 10.5. For comprehensive numerical tables of A, f, and S, including forbidden lines, see Sect. 10.22.

Experimental and theoretical methods to determine A, f, or S values as well as atomic lifetimes are discussed in Chaps. 17, 18, and 21.

Conversion relations between S and A_{ki} for the most common forbidden transitions are given in Table 10.6. Oscillator strengths f are not used for forbidden transitions, i.e., magnetic dipole (M1), electric quadrupole (E2), etc.

[Numerical example: For the $1s2p\ ^1P^\circ_1 - 1s3d\ ^1D_2$ (allowed) transition in He I at 6678.15 Å: $g_i = 3$; $g_k = 5$; $A_{ki} = 6.38 \times 10^7$ s^{-1}; $f_{ik} = 0.711$; $S = 46.9\ a_0^2\ e^2$.]

10.17.5 Relationships Between Line and Multiplet Values

The relations between the total strength and f value of a multiplet (M) and the corresponding quantities for the lines of the multiplet (allowed transitions) are

$$S_M = \sum S_{\text{line}} , \qquad (10.21)$$

$$f_M = (\bar{\lambda}\bar{g}_i)^{-1} \sum_{J_k, J_i} g_i \lambda(J_i, J_k) f(J_i, J_k) . \qquad (10.22)$$

$\bar{\lambda}$ is the weighted ("multiplet") wavelength in vacuum:

$$\bar{\lambda} = n\bar{\lambda}_{\text{air}} = hc/\overline{\Delta E} , \qquad (10.23)$$

where

$$\overline{\Delta E} = \overline{E_k} - \overline{E_i} = (\bar{g}_k)^{-1} \sum_{J_k} g_k E_k - (\bar{g}_i)^{-1} \sum_{J_i} g_i E_i , \qquad (10.24)$$

and n is the refractive index of standard air.

10.17.6 Relative Strengths for Lines of Multiplets in *LS* Coupling

Table 10.7 lists relative line strengths for frequently encountered symmetrical ($P \to P$, $D \to D$) and normal ($S \to P$, $P \to D$) multiplets in *LS* coupling. The strongest, or principal, lines are situated along the main diagonal of the table and are called x_1, x_2, etc. Their strengths normally diminish along the diagonal. The satellite lines y_n and z_n are usually weaker and deviate

Table 10.5. Wavelengths λ, upper energy levels E_k, statistical weights g_i and g_k of lower and upper levels, and transition probabilities A_{ki} for persistent spectral lines of neutral atoms. Many tabulated lines are resonance lines (marked "g"), where the lower energy level belongs to the ground term.

Spec-trum	λ^a Å	E_k (cm^{-1})	g_i	g_k	A_{ki} 10^8 s^{-1}	Accu-racy^b	Spec-trum	λ^a Å	E_k (cm^{-1})	g_i	g_k	A_{ki} 10^8 s^{-1}	Accu-racy^b
Ag	3280.7g	30 473	2	4	1.4	B	Cd	2288.0g	43 692	1	3	5.3	C
	3382.9g	29 552	2	2	1.3	B		3466.2	59 498	3	5	1.2	D
	5209.1	48 744	2	4	0.75	D		3610.5	59 516	5	7	1.3	D
	5465.5	48 764	4	6	0.86	D		5085.8	51 484	5	3	0.56	C
Al	3082.2g	32 435	2	4	0.63	C	Cl	1347.2g	74 226	4	4	4.19	C
	3092.7g	32 437	4	6	0.74	C		1351.7g	74 866	2	2	3.23	C
	3944.0g	25 348	2	2	0.493	C		4526.2	96 313	4	4	0.051	C
	3961.5g	25 348	4	2	0.98	C		7256.6	85 735	6	4	0.15	C
Ar	1048.2g	95 400	1	3	5.32	B	Co	3405.1	32 842	10	10	1.0	C+
	4158.6	117 184	5	5	0.0140	B		3453.5	32 431	10	12	1.1	C+
	4259.4	118 871	3	1	0.0398	B		3502.3	32 028	10	8	0.80	C+
	7635.1	106 238	5	5	0.245	C		3569.4	35 451	8	8	1.6	C
	7948.2	107 132	1	3	0.186	C	Cr	3578.7g	27 935	7	9	1.48	B
	8115.3	105 463	5	7	0.331	C		3593.5g	27 820	7	7	1.50	B
As	1890.4g	52 898	4	6	2.0	D		3605.3g	27 729	7	5	1.62	B
	1937.6g	51 610	4	4	2.0	D		4254.3g	23 499	7	9	0.315	B
	2288.1	54 605	6	4	2.8	D		4274.8g	23 386	7	7	0.307	B
	2349.8	53 136	4	2	3.1	D		5208.4	26 788	5	7	0.506	B
Au	2428.0g	41 174	2	4	1.99	B+	Cs	3876.1g	25 792	2	4	0.0038	C
	2676.0g	37 359	2	2	1.64	B+		4555.3g	21 946	2	4	0.0188	C
B	1825.9g	54 767	2	4	1.76	B		4593.2g	21 765	2	2	0.0080	C
	1826.4g	54 767	4	6	2.11	B		8521.1g	11 732	2	4	0.3276	AA
	2496.8g	40 040	2	2	0.864	C		8943.5g	11 178	2	2	0.287	A
	2497.7g	40 040	4	2	1.73	C	Cu	2178.9g	45 879	2	4	0.913	B
Ba	5535.5g	18 060	1	3	1.19	B		3247.5g	30 784	2	4	1.39	B
	6498.8	24 980	7	7	0.86	D		3274.0g	30 535	2	2	1.37	B
	7059.9	23 757	7	9	0.71	D		5218.2	49 942	4	6	0.75	C
	7280.3	22 947	5	7	0.53	D	F	954.83g	104 731	4	4	5.77	C
Be	2348.6g	42 565	1	3	5.547	AA		6856.0	116 987	6	8	0.494	C
	2650.6	59 696	9	9^c	4.24	AA		7398.7	115 918	6	6	0.285	C+
Bi	2228.3g	44 865	4	4	0.89	D		7754.7	117 623	4	6	0.382	C+
	2898.0	45 916	4	2	1.53	C	Fe	3581.2	34 844	11	13	1.02	B+
	2989.0	44 865	4	4	0.55	C		3719.9g	26 875	9	11	0.162	B+
	3067.7g	32 588	4	2	2.07	C		3734.9	33 695	11	11	0.902	B+
Br	1488.5g	67 184	4	4	1.2	D		3745.6g	27 395	5	7	0.115	B+
	1540.7g	64 907	4	4	1.4	D		3859.9g	25 900	9	9	0.0970	B+
	7348.5	78 512	4	6	0.12	D		4045.8	36 686	9	9	0.863	B+
C	1561.4g	64 087	5	7	1.18	A	Ga	2874.2g	34 782	2	4	1.2	C
	1657.0g	60 393	5	5	2.52	A		2943.6g	34 788	4	6	1.4	C
	1930.9	61 982	5	3	3.51	B+		4033.0g	24 789	2	2	0.49	C
	2478.6	61 982	1	3	0.340	B+		4172.0g	24 789	4	2	0.92	C
Ca	4226.7g	23 652	1	3	2.18	B+	Ge	2651.6g	37 702	1	3	0.85	C
	4302.5	38 552	5	5	1.36	C+		2709.6g	37 452	3	1	2.8	C
	5588.8	38 259	7	7	0.49	D		2754.6g	37 702	5	3	1.1	C
	6162.2	31 539	5	3	0.354	C		3039.1	40 021	5	3	2.8	C
	6439.1	35 897	7	9	0.53	D							

Table 10.5. *Continued.*

Spec-trum	λ^a Å	E_k (cm^{-1})	g_i	g_k	A_{ki} 10^8 s^{-1}	Accu-racy^b	Spec-trum	λ^a Å	E_k (cm^{-1})	g_i	g_k	A_{ki} 10^8 s^{-1}	Accu-racy^b
He	537.03g	186 209	1	3	5.66	AA	Ni	3101.6	33 112	5	7	0.63	C+
	584.33g	171 135	1	3	17.99	AA		3134.1	33 611	3	5	0.73	C+
	3888.7	185 565	3	9^c	0.09475	AA		3369.6g	29 669	9	7	0.18	C
	4026.2	193 917	9	15^c	0.1160	AA		3414.8	29 481	7	9	0.55	C
	4471.5	191 445	9	15^c	0.2458	AA		3524.5	28 569	7	5	1.0	C
	5875.7	186 102	9	15^c	0.7070	AA		3619.4	31 031	5	7	0.66	C
Hg	2536.5g	39 412	1	3	0.0800	B	O	1302.2g	76 795	5	3	3.14	A
	3125.7	71 396	3	5	0.656	B		4368.2	99 681	3	9^c	0.00727	B−
	4358.3	62 350	3	3	0.557	B		5436.9	105 019	7	5	0.0180	C+
	5460.7	62 350	5	3	0.487	B		7156.7	116 631	5	5	0.505	B
I	1782.8g	56 093	4	4	2.71	C		7771.9	86 631	5	7	0.369	A
	1830.4g	54 633	4	6	0.16	D	P	1775.0g	56 340	4	6	2.17	C
In	3039.4g	32 892	2	6	1.3	D		1782.9g	56 090	4	4	2.14	C
	3256.1g	32 915	4	4	1.3	D		2136.2	58 174	6	4	2.83	C
	4101.8g	24 373	2	2	0.56	C		2535.6	58 174	4	4	0.95	C
	4511.3g	24 373	4	2	1.02	C	Pb	2802.0g	46 329	5	7	1.6	D
K	4044.1g	24 720	2	4	0.0124	C		2833.1g	35 287	1	3	0.58	D
	4047.2g	24 701	2	2	0.0124	C		3683.5g	34 960	3	1	1.5	D
	7664.9g	13 043	2	4	0.387	B+		4057.8g	35 287	5	3	0.89	D
	7699.0g	12 985	2	2	0.382	B+	Rb	4201.8g	23 793	2	4	0.018	C
Kr	5570.3	97 919	5	3	0.021	D		4215.5g	23 715	2	2	0.015	C
	5870.9	97 945	3	5	0.018	D		7800.3g	12 817	2	4	0.370	B
	7601.5	93 123	5	5	0.31	D		7947.6g	12 579	2	2	0.340	B
	8112.9	92 294	5	7	0.36	D	S	1474.0g	67 843	5	7	1.6	D
Li	3232.7g	30 925	2	6^c	0.0117	B		1666.7	69 238	5	5	6.3	C
	4602.9	36 623	6	10^c	0.223	B		1807.3g	55 331	5	3	3.8	C
	6103.6	31 283	6	10^c	0.6860	AA		4694.1	73 921	5	7	0.0067	D
	6707.8g	14 904	2	6^c	0.3691	AA	Sc	3907.5g	25 585	4	6	1.28	C+
Mg	2025.8g	49 347	1	3	0.84	D		3911.8g	25 725	6	8	1.37	C+
	2852.1g	35 051	1	3	4.95	B		4020.4g	24 866	4	4	1.65	C+
	4703.0	56 308	3	5	0.255	C		4023.7g	25 014	6	6	1.44	C+
	5183.6	41 197	5	3	0.575	B	Si	2506.9g	39 955	3	5	0.466	C
Mn	2794.8g	35 770	6	8	3.7	C		2516.1g	39 955	5	5	1.21	C
	2798.3g	35 726	6	6	3.6	C		2881.6	40 992	5	3	1.89	C
	2801.1g	35 690	6	4	3.7	C		5006.1	60 962	3	5	0.028	D
	4030.8g	24 820	6	8	0.17	C+		5948.5	57 798	3	5	0.022	D
	4033.1g	24 788	6	6	0.165	C+	Sn	2840.0g	38 629	5	5	1.7	D
	4034.4g	24 779	6	4	0.158	C+		3034.1g	34 641	3	1	2.0	D
N	1199.6g	83 365	4	6	4.01	B+		3175.1g	34 914	5	3	1.0	D
	1492.6	86 221	6	4	3.13	B+		3262.3	39 257	5	3	2.7	D
	4935.1	106 478	4	2	0.0176	B	Sr	2428.1g	41 172	1	3	0.17	C
	7468.3	96 751	6	4	0.193	B+		4607.3g	21 698	1	3	2.01	B
	8216.3	95 532	6	6	0.223	B	Ti	3642.7g	27 615	7	9	0.774	B
Na	5890.0g	16 973	2	4	0.611	AA		3653.5g	27 750	9	11	0.754	C+
	5895.9g	16 956	2	2	0.610	AA		3998.6g	25 388	9	9	0.408	B
	5682.6	34 549	2	4	0.103	C		4981.7	26 911	11	13	0.660	C+
	8183.3	29 173	2	4	0.453	C		5210.4g	19 574	9	9	0.0357	C+
Ne	735.90g	135 889	1	3	6.11	B	Tl	2767.9g	36 118	2	4	1.26	C
	743.72g	134 459	1	3	0.486	B		3519.2g	36 200	4	6	1.24	C
	5852.5	152 971	3	1	0.682	B		3775.7g	26 478	2	2	0.625	B
	6402.2	149 657	5	7	0.514	B		5350.5g	26 478	4	2	0.705	B
	6074.3	150 917	3	1	0.603	B							

Table 10.5. *Continued.*

Spec-trum	λ^a Å	E_k (cm^{-1})	g_i	g_k	A_{ki} 10^8 s^{-1}	Accu-racy^b	Spec-trum	λ^a Å	E_k (cm^{-1})	g_i	g_k	A_{ki} 10^8 s^{-1}	Accu-racy^b
U	3566.6g	28 650	11	11	0.24	B	Xe	1192.0g	83 890	1	3	6.2	C
	3571.6	38 338	17	15	0.13	C		1295.6g	77 186	1	3	2.5	C
	3584.9g	27 887	13	15	0.18	B		1469.6g	68 046	1	3	2.8	B
								4671.2	88 470	5	7	0.010	D
								7119.6	92 445	7	9	0.066	D
V	3183.4g	31 541	6	8	2.4	C+	Zn	2138.6g	46 745	1	3	7.09	B
	4111.8	26 738	10	10	1.01	B		3302.6	62 772	3	5	1.2	B
	4379.2	25 254	10	12	1.1	C		3345.0	62 777	5	7	1.7	B
	4384.7	25 112	8	10	1.1	C		6362.3	62 459	3	5	0.474	C

^{a}A "g" following the wavelength indicates that the lower level of the transition belongs to the ground term; i.e., the line is a resonance line. Wavelengths below 2000 Å are in vacuum, and those above 2000 Å are in air.

^{b}Accuracy estimates pertain to A_{ki} values: AA, uncertainty within 1%; A, within 3%; B, within 10%; C, within 25%; D, within 50%.

^{c}The A_{ki}, λ, g_i, and g_k are multiplet values; see Eq. (10.20) and Sect. 10.17.5

Table 10.6. Conversion relations between S and A_{ki} for forbidden transitions.

	SI units^a	Numerically, in customary units^b
Electric quadrupole	$A_{ki} = \frac{16\pi^5}{15h\,\epsilon_0\lambda^5\,g_k} S$	$A_{ki} = \frac{1.1199\times10^{18}}{g_k\,\lambda^5} S$
Magnetic dipole	$A_{ki} = \frac{16\pi^3}{3h\,\lambda^3}\frac{\mu_0}{g_k} S$	$A_{ki} = \frac{2.697\times10^{13}}{g_k\,\lambda^3} S$

^{a}A in s^{-1}, λ in m. Electric quadrupole: S in m^4 C^2. Magnetic dipole: S in J^2 T^{-2}.

^{b}A in s^{-1}, λ in Å. S in atomic units: $a_0^4\,e^2 = 2.013 \times 10^{-79}$ m^4 C^2 (electric quadrupole), $e^2\,h^2/16\pi^2\,m_e^2 = \mu_B^2 = 8.601 \times 10^{-47}$ J^2 T^{-2} (magnetic dipole). μ_B is the Bohr magneton.

more from the *LS* values than the stronger diagonal lines when departures from *LS* coupling are encountered. The total multiplet strengths S_M are also listed in Table 10.7. A discussion of their normalization as well as more extensive tables are given in Ref. [30].

10.18   ATOMIC LIFETIMES

The radiative lifetime τ_k of an atomic level k is related to the sum of transition probabilities to all levels i lower in energy than k:

$$\tau_k = \left(\sum_i A_{ki}\right)^{-1} . \tag{10.25}$$

The *branching* ratio of a particular transition, say to state i', is defined as

$$A_{ki'}/\sum_i A_{ki} = A_{ki'}\,\tau_k . \tag{10.26}$$

If only one branch (i') exists (or if all other branches may be neglected), one obtains $A_{ki'}\,\tau_k = 1$, and

$$\tau_k = 1/A_{ki'} . \tag{10.27}$$

Precision lifetime measurement techniques are discussed in Chaps. 17 and 18.

10.19   REGULARITIES AND SCALING

10.19.1   Transitions in Hydrogenic (One-Electron) Species

The nonrelativistic *energy* of a hydrogenic transition [Eqs. (10.1), (10.8)] is

$$(\Delta E)_Z = (E_k - E_i)_Z = R_Z\,hc\,Z^2\,(1/n_i^2 - 1/n_k^2) . \tag{10.28}$$

Hydrogenic Z scaling. The spectroscopic quantities for a hydrogenic ion of nuclear charge Z are related to the equivalent quantities in hydrogen ($Z = 1$) as follows (neglecting small differences in the values of R_Z):

$$(\Delta E)_Z = Z^2(\Delta E)_H , \tag{10.29}$$
$$(\lambda_{\text{vac}})_Z = Z^{-2}(\lambda_{\text{vac}})_H , \tag{10.30}$$
$$S_Z = Z^{-2} S_H , \tag{10.31}$$
$$f_Z = f_H , \tag{10.32}$$
$$A_Z = Z^4 A_H . \tag{10.33}$$

For large values of Z, roughly $Z > 20$, relativistic corrections become noticeable and must be taken into account.

Table 10.7. Relative strengths for lines of multiplets in LS coupling.

Normal multiplets $S - P$, $P - D$, $D - F$, etc.

	J_m	$J_m - 1$	$J_m - 2$	$J_m - 3$	$J_m - 4$
$J_m - 1$	x_1	y_1	z_1		
$J_m - 2$		x_2	y_2	z_2	
$J_m - 3$			x_3	y_3	z_3
$J_m - 4$				x_4	y_4

Symmetrical multiplets $P - P$, $D - D$, etc.

	J_m	$J_m - 1$	$J_m - 2$	$J_m - 3$
J_m	x_1	y_1		
$J_m - 1$	y_1	x_2	y_2	
$J_m - 2$		y_2	x_3	y_3
$J_m - 3$			y_3	x_4

Normal multiplets — Multiplicity

$S - P$

	1	2	3	4	5
$S_M =$	3	6	9	12	15
x_1	3.00	4.00	5.00	6.00	7.00
y_1		2.00	3.00	4.00	5.00
z_1			1.00	2.00	3.00

$P - P$

	1	2	3	4	5
$S_M =$	9	18	27	36	45
x_1	9.00	10.00	11.25	12.60	14.00
x_2		4.00	2.25	1.60	1.25
x_3				1.00	2.25
y_1		2.00	3.75	5.40	7.00
y_2			3.00	5.00	6.75

$P - D$

	1	2	3	4	5
$S_M =$	15	30	45	60	75
x_1	15.00	18.00	21.00	24.00	27.00
x_2		10.00	11.25	12.60	14.00
x_3			5.00	5.00	5.25
y_1		2.00	3.75	5.40	7.00
y_2			3.75	6.40	8.75
y_3				5.60	6.75
z_1			.25	.60	1.00
z_2				1.00	2.25
z_3					3.00

Symmetrical multiplets — Multiplicity

$D - D$

	1	2	3	4	5
$S_M =$	25	50	75	100	125
x_1	25.00	28.00	31.11	34.29	37.50
x_2		18.00	17.36	17.29	17.50
x_3			11.25	8.00	6.25
x_4				5.00	1.25
y_1		2.00	3.89	5.71	7.50
y_2			3.75	7.00	10.00
y_3				5.00	8.75
y_4					5.00

$D - F$

	1	2	3	4	5
$S_M =$	35	70	105	140	175
x_1	35.00	40.00	45.00	50.00	55.00
x_2		28.00	31.11	34.29	37.50
x_3			21.00	22.40	24.00
x_4				14.00	14.00
x_5					7.00
y_1		2.00	3.89	5.71	7.50
y_2			3.89	7.31	10.50
y_3				5.60	10.00
y_4					7.00
z_1			.11	.29	.50
z_2				.40	1.00
z_3					1.00

f-value trends. f values for high series members (large n' values) of hydrogenic ions decrease according to

$$f(n, l \to n', l \pm 1) \propto (n')^{-3} . \qquad (10.34)$$

Data for some lines of the main spectral series of hydrogen are given in Table 10.8.

10.19.2 Systematic Trends and Regularities in Atoms and Ions with Two or More Electrons

Atomic quantities for a given state or transition in an *isoelectronic sequence* may be expressed as power series expansions in Z^{-1}:

$$Z^{-2} E = E_0 + E_1 Z^{-1} + E_2 Z^{-2} + \dots , \qquad (10.35)$$

Table 10.8. Some transitions of the main spectral series of hydrogen.

Transition	Customary name[a]	λ^b (Å)	g_i^c	g_k	A_{ki} $(10^8\ \text{s}^{-1})$
1-2	(L$_\alpha$)	1 215.67	2	8	4.699
1-3	(L$_\beta$)	1 025.73	2	18	5.575(-1)^d
1-4	(L$_\gamma$)	972.537	2	32	1.278(-1)
1-5	(L$_\delta$)	949.743	2	50	4.125(-2)
1-6	(L$_\epsilon$)	937.803	2	72	1.644(-2)
2-3	(H$_\alpha$)	6 562.80	8	18	4.410(-1)
2-4	(H$_\beta$)	4 861.32	8	32	8.419(-2)
2-5	(H$_\gamma$)	4 340.46	8	50	2.530(-2)
2-6	(H$_\delta$)	4 101.73	8	72	9.732(-3)
2-7	(H$_\epsilon$)	3 970.07	8	98	4.389(-3)
3-4	(P$_\alpha$)	18 751.0	18	32	8.986(-2)
3-5	(P$_\beta$)	12 818.1	18	50	2.201(-2)
3-6	(P$_\gamma$)	10 938.1	18	72	7.783(-3)
3-7	(P$_\delta$)	10 049.4	18	98	3.358(-3)
3-8	(P$_\epsilon$)	9 545.97	18	128	1.651(-3)

^{a}L$_\alpha$ is often called Lyman α, H$_\alpha$ = Balmer α, P$_\alpha$ = Paschen α, etc.
^{b}Wavelengths below 2000 Å are in vacuum; values above 2000 Å are in air.
^{c}For transitions in hydrogen, $g_{i(k)} = 2(n_{i(k)})^2$, where $n_{i(k)}$, is the principal quantum number of the lower (upper) electron shell.
^{d}The number in parentheses indicates the power of 10 by which the value has to be multiplied.

$$Z^2 S = S_0 + S_1 Z^{-1} + S_2 Z^{-2} + \ldots , \quad (10.36)$$

$$f = f_0 + f_1 Z^{-1} + f_2 Z^{-2} + \ldots , \quad (10.37)$$

where E_0, f_0, and S_0 are hydrogenic quantities. For transitions in which n does not change ($n_i = n_k$), $f_0 = 0$, since states i and k are degenerate.

For equivalent transitions of *homologous atoms*, f values vary gradually. Transitions to be compared in the case of the "alkalis" are [31]

$$(nl - n'l')_{\text{Li}} \rightarrow [(n+1)l - (n'+1)l']_{\text{Na}}$$
$$\rightarrow [(n+2)l - (n'+2)l']_{\text{Cu}} \rightarrow \ldots .$$

Complex atomic structures, as well as cases involving strong cancellation in the integrand of the transition integral, generally do not adhere to this regular behavior.

10.20 SPECTRAL LINE SHAPES, WIDTHS, AND SHIFTS

Observed spectral lines are always broadened, partly due to the finite resolution of the spectrometer and partly due to intrinsic physical causes. The principal physical causes of spectral line broadening are Doppler and pressure broadening. The theoretical foundations of line broadening are discussed in Chaps. 19 and 57.

10.20.1 Doppler Broadening

Doppler broadening is due to the thermal motion of the emitting atoms or ions. For a Maxwellian velocity distribution, the line shape is *Gaussian*; the full width at half maximum intensity (FWHM) is, in Å,

$$\Delta\lambda_{1/2}^D = (7.16 \times 10^{-7})\,\lambda\,(T/M)^{1/2} . \quad (10.38)$$

T is the temperature of the emitters in K, and M the atomic weight in atomic mass units (amu).

10.20.2 Pressure Broadening

Pressure broadening is due to collisions of the emitters with neighboring particles (see also Chaps. 19 and 57). Shapes are often approximately Lorentzian, i.e., $I(\lambda) \propto \{1 + [(\lambda - \lambda_0)/\Delta\lambda_{1/2}]^2\}^{-1}$. In the following formulas, all FWHM's and wavelengths are expressed in Å, particle densities N in cm^{-3}, temperatures T in K, and energies E or I in cm^{-1}.

Resonance broadening (self-broadening) occurs only between identical species and is confined to lines with the upper or lower level having an electric dipole transition (resonance line) to the ground state. The FWHM may be estimated as

$$\Delta\lambda_{1/2}^R \simeq 8.6 \times 10^{-30}(g_i/g_k)^{1/2}\lambda^2\lambda_r f_r N_i , \quad (10.39)$$

where λ is the wavelength of the observed line. f_r and λ_r are the oscillator strength and wavelength of the resonance line; g_k and g_i are the statistical weights of its upper and lower levels. N_i is the ground state number density.

For the $1s2p\ ^1P_1^\circ - 1s3d\ ^1D_2$ transition in He I [$\lambda = 6678.15$ Å; $\lambda_r\ (1s^2\ ^1S_0 - 1s2p\ ^1P_1^\circ) = 584.334$ Å; $g_i = 1$; $g_k = 3$; $f_r = 0.2762$] at $N_i = 1 \times 10^{18}$ cm^{-3}: $\Delta\lambda_{1/2}^R = 0.036$ Å.

Van der Waals broadening arises from the dipole interaction of an excited atom with the induced dipole of a ground state atom. (In the case of foreign gas broadening, both the perturber and the radiator may be in their respective ground states.) An approximate formula for the FWHM, strictly applicable to hydrogen and similar atomic structures only, is

$$\Delta\lambda_{1/2}^W \simeq 3.0 \times 10^{16}\,\lambda^2\,C_6^{2/5}(T/\mu)^{3/10}N , \quad (10.40)$$

where μ is the atom-perturber reduced mass in units of u, N the perturber density, and C_6 the interaction constant. C_6 may be roughly estimated as follows: $C_6 = C_k - C_i$,

Table 10.9. Values of Stark-broadening parameter $\alpha_{1/2}$ of the H_β line of hydrogen (4861 Å) for various temperatures and electron densities.

T(K)	N_e (cm^{-3})			
	10^{15}	10^{16}	10^{17}	10^{18}
5 000	0.0787	0.0808	0.0765	...
10 000	0.0803	0.0840	0.0851	0.0781
20 000	0.0815	0.0860	0.0902	0.0896
30 000	0.0814	0.0860	0.0919	0.0946

with $C_{i(k)} = (9.8 \times 10^{10}) \alpha_d R_{i(k)}^2$ (α_d in cm^3, R^2 in a_0^2). Mean atomic polarizability $\alpha_d \approx (6.7 \times 10^{-25})$ $(3 I_H/4 E^*)^2$ cm^3, where I_H is the ionization energy of hydrogen and E^* the energy of the first excited level of the perturber atom. $R_{i(k)}^2 \approx 2.5 \, [I_H/(I - E_{i(k)})]^2$, where I is the ionization energy of the radiator. Van der Waals broadened lines are red shifted by about one-third the size of the FWHM.

For the $1s2p \; {}^1P_1^\circ - 1s3d \; {}^1D_2$ transition in He I, and with He as perturber: $\lambda = 6678.15$ Å; $I = 198\,311$ cm^{-1}; $E^* = E_i = 171\,135$ cm^{-1}; $E_k = 186\,105$ cm^{-1}; $\mu = 2$. At $T = 15\,000$ K and $N = 1 \times 10^{18}$ cm^{-3}: $\Delta\lambda_{1/2}^W = 0.044$ Å.

Stark broadening due to charged perturbers, i.e., ions and electrons, usually dominates resonance and van der Waals broadening in discharges and plasmas. The FWHM for hydrogen lines is

$$\Delta\lambda_{1/2}^{S,H} = (2.50 \times 10^{-9}) \, \alpha_{1/2} \, N_e^{2/3} \,, \qquad (10.41)$$

where N_e is the electron density. The half-width parameter $\alpha_{1/2}$ for the H_β line at 4861 Å, widely used for plasma diagnostics, is tabulated in Table 10.9 for some typical temperatures and electron densities [30]. This reference also contains $\alpha_{1/2}$ parameters for other hydrogen lines, as well as Stark width and shift data for numerous lines of other elements, i.e., neutral atoms and singly charged ions (in the latter, Stark widths and shifts depend linearly on N_e). Other tabulations of complete hydrogen Stark profiles exist.

10.21 SPECTRAL CONTINUUM RADIATION

10.21.1 Hydrogenic Species

Precise quantum-mechanical calculations exist only for hydrogenic species. The total power ϵ_{cont} radiated (per unit source volume and per unit solid angle, and expressed in SI units) in the wavelength interval $\Delta\lambda$ is the sum of radiation due to the recombination of a free electron with a bare ion (free-bound transitions) and bremsstrahlung (free-free transitions):

$$
\begin{aligned}
\epsilon_{cont} = {} & \frac{e^6}{2\pi\epsilon_0^3 (6\pi m_e)^{3/2}} N_e N_Z Z^2 \\
& \times \frac{1}{(kT)^{1/2}} \exp\left(-\frac{hc}{\lambda kT}\right) \frac{\Delta\lambda}{\lambda^2} \\
& \times \left\{ \frac{2 Z^2 I_H}{kT} \sum_{n \geq (Z^2 I_H \lambda/hc)^{1/2}}^{n'} \frac{\gamma_{fb}}{n^3} \exp\left(\frac{Z^2 I_H}{n^2 kT}\right) \right. \\
& \left. + \bar\gamma_{fb} \left[\exp\left(\frac{Z^2 I_H}{(n'+1)^2 kT}\right) - 1\right] + \gamma_{ff} \right\}
\end{aligned}
$$

$$(10.42)$$

where N_e is the electron density, N_Z the number density of hydrogenic (bare) ions of nuclear charge Z, I_H the ionization energy of hydrogen, n' the principal quantum number of the lowest level for which adjacent levels are so close that they approach a continuum and summation over n may be replaced by an integral. (The choice of n' is rather arbitrary; n' as low as 6 is found in the literature.) γ_{fb} and γ_{ff} are the Gaunt factors, which are generally close to unity. (For the higher free-bound continua, starting with $n'+1$, an average Gaunt factor $\bar\gamma_{fb}$ is used.) For neutral hydrogen, the recombination continuum forming H^- becomes important, too [32].

In the equation above, the value of the constant factor is 6.065×10^{-55} W m^4 J$^{1/2}$ sr^{-1}. [Numerical example: For atomic hydrogen ($Z = 1$), the quantity ϵ_{cont} has the value 2.9 W m^{-3} sr^{-1} under the following conditions: $\lambda = 3 \times 10^{-7}$ m; $\Delta\lambda = 1 \times 10^{-10}$ m; $N_e (= N_{Z=1}) = 1 \times 10^{21}$ m^{-3}; $T = 12\,000$ K. The lower limit of the summation index n is 2; the upper limit n' has been taken to be 10. All Gaunt factors γ_{fb}, $\bar\gamma_{fb}$, and γ_{ff} have been assumed to be unity.]

10.21.2 Many-Electron Systems

For many-electron systems, only approximate theoretical treatments exist, based on the quantum-defect method (for results of calculations for noble gases, see, e.g., Ref. [33]). Experimental work is centered on the noble gases [34]. Modifications of the continuum by autoionization processes must also be considered.

Near the ionization limit, the f values for bound-bound transitions of a spectral series ($n' \to \infty$) make a smooth connection to the differential oscillator strength distribution $df/d\epsilon$ in the continuum [35].

10.22 SOURCES OF SPECTROSCOPIC DATA

A few data compilations have been cited above [8, 12, 30], and Sect. 10.16 includes additional references for wavelength tables. Major recent compilations of atomic transition probabilities cover C, N and O [36] and the Fe-group elements [37]. A fairly comprehensive bibliographic review of compilations of data on atomic energy levels, wavelengths, transition probabilities, and spectral line shapes and shifts was published in 1992 [38]. Extensive data from compilations and other sources, as well as bibliographies, are included in the holdings of the NIST Physics Laboratory server on the Internet [39].

REFERENCES

1. *The International System of Units (SI)*, edited by B. N. Taylor NIST Spec. Publ. 330 (U. S. Government Printing Office, Washington, 1991), p. 29.
2. E. U. Condon and G. H. Shortley, *The Theory of Atomic Spectra* (Cambridge University Press, Cambridge, 1935, reprinted 1951).
3. R. D. Cowan, *The Theory of Atomic Structure and Spectra* (University of California Press, Berkeley, 1981).
4. H. A. Bethe and E. E. Salpeter, *Quantum Mechanics of One- and Two-Electron Atoms* (Plenum, New York, 1977).
5. B. Edlén, in *Encyclopedia of Physics*, edited by S. Flügge (Springer-Verlag, Berlin, 1964). Vol. 27.
6. H. N. Russell and F. A. Saunders, Astrophys. J **61**, 38 (1925).
7. C. W. Nielson and G. F. Koster, *Spectroscopic Coefficients for the p^n, d^n, and f^n Configurations* (MIT Press, Cambridge, 1963).
8. W. C. Martin, R. Zalubas, and L. Hagan, *Atomic Energy Levels — The Rare-Earth Elements*, Nat. Stand. Ref. Data Ser., Nat. Bur. Stand. (United States Government Printing Office, Washington, 1978), No. 60.
9. A. de-Shalit and I. Talmi, *Nuclear Shell Theory* (Academic, New York, 1963).
10. H. N. Russell, A. G. Shenstone, and L. A. Turner, Phys. Rev. **33**, 900 (1929).
11. R. F. Bacher and S. Goudsmit, *Atomic Energy States* (McGraw-Hill, New York, 1932).
12. C. E. Moore, *Atomic Energy Levels*, Nat. Stand. Ref. Data Ser., Nat. Bur. Stand. (United States Government Printing Office, Washington, 1971), No. 35.
13. W. C. Martin, unpublished compilation. A more complete table including references is being prepared for the NIST Physics Laboratory WWW server.
14. B. G. Wybourne, *Spectroscopic Properties of Rare Earths* (Interscience, New York, 1965).
15. Z.-C. Yan and G. W. F. Drake, Phys. Rev. A **50**, R1980 (1994).
16. W. C. Martin, J. Opt. Soc. Am. **70**, 784 (1980).
17. G. W. F. Drake, Adv. At. Mol. Opt. Phys. **32**, 93 (1994).
18. U. Fano and W. C. Martin, in *Topics in Modern Physics, A Tribute to E. U. Condon*, edited by W. E. Brittin and H. Odabasi (Colorado Associated University Press, 1971), pp. 147–152.
19. E. R. Peck and K. Reeder, J. Opt. Soc. Am. **62**, 958 (1972).
20. T. J. Quinn, Metrologia **30**, 523 (1994).
21. A. G. Maki and J. S. Wells, *Wavenumber Calibration Tables from Heterodyne Frequency Measurements*, NIST Spec. Publ. **821** (1991).
22. C. J. Sansonetti and K.-H. Weber, J. Opt. Soc. Am. B **1**, 361 (1984).
23. B. A. Palmer, R. A. Keller, and R. Engleman, Jr., Los Alamos National Laboratory Report LA-8251-MS, UC-34a (1980).
24. B. A. Palmer and R. Engleman, Jr., Los Alamos National Laboratory Report LA-9615, UC-4 (1983).
25. S. Gerstenkorn and P. Luc, *Atlas du Spectre d'Absorption de la Molécule d'Iode entre* 14 800−20 000 cm^{-1} (Editions du CNRS, Paris, 1978); S. Gerstenkorn and P. Luc, Rev. Phys. Appl. **14**, 791 (1979).
26. V. Kaufman and B. Edlén, J. Phys. Chem. Ref. Data **3**, 825 (1974).
27. G. Norlén, Phys. Scr. **8**, 249 (1973).
28. G. Nave, S. Johansson, R. C. M. Learner, A. P. Thorne, and J. W. Brault, Astrophys. J. Suppl. Ser. **94**, 221 (1994), and references therein.
29. J. E. Sansonetti, J. Reader, C. J. Sansonetti, and N. Acquista, J. Res. Natl. Inst. Stds. Tech. **97**, 1 (1992), and references therein.
30. *Astrophysical Quantities*, edited by A. N. Cox (American Institute of Physics Press, New York, 1996), 4th ed.
31. A. W. Weiss, J. Quant. Spectrosc. Radiat. Transfer **18**, 481 (1977).
32. H. R. Griem, *Spectral Line Broadening by Plasmas* (Academic, New York, 1974).
33. J. R. Roberts and P. A. Voigt, J. Res. Natl. Bur. Stand. **75**, 291 (1971).
34. I. I. Sobelman, *Atomic Spectra and Radiative Transitions* (Springer, Berlin, 1992), 2nd ed.
35. A. T. M. Wilbers, G. M. W. Kroesen, C. J. Timmermans, and D. C. Schram, J. Quant. Spectrosc. Radiat. Transfer **45**, 1 (1991).
36. W. L. Wiese, J. R. Fuhr, and T. M. Deters, J. Phys. Chem. Ref. Data, Monograph 7 (1996).
37. J. R. Fuhr, G. A. Martin, and W. L. Wiese, J. Phys. Chem. Ref. Data **17**, Suppls. 3 and 4 (1988).
38. W. C. Martin, in *Atomic and Molecular Data for Space Astronomy*, Lecture Notes in Physics, edited by P. L. Smith and W. L. Wiese (Springer, Berlin, 1992), No. 407, p. 121.
39. NIST Physics Laboratory server, accessible through the World Wide Web, and a Web browser, such as Mosaic or Netscape. The URL is: http://physics.nist.gov. Select Physical Reference Data, and Atomic Spectroscopic Database.

11

High Precision Calculations for Helium

G. W. F. Drake

Department of Physics, University of Windsor, Windsor, Ontario N9B 3P4, Canada

11.1 INTRODUCTION

Exact analytic solutions to the Schrödinger equation are known only for atomic hydrogen, and other equivalent two-body systems (see Chap. 9). However, very high precision approximations are now available for helium, which are essentially exact for all practical purposes. This chapter summarizes the computational methods and tabulates numerical results for the ground state and several singly excited states. Similar methods can be applied to other three-body problems.

11.2 THE THREE-BODY SCHRÖDINGER EQUATION

The Schrödinger equation for a three-body system consisting of a nucleus of charge Ze, and mass M, and

two electrons of charge $-e$ and mass m_e is

$$\left[\frac{1}{2M} P_N^2 + \frac{1}{2m_e} \sum_{i=1}^{2} P_i^2 + V(\mathbf{R}_N, \mathbf{R}_i) \right] \Psi = E\Psi , \quad (11.1)$$

where $\mathbf{P}_i = (\hbar/i)\nabla_i$ and

$$V(\mathbf{R}_N, \mathbf{R}_i) = -\frac{Ze^2}{|\mathbf{R}_N - \mathbf{R}_1|} - \frac{Ze^2}{|\mathbf{R}_N - \mathbf{R}_2|} + \frac{e^2}{|\mathbf{R}_1 - \mathbf{R}_2|}$$

$$(11.2)$$

depends only on the relative particle separations. Since the center of mass (c.m.) is then an ignorable coordinate, it can be eliminated by defining the relative particle coordinates

$$\mathbf{r}_i = \mathbf{R}_i - \mathbf{R}_N$$

to obtain

$$\left[\frac{1}{2\mu}\sum_{i=1}^{2}p_i^2 + \frac{1}{M}\mathbf{p}_1\cdot\mathbf{p}_2 + V(\mathbf{r}_1,\mathbf{r}_2)\right]\Psi = E\Psi, \quad (11.3)$$

where $\mu = m_e M/(m_e + M)$ is the electron reduced mass and the term $H_{\mathrm{mp}} = \mathbf{p}_1\cdot\mathbf{p}_2/M$ is called the mass polarization operator. For computational purposes, it is usual to measure distance in units of $a_\mu = (m_e/\mu)a_0$ and energies in units of $e^2/a_\mu = 2(\mu/m_e)R_\infty$ so that Eq. (11.3) assumes the dimensionless form

$$\left[-\frac{1}{2}\sum_{i=1}^{2}\nabla_{\rho_i}^2 - \frac{\mu}{M}\nabla_{\rho_1}\cdot\nabla_{\rho_2} + V(\boldsymbol{\rho}_1,\boldsymbol{\rho}_2)\right]\Psi = \varepsilon\Psi,$$

$$(11.4)$$

where $\boldsymbol{\rho}_i = \mathbf{r}_i/a_\mu$, $\varepsilon = E/(e^2/a_\mu)$, and

$$V(\boldsymbol{\rho}_1,\boldsymbol{\rho}_2) = -\frac{Z}{\rho_1} - \frac{Z}{\rho_2} + \frac{1}{|\boldsymbol{\rho}_1-\boldsymbol{\rho}_2|}. \quad (11.5)$$

The limit $\mu/M \to 0$ defines the infinite nuclear mass problem with eigenvalue ε_0 and eigenfunction Ψ_0. If the mass polarization term is treated as a small perturbation, then the total energy assumes the form

$$E = \left[\varepsilon_0 + \frac{\mu}{M}\varepsilon_1 + \left(\frac{\mu}{M}\right)^2\varepsilon_2 + \cdots\right]\frac{\mu}{m_e}\frac{e^2}{a_0}, \quad (11.6)$$

where $\varepsilon_1 = -\langle\Psi_0|\nabla_{\rho_1}\cdot\nabla_{\rho_2}|\Psi_0\rangle$ determines the first-order specific mass shift and ε_2 is the second-order coefficient. The common $(\mu/m_e)\varepsilon_0$ mass scaling of all eigenvalues determines the normal mass shift. Since $\mu/m = 1-\mu/M$, the shift is $-(\mu/M)\varepsilon_0$.

11.2.1 Formal Mathematical Properties

Two-Particle Coalescences. The exact nonrelativistic wave function for any many-body system contains discontinuities or cusps in the spherically averaged radial derivative with respect to r_{ij} as $r_{ij} \to 0$, where $r_{ij} = |\mathbf{r}_i - \mathbf{r}_j|$ is any interparticle coordinate. If the masses and charges are m_i and q_i respectively, then the discontinuities are given by the Kato cusp condition [1]

$$\hbar^2\left(\frac{\partial\bar{\Psi}}{\partial r_{ij}}\right)_{r_{ij}=0} = \mu_{ij}q_iq_j\Psi(r_{ij}=0), \quad (11.7)$$

where $\mu_{ij} = m_im_j/(m_i + m_j)$ and $\bar{\Psi}$ denotes the wave function averaged over a sphere centered at $r_{ij} = 0$. If Ψ vanishes at $r_{ij} = 0$, then its leading dependence on r_{ij} is of the form $r_{ij}^l Y_{lm}(\mathbf{r}_{ij})$ for some integer $l > 0$ [2]. Equation (11.7) applies to any Coulombic system. The electron-nucleus cusp in the wave functions for hydrogen provides a simple example.

Three-Particle Coalescences. Three-particle coalescences are described by the Fock expansion [3], as recently discussed by Myers *et al.* [4]. For the S-states of He-like ions, the expansion has the form

$$\Psi(\mathbf{r}_1,\mathbf{r}_2) = \sum_{j=0}^{\infty}\sum_{k=0}^{[j/2]}\mathcal{R}^j(\ln\mathcal{R})^k\phi_{j,k}, \quad (11.8)$$

where [ ] denotes "greatest integer in", and $\mathcal{R} = (r_1^2 + r_2^2)^{1/2}$ is the hyperradius. The leading coefficients are

$$\phi_{0,0} = 1,$$
$$\phi_{1,0} = -\left(Zr_1 + Zr_2 - \tfrac{1}{2}r_{12}\right)/\mathcal{R},$$
$$\phi_{2,1} = -2Z\left(\frac{\pi-2}{3\pi}\right)\frac{\mathbf{r}_1\cdot\mathbf{r}_2}{\mathcal{R}^2}. \quad (11.9)$$

The next term $\phi_{2,0}$ is known in terms of a lengthy expression [4,5], but higher terms have not yet been obtained in closed form. The Fock expansion has been proved convergent for all $\mathcal{R} < \frac{1}{2}$ [6], and extended to pointwise convergence for all $\mathcal{R}$ [7,8].

Asymptotic Form. The long range behavior of many-electron wave functions has been studied from several points of view [9–11]. The basic result of Ref. [12] is that at large distances, the one-electron density behaves as

$$\rho^{1/2}(r) \sim r^{Z^*/t-1}e^{-tr}, \quad (11.10)$$

where $t = (2I_1)^{1/2}$, I_1 is the first ionization potential (in a.u.), and $Z^* = Z - N + 1$ is the screened nuclear charge seen by the outer most electron. For hydrogenic systems with principal quantum number n, $I_1 = (Z^*)^2/2n^2$.

11.3 COMPUTATIONAL METHODS

11.3.1 Variational Methods

Most high precision calculations for the bound states of three-body systems such as helium are based on the Rayleigh-Ritz variational principle. For any normalizable trial function Ψ_{tr}, the quantity

$$E_{\mathrm{tr}} = \frac{\langle\Psi_{\mathrm{tr}}|H|\Psi_{\mathrm{tr}}\rangle}{\langle\Psi_{\mathrm{tr}}|\Psi_{\mathrm{tr}}\rangle} \quad (11.11)$$

satisfies the inequality $E_{\mathrm{tr}} \geq E_1$, where E_1 is the true ground state energy. Thus E_{tr} is an upper bound to E_1. The inequality is easily proved by expanding Ψ_{tr} in the complete basis set of eigenfunctions Ψ_1, Ψ_2, Ψ_3, $\cdots$ of H with eigenvalues $E_1 < E_2 < E_3 < \cdots$, so that

$$\Psi_{\mathrm{tr}} = \sum_{i=1}^{\infty}c_i\Psi_i, \quad (11.12)$$

where the c_i are expansion coefficients. This can always be done in principle, even though the exact Ψ_i are not actually known. If Ψ_{tr} is normalized so that $\langle\Psi_{\mathrm{tr}}|\Psi_{\mathrm{tr}}\rangle = 1$, then $\sum_{i=1}^{\infty}|c_i|^2 = 1$ and

$$
\begin{aligned}
E_{\mathrm{tr}} &= |c_1|^2 E_1 + |c_2|^2 E_2 + |c_3|^2 E_3 + \cdots \\
&= E_1 + |c_2|^2 (E_2 - E_1) + |c_3|^2 (E_3 - E_1) + \cdots \\
&\geq E_1\,,
\end{aligned}
\tag{11.13}
$$

which proves the theorem.

The basic idea of variational calculations then is to write Ψ_{tr} in some arbitrarily chosen mathematical form with variational parameters (subject to normalizability and boundary conditions at the origin and infinity), and then adjust the parameters to obtain the minimum value of E_{tr}.

The minimization problem for the case of *linear* variational coefficients can be solved algebraically. For example, let

$$
\chi_p(\alpha,\beta) = r_1^i r_2^j r_{12}^k\, e^{-\alpha r_1 - \beta r_2}
\tag{11.14}
$$

denote the members of a basis set, where p is an index labeling distinct triplets of nonnegative integer values for the powers $\{i,j,k\}$, and α,β are (for the moment) fixed constants determining the distance scale.

If Ψ_{tr} is expanded in the form

$$
\Psi_{\mathrm{tr}} = \sum_{p=1}^{N} c_p \chi_p(\alpha,\beta)\,,
\tag{11.15}
$$

then the solution to the system of equations $\partial E_{\mathrm{tr}}/\partial c_p = 0$, $p = 1,\ldots,N$, is exactly equivalent to solving the N-dimensional generalized eigenvalue problem

$$
\mathbf{Hc} = \lambda \mathbf{Oc}\,,
\tag{11.16}
$$

where $\mathbf{c}$ is a column vector of coefficients c_p; and $\mathbf{H}$ and $\mathbf{O}$ have matrix elements $H_{pq} = \langle\chi_p|H|\chi_q\rangle$ and $O_{pq} = \langle\chi_p|\chi_q\rangle$. There are N eigenvalues $\lambda_1, \lambda_2, \ldots \lambda_N$, of which the lowest is an upper bound to E_1.

Extension to Excited States. By the Hylleraas-Undheim-MacDonald (HUM) theorem [13, 14], the remaining eigenvalues $\lambda_2, \lambda_3, \ldots$ are also upper bounds to the exact energies $E_2, E_3, \ldots$, provided that the spectrum is bounded from below. The HUM theorem is a consequence of the matrix eigenvalue interleaving theorem, which states that as the dimensions of $\mathbf{H}$ and $\mathbf{O}$ are progressively increased by adding an extra row and column, the N old eigenvalues λ_p fall between the $N+1$ new ones. Consequently, as illustrated in Fig. 11.1, all eigenvalues numbered from the bottom up must move inexorably downward as N is increased. Since the exact spectrum of bound states is obtained in the limit $N \to \infty$, no λ_p can cross the corresponding exact E_p on its way down. Thus $\lambda_p \geq E_p$ for every finite N.

11.3.2 Construction of Basis Sets

Since the Schrödinger equation (11.4) is not separable in the electron coordinates, basis sets which incorporate

Figure 11.1. Diagram illustrating the Hylleraas-Undheim-MacDonald Theorem. The λ_p, $p = 1,\ldots,N$ are the variational eigenvalues for an N-dimensional basis set, and the E_i are the exact eigenvalues of H. The highest λ_p lie in the continuous spectrum of H.

the $r_{12} = |\mathbf{r}_1 - \mathbf{r}_2|$ interelectron coordinate are most efficient. The necessity for r_{12} terms also follows from the Fock expansion (11.8). A basis set constructed from terms of the form (11.14) is called a Hylleraas basis set [15]. [1]

With $\chi_p(\alpha,\beta)$ defined as in (11.14), the general form for a state of total angular momentum L is

$$
\begin{aligned}
\Psi_{\mathrm{tr}} = \sum_{l_1=0}^{[L/2]} \sum_{p} C_{p,l_1} \chi_p(\alpha,\beta) r_1^{l_1} r_2^{l_2} \mathcal{Y}_{l_1\,L-l_1\,L}^{M}(\hat{\mathbf{r}}_1,\hat{\mathbf{r}}_2) \\
\pm \text{exchange}\,,
\end{aligned}
\tag{11.17}
$$

where

$$
\mathcal{Y}_{l_1 l_2 L}^{M}(\hat{\mathbf{r}}_1,\hat{\mathbf{r}}_2) = \sum_{m_1,m_2} Y_{l_1 m_1}(\hat{\mathbf{r}}_1) Y_{l_2 m_2}(\hat{\mathbf{r}}_2) \\
\times \langle l_1 l_2 m_1 m_2 | LM \rangle
\tag{11.18}
$$

is the vector coupled product of angular momenta l_1, l_2 for the two electrons. The sum over p in (11.17) typically includes all terms in (11.14) with $i + j + k \leq \Omega$, where Ω is an integer determining a so-called *Pekeris shell* of terms, and the exchange term denotes the interchange of r_1 and r_2 with $(+)$ for singlet states and $(-)$ for triplet states. Convergence is studied by progressively increasing Ω. The number of terms is

$$
N = \tfrac{1}{6}(\Omega+1)(\Omega+2)(\Omega+3)\,.
$$

---

[1] The basis set is often expressed in terms of the equivalent variables $s = r_1 + r_2$, $t = r_1 - r_2$, $u = r_{12}$.

Basis sets of this type were used by many authors, culminating in the extensive high precision calculations of Pekeris and coworkers [16] for low-lying states, using as many as 1078 terms. Their accuracy is not easily surpassed because of the rapid growth of N with Ω, and because of numerical linear dependence in the basis set for large Ω. Recently, their accuracy has been surpassed by two principal methods. The first explicitly includes powers of logarithmic and half-integral terms in χ_p, as suggested by the Fock expansion [17–19]. This is particularly effective for S-states. The second focuses directly on the multiple distance scales required for an accurate representation of the wave function by writing the trial function in terms of the double basis set [20]

$$\Psi_{\text{tr}} = \sum_{l_1=0}^{[L/2]} \sum_p \left[C_{p,l_1}^{(1)} \chi_p(\alpha_1, \beta_1) + C_{p,l_1}^{(2)} \chi_p(\alpha_2, \beta_2) \right]$$
$$\times r_1^{l_1} r_2^{l_2} \mathcal{Y}_{l_1 l_2 L}(\mathbf{r}_1, \mathbf{r}_2) \pm \text{exchange}, \qquad (11.19)$$

where each $\chi_p(\alpha, \beta)$ is of the form (11.14), but with different values for the distance scales α_1, β_1 and α_2, β_2 in the two sets of terms. They are determined by a complete minimization of E_{tr} with respect to all four parameters, producing a natural division of the basis set into an asymptotic sector and a close-range correlation sector. The method produces a dramatic improvement in accuracy for higher-lying Rydberg states (where variational methods typically deteriorate rapidly in accuracy) and is also effective for low-lying S-states [21, 22]. Nonrelativistic energies accurate to 1 part in 10^{16} are obtainable with modest computing resources.

11.3.3 Calculation of Matrix Elements

The three-body problem has the unique advantage that the full six-dimensional volume element (in the c.m. frame) can be transformed to the product of a three-dimensional angular integral (ang) and a three-dimensional radial integral (rad) over r_1, r_2, and r_{12}. The transformation is

$$\iint d\mathbf{r}_1 d\mathbf{r}_2 = \int_0^{2\pi} d\phi \int_0^{2\pi} d\varphi_1 \int_0^\pi \sin\theta_1 \, d\theta_1$$
$$\times \int_0^\infty r_1 \, dr_1 \int_0^\infty r_2 \, dr_2 \int_{|r_1-r_2|}^{r_1+r_2} r_{12} \, dr_{12}, \qquad (11.20)$$

where θ_1, φ_1 are the polar angles of $\mathbf{r}_1$ and ϕ is the angle of rotation of the triangle formed by $\mathbf{r}_1$, $\mathbf{r}_2$, and $\mathbf{r}_{12}$ about the $\mathbf{r}_1$ direction. The polar angles θ_2, φ_2 are then dependent variables. The basic angular integral is

$$\langle Y_{l_1 m_1}^*(\theta_1, \varphi_1) Y_{l_2 m_2}(\theta_2, \varphi_2) \rangle_{\text{ang}}$$
$$= 2\pi \delta_{l_1 l_2} \delta_{m_1 m_2} P_{l_1}(\cos\theta), \qquad (11.21)$$

where $\cos\theta \equiv \hat{\mathbf{r}}_1 \cdot \hat{\mathbf{r}}_2$ denotes the radial function

$$\cos\theta = \frac{r_1^2 + r_2^2 - r_{12}^2}{2 r_1 r_2}, \qquad (11.22)$$

and $P_l(\cos\theta)$ is a Legendre polynomial. The angular integral over vector-coupled spherical harmonics is [23]

$$\langle \mathcal{Y}_{l_1' l_2' L'}^{M'*}(\hat{\mathbf{r}}_1, \hat{\mathbf{r}}_2) \, \mathcal{Y}_{l_1 l_2 L}^{M}(\hat{\mathbf{r}}_1, \hat{\mathbf{r}}_2) \rangle_{\text{ang}}$$
$$= \delta_{L,L'} \delta_{M,M'} \sum_\Lambda C_\Lambda P_\Lambda(\cos\theta), \qquad (11.23)$$

where

$$C_\Lambda = \tfrac{1}{2} [(2l_1+1)(2l_1'+1)(2l_2+1)(2l_2'+1)]^{1/2}$$
$$\times (-1)^{L+\Lambda}(2\Lambda+1)$$
$$\times \begin{pmatrix} l_1' & l_1 & \Lambda \\ 0 & 0 & 0 \end{pmatrix} \begin{pmatrix} l_2' & l_2 & \Lambda \\ 0 & 0 & 0 \end{pmatrix} \begin{Bmatrix} L & l_1 & l_2 \\ \Lambda & l_2' & l_1' \end{Bmatrix},$$
$$(11.24)$$

and the sum over Λ includes all nonvanishing terms. This can be extended to general matrix elements of tensor operators by further vector coupling [23].

Radial Integrals. Table 11.1 lists formulas for the radial integrals arising from matrix elements of H, as well as those from the Breit interaction (see Sect. 21.1). Although they can all be written in closed form, some have been expressed as infinite series in order to achieve good numerical stability. The exceptions are formulas 5 and 10 in the Table, which became unstable as $\alpha \to \beta$. More elaborate techniques for these are discussed in Ref. [24]. Other cases can be derived by use of the formula

$$\langle r_1^{-1} r_2^{-1} f(r_{12}) e^{-\alpha r_1 - \beta r_2} \rangle_{\text{rad}}$$
$$= \frac{2}{\alpha^2 - \beta^2} \int_0^\infty \left(e^{-\beta r} - e^{-\alpha r} \right) r f(r) \, dr, \qquad (11.25)$$

and then differentiating or integrating with respect to α or β to raise or lower the powers of r_1 and r_2.

Total Integral. The angular integral (11.23) combined with the radial integrals from Table 11.1 yields the total integral

$$\langle \mathcal{Y}_{l_1 l_2 L}^{M*} \, \mathcal{Y}_{l_1' l_2' L}^{M} f(a, b, c; \alpha, \beta) \rangle = \sum_\Lambda C_\Lambda I_\Lambda(a, b, c; \alpha, \beta),$$
$$(11.26)$$

where

$$I_\Lambda(a, b, c; \alpha, \beta) = \langle f(a, b, c; \alpha, \beta) P_\Lambda(\cos\theta) \rangle_{\text{rad}},$$
$$f(a, b, c; \alpha, \beta) = r_1^a r_2^b r_{12}^c \, e^{-\alpha r_1 - \beta r_2}.$$

Starting from I_0 and I_1, the general I_Λ can be efficiently calculated from the recursion relations [23]

Table 11.1. Formulas for the radial integrals $I_0(a,b,c;\alpha,\beta) = \langle r_1^a r_2^b r_{12}^c e^{-\alpha r_1 - \beta r_2}\rangle_{\text{rad}}$ and $I_0^{\log}(a,b,c;\alpha,\beta) = \langle r_1^a r_2^b r_{12}^c \ln r_{12}\, e^{-\alpha r_1 - \beta r_2}\rangle_{\text{rad}}$. $\psi(n) = -\gamma + \sum_{k=1}^{n-1} k^{-1}$ is the digamma function, $_2F_1(a,b;c;z)$ is the hypergeometric function, and $s = a+b+c+5$. Except as noted, the formulas apply for $a \geq -1$, $b \geq -1$, $c \geq -1$.

---

1. $I_0(-2,-2,-1;\alpha,\beta) = \dfrac{2}{\alpha}\ln\left(\dfrac{\alpha+\beta}{\beta}\right) + \dfrac{2}{\beta}\ln\left(\dfrac{\alpha+\beta}{\alpha}\right)$

2. $I_0(a,b,c;\alpha,\beta) = \dfrac{2}{c+2}\displaystyle\sum_{i=0}^{[(c+1)/2]}\binom{c+2}{2i+1}[F_{a+2i+2,\,b+c-2i+2}(\alpha,\beta) + F_{b+2i+2,\,a+c-2i+2}(\beta,\alpha)]$

$$(c \geq -1, s \geq 0)$$

where $F_{p,q}(\alpha,\beta) = \begin{cases} \dfrac{q!}{(\alpha+\beta)^{p+1}\beta^{q+1}}\displaystyle\sum_{j=1}^{q}\dfrac{(p+j)!}{j!}\left(\dfrac{\beta}{\alpha+\beta}\right)^j & q \geq 0, p \geq 0 \\[1em] \dfrac{p!}{\alpha^{p+q+2}}\displaystyle\sum_{j=p+q+1}^{\infty}\dfrac{j!}{(j-q)!}\left(\dfrac{\alpha}{\alpha+\beta}\right)^{j+1} & q < 0, p \geq 0 \\[1em] 0\,^{\text{a}} & p < 0 \end{cases}$

3. $I_0(a,b,c;\alpha,\alpha) = \dfrac{2^{c+3}s!}{(c+2)(2\alpha)^{s+1}}\left\{\displaystyle\sum_{j=0}^{a+1}\binom{a+1}{j}(b+1)!\left[\dfrac{j!}{(j+b+2)!} - \dfrac{(j+c+2)!}{(j+b+c+4)!}\right] + (a \leftrightarrow b)\right\}$

5. $I_0(a,b,-2;\alpha,\beta) = \dfrac{(a+1)!}{\alpha^{a+2}}\displaystyle\sum_{j=0}^{a+1}\dfrac{(b+1+j)!}{j!}\left[\dfrac{\alpha^j}{(\alpha+\beta)^{b+2+j}} - \dfrac{(-\alpha)^j}{(\beta-\alpha)^{b+2+j}}\right]$

$\qquad\times\left[\ln\left(\dfrac{2\alpha}{\alpha+\beta}\right) - \psi(a+2-j) + \psi(1)\right] + \begin{pmatrix} a \leftrightarrow b \\ \alpha \leftrightarrow \beta \end{pmatrix}$

6. $I_0(a,b,-2;\alpha,\alpha) = \dfrac{2s!(a+1)!(b+1)!}{(2\alpha)^{s+1}}\left[\displaystyle\sum_{j=0}^{a+1}\dfrac{\psi(s+1-j) - \psi(a+2-j)}{j!(s-j)!} + (a \leftrightarrow b)\right]$

7. $I_0(-1,-1,-3;\alpha,\beta) = \dfrac{2(\beta\ln\beta - \alpha\ln\alpha)}{\alpha^2 - \beta^2} + \dfrac{2[\psi(2) - \ln\epsilon]}{\alpha+\beta}\,^{\text{b}}$

8. $I_0(a,b,-3;\alpha,\beta) = \left[\dfrac{(a+1)!}{\alpha^{a+1}(\alpha+\beta)^{b+2}}\displaystyle\sum_{j=0}^{a}\dfrac{(b+1+j)!}{j!(a+1-j)!}\left(\dfrac{\alpha}{\alpha+\beta}\right)^j + \begin{pmatrix} a \leftrightarrow b \\ \alpha \leftrightarrow \beta \end{pmatrix}\right] - \dfrac{s!\left[\ln(\alpha\beta\epsilon^2) - \psi(1)\right]}{(\alpha+\beta)^{s+1}}$

$\qquad - \dfrac{(a+1)!(b+1)!}{(s+1)\alpha^{a+2}\beta^{b+1}}\,_2F_1\left(a+2,1;s+2;\dfrac{\alpha-\beta}{\alpha}\right), \quad a \geq -1, b \geq -1$

9. $I_0^{\log}(-1,-1,c;\alpha,\beta) = \dfrac{2(c+1)!}{\alpha^2 - \beta^2}\left[\dfrac{\ln\alpha - \psi(c+2)}{\alpha^{c+2}} - \dfrac{\ln\beta - \psi(c+2)}{\beta^{c+2}}\right]$

10. $I_0^{\log}(a,b,c;\alpha,\beta) = \dfrac{(a+1)!}{(c+2)\alpha^{a+c+4}}\displaystyle\sum_{j=0}^{a+1}\dfrac{(b+1+j)!(a+c+3-j)!}{j!(a+1-j)!}\left[\dfrac{\alpha^j}{(\alpha+\beta)^{b+2+j}} - \dfrac{(-\alpha)^j}{(\beta-\alpha)^{b+2+j}}\right]$

$\qquad + \begin{pmatrix} a \leftrightarrow b \\ \alpha \leftrightarrow \beta \end{pmatrix}$

11. $I_0^{\log}(a,b,c;\alpha,\alpha) = \dfrac{2^{c+3}s!(b+1)!}{(c+2)(2\alpha)^{s+1}}\displaystyle\sum_{j=0}^{a+1}\binom{a+1}{j}\left\{\left[\dfrac{j!}{(j+b+2)!} - \dfrac{(j+c+2)!}{(j+b+c+4)!}\right]\left[\psi(s+1) - \ln\alpha - \dfrac{1}{c+2}\right]\right.$

$\qquad \left. + \dfrac{(a+c+3-j)!}{(s+1-j)!}[\psi(s+1-j) - \psi(a+c+4-j)]\right\} + (a \leftrightarrow b)$

---

^{a} Terms with $p < 0$ represent divergent parts which cancel from convergent differences between integrals with the same α and β.
^{b} ϵ is the radius of an infinitesimal sphere about $r_{12} = 0$ which is omitted from the range of integration.

$$I_{\Lambda+1}(a,b,c;\alpha,\beta) = \frac{2\Lambda+1}{c+2}I_\Lambda(a-1,b-1,c+2;\alpha,\beta)$$
$$- I_{\Lambda-1}(a,b,c;\alpha,\beta), \quad c \neq -2 \tag{11.27}$$

$$I_{\Lambda+1}(a,b,-2;\alpha,\beta) = (2\Lambda+1)I_\Lambda^{\log}(a-1,b-1,0;\alpha,\beta)$$
$$- I_{\Lambda-1}(a,b,-2;\alpha,\beta), \quad c = -2 \tag{11.28}$$

where

$$I_\Lambda^{\log}(a,b,c;\alpha,\beta) = \langle f(a,b,c,\alpha,\beta)\ln r_{12}P_\Lambda(\cos\theta)\rangle_{\text{rad}}.$$

The $I_\Lambda^{\log}$ integrals follow the recursion relation

$$I_{\Lambda+1}^{\log}(a,b,c;\alpha,\beta)$$
$$= \frac{(2\Lambda+1)}{c+2}\left[I_\Lambda^{\log}(a-1,b-1,c+2;\alpha,\beta)\right.$$
$$\left. - \frac{1}{c+2}I_\Lambda(a-1,b-1,c+2;\alpha,\beta)\right]$$
$$+ I_{\Lambda-1}^{\log}(a,b,c;\alpha,\beta). \tag{11.29}$$

Hamiltonian Matrix Elements. The general form of the Laplacian operator in terms of r_1, r_2, r_{12} variables is

$$\nabla_1^2 = \frac{1}{r_1^2}\frac{\partial}{\partial r_1}\left(r_1^2\frac{\partial}{\partial r_1}\right) + \frac{1}{r^2}\frac{\partial}{\partial r}\left(r^2\frac{\partial}{\partial r}\right) - \frac{l_1^2}{r_1^2}$$
$$+ \frac{2(r_1 - r_2\cos\theta)}{r}\frac{\partial^2}{\partial r_1 \partial r} - 2\left(\nabla_1^Y \cdot \mathbf{r}_2\right)\frac{1}{r}\frac{\partial}{\partial r}, \tag{11.30}$$

and similarly for ∇_2^2 with subscripts 1 and 2 interchanged. The term ∇_1^Y is understood to act only on the $\mathcal{Y}_{l_1 l_2 L}^M(\hat{\mathbf{r}}_1, \hat{\mathbf{r}}_2)$ part of the wave function. This term in ∇_1^2 can be easily evaluated by means of the effective operator replacement

$$\langle \mathcal{Y}_{l_1' l_2' L'}^{M*} \nabla_1^Y \cdot \mathbf{r}_2\, \mathcal{Y}_{l_1 l_2 L}^M\rangle_{\text{ang}} \frac{1}{r_{12}}\frac{\partial g(r_{12})}{\partial r_{12}}$$
$$\rightarrow \frac{g(r_{12})}{2r_1 r_2}\sum_\Lambda \tilde{C}_\Lambda P_\Lambda(\cos\theta) \tag{11.31}$$

for the angular part of the total integral, where

$$\tilde{C}_\Lambda = [l_1'(l_1'+1) - l_1(l_1+1) - \Lambda(\Lambda+1)]C_\Lambda. \tag{11.32}$$

The replacement (11.31) becomes an equality after radial integration with any function $g(r_{12})$ in the integrand. The matrix elements of H between arbitrary basis functions defined by

$$\chi = r_1^a r_2^b r_{12}^c\, e^{-\alpha r_1 - \beta r_2}\mathcal{Y}_{l_1 l_2 L}^M(\hat{\mathbf{r}}_1, \hat{\mathbf{r}}_2),$$
$$\chi' = r_1^{a'} r_2^{b'} r_{12}^{c'}\, e^{-\alpha' r_1 - \beta' r_2}\mathcal{Y}_{l_1' l_2' L}^M(\hat{\mathbf{r}}_1, \hat{\mathbf{r}}_2),$$

can then be written in the explicitly Hermitian form (for infinite nuclear mass)

$$\langle \chi'|H|\chi\rangle = \frac{1}{8}\sum_\Lambda C_\Lambda \sum_{i=0}^2 \left[A_i^{(1)}I_\Lambda(a_+ - i, b_+, c_+; \alpha_+, \beta_+)\right.$$
$$+ A_i^{(2)}I_\Lambda(a_+, b_+ - i, c_+; \alpha_+, \beta_+)$$
$$\left. + A_i^{(3)}I_\Lambda(a_+, b_+, c_+ - i; \alpha_+, \beta_+)\right] \tag{11.33}$$

where $a_\pm = a' \pm a$, $\alpha_\pm = \alpha' \pm \alpha$ etc., and

$$A_0^{(1)} = -\alpha_+^2 - \alpha_-^2 + 2\alpha_-\alpha_+(c_-/c_+),$$
$$A_1^{(1)} = 2\{\alpha_+(a_++2) + \alpha_-a_-$$
$$- [\alpha_+a_- + \alpha_-(a_++2)](c_-/c_+)\} - 8Z,$$
$$A_2^{(1)} = -a_+^2 - a_-^2 + 2a_+ + 2a_-(a_++1)(c_-/c_+)$$
$$+ 2l_1(l_1+1)(1 - c_-/c_+)$$
$$+ 2l_1'(l_1'+1)(1 + c_-/c_+),$$
$$A_0^{(3)} = 0, \qquad A_1^{(3)} = 8,$$
$$A_2^{(3)} = 2(c_+^2 - c_-^2),$$

with $(c_-/c_+) = 0$ for $c_+ = 0$. The $A_i^{(2)}$ are defined similarly to $A_i^{(1)}$ with the replacements $a \rightarrow b$, $\alpha \rightarrow \beta$, $l_1 \rightarrow l_2$. The overlap integral is

$$\langle \chi'|\chi\rangle = \sum_\Lambda C_\Lambda I_\Lambda(a_+, b_+, c_+; \alpha_+, \beta_+). \tag{11.34}$$

11.3.4 Other Computational Methods

Although not yet at the same level of accuracy as variational methods, certain nonvariational methods, such as finite element methods [25], solutions to the Faddeev equations [26], and the correlated-function hyperspherical-harmonic method [27], have their own advantages of flexibility and/or generality. A characteristic feature of these methods is that they provide direct numerical solutions to the three-body problem which in principle converge pointwise to the exact solution, rather than depending upon a globally optimized solution. Other methods particularly suited to doubly-excited states are discussed in Chap. 25.

11.4 VARIATIONAL EIGENVALUES

High precision variational eigenvalues are available for all states of helium up to $n = 10$ and $L = 7$ [21,22]. The nonrelativistic values of ε_0, ε_1 and ε_2 [see Eq. (11.6)] are listed in Tables 11.2 and 11.3. The ε_0 are the eigenvalues for infinite nuclear mass, and ε_1 and ε_2, together with Eq. (11.6) give the finite mass corrections for the isotopes ^{3}He and ^{4}He. The values of μ/M can be calculated from

$$\frac{\mu}{M} = \left[\frac{M_A}{5.485\,799\,015\,(130)\times 10^{-4}} - N + 1\right]^{-1} \tag{11.35}$$

Table 11.2. Nonrelativistic eigenvalue coefficients ε_0 and ε_1 for helium.

State	$\varepsilon_0(n\,^1L)$	$\varepsilon_1(n\,^1L)$	$\varepsilon_0(n\,^3L)$	$\varepsilon_1(n\,^3L)$
1 S	−2.903 724 377 034 119 5	0.159 069 475 085 84		
2 S	−2.145 974 046 054 419(6)	0.009 503 864 419 28	−2.175 229 378 236 791 30	0.007 442 130 706 04
2 P	−2.123 843 086 498 093(2)	0.046 044 524 937(1)	−2.133 164 190 779 273(5)	−0.064 572 425 024(4)
3 S	−2.061 271 989 740 911(5)	0.002 630 567 097 7(1)	−2.068 689 067 472 457 19	0.001 896 211 617 81
3 P	−2.055 146 362 091 94(3)	0.014 548 047 097(1)	−2.058 081 084 274 28(4)	−0.018 369 001 636(2)
3 D	−2.055 620 732 852 246(6)	−0.000 249 399 992 1(1)	−2.055 636 309 453 261(4)	0.000 025 322 839(1)
4 S	−2.033 586 717 030 72(1)	0.001 073 641 226 6(1)	−2.036 512 083 098 236 30(2)	0.000 742 661 516 18
4 P	−2.031 069 650 450 24(3)	0.006 254 923 554 3(1)	−2.032 324 354 296 62(2)	−0.007 555 178 98(1)
4 D	−2.031 279 846 178 687(7)	−0.000 129 175 188 7(8)	−2.031 288 847 501 795(3)	0.000 029 442 651(2)
4 F	−2.031 255 144 381 749(1)	−0.000 010 024 269 4(2)	−2.031 255 168 403 245 6(6)	−0.000 009 669 639 6
5 S	−2.021 176 851 574 363(5)	0.000 538 860 360 5(1)	−2.022 618 872 302 312 27(1)	0.000 363 697 136 49
5 P	−2.019 905 989 900 83(2)	0.003 230 021 84(2)	−2.020 551 187 256 25(1)	−0.003 810 911 035(1)
5 D	−2.020 015 836 159 984(4)	−0.000 071 883 131(6)	−2.020 021 027 446 911(5)	0.000 019 568 85(1)
5 F	−2.020 002 937 158 742 7(5)	−0.000 005 704 294 6(4)	−2.020 002 957 377 369 4(4)	−0.000 005 406 490 0(5)
5 G	−2.020 000 710 898 584 71(1)	−0.000 001 404 413 6	−2.020 000 710 925 343 92(1)	−0.000 001 404 001 3
6 S	−2.014 563 098 446 60(1)	0.000 307 704 277(1)	−2.015 377 452 992 862 19(3)	0.000 204 329 479 10
6 P	−2.013 833 979 671 73(2)	0.001 878 058 536(1)	−2.014 207 958 773 74(1)	−0.002 184 346 463(1)
6 D	−2.013 898 227 424 286(5)	−0.000 043 412 268 9(9)	−2.013 901 415 453 792(7)	0.000 012 742 22(3)
6 F	−2.013 890 683 815 549 7(3)	−0.000 003 482 257(7)	−2.013 890 698 348 532 0(2)	−0.000 003 268 458 6(8)
6 G	−2.013 889 345 387 313 22(3)	−0.000 000 898 579 9(7)	−2.013 889 345 416 952 96(3)	−0.000 000 898 123 7(7)
6 H	−2.013 889 034 754 279 72	−0.000 000 290 347 1	−2.013 889 034 754 301 55	−0.000 000 290 346 7
7 S	−2.010 625 776 210 87(2)	0.000 191 925 025(1)	−2.011 129 919 527 626 21(4)	0.000 125 981 736 89
7 P	−2.010 169 314 529 35(2)	0.001 186 152 30(1)	−2.010 404 960 007 94(2)	−0.001 366 500 8(3)
7 D	−2.010 210 028 457 98(1)	−0.000 028 027 840(2)	−2.010 212 105 955 595(2)	0.000 008 563 121(3)
7 F	−2.010 205 248 074 013(1)	−0.000 002 262 00(4)	−2.010 205 258 374 865(1)	−0.000 002 110 58(3)
7 G	−2.010 204 386 224 772 55(7)	−0.000 000 598 396 3(3)	−2.010 204 386 250 217 93(6)	−0.000 000 598 005(1)
7 H	−2.010 204 182 806 482 04(2)	−0.000 000 201 097 8	−2.010 204 182 806 512 04(1)	−0.000 000 201 097 3
7 I	−2.010 204 120 606 191 32	−0.000 000 077 775 5	−2.010 204 120 606 191 340	−0.000 000 077 775 5
8 S	−2.008 093 622 105 61(4)	0.000 127 650 436(1)	−2.008 427 122 064 721 42(6)	0.000 083 070 552 34
8 P	−2.007 789 127 133 22(2)	0.000 796 195 83(5)	−2.007 947 013 771 12(1)	−0.000 911 053 5(3)
8 D	−2.007 816 512 563 811(7)	−0.000 019 076 181(1)	−2.007 817 934 711 706(3)	0.000 005 971 123 4(3)
8 F	−2.007 813 297 115 014 1(6)	−0.000 001 545 48(1)	−2.007 813 304 535 090 8(5)	−0.000 001 436 452(2)
8 G	−2.007 812 711 494 024 1(1)	−0.000 000 415 004 0(1)	−2.007 812 711 514 424 82(9)	−0.000 000 414 690 4
8 H	−2.007 812 571 828 655 81(1)	−0.000 000 142 649 2(3)	−2.007 812 571 828 685 73(1)	−0.000 000 142 648 7(2)
8 I	−2.007 812 528 549 584 59	−0.000 000 056 935 9	−2.007 812 528 549 584 61	−0.000 000 056 935 9
8 K	−2.007 812 512 570 229 31	−0.000 000 025 111 3	−2.007 812 512 570 229 306	−0.000 000 025 111 3
9 S	−2.006 369 553 107 85(3)	0.000 089 149 638 7(7)	−2.006 601 516 715 010 67(3)	0.000 057 628 311 52
9 P	−2.006 156 384 652 86(5)	0.000 559 978 028(2)	−2.006 267 267 366 41(4)	−0.000 637 531 359(6)
9 D	−2.006 175 671 437 641(6)	−0.000 013 542 185(3)	−2.006 176 684 884 697(2)	0.000 004 306 538(6)
9 F	−2.006 173 406 897 324 6(8)	−0.000 001 099 967 1(3)	−2.006 173 412 365 043 0(7)	−0.000 001 019 651(2)
9 G	−2.006 172 991 627 586 3(2)	−0.000 000 298 267 2(1)	−2.006 172 991 643 665 0(3)	−0.000 000 298 019 8(1)
9 H	−2.006 172 891 903 619 14(2)	−0.000 000 104 002 2	−2.006 172 891 903 645 88(2)	−0.000 000 104 001 9
9 I	−2.006 172 860 732 382 57	−0.000 000 042 313 6	−2.006 172 860 732 382 60	−0.000 000 042 313 6(1)
9 K	−2.006 172 849 096 329 78	−0.000 000 019 151 6	−2.006 172 849 096 329 780	−0.000 000 019 151 6
10 S	−2.005 142 991 748 00(8)	0.000 064 697 214(3)	−2.005 310 794 915 611 3(2)	0.000 041 598 811 52
10 P	−2.004 987 983 802 22(4)	0.000 408 649 426 3	−2.005 068 805 497 8(1)	−0.000 463 433 718(8)
10 D	−2.005 002 071 654 250(6)	−0.000 009 947 506 0(6)	−2.005 002 818 080 232(8)	0.000 003 198 298(8)
10 F	−2.005 000 417 564 668 2(9)	−0.000 000 809 442(9)	−2.005 000 421 686 603 6(7)	−0.000 000 748 926 4(2)
10 G	−2.005 000 112 764 318 0(3)	−0.000 000 220 982(2)	−2.005 000 112 777 003 1(4)	−0.000 000 220 785(3)
10 H	−2.005 000 039 214 394 52(2)	−0.000 000 077 806 7	−2.005 000 039 214 417 41(2)	−0.000 000 077 806 2
10 I	−2.005 000 016 086 516 19	−0.000 000 032 059 0(1)	−2.005 000 016 086 516 22	−0.000 000 032 058 9(2)
10 K	−2.005 000 007 388 375 88	−0.000 000 014 751 4	−2.005 000 007 388 375 88	−0.000 000 014 751 4

Table 11.3. Eigenvalue coefficients ε_2 for helium.

State	$\varepsilon_2(n\,{}^1L)$	$\varepsilon_2(n\,{}^3L)$
$1\,S$	$-0.470\,391\,870(1)$	
$2\,S$	$-0.135\,276\,864(1)$	$-0.057\,495\,847\,9(2)$
$2\,P$	$-0.168\,271\,22(7)$	$-0.204\,959\,88(1)$
$3\,S$	$-0.058\,599\,3124(4)$	$-0.040\,455\,850\,5(5)$
$3\,P$	$-0.066\,047\,859(3)$	$-0.070\,292\,710(2)$
$3\,D$	$-0.057\,201\,299(9)$	$-0.054\,737\,73(1)$
$4\,S$	$-0.032\,522\,293(2)$	$-0.025\,628\,633\,8(1)$
$4\,P$	$-0.035\,159\,71(6)$	$-0.036\,129\,973(2)$
$4\,D$	$-0.032\,150\,91(2)$	$-0.030\,747\,891(7)$
$4\,F$	$-0.031\,274\,336(4)$	$-0.031\,277\,992\,1(3)$
$5\,S$	$-0.020\,647\,26(9)$	$-0.017\,322\,734\,96$
$5\,P$	$-0.021\,847\,6(3)$	$-0.022\,166\,61(9)$
$5\,D$	$-0.020\,510\,1(2)$	$-0.019\,706\,2(2)$
$5\,F$	$-0.020\,013\,498(6)$	$-0.020\,016\,561(4)$
$5\,G$	$-0.020\,003\,560\,8$	$-0.020\,003\,564\,6$
$6\,S$	$-0.014\,261\,796(4)$	$-0.012\,411\,399\,1(3)$
$6\,P$	$-0.014\,902\,86(9)$	$-0.015\,033\,58(5)$
$6\,D$	$-0.014\,199\,4(2)$	$-0.013\,707\,27(1)$
$6\,F$	$-0.013\,896\,984(2)$	$-0.013\,899\,22(3)$
$6\,G$	$-0.013\,891\,179(6)$	$-0.013\,891\,184(8)$
$6\,H$	$-0.013\,889\,619\,1$	$-0.013\,889\,619\,0$
$7\,S$	$-0.010\,438\,2(2)$	$-0.009\,304\,443\,3(3)$
$7\,P$	$-0.010\,818\,6(2)$	$-0.010\,879(2)$
$7\,D$	$-0.010\,405\,09(3)$	$-0.010\,085\,212(1)$
$7\,F$	$-0.010\,209\,2(3)$	$-0.010\,210\,7(3)$
$7\,G$	$-0.010\,205\,61(5)$	$-0.010\,205\,61(5)$
$7\,H$	$-0.010\,204\,590(2)$	$-0.010\,204\,587(2)$
$7\,I$	$-0.010\,204\,276\,7$	$-0.010\,204\,276\,8$
$8\,S$	$-0.007\,968\,944(3)$	$-0.007\,224\,770\,5(3)$
$8\,P$	$-0.008\,211\,7(5)$	$-0.008\,248\,7(6)$
$8\,D$	$-0.007\,950\,7(4)$	$-0.007\,731\,59(2)$
$8\,F$	$-0.007\,815\,9(3)$	$-0.007\,817\,0(2)$
$8\,G$	$-0.007\,813\,563(1)$	$-0.007\,813\,568(3)$
$8\,H$	$-0.007\,812\,855(4)$	$-0.007\,812\,859(5)$
$8\,I$	$-0.007\,812\,642\,9$	$-0.007\,812\,642\,9$
$8\,K$	$-0.007\,812\,563\,0$	$-0.007\,812\,563\,0$
$9\,S$	$-0.006\,282\,5136(1)$	$-0.005\,768\,028\,5(1)$
$9\,P$	$-0.006\,445\,7(2)$	$-0.006\,464\,936\,9(1)$
$9\,D$	$-0.006\,270\,99(7)$	$-0.006\,115\,2(1)$
$9\,F$	$-0.006\,175\,20(1)$	$-0.006\,176\,025\,4(7)$
$9\,G$	$-0.006\,173\,579\,6(1)$	$-0.006\,173\,592(4)$
$9\,H$	$-0.006\,173\,104(2)$	$-0.006\,173\,101(2)$
$9\,I$	$-0.006\,172\,945\,9(1)$	$-0.006\,172\,946\,0(2)$
$9\,K$	$-0.006\,172\,887\,6$	$-0.006\,172\,887\,6$
$10\,S$	$-0.005\,079\,8362(8)$	$-0.004\,709\,453\,0(1)$
$10\,P$	$-0.005\,197(1)$	$-0.005\,206\,7(1)$
$10\,D$	$-0.005\,072\,4(4)$	$-0.004\,958\,0(8)$
$10\,F$	$-0.005\,001\,76(2)$	$-0.005\,002\,386(2)$
$10\,G$	$-0.005\,000\,55(2)$	$-0.005\,000\,55(2)$
$10\,H$	$-0.005\,000\,193\,5(2)$	$-0.005\,000\,193\,5(1)$
$10\,I$	$-0.005\,000\,080\,3(4)$	$-0.005\,000\,081(1)$
$10\,K$	$-0.005\,000\,036\,9$	$-0.005\,000\,036\,8$

Table 11.4. Values of the reduced electron mass ratio μ/M.

Isotope	$\mu/M \times 10^4$
^{1}H	$5.443\,205\,67(11)$
^{2}D	$2.723\,695\,02(6)$
^{3}He	$1.819\,212\,04(3)$
^{4}He	$1.370\,745\,62(3)$
^{6}Li	$0.912\,167\,73(11)$
^{7}Li	$0.782\,020\,31(9)$
^{9}Be	$0.608\,820\,44(3)$

where M_A is the atomic mass (in amu, see Ref. [28] for a tabulation)[2] and N is the number of electrons. For ^{4}He, one can use directly the accurately known value of m_e/m_α to calculate $\mu/M = 1/(m_\alpha/m_e + 1)$. Values of μ/M for the first several isotopes are listed in Table 11.4, and the corresponding energy coefficients for the $1s^2\ {}^1S$ ground state are given in Table 11.5.

11.4.1 Expectation Values of Operators and Sum Rules

Expectation values for various powers of the radial coordinates, together with operators appearing in the Breit interaction, are listed in Table 11.6 for the ground state of helium and He-like ions. Included are all terms required to calculate $\langle V^2 \rangle$, and the oscillator strength sum rules [29]

$$S(-1) = \tfrac{2}{3}\langle (\mathbf{r}_1 + \mathbf{r}_2)^2 \rangle, \qquad (11.36a)$$

$$S(0) = 2, \qquad (11.36b)$$

$$S(1) = -\tfrac{4}{3}(\varepsilon_0 - \varepsilon_1), \qquad (11.36c)$$

$$S(2) = \frac{2\pi Z}{3}\langle \delta(\mathbf{r}_1) + \delta(\mathbf{r}_2) \rangle, \qquad (11.36d)$$

where $S(k) = \sum_n [\varepsilon_0(n^1P) - \varepsilon_0(1^1S)]^k f_{0n}$, with energies in a.u., and f_{0n} is the $1^1S - n^1P$ oscillator strength (see Sect. 11.6.1).

11.5 TOTAL ENERGIES

As discussed in Chaps. 21 and 27, relativistic and QED corrections must be added to the nonrelativistic eigenvalues of Sect. 11.4 before a meaningful comparison with measured transition frequencies can be made. The corrections are discussed in detail in Refs. [21, 30].

The terms in order of decreasing size are:
1. **Relativistic Corrections of $O(\alpha^2)$.**

---

[2]For high precision work, the helium electronic binding energy of 8.48×10^{-8} amu should be added to M_A.

Table 11.5. Nonrelativistic eigenvalues $E = \varepsilon_0 + (\mu/M)\varepsilon_1 + (\mu/M)^2\varepsilon_2$ for helium-like ions (in units of e^2/a_μ).

Atom	$\varepsilon_0(1\,^1S)$	$\varepsilon_1(1\,^1S)$	$\varepsilon_2(1\,^1S)$
H$^-$	$-0.527\,751\,016\,544\,377$	$0.032\,879\,781\,852\,30$	$-0.059\,779\,492\,64(1)$
He	$-2.903\,724\,377\,034\,119\,5$	$0.159\,069\,475\,085\,84$	$-0.470\,391\,870(1)$
Li$^+$	$-7.279\,913\,412\,669\,305\,9$	$0.288\,975\,786\,393\,99$	$-1.277\,369\,377\,6(2)$
Be^{++}	$-13.155\,566\,238\,423\,586\,7$	$0.420\,520\,303\,439\,44$	$-2.491\,572\,858\,1(1)$

Table 11.6. Expectation values of various operators for He-like ions for the case $M = \infty$ (in a.u.).

Quantity	H$^-$	He	Li$^+$	Be^{++}
$\langle r_1^2 \rangle$	$11.913\,699\,678\,05(6)$	$1.193\,482\,995\,019$	$0.446\,279\,011\,201$	$0.232\,067\,315\,531$
$\langle r_{12}^2 \rangle$	$25.202\,025\,291\,2(1)$	$2.516\,439\,312\,833$	$0.927\,064\,803\,063$	$0.477\,946\,525\,143$
$\langle \mathbf{r}_1 \cdot \mathbf{r}_2 \rangle$	$-0.687\,312\,967\,569$	$-0.064\,736\,661\,398$	$-0.017\,253\,390\,330$	$-0.006\,905\,947\,040$
$\langle r_1 \rangle$	$2.710\,178\,278\,444(1)$	$0.929\,472\,294\,874$	$0.572\,774\,149\,971$	$0.414\,283\,328\,006$
$\langle r_{12} \rangle$	$4.412\,694\,497\,992(2)$	$1.422\,070\,255\,566$	$0.862\,315\,375\,456$	$0.618\,756\,314\,066$
$\langle 1/r_1 \rangle$	$0.683\,261\,767\,652$	$1.688\,316\,800\,717$	$2.687\,924\,397\,413$	$3.687\,750\,406\,344$
$\langle 1/r_{12} \rangle$	$0.311\,021\,502\,214$	$0.945\,818\,448\,800$	$1.567\,719\,559\,137$	$2.190\,870\,773\,906$
$\langle 1/r_1^2 \rangle$	$1.116\,662\,824\,6(1)$	$6.017\,408\,867\,0(1)$	$14.927\,623\,721\,4(2)$	$27.840\,105\,671\,33(2)$
$\langle 1/r_{12}^2 \rangle$	$0.155\,104\,152\,58(3)$	$1.464\,770\,923\,350(1)$	$4.082\,232\,787\,55(2)$	$8.028\,801\,781\,824(1)$
$\langle 1/r_1 r_2 \rangle$	$0.382\,627\,890\,340$	$2.708\,655\,474\,480$	$7.011\,874\,111\,824(1)$	$13.313\,954\,940\,144(1)$
$\langle 1/r_1 r_{12} \rangle$	$0.253\,077\,567\,065$	$1.920\,943\,921\,900$	$5.069\,790\,932\,379$	$9.717\,071\,116\,528$
$\langle \delta(r_1) \rangle$	$0.164\,552\,872\,86(3)$	$1.810\,429\,318\,49(3)$	$6.852\,009\,434\,4(1)$	$17.198\,172\,544\,74(3)$
$\langle \delta(r_{12}) \rangle$	$0.002\,737\,992\,3(3)$	$0.106\,345\,371\,2(2)$	$0.533\,722\,537\,1(9)$	$1.522\,895\,354\,1(2)$
$\langle p^4 \rangle$	$2.462\,558\,614(3)$	$54.088\,067\,230(2)$	$310.547\,150\,179(6)$	$1047.278\,491\,476(2)$
$\langle H_{\text{oo}} \rangle/\alpha^2$	$-0.008\,875\,022\,10(1)$	$-0.139\,094\,690\,556(1)$	$-0.427\,991\,611\,178(9)$	$-0.878\,768\,694\,709(1)$

$$H_{\text{rel}} = H_{\text{NFS}} + H_{\text{FS}},$$
$$H_{\text{NFS}} = H_{\text{mass}} + H_{\text{D}} + H_{\text{ssc}} + H_{\text{oo}},$$
$$H_{\text{FS}} = H_{\text{so}} + H_{\text{soo}} + H_{\text{ss}}.$$

The various nonfine-structure (NFS) and fine-structure (FS) terms are defined in Sect. 21.1. The off-diagonal matrix elements of H_{FS} mix states of different spin and cause a break-down of LS-coupling.

2. Anomalous Magnetic Moment Corrections of $O(\alpha^3)$. The general FS matrix elements between states with spins S and S' due to the anomalous magnetic moment a_e are (see Sect. 27.5)

$$\langle \gamma S | H_{\text{FS}}^{\text{anom}} | \gamma' S' \rangle = 2a_e \langle \gamma S | H_{\text{so}} + \tfrac{2}{3}\delta_{S,S'} H_{\text{soo}} + \left(1 + \tfrac{1}{2}a_e\right) H_{\text{ss}} | \gamma' S' \rangle, \quad (11.37)$$

where $a_e = (g_e - 2)/2 = \alpha/(2\pi) - 0.328\,479\alpha^2 + \cdots$.

3. QED Corrections of $O(\alpha^3)$. The lowest order QED corrections (including NFS anomalous magnetic moment terms) can be written in the form $\Delta_{L,1} + \Delta_{L,2}$, where

$$\Delta E_{L,1} = \tfrac{4}{3} Z\alpha^3 \left[\ln(Z\alpha)^{-2} + \tfrac{19}{30} - \ln k_0\right] \langle \delta(\mathbf{r}_1) + \delta(\mathbf{r}_2) \rangle \quad (11.38)$$

$$\Delta E_{L,2} = \alpha^3 \left[\tfrac{89}{15} + \tfrac{14}{3}\ln\alpha - \tfrac{20}{3}\mathbf{s}_1 \cdot \mathbf{s}_2\right] \langle \delta(\mathbf{r}_{12}) \rangle - \tfrac{14}{3}\alpha^3 Q, \quad (11.39)$$

$\ln k_0$ is the two-electron Bethe logarithm defined by Eq. (27.86) and Q is the matrix element defined by Eq.

(27.83). For a highly excited $1snl$ state, $\Delta E_{L,2} \to 0$, $\langle \delta(\mathbf{r}_1) + \delta(\mathbf{r}_2) \rangle \to Z^3/\pi$, $\ln k_0 \to \ln k_0(1s) = 2.984\,128\,555$, and $\Delta E_{L,1}$ reduces to the Lamb shift of the $1s$ core state (see Sects. 28.3.4 and 28.3.5). Thus $\Delta E_{L,1}$ represents the essentially one-electron part of the QED shift with the factor of Z^3/π replaced by the correct electron density at the nucleus. Values of $\ln k_0$ for two-electron atoms are in general difficult to calculate. For a $1snl$ state with large l, the asymptotic expansion [31]

$$\ln k_0(1snl) \sim \ln k_0(1s) + \frac{1}{n^3}\left(\frac{Z-1}{Z}\right)^4 \ln k_0(nl) + 0.316\,26 Z^{-6}\langle r^{-4} \rangle_{nl} \quad (11.40)$$

becomes essentially exact. Here $\ln k_0(nl)$ is the one-electron Bethe logarithm [32] and

$$\langle r^{-4} \rangle_{nl} = \frac{16(Z-1)^4[3n^2 - l(l+1)]}{(2l-1)2l(2l+1)(2l+2)(2l+3)}. \quad (11.41)$$

For the low-lying S-states of helium [33],

$$\ln k_0(1\,^1S) = 2.983\,856\,(2), \quad (11.42a)$$
$$\ln k_0(2\,^1S) = 2.980\,058\,(2), \quad (11.42b)$$
$$\ln k_0(2\,^3S) = 2.977\,720\,(1). \quad (11.42c)$$

For higher Z, the $1/Z$ expansions [34]

Table 11.7. Total ionization energies for ^{4}He, calculated with $R_M = 3\,289\,391\,007.44$ MHz.

State	$E(n\,^1L_L)$	$E(n\,^3L_{L-1})$	$E(n\,^3L_L)$	$E(n\,^3L_{L+1})$
1 S	5 945 204 258(180)			
2 S	960 332 042.6(8.0)			1 152 842 738.0(8.0)
2 P	814 709 153.5(8.0)	876 078 647.4(8.0)	876 108 270.6(8.0)	876 110 563.6(8.0)
3 S	403 096 136.5(34.0)			451 903 467.1(41.0)
3 P	362 787 969.7(2.4)	382 109 904.0(2.4)	382 118 019.2(2.4)	382 118 678.5(2.4)
3 D	365 917 749.11(2)	366 018 892.94(2)	366 020 218.06(2)	366 020 293.39(2)
4 S	220 948 853.8(13.0)			240 193 898.6(16.0)
4 P	204 397 211.4(1.2)	212 658 041.6(1.2)	212 661 349.1(1.2)	212 661 618.8(1.2)
4 D	205 783 935.87(1)	205 842 547.90(1)	205 843 103.14(1)	205 843 139.16(1)
4 F	205 620 797.192 1(7)	205 621 029.649 2(7)	205 621 502.065 8(7)	205 621 288.021 5(7)
5 S	139 318 271.7(6.0)			148 807 319.8(8.0)
5 P	130 955 542.2(4)	135 203 443.6(4)	135 205 105.9(4)	135 205 241.4(4)
5 D	131 680 211.973(6)	131 714 043.935(6)	131 714 327.495(6)	131 714 346.715(6)
5 F	131 595 041.530 6(5)	131 595 195.264 3(6)	131 595 419.770 5(6)	131 595 327.483 4(6)
5 G	131 580 320.162 1(2)	131 580 370.975 8(2)	131 580 529.548 1(2)	131 580 446.489 9(2)
6 S	95 807 694.1(4.0)			101 166 450.4(4.0)
6 P	91 009 810.75(24)	93 472 041.87(24)	93 472 992.88(24)	93 473 070.37(24)
6 D	91 433 655.865(3)	91 454 440.606(3)	91 454 604.487(3)	91 454 615.833(3)
6 F	91 383 852.051 7(3)	91 383 954.321 5(3)	91 384 078.920 4(3)	91 384 030.814 4(3)
6 G	91 374 997.981 5(2)	91 375 027.438 1(2)	91 375 119.156 5(2)	91 375 071.134 2(2)
6 H	91 372 940.632 7(2)	91 372 961.833 9(1)	91 373 021.550 7(1)	91 372 990.247 1(1)
7 S	69 904 830.2(2.0)			73 222 276.7(2.0)
7 P	66 901 127.65(20)	68 452 586.84(20)	68 453 181.14(20)	68 453 229.54(20)
7 D	67 169 717.174(2)	67 183 264.592(2)	67 183 367.712(2)	67 183 374.936(2)
7 F	67 138 158.572 2(3)	67 138 228.573 5(2)	67 138 305.080 6(2)	67 138 276.734 8(2)
7 G	67 132 455.962 6(2)	67 132 474.536 7(3)	67 132 532.272 2(3)	67 132 502.051 9(3)
7 H	67 131 109.030 24(4)	67 131 122.381 75(4)	67 131 159.987 28(4)	67 131 140.274 18(4)
7 I	67 130 692.494 97(2)	67 130 702.504 47(2)	67 130 728.929 98(2)	67 130 715.103 36(2)
8 S	53 246 292.0(1.0)			55 440 840.7(2.0)
8 P	51 242 587.47(12)	52 282 092.17(12)	52 282 488.12(12)	52 282 520.36(12)
8 D	51 423 248.154(2)	51 432 523.250(2)	51 432 592.295(2)	51 432 597.169(2)
8 F	51 402 021.640 5(2)	51 402 071.121 6(2)	51 402 121.545 8(2)	51 402 103.381 7(2)
8 G	51 398 146.249 44(8)	51 398 158.704 3(1)	51 398 197.371 4(1)	51 398 177.136 6(1)
8 H	51 397 221.590 77(5)	51 397 230.535 37(6)	51 397 255.727 95(6)	51 397 242.521 70(6)
8 I	51 396 931.954 56(2)	51 396 938.660 18(2)	51 396 956.363 16(2)	51 396 947.100 39(2)
8 K	51 396 822.746 08(1)	51 396 827.941 01(1)	51 396 841.064 37(1)	51 396 834.214 80(1)
9 S	41 903 986.9(8)			43 430 388.7(1.0)
9 P	40 501 246.45(8)	41 231 283.33(8)	41 231 560.29(8)	41 231 582.83(8)
9 D	40 628 480.277(1)	40 635 090.450(1)	40 635 138.924(1)	40 635 142.364(1)
9 F	40 613 531.518 0(1)	40 613 567.564 1(1)	40 613 602.586 8(1)	40 613 590.219 4(1)
9 G	40 610 783.207 7(1)	40 610 791.961 35(7)	40 610 819.112 53(7)	40 610 804.906 48(7)
9 H	40 610 123.045 0(2)	40 610 129.327 1(1)	40 610 147.020 6(1)	40 610 137.745 4(1)
9 I	40 609 914.525 83(3)	40 609 919.235 42(2)	40 609 931.668 77(2)	40 609 925.163 22(2)
9 K	40 609 835.106 61(1)	40 609 838.755 19(1)	40 609 847.972 14(1)	40 609 843.161 46(1)
10 S	33 834 686.3(6)			34 938 889.0(7)
10 P	32 814 665.36(6)	33 346 784.43(6)	33 346 985.68(6)	33 347 002.06(6)
10 D	32 907 601.923(2)	32 912 470.759(1)	32 912 506.087(1)	32 912 508.602(1)
10 F	32 896 683.103 7(4)	32 896 710.074 1(4)	32 896 735.404 2(4)	32 896 726.588 7(4)
10 G	32 894 665.778 26(4)	32 894 672.163 03(4)	32 894 691.953 01(4)	32 894 681.599 81(4)
10 H	32 894 178.916 98(5)	32 894 183.496 72(5)	32 894 196.395 17(5)	32 894 189.633 59(5)
10 I	32 894 024.248 39(1)	32 894 027.681 69(1)	32 894 036.745 59(1)	32 894 032.003 04(1)
10 K	32 893 964.934 34(1)	32 893 967.594 16(1)	32 893 974.313 31(1)	32 893 970.806 33(1)

Table 11.8. QED corrections included in Table 11.7 for the S- and P-states of helium (in MHz).

State	$\Delta E_{L,1}$		$\Delta E_{L,2}$	
	Singlet	Triplet	Singlet	Triplet
$1\,S$	-45410.2		4208.06	
$2\,S$	-3135.7	-4099.2	330.359	36.883
$3\,S$	-854.1	-1035.6	91.960	8.468
$4\,S$	-336.8	-397.3	37.590	3.203
$5\,S$	-161.7	-188.8	18.880	1.544
$6\,S$	-87.8	-102.4	10.785	0.861
$7\,S$	-51.8	-60.6	6.729	0.528
$8\,S$	-32.4	-38.3	4.476	0.347
$9\,S$	-21.19	-25.3	3.126	0.240
$10\,S$	-14.32	-17.37	2.269	0.173
$2\,P$	-100.1	1214.0	62.819	45.502
$3\,P$	-33.8	347.3	19.632	12.376
$4\,P$	-14.5	143.4	8.445	5.036
$5\,P$	-7.49	72.50	4.364	2.529
$6\,P$	-4.35	41.60	2.539	1.446
$7\,P$	-2.74	26.05	1.604	0.904
$8\,P$	-1.84	17.37	1.077	0.602
$9\,P$	-1.29	12.16	0.758	0.421
$10\,P$	-0.942	8.842	0.553	0.306

$$\ln k_0(1\,^1S) = \ln[19.769\,2669\,(1 - 0.00615/Z)^2] \quad (11.43a)$$
$$\ln k_0(2\,^1S) = \ln[19.394\,2687\,(1 + 0.02040/Z)^2] \quad (11.43b)$$
$$\ln k_0(2\,^3S) = \ln[19.394\,2687\,(1 + 0.01388/Z)^2] \quad (11.43c)$$

are useful.

4. Relativistic Finite Mass Corrections of $O(\alpha^2 \mu/M)$. Relativistic finite mass corrections come from two sources. First, a transformation to relative coordinates as in Eq. (11.3) is applied to the pairwise Breit interactions among the three particles, generating the new terms [35]

$$\Delta = \Delta_{\rm oo} + \Delta_{\rm so} + 2\frac{m_e}{M} H_{\rm so}$$

where

$$\Delta_{\rm oo} = \frac{-Z\alpha^2 m_e}{2M} \sum_{i,j} \frac{1}{r_i} \left[\mathbf{p}_j \cdot \mathbf{p}_i + \hat{\mathbf{r}}_i \cdot (\hat{\mathbf{r}}_i \cdot \mathbf{p}_j)\, \mathbf{p}_i \right], \quad (11.44)$$

$$\Delta_{\rm so} = \frac{Z\alpha^2 m_e}{M} \sum_{i \neq j} \frac{1}{r_i^3} \mathbf{r}_i \times \mathbf{p}_j \cdot \mathbf{s}_i. \quad (11.45)$$

Second, the mass polarization term $H_{\rm mp}$ in Eq. (11.3) generates second-order cross-terms between $H_{\rm mp}$ and $H_{\rm rel}$. If the wave functions are calculated by solving Eq. (11.4) in scaled atomic units, the $H_{\rm mp}$ correction is then automatically included to all orders and the mass-corrected relativistic energy shift is (in units of e^2/a_0)

$$\Delta E_{\rm rel} = \left(\frac{\mu}{m_e}\right)^3 \Big\langle \left(\frac{\mu}{m_e}\right) H_{\rm mass} + H_{\rm D} + H_{\rm ssc} + H_{\rm oo}$$
$$+ \Delta_{\rm oo} + \left(1 + \frac{2m_e}{M}\right) H_{\rm so} + H_{\rm soo} + H_{\rm ss} + \Delta_{\rm so} \Big\rangle$$
$$(11.46)$$

with $\mu/m_e = 1 - \mu/M$. The difference $\Delta E_{\rm rel} - \langle H_{\rm rel} \rangle_\infty$ calculated for infinite nuclear mass is the relativistic finite mass correction.

5. Higher-order Corrections. Spin-dependent terms of $O(\alpha^4)$ are known in their entirety, and have recently been calculated to high precision [36]. Nonrelativistic operators for the spin-independent part are not known in their entirety, but the one-electron part is known from the one-electron Lamb shift to be

$$\Delta E'_{L,1} = Z\alpha^3 \left[\pi Z\alpha \left(\frac{427}{96} - 2\ln 2 \right) + 0.538\,931\frac{\alpha}{\pi} \right]$$
$$\times \langle \delta(\mathbf{r}_1) + \delta(\mathbf{r}_2) \rangle. \quad (11.47)$$

As an example, $\Delta E'_{L,1}$ contributes -50.336 MHz and -88.267 MHz to the $2\,^1P - 2\,^1S$ and $2\,^3P_J - 2\,^3S_1$ transition frequencies respectively, while the differences between theory and experiment are ~ 1 MHz and ~ -7 MHz for the two cases (see Ref. [30]). Thus, two-electron corrections (for example, relativistic corrections of relative order $Z\alpha$ to $\Delta E_{L,2}$) are evidently small.

Table 11.7 lists the calculated ionization energies for all states of helium up to $n = 10$ and $L = 7$. For the D-states and beyond, the uncertainties are sufficiently small that these states can be taken as known points of reference in the interpretation of experimental transition frequencies. However, long-range Casimir-Polder corrections [37] are not included since they still lack experimental confirmation [38]. For the S- and P-states, the QED uncertainties are much larger. The contributions from $\Delta E_{L,1} + \Delta E'_{L,1}$ and $\Delta E_{L,2}$ for these states are listed separately in Table 11.8 so that the entries in Table 11.7 can be adjusted as better calculations become available. Applications to isotope shifts and measurements of the nuclear radius are discussed in Sects. 16.3 and 86.2.

11.5.1 Quantum Defect Extrapolations

As discussed in Sect. 14.2, the ionization energies of an isolated Rydberg series of states can be expressed in the form

$$W_n = R_M (Z-1)^2/n^{*2}, \quad (11.48)$$

where $Z - 1$ is the screened nuclear charge and n^* is the effective principal quantum number defined by an iterative solution to the equation

$$n^* = n - \delta(n^*), \quad (11.49)$$

Table 11.9. Quantum defects for the total energies of helium with the ΔW_n term subtracted [see Eq. (11.52)].

δ_i	Value			
	$^1S_0$			$^3S_1$
δ_0	0.139 737 6(34)			0.296 676 7(31)
δ_2	0.025 05(39)			0.035 47(34)
δ_4	0.139(12)			0.126(10)
δ_6	−1.60(11)			−1.478(86)
	$^1P_1$	$^3P_0$	$^3P_1$	$^3P_2$
δ_0	−0.012 141 803 603(64)	0.068 328 002 51(27)	0.068 357 857 65(27)	0.068 360 283 79(23)
δ_2	0.007 519 080 4(59)	−0.018 641 975(24)	−0.018 630 462(24)	−0.018 629 228(21)
δ_4	0.013 977 80(15)	−0.012 331 65(57)	−0.012 330 40(57)	−0.012 332 75(51)
δ_6	0.004 837 3(12)	−0.007 951 5(45)	−0.007 951 2(45)	−0.007 952 7(41)
δ_8	0.001 228 3(29)	−0.005 448(10)	−0.005 450(10)	−0.005 451(9)
	$^1D_2$	$^3D_1$	$^3D_2$	$^3D_3$
δ_0	0.002 113 378 464(49)	0.002 885 580 281(22)	0.002 890 941 493(25)	0.002 891 328 825(26)
δ_2	−0.003 090 051 0(58)	−0.006 357 601 2(27)	−0.006 357 183 6(30)	−0.006 357 704 0(33)
δ_4	0.000 008 27(22)	0.000 336 67(11)	0.000 337 77(11)	0.000 336 70(13)
δ_6	−0.000 309 4(31)	0.000 839 4(16)	0.000 839 2(16)	0.000 839 5(18)
δ_8	−0.000 401(14)	0.000 379 8(72)	0.000 432 3(75)	0.000 381 1(83)
	$^1F_3$	$^3F_2$	$^3F_3$	$^3F_4$
δ_0	0.000 440 294 26(62)	0.000 444 869 89(22)	0.000 448 594 83(28)	0.000 447 379 27(21)
δ_2	−0.001 689 446(65)	−0.001 739 275(24)	−0.001 727 232(30)	−0.001 739 217(23)
δ_4	−0.000 118 3(20)	0.000 104 76(76)	0.000 152 4(9)	0.000 104 78(71)
δ_6	0.000 326(18)	0.000 033 7(69)	−0.000 248 6(83)	0.000 033 1(64)
	$^1G_4$	$^3G_3$	$^3G_4$	$^3G_5$
δ_0	0.000 124 734 490(79)	0.000 125 707 43(12)	0.000 128 713 16(10)	0.000 127 141 67(11)
δ_2	−0.000 796 230(12)	−0.000 796 498(19)	−0.000 796 246(15)	−0.000 796 484(17)
δ_4	−0.000 012 05(53)	−0.000 009 80(81)	−0.000 011 89(66)	−0.000 009 85(75)
δ_6	−0.000 013 6(69)	−0.000 019(11)	−0.000 014 1(85)	−0.000 019(10)
	$^1H_5$	$^3H_4$	$^3H_5$	$^3H_6$
δ_0	0.000 047 100 899(61)	0.000 047 797 067(43)	0.000 049 757 614(51)	0.000 048 729 846(45)
δ_2	−0.000 433 227 7(84)	−0.000 433 232 2(55)	−0.000 433 227 4(65)	−0.000 433 228 1(57)
δ_4	−0.000 008 14(26)	−0.000 008 07(16)	−0.000 008 13(19)	−0.000 008 10(16)
	$^1I_6$	$^3I_5$	$^3I_6$	$^3I_7$
δ_0	0.000 021 868 881(17)	0.000 022 390 759(20)	0.000 023 768 483(14)	0.000 023 047 609(26)
δ_2	−0.000 261 067 3(22)	−0.000 261 068 0(28)	−0.000 261 066 2(18)	−0.000 261 067 2(35)
δ_4	−0.000 004 048(67)	−0.000 004 042(87)	−0.000 004 076(58)	−0.000 004 04(11)

where $\delta(n^*)$ is the *quantum defect* defined by the Ritz expansion

$$\delta(n^*) = \delta_0 + \frac{\delta_2}{(n-\delta)^2} + \frac{\delta_4}{(n-\delta)^4} + \cdots \quad (11.50)$$

with constant coefficients δ_i. The absence of *odd* terms in this series is a special property of the eigenvalues of Hamiltonians of the form $H_C + V$, where H_C is a pure one-electron Coulomb Hamiltonian, and V is a local, short-range, spherically symmetric potential of arbitrary strength (see Ref. [39] for further discussion). For the Rydberg states of helium, odd terms must be included in the Ritz expansion (11.50) due to relativistic and mass

polarization corrections, but they can be removed again by first adjusting the energies according to

$$W_n' = W_n - \Delta W_n \quad (11.51)$$

where, to sufficient accuracy [39] [see discussion following Eq. (11.64)]

$$\Delta W_n = R_M \left\{ \frac{-3\alpha^2(Z-1)^4}{4n^4} + \left(\frac{\mu}{M}\right)^2 \frac{(Z-1)^2}{n^2} \right. $$
$$\left. \times \left[1 + \frac{5}{6}(\alpha Z)^2\right] \right\}, \quad (11.52)$$

with $Z = 2$ for helium. The quantum defect parameters listed in Table 11.9 provide accurate extrapolations to higher-lying Rydberg states, with

$$W_n = R_M/n^{*2} + \Delta W_n \,. \tag{11.53}$$

11.5.2 Asymptotic Expansions

The asymptotic expansion method [30, 40] rapidly increases in accuracy with increasing angular momentum L of the Rydberg electron, and can be used to high precision for $L \geq 7$. The method is based on a model in which:

1. the Rydberg electron, treated as a distinguishable particle, moves in the field of the core consisting of the He nucleus and a tightly bound $1s$ electron.

2. the core, as characterized by its various multipole moments, is perturbed by the electric field of the Rydberg electron.

A systematic perturbation expansion yields an asymptotic series of the form

$$\varepsilon_0^{nL} = -2 - \frac{(Z-1)^2}{2n^2} + a_0 \sum_{j=4}^{N} A_j \langle r^{-j} \rangle_{nL} \, \frac{e^2}{a_0}\,, \tag{11.54}$$

where the expectation value $\langle r^{-j} \rangle_{nL}$ is calculated with respect to the hydrogenic nL-electron wave function [41] and the series is truncated at the upper limit N where the series begins diverging. The leading coefficients A_j are

$$A_4 = -\tfrac{1}{2}\alpha_1\,, \quad A_5 = 0\,, \quad A_6 = -\tfrac{1}{2}(\alpha_2 - 6\beta_1)\,,$$

where α_k is the 2^k-pole polarizability of the hydrogenic core and β_k is a nonadiabatic correction. The exact hydrogenic values are

$$\alpha_1 = \frac{9a_0^3}{2Z^4}\,, \quad \alpha_2 = \frac{15a_0^5}{Z^6}\,, \quad \beta_1 = \frac{43a_0^5}{8Z^6}\,.$$

All terms are known up to A_{10} (see Refs. [30, 40] for detailed results). The expansions for the terms ε_0, ε_1, and ε_2 in Eq. (11.6) for helium are

$$\varepsilon_0^{nL} = -2 - \frac{1}{2n^2} - \frac{9}{64}\langle r^{-4} \rangle + \frac{69}{512}\langle r^{-6} \rangle + \frac{3833}{15\,360}\langle r^{-7} \rangle$$
$$- \left[\frac{55\,923}{65\,536} + \frac{957L(L+1)}{10\,240} \right]\langle r^{-8} \rangle - \frac{908\,185}{688\,128}\langle r^{-9} \rangle$$
$$+ \left[\frac{3\,824\,925}{1\,048\,576} + \frac{33\,275L(L+1)}{28\,672} \right]\langle r^{-10} \rangle$$
$$+ e^{(1,1)} - \frac{23}{20}e^{(1,2)}\,, \tag{11.55}$$

$$\varepsilon_1^{nL} = -\frac{9}{32}\langle r^{-4} \rangle + \frac{249}{256}\langle r^{-6} \rangle + \frac{319}{3840}\langle r^{-7} \rangle$$
$$- \left[\frac{34\,659}{16\,384} + \frac{957L(L+1)}{5120} \right]\langle r^{-8} \rangle - \frac{14\,419}{3072}\langle r^{-9} \rangle$$
$$+ \left[\frac{6\,413\,781}{262\,144} + \frac{24\,155L(L+1)}{8192} \right]\langle r^{-10} \rangle$$
$$+ 4e^{(1,1)} - \frac{53}{5}e^{(1,2)}\,, \tag{11.56}$$

$$\varepsilon_2^{nL} = -\frac{1}{2n^2} - \frac{45}{64}\langle r^{-4} \rangle + \frac{165}{512}\langle r^{-6} \rangle + \frac{2555}{3072}\langle r^{-7} \rangle$$
$$- \left[\frac{268\,485}{32\,768} + \frac{957L(L+1)}{2048} \right]\langle r^{-8} \rangle + \frac{598\,909}{172\,032}\langle r^{-9} \rangle$$
$$+ \left[\frac{3\,907\,923}{524\,288} + \frac{629\,515L(L+1)}{114\,688} \right]\langle r^{-10} \rangle$$
$$+ 14e^{(1,1)} - \frac{251}{10}e^{(1,2)}\,. \tag{11.57}$$

The terms $e^{(1,1)}$ and $e^{(1,2)}$ are second-order dipole-dipole and dipole-quadrupole perturbation corrections. Defining $f_p^L = (L + p)!/(L - p)!$, they are given by

$$e^{(i,j)} = -\frac{(2 - \delta_{j,k})2^{2i+2j+1}(2L - 2i)!(2L - 2j)!}{n^3(2L + 2i + 1)!(2L + 2j + 1)!}$$
$$\times \left[\frac{2^{2i+2j}(2L - 2i - 2j)!A^{(i,j)}}{n^{2i+2j+2}(2L + 2i + 2j + 1)!} + \frac{B^{(i,j)}}{n^{2i+2j+1}} \right] \tag{11.58}$$

with

$$A^{(1,1)} = 3n^2(3n^2 - 2f_1)(f_1 - 2)(45 + 623f_1^L + 3640f_2^L$$
$$+ 560f_3^L)\,,$$
$$B^{(1,1)} = (9n^2 - 7f_1^L)(3n^2 - f_1^L)\,,$$
$$A^{(1,2)} = -21n^6(94\,500 + 122\,850f_1^L - 1126\,125f_2^L$$
$$- 18\,931\,770f_3^L - 11\,171\,160f_4^L$$
$$- 1029\,600f_5^L - 18\,304f_6^L)$$
$$- 15n^4(94\,500 - 444\,150f_1^L + 7747\,425f_2^L$$
$$+ 337\,931\,880f_3^L + 375\,290\,190f_4^L$$
$$+ 66\,518\,760f_5^L + 2880\,416f_6^L + 29\,568f_7^L)$$
$$+ 9n^2f_1^L(90\,300 - 177\,450f_1^L + 1738\,450f_2^L$$
$$+ 133\,125\,575f_3^L + 160\,040\,870f_4^L$$
$$+ 29\,322\,216f_5^L + 1293\,600f_6^L + 13\,440f_7^L)$$
$$+ 2f_1^Lf_2^Lf_3^L(45 + 252f_1^L - 1680f_2^L - 2240f_3^L)\,,$$
$$B^{(1,2)} = 315n^6 + 125n^4(3 - 5f_1^L) - 7n^2f_1^L(43 - 39f_1^L)$$
$$- 27f_1^Lf_2^L\,.$$

The accuracy of the expansion for the ε_0, ε_1, and ε_2 can be reliably estimated to be one-half of the last $\langle r^{-j} \rangle$ term included in the sum. Formulas for the $\langle r^{-j} \rangle$ are given in Table 11.10.

The asymptotic formulas for the NFS relativistic corrections are [30, 42]

$$\langle H_{\text{mass}} + H_{\text{D}} \rangle \rightarrow -\frac{\alpha^2 Z^4}{8} + h_1(nL) + \chi_1(nL)$$
$$+ \frac{(Z\alpha)^2}{2}\left[\frac{14}{3Z^4}\langle r^{-4} \rangle - \frac{5041}{240Z^6}\langle r^{-6} \rangle \right] \tag{11.59}$$

$$\langle H_{\text{oo}} \rangle \rightarrow \frac{\alpha^2}{Z^2}\left[\langle r^{-4} \rangle + \frac{3(Z-1)}{Z^2}\langle r^{-5} \rangle \right.$$
$$\left. - \frac{3(f_1^L + 8)}{4Z^2}\langle r^{-6} \rangle \right]\,, \tag{11.60}$$

Table 11.10. Formulas for the hydrogenic expectation value $\langle r^{-j} \rangle \equiv \langle nl | r^{-j} | nl \rangle$ in terms of

$$G_p^{nl} = \frac{2^p Z^p (2l - p + 2)!}{n^{p+1}(2l + p - 1)!}, \quad f_p^l = \frac{(l+p)!}{(l-p)!}.$$

j	$\langle r^{-j} \rangle \ (a_0)$
2	$\frac{1}{2}G_2^{nl}$
3	nG_3^{nl}
4	$G_4^{nl}(3n^2 - f_1^l)$
5	$2G_5^{nl}[5n^3 - n(3f_1^l - 1)]$
6	$G_6^{nl}[35n^4 - 5n^2(6f_1^l - 5) + 3f_2^l]$
7	$2G_7^{nl}[63n^5 - 35n^3(2f_1^l - 3) + n(15f_2^l - 20f_1^l + 12)]$
8	$G_8^{nl}[462n^6 - 210n^4(3f_1^l - 7)$ $+ 42n^2(5f_2^l - 15f_1^l + 14) - 10f_3^l]$
9	$2G_9^{nl}[858n^7 - 462n^5(3f_1^l - 10) + 42n^3(15f_2^l - 75f_1^l$ $+ 101) - 2n(35f_3^l - 105f_2^l + 252f_1^l - 180)]$
10	$G_{10}^{nl}[6435n^8 - 6006n^6(2f_1^l - 9) + 1155n^4(6f_2^l - 44f_1^l$ $+ 81) - 6n^2(210f_3^l - 1365f_2^l + 4648f_1^l - 4566)$ $+ 35f_4^l]$

where

$$h_1(nL) = \frac{\alpha^2 (Z-1)^4}{2n^3}\left[\frac{3}{4n} - \frac{1}{L + \frac{1}{2}}\right] \tag{11.61}$$

is the leading one-electron Dirac energy and

$$\chi_1(nL) = \frac{\alpha^2 \alpha_1}{2}\left\{ 3\left(\frac{Z-1}{n}\right)^2 \langle r^{-4}\rangle - (Z-1)\langle r^{-5}\rangle \right.$$
$$- \frac{4(2L-2)!}{(2L+3)!}\left[4\left(\frac{Z-1}{n}\right)^6\left(n + \frac{9n^2 - 5f_1^L}{2L+1}\right)\right.$$
$$\left.\left. + (Z-1)^2\left(\frac{40f_2^L + 70f_1^L - 3}{2L+1}\right)\langle r^{-4}\rangle\right]\right\} \tag{11.62}$$

is the correction due to the dipole perturbation of the Rydberg electron. The relativistic recoil terms due to mass polarization are

$$\langle H_{\text{mass}} + H_{\text{D}}\rangle_{\text{RR}}$$
$$\to \frac{\mu}{M}\left[\frac{22(Z\alpha)^2(Z-1)}{9Z^4}\langle r^{-4}\rangle + 2(Z-1)\chi_1(nL)\right]$$
$$+ \left(\frac{\mu}{M}\right)^2\left[-\frac{5}{12}\left(\frac{\alpha Z(Z-1)}{n}\right)^2 + 4h_1(nL)\right], \tag{11.63}$$

$$\langle H_{\text{oo}}\rangle_{\text{RR}} + \langle \Delta_{\text{oo}}\rangle$$
$$\to -\frac{\alpha^2\mu}{M}\left[Z^4 + \frac{(Z-1)^4}{n^3}\left(\frac{1}{n} - \frac{3}{2L+1}\right)\right.$$
$$\left. - \frac{25[1 + 13f(Z)]}{16Z^2}\langle r^{-4}\rangle\right], \tag{11.64}$$

with $f(Z) \simeq 1 + (Z-2)/6$. The $-(5/12)[\alpha Z(Z-1)/n]^2$ term in (11.63) is the dominant contribution in helium for $L \geq 4$. It is included in Eq. (11.52) for ΔW_n, along with the leading $1/n^2$ term from (11.57), and the $1/n^4$ term from (11.61). The complete relativistic finite mass correction includes also the mass-scaling terms $-(\mu/M)\langle 4H_{\text{mass}} + 3H_{\text{D}} + 3H_{\text{oo}}\rangle$ obtained by expanding μ/m_e in Eq. (11.46). The $\langle \delta(\mathbf{r}_1)\rangle$ term is [43]

$$\pi\langle\delta(\mathbf{r}_1)\rangle \to \frac{Z^3}{2} - \frac{31}{4Z^3}\langle r^{-4}\rangle + \frac{1447}{32Z^5}\langle r^{-7}\rangle$$
$$- \frac{-31(Z-1)}{2Z^3}\left(\frac{\mu}{M}\right)\langle r^{-4}\rangle + \cdots. \tag{11.65}$$

$\langle\delta(\mathbf{r}_{12})\rangle$ vanishes exponentially as $1/n^{2L+4}$ with increasing L. The complete asymptotic expressions for the FS matrix elements are summarized by the formulas

$$\langle nL\,{}^3L_J | H_{\text{FS}} | nL\,{}^3L_J\rangle$$
$$\to T_{nL}(J)\{Z - 3 + 2S_L(J) + 2a_e[Z - 2$$
$$+ (2 + a_e)S_L(J)] + (\mu/M)[2 - 4S_L(J)]\}, \tag{11.66}$$
$$\langle nL\,{}^3L_J | H_{\text{FS}} | nL\,{}^1L_J\rangle$$
$$\to T_{nL}(L)(Z + 1 + 2a_eZ - 2\mu/M)\sqrt{L(L+1)}, \tag{11.67}$$

where

$$T_{nL}(J) = \begin{cases} -\alpha^2(L+1)\langle r^{-3}\rangle/4 & J = L-1, \\ -\alpha^2\langle r^{-3}\rangle/4 & J = L, \\ \alpha^2 L\langle r^{-3}\rangle/4 & J = L+1, \end{cases}$$

$$S_L(J) = \begin{cases} 1 & J = L, \\ \pm 1/(2J+1) & J = L \pm 1. \end{cases}$$

The asymptotic form for the QED term $\Delta E_{L,1}$ follows from Eq. (11.38) with the use of (11.65) for $\langle\delta(\mathbf{r}_1)\rangle$ and (11.40) for $\ln k_0$. The two-electron part is

$$\Delta E_{L,2} \to -\frac{7\alpha^3}{6\pi}\left(\langle r^{-3}\rangle + \frac{3}{Z^2}\langle r^{-5}\rangle\right). \tag{11.68}$$

With the use of the formulas in this section, the variationally calculated ionization energies for the K-states ($L = 7$) in Table 11.7 can be reproduced to within ± 20 Hz. For $L > 7$, the uncertainty becomes less than 1 Hz, up to the Casimir-Polder retardation effects which have not been included.

11.6 RADIATIVE TRANSITIONS

11.6.1 Basic Formulation

In a semiclassical picture, the interaction Hamiltonian with the radiation field is obtained by making the minimal coupling replacements

$$\mathbf{P}_N \to \mathbf{P}_N - \frac{Ze}{c}\mathbf{A}(\mathbf{R}_N)$$
$$\mathbf{P}_i \to \mathbf{P}_i + \frac{e}{c}\mathbf{A}_i(\mathbf{R}_i) \tag{11.69}$$

in Eq. (11.1), where

$$\mathbf{A}(\mathbf{R}) = c \left(\frac{2\pi\hbar}{\omega\mathcal{V}} \right)^{1/2} \hat{\boldsymbol{\epsilon}}\, e^{i\mathbf{k}\cdot\mathbf{R}} \qquad (11.70)$$

is the time-independent part of the vector potential $\mathbf{A}(\mathbf{r},t) = \mathbf{A}(\mathbf{r})e^{-i\omega t} + \text{c.c}$ for a photon of frequency ω, wave vector $\mathbf{k}$, and polarization $\hat{\boldsymbol{\epsilon}} \perp \mathbf{k}$ normalized to unit photon energy $\hbar\omega$ in volume $\mathcal{V}$. The linear coupling terms then yield

$$H_{\text{int}} = -\frac{Ze}{Mc}\mathbf{P}_N\cdot\mathbf{A}(\mathbf{R}_N) + \frac{e}{m_e c}\sum_{i=1}^{2} \mathbf{P}_i\cdot\mathbf{A}(\mathbf{R}_i), \quad (11.71)$$

and from Fermi's Golden Rule, the decay rate for spontaneous emission from state γ to γ' is

$$w_{\gamma,\gamma'}d\Omega = \frac{2\pi}{\hbar}|\langle\gamma|H_{\text{int}}|\gamma'\rangle|^2\rho_f, \qquad (11.72)$$

where $\rho_f = \mathcal{V}\omega^2 d\omega/(2\pi c)^3\hbar$ is the number of photon states with polarization $\hat{\boldsymbol{\epsilon}}$ per unit energy and solid angle in the normalization volume $\mathcal{V}$. In the long wavelength and electric dipole approximations, the factor $e^{i\mathbf{k}\cdot\mathbf{R}}$ in Eq. (11.70) is replaced by unity. After integrating over angles $d\Omega$ and summing over polarizations $\hat{\boldsymbol{\epsilon}}$, the decay rate reduces to

$$w_{\gamma,\gamma'} = \tfrac{4}{3}\alpha\omega_{\gamma,\gamma'}|\langle\gamma|\mathbf{Q}_p|\gamma'\rangle|^2, \qquad (11.73)$$

where $\omega_{\gamma',\gamma}$ is the transition frequency and $\mathbf{Q}_p$ is the velocity form of the transition operator

$$\mathbf{Q}_p = \frac{Z}{Mc}\mathbf{P}_N + \frac{1}{m_e c}\sum_{i=1}^{N} P_i \qquad (11.74)$$

for the general case of N electrons. From the commutator $[H_0, \mathbf{Q}_r/\hbar\omega_{\gamma,\gamma'}] = \mathbf{Q}_p$, where H_0 is the field-free Hamiltonian in Eq. (11.1), the equivalent length form is

$$\mathbf{Q}_r = -\frac{i}{c}\omega\gamma, \gamma' \left(Z\mathbf{R}_N - \sum_{i=1}^{N}\mathbf{R}_i \right). \qquad (11.75)$$

After transforming to c.m. plus relative coordinates in parallel with Eq. (11.3), the dipole transition operators become

$$\mathbf{Q}_p = \frac{Z_p}{mc}\sum_{i=1}^{N}\mathbf{p}_i, \quad \mathbf{Q}_r = \frac{i\omega_{\gamma,\gamma'}}{c}Z_r\sum_{i=1}^{N}\mathbf{r}_i, \qquad (11.76)$$

with

$$Z_p = \frac{Zm_e + M}{M}, \quad Z_r = \frac{Zm_e + M}{Nm_e + M},$$

and H_0 now contains the H_{mp} term. If Eq. (11.3) is solved exactly for the states $|\gamma\rangle$ and $|\gamma'\rangle$, then the identity

$$\langle\gamma|\mathbf{Q}_p|\gamma'\rangle = \langle\gamma|\mathbf{Q}_r|\gamma'\rangle \qquad (11.77)$$

is satisfied to all orders in m_e/M. For a neutral atom, $N = Z$ and $Z_r = 1$. If the oscillator strength is defined by

$$\begin{aligned}
f_{\gamma',\gamma} &= \frac{2m_e\omega_{\gamma',\gamma}}{3\hbar}\left(\frac{Z_r}{Z_p}\right)|\langle\gamma'|\sum_{i=1}^{N}\mathbf{r}_i|\gamma\rangle|^2 \\
&= \frac{2}{3m_e\hbar\omega_{\gamma',\gamma}}\left(\frac{Z_p}{Z_r}\right)|\langle\gamma'|\sum_{i=1}^{N}\mathbf{p}_i|\gamma\rangle|^2 \qquad (11.78)
\end{aligned}$$

then the sum rule $\sum_{\gamma'} f_{\gamma',\gamma} = N$ remains valid, independent of m_e/M. The decay rate, summed over final states and averaged over initial states, is

$$\bar{w}_{\gamma,\gamma'} = -\frac{2\alpha\hbar\omega_{\gamma,\gamma'}}{m_e c^2}Z_p Z_r \bar{f}_{\gamma,\gamma'} \qquad (11.79)$$

where $\bar{f}_{\gamma,\gamma'} = -(g_\gamma/g_{\gamma'})\bar{f}_{\gamma',\gamma}$ is the (negative) oscillator strength for photon emission, and g_γ, $g_{\gamma'}$ are the statistical weights of the states.

11.6.2 Oscillator Strength Table

Table 11.11 provides arrays of nonrelativistic oscillator strengths among various states of helium, including the effects of finite nuclear mass as a separate factor. In the absence of mass polarization, the correction factor would be $(1 + \mu/M)^{-1} \simeq 1 - \mu/M$. Mass polarization effects are particularly strong for P-states, and for transitions with $\Delta n = 0$.

The largest relativistic correction comes from singlet-triplet mixing between states with the same n, L, and J (e.g. $3\,^1D_2$ and $3\,^3D_2$) due to H_{FS}. The wave functions obtained by diagonalizing the 2×2 matrices $\mathbf{H}_0 + \mathbf{H}_{\text{NFS}} + \mathbf{H}_{\text{FS}}$ are then

$$\begin{aligned}
\Psi(n\,^1L_L) &= \Psi_0(n\,^1L_L)\cos\theta + \Psi_0(n\,^3L_L)\sin\theta \\
\Psi(n\,^3L_L) &= -\Psi_0(n\,^1L_L)\sin\theta + \Psi_0(n\,^3L_L)\cos\theta.
\end{aligned}$$

Values of $\sin\theta$ are listed in Table 11.12. The corrected oscillator strengths $\tilde{f}_{\gamma,\gamma'}$ for the singlet (s) and triplet (t) components of a $\gamma \rightarrow \gamma'$ transition can then be calculated from the values in Table 11.11 according to

$$\begin{aligned}
\tilde{f}_{\gamma,\gamma'}^{\text{ss}} &= \omega_{\gamma,\gamma'}^{\text{ss}}\left(X_{\gamma,\gamma'}^{\text{ss}}\cos\theta_\gamma\cos\theta_{\gamma'} + X_{\gamma,\gamma'}^{\text{tt}}\sin\theta_\gamma\sin\theta_{\gamma'}\right)^2, \\
\tilde{f}_{\gamma,\gamma'}^{\text{tt}} &= \omega_{\gamma,\gamma'}^{\text{tt}}\left(X_{\gamma,\gamma'}^{\text{ss}}\sin\theta_\gamma\sin\theta_{\gamma'} + X_{\gamma,\gamma'}^{\text{tt}}\cos\theta_\gamma\cos\theta_{\gamma'}\right)^2, \\
\tilde{f}_{\gamma,\gamma'}^{\text{st}} &= \omega_{\gamma,\gamma'}^{\text{st}}\left(X_{\gamma,\gamma'}^{\text{ss}}\cos\theta_\gamma\sin\theta_{\gamma'} - X_{\gamma,\gamma'}^{\text{tt}}\sin\theta_\gamma\cos\theta_{\gamma'}\right)^2, \\
\tilde{f}_{\gamma,\gamma'}^{\text{ts}} &= \omega_{\gamma,\gamma'}^{\text{ts}}\left(X_{\gamma,\gamma'}^{\text{ss}}\sin\theta_\gamma\cos\theta_{\gamma'} - X_{\gamma,\gamma'}^{\text{tt}}\cos\theta_\gamma\sin\theta_{\gamma'}\right)^2,
\end{aligned}$$

where $X_{\gamma,\gamma'}^{\text{ss}} = (f_{\gamma,\gamma'}^{\text{ss}}/\omega_{\gamma,\gamma'}^{\text{ss}})^{1/2}$, and similarly for $X_{\gamma,\gamma'}^{\text{tt}}$. From Eq. (11.78), $X_{\gamma,\gamma'}$ is proportional to the dipole *length* form of the transition operator, for which there

Table 11.11. Oscillator strengths for helium. The factor in brackets gives the finite mass correction, with $y = \mu/M$.

	$1\,^1S$	$2\,^1S$	$3\,^1S$	$4\,^1S$
$2\,^1P$	$0.276\,1647(1 - 2.282y)$	$0.376\,4403(1 + 1.255y)$	$-0.145\,4703(1 + 1.351y)$	$-0.025\,8703(1 + 0.885y)$
$3\,^1P$	$0.073\,4349(1 - 1.789y)$	$0.151\,3417(1 - 3.971y)$	$0.626\,1931(1 + 1.234y)$	$-0.307\,5074(1 + 1.097y)$
$4\,^1P$	$0.029\,8629(1 - 1.583y)$	$0.049\,1549(1 - 3.235y)$	$0.143\,8889(1 - 4.650y)$	$0.858\,0214(1 + 1.205y)$
$5\,^1P$	$0.015\,0393(1 - 1.474y)$	$0.022\,3377(1 - 2.967y)$	$0.050\,4714(1 - 3.764y)$	$0.146\,2869(1 - 5.080y)$
$6\,^1P$	$0.008\,6277(1 - 1.407y)$	$0.012\,1340(1 - 2.829y)$	$0.024\,1835(1 - 3.444y)$	$0.052\,7562(1 - 4.105y)$
$7\,^1P$	$0.005\,4054(1 - 1.362y)$	$0.007\,3596(1 - 2.75y)$	$0.013\,6794(1 - 3.279y)$	$0.025\,8918(1 - 3.75y)$
	$2\,^3S$	$3\,^3S$	$4\,^3S$	$5\,^3S$
$2\,^3P$	$0.539\,0861(1 - 3.185y)$	$-0.208\,5359(1 - 3.773y)$	$-0.031\,7208(1 - 2.819y)$	$-0.011\,3409(1 - 2.609y)$
$3\,^3P$	$0.064\,4612(1 + 5.552y)$	$0.890\,8513(1 - 2.967y)$	$-0.435\,6711(1 - 3.362y)$	$-0.067\,6073(1 - 2.359y)$
$4\,^3P$	$0.025\,7689(1 + 3.886y)$	$0.050\,0833(1 + 7.505y)$	$1.215\,2630(1 - 2.878y)$	$-0.668\,3003(1 - 3.185y)$
$5\,^3P$	$0.012\,4906(1 + 3.332y)$	$0.022\,9141(1 + 5.209y)$	$0.044\,2305(1 + 9.009y)$	$1.530\,6287(1 - 2.827y)$
$6\,^3P$	$0.006\,9822(1 + 3.063y)$	$0.011\,9933(1 + 4.460y)$	$0.021\,6301(1 + 6.198y)$	$0.041\,5177(1 + 10.215y)$
$7\,^3P$	$0.004\,2990(1 + 2.908y)$	$0.007\,0772(1 + 4.092y)$	$0.011\,7754(1 + 5.292y)$	$0.021\,1003(1 + 6.981y)$
	$2\,^1P$	$3\,^1P$	$4\,^1P$	$5\,^1P$
$3\,^1D$	$0.710\,1641(1 - 0.281y)$	$-0.021\,1401(1 + 29.947y)$	$-0.015\,3034(1 - 6.680y)$	$-0.003\,1128(1 - 6.27y)$
$4\,^1D$	$0.120\,2704(1 - 1.307y)$	$0.648\,1049(1 + 0.435y)$	$-0.040\,0610(1 + 29.183y)$	$-0.039\,2932(1 - 6.163y)$
$5\,^1D$	$0.043\,2576(1 - 1.681y)$	$0.141\,3027(1 - 0.566y)$	$0.647\,6679(1 + 0.817y)$	$-0.057\,3258(1 + 28.903y)$
$6\,^1D$	$0.020\,9485(1 - 1.866y)$	$0.056\,2766(1 - 0.936y)$	$0.152\,8104(1 - 0.170y)$	$0.669\,8361(1 + 1.056y)$
$7\,^1D$	$0.011\,8970(1 - 1.975y)$	$0.028\,8961(1 - 1.127y)$	$0.063\,5953(1 - 0.538y)$	$0.163\,0272(1 + 0.082y)$
$8\,^1D$	$0.007\,4645(1 - 2.046y)$	$0.017\,0777(1 - 1.241y)$	$0.033\,6403(1 - 0.731y)$	$0.069\,3063(1 - 0.26y)$
	$2\,^3P$	$3\,^3P$	$4\,^3P$	$5\,^3P$
$3\,^3D$	$0.610\,2252(1 - 2.029y)$	$0.112\,1004(1 + 6.653y)$	$-0.036\,9592(1 + 3.292y)$	$-0.006\,9009(1 + 2.678y)$
$4\,^3D$	$0.122\,8469(1 - 1.001y)$	$0.477\,5938(1 - 3.059y)$	$0.200\,9498(1 + 6.368y)$	$-0.088\,3017(1 + 2.939y)$
$5\,^3D$	$0.047\,0071(1 - 0.631y)$	$0.124\,5532(1 - 2.019y)$	$0.438\,3888(1 - 3.607y)$	$0.280\,0558(1 + 6.225y)$
$6\,^3D$	$0.023\,4692(1 - 0.449y)$	$0.053\,0093(1 - 1.631y)$	$0.123\,9414(1 - 2.555y)$	$0.429\,4411(1 - 3.961y)$
$7\,^3D$	$0.013\,5638(1 - 0.346y)$	$0.028\,1587(1 - 1.432y)$	$0.055\,2332(1 - 2.153y)$	$0.125\,2389(1 - 2.904y)$
$8\,^3D$	$0.008\,6047(1 - 0.280y)$	$0.016\,9809(1 - 1.315y)$	$0.030\,2853(1 - 1.94y)$	$0.057\,0589(1 - 2.498y)$
	$3\,^1D$	$4\,^1D$	$5\,^1D$	$6\,^1D$
$4\,^1F$	$1.015\,0829(1 - 1.010y)$	$0.002\,4920(1 + 3.833y)$	$-0.012\,6968(1 - 0.888y)$	$-0.002\,2631(1 - 0.890y)$
$5\,^1F$	$0.156\,8808(1 - 0.993y)$	$0.886\,1343(1 - 1.023y)$	$0.004\,6467(1 + 4.139y)$	$-0.033\,2539(1 - 0.893y)$
$6\,^1F$	$0.054\,0508(1 - 0.984y)$	$0.186\,0576(1 - 1.001y)$	$0.839\,1374(1 - 1.031y)$	$0.006\,6028(1 + 4.302y)$
$7\,^1F$	$0.025\,6799(1 - 0.978y)$	$0.072\,3229(1 - 0.994y)$	$0.196\,3692(1 - 1.014y)$	$0.826\,9464(1 - 1.039y)$
$8\,^1F$	$0.014\,4782(1 - 0.978y)$	$0.036\,6627(1 - 0.987y)$	$0.080\,7847(1 - 1.003y)$	$0.203\,1182(1 - 1.019y)$
$9\,^1F$	$0.009\,0730(1 - 0.977y)$	$0.021\,5401(1 - 0.975y)$	$0.042\,4256(1 - 1.000y)$	$0.086\,0955(1 - 1.01y)$
	$3\,^3D$	$4\,^3D$	$5\,^3D$	$6\,^3D$
$4\,^3F$	$1.014\,3389(1 - 0.997y)$	$0.003\,3992(1 - 2.166y)$	$-0.012\,8084(1 - 1.042y)$	$-0.002\,2830(1 - 1.044y)$
$5\,^3F$	$0.156\,9831(1 - 1.004y)$	$0.884\,5767(1 - 0.991y)$	$0.006\,5121(1 - 2.387y)$	$-0.033\,5369(1 - 1.043y)$
$6\,^3F$	$0.054\,1179(1 - 1.006y)$	$0.186\,0264(1 - 1.003y)$	$0.837\,0221(1 - 0.988y)$	$0.009\,3836(1 - 2.499y)$
$7\,^3F$	$0.025\,7201(1 - 1.008y)$	$0.072\,3579(1 - 1.003y)$	$0.196\,2031(1 - 0.996y)$	$0.824\,4031(1 - 0.984y)$
$8\,^3F$	$0.014\,5037(1 - 1.009y)$	$0.036\,6936(1 - 1.004y)$	$0.080\,7712(1 - 1.00y)$	$0.202\,8407(1 - 0.993y)$
$9\,^3F$	$0.009\,0903(1 - 1.008y)$	$0.021\,5632(1 - 1.011y)$	$0.042\,4344(1 - 0.99y)$	$0.086\,0373(1 - 0.99y)$

Table 11.12. Singlet-triplet mixing angles for helium.

State	$\sin\theta$	State	$\sin\theta$	State	$\sin\theta$
2 P	0.000 2783				
3 P	0.000 2558	3 D	0.015 6095		
4 P	0.000 2498	4 D	0.011 3960	4 F	0.604 1024
5 P	0.000 2473	5 D	0.010 1143	5 F	0.549 9291
6 P	0.000 2460	6 D	0.009 5289	6 F	0.518 0737
7 P	0.000 2452	7 D	0.009 2067	7 F	0.498 4184
8 P	0.000 2447	8 D	0.009 0087	8 F	0.485 5768
9 P	0.000 2444	9 D	0.008 8777	9 F	0.476 7620
10 P	0.000 2442	10 D	0.008 7862	10 F	0.470 4595
5 G	0.693 4752				
6 G	0.693 1996	6 H	0.696 2385		
7 G	0.692 9889	7 H	0.696 2377	7 I	0.697 9315
8 G	0.692 8356	8 H	0.696 2372	8 I	0.697 9315
9 G	0.692 7195	9 H	0.696 2374	9 I	0.697 9316
10 G	0.692 6329	10 H	0.696 2353	10 I	0.697 9316
8 K	0.699 1671				
9 K	0.699 1671	9 L	0.700 1089		
10 K	0.699 1671	10 L	0.700 1089	10 M	0.700 8507

are no spin-dependent relativistic corrections [44]. The mixing corrections are particularly significant for D–F and F–G transitions, where intermidiate coupling prevails. The two-state approximation becomes increasingly accurate with increasing L, but for P-states, where $\sin\theta$ is small, states with $n' \neq n$ must also be included [45].

11.7 FUTURE PERSPECTIVES

The variational calculations, together with quantum defect extrapolations for high n and asymptotic expansions for high L, provide essentially exact results for the entire singly-excited spectrum of helium. In this sense, helium joins hydrogen as a fundamental atomic system. The dominant uncertainties arise from two-electron QED effects beyond the current realm of standard atomic physics. Transition frequencies from the $1s2s\,^1S_0$ state are now known to better than ± 0.5 MHz (± 1.8 parts in 10^9) [46], and the fine structure intervals in the $1s2p\,^3P$ state have been measured to an accuracy approaching 1 kHz [47]. Transition frequencies among the $n = 10$ states are known even more accurately [38]. A complete analysis of relativistic and QED corrections of order α^4, $\alpha^5 \ln \alpha$, and α^5 a.u. will be required to match these levels of accuracy. Some progress has been made [48,49], but this remains an important challenge in fundamental atomic physics for the future.

Acknowledgments

The author is grateful to R. N. Hill and J. D. Morgan III for suggesting some of the material in Sect. 11.1. This work was supported by the Natural Sciences and Engineering Research Council of Canada.

REFERENCES

1. T. Kato, Commun. Pure Appl. Math. **10**, 151 (1957).
2. M. Hoffman-Ostenhoff, T. Hoffmann-Ostenhoff, and H. Stremnitzer, Phys. Rev. Lett. **68**, 3857 (1992).
3. V. A. Fock, Izv. Akad. Nauk SSSR, Ser. Fiz. **18**, 161 (1954) [D. Kngl. Norske Videnskab. Selsk. Forh. **31**, 138 (1958); **31**, 145 (1958)]. See also G. B. Sochilin, Int. J. Quantum Chem. **3**, 297 (1969).
4. C. R. Myers, C. J. Umriger, J. P. Sethna, and J. D. Morgan III, Phys. Rev. A **44**, 5537 (1991).
5. J. E. Gottschalk and E. N. Maslen, J. Phys. A **20**, 2781 (1987) and earlier references therein; K. McIsaac and E. N. Maslen, Int. J. Quantum Chem. **31**, 361 (1987).
6. J. H. Macek, Phys. Rev. **160**, 170 (1967).
7. J. Leray, in *Trends and Applications of Pure Mathematics to Mechanics*, Lecture Notes in Physics 195, Edited by P. G. Ciarlet and M. Roseau (Springer, Berlin, 1984), pp. 235–247.
8. J. D. Morgan III, Theoret. Chem. Acta **69**, 181 (1986).
9. P. Deift, W. Hunziker, B. Simon, and E. Vock, Commun. Math. Phys. **64**, 1 (1978).
10. J. D. Morgan III, J. Phys. A **10**, L91 (1977).
11. M. Hoffman-Ostenhoff and T. Hoffmann-Ostenhoff, Phys. Rev. A **16**, 1782 (1977).
12. R. Ahlrichs, T. Hoffman-Ostenhoff, M. Hoffmann-Ostenhoff, and J. D. Morgan III, Phys. Rev. A **23**, 2107 (1981), and earlier references therein.
13. E. A. Hylleraas and B. Undheim, Z. Phys. **65**, 759 (1930).
14. J. K. L. MacDonald, Phys. Rev. **43**, 830 (1933).
15. E. A. Hylleraas, Z. Phys. **48**, 469 (1928); **54**, 347 (1929).
16. Y. Accad, C. L. Pekeris, and B. Schiff, Phys. Rev. A **4**, 516 (1971).
17. K. Frankowski and C. L. Pekeris, Phys. Rev. **146**, 46 (1966); **150**, 366(E) (1966).
18. D. E. Freund, B. D. Huxtable, and J. D. Morgan III, Phys. Rev. A **29**, 980 (1984).
19. A. J. Thakkar and T. Koga, Phys. Rev. A **50**, 854 (1994).
20. G. W. F. Drake, Nucl. Instrum. Methods Phys. Res. Sect B **31**, 7 (1988).
21. G. W. F. Drake and Z.-C. Yan, Phys. Rev. A **46**, 2378 (1992); G. W. F. Drake, in *Long Range Casimir Forces: Theory and Recent Experiments in Atomic Systems*, Edited by F. S. Levin and D. A. Micha (Plenum Press, New York, 1993), p. 107.
22. G. W. F. Drake and Z.-C. Yan, Chem. Phys. Lett. **229**, 486 (1994).
23. G. W. F. Drake, Phys. Rev. A **18**, 820 (1978).
24. Z.-C. Yan and G. W. F. Drake, Can. J. Phys. **72**, 822 (1994).
25. J. Ackermann, Phys. Rev. A **52**, 1968 (1995), and earlier references therein.

26. C.-Y. Hu, A. A. Kvitsinsky, and J. S. Cohen, J. Phys. B **28**, 3629 (1995), and earlier references therein.
27. M. I. Haftel and V. B. Mandelzweig, Phys. Rev. A **49**, 3338 (1994), and earlier references therein.
28. A. H. Wapstra and G. Audi, Nucl. Phys. A **432**, 1 (1985).
29. A. Dalgarno and N. Lynn, Proc. Phys. Soc. (London) A **70**, 802 (1957).
30. G. W. F. Drake, Adv. At. Mol. Opt. Phys. **31**, 1 (1993).
31. S. P. Goldman and G. W. F. Drake, Phys. Rev. Lett. **68**, 1683 (1992).
32. G. W. F. Drake and R. A. Swainson, Phys. Rev. A **41**, 1243 (1990).
33. J. D. Baker, R. C. Forrey, M. Jeziorska, and J. D. Morgan III, to be published.
34. S. P. Goldman and G. W. F. Drake, J. Phys. B **16**, L183 (1983); **17**, L197 (1984).
35. A. P. Stone, Proc. Phys. Soc. (London) **77**, 786 (1961); **81**, 868 (1963).
36. Z.-C. Yan and G. W. F. Drake, Phys. Rev. Lett. **74**, 4791 (1995).
37. J. F. Babb and L. Spruch, Phys. Rev. A **38**, 13 (1988); C.-K. Au, Phys. Rev. A **39**, 2789 (1989); C.-K. Au and

M. A. Mesa, Phys. Rev. A **41**, 2848 (1990).
38. C. H. Storry, N. E. Rothery, and E. A. Hessels, Phys. Rev. Lett. **75**, 3249 (1995), and earlier references therein.
39. G. W. F. Drake, Adv. At. Mol. Opt. Phys. **32**, 93 (1994).
40. R. J. Drachman, Phys. Rev. A **47**, 694 (1993), and earlier references therein.
41. G. W. F. Drake and R. A. Swainson, Phys. Rev. A **42**, 1123 (1990).
42. E. A. Hessels, Phys. Rev. A **46**, 5389 (1992).
43. G. W. F. Drake, Phys. Rev. A **45**, 70 (1992).
44. G. W. F. Drake, Phys. Rev. A **5**, 1979 (1972); J. Phys. B **9**, L169 (1976).
45. G. W. F. Drake, Phys. Rev. A **19**, 1387 (1979).
46. C. J. Sansonetti and J. D. Gillaspy, Phys. Rev. A **45**, R1 (1992); W. Lichten, D. Shiner, and Z.-X. Zhou, Phys. Rev. A **43**, 1663 (1991).
47. D. Shiner, R. Dixson, and P. Zhao, Phys. Rev. Lett. **72**, 1802 (1994).
48. T. Zhang and G. W. F. Drake, J. Phys. B **27**, L311 (1994); T. Zhang, to be published.
49. M. I. Eides, I. B. Khriplovich, and A. I. Milstein, Phys. Lett. B **339**, 275 (1994).

12

Atomic Multipoles

W. E. Baylis

Department of Physics, University of Windsor, Windsor, Ontario N9B 3P4, Canada

12.1 INTRODUCTION

A typical atomic experiment involves the preparation of an atomic or molecular ensemble, its perturbation by a combination of collisions and external fields, and the characterization of the perturbed system through detection of emitted or scattered particles or quanta. The ensemble can be described by its density matrix, whose full specification generally requires elements between every pair of states in a complete basis set, and whose elements generally depend on spatial position, velocity, and time. The time development of such an ensemble may depend on interactions among the atoms of the ensemble with each other, with external fields, and with external perturbers. Most problems burdened by this much detail are intractable, but fortunately, significant simplifications can usually be made.

In this chapter, only ensembles of homogeneously excited, independent atoms are considered. The spatial and velocity distribution of the atoms is thus assumed to be independent of the quantum-state distribution, and the total density matrix can be factored into the product of a phase-space distribution with a state density matrix that is independent of position and velocity. The assumption is severe and prevents us from treating cases of radiation transfer (see Chap. 19) where the degree of excitation varies as a function of position or problems

of velocity-selective laser excitation. Nevertheless, many experiments in atomic physics do use ensembles that are well-described by our assumption, and even in more complex cases, our model can often serve as a starting point for analysis. Typical of the experiments for which our approach is well suited are collisionally induced polarization relaxation experiments in a gas cell, either of an excited state or of an optically pumped ground state.

In addition, we shall for the most part focus on a few isolated manifolds, usually a ground-state manifold and one or two excited-state manifolds. Each manifold, we assume, can be described by a given eigenvalue J of the total angular momentum. The qualifier "isolated" means that coherences between manifolds oscillate rapidly in time compared to other processes and can be ignored. These assumptions permit a great simplification and frequently allow a decoupling of the equations of motion, as shown below. The density matrix within each state manifold can be described by multipole moments, which are coefficients of the expansion of the density matrix in irreducible tensor operators, as described in Chap. 7. Each state multipole is associated with a physical electric or magnetic moment in the atom.

The amount of information which can be imparted to, or obtained from, an atomic ensemble is limited by the finite number of accessible multipole moments. The nature of that information may depend on the time evolution of the multipoles in external fields and their relaxation through collisions. The ideal experiment is a *complete* measurement that determines all the multipoles. Often *over-complete* experiments can be designed, but there is little reason for determining redundant information unless consistency checks are important.

Often symmetries in the experiment limit the number of nonvanishing multipoles, and frequently only populations proportional to diagonal elements of the density matrix are significant. This chapter studies the physical and geometrical significance of simple multipoles and examines whether symmetries allow a complete characterization of the ensemble with state populations. More thorough treatments can be found elsewhere [1].

12.2 POLARIZATION AND MULTIPOLES

An ensemble is said to be *polarized* if there is a non-statistical distribution of atoms in magnetic sublevels. The type of polarization can be described by the *multipoles* that exist. In many experiments, symmetries in the preparation or detection of the ensemble limit the detectable polarizations to ones caused by different sublevel populations. One can then employ the relatively simple *population* (or *diagonal*) representation. The more general case is discussed in the next section.

As an introduction to the concepts, consider an atomic manifold of total angular momentum 3/2, for example an excited alkali $P_{3/2}$ state. There are four magnetic sublevels $|m\rangle$ with $m = \pm 3/2, \pm 1/2$ so that state populations are given by a four-dimensional population-density vector

$$\mathbf{N} = \begin{pmatrix} N_{3/2} \\ N_{1/2} \\ N_{-1/2} \\ N_{-3/2} \end{pmatrix},\qquad (12.1)$$

where N_m is the density of atoms in state $|m\rangle$. If there are no external fields, the time evolution of $\mathbf{N}$ is given by

$$\dot{\mathbf{N}} = \mathcal{S} - \Gamma\mathbf{N} - \boldsymbol{\gamma}\cdot\mathbf{N},\qquad (12.2)$$

where $\mathcal{S}$ is the source vector and gives the excitation rate to the various sublevels, Γ is the radiative decay rate, assumed the same for all levels, and $\boldsymbol{\gamma}$ is the collisional matrix, whose elements $-\gamma_{mn}$ give the collisional transition rate from n to m.

The equation of motion is most easily solved if the basis used for the state space exploits the symmetry. Define the "spherical" orthonormal basis of vectors

$$\hat{\mathbf{T}}_0 = \frac{1}{2}\begin{pmatrix} 1 \\ 1 \\ 1 \\ 1 \end{pmatrix},\quad \hat{\mathbf{T}}_1 = \frac{1}{\sqrt{20}}\begin{pmatrix} 3 \\ 1 \\ -1 \\ -3 \end{pmatrix},\quad (12.3)$$

$$\hat{\mathbf{T}}_2 = \frac{1}{2}\begin{pmatrix} 1 \\ -1 \\ -1 \\ 1 \end{pmatrix},\quad \hat{\mathbf{T}}_3 = \frac{1}{\sqrt{20}}\begin{pmatrix} 1 \\ -3 \\ 3 \\ -1 \end{pmatrix}.\quad (12.4)$$

(These relations follow from the $\mathbf{T}_{LM}$ with $M = 0$. See Sects. 12.3 and 12.4.) The population vector $\mathbf{N}$ can be expanded

$$\mathbf{N} = \sum_{L=0}^{4} n_L\hat{\mathbf{T}}_L,\qquad (12.5)$$

where the coefficients $n_L = \mathbf{N}\cdot\hat{\mathbf{T}}_L$ contain all the knowable information about the population distribution. The coefficient n_L is called the $2L$ multipole moment of the ensemble. In particular, $n_0 = \frac{1}{2}\sum_m N_m$ is twice the population of the entire ensemble; $n_1 = 5^{-1/2}\sum_m mN_m = 2n_0\langle J_z\rangle/\sqrt{5}$ is the "orientation" (associated with a physical magnetic-dipole moment) of the system, $n_2 = \frac{1}{3}n_0\langle 3J_z^2 - J^2\rangle$ is the "alignment" (electric quadrupole moment), and $n_3 = (2n_0/\sqrt{45})\langle J_z(5J_z + 1 - 3\mathbf{J}^2)\rangle$ is the octupole moment of $\mathbf{N}$. The source vector $\mathcal{S}$ is readily expanded in the same basis.

If the collisions are isotropic, as is usually approximately the case in gas-cell experiments, then the collision matrix $\boldsymbol{\gamma}$ is invariant under rotations and

$$\boldsymbol{\gamma}\cdot\hat{\mathbf{T}}_L = \gamma_L\hat{\mathbf{T}}_L.\qquad (12.6)$$

The equation of motion (12.2) is separated into four uncoupled scalar equations:

$$\dot{n}_L = \mathcal{S}_L - \Gamma n_L - \gamma_L n_L.\qquad (12.7)$$

If the source term $\mathcal{S}_L = \mathcal{S}\cdot\hat{\mathbf{T}}_L$ is constant, Eqs. (12.7) have the simple solutions

$$n_L(t) = n_L(\infty) + [n_L(0) - n_L(\infty)]\exp[-(\Gamma + \gamma_L)t],\qquad (12.8)$$

with the steady-state value

$$n_L(\infty) = \mathcal{S}_L/(\Gamma + \gamma_L).\qquad (12.9)$$

The detection of such multipoles is discussed at the end of Sect. 12.5.

12.3 THE DENSITY MATRIX IN LIOUVILLE SPACE

If in the example of the last section the state basis is changed, say by rotating the axis of quantization, then populations are generally no longer sufficient to characterize the the ensemble. In the new basis, the ensemble will have been prepared in a coherent superposition of states. An adequate description for the quantum state of an ensemble in a manifold of n quantum states requires an $n \times n$ density matrix $\boldsymbol{\rho}(t)$ as discussed in Sect. 7.1.

The density matrix may be considered a vector in a space, called *Liouville space*, of n^2 dimensions. A convenient set of complex orthonormal basis vectors in this space is given by the irreducible tensor operators $\mathbf{T}_{LM}$:

$$\mathbf{T}^*_{LM} \cdot \mathbf{T}_{L'M'} := \mathrm{Tr}\left\{\mathbf{T}^\dagger_{LM}\mathbf{T}_{L'M'}\right\} = \delta_{LL'}\delta_{MM'}, \quad (12.10)$$

where the trace (Tr) expression is appropriate for the usual $n \times n$ matrix representations of the $\mathbf{T}_{LM}$. Every quantum operator on the manifold can be expanded in the basis $\{\mathbf{T}_{LM}\}$, and in particular the density matrix has an expansion

$$\boldsymbol{\rho}(t) = \sum_{LM} \rho_{LM}(t)\,\mathbf{T}_{LM}, \quad (12.11)$$

with time-dependent coefficients

$$\begin{aligned}\rho_{LM}(t) &= \mathbf{T}^*_{LM}\cdot\boldsymbol{\rho}(t) := Tr\left[\mathbf{T}^\dagger_{LM}\,\boldsymbol{\rho}(t)\right] \\ &= N\left\langle \mathbf{T}^\dagger_{LM}\right\rangle\end{aligned} \quad (12.12)$$

that are identified with the multipole moments of the ensemble. Here, $\langle\cdots\rangle$ indicates the average value over the ensemble and $N = \mathrm{Tr}\{\boldsymbol{\rho}\}$ is the normalization of the density matrix $\boldsymbol{\rho}$, which is often set equal to unity but is sometimes more conveniently set equal to the total density of atoms in the manifold.

The $\mathbf{T}_{LM}$ are defined to transform under rotations as spherical harmonics Y_{LM} [2,3]. The expansion (12.11) splits the density matrix into basis vectors $\mathbf{T}_{LM}$ which contain the geometric information and scalar coefficients (the multipole moments) $\rho_{LM}(t)$ which contain the physical information about state distributions.

The time dependence of $\boldsymbol{\rho}(t)$ is given by the Schrödinger equation and in the interaction picture takes the form

$$i\hbar\dot{\boldsymbol{\rho}} = [\mathbf{V},\,\boldsymbol{\rho}], \quad (12.13)$$

where $\mathbf{V}$ is the interaction due to external fields (including radiation) and collisions. When $\mathbf{V}$ is expanded and second-order perturbation theory is applied, the time evolution 12.13 can be expressed in a form analogous to Eq. (12.2) but with an extra term [1]:

$$\dot{\boldsymbol{\rho}} = \mathcal{S} - \boldsymbol{\Gamma}\cdot\boldsymbol{\rho} - \boldsymbol{\gamma}\cdot\boldsymbol{\rho} - i\mathbf{L}_0\cdot\boldsymbol{\rho}, \quad (12.14)$$

where $\mathbf{L}_0$ is the *Liouville operator* arising from the interaction $\mathbf{V}_0$ with slowly varying external fields. Its matrix elements with $\boldsymbol{\rho}$ are

$$(\mathbf{L}_0\cdot\boldsymbol{\rho})_{mn} = \hbar^{-1}[\mathbf{V}_0,\boldsymbol{\rho}]_{mn} = (\omega_m - \omega_n)\rho_{mn}, \quad (12.15)$$

where $\omega_m - \omega_n$ are the field-induced frequency splittings between the states.

For example, in a weak magnetic field $B_0\hat{\mathbf{z}}$ oriented along the quantization axis $\hat{\mathbf{z}}$, $\mathbf{V}_0 = \omega_0 J_z$ where $\omega_0 = g_j\mu_B B_0/\hbar$. Since $[J_z, \mathbf{T}_{LM}] = M\hbar\mathbf{T}_{LM}$ (see Chap. 7), then assuming that the collisions are isotropic and that the radiative decay rate is the same for all states of the manifold, the expansion of $\boldsymbol{\rho}$ in $\mathbf{T}_{LM}$ decouples the equation of motion (12.14) into multipoles:

$$\dot{\rho}_{LM} = \mathcal{S}_{LM} - \Gamma\rho_{LM} - \gamma_L\rho_{LM} - iM\omega_0\rho_{LM}, \quad (12.16)$$

with the solution

$$\begin{aligned}\rho_{LM}(t) &= \rho_{LM}(\infty) + [\rho_{LM}(0) - \rho_{LM}(\infty)] \\ &\quad \times \exp[-(\Gamma + \gamma_L + iM\omega_0)t]\end{aligned} \quad (12.17)$$

comprising a transient part that decays as it precesses at the angular rate ω_0 plus a steady-state part that predicts the collisionally broadened Hanle effect (see Chap. 17):

$$\rho_{LM}(\infty) = \frac{\mathcal{S}_{LM}}{\Gamma + \gamma_L + iM\omega_0}. \quad (12.18)$$

If $M = 0$, the solutions reduce to the field-independent ones given in Eq. (12.9). On the other hand, any stray field that is not aligned with an induced multipole moment can rotate the moment and possibly cause systematic measurement errors if not taken into account. Measurements of γ_L in a gas cell at temperature T are frequently used to determine thermally averaged multipole-relaxation cross sections Q_L, defined by

$$\gamma_L = N\bar{v}Q_L, \quad (12.19)$$

where N is the density of perturbers, and the mean relative velocity is $\bar{v} = 2\pi^{-1/2}(2kT/\mu)^{1/2}$, with μ the reduced mass of the polarized-atom/perturber system. Many semiclassical and full quantum calculations of such cross sections have been made, both for atoms [1, 4] and for molecules [5].

12.4 DIAGONAL REPRESENTATION: STATE POPULATIONS

An ensemble is said to be in a *coherent state* when one or more of the off-diagonal elements of the density matrix do not vanish. In a manifold of angular momentum substates of a single total angular momentum j, the ensemble has coherence if and only if it is not axially symmetric, that is, if and only if $[\boldsymbol{\rho}, J_z] \neq 0$. In an ensemble lacking coherence, or one in which any coherence that is present does not affect the observation, only the $N = 2j + 1$ diagonal elements of the $N \times N$ density matrix, those that represent state populations, need be considered. In terms of multipoles, only the N elements ρ_{L0} then play a role; the ρ_{LM} elements with $M \neq 0$ can be safely ignored. Such a *diagonal representation* is valid whenever (1) the time evolution of the system is axially symmetric and (2) either the preparation of the system or its detection is axially symmetric.

12.5 INTERACTION WITH LIGHT

A polarized system can be prepared in a variety of ways, including beam splitting in external fields, collisional excitation (including beam-foil excitation of fast beams), or radiative excitation with directed and/or polarized light. Several options for the detection of multipoles also exist, including the measurement of the anisotropy and/or polarization of scattered particles or photons. In this section, the interaction of the atomic ensemble with dipole radiation is considered.

It is convenient to define a *detection operator* for electric dipole radiation of polarization $\hat{\boldsymbol{\epsilon}}$ by

$$\mathbf{D}_\lambda (\hat{\boldsymbol{\epsilon}}) = \sum_\mu \hat{\boldsymbol{\epsilon}} \cdot \mathbf{d} \, |\lambda\mu\rangle \langle\lambda\mu| \, \mathbf{d} \cdot \hat{\boldsymbol{\epsilon}}^* \quad (12.20)$$

for decay to the state λ. Here, $\mathbf{d} = \sum_k e\mathbf{r}_k$ is the dipole-moment operator of the atom and μ is a magnetic sublevel of λ. The detection operator is a vector in Liouville space and can be expanded in irreducible tensor operators according to

$$\mathbf{D}_\lambda (\hat{\boldsymbol{\epsilon}}) = \sum_{LM} D_{\lambda LM} (\hat{\boldsymbol{\epsilon}}) \, \mathbf{T}_{LM} , \quad (12.21)$$

where

$$D_{\lambda LM} (\hat{\boldsymbol{\epsilon}}) = \mathbf{T}_{LM}^* \cdot \mathbf{D}_\lambda (\hat{\boldsymbol{\epsilon}})$$
$$= B_L (\lambda) \, \Phi_{LM} (\hat{\boldsymbol{\epsilon}}) , \quad (12.22)$$

with the dynamics contained in

$$B_L (\lambda) = (-1)^{\lambda+L+j+1} |d_{j\lambda}|^2 \begin{Bmatrix} 1 & L & 1 \\ j & \lambda & j \end{Bmatrix} , \quad (12.23)$$

and the angular dependence in

$$\Phi_{LM} (\hat{\boldsymbol{\epsilon}}) = (2L+1)^{1/2} \sum_{rs} \hat{\boldsymbol{\epsilon}} \cdot \hat{\mathbf{r}} \, (\hat{\boldsymbol{\epsilon}} \cdot \hat{\mathbf{s}})^*$$
$$\times (-1)^{1-s} \begin{pmatrix} 1 & L & 1 \\ r & \lambda & -s \end{pmatrix} . \quad (12.24)$$

Here, $\hat{\mathbf{r}}$ and $\hat{\mathbf{s}}$ range over the unit spherical tensors of rank one, namely $\pm\hat{\mathbf{1}} = \mp (\hat{\mathbf{x}} \pm i\hat{\mathbf{y}}) / \sqrt{2}$ and $\hat{\mathbf{0}} = \hat{\mathbf{z}}$ (see also Omont [6], who gives tables of Φ_{LM}), and $d_{j\lambda}$ is a reduced matrix element of $\mathbf{d}$.

The intensity of radiation of polarization $\hat{\boldsymbol{\epsilon}}'$ emitted by an excited ensemble of atoms with state density matrix ρ in its radiative decay to level λ is

$$\boldsymbol{\rho}^* \cdot \mathbf{D}_\lambda (\hat{\boldsymbol{\epsilon}}') = \sum_{m,n,\mu} \langle\lambda\mu| \mathbf{d} \cdot \hat{\boldsymbol{\epsilon}}'^* |m\rangle \, \rho_{mn} \langle n| \mathbf{d} \cdot \hat{\boldsymbol{\epsilon}}' |\lambda\mu\rangle$$
$$= \sum_{LM} \rho_{LM} D_{\lambda LM} , \quad (12.25)$$

and by selection of polarization or spatial distribution, individual multipole components $D_{\lambda LM}$ and hence ρ_{LM} can be determined. The source terms $\mathcal{S}_{LM}$ excited by electric-dipole radiation from an isotropic ground state are given by a linear combination of $D_{\lambda LM}$:

$$\mathcal{S}_{LM} = (2\pi)^2 \sum_{\lambda, \hat{\boldsymbol{\epsilon}}} u_{\hat{\boldsymbol{\epsilon}}} (\lambda) \, D_{\lambda LM} (\hat{\boldsymbol{\epsilon}}) , \quad (12.26)$$

where $u_{\hat{\boldsymbol{\epsilon}}} (\lambda)$ is the energy density of exciting radiation with polarization vector $\hat{\boldsymbol{\epsilon}}$ per unit energy. From the $3-j$ symbol in Eq. (12.24), only the $L = 0, 1$, and 2 components of ρ are observable unless the splitting of the Zeeman sublevels of λ are spectroscopically resolved. Similarly, if the ground state is unpolarized and its Zeeman sublevels unresolved, then only $L = 0, 1$, and 2 components of ρ can be excited. To excite higher-order multipoles, the ground state can be polarized by optical pumping [7].

As an example, consider the axially symmetric system discussed in Sect. 12.2 with $j = 3/2$ in the excited state and $\lambda = 1/2$ in the ground state. Only the four diagonal elements of $\mathbf{D}(\hat{\boldsymbol{\epsilon}}')$,

$$D_m (\hat{\boldsymbol{\epsilon}}') = \sum_\mu |\langle jm | \hat{\boldsymbol{\epsilon}}' \cdot \mathbf{d} | \lambda\mu\rangle|^2$$
$$= |d_{j\lambda}|^2 \sum_{\mu r} |\hat{\boldsymbol{\epsilon}}' \cdot \hat{\mathbf{r}}|^2 \begin{pmatrix} \lambda & 1 & j \\ -\mu & r & m \end{pmatrix}^2 ,$$

are needed for the $2j+1$ values of $m = -j, \ldots, j$, where μ ranges over the $2\lambda+1$ values from $-\lambda$ to λ and $r = \pm 1, 0$. One finds directly

$$\mathbf{D}(\hat{\boldsymbol{\epsilon}}') = \frac{|d_{j\lambda}|^2}{12} \left[|\hat{\boldsymbol{\epsilon}}' \cdot -\hat{\mathbf{1}}|^2 \begin{pmatrix} 3 \\ 1 \\ 0 \\ 0 \end{pmatrix} + |\hat{\boldsymbol{\epsilon}}' \cdot \hat{\mathbf{0}}|^2 \begin{pmatrix} 0 \\ 2 \\ 2 \\ 0 \end{pmatrix} \right.$$
$$\left. + |\hat{\boldsymbol{\epsilon}}' \cdot \hat{\mathbf{1}}|^2 \begin{pmatrix} 0 \\ 0 \\ 1 \\ 3 \end{pmatrix} \right] . \quad (12.27)$$

The detected signal of polarization $\hat{\boldsymbol{\epsilon}}'$ is proportional to $\mathbf{N} \cdot \mathbf{D}(\hat{\boldsymbol{\epsilon}}')$, where $\mathbf{N}$ is the population-density vector (12.1). Some of the state polarization can be monitored by looking at polarized dipole radiation. If the emitted light is observed as it propagates perpendicular to the polarization vector, the intensities are given to within an overall constant of proportionality by

$$I_{\sigma\pm} = \mathbf{D}(\pm\hat{\mathbf{1}}) \cdot \mathbf{N}, \quad I_\pi = \mathbf{D}(\hat{\mathbf{0}}) \cdot \mathbf{N}, \quad (12.28)$$
$$I_\sigma = \mathbf{D}(\hat{\boldsymbol{\xi}}) \cdot \mathbf{N} = \frac{1}{2} (I_{\sigma+} + I_{\sigma-}) , \quad (12.29)$$

where $\hat{\boldsymbol{\xi}}$ is any real unit vector (linear polarization vector) in the xy-plane. The intensities of polarized radiation may also be given in terms of Stokes parameters [8] (see also Chap. 7). By expanding $\mathbf{N}$ in the spherical basis vectors $\mathbf{T}_L$ as in Eq. (12.5), one obtains a detected signal

$$\mathbf{N} \cdot \mathbf{D}\left(\hat{\boldsymbol{\epsilon}}'\right) = \sum_{L=0}^{2j+1} n_L D_{\lambda L 0} \tag{12.30}$$

$$= \frac{|d_{j\lambda}|^2}{12} \left[2n_0 + \sqrt{5} \left(\left| \hat{\boldsymbol{\epsilon}}' \cdot \hat{\mathbf{1}} \right|^2 - \left| \hat{\boldsymbol{\epsilon}}' \cdot -\hat{\mathbf{1}} \right|^2 \right) n_1 \right. $$
$$\left. + \left(3 \left| \hat{\boldsymbol{\epsilon}}' \cdot \hat{\mathbf{0}} \right|^2 - 1 \right) n_2 \right]. \tag{12.31}$$

Thus there is no way to monitor the octupole polarization n_3 from the dipole radiation since $D_{\lambda LM} = 0$ for $L > 2$. To monitor the orientation n_1, the circular polarization must be measured

$$\frac{\sqrt{5}}{3} \frac{n_1}{n_0} = \frac{I_{\sigma+} - I_{\sigma-}}{2I_\sigma + I_\pi}, \tag{12.32}$$

since the coefficient $\mathbf{D}_{\lambda L 0}$ vanishes for any linear polarization. Conversely, from

$$\frac{n_2}{2n_0} = \frac{I_\sigma - I_\pi}{2I_\sigma + I_\pi}, \tag{12.33}$$

alignment is given by linear polarization. However, the denominator in both cases is different from that common in polarization measurements, and indeed, the ratio $(I_{\sigma+} - I_{\sigma-}) / (I_{\sigma+} + I_{\sigma-})$ is *not* proportional to the orientation but contains a contribution from the alignment, and $(I_\sigma - I_\pi) / (I_\sigma + I_\pi)$ is not linear in the alignment. The expressions for other total angular momenta j are the same except for the numerical coefficients on the LHS of Eqs. (12.32) and (12.33). Finally, if the desire is to measure the total excited-state population n_0 without any polarization component (which might rotate in stray external fields), one can choose linear polarization at an angle to make $D_{\lambda 20}$ disappear, namely at the "magic angle" $\theta = \arccos 3^{-1/2} = 54.74$ degrees. Of course, the same angle may be chosen to excite an unpolarized population.

12.6 EXTENSIONS

Although the discussion here has focused on cell experiments in which collisional perturbations are isotropic and state manifolds with a given total angular momentum j are well isolated, the concept of state multipoles is also useful in many applications where the collision symmetry is lower and where states with different j values interact, possibly due to the presence of fine and hyperfine structure [1]. Applications have been made in electron-atom collisions (Chap. 7), atom-atom collisions [9], and atom-molecule collisions [5].

REFERENCES

1. W. E. Baylis, in *Progress in Atomic Physics*, edited by W. Hanle and H. Kleinpoppen (Plenum, New York, 1977), Part. B, Chap. 28.
2. U. Fano, Rev. Mod. Phys. **29**, 74 (1957).
3. U. Fano and G. Racah, *Irreducible Tensorial Sets* (Academic, New York, 1967).
4. E. L. Lewis, Phys. Rep. **58**, 1 (1980).
5. W. E. Baylis, J. Pascale, and F. Rossi, Phys. Rev. A **36**, 4212 (1987).
6. A. Omont, Prog. Quantum Electron. **5**, 69 (1977).
7. W. Happer, Rev. Mod. Phys. **44**, 169 (1972).
8. W. E. Baylis, J. Bonenfant, J. Derbyshire, and J. Huschilt, Am. J. Phys. **61**, 534 (1993).
9. E. E. Nikitin and S. Ya. Umanski, *Theory of Slow Atomic Collisions* (Springer, Berlin, 1984).

13

Atoms in Strong Fields

S. P. Goldman

Department of Physics, The University of Western Ontario London, Ontario N6A 3K7, Canada

13.1 ELECTRON IN A UNIFORM MAGNETIC FIELD

13.1.1 Nonrelativistic Theory

The nonrelativistic Hamiltonian for an electron in an external field $\mathbf{A}$ is [1]

$$H = \frac{1}{2m}\left(\mathbf{p} - \frac{e}{c}\mathbf{A}\right)^2 - \boldsymbol{\mu}\cdot\mathbf{B} + eV, \qquad (13.1)$$

where $\mathbf{A}$ is the vector potential and V is the scalar potential. $\mathbf{A}$ and V are defined to within a gauge transformation by

$$\mathbf{A}' = \mathbf{A} + \nabla\phi, \quad V' = V - \frac{1}{c}\frac{\partial f}{\partial t}. \qquad (13.2)$$

We choose the gauge $\nabla\cdot\mathbf{A} = 0$ in which the momentum operator, $\mathbf{p} = i\hbar\nabla$, and $\mathbf{A}$ commute. $\mathbf{B}$ is a constant uniform magnetic field with vector potential

$$\mathbf{A} = \frac{1}{2}\mathbf{B}\times\mathbf{r}. \qquad (13.3)$$

μ is the magnetic moment of the electron:

$$\boldsymbol{\mu} = \frac{1}{2}g_{\mathrm{e}}\mu_B\boldsymbol{\sigma}, \qquad (13.4)$$

where the σ_i are the Pauli spin matrices, $\mu_B = e\hbar/2m_{\mathrm{e}} = 9.274\,015\,4(31)\times10^{-19}$ J/T is the Bohr magneton and $g_{\mathrm{e}} = 2.002\,319\,304\,386(20)$ is the electron g-factor which accounts for the anomalous magnetic moment of the electron (see Sect. 27.5). Consider now the case of a free electron in a constant uniform field in the z-direction with no scalar potential. This case also describes an atom in the limit of strong magnetic fields such that the Coulomb interactions are negligble. In this case we have

$$\mathbf{B} = B\hat{\mathbf{z}} \text{ with } \begin{cases} A_x = -By \\ A_y = A_z = 0 \end{cases}. \qquad (13.5)$$

For this field, the operators μ_z, p_x and p_z commute with the Hamiltonian and are therefore conserved. Calling their respective eigenvalues μ_z, p_x and p_z, with $-\infty \leq p_x, p_y \leq \infty$, the eigenstates are written as

$$\psi = e^{i(p_x x + p_z z)/\hbar}\varphi(y). \qquad (13.6)$$

Calling $y_0 = -cp_x/eB$, φ satisfies

$$-\frac{\hbar^2}{2m}\frac{d^2}{dy^2}\varphi + \frac{1}{2}m\omega_B^2\left(y - y_0\right)^2 \varphi$$
$$= \left(E + \mu_z B - \frac{p_z^2}{2m}\right)\varphi, \qquad (13.7)$$

which is the Schrödinger equation for a one-dimensional harmonic oscillator with angular frequency $\omega_B = |e|B/mc$, the cyclotron frequency of the electron. The solutions to Eq. (13.7) give the eigenstates for an electron in an external homogeneous magnetic field. They are called *Landau levels* [1] with energies given by

$$E_n = (n + \tfrac{1}{2} + m_s)\hbar\omega_B + \frac{p_z^2}{2m}, \qquad (13.8)$$

where $\hbar m_s = \pm\,\hbar/2$ is the eigenvalue of the z-component of the spin operator $\mathbf{s}$, and eigenfunctions given by

$$\varphi_n(y) = \frac{1}{\sqrt{\pi^{1/2}a_B 2^n n!}}$$
$$\times \exp\left[-\frac{(y - y_0)^2}{2a_B^2}\right] H_n\left(\frac{y - y_0}{a_B}\right), \qquad (13.9)$$

where $a_B = \sqrt{\hbar/m\omega_B}$ and the H_n are Hermite polynomials [2].

13.1.2 Relativistic Theory

The relativistic analog of Eq. (13.1) is given by the Dirac Hamiltonian

$$H_{\mathrm{D}} = c\boldsymbol{\alpha}\cdot(\mathbf{p} - \frac{e}{c}\mathbf{A}) + \beta mc^2 + eV, \qquad (13.10)$$

where α and β are the 4×4 matrices defined by Eqs. (9.62) and (9.63). The eigenfunctions of the Dirac Hamiltonian (13.10) are written as four-dimensional spinors:

$$\psi_{\mathrm{D}} = \begin{pmatrix} \psi^{(1)} \\ \psi^{(2)} \\ \psi^{(3)} \\ \psi^{(4)} \end{pmatrix} \equiv \begin{pmatrix} \phi_1 \\ \phi_{-1} \\ \chi_1 \\ \chi_{-1} \end{pmatrix} \equiv \begin{pmatrix} \phi \\ \chi \end{pmatrix}, \qquad (13.11)$$

where ϕ and χ are two-dimensional (Pauli) spinors and ϕ_i and χ_i are functions of the coordinates of the electron.

In the case $V = 0$, the lower component χ can be eliminated in the eigenvalue equation $H_{\mathrm{D}}\psi = E\psi$ to obtain a Hamiltonian equation for ϕ only, similar to Eq. (13.1) [3]. For a field (13.5) one writes ϕ in the form (as in the nonrelativistic case)

$$\phi = e^{i(p_x x + p_z z)/\hbar}\,f(y). \qquad (13.12)$$

Here, $f(y)$ satisfies the equation

$$-\frac{\hbar^2}{2m}\frac{d^2}{dy^2}\varphi + \frac{1}{2}m\omega_B^2\left(y - y_0\right)\varphi$$
$$= \left(\frac{E^2 - m^2c^4}{2mc^2} + \mu_z B - \frac{p_z^2}{2m}\right)\varphi, \qquad (13.13)$$

which reduces to the nonrelativistic form (13.7) in the limit $E \to mc^2$. In Eq. (13.13), the term involving μ is not included arbitrarily as in Eq. (13.1) but is a consequence of HF_{D}, which predicts the value $g = 2$. The value g_{e} used in Eq. (13.4) includes radiative corrections to the Dirac value.

From Eq. (13.13) we obtain

$$E_n^2 = 2mc^2\left[\frac{mc^2}{2} + (n + \tfrac{1}{2} + m_s)\hbar\omega_B + \frac{p_z^2}{2m}\right], \qquad (13.14)$$

and the eigefunctions $\psi_{m_s}^{\mathrm{D}}$ are

$$\psi_{1/2}^{\mathrm{D}} = N \begin{pmatrix} \dfrac{E + mc^2}{c}\,\psi_n \\[4pt] 0 \\[4pt] p_z\,\psi_n \\[4pt] -\dfrac{i\hbar}{a_B}\sqrt{2n}\,\psi_{n-1} \end{pmatrix},$$

$$\psi_{-1/2}^{\mathrm{D}} = N \begin{pmatrix} 0 \\[4pt] \dfrac{E + mc^2}{c}\,\psi_n \\[4pt] \dfrac{i\hbar}{a_B}\sqrt{2(n+1)}\,\psi_{n+1} \\[4pt] -p_z\,\psi_n \end{pmatrix}, \qquad (13.15)$$

where ψ_n is defined as in Eqs. (13.6) and (13.9), and N is a normalization constant.

13.2 ATOMS IN UNIFORM MAGNETIC FIELDS

13.2.1 Anomalous Zeeman Effect

Consider now the nonrelativistic Hamiltonian for a one-electron atom in the presence of an external magnetic field $\mathbf{B}$. The one-electron Hamiltonian can be written as [4]

$$H = \frac{1}{2m}\mathbf{p}^2 - \frac{1}{4\pi\epsilon_0}\frac{Ze^2}{r} + \xi(r)\mathbf{L}\cdot\mathbf{S}$$
$$+ \frac{\mu_b}{\hbar}(\mathbf{L} + g_{\mathrm{e}}\mathbf{S})\cdot\mathbf{B} + \frac{e^2}{8mc^2}(\mathbf{B} \times \mathbf{r})^2, \qquad (13.16)$$

where

$$\xi(r) = \frac{1}{2m^2c^2}\frac{Ze^2}{4\pi\epsilon_0}\frac{1}{r^3}. \qquad (13.17)$$

The anomalous Zeeman effect corresponds to the case of *weak* magnetic fields such that the magnetic interaction is small compared with the $\mathbf{L} \cdot \mathbf{S}$ spin-orbit term. The energy shifts are obtained from a perturbation of Eq. (13.16) with $\mathbf{B} = 0$. The unperturbed wave functions are eigenfunctions of $\mathbf{L}^2$, $\mathbf{S}^2$, $\mathbf{J}^2$ and J_z, with $\mathbf{J} = \mathbf{L} + \mathbf{S}$, but are not eigenfunctions of L_z or S_z. The energy levels with given values of l, s and j split in the presence of a field (13.5) according to

$$\Delta E_{m_j} = g\,\mu_B\,B\,m_j \,, \qquad (13.18)$$

where g is the *Landé splitting factor* given by

$$g = g_l\,\frac{j(j+1) - s(s+1) + l(l+1)}{2j(j+1)}$$
$$+ g_e\,\frac{j(j+1) + s(s+1) - l(l+1)}{2j(j+1)} \,, \qquad (13.19)$$

where g_e is defined by Eq. (13.4), and $g_l = 1 - m_e/M$ to lowest order in m_e/M, where m_e is the electron mass and M is the nuclear mass. To a first approximation, it is often sufficient to take $g_s = 2$ (the Dirac value) and $g_l = 1$. In the case of a many-electron atom, j, l and s are replaced by the *total* angular momenta J, L and S. In the one-electron case (neglecting corrections to g_l or g_s) the g-value is simply

$$g \approx \frac{j + \frac{1}{2}}{l + \frac{1}{2}} \,, \qquad (13.20)$$

which shows that the splitting of the $j = l + \frac{1}{2}$ levels is larger than that of the $j = l - \frac{1}{2}$ levels. The selection rules for the splitting of spectral lines are $\delta m_j = 0$ for components polarized parallel to the field (π components), and $\delta m_j = \pm 1$ for those perpendicular to the field (σ components).

13.2.2 Normal Zeeman Effect

For moderately strong fields up to $B \sim 10^4$ T, the quadratic $(\mathbf{B} \times \mathbf{r})^2$ term in Eq. (13.16) can be neglected. If the spin-orbit interaction term is also neglected, then the energies relative to the field-free eigenvalues E_n are given by

$$E_n(B) = E_n + \mu_b\,B\,(m_l + 2m_s) \,. \qquad (13.21)$$

The selection rules for transitions are $\Delta m_s = 0$ and $\Delta m_l = 0, \pm 1$. The transition energy of a spectral line is split into three components, the *Lorentz triplet*

$$\Delta E_n = \Delta E_{n,0} + (\Delta m_l)\hbar\omega_L \,, \qquad (13.22)$$

where $\Delta E_{n,0}$ is the transition energy in the absence of a field, and $\omega_L = |e|B/2m$ is the Larmor frequency. Transitions with $\Delta m_l = 0$ produce the π line at the unshifted transition energy; transitions with $\Delta m_l = \pm 1$ produce the shifted σ lines. Lorentz triplets can be observed in many-electron atoms in which the total spin is zero.

13.2.3 Paschen-Back Effect

We add now to the results of the last section the first-order perturbation caused by the spin-orbit term. The calculation can be performed in closed form for hydrogenic atoms, for which the contribution to the energy of the level n is [5]

$$\Delta E_n = 0 \,, \qquad\qquad \text{for } l = 0$$
$$\Delta E_n = \frac{\alpha^4 Z^4 \mu c^2}{2n^3}\,m_l\,m_s\,\left[l\left(l + \tfrac{1}{2}\right)(l+1)\right] \,, \quad \text{for } l \neq 0$$
$$(13.23)$$

where μ is the reduced mass, $\mu = mM/(m+M)$. A general expression of the relativistic solution for the Paschen-Back effect can be written for the one-electron case in the Pauli approximation in which the eigenstates are given in terms of ϕ in Eq. (13.11). Call $\phi_\pm^0$ the eigenfunctions with no external field corresponding to $j = l \pm \frac{1}{2}$ with unperturbed energies $E_\pm^0$ respectively. In terms of the dimensionless variables

$$\eta = \frac{\mu_B B}{E_+ - E_-} \,, \quad \epsilon = \sqrt{1 + \eta\,\frac{4m_j}{2l+1} + \eta^2} \,,$$
$$\gamma = \frac{1}{\epsilon}\left(1 + \eta\,\frac{4m_j}{2l+1}\right) \,, \qquad (13.24)$$

the energies $E_\pm$ and wave functions $\phi_\pm$ of the states, in the presence of the field B, are

$$E_\pm = \frac{E_+^0}{2}\left(1 \pm \epsilon + 2m_j\eta\right) + \frac{E_-^0}{2}\left(1 \mp \epsilon - 2m_j\eta\right) \quad (13.25)$$

and

$$\phi_\pm = \pm\sqrt{\frac{1 \pm \gamma}{2}}\,\phi_+^0 + \sqrt{\frac{1 \mp \gamma}{2}}\,\phi_-^0 \,. \qquad (13.26)$$

13.3 ATOMS IN VERY STRONG MAGNETIC FIELDS

The case with *very* strong magnetic (i.e. $B > 10^4$ T), such as those encountered at the surface of neutron stars, is also called the *quadratic Zeeman effect*, as the last term in Eq. (13.16) becomes dominant. In this range, perturbation calculations fail to yield good results as the field is too large, and even at fields of the order $B \sim 10^7$ T the Landau high B approximation of Eqs. (13.8) and (13.9) is not adequate.

Very accurate calculations have been performed using variational finite basis set techniques for both the relativistic Dirac and nonrelativistic Schrödinger Hamiltonians. The calculations use the following relativistic basis set [6] that includes nuclear size effects (R is the nuclear size) and contains both asymptotic limits, the Coulomb limit for $B = 0$ and the Landau limit for $B \to \infty$:

Table 13.1. Relativistic ground state binding energy $-E_{\text{gs}}/Z^2$ and finite nuclear size correction $\delta E_{\text{nuc}}/Z^2$ (in a.u.) of hydrogenic atoms for various magnetic fields B (in units of 2.35×10^5 T). δE_{nuc} should be added to E_{gs}.

Z	B	$-E_{\text{gs}}/Z^2$	$\delta E_{\text{nuc}}/Z^2$
1	0	0.500 006 656 597 483 75	1.55786×10^{-10}
1	10^{-5}	0.500 011 656 483 7	1.5579×10^{-10}
1	10^{-2}	0.504 981 572 360	1.5580×10^{-10}
1	10^{-1}	0.547 532 408 342 9	1.5718×10^{-10}
1	2	1.022 218 029 0	3.23×10^{-10}
1	10	1.747 800 68	1.182×10^{-9}
1	20	2.215 400 91	2.360×10^{-9}
1	200	4.727 123 3	3.032×10^{-8}
1	500	6.257 032 6	8.778×10^{-8}
20	0	0.502 691 308 407 509 8	1.3372×10^{-6}
20	1	0.503 930 867 05	1.34×10^{-6}
20	10	0.514 950 248	1.3×10^{-6}
20	100	0.612 377 94	1.4×10^{-6}
40	0	0.511 129 686 143	1.1878×10^{-5}
92	0	0.574 338 140 737 7	8.4155×10^{-4}
92	1	0.574 386 987	8.4155×10^{-4}

Table 13.2. Relativistic binding energy $-E_{2S,-1/2}$ for the $2S_{1/2}$ ($m_j = -\frac{1}{2}$) and $-E_{2P,-1/2}$ for the $2P_{1/2}$ ($m_j = -\frac{1}{2}$) excited states of hydrogen (in a.u.) in an intense magnetic field B (in units of 2.35×10^5 T).

B	$-E_{2S,-1/2}$	$-E_{2P,-1/2}$
10^{-6}	0.125 002 580 164	0.125 002 283 074
10^{-5}		0.125 006 104 950
10^{-4}	0.125 052 044 95	0.125 050 967 92
10^{-3}		0.125 499 469 4
10^{-2}	0.129 653 642 8	0.129 851 364 2
0.05	0.142 018 956	
0.1	0.148 091 738 6	0.162 411 052 4
0.2	0.148 989 58	
0.5	0.150 810 15	
1	0.160 471 07	0.260 009 34
10	0.208 955 91	0.382 663 18
100	0.256 191	0.463 664 1

$$\psi_{nl}^{(k,\nu)} = \begin{cases} r^{q_k - 1 + 2n} e^{-a_{n\nu} r^2 - \beta \rho^2} \Omega_k & r \leq R \\ b_{n\nu} r^{\gamma_\nu - 1 + n} e^{-\lambda r - \beta \rho^2} \Omega_k & r > R \end{cases} \quad (13.27)$$

with

$$\Omega_k = (\cos\theta)^{l - |m_k|} (\sin\theta)^{|m_k|} e^{im_k \phi} \omega_k , \quad (13.28)$$

where

$$n = 0, 1, \ldots, N_r, \quad k = 1,2,3,4, \quad j = 1,2,3,4,$$
$$q_1 = q_2 = k_0, \quad q_3 = q_4 = k_0',$$
$$m_k = \mu - \sigma_k/2, \quad \sigma_1 = \sigma_3 = 1, \quad \sigma_2 = \sigma_4 = -1 .$$

Here, k refers to the component $\psi^{(k)}$ in Eq. (13.11), and λ and β are variational parameters. The power of r at the origin is given by

$$k_0 = \begin{cases} |\kappa| & \text{if } \kappa < 0 \\ |\kappa| + 1 & \text{if } \kappa > 0, \end{cases} \quad (13.29)$$

$$k_0' = \begin{cases} |\kappa| + 1 & \text{if } \kappa < 0 \\ |\kappa| & \text{if } \kappa > 0. \end{cases} \quad (13.30)$$

The index ν refers to the two regular and two irregular solutions for $r > R$ that match the corresponding powers at the origin k_0 and k_0'.

$$\gamma_1 = \gamma_0, \quad \gamma_2 = \gamma_0 + 1, \quad \gamma_3 = -\gamma_0, \quad \gamma_4 = -\gamma_0 + 1, \quad (13.31)$$

$$\gamma_0 = \sqrt{\kappa^2 - (\alpha Z)^2}, \quad \kappa = \mp(j \pm \tfrac{1}{2}) + \tfrac{1}{2}, \quad (13.32)$$

$$\omega_1 = \begin{pmatrix} \vartheta_1 \\ 0 \end{pmatrix}, \quad \omega_2 = \begin{pmatrix} \vartheta_{-1} \\ 0 \end{pmatrix},$$
$$\omega_3 = \begin{pmatrix} 0 \\ i\vartheta_{-1} \end{pmatrix}, \quad \omega_4 = \begin{pmatrix} 0 \\ i\vartheta_{-1} \end{pmatrix}, \quad (13.33)$$

where ϑ_k is a two-component Pauli spinor: $\boldsymbol{\sigma}_z \vartheta_k = k\,\vartheta_i$. For even (odd) parity states, the value of l for the large components ($k = 1,2$) takes an even (odd) number greater or equal to $|m_k|$ up to $2N_\theta$ (for even parity) or $2N_\theta + 1$ (for odd parity), while for the small components ($k = 3, 4$) it takes an odd (even) number greater or equal to $|m_k|$ up to $2N_\theta + 1$ (for even parity) or $2N_\theta$ (for odd parity), since the small component has a different nonrelativistic parity than the large component. The coefficients a_{nj} and b_{nj} are determined by the continuity condition of the basis functions and their first derivatives at R. For a point nucleus, omit the section $r \leq R$; for a nonrelativistic calculation, take $\alpha = 0$ in the basis set.

Table 13.1 presents relativistic (Dirac) energies for the ground state of one-electron atoms. Values for a point nucleus and finite nuclear size corrections are given. Table 13.2 presents the relativistic energies for $n = 2$ excited states of hydrogen with the (negligible) finite nuclear size correction included. Table 13.3, which displays the relativistic corrections of the energies of the previous two tables, presents one of the most interesting relativistic results: the change in sign of the relativistic correction of the energy of the ground state at $B \sim 10^7$ T.

Table 13.3. Relativistic corrections $\delta E = (E - E_{\text{NR}})/|E_{\text{NR}}|$ to the nonrelativistic energies E_{NR} for the ground state and $n = 2$ excited states of hydrogen in an intense magnetic field B (in units of 2.35×10^5 T). The numbers between brackets denote powers of 10.

B	δE_{gs}	B	$\delta E_{2S,-1/2}$	$\delta E_{2P,-1/2}$
0.1	−1.08 [−5]	1 [−6]	−1.66 [−5]	−1.43 [−5]
1	−5.21 [−6]	1 [−4]	−1.66 [−5]	−7.86 [−6]
2	−4.03 [−6]	1 [−3]		−7.72 [−6]
3	−3.48 [−6]	1 [−2]	−1.60 [−5]	−7.30 [−6]
20	−1.09 [−6]	0.05	−1.57 [−5]	
200	4.61 [−6]	0.1	−1.74 [−6]	−6.00 [−6]
500	8.81 [−6]	1	−1.3 [−5]	−1.05 [−5]
2000	1.85 [−5]	10	−2.0 [−5]	−3.48 [−5]
5000	2.78 [−5]	100	−3.9 [−5]	−1.0 [−4]

13.4   ATOMS IN ELECTRIC FIELDS

13.4.1   Stark Ionization

An external electric field $\mathbf{F}$ introduces the perturbing potential

$$V = -\mathbf{d}\cdot\mathbf{F}, \tag{13.34}$$

where

$$\mathbf{d} = \sum_i q_i \mathbf{r}_i \tag{13.35}$$

is the dipole moment of the atom, and i runs over all electrons in the atom. In the case of strong external electric fields, bound states do not exist because the atom ionizes. Consider a hydrogenic atom in a static electric field

$$\mathbf{F} = F\hat{\mathbf{z}}. \tag{13.36}$$

The total potential acting on the electron is then

$$V_{tot}(\mathbf{r}) = -\frac{e^2 Z}{4\pi\epsilon_0}\frac{1}{r} + eFz. \tag{13.37}$$

Consider the z-dependence of this potential. Call $\rho = \sqrt{x^2 + y^2}$ and $v(z,\rho) = V(x,y,z)$. Unlike the Coulomb case in which $v_{\text{Coul}}(\pm\infty,\rho) = 0$ resulting in an infinite number of bound states, now $v(\pm\infty,\rho) = \pm\infty$ and v has a local maximum. On the z axis, this maximum occurs at $z_{max} = -\sqrt{Z|e|/(4\pi\epsilon_0 F)}$ for which $v(z_{max},0) = 0$. There is then a potential barrier through which the electron can tunnel, i.e. there are no bound states any longer but *resonances*. The potential barrier is shallower the stronger the field; the well can contain a smaller number of bound states and ionization occurs.

13.4.2   Linear Stark Effect

The electric field (13.36) produces a dipole potential

$$V_F = eFz = eF r \sqrt{\frac{4\pi}{3}} Y_{10}(\hat{\mathbf{r}}), \tag{13.38}$$

which does not preserve parity. A first-order perturbation calculation for the energy

$$E_n^{(1)} = \langle n | V_F | n \rangle \tag{13.39}$$

yields null results unless the unperturbed states are degenerate with states of opposite parity.

In the remainder of this chapter, atomic units will be used. Final results for energies can be multiplied by $2R_\infty hc$ to translate to SI or other units. The calculation can be carried out in detail for the case of hydrogenic atoms [7]. In this case it is convenient to work in parabolic coordinates: φ denotes the same angle in the xy-plane, and

$$\begin{aligned} \xi &= r + z, \\ \eta &= r - z. \end{aligned} \tag{13.40}$$

The Hamiltonian for a hydrogenic atom with a field $V_F = \frac{1}{2}F(\xi - \eta)$ from Eq. (13.38) is

$$(\xi + \eta)H = \xi h_+(\xi) + \eta h_-(\eta). \tag{13.41}$$

The wave function is written in the form

$$\Psi(\xi,\eta,\varphi) = \frac{1}{\sqrt{2\pi Z}}\, \psi_+(\xi)\psi_-(\eta)e^{im_l\varphi}, \tag{13.42}$$

with the $\psi_\pm$ satisfying

$$h_\pm(x)\psi_\pm(x) = E\psi_\pm(x), \tag{13.43}$$

where $x = \xi$ for ψ_+ and $x = \eta$ for ψ_-, and

$$h_\pm(x) = -\frac{2}{x}\frac{d}{dx}\left(x\frac{d}{dx}\right) - \frac{2Z_\pm}{x} + \frac{m_l^2}{2x^2} \mp \frac{1}{2}Fx, \tag{13.44}$$

with

$$Z = Z_+ + Z_-. \tag{13.45}$$

Using the notation

$$\begin{aligned} \epsilon &= \sqrt{-2E} \\ n_\pm &= Z_\pm/\epsilon - \tfrac{1}{2}(|m_l| + 1) \\ n &= n_+ + n_- + |m_l| + 1 \\ n_+, N_- &= 0, 1, \ldots, n - |m_l| + 1 \\ |m_l| &= 0, 1, 2, \ldots, n - 1 \\ \delta n &= n_+ - n_-, \end{aligned} \tag{13.46}$$

where n is the principal quantum number, the unperturbed eigenfunctions are

$$\psi_\pm(x) = \frac{\epsilon\, n_\pm!^{\frac{1}{2}}}{(n_\pm + |m_l|)!^{\frac{1}{2}}}\, e^{-\frac{1}{2}\epsilon x}\, (\epsilon x)^{\frac{1}{2}|m_l|}\, L_{n_\pm}^{(|m_l|)}(\epsilon x)\,,$$

$$(13.47)$$

where the $L_b^{(a)}$ are generalized Laguerre polynomials (see Sect. 9.5.2). The zero-order eigenvalue equations are

$$Z_\pm^{(0)} = \left(n_\pm + \frac{m_l + 1}{2}\right)\epsilon\,.$$

$$(13.48)$$

The first-order perturbation yields

$$Z_\pm^{(1)} = \pm\frac{1}{4}\frac{F}{\epsilon^2}\left[6n_\pm(n_\pm + m_l + 1) + m_l(m_l + 3) + 2\right].$$

$$(13.49)$$

From these

$$\epsilon = \frac{Z}{n} - \frac{3}{2\,F\left(\frac{n}{Z}\right)^2 \delta_n}\,,$$

$$(13.50)$$

and to first order in F,

$$E = -\frac{1}{2}\epsilon^2 \approx E^{(0)} + E^{(1)}\,,$$

$$E^{(0)} = -\frac{1}{2}\frac{Z^2}{n^2}\,,$$

$$E^{(1)} = \frac{3}{2}\frac{F}{Z}\,n\,\delta_n\,.$$

$$(13.51)$$

13.4.3 Quadratic Stark Effect

A perturbation linear in the field F yields no contribution to nondegenerate states (e.g. the ground state $n_+ = n_- = m = 0$; $n = 1$). In this case, the lowest order contribution comes from the *quadratic Stark effect*, the contribution of order F^2. The quadratic perturbation to a level $E_n^{(0)}$ caused by a general electric field $\mathbf{F}$ can be written in terms of the symmetric tensor α_{ij}^n as

$$E_n^{(2)} = -\frac{1}{2}\,\alpha_{ij}^n\, F_i\, F_j\,,$$

$$(13.52)$$

with

$$\alpha_{ij}^n = -2\sum_{\substack{m \\ m \neq n}} \frac{\langle n\,|d_i|\,m\rangle\,\langle m\,|d_j|\,n\rangle}{E_n - E_m}$$

$$(13.53)$$

where d_i is defined in (13.35).

For a field (13.36),

$$\Delta E_n = -\tfrac{1}{2}\,\alpha^n\, F^2\,,$$

$$(13.54)$$

where

$$\alpha^n \equiv \alpha_{zz}^n = -2e^2\sum_{\substack{m \\ m \neq n}} \frac{|\langle n\,|z|\,m\rangle|^2}{E_n - E_m}\,.$$

$$(13.55)$$

Table 13.4. Relativistic dipole polarizabilities for the ground state of hydrogenic atoms.

Z	$\alpha_{\text{rel}}^{n=1}\, Z^4/(a_0^3)$
1	4.499 751 5
5	4.493 788 3
10	4.475 164 4
20	4.400 837 6
30	4.277 562 1
40	4.106 247 4
50	3.888 179 2
60	3.625 029 5
70	3.318 865 9
80	2.972 152 4
90	2.587 720 5
100	2.168 648 3

In terms of (13.46), a general nonrelativistic expression for the *dipole polarizability* of hydrogenic ions is [8]

$$\alpha^n = \frac{a_0^3 n^4}{8 Z^4}\left[17n^2 - 3\,\delta_n^2 - 9m_l^2 + 19\right]\,.$$

$$(13.56)$$

For the ground state of hydrogenic atoms,

$$\alpha^{n=1} = \frac{9 a_0^3}{2 Z^4}\,.$$

$$(13.57)$$

Table 13.4 lists the *relativistic* values for the ground state polarizability $\alpha_{\text{rel}}^{n=1}$, obtained by calculating (13.55) using relativistic variational basis sets [9]. The values are interpolated by

$$\alpha^{n=1} = \frac{a_0^3}{Z^4}\left[\tfrac{9}{2} - \tfrac{14}{3}(\alpha Z)^2 + 0.53983(\alpha Z)^4\right]\,.$$

$$(13.58)$$

13.4.4 Other Stark Corrections

Third Order Corrections

For the energy correction cubic in the external field (13.36), one obtains [8]

$$E^{(3)} = \frac{3}{32}\, F^3 \left(\frac{n}{Z}\right)^7$$

$$\times \delta_n\left(23n^2 - \delta_n^2 + 11m_l^2 + 39\right)\,.$$

$$(13.59)$$

Relativistic Linear Stark-Shift of the Fine Structure of Hydrogen

For a Stark effect small relative to the fine structure, the degenerate levels corresponding to the same value of j split according to

$$\delta_m E_{nj} = \frac{3}{4} \sqrt{n^2 - \left(j + \tfrac{1}{2}\right)^2}\, \frac{n\,m}{j(j+1)}\, F. \qquad (13.60)$$

Other Stark Corrections in Hydrogen

The expectation value of the delta function, is, in a.u. [10],

$$2\pi \langle 1s |\, \delta(\mathbf{r}) \,| 1s \rangle = 2 - 31 F^2. \qquad (13.61)$$

For the Bethe logarithm β defined by

$$\beta_{1s} = \frac{\sum_n |\langle 1s\,|\mathbf{p}|\,n\rangle|^2 \,(E_n - E_{1s}) \ln |E_n - E_{1s}|}{\sum_n |\langle 1s\,|\mathbf{p}|\,n\rangle|^2 \,(E_n - E_{1s})}, \qquad (13.62)$$

the result is [11]

$$\beta_{1s} = 2.290\,981\,375\,205\,552\,301 + 0.316\,205(6)\,F^2. \qquad (13.63)$$

These results are useful in calculating an asymptotic expansion for the two-electron Bethe logarithm [12].

REFERENCES

1. L. D. Landau and E. M. Lifshitz, *Quantum Mechanics (Course of Theoretical Physics)* (Pergamon Press, Oxford, 1977), Vol. 3, p. 456.
2. A. Messiah, *Quantum Mechanics* (John Wiley, New York), p. 491.
3. C. Itzykson and J.-B. Zuber, *Quantum Field Theory* (McGraw-Hill, New York, 1980), p. 67.
4. H. A. Bethe and E. Salpeter, *Quantum Mechanics of One- and Two-electron Atoms* (Plenum, New York, 1977), p. 208.
5. Ref. [4], p. 211.
6. Z. Chen and S. P. Goldman, Phys. Rev. A **48**, 1107 (1993).
7. Ref. [4], p. 229.
8. Ref. [4], p. 233.
9. G. W. F. Drake and S. P. Goldman, Phys. Rev. A **23**, 2093 (1981).
10. G. W. F. Drake, Phys. Rev. A **45**, 70 (1992).
11. S. P. Goldman, Phys. Rev. A **50**, 3039 (1994).
12. S. P. Goldman and G. W. F. Drake, Phys. Rev. Lett. **68**, 1683 (1992).

14

Rydberg Atoms

T. F. Gallagher

Department of Physics, University of Virginia, Charlottesville, Virginia 22901

14.1 INTRODUCTION

Rydberg atoms are those in which the valence electron is in a state of high principal quantum number n. They are of historical interest since the observation of Rydberg series helped in the initial unraveling of atomic spectroscopy [1]. Since the 1970s they have been studied mostly for two reasons. First, Rydberg states are at the border between bound states and the continuum, and any process which can result in either excited bound states or ions and free electrons usually leads to the production of Rydberg states. Second, the exaggerated properties of Rydberg atoms allow experiments to be done which would be difficult or impossible with normal atoms.

The asymptotic expansion method and high-precision results for the Rydberg states of helium are further discussed in Sect. 11.5.2.

14.2 WAVE FUNCTIONS AND QUANTUM DEFECT THEORY

Many of the properties of Rydberg atoms can be calculated accurately using quantum defect theory, which is easily understood by starting with the H atom [2]. We shall use atomic units, as discussed in Sect. 1.3. The Schrödinger equation for the motion of the electron in a H atom in spherical co-ordinates is

$$\left(-\tfrac{1}{2}\nabla^2 - 1/r\right)\Psi(r,\theta,\phi) = E\Psi(r,\theta,\phi), \qquad (14.1)$$

where E is the energy, r is the distance between the electron and the proton, and θ and ϕ are the polar and azimuthal angles of the electron's position. Equation (14.1) can be separated, and its solution expressed as the product

$$\Psi(r,\theta,\phi) = R(r)Y_{\ell m}(\theta,\phi), \qquad (14.2)$$

where ℓ and m are the orbital and azimuthal-orbital angular momentum (i.e., magnetic) quantum numbers and $Y_{\ell m}(\theta,\phi)$ is a normalized spherical harmonic. $R(r)$ satisfies the radial equation

$$\frac{d^2 R(r)}{dr^2} + \frac{2dR(r)}{rdr} + 2ER(r) + \frac{2R(r)}{r} = \frac{\ell(\ell+1)R}{r^2}, \qquad (14.3)$$

which has the two physically interesting solutions

$$R(r) = \frac{f(\ell,E,r)}{r}, \qquad (14.4)$$

$$R(r) = \frac{g(\ell,E,r)}{r}. \qquad (14.5)$$

The f and g functions are the regular and irregular Coulomb functions which are the solutions to a variant of Eq. (14.3). As $r \to 0$ they have the forms [3]

$$f(\ell, E, r) \propto r^{\ell+1}, \qquad (14.6)$$

$$g(\ell, E, r) \propto r^{-\ell}, \qquad (14.7)$$

irrespective of whether E is positive or negative. As $r \to \infty$, for $E > 0$ the f and g functions are sine and cosine waves, i.e., there is a phase shift of $\pi/2$ between them. For $E < 0$ it is useful to introduce ν, defined by $E = -1/2\nu^2$, and for $E < 0$ as $r \to \infty$

$$f = u(\ell, \nu, r) \sin \pi\nu - v(\ell, \nu, r)e^{i\pi\nu}, \qquad (14.8)$$

$$g = -u(\ell, \nu, r) \cos \pi\nu + v(\ell, \nu, r)e^{i\pi(\nu+1/2)}, \qquad (14.9)$$

where u and v are exponentially increasing and decreasing functions of r. As $r \to \infty$, $u \to \infty$ and $v \to 0$.

Requiring that the wave function be square integrable means that as $r \to 0$ only the f function is allowed. Equation (14.8) shows that the $r \to \infty$ boundary condition requires that $\sin \pi\nu$ be zero or ν an integer n, leading to the hydrogenic Bohr formula for the energies:

$$E = -\frac{1}{2n^2}. \qquad (14.10)$$

The classical turning point of an s wave occurs at $r = 2n^2$, and the expectation values of positive powers of r reflect the location of the outer turning point, i.e.,

$$\langle r^k \rangle \approx n^{2k}. \qquad (14.11)$$

The expectation values of negative powers of r are determined by the properties of the wave function at small r. The normalization constant of the radial wave function scales as $n^{-3/2}$, so that $R(r) \propto n^{-3/2}r^{\ell+1}$ for small r. Accordingly, the expectation values of negative powers of r, except r^{-1} and any properties which depend on the small r part of the wave function, scale as n^{-3}. Using the properties of the wave function and the energies, the n-scaling of the properties of Rydberg atoms can be determined.

The primary reason for introducing the Coulomb waves instead of the more common Hermite polynominal solution for the radial function is to set the stage for single channel quantum defect theory, which enables us to calculate the wave functions and properties of one valence electron atoms such as Na. The simplest picture of an Na Rydberg atom is an electron orbiting a positively charged Na$^+$ core consisting of 10 electrons and a nucleus of charge +11. The ten electrons are assumed to be frozen in place with spherical symmetry about the nucleus, so their charge cloud is not polarized by the outer valence electron, although the valence electron can penetrate the ten-electron cloud. When the electron penetrates the charge cloud of the core electrons, it sees a potential well

Figure 14.1. Radial wave functions for H and Na showing that the Na wave function is pulled in toward the ionic core.

deeper than $-1/r$ due to the decreased shielding of the +11 nuclear charge. For Na and other alkali atoms, we assume that there is a radius r_c such that for $r < r_c$ the potential is deeper than $-1/r$, and for $r > r_c$ it is equal to $-1/r$. As a result of the deeper potential at $r < r_c$, the radial wave function is pulled into the core in Na, relative to H, as shown in Fig. 14.1. For $r \geq r_c$, the potential is a Coulomb $-1/r$ potential, and $R(r)$ is a solution of Eq. (14.3) which can be expressed as

$$R(r) = (f(\ell, \nu, r) \cos \tau_\ell - g(\ell, \nu, r) \sin \tau_\ell)/r, \qquad (14.12)$$

where τ_ℓ is the radial phase shift.

Near the ionization limit, $E \sim 0$, and as a result, the kinetic energy of the Rydberg electron is greater than $1/r_c$ (~ 10 eV) when $r < r_c$. As a result, changes in E of 0.10 eV, the $n = 10$ binding energy, do not appreciably alter the phase shift τ_ℓ, and we can assume τ_ℓ to be independent of E. The ℓ dependence of τ_ℓ arises because the centrifugal $\ell(\ell+1)/r$ term in Eq. (14.3) excludes the Rydberg electron from the region of the core in states of high ℓ. Applying the $r \to \infty$ boundary condition to the wave function of Eq. (14.12) leads to the requirement that the coefficient of u vanish, i.e.,

$$\cos \tau_\ell \sin(\pi\nu) + \sin\tau_\ell \cos(\pi\nu) = 0, \qquad (14.13)$$

which implies that $\sin(\pi\nu + \tau_\ell) = 0$ or $\nu = n - \tau_\ell/\pi$. Usually τ_ℓ/π is written as δ_ℓ and termed the quantum defect, and the energies of members of the $n\ell$ series are written as

$$E = -\frac{1}{2(n - \delta_\ell)^2} = -\frac{1}{2n^{*2}}, \qquad (14.14)$$

where $n^* = n - \delta_\ell$ is often termed the effective quantum number (see also Sect. 11.5.1).

Knowledge of the quantum defect δ_ℓ of a series of ℓ states determines their energies, and it is a straight-forward matter to calculate the Coulomb wave function

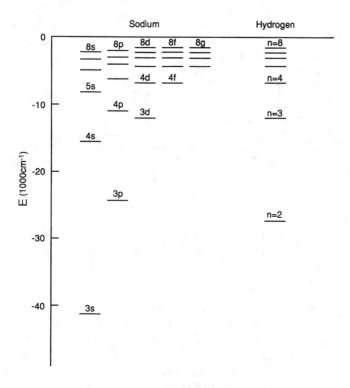

Figure 14.2. Energy levels of Na and H.

specified in Eq. (14.12) using a Numerov algorithm [4, 5]. This procedure gives wave functions valid for $r \geq r_c$, which can be used to calculate many of the properties of Rydberg atoms with great accuracy. The effect of core penetration on the energies is easily seen in the energy level diagram of Fig. 14.2. The Na $\ell \geq 2$ states have the same energies as hydrogen, while the s and p states, with quantum defects of 1.35 and 0.85 respectively, lie far below the hydrogenic energies.

Although it is impossible to discern in Fig. 14.2, the Na $\ell \geq 2$ states also lie below the hydrogenic energies. For these states it is not core penetration, but core polarization which is responsible for the shift to lower energy. Contrary to our earlier assumption that the outer electron does not affect the inner electrons if $r > r_c$, the outer electron polarizes the inner electron cloud even when $r > r_c$, and the energies of even the high ℓ states fall below the hydrogenic energies. The leading term in the polarization energy is due to the dipole polarizability of the core, α_d. For high ℓ states it gives a quantum defect of [6]

$$\delta_\ell = \frac{3\alpha_d}{4\ell^5}.\qquad(14.15)$$

Quantum defects due primarily to core polarization rarely exceed 10^{-2}, while those due to core penetration are often greater than one.

14.3 OPTICAL EXCITATION AND RADIATIVE LIFETIMES

Optical excitation of the Rydberg states from the ground state or any other low lying state is the continuation of the photoionization cross section σ_{PI} below the ionization limit. The photoionization cross section, discussed more extensively in Chap. 24, is approximately constant at the limit. Above and below the limit the average photoabsorption cross section is the same, as evidenced by the fact that a discontinuity is not evident in an absorption spectrum, i.e., it is not possible to see where the unresolved Rydberg states end and the continuum begins. Nonetheless, below the limit the cross section is structured by the Δn spacing of $1/n^3$ between adjacent members of the Rydberg series. In any experiment, there is a finite resolution $\Delta\omega$ with which the Rydberg states can be excited. It can arise, for example, from the Doppler width or a laser linewidth. This resolution determines the cross section σ_n for exciting the Rydberg state of principal quantum number n. Explicitly, σ_n is given by

$$\sigma_n = \frac{\sigma_{PI}}{n^3\Delta\omega}.\qquad(14.16)$$

A typical value for σ_{PI} is 10^{-18} cm^2. For a resolution $\Delta\omega = 1$ cm^{-1} (6×10^{-6} a.u.) the cross section for exciting an $n = 20$ atom is 3×10^{-17} cm^2.

From the wave functions of the Rydberg states, we can also derive the n^{-3} dependence of the photoexcitation cross section. The dipole matrix element from the ground state to a Rydberg state only involves the part of the Rydberg state wave function near the core. At small r, the Rydberg wave function only depends on n through the n^{-3} normalization factor, and as a result, the squared dipole matrix element between the ground state and the Rydberg state and the cross section have n^{-3} dependences.

Radiative decay, which is covered in Chap. 17, is, to some extent, the reverse of optical excitation. The general expression for the spontaneous transition rate from the $n\ell$ state to the $n'\ell'$ state is the Einstein A coefficient, given by [2]

$$A_{n\ell,n'\ell'} = \frac{4}{3}\mu_{n\ell,n'\ell'}^2\omega_{n\ell,n'\ell'}^3\frac{\alpha^3 g_>}{2g_n+1},\qquad(14.17)$$

where $\mu_{n\ell,n'\ell'}$ and $\omega_{n\ell,n',\ell'}$ are the electric dipole matrix elements and frequencies of the $n\ell \rightarrow n'\ell'$ transitions, g_n and $g_{n'}$ are the degeneracies of the $n\ell$ and $n'\ell'$ states, and $g_>$ is the greater of g_n and $g_{n'}$. The lifetime, $\tau_{n\ell}$, of the $n\ell$ state is obtained by summing the decay rates to all possible lower energy states. Explicitly,

$$\frac{1}{\tau_{n\ell}} = \sum_{n'\ell'} A_{n\ell,n'\ell'}.\qquad(14.18)$$

Due to the ω^3 factor in (14.17), the highest frequency transition usually contributes most heavily to the total radiative decay rate, and the dominant decay is likely to be the lowest lying state possible. For low-ℓ Rydberg states, the lowest lying ℓ' states are bound by orders of magnitude more than the Rydberg states, and the frequency of the decay is nearly independent of n. Only the squared dipole moment depends on n, as n^{-3}, because of the normalization of the Rydberg wave function at the core. Consequently, for low-ℓ states,

$$\tau_{n\ell} \propto n^3 . \tag{14.19}$$

As a typical example, the $10f$ state in H has a lifetime of 1.08 μs [7].

The highest ℓ states, with $\ell = n - 1$, have radiative lifetimes with a completely different n dependence. The only possible transitions are $n \to n - 1$, with frequency $1/n^3$. In this case the dipole moments reflect the large size of both the n and $n-1$ states and have the n^2 scaling of the orbital radius. Using Eq. (14.17) for $\ell = n-1$ leads to

$$\tau_{n(n-1)} \propto n^5 . \tag{14.20}$$

Another useful lifetime, τ_n, is that corresponding to the average decay rate of all ℓ, m states of the same n. It scales as $n^{4.5}$ [2, 8].

Equation (14.17) describes spontaneous decay to lower lying states driven by the vacuum. At room temperature, 300 K, there are many thermal photons at the frequencies of the $n \to n \pm 1$ transitions of Rydberg states for $n \geq 10$, and these photons drive transitions to higher and lower states [9]. A convenient way of describing blackbody radiation is in terms of the photon occupation number $\bar{n}$, given by

$$\bar{n} = \frac{1}{e^{\omega/kT} - 1} . \tag{14.21}$$

The stimulated emission or absorption rate $K_{n\ell,n'\ell'}$ from state $n\ell$ to state $n'\ell'$ is given by

$$K_{n\ell,n'\ell'} = \frac{4}{3}\mu_{n\ell,n'\ell'}\omega^3_{n\ell,n'\ell'}\frac{\alpha^3\bar{n}g_>}{2g_n+1} . \tag{14.22}$$

Summing these rates over n' and ℓ' gives the total blackbody decay rate $1/\tau_n^{bb}$. Explicitly,

$$\frac{1}{\tau_{n\ell}^{bb}} = \sum_{n'\ell'} K_{n\ell,n'\ell'} . \tag{14.23}$$

The resulting lifetime $\tau_{n\ell}^T$ at any given temperature is given by

$$\frac{1}{\tau_{n\ell}^T} = 1/\tau_{n\ell} + 1/\tau_{n\ell}^{bb} . \tag{14.24}$$

For low-ℓ states with $10 < n < 20$, blackbody radiation produces a 10% decrease in the lifetimes, but for high-ℓ

Figure 14.3. Combined Coulomb-Stark potential along the z axis when a field of 5 x 10^{-7} a.u. (2700 V/cm) is applied in the z- direction (solid). The extreme red state (R) is near the saddlepoint, and the extreme blue (B) state is held on the upfield side of the atom by an effective potential (dashed) roughly analogous to a centrifugal potential.

states of the same n, it reduces the lifetimes by a factor of ten. Since $1/\tau_{n\ell}^{bb} \propto n^{-2}$, at high n it must dominate normal spontaneous emission.

The above discussion of spontaneous and stimulated transitions is based on the implicit assumption that the atoms are in free space. If the atoms are in a cavity, which introduces structure into the blackbody and vacuum fields, the transition rates are significantly altered [10]. These alterations are described in Chap. 76. If the cavity is tuned to a resonance, it increases the transition rate by the finesse of the cavity (approximately the Q for low-order modes). On the other hand, if the cavity is tuned between resonances, the transition rate is suppressed by a similar factor.

14.4 ELECTRIC FIELDS

As a starting point, consider the H atom in a static electric field $\mathcal{E}$ in the z-direction, and focus on the states of principal quantum number n. The field couples ℓ and $\ell \pm 1$ states of the same m by the electric dipole matrix elements. Since the states all have a common zero field energy of $-1/2n^2$, and the off-diagonal Hamiltonian matrix elements are all proportional to $\mathcal{E}$, the eigenstates are field-independent linear combinations of the zero field ℓ states of the same m, and the energy shifts from $-1/2n^2$ are linear in $\mathcal{E}$. In this first-order approximation, the energies are given by [2]

$$E = -\frac{1}{2n^2} + \frac{3}{2}(n_1 - n_2)n\mathcal{E}, \qquad (14.25)$$

where n_1 and n_2 are parabolic quantum numbers (see Sect. 9.2.2) which satisfy

$$n_1 + n_2 + |m| + 1 = n. \qquad (14.26)$$

Consider the $m = 0$ states as an example. The $n_1 - n_2 = n - 1$ state is shifted up in energy by $\frac{3}{2}n(n-1)\mathcal{E}$ and is called the extreme blue Stark state, and the $n_2 - n_1 = n - 1$ state is shifted down in energy by $\frac{3}{2}n(n-1)\mathcal{E}$ and is called the extreme red Stark state. These two states have large permanent dipole moments, and in the red (blue) state the electron spends most of its time on the downfield (upfield) side of the proton as shown in Fig. 14.3, a plot of the potential along the z-axis. We have here ignored the electric dipole couplings to other n states, which introduce small second order Stark shifts to lower energy. As implied by Eq. (14.26), states of higher m have smaller shifts. In particular, the circular $m = \ell = n - 1$ state has no first order shift since there are no degenerate states to which it is coupled by the field. The Stark effect in other atoms is similar, but not identical to that observed in H. This point is shown by Fig. 14.4, a plot of the energies of the Na $m = 0$ levels near $n = 20$. The energy levels are similar to those of H in that most of the levels exhibit apparently linear Stark shifts from the zero field energy of the high-ℓ states. However, the differences are twofold. First, the levels from s and p states with nonzero quantum defects join the manifold of Stark states at some nonzero field, given approximately by [4]

$$\mathcal{E} = \frac{2\delta_\ell'}{3n^5}, \qquad (14.27)$$

where δ_ℓ' is the magnitude of the difference between δ_ℓ and the nearest integer. Second, there are avoided crossings between the blue $n = 20$ and red $n = 21$ Stark states. In H these states would cross, but in Na they do not because of the finite sized Na$^+$ core, which also leads to the nonzero quantum defects of the ns and np states. This point and other related points are described in Chap. 15.

Field ionization is both intrinsically interesting and of great practical importance for the detection of Rydberg atoms [11]. The simplest picture of field ionization can be understood with the help of Fig. 14.4. The potential along the z-axis of an atom in a field $\mathcal{E}$ in the z-direction is given by

$$V = -\frac{1}{r} - \mathcal{E}z. \qquad (14.28)$$

If an atom has an energy E relative to the zero field limit, it can ionize classically if the energy E lies above the saddle point in the potential. The required field is given by

$$\mathcal{E} = \frac{E^2}{4}. \qquad (14.29)$$

Ignoring the Stark shifts and using $E = -1/2n^2$ yields the expression

$$\mathcal{E} = \frac{1}{16n^4}. \qquad (14.30)$$

The H atom ionizes classically as described above, or by quantum mechanical tunneling which occurs at slightly lower fields. Since the tunneling rates increase exponentially with field strength, typically an order of magnitude for a 3% change in the field, specifying the classical ionization field is a good approximation to the field which gives an ionization rate of practical interest. The red and blue states of H ionize at very different fields, as shown by Fig. 14.5, a plot of the $m = 0$ Stark states out to the fields at which the ionization rates are 10^6 s^{-1} [12]. First, note the crossing of the levels of different n mentioned earlier. Second, note that the red states ionize at lower fields than do the blue states, in spite of the fact that they are lower in energy. In the red states, the electron is close to the saddle point of the potential of Fig. 14.3, and it ionizes according to Eq. (14.29). If the Stark shift of the extreme red state to lower energy is taken into account, Eq. (14.30) becomes

$$\mathcal{E} = \frac{1}{9n^4}. \qquad (14.31)$$

In the blue state the electron is held on the upfield side of the atom by an effective potential roughly analogous to a centrifugal potential, as shown by Fig. 14.3. At the same field the blue state's energy is lower relative to the saddle point of its potential, shown by the broken line of Fig. 14.3, than is the energy of the red state relative to the saddle point of its potential, given by Eq. 14.28 and shown by the solid line of Fig. 14.3. As shown by the broken line of Fig. 14.5, the classical ionization limit of Eq. (14.29) is simply a line connecting the ionization fields of the extreme red Stark states. All other states are stable above the classical ionization limit. In the Na atom, ionization of $m = 0$ states occurs in a qualitatively different fashion [12]. Due to the finite size of the Na$^+$ core, there are avoided crossings between the blue and red Stark states of different n, as is shown by Fig. 14.4. In the region above the classical ionization limit, shown by the shaded region of Fig. 14.4, the same coupling between hydrogenically stable blue states and the degenerate red

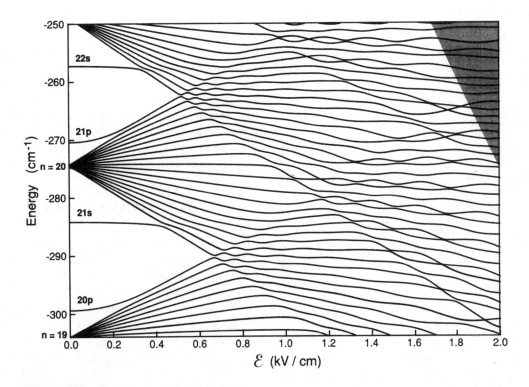

Figure 14.4. Energies of Na $m = 0$ levels of $n \approx 20$ as a function of electric field. The shaded region is above the classical ionization limit.

continua leads to autoionization of the blue states [13]. As a result, all states above the classical ionization limit ionize at experimentally significant rates. In higher m states, the core coupling is smaller, and the behavior is more similar to H.

Field ionization is commonly used to detect Rydberg atoms in a state selective manner. Experiments are most often conducted at or near zero field, and afterwards the field is increased in order to ionize the atoms. Exactly how the atoms pass from the low field to the high ionizing field is quite important. The passage can be adiabatic, diabatic or anything in between. The selectivity is best if the passage is purely adiabatic or purely diabatic, for in these two cases unique paths are followed.

In zero field, optical excitation from a ground s state leads only to final np states. In the presence of an electric field, all the Stark states are optically accessible, because they all have some p character. The fact that all the Stark states are optically accessible from the ground state allows the population of arbitrary ℓ states of nonhydrogenic atoms by a technique called Stark switching [6, 14]. In any atom other than H, the ℓ states are nondegenerate in zero field, and each of them is adiabatically connected to one, and only one, high field Stark state, as shown by Fig. 14.4. If one of the Stark states is excited with a laser and the field reduced to zero adiabatically, the atoms are left in a single zero field ℓ state.

In zero field, the photoionization cross section is structureless. However, in an electric field, it exhibits obvious structure, sometimes termed strong field mixing resonances. Specifically, when ground state s atoms are exposed to light polarized parallel to the static field, an oscillatory structure is observed in the cross section, even above the zero field ionization limit [15]. The origin of the structure can be understood with the aid of a simple classical picture [16, 17]. The electrons ejected in the downfield direction can simply leave the atom, while the electrons ejected in the upfield direction are reflected back across the ionic core and also leave the atom in the downfield direction. The wave packets corresponding to these two classical trajectories are added, and they can interfere constructively or destructively at the ionic core depending on the phase accumulation of the reflected wave packet. Since the phase depends on the energy, there is an oscillation in the photoexcitation spectrum. This model suggests that no oscillations should be observed for light polarized perpendicular to the static field, and none are. The oscillations can also be thought of as arising from the remnants of quasistable extreme blue Stark states which have been shifted above the ionization limit, and, using this approach or a WKB approach, one can show that the spacing between the oscillations at the zero field limit is $\Delta E = \mathcal{E}^{3/4}$ [18, 19].

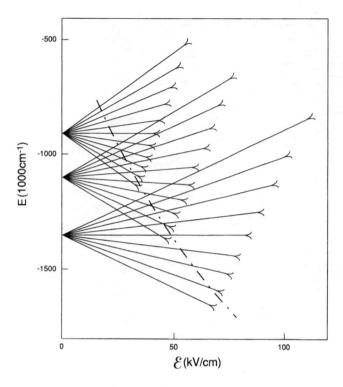

Figure 14.5. Energies of H $m = 0$ levels of $n = 9$, 10, and 11 as functions of electric field. The widths of the levels due to ionization broaden exponentially with fields, and the onset of the broadening indicated is at an ionization rate of 10^6 s^{-1}. The broken line indicating the classical ionization limit, $\mathcal{E} = E^2/4$ passes near the points at which the extreme red states ionize.

14.5   MAGNETIC FIELDS

To first order, the energy shift of a Rydberg atom due to a magnetic field B (the Zeeman effect) is proportional to the angular momentum of the atom. Since the states optically accessible from the ground state have low angular momenta, the energy shifts are the same as those of low-lying atomic states. In contrast, the second order diamagnetic energy shifts are proportional the area of the Rydberg electron's orbit and scale as B^2n^4 [20]. The diamagnetic interaction mixes the ℓ states, allowing all to be excited from the ground state, and produces large shifts to higher energies. The energy levels as a function of magnetic field are reminiscent of the Stark energy levels shown in Fig. 14.5, differing in that the energy shifts are quadratic in the magnetic field.

One of the most striking phenomena in magnetic fields is the existence of quasi-Landau resonances, spaced by $\Delta E = 3\hbar B/2$, in the photoionization cross section above the ionization limit [21]. The origin of this structure is similar to the origin of the strong field mixing resonances observed in electric fields. An electron ejected in the plane perpendicular to the B fields is

launched into a circular orbit and returns to the ionic core. The returning wave packet can be in or out of phase with the one leaving the ionic core, and thus, can interfere constructively or destructively with it. While the electron motion in the xy-plane is bound, motion in the z-direction is unaffected by the magnetic field and is unbounded above the ionization limit, leading to resonances of substantial width. The Coulomb potential does provide some binding in the z-direction and allows the existence of quasistable three-dimensional orbits [22].

14.6   MICROWAVE FIELDS

Strong microwave fields have been used to drive multiphoton transitions between Rydberg states and to ionize them. Here we restrict our attention to ionization. Ionization by both linearly and circularly polarized fields has been explored with both H and other atoms.

H atoms have been studied with linearly polarized fields of frequencies up to 36 GHz [23]. When the microwave frequency $\omega \ll 1/n^3$, ionization of $m = 0$ states occurs at a field of $\mathcal{E} = 1/9n^4(E^2/4)$, which is the field at which the extreme red Stark state is ionized by a static field. Due to the second-order Stark effect, the blue and red shifted states are not quite mirror images of each other, and when the microwave field reverses, transitions between Stark states occur. There is a rapid mixing of the Stark states of the same n and m by a microwave field, and all of them are ionized at the same microwave field amplitude, $\mathcal{E} = 1/9n^4$. Important points are that no change in n occurs and the ionization field is the same as the static field required for ionization of the extreme red Stark state. As ω approaches $1/n^3$, the field falls below $1/9n^4$ due to Δn transitions to higher lying states, allowing ionization at lower fields. This form of ionization can be well described as the transition to the classically chaotic regime [24]. For $\omega > 1/n^3$ the ionization field is more or less constant, and for $\omega > 1/2n^2$ the process becomes photoionization.

The ionization of nonhydrogenic atoms by linearly polarized fields has also been investigated at frequencies of up to 30 GHz, but the result is very different from the hydrogenic result. For $\omega \ll 1/n^3$ and low m, ionization occurs at a field of $\mathcal{E} \approx 1/3n^5$ [25]. This is the field at which the $m = 0$ extreme blue and red Stark states of principal quantum number n and $n+1$ have their avoided crossing. For $n = 20$ this field is ~ 500 V/cm, as shown by Fig. 14.4. How ionization occurs can be understood with a simple model based on a time-varying electric field. As the microwave field oscillates in time, atoms follow the Stark states of Na shown in Fig. 14.4. Even with very small field amplitudes, transitions between the Stark states of the same n are quite rapid because of the zero field avoided crossings. If the field reaches $1/3n^5$, the avoided crossing between the extreme red n and blue $n+1$

state is reached, and an atom in the blue n Stark state can make a Landau-Zener transition to the red $n+1$ Stark state. Since the analogous red-blue avoided crossings between higher lying states occur at lower fields, once an atom has made the $n \rightarrow n+1$ transition it rapidly makes a succession of transitions through higher n states to a state which is itself ionized by the field.

The Landau-Zener description given above is somewhat oversimplified in that we have ignored the coherence between field cycles. When it is included, we see that the transitions between levels are resonant multiphoton transitions. While the resonant character is obscured by the presence of many overlapping resonances, the coherence substantially increases the $n \rightarrow n+1$ transition probability even when $\mathcal{E} < 1/3n^5$. The fields required for ionization calculated using this model are lower than $1/3n^5$, in agreement with the experimental observations. Nonhydrogenic Na states of high m behave like H, because no states with significant quantum defects are included, and the $n \rightarrow n+1$ avoided crossings are vanishingly small.

Experiments on ionization of alkali atoms by circularly polarized fields of frequency ω show that for $\omega \ll 1/n^3$, a field amplitude of $\mathcal{E} = 1/16n^4$ is required for ionization [26]. This field is the same as the static field required. In a frame rotating with frequency ω, the circularly polarized field is stationary and cannot induce transitions, so this result is not surprising. On the other hand, when the problem is transformed to the rotating frame, the potential of Eq. (14.28) is replaced by

$$V = -\frac{1}{r} - \mathcal{E}x - \frac{\omega^2 \rho^2}{2}, \qquad (14.32)$$

where $\rho^2 = x^2 + y^2$, and we have assumed the field to be in the x-direction in the rotating frame. This potential has a saddle point below $\mathcal{E} = 1/16n^4$ [27]. As n or ω is raised so that $\omega \rightarrow 1/n^3$, the experimentally observed field falls below $1/16n^4$, but not so fast as implied by Eq. (14.32). Equation (14.32) is based solely on energy considerations, and ionization at the threshold field implied by Eq. (14.32) requires that the electron escape over the saddle point in the rotating frame at nearly zero velocity. For this to happen, when ω approaches $1/n^3$, more than n units of angular momentum must be transferred to the electron, which is unlikely. Models based on a restriction of the angular momentum transferred from the field to the Rydberg electron are in better agreement with the experimental results. Small deviations of a few percent from circular polarization allow ionization at fields as low as $\mathcal{E} = 1/3n^5$. This sensitivity can be understood as follows. In the rotating frame, a field with slightly elliptical polarization appears to be a large static field with a superimposed oscillating field at frequency 2ω. The oscillating field drives transitions to states of higher energy, allowing ionization at fields less than $\mathcal{E} = 1/16n^4$.

14.7 COLLISIONS

Since Rydberg atoms are large, with geometric cross sections proportional to n^4, one might expect the cross sections for collisions to be correspondingly large. In fact, such is often not the case. A useful way of understanding collisions of neutral atoms and molecules with Rydberg atoms is to imagine an atom or molecule M passing through the electron cloud of an Na Rydberg atom. There are three interactions

$$e^- - Na^+, \quad e^- - M, \quad M - Na^+. \qquad (14.33)$$

The long range $e^- - Na^+$ interaction determines the energy levels of the Na atom. The short range of the $e^- - M$ and $M - Na^+$ interactions makes it likely that only one will be important at any given time. This approximation, termed the binary encounter approximation, is described in Chap. 54. The $M - Na^+$ interaction can only lead to cross sections of $\sim 10 - 100$ Å^2. On the other hand, since the electron can be anywhere in the cloud, the cross sections due to the $e^- - M$ interaction can be as large as the geometric cross section of the Rydberg atom. Accordingly, we focus on the $e^- - M$ interaction.

Consider a thermal collision between M and an Na Rydberg atom. Typically, M passes through the electron cloud slowly compared with the velocity of the Rydberg electron, and it is the $e^- - M$ scattering which determines what happens in the $M - Na$ collision, as first pointed out by Fermi [28]. First consider the case where M is an atom. There are no energetically accessible states of atom M which can be excited by the low energy electron, so the scattering must be elastic. The electron can transfer very little kinetic energy to M, but the direction of the electron's motion can change. With this thought in mind, we can see that only the collisional mixing of nearly degenerate ℓ states of the same n has very large cross sections. The ℓ-mixing cross sections are approximately geometric at low n [29]. If the M atom comes anywhere into the Rydberg orbit, scattering into a different ℓ state occurs. At high n, the cross section decreases, because the probability distribution of the Rydberg electron becomes too dilute, and it becomes increasingly likely that the M atom will pass through the Rydberg electron's orbit without encountering the electron. The n at which the peak ℓ-mixing cross section occurs increases with the electron scattering length of the atom. While ℓ-mixing cross sections are large, n changing cross sections are small ($\sim$100 Å^2) since they cannot occur when the Rydberg electron is anywhere close to the outer turning point of its orbit [30].

If M is a molecule, there are likely to be energetically accessible vibrational and rotational transitions which can provide energy to or accept energy from the Rydberg electron, and this possibility increases the likelihood of n changing collisions with Rydberg atoms [11]. Electronic

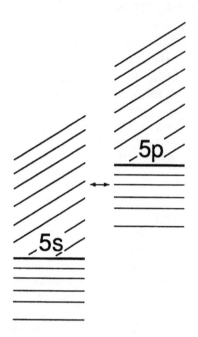

Figure 14.6. Sr^+ $5s$ and $5p$ states (—), the Rydberg states of Sr converging to these two ionic states are shown by (—), the continuum above the two ionic levels (///). The $5pn\ell$ states are coupled to the $5s\epsilon\ell'$ continua and autoionize.

energy from the Rydberg atom must be resonantly transferred to rotation or vibration in the molecule. In heavy or complex molecules, the presence of many rotational-vibrational states tends to obscure the resonant character of the transfer, but in several light systems the collisional resonances have been observed clearly [11].

Due to the range of the interactions responsible, thermal collisions with neutral atoms and molecules lead to cross sections which are necessarily smaller than the geometric cross section of the Rydberg atom. In contrast, collisions of Rydberg atoms with charged particles and resonant collisions between Rydberg atoms can have cross sections much larger than the geometric cross section of the Rydberg atom [31, 32].

14.8 AUTOIONIZING RYDBERG STATES

The bound Rydberg atoms considered thus far are formed by adding the Rydberg electron to the ground state of the ionic core. It could equally well be added to an excited state of the core [33]. Figure 14.6 shows the energy levels of the ground $5s$ state of Sr^+ and the excited $5p$ state. Adding an $n\ell$ electron to the $5s$ state yields the bound Sr $5sn\ell$ state, and adding it to the excited $5p$ state gives the doubly excited $5pn\ell$ state, which is

coupled by the Coulomb interaction to the degenerate $5s\epsilon\ell'$ continuum. The $5pn\ell$ state autoionizes at the rate $\Gamma_{n\ell}$ given by [34]

$$\Gamma_{n\ell} = 2\pi|\langle 5pn\ell|V|5s\epsilon\ell'\rangle|^2 , \qquad (14.34)$$

where V denotes the Coulomb coupling between the nominally bound $5pn\ell$ state and the $5s\epsilon\ell'$ continuum. A more general description of autoionization can be found in Chap. 25.

A simple picture, based on superelastic electron scattering from the Sr^+ $5p$ state, gives the scaling of the autoionization rates of Eq. (14.34) with n and ℓ. The $n\ell$ Rydberg electron is in an elliptical orbit, and each time it comes near the core it has an n-independent probability γ_ℓ of scattering superelastically from the Sr^+ $5p$ ion, leaving the core in the $5s$ state and gaining enough energy to escape from the Coulomb potential of the Sr^+ core. The autoionization rate of the $5pn\ell$ state is obtained by multiplying γ_ℓ by the orbital frequency of the $n\ell$ state, $1/n^3$ to obtain

$$\Gamma_{n\ell} = \frac{\gamma_\ell}{n^3} . \qquad (14.35)$$

Equation (14.35) displays the n dependence of the autoionization rate explicitly and the ℓ dependence through γ_ℓ. As ℓ increases, the closest approach of the Rydberg electron to the Sr^+ is at a larger orbital radius, so that superelastic scattering becomes progressively less probable, and γ_ℓ decreases rapidly with increasing ℓ.

To a first approximation, the Sr $5pn\ell$ states can be described by the independent electron picture used above, but in states converging to higher lying states of Sr^+, the independent electron picture fails. Consider the Sr^+ $\ell \geq 4$ states of $n > 5$. They are essentially degenerate, and the field due to an outer Rydberg electron converts the zero field ℓ states to superpositions much like Stark states. The outer electron polarizes the Sr^+ core, so that the outer electron is in a potential due to a charge and a dipole, and the resulting dipole states of the outer electron display a qualitatively different excitation spectrum than do states such as the $5pn\ell$ states, which are well described by an independent particle picture [35].

REFERENCES

1. H. E. White, *Introduction to Atomic Spectra* (McGraw-Hill, New York, 1934).
2. H. A. Bethe and E. A. Salpeter, *Quantum Mechanics of One and Two Electron Atoms* (Academic Press, New York, 1975).
3. U. Fano, Phys. Rev. A **2**, 353 (1970).
4. M. L. Zimmerman, M. G. Littman, M. M. Kash, and D. Kleppner, Phys. Rev. A **20**, 2251 (1979).
5. S. A. Bhatti, C. L. Cromer, and W. E. Cooke, Phys. Rev. A **24**, 161 (1981).

6. R. R. Freeman and D. Kleppner, Phys. Rev. A14, 1614 (1976).

7. A. Lindgard and S. E. Nielsen, At. Data Nucl. Data Tables 19, 534 (1977).

8. E. S. Chang, Phys. Rev. A 31, 495 (1985).

9. W. E. Cooke and T. F. Gallagher, Phys. Rev. A21, 588 (1980).

10. S. Haroche and J. M. Raimond, in *Advances in Atomic and Molecular Physics*, edited by D. Bates and B. Bederson (Academic Press, New York, 1985), Vol. 20.

11. F. B. Dunning and R. F. Stebbings, in *Rydberg States of Atoms and Molecules*, edited by R. F. Stebbings and F. B. Dunning (Cambridge University Press, Cambridge, 1983).

12. D. S. Bailey, J. R. Hiskes, and A. C. Riviere, Nucl. Fusion 5, 41 (1965).

13. M. G. Littman, M. M. Kash, and D. Kleppner, Phys. Rev. Lett. 41, 103 (1978).

14. R. R. Jones and T. F. Gallagher, Phys. Rev. A 38, 2946 (1988).

15. R. R. Freeman, N. P. Economou, G. C. Bjorklund, and K. T. Lu, Phys. Rev. Lett. 41, 1463 (1978).

16. W. P. Reinhardt, J. Phys. B 16, 635 (1983).

17. J. Gao, J. B. Delos, and M. C. Baruch, Phys. Rev. A 46, 1449 (1992).

18. T. F. Gallagher, *Rydberg Atoms* (Cambridge University Press, Cambridge, 1994).

19. A. R. P. Rau, J. Phys. B 12, L193 (1979).

20. D. Kleppner, M. G. Littman, and M. L. Zimmerman, in *Rydberg States of Atoms and Molecules*, edited by R. F. Stebbings and F. B. Dunning (Cambridge University Press, Cambridge, 1983).

21. W. R. S. Garton and F. S. Tomkins, Astrophys. J. 158, 839 (1969).

22. A. Holle, J. Main, G. Wiebusch, H. Rottke, and K. H. Welge, Phys. Rev. Lett. 61, 161 (1988).

23. B. E. Sauer, M. R. W. Bellerman, and P. M. Koch, Phys. Rev. Lett. 68, 1633 (1992).

24. R. V. Jensen, S. M. Susskind, and M. M. Sanders, Phys. Rept. 201, 1 (1991).

25. P. Pillet, H. B. van Linden van der Heuvell, W. W. Smith, R. Kachru, N. H. Tran, and T. F. Gallagher, Phys. Rev. A 30, 280 (1984).

26. P. Fu, T. J. Scholz, J. M. Hettema, and T. F. Gallagher, Phys. Rev. Lett. 64, 511 (1990).

27. M. Nauenberg, Phys. Rev. Lett. 64, 2731 (1990).

28. E. Fermi, Nuovo Cimento 11, 157 (1934).

29. T. F. Gallagher, S. A. Edelstein, and R. M. Hill, Phys. Rev. A 15, 1945 (1977).

30. F. Gounand and J. Berlande, in *Rydberg States of Atoms and Molecules*, edited by R. F. Stebbings and F. B. Dunning (Cambridge University Press, Cambridge, 1983).

31. T. F. Gallagher, Phys. Rept. 210, 319 (1992).

32. T. F. Gallagher, Rep. Prog. Phys. 51, 143 (1988).

33. T. F. Gallagher, J. Opt. Soc. Am. B 4, 794 (1987).

34. U. Fano, Phys. Rev. 124, 1866 (1961).

35. U. Eichmann, V. Lange, and W. Sandner, Phys. Rev. Lett. 68, 21 (1992).

15

Rydberg Atoms in Strong Static Fields

M. A. Iken and T. Uzer

School of Physics, Georgia Institute of Technology, Atlanta, Georgia 30332-0430

15.1 INTRODUCTION

Confronting classical and quantum mechanics in systems in which the classical motion is chaotic is one of the fundamental problems of this decade, as evidenced by the enormous outpouring of research on this subject [1]. In particular, the attention of many atomic physicists and nonlinear dynamicists has been focused on electron dynamics and spectroscopy in highly excited Rydberg atoms in strong external fields [2–5]. In this context, these fields are not strong by absolute measure. The relative influence of the electric field F and the magnetic field B scale with n as $n^4 F$ and $n^3 B$. Considering that the atomic unit of F is $F_0 = e^2/|q|a_0^2 = 5.14 \times 10^9$ V/cm and the atomic unit of B is $B_0 = \hbar/|q|a_0^2 = 2.35 \times 10^5$ T, for very high n, strong field conditions can be achieved even for weak laboratory fields. Atomic units will be used throughout most of this chapter. With $e^2 = q^2/4\pi\epsilon_0$, where q is the charge of the electron, and the Rydberg constant R, the unit of energy is $2R = e^2/a_0$ and the magnetic field parameter is $\gamma = B/B_0$.

The intense interest in Rydberg atoms in strong static fields stems from two principal reasons: these atoms can be prepared and manipulated in the laboratory [6] and they are amenable to theoretical treatments, often with astonishing accuracy [3]. The corresponding Hamiltonians possess deceptively simple nonlinearities which can lead to chaotic classical dynamics. The very high excitations involved imply that the detailed quantum mechanical treatment can be very tedious; on the other hand, the high excitations place the electron into a regime where reasonable accuracy can be expected from the Correspondence Principle. Thus, Rydberg atoms in external fields constitute atomic laboratories in which the quantum mechanics of classically chaotic systems can be investigated [5, 7, 8].

Research into the effects of time-dependent fields, where hydrogen or Rydberg atoms are ionized through multiphoton transitions [9–11] is treated in Sect. 72.3.8. In this chapter we consider static and uniform fields.

The different cases described here are all classically nonseparable systems and exhibit transitions from regular to irregular behavior as an external parameter (for instance the energy) is increased beyond a threshold value. Separability of the Coulomb problem, with or without external fields, in various coordinate systems is discussed by Kalnins *et al.* [12]. Possible scenarios are (i) the Stark effect (see Sect. 14.4), which is separable and therefore does not lead to chaos except when there is a core present (ii) the Quadratic Zeeman effect (QZE) [3–5, 7] (iii) parallel F and B fields (SQZE for Stark Quadratic Zeeman effect) [13, 14] or (4) crossed electric and magnetic fields [15–17].

The symmetry breaking of the Coulomb potential, induced by external fields, affects the three quantum numbers, n, l, m_l, of the Rydberg electron differently: m_l remains a good quantum number, as long as a single field direction is present; l breaks down extensively as the magnetic field is increased, whereas n breaks down only gradually. Therefore, an n-manifold of electronic energy levels does not have enough degrees of freedom for chaos (the minimum number is one degree of freedom and time, or for conservative problems, two degrees of freedom). Chaos only develops when different n-shells mix (intermanifold chaos). In contrast, the extensive symmetry breaking introduced by two misaligned (in practice, crossed) fields causes intramanifold chaos. In this case three degrees of freedom are available to the system, and the dynamics are qualitatively different [18].

The Hamiltonian for a one-electron atom in a $\hat{\mathbf{z}}$-directed magnetic field and a general static electric field is, in atomic units,

$$H = \frac{\mathbf{p}^2}{2} - \frac{1}{r} + \frac{\gamma L_z}{2} + \frac{\gamma^2 \rho^2}{8} + \mathbf{r} \cdot \mathbf{F} = E, \qquad (15.1)$$

where $\rho^2 = (x^2 + y^2)$. The third and fourth terms are the paramagnetic and diamagnetic contributions, respectively. It is useful to scale this Hamiltonian by $\mathbf{r} = \tilde{\mathbf{r}}\gamma^{2/3}$ and $t = \gamma\tilde{t}$; therefore, $\mathbf{p} = \gamma^{-1/3}\tilde{\mathbf{p}}$ [3, 4]. The Hamiltonian then becomes

$$\tilde{H} = H\gamma^{-2/3} = \frac{\tilde{\mathbf{p}}^2}{2} + \frac{\tilde{\rho}^2}{8} + \frac{\tilde{L}_z}{2} + \tilde{\mathbf{r}} \cdot \mathbf{F}\gamma^{-4/3} - \frac{1}{\tilde{r}} = W. \qquad (15.2)$$

From now on, it will be assumed that this scaling has been done, and the tildes will be dropped. The number of parameters is reduced through the use of the above magnetic scaling. The new parameters in the system are the scaled energy,

$$W = EB^{-2/3}, \qquad (15.3)$$

and the scaled field,

$$\epsilon = FB^{-4/3}. \qquad (15.4)$$

The scaling property is a distinct advantage in theoretical and experimental treatments [3, 4]. Take the case $\epsilon = 0$. Once the periodic orbits of $\tilde{H}$ are analyzed [19, 20], the results can be scaled to any γ. Similarly, for $\epsilon \neq 0$, any solution of this Hamiltonian corresponds to a one-dimensional manifold of field combinations.

15.2 QUADRATIC ZEEMAN EFFECT (QZE)

First we consider a one-electron atom in a strong magnetic field. This problem has two solvable limits: the zero-field case is the pure Coulomb limit and has spherical symmetry, the other is the free electron in a pure magnetic field which leads to the Landau levels (see Sect. 13.1.1), which are equally spaced by $\hbar\omega_c$, where ω_c is the cyclotron frequency, $\omega_c = eB/m_e$ [21]. In

the latter limit, the problem has cylindrical symmetry. The intermediate case, that is, the QZE, combines the spherical symmetry of the Coulomb term with the cylindrical symmetry of the diamagnetic term. This leaves only two constants of the motion, the magnetic quantum number and the parity. The $m_l = 0$ case is representative of other cases.

The most remarkable experimental manifestation of the QZE are Quasi-Landau (or Garton-Tomkins) oscillations, which is the name given to the resonances originally found near the ionization threshold [22]. At the ionization limit, they are separated by $\frac{3}{2}\hbar\omega_c$, independent of magnetic field strength. These correspond to the motion of the electron mainly perpendicular to the magnetic field [23].

The origin of chaos in the QZE can best be appreciated by removing the singularity at the origin by the process of regularization. Measuring time using a new time scale τ related by

$$dt/d\tau = 4r. \qquad (15.5)$$

to the old time scale t removes the difficulties associated with the Coulomb singularity [24]. When $m_l = 0$, we can write the Hamiltonian in two dimensions using semiparabolic coordinates [25],

$$\begin{aligned} \rho &= 2uv, \\ z &= u^2 - v^2, \end{aligned} \qquad (15.6)$$

and using the new time scale τ in defining the momenta results in the following Hamiltonian (which, however is not an energy but incorporates the scaled energy W as a parameter):

$$\begin{aligned} K = 4 = \frac{\mathbf{p}^2}{2} &- 4W\left(u^2 + v^2\right) \\ &+ 2u^2v^2\left(u^2 + v^2\right). \end{aligned} \qquad (15.7)$$

The corresponding eigenvalue problem ("Scaled Spectroscopy") differs considerably from conventional eigenvalue equations and is discussed, e.g., in references [2, 3].

For $W < 0$ the main term is a two dimensional isotropic oscillator of frequency $\sqrt{-2W}$; and thus, a connection is established between this problem and the dynamics of nonlinearly coupled oscillators.

The second-order correction to the energy for a general m_l is given by [26]

$$\Delta E = \frac{\gamma^2 n^2}{16}\left(n^2 + m_l^2 + 3 + \Lambda\right) \qquad (15.8)$$

which contains an adiabatic invariant (Herrick-Solov'ev, H-S invariant [15, 27]) given by

$$\Lambda = 4\mathbf{A} \cdot \mathbf{A} - 5A_z^2, \qquad (15.9)$$

where $\mathbf{A}$ is the Runge-Lenz vector,

$$\mathbf{A} = \left(-2\hbar^2 m_e E\right)^{-\frac{1}{2}} \left[\frac{1}{2}\left(\mathbf{p} \times \mathbf{L} - \mathbf{L} \times \mathbf{p}\right) - \left(\frac{m_e e^2}{r}\mathbf{r}\right)\right].$$

$$(15.10)$$

To second-order in γ, Λ is a good constant of the motion.

Investigating the chaotic behavior of the system requires locating the regions in which a transition from regular to irregular behavior occurs [3, 4, 7]. When γn^3, which is the ratio of the diamagnetic energy to the Coulomb energy, is around unity, maximal competition between the two incompatible symmetries (spherical and cylindrical) is reached. Empirically, the system makes a transition to noticeably chaotic behavior at $W \simeq -0.35$ [28].

Chaos, as it is understood in classical mechanics (and implemented in ergodic theory) does not exist in bounded, autonomous quantum systems. However, the search for unambiguous signatures of chaos in quantum mechanical systems is continuing [29]. One such signature is the change in the statistical nature of the nearest neighbor energy level separations (NNS) [29]. A NNS distribution which follows Poisson statistics,

$$P(S) = e^{-S},$$

$$(15.11)$$

(where S is the spacing between neighboring quantum levels) is often associated with regular behavior of the underlying classical system. However, the Poisson distribution does not necessarily guarantee regularity. In contrast, when the NNS statistics follow the Wigner distribution,

$$P(S) = \frac{\pi}{2}Se^{-\frac{\pi}{4}S^2},$$

$$(15.12)$$

a region of irregular behavior of the corresponding classical system can be expected [4, 29]. A change in the NNS statistics from Poisson to Wigner is seen in the QZE energy levels [3, 4, 7].

15.3 PARALLEL *F* AND *B* FIELDS: STARK QUADRATIC ZEEMAN EFFECT (SQZE)

For definitiveness, assume that the electric field is in the $\hat{\mathbf{z}}$-direction with magnitude F. The magnetic quantum number is still good because, as in the QZE, there is a unique field direction. The $m_l = 0$ case will again be used in order to reduce the problem to two dimensions. In semiparabolic coordinates,

$$K = 4 = \frac{\mathbf{p}^2}{2} + 2u^2v^2\left(u^2 + v^2\right)$$
$$- 4W\left(u^2 + v^2\right) + 4\epsilon\left(u^4 - v^4\right).$$

$$(15.13)$$

The general form for the H-S invariant in this case is [13, 14]

$$\Lambda_\beta = 4A^2 - 5\left(A_z - \beta\right)^2 + 5\beta^2,$$

$$(15.14)$$

where

$$\beta = \frac{12}{5}\frac{n^2 f}{n^4 \gamma},$$

$$(15.15)$$

and $f = F/F_0$. The chaos properties are similar to the QZE case.

15.4 CROSSED ELECTRIC AND MAGNETIC FIELDS

In arbitrarily aligned B and F fields, replacing $\mathbf{r}$ with $-\frac{3}{2}n\mathbf{A}$ in a given n-shell (the Pauli replacement [30]) gives the first-order contribution to the energy for arbitrarily oriented fields [31, 32]:

$$\Delta E = \boldsymbol{\omega}_L \cdot \mathbf{L} + \boldsymbol{\omega}_S \cdot \mathbf{A},$$

$$(15.16)$$

where $\mathbf{L}$ is the electronic angular momentum,

$$\boldsymbol{\omega}_L = \frac{e\mathbf{B}}{2m_e}$$

$$(15.17)$$

is the Larmor frequency (half of the cyclotron frequency ω_c) and

$$\boldsymbol{\omega}_S = -\frac{3}{2}\left(4\pi\epsilon_0 \frac{\hbar}{m_e e}\right)n\mathbf{F}.$$

$$(15.18)$$

An alternative expression, which brings out the symmetries involved, can be obtained by defining (in atomic units)

$$\begin{aligned}\omega_1 &= \tfrac{1}{2}\left(\mathbf{B} - 3n\mathbf{F}\right),\\ \omega_2 &= \tfrac{1}{2}\left(\mathbf{B} + 3n\mathbf{F}\right),\end{aligned}$$

$$(15.19)$$

and the generators

$$\begin{aligned}\mathbf{I}_1 &\equiv \tfrac{1}{2}\left(\mathbf{L} + \mathbf{A}\right),\\ \mathbf{I}_2 &\equiv \tfrac{1}{2}\left(\mathbf{L} - \mathbf{A}\right).\end{aligned}$$

$$(15.20)$$

With these,

$$\Delta E = \omega_1 \mathbf{I}_1 + \omega_2 \mathbf{I}_2.$$

$$(15.21)$$

In the special case of perpendicular fields, $|\omega_1| = |\omega_2|$ and the problem is degenerate.

The first-order contribution does not destroy the integrability, but the second-order contribution, which is considerably more complicated [31, 33], leads to chaos.

For definitiveness, assume that in the crossed fields case the magnetic field is in the $\hat{\mathbf{z}}$-direction and the electric field is in the $\hat{\mathbf{x}}$-direction. Since m_l is no longer a good quantum number, the regularization has to take place in four-dimensional space through the use of Kustaanheimo-Stiefel coordinates [24]:

$$x = 2\left(u_1 u_3 + u_2 u_4\right),$$
$$y = 2\left(u_1 u_2 - u_3 u_4\right),$$
$$z = u_1^2 - u_2^2 - u_3^2 + u_4^2,$$

with the time transformation

$$\frac{dt}{d\tau} = 4r, \tag{15.22}$$

and the constraint

$$u_1 p_4 - u_4 p_1 + u_3 p_2 - u_2 p_3 = 0. \tag{15.23}$$

The resultant Hamiltonian is [33]

$$K = 4 = \frac{\mathbf{p}^2}{2} + 2|\mathbf{u}|^2 \left(u_1 p_4 - u_4 p_1\right)$$
$$- 4|\mathbf{u}|^2 W + 8|u|^2 \left(u_1 u_3 + u_2 u_4\right)\epsilon$$
$$+ 2|\mathbf{u}|^2 \left(u_1^2 + u_4^2\right)\left(u_2^2 + u_3^2\right). \tag{15.24}$$

The classical and quantum treatments of this problem are currently active research subjects [16, 33, 34].

REFERENCES

1. M. C. Gutzwiller, *Chaos in Classical and Quantum Mechanics* (Springer-Verlag, New York, 1990).
2. H. Friedrich, *Theoretical Atomic Physics* (Springer-Verlag, New York, 1990).
3. H. Friedrich and D. Wintgen, Phys. Rep. **183**, 37 (1989).
4. H. Hasegawa, M. Robnik, and G. Wunner, Prog. Theor. Phys. Suppl. **98**, 198 (1989).
5. D. Delande, in *Chaos and Quantum Physics*, edited by M.-J. Giannoni, A. Voros, and J. Zinn-Justin (North-Holland, Amsterdam, 1991), p. 665.
6. T. F. Gallagher, Rep. Prog. Phys. **51**, 143 (1988).
7. D. Delande and J. C. Gay, Comments At. Mol. Phys. **19**, 35 (1986).
8. *The Hydrogen Atom*, edited by G. F. Bassani, M. Inguscio, and T. W. Hänsch (Springer-Verlag, New York, 1989).
9. J. E. Bayfield, Phys. Rep. **51**, 317 (1979).
10. G. Casati, B. V. Chirikov, D. L. Shepelyansky, and I. Guarneri, Phys. Rep. **154**, 77 (1987).
11. P. M. Koch, L. Moorman, and B. E. Sauer, Comments At. Mol. Phys. **25**, 165 (1990).
12. E. G. Kalnins, W. Miller Jr., and P. Winternitz, J. Appl. Math. **30**, 630 (1976).
13. P. Cacciani, E. Luc-Koenig, J. Pinard, C. Thomas, and S. Liberman, Phys. Rev. Lett. **56**, 1467 (1986).
14. D. Farrelly, T. Uzer, P. E. Raines, J. P. Skelton, and J. A. Milligan, Phys. Rev. A **45**, 4738 (1992).
15. E. A. Solov'ev, Zh. Eksp. Teor. Fiz. **85**, 109 (1983); Sov. Phys. JETP **58**, 63 (1983).
16. G. Wiebusch, J. Main, K. Krüger, H. Rottke, A. Holle, and K. H. Welge, Phys. Rev. Lett. **62**, 2821 (1989).
17. G. Raithel, M. Fauth, and H. Walther, Phys. Rev. A **44**, 1898 (1991).
18. J. C. Gay, Comments At. Mol. Phys. **25**, 1 (1990).
19. J. B. Delos and M.-L. Du, IEEE J. Quantum Electron. **24**, 1445 (1988).
20. J.-M. Mao, J. Shaw, and J. B. Delos, J. Stat. Phys. **68**, 51 (1992).
21. C. Cohen-Tannoudji, B. Diu, and F. Laloë, *Quantum Mechanics* (John Wiley and Sons, New York, 1977).
22. A. R. P. Rau, J. Phys. **12**, L193 (1979).
23. A. F. Starace, J. Phys. B **6**, 585 (1973).
24. E. L. Stiefel and G. Scheifele, *Linear and Regular Celestial Mechanics* (Springer-Verlag, Berlin, 1971).
25. A. R. Edmonds, J. Phys. A **22**, L673 (1989).
26. D. Delande and J. C. Gay, J. Phys. B **17**, L335 (1984).
27. D. R. Herrick, Phys. Rev. A, **26**, 323 (1982).
28. A. Harada and H. Hasegawa, J. Phys. A **16**, L259 (1983).
29. F. Haake, *Quantum Signatures of Chaos* (Springer-Verlag, Berlin, 1991).
30. W. Pauli, Z. Phys. **36**, 339 (1926).
31. E. A. Solov'ev, Zh. Eksp. Teor. Fiz. **82**, 1762 (1982); Sov. Phys. JETP, **55**, 1017 (1982).
32. J. C. Gay and D. Delande, in *Atomic Excitation and Recombination in External Fields*, edited by M. H. Nayfeh and C. W. Clark (Gordon and Breach, New York, 1985), p. 131.
33. M. J. Gourlay, T. Uzer, and D. Farrelly, Phys. Rev. A **47**, 3113 (1993).
34. J. von Milczewski, G. H. F. Diercksen, and T. Uzer, Phys. Rev. Lett. **73**, 2428 (1994).

16

Hyperfine Structure

Guy T. Emery

Bowdoin College, Brunswick, Maine 04011

16.1 INTRODUCTION

Hyperfine structure in atomic and molecular spectra is a result of the interaction between electronic degrees of freedom and nuclear properties other than the dominant one, the nuclear Coulomb field. It includes splittings of energy levels (and thus of spectral lines) from magnetic dipole and electric quadrupole interactions (and higher multipoles, on occasion). Isotope shifts are experimentally entangled with hyperfine structure, and the so-called field effect in the isotope shift can be naturally included as part of hyperfine structure. Studies of hyperfine structure can be used to probe nuclear properties, but they are an equally important probe of the structure of atomic systems, providing especially good tests of atomic wave functions near the nucleus. There are also isotope shifts owing to the mass differences between different nuclear species, and the study of these shifts provides useful atomic information, especially about correlations between electrons.

Hyperfine effects are usually small and often, but not always, it is sufficient to consider only diagonal matrix elements for the atomic or molecular system and for the nuclear system. In some cases, however, matrix elements off-diagonal in the atomic space, even though small, can be of importance; one possible result is to cause admixtures sufficient to make normally *forbidden* transitions possible.

In the diagonal case, one can picture each electron undergoing elastic scattering from the nucleus and returning to its initial bound state. As pointed out by Casimir [1], however, the *internal conversion* of nuclear gamma-ray transitions involves the inelastic down-scattering from an excited nuclear state to a lower one as an electron goes from an initial bound state to the continuum. By further conversion of bound to continuum states, one sees the connection with electron scattering from the nucleus – elastic, inelastic, and break-up. Hyperfine structure of outer-shell electronic states is at the low momentum-transfer end of this chain of related processes.

Some of the standard textbooks which discuss hyperfine structure are Refs. [2–9] and a few newer texts [10–13]. Especially relevant are Refs. [14–18] and the conference proceedings [19–21].

The study of hyperfine structure in free atoms, ions, and molecules is part of the more extensive research area of hyperfine interactions, which includes the study of atoms and molecules in matter, both at rest, for example as part of the structure of a solid, and moving, such as ions moving through condensed or gaseous matter. This more general subject also includes the ways in which atomic electrons shield the nucleus, or anti-shield it, from external or collective fields. Thus nuclear magnetic resonance, nuclear quadrupole resonance, electron-nuclear double resonance, recoilless nuclear absorption and emission, nuclear orientation, production of polarized beams, and many other widely used techniques, are intimately connected with hyperfine effects.

Though hyperfine effects are ordinarily small in electronic systems, they can become much larger in "exotic" atoms; those with a heavier lepton or hadron as the "light" particle. Hyperfine effects are typically related to light particle density at the nucleus, or to expectation values of r^{-3}, and thus scale as the cube of the light particle mass. The study of muonic atoms has contributed importantly to knowledge of the nuclear charge distribution (e.g., Refs. [22, 23]). There has been considerable interest in pionic atoms, where the strong interaction also contributes to hyperfine structure (e.g., Refs. [24, 25]), and also in kaonic, antiprotonic, and other hadronic "atoms" [26, 27].

16.2 SPLITTINGS AND INTENSITIES

16.2.1 Angular Momentum Coupling

When the nuclear system is in an isotropic environment, each nuclear state β has a definite value of nuclear angular momentum $I_\beta \hbar$, where the possible values of I_β are related to the number of nucleons (protons plus neutrons) in the same way as those for J_α are related to the number of electrons in electronic state $|\alpha\rangle$. The nuclear operators, eigenstates, and eigenvalues are related to each other in the same way as for atomic angular momentum by

$$I^2|\beta\rangle = I_\beta(I_\beta + 1)|\beta\rangle,$$
$$I_z|\beta\rangle = M_\beta|\beta\rangle, \qquad (16.1)$$

in units with $\hbar = 1$. Shift operators move the system from one M-value to another, as for the atomic system (see Chap. 2), and the operator $\mathbf{I}$ is the generator of rotations. When the combined atomic-nuclear system is

considered, in an isotropic environment, it is the total angular momentum of the combined system defined by

$$\mathbf{F} = \mathbf{J} + \mathbf{I}, \qquad (16.2)$$

that has definite values. The state of the combined system can be labeled by γ, so that

$$F^2|\gamma\rangle = F_\gamma(F_\gamma + 1)|\gamma\rangle,$$
$$F_z|\gamma\rangle = M_\gamma|\gamma\rangle. \qquad (16.3)$$

The shift operators are defined as before, and it is now $\mathbf{F}$ that is the generator of rotations of the (combined) system, or of the coordinate frame to which the system is referred.

By the rules of combining angular momenta, the possible values of the quantum number F are separated by integer steps and run from an upper limit of $J_\alpha + I_\beta$ to a lower limit of $|J_\alpha - I_\beta|$. The number of possible eigenvalues F is the smaller of $2J_\alpha + 1$ and $2I_\beta + 1$. Experimental values of the nuclear quantum number I may be found in a number of compilations [28–30].

16.2.2 Energy Splittings

Electromagnetic interactions between atomic electrons and the nucleus can be expanded in a multipole series

$$H_{\text{eN}} = \sum_k \mathbf{T}^k(\text{N}) \cdot \mathbf{T}^k(\text{e}),$$
$$\equiv \sum_{k,q} (-1)^q T_q^k(\text{N}) T_{-q}^k(\text{e}) \qquad (16.4)$$

where $\mathbf{T}^k(\text{N})$ is an irreducible tensor operator of rank k operating in the nuclear space, and similarly $\mathbf{T}^k(\text{e})$ operates in the space of the electrons. Since one is taking diagonal matrix elements (in the nuclear space, at least) in states that are to a very good approximation eigenstates of the parity operator, only even electric multipoles (E0, E2, etc.) and odd magnetic multipoles (M1, M3, etc.) contribute to the series. The effects of the parity nonconserving weak interaction are considered in Chap. 29.

The term with $k = 0$ contributes directly to the structure (and fine structure) of atomic systems, and its dominant contributions come from the external r^{-1} electrostatic field of the nucleus. The *hyperfine* Hamiltonian is defined by subtracting that term to obtain

$$H_{\text{hfs}} = \sum_i \left[\mathbf{T}^0(\text{N})\mathbf{T}^0(i) - (-Ze^2/r_i) \right]$$
$$+ \sum_{k=1} \mathbf{T}^k(\text{N}) \cdot \mathbf{T}^k(\text{e}), \qquad (16.5)$$

where Z is the nuclear charge number. The difference between the Ze^2/r term(s) and the full monopole term is called the *field effect* or *finite nuclear size effect* in the isotope shift, and the remaining terms contribute dipole ($k = 1$), quadrupole ($k = 2$), and higher multipoles in hyperfine structure.

Since the hyperfine Hamiltonian can be expressed as a multipole expansion, its contributions to the pattern of energy levels for the various F values in a given J_α, I_β multiplet in first-order perturbation theory can be described relatively simply in terms of 3-j and 6-j symbols. The contribution of the term which is the scalar product of electron and nuclear operators of multipole k is

$$\Delta E_k(JIF, JIF)$$
$$= (-)^{J+I+F} \begin{Bmatrix} J & J & k \\ I & I & F \end{Bmatrix}$$
$$\times \left[\begin{pmatrix} J & k & J \\ J & 0 & -J \end{pmatrix} \begin{pmatrix} I & k & I \\ I & 0 & -I \end{pmatrix} \right]^{-1} A_k, \quad (16.6)$$

where for $k \geq 1$,

$$A_k = \langle JJ | \mathbf{T}^k(e) | JJ \rangle \cdot \langle II | \mathbf{T}^k(N) | II \rangle. \quad (16.7)$$

The commonly used hfs coefficients A, B, etc., are related to the A_k by

$$A = A_1/IJ, \quad B = 4A_2, \quad C = A_3, \quad D = A_4. \quad (16.8)$$

The isotope shift A_0 is the matrix element of the reduced monopole operator.

The pattern of the splitting depends on the total angular momentum F wholly through the 6-j symbol. Since for $k = 0$ the value of the 6-j symbol is independent of F, the monopole term shifts all levels of the hyperfine multiplet equally, independent of the value of F.

The F-dependence of the dipole contribution can be found from the fact that the same 6-j symbol would appear for any scalar product of $k = 1$ operators, for example $\mathbf{J} \cdot \mathbf{I}$. But in this product space, with J, I, and F all good quantum numbers, the diagonal matrix elements of $\mathbf{J} \cdot \mathbf{I}$ are just

$$\langle \mathbf{J} \cdot \mathbf{I} \rangle = \tfrac{1}{2}[F(F+1) - J(J+1) - I(I+1)], \quad (16.9)$$

so that

$$\Delta E_1(JIF, JIF)$$
$$= \tfrac{1}{2}A[F(F+1) - J(J+1) - I(I+1)], \quad (16.10)$$

where, in terms of reduced matrix elements according to the convention of Brink and Satchler (Ref. [31], p. 152), (the first version given in Sect. 2.9.4)

$$A = [J(J+1)]^{-1/2}\langle J\|\mathbf{T}^1(e)\|J\rangle[I(I+1)]^{-1/2}$$
$$\times \langle I\|\mathbf{T}^1(N)\|I\rangle. \quad (16.11)$$

A is the magnetic dipole hyperfine structure constant for the atomic level J and nuclear state I. M1 hfs shows the same pattern of splittings as spin-orbit fine structure, described sometimes as the *Landé interval rule*.

Electric quadrupole hfs is described by the quadrupole hyperfine structure constant B. If we define the quantity $K = [F(F+1) - J(J+1) - I(I+1)]$, then

$$\Delta E_2(JIF, JIF)$$
$$= \frac{(\tfrac{1}{4}B)[3K(K+1)/2 - 2J(J+1)I(I+1)]}{J(2J-1)I(2I-1)}. \quad (16.12)$$

The constant B is related to the tensor operators by

$$\tfrac{1}{4}B = [J(2J-1)/(J+1)(2J+3)]^{-1/2}\langle J\|\mathbf{T}^2(e)\|J\rangle$$
$$\times [I(2I-1)/(I+1)(2I+3)]^{-1/2}\langle I\|\mathbf{T}^2(N)\|I\rangle, \quad (16.13)$$

For higher multipoles, see Ref. [32].

The multipole expansion is important because it is valid for relativistic as well as nonrelativistic situations, and for nuclear penetration effects (hyperfine anomalies discussed in Sect. 16.4.3) as well as for normal hyperfine structure. Its limitation comes from its nature as a first-order diagonal perturbation. Off-diagonal contributions, even when small, can perturb the pattern, but, more importantly, can lead to misleading values for the A_k coefficients, including the isotope shift.

16.2.3 Intensities

When hyperfine structure is observed as a splitting in an optical transition between different atomic levels, there are relations between the intensities of the components. The general rule for reduced matrix elements of a tensor operator operating in the first part of a coupled space is (Ref. [31], p. 152)

$$\langle JIF\|Q^\lambda(e)\|J'IF'\rangle$$
$$= (-1)^{\lambda+I+F'+J}(2F'+1)^{1/2} \begin{Bmatrix} F & F' & \lambda \\ J' & J & I \end{Bmatrix}$$
$$\times (2J+1)^{1/2}\langle J\|Q^\lambda(e)\|J'\rangle. \quad (16.14)$$

For a dipole transition ($\lambda = 1$) connecting atomic states J and J', with fixed nuclear spin I, the line strength $S_{FF'}$ of the hyperfine component connecting F and F' is related to the line strength $S_{JJ'}$ by

$$S_{FF'} = (2F+1)(2F'+1) \begin{Bmatrix} F & F' & 1 \\ J' & J & I \end{Bmatrix}^2 S_{JJ'}. \quad (16.15)$$

16.3 ISOTOPE SHIFTS

Two distinct mechanisms contribute to isotope shifts in atomic energy levels and transition energies. First, there are shifts due to the different mass values of the isotopes; these *mass shifts* can again be separated into two kinds, the normal mass shift and the specific mass shift. Second, there are shifts due to different nuclear charge distributions in different isotopes. Shifts of this sort are called *field shifts*, and can be considered to be the monopole part of the hyperfine interaction.

The usual convention is to describe an isotope shift in a transition as positive when the line frequency is greater for the heavier isotope.

16.3.1 Normal Mass Shift

The *normal mass shift* occurs already for one-electron atoms, where the energy scale in the c.m. frame for the electron-nucleus system is directly proportional to the reduced mass of the system, $\mu = mM/(M + m)$, where m is the mass of the electron and M that of the nucleus. The two natural limits are those of an infinitely heavy nucleus, where $\mu = m$, and positronium, where $\mu = \frac{1}{2}m$. The normal mass shift applies to all levels of all atomic systems. The fractional shift in frequency between isotopes of mass M_H and M_L is given by

$$(\nu_H - \nu_L)/\nu_H = m(M_H - M_L)/M_H(M_L + m), \quad (16.16)$$

or

$$(\nu_H - \nu_L)/\nu_L = m(M_H - M_L)/(M_H + m)M_L. \quad (16.17)$$

The normal mass shift between deuterium and protonic hydrogen is $(\nu_D - \nu_H)/\nu_H = 2.721 \times 10^{-4}$, amounting to 4.15 cm^{-1} (0.179 nm) for Balmer α. It decreases rapidly for heavy elements, with $\Delta\nu/\nu \approx (A_H - A_L)/(1823 A_H A_L)$, where the A-values are the atomic mass numbers; for the pair ^{208}Pb $-$ ^{206}Pb, $\Delta\nu/\nu$ is then 2.56×10^{-8}, corresponding to a wavelength shift of 0.000 014 nm for a line at 550 nm.

16.3.2 Specific Mass Shift

When the system has more than two particles the situation is more complicated; in particular, the center of mass of any particular electron and the nucleus is no longer the center of mass of the whole system. When c.m. motion is removed from the Hamiltonian, there remains a set of *mass polarization* terms

$$H_{\text{mp}} = \frac{1}{M} \sum_{j<k} \mathbf{p}_j \cdot \mathbf{p}_k, \quad (16.18)$$

where M is the nuclear mass and $\mathbf{p}_j$ is the momentum of the jth electron. The matrix elements of these terms can be strongly state-dependent, and the difference in their contributions for isotopes of different mass is called the

specific mass shift (SMS). For a transition $a \to b$, the lowest-order SMS between isotopes A and A' is

$$\Delta\nu(a,b; A, A') = K^{a,b}(M_A - M_{A'})/M_A M_{A'}, \quad (16.19)$$

where

$$K^{a,b} = \langle a|\sum_{j<k}\mathbf{p}_j\cdot\mathbf{p}_k|a\rangle - \langle b|\sum_{j<k}\mathbf{p}_j\cdot\mathbf{p}_k|b\rangle. \quad (16.20)$$

It was earlier thought that the SMS, like the normal mass shift, was always very small for heavy atoms, but that has turned out not to be the case.

Since the operator $\mathbf{p}_1\cdot\mathbf{p}_2$ resembles the product of two dipole transition operators, the matrix elements of H_{mp} vanish between simple product-type wave functions unless allowed by dipole selection rules. For example, in the $1sn\ell$ ^1L and ^3L states of helium, the only nonvanishing diagonal terms are the $\ell = 1$ exchange terms $\pm\langle 1s(1)np(2)|\mathbf{p}_1\cdot\mathbf{p}_2|np(2)1s(1)\rangle$, with ($-$) for singlet states and ($+$) for triplet states. The resulting Hughes-Eckart level shift [33] is positive for singlets and negative for triplets. For other states, $\langle H_{\text{mp}}\rangle$ acquires a nonzero value due to electron correlation (configuration mixing) effects, but the resulting level shifts are correspondingly smaller. Detailed tabulations for many states of helium, including second-order corrections, are given by Drake and Yan [34].

For other atoms, the diagonal matrix elements of H_{mp} within a single electron configuration are weighted by the same coefficients as those that weight the Slater exchange integral G^1. Relativistic corrections have been given by Stone [35]. Bauche and Champeau [10] provide a useful discussion of the SMS, as does King [15].

Among recent results, we mention high-precision experimental work on the difference in the splittings $2^3S_1 - 2^3P_J$ between ^{4}He and ^{3}He [36–38], which can be compared with theoretical results [38,39], and an extensive multi-isotope study in Sm II, in which isotope-shift results are used to deduce structure information for a number of levels [40].

16.3.3 Field Shift

The field shift is due to different electric monopole interactions within the nucleus for different nuclei. Since the field shift is state-dependent, it contributes to the isotope shift of electronic transition frequencies. When field shifts occur between different isomeric levels of the same nuclear species, they are called *isomer shifts*.

Following Seltzer [41], (see also the summary in Fricke *et al.* [42]), a level shift can be written as

$$\Delta E = -e^2 \int [\rho_N(A) - \rho_N(A')]$$
$$\times \left[r_N^{-1} \int \rho_e d\tau_e - \int r_e^{-1} \rho_e d\tau_e \right] d\tau_N. \quad (16.21)$$

The electronic factor inside the second bracket can be fitted as an even power series in r_N, starting with r_N^{2j+1}. The result is that the field shift for a transition a is

$$\Delta\nu(a, AA') = F^a \lambda^{AA'}, \qquad (16.22)$$

with

$$\lambda^{AA'} = \delta\langle r^2\rangle_{AA'} + (C_2/C_1)\delta\langle r^4\rangle_{AA'} \\ + (C_3/C_1)\delta\langle r^6\rangle_{AA'} + \cdots, \qquad (16.23)$$

where $\delta\langle r^k\rangle_{AA'} = \langle r^k\rangle_A - \langle r^k\rangle_{A'}$. The term F^a is an electronic factor proportional to the change in electron density at the nucleus between the initial and final electronic states; in the simplest perturbation approach F^a is equal to $-(4\pi/6)Ze^2\Delta\rho_e(0)$, where $\Delta\rho_e(0)$ is the electron density at the nucleus in the lower atomic state minus that in the upper (see Sect. 86.2). The ratios C_2/C_1 etc., which weight the higher even moments, depend only on Z and not on the particular transition; values are tabulated in Refs. [41, 43, 44], as are F-factors for K x-ray transitions. This approach has been generalized and reformulated by Blundell *et al.* [44].

Measurements of isotope shifts of optical lines have long played an important role in the determination of nuclear rms radii. A more comprehensive picture of nuclear charge distributions can be found by combining optical isotope shift results with those for x-ray lines and for transitions in muonic atoms, together with results of elastic electron scattering from nuclei; a compilation of the results of such an analysis is to be found in Fricke *et al.* [42].

16.3.4 Separation of Mass Shift and Field Shift

From the measured isotope shift for a spectral line a between a pair of isotopes A and A', one can define a *residual isotope shift* by subtracting the normal mass shift. The residual shift is the sum of two terms, each with an electronic factor and a nuclear factor

$$\Delta\nu_{a,AA'} = K^a(M_A - M_{A'})/M_A M_{A'} + F^a\lambda^{AA'}. \quad (16.24)$$

If the field shift is assumed to be negligible, the electronic factor K^a can be extracted. Similarly, if the SMS is assumed to be negligible, from the ratios $\Delta\nu(a, A_1A_2)/\Delta\nu(a, A_3A_4)$, ratios of $\lambda(A_1A_2)/\lambda(A_3A_4)$ can be obtained, and the ratios $\Delta\nu(a, A_1A_2)/\Delta\nu(b, A_1A_2)$ give ratios of the electronic factors F^a/F^b.

If both contributions have to be considered, one can gain information from a King plot [45, 15] in which a *modified residual shift*[1] $\Delta\nu'(a, AA')$ for line a is defined

---

[1] Some authors modify the whole shift, without subtracting the normal mass shift; since the isotopic mass dependence is the same for the two mass-shift contributions, the separation proceeds as before, but with an altered definition of the constant K.

as

$$\Delta\nu'(a, AA') = \Delta\nu(a, AA')M_A M_{A'}/(M_A - M_{A'}).$$

$$(16.25)$$

For each isotope pair AA', this defines a point in the King plot whose ordinate is, e.g., $\Delta\nu'(a, AA')$ and whose abscissa is $\Delta\nu'(b, AA')$. If the assumed additivity is valid, one finds a linear relationship between the modified shifts for line a and those for another line b according to

$$\Delta\nu'(a, AA') = (F^a/F^b)\Delta\nu'(b, AA') + K^a - K^b(F^a/F^b).$$

$$(16.26)$$

From the slope of the line one has the ratio of F-factors for the two lines, and from the intercept a relation between the K-factors. A sometimes useful variant of the King plot involves choosing a pair of isotopes, say B and B' as a reference pair [15, 42]. A *reduced shift* $\Delta\nu''(a, AA')$ is then defined as

$$\Delta\nu''(a, AA') = \frac{\Delta\nu(a, AA')M_A M_{A'}(M_B - M_{B'})}{(M_A - M_{A'})M_B M_{B'}},$$

$$(16.27)$$

and $\Delta\nu''(a)$ is plotted *vs.* $\Delta\nu''(b)$.

It is not in general possible to extract F- and K-values for individual transitions from optical isotope shifts alone, unless one has at least one calculated or otherwise reliably known value. Information about nuclear charge distributions, however, and isotopic changes in them, is available from other sources, including x-ray isotope shifts, muonic x-rays, and elastic electron scattering from nuclei. Calculated electronic factors for K x-ray transitions can be expected to be more reliable than those for optical transitions. The effects of finite nuclear size are much bigger in muonic than electronic atoms, and screening effects much smaller. Combined analyses of all four types of data lead to the best current knowledge of changes in nuclear charge parameters, and thus contribute to knowledge of the electronic factors F^a and K^a for optical transitions [42].

16.4 HYPERFINE STRUCTURE

We discuss first *normal* hyperfine structure, in which terms coming from penetration of electrons into the nuclear charge and current distributions can be neglected; such contributions, the *anomalous* hyperfine structure, are briefly discussed in Sect. 16.4.3. See also Sect. 21.7.2.

16.4.1 Electric Multipoles

The electric quadrupole tensor operators, when penetration is neglected, are

$$\mathbf{T}^2(N) = \rho_N(\mathbf{r}_N)r_N^2\mathbf{C}^{(2)}\,, \qquad (16.28)$$

where ρ_N is the nuclear charge density, and

$$\mathbf{T}^2(\mathbf{e}_i) = -er_i^{-3}\mathbf{C}^{(2)}(i)\,. \qquad (16.29)$$

The matrix element of $\mathbf{T}^2(N)$ that occurs in the expansion coefficient A_2 is just $e/2$ times the nuclear quadrupole moment Q:

$$eQ = 2\langle II|\mathbf{T}^2(N)|II\rangle\,. \qquad (16.30)$$

Experimental values of Q are tabulated by Raghavan [28] and more recently discussed by Pyykkö [46]. The matrix element of $\mathbf{T}^2(e)$, summed over all electrons, also has a simple semiclassical interpretation (see, e.g., Ref. [6]);

$$\langle JJ|\mathbf{T}^2(e)|JJ\rangle = (e/2)q_J$$
$$= (1/2)\langle\partial^2 V/\partial z^2\rangle_{JJ}\,, \qquad (16.31)$$

where V is the electrostatic potential at the nucleus due to the electrons. The quadrupole splitting constant can thus be expressed as

$$B = 4A_2 = e^2 Q q_J = eQ\langle\partial^2 V/\partial z^2\rangle_{JJ}\,. \qquad (16.32)$$

Because the matrix elements of spherical harmonics in lsj-coupled states are independent of l (aside from the parity requirement that $l+l'+k$ be even), the one-electron matrix elements of $\mathbf{T}^2(e)$ are [32]

$$\langle lsj\|\mathbf{T}^2(e)\|l'sj'\rangle$$
$$= -e\langle lsj\|\mathbf{C}^{(2)}\|l'sj'\rangle \int r^{-3}(gg' + ff')dr\,, \qquad (16.33)$$

where g and f are the large and small r-multiplied Dirac radial functions, and the right-hand reduced matrix element is equal to $(-1)^{k-1}C_{\frac{1}{2}0\frac{1}{2}}^{jkj'}$ with k=2 (see, e.g., Ref. [31], p. 153).

Electric hexadecapole hyperfine structure follows similar rules [32]. While there is considerable indirect evidence that many nuclear states have nonzero $E4$ moments [47], their effect on hyperfine structure has been identified only occasionally [48, 49].

16.4.2 Magnetic Multipoles

For the magnetic multipole modes, when penetration effects are neglected, and when the the nuclear current is taken to be a sum of the orbital-current and spin-current contributions of individual nucleons, the nuclear tensor operator can be written as the sum

$$\mathbf{T}^k(N) = (e\hbar/2m_p c)\sum_n [k(2k-1)]^{1/2}r_n^{k-1}$$
$$\times \left\{ 2g_{ln}(k+1)^{-1}\left[\mathbf{C}^{(k-1)}\mathbf{L}_n\right]^{(k)} \right.$$
$$\left. + g_{sn}\left[\mathbf{C}^{(k-1)}\mathbf{S}_n\right]^{(k)} \right\}\,, \qquad (16.34)$$

where $e\hbar/2m_p c$ is the nuclear magneton, and in the extreme single-nucleon model the orbital g-factors g_{ln} are 1 for protons and 0 for neutrons, while the spin g-factors g_{sn} are 5.585 695 for protons and $-3.826\,086$ for neutrons (see Chap. 1). Similarly, for electron i, the atomic Mk operator can be written

$$\mathbf{T}^k(\mathbf{e}_i) = ie[(k+1)/k]^{1/2}r_i^{-k-1}\left[\boldsymbol{\alpha}_i\mathbf{C}_i^{(k)}\right]^{(k)}\,. \qquad (16.35)$$

For the magnetic dipole case, the nuclear operator of (16.29) becomes

$$\mathbf{T}^1(N) = (e\hbar/2m_p c)\sum_n (g_{ln}\mathbf{L}_n + g_{sn}\mathbf{S}_n)\,. \qquad (16.36)$$

The nuclear currents, however, are more complicated in the strongly interacting and relatively dense environment of real nuclear systems; renormalized g-factors can be used to take some such effects into account. For the quenching of spin matrix elements, see Castel and Towner [50]; for the $M1$ mode, the effective g-factor is typically in the neighborhood of 0.5 or 0.6 of the free-space value. It is also sometimes useful to consider effective orbital g-factors, but the effect is smaller, with $g_{\mathrm{eff}}/g_{\mathrm{free}} = 1.05$– 1.1 [51]. The matrix element of $\mathbf{T}^1(N)$ is the nuclear magnetic dipole moment

$$\mu = \langle II\|\mathbf{T}^1(N)\|II\rangle\,. \qquad (16.37)$$

Experimental values of μ are tabulated by Raghavan [28].

In the semiclassical picture where $\Delta E = -\boldsymbol{\mu}\cdot\mathbf{B}(0)$, the contributions to $\mathbf{B}(0)$ from the orbital and spin magnetism of the ith electron are

$$\mathbf{B}_l(0) = -2\mu_0 r_i^{-3}\mathbf{l}_i\,, \qquad (16.38)$$

$$\mathbf{B}_s(0) = 2\mu_0 r_i^{-3}(10)^{1/2}\left[\mathbf{s}_i\mathbf{C}_i^{(2)}\right]^{(1)}\,, \qquad (16.39)$$

while the "contact" term for s-electrons is

$$\mathbf{B}_c(0) = -(16\pi/3)\mu_0\rho_i(0)\mathbf{s}\,. \qquad (16.40)$$

The one-electron matrix elements of $\mathbf{T}^1(e)$ are [32]

$$\langle lsj\|\mathbf{T}^1(e)\|l'sj'\rangle = -e(\kappa + \kappa')\langle lsj\|\mathbf{C}^{(1)}\|l'_xsj'\rangle$$
$$\times \int r^{-2}(fg' + gf')dr\,, \qquad (16.41)$$

where for a given combination lsj, $\kappa = (l-j)(2j+1)$, and l_x is the opposite-parity orbital quantum number of the small Dirac component, $l_x = l+1$ for $j = l+1/2$, $l_x = l-1$ for $j = l-1/2$.

Magnetic octupole hyperfine structure [52] follows similar rules [32, 53]. Systems studied in recent years include Eu I excited states [54].

16.4.3 Hyperfine Anomalies

When the electron density is nonzero inside the nucleus, the interaction with different kinds of nuclear current density is in general different; there is, for example, a sensitivity to differing radial distributions of spin and orbital currents. The result is that the ratio of A-values for two isotopes is not necessarily the same as the ratio of nuclear g-factors [55]. The anomaly for two nuclear species a and b is characterized by

$$^a\Delta^b = (A_a/A_b)(g_a/g_b) - 1. \qquad (16.42)$$

Δ is seldom larger than a few per cent. The theory for the $M1$ mode was worked out by Bohr and Weisskopf [56]; there is a relatively recent review of theory and experiment [57], and an especially clear exposition of the semiclassical limit [58]. Recent experimental results include the $4f^45d6s$ levels of Nd I [59] and the transitions $^9S_4 - {}^9P_5$ of Eu$^+$ [60]. Anomalies are expected to be considerably smaller for the electric mode.

REFERENCES

1. H. B. G. Casimir, *On the Interaction Between Atomic Nuclei and Electrons* (Teyler's Tweede Genootschap, Haarlem, 1936; W. H. Freeman, San Francisco, 1963).
2. L. Pauling and S. Goudsmit, *The Structure of Line Spectra* (McGraw Hill, New York, 1930).
3. G. Herzberg, *Atomic Spectra and Atomic Structure* (Prentice-Hall, New York, 1937; 2nd ed., Dover, New York, 1944).
4. G. Herzberg, *Molecular Spectra and Molecular Structure, I. Spectra of Diatomic Molecules* (Van Nostrand, Princeton, 1950), 2nd ed.
5. H. Kopfermann, *Nuclear Moments* (Academic Press, New York, 1958).
6. N. F. Ramsey, *Molecular Beams* (Clarendon, Oxford, 1956).
7. J. C. Slater, *Quantum Theory of Atomic Structure* (McGraw-Hill, New York, 1960), Vol. II.
8. I. I. Sobelman, *Introduction to the Theory of Atomic Spectra* (Pergamon Press, London, 1972).
9. H. G. Kuhn, *Atomic Spectra* (Academic Press, New York, 1962; 2nd ed. 1969).
10. A. Corney, *Atomic and Laser Spectroscopy* (Clarendon, Oxford, 1977).
11. M. Weissbluth, *Atoms and Molecules* (Academic Press, New York, 1978).
12. I. I. Sobelman, *Atomic Spectra and Radiative Transitions* (Springer Verlag, Berlin, 1979).
13. H. Haken and H. C. Wolf, *Atomic and Quantum Physics* (Springer Verlag, Berlin, 1984).
14. L. Armstrong, Jr., *Theory of the Hyperfine Structure of Free Atoms* (Wiley-Interscience, New York, 1971).
15. W. H. King, *Isotope Shifts in Atomic Spectra* (Plenum Press, New York, 1984).
16. I. Lindgren and J. Morrison, *Atomic Many-Body Theory* (Springer Verlag, Berlin, 1982).
17. J. Bauche and R.-J. Champeau, Adv. At. Mol. Opt. Phys. **12**, 39 (1976).
18. K. Heilig and A. Steudel, in *Progress in Atomic Spectroscopy, Part A*, edited by W. Hanle and H. Kleinpoppen (Plenum Press, New York, 1978), pp. 263–328.
19. *Atomic Physics*, edited by V. W. Hughes, B. Bederson, V. W. Cohen, and F. M. J. Pichanik (Plenum Press, New York, 1969), especially articles by H. M. Foley, pp. 509–522, and H. H. Stroke, pp. 523–550.
20. *AIP Conference Proceedings No. 233, Atomic Physics*, edited by J. C. Zorn, R. R. Lewis, and M. K. Weiss (American Institute of Physics, New York, 1991).
21. *AIP Conference Proceedings No. 275, Atomic Physics*, edited by H. Walther, T. W. Hänsch, and B. Neizert (American Institute of Physics, New York, 1993).
22. J. Hüfner, F. Scheck, and C. S. Wu, Muon Phys. **1**, 201 (1977); R. C. Barrett, *ibid.*, pp. 309–322.
23. Roger C. Barrett and Daphne Jackson, *Nuclear Sizes and Structure* (Oxford 1977).
24. G. Backenstoss, Ann. Rev. Nucl. Sci. **20**, 467 (1970).
25. J. Konijn, in *Pions in Nuclei*, edited by E. Oset, M. J. Vicente Vacas, and C. Garcia Recio (World Scientific, 1992), p. 303.
26. C. J. Batty, Nucl. Phys. A**508**, 89C (1990).
27. G. Backenstoss, Comtemp. Phys. **30**, 433 (1989).
28. P. Raghavan, At. Data Nucl. Data Tables **42**, 189 (1989).
29. *Nuclear Data Sheets*, edited by M. J. Martin and J. K. Tuli (Academic Press, 1995), Nos. 74–76; each issue contains a guide to the most recent compilation for each atomic mass number.
30. *Table of Isotopes*, edited by C. M. Lederer and V. S. Shirley (John Wiley & Sons, 1978), 7th ed.
31. D. M. Brink and G. R. Satchler, *Angular Momentum* (Oxford, 1968), 2nd ed.
32. C. Schwartz, Phys. Rev. **97**, 380 (1955).
33. D. S. Hughes and C. Eckart, Phys. Rev. **36**, 694 (1930).
34. G. W. F. Drake and Z.-C. Yan, Phys. Rev. A **46**, 2378 (1992).
35. A. P. Stone, Proc. Phys. Soc. (London) **77**, 786 (1961); *ibid.*, **81**, 868 (1963).
36. Ping Zhao, J. R. Lawall, and F. M. Pipkin, Phys. Rev. Lett. **66**, 592 (1991).
37. D. Shiner, R. Dixson, and V. Vedantham, Phys. Rev. Lett. **74**, 3553 (1995).
38. F. Marin, F. Minardi, F. S. Pavone, and G. W. F. Drake, Z. Phys. D **32**, 285 (1995).
39. G. W. F. Drake in *Long Range Casimir Forces: Theory and Recent Experiments on Atomic Systems*, edited by F. S. Levin and D. A. Micha (Plenum Press, New York 1993), pp. 196–9. For isotope shifts in Li$^+$, see also E. Riis, A. G. Sinclair, O. Paulsen, G. W. F. Drake, W. R. C. Rowley, and A. P. Levick, Phys. Rev. A **49**, 207 (1994).
40. P. Villemoes *et al.*, Phys. Rev. A**51**, 2838 (1995).
41. E. C. Seltzer, Phys. Rev. **188**, 1916 (1969).
42. G. Fricke *et al.*, At. Data Nucl. Data Tables **60**, 177 (1995).
43. F. Boehm and P. L. Lee, At. Data Nucl. Data Tables **37**, 455 (1987).
44. S. A. Blundell *et al.*, J. Phys. B**20**, 3663 (1987).
45. W. H. King, J. Opt. Soc. Am. **53**, 638 (1963).
46. P. Pyykkö, Z. Naturforsch. **47a**, 189 (1992).

47. R. F. Casten, *Nuclear Structure from a Simple Perspective* (Oxford University Press, 1990), pp. 296–300.

48. W. Dankwort, J. Ferch, and H. Gebauer, Z. Phys. **267**, 229 (1974).

49. O. Becker *et al.*, Phys. Rev. A**48**, 3546 (1993).

50. B. Castel and I. S. Towner, *Modern Theories of Nuclear Moments* (Oxford 1990).

51. T. Yamazaki *et al.*, Phys. Rev. Lett. **25**, 547 (1970).

52. V. Jaccarino, J. G. King, R. A. Satten, and H. H. Stroke, Phys. Rev. **94**, 1798 (1954); P. Kusch and T. G. Eck, Phys. Rev. **94**, 1799 (1954).

53. M. Mizushima, *Quantum Mechanics of Atomic Spectra and Atomic Structure* (Benjamin, New York, 1970), Sect. 9-10.

54. W. J. Childs, Phys. Rev. A**44**, 1523 (1991).

55. F. Bitter, Phys. Rev. **76**, 150 (1949).

56. A. Bohr and V. F. Weisskopf, Phys. Rev. **77**, 94 (1950).

57. S. Büttgenbach, Hyperfine Interactions **20**, 1 (1984).

58. R. A. Sorensen, Am. J. Phys. **35**, 1078 (1967).

59. P. Aufmuth *et al.*, Z. Phys. D**23**, 19 (1992).

60. G. Werth, in Ref. [21], p. 217.

17

Precision Oscillator Strength and Lifetime Measurements

L. J. Curtis

Department of Physics and Astronomy, University of Toledo, Toledo, Ohio 43606

17.1 INTRODUCTION

A spectral line arising from a radiative transition between atomic states i and k is characterized by its wavelength λ_{ik}, its intensity and its shape. In the limit of free atoms, the intensity per atom is determined by the emission transition rate A_{ik} or absorption oscillator strength f_{ki}, and the shape by the natural width $\Gamma_i = \hbar/\tau_i$, where τ_i is the lifetime of the excited state. While classical spectroscopic methods provide precise wavelength measurements (1 part in 10^8 or better; *cf.* Chap. 10), it has only recently been possible to measure oscillator strengths and lifetimes to better than a few percent, as discussed in this Chapter. Examples of applications which require an accurate knowledge of these quantities are the interpretation of astrophysical data (*cf.* Chap. 78), atmospheric physics (*cf.* Chap. 80), combustion (*cf.* Chap. 84), the modeling and diagnosis of thermonuclear plasmas (*cf.* Chap. 82), nonlinear optics (*cf.* Chap. 70), isotope separation (*cf.* Chap. 16), and the development of new types of lasers (*cf.* Chap. 69).

Precision measurements of oscillator strengths and lifetimes also provide stringent tests of atomic structure calculations. These quantities are very sensitive to the wave functions and the approximations used, particularly in cases where electron correlations (see Sect. 23.3.1) and relativistic effects (see Sect. 22.1) are significant. They provide experimental tests of fundamental theory; for example, of quantum electrodynamic corrections (see Sect. 27.3), and of the nonconservation of parity predicted by the unified electro-weak theory (see Sect. 29.1).

In applications where only modest precision is required, semi-empirical parameterizations involving quantum defects, charge screening and polarization allow a few precise measurements to be extrapolated along isoelectronic, homologous, isoionic and Rydberg sequences. Similar methods have been applied to the atomic energy levels themselves (see Sect. 10.14). They allow one to produce a very large data base of moderate precision [1].

The spontaneous transition rate A_{ik} (see Sect. 10.17.2) is the probability per unit time (s^{-1}) for an atom in any one of the g_i states of the energy level i to make a transition to any of the g_k states of the level k. The lifetime τ_i is then given by $1/\tau_i = \sum_k A_{ik}$. The branching fraction for the kth channel of the decay of level i is defined as $F_{\rm B} = \tau_i A_{ik}$, and the branching ratio (see Sect. 10.18) between two decay channels is $R_{\rm B} = A_{ik}/A_{ij}$. Emission and absorption rate constants differ by a factor of λ_{ik}^2, and the absorption oscillator strength f_{ki} (see Sect. 10.18) is defined by

$$g_k f_{ki} = C\lambda_{ik}^2 g_i A_{ik}\,, \qquad (17.1)$$

where $C = (32\pi^3\alpha a_0^2 {\rm Ry})^{-1} = 1.49919 \times 10^{-14}\,{\rm nm}^{-2}{\rm s}$. Because of this relationship, the words transition rate and oscillator strength will be used almost interchangeably.

17.2 OSCILLATOR STRENGTHS

Oscillator strengths can be determined directly through absolute emission, absorption, or dispersion measurements, or through the combined measurement of branching ratios and lifetimes. Direct measurements compare different transitions at the same time, and require sample equilibrium, a knowledge of the absolute number density, and an absolute intensity measurement. These are in contrast to time-resolved lifetime measurements, which compare relative intensities from the same transition at different times, and require no absolute measurements. However, lifetime measurements yield oscillator strengths directly only in cases where a single decay transition channel exists, such as the lowest excited level in an atom or ion. Thus, combined measurements of lifetimes and branching ratios are often used where high precision values for oscillator strengths are required. Both absorption and dispersion measurements involve the number density of the lower level of the transition, whereas emission measurements involve that of the upper level.

17.2.1 Absorption and Dispersion Measurements

Absorption measurements involve placing a sample of atoms (for example, in a gas cell, an atomic beam, an arc, a shock tube, or within the vapor column in a furnace) between a continuous light source and a spectrometer. For an isolated spectral line, the absorption cross section for a beam of photons of frequency ν passing through the sample is

$$\sigma_{ik}(\nu) = \pi\alpha(\hbar/m)g(\nu)f_{ik} \qquad (17.2)$$

where $g(\nu)$ is the spectral distribution function per unit frequency normalized so that $\int_0^\infty g(\nu)\,d\nu = 1$. If there are N atoms per unit volume, the absorption coefficient is $k_\nu = N\sigma(\nu)$. The integrated intensity lost after passing through a distance L of the sample is

$$\int_0^\infty \Delta I(\nu)\,d\nu = I_0\int_0^\infty \left(1 - \exp^{-\sigma_{ik}(\nu)NL}\right)\,d\nu$$
$$\simeq I_0\pi\alpha(\hbar/m)NLf_{ik} \qquad (17.3)$$

where the second line applies if the sample is optically thin. Otherwise, the integral can be calculated directly, if $g(\nu)$ is known, to determine f_{ik} by the curve-of-growth method.

The Furnace Method

High precision absorption measurements have been achieved by Blackwell and co-workers [2], who have used the furnace method to study the astrophysically important neutral iron spectrum. These measurements have quoted accuracies of 0.5% on a relative scale, and 2.5% on an absolute scale. This accuracy was obtained through the use of a stable and isothermal furnace, low-noise spectral intensity recording techniques, and two identical high resolution spectrometers for the simultaneous recording of pairs of absorption lines. By selecting successive line pairs of a suitable oscillator strength ratio and adjusting the temperature and vapor pressure in the furnace, a large dynamic range of oscillator strengths could be covered. Recently, corroborative studies of the uncertainties quoted in these measurements have been undertaken, including tests that are coupled to other methods that use combinations of lifetime and branching ratio measurements [2].

The Hook Method

The absorption measurements described above determine the oscillator strength from the line intensity. An alternative absorptive approach is the anomalous dispersion or "hook" method, which determines the oscillator strength from the index of refraction at wavelengths near the edge of an absorption line [3]. The advantages of this method are its large dynamic range, its insensitivity to the line shape, and the fact that it does not saturate. The absorbing gas is placed in one arm of an interferometer and a compensator is placed in the other arm. This leads to the formation of oblique interference fringes with two characteristic hooks symmetric about the center of an absorption line. The oscillator strength is determined by the wavelength separation W between the hooks. Defining $K = \lambda_{ik}N_{\rm f}$, where $N_{\rm f}$ is the number of fringes per unit wavelength, then

$$f_{ik} = \frac{\pi K W^2}{\lambda_{ik}^3 \alpha^2 a_0 N L} \qquad (17.4)$$

where a_0 is the Bohr radius.

17.2.2 Emission Measurements

Emission methods use, for example, a hollow cathode, wall-stabilized arc, or shock tube to excite the source. Recent developments in the methods of Fourier Transform Spectrometry (FTS) [4, 5] offer several advantages over grating spectroscopy, as discussed next.

Fourier Transform Spectrometry

As opposed to the dispersive nature of a grating spectrograph, FTS uses interference effects from a Michelson interferometer. All radiation admitted to the spectrometer is thus incident on the detectors at all times, but different wavelengths are distinguished by their spatial modulation frequencies. The interferogram from all sinusoidal signals is sampled at a prescribed interval of path length, and compared with a laser of known frequency following the same optical path. The spectrum is recovered from the interferogram by means of a fast Fourier transform (see Sect. 8.2.3). Thus, for an FTS instrument, the spectral range is determined by the sampling step size, whereas the resolution depends on the maximum path difference (this is in contrast to a grating spectrometer where the sampling step determines the resolution, and the scan length determines the spectral range).

The FTS method provides a number of attractive features in precision oscillator strength measurements. The axial symmetry and the replacement of the slit by a larger aperture can provide a throughput that is two orders of magnitude greater than that of a grating instrument of the same resolution. The precision and reproducibility are determined by the laser standard and the linearity of the wavenumber scale. The resolution can be increased as necessary to resolve a specific source line. The superior resolution allows blending and self-absorption to be more readily detected. The spectral range is limited only by detectors and filters. Recently, uv FTS instruments have been constructed that operate from the visible down to 175 nm. Since all wavelengths are observed at all times, errors from drifts in source conditions during scanning are reduced.

Hollow Cathode Lamps

Hollow cathode discharge lamps are used extensively for emission branching fraction measurements [4]. These lamps can generate an emission spectrum of essentially

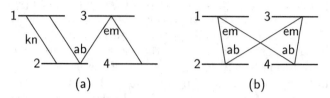

Figure 17.1. Illustration of methods for the determination of ratios of oscillator strengths.

any element, and the relatively low collision rates result in line profiles that are narrow and primarily Doppler broadened. The narrow line width is a major advantage when studying line-rich spectra. The low collision rates also imply that the discharges are far from local thermodynamic equilibrium (LTE). However, this does not affect the determination of branching fractions, where only the relative strengths of lines from a common upper level are measured. These measurements can then be put on an absolute basis if lifetime measurements are available for all of the upper levels.

17.2.3 Combined Absorption, Emission, and Lifetime Measurements

By combining measurements obtained in emission with those obtained in absorption (or dispersion) to obtain branching ratios, and then incorporating lifetime measurements, it is possible to use a scheme that requires no knowledge of level populations [3]. The scheme was originally proposed by Ladenburg in 1933, and its various modern implementations are known as "leap-frogging," "linkage" and "bow ties."

The principle of "leap-frogging" or "linkage" is illustrated in Fig. 17.1(a). The decay of level 1 is unbranched, so the $1 \to 2$ transition rate is assumed known (kn) from a lifetime measurement. This is used to specify the $2 \to 3$ oscillator strength using relative absorption (ab) measurements for the $2 \to 1 : 2 \to 3$ oscillator strength ratio (and appropriate factors of the wavelengths and degeneracies). In a similar manner, this is subsequently combined with relative emission (em) measurements of the $3 \to 2 : 3 \to 4$ branching ratio to determine the $3 \to 4$ transition rate.

The principle of "bow ties" is illustrated in Fig. 17.1(b). The two branching ratios for $1 \to 2 : 1 \to 4$ and for $3 \to 2 : 3 \to 4$ are measured in emission (em). The two oscillator strength ratios for $2 \to 1 : 2 \to 3$ and for $4 \to 1 : 4 \to 3$ are measured in absorption (ab). After correction for wavelength and degeneracy factors between f and A values, these relationships are combined into a quantity known as the "bow tie ratio," which would be unity for ideally accurate measurements. A significant deviation from unity can be used to trace the observations that are in error. Figure 17.1(b) is a "simple bow tie" connecting two upper and two lower levels with four transitions. Higher order sets of measurements can be similarly coupled; for example, a set of transitions between three lower levels and three upper levels can be coupled by nine simple bow ties.

In cases where both lifetime and complete branching ratio measurements exist for the same upper level, then the branching ratios can be normalized to branching fractions divided by the lifetime to obtain absolute transition probabilities.

17.2.4 Branching Ratios in Highly Ionized Atoms

In highly ionized atoms, many measurements of lifetimes in the 1–5% accuracy range have been made by ANDC (see Sect. 17.3.2) analysis of beam-foil measurements. However, little work exists for the precision measurement of branching ratios in highly ionized atoms, where an intensity calibration of the detection equipment is particularly difficult. In beam-foil excitation, for example, Doppler broadening and Doppler shifts, polarization due to anisotropic excitation, and the short wavelength (≤ 200 nm) nature of the radiation are not well-suited for use with standard techniques for calibrating grating spectrometer and detection systems by the use of calibrated lamps. Calibrations have been carried out using synchrotron radiation, or through the use of previously known branching ratios.

One way in which lifetime measurements are used to determine branching ratios involves precision studies of the lifetimes of the individual fine structure components of a multiplet decay. If one fine structure level has a decay channel that is not available to the other levels (for example, spin-changing transition or an autoionization mode made possible by a J-dependent intermediate coupling) then the transition rate of the extra channel can be determined by differential lifetime measurements.

The use of Si(Li) detectors in the measurement of very short wavelength radiation in highly ionized atoms also offers possibilities for branching ratio measurements. Since these devices can specify the photon energy from pulse height information without the need for a spectrometer, they can be calibrated for detection efficiency as a function of energy. This type of detection was recently used to determine the branching ratio of the magnetic dipole channel to the two-photon decay channel in the $2s\ ^2S_{\frac{1}{2}}$ state in one-electron krypton [6].

17.3 LIFETIMES

The total transition rate summed over all decay channels can be measured either through frequency-resolved studies of the level width, or through time-resolved studies of the level lifetime. In order to determine the natural linewidth in a field-free spectroscopic measurement, either the lifetime must be very short, or the Doppler, pressure, and instrumental broadenings must be made very small. Line widths have been determined using Fabry-Perot spectrometry at very low temperatures and pressures, and in beam-foil studies of radiative transitions in which the lower level decays very rapidly via autoionization. The linewidth can also be determined through the use of the phase shift method. Here modulated excitation is applied to the source, thus producing similarly modulated emitted radiation, and the width can be specified from the phase shift between the two signals. Other methods involve resonance fluorescence techniques, where sub-Doppler widths are obtained because the width of the exciting radiation selects a subset of particle motions within the sample. Resonance fluorescence techniques that can be used to determine level widths include zero-field level crossing (the Hanle effect), high-field level crossing, and double optical resonance methods, but the Hanle effect is the most common.

17.3.1 The Hanle Effect

In its most commonly used form, the Hanle effect makes use of polarized resonance radiation to excite atoms in the presence of a known variable magnetic field. The magnetic substates of the sample are anisotropically excited, and the subsequent radiation possesses a preferred angular distribution. By applying the magnetic field in a direction perpendicular to the anisotropy, the angular distribution is made to precess, producing oscillations in the radiation observed at a fixed angle. At infinite precessional frequency the intensity would be proportional to the instantaneous average angular intensity, but at finite precessional frequency it depends upon the decay that has occurred during each quarter rotation. Measured as a function of magnetic field, the emitted intensity has a Lorentzian shape centered about zero field with a width that depends on the lifetime and g-factor of the level [7, 8].

17.3.2 Time-Resolved Decay Measurements

The most direct method for the experimental determination of level lifetimes is through the time-resolved measurement of the free decay of the fluorescence radiation following a cutoff of the source of excitation. An important factor limiting the accuracy is the repopulation of the level of interest by cascade transitions from higher-lying levels. For this reason, decay curve measurements fall into two classes: those that involve selective excitation of the level of interest, thus eliminating cascading altogether; and those that use correlations between cascade connected decays to account for the effects of cascades.

Selective Excitation

Lifetime measurements accurate to within a few parts in 10^3 have been obtained through selective excitation produced when appropriately tuned laser light is incident on a gas cell or a thermal beam, or on a fast ion beam (see Sect. 18.1). With a gas cell or thermal beam, the timing is obtained by a pulsed laser and delayed coincidence detection. With a fast ion beam, time-of-flight methods are used. In either case, after removal of the background, the decay curve of intensity *vs.* time is a single exponential, and the lifetime is obtained from its semilogarithmic slope. Laser-excited time-of-flight studies were first carried out by observing the optical decay of the ion in flight following excitation using a laser beam which crossed the ion beam. In these studies, the laser light was tuned to the frequency of the desired absorption transition either through the use of a tunable dye laser, or by varying the angle of intersection to exploit the Doppler effect. Recent measurements have utilized diode laser excitation in this geometry [9]. A number of adaptations of this technique have been developed in which the laser and ion beams are made to be collinear, and are switched into and out of resonance within a segment of the beam by use of the Doppler effect [10]. The collinear geometry can provide a longer excitation region and less scattering of laser light into the detector than occurs in the crossed beam geometry. In one adaptation [11], excitation occurs within an electrostatic velocity switch, and the time resolution is obtained by physically moving the velocity switch. In another adaptation [10], the ion beam is accelerated with a spatially varying voltage ramp. The resonance region is moved relative to a fixed detector by time-sweeping either the laser tuning or the ramp voltage.

While these selective excitation methods totally eliminate the effects of cascade repopulation, they are generally limited to levels in neutral and singly ionized atoms that can be accessed from the ground state by strongly absorptive E1 transitions, and the selectivity itself is a limitation. Many very precise measurements have been made by these techniques, but they have primarily involved $\Delta n = 0$ resonance transitions in neutral alkali atoms and singly ionized alkali-like ions. A summary of these measurements is given in Table 17.1, and a comparison of these values with theoretical calculations is given in Ref. [24].

Table 17.1. Measured $np\ {}^2P_J$ lifetimes.

Atom	n	J	τ (ns)
Li	2	1/2	27.22(20)[a], 27.29(4)[b]
Na	3	1/2	16.38(8)[c], 16.40(3)[b], 16.40(5)[d]
	3	3/2	16.36(10)[c], 16.35(5)[d]
Mg$^+$	3	1/2	3.854(30)[e]
	3	3/2	3.810(40)[e]
Ca$^+$	4	1/2	7.07(7)[f], 7.098(20)[g]
	4	3/2	6.87(6)[f], 6.924(19)[g]
Cu	4	1/2	7.27(6)[h]
	4	3/2	7.17(6)[h]
Sr$^+$	5	1/2	7.47(7)[i]
	5	3/2	6.69(7)[i]
Ag	5	1/2	7.41(4)[j]
	5	3/2	6.79(3)[j]
Cd$^+$	5	1/2	3.11(4)[k]
	5	3/2	2.77(7)[k]
Cs	6	3/2	30.55(27)[l]
Ba$^+$	6	1/2	7.92(8)[i]
	6	3/2	6.312(16)[m]

[a]Ref. [12]　　[f]Ref. [13]　　[j]Ref. [14]
[b]Ref. [15]　　[g]Ref. [16]　　[k]Ref. [17]
[c]Ref. [18]　　[h]Ref. [19]　　[l]Ref. [9]
[d]Ref. [20]　　[i]Ref. [21]　　[m]Ref. [22]
[e]Ref. [23]

Nonselective Excitation

Much more general access can be obtained by nonselective excitation methods, such as pulsed electron beam bombardment of a gas cell or gas jet, or in-flight excitation of a fast ion beam by a thin foil. Pulsed electron beam excitation can be achieved either through use of a suppressor grid, or by repetitive high frequency deflection of the beam across a slit so as to chop the beam [7]. Particularly in the case of weak lines, the high frequency deflection technique offers the advantages of high current and sharp cutoff times. The high currents yield high light levels, so that high resolution spectroscopic methods can be used to eliminate the effects of line blending. Pulsed electron excitation methods are well suited to measurements in neutral and near neutral ions (although for very long lifetimes in ionized species, the decay curves can be distorted if particles escape from the viewing volume through the Coulomb explosion effect [25]). However, for highly ionized atoms, the only generally applicable method is thin foil excitation of a fast ion beam. Nonselective excitation techniques can also be applied to measurements such as the phase shift method [26] and the Hanle effect [27], in which case cascade repopulation also can become a serious problem. However, most of the attempts to eliminate or account for cascade effects have occurred in decay curve studies.

In beam-foil studies, the excitation is created in the dense solid environment of the foil, after which the ions emerge into a field free, collision free, high vacuum region downstream from the foil. A time-resolved decay curve is obtained by translating the foil upstream or downstream relative to the detection apparatus. The beam is a very tenuous plasma, which has both advantages and disadvantages. The low density avoids the effects of collisional de-excitation and radiation trapping, but also produces relatively low light levels. This requires fast optical systems with a corresponding reduction in wavelength dispersion, and care must be taken to avoid blending of these Doppler broadened lines. Methods have been developed by which grating monochromators can be refocussed to a moving light source, thus utilizing the angular dependence of the Doppler effect to narrow and enhance the lines.

In these nonselective excitation methods, the decay curve involves a sum of many exponentials, one corresponding to the primary level, and one to each level that cascades (either directly or indirectly) into it. Decay exponentials do not comprise an orthogonal set of functions, and the representation of an infinite sum by a finite sum through curve fitting methods (see Sect. 8.2.2) can lead to large errors. Fortunately, alternative methods to exponential curve fitting exist, which permit the accurate extraction of lifetimes to be made from correlated sets of nonselectively populated decay curves.

ANDC Method

Precision lifetime values have been extracted from cascade-affected decay curves by a technique known as the Arbitrarily Normalized Decay Curve (ANDC) method (see Sect. 18.2.2, [28]), which exploits dynamical correlations among the cascade-related decay curves. These correlations arise from the rate equation that connects the population of a given level to those of the levels that cascade directly into it. The instantaneous population of each level n is, to within constant factors ξ_n involving the transition probabilities and detection efficiencies, proportional to the intensity of radiation $I_n(t)$ emitted in any convenient decay branch. In terms of these intensities, the population equation for the level n can be written in terms of its direct cascades from levels i as

$$\frac{dI_n}{dt}(t_p) = \sum_i \xi_i I_i(t_p) - I_n(t_p)/\tau_n. \tag{17.5}$$

Thus, if all decay curves are measured at the same discrete intervals of time t_p, the population equation provides a separate independent linear relationship among these measured decay curves for each value of t_p, with common constant coefficients given by the lifetime τ_n and the normalization parameters ξ_i. Although the sum over cascades is formally unbounded, the dominant effects of cascading from highly excited states are often accounted for by indirect cascading through the lower states, in which case the sum can be truncated after only a few terms. ANDC analysis consists of using this equation to relate the measured $I_k(t_p)$'s (using numerical differentiation or integration) to determine τ_k and the ξ_i's through a linear regression. If all significant direct cascades have been included, the goodness-of-fit will be uniform for all time subregions, indicating reliability. If important cascades have been omitted or blends are present, the fit will vary over time subregions, indicating a failure of the analysis. Very rugged algorithms have been developed [29] that permit accurate lifetimes to be extracted even in cases where statistical fluctuations are substantial, and studies of the propagation and correlation of errors have been made. Clearly the ANDC method is most easily applied to systems for which repopulation effects are dominated by a small number of cascade channels. Further applications are discussed in Sect. 18.2.2.

17.3.3 Other Methods

Coincidence measurements provide another method for elimination of cascade effects. While the low count rates and correspondingly high accidental rates make the application of these methods difficult for optical spectra, the use of Si(Li) for the very short wavelength emission in very highly ionized systems offers new possiblilties for these measurements.

Another method of accounting for cascading involves the combined use of a laser and beam-foil excitation [30]. The beam-foil excitation provides a source of ions in excited states, and a chopped laser is used to stimulate transitions between two excited states. By subtracting the decay curves obtained with laser on and with laser off, the cascade-free, laser-produced portion of the decay curve is obtained.

17.3.4 Multiplexed Detection

Recently, the detection efficiency and reliability of beam-foil measurements have been improved through the use of position sensitive detectors (PSD), which permits measurement of decay curves as a function both of wavelength and of time since excitation. The PSD is mounted at the exit focus of the analyzing monochromator, where it records all lines within a given wavelength interval simultaneously, including reference lines with the same Doppler shifts. Decay curves can be constructed by integrating over a line profile, and that profile can be examined for exponential content to eliminate blending. The time dependent backgrounds underlying the decay curves are directly available from the neighboring channels.

The use of multiplexed detection greatly enhances the data collection efficiency, and causes many possible systematic errors to cancel in differential measurements. Effects such as fluctuations in the beam current, degradation of the foil, divergence of the beam, etc., affect all decay curves in the same way.

The most accurate measurement [31] made by beam-foil excitation ($\pm 0.26\%$) used a type of multiplexed detection in which two spectrometers simultaneously viewed the decays of the $^1P_1$ and $^3P_1$ levels of the $1s3p$ configuration in neutral helium. The $^1P_1$ emission exhibited the desired multi-exponential decay curve (with only weak cascading), whereas the $^3P_1$ emission exhibited a zero field quantum beat pattern superimposed on its decay because of the anisotropic excitation of that level. The quantum beats provided an in-beam time base calibration which permitted this high precision.

17.4 CONCLUSION

The accuracy of oscillator strength and lifetime measurements has improved greatly in the past twenty years. Nevertheless, these high accuracies have been achieved for only a restricted number of lines belonging to a few elements and ionization stages [3]. Large numbers of precision measurements must still be made as improved experimental oscillator strengths are needed, both as tests of theoretical concepts, and for diagnostics and engineering applications.

REFERENCES

1. L. J. Curtis, Phys. Scr. **43**, 137 (1991).
2. D. E. Blackwell, in *Atomic Spectra and Oscillator Strengths for Astrophysics and Fusion Research*, edited by J. E. Hansen (North-Holland, Amsterdam, 1990), p. 160.
3. M. C. E. Huber and R. J. Sandeman, Rep. Prog. Phys. **49**, 397 (1986).
4. J. E. Lawler, S. D. Bergeson, and R. C. Wamsley, Phys. Scr. **T47**, 29 (1993).
5. A. P. Thorne and R. C. M. Learner, in *Atomic Spectra and Oscillator Strengths for Astrophysics and Fusion Research*, edited by J. E. Hansen (North-Holland, Amsterdam, 1990), p. 191.
6. S. Cheng, H. G. Berry, R. W. Dunford, D. S. Gemmell, E. P. Kanter, B. J. Zabransky, A. E. Livingston, L. J. Curtis, J. Bailey, and J. A. Nolen, Jr., Phys. Rev. A **47**, 903 (1993).
7. W. L. Wiese, in *Progress in Atomic Spectroscopy B*, edited by W. Hanle and H. Kleinpoppen (Plenum, New York, 1979), p. 1101.
8. A. Corney, Adv. Electronics Electron Phys. **29**, 115 (1970).
9. C. E. Tanner, A. E. Livingston, R. J. Rafac, F. G. Serpa, K. W. Kukla, H. G. Berry, L. Young, and C. A. Kurtz, Phys. Rev. Lett. **69**, 2765 (1992).
10. J. Jin and D. A. Church, Phys. Rev. **47**, 132 (1993).
11. H. Winter and M. L. Gaillard, Z. Phys. **A281**, 311 (1977).
12. J. Carlsson and L. Sturesson, Z. Phys. **D14**, 281 (1989).
13. R. N. Gosselin, E. H. Pinnington, and W. Ansbacher, Phys. Rev. A **38**, 4887 (1988).
14. J. Carlsson, P. Jönsson, and L. Sturesson, Z. Phys. **D16**, 87 (1990).
15. A. Gaupp, P. Kuske, and H. J. Andrä, Phys. Rev. **A26**, 3351 (1982).
16. J. Jin and D. A. Church, Phys. Rev. Lett. **70**, 3213 (1993).
17. E. H. Pinnington, J. J. van Hunen, R. N. Gosselin, B. Guo, and R. W. Berends, Phys. Scr. **49**, 331 (1994).
18. J. Carlsson, Z. Phys. **D9**, 147 (1988).
19. J. Carlsson, L. Sturesson, and S. Svanberg, Z. Phys. **D11**, 287 (1989).
20. H. Schmoranzer, D. Schulze-Hagenest, and S. A. Kandela, Symposium on Atomic Spectroscopy (SAS-79), Tucson AZ 1979, Abstracts p. 195.
21. P. Kuske, N. Kirchner, W. Wittmann, H. J. Andrä, and D. Kaiser, Phys. Lett. **A64**, 377 (1978).
22. H. J. Andrä, in *Beam-Foil Spectroscopy*, edited by I. A. Sellin and D. J. Pegg (Plenum, New York, 1976) Vol. 2, p. 835.
23. W. Ansbacher, Y. Li, and E. H. Pinnington, Phys. Lett. **A139**, 165 (1989).
24. L. J. Curtis, Phys. Scr. **48**, 559 (1992).
25. L. J. Curtis and P. Erman, J. Opt. Soc. Am. **67**, 1218 (1977).
26. L. J. Curtis and W. H. Smith, Phys. Rev. **A9**, 1537 (1974).
27. M. Dufay, Nucl. Instrum. Methods **110**, 79 (1973).
28. L. J. Curtis, in *Beam-Foil Spectroscopy*, edited by S. Bashkin (Springer, Berlin, 1976), Ch. 3.
29. L. Engström, Nucl. Instrum. Meth. **202**, 369 (1982).
30. H. Harde and G. Guthöhrlein, Phys. Rev. **A10**, 1488 (1974); Y. Baudinet-Robinet, H.-P. Garnir, P.-D. Dumont, and J. Résimont, Phys. Rev. **A42**, 1080 (1990).
31. G. Astner, L. J. Curtis, L. Liljeby, S. Mannervik, and I. Martinson, Z. Phys. **279**, 1 (1976).

18

Ion Beam Spectroscopy

Eric H. Pinnington

Department of Physics, University of Alberta, Edmonton, Alberta T6G 2J1, Canada

18.1 INTRODUCTION

A beam of ionized atoms has important advantages as a spectroscopic source. Unlike arcs, sparks and high temperature plasmas, the ions in a beam can be studied in an environment that is free of electric and magnetic fields, and relatively free of interparticle collisions [1]. Standard accelerator techniques from nuclear physics may be used to produce a well-collimated, mass-analyzed beam of ions having a low velocity spread. In principle, virtually any charge state of any element, isotopically pure if required, can be obtained, given the appropriate equipment. Finally, the well-defined velocity of the beam permits the study of processes evolving in time in terms of their spatial evolution along the beam. This is particularly important in the case of lifetime measurements, discussed in Sect. 18.2.2. An experiment can involve an ion energy as low as a few keV or as high as several GeV. However, in this article the individual applications of ion beam techniques in atomic physics will be classified by the means used (foil, gas, laser etc.) to excite the ions, since this better defines the type of research in each case.

18.2 BEAM-FOIL SPECTROSCOPY

A beam of ions passing through a thin (50–200 nm) foil emerges in a range of ionization states, with the mean charge state increasing with the incident energy [2]. For example, an F^+ beam at 0.5 MeV emerges from the foil with a mean charge of about $+2e$, while an Xe^+ beam at 180 MeV emerges with a mean charge of about $+29e$. The outer electrons are distributed over many different states, most of which then decay to lower states by photon emission as the ions move away from the foil. Thin foils made from a light element, usually carbon, are used to minimize particle scattering and energy straggling. Since many different electronic states are populated in the beam-foil interaction, it is a highly *nonselective* excitation. This has advantages for spectroscopic studies (see Sect. 18.2.1), but causes cascade problems for lifetime measurements. The major disadvantage of the beam-foil light source is its low intensity, and the resulting limitation on the spectroscopic resolution that is attainable. Consequently, beam-foil spectroscopy is often carried out in conjunction with more classical, higher resolution photographic spectroscopy (see Sect. 42.1) for detailed analyses of atomic spectra. Because of the low intensity of the source, beam-foil spectroscopy is usually performed using photon-counting techniques.

18.2.1 Spectroscopic Studies

The beam-foil interaction involves an effective shielding of the outer electrons from the core as the ion passes through the foil. As the electrons emerge from the foil, some are re-captured by the core into a statistical distribution of states. Since the probability that more than one electron will be captured in an excited state is high relative to other light sources, beam-foil spectroscopy is used extensively to study doubly- and multiply-excited states [3] (see Sect. 62.1.1). The beam-foil interaction also favors the production of high-L Rydberg states [4] (see Sect. 14.1). At low incident ion energies, electron capture can give a downstream beam containing significant fractions of neutral and even negative ions. The first observation of photon emission between bound states in a negative ion was achieved using beam-foil excitation [5]. A recent example of multiple excitation in a negative ion is the identification in lithium-like He$^-$ in the $2p^3\,^4S^0$ state in which all three electrons are excited [6]. This state has also been identified in several other members of the lithium-like isoelectronic sequence.

Lifetime measurements and spectroscopic analyses often go hand-in-hand. For example, if two (or more) transitions are assigned as having the same upper state, then the shape of their intensity decay curves (see Sect. 18.2.2) should be the same within the statistical uncertainties, provided that none of the transitions is blended with another line. The time-resolution inherent in the beam-foil source can also be used to advantage in the assignment of transitions from long-lived states, such as intercombination transitions [7] (see also Sect. 10.17). Here it is useful to record the beam-foil spectrum both close to the excitation region at the foil and relatively far downstream from the foil. "Far" in this context means that the short-lived levels have had time to decay. The intercombination (or other long-lived) transitions can then often be easily identified by their relatively strong intensity in the downstream (time-delayed) spectrum.

The low intensity of the beam-foil light source inevitably results in spectra of rather modest resolution. The high velocity of the radiating ions can also increase the widths of the spectral lines (see Sect. 19.1 for Doppler line broadening). A useful trick here is to refocus the spectrometer for the moving light source by adjusting the grating-to-slit distance appropriately [8]. While this method works well over narrow spectral ranges, it requires that the spectrometer be refocused for each wavelength region. Furthermore, the linewidth is still determined by the width of the entrance slit. An ingenious suggestion by Bergkvist [9] showed that, by introducing an appropriate cylindrical lens immediately before

the spectrometer, it is actually possible to use a wide entrance slit without losing resolution, giving intensity gains of several orders of magnitude.

Beam-foil spectroscopy has also been used to make precision wavelength measurements of resonance transitions in hydrogen-, helium- and lithium-like ions in order to derive values for the Lamb Shift (see Sect. 28.3.9) [10]. In addition to the refocusing procedure mentioned above, if the measurement is made using a scanning spectrometer, it is important to investigate the possible error caused by periodic imperfections in the grating drive screw [11]. This problem is avoided if the spectrometer is used in a nonscanning mode, such as by employing a microchannel plate [12] or other detector array. It is also preferable to use calibration lines emitted by the moving ions themselves since a spectrometer that is focused for the moving ions will be out of focus for a stationary calibration source. One approach here is to use different grating orders to superimpose reference lines from other spectral regions on the spectrum containing the transitions under study [13]. Examples of beam-foil precision wavelength measurements for He-like ions [14, 15] are compared with calculations [16, 17] in Table 18.2.

18.2.2 Lifetime Measurements

Measurements Using Multi-Exponential Analysis

Since the beam-foil source is highly nonselective, the population of a given level j can be both depleted by spontaneous emission to lower levels k, and replenished by cascade transitions from higher levels i. The radiative lifetime (or mean life) τ of level j is given by the reciprocal of the sum of the transition probabilities A$_{jk}$. The population of level j is given as a function of the time t after excitation at the foil by the sum of exponential terms

$$N_j(t) = A \exp(-t/\tau) + \sum_i B_i \exp(-t/\tau_i) \qquad (18.1)$$

where the coefficients B_i can be positive or negative, and the summation includes any contributions from levels that repopulate level j, either directly or via intermediate levels. An actual measurement consists of recording a decay curve; *viz.*, the *intensity* of a transition from level j as a function of distance from the foil. The intensity is proportional to the level population. The distance x corresponds to an elapsed time of x/v, where v is the ion velocity. The measurement is usually made by stepping the foil upstream, rather than moving the observation region downstream. However, scanning the beam directly by an appropriate optical system is also possible [18]. In a typical beam-foil decay curve, the data points are spaced further apart in the tail of the curve in order to estimate the contribution of the long-lived cascade components in a reasonable total measurement time.

Fitting the observed data to Eq. (18.1) to determine the lifetime of level j must be done with care (see also Sect. 8.2.2 for further discussion of curve-fitting problems). Several computer routines have been developed, which attempt to determine the correct form of the fitting function, i.e., how many exponential terms should be included, e.g., DISCRETE [19] and HOMER [20]. A useful technique here is to record the decay curve for each of the major cascade transitions into level j, and then include the lifetimes determined from analyses of these curves as fixed parameters in fitting the decay curve from level j to Eq. (18.1).

An additional problem can arise in the measurement of very short lifetimes. These tend to be associated with short wavelength transitions for which no lenses are available to focus the beam on the spectrometer entrance slit. Consequently, the observation region extends along the beam. The effect of this averaging can be seen by integrating Eq. (18.1) from $(t - \Delta)$ to $(t + \Delta)$. Since the first data point that can be fitted to (18.1) is at a time Δ downstream, each coefficient in (18.1) becomes multiplied by a factor $\tau[1 - \exp(-2\Delta/\tau)]$, where τ is the lifetime for that component. Clearly, the short-lived components become very much weakened relative to the longer-lived cascade terms. Although this problem can be lessened at the cost of intensity by reducing the acceptance angle of the spectrometer, a preferable technique is to include the vignetting effect at the foil in the fitting function itself. Routines to do this include FITB [21] and VNET [22], both of which permit the decay curve to start on the upstream side of the foil. Lifetimes in the picosecond regime have been reported using the beam-foil technique method [23].

Measurements Using Cascade Correction (ANDC) Techniques

A more rigorous approach to the cascade problem is provided by the ANDC method [24] discussed in Sect. 17.3.2. Various methods have been developed for applying the basic ANDC principle [25–27]. The technique has been demonstrated [28, 29] to be very useful in cases where only a few cascades are important, as in atomic systems with a single active electron, (Na-like, Cu-like etc.) or with two active electrons, (Mg-like, Zn-like etc.), when it can yield reliable lifetimes with precisions of 5% or better. However, the technique tends to get results of rather lower precision when the cascade scheme is more complicated. The ANDC method is also useful when one or more strong cascades have lifetime components close to those of the primary decay, which can result in significant errors if the analysis relies on only multi-exponential curve fitting. One such example

concerns the resonance transition in Be-like ions, where the lifetimes of the $2s2p\,{}^1P^o$ resonance level and the important cascade level, $2p^2\,{}^1S$, are almost equal. This problem has been studied by Engström *et al.* [30] for the Be-like ions N IV and O V. Their results are summarized in Table 18.1, where they are compared with the weighted means of several previous beam-foil measurements and with two calculations [31, 32]. Two conclusions may be drawn from Table 18.1. First, the ANDC analyses give results in much closer accord with the calculations in this example. Second, consistency between several or many individual measurements of the same type (multi-exponential curve-fitting of beam-foil data in this instance) does not guarantee that the results are not subject to significant systematic error.

18.2.3 Other Phenomena Studied Using the Beam-Foil Interaction

Measurements of the velocity and scattering of the ions after passage through foils of known constitution and thickness have yielded data on electronic and nuclear stopping powers (see Sect. 87.2.1). Studies of the relative populations in the excited states of the ions have been used to model the ion-foil interaction itself [33]. The polarization of light emitted by ions after passage through a foil has been studied, both using linear polarization to study *alignment* of the atomic sublevels and circular polarization to study their orientation (see Sect. 12.2). Remarkably large degrees of circular polarization (up to 30%) have been observed using tilted foils, where foils have been tilted through large angles relative to the beam direction [34]. Of particular interest has been the study of quantum beats [35], in which the beam-foil decay curves show oscillations corresponding to the separation of close-lying fine- or hyperfine-structure levels that are coherently excited as a result of the very short duration ($\sim 10^{-14}$ s) of the excitation process (see Sect. 17.3.4 for further details).

18.3 BEAM-GAS SPECTROSCOPY

While the use of a gas target in place of a foil has the obvious advantage that the gas cannot "break", the loss of a tightly-localized excitation region means that the fine time-resolution of the beam-foil light source is largely lost. The beam-gas source has two main advantages, however. First, the passage of a beam of fast, highly-stripped ions through a neutral gas results in the production of recoil ions (see also Sect. 63.3) that are moving very slowly relative to the beam ions, thus avoiding the Doppler broadening problems mentioned in Sect. 18.2.1. Second, it is possible to study details of the interaction between the gas and the beam, such as the charge-exchange reactions, as they occur rather than merely observing their consequences. Differences between the emission spectrum produced by the beam-gas interaction and that produced by, say, interaction of the same ion beam with a foil, can give important clues concerning resonance processes for charge exchange or excitation exchange. Such experiments have experienced a resurgence recently with the advent of the ECR ion source [36], by which it is possible to obtain relatively high beam currents of highly-charged ions.

18.3.1 Recoil Ion Spectroscopy

When a fast beam of highly-charged ions passes through a neutral gas, it leaves a trail of ionized gas atoms in its wake. These ions recoil from the interaction region with relatively low velocities ($v/c = 10^{-4} - 10^{-5}$). Furthermore, the recoil velocities tend to be restricted to a narrow range of angles approximately perpendicular to the beam direction. Hence, observation of the radiation emitted by the recoil ions transverse to their motion; i.e., along a direction parallel to the beam, gives a very low Doppler width, $\Delta\lambda_D/\lambda$ being typically 10^{-6}, the limit imposed by the thermal motion of the target gas atoms. Recoil ion spectroscopy is therefore a useful procedure for precision wavelength measurements highly-stripped ions, such as He-like Ar^{16+} recoil ions produced by a beam of 2 GeV U^{70+} ions [37]. Some examples of wavelength measurements using recoil ions [38, 39] are compared with calculations [16, 17] in Table 18.2.

The energies and charge states of the recoil ions may be determined using standard time-of-flight techniques [43], and detecting the recoil ions in coincidence with their progenitor projectile ions yields information on the dependence of the recoil energy on the details of the ionizing collision [44]. Detecting optical emission between Rydberg states of the ions (see Sect. 14.1) in coincidence with either the projectile or recoil ions can also give information on the role played by autoionization (see Chap. 25) by an ion between its formation in the collision and its arrival at the detector, as in the measurements by Martin *et al.* using Kr^{16+} ions with an energy of 360 keV [45]. Lifetimes of metastable states in highly-stripped recoil ions can also be measured by forming a secondary beam from the recoil ions. A position-sensitive detector is then used to record in-flight x-ray emission in coincidence with the charge state of the radiating ion [46].

18.3.2 Exchange Reactions

Measurements made on the projectile ions, rather than the recoiling target ions, also yield useful information about electron-capture processes. The strength of such a process is described in terms of its cross section

(see Sect. 61.2.2). Recent work has shifted from measurements of the total cross section for electron capture in a collision, to more detailed studies of the individual capture channels [47]. Such studies provide much more stringent tests of theoretical models of the ion-atom charge transfer processes. They often involve such techniques as spectroscopy of the optical radiation or of the Auger electrons (see Sect. 25.1.1) emitted by the ions after electron exchange. This topic is covered in Chaps. 49, 62 and 63. A further example is the study by Prior *et al.* [48] of the angular distribution of Auger electrons emitted by doubly-excited states formed in hydrogen-like projectile ions with an energy of 40 keV, following double-electron capture from target helium atoms. They found that significant alignment of the magnetic substates of the projectile ions can result from electron capture. Such anisotropies in the Auger electron emission demonstrate the danger of using single-angle measurements to determine cross sections.

18.4 BEAM LASER SPECTROSCOPY

As for the beam-foil source, excitation of a beam of fast ions by a transverse tunable laser produces the localized excitation required for high temporal resolution. Now, however, the excitation is *highly selective*, permitting the population of just a single level. The laser-induced-fluorescence (LIF) signal is therefore described by a single exponential decay, for which an exact analysis with rigorous error bounds is possible. The obvious restriction to levels that can be accessed by electric dipole (E1) transitions from the ground or metastable levels may be overcome by combining laser excitation with a nonselective mode of excitation, as has been done in beam-gas-laser [49] or beam-foil-laser [50] measurements. Discussion of these precision lifetime measurements is given in Chap. 17. Here the discussion is restricted to precision atomic spectroscopy. Some recent examples of precision beam-laser wavelength measurements [40, 42] are compared with calculation [16, 41] in Table 18.2. In addition, a recent high-precision measurement of the spin-forbidden $1s2s\,^1S_0-1s2p\,^3S_1$ interval in N^{5+} [51] yields 986.321(7) cm^{-1}, in reasonable agreement with the theoretical value 986.58(30) cm^{-1} [16].

A major aim in beam-laser spectroscopy is to minimize the width of the LIF signal. The instrumental linewidth of the laser itself can be reduced to below 1 kHz, so that the width of the LIF signal is usually dominated by the velocity spread of the ions and by the divergences of the ion and laser beams. The effects of beam divergence can be minimized by using the collinear geometry, in which the ion and laser beams are parallel. If the angle between the ion and laser beams is θ, the Doppler-shifted laser frequency, as measured in the ion's rest frame, is $f_L(1 - \beta\cos\theta)$, where f_L is the laser

Table 18.1. Beam-foil lifetime measurements by Engström *et al.* [30] for the $2s2p\,{}^1P^o$ level in Be-Like N^{3+} and O^{4+} ions. Units are ns.

Ion	Experiment			Theory	
	Multi-Expl. Fitting	Other Expt.	ANDC	a	b
N^{3+}	0.50 ± 0.03 (0.35 ± 0.03^c)	0.463 ± 0.01^d	0.425 ± 0.015	0.429	0.421
O^{4+}	0.39 ± 0.02 (0.29 ± 0.02^c)	0.402 ± 0.014^e	0.338 ± 0.015	0.348	0.343

[a] Two-valence-electron model-potential calculation [31].
[b] Intermediate-coupling multiconfiguration calculation [32].
[c] Lifetime measured for the $2p^2\,{}^1S$ cascade level.
[d] Weighted mean of 4 earlier experiments quoted in Ref. [30].
[e] Weighted mean of 5 earlier experiments quoted in Ref. [30].

Table 18.2. Examples of fast-beam measurements of $\lambda_{\rm vac}$ for the $1s2s\,{}^3S - 1s2p\,{}^3P^o$ transitions in He-like Ions. This Table compares experiment and theory for some low-Z, intermediate-Z and high-Z members of the He-like isoelectronic sequence. The theoretical uncertainties (shown in brackets) reflect the leading terms omitted in the calculations by Drake [16]. Since these terms vary as Z^4, a lower experimental precision is required at higher values of Z to provide a useful test of the calculation. Thus the recoil-ion measurements at intermediate-Z and the beam-foil measurements at high-Z provide useful comparisons with theory, even though their precision is much lower than that of the beam-laser experiments, which can only be performed for relatively low Z. Units are nm.

Ion	Technique	$J-J'$	Experiment	Theory
Be^{2+}	Beam-Laser	1–0	$372.236904(5)^a$	372.23625^b, $372.23657(42)^c$
	Beam-Laser	1–1	$372.397129(4)^a$	372.39704^b, $372.39733(42)^c$
	Beam-Laser	1–2	$372.190681(4)^a$	372.19082^b, $372.19112(42)^c$
B^{3+}	Beam-Laser	1–0	$282.53674(10)^d$	$282.53616(64)^c$
	Beam-Laser	1–1	$282.66614(10)^d$	$282.66633(64)^c$
	Beam-Laser	1–2	$282.24601(7)^d$	$282.24652(64)^c$
Ne^{8+}	Recoil-Ion	1–0	$127.774(4)^e$	$127.7696(20)^c$, 127.7740^f
	Recoil-Ion	1–2	$124.8104(10)^e$	$124.8103(19)^c$, 124.8100^f
Ar^{16+}	Recoil-Ion	1–0	$66.136(4)^g$	$66.1437(57)^c$, 66.1573^f
	Recoil-Ion	1–2	$55.994(10)^g$	$55.9983(41)^c$, 55.9986^f
Ni^{26+}	Beam-Foil	1–2	$22.627(4)^h$	$22.6290(37)^c$, 22.6291^f
Cu^{27+}	Beam-Foil	1–2	$20.665(8)^i$	$20.6755(36)^c$

[a] Scholl *et al.* [40].
[b] Accad *et al.* [41].
[c] Drake [16].
[d] Dineen *et al.* [42].
[e] Hallett *et al.* [38].
[f] Plante *et al.* [17].
[g] Beyer *et al.* [39].
[h] Zacarias *et al.* [14].
[i] Buchet *et al.* [15].

frequency and $\beta = v/c$ is assumed to be much smaller than unity. The range in frequency resulting from a divergence $\Delta\theta$ is thus $f_L\beta\sin\theta\Delta\theta$, and tends to zero as θ tends to zero. One disadvantage of the collinear geometry is that, if the laser is brought into resonance with an atomic transition by adjusting the ion velocity and/or the laser frequency, the LIF signal is produced over the entire overlap region between the ion and laser beams. The resonance can be restricted to a desired region by setting the ion velocity to be slightly off-resonance. The velocity is then locally adjusted for resonance with the laser by passing the ions through a Faraday cage electrode to which a variable voltage is applied [52]. In the collinear geometry, the width of the resonance signal is usually dominated by the spread in the ion velocity $c\Delta\beta$. The width resulting from a given $\Delta\beta$ may be diminished using kinematic compression. In terms of the ion energy E and the ion mass M, the Doppler width of the LIF signal is given by

$$\Delta f_D = f_L\Delta\beta = f_L\Delta E/(2Mc^2E)^{1/2} \qquad (18.2)$$

and thus decreases as E is increased [52]. A more significant improvement in frequency resolution is made possible by including rf resonance in a laser double-resonance experiment. Here, the ions are brought into resonance with an off-resonance laser using two separate Faraday cage electrodes. The first resonance depletes the population in the ion state from which excitation occurs, thus weakening the second resonance signal. An rf field is then applied to the ions between the two electrodes. Tuning the frequency of this field over the region that corresponds to fine- or hyperfine structure intervals in the ion can repopulate the state from which laser excitation occurs, thus re-establishing the second LIF resonance signal. The width of the resonance signal is now determined by the *transit-time broadening* that results from the finite time spent by the ions in the rf-field (see also Sect. 74.6.1). For example, in the experiments of Sen *et al.* [52] with a beam of ^{131}Eu$^+$ ions at 1.35 keV, the width of the laser-rf double resonance signal is 59 kHz, compared with a width of 45 MHz obtained using a single LIF resonance.

18.5 OTHER TECHNIQUES OF ION-BEAM SPECTROSCOPY

While ion beams find uses in many other applications, possibly the three main areas involve merged beams, storage rings and studies of the ion-surface interaction at grazing incidence. Merged beam experiments usually study recombination processes involving electrons and atomic or molecular ions (see Sect. 52.1 regarding recombination processes). The advantage of using merged electron and ion beams is that the time development of the processes may be studied spatially, while maintaining a low relative velocity between the ions and electrons. This permits measurements at the low energies of importance in studies of Rydberg state formation and in some astrophysical applications. The recent advent of storage rings has taken this concept one step further, as the stored ions may be cooled using lasers to longitudinal temperatures of a few kelvins [53]. (Laser cooling is discussed in Chap. 73.) As well as permitting very precise studies of electron-ion recombination and laser-stimulated recombination, it is also possible to measure very long lifetimes [54] and to search for condensed structures of ions similar to those that have been found in ion traps (see Chap. 73). Finally, in a very different type of experiment, the ion-surface interaction is studied using a well-collimated ion beam at grazing incidence on a clean, flat surface. Such experiments have revealed very large atomic orientation [55]. This orientation can also be passed on to the nuclei of the atoms via the hyperfine interaction, thus providing a source of oriented nuclei [56].

REFERENCES

1. I. Martinson, Rep. Prog. Phys. **52**, 157 (1989).
2. K. Shima, T. Mikumo, and H. Tawara, At. Data Nucl. Data Tables **34**, 357 (1986).
3. H. G. Berry, Phys. Scr. **12**, 5 (1975); T. Andersen and S. Mannervik, Comments At. Mol. Phys. **16**, 185 (1985).
4. F. G. Serpa and A. E. Livingston, Phys. Rev. A **43**, 6447 (1991).
5. C. F. Bunge, Phys. Rev. Lett. **44**, 1450 (1980).
6. E. J. Knystautas, Phys. Rev. Lett. **69**, 2635 (1992); E. Träbert, P. H. Heckmann, J. Doerfert, and J. Granzow, J. Phys. B **25**, L353 (1993).
7. E. Träbert, Phys. Scr. **48**, 699 (1993).
8. J. O. Stoner and J. A. Leavitt, Opt. Acta **20**, 435 (1973).
9. K.-E. Bergkvist, in *Beam-Foil Spectroscopy*, edited by I. A. Sellin and D. J. Pegg (Plenum Press, New York, 1976), p. 719.
10. J. Désesquelles, Nucl. Instrum. Meth. Phys. Res. B **31**, 30 (1988); R. D. Deslattes, Phys. Scr. T **46**, 9 (1993).
11. A. E. Livingston and S. J. Hinterlong, Appl. Opt. **20**, 1727 (1981); R. DeSerio, *ibid.* **20**, 1781 (1981).
12. J. A. Demarest and R. L. Watson, Nucl. Instrum. Meth. Phys. Res. B **24/25**, 296 (1981).
13. R. DeSerio, H. G. Berry, R. L. Brooks, J. Hardis, A. E. Livingston, and S. J. Hinterlong, Phys. Rev. A **24**, 1872 (1981); J. D. Silver, A. F. McClelland, J. M. Laming, S. D. Rosner, G. C. Chandler, D. D. Dietrich, and P. O. Egan, *ibid.* **36**, 1515 (1987).
14. A. S. Zacarias, A. E. Livingston, Y. N. Lu, R. F. Ward, H. G. Berry, and R. W. Dunford, Nucl. Instrum. Meth. Phys. Res. B **31**, 41 (1988).
15. J. P. Buchet, M. C. Buchet-Poulizac, A. Denis, J. Désesquelles, M. Druetta, J. P. Grandin, X. Husson, D. Lecler, and H. F. Beyer, Nucl. Instrum. Meth. Phys. Res. B **9**, 645 (1985).
16. G. W. F. Drake, Can. J. Phys. **66**, 586 (1988).

17. D. R. Plante, W. R. Johnson, and J. Sapirstein, Phys. Rev. A **49**, 3519 (1994).

18. E. H. Pinnington, P. Weinberg, W. Verfuss, and H. O. Lutz, Z. Phys. A **281**, 325 (1977).

19. S. W. Provencher, J. Chem. Phys. **64**, 2772 (1976).

20. D. J. G. Irwin and A. E. Livingston, Comput. Phys. Commun. **7**, 95 (1974).

21. P. H. Heckmann, E. Träbert, H. Winter, F. Hannebauer, H. H. Bukow, and H. von Buttlar, Phys. Lett. A **57**, 126 (1976).

22. E. H. Pinnington, W. Ansbacher, J. A. Kernahan, Z.-Q. Ge, and A. S. Inamdar, Nucl. Instrum. Meth. Phys. Res. B **31**, 206 (1988).

23. C. L. Cocke, in *Beam-Foil Spectroscopy*, edited by I. A. Sellin and D. J. Pegg (Plenum Press, New York, 1976), p. 283; L. Engström and P. Bengtsson, Phys. Scr. **43**, 480 (1991).

24. L. J. Curtis, H. G. Berry, and J. Bromander, Phys. Scr. **2**, 216 (1970); Phys. Lett. A **34**, 169 (1971). vspace*1pt

25. E. H. Pinnington, R. N. Gosselin, J. A. O'Neill, J. A. Kernahan, K. E. Donnelly, and R. L. Brooks, Phys. Scr. **20**, 151 (1979). vspace*1pt

26. L. Engström, Nucl. Instrum. Meth. Phys. Res. **202**, 369 (1982). vspace*1pt

27. W. Ansbacher, E. H. Pinnington, J. A. Kernahan, and R. N. Gosselin, Can. J. Phys. **64**, 1365 (1986). vspace*1pt

28. E. H. Pinnington, Nucl. Instrum. Meth. Phys. Res. B **9**, 549 (1985). vspace*1pt

29. L. Engström, Phys. Scr. **40**, 17 (1989). vspace*1pt

30. L. Engström, B. Denne, J. O. Ekberg, K. W. Jones, C. Jupén, U. Litzén, W. T. Meng, A.Trigueiros, and I. Martinson, Phys. Scr. **24**, 551 (1981). vspace*1pt

31. C. Laughlin, E. R. Constantinides, and G. A. Victor, J. Phys. B **11**, 2243 (1978). vspace*1pt

32. H. Nussbaumer and P. J. Storey, Astron. Astrophys. **74**, 244 (1979).

33. T. Åberg and O. Goscinski, Phys. Rev. A **24**, 801 (1981); C. Jupén, B. Denne, J. O. Ekberg, L. Engström, U. Litzén, I. Martinson, W. Tai-Meng, A. Trigueiros, and E. Veje, *ibid.* A **26**, 2468 (1982).

34. H. G. Berry, L. J. Curtis, D. G. Ellis, and R. M. Schectman, Phys. Rev. Lett. **32**, 751 (1974); S. Huldt, L. J. Curtis, B. Denne, L. Engström, K. Ishii, and I. Martinson, Phys. Lett. A **66**, 103 (1978).

35. H. J. Andrä, Phys. Rev. Lett. **25**, 325 (1970); H. G. Berry, J. L. Subtil, E. H. Pinnington, H. J. Andrä, W. Wittmann, and A. Gaupp, Phys. Rev. A **7**, 1609 (1973).

36. R. Geller, App. Phys. Lett. **16**, 401 (1970); C. M. Lyneis and T. A. Antaya, Rev. Sci. Instrum. **61**, 221 (1990).

37. J. M. Laming and J. D. Silver, Phys. Lett. A **123**, 395 (1987).

38. W. A. Hallett, D. D. Dietrich, and J. D. Silver, Phys. Rev. A **47**, 1130 (1993).

39. H. F. Beyer, F. Folkmann, and K.-H. Schartner, Z. Phys. D **1**, 65 (1986).

40. T. J. Scholl, R. Cameron, S. D. Rosner, L. Zhang, R. A. Holt, C. J. Sansonetti, and J. D. Gillasby, Phys. Rev. Lett. **71**, 2188 (1993).

41. Y. Accad, C. L. Pekeris, and B. Schiff, Phys. Rev. A **4**, 516 (1971).

42. T. P. Dineen, N. Berrah-Mansour, H. G. Berry, L. Young, and R. C. Pardo, Phys. Rev. Lett. **66**, 2859 (1991).

43. G. P. Grandin, D. Hennecart, X. Hussin, D. Lecler, I. Lesteven-Vaisse, and D. Lisfi, Europhys. Lett. **6**, 683 (1988).

44. J. C. Levin, R. T. Short, C.-S. O, H. Cederquist, S. B. Elston, J. P. Gibbons, I. A. Sellin, and H. Schmidt-Böcking, Phys. Rev. A **36**, 1649 (1987).

45. S. Martin. A. Denis, Y. Ouerdane, and M. Carré, Phys. Lett. A **165**, 441 (1992).

46. G. Hubricht and E. Träbert, Phys. Scr. **39**, 581 (1989).

47. M. Barat, Nucl. Instrum. Meth. Phys. Res. B **9**, 364 (1985).

48. M. H. Prior, R. A. Holt, D. Schneider, K. L. Randall, and R. Hutton, Phys. Rev. A **48**, 1964 (1993).

49. D. Schutze-Hagenest, H. Harde, W. Brand, and W. Demtröder, Z. Phys. A **282**, 149 (1977); H. Schmoranzer and U. Volz, Phys. Scr. **T47**, 42 (1993).

50. Y. Baudinet-Robinet, P.-D. Dumont, H.-P. Garnir, and A. El. Himdy, Phys. Rev. A **40**, 6321 (1989).

51. E. G. Myers, J. K. Thompson, E. P. Gavathas, N. R. Clausen, J. D. Silver, and D. J. H. Howie, Phys. Rev. Lett. **75**, 3637 (1995).

52. A. Sen, W. J. Childs, and L. S. Goodman, Nucl. Instrum. Meth. Phys. Res. B **31**, 324 (1988).

53. S. Schröder *et al.*, Phys. Rev. Lett. **64**, 2901 (1990); J. S. Hangst, M. Kristensen, J. S. Nielsen, O. Poulsen J. P. Schiffer, and P. Shi, *ibid.* **67**, 1238 (1991).

54. T. Andersen, L. H. Andersen, P. Balling, H. K. Haugen, P. Hvelplund, W. W. Smith, and K. Taulbjerg, Phys. Rev. A **47**, 890 (1993).

55. H. J. Andrä, R. Fröhling, H. J. Plöhn, and J. D. Silver, Phys. Rev. Lett. **37**, 1212 (1976).

56. H. J. Andrä, R. Fröhling, H. J. Plöhn, H. Winter, and W. Wittmann, J. Phys. (Paris) **40**, C1-275 (1979).

19

Line Shapes and Radiation Transfer

Alan Gallagher

*JILA, University of Colorado and National Institute of Standards and Technology, Boulder, Colorado 80309-0440**

19.1 COLLISIONAL LINE SHAPES

The shapes of collisionally broadened atomic lines is a topic almost as old as Fraunhofer's discovery of the existence of discrete lines. Lorentz provided the first quantitative theory in 1906 [1], and Weisskopf advanced this to the *impact theory* by 1933 [2]. Holtsmark [3], Kuhn [4] and Margenau [5] meanwhile developed the *quasistatic* or *statistical* theory which describes the line wing, and Jablonski put this on a quantum mechanical footing in the context of free-free molecular radiation [6]. By the 1940s, satellite bands in the line wings, and a variety of high and low pressure line shapes and broadening rates had been measured. Initial confusion regarding the validity of the contrasting *impact* versus *static* approaches was largely resolved by unified treatments of the Fourier integral theory [7–10]. Baranger then provided a

quantum basis for the impact theory, including level degeneracies [11]. Descriptions can be found in a variety of reviews and references therein, including [2, 5, 12–16]. The broadening of molecular lines involves the additional complication of rotationally nonadiabatic collisions; this was initially addressed by Anderson [10] and later with great thoroughness by van Kranendonk [17]. This chapter and most of the above theories are concerned with neutral atomic gases, which is sometimes called *pressure broadening*. In plasmas, electron, ion, and neutral collisions all contribute to the line shapes and strengths; thus the emitted lines provide a powerful diagnostic of plasma conditions. Neither molecular nor plasma broadening will be covered here; the latter is reviewed in Refs. [14–16, 18], and in Chaps. 57 and 46.

The neutral-gas theories described above generally used phenomenological or long range forms of the atomic and molecular interactions, and most measurements were not sufficiently detailed to test the validity of these pa-

*Staff Member, Quantum Physics Division, National Institute of Standards and Technology rameterizations or the theoretical approximations. A great deal of the work since mid century has been directed towards obtaining more realistic and accurate descriptions of these interactions and of the full line shapes. In addition, many new types of observations have stimulated variations on the basic theories and descriptions. This includes topics such as collision-induced forbidden transitions, satellite shapes, spectral and polarization redistribution, orientation and alignment effects, Doppler-free spectroscopy and very low temperature collisions. Most of these topics are beyond the scope of this brief and basic description of neutral collisional line shapes.

19.1.1 Voigt Line Shape

An atomic (or molecular or ionic) line has an intrinsic Lorentzian shape that reflects the Fourier transform of the exponentially decaying spontaneous emission. For a spontaneous decay rate Γ, the fullwidth at half maximum (FWHM) is $\Delta\omega = \Gamma$. Due to the Maxwellian distribution of atomic velocities v in a thermal vapor, and the Doppler shift $\Delta\omega_D = v/\lambda$, the resonance frequencies of atoms in the laboratory frame have a Gaussian distribution. The Doppler width of this distribution is approximately $\omega_0 u/c$, where ω_0 is the resonant frequency and u is the mean thermal velocity. The full line shape is thus a convolution of the natural Lorentzian with the thermal Gaussian; a Voigt profile. In the presence of collisions, the line from each atom broadens, shifts and becomes asymmetric, and this is normally convoluted with the Gaussian velocity distribution to obtain the complete line shape. Collisions may also cause weak or dipole-forbidden transitions to become stronger as well as broader.

19.1.2 Interaction Potentials

Theories of collisional line shapes consider an ensemble average of collisional interactions. For atomic gases, the description of each individual collision or interaction generally starts with a molecular model for a pair of interacting atoms, since the Born-Oppenheimer approximation is appropriate for thermal atomic collisions (electron velocities $\gg$ nuclear velocities) [19]. The radiative transition then occurs between adiabatic electronic molecular states that separate to the atomic states of the transition under consideration. The notation $V_u(R)$ denotes the electronic energy of the upper state, and $V_g(R)$ the lower state, where R is the internuclear separation, and the total atomic energies are E_1 and E_0 with $E_1 - E_0 = h\omega_0$ (Fig. 19.1). The next simplification is to assume that the atom-pair statistically branch into adiabatic motion along each of the molecular states associated with the initial atomic state, and radiation to each of the final states is summed independently, assuming they are also

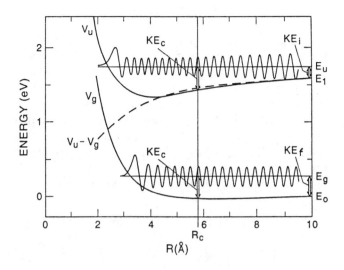

Figure 19.1. Diagrammatic representation of a free-free molecular radiative transition and the Classical Franck-Condon principle. Adiabatic potentials (V_u and V_g), the difference potential equal to the classical transition frequency (dashed), nuclear kinetic energies and wave functions $\Phi_u(R)$ and $\Phi_g(R)$ are indicated. The position R_C is the classical radiation position for the initial (E_u) and final (E_g) energies shown.

completely adiabatic. This ignores an inevitable nonadiabatic coupling between the states as their energy separation decreases to zero at large R, but as discussed below this has been shown to have a very minor effect on line broadening. The single-collision problem then reduces to calculating the spectrum of a molecular transition between each upper and lower pair of adiabatic states, for each initial state of internuclear motion and all possible final motions, and summing these weighted by the rate of initial collisional motions. However, such calculations are only necessary to elucidate particular quantum features, because the *classical oscillator approximation, impact approximation, quasistatic approximation,* and *classical Franck-Condon principle* provide major conceptual and calculational simplifications. These are presented in the context of free-free transitions in the following sections.

19.1.3 Classical Oscillator Approximation

Consider a free-free molecular radiative transition of energy $\hbar\omega = E_u - E_g$ between upper (u) and lower (g) states of total energy E_u and E_g, as shown in Fig. 19.1. Each elastic scattering state is the product of an electronic adiabatic state $\phi(r, R)$ and a nuclear motion state $\Phi(R)$ of the molecule. The electronic states u and g have effective potentials

$$V_q^e = V_q + l(l+1)/R^2 , \qquad (19.1)$$

where $q = u$ or g. Examples of such $V(R)$ and $\Phi(R)$ for the case of $l = 0$ are shown in Fig. 19.1. The electric dipole radiation operator that couples these states is normally a weak perturbation that does not alter these potentials or wave functions. For definiteness assume the atom is initially in the upper state, and that an atom pair approach with kinetic energy T_i and separate with T_f, as indicated. The intensity I, or transition probability, is proportional to the squared matrix element of the dipole operator $e\mathbf{r} = e\Sigma_i \mathbf{r}_i$ between initial and final states:

$$I \propto \left| \int d^3 R \int d^3 \mathbf{r}\, \Phi_u(R)^* \phi_u(\mathbf{r}, R)^* e\mathbf{r}\phi_g(\mathbf{r}, R)\Phi_g(R) \right|^2$$

$$= \mu^2 \left| \int d^3 R\, \Phi_u(R)^* \Phi_g(R) \right|^2 , \qquad (19.2)$$

where the last factor is the Franck-Condon factor, μ is the dipole moment, here assumed to be the atomic value independent of R, and $d^3 R = 4\pi R^2 dR$ for diatomics. The $\Phi(R)$ can be represented, in the WKB approximation, as

$$R\Phi_q = k_q^{-1/2} \exp\left[-i \int_0^\infty dR\, k_q(R) \right] , \qquad (19.3)$$

where

$$k_q = \left\{ 2M[E_q - V_q^e(R)] \right\}^{1/2} /\hbar \qquad (19.4)$$

and M is the reduced mass of the atom pair $[(\hbar^2 k^2/2M)$ is the kinetic energy in Fig. 19.1].

The integrand in the second line of Eq. (19.2) then contains $\exp[i \int d^3 R\, (k_u - k_g)]$. If one multiplies $k_u - k_g$ by $(k_u + k_g)/(k_u + k_g)$, the exponent becomes

$$\frac{2M(\omega - \omega_c)}{\hbar(k_u + k_g)} ,$$

and if one further defines a variable $t(R)$ by $dt = dR/v(R)$ with $v(R) = \hbar(k_u + k_g)/2M$, the exponent becomes $i \int \omega_c(t')dt' - i\omega t$, where

$$\omega_c(R) = [V_u(R) - V_g(R)]/\hbar \qquad (19.5)$$

and $R(t)$ is the classical trajectory in the average potential. Thus, the Franck-Condon factor reduces to the squared Fourier transform of $\exp\left[i \int^t \omega_c(t')\, dt' \right]$, where $\hbar\omega_c(R) = V_u(R) - V_g(R)$ (see Fig. 19.1). This is the *classical oscillator approximation* (COA), in which the atoms are considered to be a classical oscillator at $\omega_c(R)$ during the collisional interaction, and $R(t)$ is the classical orbit.

The $\sqrt{k}$ in the denominator of the WKB wave function leads to $1/v(R)$ in the nuclear density at R, again corresponding to classical motion. This COA is used as the starting point in many line shape calculations. It is apparent that the COA breaks down when the initial- and final-state orbits are significantly different, or when the Born-Oppenheimer or WKB approximation is not valid. However, the COA is normally a very good approximation for thermal atomic vapors when $h(\omega - \omega_0) \ll k_B T$, since most orbits are then straight lines within the R region that yields the observed radiation.

One further conceptual and mathematical simplification is to subtract ω_0 from $\omega_c(t')$ and from ω by defining $\Delta\omega_c = \omega_c - \omega_0$ and $\Delta\omega = \omega - \omega_0$. The Franck-Condon factor in Eq. (19.2) then becomes

$$I(\omega) = \left(\frac{4\mu^2 \omega^4}{3c^3} \right) \left| \int dt \exp\left[i\int^t \Delta\omega_c(t')dt' - i\Delta\omega t \right] \right|^2_{\text{av}} , \qquad (19.6)$$

where the prefactors that yield the correct intensity are explained in [10] (see also Sect. 10.7). The average is, in general, over a sequence of collisions leading to an integral over collision velocity and impact parameter, weighted by the rate of occurrence.

In the next section, the character of the broadened line core is obtained by the traditional method of evaluating Eq. (19.6) using the impact approximation.

19.1.4 Impact Approximation

Atomic collisions in thermal vapors typically occur in a time interval $\tau_c < 1$ ps, whereas the time between significant collisions is much longer at vapor pressures below a few atmospheres. Thus, one can consider the atom as radiating its unperturbed frequency ω_0 most of the time, but occasionally suffering a rapid, strong perturbation. If the duration of a collision is τ_c, then $\phi(t) = \int^t \Delta\omega_c(t')\, dt'$ in the exponent of Eq. (19.6) undergoes a net phase shift $\Delta\theta$ in a time τ_c, and is constant between collisions. If $\Delta\omega \ll 1/\tau_c$, then one can approximate this phase shift as instantaneous; this is the essential assumption of the *impact approximation*. When the factor $\exp i\Delta\theta(b, v)$, which then occurs in Eq. (19.6) is averaged over collision rates with velocity v and impact parameter b this leads to $(\gamma_c - i\delta)t$ in place of $\phi(t)$, where γ_c and δ are given by [12]

$$\gamma_c = \int_0^\infty n(v)v\, dv \int_0^\infty 2\pi b[1 - \cos\Delta\theta(b, v)]\, db ,$$

$$\delta = \int_0^\infty n(v)v\, dv \int_0^\infty 2\pi b \sin\Delta\theta(b, v)\, db .$$

Including the spontaneous emission dipole decay rate $\exp(-\Gamma t/2)$, the Fourier integral produces the normalized Lorentzian line shape

$$I(\omega) = \frac{(\Gamma/2 + \gamma_c)}{\pi\left[(\Gamma/2 + \gamma_c)^2 + (\Delta\omega - \delta)^2 \right]} . \qquad (19.7)$$

The full line shape is then a Voigt profile with a Lorentzian-component half width $\Gamma + 2\gamma_c$ and a shift of

δ from ω_0. The shift can be understood as the perturbed oscillator frequency advancing relative to ω_0 by the average value of $\sin \Delta\theta$, or alternately as indicating the direction of the frequency shift of $V_u - V_g$. The Lorentz approximation corresponds to taking $1 - \cos \Delta\theta = 1$ or 0, with the former representing a collision. This gives $\gamma = 1/\Delta t_c$, where Δt_c is the average time between strong collisions.

The b for which $\Delta\theta(b, \langle v \rangle) = 1$ is called the Weisskopf radius (R_W). The collisional line width $2\gamma_c$ can be thought of as a collisional rate $n\langle v\rangle Q$, where $Q = \pi R_W^2$. The importance of R_W is that γ_c and δ result primarily from atomic interactions in the region $R \approx R_W$. Since the size of an electronic wave function increases with increasing excitation energy, $|\Delta V(R)| = |V(R) - V(\infty)|$ is normally larger for the upper state of a transition. (High Rydberg states can be an exception.) Since $\Delta\theta = 1$ for $b = R_W$ and $\Delta\theta \sim (\Delta V/\hbar)\tau_c$, then $\Delta V(R_W) \sim \hbar/\tau_c$. For a typical $\tau_c = 1$ ps this implies $\Delta V(R_W)/hc \sim 5$ cm^{-1}, which is a relatively weak, long range interaction. Thus, reasonable approximations to the necessary $V(R)$ can often be obtained from atomic perturbation theory. Simple expressions for γ_c and δ are obtained for $V_i - V_f = A/R^n$ with $n > 2$ [20]. The case of $n = 3$ corresponds to the resonant interaction between identical atoms, while $n = 6$ corresponds to the van der Waals interaction, which holds at long range for foreign gas interactions. This result for one pair of adiabatic states must be averaged over all of the pairs that separate to each atomic state.

Higher order approximations obtain an asymmetric line, rather than a pure Lorentzian, due to the asymmetry of $\Delta V(R)$ [20]. In the line core this asymmetry appears as a multiplicative correction $1 + \Delta\omega/D$, where $|D| \sim 1/\tau_c$ [21], and with increasing detuning it increases until at $|\Delta\omega| > 1/\tau_c$ the static wing approaches the quasistatic limit described in Sect. 19.1.7. If $\Delta V(R)$ changes monotonically with decreasing R, only one side of the line has a static contribution and the other, antistatic side falls off exponentially at $|\Delta\omega| > \tau_c$ [9]. However, this situation is seldom observed, as more than one difference potential generally contributes and there is usually a static contribution on both sides of the line. Another factor that produces a small divergence from the Voigt profile is the velocity dependence of the shift and width. When combined with the higher velocities of atoms emitting or absorbing in the Doppler wing, this produces an asymmetry in the Doppler wings [22, 23].

19.1.5 Examples: Line Core

It is possible to deconvolve a Voigt line shape to separate the Doppler and Lorentzian components, and thereby deduce broadenings of considerably less than the Doppler width ([24] and references therein). However,

the broadening is most easily observed at perturber densities where the collisional broadening exceeds the Doppler broadening. Such a pressure-dependent line shape is shown in Fig. 19.2, for a range of perturber density n_P such that the broadening exceeds the Doppler width and hyperfine structure, yet the line core is described by the impact theory [25]. In Fig. 19.2 the normalized line intensity has been divided by n_P; as the line wings are proportional to n_P they are constant in such a plot, while the line center broadens and shifts with increasing n_P. For this case of fairly heavy atoms, $(2\pi c\tau_c)^{-1} = \Delta k_c$ corresponds to ~ 0.5 cm^{-1}, and the line becomes asymmetric and non-Lorentzian beyond ~ 1 cm^{-1} [Fig. 19.3(a)]; the red wing intensity falls more slowly and the blue wing more rapidly than $(\Delta\omega)^{-2}$. This behavior is typical for most heavy perturbers, and is attributed to a long range attractive V_u which dominates $V_u - V_g$.

For the lowest n_P shown in Fig. 19.2, a convolution with the Doppler, hyperfine and instrumental broadenings showed that the line is essentially a symmetric Lorentzian for $|\Delta k| < \Delta k_c$ [25]. However, at the highest density, the half-height point is beginning to fall outside of Δk_c; the impact approximation is marginally valid for describing the half width of the line at this density. Most early experiments were done at more than 10 times this density [12]; most of the line-core was at $|\Delta k| > \Delta k_c$ and describable by the static theory (see Sect. 19.1.7) rather than the impact approximation. The impact approximation was also not valid under these conditions because collisions overlap in time. These very broad lines are well represented by the multiple-perturber, static theories that assume scalarly additive perturber interactions [5, 7]. This transition between an impact and quasistatic line core, and to multiple perturber interactions, occurs at lower pressures for transitions to higher states, as the interactions have a longer range. In addition, nearby intensity peaks or satellites often occur, and strongly affect the line as pressure increases. An example calculation, based on an interpretation of measured spectra [13], is shown in Fig. 19.4. This shows how a line with a satellite feature progressively broadens and finally blends with the satellite as n_P increases.

With the advent of saturated-absorption (Doppler free) spectroscopy, collisional line broadening can be measured at much lower densities, where $2\gamma_c \ll \Delta\omega_D$. In principle, this can allow measurement of line broadenings and shifts, although with a complication that affects the line shape; the same collisions that produce optical phase shifts also change the atomic velocity. These velocity changes have a minor effect outside the Doppler envelope where high pressure measurements are normally made, but they are quite important in saturated absorption line shapes. This affects primarily the low intensity wings of the line, so it does not prevent measuring the broadening and shift of the nearly Lorentzian core. Details can

Figure 19.2. Normalized line shape of the Rb $5P_{3/2} - 5S_{1/2}$ transition broadened by Kr, for Kr densities of 4.5, 9, 18, and 27 × 10^{18}/cm^3 (top to bottom). Hyperfine structure and instrumental resolution cause ~0.3 cm^{-1} of the broadening shown.

be found in [26] and references therein. Two-photon absorption yields Doppler free lines that are not affected by velocity changing collisions. This provides the most exacting test of line shapes. These narrow lines are precisely Lorentzian, with a broadening that reflects the upper state interaction since this is usually much stronger than that of the ground state. The technique has been used to measure the broadening of two photon transitions to many excited states [27, 28].

19.1.6 Δ and γ_c Characteristics

Since 1970, neutral broadening has generally been measured in the $|\Delta\omega| < 1/\tau_c$ region where the impact approximation and Eq. (19.7) is valid. Measurements through 1982 are tabulated in Ref. [13], and through 1992 in Ref. [32]. These involve primarily metal vapor resonance lines broadened by noble gases. For collisions with the heavier (more polarizable) gases, the sizes of these measured broadening rate coefficients generally fall within a factor of 10 range, and approximately fit the prediction of Eq. (19.7) with $\Delta V_u(R) - \Delta V_g(R) = C_6 R^{-6}$ [33] and C_6 given by a simple effective quantum number formula. This occurs because the potentials are fairly close to van der Waals for $R \geq R_W$ and the broadening is insensitive to details of the potentials at $R < R_W$ since $\cos\Delta\theta$ in Eq. (19.7) averages to ~0 for the closer (strong) collisions. It also occurs because the full quantum solution for broadening by a van der Waals interaction, with Zeeman degeneracies, yields nearly the same result as the above single-level theory with an average C_6 [34].

For the heavy, more polarizable perturbers, the ex-

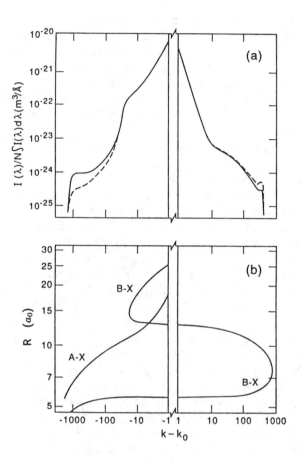

Figure 19.3. Normalized intensity in the wings of the Rb $5P_{3/2} - 5S_{1/2}$ transition broadened by Kr, in frequency units of $k = 1/\lambda$. The measured spectrum in (a) is from [29, 30, 25]. The solid line is at 310 K and the dashed line at 540 K. The difference potentials corresponding to the A,B and X states of Rb-Kr, taken from [31], are shown in (b).

cited state interactions are attractive and red shifts occur, but the measured shifts have a very poor correlation with the van der Waals prediction. As b decreases and $\Delta\theta(b, v)$ increases, $\sin\Delta\theta$ oscillates and major cancellations occur in the average of $\sin\Delta\theta$ in Eq. (19.7). The shift is therefore only a fraction of the broadening and is very sensitive to the interaction throughout the region $R \approx R_W$. This often differs considerably from the van der Waals form, at the typical ~5 cm^{-1} interaction energy at R_W. The shape of the red wing just beyond $\Delta\omega_c$ also frequently fails to fit that expected for a van der Waals interaction [25]. Thus, the often good agreement of γ_c with van der Waals numbers is not a reliable indicator of the actual $V(R)$ in the relevant R region, even for heavy noble gas perturbers.

For He and sometimes Ne perturbers, a repulsive interaction due to charge overlap normally dominates at $R \sim R_W$, causing a blue shift as well as a larger

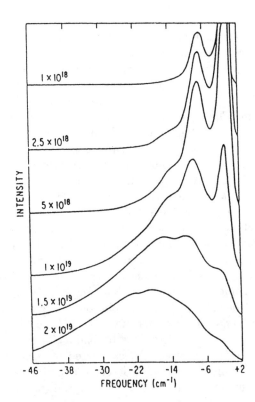

Figure 19.4. Calculated line shapes of the $Cs(9P_{1/2} - 6S_{1/2})$ line broadened by Xe at the densities indicated (from [13]). The assumed interaction is based on measured line shapes, but data corresponding to the calculated conditions are not available.

broadening than the van der Waals prediction.

19.1.7 Quasistatic Approximation

The impact approximation is valid for $|\Delta\omega| \ll 1/\tau_c$, where the $1/\tau_c$ is typically 1–10 cm^{-1}. For larger $|\Delta\omega|$ the line shape becomes asymmetric, with higher intensity on the wing corresponding to the long-range $V_u(R) - V_g(R)$. At large detunings where $\Delta\omega \gg 1/\tau_c$ a major simplification occurs. The COA describes the interacting atom pair as an oscillator of frequency $\omega_c(R) = [V_u(R) - V_g(R)]/\hbar$ when at separation R. Since R is time dependent during the classical orbit, ω_c is as well and the Fourier spectrum is broadened relative to the simple distribution of $\omega_c(t)$ that occurs during the orbit. But if the motion is sufficiently slow, the intensity at ω reduces to the probability of finding the atom pair at the appropriate $R(\omega = \omega_c)$. The spectrum then reduces, at low pressure, to the probability distribution of pair separations R, subject to Eq. (19.5) between R and ω. This is the binary quasistatic, static, or statistical spectrum, which accurately describes most line wings for $|\Delta\omega| > 1/\tau_c$. When the pressure is large enough to yield

a significant probability of one perturber at $R \leq R_c$, multiple-perturber interactions must also be considered as in [5].

This intuitive deduction of the statistical spectrum from the COA [4] can also be obtained more formally from Eq. (19.6) by expanding the exponent about the time during a collision when $\omega_c(t) = \omega$. Alternatively, it follows directly from Eq. (19.2) using WKB wave functions to evaluate free-free molecular Franck-Condon factors [6]. This result is identical to the classical Franck-Condon principle (CFCP), originally established in the context of bound-bound molecular radiation.

The CFCP yields important insights for all molecular radiation. Again consider Eq. (19.2) with the substitution of the WKB wave functions ϕ_q, given below it. Examples of ϕ_u and ϕ_g are given in Fig. 19.1. For large detunings $\omega - \omega_0$, as shown in Fig. 19.1, the integrand on the right side oscillates rapidly everywhere except at the stationary phase point R_c, where $k_u = k_g$. As a consequence, the dominant contribution to the integral occurs at R_c and one can consider the transition to be localized at R_c. Since $k_u(R_c) = k_g(R_c)$, $T_u(R_c) = T_g(R_c) = T_c$ also holds, and as can be seen in Fig. 19.1 it then follows that $\hbar\omega = V_u(R_c) - V_g(R_c)$. Thus, radiation at frequency ω "occurs" when the atoms are near R_c, where the electronic state energies differ by $\hbar\omega$. Note that this holds for all initial kinetic energies and angular momenta, as long as the conditions for validity of the Born-Oppenheimer and WKB approximations hold for the initial and final nuclear motions. This is the CFCP, which is equivalent to the classical-oscillator model for radiation at large detunings from the atomic transition.

Another insight evident from Fig. 19.1 is that the photon energy associated with the frequency difference $\omega - \omega_c$ is supplied by nuclear kinetic energy $\hbar(\omega - \omega_0) = T_i - T_f$. This transformation of nuclear into electronic energy takes place as the nuclei move from large R to R_c on one $V(R)$ and back to large R on the other.

If an absorbing or emitting atom interacts as $V_i(R)$ with a density n_P of perturbers in a vapor of temperature T, the probability of a perturber at separation $R \rightarrow R + dR$ is $n_P 4\pi R^2 \exp[-V_i(R)/k_BT]\,dR$ if the interatomic motion is in equilibrium. Inverting Eq. (19.5) for $R(\omega)$ yields $dR = d\omega/(d\omega/dR)$, and this pair of relations yields the (single perturber) quasistatic (QS) spectrum

$$I(\omega) = N n_P \Gamma 4\pi R(\omega)^2 \exp[-V_i\{\omega\}/k_BT]/[d\omega(R)/dR],$$

$$(19.8)$$

where N is the radiator density and $I(\omega)$ the radiation per unit volume and frequency interval.

Figure 19.3(a) gives an example of far wing emission line shapes versus photon energy in units of cm^{-1}, for the Kr broadened Rb D2 line for which $\Delta k_c \approx 0.7$ cm^{-1}. These data are normalized by dividing by perturber density, so they are independent of perturber density for the density region of the experiment. The excited state produces two $V_u(R)$, called the A and B states, while the ground state produces one $V_g(R)$, called the X state. Each of these potentials has a single minimum at long range and is strongly repulsive at close range, but the well depths and positions are very different [31]. This causes the complex forms of $V_u(R) - V_g(R)$ that are shown in Fig. 19.3(b). There I have plotted $\ln(R)$ vertically and $\ln(\Delta V(R)/hc)$ horizontally, where ΔV refers to $V_A - V_X$ and $V_B - V_X$. The right side of Eq. (19.8) can also be written as the exponential and constant factors times $dR^3(\omega)/d\omega$. Since $\ln[R^3(\omega)] \propto \ln[R(\omega)]$, the static spectrum at $\Delta k = \Delta\omega/2\pi c$ is proportional to the slopes of the curves in Fig. 19.3(b), divided by $|\Delta k|$ due to the $\ln(\Delta k)$ horizontal axis. One can see qualitatively that the overall spectrum follows such a relation to the lines in Fig. 19.3(b); in fact in most spectral regions this relation is quantitatively accurate.

The temperature dependence in Fig. 19.3(a) corresponds to the exponential factors in Eq. (19.8) [29]. At large R, both $\Delta V(R)$ are attractive, and this causes a large intensity on the negative Δk (red) wing. However, once $|\Delta k|$ exceeds $\sim$20 cm^{-1}, where $V_B - V_X$ reverses direction, the red wing intensity drops rapidly. This extremum in $\Delta V(R)$ causes a satellite at $\sim$−20 cm^{-1}, although it is spread out by the finite collision speed and does not cause a distinct peak in the spectrum. Satellite features are discussed in more detail in the next paragraph. The antistatic blue wing drops rapidly for several decades, then suddenly flattens beyond $\sim$10 cm^{-1} due to the positive portion of $V_B - V_X$ at small R. The remaining blue wing is the B–X band, and has a satellite at $\sim$350 cm^{-1} as $V_B - V_X$ passes through another extremum. The theory predicts this at 800 cm^{-1}, but clearly represents all the basic aspects correctly. This satellite is also spread out by finite collision speed, but a definite intensity peak remains. The red wing beyond $\sim$50 cm^{-1} is the A–X band. The feature near −1000 cm^{-1} is due to the exponential factor in Eq. (19.8), not an extremum in $\Delta V(R)$; the feature essentially disappears if the normalized intensity is extrapolated to infinite temperature.

19.1.8   Satellites

In regions of the wing where the intensity falls slowly with increasing frequency, motional broadening of the static spectrum is not noticeable and the static spectrum is a good approximation. However, if $\Delta V(R)$, or equivalently $\omega_c(R)$, has an extremum at some R_S, the denominator of Eq. (19.8) is zero at $\omega(R_S) = \omega_S$. This produces a local maximum, or satellite, in the far wing intensity, as seen in Fig. 19.3(a) at 350 cm^{-1}. If one expands $\omega(R)$ in a Taylor series about $R = R_S$ this produces in Eq. (19.8) a square root

divergence of finite area, with no intensity beyond ω_S. The area under this feature is meaningful, but not its shape; the quasistatic assumption is clearly not valid for such sharp features. The more accurate satellite shape is obtained by returning to Eq. (19.2) and expanding $V_u(R)$, $V_g(R)$ and the WKB wave functions about R_S, or using Eq. (19.6) with $\omega_c(t)$ expanded about $t(R_S)$. Sando and Wormhoudt used the former method to obtain a universal satellite shape [35]. Szudy and Baylis improved the expansion to yield a smooth transition to the quasistatic spectrum at smaller detunings [33]. This result is nearly the same as Sando *et al.* in the spectral region of the satellite, but it more accurately connects to the adjacent static line wing. Intensity undulations between the satellite and the line occur in this calculation; these arise from alternating constructive and destructive interference between two contributions to the same frequency from $R > R_S$ and $R < R_S$. This can not be seen in the low resolution of Fig. 19.3(a), but such undulations are seen near the 350 cm^{-1} satellite [30]. At antistatic detunings beyond ω_S, which are not quasistatically allowed, the calculated intensities decay exponentially. This is also observed experimentally [30] and is the same behavior predicted for the antistatic wing of a line [9].

At higher perturber densities and closer to the line core, corresponding to larger R interactions, the multiple perturber probability distribution must be included. If the interactions are additive, this leads to a secondary satellite at twice the detuning of the low pressure satellite, as seen in Fig. 19.4.

The wings of a collisionally broadened atomic line are molecular radiation. In the context of molecular bound state spectroscopy, a satellite is a "head of band heads," corresponding to a frequency region where bound-bound band heads congregate. This occurs, of course, at the classical satellite frequency and when $V_u(R) - V_g(R)$ has an extremum [36].

An extremum in $\Delta V(R)$ is the most common cause of satellites, but similar looking features can occur for other reasons. Forbidden bands often appear in the wings of forbidden atomic transitions, due to an increase in the transition dipole moment $\mu(R)$ resulting from the collisional interaction. These are described, in the QS approximation, by Eq. (19.8) with Γ replaced by $\Gamma(R)$. If $\Gamma(R)$ increases rapidly with decreasing R, the intensity increases as ω moves into the far wing until the dR^3 and exponential factors cause a net decrease at small R. This leads to forbidden bands far from the atomic frequency, such as those in Ref. [37]. In some cases, a collision-induced feature also appears at the frequency of the forbidden transition. The shapes of such features, which also include radiative collisions, in which both atoms change state, are calculated and reviewed in Ref. [38].

A variety of related line shape phenomena has been

investigated, including the relation between absorbed and emitted wavelengths (spectral redistribution), the dependence of fluorescence polarization on absorbed wavelength (polarization redistribution), and high power effects. Some references regarding these phenomena are [39–43].

19.1.9 Bound States and Other Quantum Effects

The validity of the QS spectrum requires the validity of the WKB approximation in the initial and final state, but it is not restricted to free-free molecular transitions. In fact, the equilibrium probability distribution in Eq. (19.8) must include bound states in an attractive $V_i(R)$. The QS spectrum describes the *average* behavior of bound-free and bound-bound molecular bands, as well as the free-free radiation implied by the above method of derivation. The quantum character is expressed in the discrete bound-bound lines that make up this average, and in *Condon oscillations*, where the intensity oscillates about the average $I_{QS}(\omega)$. The latter occur as oscillations in Franck-Condon factors in the bound-bound case, and as smooth oscillations in bound-free spectra and low resolution bound-bound spectra. An additional quantum feature occurs in regions of the spectrum dominated by classical turning points, usually at the far edge of a line wing where the intensity is dropping rapidly. There, quantum tunneling past the edge of the classically allowed region spreads the spectrum. Yet another is the energy $h\omega_0/2$ of the ground vibrational state, which effectively adds to $k_B T$ in Eq. (19.8) for attractive V_i. All of these quantum features become more pronounced as the reduced mass decreases; examples and details can be found in Refs. [36, 44–46].

19.1.10 Einstein A and B Coefficients

The relationship between absorption coefficient $B_{12}(\omega)$, stimulated emission coefficient $B_{21}(\omega)$ and spontaneous emission coefficient $A_{21}(\omega)$ are given by the Einstein relations; $A_{21}/B_{21} = 8\pi h\lambda^{-3}$ and $B_{21}/B_{12} = g_1/g_2$. These relations are most familiar for atomic lines, but if they are referred to the density of absorbers $dN_g/d\omega$ and emitters $dN_u/d\omega$ that emit or absorb at ω, then they also apply to the wings of lines, i.e.,

$$k(\omega) = B_{12}(\omega)\left(\frac{2\pi h}{\lambda}\right)\left(\frac{g_u}{g_g}\right)\frac{dN_g}{d\omega}$$

$$= \tfrac{1}{4}\lambda^2 A_{21}(\omega)\left(\frac{g_u}{g_g}\right)\frac{dN_g}{d\omega}, \qquad (19.9)$$

$$g(\omega) = \tfrac{1}{4}\lambda^2 A_{21}(\omega)\frac{dN_u}{d\omega}, \qquad (19.10)$$

$$I(\omega) = \hbar\omega A_{21}(\omega)\frac{dN_u}{d\omega}. \qquad (19.11)$$

Here $k(\omega)$ is the absorption coefficient due to lower state atoms, $g(\omega)$ is the stimulated emission coefficient and $I(\omega)$ the spontaneous emission due to excited state population, and g_u and g_g are the statistical weights of the atomic states. For absorbing atoms of density N_g and perturber density n_P, the QS approximation with equilibrated internuclear motion sets

$$\frac{dN_g}{d\omega} = N_g n_P 4\pi R^2 \frac{dR}{d\omega}\exp\left(-\frac{V_g}{k_B T}\right), \qquad (19.12)$$

and equivalently for a radiating atom density N_u with perturber interaction V_u. Normally most of the radiation, and $dN_g/d\omega$, is concentrated at the atomic line, so integrating over $d\omega$ near the line leads to the relations

$$\int_0^\infty k(\omega)d\omega = \frac{1}{2}\lambda^2 A_{21}N_g\left(\frac{g_u}{g_g}\right), \qquad (19.13)$$

$$\int_0^\infty I(\omega)\,d\omega = A_{21}N_u\left(\frac{h\omega}{2\pi}\right), \text{ etc.} \qquad (19.14)$$

Note that

$$\frac{g(\omega)/N_u}{k(\omega)/N_g} \propto \exp\left(-\frac{\hbar\omega}{k_B T}\right); \qquad (19.15)$$

if N_u/N_g is also in equilibrium at T, this yields the correct equilibrium relation between $k(\omega), g(\omega), I(\omega)$, and a black body spectrum. While these relations are much more general than the QS theory, the latter provides a helpful conceptual basis. The above expressions in terms of spontaneous emission thus cover all cases.

19.2 RADIATION TRAPPING

Atoms and ions efficiently absorb their own resonance radiation, and their emission can be reabsorbed before escaping a vapor. Molecules are less efficient absorbers, since each electronic transition branches into multiple-line bands, but interesting effects result if such reabsorption occurs. This emission and reabsorption process is fundamental to the formation of stellar lines, where it is called *radiation transfer*, and to confined vapors and plasmas where it is also called *radiation diffusion* or *trapping*. Fraunhofer's observation of dark lines in the stellar spectrum result from this radiation transfer process. Highly sophisticated treatments of line formation in inhomogeneous and nonequilibrium plasmas containing many species [15, 16] also apply to laboratory plasmas, but the simplifications inherent in a one- or two-element, confined plasma with cylindrical or planar symmetry leads to easier treatments. This sections discusses only a uniform density and temperature, confined atomic vapor.

The fluorescent lamp in which 254 nm mercury radiation diffuses to the walls and excites a phosphor, provides a prime example of radiation trapping. Its improvement motivated the seminal Biberman [47] and Holstein [48] papers, on which the discussion here is based. The effect of radiation trapping on the *polarization* of flourescent radiation played a major role in developing a correct understanding of the coherent response of atoms to radiation. This is reviewed in Ref. [40], and will not be covered here.

19.2.1   Holstein-Biberman Theory

An atom in a dense vapor may be excited by externally applied radiation plus the fluorescence from other excited atoms within the vapor, and it will decay by spontaneous emission (neglecting stimulated emission). This is expressed by the Holstein-Biberman equation

$$dn(\mathbf{r}, t)/dt = S(\mathbf{r}, t) + \gamma \int_{\text{vol}} K(\mathbf{r} - \mathbf{r}')n(\mathbf{r}', t)\, d^3\mathbf{r}'$$
$$- \gamma n(\mathbf{r}, t)\,, \qquad (19.16)$$

where $n(\mathbf{r}, t)$ is the excited state density at position $\mathbf{r}$, $S(\mathbf{r}, t)$ is the excitation rate due to externally applied radiation, γ is the spontaneous emission rate, the kernel $K(\mathbf{r} - \mathbf{r}')$ is the probability of a reabsorption at $\mathbf{r}$ due to fluorescence by an atom at $\mathbf{r}'$, and the integral is over the vapor filled volume [47, 48]. Since $K(\mathbf{r}, \mathbf{r}')$ is assumed the same for all excited atoms, this contains an implicit assumption that all atoms emit the same fully redistributed spectrum. The solution of this linear integral equation, subject to boundary values at the vapor boundary, can be expressed as a sum over an orthogonal set of solutions $n(\mathbf{r}, t)_i = n(\mathbf{r})_i \exp(-g_i \gamma t)$ of the homogeneous equation

$$n(\mathbf{r}, t) = \sum_{i=1}^{\infty} a(t)_i n(\mathbf{r})_i\,, \qquad (19.17)$$

where, if $S(\mathbf{r}, t) = S(\mathbf{r})f(t)$, then $a(t)_i = \bar{a}_i \int_{-\infty}^{t} f(t') \times \exp(-g_i \gamma(t - t'))dt'$ and $\bar{a}_i = \int S(\mathbf{r})n(\mathbf{r})_i d^3\mathbf{r}$. Here $n(\mathbf{r}, t)_i$ is the ith normal mode and $g_i\gamma$ is the decay rate of this mode, as it would decay without change in its shape $n(\mathbf{r})_i$ from a pulse of excitation.

Two shapes of vapor regions have been studied in detail: an infinitely long cylinder of radius R and the region between two infinite parallel plates with separation L. The first three symmetric modes of the latter slab geometry are shown with unit height in Fig. 19.5. A spatial integration over the normalized $i = 1$ or fundamental mode yields 1 and all others integrate to zero, so $a(t)_1$ equals the total excited state population. g_1 is the escape probability; i.e., the probability of photon escape averaged over the fundamental mode distribution of emitters $n(\mathbf{r})_1$. Since $n(\mathbf{r}, t)$ must be everywhere positive, the negative contributions of the higher order modes only reduce the density in some regions. The g_i can vary from 0 to 1 and increase with increasing i,

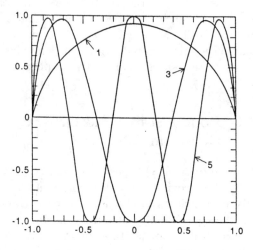

Figure 19.5. The first three symmetric eigenfunctions ($j = 0.2, 4$) of radiation trapping between slab windows, for a Doppler line profile, from Ref. [49]. The windows are at ± 1.

so that higher order modes die out faster after pulsed excitation. The ratios of decay rates is $g_i : g_3 : g_5 = 1 : 3.7 : 6.4$ for the symmetric slab modes shown in Fig. 19.5. For steady state excitation, Eq. (19.17) yields $a(t)_i = \bar{a}_i/g_i\gamma$, so the lower order modes are more heavily weighted because they decay more slowly. The fundamental mode decay rate $g_1\gamma$ is of primary interest in most situations, and we will now discuss its properties.

The kernel $K(x)$ is the probability of fluorescence transport over a distance x followed by reabsorption, averaged over the emitted frequency distribution. It is conceptually useful to express it in terms of the spectrally averaged transmission $T(x)$

$$K(x) = \frac{1}{4\pi x^2}\frac{dT(x)}{dx}\,, \qquad (19.18)$$

with

$$T(x) = \int_{0}^{\infty} \mathcal{L}(\omega)\exp(-k(\omega)x)\,d\omega\,,$$

where $\mathcal{L}(\omega)$ is the emission line shape normalized to unit area, and $x = |\mathbf{r} - \mathbf{r}'|$. If one assumes that the fluorescence frequency of an atom does not depend on the frequency it absorbed (i.e., complete spectral redistribution), this leads to $k(\omega) = \kappa \mathcal{L}(\omega)$, where $\kappa = (\lambda^2/8\pi)\,(g_u/g_g)n\Gamma$ and g_u and g_g are statistical weights. This simplification applies under most conditions and will be used here; its range of validity and more accurate treatments are discussed below.

The transmission factor $\mathcal{L}(\omega)$ and the integrand of Eq. (19.18) are shown in Fig. 19.6, for a Gaussian line shape and several values of k_0x, where k_0 is the line

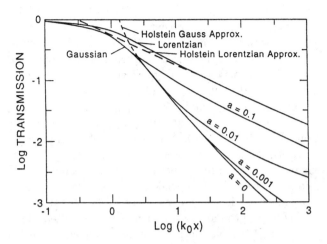

Figure 19.6. Gaussian emission spectrum $\mathcal{L}(\omega)$ (short dash line), transmissions $T(\omega)$ (long dash lines), and transmitted intensities (solid lines) for $k_0 x = 2$, 10, and 50.

Figure 19.7. Transmission $T(x)$ versus distance in units of $k_0 x$, for Voigt line shapes with the a parameters indicated, where $a = (\ln 2)^{1/2} \Delta\omega_{\text{Lor}} / \Delta\omega_{\text{Gauss}}$. The Gaussian limit corresponds to $a = 0$ and the Lorentzian limit to $a = 1$. The Holstein, large $k_0 x$, approximations are also indicated.

center absorption coefficient. At small $k_0 x$, the transmitted spectrum is similar to $\mathcal{L}(\omega)$; for these conditions $T(x) \simeq \exp(-k_{\text{av}} x)$, where $k_{\text{av}} \simeq 0.7 k_0$ is the average attenuation. For $k_0 x > 5$, the transmission is small except at the edges of the line. The transmitted radiation is then predominately in the ω region near ω_1, defined by $k_0 x \mathcal{L}(\omega_1) = 1$. Since the integrand is sharply peaked near ω_1, this leads to simple analytic forms for $T(x)$. In the large $k_0 x$ limit, $T(x) \simeq [k_0 x (\pi \ln k_0 x)^{1/2}]^{-1}$ in the Gaussian case, and $T(x) \simeq (\pi k_0 x)^{-1/2}$ for a Lorentzian line shape. These asymptotic forms of $T(x)$ are compared with the exact $T(x)$ in Fig. 19.7; $T(x)$ follows the asymptotic formulas for $k_0 x > 5$ and 10, respectively. $T(x)$ for several Voigt line shapes is also shown in Fig. 19.7; these follow the Gaussian $T(x)$ at smaller $k_0 x$, then rise above as ω_1 moves into the Lorentzian wing.

For radiative escape from a cell, transmission over distances near the cell dimension (R or L) is most important, since transport over this distance often escapes the vapor and transport over much smaller distances does not have much effect. The escape probability g_1, averaged over the fundamental mode distribution, is close to $T(R)$ or $T(L/2)$, while the higher order modes are related to the same asymptotic forms of $T(x)$ at smaller distances. Thus, in the large $k_0 L$ slab case,

$$g_i = \begin{cases} \dfrac{G_i}{k_0 L (\frac{1}{2} \ln k_0 L)^{1/2}} & \text{Gaussian line} \quad (19.19a) \\[3mm] \dfrac{G_1'}{(k_0 L)^{1/2}} & \text{Lorentzian line,} \quad (19.19b) \end{cases}$$

with $G_1 = 1.03$ and $G_1' \simeq 0.65$. For an infinite cylinder, the same equations hold with $L/2 \to R$ and slightly larger G_i values. Exact G_i and G_1' values can be found in [49].

19.2.2 Additional Factors

As noted above, the line shape of a two-level atom in a thermal vapor is a Voigt shape; a convolution of a Lorentzian of width $\Gamma + 2\gamma_c$ with a Gaussian of width $\Delta\omega_D$. In most cases, $\Delta\omega_D \gg \Gamma$, so in the absence of a buffer gas the line shape is nearly Gaussian at low density (n). As a result, ω_1 is in the Gaussian region of the line at low density and g_1 behaves similarly to the Gaussian transmission in Fig. 19.7 with x replaced by the confinement dimension. k_0 is proportional to n, so from Eq. (19.19a) g_1 is approximately inversely proportional to n for $k_0 L > 5$. As n increases, ω_1 moves further into the wing of the line, and when ω_1 reaches the Lorentzian tail of the Voigt line profile a transition to Eq. (19.19b) occurs, where k_0 corresponds to a purely Lorentzian line. (That the core of the line does not have a Lorentzian shape does not matter, since the fraction of emission well into the Lorentzian wing is nearly the same as that of a pure Lorentzian line.)

In the absence of a collision, a two level atom reradiates in its rest frame the same frequency it absorbed. Thermal motion redistributes this coherent scattering frequency within the Doppler envelope when the emission and absorption are in different directions, but it does not transfer it into the natural Lorentzian wing outside the Doppler envelope. This leads to the property that an atomic vapor will scatter frequencies in the natural wing, but will not emit in this wing unless it absorbed there or is excited by or during a collision. With line broadening collisions, a fraction $\Gamma / (\Gamma + 2\gamma_c)$ of optical attenuation is coherently scattered and a fraction $2\gamma_c / (\Gamma + 2\gamma_c)$ is redistributed into "incoherent" emission with a Lorentzian spectrum of width $\Gamma + 2\gamma_c$ centered at $\omega_0 + \delta$ in the reference frame of the moving atom. This redistributed emission can escape in the Lorentzian wing of the Voigt line. In this radiation transport problem, the consequence is that Eq. (19.19b) with $k_0 = n(\lambda^2/2\pi)(\Gamma/\gamma_c)(g_u/g_g)$, corresponding to a Lorentzian with $\Gamma_{\text{Total}} = 2\gamma_c$ not $\Gamma + 2\gamma_c$, provides the best approximation to g_1 in the density region where ω_1 is in the Lorentzian wing of the line. Since $k_0 \propto n/\gamma_c$ and in the absence of a buffer gas $\gamma_c = k_c n$, where k_c is the rate coefficient for self broadening collisions, g_1 becomes independent of n. In fact, $k_c \propto \Gamma$ as well, so g_1 is also independent of Γ. For the case of a $J = 0$ ground state and a $J = 1$ excited state, $g_1 = 0.21(\lambda/L)^{1/2}$; the broadening coefficient for other cases can be found in [50]. If the broadening is due to a buffer gas, $\gamma_c = k_c n_B$ in Eq. (19.19b) yields

$$g_1 \propto \left(\frac{n_B}{n}\right)^{1/2} ; \qquad (19.20)$$

this has been studied in [51].

Post *et al.* have numerically evaluated g_1 for all values of $k_0 L$ for slab and cylinder geometries, by integrating the radiative escape probability $g(z)$ over the fundamental mode distribution $N(z)$, where z is the position between the windows [52]. To obtain $g(z)$ they integrate over the angular distribution of the emission, using $T(x)$ from the exact line shape. Thus all features of the calculation correspond to the Holstein-Biberman theory for an isolated line without approximation. As will now be discussed real atomic vapors are generally not that simple.

Many atomic "lines" have multiple components due to hyperfine structure and isotope shifts; some components are isolated while others are separated by less than a Doppler line width and overlap. The absorption line shape then becomes a weighted sum over components, each with an equivalent Voigt shape. In a high density vapor or a plasma, collisions will usually distribute the excited state population between the isotopes and hyperfine states in proportion to their isotopic fraction and statistical weight. The emission line shape $\mathcal{L}(\omega)$ is then a similarly weighted distribution over components. Since radiation only escapes in the wings of a line component at high $k_0 L$, overlapping components act almost as a single component. If the line has M isolated components, the right-hand side of Eqs. (19.19a) and (19.19b) become sums over the fraction f_j of the intensity in the j component times the escape probability for that component. The latter is obtained, for large $k_0 L$, by replacing k_0 with $k_0 f_j$ in Eqs. (19.19a) and (19.19b). The net result, after summing over components, is an increase in g_i by a factor of $\sim M$ in the Gaussian case and $\sim M^{1/2}$ in the Lorentzian. This approximation was obtained by Holstein in the context of the Hg 254 nm radiation under conditions appropriate to the fluorescent

lamp [53]. Walsh made a more detailed study of these overlapping components [54], and the dependence of g_1 on the ratio of line separation to Doppler width is also given in [55].

19.2.3 Measurements

The overall behavior of g_1 versus n is shown in Fig. 19.8 for the Na($3P_{3/2}$) or D2 resonance line in pure Na vapor [55, 56]. In this type of experiment the fundamental mode decay rate is established by a combination of optimally exciting that spatial mode and of waiting until the fluorescence decay is exponential in time after termination of the excitation. A transition to approximately $1/n$ behavior, corresponding to Eq. (19.19a), is seen to occur at $k_0 L/2 \sim 5$. At $k_0 L/2 \sim 100$ the transition to n^0 behavior, corresponding to a self-broadened Lorentzian line in Eq. (19.19b), can be seen. The behavior at $k_0 L < 5$ fits the Milne diffusion theory [57] as well as the Post *et al.* theory shown as a solid line; this is also similar to $T(L/2)$, as seen in Fig. 19.7. For $5 < k_0 L/2 < 100$, the behavior is similar to Eq. (19.19a) (dashed line), but the Post *et al.* theory (solid line) is $\sim 20\%$ higher due to the inclusion of the Na hyperfine structure (hfs splitting $\simeq$ Doppler width). For $k_0 L/2 > 1000$, the Post theory converges to the Holstein Lorentzian-line result with $\Gamma_{\text{Total}} = 2\gamma_c$.

The experiment is complicated in the $50 < k_0 L/2 < 500$ region by fine structure mixing [56]. The $3P_{3/2}$ state was excited, but at high densities, collisions populate the $3P_{1/2}$ state, which has a smaller g_1 than the $3P_{3/2}$ state (Fig. 19.8). At low densities, $g_1^{\text{eff}} = g_1(3P_{3/2})$, and at high densities these states are statistically populated and $g_1^{\text{eff}} = \frac{1}{3}g_1(3P_{1/2}) + \frac{2}{3}g_1(3P_{3/2})$. The transition density where the fine structure mixing rate R equals Γ_{eff} is indicated in Fig. 19.8. The theory is also complicated in this intermediate $k_0 L$ region by the necessity of including incomplete frequency redistribution [52]; this leads to the dip in g_1 near $k_0 L \sim 500$. While the overall behavior of the data in Fig. 19.8 is consistent with the Post *et al.* theory, there is $\sim 30\%$ systematic discrepancy at $k_0 L/2 = 10$–100 and the dip near 500 is not seen. Part of this difference probably results from the experimental geometry, which was between a slab and a cylinder of radius $R = L/2$; g_1 for the cylinder is 17% larger than the slab value used in Fig. 19.8.

The fundamental mode decay rate has also been measured for the Hg 254 nm [59] and 185 nm [52] lines, for the Ne resonance line [60] and for the Ar resonance line [61]. The Hg measurements are complicated by multiple isotopes and hyperfine structure, producing a mixture of partially overlapping and isolated lines combined with density-dependent uncertainties in excited state populations of the various isotopes. Serious efforts to model and measure these effects have been made

Figure 19.8. Radiative escape probability g_1 for Na vapor excited to the $3P_{3/2}$ state, for a slab geometry. The Holstein approximation for the $3P_{3/2} - 3S_{1/2}$ (D2) line and the $3P_{1/2} - 3S_{1/2}$ (D1) line are indicated as dashed lines. The Post-type calculation of [58] for the D2 line is indicated as a solid line. Solid squares are data from [56], and open circles are data from [55]. The effective escape probability corresponds to the D2 line rate at low densities but a combination of D1 and D2 at high densities.

[52, 54, 59, 62]. The Ne and Ar measurements have similar complications, as will now be discussed.

In essence, g_1 behaves like the Gaussian $T(x = L/2)$ in Fig. 19.7 until n is large enough for ω_1 to approach the collision induced Lorentzian wing of the Voigt line. g_1 then decreases more slowly since the line wing does not fall off as rapidly as a Gaussian. With continued increase in n, ω_1 moves further into the Lorentzian wing, a broader spectral region escapes and g_1 reaches a minimum. Finally, when the entire escaping spectral region is Lorentzian, g_1 reaches the constant value described above. Independent and detailed treatments of this density region, including incomplete frequency redistribution, predict a dip in g_1 as seen in Fig. 19.8 [52, 61, 63, 64]. However, this has not been clearly confirmed experimentally. In Fig. 19.8 this dip occurs where fine structure mixing also occurs, and in addition the data are higher than the calculations throughout this n region. Post *et al.* [52] did observe such a dip for the Hg (149 nm) resonance line, but the data do not fit the calculation at other densities; hyperfine and isotopic structure within the line cause major complications. Payne *et al.* [61] did not observe the predicted dip for the Ar resonance line; again a minor isotope with an isolated line occurs and could be very important at these high optical depths. Phelps [60] reported such a dip for the Ne 74.3 nm resonance line, but with rather large uncertainties due to the necessity of correcting for other collisional effects. Again there are isotopes with isolated lines that may have effected the data. Thus, experiments have verified the essential aspects of the above theories, but quantitative agreement in all aspects has not yet been achieved.

The fact that the escaping radiation is concentrated in the wings of the line, near the unity optical depth point ω_1, is reflected in the emitted spectrum. Calculated examples are shown in Ref. [64]; the Gaussian case looks somewhat like the transmitted spectra in Fig. 19.6 for $x \sim L/4$. These spectra, and all results described so far, are calculated assuming no motion of the atoms. This is appropriate in the central region of the vapor, because the distance moved in an excited state lifetime ($L_v = v/\Gamma$) is much smaller than L. In fact, resonant collisions between excited and ground state atoms further limits the distance an excited atom moves in one direction before transfer of excitation. However, near the window or wall of the container, atomic motion will cause wall collisions of excited atoms and loss of radiation. This loss will be primarily within the Doppler core of the line, since these frequencies can only escape if emitted near the vapor edge. This loss depends on the excited state density in the neighborhood of the wall, and can be significant if $L_v > 1/k_0$. The excited atom density near the wall must be self consistent with the radiation transport and wall quenching. This situation has been modeled and studied experimentally ([65] and references therein).

REFERENCES

1. H. A. Lorentz, Proc. Akad. Wet. (Amsterdam) **8**, 591 (1906).
2. V. Weisskopf, Phys. Z. **34**, 1 (1933).
3. J. Holtsmark, Ann. Phys. (Leipzig) **58**, 577 (1919).
4. H. G. Kuhn, Phil. Mag. **18**, 987 (1934).
5. H. Margenau and W. W. Watson, Rev. Mod. Phys. **8**, 22 (1936).
6. A. Jablonski, Z. Physik, **70**, 723 (1931); Acta Phys. Polon. **27**, 49 (1965).
7. E. Lindholm, Ark. Fys. A **32**, 1 (1945).
8. H. M. Foley, Phys. Rev. **69**, 616 (1946); Phys. Rev. **73**, 259 (1948).
9. T. Holstein, Phys. Rev. **79**, 744 (1950).
10. P. W. Anderson, Phys. Rev. **76**, 647 (1949); Phys. Rev. **86**, 809 (1952).
11. M. Baranger, Phys. Rev. **112**, 855 (1958).
12. S-Y Chen and M. Takeo, Rev. Mod. Phys. **29**, 20 (1957).
13. N. Allard and J. Kielkopf, Rev. Mod. Phys. **54**, 1103 (1982).
14. H. Griem, *Plasma Spectroscopy* (McGraw Hill, New York, 1964); *Spectral Line Broadening by Plasmas* (Academic Press, New York, 1974).

15. J. Jeffries, *Spectral Line Formation* (Blaisdell, Waltham, 1968).
16. D. Mihalas, *Stellar Atmospheres* (W. H. Freeman, San Francisco, 1970).
17. J. van Kranendonk, Canad. J. Phys. **46**, 1173 (1968).
18. J. Cooper, Rep. Prog. Phys. **29**, 35 (1966).
19. J. Simons, *Energetic Principles of Chemical Reactions* (Jones and Bartlett, Boston, 1983).
20. J. Szudy and W. E. Baylis, J. Quant. Spectrosc. Radiat. Trans. **17**, 681 (1977).
21. R. E. Walkup, A. Spielfiedel, and D. E. Pritchard, Phys. Rev. Lett. **45**, 986 (1980).
22. J. Ward, J. Cooper, and E. W. Smith, J. Quant. Spectrosc. Radiat. Trans. **14**, 555 (1974).
23. D. G. McCarten and N. Lwin, J. Phys. B **10**, L17 (1977).
24. D. N. Stacey and R. C. Thompson, Acta Phys. Polon. A **54**, 833 (1978).
25. Ch. Ottinger, R. Scheps, G. W. York, and A. Gallagher, Phys. Rev. A **11**, 1815 (1975).
26. M. J. O'Callaghan and J. Cooper, Phys. Rev. A **39**, 6206 (1989).
27. B. P. Stoicheff and E. Weinberger, Phys. Rev. Lett. **44**, 733 (1980).
28. K. H. Weber and K. Niemax, Z Phys. A **307**, 13 (1982).
29. D. Drummond and A. Gallagher, J. Chem. Phys. **60**, 3426 (1974).
30. C. G. Carrington and A. Gallagher, Phys. Rev. A **10**, 1464 (1974).
31. J. Pascale and J. Vandeplanque, J. Chem. Phys. **60**, 2278 (1974).
32. J. R. Fuhr and A. Lesage, *Bibliography of Atomic Line Shapes and Shifts, NIST Special Publication 366* (U. S. Gov't. Printing Office, Washington, 1993), Suppl. 4.
33. J. Szudy and W. E. Baylis, J. Quant. Spectrosc. Radiat. Trans. **15**, 641 (1975).
34. D. N. Stacey and J. Cooper, J. Quant. Spectrosc. Radiat. Trans. **11**, 1271 (1971).
35. K. M. Sando and J. Wormhoudt, Phys. Rev. A **7**, 1889 (1973).
36. L. K. Lam, M. M. Hessel, and A. Gallagher, J. Chem. Phys. **66**, 3550 (1977).
37. A. Tam, G. Moe, W. Park, and W. Happer, Phys. Rev. Lett. **35**, 85 (1975).
38. T. Holstein and A. Gallagher, Phys. Rev. A **16**, 2413 (1977).
39. J. L. Carlston and A. Szoke, J. Phys. B **9**, L231 (1976).
40. A. Omont, Prog. Quant. Electr. **5**, 69 (1977).
41. J. Light and A. Szoke, Phys. Rev. A **15**, 1029 (1977).
42. K. Burnett, J. Cooper, R. J. Ballagh, and E. W. Smith, Phys. Rev. A **22**, 2005 (1980).
43. A. Streater, J. Cooper, and W. J. Sandle, J. Quant. Spectrosc. Radiat. Trans. **37**, 151 (1987).
44. J. Tellinghausen, J. Mol. Spectrosc. **103**, 455 (1984).
45. F. H. Mies, J. Chem. Phys. **48**, 482 (1968).
46. C. G. Carrington, D. Drummond, A. V. Phelps, and A. Gallagher, Chem. Phys. Lett. **22**, 511 (1973).
47. L. M. Biberman, J. Exp. Theor. Phys. U. S. S. R. **17**, 416 (1947); J. Exp. Theor. Phys. U. S. S. R. **59**, 659 (1948).
48. T. Holstein, Phys. Rev. **72**, 1212 (1947); Phys. Rev. **83**, 1159 (1951).
49. C. van Trigt, Phys. Rev. **181**, 97 (1969); Phys. Rev. A **4**, 1303 (1971); Phys. Rev. **13**, 726 (1976).
50. C. G. Carrington, D. N. Stacey, and J. Cooper, J. Phys. B **6**, 417 (1973).
51. J. Huennekins, H. J. Park, T. Colbert, and S. C. McClain, Phys. Rev. **35**, 2892 (1987).
52. H. A. Post, P. van de Weijer, and R. M. M. Cremers, Phys. Rev. A **33**, 2003 & 2017 (1986).
53. T. Holstein, Phys. Rev. **83**, 1159 (1951).
54. P. J. Walsh, Phys. Rev. **116**, 511 (1959).
55. T. Colbert and J. Huennekens, Phys. Rev. **41**, 6145 (1990).
56. J. Huennekins and A. Gallagher, Phys. Rev. A **28**, 238 (1983).
57. E. Milne, J. Math. Soc. (London) **1**, 40 (1926).
58. J. Huennekins and T. Colbert, J. Quant. Spectrosc. Radiat. Trans. **41**, 439 (1989).
59. T. Holstein, D. Alpert, and A. O. McCoubrey, Phys. Rev. **85**, 985 (1952).
60. A. V. Phelps, Phys. Rev. **114**, 1011 (1959).
61. M. G. Payne, J. E. Talmage, G. S. Hurst, and E. B. Wagner, Phys. Rev. A **9**, 1050 (1974).
62. J. B. Anderson, J. Maya, M. W. Grossman, R. Lagushenko, and J. F. Weymouth, Phys. Rev. A **3**, 2986 (1985).
63. G. J. Parker, W. N. G. Hitchon, and J. E. Lawler, J. Phys. B **26**, 4643 (1993).
64. C. van Trigt, Phys. Rev. A **1**, 1298 (1970).
65. A. Zajonc and A. V. Phelps, Phys. Rev. A **23**, 2479 (1981).

20

Thomas-Fermi and Other Density-Functional Theories

John D. Morgan III

Department of Physics and Astronomy, University of Delaware, Newark, Delaware 19716

20.1 INTRODUCTION

In the early years of quantum physics, Thomas [1] and Fermi [2] independently invented a simplified theory, subsequently known as *Thomas-Fermi theory*, to describe an atom or atomic ion with a large nuclear charge Z and a large number of electrons N. Many qualitative features of this model can be studied analytically, and the precise solution can be found by solving numerically a nonlinear ordinary differential equation. Lenz [3] demonstrated that this equation for the electrostatic potential could be derived from a variational expression for the energy as a functional of the density. Refinements to Thomas-Fermi theory include a term in the energy functional to account for electron exchange effects introduced by Dirac [4], and nonlocal gradient corrections to the kinetic energy introduced by von Weizsäcker [5].

Although the Hartree-Fock method or other more elaborate techniques of atomic structure now provide much more accurate results (see Chaps. 23, 21 and 22), Thomas-Fermi theory provides quick estimates and global insight into the total energy and other properties of a heavy atom or ion. A rigorous analysis of Thomas-Fermi theory by Lieb and Simon [6,7] showed that it is asymptotically exact in that it yields the correct leading asymptotic behavior, for both the total energy and the electronic density, in the limit as both Z and N tend to infinity, with the ratio Z/N fixed. (In a real atom, of course, relativistic and other effects become increasing important as Z increases.) However, Thomas-Fermi theory has the property that molecules do not bind, as first noted by Sheldon [8] and proved by Teller [9]. That the interatomic potential energy curve for a homonuclear diatomic molecule is purely repulsive was demonstrated

233

by Balàzs [10]. This 'no binding' property of clusters of atoms was used by Lieb and Thirring [11] to prove the *stability of matter*, in the sense that as the number of particles increases, the total energy decreases only linearly rather than as a higher power of the number of particles, as it would if electrons were bosons rather than fermions. Lieb went on to explore the mathematical structure of the modifications of the Thomas-Fermi model when gradient terms (von Weizsäcker) and/or exchange (Dirac) terms are included [12]. A review article by Spruch [13] explicates the linkage between long-developed physical intuition and the mathematically rigorous results obtained in the 1970's and 1980's. The older literature was reviewed by Gombás [14] and by March [15].

An outgrowth of Thomas-Fermi theory is the general density functional theory initiated by Hohenberg and Kohn [16] and by Kohn and Sham [17], as discussed in Sect. 20.4 of this Chapter.

20.2 THOMAS-FERMI THEORY AND ITS EXTENSIONS

The key idea in Thomas-Fermi theory and its generalizations is the replacement of complicated terms in the kinetic energy and electron-electron repulsion energy contributions to the total energy by relatively simple functionals of the electron density ρ, and perhaps $\nabla\rho$. This section first describes Thomas-Fermi theory, and then its various generalizations which attempt to correct, with varying success, some of its deficiencies.

20.2.1 Thomas-Fermi Theory

In a D-dimensional Euclidean space, the expectation value of the electronic kinetic energy operator in a quantum state ψ can be approximated by

$$\frac{\hbar^2}{m_e} 2\pi^2 \frac{D}{D+2} \left(\frac{D}{2\Omega_D}\right)^{2/D} \int \rho^{(D+2)/D}(\mathbf{r})\, d^D\mathbf{r}, \quad (20.1)$$

where

$$\Omega_D = D\pi^{D/2}/\Gamma(1 + D/2) \quad (20.2)$$

is the surface area of a unit hypersphere in D dimensions (see [13], p. 176). These expressions can easily be derived by considering the energy levels of a system of a large number of noninteracting fermions confined to a D-dimensional box. Specialization to the physically interesting case of $D = 3$ yields the well-known expression

$$\frac{\hbar^2}{m_e} 2\pi^2 \frac{3}{5} \left(\frac{3}{2\Omega_3}\right)^{2/3} \int \rho^{5/3}(\mathbf{r})\, d\mathbf{r}, \quad (20.3)$$

where

$$\Omega_3 = 3\pi^{3/2}/\Gamma(1 + 3/2) = 4\pi. \quad (20.4)$$

The electron-nucleus attraction energy in a three-dimensional space is given exactly by

$$\int \rho(\mathbf{r})V(\mathbf{r})\, d\mathbf{r}, \quad (20.5)$$

where $V(\mathbf{r})$ is the Coulomb potential due to a single nucleus ($V(\mathbf{r}) = -Z/r$) or to several nuclei ($V(\mathbf{r}) = -\sum_i Z_i/|\mathbf{r} - \mathbf{R}_i|$). The electron-electron Coulomb repulsion energy in a three-dimensional space is approximated by

$$\frac{1}{2} \int \frac{\rho(\mathbf{r})\rho(\mathbf{r}')}{|\mathbf{r} - \mathbf{r}'|}\, d\mathbf{r}\, d\mathbf{r}', \quad (20.6)$$

which tends to overestimate the actual repulsion energy because it includes the self-energy of the densities of individual electrons. This is, however, a higher-order effect for a system with a large number of electrons concentrated in a small region of space. As was suggested by Fermi and Amaldi [18], this overestimation can be approximately corrected for an atom with N electrons by multiplying this term by the ratio of the number of ordered pairs of different electrons to the total number of ordered pairs:

$$\frac{N(N-1)}{N^2} = 1 - \frac{1}{N}. \quad (20.7)$$

This is approximately correct for an atom, with many electrons concentrated close together, but it would still be an overestimate for a diffuse system, such as one composed of N electrons and N protons separated by large distances of $O(R)$, for which the ground-state electron-electron repulsion term should be proportional to $\frac{1}{2}N(N-1)/R$ rather than to N times a constant of $O(1)$. For this reason the Fermi-Amaldi correction, which complicates the mathematical analysis without eliminating the unphysical overestimation of the electron-electron repulsion term, is not usually included. It is evident that the treatment of both the electronic kinetic energy term and the electron-electron repulsion energy term depends on the assumption that the number N of electrons (actually, the number of electrons *per atom*) is large. Hence the Thomas-Fermi model is sometimes called the *statistical model* of an atom.

The three contributions to the total energy are now added together and one seeks to minimize their sum, the Lenz functional [3]

$$E[\rho] = \frac{\hbar^2}{m_e} 2\pi^2 \frac{3}{5} \left(\frac{3}{2\Omega_3}\right)^{2/3} \int \rho^{5/3}(\mathbf{r})\, d\mathbf{r}$$
$$+ \int \rho(\mathbf{r})V(\mathbf{r})\, d\mathbf{r} + \frac{1}{2} \int \frac{\rho(\mathbf{r})\rho(\mathbf{r}')}{|\mathbf{r} - \mathbf{r}'|}\, d\mathbf{r}\, d\mathbf{r}',$$

$$(20.8)$$

over all admissible densities ρ. The mathematical question now arises: what is an admissible density? The answer was provided by Lieb and Simon [6, 7]: a density for which both

$$\int \rho(\mathbf{r})\,d\mathbf{r}, \qquad (20.9)$$

the total number of electrons, and

$$\int \rho^{5/3}(\mathbf{r})\,d\mathbf{r}, \qquad (20.10)$$

which is proportional to the estimate of their kinetic energy, are finite, automatically yields finite values of the other terms in the expression for the energy. As Lieb and Simon proved, the minimization of this functional over all such densities yields a well-determined result.

Carrying out the variation of $E[\rho]$ with respect to ρ yields the Euler-Lagrange equation

$$0 = \frac{\hbar^2}{m_e}2\pi^2\Big(\frac{3}{2\Omega_3}\Big)^{2/3}\rho^{2/3}(\mathbf{r})$$
$$+ V(\mathbf{r}) + \int \frac{\rho(\mathbf{r}')}{|\mathbf{r}-\mathbf{r}'|}\,d\mathbf{r}'. \qquad (20.11)$$

The sum of the last two terms is of course the negative of the total electrostatic potential $\phi(\mathbf{r})$, so one sees that in Thomas-Fermi theory the density is proportional to the 3/2-power of the potential. To simplify subsequent manipulations, let

$$\frac{\hbar^2}{m_e}2\pi^2\Big(\frac{3}{2\Omega_3}\Big)^{2/3} = \frac{\hbar^2}{2m_e}(3\pi^2)^{2/3} = \gamma_p, \qquad (20.12)$$

so that

$$\gamma_p\rho^{2/3}(\mathbf{r}) = \phi(\mathbf{r}). \qquad (20.13)$$

By Poisson's theorem,

$$-\nabla^2\phi = 4\pi\Big(\sum_i Z_i\delta^{(3)}(\mathbf{r}-\mathbf{R}_i) - \rho(\mathbf{r})\Big), \qquad (20.14)$$

and from Eq. (20.13) one has $\rho = \gamma_p^{-3/2}\phi^{3/2}$, so from the integral equation for the electronic density ρ one obtains the differential equation

$$-\nabla^2\phi = 4\pi\Big(\sum_i Z_i\delta^{(3)}(\mathbf{r}-\mathbf{R}_i) - \gamma_p^{-3/2}\phi^{3/2}\Big) \qquad (20.15)$$

for the potential ϕ. In the case of a single nucleus, the usual separation of variables in spherical polar coordinates yields for ϕ the ordinary differential equation

$$\frac{1}{r}\frac{d^2}{dr^2}(r\phi) = 4\pi\gamma_p^{-3/2}\phi^{3/2}, \qquad (20.16)$$

whose similarity to Emden's equation, which Eddington had used to study the internal constitution of stars, was

recognized by Milne [19]. The numerical solution of this equation with the appropriate boundary conditions at $r = 0$ and $r = \infty$ was extensively discussed by Baker [20], and accurate solutions tabulated by Tal and Levy [21]. The numerical solution determines that the total energy of a neutral atom is to be

$$E = -3.678\,745\,21\ldots\,\gamma_p^{-1}Z^{7/3}$$
$$= -1.537\,490\,24\ldots\,Z^{7/3}\text{ Ry}. \qquad (20.17)$$

Another possibility is to do the constrained minization over all densities which obey

$$\int \rho(\mathbf{r})\,d\mathbf{r} = \lambda, \qquad (20.18)$$

where λ is the number of electrons, which for purposes of mathematical analysis is allowed to be nonintegral. One then introduces a Lagrange multiplier $-\mu$, the *chemical potential*, to correspond with this constraint, and thereby obtains the Euler-Lagrange equation

$$0 = \frac{\hbar^2}{m_e}2\pi^2\Big(\frac{3}{2\Omega_3}\Big)^{2/3}\rho^{2/3}(\mathbf{r})$$
$$+ V(\mathbf{r}) + \int \frac{\rho(\mathbf{r}')}{|\mathbf{r}-\mathbf{r}'|}\,d\mathbf{r}' + \mu, \qquad (20.19)$$

which holds wherever ρ is positive. As was shown by Lieb and Simon [6, 7], this procedure too is well-defined. The analogue of Eq. (20.13), the relationship between the density and the electrostatic potential for the neutral atom, is now

$$\gamma_p\rho^{2/3}(\mathbf{r}) = [\phi(\mathbf{r}) - \mu]_+, \qquad (20.20)$$

where $[f]_+ = \max(f, 0)$. The corresponding differential equation for the potential ϕ is

$$-\nabla^2\phi = 4\pi\Big(\sum_i Z_i\delta^{(3)}(\mathbf{r}-\mathbf{R}_i) - \gamma_p^{-3/2}[\phi(\mathbf{r})-\mu]_+^{3/2}\Big). \qquad (20.21)$$

Lieb and Simon rigorously proved a large number of results concerning the solution of the Thomas-Fermi model. When $V(r)$ is a sum of Coulomb potentials arising from a set of nuclei of positive charges Z_i, with $\sum_i Z_i = Z$, then the energy $E(\lambda)$ is a continuous, monotonically decreasing function of λ for $0 \le \lambda \le Z$, and its derivative $dE/d\lambda$ is the chemical potential $-\mu(\lambda)$, which vanishes at $\lambda = Z$. For λ in this range, there is a unique minimizing density ρ, whereas for $\lambda > Z$ there is no unique minimizing ρ, since one can place arbitrarily large clumps of charge with arbitrarily low energy arbitrarily far away from the nuclei. In the atomic case, with a single nucleus, $\rho(\mathbf{r})$ is a spherically symmetric monotonically decreasing function of r.

Moreover, for an atom or atomic ion, the Thomas-Fermi density obeys the *virial theorem*

$$2\langle T \rangle = -\langle V \rangle = -2E, \qquad (20.22)$$

and for a neutral atom the electronic kinetic energy, electron-nucleus attraction energy, and electron-electron repulsion energy terms in the expression for the total energy satisfy the ratios $3 : -7 : 1$.

It is straightforward to examine the behavior of the Thomas-Fermi density ρ in the limit as either $r \to 0$ or $r \to \infty$. For large r, the electron density vanishes identically outside a sphere of finite radius for a positive ion. For a neutral atom, the ordinary differential equation (20.16) for the potential ϕ can be analyzed to show that

$$\phi(r) \simeq \gamma_p \, (3\gamma_p/\pi)^2 \, r^{-4}, \qquad (20.23)$$

from which it follows that

$$\rho(r) \simeq (3\gamma_p/\pi)^3 \, r^{-6}, \qquad (20.24)$$

independently of Z, which implies that as $Z \to \infty$, a neutral atom described by the Thomas-Fermi model has a finite size defined in terms of a radius within which all but a fixed amount of electronic probability density is located. For example, if one defines the size of an atom as that value of r_a for which

$$\int_{|\mathbf{r}| \geq r_a} \rho(\mathbf{r}) \, d\mathbf{r} = \frac{1}{2}, \qquad (20.25)$$

one finds that in the large-Z limit

$$r_a = \left(\frac{8\pi}{3}\right)^{1/3} \frac{3\gamma_p}{\pi}. \qquad (20.26)$$

In atomic units, $\gamma_p = (3\pi^2)^{2/3}/2$, and

$$r_a = (9\pi)^{2/3} \, a_0 \simeq 9.3 \, a_0, \qquad (20.27)$$

which is about what one would expect for a 'real' nonrelativistic atom with a large nuclear charge Z. On the other hand, the characteristic distance scale in Thomas-Fermi theory, defined as the 'average' value of r, or in terms of a radius within which a fixed *fraction* of electronic probability density is located, is proportional to $Z^{-1/3}$, which shrinks to 0 as $Z \to \infty$. The resolution of this paradox is that outside the typical 'core' scale of distance set by $Z^{-1/3}$, within which most of the electron density is located, there resides in the 'mantle' region a fraction of electrons proportional to $Z^{2/3}/Z = Z^{-1/3}$, and almost all of these are concentrated within a sphere of radius of about 10 a_0.

Moving deeper into the core and approaching the nucleus, the $-Z/r$ singularity in the electron-nucleus Coulomb potential dominates the smeared-out electron-electron potential, so one readily finds that

$$\rho(r) \simeq \left(\frac{Z}{\gamma_p \, r}\right)^{3/2}. \qquad (20.28)$$

This singularity is integrable but unphysical, since it arises from the approximation of the local kinetic energy by $\rho^{5/3}$, which breaks down where ρ is rapidly varying on a length scale proportional to $1/Z$. In a 'real' nonrelativistic heavy atom governed by the Schrödinger equation, the actual electron density at the nucleus is finite, being proportional to Z^3. This unphysical singularity in the electron density in Thomas-Fermi theory can be eliminated by adding a gradient correction to the Thomas-Fermi kinetic energy term.

20.2.2 Thomas-Fermi-von Weizsäcker Theory

The semiclassical approximation (20.3) for the quantum kinetic energy in terms of a power of the density is capable of improvement, particularly in regions of space where the density is rapidly varying. The incorporation of such corrections leads to a gradient expansion for the kinetic energy [22]. The leading correction is of the form

$$\frac{\hbar^2}{2m} \int |(\nabla \rho^{1/2})(\mathbf{r})|^2 \, d\mathbf{r}. \qquad (20.29)$$

Addition of such a term to the Thomas-Fermi expression for the kinetic energy yields a theory which avoids many of the unphysical features of ordinary Thomas-Fermi theory at very short and moderately large distances. The more important points, as rigorously proved in Lieb's review article [12], are as follows. The leading features of the energy are unchanged; for large Z the energy $E(Z)$ of a neutral atom or atomic ion is still proportional to $Z^{7/3}$, but now there enter higher-order corrections arising from the gradient terms of order $Z^{7/3} Z^{-1/3} = Z^2$ and higher powers of $Z^{-1/3}$. The maximum number of electrons which can be bound by an atom of nuclear charge Z is no longer exactly Z, but a slightly larger number; thus Thomas-Fermi-von Weizsäcker theory allows for the formation of negatively charged atomic ions. It was further proven by Benguria and Lieb [23] that in the Thomas-Fermi-von Weizsäcker model a neutral atom can bind at most one extra electron, and that a neutral molecule can bind at most as many extra electrons as it has nuclei.

The effect on the electronic density ρ is more profound. While the general shape and properties of ρ in the 'core' and 'mantle' regions is unchanged, the fact that $\nabla \rho^{1/2}$ need not and in general does not vanish when ρ vanishes on some surface implies that for a positive ion ρ no longer vanishes outside of a sphere, as it does in the case of Thomas-Fermi theory, but instead extends over all space. For positive ions, neutral atoms, and negative ions alike, 'differential inequality' techniques (see, for example, [24]) can be used to show that $\rho(r)$ decays exponentially for large r, with the constant in the exponential proportional to $\mu^{1/2}(\lambda)$. For small r, the gradient terms dominate the energy expression, so one finds that the electronic density no longer diverges as $r \to 0$, but instead tends to a finite limit, with a first derivative which obeys a relation analogous to the Kato cusp condition [25] (see Sect. 11.2.1).

The study of molecules within the Thomas-Fermi-von Weizsäcker model involves several subtleties and pitfalls, which can lead to physical absurdities. Since two neutral atoms with different nuclear charges will in general have different chemical potentials, a pair of such atoms placed a long distance apart will *spontaneously ionize*, with a small amount of electric charge being transferred from one to the other until the chemical potentials of the positively charged ion and the negatively charged ion become equal. The result is a long-range Coulomb attraction between them [26]. This phenomenon does *not* occur in the real world, since the amount of electric charge which can be transferred is quantized in units of $-e$, and it is empirically true that the smallest atomic ionization potential exceeds the largest atomic electron affinity. For two neutral atoms with the same nuclear charges, the situation is more subtle. Nonetheless, a careful analysis shows that in this case too, though no spontaneous ionization occurs, there is a long-range attractive interaction between them arising from the overlap of the exponentially small tails of the electron clouds. Since electron correlation is not included in this model, it could not be expected to describe attractive van der Waals forces.

In summary, the Thomas-Fermi-von Weizsäcker model yields a more realistic picture of a single atom than does the Thomas-Fermi model. However, it does not provide a useful picture for understanding the interaction between atoms at large distances. These kinds of unphysical features provide a glimpse into the complicated nature of the universal density functional, which must include terms which rigorously suppress an unphysical feature like spontaneous ionization of a distant pair of heteronuclear atoms (see, for example, Refs. [27] and [28]). It is evident from the mathematical properties of Thomas-Fermi-von Weizsäcker theory and related models that a density functional which 'fixes up' the Thomas-Fermi expression simply by adding a few gradient terms and/or simple exchange terms and the like must still differ in important ways from the universal density functional, particularly for properties of extended systems.

20.2.3 Thomas-Fermi-Dirac Theory

The effect of the exchange of electrons can be approximated, following Dirac [4], by including in the Thomas-Fermi energy functional an expression of the form

$$-\frac{1}{4\pi^3}\left(3\pi^2\right)^{4/3}\int \rho^{4/3}(\mathbf{r})\,d\mathbf{r}\,. \qquad (20.30)$$

Minimization of the resulting Thomas-Fermi-Dirac energy functional over all admissible densities ρ whose integral is λ yields a well-defined $E(\lambda)$, which has the correct behavior for $\lambda \leq Z$, and it has been shown that for an atom the exchange correction to the energy is of order $Z^{5/3}$. However, this model exhibits unphysical behavior for $\lambda > Z$, because one can obtain a completely artificial lowering of the energy by placing many small clumps of electrons a large distance from the nucleus, for which the negative $\int \rho^{4/3}\,d\mathbf{r}$ term dominates the energy expression [12]. At the conclusion of his original article, Dirac clearly stated that the correction he had derived, although giving a better approximation in the interior of an atom, gives "a meaningless result for the outside of the atom" [4]. It is therefore clear that any physically reasonable theory must somehow profoundly modify this correction in the region where the electronic density is very small.

20.2.4 Thomas-Fermi-von Weizsäcker-Dirac Theory

One can also include the Dirac exchange correction in the Thomas-Fermi-von Weizsäcker energy functional. In this case, however, the mathematical foundations of the theory are still incomplete (see [12], pp. 638-9). Nonetheless, it is clear that this theory too suffers from the unphysical lowering of the energy by small clumps of electrons at large distances from the nucleus.

In summary, one can say that the inclusion of Dirac's exchange correction in its most straightforward form leads to an improvement of energies for positive ions or neutral atoms, but to unphysical behavior for systems where the charge of the electrons exceeds the nuclear charge, in line with Dirac's own observations on the limitations of his correction [4]. We see here again a manifestation of how complicated must be the behavior of the true universal density functional.

20.2.5 Thomas-Fermi Theory with Different Spin Densities

As was remarked by Lieb and Simon [7], it is possible to consider a variant of Thomas-Fermi theory with a pair of spin densities ρ_α and ρ_β for the spin-up and spin-down electrons, with the two adding together to produce the total electronic density ρ. This theory has been rigorously formulated and analyzed by Goldstein and Rieder [30]. Because the problem is nonlinear, the mathematical complications are substantial, and the theory is not a trivial extension of ordinary Thomas-Fermi theory. Goldstein and Rieder first considered the case where the total number of electrons of each type of spin is specified in advance [29]. There is no mathematical obstacle to constructing such a spin-polarized Thomas-Fermi theory, but it does not yield the kind of spontaneous spin-polarization that one observes in the ground states of many real quantum mechanical atoms and molecules, which is not surprising in view of the fact that such spin-polarization in accord with Hund's first rule arises from exchange and correlation effects not included in this simple functional. However, in the case where the electronic spins (but *not* their currents) are coupled to an external magnetic field, the ground state is naturally spin-polarized [30].

20.3   NONRELATIVISTIC ENERGIES OF HEAVY ATOMS

Thomas-Fermi theory suggests that Eq. (20.17) provides the leading term in a power series expansion for the nonrelativistic energy of a neutral atom of the form

$$E(Z) = - \left(c_7 Z^{7/3} + c_6 Z^{6/3} + c_5 Z^{5/3} + \cdots \right) \quad (20.31)$$

with $c_7 = 1.537\,490\,24\ldots$ Ry, $c_6 = -1$ Ry, and $c_5 \simeq 0.5398$ Ry. The c_6 term was first calculated by Scott [31] from the observation that it arises from the energy of the innermost electrons for which the electron-electron interaction can be neglected. The difference between the exact and Thomas-Fermi energies for this case of noninteracting electrons yields the correct c_6 [32, 13]. A mathematically nonrigorous but physically insightful justification of the Scott correction was provided in 1980 by Julian Schwinger [33]. This result has now been rigorously proved, with upper and lower bounds coinciding [34–38].

The c_5 term is much more subtle, since it arises from a combination of effects from the exchange interaction and from the bulk motion of electrons in the Thomas-Fermi potential. A general analytic procedure devised by Schwinger [39] yields the above value, in good agreement with a much earlier estimate by March and Plaskett [40].

A numerical check of these results, based on a fit to Hartree-Fock calculations for Z up to 290 with correlation corrections, yielded the values [41] $c_5 = 0.55 \pm 0.02$ Ry and $c_4 \simeq 0$. It seems likely that, because of shell structure, the terms c_4 and beyond have an oscillatory dependence on Z [42]. The oscillatory structure and other refinements of Thomas-Fermi theory are considered in a series of papers by Englert and Schwinger [43].

Recent Progress. Iantchenko, Lieb, and Siedentop [44] have recently proven Lieb's 'strong Scott conjecture' that for small r, the rescaled density for the exact quantum system converges to the sum of the densities of the bound noninteracting hydrogenic orbitals; the properties of this function were explored by Heilman and Lieb

[45] Fefferman and Seco [46] have rigorously proved the correctness of Schwinger's procedure for calculating not just the $O(Z^{6/3})$ Scott correction but also the $O(Z^{5/3})$ exchange term. Their full proof includes a demonstration that the Hartree-Fock energy agrees with the exact quantum energy through $O(Z^{5/3})$, with an error of smaller order [47]. Numerous auxiliary theorems and lemmas are published in Refs. [48–51]. Progress toward obtaining higher-order oscillatory terms is described in Refs. [52, 53]. The analytical evaluation of accurate approximations to the energy of a heavy atom, or at least of the contributions to that energy of all but the few outermost electronic orbitals, would be of particularly great value if it led to the construction of more accurate and better justified pseudopotentials [54, 55] for describing the valence orbitals.

20.4   GENERAL DENSITY FUNCTIONAL THEORY

The literature on general density functional theory and its applications is enormous, so any bibliography must be selective. The reader interested in learning more could begin by consulting a number of review articles (e.g. [56–61]), collections of articles (e.g. [62–64]), and conference proceedings (e.g. [65–75]), and the recent textbooks by Parr and Yang [76] and by Dreizler and Gross [77].

20.4.1   The Hohenberg-Kohn Theorem for the One-Electron Density

In 1964 Hohenberg and Kohn [16] argued that there exists a universal density functional $F[\rho]$, independent of the external potential $V(\mathbf{r})$, such that minimization of the sum

$$F[\rho] + \int \rho(\mathbf{r}) V(\mathbf{r}) \, d\mathbf{r} , \quad (20.32)$$

subject to the constraint

$$\int \rho(\mathbf{r}) \, d\mathbf{r} = N \text{ (a positive integer)} , \quad (20.33)$$

yields the ground state energy of a quantum-mechanical N-electron system moving in this external potential. However, Hohenberg and Kohn's 'theorem' is like a mathematical 'existence theorem'; no procedure exists to calculate explicitly this unknown universal functional, which surely is extremely complicated if it can be written down at all in closed form.[1]   Moreover, Hohenberg

---

[1]E. Bright Wilson, however, defined it, implicitly and whimsically, as follows: "Take the ground-state density and integrate it to find the total number of electrons. Find the cusps in the density to locate all nuclei, and then use the cusp condition–that the

and Kohn glossed over two problems: it is not clear *a priori* that every well-behaved ρ is derivable from a well-behaved properly antisymmetric many-electron wave function (the so-called *n*-representability problem, since *n* was used by Hohenberg and Kohn to represent the density of electrons), and it is also not clear *a priori* that every well-behaved density ρ can be derived from a quantum-mechanical many-electron wave function ψ which is the properly antisymmetric ground-state wave function for a system of electrons moving in some external potential $V(\mathbf{r})$ (the so-called *v*-representability problem, since Hohenberg and Kohn used *v* in place of V).

The *n*-representability problem was solved by Gilbert [78] and by Harriman [79], who gave a prescription for starting from an arbitrary well-behaved ρ and from it constructing a many-electron wave function ψ which generated that ρ (see also [80] and [28]). The *v*-representability problem is much more formidable, as demonstrated by the discovery that there are well-behaved densities ρ which are *not* the ground-state densities for any fermionic system in an external potential V [81, 80]. Following Percus' definition of a universal kinetic energy functional for independent fermion systems [82], Levy [83] proposed to circumvent this *v*-representability problem by modifying the definition of $F[\rho]$ so that instead of being defined in terms of densities which might not be *v*-representable, it is defined as

$$F[\rho] = \min \left(\psi, (T+V)\psi \right), \qquad (20.34)$$

with the minimum being taken over all properly antisymmetric normalized ψ's which yield that ρ.

A great deal of effort has been devoted to trying to find approximate representations of the universal functional $F[\rho]$. One route is mathematical, and features a careful exploration of the abstract properties which $F[\rho]$ must have. Another route is numerical, and can be characterized as involving the guessing of some *ansatz* with a general resemblance to Thomas-Fermi-von Weizsäcker-Dirac theory, with some flexible parameters which are determined by least-squares fitting of the energies resulting from insertion of Hartree-Fock densities into the trial functional to theoretical Hartree-Fock energies, or the like. If, however, the basic *ansatz* exhibits unphysical features in the case of negatively charged ions or heteronuclear molecules, it is not likely that the optimization of parameters in that *ansatz* will get one closer to the true universal density functional. In the opinion of this writer, significant progress in density functional theory based solely upon the one-electron density is likely to require a major revolution in our mathematical un-

derstanding of this field, with a useful procedure made explicit for constructing progressively better approximations to the universal density functional, which, like π or other transcendental numbers, probably will never be written down exactly in closed form. Moreover, the numerical solution of the highly nonlinear Euler-Lagrange equations for a very complicated density functional is likely to require large amounts of computer time, as well as problems with landing in local minima of the energy.

20.4.2 The Kohn-Sham Method for Including Exchange and Correlation Corrections

Density functional theory posed solely in terms of the one-electron density and based upon the Hohenberg-Kohn variational principle provides no general procedure for accurately calculating relatively small energy differences such as excitation energies, ionization potentials, electron affinities, or the binding energies of molecules. There is, however, a powerful method inspired by the Hohenberg-Kohn variational principle, which has been used with great success in the calculation of such quantities. This is the Kohn-Sham variational method [17].

The key idea in the Kohn-Sham variational method is to replace the nonlocal exchange term in the Hartree-Fock equations with an exchange-correlation potential, which at least in principle can be used to determine energies exactly. The oldest, simplest and most common *ansatz* used for the exchange-correlation potential involves the *local density approximation* (LDA), in which one assumes that the exchange-correlation potential for the actual system under study has the same functional form as does the exchange-correlation potential for a uniform interacting gas of electrons. If the density is not too small or not too rapidly varying, the exchange part of this potential can be approximated by $\rho^{1/3}$, which appears in Dirac's first-order approximation for exchange energies, with a systematic procedure for deriving higher-order corrections in a gradient expansion. The correlation part of this potential is accurately known from Ceperley and Alder's quantum Monte Carlo calculation of the properties of the uniform electron gas [84]. One therefore retains the important features of the quantum theory based on wave functions, with a determinantal approximation to ψ, while approximately including exchange and correlation effects through a simply computable effective potential. Higher corrections, which are important for quantitative accuracy, can be incorporated by taking account of the variation of ρ by means of a gradient expansion [22] involving the $\nabla\rho$ and higher derivatives (see, for example, [77], chapter 7), thus yielding a *generalized gradient approximation* (GGA) for the exchange-correlation potential.

The Kohn-Sham procedure has become the back-

radial derivative of the density at the cusp is minus twice the nuclear charge density at each cusp—to determine the charges on each nucleus. Finally solve Schrödinger's equation for the ground-state density or any other property that is desired" (paraphrased by B. I. Dunlap, in [71], p. 3, from J. W. D. Connolly).

bone for the vast majority of accurate calculations of the electronic structure of solids (see, for example, Refs. [60, 74]). In the 1990's, motivated by Becke's work on constructing simple gradient-corrected exchange potentials [85–88, 90, 89, 91],[2] and incorporating the Lee, Yang, and Parr (LYP) expression for the correlation potential [95] derived from Colle and Salvetti's correlation-energy formula [96], the Kohn-Sham method is finding increasing application in efficiently estimating relatively small energy differences of relevance to chemistry (see, for example, Refs. [97–99]). For definitive results, however, one must still resort to an *ab initio* theory which at least in principle converges toward the correct result.

The generation of improved generalized gradient approximations has recently become a growth industry, with increasingly many proposals of increasingly greater complexity [85, 94, 95, 100–109]. Inevitably, some expressions work better for some properties than for others. It is found that usually most of the errors in the long-range tails of the exchange and correlation potentials tend to cancel each other, thus leading to better overall energies than one could reasonably expect [110]. Under these circumstances, it is important to have benchmarks for testing the accuracy of the various approximations. Such comparisons have recently been carried out for two important sets of two-electron systems [111, 112] (see also Ref. [113]):

1. a pair of electrons moving in harmonic potential wells and coupled by the Coulomb repulsion, which yields an exactly solvable system;

2. helium-like ions of variable nuclear charge Z, for which extremely accurate energies and wave functions are available which take account of the behavior of the exact but unknown wave function in the vicinity of all two-particle coalescences and the three-particle coalescence.

The results indicate that the approximate exchange-correlation potentials differ quite considerably from the true exchange-correlation potentials, thus indicating the need for further analytical work in understanding how to design accurate exchange-correlation potentials, and for devising tests of exchange-correlation potentials for larger atoms and for molecules.

Another important way of testing the validity of various approximate exchange and correlation potentials is checking whether they obey inequalities imposed by such general properties as scaling and the Hellmann-Feynman theorem. Such general tests have been devised by Levy and his co-workers [114–123], who have found that many of the commonly used approximate potentials violate general inequalities which must be obeyed by the exact

---

[2]However, Becke's gradient-corrected exchange potential does *not* have the correct $1/r$ behavior at large r, as was observed by several authors [92–94].

potential. These abstract results are helpful in designing potentials which should be better approximations to the true potential.

20.4.3 Density Functional Theory for Excited States

The Hohenberg-Kohn theorem and the Kohn-Sham method were originally formulated in terms of the ground electronic state. These techniques can be extended to calculate the ground state of a given symmetry [124], but that leaves unresolved the issue of using density functional theory to calculate the energies of excited states for a given symmetry. Using the Rayleigh-Ritz principle for ensembles, general abstract procedures for generalizing density functional theory to excited state calculations have been formulated by Theophilou [125] and by several other workers [126–133]. Unfortunately, the errors typically seem to be much larger than for ground-state density functional theory.

20.4.4 Relativistic and Quantum Field Theoretic Density Functional Theory

At a formal level, one can discuss the development of density functional theory for a relativistic system of electrons. For an overview of this challenging subject, see the discussions by Dreizler and Gross ([77], Chap. 8) and by Dreizler [134]. Much of the formalism carries over, but no good way has yet been found of incorporating vacuum polarization corrections.

Acknowledgments

I am grateful to Cyrus Umrigar and Michael P. Teter of Cornell University for generously providing sabbatical support in the spring of 1995, when this article was written. I should also like to thank them, as well as Elliott Lieb and Mel Levy, for helpful discussions. This work was supported by my National Science Foundation grant PHY-9215442.

REFERENCES

1. L. H. Thomas, Proc. Camb. Philos. Soc. **23**, 542 (1927).
2. E. Fermi, Rend. Accad. Naz. Lincei **6**, 602-607 (1927). See also Z. Phys. **48**, 73 (1928); Z. Phys. **49**, 550 (1928).
3. W. Lenz, Z. Phys. **77**, 713 (1932).
4. P. A. M. Dirac, Proc. Camb. Philos. Soc. **26**, 376 (1930).
5. C. F. von Weizsäcker, Z. Phys. **96**, 431 (1935).
6. E. H. Lieb and B. Simon, Phys. Rev. Lett. **31**, 681 (1973).
7. E. H. Lieb and B. Simon, Adv. in Math. **23**, 22 (1977).
8. J. W. Sheldon, Phys. Rev. **99**, 1291 (1955).

9. E. Teller, Rev. Mod. Phys. **34**, 627 (1962).

10. N. Balàzs, Phys. Rev. **156**, 42 (1967).

11. E. H. Lieb and W. Thirring, Phys. Rev. Lett. **35**, 687 (1975).

12. E. H. Lieb, Rev. Mod. Phys. **53**, 603 (1981); *ibid.* **54**, 311(E) (1982).

13. L. Spruch, Rev. Mod. Phys. **63**, 151 (1991).

14. P. Gombás, *Die statistische Theorie des Atoms und ihre Anwendungen* (Springer, Berlin, 1949); *Statistische Behandlung des Atoms*, in Handbuch der Physik **36** *Atome II*, edited by S. Flügge (Springer, Berlin, 1956), 109-231.

15. N. H. March, *The Thomas-Fermi Approximation in Quantum Mechanics*, Adv. Phys. **6**, 1-98 (1957).

16. P. Hohenberg and W. Kohn, Phys. Rev. **B136**, 864 (1964).

17. W. Kohn and L. J. Sham, Phys. Rev. **A140**, 1133 (1965).

18. E. Fermi and E. Amaldi, Mem. Accad. d'Italia **6**, 119 (1934).

19. E. A. Milne, Proc. Camb. Philos. Soc. **23**, 794 (1927).

20. E. B. Baker, Phys. Rev. **36**, 630 (1930).

21. Y. Tal and M. Levy, Phys. Rev. A **23**, 408 (1981).

22. C. H. Hodges, Can. J. Phys. **51**, 1428 (1973).

23. R. Benguria and E. H. Lieb, J. Phys. B **18**, 1045 (1985).

24. R. Ahlrichs, T. Hoffmann-Ostenhof, M. Hoffmann-Ostenhof, and J. D. Morgan III, Phys. Rev. A **23**, 2107 (1981), and references therein.

25. T. Kato, Commun. Pure Appl. Math. **10**, 151 (1957).

26. J. P. Perdew, R. G. Parr, M. Levy, and J. L. Balduz, Jr., Phys. Rev. Lett. **49**, 1691 (1982).

27. M. Levy and J. P. Perdew, in [69], pp. 11–30.

28. E. H. Lieb, in [69], pp. 31–80.

29. J. A. Goldstein and G. R. Rieder, J. Math. Phys. **29**, 709 (1988).

30. J. A. Goldstein and G. R. Rieder, J. Math. Phys. **32**, 2907 (1991).

31. J. M. C. Scott, Philos. Mag. **43**, 859 (1952).

32. E. H. Lieb, Rev. Mod. Phys. **48**, 553 (1976).

33. J. Schwinger, Phys. Rev. A **22**, 1827 (1980).

34. H. Siedentop and R. Weikard, Commun. Math. Phys. **112**, 471 (1987).

35. H. Siedentop and R. Weikard, Invent. Math. **97**, 159 (1989).

36. W. Hughes, Adv. in Math. **79**, 213 (1990).

37. V. Bach, Rep. Math. Phys. **28**, 213 (1989).

38. V. J. Ivrii and I. M. Sigal, Annals of Math. **138**, 243 (1993).

39. J. Schwinger, Phys. Rev. A **24**, 2353 (1981).

40. N. H. March and J. S. Plaskett, Proc. Roy. Soc. London **A 235**, 419 (1956).

41. R. Shakeshaft, L. Spruch, and J. Mann, J. Phys. B **14**, L121 (1981).

42. R. Shakeshaft and L. Spruch, Phys. Rev. A **23**, 2118 (1981).

43. B.-G. Englert and J. Schwinger, Phys. Rev. A **26**, 2322 (1982); **29**, 2331, 2339, 2353 (1984); **32**, 28, 36, 47 (1985).

44. A. Iantchenko, E. H. Lieb, and H. Siedentop, J. Reine Angew. Math. **?**, to appear (1995?).

45. O. J. Heilman and E. H. Lieb, Phys. Rev. A **52**, 3628 (1995).

46. C. Fefferman and L. A. Seco, Bull. Amer. Math. Soc., **23**, 525 (1990).

47. C. Fefferman and L. A. Seco, Adv. in Math. **107**, 1- (1994).

48. C. Fefferman and L. A. Seco, Adv. in Math. **95**, 145 (1992).

49. C. Fefferman and L. A. Seco, Revista Mathemática Iberoamericana **9**, 409 (1993).

50. C. Fefferman and L. A. Seco, Adv. in Math. **107**, 187 (1994).

51. C. Fefferman and L. A. Seco, Adv. in Math. **108**, 263 (1994).

52. A. Cordoba, C. Fefferman, and L. A. Seco, *A Trigonometric Sum Relevant to the Nonrelativistic Theory of Atoms*, Princeton preprint (1994).

53. A. Cordoba, C. Fefferman, and L. A. Seco, *Weyl Sums and Atomic Energy Oscillations*, Princeton preprint (1995).

54. Solid State Physics **24** (1970).

55. M. C. Payne, M. P. Teter, D. C. Allan, T. A. Arias, and J. D. Joannopoulos, Rev. Mod. Phys. **64**, 1045 (1992).

56. A. K. Rajagopal, Adv. Chem. Phys. **41**, 59 (1980).

57. A. S. Bamzai and B. M. Deb, Rev. Mod. Phys. **53**, 95 (1981); **53**, 593(E) (1981). .

58. R. G. Parr, Ann. Rev. Phys. Chem. **34**, 631 (1983).

59. J. Callaway and N. H. March, Solid State Phys. **38**, 135 (1984).

60. R. O. Jones and O. Gunnarsson, Rev. Mod. Phys. **61**, 689 (1989).

61. S. B. Trickey, in *Conceptual Trends in Quantum Chemistry*, edited by E. S. Kryachko and J. L. Calais (Kluwer, Amsterdam, 1994), Vol. 1, pp. 87–100.

62. *Theory of the Inhomogeneous Electron Gas*, edited by S. Lundqvist and N. H. March (Plenum, New York, 1983).

63. *Single-Particle Density in Physics and Chemistry*, edited by N. H. March and B. M. Deb (Academic, New York, 1987).

64. *Density Functional Theory of Many-Fermion Systems*, Advances in Quantum Chemistry, edited by S. B. Trickey (Academic, New York, 1990), Vol. 21.

65. J. Keller and J. L. Gázquez, in *Density Functional Theory*, Lecture Notes in Physics No. 187 (Springer, Berlin, 1983).

66. *Many-Body Phenomena at Surfaces*, edited by D. Langreth and H. Suhl (Academic, New York, 1984).

67. *Local Density Approximations in Quantum Chemistry and Solid State Physics*, edited by J. P. Dahl and J. Avery (Plenum, New York, 1984).

68. *The Electronic Structure of Complex Systems*, NATO ASI Series B, No. 113, edited by P. Phariseau and W. M. Temmermann (Plenum, New York, 1984).

69. *Density Functional Methods in Physics*, NATO ASI Series B, No. 123, edited by R. M. Dreizler and J. da Providência (Plenum, New York, 1985).

70. *Density Matrices and Density Functionals*, edited by R. Erdahl and V. H. Smith, Jr. (Reidel, Dordrecht, 1987).

71. J. K. Labanowski and J. W. Andzelm, *Density Functional Methods in Chemistry* (Springer, New York, 1991).

72. *Density Functional Theory and its Applications*, Philos. Mag. B **69**, 725-1074 (1994).

73. E. S. Kryachko and E. V. Ludeña, *Energy Density Functional Theory of Many-Electron Systems* (Kluwer,

Dordrecht, 1990).

74. *Density Functional Theory*, NATO ASI Series B, No. 337, edited by E. K. U. Gross and R. M. Dreizler (Plenum, New York, 1995).

75. Proceedings of the CECAM Workshop, Theoret. Chim. Acta **91**, 111-266 (1995).

76. R. G. Parr and W. Yang, *Density-Functional Theory of Atoms and Molecules* (Clarendon, Oxford, 1989).

77. R. M. Dreizler and E. K. U. Gross, *Density Functional Theory: An Approach to the Quantum Many-Body Problem* (Springer, Berlin, 1990).

78. T. L. Gilbert, Phys. Rev. B **12**, 2111 (1975).

79. J. E. Harriman, Phys. Rev. A **24**, 680 (1981).

80. E. H. Lieb, in *Physics as Natural Philosophy: Essays in Honor of Laszlo Tisza on his 75th Birthday*, edited by H. Feshbach and A. Shimony (MIT, Cambridge, 1982), 111-149; reprinted in Intern. J. Quantum Chem. **24**, 243 (1983).

81. M. Levy, Phys. Rev. A **26**, 1200 (1982).

82. J. K. Percus, Intern. J. Quantum Chem. **13**, 89 (1978).

83. M. Levy, Proc. Natl. Acad. Sci. USA **76**, 6062 (1979).

84. D. M. Ceperley and B. J. Alder, Phys. Rev. Lett. **45**, 566 (1980).

85. A. D. Becke, Phys. Rev. A **38**, 3098 (1988).

86. A. D. Becke, Intern. J. Quantum Chem. Symp. **23**, 599 (1989).

87. A. D. Becke, J. Chem. Phys. **96**, 2155 (1992).

88. A. D. Becke, J. Chem. Phys. **97**, 9173 (1992).

89. A. D. Becke, J. Chem. Phys. **98**, 1372 (1993).

90. A. D. Becke, J. Chem. Phys. **98**, 5648 (1993).

91. R. M. Dickson and A. D. Becke, J. Chem. Phys. **99**, 3898 (1993).

92. G. Ortiz and P. Ballone, Phys. Rev. B **43**, 6376 (1991).

93. C. Lee and Z. Zhou, Phys. Rev. A **44**, 1536 (1991).

94. E. Engel, J. A. Chevary, L. D. Macdonald, and S. H. Vosko, Z. Phys. D **23**, 7 (1992).

95. C. Lee, W. Yang, and R. G. Parr, Phys. Rev. B **37**, 785 (1988).

96. R. Colle and O. Salvetti, Theoret. Chim. Acta **37**, 329 (1975); **53**, 55 (1979); J. Chem. Phys. **79**, 1404 (1983).

97. B. G. Johnson, P. M. W. Gill, and J. A. Pople, J. Chem. Phys. **98**, 5612 (1993).

98. C. W. Murray, N. C. Handy, and R. D. Amos, J. Chem. Phys. **98**, 7145 (1993).

99. G. J. Laming, V. Termath, and N. C. Handy, J. Chem. Phys. **99**, 8765 (1993).

100. S. H. Vosko, L. Wilk, and M. Nusair, Can. J. Phys. 1200 (1980).

101. D. C. Langreth and M. J. Mehl, Phys. Rev. Lett. **47**, 446 (1981).

102. D. C. Langreth and M. J. Mehl, Phys. Rev. B **28**, 1809 (1983); *ibid.* **29**, 2310(E) (1984).

103. J. P. Perdew and Y. Wang, Phys. Rev. B **33**, 8800 (1986).

104. J. P. Perdew, Phys. Rev. B **33**, 8822; **34**, 7406(E) (1986).

105. A. E. DePristo and J. D. Kress, J. Chem. Phys. **86**, 1425 (1987).

106. L. C. Wilson and M. Levy, Phys. Rev. B **41**, 12930 (1990).

107. J. P. Perdew, Physica B **172**, 1 (1991).

108. J. P. Perdew and Y. Wang, Phys. Rev. B **45**, 13244 (1992).

109. D. J. Lacks and R. G. Gordon, Phys. Rev. A **47**, 4681 (1993).

110. J. P. Perdew, Int. J. Quantum Chem. Symp. **27**, 93 (1993).

111. C. Filippi, C. J. Umrigar, and M. Taut, J. Chem. Phys. **100**, 1290 (1994).

112. C. J. Umrigar and X. Gonze, Phys. Rev. A **50**, 3827 (1994).

113. C. J. Umrigar, in *High Performance Computing and its Application to the Physical Sciences*, Proceedings of the Mardi Gras '93 Conference, edited by D. A. Browne, J. Callaway, J. P. Drayer, R. W. Haymaker, R. K. Kalia, J. E. Tohline, and P. Vashishta (World Scientific, Singapore, 1993), pp. 43–59.

114. M. Levy and J. P. Perdew, Phys. Rev. A **32**, 2010 (1985).

115. M. Levy and H. Ou-Yang, Phys. Rev. A **38**, 625 (1988).

116. M. Levy, in [64], pp. 69–95.

117. M. Levy, in [71], pp. 175–193.

118. A. Görling and M. Levy, Phys. Rev. A **45**, 1509 (1992).

119. Q. Zhao, M. Levy, and R. G. Parr, Phys. Rev. A **47**, 918 (1993).

120. A. Görling and M. Levy, Phys. Rev. B **47**, 13105 (1993).

121. M. Levy and J. P. Perdew, Phys. Rev. B **48**, 11638 (1993).

122. A. Görling and M. Levy, Phys. Rev. A **50**, 196 (1994).

123. M. Levy and A. Görling, Philos. Mag. B **69**, 763 (1994).

124. O. Gunnarsson and B. J. Lundqvist, Phys. Rev. B **13**, 4274 (1976).

125. A. K. Theophilou, J. Phys. C **12**, 5419 (1979); Philos. Mag. B **69**, 771 (1994).

126. J. Katriel, J. Phys. C **13**, L375 (1980).

127. N. Hadjisavvas and A. Theophilou, Phys. Rev. A **32**, 720 (1985).

128. W. Kohn, Phys. Rev. A **34**, 737 (1986).

129. E. K. U. Gross, L. N. Oliveira, and W. Kohn, Phys. Rev. A **37**, 2805, 2809 (1988).

130. L. N. Oliveira, E. K. U. Gross, and W. Kohn, Phys. Rev. A **37**, 2821 (1988), and earlier references therein.

131. H. Englisch, H. Fieseler, and A. Haufe, Phys. Rev. A **37**, 4570 (1988).

132. L. N. Oliveira, in [64], pp. 135–154.

133. L. N. Oliveira, E. K. U. Gross, and W. Kohn, Int. J. Quantum Chem. Symp. **24**, 707 (1990).

134. R. M. Dreizler, Phys. Scr. **T46**, 167 (1993).

21

Atomic Structure: Multiconfiguration Hartree-Fock Theories

Charlotte F. Fischer

Department of Computer Science, Box 1679B, Vanderbilt University, Nashville, Tennessee 37235

21.1 HAMILTONIANS: SCHRÖDINGER AND BREIT-PAULI

The state of a many-electron system is described by a wave function Ψ that is the solution of a partial differential equation (called the wave equation),

$$(H - E)\Psi = 0, \qquad (21.1)$$

where H is the Hamiltonian operator for the system and E the total energy. The operator H depends on the system (atomic, molecular, solid-state, etc.) as well as the underlying quantum mechanical formalism (nonrelativistic, Breit-Pauli, Dirac-Coulomb, or Dirac-Breit, etc.). In atomic systems, the Hamiltonian of the nonrelativistic Schrödinger equation is (in atomic units)

$$H_{\mathrm{nr}} = -\frac{1}{2}\sum_{i=1}^{N}\left(\nabla_i^2 + \frac{2Z}{r_i}\right) + \sum_{i<j}\frac{1}{r_{ij}}. \qquad (21.2)$$

Here Z is the nuclear charge of the atom with N electrons, r_i is the distance of electron i from the nucleus, and r_{ij} is the distance between electron i and electron j. This equation was derived under the assumption of a point-nucleus of infinite mass. The term $2Z/r$ represents the nuclear attraction and $1/r_{ij}$ the inter-electron repulsion. The operator H_{NR} has both a discrete and continuous spectrum: for the former, $\Psi(\mathbf{r}_1, \mathbf{r}_2, \ldots, \mathbf{r}_N)$ has a probability interpretation and consequently must be square integrable. In the Breit-Pauli approximation, the Hamiltonian is extended to include relativistic corrections up to relative order $(\alpha Z)^2$. It is convenient to write the Breit-Pauli Hamiltonian as the sum [1]

$$H_{\mathrm{BP}} = H_{\mathrm{nr}} + H_{\mathrm{rel}}, \qquad (21.3)$$

where H_{rel} represents the relativistic contributions. The latter may again be subdivided into nonfine-structure (NFS) and fine structure (FS) contributions:

$$H_{\mathrm{rel}} = H_{\mathrm{NFS}} + H_{\mathrm{FS}}. \qquad (21.4)$$

The NFS contributions

$$H_{\mathrm{NFS}} = H_{\mathrm{mass}} + H_{\mathrm{D}} + H_{\mathrm{ssc}} + H_{\mathrm{oo}} \qquad (21.5)$$

shift nonrelativistic energy levels without splitting the levels. The mass-velocity term

$$H_{\mathrm{mass}} = -\frac{\alpha^2}{8} \sum_i \nabla_i^4 \qquad (21.6)$$

corrects for the variation of mass with velocity; the one- and two-body Darwin terms

$$H_{\mathrm{D}} = -\frac{\alpha^2 Z}{8} \sum_i \nabla_i^2 r_i^{-1} + \frac{\alpha^2}{4} \sum_{i<j} \nabla_i^2 r_{ij}^{-1} \qquad (21.7)$$

are the corrections of the one-electron Dirac equation due to the retardation of the electromagnetic field produced by an electron; the spin-spin contact term

$$H_{\mathrm{ssc}} = -\frac{8\pi\alpha^2}{3} \sum_{i<j} (\mathbf{s}_i \cdot \mathbf{s}_j) \delta(\mathbf{r}_{ij}) \qquad (21.8)$$

accounts for the interaction of the spin magnetic moments of two electrons occupying the same space; the orbit-orbit interaction

$$H_{\mathrm{oo}} = -\frac{\alpha^2}{2} \sum_{i<j} \left\{ \frac{\mathbf{p}_i \cdot \mathbf{p}_j}{r_{ij}} + \frac{\mathbf{r}_{ij}(\mathbf{r}_{ij} \cdot \mathbf{p}_i) \cdot \mathbf{p}_j}{r_{ij}^3} \right\} \qquad (21.9)$$

accounts for the interaction of two orbital moments.

The FS contributions

$$H_{\mathrm{FS}} = H_{\mathrm{so}} + H_{\mathrm{soo}} + H_{\mathrm{ss}}, \qquad (21.10)$$

split the nonrelativistic energy levels into a series of closely-spaced fine structure levels. The nuclear spin-orbit interaction

$$H_{\mathrm{so}} = \frac{\alpha^2 Z}{2} \sum_i \frac{1}{r_i^3} (\mathbf{l}_i \cdot \mathbf{s}_i), \qquad (21.11)$$

represents the interaction of the spin and angular magnetic moments of an electron in the field of the nucleus. The spin-other-orbit term

$$H_{\mathrm{soo}} = -\frac{\alpha^2}{2} \sum_{i \neq j} \left(\frac{\mathbf{r}_{ij}}{r_{ij}^3} \times \mathbf{p}_i \right) \cdot (\mathbf{s}_i + 2\mathbf{s}_j), \qquad (21.12)$$

and the spin-spin term

$$H_{\mathrm{ss}} = \alpha^2 \sum_{i<j} \frac{1}{r_{ij}^3} \left\{ \mathbf{s}_i \cdot \mathbf{s}_j - \frac{3}{r_{ij}^2} (\mathbf{s}_i \cdot \mathbf{r}_{ij})(\mathbf{s}_j \cdot \mathbf{r}_{ij}) \right\}, \qquad (21.13)$$

arise from spin-dependent interactions with the other electrons in the system.

21.2 WAVE FUNCTIONS: *LS* AND *LSJ* COUPLING

In the configuration interaction model, the approximate wave function Ψ for a many-electron system is expanded in terms of configuration state functions (CSF).

The assignment of nl quantum numbers to electrons specifies a configuration, often written as $(n_1 l_1)^{q_1} (n_2 l_2)^{q_2} \cdots (n_m l_m)^{q_m}$, where q_i is the occupation of subshell $(n_i l_i)$. Associated with each subshell are one-electron spin-orbitals

$$\phi(r, \theta, \varphi, \sigma) = (1/r) P_{nl}(r) Y_{lm_l}(\theta, \varphi) \chi_{m_s}(\sigma),$$

where $P_{nl}(r)$ is the radial function, $Y_{lm_l}(\theta, \varphi)$ a spherical harmonic, and $\chi_{m_s}(\sigma)$ a spinor. Each CSF is a linear combination of products of one-electron spin-orbitals, one for each electron in the system, such that the sum is an eigenfunction of the total angular momenta operators L^2, L_z and the total spin operators S^2, S_z. It can be considered to be a product of radial factors, one for each electron, an angular and a spin factor obtained by vector coupling methods. It also is required to be antisymmetric with respect to the interchange of any pair of electron coordinates. Often, the specification of the configuration and the final LS quantum numbers is sufficient to define the configuration state, but this is not always the case. Additional information about the order of coupling or the seniority of a subshell of equivalent electrons is needed. Let γ specify the configuration information and any additional information about coupling to uniquely specify the configuration state function denoted by $\Phi(\gamma LS)$.

The wave function for a many-electron system is usually labeled in the same manner as a CSF and generally designates the largest component. Thus, in the LS approximation,

$$\Psi(\gamma LS) = \sum_{\alpha=1}^{M} c_\alpha \Phi(\gamma_\alpha LS).$$

However, cases are known where the configuration states are so highly mixed that no dominant component can be found. Then the assignment is made using other criteria. Clearly no two states should have the same label.

In the LSJ scheme, the angular and spin momenta are coupled to form an eigenstate of the total momenta J^2, J_z. The label often still includes an LS designation, as in $2p^2\,{}^3P_2$, but only the subscript J is a good quantum number. Thus,

$$\Psi(\gamma LSJ) = \sum_{\alpha=1}^{M} c_\alpha \Phi(\gamma_\alpha L_\alpha S_\alpha J). \qquad (21.14)$$

21.3 VARIATIONAL PRINCIPLE

Variational theory shows the equivalence between solutions of the wave equation, $(H - E)\psi = 0$, and stationary solutions of a functional. For bound states where approximate solutions Ψ are restricted to a square integrable subspace, say $\tilde{\mathcal{H}}$, the best solutions are those for which the energy functional

$$\mathcal{E}(\Psi) = \langle \Psi | H | \Psi \rangle / \langle \Psi | \Psi \rangle \qquad (21.15)$$

is stationary. The condition $\delta\mathcal{E}(\Psi) = 0$ leads to

$$\langle \delta\Psi | H - E | \Psi \rangle = 0, \quad \forall\, \delta\Psi \in \tilde{\mathcal{H}}, \ E = \mathcal{E}(\Psi). \qquad (21.16)$$

Several results readily follow. The eigenvalues of H are bounded from below. Let $E_0 \le E_1 \le \cdots$. Then

$$E_0 \le \mathcal{E}(\Psi), \quad \forall\, \Psi \in \tilde{\mathcal{H}}. \qquad (21.17)$$

Consequently, for any approximate wave function, the computed energy is an upper bound to the exact lowest eigenvalue. By the Hylleraas-Undheim-MacDonald theorem (see Sect. 11.3.1), the computed excited states are also upper bounds to the exact eigenvalues, provided that the correct number of states lies below.

21.4 HARTREE-FOCK THEORY

In the Hartree-Fock (HF) approximation, the approximate wave function consists of only one configuration state function. The radial function of each spin-orbital is assumed to depend only on the nl quantum numbers.

These are determined using the variational principle and the nonrelativistic Schrödinger Hamiltonian.

The energy functional can be written as an energy expression for the matrix element $\langle \Phi(\gamma LS) | H | \Phi(\gamma LS) \rangle$. Racah algebra may be used to evaluate the spin-angular contributions, resulting in two types of radial integrals:

One-body. Let $\mathcal{L}$ be the differential operator

$$\mathcal{L} = \frac{d^2}{dr^2} + \frac{2Z}{r} - \frac{\ell(\ell+1)}{r^2}. \qquad (21.18)$$

Then,

$$I(nl, n'l') = -\frac{1}{2}\int_0^\infty P(nl;r)\mathcal{L}P(n'l';r)\,dr. \qquad (21.19)$$

Two-body. The other integrals arise from the multipole expansion of the two-electron part

$$\frac{1}{r_{12}} = \sum_k \frac{r_<^k}{r_>^{k+1}} P^k(\cos\theta), \qquad (21.20)$$

where θ is the angle between the vectors $\mathbf{r}_1$ and $\mathbf{r}_2$, and $r_<$, $r_>$ are the lesser and greater of r_1, r_2, respectively. In general, let a, b, c, d be four nl quantum numbers, two from the left (bra) and two from the right (ket) CSF. Then

$$R^k(ab, cd) = \int_0^\infty \int_0^\infty P(a;r_1)P(b;r_2)$$
$$\times \frac{r_<^k}{r_>^{k+1}} P(c;r_1)P(d;r_2)\,dr_1\,dr_2, \qquad (21.21)$$

which is called a Slater integral. It has the symmetries

$$R^k(ad, cb) \equiv R^k(cb, ad) \equiv R^k(cd, ab) \equiv R^k(ab, cd).$$

In the Hartree-Fock approximation, the Slater integrals that occur depend on only two sets of quantum numbers. These special cases are denoted separately as

$$F^k(a, b) \equiv R^k(ab, ab) \text{ and } G^k(a, b) \equiv R^k(ab, ba). \qquad (21.22)$$

The former is the direct interaction between a pair of orbitals whereas the latter arises from the exchange operator.

The energy expression may be written as

$$\mathcal{E}(\gamma LS) = \sum_i^m q_i \left[I(n_il_i, n_il_i) \right.$$
$$\left. + \frac{q_i - 1}{2} \sum_{k=0}^{2l_i} f_k(i) F^k(n_il_i, n_il_i) \right]$$
$$+ \sum_{j<i} q_i q_j \left[F^0(n_il_i, n_jl_j) \right.$$
$$\left. + \sum_{k=|l_i-l_j|}^{l_i+l_j} g_k(i, j) G^k(n_il_i, n_jl_j) \right]. \qquad (21.23)$$

In general, the coefficients $f_k(i)$ and $g_k(i,j)$ depend not only on the configuration, but also on the coupling. An extremely useful concept, introduced by Slater, is the "average energy of a configuration", $\mathcal{E}(\mathrm{av})$. This is a weighted average of all possible LS terms, where the weighting factor is $(2L+1)(2S+1)$. In this case the coefficients have simple formulas, independent of the coupling:

$$f_k(i) = -\left(\frac{2l_i+1}{4l_i+1}\right)\begin{pmatrix} l_i & k & l_i \\ 0 & 0 & 0 \end{pmatrix}^2,$$
$$(21.24)$$
$$g_k(i,j) = -\left(\frac{2l_i+1}{4l_i+1}\right)\begin{pmatrix} l_i & k & l_j \\ 0 & 0 & 0 \end{pmatrix}^2.$$

Let $\mathcal{I}(P_a)$ be an integral that depends on P_a. Then $\delta\mathcal{I}$ is defined as the first-order term of $\mathcal{I}(P_a + \delta P_a) - \mathcal{I}(P_a)$. To derive the first-order variation of the F^k and G^k integrals (and R^k in general) it is convenient to replace the variables (r_1, r_2) by (r, s) and introduce the functions

$$Y^k(ab;r) = r\int_0^\infty \frac{r_<^k}{r_>^{k+1}} P(a;s)P(b;s)\,ds$$
$$= \int_0^r \left(\frac{s}{r}\right)^k P(a;s)P(b;s)\,ds$$
$$+ \int_r^\infty \left(\frac{r}{s}\right)^{k+1} P(a;s)P(b;s)\,ds \quad (21.25)$$

so that

$$F^k(a,b) = \int_0^\infty P^2(a;r)\left(\frac{1}{r}\right)Y^k(bb;r)\,dr \quad (21.26)$$

and

$$G^k(a,b) = \int_0^\infty P(a;r)P(b;r)\left(\frac{1}{r}\right)Y^k(ab;r)\,dr. \quad (21.27)$$

Then the variations are

$$\delta I(a,b) = -\frac{1}{2}(1+\delta_{a,b})\int_0^\infty \delta P(b;r)\mathcal{L}P(b;r)\,dr,$$
$$\delta F^k(a,b) = 2(1+\delta_{a,b})\int_0^\infty \delta P(a;r)P(a;r)$$
$$\times\left(\frac{1}{r}\right)Y^k(bb;r)\,dr, \quad (21.28)$$
$$\delta G^k(a,b) = 2\int_0^\infty \delta P(a;r)P(b;r)\left(\frac{1}{r}\right)Y^k(ab;r)\,dr.$$

The part of the expression that depends on $P(n_il_i;r)$, for example, is the negative of the removal energy of the entire $(n_il_i)^{q_i}$ subshell, say $-\bar{\mathcal{E}}((n_il_i)^{q_i})$. The stationary condition for a Hartree-Fock solution applies to this expression, but since the variations must be constrained in order to satisfy orthonormality assumptions, Lagrange multipliers λ_{ij} need to be introduced. The stationary condition applies to the functional

$$\mathcal{F}(P(n_il_i)) = -\bar{\mathcal{E}}((n_il_i)^{q_i})$$
$$+ \sum_j \delta_{l_i,l_j}\lambda_{ij}\langle P(n_il_i)|P(n_jl_j)\rangle. \quad (21.29)$$

Applying the variational conditions to each of the integrals, and dividing by $-q_i$, we get the equation

$$\left(\frac{d^2}{dr^2} + \frac{2}{r}[Z - Y(n_il_i;r)] - \frac{l(l+1)}{r^2} - \varepsilon_{ii}\right)P(n_il_i;r)$$
$$= \frac{2}{r}X(n_il_i;r) + \sum_{j\neq i}\delta_{l_i,l_j}\varepsilon_{ij}P(n_jl_j;r), \quad (21.30)$$

where

$$Y(n_il_i;r) = 2(q_i-1)\sum_k f_k(i)Y^k(n_il_in_il_i;r)$$
$$+ \sum_{j\neq i}q_jY^0(n_jl_jn_jl_j;r), \quad (21.31)$$
$$X(n_il_i;r) = \sum_{j\neq i}q_j\sum_k g_k(i,j)Y^k(n_il_in_jl_j;r)P(n_jl_j;r).$$

21.4.1 Diagonal Energy Parameters and Koopmans' Theorem

The diagonal (ε_{ii}) and off-diagonal (ε_{ij}) energy parameters are related to the Lagrange multipliers by $\varepsilon_{ii} = 2\lambda_{ii}/q_i$ and $\varepsilon_{ij} = \lambda_{ij}/q_i$. In fact,

$$\varepsilon_{ii} = \frac{2}{q_i}\left(\bar{E}[(n_il_i)^{q_i}] - 2(q_i-1)\sum_k f_k(i)F^k(n_il_i, n_il_i)\right)$$
$$(21.32)$$

where $\bar{E}[(n_il_i)^{q_i}]$ is the Hartree-Fock value of the removal energy functional $\bar{\mathcal{E}}[(n_il_i)^{q_i}]$. In the special case where $q_i = 1$, ε_{ii} is twice the removal energy, or ionization energy. This is often referred to as Koopmans' Theorem; but, as discussed in Sect. 21.4.3, if a rotation of the radial basis leaves the wave function unchanged while transforming the matrix of energy parameters (ε_{ij}), the removal energies are extreme values obtained by setting the off-diagonal energy parameters to zero. For multiply occupied shells, $\varepsilon_{ii}/2$ can be interpreted as an average removal energy, with a correction arising from the self-interaction.

21.4.2 The Fixed-Core Hartree-Fock Approximation

The above derivation has assumed that the solution is stationary with respect to *all* allowed variations. In practice, it may be convenient to assume that certain radial functions are "fixed" or "frozen". In other words, these radial functions are assumed to be given. Such approximations are often made for core orbitals and so, this is called a fixed-core HF approximation.

21.4.3 Brillouin's Theorem

The Hartree-Fock approximation has some special properties not possessed by other single configuration approximations. One such property is referred to as satisfying Brillouin's theorem, though, in complex systems with multiple open shells of the same symmetry, Brillouin's theorem is not always obeyed.

Let $\Phi^{\mathrm{HF}}(\gamma LS)$ be a Hartree-Fock configuration state, where γ denotes the configuration and coupling scheme. With $\Phi^{\mathrm{HF}}(\gamma LS)$ are associated the m Hartree-Fock radial functions $P^{\mathrm{HF}}(n_1 l_1; r)$, $P^{\mathrm{HF}}(n_2 l_2; r)$, ..., $P^{\mathrm{HF}}(n_m l_m; r)$. These radial functions define the occupied orbitals. To this set may be added virtual orbitals that maintain the necessary orthonormality conditions. Let one of the radial functions (nl) be replaced by another $(n'l)$, either occupied or virtual, without any change in the coupling of the spin-angular factor. Let the resulting function be denoted by $F(nl \rightarrow n'l)$.

The perturbation of the Hartree-Fock radial function, $P(nl; r) \rightarrow P^{\mathrm{HF}}(nl; r) + \epsilon P(n'l; r)$ induces a perturbation $\Phi^{\mathrm{HF}}(\gamma LS) \rightarrow \Phi^{\mathrm{HF}}(\gamma LS) + \epsilon F(nl \rightarrow n'l)$. But the Hartree-Fock energy is stationary with respect to such variations and so,

$$\langle \Phi^{\mathrm{HF}}(\gamma LS) \mid H \mid F(nl \rightarrow n'l)\rangle = 0. \quad (21.33)$$

If the function $F(nl \rightarrow n'l)$ is a CSF for a configuration γ^*, or proportional to one, then Brillouin's theorem is said to hold between the two configuration states. When $n'l$ is a virtual orbital, it may happen that $F(nl \rightarrow n'l)$ is a linear combination of CSFs, as in the $2p \rightarrow 3p$ replacement from $2p^3\,{}^2P$, yielding a linear combination of $\{2p^2({}^1S)3p, 2p^2({}^3P)3p, 2p({}^1D)3p\}$, the linear combination being determined by coefficients of fractional parentage. Thus, Brillouin's theorem will not hold for any of the three individual configuration states in the above equation, only for the linear combination.

When perturbations are constrained by orthogonality conditions between occupied orbitals, the perturbation is of the form of a rotation, where both are perturbed simultaneously,

$$\begin{aligned} P(nl; r) &\rightarrow P(nl; r) + \epsilon P(n'l; r), \\ P(n'l; r) &\rightarrow P(n'l; r) - \epsilon P(nl; r). \end{aligned} \quad (21.34)$$

Then the perturbation has the form $F(nl \rightarrow n'l, n'l \rightarrow -nl)$. For $1s^2 2s\,{}^2S$, the simultaneous perturbations, $F(1s \rightarrow 2s, 2s \rightarrow -1s)$, lead to a linear combination of $\{1s2s^2\,{}^2S, 1s^3\,{}^2S\}$. The CSF for the $1s^3\,{}^2S$ is identically zero by antisymmetry, and so Brillouin's theorem holds for the lithium-like ground state. In the $1s2s\,{}^3S$ state, neither the $1s \rightarrow 2s$ nor the $2s \rightarrow 1s$ substitutions are allowed; in fact, it can be shown that for these states, Brillouin's theorem holds for all mono-excited configurations. The same is not true for $1s2s\,{}^1S$ where the simultaneous perturbations lead to the condition

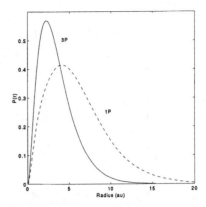

Figure 21.1. A comparison of the $2p$ Hartree-Fock radial functions for the $1s2p\,{}^{1,3}P$ states of Be.

$$\langle \Phi^{\mathrm{HF}}(1s2s\,{}^1S) \mid H \mid [\Phi(1s^2) - \Phi(2s^2)]\rangle = 0. \quad (21.35)$$

Thus, Brillouin's theorem is not obeyed for either the $\Phi(1s^2)$ or $\Phi(2s^2)$ CSF in an HF calculation for $1s2s$.

The importance of Brillouin's theorem lies in the fact that certain interactions have already been included to first order. This has the consequence that certain classes of diagrams can be omitted in many-body perturbation theory [2].

21.4.4 Properties of Hartree-Fock Functions

Term Dependence. The radial distribution for a given nl orbital may depend significantly on the LS term. A well known example is the $1s^2 2s2p$ configuration in Be which may couple to form either a ${}^3P$ or ${}^1P$ term. The energy expression differs only in the exchange interaction, $\pm(1/3)G^1(2s, 2p)$, where the $+$ refers to ${}^1P$ and the $-$ to ${}^3P$. Clearly, the energies of these two terms differ. What is not quite as obvious is the extent to which the $P(2p)$ radial functions differ for the two states. The most affected orbital is the one that is least tightly bound, which in this case is the $2p$ orbital. Figure 21.1 shows the two radial functions. The ${}^1P$ orbital is far more diffuse (not as localized) as the one for ${}^3P$. Such a change in an orbital is called LS term dependence.

Orbital Collapse. Another phenomenon, called *orbital collapse*, occurs when an orbital rapidly contracts as a function of the energy. This could be an LS dependent effect, but it can also occur along an isoelectronic sequence. This effect is most noticeable in the high-l orbitals. In hydrogen, the mean radius of an orbital is $\langle r \rangle = (1/2)[3n^2 - l(l+1)]\,a_0$. Thus, the higher-$l$ orbitals are more contracted; but in neutral systems, the high-l orbitals have a higher energy and are more diffuse. This is due, in part, to the $l(l+1)/r^2$ angular momentum barrier that appears in the definition of the $\mathcal{L}$ operator. In

Table 21.1. The effective quantum number and quantum defect parameters of the $2snd$ Rydberg series in Be.

| n | ^3D | | ^1D | |
	n^*	$\delta(nl)$	n^*	$\delta(nl)$
3	2.968	0.032	3.014	-0.014
4	3.960	0.040	4.012	-0.012
5	4.957	0.043	5.013	-0.013
6	5.955	0.045	6.013	-0.013

the Hartree model, it is possible for $V(r) + l(l + 1)/r^2$ to have two wells: an inner well and an outer shallow well. As the lowest eigenfunction changes rapidly from the outer well to the inner well, as Z changes, orbital collapse is said to occur.

Quantum Defects and Rydberg Series. Spectra of atoms often exhibit phenomena associated with a Rydberg series of states where one of the electrons is in an nl orbital, with n assuming a sequence of values. An example is the $1s^2 2snd$ ^3D series in Be, $n = 3, 4, 5, \dots$. For such a series, a useful concept is that of a quantum defect parameter δ. In hydrogen, the ionization energy (I_P) in atomic units is $1/(2n^2)$. In complex neutral systems, the effective charge would be the same at large r. As n increases, the mean radius becomes larger and the probability of the electron being in the hydrogen-like potential increases. Thus, one could define an effective quantum number, $n^* = n - \delta$, such that

$$I_P(nl) = (1/2)/(n - \delta)^2 . \qquad (21.36)$$

For ionized systems, the equation must be modified to $I_P(nl) = (1/2)[(Z - N + 1)/(n - \delta)]^2$.

Often, this parameter is defined with respect to observed data, but it can also be used to evaluate Hartree-Fock energies, where $I_P = \varepsilon_{nl,nl}/2$, so that $\varepsilon_{nl,nl} = [(Z - N + 1)/(n - \delta)]^2$. Table 21.1 shows the effective quantum number and quantum defect for the Hartree-Fock $2snd$ ^3D and ^1D orbitals in Be as a function of n. For the triplet part the quantum defect is positive whereas for the singlet it is negative. This is the effect of exchange. Note that as n increases the quantum defect becomes constant. This observation is often the basis for determining ionization potentials from observed data.

21.5 MULTICONFIGURATION HARTREE-FOCK THEORY

The Hartree-Fock method predicts many atomic properties remarkably well; but when analyzed carefully, systematic discrepancies can be observed. Consider the ionization potentials tabulated in Table 21.2 compared with the observed values. In these calculations, the energy of the ion was computed using the same radial func-

Table 21.2. Observed and Hartree-Fock ionization potentials for the ground states of neutral atoms, in eV. (See also Table 10.3.)

Atom	Obs.	HF	Diff.
Li	5.39	5.34	0.05
Be	9.32	8.42	0.90
B	8.30	8.43	-0.13
C	11.26	11.79	-0.53
N	14.53	15.44	-0.91
O	13.62	14.45	-0.85
F	17.42	18.62	-1.20
Ne	21.56	23.14	-1.58

tions as for the atom. Thus, no "relaxation" effects were included.

The observed data include other effects as well, such as relativistic effects, finite mass and volume of the nucleus, but these are small for light atoms. For these systems, the largest source of discrepancy arises from the fact that the Hartree-Fock solution is an independent particle approximation to the exact solution of Schrödinger's equation. Neglected entirely is the notion of "correlation in the motion of the electrons"; each electron is assumed to move independently in a field determined by the other electrons. For this reason, the error in the energy was defined by Löwdin [3], to be the correlation energy, that is,

$$E^{\mathrm{corr}} = E^{\mathrm{exact}} - E^{\mathrm{HF}} . \qquad (21.37)$$

In this definition, E^{exact} is not the observed energy – it is the exact solution of Schrödinger's equation which itself is based on a number of assumptions.

21.5.1 Z-dependent Theory

An indication of the important correlation corrections can be obtained from a perturbation theory study of the exact wave function. In the following section, we follow closely the approach taken by Layzer et al. [4] in the study of the Z-dependent structure of the total energy.

Let us introduce a new scaled length, $\rho = Zr$. Then the Hamiltonian becomes

$$H = Z^2 \left(H_0 + Z^{-1}V \right) , \qquad (21.38)$$

where

$$H_0 = -\frac{1}{2} \sum_i \left(\nabla_i^2 + \frac{2}{\rho} \right) , \qquad (21.39)$$

$$V = \sum_{i>j} \frac{1}{\rho_{ij}} , \qquad (21.40)$$

and Schrödinger's equation becomes

$$\left(H_0 + Z^{-1}V\right)\psi = \left(Z^{-2}E\right)\psi. \qquad (21.41)$$

With $Z^{-1}V$ regarded as a perturbation, the expansions of ψ and E in the powers of Z^{-1} are

$$\psi = \psi_0 + Z^{-1}\psi_1 + Z^{-2}\psi_2 + \cdots, \qquad (21.42)$$

in the ρ unit of length, and

$$E = Z^2 \left(E_0 + Z^{-1}E_1 + Z^{-2}E_2 + Z^{-3}E_3 + \cdots\right). \qquad (21.43)$$

The zero-order equation is

$$H_0\psi_0 = E_0\psi_0. \qquad (21.44)$$

The solutions of this equation are products of hydrogenic orbitals.

Let $|(nl)\nu LS\rangle$ be a configuration state function constructed by vector coupling methods from products of hydrogenic orbitals. Here (nl) represents a set of N quantum numbers $(n_1l_1, n_2l_2, \ldots, n_Nl_N)$ and ν any additional quantum numbers such as the coupling scheme or seniority needed to distinguish the different configuration states. Then,

$$H_0|(nl)\nu LS\rangle = E_0|(nl)\nu LS\rangle, \qquad (21.45)$$

$$E_0 = -\frac{1}{2}\sum_i \frac{1}{n_i^2}. \qquad (21.46)$$

Since E_0 is independent of the l_i, different configurations may have the same E_0; that is, E_0 is degenerate. According to first-order perturbation theory for degenerate states, ψ_0 then is a linear combination of the degenerate configuration state functions $|(nl')\nu'LS\rangle$ with the same set of principal quantum numbers n_i and parity π. The coefficients are components of an eigenvector of the interaction matrix $\langle(nl')\nu'LS|V|(nl)\nu LS\rangle$, and E_1 is the corresponding eigenvalue. This is the set of configurations referred to as the *complex* by Layzer [5] and denoted by the quantum numbers $(n)\pi LS$.

The zero-order wave function ψ_0 describes the many-electron system in a general way. It can be shown that the square of the expansion coefficients of ψ_0 over the degenerate set of configuration states can be interpreted as a probability that the many-electron system is in that configuration state, that ψ_1 is then a weighted linear combination of first-order corrections to each such configuration state. Let us now assume the nondegenerate case where $\psi_0 = \Phi(\gamma LS)$. The configurations interacting with γLS are of two types: those that differ by a single electron (single substitution S) and those that differ by two electrons (double substitution D). The former can be further subdivided into three categories:
(i) Those that differ from γLS by one principal quantum number but retain the same spin-angular coupling. These configuration states are part of *radial correlation*.

(ii) Those that differ by one principal quantum number but differ in their coupling. If the only change is the coupling of the spins, the configuration states are part of *spin-polarization*.
(iii) Those that differ in the angular momentum of one electron and are accompanied by a change in angular coupling of the configuration state and possibly also the spin coupling. The latter represent *orbital-polarization*.

The sums over intermediate states involve infinite sums. In practice, the set of orbitals is finite. In the nondegenerate case, these orbitals can be divided into occupied orbitals and unoccupied, or virtual, orbitals depending on whether or not they occur in the *reference configuration* that defines γLS. Single and double (SD) replacements of occupied orbitals by other occupied or virtual orbitals generate the set of configurations that interact with ψ_0. Consider the $1s^22s$ ground state of Li and the $\{1s, 2s, 3s, 4s, 2p, 3p, 4p, 3d, 4d, 4f\}$ set of orbitals. The $1s$ and $2s$ orbitals are occupied and all the other orbitals are virtual orbitals, vl. The set of replacements can then be classified as follows:

Replacement	Configuration	Type of correlation
$1s \rightarrow 2s$	$1s2s^2$	Radial and Spin-polarization
$2s \rightarrow vs$	$1s^2vs$	Radial
$1s \rightarrow vs$	$1svs(^1S)2s$	Radial
	$1svs(^3S)2s$	Spin-polarization
$1s2s \rightarrow vlv'l$	$1svlv'l$	Core-polarization
$1s^2 \rightarrow vlv'l$	$2svlv'l$	Core

The above discussion has considered only the Z-dependence of the wave function, but the notion can readily be extended to other properties. For example, in transition studies, the dipole transition matrix element decreases as $1/Z$, whereas the transition energy increases linearly with Z for $\Delta n = 0$ transitions, and quadratically as Z^2 otherwise. A first-order theory for oscillator strengths (FOTOS) [6] is based on similar concepts.

21.5.2 The MCHF Approximation

In the multiconfiguration Hartree-Fock (MCHF) method, the wave function is approximated by a linear combination of orthogonal configuration states so that

$$\Psi(\gamma LS) = \sum_i^m c_i\Phi(\gamma_i LS), \text{ where } \sum_i^m c_i^2 = 1. \quad (21.47)$$

Then the energy expression becomes

$$\mathcal{E}(\Psi(\gamma LS)) = \sum_i^m \sum_j^m c_ic_jH_{ij} \qquad (21.48)$$

where

$$H_{ij} = \langle \Phi(\gamma_i\, LS) \mid H \mid \Phi(\gamma_j\, LS) \rangle. \qquad (21.49)$$

Because $H_{ij} = H_{ji}$, the sum over i, j may be limited to the diagonals and the lower part of the matrix $\mathbf{H} = (H_{ij})$, called the *interaction* matrix. Let $\mathbf{c} = (c_i)$ be a column vector of the expansion coefficients, also called *mixing coefficients*. Then the energy of the system is

$$E = \mathbf{c}^{\mathrm{T}}\mathbf{H}\mathbf{c}. \qquad (21.50)$$

Let $\mathbf{P}$ be the column vector of radial functions, $[P_a, P_b, \ldots]^{\mathrm{T}}$. Since the interaction matrix elements depend on the radial functions, it is clear that the energy functional depends on both $\mathbf{P}$ and $\mathbf{c}$.

In deriving the MCHF equations, the energy needs to be expressed in terms of the radial functions and $\mathbf{c}$. From the theory of angular momenta, it follows that

$$H_{ij} = \sum_{ab} q_{ab}^{ij} I(a,b) + \sum_{abcd;k} v_{abcd;k}^{ij} R^k(ab, cd), \quad (21.51)$$

where the sum over ab or $abcd$ covers the occupied orbitals of configuration states i and j.

Substituting into the energy expression (21.48), and interchanging the order of summation, we get

$$\mathcal{E}(\Psi) = \sum_{ab} q_{ab} I(a,b) + \sum_{abcd;k} v_{abcd;k} R^k(ab, cd), \quad (21.52)$$

where

$$q_{ab} = \sum_i \sum_j c_i c_j q_{ab}^{ij} \ \text{and}\ v_{abcd;k} = \sum_i \sum_j c_i c_j v_{abcd;k}^{ij}.$$

In this form, the energy is expressed as a list of integrals and their contribution to the energy – a form suitable for the derivation of the MCHF radial equations.

As in the derivation of the Hartree-Fock equations, the stationary principle must be applied to a functional that includes Lagrange multipliers for all the constraints. Thus,

$$\mathcal{F}(\mathbf{P}, \mathbf{c}) = \mathcal{E}(\Psi) + \sum_{a<b} \delta_{\ell_a, \ell_b} \lambda_{ab} \langle a|b \rangle - E \sum_i c_i^2. \quad (21.53)$$

In deriving the stationary conditions with respect to variations in c_i, the most convenient form for $\mathcal{E}(\Psi)$ is Eq. (21.48), which leads to the secular equation

$$\mathbf{H}\mathbf{c} = E\mathbf{c}. \qquad (21.54)$$

Thus, the Lagrange multiplier E is the total energy of the system.

The requirement of a stationary condition with respect to variations in the radial functions leads to a system of equations with exactly the same form as the Hartree-Fock equations except: (i) the occupation numbers q_{aa} are not integers but rather *expected* occupation numbers, and (ii) the function $X(r)$ arises not only from the exchange of electrons within a configuration state, but also from interactions between configuration states.

A solution of the MCHF problem requires the simultaneous solution of the secular equation and the variational radial equations. When the latter are assumed to be given, then only the secular problem needs to be solved and the problem is called a *configuration interaction* (CI) problem. If any radial function is optimized, the calculation is called a *Multi-configuration Hartree Fock* (MCHF) calculation. The iterative procedure for its solution is the MCHF-SCF method, details of which can be found in [7].

21.5.3 Systematic Methods

For MCHF calculations with only a few configuration states, the latter must be carefully chosen. According to first-order perturbation theory, the expansion coefficient $c_i = \langle \Phi_i | H | \Phi_0 \rangle / (E_0 - E_i)$. Clearly, configurations near in energy to the state under investigation are candidates for a strong interaction, but it is also important to remember the numerator in this expression. Generally, the latter is large if the electrons in the two configurations occupy the same region of space. Thus in the ground state of beryllium, the strongest mixing is with $2p^2\,{}^1S$, even though the energy of the latter is far removed from the energy of $2s^2\,{}^1S$. In highly ionized systems, the complex identifies the configurations that might interact strongly, but for neutral systems, there are many exceptions, particularly in atoms such as the transition metals where the $4s$ and $4p$ subshells may be filled before the $3d$ subshell.

For large scale computation, it is desirable to define the configuration states systematically in terms of an active set (AS) of orbitals. Several notations are commonly used for this set. A very simple designation is "the $n = 4$ active set" in that all orbitals up to and including those with principal quantum number $n = 4$ are included. Another notation common in quantum chemistry is one where the number of orbitals of each symmetry is specified, as in $4s3p2d1f$.

Closely associated with this set is the set of configuration states that can be generated. The latter is referred to as the complete active space (CAS). Often the CAS is defined relative to a set of closed shells (or subshells) common to all configurations. For example, the complex for the $3s^2 3p^2\,{}^3P$ ground state of Si, is a neon-like core, $1s^2 2s^2 2p^6$, coupled to the $\{N = 4, \pi = \text{even}, {}^3P\}$ CAS of the active set $\{3s, 3p, 3d\}$. If the AS is extended to $\{3s, 3p, 3d, 4s, 4p, 4d, 4f\}$, then all single, double, triple, quadruple (SDTQ) replacements from the outer four electrons are generated. Of course, the number of CSF increase rapidly, both with the size of the active set and with the number of electrons N defining the CAS. For this reason, other models may be used, such as multi-reference singles and doubles (MR-SD) modeled on the results of Z-dependent perturbation theory. These multi-reference functions may be extended to include all important contributors to a wave function. Calculations in which correlation is restricted to a few outer electrons is called *valence correlation*.

A consequence of a CAS expansion is that the MCHF problem is over determined: a rotation of radial functions of the same symmetry merely transforms the CSFs and the expansion coefficients. The wave function and energy do not change. When similar situations occur in Hartree-Fock theory, Koopmans' theorem requires that the lagrange multiplier associated with the rotation be set to zero. In MCHF calculations, the degree of freedom is usually used to eliminate a CSF or, more precisely, to determine that solution for which a specific CSF has a zero expansion coefficient. A generalized Brillouin's Theorem (GBT) then holds. In the case of the helium ground state (or any ns^2 pair function), applying GBT leads to the *natural orbital expansion* of the form $\{1s^2, 2s^2, 3s^2, \ldots, 2p^2, 3p^2, \ldots, 3d^2, \ldots\}$ in which all CSF differ by two electrons. For other symmetries, such as $1s2p\,^3P$, a *reduced form* [8] can also be defined in which all CSF differ by two electrons, but now involves different sets of orthonormal radial functions for the different partial waves as in $\{1s2p_1, 2s3p_1, 3s4p_1, \ldots, 2p_23d_1, 3p_24d_1, \ldots\}$. Such expansions yield the fastest rate of convergence, but are difficult to apply in complex systems. For a history of Brillouin's theorem and its use in solving multiconfiguration self-consistent field problems see Ref. [9].

21.5.4 Excited States

For ground states or atomic states lowest in their symmetry, the variational procedure is a minimization procedure, and consquently any approximate energy is an upperbound. For all others the energies are stationary. Such calculations may be difficult. One of the most difficult has been the HF calculation for $1s2s\,^1S$ and, once obtained is disappointing since the energy is too low. Excited state calculations become minimization problems through the use of the Hylleraas-Undheim-MacDonald theorem (Sect. 11.3.1). Consider a CAS calculation over the active set $\{1s, 2s\}$ for which the CSFs are $\{1s^2, 1s2s, 2s^2\}$. In determining orbitals, we have a degree of freedom so one CSF may be removed. Selecting $2s^2$ has the consequence that the eigenvalue for the $1s2s$ state to be determined is now the second eigenvalue, and hence an upperbound to the exact.

21.5.5 Autoionizing States

The MCHF variational method may be applied to core excited states imbedded in the continuum, provided that certain CSFs with filled shells are omitted. An example is $1s2s2p\,^2P$ of Lithium. An MCHF calculation omitting $1s^2np$ configuration states represents the localized charge distribution of a state in the continuum. However, the resulting energy is not guaranteed to be an upper bound to the exact energy.

The saddle-point variational method, as described in Chap. 25 is a minimax method for such states.

21.6 CONFIGURATION INTERACTION METHODS

Configuration interaction (CI) methods differ from variational methods like MCHF in that the radial functions are assumed to be fixed, and hence known in advance. There are several situations where these methods can be used effectively:

MCHF with Breit-Pauli. Since the Breit-Pauli operators are valid only as first order perturbations, variational calculations for the Breit-Pauli Hamiltonian are not justified. Instead, MCHF calculations are performed to provide a basis for LSJ wave functions of Eq. 21.14. The expansion coefficients are obtained as a CI calculation.

Full-Core Methods. The full-core plus correlation method (FCPC) [10] is a configuration interaction method. It is mostly used for an atomic system in which a well defined "core" is present. For example, in a system such as the three-electron $1s^2nl$ or the four-electron $1s^2nln'l'$, the system has a $1s1s$ core. In this case, the wave function for an N-electron system is written as

$$\Psi(1, 2, 3, \ldots, N) = \mathcal{A}\phi_{1s1s}(1, 2) \sum_i C_i \Phi_i(3, 4, \ldots, N)$$
$$+ \mathcal{A} \sum_j D_j \psi_j(1, 2, 3, \ldots, N),$$

where $\phi_{1s1s}(1, 2)$, the wave function of the core, is used as a single term, and $\mathcal{A}$ denotes antisymmetrization. The relaxation of the core and other correlation effects are accounted for by the last term in this equation. The correlation effect of the $1s1s$ core is, in general, very strong and it is difficult to fully account for this effect in a conventional N-electron wave function. If inaccurate results are obtained from such a wave function, it may be difficult to distinguish between errors coming from the core part or from the other parts. In the FCPC, the correlation effect in $\phi_{1s1s}(1, 2)$ can be precalculated to a desired accuracy such that ψ_j no longer contains the contribution from the unperturbed core. In physical processes such as ionization, optical transitions and others, the $1s1s$ core wave function in the final state does not change much from that of the initial state. Using the same $\phi_{1s1s}(1, 2)$ for the core may minimize possible errors due to the inaccuracy of the core wave function. The application of this method is not limited to systems with

Table 21.3. Comparison of theoretical and experimental energies for Be $1s^2 2s^2 \, ^1S$ in hartrees. All theoretical values include some form of extrapolation.

Ref.	Method	$-E_{\mathrm{NR}}$	$-E_{\mathrm{rel}}$
[12]	FEM MCHF	14.66737	14.66967
[13]	MCHF	14.667315	
[11]	Full-core CI	14.6673492	14.6696774
[14]	Semi-empirical	14.667353	
[15]	Experiment		14.6696759

$1s1s$ cores. For example, for the lithium-like $1s2snp \, ^4P^o$, for $n \geq 4$, the $1s2s \, ^3S$ can also be considered as an appropriate "core".

From a computational view point, the FCPC wave function has the advantage that it reduces the matrix size of the secular equation substantially. This drastically reduces the memory and CPU time needed on a computer. This method has been quite successful in getting accurate results for four-electron systems as shown in Table 21.3. See Ref. [11] for more details.

Slater Type Orbitals. The essential characteristic of a radial function can be well represented by the expansion

$$P_{nl}(r) = \sum_j c_{jnl} \phi_{jnl}(r),$$

where

$$\phi_{jnl}(r) = \left[\frac{(2\zeta_{jnl})^{2I_{jnl}+1}}{(2I_{jnl})} \right]^{1/2} r^{I_{jnl}} \exp(-\zeta_{jnl} r)$$

is a *Slater Type Orbital* (STO). Optimized sets of parameters have been tabulated for many Hartree-Fock wave functions and these may be used to represent the core [16]. Others may be added and selected orbitals *exponent optimized* (only the exponent is varied) so as to augment the basis. This method has been used effectively by A. Hibbert as implemented in the *Configuration Interaction Version 3* program [17].

Spline Basis. The analytic basis methods described in the previous section have some similarities with MCHF in that linear combinations of STOs first represent orbitals which then define the CSFs. These bases result in extensive cancellation as shown by Hansen *et al.* [18]. An expansion in B-spline basis functions, $B_{i,k}(r)$, which are a basis for a piecewise polynomial subspace (see Chap. 8), provide a more flexible basis with very little cancellation in this representation.

By solving a radial equation with a well chosen potential using spline basis methods, a complete set of orbitals for a piecewise polynomial space can be obtained through diagonalization of the interaction matrix. The resulting basis may then be used in CI calculations. Such

methods have been reviewed recently by Hansen *et al.* [18] and applied to Rydberg series [19].

21.7 ATOMIC PROPERTIES

The discussion so far has concentrated entirely on determining accurate wave functions based on expressions for the energy. Energy levels are a by-product of such calculations, but once a wave function is known a number of atomic properties can be evaluated.

21.7.1 Isotope Effects

The isotope shift observed in atomic transitions can be separated into a mass shift and a field shift. The mass shift is due to differences in the nuclear mass of the isotopes, and is the dominant effect for light atoms. The volume shift arises from the finite volume of the nuclear charge distribution, and is important for heavy atoms. From a physical point of view the field shift is the more interesting, since it yields information about differences in the nuclear charge distribution between the isotopes.

The isotope shift in a transition is given as the difference between the shift for the upper and lower level. The individual shifts are often large, but cancel, and therefore it is necessary to calculate them very accurately in order to get a reliable value for the difference. The energy shifts are evaluated in first-order perturbation theory with wave functions obtained from the zero-order Schrödinger Hamiltonian, H_0.

Mass Shift. Up to this point, the nuclear mass has been taken to be infinite. If instead it has a finite value M, an N-electron atom turns into an $(N+1)$-particle dynamical system. A transformation to center of mass $\mathbf{R}$ plus relative coordinates $\mathbf{r}_i' = \mathbf{r}_i - \mathbf{r}_{nuc}$ yields the transformed Hamiltonian [20]

$$H = \frac{P^2}{2M_{\mathrm{tot}}} + \sum_{i=1}^{N} \frac{p_i'^2}{2\mu} + \sum_{i<j}^{N} \frac{\mathbf{p}_i' \cdot \mathbf{p}_j'}{M} + V(\mathbf{r}_i'), \quad (21.55)$$

where $M_{\mathrm{tot}} = M + Nm$ is the total mass and $\mu = Mm/(M+m)$ is the reduced mass. The first term is the kinetic energy of the center of mass, which can be neglected if $\mathbf{R}$ is an ignorable coordinate. For the remaining terms, introduce the scaled distances $\rho_i = \mathbf{r}_i'/a_\mu$, where $a_\mu = (m/\mu)a_0$ is the scaled Bohr radius. If the potential is entirely Coulombic, then H assumes the form

$$H = -\frac{1}{2} \sum_{i=1}^{N} \nabla_{\rho_i}^2 - \frac{\mu}{M} \sum_{i<j}^{N} \nabla_{\rho_i} \cdot \nabla_{\rho_j} + V(\rho_i) \quad (21.56)$$

in units of e^2/a_μ. This can be written in the form $H = H_0 + H_{\mathrm{MP}}$, where H_0 includes the first and last terms of (21.56), and

$$H_{MP} = -\frac{\mu}{M} \sum_{i<j}^{N} \nabla_{\rho_i} \cdot \nabla_{\rho_j} \quad (e^2/a_\mu) \qquad (21.57)$$

is the additional *mass polarization* term.

Equation (21.56) gives rise to two kinds of mass shifts. First, since H is identical to the infinite mass Hamiltonian, all its eigenvalues E_0 are multiplied by $a_0/a_\mu = \mu/m$, resulting in the *normal mass shift* (nms)

$$\Delta E_{nms} = -(\mu/M)E_0. \qquad (21.58)$$

Second, if H_{MP} is treated as a first order perturbation, then the resulting *specific mass shift* (sms) is

$$\Delta E_{sms} = \langle \Psi_0 \mid H_{MP} \mid \Psi_0 \rangle \quad (e^2/a_\mu) \qquad (21.59)$$

where Ψ_0 is an eigenvector of H_0. The *specific mass shift parameter* S is defined by

$$S = \langle \Psi_0 \mid -\sum_{i<j}^{N} \nabla_{\rho_i} \cdot \nabla_{\rho_j} \mid \Psi_0 \rangle. \qquad (21.60)$$

Field Shift. The field shift of an atomic energy level is due to the extended nuclear charge distribution. The field inside the nucleus deviates from the Coulomb field of a point charge, and this is reflected in the calculated levels. For light atoms, the resulting energy correction to the level E_0 is expressed in terms of the nonrelativistic electron probability at the origin, $|\Psi(0)|^2$:

$$E_{fs} = \frac{2\pi}{3} Z \langle r_N^2 \rangle |\Psi(0)|^2, \qquad (21.61)$$

where $\langle r_N^2 \rangle$ is the mean square radius of the nucleus. For heavier atoms ($Z > 10$) it becomes necessary to include a relativistic correction factor. Numerical values of this correction factor can be found in Ref. [21].

Table 21.4 shows the convergence of an MCHF calculation for the specific mass shift parameter and the electron density at the nucleus of the ground state of Boron II as the active set increases. The calculated $^{11}B - {}^{10}B$ isotope shift is 13.3 mÅ with an estimated uncertainty of 1%. The size of the isotope shift is similar to the limit of resolution of the Goddard High Resolution Spectrograph aboard the Hubble Space Telescope [22].

21.7.2 Hyperfine Effects

The magnetic hyperfine structure (hfs) is due to an interaction between the magnetic field generated by the electrons and the nuclear magnetic dipole moment. The interaction couples the nuclear and electronic angular momenta to a total momentum $\mathbf{F} = \mathbf{I} + \mathbf{J}$, and the interaction energy can be written as the expectation value of a scalar product between an electronic and a nuclear tensor operator [23] (see Chap. 16)

Table 21.4. Specific mass shift parameter and electron density at the nucleus as a function of the Active Set.

Active set	S (a.u.)	$|\psi(0)|^2$
HF	0.000 00	72.629
2s1p	−0.020 17	72.452
3s2p1d	0.625 18	72.490
4s3p2d1f	0.624 81	72.497
5s4p3d2f1g	0.601 69	72.501
6s5p4d3f2g1h	0.598 03	72.503
7s6p5d4f3g2h1i	0.597 09	72.504

$$W_{M1}(J) = \langle \gamma_I \gamma_J IJFM_F \mid \mathbf{T}^{(1)} \cdot \mathbf{M}^{(1)} \mid \gamma_I \gamma_J IJFM_F \rangle. \qquad (21.62)$$

The nuclear operator $\mathbf{M}^{(1)}$ is related to the scalar magnetic dipole moment, μ_I, according to

$$\langle \gamma_I II \mid M_0^{(1)} \mid \gamma_I II \rangle = \mu_I. \qquad (21.63)$$

The magnetic dipole moments are known quantities, obtained with high accuracy from experiments. For a recent tabulation see Ref. [24].

The electronic operator is

$$\mathbf{T}^{(1)} = \frac{\alpha^2}{2} \sum_{i=1}^{N} \left\{ 2\,\mathbf{l}^{(1)}(i) r_i^{-3} - g_s \sqrt{10}\,[\mathbf{C}^{(2)}(i) \times \mathbf{s}^{(1)}(i)]^{(1)} r_i^{-3} \right. $$
$$\left. + g_s \frac{8}{3} \pi \delta(\mathbf{r}_i) \mathbf{s}^{(1)}(i) \right\}, \qquad (21.64)$$

where $g_s = 2.002 319 3$ is the electron spin g-factor, $\delta(\mathbf{r})$ the three-dimensional delta function and $\mathbf{C}_q^{(k)} = \sqrt{4\pi/(2k+1)} Y_{kq}$, with Y_{kq} being a normalized spherical harmonic. The first term of the electronic operator represents the magnetic field generated by the orbiting electric charges and is called the orbital term. The second term represents the field generated by the orbiting magnetic dipole moments, which are coupled to the spin of the electrons. This is the spin-dipole term. The last term represents the contact interaction between the nuclear magnetic dipole moment and the electron magnetic moment. It is called the Fermi contact term and contributes only for s-electrons.

By recoupling I and J, the interaction energy can be rewritten as

$$W_{M1}(J) = (-1)^{I+J-F} W(IJ\,IJ; F1)$$
$$\times \langle \gamma_J J \| \mathbf{T}^{(1)} \| \gamma_J J \rangle \langle \gamma_I I \| \mathbf{M}^{(1)} \| \gamma_I I \rangle, \quad (21.65)$$

where $W(IJ\,IJ; F1)$ is a W coefficient of Racah. When the magnetic dipole interaction constant

$$A_J = \frac{\mu_I}{I} \frac{1}{[J(J+1)(2J+1)]^{\frac{1}{2}}} \langle \gamma_J J \| \mathbf{T}^{(1)} \| \gamma_J J \rangle \quad (21.66)$$

is introduced, the energy is given by

$$W_{M1}(J) = \frac{1}{2} A_J C, \qquad (21.67)$$

where $C = F(F + 1) - J(J + 1) - I(I + 1)$. In theoretical studies, the A-factor is often given as a linear combination of the hyperfine parameters

$$a_l = \langle \gamma LSM_LM_S \mid \sum_i l_0^{(1)}(i)r_i^{-3} \mid \gamma LSM_LM_S \rangle, \quad (21.68)$$

$$a_d = \langle \gamma LSM_LM_S \mid \sum_i 2C_0^{(2)}(i)s_0^{(1)}(i)r_i^{-3} \mid \gamma LSM_LM_S \rangle,$$
$$(21.69)$$

$$a_c = \langle \gamma LSM_LM_S \mid \sum_i 2s_0^{(1)}(i)r_i^{-2}\delta(r_i) \mid \gamma LSM_LM_S \rangle,$$
$$(21.70)$$

where $M_L = L$ and $M_S = S$. These parameters correspond to the orbital, spin-dipole and Fermi contact term of the electronic operator.

The electric hyperfine structure is due to the interaction between the electric field gradient produced by the electrons and the nonspherical charge distribution of the nucleus. The interaction energy is

$$W_{E2}(J) \qquad (21.71)$$
$$= \langle \gamma_I \gamma_J IJFM_F \mid \mathbf{T}^{(2)} \cdot \mathbf{M}^{(2)} \mid \gamma_I \gamma_J IJFM_F \rangle,$$

where the nuclear operator $\mathbf{M}^{(2)}$ is related to the scalar electric quadrupole moment, Q, according to

$$\langle \gamma_I II \mid M_0^{(2)} \mid \gamma_I II \rangle = \frac{Q}{2}. \qquad (21.72)$$

The electronic operator is

$$\mathbf{T}^{(2)} = -\sum_{i=1}^{N} \mathbf{C}^{(2)}(i)r_i^{-3}, \qquad (21.73)$$

and represents the electric field gradient. By introducing the electric quadrupole interaction constant B,

$$B_J = 2Q \left[\frac{J(2J-1)}{(J+1)(2J+1)(2J+3)} \right]^{\frac{1}{2}} \langle \gamma_J J \| \mathbf{T}^{(2)} \| \gamma_J J \rangle, \qquad (21.74)$$

the interaction energy can be written as

$$W_{E2}(J) = B_J \frac{\frac{3}{4}C(C+1) - I(I+1)J(J+1)}{2I(2I-1)J(2J-1)}. \qquad (21.75)$$

In many cases the electric hyperfine interaction is weaker than the magnetic and manifests itself as a small deviation from the Landé interval rule for the magnetic hfs. If the electronic part of the interaction can be calculated

Table 21.5. MCHF Hyperfine constants (in MHz) for the $1s^2 2s 2p$ ^1P state of B II.

Active set	^{10}B		^{11}B	
	A_1	B_1	A_1	B_1
HF	60.06	8.338	179.36	4.001
$2s1p$	60.22	8.360	179.83	4.011
$3s2p1d$	60.98	8.193	182.11	3.932
$4s3p2d1f$	60.05	7.677	179.34	3.684
$5s4p3d2f1g$	60.48	7.764	180.62	3.725
$6s5p4d3f2g1h$	60.85	8.052	181.71	3.864
$7s6p5d4f3g2h1i$	60.81	8.002	181.60	3.840

accurately, a value of the electric quadrupole moment Q, which is a difficult quantity to measure with direct nuclear techniques, can be deduced from the measured B-factor. A recent tabulation of nuclear quadrupole moments is given in Ref. [25].

Table 21.5 shows the convergence of an MCHF active space calculation for two different isotopes for the $1s^2 2s 2p$ ^1P state of B II [22]. Some oscillations are observed since each new "layer" of orbitals may localize in different regions of space.

21.7.3 Metastable States and Lifetimes

States above an ionization threshold may decay via a radiationless transition to a continuum. When the interaction with the continuum is spin-forbidden the state is metastable. The $nsnp^2$ ^4P of negative ions, for example, decay through Breit-Pauli interactions with $nsnp^2$ doublets. The latter, in turn interact with continuum states, thus opening a decay channel. Such metastable states may be treated as bound states.

The foundation for the theory of autoionization was laid down by Fano [26], where he developed a configuration interaction (CI) theory for autoionization. Let $\Psi_b(N; \gamma LS)$ be a normalized, multiconfiguration component of a discrete perturbor for an N-electron system, in which all orbitals are bound orbitals, decaying exponentially for large r. Let Ψ_k be an asymptotically normalized continuum component of the wave function, also for an N-electron system, at energy E, of the form,

$$\Psi_k(N; E\gamma'LS) = |\Psi_b(N-1; \beta\tilde{L}\tilde{S}) \bullet \phi(kl)LS\rangle, \quad (21.76)$$

where $\Psi_b(N-1; \beta\tilde{L}\tilde{S})$ is a bound MCHF wave function for the $(N-1)$-electron target system. Then, in the LS-coupling scheme, the width of the autoionizing state is given by the "Golden Rule"

$$\Gamma = 2\pi V_{E_0}^2, \qquad (21.77)$$

where

$$V_{E_0} = \langle \Psi_b(N; \gamma LS)|H - E_0|\Psi_k(N; E_0\gamma'LS)\rangle . \quad (21.78)$$

A similar formula can be derived for the LSJ-scheme and the Breit-Pauli Hamiltonian. In the above equation, $E_0 = \langle \Psi_b(N; \gamma LS)|H|\Psi_b(N; \gamma LS)\rangle$. The energy of the core, or target, is $E_{\text{target}} = \langle \Psi_b(N - 1; \beta \tilde{L}\tilde{S})|H|\Psi_b(N -1; \beta \tilde{L}\tilde{S})\rangle$, where H in the latter equation is the Hamiltonian for an $(N - 1)$-electron system.

The wave function for the continuum component is assumed to have only one unknown, namely $\phi(kl) = P_{kl}(r)|ls\rangle$, the one-particle continuum function, where $|ls\rangle$ is the known spin-angular part. The radial equation for $P_{kl}(r)$ has exactly the same form as Eq. (21.26), except that $\varepsilon_{nl,nl} = -k^2$, where $E_0 = E_{\text{target}} + k^2/2$. The radial function can be obtained iteratively using an SCF procedure from outward integration.

One of the more accurate calculations of a lifetime is that for He$^-$ $1s2s2p\,^4P_{5/2}$ by Miecznik *et al.* [27]. In this case, a lifetime of 345±10 μs was found and compared with a recent experimental value of 350±15 μs [28]. In this LSJ state, the ^4P interacts with the $1s^2kf\,^2F$. It was found that correlation in the target of the continuum orbital modified the lifetime. In calculations like these, it is always a question whether orthogonality conditions should be applied (as in projection operator formalism (see Chap. 25)). Some theorems relating to this question have been published by Brage, Froese Fischer, and Vaeck [29].

21.7.4   Transition Probabilities

The most fundamental quantity for the probability of a transition from an initial state i to a final state f is the reduced matrix element related to the line strength by

$$S^{1/2} = \langle \Psi_i||O||\Psi_f\rangle , \quad (21.79)$$

where O is the transition operator. In the case of the electric dipole transition, there are two frequently used forms: the length form $O = \sum_j \mathbf{r}_j$, and the velocity form, $O = \sum_j \nabla_j/E_{if}$, where $E_{if} = E_f - E_i$. For exact wave functions, the two forms are equivalent, but for approximate wave functions, the matrix elements in general differ. Thus, the computation of the line strength and the oscillator strength, or f-value, where

$$f = (2/3)E_{if}S/((2S_i + 1)(2L_i + 1)),$$

forms a critical test of the wave function and also the model describing a many-electron system.

Some well-known discrepancies between theory and experiment have existed now for more than a decade for the resonance transitions of Li and Na. For the nonrelativistic $2s - 2p$ transition in Li, a full-core CI [30] calculation produced f-values of (0.74704, 0.74704, 0.75378) for the length-, velocity-, and acceleration form,

Table 21.6. Convergence of transition data for the $1s^22s^22p\,^2P^o \rightarrow 1s^22s2p^2\,^2D$ transition in Boron with increasing active set.

N	gf_l	gf_v	S_l	ΔE (cm^{-1})
3	0.6876	0.8156	2.5534	53197
4	0.2456	0.2696	0.9959	48720
5	0.2625	0.2695	1.0705	48440
6	0.2891	0.2866	1.1868	48125
7	0.2928	0.2900	1.2036	48051
Expt. [a]	0.28(02)			47857

[a] Ref. [33]

respectively. When relativistic corrections were included, the value changed to 0.74715, in agreement with a number of theories, tabulated to only four decimal places. A fast beam-laser experiment by Gaupp *et al.* [31], yielded a value of 0.7416±0.0012: the most accurate theory is well outside the experimental uncertainty. A comparison of theory and experiment is included in [30].

For MCHF calculations of transition data, an important consideration is that the matrix element is between two different states. For independently optimized wave functions, the orbitals of the initial and final states are not orthonormal, as assumed when Racah algebra techniques are used to evaluate the transition matrix element. Through the use of biorthogonal transformations, the orbitals and the coefficients of expansion of the wave functions of the initial and final state can be transformed efficiently so that standard Racah algebra techniques may be applied [32]. Table 21.6 shows the convergence of an MCHF calculation for the ground state of Boron from independently optimized wave functions.

21.7.5   Electron Affinities

By definition, the electron affinity is $A_e = E(A^-) -E(A)$. Thus it is the energy difference between Hamiltonians differing by one electron. Correlation plays a very important role in the binding of the extra electron in the negative ion. It has been known for a long time that the alkali metals have a positive electron affinity, but only recently has it been found, theoretically [34] and experimentally [35], that some of the alkaline earths may also be able to bind an extra electron. The d-electrons need to play a strong role, so Be and Mg, do not have a positive A_e, but according to the most recent experimental measurement, the electron affinity for Ca is 18 meV [36]. A calculation based on the spline methods and using a model potential with adjustable parameters to describe the core, obtained a value of 17.7 meV [37] in close agreement with experiment.

Atomic systems such as Ca are often thought of as

two-electron systems and indeed, for qualitative descriptions, many observations can be explained. A number of physical effects need to be considered when predicting such electron affinities:

(i) Valence Correlation is crucial for obtaining binding.

(ii) Intershell correlation between the valence electrons and the core (core polarization) is also important. The first electron may polarize the core considerably, but this is reduced by the second electron since the two avoid each other dynamically and prefer to be on opposite sides of the core.

(iii) Core rearrangement, which occurs because of the presence of one or more outer electrons, and is particularly large if any of these penetrate the core. In the case of Ca^+, the fixed-core Hartree-Fock energy of $3d\,^2D$ state is 300 meV higher if computed in the fixed potential the Ca^{+2} ion compared with a fully variational calculation!

(iv) Intracore exclusion effects due to the presence of an extra valence electron which reduces the correlation of the core.

(v) Relativistic shift effects, which are present in observed levels and are particularly important for s-electrons.

Model potential methods attempt to capture all but valence correlation in a potential so that calculations for Calcium, for example, can proceed as though for a two-electron system. A review of various theoretical approaches, many of which include different effects, may be found in Ref. [38].

For small systems such as Li, the electron affinity has been computed [39] to experimental accuracy [40] of 0.6176 ± 0.0002 eV.

21.8 SUMMARY

More comprehensive treatments of atomic structure may be found [41,42]. An atomic structure package is available for many of the calculations described here [43]. A review of the application of systematic procedures to the prediction of atomic properties has been published [44].

REFERENCES

1. R. Glass and A. Hibbert, Comput. Phys. Commun. **16**, 19 (1978).
2. I. Lindgren and J. Morrison, *Atomic Many-Body Theory*, (Springer-Verlag, New York, 1982).
3. P.-O. Löwdin, Phys. Rev. **97**, 1509 (1955).
4. D. Layzer, Z. Horák, M. N. Lewis, and D. P. Thompson, Ann. Phys. (N. Y.) **29**, 101 (1964).
5. D. Layzer, Ann. Phys. (N. Y.) **8**, 271 (1959).
6. C. A. Nicolaides and D. R. Beck, Chem. Phys. Lett. **35**, 79 (1975).
7. C. Froese Fischer, Comput. Phys. Rep. **3**, 273 (1986).
8. C. Froese Fischer, J. Comput. Phys. **13**, 502 (1973).
9. M. R. Godefroid, J. Lievin, and J.-Y. Metz, Int. J. Quantum Chem. XL, 243 (1991).
10. K. T. Chung, *Many-Body Theory of Atomic Structure and Photoionization*, edited by T. N. Chang (World Scientific, Singapore, 1993), p. 83.
11. K. T. Chung, X.-W. Zhu, and Z.-W. Wang, Phys. Rev. A **47**, 1740 (1993).
12. A. Mårtensson-Pendrill, S. A. Alexander, J. Olsen, P. Öster, H. M. Quiney, S. Salomonson, and D. Sundholm, Phys. Rev. A **43**, 3355 (1991).
13. C. Froese Fischer, J. Phys. B **26**, 855 (1993).
14. E. Lindroth, H. Persson, S. Salomonson, and A.-M. Mårtensson-Pendrill, Phys. Rev. A **45**, 1493 (1992).
15. R. L. Keely, J. Phys. Chem. Ref. Data, **16**, Suppl. 1 (1987).
16. E. Clementi and C. Roetti, At. Data Nucl. Data Tables, **14**, 177 (1974).
17. A. Hibbert, Comput. Phys. Commun. **9**, 141 (1975).
18. J. E. Hansen, M. Bentley, H. W. van der Hart, M. Landtman, G. M. S. Lister, Y.-T. Shen, and N. Vaeck, Phys. Scr. **T47**, 7 (1993).
19. T. Brage and C. Froese Fischer, Phys. Scr. **49**, 651 (1994).
20. D. S. Hughes and C. Eckart, Phys. Rev. **36**, 694 (1930).
21. W. H. King, *Isotope Shifts in Atomic Spectra* (Plenum, New York, 1984).
22. P. Jönsson, S. G. Johansson, and C. Froese Fischer, Astrophys. J. **429**, L45, (1994).
23. I. Lindgren and A. Rosén, Case Stud. At. Phys. 1772 **4**, 93 (1974).
24. P. Raghavan, At. Data Nucl. Data Tables **42**, 189 (1989).
25. P. Pyykkö, Z. Naturforsch. **47a**, 189 (1992).
26. U. Fano, Phys. Rev. **129**, 1866 (1961).
27. G. Miecznik, T. Brage, and C. Froese Fischer, Phys. Rev. A **47**, 3718 (1993).
28. T. Andersen, L. H. Andersen, P. Balling, H. K. Haugen, P. Hvelplund, W. W. Smith, and K. Taulbjerg, Phys. Rev. A **47**, 890 (1993).
29. T. Brage, C. Froese Fischer, and N. Vaeck, J. Phys. B **26**, 621 (1993).
30. K. T. Chung, in *Proceedings of the International Conference on Highly Charged Ions, Manhattan Kansas, 1992* (American Institute of Physics, 1993).
31. A. Gaupp, P. Kuske, and H. J. Andrä, Phys. Rev. A **26**, 3351 (1982).
32. J. Olsen, M. R. Godefroid, P. Jönsson, P. Å. Malmqvist, and C. Froese Fischer, Phys. Rev. E **52**, 449 (1995).
33. T. R. O'Brian and J. E. Lawler, Astron. Astrophys. **255**, 420 (1992).
34. C. Froese Fischer, J. B. Lagowski, and S. H. Vosko, Phys. Rev. Lett. **59**, 2263 (1987).
35. D. J. Pegg, J. S. Thompson, R. N. Compton, and G. D. Alton, Phys. Rev. Lett. **59**, 2267 (1989).
36. K. W. McLaughlin and D. W. Duquette, Phys. Rev. Lett. **72**, 1176 (1994).
37. H. W. van der Hart, C. Laughlin, and J. E. Hansen, Phys. Rev. Lett. **71**, 1506 (1993).
38. C. Froese Fischer and T. Brage, Can. J. Phys. **70**, 1283 (1992).
39. C. Froese Fischer, J. Phys. B **26**, 855 (1993).
40. J. Dellwo, Y. Liu, D. J. Pegg, and G. D. Alton, Phys. Rev. A **45**, 1544 (1992).

41. I. I. Sobel'man, *Introduction to the Theory of Atomic Spectra* (Pergamon Press, Oxford, 1972).

42. R. D. Cowan, *The Theory of Atomic Structure and Spectra* (University of California Press, Berkeley, 1981).

43. C. Froese Fischer, Computer Phys. Commun. **64**, 369 (1991).

44. C. Froese Fischer and P. Jönsson, Comput. Phys. Commun. **84**, 37 (1994).

22

Relativistic Atomic Structure

I. P. Grant

University of Oxford, Mathematical Institute, 24/29 St. Giles', Oxford OX1 3LB, England

22.1 INTRODUCTION

Relativistic quantum mechanics is required for the description of atoms and molecules whenever their orbital electrons probe regions of space with high potential energy near the atomic nuclei. Primary effects of a relativistic description include changes to spatial and momentum distributions; spin-orbit interactions; quantum electrodynamic corrections such as the Lamb shift; and vacuum polarization. Secondary effects in many-electron systems arise from shielding of the outer electrons by the distributions of electrons in penetrating orbitals; they change orbital binding energies and dimensions and so modify the order in which atomic shells are filled in the lower rows of the Periodic Table.

Relativistic atomic and molecular structure theory can be regarded as a simplification of the fundamental description provided by quantum electrodynamics (QED). This treats the atom or molecule as an assembly of electrons and atomic nuclei interacting primarily by exchanging photons. This model is far too difficult and general for most purposes, and simplifications are required. The most important of these is the representation of the nuclei as classical charge distributions, or even as point particles. Since the motion of the nuclei is usually slow relative to the electrons, it is often adequate to treat the nuclear motion nonrelativistically, or even to start from nuclei in fixed positions, correcting subsequently for nuclear motion.

The emphasis in this chapter is on relativistic methods for the calculation of atomic structure for general many-electron atoms based on an effective Hamiltonian derived from QED in the manner sketched in Sect. 22.3 below. An understanding of the Dirac equation, its solutions and their numerical approximation, is essential material for studying many-electron systems, just as the corresponding properties of the Schrödinger equation underpin Chap. 21.

We shall use atomic units throughout. Where it aids interpretation we shall, however, insert factors of c, m_e and $\hbar$. In these units, the velocity of light, c, has the numerical value $\alpha^{-1} = 137.035\,989\,5(61)$, where α is the fine structure constant.

22.1.1 Relativistic Notation: Minkowski Space-Time

An *event* in Minkowski space-time is defined by a 4-vector $x = \{x^\mu\}$ $(\mu = 0, 1, 2, 3)$ where $x^0 = ct$ is the time coordinate and x^1, x^2, x^3 are Cartesian coordinates in 3-space. The bilinear form[1]

$$(x, y) = x^\mu g_{\mu\nu} y^\nu \,, \tag{22.1}$$

in which

$$g = (g_{\mu\nu}) = (g^{\mu\nu}) = \begin{pmatrix} 1 & 0 & 0 & 0 \\ 0 & -1 & 0 & 0 \\ 0 & 0 & -1 & 0 \\ 0 & 0 & 0 & -1 \end{pmatrix} \tag{22.2}$$

are called *metric coefficients*, defines the metric of Minkowski space.

22.1.2 Lorentz Transformations

Lorentz transformations are defined as linear mappings Λ such that

$$(\Lambda x, \Lambda y) = (x, y) \tag{22.3}$$

so that

$$g_{\mu\nu} = \Lambda^\rho{}_\mu \, g_{\rho\sigma} \, \Lambda^\sigma{}_\nu \,. \tag{22.4}$$

This furnishes 10 equations connecting the 16 components of Λ; at most 6 components can be regarded as independent parameters. The (infinite) set of Λ matrices forms a regular matrix group (with respect to matrix multiplication) called the Lorentz group, $\mathcal{L}$, designated O(3,1) [1, 2].

22.1.3 Classification of Lorentz Transformations

Rotations

Lorentz transformations with matrices of the form

---

[1]The Einstein suffix convention, in which repeated pairs of Greek subscripts are assumed to be summed over all values 0,1,2,3, will be used where necessary in this chapter.

$$\Lambda = \begin{pmatrix} 1 & \mathbf{0}^\top \\ \mathbf{0} & \mathbf{R} \end{pmatrix}, \qquad (22.5)$$

where $\mathbf{R} \in SO(3)$ is an orthogonal 3×3 matrix with determinant $+1$, and $\mathbf{0}$ is a null three dimensional column vector, correspond to three-dimensional space rotations. They form a group isomorphic to $SO(3)$.

Boosts

Lorentz transformations with matrices of the form

$$\Lambda = \begin{pmatrix} \gamma(v) & \gamma(v)\mathbf{v}^\top \\ \gamma(v)\mathbf{v} & \mathbf{I}_3 + (\gamma(v) - 1)\mathbf{n}\mathbf{n}^\top \end{pmatrix}, \qquad (22.6)$$

with $\mathbf{v} = v\mathbf{n}$ a three dimensional column vector, $|\mathbf{n}| = 1$, $v = |\mathbf{v}|$ and $\gamma(v) = (1 - v^2/c^2)^{-1/2}$, are called *boosts*. The matrix Λ describes an 'active' transformation from an inertial frame in which a free classical particle is at rest to another inertial frame in which its velocity is $\mathbf{v}$.

Boosts form a submanifold of $\mathcal{L}$ though they do not in general form a subgroup. However, the set of boosts in a fixed direction $\mathbf{n}$ form a one-parameter subgroup.

Discrete Transformations

The matrices

$$P = \begin{pmatrix} 1 & \mathbf{0}^\top \\ \mathbf{0} & -\mathbf{I}_3 \end{pmatrix}, \ T = \begin{pmatrix} -1 & \mathbf{0}^\top \\ \mathbf{0} & \mathbf{I}_3 \end{pmatrix} \text{with } PT = -\mathbf{I}_4$$

$$(22.7)$$

are called discrete Lorentz transformations and form a subgroup of the Lorentz group along with the identity $\mathbf{I}_4$. The matrix P performs *space* or *parity inversion*; the matrix T performs *time reversal*.

Infinitesimal Lorentz Transformations

The proper Lorentz transformations close to the identity are of particular importance: they have the form

$$\Lambda^\mu{}_\nu = \delta^\mu{}_\nu + \epsilon\omega^\mu{}_\nu + \cdots,$$
$$(\Lambda^{-1})^\mu{}_\nu = \delta^\mu{}_\nu - \epsilon\omega^\mu{}_\nu + \cdots, \qquad (22.8)$$
$$\text{where}$$
$$\omega_{\mu\nu} = -\omega_{\nu\mu}$$

and ϵ is infinitesimal. The *infinitesimal generators*, components $\omega_{\mu\nu}$, can be treated as quantum mechanical observables: see Sect. 22.2.1.

The Lorentz Group

The Lorentz group $\mathcal{L}$ is a Lie group with a six-dimensional group manifold which has four connected components, namely

$$\mathcal{L}^\uparrow_+ \equiv \left\{ \Lambda \in \mathcal{L} \,|\, \Lambda^0{}_0 \geq 1, \det\Lambda = +1 \right\}, \qquad (22.9)$$
$$\mathcal{L}^\uparrow_- \equiv \left\{ \Lambda \in \mathcal{L} \,|\, \Lambda^0{}_0 \geq 1, \det\Lambda = -1 \right\} = P\mathcal{L}^\uparrow_+, \qquad (22.10)$$
$$\mathcal{L}^\downarrow_+ \equiv \left\{ \Lambda \in \mathcal{L} \,|\, \Lambda^0{}_0 \leq 1, \det\Lambda = -1 \right\} = T\mathcal{L}^\uparrow_+, \qquad (22.11)$$
$$\mathcal{L}^\downarrow_+ \equiv \left\{ \Lambda \in \mathcal{L} \,|\, \Lambda^0{}_0 \leq 1, \det\Lambda = +1 \right\} = PT\mathcal{L}^\uparrow_+. \qquad (22.12)$$

The connected component $\mathcal{L}^\uparrow_+$ containing the identity is a Lie subgroup of $\mathcal{L}$ called the *proper Lorentz group*. All its group elements can be obtained from boosts and rotations. It is not simply connected because the subgroup of rotations is not simply connected. The group is also noncompact as the subset of boosts is homeomorphic to $\mathcal{R}^3$.

These topological properties of $\mathcal{L}^\uparrow_+$ are essential for understanding the properties of relativistic wave equations. In particular the multiple connectedness forces the introduction of spinor representations, and to the appearance of half-integer angular momenta or spin.

22.1.4 Contravariant and Covariant Vectors

Contravariant 4-vectors (such as events x) transform according to the rule

$$a^\mu \mapsto a^{\mu\prime} = \Lambda^\mu{}_\nu a^\nu. \qquad (22.13)$$

Covariant 4-vectors can be formed by writing

$$a_\mu = g_{\mu\nu}a^\nu, \qquad (22.14)$$

so that

$$a^\mu a_\mu = a^\mu g_{\mu\nu}a^\nu = (a, a) \qquad (22.15)$$

is invariant with respect to Lorentz transformations. Similarly, we can construct a contravariant 4-vector from a covariant one by writing

$$a^\mu = g^{\mu\nu}a_\nu. \qquad (22.16)$$

The transformation law for covariant vectors is therefore

$$a_\mu \mapsto a'_\mu = [\Lambda^{-1}]^\nu{}_\mu a_\nu. \qquad (22.17)$$

The most important example of a covariant vector is the 4-momentum operator

$$p_\mu = i\frac{\partial}{\partial x^\nu} \quad \mu = 0, 1, 2, 3. \qquad (22.18)$$

From this we derive the *contravariant* 4-momentum operator with components p^μ by writing

$$p^\mu = g^{\mu\nu} p_\nu$$
$$= \left(i\frac{\partial}{\partial x^0}, -i\frac{\partial}{\partial x^1}, -i\frac{\partial}{\partial x^2}, -i\frac{\partial}{\partial x^3} \right), \quad (22.19)$$

in agreement with nonrelativistic expressions.

22.1.5 Poincaré Transformations

More generally, a Poincaré transformation is obtained by combining Lorentz transformations and space-time translations:

$$\Pi(x) = \Lambda x + a. \quad (22.20)$$

The set of all Poincaré transformations, $\Pi = (a, \Lambda)$, with the composition law

$$(a_1, \Lambda_1)(a_2, \Lambda_2) = (a_1 + \Lambda_1 a_2, \Lambda_1 \Lambda_2), \quad (22.21)$$

also forms a group, $\mathcal{P}$.

Properties of the Lorentz and Poincaré groups will be introduced as needed. For a concise account of their properties see Ref. [3]. For more detail on relativistic quantum mechanics in general see textbooks such as [3,4].

22.2 DIRAC'S EQUATION

We present Dirac's equation for an electron in a classical electromagnetic field defined by the 4-potential $A_\mu(x)$:

Covariant Form.
$$\{\gamma^\mu(p_\mu - eA_\mu(x)) - m_{\mathrm{e}}c\}\, \psi(x) = 0. \quad (22.22)$$

where

γ^μ $(\mu = 0,1,2,3)$, are 4×4 matrices.

$\psi(x)$ is a 4-component spinor wave function.

Here, and elsewhere in this chapter, identity matrices are omitted when it is safe to do so.

Dirac Gamma Matrices.

Anticommutation relations:
$$\gamma^\mu \gamma^\nu + \gamma^\nu \gamma^\mu = 2g^{\mu\nu}.$$

Standard representation:

$$\gamma^0 = \begin{pmatrix} 1 & 0 \\ 0 & -1 \end{pmatrix}$$

$$\gamma^i = \begin{pmatrix} 0 & \sigma^i \\ -\sigma^i & 0 \end{pmatrix} \quad i = 1,2,3,$$

where σ_i $i = 1,2,3$ are Pauli matrices [1–4].

Noncovariant Form. In the majority of atomic structure calculations, a frame of reference is chosen in which the nuclear center is taken to be fixed at the origin. In this case it is convenient to write Dirac's equation in noncovariant form. Then functions of

$$x = (x^0, \mathbf{x}),$$

where $x^0 = ct$, can be regarded as functions of the time t and the position 3-vector $\mathbf{x}$, so that (22.22) is replaced by

$$i\frac{\partial}{\partial t}\psi(\mathbf{x}, t) = \hat{\mathbf{h}}_D \psi(\mathbf{x}, t) \quad (22.23)$$

where the scalar and 3-vector potentials are defined by

$$\phi(\mathbf{x}, t) = cA^0(x),$$
$$\mathbf{A}(\mathbf{x}, t) = (A^1(x), A^2(x), A^3(x)), \quad (22.24)$$

and

$$\hat{\mathbf{h}}_D = \left\{ c\boldsymbol{\alpha} \cdot (\mathbf{p} - e\mathbf{A}(\mathbf{x}, t)) + e\phi(\mathbf{x}, t) + \beta m_{\mathrm{e}}c^2 \right\} \quad (22.25)$$

defines the Dirac Hamiltonian. The matrices $\boldsymbol{\alpha}$, with Cartesian components $(\alpha^1, \alpha^2, \alpha^3)$, and β, have the standard representation

$$\beta = \gamma^0 = \begin{pmatrix} 1 & 0 \\ 0 & -1 \end{pmatrix} \quad (22.26)$$

$$\alpha^i = \gamma^0 \gamma^i = \begin{pmatrix} 0 & \sigma^i \\ \sigma^i & 0 \end{pmatrix} \quad i = 1,2,3. \quad (22.27)$$

22.2.1 Characterization of Dirac States

The solutions of Dirac's equation span representations of the Lorentz and Poincaré groups, whose infinitesimal generators can be identified with physical observables. The Lorentz group algebra has 10 independent self-adjoint infinitesimal generators: these can be taken to be the components p^μ of the four-momentum (which generate displacements in each of the four coordinate directions); the three generators, J_i, of rotations about each coordinate axis in space; and the pseudovector w_μ. The irreducible representations can be characterized by invariants

$$(p, p) = m_e^2 c^2, \quad (22.28)$$
$$(w, w) = -m_e^2 c^2 \mathbf{s}^2 = -\tfrac{3}{4} m_e^2 c^2, \quad (22.29)$$

where p is the momentum four-vector and $\mathbf{s}$ is a 3-vector defined in terms of Pauli matrices by

$$s_i = \tfrac{1}{2}\sigma_i, \quad i = 1,2,3.$$

which can be interpreted as the electronic angular momentum (intrinsic spin) in its rest frame. For more detail see [3] and the original papers [5,6].

22.2.2   The Charge-Current 4-Vector

Dirac's equation (22.22) is covariant with respect to Lorentz (22.3) and Poincaré (22.20) transformations, provided that there exists a nonsingular 4×4 matrix $S(\Lambda)$ with the property

$$\psi'(x) = S(\Lambda)\psi(\Lambda^{-1}(x - a)). \qquad (22.30)$$

The matrices $S(\Lambda)$ are characterized by the equation

$$S^{-1}(\Lambda)\gamma^\lambda S(\Lambda) = \Lambda^\lambda{}_\mu \gamma^\mu. \qquad (22.31)$$

The most important observable expression required in this chapter is the *charge-current four-vector*

$$j^\mu = ec\overline{\psi}(x)\gamma^\mu\psi(x), \qquad (22.32)$$

where the Dirac adjoint is defined by

$$\overline{\psi}(x) = \psi^\dagger(x)\gamma^0, \qquad (22.33)$$

and the dagger denotes spinor conjugation and transposition. Since

$$\begin{aligned}\overline{\psi}'(x) &= \overline{\psi}(\Lambda^{-1}(x-a))\gamma^0 S(\Lambda)^\dagger \gamma^0 \\ &= \overline{\psi}(\Lambda^{-1}(x-a))S^{-1}(\Lambda),\end{aligned}$$

$j^\mu(x)$ transforms as a 4-vector

$$j^{\mu\prime}(x) = \Lambda^\mu{}_\nu j^\nu(x)$$

by virtue of (22.31). The component $j^0(x)$ can be interpreted as a multiple of the *charge density* $\rho(x)$,

$$j^0(x) = ec\rho(x) = ec\overline{\psi}(x)\gamma^0\psi(x) = ec\psi^\dagger(x)\psi(x) \qquad (22.34)$$

and the space-like components as the *current density*

$$j^i(x) = ec\overline{\psi}(x)\gamma^i\psi(x) = ec\psi^\dagger(x)\alpha^i\psi(x). \qquad (22.35)$$

The charge-current density satisfies a continuity equation, which in noncovariant form reads

$$\frac{\partial\rho(x)}{\partial t} + \sum_{i=1}^{3}\frac{\partial j^i(x)}{\partial x^i} = 0,$$

or, in covariant notation,

$$\partial_\mu j^\mu = 0. \qquad (22.36)$$

This is readily proved by using the Dirac equation (22.22) and its Dirac adjoint. Equation (22.36) is clearly invariant under Poincaré transformations, and this yields the important property that *electric charge* is conserved in Dirac theory.

22.3   QED: RELATIVISTIC ATOMIC AND MOLECULAR STRUCTURE

22.3.1   The QED Equations of Motion

The conventional starting point [7–10] for deriving equations of motion in quantum electrodynamics (QED) is a Lagrangian density of the form

$$\mathcal{L}(x) = \mathcal{L}_{\text{em}}(x) + \mathcal{L}_{\text{e}}(x) + \mathcal{L}_{\text{int}}(x). \qquad (22.37)$$

The first term is the Lagrangian density for the free electromagnetic field, $F^{\mu\nu}(x)$,

$$\mathcal{L}_{\text{em}}(x) = -\tfrac{1}{4}F^{\mu\nu}F_{\mu\nu}, \qquad (22.38)$$

the second term is the Lagrangian density for the electron-positron field in the presence of the external potential $A_{\text{ext}}^\mu(x)$,

$$\mathcal{L}_{\text{e}}(x) = \overline{\psi}(x)\{\gamma_\mu(p^\mu - eA_{\text{ext}}^\mu(x)) - m_{\text{e}}c\}\psi(x). \qquad (22.39)$$

We assume that the electromagnetic fields are expressible in terms of the four-potentials by

$$F^{\mu\nu} = \partial^\mu A_{\text{tot}}^\nu - \partial^\nu A_{\text{tot}}^\mu,$$

where

$$A_{\text{tot}}^\mu(x) = A_{\text{ext}}^\mu(x) + A^\mu(x)$$

is the sum of a four-potential $A_{\text{ext}}^\mu(x)$ describing the fields generated by classical external charge-current distributions, and a quantized field $A^\mu(x)$ which through

$$\mathcal{L}_{\text{int}}(x) = -j_\mu(x)A^\mu(x), \qquad (22.40)$$

accounts for the interaction between the uncoupled electrons and the radiation field. The field equations deduced from (22.37) are

$$\begin{aligned}\{\gamma_\mu(p^\mu - A_{\text{ext}}^\mu(x)) - m_{\text{e}}c\}\,\psi(x) &= j_\mu(x)A^\mu(x) \\ \partial_\mu F^{\mu\nu}(x) &= j^\nu(x),\end{aligned} \qquad (22.41)$$

and clearly exhibit the coupling between the fields.

Quantum electrodynamics requires the solution of the system (22.41) when $A^\mu(x)$, $\psi(x)$ and its adjoint $\overline{\psi}(x)$ are quantized fields. This formulation is purely formal: it ignores all questions of zero-point energies, normal ordering of operators, choice of gauge associated with the quantized photon field, or the need to include (infinite) counterterms to render the theory finite.

22.3.2   The Quantized Electron-Positron Field

Furry's bound interaction picture of QED [11, 7] exploits the fact that a one-electron model is often a good starting point for a more accurate calculation of atomic or molecular properties. The electrons are described by a field operator

$$\psi(x) = \sum_{E_m > E_F} a_m \psi_m(x) + \sum_{E_n < E_F} b_n^\dagger \psi_n(x), \quad (22.42)$$

where $E_F \geq -mc^2$ is a "Fermi level" separating the states describing electrons (bound and continuum) from the positron states (lower continuum) in the chosen time-independent model potential $V(\mathbf{r})$. Equation (22.42) is written as if the spectrum were entirely discrete, as in finite matrix models; more generally, this must be replaced by integrals over the continuum states together with a sum over the bound states. We assume that the amplitudes $\psi_m(x)$ are orthonormalized (which can be achieved, for example, by enclosing the system in a finite box). The operators a_m and $a_m^\dagger$ respectively annihilate and create electrons, and b_n and $b_n^\dagger$ perform the same role for vacancies in the "negative energy" states, which we interpret as antiparticles (positrons). These operators satisfy the *anticommutation* rules (see Sect. 6.1.1)

$$\{a_m, a_{m'}^\dagger\} = \delta_{m,m'},$$
$$\{b_n, b_{n'}^\dagger\} = \delta_{n,n'}, \quad (22.43)$$

where $\{a, b\} = ab + ba$. All other anticommutators vanish. The operator representing the number of electrons in state m is then

$$N_m = a_m^\dagger a_m, \quad (22.44)$$

having the eigenvalues 0 or 1; the states of a system of noninteracting electrons and positrons can therefore be labeled by listing the occupation numbers, 0 or 1 of the one-electron states participating.

We define the *vacuum state* as the (reference) state $|0\rangle$ in which $N_m = N_n = 0$ for all m, n, so that

$$a_m|0\rangle = b_n|0\rangle = 0. \quad (22.45)$$

The operator representing the total number of particles is given by

$$N = \int \psi^\dagger(x)\psi(x)\,d\mathbf{x} = \sum_{E_m > E_F} N_m + \sum_{E_n < E_F}(1 - N_n).$$

This is not quite satisfactory: $N = \sum_{E_n < E_F} 1$ is infinite for the vacuum state, as are the total charge and energy of the vacuum.

These infinite "zero-point" values can be eliminated by introducing *normal ordered operators*. A product of annihilation and creation operators is in normal order if it is rearranged so that all annihilation operators are to the right of all creation operators. Such a product has a null value in the vacuum state. In performing the rearrangement, each anticommutator is treated as if it

were zero. We denote normal ordering by placing the operators between colons. Thus $: a_m^\dagger a_m := a_m^\dagger a_m$ whilst $: b_n b_n^\dagger := -b_n^\dagger b_n$. This means that if we *redefine* N by

$$N = \int : \psi^\dagger(x)\psi(x) : d\mathbf{x} = \sum_{E_m > E_F} N_m - \sum_{E_n < E_F} N_n, \quad (22.46)$$

then $\langle 0|N|0\rangle = 0$. The same trick eliminates the infinity from the total energy of the vacuum;

$$H_0 = \int : \psi^\dagger(x)\hat{\mathbf{h}}_D\psi(x) : d\mathbf{x}$$
$$= \sum_{E_m > E_F} N_m E_m - \sum_{E_n < E_F} N_n E_n, \quad (22.47)$$

so that $\langle 0|H_0|0\rangle = 0$.

The current density operator is given by the commutator of two field variables

$$j^\mu(x) = -\tfrac{1}{2}ec[\bar{\psi}(x)\gamma^\mu, \psi(x)], \quad (22.48)$$

where the Dirac adjoint, $\bar{\psi}(x)$ is defined by (22.33). The definition (22.48) differs from (22.32) by expressing the total current as the difference between the electron (negatively charged) and positron (positively charged) currents. We can write

$$j^\mu(x) =: j^\mu(x) : + ec\,\mathrm{Tr}[\gamma^\mu S_F(x, x)], \quad (22.49)$$

where $S_F(x, y)$ is the Feynman causal propagator, defined below. Since $\langle 0| : j^\mu(x) : |0\rangle = 0$, the last term in (22.49) is the *vacuum polarization current* due to the asymmetry between positive and negative energy states induced by the external field. From this, the net charge of the system is

$$Q = \frac{1}{c}\int j^0(x)\,d^3x \quad (22.50)$$
$$= -e\left(\sum_{E_m > E_F} N_m - \sum_{E_n < E_F} N_n\right) + Q_{\mathrm{vac}}.$$

Q_{vac} is the total charge of the vacuum, which vanishes for free electrons, but is finite in the presence of an external field (the phenomenon of *vacuum polarization*). Note that whilst Q is conserved for all processes, the total number of particles need not be; it is always possible to add virtual states incorporating electron-positron pairs without changing Q.

22.3.3 Quantized Electromagnetic Field

The four-potential of the quantized electromagnetic field can be expressed in terms of a spectral expansion over the field modes in, for example, plane waves

$$A_\mu(x) = \int \frac{d^3k}{2(2\pi)^3 k_0} \sum_{\lambda=0}^{3} [q^{(\lambda)}(k)\epsilon_\mu^{(\lambda)}(k)e^{-ik\cdot x}$$
$$+ q^{(\lambda)\dagger}(k)\epsilon_\mu^{(\lambda)*}(k)e^{ik\cdot x}]. \qquad (22.51)$$

The vectors $\epsilon_\mu^{(\lambda)}(k)$ describe the polarization modes; there are four linearly independent vectors, which may be assumed real, for each k on the positive light cone. Two of these ($\lambda = 1, 2$) can be chosen perpendicular to the photon momentum **k** (*transverse polarization*); one component ($\lambda = 3$) along **k** (*longitudinal polarization*); and the final component ($\lambda = 0$) is time-like (*scalar polarization*). The operators $q^{(\lambda)}(k)$ and $q^{(\lambda)\dagger}(k)$ describe respectively photon absorption and emission. They satisfy commutation relations

$$[q^{(\lambda)}(k), q^{(\lambda')\dagger}(k')] = \delta_{\lambda,\lambda'}\delta_{k,k'}, \quad \lambda, \lambda' = 1, 2, 3$$
$$[q^{(0)}(k), q^{(0)\dagger}(k')] = -\delta_{k,k'}; \qquad (22.52)$$

all other commutators vanish. The photon vacuum state, $|0\rangle_\gamma$, is such that

$$q^{(\lambda)}(k)|0\rangle_\gamma = 0. \qquad (22.53)$$

Further details may be found in the texts [7–10].

22.3.4 QED Perturbation Theory

The textbook perturbation theory of QED, see for example [7, 8, 10, 12] and other works, has been adapted for applications to relativistic atomic and molecular structure and is also the source of methods of nonrelativistic many-body perturbation theory (MBPT). We offer a brief sketch emphasizing details not found in the standard texts.

The Perturbation Expansion

The Lagrangian approach leads to an interaction Hamiltonian

$$H_I = -\int j^\mu(x)A_\mu(x)\,d\mathbf{x}. \qquad (22.54)$$

In the interaction representation, this gives an equation of motion

$$i\hbar\partial_t|\Phi(t)\rangle = H_I(t)|\Phi(t)\rangle, \qquad (22.55)$$

where $|\Phi(t)\rangle$ is the QED state vector, and

$$H_I(t) = \exp(iH_0 t/\hbar)H_I\exp(-iH_0 t/\hbar),$$

where $H_0 = H_{em} + H_e$ is the Hamiltonian for the uncoupled photon and electron-positron fields. If $S(t, t')$ is the *time development* operator such that

$$|\Phi(t)\rangle = S(t, t')|\Phi(t')\rangle,$$

then

$$i\hbar\partial_t S(t, t') = H_I(t)S(t, t').$$

The equivalent integral equation, incorporating the inital condition $S(t, t) = 1$,

$$S(t, t_0) = 1 - \frac{i}{\hbar}\int_{t_0}^{t} H_I(t_1)S(t_1, t_0)d_1 t, \qquad (22.56)$$

can be solved iteratively, giving

$$S(t, t_0) = 1 + \sum_{n=1}^{\infty} S^{(n)}(t, t_0), \qquad (22.57)$$

where

$$S^{(n)}(t, t_0) = (-i/\hbar)^n \int_{t_0}^{t} dt_1 \int_{t_0}^{t_1} dt_2 \cdots \int_{t_0}^{t_{n-1}} dt_n$$
$$\times H_I(t_1)H_I(t_2)\cdots H_I(t_n).$$

This can be put into a more symmetric form by using *time-ordered* operators. Define the T-product of two operators by

$$T(A(t_1)B(t_2)) = \begin{cases} A(t_1)B(t_2), & t_1 > t_2 \\ \pm B(t_2)A(t_1), & t_2 > t_1 \end{cases} \qquad (22.58)$$

where the *positive* sign refers to the product of photon operators and the *negative* sign to electrons. Then

$$S^{(n)}(t, t_0) = \frac{(-i/\hbar)^n}{n!}\int_{t_0}^{t} dt_1 \int_{t_0}^{t} dt_2 \cdots \int_{t_0}^{t} dt_n$$
$$\times T(H_I(t_1)H_I(t_2)\cdots H_I(t_n)). \qquad (22.59)$$

The operator $S(t, t')$ relates the state vector at time t to the state vector at some earlier time $t' < t$. Its matrix elements therefore give the transition amplitudes for different processes, for example the emission or absorption of radiation by a system, or the outcome of scattering of a projectile from a target. The techniques for extracting cross-sections and other observable quantities from the S-operator are described at length in the texts [7, 8, 10, 12].

Although the use of normal ordering means that the charge and mass of the reference state, the vacuum, is zero, it fails to remove other infinities due to the occurrence of divergent integrals. The method of extracting finite quantities from this theory involves *renormalization* of the charge and mass of the electron. We shall refer especially to [10, Chap. 8] for a detailed discussion. The most difficult technical problems are posed by *mass renormalization*. Formally, we modify the interaction Hamiltonian to read

$$j^\mu(x)A_\mu(x) - \delta M(x),$$

where $\delta M(x)$ is the *mass renormalization operator*

$$\delta M(x) = \tfrac{1}{2}\delta m[\bar{\psi}(x), \psi(x)]$$

where δm is infinite.

A further problem is that electrons in a many-electron atom or molecule move in a potential which is quite unlike that of the bare nucleus. It is therefore useful to introduce a local *mean field* potential, say $U(x)$, representing some sort of average interaction with the rest of the electron charge distribution, so that the zero-order orbitals satisfy

$$\left\{ c\boldsymbol{\alpha}\cdot\mathbf{p} + \beta c^2 + V_{\text{nuc}}(\mathbf{x}) + U(\mathbf{x}) - E_m \right\} \psi_m(\mathbf{x}) = 0 \,.$$
$$(22.60)$$

With this starting point, the interaction Hamiltonian becomes

$$H_I(x) = H_I^{(1)}(x) + H_I^{(2)}(x) \,, \qquad (22.61)$$

where

$$H_I^{(1)}(x) = -U(x), \quad H_I^{(2)}(x) = j^\mu(x)A_\mu(x) - \delta M(x) \,,$$

and the electron current is defined in terms of the mean field orbitals of (22.60). The expression $H_I^{(2)}(x)$ is sometimes referred to as a *fluctuation potential*. The term $j^\mu(x)A_\mu(x)$ is proportional to the electron charge, e, which serves as an ordering parameter for perturbation expansions.

Effective Interactions

Although the S-matrix formalism provides in principle a complete computational scheme for many-electron systems, it is generally too cumbersome for practical use, and approximations are necessary. Usually, this is a matter of selecting a subset of dominant contributions to the perturbation series depending on the application. We are faced with the evaluation of T-products of the form

$$T(\phi(t_1)\phi(t_2)\cdots\phi(t_n))$$

which is done using Wick's Theorem [10, p. 25].

In the simplest case,

$$T(\phi(t_1)\phi(t_2)) =: \phi(t_1)\phi(t_2) : + \langle 0 \mid T(\phi(t_1)\phi(t_2)) \mid 0 \rangle \,.$$
$$(22.62)$$

The vacuum expectation value is called a *contraction*. More generally, we have

$$T(\phi(t_1)\phi(t_2)\cdots\phi(t_n))$$
$$=: \phi(t_1)\phi(t_2)\cdots\phi(t_n) :$$
$$+ [\langle 0|T(\phi(t_1)\phi(t_2))|0\rangle : \phi(t_3)\cdots\phi(t_n) : +\text{permutations}]$$
$$+ [\langle 0|T(\phi(t_1)\phi(t_2))|0\rangle \langle 0|T(\phi(t_3)\phi(t_4))|0\rangle$$
$$\times : \phi(t_5)\cdots\phi(t_n) : +\text{permutations}]$$

This result has the effect that a T-product with an odd number of factors vanishes. A rigorous statement can be found in all standard texts; each term in the expansion gives rise to a Feynman diagram which can be interpreted as the amplitude of a physical process. As an example, consider the simple but important case

$$S^{(2)} = \frac{(ie)^2}{2!} T(j^\mu(x)A_\mu(x).j^\nu(y)A_\nu(y)) \,. \qquad (22.63)$$

One of the terms (there are others) found by using Wick's Theorem is

$$j^\mu(x)j^\nu(y)\langle 0|T(A_\mu(x)A_\nu(y))|0\rangle \,.$$

We see that this involves the contraction of two photon amplitudes

$$-\tfrac{1}{2}D_{F\mu\nu}(x-y) = \langle 0 \mid T(A_\mu(x)A_\nu(y)) \mid 0 \rangle \,,$$

which plays the role of a *propagator* [see Eq. (22.69)]: it relates the photon amplitudes at two space-time points x, y. With the introduction of a spectral expansion for the electron current (22.48), the contribution to the energy of the system becomes

$$\frac{1}{2}\sum_{pqrs} : a_p^\dagger a_q^\dagger a_s a_r : \langle pq|V|rs\rangle \,, \qquad (22.64)$$

which can be interpreted, in the familiar language of ordinary quantum mechanics, as the energy of two electrons due to the electron-electron interaction V which is directly related to the photon propagator.

22.3.5 Propagators

Propagators relate field variables at different space-time points. Here we briefly define those most often needed in atomic and molecular physics.

Electrons

Define Feynman's causal propagator for the electron-positron field by the contraction

$$S_F(x_2, x_1) = \langle 0|T(\psi(x_2)\bar{\psi}(x_1))|0\rangle \,. \qquad (22.65)$$

This has a spectral decomposition of the form

$$S_F(x_2, x_1) = \begin{cases} \sum_{E_m > E_F} \psi_m(x_2)\bar{\psi}_m(x_1) & t_1 > t_2 \,, \\ -\sum_{E_n < E_F} \psi_n(x_2)\bar{\psi}_n(x_1) & t_1 < t_2 \,, \end{cases}$$
$$(22.66)$$

which ensures that positive energy solutions are propagated forwards in time, and negative energy solutions backwards in time in accordance with the antiparticle interpretation of the negative energy states. By noting that the stationary state solutions $\psi_m(x)$ have time dependence $\exp(-iE_m t)$, we can write (22.66) in the form

$$
\begin{aligned}
S_F(x_2, x_1) &= \frac{1}{2\pi i} \int_{-\infty}^{\infty} \sum_n \frac{\psi_n(\mathbf{x}_2)\bar{\psi}_n(\mathbf{x}_1)}{E_n - z(1 + i\delta)} e^{-iz(t_2 - t_1)} \, dz \\
&= \frac{1}{2\pi i} \int_{-\infty}^{\infty} G(\mathbf{x}_2, \mathbf{x}_1, z)\gamma^0 e^{-iz(t_2 - t_1)} \, dz \,,
\end{aligned}
$$

(22.67)

where δ is a small positive number, the sum over n includes *the whole spectrum*, and where the Green's function $G(\mathbf{x}_2, \mathbf{x}_1, z)$, in the specific case in which the potential of the external field $a^\mu(x)$ has only a scalar time-independent part, $V_{\text{nuc}}(\mathbf{x})$, satisfies

$$
\begin{aligned}
\{ c\boldsymbol{\alpha} \cdot \mathbf{p} + \beta c^2 &+ (V_{\text{nuc}}\mathbf{x}) - z \} G(\mathbf{x}, \mathbf{y}, z) \\
&= \delta^{(3)}(\mathbf{x}_2 - \mathbf{x}_1) \,.
\end{aligned}
$$

(22.68)

$G(\mathbf{x}_2, \mathbf{x}_1, z)$ is a meromorphic function of the complex variable z with branch points at $z = \pm c^2$, and cuts along the real axis (c^2, ∞) and $(-\infty, -c^2)$. The poles lie on the segment $(-c^2, c^2)$ at the bound eigenvalues of the Dirac Hamiltonian for this potential.

Photons

The photon propagator $D_{F\mu\nu}(x_2 - x_1)$ is constructed in a similar manner:

$$
-\tfrac{1}{2} D_{F\mu\nu}(x_2 - x_1) = \langle 0 | T(A_\mu(x_2)A_\nu(x_1)) | 0 \rangle, \quad (22.69)
$$

where μ_0 is the permeability of the vacuum. This has the integral representation

$$
\begin{aligned}
D_{F\mu\nu}(x_2 - x_1) &= g_{\mu\nu} D_F(x_2 - x_1) \\
&= -g_{\mu\nu} \frac{i}{(2\pi)^4} \int d^4 q \, D(q^2),
\end{aligned}
$$

(22.70)

where

$$
D(q^2) = \frac{1}{q^2 + i\delta} \,,
$$

and δ is a small positive number. This is not unique, as the four-potentials depend on the choice of gauge; for details see [8, Sect. 77]. The various forms for the electron-electron interaction given below express such gauge choices.

22.3.6 Effective Interaction of Electrons

The expression (22.63) can be viewed in several ways: it is the interaction of the current density $j^\mu(x)$ at the space-time point x with the four-potential due to the current $j^\mu(y)$; the interaction of the current density $j^\mu(y)$

with the four-potential due to the current $j^\mu(x)$; or, as is commonly assumed in nonrelativistic atomic theory, the effective interaction between two charge density distributions, as represented by (22.64). In terms of the corresponding Feynman diagram, it can be thought of as the energy due to the exchange of a virtual photon.

The form of V depends on the choice of gauge potential, as follows.

Feynman gauge

$$
\begin{aligned}
&\langle pq|V|rs \rangle \\
&= \int\int \psi_p^\dagger(\mathbf{x})\psi_r(\mathbf{x}) v_{sq}^F(\mathbf{x}, \mathbf{y})\psi_q^\dagger(\mathbf{y})\psi_s(\mathbf{y}) d^3x \, d^3y \quad (22.71)
\end{aligned}
$$

where

$$
v_{sq}^F(\mathbf{x}, \mathbf{y}) = \frac{e^{i\omega_{sq} R}}{R}(1 - \boldsymbol{\alpha_x} \cdot \boldsymbol{\alpha_y}), \quad (22.72)
$$

with

$$
\mathbf{R} = \mathbf{x} - \mathbf{y}, \quad R = |\mathbf{R}|, \quad \omega_{sq} = \frac{E_s - E_q}{c} \,.
$$

This interaction gives both a real and an imaginary contribution to the energy; only the former is usually taken into account in structure calculations. Since the orbital indices are dummy variables, it is usual to symmetrize the interaction kernel by writing

$$
\bar{v}^F(\mathbf{x}, \mathbf{y}) = \frac{1}{2}\left(v_{sq}^F(\mathbf{x}, \mathbf{y}) + v_{rp}^F(\mathbf{x}, \mathbf{y}) \right), \quad (22.73)
$$

which places the orbitals on an equal footing.

Coulomb gauge

Here the Feynman propagator is replaced by that for the Coulomb gauge, giving

$$
\begin{aligned}
v_{sq}^T(\mathbf{x}, \mathbf{y}) = \frac{e^{i\omega_{sq} R}}{R} &- \left(\boldsymbol{\alpha_x} \cdot \boldsymbol{\alpha_y} \frac{e^{i\omega_{sq} R}}{R} \right. \\
&\left. + (\boldsymbol{\alpha_x} \cdot \boldsymbol{\nabla})(\boldsymbol{\alpha_y} \cdot \boldsymbol{\nabla}) \frac{e^{i\omega_{rp} R} - 1}{\omega_{rp}^2 R} \right)
\end{aligned}
$$

in which the operator $\boldsymbol{\nabla}$ involves differentiation with respect to $\mathbf{R}$.

Symmetrization is also used with this interaction.

Breit Operator

The *low frequency limit*, $\omega_{rp} \to 0$, $\omega_{sq} \to 0$, is known as the Breit interaction:

$$
\lim_{\omega_{rp}, \omega_{sq} \to 0} v_{sq}^T(\mathbf{x}, \mathbf{y}) = \frac{1}{R} + v^B(\mathbf{R}), \quad (22.74)
$$

where

$$
v^B(\mathbf{R}) = -\frac{1}{2R}\left(\boldsymbol{\alpha_x} \cdot \boldsymbol{\alpha_y} + \frac{\boldsymbol{\alpha_x} \cdot \mathbf{R} \, \boldsymbol{\alpha_y} \cdot \mathbf{R}}{R^2} \right).
$$

Gaunt Operator

This is a further approximation in which $v^B(\mathbf{R})$ is replaced by

$$v^G(\mathbf{R}) = -\frac{\boldsymbol{\alpha}_x \cdot \boldsymbol{\alpha}_y}{R}, \qquad (22.75)$$

the residual part of the Breit interaction being neglected.

Comments

The choice of gauge should not influence the predictions of QED for atomic and molecular structure when the perturbation series is summed to convergence, so that it should not matter if the unapproximated effective operators are taken in Feynman or Coulomb gauge. However, this need not be true at each order of perturbation. It has been shown that the results are equivalent, order by order, if the orbitals have been defined in a *local* potential, but not otherwise. There have also been suggestions that the Feynman operator introduces spurious terms in lower orders of perturbation that are canceled in higher orders [13]. For this reason, most structure calculations have used Coulomb gauge.

It is often argued, following Bethe and Salpeter [14, Sect. 38], that the Breit interaction should only be used in first order perturbation theory. The reason is the approximation $\omega \to 0$; however, this approximation is quite adequate for many applications in which the dominant interactions involve only small energy differences.

22.4 MANY-BODY THEORY FOR ATOMS

The relativistic theory of atomic structure can be viewed as a simplification of the QED approach using an effective Hamiltonian operator in which the Dirac electrons interact through the effective electron-electron interaction of Sect. 22.3.6. This approach retains the dominant terms from the perturbation solution; those that are omitted are small and can, with sufficient trouble, be taken into account perturbatively [15, 10]. In particular, radiative correction terms requiring renormalization are explicitly omitted, and their effects incorporated at a later stage. Once a model has been chosen, the techniques and methods used for practical calculations acquire a close resemblance to those of the nonrelativistic theory described, for example in Chap. 21.

22.4.1 Effective Hamiltonians

The models which are closest to QED are those in which the full electron-electron interaction is included,

usually in Coulomb gauge. We define a Fock space Hamiltonian

$$H_{DCB} = H_0 + H_1 + H_2 \qquad (22.76)$$

where, as in Eq. (22.47),

$$H_0 = \sum_{E_m > E_F} N_m E_m - \sum_{E_n < E_F} N_n E_n, \qquad (22.77)$$

in which the states are those determined with respect to a mean-field central potential $U(\mathbf{x})$ as in (22.60)

$$\left\{ c\boldsymbol{\alpha} \cdot \mathbf{p} + \beta c^2 + V_{\text{nuc}}(\mathbf{x}) + U(\mathbf{x}) - E_m \right\} \psi_m(\mathbf{x}) = 0,$$

and

$$H_1 = -\sum_{pq} : a_p^\dagger a_q : \langle p|U(\mathbf{x})|q \rangle$$

$$H_2 = \frac{1}{2} \sum_{pqrs} : a_p^\dagger a_q^\dagger a_r a_s : \langle pq|V|sr \rangle.$$

Here the sums run only over states p with $E_p > E_F$; this means that states with $E_p < E_F$ are treated as inert.

The models are named according to the choice of V from Sect. 22.5.3:

Dirac-Coulomb-Breit Models. These incorporate the *full* Coulomb gauge operator (22.74) or the less accurate Breit operator (22.74). The fully retarded operator is usually taken in the symmetrized form. The Gaunt operator (22.75) is sometimes considered as an approximation to the Breit operator.

Dirac-Coulomb Models. The electron-electron interaction is simply taken to be the static $1/R$ potential. Note that although the equations are relativistic, the choices of electron-nucleus interaction all implicitly restrict these models to a frame in which the nuclei are fixed in space. The full electron-electron interaction is gauge invariant; however, it is common to start from the Dirac-Coulomb operator, in which case the gauge invariance is lost. Since radiative transition rates are sensitive to loss of gauge invariance [16] the choice of potential in (22.76) can make a big difference. Such choices may also affect the rate of convergence in correlation calculations in which the relativistic parts of the electron-electron interaction are treated as a second, independent, perturbation.

22.4.2 Nonrelativistic Limit: Breit-Pauli Hamiltonian

The nonrelativistic limit of the Dirac-Coulomb-Breit Hamiltonian is described in Chap. 21. The derivation is given in many texts, for example [8, 10, 14], and in principle involves the following steps:

1. Express the relativistic 4-spinor in terms of nonrelativistic Pauli 2-spinors of the form (see Sect. 21.2)

$$\phi_{nlm_l,m_s}(\mathbf{x}) = \text{const.} \frac{P_{nl}(r)}{r} Y_{lm_l}(\theta, \phi) \chi_{m_s}(\sigma) \,,$$

where χ_{m_s} is a 2-component eigenvector of the spin operator $\mathbf{s}$ to lowest order in $1/c$.

2. Extract effective operators to order $1/c^2$.

Thus the Breit-Pauli Hamiltonian is written as the sum of terms of Sect. 21.2 which can be correlated with specific parts of the parent relativistic operator:

1. One-body terms originate from the Dirac Hamiltonian: they are H_{mass} (21.5), the one-body part of H_{Darwin} (21.7) and the spin-orbit couplings H_{so} (21.11) and H_{soo} (21.12). The forms given in these equations assume that the electron interacts with a point-charge nucleus and only require the Coulomb part of the electron-electron interaction.

2. Two-body terms, including the two-body parts of H_{Darwin} (21.7), the spin-spin contact term H_{ssc} (21.8), the orbit-orbit term H_{oo} (Eq. 21.9) and the spin-spin term H_{ss} (21.13) originate from the Breit interaction.

22.4.3 Perturbation Theory: Nondegenerate Case

We give a brief resumé of the Rayleigh-Schrödinger perturbation theory following Lindgren [17]. The material presented here supplements the general discussion of perturbation theory in Chap. 5. First consider the simplest case with a nondegenerate reference state Φ belonging to the Hilbert space $\mathcal{H}$ satisfying

$$H_0|\Phi\rangle = E_0|\Phi\rangle \,, \tag{22.78}$$

which is a first approximation to the solution of the full problem

$$H|\Psi\rangle = E|\Psi\rangle \,, \quad H = H_0 + V \,. \tag{22.79}$$

Next, introduce a projection operator P such that

$$P = |\Phi\rangle\langle\Phi|, \quad P|\Phi\rangle = |\Phi\rangle \,,$$

and its complement $Q = 1 - P$, projecting onto the complementary subspace $\mathcal{H}\backslash\{\Phi\}$. With the *intermediate normalization*

$$\langle\Phi|\Psi\rangle = \langle\Phi|\Phi\rangle = 1 \,,$$

it follows that

$$P|\Psi\rangle = |\Phi\rangle\langle\Phi|\Psi\rangle = |\Phi\rangle \,,$$

so that the perturbed wave function can be decomposed into two parts:

$$|\Psi\rangle = (P + Q)|\Psi\rangle = |\Phi\rangle + Q|\Psi\rangle \,.$$

Thus, with intermediate normalization,

$$E = \langle\Phi|H|\Psi\rangle = E_0 + \langle\Phi|V|\Psi\rangle \,.$$

We now use this decomposition to write (22.79) in the form

$$(E_0 - H_0)|\Psi\rangle = (V - \Delta E)|\Psi\rangle \,, \tag{22.80}$$

where $\Delta E = \langle\Phi|V|\Psi\rangle$. Thus

$$(E_0 - H_0)Q|\Psi\rangle = Q(V - \Delta E)|\Psi\rangle \,. \tag{22.81}$$

Introduce the *resolvent operator*

$$R = \frac{Q}{E_0 - H_0} \,, \tag{22.82}$$

which is well-defined except on $\{\Phi\}$. Then the perturbation contribution to the wave function is

$$Q|\Psi\rangle = R(V - \Delta E)|\Psi\rangle = R(V|\Psi\rangle - |\Psi\rangle\langle\Phi|V|\Psi\rangle) \,.$$

The *Rayleigh-Schrödinger* perturbation expansion can now be written

$$|\Psi\rangle = |\Phi\rangle + |\Psi^{(1)}\rangle + |\Psi^{(2)}\rangle + \cdots$$
$$E = E_0 + E^{(1)} + E^{(2)} + \cdots$$

The contributions are ordered by the number of occurrences of V, the leading terms being

$$|\Psi^{(1)}\rangle = RV|\Phi\rangle \,,$$
$$|\Psi^{(2)}\rangle = (RVRV - R^2VPV)|\Phi\rangle \,,$$

and so on. The corresponding contribution to the energy can then be found from

$$E^{(n)} = \langle\Phi|V|\Psi^{(n-1)}\rangle \,.$$

22.4.4 Perturbation Theory: Open-Shell Case

Consider now the case in which there are several unperturbed states, $|\Phi^{(a)}\rangle$, $a = 1, 2, \ldots, d$, having the same energy E_0, which span a d-dimensional linear subspace (the *model space*) $\mathcal{M} \subset \mathcal{H}$, so that

$$H_0|\Phi^{(a)}\rangle = E_0|\Phi^{(a)}\rangle, \quad a = 1, 2, \ldots, d \,.$$

Let P be the projector onto $\mathcal{M}$, and Q onto the orthogonal subspace $\mathcal{M}^\perp$.

The perturbed states $|\Psi^{(a)}\rangle, a = 1, 2, \ldots, d$ are related to the unperturbed states by the *wave operator* Ω,

$$|\Psi^{(a)}\rangle = \Omega|\Phi^{(a)}\rangle, \quad a = 1, 2, \ldots, d.$$

The *effective Hamiltonian*, H_{eff}, is defined so that

$$H_{\text{eff}}|\Phi^{(a)}\rangle = E^{(a)}|\Phi^{(a)}\rangle,$$

and thus

$$\Omega H_{\text{eff}}|\Phi^{(a)}\rangle = E^{(a)}|\Psi^{(a)}\rangle = H\Omega|\Phi^{(a)}\rangle.$$

Thus on the domain $\mathcal{M}$ we can write an operator equation

$$\Omega H_{\text{eff}} P = H\Omega P, \tag{22.83}$$

known as the Bloch equation. We now partition H_{eff} so that

$$H_{\text{eff}} P = (H_0 + V_{\text{eff}})P,$$

enabling a reformulation of (22.83) as the commutator equation

$$[\Omega, H_0]P = (V\Omega - \Omega V_{\text{eff}})P. \tag{22.84}$$

With the intermediate normalization convention of Sect. 22.4.3, this becomes

$$|\Psi^{(a)}\rangle = P|\Phi^{(a)}\rangle$$

so that $P\Omega P = P$ and

$$H_{\text{eff}} P = PH\Omega P, \quad V_{\text{eff}} P = PV\Omega P.$$

Then (22.84) can be put in the final form

$$[\Omega, H_0]P = (V\Omega - \Omega PV\Omega)P. \tag{22.85}$$

The general Rayleigh-Schrödinger perturbation expansion can now be generated by expanding the wave operator order by order

$$\Omega = 1 + \Omega^{(1)} + \Omega^{(2)} + \cdots,$$

and inserting into (22.85), resulting in a hierarchy of equations

$$[\Omega^{(1)}, H_0]P = (V - PV)P = QVP,$$
$$[\Omega^{(2)}, H_0]P = (QV\Omega^{(1)} - \Omega^{(1)}PV)P,$$

and so on, with $H_{\text{eff}}^{(n)} = PV\Omega^{(n-1)}$.

22.4.5 Perturbation Theory: Algorithms

The techniques of QED perturbation theory of Sect. 22.3.4 can be utilized to give computable expressions for perturbation calculations order by order. They exploit the second quantized representation of operators of Sect.

22.4.1 along with the use of diagrams to express the contributions to the wave operator and the energy as sums over virtual states. The use of Wick's theorem to reduce products of normally-ordered operators, and the *linked-diagram* or *linked-cluster* theorem are explained in Lindgren's article [17] and Chap. 5. Further references and discussion of features which can exploit vector-processing and parallel-processing computer architectures may be found in [18].

The theory can also be recast so as to sum certain classes of terms to completion. This depends on the possibility of expressing the wave operator as a normally ordered exponential operator

$$\Omega = \{\exp S\} = 1 + \{S\} + \frac{1}{2!}\{S^2\} + \cdots,$$

where the normally ordered operator S is known as the *cluster* operator. Expanding S order by order leads to the *coupled cluster expansion* (see also Chaps. 5 and 27).

22.5 SPHERICAL SYMMETRY

A popular starting point for most calculations in atomic and molecular structure is the independent particle central field approximation. This assumes that the electrons move independently in a potential field of the form

$$A_0(x) = \frac{1}{c}\phi(r), \quad r = |\mathbf{x}|;$$
$$A_i(x) = 0, \quad i = 1, 2, 3. \tag{22.86}$$

Clearly $\phi(r)$ is left unchanged by any rotation about the origin, $r = 0$, but transforms as the component $A_0(x)$ of a 4-vector under other types of Lorentz and Poincaré transformation such as boosts or translations. However, solutions in central potentials of this form have a simple form which is convenient for further calculation.

With this restriction on the 4-potential, Dirac's Hamiltonian becomes

$$\hat{\mathbf{h}}_D = \{c\boldsymbol{\alpha}\cdot\mathbf{p} + e\phi(r) + \beta m_e c^2\}. \tag{22.87}$$

Consider stationary solutions with energy E satisfying

$$\hat{\mathbf{h}}_D\psi_E(x) = E\psi_E(x).$$

Since $\hat{\mathbf{h}}_D$ is invariant with respect to rotation about $r = 0$, it commutes with the operators J_1, J_2, J_3 introduced in (22.2), which can be interpreted as the components of the total angular momentum $\mathbf{j}$ of the electron, usually decomposed into an orbital part $\mathbf{l}$ and a spin part $\mathbf{s}$,

$$\mathbf{j} = \mathbf{l} + \mathbf{s} \tag{22.88}$$

where

$$l_j = i\epsilon_{jkl}x_k\partial_l, \quad j = 1, 2, 3$$
$$s_j = \tfrac{1}{2}\epsilon_{jkl}\sigma_{kl}, \quad j = 1, 2, 3.$$

22.5.1 Eigenstates of Angular Momentum

We can construct simultaneous eigenstates of the operators $\mathbf{j}^2$ and j_3 by using the product representation $\mathcal{D}^{(l)} \times \mathcal{D}^{(1/2)}$ of the rotation group SO(3), which is reducible to the Clebsch-Gordan sum of two irreps

$$\mathcal{D}^{(l+1/2)} \oplus \mathcal{D}^{(l-1/2)} . \tag{22.89}$$

We construct a basis for each irrep from products of basis vectors for $\mathcal{D}^{(1/2)}$ and $\mathcal{D}^{(l)}$ respectively. $\mathcal{D}^{(1/2)}$ is a 2-dimensional representation spanned by the simultaneous eigenstates ϕ_σ of $\mathbf{s}^2$ and s_3

$$\mathbf{s}^2 \phi_\sigma = \tfrac{3}{4}\phi_\sigma, \quad s_3 \phi_\sigma = \sigma\phi_\sigma, \quad \sigma = \pm\tfrac{1}{2},$$

for which we can use 2-rowed vectors

$$\phi_{1/2} = \begin{pmatrix} 1 \\ 0 \end{pmatrix}, \qquad \phi_{-1/2} = \begin{pmatrix} 0 \\ 1 \end{pmatrix} .$$

The representation $\mathcal{D}^{(l)}$ is $(2l+1)$-dimensional; its basis vectors can be taken to be the spherical harmonics

$$\{Y_l^m(\theta,\varphi) \,|\, m = -l, -l+1, \ldots, l\},$$

so that

$$\mathbf{l}^2 Y_l^m(\theta,\varphi) = l(l+1)\hbar^2 Y_l^m(\theta,\varphi),$$
$$l_3 Y_l^m(\theta,\varphi) = m\hbar Y_l^m(\theta,\varphi).$$

We shall assume that spherical harmonics satisfy the standard relations

$$l_\pm Y_l^m(\theta,\varphi) = [l(l+1) - m(m\pm 1)]^{1/2}\hbar Y_l^{m\pm 1}(\theta,\varphi),$$

where $l_\pm = l_1 \pm l_2$, so that

$$Y_l^m(\theta,\varphi) = \left(\frac{2l+1}{4\pi}\right)^{1/2} C_l^m(\theta,\varphi), \tag{22.90}$$

$$C_l^m(\theta,\varphi) = (-1)^m \left[\frac{(l-m)!}{(l+m)!}\right]^{1/2} P_l^m(\theta)e^{im\varphi} \text{ if } m \geq 0,$$
$$C_l^{-m}(\theta,\varphi) = (-1)^m C_l^m(\theta,\varphi)^* .$$

Basis functions for the representations $\mathcal{D}^j$ with $j = l \pm \tfrac{1}{2}$ have the form [2]

$$\chi_{j,m,a}(\theta,\varphi)$$
$$= \sum_\sigma \langle l, m-\sigma, \tfrac{1}{2}, \sigma \,|\, l, \tfrac{1}{2}, j, m \rangle Y_l^{m-\sigma}(\theta,\varphi)\phi_\sigma \tag{22.91}$$

where $\langle l, m-\sigma, \tfrac{1}{2}, \sigma \,|\, l, \tfrac{1}{2}, j, m \rangle$ is a Clebsch-Gordan coefficient with

---

[2] The order of coupling is significant, and great confusion results from a mixing of conventions. Here we couple in the order l, s, j. The same spin-angle functions are obtained if we use the order s, l, j but there is a phase difference $(-1)^{l-j+1/2} = (-1)^{(1-a)/2}$. You have been warned!

$$l = j - \tfrac{1}{2}a, \quad a = \pm 1, \quad m = -j, -j+1, \ldots, j-1, j.$$

Inserting explicit expressions for the Clebsch-Gordan coefficients gives

$$\chi_{j,m,-1}(\theta,\varphi) = \begin{bmatrix} -\left(\dfrac{j+1-m}{2j+2}\right)^{1/2} Y_{j+1/2}^{m-1/2}(\theta,\varphi) \\ \left(\dfrac{j+1+m}{2j+2}\right)^{1/2} Y_{j+1/2}^{m+1/2}(\theta,\varphi) \end{bmatrix},$$

$$\tag{22.92}$$

$$\chi_{j,m,1}(\theta,\varphi) = \begin{bmatrix} \left(\dfrac{j+m}{2j}\right)^{1/2} Y_{j-1/2}^{m-1/2}(\theta,\varphi) \\ -\left(\dfrac{j-m}{2j}\right)^{1/2} Y_{j-1/2}^{m+1/2}(\theta,\varphi) \end{bmatrix}.$$

The vectors (22.92) satisfy

$$\mathbf{j}^2\chi_{j,m,a} = j(j+1)\chi_{j,m,a}, \quad \mathbf{s}^2\chi_{j,m,a} = \tfrac{3}{4}\chi_{j,m,a},$$
$$\mathbf{l}^2\chi_{j,m,a} = l(l+1)\chi_{j,m,a}, \quad l = j - \tfrac{1}{2}a, \quad a = \pm 1. \tag{22.93}$$

The parity of the angular part is given by $(-1)^l$, with the two possibilities distinguished by means of the operator

$$K' = -(\mathbf{j}^2 - \mathbf{l}^2 - \mathbf{s}^2 + 1) = -(2\mathbf{s}\cdot\mathbf{l} + 1) \tag{22.94}$$

so that

$$K'\chi_{j,m,a} = k'\chi_{j,m,a}, \quad k' = -(j+\tfrac{1}{2})a, \quad a = \pm 1.$$

The basis vectors are orthonormal on the unit sphere with respect to the inner product

$$(\chi_{j',m',a'} \,|\, \chi_{jma})$$
$$= \iint \chi_{j',m',a'}^\dagger(\theta,\varphi)\chi_{j,m,a}(\theta,\varphi) \sin\theta d\theta d\varphi$$
$$= \delta_{j',j}\delta_{m',m}\delta_{a',a} . \tag{22.95}$$

22.5.2 Eigenstates of Dirac Hamiltonian in Spherical Coordinates

Eigenstates of Dirac's Hamiltonian (22.87) in spherical coordinates with a spherically symmetric potential $V(r) = e\phi(r)$,

$$\hat{\mathbf{h}}_D\psi_E(\mathbf{r}) = E\psi_E(\mathbf{r}), \tag{22.96}$$

are also simultaneous eigenstates of $\mathbf{j}^2$, of j_3 and of the operator

$$K = \begin{bmatrix} K' & 0 \\ 0 & -K' \end{bmatrix}, \tag{22.97}$$

where K' is defined in (22.94) above. Denote the corresponding eigenvalues by j, m and κ, where

$$\kappa = \pm \left(j + \frac{1}{2} \right) . \qquad (22.98)$$

Then the simultaneous eigenstates take the form

$$\psi_{E\kappa m}(\mathbf{r}) = \frac{1}{r} \left[\begin{array}{c} P_{E\kappa}(r)\chi_{\kappa,m}(\theta,\varphi) \\ iQ_{E\kappa}(r)\chi_{-\kappa,m}(\theta,\varphi) \end{array} \right], \qquad (22.99)$$

where $\kappa = -(j+1/2)a$ is the eigenvalue of K', and the notation $\chi_{\kappa,m}$ replaces the notation $\chi_{j,m,a}$ used previously in (22.91). The factor i in the lower component ensures that, at least for bound states, the radial amplitudes $P_{E\kappa}(r), Q_{E\kappa}(r)$ can be chosen to be real. This decomposition into radial and angular factors exploits the identity

$$\boldsymbol{\sigma}\cdot\mathbf{p} \left\{ \frac{F(r)}{r} \chi_{\kappa,m}(\theta,\varphi) \right\}$$
$$= i\hbar \frac{1}{r} \left\{ \frac{dF}{dr} + \frac{\kappa F}{r} \right\} \chi_{-\kappa,m}(\theta,\varphi) \qquad (22.100)$$

and gives a reduced eigenvalue equation

$$\left[\begin{array}{cc} mc^2 - E + V & -c\left(\frac{d}{dr} - \frac{\kappa}{r}\right) \\ c\left(\frac{d}{dr} + \frac{\kappa}{r}\right) & -mc^2 - E + V \end{array} \right] \left[\begin{array}{c} P_{E\kappa}(r) \\ Q_{E\kappa}(r) \end{array} \right] = 0. \qquad (22.101)$$

Angular Density Distributions

It is a remarkable fact that the angular density distribution

$$A_{\kappa,m}(\theta,\varphi) = \chi_{\kappa,m}(\theta,\varphi)^{\dagger}\chi_{\kappa,m}(\theta,\varphi), \qquad (22.102)$$

where $m = -j, -j+1, \ldots, j-1, j$, is *independent of the sign of* κ; the equivalence of

$$A_{j+1/2,m}(\theta,\varphi) = \frac{1}{4\pi} \frac{(j-m)!}{(j+m)!}$$
$$\times \left\{ (j-m+1)^2 |P_{j+1/2}^{m-1/2}(\mu)|^2 + |P_{j+1/2}^{m+1/2}(\mu)|^2 \right\},$$

and

$$A_{-(j+1/2),m}(\theta,\varphi) = \frac{1}{4\pi} \frac{(j-m)!}{(j+m)!}$$
$$\times \left\{ (j+m)^2 |P_{j-1/2}^{m-1/2}(\mu)|^2 + |P_{j-1/2}^{m+1/2}(\mu)|^2 \right\},$$

where $\mu = \cos\theta$, was demonstrated by Hartree [19].

Angular densities for the lowest $|\kappa|$ values are given in Table 22.1. The corresponding nonrelativistic angular densities

$$A_{l,m}(\theta,\varphi)_{nr} = |Y_l^m(\theta,\varphi)|^2 = \frac{2l+1}{4\pi} \frac{(l-m)!}{(l+m)!} |P_l^{|m|}(\mu)|^2 ;$$

are listed in Table 22.2.

Table 22.1. Relativistic angular density functions.

$|\kappa|$	$|m|$	$4\pi . A_{|\kappa|,m}(\theta,\varphi)$
1	$\frac{1}{2}$	1
2	$\frac{3}{2}$	$\frac{3}{2}\sin^2\theta$
	$\frac{1}{2}$	$\frac{1}{2}(1+3\cos^2\theta)$
3	$\frac{5}{2}$	$\frac{15}{8}\sin^4\theta$
	$\frac{3}{2}$	$\frac{3}{8}\sin^2\theta(1+15\cos^2\theta)$
	$\frac{1}{2}$	$\frac{3}{4}(3\cos^2\theta-1)^2+3\sin^2\theta\cos^2\theta$

Radial Density Distributions

The probability density distribution $\rho_{E\kappa m}(\mathbf{r})$ associated with the stationary state (22.99) is given by

$$\rho_{E,\kappa,m}(\mathbf{r}) = \frac{1}{r^2} \left\{ |P_{E,\kappa}(r)|^2 A_{\kappa,m}(\theta,\varphi) \right.$$
$$\left. + |Q_{E,\kappa}(r)|^2 A_{-\kappa,m}(\theta,\varphi) \right\} . \qquad (22.103)$$

Since $A_{\kappa,m}$ does not depend on the sign of κ, the angular part can be factored so that

$$\rho_{E,\kappa,m}(\mathbf{r}) = \frac{D_{E,\kappa}(r)}{r^2} A_{|\kappa|,m}(\theta,\varphi),$$

where

$$D_{E,\kappa}(r) = \left\{ |P_{E,\kappa}(r)|^2 + |Q_{E,\kappa}(r)|^2 \right\} \qquad (22.104)$$

defines the *radial density* distribution.

Subshells in $j - j$ Coupling

The notion of a *subshell* depends on the observation that the set $\{\psi_{E,\kappa,m}, m = -j,\ldots,j\}$ have a common radial density distribution. The simplest atomic model is one in which the electrons move independently in a mean field central potential. Since

Table 22.2. Nonrelativistic angular density functions.

l	$|m|$	$4\pi . A_{l,m}(\theta,\varphi)_{nr}$
0	0	1
1	1	$\frac{3}{2}\sin^2\theta$
	0	$3\cos^2\theta$
2	2	$\frac{15}{8}\sin^4\theta$
	1	$\frac{15}{2}\sin^2\theta\cos^2\theta$
	0	$\frac{5}{4}(3\cos^2\theta-1)^2$

Table 22.3. Spectroscopic labels and angular quantum numbers.

Label:	s	$\bar{p}$	p	$\bar{d}$	d	$\bar{f}$	f
$\kappa = -(j + \frac{1}{2})a$	-1	$+1$	-2	$+2$	-3	$+3$	-4
$j = l + \frac{1}{2}a$	$\frac{1}{2}$	$\frac{1}{2}$	$\frac{3}{2}$	$\frac{3}{2}$	$\frac{5}{2}$	$\frac{5}{2}$	$\frac{7}{2}$
a	1	-1	1	-1	1	-1	1
$l = j - \frac{1}{2}a$	0	1	1	2	2	3	3
$\bar{l} = j + \frac{1}{2}a$	1	0	2	1	3	2	4

$$\sum_{m=-j}^{j} \rho_{E,\kappa,m}(\mathbf{r}) = \frac{2j + 1}{4\pi} \frac{D_{E,\kappa}(r)}{r^2}, \qquad (22.105)$$

a state of $2j + 1$ independent electrons, with one in each member of the set $\{\psi_{E,\kappa,m}, m = -j, \ldots, j\}$, has a spherically symmetric probability density. If E belongs to the point spectrum of the Hamiltonian, then (22.105) gives a distribution localized in r, and we refer to the states $\{E, \kappa, m\}, m = -j, \ldots, j$ as belonging to the subshell $\{E, \kappa\}$.

The notations in use for Dirac central field states are set out in Table 22.3. Here l is associated with the orbital angular quantum number of the upper pair of components and $\bar{l}$ with the lower pair. Note the useful equivalence

$$\kappa(\kappa + 1) = l(l + 1).$$

Defining $\bar{\kappa} := -\kappa$ we have also $\bar{\kappa}(\bar{\kappa} + 1) = \bar{l}(\bar{l} + 1)$.

22.5.3 Radial Amplitudes

Textbooks on quantum electrodynamics usually contain extensive discussions of the formalism associated with the Dirac equation but rarely go beyond the treatment of the hydrogen atom (see Chap. 10). Greiner's textbook [4] is an honorable exception, with many worked examples. A more exhaustive list of problems in which exact solutions are known is contained in [20]; it is particularly rich in detail about equations of motion and Green's functions in external electromagnetic fields of various configurations; coherent states of relativistic particles; charged particles in quantized plane wave fields. It also incorporates discussion of extensions of the Dirac equations due to Pauli which include explicit interaction terms arising from anomalous magnetic or electric moments.

Atoms and molecules with more than one electron are not soluble analytically so that numerical models are needed to make predictions. The solutions are sensitive to boundary conditions on which we focus in this section. For large r, solutions of (22.101) can be found proportional to $\exp(\pm\lambda r)$, where

$$\lambda = +\sqrt{c^2 - E^2/c^2}. \qquad (22.106)$$

Thus λ is *real* when $-c^2 \le E \le c^2$, and *pure imaginary* otherwise.

Singular Point at $r = 0$

Singularities of the nuclear potential near $r = 0$ have a major influence on the nature of solutions of the Dirac equation. Suppose that the potential has the form

$$V(r) = -\frac{Z(r)}{r}, \qquad (22.107)$$

so that $Z(r)$ is the effective charge seen by an electron at radius r from the nuclear center. The dependence of $Z(r)$ on r may reflect the finite size of the nuclear charge distribution, so far treated as a point, or the screening due to the environment. Assume that $Z(r)$ can be expanded in a power series of the form

$$Z(r) = Z_0 + Z_1 r + Z_2 r^2 + \cdots \qquad (22.108)$$

in a neighborhood of $r = 0$. This property characterizes a number of well-used models

1. *Point nucleus*: $Z_0 \ne 0$; $Z_n = 0$, $n > 0$.

2. *Uniform nuclear charge distribution*:

$$V(r) = \begin{cases} -\dfrac{3Z}{2a}\left(1 - \dfrac{r^2}{3a^2}\right), & 0 \le r \le a, \\ -\dfrac{Z}{r}, & r > a. \end{cases} \qquad (22.109)$$

This gives the expansion $Z_0 = -3Z/2a$, $Z_1 = 0$, $Z_2 = +Z/2a^3$, $Z_n = 0$ for $n > 2$ when $r \le a$.

3. *Fermi distribution*: The nuclear charge density has the form

$$\rho_{\text{nuc}}(r) = \frac{\rho_0}{1 + \exp[(r - a)/d]},$$

where ρ_0 is chosen so that the total charge on the nucleus is Z.

Other nuclear models, reflecting the density distributions deduced from nuclear scattering experiments, can be found in the literature.

Series Solutions Near $r = 0$

Any solution for the radial amplitudes of Dirac's equation in a central potential

$$u(r) = \begin{bmatrix} P(r) \\ Q(r) \end{bmatrix}, \qquad (22.110)$$

with radial density

$$D(r) = P^2(r) + Q^2(r),$$

can be expanded in a power series near the singular point at $r = 0$ in the form

$$u(r) = r^\gamma[u_0 + u_1 r + u_2 r^2 + \cdots], \qquad (22.111)$$

where

$$u_k = \begin{bmatrix} p_k \\ q_k \end{bmatrix}, \qquad k = 1, 2, \ldots$$

and γ, p_k, q_k are constants which depend on the nuclear potential model.

Point Nuclear Models

For a Coulomb singularity, $Z_0 \neq 0$, the leading coefficients satisfy

$$-Z_0 p_0 + c(\kappa - \gamma)q_0 = 0,$$
$$c(\kappa + \gamma)p_0 - Z_0 q_0 = 0, \qquad (22.112)$$

so that

$$\gamma = \pm\sqrt{\kappa^2 - \frac{Z_0^2}{c^2}}, \qquad (22.113)$$
$$\frac{q_0}{p_0} = \frac{Z_0}{c(\kappa - \gamma)} = \frac{c(\kappa + \gamma)}{Z_0}.$$

Finite Nuclear Models

Finite nuclear models, for which $Z_0 = 0$, have no singularity in the potential at $r = 0$. The indicial equation (22.113) reduces to $\gamma = \pm|\kappa|$, so that for $\kappa < 0$,

$$P(r) = p_0 r^{l+1} + O(r^{l+3}), \qquad (22.114)$$
$$Q(r) = q_1 r^{l+2} + O(r^{l+4}), \qquad (22.115)$$

with

$$q_1/p_0 = (E - mc^2 + Z_1)/[c(2l + 3)], \qquad q_0 = p_1 = 0,$$

and for $\kappa \geq 0$,

$$P(r) = p_1 r^{l+1} + O(r^{l+3}), \qquad (22.116)$$
$$Q(r) = q_0 r^l + O(r^{l+4}), \qquad (22.117)$$

with

$$p_1/q_0 = -(E - mc^2 + Z_1)/[c(2l + 1)], \qquad p_0 = q_1 = 0.$$

In both cases the solutions consist of either purely *even* powers or purely *odd* powers of r, contrasting strongly with the point nucleus case, where both even and odd powers are present in the series expansion.

The Nonrelativistic Limit

For a solution linked to a nonrelativistic state with orbital angular momentum l, one expects the nonrelativistic limit

$$P(r) = O(r^{l+1}), \qquad c \to \infty.$$

The limiting behavior reveals some significant features.

Finite Nuclear Models. The behavior is entirely regular:

$$P(r) = O(r^{l+1}), \qquad Q(r) = O(c^{-1}) \to 0.$$

Point Nuclear Models. Since

$$\gamma = |\kappa| - \frac{Z^2}{2c^2|\kappa|} + \cdots,$$

Eq. (22.113) shows that the leading coefficient p_0 *vanishes* in the limit so that,

$$P(r) \approx p_1 r^{l+1}(1 + O(r^2)), \qquad \text{when } \kappa \geq 1, \ l = \kappa.$$
$$(22.118)$$

All higher powers of *odd* relative order vanish in the limit for both components. The behavior in the case $\kappa < 0$ is entirely regular.

22.5.4 Square Integrable Solutions

Square integrable solutions require $\int D_{E,\kappa}(r)dr$ to be finite; since the solutions are smooth, except possibly near the singular endpoints $r \to 0$ and $r \to \infty$, we focus on the behavior at the endpoints:

$r \to \infty$. For *real* values of λ the condition

$$\int_R^\infty D_{E,\kappa}(r)dr < \infty, \qquad 0 < R < \infty,$$

requires that $P_{E\kappa}(r)$, $Q_{E\kappa}(r)$ are proportional to $\exp(-\lambda r)$ with $\lambda > 0$.

This means that bound states can only exist when E lies in the interval $-c^2 \leq E \leq c^2$. Outside this interval solutions are necessarily of scattering type and so

$$\int_R^\infty D_{E,\kappa}(r)dr$$

diverges when $|E| > c^2$.

$r \to 0$. This limit requires

$$\int_0^{R'} D_{E,\kappa}(r)dr < \infty, \qquad R' > 0.$$

Since $D_{E,\kappa}(r) \sim r^{\pm 2\gamma}$ as $r \to 0$, this condition holds when $\pm\gamma > -\frac{1}{2}$. Only the solution with $\gamma > 0$ satisfies the condition when $|\gamma| > \frac{1}{2}$, or $Z < \alpha^{-1}\sqrt{\kappa^2 - \frac{1}{4}}$, and the solution with $\gamma < 0$ must be disregarded. This corresponds to the *limit point case* of a second-order differential operator [21]. In the special case $|\kappa| = 1$ or $j = \frac{1}{2}$ this limits Z to be smaller than $c\sqrt{3}/2 \approx 118.6$. For $Z > c\sqrt{3}/2$, both solutions are square integrable near the origin (the *limit circle case*) and the differential operator is no longer essentially self-adjoint.

The Coulomb potential must have a finite expectation for any physically acceptable solution, so that we also require

$$\int_0^{R'} D_{E,\kappa}(r) \frac{dr}{r} < \infty, \quad R' > 0 \,.$$

This is always satisfied by the solution with $\gamma > 0$ for all $|Z| < \alpha^{-1}|\kappa|$, but not by the solution with $\gamma < 0$. Imposing this condition restores essential self-adjointness (on a restricted domain) for $118 < Z \leq 137$.

22.5.5 Hydrogenic Solutions

The wave functions for hydrogenic solutions of Dirac's equation are presented in Sect. 9.3. Here we note some properties of hydrogenic solutions that reveal dynamical effects of relativity in the absence of screening by orbital electrons. In this case $Z_0 = Z$, $Z_n = 0$, $n > 0$. When $-c^2 < E < c^2$ we have bound states. The parameter λ, (22.106), can conveniently be written

$$\lambda = Z/N \,, \tag{22.119}$$

so that rearranging (22.106) gives

$$E = +c^2 \sqrt{1 - \frac{Z^2}{N^2 c^2}} \,, \tag{22.120}$$

essentially equivalent to Sommerfeld's fine structure formula. In the formal nonrelativistic limit, $c \to \infty$, we have

$$E = c^2 - \frac{Z^2}{2N^2} + O(1/c^2)$$

so that N is closely related to the principal quantum number, n, appearing in the Rydberg formula. As in Sect. 9.3, we write $\rho = 2\lambda r$.

Define the *inner quantum number*

$$n_r = -a = -\gamma + \frac{NE}{c^2}, \qquad n_r = 0, 1, 2, \ldots \,.$$

Substitute for E from (22.120) to get

$$N = [(n_r + \gamma)^2 + \alpha^2 Z^2]^{1/2}$$
$$= [n^2 - 2n_r(|\kappa| - \gamma)]^{1/2} \,, \tag{22.121}$$

where $n = n_r + |\kappa|$ is the *principal quantum number*, the exact equivalent of the principal quantum number

of the nonrelativistic state to which the Dirac solution reduces in the limit $c \to \infty$. With this notation, the radial amplitudes for bound hydrogenic states are

$$P_{E\kappa}(r)$$
$$= \mathcal{N}_{E\kappa}(c + E/c)^{1/2} \rho^\gamma e^{-\rho/2} [-n_r M(-n_r + 1, 2\gamma + 1; \rho)$$
$$+ (N - \kappa) M(-n_r, 2\gamma + 1; \rho)] \,, \tag{22.122}$$

$$Q_{E\kappa}(r)$$
$$= \mathcal{N}_{E\kappa}(c - E/c)^{1/2} \rho^\gamma e^{-\rho/2} [-n_r M(-n_r + 1, 2\gamma + 1; \rho)$$
$$- (N - \kappa) M(-n_r, 2\gamma + 1; \rho)] \,, \tag{22.123}$$

where

$$\mathcal{N}_{E\kappa} = \left\{ \frac{\alpha Z}{2N^2(N - \kappa)} \cdot \frac{\Gamma(2\gamma + n_r + 1)}{n_r [\Gamma(2\gamma + 1)]^2} \right\}^{1/2}$$

is the normalization constant. For definitions of the confluent hypergeometric functions $M(a, b; c; z)$ see [22, Sect. 13.1]

Table 22.4 lists expectation values of simple powers of the radial variable $\rho = 2Zr/N$ from [23] and [24]. Simple algebra, using the inequalities $\gamma < |\kappa|$ and $N < n$, yields the inequality

$$\langle \rho^s \rangle_{n\kappa} < \langle \rho^s \rangle_{nl}, \quad s > 0 \,;$$

the inequality is reversed for $s < 0$. In the same way, it is easy to deduce that relativistic hydrogenic eigenvalues lie below the nonrelativistic eigenvalues

$$\epsilon_{n\kappa} < \epsilon_{nl} \,.$$

Thus, in the absence of screening, Dirac orbitals both *contract* and are *stabilized* with respect to their nonrelativistic counterparts. The relativistic and nonrelativistic expectation values approach each other as the *relativistic coupling constant, $Z/c = \alpha Z \to 0$.* This formal *nonrelativistic limit* is approached as $\alpha \to 0$ or $c \to \infty$, in which the speed of light is regarded as infinite.

22.5.6 The Free Electron Problem in Spherical Coordinates

The radial equation (22.101) for the free electron ($V(r) = 0$) gives a pair of first order ordinary differential equations

$$(mc^2 - E)P_{E\kappa}(r) = c\left(\frac{d}{dr} - \frac{\kappa}{r}\right)Q_{E\kappa}(r),$$
$$c\left(\frac{d}{dr} + \frac{\kappa}{r}\right)P_{E\kappa}(r) = (mc^2 + E)Q_{E\kappa}(r), \tag{22.124}$$

from which we deduce that

Table 22.4. Radial moments $\langle \rho^s \rangle$.

s	Nonrelativistic	Relativistic [a]
2	$2n^2[5n^2 + 1 \\ -3l(l+1)]$	$2\left\{ N^2(5N^2 - 2\kappa^2)R^2(N) \\ +N^2(1-\gamma^2) - 3\kappa N^2 R(N) \right\}$
1	$3n^2 - l(l+1)$	$-\kappa + (3N^2 - \kappa^2)R(N)$
0	1	1
-1	$\dfrac{1}{2n^2}$	$\dfrac{n\gamma + (|\kappa| - \gamma)|\kappa|}{2\gamma N^3}$
-2	$\dfrac{1}{2n^3(2l+1)}$	$\dfrac{\kappa^2 R(N)}{2\gamma^2 N^3(2\gamma - \mathrm{sgn}\kappa)}$
-3	$\dfrac{1}{4n^3 l(l+1)(2l+1)}$	$\dfrac{N^2 + 2\gamma^2\kappa^2 - 3N^2\kappa R(N)}{4N^5\gamma(\gamma^2 - 1)(4\gamma^2 - 1)}$

[a] $R(N) = \sqrt{1 - Z^2/N^2c^2}$.

$$\frac{d^2 P_{E\kappa}(r)}{dr^2} + \left(p^2 - \frac{\kappa(\kappa+1)}{r^2} \right) P_{E\kappa}(r) = 0\,,$$

$$\frac{d^2 Q_{E\kappa}(r)}{dr^2} + \left(p^2 - \frac{\bar\kappa(\bar\kappa+1)}{r^2} \right) Q_{E\kappa}(r) = 0\,, \tag{22.125}$$

where $p^2 = m^2c^2 - E^2/c^2 = \mathbf{p}.\mathbf{p}$ and the angular quantum numbers κ and $\bar\kappa$ are associated respectively with the upper and lower components. These are defining equations of Riccati-Bessel functions [22, Sect. 10.3] of orders l and $\bar l$ respectively, where

$$\kappa(\kappa+1) = l(l+1)\,, \quad \bar\kappa(\bar\kappa+1) = \bar l(\bar l+1)\,.$$

Thus the solutions of (22.125) are functions of the variable $x = pr$ of the form

$$P_{E\kappa}(r) = Axf_l(x)\,, \quad Q_{E\kappa}(r) = Bxf_{\bar l}(x)\,,$$

where the ratio of A and B is determined by equations (22.124) and where $f_l(x)$ is a spherical Bessel function of the first, second or third kind [22, Sect. 10.1.1]. Thus

$$P_{E\kappa}(r) = \mathcal{N} \left(\frac{E + mc^2}{\pi E} \right)^{1/2} xf_l(x)\,,$$

$$Q_{E\kappa}(r) = \mathcal{N}\mathrm{sgn}\,(\kappa) \left(\frac{E - mc^2}{\pi E} \right)^{1/2} xf_{\bar l}(x)\,. \tag{22.126}$$

Equations (22.124) require that Riccati-Bessel solutions of the same type be chosen for both components. The possibilities are:

Standing Waves. The two solutions of the same type are $f_l(x) = j_l(x)$, $f_l(x) = y_l(x)$. The $j_l(x)$ are bounded everywhere, including the singular points $x = 0$, $x \to \infty$ and have zeros of order l at $x = 0$. The $y_l(x)$ are bounded at infinity but have poles of order $l+1$ at $x = 0$.

Progressive Waves. The spherical Hankel functions (functions of the third kind) are linear combinations

$$h_l^{(1)}(x) = j_l(x) + iy_l(x)\,, \quad h_l^{(2)}(x) = j_l(x) - iy_l(x)\,.$$

Recalling that p is *real* if and only if $|E| > mc^2$, we see that $h_l^{(1)}(x), h_l^{(2)}(x)$ are bounded as $x \to \infty$ and have poles of order $l+1$ at $x = 0$. Notice that when $|E| < mc^2$, which does not occur for a free particle, p becomes pure imaginary and no solution exists which is finite at both singular points.

The normalization constant $\mathcal{N}$ can be determined by using the well-known result

$$\int_0^\infty j_l(pr)j_l(p'r)r^2\,dr = \frac{\pi}{2p^2}\delta(p - p')\,.$$

The choice $\mathcal{N} = 1$ ensures that

$$\int_0^\infty \left(P_{E\kappa}^\dagger(r)P_{E'\kappa}(r) + Q_{E\kappa}^\dagger(r)Q_{E'\kappa}(r) \right) dr = \delta(p - p')\,.$$

Noting that

$$\delta(E - E') = \left| \frac{dp}{dE} \right| \delta(p - p')\,,$$

and $dp/dE = c^2p/E$ gives

$$\int_0^\infty \left(P_{E\kappa}^\dagger(r)P_{E'\kappa}(r) + Q_{E\kappa}^\dagger(r)Q_{E'\kappa}(r) \right) dr = \delta(E - E')\,.$$

when $\mathcal{N} = \left(|E|/c^2p \right)^{1/2}$.

22.6 NUMERICAL APPROXIMATION OF CENTRAL FIELD DIRAC EQUATIONS

The main drive for understanding methods of numerical approximation of solutions of Dirac's equation comes from their application to many-electron systems. Approximate wave functions for atomic or molecular states are usually constructed from products of one-electron orbitals, and their determination exploits knowledge gained from the treatment of one-electron problems. Whilst the numerical methods described here are strictly one-electron in character, extension to many-electron problems is relatively straightforward.

22.6.1 Finite Differences

The numerical approximation of eigensolutions of the first order system of differential equations (22.101)

$$E \begin{bmatrix} P_{E\kappa}(r) \\ Q_{E\kappa}(r) \end{bmatrix}$$

$$= \begin{bmatrix} mc^2 + V(r) & -c\left(\frac{d}{dr} - \frac{\kappa}{r}\right) \\ c\left(\frac{d}{dr} + \frac{\kappa}{r}\right) & -mc^2 + V(r) \end{bmatrix} \begin{bmatrix} P_{E\kappa}(r) \\ Q_{E\kappa}(r) \end{bmatrix}$$

$$(22.127)$$

can be achieved by more or less standard finite difference methods given in texts such as [25]. For states in either continuum, $E > mc^2$ or $E < -mc^2$, the calculation is completely specified as an *initial value problem* for a prescribed value of E starting from power series solutions in the neighborhood of $r = 0$. Solutions of this sort exist for all values of (complex) E except at the bound eigensolutions in the gap $-mc^2 < E < mc^2$. For bound states, the calculation becomes that of a *two-point boundary value problem* in which the eigenvalue E has to be determined iteratively along with the numerical solution. We concentrate on the latter, which is more involved.

It is convenient to write

$$\epsilon_{n\kappa} = E_{n\kappa} - mc^2, \qquad (22.128)$$

so that ϵ approaches the nonrelativistic eigenvalue in the limit $c \to \infty$. For the one-electron problem, (22.101) can be written in the general form

$$\mathbf{J}\frac{du}{ds} + \frac{1}{c}\frac{dr}{ds}(r\epsilon + \mathbf{W}(s))u(s) = \chi(s)\frac{dr}{ds}, \qquad (22.129)$$

where $u(s)$ and $\chi(s)$ are two-component vectors, such that

$$u(s) = \begin{pmatrix} P(s) \\ Q(s) \end{pmatrix}, \quad \mathbf{J} = \begin{pmatrix} 0 & 1 \\ -1 & 0 \end{pmatrix},$$

$$\mathbf{W}(s) = \begin{pmatrix} -rV(r) & -c\kappa \\ -c\kappa & 2rc^2 - rV(r) \end{pmatrix},$$

and $r(s)$ is a smooth differentiable function of a new independent variable s. This facilitates the use of a uniform grid for s mapping onto a suitable nonuniform grid for r. Common choices are

$$r_n = r_0 e^{s_n}, \quad s_n = nh, \quad n = 0, 1, 2, \ldots, N,$$

for suitable values of the parameters r_0 and h, and

$$Ar_n + \log\left(1 + \frac{r_n}{r_0}\right) = s_n, \quad n = 0, 1, 2, \ldots, N,$$

where A is a constant, chosen so that the spacing in r_n increases exponentially for small values of n and approaches a constant for large values of n. The exponentially increasing spacing is appropriate for tightly bound solutions, but a nearly linear spacing is advisable

to ensure numerical stability in the tails of extended and continuum solutions.

The most convenient numerical algorithm involves *double shooting* from $s_0 = 0$ and $s_N = Nh$ towards an intermediate *join* point $s = Jh$, adjusting ϵ until the trial solutions have the right number of nodes and have left- and right-limits at $s = Jh$ which agree to a pre-set tolerance (commonly about 1 part in 10^8).

The *deferred correction* method [26, 27] allows the precision of the numerical approximation to be improved as the iteration converges. Consider the simplest implicit linear difference scheme for the first order system

$$\frac{dy}{ds} = F(y(s), s),$$

based on the trapezoidal rule of quadrature, is

$$z_{j+1} - z_j = \frac{1}{2}h(F_{j+1} + F_j), \qquad (22.130)$$

which has a local truncation error $O(h^2)$. The precision can be improved, at the expense of increasing the computational cost per iterative cycle, by adding higher order difference terms to the right-hand side in Eq. (22.130). Use of the trial solution from the previous cycle leaves the stability properties of (22.130) are unaltered, but the converged solution has much higher accuracy.

To apply this to the Dirac system, write $f(s) = dr/ds$ and

$$\mathbf{A}_j^{\pm} = \mathbf{J} \pm \frac{h}{2c}f(s_j)\mathbf{W}(s_j).$$

Also consider a slightly generalized problem in which $V(r)$ is replaced by a discretized potential $V_j^{(\nu)}$ that may change from one iteration to the next, as in a self-consistent field calculation. The first iteration is

$$\mathbf{A}_{j+1}^{+(0)} U_{j+1}^{(1)} - \mathbf{A}_j^{-(0)} U_j^{(1)}$$
$$+ \epsilon^{(0)}\frac{h}{2c}\left(r_{j+1}f(s_{j+1})U_{j+1}^{(1)} + r_j f(s_j)U_j^{(1)}\right)$$
$$= \frac{1}{2}h\left(f(s_{j+1})\chi(s_{j+1})^{(0)} + f(s_j)\chi(s_j)^{(0)}\right), (22.131)$$

where superscript 0 refers to initial estimates and superscript 1 to the result of the first iteration. On the $(\nu + 1)$-th iteration, we solve

$$\mathbf{A}_{j+1}^{+(\nu)} U_{j+1}^{(\nu+1)} - \mathbf{A}_j^{-(\nu)} U_j^{(\nu+1)}$$
$$+ \epsilon^{(\nu)}\frac{h}{2c}\left(r_{j+1}f(s_{j+1})U_{j+1}^{(\nu+1)} + r_j f(s_j)U_j^{(\nu+1)}\right)$$
$$= \frac{1}{2}h\left(f(s_{j+1})\chi(s_{j+1})^{(\nu)} + f(s_j)\chi(s_j)^{(\nu)}\right)$$
$$+ \frac{1}{12}\delta^3 U_{j+1/2}^{(\nu)} + \cdots, \qquad (22.132)$$

where $\delta^3 U_{j+1/2}^{(\nu)}$ is the central-difference correction of order 3 [22, Sect. 25.1.2]. Higher order difference corrections (at least to order 5) are included in modern codes to improve the accuracy and numerical stability of weakly bound solutions. This deferred correction algorithm can be shown to converge asymptotically to the required solution of the differential system with a local truncation error of order $O(h^{2p+2})$ when difference corrections of order $2p + 1$ are employed [28].

22.6.2 Expansion Methods

Methods of solving the Dirac equation which represent the one-electron wave function as a linear combination of sets of square integrable functions (basis sets) have become popular in the last 10 years. Simple and rigorous criteria for choosing effective basis sets for this purpose are now available, and classes of functions that satisfy these criteria are known. Consequently, cheap and accurate calculations of the electronic structure of atoms and molecules are now a practical possibility.

Finite difference algorithms generate eigensolutions one at a time. Basis set methods replace the differential operator $\hat{\mathbf{h}}_D$ of Eq. (22.87) with a finite symmetric (in some cases complex Hermitian) matrix of dimension $2N$. The spectrum of this operator, which is of course a pure point spectrum, consists of three pieces: N eigensolutions with $E < -mc^2$ ($\epsilon < -2mc^2$) representing the eigenstates of the lower continuum; $N_b < N$ eigensolutions in the gap $-mc^2 < E < mc^2$ ($-2mc^2 < \epsilon < 0$) corresponding to bound states; and $N - N_b$ eigensolutions with $E > mc^2$ ($\epsilon > 0$) representing the eigenstates of the upper continuum. For properly chosen basis sets, the approximation properties of bound state eigensolutions are similar to those of the equivalent nonrelativistic eigensolutions. Solutions at continuum energies have the correct behavior near $r = 0$, but their amplitudes decrease expoentially like bound state solutions at large values of r. The criteria on which this description rests are as follows:

A. The eigenstates of $\hat{\mathbf{h}}_D$ are 4-component central field spinors whose components are coupled. The basis functions should therefore also consist of 4-component spinors of the form

$$\Phi_{\kappa m}(\mathbf{r}) = \frac{1}{r} \left[\begin{array}{c} f_\kappa^L(r) \chi_{\kappa,m}(\theta, \varphi) \\ i f_\kappa^S(r) \chi_{-\kappa,m}(\theta, \varphi) \end{array} \right]. \quad (22.133)$$

B. The spinor basis functions should, as far as practicable, satisfy the boundary conditions near $r = 0$ of Sect. 22.5.3. They should also be square integrable at infinity.

C. Acceptable spinor basis functions should satisfy the relation

$$i\hbar \frac{f_\kappa^S(r)}{r} \chi_{-\kappa,m}(\theta, \varphi) \to \boldsymbol{\sigma} \cdot \mathbf{p} \frac{f_\kappa^L(r)}{r} \chi_{\kappa,m}(\theta, \varphi) \quad (22.134)$$

in the *nonrelativistic limit*, $c \to \infty$.

D. Acceptable spinor basis functions must have *finite expectation values* of component operators of $\hat{\mathbf{h}}_D$, namely $\boldsymbol{\alpha} \cdot \mathbf{p}$, β and $V(r)$.

Finite Basis Set Formalism

Assume that each solution of the target problem is approximated as a linear combination

$$\psi_{\kappa m}(\mathbf{r}) = \frac{1}{r} \left[\begin{array}{c} \sum_j c_{\kappa j}^L f_{\kappa j}^L(r) \chi_{\kappa,m}(\theta, \varphi) \\ i \sum_j c_{\kappa j}^S f_{\kappa j}^S(r) \chi_{-\kappa,m}(\theta, \varphi) \end{array} \right], \quad (22.135)$$

where $c_{\kappa j}^L, c_{\kappa j}^S \quad j = 1 \cdots N$, are arbitrary constants, so that each j-term on the right-hand side has the form (22.133). This enables us to construct a *Rayleigh quotient*

$$W[\psi] = \frac{\langle \psi | \hat{\mathbf{h}}_D | \psi \rangle}{\langle \psi | \psi \rangle}, \quad (22.136)$$

where both $\langle \psi | \hat{\mathbf{h}}_D | \psi \rangle$ and $\langle \psi | \psi \rangle$ are quadratic expressions in the expansion coefficients c_j^L, c_j^S. By requiring that $W[\psi]$ shall be stationary with respect to arbitrary variations in the expansion coefficients, we arrive at the matrix eigenvalue equation

$$\mathbf{f}_\kappa \left[\begin{array}{c} \mathbf{c}_\kappa^L \\ \mathbf{c}_\kappa^S \end{array} \right] = \epsilon \left[\begin{array}{cc} \mathbf{S}_\kappa^{LL} & 0 \\ 0 & \mathbf{S}_\kappa^{SS} \end{array} \right] \left[\begin{array}{c} \mathbf{c}_\kappa^L \\ \mathbf{c}_\kappa^S \end{array} \right], \quad (22.137)$$

where the matrix Hamiltonian is denoted by

$$\mathbf{f}_\kappa = \left[\begin{array}{cc} \mathbf{V}_\kappa^{LL} & c\boldsymbol{\Pi}_\kappa^{LS} \\ c\boldsymbol{\Pi}_\kappa^{SL} & \mathbf{V}_\kappa^{SS} - 2mc^2 \mathbf{S}_\kappa^{SS} \end{array} \right],$$

$\mathbf{c}_\kappa^L$, $\mathbf{c}_\kappa^S$ are N-vectors, and $\mathbf{V}_\kappa^{LL}, \mathbf{V}_\kappa^{SS}, \mathbf{S}_\kappa^{LL}, \mathbf{S}_\kappa^{SS}, \boldsymbol{\Pi}_\kappa^{LS}$ and $\boldsymbol{\Pi}_\kappa^{SL}$ are all $N \times N$ matrices. Using superscripts T to denote either of the letters L, S, the elements of the matrices are defined by

$$S_{\kappa ij}^{TT} = \int_0^\infty f_{i\kappa}^{T*}(r) f_{j\kappa}^T(r) dr, \quad (22.138)$$

$$V_{\kappa ij}^{TT} = \int_0^\infty f_{i\kappa}^{T*}(r) V(r) f_{j\kappa}^T(r) dr, \quad (22.139)$$

and

$$\Pi_{\kappa ij}^{LS} = \int_0^\infty f_{i\kappa}^{L*}(r) \left(-\frac{d}{dr} + \frac{\kappa}{r} \right) f_{j\kappa}^S(r) dr, \quad (22.140)$$

$$\Pi_{\kappa ij}^{SL} = \int_0^\infty f_{i\kappa}^{S*}(r) \left(\frac{d}{dr} + \frac{\kappa}{r} \right) f_{j\kappa}^L(r) dr. \quad (22.141)$$

If $f_{i\kappa}^L(r)$ and $f_{i\kappa}^S(r)$ vanish at both $r = 0$ and $r \to \infty$, then a simple integration by parts shows that $\boldsymbol{\Pi}_\kappa^{LS}$ and $\boldsymbol{\Pi}_\kappa^{SL}$ are Hermitian conjugate matrices.

Physically Acceptable Basis Sets

The four criteria described above are exploited in the following way:

A. The structure of (22.133) ensures (i) that the upper and lower components have properly matched angular behavior. It also emphasizes that the radial parts are part of a spinor structure which should be kept intact when making approximations.

B. The nuclear singularity drives the dynamics of the electronic motion. It is therefore important that approximate trial solutions should have the correct analytic character as defined in Sects. 22.5.3 and 22.5.4. An expansion of $f_{i\kappa}^L(r)$ and $f_{i\kappa}^S(r)$ at $r = 0$ must reproduce this analytic behavior *exactly* if the approximation is to be physically reliable. The boundary conditions are part of the definition of a quantum mechanical operator; changing them gives a different operator with a different eigenvalue spectrum, so that trial functions which do not satisfy the boundary conditions of the physical problem cannot reproduce the physical solution.

The behavior as $r \to \infty$ is less crucial. Provided a bound wavefunction is well approximated over the region containing most of the electron density, the results are insensitive to many choices.

C. The correct reduction of the Dirac equation to Schrödinger's equation in the nonrelativistic limit (for example, see Ref. [4, p. 97]) depends upon the operator identity

$$\mathbf{p}^2 = (\boldsymbol{\sigma}\cdot\mathbf{p})(\boldsymbol{\sigma}\cdot\mathbf{p}).$$

In the basis set formalism, the matrix equivalent of this equation is

$$T_{lij} = \frac{1}{2}\sum_{k=1}^N \Pi_{\kappa i k}^{LS}\Pi_{\kappa k j}^{SL}, \qquad (22.142)$$

where

$$T_{lij} = \int_0^\infty f_{i\kappa}^{L*}(r)\frac{1}{2}\left\{-\frac{d^2}{dr^2} + \frac{l(l+1)}{r^2}\right\}f_{j\kappa}^L(r)\,dr$$

is the ij-element of the nonrelativistic radial kinetic energy matrix. *This is not true in general unless Criterion C holds* [29,30]. The criterion can only be satisfied by matched pairs of functions $f_{i\kappa}^L(r), f_{i\kappa}^S(r)$, ruling out all choices of basis set in which large and small components are not matched in pairs.

Another way of viewing this result is to observe that for a general basis set, the sum over intermediate states in (22.142) is necessarily incomplete. The Hermitian conjugacy property, $\Pi_{\kappa ij}^{LS} = \Pi_{\kappa ji}^{SL}$ ensures that the omitted terms give real and non-negative contributions. Thus all other choices of basis set cause (22.142) to

underestimate the nonrelativistic kinetic energy [29] and to give spuriously large relativistic energy corrections.

We emphasize that (22.134) need only be true in the limit $c \to \infty$; however, basis sets used for finite values of c should be smooth functions of c^{-1} as $c \to \infty$ so that the equality

$$i\hbar\frac{f_\kappa^S(r)}{r}\chi_{-\kappa,m}(\theta,\varphi) = \boldsymbol{\sigma}\cdot\mathbf{p}\,\frac{f_\kappa^L(r)}{r}\chi_{\kappa,m}(\theta,\varphi)$$

$$(22.143)$$

holds *in the limit*.

D. This ensures that the basis functions are in the domain of the Dirac operator; the meaning of this statement can be made precise in a functional analytic discussion such as in [3]. Some implications for the finite basis set approach are given in the author's paper [15, pp. 235–253], which discusses the convergence of expectation values of operators for approximate Dirac wavefunctions obtained by this method.

Here the main importance is that a (possibly singular) multiplicative operator $V(r)$ (say, $-Z/r$) has $N \times N$ matrices $\mathbf{V}_\kappa^{LL}, \mathbf{V}_\kappa^{SS}$ with *finite* elements. This must be true both for exact solutions and for approximations if the wave functions are to represent physical states. In particular, both matrices must have a *lowest eigenvalue* $V_{\min}^{(N)}$ say. Consider now the quantity $(\psi|\hat{\mathbf{h}}_D(\lambda)|\psi)$, where

$$\hat{\mathbf{h}}_D(\lambda) = \left\{c\boldsymbol{\alpha}\cdot\mathbf{p} + \beta m_e c^2\right\} + \lambda V(r).$$

With $\lambda = 0$ we have a free Dirac particle with a two-branched continuous spectrum $E > mc^2$ and $E < -mc^2$. A negative definite $V(r)$ has always $(\psi|V(r)|\psi) > V_{\min}$; clearly,

$$V_{\min}^{(N)} \geq V_{\min} > -2mc^2, \qquad (22.144)$$

for all values of N. So if we increase λ from 0 to 1, the eigenvalues of trial solutions corresponding to eigenvalues in the upper continuum at $\lambda = 0$ will be smoothly decreasing functions of λ bounded below by $V_{\min}$ for all values of N. It follows that *the upper set of eigenvalues has a fixed lower bound* in the gap $(-mc^2, mc^2)$ for each finite matrix approximation.

If the basis set satisfies suitable completeness criteria in an appropriate Hilbert space as $N \to \infty$ (see [15, pp. 235–253] for more details) we see that, if (22.144) holds for all values of N, the infinite sequence $\{E_{N+i}^{(N)}, \quad N = N_0, N_0 + 1, \ldots\}$ of eigenvalues approximating the ith bound state has a finite lower bound, and therefore, by the completeness of the real numbers, it must have a limit point E_i in the bound state gap $(-mc^2, mc^2)$. Thus Rayleigh-Ritz approximations for Dirac's Hamiltonian converge in the same fashion as the corresponding nonrelativistic Rayleigh-Ritz approximations [30].

22.6.3 Catalogue of Basis Sets for Atomic Calculations

A. L-Spinors. L-spinors are the relativistic analogue of Coulomb Sturmian functions [31, 32] and are related analogously to the Dirac hydrogenic solutions. They satisfy equations of the form

$$\Lambda S_{n_r\kappa}(x) := \Lambda \begin{bmatrix} A_L f^{(L)}(x) \\ A_S f^{(S)}(x) \end{bmatrix} = 0 \qquad (22.145)$$

where

$$\Lambda := \begin{bmatrix} c^2\sigma^-(\lambda) - \dfrac{\lambda N_{n_r\kappa}}{x} & c\left(-\dfrac{d}{dx} + \dfrac{\kappa}{x}\right) \\ c\left(\dfrac{d}{dx} + \dfrac{\kappa}{x}\right) & -c^2\sigma^+(\lambda) - \dfrac{\lambda N_{n_r\kappa}}{x} \end{bmatrix},$$

λ is an arbitrary positive constant,

$$\sigma^-(\lambda) = \left[1 - \sqrt{1 - \lambda^2/c^2}\right],$$
$$\sigma^+(\lambda) = \left[1 + \sqrt{1 - \lambda^2/c^2}\right],$$

and $N_{n_r,\kappa} = \sqrt{n_r^2 + 2n_r\gamma + \kappa^2}$ is the Dirac apparent principal quantum number of Sect. 22.3. The *exact* Dirac Coulomb solution is obtained by choosing λ such that $\lambda = Z/N_{n_r\kappa}$. The explicit form for L-spinors, in terms of Laguerre polynomials (see Sect. 9.4.2), $L_{n_r}^{(2\gamma)}(2\lambda x)$, is

$$f_{\kappa,n_r}^{(L)}(x) = \mathcal{N}_{n_r,\kappa} x^\gamma e^{-\lambda x} \left[-(1 - \delta_{n_r,0}) L_{n_r-1}^{(2\gamma)}(2\lambda x)\right.$$
$$\left. + \frac{(N_{n_r,\kappa} - \kappa)}{(n_r + 2\gamma)} L_{n_r}^{(2\gamma)}(2\lambda x)\right]. \qquad (22.146)$$

$$f_{\kappa,n_r}^{(S)}(x) = \mathcal{N}_{n_r,\kappa} x^\gamma e^{-\lambda x} \left[-(1 - \delta_{n_r,0}) L_{n_r-1}^{(2\gamma)}(2\lambda x)\right.$$
$$\left. - \frac{(N_{n_r,\kappa} - \kappa)}{(n_r + 2\gamma)} L_{n_r}^{(2\gamma)}(2\lambda x)\right]. \qquad (22.147)$$

The constants A_T, $T = L, S$, account for the difference in normalization of the hydrogenic functions (22.123) from the L-spinors. In the limit $c \to \infty$, we see that

$$f^{(L)}(x) \to \text{const. } x^{l+1} e^{-\lambda x} L_{n_r}^{(2l+1)}(2\lambda x), \qquad (22.148)$$

which is the corresponding nonrelativistic Coulomb Sturmian function. L-spinors with different values of n_r are orthogonal with respect to the weight function $1/x$ since

$$(N_{n,\kappa} - N_{n',\kappa}) \int_0^\infty S_{n',\kappa}^\dagger(x) S_{n,\kappa}(x) \frac{dx}{x} = 0. \qquad (22.149)$$

The method of construction ensures that all criteria A – D are satisfied. Also, L-spinor component sets are complete in the Hilbert spaces $L^2(\mathcal{R}_+)$, $W_2^{(1)}(\mathcal{R}_+)$ and $W_2^{(2)}(\mathcal{R}_+)$ ensuring convergence of bound state solutions obtained with the Rayleigh-Ritz method.

In practice it is convenient to normalize the components independently, since this facilitates the construction of the blocks of the matrix Hamiltonian (22.137), which are banded when the operators are powers r^n, $n > -1$. The Coulomb potential is, of course, diagonal, by virtue of (22.149).

L-spinors are most useful for hydrogenic problems, either for isolated atoms or for atoms in static electromagnetic fields (see Chap. 13). The equivalent nonrelativistic Coulomb Sturmians have for a long time been used to study the Zeeman effect on high Rydberg levels, especially in the region where chaotic behavior is expected [33] (see Chap. 15).

B. S-Spinors. S-spinors have the functional form of *the most nearly nodeless* L-spinors characterized by the minimal value of n_r, and can be viewed as the relativistic analogues of Slater functions (STOs). When κ is negative, take $n_r = 0$, so that (22.146)

$$f_{\kappa,0}^{(L)}(x) = -f_{\kappa,0}^{(S)}(x)$$
$$= \mathcal{N}_{\kappa,0} x^\gamma \exp(-\lambda x) \frac{N_{0,\kappa} - \kappa}{2\gamma} L_0^{(2\gamma)}(2\lambda x).$$

When κ is positive, we must take $n_r = 1$, and then

$$f_{\kappa,1}^{(L)}(x) = \mathcal{N}_{1,\kappa} x^\gamma e^{-\lambda x}$$
$$\times \left[-L_0^{(2\gamma)}(2\lambda x) + \frac{N_{1,\kappa} - \kappa}{1 + 2\gamma} L_1^{(2\gamma)}(2\lambda x)\right],$$
$$f_{\kappa,1}^{(S)}(x) = \mathcal{N}_{1,\kappa} x^\gamma e^{-\lambda x}$$
$$\times \left[-L_0^{(2\gamma)}(2\lambda x) - \frac{N_{1,\kappa} - \kappa}{1 + 2\gamma} L_1^{(2\gamma)}(2\lambda x)\right].$$

These can be simplified by inserting the explicit expressions $L_0^{(2\gamma)}(2\lambda x) = 1$, $L_1^{(2\gamma)}(2\lambda x) = 2\gamma + 1 - 2\lambda x$. We define a *set of S-spinors* with exponents $\{\lambda_m, m = 1, 2, \ldots, N\}$ by rewriting the above in the form

$$f_m^{(T)}(r) = A^T g_m(\gamma, r) + B^T g_m(\gamma + 1, r), \qquad (22.150)$$

where, as in Eq. (22.145), $T = L, S$, $g_m(\theta, r) = r^\theta e^{-\lambda_m r}$,

$$A^L = A^S = 1, \quad B^L = B^S = 0 \quad \text{for } \kappa < 0,$$

$$\left.\begin{array}{l} A^L = \dfrac{(\kappa + 1 - N_{1,\kappa})(2\gamma + 1)}{2(N_{1,\kappa} - \kappa)} \\[2mm] A^S = \dfrac{(\kappa - 1 - N_{1,\kappa})(2\gamma + 1)}{2(N_{1,\kappa} - \kappa)} \\[2mm] B^L = B^S = 1 \end{array}\right\} \quad \text{for } \kappa > 0, \qquad (22.151)$$

and

$$\gamma = \sqrt{\kappa^2 - Z^2/c^2}, \quad N_{1,\kappa} = \sqrt{\kappa^2 + 2\gamma + 1}.$$

The choice of the set of positive real exponents $\{\lambda_m, m = 1, 2, \ldots, N\}$, must be such as to assure Rayleigh-Ritz convergence [15, pp. 235–253] and to maximize the rate at which it is attained. In particular, if *one* particular exponent is chosen to have the value $\lambda_m = Z/N_{r,\kappa}$, then the corresponding S-spinor is a true hydrogenic solution. In this case the trial solution is exact. Clearly, S-spinors inherit desirable properties of L-spinors and, in particular, satisfy criteria A – D.

All elements of the matrix Hamiltonian of the Dirac hydrogenic problem can be expressed in terms of Euler's integral for the gamma function [22, Sect. 6.1.1]:

$$\Gamma(z) = k^z \int_0^\infty t^{z-1} e^{-kt}\,dt, \quad (\Re z > 0, \Re k > 0)$$

and are therefore readily written down and evaluated. The effectiveness of this method depends upon the choice of exponent set: see D below. We refer to calculations using this scheme for many-electron systems in Sect. 22.7.

C. G-Spinors. The G-spinors are the relativistic analogues of nonrelativistic spherical Gaussians (SGTO), popular in quantum chemistry for studying the electronic structure of atoms and molecules. They satisfy the relativistic boundary conditions for a *finite size* nuclear charge density distribution at $r = 0$, and are therefore the most convenient for relativistic molecular electronic structure calculations. They are defined so that (22.143) holds for finite c as well as in the nonrelativistic limit, which is equivalent to

$$f_m^{(S)}(r) = \text{const.} \left(\frac{d}{dr} + \frac{\kappa}{r}\right) f_m^{(L)}(r). \quad (22.152)$$

Thus if

$$f_m^{(L)}(r) = \mathcal{N}_{l,m}^{(L)} r^{l+1} e^{-\lambda_m r^2}, \quad (22.153)$$

$$f_m^{(S)}(r) = \mathcal{N}_{l,m}^{(S)} \left[(\kappa + l + 1)r^l - 2\lambda_m r^{l+2}\right] e^{-\lambda_m r^2}. \quad (22.154)$$

Note that the leading term in (22.154) vanishes when $\kappa < 0$, so that the radial amplitude $r^{-1} f_m^{(S)}(r)$ is never singular, even in the s-state case when $\kappa = -1, l = 0$.

D. Exponent Sets for S- and G-Spinors. Quantum chemists are familiar with the use of nonrelativistic STO and GTO basis sets, and there are extensive collections of optimized exponents which permit economical calculations for atomic and molecular calculations [34]. These sets are a good starting point for relativistic calculations also. By and large, the compilations ignore mathematical completeness, which although desirable is in practice unattainable. However, basis sets can almost always be constructed to give adequate numerical precision for most purposes.

An effective alternative to optimization, especially for atoms, is to use geometrical sequences $\{\lambda_m\}$ of the form

$$\lambda_m = \alpha_N \beta_N^{m-1}, \quad m = 1, 2, \ldots, N, \quad (22.155)$$

which depend upon just two parameters α_N, β_N. A convenient way to do this is to find a pair $\alpha_{N_0}, \beta_{N_0}$ for small

N_0, say $N_0 = 9$, in a cheap and simple nonrelativistic calculation and then to increase N systematically using relations such as

$$\frac{\alpha_N}{\alpha_{N_0}} = \left(\frac{\beta_N - 1}{\beta_{N_0} - 1}\right)^a, \quad \text{or} \quad \frac{\ln \beta_N}{\ln \beta_{N_0}} = \left(\frac{N_0}{N}\right)^b,$$

where a, b are positive constants. Experience shows that no linear dependence problems (caused by ill-conditioning of the $\mathbf{S}^T$ matrices) are encountered when $\beta_N > 1.2$ for S-spinors, with N up to about 30, or $\beta_N > 1.5$ for G-spinors with N up to about 50.

E. Other Types of Analytic Basis Set; Variational Collapse. The earliest work with atoms [35] used STO functions of the form $\{r^\gamma \exp(-\lambda_m r), \quad m = 1, \ldots N\}$ for both large and small components, whilst Kagawa [36] used integer powers instead of the noninteger γ. Drake and Goldman [37] used functions of the form $\{r^{\gamma+i} \exp(-\lambda r), \quad i = 0, \ldots N-1\}$. For hydrogenic problems, these worked well for negative κ states, but gave a single spurious eigenvalue for positive κ, which could be simply deleted from the basis set. Various test calculations are included in the review article [38, Sect. IV]. Other attempts to use GTOs in the early 1980's led to problems interpreted as failure of the Rayleigh-Ritz methods because of the presence of "negative energy states" with a spectrum unbounded below: so-called "variational collapse". It is clear that all these approaches fail to observe three, and sometimes all, of the four criteria for acceptable basis sets. They are incapable of satisfying the physical boundary conditions, and it is therefore hardly surprising that they give unphysical spectra.

Several procedures have been advocated to overcome the problem, of which the two most popular are "kinetic balance" and projection operators. Kinetic balance, suggested by Lee and McLean [39], advocates augmenting a GTO basis, common to both large and small components, with additional functions to "balance the set kinetically". This appears to "fix up" the problem for the upper spectrum, but introduces spurious states, mainly in the lower part of the spectrum, as well as increasing the size of the small component basis set. There is no rigorous nonrelativistic limit, and no mathematical proof of convergence such as that guaranteed by Criteria A – D. A model with spurious negative-energy states cannot furnish a consistent physical interpretation of negative energy solutions as positron states, expected of a proper relativistic theory.

If "variational collapse" is attributed to the absence of a lower bound to the Dirac spectrum as whole, the idea of introducing a projection operator to eliminate collapse seems attractive. This is easy to do for free electrons, where the operators

$$\Lambda^\pm = \frac{\hat{\mathbf{h}}_D \pm E}{2|E|},$$

select positive/negative energy solutions. Unfortunately, this cannot be done in the presence of a potential except by an approximation which complicates calculations and reduces the efficiency of algorithms. The "negative energy sea" also depends upon the choice of potential; perturbing the potential (as long as it does not change the domain of the Hamiltonian) induces a unitary transformation taking one set of eigenstates into another which inevitably mixes the old positive and negative energy states. For example a relativistic calculation on a hydrogenic atom in which the nuclear charge is perturbed gives incorrect answers if the negative energy contribution to the perturbation series is omitted [40].

In any event, the finite matrix eigensolutions include both positive energy and negative energy states. It is therefore a simple matter to exclude the negative energy states if their contribution is expected to be negligible; this is the *no virtual-pair approximation*. The negative energy solutions are inert spectators for most atomic processes, just as are those positive energy solutions which lie deep in the atomic core. It is easy to go beyond the no virtual-pair approximation if the physical problem demands it.

F. Finite Element Methods. The use of B-splines in relativistic atomic calculations was popularized by Johnson *et al.* [41] following earlier work on relativistic ion-ion collisions by Bottcher and Strayer [42]. See Sect. 8.2.1; also Sect. 21.6 for applications in nonrelativistic atomic structure.

Whilst splines have proved very effective in calculations of spectroscopic properties of heavy ions using relativistic many-body theory (see Chap. 27), there are difficulties in meeting the criteria relating to boundary conditions, both at $r = 0$ and as $r \to \infty$. A discussion of this point is outside the scope of this chapter.

22.7 MANY-BODY CALCULATIONS

22.7.1 Atomic States

The construction of atomic many-electron wavefunctions from products of central field Dirac orbitals is employed to simplify the algorithms for calculating electronic structures and properties. This can be either in the context of expansions in Slater determinants of the traditional type, or by use of Racah algebra. A complete description of the methods of the latter sort used in popular computer codes is found in [27, Sect. 2].

22.7.2 Slater Determinants

An antisymmetric state of N independent electrons in configuration space can be constructed in the form

$$\{\alpha_1, \alpha_2, \ldots, \alpha_N\}$$

$$= \langle \mathbf{x}_1, \mathbf{x}_2, \ldots, \mathbf{x}_n | a_{\alpha_1}^\dagger a_{\alpha_2}^\dagger \cdots a_{\alpha_N}^\dagger | 0 \rangle$$

$$= \frac{1}{N!} \begin{vmatrix} \psi_{\alpha_1}(\mathbf{x}_1) & \psi_{\alpha_2}(\mathbf{x}_1) & \cdots & \psi_{\alpha_N}(\mathbf{x}_1) \\ \psi_{\alpha_1}(\mathbf{x}_2) & \psi_{\alpha_2}(\mathbf{x}_2) & \cdots & \psi_{\alpha_N}(\mathbf{x}_2) \\ \cdots\cdots\cdots\cdots\cdots\cdots\cdots\cdots\cdots\cdots\cdots\cdots\cdots \\ \psi_{\alpha_1}(\mathbf{x}_N) & \psi_{\alpha_2}(\mathbf{x}_N) & \cdots & \psi_{\alpha_N}(\mathbf{x}_N) \end{vmatrix}.$$

$$(22.156)$$

This *Slater determinant* is an antisymmetric eigenfunction of H_0 corresponding to the energy $\sum E_{\alpha_n}$ and of the angular momentum projection $J_3 = \sum j_{3,\alpha_n}$ corresponding to the eigenvalue $M_3 = \sum m_{3,\alpha_n}$. Defining the parity of a Dirac electron orbital as that of its upper component, $(-1)^{l_{\alpha_n}}$, we see that this has parity $\prod (-1)^{l_{\alpha_n}}$.

22.7.3 Configurational States

Configurational state functions (CSF) having specified total angular momentum J and parity Π can be constructed by vector addition of the individual angular momenta: $\mathbf{J} = \sum \mathbf{j}_{\alpha_n}$. We write such states as

$$\phi(\gamma J M)$$
$$= \sum_{\{m_{\alpha_n}\}} \langle \gamma J M | m_{\alpha_1}, m_{\alpha_2} \ldots, m_{\alpha_N} \rangle \{\alpha_1, \alpha_2, \ldots, \alpha_N\},$$

$$(22.157)$$

where $\langle \gamma J M | m_{\alpha_1}, m_{\alpha_2} \ldots, m_{\alpha_N} \rangle$ is a generalized Clebsch-Gordon coefficient, and γ defines the angular momentum coupling scheme.

A list of orbital quantum numbers, $\{\alpha_1, \alpha_2, \ldots, \alpha_N\}$ defines an electron *configuration*. If the configuration belongs to a single subshell, then the states share a common set of labels $\{n, \kappa\}$ where n is the principal quantum number. In $j-j$ coupling, the $\alpha-subshell\ states$ of N_α equivalent electrons can therefore be identified (we can suppress the projection M_α and the parity Π_α) by the labeling α^{N_α}, γ_α, J_α, where γ_α distinguishes degenerate states of the same J_α. For $j-j$ coupling, such labels are needed only for $j \geq \frac{5}{2}$; the *seniority scheme*, [27, Sects. 2.3, 2.4], provides a complete classification for $j < \frac{9}{2}$. A list of states of configurations j^N, classified in terms of the *seniority number* v and of *total angular momentum* J, appears in Table 22.5.

22.7.4 CSF Expansion

Atomic state functions (ASF) are linear superpositions of CSF's, of the form

$$\Psi(\gamma \Pi J) = \sum_{\alpha=1}^{N} c_\alpha \phi(\gamma_\alpha J), \qquad (22.158)$$

where c_α are a set of (normally) real coefficients. These coefficients are usually chosen so that $\Psi(\gamma \Pi J)$ is an eigenstate of the many-electron Hamiltonian matrix in a finite subspace of CSF's.

22.7.5 Matrix Element Construction

A full presentation of the reduction of matrix elements between CSF's to computable form is beyond the scope of this chapter. There are two approaches: one is based on expanding all CSF's and ASF's in Slater determinants, whilst the other exploits the properties of central field orbital spinors. The principles underlying the first are straightforward and may be found in atomic physics texts and review articles such as [26, 27].

The use of second quantization and diagrammatic methods of the quantum theory of angular momentum provides a powerful means of reducing matrix elements between atomic CSF's to a linear combination of radial integrals in a systematic way. The method, which is fully explained in [27], leads to a complete classification of matrix element expressions for all the one- and two-electron operators treated in this chapter. A full implementation within the $j-j$ coupling seniority scheme is available in various versions of the GRASP code [43–45].

22.7.6 Dirac-Hartree-Fock and Other Theories

The notation above echoes that of the nonrelativistic theory of Chap. 21, and it is possible to proceed along similar lines.

Dirac-Hartree-Fock Theory

Dirac-Hartree-Fock theory works exactly as described in Sect. 21.4; relativistic counterparts of Koopmans' Theorem, fixed-core approximations, Brillouin's Theorem are easy to obtain. The properties of Dirac-Hartree-Fock functions closely resemble those of Hartree-Fock functions, though allowance must be made for the fact that, for example, np orbitals (with $\kappa = -2$, $j = \frac{3}{2}$) and $n\bar{p}$ orbitals (with $\kappa = +1$, $j = \frac{1}{2}$) have different spatial distributions as a consequence of the dynamical and indirect effects of relativity. For further insight see [23, 26].

Most such calculations are currently made with updated versions of the codes of Desclaux [46] or Grant [43–45] which rely on finite difference methods resting on the techniques of Sect. 22.6.1. Further details may be found in the code descriptions.

Finite Matrix Methods for Atoms and Molecules

In view of the rapid pace of development of finite matrix methods, especially for the treatment of relativistic molecular electronic structure in the Born-Oppenheimer (fixed nucleus) approximation, it seems appropriate to give a brief outline of the extension of the one-electron equations of Sect. 22.4.2 to the many-electron case.

The method of approximation generalizes the one-body approximation scheme of Sect. 22.6.2 to the many-body problem based on the effective Hamiltonian of Sect. 22.4.1. This leads to an energy functional of the form

$$E = E_0 + E_1 \qquad (22.159)$$

where E_0 is the expected value of H_0 (22.76) and E_1 the expected value of H_1 for the finite basis many-body trial function. This leads to matrix Dirac-Fock equations of the form

$$\mathbf{FX} = \mathbf{ESX}. \qquad (22.160)$$

In general, the Fock matrix $\mathbf{F}$ is a sum of several matrices

$$\mathbf{F} = \mathbf{f} + \mathbf{g} + \mathbf{b}, \qquad (22.161)$$

Table 22.5. j^N configurational states in the seniority scheme. The multiplicity of each unresolved degenerate state is indicated by a superscript.

j	N	v	J	j	N	v	J
$\frac{1}{2}$	0,2	0	0	$\frac{9}{2}$	0,10	0	0
	1	0	$\frac{1}{2}$		1,9	1	$\frac{9}{2}$
$\frac{3}{2}$	0,4	0	0		2,8	0	0
	1,3	1	$\frac{3}{2}$			2	2,4,6,8
	2	0	0		3,7	1	$\frac{9}{2}$
	2	2	2			3	$\frac{3}{2},\frac{5}{2},\frac{7}{2},\frac{9}{2},\frac{11}{2},\frac{13}{2},\frac{15}{2},\frac{17}{2},\frac{21}{2}$
$\frac{5}{2}$	0,6	0	0		4,6	0	0
	1,5	1	$\frac{5}{2}$			2	2,4,6,8
	2,4	0	0			4	$0,2,3,4^2,5,6^2,7,8,9,10,12$
		2	2,4		5	1	$\frac{9}{2}$
	3	1	$\frac{5}{2}$			3	$\frac{3}{2},\frac{5}{2},\frac{7}{2},\frac{9}{2},\frac{11}{2},\frac{13}{2},\frac{15}{2},\frac{17}{2},\frac{21}{2}$
		3	$\frac{3}{2},\frac{9}{2}$	$\frac{7}{2}$	5		$\frac{3}{2},\frac{5}{2},\frac{7}{2},\frac{9}{2},\frac{11}{2},\frac{13}{2},\frac{15}{2},\frac{17}{2},\frac{19}{2},\frac{25}{2}$
$\frac{7}{2}$	0,8	0	0				
	1,7	1					
	2,6	0	0				
		2	2,4,6				
	3,5	1	$\frac{7}{2}$				
		3	$\frac{3}{2},\frac{5}{2},\frac{9}{2},\frac{11}{2},\frac{15}{2}$				
	4	0	0				
		2	2,4,6				
		4	2,4,5,8				

where, for each symmetry κ and nuclear center, A, of the molecule, $\mathbf{f}$ can be partitioned into blocks

$$\mathbf{f}_{A\kappa} = \begin{bmatrix} \mathbf{V}_{A\kappa}^{LL} & c\boldsymbol{\Pi}_{A\kappa}^{LS} \\ c\boldsymbol{\Pi}_{A\kappa}^{SL} & \mathbf{V}_{A\kappa}^{SS} - 2mc^2\mathbf{S}_{A\kappa}^{SS} \end{bmatrix}. \qquad (22.162)$$

The matrix

$$\mathbf{g} = \begin{bmatrix} \mathbf{J}^{LL} - \mathbf{K}^{LL} & -\mathbf{K}^{LS} \\ -\mathbf{K}^{SL} & \mathbf{J}^{SS} - \mathbf{K}^{SS} \end{bmatrix} \qquad (22.163)$$

is the matrix of the Coulomb repulsion part of the electron-electron interaction and

$$\mathbf{b} = \begin{bmatrix} \mathbf{B}^{LL} & \mathbf{B}^{LS} \\ \mathbf{B}^{SL} & \mathbf{B}^{SS} \end{bmatrix}. \qquad (22.164)$$

is the matrix of the Breit interaction.

In the atomic (one nuclear center) case, following [47], these matrices can also be blocked by symmetry κ. Using superscripts T to label the L or S components, and the notation $\bar{T}$ to denote the complementary label: $\bar{T} = S$ when $T = L$ or $\bar{T} = L$ when $T = S$, then the direct Coulomb part $\mathbf{J}_\kappa^{TT}$ has matrix elements

$$J_{\kappa pq}^{TT} = \sum_{\kappa' rs}(2j'+1)\left[D_{\kappa' rs}^{TT}J_{\kappa pq,\kappa' rs}^{0,TTTT} + D_{\kappa' rs}^{\bar{T}\bar{T}}J_{\kappa pq,\kappa' rs}^{0,TT\bar{T}\bar{T}}\right],$$

$$(22.165)$$

whilst the exchange part $\mathbf{K}_\kappa^{TT'}$ has the form

$$K_{\kappa pq}^{TT'} = \sum_{\kappa' rs}\sum_\nu (2j'+1)b_\nu(jj')D_{\kappa' rs}^{TT'}K_{\kappa pq,\kappa' rs}^{\nu,TT'TT'},$$

$$(22.166)$$

where TT' denotes any combination of component labels. Here $\mathbf{D}_\kappa^{TT'}$ is a density matrix with elements

$$D_{\kappa pq}^{TT'} = c_{\kappa p}^{T*}c_{\kappa q}^{T'}, \qquad (22.167)$$

where $c_{\kappa p}^T$ are the expansion coefficients. The Breit interaction matrices have the similar form

$$B_{\kappa pq}^{TT} = \sum_{\kappa' rs}\sum_\nu (2j'+1)e_\nu(jj')D_{\kappa' rs}^{\bar{T}\bar{T}}K_{\kappa pq,\kappa' rs}^{\nu,TT\bar{T}\bar{T}},$$

$$(22.168)$$

and

$$B_{\kappa pq}^{T\bar{T}} = \sum_{\kappa' rs}\sum_\nu (2j'+1)D_{\kappa' rs}^{\bar{T}T}$$
$$\times \left[d_\nu(\kappa\kappa')K_{\kappa pq,\kappa' rs}^{\nu,T\bar{T}\bar{T}T} + g_\nu(\kappa\kappa')M_{\kappa pq,\kappa' rs}^{\nu,T\bar{T}\bar{T}T}\right].$$

$$(22.169)$$

The matrix elements are constructed from standard radial integrals

$$J_{\kappa pq,\kappa' rs}^{\nu,TTT'T'} = \int_0^\infty \int_0^\infty f_{\kappa p}^T(r_1)f_{\kappa q}^T(r_1)U_\nu(r_1,r_2)$$
$$\times f_{\kappa' r}^{T'}(r_2)f_{\kappa' s}^{T'}(r_2)dr_2\,dr_1 \qquad (22.170)$$

where

$$U_\nu(r_1,r_2) = \begin{cases} r_1^\nu/r_2^{\nu+1} & \text{for } r_1 < r_2\,, \\ r_2^\nu/r_1^{\nu+1} & \text{for } r_1 > r_2\,. \end{cases}$$

Similarly

$$K_{\kappa pq,\kappa' rs}^{\nu,TT'TT'} = J_{\kappa p,\kappa' r,\kappa q,\kappa' s}^{\nu,TTT'T'} \qquad (22.171)$$

and

$$M_{\kappa pq,\kappa' rs}^{\nu,T\bar{T}\bar{T}T} = \int_0^\infty \int_{r_1}^\infty f_{\kappa p}^T(r_1)f_{\kappa' r}^{\bar{T}}(r_1)U_\nu(r_1,r_2)$$
$$\times f_{\kappa q}^{\bar{T}}(r_2)f_{\kappa' s}^T(r_2)dr_2\,dr_1\,. \qquad (22.172)$$

Further details about the coefficients $b_\nu(jj')$, $e_\nu(jj')$, $d_\nu(\kappa\kappa')$ and $g_\nu(\kappa\kappa')$ may be found in [47].

This formalism has been implemented for closed shell atoms with both S-spinors and G-spinors [47]. Computational aspects of calculating the radial integrals using S-spinors are discussed in [48], and can be adapted with relatively small modifications to G-spinor basis sets. As yet, there have been relatively few applications by comparison with codes based on finite difference methods, but the potential can be gauged from papers such as [47, 49–52], which deal with Dirac-Fock and Dirac-Fock-Breit calculations, many-body perturbation theory and coupled-cluster schemes.

G-spinor basis sets provide the most promising technique for application to the electronic structure of molecules; computer codes are under active development.

Electron Correlation in Atomic Calculations

Here we use the term *correlation* to denote methods which go beyond the single determinant approximation of Dirac-Hartree-Fock theory. These include *configuration interaction* schemes, in which each ASF is represented as a linear combination of CSF's built from previously determined orbital spinors and *multiconfiguration Dirac-Fock* calculations in which the orbitals are optimized simultaneously. Calculations representative of state of the art techniques will be found in [53, 54].

Many-body perturbation theory calculations and coupled-cluster calculations are not well suited to calculations with finite difference codes, because of the expense of calculating more than a limited orbital basis and all the matrix elements required. Calculations based on finite matrix methods enable this sort of calculation to be done more economically. Some justification for the use of finite matrix methods in relativistic many-body theory is given in [15, pp. 235–253].

The relativistic version of quantum defect theory [55] also gives insight into the competing roles of relativistic dynamics and screening in atoms. Compared with nonrelativistic quantum defect theory, it has been under-used.

22.7.7 Radiative Corrections

The term "radiative corrections" is usually interpreted to mean QED contributions to energies, expectation values or rates of atomic or molecular processes that arise from interaction between the electron-positron and photon fields, apart from those directly attributable to the nonrelativistic Coulomb interaction. This includes the relativistic and retardation effects embodied in the effective interaction between electrons as well as contributions from processes that are not so included. We consider two such processes, the electron self-energy and the vacuum polarization, which involve interactions of the same formal order as those giving rise to the covariant electron-electron interaction discussed above, but which are formally infinite. These are the lowest order processes requiring renormalization. See [7, 8, 10, 15] for more detail.

Electron Self-Energy

For a one-electron system, the renormalized expression for the self-energy of an electron in the state a in Feynman gauge is

$$\Delta E_a = \lim_{\Lambda \to \infty} \Re \left\{ -i\alpha\pi.mc^2 \int \bar{\psi}_a(x_2)\gamma^\mu S_F(x_2, x_1) \right.$$
$$\times \gamma^\nu \psi_a(x_1) g_{\mu\nu} D_F^\Lambda(x_2 - x_1) d^3x_2 d^3x_1 d(t_2 - t_1)$$
$$\left. - \delta m(\Lambda)\langle\psi_a|\beta|\psi_a\rangle \right\}, \qquad (22.173)$$

where

$$\delta m(\Lambda) = \frac{\alpha}{\pi}mc^2\left[\frac{3}{4}\ln(\Lambda^2) + \frac{3}{8}\right].$$

This represents the contribution from virtual processes involving the exchange of a single photon. The photon propagator has been modified to give the photon and effective mass Λ, so that the denominator of $D(q^2)$ (22.69) becomes $q^2 - \Lambda^2 + i\delta$. The two parts of this formula diverge as $\Lambda \to \infty$, though the limit of their difference is finite. This makes calculation difficult and expensive. There are several approaches:

1. For atomic number $Z \lesssim 20$, an expansion in powers of the electron-nucleus coupling parameter $\alpha Z = Z/c$ is satisfactory.

2. At larger atomic numbers an expansion in αZ evidently fails to converge, and nonperturbative methods

must be sought. This too is computationally difficult and expensive. The results for hydrogenic ions have been tabulated [56] for atomic numbers in the range $1 \leq Z \leq 100$. See [10, Chap. 2] for an up-to-date summary biased towards applications to the spectroscopy of highly-ionized atoms.

3. Processes involving more than one virtual photon are hard to calculate, and have mostly been ignored. See [10] for references.

Vacuum Polarization

The contribution of vacuum polarization is next in order of importance in the list of radiative corrections in atoms. As shown by Eq. (22.49), the nuclear potential generates a current in the vacuum that is responsible for a short-range screening of the nuclear charge. This can be represented as a local perturbing potential which is easy to take into account [57].

22.7.8 Radiative Processes

The operator $j^\mu(x)A_\mu(x)$ which occurs in the interaction Hamiltonian (22.61) describes processes in which the number of photons present can increase or decrease by one. The Fock space operator may be written

$$H_{int} = \sum_{a,b}\sum_\rho \left[a_a^\dagger a_b q_\rho^\dagger M_{ab}^{(\rho)\dagger}(t) + a_a^\dagger a_b q_\rho M_{ab}^{(\rho)}(t) \right],$$
$$(22.174)$$

where the first set of terms in the sum represents emission of a photon in the mode labeled ρ and the second to absorption of a photon by the same initial state. The operators a_a and $a_a^\dagger$ are anticommuting annihilation and creation operators of electrons, whilst q_ρ and $q_\rho^\dagger$ are *commuting* annihilation and creation operators of photons. If ω denotes photon frequency, then

$$M_{ab}^{(\rho)\dagger}(t) = M_{ab}^{(\rho)}e^{i(E_a - E_b + \omega)t},$$
$$M_{ab}^{(\rho)}(t) = M_{ab}^{(\rho)}e^{i(E_a - E_b - \omega)t},$$

where

$$M_{ab}^{(\rho)} = \left(\frac{\omega}{\pi c}\right)^{1/2}\int \psi_a^\dagger(\mathbf{x})[\Phi^{(\rho)}(\mathbf{x}) + c\boldsymbol{\alpha}\cdot\mathbf{A}^{(\rho)}(\mathbf{x})]\psi_b(\mathbf{x})d^3x$$

is the transition amplitude. For a discussion of this expression including the effect of gauge transformations on the computed amplitudes, the elimination of angular coordinates for atomic central field orbitals and connection with the nonrelativistic limit, see [27, 10, 16].

REFERENCES

1. B. G. Wybourne, *Classical Groups for Physicists* (John Wiley, New York, 1974).
2. J. P. Elliott and P. G. Dawber, *Symmetry in Physics* (Macmillan, Basingstoke, and London, 1979).
3. B. Thaller, *The Dirac Equation* (Springer-Verlag, Berlin, 1992).
4. W. Greiner, *Relativistic Quantum Mechanics* (Springer-Verlag, Berlin, 1990).
5. L. L. Foldy, Phys. Rev. **102**, 568 (1956).
6. Yu. M. Shirokov, Sov. Phys. JETP **6**, 664, 919, 929 (1958).
7. S. S. Schweber, *Introduction to Relativistic Quantum Field Theory* (Harper and Row, New York, 1964).
8. V. B. Berestetskii, E. M. Lifshitz, and L. P. Pitaevskii, *Relativistic Quantum Theory* (Pergamon Press, Oxford, 1971).
9. C. Itzykson and J.-B. Zuber, *Quantum Field Theory* (McGraw-Hill, New York, 1980).
10. L. N. Labzowsky, G. L. Klimchitskaya, and Yu. Yu. Dmitriev, *Relativistic Effects in the Spectra of Atomic Systems* (Institute of Physics Publishing, Bristol & Philadelphia, 1993).
11. W. H. Furry, Phys. Rev. **81**, 115 (1951).
12. I. P. Grant and H. M. Quiney, Adv. At. Mol. Phys. **23**, 37 (1988).
13. I. Lindgren, J. Phys. B **23**, 1085 (1990).
14. H. A. Bethe and E. E. Salpeter, *Quantum Mechanics of One-, and Two-Electron Sysems* (Berlin, Springer-Verlag, 1957).
15. *Relativistic, Quantum Electrodynamic, and Weak Interaction Effects in Atoms*, edited by W. R. Johnson, P. J. Mohr, and J. Sucher (American Insitute of Physics, New York, 1989).
16. I. P. Grant, J. Phys. B **7**, 1458 (1974).
17. I. Lindgren in [15] [pp. 3–27].
18. D. J. Baker, D. Moncrieff, and S. Wilson, in *Supercomputational Science*, edited by R. G. Evans and S. Wilson (Plenum Press, New York, 1990), pp. 201–209.
19. D. R. Hartree, Proc. Camb. Phil. Soc. **25**, 225 (1929).
20. V. G. Bagrov and D. M. Gitman, *Exact Solutions of Relativistic Wave Equations* (Kluwer Academic Publishers, Dordrecht, 1990).
21. E. A. Coddington and N. Levinson, *Theory of Ordinary Differential Equations* (McGraw-Hill, New York, 1955).
22. M. Abramowitz and I. A. Stegun, *Handbook of Mathematical Functions* (Dover, New York, 1965).
23. V. M. Burke and I. P. Grant, Proc. Phys. Soc. **90**, 297 (1967).
24. J. Kobus, J. Karwowski, and W. Jaskolski, Phys. Rev. A **20**, 3347 (1987).
25. G. Hall and J. M. Watt, *Modern Numerical Methods for Ordinary Differential Equations* (Clarendon Press, Oxford, 1976).
26. I. P. Grant, Adv. Phys. **19**, 747 (1970).
27. I. P. Grant, *Methods in Computational Molecular Physics*, edited by S. Wilson and G. H. F. Diercksen (Plenum Press, New York, 1992), Vol. 2, p. 1.
28. I. P. Grant, Phys. Scr. **21**, 443 (1980).
29. K. G. Dyall, I. P. Grant, and S. Wilson, J. Phys. B **17**, 493 (1984).
30. I. P. Grant, J. Phys. B **19**, 3187 (1986).
31. M. Rotenberg, Ann. Phys. **19**, 262 (1962).
32. M. Rotenberg, Adv. At. Mol. Phys. **6**, 233 (1970).
33. C. W. Clark and K. T. Taylor, J. Phys. B **15**, 1175 (1982).
34. P. Čarsky and M. Urban, *Ab Initio Calculations. Methods, and Applications in Chemistry* (Springer-Verlag, Berlin, 1980); *Gaussian Basis Sets for Molecular Calculations*, edited by S. Huzinaga, J. Andzelm, M. Klobukowski, E. Radzio-Andselm, Y. Sakai, and H. Tatewaki (Elsevier, Amsterdam, 1984); R. Poirier, R. Kari, and I. G. Csizmadia *Handbook of Gaussian Basis Sets* (Elsevier, Amsterdam, 1985).
35. Y.-K. Kim, Phys. Rev. **154**, 17 (1967); Phys. Rev. A **159**, 190 (1967).
36. T. Kagawa, Phys. Rev. A **12**, 2245 (1975); Phys. Rev. **22**, 2340 (1980).
37. G. W. F. Drake and S. P. Goldman, Phys. Rev. A **23**, 2093 (1981).
38. G. W. F. Drake and S. P. Goldman, Adv. At. Mol. Phys. **25**, 393 (1988).
39. Y. S. Lee and A. D. McLean, J. Chem. Phys. **76**, 735 (1982).
40. H. M. Quiney, I. P. Grant, and S. Wilson, J. Phys. B **18**, 2805 (1985).
41. W. R. Johnson, S. A. Blundell, and J. Sapirstein, Phys. Rev. A **37**, 307 (1988); Phys. Rev. A **40**, 2233 (1988).
42. C. Bottcher and M. R. Strayer, in *AIP Conference Proceedings # 136* (American Insitute of Physics, New York, 1987), p. 268.
43. I. P. Grant, B. J. McKenzie, P. H. Norrington, D. F. Mayers, and N. C. Pyper, Comput. Phys. Commun. **21**, 207 (1980).
44. K. G. Dyall, I. P. Grant, C. T. Johnson, F. A. Parpia, and E. P. Plummer, Comput. Phys. Commun. **55**, 425 (1989).
45. F. A. Parpia, C. F. Fischer, and I. P. Grant, Comput. Phys. Commun., (in press); CPC Program Library Catalog: ADCU.
46. J. P. Desclaux, Comput. Phys. Commun. **9**, 31 (1975); see also *ibid.* **13**, 71 (1977).
47. H. M. Quiney, I. P. Grant, and S. Wilson, J. Phys. B **20**, 1413 (1987).
48. H. M. Quiney, in *Supercomputational Science*, edited by R. G. Evans and S. Wilson (Plenum Press, New York, 1990), pp. 159–200.
49. S. Wilson, in *Methods in Computational Chemistry*, edited by S. Wilson (Plenum Press, New York, 1988), Vol. 2, p. 73.
50. H. M. Quiney, in *Methods in Computational Chemistry*, edited by S. Wilson (Plenum Press, New York, 1988), Vol. 2, p. 227.
51. H. M. Quiney, I. P. Grant, and S. Wilson, in *Lecture Notes in Chemistry # 52*, edited by U. Kaldor (Springer-Verlag, Berlin, 1989), p. 307.
52. H. M. Quiney, I. P. Grant, and S. Wilson, J. Phys. B **23**, L271 (1990).
53. S. Fritzsche and I. P. Grant, Phys. Scr. **50**, 473 (1994).
54. A. Ynnerman and C. F. Fischer, Phys. Rev. A **51**, 2020 (1995).

55. W. R. Johnson and K. T. Cheng, J. Phys. B **12**, 863 (1979); see also C. M. Lee and W. R. Johnson, Phys. Rev. A **22**, 979 (1980).

56. W. R. Johnson and G. Soff, At. Data Nucl. Data Tables **33**, 405 (1985).

57. E. A. Uehling, Phys. Rev. **48**, 55 (1935); see also E. H. Wichmann and N. M. Kroll, Phys. Rev. **101**, 843 (1956); and L. W. Fullerton and G. A. Rinker, Phys. Rev. A **13**, 1283 (1976).

23

Many-Body Theory of Atomic Structure and Processes

M. Ya. Amusia

Argonne National Laboratory, Argonne, Illinois 60439,
A. F. Ioffe Physical-Technical Institute, St. Petersburg 194021, Russia

23.1 INTRODUCTION

All atoms except hydrogen are many-body systems in which the interelectron interaction plays an important or even decisive role. The aim of this chapter is to describe a consistent method for calculating the structure of atoms and the characteristics of different atomic processes, by applying perturbation theory to take into account the interelectron interaction. This method involves drawing a characteristic diagram based on the structure or process. This is then used to create an analytical expression to the lowest order in the interelectron interaction. Higher-order corrections are subsequently generated.

This technique was invented about half a century ago in quantum electrodynamics by Feynman [1], then mod-

ified and adjusted for multiparticle systems by a number of authors. Its application to atomic structure and atomic processes required further modifications initiated at the end of the fifties (see e.g. Ref. [2]) and later. The corresponding technique was successfully applied to the calculation of a wide variety of characteristics and processes in many papers and several review articles [3, 4]. The increasing amount of experimental data available has led to improved accuracy for this technique, so that it can be applied to current problems [5].

The elements of the diagrammatic technique, which form a convenient and simple "language", are given together with the rules for creating "sentences" using basic "words". A kind of "dictionary" helps to translate diagrammatic "sentences" into analytical expressions suitable for calculations.

An essential part of the program is to learn how the simplest approximation can be improved, and what are the mechanisms and processes connected with and responsible for higher-order corrections.

When the diagrammatic technique of many-body theory is used, it is unnecessary to be restricted to a finite number of lowest-order terms in the interelectron interaction. On the contrary, some infinite sequences may be taken into account. The sum of all many-body diagrams is completely equivalent to the many-particle Schrödinger equation. Therefore taking all of them into account is just as complicated as solving the corresponding equation. Compared with other approaches, the diagrammatic technique can easily uncover hidden approximations and transparently demonstrate possible sources of corrections.

23.2 DIAGRAMMATIC TECHNIQUE

23.2.1 Basic Elements

Each physical atomic process involves an electronic interaction with a projectile or external field, in general time-dependent, or a mutual interelectronic interaction. By convention, the ground state of the atom (if it is not degenerate) is regarded as the vacuum state. Then the simplest process in this target is excitation of an electron to an unoccupied level, leaving behind a vacancy. The basic elements of a diagram are

$$(23.1)$$

where (a) with an arrow directed to the right represents an electron excited to a vacant level; (b) with an arrow directed to the left represents a vacancy; (c) with a cross represents the static Coulomb interaction; (d) represents the interelectron Coulomb interaction; (e) represents interaction with a time-dependent external field, usually electromagnetic; and (f) represents the very act of interaction.

The elements (23.1a–f) in combination can describe the following real or virtual basic processes

$$(23.2)$$

which represent (a) photon absorption by the vacuum with electron-vacancy pair creation; (b) electron excitation; and (c) vacancy excitation. Diagrams (23.2) depict processes as developing in time, shown increasing from left to right. A vacancy can be thought of as an antiparticle to the electron, moving backward in time. The time-reverse of processes (23.2) represent processes of photon emission due to annihilation of an electron-vacancy pair,

vacancy transition, and electron inelastic scattering, respectively.

A static [for example, Coulombic (23.1c)] field can virtually create an electron-vacancy pair, or affect the moving electron or vacancy, as shown in the following diagrams:

$$(23.3)$$

Just as for (23.2), diagrams (23.3) have their time-reversed counterparts.

Inclusion of interelectron interaction leads to a number of processes of which some examples are

$$(23.4)$$

Here, (23.4a) describes creation of two electron-vacancy pairs, (23.4b) represents the simplest picture of electron inelastic scattering, (23.4c) depicts vacancy decay with electron-vacancy pair creation, (23.4d) stands for electron-electron scattering, (23.4e) represents a process which can be called electron-vacancy annihilation and creation, while (23.4f) shows electron-vacancy scattering.

23.2.2 Construction Principles for Diagrams

The foundations of diagrammatic techniques are discussed in a number of books such as Ref. [6]. This chapter presents recipes for the construction and evaluation of diagrams corresponding to various atomic processes [7].

The basic procedure is to connect the initial and final states of the atom, drawn at the left and right sides of the diagram, using any of the elements in (23.1). In doing so, the following rules apply:

1. At each dot (23.1f), only three lines can meet: wavy (or dashed) and electron-vacancy.

2. A vacancy cannot be transformed into an electron or vice versa.

3. Electrons and vacancies can be created only pairwise from the vacuum.

4. Only linked diagrams are allowed; i.e. only those having no parts entirely disconnected from one another.

The simplest or initial approximation to a process is represented by a diagram which includes the lowest possible number of elements (23.1)–(23.4). Higher-order corrections can be derived by including additional elements of interaction with the static field of the nucleus (23.1c) and between electrons and/or vacancies (23.1d).

As an illustration of the method, consider the following three processes:

1. one-electron photoionization — the initial state is a photon while the simplest final state is an electron-vacancy pair. They can be combined together giving the basic diagram (23.2a).

2. elastic electron scattering — the initial and final states are single electrons. To describe the simplest scattering process, the interaction with the Coulomb field must be taken into account, leading to (23.3b). To account for interelectron interaction, the simplest element

$$(23.5)$$

must be introduced. It is a modification of (23.4b) accounting for the interaction of an incoming electron with all target electrons individually, not altering their states. This is emphasized by the loop in which the same vacancy leaving the lower dot reenters it. Indistinguishability of all electrons as fermions is taken into account by permutation of the electron (vacancy) line ends, as illustrated in the diagrams:

$$(23.6)$$

Diagram (23.6a) is obtained from (23.5) by permutation of the electron lines on one side of the interelectron interaction. Diagram (23.6b) is equivalent to (23.6a), but is simpler to draw.

3. inelastic electron scattering — the initial state is a single electron. For the final state we choose one with two electrons and therefore a single vacancy. The simplest diagram in this case is given by (23.4b).

To illustrate the description of the ground state characteristics, consider the contributions to the ground state energy of an atom. If this state is not degenerate, its potential energy is given by vacuum diagrams which have no free lines in the initial or final states. The simplest vacuum diagrams are

$$(a) \qquad (b) \qquad (c) \qquad (23.7)$$

Higher-order corrections to all these diagrams can be obtained by adding elements such as a static external field (23.3) or interelectron interaction (23.4) without

changing the initial and final states of the processes. The lowest-order processes are represented by (23.2a), (23.5), (23.6), (23.4b), and (23.7). There are many corrections even in the next order of interaction, either with an external field or with electrons or vacancies. To illustrate, only one correction to each process will be presented:

1. Simple photoionization (23.2a) may be combined with (23.4e) to obtain

$$(23.8)$$

This describes the effect of the creation of another electron-vacancy pair, after annihilation of the first one formed by absorption of the initial photon.

2. Simple elastic electron scattering (23.5) can be combined with an extra interaction term (23.4f) between the incoming electron and the vacancy of the loop (23.5), to obtain

$$(23.9)$$

3. Simple inelastic electron scattering (23.4b) can also be combined with (23.4f), accounting for the interaction of an electron and vacancy created in the lowest-order process, to obtain

$$(23.10)$$

4. The ground state energy term (23.7b) can be combined with the element (23.3c) of the interaction between the vacancy in the (23.7b) loops and the static field to obtain

$$(23.11)$$

Higher-order corrections can be constructed step by step by introducing further elements of interaction. In some cases, classes of diagrams may be taken into account up to infinite order by solving closed systems of integral or differential equations.

23.2.3 Correspondence Rules

These rules describe how to obtain an analytical expression corresponding to a given diagram. One starts by choosing a zero-order approximation which can be that of independent electrons moving in the Coulomb field of an atomic nucleus. Atoms with completely occupied shells, or subshells having a nondegenerate ground state, can be chosen as the vacuum. Electron (vacancy) states are characterized in this case by the quantum numbers n, ℓ, m_ℓ, and $\sigma = \pm 1/2$.

The first correspondence rule is to substitute a matrix element for each interaction:

$$
\begin{aligned}
&\text{Diagram } (23.2) \rightarrow \langle p|W|q\rangle\,, \\
&\text{Diagram } (23.3) \rightarrow \langle p|U|q\rangle\,, \quad (23.12)\\
&\text{Diagram } (23.4) \rightarrow \langle pt|V|qs\rangle\,,
\end{aligned}
$$

where W is the interaction potential of an electron with the external time-dependent field, U is the interaction potential of an electron (vacancy) with an external static field, for example that of the nucleus, and V is the Coulombic interelectron interaction. Each of the letters p, q, t, s represents a full set of n, ℓ, m_ℓ, σ quantum numbers. Vacancy states are below (and include) the highest occupied energy level, called the Fermi level, so that $p \leq F$. Electron states are above the Fermi level so that $q > F$. Thus diagram (23.2a) is represented by $\langle p|W|q\rangle$ with $p \leq F$ and $q > F$.

Apart from initial and final states, each diagram can have sections, i.e., intervals between successive interactions. For instance (23.9) and (23.10) each have one section. Each section is represented by an inverse energy denominator ε_d^{-1}. It includes the sum over all vacancy energies $\sum_{\text{vac}} \varepsilon_i$ minus the sum of the electron energies $\sum_{\text{el}} \varepsilon_n$ to which the entrance energy E of the diagram (e.g. $\hbar\omega$ for a time-dependent field) must be added:

$$
\varepsilon_d^{-1} = \left(\sum_{\text{vac}} \varepsilon_i - \sum_{\text{el}} \varepsilon_n + E \right)^{-1}. \quad (23.13)
$$

The second correspondence rule is to identify sections and write down their energy denominators. After attributing to each electron (vacancy) line a letter, denoting its state, the analytical expression for a diagram is given by

ANALYTICAL EXPRESSION

$= $ (THE PRODUCT OF ALL INTERACTION

  MATRIX ELEMENTS)

$\times$ (ALL ENERGY DENOMINATORS)$^{-1}$ (23.14)

$\times (-1)^L$ SUMMED OVER ALL INTERMEDIATE

  ELECTRON AND VACANCY STATES,

where L is equal to the sum of the total number of vacancy lines and closed vacancy or electron-vacancy loops.

Although electrons are fermions, the summation in (23.14) has no additional restrictions caused by the Pauli principle. It runs over all electron ($> F$) and vacancy ($\leq F$) states, including those where two or more electrons

(or vacancies) are in the same state. The correspondence rules (23.12), (23.13), and (23.14) can be illustrated by giving as examples the analytical expressions of two diagrams (23.8) and (23.9).

Attributing letters denoting electron and vacancy states, diagram (23.8) becomes

$$(23.15)$$

According to (23.12)–(23.14), the analytical formula

$$
A_{if}(\omega) = \sum_{r > F, t \leq F} \frac{\langle t|W|r\rangle \langle ri|V|tf\rangle}{\varepsilon_t - \varepsilon_r + \omega}(-1)^{2+1} \quad (23.16)
$$

is obtained. The symbol $\sum$ includes summation over discrete levels and integration. In (23.16), the intermediate state is $r > F$, $t \leq F$ and the diagram has two vacancies (t and i) and one loop rt. Integration must be performed over those states r which belong to the continuum.

Assigning letters denoting states, (23.9) appears as

$$(23.17)$$

where $p, q, p', r > F$, while $t \leq F$. According to (23.14),

$$
\Delta E = \sum_{r, q > F; t \leq F} \frac{\langle pt|V|qr\rangle \langle qr|V|p't\rangle}{\varepsilon_t - \varepsilon_q - \varepsilon_r + \varepsilon_p}(-1)^{1+1}, \quad (23.18)
$$

where the intermediate states are $q, r > F$ and $t \leq F$. It has one vacancy and one electron-vacancy loop rt.

An intermediate state in a diagram can be real or virtual. It is real if the energy conservation law can be fulfilled, i.e. if for some values of the section energy the following relation holds:

$$
E = \sum_{\text{el}} \varepsilon_n - \sum_{\text{vac}} \varepsilon_i\,. \quad (23.19)
$$

If (23.19) can be fulfilled, a prescription for avoiding the singularity in (23.13) is to substitute the expression

$$
\begin{aligned}
Q &= \lim_{\eta \to 0} \left(E - \sum_{\text{el}} \varepsilon_n + \sum_{\text{vac}} \varepsilon_i + i\eta \right)^{-1} \\
&= P \left(E - \sum_{\text{el}} \varepsilon_n + \sum_{\text{vac}} \varepsilon_i \right)^{-1} \\
&\quad - i\pi\delta \left(E - \sum_{\text{el}} \varepsilon_n + \sum_{\text{vac}} \varepsilon_i \right). \quad (23.20)
\end{aligned}
$$

for ε_d^{-1}. Here P denotes that the principal value is to be taken on integration over intermediate state energies. The result of (23.20) can thus be complex.

An intermediate state is virtual if the energy conservation law (23.19) is violated for all values of the section energy. In general, the bigger the virtuality, i.e. the difference $E - \sum_{\text{el}} \varepsilon_n + \sum_{\text{vac}} \varepsilon_i$, the smaller the contribution.

23.2.4 Higher-Order Corrections and Summation of Sequences

An important feature of the diagrammatic technique is the convenience in constructing higher-order corrections and in the summation of infinite sequences of diagrams. According to (23.13), each new interaction line leads to an additional interaction matrix element, extra energy denominator and summation over new intermediate states.

An important example of infinite summation is that of determining the one-electron states. The interaction with the nucleus (23.3b) and with atomic electrons (23.5) and (23.6) is not small and must be taken into account nonperturbatively; i.e., these elements must be iterated infinitely. To simplify the drawing, only the element (23.5) is repeated, leading to the diagrammatic equation:

$$(23.21)$$

Indeed, everything in the infinite sum which is in front of the dashed line repeats the infinite sum itself, thus leading to a closed equation of the form

$$\langle \tilde{p} | = \langle p | + \sum_{q > F} \langle \tilde{p}i|V|qi\rangle \frac{1}{-\varepsilon_q + \varepsilon_p} \langle q | . \qquad (23.22)$$

The two interactions leading to (23.21) can be permuted, so that the interaction 1 can be after 2. This leads to extension of the sum to include states with $q \leq F$. As a result, the summation in (23.22) must be performed over all states q.

Interaction with the nucleus and the other electrons affects also the occupied (or vacancy) states i in (23.21) and therefore the latter must be modified by inserting the elements (23.3b), (23.5), and (23.6) into them. Here again, the vacancy line in (23.5) and (23.6) must be modified by including the corrections (23.3b), (23.5), (23.6) and so on. Finally, the diagrammatic equation

$$(23.23)$$

is obtained. The doubled line for i emphasizes that the vacancy wave function is determined by an equation similar to (23.21). The corresponding analytical equation looks like (23.22), but includes also the Coulombic interaction with the nucleus and the exchange interaction with other atomic electrons. The summation over q in this equation is extended over all q, not only $q > F$. Multiplying the corresponding equation by $(\hat{H}_0 - \varepsilon_p)$ from the right[1], where $\hat{H}_0 = -\nabla^2/2$, and using the completeness of the functions $\sum_q |q\rangle\langle q| = \delta(\mathbf{r} - \mathbf{r}')$, results in the equation

$$\left[-\frac{\nabla^2}{2} - \frac{Z}{r} + \sum_{i \leq F} \int \frac{d\mathbf{r}'}{|\mathbf{r}' - \mathbf{r}|} |\phi_i(\mathbf{r}')|^2 - \varepsilon_p \right] \phi_p(\mathbf{r})$$
$$= \sum_{i \leq F} \int \frac{d\mathbf{r}'}{|\mathbf{r}' - \mathbf{r}|} \phi_i^*(\mathbf{r}')\phi_p(\mathbf{r}')\phi_i(\mathbf{r}) \qquad (23.24)$$

for the electron wave function $\phi_p(\mathbf{r})$. Here $\phi_i(\mathbf{r})$ are wave functions determined by equations similar to (23.24). These are the Hartree-Fock (HF) equations.

HF includes a part for interelectron interaction matrix elements, namely that given by (23.5) and (23.6). The rest is called the residual interaction, and its inclusion leads beyond the HF frame, accounting for correlations.

When a perturbative approach is used, it is essential to define the zero-order approximation. In this chapter, and very often in the literature, the Hartree-Fock approximation is used in this role. To simplify the drawing of diagrams, from now on single (rather than double) lines will represent electrons (vacancies), whose wave functions are determined in the HF approximation by Eq. (23.24). Obviously, in this case elements (23.3a), (23.5) and (23.6) should not be added to any other diagrams.

The procedure used in deriving (23.21) and (23.23) is in fact more general. Let us separate all diagrams describing elastic scattering which do not include a single one-electron or one-vacancy state as intermediate. Depicting their total contribution by a square, the precise one-particle state is determined by an infinite sequence of iterative diagrams which can be summed, similarly to (23.21), by

---

[1] Atomic units are used in this chapter: $e = m_e = \hbar = 1$

$$\xRightarrow{p}\bullet = \xrightarrow{p}\bullet + \xrightarrow{p}\boxed{\hat{\Sigma}}\xrightarrow{q}\bullet + \xrightarrow{p}\boxed{\hat{\Sigma}}\xrightarrow{q'}\boxed{\hat{\Sigma}}\xrightarrow{q}\bullet$$

$$+ \cdots + \xrightarrow{p}\boxed{\hat{\Sigma}}\bigg\downarrow_q + \cdots$$

$$= \xrightarrow{p}\bullet + \xRightarrow{p}\boxed{\hat{\Sigma}}\xrightarrow{q}\bullet + \xRightarrow{p}\boxed{\hat{\Sigma}}\bigg\downarrow_q . \qquad (23.25)$$

Here the single line stands for an HF state. Using the correspondence rule (23.14), an analytical equation similar to the Schrödinger equation can be derived with the operator $\hat{\boldsymbol{\Sigma}}$ playing the role of an external potential. The essential difference is, however, that this "potential" depends in principle upon the energy and state of the particle. The same kind of iterative procedure leading to (23.21) or (23.23) will be used several times in this chapter.

Other zero-order approximations can be chosen. But then diagrams with corrections of the type (23.3a) must be included, with the external static field potential equal to the difference between the HF and the chosen one.

23.3 CALCULATION OF ATOMIC PROPERTIES

23.3.1 Electron Correlations in Ground State Properties

A major advantage of the diagrammatic technique in many-body theory is that it is usually unnecessary to know the total wave function of the atom. On the contrary, only actively participating electrons or vacancies appear in a diagram. The HF zero-order approximation for one-electron and one-vacancy wave functions is used in what follows. All atomic characteristics and cross sections for atomic processes calculated with HF form the one-electron approximation. Everything beyond the HF frame, i.e., caused by residual interaction, are called correlation corrections or correlations. They can be calculated using the many-body perturbation theory (MBPT) [3], random phase approximation (RPA) [8] and random phase approximation with exchange (RPAE) or its generalized version GRPAE [7].

The simplest diagrammatic expression for the correlation energy is given by the two diagrams

$$(23.26)$$

The analytical expression $\Delta E_{\text{corr}}^{(2)}$ for (23.26a) is

$$\Delta E_{\text{corr}}^{(2)} = \sum_{k,n > F; i,j \leq F} \frac{\langle ij|V|kn\rangle \langle kn|V|ij\rangle}{\varepsilon_i + \varepsilon_j - \varepsilon_k - \varepsilon_n} . \qquad (23.27)$$

The analytical expression for (23.26b) differs from (23.27) by the sign and an exchange matrix element $\langle kn|V|ji\rangle$ instead of a direct $\langle kn|V|ij\rangle$ one. The contribution (23.26) overestimates the correlation energy by $\sim 10\%$.

Diagrams (23.26) can also be used to describe the interaction potential of two atoms, designated A and B. Let the ki states belong to A and nj to B. At large distances R between the atoms, the contribution of (23.26b) is exponentially small. Because the vacancies i and j are located inside atoms A and B respectively, the interelectron potential $V = |\mathbf{r_A} - \mathbf{r_B} + \mathbf{R}|^{-1}$ at large distances $R \gg R_{A,B}$, ($R_{A,B}$ are atomic radii), can be expanded as a series in powers of R^{-1}. The first term giving a nonvanishing contribution to (23.27) is $V \simeq R^{-3}[(\mathbf{r_A} \cdot \mathbf{r_B}) - 3(\mathbf{r_A} \cdot \mathbf{n})(\mathbf{r_B} \cdot \mathbf{n})]$, $\mathbf{n}$ being the unit vector in the direction of $\mathbf{R}$. Substituted into (23.26), this potential leads to the expression

$$U(R) = -\frac{C_6}{R^6} \qquad (23.28)$$

for the interatomic potential [9], where

$$C_6 \sim \sum_{k,n > F; i,j \leq F} \frac{|\langle i|\mathbf{r}|k\rangle|^2 |\langle j|\mathbf{r}|n\rangle|^2}{(\varepsilon_i + \varepsilon_j - \varepsilon_k - \varepsilon_n)} . \qquad (23.29)$$

Calculations (see e.g. Ref. [10]) show that the inclusion of higher-order corrections is important for obtaining accurate values for ΔE_{corr} and C_6. However, to improve accuracy by taking into account the corrections to diagrams (23.26) requires considerable effort. Indeed, there are several types of corrections to (23.26) such as (i) screening of the Coulombic interelectron interaction by the electron-vacancy excitations; (ii) interaction between vacancies ij; (iii) interaction between electrons and vacancies $ki(nj)$ $(kj(ni))$; and (iv) interaction between electrons kn. Corrections to the HF field itself which acts upon electrons k, n and vacancies i, j are discussed in Sect. 23.3.3.

Screening of the Coulomb interelectron interaction is very important, and in many cases must be taken into account nonperturbatively. The simplest way to do this is to use RPA, which defines the effective interelectron interaction $\tilde{\Gamma}$ as a solution of an integral equation, shown diagrammatically by

$$\tilde{\Gamma}\;\reflectbox{\rotatebox{90}{$\}$}} = \big\} + \overset{k}{\underset{i}{\circlearrowleft}}\;\tilde{\Gamma} \qquad (23.30)$$

If V in (23.27) is replaced by $\tilde{\Gamma}$, an expression for ΔE_{corr} in RPA can be derived.

Exchange is very important in atoms and molecules, so diagram (23.30) can be modified to include this effect, thus leading to the effective interaction Γ in RPAE [10]:

$$\tilde{\Gamma}\;\reflectbox{\rotatebox{90}{$\}$}} = \big\} + \overset{k}{\underset{i}{\circlearrowleft}}\;\tilde{\Gamma} + \overset{k}{\underset{i}{\times}}\;\tilde{\Gamma} \qquad (23.31)$$

Replacing V in (23.27) by Γ gives an expression for ΔE_{corr} in RPAE. Taking into account screening also affects the long-range interatomic interaction considerably by altering the constant C_6 in (23.28).

The ground state energy of an atom or molecule is modified by an external field. For a not too intense electromagnetic field, the simplest correction to the ground state energy is given by the diagrams

$$\cdots\!\!\!\!\raisebox{-1ex}{}\cdots + \quad \qquad (23.32)$$

Considering a dipole external field, its interaction with the atomic electrons is given by $W = \sum_{i\leq F} \mathbf{E}\cdot\mathbf{r}_i$, $\mathbf{E}$ being the strength of the field. The ground state energy shift is given by $\Delta E = -\alpha(\omega)E^2/2$, where $\alpha(\omega)$ is the dynamical dipole polarizability and ω is the frequency of the field. According to (23.32), $\alpha(\omega)$ is determined by

$$\alpha(\omega) = \sum_{k>F;i\leq F} \frac{2|\langle i|z|k\rangle|^2(\varepsilon_k - \varepsilon_i)}{(\varepsilon_k - \varepsilon_i)^2 - \omega^2}, \qquad (23.33)$$

where z is a component of the vector $\mathbf{r}$.

RPAE corrections to $\alpha(\omega)$ are discussed in Sect. 23.3.5 in connection with the photoionization process.

Nondipole polarizabilities of other multipolarities can be obtained in the lowest order of interelectron interaction using (23.32) with a properly chosen interaction operator between the electromagnetic field and an electron, instead of $W = \sum_{i\leq F} \mathbf{E}\cdot\mathbf{r}_i$.

23.3.2 Characteristics of One-Particle States

A single vacancy or electron can propagate from one instant of interaction to another, as described to zero order by elements (23.1b) [or (23.1a)] with dots (23.1f) at the ends. This line represents an HF one-particle state with a given angular momentum, spin, and total momentum. Accounting for virtual or real atomic excitations leads, for a vacancy, to a diagram similar to (23.25) but with oppositely directed arrows. Because the interaction with these excitations is usually much smaller than the energy distance between shells, in the sum over q only the term $q = i$, i being the considered vacancy state, need be taken into account. Interaction with the vacuum leaves the angular momentum, spin, and total momentum unaltered. It can however change the energy, and lead to a finite lifetime for a vacancy state.

Analytically, the vacancy propagation in the HF approximation is described by the one-particle HF Green's function G^{HF}:

$$G_i^{\mathrm{HF}}(E) = 1/(\varepsilon_i - E). \qquad (23.34)$$

Solving equation (23.25) for a vacancy i with only Σ_{ii} terms included gives

$$G_i(E) = 1/[\varepsilon_i + \Sigma_{ii}(E) - E]. \qquad (23.35)$$

The pole in $G(E)$ which determines the vacancy energy is shifted from $E = \varepsilon_i$ to $E_i = \varepsilon_i + \Sigma_{ii}(E_i)$. The quantity $\Sigma_{ii}(E)$ is called the self-energy, and is in general a complex function of energy, its imaginary part determining the lifetime of the vacancy i. Near E_i, Eq. (23.35) can be written in the form

$$G_i(E) \approx F_i/[\varepsilon_i + \Sigma_{ii}(E_i) - E] = F_i/(E_i - E), \qquad (23.36)$$

where

$$F_i = \left[1 - \frac{\partial \Sigma_{ii}(E)}{\partial E}\bigg|_{E=E_i}\right]^{-1} \qquad (23.37)$$

is called the *spectroscopic factor*. It characterizes the probability for more complicated configurations to be admixed into a single vacancy state i [7].

An important problem is to calculate the self-energy part $\Sigma(E)$. The first nonzero contributions are

$$+ \text{ exchange terms.} \qquad (23.38)$$

Specific calculations [10] demonstrate that if the intermediate electron states n [in (23.38a)] and kn [in (23.38b)] are found in the field of vacancies jj' and $ii'j$, the diagrams (23.38) are able to reproduce the values of the correlation energy shift with about 5% accuracy. For outer subshell vacancies, the contributions (23.38a) and (23.38b) are almost equally important, to a large extent canceling each other. For example, (23.38a) shifts the outer $3p$ vacancy in Ar to lower binding energies by 0.1 Ry, while the contribution of (23.38b) is -0.074 Ry. The total value 0.026 Ry is small and close to the experimental one, which is 0.01 Ry. For inner vacancies, (23.38a) is dominant because the intermediate states in (23.38b) have large virtualities and are therefore small. The main contribution to the sum over j' comes from the term $j' = i' = i$, which gives for the energy shift of level i

$$\Delta\varepsilon_i = \Sigma_{ii}^{(2)}(\varepsilon_i) = \sum_{n>F;j\leq F} \frac{|\langle n|r^{-1}|j\rangle|^2}{\varepsilon_n - \varepsilon_j}. \qquad (23.39)$$

The value (23.39) is positive. Most important higher-order corrections will be included if V in (23.38a) is replaced by Γ from (23.31).

The physical meaning of diagram (23.38a) is transparent: it accounts for configuration mixing of one vacancy i and "two vacancies jj'—one electron n" states in the lowest order in the interelectron interaction. Diagram (23.38b) is not as transparent, and for $i = i'$ its intermediate state appears to violate the Pauli principle. However, as noted in connection with (23.14), the Pauli principle should not be considered as a restriction in constructing intermediate states.

Diagram (23.38a) and its exchange can have an imaginary part, which gives the probability of Auger decay $\gamma_i^{(A)}$, calculated to the lowest order in the interelectron interaction. For (23.38a) one has

$$\gamma_i^{(A)} = \Im\left[\Sigma_{ii}(E_i)\right]$$
$$= 2\pi \sum_{j,j' \leq F; n > F} |\langle in|V|j'j\rangle|^2 \delta(\varepsilon_j + \varepsilon_{j'} - \varepsilon_n - \varepsilon_i).$$

$$(23.40)$$

The width $\gamma_i^{(A)}$ is usually much smaller than $\Re[\Sigma_{ii}(E_i)]$, but there are several exceptional cases with abnormally large Auger widths, among which the most impressive is the $4p$-vacancy in Xe with its $\gamma_{4p} \approx 10$ eV.

Higher-order corrections include those which are taken into account when V in (23.40) is replaced by Γ from (23.31). The others include jj' vacancy-vacancy interaction, the interaction between vacancies jj' and the electron n and so on. As noted above, all of these can be obtained step by step by inserting the elements (23.4) into (23.38). To select the most important corrections, a physical idea and/or experience are necessary. For instance, if the energy transferred in the decay process $\Delta\varepsilon = \varepsilon_{j'} - \varepsilon_j$ is close to some threshold energies of atomic intermediate or outer shells, corrections which include virtual excitation of this shell must be taken into account.

The contribution to the spectroscopic factor from (23.38a) is given according to (23.37) by

$$F_i^{(2)} \equiv \left[1 + \sum_{j,j' \leq F; n > F} \frac{|\langle in|V|j'j\rangle|^2}{(\varepsilon_j + \varepsilon_{j'} - \varepsilon_n - \varepsilon_i)^2}\right]^{-1}.$$

$$(23.41)$$

Generally, for any Fermi particle, $F_i \leq 1$ [11] because there cannot be more than one particle in a given state. Note that the integrand in (23.41) is the lowest-order admixture of the $jj'n$ state to a pure one-vacancy state i. A small F value means strong mixing. For atoms, F_i is usually close to 1, but there are exceptions where F is small. For example, F_{5s} in Xe is about 0.33 [7].

The operator $\Sigma_{ij(n)}(\varepsilon)$ has nondiagonal matrix elements, which leads to admixture of other one-vacancy j or one-electron n states to the vacancy i. A measure of this admixture is given by the ratio $\Sigma_{ij(n)}(\varepsilon_i)/(\varepsilon_{j(n)} - \varepsilon_i)$.

In higher orders, decay processes more complex than those described by the imaginary part of (23.38a) become possible. For example, this could be a two-electron Auger decay in which the transition energy is distributed

between two outgoing electrons. An example of the lowest-order diagram for this process is

$$(23.42)$$

This is one of those diagrams which describe the mixing of a pure one-vacancy state with a quite complex configuration $jj_1j_2\varepsilon_1\varepsilon_2$.

23.3.3 Electron Scattering

Propagation of an electron in a discrete level or in a scattering state can be described in the same way as for a vacancy. The electron wave function is determined by equation (23.25). Using the correspondence rule (23.14), equation (23.25) can be expressed analytically in the form

$$\left[-\frac{\nabla^2}{2} - \frac{Z}{r} + \sum_{i \leq F} \int \frac{dr'}{|\mathbf{r}' - \mathbf{r}|} |\phi_i(\mathbf{r}')|^2 - \varepsilon\right]\psi_\varepsilon(\mathbf{r})$$
$$= \sum_{i \leq F} \int \frac{dr'}{|\mathbf{r}' - \mathbf{r}|} \phi_i^*(\mathbf{r}')\psi_\varepsilon(\mathbf{r}')\phi_i(\mathbf{r})$$
$$+ \int \hat{\Sigma}(\mathbf{r}, \mathbf{r}', \varepsilon)\psi_\varepsilon(\mathbf{r}')d\mathbf{r}'.$$

$$(23.43)$$

The terms with the Coulombic interelectron interaction $|\mathbf{r}' - \mathbf{r}|^{-1}$ determine the Hartree-Fock self-consistent potential. The last term in (23.43) represents the nonlocal energy dependent polarizational interaction of the continuous spectrum electron with the target atom. Although (23.43) resembles the ordinary Schrödinger equation, because of the energy dependence of the self-energy part $\hat{\Sigma}(\mathbf{r}, \mathbf{r}', \varepsilon)$, it is not the same. Consequently, the wave function $\psi_\varepsilon(\mathbf{r})$, often called a Dyson orbital, must be normalized according to the condition

$$(\psi_{\varepsilon'}|\psi_\varepsilon) = F_\varepsilon \delta(\varepsilon' - \varepsilon),$$

$$(23.44)$$

where

$$F_\varepsilon = \left[1 - \left(\varepsilon \left|\frac{\partial\hat{\Sigma}(\mathbf{r}, \mathbf{r}', E)}{\partial E}\right| \varepsilon\right)\Bigg|_{E=\varepsilon}\right]^{-1}.$$

$$(23.45)$$

This is different from that for ordinary wave functions by the factor $F_\varepsilon < 1$. In F_ε, the matrix element of the operator $\partial\hat{\Sigma}(\mathbf{r}, \mathbf{r}', E)/\partial E$ is calculated between states $\psi_\varepsilon(\mathbf{r})$. It is seen that F_ε is the same spectroscopic factor as determined by (23.37), but for a continuous spectrum or an excited electron state.

The object described by the wave function $\psi_\varepsilon(\mathbf{r})$ differs from an individual electron because it can be unstable, and its state is mixed with those of more complicated configurations, such as "two electron $k'k''$—one vacancy j''". This object is called a *quasi-electron*.

Equation (23.43) also determines the energies of electrons in discrete levels which are shifted from their HF values. Contrary to the case of deep vacancies, it is impossible to predict the sign of the energy shift without detailed calculations.

For incoming electron energies ε higher than the target ionization threshold, the operator $\hat{\Sigma}(\mathbf{r}, \mathbf{r}', \varepsilon)$ acquires an imaginary part, thus becoming an optical potential for the projectile. The additional elastic scattering phase shifts from their HF values can be expressed via matrix elements of $\hat{\Sigma}(r, r', \varepsilon)$ between the wave functions $\phi_\varepsilon^*(r)$ and $\psi_\varepsilon(r)$; but to find numbers for these phase shifts, the self-energy part or polarizational interaction $\hat{\Sigma}(\mathbf{r}, \mathbf{r}', \varepsilon)$ must be calculated. It appears that the second-order projectile-target interactions

$$\text{(a)} \qquad + \qquad \text{(b)} \qquad + \text{ exchange terms} \qquad (23.46)$$

provide a reasonably good approximation [10, 12].

The expression for $\hat{\Sigma}(\mathbf{r}, \mathbf{r}', \varepsilon)$ simplifies at distances far from the target. Only (23.46a) contributes in this region, while other terms are exponentially small. Expanding the Coulomb interelectron interactions in (23.46a) in powers of $r_{1'}/r_1 \ll 1$, $r_{2'}/r_2 \ll 1$ gives

$$\hat{\Sigma}(\mathbf{r}, \mathbf{r}', \varepsilon) = -\delta(\mathbf{r} - \mathbf{r}')\frac{\alpha^{\mathrm{HF}}(0)}{2r^4}, \quad r, r' \to \infty \qquad (23.47)$$

where $\alpha^{\mathrm{HF}}(0)$ is the static ($\omega = 0$) dipole polarizability determined by (23.33).

Accounting for each additional interaction between the projectile and target increases the power of r in the denominator of (23.47). Most important is the interaction between target electrons. By including this, the asymptotic expression (23.47) is also modified, where instead of $\alpha^{\mathrm{HF}}(0)$, the RPAE polarizability $\alpha^{\mathrm{RPAE}}(0)$ appears. If the target-projectile interelectron interaction is taken into account to second order as in (23.46) while all the rest is included exactly, the expression for $\hat{\Sigma}(\mathbf{r}, \mathbf{r}', \varepsilon)$ for $r, r' \to \infty$ is still given by (23.46), but with the exact static polarizability.

Many experimental results for low energy electron scattering by noble gases, alkalis, and alkaline earths agree well with calculations of elastic scattering phase shifts obtained by solving equation (23.42) in which $\hat{\Sigma}(\mathbf{r}, \mathbf{r}', \varepsilon)$ is given by (23.46) with RPAE corrections taken into account. Total cross sections are reproduced with an accuracy as high as several percent, including the Ramsauer minimum region. This is illustrated in Fig. 23.1 for the e^- + Xe case [12]. This approximation is also reasonably good in describing the angular

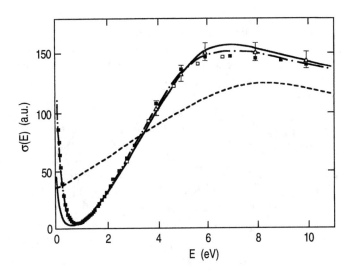

Figure 23.1. Electron-Xe atom elastic scattering cross section [12]. Solid line: including polarization interaction; dashed line: HF; dash-dotted line: experiment.

distributions. As the projectile energy ε increases, the contribution of (23.46) decreases rapidly.

The approach presented here applies to other incoming particles, for instance positrons e^+. The exchange between e^+ and target electrons must be omitted by discarding (23.6) and all but (23.46a) terms in the polarization interaction $\hat{\Sigma}(\mathbf{r}, \mathbf{r}', \varepsilon)$. The incoming positron in its intermediate state k' [see (23.46a)] interacts strongly with the virtually excited electron k'', forming a positronium-like object. Corresponding diagrams are obtained by inserting elements (23.4d) for the $e^+ - e^-$ interaction into (23.46a). Summation of the infinite sequence of such diagrams corresponds to substituting the product $\phi_{k'}(\mathbf{r_1})\phi_{k''}(\mathbf{r_2})$ in the intermediate state of (23.46a) by the exact e^+e^- wave function in the field of a target with a vacancy j. Such a program is very complicated, and a simplification has proved to be satisfactory. Only the positronium (Ps) binding is taken into account by subtracting its binding energy in the denominator of (23.13). This is equivalent to adding the Ps ionization potential I_{Ps} to ε in (23.46) [13]. It enhances the polarization interaction and leads to an interesting qualitative feature: the possibility of alteration of the sign of the interaction (23.46). Indeed, instead of $\alpha(0)$, the corresponding expression for e^+-atom collision includes $\alpha(I_{\mathrm{Ps}})$. For alkalis, the binding energy is less than I_{Ps}, and at the energy $\hbar\omega = I_{\mathrm{Ps}}$ the polarizability (23.33) is negative, while $\alpha(0) > 0$. This leads to a repulsive polarizational interaction, instead of the usual attractive one. This difference affects the cross section qualitatively [14].

According to diagram (23.23), Eq. (23.43) describes the target with an additional electron. If the target is a neutral atom, solution of (23.43) with discrete energy values describes negative ion states; both ground and excited.

Again as in Sect. 23.3.2, diagrams (23.46) with RPAE corrections form a reasonably good starting point for calculating the negative ion binding energies, even in cases when this binding is comparatively strong, as in alkaline earth negative ions [15]. The inclusion of only the outer shell polarizability (j is a vacancy in the outer shell) leads to overbinding of the additional electron forming the negative ion. Only the inclusion of screening due to inner shell excitations yields good agreement with experiment. For instance, recent measurements of Ca$^-$ affinity [16] give more than 20 meV, while the calculations without inner shell excitations give about 50 meV [15]. Their inclusion must considerably reduce the theoretical value.

23.3.4 Two-Electron and Two-Vacancy States

One can construct a diagrammatic equation for the wave function of a two-electron or a two-vacancy state by separating all diagrams describing two-electron (two-vacancy) scattering which do not include these states as intermediate, and denoting their total contribution by a circle. Then the exact two-electron (vacancy) state is determined by the infinite sequence of diagrams

The analytic equation for two electrons in an atom interacting with each other can be written in the form

$$\left(\hat{H}_1^{\mathrm{HF}} + \hat{\Sigma}_1 + \hat{H}_2^{\mathrm{HF}} + \hat{\Sigma}_2 + \hat{Q}\hat{\Pi}_{12} - \varepsilon \right) \psi_{12}(\mathbf{r_1 \cdot r_2}) = 0 .$$

(23.49)

Here $\hat{H}_{1(2)}^{\mathrm{HF}}$ is the HF part of the one-particle Hamiltonian in (23.24), $\hat{\Pi}_{12}$ is the effective interelectron interaction and $\hat{Q}$ is the projection operator

$$\hat{Q} = 1 - n_1 - n_2 ,$$

(23.50)

with $n_{1(2)}$ being the Fermi step function, $n_{1(2)} = 1$ for $1(2) \leq F$ and $n_{1(2)} = 0$ for $1(2) > F$. The function $n_{1(2)}$ thus eliminates contributions of vacant states. The operator $\hat{Q}$ takes into account the fact that propagation

of two electrons (or, more precisely, quasi-electrons due to the presence of $\hat{\Sigma}$) takes place in a system of other particles which occupy all levels up to the Fermi level. The presence of $\hat{Q}$ makes (23.49) essentially different from a simple two-electron Schrödinger equation: $\hat{Q}$ requires that after each act of interaction described by $\hat{\Pi}_{12}$, both participants either remain electrons or become vacancies.

Diagrams (23.48) describe the states of two electrons outside the closed shell core, or electron scattering by an atom with one electron outside the closed shells. In general, $\hat{\Pi}_{12}$ is nonlocal, dependent upon ε, and can have an imaginary part. The lowest-order approximation to $\hat{\Pi}_{12}$ is $V = |\mathbf{r}_1 - \mathbf{r}_2|^{-1}$. Equation (23.49) must be solved in order to obtain, for example, the excitation spectrum of two electrons in atoms with two electrons outside closed subshells, such as Ca. Instead of V, the interaction Γ from (23.31) can be used, which would account for screening due to virtual excitation of inner shell electrons. For low level two-electron excitations, the energy dependence of $\hat{\Pi}_{12}$ can be neglected.

The same type of equation can be obtained for two-vacancy states, describing their energies, decay widths, and structure due to configuration mixing with more complex states. For inner and intermediate shell vacancies, however, the corresponding corrections can be taken into account perturbatively, and the screening by outer shells is not essential. The admixture of outer shell excitation with inner vacancies is important, leading to satellites of the main spectral lines.

The interaction between vacancies leads to correlation two-vacancy decay processes in which the energy is carried away by a single electron or photon. Some diagrams exemplifying these processes in the lowest possible order of interaction are

Even in this order, there are several diagrams giving together the amplitude of the correlation Auger (23.51a) and radiative (23.51b) decay. For inner shells, a specific feature of such processes is that the released energy is about twice that for a single vacancy decay. Of course, in these cases the decay probability is relatively small. It is not, however, necessarily much smaller than that of individual vacancies if the energies of the intermediate and initial states are close to each other.

The presence of residual two-body forces leads to effective multiparticle interactions. The simplest diagram presenting a three-electron interaction is given by

(23.52)

The role of multiparticle interactions in atomic structure is far from being clear. Diagrams similar to (23.52) are important if it is of interest to calculate the energy levels of atoms with three or more electrons (or vacancies) outside of closed shells. This is a very complicated calculation because even two electrons in the Coulomb field of the nucleus is a difficult three-body problem.

23.3.5 Electron-Vacancy States

The one-electron—one-vacancy state is the simplest excitation of a closed shell system under the action of an external time-dependent field, represented by (23.2a). Beginning with electrons and vacancies described in the HF approximation, the result of residual interactions leads to excitations of more complex states, including those with two or more electron-vacancy pairs. The interaction can also lead to a single electron-vacancy pair. Let us concentrate on the latter case and separate all diagrams describing electron-vacancy interaction which do not include these states as intermediate, and denote them by a circle. Then the exact electron-vacancy state is determined by the infinite sequence of diagrams

$$ \tag{23.53} $$

Contrary to the electron-electron case in Sect. 23.3.4, the analytical expression corresponding to (23.53) cannot be represented as a Schrödinger-type equation. Indeed, being symmetric under time reversal, (23.53) leads to an equation depending, unlike (23.49), upon the second power of the electron-vacancy energy ω. Note that R in (23.53) and $\hat{\Pi}$ in (23.46) are different.

It is necessary to solve (23.53) when calculating the photoabsorption amplitude, which can then be represented as

$$ \omega \text{----} \qquad \tag{23.54} $$

The amplitude for elastic photon scattering is also expressed via the exact electron-vacancy state, determined by (23.53) to be

$$ \omega \text{----} \qquad \text{----} \; \omega \tag{23.55} $$

The other case where it is necessary to solve (23.53) is the scattering of electrons, both elastic and inelastic, by atoms with a vacancy in their outer shell, such as the halogens.

The simplest approximation to R is given by the Coulomb interaction to lowest order (23.4e) and (23.4f). With such R, the sequence of diagrams (23.53) is the same as (23.31) and thus forms the RPAE, which is often used to describe photoionization and other atomic processes. Both terms (23.4e) and (23.4f) contribute to the electron-vacancy in (23.53) only if the external field is spin independent, such as an ordinary photon. For a magnetic interaction, which is proportional to spin, the term (23.4e) does not contribute.

23.3.6 Photoionization in RPAE and Beyond

In RPAE, the photoionization amplitude $\langle k|D(\omega)|i\rangle$ is determined by solving an integral equation obtained from (23.53) and (23.54) using the correspondence rule (23.14) [7]:

$$
\langle k|D(\omega)|i\rangle = \langle k|d|i\rangle
$$
$$
+ \sum_{k' > F; \, i' \leq F} \left[\frac{\langle k'|D(\omega)|i'\rangle \langle ki'|V|ik' - k'i\rangle}{\omega - \varepsilon_{k'} + \varepsilon_{i'} + i\eta} \right.
$$
$$
\left. - \frac{\langle i'|D(\omega)|k'\rangle \langle kk'|V|ii' - i'i\rangle}{\omega - \varepsilon_{k'} - \varepsilon_{i'} - i\eta} \right], \tag{23.56}
$$

where d is the dipole operator describing the photon-electron interaction, and $V = |\mathbf{r}_1 - \mathbf{r}_2|^{-1}$. To obtain the RPAE photoionization cross section, the usual expression (24.19) must be multiplied by the square modulus of the ratio $\langle k|D(\omega)|i\rangle / \langle k|d|i\rangle$, with $\omega = \varepsilon_k + I_i$. The RPAE corrections described by the term in square brackets in (23.56) are very large for outer and intermediate electron shells. In the sum over i', if only terms with the same energies $\varepsilon_{i'} = \varepsilon_i$ are included, then equation (23.56) accounts for intrashell correlations. By adding terms with $\varepsilon_{i'} \neq \varepsilon_i$, the effect of intershell correlations is taken into account.

Intrashell correlations are most important for multi-electron shells with large photoionization cross sections. Their inclusion leads to a description of qualitatively huge maxima in the cross section called giant resonances. An example is the $4d^{10}$ photoionization cross section of Xe shown in Fig. 23.2 [7], where satisfactory agreement with experiment is demonstrated. It appears that all RPAE intrashell time-forward diagrams, such as that on the first line in (23.53), and the first term in brackets in (23.56) with $\varepsilon_{i'} = \varepsilon_i$, may be taken into account by the matrix element $\langle \tilde{k}|d|i\rangle$. This is the one electron approximation, but with the function $\tilde{\phi}_k(\mathbf{r})$ calculated in the term-dependent HF approximation [7].

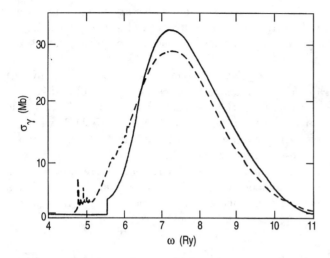

Figure 23.2. Photoabsorption in the vicinity of the $4d^{10}$ subshell threshold in Xe [7]. Solid line: RPAE; dashed line: experiment.

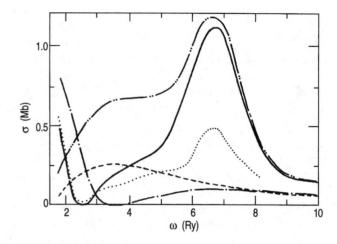

Figure 23.3. Photoionization of $5s^2$ electrons in Xe [7]. Solid line: RPAE with effects of $5p^6$ and $4d^{10}$ included; dashed line: $5s^2$ electrons only; dash-dotted line: with effect of $5p^6$ electrons; dash-double-dotted line: with effect of $4d^{10}$ electrons; dotted line: experiment.

Term-dependency means that only the total angular momentum and spin and their projections for the electron-vacancy pair are conserved, being equal to that of the incoming photon. The individual values for the electron and vacancy angular momentum and spin are not considered to be good quantum numbers. Thus the term-dependent HF includes a large fraction of RPAE correlations.

Subshells with small photoionization cross sections are strongly influenced by their multielectron neighbors. This is a manifestation of intershell correlations. These are taken into account if the sum over i' in (23.56) includes terms with $\varepsilon_{i'} \neq \varepsilon_i$. An example is the $5s^2$ subshell in Xe, which is strongly affected by the outer $5p^6$ and inner $4d^{10}$ neighboring electrons. Due to this interaction, the $5s^2$ cross section is completely altered as illustrated in Fig. 23.3 [7]. The RPAE results predict a qualitative feature of the experimental data, namely the formation of the correlation minimum in the cross section. Note that the amplitude of electron elastic scattering on an atom with a vacancy is expressed in RPAE via Γ given by (23.31) [17].

For the inner or deep intermediate shells, RPAE proves to be insufficient. First, screening of the Coulomb interaction between the outgoing or virtually excited electron and the vacancy [see (23.4f)] must be taken into account. This can be done by replacing V by Γ from (23.31). The ionization potential (or the energy of the vacancy i) must also be corrected, which requires inclusion of at least the contribution from the first term of (23.38). It has been demonstrated [18] that the screening of the electron-vacancy interaction can be taken into account by calculating the wave function of the virtually excited or outgoing electron in the self-consistent HF

field of an ion instead of that of a neutral atom. A method which uses only these one-particle wave functions in (23.56) is called the generalized RPAE or GRPAE. The use of this approximation considerably improves the agreement with experiment near the intermediate shell thresholds, decreasing there the cross section value and shifting its maximum to higher energies.

Close to inner shell thresholds, the Auger decay of a deep vacancy must be taken into account. Due to decay, the photoelectron instantly finds itself in the field of at least two vacancies instead of one, leading to considerable growth of the threshold cross section. Diagrammatically, the effect of decay may be described by

$$ \text{(23.57)} $$

Here, the double line emphasizes that starting from the moment of decay, the photoelectron moves in the double field $j_1 j_2$ instead of a single i vacancy field. For inner vacancies, this is a strong effect which can lead even to recapture of the photoelectron into some of the discrete levels in the field of the double vacancy $j_1 j_2$.

So-called post-collision interactions in photoionization can be taken into account by this diagram when the Auger electron is much faster than the photoelectron [19]. If their speed is of the same order, their mutual Coulomb repulsion must be accounted for, leading to additional alteration of energy and angular redistributions.

A photoelectron can excite or knock out another atomic electron. To lowest order in the residual interelectron interaction, this process can be represented by

(23.58)

While the formation of an initial electron-vacancy ki_1 pair requires RPAE or GRPAE for its description, the second step of (23.58) can be reasonably well reproduced by the lowest-order term in V. It appears that process (23.58) has a high probability for not too fast photoelectrons, changing considerably the cross section [20] for ionization by creating a vacancy i_1.

A comparatively simple diagram

(23.59)

describes the photoionization process in which a more complicated state is created in the ion than a single vacancy, e.g. a state with two vacancies and one electron. Here the doubled arrow indicates that the electron is in a discrete level n.

Diagrams (23.57)–(23.59) present some corrections which mix electron-vacancy and two-electron–two-vacancy configurations. Each additional interaction line increases the number of possible physical processes considerably. With growth of the number of particles actively participating in a process, the calculational difficulties increase enormously. However, this is not a shortcoming of the diagrammatic approach, but a specific feature of more and more complex physical processes.

23.3.7 Photon Emission and Bremsstrahlung

The amplitude of photon emission in lowest order is given by the time-reverse of (23.2c):

$$\xrightarrow{\quad k \quad} \bullet \xrightarrow{\quad k' \quad}$$

(23.60)

This diagram represents ordinary Bremsstrahlung; i.e. a process of projectile deceleration in the field of the target. If the target has internal structure as atoms do, it can be really or virtually excited during the collision process. The simplest excitation means creation of an electron-vacancy pair. The annihilation of this pair results via the time-reverse of (23.2a) in photon emission. The process thus looks like

(23.61)

To obtain the total Bremsstrahlung amplitude, the terms (23.60) and (23.61) must be summed. The polarization radiation (PR) created by the mechanism (23.61) has a number of features which are different from the ordinary Bremsstrahlung (OB) represented by (23.60). The intensity is proportional to $1/M_p^2$ for OB, where M_p is the projectile mass, and the spectrum, at least for high ε_k, is proportional to $1/\omega$. On the other hand, the PR intensity is almost completely independent of M_p and its frequency dependence is quite complex, being determined by the target polarizability $\alpha(\omega)$ [21]. PR is most important for frequencies ω of the order of and higher than the target's ionization potential. At sufficiently large distances and for neutral targets, PR starts to predominate over OB. Close to discrete excitations of the k' electron, the contribution (23.61) becomes resonantly enhanced.

Higher-order corrections are important in the PR amplitude. First, the Coulomb interaction V in (23.61) must be replaced by Γ from (23.31).

The analytical expression for the total Bremsstrahlung amplitude, including RPAE corrections to (23.61), is given by the expression

$$\langle k|A(\omega)|k'\rangle = \langle k|d|k'\rangle + \sum_{k''>F;\, i\leq F} \langle ki|V|k'k''\rangle$$

$$\times \lim_{\eta\to+0} \frac{2(\varepsilon_{k''}-\varepsilon_i)}{\omega^2-(\varepsilon_{k''}-\varepsilon_i)^2+i\omega\eta}\langle i|D(\omega)|k''\rangle . \quad (23.62)$$

To derive the Bremsstrahlung spectrum, the usual general expression must be multiplied by the square modulus of the ratio $\langle k|A(\omega)|k'\rangle/\langle k|d|k'\rangle$. If the incoming electron is slow, corrections (23.46) also become important. The intermediate state in (23.61) includes two electrons k' and k, and a vacancy i. The extent of interaction between them could be considerable.

An important feature of PR is that it is nonzero even if the projectile is neutral, but is able to polarize the target. For example, it leads to emission of continuous spectrum radiation in atom-atom collisions, whose intensity for frequencies of the order of the ionization potentials is close to that in electron-atom collisions.

The second and higher orders in the residual interaction involve processes more complicated than (23.61), for instance those which include simultaneous photon emission and target excitation (ionization) [22].

23.4 CONCLUDING REMARKS

It is most convenient to apply diagrammatic techniques to closed shell atoms whose ground state is nondegenerate. Degeneracy means that some of the energy denominators (23.13) become zero with nonzero statistical weight. All such contributions must be summed to eliminate this degeneracy. This leads to strong mixing of some states. For example, the energy required for electron-vacancy transitions jn [see (23.38)] within an open shell is zero, thus leading to strong mixing of i and ijn states. If a pair with zero excitation energy has nonzero angular momentum, taking into account the mixing within such a pair destroys angular momentum as a characteristic of a one-vacancy state. This makes all calculations much more complicated, reflecting a specific feature of the degenerate physical system.

In using the diagrams and formulas presented above, it is essential that the interelectron and electron-nucleus interactions be purely potential. Inclusion of retardation and spin-dependence in the interparticle interaction makes the calculations much more complicated. These parts of the interaction appear as relativistic corrections. They are comparatively small in all but the heaviest atoms, and can be taken into account perturbatively. Beyond lowest order, these additional interactions are strongly altered when virtual excitations of electron-vacancy pairs and the Coulomb interaction between them is taken into account. An example is given by the sequence of diagrams

$$(23.63)$$

where the heavy dashed line stands for the spin-dependent interelectron interaction. Note that here there are no electron-vacancy loops as in (23.28) because the Coulomb interaction is unable to affect the electron spin and thus to transfer spin excitations.

The same kind of diagrams describes the one-particle field acting upon an electron or vacancy due to the presence of spin-orbit interaction or weak interaction between electrons and the nucleus. For instance, the effective weak potential includes contributions from the sequence

$$(23.64)$$

This is another example demonstrating how flexible and convenient is the many-body approach in considering different processes and interactions.

Acknowledgment

This work was supported by the U. S. Department of Energy, Office of Basic Energy Sciences, under Contract W-31-109-Eng-38.

REFERENCES

1. R. P. Feynman, *Quantum Electrodynamics* (W. A. Benjamin, New York, 1961).
2. H. A. Bethe and J. Goldstone, Proc. R. Soc. London A **238**, 551 (1957).
3. H. P. Kelly, in *Advances in Theoretical Physics*, edited by Keith A. Brueckner (Academic Press, New York, 1968), Vol. 2, pp. 75–169.
4. M. Ya. Amusia, *Many-body Effects in Electron Atomic Shells* (A. F. Ioffe Physical-Technical Institute Publications, Leningrad, 1968), [in Russian], pp. 1–144.
5. *Proceedings of the 10th Conference of Vacuum Ultraviolet Radiation Physics, Paris, 1992*, edited by F. J. Wuilleumier, Y. Petroff, and I. Nenner (World Scientific, Singapore, 1993).
6. N. H. March, W. H. Young, and S. Sampanthar, *The Many-Body Problem in Quantum Mechanics* (Cambridge University Press, Cambridge, 1967).
7. M. Ya. Amusia, *Atomic Photoeffect* (Plenum Press, 1990).
8. D. Pines, *The Many-Body Problem* (W. A. Benjamin, New York, 1961).
9. L. D. Landau and E. M. Lifshits, *Quantum Mechanics* (Pergamon Press, 1965), pp. 319–322.
10. M. Ya. Amusia and N. A. Cherepkov, Case Studies At. Phys. **5**, 47 (1975).
11. A. B. Migdal, *Theory of Finite Fermi-Systems and Applications to Atomic Nuclei* (Interscience, New York, 1967).
12. W. R. Johnson and C. Guet, Phys. Rev. A **49**, 1041 (1994).
13. M. Ya. Amusia, N. A. Cherepkov, L. V. Chernysheva, and S. G. Shapiro, J. Phys. B **9**, 16, L469 (1976).
14. S. Zhou, S. P. Parikh, W. E. Kauppila, C. K. Kwan, D. Lin, A. Surdutovich, and T. S. Stein, Phys. Rev. Lett. **73**, 2, 236 (1994).
15. L. V. Chernysheva, G. F. Gribakin, V. K. Ivanov, and M. Yu. Kuchiev, J. Phys. B **21**, L419 (1988).
16. C. W. Walter and J. R. Peterson, Phys. Rev. Lett. **68**, 2281 (1992).
17. M. Ya. Amusia, V. A. Sosnivker, N. A. Cherepkov, L. V. Chernysheva, and S. I. Sheftel, J. Tech. Phys. (USSR Acad. Sci.) **60**, 1 (1990) [in Russian].
18. M. Ya. Amusia, in *Photoionization in VUV and Soft X-ray Frequency Regions*, edited by U. Becker and D. Shirley (Plenum Press, in press).
19. M. Ya. Amusia, M. Yu. Kuchiev, and S. A. Sheinerman, J. Exp. Theor. Phys. **76**, 2, 470 (1979).
20. M. Ya. Amusia, G. F. Gribakin, K. L. Tsemekhaman, and V. L. Tsemekhaman, J. Phys. B **23**, 393 (1990).
21. M. Ya. Amusia, Phys. Rep. **162**, 5, 249 (1988).
22. M. Ya. Amusia, in *Polarizational Radiation of Particles and Atoms*, edited by V. N. Tsitovich and I. M. Oiringel (Plenum Press, 1992).

24

Photoionization of Atoms

Anthony F. Starace

Department of Physics and Astronomy, The University of Nebraska, Lincoln, Nebraska 68588-0111

24.1 INTRODUCTION

This chapter outlines the theory of atomic photoionization, and the dynamics of the photon-atom collision process. Those kinds of electron correlation that are most important in photoionization are emphasized, although many qualitative features can be understood within a central field model. The particle-hole type of electron correlations are discussed, as they are by far the most important for describing the single photoionization of atoms near ionization thresholds. Detailed reviews of atomic photoionization are presented in Refs. [1] and [2]. Current activities and interests are well-described in two recent books [3, 4]. Other related topics covered in this volume are, experimental studies of photon interactions at both low and high energies in Chaps. 59 and 60, photodetachment in Chap. 58, and theoretical descriptions of electron correlations in Chap. 23, autoionization in Chap. 25, and multiphoton processes in Chap. 72.

24.2 GENERAL CONSIDERATIONS

24.2.1 The Interaction Hamiltonian

Consider an N-electron atom with nuclear charge Z. In nonrelativistic approximation, it is described by the Hamiltonian

301

$$H = \sum_{i=1}^{N} \left(\frac{p_i^2}{2m} - \frac{Ze^2}{r_i} \right) + \sum_{i>j=1}^{N} \frac{e^2}{|\mathbf{r}_i - \mathbf{r}_j|} . \quad (24.1)$$

The one-electron terms in brackets describe the kinetic and potential energy of each electron in the Coulomb field of the nucleus; the second set of terms describe the repulsive electrostatic potential energy between electron pairs. The interaction of this atom with external electromagnetic radiation is described by the additional terms obtained upon replacing $\mathbf{p}_i$ by $\mathbf{p}_i + (|e|/c)\mathbf{A}(\mathbf{r}_i, t)$, where $\mathbf{A}(\mathbf{r}_i, t)$ is the vector potential for the radiation. The interaction Hamiltonian is thus

$$H_{\text{int}} = \sum_{i=1}^{N} \left\{ \frac{+|e|}{2mc} [\mathbf{p}_i \cdot \mathbf{A}(\mathbf{r}_i, t) + \mathbf{A}(\mathbf{r}_i, t) \cdot \mathbf{p}_i] \right.$$
$$\left. + \frac{e^2}{2mc^2} |\mathbf{A}(\mathbf{r}_i, t)|^2 \right\} . \quad (24.2)$$

Under the most common circumstance of single-photon ionization of an outer-subshell electron, the interaction Hamiltonian in (24.2) may be simplified considerably. First, the third term in (24.2) may be dropped, as it introduces two-photon processes (since it is of second order in $\mathbf{A}$). In any case, it is small compared with single photon processes since it is of second order in the coupling constant $|e|/c$. Second, we choose the Coulomb gauge for $\mathbf{A}$, which fixes the divergence of $\mathbf{A}$ as $\nabla \cdot \mathbf{A} = 0$. $\mathbf{A}$ thus describes a transverse radiation field. Furthermore $\mathbf{p}$ and $\mathbf{A}$ now commute and hence the first and second terms in (24.2) may be combined. Third, we introduce the following form for $\mathbf{A}$:

$$\mathbf{A}(\mathbf{r}_i, t) = \left(\frac{2\pi c^2 \hbar}{\omega V} \right)^{\frac{1}{2}} \hat{\boldsymbol{\epsilon}} e^{i(\mathbf{k} \cdot \mathbf{r}_i - \omega t)} . \quad (24.3)$$

This classical expression for $\mathbf{A}$ may be shown [5] to give photoabsorption transition rates that are in agreement with those obtained using the quantum theory of radiation. Here $\mathbf{k}$ and ω are the wave vector and angular frequency of the incident radiation, $\hat{\boldsymbol{\epsilon}}$ is its polarization unit vector, and V is the spatial volume. Fourth, the *electric dipole* (E1) approximation, in which exp $i(\mathbf{k} \cdot \mathbf{r}_i)$ is replaced by unity, is usually appropriate. The radii r_i of the atomic electrons are usually of order 1Å. Thus for $\lambda \gg 100$Å, $|\mathbf{k} \cdot \mathbf{r}_i| \ll 1$. Now $\lambda \gg 100$Å corresponds to photon energies $\hbar\omega \ll 124$eV. For outer atomic subshells, most of the photoabsorption occurs for much smaller photon energies, thus validating the use of the E1 approximation.[1] Use of all of the above conventions

---

[1] This approximation cannot be used uncritically, however. For example, photoionization of excited atoms (which have large radii), photoionization of inner subshells (which requires the use of short wavelength radiation), and calculation of differential cross sections or other measurable quantities that are sensitive to the overlap of electric dipole and higher multipole amplitudes all require that the validity of the electric dipole approximation be checked.

and approximations allows the reduction of H_{int} in Eq. (24.2) to the simplified form

$$H_{\text{int}} = \frac{+|e|}{mc} \left(\frac{2\pi c^2 \hbar}{\omega V} \right)^{\frac{1}{2}} \sum_{i=1}^{N} \hat{\boldsymbol{\epsilon}} \cdot \mathbf{p}_i \exp(-i\omega t) . \quad (24.4)$$

H_{int} thus has the form of a harmonically time-dependent perturbation. According to time-dependent perturbation theory, the photoionization cross section is proportional to the absolute square of the matrix element of (24.4) between the initial and final electronic states described by the atomic Hamiltonian in (24.1). Atomic units, in which $|e| = m = \hbar = 1$, are used in what follows.

24.2.2 Alternative Forms for the Transition Matrix Element

The matrix element of (24.4) is proportional to the matrix element of the momentum operator $\sum_i \mathbf{p}_i$. Alternative expressions for this matrix element may be obtained from the following operator equations involving commutators of the exact atomic Hamiltonian in (24.1):

$$\sum_{i=1}^{N} \mathbf{p}_i = -i \left[\sum_{i=1}^{N} \mathbf{r}_i, H \right] , \quad (24.5)$$

$$\left[\sum_{i=1}^{N} \mathbf{p}_i, H \right] = -i \sum_{i=1}^{N} \frac{Z\mathbf{r}_i}{r_i^3} . \quad (24.6)$$

Matrix elements of (24.5) and (24.6) between eigenstates $\langle \psi_0 |$ and $|\psi_f \rangle$ of H having energies E_0 and E_f respectively give

$$\langle \psi_0 | \sum_{i=1}^{N} \mathbf{p}_i |\psi_f \rangle = -i\omega \langle \psi_0 | \sum_{i=1}^{N} \mathbf{r}_i |\psi_f \rangle , \quad (24.7)$$

$$\langle \psi_0 | \sum_{i=1}^{N} \mathbf{p}_i |\psi_f \rangle = \frac{-i}{\omega} \langle \psi_0 | \sum_{i=1}^{N} \frac{Z\mathbf{r}_i}{r_i^3} |\psi_f \rangle , \quad (24.8)$$

where $\omega = E_f - E_0$. Matrix elements of $\sum_{i=1}^{N} \mathbf{p}_i$, $\sum_{i=1}^{N} \mathbf{r}_i$, and $\sum_{i=1}^{N} Z\mathbf{r}_i/r_i^3$ are known as the "velocity," "length," and "acceleration" forms of the E1 matrix element.

Equality of the matrix elements in (24.7) and (24.8) does not hold when approximate eigenstates of H are used [6]. In such a case, qualitative considerations may help to determine which form is most reliable. For example, the length form tends to emphasize the large r part of the approximate wave functions, the acceleration form tends to emphasize the small r part of the wave functions, and the velocity form tends to emphasize intermediate values of r.

If instead of employing approximate eigenstates of the exact H, one employs exact eigenstates of an approximate N-electron Hamiltonian, then inequality of the matrix elements in (24.7) and (24.8) is a measure of the non-locality of the potential in the approximate Hamiltonian

[7]. The exchange part of the Hartree-Fock potential is an example of such a nonlocal potential. Nonlocal potentials are also implicitly introduced in configuration interaction calculations employing a finite number of configurations [7]. One may eliminate the ambiguity of which form of the E1 transition operator to use by requiring that the Schrödinger equation be gauge invariant. Only the length form is consistent with such gauge invariance [7].

However, equality of the alternative forms of the transition operator does not necessarily imply high accuracy. For example, they are exactly equal when one uses an approximate local potential to describe the N-electron atom, as in a central potential model, even though the accuracy is often poor. The length and velocity forms are also exactly equal in the random phase approximation [8], which does generally give accurate cross sections for single photoionization of closed shell atoms. No general prescription exists, however, for ensuring that the length and velocity matrix elements are equal at each level of approximation to the N-electron Hamiltonian.

24.2.3 Selection Rules for Electric Dipole Transitions

If one ignores relativistic interactions, then a general atomic photoionization process may be described in LS-coupling as follows:

$$\mathcal{A}(L, S, M_L, M_S, \pi_{\mathcal{A}}) + \gamma(\pi_\gamma, \ell_\gamma, m_\gamma)$$
$$\longrightarrow \mathcal{A}^+(\bar{L}\bar{S}\pi_{\mathcal{A}^+})\varepsilon\ell(L', S', M_{L'}, M_{S'}). \quad (24.9)$$

Here the atom $\mathcal{A}$ is ionized by the photon γ to produce a photoelectron with kinetic energy ε and orbital angular momentum ℓ. The photoelectron is coupled to the ion $\mathcal{A}^+$ with total orbital and spin angular momenta L' and S'. In the electric dipole approximation, the photon may be regarded as having odd parity, i.e., $\pi_\gamma = -1$, and unit angular momentum, i.e., $\ell_\gamma = 1$. This is obvious from Eqs. (24.7) and (24.8), where the E1 operator is seen to be a vector operator. The component m_γ of the photon in the E1 approximation is ± 1 for right or left circularly polarized light and 0 for linearly polarized light.[2] Angular momentum and parity selection rules for the E1 transition in (24.9) imply the following relations between the initial and final state quantum numbers:

$$L' = L \oplus 1 = \bar{L} \oplus \ell, \quad (24.10)$$

$$M_{L'} = M_L + m_\gamma = M_{\bar{L}} + m_\ell, \quad (24.11)$$

$$S' = S = \bar{S} \oplus \tfrac{1}{2}, \quad (24.12)$$

$$M_{S'} = M_S = M_{\bar{S}} + m_s, \quad (24.13)$$

$$\pi_{\mathcal{A}}\pi_{\mathcal{A}^+} = (-1)^{\ell+1}. \quad (24.14)$$

---

[2]The z axis is taken as $\hat{\mathbf{k}}$ in the case of circularly polarized light and as $\hat{\boldsymbol{\epsilon}}$ in the case of linearly polarized light, where $\mathbf{k}$ and $\hat{\boldsymbol{\epsilon}}$ are defined in Eq. (24.3).

Equation (24.14) follows from the parity $(-1)^\ell$ of the photoelectron. The direct sum symbol $\oplus$ denotes the vector addition of $\mathbf{A}$ and $\mathbf{B}$ i.e, $\mathbf{A} \oplus \mathbf{B} = A + B, A + B - 1, \dots, |A - B|$.

In (24.9), the quantum numbers $\alpha \equiv \bar{L}, \bar{S}, \pi_{\mathcal{A}^+}, \ell, L', S', M_{L'}, M_{S'}$ (plus any other quantum numbers needed to specify uniquely the state of the ion $\mathcal{A}^+$) define a final state *channel*. All final states that differ only in the photoelectron energy ε belong to the same channel. The quantum numbers $L', S', M_{L'}, M_{S'}$, and $\pi_{\text{tot}} = (-1)^\ell \pi_{\mathcal{A}^+}$ are the only good quantum numbers for the final states. Thus the Hamiltonian (24.1) mixes final state channels having the same angular momentum and parity quantum numbers but differing quantum numbers for the ion and the photoelectron; i.e., differing $\bar{L}, \bar{S}, \pi_{\mathcal{A}^+}$, and ℓ but the same $L', S', M_{L'}, M_{S'}$ and $(-1)^\ell \pi_{\mathcal{A}^+}$.

24.2.4 Boundary Conditions on the Final State Wave Function

Photoionization calculations obtain final state wave functions statisfying the asymptotic boundary condition that the photoelectron is ionized in channel α. This boundary condition is expressed as

$$\psi_{\alpha E}^-(\mathbf{r}_1 s_1, \dots, \mathbf{r}_N s_N)$$
$$\xrightarrow[r_N \to \infty]{} \theta_\alpha(\mathbf{r}_1 s_1, \dots, \hat{\mathbf{r}}_N \mathbf{s}_N) \frac{1}{i(2\pi k_\alpha)^{\frac{1}{2}}} \frac{1}{r_N} e^{i\Delta_\alpha}$$
$$- \sum_{\alpha'} \theta_{\alpha'}(\mathbf{r}_1 s_1, \dots, \hat{\mathbf{r}}_N \mathbf{s}_N) \frac{1}{i(2\pi k_{\alpha'})^{\frac{1}{2}}} \frac{1}{r_N} e^{-i\Delta_{\alpha'}} S_{\alpha'\alpha}^\dagger,$$
$$(24.15)$$

where the phase appropriate for a Coulomb field is

$$\triangle_\alpha \equiv k_\alpha r_N - \tfrac{1}{2}\pi\ell_\alpha + \frac{1}{k_\alpha} \log 2k_\alpha r_N + \sigma_{\ell_\alpha}. \quad (24.16)$$

The minus superscript on the wave function in (24.15) indicates an "incoming wave" normalization: i.e., asymptotically $\psi_{\alpha E}^-$ has outgoing spherical Coulomb waves only in channel α, while there are incoming spherical Coulomb waves in all channels. $S_{\alpha'\alpha}^\dagger$ is the Hermitian conjugate of the S-matrix of scattering theory, θ_α indicates the coupled wave function of the ion and the angular and spin parts of the photoelectron wave function, k_α is the photoelectron momentum in channel α and ℓ_α is its orbital angular momentum, and σ_{ℓ_α} in (24.16) is the Coulomb phase shift.

While one calculates channel functions $\psi_{\alpha E}^-$, experimentally one measures photoelectrons which asymptotically have well-defined linear momenta $\mathbf{k}_\alpha$ and well-defined spin states $m_{\frac{1}{2}}$, and ions in well-defined states $\bar{\alpha} \equiv \bar{L}\bar{S}M_{\bar{L}}M_{\bar{S}}$. The wave function appropriate for this experimental situation is related to the channel functions by uncoupling the ionic and electronic orbital and spin angular momenta and projecting the photoelectron angular momentum states ℓ_α, m_α onto the direction $\hat{\mathbf{k}}_\alpha$ by means of the spherical harmonic $Y^\star_{\ell_\alpha m_\alpha}(k_\alpha)$. This relation is [1]:

$$
\psi^-_{\bar{\alpha} k_\alpha}(\mathbf{r}_1 \mathbf{s}_1, \ldots, \mathbf{r}_N \mathbf{s}_N)
$$

$$
= \sum_{\ell_\alpha m_\alpha} \frac{i^{\ell_\alpha} \exp(-i\sigma_{\ell_\alpha})}{k_\alpha^{\frac{1}{2}}} Y^\star_{\ell_\alpha m_\alpha}(\hat{\mathbf{k}}_\alpha) \sum_{L M_L S M_S} \langle \bar{L} M_{\bar{L}} \ell_\alpha m_\alpha | L M_L \rangle
$$

$$
\times \langle \bar{S} M_{\bar{S}} \tfrac{1}{2} m_{\frac{1}{2}} | S M_S \rangle \psi^-_{\alpha E}(\mathbf{r}_1 \mathbf{s}_1, \ldots, \mathbf{r}_N \mathbf{s}_N), \qquad (24.17)
$$

where the coefficients in brackets are Clebsch-Gordan coefficients. This wave function is normalized to a delta function in momentum space, i.e.,

$$
\int (\psi^-_{\bar{\alpha} \mathbf{k}_\alpha})^\dagger \psi^-_{\bar{\alpha}' \mathbf{k}_{\alpha'}} d^3 r = \delta_{\bar{\alpha}\bar{\alpha}'} \delta(\mathbf{k}_\alpha - \mathbf{k}_{\alpha'}). \qquad (24.18)
$$

The factors $i^{\ell_\alpha} \exp(-i\sigma_{\ell_\alpha}) k_\alpha^{-\frac{1}{2}}$ ensure that for large r_N Eq. (24.17) represents a Coulomb wave (with momentum $\mathbf{k}_\alpha$) times the ionic wave function for the state $\bar{\alpha}$ plus a sum of terms representing incoming spherical waves. Thus only the ionic term $\bar{\alpha}$ has an outgoing wave. One uses the wave function in (24.17) to calculate the angular distribution of photoelectrons.

24.2.5 Photoionization Cross Sections

If one writes H_{int} in (24.4) as $H_{\text{int}}(t) = H_{\text{int}}(0)^{-i\omega t}$, then from first order time-dependent perturbation theory, the transition rate for transition from an initial state with energy E_0 and wave function ψ_0 to a final state with total energy E_f and wave function $\psi^-_{\bar{\alpha} k_\alpha}$ is

$$
dW_{\mathbf{k}_\alpha} = 2\pi |\langle \psi_0 | H_{\text{int}}(0) | \psi^-_{\bar{\alpha} \mathbf{k}_\alpha} \rangle|^2
$$

$$
\times \delta(E_f - E_0 - \omega) k_\alpha^2 dk_\alpha d\Omega(\hat{\mathbf{k}}_\alpha). \qquad (24.19)
$$

The delta function expresses energy conservation and the last factors on the right are the phase space factors for the photoelectron. Dividing the transition rate by the incident photon current density c/V, integrating over dk_α, and inserting $H_{\text{int}}(0)$, the differential photoionization cross section is

$$
\frac{d\sigma_{\bar{\alpha}}}{d\Omega} = \frac{4\pi^2}{c} \frac{k_\alpha}{\omega} |\hat{\boldsymbol{\epsilon}} \cdot \langle \psi_0 | \sum_{i=1}^N \mathbf{p}_i | \psi^-_{\bar{\alpha} \mathbf{k}_\alpha} \rangle|^2. \qquad (24.20)
$$

Implicit in Eqs. (24.19) and (24.20) is an average over initial magnetic quantum numbers $M_{L_0} M_{S_0}$ and a sum over final magnetic quantum numbers $M_{\bar{L}} M_{\bar{S}} m_{\frac{1}{2}}$. The length form of Eq. (24.20) is obtained by replacing each $\mathbf{p}_i$ by $\omega \mathbf{r}_i$ [*cf.* Eq. (24.7)].

Substitution of the final state wave function (24.17) in Eq. (24.20) permits one to carry out the numerous summations over magnetic quantum numbers and obtain the form

$$
\frac{d\sigma_{\bar{\alpha}}}{d\Omega} = \frac{\sigma_{\bar{\alpha}}}{4\pi} [1 + \beta P_2(\cos\theta)] \qquad (24.21)
$$

for the differential cross section [9]. Here $\sigma_{\bar{\alpha}}$ is the partial cross section for leaving the ion in the state $\bar{\alpha}$, β is the asymmetry parameter [10], $P_2(\cos\theta) = \frac{3}{2}\cos^2\theta - \frac{1}{2}$, and θ indicates the direction of the outgoing photoelectron with respect to the polarization vector $\hat{\boldsymbol{\epsilon}}$ of the incident light. The form of (24.21) follows in the electric dipole approximation from general symmetry principles, provided that the target atom is unpolarized [11]. The partial cross section is given in terms of reduced E1 matrix elements involving the channel functions in (24.15) by

$$
\sigma_{\bar{\alpha}} = \frac{4\pi^2}{3c} \omega [L]^{-1} \sum_{\ell_\alpha L'} |\langle \psi_0 \| \sum_{i=1}^N r_i^{[1]} \| \psi^-_{\alpha E} \rangle|^2. \qquad (24.22)
$$

The β parameter has a much more complicated expression involving interference between different reduced dipole amplitudes [10]. Thus measurement of β provides information on the relative phases of the alternative final state channel wave functions, whereas the partial cross-section in Eq. (24.22) does not. From the requirement that the differential cross section in (24.21) be positive, one sees that $-1 \le \beta \le +2$.

24.3 AN INDEPENDENT ELECTRON MODEL

The many-body wave functions ψ_0 and $\psi^-_{\alpha E}$ are usually expressed in terms of a basis of independent electron wave functions. Key qualitative features of photoionization cross sections can often be interpreted in terms of the overlaps of initial and final state one electron radial wave functions [12]. The simplest independent electron representation of the atom, the central potential model, proves useful for this purpose.

24.3.1 Central Potential Model

In the central potential (CP) model the exact H in (24.1) is approximated by a sum of single-particle terms describing the independent motion of each electron in a central potential $V(r)$:

$$
H_{\text{CP}} = \sum_{i=1}^N \left[\frac{p_i^2}{2m} + V(r_i) \right]. \qquad (24.23)
$$

The potential $V(r)$ must describe the nuclear attraction and the electron-electron repulsion as well as possible and must satisfy the boundary conditions

$$
V(r) \xrightarrow[r \to 0]{} -Z/r \quad \text{and} \quad V(r) \xrightarrow[r \to \infty]{} -1/r \qquad (24.24)
$$

in the case of a neutral atom. $H_{\rm CP}$ is separable in spherical coordinates and its eigenstates can be written as Slater determinants of one-electron orbitals of the form $r^{-1}P_{n\ell}Y_{\ell m}(\Omega)$ for bound orbitals and of the form $r^{-1}P_{\varepsilon\ell}(r)Y_{\ell m}(\Omega)$ for continuum orbitals. The one-electron radial wave functions satisfy

$$\frac{d^2 P_{\varepsilon\ell}(r)}{dr^2} + 2\left[\varepsilon - V(r) - \frac{\ell(\ell+1)}{2r^2}\right]P_{\varepsilon\ell}(r) = 0, \quad (24.25)$$

subject to the boundary condition $P_{\varepsilon\ell}(0) = 0$, and similarly for the discrete orbitals $P_{n\ell}(r)$. Hermann and Skillman [13] have tabulated a widely used central potential for each element in the periodic table as well as radial wave functions for each occupied orbital in the ground state of each element.

24.3.2 High Energy Behavior

The hydrogen atom cross section, which is nonzero at threshold and decreases monotonically with increasing photon energy, serves as a model for inner-shell photoionization cross sections in the x-ray photon energy range. A sharp onset at threshold followed by a monotonic decrease above threshold is precisely the behavior seen in x-ray photoabsorption measurements. A simple hydrogenic approximation at high energies may be justified theoretically as follows: (1) Since a free electron cannot absorb a photon (because of kinematical considerations), at high photon energies one expects the more strongly bound inner electrons to be preferentially ionized as compared with the outer electrons. (2) Since the $P_{n\ell}(r)$ for an inner electron is concentrated in a very small range of r, one expects the integrand of the radial dipole matrix element to be negligible except for those values of r where $P_{n\ell}(r)$ is greatest. (3) Thus it is only necessary to approximate the atomic potential locally, e.g., by means of a screened Coulomb potential

$$V_{n\ell}(r) = -\left(\frac{Z - s_{n\ell}}{r}\right) + V_{n\ell}{}^{\circ} \quad (24.26)$$

appropriate for the $n\ell$ orbital. Here $s_{n\ell}$ is the "inner-screening" parameter, which accounts for the screening of the nuclear charge by the other atomic electrons, and $V_{n\ell}^{\circ}$ is the "outer-screening" parameter, which accounts for the lowering of the $n\ell$ electrons' binding energy due to repulsion between the outer electrons and the photoelectron as the latter leaves the atom. The potential in (24.26) predicts hydrogen-like photoionization cross sections for inner-shell electrons with onsets determined by the outer-screening parameters $V_{n\ell}^{\circ}$.

Use of more accurate atomic central potentials in place of the screened hydrogenic potential in (24.26) generally enables one to obtain photoionization cross sections in the keV photon energy region to within 10%

of the experimental results [14]. For high but still nonrelativistic photon energies, i.e., $\omega \ll mc^2$, the energy dependence of the cross section for the $n\ell$ subshell is [15]

$$\sigma_{n\ell} \sim \omega^{-\ell - \frac{7}{2}}. \quad (24.27)$$

24.3.3 Near Threshold Behavior

For photons in the vuv energy region, i.e., near the outer-subshell ionization thresholds, the photoionization cross sections for subshells with $\ell \geq 1$ frequently have distinctly nonhydrogenic behavior. The cross section, instead of decreasing monotonically as for hydrogen, rises above threshold to a maximum (the so called *delayed maximum* above threshold). Then it decreases to a minimum (the *Cooper minimum* [16,17]) and rises to a second maximum. Finally the cross section decreases monotonically at high energies in acordance with hydrogenic behavior. Such nonhydrogenic behavior may be interpreted as due either to an effective potential barrier or to a zero in the radial dipole matrix element. We examine each of these effects in turn.

The *delayed maximum* above outer subshell ionization thresholds of heavy atoms (i.e., $Z \gtrsim 18$) is due to an effective potential barrier seen by $\ell = 2$ and $\ell = 3$ photoelectrons in the region of the outer edge of the atom [*cf.* Eq. (24.25)]. This effective potential lowers the probability of photoelectron escape until the photoelectrons have enough excess energy to surmount the barrier. Such behavior is nonhydrogenic. Furthermore, in cases where an inner subshell with $\ell = 2$ or 3 is being filled as Z increases (as in the transition metals, the lanthanides and the actinides) there is a double well potential. This double well has profound effects on the $3p$-subshell spectra of the transition metals, the $4d$-subshell spectra of the lanthanides, and the $5d$-subshell spectra of the actinides, as well as on atoms with Z just below those of these series of elements [18,19].

Cross section minima arise due to a change in sign of the radial dipole transition matrix element in a particular channel [20,21]. Rules for predicting their occurrence were developed by Cooper [16,17]. Studies of their occurrence in photoionization from excited states [22], in high Z atoms [23], and in relativistic approximation [24] have been carried out. Only recently has a proof been given [25] that such minima do not occur in atomic hydrogen spectra. For other elements, there are further rules on when and how many minima may occur [26].

Often within such minima, one can observe effects of weak interactions that are otherwise obscured. Relativistic and weak correlation effects on the asymmetry parameter β for s-subshells is a notable example [27]. Wang et al. [28] have also emphasized that near such minima in the E1 amplitudes, one cannot ignore the effects of quadrupole and higher corrections to the differential cross section. Central potential model calculations [28] show that quadrupole corrections can be as large as 10% of the E1 cross section at such cross section minima, even for low photon energies.

24.4 PARTICLE-HOLE INTERACTION EFFECTS

The experimental photoionization cross sections for the outer subshells of the noble gases[3] near the ionization thresholds can be understood in terms of interactions between the photoelectron, the residual ion, and the photon field which are called, in many-body theory language, "particle-hole" interactions (see Chap. 23). These may be described as interactions in which two electrons either excite or de-excite each other out of or into their initial subshell locations in the unexcited atom. To analyze the effects of these interactions on the cross section, it is convenient to classify them into three categories: intrachannel, virtual double excitation, and interchannel. These alternative kinds of particle-hole interactions are illustrated in Fig. 24.1 using both many-body perturbation theory (MBPT) diagrams and more "physical" scattering pictures. We discuss each of these types of interaction in turn.

24.4.1 Intrachannel Interactions

The MBPT diagram for this interaction is shown on the left in Fig. 24.1(a); on the right a slightly more pictorial description of this interaction is shown. The wiggly line indicates a photon, which is absorbed by the atom in such a way that an electron is excited out of the $n\ell$ subshell. During the escape of this excited electron, it collides or interacts with another electron from the same subshell in such a way that the second electron absorbs all the energy imparted to the atom by the photon; the first electron is de-excited back to its original location in the $n\ell$ subshell. For closed-shell atoms, the photoionization process leads to a $^1P_1$ final state in which the intrachannel interaction is strongly repulsive. This interaction tends to broaden cross section maxima and push them to higher photon energies as compared with the results of central potential model calculations.

Intrachannel interaction effects are taken into account automatically when the correct Hartree-Fock (HF) basis set is employed in which the photoelectron sees a net Coulomb field due to the residual ion and is coupled to the ion to form the appropriate total orbital L and spin

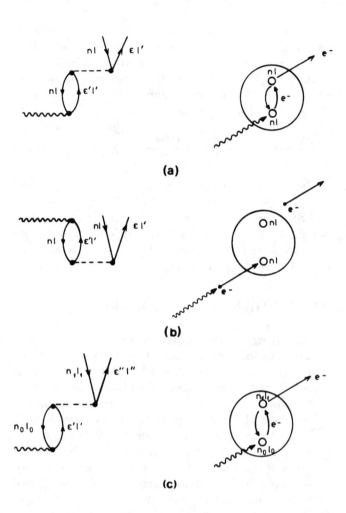

Figure 24.1. MBPT diagrams (left) and scattering pictures (right) for three kinds of particle-hole interaction: (a) intrachannel scattering following photoabsorption; (b) photoabsorption by a virtual doubly-excited state; (c) interchannel scattering following photoabsorption.

S angular momenta. Any other basis set requires explicit treatment of intrachannel interactions.

24.4.2 Virtual Double Excitations

The MBPT diagram for this type of interaction is shown on the left in Fig. 24.1(b). Topologically, this diagram is the same as that on the left in Fig. 24.1(a). In fact, the radial parts of the two matrix elements are identical; only the angular factors differ. A more pictorial description of this interaction is shown on the right of Fig. 24.1(b). The ground state of the atom before photoabsorption is shown to have two electrons virtually excited out of the $n\ell$ subshell. In absorbing the photon, one of these electrons is de-excited to its original location in the $n\ell$ subshell, while the other electron in

Figure 24.2. Photoionization cross section for the $3p$ and $3s$ subshells of Ar. HFL and HFV indicate the length and velocity results obtained using HF orbitals calculated in a $^1P_1$ potential. Dot-dash and dashed lines represent the length and velocity results of the MBPT calculation of Kelly and Simons [29]. Only the four lowest $3s \to np$ resonances are shown; the series converges to the $3s$ threshold at 29.24 eV. Experimental results are those of Samson[30] above 37 eV and of Madden *et al.* [31] below 37 eV (From Ref. [29]).

ionized. These virtual double excitations imply a more diffuse atom than in central-potential or HF models, with the effect that the overly repulsive intrachannel interactions are weakened, leading to cross sections for noble gas atoms that are in very good agreement with experiment with the exception that resonance features are not predicted.

24.4.3 Interchannel Interactions

The interchannel interaction shown in Fig. 24.1(c) is important, particularly for s subshells. This interaction has the same form as the intrachannel interaction shown in Fig. 24.1(a), except now when an electron is photoexcited out of the $n_0 \ell_0$ subshell, it collides or interacts with an electron in a different subshell—the $n_1 \ell_1$ subshell. This interaction causes the second electron to be ionized, and the first electron to fall back into its original location in the $n_0 \ell_0$ subshell.

Interchannel interaction effects are usually very conspicuous features of photoionization cross sections. When the interacting channels have partial photoionization cross sections which differ greatly in magnitude, one finds that the calculated cross section for the weaker channel is completely dominated by its interaction with the stronger channel. At the same time, it is often a safe approximation to ignore the effect of weak channels on stronger

channels. In addition, when the interacting channels have differing binding energies, their interchannel interactions lead to resonance structure in the channel with lower binding energy (arising from its coupling to the Rydberg series in the channel with higher binding energy).

24.4.4 Photoionization of Ar

An example of both the qualitative features exhibited by photoionization cross sections in the vuv energy region and of the ability of theory to calculate photoionization cross reactions is provided by photoionization of the $n = 3$ subshell of argon, i.e.,

$$\text{Ar}3s^2 3p^6 + \gamma \to \text{Ar}^+ 3s^2 3p^5 + e^-$$
$$\to \text{Ar}^+ 3s 3p^6 + e^- . \quad (24.28)$$

Figure 24.2 shows the MBPT calculation of Kelly and Simons [29], which includes both intrachannel and interchannel interactions as well as the effect of virtual double excitations. The cross section is in excellent agreement with experiment [30,31], even to the extent of describing the resonance behavior due to discrete members of the $3s \to \varepsilon p$ channel.

Figure 24.2 illustrates most of the features of photoionization cross sections described so far. First, the cross section rises to a delayed maximum just above the threshold because of the potential barrier seen by photoelectrons from the $3p$ subshell having $\ell = 2$. For photon energies in the range of 45 eV–50 eV, the calculated cross section goes through a minimum because of a change in sign of the $3p \to \varepsilon d$ radial dipole amplitude. The HFL and HFV calculations include the strongly repulsive intrachannel interactions in the ^1P final-state channels and calculate the transition amplitude using the length (L) and velocity (V) form respectively for the electric dipole transition operator [*cf.* Eq. (24.7)]. With respect to the results of central potential model calculations, the HFL and HFV results have lower and broader maxima at higher energies. They also disagree with each other by a factor of two! Inclusion of virtual double excitations results in length and velocity results that agree to within 10% with each other and with experiment, except that the resonance structures are not reproduced. Finally, taking into account the interchannel interactions, one obtains the length and velocity form results shown in Fig. 24.2 by dash-dot and dashed curves respectively. Agreement with experiment is excellent and the observed resonances are well-reproduced.

24.5 THEORETICAL METHODS FOR PHOTOIONIZATION

24.5.1 Calculational Methods

Most of the *ab initio* methods for the calculation of photoionization cross sections (e.g., the MBPT method [32], the close-coupling (CC) method [33], the R-matrix method [34], the random phase approximation (RPA) method [8], the relativistic RPA method [35], the transition matrix method [36], the multiconfiguration Hartree-Fock (MCHF) method [37, 38], etc.) have successfully calculated outer p-subshell photoionization cross sections of the noble gases by treating in their alternative ways the key interactions described above, i.e., the particle-hole interactions. In general, these methods all treat both intrachannel and interchannel interactions to infinite order and differ only in their treatment of ground state correlations. (The exception is MBPT, which often treats interchannel interactions between weak and strong channels only to first or second order.) These methods therefore stand in contrast to central potential model calculations, which do not treat any of the particle-hole interactions, and single-channel term-dependent HF calculations, which treat only the intrachannel interactions. The key point is that selection of the interactions that are included in a particular calculation is more important than the method by which such interactions are handled.

Treatment of photoionization of atoms other than the noble gases presents additional challenges for theory. For example, elements such as the alkaline earths, which have s^2 outer subshells, require careful treatment of electron pair excitations in both initial and final states. Open shell atoms have many more ionization thresholds than do the noble gases. Treatment of the resultant rich resonance structures typically relies heavily on quantum defect theory [38] (see Chap. 45). All the methods listed above can be used to treat elements other than the noble gases, but a method which has come to prominence because of the excellent results it obtains for both alkaline earth and open-shell atoms is the eigenchannel R-matrix method [39].

24.5.2 Other Interaction Effects

A number of interactions, not of the particle-hole type, lead to conspicuous effects in localized energy regions. When treating photoionization in such energy regions, one must be careful to choose a theoretical method which is appropriate. Among the interactions which may be important are the following:

Relativistic and Spin-Dependent Interactions. The fact that $j = \ell - \frac{1}{2}$ electrons are contracted more than $j = \ell + \frac{1}{2}$ electrons at small distances has an enormous effect on the location of cross section minima in heavy elements [14, 40]. It may explain the large observed differences in the profiles of a resonance decaying to final states that differ only in their fine structure quantum numbers [41].

Inner-Shell Vacancy Rearrangement. Inner-shell vacancies often result in significant production of satellite structures in photoelectron spectra. Calculations for inner subshell partial photoionization cross sections are often substantially larger than results of photoelectron measurements [42–44]. This difference is attributed to such satellite production, which is often not treated in theoretical calculations.

Polarization and Relaxation Effects. Negative ion photodetachment cross sections often exhibit strong effects of core polarization near threshold. These effects can be treated semi-empirically, resulting in excellent agreement between theory and experiment [45]. Even for inner shell photoionization cross sections of heavy elements, *ab initio* theories do not reproduce measurements near threshold without the inclusion of polarization and relaxation effects [46, 47].

An Example. The calculation of the energy dependence of the asymmetry parameter β for the 5s subshell of xenon requires the theoretical treatment of all of the above effects. In the absence of relativistic interactions, β for Xe5s would have the energy-independent value of two. Deviations of β from two are therefore an indication of the presence of these relativistic interactions. The greatest deviation of β from two occurs in the localized energy region where the partial photoionization cross section for the 5s subshell has a minimum. In this region, however, relativistic calculations show larger deviations from two than are observed experimentally. Inner shell rearrangement and relaxation effects play an important role [48, 49] and must be included to achieve good agreement with experiment.

24.6 FUTURE DIRECTIONS

The construction of high brightness synchrotron light sources and the increasing use of lasers are providing the means to study atomic photoionization processes at an unparalleled level of detail. The synchrotrons generally produce photons in the soft x-ray and x-ray regions. Thus, inner shell vacancy production and decay, satellite production, and multiple ionization phenomena are all being increasingly studied. Laser sources are allowing production of atoms in tailored initial states. Thus, photoionization of excited atoms and, in particular, complete measurements of particular photoionization processes, are now possible. Recent collections of short review papers provide references to these topics [3, 4]. In addition, two recent reviews of experimental results for noble gas atom photoionization [50] and for metal atom photoionization [51] provide also valuable information on the current state of the corresponding theoretical results.

REFERENCES

1. A. F. Starace, in *Handbuch der Physik*, edited by W. Mehlhorn (Springer, Berlin, 1982), Vol XXXI, pp. 1–121.
2. M. Ya. Amusia, *Atomic Photoeffect* (Plenum, New York, 1990).
3. *Many-Body Theory of Atomic Structure and Photoionization*, edited by T. N. Chang (World Scientific, Singapore, 1993).
4. *VUV and Soft X-Ray Photoionization Studies*, edited by U. Becker and D. A. Shirley (in press) (Plenum, New York, 1996).
5. J. J. Sakurai, *Advanced Quantum Mechanics* (Addison-Wesley, Reading, 1967), p. 39.
6. S. Chandrasekhar, Astrophys. J. **102**, 223 (1945).
7. A. F. Starace, Phys. Rev. A **3**, 1242 (1971); **8**, 1141 (1973).
8. M. Ya. Amusia and N. A. Cherepkov, Case Studies in At. Phys. **5**, 47 (1975).
9. J. M. Blatt and L. C. Biedenharn, Rev. Mod. Phys. **24**, 258 (1952).
10. *cf.* Sect. 7 of Ref. [1].
11. C. N. Yang, Phys. Rev. **74**, 764 (1948).
12. See, e.g., Eqs. (9.6)–(9.15) of [1] for expressions for the reduced matrix elements in Eq. (24.22) in which all angular integrations have been carried out and the results are expressed in terms of one-electron radial dipole matrix elements.
13. F. Hermann and S. Skillman, *Atomic Structure Calculations* (Prentice-Hall, Englewood Cliffs, 1963).
14. R. H. Pratt, A. Ron, and H. K. Tseng, Rev. Mod. Phys. **45**, 273 (1973).
15. H. A. Bethe and E. E. Salpeter, *Quantum Mechanics of One- and Two-Electron Atoms* (Springer, Berlin, 1957), Sects. 69–71.
16. J. W. Cooper, Phys. Rev. **128**, 681 (1962).
17. U. Fano and J. W. Cooper, Rev. Mod. Phys. **40**, 441 (1968).
18. Ref. [1], pp. 50–55, and references therein.
19. *Giant Resonances in Atoms, Molecules, and Solids*, edited by J. P. Connerade, J. M. Estiva, and R. C. Karnatak (Plenum, New York, 1987), and references therein.
20. D. R. Bates, Mon. Not. R. Astron. Soc. **106**, 432 (1946).
21. M. J. Seaton, Proc. Roy. Soc. A **208**, 418 (1951).
22. A. Z. Msezane and S. T. Manson, Phys. Rev. Lett. **48**, 473 (1982).
23. Y. S. Kim, R. H. Pratt, and A. Ron, Phys. Rev. A **24**, 1626 (1981).
24. Y. S. Kim, A. Ron, R. H. Pratt, B. R. Tambe, and S. T. Manson, Phys. Rev. Lett. **46**, 1326 (1981).
25. S. D. Oh and R. H. Pratt, Phys. Rev. A **34**, 2486 (1986).
26. R. H. Pratt, R. Y. Yin, and X. Liang, Phys. Rev. A **35**, 1450 (1987); R. Y. Yin and R. H. Pratt, Phys. Rev. A **35**, 1149 (1987); **35**, 1154 (1987).
27. S. T. Manson and A. F. Starace, Rev. Mod. Phys. **54**, 389 (1982).
28. M. S. Wang, Y. S. Kim, R. H. Pratt, and A. Ron, Phys. Rev. A **25**, 857 (1982).
29. H. P. Kelly and R. L. Simons, Phys. Rev. Lett. **30**, 529 (1973).
30. J. A. R. Samson, Adv. At. Mol. Phys. **2**, 177 (1966).
31. R. P. Madden, D. L. Ederer, and K. Codling, Phys. Rev. **177**, 136 (1969).
32. H. P. Kelly in *Photoionization and Other Probes of Many-Electron Interactions*, edited by F. J. Wuilleumier (Plenum, New York, 1976), pp. 83–109.
33. P. G. Burke and M. J. Seaton, Methods Comput. Phys. **10**, 1 (1971).
34. P. G. Burke and W. D. Robb, Adv. At. Mol. Phys. **11**, 143 (1975); *Electronic and Atomic Collisions: Invited Papers and Progress Reports*, edited by G. Watel (North Holland, Amsterdam, 1978), pp. 201–280.
35. W. R. Johnson, C. D. Lin, K. T. Cheng, and C. M. Lee, Phys. Scr. **21**, 409 (1980).
36. T. N. Chang and U. Fano, Phys. Rev. A **13**, 263 (1976); **13**, 282 (1976).
37. J. R. Swanson and L. Armstrong, Jr., Phys. Rev. A **15**, 661 (1977); **16**, 1117 (1977).
38. M. J. Seaton, Rep. Prog. Phys. **46**, 167 (1983).
39. C. H. Greene, in *Fundamental Processes of Atomic Dynamics*, edited by J. S. Briggs, H. Kleinpoppen, and H. O. Lutz (Plenum, New York, 1988), pp. 105–127, and references therein.
40. D. W. Lindle, T. A. Ferrett, P. A. Heiman, and D. A. Shirley, Phys. Rev. A **37**, 3808 (1988).
41. M. Krause, F. Cerrina, A. Fahlman, and T. A. Carlson, Phys. Rev. Lett. **51**, 2093 (1983).
42. J. B. West, P. R. Woodruff, K. Codling, and R. G. Houlgate, J. Phys. B **9**, 407 (1976).
43. M. Y. Adam, F. Wuilleumier, N. Sandner, V. Schmidt, and G. Wendin, J. Phys. (Paris) **39**, 129 (1978).
44. U. Becker, T. Prescher, E. Schmidt, B. Sonntag, and H.-E. Wetzel, Phys. Rev. A **33**, 3891 (1986).
45. See, e.g., K. T. Taylor, and D. W. Norcross, Phys. Rev. A **34**, 3878 (1986).
46. M. Ya. Amusia, in *Atomic Physics 5*, edited by R. Marrus, M. Prior, and H. Shugart (Plenum, New York, 1977), pp. 537–565.
47. W. Jitschin, U. Werner, G. Materlik, and G. D. Doolen, Phys. Rev. A **35**, 5038 (1987).
48. G. Wendin and A. F. Starace, Phys. Rev. A **28**, 3143 (1983).
49. J. Tulkki, Phys. Rev. Lett. **62**, 2817 (1989).
50. V. Schmidt, Rep. Prog. Phys. **55**, 1483 (1992).
51. B. Sonntag and P. Zimmerman, Rep. Prog. Phys. **55**, 911 (1992).

25

Autoionization

A. Temkin

Laboratory for Astronomy and Solar Physics, Goddard Space Flight Center, National Aeronautics and Space Administration, Greenbelt, Maryland 20771

25.1 INTRODUCTION

The phenomenon of autoionization, or more particularly the autoionization state itself, is treated for the most part in this Chapter as a bound state. The process is rigorously a part of the scattering continuum (*cf.* Chap. 45), but, due mostly to the work of Feshbach [1], a rigorous formulation can be established whereby the main element of the theory can be made into a bound state problem with the scattering elements built around it. The major constituent of both these features is accomplished with projection operators. A brief description of the above elements of the theory, centered around projection operators, is the aim of this chapter [2]. Energies are in Rydbergs unless otherwise noted.

25.1.1 Auger Effect

Autoionization falls within the general class of phenomena known as the Auger effect. In the Auger effect an atomic system "seemingly" [*cf.* Sect. 25.1.3] spontaneously decays into a partition of its constituent parts.

25.1.2 Autoionization, Autodetachment, and Radiative Decay

If the initial, composite system is neutral, or positively charged, and its constituent decay particles are an electron and the residual ion, then the process is called *autoionization*. If the original system is a negative ion, so that the residual heavy particle system is neutral, then the process is technically called *autodetachment*; for the most part, the physics and the mathematical treatment are the same.

It is also possible, before electron emission takes place, that the system will alternatively decay radiatively to an autoionization state of lower energy, or a true bound state of the composite system. The latter process is called *radiative stabilization* [which is a basic part of dielectronic recombination (*cf.* Chap. 53)].

25.1.3 Formation, Scattering, and Resonances

Autoionizing states are formed by scattering processes and photoabsorption. These are the inverses of the autoionization and photon emission processes by which they can decay. In the scattering process, formation of the autoionization state corresponds to a resonance in the scattering cross section (*cf.* Chap. 45). Autoionization is the process which corresponds to the decay of the resonance. The decay of the resonance (autoionization) is then seen to be the last half of the resonant scattering process. Strictly speaking therefore, the resonant or autoionization state, although it may be long-lived, is not completely stationary, and that is the reason that the word "seemingly" was used to describe the Auger process. The compound system can also be formed by absorption of photons impinging on a bound state (usually the ground state) of the compound system, in which case the autionization state shows up as a "line" in the absorption spectrum.

25.2 THE PROJECTION OPERATOR FORMALISM

In the energy domain where the Schrödinger equation (SE)

$$H\Psi = E\Psi \tag{25.1}$$

describes scattering, the wave function does not vanish at infinity; *i.e.*

$$\lim_{r_i \to \infty} \Psi \neq 0, \tag{25.2}$$

where r_i is the radial coordinate of ith electron.

The basic idea of the projection operator formalism [1] is to define projection operators P and Q which separate Ψ into scattering-like ($P\Psi$) and quadratically integrable ($Q\Psi$) parts:

$$\Psi = P\Psi + Q\Psi. \tag{25.3}$$

Implicit in Eq. (25.3) are

$$\begin{aligned} \text{completeness}&: P + Q = 1, \\ \text{idempotency}&: P^2 = P, \quad Q^2 = Q, \\ \text{orthogonality}&: PQ = 0, \end{aligned} \tag{25.4}$$

and the asymptotic properties

$$\lim_{r_i \to \infty} \begin{cases} P\Psi = \displaystyle\lim_{r_i \to \infty} \Psi \\ Q\Psi = 0 \end{cases}. \tag{25.5}$$

25.2.1 The Optical Potential

Straightforward manipulation of Eqs. (25.1) and (25.3) leads to an important relation between $Q\Psi$ and $P\Psi$:

$$Q\Psi = (E - QHQ)^{-1} QHP\Psi. \tag{25.6}$$

From Eq. (25.6) a basic equation for $P\Psi$ [which, by virtue of Eq. (25.5), contains all the scattering information] emerges:

$$(PHP + V_{\text{op}} - E)P\Psi = 0. \tag{25.7}$$

The most significant part of Eq. (25.7) is the optical potential V_{op} given by

$$V_{\text{op}} = PHQ(E - QHQ)^{-1}QHP. \tag{25.8}$$

V_{op} is a nonlocal potential; the most incisive way to give it meaning is to the define the QHQ problem.

25.2.2 Expansion of V_{op}: The QHQ Problem

The following eigenvalue problem constitutes the heart of the projection operator formalism:

$$QHQ\Phi_n = \mathcal{E}_n\Phi_n. \tag{25.9}$$

For calculational purposes, it is best to recast Eq. (25.9) in the variational form

$$\delta\left[\frac{\langle\Phi QHQ\Phi\rangle}{\langle\Phi Q\Phi\rangle}\right] = 0. \tag{25.10}$$

This equation may yield a discrete plus a continuous spectrum in an energy domain where the SE has only a continuous spectrum. Moreover, if the Q operator is appropriately chosen, then the discrete eigenvalues $\mathcal{E}_n$ are close to the desired class of many-body resonances, called variously Feshbach resonances core-excited [3] or doubly-excited states. In terms of these solutions, V_{op} has the expansion

$$V_{\text{op}} = \sum_n PHQ\,\Phi_n\rangle(E - \mathcal{E}_n)^{-1}\langle\Phi_n\,QHP. \tag{25.11}$$

Table 25.1. Test of sum rule (25.15) for the lowest He$^-$ ($1s2s^2\,^2S$) autodetachment state (*cf.* [6]). Projection operators are based on Hylleraas ϕ_0 and the variational form of v_α given below:

$$\phi_0 \propto (1 + C_1 r_1 + C_2 r_2 + C_{12} r_{12}) \exp(-\gamma_1 r_1 - \gamma_2 r_2)$$
$$+ (r_1 \leftrightarrow r_2),$$
$$v_\alpha \propto (c_{11}^{(\alpha)} + c_{12}^{(\alpha)} r) \exp(-\gamma_1 r)$$
$$+ (c_{21}^{(\alpha)} + c_{22}^{(\alpha)} r) \exp(-\gamma_2 r),$$

and the variational eigenvalues are [6]

$$\lambda_1 = 1.009\,453 \quad \lambda_2 = 232.8540$$
$$\lambda_3 = 80\,101.08 \quad \lambda_4 = 4\,817\,341.$$

Summation	Value (Ry^{-1})
$\displaystyle\sum_{\alpha=1}^{4}(\lambda_\alpha)^{-1}$	0.994 9425
$\langle K(r\vert r)\rangle_r$	0.994 9514

25.3 FORMS OF P AND Q

25.3.1 The Feshbach Form

Projection operators are not unique. Feshbach [1] has sketched a derivation of a "robust" projection operator for the general N-electron target system. Robust means that $Q\Phi$ is devoid of open channels. The complete expression for P, including inelastic channels [5], is

$$P = \sum_{i=1}^{N+1} \Big[\boldsymbol{\psi}(r^{(i)}))\cdot\langle\boldsymbol{\psi}(r^{(i)})$$
$$+ \sideset{}{'}\sum_{\lambda_\alpha} \frac{\boldsymbol{v}_\alpha(r_i)\cdot\boldsymbol{\psi}(r^{(i)}))\langle\boldsymbol{v}_\alpha(r_i)\cdot\boldsymbol{\psi}(r^{(i)})}{\lambda_\alpha - 1} \Big], \quad (25.12)$$

where the prime on the second summation means that terms with $\lambda_\alpha = 1$ are to be omitted, r_i denotes the radial coordinate of electron i, $r^{(i)}$ stands for the angular and spin coordinates of electron i plus all coordinates of the remaining N electrons. Thus $r^{(i)}$ indicates the totality of all coordinates of the $(N+1)$ electrons *except* the radial coordinate of the ith electron r_i. $\boldsymbol{\psi}(r^{(i)})$ is the vector channel wave function in which the angular momentum and spin of the electron i are coupled to the target in state ν ($\nu = 0, \ldots, \nu_{\max}$). A component of a vector labels the inelastic channel, and dot products represent sums over channels. The definition $Q = 1 - P$ is then made explicit by completeness.

The α-indexed quantities in Eq. (25.12) arise from exchange; they are the eigensolutions of the integral eigenvalue problem

$$\boldsymbol{v}_\alpha(r_i) = \lambda_\alpha\langle\mathbf{K}(r_i\vert r_j)\cdot\boldsymbol{v}_\alpha(r_j)\rangle_{r_j}. \quad (25.13)$$

Table 25.2. Comparison of methods for calculating the energy of the lowest He$^-$ ($1s2s^2\,^2S$) autodetachment state. The QHQ results are denoted $(\hat{\mathcal{E}} - E_0)_{\text{Quasi}}$ for the quasi-projection method [7] and $(\mathcal{E} - E_0)_{\text{Complete}}$ for the complete projection method [4]. The entries labeled "Other results" give the full resonant renergy. Units are eV.

Target	$(\hat{\mathcal{E}} - E_0)_{\text{Quasi}}$	$(\mathcal{E} - E_0)^a_{\text{Complete}}$
closed shell	19.294	19.593
open shell	19.379	19.666
$\vert 1s1s'\vert + 2p^2$	19.382	19.615
4 term Hylleraas	19.375	19.496
10 term Hylleraas	19.373	19.504

Other results 19.402[b], 19.376[c], 19.367[d], 19.367 ± 0.007[e]

[a]Values obtained using $R_\infty = 13.605\,698$ eV and $E_0 = -79.0151$ eV [8].
[b]Complex rotation method; Junker and Huang [9].
[c]Hole-projection complex-rotation; Davis and Chung [10].
[d]Hermitian-representation complex-rotation; [11].
[e]Experiment; Brunt *et al.* [12].

Here, $\mathbf{K}(r_i\vert r_j)$ is a matrix with components

$$K_{\mu\nu}(r_i\vert r_j) \propto N\langle\psi_\mu(r^{(i)})\psi_\nu(r^{(j)})\rangle_{r^{(ij)}}, \quad (25.14)$$

and the final subscript indicates the variables being integrated over. In the inelastic regime, therefore, the $[\boldsymbol{v}_\alpha, \lambda_\alpha]$ are not associated with specific channels, but rather with the totality of open channels. This means that every component of $\mathbf{v}_\alpha$ is associated with all inelastic channels. They are orthogonal, and can be normalized so that $\langle\mathbf{v}_\alpha\cdot\mathbf{v}_\beta\rangle = \delta_{\alpha\beta}$. They obey several sum rules [6], of which the most useful is

$$\sum_{\alpha=1}^{n_\lambda}(\lambda_\alpha)^{-1} = \sum_{\nu=0}^{\nu_{\max}}\langle K_{\nu\nu}(r\vert r)\rangle_r, \quad (25.15)$$

where n_λ is the number of eigenvalues of Eq. (25.13). The v_α can be accurately calculated by use of a variational principle [6]. A test of Eq. (25.15) for the lowest He$^-$(^2S) resonance, using Hylleraas functions to evaluate QHQ, is shown in Table 25.1 (*cf.* Ref. [6]), and results for the resonance position are compared in Table 25.2.

25.3.2 Reduction for the $N = 1$ Target

Explicit rigorous P and Q operators of the above type are only possible for $N = 1$ (i.e. hydrogenic) targets. In that case, spin can be easily eliminated by using spatially symmetric or antisymmetric wave functions. In the elastic region, P and Q reduce to [13, 7]

$$P = P_1 + P_2 - P_1 P_2; \quad Q = 1 - P_1 - P_2 + P_1 P_2.$$
$$(25.16)$$

Table 25.3. Energies $\mathcal{E}_s$ of the He($2s2p\,^1P^o$) autoionization states below He$^+$ ($n = 2$) threshold from the variational calculations of O'Malley and Geltman [14]. Units are Ry. N is the number of terms in the trial function.

N	$s = 1$	$s = 2$	$s = 3$
9	−1.377 08	−1.178 92	−1.097 16
15	−1.380 44	−1.183 12	−1.083 20
20	−1.382 16	−1.183 48	−1.108 28
25	−1.383 16	−1.190 00	−1.111 88

Here the $P_i = \phi(r_i)\rangle\langle\phi(r_i)$ are purely spatial projectors. Forms for the inelastic continuum are easily generalized [14].

There have been many calculations of QHQ for two-electron systems ($N=1$), starting with fundamental work of O'Malley and Geltman [14]. A small sample is given in Table 25.3. They illustrate convergence to well defined values in the continuum, all eigenvalues corresponding to resonances.

25.3.3 Alternative Projection and Projection-Like Operators

Two alternative methods based on the idea of projection are available: quasi-projectors and hole projection operators.

Quasi-projectors [15] relax the condition of idempotency, but still maintain a discrete spectrum, which is in a one-to-one correspondence with resonances, with a predeterminable number of exceptions [2].

Hole projection operators have proven to be a more practical and effective approach [16]. The method uses one-particle (say, hydrogenic) orbitals, $\phi_n(q; \mathbf{r})$, to build holes via projectors, $[1 - \sum_n \phi_n\rangle\langle\phi_n]$, operating on the ($N + 1$)-particle wave function. The Rayleigh-Ritz functional is minimized with respect to the parameters in Φ, but it is maximized with respect to q, the nonlinear parameter in all the ϕ_n. In a model case, this *minimax* procedure has been shown [16] to optimize the eigenvalues to describe resonance energies; many calculations since then [17] have verified the minimax criteria in many-electron systems. More recently, the technique has been combined with complex rotation, so as to enable calculation of other resonant quantities [18]. Remarkably accurate results have been obtained (*cf.* Table 25.4).

Finally, hole projectors are ideally suited for inner-shell vacancy states of many-electron systems if high accuracy is required [2]. Effectively, this amounts to a reliable method for optimizing parameters of a hole orbital to be used in an Auger transition integral for filling such a vacancy, although that method has apparently never been used (*cf.* Chap. 60).

25.4 WIDTH, SHIFT, AND SHAPE PARAMETER

25.4.1 Width and Shift

Here one requires $P\Psi$ as well as $Q\Psi$. The former is obtained from a "nonresonant continuum," defined as the scattering solution of

$$[PHP + V_s^{(\mathrm{nr})} - E]P\Psi_s^{(\mathrm{nr})} = 0,\qquad(25.17)$$

where the nonresonant potential

$$V_s^{(\mathrm{nr})} = V_{\mathrm{op}} - PHQ\Phi_s\rangle(E - \mathcal{E}_s)^{-1}\langle\Phi_s QHP\qquad(25.18)$$

excludes the resonant state s from the optical potential. In terms of $P\Psi_s^{(\mathrm{nr})}$, whose phase shift, η_0, is smooth in the vicinity of $E \approx \mathcal{E}_s$, a solution of the complete problem, Eq. (25.7), can be constructed [14] with a phase shift $\eta_0 + \eta_r$, where

$$\eta_r = \arctan\left[\frac{\Gamma/2}{(\mathcal{E}_s + \Delta_s) - E}\right]\qquad(25.19)$$

exhibits typical resonant behavior ($0 < \eta_r < \pi$). From Eq. (25.19) it is clear that the "true" position of the resonance is

$$E_s(E) = \mathcal{E}_s + \Delta_s(E).\qquad(25.20)$$

The width Γ_s and shift Δ_s are [14]

$$\Gamma_s(E) = 2k|\langle\Psi_s^{(\mathrm{nr})}(E)\,PHQ\,\Phi_s\rangle|^2,\qquad(25.21)$$

$$\Delta_s(E) = \langle\Phi_s\,QHP\,G_P(E)\,PHQ\Phi_s\rangle,\qquad(25.22)$$

where k is the scattering momentum (i.e. $k^2 = E - E_0$), and G_P is the Green's function associated with Eq. (25.17); G_P can be simplified from the form given in [14] (*cf.* Eq. (2.28) of first article of [2]).

Equation (25.20) is an implicit equation for the energy at which the resonance occurs. It can be solved graphically [24], and that energy defines the Feshbach resonant energy E_{F}, which differs (very slightly) from the Breit-Wigner energy (*cf.* Sect. 25.4.3 and Fig. 25.1).

25.4.2 Shape Parameter

The shape of an isolated radiative transition between an autoionization state and some other state can be described by Γ, E, and an additional parameter q_s, often called the shape parameter [25]. With the radiative transition operator labeled T, the ratio of transition probabality (in, say, absorption from an initial state labeled $i\rangle$) through the resonant state to its nonresonant value is parametrized in its Fano form [25] on the left-hand side of the equation

Table 25.4. Comparison of high precision calculations with experiment for the resonance parameters of the He($^1P^o$) resonances below the $n = 2$ threshold. For photoabsorption, the appropriate Rydberg constant is $R_M = R_\infty(1+m/M)$ [19]. The value used here is $R_M = 13.603\,833$ eV, together with $E_0 = -79.0151$ eV [8].

| Quantity | Units | Calculation[a] | | Experiment | |
		$s = 1$	$s = 2$	$s = 1$	$s = 2$
$\mathcal{E}_s$	Ry	−1.385 7895	−1.194 782		
Δ	eV	−0.007 13	0.000 6202		
$E_s - E_0$	eV	60.144 4	62.758 7	60.133±0.015[b]	62.756±0.01[b]
				60.151±0.0103[c]	
Γ	eV	0.0369[d]	0.000 1165	0.038±0.004[b]	
				0.038±0.002[c]	
q_s		−2.849[e]	−4.606[e]	−2.80±0.025[b]	
				−2.55±0.016[c]	

[a]Bhatia and Temkin [19], except as noted.
[b]Madden and Codling [20].
[c]Morgan and Ederer [21].
[d]Bhatia, Burke, and Temkin [22].
[e]Bhatia and Temkin [23].

$$\frac{(e_s + q_s)^2}{1 + e_s^2} = \frac{|\langle P\Psi + Q\Psi|T|i\rangle|^2}{|\langle P\Psi_s^{(\mathrm{nr})} + Q\Psi_s^{(\mathrm{nr})}|T|i\rangle|^2}, \quad (25.23)$$

where

$$e_s = (E - E_s)/(\Gamma_s/2) \quad (25.24)$$

is the scaled energy.

To analyze (25.23) in the Feshbach formalism, the key point [14] is to recognize that the bras on the RHS must include P as well as Q parts of the respective wave functions, as is indicated in (25.23). That is because the T operator is a perturbation and not part of the dynamical problem. With T in length form

$$T = \sum_{j=1}^{N+1} z_j, \quad (25.25)$$

$P\Psi_s$ and $P\Psi_s^{(\mathrm{nr})}$ are, in principle, determined by (25.7) and (25.17). $Q\Psi_s$ is then derived from $P\Psi_s$ using (25.6). But $Q\Psi_s^{(\mathrm{nr})}$ excludes the sth term from the RHS of (25.6) so that

$$Q\Psi_s^{(\mathrm{nr})} = \sum_{n\neq s} \frac{\Phi_n\rangle\langle\Phi_n QHP\Phi^{(\mathrm{nr})}\rangle}{E - \mathcal{E}_n}. \quad (25.26)$$

An explicit formula for q_s was derived in Ref. [2]; it is summarized by

$$q_s = \frac{\langle\Phi_s\tilde{Q}|T|i\rangle}{k\langle i|T|\Psi_s^{(\mathrm{nr})}\rangle\langle\Psi_s^{(\mathrm{nr})}|PHQ\Phi_s\rangle}, \quad (25.27)$$

where

$$\tilde{Q} = Q + QHPG + \sum_{n\neq s} \frac{QHPGHQ|\Phi_n\rangle\langle\Phi_n}{E - \mathcal{E}_n}. \quad (25.28)$$

The Green's function G in (25.28) is the one associated with Eq. (25.17) It can be expanded in terms of the eigensolutions of (25.17), but its spectrum may have a discrete as well as continuous part, in which case,

$$G = \sum_\nu \frac{P\Psi_\nu^{(\mathrm{nr})}\rangle\langle P\Psi_\nu^{(\mathrm{nr})}}{E_s - E_\nu}$$
$$+ \frac{\wp}{\pi} \int \frac{P\Psi_s^{(\mathrm{nr})}(E')\rangle\langle P\Psi_s^{(\mathrm{nr})}(E')\sqrt{E'}\,dE'}{E_s - E'}. \quad (25.29)$$

The sum over ν refers to the discrete part of the spectrum of Eq. (25.17) (if there is one), and $\wp$ denotes a principal value integral over the continuum solutions. The latter are always assumed to be normalized as plane waves (not energy normalized) throughout this chapter.

Equation (25.27) is a nontrivial example of what can be done with the projection operator technique. Not only does it allow very accurate calculations (*cf.* Refs. [18, 19] and footnote e of Table 25.2), but it provides a theoretical incisiveness which far exceeds all previous resonance formalisms.

A formula for the resonant scattering cross section can be derived which is of the same form as the left-hand side of Eq. (25.23); however, in that case, the parameter corresponding to q_s is related to the nonresonant phase shift [26], and has no quantitative relationship to the above shape parameter (q_s).

Figure 25.1. Precision calculation of $H^-(^1S)$ resonance. Solid curve: $E+1$ *vs.* k^2, where $E = \mathcal{E} + \Delta$, Eq. (25.20). k_F^2 is that value of k^2 at which $E+1 = k^2$. Dashed curve is Γ *vs.* k^2 from Eq. (25.21), and $\Gamma_F = \Gamma(k_F^2)$. Curves are from calculations of Ref. [24]; results are $E_F = 9.55735$ eV and $\Gamma_F = 0.0470605$ eV. Applying corrections Eqs. (25.31), give finally $E_{BW} = E_F + O(10^{-6})$, $\Gamma_{BW} = 0.04717$ eV.

25.4.3 Relation to Breit-Wigner Parameters

Inferring resonance parameters from experimental data is generally done by fitting to resonance formulae in which the resonance parameters are assumed to be energy independent. A phase shift for example would be inferred by assuming

$$\eta(E) = \delta_0(E) + \arctan\left(\frac{\Gamma_{BW}/2}{E_{BW} - E}\right). \quad (25.30)$$

The relation between the above Breit-Wigner parameters and those of the Feshbach theory has been derived in lowest order by Drachman [27]:

$$E_{BW} = E_F - (1/4)\Gamma_F[d\Gamma_F/dE]_{E=E_F},$$
$$\Gamma_{BW} = \Gamma_F[1 + d\Delta_F/dE]_{E=E_F}, \quad (25.31)$$
$$\delta_0(E) = \eta_0(E) - (1/2)d\Gamma_F/dE.$$

In only one precision calculation (for the lowest ^1S resonance in electron-hydrogen scattering) have these differences, thus far, been evaluated [24]. A précis is given in Fig. 25.1.

25.5 RELATED TOPICS

This chapter is necessarily of limited scope. Within the projection operator formalism, overlapping resonance theory [28] is discussed in Sect. 45.2.3. Recent calculations [29] have shown that such effects, when present,

can induce significant alteration of isolated resonance results. Other prominent items not included are stabilization methods [30], hyperspherical coordinate methods, and complex rotation (*cf.* Chap. 21). Complex rotation is based on a theorem of Balslev and Combes [31]; it states that if the electronic coordinates of a continuum problem are rotated in the complex plane, then a variational calculation will uncover a set of poles whose real parts are resonance positions (which include the shifts) and whose imaginary parts are one-half of the widths. Calculations for two-electron systems have led to impressive accuracy [32]. Hyperspherical methods have the appealing property of presenting energies as a function of the hyperradius, $R = [\sum_i r_i^2]$, which look like potential energy curves of diatomic molecules. They have also engendered additional quantum labels [33] corresponding to new (approximate) symmetries, again reminiscent of molecular structure. They thus have a global character not present in the foregoing methods. On a purely quantitative level, however, they are not generally as accurate as methods based on the projection operator formalism.

There are many other areas involving autoionization, such as satellite lines [34], inner-shell ionization [35], to mention only a few (*cf.* Chap. 60 and Ref. [36]). In addition, significant application of the phenomena associated with autoionization to diagnostics of astrophysical [37] and fusion [38] plasmas, for example, shows that autoionization has considerable applied utility.

REFERENCES

1. H. Feshbach, Ann. Phys. (N. Y.) **19**, 287 (1962).
2. For a more complete review of much of this material, *cf.* A. Temkin and A. K. Bhatia in *Autoionization, Recent Developments and Applications*, edited by A. Temkin (Plenum, New York, 1985), pp. 1 *ff*, 35 *ff*.
3. H. S. Taylor, Adv. Chem. Phys. **18**, 91 (1970).
4. A. Berk, A. K. Bhatia, B. R. Junker, and A. Temkin, Phys. Rev. A **34**, 4591 (1986).
5. A. Temkin and A. K. Bhatia, Phys. Rev. A **31**, 1259 (1985).
6. A. Berk and A. Temkin, Phys. Rev. A **32**, 3196 (1985).
7. Y. Hahn, Ann. Phys. (N. Y.) **58**, 137 (1960).
8. C. L. Pekeris, Phys. Rev. **112**, 1649 (1958).
9. B. R. Junker and C. L. Huang, Phys. Rev. A **18**, 313 (1978).
10. B. F. Davis and K. Chung, Phys. Rev. A **29**, 313, 2437 (1978).
11. M. Bylicki, J. Phys. B **24**, 413 (1991).
12. J. Brunt, G. King, and F. Read, J. Phys. B **10**, 433 (1977).
13. Y. Hahn, T. F. O'Malley, and L. Spruch, Phys. Rev. **128**, 932 (1962).
14. T. F. O'Malley and S. Geltman, Phys. Rev. A **137**, 1344 (1965).
15. A. Temkin, A. K. Bhatia, and J. N. Bardsley, Phys. Rev. A **7**, 1633 (1972).

16. K. T. Chung, Phys. Rev. A **20**, 1743 (1979).
17. K. T. Chung and B. F. Davis in *Autoionization, Recent Developments and Applications*, edited by A. Temkin (Plenum, New York, 1985), p. 73.
18. K. Chung and B. F. Davis, Phys. Rev. A **26**, 3278 (1982).
19. A. K. Bhatia and A. Temkin, Phys. Rev. A **11**, 2018 (1975).
20. R. P. Madden and K. Codling, Astrophs. J. **141**, 364 (1965).
21. H. Morgan and D. Ederer, Phys. Rev. A **29**, 1901 (1984).
22. A. K. Bhatia, P. G. Burke, and A. Temkin, Phys. Rev. A **8**, 21 (1973).
23. A. K. Bhatia and A. Temkin, Phys. Rev. A **29**, 1895 (1984).
24. Y.-K. Ho, A. K. Bhatia, and A. Temkin, Phys. Rev. A **15**, 1432 (1977).
25. U. Fano, Phys. Rev. **124**, 1866 (1960).
26. K. Smith, *The Calculation of Atomic Collision Processes* (Wiley-Interscience, New York, 1971), p. 47.
27. R. J. Drachman, Phys. Rev. A **15**, 1432 (1977).
28. H. Feshbach, Ann. Phys. (N. Y.) **43**, 410 (1967).
29. D. C.Griffin, M. S. Pindzola, F. Robicheaux, T. W. Gorczyca, and N. R. Badnell, Phys. Rev. Lett. **72**, 3491 (1994).
30. V. A. Mandelshtam, T. R. Ravuri, and H. S. Taylor, J. Chem. Phys. **101**, 8792 (1994).
31. E. B. Balslev and J. W. Combes, Comm. Math. Phys. **22**, 280 (1971).
32. Y.-K. Ho, Phys. Rep. **99**, 1 (1983).
33. C.-D. Lin in *Review of Fundamental Processes and Applications of Atoms and Ions*, edited by C.-D. Lin (World Scientific, 1993), p. 24.
34. F. Bely-Dabau, A. H. Gabriel, and S. Valonte, Mon. Not. Roy. Astron. Soc. **186**, 305 (1979).
35. C. J. Powell in *Electron Impact Ionization*, edited by T. D. Mark and G. H. Dunn, (Springer Verlag, 1985), Chap. 6.
36. R. D. Cowan, *The Theory of Atomic Structure and Spectra* (University of California Press, 1981).
37. G. Doschek in *Autoionization (op. cit.)*, p. 171.
38. M. Finkenthal in *AIP Conference Proceedings # 206; Atomic Processes in Plasmas, 1990*, edited by Y.-K. Kim and R. E. Elton (American Institute of Physics, New York, 1990), p. 95.

26

Green's Functions of Field Theory

Gordon Feldman and Thomas Fulton

The Henry A. Rowland Department of Physics and Astronomy, The Johns Hopkins University, Baltimore, Maryland 21218

26.1 INTRODUCTION

The discussion in this chapter is restricted to Green's function techniques as applied to problems in atomic physics, specifically to the calculation of higher order (correlation, Breit, as well as radiative) corrections to energy levels, and also of transition amplitudes for radiative transitions of atoms which are gauge invariant (GI) at every level of approximation.

Green's function techniques were first applied to many-electron atoms in 1971 as specific instances of the use of field theory techniques in many-particle problems [1, 2]. They initially provided alternative derivations of known approximations such as the Hartree-Fock (HF) approximation and random phase approximation (RPA). The starting point for the derivations was a nonrelativistic (NR) field-theoretical effective Hamiltonian for the system, which involved the nucleus-electron potential and only Coulomb interactions between electrons.

It subsequently became apparent that it was possible to formulate the problem more generally, with the full QED Hamiltonian as a starting point [3, 4]. Thus, the theory contains both relativity, and virtual and real transverse photons, as well as the Coulomb interaction between the electrons. The relativistic (R) approximate equations, such as the Dirac-Fock (DF) and relativistic random phase approximation (RRPA) are the natural outcomes of the formalism, and the nonrelativistic (NR)

results can be viewed as approximations to these R cases. Moreover, the Green's function approach (GFA) now provides a means of carrying out programs involving systematic approximations of successively higher and higher accuracy. The GFA provides a framework which allows one to make corrections to results obtained in the DF approximation or in the coupled cluster approximation (CCA) and many-body perturbation theory (MBPT), including magnetic (Breit type) interactions [5, 6]. It ensures that there is neither double counting nor omission of contributions. For radiative transitions, the formalism allows for a systematic treatment of such effects which is gauge independent (GI) at any given level of approximation [7–10]. Finally, the GFA is numerically implementable. Renormalization can be carried out for radiative corrections, resulting in finite and calculable expressions [11–13]. The integro-differential equations (i.e. the Dyson equations) needed to calculate energies or transition amplitudes in nontrivial approximations are also soluble [8, 9, 14].

In quantum field theory, Green's functions are defined in terms of vacuum expectation values of products of field operators. While this restriction can be relaxed, expectation values must still be taken for a nondegenerate state. As a practical matter, this requirement ultimately restricts one to consider atoms with electron numbers associated only with closed shells or subshells, and those with closed shells or subshells plus or minus one or two electrons. The corresponding Green's functions considered here are the two- and four-point functions for energy levels, and the three- and five-point functions for transition amplitudes (leading to oscillator strengths). The restriction in electron numbers is clearly not a severe one. It allows one to cover many atomic species.

Starting from relativistic QED, the electron field operator $\psi(\mathbf{r}, t)$ written in the Heisenberg picture, satisfies equal-time anticommutation relations

$$\{\psi(\mathbf{r}, t), \psi(\mathbf{r}', t)\} = \{\psi^\dagger(\mathbf{r}, t), \psi^\dagger(\mathbf{r}', t)\} = 0, \, (26.1)$$

$$\{\psi(\mathbf{r}, t), \psi^\dagger(\mathbf{r}', t)\} = \delta^3(\mathbf{r} - \mathbf{r}'). \quad (26.2)$$

(Spinor labels are suppressed.)

The time-translation operator, acting on any Heisenberg operator $\mathcal{O}(\mathbf{r}, t) = \{\psi(\mathbf{r}, t), \, \psi^\dagger(\mathbf{r}, t), \, j_\mu(\mathbf{r}, t), \dots\}$, is

$$\mathcal{O}(\mathbf{r}, t) = e^{i\mathcal{H}t} \mathcal{O}(\mathbf{r}, 0) e^{-i\mathcal{H}t}, \quad (26.3)$$

where $\mathcal{H}$ is the full QED Hamiltonian.

Notations and Definitions. For brevity, plainface numbers are used to denote both a coordinate and time, while boldface numbers denote a coordinate vector alone. For example,

$$1 \equiv (\mathbf{1}, t_1) \equiv (\mathbf{r}_1, t_1),$$
$$d^3\mathbf{1} \equiv d^3\mathbf{r}_1, \quad (26.4)$$
$$d^4\mathbf{1} \equiv d^3\mathbf{r}_1 dt_1.$$

For radiative transitions $\mathcal{M}_{fi}^{N'}(k_0)$ denotes the transition amplitude for the emission of a single photon of energy k_0 for an N'-electron atom from an initial energy $E_i^{N'}$ to a final energy $E_f^{N'}$, where

$$\mathcal{M}_{fi}^{N'}(k_0) \equiv (2\pi)^4 \delta(E_f^{N'} - E_i^{N'} - k_0) M_{fi}^{N'}(k_0). \quad (26.5)$$

The notation on the rhs of Eq. (26.5) is somewhat redundant, since k_0 is taken to be $k_0 = E_f^{N'} - E_i^{N'}$.

The current density operators at $t = 0$ are given respectively by

$$j_\mu(\mathbf{r}, 0)$$
$$= \begin{cases} \frac{1}{2} \int d^3\mathbf{1} d^3\mathbf{2} \left[\psi^\dagger(\mathbf{1}, 0), u_\mu^{\mathrm{R}}(\mathbf{12}; \mathbf{r})\psi(\mathbf{2}, 0) \right] & \text{for R} \\ \int d^3\mathbf{1} d^3\mathbf{2} \psi^\dagger(\mathbf{1}, 0) u_\mu^{\mathrm{NR}}(\mathbf{12}; \mathbf{r})\psi(\mathbf{2}, 0) & \text{for NR} \end{cases}$$
$$(26.6)$$

with

$$u_\mu(\mathbf{12}; \mathbf{r})$$
$$= \begin{cases} e \, \alpha_\mu \delta^3(\mathbf{1} - \mathbf{r})\delta^3(\mathbf{2} - \mathbf{r}) & \text{for R} \\ e \left[1, \left(\frac{1}{2im}\right)(\nabla_2 - \nabla_1) \right] \delta^3(\mathbf{1} - \mathbf{r})\delta^3(\mathbf{2} - \mathbf{r}) & \text{for NR,} \end{cases}$$
$$(26.7)$$

where $\alpha_\mu = (\boldsymbol{\alpha}, 1)$, and the components of $\boldsymbol{\alpha}$ are the usual Dirac matrices.

The corresponding charge operators are defined as

$$Q = \begin{cases} \frac{1}{2}e \int d^3\mathbf{1} \left[\psi^\dagger(\mathbf{1}, 0), \psi(\mathbf{1}, 0) \right] & \text{for R} \\ e \int d^3\mathbf{1} \psi^\dagger(\mathbf{1}, 0)\psi(\mathbf{1}, 0) & \text{for NR} \end{cases}. \quad (26.8)$$

The transition amplitude for a photon of polarization four-vector $\epsilon^\mu(k)$, momentum $\mathbf{k}$, energy $k_0(k_0 = |\mathbf{k}|)$, and photon attachment point $\mathbf{r}$ is

$$M_{fi}^{N'}(k_0) = \int d^3\mathbf{r} \frac{e^{i\mathbf{k}\cdot\mathbf{r}}}{\sqrt{2k_0}} M_{fi}^{N'}(\mathbf{r}; k_0). \quad (26.9)$$

In terms of the current density operator, $M_{fi}^{N'}(\mathbf{r}; k_0)$ can be written as

$$M_{fi}^{N'}(\mathbf{r}; k_0) = \left\langle \begin{matrix} N' \\ f \end{matrix} \right| \epsilon^\mu(k) j_\mu(\mathbf{r}, 0) \left| \begin{matrix} N' \\ i \end{matrix} \right\rangle$$
$$\equiv \left\langle \begin{matrix} N' \\ f \end{matrix} \right| j^k(\mathbf{r}, 0) \left| \begin{matrix} N' \\ i \end{matrix} \right\rangle, \quad (26.10)$$

where $\left| \begin{matrix} N' \\ n \end{matrix} \right\rangle$ is a state of leptonic charge number N', with N' corresponding to an atom of N' electrons, with total energy $E_n^{N'}$. The term *lepton charge* is used to refer to the charge of electrons and positrons.

In the dipole approximation, $e^{i\mathbf{k}\cdot\mathbf{r}} \approx 1$, $\epsilon^\mu j_\mu$ contains the quantity

$$\epsilon^\mu(k) u_\mu(\mathbf{12}; \mathbf{r}) \equiv \lambda_k(\mathbf{r})\delta(\mathbf{1} - \mathbf{r})\delta(\mathbf{2} - \mathbf{r}), \quad (26.11)$$

where, in the radiation gauge,

$$\lambda_k(\mathbf{r}) = \begin{cases} e\boldsymbol{\epsilon}(k)\cdot\boldsymbol{\alpha} & \text{velocity form} \\ iek_0\boldsymbol{\epsilon}(k)\cdot\mathbf{r} & \text{length form} \end{cases}. \quad (26.12)$$

26.2 THE TWO-POINT GREEN'S FUNCTION

The two-point Green's function, or one-body propagator [3, 4], for a system of lepton charge Ne is defined as the expectation value of a time-ordered product

$$G^N(1, 1') \equiv -i \left\langle \begin{matrix} N \\ 0 \end{matrix} \right| T\left[\psi(1)\psi^\dagger(1')\right] \left| \begin{matrix} N \\ 0 \end{matrix} \right\rangle, \quad (26.13)$$

where $\left| \begin{matrix} N \\ 0 \end{matrix} \right\rangle$ is the ground state of leptonic charge number N, with N corresponding to an atom with electron number N in a filled shell or subshell (N=2, 4, 8, $\dots$).

The relative time $(t_1 - t_1')$ Fourier transform of Eq. (26.13) yields the spectral representation of $G^N(1, 1')$:

$$G_\omega^N = \sum_j \frac{|u^j\rangle\langle u^j|}{\omega - \varepsilon^j + i\eta} + \sum_\zeta \frac{|v^\zeta\rangle\langle v^\zeta|}{\omega - \varepsilon^\zeta + i\eta}, \quad \eta = 0^+, \quad (26.14)$$

where

$$G_\omega^N(1, 1') \equiv \langle 1| G_\omega^N |1'\rangle, \quad u^j(\mathbf{r}) \equiv \langle \mathbf{r} | u^j \rangle, \quad \text{etc.}, \quad (26.15)$$

and where the symbol $\sum\!\!\!\!\!\int$ denotes a summation over discrete and integration over continuous states. The two terms in Eq. (26.14) are obtained by counting each time-ordering separately in Eq. (26.13), introducing a complete set of intermediate states and using the time translation operator [Eq. (26.3)]. The functions $u^j(\mathbf{r})$ and $v^\zeta(\mathbf{r})$ are defined by

$$u^j(\mathbf{r}) \equiv \left\langle {}^N_0 \left| \psi(\mathbf{r},0) \right| {}^{N+1}_j \right\rangle ,$$

$$\left\langle {}^N_0 \left| \psi(\mathbf{r},t) \right| {}^{N+1}_j \right\rangle = e^{i(E^N_0 - E^{N+1}_j)t} u^j(\mathbf{r}) , \quad (26.16)$$

and

$$v^\zeta(\mathbf{r}) \equiv \left\langle {}^{N-1}_\zeta \left| \psi(\mathbf{r},0) \right| {}^N_0 \right\rangle ,$$

$$\left\langle {}^{N-1}_\zeta \left| \psi(\mathbf{r},t) \right| {}^N_0 \right\rangle = e^{i(E^{N-1}_\zeta - E^N_0)t} v^\zeta(\mathbf{r}) . \quad (26.17)$$

Here, $\left| {}^N_0 \right\rangle$ is the ground state of an atom of lepton charge number N (corresponding to a nondegenerate state — a closed shell or subshell) of energy E^N_0, $\left| {}^{N+1}_j \right\rangle$ an atomic state of total energy E^{N+1}_j and leptonic charge number $N+1$, and $\left| {}^{N-1}_\zeta \right\rangle$ is the state of energy E^{N-1}_ζ and leptonic charge number $N-1$. These several states satisfy the eigenvalue equations

$$\mathcal{H} \left| {}^N_0 \right\rangle = E^N_0 \left| {}^N_0 \right\rangle ,$$

$$\mathcal{H} \left| {}^{N+1}_j \right\rangle = E^{N+1}_j \left| {}^{N+1}_j \right\rangle , \quad (26.18)$$

$$\mathcal{H} \left| {}^{N-1}_\zeta \right\rangle = E^{N-1}_\zeta \left| {}^{N-1}_\zeta \right\rangle .$$

The energy parameters ε^j and ε^ζ are defined by

$$\varepsilon^j \equiv E^{N+1}_j - E^N_0 , \quad (26.19)$$

$$\varepsilon^\zeta \equiv E^N_0 - E^{N-1}_\zeta . \quad (26.20)$$

Equations (26.19) and (26.20) are generalizations of Koopmans's theorem [15] (see Sect. 21.4.1). In the DF approximation, the state $\left| {}^{N+1}_j \right\rangle$ can be thought of as N effective particles (electrons) making up the core, plus one valence electron, with energy label j. The atom can also be an isoelectronic ion with nuclear charge number Z. In this (DF) approximation, the state $\left| {}^{N-1}_\zeta \right\rangle$ is one of two possible types. It can have N independent electrons making up the core, with one of the core electrons missing, or equivalently the N electron core with one electron-hole, with energy label ζ. There is a finite number N of such hole states, which shall be labeled a. There are no other states in the NR (HF) case. In the R (DF) case, the second group of states $\left| {}^{N-1}_\zeta \right\rangle$ can also describe an atom with a core of N electrons, plus one positron. Its continuum energy label, will be taken as $\zeta = \bar{\ell}$. This energy will appear with a negative sign in Eq. (26.14).

The second step in leading to an explicit $G^N(1, 1')$ is the generation of a Dyson equation which it satisfies.

In both the DF and HF approximations, the Dyson equation can be obtained in a successive series of steps [11]. Working in the Coulomb gauge (see Sect. 27.3), begin with the Coulomb interaction between electrons and neglect the exchange of transverse virtual photons. The Dyson equation for the resulting two-point function is then expanded as a power series in the electron-electron (ee) interaction

$$V_{ee} \equiv V = \frac{\alpha}{|\mathbf{x} - \mathbf{y}|} , \quad (26.21)$$

resulting in an infinite set of Feynman diagrams ($\alpha = e^2/4\pi$ in rationalized mks units). As a next approximation, consider only diagrams involving single Coulomb exchanges and their iterates ("ladders"). That is, set aside for later consideration nonladder Feynman diagrams of two or more Coulomb photons and their iterates (e.g., two crossed Coulomb photon lines). A summation of the infinite set of these remaining terms generates an equation for a propagator labeled $G^N_{\omega,\text{Coul}}$ which contains Coulomb radiative corrections in its kernel. Finally, modify the kernel by isolating these radiative corrections (self energy and vacuum polarization) by constructing a spectral representation of $G^{\text{``0''}}_\omega$ which mimics that for the usual QED propagator, which is a vacuum (rather than an N-lepton ground state) expectation value. This requires shifting the poles corresponding to core energies of the atom from the upper to the lower complex ω plane. The shifting of poles is accomplished by means of the equation

$$\frac{1}{x - i\eta} - \frac{1}{x + i\eta} = 2\pi i \delta(x) , \quad (26.22)$$

which, when used to shift the poles of core electrons in the ω plane, gives

$$G^N_{\omega,\text{Coul}} = 2\pi i \sum_{a=1}^{N} |v^a\rangle \langle v^a| \, \delta(\omega - \varepsilon^a) + G^{\text{``0''}}_\omega . \quad (26.23)$$

It is the first term on the right-hand side of this equation which occurs in the kernel of Eq. (26.24) below and generates the DF approximation.

Some of the terms set aside in the course of this sequence of approximations are reconsidered in Sect. 26.5 to obtain more accurate energy results. What remains at the end of the sequence is an approximation to G^N_ω of Eq. (26.14) designated as $G^N_{\omega,\text{DF}}$, and which satisfies the self-consistent Dyson equation

$$G^N_{\omega,\text{DF}} = g_\omega + g_\omega \Sigma^N_{\text{DF}} G^N_{\omega,\text{DF}} , \quad (26.24)$$

where $\sum^N_{\text{DF}}$ is the kernel defined in Eq. (26.27). For the R (DF) case,

$$g^{-1}_\omega = \omega - h^R(\mathbf{r}), \quad h^R(\mathbf{r}) = \boldsymbol{\alpha}\cdot\mathbf{p} + m\beta - Z\alpha/r, \quad (26.25)$$

and the corresponding h^{NR}, used in the HF case, is

$$h^{\mathrm{NR}}(\mathbf{r}) = \frac{\mathbf{p}^2}{2m} - \frac{Z\alpha}{r}. \qquad (26.26)$$

The function g_ω in Eq. (26.25) is the R or NR Coulomb Green's function. It is the solution of the corresponding inhomogeneous c-number Schrödinger or Dirac equation. It is a single-particle equation which has had a long history of specific treatments (see Chap. 9 and Refs. [16–18]). It is the semiclassical limit of the two-point propagator we consider here when only the c-number nuclear Coulomb potential is kept and all other (q-number) interactions are turned off.

In the DF and HF approximations in Eq. (26.14), we replace ε^j by $e_j > 0$, $|u^j\rangle$ by $|j\rangle$ (the valence energies and states of an atom with $N+1$ electrons and a frozen relaxed core of N electrons), ε^ζ by $e_a > 0$, $|u^\zeta\rangle$ by $|a\rangle$, for the N discrete electron core states and ε^ζ by $e_{\bar\ell} < 0$, $|u^\zeta\rangle$ by $|\bar\ell\rangle$, for the continuum of negative energy states (which do not appear in the HF approximation). The kernel Σ_{DF}^N contains *only* core states $|a\rangle$, in *both* the DF and HF approximations, and is given by the sum over states

$$\begin{aligned}
\langle m|\Sigma_{\mathrm{DF}}^N|n\rangle &= \alpha \sum_{a=1}^N \int d^3x\,d^3y\,\frac{1}{|\mathbf{x}-\mathbf{y}|}\,\langle m|\mathbf{x}\rangle\langle a|\mathbf{y}\rangle \\
&\quad \times \{\langle\mathbf{x}|n\rangle\langle\mathbf{y}|a\rangle - \langle\mathbf{x}|a\rangle\langle\mathbf{y}|n\rangle\} \\
&\equiv \sum_{a=1}^N \{({}^m_a|V|{}^n_a) - ({}^m_a|V|{}^a_n)\} \\
&\equiv \sum_{a=1}^N [{}^m_a|V|{}^n_a] \qquad (26.27)
\end{aligned}$$

for arbitrary states m and n. The first term in brackets is the Hartree term and the second is the electron exchange term in the DF (HF) approximation. The DF (HF) equation is the homogeneous equation corresponding to the inhomogeneous Eq. (26.24). Thus,

$$(e_n - h - \Sigma_{\mathrm{DF}}^N)|n\rangle = 0, \qquad \langle n|n'\rangle = \delta_{nn'}. \qquad (26.28)$$

The states $|n\rangle$ (valence, core, and negative energy states) are orthonormal and complete. In the coordinate basis, Eq. (26.28) takes the more familiar form

$$[e_n - h(\mathbf{x})]\langle\mathbf{x}|n\rangle - \langle\mathbf{x}|\Sigma_{\mathrm{DF}}^N|n\rangle = 0, \qquad (26.29)$$

where

$$\begin{aligned}
\langle\mathbf{x}|\Sigma_{\mathrm{DF}}^N|n\rangle &= \sum_m \sum_{a=1}^N \langle\mathbf{x}|m\rangle [{}^m_a|V|{}^n_a] \equiv \sum_{a=1}^N [{}^x_a|V|{}^n_a] \\
&\equiv \int V_{\mathrm{DF}}(\mathbf{x},\mathbf{y})d^3y\langle\mathbf{y}|n\rangle. \qquad (26.30)
\end{aligned}$$

One can also generate equations corresponding to higher approximations than DF, using the same approach. For example, one can obtain a Brueckner equation [19],

$$[e_n - h - \Sigma_{\mathrm{DF}}^N - \Sigma_{\mathrm{B}}^N(e_n)]|n\rangle = 0, \qquad (26.31)$$

and the states now satisfy the orthonormality condition

$$\begin{aligned}
&\lim_{\omega\to e_n} \int d^3x\,d^3y\,\langle n|\mathbf{x}\rangle\langle\mathbf{y}|n'\rangle \\
&\times \left(\frac{[(\omega - e_n)\delta^3(\mathbf{x}-\mathbf{y}) - \langle\mathbf{x}|\{\Sigma_{\mathrm{B}}^N(\omega) - \Sigma_{\mathrm{B}}^N(e_n)\}|\mathbf{y}\rangle]}{\omega - e_n}\right) \\
&= \delta_{nn'}, \qquad (26.32)
\end{aligned}$$

where the energy-dependent kernel $\Sigma_{\mathrm{B}}(e_n)$ arises from irreducible Feynman diagrams involving two Coulomb photons. The kernel is given by

$$\begin{aligned}
&\langle m|\Sigma_{\mathrm{B}}^N(e_n)|n\rangle \\
&= \sum_{a,i,j} \frac{1/2}{e_n + e_a - e_i - e_j} [{}^m_a|V|{}^i_j][{}^i_j|V|{}^n_a] \\
&\quad + \sum_{a,b,i} \frac{1/2}{e_n + e_i - e_a - e_b} [{}^m_i|V|{}^a_b][{}^a_b|V|{}^n_i]. \qquad (26.33)
\end{aligned}$$

Σ involves summations over core states (a,b), and summation and integration over valence states (i,j). In perturbation theory, the lowest order contribution of $\Sigma_{\mathrm{B}}(e_n)$ is an ee correlation term.

26.3 THE FOUR-POINT GREEN'S FUNCTION

The four-point Green's function [4, 8, 11], or two-body propagator, is defined as

$$G^N(12, 1'2') \equiv -\langle{}^N_0|T[\psi(1)\psi^\dagger(1')\psi(2)\psi^\dagger(2')]|{}^N_0\rangle. \qquad (26.34)$$

In order to avoid unnecessary complication, only simple ladders of $V_{\mathrm{ee}} \equiv V$ are considered. There are 4! possible time orderings. Of these, there are four groups of four with $t_1, t_{1'} > t_2, t_{2'}$, $t_1, t_{1'} < t_2, t_{2'}$, $t_1, t_2 > t_{1'}, t_{2'}$, and $t_1, t_2 < t_{1'}, t_{2'}$, corresponding to the particle-hole (PH/HP) and the two-particle/two-hole (2P/2H) cases for the first eight and last eight time orderings respectively. For each of these cases, introduce a total time and relative time variable defined by

$$\begin{aligned}
T &= \frac{1}{2}(t_1 + t_{1'}), \quad T' = \frac{1}{2}(t_2 + t_{2'}), \\
t &= t_1 - t_{1'}, \qquad t' = t_2 - t_{2'}, \qquad (26.35)
\end{aligned}$$

for the first eight cases (PH/HP) and

$$\begin{aligned}
T &= \frac{1}{2}(t_1 + t_2), \quad T' = \frac{1}{2}(t_{1'} + t_{2'}), \\
t &= t_1 - t_2, \qquad t' = t_{1'} - t_{2'}, \qquad (26.36)
\end{aligned}$$

for the last eight cases (2P/2H). For a particular set of eight time orderings, a time translation with respect to the relevant c.m. time t or t', followed by a separate Fourier transformation for each case with respect to $T - T'$ (with integration variable $d\Omega$), yields contributions with poles in the separately defined Ω-planes at

$$\omega_{(i\alpha)} = E^N_{(i\alpha)} - E^N_0, \quad \text{for PH} \quad (26.37a)$$

$$-\omega_{(i\alpha)} = E^N_0 - E^N_{(i\alpha)}, \quad \text{for HP} \quad (26.37b)$$

$$\omega_{(ij)} = E^{N+2}_{(ij)} - E^N_0, \quad \text{for 2P} \quad (26.38a)$$

$$\omega_{(ab)} = E^N_0 - E^{N-2}_{(ab)}, \quad \text{for 2H}. \quad (26.38b)$$

Equations (26.37ab) parallel Eqs. (26.19) and (26.20) as generalizations of Koopmans' theorem. The spectral representation is of a form similar to Eq. (26.14), with wave functions corresponding to Eqs. (26.16) and (26.17) given by

$$\chi_{(ia)}(11') \equiv \left\langle \mathbf{1} \left| \chi_{(ia)}(t) \right| \mathbf{1}' \right\rangle e^{-i\omega_{(ia)}T}$$
$$= \left\langle {}^N_0 \left| T\left[\psi(1)\psi^\dagger(1')\right] \right| {}^N_{(ia)} \right\rangle \quad (26.39)$$

for the PH/HP case and

$$\varphi_{(ij)}(12) \equiv \left\langle \mathbf{12} \left| \varphi_{(ij)}(t) \right\rangle e^{-i\omega_{(ij)}T} \right.$$
$$= \left\langle {}^N_0 \left| T\left[\psi(1)\psi(2)\right] \right| {}^{N+2}_{(ij)} \right\rangle, \quad (26.40)$$

$$\gamma_{(ab)}(1'2') \equiv \left\langle \mathbf{1}'\mathbf{2}' \left| \gamma_{(ab)}(t') \right\rangle e^{-i\omega_{(ab)}T'} \right.$$
$$= \left\langle {}^{N-2}_{(ab)} \left| T\left[\psi(1')\psi(2')\right] \right| {}^N_0 \right\rangle, \quad (26.41)$$

for the 2P and 2H cases, respectively. The antisymmetry under exchange follows from the definitions (26.40) and (26.41):

$$\zeta_{(\tau)}(12) = -\zeta_{(\tau)}(21), \quad \zeta_{(\tau)} = \varphi_{(ij)}, \gamma_{(ab)}, \quad (26.42)$$

where it is understood that the suppressed spinor indices are interchanged as well as the coordinate and time variables.

The three amplitudes defined above satisfy Bethe-Salpeter (BS) equations. The PH/HP case is the analog of the positronium atom and the 2P case is analogous to He. For the case of Coulomb ladder exchanges, to which we have restricted ourselves, these BS equations can be reduced to simpler ones, with the relative time set equal to zero (the Salpeter equation [20]). The corresponding BS wave functions in the DF (HF) basis (rather than the coordinate basis), are for PH/HP,

$$\left\langle m \left| \bar\chi_{(ia)} \right| n \right\rangle \equiv \int d^3\mathbf{1} \, d^3\mathbf{1}' \langle m|\mathbf{1}\rangle\langle\mathbf{1}|\chi_{(ia)}(0)|\mathbf{1}'\rangle\langle\mathbf{1}'|n\rangle, \quad (26.43)$$

with $m = k$ and $n = c$, or $m = c$ and $n = k$, and for 2P/2H,

$$\left\langle mn \left| \bar\zeta_{(\tau)} \right\rangle \equiv \int d^3\mathbf{1} \, d^3\mathbf{2}\langle m|\mathbf{1}\rangle\langle n|\mathbf{2}\rangle \left\langle \mathbf{12} \left| \zeta_{(\tau)}(0) \right\rangle \right. , \quad (26.44)$$

where $\zeta_{(\tau)} = \varphi_{(ij)}, \gamma_{(ab)}$. The states $|i\rangle$, $|j\rangle$, $|a\rangle$, $|b\rangle$, $|m\rangle$, and $|n\rangle$ label one-particle DF (HF) eigenkets and e_i, e_j, e_a, e_b, e_m, and e_n, the corresponding eigenvalues.

In the PH/HP case, the BS equation is

$$-(\omega_{(ia)} - e_k + e_c) \left\langle i \left| \bar\chi_{(ia)} \right| a \right\rangle$$
$$= \sum_{j,b} \left\{ \left[{}^k_b \left| V \right| {}^j_c\right] \left\langle j \left| \bar\chi_{(ia)} \right| b \right\rangle + \left[{}^k_j \left| V \right| {}^b_c\right] \left\langle b \left| \bar\chi_{(ia)} \right| j \right\rangle \right\}, \quad (26.45)$$

$$(\omega_{(ia)} - e_c + e_k) \left\langle a \left| \bar\chi_{(ia)} \right| i \right\rangle$$
$$= \sum_{j,b} \left\{ \left[{}^c_b \left| V \right| {}^j_k\right] \left\langle j \left| \bar\chi_{(ia)} \right| b \right\rangle + \left[{}^c_j \left| V \right| {}^b_k\right] \left\langle b \left| \bar\chi_{(ia)} \right| j \right\rangle \right\}. $$

In the 2P/2H case, the coupled pairs of BS equations are

$$-(\omega_{(\tau)} - e_c - e_d) \left\langle cd \left| \bar\zeta_{(\tau)} \right\rangle \right. = \sum_{m>n} \left[{}^c_d \left| V \right| {}^m_n\right] \left\langle mn \left| \bar\zeta_{(\tau)} \right\rangle \right. , \quad (26.46)$$

$$(\omega_{(\tau)} - e_k - e_\ell) \left\langle k\ell \left| \bar\zeta_{(\tau)} \right\rangle \right. = \sum_{m>n} \left[{}^k_\ell \left| V \right| {}^m_n\right] \left\langle mn \left| \bar\zeta_{(\tau)} \right\rangle \right. , $$

with m and n in these equations labeling either both core, or both valence states.

The wave functions also satisfy the additional conditions:

$$\left\langle c \left| \bar\chi_{(ia)} \right| d \right\rangle = \left\langle k \left| \bar\chi_{(ia)} \right| \ell \right\rangle = 0,$$
$$\text{PH/HP (no 2P/2H terms)} \quad (26.47)$$

$$\left\langle ck \left| \bar\varphi_{(ij)} \right\rangle = \left\langle ck \left| \bar\gamma_{(ab)} \right\rangle = 0, \right.$$
$$\text{2P/2H (no PH/HP terms)} \quad (26.48)$$

The single indices a, b, c, d refer to core and i, j, k, l to valence DF (HF) states. The BS wave functions satisfy the orthonormality conditions

$$\sum_{k,c} \left\{ \left\langle k \left| \bar\chi_{(ia)} \right| c \right\rangle \left\langle c \left| \bar\chi_{(ia)'} \right| k \right\rangle \right.$$
$$+ \left\langle c \left| \bar\chi_{(ia)} \right| k \right\rangle \left\langle k \left| \bar\chi_{(ia)'} \right| c \right\rangle \right\} = \delta_{(ia)(ia)'}, \quad (26.49)$$

for the PH/HP case, and

$$\pm \left\{ \sum_{c>d} \left\langle \bar\zeta_{(\tau)} \left| cd \right\rangle \left\langle cd \left| \bar\zeta_{(\tau')} \right\rangle \right. \right.$$
$$\left. - \sum_{k>\ell} \left\langle \bar\zeta_{(\tau)} \left| k\ell \right\rangle \left\langle k\ell \left| \bar\zeta_{(\tau')} \right\rangle \right. \right\} = \delta_{(\tau)(\tau')}, \quad (26.50)$$

where $+(-)$ corresponds to the 2H(2P) case and $(\tau) = (ab)$ or (ij).

The PH/HP case in the GFA, involving Coulomb ladders for the ee interaction, is just the R and NR RPA [2]. The labels c, d should also refer to antiparticles, but the contributions of the integrals from these terms to Eqs. (26.45) and (26.46) are negligible.

26.4   RADIATIVE TRANSITIONS

For the majority of the applications, one begins with the function $\Gamma^N(12;3)$, the reducible three-point vertex:

$$\Gamma^N(12;3) = -\left\langle {}^N_0 \left| T\left[\psi(1)j^k(3)\psi^\dagger(2)\right] \right| {}^N_0 \right\rangle . \quad (26.51)$$

Our usual strategy is followed. The spectral representation serves to identify the functions of ultimate interest. (Energies are not relevant in this case.) One then generates Dyson equations in the chosen approximation by summing an infinite series of perturbation terms.

There are 3! time orderings in Eq. (26.51). As with the four-point function, not all of them are subsequently useful. The two useful cases are $t_1 > t_3 > t_2$, and $t_2 > t_3 > t_1$. To obtain a spectral representation for these two cases, one first carries out a time translation of t_3, using the operator $\exp(i\mathcal{H}t_3)$ of Eq. (26.3), so that $t_1 \to \tau_1 = t_1 - t_3$, $t_2 \to \tau_2 = t_2 - t_3$. One then introduces complete sets of intermediate states. The functions defined in connection with the two-point function in Eqs. (26.16) and (26.17) will now appear, as well as the radiative transition amplitude, defined in Eq. (26.10). If one next carries out a separate time translation of τ_1 and τ_2 and Fourier-transforms the resulting expressions, one obtains (with **3** replaced by **r**)

$$\Gamma^N(\mathbf{r};\omega_1\omega_2) = \sum_{j\ell}\!\!\!\!\!\!\!\!\!\!\!\!\!\! \frac{|u^j\rangle M^{N+1}_{j\ell}(\mathbf{r})\langle u^\ell|}{(\omega_1 - \varepsilon_j + i\eta)(\omega_2 - \varepsilon_\ell + i\eta)}$$
$$+ \sum_{\zeta\chi}\!\!\!\!\!\!\!\!\!\!\!\!\!\! \frac{|v^\zeta\rangle M^{N-1}_{\zeta\chi}(\mathbf{r})\langle v^\chi|}{(\omega_1 - \varepsilon_\zeta - i\eta)(\omega_2 - \varepsilon_\chi - i\eta)} + \cdots .$$

$$(26.52)$$

We next define the three-point irreducible electron vertex Λ^N from the reducible vertex Γ^N (the electron "legs" in Γ^N are missing in Λ^N) as

$$\langle \mathbf{1} | \Gamma^N(\mathbf{r};\omega_1\omega_2) | \mathbf{2} \rangle$$
$$\equiv \iint d^3\mathbf{1}'d^3\mathbf{2}' \langle \mathbf{1} | G^N_{\omega_1} | \mathbf{1}' \rangle \langle \mathbf{1}' | \Lambda^N(\mathbf{r};\omega_1\omega_2) | \mathbf{2}' \rangle$$
$$\times \langle \mathbf{2}' | G^N_{\omega_2} | \mathbf{2} \rangle , \quad (26.53)$$

and pick out the residues of the ω_1 and ω_2 poles at specific energies ε_m and ε_n. [$\Lambda^N(\mathbf{r};\omega_1\omega_2)$ has no such poles.] With the equivalent of the DF(HF) approximation to the Dyson equations [see Eq. (26.57) below], the corresponding kets form an orthonormal set, and scalar products

calculated with respect to $\langle m|$ and $|n\rangle$. The second term on the right-hand side of Eq. (26.52) refers to hole states and is of less interest than the first term describing 1P-1P transitions. From the first term, the transition matrix element in terms of Λ^N is

$$M^{N+1}_{fi}(\mathbf{r};k_0) = \left\langle f \left| \Lambda^N(\mathbf{r};e_f e_i) \right| i \right\rangle . \quad (26.54)$$

Generation of a Dyson Equation. The approximation of only Coulomb "ladder" ee interactions in the two-point Green's function produces an infinite set of Feynman diagrams to which one end of a single transverse photon line is attached in all possible ways. A resummation of these diagrams generates the $G^N_{\omega,\text{Coul}}$ functions, which are approximated by the DF (HF) propagators written in their spectral form. The scalar products leading to Eq. (26.54) can then be taken.

A further simplification results because of the Coulomb ladder approximation, which is the same as in the four-point Green's function case: the relative time can be set equal to zero, which corresponds to integrating over the relative frequency, $\omega = \omega_1 - \omega_2$. With the total frequency defined as $\Omega = \frac{1}{2}(\omega_1 + \omega_2)$, and the definitions

$$\bar\Lambda^N(\mathbf{r},\Omega) = \frac{1}{2\pi}\int d\omega \Lambda^N(\mathbf{r};\omega + \tfrac{1}{2}\Omega, \omega - \tfrac{1}{2}\Omega),$$

$$\langle m | \bar\Lambda^N(\Omega) | n \rangle = \int d^3\mathbf{r} \frac{e^{i\mathbf{k}\cdot\mathbf{r}}}{\sqrt{2k_0}} \langle m | \bar\Lambda^N(\mathbf{r},\Omega) | n \rangle , \quad (26.55)$$
$$\Omega_{mn} = e_n - e_m = k_0 ,$$

with a relation similar to Eq. (26.55) for the quantity $\lambda_k(r)$ defined in Eq. (26.11); i.e.,

$$\langle m | \lambda(k_0) | n \rangle = \int d^3\mathbf{r} \frac{e^{i\mathbf{k}\cdot\mathbf{r}}}{\sqrt{2k_0}} \langle m | \lambda_k(r) | n \rangle , \quad (26.56)$$

the matrix elements in the DF (HF) basis are [14, 19]

$$\langle m | \bar\Lambda^N(k_0) | n \rangle$$
$$= \langle m | \lambda(k_0) | n \rangle$$
$$+ \sum_{a j}\left\{ \frac{1}{e_a - e_j + k_0} \left[{}^m_j | V | {}^n_a\right] \left\langle a | \bar\Lambda^N(k_0) | j \right\rangle \right.$$
$$\left. + \frac{1}{e_a - e_j - k_0} \left[{}^m_a | V | {}^n_j\right] \left\langle j | \bar\Lambda^N(k_0) | a \right\rangle \right\} . \quad (26.57)$$

As discussed in Sect. 26.2, the label a should include not just hole states but also negative energy states, which have been neglected in Eq. (26.57). Note also that only PH or HP matrix elements appear on the right-hand side of Eq. (26.57). Therefore, a closed set of inhomogeneous linear algebraic equations for $\langle b | \bar\Lambda^N(k_0) | \ell \rangle$ and $\langle \ell | \bar\Lambda^N(k_0) | b \rangle$ results from setting $m = b$, $n = \ell$ or $m = \ell$, $n = b$, respectively, in Eq. (26.57). These equations can be solved numerically and the resulting $\langle a | \bar\Lambda^N(k_0) | j \rangle$ and $\langle j | \bar\Lambda^N(k_0) | a \rangle$ substituted in Eq. (26.57) to obtain the final result

$$M_{fi}^{N+1}(k_0) \equiv \langle f | \bar\Lambda^N(k_0) | i \rangle . \tag{26.58}$$

An integro-differential equation form, which provides the option of choosing alternative numerical techniques [14] is obtained from Eq. (26.57), after some rearrangement and passage to a coordinate basis. Defining $m_{fi}(k_0)$ from $\lambda(k_0)$ in analogy with the definition of $M_{fi}^{N+1}(k_0)$ from $\bar\Lambda^N(k_0)$ in Eq. (26.58), one has

$$M_{fi}^{N+1}(k_0) = m_{fi}(k_0) + \sum_a \{ \langle a | \lambda(k_0) | A^- \rangle$$
$$+ \langle A^+ | \lambda(k_0) | a \rangle \}, \tag{26.59}$$

where

$$[h(\mathbf{r}) \mp k_0 - e_a] \langle \mathbf{r} | A^\pm \rangle + \sum_a \begin{bmatrix} \mathbf{r} \\ a \end{bmatrix} V \begin{vmatrix} A^\pm \\ a \end{vmatrix}$$
$$= -\sum_j \langle \mathbf{r} | j \rangle \langle j | [v_\pm + \mathcal{V}_\pm] | a \rangle , \tag{26.60}$$

$$\langle j | v_+ | a \rangle = \begin{bmatrix} j \\ i \end{bmatrix} V \begin{vmatrix} a \\ f \end{vmatrix}, \quad \langle j | v_- | a \rangle = \begin{bmatrix} j \\ f \end{bmatrix} V \begin{vmatrix} a \\ i \end{vmatrix}, \tag{26.61}$$

$$\langle j | \mathcal{V}_\pm | a \rangle = \sum_b \left\{ \begin{bmatrix} j \\ b \end{bmatrix} V \begin{vmatrix} a \\ B_\pm \end{vmatrix} + \begin{bmatrix} j \\ B_\mp \end{bmatrix} V \begin{vmatrix} a \\ b \end{vmatrix} \right\} . \tag{26.62}$$

The three-point Green's function, as can be seen from this summary, describes transition amplitudes for radiative transitions between two valence states of atoms with closed shells (subshells) plus one electron (the 1P case). In the Coulomb ladder approximation, closely related to the DF equation, the photon vertex $\langle \mathbf{1} | \Lambda^N(\mathbf{r}; \omega_1\omega_2) | \mathbf{2} \rangle$ of Eq. (26.53), is nonlocal in space, rather than the local vertex $\langle \mathbf{1} | \lambda_k(\mathbf{r}) | \mathbf{2} \rangle$ of Eq. (26.11), (as follows from the factor of $\delta^3(\mathbf{1} - \mathbf{r})\delta^3(\mathbf{2} - \mathbf{r})$). The presence of these additional nonlocal contributions to $M_{fi}^{N+1}(k_0)$ is made apparent in Eq. (26.59). The $m_{fi}(k_0)$ term is the contribution of the local vertex. Of course, if one knew the *exact* $N + 1$ electron wave functions and energies, only the local vertex would enter and a GI result would be obtained. However, the length and velocity versions of Eq. (26.12) are equal and GI even in the approximation just discussed [3, 7]. Gauge invariance is an *essential* constraint on radiative transition amplitudes. It has been proven for general gauges not only in the present approximation, but also in the ones discussed below [8, 10], all of them arising in the GFA. Since the effective potential in the DF (or HF) approximation is nonlocal, the effective current must also be nonlocal in order to maintain the GI.

The somewhat more complicated Dyson equation satisfied by the nonlocal vertex corresponding to the Brueckner approximation has also been generated [21], and put in a numerically implementable form.

An alternative approximation [3, 7] for transition matrix elements, proposed earlier [2] than the one just discussed, is based on the RPA [our PH/HP case, with wave function $\bar\chi_{(fa)}$ the solution of Eq. (26.45).]

Reference to Eqs. (26.6), (26.7), (26.10) and (26.11) gives immediately

$$M_{fa}^N(\mathbf{r}; \omega_{(fa)}) = -\lambda_k(\mathbf{r}) \langle \mathbf{r} | \bar\chi_{(fa)} | \mathbf{r} \rangle , \tag{26.63}$$

where $k_0 = k = \omega_{(fa)}$ and N correspond to atoms with a closed shell or subshell of electrons.

Aside from being a different approximation to transition amplitudes than Eq. (26.54), (26.63) covers a different set of cases than does (26.54), since the corresponding initial states i in (26.54) are restricted to core states, a, in (26.63). Thus, for example, the case $N = 2$ in (26.54) can describe transitions between any two valence states of Li. The corresponding case for (26.63) is $N = 4$, but only transitions to a higher level, originating in the $1s$ or $2s$ level of Li, can be described by the formalism.

Finally, we shall discuss radiative transitions [8] for 2P atoms (closed shell/subshell plus two valence electrons). The general case involves a five-point nonlocal vertex. However, in the ladder approximation, this reduces to transition matrix elements which contain combinations of DF, 2P BS wave functions, and three-point functions. The final expressions are

$$M_{fi}^{N+2}(\mathbf{r}; k_0)$$
$$= \left\langle \bar\zeta_{(fp)} \left| \left(\frac{\bar\Lambda^N(\mathbf{r}, k_0)}{k_0 - \Delta H} V - V \frac{\bar\Lambda^N(\mathbf{r}, k_0)}{k_0 - \Delta H} \right) \right| \bar\zeta_{(iq)} \right\rangle \tag{26.64}$$

The expressions appearing in this equation are, in more detail, in the DF basis:

$$\left\langle \begin{matrix} m \\ n \end{matrix} \left| \frac{\bar\Lambda^N(\mathbf{r}, k_0)}{k_0 - \Delta H} \right| \begin{matrix} q \\ s \end{matrix} \right\rangle \equiv \frac{\langle m | \bar\Lambda^N(\mathbf{r}, k_0) | q \rangle \delta_{ns}}{k_0 + e_q - e_m} ,$$
$$\tag{26.65}$$

where m, n, q, s label DF states (P or H), $|{}^q_s\rangle \equiv |qs\rangle \equiv |q\rangle|s\rangle$, Eq. (26.65) serves to define ΔH, as a difference of two DF Hamiltonians, the argument $\mathbf{r}$ of $\bar\Lambda$ refers to the electron which emits the photon, and

$$\langle {}^m_n | V | \bar\zeta_{(\tau)} \rangle = \sum_{cd} ({}^m_n | V | {}^c_d) \langle cd | \bar\zeta_{(\tau)} \rangle$$
$$+ \sum_{k\ell} ({}^m_n | V | {}^k_\ell) \langle k\ell | \bar\zeta_{(\tau)} \rangle . \tag{26.66}$$

In Eq. (26.66), we used the fact that in the DF basis, $|\bar\zeta_{(\tau)}\rangle$, only has 2P or 2H components [see (26.46) and (26.48)].

26.5 RADIATIVE CORRECTIONS

This section summarizes radiative corrections for 1P/1H atoms, starting from the two-point Green's function. The Dyson equations are generated and solved perturbatively for the energy to order $\alpha^5 m$ (α^3 a.u.). The perturbation theory starts from the DF solution as the zero-order one.

As done for the three-point function in Sect. 26.4, the Dyson equation for radiative corrections involving a single transverse photon is generated by expanding the two-point propagator to all orders in the Coulomb ladder approximation, inserting a transverse virtual photon in all possible ways, and then resumming. The resulting integral equation [22] is quite complicated, containing even the three-point vertex in its inhomogeneous term, as well as a mass counter term to eliminate divergences. It is sufficient for the present purposes to expand these vertices through first order matrix elements of $V_{ee} \equiv V$, but excluding matrix elements of type $\begin{bmatrix} m \\ a \end{bmatrix} |V| \begin{bmatrix} n \\ a \end{bmatrix}$ (where a is the label for core states), since these are already included in the DF approximation (see Eq. (26.27)).

Other radiative corrections are also generated, which include Coulomb photons. Two of these corrections, involving only such photons, have already been referred to in the text between Eqs. (26.21) and (26.23). Finally, to order $\alpha^5 m$, corrections involving two transverse photons must be included.

Systematic application of the pole-shifting process in Eq. (26.23) to the one- and two-transverse-photon and the Coulomb photon-expressions yields two types of terms: first, photon exchange (of one transverse photon or of two-photons of either kind, Coulomb or transverse) between core and valence states; and second, self-energy and vacuum polarization contributions. Among the photon exchange terms, one can identify those contributions included in other approaches (at least for a few special cases, such as those arising from a single transverse photon interaction, and from two Coulomb interactions and Coulomb-Breit interactions with positive energy intermediate electron states). These are electron correlation terms which are included as part of a MBPT calculation [5], or one which involves consideration of an infinite subset of MBPT terms [6]. They need not be re-evaluated. The terms not covered by calculations of the type in Refs. [5,6], which are of $O(m\alpha^5 c^2)$, involve retardation in Coulomb-transverse photon exchange, negative energy intermediate electron states for two Coulomb and Coulomb-transverse photon exchanges, two transverse photon exchange, self-energy and vacuum polarization terms, as well as anomalous magnetic moment corrections.

After lengthy calculation, one obtains final results in finite analytic form which are numerically implementable. The ones given below are obtained after further approximations. First, the remnants of the original integral equation are solved iteratively. Second, the self-energy terms are calculated in a joint expansion [23, 24] in α and αZ, and are thus only valid for low Z in isoelectronic sequences. Finally, there are characteristic logarithmic terms generated by low virtual photon momenta in the self-energy contributions. A standard approach

is to scale these with a factor of $(\alpha Z)^2$ to obtain Bethe log (BL) terms as constants independent of α, together with constant log (CL) terms of the form $\ln(\alpha Z)^2$. The BL terms are independent of Z for hydrogenic ions, and remain nearly so for other atoms. BL and CL terms are associated with both nuclear and ee Coulomb potentials. The ee CL terms arising from core and valence self-energy terms are canceled exactly by those coming from Coulomb and transverse photon exchange. Thus, the numerous and rather complicated corresponding ee BL terms, which also are *individually* small compared to their associated CL terms, should additionally almost cancel, and are therefore neglected.

One finally obtains [with all state labels denoting principal and orbital quantum numbers as well as spin indices: $n \equiv (n, \ell; m_s)$]

$$\Delta E(n, \ell) = Z |\langle 0|n \rangle|^2 F(n, \ell), \qquad (26.67)$$

in a.u., where $\langle 0|n \rangle \equiv \langle 0|n, 0 \rangle$ is the s-state HF wave function[1] at the origin of coordinates, and serves to define an effective (shielded) nuclear charge $Z_{n,\text{eff}}$:

$$\langle \mathbf{r}|n, \ell \rangle \equiv \Psi_{n,\ell}(\mathbf{r}), |\langle 0|n \rangle|^2 \equiv \frac{1}{\pi n^3} (Z_{n,\text{eff}} \, \alpha)^3. \quad (26.68)$$

$F(n, \ell)$ consists of a valence or hole contribution, labeled $F_{\text{v,h}}$, a core term and one due to photon exchange between electrons:

$$F(n, \ell) = F_{\text{v,h}}(n, \ell) + F_{\text{core}}(n, \ell) + F_{\text{ee}}(n, \ell), \quad (26.69)$$

$$F_{\text{v,h}}(n, \ell) = \frac{4}{3} \left[N(Z)\delta_{\ell 0} + L(n\ell) + U_{\text{v,h}}^{\text{so}}(n\ell) \right], \quad (26.70)$$

$$F_{\text{core}}(n, \ell) = \frac{4}{3Z} [N(Z)\rho(n\ell) + L_{\text{core}}(n\ell) + U(n\ell) + U_{\text{core}}^{\text{so}}(n\ell)], \quad (26.71)$$

$$F_{\text{ee}}(n, \ell) = \frac{4}{3Z} \left[E(Z)K^C(n\ell) + K^L(n\ell) \right], \quad (26.72)$$

with $U_{\text{v,h}}^{\text{so}}(n\ell)$, $U_{\text{core}}^{\text{so}}(n\ell)$ being the spin-orbit terms, and where

$$N(Z) = \ln \frac{1}{(Z\alpha)^2} + \frac{19}{30} + Z\alpha C_5, \qquad (26.73)$$

$$C_5 = 3\pi \left(1 + \frac{11}{128} - \frac{1}{2} \ln 2 \right), \qquad (26.74)$$

$$E(Z) = -\frac{7}{2} \ln \frac{1}{Z\alpha} + \frac{59}{20} - \frac{9}{8}\pi, \qquad (26.75)$$

$$L(n\ell) = \frac{i/4\pi}{|\langle 0|n \rangle|^2} \left\langle n, \ell \left| \left[\mathbf{p} \cdot \mathcal{L}_{\text{B}}(n\ell) \frac{\mathbf{r}}{r^3} - \frac{\mathbf{r}}{r^3} \mathcal{L}_{\text{B}}(n\ell) \cdot \mathbf{p} \right] \right| n, \ell \right\rangle, \quad (26.76)$$

$$\mathcal{L}_{\text{B}}(n\ell) \equiv \ln \frac{Z^2}{2|e_n - H|}, \qquad (26.77)$$

---

[1] Just as in hydrogen, for which the final expressions for the radiative corrections require NR and not R wave functions and energies, so in this case we use HF (NR) and not DF (R) quantities.

H is the HF Hamiltonian, and $e_n \equiv e(n\ell)$ is the HF energy, and

$$U^{\rm so}_{\rm v,h}(n\ell) = \frac{3/16\pi}{|\langle \mathbf{0}|n\rangle|^2} \left\langle n \left| \frac{1}{r^3} \right| n \right\rangle C_\ell \tag{26.78}$$

where

$$C_\ell = \begin{cases} \ell & j = \ell + \frac{1}{2} \\ -(\ell+1) & j = \ell - \frac{1}{2} \end{cases}, \tag{26.79}$$

$$\rho(n\ell) = \frac{i/4\pi}{|\langle \mathbf{0}|n\rangle|^2} \sum_a \sum_{jk} \left\{ \frac{(e_k - e_j)(e_a - 2e_k + e_j)}{e_a - e_j} \left[{}^n_a \right| V \left| {}^n_j \right] \langle j|\mathbf{r}|k\rangle \cdot \langle k|\mathbf{p}|a\rangle + {\rm c.c.} \right\}, \tag{26.80}$$

$$L_{\rm core}(n\ell) = \frac{i/4\pi}{|\langle \mathbf{0}|n\rangle|^2} \sum_a \sum_{jk} \left\{ \frac{(e_k - e_j)(e_a - 2e_k + e_j)}{e_a - e_j} \left[{}^n_a \right| V \left| {}^n_j \right] \ln \frac{Z^2}{2|e_a - e_k|} \langle j|\mathbf{r}|k\rangle \cdot \langle k|\mathbf{p}|a\rangle + {\rm c.c.} \right\}, \tag{26.81}$$

$$U_{(1)}(n\ell) = \frac{1/2\pi}{|\langle \mathbf{0}|n\rangle|^2} \sum_a \left\{ \langle a \left| \mathbf{p}^2 \right| a \rangle \left[{}^n_a \right| V \left| {}^n_a \right] - \left[{}^{n}_{(a)'} \right| V \left| {}^{n}_{(a)'} \right] - \left(\left[{}^{(b)'}_a \right| V \left| {}^{a}_{(b)'} \right] + {\rm c.c.} \right) \delta_{nb} \right\}, \tag{26.82}$$

$$U_{(2)}(n\ell) = \frac{i/2\pi}{|\langle \mathbf{0}|n\rangle|^2} \sum_a \sum_{k<q} \left\{ \frac{(2e_a - e_k - e_q)}{e_q - e_k} \left[{}^n_q \right| V \left| {}^n_k \right] \left(\ln \left| \frac{e_a - e_q}{e_a - e_k} \right| + 2 \right) \langle a|\mathbf{p}|q\rangle \cdot \langle k|\mathbf{p}|a\rangle + {\rm c.c.} \right\}, \tag{26.83}$$

$$U^{\rm so}_{\rm core}(n\ell) = \frac{3/16\pi}{|\langle \mathbf{0}|n\rangle|^2} \sum_{aj} \left\{ \frac{\delta_{\ell_a \ell_j}}{e_a - e_j} \left[{}^n_a \right| V \left| {}^n_j \right] \left\langle j \left| \frac{1}{r^3} \right| a \right\rangle C_\ell + {\rm c.c.} \right\}, \tag{26.84}$$

$$K^C(n\ell) = \frac{1}{|\langle \mathbf{0}|n\rangle|^2} \sum_a \left[{}^n_a \left| \delta^3(\mathbf{x} - \mathbf{y}) \right| {}^n_a \right], \tag{26.85}$$

$$K^L(n\ell) = \frac{7/2}{|\langle \mathbf{0}|n\rangle|^2|} \sum_a \left[{}^n_a \left| \frac{1}{4\pi} \nabla^2 \left(\frac{\ell n(Z|\mathbf{x} - \mathbf{y}|) + \gamma}{|\mathbf{x} - \mathbf{y}|} \right) \right| {}^n_a \right], \tag{26.86}$$

where

$$U(n\ell) \equiv U_{(1)}(n\ell) + U_{(2)}(n\ell), \tag{26.87}$$

$$\langle \mathbf{r}|(a)'\rangle \equiv \frac{d}{dr} \langle \mathbf{r}|a\rangle, \tag{26.88}$$

and γ is the Euler-Mascheroni constant. As in Eq. (26.27), x refers to the top row and y to the bottom row in the two-row expression in Eq. (26.86).

$L(n\ell)$, which is associated with the valence electron and the nuclear Coulomb potential, has the form of a hydrogenic BL, except that it is calculated with HF wave functions and energies. $L_{\rm core}(n\ell)$ comes from BL terms associated with core electrons.

The expressions appearing in Eqs. (26.80), (26.81) and (26.83) originate from a double commutator, $[\mathbf{p} \cdot [\mathbf{p}, H]]$, which is then approximated by using the commutator $[\mathbf{r}, H] \approx i\mathbf{p}$. This is not an exact result because the ee exchange term in the HF potential is neglected. One thus obtains [13]

$$\langle m \left| [\mathbf{p} \cdot [\mathbf{p}, H]] \right| n \rangle \approx i(e_n - 2e_k + e_m)(e_k - e_n) \langle m \left| \mathbf{p} \right| k \rangle \cdot \langle k \left| \mathbf{r} \right| n \rangle. \tag{26.89}$$

A less accurate approximation is [12]

$$4\pi Z \delta^3(\mathbf{r}) \approx [\mathbf{p} \cdot [\mathbf{p}, H]]. \tag{26.90}$$

Using the left-hand side of this equation instead of the approximation (26.89) to the right-hand side would lead to only s-state a, j and p-state k contributions in Eq. (26.80) and alternative forms of Eqs. (26.81) and (26.83).

The major part of the contribution to the energy due to radiative corrections comes from $F_{\rm v,h}(n,0)$ (s-states), and numerical tests indicate that the principal effect in this term comes from the renormalization of the electron density at the coordinate origin due to electron shielding. This reduction occurs because the shielded wave function is more spread out than the unshielded one. Thus, one can extend the above results semi-empirically to high Z (for which $F_{\rm core}(n,\ell)$ and $F_{\rm ee}(n,\ell)$ play a much smaller role because of their $1/Z$ dependence) by using the hydrogenic results

[25] for high Z, i.e., replacing Z^4 for hydrogen by $Z(Z_{n,\text{eff}})^3$ for 1P/1H atoms [where $Z_{n,\text{eff}}$ is the shielded nuclear charge number, defined in Eq. (26.68), and Z is the unshielded nuclear charge number]. The results obtained in this way are competitive with other evaluations of ΔE_n [26].

Finally, there is a correction to $F(n, \ell)$ for smaller Z when the integral equations for the self-energy and vacuum polarization contributions are solved more accurately than by iteration. These corrections play little role in s-state energies, but are somewhat more important in p-states for the hard core case (closed shells, $N = 2, 10$, etc.). They are expected to provide a much more significant contribution in the soft core case (closed *sub*shells, $N = 4, 12$, etc.).

The correction is based on the simpler approximation given in Eq. (26.90) of the more accurate Eq. (26.89). It is given by

$$\delta F(n, \ell) = \frac{1}{Z\langle \mathbf{0}|n\rangle^2} \sum_a \sum_j \left\{ \frac{\left[\begin{smallmatrix} n \\ a \end{smallmatrix}\middle| V \middle|\begin{smallmatrix} n \\ j \end{smallmatrix}\right]}{e_a - e_j} [\langle j|\mathcal{E}|a\rangle \right.$$
$$\left. - Z\mathcal{F}_{\text{v,h}}^{(0)} \langle j|\mathbf{0}\rangle \langle \mathbf{0}|a\rangle] + \text{c.c.} \right\}, \qquad (26.91)$$

where

$$\langle j|\mathcal{E}|a\rangle = Z\mathcal{F}_{\text{v,h}}^{(0)} \langle j|\mathbf{0}\rangle \langle \mathbf{0}|a\rangle$$
$$+ \sum_b \sum_k \left(\frac{\left[\begin{smallmatrix} j \\ b \end{smallmatrix}\middle| V \middle|\begin{smallmatrix} a \\ k \end{smallmatrix}\right]}{e_b - e_k} [\langle k|\mathcal{E}|b\rangle + \text{c.c.}] \right), \quad (26.92)$$

$$\mathcal{F}_{\text{v,h}}^{(o)} = \frac{4}{3}\{N(Z) + L\}. \qquad (26.93)$$

The inhomogeneous term in Eq. (26.92) appears only for states $|j\rangle$ and $|a\rangle$ which are s-states. L in Eq. (26.93) is taken to be a constant, an approximation sufficient for the desired accuracy, and reflects the fact that $L(n, 0)$ for s-states is essentially constant as a function of radial quantum number, and is approximately the same for hydrogen and the HF approximation.

In order to obtain the correction $\delta F(n, \ell)$ of Eq. (26.91), it is necessary first to solve the coupled inhomogeneous linear equations, Eq. (26.92). While the sum $\sum_a$ over core states is always over a finite number of discrete states, the symbol $\sum$ denotes an infinite sum over discrete bound valence states and an integral over the continuum of such states. Indeed, expressions of this type occur throughout the GFA. They can be dealt with by the use of finite basis techniques, for example the B-spline approach [27, 28].

REFERENCES

1. A. L. Fetter and J. D. Walecka, *Quantum Theory of Many Particle Systems* (McGraw-Hill, New York, 1971).
2. Gy. Csanak, H. S. Taylor, and R. Yaris, Adv. Atom. Mol. Phys. **7**, 287 (1971).
3. G. Feldman and T. Fulton, Ann. Phys. (N. Y.) **172**, 40 (1986).
4. G. Feldman and T. Fulton, Ann. Phys. (N. Y.) **179**, 20 (1987).
5. W. R. Johnson, S. A. Blundell, and J. Sapirstein, Phys. Rev. A **37**, 2764 (1988).
6. W. R. Johnson, S. A. Blundell, Z. W. Liu, and J. Sapirstein, Phys. Rev. A **40**, 2233 (1989).
7. G. Feldman and T. Fulton, Ann. Phys. (N. Y.) **152**, 376 (1984).
8. S.-S. Liaw, G. Feldman, and T. Fulton, Phys. Rev. A **38**, 5985 (1988).
9. S.-S. Liaw, Phys. Rev. A **47**, 1726 (1993).
10. S.-S. Liaw and F.-Y. Chiou, Phys. Rev. A **49**, 2435 (1994).
11. G. Feldman and T. Fulton, Ann. Phys. (N. Y.) **201**, 193 (1990).
12. G. Feldman, T. Fulton, and J. Ingham, Ann. Phys. (N. Y.) **219**, 1 (1992).
13. A. Devoto, G. Feldman, and T. Fulton, Ann. Phys. (N. Y.) **232**, 88 (1994).
14. T. Fulton and W. R. Johnson, Phys. Rev. **A34**, 1686 (1986).
15. T. H. Koopmans, Physica **1**, 104 (1933).
16. L. Hostler, J. Math. Phys. **5**, 1234 (1964).
17. J. Schwinger, J. Math. Phys. **5**, 1606 (1964).
18. R. A. Swainson and G. W. F. Drake, J. Phys. A: Math. Gen. **24**, 95, 1801 (1991).
19. S.-S. Liaw, Phys. Rev. A **48**, 3555 (1993).
20. E. E. Salpeter, Phys. Rev. **87**, 328 (1952).
21. Ref. [10], Eq. (15).
22. Ref. [11], Fig. 6.
23. J. B. French and V. F. Weisskopf, Phys. Rev. **75**, 1240 (1949).
24. G. W. Erickson and D. R. Yennie, Ann. Phys. (N. Y.) **35**, 271, 447 (1965).
25. W. R. Johnson and G. Soff, At. Data Nucl. Data Tables **33**, 405 (1985).
26. K. T. Cheng, W. R. Johnson, and J. Sapirstein, Phys. Rev. Lett. **66**, 2960 (1991).
27. C. de Boor, *A Practical Guide to Splines* (Springer, N. Y., 1987).
28. W. R. Johnson and J. Sapirstein, Phys. Rev. Lett. **57**, 1126 (1986).

27

Quantum Electrodynamics

J. Sapirstein

Department of Physics, University of Notre Dame, Notre Dame, Indiana 46556

27.1 INTRODUCTION

Quantum Electrodynamics (QED) is the underlying theory of atomic and molecular physics. Despite this generality, it is not necessary to use the full theory in most atomic physics problems. This is because in the nonrelativistic limit, QED reduces to the Schrödinger equation, and the extra physics in QED in general is quite small, being suppressed by powers of the fine structure constant α. Given the difficulty of solving the Schrödinger equation with high accuracy in most atomic physics situations, these small corrections can usually be neglected. The theory is however needed to explain small deviations from the solution to the Schrödinger equation in simple systems, in particular a single electron in a constant magnetic field, one-electron atoms, and helium. Larger deviations occur for highly charged ions, and also for high-energy scattering of electrons and photons. We note that a rather extensive review of QED is available [1], and refer the reader interested in more details to that

work. In addition, comparison with experiment is made by P. J. Mohr in Chap. 28, and thus is not done here.

27.2 COVARIANT PERTURBATION THEORY

QED combines relativity, electromagnetism, and quantum mechanics. As the first two theories were well understood when quantum mechanics was developed, the development of the the fundamental equations of QED (after Dirac's introduction of his relativistic equation for the electron) took place rapidly, being in place in 1928 [2]. However, it was recognized almost immediately that when higher order perturbation theory was considered, infinities associated with short wavelengths, known as ultraviolet divergences, were present, and that this apparently predicted infinite shifts in spectral lines. These difficulties were not overcome for two decades, but at that time improvements in calculational technology coupled with an understanding that the infinities could be grouped into renormalizations of the electron's mass, charge, and wave function and the photon's wave function, led to the modern form of QED. A central object in this form of the theory is the S-matrix. To introduce it, we start with the Schrödinger equation,

$$i\hbar\frac{\partial}{\partial t}\Psi(t) = (H_0 + H_I)\Psi(t)\,, \qquad (27.1)$$

where H_0 is the Hamiltonian of free electrons and photons (although this can be easily generalized to include external potentials such as a nuclear Coulomb field), and H_I the electromagnetic interaction between them. These Hamiltonians follow from the Lagrangian density $\mathcal{L} = \mathcal{L}_0 + \mathcal{L}_1$, where

$$\begin{aligned}\mathcal{L}_0 = \bar{\psi}_0(\mathbf{x},t)(\gamma_\mu p^\mu - m_0)\psi_0(\mathbf{x},t) \\ - \tfrac{1}{4}F_{0\mu\nu}(\mathbf{x},t)F_0^{\mu\nu}(\mathbf{x},t)\end{aligned} \qquad (27.2)$$

and

$$\mathcal{L}_I = -e_0\bar{\psi}_0(\mathbf{x},t)\gamma_\mu\psi_0(\mathbf{x},t)A_0^\mu(\mathbf{x},t)\,. \qquad (27.3)$$

The 0 subscripts in the above emphasize that the fields and couplings are unrenormalized: renormalization will be discussed in the next section. In addition $\bar{\psi}p^\mu\psi$ is understood to represent $\frac{1}{2}i\bar{\psi}(\partial^\mu\psi) - \frac{1}{2}i(\partial^\mu\bar{\psi})\psi$, and gauge fixing terms have been suppressed. By making the unitary transformation

$$\Psi(t) \equiv e^{-iH_0t/\hbar}\Phi(t)\,, \qquad (27.4)$$

which transforms from the Schrödinger to the interaction representation, and further defining the U matrix through

$$\Phi(t) = U(t,-\infty)\Phi(-\infty)\,, \qquad (27.5)$$

one finds an equation for this matrix

$$i\hbar\frac{\partial}{\partial t}U(t,-\infty) = \hat{\mathbf{H}}_I(t)U(t,-\infty)\,, \qquad (27.6)$$

where

$$\hat{\mathbf{H}}_I(t) = e^{iH_0t/\hbar}H_Ie^{-iH_0t/\hbar}\,. \qquad (27.7)$$

Solving this equation iteratively then gives for the S-matrix, defined as $S = U(\infty, -\infty)$,

$$\begin{aligned}S = \sum_{n=0}^{\infty}(-i)^n\frac{1}{n!}\int_{-\infty}^{\infty}dx_1\cdots\int_{-\infty}^{\infty}dx_n \\ \times T[\hat{\mathbf{H}}_1(x_1),\ldots\hat{\mathbf{H}}_1(x_n)]\,,\end{aligned} \qquad (27.8)$$

where T is the time ordering operator. An initial state consisting of free electrons, positrons, and photons will then have an amplitude to scatter into a final state with different momenta and perhaps different numbers of particles given by the S-matrix. This amplitude can be calculated using Wick's theorem, and the result conveniently represented by Feynman diagrams. Lowest order results of this procedure (tree approximation) describe processes such as electron scattering, electron-positron annihilation, etc. to fairly high precision. However, as mentioned above, when higher terms in the perturbation expansion are considered, diagrams containing closed loops are encountered that are formally infinite, and a renormalization program must be introduced.

27.3 RENORMALIZATION THEORY AND GAUGE CHOICES

Before we discuss renormalization theory, we mention that QED has the same freedom to choose gauge as classical electromagnetism. We will discuss four gauges that have been used in QED calculations, though there is of course an arbitrary number. All of these gauges can be defined through the photon propagator in momentum space. If this is defined by

$$\int d^4x e^{-ik\cdot x}\langle 0|T(A_\mu(x)A_\nu(0))|0\rangle \equiv -i\frac{G_{\mu\nu}(k)}{k^2}\,, \quad (27.9)$$

the Coulomb gauge is defined by

$$\begin{aligned}G_{00} &= -\frac{k^2}{\mathbf{k}^2}\,, \\ G_{ij} &= -\left(\delta_{ij} - \frac{k_ik_j}{\mathbf{k}^2}\right)\,, \qquad (27.10) \\ G_{i0} &= G_{0i} = 0\,.\end{aligned}$$

While this gauge is particularly physical, with the G_{00} part directly corresponding to the instantaneous Coulomb interaction and the G_{ij} to magnetic interactions, it is relatively difficult to work with. For ease of calculation, the covariant gauges, defined by

$$G_{\mu\nu} = g_{\mu\nu} + \beta\frac{k_\mu k_\nu}{k^2} \qquad (27.11)$$

are useful, particularly the case $\beta = 0$, the Feynman gauge. Other values of β are $\beta = 2$, the Yennie gauge [3], and $\beta = -1$, the Landau gauge. The former has the advantage of controlling infrared divergences, and the latter of controlling ultraviolet divergences.

Two of these infinities are first encountered when the self-energy of a free electron is calculated in second order perturbation theory. In order to deal with finite quantities, we first must introduce a device to regulate the high-frequency range of the integrations. This can be done in a number of ways, among them Pauli-Villars regularization [4]. In this method one modifies the photon propagator to

Figure 27.1. Ultraviolet divergent one-loop Feynman diagrams.

$$1/q^2 \to 1/(q^2 - \lambda^2) - 1/(q^2 - \Lambda^2)\,, \qquad (27.12)$$

where λ is a photon mass that regulates infrared divergences and Λ an ultraviolet cutoff mass. In this case the self-energy operator is represented by the Feynman diagram of Fig. 27.1a, and is, using Feynman gauge,

$$\Sigma(p) = -ie^2 \int \frac{d^4k}{(2\pi)^4} \gamma_\mu \frac{1}{\slashed{p} - \slashed{k} - m_0} \gamma^\mu$$
$$\times \left(\frac{1}{k^2 - \lambda^2} - \frac{1}{k^2 - \Lambda^2} \right)\,. \qquad (27.13)$$

Combining the two denominators together with a Feynman parameter and carrying out the integration over k then leads to

$$\Sigma(p) = \delta m^{(2)} + B^{(2)}(\slashed{p} - m_0) + \Sigma_F(p)\,, \qquad (27.14)$$

where

$$\delta m^{(2)} = \frac{3\alpha m_0}{2\pi} \left[\ln(\Lambda/m_0) + 1/4 \right] \qquad (27.15)$$

and

$$B^{(2)} = -\frac{\alpha}{2\pi} [\ln(\Lambda/m_0) + 2\ln(\lambda/m_0) + 9/4]\,. \qquad (27.16)$$

Σ_F will not be given here, but the important point is that it is ultraviolet finite. Thus the ultraviolet infinities of QED are isolated in the first two terms, which are of a very simple structure.

The next infinity is connected with the vertex function of Fig. 27.1b. This is defined by the equation

$$\Gamma_\mu = -ie_0^3 \int \frac{d^4k}{(2\pi)^4} \gamma_\rho \frac{1}{\slashed{p} - \slashed{k} - m} \gamma_\mu$$
$$\times \frac{1}{\slashed{p}' - \slashed{k} - m} \gamma^\rho \left(\frac{1}{k^2 - \lambda^2} - \frac{1}{k^2 - \Lambda^2} \right)\,. \quad (27.17)$$

While Γ_μ is a fairly complicated function, its ultraviolet divergent part is simply a multiple of γ_μ. When the electron momenta p and p' are equal and on shell, this integral can be evaluated to be

$$\Gamma_\mu = e_0 \gamma_\mu L^{(2)} \qquad (27.18)$$

where

$$L^{(2)} = \frac{\alpha}{2\pi} [\ln(\Lambda/m) + 2\ln(\lambda/m) + 9/4]\,. \qquad (27.19)$$

The fact that $L^{(2)} = -B^{(2)}$ is a consequence of the Ward identity [5].

The final infinity of second-order QED arises from the vacuum polarization diagram of Fig. 27.1c. The associated integral is

$$\Pi_{\mu\nu}(p) = -ie^2 \int \frac{d^4k}{(2\pi)^4} \mathrm{Tr} \left[\gamma_\mu \frac{1}{\slashed{p} + \slashed{k} - m_0} \gamma_\nu \frac{1}{\slashed{k} - m_0} \right]\,. \qquad (27.20)$$

In this case the integral cannot be regulated as before, since there is no photon propagator to modify. Instead one can subtract a similar integral with the electron mass replaced by M. Vacuum polarization is particularly sensitive to ultraviolet divergences, since the nominal order of the divergence is quadratic. However, gauge invariance requires $\Pi_{\mu\nu}(p)$ to have the structure $(p^2 g_{\mu\nu} - p_\mu p_\nu)\Pi(p^2)$, and if one considers only the $p_\mu p_\nu$ part of the vacuum polarization integral, the ultraviolet divergence is only logarithmic. This divergence is independent of the photon momentum p, and one can write

$$\Pi(p^2) = C^{(2)} + \Pi_{\mathrm{finite}}(p^2)\,, \qquad (27.21)$$

where

$$C^{(2)} = \frac{\alpha}{3\pi} \ln(M^2/m_0^2) \qquad (27.22)$$

and

$$\Pi_{\mathrm{finite}}(p^2) = -\frac{2\alpha}{\pi} \int_0^1 dx\, x(1-x) \ln[1 - x(1-x)p^2/m_0^2]\,. \qquad (27.23)$$

The four infinite quantities encountered in second order perturbation theory are modified by higher order corrections, but no new divergent structures arise. The basic idea of renormalization is to note that these structures are already present in the lowest order Lagrangian. We now make the following definitions:

$$m = m_0 + \delta m\,, \qquad (27.24)$$

$$\psi(x) = Z_2^{-1/2} \psi_0(x)\,, \qquad (27.25)$$

$$A^\mu(x) = Z_3^{-1/2} A_0^\mu(x)\,, \qquad (27.26)$$

and

$$e = e_0 Z_1^{-1} Z_2 Z_3^{1/2}\,. \qquad (27.27)$$

These correspond to an additive renormalization of the electron mass and multiplicative renormalizations of the electron and photon wave functions and the electron charge. Rewriting the original bare Lagrangian in terms of these renormalized quantities then gives that Lagrangian without the 0 subscripts, plus the following counterterms:

$$\mathcal{L}_{CT1} = Z_2 \delta m \bar{\psi}(\mathbf{x}, t) \psi(\mathbf{x}, t), \tag{27.28}$$

$$\mathcal{L}_{CT2} = -e(Z_1 - 1)\bar{\psi}(\mathbf{x}, t)\gamma_\mu \psi(\mathbf{x}, t) A^\mu(\mathbf{x}, t), \tag{27.29}$$

$$\mathcal{L}_{CT3} = -\tfrac{1}{4}(Z_3 - 1)F_{\mu\nu}(\mathbf{x}, t)F^{\mu\nu}(\mathbf{x}, t), \tag{27.30}$$

and

$$\mathcal{L}_{CT4} = (Z_2 - 1)\bar{\psi}(\mathbf{x}, t)(\gamma_\mu p^\mu - m)\psi(\mathbf{x}, t). \tag{27.31}$$

By choosing $Z_2 = 1 + B^{(2)}$, $Z_1 = 1 - L^{(2)}$, $\delta m = \delta m^{(2)}$, and $Z_3 = 1 - C^{(2)}$, these counterterms will precisely cancel the previously encountered divergences in second order. At this point we identify m and e as the experimentally determined mass and charge of the electron: as long as these are used, the radiative corrections discussed above have no effect for free electrons. However, when an electron undergoes scattering or is in the presence of an external magnetic or nuclear Coulomb field, the finite terms no longer vanish, and give rise to small corrections. We now turn to a discussion of these corrections.

27.4 TESTS OF QED IN LEPTON SCATTERING

The highest energy tests of QED come from scattering experiments at accelerators. While the dominant part of QED corrections for all the other tests discussed in this chapter involves electron and photon propagators close to the mass shell, scattering experiments involve propagators very far off the mass shell, which allows tests of the theory at very small distances. It is standard to parameterize possible deviations from the predictions of QED at these small distances by the introduction of form factors of the form

$$F(q^2) = 1 - \frac{q^2}{q^2 - \Lambda^2}, \tag{27.32}$$

where q is photon momentum at an electron-photon vertex. In QED Λ is infinite and this form factor is unity even at very high q^2, but this can be tested in various scattering experiments. For example, Bhabha scattering, $e^+e^- \to e^+e^-$, has been accurately measured at high center of mass energy, $\sqrt{s} = 34.8$ GeV, at TASSO [6]. To compare with QED, very sizable radiative corrections must be carefully calculated, and at these energies electroweak effects involving the Z boson, while small, must also be considered. Although the accuracy of the experiments is not high compared with atomic physics

measurements, being at the percent level, the good agreement with QED that is found allows lower limits on the cutoff $\Lambda > 500$ GeV to be placed. This corresponds to distances of under 10^{-16} cm. It is of interest to compare this sensitivity with that available from atomic physics tests. The change in the photon propagator given above corresponds to a potential $e^{-\Lambda r}/r$. This would lead to an energy shift of a $2s$ electron in hydrogen of $46/\Lambda^2$ kHz with Λ in units of GeV. Thus even 1 kHz accuracy in the Lamb shift would only restrict $\Lambda > 7$ GeV.

27.5 ELECTRON AND MUON G FACTORS

One of the successes of the Dirac equation is the prediction $g = 2$ for the electron. The leading correction to this result coming from QED is the Schwinger correction [7],

$$g = 2\left(1 + \frac{\alpha}{2\pi}\right). \tag{27.33}$$

While in principle this is an external field problem, because of the weakness of laboratory magnetic fields, the correction can be related to Feynman diagrams with free propagators. To see the weakness we note that $eB/m_e^2 = 2.3 \times 10^{-14} B(\text{gauss})$. In extremely intense magnetic fields such as can be encountered in astrophysical situations, a bound state approach [8] should be used both for calculating energy shifts and the imaginary part of these shifts, which describe synchrotron radiation. An example of the more precise approach is Demeur's formula for the (real) energy shift of an electron in the lowest energy level

$$\Delta E = \frac{m\alpha}{2\pi}\left\{ -\frac{eB}{2m^2} + \left(\frac{eB}{m^2}\right)^2\left(\frac{4}{3}\ln\frac{m^2}{2eB} - \frac{13}{18}\right) \right.$$
$$\left. + \left(\frac{eB}{m^2}\right)^3\left(\frac{14}{3}\ln\frac{m^2}{2eB} - \frac{32}{5}\ln 2 + \frac{83}{90}\right) + \cdots \right\}. \tag{27.34}$$

The second term could actually be seen at the present level of precision, but is spin-independent, and the third term is negligible. An interesting feature of the experiment is the effect of the conducting cavity, which must be understood to extract the correct value of $g - 2$ [9].

After the initial verification of the Schwinger correction, experiments of increasing precision culminating in the Penning trap measurements in Washington [10] have stimulated advances in theoretical calculations. These involve the evaluation of constants C_i defined by

$$a_e = C_1\frac{\alpha}{\pi} + C_2\left(\frac{\alpha}{\pi}\right)^2 + C_3\left(\frac{\alpha}{\pi}\right)^3 + C_4\left(\frac{\alpha}{\pi}\right)^4 + \cdots, \tag{27.35}$$

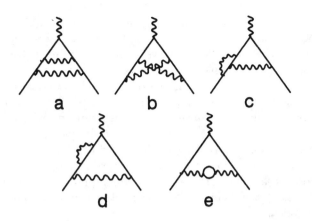

Figure 27.2. Two-loop Feynman diagrams contributing to g-2.

Table 27.1. Contributions to C_2 in Yennie gauge.

Graph	Value
a	$-\frac{1}{2}B^{(2)} - \frac{3}{16}$
b	$\frac{5}{4}\zeta(3) - \frac{5}{6}\pi^2 \ln 2 + \frac{5}{12}\pi^2 + \frac{7}{12}$
c	$-B^{(2)} - \frac{1}{2}\zeta(3) + \frac{1}{3}\pi^2 \ln 2 - \frac{29}{24}$
d	$B^{(2)} - \frac{9}{8}$
e	$-\frac{1}{3}\pi^2 + \frac{119}{36}$
counterterm	$\frac{1}{2}B^{(2)}$

where $a_{\mathrm{e}} = (g-2)/2$ is the anomalous magnetic moment of the electron. The computational effort involved in computing the coefficients C_i increases very rapidly with i, and the four loop calculation is the largest QED calculation ever carried out. The situation with regard to these calculations is as follows. After the calculation of the Schwinger correction, the next step was the evaluation of the seven Feynman diagrams of Fig. 27.2. A feature of the one-loop calculation, that it is ultraviolet finite, is no longer present at this level, and renormalizations of the self-mass, vertex, and wave function must be performed, although the latter two cancel by Ward's identity. When the calculation is carried out in Feynman gauge, each graph has an infrared divergence that must be regulated in some fashion, for example by giving the photon a small mass λ. This calculation was first correctly carried out by Sommerfield [11] and Petermann [12]. The result is

$$C_2 = \frac{197}{144} + \frac{\pi^2}{6}(0.5 - 3\ln 2) + 0.75\zeta(3),\qquad (27.36)$$

where $\zeta(3) = 1.20205\cdots$ is the Riemann zeta function of argument 3. While each individual diagram is infrared divergent in Feynman gauge, Adkins [13] has recently used Yennie gauge [3] to recalculate the effect, free of infrared divergent terms. His results are given in Table 27.1.

The vacuum polarization graph of Fig. 27.2e plays an interesting role. While the result for C_2 given above includes only the case where the intermediate particles in the loop are electrons, that loop can also involve any charged particle, such as the muon, pion, or tau. However, because all these particles are much heavier than the electron, which sets the energy scale of the Feynman integral, their effect is suppressed by the square of the mass ratio of the electron to their mass, and are thus quite small. Specifically, in units of 10^{-12}, the muon

loop contributes 2.80, the tau loop 0.01, and hadrons 1.6(2). These act to increase C_2 by 0.000 000 82. We note in passing that the effect of the weak interactions enters in one loop, and contributes 0.05 10^{-12} to the electron $g-2$ value and 195(10) 10^{-11} for the muon.

When the anomalous magnetic moment of the muon is considered the situation changes significantly. Firstly, the contribution of electrons in the loop is enhanced, since they are now relatively light particles, and the vacuum polarization loop behaves as a logarithm of the mass ratio. Specifically, the difference between the muon and electron $g-2$ factors behaves as

$$a_\mu - a_{\mathrm{e}} = 1.094\left(\frac{\alpha}{\pi}\right)^2 + 22.9\left(\frac{\alpha}{\pi}\right)^3 + 127.0\left(\frac{\alpha}{\pi}\right)^4 + \cdots.$$
$$(27.37)$$

These large coefficients arise primarily because the vacuum polarization loops involving electrons change the effective value of α to $\alpha/[1 - (2\alpha/3\pi)\ln m_\mu/m_{\mathrm{e}}]$. Note however that such logarithmic terms also arise from other sources, most notably the light-by-light scattering graphs that enter first in third order.

The second important change in the muon case is the significant role of strongly interacting particles in the loop. Fortunately, while our present inability to carry out high accuracy calculations of the strong interactions could in principle interfere with the interpretation of the muon $g-2$ as a QED test, the bulk of this contribution can be related to the experimentally available cross-section for e^+e^- annihilation into hadrons [14]. Nevertheless, the experimental uncertainties still are the dominant ones for muon $g-2$.

The calculation of C_3 involves 72 Feynman graphs, although they can be grouped together into a smaller number of gauge invariant sets. There has been considerable progress in the analytic evaluation of these graphs [15], and at present only a few have to be calculated numerically. The evaluation of such high-order graphs requires an intricate set of subtractions to lead to finite answers, and provides a practical demonstration of the renormalizability of QED. The present result [16] for C_3 is

$$C_3 = 1.176\ 11(42).\qquad (27.38)$$

Finally, the very large scale calculation of C_4, which is almost completely numerical, has been carried out by Kinoshita and Lindquist [17]. When their result,

$$C_4 = -1.434(138), (27.39)$$

is compared with experiment, agreement is found, but the largest source of error is the uncertainty in the fine structure constant as determined from solid state physics. However, if one instead assumes the validity of QED, the situation can be turned around to determine a QED value of the fine structure constant. This is done by combining the experimental result [10]

$$a_{e^-} = 1\,159\,652\,188.4(4.3) \times 10^{-12}, (27.40)$$
$$a_{e^+} = 1\,159\,652\,187.9(4.3) \times 10^{-12},$$

with the previous formulas for the C coefficients (including the small vacuum polarization corrections for C_2 discussed above). The result is

$$\alpha_{\text{QED}}^{-1} = 137.035\,992\,22(51)(63)(48), (27.41)$$

where the errors come from experiment, C_3, and C_4 respectively. It would be of great interest to have another QED determination of the fine structure constant of comparable accuracy: the most promising system for this purpose appears to be the fine structure of helium, as will be discussed below.

27.6 RECOIL CORRECTIONS

Because of the smallness of the ratio of the electron mass to most nuclear masses, a reasonable approximation to atoms is to treat the nucleus as a source of a classical Coulomb and magnetic dipole field, which corresponds to the use of Furry representation [18]. In addition, the leading correction to this approximation can be accounted for by using the reduced mass in place of the electron mass. However, to calculate higher order terms consistently, one must treat a one-electron atom as a two-body system and an N-electron atom as a $N+1$-body system. Shortly after the development of modern QED, a number of workers [19–21] developed two-body equations by considering the Green's function for electron-nucleus scattering. Considered as a function of the total energy in the c.m. frame, this Green's function has poles at bound state energies. In practice an approximate Green's function is considered that has poles at either the Schrödinger or the Dirac energies, with reduced mass built in to some degree. Then a perturbation theory is set up that allows corrections of higher order in α and m_e/m_N to be calculated in a systematic way. Here m_N is the mass of the positive particle in the atom, a proton, nucleus, muon, or positron. In the latter case of course corrections of order m_e/m_N, known as recoil

corrections, have no meaning. A distinguishing feature of this approach to the bound state problem is its flexibility: there is no one 'correct' starting equation, and in fact there are many different ways to set up the formalism [22, 23]. For example, as mentioned above one can build in either the physics of the Schrödinger or Dirac equation in the lowest order approximation, with the former approach giving an incorrect fine structure in order α^2 a.u. However, upon inclusion of first order perturbation theory, the fine structure to this order will be correctly included. The importance of the starting formalism is more its ease of use, since an awkward approach can make the calculation of higher-order corrections unmanageably complicated.

To introduce the Bethe-Salpeter equation, we consider the truncated four-point Green's function in the c.m. frame, which satisfies

$$G_T(P, p, p') = K(P, p, p')$$
$$+ \int \frac{d^4k}{(2\pi)^4} K(P, p, k) S(P, k) G_T(P, k, p'),$$
$$(27.42)$$

where $P = (E, \mathbf{0})$, and G has poles when E is at a bound state energy. For simplicity this equation can also be written formally as $G_T = K + KSG_T$. K is known as the irreducible kernel, where irreducible means that K cannot be separated into two parts by cutting two intermediate fermion propagators. It is dominated by one-photon exchange. The function S represents the propagation of the electron and nucleus as free particles, and is given by (in the following we label the electron as particle 1, $m_1 = m_e$, the nucleus as particle 2, $m_2 = m_N$, and define $M = m_1 + m_2$)

$$S(P, k) = \left[\frac{i}{\not{P}_1 + \not{k} - m_1} \right]_1 \left[\frac{i}{\not{P}_2 - \not{k} - m_2} \right]_2, (27.43)$$

where $P_1 = (\tau_1 E, \mathbf{0})$ and $P_2 = (\tau_2 E, \mathbf{0})$, with $\tau_1 = m_1/M$ and $\tau_2 = m_2/M$. The original Bethe-Salpeter equation is obtained by considering the related untruncated Green's function, defined by $G = S + SG_T S$, which can easily be shown to satisfy $G = S + SKG$. To get to a bound state equation, we require that near a bound state pole G behaves as

$$G = \frac{\psi_n(P, p)\bar{\psi}_n(P, p')}{E - E_n} + \text{terms finite as } E \to E_n. (27.44)$$

Because S has no singularity at this point, by multiplying by $E - E_n$ and taking the limit, followed by multiplying by S^{-1}, one arrives at the Bethe-Salpeter equation in momentum space,

$$[\not{P}_1 + \not{p} - m_1]_1 [\not{P}_2 - \not{p} - m_2]_2 \psi_n(P, p)$$
$$= - \int \frac{d^4k}{(2\pi)^4} K(P, p, k) \psi_n(P, k). (27.45)$$

If the kernel is approximated by

$$K_0(P, p, p') = \frac{-4\pi i Z\alpha}{|\mathbf{p} - \mathbf{p}'|^2} [\gamma_0]_1 [\gamma_0]_2 , \qquad (27.46)$$

the Bethe-Salpeter equation can be simplified by multiplying both sides by S and integrating over all values of p_0, further defining $\phi_n(P, \mathbf{p}) \equiv \int dp_0 \psi(P, p)$. Because K_0 is independent of p_0 and k_0, it is straightforward to show, using

$$\frac{1}{\not{p} - m} = \frac{\Lambda_+(\mathbf{p})\gamma_0}{p_0 - E_p + i\epsilon} + \frac{\Lambda_-(\mathbf{p})\gamma_0}{p_0 + E_p - i\epsilon} , \qquad (27.47)$$

where $E_p \equiv \sqrt{\mathbf{p}^2 + m^2}$, that

$$\phi_n(P, \mathbf{p}) = \frac{1}{E - H_1 - H_2} \Big[\Lambda_+^{(1)}(\mathbf{p})\gamma_0 \Lambda_+^{(2)}(-\mathbf{p})\gamma_0$$
$$- \Lambda_-^{(1)}(\mathbf{p})\gamma_0 \Lambda_-^{(2)}(-\mathbf{p})\gamma_0 \Big] \int \frac{d^3k}{(2\pi)^3} \frac{4\pi Z\alpha}{|\mathbf{p} - \mathbf{k}|^2}$$
$$\times \phi_n(P, \mathbf{k}) . \qquad (27.48)$$

Here $H_i \equiv \boldsymbol{\alpha}_i \cdot \mathbf{p}_i + \beta_i m_i$ and the Λ factors are positive and negative energy projection operators, defined by

$$\Lambda_\pm(\mathbf{p}) = \frac{E_p \pm (\boldsymbol{\alpha} \cdot \mathbf{p} + m\gamma_0)}{2E_p} . \qquad (27.49)$$

This equation is called the Salpeter equation [21]. It has an interesting relation to the Breit equation [24], an early attempt to establish a two-body equation. The Breit equation can be obtained from the Salpeter equation by replacing the expression involving projection operators with unity. However, the projection operators correctly incorporate hole theory, so this procedure is not valid. This can be seen most clearly by writing

$$\Lambda_+^{(1)}\Lambda_+^{(2)} - \Lambda_-^{(1)}\Lambda_-^{(2)} = \big[(\Lambda_+^{(1)} + \Lambda_-^{(1)})(\Lambda_+^{(2)} + \Lambda_-^{(2)})$$
$$- 2\Lambda_-^{(1)}\Lambda_-^{(2)} - \Lambda_+^{(1)}\Lambda_-^{(2)} - \Lambda_-^{(1)}\Lambda_+^{(2)} \big] . \quad (27.50)$$

The first term on the right-hand side corresponds to the Breit equation. However, the other terms on the right-hand side are inconsistent with hole theory, and give rise to several problems. One is that the ultraviolet behavior is insufficiently damped. To see this, recall that this expression is multiplied by $1/(E - H_1 - H_2)$. Using $H_i\Lambda_\pm = \pm E_p\Lambda_\pm$, one can see that the left-hand side falls as $1/p$ for large p. However, the last two terms on the right-hand side have denominators such as $E + \sqrt{\mathbf{p}^2 + m_1^2} - \sqrt{\mathbf{p}^2 + m_2^2}$, which provide no falloff at large p. Because of this, when the Breit equation is used in perturbation theory unphysical ultraviolet divergences are encountered. This is related to the fact that when the Breit equation is applied to helium, a spurious contribution to the fine structure arises that has to be discarded in an *ad hoc* manner. This same issue arises in applying

QED to many electron atoms, and will be discussed in the last section.

While the Salpeter equation provides a consistent starting point for recoil corrections, it is difficult to carry out high order calculations with it. As discussed above, more tractable equations can be derived by replacing the propagator S with approximations that incorporate the dominant nonrelativistic physics of atoms. An example would be to replace S with $S_0 = 2\pi i\delta(k_0)s_0(P, \mathbf{k})$, where

$$s_0(P, \mathbf{k}) = 2m_r\Lambda^{(1)}(\mathbf{k})\gamma_0\Lambda^{(2)}(-\mathbf{k})\gamma_0 \frac{1}{\mathbf{k}^2 + \gamma^2} , \quad (27.51)$$

and γ is related to E by $\gamma^2 = 2m_r(m_1 + m_2 - E)$. Because of the delta function, one can turn the equation

$$G_T(P, p, p') = \bar{K}(P, p, p')$$
$$+ \int \frac{d^4k}{(2\pi)^4} \bar{K}(P, p, k)S_0(P, k)G_T(P, k, p')$$
$$\qquad (27.52)$$

into an effective three dimensional form by setting $p_0 = p_0' = 0$. $\bar{K}$ can be determined perturbatively from the equation $\bar{K} = K + K(S - S_0)\bar{K}$. Following the same steps as before, one can introduce the untruncated Green's function and find a bound state equation. However, with simpler forms for S_0, these equations can be solved analytically when the kernel is K_0, unlike the case of the Salpeter equation. These three-dimensional forms have allowed corrections of quite high order to be calculated, as will be discussed below.

27.7   FINE STRUCTURE

The fine structure of hydrogenic atoms in the nonrecoil limit is correctly described by the Dirac equation, which gives for a state of principal quantum number n and spin j $E_{nj} = m_e f(n, j)$, with

$$f(n, j) = \left[1 + \frac{(Z\alpha)^2}{(n - \beta)^2} \right]^{-1/2} . \qquad (27.53)$$

Here

$$\beta \equiv j + \frac{1}{2} - \sqrt{(j + \tfrac{1}{2})^2 - (Z\alpha)^2} , \qquad (27.54)$$

which gives the expansion

$$E_{nj} = m \left\{ 1 - \frac{(Z\alpha)^2}{2n^2} - \frac{(Z\alpha)^4}{2n^3} \left[\frac{1}{j + \frac{1}{2}} - \frac{3}{4n} \right] + \cdots \right\} . \qquad (27.55)$$

With the use of any of the Bethe-Salpeter formalisms discussed in the previous section one can include recoil corrections exactly to order $m(Z\alpha)^4$, and to this order

$$E = M + m_r[f(n,j) - 1] - \frac{m_r^2}{2M}[f(n,j) - 1]^2$$
$$+ \frac{(Z\alpha)^4 m_r^3}{2n^3 m_N^2}\left[\frac{1}{j + \frac{1}{2}} - \frac{1}{l + \frac{1}{2}}\right](1 - \delta_{l0}). \quad (27.56)$$

The last term in this expression leads to a slight breaking of the degeneracy of the $2s_{1/2}$ and $2p_{1/2}$ states: this splitting is referred to as the Lamb shift. A larger breaking arises from the finite size of the nucleus, which in the nonrelativistic limit shifts s-states by

$$\Delta E_n(\text{finite size}) = \frac{2}{3n^3}(Z\alpha)^4 m_r^3 \langle r^2 \rangle. \quad (27.57)$$

A major issue for the Lamb shift in hydrogen is the disagreement between measurements at Stanford [25] and Mainz [26] of the charge radius of the proton. As discussed in Mohr's chapter, the 18 kHz difference creates difficulties in interpreting the experimental status of the Lamb shift. While one may determine the proton charge radius independently by assuming the validity of QED and completing all calculations that enter at the few kHz level, there is a clear need for a definitive electron-scattering experiment.

The largest correction to the Lamb shift comes from the self-energy and vacuum polarization graphs previously introduced in Sect. 27.3, with the understanding that the free electron propagators are replaced by Dirac-Coulomb propagators. The self-energy diagram in the Feynman gauge gives an energy shift

$$\Delta E_n(\text{SE}) = -ie^2 \int d^3r\, d^3r' \int \frac{d^4k}{(2\pi)^4} \frac{e^{i\mathbf{k}\cdot(\mathbf{r}-\mathbf{r}')}}{k^2 + i\epsilon} \bar{\psi}_n(\mathbf{r})$$
$$\times \gamma_\mu S_F(\mathbf{r}, \mathbf{r}'; E_n - k_0)\gamma^\mu \psi_n(\mathbf{r}'), \quad (27.58)$$

and a vacuum polarization shift

$$\Delta E_n(\text{VP}) = -ie^2 \int d^3r \int d^3r' \int \frac{d^3k}{(2\pi)^3} \int \frac{dE}{2\pi} \bar{\psi}_n(\mathbf{r})\gamma_0$$
$$\times \psi_n(\mathbf{r}) \frac{e^{i\mathbf{k}\cdot(\mathbf{r}-\mathbf{r}')}}{\mathbf{k}^2} \text{Tr}[\gamma^0 S_F(\mathbf{r}', \mathbf{r}'; E)]. \quad (27.59)$$

In both cases the counterterms discussed above are understood to be added. It is conventional to pull out the overall behavior in Z, α, and the principal quantum number n as follows:

$$\Delta E_n(\text{SE}) + \Delta E_n(\text{VP}) \equiv m\alpha \frac{(Z\alpha)^4}{\pi n^3} F_n(Z\alpha). \quad (27.60)$$

The evaluation of the function $F_n(Z\alpha)$ is quite difficult. At low Z, a perturbative expansion to order $(Z\alpha)^2$ is available, and is expressed as

$$F_n(Z\alpha) = A_{40} + A_{41}\ln(Z\alpha)^{-2} + (Z\alpha)A_{50}$$
$$+ (Z\alpha)^2[A_{62}\ln^2(Z\alpha)^{-2} + A_{61}\ln(Z\alpha)^{-2}$$
$$+ A_{60}] + \cdots. \quad (27.61)$$

The constants A can be found in Mohr's Chap. 28. A_{40} contains a constant that must be obtained numerically, known as the Bethe logarithm. It can be defined from a nonrelativistic limit of the Coulomb gauge version of Eq. (27.58), in which one carries out the integral over k_0 with Cauchy's theorem, keeping only the photon pole, makes the dipole approximation, and replaces α with $\mathbf{p}/m$. In that limit one has

$$\Delta E_n^{\text{NR}} = \frac{2e^2}{3\text{m}^2} \int_0^\Lambda \frac{d^3k}{2\omega(2\pi)^3} \sum_m \frac{|\mathbf{p}_{nm}|^2}{E_n - \omega - E_m}, \quad (27.62)$$

where $\mathbf{p}_{nm} \equiv \langle n|\mathbf{p}|m\rangle$. Because negative energy states are not present, this expression diverges linearly with Λ. The Bethe logarithm is defined by subtracting $-1/\omega$ from the above denominator and carrying out the d^3k integration, giving

$$\Delta E_n^{\text{NR}} = \frac{e^2}{6\text{m}^2\pi^2} \sum_m |\mathbf{p}_{nm}|^2(E_m - E_n)\ln\frac{\Lambda}{|E_m - E_n|}$$
$$\equiv \frac{4\text{m}\alpha(Z\alpha)^4}{3\pi n^3}\ln\left[\frac{\Lambda}{k_0(n)\text{Ry}}\right]. \quad (27.63)$$

Elegant high accuracy determinations of these quantities using Schwinger's Coulomb Green's function have been made [27, 28]. Reference [29] contains an extensive tabulation.

We next discuss recoil corrections to the Lamb shift. One recoil effect is the modification of the A constants by a factor $(m_r/m_e)^3$ (although part of A_{40} is modified instead by $(m_r/m_e)^2$). Another correction, similar to the Lamb shift but involving a photon exchanged between the electron and nucleus was first derived by Salpeter [21], and is given by

$$\Delta E_n = \frac{m_r^3}{m_e m_N}\frac{(Z\alpha)^5}{\pi n^3}\left\{\frac{2}{3}\delta_{l0}\ln\left(\frac{1}{Z\alpha}\right) - \frac{8}{3}\ln k_0(n)\right.$$
$$- \frac{7}{3}a_n - \frac{1}{9}\delta_{l0} - \frac{2}{m_N^2 - m_e^2}\delta_{l0}$$
$$\times \left.\left[m_N^2 \ln\frac{m_e}{m_r} - m_e^2 \ln\frac{m_N}{m_r}\right]\right\}, \quad (27.64)$$

where $a_{1s} = -3 - \ln 4$, $a_{2s} = -9/2$, and $a_{2p} = -13/3$. Present research is concerned with corrections of order $\alpha^6 m_e^2/m_N$.

Finally we discuss the two-loop Lamb shift. The leading contribution is [30]

$$\Delta E_n(\text{two - loop}) = \frac{m\alpha^2(Z\alpha)^4}{\pi^2 n^3}H_n(Z\alpha), \quad (27.65)$$

where for s-states

$$H_n = -\frac{4358}{1296} - \frac{10}{27}\pi^2 + \frac{3}{2}\pi^2\ln 2 - \frac{9}{4}\zeta(3), \quad (27.66)$$

and for non-s states

$$H_n = \left[\frac{197}{72} + \frac{\pi^2}{6} - \pi^2 \ln 2 + \frac{3}{2}\zeta(3)\right]\frac{C_{lj}}{2(2l+1)}, \quad (27.67)$$

where $C_{lj} = 2(j-l)/(j+1/2)$. There has been recent work on $Z\alpha$ corrections to this quantity [31,32], but the calculations disagree, and another calculation will be required to settle the issue [33].

27.8 HYPERFINE STRUCTURE

27.8.1 Muonium Hyperfine Splitting

After the electron $g-2$, the best test of QED is afforded by ground state muonium hyperfine splitting. This is dominated by the Fermi splitting, which is given by

$$E_{\rm F} = \frac{16}{3}\alpha^2 \frac{m_r^3}{m_e^2 m_\mu} cR_\infty, \quad (27.68)$$

where m_r is the reduced mass. The leading correction to the Fermi splitting involves the electron and muon anomalous magnetic moment, which act to increase the splitting by a factor $(1+a_e)(1+a_\mu)$. It is convenient to split the remaining corrections into nonrecoil and recoil parts. The nonrecoil terms are given by

$$\Delta\nu_{\rm nonrecoil} = (1+a_\mu)\left\{1 + a_e + \frac{3}{2}(Z\alpha)^2\right.$$
$$+ \alpha(Z\alpha)(\ln 2 - \tfrac{5}{2})$$
$$- \frac{8\alpha(Z\alpha)^2}{3\pi}\ln(Z\alpha)\left[\ln(Z\alpha) - \ln 4 + \frac{281}{480}\right]$$
$$+ \frac{\alpha(Z\alpha)^2}{\pi}[14.88\,(0.29)]$$
$$\left. + \frac{\alpha^2(Z\alpha)}{\pi}D_1\right\}E_{\rm F}. \quad (27.69)$$

A recent development has been the complete evaluation of D_1, which is a binding correction to the two-loop $g-2$ contribution, by Kinoshita and Nio [34], who found $D_1 = 0.82(4)$. Note also the constant 14.88 has changed from the previous value of 15.38 because of a new calculation of the vacuum polarization component, as discussed further in Mohr's chapter. Subtracting $\Delta\nu_{\rm nonrecoil}$ from experiment leaves 795 kHz to be accounted for by recoil. The present state of recoil corrections is given by

$$\Delta\nu_{\rm recoil} = \left\{-\frac{3Z\alpha}{\pi}\frac{m_e m_\mu}{m_\mu^2 - m_e^2}\ln\frac{m_\mu}{m_e} + \frac{(Z\alpha)^2 m_r^2}{m_e m_\mu}\right.$$
$$\times\left[-2\ln(Z\alpha) - 8\ln 2 + \frac{65}{18}\right]$$
$$+ \frac{\alpha(Z\alpha)}{\pi^2}\frac{m_e}{m_\mu}\left[-2\ln^2\frac{m_\mu}{m_e} + \frac{13}{12}\ln\frac{m_\mu}{m_e}\right.$$
$$\left.+ \frac{21}{2}\zeta(3) + \frac{\pi^2}{6} + \frac{35}{9} + 2.15(14)\right.$$

$$+ \frac{\alpha}{\pi}\left(-\frac{4}{3}\ln^3\frac{m_\mu}{m_e} + \frac{4}{3}\ln^2\frac{m_\mu}{m_e}\right)\right]\right\}E_{\rm F}. \quad (27.70)$$

The full power of modern forms of the Bethe-Salpeter equation was needed for the evaluation of the second term [35]. While these terms account for just the needed 795 kHz mentioned above, the relatively large uncertainty in the muon mass leads to an uncertainty in the Fermi splitting of 1.3 kHz, and further progress in muonium hyperfine splitting will need this uncertainty to be reduced.

27.8.2 Hydrogen Hyperfine Splitting

The situation in hydrogen hyperfine splitting is quite different from the muonium case because of the structure of the proton. The recoil corrections in Eq. (27.70) that involve $\ln m_\mu/m_e$ involve the offshell muon propagator. Because the muon is a pointlike particle, there is no uncertainty in the calculation. However, when it is replaced by a proton, two strong-interaction problems arise that limit the theoretical accuracy that can be attained. The first arises from the fact that the charge distribution of the proton modifies the hyperfine splitting. The fractional correction can be expressed in terms of the electric and magnetic form factors of the proton as follows:

$$\delta_p(\text{Zemach}) = \frac{2\alpha m_e}{\pi^2}\int\frac{d^3p}{p^4}\left[\frac{G_E(-p^2)G_M(-p^2)}{1+\kappa} - 1\right], \quad (27.71)$$

where κ is the anomalous magnetic moment of the proton. It contributes -38.72 ppm, with an error of 0.5 ppm coming from the uncertainty in the form factors. The second correction, $\delta_{\rm recoil}$, replaces the first term in $\Delta\nu_{\rm recoil}$ discussed in muonium hfs. The logarithm of the ratio of the muon and electron masses in that expression arises from high internal momenta. In the case of hydrogen hfs, this is sensitive to details of the proton structure. The most recent evaluation of this quantity [36] gives a 5.68 ppm effect, which is smaller than the Zemach correction only because of cancellation of individual terms of comparable size. When these corrections are added to the other QED corrections, theory agrees with experiment at under the 1 ppm level, with hadronic uncertainties of about 0.5 ppm. This limits the size of hadronic effects arising from the polarizability of the proton, which is known to be bounded in magnitude by 4 ppm from inelastic electron scattering data [37]. As with the case of the Lamb shift in hydrogen, further progress will require better understanding from either an experimental or theoretical approach of properties of the proton.

27.9 ORTHOPOSITRONIUM DECAY RATE

The decay rate of orthopositronium has been measured with increasing accuracy over the years, and has reached the level where corrections of order α^2 to the lowest order decay rate need to be considered. This lowest order rate is

$$\Gamma^0 = 2\frac{\pi^2 - 9}{9\pi}m\alpha^6 \, . \qquad (27.72)$$

The first order correction to this rate was first correctly calculated in Ref. [38]. Analytic and numeric improvements in this work [39] have led to the present value

$$\Gamma^1 = \frac{\alpha}{\pi}[-10.2866 \pm 0.0006]\Gamma^0 \, . \qquad (27.73)$$

The large coefficient of this correction is notable. In particular, a naive estimate of the as-yet uncalculated order α^2 corrections would suggest a coefficient of order 100. In fact, such a coefficient is required to account for the present discrepancy between theory and experiment (see Chap. 28). Recently parts of this very large scale calculation have been carried out, and relatively large coefficients found. We refer to two calculations in particular. The first is the determination by Labelle, Lepage, and Magnea [40] in the framework of nonrelativistic QED [41], of a part of the correction

$$\Gamma^2_{\mathrm{NR}} = 24.6(\alpha/\pi)^2\Gamma^0 \, . \qquad (27.74)$$

However, this result is incomplete, and must be augmented by a calculation of a large set of Feynman diagrams. Part of this calculation has been carried out by Burichenko [42], with a result

$$\Gamma^2_{\mathrm{R}} = 28(\alpha/\pi)^2\Gamma^0 \, . \qquad (27.75)$$

While nothing definite can be said until the complete calculation is carried out, these large results leave open the possibility that QED will correctly describe the decay rate of orthopositronium.

27.10 PRECISION TESTS OF QED IN NEUTRAL HELIUM

The Bethe-Salpeter equation discussed above was developed in terms of an expansion of free electron and nuclear propagators. However, if these propagators are replaced with two electron propagators in an external Coulomb field, the formalism, while now more complicated, allows a rigorous treatment of helium in the nonrecoil limit. This was carried out by Araki [43] and Sucher [44], and allowed theoretical predictions up to order $m\alpha^5$. However, the actual calculations are far more difficult because of the more complicated propagators. Firstly, the

Schrödinger equation cannot be solved analytically, but rather numerically. However, the use of sophisticated Hylleras basis sets [45, 46], allows accuracies far beyond the experimental precision. Secondly, the fine structure cannot be calculated analytically, but instead involves expectation values of operators that are somewhat singular: again the high-quality basis sets just mentioned allow the precision evaluation of this quantity [46]. A third difficulty is the evaluation of the Lamb shift: in helium, the analog of the Bethe logarithm discussed in connection with hydrogen is much more difficult to evaluate.

Most of the leading QED corrections in helium can be expressed in terms of expectation values of various operators $\langle O \rangle$, where

$$\langle O \rangle \equiv \int d^3r_1 d^3r_2 \phi_0^*(\mathbf{r}_1, \mathbf{r}_2) O(\mathbf{r}_1, \mathbf{r}_2) \phi_0(\mathbf{r}_1, \mathbf{r}_2) \, . \qquad (27.76)$$

Explicitly, one has

$$
\begin{aligned}
H^{(4)} = {} & \frac{\alpha}{\pi}\langle H_{\mathrm{so}} + H'_{\mathrm{soo}} + \tfrac{1}{2}H''_{\mathrm{soo}} + H_{\mathrm{ss}} + H_{\mathrm{ssc}}\rangle \\
& + \left[\frac{89}{15} + \frac{14}{3}\ln\alpha - 4\mathbf{s}_1\!\cdot\!\mathbf{s}_2\right]\frac{\alpha^2}{m^2}\langle\delta(\mathbf{r}_{12})\rangle \\
& + \left[\frac{19}{30} + \ln(Z\alpha)^{-2}\right]\frac{4\alpha^2 Z}{3m^2}\langle\delta(\mathbf{r}_1) + \delta(\mathbf{r}_2)\rangle \\
& - \frac{14}{3}\frac{\alpha^2}{m^2}Q - \frac{2\alpha}{3\pi m^2}M \, .
\end{aligned}
\qquad (27.77)
$$

Here

$$H_{\mathrm{so}} = \frac{\alpha Z}{2m^2}\left[\frac{\mathbf{s}_1\!\cdot\!\mathbf{L}_1}{r_1^3} + \frac{\mathbf{s}_2\!\cdot\!\mathbf{L}_2}{r_2^3}\right] , \qquad (27.78)$$

$$H'_{\mathrm{soo}} = \frac{-\alpha}{2m^2 r_{12}^3}[\mathbf{s}_1\!\cdot\!\mathbf{r}_{12}\times\mathbf{p}_1 + \mathbf{s}_2\!\cdot\!\mathbf{r}_{21}\times\mathbf{p}_2] , \qquad (27.79)$$

$$H''_{\mathrm{soo}} = \frac{-\alpha}{m^2 r_{12}^3}[\mathbf{s}_2\!\cdot\!\mathbf{r}_{12}\times\mathbf{p}_1 + \mathbf{s}_1\!\cdot\!\mathbf{r}_{21}\times\mathbf{p}_2] , \qquad (27.80)$$

$$H_{\mathrm{ss}} = \frac{\alpha}{m^2 r_{12}^3}[\mathbf{s}_1\!\cdot\!\mathbf{s}_2 - 3(\mathbf{s}_1\!\cdot\!\mathbf{r}_{12})(\mathbf{s}_2\!\cdot\!\mathbf{r}_{12})] \qquad (27.81)$$

$$H_{\mathrm{ssc}} = \frac{-8\pi\alpha}{3m^2}\mathbf{s}_1\!\cdot\!\mathbf{s}_2\delta(\mathbf{r}_{12}) , \qquad (27.82)$$

and $\mathbf{r}_{12} = \mathbf{r}_1 - \mathbf{r}_2$. The sum $H'_{\mathrm{soo}} + H''_{\mathrm{soo}} = H_{\mathrm{soo}}$ is the total spin-other-orbit interaction defined in Eq. 21.12. In the above, units in which $\hbar = c = 1$ are used. Then p and $1/r$ are of order $m\alpha$, and $H^{(4)}$ is of order $m\alpha^5$. The quantity Q is defined by

$$Q = \frac{1}{4\pi}\lim_{a\to 0}\left\langle\frac{1}{r_{12}^3(a)} + 4\pi\delta(\mathbf{r}_{12})(\gamma + \ln a)\right\rangle , \qquad (27.83)$$

where $r_{12}(a)$ vanishes when $r_{12} < a$. M is given by

$$M = \sum_n (E_n^0 - E^0)|\mathbf{P}_n|^2 \ln\frac{(E_n^0 - E^0)}{Z^2 m\alpha^2/2} , \qquad (27.84)$$

where

$$\mathbf{P}_n = \int d^3r_1 d^3r_2 \phi_n^*(\mathbf{r}_1, \mathbf{r}_2)(\mathbf{p}_1 + \mathbf{p}_2)\phi_0(\mathbf{r}_1, \mathbf{r}_2). \quad (27.85)$$

M can be written in terms of an expectation value by defining the Bethe logarithm $\ln k_0$ through

$$M \equiv \sum_n (E_n^0 - E^0)|\mathbf{P}_n|^2 \ln k_0$$
$$= 2\pi\alpha Z \langle \delta(\mathbf{r}_1) + \delta(\mathbf{r}_2) \rangle \ln k_0, \quad (27.86)$$

where the oscillator strength sum rule 11.36d has been used. The evaluation of this Bethe logarithm is significantly more difficult than in the hydrogenic case, though $1/Z$ expansions for low-L states and asymptotic expansions for high-L states are available [29, 47]. An important recent development has been the direct evaluation of this quantity [48] for S-states. A complete calculation of the next order corrections, that is, those of order $m\alpha^6$, has not yet been carried out in general, although Douglas and Kroll [49] have carried out this calculation for P-states. An outstanding problem in QED is the completion of the $m\alpha^6$ calculation for all states and the extension of the fine structure calculation to order $m\alpha^7$. This latter calculation should allow an extraction of the fine structure constant from helium competitive to that obtained from electron $g - 2$ [50].

27.11 QED IN HIGHLY CHARGED ONE-ELECTRON IONS

Because the Lamb shift scales as Z^4 and energy levels as Z^2, the relative importance of this effect increases as one goes out along an isoelectronic sequence. In addition, the approach used in the previous section, which relies on an expansion in powers of $Z\alpha$, becomes inappropriate, and requires methods that do not use such an expansion. Such methods were introduced by Wichmann and Kroll [51] for vacuum polarization. They were extended to the more difficult self-energy calculation by Brown, Langer and Schaefer [52], and the first correct calculations using the method were carried out by Desiderio and Johnson [53]. The basic idea is first to carry out the d^3k integration in Eq. (27.58) analytically, leaving an integration over d^3r, d^3r', and k_0 of a product of the electron and photon propagators in coordinate space. If one then makes a partial wave expansion of these propagators, the angle integrations can be carried out, and the self-energy becomes a sum over partial waves of an integral over two radial coordinates and the photon energy, which integral is carried out numerically. While simple in principle, this method is numerically awkward because the parts of the self-energy that are ultraviolet divergent are not well described in coordinate space. To solve this problem, the parts of the self-energy that have this sensitivity, which are associated with the part of the electron propagator in which the electron propagates

freely, or with a single interaction, can be separated out. These subtractions allowed the first high-accuracy calculation of the self-energy in Coulomb potentials by Mohr [54]. Later Blundell and Snyderman [55] used such subtractions in a purely numeric approach that allowed the treatment of non-Coulomb potentials, which must be used for the many-electron ions discussed in the next section. While the most accurate calculations have been carried out for the case of the Coulomb potential, the finite size of the nucleus cannot be neglected, particularly at high Z. This issue has recently been studied by several groups [56, 57], and is now well understood. The experimental status of QED in one-electron ions is discussed by Mohr in this volume.

27.12 QED IN HIGHLY CHARGED MANY-ELECTRON IONS

The previous applications of QED have been concentrated on simple systems, the most complex being helium. However, slight modifications of the S-matrix approach to QED described in Sect. 27.2 allow the treatment of ions with any number of electrons. In practice, this does not necessarily allow progress, since diagrams of arbitrarily high order contribute at the level of 1 a.u. in the many-body problem. However, when the nuclear Coulomb field dominates the Coulomb fields of the electrons, a small parameter is available, the quantity $1/Z$. Physically this simply reflects the fact that the energy of repulsion of any pair of electrons is of the order of Z a.u., while the energy of attraction to the nucleus scales as Z^2 a.u. This $1/Z$ expansion has been studied empirically in the context of many-body perturbation theory, and it can be seen that while diagrams involving two-photon interactions are important, those involving three photons are highly suppressed. Thus the possibility exists of putting the QED of highly charged ions on the same footing as, for example, the electron $g - 2$, in the sense that evaluation of a limited number of Feynman diagrams will allow precision tests of QED. However, while in principle one can start with the original Furry representation, in practice, particularly when larger numbers of electrons are present, it is better to start with non-Coulomb potentials that build in a major part of the electron screening present in the ion. For example, if one considers sodiumlike platinum ($Z = 78$), the valence electron should see a nuclear charge of around 68 because of the ten 'core' electrons. One potential that builds in this property is the so-called 'core-Hartree' potential [57], which is defined so that

$$V_{\mathrm{CH}}(r) = V_{\mathrm{nuc}}(r) + \sum_a (2j_a + 1)v_0(a, a; r), \quad (27.87)$$

where

$$v_0(a, a, : r) \equiv \int_0^\infty dr' \frac{1}{r_>} [g_a^2(r') + f_a^2(r')], \quad (27.88)$$

with g_a and f_a the upper and lower radial components of the Dirac wave function for core state a. For sodium-like systems, a ranges over the $1s$, $2s$, $2p_{1/2}$, and $2p_{3/2}$ states. These states are solved self consistently. The resulting potential gives results close to a full Dirac-Fock potential, but has the advantage of being local, which makes the connection with QED transparent. Because as $r \to \infty$ $v_0(a, a; r) \to 1/r$, the long range behavior of this potential has the physically expected limit. It is very simple to modify the Furry representation to include this potential: one simply writes

$$H_0 = \int d^3 r \psi^\dagger(\mathbf{r})[\boldsymbol{\alpha} \cdot \mathbf{p} + \beta m + V_{\mathrm{CH}}(r)]\psi(\mathbf{r}) \quad (27.89)$$

and

$$H_I = \int d^3 r \psi^\dagger(\mathbf{r})[V_{\mathrm{nuc}}(r) - V_{\mathrm{CH}}(r)]\psi(\mathbf{r})$$
$$- e \int d^3 r \psi^\dagger(\mathbf{r})\boldsymbol{\alpha} \cdot \mathbf{A}(\mathbf{r})\psi(\mathbf{r})$$
$$+ \frac{\alpha}{2} \int \frac{d^3 r d^3 r'}{|\mathbf{r} - \mathbf{r}'|} \psi^\dagger(\mathbf{r})\psi(\mathbf{r})\psi^\dagger(\mathbf{r}')\psi(\mathbf{r}'). \quad (27.90)$$

The only difference with the QED discussed in the previous section is the first term of H_I, which acts in a manner similar to the self-mass counterterm, though it is of course finite. However, when more than one electron is present, new types of Feynman diagram are encountered: a representative set of them is given in Fig. 27.3. There is an interesting connection between these diagrams and the many-body perturbation theory (MBPT) method of solving the many-electron Schrödinger equation. In this latter method the N-electron Hamiltonian is written $H = H_0 + V_C$, where

$$H_0 = \sum_{i=1}^{i=N} \left[\boldsymbol{\alpha}_i \cdot \mathbf{p}_i + \beta_i m - \frac{Z\alpha}{r_i} + U(r_i) \right] \quad (27.91)$$

and

$$V_C = \frac{\alpha}{2} \sum_{ij} \frac{1}{|\mathbf{r}_i - \mathbf{r}_j|} - \sum_{i=1}^{i=N} U(r_i). \quad (27.92)$$

Care must be taken when working with the relativistic version of the N-electron Schrödinger equation because of negative energy states, a point that has been emphasized by Sucher [58]. This problem shows up in MBPT in second order, where the associated energy shift for the ground state of an alkali atom with valence electron v in the case when U is the Hartree-Fock potential is

$$E^{(2)} = - \sum_{amn} \frac{g_{vamn}(g_{mnva} - g_{mnav})}{\epsilon_m + \epsilon_n - \epsilon_a - \epsilon_v}$$
$$+ \sum_{abm} \frac{g_{abmv}(g_{mvab} - g_{mvba})}{\epsilon_m + \epsilon_v - \epsilon_a - \epsilon_b}. \quad (27.93)$$

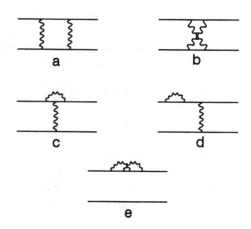

Figure 27.3. Two-photon diagrams contributing to energy levels of highly-charged ions.

Here a and b range over the occupied core states, and m and n over excited states including in principle negative energy states. However, if they are included, the energy denominator in the first term of the above expression can vanish when m is a positive energy state and n negative energy, and vice versa. This problem can be avoided by restricting sums over excited states to positive energy states, but this procedure must be justified in the framework of QED. This can indeed be done [59], with the result that the use of MBPT with the positive energy restriction is justified, but negative energy state effects enter in a well-defined way, giving effects on the order of the Lamb shift.

To illustrate the present status of QED in highly charged many-electron ions, we consider the spectrum of sodium-like platinum. The transition energy between $3p_{3/2}$ and $3s_{1/2}$ has been measured [60] to be 653.44(7) eV. When MBPT is applied to this transition through second order, the answer depends slightly on the starting potential. The Hartree Fock potential gives 659.56 eV, while the core-Hartree potential described above gives 659.59 eV. A modification of the core-Hartree potential in which the factor $2j_a + 1$ is reduced by 1 for the last core state give 659.57 eV. This spread of 0.03 eV gives a measure of the convergence of MBPT: it should be compared with a spread of 12 eV in lowest order and 0.4 eV in first order. Thus the difference with experiment of 6.1 eV that is to be explained by QED is reliable at the 0.1 eV level.

The QED calculations in highly charged ions are not as advanced as in the QED tests in simple neutral atoms. As discussed above, only recently have the calculations of one-loop radiative corrections in non-Coulomb potentials become possible. In the case of the core-Hartree potential they give exactly the 6.1 eV required to explain the difference between MBPT and experiment. This is actually somewhat unfortunate, since there is a set of interesting two-photon Feynman diagrams, some of which are shown in Fig. 27.3, that have yet to be completely evaluated. These are expected to contribute at the few tenths of an eV level, and it is likely the good agreement with experiment found above involves fortuitous calculations. The graph of Fig. 27.3a plays a particularly interesting role, since it can be shown to 'contain' $E^{(2)}$. Specifically, when both photons are Coulomb photons and the electron propagators are written in terms of a spectral representation, the fourth component of the internal momentum integration can be carried out with contour methods. When both propagators involve positive energy states, $E^{(2)}$ from MBPT is precisely reproduced, including of course the previously introduced *ad hoc* rule of having only positive energy states in intermediate sums. When one is positive and the other negative the undefined term discussed above could arise, but since in this case both poles are on the same side of the axis, the contribution vanishes. However, when both are negative, a contribution outside of MBPT, part of the QED effect, arises. It is of order $Z^3 \alpha^3$ a.u., in other words, $1/Z$ of the leading Lamb shift. For sodium-like platinum, this should enter at the 0.1 eV level. For this reason, to be sensitive to corrections beyond the one-loop Lamb shift, ever more precise experiments along with more sophisticated MBPT calculations are required. As this progress is made, the interesting prospect arises of extending the high-accuracy QED tests in simple neutral systems discussed in this review to more complicated many-electron ions, and eventually to all the neutral atoms in the periodic table.

REFERENCES

1. *Quantum Electrodynamics*, edited by T. Kinoshita (World Scientific, Singapore, 1990).
2. An excellent collection of the early papers of QED is *Selected Papers on Quantum Electrodynamics*, edited by J. Schwinger (Dover Publications, New York, 1958). See also *QED and the Men Who Made It*, by Silvan S. Schweber, (Princeton University Press, Princeton, New Jersey, 1994).
3. H. Fried and D. R. Yennie, Phys. Rev. **112**, 1391 (1958).
4. W. Pauli and F. Villars, Rev. Mod. Phys. **21**, 434 (1949).
5. J. C. Ward, Phys. Rev. **73**, 416 (1948).
6. This experiment and others are discussed by Hans-Ulrich Martyn in Ref. 1.
7. J. Schwinger, Phys. Rev. **73**, 416 (1948).
8. W.-Y. Tsai and A. Yildiz, Phys. Rev. D **8**, 3446 (1973).
9. G. Gabrielse, J. Tan, and L. S. Brown, in Ref. 1.
10. R. S. Van Dyck Jr., P. B. Schwinberg, and H. G. Dehmelt, Phys. Rev. Lett. **59**, 26 (1987).
11. C. M. Sommerfield, Ann. Phys. (N. Y.) **5**, 26 (1958).
12. A. Petermann, Helv. Phys. Acta **30**, 407 (1957).
13. G. S. Adkins, Phys. Rev. D **39**, 3798 (1989).
14. T. Kinoshita, B. Nizic, and Y. Okamoto, Phys. Rev. D **31**, 2108 (1985).
15. R. Z. Roskies, E. Remiddi, and M. J. Levine, in Ref. 1.
16. P. Cvitanovic and T. Kinoshita, Phys. Rev. D **10**, 4007 (1974): an updated version of this calculation is given by T. Kinoshita in Ref. 1.
17. T. Kinoshita and W. B. Lindquist, Phys. Rev. D **42**, 636 (1990).
18. W. Furry, Phys. Rev. **107**, 1448 (1957).
19. J. Schwinger, Proc. Nat. Acad. Sci. USA, **37**, 452, 455 (1951).
20. E. E. Salpeter and H. A. Bethe, Phys. Rev. **84**, 1232 (1951).
21. E. E. Salpeter, Phys. Rev. **87**, 328 (1952).
22. A. Klein and T.-S. H. Lee, Phys. Rev. D **10**, 4308 (1974).
23. G. P. Lepage, Phys. Rev. A **16**, 863 (1977).
24. G. Breit, Phys. Rev. **29**, 553 (1929).
25. L. N. Hand, D. J. Miller, and R. Wilson, Rev. Mod. Phys. **35**, 335 (1963).
26. G. G. Simon, Ch. Schmidt, F. Borkowski, and V. H. Walther, Nucl. Phys. A **333**, 381 (1980).
27. M. Lieber, Phys. Rev. **174**, 2037 (1968).
28. R. W. Huff, Phys. Rev. **186**, 1367 (1969).
29. G. W. F. Drake and R. A. Swainson, Phys. Rev. A**41**, 1243 (1990).
30. T. W. Appelquist and S. J. Brodsky, Phys. Rev. A **2**, 2293 (1970); B. E. Lautrup, A. Peterman, and E. deRafael, Phys. Lett. B**31**, 577 (1970); R. Barbieri, J. A. Mignaco, and E. Remiddi, Nuovo Cimento Lett. **3**, 588 (1970).
31. C. Pachuki, Phys. Rev. Lett. **72**, 3154 (1994).
32. T. Ward (private communication).
33. After the preparation of this chapter, another calculation in agreement with Pachuki was presented by M. I. Eides, and V. A. Shelyuto, JETP Letters **61**, 478 (1995).
34. T. Kinoshita and M. Nio, Phys. Rev. Lett. **72**, 3803 (1994).
35. G. T. Bodwin, D. R. Yennie, and M. A. Gregorio, Rev. Mod. Phys. **57**, 723 (1985).
36. G. T. Bodwin and D. R. Yennie, Phys. Rev. D **37**, 498 (1988).
37. V. W. Hughes and J. Kuti, Ann. Rev. Nucl. Part. Sci. **33**, 611 (1983).
38. W. E. Caswell, G. P. Lepage, and J. Sapirstein, Phys. Rev. Lett. **38**, 488 (1977).
39. Gregory S. Adkins, Ali A. Salahuddin, and Koenraad E. Schalm, Phys. Rev. A **45**, 7774 (1992).
40. P. Labelle, G. P. Lepage, and U. Magnea, Phys. Rev. Lett. **72**, 2006 (1994).
41. W. E. Caswell and G. P. Lepage, Phys. Lett. B **167**, 437 (1986).
42. A. P. Burichenko, Phys. At. Nucl. **56** (5), 640 (1993) [Yad. Fiz. **56**, 123 (1993)].
43. H. Araki, Prog. Theor. Phys. **17**, 619 (1957).
44. J. Sucher, Phys. Rev. **109**, 1010 (1958).
45. J. D. Morgan III, in *AIP Conference Proceedings #189, Relativistic, Quantum Electrodynamic, and Weak Interaction Effects in Atoms, 1988*, edited by W. Johnson, P. Mohr, and J. Sucher (American Institute of Physics, New York, 1988), p. 123.
46. G. W. F. Drake and Z.-C. Yan, Phys. Rev. A **46**, 2378 (1992); G. W. F. Drake, Adv. Mol. Opt. Phys. **31**, 1

(1993).

47. S. P. Goldman and G. W. F. Drake, Phys. Rev. Lett. **68**, 1683 (1992).

48. J. D. Baker, R. C. Forrey, J. D. Morgan III, R. N. Hill, M. Jesiorska, and J. Schertzer, Bull. Am. Phys. Soc. **38**, No. 3 (1993).

49. M. Douglas and N. Kroll, Ann. Phys. (N. Y.) **82**, 89 (1974).

50. Z.-C. Yan and G. W. F. Drake, Phys. Rev. Lett. **74**, 4791 (1995).

51. E. H. Wichmann and N. M. Kroll, Phys. Rev. **101**, 843 (1956).

52. G. E. Brown, J. S. Langer, and G. W. Schaefer, Proc. R. Soc. London Ser. A **251**, 92 (1959).

53. A. M. Desiderio and W. R. Johnson, Phys. Rev. A **3**, 1267 (1971).

54. P. J. Mohr, Ann. Phys. (N. Y.) **88**, 26, 52 (1974): higher accuracy results are given for $n = 1$ and $n = 2$ states by P. J. Mohr, Phys. Rev. A **46**, 4421 (1992), and for $n = 3$, 4, and 5 states by P. J. Mohr and Y. K. Kim, Phys. Rev. A **45**, 2727 (1992).

55. S. A. Blundell and N. Snyderman, Phys. Rev. A **44**, R1427 (1991).

56. P. J. Mohr and G. Soff, Phys. Rev. Lett. **70**, 158, 1993.

57. K. T. Cheng, W. R. Johnson, and J. Sapirstein, Phys. Rev. A **47**, 1817 (1993).

58. J. Sucher, Phys. Rev. A **22**, 348 (1980).

59. J. Sapirstein, Phys. Scr. T**46**, 52 (1993).

60. T. E. Cowan, C. L. Bennett, D. D. Dietrich, J. Bixler, C. J. Hailey, J. R. Henderson, D. A. Knapp, M. A. Levine, R. E. Marrs, and M. B. Schneider, Phys. Rev. Lett. **66**, 1150 (1991).

Tests of Fundamental Physics

Peter J. Mohr

National Institute of Standards and Technology, Gaithersburg, Maryland 20899

28.1 INTRODUCTION

This Chapter describes comparisons of precise measurements and theoretical predictions that provide tests of our knowledge of fundamental physics. The focus is on several quantitative tests of quantum electrodynamics (QED).

The basic formulation of the theory of QED and calculational methods are discussed in Chap. 27. Here, only the end results of calculations are collected, numerically evaluated, and compared with the corresponding experiments. (All quoted uncertainties are meant to be approximately at the one standard deviation level.)

It should be remarked that QED theory and the fundamental constants that are employed in evaluating the theoretical expressions are intimately linked. Fundamental constants are discussed in Chap. 1. Values of the constants needed for comparison of theory and experiment are generally determined by other comparisons of theory and experiment, so that only the consistency of a set of tests is checked. However, the fact that this overall consistency is maintained at a high level of precision and over a broad range of phenomena provides confidence that QED is sound despite mathematical shortcomings in its formulation.

28.2 *g*-FACTOR ANOMALIES

One of the most precise tests of QED is the comparison of theory and experiment for the anomalous magnetic moment of the electron. The current status of this comparison is given in the next section, and the corresponding results for the muon anomalous magnetic moment are discussed in the subsequent section.

28.2.1 Electron *g*-Factor Anomaly

The electron magnetic moment is proportional to

$$g_e = 2(1 + a_e), \quad (28.1)$$

where the anomaly a_e characterizes the deviation of the *g*-factor from the Dirac value of $g_e(\text{Dirac}) = 2$. Measurement of the anomaly by the University of Washington group has yielded values for the electron and positron given respectively by [1]

$$a_{e^-}(\text{exp}) = 1\,159\,652\,188.4(4.3) \times 10^{-12}, \quad (28.2)$$
$$a_{e^+}(\text{exp}) = 1\,159\,652\,187.9(4.3) \times 10^{-12}. \quad (28.3)$$

The same experiment has provided a precision comparison of the electron and positron *g*-factors

$$\frac{g_{e^-}}{g_{e^+}} = 1 + (0.5 \pm 2.1) \times 10^{-12}, \quad (28.4)$$

which provides a test of CPT invariance of the electron-positron system.

The QED theoretical value of the electron magnetic moment anomaly is given by

$$a_e(\text{QED}) = A_1 + A_2\,(m_e/m_\mu) + A_2\,(m_e/m_\tau)$$
$$+ A_3\,(m_e/m_\mu, m_e/m_\tau), \quad (28.5)$$

where the term A_1 arises from Feynman diagrams with only electron lines, A_2 corresponds to diagrams with one or more closed loops of one type of heavier lepton (μ or τ), and A_3 results from diagrams with both heavier particles in closed loops. Each of these terms is calculated as a power series in the fine-structure constant α

$$A_i = A_i^{(2)} \left(\frac{\alpha}{\pi}\right) + A_i^{(4)} \left(\frac{\alpha}{\pi}\right)^2$$
$$+ A_i^{(6)} \left(\frac{\alpha}{\pi}\right)^3 + A_i^{(8)} \left(\frac{\alpha}{\pi}\right)^4 + \cdots, \quad (28.6)$$

where $i = 1, 2, 3$.

For the purely electron diagrams, the leading term is the Schwinger term [2]

$$A_1^{(2)} = \tfrac{1}{2}, \quad (28.7)$$

and the fourth-order term is also known in analytic form to be [3–6]

$$A_1^{(4)} = \tfrac{197}{144} + \left(\tfrac{1}{2} - 3\ln 2\right)\zeta(2) + \tfrac{3}{4}\zeta(3)$$
$$= -0.328\,478\,965 \cdots. \quad (28.8)$$

The sixth-order term has been evaluated by a combination of analytic and numerical methods [7–10]

$$A_1^{(6)} = 1.181\,259(40). \quad (28.9)$$

Recent improved numerical calculations have revealed that earlier evaluations of this term suffered from a slight mismatch in combining analytic and numerical contributions, so the value has shifted by a small but significant amount [10]. A numerical evaluation of the eighth-order coefficient gives [7,9]

$$A_1^{(8)} = -1.557(70), \quad (28.10)$$

which results from the evaluation of 891 Feynman diagrams.

In addition to the QED contributions listed above, there are small corrections for muon and tauon vacuum polarization loops, hadronic vacuum polarization loops (estimated from e^+–e^- annihilation into hadrons data), and weak interactions. For the muon and tauon contributions, we have [11]

$$A_2^{(2)}(x) = 0,$$
$$A_2^{(4)}(x) = \tfrac{1}{45}x^2 + \mathcal{O}(x^4 \ln x), \quad (28.11)$$

which, together with [12]

$$\frac{m_\mu}{m_e} = 206.768\,262(30) \quad (28.12)$$

and [13]

$$m_\tau = 1771.1(5)\text{MeV} \quad \text{or} \quad \frac{m_\tau}{m_e} = 3478(1) \quad (28.13)$$

give

$$A_2(m_e/m_\mu) = 2.804 \times 10^{-12}, \quad (28.14)$$
$$A_2(m_e/m_\tau) = 0.010 \times 10^{-12}. \quad (28.15)$$

The hadronic and weak interaction contributions are [14]

$$a_e(\text{had}) = 1.6(2) \times 10^{-12} \quad (28.16)$$

and [14–16]

$$a_e(\text{weak}) = 0.04 \times 10^{-12}. \quad (28.17)$$

With the 1986 CODATA recommended value for the fine-structure constant [12]

$$\alpha^{-1} = 137.035\,989\,5(61), \quad (28.18)$$

the total calculated value for the anomaly is

$$a_e(\text{th}) = 1\,159\,652\,272(52) \times 10^{-12}, \quad (28.19)$$

where the number in parentheses is the combined uncertainty from Eqs. (28.9), (28.10), (28.16), and (28.18). The dominant source of uncertainty is the estimated precision in the measurement of the fine-structure constant. With the value of α determined by the quantized Hall effect measurement [17]

$$\alpha^{-1}(\text{QHE}) = 137.035\,997\,9(32)\,, \qquad (28.20)$$

the electron anomaly is

$$a_{\text{e}}(\text{th}) = 1\,159\,652\,201(27) \times 10^{-12}\,. \qquad (28.21)$$

The value of the fine-structure constant for which the theoretical and experimental values of the electron anomalous magnetic moment are equal is given by

$$\alpha^{-1}(g-2) = 137.035\,999\,44(57)\,, \qquad (28.22)$$

which has the smallest estimated uncertainty among the various methods of determining α.

28.2.2 Muon *g*-Factor Anomaly

The measured values of the muon *g*-factor anomaly are

$$a_{\mu^-}(\text{exp}) = 1\,165\,936(12) \times 10^{-9}\,, \qquad (28.23)$$
$$a_{\mu^+}(\text{exp}) = 1\,165\,910(11) \times 10^{-9}\,, \qquad (28.24)$$

for the negative and positive muon respectively [18, 19]. For the theoretical value, the QED part is given by

$$a_{\mu}(\text{QED}) = A_1 + A_2\,(m_{\mu}/m_{\text{e}}) + A_2\,(m_{\mu}/m_{\tau})$$
$$+ A_3\,(m_{\mu}/m_{\text{e}}, m_{\mu}/m_{\tau})\,, \qquad (28.25)$$

where the A_i are the same functions as in the previous section evaluated at the indicated mass ratios. The mass-dependent terms give a relatively greater contribution to the total for the muon compared with the electron, primarily due to the relatively larger effect of vacuum polarization corrections from a lighter lepton. The leading fourth-order terms are given by [20–25]

$$A_2^{(4)}(x) = \tfrac{1}{3}\ln x - \tfrac{25}{36} + \tfrac{3}{2}\zeta(2)x^{-1} - 4x^{-2}\ln x$$
$$+ 3x^{-2} - \tfrac{15}{2}\zeta(2)x^{-3} + \mathcal{O}(x^{-4}\ln^2 x)\,, \qquad (28.26)$$

for large x, so that

$$A_2^{(4)}\,(m_{\mu}/m_{\text{e}}) = 1.094\,258\cdots\,, \qquad (28.27)$$

and [25, 11]

$$A_2^{(4)}(x) = \tfrac{1}{45}x^2 + \tfrac{1}{70}x^4 \ln x + \mathcal{O}(x^4) \qquad (28.28)$$

for small x, so

$$A_2^{(4)}\,(m_{\mu}/m_{\tau}) = 7.8 \times 10^{-5}\,. \qquad (28.29)$$

The sixth-order terms are known partially analytically and partially numerically. Current values are [26–28]

$$A_2^{(6)}\,(m_{\mu}/m_{\text{e}}) = 22.868\,379\cdots\,,$$
$$A_2^{(6)}\,(m_{\mu}/m_{\tau}) = 3.55(10) \times 10^{-4}\,, \qquad (28.30)$$

and the leading term of A_3 ($A_3^{(2)} = A_3^{(4)} = 0$) is [26–28]

$$A_3^{(6)}\,(m_{\mu}/m_{\text{e}}, m_{\mu}/m_{\tau}) = 5.24(1) \times 10^{-4}\,. \qquad (28.31)$$

Higher-order terms are [26, 29]

$$A_2^{(8)}\,(m_{\mu}/m_{\text{e}}) = 127.49(41)\,,$$
$$A_3^{(8)}\,(m_{\mu}/m_{\text{e}}, m_{\mu}/m_{\tau}) = 0.079(3)\,, \qquad (28.32)$$

and [26]

$$A_2^{(10)}\,(m_{\mu}/m_{\text{e}}) = 570(140)\,. \qquad (28.33)$$

A sizable correction arises from the hadronic vacuum polarization correction [30]

$$a_{\mu}(\text{had}) = 70.52(76) \times 10^{-9}\,. \qquad (28.34)$$

Weak interaction effects contribute [31]

$$a_{\mu}(\text{weak}) = 1.95(10) \times 10^{-9}\,. \qquad (28.35)$$

Together, these contributions give

$$a_{\mu}(\text{th}) = 1\,165\,919.5(8) \times 10^{-9} \qquad (28.36)$$

for the total theoretical anomaly.

28.3 HYDROGEN-LIKE ATOMS

Theoretical energy levels of hydrogen-like atoms are determined mainly by the Dirac eigenvalue, quantum electrodynamic effects such as the self energy and vacuum polarization, finite-nuclear-size corrections, and nuclear motion effects. Earlier tabulations of these effects are given by [32] and [33].

28.3.1 Dirac Eigenvalue

The binding energy of an electron in a static Coulomb field is determined predominantly by the Dirac eigenvalue

$$E_{\text{D}} = \left[1 + \frac{(Z\alpha)^2}{(n-\delta)^2}\right]^{-1/2} m_{\text{e}}c^2\,, \qquad (28.37)$$

where

$$\delta = |\kappa| - \left[\kappa^2 - (Z\alpha)^2\right]^{1/2}\,, \qquad (28.38)$$

n is the principal quantum number, κ is the angular momentum-parity quantum number ($\kappa = -1, 1, -2$ for $S_{1/2}$, $P_{1/2}$, $P_{3/2}$ states, respectively), and $m_{\text{e}}c^2$ is the electron rest energy. States with the same principal quantum number n and angular momentum quantum number $j = |\kappa| - \tfrac{1}{2}$ have degenerate eigenvalues. The nonrelativistic orbital angular momentum is given by $l = |\kappa + \tfrac{1}{2}| - \tfrac{1}{2}$.

28.3.2 Relativistic Recoil

Corrections to the Dirac eigenvalue that take into account the finite mass of the nucleus m_N are included in the more general expression for atomic energy levels [34, 35]

$$E_M = Mc^2 + [f(n,j)-1]m_r c^2 - [f(n,j)-1]^2 \frac{m_r^2 c^2}{2M}$$
$$+ \frac{1-\delta_{l0}}{\kappa(2l+1)} \frac{(Z\alpha)^4 m_r^3 c^2}{2n^3 m_N^2} + \cdots, \quad (28.39)$$

where

$$f(n,j) = \left[1 + \frac{(Z\alpha)^2}{(n-\delta)^2}\right]^{-1/2}, \quad (28.40)$$

$M = m_e + m_N$, and $m_r = m_e m_N/(m_e+m_N)$.

Relativistic corrections to this expression are considered relativistic-recoil corrections. The leading term has been calculated by Salpeter [36]

$$E_S = \frac{m_r^3}{m_e^2 m_N} \frac{(Z\alpha)^5}{\pi n^3} m_e c^2$$
$$\times \left\{ \frac{1}{3}\delta_{l0}\ln(Z\alpha)^{-2} - \frac{8}{3}\ln k_0(n,l) - \frac{1}{9}\delta_{l0} - \frac{7}{3}a_n \right.$$
$$\left. - \frac{2}{m_N^2 - m_e^2}\delta_{l0}\left[m_N^2 \ln\left(\frac{m_e}{m_r}\right) - m_e^2 \ln\left(\frac{m_N}{m_r}\right)\right]\right\}, \quad (28.41)$$

where [37]

$$a_n = -2\left[\ln\left(\frac{2}{n}\right) + \sum_{i=1}^n \frac{1}{i} + 1 - \frac{1}{2n}\right]\delta_{l0}$$
$$+ \frac{1-\delta_{l0}}{l(l+1)(2l+1)}. \quad (28.42)$$

The next-order correction has been investigated recently with the result [38–41]

$$E_R = \frac{m_e}{m_N} \frac{(Z\alpha)^6}{n^3} m_e c^2 D_{60} \quad (28.43)$$

where

$$D_{60} = \begin{cases} 4\ln 2 - \frac{7}{2} & nS_{1/2}, \\ \frac{1}{3} & 2P_{1/2}, \\ \frac{1}{3} & 2P_{3/2}. \end{cases} \quad (28.44)$$

28.3.3 Nuclear Polarization

Nuclear polarization, which arises from interactions between the electron and nucleus in which excited states of the nucleus are included in the intermediate states, has been calculated for hydrogen-like lead and uranium in Ref. [42]. The results appear in Table 28.1.

This effect is negligible in hydrogen at present, but it gives the $1S$ level an estimated shift of -19 kHz in deuterium [43].

Table 28.1. Nuclear polarization effect (in meV).

Z	$1S_{1/2}$	$2S_{1/2}$	$2P_{1/2}$
82	-101.3	-17.4	-1.4
92	-1079.2	-202.7	-23.4

28.3.4 Self Energy

The second-order self-energy level shift is given by

$$E_{SE}^{(2)} = \frac{\alpha}{\pi}\frac{(Z\alpha)^4}{n^3} F(Z\alpha)\, m_e c^2, \quad (28.45)$$

where

$$F(Z\alpha) = A_{41}\ln(Z\alpha)^{-2} + A_{40} + A_{50}(Z\alpha)$$
$$+ A_{62}(Z\alpha)^2\ln^2(Z\alpha)^{-2} + A_{61}(Z\alpha)^2\ln(Z\alpha)^{-2}$$
$$+ G_{SE}(Z\alpha)(Z\alpha)^2, \quad (28.46)$$

with [44]

$$A_{41} = \frac{4}{3}\delta_{l0},$$
$$A_{40} = -\frac{4}{3}\ln k_0(n,l) + \frac{11}{18}\delta_{l0} - \frac{1}{2\kappa(2l+1)},$$
$$A_{50} = \left(\frac{139}{32} - 2\ln 2\right)\pi\delta_{l0},$$
$$A_{62} = -\delta_{l0},$$
$$A_{61} = \begin{cases} \frac{28}{3}\ln 2 - \frac{21}{20} & 1S_{1/2}, \quad (28.47)\\ \frac{16}{3}\ln 2 + \frac{67}{30} & 2S_{1/2}, \\ \frac{4}{3}\ln 2 + \frac{391}{80} & 4S_{1/2}, \; . \\ \frac{103}{180} & 2P_{1/2}, \\ \frac{29}{90} & 2P_{3/2}. \end{cases}$$

Bethe logarithms appearing in Eq. (28.47) are given by [45]

$$\ln k_0(1,0) = 2.984\,128\,555\cdots,$$
$$\ln k_0(2,0) = 2.811\,769\,893\cdots,$$
$$\ln k_0(4,0) = 2.749\,811\,840\cdots, \quad (28.48)$$
$$\ln k_0(2,1) = -0.030\,016\,708\cdots.$$

The function $G_{SE}(Z\alpha)$ in (28.46) gives the higher-order contributions to the self energy. The low-Z limit of this function $G_{SE}(0) = A_{60}$ has been calculated by Pachucki [46]:

$$A_{60} = \begin{cases} -30.928\,90(1) & 1S_{1/2} \\ -31.840\,47(1) & 2S_{1/2} \end{cases}. \quad (28.49)$$

Values for the function $G_{SE}(Z\alpha)$, based on extrapolation to low Z of numerical values calculated at higher Z, are given in Table 28.2 [47–49]. The extrapolated values of

Table 28.2. The function $G_{SE}(Z\alpha)$.

Z	$1S_{1/2}$	$2S_{1/2}$	$4S_{1/2}$	$2P_{1/2}$	$2P_{3/2}$
1	$-30.31(4)$	$-31.23(4)$	$-30.9(2)$	$-0.98(4)$	$-0.49(2)$
2	$-29.79(4)$	$-30.68(4)$	$-30.4(2)$	$-0.96(4)$	$-0.48(2)$

Table 28.3. The function $F(Z\alpha)$.

Z	$1S_{1/2}$	$2S_{1/2}$	$2P_{1/2}$	$2P_{3/2}$
15	3.801 41	4.050 88	$-0.104\,55$	0.136 57
16	3.672 90	3.924 42	$-0.102\,27$	0.137 95
17	3.554 20	3.807 84	$-0.099\,92$	0.139 36
18	3.444 15	3.699 98	$-0.097\,51$	0.140 82
26	2.783 94	3.059 47	$-0.076\,22$	0.153 62
36	2.279 70	2.585 36	$-0.044\,99$	0.171 81
54	1.783 14	2.162 05	0.025 64	0.208 57
82	1.495 69	2.077 34	0.206 32	0.271 86
92	1.490 92	2.199 49	0.319 34	0.295 38

Table 28.4. The function $G_{VP}^{(1)}(Z\alpha)$.

Z	$1S_{1/2}$	$2S_{1/2}$	$4S_{1/2}$	$2P_{1/2}$	$2P_{3/2}$
1	-0.6187	-0.8089	-0.8066	-0.0640	-0.0141
2	-0.6077	-0.7961	-0.7938	-0.0638	-0.0140

Table 28.5. The function $H^{(1)}(Z\alpha)$.

Z	$1S_{1/2}$	$2S_{1/2}$	$2P_{1/2}$	$2P_{3/2}$
15	$-0.244\,31$	$-0.246\,40$	$-0.000\,76$	$-0.000\,15$
16	$-0.243\,50$	$-0.245\,87$	$-0.000\,87$	$-0.000\,17$
17	$-0.242\,75$	$-0.245\,42$	$-0.000\,98$	$-0.000\,19$
18	$-0.242\,08$	$-0.245\,06$	$-0.001\,10$	$-0.000\,21$
26	$-0.238\,89$	$-0.245\,06$	$-0.002\,38$	$-0.000\,41$
36	$-0.240\,05$	$-0.252\,11$	$-0.004\,93$	$-0.000\,75$
54	$-0.256\,87$	$-0.287\,12$	$-0.013\,78$	$-0.001\,55$
82	$-0.340\,88$	$-0.445\,58$	$-0.057\,50$	$-0.003\,35$
92	$-0.406\,41$	$-0.574\,27$	$-0.099\,38$	$-0.004\,20$

$G_{SE}(Z\alpha)$ appear to be consistent with the limits at $Z = 0$ within the quoted uncertainties. Karshenboĭm has done similar extrapolations for $G_{SE}(\alpha)$ with slightly different numerical results [50].

At intermediate and high Z, the series in (28.46) converges slowly numerically, so it is preferable to work directly with numerical values for $F(Z\alpha)$. Results are shown in Table 28.3 [48]. In that table, the values at $Z = 16, 17, 18$ are obtained by interpolation of the published numbers, as discussed in Ref. [32].

28.3.5 Vacuum Polarization

The vacuum-polarization level shift is

$$E_{VP}^{(2)} = \frac{\alpha}{\pi} \frac{(Z\alpha)^4}{n^3} H(Z\alpha) m_e c^2, \quad (28.50)$$

where the function $H(Z\alpha)$ is divided into the part corresponding to the Uehling potential, denoted here by $H^{(1)}(Z\alpha)$, and the higher-order remainder $H^{(3+)}(Z\alpha)$ $= H^{(3)}(Z\alpha) + H^{(5)}(Z\alpha) + \cdots$, where the index denotes the order in powers of the external field. The individual terms are expanded in power series in $Z\alpha$ as

$$H^{(1)}(Z\alpha) = C_{40} + C_{50}(Z\alpha) + C_{61}(Z\alpha)^2 \ln(Z\alpha)^{-2}$$
$$+ G_{VP}^{(1)}(Z\alpha)(Z\alpha)^2, \quad (28.51)$$
$$H^{(3+)}(Z\alpha) = G_{VP}^{(3+)}(Z\alpha)(Z\alpha)^2, \quad (28.52)$$

with

$$\begin{aligned} C_{40} &= -\tfrac{4}{15}\delta_{l0}, \\ C_{50} &= \tfrac{5}{48}\pi\,\delta_{l0}, \\ C_{61} &= -\tfrac{2}{15}\delta_{l0}. \end{aligned} \quad (28.53)$$

Table 28.6. The function $H^{(3)}(Z\alpha)$.

Z	$1S_{1/2}$	$2S_{1/2}$	$2P_{1/2}$	$2P_{3/2}$
15	0.0006	0.0006	0.0000	0.0000
16	0.0006	0.0006	0.0000	0.0000
17	0.0007	0.0007	0.0000	0.0000
18	0.0008	0.0008	0.0000	0.0000
26	0.0015	0.0015	0.0000	0.0000
36	0.0027	0.0028	0.0001	0.0000
54	0.0058	0.0063	0.0004	0.0001
82	0.0142	0.0176	0.0033	0.0005
92	0.0194	0.0257	0.0063	0.0007

The part $G_{VP}^{(1)}(Z\alpha)$ arises from the Uehling potential, and is readily calculated numerically [51]. Values appear in Table 28.4. The higher-order remainder $G_{VP}^{(3+)}(Z\alpha)$ has been considered by Wichmann and Kroll, and the low-Z limit is given by [32, 52, 53]

$$G_{VP}^{(3)}(0) = \left(\tfrac{19}{45} - \tfrac{1}{27}\pi^2\right)\delta_{l0}. \quad (28.54)$$

Complete numerical calculations have been done for higher Z, as discussed below. Results for $H^{(3)}(Z\alpha)$ are listed in Table 28.6 [33].

28.3.6 Higher-Order Radiative Corrections

Radiative corrections from two virtual photons, of order α^2, have been calculated to the lowest two orders in $Z\alpha$. The lowest order level shifts are characterized as a self-energy correction [54–56]

$$E_{\text{SE}}^{(4)} = \left(\frac{\alpha}{\pi}\right)^2 \frac{(Z\alpha)^4}{n^3} m_e c^2$$
$$\times \left[2\pi^2 \ln 2 - \tfrac{49}{108}\pi^2 - \tfrac{4819}{1296} - 3\zeta(3)\right]\delta_{l0} , \quad (28.55)$$

a magnetic-moment correction [4, 5]

$$E_{\text{MM}}^{(4)} = \left(\frac{\alpha}{\pi}\right)^2 \frac{(Z\alpha)^4}{n^3} m_e c^2$$
$$\times \left[\tfrac{1}{2}\pi^2 \ln 2 - \tfrac{1}{12}\pi^2 - \tfrac{197}{144} - \tfrac{3}{4}\zeta(3)\right]\frac{1}{\kappa(2l+1)} , \quad (28.56)$$

and a vacuum-polarization correction [57, 58]

$$E_{\text{VP}}^{(4)} = \left(\frac{\alpha}{\pi}\right)^2 \frac{(Z\alpha)^4}{n^3} m_e c^2 \left[-\tfrac{82}{81}\right]\delta_{l0} , \quad (28.57)$$

where ζ is the Riemann zeta function.

Terms of relative order $Z\alpha$ can be divided into two classes depending on whether the corresponding Feynman diagrams do or do not have closed electron loops. The former category gives [59–63]

$$E_{\text{EL}}^{(4)} = \left(\frac{\alpha}{\pi}\right)^2 \frac{(Z\alpha)^5}{n^3} m_e c^2 \left[2.7106\cdots\right]\delta_{l0} , \quad (28.58)$$

while the latter category gives [64–66]

$$E_{\text{NL}}^{(4)} = \left(\frac{\alpha}{\pi}\right)^2 \frac{(Z\alpha)^5}{n^3} m_e c^2 \left[-24.269(3)\right]\delta_{l0} . \quad (28.59)$$

Expressing the complete fourth-order level shift as

$$E^{(4)} = \left(\frac{\alpha}{\pi}\right)^2 \frac{(Z\alpha)^4}{n^3} m_e c^2 \left[B_{40} + B_{50}(Z\alpha) + \cdots\right] , \quad (28.60)$$

we have

$$B_{40} = \left[2\pi^2 \ln 2 - \tfrac{49}{108}\pi^2 - \tfrac{6131}{1296} - 3\zeta(3)\right]\delta_{l0}$$
$$+ \left[\tfrac{1}{2}\pi^2 \ln 2 - \tfrac{1}{12}\pi^2 - \tfrac{197}{144} - \tfrac{3}{4}\zeta(3)\right]\frac{1}{\kappa(2l+1)} ,$$
$$B_{50} = -21.558(3)\delta_{l0} . \quad (28.61)$$

Estimates of higher-order terms have been made [50, 67].

The unexpectedly large value of B_{50}, $|B_{50}/B_{40}| \approx 40$ for S states, causes some theoretical predictions to shift beyond prior uncertainty estimates. Revised numerical results are given in Sect. 28.3.9.

Calculations of fourth-order corrections that include all orders in $Z\alpha$ have yielded partial results, but are still in the early stages of development [68, 69].

28.3.7 Reduced Mass and Radiative-Recoil Corrections

The dominant effect of nuclear motion on the radiative corrections is taken into account by multiplying the self energy and vacuum polarization by the reduced-mass factor $(m_r/m_e)^3$, except the magnetic moment term which is multiplied by the factor $(m_r/m_e)^2$. In addition, the argument of the logarithms in Eqs. (28.46) and (28.51) gains a factor of (m_e/m_r) [35]. This gives a modification

$$F(Z\alpha) + H(Z\alpha)$$
$$\rightarrow \left(\frac{m_r}{m_e}\right)^3 \left\{\left[\frac{4}{3}\ln\left(\frac{m_e}{m_r(Z\alpha)^2}\right) - \frac{4}{15} + \frac{10}{9}\right]\delta_{l0}\right.$$
$$\left. - \frac{4}{3}\ln k_0(n,l)\right\} - \left(\frac{m_r}{m_e}\right)^2 \frac{1-\delta_{l0}}{2\kappa(2l+1)}$$
$$+ \left(\frac{m_r}{m_e}\right)^3 \left(\frac{139}{32} + \frac{5}{48} - 2\ln 2\right)\delta_{l0}\pi(Z\alpha)$$
$$+ \cdots \quad (28.62)$$

to Eqs. (28.45) and (28.50).

Corrections to this prescription, radiative-recoil effects of order

$$\alpha\frac{(Z\alpha)^5 m_e}{n^3 m_N} m_e c^2 , \quad (28.63)$$

are the subject of recent and current study [70–72].

28.3.8 Nuclear Finite Size

At low Z, the effect of the finite nuclear size on level energies is given by

$$E_{\text{NS}} = \frac{2}{3}\frac{(Z\alpha)^4}{n^3}\left(\frac{R}{\lambda_C}\right)^2 \delta_{l0}\, m_e c^2 , \quad (28.64)$$

where R is the root-mean-square (rms) charge radius of the nucleus, and $\lambda_C = \alpha a_0$ is the Compton wavelength.

For intermediate values of Z (≤ 40), more accurate expressions that neglect higher-order terms of order $(Z\alpha)^4$ and $Z\alpha R/\lambda_C$ relative to the lowest-order S-state shift are given by [32]

$$E_{\text{NS}}\left(1S_{1/2}\right) = \left[1 + 0.50(Z\alpha)^2\right]\frac{2(Z\alpha)^2}{3}\left(\frac{Z\alpha R}{\lambda_C}\right)^{2s} m_e c^2 ,$$

$$E_{\text{NS}}\left(2S_{1/2}\right) = \left[1 + 1.38(Z\alpha)^2\right]\frac{(Z\alpha)^2}{12}\left(\frac{Z\alpha R}{\lambda_C}\right)^{2s} m_e c^2 ,$$

$$E_{\text{NS}}\left(2P_{1/2}\right) = \frac{(Z\alpha)^4}{64}\left(\frac{Z\alpha R}{\lambda_C}\right)^{2s} m_e c^2 ,$$

$$E_{\text{NS}}\left(2P_{3/2}\right) = 0 , \quad (28.65)$$

where $s = [1 - (Z\alpha)^2]^{1/2}$. For higher Z, nuclear-size corrections are obtained by numerical calculation of the Dirac eigenvalue [73], or by approximate analytic calculation [74].

At high Z, there are additional corrections that arise from the effect of the finite size of the nucleus on the

Table 28.7. The function $\Delta F(Z\alpha, R)$.

Z	R (fm)	$1S_{1/2}$	$2S_{1/2}$	$2P_{1/2}$
26	3.730	−0.000 17	−0.000 18	−0.000 00
36	4.230	−0.000 38	−0.000 40	−0.000 00
54	4.826	−0.001 28	−0.001 46	−0.000 02
82	5.505	−0.008 43	−0.012 02	−0.000 71
92	5.863	−0.018 49	−0.029 09	−0.002 48

Table 28.9. The function $H^{(3+)}(Z\alpha, R)$.

Z	R (fm)	$1S_{1/2}$	$2S_{1/2}$	$2P_{1/2}$
36	4.230	0.002 74	0.002 83	0.000 08
54	4.826	0.005 92	0.006 42	0.000 45
82	5.505	0.015 05	0.018 58	0.003 47
92	5.860	0.020 68	0.027 25	0.006 82

radiative corrections. The main change is in the lowest-order self energy and vacuum polarization [75–78].

For the self energy, the level shift in the field of a finite nucleus is given by Eq. (28.45) with $F(Z\alpha)$ generalized to $F(Z\alpha, R)$, which includes the effect of the finite size of the nucleus. The nuclear-size correction is proportional to $\Delta F(Z\alpha, R) = F(Z\alpha, R) - F(Z\alpha, 0)$, and the level shift ΔE_{SE} is

$$\Delta E_{\mathrm{SE}} = \frac{\alpha}{\pi} \frac{(Z\alpha)^4}{n^3} \Delta F(Z\alpha, R)\, m_e c^2 . \qquad (28.66)$$

Calculated values of the size effect are listed in Table 28.7 [75]. The values in this table are based on a nuclear model in which the charge is uniformly distributed in a sphere. The model dependence is expected to be less than 0.5% for any of the listed cases.

The analogous correction to the vacuum polarization, replacement of $H(Z\alpha)$ by $H(Z\alpha, R)$, leads to a correction of the form

$$\Delta E_{\mathrm{VP}} = \frac{\alpha}{\pi} \frac{(Z\alpha)^4}{n^3} \Delta H(Z\alpha, R)\, m_e c^2 , \qquad (28.67)$$

where $\Delta H(Z\alpha, R) = H(Z\alpha, R) - H(Z\alpha, 0)$. The Uehling potential correction $\Delta H^{(1)}(Z\alpha, R)$ is listed in Table 28.8 [33, 78]. There is remarkable cancellation between the nuclear-size corrections to the self energy and the Uehling potential, so the net effect is small. Table 28.9 lists the higher-order vacuum polarization $H^{(3+)}(Z\alpha, R)$, including the nuclear-size effect [77, 78]. A large part of the numbers in this table is accounted for in $H^{(3)}(Z\alpha)$ in Table 28.6.

28.3.9 Comparison of Experiment and Theory

In this section, theoretical predictions of the formulas described in the preceding sections are compared with experiment.

A summary of theoretical contributions to the Lamb shift $\mathcal{S} = E(2S_{1/2}) - E(2P_{1/2})$ in hydrogen is given in Table 28.10. The nuclear-size effect corresponds to the rms charge radius of the proton $R = 0.862(12)$ fm, measured by Simon *et al.* [79]. An earlier measurement, $R = 0.805(11)$ [80], yields a theoretical Lamb shift of $\mathcal{S} = 1057.820(6)$ MHz. In Table 28.10, the small contribution of the term of order $(m_e/m_N)^2(Z\alpha)^4 m_e c^2$ from Eq. (28.39) is included in the value of the relativistic recoil effect. Recent measurements of the Lamb shift are listed in Table 28.11. In that table, the Hagley and Pipkin value is based on a measurement of the interval $E(2P_{3/2}) - E(2S_{1/2}) = \Delta E - \mathcal{S}$. The corresponding Lamb shift is calculated with the theoretical value of the fine-structure splitting in hydrogen $\Delta E = E(2P_{3/2}) - E(2P_{1/2}) = 10\,969.041\,4(6)$ MHz, based on the quantized Hall effect value of alpha [17].

Measurements and theory for the ground-state (1S-state) radiative level shift in hydrogen, often also called the Lamb shift, which by convention includes all contributions to the energy level except hyperfine structure and the first two terms of Eq. (28.39), $Mc^2 + [f(n,j) - 1]m_r c^2$, are listed in Table 28.12. The experimental values are deduced from measurements that compare different transitions in hydrogen in order to eliminate uncertainty from

Table 28.8. The function $\Delta H^{(1)}(Z\alpha, R)$.

Z	R (fm)	$1S_{1/2}$	$2S_{1/2}$	$2P_{1/2}$
36	4.230	0.000 22	0.000 25	0.000 01
54	4.826	0.000 93	0.001 06	0.000 03
82	5.505	0.007 73	0.010 82	0.000 82
92	5.860	0.018 29	0.028 14	0.002 98

Table 28.10. Theoretical contributions to the Lamb shift in hydrogen.

Contribution	Order ($m_e c^2$)	Value (MHz)
Self energy	$\alpha(Z\alpha)^4$	1085.817
Vacuum polarization	$\alpha(Z\alpha)^4$	−26.897
Fourth order	$\alpha^2(Z\alpha)^4$	0.064(4)
Reduced mass	$(m_e/m_N)\alpha(Z\alpha)^4$	−1.646(3)
Relativistic recoil	$(m_e/m_N)(Z\alpha)^5$	0.355
Nuclear size	$(Z\alpha)^4(R/\lambda_C)^2$	0.145(4)
Total		1057.839(6)

Table 28.11. Lamb shift in hydrogen.

Reference	$\mathcal{S}$ (MHz)
Newton Andrews & Unsworth [81]	1057.862(20)
Pal'chikov Sokolov & Yakovlev [82]	1057.8514(19)
Lundeen & Pipkin [83]	1057.845(9)
Hagley & Pipkin [84]†	1057.841(12)
Theory	1057.839(6)

† Indirect measurement

Table 28.12. $1S$ Lamb shift in hydrogen.

Reference	$L(1S)$ (MHz)
Nez *et al.* [85]	8172.815(70)
Weitz *et al.* [86]	8172.86(6)
Berkeland Hinds & Boshier [87]	8172.827(51)
Theory	8172.80(5)

Table 28.13. Lamb shift in He$^+$.

Reference	$\mathcal{S}$ (MHz)
Dewey & Dunford [89]	14 042.0(1.2)
van Wijngaarden Kwela & Drake [90]	14 042.52(16)
Theory	14 041.3(2)

the Rydberg constant.

Measurements of the Lamb shift in hydrogen-like ions test the theory in the presence of a strong binding field. Table 28.13 shows recent experiments and theory for the Lamb shift in He$^+$. Lamb shift measurements on one-electron atoms with Z in the range 15–18 are listed in Table 28.14. The higher-order binding correction to the two-photon QED effects brings theory into disagreement with the latest measurement in He$^+$, but an apparent discrepancy with experiments on hydrogenic phosphorus and sulfur is reduced by the same correction. The theoretical uncertainty in the mid- and high-Z range is increased over earlier estimates [32, 33, 88], because the higher-order correction grows more rapidly with Z than was anticipated, and still higher-order corrections could change the predictions by a comparable amount.

Experiments on levels in high-Z one-electron atoms such as hydrogen-like uranium are currently in a stage of rapid development. For these systems, nuclear-size effects play a relatively larger role than in low-Z atoms. Experiments and theory are compared in Table 28.15. The theoretical contributions are listed in more detail in Table 28.16. The large uncertainty of the fourth-order QED term is due to the fact that it is based on a truncated power series. Calculations of QED effects from two-photon diagrams that include all orders in $Z\alpha$ will eliminate this source of uncertainty.

Table 28.14. Lamb shift in mid-Z hydrogen-like ions.

Z	$\mathcal{S}$(th) (THz)	$\mathcal{S}$(exp) (THz)	Reference
15	20.23(2)	20.13(20)	Pellegrin *et al.* [91]†
15		20.188(29)	Pross *et al.* [92]†
16	25.34(3)	25.14(24)	Zacek *et al.* [93]
16		25.266(63)	Georgiadis *et al.* [94]†
17	31.30(4)	31.19(22)	Wood *et al.* [95]
18	38.19(6)	37.89(38)	Gould & Marrus [96]

† Indirect measurement

Table 28.15. $1S$ Lamb shift in hydrogen-like uranium.

Reference	$L(1S)$ (eV)
Briand *et al.* [97]	520(130)
Stöhlker *et al.* [98]	429(63)
Lupton *et al.* [99]	509(92)
Beyer *et al.* [100]	470(16)
Theory	465(1)

Table 28.16. $1S$ Lamb shift in hydrogen-like uranium

Contribution	$L(1S)$ (eV)
Self energy	359.5
Self-energy size	−4.5
Uehling vac pol	−98.0
Uehling vac pol size	4.4
Wichmann Kroll	5.0
Fourth order	0.3(1.0)
Relativistic recoil	0.1(1)
Nuclear polarization	−1.1(3)
Nuclear size	198.8(4)
Total	465(1)

28.4 MUONIUM ATOM

Muonium is the atom consisting of a positive muon and an electron. From the theoretical perspective, this atom has the advantage over hydrogen that the muon is a Dirac particle which is better understood than the proton. On the other hand, the muon is lighter than the proton, $m_\mu \approx 207 m_e$, so that higher-order recoil effects must be calculated to reach the same level of precision in the theory.

28.4.1 Hyperfine Structure

An intensively studied property of muonium is the ground-state hyperfine splitting. Comparison of theory and experiment for this splitting provides a link between the mass of the muon and the value of the fine-structure constant.

The experimental value of the hyperfine splitting of the ground state of muonium is [101]

$$\Delta\nu(\text{exp}) = 4\,463\,302.88(16) \text{ kHz}, \qquad (28.68)$$

which represents an uncertainty of 0.036 ppm.

The dominant part of the theoretical value is given by the Fermi formula [102]

$$\nu_{\text{F}} = \frac{16}{3} Z^3 \alpha^2 c R_\infty \frac{m_e}{m_\mu} \left[1 + \frac{m_e}{m_\mu}\right]^{-3}. \qquad (28.69)$$

The various corrections to this value are grouped into categories according to [35, 103]

$$\Delta\nu(\text{th}) = \Delta\nu_{\text{rad}} + \Delta\nu_{\text{rec}} + \Delta\nu_{\text{r-r}}$$
$$+ \Delta\nu_{\text{had}} + \Delta\nu_{\text{weak}}, \qquad (28.70)$$

where the terms labeled rad, rec, r-r, had, and weak take into account the radiative, recoil, radiative-recoil, hadronic, and weak interaction corrections to the hyperfine splitting, respectively.

Radiative corrections are included together with the other nonrecoil contributions in

$$\Delta\nu_{\text{rad}} = \nu_{\text{F}}(1 + a_\mu)\Bigg\{1 + \frac{3}{2}(Z\alpha)^2 + \frac{17}{8}(Z\alpha)^4 + a_e$$
$$+ \alpha(Z\alpha)\left(\ln 2 - \frac{5}{2}\right) + \frac{\alpha(Z\alpha)^2}{\pi}\left[-\frac{2}{3}\ln^2(Z\alpha)^{-2}\right.$$
$$+ \left(\frac{281}{360} - \frac{8}{3}\ln 2\right)\ln(Z\alpha)^{-2} + 14.88(29)\Bigg]$$
$$+ \alpha(Z\alpha)^3 C_\alpha \ln(Z\alpha)^{-2} + \frac{\alpha^2(Z\alpha)}{\pi} 0.7717(4)$$
$$+ \frac{\alpha^2(Z\alpha)^2}{\pi^2}\left[-\frac{1}{3}\ln^2(Z\alpha)^{-2} + C_{\beta\epsilon}\ln(Z\alpha)^{-2}\right]\Bigg\}$$
$$+ \cdots, \qquad (28.71)$$

where a_e and a_μ are the anomalous magnetic moments of the electron and muon, respectively. The constant 14.88(29) in the term proportional to $\alpha(Z\alpha)^2$ is made up of a contribution 15.10(29) from the self energy [104] and a contribution $\frac{34}{225} - \frac{8}{15}\ln 2$ from vacuum polarization [105, 106]. Earlier work is reviewed in Ref. [35], and the coefficient of the term proportional to $\alpha^2(Z\alpha)$ is the result of a recent series of calculations of two-photon Feynman diagrams [64, 66, 107–110]. The coefficient of the term proportional to $\alpha^2(Z\alpha)^2 \ln^2(Z\alpha)^{-2}$ is given in Ref. [67], and the contributions of the terms proportional to C_α [111] and $C_{\beta\epsilon}$ are given numerically below from Ref. [110].

Recoil corrections are given by [35, 67, 110]

$$\Delta\nu_{\text{rec}} = \nu_{\text{F}}\Bigg\{-\frac{3Z\alpha}{\pi}\frac{m_e m_\mu}{m_\mu^2 - m_e^2}\ln\left(\frac{m_\mu}{m_e}\right)$$
$$+ \frac{(Z\alpha)^2 m_r^2}{m_e m_\mu}\left[\ln(Z\alpha)^{-2} - 8\ln 2 + \frac{65}{18}\right]$$
$$+ \frac{(Z\alpha)^3}{\pi}\frac{m_e}{m_\mu}\left[C_\delta \ln\left(\frac{m_\mu}{m_e}\right)\ln(Z\alpha)^{-2}\right.$$
$$\left.- \frac{1}{6}\ln^2(Z\alpha)^{-2} + C_\gamma \ln(Z\alpha)^{-2}\right]\Bigg\}$$
$$+ \cdots, \qquad (28.72)$$

where $m_r = m_e m_\mu/(m_e + m_\mu)$.

The radiative-recoil corrections are [35, 67, 112, 113]

$$\Delta\nu_{\text{r-r}} = \nu_{\text{F}}\frac{\alpha}{\pi}\frac{m_e}{m_\mu}\Bigg\{\frac{\alpha}{\pi}\left[-2\ln^2\left(\frac{m_\mu}{m_e}\right) + \frac{13}{12}\ln\left(\frac{m_\mu}{m_e}\right)\right.$$
$$\left.+ \frac{21}{2}\zeta(3) + \zeta(2) + \frac{35}{9}\right] + \frac{4}{3}\alpha^2 \ln^2 \alpha^{-2}$$
$$+ \frac{\alpha^2}{\pi^2}\left[-\frac{4}{3}\ln^3\left(\frac{m_\mu}{m_e}\right) + \frac{4}{3}\ln^2\left(\frac{m_\mu}{m_e}\right)\right]\Bigg\}$$
$$+ \cdots. \qquad (28.73)$$

The terms in Eqs. (28.71) and (28.72) corresponding to the coefficients C_α, $C_{\beta\epsilon}$, C_γ, and C_δ, contribute -0.381 kHz, -0.003 kHz, -0.189 kHz, and -0.210 kHz, respectively, to the hyperfine splitting for a total of [110]

$$\Delta\nu_{\alpha-\epsilon} = -0.783(50) \text{ kHz}. \qquad (28.74)$$

Hadronic [114] and weak interaction [115] corrections are given respectively by

$$\Delta\nu_{\text{had}} = 0.250(16) \text{ kHz}, \qquad (28.75)$$
$$\Delta\nu_{\text{weak}} \approx -0.065 \text{ kHz}. \qquad (28.76)$$

With the value of the fine-structure constant deduced from the quantized Hall effect measurement [17] and values for R_∞ and m_μ/m_e from Refs. [85] and [101]

$$R_\infty = 10\,973\,731.568\,34(24) \text{ m}^{-1},$$
$$\frac{m_\mu}{m_e} = 206.768\,259(62), \qquad (28.77)$$

the theoretical value

$$\Delta\nu(\text{th}) = 4\,463\,302.2(1.4) \text{ kHz} \qquad (28.78)$$

is obtained. The uncertainty in Eq. (28.78) is due primarily to the uncertainty in the muon-electron mass ratio, with uncertainty in the theory and in alpha contributing 0.17 kHz and 0.21 kHz, respectively.

Since the uncertainty in the muon-electron mass ratio is the dominant component of the uncertainty in Eq. (28.78), one can determine this quantity with a smaller estimated uncertainty as the value for which theory and experiment are equal:

$$\frac{m_\mu}{m_e} = 206.768\,227(14). \qquad (28.79)$$

Acknowledgments

The author gratefully acknowledges helpful conversations with Dr. Barry N. Taylor, Prof. Michael Eides, Prof. Toichiro Kinoshita, Dr. Makiko Nio, and Prof. Jonathan Sapirstein.

REFERENCES

1. R. S. Van Dyck, Jr., P. B. Schwinberg, and H. G. Dehmelt, Phys. Rev. Lett. **59**, 26 (1987).
2. J. Schwinger, Phys. Rev. **73**, 416 (1948).
3. R. Karplus and N. M. Kroll, Phys. Rev. **77**, 536 (1950).
4. C. M. Sommerfield, Phys. Rev. **107**, 328 (1957).
5. A. Petermann, Helv. Phys. Acta **30**, 407 (1957).
6. G. S. Adkins, Phys. Rev. D **39**, 3798 (1989).
7. T. Kinoshita and W. B. Lindquist, Phys. Rev. D **42**, 636 (1990).
8. S. Laporta and E. Remiddi, Phys. Lett. B **265**, 182 (1991).
9. T. Kinoshita, IEEE Trans. Instrum. Meas. **44**, 498 (1995).
10. T. Kinoshita, Phys. Rev. Lett. **75**, 4728 (1995).
11. B. E. Lautrup and E. de Rafael, Phys. Rev. **174**, 1835 (1968).
12. E. R. Cohen and B. N. Taylor, Rev. Mod. Phys. **59**, 1121 (1987).
13. L. Montanet *et al.*, Phys. Rev. D **50**, 1173 (1994).
14. T. Kinoshita, in *New Frontiers in High Energy Physics*, edited by B. Kursunoglu, A. Perlmutter, and L. F. Scott (Plenum, New York, 1978), p. 127.
15. R. Jackiw and S. Weinberg, Phys. Rev. D **5**, 2396 (1972).
16. K. B. Samuel and M. A. Samuel, Can. J. Phys. **68**, 1359 (1990).
17. M. E. Cage *et al.*, IEEE Trans. Instrum. Meas. **38**, 284 (1989).
18. J. Bailey *et al.*, Phys. Lett. B **68**, 191 (1977).
19. F. J. M. Farley and E. Picasso, in *Quantum Electrodynamics*, edited by T. Kinoshita (World Scientific, Singapore, 1990), Chap. 11, p. 479.
20. H. Suura and E. Wichmann, Phys. Rev. **105**, 1930 (1957).
21. A. Petermann, Phys. Rev. **105**, 1931 (1957).
22. H. H. Elend, Phys. Lett. **20**, 682 (1966).
23. H. H. Elend, Phys. Lett. **21**, 720 (1966).
24. G. W. Erickson and H. T. Liu, UCD-CNL-81 report (1968).
25. G. Li, R. Mendel, and M. A. Samuel, Phys. Rev. D **47**, 1723 (1993).
26. T. Kinoshita, B. Nižić, and Y. Okamoto, Phys. Rev. D **41**, 593 (1990).
27. M. A. Samuel and G. Li, Phys. Rev. D **44**, 3935 (1991).
28. S. Laporta and E. Remiddi, Phys. Lett. B **301**, 440 (1993).
29. S. Laporta, Phys. Lett. B **312**, 495 (1993).
30. Ľ. Martinovič and S. Dubnička, Phys. Rev. D **42**, 884 (1990).
31. T. Kinoshita and W. J. Marciano, in *Quantum Electrodynamics*, edited by T. Kinoshita (World Scientific, Singapore, 1990), Chap. 10, p. 419.
32. P. J. Mohr, At. Data Nucl. Data Tables **29**, 453 (1983).
33. W. R. Johnson and G. Soff, At. Data Nucl. Data Tables **33**, 405 (1985).
34. W. A. Barker and F. N. Glover, Phys. Rev. **99**, 317 (1955).
35. J. R. Sapirstein and D. R. Yennie, in *Quantum Electrodynamics*, edited by T. Kinoshita (World Scientific, Singapore, 1990), Chap. 12, p. 560.
36. E. E. Salpeter, Phys. Rev. **87**, 328 (1952).
37. G. W. Erickson, J. Phys. Chem. Ref. Data **6**, 831 (1977).
38. M. Doncheski, H. Grotch, and G. W. Erickson, Phys. Rev. A **43**, 2152 (1991).
39. R. N. Fell, I. B. Khriplovich, A. I. Milstein, and A. S. Yelkhovsky, Phys. Lett. A **181**, 172 (1993).
40. K. Pachucki and H. Grotch, Phys. Rev. A **51**, 1854 (1995).
41. E. A. Golosov, A. S. Elkhovskiĭ, A. I. Mil'shteĭn, and I. B. Khriplovich, Zh. Eksp. Teor. Fiz. **107**, 393 (1995), [JETP **80**, 208 (1995)].
42. G. Plunien, B. Müller, W. Greiner, and G. Soff, Phys. Rev. A **43**, 5853 (1991).
43. K. Pachucki, D. Leibfried, and T. W. Hänsch, Phys. Rev. A **48**, 1 (1993).
44. G. W. Erickson and D. R. Yennie, Ann. Phys. (N. Y.) **35**, 271, 447 (1965).
45. G. W. F. Drake and R. A. Swainson, Phys. Rev. A **41**, 1243 (1990).
46. K. Pachucki, Ann. Phys. (N. Y.) **226**, 1 (1993).
47. P. J. Mohr, Phys. Rev. Lett. **34**, 1050 (1975).
48. P. J. Mohr, Phys. Rev. A **46**, 4421 (1992).
49. P. J. Mohr and Y.-K. Kim, Phys. Rev. A **45**, 2727 (1992).
50. S. G. Karshenboĭm, Zh. Eksp. Teor. Fiz. **106**, 414 (1994), [JETP **79**, 230 (1994)].
51. P. J. Mohr, Phys. Rev. A **26**, 2338 (1982).
52. E. H. Wichmann and N. M. Kroll, Phys. Rev. **101**, 843 (1956).
53. P. J. Mohr, in *Beam-Foil Spectroscopy*, edited by I. A. Sellin and D. J. Pegg (Plenum Press, New York, 1976), Vol. 1, p. 89.
54. B. E. Lautrup, A. Peterman, and E. de Rafael, Phys. Lett. B **31**, 577 (1970).
55. R. Barbieri, J. A. Mignaco, and E. Remiddi, Lett. Nuovo Cimento **3**, 588 (1970).
56. A. Peterman, Phys. Lett. B **35**, 325 (1971).
57. M. Baranger, F. J. Dyson, and E. E. Salpeter, Phys. Rev. **88**, 680 (1952).
58. G. Källén and A. Sabry, Dan. Mat. Fys. Medd. **29**, No. 17 (1955).
59. M. I. Eides, H. Grotch, and D. A. Owen, Phys. Lett. B **294**, 115 (1992).
60. M. I. Eides and H. Grotch, Phys. Lett. B **301**, 127 (1993).
61. M. I. Eides and H. Grotch, Phys. Lett. B **308**, 389 (1993).
62. M. I. Eides, H. Grotch, and P. Pebler, Phys. Rev. A **50**, 144 (1994).
63. K. Pachucki, Phys. Rev. A **48**, 2609 (1993).

64. M. I. Eides, S. G. Karshenboim, and V. A. Shelyuto, Phys. Lett. B **312**, 358 (1993).
65. K. Pachucki, Phys. Rev. Lett. **72**, 3154 (1994).
66. M. I. Eides and V. A. Shelyuto, Pis'ma Zh. Eksp. Teor. Fiz. **61**, 465 (1995) [JETP Lett. **61**, 478 (1995)].
67. S. G. Karshenboïm, Zh. Eksp. Teor. Fiz. **103**, 1105 (1993), [JETP **76**, 541 (1993)].
68. T. Beier and G. Soff, Z. Phys. D **8**, 129 (1988).
69. I. Lindgren *et al.*, J. Phys. B **26**, L503 (1993).
70. G. Bhatt and H. Grotch, Phys. Rev. A **31**, 2794 (1985).
71. G. Bhatt and H. Grotch, Phys. Rev. Lett. **58**, 471 (1987).
72. G. C. Bhatt and H. Grotch, Ann. Phys. (N. Y.) **178**, 1 (1987).
73. T. Franosch and G. Soff, Z. Phys. D **18**, 219 (1991).
74. V. M. Shabaev, J. Phys. B **26**, 1103 (1993).
75. P. J. Mohr and G. Soff, Phys. Rev. Lett. **70**, 158 (1993).
76. K. T. Cheng, W. R. Johnson, and J. Sapirstein, Phys. Rev. A **47**, 1817 (1993).
77. G. Soff and P. J. Mohr, Phys. Rev. A **38**, 5066 (1988).
78. H. Persson, I. Lindgren, S. Salomonson, and P. Sunnergren, Phys. Rev. A **48**, 2772 (1993).
79. G. G. Simon, Ch. Schmitt, F. Borkowski, and V. H. Walther, Nucl. Phys. A **333**, 381 (1980).
80. L. N. Hand, D. G. Miller, and R. Wilson, Rev. Mod. Phys. **35**, 335 (1963).
81. G. Newton, D. A. Andrews, and P. J. Unsworth, Philos. Trans. R. Soc. London **290**, 373 (1979).
82. V. G. Pal'chikov, Yu. L. Sokolov, and V. P. Yakovlev, Pis'ma Zh. Eksp. Teor. Fiz. **38**, 347 (1983), [JETP Lett. **38**, 418 (1983)].
83. S. R. Lundeen and F. M. Pipkin, Metrologia **22**, 9 (1986).
84. E. W. Hagley and F. M. Pipkin, Phys. Rev. Lett. **72**, 1172 (1994).
85. F. Nez *et al.*, Europhys. Lett. **24**, 635 (1993).
86. M. Weitz *et al.*, Phys. Rev. Lett. **72**, 328 (1994).
87. M. Boshier, private communication (1995).
88. P. J. Mohr, Nucl. Instrum. Meth. B **87**, 232 (1994).
89. M. S. Dewey and R. W. Dunford, Phys. Rev. Lett. **60**, 2014 (1988).
90. A. van Wijngaarden, J. Kwela, and G. W. F. Drake, Phys. Rev. A **43**, 3325 (1991).
91. P. Pellegrin, Y. El Masri, and L. Palffy, Phys. Rev. A **31**, 5 (1985).
92. H.-J. Pross *et al.*, Phys. Rev. A **48**, 1875 (1993).
93. V. Zacek *et al.*, Z. Phys. A **318**, 7 (1984).
94. A. P. Georgiadis *et al.*, Phys. Lett. A **115**, 108 (1986).
95. O. R. Wood, II *et al.*, Phys. Rev. Lett. **48**, 398 (1982).
96. H. Gould and R. Marrus, Phys. Rev. A **28**, 2001 (1983).
97. J. P. Briand *et al.*, Phys. Rev. Lett. **65**, 2761 (1990).
98. Th. Stöhlker *et al.*, Phys. Rev. Lett. **71**, 2184 (1993).
99. J. H. Lupton *et al.*, Phys. Rev. A **50**, 2150 (1994).
100. H. F. Beyer, IEEE Trans. Instrum. Meas. **44**, 510 (1995).
101. F. G. Mariam *et al.*, Phys. Rev. Lett. **49**, 993 (1982).
102. E. Fermi, Z. Phys. **60**, 320 (1930).
103. G. T. Bodwin, D. R. Yennie, and M. A. Gregorio, Rev. Mod. Phys. **57**, 723 (1985).
104. J. R. Sapirstein, Phys. Rev. Lett. **51**, 985 (1983).
105. S. J. Brodsky and G. W. Erickson, Phys. Rev. **148**, 26 (1966).
106. S. M. Schneider, W. Greiner, and G. Soff, Phys. Rev. A **50**, 118 (1994).
107. M. I. Eides, S. G. Karshenboim, and V. A. Shelyuto, Phys. Lett. B **229**, 285 (1989).
108. M. I. Eides, S. G. Karshenboim, and V. A. Shelyuto, Phys. Lett. B **249**, 519 (1990).
109. M. I. Eides, S. G. Karshenboim, and V. A. Shelyuto, Phys. Lett. B **268**, 433 (1991), **316**, E631 (1993), **319**, E545 (1993).
110. T. Kinoshita and M. Nio, Phys. Rev. Lett. **72**, 3803 (1994).
111. G. P. Lepage, quoted in [110].
112. M. I. Eides and V. A. Shelyuto, Phys. Lett. B **146**, 241 (1984).
113. M. I. Eides, S. G. Karshenboim, and V. A. Shelyuto, Phys. Lett. B **216**, 405 (1989).
114. A. Karimkhodzhaev and R. N. Faustov, Yad. Fiz. **53**, 1012 (1991), [Sov. J. Nucl. Phys. **53**, 626 (1991)].
115. M. A. B. Bég and G. Feinberg, Phys. Rev. Lett. **33**, 606 (1974); **35**, 130 (1975).

29

Parity Nonconserving Effects in Atoms

J. Sapirstein

Department of Physics, University of Notre Dame, Notre Dame, Indiana 46556

29.1 INTRODUCTION

Until 1957 the invariance of the laws of physics under the process of parity inversion was assumed to hold. The concept of this invariance has its origin in atomic physics, where Laporte introduced it to explain certain aspects of the iron spectrum. The underlying theory of atomic structure, Quantum Electrodynamics (QED), is an example of a theory that has this invariance. However, it was shown in 1957 that weak interaction processes are not invariant under parity inversion. At that time the only known weak interactions were charge changing, and thus did not affect the spectrum of a stable atom. However, if charge *conserving* neutral weak interactions exist, these can lead to parity nonconserving (PNC) processes in atoms. These include electric dipole transitions between states of the same parity, and optical rotation. It is now well established that such interactions do exist, and are mediated by the Z boson. Unfortunately, because of the extremely small ratio of the energy scale of atoms to the Z mass, PNC processes in atoms must be correspondingly small. However, in 1974 the Bouchiats [1], in a paper that laid the foundation for the field, showed that PNC transitions in heavy atoms with atomic number Z were enhanced by a factor of Z^3. While still very small, this effect has been observed in a variety of heavy atoms, specifically cesium ($Z = 55$) [2, 3], thallium ($Z = 81$) [4, 5], lead ($Z = 82$) [6], and bismuth ($Z = 83$)

[7]. The accurate calculation of the electronic structure of such atoms is, of course, a very challenging atomic physics problem, and such calculations must be carried out before the experiments can be interpreted in terms of particle physics. This problem is not present for hydrogen, but experimental problems have impeded progress in this direction.

At the time of the Bouchiats's paper, the question of whether or not neutral currents existed was an important problem in weak interaction physics, and even a qualitative observation of atomic PNC would have been of great interest. However, now that neutral currents are well established, what has become the most important aspect of atomic PNC is the precise measurement of this effect. This is because of the present intense interest in possible modifications of the Standard Model. While this model, described in more detail in the next section, is extremely successful, it is incomplete in some ways, in particular in the way particle masses are treated. For example, in the Standard Model, the mystery of the fermion mass spectrum is simply transferred to another mystery of the sizes of Yukawa coupling constants that couple Higgs fields to fermions. It is hoped that some deviation from the Standard Model will be uncovered, generally referred to as *new physics*, that will provide a hint as to a more satisfactory theory. Because, at the present time, all Standard Model predictions are in agreement with experiment at the few percent level, the hoped for deviation will be small, and for this reason high precision experiments, combined with calculations of similar accuracy, are needed. For this reason, atomic PNC, while apparently a somewhat esoteric branch of atomic physics, actually is involved with one of the most central problems of the field, namely, the accurate solution of the Schrödinger equation for many-electron atoms. We will describe an approach to this fundamental problem in Sect. 29.4.

Another weak interaction effect important in atomic physics has to do with CP noninvariance, which refers to noninvariance under the simultaneous operations of parity inversion (P) and charge conjugation (C). While the weak interactions are not invariant under P, until 1964 it appeared that they *were* invariant under CP. However, in 1964, measurements of the neutral K meson (kaon) system showed that CP invariance is slightly broken. While this does not directly affect atomic physics, it has been suggested that this still incompletely explained breaking in the quark sector may have a counterpart in the lepton sector, which could have as a consequence a nonvanishing electron electric dipole moment (edm). This opens the possibility of using precise atomic experiments to search for such an effect. In this case, as with 'ordinary' PNC, a calculation of atomic structure is necessary. The requirements of precision are not as important in this case, as the simple detection of a nonzero atomic edm would be a discovery of importance comparable to the detection of CP noninvariance in the kaon system. Because of limitations of space, we do not discuss this interesting field further, but instead refer the interested reader to a recent review [8].

29.2 THE STANDARD MODEL

The Standard Model is the present theory of the strong, weak, and electromagnetic interactions. We are concerned with the part of the Standard Model that unifies the weak and electromagnetic interactions, which we refer to as the electroweak theory. This theory involves the interactions of leptons and quarks with four vector bosons, the photon (A), a heavy neutral partner of the photon (Z), and two charged bosons (W$^+$, W$^-$). Interactions involving the photon conserve parity, while interactions involving the other bosons do not. The neutral currents affecting atomic PNC are described by the Lagrangian

$$L = -e \sum_i q_i \bar{\psi}_i \gamma_\mu \psi_i A^\mu$$
$$- \frac{e}{\sin 2\theta_W} \sum_i \bar{\psi}_i \gamma_\mu (V_i - A_i \gamma_5) \psi_i Z^\mu , \quad (29.1)$$

where i ranges over three types of fermions, the electron ($q_e = -1$, $V_e = -\frac{1}{2} - 2\sin^2\theta_W$, $A_e = -\frac{1}{2}$), the up quark ($q_u = \frac{2}{3}$, $V_u = \frac{1}{2} - \frac{4}{3}\sin^2\theta_W$, $A_u = \frac{1}{2}$), and the down quark ($q_d = -\frac{1}{3}$, $V_d = -\frac{1}{2} + \frac{2}{3}\sin^2\theta_W$, $A_d = -\frac{1}{2}$). θ_W is a fundamental parameter of the Standard Model known as the *weak angle*, which we discuss further below.

The physics described by the above Lagrangian, which leads to atomic PNC, is the exchange of a virtual Z boson, either between a quark in the nucleus and an electron, or between two electrons, though calculations of the latter effect show it to be negligible [9]. PNC arises when the Z matrix element acts as a polar vector on the nucleus and as an axial vector on the electron ($V_N A_e$), or vice versa ($A_N V_e$). The dominant PNC contribution comes from the former case because all the quarks contribute coherently. The latter case however cannot be neglected, and is discussed in Sects. 29.3 and 29.5. Because the nuclear current is of a vector nature, one can introduce a conserved charge, the *weak charge*,

$$Q_W = 2Z(2V_u + V_d) + 2N(V_u + 2V_d) . \quad (29.2)$$

Here, Z is the number of protons and N the number of neutrons. Putting in the above values for V_u and V_d gives

$$Q_W = Z(1 - 4\sin^2\theta_W) - N . \quad (29.3)$$

It would appear that the dependence of Q_W on the weak angle would allow it to be determined from atomic PNC. While this can be done, the above discussion has been only at the tree level, i.e., Feynman diagrams without loops. One of the most important features of the electroweak theory is that radiative corrections can be calculated. These corrections enter at the percent level, and must be included for accurate interpretation of experiments sensitive to electroweak effects. Until recently, of the parameters affecting the electroweak theory, only the fine structure constant and the muon decay rate were known with high precision, which left θ_W as a free parameter. However, now that the Z mass has been measured with very high precision, it is possible to use that measurement to determine θ_W, and then to predict a large set of radiatively corrected electroweak processes, including atomic PNC [10]. While there is still some sensitivity to the top quark and Higgs masses, as discussed in Sect. 29.6, using the Z mass allows a prediction for Cs of

$$Q_W = -73.2(0.2) . \quad (29.4)$$

Therefore, the importance of atomic PNC for fundamental physics is not so much its ability to determine θ_W, but rather the fact that the Standard Model makes a definite prediction for the size of the effect. Thus, any disagreement indicates new physics not contained in the Standard Model, and conversely, agreement to within a given precision puts limits on possible deviations from the Standard Model.

29.3 PNC IN CESIUM

We concentrate on PNC in Cs because it is the simplest atom in which the effect has been measured. This is because it consists of a single electron outside a closed Xe-like core which is relatively unpolarizable. This should be contrasted to, for example, Tl. While thallium nominally also consists of one $6p_{1/2}$ electron outside a closed core, part of that core is a filled $6s_{1/2}^2$ subshell. It is quite easy to polarize the outer subshell, so that one really has three electrons outside a closed core. This leads to distinctly poorer convergence properties of many-body perturbation theory (MBPT), the theoretical method used for these calculations, and consequently, less accurate atomic theory predictions. Similar considerations apply to Pb and Bi. The PNC transition in Cs that has been studied is $6s_{1/2} \to 7s_{1/2}$, and accurate measurements have been made by the Bouchiats [2] and by Wieman, *et al.* [3]. Cs is a 55-electron atom with a nucleus consisting of 78 neutrons and 55 protons with nuclear spin $I = 7/2$. The total angular momentum of atomic *s*-states is then $F = 3$ or $F = 4$. Both of the transitions, $6s_{1/2}(F = 4) \to 7s_{1/2}(F = 3)$ and $6s_{1/2}(F = 3) \to 7s_{1/2}(F = 4)$ have been measured, allowing the isolation of PNC effects that depend on the spin of the nucleus. The bulk of atomic PNC comes from the timelike contribution of the $(V_N A_e)$ exchange discussed in the previous section, and can be described by the effective Hamiltonian

$$H_W = \frac{G_F}{\sqrt{8}} Q_W \rho_{\text{nuc}}(\mathbf{r}) \gamma_5 . \tag{29.5}$$

Here, $\rho_{\text{nuc}}(\mathbf{r})$ is a weighted average of the neutron and proton distributions in the nucleus, which leads to nuclear structure uncertainties that will be discussed in Sect. 29.5. Using this Hamiltonian, a large scale calculation [9] leads to a prediction for the nuclear-spin-independent part of the PNC transition

$$E_{\text{PNC}} = -0.905(9) \times 10^{-11} i |e| a_0 (-Q_W/N) . \tag{29.6}$$

Here, the unknown Q_W has been factored out and divided by its approximate value $-N$. This result is in good agreement with independent MBPT calculations of the Novosibirsk [11] and Göteborg [12] groups. When this is compared with the experimental measurement [3]

$$E_{\text{PNC}}^{\text{exp}} = -0.8252(184)[61] \times 10^{-11} i |e| a_0 , \tag{29.7}$$

there results a prediction for Q_W of

$$Q_W = -71.04(1.58)[0.88] , \tag{29.8}$$

where the first error is experimental and the second theoretical. The spacelike part of $(V_N A_e)$ exchange and the timelike part of $(A_N V_e)$ exchange are negligible, but the spacelike part of the latter gives a nuclear-spin-dependent effect that will be discussed in Sect. 29.5. Also discussed below is an interesting nuclear physics source of PNC known as the *anapole* moment [13], which arises from photon exchange with weak radiative corrections on the nuclear vertex. This effect enters at the several percent level, but in a way that can be subtracted out, as will be described in Sect. 29.5.

29.4 MANY-BODY PERTURBATION THEORY

While there are several ways of solving the many-electron Schrödinger equation, the most accurate treatments of Cs PNC are based on many-body perturbation theory (MBPT). In MBPT, the Hamiltonian is broken up into $H = H_0 + V_C$, where

$$H_0 = \sum_{i=1}^{N} [\boldsymbol{\alpha}_i \cdot \mathbf{p}_i + \beta_i m + U(r_i)], \tag{29.9}$$

$$V_C = \frac{1}{2} \sum_{ij} \frac{\alpha}{|\mathbf{r}_i - \mathbf{r}_j|} - \sum_{i=1}^{N} \left[\frac{Z\alpha}{r_i} + U(r_i) \right] . \tag{29.10}$$

The starting potential $U(r)$ is generally chosen to be a frozen-core Hartree-Fock (HF) potential. This model of the atom is rather inaccurate; valence removal energies disagree with experiment by about 10%, and matrix elements of the hyperfine operator by about 40%. Thus it is essential for accurate calculations to include the effects of V_C as fully as possible. MBPT proceeds by expanding the many-body wave function and the energy in powers of V_C. While going to second-order in V_C improves agreement with experiment to the percent level for energies and few percent level for matrix elements [14], more powerful methods that sum infinite classes of MBPT contributions are needed for higher accuracy, and thus we turn to a description of these *all-order* methods.

If the frozen-core HF wave function is denoted by Ψ_0, all-order *singles-doubles* methods add to it a correction $\Delta \Psi$ with either one or two excited orbitals in the HF wave function:

$$\Delta\Psi = \left[\sum_{am} \rho_{ma} a_m^\dagger a_a + \sum_{abmn} \rho_{mnab} a_m^\dagger a_n^\dagger a_a a_b \right.$$
$$\left. + \sum_{m} \rho_{mv} a_m^\dagger a_v + \sum_{amn} \rho_{mnav} a_m^\dagger a_n^\dagger a_a a_v \right] \Psi_0.$$
$$\tag{29.11}$$

The first two terms of Eq. (29.11) describe single and double excitations of the closed core, while the last two describe single and double excitations of the atom where the valence orbital is also excited. Substituting Eq. (29.11) into the Schrödinger equation, one obtains a set of coupled equations for the expansion coefficients that can be found in Ref. [15]. The first and second iterations of the equations for the expansion coefficients lead to results that are identical to first- and second-order perturbation theory. In third-order perturbation theory, terms associated with triple excitations contribute to the energy. These terms have no counterpart in the iterative solution to the equations under consideration.

To account for such terms, it is necessary to add to $\Delta\Psi$ a triple-excitation correction of the specific form

$$\delta\Psi = \left(\sum_{abcmnr} \rho_{mnrabc} a_n^\dagger a_m^\dagger a_r^\dagger a_a a_b a_c \right.$$
$$\left. + \sum_{abmnr} \rho_{mnrabv} a_n^\dagger a_m^\dagger a_r^\dagger a_a a_b a_v \right) \Psi_0 . \quad (29.12)$$

Such a term enters in two ways. First, there is a set of equations giving the triple-excitation coefficients in terms of the single-, double-, and triple-excitation coefficients. Second, the triple-excitation coefficients enter on the right-hand side of equations for the single- and double-excitation coefficients. Solving for the triple-excitation coefficients in terms of the singles and doubles (ignoring the triples on the right-hand sides of these equations), one can use them on the right-hand sides of the equations for the singles. This procedure leads to equations which, when iterated to third-order, include all of the terms from MBPT. However, the triples also modify the doubles equation, and this more computationally demanding step has not yet been implemented. As the effects of this modification enter first in fourth-order MBPT, the calculation is complete through third-order, but still misses some fourth-order contributions. Nevertheless, the method is accurate enough to predict PNC to 1%. Greater accuracy is expected when the doubles equation is modified, which should lead to a calculation complete through fourth-order. Work on this problem is in progress.

29.5 PNC CALCULATIONS

The above discussion of MBPT allows the calculation on Cs to be accurate to a few tenths of a percent for energies and under 1% for 'ordinary' matrix elements [15], and thus should allow a calculation of similar accuracy for PNC. To calculate such transitions, one modifies H_0 by adding the weak interaction operator H_W to the HF potential. This approach leads to a generalization of the single-particle states in which each state acquires an opposite-parity admixture, that is

$$\phi_k \rightarrow \phi_k + \tilde{\phi}_k . \quad (29.13)$$

Thus, for example, each $s_{1/2}$ orbital will pick up a small $p_{1/2}$ state admixture. One can then calculate a PNC

transition along the lines of a parity allowed transition calculation, simply replacing each orbital occurring in the MBPT formulas in turn with its opposite-parity admixture. Details of this approach can be found in Ref. [9].

There are several smaller effects that must be considered when the 1% level of accuracy is reached. The first stems from the fact that the Z couples to the density of up and down quarks, which in turn is related to the density of protons and neutrons in the Cs nucleus. While experimental data are available on the charge distribution [16], the neutron distribution is available only theoretically. While this uncertainty can be shown to be unimportant for Cs^{133} [9], there has been interest in looking at PNC in several different isotopes with the aim of taking ratios and greatly reducing atomic physics uncertainties. However, the neutron distribution in different nuclei is more uncertain, and taking the ratio enhances the nuclear physics uncertainty. This issue has been addressed recently by Chen and Vogel [17], who estimate the nuclear physics uncertainties to be smaller than the anticipated experimental error.

The next small effect is the interaction between the nuclear axial-vector current and the electron vector current from Z exchange. In the limit of nonrelativistic nucleon motion, this interaction is given by the spin-dependent correction to H_W

$$H_W^{(2)} = -\frac{G}{\sqrt{2}} K_2 \frac{\kappa - 1/2}{I(I+1)} \boldsymbol{\alpha}\cdot\mathbf{I}\rho(r) . \quad (29.14)$$

Here, $\kappa = 4$, $I = 7/2$ and $K_2 \cong -0.05$ for the valence proton of ^{133}Cs. Additionally, parity violation in the nucleus leads to a parity-violating nuclear moment, the anapole moment mentioned above, that couples electromagnetically to the atomic electrons. The anapole-electron interaction is described by a Hamiltonian similar to (29.14), with K_2 replaced by $-\kappa/(\kappa - \frac{1}{2})K_a$, where $K_a = 0.24 - 0.33$ is determined from nuclear model calculations [18]. Linear combinations of amplitudes for different $F \rightarrow F'$ transitions can be used to isolate either the spin-dependent or spin-independent parts of the interaction. The final smaller effects are Z exchange between the electrons, which as mentioned above turns out to be negligible, and the inclusion of the Breit interaction, which gives a 0.2% effect included in Eq. (29.6).

29.6 COMPARISON WITH EXPERIMENT

We can now make use of the above analysis to extract the value of the weak charge Q_W from experiment. The PNC amplitudes measured by Noecker, Masterson, and Wieman [3] are

$$\Im(E_{\rm PNC})/\beta = \begin{cases} -1.639(47)(08) & F = 4 \rightarrow 3 \\ -1.513(49)(08) & F = 3 \rightarrow 4 \end{cases} \quad (29.15)$$

in units of mV/cm. The quantity β is the vector part of the Stark-induced polarizability for the $6s \rightarrow 7s$ transition in Cs. This quantity has also been calculated with an accuracy of better than 1% using the all-order techniques outlined above, giving $\beta = 27.00(20)\,a_0^3$. Eliminating the spin-dependence from (29.15) by taking an appropriate linear combination and using the theoretical value for β, one finds

$$\Im(E_{\mathrm{PNC}}^{\mathrm{exp}}) = -0.8252(184)(61) \times 10^{-11}\,|e|a_0\,, \quad (29.16)$$

where the first error is from experiment and the second from theory. Combining this result with the calculation of the spin-independent amplitude given in Eq. (29.6), we obtain

$$Q_W = -71.04(1.58)(0.88)\,. \quad (29.17)$$

Alternatively, taking the opposite linear combination to eliminate the spin-independent terms in (29.15), we obtain the value

$$K_a - (\kappa - \tfrac{1}{2})/\kappa K_2 = 0.83 \pm 0.46 \quad (29.18)$$

for the constant governing the spin-dependent interaction.

Radiative corrections to Q_W incorporating a parametrization of new physics beyond the Standard Model have been worked out by Marciano and Rosner [10], who found that for ^{133}Cs$_{55}$,

$$Q_W = -73.20 - 0.8S - 0.005T \pm 0.13\,, \quad (29.19)$$

assuming the values $m_t = 140$ GeV for the top quark mass and $m_H = 100$ GeV for the Higgs particle mass. The parameters S and T in Eq. (29.19) are associated partly with deviations of the top quark and Higgs mass from their assumed values and partly with new physics beyond the Standard Model. The small factor multiplying T makes this prediction very insensitive to the top quark mass in the absence of new physics. Unfortunately, both the experimental and theoretical errors are presently too large to make atomic PNC in Cs a precision test of the Standard Model. However, there are two features of Cs PNC that even at the present accuracy lead to particle physics implications. The first is the possibility of a large positive value of S in technicolor theories [19]. A value of $S = 2$ moves the theoretical prediction for Q_W more than 2 experimental standard deviations away from the experimental value. The second is the effect of extra Z bosons, which is not accounted for in Eq. (29.19). Exchange of new Z's can be shown to be strongly constrained by atomic PNC [20]. Perhaps more interesting is the possibility of having entirely new physics

that has not been thought of. Since new physics affects different weak interaction tests differently, it is important to have as many such tests as possible. The value of atomic PNC tests will increase when the next stage of accuracy is reached, at which time atomic physics will have a significant role in precision tests of the Standard Model.

Acknowledgments

The work described here was carried out in collaboration with S. A. Blundell and W. R. Johnson, and was supported in part by NSF grant PHY-92-04089.

REFERENCES

1. M. A. Bouchiat and C. C. Bouchiat, J. Phys. (Paris) **35**, 899 (1974).
2. M. A. Bouchiat, *et al.*, J. Phys. (Paris) **47**, 1709 (1986).
3. M. C. Noecker, B. P. Masterson, and C. E. Wieman, Phys. Rev. Lett. **61**, 310 (1988).
4. T. M. Wolfenden, P. E. G. Baird, and P. G. H. Sandars, Europhys. Lett. **15**, 731 (1991).
5. P. A. Vetter, D. M. Meekhof, P. K. Majumder, S. K. Lamoreaux, and E. N. Fortson, Phys. Rev. Lett. **74**, 2658 (1995).
6. D. M. Meekof, P. Vetter, P. K. Majumder, S. K. Lamoreaux, and E. M. Fortson Phys. Rev. Lett. **71**, 3442 (1993).
7. M. J. D. Macpherson, K. P. Zetie, R. B. Warrington, D. N. Stacey, and J. P. Hoare, Phys. Rev. Lett. **67**, 2784 (1991).
8. W. Bernreuther and M. Suzuki, Rev. Mod. Phys. **63**, 313 (1991).
9. S. A. Blundell, J. Sapirstein, and W. R. Johnson, Phys. Rev. D **45**, 1602 (1992).
10. W. Marciano and J. Rosner, Phys. Rev. Lett. **65**, 2963 (1990).
11. V. A. Dzuba, V. V. Flambaum, and O. P. Sushkov, Phys. Lett. A **141**, 147 (1989).
12. A. C. Hartley, E. Lindroth, and A.-M. Mårtensson-Pendrill, J. Phys. B **23**, 3417 (1990).
13. Ya. Zel'dovich, Zh. Eksp. Teor. Fiz. **33**, 1531 (1957); Sov. Phys. JETP **6**, 1184 (1958).
14. W. R. Johnson, M. Idrees, and J. Sapirstein, Phys. Rev. A **35**, 3218 (1987).
15. S. A. Blundell, W. R. Johnson, and J. Sapirstein, Phys. Rev. A **43**, 3407 (1991).
16. R. Engfer, *et al.*, At. Data Nucl. Data Tables **14**, 479 (1974).
17. B. Q. Chen and P. Vogel, Phys. Rev. C **48**, 1392 (1993).
18. V. V. Flambaum, I. B. Khriplovich, and O. P. Sushkov, Phys. Lett. B **146**, 367 (1984).
19. M. E. Peskin and T. Takeuchi, Phys. Rev. Lett. **65**, 964 (1990).
20. P. Langacker and M. Luo, Phys. Rev. D **45**, 278 (1992).

30

Molecular Structure

David R. Yarkony

Department of Chemistry, The Johns Hopkins University, Baltimore, Maryland 21218

30.1 MOLECULAR STRUCTURE: INTRODUCTION

Molecular structure is a reflection of the Born-Oppenheimer separation of electronic and nuclear motion, which is in turn a consequence of the large difference between the electron and nuclear masses. One consequence of this separation is the concept of a potential energy surface for nuclear motion created by the faster moving electrons. Corollaries include equilibrium structures, transition states and reaction paths which are the foundation of the description of molecular structure and reactivity. However the Born-Oppenheimer approximation is not uniformly applicable and its breakdown results in perturbations in molecular spectra, radiationless decay and nonadiabatic chemical reactions.

There are many issues that can be addressed in a discussion of molecular structure, including the structure and bonding of individual classes of molecules, computational and/or experimental techniques used to determine or infer molecular structure, the accuracy of those methods, etc. In an effort to provide a broad view of the essential aspects of molecular structure, this Chapter considers issues in molecular structure from a theoretical/computational perspective using the Born-Oppenheimer approximation as the point of origin. Rather than providing a compendium of results, this chapter will explain how issues in molecular structure are investigated and how the questions that can be addressed reflect the available methodology. Even with these restrictions the scope of this topic remains enormous and precludes a detailed presentation of any one issue. Thus the abbreviated discussions in this work are supplemented by ample references to the current literature.

Several aspects of potential energy surfaces and their relation to molecular structure will be considered: (i) the electronic structure techniques used to determine a single point on a potential energy surface, and the interactions that couple the electronic states in question, (ii) the local properties of potential energy surfaces, in particular equilibrium structures and rovibrational levels that provide the link to experimental inferences concerning molecular structure, (iii) global chemistry deduced from potential energy surfaces including reaction mechanisms and reaction paths and (iv) phenomena resulting from the nonadiabatic interactions that couple potential energy surfaces.

30.2   CONCEPTS

30.2.1   Nonadiabatic Ansatz: Born-Oppenheimer Approximation

Basic Quantities

The total Hamiltonian for electronic and nuclear motion in the space fixed coordinate frame, in atomic units, is

$$H^{\mathrm{eN}}(\mathbf{r}, \mathbf{R}) = \sum_{\alpha=1}^{N} \frac{-1}{2M_\alpha} \nabla_\alpha^2 + H^{\mathrm{e}}(\mathbf{r}; \mathbf{R})$$
$$\equiv T^{\mathrm{nuc}} + H^{\mathrm{e}}(\mathbf{r}; \mathbf{R}), \qquad (30.1)$$

where $\mathbf{R}$ denotes the $3N$ nuclear coordinates, with $\mathbf{R} = (\mathbf{R}_1, \mathbf{R}_2, \ldots, \mathbf{R}_N)$, $\mathbf{R}_i = (X_i, Y_i, Z_i)$, $\mathbf{r}$ denotes the $3M$ electronic coordinates using similar conventions, T^{nuc} is the nuclear kinetic energy operator and $H^{\mathrm{e}}(\mathbf{r}, \mathbf{R})$ is the total electronic Hamiltonian taken as

$$H^{\mathrm{e}}(\mathbf{r}; \mathbf{R}) = H^0(\mathbf{r}; \mathbf{R}) + H^{\mathrm{rel}}(\mathbf{r}; \mathbf{R}). \qquad (30.2)$$

Here $H^0(\mathbf{r}; \mathbf{R})$ is the nonrelativistic Born-Oppenheimer Hamiltonian

$$H^0(\mathbf{r}; \mathbf{R}) = -\frac{1}{2} \sum_{i=1}^{M} \nabla_i^2 - \sum_{K,i} \frac{Z_K}{|\mathbf{R}_k - \mathbf{r}_i|}$$
$$+ \frac{1}{2} \sum_{i \neq j}^{M} \frac{1}{|\mathbf{r}_i - \mathbf{r}_j|} + \frac{1}{2} \sum_{K \neq L}^{N} \frac{Z_K Z_L}{|\mathbf{R}_K - \mathbf{R}_L|}, \qquad (30.3)$$

and $H^{\mathrm{rel}}(\mathbf{r}; \mathbf{R})$ is the relativistic contribution to the electronic Hamiltonian, for light atoms conventionally treated within the Breit-Pauli approximation [1] and discussed further in Sect. 30.4.2. Note that in Eq. (30.3) the nuclear kinetic energy is absent, so that only the $\mathbf{r}_i$ are dynamical variables in Eqs. (30.2) and (30.3).

Using the Born-Huang approach [2] to the Born-Oppenheimer approximation, the total wave function $\Psi_L^{\mathrm{eN}}(\mathbf{r}, \mathbf{R})$ is expanded in a basis of electronic states, $\Psi_I^{\mathrm{e}}(\mathbf{r}; \mathbf{R})$. The total wave function for the system thus has the form

$$\Psi_L^{\mathrm{eN}}(\mathbf{r}, \mathbf{R}) = \sum_I \Psi_I^{\mathrm{e}}(\mathbf{r}; \mathbf{R}) \beta_I^L(\mathbf{R}). \qquad (30.4)$$

Equation (30.4) is valid for any complete set of electronic states depending parametrically on nuclear coordinates. The parametrical dependence of $\Psi_I^{\mathrm{e}}(\mathbf{r}; \mathbf{R})$ on $\mathbf{R}$, denoted by the semicolon, reflects the fact that the $\mathbf{R}$ are not dynamical variables in Eqs. (30.2) and (30.3). As a practical matter it is necessary to make a particular choice of $\Psi_I^{\mathrm{e}}(\mathbf{r}; \mathbf{R})$ in order to limit the size of the expansion in electronic states. An adiabatic electronic state is an eigenfunction of $H^{\mathrm{e}}(\mathbf{r}; \mathbf{R})$ for fixed $\mathbf{R}$. A particularly useful choice of adiabatic state, denoted by $\Psi_I^0(\mathbf{r}; \mathbf{R})$, employs $H^0(\mathbf{r}; \mathbf{R})$ rather than the full $H^{\mathrm{e}}(\mathbf{r}, \mathbf{R})$. These electronic wave functions satisfy

$$H^0(\mathbf{r}; \mathbf{R}) \Psi_I^0(\mathbf{r}; \mathbf{R}) = E_I^0(\mathbf{R}) \Psi_I^0(\mathbf{r}; \mathbf{R}). \qquad (30.5)$$

The $\Psi_I^0(\mathbf{r}; \mathbf{R})$ are determined up to a geometry dependent phase. This phase is usually chosen such that the $\Psi_I^0(\mathbf{r}; \mathbf{R})$ are real. This assumption is acceptable *except* in the situation where there is a conical intersection on the potential energy surface in question. In that case, the real-valued $\Psi_I^0(\mathbf{r}; \mathbf{R})$ changes sign when a closed loop surrounding the conical intersection point is traversed, that is, $\Psi_I^0(\mathbf{r}; \mathbf{R})$ is not single-valued [3]. This geometric or Berry [4] phase condition has consequences in such phenomena as the dynamic Jahn-Teller effect [5,6] but will not be addressed in detail in this review.

As a consequence of the parametric dependence of $\Psi_I^0(\mathbf{r}; \mathbf{R})$ on $\mathbf{R}$, $E_I^0(\mathbf{R})$ becomes a function of $\mathbf{R}$ and is referred to as the nonrelativistic Born Oppenheimer potential energy surface for reasons discussed below. Approaches based on Eq. (30.5) are most appropriate for molecular systems with only light atoms. However with the use of pseudopotential techniques [7], formally equivalent approaches can be developed for heavier systems (see Sect. 30.4.2).

Inserting Eq. (30.4) into the time independent Schrödinger equation $[H^{\mathrm{eN}}(\mathbf{r};\mathbf{R})-E]\Psi^{\mathrm{eN}}(\mathbf{r};\mathbf{R})=0$ and taking the dot product with the electronic basis state $\Psi_I^0(\mathbf{r};\mathbf{R})$ gives the following system of coupled equations for the rovibronic functions $\beta_I^L(\mathbf{R})$ [8,9]:

$$[T^{\mathrm{nuc}} + E_I^0(\mathbf{R}) - E]\beta_I(\mathbf{R})$$
$$= -\sum_J \Big[\tilde{K}^{IJ}(\mathbf{R}) + H_{IJ}^{\mathrm{BP}}(\mathbf{R})$$
$$- \sum_{\alpha=1}^N \Big\{ \frac{1}{2M_\alpha}[\mathbf{f}_\alpha^{IJ}(\mathbf{R})\cdot\nabla_\alpha + \nabla_\alpha\cdot\mathbf{f}_\alpha^{IJ}(\mathbf{R})] \Big\} \Big]\beta_J(\mathbf{R}),$$

(30.6)

where the state label L on β_I^L has been suppressed, and

$$\mathbf{f}_\alpha^{IJ}(\mathbf{R}) \equiv [f_{X_\alpha}^{IJ}(\mathbf{R}), f_{Y_\alpha}^{IJ}(\mathbf{R}), f_{Z_\alpha}^{IJ}(\mathbf{R})], \quad (30.7)$$

$$\tilde{K}^{JI}(\mathbf{R}) = \sum_{W,\alpha} \frac{1}{2M_\alpha} \tilde{k}_{W_\alpha W_\alpha}^{JI}(\mathbf{R}), \quad (30.8)$$

$$f_{W_\alpha}^{JI}(\mathbf{R}) = \left\langle \Psi_J^0(\mathbf{r};\mathbf{R}) \left| \frac{\partial}{\partial W_\alpha}\Psi_I^0(\mathbf{r};\mathbf{R}) \right\rangle_r, \quad (30.9)$$

$$\tilde{k}_{W_\alpha W_\beta'}^{JI} = \left\langle \frac{\partial}{\partial W_\alpha}\Psi_J^0(\mathbf{r};\mathbf{R}) \left| \frac{\partial}{\partial W_\beta'}\Psi_I^0(\mathbf{r};\mathbf{R}) \right\rangle_r, \quad (30.10)$$

with $W = X, Y, Z$, $\alpha = 1\cdots N$. The subscript r on the matrix elements in Eqs. (30.9)–(30.10) denotes integration over all electronic coordinates and the final ∇_α in Eq. (30.6) acts on both $\mathbf{f}_\alpha^{IJ}(\mathbf{R})$ and $\beta_J(\mathbf{R})$. While this representation may not be optimal for a treatment of the nuclear dynamics, since it employs a space fixed frame representation, it is adequate for the present development as it clearly displays the origins the Born-Oppenheimer approximation and its breakdown.

Modifications to the Nonrelativistic Born-Oppenheimer Potential Energy Surface

When the interstate couplings on the right-hand side of Eq. (30.6) can be neglected, the nuclear motion is governed by the effective potential given by $V^I(\mathbf{R}) \equiv E_I^e(\mathbf{R}) + \tilde{K}^{II}(\mathbf{R}) \equiv E_I^0(\mathbf{R}) + H_{II}^{BP}(\mathbf{R}) + \tilde{K}^{II}(\mathbf{R})$ and the Born-Oppenheimer approximation is valid. Here it is assumed that $\mathbf{f}_\alpha^{II}(\mathbf{R}) = 0$, a sufficient condition for which is that the $\Psi_I^0(\mathbf{r};\mathbf{R})$ are chosen real-valued, see discussion following Eq. (30.5). $E_I^0(\mathbf{R})$ is generally the principal $\mathbf{R}$-dependent contribution to $V^I(\mathbf{R})$. $H_{II}^{\mathrm{BP}}(\mathbf{R})$ is referred to as the relativistic contribution to the nonrelativistic Born-Oppenheimer potential energy surface and is discussed further in Sect. 30.4.2. The contribution $\tilde{K}^{II}(\mathbf{R})$ is referred to as the *adiabatic correction*. Unlike $E_I^0(\mathbf{R})$

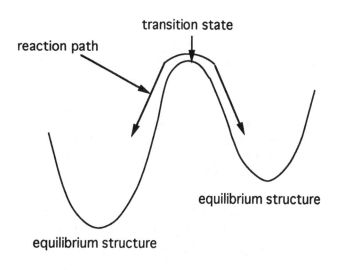

Figure 30.1. Schematic representation of a transition state and reaction path connecting two minima.

and $H_{IJ}^{\mathrm{BP}}(\mathbf{R})$, which are independent of mass, $\tilde{K}^{II}(\mathbf{R})$ is mass dependent [see Eqs. (30.8) and (30.10)]. It has been computed from first principles [10–12] and inferred from experiments [13,14] principally for diatomic systems.

30.2.2 Born-Oppenheimer Potential Energy Surfaces and Their Topology

$E_I^0(\mathbf{R})$ is the main focus of this chapter. There have been several recent reviews of ground state ($I = 1$) potential energy surfaces [15,16].

Equilibrium structures (stable or metastable species) represent local minima on $E_I^0(\mathbf{R})$, that is $g_{W_\alpha}^I(\mathbf{R}) \equiv \partial E_I^0(\mathbf{R})/\partial W_\alpha = 0$ for all W_α at $\mathbf{R} = \mathbf{R}^e$. There may be several equilibrium structures for a given set of nuclei; for example the atoms H, C and N form stable molecules HCN and HNC. Saddle points or transition states, extrema on $E_I^0(\mathbf{R})$ with one negative eigenvalue of the *Hessian matrix* $\mathbf{F}^I(\mathbf{R})$, whose elements are defined by $F_{W_\alpha W_\beta}^I(\mathbf{R}) \equiv \partial^2 E_I^0(\mathbf{R})/\partial W_\alpha \partial W_\beta$, represent mountain passes separating the various equilibrium structures and the asymptotes — values of $\mathbf{R}$ corresponding to isolated molecular fragments. This situation is illustrated in Fig. 30.1.

Reaction paths [17] connect the asymptotes, minima and saddle points. They can best be defined as the steepest descent paths from a transition state structure down to local minima [17], although methods for walking uphill (shallowest ascent path) [18] also exist. Unlike equilibrium structures and saddle points which are independent of the choice of coordinate system used to represent $\mathbf{R}$, the reaction path is coordinate system dependent. The intrinsic reaction coordinate [17], the reaction coordinate in mass weighted cartesian coordinates $\mathbf{R}_i \rightarrow \mathbf{R}_i M_i$, is most frequently used (see Sect. 30.5.4)

These features of the potential energy surfaces and their determination will be discussed further below. Their determination consists of two parts: (i) the level of treatment, i.e. the electronic structure technique used to determine $E_I^0(\mathbf{R})$ and (ii) the characterization of the features of $E_I^0(\mathbf{R})$, i.e. minima, saddle points, and reaction paths [19,20] at the level of treatment chosen. In this chapter, only fully *ab initio* levels of treatment [21,22] are considered.

30.2.3 Classification of Interstate Couplings: Adiabatic and Diabatic Bases

Intersurface couplings, shown on the right-hand side of Eq. (30.6), result in the breakdown of the single potential energy surface Born-Oppenheimer approximation. Within the Born-Huang *ansatz* such a nonadiabatic process is interpreted in terms of motion on more than one Born-Oppenheimer potential energy surface [8]. From Eq. (30.6), it is seen that there are three types of matrix elements coupling the electronic states, $\tilde{K}^{IJ}(\mathbf{R})$, $H_{IJ}^{\mathrm{BP}}(\mathbf{R})$ and $\mathbf{f}^{IJ}(\mathbf{R})$. Two of these couplings, $\tilde{K}^{IJ}(\mathbf{R})$ and $\mathbf{f}^{IJ}(\mathbf{R})$ arise from the nuclear kinetic energy operator T^{nuc}, while the third coupling arises from the Breit-Pauli interaction H^{BP}. This classification of the intersurface interactions is a consequence of the choice of the adiabatic electronic states through Eq. (30.5). Other choices of $\Psi_I^0(\mathbf{r};\mathbf{R})$, most notably the diabatic electronic states [23–27], are possible.

In the adiabatic state basis, $H^0(\mathbf{r};\mathbf{R})$ is diagonal and in the absence of relativistic effects, intersurface couplings originate exclusively from the derivative coupling terms in Eqs. (30.9) and (30.10). The diabatic basis seeks to 'transfer' the intersurface coupling from these derivative operators to a potential term analogous in form to $H_{IJ}^{\mathrm{BP}}(\mathbf{R})$. The diabatic basis, $\Psi_I^{0-d}(\mathbf{r};\mathbf{R})$, is a unitary transformation of the adiabatic electronic basis, defined such that [24]

$$f_{R_\alpha}^{JI,d} \equiv \left\langle \Psi_J^{0-d}(\mathbf{r};\mathbf{R}) \left| \frac{\partial}{\partial R_\alpha} \Psi_I^{0-d}(\mathbf{r};\mathbf{R}) \right. \right\rangle_r = 0 \quad (30.11)$$

where R_α is an internal coordinate. For polyatomic systems ($N > 2$), rigorous diabatic bases do not exist [27] and approximate diabatic bases are sought [27–30]. A discussion of this issue can be found in Ref. [31].

30.2.4 Surfaces of Intersection of Potential Energy Surfaces

The intersurface couplings are most effective in promoting a nonadiabatic process when $|V^I(\mathbf{R}) - V^J(\mathbf{R})|$ is small for some J on the right-hand side of Eq. (30.6), so that

$$E_I^e(\mathbf{R}) + \tilde{K}^{II}(\mathbf{R}) - [E_J^e(\mathbf{R}) + \tilde{K}^{JJ}(\mathbf{R})]$$
$$\equiv \Delta E_{IJ}^0(\mathbf{R}) + \Delta E_{IJ}^{\mathrm{BP}}(\mathbf{R}) + \Delta K_{IJ}(\mathbf{R}) \quad (30.12)$$

is small. Here

$$\Delta E_{IJ}^0(\mathbf{R}) \equiv E_I^0(\mathbf{R}) - E_J^0(\mathbf{R}), \quad (30.13)$$

$$\Delta K_{IJ}(\mathbf{R}) \equiv \tilde{K}^{II}(\mathbf{R}) - \tilde{K}^{JJ}(\mathbf{R}), \quad (30.14)$$

$$\Delta E_{IJ}^{\mathrm{BP}}(\mathbf{R}) \equiv H_{II}^{\mathrm{BP}}(\mathbf{R}) - H_{JJ}^{\mathrm{BP}}(\mathbf{R}). \quad (30.15)$$

Since in general $\Delta K_{IJ}(\mathbf{R})$ is quite small, and for the low atomic number systems considered here H_{II}^{BP} is itself small, we are led to consider regions of nuclear coordinate space for which the magnitude of $\Delta E_{IJ}^0(\mathbf{R})$ is small, referred to as avoided intersections when $\Delta E_{IJ}^0(\mathbf{R}) \neq 0$ and as conical and Renner-type [32] intersections when $\Delta E_{IJ}^0(\mathbf{R}) = 0$.

The set of $\mathbf{R}$ for which $\Delta E_{IJ}^0(\mathbf{R}) = 0$, the *noncrossing rule*, was first discussed in 1929 by von Neumann and Wigner [33]. The rule states that for diatomic systems, crossings between two potential energy curves of the same symmetry are not possible (actually, are extremely rare), whereas for polyatomic systems, potential energy surface intersections are allowed. Mathematically the noncrossing rule is expressed in terms of the dimensions of a surface, the surface of intersection, on which a set of conditions can be satisfied [3,34]. For electronic states of the same symmetry (neglecting electron spin degeneracy), the dimension of the surface of conical intersection is $K - 2$ where K is the number of internal nuclear degrees of freedom. For states of different spin symmetry, the dimension of the surface of intersection is $K' - 1$.

30.3 CHARACTERIZATION OF POTENTIAL ENERGY SURFACES

This section is concerned with the determination of the approximate solution of Eq. (30.5). For this purpose it is frequently convenient to re-express $H^0(\mathbf{r};\mathbf{R})$ in second quantized [35] form as

$$H^0(\mathbf{r};\mathbf{R}) = \sum_{i,j} h_{ij} a_i^\dagger a_j + \frac{1}{2} \sum_{i,j,k,l} (il|jk) a_i^\dagger a_j^\dagger a_k a_l,$$
$$(30.16)$$

where h_{ij} and $(il|jk)$ are standard abbreviations [36] for the one electron (kinetic energy and nuclear–electron attraction) and two electron (electron–electron repulsion) integrals respectively, and $a_i^\dagger$ and a_i are the fermion creation and destruction operators. The integrals and the creation and destruction operators are defined in terms of a basis of one electron functions, usually molecular spin-orbitals $\phi_i\gamma_i$, $\gamma_i = \alpha$ or β. Thus

$$h_{ij}(\mathbf{R}) = \left\langle \phi_i(\mathbf{r}_k)\gamma_i \middle| -\tfrac{1}{2}\nabla_k^2 \right.$$
$$\left. - \sum_K \frac{Z_K}{|\mathbf{R}_K - \mathbf{r}_k|} \middle| \phi_j(\mathbf{r}_k)\gamma_j \right\rangle_{\boldsymbol{r}_k} \quad (30.17)$$

$$(ij|kl) = \left\langle \phi_i(\mathbf{r}_m)\gamma_i\phi_j(\mathbf{r}_m)\gamma_j \middle| \frac{1}{|\mathbf{r}_m - \mathbf{r}_n|} \right.$$
$$\left. \times \middle| \phi_k(\mathbf{r}_n)\gamma_k\phi_l(\mathbf{r}_n)\gamma_l \right\rangle_{\boldsymbol{r}_m,\boldsymbol{r}_n}. \quad (30.18)$$

Equation (30.18) is referred to as the Mulliken notation [37] for a two-electron integral. It differs from the frequently used [36] bra-ket representation $\langle ij|kl \rangle = (ik|jl)$. The molecular orbitals $\phi_i(\mathbf{r}_j; \mathbf{R})$ are in turn expanded in terms of a basis set of atom-centered functions [38], $\chi(\mathbf{r}_j; \mathbf{R})$, with $\phi_i(\mathbf{r}_j; \mathbf{R}) = \sum_{P=1}^L \chi_P(\mathbf{r}_j; \mathbf{R})T_{P,i}(\mathbf{R})$. Note that here we have distinguished between a spin-orbital $\phi_i(\mathbf{r}_j; \mathbf{R})\gamma_i$ and a molecular orbital $\phi_i(\mathbf{r}_j; \mathbf{R})$, but we will follow the usual convention of allowing the functions in Eqs. (30.17) and (30.18) to be molecular orbitals or spin-orbitals with the use being clear from the context.

Since $H^0(\mathbf{r}; \mathbf{R})$ is independent of electron spin, it can also be written in terms of the $\tilde{E}_{pq}$, spin-averaged excitation operators [39], where $\tilde{E}_{pq} = \sum_\gamma a_{p\gamma}^\dagger a_{q\gamma}$, p, q label orbitals and $\gamma = \alpha, \beta$ so that

$$H^0(\mathbf{r}; \mathbf{R}) = \sum_{i,j} h_{ij}\tilde{E}_{ij} + \tfrac{1}{2}\sum_{i,j,k,l}(ij|kl)(\tilde{E}_{ij}\tilde{E}_{kl} - \delta_{jk}\tilde{E}_{il}).$$
$$(30.19)$$

The $\tilde{E}_{pq}$ satisfy $[\tilde{E}_{ij}, \tilde{E}_{kl}] = \tilde{E}_{il}\delta_{jk} - \tilde{E}_{kj}\delta_{il}$ as a consequence of the commutation relations for the $a_i^\dagger$ and a_i and are generators of the unitary group [40] (see Chap. 4). This observation allows the powerful machinery of the unitary group to be applied to the evaluation of $E_I^0(\mathbf{R})$ [41] (see Sect. 30.2.2).

The determination of $E_I^0(\mathbf{R})$ involves two steps; (i) the determination of the molecular orbitals $\phi_i(\mathbf{r}_j; \mathbf{R})$, and (ii) sometimes simultaneously, the determination of the approximate wave function $\Psi_I^0(\mathbf{r}; \mathbf{R})$.

Wave functions discussed in this section are constructed from functions that are antisymmetrized product of spin-orbitals $\phi_i(\mathbf{r}_j; \mathbf{R})\gamma_i$

$$\Psi_I^0(\mathbf{r}; \mathbf{R}) = \mathcal{A}\prod_{i=1}^M \phi_i(\mathbf{r}_j; \mathbf{R})\gamma_i, \quad (30.20)$$

where $\gamma_i = \alpha$ or β. The antisymmetrizer $\mathcal{A}$ is required to take into account the Fermi statistics of the electrons. The wave function in Eq. (30.20) is referred to as a Slater determinant and is an eigenfunction of M_s, the z-component of total electron spin. A linear combination of Slater determinants that is also an eigenfunction of S^2 is referred to as a configuration state function (CSF) [42].

Equation (30.20) corresponds to a particular distribution of electrons in the orbitals $\phi_i(\mathbf{r}_j; \mathbf{R})$. This distribution is referred to as an electron configuration.

In principle, it is possible to learn everything about the topology of a potential energy surface from the point-wise evaluation of $E_I^0(\mathbf{R})$. In practice however, the determination of topological features, including location of equilibrium structures, saddle points and reaction paths, characterization of quadratic and anharmonic force fields [43], and even the evaluation of derivative couplings [44], has benefitted immensely from techniques in which the energy gradient $\mathbf{g}^I(\mathbf{R})$ or higher derivatives are determined directly from knowledge of the wave function at the $\mathbf{R}$ in question. These analytic derivative techniques, i.e. techniques that do not use divided difference differentiation, have been actively developed since their first introduction [45].

The techniques most commonly used to calculate $E_I^0(\mathbf{R})$ and its derivatives are described below.

30.3.1 The Self-Consistent Field (SCF) Method

The SCF Energy

In this most basic treatment, the electronic wave function $\Psi_I^{0-\text{SCF}}$ is taken as a single Slater determinant or a single CSF. The spin-orbitals are determined to give an extremum of the energy functional $\langle \Psi_I^{0-\text{SCF}}(\mathbf{r}; \mathbf{R})|H^0(\mathbf{r}; \mathbf{R})|\Psi_I^{0-\text{SCF}}(\mathbf{r}; \mathbf{R})\rangle_r$. This gives rise to the self-consistent field conditions. For restricted closed shell wave functions [46], defined by the conditions

$$\Psi_I^{0-\text{RSCF}}(\mathbf{r}; \mathbf{R}) = \mathcal{A}\prod_{i=1}^L \phi_{2i}(\mathbf{r}_{2i}; \mathbf{R})\alpha\phi_{2i+1}(\mathbf{r}_{2i+1}; \mathbf{R})\beta$$
$$(30.21)$$

with $\phi_{2i}(\mathbf{r}_j; \mathbf{R}) = \phi_{2i-1}(\mathbf{r}_j; \mathbf{R})$, $i = 1, \ldots, L$ and $2L = M$, the SCF equations are

$$\varepsilon_{ri} = h_{ri} + \sum_{j \in \{j_{\text{occ}}\}} 2(ir|jj) - (ij|rj) = 0 \quad (30.22)$$

for ϕ_i, ϕ_r occupied and virtual orbitals, respectively. Here, occupied orbitals $\{j_{\text{occ}}\}$ are those occurring in the orbital product in Eq. (30.21), and the remainder of the orbitals are referred to as virtual orbitals. The solution to Eq. (30.22) is referred to as the canonical SCF solution, provided the orbitals satisfy the SCF equations

$$F\phi_i = \lambda_i\phi_i, \quad (30.23)$$

where the Fock operator F is defined by $F_{ri} = \varepsilon_{ri}$ and $\lambda_i = \varepsilon_{ii}$. The corresponding electronic energy, $E_I^0(\mathbf{R}) \equiv E_I^{0-\text{SCF}}(\mathbf{R}) = \sum_{j \in \{j_{\text{occ}}\}}(\varepsilon_{jj} + h_{jj})$.

SCF Energy Derivatives

The derivative of $E_i^{0-\text{SCF}}(\mathbf{R})$ can be expanded through third order using analytic gradient techniques [47, 48], and through fourth order using divided differences and more recently using analytic derivative techniques [43]. The ability to evaluate the energy derivatives or force field through fourth order is important in the determination of vibrational properties of molecules, as discussed in Sect. 30.5.2.

Direct SCF

The range of molecular systems accessible to treatment at the SCF level has been expanded considerably by the introduction of the direct SCF methods [49]. In this method, none of the two-electron integrals in the χ basis $(pq|rs)$ are stored during the iterative solution of Eqs. (30.22) and (30.23). Since the number of such integrals grows as $L^4/8$, direct SCF procedures avoid the 'L^4-storage bottleneck', enabling SCF wave functions to be determined for systems with more than 100 atoms and basis sets with more than 1000 functions. (If all the $L^4/8$ integrals were stored on disk this basis would require $\sim$1000 gigabytes).

30.3.2 Electron Correlation: Wave Function Based Methods

Wave functions more accurate than SCF wave functions are obtained by including the effects of electron correlation. The correlation energy is defined as $E_I^{\text{corr}}(\mathbf{R}) \equiv E_I^0(\mathbf{R}) - E_I^{0-\text{SCF}}(\mathbf{R})$. Methods for the determination of $E_I^{\text{corr}}(\mathbf{R})$ are commonly classified as single reference or mulitreference methods. In single reference methods, an SCF wave function $\Psi_I^{0-\text{SCF}}(\mathbf{r}; \mathbf{R})$ given by Eq. (30.20) is improved. In multireference methods, the starting point is the space spanned by a set of terms like Eq. (30.20). This space is referred to as the reference space and the functions in the space as the reference configurations. Sometimes the molecular orbitals $\phi(\mathbf{r}_i; \mathbf{R})$ for the reference space are the SCF orbitals of a single reference configuration calculation, but more often the $\phi(\mathbf{r}_i; \mathbf{R})$ are chosen to satisfy multiconfigurational self-consistent field (MCSCF) equations [50–53].

Single Reference Methods

Second Order Møller-Plesset Perturbation Theory (MP2). In this approach [54], the solutions of the SCF equations are used to determine $E_I^0(\mathbf{R})$ to second order as

$$E_I^{0-\text{MP2}}(\mathbf{R}) = E_I^{0-\text{SCF}}(\mathbf{R}) - \frac{1}{4} \sum_{a,b,i,j} \frac{(ab||ij)^2}{\lambda_a + \lambda_b - \lambda_i - \lambda_j},$$

(30.24)

where i, j denote occupied orbitals, a, b denote virtual orbitals, and

$$(ab||ij) = (ai|bj) - (aj|bi) = \langle ab|ij \rangle - \langle ab|ji \rangle.$$

Both gradients of $E_I^{0-\text{MP2}}(\mathbf{R})$ and direct implementations of MP2 [37, 55] are currently available. Higher order perturbation theories, also in common use, are discussed in the context of coupled cluster methods which are considered next.

Coupled Cluster Method. Perhaps the most reliable method currently available for characterizing near equilibrium properties of moderately sized molecules is the coupled cluster approach [56–60] (*cf.* Chap. 5). In this approach, the exact wave function is written as a unitary transformation of a reference wave function $\Phi_0(\mathbf{r}; \mathbf{R})$ [usually, but not necessarily, the SCF wave function $\Phi_0(\mathbf{r}; \mathbf{R}) = \Psi_I^{0-\text{SCF}}(\mathbf{r}; \mathbf{R})$],

$$\Psi_I^{0-\text{CC}} = \exp(T)\Phi_0.$$

(30.25)

where T is an excitation operator defined as

$$T = \sum_p T_p,$$

(30.26)

with

$$T_p = \frac{1}{p!^2} \sum_{i,j,k,\dots,a,b,c} t_{ijk\dots}^{abc\dots} a_a^\dagger a_b^\dagger a_c^\dagger a_i a_j a_k,$$

(30.27)

where i, j, k denote occupied orbitals in Φ_0 and a, b, c denote virtual orbitals. Thus

$$\Psi_I^{0-\text{CC}} = \Phi_0 + \sum_{i,a} t_i^a \Phi_i^a + \frac{1}{4} \sum_{i,j,a,b} t_{ij}^{ab} \Phi_{ij}^{ab} + \cdots,$$

(30.28)

where Φ_i^a, Φ_{ij}^{ab} are single and double excitations from Φ_0, and it is the amplitudes t that must be determined.

Wave functions of the form in Eqs. (30.25)–(30.28) have the important property referred to as size consistency [60] or size extensivity [59]. Consider N noninteracting helium atoms. The electronic wave function for the ith helium atom in the natural orbital basis [61] can be written as $\Psi_i = [1 + T_2(i)]\phi_0(i)$, so that the N atom electronic wave function becomes

$$\Psi = \prod_i \Psi_i = \prod_i [1 + T_2(i)]\Phi_0$$

$$= \left[\exp \sum_i T_2(i) \right] \Phi_0 = \exp(T_2)\Phi_0$$

(30.29)

Thus this exponential type of solution scales properly with the number of particles.

To obtain the amplitudes $\mathbf{t}$, define $H_N = H^0 - \langle\Phi_0|H^0|\Phi_0\rangle$, $E_I^{0-\mathrm{CC}} = \Delta E_I^{0-\mathrm{CC}} + \langle\Phi_0|H^0|\Phi_0\rangle \equiv \Delta E_I^{0-\mathrm{CC}} + E_I^{0-0}$, $P = |\Phi_0\rangle\langle\Phi_0|$ and $Q = 1 - P$. Then, inserting Eqs. (30.25)–(30.28) into Eq. (30.5) gives the energy (30.30) and amplitude (30.31) equations, which define the coupled cluster approach:

$$\Delta E_I^{0-\mathrm{CC}} = \langle\Phi_0|\tilde{H}_N|\Phi_0\rangle, \tag{30.30}$$

$$\langle\Phi_{ijk\cdots}^{abc\cdots}|\tilde{H}_N|\Phi_0\rangle = 0 \text{ for all } \Phi_{ijk\cdots}^{abc\cdots}, \tag{30.31}$$

where

$$\begin{aligned}
\tilde{H}_N &= \exp(-T)H_N\exp(T) \\
&= -E_I^{0-0} + \exp(-T)H^0\exp(T). \tag{30.32}
\end{aligned}$$

To appreciate the nature of these equations, consider the approximation $T = T_2$ constructed from SCF orbitals [36,59], referred to as the coupled cluster doubles (CCD) level. At this level the energy equation becomes

$$\begin{aligned}
E_I^{0-\mathrm{CCD}} &= \langle\Phi_0|\exp(-T_2)H^0\exp(T_2)|\Phi_0\rangle \\
&= \langle\Phi_0|H^0T_2|\Phi_0\rangle + E_I^{0-0} \tag{30.33} \\
&= E_I^{0-0} + \tfrac{1}{4}\sum_{i,j,r,s}\langle\Phi_0|H^0|a_r^\dagger a_s^\dagger a_i a_j\Phi_0\rangle t_{ij}^{rs} \\
&= E_I^{0-0} + \tfrac{1}{4}\sum_{i,j,r,s}(rs||ij)t_{ij}^{rs}
\end{aligned}$$

for i,j occupied and r,s virtual orbitals. The amplitude equations become

$$\begin{aligned}
&\langle\Phi_{ij}^{rs}|\exp(-T_2)H^0\exp(T_2)|\Phi_0\rangle \\
&= \langle\Phi_{ij}^{rs}|\left(1 - T_2 + \tfrac{1}{2}T_2^2\right)H^0\left(1 + T_2 + \tfrac{1}{2}T_2^2\right)|\Phi_0\rangle \\
&= 0 \tag{30.34}
\end{aligned}$$

which reduces (after considerable commutator algebra) [36] to

$$\begin{aligned}
&(\lambda_r + \lambda_s - \lambda_i - \lambda_j)t_{ij}^{rs} \\
&= -(rs||ij) - \sum_{p>q}(rs||pq)t_{ij}^{pq} - \sum_{k>l}(kl||ij)t_{kl}^{rs} \\
&\quad + \sum_{k,p}\big[(ks||jp)t_{ik}^{rp} - (kr||jp)t_{ik}^{sp} \\
&\quad - (ks||ip)t_{jk}^{rp} + (kr||ip)t_{jk}^{sp}\big] \\
&\quad - \sum_{k>l;p>q}(kl||pq)\big[t_{ij}^{pq}t_{kl}^{rs} - 2(t_{ij}^{rp}t_{kl}^{sq} + t_{ij}^{sq}t_{kl}^{rp}) \\
&\quad - 2(t_{ik}^{rs}t_{jl}^{pq} + t_{ik}^{pq}t_{jl}^{rs}) + 4(t_{ik}^{rp}t_{jl}^{sq} + t_{ik}^{sq}t_{jl}^{rp})\big]. \tag{30.35}
\end{aligned}$$

The solution to Eq. (30.35) is obtained iteratively. The first iteration gives $t_{ij}^{rs} = -(rs||ij)/(\lambda_r + \lambda_s - \lambda_i - \lambda_j)$, which when inserted into Eq. (30.33) gives Eq. (30.24), i.e. the MP2 result. If the quadratic terms in Eq. (30.35) are neglected, then the second iteration gives the third order Møller-Plesset (MP3) result [36]. The result of iterating Eq. (30.35) to convergence gives the CCD result.

Of the levels of coupled cluster treatments in current use, those including T_1, T_2, T_3 and T_4 in Eq. (30.25) provide the most reliable results [62–66].

Multireference Methods

Single reference methods provide probably the most powerful tools for treating near-equilibrium properties of ground electronic state systems. In other instances, such as the study of electronically excited states, determination of global potential energy surfaces and for systems with multiple open shells such as diradicals, multireference techniques are found to be extremely useful. In the multireference techniques discussed below, the wave function is written as

$$\Psi_I^{0-\mathrm{MRF}}(\mathbf{r};\mathbf{R}) = \sum_\alpha c_\alpha^I\psi_\alpha(\mathbf{r};\mathbf{R}) \tag{30.36}$$

where $\psi_\alpha(\mathbf{r};\mathbf{R})$ is a CSF. This expansion is usually referred to as a configuration interaction (CI) expansion, and the coefficients $\mathbf{c}^I(\mathbf{R})$ are referred to as CI coefficients [42]. Since Eq. (30.36) does not involve the exponential *ansatz*, it is not automatically size extensive. In the description of these wave functions, it is useful to generalize the notion of occupied and virtual orbitals to inactive, active, and virtual orbitals, where, referring to the CSF's defining the reference space, inactive orbitals are fully occupied in all CSFs, virtual orbitals are not occupied in any CSF and the active orbitals are (partially) occupied in at least one CSF.

Most multireference techniques begin with the determination of a multiconfigurational self-consistent field (MCSCF) wave function or state averaged MCSCF (SA-MCSCF) wave function [50–53,67]. This approach is capable of describing the internal or static correlation energy, the part of the correlation energy that leads to sizeable separation of two electrons in a pair, and near-degeneracy effects. A particularly robust type of MCSCF wave function, the complete active space (CAS) [68–70] wave function, includes in Eq. (30.36) all CSFs arising from the distribution of the available electrons among the active orbitals. Note that a CAS wave function is size consistent.

The remaining part of the correlation energy, the dynamic correlation, describes the two-electron cusp i.e. the regions of space for which two electrons experience the singularity in the Coulomb potential. Empirically, it has been shown [71] that wave functions capable of providing chemically accurate descriptions of the dynamic correlation can be obtained by augmenting the MCSCF or reference wave function with all CSFs that differ by at most two molecular orbitals, double excitations, [see Eq. (30.28)] from those of the reference space. Such wave functions are generally referred to as multireference single and double excitation configuration interaction (MR-SDCI) wave functions. First (second) order wave functions [72] include all single (single and double) excitations relative to a CAS reference space.

Multiconfigurational Self-Consistent Field (MCSCF) Theory. In the MCSCF approximation, a wave function $\Psi_I^{0-\text{MCSCF}}(\mathbf{r};\mathbf{R}) = \sum_\alpha c_\alpha^I \psi_\alpha(\mathbf{r};\mathbf{R})$ of the form in Eq. (30.36) is to be determined. In this procedure, both the molecular orbitals $\phi(\mathbf{r}_j;\mathbf{R})$ and $\mathbf{c}^I(\mathbf{R})$ are determined from the requirement that

$$E_I^{0-\text{MCSCF}} = \langle \Psi_I^{0-\text{MCSCF}}(\mathbf{r};\mathbf{R}) | H^0 | \Psi_I^{0-\text{MCSCF}}(\mathbf{r};\mathbf{R}) \rangle_\mathbf{r}, \tag{30.37}$$

be a minimum. In a popular variant, the state averaged MCSCF (SA-MCSCF) [50–53] procedure, the average energy functional

$$\begin{aligned} E^{0-\text{SAMCSCF}} = \sum_{I=2}^{K} w_I \langle \Psi_I^{0-\text{MCSCF}}(\mathbf{r};\mathbf{R}) | H^0 | \\ \times \Psi_I^{0-\text{MCSCF}}(\mathbf{r};\mathbf{R}) \rangle_\mathbf{r}, \end{aligned} \tag{30.38}$$

where the weight vector $\mathbf{w} = (w_1, w_2, \dots, w_K)$ has only positive elements, is minimized. This procedure should be compared with the multireference CI approach described below in which a predetermined set of molecular orbitals is used.

The optimum molecular orbitals and CI coeffients can be written as unitary transformations of an initial set of such quantities, i.e.

$$\Psi_I^{0-\text{MCSCF}} = \exp(\gamma) \exp(\Delta) \Phi_I^0 \tag{30.39}$$

where

$$\gamma = \sum_{i,s} \gamma_{s,i} a_s^\dagger a_i, \quad \Delta = \sum_{n,l} \Delta_{l,n} |\Phi_l^0\rangle \langle \Phi_n^0|, \tag{30.40}$$

and γ, Δ are general anti-Hermitian matrices. Since $m_{i,j} = -m_{j,i} \equiv m_{ij}$ for $\mathbf{m} = \gamma$, Δ, the upper triangle of $\mathbf{m}$ forms a vector $\vec{m}$ that enumerates the independent parameters of $\mathbf{m}$. The MCSCF or SA-MCSCF equations can be succinctly formulated by inserting Eq. (30.39) into Eq. (30.37) or Eq. (30.38), expanding the commutators to second order and requiring $\partial E/\partial m_{ij} = 0$ [36]. The result provides a system of Newton-Raphson equations that can be solved iteratively for the γ, Δ [52, 53, 67, 73].

Multireference Configuration Interaction Theory: The MR-SDCI Method. In multireference configuration interaction theory, the wave function is again of the form $\Psi_I^{0-\text{MRCI}}(\mathbf{r};\mathbf{R}) = \sum_\alpha c_\alpha^I(\mathbf{R}) \psi_\alpha(\mathbf{r};\mathbf{R})$, but now CSFs involving the large space of virtual orbitals are included. In this approach the CI coefficients are found for a predetermined set of molecular orbitals. The CSF expansions at the MR-SDCI level become quite large (1-10 million CSFs is routine), and even larger expansions are tractable using specialized methods. The $\mathbf{c}^I$ satisfy the usual matrix equation

$$\left[\mathbf{H}^0 - E_I^0 \right] \mathbf{c}^I = 0 \tag{30.41}$$

where

$$\begin{aligned} H_{\alpha\beta}^0 &= \langle \psi_\alpha | H^0 | \psi_\beta \rangle \\ &= \sum_{i,j} A_{ij}^{\alpha\beta} h_{ij} + \sum_{i,j,k,l} A_{ijkl}^{\alpha\beta} (ij|kl). \end{aligned} \tag{30.42}$$

It is the computationally elegant solution of Eq. (30.41) for large expansions that is the essence of modern MR-SDCI methods.

Because of the large dimension of the CSF space, it is not possible (or even desirable) to find all the solutions of Eq. (30.41). The few lowest eigenstates and eigenvalues can be found [74] using an iterative direct CI procedure [75] in which a subspace is generated sequentially from the residual defined at the kth iteration by

$$\sigma_\mu^{(k)} = \sum_\nu \left[H_{\mu\nu}^0 - E^{(k-1)} \delta_{\mu\nu} \right] c_\nu^{(k-1)}, \tag{30.43}$$

where for simplicity of notation, the state index I is suppressed. The computationally demanding step in this procedure is the efficient evaluation of the $A_{ijkl}^{\alpha\beta} = \langle \psi_\alpha | \tilde{E}_{ij} \tilde{E}_{kl} - \delta_{jk} \tilde{E}_{il} | \psi_\beta \rangle$. Key to the efficiency of this evaluation is the factorization formally achieved by

$$\langle \psi_\alpha | \tilde{E}_{ij} \tilde{E}_{kl} | \psi_\beta \rangle = \sum_m \langle \psi_\alpha | \tilde{E}_{ij} | \psi_m \rangle \langle \psi_m | \tilde{E}_{kl} | \psi_\beta \rangle. \tag{30.44}$$

Using unitary group techniques, this apparently intractable summation can be used to express the $A_{ijkl}^{\mu\nu}$ as a simple finite product [76].

Contracted CI and Complete Active Space Perturbation Theory (CASPT2). The direct approach outlined above makes treatment of large MR-SDCI expansions possible. However as the size of the reference space grows, the CSF space in the MR-SDCI expansion may become intractably large, particularly if the full second-order wave function is used. To avoid this bottleneck, the reference CSFs may be selected from the active space, and perturbation theory may be used to select CSFs involving orbitals in the virtual space [42]. The use of selection procedures complicates the implementation of 'direct' techniques although recently progress in selected direct CI procedures has been reported [77–79]. Alternatively, new techniques have been developed that avoid this selection procedure. In these approaches, the MCSCF wave function itself is used as the reference wave function for CI or pertubation theory techniques. The use of a reference wave function rather than a reference space considerably reduces the size of the CSF space to be handled. In this approach, one of the principal computational complications is that the excited functions are not necessarily mutually orthogonal. Two computational procedures currently in wide use, known as contracted CI [80–83] and CASPT2 [84, 85], are based on this approach.

The CASPT2 method is a computationally efficient variant of second order perturbation theory in which the reference wave function is a CAS-MCSCF wave function and thus may itself contain tens of thousands of CSFs. In this case the full multireference CI problem would be intractable owing to the large space of double excitations. A similar approach is adopted in the contracted CI method, in that the excitations are defined in terms of a general MCSCF reference wave function $\Psi_I^{0-\text{MRF}}$ rather than the reference space as in the MR-SDCI methods described above.

30.3.3 Electron Correlation: Density Functional Theory

The approaches in Sects. 30.2.1 and 30.2.2 can be referred to as wave function based approaches in the sense that determination of $E_I^0(\mathbf{R})$ is accompanied by the determination of the corresponding electronic wave function $\Psi_I^0(\mathbf{r};\mathbf{R})$. An alternative approach is the known as density functional theory (DFT) [86]. The ultimate goal of DFT is the determination of total densities and energies without the determination of wave functions, as in the Thomas-Fermi approximation. DFT is based on the Hohenberg-Kohn Theorem [87], which states that the total electronic density can be considered to be the independent variable in a multi-electron theory. Computationally viable approaches exploit the Kohn and Sham formulation [88], which introduces molecular orbitals as an intermediate device.

The essential features of the Kohn-Sham (KS) theory [86], are as follows [89]. Assume that the real N-electron system for a particular arrangement of the nuclei $\mathbf{R}$ has a total electron density $\rho(\mathbf{r};\mathbf{R})$. Consider a system of N independent noninteracting electrons subject to a one-body potential V_0 with total density $\rho^0(\mathbf{r};\mathbf{R})$ such that $\rho(\mathbf{r};\mathbf{R}) = \rho^0(\mathbf{r};\mathbf{R})$. The corresponding independent particle orbitals, the Kohn-Sham orbitals $\phi_i^{\text{KS}}(\mathbf{r}_j;\mathbf{R})$, satisfy a Hartree-Fock-like equation

$$\left(-\tfrac{1}{2}\nabla^2 + V_0 - \varepsilon_i\right)\phi_i^{\text{KS}} = 0 \qquad (30.45)$$

with

$$\rho^0(\mathbf{r};\mathbf{R}) = \sum_i |\phi_i^{\text{KS}}|^2 . \qquad (30.46)$$

The relation between the energy of the ideal system and that of the true system $E^{0-\text{DFT}}(\mathbf{R})$ is obtained from the adiabatic connection formula [89].

In order to determine $E^{0-\text{DFT}}(\mathbf{R})$, functions ϕ_i^{KS} are required, which in turn means that V_0, the Kohn-Sham noninteracting one-body potential, must be determined. V_0 is written as [*cf.* Eq. (30.3)]

$$V_0 = V_{\text{e-N}} + V_{\text{Coul}} + V_{\text{xc}} \qquad (30.47)$$

where V_{Coul} is the Coulomb interaction corresponding to the electron density, $V_{\text{e-N}}$ is the electron-nuclear

attraction interaction and V_{xc} is the exchange-correlation density. The effect of Eq. (30.47) is to isolate from V_0 the straightforward contributions V_{Coul} and $V_{\text{e-N}}$, and transfers our ignorance to the remaining portion V_{xc}. Although, by the Hohenberg-Kohn theorem V_{xc} must exist, its determination remains the challenge of modern density functional theory which currently uses approximate functional forms. These approximate treatments of V_{xc} are quite useful in practice, and for large systems, DFT offers a promising alternative to wave function based methods.

Through Eq. (30.45), the KS approach is formally similar to SCF theory, although the Kohn-Sham theory is in principle exact. This formal similarity is exploited in the evaluation of the derivative $E^{0-\text{DFT}}(\mathbf{R})$ [90].

30.3.4 Weakly Interacting Systems

When attempting to describe weak chemical interactions such as van der Waals or dispersion forces, the techniques outlined in Sect. 30.3.2 must be modified somewhat. These modifications arise from the finiteness, and hence incompleteness, of the basis (the set χ) used to describe the molecular orbitals. Assume that the interaction of two molecules A and B is to be determined. Consider the description of molecule A as the distance between A and B decreases from infinity. The atom centered basis functions on molecule B augment those on molecule A, lowering its energy, independent of any physical interaction. This computational artifact serves to overestimate the interaction energy, and is known as the basis set superposition error [91]. In chemically bonded systems where interaction energies are large, it is of negligible importance. However in weakly bonded systems for which the interaction energies may be on the order of 10 to 100 cm^{-1}, the basis set superpostion error can be significant.

The basis set superposition error can be reduced by the counterpoise correction [92, 93]. In this approach, the interaction energy is evaluated directly as

$$E_I^{0-\text{int}}(\mathbf{R}) = E_I^0(\mathbf{R}) - \left[E_I^{0-A}(\mathbf{R}) + E_I^{0-B}(\mathbf{R})\right], \quad (30.48)$$

where $E_I^{0-A}(\mathbf{R})$ and $E_I^{0-B}(\mathbf{R})$, the energies of A and B respectively, are evaluated in the full basis.

30.4 INTERSURFACE INTERACTIONS: PERTURBATIONS

The existence of interstate interactions can lead to 'unexpected' shifts in spectral lines as well as predissociation of the states themselves [94]. These situations are illustrated in Figs. 30.2a and 30.2b, which present the

Figure 30.2. (a) Adiabatic potential energy curves for the $1,2,3\,^3\Pi_g$ states of Al_2 from Ref. [95]. (b) Derivative couplings $f^{IJ}(\mathbf{R})$ for $(I,J) = 1,3\,^3\Pi_g$, $1,2\,^3\Pi_g$ and $2,3\,^3\Pi_g$ from Ref. [95].

$1,2,3\,^3\Pi_g$ potential energy curves for Al_2 and the corresponding derivative couplings $\mathbf{f}^{IJ}(\mathbf{R})$ respectively. The derivative couplings were evaluated using the method described in Sect. 30.4.1. In this molecule, derivative couplings between the $2,3\,^3\Pi_g$ states are responsible for the perturbations in the vibrational levels of the bound $2,3\,^3\Pi_g$ states. Derivative couplings of these states with the $1\,^3\Pi_g$ state causes predissociation of all levels in the $2,3\,^3\Pi_g$ manifold [95].

Sect. 30.1 describes two classes of interstate matrix elements that can lead to these nonadiabatic phenomena, the derivative coupling matrix elements in Eqs. (30.7)–(30.10) and H_{IJ}^{rel}, which is usually treated for low Z systems within the Breit-Pauli approximation. An illustration of a nonadiabatic process induced by H^{rel} is provided in Sect. 30.6. A key issue in the treatment of the electronic structure aspects of these phenomena is the reliable evaluation of the interstate matrix elements. The interstate interactions are usually of most interest in regions of nuclear coordinate space far removed from the equilibrium nuclear configuration. Thus it is desirable to evaluate these interactions using multireference CI wave functions which (see Sect. 30.2) are well suited for use

in these regions of coordinate space. The evaluation of nonadiabatic interactions, based on SA-MCSCF/CI wave functions, is discussed below.

30.4.1 Derivative Couplings

From Eqs. (30.9)–(30.11), two classes of matrix elements are required, $\tilde{k}_{W_\alpha W_\beta}^{JI}(\mathbf{R})$ and $f_{W_\alpha}^{JI}(\mathbf{R})$. In fact, techniques to evaluate both of these exist [44]. However, it is common to approximate $\tilde{k}_{W_\alpha W_\beta}^{JI}(\mathbf{R})$ by

$$\tilde{k}_{W_\alpha W'_\beta}^{JI}(\mathbf{R}) = \sum_M \left\langle \frac{\partial}{\partial W_\alpha} \Psi_J^0(\mathbf{r};\mathbf{R}) \middle| \Psi_M^0(\mathbf{r};\mathbf{R}) \right\rangle_r$$
$$\times \left\langle \Psi_M^0(\mathbf{r};\mathbf{R}) \middle| \frac{\partial}{\partial W'_\beta} \Psi_I^0(\mathbf{r};\mathbf{R}) \right\rangle_r$$
$$= \sum_M f_{W_\alpha}^{MJ} f_{W'_\beta}^{MI}. \qquad (30.49)$$

with the (in principle infinite) summation over states truncated to reflect only the states explicitly treated in the nonadiabatic dynamics. Thus it is sufficient to discuss the determination of $f_{W_\alpha}^{JI}(\mathbf{R})$.

The use of analytic derivative theory [96, 97] greatly improves the computational efficiency of evaluating $f_{W_\alpha}^{JI}(\mathbf{R})$ relative to the earlier divided difference techniques [98]. The key ideas are given below. In this presentation, the standard, real-valued normalization is used. Additional contributions owing to the geometric phase, if required, must be evaluated separately as they do not follow from the electronic Schrödinger equation at a single point [99].

Differentiation of the $\Psi_I^0(\mathbf{r};\mathbf{R})$ defined in Eq. (30.36) gives:

$$\frac{\partial}{\partial W_\alpha} \Psi_I^0(\mathbf{r};\mathbf{R}) = \sum_\lambda \left\{ \left[\frac{\partial}{\partial W_\alpha} c_\lambda^I(\mathbf{R}) \right] \psi_\lambda(\mathbf{r};\mathbf{R}) \right.$$
$$\left. + c_\lambda^I(\mathbf{R}) \left[\frac{\partial}{\partial W_\alpha} \psi_\lambda(\mathbf{r};\mathbf{R}) \right] \right\}. \qquad (30.50)$$

Thus $f_{W_\alpha}^{JI}(\mathbf{R})$, consists of two terms

$$f_{W_\alpha}^{JI}(\mathbf{R}) = {}^{\mathrm{CI}}f_{W_\alpha}^{JI}(\mathbf{R}) + {}^{\mathrm{CSF}}f_{W_\alpha}^{JI}(\mathbf{R}) \qquad (30.51)$$

where the CI contribution is given by

$${}^{\mathrm{CI}}f_{W_\alpha}^{JI}(\mathbf{R}) = \sum_\lambda c_\lambda^J(\mathbf{R}) \left[\frac{\partial}{\partial W_\alpha} c_\lambda^I(\mathbf{R}) \right], \qquad (30.52)$$

and the CSF contribution has the form

$${}^{\mathrm{CSF}}f_{W_\alpha}^{JI}(\mathbf{R})$$
$$= \sum_{\lambda,\mu} c_\lambda^J(\mathbf{R}) \left\langle \psi_\lambda(\mathbf{r};\mathbf{R}) \middle| \frac{\partial}{\partial W_\alpha} \psi_\mu^I(\mathbf{r};\mathbf{R}) \right\rangle_r c_\mu^I(\mathbf{R}).$$
$$(30.53)$$

Evaluation of $^{\text{CSF}}f_{W_\alpha}^{JI}(\mathbf{R})$ is straightforward using analytical derivative techniques [100], so that the remainder of the discussion focusses on $^{\text{CI}}f_{W_\alpha}^{JI}(\mathbf{R})$. From Eq. (30.52), it would appear that the derivative of the CI coefficients $\partial/\partial W_\alpha c_\lambda^I(\mathbf{R}) \equiv V_{W_\alpha,\lambda}^I(\mathbf{R})$ would be required to evaluate $^{\text{CI}}f_{W_\alpha}^{JI}(\mathbf{R})$. This is quite costly and is in fact not necessary, since only the projection onto the state $\Psi_J^0(\mathbf{r};\mathbf{R})$ is required. Eq. (30.52) for $^{\text{CI}}f_{W_\alpha}^{JI}(\mathbf{R})$ can be recast in a form similar to that of $\mathbf{g}^I(\mathbf{R})$. This transformation of Eq. (30.52) enables the explicit determination of $\mathbf{V}_{W_\alpha}^I(\mathbf{R})$ to be avoided, and is the key to the efficient use of analytic gradient techniques in the evaluation of $\mathbf{f}^{JI}(\mathbf{R})$.

Differentiating Eq. (30.41) with respect to W_α gives

$$\left[\mathbf{H}^0 - E_I^0(\mathbf{R})\right]\mathbf{V}_{W_\alpha}^I(\mathbf{R})$$
$$= -\left\{\frac{\partial}{\partial W_\alpha}\left[\mathbf{H}^0 - E_I^0(\mathbf{R})\right]\right\}\mathbf{c}^I(\mathbf{R}). \tag{30.54}$$

Taking the dot product of Eq. (30.54) with $\mathbf{c}^J(\mathbf{R})$ gives

$$^{\text{CI}}f_{W_\alpha}^{JI}(\mathbf{R}) \equiv \mathbf{c}^J(\mathbf{R})^\dagger \mathbf{V}_{W_\alpha}^I(\mathbf{R}) \tag{30.55}$$
$$= \Delta E_{IJ}^0(\mathbf{R})^{-1}\mathbf{c}^J(\mathbf{R})^\dagger\frac{\partial\mathbf{H}^0(\mathbf{R})}{\partial W_\alpha}\mathbf{c}^I(\mathbf{R}) \tag{30.56}$$
$$\equiv \Delta E_{IJ}^0(\mathbf{R})^{-1}h_{W_\alpha}^{JI}(\mathbf{R}). \tag{30.57}$$

Observe that Eqs. (30.56) and (30.57) are not the Hellmann-Feynman theorem [100, 101] (see Chap. 49), to which they bear a formal resemblance, since it is not the Hamiltonian operator $H^0(\mathbf{r};\mathbf{R})$ but rather the Hamiltonian matrix $\mathbf{H}^0(\mathbf{R})$ that is being differentiated. Since the energy gradient has the form [44]

$$\frac{\partial}{\partial W_\alpha}E_I^0(\mathbf{R}) = \mathbf{c}^I(\mathbf{R})^\dagger\frac{\partial\mathbf{H}^0(\mathbf{R})}{\partial W_\alpha}\mathbf{c}^I(\mathbf{R}), \tag{30.58}$$

its relation to $^{\text{CI}}f_{W_\alpha}^{JI}(\mathbf{R})$ is clear. This identification is the key step in the evaluation of $^{\text{CI}}f_{W_\alpha}^{JI}(\mathbf{R})$ using analytic gradient techniques [96, 97].

30.4.2 Breit-Pauli Interactions

For light systems, it is possible to introduce relativistic effects using the Breit-Pauli approximation [1], in which the four component Dirac description of a single electron is replaced by a two component (α, β) description. The $H^e(\mathbf{r};\mathbf{R})$ becomes [1, 102]

$$H^e(\mathbf{r};\mathbf{R}) = H^0(\mathbf{r};\mathbf{R}) + H^{\text{rel}}(\mathbf{r};\mathbf{R}) \tag{30.59}$$

where, in parallel with Eqs. (21.4)–(21.13) for the atomic case, the relativistic correction $H^{\text{rel}} = \sum_{k=1}^3 H^k(\mathbf{r};\mathbf{R})$ can be divided into spin-dependent (SD) and spin-independent (SI) parts, plus an external field interaction term H^{ext}. These are given by

$$H^1 = H^{\text{SD}} \equiv H^{\text{so}} + H^{\text{soo}} + H^{\text{ss}}, \tag{30.60}$$
$$H^2 = H^{\text{SI}} \equiv H^{\text{mass}} + H^{\text{D}} + H^{\text{ssc}} + H^{\text{oo}}, \tag{30.61}$$
$$H^3 = H^{\text{ext}}, \tag{30.62}$$

where, in atomic units,

$$H^{\text{so}} = \frac{\alpha^2}{2}\sum_{K,i}\frac{Z_K(\mathbf{r}_i - \mathbf{R}_K)\times\mathbf{p}_i\cdot\mathbf{s}_i}{|\mathbf{r}_i - \mathbf{R}_K|^3}, \tag{30.63}$$

$$H^{\text{soo}} = \frac{-\alpha^2}{2}\sum_{i\neq j}\frac{\mathbf{r}_{ij}\times\mathbf{p}_i\cdot(\mathbf{s}_i + 2\mathbf{s}_j)}{|\mathbf{r}_{ij}|^3}, \tag{30.64}$$

$$H^{\text{ss}} = \alpha^2\sum_{i<j}\left[\frac{\mathbf{s}_i\cdot\mathbf{s}_j}{|\mathbf{r}_{ij}|^3} - \frac{3(\mathbf{r}_{ij}\cdot\mathbf{s}_i)(\mathbf{r}_{ij}\cdot\mathbf{s}_j)}{|\mathbf{r}_{ij}|^5}\right], \tag{30.65}$$

$$H^{\text{mass}} = \frac{-\alpha^2}{8}\sum_i p_i^4, \tag{30.66}$$

$$H^{\text{D}} = \frac{-\alpha^2}{8}\left[\sum_{K,i}z_K\nabla_i^2|\mathbf{r}_i - \mathbf{R}_K|^{-1}\right.$$
$$\left. - \sum_{i\neq j}\nabla_i^2|\mathbf{r}_{ij}|^{-1}\right], \tag{30.67}$$

$$H^{\text{ssc}} = \frac{-8\pi\alpha^2}{3}\sum_{i<j}(\mathbf{s}_i\cdot\mathbf{s}_j)\delta(\mathbf{r}_{ij}), \tag{30.68}$$

$$H^{\text{oo}} = \frac{-\alpha^2}{4}\left[\sum_{j\neq i}\frac{\mathbf{p}_i\cdot\mathbf{p}_j}{|\mathbf{r}_{ij}|} - \frac{\mathbf{r}_{ij}\cdot[\mathbf{r}_{ij}\cdot\mathbf{p}_j]\mathbf{p}_i}{|\mathbf{r}_{ij}|^3}\right], \tag{30.69}$$

$$H^{\text{ext}} = \frac{\alpha^2}{2}\sum_i\mathbf{E}(\mathbf{r}_i)\times\mathbf{p}_i\cdot\mathbf{s}_i + \frac{i\alpha^2}{4}\sum_i\mathbf{E}(\mathbf{r}_i)\cdot\mathbf{p}_i$$
$$+ 2\mu\sum_i\boldsymbol{\mathcal{H}}(\mathbf{r}_i)\cdot\mathbf{s}_i, \tag{30.70}$$

$\mathbf{E}(\mathbf{r})$ and $\boldsymbol{\mathcal{H}}(r)$ are electric and magnetic fields, and $\mu = e\hbar/(2me)$ is the Bohr magneton. The physical significance of those terms is the same as discussed is Sect. 21.1.

One of the most important consequences of these relativistic effects is that total electron spin is no longer a good quantum number as it is in $H^0(\mathbf{r};\mathbf{R})$. The term H^{SD} couple states corresponding to distinct eigenvalues of S^2 and lead to the nonadiabatic effects that are the subject of this section.

The Breit-Pauli approximation is most useful for light atoms, but the approximation breaks down when Z becomes large [103–105]. One of the principal effects omitted in a treatment which includes only H^{SD} is the relativistic contraction of the molecular orbitals [106] due to the mass-velocity operator (H^{mass}), an effect whose importance increases with Z. Several approaches exist which attempt to correct this situation while retaining the spirit and simplifications of the Breit-Pauli approximation. The first of these is the relativistic effective core potential (ECP) approximation [107–110].

In this approach, the results of an atomic Dirac-Fock calculation [111] are used to replace innermost or core electrons of a given atom with (i) an effective one electron potential that modifies the electron-nuclear attraction term in H^0 and (ii) an effective one electron spin-orbit operator, so that

$$H^e(\mathbf{r}; \mathbf{R}) \to H^{e-\mathrm{ECP}}(\mathbf{r}; \mathbf{R})$$
$$= H^{0-\mathrm{ECP}}(\mathbf{r}; \mathbf{R}) + H^{\mathrm{so-ECP}}(\mathbf{r}; \mathbf{R}). \quad (30.71)$$

The formal similarity between $H^{e-\mathrm{ECP}}$ and $H^e(\mathbf{r}; \mathbf{R})$ results in a similar phenomenological interpretation of relativistically induced nonadiabatic processes. Applications of this approach have been recently reviewed [106, 112].

A second approach includes all electrons explicitly, and uses H^{SD} as defined above. The relativistic contraction of the core electrons is included by using a variational one-component spin-free approximation [113, 114] to the no-pair Hamiltonian [115] at the orbital optimization stage. The variational nature of the approximation provides advantages over the use of the H^{mass} term. Recent applications of this approach to the spectra of CuH and NiH have been reported [116, 117].

In order to evaluate $H_{IJ}^{\mathrm{SD}}(\mathbf{R})$, it is necessary to specify the molecular orbitals to be used to construct the $\Psi_I^0(\mathbf{r}; \mathbf{R})$. The choice of molecular orbitals is dictated by the following considerations. Matrix elements of H^{SD} between different states are required. The molecular orbitals appropriate for one state may not be appropriate for the description of the second state. Two approaches are available to handle this situation. In one approach, distinct sets of (mutually nonorthogonal) molecular orbitals are used to describe each state [118]. This permits a more compact description of the spaces in question. However in this case one is required to evaluate the matrix element of a two electron operator H^{soo} and/or H^{ss} in a nonorthogonal molecular orbital basis, an imposing computational task. This significantly limits the size of the CSF space which is tractable. The alternative approach is to use a common orthonormal basis balanced between the two spaces in question, and to use larger CSF spaces [119–122]. The use of a common orthonormal basis decreases significantly the computational effort required to evaluate the matrix elements. A symbolic matrix element method [123] has been applied to H^{SD}, as described in a recent review [124].

30.4.3 Surfaces of Intersection

The preceding subsections have considered what must be calculated in order to characterize an electronically nonadiabatic process. As noted in Sect. 30.2, it is also necessary to consider where in nuclear coordinate space electronic nonadiabaticity is important. Nonadiabatic processes are important in regions of close approach of

the potential energy surfaces with regions of surface intersections being of preeminent interest. Until quite recently, these surfaces of intersection were determined by indirect methods, i.e., the potential energy surfaces were characterized, and then the surface of intersection was determined. This made the determination of these surfaces of intersection a computationally daunting task. However recent computational advances now make it possible to determine these surfaces of intersection directly, i.e., without prior determination of the individual potential energy surfaces. This point is discussed next.

A point on the surface of conical intersection of two states of the same symmetry, subject to a set of geometric equality constraints of the form $C_i(\mathbf{R}) = 0$, $i = 1, m$, is determined from the Newton-Raphson equations [125]

$$-\begin{bmatrix} \mathbf{Q}^{IJ}(\mathbf{R}, \boldsymbol{\xi}, \boldsymbol{\lambda}) & \mathbf{g}^{IJ}(\mathbf{R}) & \mathbf{h}^{IJ}(\mathbf{R}) & \mathbf{k}(\mathbf{R}) \\ \mathbf{g}^{IJ}(\mathbf{R})^\dagger & 0 & 0 & 0 \\ \mathbf{h}^{IJ}(\mathbf{R})^\dagger & 0 & 0 & 0 \\ \mathbf{k}(\mathbf{R})^\dagger & \mathbf{0}^\dagger & \mathbf{0}^\dagger & 0 \end{bmatrix} \begin{bmatrix} \delta\mathbf{R} \\ \delta\xi_1 \\ \delta\xi_2 \\ \delta\boldsymbol{\lambda} \end{bmatrix}$$

$$= \begin{bmatrix} \mathbf{g}^I(\mathbf{R}) + \xi_1 \mathbf{g}^{IJ}(\mathbf{R}) + \xi_2 \mathbf{h}^{IJ}(\mathbf{R}) + \sum_{i=1}^m \lambda_i \mathbf{k}^i(\mathbf{R}) \\ \Delta E_{IJ}(\mathbf{R}) \\ 0 \\ \mathbf{C}(\mathbf{R}) \end{bmatrix}$$
$$(30.72)$$

where $\delta\mathbf{R} = \mathbf{R}' - \mathbf{R}$, $\delta\boldsymbol{\lambda} = \boldsymbol{\lambda}' - \boldsymbol{\lambda}$, $\delta\boldsymbol{\xi} = \boldsymbol{\xi}' - \boldsymbol{\xi}$, $g_\alpha^{IJ}(\mathbf{R}) \equiv \partial\Delta E_{IJ}(\mathbf{R})/\partial R_\alpha$, $k_\alpha^i(\mathbf{R}) \equiv \partial C_i(\mathbf{R})/\partial R_\alpha$, $h_\alpha^{IJ}(\mathbf{R}) \equiv \mathbf{c}^{I\dagger}(\mathbf{R})\partial\mathbf{H}^0(\mathbf{R})/\partial R_\alpha \mathbf{c}^J(\mathbf{R})$, $\boldsymbol{\xi}$ and $\boldsymbol{\lambda}$ are Lagrange multipliers, and $\mathbf{Q}^{IJ}(\mathbf{R}, \boldsymbol{\xi}, \boldsymbol{\lambda})$ is a matrix of second derivatives [125]. For two states of different symmetry, the analogue of Eq. (30.72) is used with the terms related to ξ_2 omitted [126]. The excellent performance of this algorithm has been documented [125, 127].

Equations (30.72) can be motivated as follows. $\mathbf{R}^c$ is sought so that $E_I^0(\mathbf{R})$ is minimized subject to the constraints $E_I^0(\mathbf{R}) = E_J^0(\mathbf{R})$ and $\mathbf{C}(\mathbf{R}) = 0$. The key is to impose the first of these constraints, noting that at each step in the Newton-Raphson procedure, $\mathbf{c}^I(\mathbf{R})$ and $\mathbf{c}^J(\mathbf{R})$ are eigenvectors. Equations (30.72) are the Newton-Raphson equations corresponding to the Lagrangian function [128]

$$L_{IJ}(\mathbf{R}, \boldsymbol{\xi}, \boldsymbol{\lambda}) = E_I(\mathbf{R}) + \xi_1 \Delta E_{IJ}(\mathbf{R})$$
$$+ \xi_2 H_{IJ}(\mathbf{R}) + \sum_{k=1}^M \lambda_k C_k(\mathbf{R}), \quad (30.73)$$

provided that the gradient of $H_{IJ}^0(\mathbf{R})$ is interpreted as a change in $\mathbf{H}^0(\mathbf{R})$ within the subspace spanned by $\mathbf{c}^I(\mathbf{R})$ and $\mathbf{c}^J(\mathbf{R})$. This can be derived from quasidegenerate perturbation theory and understood as follows. Assume for convenience that $m = 0$, i.e. there are no geometrical constraints.

Consider an $\mathbf{R}$ for which Eq. (30.72) is not satisfied. The 2×2 matrix $\mathbf{H}(\mathbf{R})$ with matrix elements $\mathbf{H}_{KL}(\mathbf{R})$, $K, L \in \{I, J\}$ becomes at $\mathbf{R} + \delta\mathbf{R}$:

$$\mathbf{H}_{IJ}(\mathbf{R} + \delta\mathbf{R})$$
$$= \mathbf{c}^{I\dagger}(\mathbf{R}) \left[\mathbf{H}(\mathbf{R}) + \sum_\alpha \frac{\partial \mathbf{H}(\mathbf{R})}{\partial R_\alpha} \delta R_\alpha \right] \mathbf{c}^J(\mathbf{R}) \quad (30.74)$$

$\mathbf{H}(\mathbf{R})$ is diagonal but nondegenerate at $\mathbf{R}$, i.e.,

$$\mathbf{H}(\mathbf{R}) = \begin{bmatrix} E_I^0(\mathbf{R}) - E_J^0(\mathbf{R}) & 0 \\ 0 & 0 \end{bmatrix}. \quad (30.75)$$

From Eq. (30.74) at $\mathbf{R}^c = \mathbf{R} + \delta\mathbf{R}$, $\mathbf{H}(\mathbf{R}^c)$ becomes, to first order, ignoring an irrelevant uniform shift of the diagonal elements,

$$\mathbf{H}(\mathbf{R}^c) = \begin{bmatrix} \Delta E_{IJ}(\mathbf{R}) + \mathbf{g}^{IJ\dagger} \cdot \delta\mathbf{R} & \mathbf{h}^{IJ\dagger} \cdot \delta\mathbf{R} \\ \mathbf{h}^{IJ\dagger} \cdot \delta\mathbf{R} & 0 \end{bmatrix}. \quad (30.76)$$

Thus Eq. (30.72) is seen to be the requirement that $\mathbf{H}(\mathbf{R}^c)$ has degenerate eigenvalues, with eigenvectors $\mathbf{c}^I(\mathbf{R})$ and $\mathbf{c}^J(\mathbf{R})$.

When Eqs. (30.72) have been solved, $\mathbf{H}_{IJ}(\mathbf{R}^c)$ is diagonal and degenerate, $E_I(\mathbf{R}^c)$ has been minimized, and $\mathbf{g}^I(\mathbf{R}^c) = 0$, except along directions contained in the two dimensional subspace spanned by $\mathbf{g}^{IJ}(\mathbf{R}^c)$ and $\mathbf{h}^{IJ}(\mathbf{R}^c)$. From Eq. (30.76), it is these two directions which lift the degeneracy of $\mathbf{H}(\mathbf{R}^c)$. Equation (30.72) is also relevant to the reaction path in a nonadiabatic process as discussed in Sect. 30.5.4. A discussion of these points from an alternative perspective has been presented by Radazos *et al.* [129].

A solution to Eq. (30.72) is referred to as a conical intersection, although rigorously degenerate states cannot be obtained from a numerical procedure. If required, the existence of a conical intersection can be rigorously established by showing that the adiabatic wave functions undergo a change of sign when transported around a closed loop containing $\mathbf{R}^c$ in the plane defined by $\mathbf{g}^{IJ}(\mathbf{R}^c)$ and $\mathbf{h}^{IJ}(\mathbf{R}^c)$. This is the geometric or Berry phase criterion [130,131]. It has been explicitly demonstrated for a conical intersection in O_3 by Ruedenberg and coworkers [132].

30.5 NUCLEAR MOTION

30.5.1 General Considerations

The determination of the rovibrational spectrum of polyatomic systems from first principles is a problem of primary importance since it allows the determination of molecular forces and structure from spectral data. In the adiabatic case, the solution of [see Eq. (30.6)]

$$[T^{\mathrm{nuc}}(\mathbf{R}) + E_I^0(\mathbf{Q}) - E_K]\beta_K(\mathbf{R}) = 0 \quad (30.77)$$

is sought, where $\mathbf{Q}$ denotes a set of internal nuclear coordinates. The reliability of the solution of Eq. (30.77) reflects the accuracy of the Born-Oppenheimer potential energy surface $E_I^0(\mathbf{Q})$ appearing in that equation. The methods for determining $E_I^0(\mathbf{Q})$ were discussed in Sect. 30.3. Conversely, the determination of the E_K from spectroscopic measurements can be used to infer information concerning $E_I^0(\mathbf{Q})$. In either case, the accurate solution of Eq. (30.77) is requisite and this section is concerned with its solution.

The operator in Eq. (30.77) has a continuous spectrum since T^{nuc} includes translations of the nuclear center of mass (c.m.). An operator with a discrete spectrum is obtained by replacing the Hamiltonian in Eq. (30.77) with one in which the translation of the nuclear c.m. has been eliminated by transforming to the c.m. frame. In the center of mass frame, Eq. (30.77) has the general form [133]

$$[T^{\mathrm{vr}}(\boldsymbol{\Omega}, \mathbf{Q}) + T^{\mathrm{vib}}(\mathbf{Q}) + E_I^0(\mathbf{Q}) - E_K]\beta(\boldsymbol{\Omega}, \mathbf{Q}) = 0 \quad (30.78)$$

where $\boldsymbol{\Omega}$ are three rotational coordinates and T^{vr} and T^{vib} are the rotational and vibrational kinetic energy operators. In $T^{\mathrm{vr}}(\boldsymbol{\Omega}, \mathbf{Q})$ all the complexity associated with the coupling of nuclear and electronic angular momentum is buried. The determination of the appropriate form for Eq. (30.78) is by no means straightforward [134, 135], and treatment of the effects of multiple angular momentum is a complex problem in angular momentum algebra.

In diatomic systems, the vibrational problem involves only one internal coordinate and is straightforward. Angular momentum coupling is usually treated using Hund's case (a), (b), etc., or an intermediate case approach [94] with Van Vleck's reversed angular momentum commutation relations being helpful in analyzing the coupled angular momentum problem [134].

Polyatomic systems introduce new complications, since in addition to the increased dimensionality of the vibrational problem, internal and rotational coordinates interconvert for colinear arrangements of the nuclei; this is particularly relevant in triatomic systems. In addition, since the $C_{\infty v}$ point group has doubly degenerate representations at collinear geometries, electronic state degeneracies may arise (the Renner-Teller effect) [32, 136, 137], further complicating the analysis. Techniques associated with the description of triatomic systems involving coupled angular momentum and electronic degeneracy are illustrated in Sect. 30.5.3.

More generally, Eq. (30.78) can be rewritten as:

$$[T^{\mathrm{vr}}(\boldsymbol{\Omega}, \mathbf{Q}^{\mathrm{e}}) + T^{\mathrm{vib}}(\mathbf{Q}) + E_I^{\mathrm{sep}}(\mathbf{Q}) - E_K]\beta(\mathbf{R})$$
$$= -\Delta T^{\mathrm{vr}}(\boldsymbol{\Omega}, \mathbf{Q})\beta(\mathbf{R}) - \Delta E_I^0(\mathbf{Q}), \quad (30.79)$$

where

$$\Delta T^{\mathrm{vr}}(\mathbf{\Omega}, \mathbf{Q}) \equiv T^{\mathrm{vr}}(\mathbf{\Omega}, \mathbf{Q}) - T^{\mathrm{vr}}(\mathbf{\Omega}, \mathbf{Q}^{\mathrm{e}}), \qquad (30.80)$$

$$\Delta E_I^0(\mathbf{Q}) \equiv E_I^0(\mathbf{Q}) - E_I^{\mathrm{sep}}(\mathbf{Q}), \qquad (30.81)$$

$\mathbf{Q}^{\mathrm{e}}$ is an equilibrium structure, and $E_I^{\mathrm{sep}}(\mathbf{Q})$ is a separable function of the normal coordinates. In this case, the solution to the left-hand side of Eq. (30.79) can be factorized into a rotational part and a vibrational part [138] according to

$$\psi_{\boldsymbol{l}}^J = \sum_K k_K^J D_{K0}^J(\theta, \chi) \prod_i \alpha_{l_i}(Q_i) \qquad (30.82)$$

where D_{K0}^J is a symmetric top wave function [139], and $\alpha_j(Q_i)$ is a pure vibrational wave function, e.g. harmonic oscillator function, for the ith internal normal coordinate. It is in this approximation that the notion of a molecule rotating and vibrating about a fixed molecular structure is achieved. In this case, E_K is just the sum of the pure rotational and vibrational energies. These rigid rotator-vibrator solutions then form the basis for the inclusion of the effects of the right-hand side using for example perturbation theory [140]. They can also be used as a basis for a nonperturbative treatment (see Sect. 30.5.2).

The classic treatments of Eq. (30.77) [141–144] have been periodically revisited [134, 145]. In these investigations, the contributions of electronic angular momentum to this Hamiltonian were frequently suppressed [144] so that these treatments are appropriate to totally symmetric electronic states. Van Vleck [134] has shown how the effects of electronic angular momentum can be incorporated into these treatments.

For nonlinear polyatomic systems, the Hamiltonian in Eq. (30.78) can be transformed to the Eckart or body fixed frame in which the reference axis is a body fixed axis oriented along the principal moments of inertia [146], although other choices of the body fixed axes are possible [147]. The principal moments of inertia represent the eigenvectors of the inertial tensor matrix [144], I_{ij} $i, j = x, y, z$ where

$$I_{xx} = \sum_i M_i(\mathbf{R}_i^{\mathrm{e}} \cdot \mathbf{R}_i^{\mathrm{e}} - X_i^{\mathrm{e}} X_i^{\mathrm{e}}) \qquad (30.83)$$

$$I_{xy} = -\sum_i M_i(X_i^{\mathrm{e}} Y_i^{\mathrm{e}}) \qquad (30.84)$$

and cyclic permuations, and $\mathbf{R}^{\mathrm{e}}$ denotes an equilibrium structure. The Hamiltonian determined by this procedure is referred to as the Watson Hamiltonian [145], and is widely used in discussing the rovibrational spectrum of nonlinear polyatomic molecules in singlet electronic states. Treatments of the Watson Hamiltonian are mentioned in Sect. 30.5.2. For linear molecules, an alternative treatment is required since there is one fewer overall rotational coordinate and one more internal coordinate [138].

30.5.2 Rotational-Vibrational Structure

Various approaches to the solution of Eq. (30.79) exist. The principal issues, which are interrelated, are (i) the range of nuclear configurations over which $E_I^0(\mathbf{Q})$ is known, (ii) the coordinate system used to express the internal coordinates, and (iii) particularly in larger systems, the number of modes or internal coordinates retained in the calculations (see Sect. 30.5.3). With regard to point (i), two approaches are currently in use. The force field method uses a power series expansion of $E_I^0(\mathbf{Q})$ about $E_I^0(\mathbf{Q}^{\mathrm{e}})$, that is (using the Einstein summation convention)

$$E_I^0(\mathbf{Q}) = E_I^0(\mathbf{Q}^{\mathrm{e}}) + \frac{1}{2} \frac{\partial^2 E_I^0(\mathbf{Q}^{\mathrm{e}})}{\partial Q_i \partial Q_j} \Delta Q_i \Delta Q_j$$
$$+ \frac{1}{6} \frac{\partial^3 E_I^0(\mathbf{Q}^{\mathrm{e}})}{\partial Q_i \partial Q_j \partial Q_k} \Delta Q_i \Delta Q_j \Delta Q_k \quad (30.85)$$

together with perturbation theory to determine spectroscopic constants. In this approach, the forces fields [the partial derivatives in Eq. (30.85)] are usually evaluated directly with the aid of analytic gradient techniques (see Sect. 30.3). Since the expansion of $E_I^0(\mathbf{Q})$ is truncated, the results are not independent of the coordinate system used. For example, significant differences in the description of Fermi-resonance parameters [148, 149] in rectilinear and curvilinear coordinates [150] have been reported.

Alternatively, $E_I^0(\mathbf{Q})$ can be represented by a grid of points around $\mathbf{Q}^{\mathrm{e}}$. In the most reliable calculations reported to date, $E_I^0(\mathbf{Q})$ is determined using the coupled cluster techniques discussed in Sect. 30.3. Then Eq. (30.79) can be solved in a basis analogous to $\psi_{\boldsymbol{l}}^J$ in Eq. (30.82). This approach is frequently referred to as the vibrational CI problem [151]. The reliability of the results depends to a considerable extent on the basis functions and coordinate system used to describe the problem. Considerable success has been reported for a technique in which the Watson Hamiltonian and harmonic oscillator functions are used to solve Eq. (30.79) [151, 152].

30.5.3 Coupling of Electronic and Rotational Angular Momentum in Weakly Interacting Systems

An understanding of the molecular structure of the weakly bound compounds of noble gas atoms and diatomic molecules provides important insights into the nature of chemical bonding. The inference of structural data from spectroscopic observations is an important aspect of this problem. In this subsection, a theoretical framework for understanding the spectroscopy of these systems is outlined as an illustration of the methods used to treat coupling of electronic and nuclear angular momentum in weakly interacting triatomic molecular systems. Detailed discussion of this class of problems can be found in Refs. [153–155].

As an example, consider the rovibronic structure of a noble gas-diatom complex Rg-AB, in either its $^1\Sigma^+$ or $^1\Pi$ states (which may be closely spaced). Note that degeneracy of the $^1\Pi$ state will only persist for collinear geometries of the triatom system. The rovibrational wave functions can be expanded in a product basis of functions describing (i) the rovibronic structure of AB and (ii) the relative motion (vibrational and end-over-end rotation) of Rg and AB. Since total angular momentum J and its space fixed projections M are good quantum numbers, the rovibronic wave function can be expanded as

$$\Psi^{JM} = \frac{1}{R} \sum_{vj\Omega l\varepsilon} C^{JM}_{j\Omega l\varepsilon v}(R)|\psi_{\mathrm{Rg}}\rangle|v_{Ij}\rangle|Ij\Omega\varepsilon, l; JM\rangle$$

(30.86)

where ψ_{Rg} is the ground state wave function for the noble gas atom, l denotes the angular momentum associated with the end-over-end motion, and v, j, ε and I denote respectively the vibrational, angular momentum (electronic + rotational), e/f symmetry index and state label of the electronic state of the AB molecule in Hund's case (a) basis. The angular momentum coupling algebra noted above is reflected in the definition of $|Ij\Omega\varepsilon, l; JM\rangle$, which is

$$|Ij\Omega\varepsilon, l; JM\rangle$$
$$= \sum_{m_j m_l} \langle jm_j lm_l|JM\rangle Y_{lm_l}(\theta, \phi)\psi_{Ijm_j\Omega\varepsilon}(\beta, \alpha)$$

(30.87)

where θ, ϕ and β, α are the polar and azimuthal angles for the line connecting the noble gas atom to the center of mass of the diatom and the diatom axis, respectively, and $\langle \cdots|\cdots\rangle$ is a Clebsch-Gordan coefficient [156] (see Chap. 2). The $\psi_{Ijm_j\Omega\varepsilon}$ are defined in turn by [94]

$$\psi_{Ijm_j\Omega\varepsilon} = \left[\frac{2J+1}{8\pi}\right]^{1/2} \left[D^j_{m_j,\Omega}(\alpha,\beta,0)^*|I^{2S+1}\Lambda^+(\Sigma)\rangle\right.$$
$$\left. + \varepsilon D^j_{m_j,-\Omega}(\alpha,\beta,0)^*|I^{2S+1}\Lambda^-(-\Sigma)\rangle\right] \quad (30.88)$$

or

$$\psi_{Ijm_j\Omega\varepsilon} = \left[\frac{2J+1}{4\pi}\right]^{1/2} D^j_{m_j,0}(\alpha,\beta,0)^*|I^{2S+1}\Lambda(\Sigma)\rangle$$

(30.89)

when $\Lambda = \Sigma = 0$, where the electronic state has term symbol $I^{2S+1}\Lambda$ with $\Omega = \Lambda + \Sigma$.

The $\mathbf{C}^{JM}(\mathbf{R})$ satisfy the usual close coupled equations [157, 158] (see Sect. 45.2.1)

$$\left\{\left[\frac{-1}{2\mu}\frac{d^2}{dR^2} + \frac{l(l+1)}{2\mu R^2}\right]\mathbf{I} - \frac{\mathbf{k}^2}{2\mu} + \mathbf{V}(r)\right\}\mathbf{C}^{JM}(r) = 0.$$

(30.90)

where $l(l+1)$ and $\mathbf{k}^2/2\mu$ designate the diagonal matrices of orbital angular momentum and asymptotic net scattering energy of individual channels. $\mathbf{V}(R)$ represents the matrix elements of $H^e(\mathbf{r};\mathbf{R})$ in the vibronic basis defined in Eqs. (30.86)–(30.89). It is built from terms of the form

$$\left\langle Y_{lm}D^{j\,*}_{m_j\Omega}v_{Ij}\right|\left\langle \psi_{\mathrm{Rg}}I^{2S+1}\Lambda\right|H^e(\mathbf{r};\mathbf{R})$$
$$\times \left|\psi_{\mathrm{Rg}}I'^{2S'+1}\Lambda'\right\rangle_r \left|v_{I'j'}Y_{l'm'}D^{j'\,*}_{m'_j\Omega'}\right\rangle$$
$$\equiv \left\langle Y_{lm}D^{j\,*}_{m_j\Omega}v_{Ij}\right|H^e_{II'}(R,\tilde{\beta},r)\left|Y_{l'm'}D^{j'\,*}_{m'_j\Omega'}v_{I'j'}\right\rangle$$

(30.91)

The angle $\tilde{\beta}$ is the polar angle of the diatom with respect to the atom-diatom axis. The angular integrations on the right-hand side of Eq. (30.91) are accomplished by expanding the angular dependence of $H^e_{II'}$ in d-matrices and using angular momentum coupling algebra [158]. $H^e_{II'}$ gives the interaction potential of the ground state noble gas atom and the $I^{2S+1}\Lambda$ state of AB. It represents the potential coupling in the diabatic basis $\psi_{\mathrm{Rg}}|I^{2S+1}\Lambda\rangle$. These matrix elements are derived from the corresponding adiabatic potential energy surfaces, determined using the techniques of Sect. 30.3. Since the van der Waals interaction is weak, the counterpoised method discussed in Sect. 30.3.4 should be used. When only one electronic state of a given symmetry is involved (point group C_s in the case of a triatom), the adiabatic and diabatic potentials are taken as equal. This is the case if for example only the $^1\Pi$ state of AB is considered. When more than one electronic state of a given sysmmetry is involved, for example when both $^1\Pi$ and $^1\Sigma^+$ electronic states of AB are included, the adiabatic potentials are used to determine the diabatic state potentials using an approximate adiabatic → diabatic state transformation [159, 160].

The solution to Eq. (30.90) can be obtained by either variational methods or by direct integration [153]. Results for Ar-OH [161], Ar-BH [155] and Ar-CH [162], based on potential energy surfaces determined from multireference contracted CI wave functions, are encouraging. In these studies, the r-dependence of the potential energy surfaces was neglected.

30.5.4 Reaction Path

The rovibrational spectrum reflects the molecular structure in the vicinity of an equilibrium structure of the molecule. However a chemical reaction samples a much broader range of molecular structures. The evolution of molecular structure can be characterized by the reaction path, which represents a minimum energy path along the potential energy surface connecting the reactants and products. The reaction path Hamiltonian [163, 164] enables dynamical aspects of a chemical reaction to be inferred from a characterization of the reaction path.

The reaction path $\mathbf{R}(s)$ on potential energy surface $E_I^0(\mathbf{R})$ is the curve in mass weighted Cartesian coordinates given parametrically in terms of the arc length s and is defined as that solution to the differential equation

$$\frac{d\mathbf{R}}{ds} = \mathbf{g}^I / \sqrt{\mathbf{g}^{I\dagger}\mathbf{g}^I} \equiv \mathbf{R}'(s) \qquad (30.92)$$

which approaches the saddle point from below [17, 165], where the arc length is given by

$$ds^2 = \sum_{i=1}^N dR_i^2 . \qquad (30.93)$$

At each point s, $\mathbf{R}(s)$ represents the molecular structure in terms of $3N$ mass weighted Cartesian coordinates. (In this subsection, the notation suppresses the difference between Cartesian and mass weighted Cartesian coordinates.)

The reaction path can be obtained by integrating Eq. (30.92). Integration begins at the saddle point along the direction of the single negative eigenvalue of $\mathbf{F}^I$, which is related to the reaction path curvature $d\mathbf{R}^2/ds^2$, by

$$\frac{d\mathbf{R}^2}{ds^2} \equiv \frac{d\mathbf{R}'}{ds} = [\mathbf{F}^I \mathbf{R}' - (\mathbf{R}'^\dagger \mathbf{F}^I \mathbf{R}')\mathbf{R}'] / \sqrt{\mathbf{g}^{I\dagger}\mathbf{g}^I} .$$
$$(30.94)$$

Thus integration of Eq. (30.92) gives the steepest descent path from the saddle point. This integration requires in principle only knowledge of the energy gradient, $\mathbf{g}^I(\mathbf{R})$, which is readily available using analytic gradient techniques (see Sect. 30.3). This approach works quite well when the reaction path is used only to investigate the topology of a potential energy surface. However if quantities such as the curvature are desired, procedures that follow the reaction path more closely are required. Such procedures require higher derivatives of the energy, as fully discussed in Ref. [165].

In a nonadiabatic process, the reaction path necessarily involves more than one potential energy surface. In the case of nonadiabatic reactions involving avoided intersections, the reaction path is difficult to define since the propensity for an intersurface transition spreads out over a range of nuclear coordinates. However, for nonadiabatic reactions that procede though a surface of intersection of two potential energy surfaces, the reaction path passes through the minimum energy point on the surface of intersection of the two potential energy surfaces [166]. An example of such a process is provided in Sect. 30.6. The minimum energy point on the surface of intersection satisfies Eq. (30.72) with no geometric constraints. For a spin-forbidden reaction, the solution of Eq. (30.72) yields

$\mathbf{g}^I \mathbf{R} + \xi_1 \mathbf{g}^{IJ}(\mathbf{R}) = 0$, so that $\mathbf{g}^I(\mathbf{R}) = \xi_1/(1+\xi_1)\mathbf{g}^J(\mathbf{R})$, i.e., the gradients on the two surfaces are either parallel or anti-parallel depending on the value of ξ_1. The relationship between the gradients in the vicinity of an intersection of two states of the same symmetry is not as simple. The wave functions are not differentiable at a point of intersection. In the vicinity of the point of intersection, the condition $\mathbf{g}^I(\mathbf{R}) + \xi_1 \mathbf{g}^{IJ}(\mathbf{R}) + \xi_2 \mathbf{h}^{IJ}(\mathbf{R}) \approx 0$ holds, so that $\mathbf{g}^I(\mathbf{R})$ and $\mathbf{g}^J(\mathbf{R})$ will not be simply related. Finally, note that the adiabatic correction, defined in Sect. 30.2, is large (and positive) in the vicinity of the conical intersection, so that the nuclear wave function exhibits a node at this point.

30.6 REACTION MECHANISMS: A SPIN-FORBIDDEN CHEMICAL REACTION

This section presents an example to illustrate how the ideas outlined in the preceding sections can be used to obtain a mechanistic description of a chemical reaction which could not be deduced on intuitive grounds. Considered here is the ground state reaction

$$\begin{aligned} \mathrm{CH}(X^2\Pi) + \mathrm{N_2}(X^1\Sigma_g^+) &\to \mathrm{HCN_2}(1\,^2A'') \\ &\to \mathrm{HCN_2}(1\,^4A'') \to \mathrm{HCN}(X^1\Sigma^+) + \mathrm{N}(^4S) \end{aligned} \qquad (30.95)$$

which is of considerable importance in the chemistry of planetary atmospheres [167, 168] and hydrocarbon flames, having been suggested as the initial step in the production of prompt or Fenimore NO [168–171]. However because the reaction is spin-forbidden, its importance has been questioned.

The reaction will be interpreted in terms of the intermediate complex model [172, 173]. In this model, the reaction can occur despite a small probability for intersystem crossing by repeatedly traversing the $^2A'' - {}^4A''$ surface of intersection. The repeated traversals result from a local minimum on the $^2A''$ potential energy surface. This mechanism is operative for the spin-forbidden oxygen quenching reactions [172, 173]

$$\begin{aligned} \mathrm{O}(^1D) + \mathrm{N_2}(X^1\Sigma_g^+) &\to \mathrm{N_2O}^* \\ &\to \mathrm{O}(^3P) + \mathrm{N_2}(X^1\Sigma_g^+) \end{aligned} \qquad (30.96)$$

$$\begin{aligned} \mathrm{O}(^1D) + \mathrm{CO}(X^1\Sigma^+) &\to \mathrm{CO_2}^* \\ &\to \mathrm{O}(^3P) + \mathrm{CO}(X^1\Sigma^+) \end{aligned} \qquad (30.97)$$

which are important in the chemistry of the atmosphere.

To demonstrate the feasibility of reaction (30.95) as an intermediate complex assisted chemical reaction it is necessary to (i) determine the energetically relevant portion of the doublet-quartet surface of intersection, (ii) determine the spin-orbit interaction coupling the doublet and quartet surfaces, (iii) characterize a local minimum on the doublet surface from which the surface of intersection is accessible, (iv) characterize a path from the reactant channel to this region and (v) determine the exit channel path on the quartet surface. The results of electronic structure calculations at the MR-CI level, addressing these points [174–178] are summarized below.

Local Extrema on the $^2A'' - {}^4A''$ Surface of Intersection. Points in the $^2A'' - {}^4A''$ surface of intersection were determined from the solution to Eq. (30.72) [126, 174]. The minimum energy point on the surface of intersection, labelled $C_{mex}^{C_{2v}}$, has approximate C_{2v} symmetry, and is shown in Fig. 30.3. It is the region of this crossing point that must be accessed from the doublet surface.

The Spin–Orbit Interaction. In the double group corresponding to C_s symmetry, the $^4A''$ and $^2A''$ wave functions each carry degenerate irreducible representations, Kramer's doublets [179]. The following pairs of nonrelativistic zeroth-order wave functions can be used to span these degenerate representations (i) $\Psi^0[^2A''(\frac{1}{2})]$, $\Psi^0[^2A''(-\frac{1}{2})]$, (ii) $i\Psi^0[^4A''(\frac{1}{2})]$, $i\Psi^0[^4A''(-\frac{1}{2})]$, and (iii) $i\Psi^0[^4A''(\frac{3}{2})]$, $i\Psi0[^4A''(-\frac{3}{2})]$, where the M_s value has been given parenthetically. In this case, all nonvanishing matrix elements connecting the components of the $^4A''$ and $^2A''$ states can be expressed in terms of the single real-valued matrix element [139]

$$H^{so}(^4A'', {}^2A'') \equiv \langle i\Psi^0[^4A''(\tfrac{3}{2})]|H^{so}|\Psi^0[^2A''(\tfrac{1}{2})]\rangle$$

$$(30.98)$$

which was found to be ≈ 12.5 cm^{-1} [126, 174] in the vicinity of the minimum energy crossing point determined above.

Local Extrema on the $^2A''$ Potential Energy Surface. Two local minima on the $^2A''$ potential energy surface have been found [174–176]. The local minimum configuration labeled C_{min}^{dative} pictured in Fig. 30.3 represents a datively bonded structure in which N_2 donates a pair of electrons to the empty CH(1π) orbital. This can be thought of as a reactant channel structure, since the N-N and C-H bond lengths are similar to those in the isolated molecules. This point on the $^2A''$ potential energy surface is stable by 20.2 kcal/mol relative to the doublet asymptote in the reactant channel. The second minimum on the $^2A''$ potential energy surface is also pictured in Fig. 30.3 and is seen to have C_{2v} symmetry. This configuration point, denoted $C_{min}^{C_{2v}}$, is stable by 22.3 kcal/mol relative to the doublet asymptote in the reactant channel. Note that $C_{min}^{C_{2v}}$ and $C_{mex}^{C_{2v}}$ differ only in the length of the N-N bond.

Reaction Path. From the above data it might appear that both the $C_{min}^{C_{2v}}$ and C_{min}^{dative} structures are involved in the mechanism of reaction (30.95). However the energy of the transition state connecting these two minima is quite large [176], so that the reaction path avoids C_{min}^{dative} [177]. The true reaction path involves $C_{min}^{C_{2v}}$,

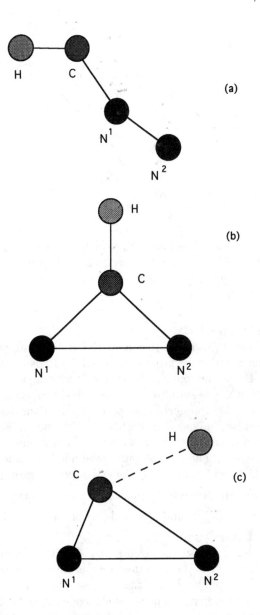

Figure 30.3. Key bond distances in Å for (a) $R_{min}^{dative}(\mathrm{CH}) = 1.082$, $R_{min}^{dative}(\mathrm{CN}^1) = 1.340$, $R_{min}^{dative}(\mathrm{N}^1\mathrm{N}^2) = 1.143$; (b) $R_{min}^{C_{2v}}(\mathrm{CH}) = 1.071$, $R_{min}^{C_{2v}}(\mathrm{CN}^1) = 1.311$; $R_{min}^{C_{2v}}(\mathrm{N}^1\mathrm{N}^2) = 1.735$; $R_{mex}^{C_{2v}}(\mathrm{CH}) = 1.076$, $R_{mrex}^{C_{2v}}(\mathrm{CN}^1) = 1.3$, $R_{mex}^{C_{2v}}(\mathrm{N}^1\mathrm{N}^2) = 2.221$; (c) $R_{TS}^c(\mathrm{CH}) = 1.093$, $R_{TS}^c(\mathrm{CN}^1) = 1.515$, $R_{TS}^c(\mathrm{N}^1\mathrm{N}^2) = 1.214$. The notation is as defined in Sect. 30.6. The dashed line in structure (c) indicates that H is out of the plane of the paper. (a), (b) from Ref. [124], (c) from Ref. [177].

$C_{mex}^{C_{2v}}$ and a reactant channel transition state C_{TS}^c that connects $C_{min}^{C_{2v}}$ and the reactant channel species CH + N_2. C_{TS}^c is also pictured in Fig. 30.3. It involves the nonplanar cis- approach of HC to N_2 [177]. The total reaction path is presented in Fig. 30.4 [178]. Note that following intersystem crossing, the system encounters a final transition state before proceding to the spin-

Figure 30.4. Reaction path for $CH(X^2\Pi) + N_2(X^1\Sigma_g^+)$ $\rightarrow HCN(X^1\Sigma^+) + N(^4S)$ after Ref. [177]. Energy scale, which is approximate, is in kcal/mol.

forbidden products.

From this qualitative analysis, a bimodal picture of the reaction of ground state CH with N_2 emerges. $CH(^2\Pi)$ can be removed by either (i) forming a complex $\mathcal{C}_{min}^{dative}$ which is stabilized by a third body collision or (ii) forming the spin-forbidden products with the reaction mechanism indicated in Fig. 30.4. This mechanism has been deduced entirely from the computational results cited above. The key computational findings responsible for this mechanism are (i) the existence and geometrical relationship between $\mathcal{C}_{min}^{C2v}$ and $\mathcal{C}_{mex}^{C2v}$, (ii) the large barrier separating $\mathcal{C}_{min}^{dative}$ and $\mathcal{C}_{min}^{C2v}$ and (iii) the entrance channel path across $\mathcal{C}_{TS}^{c}$ to $\mathcal{C}_{min}^{C2v}$.

Since there is apparently no or little barrier to forming $\mathcal{C}_{min}^{dative}$ [176], at low temperatures CH will be removed exclusively by this process at a rate with a small or negative temperature dependence, indicating a barrierless reaction. At higher temperatures, the rate constant is expected to exhibit a more usual Arrhenius type behavior as the spin-forbidden channel can be accessed [180]. These expectations are consistent with previous low temperature measurements [168,181] and high temperature shock tube results [171,182,183]. The above reaction path for the spin-forbidden reaction has been used in a two internal coordinate model for reaction (30.95) [180] that also supports the proposed description of this reaction.

Acknowledgment

This work has been supported in part by NSF grant CHE 94-04193, AFOSR grant F49620-93-1-0067 and DOE grant DE-FG02-91ER14189

REFERENCES

1. H. A. Bethe and E. E. Salpeter, *Quantum Mechanics of One and Two Electron Atoms* (Plenum/Rosetta, New York, 1977).
2. M. Born and K. Huang, *Dynamical Theory of Crystal Lattices* (Oxford University Press, 1954).
3. G. Herzberg and H. C. Longuet-Higgins, Disc. Faraday Soc. **35**, 77 (1963).
4. M. V. Berry, Proc. R. Soc. London, Sect. A **392**, 45 (1984).
5. H. C. Longuet-Higgins, Adv. Spectr. **2**, 429 (1961).
6. W. H. Gerber and E. Schumacher, J. Chem. Phys. **69**, 1692 (1978).
7. L. R. Kahn, P. Baybutt, and D. G. Truhlar, J. Chem. Phys. **65**, 3826 (1976).
8. J. C. Tully, in *Modern Theoretical Chemistry*, edited by W. H. Miller (Plenum, New York, 1976), Vol. 2, p. 217.
9. B. C. Garrett and D. G. Truhlar, *Theoretical Chemistry Advances and Perspectives* (Academic Press, New York, 1981).
10. D. M. Bishop and L. M. Cheung, J. Chem. Phys. **78**, 1396 (1983).
11. D. M. Bishop and L. M. Cheung, J. Chem. Phys. **80**, 4341 (1984).
12. J. O. Jensen and D. R. Yarkony, J. Chem. Phys. **89**, 3853 (1988).
13. C. R. Vidal and W. C. Stwalley, J. Chem. Phys. **77**, 883 (1982).
14. Y. C. Chen, D. R. Harding, W. C. Stwalley, and C. R. Vidal, J. Chem. Phys. **85**, 2436 (1986).
15. J. N. Murrell, S. Carter, S. C. Farantos, P. Huxley, and A. J. C. Varandas, *Molecular Potential Energy Surfaces* (Wiley, New York, 1984).
16. D. G. Truhlar, R. Steckler, and M. S. Gordon, Chem. Rev. **87**, 217 (1987).
17. K. Fukui, Acc. Chem. Res. **14**, 363 (1981).
18. C. J. Cerjan and W. H. Miller, J. Chem. Phys. **75**, 2800 (1981).
19. H. B. Schlegel, Adv. Chem. Phys. **67**, 249 (1987).
20. H. B. Schlegel, in *Modern Electronic Structure Theory*, edited by D. R. Yarkony (World Scientific, Singapore, 1995), Part 1.
21. *Modern Theoretical Chemistry*, edited by H. F. Schaefer (Plenum Press, New York, 1976).
22. *Modern Electronic Structure Theory*, edited by D. R. Yarkony (World Scientific, Singapore, 1995).
23. A. D. McLachlan, Mol. Phys. **4**, 417 (1961).
24. F. T. Smith, Phys. Rev. **179**, 111 (1969).
25. W. Lichten, Phys. Rev. **131**, 339 (1963).
26. W. Lichten, Phys. Rev. **164**, 131 (1967).
27. C. A. Mead and D. G. Truhlar, J. Chem. Phys. **77**, 6090 (1982).
28. M. Baer, Chem. Phys. Lett. **35**, 112 (1975).
29. M. Baer, Chem. Phys. **15**, 49 (1976).
30. T. Pacher, C. A. Mead, L. S. Cederbaum, and H. Koppel, J. Chem. Phys. **91**, 7057 (1989).
31. V. Sidis, in *State-Selected and State-to-State Ion-Molecule Reaction Dynamics Part 2, Theory*, edited by M. Baer and C.-Y. Ng (John Wiley and Sons, New York, 1992), Vol. 82, p. 73.

32. R. Renner, Z. Phys. **92**, 172 (1934).
33. J. von Neumann and E. Wigner, Phys. Z. **30**, 467 (1929).
34. C. A. Mead, J. Chem. Phys. **70**, 2276 (1979).
35. A. Szabo and N. S. Ostlund, *Modern Quantum Chemistry: Introduction to Advanced Electronic Structure Theory* (McMillan, New York, 1982).
36. P. Jørgensen and J. Simons, *Second Quantization Based Methods in Quantum Chemistry* (Academic Press, New York, 1981).
37. J. Almlöf, in *Modern Electronic Structure Theory*, edited by D. R. Yarkony (World Scientific, Singapore, 1995), Part 1.
38. T. H. Dunning and P. J. Hay, in *Modern Theoretical Chemistry*, edited by H. F. Schaefer (Plenum, New York, 1976), Vol. 3.
39. B. O. Roos, P. Linse, P. E. M. Siegbahn, and M. R. A. Blomberg, Chem. Phys. **66**, 197 (1982).
40. M. Moshinsky, *Group Theory and the Many Body Problem* (Gordon and Breach, New York, 1968).
41. *The Unitary Group*, in *Lecture Notes in Chemistry*, edited by J. Hinze (Springer-Verlag, Berlin, 1979), Vol. 22.
42. I. Shavitt, in *Modern Theoretical Chemistry*, edited by H. F. Schaefer (Plenum Press, New York, 1976), Vol. 3, p. 189.
43. Y. Yamaguchi, Y. Osamura, J. D. Goddard, and H. F. Schaefer, *A New Dimension to Quantum Chemistry: Analytic Derivative Methods in ab initio Molecular Electronic Structure Theory* (Oxford University Press, Oxford, 1994).
44. B. H. Lengsfield and D. R. Yarkony, in *State-Selected and State to State Ion-Molecule Reaction Dynamics: Part 2 Theory*, edited by M. Baer and C. Ng (John Wiley and Sons, New York, 1992), Vol. 82.
45. P. Pulay, Mol. Phys. **17**, 197 (1969).
46. C. C. J. Roothaan, Rev. Mod. Phys. **23**, 69 (1951).
47. J. F. Gaw, Y. Yamaguchi, and H. F. Schaefer, J. Chem. Phys. **81**, 6395 (1984).
48. J. F. Gaw, Y. Yamaguchi, H. F. Schaefer, and N. C. Handy, J. Chem. Phys. **85**, 5132 (1986).
49. J. Almlöf, J. Comput. Chem. **3**, 385 (1982).
50. K. Docken and J. Hinze, J. Chem. Phys. **57**, 4928 (1972).
51. J. Hinze, J. Chem. Phys. **59**, 6424 (1973).
52. H. Werner and W. Meyer, J. Chem. Phys. **74**, 5794 (1981).
53. B. H. Lengsfield, J. Chem. Phys. **77**, 4073 (1982).
54. C. Møller and M. S. Plesset, Phys. Rev. **46**, 618 (1934).
55. J. Almlöf, Chem. Phys. Lett. **181**, 319 (1991).
56. J. Cizek, J. Chem. Phys. **45**, 4256 (1966).
57. J. Cizek, Adv. Chem. Phys. **14**, 35 (1969).
58. J. Paldus and J. Cizek, in *Energy Structure and Reactivity*, edited by D. Smith and W. B. McRae (Wiley, New York, 1973), p. 35.
59. R. J. Bartlett and G. D. Purvis III, Int. J. Quantum Chem. Symp. **14**, 561 (1978).
60. J. A. Pople, R. Krishnan, H. B. Schlegel, and J. S. Binkley, Int. J. Quantum Chem. **14**, 545 (1978).
61. P. O. Löwdin, Phys. Rev. **97**, 1474 (1955).
62. R. J. Bartlett, J. D. Watts, S. A. Kucharski, and J. Noga, Chem. Phys. Lett. **165**, 513 (1990).
63. S. A. Kucharski and R. J. Bartlett, Chem. Phys. Lett. **158**, 550 (1989).
64. K. Raghavachari, G. W. Trucks, J. A. Pople, and M. Head-Gordon, Chem. Phys. Lett. **157**, 479 (1989).
65. J. D. Watts, J. Gauss, and R. J. Bartlett, J. Chem. Phys. **98**, 8718 (1993).
66. R. J. Bartlett, in *Modern Electronic Structure Theory*, edited by D. R. Yarkony (World Scientific, Singapore, 1995), Part 2.
67. R. N. Diffenderfer and D. R. Yarkony, J. Phys. Chem. **86**, 5098 (1982).
68. B. O. Roos, P. R. Taylor, and P. E. M. Siegbahn, Chem. Phys. **48**, 157 (1980).
69. B. O. Roos, Int. J. Quantum Chem. Symp. **14**, 175 (1980).
70. P. Siegbahn, A. Heiberg, B. Roos, and B. Levy, Phys. Scr. **21**, 323 (1980).
71. C. W. Bauschlicher, S. R. Langhoff, and P. R. Taylor, Adv. Chem. Phys. **77**, 103 (1990).
72. H. J. Silverstone and O. Sinanoglu, J. Chem. Phys. 44, 1899 (1966).
73. E. Dalgaard and P. Jørgensen, J. Chem. Phys. **69**, 3833 (1978).
74. E. R. Davidson, J. Comput. Phys. **17**, 87 (1975).
75. B. O. Roos, Chem. Phys. Lett. **15**, 153 (1972).
76. P. E. M. Siegbahn, J. Chem. Phys. **72**, 1647 (1980).
77. R. J. Harrison, J. Chem. Phys. **94**, 5021 (1991).
78. R. Caballol and J. P. Malrieu, Chem. Phys. Lett. **188**, 543 (1992).
79. A. Povill, J. Rubio, and F. Illas, Theo. Chim. Acta **82**, 229 (1992).
80. P. J. Knowles and H. Werner, Chem. Phys. Lett. **115**, 259 (1985).
81. P. J. Knowles and H. Werner, Chem. Phys. Lett. **145**, 514 (1988).
82. H. Werner and P. J. Knowles, J. Chem. Phys. **82**, 5053 (1985).
83. H. Werner and P. J. Knowles, J. Chem. Phys. **89**, 5803 (1988).
84. K. Andersson, P. Å. Malmqvist, B. O. Roos, A. J. Sadlej, and K. Wolinski, J. Phys. Chem. **94**, 5483 (1990).
85. K. Andersson, P. Å. Malmqvist, and B. O. Roos, J. Chem. Phys. **96**, 1218 (1992).
86. R. G. Parr and W. Yang, *Density-Functional Theory of Atoms and Molecules* (Oxford, New York, 1989).
87. P. Hohenberg and W. Kohn, Phys. Rev. B **136**, 864 (1964).
88. W. Kohn and L. J. Sham, Phys. Rev. A **140**, 1133 (1965).
89. A. D. Becke, in *Modern Electronic Structure Theory*, edited by D. R. Yarkony (World Scientific, Singapore, 1995), Part 2.
90. P. Pulay, in *Modern Electronic Structure Theory*, edited by D. R. Yarkony (World Scientific, Singapore, 1995), Part 2.
91. B. Liu and A. D. McLean, J. Chem. Phys. **59**, 4557 (1973); **91**, 2348 (1989).
92. S. F. Boys and F. Bernardi, Mol. Phys. **19**, 553 (1970).
93. M. Gutowski, J. van Lenthe, J. C. G. M. van Duijneveldt-van de Rijdt, and F. B. van Duijneveldt, J. Chem. Phys. **98**, 4728 (1993).
94. H. Lefebvre-Brion and R. W. Field, *Perturbations in the Spectra of Diatomic Molecules* (Academic Press, New

York, 1986).

95. S. Han, H. Hettema, and D. R. Yarkony, J. Chem. Phys. **102**, 1955 (1995).

96. B. H. Lengsfield, P. Saxe, and D. R. Yarkony, J. Chem. Phys. **81**, 4549 (1984).

97. P. Saxe, B. H. Lengsfield, and D. R. Yarkony, Chem. Phys. Lett. **113**, 159 (1985).

98. R. J. Buenker, G. Hirsch, S. D. Peyerimhoff, P. J. Bruna, J. Romelt, M. Bettendorff, and C. Petrongolo, in *Current Aspects of Quantum Chemistry 1981*, edited by R. Carbo (Elsevier, New York, 1982), Vol. 21, p. 81.

99. C. A. Mead, J. Chem. Phys. **78**, 807 (1993).

100. J. Hellmann, *Einführung in die Quantenchemie* (Deuticke, Leipzig, 1937).

101. R. P. Feynman, Phys. Rev. **56**, 340 (1939).

102. S. R. Langhoff and C. W. Kern, in *Modern Theoretical Chemistry*, edited by H. F. Schaefer (Plenum, New York, 1977), Vol. 4, p. 381.

103. M. Blume and R. E. Watson, Proc. R. Soc. London A **270**, 127 (1962).

104. M. Blume and R. E. Watson, Proc. R. Soc. London A **271**, 565 (1963).

105. S. Fraga, K. M. S. Saxena, and B. W. N. Lo, At. Data **3**, 323 (1971).

106. W. C. Ermler, R. B. Ross, and P. A. Christiansen, in *Advances in Quantum Chemistry* (Academic Press, New York, 1988), Vol. 19, p. 139.

107. W. C. Ermler, Y. S. Lee, P. A. Christiansen, and K. S. Pitzer, Chem. Phys. Lett. **81**, 70 (1981).

108. Y. S. Lee, W. C. Erlmer, and K. S. Pitzer, J. Chem. Phys. **67**, 5861 (1977).

109. R. M. Pitzer and N. W. Winter, J. Phys. Chem. **92**, 3061 (1988).

110. W. J. Stevens and M. Krauss, J. Chem. Phys. **76**, 3834 (1982).

111. J. P. Desclaux, Comp. Phys. Comm. **9**, 31 (1975).

112. K. Balasubramanian, J. Phys. Chem. **93**, 6585 (1989).

113. B. A. Hess, Phys. Rev. A **33**, 3742 (1986).

114. B. A. Hess and P. Chandra, Phys. Scr. **36**, 412 (1987).

115. J. Sucher, Phys. Rev. A **22**, 348 (1980).

116. C. M. Marian, J. Chem. Phys. **93**, 1176 (1990).

117. C. M. Marian, J. Chem. Phys. **94**, 5574 (1991).

118. T. R. Furlani and H. F. King, J. Chem. Phys. **82**, 5577 (1985).

119. B. A. Hess, R. J. Buenker, C. M. Marian, and S. D. Peyerimhoff, Chem. Phys. 71, 79 (1982).

120. B. A. Hess, R. J. Buenker, C. M. Marian, and S. D. Peyerimhoff, Chem. Phys. **71**, 79 (1982).

121. D. R. Yarkony, J. Chem. Phys. **84**, 2075 (1986).

122. J. O. Jensen and D. R. Yarkony, Chem. Phys. Lett. **141**, 391 (1987).

123. B. Liu and M. Yoshimine, J. Chem. Phys. **74**, 612 (1981).

124. D. R. Yarkony, Int. Rev. Phys. Chem. **11**, 195 (1992).

125. M. R. Manaa and D. R. Yarkony, J. Chem. Phys. **99**, 5251 (1993).

126. D. R. Yarkony, J. Phys. Chem. **97**, 4407 (1993).

127. D. R. Yarkony, J. Chem. Phys. **100**, 3639 (1994).

128. R. Fletcher, *Practical Methods of Optimization* (John Wiley and Sons, New York, 1981), Vol. 2.

129. I. N. Radazos, M. A. Robb, M. A. Bernardi, and M.

Olivucci, Chem. Phys. Lett. **197**, 217 (1992).

130. G. Herzberg and H. C. Longuet-Higgins, Disc. Faraday Soc. **35**, 77 (1963).

131. M. V. Berry, Proc. R. Soc. London A **392**, 45 (1984).

132. S. Xantheas, S. T. Elbert, and K. Ruedenberg, J. Chem. Phys. **93**, 7519 (1990).

133. J. Tennyson, S. Miller, and B. T. Sutcliffe, J. Chem. Soc. Faraday Trans. **84**, 1295 (1988).

134. J. H. Van Vleck, Rev. Mod. Phys. **23**, 213 (1951).

135. M. Mizushima, *The Theory of Rotating Diatomic Molecules* (John Wiley and Sons, New York, 1975).

136. T. J. Lee, D. J. Fox, H. F. Schaefer, and R. M. Pitzer, J. Chem. Phys. **81**, 356 (1984).

137. R. N. Dixon, Mol. Phys. **54**, 333 (1985).

138. R. J. Whitehead and N. C. Handy, J. Mol. Spec. **55**, 356 (1975).

139. M. E. Rose, *Elementary Theory of Angular Momentum* (John Wiley and Sons, New York, 1957).

140. I. M. Mills, in *Molecular Spectroscopy: Modern Research*, edited by K. N. Rao and C. W. Mathews (Academic Press, 1970), p. 115.

141. C. Eckart, Phys. Rev. **47**, 552 (1935).

142. E. B. Wilson, Jr. and J. B. Howard, J. Chem. Phys. **4**, 260 (1936).

143. B. T. Darling and D. M. Dennison, Phys. Rev. **57**, 128 (1940).

144. E. B. Wilson, J. C. Decius, and P. C. Cross, *Molecular Vibrations* (McGraw-Hill, New York, 1955).

145. J. K. G. Watson, Mol. Phys. **15**, 479 (1968).

146. R. N. Zare, *Angular Momentum* (John Wiley and Sons, New York, 1988).

147. B. T. Sutcliffe and J. Tennyson, Mol. Phys. **58**, 1053 (1986).

148. J. Segall, R. N. Zare, H. R. Dubal, M. Lewerenz, and M. Quack, J. Chem. Phys. **86**, 634 (1987).

149. W. H. Green, W. D. Lawrence, and C. B. Moore, J. Chem. Phys. **86**, 6000 (1987).

150. R. D. Amos, J. G. Gaw, N. C. Handy, and S. Carter, J. Chem. Soc. Faraday Trans. **84**, 1247 (1988).

151. P. Botschwina, Chem. Phys. **68**, 41 (1982).

152. P. Botschwina, J. Chem. Soc. Faraday Trans. **84**, 1263 (1988).

153. J. M. Hutson, in *Advances in Molecular Vibrations and Collision Dynamics*, edited by J. M. Bowman and M. A. Ratner (JAI Press, Greenwich, 1991), Vol. 1A.

154. M. L. Dubernet, D. Flower, and J. M. Hutson, J. Chem. Phys. **94**, 7602 (1991).

155. M. H. Alexander, S. Gregurick, and P. J. Dagdigian, J. Chem. Phys. **101**, 2887 (1994).

156. D. M. Brink and G. R. Satchler, *Angular Momentum* (Clarendon, Oxford, 1968).

157. A. Arthurs and A. Dalgarno, Proc. Roy. Soc. London A **256**, 540 (1960).

158. M. Alexander, Chem. Phys. **92**, 337 (1985).

159. H. Werner, B. Follmeg, and M. H. Alexander, J. Chem. Phys. **89**, 3139 (1988).

160. H. Werner, B. Follmeg, M. H. Alexander, and D. Lemoine, J. Chem. Phys. **91**, 5425 (1989).

161. C. Chakravarty and D. Clary, J. Chem. Phys. **94**, 4149 (1991).

162. M. H. Alexander, S. Gregurick, P. J. Dagdigian, G. W.

Lemire, M. J. McQuaid, and R. C. Sausa, J. Chem. Phys. **101**, 4547 (1994).

163. W. H. Miller, in *The Theory of Chemical Reaction Dynamics*, edited by D. C. Clary (Reidel, Dordrecht, 1986).

164. W. H. Miller, N. Handy, and J. Adams, J. Chem. Phys. **72**, 99 (1980).

165. M. Page and J. W. McIver, J. Chem. Phys. **88**, 922 (1988).

166. S. Kato, R. L. Jaffe, A. Komornicki, and K. Morokuma, J. Chem. Phys. **78**, 4567 (1983).

167. D. F. Strobel, Planet. Space Sci. **30**, 839 (1982).

168. M. R. Berman and M. C. Lin, J. Phys. Chem. **87**, 3933 (1983).

169. C. P. Fenimore, in *13th International Symposium on Combustion* (The Combustion Institute, Pittsburgh, 1971), p. 373.

170. J. Blauwens, B. Smets, and J. Peeters, in *16th International Symposium on Combustion* (The Combustion Institute, Pittsburgh, 1977), p. 1055.

171. A. J. Dean, R. K. Hanson, and C. T. Bowman, in *23rd International Symposium on Combustion* (The Combustion Institute, Pittsburgh, 1990), p. 259.

172. J. C. Tully, J. Chem. Phys. **61**, 61 (1974).

173. G. E. Zahr, R. K. Preston, and W. H. Miller, J. Chem. Phys. **62**, 1127 (1975).

174. M. R. Manaa and D. R. Yarkony, J. Chem. Phys. **95**, 1808 (1991).

175. M. R. Manaa and D. R. Yarkony, Chem. Phys. Lett. **188**, 352 (1992).

176. J. M. L. Martin and P. R. Taylor, Chem. Phys. Lett. **209**, 143 (1993).

177. S. P. Walch, Chem. Phys. Lett. **208**, 214 (1993).

178. T. Seideman and S. P. Walch, J. Chem. Phys. **101**, 3656 (1994).

179. M. Tinkham, *Group Theory and Quantum Mechanics* (McGraw-Hill, New York, 1964).

180. T. Seideman, J. Chem. Phys. **101**, 3662 (1994).

181. K. H. Becker, B. Engelhardt, H. Geiger, R. Kurtenbach, G. Schrey, and P. Wissen, Chem. Phys. Lett. **154**, 342 (1992).

182. D. Lindackers, M. Burmeister, and P. Roth, in *23rd International Symposium on Combustion* (Combustion Institute, Pittsburgh, 1990), p. 251.

183. D. Lindackers, M. Burmeister, and P. Roth, Combustion, and Flame **81**, 251 (1990).

31

Molecular Symmetry and Dynamics

William G. Harter

Department of Physics, University of Arkansas, Fayetteville, Arkansas 72701

31.1 DYNAMICS AND SPECTRA OF MOLECULAR ROTORS

Molecules are aggregates of two or more nuclei bound by at least one electron. The nuclei of most stable molecules can be imagined to be points in a more or less rigid body whose relative positions are constrained by an electronic bonding potential. This potential depends strongly upon the electronic state as described in Chap. 30. For this discussion, we consider only stable molecules in their electronic ground state.

Motions which stretch or compress the bonds are called *vibrational* motions, and give rise to spectral resonances in the infrared region of the spectrum. Typical vibrational quanta (v_0) lie between 80 cm^{-1} (the lowest GeBr$_4$ mode) and 3020 cm^{-1} (the highest CH$_4$ mode). (A wave number of 1000 cm^{-1} corresponds to a wavelength of 10 μm and a frequency of 29.979 2458 THz.) Vibrational amplitudes are usually tiny, since zero-point motions or vibrations involving one or two quanta ($v = 0, 1, 2, \ldots$) are constrained by the steep bonding potential to less than a few percent of bond lengths.

Overall rotation of molecules in free space is unconstrained, and gives rise to far-infrared or microwave pure rotational transitions or sidebands on top of vibrational spectra. Typical rotational quanta lie between 0.18 cm^{-1} (5.4 GHz) for SF$_6$ and 10.6 cm^{-1} for CH$_4$. Individual molecules are free to rotate or translate as a whole while undergoing their tiny but rapid vibrations.

A vibrating molecule can be thought of as a tumbling collection of masses held together by 'springs' (the electronic vibrational potential or force field) and is called a *semi-rigid rotor*. The coupling of rotational and vibrational motion is called *rovibrational* coupling, and includes *centrifugal* and *Coriolis* coupling discussed in Sect. 31.4.

This discussion of molecular dynamics and spectra mainly involves molecular rotation and properties of rotationally excited molecules, particularly those with high rotational quantum number $J = 10$–200. However, the discussion also applies to molecules in excited vibrational states, and even to certain cases of molecules in excited electronic states. The analysis of *vibronic* (vibrational-electronic), *rovibrational* (rotational-vibrational), or *rovibronic* (all three) types of excitation can be very complicated [1–5] and is beyond the scope of this chapter, but consideration of these problems can all benefit from the elementary treatment given here.

31.1.1 Rigid Rotors

As a first approximation, and for the purposes of discussing basic molecular dynamics and spectra, one may ignore vibrations and model stable molecules as 'stick-and-ball' structures or *rigid rotors*. Then the Hamiltonian just has the following three terms

$$H = AJ_x^2 + BJ_y^2 + CJ_z^2. \quad (31.1)$$

Here, J_x, J_y, J_z are rotational angular momentum operators, and the rotational constants are

$$A = \tfrac{1}{2}(I_x), \quad B = \tfrac{1}{2}(I_y), \quad C = \tfrac{1}{2}(I_z). \quad (31.2)$$

where the I_α are the principal moments of inertia of the body. This implies that the J-coordinate system being used is a special one fixed to the rotor and aligned with its principal axes.

Many molecules, particularly all diatomic molecules, have two of these rotational constants equal, say $A = B$. Such rotors are called *symmetric tops*, and their Hamiltonian can be written in terms of the square of the total angular momentum $\mathbf{J}\cdot\mathbf{J}$ and one other component J_z as

$$H = BJ_x^2 + BJ_y^2 + CJ_z^2 = B\mathbf{J}\cdot\mathbf{J} + (C - B)J_z^2. \quad (31.3)$$

This gives a simple formula for the symmetric top rotational energy levels in terms of the quantum numbers J for total angular momentum and K for the body z-component

$$E(J,K) = BJ(J + 1) + (C - B)K^2. \quad (31.4)$$

However, this simple eigenvalue formula hides the structure of the eigenstates or eigenfunctions. Indeed, the full Schrödinger angular differential equation based upon Hamiltonian (31.1) is much more lengthy. It is important to remember that H is written in a rotating *body* coordinate system, and must be connected to a star-fixed or *laboratory* frame to get the full theory.

31.1.2 Molecular States Inside and Out

Rotor angular momentum eigenfunctions can be expressed as a continuous linear combination of rotor angular position states $|\alpha\beta\gamma\rangle$ defined by the Euler angles of laboratory azimuth α, polar angle β of body z-axis, and body azimuth or 'gauge twist' γ

$$\left| \begin{matrix} J \\ MK \end{matrix} \right\rangle = \frac{\sqrt{2J+1}}{8\pi^2} \int_0^{2\pi} d\alpha \int_0^\pi \sin\beta\, d\beta$$
$$\times \int_0^{2\pi} d\gamma\, D_{MK}^{J*}(\alpha\beta\gamma)|\alpha\beta\gamma\rangle, \quad (31.5)$$

where the rotor wave functions D_{MK}^{J*} are just the conjugates of the Wigner rotation matrices described in Sect. 2.4.1, and row and column indices M and K are the lab and body components of the angular momentum, respectively [5–7].

An important feature of polyatomic molecules is that their angular momentum states have two kinds of azimuthal quantum numbers. In addition to the usual lab component M associated with the coordinate α (α and β are usually labeled ϕ and ϑ), there is a body component K associated with the Euler coordinate γ, the body azimuthal angle for the body z-axis relative to the laboratory Z-axis.

The physics of atomic or diatomic angular momentum states has no internal or 'body' structure, so that the quantum number K is always zero. Unless one sets $K = 0$, the energy formula (31.4) blows up for a point particle because z-inertia for a point is zero and C is infinite. Also, the dimension of the angular momentum state multiplet of a given J is larger than the usual $(2J + 1)$ found in atomic or diatomic molecular physics. In polyatomic rotors, the number of states for each J is $(2J + 1)^2$ since both M and K quantum numbers range from $-J$ to $+J$.

A futher important feature is that the molecular rotor wave functions contain as a special $K = 0$ case all the usual atomic spherical harmonics Y_m^ℓ complete with correct normalization and phase,

$$\sqrt{4\pi}Y_m^\ell(\phi\vartheta) = D_{m0}^{\ell*}(\phi\vartheta)\sqrt{2\ell + 1}. \quad (31.6)$$

This is part of a powerful symmetry principle: group representations are quantum wave functions, and symmetry analysis is an extension of Fourier analysis — not just for translations as in standard Fourier transforms, but for *any* group of symmetry operations. The usual Fourier coefficients $\exp(ikx)$ are replaced by the D functions in the rotational Fourier transform embodied by Eq. (31.5).

Molecular rotational analysis displays another important aspect of symmetry analysis in general. For every group of symmetry operations, such as external lab-based rotations, there is an independent dual group of internal or body-based operations. The external symmetry of the environment is independent of the internal symmetry of the molecular body, and all the operations of one commute with all those of the other. The molecular rotation group is written as an outer product $R(3)_{\text{lab}} \otimes R(3)_{\text{body}}$ of the external and internal parts. Thus the degeneracy associated with the group representations for a single J is $(2J + 1)^2$.

The inversion or parity operator $\mathbf{I}(\mathbf{r} \rightarrow -\mathbf{r})$ can be defined to be the same for both lab and body frames. Including $\mathbf{I}$ with the rotational group $R(3)$ gives the orthogonal group $O(3) = R(3) \otimes \{\mathbf{1}, \mathbf{I}\}$. If parity is conserved (e.g., no weak neutral currents), the fundamental molecular orthogonal group is $O(3)_{\text{lab}} \otimes O(3)_{\text{body}}$.

How this symmetry breaks down and which levels split depends upon both the perturbative laboratory environment and the internal molecular structure. A *spherical top* Hamiltonian is Eq. (31.1) with $A = B = C$. This has a full $O(3)_{\text{body}}$ (spherical) symmetry since it is just $B\mathbf{J}\cdot\mathbf{J}$. Given that the rotor is in an $O(3)_{\text{lab}}$ laboratory (empty space), the original symmetry $O(3)_{\text{lab}} \otimes O(3)_{\text{body}}$ remains intact, and the $(2J + 1)^2$ degeneracy is to be expected. However, a *symmetric rotor* in a lab vacuum has its internal symmetry broken to $O(2)_{\text{body}}$ if $A = B \neq C$, and the energies given by Eq. (31.4) result in internal quantum singlets for $K = 0$ and $\pm K$ doublets for $K \neq 0$. But each of these levels still has a lab degeneracy of $(2J + 1)$ if $O(3)_{\text{lab}}$ is still in effect. So the $(2J + 1)^2$ level degeneracies are each split into multiplets of degeneracy $(2J+1)$ and $2(2J+1)$ for $K = 0$ and $K \neq 0$ respectively. The resulting levels are often labeled $\Sigma, \Pi, \Delta, \Phi, \Gamma, \ldots$ in analogy with the atomic $s, p, d, f, g, \ldots$ labels of Bohr model orbitals.

Only by perturbing the lab environment can one reduce the $O(3)_{\text{lab}}$ symmetry and split the M degeneracies. For example, a uniform electric field would reduce the $O(3)_{\text{lab}}$ to $O(2)_{\text{lab}}$, giving Stark splittings which consist of external quantum singlets for $M = 0$ and $\pm M$ doublets for $M \neq 0$. A uniform magnetic field would reduce the $O(3)_{\text{lab}}$ to $R(2)_{\text{lab}}$, giving Zeeman splittings into external quantum singlets for each M. The analogy between atomic external field splitting and internal molecular rotational structure is sometimes a useful one, as used in Sects. 31.4 and 31.5.2.

31.1.3 Rigid Asymmetric Rotor Eigensolutions and Dynamics

The general case for the rigid rotor Hamiltonian (31.1) has three unequal principal moments of inertia, so that $A \neq B \neq C$. This is called the rigid *asymmetric top* Hamiltonian, and provides a first approximation for modeling rotation of low symmetry molecules such as H_2O. Also, several properties of its eigensolutions are shared by more complicated systems. The dynamics of an asymmetric top is quite remarkable, as demonstrated by tossing a tennis racquet in the air, flat side up. The corresponding behavior of molecules needs to be studied carefully in order to be well understood.

Given the total angular momentum J, a $(2J + 1)$-dimensional matrix representation of H may be constructed using standard matrix elements of angular momentum operators J_x, J_y and J_z as given in Chap. 2. The H matrix connects states with $(2J + 1)$ different body quantum numbers K $(-J \leq K \leq J)$, but the matrix is independent of the lab quantum numbers M, so that there are $(2J+1)$ identical H matrices, one for each value of the lab quantum number M $(-J \leq M \leq J)$.

A plot of the 21 eigenvalues of (31.1) for $J = 10$ is shown in Fig. 31.1. Here, the constants are set to $A = 0.2$ cm^{-1} and $C = 0.6$ cm^{-1} while B is varied between $B = A$, which corresponds to a *prolate* symmetric top (elongated cylindrical object) and $B = C$, which corresponds to an *oblate* symmetric top (flattened cylindrical object). For all B values between those of A and C, the object is asymmetric.

The left-hand end ($A = B = 0.2$, $C = 0.6$ cm^{-1}) of the plot in Fig. 31.1 corresponds to a prolate symmetric top. The symmetric top level spectrum is given by Eq. (31.4). It consists of a lowest singlet state corresponding to $K = 0$ and an ascending quadratic ladder of doublets corresponding to $K = \pm 1, \pm 2, \ldots, \pm J$. The right-hand end ($A = 0.2$, $B = C = 0.6$ cm^{-1}) of the plot corresponds to an oblate symmetric top with a descending quadratic ladder of levels, and the $K = 0$ level is highest. Also, the internal K-axis of quantization switches from the body z-axis for ($A = B = 0.2$, $C = 0.6$) to the body x-axis for ($A = 0.2$, $B = C = 0.6$). The lab M-degeneracy is invisible here, but exists nevertheless.

For intermediate values of B, one has an asymmetric top level structure and, strictly speaking, no single axis of quantization. As a result, the eigenlevel spectrum is quite different. A detailed display of asymmetric top levels for the case ($A = 0.2$, $B = 0.4$, $C = 0.6$) is given at the bottom of Fig. 31.2. They are shown to correspond to semiclassical orbits in Sect. 31.2. This example is the most asymmetric top, since parameter B has a value midway between those of the symmetric top limits $B = A$ and $B = C$.

The twenty-one $J = 10$ asymmetric top levels are

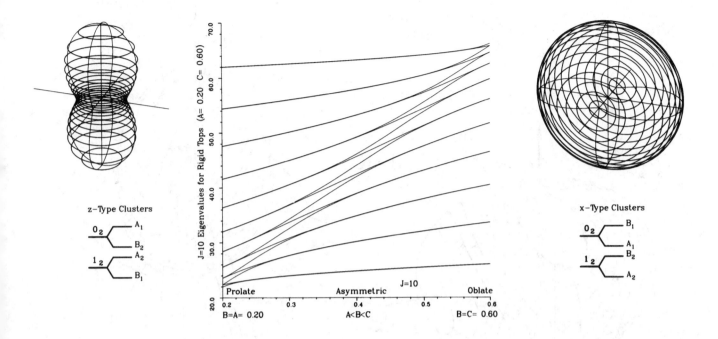

z–Type Clusters

0_2 — A_1, B_2

1_2 — A_2, B_1

x–Type Clusters

0_2 — B_1, A_1

1_2 — B_2, A_2

Figure 31.1. $J = 10$ eigenvalue plot for symmetric and asymmetric rigid rotors. ($A = 0.2$, $C = 0.6$ cm^{-1}, $A < B < C$). Prolate and oblate RE surfaces are shown.

arranged into roughly ten doublets and one singlet. This is like the symmetric top spectrum except that the doublets are all split by varying amounts, and the singlet is well separated from other levels in the middle of the band instead of being crowded onto the top or bottom. The doublet splittings are magnified in circles drawn next to the levels, and these indicate that the splitting decreases quasi-exponentially with each doublet's separation from the central singlet.

The rotor doublet splitting is called *superfine structure* and can be viewed as the result of a dynamic tunneling process in a semiclassical model of rotation [8–10]. Such a model is useful for clarifying the classical-quantum correspondence for polyatomic rovibrational dynamics in general. It can also help to derive simple approximations for eigenvalues and eigenvectors.

31.2 ROTATIONAL ENERGY SURFACES AND SEMICLASSICAL ROTATIONAL DYNAMICS

A semiclassical model of molecular rotation can be based upon what is called a *rotational energy surface* (RE) [7–14]. Examples of RE surfaces for an asymmetric top are shown in Fig. 31.2 and for prolate and oblate

symmetric tops in Fig. 31.1. Each surface is a radial plot of the classical energy derived from the Hamiltonian (31.1) as a function of the polar *directions* of the classical angular momentum **J**-vector in the *body* frame. The magnitude of **J** is fixed for each surface. Note that the **J**-vector in the *lab* frame is a classical constant of the motion if there are no external perturbations. However, **J** may gyrate considerably in the moving *body* frame, but its magnitude stays the same in all frames for free rotation.

The RE surface differs from what is called a *constant energy* (CE) surface which is obtained by simply plotting $E = H = $ const. in **J**-space using Eq. (31.1). A rigid rotor CE surface is an ellipsoid covering a range of $|\mathbf{J}|$ values at a single energy. On the other hand, an RE surface, is a spherical harmonic plot at a single $|\mathbf{J}|$-value for a range of energies. The latter is more appropriate for spectroscopic studies of fine structure, since one value of J corresponds to a multiplet of energy levels or transitions. An RE surface also shows loci of high and low energy rotations. Also, it has roughly the same shape as the body it represents, i.e., it is long in the direction in which the body is long.

For a freely rotating molecule, the laboratory components of the classical total angular momentum **J** are constant. If **J** is used to define the lab z-axis, then the direction of the **J**-vector in the body frame is given by

Figure 31.2. $J = 10$ rotational energy surface and related level spectrum for an asymmetric rigid rotator $(A = 0.2,\ B = 0.4,\ C = 0.6\ \mathrm{cm}^{-1}\ )$.

polar and body azimuthal coordinates β and γ, which are the second and third Euler angles, respectively. (It is conventional to use the *negatives* $-\beta$ and $-\gamma$ as polar coordinates, but this will not be necessary here.) Then the body components of the **J**-vector are written as

$$
\begin{aligned}
J_x &= |J| \sin\beta \cos\gamma\,, \\
J_y &= |J| \sin\beta \sin\gamma\,, \\
J_z &= |J| \cos\beta\,,
\end{aligned}
\tag{31.7}
$$

where the magnitude is $|J| = \sqrt{J(J+1)} \cong J + \tfrac{1}{2}$.

Substituting this into Eq. (31.1) gives an expression

$$
E(\beta,\gamma) = \langle H \rangle = J(J+1)\big[\sin^2\beta\,(A\cos^2\gamma + B\sin^2\gamma) + C\cos^2\beta\big]
\tag{31.8}
$$

for the general rigid rotor RE surface radius in polar coordinates. The prolate symmetric top ($A = B < C$) expression

$$
E(\beta) = \langle H \rangle = J(J+1)\big[B + (C - B)\cos^2\beta\big]
\tag{31.9}
$$

is independent of azimuth γ. The 3-dimensional plots of these expressions are shown in Figs. 31.1 and 31.2.

The RE surfaces have topography lines of constant energy ($E = \text{const.}$). These lines are the intersection of the RE surface (constant $|J|$) with spheres of constant energy. The topography lines are allowed classical paths of the **J**-vector in the body frame, since these paths conserve both energy and momentum.

The trajectories in these figures are special ones. They are the quantizing trajectories for total angular momentum $J = 10$. For the prolate symmetric top, the quantizing trajectories have integral values for the body z-component K. According to the Dirac vector model, angular momentum vectors trace out a cone of altitude K and height $|J| = \sqrt{J(J+1)}$. The quantizing polar angles Θ_K^J are given by

$$
\Theta_K^J = \cos^{-1} \frac{K}{\sqrt{J(J+1)}}\,, \quad (K = J, J-1, \ldots, -J)
\tag{31.10}
$$

and these are the latitude angles of the paths on the RE surfaces in Fig. 31.1 for $K = 10, 9, \ldots, -10$. (For the oblate RE surface, the angles are relative to the x-axis.) If $\beta = \Theta_K^J$ is substituted into the symmetric top RE surface equation (31.9), the result is

$$
E(\Theta_K^J) = J(J+1)B + (C - B)K^2\,,
\tag{31.11}
$$

which is precisely the symmetric top eigenvalue equation (31.4). The quantizing paths are circles which lie at the intersections of the Dirac angular momentum cones and the RE surfaces. The angle Θ_K^J is a measure of the angular momentum uncertainty ΔJ_x or ΔJ_y transverse

to the z-axis of quantization. Clearly, $K = J$ states have minimum uncertainty.

For the asymmetric top, the classical paths which conserve both $|J|$ and E fall into one of two types. First, there are those pairs of equal-energy orbits which go around the hills on the plus or minus end of the body z-axis and correspond to the $\pm K$ pairs of levels in the upper half of the level spectrum drawn in Fig. 31.2. Then there are the pairs of levels belonging to the equal-energy orbits in either of the two valleys surrounding the body x-axis and associated with the pairs of levels in the lower half of the level spectrum.

The upper pairs of paths are seen to be distorted versions of the prolate top orbits seen on the left of Fig. 31.1, while the lower pairs are distorted versions of the oblate top orbits seen on the right of Fig. 31.1. The distortion makes J_z deviate from a constant K-value, and corresponds to K-mixing in the quantum states. This also shows that more than one axis of quantization must be considered; the prolate-like paths are based on the z-axis, while the oblate-like paths belong to the body x-axis.

The two types of orbit are separated by what is called a *separatrix* curve which crosses the saddle points on either side of the body y-axis. In the example shown in Fig. 31.2, the separatrix is associated with a single level which separates the upper and lower energy doublets. The doublets that are closer to the separatrix level are split more than those which are further away. Apart from the splitting, the energy levels can be obtained by generalized Bohr quantization of the classical paths on the RE surface. The quantization condition is

$$
\oint J_z\, d\gamma = K\,,
\tag{31.12}
$$

where

$$
J_z = \sqrt{\frac{J(J+1)(C\cos^2\gamma + B\sin^2\gamma) - E}{(C\cos^2\gamma + B\sin^2\gamma) - A}}
\tag{31.13}
$$

follows from Eqs. (31.7) and (31.8). The resulting E_K-values are obtained by iteration.

The doublet or superfine splitting is a quantum effect which may be associated with tunneling between orbits that would have had equal energies E_K in the purely classical or semiclassical model. Most of the tunneling can be obtained from an integral over the saddle point between each pair of equal-energy quantizing paths. To a good approximation, the Kth tunneling amplitude is given by

$$
S_K = \nu_K \exp(-P_K)
\tag{31.14}
$$

where

$$
P_K = i \int_{\gamma_-}^{\gamma_+} d\gamma \sqrt{\frac{J(J+1)(C\cos^2\gamma + B\sin^2\gamma) - E_K}{(C\cos^2\gamma + B\sin^2\gamma) - A}}
\tag{31.15}
$$

Table 31.1.

Eigenvectors	$|z\rangle$	$|\bar{z}\rangle$	Eigenvalues
$|A\rangle$	1	1	$E^A(K) = E_K + 2S_K$
$|B\rangle$	1	–1	$E^B(K) = E_K - 2S_K$

is the saddle path integral between closest approach points $\gamma+$ and $\gamma-$, and ν_K is the classical precession frequency or quantum level spacing around energy level E_K. Each amplitude S_K is doubled, since there are two tunneling paths, and it appears in a tunneling Hamiltonian matrix for the Kth semiclassical doublet of z and $-z = \bar{z}$ paths

$$\langle H \rangle = \begin{pmatrix} E_K & 2S_K \\ 2S_K & E_K \end{pmatrix} \begin{matrix} |z\rangle \\ |\bar{z}\rangle \end{matrix} . \qquad (31.16)$$

The resulting tunneling energy eigensolutions are given in Table 31.1. A- or B-states correspond to symmetric and antisymmetric combinations of waves localized in the two semiclassical paths. The symmetry of rotational states is considered in Sect. 31.3.

The total doublet splitting is $4S_K$, and decreases exponentially with the saddle point integral (31.15). The superfine $A - B$ splittings in Fig. 31.2 are seen to range from several GHz near the separatrix down to only 26 kHz for the highest-K doublets at the band edges.

Meanwhile, the typical interdoublet level spacing or classical precession frequency is about 150 GHz for the $J = 10$ levels shown in Fig. 31.2. This K-level spacing is labeled *rotational fine structure* splitting, and would also be present in the symmetric top case. The superfine splitting of the symmetric top doublets goes exactly to zero, because the rotor has $O(2)_{\text{body}}$ symmetry if $A = B$ or $B = C$. Then all tunneling amplitudes cancel.

Finally, consider the spacing between adjacent J-levels called *rotational structure* in a spectrum. This spacing is

$$E(J, K) - E(J - 1, K) = 2BJ, \qquad (31.17)$$

according to the symmetric top energy formula (31.4). For the example just treated, this is about 10 cm^{-1} or 300 GHz. It corresponds to the actual rotation frequency of the body. It is the only kind of roatational dynamics or spectrum possible for a diatomic rotor. However, a diatomic molecule can have internal electronic or nuclear spin rotation, which gives an additional fine structure [1, 6, 15].

To summarize, polyatomic molecules exhibit all three types of rotational motion (from faster to slower): *rotational*, *precessional*, and precessional *tunneling*. These are related to three kinds of spectral structure (from coarser to finer): *rotational* structure, *fine* structure, and *superfine* structure, respectively. Again, this neglects in-ternal rotational and spin effects which can have abnormally strong rotational resonance coupling due to the superfine structure [9, 16]. Examples of this are discussed in Sect. 31.5.2.

31.3 SYMMETRY OF MOLECULAR ROTORS

Molecular rotational symmetry is most easily introduced using examples of rigid rotors. Molecular rotor structure may have more or less internal molecular symmetry, depending on how the nuclei are positioned relative to one another in the body frame. Molecular rotational symmetry is described by one of the elementary rotational point symmetry groups. These are the n-fold axial cyclic groups C_n and the polygonal dihedral groups D_n $(n = 1, 2, \dots)$, the tetrahedral group T, the cubic-octahedral group O, or the icosahedral group Y. All other point groups such as C_{nv}, T_d, and O_h are combinations of an elementary group with the inversion operation $\mathbf{I}(\mathbf{r} \rightarrow -\mathbf{r})$. Each of these groups consists of operations which leave at least one point (the origin) of a structure fixed while mapping identical atoms or nuclei into each other in such a way that the appearance of the structure is unchanged. The point groups are subgroups of the nuclear permutation groups [17].

In other words, molecular symmetry is based upon the absolute identity of all atoms or, more precisely, nuclei of a given atomic number Z and mass number A. It is the identity of the so-called 'elementary' electronic and nucleonic constituent particles which underlies the symmetry.

The Pauli principle states that *all half-integer spin particles are antisymmetrized with every other one of their kind in the universe*. The Pauli exclusion principle and the related Bose symmetrization principle determine much of molecular symmetry and dynamics, just as the Pauli principle is fundamental to electronic structure.

31.3.1 Asymmetric Rotor Symmetry Analysis

For an asymmetric rigid rotor, any rotation which interchanges x, y, or z-axes of the body cannot possibly be a symmetry, since all three axes are assumed to have differing inertial constants. This restricts consideration to 180° rotations about the body axes which are elements of the groups C_2 and D_2.

The two symmetry types for C_2 are *even* (denoted A or 0_2) and *odd* (denoted B or 1_2) with respect to a 180° rotation. For D_2, which is just $C_2 \otimes C_2$, the four symmetry types are *even-even* (denoted A_1), *even-odd* (denoted A_2), *odd-even* (denoted B_1), and *odd-odd* (denoted B_2) with respect to 180° rotations about the y-

Table 31.2. Character table for symmetry group C_2.

C_2	1	R
A	1	1
B	1	–1

and x-axes, repectively. (The z-symmetry is determined by a product of the other two, since $\mathbf{R}_z = \mathbf{R}_x \, \mathbf{R}_y$.) This is summarized in the character Tables 31.2 and 31.3. The rotational energy surface for a rigid rotor shown in Fig. 31.2 is invariant under 180° rotations about each of the three body axes. Therefore, its Hamiltonian symmetry is D_2 and its quantum eigenlevels must correspond to one of the four types in Table 31.3. The D_2 symmetry labels are called *rotational* (or in general *rovibronic*) *species* of the molecular state. The species label the symmetry of a quantum wave function associated with a pair of C_2 symmetric semiclassical paths.

The classical **J**-paths come in D_2 symmetric pairs, but each individual classical **J**-path on the rigid rotor RE surface has a C_2 symmetry which is a subgroup of D_2. Each path in the valley around the x-axis is invariant under just the 180° rotation around the x-axis. This is $C_2(x)$ symmetry. The other member of its pair which goes around the negative x-axis also has this local $C_2(x)$ symmetry. The combined pair of paths has full D_2 symmetry, but classical mechanics does not permit occupation of two separate paths. *Multiple path occupation* is a quantum effect.

Similarly, each individual **J**-path in the hill around the z-axis is invariant under just the 180° rotation around the z-axis, so it has $C_2(z)$ symmetry, as does the equivalent path around the negative z-axis. Only the separatrix has the full D_2 symmetry, since its pairs are linked up on the y-axis to form the boundary between x and z paths. No **J**-paths encircle the unstable y-axis, since it is a saddle point.

Each classical **J**-path near the x- or z-axes belongs to a particular K-value through semiclassical quantization conditions (31.12). Depending upon whether the K-value is even (0_2) or odd (1_2), the corresponding K-doublet is correlated with a pair of D_2 species as shown in the

correlation tables in Fig. 31.3. These three correlation tables give the axial 180° rotational symmetry of each D_2 species for rotation near each body axis x, y, and z, respectively. Only the stable rotation axes x and z need to be considered here.

For example, consider $K = 10$ paths which lie lowest in the x-axis valleys. Since $K = 10$ is even (0_2), it is correlated with an A_1 and B_1 superfine doublet [see the 0_2 column in the $C_2(x)$ table]. On the high end near the z-axis hill top, $K = 10$ gives rise to an A_1 and B_2 doublet [see the 0_2 column in the $C_2(z)$ table]. All the doublets in Fig. 31.2 may be assigned in this way.

31.4 TETRAHEDRAL-OCTAHEDRAL ROTATIONAL DYNAMICS AND SPECTRA

The highest symmetry rigid rotor is the *spherical top* for which the three inertial constants are equal ($A = B = C$). The spherical top Hamiltonian

$$H = B\mathbf{J}{\cdot}\mathbf{J} \qquad (31.18)$$

has the full $R(3)_{\text{lab}} \otimes R(3)_{\text{body}}$ symmetry. With inversion parity, the symmetry is $O(3)_{\text{lab}} \otimes O(3)_{\text{body}}$. In any case, the J-levels are $(2J + 1)^2$-fold degenerate. The resulting $BJ(J+1)$ energy expression is the first approximation for molecules which have the regular polyhedral symmetry of, for example, a tetrahedron (CF_4), cube (C_6H_6), octahedron (SF_6), dodecahedron or icosahedron ($C_{20}H_{20}$, $B_{12}H_{12}$, or C_{60}). Rigid regular polyhedra have isotropic or equal inertial constants, and rotate as if their mass distributions were perfectly spherical.

However, no molecule can really have spherical $O(3)_{\text{body}}$ symmetry. Even molecules of highest symmetry contain nuclear mass points, and therefore have a finite internal point symmetry. Evidence of octahedral or tetrahedral symmetry shows up in fine structure splittings analogous to those discussed previously for asymmetric tops. However, spherical top fine structure is due to symmetry breaking caused by anisotropic or tensor rotational distortion. To discuss this, one needs to consider what are called *semirigid* rotors.

31.4.1 Semirigid Octahedral Rotors and Centrifugal Tensor Hamiltonians

The lowest order tensor centrifugal distortion perturbation has the same form for both tetrahedral and octahedral molecules. It is simply a sum of fourth powers of angular momentum operators given in the third term below. The first two terms are scalar rotor energy (31.18) and scalar centrifugal energy.

Table 31.3. Character table for symmetry group D_2.

D_2	1	$\mathbf{R}_x$	$\mathbf{R}_y$	$\mathbf{R}_z$
A_1	1	1	1	1
A_2	1	–1	1	–1
B_1	1	1	–1	–1
B_2	1	–1	–1	1

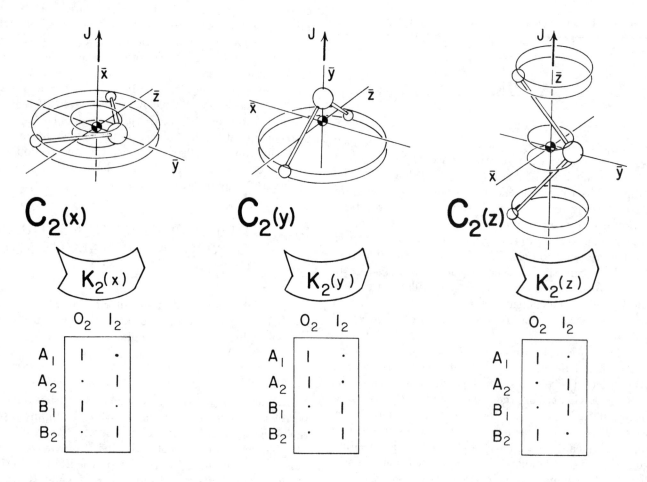

Figure 31.3. Tables of correlations between D_2 symmetry species and the even (0_2) and odd (I_2) symmetry species of subgroups $C_2(x)$, $C_2(y)$, and $C_2(z)$.

$$H = B|J|^2 + D||J|^4 + 10t_{044}\left[J_x^4 + J_y^4 + J_z^4 - (3/5)J^4\right]$$
$$(31.19)$$

The tensor term includes the scalar $(3/5)J^4$ to preserve the center of gravity of the tensor level splitting. This type of semirigid rotor Hamiltonian was first used in the study of methane (CH_4) spectra [18].

The scalar terms do not reduce the symmetry or split the levels. The tensor (t_{044}) term breaks the molecular symmetry from $O(3)_{lab} \otimes O(3)_{body}$ to a lower symmetry subgroup $O(3)_{lab} \otimes T_{d\,body}$ or $O(3)_{lab} \otimes O_{h\,body}$ and splits the $(2J+1)^2$-fold degeneracy into intricate fine structure patterns which are analogous to cubic crystal field splitting of atomic orbitals. The first calculations of the tensor spectrum were done by direct numerical diagonalization [18–21]. As a result, many of the subtle properties of the eigensolutions were missed. A semiclassical symmetry analysis [22] described in the following Sections exposes these properties.

31.4.2 Octahedral and Tetrahedral Rotational Energy Surfaces

By substituting (31.7) and plotting energy as a function of body polar angles β and γ, a rotational energy surface is obtained, two views of which are shown in Fig. 31.4. Here the tensor term is exaggerated in order to exhibit the topography clearly. (In SF_6, the t_{044} coefficient is only about 5.44 Hz, while the rotational constant is $B = 0.09$ cm^{-1}.)

A positive tensor coefficient $(t_{044} > 0)$ gives the octahedral shaped RE surface shown in Fig. 31.4. This is appropriate for octahedral molecules, since they are least susceptible to distortion by rotations around the x-, y-, and z-axes containing the strong radial bonds. Therefore, the rotational energy is highest for the **J**-vector in any of the six axial directions $(\pm1,0,0)$, $(0,\pm1,0)$, $(0,0,\pm1)$. This accounts for the six hills on the RE surface in Fig. 31.4.

However, if the **J**-vector is set in any of the eight interaxial directions $(\pm1,\pm1,\pm1)$, the centrifugal force will more easily bend the weaker angular bonds, raise the molecular inertia, and lower the rotational energy.

This accounts for the eight valleys in Fig. 31.4.

A negative tensor coefficient ($t_{044} < 0$) gives a cubic shaped RE surface. Usually, this is appropriate for cubic and tetrahedral molecules, since they are most susceptible to distortion by rotations around the x-, y-, and z-axes which lie between the strong radial bonds on the cube diagonals. Instead of six hills, one finds six valleys and eight hills on the cubic RE surface.

Note that a semirigid tetrahedral rotor may have the same form of rotational Hamiltonian and RE surface as a cubic rotor. The four tetrahedral atomic sites are in the same directions as four of the eight cubic sites. The other four cubic sites form an inverted tetrahedron which is indistinguishable from the original tetrahedron.

If only tetrahedral symmetry were required, the Hamiltonian could contain a third order tensor of the form $J_x J_y J_z$. However, pure rotational Hamiltonians must also satisfy time-reversal symmetry: the energy for each **J** must be the same as for $-$**J**, and rotational sense should not matter. This excludes all odd powers of **J**. Rotational energy surfaces must have inversion symmetry even if the corresponding molecules do not. Hence the tetrahedral RE surface must have octahedral symmetry.

31.4.3 Octahedral and Tetrahedral Rotational Fine Structure

An example of rotational fine structure for angular momentum quantum number $J = 30$ is shown in Fig. 31.4. The levels are grouped mainly in clusters belonging to the octahedral symmetry species A_1, A_2, E, T_1, or T_2. The characters of these species are given in Table 31.4, (which also applies to the group T_d, where T_1 and T_2 are sometimes labeled F_1 and F_2). The first column gives the dimension or degeneracy of each species; A_1 and A_2 are singlets, E is a doublet, while T_1 and T_2 are triplets. These species form two clusters (A_1, T_1, T_2, A_2) and (T_2, E, T_1) in the low end of the spectrum and six clusters (T_1, T_2), (A_2, T_2, E), (T_1, T_2), (E, T_1, A_1), (T_1, T_2), and (A_2, T_2, E) in the upper part of the spectrum. (See the right-hand side of Fig. 31.4). Note that the total dimension or (near) degeneracy for each of the two lower clusters is eight: $(1+3+3+1)$ and $(3+2+3)$, while the upper clusters each have sixfold (near) degeneracy: $(3+3)$, $(1+3+2)$, etc.

Each of the two lower eightfold clusters can be associated with semiclassical quantizing paths in an RE surface valley as shown in Fig. 31.4. The eightfold dimension or (near) degeneracy occurs because each quantizing path is repeated eight times; once in each of the eight identical valleys. Similarly, the sixfold cluster dimension occurs because there are six identical hills and each quantizing path is repeated six times around the surface.

The majority of the paths lie on the hills because the

Table 31.4. Character table for symmetry group O.

O	0°	120°	180°	90°	180°
A_1	1	1	1	1	1
A_2	1	1	1	-1	-1
E	2	-1	2	0	0
T_1	3	0	-1	1	-1
T_2	3	0	-1	-1	1

hills are bigger than the valleys. The hills subtend a half angle of 35.3° at the separatrix, while the valleys only have 19.5°. To estimate the number of paths or clusters in hills or valleys, the angular momentum cone angles for $J = 30$ may be calculated using Eq. (31.10). The results are displayed in Fig. 31.5. The result is consistent with the spectrum in Fig. 31.4. Only the two highest K-values of $K = 29, 30$ have cones small enough to fit in the valleys, while the six states $K = 25$ to 30 can all fit together on the hills.

The angular momentum cone formula also provides an estimate for each level cluster energy. The estimates become more and more accurate as K increases and approaches J and the uncertainty angle Θ_K^J decreases. Paths for higher K are more nearly circular and therefore more nearly correspond to symmetric top quantum states of pure K. The paths on octahedral surfaces are more nearly circular for a given K than are those on the asymmetric top RE surface, and so the octahedral rotor states can be better approximated with those of the symmetric top.

31.4.4 Octahedral Superfine Structure

The octahedral RE surface has many more local hills and valleys and corresponding types of semiclassical paths than are found on the rigid asymmetric top RE surface. The tunneling between multiple paths gives an octahedral surface superfine structure which can be much more complicated than the simple asymmetric top doublet splittings. Still, the same basic symmetry correlations and tunneling mechanisms that work for the asymmetric top may be used.

First, the octahedral symmetry must be correlated with the local symmetry of the paths on the hills and in the valleys. The hill paths have C_4 symmetry, while valley paths have a local C_3 symmetry. This is seen most clearly for the low-K paths near the separatrix which are least circular. The C_3 and C_4 correlations are given in Fig. 31.6 along with a sketch of the corresponding molecular rotation for each type of path.

To find the octahedral species associated with the $K_3 = 30$ path in a C_3 valley, one notes that 30 is 0 modulo 3. Hence the desired species are found in the 0_3 column of

388 / *Atomic, Molecular, & Optical Physics Handbook*

Figure 31.4. $J = 10$ rotational energy surface and related level spectrum for a semirigid octahedral or tetrahedral rotor.

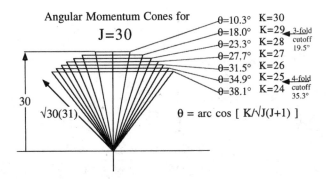

Figure 31.5. $J = 30$ angular momentum cone half angles and octahedral cutoffs.

the C_3 correlation table: (A_1, A_2, T_1, T_2). This is what appears (not necessarily in the same order) in the lower left corner of Fig. 31.4. Similarly, the species (A_2, E, T_2) for a $K_4 = 30$ path on top of a C_4 hill are found in the 2_4 column of the C_4 correlation table since 30 is 2 modulo 4; these appear on the other side of Fig. 31.4. Clusters (T_1, T_2) for $K_4 = 29$ and (A_1, E, T_1) for $K_4 = 28$ are found similarly.

A multiple path tunneling calculation analogous to the one for rigid rotors can be applied to approximate octahedral superfine splittings. Consider the cluster (A_1, E, T_1) for $K_4 = 28$, for example. Six C_4-symmetric paths located on octahedral vertices on opposite sides of the x-, y-, and z-axes may be labeled $\{|x\rangle, |\bar{x}\rangle, |y\rangle, |\bar{y}\rangle, |z\rangle, |\bar{z}\rangle\}$. A tunneling matrix between the six paths is the following:

$$\langle H \rangle_{K_4=28} = \begin{pmatrix} H & 0 & S & S & S & S \\ 0 & H & S & S & S & S \\ S & S & H & 0 & S & S \\ S & S & S & S & H & 0 \\ S & S & S & S & 0 & H \end{pmatrix} \begin{matrix} |x\rangle \\ |\bar{x}\rangle \\ |y\rangle \\ |\bar{y}\rangle \\ |z\rangle \\ |\bar{z}\rangle \end{matrix} \quad (31.20)$$

where the tunneling amplitude between nearest neighbor octahedral vertices is S, but is assumed to be zero between antipodal vertices. The eigenvectors and eigenvalues for this matrix are given in Table 31.5. This predicts that the triplet (T_1) level should fall between the singlet (A_1) and the doublet (E) levels and the singlet–triplet spacing $(4S)$ should be twice the splitting $(-2S)$ between triplet and doublet. This 2:1 ratio is observed in (E, T_1, A_1) and (A_2, T_2, E) clusters which can be resolved and also in numerical calculations [18–21].

The tunneling amplitudes can be calculated by a separatrix path integral analogous to the asymmetric top formula (31.14) [10,11]. As shown in Fig. 31.4, the tunneling rates or superfine splittings near the separatrix are ~ 1 MHz, which is only a little less than the classical

precession frequency. But as K approaches J on the hill tops, the tunneling rate slows down to a few Hz.

31.5 HIGH RESOLUTION ROVIBRATIONAL STRUCTURE

A display of spectral hierarchy for higher and higher resolution is shown in Fig. 31.7 for the 630 cm^{-1} or 16 μm bands of CF$_4$. This will serve to summarize the possible rovibrational spectral structures and place them in a larger context. The ν_4 resonance in part (a) corresponds to a dipole-active $n_4 = 0 \rightarrow 1$ vibrational transition, and is just one of many vibrational structures to study. The $P(54)$ sideband resonance in part (b) corresponds to a $(J = 54) \rightarrow (J - 1)$ rotational transition, and is just one of a hundred or so rotational structures to study within the ν_4 bands.

Each band is something like a Russian doll; it contains structure within structure within structure down to a resolution of a few tens of Hz. Examples of the rotational fine and superfine structures discussed in Sect. 31.4 are shown in Fig. 31.7(c) and (d), but even more resolution is needed to see the hyperfine structure in (e). Such extremely high resolution has been reached with a CO$_2$ saturation absorption spectrometer [23,24]. They have studied the 10 μm bands of SF$_6$ and SiF$_4$, the latter being very simmilar to CF$_4$ [25].

31.5.1 Tetrahedral Nuclear Hyperfine Structure

High resolution spectral studies of SiF$_4$ show unanticipated effects involving the four fluorine nuclear spins and magnetic moments and their associated hyperfine states. First, the Pauli principle restricts the nuclear spin multiplicity associated with each of the rotational symmetry species in much the same way that atomic $L - S$ coupled states ^{2S+1}L have certain spin multiplicities $(2S+1)$ allowed for a given orbital L species involving two or more equivalent electrons. Second, since superfine split-

Table 31.5. Eigenvectors and eigenvalues of the tunneling matrix for the (A_1, E, T_1) cluster with $K = 28$

Eigenvector	$|x\rangle$	$|\bar{x}\rangle$	$|y\rangle$	$|\bar{y}\rangle$	$|z\rangle$	$|\bar{z}\rangle$	Eigenvalue
$\sqrt{6}|A_1\rangle =$	1	1	1	1	1	1	$E^{A_1} = H + 4S$
$\sqrt{12}|E,1\rangle =$	2	2	−1	−1	−1	−1	$E^E = H - 2S$
$2|E,2\rangle =$	0	0	1	1	−1	−1	
$\sqrt{2}|T_1,1\rangle =$	1	−1	0	0	0	0	$E^{T_1} = H$
$\sqrt{2}|T_1,2\rangle =$	0	0	1	−1	0	0	
$\sqrt{2}|T_1,3\rangle =$	0	0	0	0	1	−1	

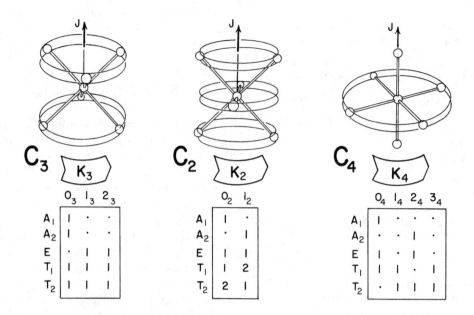

	0_3	1_3	2_3
A_1	1	·	·
A_2	1	·	·
E	·	1	1
T_1	1	1	1
T_2	1	1	1

	0_2	1_2
A_1	1	·
A_2	·	1
E	1	1
T_1	1	2
T_2	2	1

	0_4	1_4	2_4	3_4
A_1	1	·	·	·
A_2	·	·	1	·
E	1	·	1	·
T_1	1	1	·	1
T_2	·	1	1	1

Figure 31.6. Tables of correlations between O symmetry species and the cyclic axial symmetry species (K_p means K mod p) of subgroups C_3, C_2 and C_4.

tings can easily be tiny, different spin species can end up so close that hyperfine interactions, however small, can cause strongly resonant mixing of normally inviolate species. Finally, a very pure and simple form of spontaneous symmetry breaking is observed in which otherwise equivalent nuclei are separated into different subsets by the rotational dynamics.

The nuclear spin to rotational species correlation is easily found by correlating the full permutation symmetry (S_n for XY_n molecules) with the full molecular rotation and parity symmetry [$O(3)_{\text{lab}} \otimes T_{d\,\text{body}}$ for CF_4 or $O(3)_{\text{lab}} \otimes O_{h\,\text{body}}$ for SF_6]. For spin-$\frac{1}{2}$ nuclei, the Pauli principle allows a spin of $I = 2$ and a spin multiplicity of $2I + 1 = 5$ for (J^+, A_2) or (J^-, A_1) species, but excludes (J^-, A_2) or (J^+, A_1) species altogether. The Pauli allowed spin for (J^+, T_1) or (J^-, T_2) species is $I = 1$ with multiplicity 3, but there are no allowed (J^+, T_2) or (J^-, T_1) species. Finally, both (J^+, E) and (J^-, E) belong to singlet spin $I = 0$ and are singlet partners to an *inversion doublet*. (None of the other species can have both + and − parity.)

The E inversion doublet is analogous to the doublet in NH_3 which is responsible for the ammonia maser. However, NH_3-type inversion is not feasible in CF_4 or SiF_4, and so the splitting of the E doublet in these molecules is due to hyperfine resonance [9, 16, 29].

The Pauli analysis gives the number of hyperfine lines that each species would exhibit if it were isolated and resolved, as shown in the center of Fig. 31.7(e). The rotational singlets A_1 and A_2 have 5 lines each, the rotational triplets T_1 and T_2 are spin triplets, and the rotational doublet E is a spin singlet but an inversion

doublet. If the hyperfine structure of a given species A_1, A_2, T_1, T_2, or E is not resolved, then their line heights are proportional to their total spin weights of 5, 5, 3, 3, and 2, respectively.

If the unresolved species are clustered, then the total spin weights of each add to give a characteristic cluster height. The line heights of the C_4 clusters (T_1, T_2), (A_2, T_2, E), (T_1, T_2), (E, T_1, A_1) are 6, 10, 6, 10, respectively. The line heights in the C_3 clusters (A_1, T_1, T_2, A_2), (T_1, E, T_2), (T_1, E, T_2) are 16, 8, 8, respectively. This is roughly what is seen in the $P(54)$ spectrum in Fig. 31.7(c).

31.5.2 Superfine Structure and Spontaneous Symmetry Breaking

The superfine cluster splittings ($2S$, $4S$, etc.) are proportional to the **J**-precessional tunneling or 'tumbling' rates between equivalent C_3 or C_4 symmetry axes, and they decrease with increasing K_3 or K_4. At some point, the superfine splittings decrease to less than the hyperfine splittings which are actually increasing with K. The resulting collision of superfine and hyperfine structure has been called *superhyperfine* structure or *Case 2* clusters. The following is a rough sketch of the phenomenology of this very complex effect, using the results of Pfister [25].

As long as the tunneling rates are > 1 MHz, the nuclear spins tend to average over spherical top motion. The spins couple into states of good total nuclear spin I, which in turn couple weakly with the overall angular momentum and with well defined rovibrational species A_1,

(a) CF₄ Vibrational Structure

$\nu_2 = 435.0\ cm^{-1}$

$\nu_4 = 631.2\ cm^{-1}$

$\nu_1 = 908.5\ cm^{-1}$

$\nu_3 = 1283.0\ cm^{-1}$

$\nu_3(^{13}C)$ $2\nu_4$

(b) ν_4 Rotational Structure

P(50) P(40) P(30) P(20) P(10) Q R(10) R(20) R(30) R(40) R(50) R(60)

600 cm⁻¹ 610 620 630 640 650 660

(c) P(54) Fine (Centrifugal) Structure

FASTER 4-FOLD ROTATION

2-FOLD TUMBLING

FASTER 3-FOLD ROTATION

54 53 52 50 51 48 49 47 46 45 44 51 52 53 54

P(54)

3.0 Ghz

603.3 / 603.4 / 603.5 / 603.6 cm⁻¹

(d) Superfine ("Tumbling") Structure

2.5Mhz 120Mhz 9.9Mhz 0.97Mhz 48 Khz

A₁ F E F₂ F₁ F₁ E F₂ F₁ E F₂ A₂F₂F₁A₁

20 Khz F₂ F₁

2.7 Khz E F₂ A₂

3.5 millihz 9.7 millihz 9.4 hz 110 hz

E F₂ A₂ F₂ F₁ A₁ F₁ E F₂ F₁

(e) Hyperfine (Nuclear Spin) Structure

~50 Khz

~1-100Khz

CASE (2) ‖ CASE(1)

Figure 31.7. Rovibrational structure in the 630 cm⁻¹ or 16 μm bands of CF₄ [16]. (a) Vibrational resonances and band profiles. (Raman spectra from Ref. [26]). (b) Rotational P, Q, and R band structure corresponding to $J \to J-1$, $J \to J$, and $J \to J+1$ transitions. (FTIR spectra from Ref. [27]). (c) $P(54)$ rotational fine structure due to rotation-vibration coupling and angular momentum precessional motion. (Laser diode spectra from Ref. [28]). (d) Superfine structure due to precessional tunneling [25]. (e) Hyperfine structure due to nuclear spin precession [25].

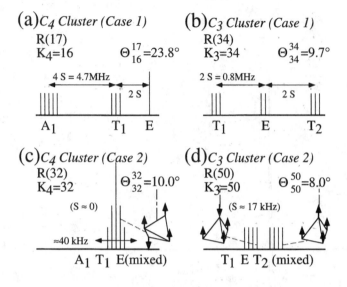

(a) *C_4 Cluster (Case 1)*
R(17)
$K_4=16$ $\Theta_{16}^{17}=23.8°$

$4\,S=4.7\,\text{MHz}$ $2\,S$

A_1 T_1 E

(b) *C_3 Cluster (Case 1)*
R(34)
$K_3=34$ $\Theta_{34}^{34}=9.7°$

$2\,S=0.8\,\text{MHz}$ $2\,S$

T_1 E T_2

(c) *C_4 Cluster (Case 2)*
R(32)
$K_4=32$ $\Theta_{32}^{32}=10.0°$

$(S\approx 0)$

$\approx 40\,\text{kHz}$

A_1 T_1 E(mixed)

(d) *C_3 Cluster (Case 2)*
R(50)
$K_3=50$ $\Theta_{50}^{50}=8.0°$

$(S\approx 17\,\text{kHz})$

T_1 E T_2 (mixed)

Figure 31.8. Stick sketches for examples of superfine and hyperfine spectral structure found by Pfister [25]. (a),(b) Case 1 clusters (high tunneling amplitude S). (c),(d) Case 2 clusters (low tunneling amplitude S).

A_2, T_1, T_2, or E as described above. The resulting is called *Case 1*, and is analogous to *LS* coupling in atoms.

Stick figures for two examples of spectra observed by Pfister [25] are shown in Fig. 31.8(a) and (b). The first Case 1 cluster in (a) is a C_4-type (0_4) cluster (A_1,T_1,E) which was solved in Table 31.5. The other Case 1 cluster in (b) is a C_3-type $(\pm 1_3)$ cluster (T_1,E,T_2) (recall C_3 correlations in Fig. 31.3). They are similar to the corresponding sketches shown in Fig. 31.7(e). One notable difference is that the inversion doublet shows little or no splitting in the (A_1,T_1,E) cluster, but does split in the (T_1,E,T_2) cluster.

When the tunneling rates fall below 10 or 20 kHz, the angular momentum can remain near a particular C_3 or C_4 symmetry axis for a time longer than the nuclear spin precession periods. Spin precession rates and hyperfine splittings are ~ 50 kHz, and increase with K. Hence there is plenty of time for each of the nuclear spins to align or antialign with the C_3 or C_4 symmetry axes of rotation. This is Case 2 coupling, and the resulting spectrum resembles that of an NMR scan of the nuclei, but the magnetic field here is provided by the molecule itself through its own body frame rotation.

If SiF$_4$ rotates uniformly about one C_4 symmetry axis, then all four F nuclei occupy equivalent positions at the same average distance from the rotation axis and experience the same local magnetic fields. The molecule can be thought of as a pair of diatomic F-F rotors with each one symmetrized or antisymmetrized so as to make the whole state symmetric. Table 31.6 shows the spin-$\frac{1}{2}$ base states arranged horizontally according

Table 31.6. Spin-$\frac{1}{2}$ basis states for SiF$_4$ rotating about a C$_4$ symmetry axis.

$I_z=2$	$I_z=1$	$I_z=0$	$I_z=-1$	$I_z=-2$
		$\left\lvert\begin{smallmatrix}\uparrow&\uparrow\\\downarrow&\downarrow\end{smallmatrix}\right\rangle$		
		$\lvert\uparrow\downarrow\ \uparrow\downarrow\rangle$		
	$\lvert\uparrow\downarrow\ \uparrow\uparrow\rangle$	$\lvert\downarrow\downarrow\ \uparrow\uparrow\rangle$	$\lvert\downarrow\downarrow\ \uparrow\downarrow\rangle$	
$\lvert\uparrow\uparrow\ \uparrow\uparrow\rangle$	$\lvert\uparrow\uparrow\ \uparrow\downarrow\rangle$	$\lvert\uparrow\uparrow\ \downarrow\downarrow\rangle$	$\lvert\uparrow\downarrow\ \downarrow\downarrow\rangle$	$\lvert\downarrow\downarrow\ \downarrow\downarrow\rangle$

to the total projection I_z of nuclear spins on the C_4 axis. Horizontal arrays $(\uparrow\downarrow)$ of spins denote symmetric states, while vertical arrays $\left(\begin{smallmatrix}\uparrow\\\downarrow\end{smallmatrix}\right)$ denote antisymmetric spin states. The hyperfine energy is approximately proportional to the projection I_z. The resulting spectrum is a (1,2,4,2,1)-degenerate pyramid of equally spaced lines as shown in Fig. 31.8(c). Four spin-$\frac{1}{2}$ states without symmetry restrictions would give the standard binomial (1,4,6,4,1)-degeneracy seen in NMR spectra.

If the molecule settles upon C_3 symmetry axes of rotation, the situation is markedly different. The four nuclei no longer occupy equivalent positions. One nucleus sits on the rotation axis, while the other three occupy equivalent off-axis positions. The off-axis nuclei experience a different local magnetic field than the lone on-axis nucleus [see Fig. 31.8(d)]. From the spectrum, it appears that the spin-up to spin-down energy difference is much greater for the on-axis nucleus than for the three equatorial nuclei, whose spin states form an energy quartet $\{\lvert\uparrow\uparrow\uparrow\rangle, \lvert\uparrow\uparrow\downarrow\rangle, \lvert\uparrow\downarrow\downarrow\rangle, \lvert\downarrow\downarrow\downarrow\rangle\}$. The on-axis nucleus has an energy doublet with a large splitting, so that the four nuclei together give a doublet of quartets as shown in the figure.

If the off-axis nuclei had experienced the greatest splitting, then the spectrum would be a quartet of doublets instead of a doublet of quartets. This does occur in the SF$_6$ superhyperfine structure, which shows a quintet of triplets for a Case 2 C_4 symmetry cluster. For either one of these molecules, it is remarkable how different the rovibrational 'chemical shifts' can become for equivalent symmetry sites. The result is a microscopic example of spontaneous symmetry breaking.

31.5.3 Extreme Molecular Symmetry Effects

The most common high-symmetry molecules belong to either the tetrahedral T_d or cubic-octahedral O groups. Until the recent discovery of fullerenes and of the structure of virus coats, the occurrence of molecular point groups of icosahedral groups was thought to be rare or nonexistent in nature [30, 31].

For an extreme example of symmetry breaking effects, consider the buckminsterfullerene or Buckyball molecule C_{60} which has the highest possible molecular point symmetry Y_h. A semiclassical approach to rotational symmetry and dynamics is useful here, since the rotational quantum constant is so small for fullerenes (for C_{60}, $2B = 0.0056$ cm^{-1} or 168 MHz [32, 33]).

Since there are two isotopes ^{12}C (nuclear spin 0) and ^{13}C (nuclear spin $\frac{1}{2}$), it is possible to have a Bose-symmetric molecule $^{12}C_{60}$, a Fermi-symmetric molecule $^{13}C_{60}$, or many broken-symmetry combinations $^{12}C_x\,^{13}C_{60-x}$. The most likely combination is $^{12}C_{59}\,^{13}C$, which has no rotational symmetry at all, and only one reflection plane. This may be the most extreme example of molecular isotopic symmetry breaking; it goes from the highest possible symmetry Y_h to the lowest C_h.

The Fermi-symmetric $^{13}C_{60}$ molecule has ten times as many rotating spin-$\frac{1}{2}$ nuclei as SF$_6$, and 2^{10} times as many hyperfine states, giving about 1.15×10^{18} spin states distributed among 10 symmetry species [34]. In contrast, the Bose-symmetric $^{12}C_{60}$ has only one spin symmetry species allowed by the Bose exclusion principle. It provides an even more extreme example of Bose exclusion than the Os^{16}O$_4$ molecule. In all, 119 of the 120 Y_h rovibrational symmetry states are Bose excluded, giving $^{12}C_{60}$ an extraordinarily sparse rotational structure. However, it only takes the addition of one extra neutron to make $^{12}C_{59}\,^{13}C$. Then all the excluded rovibrational states must return!

REFERENCES

1. G.Herzberg, *Molecular Spectra and Structure: Spectra of Diatomic Molecules* (Van Nostrand-Reinhold, New York, 1950), Vol. I.
2. G.Herzberg, *Molecular Spectra and Structure: Infrared and Raman Spectra of Polyatomic Molecules* (Van Nostrand-Reinhold, New York, 1945), Vol. II.
3. G.Herzberg, *Molecular Spectra and Structure: Electronic Structure of Polyatomic Molecules* (Van Nostrand-Reinhold, New York, 1966), Vol. III.
4. E. B. Wilson, J. C. Decius, and P. C. Cross, *Molecular Vibrations* (McGraw Hill, New York, 1955).
5. D. Papousek and M. R. Aliev, *Studies in Physical and Theoretical Chemistry: Molecular Vibrational-Rotational Spectra* (Elsevier, Amsterdam, 1982), Vol. 17.
6. R. N. Zare, *Angular Momentum: Understanding Spatial Aspects in Chemistry and Physics* (Wiley Interscience, New York, 1988).
7. W. G. Harter, *Principles of Symmetry, Dynamics, and Spectroscopy* (Wiley Interscience, New York, 1993).
8. W. G. Harter and C. W. Patterson, J. Math. Phys. **20**, 1453 (1979).
9. W. G. Harter, Phys. Rev. A**24**, 192 (1981).
10. W. G. Harter and C. W. Patterson, J. Chem. Phys. **80**, 4241 (1984).
11. W. G. Harter, Comp. Phys. Rep. **8**, 319 (1988).
12. D. A. Sadovskiĭ and B. I. Zhilinskiĭ, Mol. Phys. **65**, 109 (1988).
13. D. A. Sadovskiĭ and B. I. Zhilinskiĭ, Phys. Rev. A**47**, 2653 (1993).
14. L. M. Pavlichenkov, Phys. Rep. **226**, 173 (1993).
15. W. G. Harter, C. W. Patterson, and F. J. da Paixao, Rev. Mod. Phys. **50**, 37 (1978).
16. W. G. Harter and C. W. Patterson, Phys. Rev. A **19**, 2277 (1979).
17. P. R. Bunker, *Molecular Symmetry and Spectroscopy* (Academic Press, New York, 1979).
18. K. T. Hecht, J. Mol. Spectrosc. **5**, 535 (1960).
19. K. R. Lea, M. J. M. Leask, and W. P. Wolf, J. Phys. Chem. Solids **23**, 1381 (1962).
20. A. J. Dorney and J. K. G. Watson, J. Mol. Spectrosc. **42**, 1 (1972).
21. K. Fox, H. W. Galbraith, B. J. Krohn, and J. D. Louck, Phys. Rev. A **15**, 1363 (1977).
22. W. G. Harter and C. W. Patterson, J. Chem. Phys. **66**, 4872 (1977).
23. J. Bordé and Ch. J. Bordé, J. Chem. Phys. **71**, 417 (1982).
24. Ch. Bordé, J. Bordé, Ch. Breant, Ch. Chardonnet, A. Vanderberghe, and Ch. Salomon in *Laser Spectroscopy VII* **108**, 95 (Springer, Berlin, 1985).
25. O. Pfister, thesis, Université Paris-Nord, 1993.
26. B. Montostari and A. Weber, J. Chem. Phys. **33**, 1867 (1960).
27. A. Stein, P. Rabinowitz, and A. Kaldor, Opt. Lett. **3**, 97 (1978).
28. R. S. McDowell, C. W. Patterson, C. R. Jones, M. I. Buchwald, and J. M. Telle, Proc. Soc. Photo-Opt. Instrum. Eng. **190**, 263 (1979).
29. R. J. Butcher, Ch. Chardonnet, and Ch. Bordé, Phys. Rev. Lett. **70**, 2698 (1993).
30. H. W. Kroto, J. R. Heath, S. C. O'Brian, R. F. Curl, and R. E. Smalley, Nature **318**, 162 (1985).
31. W. Kratschmer, W. D. Lamb, K. Fostiropoulos, and D. R. Huffman, Nature **347**, 354 (1990).
32. W. G. Harter and D. E. Weeks, Chem Phys. Lett. **132**, 187 (1986).
33. D. E. Weeks and W. G. Harter, Chem. Phys. Lett. **144**, 366 (1988); **176**, 209 (1991).
34. W. G. Harter and T. C. Reimer, Chem. Phys. Lett. **194**, 230 (1992).

32

Radiative Transition Probabilities

David L. Huestis

Molecular Physics Laboratory, SRI International, Menlo Park, California 94025

32.1 OVERVIEW

32.1.1 Intensity Versus Line-Position Spectroscopy

The fact that atoms and molecules absorb and emit radiation with propensities that vary with wavelength is the origin of the field called spectroscopy. The relatively sharp intensity maxima are interpreted as corresponding to transitions between discrete states or energy levels. The frequencies or energies of these transitions are used as the primary source of information about the internal structure of the atom or molecule. Line positions can be measured with very high precision (1 ppm or better). Excellent calibration standards have been developed. The quality of these experimental data has attracted extensive analytical and theoretical effort. Sophisticated parametrized models have been developed in which the smallest shifts from the expected line positions can be used to identify perturbations or other subtle effects.

For molecules, knowledge of the strengths of these transitions is far less well developed. One reason is that quantitative experimental data on rotationally-resolved absorption cross sections and emission intensities are much rarer and the experiments themselves are much more difficult to calibrate. Few measurements claim a precision better than 1% and agreement within 10% of measurements in different laboratories is typically viewed as good. This situation is undesirable because most applications of molecular spectroscopy are in fact measurements of intensity. In many cases the strengths of absorptions or emissions are used to infer gas composition, temperature, time evolution, or other environmental conditions. In other examples the actual absorption and emission is the primary interest. Among the most important of these are atmospheric absorption of solar radiation and the greenhouse effect.

32.1.2 Scope of Chapter and Approach

This Chapter summarizes the theory of radiative transition probabilities or intensities for rotationally-resolved (high-resolution) molecular spectra. A combined treatment of diatomic, linear, symmetric-top, and asymmetric-top molecules is based on angular momentum relations. Generality and symmetry relations are emphasized. The *energy-intensity* model is founded in a rotating-frame basis-set expansion of the wave functions, Hamiltonians, and transition operators. The intensities of the various rotational branches are calculated from a small number of transition-moment matrix elements, whose relative values can be assumed from the supposed nature of the transition, or inferred by fitting experimental intensities.

32.2 MOLECULAR WAVE FUNCTIONS IN THE ROTATING FRAME

32.2.1 Symmetries of the Exact Wave Function

The exact total wave function for any isolated molecule with well-defined energy and total angular momentum can be expressed in a basis-set expansion over configurations with well-defined internal quantum numbers,

$$\Phi_{\text{exact}} = \Phi_{\text{trans}} \sum_{\alpha\beta\gamma\delta\epsilon} C_{\alpha\beta\gamma\delta\epsilon} \Phi^{\alpha}_{\text{rot}} \Phi^{\beta}_{\text{vib}} \Phi^{\gamma}_{\text{elec}} \Phi^{\delta}_{\text{espin}} \Phi^{\epsilon}_{\text{nspin}} .$$

(32.1)

In principle, the coefficients $C_{\alpha\beta\gamma\delta\epsilon}$ can be found only by diagonalizing the exact Hamiltonian. In practice one attempts to find a sufficiently good approximation, containing only a few terms, with coefficients chosen by diagonalizing an approximate or model Hamiltonian. This is the basis of the energy-intensity model developed in Sect. 32.3. As discussed by Longuet-Higgins [1] and Bunker [2], there are only six true symmetries of the exact Hamiltonian of an isolated molecule:

1. translation of the center of mass;

2. permutation of electrons;

3. permutation of identical nuclei;

4. time reversal or momentum reversal;

5. inversion of all particles through the center of mass;

6. rotation about space-fixed axes.

Of these, only the symmetries numbered 5 and 6 give quantum numbers (parity and the total angular momentum F) that are both rigorous and useful spectroscopic labels of the states of the molecule. The other symmetries are convenient for simplifying the description of the molecular wave function, for the evaluation of relations between matrix elements, and for classification of molecular states according to approximate symmetries.

The first symmetry, translation of the center of mass, allows the choice of a coordinate system referenced to the center of mass, and suppression of the portion of the wave function describing motion through space (as long as the molecule does not dissociate).

Symmetry number 2, exchange of electrons, does not directly provide any labels or quantum numbers, since the Fermi-Dirac statistics of electrons require that all wave functions must be antisymmetric. However, it provides considerable information about the probable electronic states since it controls whether molecular orbitals can be doubly or only singly occupied. For most (low-Z) molecules, each state will have a nearly well-defined value of electron spin: singlet or triplet for example. Admixture of other spin values usually can be treated as a perturbation. These points will be elaborated in Sects. 32.4.4 and 32.7.4.

Permutation of identical nuclei, symmetry number 3, also gives an identical quantum number to all the states of the molecule (± 1 depending on the character of the permutation and on whether nuclei with integral or half-integral spin are being permuted). It supplies little direct information about the energy separations between the states of the molecule. On the other hand, many molecules have identical nuclei in geometrically or dynamically equivalent positions. The existence of spatial symmetry, for nonplanar molecules, is really the same thing as permutational symmetry. Consequently, nuclear permutation, combined with inversion (symmetry number 5), is the basis for naming the states according to the approximate spatial symmetry group of the molecular frame and vibrational motion. These concepts will be explored in Sect. 32.4.3.

Symmetry number 4, time reversal, is both subtle and simple. In the absence of external magnetic fields the Hamiltonian for a molecule will contain only even combinations of angular momentum operators, e.g. $F_\alpha F_\beta$, $F_\alpha L_\beta$, or $F_\alpha S_\beta$. Thus changing the signs of all the angular momenta should result in an equivalent wave function. This will require that matrix elements retain the same absolute value when the angular quantum numbers are reversed, leading in general to complex conjugation [3].

Spatial inversion, symmetry number 5, is always an allowed operation for any molecule, even if it appears to lack internal inversion symmetry. This operation can be considered as a symmetry of the spherically-symmetric laboratory in which the molecule resides. If the molecule is linear, triatomic, rigid with a plane of symmetry, or is nonrigid with accessible vibrational or tunneling modes that correspond to plane reflections, inversion symmetry divides the states of the molecule into two classes, called parities. Perturbations can occur only between states of the same parity. For optical transitions, the change in parity of the states must match the parity of the operator. Otherwise, reflection of the molecule in a plane will interchange inconvertible optical isomers. Such optical isomers are energetically degenerate, so in all cases, inversion through the center of mass remains a valid symmetry of the rotating molecule. However, the separation of the states into two kinds does not provide any selection rules. The two parity classes are perfectly degenerate, thus there is always an allowed level with the correct parity either for perturbations or for optical transitions.

32.2.2 Rotation Matrices

The final symmetry, rotation about the center of mass, restricts the discussion to states with well-defined laboratory angular momentum, and to re-expression of the exact wave function by changing variables from *laboratory* coordinates to *body-fixed* or internal coordinates, and introducing the Euler angles relating these two coordinate systems,

$$\Phi_{\text{exact}}^{F,M_F}(\text{lab}) = \sum_{K_F} \Phi_{\text{rot}}^{F,M_F K_F}(\text{Euler angles})$$
$$\times \Phi_{\text{vesn}}^{(F,K_F)}(\text{internal, spins}). \quad (32.2)$$

Here F is the total angular momentum of the molecule, including vibrational, mechanical-rotation, electron-orbital, electron-spin, and nuclear-spin contributions. M_F and K_F are the projections of F in the laboratory and body-fixed frames, respectively. In the majority of cases, the magnitude of nuclear hyperfine interactions is sufficiently small that its influence can be ignored when analyzing wave functions and computing energies. Thus the quantum numbers J, M_J, and K_J, or just J, M, and K can be used.

Explanation is postponed of how the body-fixed frame is to be selected, but for any choice, the wave function for rotation of the entire molecule can be expressed using a rotation matrix [4, 5]

$$\Phi_{\text{rot}}^{F,M_F K_F}(\text{Euler angles}) = \left[\frac{(2F+1)}{8\pi^2}\right]^{\frac{1}{2}} D_{M_F K_F}^{*F}(\phi, \theta, \chi).$$
$$(32.3)$$

For diatomics, Zare [5] suggests multiplying by $(2\pi)^{1/2}$ and setting $\chi = 0$. The internal wave function for the vibrational, electronic, electron-spin, and nuclear-spin degrees of freedom [thus the label (vesn)] can be thought of as the partial summation

$$\Phi_{\text{vesn}}^{(F,K_F)}(\text{internal, spins}) = \sum_{\beta\gamma\delta\epsilon} C_{(FK_F)\beta\gamma\delta\epsilon}$$
$$\times \Phi_{\text{vib}}^{\beta} \Phi_{\text{elec}}^{\gamma} \Phi_{\text{espin}}^{\delta} \Phi_{\text{nspin}}^{\epsilon}, \quad (32.4)$$

expressed in the internal or rotated coordinate system. Note that the FK_F designation is only a parametric label. The rotational wave function has been absorbed into the rotation matrix.

32.2.3 Transformation of Ordinary Objects into the Rotating Frame

The assumption of rotational symmetry allows re-expression of matrix elements between total wave functions as a sum of matrix elements between internal wave functions. For example, the tensor operator $T^{(L)}$ belonging to the L representation of the rotation group, can be written in the rotating frame as [6, 7, 5, 8]

$$T_p^{(L)}(\text{lab}) = \sum_q D_{pq}^{*L}(\phi\theta\chi)T_q^{(L)}(\text{body})\,, \qquad (32.5)$$

and evaluate matrix elements that might represent radiative transitions:

$$\langle \psi^{F',M_F''+p}|T_p^{(L)}(\text{lab})|\Phi^{F'',M_F''}\rangle$$

$$= \left[\frac{(2F''+1)}{(2F'+1)}\right]^{\frac{1}{2}} \langle F''M_F'', Lp|F'M_F''+p\rangle$$

$$\times \sum_{qK_F''} \langle F''K_F'', Lq|F'K_F''+q\rangle$$

$$\times \langle \psi_{\text{vesn}}^{(F',K_F''+q)}|T_q^{(L)}(\text{body})|\Phi_{\text{vesn}}^{(F'',K_F'')}\rangle\,, \qquad (32.6)$$

where $\langle F''M_F'', Lp|F'M_F''+p\rangle$ and $\langle F''K_F'', Lq|F'K_F''+q\rangle$ are Clebsch-Gordan coefficients that vanish if $|F'-F''| > L$, $|M_F''+p| > F'$, or $|K_F''+q| > F'$.

This transformation forms the basis for derivation of rotational branch strengths (Sect. 32.7.2) and for description of electron motions that are weakly coupled to the molecular frame (Sect. 32.7.4).

32.3 THE ENERGY-INTENSITY MODEL

32.3.1 States, Levels, and Components

The previous section introduced the concept of representing the wave function of a molecule as a product of five simpler wave functions:

$$\psi \sim \psi_{\text{elec}} \cdot \psi_{\text{vib}} \cdot \psi_{\text{rot}} \cdot \psi_{\text{espin}} \cdot \psi_{\text{nspin}}\,. \qquad (32.7)$$

This construction yields a similar separation of the Hamiltonian,

$$H \sim H_{\text{elec}} + H_{\text{vib}} + H_{\text{rot-fs}} + H_{\text{hf}}\,, \qquad (32.8)$$

and representations of the energies as sums of contributions,

$$E \sim T_{\text{e}} + G_v + F_c(J)\,, \qquad (32.9)$$

and absorption or emission transition strengths as products,

$$I \sim I_{\text{elec}} \cdot I_{\text{vib}} \cdot I_{\text{rot-fs}} \cdot I_{\text{hf}}\,. \qquad (32.10)$$

Whatever theoretical arguments might favor such a separation, the real impetus is the empirical observation that most molecular absorption and emission spectra exhibit recognizable patterns arising from the dissimilar magnitudes of the energies associated with these five degrees of freedom. Separation of the wave function and the Hamiltonian into these four or five contributions facilitates the assignment of molecular spectra, in addition

to suggesting models with parameters that can be adjusted to quantitatively represent the observed spectra.

Most states of molecules are dominated by a single set of electronic and vibrational quantum numbers. Electronic states are often well separated. With each electronic state is associated a potential energy surface, the energy at the minimum being labeled T_{e}. Motion of nuclei within this potential generates various levels corresponding to different vibrational quantum numbers, following regular patterns or progressions in energy, summarized by a small number of parameters called vibrational frequencies. The quantity G_v represents the energy of the the vibrational level above the potential minimum. For each vibrational level, a progression of rotational levels is expected. For linear molecules in electronic states without electronic angular momentum (i.e. $^1\Sigma$ states) the rotational energies are also reproduced by a few rotational constants.

For more complicated molecules and electronic states, i.e. most cases, there are multiple energetically distinct levels with the same value of J (in addition to the $2J+1$ orientational degeneracy of each level). These multiple levels all share the same nominal quantum numbers (additional analysis may subdivide them into parity or permutational symmetry types). These sublevels are called "components" with energies expressed by the notation $F_c(J)$. The quantity N_c, the number of components expected, reflects the assignment of the nature of the vibronic state. For linear molecules there is a limited number of components corresponding to the various orientations of electron spin and orbital angular momentum. For example, a $^2\Pi$ electronic state will have four components for each value of J (except for $J = \frac{1}{2}$, where there are only two components). For nonlinear molecules the number of components increases with J, proportional to $2J+1$, corresponding to various possible projections of the total angular momentum onto the tumbling molecular frame.

The conclusion of this analysis is that a basis set be chosen, over which a model rotational and fine-structure Hamiltonian can be expressed. The wave functions then become vectors of numbers. *A priori*, only the form of the matrix elements and their dependence on J and body-frame projection (K or Ω) are known. Little is known in advance about how strong the interactions are in any given molecule. Thus one tends to write the Hamiltonian with parameters that are to be determined by fitting the observed energy levels.

Similarly, the choice of the basis sets for the upper and lower states specifies the overall form of the matrix of transition moments between the basis functions. The transition can be chosen to be of a simple standard form, for example, parallel or perpendicular, with only one unknown parameter representing the overall strength of the transition. Alternatively, the transition matrix elements can be considered to be independently adjustable, within the symmetry restrictions that are required (time reversal) or assumed (spatial symmetry).

32.3.2 The Basis Set and Matrix Hamiltonian

For linear molecules it is convenient to choose a basis set labeled by the projections of orbital and spin angular momenta in the body-frame coordinate system, represented symbolically by

$$|\Lambda\Sigma; JM\Omega\rangle = \left[\frac{(2J+1)}{8\pi^2}\right]^{1/2} D_{M\Omega}^{*J}|\Lambda\Sigma\rangle, \quad (32.11)$$

where $\Omega = \Lambda + \Sigma$. This is called the Hund's case (a) basis set, which is an accurate representation in a single term if the body-frame angular momenta are nearly conserved. This is true if the spin-orbit interaction is larger than the separation between rotational levels. Under all circumstances, this basis set facilitates construction of the matrix Hamiltonian and representation of sources of transition probability [9].

One parametrization for the spin-rotation Hamiltonian is provided by Brown *et al.* [10–12]:

$$\begin{aligned}
\mathbf{H}_{\text{spin}-\text{rot}} = {} & T_e + G_v + B_v\mathbf{N}^2 - D_v\mathbf{N}^4 \\
& + \tfrac{1}{2}[A_v + A_{D_v}\mathbf{N}^2, L_zS_z]_+ \\
& + (\gamma_v + \gamma_{D_v}\mathbf{N}^2)\mathbf{N}\cdot\mathbf{S} \\
& + \tfrac{1}{3}[\lambda_v + \lambda_{D_v}\mathbf{N}^2, 3S_z^2 - \mathbf{S}^2]_+ \\
& + \eta_v L_zS_z[S_z^2 - \tfrac{1}{5}(3\mathbf{S}^2 - 1)] \\
& - \tfrac{1}{4}[o_v + o_{D_v}\mathbf{N}^2, \Lambda_+^2 S_-^2 + \Lambda_-^2 S_+^2]_+ \\
& + \tfrac{1}{4}[p_v + p_{D_v}\mathbf{N}^2, \Lambda_+^2 S_- N_- + \Lambda_-^2 S_+ N_+]_+ \\
& + \tfrac{1}{4}[q_v + q_{D_v}\mathbf{N}^2, \Lambda_+^2 N_-^2 + \Lambda_-^2 N_+^2]_+,
\end{aligned}$$

$$(32.12)$$

where $[x, y]_+$ is the anticommutator $(xy + yx)$, and $\mathbf{N} = \mathbf{J} - \mathbf{S}$. Zare *et al.* [13] provide an alternative parametrization, with different interpretations of the *spectroscopic constants* (B, D, A, γ, λ, etc.) because they multiply different symbolic operators. One significant difference is that Zare *et al.* use the "mechanical angular momentum" $\mathbf{R} = \mathbf{J} - \mathbf{L} - \mathbf{S}$ as the expansion operator, rather than $\mathbf{N}$, which might be called the "spinless angular momentum." These differences mean that care must be taken in attempting to construct simulated spectra from published constants. In spite of much discussion in the literature, there is little theoretical foundation for preferring one parametrization over another, as long as the observed levels are accurately fit. In a number of cases naive assumptions about the origin of certain types of interactions have been overturned. For example, the spin-spin interaction, represented by the constant λ, is often dominated by

level shifts due to off-diagonal spin-orbit perturbations [14].

For polyatomic molecules, a suitable basis set for expansion can be chosen to have a similar form [15, 16, 8]

$$|l\Lambda\Sigma; JMK\rangle = \left[\frac{(2J+1)}{8\pi^2}\right]^{1/2} D_{MK}^{*J}|l\Lambda\Sigma\rangle, \quad (32.13)$$

where Λ and Σ represent the projections of the electron-orbital (L) and spin (S) angular momenta, and l represents the projection of the vibrational angular momentum (p for degenerate vibrational modes). This is the symmetric top basis set. Generalizing the work of Watson [17, 18], the parametrized Hamiltonian might be written in a form such as

$$\begin{aligned}
\mathbf{H}_{\text{rot}} = \sum h_{\zeta\eta\theta l}^{\alpha\beta\gamma\delta\epsilon} \Big\{ & (\mathbf{J}^2)^\alpha (J_z)^{2\beta}(\mathbf{J}_+^{2\gamma} + \mathbf{J}_-^{2\gamma}) \\
& \times (\mathbf{J}\cdot\mathbf{p})^\delta(\mathbf{J}\cdot\mathbf{L})^\epsilon(\mathbf{J}\cdot\mathbf{S})^\zeta \\
& \times (\mathbf{p}\cdot\mathbf{L})^\eta(\mathbf{p}\cdot\mathbf{S})^\theta(\mathbf{L}\cdot\mathbf{S})^l \Big\},
\end{aligned}$$

$$(32.14)$$

where the {} indicates that an appropriately symmetric combination be constructed with anticommutators.

For both linear and nonlinear molecules, it is convenient to use the Wang transformation [19] to combine basis functions with opposite sense of rotation: for diatomics

$$\frac{1}{\sqrt{2}}\left[||\Lambda|, \Sigma; J, M, \Omega\rangle \pm |-|\Lambda|, -\Sigma; J, M, -\Omega\rangle\right]; \quad (32.15)$$

and for polyatomics

$$\frac{1}{\sqrt{2}}\left[|l, \Lambda, \Sigma; J, M, K\rangle \pm |-l, -\Lambda, -\Sigma; J, M, -K\rangle\right]. \quad (32.16)$$

For diatomic molecules, these combinations can be assigned the parity $\pm(-1)^{J+S+s}$, where $s = 1$ for Σ^- states, and 0 otherwise [13, 20]. For symmetric top molecules, each term is to be accompanied by the appropriate hidden nuclear-spin basis function [21, 8].

For asymmetric top molecules, the Wang transformation divides the basis functions into four symmetry classes $E^\pm$ and $O^\pm$ according to the combining sign and whether K is even or odd. The eigenstates are often labeled by two projection quantum numbers called K_{-1} and K_1. Assuming that $A \geq B \geq C$, the asymmetry parameter

$$\kappa = \frac{(2B - A - C)}{(A - C)} \quad (32.17)$$

ranges from -1 for a prolate symmetric top ($B = C$) to 1 for an oblate symmetric top ($B = A$). A, B, and C are the rotational constants, or reciprocals of the moments of inertia, about the three principal top axes. Each asymmetric top level can be correlated with specific symmetric top levels (i.e. K-values) in the two limits. The prolate limiting K-value is called K_{-1} and the oblate limit is called K_1 (i.e. $\kappa = \pm 1$). Note that the symmetric-top principal axes rotate by 90° during this correlation. The eigenstates are given additional symmetry names (ee,eo,oe,oo) according to whether K_{-1} and K_1 are even or odd. Papoušek and Aliev [18] discuss the relations between the $(E^\pm, O^\pm)$ and (ee,eo,oe,oo) labeling schemes.

32.3.3 Fitting Experimental Energies

Having chosen a basis set and model Hamiltonian for both the upper and lower levels, the observed transition energies can be used to infer the numerical values of the constants that best fit the spectrum. The following quotation provides a good description of the process:

> The calculational procedure logically divides into three steps: (1) The matrix elements of the upper and lower state Hamiltonians are calculated for each J value using initial values of the adjustable molecular constants; (2) both Hamiltonians are numerically diagonalized and the resulting sets of eigenvalues are used to construct a set of calculated line positions; and (3) from a least-squares fit of the calculated to the observed line positions, an improved set of molecular constants is generated. This nonlinear least-squares procedure is repeated until a satisfactory set of molecular constants is obtained.

This quotation is taken from the article by Zare *et al.* [13] in which they describe the basis for the LINFIT computer program, one of the first to accomplish direct extraction of constants from diatomic spectral line positions based on numerically diagonalized Hamiltonians.

32.3.4 The Transition Moment Matrix

Diagonalization of the model Hamiltonians for the upper and lower states yields vector wave functions that can be used for calculating matrix elements, especially those needed to evaluate radiative transition probabilities. The wave functions for diatomic molecules have the form

$$\psi'_{J'M'c'} = \sum_{\Lambda'\Sigma'} b^{J'c'}_{\Lambda'\Sigma'} |\Lambda'\Sigma'; J'M'\Omega'\rangle',$$
$$\Phi''_{J''M''c''} = \sum_{\Lambda''\Sigma''} a^{J''c''}_{\Lambda''\Sigma''} |\Lambda''\Sigma''; J''M''\Omega''\rangle''. \tag{32.18}$$

Sect. 32.2.3 expresses matrix elements of laboratory-frame operators in terms of matrix elements in the rotating body-fixed frame. Terms of the form

$$\mu_{K'K''} = \langle \psi^{(J'K')} | T_q^{(L)}(\text{body}) | \Phi^{(J''K'')} \rangle \theta(-q) \tag{32.19}$$

need to be evaluated. These terms are multiplied by zero if $K' \neq K'' + q$. In the diatomic basis set these become

$$\mu_{\Lambda'\Sigma'\Lambda''\Sigma''} = {}'\langle \Lambda'\Sigma' | T_q^{(L)}(\text{body}) | \Lambda''\Sigma'' \rangle'' \theta(-q), \tag{32.20}$$

where $\theta(-q)$ is a phase factor described in Sect. 32.7.2. Only a few of these matrix elements are independent and nonzero. For electric dipole transitions ($L = 1$), time-reversal and inversion-symmetry can be used to establish the relation [9, 22, 23]

$$\mu_{-\Omega'-\Omega''} = \eta(-1)^{\Omega'-\Omega''} \mu_{\Omega'\Omega''}. \tag{32.21}$$

The sign of $\eta = \pm 1$ is determined by the overall character of the electronic transition, and is related to the classification of levels into e- and f-parity types and to the determination of which components are involved in the rotational branches (P, Q, and R). These concepts are elaborated in Sect. 32.4.

32.3.5 Fitting Experimental Intensities

For allowed transitions in linear molecules and symmetric tops, only one independent parameter is normally expected in the transition moment matrix. Thus no additional information is available from fitting the experimental rotational branch strengths (assuming the energy-intensity model is adequate). In diatomic molecules, the intensities of different vibrational bands can be used to infer the internuclear-distance dependence of the electronic transition moment (for example, see Luque and Crosley [24]).

For forbidden transitions and allowed transitions in asymmetric tops, more than one independent parameter is expected. The intensity of a single given rotational line can be expressed in the form

$$I^{\text{line}} = \left| \sum_{K'K''} \mu_{K'K''} Z_{K'K''}(\text{line}) \right|^2, \tag{32.22}$$

where $Z_{K'K''}(\text{line})$ can be calculated in advance from the energies (wave functions) and quantum numbers alone, using the formulas in Sect. 32.7.2. Nonlinear least-squares fitting can be used to derive the best intensity parameters [23, 25–28], analysis of which can help characterize the nature of the transition, and identify the sources of transition probability.

32.4 SELECTION RULES

32.4.1 Symmetry Types

Selection rules are guidelines for identifying which transitions are expected to be strong and which are expected to be weak. These rules are based on classifying rovibronic levels into labeled symmetry types. Some symmetry distinctions are effectively exact: such as total angular momentum F, or laboratory-inversion parity. Others are approximate, derived from estimates that certain matrix elements are expected to be much larger than others. The most important of these are based on electron spin (for light molecules) and geometrical point-group symmetry (for relatively rigid polyatomics). In actual fact, no transition is completely forbidden. The multipole nature of electromagnetic radiation (electric-dipole, magnetic-dipole, electric-quadrupole, etc.) implies that any change in angular momentum or parity is possible in principle. Practical interest emphasizes identification of the origin of the strongest source of transition probability, and estimation of the strengths of the weak transitions relative to the stronger ones. The result is a collection of *propensity rules* using *selection rules* as tools of estimation.

Basis functions for expansion of the wave functions for the upper and lower states were chosen in Sect. 32.3.2. The first step in the symmetry classification of rovibronic levels consists of identifying various linear combinations of basis functions that block-diagonalize the exact or approximate Hamiltonians. Symmetry-type names are then assigned to these linear combinations based on the value of F or J and knowledge of the symmetry properties of the underlying vibrational and electronic states. Thus each eigenfunction or rovibronic level consists of an expansion over only one of the kinds of linear combination, and the level can be assigned a specific symmetry type.

Similarly, the basis-set expansion leads to a matrix representation of the possible transitions. Spin- and spatial-symmetry arguments establish relationships between these transition matrix elements, and provide estimates of which are much smaller than the others. Each combination of upper- and lower-state symmetry types results in a specific pattern of rotational branches. The most important patterns are ΔJ even (Q-branches) or odd (P- and R-branches), and intensity alternation for consecutive values of J (nuclear spin statistics).

32.4.2 Rotational Branches and Parity

The symmetry of time or momentum reversal implies that changing the signs of all the angular momenta should result in an equivalent wave function. For example, the phase convention

$$\Phi_{\text{vesn}}^{(F,-K_F)}(\text{internal, spins})$$
$$= (-1)^{-F+K_F} \Phi_{\text{vesn}}^{*(F,K_F)}(\text{internal}, -\text{spins}) \quad (32.23)$$

can be chosen to establish that the relative phases of matrix elements of the Hamiltonian can be taken as

$$\langle D_{M_F-K_F'}^{*F} \Phi_{\text{vesn}}^{(F,-K_F')} | H | D_{M_F-K_F''}^{*F} \Phi_{\text{vesn}}^{(F,-K_F'')} \rangle$$
$$= \langle D_{M_F K_F'}^{*F} \Phi_{\text{vesn}}^{(F,K_F')} | H | D_{M_F K_F''}^{*F} \Phi_{\text{vesn}}^{(F,K_F'')} \rangle^* . \quad (32.24)$$

The formula for matrix elements of optical transition operators can also be reanalyzed,

$$\langle \psi^{F',M_F''+p} | T_p^{(L)}(\text{lab}) | \Phi^{F'',M_F''} \rangle$$
$$= \left[\frac{(2F''+1)}{(2F'+1)} \right]^{1/2} \langle F''M_F'', Lp | F'M_F''+p \rangle$$
$$\times \frac{1}{2} \sum_{qK''} \langle F''K_F'', Lq | F'K_F''+q \rangle$$
$$\times \left\{ \langle \psi_{\text{vesn}}^{(F',K_F''+q)} | T_q^{(L)}(\text{body}) | \Phi_{\text{vesn}}^{(F'',K_F'')} \rangle \right.$$
$$\left. + (-1)^{F'+L-F''} \langle \psi_{\text{vesn}}^{(F',-K_F''-q)} | T_{-q}^{(L)}(\text{body}) | \Phi_{\text{vesn}}^{(F'',-K_F'')} \rangle \right\} , \quad (32.25)$$

to establish that all contributions are purely real or purely imaginary [29]. Since such transition matrix elements will be used as absolute squares, they can be treated as if they were purely real.

Parity classification of molecular states according to inversion through the center of mass is important for establishing which transitions are electric-dipole allowed and which states can perturb each other. As discussed by Larsson [4] such classification is not without subtlety and opportunity for confusion. Inversion of the laboratory spatial coordinates (i_{sp}, also called E^* [2]) is equivalent to a reflection σ of the molecule-fixed electronic and nuclear coordinates in an arbitrary plane followed by rotation of the molecular frame by 180° about an axis through the origin and perpendicular to the reflection plane (if F is half-integral special care must be taken about the sense of rotation). It follows that

$$i_{\text{sp}} \Phi_{\text{exact}}^{F,M_F}(\text{lab})$$
$$= E^* \Phi_{\text{exact}}^{F,M_F}(\text{lab})$$
$$= \eta(-1)^{F-\gamma} \Phi_{\text{exact}}^{F,M_F}(\text{lab})$$
$$= \left[\frac{(2F+1)}{8\pi^2} \right]^{1/2} \sum_{K_F} (-1)^{F-K_F} D_{M_F-K_F}^{*F} \sigma_{xz} \Phi_{\text{vesn}}^{(F,K_F)} \quad (32.26)$$

and

$$\sigma_{xz} \Phi_{\text{vesn}}^{(F,K_F)} = \eta(-1)^{K_F-\gamma} \Phi_{\text{vesn}}^{(F,-K_F)} , \quad (32.27)$$

where $\gamma = 0$ or $\frac{1}{2}$ for integral or half-integral F, respectively, and η is the parity label for the state having values of ± 1. In linear molecules, levels with $\eta = +1$ are called e-levels while those with $\eta = -1$ are called f-levels [30, 31].

Inversion symmetry can be combined with time reversal to establish that all matrix elements of the Hamiltonian can be taken to be real [29]. The wave function can also be expressed in the form of the Wang transformation [19], uniting the $\pm K_F$ components,

$$\Phi_{\text{exact}}^{F,M_F}(\text{lab}) = \left[\frac{(2F+1)}{8\pi^2} \right]^{\frac{1}{2}} \sum_{K_F \geq 0} \left\{ D_{M_F K_F}^{*F} \Phi_{\text{vesn}}^{(F,K_F)} \right.$$

$$+ \eta(-1)^{-K_F+\gamma} D^{*F}_{M_F-K_F} \sigma_{xz} \Phi^{(F,K_F)}_{\text{vesn}} \Big\} [2(1+\delta_{K_F0})]^{-\frac{1}{2}} .$$

$$(32.28)$$

If the molecule is rigid and has a plane of symmetry, or is nonrigid with accessible vibrational or tunneling modes that correspond to plane reflections, inversion symmetry divides the states of the molecule into two classes, according to the sign of η. Perturbations can occur only for $\Delta F = 0$ and $\eta'\eta'' = +1$. For optical transitions, the change in parity of the states must match the parity of the operator. Odd operators (e.g. electric-dipole) require $\eta'\eta''(-1)^{\Delta F} = -1$. Even operators (e.g. magnetic-dipole and electric-quadrupole) require $\eta'\eta''(-1)^{\Delta F} = +1$.

32.4.3 Nuclear Spin, Spatial Symmetry, and Statistics

For most molecules, the coupling of nuclear spin with the electron-spin, electron-orbital, and frame-rotational angular momenta is sufficiently weak that treatment of the energetics of hyperfine interactions can be postponed. The first-order effect of nuclear spin is that rovibronic wave functions for molecules containing identical nuclei must be combined with appropriate nuclear spin wave functions in order to obtain the necessary Fermi-Dirac or Bose-Einstein nuclear permutation symmetry. For many molecules, there exist combinations of nuclear permutations that correspond to combinations of frame rotations, laboratory inversions, and feasible vibrational motions (the rotational wave function makes a contribution because renumbering the nuclei requires a reanalysis of the Euler angles). For rigid molecules, these permutations (possibly including inversion) can be used to generate the point symmetry group of the molecule. For fluxional molecules, with multiple energetically equivalent nuclear configurations, a rather large "molecular symmetry group" can result, one that may not correspond to any ordinary point group [1, 2].

In the discussion immediately following, consider the case of N occurrences of one kind of nucleus, the others being unique (e.g. PD$_3$). The treatment can easily be extended to the case of multiple kinds of identical nuclei (e.g. C$_2$H$_6$). The exact wave function can be rearranged into a sum over products of the form

$$\Phi_{\text{exact}} = \sum_{a,b} \Phi^{(a)}_{\text{rves}} \Phi^{(b)}_{\text{nspin}} , \qquad (32.29)$$

where $\Phi^{(a)}_{\text{rves}}$ is a rovibronic wave function belonging to the $\Gamma(a)$ representation of the symmetric group S_N of permutations over N objects, and $\Phi^{(b)}_{\text{nspin}}$ is a nuclear spin wave function, belonging to the $\Gamma(b)$ representation of S_N. In order to obtain the correct permutation

symmetries for the overall wave function, the only terms that can appear in this sum are those for which the direct product $\Gamma(a) \otimes \Gamma(b)$ contains the symmetric or antisymmetric representation, for bosons or fermions respectively.

Assumption of negligible hyperfine interactions allows evaluation of matrix elements of the form

$$\langle \Phi^{(c)}_{\text{rves}} \Phi^{(d)}_{\text{nspin}} | H | \Phi^{(a)}_{\text{rves}} \Phi^{(b)}_{\text{nspin}} \rangle$$
$$= \langle \Phi^{(c)}_{\text{rves}} | H | \Phi^{(a)}_{\text{rves}} \rangle \langle \Phi^{(d)}_{\text{nspin}} | \Phi^{(b)}_{\text{nspin}} \rangle , \qquad (32.30)$$

which vanishes unless $\Gamma(a) = \Gamma(c)$ and $\Gamma(b) = \Gamma(d)$. Thus the nearly-exact wave function can be written as a sum over products where the rovibronic and nuclear-spin factors correspond to basis functions from single known representations:

$$\Phi_{\text{exact}} \approx \sum_{a,b} \Phi^{(a)}_{\text{rves}} \Phi^{(b)}_{\text{nspin}} , \qquad (32.31)$$

with $\Gamma(a) = \Gamma_{\text{rves}}$, and $\Gamma(b) = \Gamma_{\text{nspin}}$. This divides the states of the molecule into a number of noninteracting symmetry classes, labeled by the representations of the symmetric group. In the absence of hyperfine interactions, optical transitions are possible only within a certain symmetry class.

Thus the existence of spatial or dynamical symmetry implies that each rovibronic wave function transforms according to a particular representation of a subgroup of the permutation-inversion group (called CNPI by Bunker [2]). Each representation includes only specific values of nuclear spin, corresponding to the permutational properties of the nuclear spin wave functions. The most important effect of this analysis is to assign statistical weights or relative intensities to the different symmetry types. For example, the symmetry group for NH$_3$ is D_{3h} (including umbrella inversion), with representations A'_1, A'_2, A''_1, A''_2, E', and E''. The A'_1 and A''_1 representations must be combined with the (nonexistent) antisymmetric spin function, yielding a statistical weight of 0. Similarly, A'_2 and A''_2 combine with the symmetric $I = \frac{3}{2}$ spin function, with a statistical weight of 4 (i.e. $2I+1$). E' and E'' combine with the nonsymmetric $I = \frac{1}{2}$ spin functions, with a statistical weight of 2. This material is discussed from various viewpoints in numerous articles and text books, of which only a few can be cited here [1, 2, 8, 18, 21, 32–38]. See Chap. 31 for additional details and examples.

Although this analysis appears rather complicated, the selection rules that result are actually the same, at least in simple cases, as the ones that are derivable from simpler ideas. For example, for a $^1\Sigma^+_g$ lower state, even J levels are permutation symmetric and have parity $+1$, while odd J levels are permutation antisymmetric and have parity -1. For a $^1\Sigma^+_u$ upper state, even J levels are permutation antisymmetric and have parity $+1$, while odd J levels are permutation symmetric and have parity -1. Both the parity selection rule and permutation-symmetry selection rule independently require that $\Delta J = \pm 1$ for electric-dipole transitions. Similarly, that the permanent dipole moment of a symmetric-top molecule must lie along the body-fixed axis replicates the $\Delta K = 0$ selection rule for pure-rotation transitions provided by permutational symmetry arguments. This means that when simulating absorption and emission spectra, the nuclear-spin wave function can usually be ignored. The intensity alternation imposed by spin-statistics can be represented by multiplying each wave function by the appropriate factor, for example, 0 or $[(2I+1)/(2i+1)^N]^{1/2}$, where I is the total nuclear spin, and i is the spin of one of the N equivalent nuclei.

Group theory remains vital for understanding the relative strengths of vibrational transitions in polyatomics (see Cotton [39], for example, and Sect. 32.6.2) and becomes very interesting as interaction between vibration and rotation increases. For the purposes of this discussion, the most important issue is identification of which transition-moment matrix elements $\mu_{K'K''}$ vanish and which are related by symmetry.

32.4.4 Electron Orbital and Spin Angular Momenta

For all molecules, the strongest transitions tend to be those that conserve electron spin. The zero-order transition-moment matrix is diagonal both in total spin and in spin-projection onto the body-frame axis. In the $|\Lambda\Sigma\rangle$ basis set for linear molecules this is expressed by

$$\mu_{\Lambda'\Sigma'\Lambda''\Sigma''} = \mu_{\Lambda'\Lambda''}\delta_{\Sigma'\Sigma''} \,. \tag{32.32}$$

The transition-moment tensor operator $T_q^{(L)}$(body) can connect basis functions that differ in Λ by at most L. Allowed electric-dipole transitions thus satisfy

$$\mu_{\Lambda'\Lambda''} = 0, \quad \text{for} \quad |\Lambda' - \Lambda''| > 1 \,. \tag{32.33}$$

In the usual case that the upper and lower states each consist of only a single value of $|\Lambda|$, there is only one independent, nonzero, matrix element $\mu_{|\Lambda'|,|\Lambda''|}$ with

$$\mu_{-|\Lambda'|,-|\Lambda''|} = \eta(-1)^{|\Lambda'|-|\Lambda''|}\mu_{|\Lambda'|,|\Lambda''|} \,. \tag{32.34}$$

See Sect. 32.8 for a discussion of spin-forbidden and orbitally-forbidden transitions. Similar arguments and phase relationships can be developed for polyatomic molecules with nonzero electron spin or degenerate vibrational or electronic states.

32.5 ABSORPTION CROSS SECTIONS AND RADIATIVE LIFETIMES

32.5.1 Radiation Relations

Among the most important radiation relations is the connection between the absorption cross section and the rate of spontaneous emission. Einstein [40] introduced his A and B coefficients to describe the rates of absorption and emission of radiation of a collection of two-level atoms or molecules in equilibrium with a radiation field at the same temperature. The discussion here follows that of Condon and Shortley [41], Penner [42], Thorne [43], and Steinfeld [44] (with corrections). Also see Chaps. 10, 17, and 66 of this Handbook. The number of absorption events per unit volume per unit time is written as

$$N_l B_{lu}\rho(\nu) \,. \tag{32.35}$$

While the rate of emission is

$$N_u B_{ul}\rho(\nu) + N_u A_{ul} \,. \tag{32.36}$$

In thermal equilibrium, the radiative energy density is given by the Planck blackbody law

$$\rho(\nu) = \left(\frac{8\pi h\nu^3}{c^3}\right) [e^{h\nu/kT} - 1]^{-1} \,, \tag{32.37}$$

and the ground and excited state densities satisfy a Boltzmann relationship,

$$\frac{N_u}{N_l} = \left(\frac{g_u}{g_l}\right) e^{-h\nu/kT} \,, \tag{32.38}$$

where g_u and g_l are the degeneracies of the upper and lower states. The requirement that the rates of absorption and emission must be equal leads to the relations

$$A_{ul} = \left(\frac{8\pi h\nu^3}{c^3}\right) B_{ul} = \left(\frac{8\pi h\nu^3}{c^3}\right)\left(\frac{g_l}{g_u}\right) B_{lu} \,. \tag{32.39}$$

Numerical values of the B coefficients can be derived from the optical absorption cross section, and thus

$$A_{ul} = \left(\frac{8\pi\nu^2}{c^2}\right)\left(\frac{g_l}{g_u}\right) \int \sigma_{\text{abs}}(\nu)\,d\nu$$

$$= \left(\frac{8\pi\nu^2}{c^2}\right) \int \sigma_{\text{se}}(\nu)\,d\nu \,. \tag{32.40}$$

Finally, the expression for the absorption oscillator strength is

$$f_{\text{abs}} = (4\pi\epsilon_0)\left(\frac{mc^2}{8\pi^2\nu^2 e^2}\right)\left(\frac{g_u}{g_l}\right) A_{ul}$$

$$= (4\pi\epsilon_0)\left(\frac{mc}{\pi e^2}\right) \int \sigma_{\text{abs}}(\nu)\,d\nu \,. \tag{32.41}$$

The emission oscillator strength is simply related to that for absorption: $f_{\text{em}} = -(g_l/g_u)f_{\text{abs}}$. The oscillator strength offers considerable advantages as a means of reporting and comparing the strengths of radiative transitions. It is dimensionless, obeys the simple sum rule (for electric-dipole transitions)

$$\sum_u f_{ul} = \text{number of electrons}, \qquad (32.42)$$

and is directly derivable from an experimental absorption cross section even before the assignment of the upper level has been determined (i.e. before its degeneracy is known).

32.5.2 Transition Moments

In many cases, the intention is to construct model quantum mechanical wave functions for the two states involved in the transition under study. In addition, *ab initio* electronic wave functions and matrix elements may be available (see Chap. 30). Quantum mechanics suggests the following expression for the Einstein A coefficient (see Sect. 11.6.1):

$$A_{ul} = \left(\frac{64\pi^4\nu^3}{3hc^3}\right)\left(\frac{1}{4\pi\epsilon_0}\right)\left(\frac{1}{g_u}\right)\sum_{u',l'',p} |\langle\psi'_{u'}|er_p|\psi''_{l''}\rangle|^2. \qquad (32.43)$$

The summation is over all three optical polarization directions p (i.e. r_p runs over x, y, and z in the lab frame), all degenerate components l'' of the lower state (i.e. g_l of them), and all degenerate components u' of the upper state (i.e. g_u of them). This triple sum is also called the *line strength* S_{ul}. Division by the upper-level degeneracy corrects for the fact that the transitions should be averaged rather than summed over the initial levels.

In practice, choosing the appropriate degeneracy to divide by is a question of some ambiguity. For atoms, it is sufficient to understand how the individual matrix elements and the line strength were calculated. For example, Bethe and Salpeter [45] use a degeneracy of $(2L+1)$ for Schrödinger wave functions for the hydrogen atom, and $(2J+1)$ for Dirac wave functions.

For molecules with internal angular momentum, i.e. everything other than $^1\Sigma$ states of linear molecules, the situation is much more complicated. For electric-dipole allowed transitions in light molecules, *ab initio* transition moments are calculated in a body-fixed coordinate system, ignoring spin, and not summed over anything.

For diatomic molecules, following the work of Whiting *et al.* [22,46], the transition probability from a single upper-state component $(J'c')$ to a single lower-state component $(J''c'')$ is written as

$$A_{J'c'J''c''} = \left(\frac{64\pi^4\nu^3}{3hc^3}\right)\left(\frac{1}{4\pi\epsilon_0}\right)$$
$$\times \left(\frac{1}{2J'+1}\right)q_{v'v''}|R_e|^2 S_{J'J''}^{c'c''}. \qquad (32.44)$$

In this formula $q_{v'v''}|R_e|^2$ represents the rotationless contribution to the transition moment, symbolically represented as a product of a vibrational overlap ($q_{v'v''}$, i.e. a Franck-Condon factor) and an electronic-only component $|R_e|^2$ (see Sect. 32.6.1). All of the rotational complexity is absorbed into the rotational-branch strength factor $S_{J'J''}^{c'c''}$ (see Sects. 32.7.2 and 32.7.3). The issue to be addressed here is how to divide numerical factors between $|R_e|^2$ and $S_{J'J''}^{c'c''}$. One approach is to construct an estimate for the rotationless transition probability

$$A_{v'v''} = \left(\frac{1}{N_c'}\right)\sum_{c'J''c''} A_{J'c'J''c''}, \qquad (32.45)$$

where N_c' is the number of internal spin-orbit components of the upper state. Whiting *et al.* suggest that $S_{J'J''}^{c'c''}$ be normalized such that for spin-allowed transitions

$$\sum_{c'J''c''} S_{J'J''}^{c'c''} = (2-\delta_{0,\Lambda'}\delta_{0,\Lambda''})(2S+1)(2J'+1). \qquad (32.46)$$

The first factor is 1 for Σ–Σ transitions, and 2 for all others. The final factor is replaced by $(2J''+1)$ if the sum is over J' instead of J''. For spin-forbidden transitions the following is a plausible extension of this sum rule,

$$\sum_{c'J''c''} S_{J'J''}^{c'c''} = \max(N_c', N_c'')(2J'+1). \qquad (32.47)$$

Sect. 32.7.3 provides a corresponding sum rule for polyatomic molecules. This normalization yields

$$A_{v'v''} = \left(\frac{64\pi^4\nu^3}{3hc^3}\right)\left(\frac{1}{4\pi\epsilon_0}\right)\left(\frac{\max(N_c', N_c'')}{N_c'}\right)q_{v'v''}|R_e|^2 \qquad (32.48)$$

and for spin-allowed transitions, the simple spin-free expressions for the electronic transition moments:

$$|R_e|^2 = |\langle\Lambda|ez|\Lambda\rangle|^2 \qquad (32.49)$$

for parallel transitions and

$$|R_e|^2 = \left|\langle\Lambda+1|e\tfrac{1}{\sqrt{2}}(x+iy)|\Lambda\rangle\right|^2 \qquad (32.50)$$

for perpendicular transitions.

32.6 VIBRATIONAL BAND STRENGTHS

32.6.1 Franck-Condon Factors

The Born-Oppenheimer separation of electron and nuclear motion suggests that during an optical transition between different electronic states the nuclei should change neither their position nor momentum. This concept was developed from semiclassical arguments by Franck [47] and justified quantum mechanically by Condon [48]. Following Herzberg [49] and Steinfeld [44] the vibronic transition moment can be written as

$$
\begin{aligned}
\mu_{v'v''} &= \langle \psi_e' \psi_{v'}' | \mu | \psi_e'' \psi_{v''}'' \rangle, \\
&= \int dR \, \psi_{v'}^{*\prime}(R) \psi_{v''}''(R) \int dr \, \psi_e^{*\prime}(r,R) \psi_e''(r,R), \\
&= \int dR \, \psi_{v'}^{*\prime}(R) \psi_{v''}''(R) \mu(R).
\end{aligned}
\tag{32.51}
$$

If the R-dependence of $\mu(R)$ is sufficiently weak, it can be factored out to obtain

$$
\mu_{v'v''} = R_e \int dR \, \psi_{v'}^{*\prime}(R) \psi_{v''}''(R), \tag{32.52}
$$

where R_e is called the electronic transition moment. The transition probability is proportional to the square of the above, which is usually written as

$$
I \sim q_{v'v''} R_e^2, \tag{32.53}
$$

where

$$
q_{v'v''} = \left| \int dR \, \psi_{v'}^{*\prime}(R) \psi_{v''}''(R) \right|^2, \tag{32.54}
$$

the square of the overlap between initial and final vibrational wave functions, is called the Franck-Condon factor.

The Franck-Condon factors satisfy the sum rule

$$
\sum_{v'} q_{v'v''} = \sum_{v''} q_{v'v''} = 1 \tag{32.55}
$$

provided the summations include the continuum vibrational wave functions above the dissociation limits. The Franck-Condon approach also can be used to calculate intensities and cross sections for bound-free [50] and free-free [51] emission and absorption.

In some cases the variation of $\mu(R)$ is significant. Calculation of the effect on the intensities can usually be handled by the *r-centroid* method in which the expression

$$
\bar{r}_{v'v''} = \frac{\langle \psi_{v'}' | R | \psi_{v''}'' \rangle}{\langle \psi_{v'}' | \psi_{v''}'' \rangle} \tag{32.56}
$$

is used to calculate an effective internuclear distance for the transition. The transition strength is then proportional to $q_{v'v'} |\mu(\bar{r}_{v'v''})|^2$. An advantage of this formulation is that the vibrational overlaps can be calculated from energy information only, before the transition moment function is known and before the transition strengths are investigated experimentally. The quantitative accuracy of the r-centroid method for transition

moments that are not linear in the internuclear distance, has been addressed by a considerable literature, which has been summarized by McCallum [52].

A second complication arises from the fact that the vibrational wave functions themselves depend parametrically on the rotational angular momentum. Calculation of rotationally-dependent Franck-Condon factors is described by Dwivedi *et al.* [53] who also discuss the r-centroid method.

32.6.2 Vibrational Transitions

Vibrational transitions derive their strength from variation of the "permanent" dipole moment of the molecule as a function of geometry or internuclear coordinates. As described by several authors [33, 34, 36, 54] one can expand the dipole moment as a power series in the internal-Cartesian or normal-mode coordinates

$$
M_{xyz}(Q) = M_{xyz}^0 + \sum_i \left(\frac{\partial M_{xyz}}{\partial Q_i} \right) Q_i + \cdots \tag{32.57}
$$

and calculate intensities from a formula like

$$
I \sim |\langle \psi_{v_1'} \psi_{v_2'} \cdots | M(\mathbf{Q}) | \psi_{v_1''} \psi_{v_2''} \cdots \rangle|^2. \tag{32.58}
$$

For homonuclear diatomic molecules, the dipole moment vanishes identically, so there is no rovibrational spectrum. The dipole moment for heteronuclear diatomics is often close to linear in the internuclear distance. The harmonic oscillator model suggests that transitions with $\Delta v = \pm 1$ are the strongest, with intensities approximated by

$$
I_{v+1,v} \sim \left| \frac{dM}{dR} \right|^2 (v+1). \tag{32.59}
$$

Overtone bands, i.e. with $|\Delta v| > 1$, are observed, as dramatically illustrated by the $\Delta v = 4, 5$ emissions from the OH radical observed from the Earth's night sky [55].

For polyatomic molecules, overtone and combination bands are often quite strong. The presence or absence of which is used to establish the symmetries of the vibrational modes. In general, it is difficult to construct quantitative vibrational intensity formulas with only a few parameters that can be inferred experimentally.

32.7 ROTATIONAL BRANCH STRENGTHS

32.7.1 Branch Structure and Transition Type

The overall rotational structure of a molecular transition is determined by the relative values and phases of the body-frame transition-moment matrix elements, the relative values and phases of coefficients in the expansion of the upper-state and lower-state component wave functions over the angular-momentum-projection basis functions, the energy separations between the components, and the relative values and phases of the vector-coupling coefficients. In simple cases, each lower component might be connected to only a single upper component [Hund's case (b) or symmetric tops], or $\Delta J = \pm 1$ (P- and R-branches) may dominate over $\Delta J = 0$ (Q-branches).

For diatomic molecules, symmetry arguments are used to divide the components into the two parity classes e and f. For electric dipole transitions, the selection rules from Sect. 32.4.2 imply that

$$
\begin{array}{ll}
(N'_e N''_e + N'_f N''_f) & P\text{- and } R\text{-branches} \\
(N'_e N''_f + N'_f N''_e) & Q\text{-branches}
\end{array}
\tag{32.60}
$$

are expected, where N_e and N_f indicate the number of components of each parity class (N_e and N_f differ by no more than one). Rotational branches are labeled with the notation $^{\Delta R}\Delta J_{c'c''}$, using symbols P for -1, Q for 0, and R for $+1$. ΔR indicates the "apparent" change in mechanical rotational angular momentum (i.e. energy) [see Hund's case (b) in Sect. 32.7.4] and ΔJ indicates the "actual" change (i.e. quantum mechanical). The labels $c'c''$ indicate the components involved. Thus a $^{P}Q_{21}$ branch is expected to be "red shaded" ($\Delta R = -1$) with $\Delta J = 0$ involving the second component of the upper state and the first (lowest) component of the lower state. The notation $^{P}Q_{21}(J)$ [or sometimes $^{P}Q_{21}(R = N)$ for Σ lower states] identifies an individual rotational line and specifies the rotational quantum number of the lower state involved in the transition. If the upper- and lower-state component numbers are the same, one of them may be dropped. Thus P_{22} is sometimes written as P_2.

For symmetric top molecules, the rotational branches are labeled ΔJ_K (e.g. P_1). For asymmetric tops, the branches are labeled by $\Delta J_{\Delta K_{-1}, \Delta K_1}$, where K_{-1} and K_1 are the prolate- and oblate-limit angular momenta projections (described in Sect. 32.3.2).

The transition dipole moment commonly lies parallel or perpendicular to the body-frame axis. In the former case, $\mu_{K'K''}$ vanishes for $K' \neq K''$, and in the latter for $K' = K''$. Thus parallel bands correspond to $\Delta K = 0$ transitions, while perpendicular bands have $\Delta K = \pm 1$. As enforced by the vector-coupling coefficients or Hönl-London factors described below, for low values of K (e.g. diatomics), $\Delta K = 0$ implies strong $\Delta J = \pm 1$ (P and R) branches and weak $\Delta J = 0$ (Q) branches. On the other hand, $\Delta K = \pm 1$ leads to Q-branches that are approximately twice as strong as either the P- or R-branches.

32.7.2 Hönl-London Factors

The matrix model Hamiltonians for the upper and lower states have been diagonalized, yielding the wave functions

$$
\psi_{c'}^{J'M'} = \sum_{K'} b_{J'c'}^{K'} \left[\frac{(2J'+1)}{8\pi^2} \right]^{\frac{1}{2}} D_{M'K'}^{*J'}(\text{lab}) \chi_{K'}(\text{body})
\tag{32.61}
$$

and

$$
\Phi_{c''}^{J''M''} = \sum_{K''} a_{J''c''}^{K''} \left[\frac{(2J''+1)}{8\pi^2} \right]^{\frac{1}{2}}
$$
$$
\times D_{M''K''}^{*J''}(\text{lab}) \xi_{K''}(\text{body}) .
\tag{32.62}
$$

In these expressions, the designations K' and K'' are slightly symbolic. They represent the body-frame projection of the total angular momentum and also a running index over basis functions. For complicated cases, more than one basis function can have a given value of K.

Following Sect. 32.2.3, the rotational branch strength is then written as

$$
S_{J'J''}^{c'c''}
$$
$$
= \sum_{pM'M''} |\langle \psi_{c'}^{J'M'} | T_p^{(L)}(\text{lab}) | \Phi_{c''}^{J''M''} \rangle|^2
$$
$$
= (2J''+1) \left| \sum_{q'K'K''} b_{J'c'}^{*K'} a_{J''c''}^{K''} \langle \chi_{K'} | T_q^{(L)}(\text{body}) | \xi_{K''} \rangle \right.
$$
$$
\left. \times \langle J''K'', Lq | J'K' \rangle \right|^2
\tag{32.63}
$$

or

$$
S_{J'J''}^{c'c''} = \left| \sum_{K'K''} b_{J'c'}^{*K'} a_{J''c''}^{K''} \mu_{K'K''} \zeta(J', K', J'', K'') \right|^2 ,
\tag{32.64}
$$

where

$$
\mu_{K'K''} = \langle \chi_{K'} | T_q^{(L)}(\text{body}) | \xi_{K''} \rangle \theta(K'' - K')
\tag{32.65}
$$

is the body-frame transition-moment matrix introduced in Sect. 32.3.4, with relative values that are hypothesized based on interpretation of the nature of the transition, calculated from *ab initio* wave functions, or inferred by fitting the observed rotational branch strengths. The Clebsch-Gordan expression

$$
\zeta(J', K', J'', K'')
$$
$$
= (2J''+1)^{1/2} \langle J''K'', LK' - K'' | J'K' \rangle
$$
$$
\times \theta(K'' - K') \theta(J' - J'')
\tag{32.66}
$$

represents the transformation of the radiation field from the laboratory-frame to the body-frame, also related to the "direction cosines" used by many authors. The additional phase factors

$$\theta(k) = \operatorname{sgn}(k) = \begin{cases} +1 & k \geq 0 \\ -1 & k < 0 \end{cases} \qquad (32.67)$$

have been included here to make the signs and symmetry relations of $\mu_{K'K''}$ and $\zeta(J', K', J'', K'')$ agree with those already in use [9, 22, 56, 57]. They are related to the choice of the leading signs when T_+ and T_- are expressed as $\pm(T_x + iT_y)$ and $\pm(T_x - iT_y)$. Their inclusion has no effect for spin-allowed transitions with only one source of transition probability, e.g. purely parallel or perpendicular.

In the usual case of electric-dipole (or magnetic-dipole) radiation (i.e. $L = 1$), ζ^2 is the well-known Hönl-London Factor [58]. $\zeta(J', K', J'', K'')$ is a real, signed quantity: negative for $\Delta J \Delta K > 0$; or $\Delta J = \Delta K = 0$ and $K < 0$; otherwise positive [9].

Setting $L = 2$ provides intensity formulas for electric-quadrupole [59], Raman [60], and two-photon [61, 62] transitions. Additional Rayleigh-like terms can appear for $K' = K'' \neq 0$. Halpern *et al.* [62] also give formulas for three-photon transitions in diatomics, expressed in terms of Clebsch-Gordan coefficients with $L = 3$ and $L = 1$ (for $|\Delta\Omega| \leq 1$).

32.7.3 Sum Rules

The orthonormality relations for component eigenvectors

$$\sum_{c'} b^{*K'}_{J'c'} b^K_{J'c'} = \delta_{K',K}, \qquad (32.68)$$

$$\sum_{c''} a^{*K''}_{J''c''} a^K_{J''c''} = \delta_{K'',K} \qquad (32.69)$$

can be used to construct the sum rule

$$\sum_{c'c''} S^{c'c''}_{J'J''} = \sum_{K'K''} |\mu_{K'K''} \zeta(J', K', J'', K'')|^2. \qquad (32.70)$$

Finally, the orthonormality relations of the Clebsch-Gordan coefficients result in

$$\sum_{J'c'c''} S^{c'c''}_{J'J''} = (2J'' + 1) \sum_{K'K''} |\mu_{K'K''}|^2, \qquad (32.71)$$

$$\sum_{J''c'c''} S^{c'c''}_{J'J''} = (2J' + 1) \sum_{K'K''} |\mu_{K'K''}|^2. \qquad (32.72)$$

As discussed in Sect. 32.5.2, it is convenient to have the $\mu_{K'K''}$ matrix elements consist of numbers that represent the nature of the transition but not its strength, the latter being expressed by the "vibrational" ($q_{v'v''}$) and "electronic" (R_e) contributions. Following Sect. 32.5.2,

for diatomic molecules, the "orientational" part $\mu_{K'K''}$ is taken to have a fixed sum rule

$$\sum_{K'K''} |\mu_{K'K''}|^2 = \max(N'_c, N''_c), \qquad (32.73)$$

where N'_c is the number of components (K' values, or basis functions) for the upper state, and N''_c is the number of components in the lower state. For polyatomic molecules, the sum rule can be written as

$$(2J + 1)^2 (\mu_a^2 + \mu_b^2 + \mu_c^2) \max(N'_c, N''_c), \qquad (32.74)$$

where

$$\mu_a^2 + \mu_b^2 + \mu_c^2 = |\mu_+|^2 + |\mu_-|^2 + |\mu_0|^2 = 1 \qquad (32.75)$$

and N'_c and N''_c are the numbers of spin-electronic-vibrational components in the upper and lower states, respectively. Also see Whiting *et al.* [22, 46] and Brown *et al.* [7].

32.7.4 Hund's Cases

In diatomic molecules, several limiting cases are useful as short-hand or first-approximation concepts for classification of energy levels and rotational branch strengths. These are called the Hund's cases [63]. They are distinguished by the extent to which the electron orbital and spin angular momenta are rigidly attached to the tumbling molecular frame, i.e. whether Λ, Σ, and S are good quantum numbers. Hund's cases are discussed in many journal articles and in every textbook dealing with the rotational structure of diatomic spectra. An appealing recent description is provided by Nikitin and Zare [64].

In most works, the emphasis has been on finding a favorable zero-order approximation for perturbation expansion of energy levels. The advance of precision measurement of transition energies and the availability of sophisticated parametrical matrix models and fast computers on which to realize them, has reduced the importance of Hund's cases for actual computations. In particular, the need to derive and implement numerous explicit energy and intensity formulas leads to unfortunate transcription errors. Nevertheless, they remain of value for qualitative and pedagogical understanding, especially for estimates of the relative intensities of rotational branches.

Hund's case (a) describes the situation in which Λ and Σ are separately well-defined. This is a common case in which the separation between electronic states, i.e. different values of $|\Lambda|$, is larger than the spin-orbit interaction, which in turn, is larger than the separation between rotational levels. At low J, there are $(2S + 1)$ pairs of nearly-degenerate energy levels separated from each other by the spin orbit constant: $E \sim A\Lambda\Sigma + BJ(J + 1)$. The wave functions are of the form

$$\psi_{\Omega\pm}^{JM} = \frac{1}{\sqrt{2}} \left[\frac{(2J+1)}{8\pi^2} \right]^{1/2}$$
$$\times \left[D_{M\Lambda+\Sigma}^{*J} |\Lambda, \Sigma\rangle \pm D_{M-\Lambda-\Sigma}^{*J} | -\Lambda, -\Sigma\rangle \right]$$
$$= \frac{1}{\sqrt{2}} \left[|\Lambda, \Sigma; J, M, \Omega\rangle \pm | -\Lambda, -\Sigma; J, M, -\Omega\rangle \right],$$
$$(32.76)$$

with only two nonzero expansion coefficients a_{Jc}^K. If both the upper and lower states are well described by Hund's case (a), then each lower component is optically connected only to the upper components with the same value of $|\Sigma|$. Then

$$S_{J'J''}^{\Omega'\Omega''} = (2J''+1)|\langle J''\Omega'', 1\Omega'-\Omega|J'\Omega'\rangle|^2, \quad (32.77)$$

with $\Omega'' = \Lambda'' + \Sigma$ and $\Omega' = \Lambda' + \Sigma$.

Hund's case (b) indicates that Λ is well-defined, but spin-orbit coupling is weak. The components correspond to well-defined values of $\mathbf{N} = \mathbf{J} - \mathbf{S}$, ranging from $|J - S|$ to $(J + S)$, with energies approximated by $E \sim BN(N+1)$, and wave functions of the form

$$\psi_{N\pm}^{JM} = \sum_\Sigma (-1)^{S+\Sigma} \langle J -\Omega, S\Sigma|N-\Lambda\rangle$$
$$\times \frac{1}{\sqrt{2}} \left[|\Lambda, \Sigma; J, M, \Omega\rangle \pm | -\Lambda, -\Sigma; J, M, -\Omega\rangle \right].$$
$$(32.78)$$

This equation is derivable from Mizushima's equation (2-3-26) [14] and Zare's equations (2.8), (2.26), and (3.105) [5], using the lab-to-body transformation

$$|SM_S\rangle(\text{lab}) = \sum_\Sigma D_{M_S\Sigma}^{*S}(\phi\theta\chi)|S\Sigma\rangle(\text{body}). \quad (32.79)$$

It disagrees with Judd's problem 9.1 [65] by a phase factor $(-1)^{J+2S+\Sigma-N}$ but agrees with Mizushima's expansion of a $^3\Pi$ state (page 287 [14]) if the Clebsch-Gordan coefficients are taken from Condon and Shortley, page 76 [41].

If both the upper and lower states are well described by Hund's case (b), these wave functions can be substituted into the general rotational-branch strength equations above. Following Edmonds [66] [Eqs. (6.2.8) and (6.2.13), and Table 5] yields the square of a product of Clebsch-Gordan and Racah coefficients

$$S_{J'J''}^{N'N''} = (2J''+1)(2J'+1)(2N''+1)$$
$$\times |\langle N''\Lambda'', 1\Lambda'-\Lambda''|N'\Lambda'\rangle$$
$$\times W(N', J'', N'', J'; S, 1)|^2. \quad (32.80)$$

The Clebsch-Gordan coefficient enforces the case (b) selection rule $\Delta N = 0, \pm 1$, while the Racah coefficient provides the $\Delta J = \Delta N$ propensity rule, which becomes more precise as N increases. A similar propensity rule, $\Delta F = \Delta J$, is common for transitions between hyperfine components (see also Femenias [67]).

Hund's case (c) corresponds to the situation in which spin-orbit coupling is so strong that each level described by the projection Ω actually consists of multiple values of $|\Lambda|$ (e.g. mixing of Σ and Π states) or multiple values of S (e.g. mixing of singlet and triplet spins). This limiting case is formally similar to Hund's case (a), but no assumptions can be made about the relative magnitudes of transition-moment matrix elements $\mu_{\Omega'\Omega''}$. Any of which can be nonzero for $|\Delta\Omega| \leq 1$, for example

$$S_{J'J''}^{\Omega'_+\Omega''_\pm} = \frac{1}{4}|\mu_{\Omega'\Omega''}\zeta(J',\Omega',J'',\Omega'')$$
$$\pm \mu_{\Omega'-\Omega''}\zeta(J',\Omega',J'',-\Omega'')$$
$$+ \mu_{-\Omega'\Omega''}\zeta(J',-\Omega',J'',\Omega'')$$
$$\pm \mu_{-\Omega'-\Omega''}\zeta(J',-\Omega',J'',-\Omega'')|^2. \quad (32.81)$$

The symmetry (sign) relations between $\mu_{\Omega'\Omega''}\zeta(J',\Omega',J'',\Omega'')$ and $\mu_{-\Omega'-\Omega''}\zeta(J',-\Omega',J'',-\Omega'')$ determine whether this transition occurs only for $\Delta J = \pm 1$ (P- and R-branches) or only for $\Delta J = 0$ (Q-branches).

The interest and complexity of Hund's case (c) were exemplified by a seminal work by Kopp and Hougen [68], who considered $\Omega' = \frac{1}{2}$, $\Omega'' = \frac{1}{2}$ transitions, under the assumption that both states could consist of arbitrary mixtures of $^2\Sigma_{1/2}$ and $^2\Pi_{1/2}$ character. Each of the six rotational branches shows constructive or destructive interference of parallel ($\Delta\Omega = 0$) and perpendicular ($\Delta\Omega = \pm 1$) contributions. Hund's case (c) also describes spin-orbit mixing collisions [69] or dissociation to specific spin-orbit limits [70–72].

Hund's case (d) arises in the investigation of Rydberg series [73], in which the separation between Σ and Π from the same orbital configuration approach each other as the principal quantum number (n) increases. Spin-orbit coupling between these projections also diminishes. The eigenfunction components correspond to well-defined values of $\mathbf{R} = \mathbf{J} - \mathbf{L}$, ranging from $|J - L|$ to $(J + L)$, with energies approximated by $E \sim BR(R+1)$, and wave functions of the form

$$\psi_{RL}^{JM} = \sum_\Lambda (-1)^{L+\Lambda} \langle J -\Lambda, L\Lambda|R0\rangle|L\Lambda; JM\Lambda\rangle. \quad (32.82)$$

Carroll [73] used intensity formulas provided by Kovacs [59] to analyze, by spectral simulation, the $4p$–$15p$ ($^1\Sigma_u^+$ and $^1\Pi_u$) Rydberg states of N_2 excited from the ground $X^1\Sigma_g^+$. Three strong Q-form branches survive, corresponding to $R' = J''$, two arising from $\Pi \leftarrow \Sigma$ and one from $\Sigma \leftarrow \Sigma$ case (a) branches. The remaining O-form ($R' = J'' - 2$) and S-form ($R' = J'' + 2$) branches fade rapidly as n increases. With the phase conventions used here, this situation corresponds to body-frame transition matrix elements satisfying

$$\mu_{00} = \mu_{-10} = -\mu_{10}. \quad (32.83)$$

In the opposite case, corresponding to a parity change of the parent-ion core [72], two of the Q-form branches are extinguished, while one Q-form, one O-form, and one S-form branch remain. The transition matrix elements would satisfy

$$\mu_{00} = 0\,, \quad \mu_{-10} = \mu_{10}\,. \qquad (32.84)$$

Hund's case (d) polyatomics are also known [8].

Hund's case (e) would correspond to a situation in which **L** and **S** are strongly coupled to each other, but neither is strongly coupled to the internuclear axis. No examples are known for bound states of molecules.

32.7.5 Symmetric Tops

For transitions between nondegenerate vibronic states, the transition moment must lie along the principal top axis, leading to the selection rule $\Delta K = 0$. Otherwise, Hougen's convenient quantum number [35] $G = \Lambda + l - K$, provides the selection rule $\Delta G = 0, \pm n$ (for an n-fold major symmetry axis) (see Sects. 32.3.2, 32.4.3, and Chap. 31). Transitions with $\Delta G = \pm n$ are much weaker than those with $\Delta G = 0$ and are not calculable from a simple formula. Branch intensities can be calculated with the Hönl-London formulas of Sect. 32.7.2.

32.7.6 Asymmetric Tops

In general, no assumptions can be made about the orientation of the transition moment. The vector representations (μ_x, μ_y, μ_z), (μ_a, μ_b, μ_c), and $(\mu_0, \mu_{+1}, \mu_{-1})$ can have any combination of independent nonzero values. It is common that one of the (μ_a, μ_b, μ_c) values is significantly larger than the others, especially for planar molecules with a two-fold symmetry axis. In this case one obtains a type A, B, or C band, if μ_a, μ_b, μ_c dominates, respectively [74, 33]. The tradition of analytic calculation of line strengths from explicit representations of wave functions and transition moments leads to formulas of considerable complexity, with somewhat restrictive assumptions [37, 38, 75]. In the more general notation of Sect. 32.7.2, the rotational line strength can be written as

$$S^{\tau'\tau''}_{J'J''} = \left| \sum_{K'K''} b^{*K'}_{J'\tau'} a^{K''}_{J''\tau''} (\mu_0 \delta_{K'K''} + \mu_+ \delta_{K'K''+1} \right.$$
$$\left. + \mu_- \delta_{K'K''-1}) \zeta(J', K', J'', K'') \right|^2, \qquad (32.85)$$

where

$$\mu_0 = \mu_c\,, \quad |\mu_+| = |\mu_-|\,,$$
$$|\mu_+|^2 + |\mu_-|^2 = \mu_a^2 + \mu_b^2\,. \qquad (32.86)$$

Papousek and Aliev [18] and Zare [5] follow the present formulation, but with somewhat less generality with respect to wave function expansion coefficients or transition moment components.

32.8 FORBIDDEN TRANSITIONS

32.8.1 Spin-Changing Transitions

The formalism presented above permits simulation of any allowed transition or forbidden transitions mediated by spin-orbit or spin-spin perturbations, or any perturbation that is diagonal in Ω. For spin-allowed transitions, the transition moment matrix is taken to be diagonal in and independent of the spin projection, so that

$$\mu_{\Lambda'\Sigma'\Lambda''\Sigma''} = \mu_{\Lambda'\Lambda''}\delta_{\Sigma'\Sigma''}\,. \qquad (32.87)$$

For forbidden transitions, or complicated Hund's case (c) mixings, the transition moment matrix elements can be considered as independent variable parameters, limited only by the symmetry constraint

$$\mu_{-\Omega'-\Omega''} = \eta(-1)^{\Omega'-\Omega''}\mu_{\Omega'\Omega''}\,, \qquad (32.88)$$

and the fact that terms with $|\Delta\Omega| > 1$ will be multiplied by zero. Alternatively, a specific set of candidate perturbers can be selected, and the Λ- and Σ-dependence of their contributions to the transition-moment matrix evaluated explicitly. For example, first-order spin-orbit mixing would lead to terms of the form

$$\mu^{S'S''}_{\Lambda'\Sigma'\Lambda''\Sigma''} = \sum_{\Lambda} \left(\mu^{S'S'}_{\Lambda'\Sigma'\Lambda\Sigma'} \frac{\langle S'\Lambda\Sigma'|H_{SO}|S''\Lambda''\Sigma''\rangle}{\Delta E''} \right.$$
$$\left. + \left(\frac{\langle S'\Lambda'\Sigma'|H_{SO}|S''\Lambda\Sigma''\rangle}{\Delta E'} \right) \mu^{S''S''}_{\Lambda\Sigma''\Lambda''\Sigma''} \right)\,. \qquad (32.89)$$

However, care must be taken in reducing these matrix elements using the Wigner-Eckart theorem, for example, following Lefebvre-Brion and Field [56], in order to satisfy the $\Delta\Omega = 0$ requirement for matrix elements of the rotationless Hamiltonian.

32.8.2 Orbitally-Forbidden Transitions

Even if the upper and lower states share the same value of electron spin, the transition may still be forbidden. The change in orbital angular momentum may be too large, $|\Delta\Lambda| > 1$, or a change in reflection parity, $\Sigma^- \rightarrow \Sigma^+$, may cause the zero-order transition matrix elements to vanish. Spin-orbit mixing with other $^{2S+1}\Lambda$ states, as described above, is usually the largest source of transition probability. In addition, terms in the Hamiltonian of the form **J·L** lead to contributions to the transition strength that increase with J, and that may mix-in higher values of Ω than were present in the zero-order $\Lambda\Sigma$ basis set for the upper and lower states. This situation can be represented by generalizing the formula from Sect. 32.7.2, following Huestis *et al.* [23],

$$S_{J'J''}^{c'c''} = \left| \sum_{K'K''} b_{J'c'}^{*K'} a_{J''c''}^{K''} \right.$$
$$\left. \times \sum_{i=-1}^{1} \mu_{K'K''}^{(i)} \zeta^{(i)}(J', K', J'', K'') \right|^2, \quad (32.90)$$

where $\mu_{K'K''}^{(0)}$ is the rotationless contribution ($\mu_{K'K''}$ from Sect. 32.7.2) and $\mu_{K'K''}^{(\pm 1)}$ are the new rotation-assisted terms. The new reflection-symmetry rule is

$$\mu_{-K'-K''}^{(i)} = \eta(-1)^{K'-K''+i} \mu_{K'K''}^{(-i)}. \quad (32.91)$$

The revised square-root Hönl-London factors are

$$\zeta^{(0)}(J', K', J'', K'') = \zeta(J', K', J'', K'') \quad (32.92)$$

(from Sect. 32.7.2) and

$$\zeta^{(\pm 1)}(J', K', J'', K'') = \tfrac{1}{2} \left([J'(J'+1) - K'(K' \mp 1)]^{\frac{1}{2}} \right.$$
$$\times \zeta(J', K' \mp 1, J'', K'')$$
$$+ [J''(J''+1) - K''(K'' \pm 1)]^{\frac{1}{2}}$$
$$\left. \times \zeta(J', K', J'', K'' \pm 1) \right).$$
$$(32.93)$$

As in Sect. 32.7.2 the symbols K' and K'' represent $\Lambda'\Sigma'$ and $\Lambda''\Sigma''$ when used as labels, and Ω' and Ω'' when used as numbers (a distinction that is relevant only when $S \geq |\Lambda|$ and $\Lambda \neq 0$). This formulation is more symmetric than that proposed by Huestis *et al.* [23], in that it explicitly allows for either the upper or lower state to be mixed by rotation (of significance only for low J and $\Delta\Lambda > 1$).

Acknowledgments

This work was supported by the NSF Atmospheric Chemistry Program, the NASA Stratospheric Chemistry Section, and the NASA Space Physics Division.

REFERENCES

1. H. C. Longuet-Higgins, Mol. Phys. **6**, 445 (1963).
2. P. R. Bunker, *Molecular Symmetry and Spectroscopy* (Academic Press, New York, 1979).
3. E. P. Wigner, *Group Theory* (Academic Press, New York, 1959).
4. M. Larsson, Phys. Scr. **23**, 835 (1981).
5. R. N. Zare, *Angular Momentum* (John Wiley & Sons, New York, 1988).
6. E. S. Chang and U. Fano, Phys. Rev. A **6**, 173 (1972).
7. J. M. Brown, B. J. Howard, and C. M. L. Kerr, J. Mol. Spectrosc. **60**, 433 (1976).
8. H. Helm, L. J. Lembo, P. C. Cosby, and D. L. Huestis, in *Fundamentals of Laser Interactions II*, edited by F. Ehlotzky (Springer Verlag, Berlin, 1989).
9. J. T. Hougen, in *The Calculation of Rotational Energies and Line Intensities in Diatomic Molecules*, NBS Monograph 115 (U. S. Government Printing Office, Washington DC, 1970).
10. J. M. Brown, E. A. Colbourn, J. K. G. Watson, and F. D. Wayne, J. Mol. Spectrosc. **74**, 294 (1979).
11. J. M. Brown and A. J. Merer, J. Mol. Spectrosc. **74**, 488 (1979).
12. J. M. Brown, D. J. Milton, J. K. G. Watson, R. N. Zare, D. L. Albritton, M. Horani, and J. Rostas, J. Mol. Spectrosc. **90**, 139 (1981).
13. R. N. Zare, A. L. Schmeltekopf, W. J. Harrop, and D. L. Albritton, J. Mol. Spectrosc. **46**, 37 (1973).
14. M. Mizushima, *The Theory of Rotating Diatomic Molecules* (John Wiley & Sons, New York, 1975).
15. G. Herzberg, *Molecular Spectra and Molecular Structure III. Electronic Spectra and Electronic Structure of Polyatomic Molecules* (Van Nostrand Reinhold, New York, 1966).
16. L. J. Lembo, H. Helm, and D. L. Huestis, J. Chem. Phys. **90**, 5299 (1989).
17. J. K. G. Watson, in *Vibrational Spectra and Structure*, edited by J. Durig (Elsevier, Amsterdam, 1977), Vol. 6.
18. D. Papousek and M. R. Aliev, *Molecular Vibrational-Rotational Spectra* (Elsevier, Amsterdam, 1982).
19. S. C. Wang, Phys. Rev. **34**, 243 (1929).
20. H. Helm, P. C. Cosby, R. P. Saxon, and D. L. Huestis, J. Chem. Phys. **76**, 2516 (1982).
21. C. H. Townes and A. L. Schawlow, *Microwave Spectroscopy* (McGraw Hill, New York, 1955) [reprinted, Dover, New York, 1975].
22. E. E. Whiting and R. W. Nicholls, Astrophys. J. Suppl. Series **235**, 27 (1974).
23. D. L. Huestis, R. A. Copeland, K. Knutsen, T. G. Slanger, R. T. Jongma, M. G. H. Boogaarts, and G. Meijer, Can. J. Phys. **72**, 1109 (1994).
24. J. Luque and D. R. Crosley, J. Quant. Spectrosc. Radiat. Transfer **53**, 189 (1995).
25. T. G. Slanger and D. L. Huestis, J. Chem. Phys. **78**, 2274 (1983).
26. C. M. L. Kerr and J. K. G. Watson, Can. J. Phys. **64**, 36 (1986).
27. M. J. Dyer, G. W. Faris, P. C. Cosby, D. L. Huestis, and T. G. Slanger, Chem. Phys. **171**, 237 (1993).
28. S. A. McFarlane and D. L. Huestis (to be published, 1995).
29. C. Di Lauro, F. Lattanzi, and G. Graner, Molec. Phys. **71**, 1285 (1990).
30. R. L. Kronig, *Band Spectra and Molecular Structure* (Cambridge University, London, 1930).
31. J. M. Brown, J. T. Hougen, K.-P. Huber, J. W. C. Johns, I. Kopp, H. Lefebvre-Brion, A. J. Merer, D. A. Ramsey, J. Rostas, and R. N. Zare, J. Mol. Spectrosc. **55**, 500 (1975).

32. E. B. Wilson, Jr., J. Chem. Phys. **3**, 276 (1935).
33. G. Herzberg, *Molecular Spectra and Molecular Structure II. Infrared and Raman Spectra of Polyatomic Molecules* (Van Nostrand Reinhold, New York, 1945).
34. E. B. Wilson, Jr., J. C. Decius, and P. C. Cross, *Molecular Vibrations* (McGraw Hill, New York, 1955).
35. J. T. Hougen, J. Chem. Phys. **37**, 1433 (1962).
36. H. C. Allen and P. C. Cross, *Molecular Vib-Rotors* (John Wiley & Sons, New York, 1963).
37. J. E. Wollrab, *Rotational Spectra and Molecular Structure* (Academic Press, New York, 1967).
38. H. W. Kroto, *Molecular Rotation Spectra* (John Wiley, London, 1975) [reprinted, Dover, New York, 1992].
39. F. A. Cotton, *Chemical Applications of Group Theory, 2nd Ed.* (John Wiley & Sons, New York, 1971).
40. A. Einstein, Physik. Z. **18**, 121 (1917).
41. E. U. Condon and G. H. Shortley, *The Theory of Atomic Spectra* (Cambridge University, London, 1935) [reprinted 1967].
42. S. S. Penner, *Quantitative Molecular Spectroscopy and Gas Emissivities* (Addison-Wesley, Reading, 1995).
43. A. P. Thorne, *Spectrophysics* (Chapman and Hall, London, 1974).
44. J. I. Steinfeld, *Molecules and Radiation* (MIT, Cambridge, 1978).
45. H. A. Bethe and E. E. Salpeter, *Quantum Mechanics of One- and Two-Electron Atoms* (Springer-Verlag, Berlin, 1957).
46. E. E. Whiting, A. Schadee, J. B. Tatum, J. T. Hougen, and R. W. Nicholls, J. Mol. Spectrosc. **80**, 249 (1980).
47. J. Franck, Trans. Faraday Soc. **21**, 536 (1925).
48. E. U. Condon, Phys. Rev. **32**, 858 (1928).
49. G. Herzberg, *Molecular Spectra and Molecular Structure I. Spectra of Diatomic Molecules* (Van Nostrand Reinhold, New York, 1950).
50. J. G. Winans and E. C. G. Stueckelberg, Proc. Nat. Acad. Sci. **14**, 867 (1928).
51. R. E. M. Hedges, D. L. Drummond, and A. Gallagher, Phys. Rev. A **6**, 1519 (1972).
52. J. C. McCallum, J. Quant. Spectrosc. Radiat. Transfer **21**, 563 (1979).
53. P. H. Dwivedi, D. Branch, J. H. Huffaker, and R. A. Bell, Astrophys. J. Suppl. **36**, 573 (1978).
54. G. W. King, *Spectroscopy and Molecular Structure* (Holt, Rinehart and Winston, New York, 1964).
55. A. B. Meinel, Astrophys. J. **111**, 555 (1950).
56. H. Lefebvre-Brion and R. W. Field, *Perturbations in the Spectra of Diatomic Molecules* (Academic Press, New York, 1986).
57. D. L. Huestis, DIATOM Spectral Simulation Computer Program, Version 7.0 (SRI International, Menlo Park, CA, 1994).
58. H. Hönl and F. London, Z. Phys. **33**, 803 (1925).
59. I. Kovacs, *Rotational Structure in the Spectra of Diatomic Molecules* (Am. Elsevier, New York, 1969).
60. A. Weber, in *The Raman Effect*, edited by A. Anderson (M. Dekker, New York, 1973), Vol. 2.
61. K. Chen and E. S. Yeung, J. Chem. Phys. **69**, 43 (1978).
62. J. B. Halpern, H. Zacharias, and R. Wallenstein, J. Mol. Spectrosc. 79, 1 (1980).
63. F. Hund, Z. Physik **36**, 657 (1926); **40**, 742 (1927); **42**, 93 (1927).
64. E. E. Nikitin and R. N. Zare, Mol. Phys. **82**, 85 (1994).
65. B. R. Judd, *Angular Momentum Theory for Diatomic Molecules* (Academic Press, New York, 1975).
66. A. R. Edmonds, *Angular Momentum in Quantum Mechanics, 2nd Ed.* (Princeton University, Princeton, 1960) [reprinted with corrections, 1974].
67. J. L. Femenias, Phys. Rev. A **15**, 1625 (1977).
68. I. Kopp and J. T. Hougen, Can. J. Phys. **45**, 2581 (1967).
69. A. P. Hickman, D. L. Huestis, and R. P. Saxon, J. Chem. Phys. **96**, 2099 (1992).
70. H. Helm, P. C. Cosby, and D. L. Huestis, J. Chem. Phys. **73**, 2629 (1980).
71. H. Helm, P. C. Cosby, and D. L. Huestis, Phys. Rev. A **30**, 851 (1984).
72. P. C. Cosby and D. L. Huestis, J. Chem. Phys. **97**, 6108 (1992).
73. P. K. Carroll, J. Chem. Phys. **58**, 3597 (1993).
74. H. H. Nielsen, Phys. Rev. **38**, 1432 (1931).
75. P. C. Cross, R. M. Hainer, and G. W. King, J. Chem. Phys. **12**, 210 (1944).

33

Molecular Photodissociation

Abigail J. Dobbyn, David H. Mordaunt, and Reinhard Schinke

Max-Planck-Institut für Strömungsforschung, 37073 Göttingen, Germany

33.1 INTRODUCTION

Molecular photodissociation is the photoinitiated fragmentation of a bound molecule [1]. The purpose of this chapter is to outline the ways in which molecular photodissociation is studied in the gas phase [2]. The results are particularly relevant to the investigation of the species involved in combustion and atmospheric reactions [3].

Conceptually the photodissociation process can be divided into three stages. During the first stage the molecule absorbs a photon and is promoted to an excited state. This is generally an excited electronic state, but can be a highly excited vibrational state in the ground electronic state. In the second stage, the transient complex evolves through a series of *transition states*, until finally, in the third stage, the molecule enters the *exit channel* and dissociates into the products. Schematically, this might be represented, for a triatomic molecule ABC, as

$$ABC + \hbar\omega \rightarrow (ABC)^{\ddagger} \rightarrow AB(v, j) + C. \quad (33.1)$$

In the case of the triatomic molecule represented here, the dissociation involves the transformation of one of the vibrational modes to a translational, or dissociative, mode, another vibrational mode (the bending) to rotational motion of the products (j), whilst the third vibrational mode is preserved (v).

When the molecule is promoted to an electronic state which has a purely repulsive potential energy surface (PES), it undergoes very rapid dissociation, often in less than one vibrational period. This is called *direct dissociation*. However, the dissociation of the transient complex can be delayed, taking place over many vibrational periods. This is called *indirect dissociation*, or predissociation, and has been divided into three different categories [4], though as with the division between direct and indirect, this is sometimes somewhat arbitrary.

Vibrational Predissociation (Herzberg type II). In this case, the transient complex is on a vibrationally adiabatic potential energy surface (this is an effective potential for the molecule when it is in a particular vibrational state v) which is not dissociative, or which has a barrier to dissociation. Therefore, to dissociate it

411

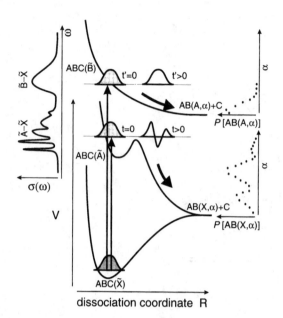

Figure 33.1. Schematic representation of the uv photodissociation of a triatomic molecule ABC into products AB(α) and C, illustrating the total absorption cross section $\sigma(\omega)$, evolution of the molecular wavepacket, and asymptotic product state distributions $P(\alpha)$, for direct and indirect dissociations on $\tilde{B}$ and $\tilde{A}$ state PESs respectively.

must either tunnel through the barrier, which is the only possibility for $v = 0$, or undergo a nonadiabatic transition to a lower vibrational state, thereby transferring energy from a vibrational degree of freedom to the dissociative mode. This process of energy exchange is commonly called the intramolecular redistribution of vibrational energy (IVR).

Rotational Predissociation (Herzberg type III). In this case, the transient complex is on a nondissociative rotationally adiabatic potential energy surface. Therefore, in a similar manner to vibrational predissociation, if it is to dissociate it must undergo a nonadiabatic transition to a lower rotational state, thereby transferring energy from rotation to the dissociative mode.

Electronic Predissociation (Herzberg type I). In this case, the PES of the electronic state of the transient complex is not dissociative at the given energy, and in order to dissociate the molecule must undergo a nonadiabatic transition to a second dissociative electronic state. This involves the coupling of nuclear and electronic motion and therefore leads to a break down of the Born-Oppenheimer (BO) approximation. There are two main types of electronic predissociation. In the first case, there is only a very small coupling, and no actual crossing, between two different electronic states, and the transition between the two is driven by the very high density of vibrational states on the second electronic state. This is called *internal conversion* for spin-allowed processes, and *intersystem crossing* for spin-forbidden processes. In the second case, the transition between the electronic states is driven by strong coupling. This coupling can be vibronic (vibrational-electronic) in nature, e.g., for the *Renner-Teller* and *Jahn-Teller* effects, or purely electronic, as in the case of a *conical intersection*.

Selection Rules. Two sets of selection rules apply to photodissociation. The first set governs the allowed states to which the molecule can be promoted by the photon. These selection rules are simply those for bound-state spectroscopy (see Sect. 32.4). Note in particular the selection rule $\Delta J = 0, \pm 1$. This has important practical implications since it means that a molecule which is initially rotationally cold remains so after absorption of a photon. Thus, those observables which are averaged over J will have a clear structure experimentally, and will be easier to calculate theoretically. This is in contrast to scattering experiments which in general involve a summation over many J states (see Chap. 35).

The second set of selection rules governs the dissociation process. The transient complex, or prepared (p) state, will undergo transitions to a final (f) vibrational, rotational or electronic state in order to dissociate; these transitions have their own set of selection rules. As for all selection rules, these are determined on the basis of symmetry. For a total wave function Ψ and a perturbation function W, which consists of the coupling terms or neglected terms in the Hamiltonian, $\int \Psi_p^* W \Psi_f \, d\tau$ must be nonzero for a transition to take place. As W forms a part of the Hamiltonian, it is totally symmetric, and therefore the integral is nonzero only if the prepared and final state irreducible representations are equal, $\Gamma_p = \Gamma_f$. If there is a transition to an excited electronic state, the point groups of the initial and final states are often not the same, in which case the point group formed by the joint elements of symmetry is used, or, in the case where there is no stable geometry for one of the states, the symmetry of the potential is used. For a diatomic molecule these selection rules are given in Table 47.1.

If the motion can be separated into vibrational, rotational, and electronic parts, so that $\Psi = \Psi^v \Psi^r \Psi^e$ and $W = W^v + W^r + W^e$, it is then possible to derive three separate selection rules: $\Gamma_p^r = \Gamma_f^r$ for the rotational motion, i.e. conservation of internal angular momentum; $\Gamma_p^e = \Gamma_f^e$ for the electronic motion; and $\Gamma_p^v = \Gamma_f^v$ for the vibrational motion. Since the final vibrational state is in the continuum, in practice all vibrational species (Γ_f^v) are available at a given energy, so that the vibrational selection rule is not significantly restrictive. This separation is not possible in the case of electronic predissociation occurring through the Renner-Teller or

Jahn-Teller effect, where it is necessary to consider the vibronic species of the initial and final states.

33.2 OBSERVABLES

Fundamental to any study of photodissociation is the measurement or calculation of the characteristic properties, or observables, of the reaction, from which the underlying dynamics of the fragmentation process can be inferred.

33.2.1 Scalar Properties

The *absorption cross section* $\sigma(\omega)$ is a measure of the capacity of the molecule to absorb photons with frequency ω. It is analogous to the line intensity in bound-state spectroscopy. Assuming that the light-matter interaction is weak (Chap. 66), and that the light pulse is on for a long time, the absorption cross section is given by

$$\sigma(\omega) \propto \omega_{fi} \left| \langle \Psi_f | \mathbf{E} \cdot \hat{\boldsymbol{\mu}} | \Psi_i \rangle \right|^2, \qquad (33.2)$$

where Ψ_i and Ψ_f represent the initial and final states, whose energies differ by $\hbar \omega_{fi}$; $\mathbf{E}$ is a unit vector in the direction of the polarization of the electric field, and $\hat{\boldsymbol{\mu}}$ is the electric dipole operator of the molecule. Assuming the Born-Oppenheimer separation of electronic and nuclear motion, Eq. (33.2) can be rewritten as

$$\sigma(\omega) \propto \omega_{fi} \left| \langle \Psi_f^{\mathrm{rv}} | \mu_{fi} | \Psi_i^{\mathrm{rv}} \rangle \right|^2, \qquad (33.3)$$

where the electronic transition dipole moment μ_{fi} equals $\langle \Psi_f^{\mathrm{e}} | \mathbf{E} \cdot \hat{\boldsymbol{\mu}} | \Psi_i^{\mathrm{e}} \rangle$, and is in general dependent on the internal coordinates of the molecule. The superscripts (r,v) will be dropped from now on, and Ψ will refer to the wave function for the internal coordinates of the molecule.

The absorption cross section reflects not only the nature of the transient complex but also its evolution through the transition states. For direct dissociation the absorption cross section is usually very broad and structureless. In contrast, the absorption cross section for predissociation is structured, containing lines which are normally Lorentzian in shape, and whose widths Γ are related to the lifetime of the transient complex at that energy by $\tau = \hbar / \Gamma$.

The *partial photodissociation cross sections* $\sigma(\omega, \alpha)$ are a measure of the capacity of the molecule to absorb photons with frequency ω *and* to yield products in quantum state α. They are defined by

$$\sigma(\omega, \alpha) \propto \omega_{fi} \left| \langle \Psi_f^{\alpha} | \mu_{fi} | \Psi_i \rangle \right|^2, \qquad (33.4)$$

where Ψ_f^{α} is the final wave function for the products in the quantum state α. The partial cross sections

for direct dissociation are broad and featureless. For predissociation, similar structures are seen in the partial cross sections as in the absorption cross section. The absorption or *total* cross section is given by the sum of the partial cross sections over all final product states:

$$\sigma(\omega) = \sum_{\alpha} \sigma(\omega, \alpha). \qquad (33.5)$$

The *rotational and vibrational product distributions* $P(\omega, \alpha)$ provide information about the amount of product formed by a photon with frequency ω in a particular rotational or vibrational state α. These are related to the partial cross sections by $P(\omega, \alpha) = \sigma(\omega, \alpha)/\sigma(\omega)$. The rotational and vibrational product distributions reflect the nature of the transient complex as it enters the exit channel, as well as the dynamics in the exit channel.

The *branching ratios* for different chemical species produced in photodissociation are defined as the fraction of the total number of parent molecules that produce the particular species of interest. In Eq. (33.1), the molecule ABC dissociated into AB + C. It might equally well have dissociated to A + BC, or indeed A + B + C. It is clear then that there may be several different reaction schemes, or channels, for the photodissociation of one particular molecule. Thus, the branching ratio for forming AB is the yield of this first channel divided by the total dissociation yield into all possible channels. Further, it would sometimes be possible to produce AB, or any of the other chemical species, in various electronic, vibrational or/and rotational quantum states. In this case, the branching ratio for forming AB(α) is the yield of AB in the specific quantum state α divided by the total yield of AB; however, this describes the branching into only this particular reaction channel.

For the reaction scheme represented in Eq. (33.1), the *quantum yields* of the products AB and C are the same. However, for example, in the reaction

$$\mathrm{ABC}_2 + \hbar\omega \rightarrow (\mathrm{ABC}_2)^{\ddagger} \rightarrow \mathrm{AB}(v, j) + 2\mathrm{C}, \qquad (33.6)$$

the quantum yield of C is twice that for AB. In general, the quantum yield of a particular product fragment for one reaction channel is the ratio of the number of fragments formed to the number of photons absorbed. However, it is again possible for a molecule to dissociate into various different reaction channels. In such a case, to obtain the *overall quantum yield* for a particular product, the quantum yield for each reaction channel must be summed over all the available reaction channels, taking into account the branching ratios for the channels.

33.2.2 Vector Correlations

Photodissociation is by its very nature an anisotropic process, as can be seen from Eq. (33.2). The operator

$\hat{\boldsymbol{\mu}}$ defines a specific axis in the *molecular body-fixed* frame of reference. At the instant of photoexcitation, $\hat{\boldsymbol{\mu}}$ is preferentially aligned parallel to the polarization of the electric field **E** in the *external laboratory space-fixed* frame of reference. Hence, **E** defines a specific axis, and thus cylindrical symmetry in the body-fixed frame. If fragmentation occurs on a time scale which is short compared with overall rotation of the excited complex, this correlation persists between the body-fixed frame and the space-fixed frame, and a wealth of information can be obtained. However, rotation of the transient complex prior to fragmentation serves to degrade this symmetry in the external body-fixed frame.

Three vectors fully describe the photodissociation process for both the parent molecule and the products: (i) $\hat{\boldsymbol{\mu}}$ in the body-fixed frame (and hence **E**, in the space-fixed frame, at the instant of photoexcitation); (ii) **v**, the recoil velocity of the products; and (iii) **j**, the rotational angular momenta of the fragments. Vector correlations can exist between all of these vectors [5].

The most commonly observed is the angular distribution of the photofragments $I(\theta, \alpha)$, i.e. the relation

$$I(\theta, \alpha) \propto \frac{1}{4\pi}\left[1 + \beta(\alpha)\mathrm{P}_2(\cos\theta)\right]. \qquad (33.7)$$

between **v** and **E**. $P_2(x)$ is the second-order Legendre polynomial and θ is the angle between **v** and **E**. The *anisotropy parameter* $\beta(\alpha)$ ranges between -1 for a perpendicular transition and $+2$ for a parallel transition. Thus, measuring the angular distribution of the fragments provides information about the type of electronic transition and hence the electronic symmetry of the excited state [6]. If the alignment between the body-fixed and space-fixed frames is destroyed, the angular distribution becomes *isotropic* and $\beta(\alpha) = 0$. The anisotropy parameter depends on the product channel α.

A second vector correlation concerns the direction of **j** with respect to **E**. Fragmentation generates rotational motion in the nuclear plane: for a perpendicular transition this is perpendicular to the plane containing the atoms, leading to the projection of **j** being preferentially aligned parallel to $\hat{\boldsymbol{\mu}}$, and thus **E** in the space-fixed frame. For a parallel transition, the opposite would be true, i.e. **j** would be aligned in the plane perpendicular to $\hat{\boldsymbol{\mu}}$. The alignment of **j** leads to polarized emission/absorption depending on whether molecules are created in an electronically excited/ground state. Therefore the orientation of the product polarization with respect to the original photolysis polarization **E**, also yields information about the symmetry of the electronic states involved in dissociation.

The final association in this series, is independent of the space-fixed frame, since **v** and **j** are both defined in the body-fixed frame. Unlike the two previous correlations, a long lifetime does not destroy the alignment, as it is not established until the bond breaks and the two fragments recoil. For a tetratomic (or larger) molecule

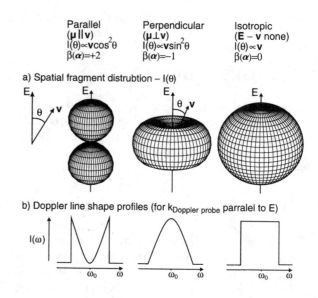

Figure 33.2. Spatial recoil anisotropies and Doppler line shape profiles, for parallel and perpendicular transitions compared with an isotropic distribution.

there are, in principle, two possible sources of product rotational excitation: bending motion in a plane of the molecule producing fragments with **v** perpendicular to **j**; or torsional motion leading to fragment rotation out of the plane. A prime example of this is the distinction between frisbee and propeller type motion of the two OH fragments in the dissociation of hydrogen peroxide [7]. Measurement of only the scalar properties *cannot* discriminate between these two possibilities, highlighting the additional information that can be gained about the bond rupture and the exit channel dynamics by the study of vector correlations.

33.3 EXPERIMENTAL TECHNIQUES

Early photochemical experiments used broad white-light continuum sources and large diffractometers [4]. However, it has been the development of lasers in combination with molecular-beam techniques that has dramatically increased the understanding of photodissociation processes. The ever increasing spectral and time resolutions of lasers, in addition to the power and range of wavelengths available, have made it possible to excite molecules selectively and with high efficiency. This has enabled state-specific preparation of the parent molecule, study of time evolution, as well as the measurement of the scalar and vector properties of the asymptotic products [8].

Specification of the Initial State. A room temperature sample of a gas will have a Boltzmann distribution over rotational states. Molecular beam techniques provide an improved specification of the initial angular momenta in the parent molecule by isentropically cooling its internal rotational energy [9]. Full quantum state specification can be achieved by various two-photon excitation schemes, e.g., stimulated emission pump (SEP) spectroscopy and vibrationally mediated dissociation. SEP is commonly used to study dissociation on the ground state PES; the molecules are excited to a stable upper electronic state, stimulated emission back to the ground state prepares a *single* quasibound state. Vibrationally mediated dissociation provides information about both ground and upper electronic states; the molecules are excited to a stable intermediate vibrational level on the ground state and further excitation promotes this fully defined wave function to an upper dissociative electronic state [10].

Detailed Measurement of the Absorption Cross Section. uv and vuv electronic spectroscopy has proven a very powerful tool for examining the interaction of a photon with a parent molecule (see Chap. 67). Absorption cross sections are typically measured by scanning the frequency domain and monitoring either the intensity of radiation absorbed or the flux of product molecules produced. State specific detection of the product flux yields the partial cross section $\sigma(\omega, \alpha)$. Explicit measurement of $\sigma(\omega)$ is a direct application of the Beer-Lambert law (see Sect. 67.2) and thus depends on the length of the optical cavity: cavity ring-down spectroscopy with multiple passes through a cell provides an effective cell length of several tens of kilometers.

Evolution of the Transient Complex. The evolution of the molecular wavepacket can be probed by time-resolved spectroscopy, as discussed in Chap. 34 and Ref. [11]. Real-time analysis of the molecular wavepacket provides a direct insight into the forces acting during molecular photodissociation. This type of time-resolved spectroscopy and the energy-resolved spectroscopy described above are mutually exclusive due to the time-energy uncertainty principle (see Sect. 77.3.1).

Asymptotic Properties. The vast majority of photodissociation studies determine asymptotic properties of the dissociation process, measuring either internal energy, recoil velocity, or angular distributions of the dissociation products. The product state distributions are usually explicitly probed by laser-induced fluorescence (LIF), resonance-enhanced multiphoton ionization (REMPI) spectroscopy, or coherent Raman scattering with the relative populations of the products obtained via line intensities.

The distribution and anisotropy of the recoil velocities are measured using Doppler spectroscopy or time-of-flight (TOF) techniques [12]. Doppler spectroscopy uses the Doppler-broadening of lines in the LIF or REMPI excitation spectra (see Sect. 67.6); the profile of the line shape also depends on the recoil anisotropy of the probed species (see Fig. 33.2). However, many important molecular fragments are not amenable to spectroscopic detection. Thus, though lacking the ultimate state-specificity of spectroscopy, TOF techniques by virtue of their general applicability provide an appealing alternative route to determine the product state distributions. In TOF techniques, the time is recorded for photofragments to recoil a known distance from the interaction region to a detector. Due to total energy and momentum conservation, the translational energy distribution of a fragment state specifically detected directly implies the internal energy distribution of the other partner product. In less favorable cases where this is not possible, *coincidence* detection schemes are employed to define the partition between translational and internal energies. Rotation of the detection axis with respect to the polarization of the photolysis laser yields the recoil anisotropy. Advances in spatially resolved detection schemes are now providing an improved measure of vector correlations.

Mass spectrometry can be used to measure branching ratios and quantum yields, which can also be obtained from the techniques described above.

33.4 THEORETICAL TECHNIQUES

The calculation of the observables of photodissociation can be carried out using quantum mechanics either in the time-independent or the time-dependent frame, as well as using classical mechanics [1]. Theoretical studies have contributed greatly to the understanding of photodissociation processes, as they provide the ability not only to calculate the observables, but also, through the knowledge of the wave function, to view directly the dissociation dynamics. This has enabled the inference of the underlying dynamics from the observables of the reaction to be more precisely established.

Due to computer limitations, the majority of quantum mechanical studies currently treat fully only three degrees of freedom, and thus have mainly concentrated on triatomic molecules. Jacobi coordinates are usually used, with the appropriate set for the dissociation of ABC into AB and C as follows: R, denoting the distance between the atom C and the centre of mass of the AB fragment; r, denoting the internal vibration coordinate of AB; and γ, denoting the bending angle between R and r.

The initial state Ψ_i is generally taken to be a single bound state. It is obtained either by the solution of the Schrödinger equation at a particular energy (see Sect. 30.1), or simply by taking a product of three Gaussians in the three coordinates, with the parameters of the Gaussians being determined from spectroscopic information on the ground state.

Further, to calculate the observables it is also necessary to have information about μ_{fi}. However, this is often assumed to be a constant, i.e. independent of the internal coordinates of the molecule. The *Franck-Condon principle* assumes that the nuclear geometry changes after the electronic transition, and not during it. Therefore a molecule, with a particular geometry, will, when promoted by the photon to the excited electronic state, be centred around the same geometry, which is thus referred to as the Franck-Condon region, or point.

To carry out any dynamical calculations it is necessary to have PESs for the electronic states involved. These are usually obtained from *ab initio* calculations which are described in Chap. 30. The accuracy of the PES surface largely determines the accuracy of the results obtained, as the PES essentially determines the dynamics of the fragmentation process.

In the *time-independent* approach, a solution of the time-independent Schrödinger equation

$$(\hat{\mathbf{H}} - E)\Psi^\alpha = 0, \qquad (33.8)$$

is sought for a specific total energy E subject to appropriate boundary conditions at infinite product separation. There are many different approaches to solving this problem, but they can be broadly separated into two groups: scattering methods and $\mathcal{L}^2$ methods. The scattering methods involve the solution of the coupled channel equations described in Chap. 35. These can be solved directly to yield the wave functions, which can then be used to calculate the observables, or can be solved indirectly to provide similar information. The use of $\mathcal{L}^2$ methods, which attempt to expand Ψ in a finite basis set, is not directly applicable since the wave functions are in the continuum and spread out to infinite distances in the R coordinate. Thus various modifications have been introduced in order to take this into account. The most important of these use variational principles [13], such as that due to Kohn [14]. In the Kohn variational principle the wave function in the *inner* or interaction region is expanded in a finite $\mathcal{L}^2$ basis. However, in the *outer* region, the wave function is expanded in an energy dependent basis of outgoing and incoming waves, which are approximate solutions of the coupled channel equations. Other methods which can sometimes be used to indirectly extract information about the observables are stabilization [15] and complex scaling [16].

In the *time-dependent* approach [17], one solves the time-dependent Schrödinger equation

$$i\hbar\frac{\partial}{\partial t}\Phi(t) = \hat{\mathbf{H}}\Phi(t) \qquad (33.9)$$

for the wavepacket $\Phi(t)$ with initial condition $\Phi(0) = \Psi_i$, i.e. it is assumed that the molecule is *vertically* promoted by an infinitely short pulse to the electronic state under consideration. The wavepacket is a coherent superposition of stationary wave functions Ψ^α (see Sect. 34.1), and since it comprises many of the stationary states, it contains all the information necessary to characterize the dissociation.

The total absorption cross section is given by

$$\sigma(\omega) \propto \int_{-\infty}^{+\infty} dt\, S(t)\, e^{-i\omega t}, \qquad (33.10)$$

where the autocorrelation function $S(t)$ is defined as

$$S(t) = \langle\Phi(0)\,|\,\Phi(t)\rangle. \qquad (33.11)$$

The autocorrelation function reflects the motion of the wavepacket, and therefore is a convenient means for visualizing the molecular dynamics. The individual partial cross sections can be obtained in the limit $t \to \infty$ by projection of the wavepacket onto the stationary wave functions of the products, i.e. plane waves in the dissociation coordinate and vibrational-rotational wave functions for the free products.

33.5 CONCEPTS IN DISSOCIATION

There has been substantial experimental and theoretical work to elucidate the processes involved in photodissociation from knowledge of the observables, and much progress has been made. In this section, an attempt is made to present some of the simpler ideas which have emerged [1].

33.5.1 Direct Dissociation

Direct dissociation is the very fast rupture of a bond after a molecule has been promoted to an electronic state which has a purely repulsive PES. A very clear picture of this process can be obtained from wavepacket calculations: the wavepacket which is placed on the repulsive surface moves directly down the PES and into the exit channel. The autocorrelation function decays from one to zero in a short time and does not show any recurrences, i.e. oscillations in the autocorrelation function. The absorption cross section $\sigma(\omega)$, which is the Fourier transform of the autocorrelation function [Eq. (33.10)], is therefore a very broad Gaussian with no structure. The breadth of $\sigma(\omega)$ is inversely proportional to the width of the autocorrelation function and can, using simple classical pictures, be taken to be approximately proportional to the steepness of the potential at the Franck-Condon point. The partial cross sections have a similar structure to the total cross section, though they have differing intensities and are shifted relative to each other on the energy scale.

The product distributions can be predicted using simple classical pictures. These methods can be divided into two groups, depending on the extent of the excitation/de-excitation, or coupling, in the exit channel. If there is very little excitation/de-excitation in the exit channel, the rotational and vibrational product distributions are best described using Franck-Condon mapping. Another

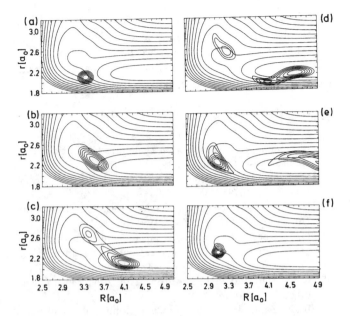

Figure 33.3. Time evolution of a wavepacket in the dissociation of FNO in the S_1 state.

model which gives good results for the rotational distributions is the impact parameter, or impulsive, model. If excitation/de-excitation in the exit channel is not negligible, the product distributions are best described using the *reflection principle*. This relates the distribution of the initial wavepacket in γ to the final rotational distributions through the classical excitation function. Similarly, the distribution of the initial wavepacket in space is related to the final vibrational distributions through another classical excitation function.

Even for a purely repulsive PES, the potential may be very flat in the Franck-Condon region so that the molecule may be able to undergo one internal vibration before it dissociates. In this case, there is a diffuse structure in $\sigma(\omega)$, associated with recurrences in the autocorrelation function. The spacing of the structures in $\sigma(\omega)$ are related to the period of the internal vibrations by $\Delta E = 2\pi\hbar/T$. Diffuse structures in $\sigma(\omega)$ have also been linked to unstable periodic orbits.

33.5.2 Vibrational Predissociation

Vibrational predissociation is dissociation delayed due to the trapping of the energy of the molecule in modes orthogonal to the dissociation coordinate. It can be explained very clearly in the time-independent picture as resonances (sometimes known as Feshbach resonances), which are simply extensions of the bound states into the continuum. $\sigma(\omega)$ in this case consists of a series of Lorentzian lines, whose width is inversely proportional to the lifetime of the resonance. In the case where the internal modes of the molecule are not strongly

coupled to each other or to the dissociation mode, these resonance states can often be assigned, with the number of quanta in each mode being specified. In this case, the lines in $\sigma(\omega)$ form a series of progressions. The widths of these lines, and thus the lifetimes of the resonance states, often show trends relating the lifetimes to the assignment. This is called *mode-specificity*. In the case that the system is strongly mixed, it is not possible to make an assignment of the resonances, and the lifetimes show strong fluctuations. This is called *statistical state-specificity* [18].

The resonances can also be seen in the time-dependent picture, where the autocorrelation show many recurrences with periods T, depending on the fundamental frequencies of the internal modes $\omega = 2\pi/T$ (see Sect. 34.1).

The partial cross sections for vibrational predissociation also consist of Lorentzian lines, with positions and widths exactly as for the total cross sections, but with differing intensities. The partial widths, which describe the rate of dissociation into each product channel, are given by

$$\Gamma_\alpha = \Gamma \frac{\sigma(\omega, \alpha)}{\sigma_{\text{tot}}(\omega)}. \qquad (33.12)$$

In the weak coupling case, simple pictures can be used to describe the product distributions. The rotational product distributions can be explained using again the reflection principle; but in this case, instead of considering the distribution of the initial wave function in γ, the distribution of the wave function at the transition state is used. The vibrational product distributions can often be well described by examining vibrationally adiabatic curves.

In the case that the modes are strongly coupled, the simple models break down. It is then sometimes possible to use statistical models to describe both the rates and the product distributions. One example of these unimolecular-statistical theories is the Rampsberger-Rice-Karplus-Marcus (RRKM) theory, which is widely used for the description of unimolecular dissociations [19]. Another example is phase space theory (PST) [20] which is often used to calculate the product distributions for reactions which have no barrier. The quantum mechanical results fluctuate about these average values. These fluctuations, which can be considered as being independent of the system, can be described well by the predications of random matrix theory [21].

33.5.3 Electronic Predissociation

Nonadiabatic transitions between two or more electronic states are a common phenomenon in photodissociation [22] as well as in other chemical reactions (see

Chap. 47). Such transitions can result in the production of both electronically excited and ground state fragments.

Adiabatic molecular PESs can vary in complex fashions. Many of these contortions arise from avoided and real crossings of the surfaces, and in all such cases, the physical and chemical understanding is greatly facilitated by expressing the adiabats in terms of the diabatic states (see Chap. 30). The electronic diabatic states are chosen to simplify the structure of the electronic wave functions by incorporating the off-diagonal or coupling elements as a pure potential energy term, rather than as a kinetic term, or a mixture of both.

Under these conditions, the BO approximation is inadequate since there is coupling between the different adiabatic states, and the electronic and nuclear motion cannot be separated. Therefore the solution of the time-dependent Schrödinger equation (see Sect. 33.4) requires the set of coupled equations

$$i\hbar \frac{\partial}{\partial t} \begin{pmatrix} \Psi_1 \\ \Psi_2 \end{pmatrix} = \begin{pmatrix} V_{11} + T_{11} & V_{12} \\ V_{12} & V_{22} + T_{22} \end{pmatrix} \begin{pmatrix} \Psi_1 \\ \Psi_2 \end{pmatrix},$$

(33.13)

to be propagated, where the coupling between the diabatic surfaces is in the potential (V) and not the kinetic (T) terms. The wavepacket evolves on both diabatic (or adiabatic) surfaces, and shows a complicated motion in moving between the two surfaces. The coupling between the nuclear and electronic motion can be thought of as resulting in the nuclear motion forcing the transfer of a valence electron to another molecular orbital. Since the efficiency of this transfer is greatest when the orbitals are degenerate, the crossings of the wavepacket between the PESs are generally localized around their degeneracies. Finally, the wavepacket moves out on the adiabatic surfaces towards the products, with which they are correlated.

33.6 SUMMARY

Photodissociation of polyatomic molecules is an ideal field for studying the details of molecular dynamics. The primary goal of the experimental and theoretical approaches (see Sects. 33.3, 33.4) is to understand the connection between the observables (see Sect. 33.2) and the underlying chemical dynamics (see Sect. 33.5). Once this connection has been established, it is possible to have a detailed understanding of the dissociation

dynamics, transition states geometries, and the PESs which ultimately govern the molecular chemical reactivity. The interplay between powerful experimental and theoretical techniques has enabled this goal to be realized for many photodissociation reactions.

REFERENCES

1. R. Schinke, *Photodissociation Dynamics* (Cambridge University Press, Cambridge, 1993).
2. H. Okabe, *Photochemistry of Small Molecules* (Wiley, New York, 1978).
3. R. P. Wayne, *Chemistry of Atmospheres*, 2nd ed., (Oxford University Press, New York, 1991).
4. G. Herzberg, *Molecular Spectra and Molecular Structure III, Electronic Spectra and Electronic Structure of Polyatomic Molecules* (D. Van Nostrand, New York, 1967).
5. G. E. Hall and P. L. Houston, Ann. Rev. Phys. Chem. **40**, 375 (1989).
6. R. N. Zare, *Angular Momentum* (John Wiley and Sons, New York, 1988).
7. A. U. Grunewald, K.-H. Gericke, and F. J. Comes, J. Chem. Phys. **87**, 5709 (1987); **89**, 345 (1988).
8. *Molecular Photodissociation Dynamics*, edited by M. N. R. Ashfold and J. E. Baggott (Royal Society of Chemistry, London, 1987).
9. *Atomic and Molecular Beams Methods*, edited by G. Scoles (Oxford University Press, New York, 1988).
10. F. F. Crim, Ann. Rev. Phys. Chem. **44**, 397 (1993).
11. *Femtosecond Chemistry*, edited by J. Manz and L. Wöste (VCH, Weinheim, 1995).
12. M. N. R. Ashfold, I. R. Lambert, D. H. Mordaunt, G. P. Morley, and C. M. Western, J. Phys. Chem. **96**, 2938 (1992).
13. R. K. Nesbet, *Variational Methods in Electron-Atom Scattering Theory* (Plenum, New York, 1980).
14. J. Z. H. Zhang and W. H. Miller, J. Chem. Phys. **92**, 1811 (1990).
15. R. Lefebvre, J. Phys. Chem. **89**, 4201 (1985).
16. W. P. Reinhardt, Ann. Rev. Phys. Chem. **33**, 223 (1982).
17. R. Kosloff, J. Phys. Chem. **92**, 2087 (1988).
18. W. L. Hase, S. Cho, D. Lu, and K. N. Swamy, Chem. Phys. **139**, 1 (1989).
19. P. J. Robinson and K. A. Holbrook, *Unimolecular Reactions* (Wiley-Interscience, London, 1972).
20. P. Pechukas and J. C. Light, J. Chem. Phys. **42**, 3285 (1965); P. Pechukas, J. C. Light, and C. Rankin, *ibid.* **44**, 794 (1966).
21. T. A. Brody, J. Flores, J. B. French, P. A. Mello, A. Pandey, and S. S. M. Wong, Rev. Mod. Phys. **53**, 385 (1981).
22. R. N. Dixon, Chem. Soc. Rev. **23**, 375 (1994).

34

Time-Resolved Molecular Dynamics

V. Engel

Institut für Physikalische Chemie, Universität Würzburg, Marcusstraße 9-11, 97070 Würzburg, Germany

34.1 INTRODUCTION

Time-resolved experiments have been performed on a diversity of molecular systems. Applications of spectroscopic techniques which work in the time-domain range from the detection of simple vibrational motion of a diatomic molecule to the direct determination of relaxation times in polyatomic molecules in a liquid environment, or the recording of isomerization processes in biomolecules. Although very different molecular systems have been investigated with various experimental techniques, the principles of the experiments are more or less the same. A brief description of the basic ideas is given, with the emphasis on gas phase molecules under collision-free conditions as usually provided in a molecular beam.

For the development of ultrashort pulse techniques (see Chap. 68) and their application to areas as different as optical engineering, solid state physics or biology, see the series of conference proceedings, Refs. [1] and [2]. For special applications to chemistry, consult the books [3, 4] and the special issue of the *Journal of Physical Chemistry* [5] on femtochemistry.

Besides an overall rotation and the translational motion of the molecular c.m., the nuclei within a molecule possess vibrational degrees of freedom (see Chap. 30). The detection of molecular motion in real time with modern spectroscopic tools is the issue of the present chapter. The internal vibrational dynamics is discussed, but the considerations apply equally well to the case of a fragmentation process (see Chap. 72) where the dissociation dynamics of the fragments is studied.

Classically, a molecule can be imagined to consist of atoms connected by springs, each spring representing a chemical bond. The atoms vibrate, performing a periodic oscillation around an equilibrium position. Quantum mechanically, the molecule has discrete vibrational states φ_n, where n represents a set of vibrational quantum numbers and ϵ_n are the corresponding eigenenergies. A vibrational period T_{vib} can be obtained from the energy separation of the energies ϵ_n:

$$T_{\text{vib}} = \frac{2\pi\hbar}{\epsilon_{n+1} - \epsilon_n}. \qquad (34.1)$$

Typical vibrational periods for smaller molecules are of the order of several hundred femtoseconds.

Traditional high resolution spectroscopy uses laser pulses with a temporal width T_P much larger than any of the time scales on which the internal molecular motion takes place:

$$T_P \gg T_{\text{vib}}. \qquad (34.2)$$

In the energy domain, this means that the spectral width of the laser pulse is small enough to excite a single eigenstate φ_n. The time-evolution of this state is that of a stationary state:

$$\psi(t) = a_n e^{-i\epsilon_n t/\hbar} \varphi_n, \qquad (34.3)$$

where a_n is a complex number.

To detect molecular motion within a time-resolved measurement, pulses with a width T_P smaller than the vibrational period have to be used:

$$T_P < T_{vib}. \qquad (34.4)$$

Thus, to resolve vibrational dynamics of smaller molecules, one needs ultrashort pulses in the femtosecond regime. Since a short pulse has a broad spectral distribution, it is possible to excite several eigenstates simultaneously: a coherent superposition of the functions φ_n is prepared. The resulting wave function is a vibrational wave packet:

$$\psi(t) = \sum_n a_n e^{-i\epsilon_n t/\hbar} \varphi_n. \qquad (34.5)$$

The packet moves spatially and its motion is determined by the topology of the potential energy surface of the electronic state in which the molecule is prepared. The principle of time-resolved molecular dynamics is to create such wave packets within a molecule and follow their motion in time.

Pulses in the femtosecond regime are available (see Chap. 68). Schemes have been developed to use these pulses to detect wave-packet dynamics in real time. This is of course not limited to internal vibrational dynamics but also to fragmentation processes. In this case the sum in Eq. (34.5) has to be replaced by an integral over continuum functions. The condition (34.4) then reads $T_P < T_F$, where T_F is the time during which the atoms separate and do not interact with each other afterwards.

Typical experimental schemes are described in Sect. 34.2, and a theoretical treatment which closely follows the experimental procedure is outlined in Sect. 34.3. Applications are listed in Sect. 34.4, and Sect. 34.5 contains a summary.

34.2   PUMP/PROBE EXPERIMENTS

The experimental setup for a time-resolved measurement commonly uses two ultra-short laser pulses which are delayed with respect to each other. A single pulse is produced and split. The resulting two pulses are delayed by sending them along different paths until they reach the molecular sample. The temporal difference can be adjusted by variation of the pathlength. The first pulse (pump) excites the molecule and prepares a wavepacket, as in Eq. (34.5). The second pulse (probe) interacts with the molecular sample after a defined delay-time and prepares the system in yet another state. A signal is then measured as a function of the delay-time between the pulses. The idea is that, since the wave packet is located in different spatial regions at different times, the signal in general depends on the delay time.

There are various detection schemes. If the probe pulse prepares the molecule in an energetically higher electronic state, the fluorescence to a lower electronic state can be measured. The pump/probe signal consists of the total fluorescence signal, recorded as a function of the delay time between the pulses. Since this signal is proportional to the population created in the state from which the radiative decay takes place, the signal reflects how effective the probe-pulse excitation occurs which, in turn, depends on the wave packet dynamics in the intermediate state accessed by the pump-pulse excitation. Equivalently one may use the probe-pulse to induce a stimulated (direct or Raman) emission process and measure the emission signal. Also, it is possible to detect the transient absorption spectrum of the system which is subject to the probe-pulse excitation. Another possibility is that the parameters of the probe-pulse are chosen such that the molecule is ionized. Then, an ion signal can be detected as a function of the pulse delay. This pump/probe ionization spectroscopy often involves multiphoton transitions induced by one or both of the pulses. An interferometric method, where a molecule is exposed to two identical and time-delayed pulses has been used [6]. The experiment includes the measurement of an interference population resulting from photon absorption from two identical pulses.

34.3   THEORETICAL DESCRIPTION

There are several theoretical approaches to describe the experiments which have been outlined in Sect. 34.2. They are based on classical, quantum mechanical or semiclassical descriptions [3, 7]. In more complex situations such as real-time measurements in a liquid enviroment, a quantum mechanical treatment based on density matrices (see Chap. 7) is of advantage.

The direct quantum mechanical approach is to solve the time-dependent Schrödinger equation. Suppose that three electronic states $|i\rangle$ $(i=1,2,3)$ participate in the excitation process. Then the nuclear wave function ψ has three components ψ_i. The equation of motion is

$$\begin{pmatrix} H_1 & W_{12} & 0 \\ W_{12} & H_2 & W_{23} \\ 0 & W_{23} & H_3 \end{pmatrix} \psi(r,t) = i\hbar \frac{\partial}{\partial t} \psi(r,t), \qquad (34.6)$$

with the field-matter interaction

$$W_{ji}(r,t) = E_0\mu_{ji}(r)\{f(t-T_1)\cos[\omega_1(t-T_1)]$$
$$+ f(t-T_2)\cos[\omega_2(t-T_2)]\}. \quad (34.7)$$

Here H_i denotes the nuclear Hamiltonian within the electronic state $|i\rangle$. The components $\psi_i(r,t)$ are coupled by the dipole interaction terms $W_{ji}(r,t)$ (see Chap. 66) which contain the projection of the (classical) electric field vector of magnitude E_0 on the electric transition dipole moment $\mu_{ji}(r)$ between the states $|j\rangle$ and $|i\rangle$. $f(t)$ describes the envelope function of the laser pulses and ω_1, ω_2 are the frequencies of pump- and probe-pulse. The pulses interact with the molecule around the times T_1, T_2 so that the time-delay is $T = T_2 - T_1$. The nuclear coordinates are collectively denoted by r.

Since in most cases the experimental signal is proportional to the population in the state prepared by the probe-pulse, it is sufficient to calculate the wave function $\psi_3(r,t)$. In principle it is possible to integrate equation (34.6) numerically with a given initial condition. This indeed is necessary if the pulses used are of high intensity. In most cases however, when the interest is in the molecular dynamics rather than in nonlinear effects induced by high power laser pulses, a perturbative approach 5.3 is of conceptual and technical advantage.

Except in the case where the two pulses have temporal overlap, the pump and probe processes may be separated. In first-order perturbation theory, the wave function created by the pump pulse in state $|2\rangle$ is

$$\psi_2(r,t) = \frac{1}{i\hbar}\int_0^t dt' U_2(t-t')W_{21}(r,t')U_1(t')\varphi_{1,n}(r),$$

$$(34.8)$$

where $\varphi_{1,n}$ is the initial (stationary) vibrational state of the molecule with the vibrational quantum numbers n and eigenenergy ϵ_n. U_i is the propagator in the respective electronic state $|i\rangle$.

If only discrete states are excited, one may expand ψ_2 in the set of vibrational functions $\{\varphi_{2,m}\}$ of the electronic state $|2\rangle$:

$$\psi_2(r,t) = \sum_m e^{-iE_m t/\hbar}\varphi_{2,m}(r)c_{mn}I_{mn}(\omega_1,t), \quad (34.9)$$

with the Franck-Condon factors (see Chap. 32)

$$c_{mn} = \int_0^\infty dr\varphi_{2,m}\mu_{21}(r)\varphi_{1,n}, \quad (34.10)$$

and the time-integrals

$$I_{mn}(\omega_1,t) = \int_0^t dt' f(t')e^{i(E_m-\epsilon_n-\hbar\omega_1)t'/\hbar}. \quad (34.11)$$

Here E_m denotes the vibrational energy corresponding to the vibrational state $\varphi_{2,m}$. From Eqs. (34.9)–(34.11) it is

clear that ψ_2 is a vibrational wave packet. The weights of the states which build the packet are determined by the Franck-Condon factors and the Fourier-transform of the pulse envelope $f(t)$ taken with respect to the energy $E_m - \epsilon_n - \hbar\omega$. In the limit of an infinite pulse, I_{mn} becomes proportional to a δ-function and, for resonance excitation, only one vibrational state is excited. Once the pump pulse no longer interacts with the system, the packet propagates unperturbed. This motion is detected in a time-resolved experiment by exposing the molecule to a probe pulse at time T. As in Eq. (34.9), the wave function in state $|3\rangle$ can be written as

$$\psi_3(r,t) = \frac{1}{i\hbar}\int_T^t dt' U_3(t-(t'-T))W_{32}(r,t')$$
$$\times U_2(t'-T)\psi_2(r,T). \quad (34.12)$$

However, there is an essential difference between the wave functions ψ_2 and ψ_3. The wave packet created by the pump-pulse results from a stationary initial state, whereas ψ_3, prepared by the probe-pulse, contains the wave packet ψ_2 as initial state. The packet changes its position and is centered around different distances r for different delay times T.

Since, during the short time the probe-pulse interacts with the molecule, the heavy nuclei do not move, one may neglect the kinetic energy part of the propagators U_i in Eq. (34.12), which yields the approximate wave function

$$\psi_3(r,t) \sim \mu_{32}(r)\psi_2(r,T)\int_T^t dt' f(t')$$
$$\times \exp[i(V_3(r)-V_2(r)-\hbar\omega_2)t'/\hbar], \quad (34.13)$$

where V_3, V_2 are the potentials in the respective electronic state. The time integral has its largest amplitude at the point $r = r_M$, satisfying the equation

$$V_3(r) - V_2(r) = \hbar\omega_2. \quad (34.14)$$

Assuming that the wave packet is localized, this implies that the norm of ψ_3, and thus the population in state $|3\rangle$, is maximal at those times T when the center of the moving wave packet ψ_2 is located around the position r_M, for which the difference potential $V_3(r)-V_2(r)$ equals the laser energy $\hbar\omega_2$. This consideration shows the central idea of a pump/probe scheme; for a fixed frequency ω_2, it is possible to detect the wave packet each time it reaches the area around the position r_M since then the pump/probe signal exhibits a maximum. If ω_2 is changed, the packet is detected when it visits a different position r_M.

34.4   APPLICATIONS

Numerous studies of molecular dynamics in the time-domain can be found in Ref. [1]. Selected examples are listed below to show possible and actual applications of pump/probe spectroscopy. The emphasis is on wave packet motion rather than on the determination of relaxation times or the time-scales for energy transfer between internal modes of a molecule or between a molecule and its surroundings.

34.4.1 Internal Vibrational Dynamics of Small Molecules in the Gas Phase

Prototype experiments investigated the simplest case of a molecular motion: the vibration of a diatomic molecule [3,8]. The change of the bond length within an electronically excited state can be detected in real-time for molecules such as I_2, Na_2, Bi_2, Cs_2. Not only the vibrational periods but also the Born-Oppenheimer bound-state potentials (see Chap. 30) can be constructed from the data [8]. Besides this information, the competition of multiphoton ionization, autoionization (Chap. 72) and fragmentation can be investigated with time-resolved measurements [9].

Gas phase vibrational dynamics in three dimensions was investigated in the Na_3 molecule [9]. A wave packet moving on a three dimensional potential energy surface was prepared. From the pump/probe signal it was possible to extract vibrational frequencies corresponding to different vibrational modes of the molecule.

34.4.2 Elementary Chemical Reactions

The study of gas phase chemical reactions in real time has been pioneered by Zewail and coworkers [8]. It is of primary interest to chemistry to monitor how chemical bonds are formed and broken.

In the case of a direct bond rupture, the time scale to be resolved is much shorter than in cases where long lived resonances exist. In a first femtosecond experiment, the ICN molecule was prepared in an electronically excited state with a repulsive potential energy surface. Free CN was detected in a pump/probe arrangement [10]. By changing the probe-laser frequency the atom-molecule separation could be recorded in time and for different internuclear distances r_M.

The decay of a quasibound complex via electronic predissociation was monitored. From the time signal, the nonadiabatic coupling between a bound and a dissociative electronic state could be extracted for the NaI molecule [11]. Rotational/vibrational predissociation of van der Waals complexes has been investigated [12]. In this case the complex is prepared in an unstable state and the coupling of internal degrees of freedom to a dissociative reaction coordinate yields a finite lifetime. Another type of indirect decay was studied in the OClO molecule

which decays on different time scales [13]. These time scales are determined by the coupling of internal nuclear degrees of freedom on the one hand, and the coupling between electronic and nuclear motion on the other. The *transition state*, which separates chemical products from educts, could be observed by photochemically preparing a wave packet consisting of states with energies close to the reaction threshold and detecting the separation from the transition region.

34.4.3 Molecular Dynamics in Liquids

Besides the gas-phase studies, one main application of time-resolved measurements is the study of molecular dynamics in a liquid surrounding. Experiments have been carried out to determine the transfer of vibrational energy in liquids [2]. It is now possible to determine directly the time scales for these processes.

Another important phenomenon is that of the *caging* of a molecule in a liquid. This effect describes the fragmentation and recombination of a molecule, the latter being induced by the cage consisting of the surrounding molecules. This effect can be detected within a time-resolved measurement [8]. Coherent vibrational motion of the photoproduct I_2^- originating from photodissociation of I_3^- was detected [14]. Also, the experiments showed that the coherence of the internal vibrational dynamics of the mother molecule persists several picoseconds in an alcohol solution.

34.4.4 Cluster Dynamics

Clusters (see Chap. 38) are normally detected by mass spectrometry, which requires ionization of the cluster to be detected. The use of nanosecond pulses carries an inherent problem: the detection laser not only ionizes the parent cluster, but also induces fragmentation of neutral clusters with different masses. The fragments are ionized as well so that it is not possible to distinguish fragment ions from the parent ions. This is different if short pulses are used; if the interaction time is on the femtosecond scale, the neutral clusters do not fragment since the nuclei are basically frozen during the ionization process. This made it possible to obtain new results on the ionization of larger clusters. For metal clusters (Na_n), it could be shown that states of different energy and lifetime exist in intermediate electronic states of the neutral molecule [3].

34.5 SUMMARY

The detection of molecular motion, e.g. temporal changes in bond lengths or bond angles in real time, is possible with pump/probe experiments using ultrashort laser pulses. A first pulse prepares a linear combination of molecular eigenstates which is a moving wave packet. The internal motion or the fragmentation dynamics is detected by excitation of the molecule with a second, time-delayed pulse. The wave packet serves as initial state for the second excitation process. It is localized in different areas of space at different times, which leads to variations in the experimental signal as a function of the delay time. Thus the signal directly reflects the molecular motion. For different probe laser frequencies ω_2, the wave packet dynamics is probed at different bondlengths.

Many molecular systems have been investigated by now. Using time-resolved experiments, it is possible to obtain information on vibrational/rotational periods, potential energy surfaces, the course of a chemical reaction, nonadiabatic couplings between internal nuclear or nuclear and electronic degrees of freedom, relaxation and recombination times, and competing processes in the decay of neutral and ionic states of a molecule.

REFERENCES

1. *Picosend Phenomena, Ultrafast Phenomena I-VIII* (Springer, Heidelberg, 1978-1993), Nos. 4, 14, 23, 38, 46, 48, 53, 55.

2. *Ultrashort Laser Pulses*, edited by W. Kaiser (Springer, Heidelberg, 1993).

3. *Femtosecond Chemistry*, edited by L. Wöste and J. Manz (VCH, Heidelberg, 1995).

4. A. H. Zewail, *Femtochemistry* (World Scientific, 1994), Vols. 1 & 2.

5. *Femtochemistry*, special issue of J. Phys. Chem. **97** (1993).

6. N. F. Scherer, R. J. Carlson, A. Matro, M. Du, A. J. Ruggiero, V. Romero-Rochin, J. A. Cina, G. R. Fleming, and S. A. Rice, J. Chem. Phys. **95**, 1487 (1991).

7. S. Mukamel, Ann. Rev. Phys. Chem. **41**, 7806 (1991); S. H. Lin, B. Fain, and N. Hamer, Adv. Chem. Phys. **79**, 133 (1990).

8. A. H. Zewail in Ref. [5]

9. Th. Baumert and G. Gerber in Ref. [3].

10. M. Dantus, M. J. Rosker, and A. H. Zewail, J. Chem. Phys. **87**, 2395 (1987); **89**, 6128 (1988).

11. T. S. Rose, M. J. Rosker, and A. H. Zewail, J. Chem. Phys. **88**, 6672 (1987); **97**, 7415 (1989).

12. M. Gutmann, D. M. Willberg, and A. H. Zewail, J. Chem. Phys. **97**, 8048 (1992).

13. T. Baumert, J. L. Herek, and A. H. Zewail, J. Chem. Phys. **99**, 4430 (1993).

14. U. Banin, A. Waldman, and S. Ruhman, in Ref. [1], No. 55 (1993).

35

Nonreactive Scattering

D. R. Flower

Department of Physics, University of Durham, Durham DH1 3LE, United Kingdom

35.1 DEFINITIONS

The cross section σ for a transition from state j' to j is defined classically as

$$\sigma(j \leftarrow j') = 2\pi \int_0^\infty P_b(j \leftarrow j')b\,db, \qquad (35.1)$$

where P_b is the transition probability for impact parameter b (see Fig. 35.1). The impact parameter is related to the relative angular momentum quantum number, ℓ, by

$$2\mu Eb^2 = \ell(\ell+1)$$
$$= k^2 b^2, \qquad (35.2)$$

where k is the wave number at relative collision energy E and μ is the reduced mass; atomic units ($e = m_e = \hbar = 1$) are used throughout. Differentiating (35.2) and setting $d\ell = 1$ in the quantal limit,

$$b\,db = \frac{2\ell+1}{2k^2},$$

whence the quantum mechanical equivalent of (35.1) may be obtained

$$\sigma(j \leftarrow j') = \frac{\pi}{k_{j'}^2} \sum_\ell (2\ell+1)P_\ell(j \leftarrow j'), \qquad (35.3)$$

$k_{j'}$ being the wave number in the initial channel. If the initial state is degenerate and the $\omega_{j'}$ degenerate substates are labeled by Ω', then

$$P_\ell(j \leftarrow j') = \frac{1}{\omega_{j'}} \sum_{\Omega',\Omega} |T_\ell(j\Omega, j'\Omega')|^2, \qquad (35.4)$$

where T_ℓ is an element of the transmission matrix $\mathbf{T}$, which contains all the information on the scattering event. The scattering matrix $\mathbf{S}$ is related to the $\mathbf{T}$ matrix by

$$\mathbf{S} = \mathbf{1} - \mathbf{T}, \qquad (35.5)$$

and thence to the reactance matrix $\mathbf{K}$ through

$$\mathbf{S} = \frac{1+i\mathbf{K}}{1-i\mathbf{K}}. \qquad (35.6)$$

The elements of the $\mathbf{K}$ matrix are real. Conservation of the incident flux of particles requires that

$$\sum_{j\Omega} |S(j\Omega, j'\Omega')|^2 = 1. \qquad (35.7)$$

Microreversibility (time-reversal symmetry) implies that the $\mathbf{S}$ matrix is symmetric,

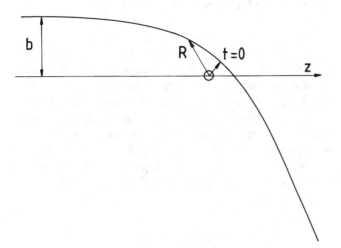

Figure 35.1. Classical scattering by a fixed scattering centre. The trajectory is symmetric about the point of closest approach, which is the time origin.

$$S(j\Omega, j'\Omega') = S(j'\Omega', j\Omega),$$

and hence

$$\sigma(j \leftarrow j')k_{j'}^2\omega_{j'} = \sigma(j' \leftarrow j)k_j^2\omega_j. \qquad (35.8)$$

The thermally averaged (Maxwellian) rate coefficient is

$$\langle \sigma v \rangle_{j \leftarrow j'} = \left(\frac{8\kappa T}{\pi\mu}\right)^{\frac{1}{2}} \int_0^\infty x_{j'}\sigma(j \leftarrow j')e^{-x_{j'}}\,dx_{j'}, \quad (35.9)$$

where $x_{j'} = \mu v_{j'}^2/2\kappa T$, $v_{j'}$ being the relative collision velocity, κ is Boltzmann's constant, and $(8\kappa T/\pi\mu)^{\frac{1}{2}}$ may be identified with the mean thermal velocity at temperature T. From (35.8) and (35.9),

$$\langle \sigma v \rangle_{j \leftarrow j'}\,\omega_{j'}\exp(-\epsilon_{j'}/\kappa T) = \langle \sigma v \rangle_{j' \leftarrow j}\,\omega_j\exp(-\epsilon_j/\kappa T),$$
$$(35.10)$$

where $\epsilon_{j'}$, ϵ_j denote the energies of the states j', j with respect to the reference level. Equation (35.10) relates the rate coefficients of inverse transitions to their relative degeneracies and excitation energies.

35.2 SEMICLASSICAL METHOD

If the collision dynamics is formulated using elements of both classical and quantum mechanics, then the method is called *semiclassical*. Thus, the relative motion might be described by a classical trajectory, whilst internal degrees of freedom (rotation, vibration, ...) are quantized. The de Broglie wavelength associated with the linear motion of a proton attains the Bohr radius at a collision energy of 0.29 eV, and a classical trajectory is a good approximation at much higher energies.

The introduction of a classical trajectory leads to a time-dependent Hamiltonian, and Schrödinger's equation may be reduced to a set of coupled, first-order differential equations of the form

$$i\dot{\mathbf{a}} = \mathbf{V}\mathbf{a}, \qquad (35.11)$$

where $\mathbf{a}$ is a column vector of transition amplitudes and $\dot{\mathbf{a}}$ is the vector of their time derivatives. The square matrix $\mathbf{V}$ is the interaction matrix, whose elements incorporate oscillatory factors of the type $\exp[(\epsilon_{j'} - \epsilon_j)t]$. The trajectory is taken to be symmetric about the point of closest approach, $t = 0$, and $\mathbf{a}$ is initialized through

$$a_j \to \delta_{j'j} \quad \text{as } t \to -\infty.$$

The differential equations (35.11) may be integrated numerically, and the transition probabilities derived from

$$P_b(j \leftarrow j') \to |a_j(b)|^2 \quad \text{as } t \to \infty.$$

The cross sections are given by Eq. (35.1).

35.3 QUANTAL METHOD

The usual approach to nonreactive scattering [1] is based on the Born-Oppenheimer approximation. The interaction potential V is taken to be a known function of the *nuclear* coordinates, the electronic energy being calculated for each geometry and separation of target and projectile. The total Hamiltonian may then be written

$$H = -\frac{1}{2\mu}\nabla_R^2 + h_1(\mathbf{x}_1) + h_2(\mathbf{x}_2) + V(\mathbf{R}, \mathbf{x}_1, \mathbf{x}_2),$$
$$(35.12)$$

where the first term represents the relative kinetic energy of the target and projectile, and h_1, h_2 are the Hamiltonians of the isolated molecules, functions of the intramolecular nuclear coordinates $\mathbf{x}_1, \mathbf{x}_2$. The eigenfunctions ψ_1, ψ_2 of h_1, h_2 describe the vibrational and rotational motions of the isolated molecules and form a basis for expanding the total wave function Ψ of the system:

$$\Psi(\mathbf{R}, \mathbf{x}_1, \mathbf{x}_2) = \sum_{\alpha_1\alpha_2\ell m} \frac{F(\alpha_1\alpha_2\ell m|R)}{R} Y_{\ell m}(\Theta, \Phi)$$
$$\times \psi_1(\alpha_1|\mathbf{x}_1)\psi_2(\alpha_2|\mathbf{x}_2). \qquad (35.13)$$

In (35.13), α_1, α_2 denote the sets of quantum numbers required to specify the states of the isolated molecules, and $Y_{\ell m}(\Theta, \Phi)$ is a spherical harmonic function of the angular coordinates of the intermolecular vector $\mathbf{R}$. $F(\alpha_1\alpha_2\ell m|R)$ are R-dependent expansion coefficients which, as will be shown below, are determined by solving the Schrödinger equation

$$(H - E)\Psi = 0. \qquad (35.14)$$

These solutions may be arranged as the columns of a square matrix $\mathbf{F}(R)$ in which each column is labeled by a different initial scattering state. The radial functions must satisfy the physical boundary conditions

$$\mathbf{F}(R) \rightarrow \mathbf{0} \qquad \text{as } R \rightarrow 0$$
$$\mathbf{F}(R) \rightarrow \mathbf{J}(R)\mathbf{A} - \mathbf{N}(R)\mathbf{B} \quad \text{as } R \rightarrow \infty,$$

where $\mathbf{J}$ and $\mathbf{N}$ are diagonal matrices whose nonvanishing elements are given by

$$J_{ii} = k_i^{\frac{1}{2}} R j_\ell(k_i R), \qquad (35.15)$$

$$N_{ii} = k_i^{\frac{1}{2}} R n_\ell(k_i R), \qquad (35.16)$$

j_ℓ, and n_ℓ are spherical Bessel functions of the first and second kinds, and k_i is the wave number in channel i (a given set of values of the quantum numbers $\alpha_1 \alpha_2 \ell m$). The reactance matrix

$$\mathbf{K} = \mathbf{B}\mathbf{A}^{-1} \qquad (35.17)$$

yields the state-to-state cross sections.

35.4 SYMMETRIES AND CONSERVATION LAWS

The expansion of the total wave function, Eq. (35.13), does not explicitly incorporate the invariance of the Hamiltonian under an arbitrary rotation in space or reflection in the coordinate origin. The Hamiltonian commutes with $\mathbf{J}^2$, where $\mathbf{J}$ is the total angular momentum, and any component J_x, J_y, or J_z of $\mathbf{J}$, although these components do not commute amongst themselves. Eigenfunctions of H may, therefore, be chosen to be simultaneous eigenfunctions of $\mathbf{J}^2$ and (conventionally) J_z, with eigenvalues $J(J+1)$ and M. If $\mathbf{j}_1$ and $\mathbf{j}_2$ are the angular momenta of the isolated molecules, then

$$\mathbf{j}_{12} = \mathbf{j}_1 + \mathbf{j}_2$$

is their resultant and

$$\mathbf{J} = \mathbf{j}_{12} + \boldsymbol{\ell},$$

where $\boldsymbol{\ell}$ denotes the relative angular momentum of the two molecules. The coupling of the angular momenta to the multipoles of the interaction potential [see Eq. (35.30) below] leads to restrictions on the changes in the angular momenta that can occur.

The parity operator, P, reflects the coordinates in the origin. Because two successive operations with P restore the original values of the coordinates, the corresponding eigenvalue p satisfies the equation $p^2 = 1$, or $p = \pm 1$. For electromagnetic interactions, the commutation of P and H implies conservation of the parity of the system.

If one takes advantage of the conservation laws associated with these symmetries of the system, substantial savings in computing time can be made: only one value of the total angular momentum and of the parity need to be considered simultaneously.

35.5 COORDINATE SYSTEMS

The natural choice of coordinate system in which to express the interaction potential $V(\mathbf{R}, \mathbf{x}_1, \mathbf{x}_2)$ is a body-fixed (BF) system in which the z-axis coincides with the direction of the intermolecular vector $\mathbf{R} = (R, \Theta, \Phi)$. A rotation of the space-fixed (SF) coordinate system through the Euler angles $(\Phi, \Theta, 0)$ generates such a BF frame. The intramolecular coordinates $\mathbf{x}_1, \mathbf{x}_2$ must then be expressed relative to the BF frame, as must the Laplacian operator ∇_R^2 which appears in the expression for the total Hamiltonian, Eq. (35.12). The latter may be written as

$$\begin{aligned}
\nabla_R^2 &= \frac{1}{R}\frac{\partial^2}{\partial R^2}R - \frac{\boldsymbol{\ell}^2}{R^2} \\
&= \frac{1}{R}\frac{\partial^2}{\partial R^2}R - \frac{(\mathbf{J}-\mathbf{j}_{12})^2}{R^2}, \qquad (35.18)
\end{aligned}$$

which are the forms suitable for calculations in SF and BF coordinates, respectively.

A unitary transformation relates the normalized eigenfunctions of a given parity in SF and BF coordinates. In Dirac notation,

$$|j_{12}\ell\epsilon JM\rangle_{\text{SF}} = \sum_{\bar{\Omega}} |j_{12}\bar{\Omega}\epsilon JM\rangle_{\text{BF}} \langle j_{12}\bar{\Omega}\epsilon JM|j_{12}\ell JM\rangle,$$

$$(35.19)$$

where $\bar{\Omega} = |\Omega|$ and Ω is the projection of $\mathbf{J}$ on the BF z-axis. As the projection of $\boldsymbol{\ell}$ on the intermolecular axis is zero, Ω is also the projection of $\mathbf{j}_{12}$ on the BF z-axis. The absolute value of Ω appears in the transformation because $|j_{12}\pm\Omega JM\rangle$ are *not* eigenfunctions of the parity operator P, whereas the linear combinations

$$|j_{12}\bar{\Omega}\epsilon JM\rangle = \frac{|j_{12}\Omega JM\rangle + \epsilon|j_{12}-\Omega JM\rangle}{[2(1+\delta_{\bar{\Omega}0})]^{\frac{1}{2}}}$$

$(\epsilon = \pm 1)$ are eigenfunctions of P. The factor $[2(1+\delta_{\bar{\Omega}0})]^{\frac{1}{2}}$ ensures the correct normalization of these functions. The elements of the matrix which perform the unitary transformation (35.19) are

$$\langle j_{12}\bar{\Omega}\epsilon JM|j_{12}\ell\epsilon JM\rangle = \left[\frac{2(2\ell+1)}{(1+\delta_{\bar{\Omega}0})(2J+1)}\right]^{\frac{1}{2}} C_{\bar{\Omega}0\bar{\Omega}}^{j_{12}\ell J},$$

$$(35.20)$$

where $C_{\bar{\Omega}0\bar{\Omega}}^{j_{12}\ell J}$ is a Clebsch-Gordan coefficient.

35.6 SCATTERING EQUATIONS

Using the orthogonality property of the eigenfunctions ψ_1, ψ_2 and the spherical harmonics $Y_{\ell m}$, Schrödinger's equation for the scattering system may be reduced to a set of coupled, ordinary, second-order differential equations for the radial functions $F(R)$. Expressed in matrix form, they become

$$\left[1\frac{d^2}{dR^2} + \mathbf{W}(R)\right]\mathbf{F}(R) = 0\,. \qquad (35.21)$$

There exists a set of equations (35.21) for each value of the total angular momentum J and parity p (*cf.* Sect. 35.4). The matrix $\mathbf{W}$ may be written

$$\mathbf{W}(R) = \mathbf{k}^2 - 2\mu\mathbf{V}_{\text{eff}}\,, \qquad (35.22)$$

where $\mathbf{k}^2$ is a diagonal matrix whose nonvanishing elements are

$$k^2_{\alpha_1\alpha_2} = 2\mu(E - \epsilon_{\alpha_1} - \epsilon_{\alpha_2})\,. \qquad (35.23)$$

$k_{\alpha_1\alpha_2}$ is the wave number at infinite separation $(R \to \infty)$ when the molecules are in eigenstates α_1, α_2 with eigenenergies $\epsilon_{\alpha_1}, \epsilon_{\alpha_2}$. $\mathbf{V}_{\text{eff}}$ is the matrix of the effective potential

$$V_{\text{eff}} = V(\mathbf{R}, \mathbf{x}_1, \mathbf{x}_2) + \frac{\ell^2}{2\mu R^2}\,, \qquad (35.24)$$

in which V is the interaction potential and $\ell^2/(2\mu R^2)$ may be identified with the centrifugal potential. There exists a standard computer code for solving equations of the form (35.21) [2].

35.7 MATRIX ELEMENTS

35.7.1 Centrifugal Potential

When evaluated in the SF frame, the matrix of the centrifugal potential is diagonal, with nonvanishing elements $\ell(\ell+1)/(2\mu R^2)$. In the BF frame, the diagonal elements are [*cf.* Eq. (35.18)]

$$\langle j_{12}\bar{\Omega}\epsilon JM|(\mathbf{J} - \mathbf{j}_{12})^2/2\mu R^2|j_{12}\bar{\Omega}\epsilon JM\rangle$$
$$= [J(J+1) + j_{12}(j_{12}+1) - 2\bar{\Omega}^2]/(2\mu R^2)\,, \qquad (35.25)$$

and, in addition, there are off-diagonal elements

$$\langle j_{12}\bar{\Omega}\epsilon JM|(\mathbf{J} - \mathbf{j}_{12})^2/2\mu R^2|j_{12}\bar{\Omega}\pm 1\epsilon JM\rangle$$
$$= -\left\{(1+\delta_{\bar{\Omega}0})(1+\delta_{\bar{\Omega}\pm1,0})[J(J+1) - \bar{\Omega}(\bar{\Omega}\pm1)]\right.$$
$$\left.\times [j_{12}(j_{12}+1) - \bar{\Omega}(\bar{\Omega}\pm1)]\right\}^{\frac{1}{2}}/(2\mu R^2)\,. \qquad (35.26)$$

The matrix elements (35.26), which are off-diagonal in $\bar{\Omega}$, are associated with the rotation in space of the BF coordinate system (Coriolis coupling). In the *coupled states* approximation [3], the off-diagonal elements (35.26) are neglected, and the diagonal elements (35.25) are often replaced by their SF equivalent form $\ell(\ell+1)/(2\mu R^2)$. The matrix of the interaction potential, on the other hand, continues to be evaluated in BF coordinates. The net effect of these approximations is to ignore the rotation of the BF frame in the course of the collision, and associated dynamical terms.

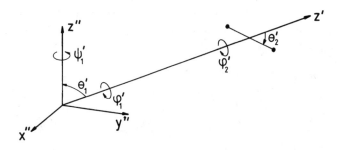

Figure 35.2. The body fixed coordinate system: scattering of a rigid rotor and a symmetric top.

35.7.2 Interaction Potential

The interaction potential is usually expressed and computed in BF coordinates. For the purposes of the analysis, it is convenient to derive a multipolar expansion of the potential from a least-squares fit to the original data points. If the collision calculations are being done in the BF frame, the potential matrix elements may be evaluated directly. However, if the SF frame is to be used, the potential expansion must first be transformed into SF form.

Consider the interaction between a symmetric top and a rigid rotor [4,5]. This complicated example serves as a paradigm, and simpler cases may be derived by reduction. In the BF frame, the potential can be expanded as

$$V(R, \hat{\boldsymbol{\omega}}'_1, \hat{\mathbf{r}}'_2) = \sum_{\substack{\lambda_1\lambda_2 \\ \mu\nu}} v_{\lambda_1\lambda_2\mu\nu}(R)\mathcal{D}^{\lambda_1}_{\mu\nu}(\hat{\boldsymbol{\omega}}'_1)Y_{\lambda_2-\nu}(\hat{\mathbf{r}}'_2)\,, \qquad (35.27)$$

where the Euler angles $\hat{\boldsymbol{\omega}}'_1 = (\phi'_1, \theta'_1, \psi'_1)$ determine the orientation of the principal axes of the top and $\hat{\mathbf{r}}'_2 = (\theta'_2, \phi'_2)$ the orientation of the axis of the rotor (Fig. 35.2). The expansion coefficients $v_{\lambda_1\lambda_2\mu\nu}(R)$ derive from the least-squares fit. Expanding the rotation matrix element $\mathcal{D}^{\lambda_1}_{\mu\nu}$ and the spherical harmonic $Y_{\lambda_2-\nu}$, Eq. (35.27) becomes

$$V(R, \hat{\boldsymbol{\omega}}'_1, \hat{\mathbf{r}}'_2) = \sum_{\substack{\lambda_1\lambda_2 \\ \mu\nu}} v_{\lambda_1\lambda_2\mu\nu}(R)\left(\frac{2\lambda_2+1}{4\pi}\right)^{\frac{1}{2}} e^{i\mu\psi'_1}d^{\lambda_1}_{\mu\nu}(\theta'_1)$$
$$\times e^{i\nu\phi'_1}d^{\lambda_2}_{0-\nu}(\theta'_2)e^{-i\nu\phi'_2} \qquad (35.28)$$

where

$$d^j_{m'm}(\beta) = \langle jm'|\exp(i\beta J_y)|jm\rangle \qquad (35.29)$$

according to the definition in [6]. Note that the $d^j_{m'm}$ are real. The combination of positive and negative values of the index ν in Eq. (35.28) reflects the invariance of the potential under rotations about the intermolecular axis (alternatively, its dependence on the *difference* between ϕ_1' and ϕ_2'). In SF coordinates, the potential becomes

$$V(\mathbf{R}, \hat{\boldsymbol{\omega}}_1, \hat{\mathbf{r}}_2) = \sum_{\substack{\lambda_1 \lambda_2 \\ \lambda \mu}} v_{\lambda_1 \lambda_2 \lambda \mu}(R) \sum_{m_1 m_2} C_{m_1 m_2 m}^{\lambda_1 \lambda_2 \lambda} \mathcal{D}_{\mu m_1}^{\lambda_1}(\hat{\boldsymbol{\omega}}_1)$$
$$\times Y_{\lambda_2 m_2}(\hat{\mathbf{r}}_2) Y_{\lambda m}^*(\hat{\mathbf{R}}). \quad (35.30)$$

The expansion coefficients in the SF and BF frames are related by

$$v_{\lambda_1 \lambda_2 \lambda \mu}(R) = \left(\frac{4\pi}{2\lambda + 1}\right)^{\frac{1}{2}} \sum_{\nu \geq 0} C_{\nu -\nu 0}^{\lambda_1 \lambda_2 \lambda}(1 + \delta_{\nu 0})^{-1}$$
$$\times [v_{\lambda_1 \lambda_2 \mu \nu}(R) + (-1)^{\lambda_1 + \lambda_2 + \lambda} v_{\lambda_1 \lambda_2 \mu -\nu}(R)]. \quad (35.31)$$

The matrix elements of the potential given by Eq. (35.30) are

$$\langle j_1 k j_2 j_{12} \ell J M | V(\mathbf{R}, \hat{\boldsymbol{\omega}}_1, \hat{\mathbf{r}}_2) | j_1' k' j_2' j_{12}' \ell' J M \rangle$$
$$= \sum_{\substack{\lambda_1 \lambda_2 \\ \lambda \mu}} v_{\lambda_1 \lambda_2 \lambda \mu}(R)(-1)^{j_1' + j_2' + j_{12} + k' - J}\left(\frac{2\lambda + 1}{4\pi}\right)$$
$$\times [(2j_1 + 1)(2j_2 + 1)(2j_{12} + 1)(2\ell + 1)(2\lambda_2 + 1)$$
$$\times (2\ell' + 1)(2j_{12}' + 1)(2j_2' + 1)(2j_1' + 1)]^{\frac{1}{2}}$$
$$\times \begin{pmatrix} \lambda & \ell & \ell' \\ 0 & 0 & 0 \end{pmatrix} \begin{pmatrix} \lambda_2 & j_2 & j_2' \\ 0 & 0 & 0 \end{pmatrix} \begin{pmatrix} \lambda_1 & j_1 & j_1' \\ \mu & k & -k' \end{pmatrix}$$
$$\times \begin{Bmatrix} \ell' & \ell & \lambda \\ j_{12} & j_{12}' & J \end{Bmatrix} \begin{Bmatrix} j_{12} & j_2 & j_1 \\ j_{12}' & j_2' & j_1' \\ \lambda & \lambda_2 & \lambda_1 \end{Bmatrix}, \quad (35.32)$$

where k, k' denote the projection of j_1, j_1' on the symmetry axis of the symmetric top; the quantities on the last two lines of the equation are 3-j, 6-j, and 9-j symbols. Note that the properties of the 3-j coefficients imply that the matrix elements corresponding to odd values of $\lambda + \ell + \ell'$ or $\lambda_2 + j_2 + j_2'$ vanish identically. Thus, if the the rigid rotor is a homonuclear molecule, and hence λ_2 is

even, transitions cannot be induced between even and odd values of the angular momentum of the rotor.

The potential matrix elements for other important cases may be derived from Eq. (35.32). Rigid rotor-rigid rotor [7] obtains when $k = \mu = 0$, and atom-rigid rotor when, additionally, $j_1 = \lambda_1 = 0$. References to numerical results for specific systems have been compiled in Appendix 2 of [8].

In order to exploit conservation of the parity of the total system, it is necessary to use symmetry adapted rotational eigenfunctions

$$|j_1 \bar{k} m \eta\rangle = \frac{|j_1 k m\rangle + \eta |j_1 - k m\rangle}{[2(1 + \delta_{\bar{k} 0})]^{\frac{1}{2}}},$$

where $\eta = \pm 1$ and $\bar{k}$ is the absolute magnitude of the projection of $\mathbf{j}_1$ on the symmetry axis of the top. The parity of the total wave function $|j_1 \bar{k} \eta j_2 j_{12} \ell J M\rangle$ is then $p = \eta(-1)^{j_1 + j_2 + \ell + \bar{k}}$, and is conserved during the collision. This fact enables the coupled equations to be separated into two noninteracting parity blocks. Because the orientation of the SF z-axis is arbitrary, the matrix elements (35.32) must be independent of M.

REFERENCES

1. R. B. Bernstein, *Atom-molecule Collision Theory* (Plenum, New York, 1979).
2. J. M. Hutson and S. Green, MOLSCAT Computer Code Version 12, distributed by CCP6 (UK Science and Engineering Research Council, Daresbury Laboratory, 1993).
3. A. S. Dickinson, Computer Phys. Commun. **17**, 51 (1979).
4. A. Offer and D. R. Flower, J. Chem. Soc. Faraday Trans. **86**, 1659 (1990).
5. C. Rist, M. H. Alexander, and P. Valiron, J. Chem. Phys. **98**, 4662 (1993).
6. A. R. Edmonds, *Angular Momentum in Quantum Mechanics* (Princeton University Press, Princeton, 1974).
7. J. Schaefer and W. Meyer, J. Chem. Phys. **70**, 344 (1979).
8. D. R. Flower, *Molecular Collisions in the Interstellar Medium* (Cambridge University Press, Cambridge, 1990).

36

Gas Phase Reactions

Eric Herbst

Department of Physics and Department of Astronomy, The Ohio State University, Columbus, Ohio 43210

36.1 INTRODUCTION

The rates of gas phase chemical processes can be described by rate laws in which the rate of formation of products or disappearance of reactants is related to the product of the concentrations of reactants raised to various powers [1]. As an example, consider the rate of disappearance of reactant A in a gas mixture containing species A, B, and C. The rate law for this disappearance can be expressed by the equation

$$d[A]/dt = -k[A]^a[B]^b[C]^c, \qquad (36.1)$$

where the symbols [ ] refer to concentration, and the rate coefficient k is dependent on temperature, and possibly other parameters such as the total gas density. The above law is not the most general that can be envisaged. For example, if species A reacts via more than one set of processes, more than one negative term on the right-hand side of the equation will be needed. In addition, if the backward reaction to form A from products is appreciable, a positive term must also be included. At equilibrium, the rate of change of reactant must be zero.

The relation above does not necessarily refer to one

chemical reaction, but to a succession of elementary reactions known collectively as a *mechanism*. The most elementary reaction is a simple bimolecular process with a second order rate law of the type

$$d[C]/dt = -d[A]/dt = k[A][B], \qquad (36.2)$$

where two species A and B collide to form products, one of which can be labeled C. In this law, the rate coefficient k is related simply to the reaction cross section σ via the equation

$$k = \sigma v, \qquad (36.3)$$

where v is the relative velocity of reactants. The rate coefficient in a bimolecular process has units of volume per time, typically $cm^3 \, s^{-1}$. In a thermal system, the rate coefficient $k(T)$ is averaged over all degrees of freedom of the reactants, both internal and translational. The most specific rate coefficient is termed a *state-to-state coefficient* and refers to a process in which reactant A in quantum state a reacts with reactant B in quantum state b at a specific translational energy $\mathcal{T}$ to form products in specific states separating with a specific translational energy [2].

Although normal bimolecular processes produce more than one product, it is also possible for the two reactants A and B to stick together if sufficient energy is released in the form of a photon:

$$A + B \longrightarrow AB + h\nu. \qquad (36.4)$$

Such a process is called *radiative association* [3], and has mainly been studied for ion-molecule systems, i.e., reactions in which one of the two reactants is an ion. The process of radiative association is normally thought to occur in two steps. The first step produces a *collision complex AB^**, which is a molecule existing in a transitory fashion above its dissociation limit:

$$A + B \longrightarrow AB^*. \qquad (36.5)$$

Once formed, the complex, which is often thought of as an ergodic entity retaining memory only of its total energy and angular momentum, can either redissociate into reactants, or emit a photon of sufficient energy to stabilize itself:

$$AB^* \longrightarrow AB + h\nu. \qquad (36.6)$$

Since redissociation of the complex is generally more rapid than radiative emission, radiative association rate coefficients are normally small. Emission of one infrared photon is sufficient to achieve stabilization, but stabilization of a complex via electronic emission may be a more rapid mechanism if suitable electronic states exist [3, 4]. The lifetime of the complex against redissociation into reactants is a strong direct function of the binding energy of the species and the number of atoms it possesses.

In addition to bimolecular processes, two other types of reactions often referred to as elementary are unimolecular and termolecular reactions. Although complex reaction mechanisms are sometimes divided into unimolecular and termolecular steps, these are not strictly elementary because they can be subdivided into a series of bimolecular steps. In a unimolecular reaction [5, 6], a molecule A is destroyed at sufficiently high gas density by a process that has the seemingly simple first order rate law

$$d[A]/dt = -k[A]. \qquad (36.7)$$

At low pressures, on the other hand, the rate law is second order. A simplified series of events, called the *Lindemann mechanism*, explains these limiting cases by invoking the activation of species A by strong inelastic collisions with bath gas M to form an activated complex A^* which can either be deactivated by inelastic collisions or spontaneously decompose, since it possesses sufficient energy to do so. The steps are written as

$$A + M \rightleftharpoons A^* + M, \qquad (36.8)$$
$$A^* \longrightarrow B + C, \qquad (36.9)$$

where the spontaneous destruction of A^* can be thought of as a truly elementary unimolecular reaction, akin to spontaneous emission of radiation. Studies of spontaneous dissociation occupy an important place in gas phase reaction theory, and are discussed in Sects. 36.2.2 and 36.2.4 in the context of dissociation of intermediate complexes. If the rate coefficients for the forward and reverse processes in Eq. (36.8) are labeled k_1 and k_{-1} respectively, and the rate coefficient for spontaneous dissociation of A^* into products is labeled k_2 (s^{-1}), application of the steady-state principle to the concentration of the activated complex, namely,

$$d[A^*]/dt = k_1[A][M] - k_{-1}[A^*][M] - k_2[A^*] \approx 0 \qquad (36.10)$$

leads to the rate law

$$-d[A]/dt = k'[A][M], \qquad (36.11)$$

where

$$k' = k_1 k_2 / (k_{-1}[M] + k_2). \qquad (36.12)$$

At low pressures, $k_{-1}[M] \ll k_2$ and $k' \approx k_1$ so that a second order rate law prevails. At high pressures, $k_{-1}[M] \gg k_2$ and $k' \approx k_1 k_2/k_{-1}[M]$ so that the rate law becomes first order

$$-d[A]/dt \approx (k_1/k_{-1})k_2[A]. \qquad (36.13)$$

Since both the activation and deactivation of complexes occur stepwise, rather than in single strong collisions, reality is far more complex than the Lindemann mechanism [7], and, at the highest degree of detail, master equation treatments are needed, especially for intermediate pressures [8].

In a termolecular reaction, which is actually the inverse of a unimolecular process [8], two species B and C collide to form a collision complex A^*, which can be regarded as the activated complex of stable species A. The collision complex can be stabilized by subsequent strong collisions with other species:

$$A^* + M \longrightarrow A + M, \qquad (36.14)$$

or can redissociate into reactants. If the complex formation step occurs with a bimolecular rate coefficient k_{-2}, and k_{-1} and k_2 refer to complex stabilization and dissociation as in the unimolecular case, the rate law for the termolecular process can be written

$$d[A]/dt = k'[B][C][M], \qquad (36.15)$$

where

$$k' = k_{-2}k_{-1}/(k_{-1}[M] + k_2). \qquad (36.16)$$

At low pressures, $k' \approx k_{-2}k_{-1}/k_2$ and the rate law is third order. At high pressures, when every activated complex is deactivated, the rate law becomes second order since $k' \approx k_{-2}/[M]$. Actually, at very low pressures radiative stabilization of the complex dominates and the rate law once again becomes second order. As in the unimolecular case, the complex does not really undergo strong inelastic collisions, so that reality is once again more complex than pictured here [8]. For detailed theories of thermal reactions, there is the additional problem in Eqs. (36.12) and (36.16) of deciding when thermalization of the partial rate coefficients should be undertaken. In reality, the partial rate coefficients refer to reactions with specific amounts of energy and angular momentum, and should not be thermally averaged before incorporation into the equations for k'.

36.2 NORMAL BIMOLECULAR REACTIONS

The rates of elementary (bimolecular) chemical reactions are governed by Born-Oppenheimer potential surfaces which contain electronic energies and nuclear-nuclear repulsions. Reactions can be classified as exoergic or endoergic depending upon whether the 0 K energies of the products lie below or above those of the reactants, respectively. In the simplest types of exoergic chemical reactions, the potential energy flows downhill from reactants to products, or flows downhill from reactants to a global minimum (the reaction complex) after which it flows uphill to products. More commonly, the morphology of the potential surface is such that after some long-range attraction, the potential rises as old chemical bonds are broken before falling as new bonds are formed. Generally, there is a minimum energy pathway through the region of large potential referred to as the *reaction coordinate*. The system is said to traverse a *transition state barrier* which refers to the configuration of atoms at which a potential saddle point occurs. The height of this transition state barrier is related to the activation energy barrier E_a in the classical Arrhenius rate law

$$k(T) = A(T) \exp(-E_a/[k_B T]), \qquad (36.17)$$

for rate coefficients of bimolecular reactions which contain short-range barriers [1]. In the Arrhenius expression, k_B is the Boltzmann constant, and the pre-exponential factor $A(T)$ can be related to the form of the long-range potential, or to an equilibrium coefficient between the transition state and reactants (see the discussion of activated complex theory in Sect. 36.2.3). Although fits of experimental data over short temperature ranges often assume the pre-exponential factor to be totally independent of temperature, theories show that this is not strictly true in most instances. A more serious problem with the expression undoubtedly occurs at low temperatures since tunneling will clearly lead to deviations.

Although, in principle, it is possible to calculate reaction cross sections and rate coefficients via the quantum theory of scattering, in practice few systems have been studied by this technique given the immense computational effort required [9, 10]. Another technique which is useful for systems with a small number of atoms is the quasiclassical trajectory approach (see Chap. 56), in which atoms move classically on quantum mechanically generated Born-Oppenheimer potential surfaces. For most reactions, however, neither technique is applicable, and a variety of simpler approaches has been developed, using capture and statistical approximations; these approaches will be emphasized in this section. Indeed, the use of an ergodic complex in our preliminary discussion of association and unimolecular reactions above presages the use of statistical approximations. An excellent high-level review article on many of the topics covered here has been written by Troe [11].

36.2.1 Capture Theories

For reactions that do not possess a potential energy barrier at short-range, it is tempting to apply long-range capture theories between structureless particles to calculate the reaction rate coefficient. Such theories assume that (a) all hard collisions lead to reaction, and (b) hard collisions occur for all partial waves up to a maximum impact parameter $b_{\max}$ or relative angular momentum quantum number $L_{\max}$. This is normally defined so that the reactant translational energy $\mathcal{T}_{AB}$ is just sufficient to overcome a centrifugal barrier. The centrifugal barrier produces a long-range maximum in the effective potential energy function V_{eff} given by the relation

$$V_{\mathrm{eff}}(r, b) = V(r) + \mathcal{T}_{AB} b^2/r^2, \qquad (36.18)$$

where r is the separation between reactants and $\mathcal{T}_{AB} b^2/r^2$ is the angular kinetic energy. If the reactants overcome the centrifugal barrier, they will spiral in towards each other in the absence of short-range repulsive forces. The approach has had its most notable success for exoergic reactions between ions and nonpolar neutral molecules [12]. The long-range potential in this situation is simply (in cgs-esu units)

$$V(r) = -e^2 \alpha_d/2r^4, \qquad (36.19)$$

where α_d is the polarizability of the neutral reactant. This potential leads to the *Langevin rate coefficient*

$$k = v\pi b_{\max}^2 = 2\pi e(\alpha_d/\mu)^{1/2}, \qquad (36.20)$$

where μ is the reduced mass of the reactants. The theory predicts a temperature independent rate coefficient with magnitude $\approx 10^{-9} \mathrm{cm}^3 \mathrm{s}^{-1}$. Numerous experiments from above room temperature to below 30 K confirm its validity for the majority of ion-molecule reactions, which appear rarely to possess potential energy surfaces with short-range barriers [13, 14].

An analogous central force potential for structureless neutral-neutral reactants is the van der Waals or Lennard-Jones attraction

$$V(r) = -C_6/r^6 , \qquad (36.21)$$

where C_6 can be defined simply in terms of the ionization potentials and polarizabilities of the reactants [15, 16]. The rate coefficient obtained using a capture theory with this long-range potential is given, after translational thermal averaging, by the equation [16]

$$k(T) = 8.56 \, C_6^{1/3} \mu^{-1/2} (k_B T)^{1/6} , \qquad (36.22)$$

where all quantities are in cgs units. Equation (36.22) leads to estimates for the rate coefficient at room temperature in the range $\approx 10^{-10} - 10^{-9}$ cm^3 s^{-1}. Unlike the situation for ion-molecule reactions, this estimate has not received much attention mainly because most neutral-neutral reactions involve activation energy. Even for those systems without activation energy, the approximation appears to lead to rate coefficients that are too large by at least a factor of a few [16]. In place of the result of Eq. (36.22), kineticists often use the simple hard-sphere model with atomic dimensions for the reaction cross section.

Both long-range potentials considered above are isotropic in nature. A variety of capture theories [17, 18] has been developed which takes angular degrees of freedom into account. For ion-molecule systems in which the neutral species has a permanent dipole moment μ_d, the long-range potential becomes

$$V(r, \theta) = -\frac{e^2 \alpha_d}{2r^4} - \frac{e \mu_d \cos \theta}{r^2} , \qquad (36.23)$$

where θ is the angle between the radius vector from the charged species to the c.m. of the dipolar species and the dipole vector. Adiabatic effective (centrifugal) potential energy curves can be defined at any given fixed intermolecular separation r by diagonalizing angular kinetic energy and potential energy matrices using a suitable basis set. If an atomic ion is reacting with a linear neutral, a suitable basis set would consist of spherical harmonics for the rotation of the linear molecule (angular momentum j) as well as rotation matrices for the relative motion (angular momentum L) between the species. The eigenvalues of a matrix with fixed total angular momentum J then correspond to the effective radial potential energy functions V_{eff} at each r. The L, j labeling of the potential curves can be ascertained by starting the calculation at sufficiently large r so that L and j are reasonably good quantum numbers. Additional approximations, such as the centrifugal sudden approximation,

can be made to facilitate the calculation [19]. Making the adiabatic assumption that transitions between these potentials are not allowed at long range, one obtains an adiabatic capture rate coefficient for each initial value of j and translational (radial kinetic) energy T_{AB} from the criterion

$$V_{\text{eff}}(j, L_{\max}, R) = T_{AB} , \qquad (36.24)$$

where R is the separation corresponding to the maximum effective potential. Unlike the Langevin approach, the rate coefficients thermalized over translation for each j show an inverse dependence on temperature, with the $j = 0$ state showing the most severe dependence since the dipole is essentially "locked" for this state. Thermal averaging over j typically leads to rate coefficients with an inverse dependence on temperature between $T^{-1/2}$ and T^{-1} [20]. Rate coefficients as large as 10^{-7} cm^3 s^{-1} can be obtained as the temperature approaches 10 K. Although most ion-dipole reactions seem to obey capture theory models at low temperature, there are exceptions.

An alternative approach to ion-dipole reactions using the classical concept of adiabatic invariants has been developed [21]. In addition, variants to the capture theory discussed here have been formulated. A statistical adiabatic approach has been developed [17], in which adiabatic effective potential maxima are not used to define capture cross sections, but to define rate coefficients via the activated complex (transition state) theory discussed in Sect. 36.2.3, which posits thermal equilibrium between reactants and the species existing at the potential maxima. One advantage of this simplification is that it permits the adiabatic treatment to be more easily extended to complex geometries, especially if perturbation approximations are utilized. Other approaches involve the variational principle; an upper bound to the capture rate coefficient can be determined within the transition state, or *bottleneck* approach, which defines the transition state through minimization of the number of available vibrational/rotational states at the bottleneck [22] (see Sect. 36.2.3).

Nonspherical capture theories have also been used [23, 24] to study neutral-neutral reactions. The role of atomic fine structure at low temperatures is an especially interesting application; reactions involving atomic C and O with a variety of reactants have been studied. Use of an electrostatic potential shows that the reactivity of C or O atoms in their $^3P_0$ states with dipolar species is minimal. This is particularly important for atomic carbon since at low temperatures it lies primarily in its ground $^3P_0$ state. The choice of an electrostatic potential has been disputed [16] because experimental results for C–hydrocarbon reactions at room temperature are best understood if the long-range potential is dispersive (Lennard-Jones) in character rather than electrostatic.

Reactions between radicals in $^2\Pi$ states (e.g. OH) and $^2\Sigma$ states (e.g. CN) and stable molecules have also been considered, especially at low temperature, with long range potentials that contain both electrostatic and dispersion terms. The results can be compared with new low temperature experimental results on the CN-O_2 reaction, but the temperature dependence is not matched by theory if the dispersion term is included [24,25].

36.2.2 Phase Space Theories

Capture theories tell us neither the products of reaction, if several sets of exoergic products are available, nor the distributions of quantum states of the products. The simplest approach to these questions for reactions with barrierless potentials is to make a statistical approximation — all detailed outcomes being equally probable as long as energy and angular momentum are conserved. Such a result requires strong coupling at short range. The most prominent treatment along these lines is referred to as *phase space theory* [26]. In this theory, the cross section σ for a reaction between two species A and B with angular momentum quantum numbers J_A and J_B and in specific vibrational-electronic states colliding with asymptotic translational kinetic energy T_{AB} to form products C and D in specific vibrational-electronic states with angular momentum quantum numbers J_C and J_D is

$$\sigma(J_A, J_B \to J_C, J_D) = \frac{\pi\hbar^2}{2\mu T_{AB}} \sum_{L_i, J} (2L_i + 1)$$
$$\times P(J_A, J_B, L_i \to J)P'(J \to J_C, J_D), \qquad (36.25)$$

where J is the total angular momentum of the combined system, L_i is the initial relative angular momentum of reactants, P is the probability that the angular momenta of the reactants add vectorially to form J, P' is the probability that the combined system with angular momentum J dissociates into the particular final state of C and D, μ is the reduced mass of reactants, and the summation is over the allowable ranges of initial relative angular momentum and total angular momentum quantum numbers. P' is equal to the sum over the final relative angular momentum L_f, of angular momentum allowed $(J \to L_f, J_C, J_D)$ combinations leading to the specific product state divided by the sum of like combinations for all energetically accessible product and reactant states. The ranges of initial and final relative angular momenta are given by appropriate capture models (e.g. Langevin, ion-dipole, Lennard-Jones) as well as angular momentum triangular rules. This procedure involves the implicit assumption that strong coupling does not occur at long range; rather, adiabatic effective potentials can be assumed for initial and final states. The state-to-state rate coefficient is simply the cross section multiplied by the relative velocity of the two reactants. Summation over

all product states, as well as thermal averaging over the reactant state distributions and the translational energy distribution, can all be undertaken. Strategies for summing/integrating over rotational/vibrational degrees of freedom have been given [27]. As can be seen by writing out the expression for P', the phase space formula for the state-to-state cross section in Eq. (36.25) obeys microscopic reversibility:

$$2\mu T_{AB}\sigma(J_A, J_B \to J_C, J_D)$$
$$= 2\mu' T'_{AB}\sigma(J_A, J_B \leftarrow J_C, J_D) \qquad (36.26)$$

where the reduced mass and asymptotic translational energy of the products are denoted by primes. Thus, thermalization of the forward and backward rates leads to detailed balancing.

As a detailed theory for prediction of reaction products and their state distributions, phase space theory and its variants have had mixed success. It is true that the theory correctly predicts that exoergic reactions occurring on barrierless potential surfaces proceed on every strong collision. It is also true, however, that the theory is generally not useful in predicting the branching ratios among several sets of exoergic products because potential surfaces do not often show barrierless pathways for more than one set of products. When applied to product state distributions, the theory yields statistical results, so that population inversions in nondegenerate degrees of freedom (such as the vibrations of a diatomic molecule) are not replicated. With respect to more thermalized entities, such as total cross sections vs. collision energy for endoergic reactions [28], the theory can be quite successful, especially when the potential surface involves a deep minimum or intermediate complex. In this instance, the strong coupling hypothesis comes closest to actuality. A deep well is also associated with a high density of vibrational states in the quasicontinuum above the dissociation limit of the molecular structure defined by the potential minimum [1,29]. Such a high density renders direct (specific) dynamical processess less likely.

A useful variant of phase space theory if complex lifetimes are needed [30] (see Sects. 36.2.4 and 36.3) is based on a unimolecular decay theory of Klots [31]. The reaction is considered to proceed via a capture cross section to form the intermediate complex, which then can dissociate into all available reactant and product states consistent with conservation of energy and angular momentum. The complex dissociation rate k_{uni} into a specific state can be obtained via the principle of microscopic reversibility in terms of the capture cross section from that state, obtained with $P' = 1$ in Eq. (36.25), to form the complex. In particular, if a complex with total angular momentum J can dissociate into one state of reactants A and B separating with translational energy T_{AB}, then

$$k_{\text{uni}} = \rho_{\text{vib}}^{-1} g(J)^{-1} g(J_A) g(J_B) 2\mu \mathcal{T}_{AB} \sigma(J_A, J_B \to J),$$

$$(36.27)$$

where ρ_{vib} is the density of complex vibrational states obtained via a variety of prescriptions [1, 29], and g is the rotational degeneracy. A cross section analogous to that in Eq. (36.25) can be formulated in terms of capture to form the complex multiplied by the complex dissociation rate into a particular state divided by the total (summed) dissociation rate. Note that the dissociation rate of the complex is proportional to ρ_{vib}^{-1}. Since ρ_{vib} is a strong function of the well depth, long-lived complexes are associated with large well depths (>1 eV). The Klots form for k_{uni} is especially useful for ion-molecule systems, where the cross section for complex formation can be assumed to be Langevin or ion-dipole. The concept of a long-range potential is less useful for most neutral-neutral systems, so that unimolecular rate coefficients for unstable entities are normally obtained quite differently in terms of ρ_{vib} at the transition state (see Sect. 36.2.4).

For smaller reaction systems, especially those involving ions, it is quite common for the electronic states of reactants to correlate with more than one potential surface of the combined system, although typically only one surface leads to energetically accessible products. In such instances, it is normal to assume statistical partitioning, although fine structure effects can complicate this picture. A generalization of phase space theory to account for multiple potential surfaces has been proposed for the $C^+ + H_2$ reaction [32].

Phase space theory has been used to predict product branching ratios for dissociative recombination reactions between polyatomic positive ions and electrons [33]; its success here has been limited at best.

36.2.3 Short-Range Barriers

Most reactions involving neutral molecules, as well as a minority of ion-molecule reactions, possess potential surfaces with short-range barriers. The experimental rate coefficients of these systems over selected temperature ranges are normally fit to the Arrhenius expression (36.17).

The simplest theoretical method of taking short-range potential barriers into account is the *line-of-centers* approach, which resembles the capture theories previously discussed [2]. In this crude approximation, it is assumed that structureless reactants colliding with impact parameter b along a repulsive potential must reach some minimum distance, d, for reaction to occur. If the potential energy in the absence of angular momentum at d is E_0, this condition implies that the asymptotic translational energy of reactants $\mathcal{T}_{AB}$ must exceed the sum of E_0 and the centrifugal energy $\mathcal{T}_{AB} b^2 / d^2$, which in turn yields a maximum impact parameter b_{max}. Thermal

averaging of the rate coefficient $k = v\pi b_{\text{max}}^2$ over a Maxwell-Boltzmann distribution then yields

$$k(T) = \pi d^2 [8k_B T / \pi \mu]^{1/2} \exp(-E_0 / k_B T), \quad (36.28)$$

which bears some resemblance to the Arrhenius expression if one equates the parameter E_0 with the activation energy. Assuming $T = 300$ K, $\mu = 10$ u, and $d = 1$ Å, the pre-exponential factor is 2.5×10^{-11} cm^3 s^{-1}, which lies in a typical range.

The simple line-of-centers approach reduces the problem to that of structureless particles. The standard method of including all degrees of freedom is to use canonical ensemble statistical mechanics and to imagine that the transition state is in equilibrium with reactants. In the transition state, one of the vibrational degrees of freedom of a normal polyatomic molecule is replaced by a degree of freedom (the reaction coordinate) along which the potential is a maximum (with a corresponding imaginary frequency of vibration). The reaction coordinate is treated as a separable translation, so that the reaction rate coefficient can be envisaged as the equilibrium coefficient between transition state (minus one coordinate) and reactants multiplied by the (average) speed of the transition state structure over the saddle point in the potential energy surface. The canonical result for $k(T)$ is given by [1, 2]

$$k(T) = \frac{k_B T}{h} \frac{q_{AB\dagger}}{q_A q_B} \exp(-E_0 / k_B T) \quad (36.29)$$

for reactants A and B, where E_0 is the energy difference between the transition state and the reactants referred to zero-point levels, the $\dagger$ refers to the transition state, and the q are partition functions per unit volume. The partition functions can be factored into products representing electronic, vibrational, rotational, and translational degrees of freedom [34]. This formulation for $k(T)$ is known as the *activated complex theory* (ACT); a more appropriate name would be transition state theory since the term activated complex is also used to refer to an unstable state of a molecule in a deep potential well. The rate coefficient can also be written in terms of the thermodynamic parameters $\Delta H^\dagger$ and $\Delta S^\dagger$ [1].

Both the size and temperature dependence of the pre-exponential factor depend critically on the characteristics of transition state and reactants. Originally, ACT theory was used mainly to fit transition state characteristics to rate data. Increasingly accurate *ab initio* calculations of potential surfaces now allow purely theoretical determinations of $k(T)$ [35].

In addition to the assumption of a separable reaction coordinate that can be treated as a translation, several other assumptions have gone into the derivation of the ACT rate coefficient. First, molecules at the saddle point configuration (the transition state) have been arbitrarily chosen to be in equilibrium with reactants.

More recently, the transition state assumption has been generalized to refer to that portion of the potential surface in which the reaction flux or number of states is a minimum [36], whether or not this occurs at a well-defined saddle point. *Loose* transition states can thus be defined even if there is no barrier along the reaction coordinate [37]. The procedure can be undertaken for transition states as a function of temperature, energy, or, in its most detailed version, energy and angular momentum. The variational theorem shows rate coefficients determined in this way to be upper bounds to the true rate coefficients, the more detailed procedures leading to the better bounds [17, 36, 37]. Secondly, the implicit assumption is made that translation along the reaction coordinate at the transition state structure invariably leads to products. Sometimes a transmission coefficient κ is used as an unknown factor in the expression for $k(T)$ to account for the possibility that translation leads back to reactants instead. Thirdly, the assumption is made that two-body collisions can lead to canonical statistical equilibrium. This assumption becomes worse as temperature declines, because at low temperatures the long-range centrifugal constraints on angular momentum become more significant [3]. The influence of the long-range potential on ACT is contained in the statistical adiabatic theory discussed earlier [17]. Here loose transition states are defined as the maxima of effective adiabatic potential curves, and Eq. (36.29) must be modified since, as in detailed variational transition state theory, the transition states are themselves dependent on energy and angular momentum. A related problem is how to treat angular momentum constraints at low temperatures for potential surfaces which also contain a large short-range barrier or *tight* transition state. One approach is discussed in Sect. 36.2.4.

Rate coefficients derived from ACT do not indicate what the reaction products or sets of products might be. This information can be obtained from *ab initio* studies of the potential surface if different transition states lead to different products.

36.2.4 Complexes Followed by Barriers

Since long-range forces are always attractive, it makes sense to consider theories in which attraction and repulsion occur sequentially. For ion-molecule systems, this is especially important because many potential surfaces are monotonically attractive from long range to formation of a deep minimum at short range, but possess transition state barriers in their exit channels which are not large enough to prevent reaction but which affect the reaction dynamics. In addition, there are ion-molecule systems with potential surfaces closer to the norm for neutral-neutral species; in these systems there is only a weak long-range minimum followed by a short-range transi-

tion state barrier with energy above that of reactants. For the former type and, more arguably, for the latter type of potential surface, one can assume that the reaction proceeds through initial formation of an ergodic complex, followed by dissociation of the complex back into reactants or over the transition state barrier. If the reactants are labeled A and B, the complex AB^*, and the transition state $AB^\dagger$, the mechanism is

$$A + B \rightleftharpoons AB^* \longrightarrow AB^\dagger \longrightarrow \text{Product}, \qquad (36.30)$$

which leads to the steady-state rate law

$$d[A]/dt = -k[A][B], \qquad (36.31)$$

$$k = \frac{k_{\text{cf}}}{k_{\text{cd}} + k_{\text{cd}'}} k_{\text{cd}'}, \qquad (36.32)$$

where k_{cf}, k_{cd}, and $k_{\text{cd}'}$ are the rate coefficients for complex formation, redissociation into reactants, and dissociation into products over the transition state, respectively. One statistical approach to such systems is to use a capture theory for complex formation, to use the Klots formulation of complex redissociation into reactants, and to use a different theory for the unimolecular dissociation of the complex into products. Transition state theories of unimolecular complex decay have been studied for many years [1]. The current version is called RRKM theory, after the four authors Rice, Ramsperger, Kassel, and Marcus [5–8]. This theory builds upon the earlier RRK approach, in which random intramolecular vibrational energy transfer leads to large amounts of energy in the bond to be broken. An alternative hypothesis, in which amplitudes of well-defined normal modes add up to extend the bond to be broken past a certain amount, is now discredited [38]. In the RRKM approach, which is perfectly analogous to ACT theory, an equilibrium is envisaged between the activated complex and transition state species. The main use of the theory is in thermal unimolecular decay where it explains complex activation, deactivation, and random unimolecular dissociation as a function of gas density. At high density in a thermal environment, an equilibrium among complex, activated complex, and transition state leads to a thermal unimolecular decay rate [see Eq. (36.13)]

$$k_{\text{RRKM}}(T) = \frac{k_{\text{B}}T}{h} \frac{q_{AB^\dagger}}{q_{AB}} \exp(-E_0/k_{\text{B}}T), \qquad (36.33)$$

which is perfectly analogous to the ACT result. At lower pressures, the activated complex is not in thermal equilibrium with the stable complex and the details of complex activation and deactivation are important [7, 8].

In the microcanonical formulation of RRKM theory, the dissociation rate coefficient k_{RRKM} as a function of (activated) complex total energy E and angular momentum J is given by

$$k_{\text{RRKM}}(E, J) = \frac{N^\dagger(E - E_0 - E_{\text{rot}}(J))}{h\rho^*(E - E_{\text{rot}}(J))}, \qquad (36.34)$$

where $N^\dagger$ refers to the total number of vibrational states of the transition state from its minimum allowable saddle point energy E_0 through E, and ρ^* refers to the density of vibrational states of the complex at energy E. For both the transition state and the complex, the available vibrational energy is the total available energy minus the rotational energy $E_{\rm rot}(J)$, which is a function of the angular momentum. The energy not used for vibration and rotation in the transition state is considered to belong to the separable reaction coordinate. The most common expressions for the number and density of vibrational states are the semiclassical empirical values [1, 29]; direct counting schemes also exist [39]. Both empirical and direct counting refer to states representing a bath of harmonic oscillators; anharmonic effects are rarely treated.

A theory of reaction rates for the mechanism in Eq. (36.30) incorporating a capture theory, and the Klots and RRKM unimolecular decays has been applied successfully to a variety of ion-molecule reactions in competition with association channels for which the reaction proceeds via a deep well and an exit channel barrier (see Sect. 36.3.3) [40–42]. Some authors prefer to use RRKM theory to describe redissociation of the complex into reactants [8, 43] by means of a loose transition state [44], which can be defined via a variety of variational and other methods [37]. Some recent work on unimolecular decay of polyatomic ions and neutrals following excitation indicates the superiority of the variational RRKM approach to unimolecular decay [45], especially if there is a significant amount of energy. Both the Klots and RRKM approaches to complex redissociation have problems associated with them: the Klots expression obeys microscopic reversibility but requires both a capture cross section and the assumption that the phase space formulation of the probability of individual quantum states is accurate; the RRKM expression does not obviously obey microscopic reversibility and, unless a variational calculation is performed, can create a somewhat artificial transition state.

Reaction mechanisms with more than one potential barrier can also be treated via a combination of ACT, capture, and unimolecular approaches [46, 47].

36.2.5 The Role of Tunneling

The role of tunneling in bimolecular and unimolecular reactions grows in importance as the temperature is lowered and hopping over potential barriers becomes more difficult. A simple one-dimensional correction for tunneling in the ACT expression for bimolecular rate coefficients, obtained by Wigner [1, 48], appears to be reasonable if the tunneling mechanism is not dominant. This correction $\Gamma > 1$ is simply the quantum correction to the partition function for the reaction coordinate:

$$\Gamma \approx 1 - \frac{(h\nu_i)^2}{24(k_{\rm B}T)^2} + \cdots, \qquad (36.35)$$

where ν_i is the (imaginary) harmonic frequency at the saddle point. Improved corrections have also been developed [49] and, for the H + H$_2$ reaction, tested versus accurate quantum calculations [50]. At very low temperatures, such as those found in the interstellar medium, tunneling corrections are likely to be very large and to require proper multidimensional treatments.

A one-dimensional tunneling correction to the RRKM expression for the microcanonical unimolecular decay rate coefficient also exists [51]. The effective potential representing the reaction coordinate at the saddle point is assumed to be an Eckart barrier. The probability of tunneling for each vibrational state of the transition state is computed, and this probability takes the place of simply counting the state in the standard formula for $k_{\rm RRKM}$. In particular, $N^\dagger(E - E_0 - E_{\rm rot})$ in the equation for $k_{\rm RRKM}(E, J)$ is replaced by

$$N_{QM}^\dagger(E - E_0 - E_{\rm rot}) = \sum_n P(E - E_0 - E_{\rm rot} - \epsilon_n),$$
$$(36.36)$$

where ϵ_n is the vibrational energy of state n of the transition state with respect to its zero point energy, and the sum is over all states n for which the energy in the reaction coordinate (the energy in parentheses on the right-hand-side) is negative, but not so negative that the classical energy in the reaction coordinate lies below the minimum of the complex potential well. In general, only a few vibrational states of the transition state need be considered for the tunneling correction.

This tunneling correction has been incorporated into statistical theories for the rate of reactions that proceed via complexes and transition state barriers [47]. It has been used recently in a calculation of the slow ion-molecule reaction

$$\mathrm{NH_3^+ + H_2 \longrightarrow NH_4^+ + H}, \qquad (36.37)$$

which successfully reproduces the observation that, despite initially decreasing as the temperature is reduced below 300 K, the reaction rate coefficient begins to increase as the temperature is decreased below 100 K [52]. The parameters for the calculation were obtained from an *ab initio* calculation of the potential surface, which shows the system to proceed through a weakly-bound long-range complex before encountering a transition state with a rather small barrier of ≈ 0.25 eV. Although the dynamical theory is not quantitative (presumably because of the one-dimensional tunneling approximation), it does reproduce the isotope effects seen when the reactants $\mathrm{NH_3^+ + D_2}$ and $\mathrm{ND_3^+ + H_2}$ are used, definitively showing that tunneling is the cause. The actual increase in rate at very low temperature comes from the fact that the tunneling rate is less dependent on temperature than the dissociation rate of the complex into reactants. Similar calculations have been performed for the analogous ion-molecule reaction

$$C_2H_2^+ + H_2 \longrightarrow C_2H_3^+ + H \qquad (36.38)$$

to explain a similar observation, although in this latter instance there is still a controversy concerning whether or not the reaction is truly exoergic [53]. It is interesting to speculate on whether similar effects can be detected for analogous neutral radical–H_2 reactants with moderate activation energy barriers, such as

$$CCH + H_2 \longrightarrow C_2H_2 + H. \qquad (36.39)$$

This system possesses a transition state of rather moderate energy [35], but the only long-range complex is presumably the very weakly-bound van der Waals structure. Unlike the ion-molecule case, there are very few experimental studies of low temperature neutral-neutral reactions. Studies of pressure broadening at very low temperatures do reveal, however, the strong influence of the van der Waals bond [54].

36.3   ASSOCIATION REACTIONS

Association reactions have been studied for both ion-molecule and neutral-neutral systems. For the ion-molecule case, association processes can be investigated in great detail at low densities near or in the radiative association regime because ions can be stored in low pressure traps. Recent review articles [55, 56] describe such experiments for small and large ions, respectively. The results of many higher pressure studies of termolecular ion-molecule association, especially those undertaken in the SIFT apparatus, are also available [57]. In general, below the high pressure limit, the rate of association reactions without activation energy is known to increase with (a) an increase in the size of the reactants, (b) an increase in the bond energy of the product species, and (c) a decrease in temperature. These trends are all reproduced by statistical theories in which association proceeds by the formation of a complex followed by radiative and/or collisional stabilization [3]. More limited data for termolecular neutral-neutral reactions, with and without short-range potential barriers, have also been reviewed [58]. Although radiative stabilization can proceed via a single photon, collisional stabilization proceeds most probably in a stepwise fashion rather than in a single strong collision. The most detailed statistical theories incorporating multistep collisional stabilization use master equation techniques; in general it is preferable to solve the inverse problem of unimolecular dissociation via detailed RRKM theory and then invoke detailed balance [8, 43]. Such theories are more successful than strong

collision approaches, especially in considering the dependence of ternary association rates on pressure. This dependence can be especially difficult to treat when association competes with an exoergic channel which does not necessarily dominate because of a barrier in the exit channel.

36.3.1   Radiative Stabilization

The problem of radiative association in the absence of competitive exoergic channels is in principle much more simple. In addition to the rates of complex formation and dissociation, one needs only the rate of stabilization via spontaneous emission. If the complex abundance is at steady state, the rate law for radiative association of species A and B is simply

$$d[A]/dt = -k_{ra}[A][B], \qquad (36.40)$$

$$k_{ra} = \frac{k_{cf}}{k_{cd} + k_{rad}} k_{rad}, \qquad (36.41)$$

where k_{ra} represents the rate coefficient for radiative association and k_{rad} refers to the rate coefficient for radiative stabilization. If ternary association is considered in addition, the rate coefficient for radiative association is the low pressure limit of a more complex rate expression.

Radiative stabilization, which is normally considered to proceed via emission of a single infrared photon, has been studied in some detail [59–61]. If the unstable complex is imagined to be a collection of harmonic oscillators, each vibrational state i, which is best regarded as a Feshbach resonance, can be expressed as a set of occupation numbers $n_1^i, n_2^i, \ldots, [n^i]$ for the assorted modes. If it is further assumed that upon formation, the complex has a probability P_i of being formed in state i, the rate of single-photon vibrational emission k_{rad} (s^{-1}) is then given by

$$k_{rad} = \sum_{i,j} P_i A_{i->j} = \sum_{[n^i],[n^j]} P_{[n^i]} A_{[n^i]->[n^j]}, \qquad (36.42)$$

where the sum is over initial states i and final states j, and A is the Einstein coefficient for spontaneous emission. With some algebraic manipulation and the assumption that dipole selection rules apply, the result simplifies to

$$k_{rad} = \sum_{k=1}^{s} \sum_{n_k} P_{n_k}^k n_k A_{1->0}^k, \qquad (36.43)$$

where the index k refers to each of the s normal modes, the index n_k to the occupation number of mode k, and the Einstein A coefficient, which here refers to the fundamental transition of mode k, can be expressed in terms of the absorption intensity (see Chap. 10). The probability $P_{n_k}^k$ that mode k is excited to state n_k is given by the statistical formula

$$P^k_{n_k} = \rho'_{\text{vib}}(E_{\text{vib}} - n_k h\nu_k)/\rho_{\text{vib}}(E_{\text{vib}}), \qquad (36.44)$$

where ρ_{vib} is the vibrational density of states and ρ'_{vib} is the reduced density of states when n_k quanta are assigned to mode k. The role of overtone and combination contributions to the radiative emission rate has also been investigated [60], as have a canonical approximation to the microcanonical formulation discussed here [61] and the small additional effect of stabilization by sequential emission [59]. With the use of the classical approximation to the standard vibrational density of state functions [29], Eq. (36.43) for k_{rad} reduces to a linear function of the vibrational energy:

$$k_{\text{rad}} = \frac{E_{\text{vib}}}{s} \sum_{k=1}^{s} \frac{A^k_{1->0}}{h\nu_k}. \qquad (36.45)$$

The constant of proportionality between k_{rad} and E_{vib} depends strongly on the fundamental intensities; many of these can be obtained from infrared absorption spectra, although there is normally insufficient information for all of the active modes in polyatomic molecules, especially ions. Reasonable estimates, as well as limited experimental data, show that k_{rad} varies between 10 and 1000 s^{-1} for vibrational energies up to a few eV [55, 56, 59–61]. The radiative stabilization rate typically decreases slightly with increasing molecular size due to the inclusion of infrared inactive modes of vibration. There also appears to be a distinction between ions and neutral species, with rates for ions apparently faster.

Since spontaneous emission rates between electronic states are often far more rapid than infrared vibrational rates, the possibility that radiative stabilization can proceed via electronic emission has been examined [4]. The well-studied association between C$^+$ and H$_2$ appears to proceed via electronic relaxation, with a radiative stabilization rate one to two orders of magnitude larger than would be the case for vibrational emission. Whether or not the association between CH$_3^+$ and H$_2$ proceeds via an electronically excited complex has been debated for some time; the answer is apparently negative [55].

36.3.2 Complex Formation and Dissociation

The formation and dissociation of the complex can be studied with a variety of statistical approximations other than what is discussed in this Section. In particular, complex dissociation has been treated by a thermal RRKM approach [56, 61], an energy but not angular momentum specific RRKM approach [59], variational transition state theory, statistical adiabatic theory, and flexible transition state theory [58]. We first consider a simple thermal approximation for systems without activation energy incorporating microscopic reversibility

[3]. In the limit that the complex redissociation into reactants is much more rapid than radiative stabilization, the rate coefficient k_{ra} for radiative association in the thermal model is

$$k_{\text{ra}} = K(T)k_{\text{rad}}, \qquad (36.46)$$

where $K(T)$ is the ratio between k_{cf} and k_{cd} in a canonical ensemble, and the radiative stabilization rate has been assumed to be independent of temperature. The equilibrium coefficient $K(T)$ between complex and reactants can be rewritten in terms of partition functions in the standard way [34]. However, the partition function of the complex is best calculated via the equation

$$q_{AB^*} = \int_0^\infty \rho_{\text{vr}}(E + D_0)\exp(-E/k_BT)\,dE,$$
$$\approx \rho_{\text{vr}}(D_0)k_BT, \quad D_0 \gg k_BT, \qquad (36.47)$$

where ρ_{vr} is the density of vibrational-rotational states [29], E is the energy of reactants, and D_0 is the bond energy of the complex. One immediate prediction of this approach (deriving mainly from the rotational partition functions of the reactants) is that k_{ra} possesses a strong inverse dependence on temperature since

$$K(T) \propto T^{-(r_A+r_B+1)/2}, \qquad (36.48)$$

where r refers to the number of rotational degrees of freedom (two for a linear molecule and three for a nonlinear one). It also predicts a strong dependence on well depth and size of the complex since the density of complex states is a strong function of both these parameters. Although the thermal model agrees with the strong inverse temperature dependence of the coefficients for both radiative and, more commonly, ternary ion-molecule association [3, 57] (in which case, the thermal model with strong collisions replaces k_{rad} with a collisional rate coefficient), the absolute rate coefficients calculated tend to range up to an order of magnitude too high. Given the large range of values exhibited by radiative association rate coefficients (10^{-20}–10^{-9} cm^3 s^{-1}), this might not seem too large a problem. It has been shown, however, that the thermal model is deficient, especially at low temperatures because thermal equilibrium cannot be achieved by two-body collisions [3]. A revised approach, called the *modified thermal theory*, replaces the rovibrational density of complex states in the thermal model with a vibrational density of states coupled with sums over the allowable ranges of complex angular momentum achievable from the specific capture model assumed [3]. The result is a somewhat lessened inverse temperature dependence and somewhat smaller rate coefficients, especially at low temperatures. Both modifications result in better agreement with experiment [3].

The modified thermal approach considers structureless reactants; full consideration of the internal states of the reactants is achieved via the Klots version of phase space theory, in which k_{cf} and k_{cd} in Eq. (36.41) are the capture and unimolecular rates discussed in Sect. 36.2.2. The phase space treatment reduces to the modified thermal treatment for small reactant angular momentum if the possibility of saturation ($k_{cd}(E, J) < k_{rad}$) may be ignored. Both conditions are normally met. The phase space approach has also been used for ion-molecule ternary association reactions, but here is coupled with the strong collision hypothesis and must be regarded as inferior to the more detailed RRKM calculations with master equation treatments for collisional stabilization [8, 43].

36.3.3 Competition with Exoergic Channels

There have been numerous reactions reported (especially ion-molecule reactions) in which exoergic reaction channels compete with association channels, both radiative (at low pressure) and ternary (at higher pressures). One view of such competition is that it occurs via a sequential mechanism in which the reactants form a long-lived collision complex which either redissociates, dissociates into exoergic products, or is stabilized [40, 41]. The rate coefficient for competitive radiative association is then given by the steady-state expression

$$k_{ra} = \frac{k_{cf}}{k_{cd} + k_{cd'} + k_{rad}} k_{rad}, \qquad (36.49)$$

where $k_{cd'}$ refers to complex dissociation into products. The additional and normally large $k_{cd'}$ term in the denominator means that the thermal and modified thermal theories cannot, in general, be used. The phase space treatment for this mechanism [40, 41, 62, 63] shows that association cannot compete with exoergic channels unless there is a significant barrier in the exit channel which considerably slows the dissociation rate of the complex into products, especially for partial waves of high angular momentum. Such barriers tend to be large enough to slow dissociation down, but not large enough to require tunneling. Although the results of phase space calculations are in good agreement with experiment for a variety of competitive systems [55, 62], they are once again inferior to RRKM calculations with master equation collisional deactivation when the competition involves collisional stabilization of the complex [8, 40, 41, 43].

Another mechanism exists for competition between association and normal exoergic channels [55]. *Ab initio* studies show that there is a parallel type of competition in which the product and association channels occur on different portions of the potential surface, with a branching at long range in the entrance channel. For example, the competing reactions [55, 64]

$$C_2H_2^+ + H_2 \longrightarrow C_2H_3^+ + H, \qquad (36.50)$$
$$C_2H_2^+ + H_2 \longrightarrow C_2H_4^+, \qquad (36.51)$$

occur via distinct pathways. The former reaction is a possibly endoergic direct process in which the molecular hydrogen attacks perpendicularly, leading to the cyclic form of the $C_2H_3^+$ ion, whereas the association reaction occurs via the deep well of the ethylene ion. The competition in the analogous $C_3H^+ + H_2$ system is not currently well understood, with both parallel and series mechanisms suggested [55, 65].

REFERENCES

1. H. S. Johnston, *Gas Phase Reaction Rate Theory* (Ronald, New York, 1966).
2. R. D. Levine and R. B. Bernstein, *Molecular Reaction Dynamics* (Oxford Univ. Press, New York, 1974).
3. D. R. Bates and E. Herbst, in *Rate Coefficients in Astrochemistry*, edited by T. J. Millar and D. A. Williams (Kluwer, Dordrecht, 1988), p. 17.
4. E. Herbst and D. R. Bates, Astrophys. J. **329**, 410 (1988).
5. P. J. Robinson and K. A. Holbrook, *Unimolecular Reactions* (Wiley-Interscience, New York, 1972).
6. W. Forst, *Theory of Unimolecular Reactions* (Academic, New York, 1973).
7. J. Troe, J. Phys. Chem. **83**, 114 (1979).
8. R. G. Gilbert and S. C. Smith, *Theory of Unimolecular and Recombination Reactions* (Blackwell, Oxford, 1990).
9. W. Jakubetz, D. Sokolovski, J. N. L. Connor, and G. C. Schatz, J. Chem. Phys. **97**, 6451 (1992).
10. S. M. Auerbach and W. H. Miller, J. Chem. Phys. **98**, 6917 (1993).
11. J. Troe, Adv. Chem. Phys. LXXXII, Part 2, 485 (1992).
12. T. Su and M. T. Bowers, in *Gas Phase Ion Chemistry*, edited by M. T. Bowers (Academic, New York, 1979), Vol. 1, p. 83.
13. B. R. Rowe, in *Rate Coefficients in Astrochemistry*, edited by T. J. Millar and D. A. Williams (Kluwer, Dordrecht, 1988), p. 135.
14. V. G. Anicich and W. T. Huntress, Astrophys. J. Suppl. **62**, 553 (1986).
15. J. O. Hirschfelder, C. F. Curtiss, and R. B. Bird, *Molecular Theory of Gases and Liquids* (Wiley, New York, 1954).
16. D. C. Clary, N. Halder, D. Husain, and M. Kabir, Astrophys. J. **422**, 416 (1994).
17. S. C. Smith and J. Troe, J. Chem. Phys. **97**, 8820 (1992).
18. D. C. Clary, in *Rate Coefficients in Astrochemistry*, edited by T. J. Millar and D. A. Williams (Kluwer, Dordrecht, 1988), p. 1.
19. D. C. Clary, J. Phys. Chem. **91**, 1718 (1987).
20. N. G. Adams, D. Smith, and D. C. Clary, Astrophys. J. **296**, L31 (1985).
21. D. R. Bates and W. L. Morgan, J. Chem. Phys. **87**, 2611 (1987).
22. S. C. Smith, M. J. McEwan, and R. G. Gilbert, J. Phys. Chem. **93**, 8142 (1989).
23. M. M. Graff, Astrophys. J. **339**, 239 (1989).

24. D. C. Clary, T. S. Stoecklin, and A. G. Wickham, J. Chem. Soc. Faraday Trans. **89**, 2185 (1993).

25. B. R. Rowe, A. Canosa, and I. R. Sims, J. Chem. Soc. Faraday Trans. **89**, 2193 (1993).

26. J. C. Light, Disc. Faraday Soc. **44**, 14 (1967).

27. W. J. Chesnavich and M. T. Bowers, in *Gas Phase Ion Chemistry*, edited by M. T. Bowers (Academic, New York, 1979), p. 119.

28. D. Gerlich, J. Chem. Phys. **90**, 3574 (1989).

29. G. Z. Whitten and B. S. Rabinovitch, J. Chem. Phys. **38**, 2466 (1963); *ibid.*, **41**, 1883 (1964).

30. L. M. Bass, W. J. Chesnavich, and M. T. Bowers, J. Am. Chem. Soc. **101**, 5493 (1979).

31. C. E. Klots, J. Phys. Chem. **75**, 1526 (1971).

32. E. Herbst and S. K. Knudson, Chem. Phys. **55**, 293 (1981).

33. E. T. Galloway and E. Herbst, Astrophys. J. **376**, 531 (1991).

34. T. L. Hill, *An Introduction to Statistical Thermodynamics* (Addison-Wesley, Reading, 1960).

35. L. B. Harding, G. C. Schatz, and R. A. Chiles, J. Chem. Phys. **76**, 5172 (1982).

36. D. G. Truhlar and B. C. Garrett, Acc. Chem. Res. **13**, 440 (1980).

37. W. L. Hase and D. M. Wardlaw, in *Bimolecular Collisions*, edited by M. N. R. Ashfold and J. E. Baggott (Royal Society of Chemistry, London, 1989), p. 171.

38. N. B. Slater, *Theory of Unimolecular Reactions* (Cornell University Press, Ithaca, 1959).

39. S. E. Stein and B. S. Rabinovitch, J. Chem. Phys. **58**, 2438 (1973).

40. L. M. Bass, R. D. Cates, M. F. Jarrold, N. J. Kirchner, and M. T. Bowers, J. Am. Chem. Soc. **105**, 7024 (1983).

41. E. Herbst, J. Chem. Phys. **82**, 4017 (1985).

42. S. Wlodek, D. K. Bohme, and E. Herbst, Mon. Not. R. Astron. Soc. **242**, 674 (1990).

43. S. C. Smith, P. F. Wilson, P. Sudkeaw, R. G. A. R. MacLagan, M. J. McEwan, V. G. Anicich, and W. T. Huntress, J. Chem. Phys. **98**, 1944 (1993).

44. S. C. Smith, J. Chem. Phys. **97**, 2406 (1993).

45. S. J. Klippenstein, J. D. Faulk, and R. C. Dunbar, J. Chem. Phys. **98**, 243 (1993).

46. Y. Chen, A. Rauk, and E. Tschuikow-Roux, J. Phys. Chem. **95**, 9900 (1991).

47. M. J. Frost, P. Sharkey, and I. W. M. Smith, J. Phys. Chem. **97**, 12254 (1993).

48. E. P. Wigner, Z. Phys. Chem. **B19**, 203 (1932).

49. T. N. Truong and D. G. Truhlar, J. Chem. Phys. **93**, 1761 (1990); *ibid.*, **97**, 8820 (1993).

50. T. Takayanagi, N. Masaki, K. Nakamura, M. Okamota, S. Sato, and G. C. Schatz, J. Chem. Phys. **86**, 6133 (1987).

51. W. H. Miller, J. Am. Chem. Soc. **101**, 6810 (1979).

52. E. Herbst, D. J. DeFrees, D. Talbi, F. Pauzat, W. Koch, and A. D. McLean, J. Chem. Phys. **94**, 7842 (1991).

53. D. Smith, J. Glosik, V. Skalsky, P. Spanel, and W. Lindinger, Int. J. Mass Spectrom. Ion Proc. **129**, 145 (1993).

54. D. C. Flatin, J. J. Holton, M. M. Beaky, T. M. Goyette, and F. C. De Lucia, J. Mol. Spectrosc. **164**, 425 (1994).

55. D. Gerlich and S. Horning, Chem. Rev. **92**, 1509 (1992).

56. R. C. Dunbar, in *Unimolecular and Bimolecular Ion-Molecule Reaction Dynamics*, edited by C. Y. Ng, T. Baer, and I. Powis (Wiley, New York, 1994), p. 270.

57. N. G. Adams and D. Smith, in *Reactions of Small Transient Species: Kinetics and Energetics*, edited by A. Fonijn and M. A. A. Clyne (Academic, New York, 1983), p. 311.

58. J. W. Davies and M. J. Pilling, in *Bimolecular Collisions*, edited by M. N. R. Ashfold and J. E. Baggott (Royal Society of Chemistry, London, 1989), p. 105.

59. J. R. Barker, J. Phys. Chem. **96**, 7361 (1992).

60. E. Herbst, Chem. Phys. **65**, 185 (1982).

61. R. C. Dunbar, Int. J. Mass Spectrom. Ion Proc. **100**, 423 (1990).

62. E. Herbst and M. J. McEwan, Astron. Astrophys. **229**, 201 (1990).

63. E. Herbst and R. C. Dunbar, Mon. Not. R. Astron. Soc. **253**, 341 (1991).

64. S. A. Maluendes, A. D. McLean, and E. Herbst, Chem. Phys. Lett. **217**, 571 (1994).

65. S. A. Maluendes, A. D. McLean, K. Yamashita, and E. Herbst, J. Chem. Phys. **99**, 2812 (1993).

37

Gas Phase Ionic Reactions

Nigel G. Adams

Department of Chemistry, University of Georgia, Athens, Georgia 30602

37.1 INTRODUCTION

Ionic reactions in the gas phase is a broad field encompassing a multitude of interactions between ions (both positively and negatively charged), electrons and neutrals. These can be as simple as the transfer of an electron between molecules, or can be complex with considerable bond breaking, reforming and rearrangement. Reactions of the general type

$$A^{+,-} + B^{-,n} \rightarrow \text{products}, \qquad (37.1)$$

illustrating this diversity, are given in Table 37.1. The reaction process can be considered as consisting of three parts: (a) the initial interaction, in which the colliding particles, $A^{+,-}$ and $B^{-,n}$ are drawn together by an attractive potential, (b) the reaction intermediate and transition state, in which reactants are transformed into products and (c) the weakening interaction as the product particles separate. Interactions can also occur where only elastic scattering is involved and these are considered in Chap. 65. Part (a) of the interaction is readily studied theoretically by classical mechanics in terms of reactant particle motions controlled by the interaction

potential $u(r)$ between the particles, where r is the interparticle separation. For ion-neutral interactions, this can take the form (i) of an attractive ion-polarization or induced dipole potential [1] $u(r) = -\alpha_\mathrm{d} q^2/2r^4$, where α_d is the polarizability of the neutral and q is the charge on the ion; anisotropy can also be taken into account in the polarizability [2], (ii) of an ion-permanent dipole potential [3] with $u(r) = -q\mu_D \cos\theta/r^2$, where μ_D is the permanent dipole of the neutral and θ is the angle that the dipole makes with r, and (iii) of an ion-quadrupole potential [4] with $u(r) = -Qq(3\cos^2\theta - 1)/2r^3$, where Q is the quadrupole moment and θ is the angle the quadrupole axis makes with r. Such capture theories are considered in Chap. 36. Other interaction potentials can be considered, but these are of lesser significance. Coulombic interaction potentials $u(r) = -q_1 q_2/r$, where q_1 and q_2 are the charges on the interacting particles, are appropriate for positive ion–negative ion recombination (more correctly termed mutual neutralization) and electron-ion recombination. For processes involving electrons, i.e., electron attachment and electron-ion recombination, the wave nature of the electron also has to be considered [5, 6].

In addition to providing a means of bringing the reactant particles together, the attractive interaction

Table 37.1. Examples illustrating the range of ionic reactions that can occur in the gas phase.

Reaction Process	Reaction Type
$Ar^+ + O_2 \rightarrow O_2^+ + Ar$ $O_2^- + NO_2 \rightarrow NO_2^- + O_2$	Nondissociative charge transfer/charge exchange
$He^+ + O_2 \rightarrow O^+ + O + He$	Dissociative charge transfer
$O^+ + N_2 \rightarrow NO^+ + N$	Ion/atom interchange or atom abstraction
$O^- + CH_4 \rightarrow OH^- + CH_3$	Atom abstraction
$H_3O^+ + HCN \rightarrow H_2CN^+ + H_2O$ $OH^- + HCN \rightarrow CN^- + H_2O$	Proton transfer
$OH^- + H \rightarrow H_2O + e^-$	Associative detachment
$CH_3^+ + H_2 + M \rightarrow CH_5^+ + M$ $Cl^- + BCl_3 + M \rightarrow BCl_4^- + M$	Ternary collisional association
$He_2^+ + e^- \rightarrow He_2(2He) + h\nu$	Radiative electron-ion recombination
$HCO^+ + e^- \rightarrow H + CO$	Dissociative electron-ion recombination
$NO^+ + NO_2^- \rightarrow NO + NO_2$	Ion-ion recombination (mutual neutralization)
$e^- + CCl_4 \rightarrow Cl^- + CCl_3$	Dissociative electron attachment
$e^- + C_7F_{14} \rightarrow C_7F_{14}^-$	Nondissociative electron attachment
$e^- + O_2 + M \rightarrow O_2^- + M$	Ternary collision stabilized attachment
$^{13}C^+ + {}^{12}CO \rightleftharpoons {}^{12}C^+ + {}^{13}CO$ $OD^- + NH_3 \rightleftharpoons OH^- + NH_2D$	Isotope exchange

potential has another function. In part (b) of the reaction mechanism, where considerable rearrangement of the atoms in the colliding species may occur (i.e., in the intermediate complex and transition state), there may be barriers to reaction. Here, the kinetic energy gained from the interaction potential is available in the intermediate complex for overcoming such energy barriers (very much less interaction energy is available in neutral-neutral reactions and the effects of energy barriers are very much more evident [7]). Of course, this energy has to be reconverted to potential energy as the product particles separate, and is therefore not available to drive the overall reaction. However, the amount of energy returned may differ from that initially converted to kinetic energy since the polarizability/dipole moment of the product neutral(s) may differ from those of the reactant neutrals. Generally, the final part (c) of the interaction has little influence on the magnitude of the reaction rate coefficient or on the product distribution once the products have significantly separated (i.e., beyond the range at which processes such as long range electron transfer can occur). Thus, to a major degree, part (b) of the mechanism and the reaction energetics determine the products of the reaction. Theories which address this part of the interaction are also discussed in Chap. 36 and these have met with mixed success. Thus,

at present, experimental measurements are providing a more definitive understanding of reactions and their mechanisms.

Also, the situation is not in general clear as to the form of the thermodynamic energy that governs whether a reaction will proceed spontaneously, i.e., whether it is controlled by the enthalpy change ΔH in the reaction or by the Gibbs' free energy change $\Delta G = \Delta H - T\Delta S$, where ΔS is the entropy change in the reaction [8]. This is discussed in more detail in Sect. 37.2.

Rate coefficients and product distributions for these reaction processes are important in all ionized media where molecular species exist, such as interstellar gas clouds [9], planetary atmospheres [10] (including that of the Earth [11, 12]), comets (Chap. 79), the space shuttle environment [13], laser plasmas [14], and plasmas used to etch semiconductors [15], etc.

37.2 REACTION ENERGETICS

The availability of sufficient energy is a primary consideration for determining whether a reaction can proceed spontaneously. Criteria for determining whether energy is given out or absorbed in a reaction are (i) exo- or endoergicity, ΔE representing the internal energy change involved in a single interaction, (ii) exo- or endothermicity for an ensemble of particles in thermal equilibrium as defined by the enthalpy change per mole, ΔH, in the reaction and (iii) exergonic or endergonic as defined by the Gibbs' free energy change, ΔG [16]. ΔE is related to ΔH at temperature T by

$$\Delta H_T \approx -N\Delta E_T, \tag{37.2}$$

where N is Avogadro's number. ΔE is usually deduced from bond energies [17–19], ionization potentials [20], electron affinities [21,22], proton affinities [21,23,24], gas phase basicities [21,23–25], etc. ΔH_T° is determined from the heats of formation $H_{f,T}^\circ$ by [26]

$$\Delta H_T^\circ = \sum_{products} H_{f,T}^\circ - \sum_{reactants} H_{f,T}^\circ, \tag{37.3}$$

where the superscript o refers to the standard state [16] of the reactants and products [for example, see Eq. (37.1)]. ΔG_T° is determined from ΔH_T° and the entropy change ΔS_T° by

$$\Delta G_T^\circ = \Delta H_T^\circ - T\Delta S_T^\circ, \tag{37.4}$$

where

$$\Delta S_T^\circ = \sum_{products} S_T^\circ - \sum_{reactants} S_T^\circ, \tag{37.5}$$

and the S_T° are the standard entropies [16]. In cases where all of the $H_{f,T}^\circ$ and S_T° are not available, they can often be deduced by constructing other reactions involving the species of interest and other species for which the required thermodynamic information is known (Ref. [16] discusses the details of ways in which this can be achieved). Alternatively, the magnitude of a thermodynamic parameter can be calculated using equilibrium statistical thermodynamics, if the energies of all of the occupied molecular energy levels are known [27]. Thus, for studies where all reacting particles of a given type have the same energy, such as in beam/beam interactions where cross sections $\sigma(E)$ are measured, then ΔE is most appropriate. Alternatively, for reactions involving an ensemble of particles in thermal equilibrium at a temperature T, such as those studied in high pressure mass spectrometer ion sources or afterglows where rate coefficients $k(T)$ are measured, ΔH and ΔG are more appropriate. (The terms "rate coefficient" and "rate constant" are used interchangeably in the literature.) $\sigma(E)$ and $k(T)$ are directly related via

$$k(T) = \int_0^\infty v(E)f(E)\sigma(E)\,dE, \tag{37.6}$$

where $v(E)$ is the relative velocity of the reactants and $f(E)$ is the Maxwell-Boltzmann energy distribution. Whether ΔH or ΔG is more important for determining

reaction spontaneity depends on the degree of interaction of the reacting systems with the surroundings during the course of the whole reaction process. For example, if the reactions occur at low pressure such that the reaction time is much less than the collision time with the background gas, there will be no interactions with the surroundings, and the only energy available in the reactions will be ΔH. Here, ΔS can only determine the probability that the intermediate complex dissociates forward to products or back to reactants. At the other extreme, if the reactions are conducted at high pressure such that the reaction species are always in thermal equilibrium with the surroundings (the limiting case of this is reaction in solution), then the additional energy $T\Delta S$ is available, and ΔG determines whether the reactions are spontaneous. Obviously, for intermediate pressures, there is a varying degree of contact with the surroundings, and which of ΔH and ΔG is most applicable is more obscure. The definition of surrounding is also somewhat loose, since in the present context it represents anything that can provide a source of energy during the reaction (for example, vibrational modes of the reacting species that are not involved in the interaction and which could be cooled as the reaction proceeds). Usually, the $T\Delta S$ term is not sufficiently large that ΔH and ΔG have different signs so that confusion about the spontaneity of reactions often does not arise. Recently, however, reactions have been discovered which appear to proceed spontaneously even though ΔH is positive, and there has been considerable discussion of this [8].

Eventually, all reaction processes will reach equilibrium as defined by

$$\Delta G = -RT \ln K, \tag{37.7}$$

where

$$K = (p_C^c\, p_D^d)/(p_A^a\, p_B^b) \tag{37.8}$$

is the equilibrium constant for the reaction

$$aA + bB \rightleftharpoons cC + dD. \tag{37.9}$$

The subscripts to the pressures p denote the components A to D, and the superscripts represent the stoichiometries [16,27]. The equilibrium constant K also equals k_f/k_r, the ratio of the forward to reverse rate coefficients, obtained for thermalized particles at temperature T. Thus, from measurements of k or the equilibrium constant at a series of temperatures, ΔH and ΔS of the reaction can be obtained.

37.3 CHEMICAL KINETICS

The rate coefficient k for the reaction process

$$A + B \rightarrow C + D \qquad (37.10)$$

is defined by the rate equation [28]

$$\frac{d[A]}{dt} = \frac{d[B]}{dt} = -k[A][B], \qquad (37.11)$$

where the square parentheses represent the concentrations of the enclosed species, and the units of k are typically cm^3 $molec^{-1}$ s^{-1} (often $molec^{-1}$ is not written explicitly in the units). These k are deduced from the time variation of specific concentrations of thermalized reactants under a variety of conditions. For ion-neutral reactions, with A being the positive or negative ion species, the situation can usually be achieved in the laboratory where $[B] \gg [A]$, i.e., a pseudo first-order reaction. Simple integration yields

$$[A]_t = [A]_0 \exp(-k[B]_0 t), \qquad (37.12)$$

where the subscript 0 indicates that this concentration is time invariant. Similar circumstances apply to binary electron attachment reactions, where $[A] = [e]$, the electron number density (the symbol β is often used here in place of k). If the reaction proceeds by association that is stabilized by collision with a third body M then

$$A + B + M \rightarrow AB + M, \qquad (37.13)$$

and the solution is

$$[A]_t = [A]_0 \exp(-k[B]_0[M]_0 t) \qquad (37.14)$$

with k now having units of cm^6 $molec^{-2}$ s^{-1}. Note that situations exist where binary and ternary reactions can occur simultaneously.

If $[B] \ggg [A]$, the solution of Eq. (37.11) is different. For the special case $[B] = [A]$, integration yields

$$1/[A]_t - 1/[A]_0 = kt, \qquad (37.15)$$

and similarly for $[B]$. This situation is usually achieved in the laboratory for the determination of k for electron-ion and ion-ion recombination (the symbols α_e and α_i respectively are often used to replace k here). In many applications where several processes contribute to $[A]$ and $[B]$, and where the simple limits are not applicable, the situation has to be analyzed numerically.

37.4 REACTION PROCESSES

The wide variety of possible reaction processes is listed in Table 37.1 by specific examples. However, the reaction mechanisms illustrated are completely general. Magnitudes of the rate coefficients for binary interactions between charged and neutral particles, vary from 10^{-7} cm^3 s^{-1} for electron attachment [29,30] to 10^{-8} or 10^{-9} cm^3 s^{-1} for unit efficiency ion-molecule reactions

[31,32]. The efficiency is defined as the ratio of the measured k to the theoretical collisional value, i.e., that determined using the appropriate interaction potential. Ion-molecule reactions involving molecules with large permanent dipoles (for example, HCN at 2.98 Debye and HCl at 1.08 Debye [33]), can have $k > 10^{-7}$ cm^3 s^{-1} at low T [33–35] due to locking of the dipole along the line joining the reactant species, thereby maximizing the strength of the interaction. Rate coefficients can be much smaller (by orders of magnitude) than these upper limits if the efficiency of part (b) of the mechanism is small.

For dissociative electron-ion recombination, the upper limits on k are generally much larger as a consequence of the long range Coulombic interaction potential, varying from $\sim 10^{-7}$ cm^3 s^{-1} for diatomic ions to $> 10^{-6}$ cm^3 s^{-1} for more polyatomic species [36–38]. For ion-ion recombination, the rate coefficients are about an order of magnitude smaller than these values because of the larger mass of the negative ion relative to the electron, and thus the smaller interaction velocity (the σ of the two processes are, in fact, similar) [39]. Electron-ion recombinations that are radiatively stabilized generally have a small k, as small as $\sim 10^{-12}$ cm^3 s^{-1} [40], because of the large magnitude of the radiative lifetime of the reaction intermediate compared with the time for autoionization. For most reaction types, experimental rate coefficients are available at room temperature, and for some, the temperature and/or energy dependencies have also been determined [29,30,37–39,41]. Generally, only the ion products are identified using mass spectrometry, and the product neutrals are not determined (unless the energetics allow only one neutral product), and the states of excitation of the products are not identified.

For ternary processes involving association, k varies from totally saturated at the binary collision limit to very small ($\sim 10^{-32}$ cm^6 s^{-1}) for ion-neutral reactions [31,32] and similarly for electron attachment [42]. Little is known about collision stabilized recombination [43]. Each type of reaction process has its own characteristic behavior and dependence on temperature. Rate coefficients, temperature dependencies, and product distributions (where available) have been tabulated for ion-neutral reactions [31,32,44] and electron attachment [29,30]. Less data are available for electron-ion and ion-ion recombination, so no attempts have been made to compile these. Data are available in Refs. [6,36–39].

37.4.1 Binary Ion-Neutral Reactions

Charge Transfer and Charge Exchange

These processes involve the exchange of an electron which can occur at relatively large interparticle separations (i.e., up to ~ 6 Å [45]). Thus, in principle, k values can be larger than the collision limiting value. Such reactions are, in general, relatively fast, although there are some notable exceptions (for example, $He^+ + H_2$, $k = 1 \times 10^{-13}$ cm^3 s^{-1}; $Ne^+ + H_2$, $k < 2 \times 10^{-14}$ and $Ne^+ + N_2$, $k = 1.1 \times 10^{-13}$ [31]). Attempts have been made to relate the efficiency of charge transfer to the Franck Condon overlap between the neutral reactant and the product ion, with mixed success [46]. More energy is generally available in the positive ion reactions than the negative ion reactions since ionization potentials are much larger than electron affinities, and thus more dissociative products would be expected, as is usually observed. The reaction

$$He^+ + N_2 \rightarrow \begin{cases} N^+ + N + He + 0.28 \text{ eV} \\ N_2^+ + He + 9.00 \text{ eV}, \end{cases}$$

is particularly interesting and has been studied in considerable detail. The product distribution slightly favors N^+ (60%) rather than N_2^+ [47], with the N^+ channel becoming even more important with increasing vibrational temperature of the N_2 [48]. Spectroscopic emission studies [49] have shown that a significant fraction ($\geq 5\%$) of the reactions proceed by charge transfer into the N_2^+ ($C\,^2\Sigma_g^+$) state followed by the radiative decay

$$N_2^+(C\,^2\Sigma_u^+) \rightarrow N_2^+(X\,^2\Sigma_g^+) + h\nu. \qquad (37.16)$$

This channel competes with the predissociation

$$N_2^+(C\,^2\Sigma_u^+) \rightarrow N^+ + N. \qquad (37.17)$$

Proton Transfer

Where proton transfer is significantly exoergic, as determined by the difference in the proton affinities of the reactant and product neutrals, reaction usually proceeds at the collisional rate [50]. If the reaction is close to thermoneutral, then the amount of phase space available when the intermediate complex dissociates to products is similar to that available when it dissociates back to the reactants, and the k approximates to one-half the collisional value. For proton transfer reactions which are not highly exo- or endoergic, k_f and k_r can be measured and thus ΔG determined [using Eq. (37.7)]. If ΔS can be determined in some way, or if k_f/k_r can be determined as a function of T, then ΔH can be deduced and used to obtain the proton affinity difference (ΔS can also be determined in the latter case). This has been used to construct proton affinity scales [23, 24]. Care is required in such studies to ensure that no vibrational excitation remains in the reactant ion due to its formation, and that the identities (i.e., isomeric forms) of the reactant and product ions are known (e.g. whether the ion is HCO^+ or the higher energy form COH^+ [51]).

Often the individual reaction processes do not occur in isolation. For example, in the reactions of CH_4^+ with COS, H_2S, NH_3, H_2CO and CH_3OH, both charge transfer and proton transfer are energetically possible, and both channels are observed. Here, the least exoergic channel is favored in all cases [52]. Proton transfer also occurs in negative ion reactions, for example those of OH^- and NH_2^- giving H_2O and NH_3 respectively (e.g. see Table 37.1), and these types of reactions are production sources for many negative ions [53].

Ion-Atom Interchange and Atom Abstraction

In many simple cases, these two processes are the same, for example

$$Ar^+ + H_2 \rightarrow ArH^+ + H, \qquad (37.18)$$

but not so for the more complicated molecules (e.g., O^+ + CH_4; see Table 37.1). Such reactions, when exoergic, usually occur close to the collisional rate although somewhat slower. Series of such reactions with H_2 occur in the interstellar medium and are responsible for producing the hydrogenation in many of the species observed there, for example, in CH_3^+ production from CH^+, NH_4^+ from NH^+ and H_3O^+ from OH^+ (see Chap. 78). The reaction,

$$NH_3^+ + H_2 \rightarrow NH_4^+ + H, \qquad (37.19)$$

is particularly interesting. At temperatures greater than 300 K, the reaction shows an activation energy barrier of 2 kcal mol^{-1}. The rate coefficient k decreases with decreasing T, reaching a minimum of 2×10^{-13} cm^3 s^{-1} at 80 K and then increases at lower T due to tunneling through the barrier [41]. This behavior occurs because, at the higher temperatures, the lifetime of the intermediate complex is not sufficient for significant tunneling to occur, but there is sufficient energy in the reacting species to overcome the barrier. At lower temperatures, there is not sufficient energy available to overcome the barrier, but the lifetime of the intermediate complex becomes long enough for significant tunneling to occur [54]. This explanation has been substantiated by the isotopic studies of the reactions $NH_3^+ + D_2$, and ND_3^+ $+H_2$ and D_2 [55].

For some reactions, isotopic labeling has been used to identify the reaction mechanism. For example, the reaction

$$O^+ + O_2 \rightarrow O_2^+ + O \qquad (37.20)$$

has been shown to be predominantly charge transfer, rather than ion-atom interchange, by labeling the ion as $^{18}O^+$ [56, 57].

Associative Detachment

For negative ions, an additional process is possible for which there is no equivalent positive ion analog (although there is the related process of chemi-ionization, i.e., $AB + C \rightarrow ABC^+ + e^-$). In this process, the negative ion and the neutral associate, and the association is stabilized by the ejection of the electron (see Table 37.1). Such reactions can only occur when the electron detachment energy is less than the energy of the bond that is produced. Therefore, these reactions usually involve radical species which produce stable molecules. k values are usually an appreciable fraction of the collisional values [58,59]. Infrared emissions have been detected from a series of these reactions [60], for example O^- with CO and F^-, Cl^- and CN^- with H, and show that the reactions populate the highest vibrational levels that are energetically accessible.

Other Binary Ion-Molecule Reaction Channels

As reactant species become more polyatomic, there is a greater variety of reaction processes that can occur, and these often occur in parallel. The processes that occur are too numerous to list so a few examples will have to suffice. Reactions that occur in isolation are insertion reactions of the type [61]

$$C^+ + C_nH_m \rightarrow C_{n+1}H_{m-x}^+ + xH \qquad (37.21)$$

and

$$S^+ + C_nH_m \rightarrow H_{m-1}C_nS^+ + H. \qquad (37.22)$$

Multiple channels are very evident in reactions of ions produced from species with large ionization potentials and small proton affinities, for example, NH^+ [62],

$$NH^+ + CH_3NH_2 \rightarrow \begin{cases} H_4CN^+ + NH_2 \\ CH_3NH_2^+ + NH & \text{(charge tsfr.)} \\ CH_3NH_3^+ + N & \text{(proton tsfr.)} \\ H_2CN^+ + (NH_3 + H) \\ H_3CN^+ + NH_3. \end{cases}$$

In addition to charge and proton transfer, other channels requiring more rearrangement are evident.

Isotopically Labeled Reactants

A great deal can be learned about reaction mechanisms in significantly exoergic ion-neutral reactions by isotopic labeling. Some such reactions have been mentioned above, but a particularly graphic example is the reaction

$$CH_4^+ + CH_4 \rightarrow CH_5^+ + CH_3, \qquad (37.23)$$

which was assumed to be either proton transfer and/or H-atom abstraction, depending on the interaction energy. By studying the reaction with both the ion and the neutral reactants separately deuterated, the reaction was shown to be much more complex, with the product ions being CH_4D^+ (10%), $CH_3D_2^+$ (22%), $CH_2D_3^+$ (43%) and CHD_4^+ (25%) for the reaction of CH_4^+ with CD_4 [63]. The reaction clearly proceeds via a long-lived intermediate in which there is a large degree of isotope mixing before unimolecular decomposition to products.

A further class of isotopic reactions are those for which the exoergicity is provided only by the different zero point energies of the reactants and products. Many such reactions have been studied. Examples of the various types are: (i) The symmetrical charge transfer

$$^{15}N_2^+ + {}^{14}N_2 \rightarrow {}^{14}N_2^+ + {}^{15}N_2, \qquad (37.24)$$

which proceeds in both directions at more than half the collisional rate. This implies that the reaction proceeds via long-range charge transfer, a conclusion substantiated by the fact that no mixed product $^{14}N\,^{15}N^+$ is produced [56]. (ii) Proton transfer reactions exemplified by $^{14}N_2H^+ + {}^{15}N_2$ and $H^{12}CO^+ + {}^{13}CO$, which are both exoergic by about 1 meV [41]. Some isotopic scrambling has been observed in the latter reaction by using the double isotopic substitution [64],

$$H\,^{12}C\,^{18}O^+ + {}^{13}C\,^{16}O \rightarrow \begin{cases} H\,^{13}C\,^{16}O^+ + {}^{12}C\,^{18}O \\ \qquad\qquad (\geq 90\%) \\ H\,^{13}C\,^{18}O^+ + {}^{12}C\,^{16}O \\ \qquad\qquad (\leq 10\%). \end{cases}$$

Other examples where considerable isotope scrambling occurs are H/D exchange in the reaction systems $H_3^+ + H_2$, $CH_3^+ + H_2$, $H_3O^+ + H_2O$, $CH_5^+ + CH_4$, etc. [41]. Such scrambling reactions are believed to proceed via proton bound dimer intermediates (e.g., $H_2O\cdots H^+ \cdots D_2O$). (iii) Isotope exchange reactions such as $^{13}C^+ + {}^{12}CO \rightarrow {}^{12}C^+ + {}^{13}CO$ have also been studied. For this reaction, both k_f and k_r increase with decreasing T due to an increase in the intermediate complex lifetime. k_r decreases at lower T due to the endoergicity in this reaction direction [63].

Isotope exchange is also observed to occur with negative ions, *viz.*

$$DO^- + MH \rightarrow HO^- + MD, \qquad (37.25)$$

and is a common process. Particularly significant is the comparison where MH is H_2 and NH_3. These species have very similar gas phase acidities, and thus the exoergicities of the reactions are similar. However, NH_3 has a larger polarizability than H_2 and, in addition, has a permanent dipole moment. The observation that the k for the NH_3 reaction is a factor of ten larger than that for the H_2 reaction is explained as being due to the stronger interaction potential which makes more energy available in the intermediate complex and facilitates H/D exchange [53].

Temperature Dependencies of Binary Reactions

When the k values at room temperature are close to the collisional value, appreciably exoergic reactions generally exhibit little temperature dependence. For slower reactions, a significant inverse temperature dependence is observed, being $\sim T^{-1/2}$ for several reactions. This behavior is rationalized as a decreasing lifetime of the intermediate complex with increasing temperature, thus allowing less time for reaction [41].

37.4.2 Ternary Ion-Molecule Reactions

In cases where binary channels are not energetically possible, association reactions can still occur, both for positive and negative ions. The reactions can be as simple as

$$He^+ + He + He \rightarrow He_2^+ + He, \tag{37.26}$$

to associations more complex than those listed in Table 37.1. Rate coefficients k_3 at room temperature vary from 1×10^{-31} cm^6 s^{-1} for reactions such as (37.26) to in excess of 1×10^{-25} cm^6 s^{-1}, this upper limit being due to experimental constraints rather than the reaction itself. The mechanism usually postulated is

$$A^\pm + B \rightleftharpoons (AB^\pm)^*, \tag{37.27}$$
$$(AB^\pm)^* + M \rightarrow AB^\pm + M^* \tag{37.28}$$

where M stabilizes the excited intermediate, $(AB^\pm)^*$ by removing energy in the collision. For this type of reaction, there is a considerable dependence of k on T. Statistical theory predicts (see Chap. 36),

$$k_3 \propto T^{-(\ell/2+\delta)}, \tag{37.29}$$

where ℓ is the number of rotational degrees of freedom in the separated reactants, and δ is a parameter attributed to the temperature variation of the collision efficiency of the stabilizing third body M. Experimentally, many association reactions exhibit such power law dependencies of the k_3 [41,44] with δ being small (~ 0.2 or 0.3) for a helium third body. Also, some evidence exists for a contribution of vibrational degrees of freedom to the temperature coefficient in cases where the vibrational levels in the reactant neutral are closely spaced [65].

For complex reactant species, lifetimes of intermediate complexes become very long so that, in the higher pressure experiments (see below), all intermediate complexes are stabilized, and thus the reaction is independent of the pressure of M. This form of "saturation" can be eliminated in many cases by using low pressure experiments, so that the time between collisions is long, and normal ternary kinetics [see Eq. (37.14)] are restored. Under these conditions, sometimes there is still a pressure

independent component to the k which is postulated as being due to radiative stabilization of the intermediate, *viz.*

$$(AB^\pm)^* \rightarrow AB^\pm + h\nu. \tag{37.30}$$

An example of this is the reaction [66]

$$CH_3^+ + CH_3CN \rightarrow CH_3^+ \cdot CH_3CN + h\nu. \tag{37.31}$$

The stabilizing photon is often considered to be in the infrared due to a vibrational transition within the ground electronic state (with a radiative lifetime of $\sim 10^{-2}$ to 10^{-3} s). Note that only a photon with $h\nu > \frac{3}{2}k_B T$ is required for the complex (although still vibrationally excited) to be stable against unimolecular dissociation. However, there are cases where it is believed that an electronically excited state is accessed [67,68]. Examples are

$$C^+ + H_2 \rightarrow CH_2^+ + h\nu \tag{37.32}$$

and

$$Cl^- + BCl_3 \rightarrow BCl_4^- + h\nu. \tag{37.33}$$

As yet, radiative association has not been observed directly (i.e., by the detection of the emitted photon), however, the kinetic evidence is strong for the existence of this process.

For the more rapid collisional associations, competition with binary channels is possible, for example

$$CH_3^+ + NH_3 \rightarrow \begin{cases} H_4CN^+ + H_2 & (70\%) \\ NH_4^+ + CH_2 & (10\%) \\ \overset{He}{\rightarrow} CH_3^+ \cdot NH_3 + He & (20\%) \end{cases}$$

where the percentages refer to product abundances at a He pressure of 0.2 Torr [69].

37.5 ELECTRON ATTACHMENT

These reactions have generally been studied by different techniques from those used to examine ion-neutral reactions and also by different workers [29,30]. Analogous to the ion-neutral situation, there are dissociative processes,

$$e^- + CCl_4 \rightarrow Cl^- + CCl_3, \tag{37.34}$$

and nondissociative processes,

$$e^- + C_6F_6 \rightarrow C_6F_6^-, \tag{37.35}$$

the former being stabilized by dissociation, and the latter occurring because of the long lifetime of the intermediate against autodetachment with eventual collisional stabilization (*cf.* saturation in ion-neutral association reactions; Sect. 37.4.2). Measured rate coefficients vary in the range 1×10^{-7} to the smallest measurable value of $\sim 1 \times 10^{-12}$ cm^3 s^{-1} [29,30]. The upper limit on k for this process is determined, from a consideration of the electron de Broglie wavelength $\lambda = \lambda/2\pi$, to be $\sim 5 \times 10^{-7}(300/T)^{1/2}$ cm^3 s^{-1} [70], or, more rigorously,

$$k(E) = \sum_i (\sigma/\pi\lambda^2)_i/[h\rho(E)], \qquad (37.36)$$

where $\rho(E)$ is the density of states and

$$\sigma/\pi\lambda^2 \sim 1 - \exp[-(4\gamma^2 \mathcal{T})^{1/2}], \qquad (37.37)$$

with $\gamma^2 R_\infty = (2\mu/m)^2 \alpha_d/a_0^3$ [5]. σ is the collision cross section, $\mathcal{T}$ the kinetic energy, α_d the polarizability, a_0 the Bohr radius, μ the reduced mass, and i represents the available product channels. Where k is less than the upper limit value, this is usually due to activation energy barriers, and k shows Arrhenius behavior

$$k(T) = k_0 \exp\left(-\Delta E/k_B T\right), \qquad (37.38)$$

where k_B is the Boltzmann constant, with the barrier height ΔE being in the range 0 to ~ 300 meV [30]. If the attachment is nondissociative, and the lifetime of the intermediate complex against autoionization is small compared with the collision time (as occurs for less complex species), the reaction exhibits normal ternary behavior, for example,

$$e^- + O_2 + M \rightarrow O_2^- + M. \qquad (37.39)$$

Some interesting mechanisms have been identified for attachment. Different product ions have been observed in experiments conducted at very different pressures, for example, in the reaction

$$C_6F_5I + e^- \rightarrow C_6F_5^- + I \quad (\geq 95\%) \qquad (37.40)$$
$$\rightarrow C_6F_5I^- \quad (\leq 5\%). \qquad (37.41)$$

at high pressures (~ 1 Torr), the products, as indicated, were observed with an association channel [see Eq. (37.41)] being detected [71]. This association channel was not seen in low pressure experiments [72]. Such behavior is explained in terms of the relative magnitudes of the autodetachment lifetime and the time between collisions. When the former time is larger, the association product will dominate, whereas if the converse is true, dissociation products dominate. Thus, the different product distributions, rather than indicating discrepancies, yield information about the autodetachment lifetime.

In attachments to the Br-containing compounds, CF_2Br_2, $CFBr_3$, and CF_2BrCF_2Br and CH_2BrCH_2Br, in which the Br atoms are on different carbon atoms, up to a 20% Br_2^- product is observed [73]. This shows that atoms in product molecules can come from spatially separated parts of the reactant molecule. Thus, considerable distortion and rearrangement must occur in the intermediate complex.

37.6 RECOMBINATION

This process has been much less studied than ion-neutral reactions with only ~ 100 reactions being studied [38] compared to about 10,000 in the former case [31]. Also, very little information is available concerning the products since these are neutral and very much more difficult to detect than ions.

37.6.1 Electron-Ion Recombination

Rate coefficients for dissociative electron-ion recombination at room temperature vary from $>1 \times 10^{-6}$ for polyatomic ions to $\sim 1 \times 10^{-7}$ cm^3 s^{-1} for diatomic ions [36,37]. For large k, there is little temperature dependence, temperature dependencies being more marked for the slower reactions. Two mechanisms have been proposed: (i) the direct mechanism [74] where the neutralized ion undergoes a radiationless transition to a repulsive potential curve on which dissociation to products occurs, and (ii) the indirect mechanism [75] where the neutralized ion initially transfers to a Rydberg state and then undergoes a radiationless transition to the repulsive curve. For the former process, the theoretical temperature dependence is $T_e^{-0.5}$, while it is $T_v^{-1} T_e^{0.5}$ for the latter, where the subscripts v and e refer to the ion vibrational and electron temperatures respectively [6]. Experimental T-dependencies for reactions which are not close to the collisional limit have power law dependencies of ~ 0.7 to 1 over the temperature range 80 to 600 K [37], i.e., between the theoretical predictions. Less is known about the products. Detailed theory carried out for the diatomic ions O_2^+, NO^+ and N_2^+, and to a large degree HCO^+, are in general agreement with experiment [76,77]. For the more polyatomic ions, the theory is not yet sufficiently quantitative [78,79] and reliance is placed on experiment. H-atom contributions to the product distributions for ten ions (N_2H^+, HCO^+, HCO_2^+, N_2OH^+, $OCSH^+$, H_2CN^+, H_3O^+, H_3S^+, NH_4^+ and CH_5^+) vary from $\sim 20\%$ for $OCSH^+$ to 120% for CH_5^+ [80]. OH is a substantial product (30 to 65%) in the above reactions where it is energetically possible [81] (i.e., excluding HCO^+), except in the case of $OCSH^+$, perhaps indicating that the proton is exclusively on the S-atom rather than the O-atom (i.e., in the lowest energy form [82,83]). Electron-ion recombination is an energetic process and electronically excited states can also be populated. Vibrational population distributions for these states can readily be determined if Einstein A coefficients are known for the observed transitions ($I \propto A[*]$, where I is the photon intensity, and $[*]$ is the number density of the excited state). The N_2 ($B\,^3\Pi_g$) state and the CO ($a\,^3\Pi_r$) state vibrational population produced in the recombinations of N_2H^+, HCO^+, HCO_2^+ and CO_2^+ have been determined [84,85]. Recent theory [86] considers (i) the impulsive force on the molecular fragments as the neutralized ion rapidly dissociates, and (ii) the Franck Condon overlap between the wave functions of the molecular products (in various vibrational states) and the wave function of this particular fragment in the neutralized ion before it dissociates. Theory underestimates the populations of the higher levels, but does predict the small observed oscillation in the occupancy of the various vibrational levels in the CO ($a\,^3\Pi_r$) state generated in the recombination of HCO^+ [87].

37.6.2 Ion-Ion Recombination (Mutual Neutralization)

Less information is available on this process than electron-ion recombination. k values vary from about 4×10^{-8} to 1×10^{-7} cm^3 s^{-1} at room temperature [38,39] with power law temperature dependencies of ~0.4 for the only two systems that have been studied as a function of T ($NO^+ + NO_2^-$ and $NH_4^+ + Cl^-$) [39]. This is consistent with theory [88] in which the Landau-Zener approximation is used to determine the probability of a crossing between the Coulombic ion-ion attractive potential and the potential of the neutral products. This transition occurs by an electron transfer (the optimum distance for such a transfer when a favorable crossing exists is about 10 Å). Theory gives that

$$k \sim 2(v^2 Q)(\mu/2\pi k_B T)^{1/2}, \qquad (37.42)$$

where Q is the cross section

$$Q = \pi R_c^2 [1 + (R_c E)^{-1}]. \qquad (37.43)$$

R_c is the crossing distance, E is the interaction energy and v is the relative velocity at the crossing point [88]. Once neutralized by electron transfer, the products continue undeflected with the velocity gained from the Coulombic field. The only product that has been detected from ion-ion recombination is the NO ($A\,^2\Sigma^+$) state product from $NO^+ + NO_2^-$, and this is consistent with the electron transfer mechanism [89]. Few data are also available on the effects of pressure. For the reactions $SF_3^+ + SF_5^-$ and $NO^+ + NO_2^-$, k increases by about a factor of 4 between 1 and 8 Torr [39,43].

Acknowledgment

The support of the National Science Foundation and the Petroleum Research Fund-ACS for my experimental research program is gratefully acknowledged.

REFERENCES

1. G. Gioumousis and D. P. Stevenson, J. Chem. Phys. **29**, 294 (1958).
2. L. Bass, T. Su, and M. T. Bowers, Int. J. Mass Spectrom. Ion. Phys. **28**, 389 (1978).
3. T. Su and M. T. Bowers, in *Gas Phase Ion Chemistry*, edited by M. T. Bowers (Academic, New York, 1979), Vol. 1, p. 83.
4. T. Su and M. T. Bowers, Int. J. Mass Spectrom. Ion Phys. **25**, 1 (1975).
5. C. E. Klots, Chem. Phys. Lett. **38**, 61 (1976).
6. J. N. Bardsley and M. A. Biondi, Adv. At. Mol. Phys. **6**, 1 (1970).
7. I. W. M. Smith, in *Rate Coefficients in Astrochemistry*, edited by T. J. Millar and D. A. Williams (Kluwer, Dordrecht, 1988), p. 103.
8. M. J. Henchman, in *Structure/Reactivity and Thermochemistry*, edited by P. Ausloos and S. G. Lias (Reidel, Dordrecht, 1987), p. 381.
9. A. Dalgarno, J. Chem. Soc. Farad. Trans. **89**, 2111 (1993).
10. S. K. Atreya, *Atmospheres and Ionospheres of the Outer Planets and their Satellites* (Springer-Verlag, Berlin, 1986).
11. E. E. Ferguson, F. C. Fehsenfeld, and D. L. Albritton, in *Gas Phase Ion Chemistry*, edited by M. T. Bowers (Academic, New York, 1979), Vol. 1, p. 45.
12. D. Smith and N. G. Adams, Topics Curr. Chem. **89**, 1 (1980).
13. E. Murad and S. Lai, J. Geophys. Res. **91**, 13745 (1986).
14. E. W. McDaniel and W. L. Nighan, *Applied Atomic Collision Physics* (Academic, New York, 1982).
15. G. Turban, Pure Appl. Chem. **56**, 215 (1984).
16. I. M. Klotz and R. M. Rosenberg, *Chemical Thermodynamics* (Krieger, Malabar, 1991).
17. *CRC Handbook of Chemistry and Physics*, edited by R. C. Weast (CRC Press, Boca Raton, 1988), 69th ed.
18. K. P. Huber and G. Herzberg, *Molecular Spectra and Molecular Structure, IV Constants of Diatomic Molecules* (Van Nostrand Reinhold, New York, 1979).
19. G. Herzberg, *Molecular Spectra and Molecular Structure, III Electronic Spectra and Electronic Structure of Polyatomic Molecules* (Van Nostrand, New Jersey, 1967).
20. H. M. Rosenstock, K. Draxl, B. W. Steiner, and J. T. Herron, J. Phys. Chem. Ref. Data **6**, I1 (1977).
21. NIST Positive and Negative Ion Energetics Database, Standard Reference Data Program (Gaithersburg, Maryland, 1990).
22. B. K. Janousek and J. I. Brauman, in *Gas Phase Ion Chemistry*, edited by M. T. Bowers (Academic, New York, 1979), Vol. 2, p. 53.
23. S. G. Lias, J. F. Liebman, and R. D. Levin, J. Phys. Chem. Ref. Data **13**, 695 (1984).
24. S. G. Lias, J. E. Bartmess, J. F. Liebman, J. L. Holmes, R. D. Levin, and W. G. Mallard, J. Phys. Chem. Ref. Data **17**, Suppl. 1 (1988).
25. J. E. Bartmess and R. T. McIver, in *Gas Phase Ion Chemistry*, edited by M. T. Bowers (Academic, New York, 1979), Vol. 2, p. 88.
26. The NBS Tables of Chemical Thermodynamic Properties, J. Chem. Phys. Ref. Data **11**, Suppl. 2 (1982).
27. D. A. McQuarrie, *Statistical Mechanics* (Harper-Collins, New York, 1976).
28. J. I. Steinfeld, J. S. Francisco, and W. L. Hase, *Chemical*

Kinetics and Dynamics (Prentice-Hall, Englewood Cliffs, 1989), p. 1.

29. L. G. Christophorou, D. L. McCorkle, and A. A. Christodoulides, in *Electron-Molecule Interactions and their Applications,* edited by L. G. Christophorou (Academic, Orlando, 1984), Vol. 1.

30. D. Smith and N. G. Adams, in *Swarms of Ions and Electrons in Gases,* edited by W. Lindinger, T. D. Mark, and F. Howorka (Springer-Verlag, Vienna, 1984), p. 284.

31. Y. Ikezoe, S. Matsuoka, M. Takebe, and A. A. Viggiano, *Gas Phase Ion-Molecule Reaction Rate Constants through 1986* (Maruzen, Tokyo, 1987).

32. V. G. Anicich, J. Phys. Chem. Ref. Data **22**, 1469 (1993).

33. D. C. Clary, D. Smith, and N. G. Adams, Chem. Phys. Lett. **119**, 320 (1985).

34. D. C. Clary, Mol. Phys. **54**, 605 (1985).

35. N. G. Adams, D. Smith, and D. C. Clary, Ap. J. **296**, L31 (1985).

36. R. Johnsen, Int. J. Mass Specrom. Ion Proc. **81**, 67 (1987).

37. N. G. Adams and D. Smith, in *Rate Coefficients in Astrochemistry*, edited by T. J. Millar and D. A. Williams (Kluwer, Dordrecht, 1988), p. 173.

38. J. B. A. Mitchell, Phys. Rep. **186**, 216 (1990).

39. D. Smith and N. G. Adams, in *Physics of Ion-Ion and Electron-Ion Collisions*, edited by F. Brouillard and J. W. McGowan (Plenum, New York, 1983), p. 501.

40. D. R. Bates and A. Dalgarno, in *Atomic and Molecular Processes*, edited by D. R. Bates (Academic, London, 1962), p. 245.

41. N. G. Adams and D. Smith, in *Reactions of Small Transient Species*, edited by A. Fontijn (Academic Press, London, 1983), p. 311.

42. H. W. S. Massey, *Negative Ions* (Cambridge University Press, Cambridge, 1976).

43. D. Smith and N. G. Adams, Geophys. Res. Lett. **9**, 1085 (1982).

44. M. Meot-Ner, in *Gas Phase Ion Chemistry* , edited by M. T. Bowers (Academic Press, New York, 1979), Vol. 1, p. 198.

45. D. Smith, N. G. Adams, E. Alge, H. Villinger, and W. Lindinger, J. Phys. B **13**, 2787 (1980).

46. J. B. Lauderslager, W. T. Huntress, and M. T. Bowers, J. Chem. Phys. **61**, 4600 (1974).

47. N. G. Adams and D. Smith, J. Phys. B **9**, 1439 (1976).

48. A. L. Schmeltekopf, E. E. Ferguson, and F. C. Fehsenfeld, J. Chem. Phys. **48**, 2966 (1968).

49. E. C. Inn, Planet. Space Sci. **15**, 19 (1967).

50. D. K. Bohme, in *Interactions between Ions and Molecules*, edited by P. Ausloos (Plenum, New York, 1975), p. 489.

51. M. J. McEwan, in *Advances in Gas Phase Ion Chemistry*, edited by N. G. Adams and L. M. Babcock (JAI Press, Greenwich, 1992), Vol. 1, p. 1.

52. N. G. Adams and D. Smith, Chem. Phys. Lett. **54**, 530 (1978).

53. C. H. DePuy, in *Ionic Processes in the Gas Phase*, edited by M. A. Almoster Ferreira (Reidel, Dortrecht, 1984), p. 227.

54. S. E. Barlow and G. H. Dunn, Int. J. Mass Spectrom. Ion Proc. **80**, 227 (1987).

55. N. G. Adams and D. Smith, Int. J. Mass Spectrom. Ion Proc. **61**, 133 (1984).

56. F. C. Fehsenfeld, D. L. Albritton, D. L. Bush, P. G. Fornier, T. R. Govers, and J. Fourier, J. Chem. Phys. **61**, 2150 (1974).

57. I. Dotan, Chem. Phys. Lett. **75**, 509 (1980).

58. F. C. Fehsenfeld, in *Interactions between Ions and Molecules*, edited by P. Ausloos (Plenum, New York, 1975), p. 387.

59. A. A. Viggiano and J. F. Paulson, in *Swarms of Ions and Electrons in Gases*, edited by W. Lindinger, T. D. Mark, and F. Howorka (Springer-Verlag, Vienna, 1984), p. 218.

60. V. M. Bierbaum, G. B. Ellison, and S. R. Leone, in *Gas Phase Ion Chemistry* , edited by M. T. Bowers (Academic Press, New York, 1984), Vol. 3, p. 1.

61. D. Smith, N. G. Adams, and E. E. Ferguson, in *Molecular Astrophysics*, edited by T. W. Hartquist (Cambridge University Press, Cambridge, 1990), p. 181.

62. N. G. Adams, D. Smith, and J. F. Paulson, J. Chem. Phys. **72**, 288 (1980).

63. D. Smith and N. G. Adams, in *Ionic Processes in the Gas Phase*, edited by M. A. Almoster Ferreira (Reidel, Dortrecht, 1984), p. 41.

64. D. Smith and N. G. Adams, Ap. J. **242**, 424 (1980).

65. A. A. Viggiano and J. F. Paulson, J. Phys. Chem. **95**, 10719 (1991).

66. E. Herbst and M. J. McEwan, Astron. Astrophys. **229**, 201 (1990).

67. E. Herbst, J. G. Schubert, and P. R. Certain, Ap. J. **213**, 696 (1977).

68. C. R. Herd and L. M. Babcock, J. Phys. Chem. **93**, 245 (1989).

69. D. Smith and N. G. Adams, Chem. Phys. Lett. **54**, 535 (1978).

70. J. M. Warman and M. C. Sauer, Int. J. Radiat. Chem. **3**, 273 (1971).

71. C. R. Herd, N. G. Adams, and D. Smith, Int. J. Mass Spectrom. Ion Proc. **87**, 331 (1989).

72. W. T. Naff, R. N. Compton, and C. D. Cooper, J. Chem. Phys. **54**, 212 (1971).

73. D. Smith, C. R. Herd, N. G. Adams, and J. F. Paulson, Int. J. Mass Spectrom. Ion Proc. **96**, 341 (1990).

74. D. R. Bates, Phys. Rev. **78**, 492 (1950).

75. J. N. Bardsley, J. Phys. B **1**, 349, 365 (1968).

76. S. L. Guberman, in *Dissociative Recombination: Theory, Experiment and Applications*, edited by B. R. Rowe, J. B. A. Mitchell, and A. Canosa (Plenum, New York, 1993), p. 47.

77. D. Talbi and Y. Ellinger, in *Dissociative Recombination: Theory, Experiment and Applications*, edited by B. R. Rowe, J. B. A. Mitchell, and A. Canosa (Plenum, New York, 1993), p. 59.

78. D. R. Bates, Ap. J. **344**, 531 (1989).

79. E. T. Galloway and E. Herbst, Ap. J. **376**, 531 (1991).

80. N. G. Adams, C. R. Herd, M. Geoghegan, D. Smith, A. Canosa, B. R. Rowe, J. L. Queffelec, and M. Morlais, J. Chem. Phys. **94**, 4852 (1991).

81. C. R. Herd, N. G. Adams, and D. Smith, Ap. J. **349**, 388 (1990).

82. P. G. Jasien and W. J. Stevens, J. Chem. Phys. **83**, 2984 (1985).

83. M. Scarlett and P. R. Taylor, Chem. Phys. **101**, 17 (1986).

84. N. G. Adams and L. M. Babcock, J. Phys. Chem. **98**, 4564 (1994).
85. B. L. Foley and N. G. Adams (to be published).
86. D. R. Bates, Mon. Not. R. Astron. Soc. **263**, 369 (1993).
87. N. G. Adams and L. M. Babcock, Ap. J. **434**, 184 (1994).
88. R. E. Olsen, J. Chem. Phys. **56**, 2979 (1972).
89. D. Smith, N. G. Adams, and M. J. Church, J. Phys. B **11**, 4041 (1978).

38

Clusters

M. L. Mandich

Bell Laboratories, Lucent Technologies Inc., Murray Hill, New Jersey 07974

38.1 INTRODUCTION

Clusters are small aggregates of atoms or molecules which are transitional forms of matter between atoms or molecules and their corresponding bulk forms. Just as this definition spans an incredibly broad range of clusters from, say, He_2 to $Na_{10\,000}$, so do the properties of these clusters span a broad range. This chapter attempts to bring order to this diverse cluster kingdom by first sorting them into six general categories. Within each category, the physics and chemistry of the more or less similar cluster species are described. Particular emphasis is placed on the unique properties of clusters owing to their finite size and finite lattice.

This chapter summarizes one of the youngest topics in this volume. Much of what is known is highly qualitative and has not yet been assembled into overarching tables or equations. Thus, this review is best regarded as a progress report on the current knowledge in this rapidly advancing field. Many of the concepts and the language used to discuss clusters are derived from condensed matter physics. The nature of these clusters impels such descriptions.

Clusters discussed in this chapter are isolated species composed primarily of a single type of atom or molecule. Most of these clusters are highly reactive and can only be made and studied under rarified conditions such as in a molecular beam. In keeping with the definition of clusters given above, this chapter will not attempt to cover the truly vast literature on atomic and molecular dimers and trimers. A great many of these species have been thoroughly characterized, however they are better described as molecules rather than "tiny clusters". Stabilized clusters in the form of Zintl ions, colloids, and nanoparticles have also been made. While it is impossible to resist mentioning these clusters throughout this chapter, they are more appropriate to condensed matter physics. Finally, experimental and theoretical techniques for forming and studying clusters will not be covered; excellent reviews of these methods are available elsewhere [1–6].

38.2 METAL CLUSTERS

Clusters of a wide variety of refractory and nonrefractory metals have been made and studied. These include clusters of alkali metals, transition metals, coinage metals, main group metals, and lanthanides. Clusters of alkali metals, particularly sodium, are the most well studied and understood. The emerging picture of alkali clusters is that they behave much like quasifree electron metal spheres. Thus, approximations such as the jellium model used for describing bulk metals explain a large number of alkali metal properties quite well [2]. Other metallic clusters, such as some main group metal and noble metal clusters, can also be understood within the jellium model. Nonetheless, there are examples of metallic clusters which deviate significantly from these simple models, particularly among the transition metal clusters.

38.2.1 Geometric Structures

The ground state geometry is known accurately for only a handful of small metal clusters. Examples include lithium and sodium clusters containing up to nine atoms, where electronic spectra are compared with accurate *ab initio* quantum chemical calculations to de-

duce their structures [5–7]. Such an approach is not a general one, either experimentally or theoretically. As the size and/or atomic valency of the metal cluster increases, the number of possible ground state structures grows enormously. Interpretation of experimental spectra requires theoretical guidance, which means that each one of these structures must be calculated using an accurate electronic structure calculational method. Thorough theoretical investigation is only practical with current computational tools for clusters containing up to about thirteen atoms. Methods such as molecular dynamics combined with density functional calculations have been used to speed up the process of finding and comparing various isomers [4, 8]. Such approaches are still restricted by available computational power to clusters with relatively few atoms and valence electrons per atom. Metal clusters of atoms with higher valency, e.g., mid-row transition metals, have proved to be quite difficult to treat accurately [9]. The large electron correlation problems inherent in these clusters must be treated semi-empirically, leading to large uncertainties in the relative energies between isomers of different spin and geometry.

Despite these difficulties, approximate geometries are known for many metal clusters. For cluster sizes greater than, roughly, several tens of atoms, there is strong experimental evidence that spherical close packed geometries, particularly the Mackay filled icosahedra, predominate the geometric structures [7, 10, 11] (The Mackay filled icosahedra contain concentric closed icosahedral shells of atoms plus one central atom. These structures are pentagonally symmetric and occur every n atoms by n = 13, 55, ⋯ 1/3(2n+1)(5n²+5n+3). [4]). These geometries have been deduced by a variety of means. For example, abundance mass spectra of metal clusters up to sizes containing thousands of atoms show intensity enhancements, termed *magic numbers*, at each cluster size which corresponds to a complete Mackay icosahedron. Saturation coverages of transition metal clusters with various reagents can be explained in terms of covering the icosahedral faces of these clusters. The persistence of icosahedral structures to large cluster sizes raises the question about where the crossover to the metallic packing occurs. For example, a spherical piece of an fcc metal would have cuboctahedral symmetry rather than noncrystalline icosahedral symmetry. Ultrafine metal particles containing 10^4 to 10^5 atoms typically have bulk crystalline geometries [12]. Little theoretical or experimental data currently exist to resolve the question of when the bulk crystalline structure emerges [10]. The point of crossover must involve kinetic as well as energetic factors. Although icosahedral packing is seen in beam experiments on metal clusters containing hundreds to thousands of atoms, high resolution electron microscopy experiments on supported metal particles in this size range show fluctuating structures that can rapidly evolve between icosahedral, cuboctahedral, and other crystalline arrangements, depending on cluster temperature [13].

38.2.2 Electronic and Magnetic Properties

A number of electronic properties have been measured for metallic clusters, particularly alkali metal and noble metal clusters. These include ionization potentials, electronic affinities, polarizabilities, and photoabsorption cross sections.

The spherical jellium approximation is generally a good model for these properties in clusters of alkali and noble metals [2,7]. This model treats the valence electrons of the cluster as a delocalized sea of electrons smeared over a uniform spherical background of ionic cores. The energy levels of such a jellium sphere can be calculated by confining the jellium electrons to a three dimensional potential. A suitable form of this potential yields level spacings that are given by principal and angular momentum quantum numbers (n) and (l). There are no restrictions on l for a given n, and the degeneracy of a given l level is $2(2l+1)$. The levels order as $1s$, $1p$, $1d$, $2s$, $1f$, $2p$, $1g$, $2d$, $3s$, $\cdots$. The electron configuration of a cluster is given by filling successive energy levels (termed shells) with the available valence electrons. Special stability occurs for those clusters with a closed shell configuration. This stabilization can be seen in the experimental measurements of the dependences of total binding energies, ionization potentials, electron affinities, and polarizabilities on cluster size [2,7,10,11,13,14]. The potential used to calculate the energy levels within the jellium model has been formulated to include elliptical distortions of open shell clusters [7]. This refinement has been successful in describing the fine structures of these trends. Despite the usefulness of the jellium model, it is not applicable for very small or very large clusters. In many smaller clusters containing up to a few tens of atoms, electron-ionic lattice interactions cannot be neglected, and the spherical approximation is poor. For these clusters, accurate quantum chemical calculations must be used to determine the electronic structure. Alternatively, for clusters larger than roughly 500 to 3000 atoms (depending on which clusters and the cluster temperature), the level spacings become increasingly continuous, blurring the shell structure. When this occurs, other effects are seen to dominate the trends in cluster stability, particularly the stability arising from completing a geometric shell of atoms on the cluster [7,10,11,15].

The convergence of cluster electronic structure to bulk metal electronic structure has been seen in the evolution of metal cluster ionization potentials (I_P) and electron affinities (A_e) as a function of cluster size. A good approximation for the overall trend in the I_P as a function of cluster size (N is the number of atoms) is given by the electrostatic model for the work function of a classical conducting sphere [2]

$$I_{P,N} = W_B + A\frac{e^2}{R_N},\qquad(38.1)$$

where W_B is the work function of the bulk metal, e is the electron charge, and R is the cluster radius, which is often set proportional to $N^{1/3}$. A is a constant which is found experimentally to have values of about 0.3 to 0.5; the theoretically derived value for A depends on the model used [2,11]. A similar form of Eq. (38.1), where A is replaced by $(A-1)$, describes the A_e of a negatively charged cluster. Experimentally, the I_P and A_e of alkali metal, noble metal, and some main group metal clusters behave as described by Eq. (38.1) with shell structure superimposed on the overall trend [2,6,14]. Equation (38.1) predicts a smooth convergence of the work function to the bulk value with increasing cluster size. This has been seen experimentally, such as in copper cluster valence and inner shell A_e which extrapolate smoothly to the corresponding bulk values [16]. In mercury clusters, the transition to bulk metallic behavior occurs more abruptly and clusters with less than ~17 atoms appear to behave as nonmetals [17]. Transition metal clusters often deviate strongly from the trend in Eq. (38.1) [10].

Electronic spectra are available for a number of metal clusters including alkali, noble metal, transition metal, and aluminum clusters [7,10,11]. Small alkali clusters exhibit rich spectra in the visible. Many of these spectra have been assigned using accurate electronic structure calculations as described above. Visible spectra of larger alkali metal clusters and other metal clusters are typified by giant resonances with cross sections reaching values as large as 2000 Å^2 [6,7]. In the absence of detailed electronic structure information, these spectra have been assigned using comparisons with bulk metal spectra. In particular, the giant resonances are assigned to collective excitations of the cluster valence electrons, in an analogy to bulk metal plasmon resonances. Theoretical treatments of these giant resonances for clusters have been derived from classical treatments of conducting spheres driven by an external electromagnetic field [7,10]. While the blue shift of the resonance frequency with increasing cluster size is well predicted by the classical models, other details are less well described, such as the magnitude of the shift and the width of the resonance. Clusters which are known to be nonspherical from other measurements exhibit multiple resonance peaks; these have yet to be quantitatively described by theory. Transition metal clusters also show evidence of collective excitations; however, the magnitudes of these absorptions are 2 to 5 times larger than predicted by classical models [18]. An intriguing *non sequitur* occurs when the classical models are extended down to describe the resonances of small alkali clusters. This is illustrated by the example of Na$_8$, which exhibits a large single resonance in the photoabsorption spectrum. At first glance, the classical jellium model does fine: it predicts a spherical closed shell cluster which should exhibit a single resonance. Yet, a more thorough examination of the time dependence of this feature has revealed that it consists of four overlapping absorptions [10, 19]. These multiple absorptions clearly must arise from one-electron excitations, not a collective all-electron excitation [5]. Further work is needed to weave together the two disparate pictures of collective versus one electron excitations in metal clusters.

Inner shell electrons of some metal clusters have been probed spectroscopically. Whereas excitations of the delocalized valence electrons primarily reflect the entire cluster environment, excitations of the localized inner shell electrons reflect the atomic environment. In mercury clusters, excitation of the $5d$ core electrons reveals a transition from insulating clusters at small sizes to more metal-like clusters with increasing s-p hybridization typical of bulk mercury [20]. Inner core electron spectra of copper and antimony clusters also reveal details regarding the evolution of the cluster lattice structure as a function of size [11, 16]. Such valuable information can be obtained from inner shell electron spectra of metal clusters that more experiments are warranted.

Magnetic moments have been measured and described theoretically for a range of transition metal, rare earth, and Group III metal clusters [5, 7, 10, 21, 22]. Examples of clusters which exhibit a positive magnetic moment include cobalt, iron, nickel, gadolinium, and terbium clusters. Clusters of metals such as vanadium, palladium, chromium, and aluminum are observed to be diamagnetic. The effective moment per atom, μ, in magnetic clusters is greater than in the bulk because of the lower average coordination number in clusters [22]. As the cluster size increases, the surface to volume atom ratio decreases and μ converges to the bulk value. The size of these magnetic clusters is smaller than the critical domain size for bulk ferromagnetism, thus they are best described as paramagnetic with a single moment given by $N\mu$, where N is the number of atoms in the cluster. Stern-Gerlach deflection data have been obtained to measure μ. A critical parameter in interpreting these data is the so-called blocking temperature T_B, the temperature at which the cluster moment unlocks from the cluster axes and orients thermally in an external field. For clusters with temperatures $T > T_B$, the observed effective moment μ_{eff} for paramagnetic clusters is given by [22]

$$\mu_{\text{eff}} = \mu \left[\coth\left(\frac{N\mu H}{k_B T}\right) - \frac{k_B T}{N\mu H} \right], \qquad (38.2)$$

where k_B is the Boltzmann constant, and H is the magnetic field strength. Medium size iron and cobalt clusters behave well according to this classically derived expression. Deviations from this model are observed in internally cold clusters and rare earth clusters which

can be explained by partial and complete locking of the magnetic moment to the cluster lattice. Also, in very large clusters, $N\mu$ becomes sufficiently large that alignment overwhelms thermal statistical behavior.

38.2.3 Chemical Properties

The chemistry of both charged and neutral metal clusters has been studied, particularly for clusters of transition metals and Group III metals [6, 7, 23, 24]. Both chemisorption and physisorption are seen, depending on the type of metal cluster and reagent. In many cases, the observed chemistry is quite similar to that observed for the corresponding bulk metal. For example, platinum clusters dehydrogenate hydrocarbons, hydrogen chemisorbs on transition metal clusters with the exception of coinage metal clusters, and oxygen reacts readily with aluminum and iron clusters.

Chemisorption reactions of metal clusters have been seen with a wide variety of reagents. Products, reaction rates, and activation barriers have been measured as a function of cluster size. With a few notable exceptions, such as the reactions of hydrogen with nickel and aluminum clusters, none of these reactions are understood at the microscopic level [10, 11]. Macroscopically, some correlations have been seen between reaction rates and other measured cluster properties. For example, some cluster reactions show evidence that the shell structure affects reaction rates, with open shell clusters having much higher reaction rate and lower activation barriers than clusters with closed shells. In other cases, the pattern of cluster reaction rates with cluster size correlates with the cluster I_P's. These correlations have been used to infer the essential cluster-reagent interaction which governs the reactivity for a given set of clusters with a given reagent. For example, open shell clusters and clusters with low I_P's favor reactions which involve electron donation from the cluster to the reagent at a critical point along the reaction coordinate. Such generalities must be made with caution, however, since they do not hold up well over a broad range of cluster sizes, compositions, and reagents. Even a single cluster size can exhibit complex reactivity: isomers with different reactivity have been observed for niobium clusters and, in some cases, these isomers have been interconverted by annealing. Of course, complexity is to be expected, given the richness and diversity of chemistry that is known to occur for different metal systems in various electronic and geometric environments.

Geometric structures of transition metal clusters have been studied using physisorption reactions. Weakly bound adsorbates such as hydrogen, water, and ammonia show strong saturation behavior in their uptake by metal clusters [10, 11, 25]. Trends in the saturation coverage with cluster size yield information about the type of binding site and total number of binding sites for a given cluster size. Additional information regarding the nature of the adsorbate site is often available from studies of the corresponding physisorption on bulk metal surfaces. This knowledge is used to sort through possible cluster structures and deduce which geometries exhibit the correct number and type of adsorbate binding sites. For example, saturation coverages of iron, cobalt, and nickel clusters correlate with icosahedrally packed geometries over a wide range of sizes. Adsorbate binding energies have also been determined in some cases [25].

38.2.4 Stable Metal Cluster Molecules and Metallocarbohedrenes

Thus far, this section has focussed entirely on the properties of metal clusters isolated in the gas phase. This discussion would not be complete, however, without mentioning that a number of metal cluster molecules have been made which are sufficiently stable that they can be made in quantity and bottled [7, 10, 11, 26]! Also, recently a new class of metal-carbon clusters has been discovered, termed metallocarbohedrenes, that are believed to be sufficiently stable and abundant that they can be made in bulk. The brief overview of these isolatable clusters given below is not meant to be complete but is intended to introduce the large and impressive body of work in this field. Interested readers are encouraged to consult the reviews cited and references therein.

Metal cluster molecules consist of a metal cluster core surrounded by a stabilizing ligand shell. A large number of such metal cluster molecules has been made, including some with metal cores as large as 300 platinum or 561 palladium atoms. In many cases, crystals of single size clusters have been made with an exactly known number of metal atoms in the core. The availability of macroscopic samples of these clusters has made it possible to measure a number of their properties. Exact structures are known for many metal cluster molecules from x-ray crystallography. Electronic and magnetic properties have been determined which reveal the development of metallic behavior within the metal core. The ligand shell is found to interact strongly with the metal atoms on the surface of the metal core. This outer shell of metal atoms does not behave as a surface of metal atoms with free valence electrons such as is found on the surface of a bulk metal. Undoubtedly, such strong interactions between surface metal atoms and coordinating ligands are necessary to make a cluster which is sufficiently stable to be isolated and crystallized without coalescing. The core of atoms inside the outer metal shell of atoms does not appear to be greatly perturbed by the coordinating ligands. Studies of the electronic and magnetic properties of the core show the onset of metallic properties as a function of particle size and atomic packing.

A distinct class of transition metal-carbon clusters has been recently been found which have been termed metallocarbohedrenes [27]. Within this class, the clusters M_8C_{12} (M = Ti, V, and Zr) are particularly abundant and stable. Extensive electronic structure calculations show that the structure of M_8C_{12} is metallic and should be viewed as a distorted M_8 cube where each face is decorated with a C_2 dimer [28]. This structure differs greatly from the corresponding bulk metal carbides, which have cubic rock salt crystalline forms. Furthermore, the metal carbide cluster formation conditions can be adjusted to yield cubic fragments of the bulk. The related metal nitride clusters are only observed to form cubic structures. Theoretical calculations on the metallocarbohedrene and cubic forms of these metal carbides and nitrides show that they are comparable in energy for the carbides, but that the cubic structures are much more stable in the metal nitride clusters [29]. Despite their apparent stability, none of the metallocarbohedrenes has yet been isolated and purified.

38.3 CARBON CLUSTERS

The discovery of the especially stable, spherical cluster of sixty carbon atoms, named buckminsterfullerene, has ignited an intense research effort in carbon clusters [7, 8, 10, 11, 30–35]. The family of pure carbon clusters that have been made extends from the dimer all the way up to tubular shaped clusters containing thousands of atoms. Several of these clusters, most notably C_{60} and C_{70}, have been isolated as a single size in macroscopic quantities [36]. The fascinating properties of bulk materials made of pure and doped C_{60} and C_{70} are outside the scope of this review [8, 10, 11, 33, 35, 37].

Carbon clusters can be roughly grouped into three distinct classes. The first consists primarily of small carbon clusters which have linear or ring geometries. Hollow spheres (termed fullerenes) appear at about 28 carbon atoms and persist up to at least several hundred atoms. Very large carbon clusters containing hundreds to thousands of atoms assume various forms, including onion-like structures of concentric spheres and hollow tubes.

38.3.1 Small Carbon Clusters

Theoretical and experimental studies have found that rings and linear chains are the most stable configurations for small carbon clusters. Below about ten atoms, linear cumulenes are the most stable structures for odd numbered carbon clusters and ionic carbon clusters [13, 24, 38, 39]. For the even numbered neutral carbon clusters, C_4, C_6, and C_8, however, cyclic and linear geometries are nearly isoenergetic [7, 11, 13, 38]. Various theoretical and experimental studies have yielded conflicting results as to which geometry is more stable for each of these three clusters [7, 38, 40]. Much of the controversy over the experimental evidence probably arises because the two structures are so close in energy that either isomer or both may be present, depending on the preparation conditions. By way of proof, direct experimental data have been found for coexisting linear and cyclic isomers of C_{7-9}^+ and C_{11} [24, 40]. Also, some experimental probes are not equally sensitive to the signature from a linear versus a cyclic geometry.

At ten carbon atoms, a distinct transition occurs from linear to cyclic structures [24, 39, 40]. Starting at this size, the additional bonding stabilization accrued by joining the two ends of the linear chain overcomes the strain energy resulting from ring closure. Stable monocyclic ring structures are observed for carbon clusters over a surprisingly wide range of sizes, even persisting into the size range where three dimensional fullerenes appear. Bicyclic rings, higher order polycyclic rings, and graphitic fragments also occur in this size range but many of these configurations appear to be metastable [39, 41].

A_e's and I_P's have been measured for these small carbon clusters. Dramatic effects appear in the size dependence of the A_e's as a result of the changeover from linear to ring structures [40]. A_e's of the chain structures are noticeably higher than for ring structures for carbon clusters containing similar numbers of atoms. Distinct odd-even alternations in the A_e's are also seen, plus there is evidence for aromatic stabilization according to the $4n + 2$ rule. No break in the I_P's is seen across the structural transition, but the trend in the I_P's also follows the $4n + 2$ rule expected for aromaticity.

Chemical reactions have been observed for both neutral and charged small carbon clusters, particularly for the cations [7, 24, 39]. In general, the linear cationic clusters, containing fewer than ten atoms, react readily with a variety of reagents and exhibit reactivity typical of carbenes. The ring shaped cationic clusters with ten or more atoms are much less reactive, and often show no detectable reaction with reagents that react efficiently with the smaller carbon clusters. It is this differential reactivity which has revealed the presence of coexisting linear and cyclic isomers. Evidence for polyacetylene versus cumulene structure in the small linear chains can also be seen in the cluster reaction patterns.

38.3.2 Fullerenes

Fullerenes make up a class of carbon clusters with closed hollow carbon atom cage morphologies. The term *"fullerene"* was inspired by the geodesic domes of architect R. Buckminster Fuller, and has come into widespread usage despite its nonstandard nomenclature. The fullerene cage network is composed of interlocking rings of sp^2 hybridized carbon atoms, where every carbon atom is bonded to three other carbon atoms. Five and six member rings predominate in the cages. Highly strained four and smaller membered rings are unfavorable. Seven and higher membered rings usually can facilely rearrange within the network to form five and six membered rings. Delocalization of the π electrons over the cage contributes significantly to the stabilization of the fullerenes.

Assuming at least five-membered rings, the smallest possible fullerene is C_{20}, consisting of twelve pentagons. However, the observed fullerenes all contain about 30 atoms or more. Larger fullerenes are formed by joining together pentagons, hexagons, and heptagons, with the pentagons providing the curvature necessary to close the cage. In fullerene structures composed entirely of pentagons, hexagons, and heptagons, it can be rigorously shown from Euler's theorem that, $N_5 = N_7 + 12$, where N_5 is the number of pentagons and N_7 the number of heptagons. By far the most abundant fullerene that is seen is C_{60}. There are 1812 possible isomers for C_{60} but only one forms in great abundance [33], which has each of its 12 pentagons isolated and surrounded by its 20 hexagons. The resulting molecule, resembling a soccer ball, belongs to the highest point symmetry group I_h, where all the carbon atoms are equivalent [30–33]. The *isolated pentagon rule* arises out of minimizing strain and maximizing resonance energy and is a powerful tool for predicting the most stable fullerene isomers [33]. Besides C_{60}, a number of other fullerenes have been seen, most notably C_{70} which is the next larger fullerene that can be constructed using the isolated pentagon rule. The largest proven fullerene structure contains 84 carbon atoms. It is possible to construct increasingly larger fullerene cages containing hundreds of carbon atoms. Such giant fullerenes have yet to be conclusively verified in experiments, however evidence for their existence has been seen in bulk samples containing fullerene mixtures and in gas phase abundances of carbon clusters [32, 33, 35].

A number of the fullerenes has been made and purified as a single size in macroscopic quantities [33, 35, 36]. This has enabled their geometries to be determined quite accurately using a wide variety of spectroscopic and theoretical techniques. Even structures of fullerenes such as C_{76}, C_{78}, and C_{84} which occur as isomeric mixtures have been elucidated. The electronic structures of many of these fullerenes have been calculated and compared with experiment [33, 34].

Yet to be resolved is the issue of how these low entropy fullerene structures can form so efficiently in carbon vapor. One proposal is that cup-like prefullerenes form first which add new carbon moieties to close the cage. Other suggestions include coalescence of small carbon rings such as C_6 or C_{10}, or stacking of intermediate size rings on a small "seed" ring such as C_{10}, or folding up of large defective graphitic fragments to form closed hollow spheroids [8, 10, 32, 33]. Some intriguing insights into this open question are provided by experiments on nonfullerene metastable forms of carbon clusters in the fullerene size regime [39]. These species can be made in the gas phase and are found to have bicyclic, graphitic, and other polycyclic ring shapes which are not in the form of partial fullerene cages. Upon annealing, some of these ring forms are seen to convert into the fullerenes, which suggests that they are important intermediates for forming the fullerene cage. Similar polycyclic ring shapes are observed during the initial "melting" of fullerenes in molecular dynamics simulations [42].

Since the fullerenes are hollow, much attention has focussed on putting something inside. A wide variety of noble gas atoms and metal atoms has been successfully loaded into fullerenes, forming a family of so-called "endohedral" complexes [32, 33, 35, 43–47]. Noble gas atoms encapsulated in fullerenes appear to assume central positions within the cage and do not perturb the overall electronic structure of the fullerene [11, 43, 48]. Endohedral complexes where the encapsulated species is a metal atom, called *metallofullerenes*, encompass a wide range of metals (M) and fullerene sizes from $U@C_{28}$ to $Sc_3@C_{82}$ (@ denotes that the metal atom is inside the fullerene). Separation and purification of a macroscopic amount has been achieved for several of these [44, 45, 47]. Theoretical and experimental data indicate that the endohedral metallofullerenes have considerable charge transfer from metal atom(s) to the fullerene cage [33, 43, 44, 47, 48]. This interaction affects the oxidation states of the metal atom(s) and the cage and, in some cases, causes the metal atom(s) to locate off-center. It is yet to be generally understood why, within the range of known fullerene sizes, only some form endohedral complexes, particularly C_{28}, C_{60}, C_{70}, C_{74}, C_{80}, C_{82}, and C_{84}. The $M@C_{28}$ endohedral complex is especially intriguing because it appears that C_{28} itself does not form as an empty fullerene.

The unsaturated surfaces of fullerenes undergo a broad range of chemical reactions. The availability of macroscopic samples of fullerenes, especially C_{60} and C_{70}, has enabled the preparation of numerous fullerene derivatives. With their high electron affinities, fullerenes readily form a wide variety of charge transfer compounds. These include the so-called *exohedral* complexes, where metal atoms are attached to the outside of the cage. Particularly stable exohedral metal-fullerene clusters have been observed which have one alkaline earth atom decorating each ring of the cage [49]. The unsaturated double bonds of the fullerenes can be functionalized with reagents such as halogens, aromatics, and alcohols [33, 35]. Some of the bulk forms of these complexes and derivatives exhibit amazing properties such as superconductivity at temperatures as high as 33 K seen in C_{60} films doped with alkali or alkaline earth metals [47]. References [33, 35, 47] contain recent reviews.

38.3.3 Giant Carbon Clusters: Tubes, Capsules, Onions, Russian Dolls, Papier Mâché ...

Giant clusters of pure carbon have been found which have tubular, capsular, and spherical shapes [8, 10, 33, 47, 50]. These clusters occur both as single entities and with multiple concentric layers. The basic structure consists of a spiralled or rolled graphitic sheet made up of hexagonal rings of sp^2 hybridized carbon atoms. Just as in the fullerenes, pentagonal rings provide the curvature required to form a ball or to cap the ends of tube shapes. Negative curvature has also been seen which is believed to result from heptagonal rings. Carbon tubes and capsules are formed prolifically in the same carbon arcs which produce fullerenes. Typical diameters range from about 5 to 100 nm, depending on the inner diameter and number of layers, and lengths of up to several microns are seen. For concentric or spiralled structures, the average interlayer spacing is 3.4 to 3.5 Å which is slightly larger than in crystalline graphite. Although tubes and capsules are the most commonly seen morphologies, these are observed to convert to layered spherical "onion" structures under intense irradiation [8]. This suggests that collapsed onion structures are more stable. Currently, there exists some debate over whether the multilayer tubes or capsules consist of layers of complete shells within shells, such as in a "Russian doll," or whether these layers are so highly defective that the overall structures are best described as "papier-mâché" consisting of numerous overlying graphitic fragments [50].

Little experimental detail is available on the electronic or mechanical properties of these tubes. Theoretical calculations have been performed on the electronic properties of perfect tubes constructed in various ways [33, 47]. Numerous configurations of tubes are possible, depending on the tube diameter and the overall screw axis formed by the rows of carbon hexagons wound around the tube waist. The geometrical arrangement strongly affects the electronic properties of a tube. Appropriate choices of diameter and tilt angle of the screw axis yield tubes which are metallic or semiconducting. Tubes and capsules with perfect lattice arrangements are calculated to be extremely stiff, forming the strongest carbon fibers known. Almost no data exist on the chemical properties of carbon tubes, capsules, and onions. While they are stable enough to isolate in air, reaction occurs with O_2 and CO_2 at high temperatures which destroy the tubes [47]. Finally, as for the fullerenes, tubes, capsules, and onions have a hollow cavity which can be filled. These "nanocapsules" have been successfully loaded with lead and gold atoms, as well as crystalline metal carbide particles [47].

38.4   IONIC CLUSTERS

A growing body of experimental and theoretical evidence on alkali halide and alkaline earth oxide clusters shows that these tiny clusters of ionic materials are ionically bound. Most of these clusters have highly ordered crystal structures even at very small sizes. Thus, these clusters offer excellent systems for studying electron localization on finite-sized crystalline lattices.

38.4.1   Geometric Structures

Most ionic clusters of alkali halides and alkaline earth oxides assume the cubic rock salt lattice typical of bulk sodium chloride [6, 7, 10, 13, 51]. This lattice is favored not only for clusters of rock salt cubic solids such as NaCl and NaF, but also for other clusters such as $Cs_x I_y$ where the bulk form has the cesium chloride cubic crystal structure. Exceptions occur when the cluster size is smaller than, roughly, a unit cube, or when the cluster anions and cations differ greatly in ionic radius, such as in lithium bromide clusters (even though solid LiBr has the NaCl crystal structure) [10]. The bulk rock salt lattice is retained even for clusters where the number of anions and cations are unequal, as long as the deviation from stoichiometry is not too great. In clusters with a large excess of alkali atoms, the extra metal atoms segregate to a face of the cluster and form a metallic overlayer [51].

Maximization of ionic interactions and minimization of surface energy leads to cuboid crystal morphologies for these clusters which have as many $(1\,0\,0)$ faces as possible. Particularly stable clusters occur for sizes where the cuboid lattice is completely filled and has nearly equal numbers of ions on all sides [51]. Excess electrons or holes also play a crucial role in the stability of ionic clusters. For example, while the $[Na_{14}Cl_{13}]^+$ cluster forms a particularly stable cube (with 9 atoms on each square face and one atom in the center), its counterpart $[Na_{13}Cl_{14}]^+$ does not because it has fewer holes than the available electrons; however, the anion $[Na_{13}Cl_{14}]^-$ does form a particularly stable cube [11, 51].

38.4.2   Electronic and Chemical Properties

The I_P's, A_e's, photoabsorption spectra and photoabsorption cross sections have all been measured for ionic clusters, particularly alkali halide $[M_j X_k]$ clusters. Understanding the role of the coulombic interactions, particularly excess electrons, in these clusters is key to understanding the trends in their electronic properties with cluster size, composition, and overall charge. Excess electrons are known to localize in at least four distinct ways: at anion vacancies, in weakly bound surface states, on

specific alkali metal ions, or in cation-anion pair dipole fields [11, 51, 52].

First, there is a large class of ionic clusters which contain equal numbers of electrons and holes, such as clusters in the series $[M_{j+1}X_j]^+$. I_P and A_e of clusters within this class reflect the overall stability associated with forming the *perfect* filled cuboid lattices described in Sect. 38.4.1. Such perfect clusters do not chemisorb polar molecules, whereas the *imperfect* clusters readily do [51]. The lowest energy absorptions for clusters in this class result from charge transfer excitations, just as in perfect ionic solids. Spectra obtained for $[Cs_{j+1}I_j]^+$ clusters show large cross sections for absorption, consistent with charge transfer, and features which converge towards the bulk for the perfectly cubic cluster, $[Cs_{14}I_{13}]^+$ [53].

Other ionic clusters do not contain equal numbers of electrons and holes; clusters with excess electrons have been the most studied. One such class consists of clusters with at least one anionic vacancy and one or more excess electrons. In this class, e.g., $[Na_{14}Cl_{12}]^+$, the excess electron localizes at the lattice site of the missing anion. This is conceptually similar to an F-center in an ionic bulk crystal. Enhanced stabilization is observed for those size clusters within this class where the excess electron sitting at an anionic site yields a filled cuboid lattice. Nonetheless, because the localized electron has a large zero-point energy, its binding energy is much less than an anion at the same site. This is reflected in both the first and second I_P's, as well as the A_e's of this class of clusters [51]. Just as for F-centers in ionic crystals, clusters with an excess electron localized on an anionic vacancy have strong optical absorptions at energies well below the charge transfer bands [51].

The next class of ionic clusters with excess electrons consists of perfect cuboid ionic clusters which contain one or more excess electrons, but do not have a defect binding site such as an anionic vacancy or an excess metal atom. In these clusters, e.g., $Na_{14}Cl_{13}$, the electron is quite weakly bound in a surface state which primarily involves the surface metal cations. These clusters have particularly low electron binding energies; for example, the I_P of ~ 1.9 eV determined for $Na_{14}F_{13}$ is the lowest I_P measured for any compound [51].

In the other two known classes of ionic clusters with excess electrons, the extra electrons have been calculated to localize on a metal cation or in a dipole potential well [54]. Some evidence for these forms of localization has been seen in photoabsorption, I_P's and A_e's [10, 11, 51]. For example, the photoelectron spectra of some $[Na_{j+1}Cl_j]^-$ clusters show that the two excess electrons are singlet-coupled, and localized either at an anion vacancy, or at a single Na site so that they behave as a Na^- anion loosely bound to a neutral $[Na_j Cl_j]$ cluster. The spectral behavior of other clusters within this same series, however, suggests they have the two excess electrons in a triplet coupled state, forming the analog of a bipolaron in a solid [10].

38.5   SEMICONDUCTOR CLUSTERS

Semiconductor clusters make up a class of clusters where, by analogy with bulk semiconductors, covalent forces are expected to dominate electronic and geometric structure. Silicon clusters are by far the most studied of the semiconductor clusters. Some information is available on germanium clusters and compound clusters made of Groups III and V atoms or Groups II and IV atoms. Also, there are the well known stable molecules of bulk semiconductors such as the P_4 tetrahedron and various sulfur rings; these have been reviewed in detail elsewhere [3]. Clusters of other possible semiconductors have been made, but little data beyond their nascent distributions are available [3]. There is also a growing body of data on silicon, III-V, and, especially II-VI semiconductor clusters in the nanometer size regime where bulk samples of stabilized forms of these clusters have been made and isolated. The crystalline and electronic structure properties, particularly quantum confinement effects, are reviewed in Refs. [10, 11, 55].

38.5.1   Silicon and Germanium Clusters

The geometries of small silicon clusters depart radically from microcrystalline fragments of the bulk silicon diamond lattice. The structures of silicon clusters containing up to about 13 atoms have been studied extensively both experimentally and theoretically [56]. These structures are more compact and, starting at Si_7 which has a pentagonal bipyramidal structure, have higher coordination than silicon in the bulk lattice. Multiple isomers of similar energy also appear starting at about Si_{10}. For example, the tetracapped trigonal prism and the symmetric tetracapped octahedron structures of Si_{10} are nearly isoenergetic; microcrystalline fragments such as the adamantane form of Si_{10} are much higher in energy. The geometric structures of larger silicon clusters are less well known. The gross shapes of clusters containing up to about 60 silicon atoms have been found experimentally to undergo a transition between Si_{20} and Si_{30} from increasingly elongated structures to more spherical structures [24]. In the transition region, both prolate and oblate isomers are observed for a single cluster size. Multiple isomers over a wide range of silicon cluster sizes have been repeatedly observed in various experiments; it appears that somewhat different sets of isomers can be produced depending on the cluster formation conditions [24]. Elucidating the ground state structures of these larger clusters theoretically is, in general, an intractable problem and requires simplifying approaches. Use of semiempirical quantum mechanical techniques or silicon interaction potentials derived from bulk silicon, however, has led

to unsatisfactory results, which suggests that the silicon atoms in silicon clusters are strongly reconstructed away from the usual sp^3 silicon atom environment [57, 58]. Some consensus has recently emerged that larger silicon clusters consist of internal silicon atoms strongly interconnected to a surrounding cage that has been described as a buckled fullerene [57, 59]. Such a silicon cage must not be construed as a true fullerene, however, since silicon does not form the strong double bonds which stabilize the interconnected carbon rings of the fullerenes. Within this scenario, the shape change is believed to occur at the point where the cage begins to contain one or more internal atoms which provide the additional bonding needed to stabilize a spherical geometry.

I_P's and A_e's, cohesive energies, and photoabsorption spectra have been measured for silicon clusters containing up to several hundred atoms. Silicon cluster I_P's start near the Si atom I_P, fall abruptly between 20 and 22 silicon atoms where the shape change has been observed, and then slowly converge towards the bulk work function. This convergence is apparently quite gradual since little change is seen between Si_{100} and Si_{200} [60]. A_e's are only known for silicon clusters up to ~15 atoms and reflect the large structural changes which occur in this small size regime [3]. Indication of a structural shape change is not observed in either the trends in the cohesive energies or the electronic spectra for silicon clusters. The cohesive energies increase smoothly with increasing size and exceed the cohesive energy of bulk silicon by 10-20% [61]. Silicon clusters exhibit strong sharp absorption spectra in the near uv. A common set of absorption features appears at ~Si_{15} and persists up to at least Si_{70} [62]. This suggests that the strong absorptions arise from localized Si-Si bond excitations, and that these obscure the delocalized excitations which are more sensitive to silicon cluster structure. The most unusual aspect of these spectra is that the common signature for these minute clusters which are strongly reconstructed from the bulk is, nonetheless, strikingly similar to the spectrum of bulk diamond-lattice crystalline silicon.

Silicon cluster ions are observed to react with a variety of reagents [3, 24, 61]. For example, silicon clusters chemisorb small organic molecules such as ethylene, and inorganic molecules such as O_2, NH_3, and XeF_2. Some of these reactions are sufficiently exothermic that they cause cluster fragmentation, or loss of small neutral fragments. The chemistry of silicon clusters often bears a close relationship to that known for bulk silicon surfaces; however, the clusters are often much less reactive by as much as several orders of magnitude. Silicon cluster reactions reveal the presence of numerous isomers which differ in their reaction rate for a given reagent. Isomerization has been induced by adding sufficient thermal energy to some clusters; the resulting more stable cluster form may or may not be more reactive than the higher energy isomer. This is particularly well illustrated in reactions of oblate versus prolate isomers where the more reactive isomer varies depending on cluster size [24]. In general, differential chemistry is not a useful predictor of cluster geometry *per se*, and appears to correlate more readily with other factors such as the number and type of dangling bonds available for reaction.

Much less is known about germanium clusters. The most stable structures that have been calculated for germanium clusters are quite similar to those for silicon clusters, although the overall binding energies are lower for germanium clusters [3, 7]. Available data such as A_e's and photoelectron spectra for germanium clusters also underscore their similarities to silicon clusters in general [3]. The chemistry of germanium clusters has not been reported.

38.5.2 Group III-V and Group II-VI Semiconductor Clusters

Geometric structures have been calculated for some of the smaller III-V and II-VI clusters, particularly aluminum phosphide, gallium arsenide, and magnesium sulfide clusters [3, 11, 63]. As for silicon clusters, these clusters differ significantly from microcrystalline fragments snipped from their corresponding bulk crystalline forms. Electronegativity differences between the two constituent atoms play a major role in determining and stabilizing these structures. This is manifested in several ways. The bonding arrangements in the most stable structures have, in general, alternating electropositive and electronegative atoms. In those clusters such as III-V clusters where electronegativity differences are smaller, covalent interactions predominate ionic interactions and the energetics of the various geometric structures are similar to those of the covalent silicon clusters. In II-VI clusters, electronegativity differences are much larger, and these clusters have structures more comparable to those seen in ionic clusters where ionic interactions are maximized.

Some data are available on the electronic structures of III-V clusters. A strong even-odd alternation occurs in both the I_P's and A_e's of gallium arsenide clusters where those clusters having a total even number of atoms have higher I_P's and lower A_e's than neighboring odd-numbered clusters [3, 64]. This has been explained by electronic structure calculations which find that, in general, the odd clusters are triplets while the even clusters of gallium arsenide are singlets [63]. Electronic absorption spectra have also been recorded for indium phosphide clusters which exhibit strong differences with cluster size and stoichiometry [65]. These spectra are also consistent with odd-numbered In_xP_y clusters having open-shell configurations, and even-numbered In_xP_y clusters having closed-shell singlet ground states, even for clusters which are considerably off-stoichiometry. A strikingly similar strong continuum-like absorption appears in the blue end of spectra for the even-numbered In_xP_y clusters. The onset of these absorptions lies close to the bulk indium phosphide band gap. The overall spectral behavior, however, is quantitatively similar to the absorptions seen in semiconductor glasses.

Little is known about the chemical properties of these clusters, with gallium arsenide clusters being the only ones studied. Hydrogen chloride is observed ubiquitously to etch $[Ga_xAs_y]^-$; however, multiple isomers are seen with differing degrees of reactivity [3]. Chemisorption of ammonia on $[Ga_xAs_y]^+$ has also been seen with the highest rates of reactivity occurring for the stoichiometric $(x = y)$ clusters [66].

38.6 NOBLE GAS CLUSTERS

Clusters have been made of all of the stable noble gases including helium. As a general class, these clusters are the most weakly bound of all clusters, and are held together only by van der Waals forces. These interactions are well understood theoretically, and thus noble gas clusters are excellent model systems for studying a variety of structural and electronic effects at finite sizes. Helium clusters behave much as quantum liquids, and are treated separately in the discussion below.

38.6.1 Geometric Structures

The overall evolution of noble gas cluster structures as a function of size is well established [4, 6, 11, 67, 68]. Icosahedral packing dominates at small cluster sizes. This packing involves polyicosahedral structures for the small cluster sizes, and different noble gas clusters exhibit somewhat different preferred arrangements. Starting at 100 to 250 atoms (depending on which noble gas), formation of closed-shell Mackay icosahedra dominates the structures [69]. Finally, at somewhere in the range 600 to 6000 atoms, the structures cross over to the close-packed fcc arrangement of the noble gas solids.

The increased coordination characteristic of the Mackay icosahedra explains why this morphology is adopted at small cluster sizes. With the addition of each successive icosahedral shell however, this five-fold symmetric packing becomes increasingly strained as a result of both atom-atom radial compression and tangential dilation. At some size, icosahedral structures are no longer more stable than fcc structures, and the noble gas clusters undergo a phase change. Currently, the critical size required for fcc packing is a matter of some dispute. Experimental evidence for this transformation at ~600 atoms comes from electron diffraction data [67]. The spectral signatures of molecules doped in noble gas clusters suggest however, that this transition occurs at ~2000 atoms [68]. Theoretical studies find a range of critical sizes, depending on the treatment used, but typically favor an even larger size regime [4, 11, 13, 68].

Knowledge of accurate pair-wise interaction potentials for noble gases has enabled extensive simulations of the physical properties of noble gas clusters [4, 6, 7, 70]. Phenomena such as specific heat, bond length and coordination number fluctuations, phase equilibria, dynamical freezing and melting, isomerization, and solvation have been explored using various molecular dynamics simulations. Numerous effects peculiar to the finite sizes and proliferation of isomers in small noble gas clusters are observed. For example, the melting temperature decreases significantly with an overall decrease in cluster size. Internal diffusion rates depend strongly on cluster size and the stability of certain favored structures such as the complete Mackay icosahedra. A well-studied finite size effect is the dependence of the mean energy per particle in a cluster versus the particle temperature (the so-called *caloric curve*) in the region where the particle is observed to "melt." Unlike in the bulk, calculations find hysteresis in the melting transition of small noble gas clusters where solid-like and liquid-like forms are observed to coexist. Such coexistence is quite size-dependent. It becomes more pronounced when the clusters are rapidly heated or cooled, and appears to be a finite time averaging effect [4, 6, 7].

38.6.2 Electronic Properties

The fragility of noble gas clusters has hampered measurements of their I_P's and A_e's. Extensive fragmentation is known to accompany ionization, making it difficult to establish the size of the ionized parent cluster. Nonetheless, I_P's have been measured for cluster sizes containing up to several tens of atoms [71]. Measured I_P's agree with theoretical predictions which are based on the assumption that the ionizing state within the cluster is a dimer, trimer, or higher n-mer cationic core [6, 71]. Absorption profiles of noble gas cations provide supporting evidence for the delocalization of the positive charge over a n-mer unit within the cluster, although consensus has yet to be reached on the size of this cationic cluster within the cluster [6, 11].

Several distinct types of electronic excitations are observed in absorption spectra of neutral noble gas clusters [6, 7, 11, 72]. Clusters containing less than ~30 atoms exhibit broad absorptions near the atomic resonance lines. These absorptions are molecular-like, but have not been described in detail above the dimer. In addition, broad continuum absorptions assigned to Rydberg excitations are observed at these small sizes. Bulk-like excitations corresponding to surface and bulk excitons in solid noble gases emerge for clusters larger than ~50 atoms. The profiles of these excitonic absorptions indicate that they arise from delocalized excitations analogous to Wannier

excitons in bulk solids. These cluster exciton absorptions are blue-shifted relative to the bulk as a result of quantum confinement of the exciton within the small cluster. Another type of excitonic excitation has also been reported which appears to be a localized excitonic state whose character is sensitive to the cluster structure. Relaxation of cluster excited states is accompanied by extensive fragmentation [7]. Details of the relaxation processes differ according to noble gas, cluster size and cluster structure. Again, these differences can be traced to finite size effects in these noble gas clusters [4].

38.6.3 Doped Noble Gas Clusters

Various chromophores have been added to noble gas clusters as a microscopic variation of matrix isolation [3, 4, 7, 10, 11, 68, 73]. A wide range of chromophores have been studied including other noble gas atoms, metal atoms, small polyatomic molecules such as SF_6, CH_3F, and HCl, and a variety of organic molecules such as benzene, carbazole, and naphthalene. Comparisons of the electronic and vibrational spectra of these guest molecules with their spectra in noble gas liquids or solid matrices have revealed information such as noble gas cluster structure, solvation effects, and solute diffusion.

A crucial issue in understanding the spectral signatures of doped noble gas clusters is the location of the solute in or on the noble gas cluster. Chromophores well-embedded in a noble gas cluster behave as molecules surrounded by a dielectric medium, which is most correctly viewed as both imperfect and finite sized [4, 7, 73, 74]. The evolution of the spectral changes with cluster size reflect the interplay of repulsive forces and collective dielectric effects as the cluster builds the solvent shell around the solute core. Both theoretical and experimental data illustrate this evolution and show how, once the first few solvent shells are established, the spectrum converges asymptotically to the matrix isolation value. For smaller clusters, the effects of too few or incomplete solvation shells appear as shifts and/or broadening of the spectral lines relative to the bulk [3, 4]. Differential solvation of the ground and excited states causes spectral shifts which are typically to lower energy since the excited state is usually more stabilized than is the ground state. Spectral broadening results from multiple isomers. Broadened electronic spectra have also been explained as a signature of cluster melting, although this interpretation has been questioned [3, 4, 73, 74]. Some infrared spectra of solute molecules have also yielded information on the nature of the solute binding site, revealing that the solvating noble gas cluster undergoes an icosahedral to fcc phase transition in a critical size regime [68].

Not all dopant species are found well-solvated, or *wetted* in the cluster interior. Both electronic and infrared dopant spectra indicate the existence of chromophores bound to the cluster surface. In some cases, the chromophore is ubiquitously *nonwetting*, such as for SF_6-Xe_n and SiF_4-Ar_n, and resides exclusively on the cluster surface [68]. In other systems, e.g., carbazole-Ar_n and CF_3Cl-Ar_n, both wetting and nonwetting are observed [7, 68, 73]. Systems such as these have enabled measurements of solute diffusion rates into or out of the cluster. Finally, wetting-nonwetting transitions have been observed which depend on cluster size and/or cluster temperature [7, 73]. Understanding this wide variety of behavior requires modeling cluster solute-solvent structures (in both ground and excited states in order to interpret electronic spectra) as a function of size and temperature. To date, most of the theoretical simulations have focussed on smaller clusters at a given temperature [4, 7, 68, 73, 74]. Nonetheless, theoretical models have provided qualitative and quantitative explanations for some of the observed spectral shifts, and have shown that the propensity for wetting/nonwetting can be related to the degree to which maximum possible coordination within the solvent is attainable.

38.6.4 Helium Clusters

Quantum effects play a dominant role in helium clusters since they consist of very weakly interacting small mass particles. Statistical effects are also expected since ^3He is a fermion and ^4He is a boson. Overall, helium clusters are believed to be in fluid-like or superfluid-like states with highly fluctional structures [75].

Neutral helium clusters have been made experimentally in sizes ranging from the dimer up to 10^6 atoms [4, 11, 68]. The diatomic He_2, has just been detected recently and is found to have a binding energy of only $\sim 10^{-7}$ eV with an average internuclear distance of ~ 55 Å [76]. Evidence for magic numbers in the helium cluster abundances is seen; however, the sizes of these especially stable clusters differ substantially from what is seen in the other noble gas clusters [4]. Helium cluster structures have been theoretically investigated using fully quantum mechanical treatments [4, 75]. These studies find that ^4He_n clusters should be bound at all sizes, but that a minimum number of atoms, ~ 30, are required for ^3He_n clusters to be stable. The latter prediction has not yet been verified experimentally, owing to the expense of ^3He, and the difficulties of ascertaining the true size of helium clusters since they easily boil off atoms upon ionization. Calculations reveal that the packing of helium clusters is highly delocalized with no evidence of icosahedral morphology. The structures derived have been found to be extremely sensitive to temperature and total angular momentum. Cluster binding energies and densities increase smoothly and monotonically with increasing size, approaching bulk behavior at about 300 atoms. Thus, helium clusters do not exhibit any especially stable sizes

(magic numbers) such as the heavier noble gases do. In short, helium clusters behave as liquid-like quantum fluids, and may even be superfluid-like at the temperatures required to stabilize them in beam experiments. Based on these calculations, it appears that the magic numbers observed experimentally in abundance spectra pertain to the more strongly bound ionized helium clusters, rather than the neutral clusters which are the subject of this discussion.

Electronic absorption spectra have recently been recorded for clusters containing ~ 50–10^6 helium atoms [77]. Broad strong absorption bands are observed which do not behave like the Wannier exciton bands seen for the heavier noble gas clusters, nor are they well-described by the Frenkel excitonic model. Note that although the Wannier and Frenkel exciton models were originally developed for solids with translational symmetry, they are also good descriptions for excitations in liquid noble gases [4]. At this writing, the helium cluster electronic spectra cannot be compared with absorption spectra of liquid or solid helium because these latter spectra have not yet been measured! With this limitation and a current lack of sufficient theoretical guidance, the helium cluster spectra have not yet been thoroughly interpreted. Theoretical calculations are available which describe the collective excitations in helium clusters on the electronic ground state surface [75]. Spectra in this energy regime have not yet been recorded.

Experimentally, it has been quite easy to dope helium clusters with various atomic and molecular species such as other noble gas atoms, oxygen, and SF_6 [4, 11, 68]. Electrons, however, are not well-solvated and negatively charged helium clusters only appear for sizes containing more than 10^5–10^6 atoms [11]. Impurity species provide a spectroscopic probe for studying the properties of the solvating helium cluster. Infrared spectra of helium clusters doped with SF_6 show that the impurity molecule resides on the cluster surface, in contradiction of theoretical calculations which predict that it should be found inside a helium cluster. Understanding this discrepancy has stimulated further theoretical investigations that reveal the dramatic structural effects which can occur when angular momentum is added to these very fragile clusters during dopant pick-up [78].

38.7 MOLECULAR CLUSTERS

Molecular clusters are in many ways similar to noble gas clusters. Both are weakly bound, and the interactions between the particles which make up the cluster can usually be described to a good approximation as the sum of pairwise interactions. The molecule constituency adds complexity, however, which is reflected in the diversity seen amongst these clusters. Molecular clusters provide model systems for studying solvation, including electron solvation, nucleation, and phase transitions at finite sizes. Furthermore, properties which are often difficult to study in the bulk, such as phase transition dynamics, are much more amenable to study in molecular clusters.

Numerous molecular clusters have been made and studied. By far the most work has focussed on the smaller molecular clusters such as molecular dimers, trimers and tetramers. A number of outstanding reviews are available on this subset of molecular clusters [3, 4, 6, 67, 79].

38.7.1 Geometric Structures and Phase Dynamics

In contrast to noble gas clusters, molecular clusters typically assume bulk phase structures at relatively small sizes. Thus, the range of condensed phases of molecular solids is reflected in the variety of packing structures found in molecular clusters. At one extreme are clusters of small nearly-spherical molecules such CH_4 and N_2 which form clusters closely connected to noble gas clusters [67]. Small clusters of these molecules are packed in polyicosahedral arrangements at small sizes, and cross over to fcc structures at sizes containing several thousand atoms. At another extreme are clusters such as water clusters where stronger hydrogen bonding forces lead to the formation of crystalline networks such as in a diamond cubic phase [67]. Some molecular clusters appear to be liquid-like with thermal motion wiping out any persistent periodic order. The liquid-like structure observed in clusters of molecules such as benzene and various hydrocarbons may freeze into a crystalline or amorphous state at sufficiently low temperatures.

Some molecular clusters have been found to exhibit a panoply of phases, depending on size and temperature. For example, not only do clusters of TeF_6 assemble into the bulk body-centered cubic and orthorhombic phases, they can also be made in other phases including two trigonal forms, a rhombohedral phase, and two monoclinic phases [80]. Similarly complex phase formation has been found in other clusters of other hexafluoride molecules such as SF_6, SeF_6, MoF_6, and WF_6.

Phase transitions in molecular clusters have been examined both experimentally and theoretically using molecular dynamics. Transition temperatures are found to depend on cluster size, and span a much broader range than in the corresponding bulk phase transitions. Large rates for nucleation and phase changes are seen in molecular clusters. Once the critical nucleus, which initiates the phase change, forms in the interior of the cluster, the remainder of the cluster transforms to the new phase extremely rapidly. Critical nucleus sizes depend on the cluster size and are typically quite small, e.g., for TeF_6 clusters, they contain only a few dozen molecules for clusters consisting of a few hundred molecules [10, 80]. A key finding is that phase transitions

in molecular clusters appear to violate *Ostwald's step rule*, which states that equilibrium systems must pass through all intermediate-energy stable phases during a transition from a higher energy to a lower energy phase [10]. This noncompliance apparently occurs because of the speed at which molecular clusters undergo phase transitions: intermediate phases simply do not have time to form.

38.7.2 Electronic Properties: Charge Solvation

I_P's, A_e's, and some spectral data are available for larger molecular clusters. Nearly all of the spectra measured pertain to clusters containing less than ten molecules [3, 4, 6, 67, 79]. Precise molecular orientations within many of these clusters have been derived from detailed analyses of these spectra. Such information has not yet emerged from the few spectra that have been measured for larger molecular clusters [3, 6]. Alternatively, measurements of electronic properties of larger clusters have generally targeted a different issue: solvation of excess positive or negative charge in a restricted bulk-like system.

Charged molecular clusters provide model systems for studying the mechanism of charge stabilization within the confines of a finite system. Excess positive charge appears to be highly localized in molecular clusters, residing on a small unit containing at most a few monomers. This positive core is surrounded and stabilized by overlying shells of molecules. For example, $(CO_2)_n^+$ clusters behave as the dimer cation $(CO_2)_2^+$, surrounded by the remaining CO_2 molecules [4]. Positively charged solutes such as alkali cations have also been introduced to study solvation of positive charge within molecular clusters [10]. For those dopants which have much lower ionization potentials than the solvating molecules, the positive charge remains strongly localized on the impurity. The cluster molecules are observed to build up solvation shells around this central impurity. The first few solvation shells are strongly affected by the positive charge, which has a decreasing influence with each successive layer. The structure of such a cluster reflects the accommodation between the geometry of the molecules influenced by the central positive charge, and those in the outer solvation layers where intermolecular forces predominate [10, 81]. Evidence for intracluster reactions in positively charged clusters has also been seen, such as proton transfer in cationic water or ammonia clusters [3, 11].

Whereas virtually every molecular cluster containing as little as two molecules exists stably as a positive ion, the same cannot be said for the negative cluster ions. Measurements of the minimum number of molecules required within the cluster to support an additional electron show that, while the dimeric $(HCl)_2^-$, $(SO_2)_2^-$, and $(H_2O)_2^-$ clusters are stable, ~35 or ~41 molecule-clusters of ND_3 or NH_3, respectively, are required to stabilize an electron [4]. Spectroscopic studies show evidence for internal and external solvation of the excess electron in these anionic clusters.

Charge solvation in water clusters has been examined in some detail. Of particular interest is the $(H_2O)_{20}$ cluster which, from experimental and theoretical data, appears to form an especially stable, well-defined clathrate cage. The interior of this cage is large enough that it can and does hold various cations such as NH_4^+, H_3O^+, and alkali ions [10]. This cage is not observed to surround negative ions or electrons, which instead, reside on the surface [10]. This is a general result for excess negative charge in water clusters in this size regime where excess charge can be external to the cluster. Excess electrons in water clusters are found on the cluster surface for small cluster sizes up to 60 to 70 water molecules. Above this size, the electron resides in the cluster interior, and behaves analogously to hydrated electrons in bulk water [4, 7].

REFERENCES

1. Special issue on Gas Phase Clusters, Chem. Rev. **86** (1986).
2. W. A. deHeer, W. D. Knight, M. Y. Chou, and M. L. Cohen, in *Solid State Physics*, edited by H. Ehrenreich and D. Turnbull (Academic Press, New York, 1987), Vol. 40.
3. *Atomic and Molecular Clusters*, edited by E. R. Bernstein (Elsevier, New York, 1990).
4. *The Chemical Physics of Atomic and Molecular Clusters*, edited by G. Scoles (North-Holland, New York, 1990).
5. V. Bonacic-Koutecky, P. Fantucci, and J. Koutecky, Chem. Rev. **91**, 1035 (1991).
6. *Clusters of Atoms and Molecules*, edited by H. Haberland (Springer Verlag, Berlin, 1994).
7. Proceedings of the 5th International Meeting on Small Particles and Inorganic Clusters, Z. Phys. D. **19–20** (1991).
8. *Clusters and Fullerenes*, edited by V. Kumar, T. P. Martin, and E. Tosatti (World Scientific, River Edge, 1993).
9. M. D. Morse, Chem. Rev. **86**, 1049 (1986).
10. Proceedings of the 6th International Meeting on Small Particles and Inorganic Clusters, Z. Phys. D. **26-26S** (1993).
11. *Physics and Chemistry of Finite Systems: From Clusters to Crystals*, edited by P. Jena, S. N. Khanna, and B. K. Rao (Kluwer Academic Publishers, Netherlands, 1992), Vol. 1 & 2.
12. C. G. Granqvist and R. A. Buhrman, J. Appl. Phys. **47**, 2200 (1976).
13. Proceedings of the 4th International Meeting on Small Particles and Inorganic Clusters, Z. Phys. D. **12** (1989).
14. *Advances in Metal and Semiconductor Clusters*, edited by M. A. Duncan (JAI Press, Greenwich, 1994), Vol. 2.
15. S. Bjornholm, J. Borggreen, O. Echt, K. Hansen, J. Pedersen, and H. D. Rasmussen, Phys. Rev. Lett. **65**, 1627 (1990); T. P. Martin, T. Bergmann, H. Gohlich, and T. Lange, Chem. Phys. Lett. **172**, 209 (1990).
16. O. Cheshnovsky, K. J. Taylor, J. Conceicao, and R. E. Smalley, Phys. Rev. Lett. **64**, 1785 (1990).
17. K. Rademann, Ber. Bunsenges. Phys. Chem. **93**, 653 (1989).
18. M. B. Knickelbein and W. J. C. Menezes, Phys. Rev. Lett. **69**, 1046 (1992).
19. T. Baumert, C. Rottgermann, C. Rothenfusser, R. Thalweiser, V. Weiss, and G. Gerber, Phys. Rev. Lett. **69**, 1512 (1992).
20. C. Brechignac *et al.*, Phys. Rev. Lett. **60**, 275 (1988).
21. J. P. Bucher, D. C. Douglass, and L. A. Bloomfield, Phys. Rev. Lett. **66**, 3052 (1991); I. M. L. Billas, J. A. Becker, A. Chatelain, and W. A. deHeer, Phys. Rev. Lett. **71**, 4067 (1993).
22. S. N. Khanna and S. Linderoth, Phys. Rev. Lett. **67**, 742 (1991).
23. D. J. Trevor and A. Kaldor in *High-Energy Processes in Organometallic Chemistry*, edited by K. S. Suslick (ACS Symposium Series #333, 1987).
24. D. C. Parent and S. L. Anderson, Chem. Rev. **92**, 1541 (1992).
25. E. K. Parks, T. D. Klots, B. J. Winter, and S. J. Riley, J. Chem. Phys. **99**, 5831 (1993); E. K. Parks and S. J. Riley, J. Chem. Phys. **99**, 5898 (1993).
26. T. Schmid, Chem. Rev. **92**, 1709 (1992); L. J. de Jongh *et al.*, in *Cluster Models For Surface and Bulk Phenomena*, edited by G. Pacchioni and P. G. Bagus (Plenum, New York, 1991).
27. B. C. Buo, K. P. Kerns, and A. W. Castleman, Jr., Science **255**, 1411 (1992).
28. H. Chen, M. Feyereisen, X. P. Long, and G. Fitzgerald, Phys. Rev. Lett. **71**, 1732 (1993).
29. B. V. Reddy and S. N. Khanna, Chem. Phys. Lett. **209**, 104 (1993).
30. H. W. Kroto, J. R. Heath, S. C. O'Brien, R. F. Curl, and R. E. Smalley, Nature, **318**, 162 (1985).
31. W. R. Creasy, Fullerene Sci. Tech. **1**, 23 (1993).
32. H. W. Kroto, A. W. Allaf, and S. P. Balm, Chem. Rev. **91**, 1213 (1991).
33. *Buckminsterfullerenes*, edited by W. E. Billups and M. A. Ciufolini (VCH Publishers, New York, 1993).
34. J. Cioslowski in *Electronic Structures of Fullerenes and Their Derivatives* (Oxford University Press, Oxford, 1994).
35. Special issue on Buckminsterfullerenes, Acc. Chem. Res. **25** (1992).
36. W. Kratschmer, L. D. Lamb, K. Fostiropoulos, and D. R. Huffman, Nature **347**, 354 (1990).
37. A. F. Hebard *et al.*, Nature **350**, 600 (1991).
38. J. R. Heath and R. J. Saykally in *On Clusters and Clustering*, edited by P. J. Reynolds (North-Holland, Amsterdam, 1993), p. 7.
39. G. von Helden, M.-T. Hsu, N. Gotts, and M. T. Bowers, J. Phys. Chem. **97**, 8182 (1993).
40. W. Weltner and R. J. Van Zee, Chem. Rev. **89**, 1713 (1989).
41. Special issue on Fullerenes, Carbon, and Metal-Carbon

Clusters, Int. J. Mass Spectrom. Ion Processes **138**, (1994).

42. S. G. Kim and D. Tomanek, Phys. Rev. Lett. **72**, 2418 (1994).

43. D. S. Bethune, R. D. Johnson, R. J. Salem, M. S. de Vries, and C. S. Yannoni, Nature **366**, 123 (1993).

44. X.-D. Wang, T. Hazhizume, Q. Xue, H. Shinohara, Y. Saito, Y. Nishina, and T. Sakurai, Japan. J. Appl. Phys. **32**, L147 (1993).

45. S. Hino, H. Takahashi, K. Iwasaki, K. Matsumoto, T. Miyazaki, S. Hasegawa, K. Kikuchi, and Y. Achiba, Phys. Rev. Lett. **71**, 4261 (1993).

46. M. Saunders, H. A. Jimenez-Vazquez, R. J. Cross, S. Mroczkowski, M. L. Gross, D. E. Giblin, and R. J. Poreda, J. Am. Chem. Soc. **116**, 2193 (1994).

47. *Fullerenes* in Solid State Physics, edited by H. Ehrenreich and F. Spaepen (Academic Press, New York, 1994), Vol. 48.

48. J. Cioslowski, in *Spectroscopic and Computational Studies of Supramolecular Systems*, edited by J. E. D. Davies (Kluwer Academic Publishers, Netherlands, 1992).

49. U. Zimmermann, N. Malinowski, U. Naher, S. Frank, and T. P. Martin, Phys. Rev. Lett. **72**, 3542 (1994).

50. O. Zhou, R. M. Fleming, D. W. Murphy, C. H. Chen, R. C. Haddon, A. P. Ramirez, and S. H. Glarum, Science **263**, 1744 (1994).

51. R. L. Whetten, Acc. Chem. Res. **26**, 49 (1993).

52. R. N. Barnett, U. Landman, D. Scharf, and J. Jortner, Acc. Chem. Res. **22**, 350 (1989).

53. X. Li and R. L. Whetten, J. Chem. Phys. **98**, 6170 (1993).

54. G. Rajagopal, R. N. Barnett, and U. Landman, Phys. Rev. Lett. **67**, 727 (1991).

55. M. L. Steigerwald and L. E. Brus, Acc. Chem. Res. **23**, 183 (1990); Y. Wang, Acc. Chem. Res. **24**, 133 (1991); L. Brus, Adv. Mater. **5**, 286 (1993).

56. K. Raghavachari and L. A. Curtiss in *Quantum Mechanical Electronic Structure Calculations With Chemical Accuracy*, edited by R. S. Langhoff (Kluwer Academic Publishers, Netherlands, 1994).

57. K. Raghavachari, Phase Transitions **24-26**, 61 (1990).

58. N. Bingelli, J. L. Martins, and J. R. Chelikowsky, Phys. Rev. Lett. **68**, 2956 (1992).

59. E. Kaxiras and K. Jackson, Phys. Rev. Lett. **71**, 727 (1993); U. Röthlisberger, W. Andreoni, and M. Parrinello, Phys. Rev. Lett. **72**, 665 (1994).

60. K. Fuke, K. Tsukamoto, F. Misaizu, and M. Sanekata, J. Chem. Phys. **99**, 7807 (1993).

61. M. F. Jarrold, Science **252**, 1085 (1991).

62. K.-D. Rinnen and M. L. Mandich, Phys. Rev. Lett. **69**, 1823 (1992).

63. M. A. Al-Laham and K. Raghavachari, J. Chem. Phys. **98**, 8770 (1993); L. Lou, P. Nordlander, and R. E. Smalley, J. Chem. Phys. **97**, 1858 (1992); R. M. Graves and G. E. Scuseria, J. Chem. Phys. **95**, 6602 (1991).

64. C. Jin, K. J. Taylor, J. Conceicao, and R. E. Smalley, Chem. Phys. Lett. **175**, 17 (1990).

65. K.-D. Rinnen, K. D. Kolenbrander, A. M. DeSantolo, and M. L. Mandich, J. Chem. Phys. **96**, 4088 (1992).

66. L. Wang, L. P. F. Chibante, F. K. Tittel, R. F. Curl, and R. E. Smalley, Chem. Phys. Lett. **172**, 335 (1990).

67. Special issue on Gas Phase Clusters, Chem. Rev. **86** (1986).

68. S. Goyal, D. L. Schutt, and G. Scoles, Acc. Chem. Res. **26**, 123 (1993).

69. W. Miehle, O. Kandler, and T. Leisner, J. Chem. Phys. **91**, 5940 (1989).

70. D. J. Wales, Mol. Phys. **78**, 151 (1993); R. S. Berry, Chem. Rev. **93**, 2379 (1993); H. Matsuoka, T. Hirokawa, M. Matsui, and M. Doyama, Phys. Rev. Lett. **69**, 297 (1992).

71. W. Kamke, J. de Vries, J. Krauss, E. Kaiser, B. Kamke, and I. V. Hertel, Z. Phys. D **14**, 339 (1989).

72. J. Wormer, M. Joppien, G. Zimmerer, and T. Moller, Phys. Rev. Lett. **67**, 2053 (1991).

73. S. Leutwyler and J. Bosiger, Chem. Rev. **90**, 489 (1990).

74. J. E. Adams and R. M. Stratt, J. Chem. Phys. **99**, 789 (1993).

75. K. B. Whaley, Int. Rev. Phys. Chem. **13**, 41 (1994).

76. F. Luo, G. C. McBane, G. Kim, C. F. Giese, and W. R. Gentry, J. Chem. Phys. **98**, 3564 (1993); J. A. Anderson, C. A. Traynor, and B. M. Boghosian, J. Chem. Phys. **99**, 345 (1993).

77. M. Joppien, R. Karbach, and T. Moller, Phys. Rev. Lett. **71**, 2654 (1993).

78. M. A. McMahon, R. N. Barnett, and K. B. Whaley, J. Chem. Phys. **99**, 8816 (1993).

79. E. J. Bieske and J. P. Maier, Chem. Rev. **93**, 2603 (1993); A. McIlroy and D. J. Nesbitt, Adv. Mol. Vib. Collision Dyn. **A1**, 109 (1991).

80. S. Xu and L. S. Bartell, J. Phys. Chem. **97**, 13544 (1993); S. Xu and L. S. Bartell, J. Phys. Chem. **97**, 13550 (1993).

81. T. J. Selegue, N. Moe, J. A. Draves, and J. M. Lisy, J. Chem. Phys. **96**, 7268 (1992).

39

Infrared Spectroscopy

Henry Buijs

BOMEM, Hartmann and Braun, 450 Avenue St.-Jean-Baptiste, Québec, Québec G2E 5S5, Canada

39.1 INFRARED SPECTROSCOPY

Infrared spectroscopy consists of the measurement of interactions of waves of the infrared (ir) part of the electromagnetic spectrum with matter. The ir spectrum starts just beyond the red part of the visible spectrum at a wavelength $\lambda = 700$ nm and extends to the microwave region at $\lambda = 0.1$ cm. Electromagnetic waves are generally described in terms of their frequency ν in Hz. In ir spectroscopy it is common practice however to use the spatial frequency $\sigma = \nu/c$. These are called wavenumbers or Kaisers and have units of cm^{-1}. In this way the near, mid and far ir spectrum spans the frequencies from 14300 cm^{-1} to 10 cm^{-1}. The interactions observed in the ir spectrum involve principally the energies associated with molecular structure change. Infrared spectroscopy is therefore useful for molecular structure elucidation and the identification and quantification of different molecular species in a sample [1].

The most common ir analysis of a sample is by ir *absorption* spectroscopy. This involves transmitting a beam of intense ir radiation through the sample and observing the distribution of wavenumbers absorbed by the molecules. Molecules in a sample may also be studied by ir *emission* spectroscopy simply by observing specific wavenumbers being emitted by virtue of the nonzero absolute temperature of the sample. Finally, radiation *reflected* from a smooth surface of a solid sample also provides information about the molecular structure of the material by virtue of the anomalous dispersion associated with absorption bands.

39.2 INTENSITIES OF INFRARED RADIATION

For strong interactions of electromagnetic waves with matter, the emitted and absorbed intensities are governed by Planck's radiation law in addition to the emissivity and absorptivity of the material. Planck's radiation law for thermal radiation from an ideal black body is

$$P_{\text{bb}}(\sigma)\, d\sigma = \frac{c_1 \sigma^3 d\sigma}{\exp(h\sigma/k_{\text{B}}T) + 1}, \tag{39.1}$$

where c_1 is a proportionality constant. Depending on the definition of c_1, $P_{bb}(\sigma)$ may represent a radiation density per unit spectral interval (cm^{-1}) in a cavity at temperature T in ergs/cm^3, or an energy flux emitted from a surface in W/cm^2 steradians. At frequencies low compared with h/k_BT, the energy distribution increases with σ^2 and is approximately proportional to T at a given σ. At high frequency, the energy distribution falls off exponentially. In the near ir, a high temperature is required to emit radiation. Room temperature objects emit strongest in the 300 to 600 cm^{-1} region, and emit negligible energy above 3000 cm^{-1}. Material cooled to liquid nitrogen temperature (77 K) only emit below 100 cm^{-1}, while material cooled to liquid helium temperature (4.2 K) only emit below 20 cm^{-1}. In contrast to visible spectroscopy, ir absorption spectroscopy is complicated by emission of ir radiation from the sample and the surrounding environment.

39.3 SOURCES FOR IR ABSORPTION SPECTROSCOPY

A silicon carbide element electrically heated to 1400 K provides a strong continuum of ir radiation over a major part of the ir spectrum. It is commonly used as a source of radiation for ir absorption spectroscopy. For near ir spectroscopy, a tungsten filament lamp operated at 2800 K provides a strong continuum all the way up to the visible part of the spectrum; it is not useful below 3000 cm^{-1} because of absorption by the glass or quartz envelope. Various electrically heated ceramic elements, such as the Nernst glower and high temperature carbon rods, have been devised to serve as ir sources.

39.4 SOURCE, SPECTROMETER, SAMPLE AND DETECTOR RELATIONSHIP

Since a sample at room temperature emits ir radiation in the mid ir, it is important to distinguish between transmitted radiation used in the determination of its absorption spectrum and its emission spectrum. By employing an intense ir beam, the effect of emission is minimized. Further distinction is achieved by encoding the ir beam before it impinges on the sample.

With classical grating or prism spectrometers, the source radiation is chopped by means of a mechanical chopper before it passes through the sample. The ir detector is provided with a means of synchronously decoding the chopped signal, thereby eliminating the emitted spectrum. Often the chopper is arranged such that it alternately switches between an empty reference beam and the sample. The logarithm of the ratio of the demodulated sample and reference spectra provides the absorption spectrum directly.

In *Fourier Transform Spectroscopy*, (FT-IR), a scanning Michelson interferometer provides the encoding function directly and no chopper is required. The interferometer is commonly placed before the sample so that it does not encode the thermally emitted radiation of the sample. Ratio recording of a sample against an empty beam is not common in FT-IR. Instead the absorption spectrum is obtained by sequentially recording the spectra of the sample and of the empty beam (sample removed), and computing the logarithm of the ratio numerically.

If it is not convenient to place the sample after the scanning Michelson interferometer, the absorption spectrum of a sample placed in front of the interferometer can be deduced by subtracting the separately recorded emission spectrum from the combined transmission plus emission spectrum.

IR emission and reflection spectroscopy forms the basis for remote sensing. Solid and gaseous (cloud) objects may be identified and quantified by direct observation of their ir spectra at a distance. Gaseous clouds reflect poorly, providing only transmitted or emitted ir radiation. Their emission spectrum is contrasted directly with the spectrum of the scene or object beyond the cloud. With a background at lower temperature than the gas cloud, the gas spectrum appears in emission, while with a warmer background the gas spectrum appears in absorption.

Only a few solid materials transmit ir radiation over a substantial thickness. Remote sensing of solid objects relates therefore to surface emission and (diffuse) reflection of ir radiation from the surrounding environment.

39.5 SIMPLIFIED PRINCIPLE OF FT-IR SPECTROSCOPY

In FT-IR spectroscopy, the spectrum of a beam of incident ir radiation is obtained by first generating and recording an interferogram with a scanning Michelson interferometer. Subsequently the interferogram is inverted by means of a cosine Fourier transform into the spectrum.

39.5.1 Interferogram Generation: The Michelson Interferometer

The scanning Michelson interferometer shown in Fig. 39.1 consists of a beam splitter, which is a substrate with a dielectric coating such that 50% of an incident beam is reflected and the remaining 50% is transmitted, and two plane mirrors (M$_1$ and M$_2$), one or both of which are translated along the direction of the beam. After splitting, the two equal amplitude wave fronts are prop-

Figure 39.1. The Michelson interferometer.

agated along different optical paths. The mirrors at the end of each path return the wavefronts to the beamsplitter, which then acts as a wavefront combiner. Because of their common coherent origin, the wavefronts interfere with one another when they combine. The state of interference is varied by scanning one or both of the mirrors such that there is a variable time delay between the two separated beams. The resulting intensity variation of the combined output beam as a function of relative time delay is the interferogram.

39.5.2 Description of Wavefront Interference with Time Delay

The intensity $I_0(\nu)$ at frequency ν of a plane wave in space is given by the expectation value of its electric field vector $\mathbf{E}(\nu, t) = \mathcal{E}(\nu)\exp(i2\pi\nu t)$ according to

$$I_0(\nu) = \langle \mathbf{E}(\nu, t)|\mathbf{E}(\nu, t)\rangle = \mathcal{E}(\nu)^2 . \quad (39.2)$$

The intensity at the output of the interferometer due to an incident intensity $I_0(\nu)$ is given by $I(\nu, \delta)$, where δ is the time delay between the two wavefronts which have been propagated along two different paths, and

$$\begin{aligned}
I(\nu, \delta) &= \langle \mathbf{E}(\nu, t) + \mathbf{E}(\nu, t + \delta) \,|\, \mathbf{E}(\nu, t) + \mathbf{E}(\nu, t + \delta)\rangle \\
&= \tfrac{1}{4}\mathcal{E}(\nu)^2[2 + e^{-i2\pi\nu\delta} + e^{i2\pi\nu\delta}] \\
&= \tfrac{1}{2}I_0(\nu)[1 + \cos(2\pi\nu\delta)] .
\end{aligned} \quad (39.3)$$

As can be seen, the output intensity of a single frequency source at the output of an ideal scanning Michelson interferometer fluctuates sinusoidally between zero and the input intensity $I_0(\nu)$ as the time delay δ between the

separated wavefronts is varied by means of scanning one of the mirrors.

The quantity δ is related to mirror displacement x with respect to equal distance of the mirrors from the beamsplitter by

$$\delta = 2x\cos\theta/c , \quad (39.4)$$

where θ is the angle between the wavefront and the optical axis of the spectrometer, and the optical axis is the normal to each plane mirror M_1 and M_2. From this, Eq. (39.3) becomes

$$I_0(\nu, x) = \tfrac{1}{2}I_0(\nu)\{1 + \cos[2\pi\nu(2x\cos\theta/c)]\} , \quad (39.5)$$

or, using $\sigma = \nu/c$,

$$I_0(\sigma, x) = \tfrac{1}{2}I_0(\sigma)\{1 + \cos[2\pi\sigma(2x\cos\theta)]\} . \quad (39.6)$$

Thus the output intensity fluctuates at frequency $\sigma' = 2\sigma\cos\theta$ as a function of the mirror displacement x.

The incident intensity generally consists of a distribution of intensities over many frequencies $S(\nu)\,d\nu$ with integrated intensity

$$I_0 = \int S(\nu)\,d\nu . \quad (39.7)$$

For this case, the output intensity of an ideal scanning Michelson interferometer is given by

$$I_0(\delta) = \frac{1}{2}\int S(\nu)[1 + \cos(2\pi\nu\delta)]\,d\nu . \quad (39.8)$$

The second term on the right side of Eq. (39.8) has the form of the cosine Fourier transform of the spectrum. By rearrangement of (39.8), it is given by

$$\int S(\nu)\cos(2\pi\nu\delta)\,d\nu = 2I_0(\delta) - I_0 . \quad (39.9)$$

The constant term I_0 provides no useful information about the spectrum. The inverse cosine Fourier transform of $2I_0(\delta)$ results in the spectrum $S(\nu)$ according to

$$S(\nu) = \int 2I_0(\delta)\cos(2\pi\nu\delta)\,d\delta , \quad (39.10)$$

$$S(\sigma) = \int 2I_0(x)\cos(2\pi\sigma'x)\,dx . \quad (39.11)$$

Contrary to classical spectrometers, where the spectrum is sequentially scanned, there is no segregation of frequencies of the input intensity. All frequencies in the source are modulated simultaneously by the scanning Michelson interferometer into a single interferogram signal. This multiplex mode of spectrum determination was first exploited by Felgett [2]. It contributes to a large advantage in sensitivity compared with other spectrometers, and is referred to as the Felgett or multiplex advantage.

39.5.3 The Operation of Spectrum Determination

The output interferogram is detected by an ir detector which converts the intensity variations $I_0(x)$ as a function of different mirror positions x into an electrical signal. Continuous determination of the inverse cosine Fourier transform of the evolving interferogram requires continuous multiplication of the signal by cosine functions with all the different frequencies of the spectrum and integrating these products. Continuous Fourier analysis with a bank of narrow band filters has been implemented both in analog and digital form in early versions of FT-IRs [3].

It is however by far more practical to capture the interferogram signal in numeric form, using an analog to digital converter, store it into a computer memory, and compute the Fourier transform numerically after the mirror displacement range has been covered.

The numerical representation of the interferogram is determined at known intervals of mirror displacement Δx. The computed spectrum is then determined at regular intervals of frequency $\Delta \sigma$ by the discrete cosine Fourier transform

$$S(j\Delta\sigma) = \sum_n 2I_0(n\Delta x)\cos(2\pi j\Delta\sigma n\Delta x). \quad (39.12)$$

The sampling interval Δx of optical path difference determines the extent of the numerically computed spectrum. A higher density of sampling permits a wider spectral range to be determined, up to

$$\sigma_{\max} = \frac{1}{2\Delta x}. \quad (39.13)$$

Beyond $\sigma_{\max}$, the spectrum repeats in reverse order, and beyond $2\sigma_{\max}$, the spectrum repeats as is. This is called spectral aliasing, and results from the incomplete knowledge of the full interferogram function between the discrete numeric representation. In order to insure that the numeric representation of the interferogram describes the continuous function uniquely, it is important to band limit the interferogram information to the range 0 to $\sigma_{\max}$ by means of optical and electrical filtering.

Conversely, the higher the density of sampling in the spectral domain, the longer the interferogram needs to be. For a wavenumber range $\Delta\sigma$, the length is

$$x_{\max} = \frac{1}{2\Delta\sigma}. \quad (39.14)$$

From the orthogonality property of the discrete cosine Fourier transform, unique linearly independent information can occur only at spectral intervals equal to or greater than the sampling interval. Hence the sampling interval in the spectrum is related to the achievable resolution. The full width at half maximum of the representation of a single frequency in the spectrum is

$1.2\Delta\sigma$. This factor applies for the case where the single cosine wave interferogram has been abruptly truncated at the end of the mirror scan. The lineshape function for this case is quite oscillatory due to the abrupt termination of the interferogram signal at the end of the scan. This is not always satisfactory, and frequently the interferogram is modified by a windowing or apodization function to make the lineshape more localized and monotonic. Apodization always results in an increase in the full width at half maximum.

A particularly convenient and accurate way to establish the sampling intervals of the interferogram is the use of a single frequency laser directed coaxially with the source radiation through the scanning Michelson interferometer. The intensity of the laser at the output of the interferometer is a highly consistent cosine wave with one cycle per change in mirror movement of one half wavelength of the laser light.

39.6 OPTICAL ASPECTS OF FT-IR TECHNOLOGY

The description of wavefront interference developed in Sect. 39.5.2 applies to interference of plane wave fronts only. A plane wave of ir radiation is obtained at the output of a collimator optics having an ir point source at its focus. In practice a point source has insignificant intensity. A finite size source, which may be represented by a distribution of point sources in the focal plane of the collimator, provides a distribution of plane wavefronts with different angles of propagation through the scanning Michelson interferometer.

As shown in Eq. (39.6), this distribution of angles results in a distribution of modulation frequencies as a function of mirror displacement x of the output intensity for a given ir wavenumber. This is illustrated in Fig. 39.2 which shows, for a single frequency ir source, the distribution of output intensity modulation frequencies for an ideal point source on the optical axis, a circularly symmetric distribution of uniform intensity about the optical axis, and finally a circularly symmetric distribution positioned slightly off the axis of the collimator.

As a result of illuminating the scanning Michelson interferometer with a finite size source, a single ir frequency source is observed having a distribution of modulation frequencies in its interferogram signal. This distribution limits the ability to resolve two closely spaced ir frequencies and determines the optical resolution limit of the FT-IR: The larger the extent of the source, the more restricted the resolution becomes. The ratio $\sigma/\Delta\sigma$ is the resolving power R of the scanning Michelson interferometer and, based on Eq. (39.6), is given by

$$R = 1/(1 - \cos\theta_{\mathrm{m}}), \quad (39.15)$$

where θ_{m} is the maximum off-axis angle of illumination.

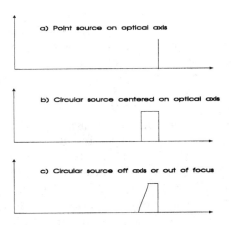

Figure 39.2. Distribution of interferogram modulation frequencies for a single optical frequency source.

For small θ_m, $\cos\theta_m \approx (1 - \frac{1}{2}\theta_m^2)$, so that

$$R \sim 2/\theta_m^2 . \tag{39.16}$$

From Eq. 39.14, a resolution limit is also imposed by the maximum length of the interferogram recorded. The interferogram length dependent resolution is constant for all spectral regions, while the optical resolution is proportional to the spectral frequency. Both resolution limits combine to give the overall resolution of an FT-IR. At low resolution, the available throughput is so high that the optical resolution is often negligible compared with the length resolution. At high resolution, throughput is at a premium and the optical resolution is often closely matched to the length resolution at the frequency of interest.

To insure a symmetrical frequency distribution, the area integrated illumination must increase as $\sin\theta$, which is approximately linear for small angles, up to its maximum θ_m, as shown with the centered circular illumination in Fig. 39.2. Any deviation from this, such as off-axis positioning of the circle, noncircular shapes, or a poorly focused centered symmetrical circle, will result in a gradual roll-off of the distribution on the low frequency side only [4]. This results in an asymmetric spectral line shape.

For a collimator of given focal length and for a given resolving power, it is easily shown that the area of the source, or the stop that delineates it, is much larger than the slit area for classical grating spectrometers. This is particularly the case at high resolving powers.

This throughput advantage was first pointed out by Jacquinot [5], and plays an important role in the large sensitivity advantage of FT-IR. The stop that delineates the source extent for a scanning Michelson interferometer is often referred to as the Jacquinot stop or the field of view stop.

39.7 THE SCANNING MICHELSON INTERFEROMETER

Optical throughput is not only determined by the area of the field of view stop, but also the solid angle subtended by the rays traversing this area. The solid angle Ω of rays traversing a Jacquinot stop positioned in the focal plane of an input collimator is given by the ratio of the interferometer beam area divided by the square of the focal length of the collimator.

For a given collimator focal length, the area of the Jacquinot stop is inversely proportional to the resolving power. To maintain equal throughput, the area of the interferometer optics should be increased as resolving power is increased in order to offset the decrease in the Jacquinot stop area.

It is common to construct interferometers with 2.5 cm diameter optics for resolving powers up to 5 000, 5.0 cm diameter optics for resolving powers up to 40 000 and 7.5 cm diameter optics for resolving powers up to 1 000 000.

In order to obtain a uniform state of interference across the entire beam of the interferometer, the beamsplitter substrate and the two mirrors must be flat to within a small fraction of the wavelength used. Also, these elements must be oriented correctly so that the optical path difference error across the beam is less than a small fraction of the wavelength.

Figure 39.1 shows two substrates at the beamsplitter position. One of the substrates supports the beamsplitting coating, while the companion substrate of precisely the same thickness acts as a compensating element to insure identical optical paths through the two arms of the interferometer. To avoid secondary interference effects, both beamsplitter and compensator substrates are normally wedged. The direction of the wedges of the two substrates must be aligned again to insure symmetry in both arms of the interferometer.

The maintenance of a very close orientational alignment tolerance of the two mirrors with respect to the beamsplitter in a stable manner over time and while scanning one of the mirrors is the greatest challenge of interferometer design and is also the greatest weakness of FT-IR.

In early models of FT-IRs, alignment was maintained by means of a stable mechanical structure and a highly precise linear air bearing for the scanning mirror. Satisfactory operation required a stable environment and frequent alignment tuning and could be achieved for mirror displacements of only several centimeters, thus limiting the maximum resolution.

Different techniques have been developed to overcome this weakness in FT-IR. The two most prominent are (1) Dynamic alignment of the interferometer, where optical alignment is servo-controlled using the reference laser not only for mirror displacement control but also for mirror orientation control, and (2) the use of cube corner mirrors in place of the flat mirrors in the interferometer. Dynamic alignment has the advantage of retaining a high degree of simplicity in the optical design of the interferometer. On the other hand it is more complex electronically. Cube corners have the property of always reflecting light 180° to incident light independent of orientation. Cube corners always insure wavefront parallelism at the point of recombination of the two beams in the interferometer. Cube corners lack a defined optical axis. In a cube corner interferometer, the optical axis is defined as the direction in which the wavefronts undergo zero shear.

The scanning Michelson Interferometer is normally provided with a drive mechanism to displace one of the mirrors precisely parallel to its initial position and at uniform velocity. The uniform velocity translates the mirror displacement dependent intensities into time dependent intensities. This facilitates signal processing electronics. In some measurement scenarios however, where the sample spectrum may vary with time, it is undesirable to deal with the multifrequency time-varying intensities of the interferogram signal. In this case it is preferable to scan the moving mirror in a stepwise mode, where the mirror is momentarily stationary at the time of signal measurement and then advanced rapidly to the next position. The mirror scan velocity v can be varied so that electrical signals can have different frequencies for the same optical frequencies:

$$f = \sigma/2v .$$ (39.17)

The scan velocity is normally selected to provide the most favorable frequency regime for the detector and electronics used and for the mechanical capabilities of the mirror drive. Typically the velocity range is from 0.05 cm/s to 4 cm/s, putting the frequencies in the audio range. At these velocities, the measurement scan may be completed in a shorter time than is desired for signal averaging purposes. In this case it is common to repeat the scan a number of times and add the results together: this is called co-adding of scans.

39.8   CONCLUSION

The modern technique of Fourier transform ir spectroscopy has evolved rapidly from its beginnings in the early 1950s to being the dominant technique of ir spectroscopy in many diverse disciplines. The Aspen Conference on Fourier Spectroscopy in 1970 was the watershed for FT-IR, where many fundamental issues of the technique were treated [6]. A practical book by Griffith and de Haseth gives many examples of applications of FT-IR [7].

FT-IR is a powerful technique for infrared spectroscopy. It has a large sensitivity advantage over conventional dispersive spectrometers because of the efficient multiplexing of all the spectral elements and the greater throughput of the Jacquinot stop. It can be used efficiently for resolving powers from less than 1 000 up to 1 000 000. FT-IR combines techniques of optical interferometry, laser metrology and digital signal processing.

The computation of the cosine Fourier transform was initially a daunting task. Today, with highly efficient factoring algorithms brought to FT-IR by Forman [8], and the widespread availability of inexpensive high performance personal computers, the computation time for a Fourier transform is only a fraction of a second per 1000 data points in the interferogram.

A traditional weakness of FT-IR is the severe demand of alignment stability and accuracy of the scanning Michelson interferometer. Newer optical designs and control procedures have largely overcome this weakness. Now FT-IR can be justified not only for its high sensitivity but also for its high degree of reproducibility and stability, permitting demanding applications in quantitative ir analysis.

REFERENCES

1. N. B. Colthup, L. H. Daly, and S. E. Wiberley, *Introduction to Infrared and Raman Spectroscopy* (Academic Press, New York, 1990).
2. P. B. Fellgett, J. Phys. Radium **19**, 187 and 237 (1958).
3. J. E. Hoffman and G. A. Vanasse, J. Phys. (Paris) **28**, C2:79 (1967).
4. P. Saarinen and J. Kauppinnen, Appl. Opt. **31**, 2353 (1992).
5. P. Jacquinot, J. Opt. Soc. Am. **44**, 761 (1954).
6. Aspen International Conference on Fourier Spectroscopy, 1970 edited by G. A. Vanasse, A. T. Stair, and D. J. Baker, AFCRL-71-0019.
7. P. R. Griffith and J. A. de Haseth, *Fourier Transform Infrared Spectrometry* (Wiley Interscience, New York, 1986), (Chemical Analysis Vol. 83).
8. M. L. Forman, J. Opt. Soc. Am. **56**, 978 (1966).

40

Laser Spectroscopy in the Submillimeter and Far-Infrared Region

K. M. Evenson

Time and Frequency Division, National Institute of Standards and Technology, Boulder, Colorado 80303

40.1 INTRODUCTION

Research in the submillimeter and far-infrared (SM-FIR) regions of the electromagnetic spectrum (1000 to 150 μm, 0.3 to 2.0 THz; and 150 to 20 μm, 2.0 to 15 THz, respectively) had been relatively inactive until about 25 years ago. Three events were responsible for enhanced activity in this part of the electromagnetic spectrum: the discovery of far-infrared (FIR) lasers [1], the development of background-limited detectors [2], and the invention of the FIR Fourier transform (FT) spectrometer [3]. Following these developments major discoveries have taken place in laboratory spectroscopic studies [4,5], in astronomical observations [6], and in spectroscopic studies of our upper atmosphere [7].

Rotational transition frequencies of light molecules (such as hydrides) lie in this region, and the associated electric dipole transitions are especially strong at these frequencies. In fact they are 10 000 times stronger than at microwave frequencies because they are 100 times typical microwave frequencies and their peak absorptivities depend approximately on the square of the frequency. Fine structure transitions of atoms and molecules also

lie in this region; however, they are much weaker magnetic dipole transitions. The observation of fine structure spectra is very important in determining atomic concentrations in astronomical and atmospheric sources and for determining the local physical conditions. Bending frequencies of larger molecules also lie in this region, but their transitions are not as strong as rotational transitions.

Spectral accuracy has been increased by several orders of magnitude with the extension of direct frequency measurement metrology into the SM-FIR region [8]. Transitions whose frequencies have been measured (including absorptions and laser emissions) are useful wavelength calibration sources ($\lambda_{\text{vac}} = c/\nu$) for FT spectrometers. FIR spectra of a series of rotational transitions have been measured in CO [9], HCl [10], HF [11], and CH_3OH [12] to be used for FT calibration standards. These lines are ten to a hundred times more accurately measured than can be realized in present state-of-the-art Fourier transform spectrometers; thus, they are excellent calibration standards. High-accuracy and high-resolution spectroscopy has permitted the spectroscopic assignment of the SM-FIR lasing transitions themselves [13] resulting in a much better understanding of the lasing process.

Astronomical spectroscopy in this region [6] may be in emission or absorption and is performed using either interferometric [14] (wavelength-based), or heterodyne (frequency-based) [15] techniques to resolve the individual spectral features.

Most high resolution spectra of our upper atmosphere have been taken with FT spectrometers flown above the heavily absorbing water vapor region in our lower atmosphere. Emission lines are generally observed in these spectrometers [7].

40.2 EXPERIMENTAL TECHNIQUES USING COHERENT SM-FIR RADIATION

The earliest sources of coherent SM radiation came from harmonics of klystron radiation generated in point-contact semiconductor diodes [16]. Spectroscopically useful powers to about one terahertz are produced. This technique is being replaced by electronic oscillators which oscillate to over one terahertz [17]. Spectroscopy above this frequency generally is performed with either lasers or FT spectrometers (see Chap. 39).

Laser techniques use either tunable radiation synthesized from the radiation of other lasers, or fixed-frequency SM-FIR laser radiation and tuning of the transition frequency of the species by an electric or magnetic field. Spectroscopy with tunable far-infrared radiation is called TuFIR spectroscopy. Spectroscopy with fixed frequency lasers is called either laser electric resonance (LER) or laser magnetic resonance (LMR). LMR is applicable only to paramagnetic species and is noteworthy for its extreme sensitivity. LMR spectroscopy has been more widely applied than LER, and is discussed in this chapter.

Tunable SM-FIR radiation has been generated either by adding microwave sidebands to radiation from a SM-FIR laser [18] or by using a pair of higher frequency lasers and generating the frequency difference [19]. The sideband technique was first reported by Dymanus [18] and uses a Schottky diode as the mixing element. It has been used up to 4.25 THz and produces a few microwatts [20]. The CO_2 laser frequency-difference technique with difference generation in the metal-insulator-metal (MIM) diode covers the FIR region out to over 6 THz and produces about 0.1 μW. This technique uses fluorescence-stabilized CO_2 lasers whose frequencies have been directly measured, and it is about two orders of magnitude more accurate than the sideband technique. However, it is somewhat less sensitive because of the decreased power available. There are several review articles on the laser sideband technique [20, 21], and only the laser difference technique is described here.

40.2.1 Tunable FIR Spectroscopy with CO_2 Laser Difference Generation in a MIM Diode

There are two different ways of generating FIR radiation using a pair of CO_2 lasers and the MIM diode. One is by second-order generation, in which tunability is achieved by using a tunable waveguide CO_2 laser as one of the CO_2 lasers; it is operated at about 8 kPa (60 Torr) and is tunable by about ±120 MHz. The second technique uses third-order generation, in which tunable microwave sidebands are added to the difference frequency of the two CO_2 lasers. The complete spectrometer which can be operated in either second or third order is shown in Fig. 40.1.

The FIR frequencies generated are:

$$\text{second-order: } \nu_{\text{fir}} = |\nu_{1,\text{CO}_2} - \nu_{\text{W,CO}_2}| \tag{40.1a}$$

$$\text{third-order: } \nu_{\text{fir}} = |\nu_{1,\text{CO}_2} - \nu_{\text{W,CO}_2}| \pm \nu_{\text{mw}} \tag{40.1b}$$

where ν_{fir} is the tunable FIR radiation, ν_{1,CO_2} and $\nu_{\text{W,CO}_2}$ are the CO_2 laser frequencies, and ν_{mw} is the microwave frequency.

Different MIM diodes are used for the two different orders: in second-order, a tungsten whisker contacts a nickel base and the normal oxide layer on nickel serves as the insulating barrier; in third-order, a cobalt base with its natural cobalt oxide layer is substituted for the nickel base. The third-order cobalt diodes produce about one third as much FIR radiation in each sideband as there is in the second-order difference; hence third-order generation is not quite as sensitive as second-order. The much larger tunability, however, makes it considerably easier to use.

The common isotope of CO_2 is used in both the waveguide laser and in laser 2; in laser 1, one of four isotopic species is used. Ninety percent of all frequencies from 0.3 to 4.5 THz can be synthesized, and the coverage between 4.5 and 6.3 THz decreases. Ninety megahertz acoustooptic modulators (AOMs) are used in the output beams of the two CO_2 lasers which irradiate the MIM diode; they increase the frequency coverage by an additional 180 MHz and isolate the CO_2 lasers from the MIM diode. This isolation decreases amplitude noise in the generated FIR radiation, caused by the feedback to the CO_2 laser from the MIM diode, by an order of magnitude; hence, the spectrometer sensitivity increases by an order of magnitude.

The radiations from laser 1 and the waveguide laser are focused on the MIM diode. Laser 2 serves as a frequency reference for the waveguide laser; the two lasers beat with each other in the HgCdTe detector and a servosystem offset-locks the waveguide laser to laser 2. Lasers 1 and 2 are frequency modulated using piezoelectric drivers on the end mirrors, and they are servoed to the line center of 4.3 μm saturated fluorescence

Figure 40.1. Tunable far-infrared spectrometer for second or third order operation using CO_2 laser difference frequency generation in the MIM diode.

signals obtained from the external low-pressure CO_2 reference cells. In both second and third order, the CO_2 reference lasers are stabilized to line center with an uncertainty of 10 kHz. The overall uncertainty in the FIR frequency due to two lasers is thus $\sqrt{2} \times 10$ or 14 kHz. This number was determined experimentally in a measurement of the rotational frequencies of CO out to $J = 38$ (4.3 THz), with the analysis of the data set determining the molecular constants [9].

The CO_2 radiation is focused by a 25 mm focal length lens onto the conically sharpened tip of the 25 μm diameter tungsten whisker. The FIR radiation is emitted from the 0.1 to 3 mm long whisker in a conical long-wire antenna pattern. Then it is collimated to a polarized beam by a corner reflector [22] and a 30 mm focal length off-axis segment of a parabolic mirror.

The largest uncertainty in the measurement of a transition frequency comes from finding the centers of the Doppler broadened lines. This is about 0.05 of the line width for lines observed with a signal-to-noise ratio of 50 or better and coresponds to about 0.05 to 0.5 MHz. A linefitting program [23] improves the line center determinations by nearly an order of magnitude.

The FIR radiation is frequency modulated due to a 0.5 to 6 MHz frequency modulation of CO_2 laser 1; this modulation is at a rate of 1 kHz. The FIR detector and lock-in amplifier detect at this modulation rate; hence the derivatives of the absorptions are recorded.

Absorption cells from 0.5 to 3.5 m in length with diameters ranging from 19 to 30 mm have been used in the spectrometer. The cells have either glass, copper, or Teflon walls and have polyethylene or polypropylene windows at each end. These long absorption cells naturally lend themselves to electrical discharges for the study of molecular ions. A measurement of the HCO^+ line at 1 THz exhibited a signal-to-noise ratio of 100:1 using a 1 s time constant; this is the same signal-to-noise ratio which was obtained using the laser sideband technique.

The instrumental resolution of the spectrometer is limited by the combined frequency fluctuation from each CO_2 laser (about 15 kHz). This is less than any Doppler-limited line width and, therefore, does not limit the resolution. The entire data system is computerized to facilitate the data recording and optimize the data handling.

Improvements in this TuFIR technique may come from either better diodes or detection schemes. The nonlinearity of the MIM diode is extremely small, and conversion efficiencies could be much larger if a more efficient diode is discovered. Differential detection (which requires more sensitive detectors) could also significantly improve the sensitivity and permit the detection of weaker lines.

40.2.2 Laser Magnetic Resonance

Laser magnetic resonance (LMR) is performed by magnetically tuning the transition frequencies of paramagnetic species into coincidence with the fixed frequency radiation from a laser. LMR is a type of Zeeman spectroscopy [24], and its chief forte is its extreme sensitivity. Approximately one hundred species have been observed in the SM-FIR region. These include atoms, diatomics (especially hydrides), polyatomics, ions, metastables, metastable ions [25], and many "first observations" of free radicals. These observations are summarized in several review articles on LMR [26, 27].

The FIR LMR spectrometer at NIST, Boulder is shown in Fig. 40.2. This spectrometer has an intracavity paramagnetic sample in a variable magnetic field which is labeled "sample volume" in the figure. A regular 38 cm EPR magnet with a 7.5 cm air gap is used. The laser cavity is divided into two parts by a Brewster angle polypropylene beam splitter about 12 μm thick. A CO_2 laser optically pumps the FIR laser gas. This pump beam makes many nearly perpendicular passes reflected by the walls of the gold-lined Pyrex tube. A 45° coupler serves to couple out the FIR radiation from the cavity to the helium-cooled detector. Mirrors each with an 89 cm radius of curvature are used in the nearly confocal geometry of the 94 cm cavity. One of the gold-coated end mirrors is moved with a micrometer to tune the cavity to resonance and also to determine the oscillating wavelength by moving the end mirror several half-wavelengths. The beam splitter is rotatable so that the polarization of the laser can be varied with respect to the magnetic field. Quartz spacers are used to minimize the thermal expansion of the cavity. This LMR spectrometer oscillates at wavelengths between 40 and 1000 μm. The technology of FIR lasers is reviewed by Nigel Douglas [28].

The LMR technique requires a close coincidence (typically within 20 GHz) between the absorption line and the frequency of the laser. Zero field frequencies which are 100 times more accurate than those obtained from optical spectra are obtained from the analysis of the Zeeman spectra observed (i.e. using the laser frequency and the magnetic field values). The accuracy is within 1 or 2 MHz and has permitted the far infrared astronomical

observation of many of these species.

Atomic FIR LMR spectra are due to fine structure transitions and are magnetic dipole transitions; hence, they are several orders of magnitude weaker than electric dipole rotational spectra of molecules. The production of a sufficient atomic number density can be difficult; however, "atomic flames" have been very effective sources, and the high sensitivity of LMR has been the most successful spectroscopic technique for measuring these transitions. The atoms O, C, metastable Mg, S, Si, Fe, Al, N^+, C^+, and F^+ have been measured by FIR LMR. Atomic Zeeman spectra are relatively simple to analyze, and fine structure frequencies accurate to within 1 MHz can be determined.

In a FIR LMR spectrometer the sample is inside the laser cavity; hence, sub-Doppler line widths can be observed by operating the sample at low pressure and observing saturation dips in the signals. This has permitted the resolution of hyperfine structure in a number of hydrides. The observation of the resolved hyperfine structure is useful in the identification of the species involved and also yields accurate values of the hyperfine splittings.

Rotational spectra of most hydrides lie in this spectral region and many of the hydrides, such as OH, CH, SH, and NH have been observed. Others, such as CaH, MnH, TiH, and ZnH are excellent candidates for LMR studies. Ions are much more difficult because of their low concentrations; however, the use of a special microwave discharge operating in the magnetic field has proved to be an extremely productive source of ions for LMR studies [25].

The spectra of a number of polyatomic species has been observed by LMR: NO_2, HO_2, HCO, PH_2, CH_2, CH_3O, and CCH. The spectra of these polyatomic species are more difficult to analyze, resulting in a somewhat less accurate prediction of the zero-field frequencies. Spectra of the extremely elusive CH_2 and CD_2 have finally yielded to analysis with the simultaneous observation of FIR LMR spectra [29] and IR LMR spectra [30]. The data yield rotational constants which predict the ground state rotational transition frequencies and permit the determination of the singlet-triplet splitting in that molecule [31].

The sensitivity of TuFIR is only about 1% of that of laser magnetic resonance, but it is difficult to compare the two because the sample can be very large in the TuFIR spectrometer and is limited to about 2 cm in the LMR spectrometer.

40.2.3 TuFIR and LMR Detectors

Four different detectors have been used in the TuFIR and LMR spectrometers: (1) an InSb 4 K, liquid ^{4}He-cooled, hot-electron bolometer, operating from 0.3

Figure 40.2. Far-infrared laser magnetic resonance spectrometer using an optically pumped FIR laser.

to 0.6 THz with a noise equivalent power (NEP) of about 10^{-13} W/$\sqrt{\text{Hz}}$; (2) a gallium doped germanium bolometer, cooled to the lambda point of liquid-helium, operating from 0.6 to 8.5 THz with an NEP of about 10^{-13} W/$\sqrt{\text{Hz}}$; (3) a similar, but liquid ^{3}He-cooled bolometer, with a NEP two orders of magnitude smaller; and (4) an unstressed Ge:Ga photoconductor, cooled to 4 K, with an NEP of 10^{-14} W/$\sqrt{\text{Hz}}$, operating from 2.5 to 8.5 THz. The optimization of detector systems for this new technique has been responsible for a significant improvement in sensitivity of these spectrometers.

40.3 SUBMILLIMETER AND FIR ASTRONOMY

Submillimeter astronomy is reviewed in Ref. [6]. Many atomic species (both neutrals and ions) and more than 96 molecules have been detected in interstellar space; many were first observed in the submillimeter and FIR region. In the microwave and submillimeter region, radio (heterodyne) techniques are employed in the receivers. At higher frequencies, interferometric techniques are used [14]; however, heterodyne techniques recently have been employed at these higher frequencies and have resulted in the detection of CO [32] and OH [33] at 2.5 THz and methylene [34] at 30 THz. Submillimeter radio astronomy observatories (at high altitudes above much of the water vapor absorptions in our atmosphere)

use laboratory determined frequencies. For example, the Kuiper Airborne Observatory and the submillimeter and FIR telescopes on Mauna Kea, Hawaii require frequencies accurate to about 1 MHz in their heterodyne receivers. For heterodyne detection, fixed frequency FIR gas lasers generally serve as local oscillators, and the frequencies of these lasers must be known. Nearly 1000 FIR laser lines have been directly frequency measured and are summarized in a review article [35].

There are four main goals of laboratory SM-FIR spectroscopy which serve the needs of the SM-FIR radio astronomy field:

1. to provide accurate frequencies of SM-FIR species for their detection,

2. to find new far infrared active species,

3. to measure the frequencies of FIR species which can be used to calibrate Fourier transform spectrometers, and

4. to measure frequencies of far-infrared lasers for use as local oscillators in radio astronomy receivers (and to be used in the analysis of laboratory LMR data).

40.4 UPPER ATMOSPHERIC STUDIES

A very impressive set of SM-FIR spectra of our upper atmosphere has been observed using balloon-borne

FT spectrometers flown at altitudes where the lines are narrow and the spectrometer is above the "black," heavily absorbing water vapor transitions [7]. A number of very important species with very strong SM-FIR spectra have been observed. They include: O_2, H_2O, NO, ClO, OH, HO_2, O_3, and O. Numerous lines have not yet been identified. It is difficult to calibrate these instruments absolutely because the spectra come from the species emitting at temperatures of about 200 K, and from an indeterminate path length. The high sensitivity of LMR might provide an alternate way of measuring the paramagnetic species in our upper atmosphere by flying an LMR spectrometer to high altitudes. A light-weight solenoid magnet could be used to increase the path length. Absolute concentrations of the paramagnetic species such as OH, HO_2, NO, and O could be obtained.

Acknowledgments

I have benefitted invaluably from the help and collaboration of John Brown, Oxford University, England for his theoretical and experimental work over many years of collaboration, I.G. Nolt of NASA, Langley for his assistance with the detector technology, and Kelly Chance for his line shape fitting program and his assistance with our studies of upper atmospheric species.

REFERENCES

1. A. Crocker, H. A. Gebbie, M. F. Kimmitt, and L. E. S. Mathias, Nature **201**, 250 (1964).
2. E. E. Haller, Infrared Phys. Technol. **35**, 127-146 (1994).
3. J. Strong and G. Vanasse, J. Opt. Soc. Amer. **49**, 884 (1959).
4. K. M. Evenson, H. P. Broida, J. S. Wells, and R. J. Mahler, Phys. Rev. Lett. **21**, 1038 (1968).
5. D. D. Bícaníc, B. F. J. Zuidberg, and A. Dymanus, Appl. Phys. Lett. **32**, 367 (1978).
6. T. G. Phillips, *Millimetre and Submillimetre Astronomy*, edited by W. B. Burton (Kluwer academic Publishers, Doordrecht 1988), p. 1.
7. W. A. Traub, K. V. Chance, D. G. Johnson, and K. W. Jucks, Proc. Soc. Photo-Opt. Instrum. Engrs. **1491**, 298 (1991).
8. D. A. Jennings, K. M. Evenson, and D. J. E. Knight, Proc. IEEE **74**, 168 (1986).
9. T. D. Varberg and K. M. Evenson, IEEE Trans. Instrum. Meas. **42**, 412 (1993).
10. I. G. Nolt, J. V. Radostitz, G. DiLonardo, K. M. Evenson, D. A. Jennings, K. R. Leopold, M. D. Vanek, L. R. Zink, A. Hinz, and K. V. Chance, J. Mol. Spectrosc. **125**, 274-287 (1987).
11. D. A. Jennings, K. M. Evenson, L. R. Zink, C. Demuynck, J. L. Destombes, and B. Lemoine, J. Mol. Spectrosc. **122**, 477 (1987).
12. F. Matsushima, K. M. Evenson, and L. R. Zink, J. Mol. Spectrosc. **164**, 517 (1994).
13. L.-H. Xu, R. M. Lees, K. M. Evenson, C.-C. Chou, J.-T. Shy, and E. C. C. Vasconcellos, Can. J. Phys. **72**, 1155 (1994).
14. J. W. V. Storey, D. M. Watson, and C. H. Townes, Int. J. Infrared Millimeter Waves **1**, 15 (1980).
15. H.-P. Röser, Infrared Phys. **32**, 385 (1991).
16. P. Helminger, J. K. Messer, and F. C. DeLucia, Appl. Phys. Lett. **42**, 309 (1983).
17. S. P. Belov, L. I. Gershstein, A. F. Krupnov, A. V. Maslovsky, V. Spirko, and D. Papousek, J. Mol. Spectrosc. **84**, 288 (1980).
18. D. D. Bícaníc, B. F. J. Zuidberg, and A. Dymanus, Appl. Phys. Lett. **32**, 367 (1978).
19. K. M. Evenson, D. A. Jennings, and F. R. Petersen, Appl. Phys. Lett. **44**, 576 (1984).
20. P. Verhoeve, E. Zwart, M. Versluis, M. Drabbels, J. J. ter Meulen, W. Leo Meerts, and A. Dymanus, Rev. Sci. Instrum. **61**, 1612 (1990).
21. G. A. Blake, K. B. Laughlin, R. C. Cohen, K. L. Busarow, D.-H. Gwo, C. A. Schmuttenmaer, D. W. Steyert, and R. J. Saykally, Rev. Sci. Instrum. **62**, 1693 (1991); **62**, 1701 (1991).
22. E. N. Grossman, Infrared Phys. **29**, 875 (1989).
23. K. V. Chance, D. A. Jennings, K. M. Evenson, M. D. Vanek, I. G. Nolt, J. V. Radostitz, and K. Park, J. Mol. Spectrosc. **146**, 375 (1991).
24. J. S. Wells and K. M. Evenson, Rev. Sci. Instrum. **41**, 226 (1970).
25. T. D. Varberg, K. M. Evenson, and J. M. Brown, J. Chem. Phys. **100**, 2487 (1994).
26. K. M. Evenson, Faraday Discussions of The Royal Society of Chemistry **71**, 7 (1981).
27. D. K. Russell, in *Electron Spin Resonance*, edited by M. Symons (Specialist Periodical Reports), The Royal Sociey of Chemistry, London, **16**, 64 (1990).
28. N. G. Douglas, *Millimetre and Submillimetre Wavelength Lasers* (Springer-Verlag, Berlin, Heidelberg, 1990).
29. T. J. Sears, P. R. Bunker, A. R. W. McKellar, K. M. Evenson, D. A. Jennings, and J. M. Brown, J. Chem. Phys. **77**, 5348 (1982).
30. T. J. Sears, P. R. Bunker, and A. R. W. McKellar, J. Chem. Phys. **75**, 4731 (1981).
31. A. R. W. McKellar, P. R. Bunker, T. J. Sears, K. M. Evenson, R. J. Saykally, and S. R. Langhoff, J. Chem. Phys. **79**, 5251 (1983).
32. R. T. Boreiko and A. L. Betz, Astrophys. J. **346**, L97 (1989).
33. A. L. Betz and R. T. Boreiko, Astrophys. J. **346**, L101 (1989).
34. D. M. Goldhaber, A. L. Betz, and J. J. Ottusch, Astrophys. J. **314**, 356 (1987).
35. M. Inguscio, G. Moruzzi, K. M. Evenson, and D. A. Jennings, J. Appl. Phys. **60**, R161 (1986), Table 5.

41

Spectroscopic Techniques: Lasers

Paul Engelking

Department of Chemistry and Chemical Physics Institute, University of Oregon, Eugene, Oregon 97403-1253

41.1 LASER BASICS

41.1.1 Stimulated Emission

A cross section σ_{21} for absorption of radiation by a lower state 1 engenders a balancing cross section σ_{12} for emission stimulated by radiation interacting with an upper state 2. Detailed balance relates these two cross sections according to

$$g_2\sigma_{12} = g_1\sigma_{21}\,, \tag{41.1}$$

where g_1 and g_2 are the statistical degeneracies of their respective states [1].

For a collection of emitting and absorbing states with densities n_2 and n_1, amplification may occur when $n_2\sigma_{12} > n_1\sigma_{21}$, which leads to a requirement for an *inversion* of the state populations:

$$n_2/n_1 > g_2/g_1\,. \tag{41.2}$$

The rate of spontaneous emission at frequency ν can be modeled itself by stimulated emission induced by a noise source of the magnitude of the density of states $\rho(\nu)$

$$\gamma_{12}(\nu) = \sigma_{12}(\nu)\rho(\nu)/c = \sigma_{12}(\nu)8\pi\lambda^{-2}\,. \tag{41.3}$$

41.1.2 Laser Configurations

A practical laser combines a population inversion with a means for controlling the radiation.

The basic laser source is the laser oscillator, an amplifier possessing positive feedback. The usual form is simply a piece of active gain medium placed inside a resonant optical cavity (Fig. 41.1). Tunability is produced if the resonant cavity is frequency selective and adjustable. Many laser sources use an amplifier after the oscillator.

479

Figure 41.1. Simple laser oscillator and beam parameters. Distances z are generally measured from the minimum beam waist w_0. The beam appears with a far-field divergence angle θ_0.

41.1.3 Gain

The fundamental gain per pass is given by

$$G = J/J_0 = \exp[(\kappa - \mu)L],\qquad(41.4)$$

where J/J_0 is the ratio of light output to input, κ is the gain coefficient, μ is the nonradiative loss rate, and L is the path length. The gain coefficient

$$\kappa = n^*\sigma_{12}\qquad(41.5)$$

depends upon the net inversion n^*

$$n^* = n_2 - (g_1/g_2)n_1\,.\qquad(41.6)$$

If λ is the wavelength of radiation, F_{12} is the emission line shape function normalized over frequency ν, τ_2 is the lifetime of the transition, and f_{12} is the branching ratio for the upper state to undergo this transition, then the stimulated emission cross section is

$$\sigma_{12} = \frac{\lambda^2 f_{12} F_{12}(\nu)}{8\pi\tau_2}\qquad(41.7)$$

For a Lorentzian lifetime-broadened line, the cross section for stimulated emission at the line center becomes

$$\sigma_{12} = \frac{\lambda^2 f_{12}}{4\pi^2\Gamma_{12}\tau_2}\,,\qquad(41.8)$$

where Γ_{12} is the full width at half maximum of the line.

41.1.4 Laser Light

Lasers are inherently bright sources of radiation: the radiation field within a practical laser must be high enough for stimulated emission to compete with spontaneous emission. The effective source of spontaneous fluctuations approximates that of the density of states.

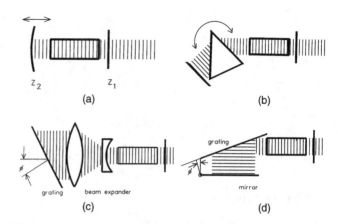

Figure 41.2. Tunable laser oscillator geometries. (a) Fabry-Perot: tuning is usually by changing cavity length, although changing the index of refraction by changing temperature or current is common with solid state laser diodes. (b) Littrow prism line selector: typical of atomic ion lasers capable of multiple line output. (c) Littrow grating tuning: common in pulsed dye lasers with high gain (> 10) per pass. Telescope increases resolution by filling, and reducing angular divergence at the grating. (d) Grazing incidence, mirror tuned, grating mount.

In terms of the beam parameters photon flux J per solid angle Ω, and frequency ν, this is

$$\frac{d^2 J}{d\Omega d\nu} = \frac{2\nu^2}{\epsilon_r c^2}\,.\qquad(41.9)$$

Beam *quality* is given by the product of the angular divergence times the beam width. Highest beam quality is associated with diffraction limited light emitted from a Gaussian spot. For circular laser beams traveling in the z-direction, this corresponds to a solution of the electromagnetic wave equation

$$u(r,\phi,z) = \psi(r,z)\exp(-ikz)\,,\qquad(41.10)$$

where u is a polarization component of the field. For high values of $k = 2\pi/\lambda$, corresponding to short wavelength, the adiabatic radial solution is also Gaussian:

$$\psi(r,z) = \exp\{-i[P + kr^2/(2q)]\}\,,\qquad(41.11)$$

where the complex phase shift P, beam parameter q, and beam radius w are functions of z:

$$P(z) = -i\ln[1 - i\lambda z/(\pi w_0^2)]$$
$$= -i\ln\sqrt{1 + [\lambda z/(\pi w_0^2)]^2}$$
$$\quad - \tan^{-1}[\lambda z/(\pi w_0^2)]\,,\qquad(41.12)$$
$$q(z) = i\pi w_0^2/\lambda + z\,,\qquad(41.13)$$
$$w^2(z) = w_0^2\left[1 + \left(\frac{\lambda z}{\pi w_0^2}\right)^2\right]\,.\qquad(41.14)$$

Here, w_0 is the beam waist parameter, the minimum width of the Gaussian beam at a focused spot. For Gaussian beams, the product of the minimum beam waist and beam divergence angle θ_0 is given by

$$\theta_0 w_0 = \lambda/\pi . \qquad (41.15)$$

The beam waist and divergence follow optical imaging according to paraxial ray theory.

Higher order circular modes with p radial nodes and l angular node planes are specified by multiplying Eqs. (41.11) and (41.12) by angular and radial factors to obtain

$$\psi_{pl}(r,\phi,z) = \left[\sqrt{2}r/w\right]^l L_p^l(2r^2/w^2) e^{il\phi} \psi(r,z) , \qquad (41.16)$$

$$P_{pl}(z) = (2p + l + 1)P(z) . \qquad (41.17)$$

Here, the functions $L_p^l(x)$ are the Laguerre polynomials as defined in Eq. (9.187). The radial phase shifts produce a wave front curvature of effective radius

$$R = z + \left[(2p + l + 1)/2 + w_0^2 \pi/\lambda^2\right]w_0^2 \pi/z . \qquad (41.18)$$

Modes with the same values of $2p + l$ have identical axial and radial phase shifts. The two polarization components of the electromagnetic field double the degeneracies of all modes considered here. Often these degeneracies are split in practice by optical inhomogeneities of the medium through which they pass. More details can be found in the summary of Kogelnick and Li [2], or in the texts by Verdeyen [3] or Svelto [4].

Some applications require knowledge of the electric field in addition to the flux density J. For purely sinusoidal single mode beams, the rms field is

$$\langle E \rangle = \left(\frac{h\nu J}{c\epsilon_0}\right)^{1/2} . \qquad (41.19)$$

Nonlinear effects are often expressed in terms of powers of the field by

$$\langle E^n \rangle = 2^{(n-1)/2} \langle E \rangle^n \qquad (41.20)$$

for single mode and multimode radiation of random frequency spacings. For m equally spaced modes, this is increased by $m!/(m - n)!$.

41.2   LASER DESIGNS

41.2.1   Cavities

The simple Fabry-Perot cavity consists of two spherical mirrors facing one another. The surfaces are chosen to be constant phase surfaces for the desired modes [Eq. (41.18)]. Stability criteria are shown in Fig. 41.3. A cavity is stable when initial angles θ and displacements

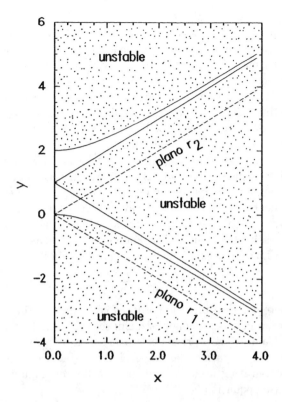

Figure 41.3. Stability parameters for simple two-mirror laser cavities of length L and mirror radii of curvature R_1 and R_2. Here, $x = \frac{1}{2}(L/R_1 - L/R_2)$ is the mean curvature difference of the two mirrors; $y = \frac{1}{2}(L/R_1 + L/R_2)$ is the mean curvature of the two mirrors. Cavities with parameters in the unshaded region are stable.

r of paraxial rays transform during a round trip into θ' and r' satisfying

$$-2 < \frac{\partial\theta'}{\partial\theta} + \frac{\partial r'}{\partial r} < 2 . \qquad (41.21)$$

At frequencies for which the round-trip phase change per passage

$$\delta(\nu) = 2\pi(z_2 - z_1)\nu/c + 2[P_{pl}(z_2) - P_{pl}(z_1)]$$
$$\approx 2\pi(z_2 - z_1)/\lambda + \pi(2p + l + 1) \qquad (41.22)$$

is an integer multiple of 2π, the phases from different passages interfere constructively, giving longitudinal modes. [Here $P_{pl}(z_i)$ gives the additional phase shift for higher transverse modes at mirrors $i = 1, 2$].

For a particular radial mode structure in an empty Fabry-Perot cavity, the ratio of the maximum cavity decay time for these standing waves to the minimum cavity decay time for frequencies between longitudinal modes is

$$(1 + r)^2/(1 - r)^2 . \qquad (41.23)$$

Here r is the reflectivity of the end mirrors; for cavities with mirrors having different reflectivities, one may use the square root of the product of their reflectivities. A simple Fabry-Perot cavity may be tuned by changing the cavity length or by changing the index of refraction of the cavity material. Since both of these may be properties of temperature, temperature tuning may be possible. The index of refraction of a material also may be sensitive to the intensity of excitation. Diode lasers consisting of a semiconductor die with polished, reflecting faces are often tuned by changing temperature and pumping current.

For lasers with a dispersive optical element within the cavity, highest selectivity is obtained when the light has low angular divergence and impinges upon the dispersive element as nearly plane waves. A beam expander may reduce the angular spread while simultaneously increasing the beam width. For a grating used as a mirror in a Littrow mount, the dispersion equation is

$$\Delta\lambda = (d/n)\cos\phi\,\Delta\phi\,, \qquad (41.24)$$

where n is the diffraction order, d is the distance between lines, and ϕ is the angle of incidence off normal.

Cavities with prisms or gratings can be conveniently tuned by rotating the angle of the dispersive element. For a grating used as a mirror in a Littrow mount, the grating equation is

$$\lambda = (2d/n)\sin\phi\,. \qquad (41.25)$$

Often, more than one longitudinal cavity mode operates within the selected frequency band, and tuning consists of "hopping" from mode to mode, rather than smoothly sweeping a single line across a band of frequencies. Smooth frequency tuning can be achieved, for example, in a design by Wallenstein and Hänsch [5], in which the grating and Fabry-Perot cavity are placed together inside a chamber. The whole laser is then tuned by changing the index of refraction of the gas inside by varying its pressure.

The current trend with pulsed lasers is to use a very lossy, short oscillator cavity in which the longitudinal modes are nearly absent, making up for the cavity losses with a very high gain lasing medium. The front, output mirror of such a cavity may actually be *anti*reflection coated, with a reflectivity of only a percent or less.

Low gain, cw lasers often use a combination of cavity length tuning along with dispersive element rotation, such as a prism or Lyot filter. Often, lasing on a traveling wave in a ring configuration is used to avoid longitudinal modes, as illustrated in Fig. 41.4. Some commonly used gain media are listed in Tables 41.1 and 41.2.

41.2.2 Pumping

Many methods, including electrical discharges and flashlamps, have been used to pump the gain media of lasers. Generally, the best pump is another laser.

Two notable pump lasers have dominated the field of tunable, visible lasers: pulsed neodymium YAG,

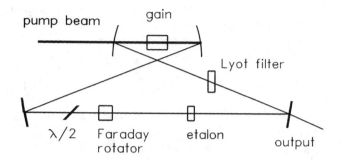

Figure 41.4. Ring laser. The Faraday rotator and half-wave plate permit circulation of the cavity fields in only one direction.

frequency doubled to ~503 nm, and cw Ar II ion at 514.5 nm. Both are extremely effective at exciting the highly efficient rhodamine class dyes in the red–yellow portion of the visible spectrum. The typical pump beam of a Nd/YAG laser enters the amplifying dye cell transversely along one on the faces of the cell. The typical cw pump beam enters the dye almost collinearly with the laser axis.

By temporarily "spoiling" the Q by making the laser cavity lossy, lasing can be held off until the gain medium stores a greater energy density than the minimum required for lasing. Rapidly switching off the loss mechanism releases this energy in one giant pulse. An electronic optical shutter, such as a Pockels cell, or a saturable dye inside the laser cavity, designed to photobleach from the spontaneous emission just before the laser reaches threshold, are commonly used.

Periodically spoiling the laser gain or the cavity Q at the period of a round trip produces intense, short pulses. Viewed from the frequency domain, the phases of individual longitudinal modes are "locked" together to produce a light packet circulating at the frequency of the reciprocal of the mode spacing. Extremely short pulses (<1 ps) can be produced by mode locking. Practically, mode locking can be achieved by using a thin, saturable absorber near one of the cavity mirrors [6], or by acoustically modulating an optical element in the cavity — even an end mirror itself. One of the best ways of mode locking is to pump a short lifetime gain medium, such as a dye, with mode locked laser light, such as from a mode locked argon ion laser [7].

41.3 INTERACTION OF LASER LIGHT WITH MATTER

41.3.1 Linear Absorption

The absorption cross section σ_{fi} for transition to the final state $|f\rangle$, when integrated over frequency ν, is given

Table 41.1. Fixed frequency lasers.

Laser	Wavelength (nm)	Excitation method
ArCl	~170 (band)	pulsed, gas discharge
ArF	~193	pulsed, gas discharge
KrCl	~222	pulsed, gas discharge
KrF	~248	pulsed, gas discharge
XeBr	~282	pulsed, gas discharge
XeCl	~308	pulsed, gas discharge
XeF	~351, 353	pulsed, gas discharge
N_2	~337.1 (other bands)	pulsed (1-10 ns), gas discharge
Ar^+	488.0, 514.5 (454.4, 457.9, 465.8, 472.7, 476.5, 501.7, 528.7)	cw, gas discharge
Ar^{+2}	351.1, 363.8	cw, gas discharge, high magnetic field
Kr^+	568.2, 647.1, (476.2, 520.8, 530.9)	cw, gas discharge
Kr^{+2}	350.7, 356.4, 406.7	cw, gas discharge, high magnetic field
Ne/He	632.8, 1152.3, 3390 (others)	cw, gas discharge
Cr^{+3}-Ruby	694.3	pulsed, flashlamp
Nd^{+3}-YAG	~1065, ~1300	pulsed: flashlamp; cw: lamp, LED or laser diode
Nd^{+3}-glass(various)	~920, ~1060, ~1370	pulsed, flashlamp
Yb^{+3}-glass	~1060	"
Er^{+3}-glass(various)	~1540, ~1536, ~1543, ~1550	pulsed: flashlamp; cw: lamp, LED or laser diode
Xe	3507	cw gas discharge
CO_2	~10600 (lines)	cw: gas discharge or gas dynamic
Molecular vibration and rotation	1×10^5–3×10^6 (numerous lines)	ir laser (CO_2, CO, $\cdots$)

theoretically by the leading first-order perturbation of the initial state $|i\rangle$ in the electric dipole approximation

$$\int \sigma_{fi}(\nu)\,d\nu = 4\pi^2 \alpha \bar{\nu} \Big| \langle f| \sum_e \mathbf{r}_e \cdot \hat{\boldsymbol{\epsilon}} |i\rangle \Big|^2, \qquad (41.26)$$

where the sum is over all charges e at distances r_e, $\hat{\boldsymbol{\epsilon}}$ is a unit polarization vector, and $\bar{\nu}$ is the averaged transition energy. Averaged over orientations and summed over possible final states, each electron contributes to the total integrated absorption cross section one electron oscillator, $\pi r_0 c \approx 0.03$ cm^{2}s^{-1}; here, r_0 is the classical electron radius. Electronic absorption bands typically contain an oscillator strength, $f = 0.01$–0.5 of an "electron oscillator," while weaker vibrational transitions have $f = 10^{-6}$–10^{-4} in each band.

For narrow lines with radiation of broader band width $\Delta\nu$ at flux density J, the linear absorption rate constant can be usefully estimated as $(\pi r_0 c) f J / \Delta\nu$.

41.3.2 Multiphoton Absorption

Second-order perturbation theory gives the theoretical two-photon contribution to the absorption. An absorption cross section $\sigma^{(2)} J_1$ for a photon of frequency ν_2 is induced by an off-resonance monochromatic field of frequency ν_1 and photon flux density J_1. When integrated over the frequency of the second photon, the second-order cross section can be related to the dipole matrix elements

$$\int \sigma_{fi}^{(2)}(\nu_1, \nu_2) J_1 \, d\nu_2 = \frac{4\pi^2 \alpha^2}{\nu_1 \bar{\nu}_2} J_1$$

$$\times \Big| \sum_m \frac{\nu_{fm}\nu_{mi}\langle r_{fm}\rangle\langle r_{mi}\rangle}{\nu_{mi} - \nu_1} \Big|^2$$

$$(41.27)$$

where

$$\langle r_{fm}\rangle = \langle f| \sum_e \mathbf{r}_e \cdot \hat{\boldsymbol{\epsilon}}_2 |m\rangle,$$

$$\langle r_{mi}\rangle = \langle m| \sum_e \mathbf{r}_e \cdot \hat{\boldsymbol{\epsilon}}_1 |i\rangle, \qquad (41.28)$$

and the sum is taken over intermediate states $|m\rangle$ having frequencies ν_{mi} and ν_{fm} for transitions to the initial and final states. The energy for the overall transition comes from two photons; hence the two-photon resonance condition $\nu_{fi} = \nu_1 + \nu_2$. A special case often occurs when only one radiation frequency is used: then $\nu_1 = \nu_2$, and two-photon resonance is achieved when the energy of the transition corresponds to twice the frequency of the radiation field.

Table 41.2. Approximate tuning ranges for tunable lasers.

Laser	Wavelengths (nm)	Notes
Dye solution	<330 – >1200 [~10 each dye]	Pulsed: laser, flashlamp; cw: laser
Alexandrite	700–820	Pulsed: lamp, laser; cw: laser
Ti^{+3}/Sapphire	680–1100	Pulsed: laser, lamp; cw: laser
GaAs	840–900	pn junction
InGaAlP/GaAs	630–700 [1–10]	pn junction
GaAsP	550–880	pn junction
AlGaAs, AlGaAs/GaAs	~820 [1–20], 720–880	pn junction, temp. & current tuning
InP	~900	"
GaInAs	906–3100	"
InGaAlAs/GaAs	800–1100	"
InPAs	900–4000 [1–50]	"
InGaAsP/InP	1200–1650	"
InAs	~3100	"
InSb	~5200	"
PbS	~4300	"
PbTe	~6500	"
PbSe	~8500	"
PbSnSeTe "Lead Salt Diode"	4500–15000 [1–50]	"
Color Centers: (F2+)/LiF	800–1040	cw: laser, Ar, Kr
/NaF	900–1050	"
/KCl:Tl	1400–1700	cw: laser, Ar, Kr, Nd/YAG
/KF	1300–1400	"
/NaCl	1400–1600	"
/KCl	1600–1700	"
/KBr	1700–1900	"
/KCl:Li	2500–2900	"
/KCl:Na	1600–1950	"

This integrated, induced cross section is roughly of the order $\alpha(\pi cr_0)^2 J_1/(\nu_1\bar{\nu}_2)$. Two-photon absorption becomes comparable to the one-photon absorption with off-resonance fields on the order of

$$J_1 \approx \frac{\nu_1\nu_2}{\alpha\pi cr_0} \qquad (41.29)$$

or about 10^{21} photons s^{-1}cm^{-2} (10^{17} Wm^{-2}s^{-1} for typical green light).

Typical dipole-allowed molecular multiphoton absorption cross sections are ~10^{-58} m^{4}s for two photons and 10^{-94} m^{6}s^2 for three photons.

Multiphoton absorption was one of the first effects explored with lasers. Two-photon absorption was first reported for inorganic crystals containing europium ions [8]. Blue and ultraviolet fluorescence appeared in the interaction of red ruby laser light with organic compounds [9–12]. Others observed two-photon absorption directly [13,14]. Selection rules for multiphoton absorption are summarized by McClain [15]; recent work is reviewed by Ashfold and Howe [16].

Highly excited states may subsequently ionize in the intense fields in a multiphoton ionization (MPI) process.

The ionization signal is often detected in a proportional ionization cell. A low pressure cell containing the vapor of a transition metal organometallic, such as iron carbonyl or ferrocene which photodissociates to give the metal atom, may be used for wavelength calibration by MPI.

41.3.3 Level Shifts

High radiation power causes resonant frequencies to shift, responding to the ac Stark effect [17,18]. Even moderate fields, tuned near resonance, interact strongly with an atom or small molecule, which undergoes rapid excitation and stimulated emission at the characteristic power dependent Rabi frequency,

$$\nu_{\text{Rabi}} = \frac{2\pi g_1}{g_1 + g_2}\sigma_{21}J. \qquad (41.30)$$

Fluorescence from such an interacting system has a characteristic "head and shoulders" spectrum best understood as radiation at the fundamental frequency amplitude modulated at the Rabi frequency [19].

41.3.4 Hole Burning

Radiation at a particular frequency generally moves population out of states that absorb that radiation. Molecules that interact resonantly with the radiation may also spontaneously emit at different frequencies, thereby ending in nonabsorbing states. This optical pumping effect can make resonance features in an absorption spectrum disappear at high power levels [20]. Depletion may appear as a "hole" in the absorption spectrum [21]. Recently, interest has shifted to permanent "hole burning" as a method of information storage in materials.

Hole burning is the basis for Doppler-free Lamb-dip spectroscopy, in which only absorbing atoms or molecules having little or no velocity component along the axis of two counterpropagating beams are temporarily depleted [22]. This technique is commonly used for laser frequency stabilization, such as with the iodine-locked He-Ne laser.

41.3.5 Nonlinear Optics

Multiplying

Nonlinear susceptibility of an optical medium can generate radiation at frequencies which are multiples of the frequency of laser radiation passing through. Phenomenologically, the second order polarization

$$\mathbf{P}_{2\nu} = \epsilon_0 \chi^{[2]} \mathbf{E}_\nu^2 \tag{41.31}$$

is given in terms of the second order nonlinear susceptibility $\chi^{[2]}$, a third rank tensor, and the electric field at the fundamental frequency (see Chap. 70). This nonlinear susceptibility can range from 0.5–5 pm/V for typical materials used for frequency doubling. Typical materials and their use are reviewed by Bordui and Fejer [23].

For a nonlinear process occurring over a length l in a cylindrical region with Gaussian waist w_0 with polarization P_0 on axis, the far field flux is given by

$$J(R,\theta) = \frac{\pi^4 \nu^3 n^3 w_0^4 P_0^2}{4hc^3 \epsilon_0 R^2} \left[\frac{\sin^2[(k_0 - k\cos\theta)l/2]}{(k_0 - k\cos\theta)^2} \right]$$
$$\times \exp[-k^2 w_0^2 \sin^2(\theta)/2] . \tag{41.32}$$

Here, n is the index of refraction and k the propagation constant for the induced radiation, while k_0 is that for the induced polarization, the vector sum of those of the original radiation. When phase matched, $k_0 - k\cos\theta = 0$, and the term in the brackets maximizes to $(l/2)^2$.

The greatest difficulty is in selecting materials which can be phase matched such that the relative phases of the fundamental and overtone radiation propagate together through the material; otherwise radiation at the higher frequency generated at different places inside the material destructively interfere. Phase matching is usually achieved by either angle tuning of a birefringent crystal, or by temperature tuning.

Only materials without a center of inversion in their crystal structures have a second order nonlinear susceptibility. All materials, including gases, will have a third order nonlinear susceptibility $\chi^{[3]}$. This can be used to generate third harmonics, especially in the vuv, where doubling materials are not available [24, 25].

Mixing

The same materials that permit frequency doubling and tripling also allow 3-wave and 4-wave frequency mixing. The frequency matching conditions are, respectively,

$$\nu_1 \pm \nu_2 = \nu_3 ,$$
$$\nu_1 \pm \nu_2 \pm \nu_3 = \nu_4 . \tag{41.33}$$

Tunable uv radiation may be generated by adding the frequencies of a fixed and tunable visible outputs. Tunable ir has been obtained by differencing fixed and tunable visible lasers. Mixing of radiation from an Ar ion laser with that from a tuned R6G dye laser in lithium iodate to produce tunable 2200–4600 nm radiation is noteworthy.

Optical Parametric Oscillator

It is possible to reverse 3-wave mixing, generating two frequencies whose sum is that of the input radiation. In parametric generation, the output frequencies are given by the phase match conditions. Both the desired frequency and the secondary "idler" frequency must be allowed to build up in the nonlinear medium. The idler radiation is not present initially, but results from the frequency mixing process itself. The process has many of the characteristics of a laser oscillator, including that of a gain threshold. This makes tuning of an optical parametric oscillator similar to that of a laser, but with more degrees of freedom: now oscillation at two frequencies must be attained simultaneously, along with the correct *phase matching* of the nonlinear material [26].

41.3.6 Raman Scattering

It is possible to have one or more of the fields in a mixing process belong to just polarization, rather than radiation. The frequency additive case of multiphoton absorption has already been consisdered; the frequency subtractive case is Raman scattering.

Incoherent Raman Scattering

Radiation at a higher frequency can excite a lower frequency vibration or rotation within a material, with appearance of radiation at the frequency of the incident radiation minus that of the absorption. The integrated cross section for this effect is given by Eq. (41.27). However, in this case the cross section is for *emission* of the second photon. The rate of spontaneous Raman emission is obtained by multiplying Eq. (41.27) by the spectral flux density of the zero-point field, $8\pi\nu^2/\epsilon_r c$. Typical vibrational Raman cross sections for transparent molecules are about 10^{-34} $m^2\,sr^{-1}$ in the blue-green (488 nm) [27].

Alternatively, energy can be extracted from an excited state, with the inelastically scattered photon departing with the sum of the incident frequency and that of the deexcitation. The term "anti-Stokes" distinguishes it from the more usual "Stokes" type of Raman scattering.

Coherent Raman Emission

The integrated cross section for emission stimulated at the Raman frequency is given directly by Eq. (41.27).

A fourth-order mixing process that is less susceptible to saturation involves coherent anti-Stokes Raman scattering (CARS). Two beams excite the material: the difference of their frequencies corresponds to an excitation of the material. Stimulated Raman scattering excites the material, which then deexcites through an anti-Stokes process, giving rise to a third, higher frequency radiation field. The phase is determinate, and the radiation leaves the region of scattering as a beam [28].

If the incident radiation induces a Raman process over a sufficiently long path, the stimulated Raman process can be used for gain at both the Stokes and anti-Stokes frequencies. Since spontaneous Raman processes are proportional to the integrated cross section, while gain in the stimulated Raman process is proportional to the peak cross section, simple materials with sharp, simple line spectra are most suitable for Raman gain media. While Raman lasers have been produced using vibrational excitation of organic liquids, currently the most important technical application is for Raman shifting the output of lasers, tunable or otherwise, to frequencies which otherwise would be inaccessible. The high pressure H_2 Raman shifter produces, at low powers, beams consisting almost entirely of well separated, sharp lines shifted by 4160 cm^{-1} from the pump beam; at high powers, a series of Stokes and anti-Stokes bands appear, each separated by 4160 cm^{-1} from each other.

REFERENCES

1. A. Einstein, Phys. Z. **18**, 121 (1917).
2. H. Kogelnik and T. Li, Proc. IEEE **54**, 1312 (1966).
3. J. T. Verdeyen *Laser Electronics* (Prentice Hall, Englewood Cliffs, 1994), 3rd ed.
4. O. Svelto, *Principles of Lasers* (Plenum, New York 1976).
5. R. Wallenstein and T. W. Hänsch, Opt. Commun. **14**, 353 (1975).
6. E. P. Ippen, C. V. Shank, and A. Dienes, Appl. Phys. Lett. **21**, 348 (1972).
7. R. K. Jain and C. P. Ausschnitt, Opt. Lett. **2**, 117 (1978).
8. W. Kaiser and C. G. B. Garrett, Phys. Rev. Lett. **7**, 229 (1961).
9. W. L. Peticolas, J. P. Goldsborough, and K. E. Rieckhoff, Phys. Rev. Lett. **10**, 4345 (1963).
10. W. L. Peticolas and K. E. Reickhoff, J. Chem. Phys. **39**, 1347 (1963).
11. S. Singh and B. P. Stoicheff, J. Chem. Phys. **38**, 2032 (1963).
12. S. Z. Weisz, A. B. Zahlen, J. Gilreath, R. C. Jarnagin, and M. Silver, J. Chem. Phys. **41**, 3491 (1964).
13. J. J. Hopfield, J. M Warlock, and K. Park, Phys. Rev. Lett. **11**, 414 (1963).
14. B. Staginnus, D. Frölich, and T. Caps, Rev. Sci. Instrum. **39**, 1129 (1968).
15. W. M. McClain, Acc. Chem. Res. **7**, 129 (1974).
16. M. N. Ashfold and J. D. Howe, Ann. Rev. Phys. Chem. **45**, 57 (1994).
17. H. J. Carmichael and D. F. Walls, J. Phys. B **9**, 1199 (1976).
18. L.-P. Li, B.-X. Yang, and P. M. Johnson, J. Opt. Soc. Am. B**2**, 748 (1985).
19. B. R. Mallow, Phys. Rev. **188**, 1969 (1969).
20. L. J. Rothberg, D. P. Gerrity, and V. Vaida, J. Chem. Phys. **75**, 4403 (1981).
21. S. Völker, Ann. Rev. Phys. Chem. **40**, 499 (1989).
22. W. R. Bennett Jr., Phys. Rev. **126**, 580 (1962).
23. P. F. Bordui and M. F. Martin, Ann. Rev. Mater. Sci. **23**, 321 (1993).
24. A. Kung, Opt. Lett. **8**, 24 (1983).
25. J. Bokor, P. Bucksbaum, and R. Freeman, Opt. Lett. **8**, 217 (1983).
26. J. Falk, J. M. Yarbourough, and E. O. Ammann, IEEE J. Quantum Elec. **7**, 359 (1971).
27. C. M. Penney, R. L. St. Peters, and M. Lapp, J. Opt. Soc. Am. **64**, 712 (1974).
28. P. R. Regnier and J. P. Taran, Appl. Phys. Lett. **23**, 240 (1973).

42

Spectroscopic Techniques: Ultraviolet

G. Stark* and Peter L. Smith[†]

*Department of Physics, Wellesley College, Wellesley, Massachusetts 02181
[†]Harvard-Smithsonian Center for Astrophysics, Cambridge, Massachusetts 02138

42.1 INTRODUCTION

The ultraviolet (uv) spectral region extends from the short wavelength side of the visible, about 400 nm, to the long wavelength side of the x-ray region at approximately 10 nm. Ultraviolet photon energies range from about 3 to 120 eV. Because laboratory air at STP does not transmit below about 200 nm, the uv region is conventionally subdivided into the *near ultraviolet*, wavelengths $\lambda > 200$ nm, and the *vacuum ultraviolet* (vuv), $\lambda < 200$ nm. Further subdivisions are widespread, but not uniformly adopted. The term *middle ultraviolet* (muv) is sometimes used to designate the 200 to 300 nm wavelength region, with the term *near uv* used only for 300 to 400 nm. Vuv wavelengths are often divided into the *far ultraviolet* (fuv), 100 to 200 nm, and the *extreme ultraviolet* (euv), 10 to 100 nm. In this review, we will simply refer to the uv (10 to 400 nm), the near uv (200 to 400 nm), and the vuv (10 to 200 nm).

The absorption of uv radiation by atoms and molecules involves transitions to highly excited discrete and continuum levels. The relevant atomic and molecular physics of highly excited states is discussed in Chaps. 14, 25, 24, and 59. Ultraviolet absorption initiates many important photochemical reactions; applications in astrophysics, aeronomy, and combustion are reviewed in Chaps. 78, 80, and 84. The experimental techniques used at ultraviolet wavelengths include high-resolution spectroscopic measurements of wavelengths and line shapes (Chap. 10), absolute photoabsorption oscillator strength and radiative lifetime measurements (Chap. 17), fluorescence spectroscopy, and energy, angle, and spin-resolved photoelectron spectroscopy (Chap. 59).

The design of uv spectroscopic experiments and apparatus must conform to the constraints associated with producing, dispersing, and detecting uv photons. In this chapter, we review the instrumentation available for uv spectroscopy, concentrating on the vuv, where special instrumentation is necessary. Recent advances are stressed, particularly in the areas of synchrotron radiation and the production of vuv laser light.

The classic authoritative review of vuv instrumentation is Ref. [1], published in 1967. There are many comprehensive reviews of vuv light sources, spectrometers, and detectors, which are referenced in the appropriate sections of this chapter.

42.2 LIGHT SOURCES

42.2.1 Synchrotron Radiation

Synchrotron radiation (SR) facilities provide intense continuum radiation in the visible, uv, and soft x-ray spectral regions. The radiation is emitted by relativistic electrons (or positrons) accelerated by the Lorentz force of a magnetic field. In addition to its continuum nature, synchrotron radiation is characterized by high degrees of collimation and polarization. Recent advances in magnet design have produced *wigglers* and *undulators* that dramatically increase spectral brightness at selectable wavelengths and are stimulating significant growth in synchrotron-radiation-based spectroscopic research. References [2–9] give general reviews on the nature of synchrotron radiation and on specific SR-based experimental techniques.

SR facilities use either a linear accelerator, a microtron, or a synchrotron to accelerate electrons or positrons to relativistic energies (0.1 to 10 GeV). The particles are then injected into a *storage ring*, where a series of bending magnets steers them in a closed orbit. The particles in the storage ring radiate energy, which is replaced by radio frequency (rf) accelerating cavities that are part of the ring. These accelerating systems produce well-defined, regularly spaced *bunches* of stored particles, so that radiation from SR facilities is characterized by sub-nanosecond pulses at 1 to 10 MHz repetition rates.

Typical storage ring currents are on the order of 100 to 500 mA. The current decreases continuously because of collisions with residual gas molecules and electron-electron (or positron-positron) scattering. Beam lifetimes in most facilities range from a few to about 20 hours. The longest lifetimes are achieved by positron beams, which repel any ions produced in residual gas collisions. Storage ring pressures must be in the 10^{-10} Torr range in order to reduce collisions to acceptable levels; such pressure requirements can restrict the types of measurements performed at SR facilities.

The spectral properties of synchrotron radiation can be derived from Larmor's formula for an accelerating relativistic charged particle [4, 10–14]. The total power P radiated by a beam of current I and energy E in a bending magnet field B is

$$P = 0.0265 E^3 I B \text{ kW}, \qquad (42.1)$$

where, throughout this chapter, I is in mA, E is in GeV, and B is in T. For example, a 300 mA beam of

1 GeV electrons radiates about 8 kW when traversing the field of a typical 1.0 T bending magnet. One-half of the total power is radiated above the *critical* wavelength, λ_c, where

$$\lambda_c = 1.9/(BE^2) \text{ nm}. \qquad (42.2)$$

For the example just considered, $\lambda_c = 1.9$ nm.

The power radiated by, and the flux from, synchrotron radiation sources is often expressed per fractional (or percent) bandwidth. For example the flux Φ, in units of photons$(\text{cm}^2\text{s})^{-1}$, at 10 nm (124 eV) per 1 % bandwidth is the flux in a 0.1 nm band at 10 nm, which is the same as that in a 1.24 eV energy band. The radiated power per fractional bandwidth peaks at $\lambda \approx 0.75\,\lambda_c$.

The flux per fractional bandwidth reaches a broad maximum at $\lambda \approx 3.3\,\lambda_c$ and decreases slowly at longer wavelengths, with a limiting behavior that is proportional to $\lambda^{-1/3}$ for $\lambda \gg \lambda_c$. Thus, although the photon flux typically peaks in the soft x-ray region, a high flux is also present throughout the vuv wavelength range. Below λ_c, the flux drops rapidly, with little usable radiation at wavelengths below about $0.1\lambda_c$.

The angular distribution of the radiated light is sharply peaked in the instantaneous direction of the beam of radiating particles. Practically all the radiation is emitted into an opening half-angle θ that at $\lambda = \lambda_c$ is equal to $1/\gamma$, where $\gamma = E/mc^2$; i.e.,

$$\theta = 1/\gamma = 5.11 \times 10^{-4}/E \text{ radians}. \qquad (42.3)$$

Radiation from beams with energies of a few GeV has a divergence somewhat less than 1 mrad at λ_c. The divergence increases at longer wavelengths, with

$$\theta \approx (\gamma)^{-1}(\lambda/\lambda_c)^{1/3} \text{ radians}. \qquad (42.4)$$

Radiation from bending magnets maintains this narrow divergence only in the vertical plane; in the orbital plane, the radiation is emitted in an opening angle equal to the bending angle of the magnet. Typical beam cross sections are 0.01 to 1 mm^2. Thus, the extremely high spectral brightness, i.e., photon flux per unit solid angle and % bandwidth, of synchrotron sources reflects both the very low divergence of the radiation and the small source size. A typical brightness associated with bending magnet radiation is on the order of 10^{12} photons $(\text{s mm}^2 \text{ mrad}^2 \, 0.1\% \text{ bandwidth})^{-1}$.

Synchrotron radiation has well-characterized polarization properties. Within the orbital plane the radiation is completely linearly polarized with the electric field vector in the orbital plane; above and below the orbital plane the radiation is elliptically polarized, with the degree of polarization dependent on both wavelength and viewing angle. The practical use of the elliptically polarized radiation is limited by the rapid decrease in intensity above and below the orbital plane.

The pulsed nature of the radiation at synchrotron facilities is ideal for sub-nanosecond time-resolved spectroscopies [15–17]. Electron bunch lengths l_b are typically a few centimeters; radiation is observed from each bunch for a time interval approximately equal to $l_b/c \approx 100$ ps. The time for one full orbit is determined by the size of the storage ring and varies from about 20 ns (SURF II facility at NIST) to 500 to 1000 ns at the larger rings. The number of bunches is typically on the order of a few hundred, resulting in pulse repetition rates from a few MHz to about 500 MHz.

Many synchrotron facilities incorporate straight sections between bending magnets to accommodate insertion devices such as *wigglers* and *undulators*. These are linear arrays of magnets with alternating polarities that cause the beam trajectory to oscillate, although no net displacement of the beam occurs. The oscillations produce synchrotron radiation that is characterized by extremely high brightness and that can be spectrally tailored to individual experiments [4, 13, 18, 19].

Wigglers and undulators are distinguished by the value of their *deflection parameter* K, which is a measure of the maximum bending angle of the stored particle beam in units of θ, the angular divergence of the emitted radiation. The deflection parameter K is determined by the peak strength of the alternating magnetic field B_0 and its spatial period x_B by

$$K = 0.934 B_0 x_B \,, \qquad (42.5)$$

with x_B in cm.

In the wiggler regime, $K \gg 1$; i.e., the angle at which the magnets deflect the electron beam is large compared with θ. The resulting spectrum resembles that from a bending magnet, but is brighter by a factor of N_m, the number of magnet periods (usually 10 to 100). Wigglers are generally used as wavelength shifters; magnetic fields larger than those available in conventional bending magnets decrease the critical wavelength of the spectral distribution and shift the overall spectrum to shorter wavelengths.

In the undulator regime, $K \ll 1$; i.e., the angle at which the electrons are deflected is close to θ and radiation from individual oscillations adds coherently at certain resonant wavelengths, producing a gain of N_m^2. Undulator radiation is characterized by sharp peaks at a fundamental wavelength λ_1 and its odd harmonics. λ_1 is determined by the beam energy and x_B:

$$\lambda_1 \approx 0.1\, x_B/(2\gamma^2) = 1.3\, x_B E^{-2} \text{ nm}, \qquad (42.6)$$

where x_B is in cm. For example, the fundamental wavelength is 6.5 nm for a 1 GeV beam traversing an undulator with a spatial period of 5 cm. The fractional bandwidth of the peaks in an undulator spectrum is $\sim 1/N_m$. Unlike the radiation from a bending magnet, the angular divergence of undulator radiation is sharply

peaked both vertically and in the orbital plane; the brightness of undulator radiation is $\sim 10^{18}$ to 10^{19} photons/(s mm^2 mrad^2 0.1% bandwidth), about six orders of magnitude greater than bending magnet radiation. The wavelength of the fundamental is tunable; the strength of the peak magnetic field is altered via changes in the size of the gaps between the poles of the magnets.

42.2.2   Laser-Produced Plasmas

When the output of a high-power pulsed laser is focused onto a solid target, a short-lived, high-temperature ($T \approx 50$ to 100 eV), high-density ($n_e \approx 10^{21}$ cm^{-3}) plasma is created. The radiation from certain target materials, particularly the rare earths ($57 \leq Z \leq 71$) and neighboring metals on the periodic table, produces a strong vuv quasicontinuum that is essentially free of discrete lines [20, 21]. The continua are most intense in the 4 to 30 nm region but often extend to about 180 nm.

The primary mechanisms responsible for the continua are recombination radiation (ionization stages up to ~ 16 are attained) and bremsstrahlung. The absence of discrete features in rare earth plasmas is thought to be caused by the extreme complexity of rare earth atomic energy level structures; individual emission lines are blended into an apparent continuum [22]. The necessary peak laser powers, which are in the range of 10^{10} to 10^{11} W cm^{-2} [23, 24], are easily achieved with most commercially available Q-switched lasers. The continuum pulse duration is comparable to the duration of the laser pulse. Peak output intensities are significantly greater than those from other pulsed *table-top* continuum sources such as spark discharges; synchrotron sources produce much higher average intensities [24]. The target material is usually a cylindrical rod that is rotated to present a fresh surface for each laser shot. Best results are reported for laser incidence and viewing angles of 45° [25]. The continuum intensity is unaffected by ambient gas pressures up to a few Torr [21]. References [23, 26, 27] present spectra of the plasmas of nine elements; the cleanest continua being produced by the rare earths Yb and Sm and by W, Pb, and Hf.

42.2.3   Arcs, Sparks, and Discharges

Several laboratory sources of vuv continuum radiation are based on gas discharges, high-pressure arcs, and low-pressure and vacuum sparks. While these well-established radiation sources are not as intense as synchrotron radiation and laser-produced plasmas, the traditional sources have the advantages of being portable and inexpensive.

H$_2$ and D$_2$ Discharges.   Direct current discharges (approximately 100 to 500 mA) through 1 to 2 Torr of H$_2$ or D$_2$ generate continuum radiation from about 165 to

350 nm [1, 28]. The continuum is produced by transitions from the bound $1s\sigma\ 2s\sigma\ a^3\Sigma_g^+$ state to the repulsive $1s\sigma\ 2p\sigma\ b^3\Sigma_u^+$ state [29]; the very steep potential curve of the lower state results in the extended nature of the continuum. The photon flux per unit wavelength from such lamps peaks at about 185 nm; below 165 nm the many-line spectra of H_2 or D_2 begin to dominate.

Ar Mini-Arc. A high-current (20 to 50 A) *wall-stabilized* arc discharge through 1 to 2 atm of Ar, known as a mini-arc, produces a continuum associated with recombination radiation. The continuum extends from the near uv down to approximately 110 nm [28,30]. The stability and reproducibility of the output of the mini-arc has led to its as a secondary radiometric standard in the vuv [30].

Noble Gas Discharges. High voltage (10 kV), mildly condensed, repetitive (5 kHz) discharges through the noble gases produce continua associated with molecular transitions between bound excited states and the very weakly bound ground states of the noble gas dimers [1,31,32]. These continua begin at the first resonance line of the atomic species and extend a few tens of nanometers to longer wavelengths. For example, the onset of the He continuum is approximately 59 nm, while the useful continuum extends from 65 to 95 nm with a peak intensity at about 80 nm. Discharges through He, Ar, Kr, and Xe cover the 65 to 180 nm region. Optimum discharge pressures vary from about 40 Torr for He up to a few hundred Torr for Kr and Xe. Differential pumping is required when using these discharge sources at wavelengths shortward of 105 nm, the short wavelength transmission limit of window materials (see Sect. 42.6).

Flash Discharges and Vacuum Sparks. Flash discharges of approximately 1 μs duration through low-pressure (approximately 0.02 Torr) gases in ceramic or glass capillaries yield useful continua down to about 30 nm [1,33]. The continua are produced by passage of the discharge through sputtered wall materials and are independent of the carrier gas. The BRV source [34], a pulsed vacuum spark with extremely low inductance, utilizes a high-Z anode (e.g., W or U) to generate a smooth continuum down to about 10 nm. The short pulse duration (approximately 50 ns) [35] and well-defined triggering make this source useful for transient absorption studies at short wavelengths.

Line Radiation Sources. Hollow cathode discharges [36], which are easy and inexpensive to construct and operate, are the most common line-emission sources used in uv spectroscopy. Low power (approximately 5 W) lamps designed to produce the spectra of about 70 elements are commercially available for spectrochemical and atomic absorption spectroscopy. Line widths are narrow, so blends are avoided. Most manufacturers can provide lamps with special windows for use at vuv wavelengths (see Sect. 42.6).

High-current, differentially-pumped hollow cathodes

sources, which have been developed for use as radiometric standards in the 40 to 125 nm wavelength range [37], can also be used as line sources. At shorter wavelengths, e.g., 5 to 40 nm, the Penning discharge has proved to be a useful source of line radiation [38].

42.2.4 Supercontinuum Radiation

Continua extending throughout the near uv wavelength range to below 200 nm can be produced by focusing femtosecond laser pulses in both liquids and noble gases. This *supercontinuum* generation [39] allows for time-resolved, sub-picosecond spectroscopy of photochemical reactions [40,41]. The supercontinuum arises from a number of competing nonlinear processes [42–44]. Access to the uv has been demonstrated both by focusing 308 nm laser pulses in water [45] and by focusing 248 nm laser pulses in high pressure (10 to 40 atm) cells of Ne, Ar, and Kr [46]. The continuum produced using H_2O is particularly broad, extending from 200 to 600 nm. The neon continuum extends into the vuv to at least 187 nm.

42.3 PULSED LASERS

Primary Lasers. Pulsed lasing at vuv wavelengths has been achieved in a number of media, including the molecular gases H_2, CO, and F_2, excimer systems (Ar_2, Kr_2, Xe_2, ArF), and Auger-pumped noble gases (Xe, Kr) [47]. The fundamental barrier to producing stimulated emission in the vuv is the ν^3 dependence of the spontaneous emission rate, which necessitates very high pump powers to create and maintain population inversions. Existing vuv lasers have generally not been widely used in spectroscopic applications because of either a lack of tunability or unacceptably broad line widths.

Pulsed lasing in CO, H_2, and F_2 occurs between rovibronic levels of a high-lying electronic state and excited vibrational levels of the ground electronic state. The lasing output consists of discrete lines and is not tunable. The CO and H_2 outputs [48,49] span relatively broad regions (181 to 197 nm and 110 to 164 nm, respectively), while the F_2 laser output consists of two or three closely-spaced lines at 157 nm [50,51]. Pulse energies are in the 1 to 100 μJ range for CO and H_2 lasers. Commercially available F_2 lasers offer pulse energies of about 50 mJ at 50 Hz repetition rates.

Noble-gas and noble-gas-halide excimer lasers operate via bound-upper-level to continuum-lower-level transitions. The resulting gain profiles are broad and the outputs are typically tunable over about 2 nm. ArF excimer systems (193 nm) are pumped by high pressure gas discharges; lasers with pulse energies of $\sim$20 to 40 mJ are commercially available. Xe_2 (173 nm), Kr_2 (146 nm), and Ar_2 (126 nm) lasers [52–54] produce comparable pulse energies but require electron beam excitation. Auger-pumped lasers achieve a population inversion through the ejection of an inner-shell electron from a neutral atom to produce a highly excited singly ionized species, followed by Auger decay to excited states of the doubly ionized species. The soft x-ray (approximately 100 eV) pump photons often originate from a laser-produced plasma. The lasing output is nontunable. Lasing in Xe (108.9 nm) and Kr (90.7 nm) [55, 56] has been achieved with output pulses in the range of a few μJ, and Auger-pumped lasing in Zn (at approximately 130 nm) has also been reported [57] .

Nonlinear Techniques. In the past twenty years, developments in the techniques of nonlinear optics have extended the range of tunable, pulsed coherent radiation to vuv wavelengths. Nonlinear frequency conversion techniques, such as stimulated anti-Stokes Raman scattering, harmonic generation, and sum- and difference-frequency mixing (see Chap. 70), can produce narrow bandwidth radiation spanning the vuv down to 60 nm with readily accessible laser instrumentation. The generated vuv radiation has the spectral and spatial characteristics of the input laser radiation. Pulses of 10^{10} to 10^{12} photons are commonly generated with bandwidths on the order of 0.1 cm^{-1} (2×10^{-4} nm). Many comprehensive reviews are available on this subject [47, 58–64]. The standard nonlinear techniques for producing tunable coherent radiation with $\lambda < 200$ nm are summarized below.

Second harmonic generation in nonlinear optical crystals is a well-established method for generating tunable laser light in the near uv (see Chap. 41). At the shortest wavelengths, second harmonic generation in β-barium borate (BBO) produces usable outputs down to approximately 205 nm, below which the phase matching requirement between the fundamental and the second harmonic cannot be met. Sum-frequency mixing ($\omega = \omega_1 + \omega_2$) in BBO [65], where ω_1 is typically the Nd:YAG fundamental (1064 nm) and ω_2 is tunable uv light, extends the useful range of BBO to approximately 190 nm, where the crystal begins to exhibit strong absorption.

Stimulated anti-Stokes Raman scattering of uv laser light in molecular gases (e.g., H_2, N_2, CH_4) is used to generate coherent radiation down to about 120 nm, although low conversion efficiencies often restrict the usable output to $\lambda > 160$ nm. The scattering is a four-wave mixing process resulting in a series of output frequencies shifted from the pump laser frequency by multiples of the vibrational splitting of the ground electronic state of the gas [66, 67]. Because high anti-Stokes orders are not efficiently generated, H_2, with its large ground state vibrational splitting of 4155 cm^{-1}, is often the gas of choice. The experimental requirements are relatively minimal; a tunable visible or uv source (e.g., a frequency-doubled dye laser) and a high pressure gas cell (about 2 to 10 atm of H_2 in a 1 m cell).

Conversion efficiencies of a few percent are reported for the first few anti-Stokes orders in H_2 with pump laser energies of about 10 to 50 mJ [68, 69]; efficiencies then drop continuously, reaching about 10^{-5} to 10^{-6} for the highest orders ($n = 9$ to 13) reported.

Third harmonic generation ($\omega = 3\omega_1$), sum-frequency mixing ($\omega = 2\omega_1 + \omega_2$), and difference-frequency mixing ($\omega = 2\omega_1 - \omega_2$) in appropriate noble gases and metal vapors are the most broadly applicable methods for producing tunable vuv light. These processes result from the presence of the third-order nonlinear term in the expansion of the induced macroscopic polarization of the gas as a power series in the electric field (see Chap. 70). (The second-order term, which is responsible for frequency doubling in crystals, is zero for isotropic media such as gases.) The power generated in third-order effects is proportional to

$$N^2 |\chi^{(3)}|^2 P_1^2 P_2 F , \qquad (42.7)$$

where N is the number density of the gas, $\chi^{(3)}$ is the third-order nonlinear susceptibility, P_1 and P_2 are the input laser powers at ω_1 and ω_2, and the factor F describes the phase matching between the generated vuv light and the induced polarization [59, 60]. For third harmonic generation, $P_1^2 P_2$ is replaced by P_1^3.

The most critical constraint in third-order frequency conversion is the phase matching requirement; a comprehensive treatment is presented in [70]. The factor F is a function of the product $b\Delta k$, where b is the confocal beam parameter of the focused input radiation and Δk is the phase mismatch between the generated vuv light and the input radiation:

$$\Delta k = k - (2k_1 + k_2) . \qquad (42.8)$$

where k_i is the wave vector of the radiation with frequency ω_i. In the standard case of tight focusing ($b \ll L$, where L is the linear dimension of the gas cell) for sum-frequency mixing and third harmonic generation processes, F is nonzero only for $\Delta k < 0$. Therefore, these techniques are applicable only in spectral regions where the gas exhibits negative dispersion. In contrast, for difference-frequency mixing, the factor F is nonzero in regions of both positive and negative dispersion, and wider tuning ranges are generally possible.

Conversion efficiencies for sum- and difference-frequency mixing schemes are usually in the range of 10^{-7} to 10^{-4} with input peak laser powers of 1 to 10 MW. Resonant enhancement of $\chi^{(3)}$, achieved by tuning the input radiation to transition frequencies in the gas, dramatically increases conversion efficiencies (by factors of about 10^4) and allows for much lower (~ 1 kW) input laser powers. Most commonly, one of the incident frequencies is tuned to an allowed two-photon transition, with the second input frequency providing subsequent tunability in the vuv; resonant methods therefore require two tunable inputs.

Metal vapors (Mg, Zn, Hg) are negatively dispersive over fairly broad regions of the vuv between 85 and 200 nm and are consequently used for sum-frequency mixing and third harmonic generation. Conversion efficiencies are further enhanced in these vapors by three photon resonances ($2\omega_1 + \omega_2$) with the ionization continuum or broad autoionizing features [62]. One experimental drawback is the complexity associated with generating the metal vapors in ovens or heat pipes.

Noble gases are generally less suited for sum-frequency mixing because they exhibit negative dispersion over fairly limited spectral ranges in the vuv. However, they do provide an experimentally simple medium for difference-frequency mixing schemes. In particular, Xe and Kr have been used in resonant and nonresonant difference-frequency mixing schemes to produce tunable radiation over the 120 to 200 nm region [71, 72]. The noble gases are used for sum-frequency mixing and third harmonic generation over limited spectral ranges. With these techniques, Kr, Ar, and Ne have proved useful for the generation of tunable radiation down to 72 nm [71, 73, 74]. Below the LiF transmission cutoff (105 nm) the gases are introduced as pulsed jets. [75] describe an ultranarrow bandwidth ($\sim$0.007 cm^{-1}) system, utilizing sum-frequency mixing and third harmonic generation schemes in Xe, Ar, and CO, capable of producing coherent radiation between 74 and 124 nm. A representative summary of third-order frequency conversion schemes is presented in Table 42.1.

Higher-order frequency conversion techniques have been used to generate both fixed frequency and tunable coherent radiation shortward of 70 nm. For example, radiation at 53.2 nm and 38.0 nm has been produced via fifth and seventh harmonic generation in He, using the 266.1 nm Nd:YAG fourth harmonic [83]. Tunable radiation at 58 nm has been produced through fifth harmonic generation in C_2H_2 using the frequency-doubled output of a dye laser [84].

42.4 SPECTROMETERS

42.4.1 Grating Spectrometers

The design and characteristics of vuv grating spectrometers and monochromators are reviewed by many authors [1, 33, 85–87]. Two basic types of spectrometer are used in the vuv; normal incidence instruments for 200 nm $> \lambda >$ 30 nm and grazing incidence instruments for 50 nm $> \lambda >$ 2 nm. The particulars of these two types are largely dictated by the low reflectivities of both metal and dielectric surfaces in the vuv.

Concave gratings are used almost exclusively in vuv spectrometers. Such gratings provide both dispersion

Table 42.1. Representative third-order frequency conversion schemes for generation of tunable coherent vuv light.

Medium	λ (nm)	Process	Ref.
Sr	178–196	res. sum mixing	[76]
Mg	121–174	res. sum mixing	[62]
Zn	106–140	res. sum mixing	[77]
Hg	142–182	nonres. tripling	[59]
	117–122	res. sum mixing	[78]
	85–125	res. sum mixing	[79]
Xe	160–206	nonres. diff. mixing	[72]
	140–147	nonres. tripling	[80]
	113–119	nonres. sum mixing	[72]
Kr	117–150	res. diff. mixing	[81]
	110–116	nonres. sum mixing	[72]
	120–124	nonres. tripling	[82]
	127–180	res. diff. mixing	[71]
	72–83	res. sum mixing	[71]
Ar	97–105	nonres. tripling	[73]
Ne	72–74	nonres. tripling	[74]

and focusing, thus eliminating the need for additional mirrors and their associated reflection losses. Most vuv spectrometers make use of the focusing properties of the *Rowland circle*, which is tangent to the grating at its center, lies in a plane perpendicular to the grating grooves, and has a diameter equal to the radius of curvature of the grating [1, 88]. A source on a horizontal Rowland circle is focused horizontally by the grating to a location also on the circle. The dispersion introduced by the grating results in a focused, diffracted spectrum lying on the Rowland circle. Almost all normal incidence and grazing incidence vuv instruments are designed with the entrance and exit slits (or photographic plate) lying on, or nearly on, the Rowland circle.

The image formed by a concave grating is not stigmatic, i.e., the vertical focus does not coincide with the horizontal focus. Hence, a point source is imaged into a vertical line on the Rowland circle [1, 87]. Astigmatism is particularly severe at grazing incidence angles, resulting in both loss of signal (the image of the entrance slit being larger in extent than the exit slit) and loss of resolution (the image of the entrance slit is curved in the dispersion direction). Aspherical concave gratings (e.g., toroidal gratings) reduce the astigmatism associated with conventional spherical gratings (see references in [85, 89]). Perhaps the most important recent advance in concave grating production is the use of interference techniques to produce gratings called holographic gratings [89, 90]. Interference techniques eliminate the periodic irregularities in conventionally ruled gratings that lead to spectroscopic "ghosts", reduce the level of scattered light by significant amounts, allow for very high groove densities (e.g., 4800 mm^{-1}), and can be relatively easily applied to aspheric surfaces.

Normal incidence grating designs are appropriate for wavelengths greater than about 30 to 40 nm. The most common types include the *Eagle*, *Wadsworth*, and *Seya-Namioka* designs [85, 87]. Eagle mounts, with entrance and exit slits on the Rowland circle and approximately equal angles of incidence and reflection ($< 10°$), can be either in-plane or off-plane. Photographic resolving powers $\lambda/\delta\lambda$ of more than 2.5×10^5 and photoelectric resolving powers of $\sim$1 to 2×10^5 have been achieved in the 100 nm region with 6.65 m Eagle mount instruments [91–93]. In the Wadsworth mount, the light source is at a large distance from the grating and no entrance slit is required. This design is appropriate for collimated light sources such as synchrotron radiation. It has the advantages of high throughput and minimal astigmatism [1, 86]. The Seya-Namioka mount is a Rowland circle instrument in which the angle subtended by the fixed entrance and exit slits relative to the center of the grating is approximately 70°. The spectrometer remains in good focus for small grating rotations, resulting in a simple scanning mechanism and thus an inexpensive design [1]. Seya-Namioka instruments provide high throughput at moderate spectral resolutions (typically 0.02 to 0.05 nm) [86].

The normal incidence reflectance of all standard metal coatings is no greater than a few percent for $\lambda < 30$ nm [87]; in this wavelength region grazing incidence instruments take advantage of the total reflection of photons at extreme grazing angles. There is a sharp reflectance cutoff at photon energies above a characteristic energy that typically limits the use of grazing incidence optics to $\lambda \gtrsim 1$ nm. Grazing incidence designs, tailored to the constraints of synchrotron radiation facilities, are described in [4, 86]. Astigmatism becomes severe at grazing angles; aspherical gratings, produced interferometrically, are often used to reduce this aberration [86]. The resolving power of a grazing incidence instrument is generally lower than that of a comparably sized normal incidence instrument, in large part because of the decreased effective width of the grating at grazing incidence [1]. The largest grazing incidence instruments achieve resolving powers of about 10^5 in the 5 to 40 nm region [94, 95].

A number of nonstandard instrument designs are reported in the literature, often for use in vuv astronomy and aeronomy applications. These novel designs take advantage of the properties of conical diffraction [96, 97], variable line-space gratings [98], and dual-grating crossed dispersion or echelle mounts [99, 100] to reduce the effects of aberrations. Ref. [85] reviews many specialized vuv spectrometer designs under development as of 1985.

42.4.2 Fourier Transform Spectrometers

Fourier transform spectroscopy (FTS) is a well-established technique for high-resolution emission and absorption spectroscopy at infrared and visible wavelengths. Recently, the technique has been extended into the near uv and vuv regions, where FTS is characterized by the large optical throughput, high spectral resolution, and accurate linear wavenumber scale available at longer wavelengths [101]. However, the multiplex advantage of FTS is not realized at uv wavelengths because the signal-to-noise ratio is photon-noise limited rather than detector limited.

A scanning Michelson interferometer with a fused silica beamsplitter has achieved a resolving power of 1.8×10^6 at wavelengths down to 178 nm [101]. The same interferometer design, with a MgF$_2$ beamsplitter, has achieved a resolving power of 8.5×10^5 at wavelengths shorter than 140 nm [102]. This spectral resolution is significantly better than that realized by the best vuv grating instruments.

At shorter wavelengths, diffraction gratings can be used as division-of-amplitude beamsplitters to create all-reflecting interference spectrometers for vuv wavelengths. This technique has been used to create *spatial heterodyned spectrometers* for astronomical applications. [103–105]. The interferometer mirrors are not moved, in such instruments; the interferogram is viewed by an array detector.

42.5 DETECTORS

There is a wide variety of photon detectors with useful response and sensitivity at vuv wavelengths. With the exception of those involving the photo-ionization of gases, vuv detectors are based on the same underlying principles as their counterparts in the visible and infrared regions — the common detection schemes are initiated by surface photoemission, electron-hole pair creation in semiconducting materials, or chemical changes in photographic emulsions. The details of the design of vuv detectors, and the constraints on their use, are most often determined by the low levels of transmission of vuv light through suitable window and semiconducting materials. Nevertheless, vuv detectors with single photon counting sensitivities and/or imaging capabilities are widely available. Reviews of vuv detectors can be found in [1, 33, 106–109]. Reference [110] reviews the operating characteristics and calibration of vuv detector standards. A more general summary of photon detector technology is presented in [111]. Below we summarize and compare the most common vuv detection schemes, including the use of photographic plates, photomultiplier tubes and vacuum photodiodes, multichannel plate detectors, silicon photodiodes, charge-coupled devices, and ionization chambers.

Photographic Plates. Photographic plates have long been used at vuv wavelengths for spectroscopic measurements, and they are still often the detector of choice for both spectroscopic surveys and high resolution wavelength measurements. They have the advantages of being imaging detectors with very high spatial resolution; spectral line positions can be determined with an uncertainty of about 1 μm. Plates also have multiplexing and very flexible integrating capabilities, being insensitive to fluctuations in signal intensities during the period of exposure. However, when compared with photoelectric detectors, photographic plates have the disadvantages of limited dynamic range and a nonlinear response, which makes radiometric measurements very problematic when plates are used. The gelatin base used in standard photographic emulsions strongly absorbs vuv radiation, so special plates with no gelatin base (e.g., Kodak SWR or 101-01) are required for vuv work [1].

Photomultiplier Tubes. Photomultiplier tubes (PMTs) [112] are used through the vuv for single-photon-counting, nonimaging, applications. Dark count rates are low (about 1 s^{-1}) for solar-blind tubes with 1 cm^2 photocathodes; pulse rise times are about 5 to 20 ns. In the 105 to 200 nm region, useful PMT window materials are fused silica, MgF$_2$, and LiF; sapphire is also used in environments where ionizing radiation, which causes sapphire to fluoresce, is not present. In the windowless region of the vuv wavelength range (2 nm $\leq \lambda \leq$ 105 nm), two options are available: the PMT may be operated bare, i.e., without a window, or a fluorescent coating can be deposited on the window to down-convert the vuv light to longer wavelengths.

Peak quantum efficiencies for vuv photomultiplier tubes are about 15 to 20%. Some of the most useful coatings for photocathodes in the vuv are CsI, KBr, and other alkali halides. These materials combine high vuv quantum efficiencies with a solar blind response — their relatively high photoelectric work functions result in long wavelength cutoffs between 150 and 300 nm. Solar blind PMTs have low dark count rates and have minimal response to stray ambient light.

Bare PMTs, although used in some applications, are constrained by the degradation of many photocathode and dynode surfaces upon exposure to air and humidity. Metal surfaces such as tungsten and aluminum/Al$_2$O$_3$ [108] must be used rather than the higher efficiency alkali halide coated surfaces. Sodium salicylate is the most commonly used vuv-visible conversion phosphor; others include liumogen, terphenyl, and coronene [1, 113, 114]. The conversion efficiency of sodium salicylate (peak fluorescence about 430 nm) is relatively constant for vuv light from 30 to 200 nm. It displays some aging effects, which may be associated with contaminants such as oil vapor in the associated vacuum system [1]. The fluorescent decay time of sodium salicylate is ~10 ns [1].

Vacuum Photodiodes. Vacuum photodiodes consist of a photocathode and an anode, with no amplifying dynode chain. The current leaving the photocathode is measured. Because of their stability, spatial uniformity, and portability, vacuum photodiodes are used as radiometric transfer-standard detectors throughout much of the vuv [108, 115, 116]. There is no gain in vacuum photodiodes, so minimum signal levels that can be accurately measured are on the order of 10^6 photons/s. For fast timing applications, vacuum photodiodes have been designed with risetimes as short as about 60 ps [106].

Microchannel Plates. The microchannel plate (MCP) detector is a photo-emissive array detector that combines the single-photon counting sensitivity of a PMT with high resolution imaging capability [107–109, 117]. An MCP consists of an array of semiconducting glass channels with diameters of about 10 to 25 μm and length-to-diameter ratios of about 50:1. Electrons ejected from the front surface of an MCP via the photoelectric effect are accelerated through the channels; repeated collisions with the channel walls result in an amplification of the charge by about 10^6. The exiting charge clouds are detected by a variety of position-sensing anode structures.

Bare MCPs have quantum efficiencies of about 10% for $\lambda \leq$ 100 nm and a long wavelength cutoff about 120 nm [118]. Alkali halide coatings increase the quantum efficiency to about 20% and extend MCP sensitivities toward longer wavelengths [118–120]. Feedback instabilities produced by positive ions created in the channels during the electron cloud amplification are minimized through two common channel geometries; the chevron configuration, where two or more straight-channel MCPs oriented at different angles are cascaded [117], or a configuration with one set of curved channels [121, 122].

Readout schemes for determining the position of individual detected photons rely on either direct detection of the resulting electron cloud, or conversion of the electron cloud into visible photons via a phosphor to produce an optical image [123]. Direct detection schemes use centroid-detecting anodes such as the *wedge and strip* design [124, 118], or discrete anode arrays that locate event positions digitally, e.g., the MAMA (multianode microchannel array) [125, 126] or the CODACON systems [127]. Spatial resolutions of 15 μm are possible.

Silicon Photodiodes. Broad use of silicon photodiodes at vuv wavelengths has traditionally been limited by the strong absorption of vuv photons in the outer SiO$_2$ passivation layer that covers the $p - n$ junction of these devices. Standard silicon detectors are sensitive throughout the infrared and visible regions and also in the soft x-ray ($\lambda <$ 2 nm) and x-ray regions. Significant improvements in silicon photodiode sensitivities in the vuv have recently been realized by thinning the SiO$_2$ passivation region to thicknesses of about 5 to 10 nm. Refs.

[128–131] report devices with a quantum efficiency of 120% (electron-hole pairs per incident photon) at 100 nm. The development of such photodiodes with appreciable vuv sensitivity and good temporal stability has led to their recent use as radiometric transfer standards. Because silicon photodiodes respond strongly to radiation throughout the visible, ir, and x-ray regions, a method for rejection of stray light is essential for their effective use in the vuv. Ref. [132] reports on the use of thin-film filters, deposited on the photodiode surface, to restrict the bandpass of the radiation impinging on the diode to selected vuv wavelengths.

Charge-Coupled Devices. CCDs are widely used for low light level imaging applications throughout the visible and near-ir regions. In the vuv, charge-coupled devices suffer from strong absorption both by surface gate structures and by the inactive passivation layer. Two approaches are employed to overcome these limitations: the CCD front surface is overcoated with a photon down-converting phosphor [133], or a thinned and surface-treated CCD is back-illuminated with vuv light [134, 135]. Such techniques have resulted in CCD vuv quantum efficiencies that rival those of photoemissive devices such as microchannel plates. A comparison of CCD and MCP performance in the vuv is given by [118]; they conclude that MCPs are most appropriate for low light-level imaging where photon counting is applicable, while the larger dynamic range and readout capabilities of CCDs make them suitable for higher light level applications.

Ionization Chambers. In gas ionization chambers, the impact of vuv photons with gas atoms or molecules produce electron-ion pairs which are then collected by the application of an appropriate voltage. The photoionization efficiency and the detection efficiency approach 100% for photon energies above the ionization threshold of the gas [1, 108]. With appropriate gain, ionization chambers can detect individual photons (e.g., the Geiger counter); these chambers require relatively high gas pressures (100 to 2000 Torr) and therefore cannot be used in the windowless region of the vuv. The double ionization chamber described by [136] serves as a primary standard detector from 5 to 100 nm [108].

42.6 OPTICAL MATERIALS

The design of vuv instrumentation is dictated in large part by the constraints of vuv optical materials. Transmission through bulk materials is limited to $\lambda \gtrsim 105$ nm, the short wavelength transmission limit of LiF. Normal incidence reflectance from metal surfaces and coatings decreases dramatically at short wavelengths; polarizers and narrow-band interference filters are relatively difficult to produce because of the lack of materials with suitable optical constants. Below we briefly summarize the properties and uses of vuv optical materials. Comprehensive discussions can be found in [1, 33].

Windows. Few bulk materials transmit light below 200 nm; none transmit light from about 2 nm to 105 nm. In the 105 to 200 nm wavelength region, the most common window materials are synthetic fused quartz or suprasil (with a short λ transmission cutoff about 160 nm), sapphire (145 nm), CaF_2 (125 nm) MgF_2 (112 nm), and LiF (105 nm). Compilations of the transmission characteristics of these and other materials are presented in [1, 33, 87, 137]. Single crystal windows with dimensions up to 10 cm can be obtained for most of these materials. Sapphire fluoresces when struck by ionizing radiation and is therefore not suitable for certain environments.

Capillaries. In the windowless region, 2 to 105 nm, it is standard practice to use differentially pumped slits or circular apertures to isolate different pressure regions, e.g., to isolate an absorption cell from a vacuum spectrometer. This technique has the drawback of requiring large pumping capacities and limiting optical throughputs. Thin films (see below) can be used below 105 nm as windows, but they provide limited transmission in narrow bandpasses and cannot support large pressure differentials. Ref. [138] describes a differentially pumped MCP capillary array as an alternative window in the vuv. The capillaries are typically ~3 mm in length and have diameters of 10 to 100 μm. Measured optical transmissions are in the 20 to 50% range throughout the vuv; tradeoffs must be considered between optical throughput and gas conductance. The main limitation to an MCP window is its small angular aperture.

Thin Films. Below 105 nm, thin (about 10 to 200 nm) metallic films are used as transmission filters with bandpasses of approximately 10 to 50 nm. Details of filter properties are given in [139]. A review of thin film filter preparation techniques is presented in [140]. Ref. [141] describes composite metallic filters (e.g., Al/Ti/C, Ti/Sb/Al), developed for the Extreme Ultraviolet Explorer satellite, with transmission bandpasses of about 10 to 20 nm and high rejection at wavelengths of strong vuv geocoronal lines.

Coatings. Above 120 nm, the principal broadband reflector for vuv wavelengths is Al with a thin protective overcoat of MgF_2, having a normal incidence reflectance of ~80 to 85% [1]. The normal incidence reflectivities of all materials drop dramatically below about 100 nm. Materials with the highest reflectivities include Os, Pt, Au, and Ir, with reflectivities of 15 to 30% from 30 to 110 nm [142, 1]. Ref. [143] reviews the preparation of vuv reflectance coatings for diffraction gratings. Ref. [144] reports grazing incidence reflection coefficients for Rh, Os, Pt, and Au from 5 to 30 nm; Rh has the highest reflectivity in this region.

Interference Filters and Multilayer Coatings. The development of interference filters for vuv wavelengths has been limited by the availability of coating materials with both high transmission and appreciable range of refractive differentials [145]. Multilayer dielectric reflectors thus require many layers to compensate for the small reflectivities at the material interfaces. Ref. [142] evaluates the commercially available vuv reflectance filters, antireflection coatings, and neutral density filters as of 1983. More recent developments are described by [146–149].

The development of multilayer reflection coatings at wavelengths below 30 nm has led to an increasing use of normal incidence optics in the nominally grazing incidence spectral region [150–152]. Reference [153] summarizes synthesis procedures and models of multilayer structures. References [154] and [155] describe normal incidence gratings coated with Ir/Si and Mo/Si multilayers. The performance of Si/Mo multilayer mirrors is described by [156,157]. Peak reflectivities of ~50% are reported with bandpasses of ~2 nm.

Polarizers. The production and detection of linearly polarized light in the vuv is discussed in Refs. [1, 87, 158–160]. Above about 112 nm, transmission polarizers, based on the birefringence of MgF_2, are employed [161, 162]. Reflection polarizers must be used below 112 nm. An analysis of double-reflection circular polarizers is given in Ref. [163].

REFERENCES

1. J. A. R. Samson, *Techniques of Vacuum Ultraviolet Spectroscopy* (Wiley, New York, 1967).
2. *Handbook on Synchrotron Radiation*, edited by G. S. Brown and D. E. Moncton (North Holland, Amsterdam, 1991), Vol. 3.
3. *Handbook on Synchrotron Radiation*, edited by E.-E. Koch (North Holland, Amsterdam, 1983), Vol. 1.
4. G. Margaritondo, *Introduction to Synchrotron Radiation* (Oxford University Press, New York, 1988).
5. *Handbook on Synchrotron Radiation*, edited by G. V. Marr (North Holland, Amsterdam, 1987), Vol. 2.
6. V. Schmidt, Rep. Prog. Phys. **55**, 1483 (1992).
7. *Synchrotron Radiation Research*, edited by H. Winick and S. Doniach (Plenum, New York, 1980).
8. H. Winick, in *Physics of Particle Accelerators* (American Institute of Physics, New York, 1989) p. 2138.
9. H. Winick, *Synchrotron Radiation Sources: A Primer* (World Scientific, Singapore, 1994).
10. E.-E. Koch, D. E. Eastman, and Y. Farge, in *Handbook on Synchrotron Radiation*, edited by E.-E. Koch (North Holland, Amsterdam, 1983), Vol. 1, p. 1.
11. S. Krinsky, M. L. Perlman, and R. E. Watson, in *Handbook on Synchrotron Radiation*, edited by E.-E. Koch (North Holland, Amsterdam, 1983), Vol. 1, p. 65.
12. I. H. Munro and G. V. Marr, in *Handbook on Synchrotron Radiation*, edited by G. V. Marr (North Holland, Amsterdam, 1987), Vol. 2, p. 1.
13. K. Wille, Rep. Prog. Phys. **54**, 1005 (1991).
14. H. Winick, in *Synchrotron Radiation Research*, edited by H. Winick and S. Doniach (Plenum, New York, 1980), p. 11.
15. T. Möller and G. Zimmerer, Phys. Scr. T **17**, 177 (1987).
16. I. H. Munro and N. Schwentner, Nucl. Instrum. Methods **208**, 819 (1983).
17. V. Rehn, Nucl. Instrum. Methods **177**, 193 (1980).
18. G. S. Brown and W. Lavender, in *Handbook on Synchrotron Radiation*, edited by G. S. Brown and D. E. Moncton (North Holland, Amsterdam, 1991), Vol. 3, p. 37.
19. P. Elleaume, Rev. Sci. Instrum. **63**, 321 (1992).
20. P. K. Carroll, E. T. Kennedy, and G. O'Sullivan, Opt. Lett. **2**, 72 (1978).
21. P. K. Carroll, E. T. Kennedy, and G. O'Sullivan, Appl. Opt. **19**, 1454 (1980).
22. G. O'Sullivan, J. Phys. B **16**, 3291 (1983).
23. J. M. Bridges, C. L. Cromer, and T. J. McIlrath, Appl. Opt. **25**, 2208 (1986).
24. J. T. Costello, J. P. Mosnier, E. T. Kennedy, P. K. Carroll, and G. O'Sullivan, Phys. Scr. T**34**, 77 (1991).
25. P. K. Carroll, E. T. Kennedy, and G. O'Sullivan, IEEE. J. Quantum Electron. **19**, 1807 (1983).
26. P. Gohil, V. Kaufman, and T. J. McIlrath, Appl. Opt. **25**, 2039 (1986).
27. F. B. Orth, K. Ueda, T. J. McIlrath, and M. L. Ginter, Appl. Opt. **25**, 2215 (1986).
28. A. P. Thorne, *Spectrophysics* (Chapman and Hall, London, 1988), 2nd. ed.
29. K. P. Huber and G. Herzberg, *Constants of Diatomic Molecules* (Van Nostrand Reinhold, New York, 1979).
30. J. M. Bridges and W. R. Ott., Appl. Opt. **16**, 367 (1977).
31. Y. Tanaka, J. Opt. Soc. Am. **45**, 710 (1955).
32. Y. Tanaka, A. S. Jursa, F. J. LeBlanc, J. Opt. Soc. Am. **48**, 304 (1958).
33. A. N. Zaidel and E. Ya. Schreider, *Vacuum Ultraviolet Spectroscopy* (Humphrey, Ann Arbor, 1970).
34. G. Balloffet, J. Romand, and B. Vodar, C. R. Acad. Sci. **252**, 4139 (1961).
35. T. B. Lucatorto, T. J. McIlrath, and G. Mehlman, Appl. Opt. **18**, 2916 (1979).
36. M. E. Pillow, Spectrochem. Acta. B**36**, 821 (1981).
37. J. Hollandt, M. Kühne, and B. Wende, Appl. Opt. **33**, 68 (1994).
38. D. S. Finley, S. Bowyer, F. Paresce, and R. F. Malina, Appl. Opt. **18**, 649 (1979).
39. R. R. Alfano and S. L. Shapiro, Phys. Rev. Lett. **24**, 592 (1970).
40. J. H. Glownia, J. A. Misewich, and P. P. Sorokin, in *The Supercontinuum Laser Source*, edited by R. R. Alfano (Springer-Verlag, New York, 1989), p. 337.
41. J. H. Glownia, J. A. Misewich, and P. P. Sorokin, J. Chem. Phys. **92**, 3335 (1990).
42. *The Supercontinuum Laser Source*, edited by R. R. Alfano (Springer-Verlag, New York, 1989).
43. P. B. Corkum and C. Rolland, in *The Supercontinuum Laser Source*, edited by R. R. Alfano (Springer-Verlag, New York, 1989), p. 318.
44. Y. R. Shen and G. Yang, in *The Supercontinuum Laser Source*, edited by R. R. Alfano (Springer-Verlag, New York, 1989), p. 1.

45. G. Rodriguez, J. P. Roberts, and A. J. Taylor, Opt. Lett. **19**, 1146 (1994).

46. T. R. Gosnell, A. J. Taylor, and D. P. Greene, Opt. Lett. **15**, 130 (1990).

47. S. M. Hooker and C. E. Webb, Prog. Quant. Electron. **18**, 227 (1994).

48. R. W. Dreyfus and R. T. Hodgson, Phys. Rev. A **9**, 2635 (1974).

49. R. T. Hodgson, J. Chem. Phys. **55**, 5378 (1971).

50. J. K. Rice, A. K. Hays, and J. R. Woodworth, Appl. Phys. Lett. **31**, 31 (1977).

51. J. R. Woodworth and J. K. Rice, J. Chem. Phys. **69**, 2500 (1978).

52. P. W. Hoff, J. C. Swingle, and C. K. Rhodes, Appl. Phys. Lett. **23**, 245 (1973).

53. W. M. Hughes, J. Shannon, and R. Hunter, Appl. Phys. Lett. **24**, 488 (1974).

54. M. H. R. Hutchinson, Appl. Opt. **19**, 3883 (1980).

55. H. C. Kapteyn, R. W. Lee, and R. W. Falcone, Phys. Rev. Lett. **57**, 2939 (1986).

56. H. C. Kapteyn and R. W. Falcone, Phys. Rev. A **37**, 2033 (1988).

57. D. J. Walker, C. P. J. Barty, G. Y. Yin, J. F. Young, and S. E. Harris, Opt. Lett. **12**, 894 (1987).

58. M. N. R. Ashfold and J. D. Prince, Contemp. Phys. **29**, 125 (1988).

59. R. Hilbig, G. Hilber, A. Lago, B. Wolff, and R. Wallenstein, Comments At. Mol. Phys. **18**, 157 (1986).

60. W. Jamroz and B. P. Stoicheff, in *Progress in Optics*, edited by E. Wolf (North-Holland, Amsterdam, 1983), Vol. XX, p. 327.

61. B. P. Stoicheff, in *Frontiers of Laser Spectroscopy of Gases*, edited by A. C. Alves, J. M. Brown, and J. M. Hollas (Kluwer, Dordrecht, 1988), p. 63.

62. B. P. Stoicheff, in *Frontiers in Laser Spectroscopy*, edited by T. W. Hänsch and M. Inguscio (North Holland, Amsterdam, 1994), p. 105.

63. C. R. Vidal, in *Tunable Lasers*, edited by L. F. Mollenauer and J. C. White (Springer-Verlag, Heidelberg, 1987), p. 57.

64. S. C. Wallace, in *Photophysics and Photochemistry in the Vacuum Ultraviolet*, edited by S. P. McGlynn, G. L. Findley, and R. H. Huebner (D. Reidel, Dordrecht, 1985), p. 105.

65. W. Mückenheim, P. Lokai, B. Burghardt, and D. Basting, Appl. Phys. B **45**, 259 (1988).

66. A. Penzkofer, A. Laubereau, and W. Kaiser, Prog. Quant. Electron. **6**, 55 (1979).

67. Y. R. Shen, *The Principles of Nonlinear Optics* (Wiley-Interscience, New York, 1984).

68. Y. Huo, K. Shimizu, and T. Yagi, J. Appl. Phys. **72**, 3258 (1992).

69. H. Schomberg, H. F. Döbele, and B. Rückle, Appl. Phys. B **30**, 131 (1983).

70. G. J. Bjorklund, IEEE J. Quantum Electron. **11**, 289 (1975).

71. G. Hilber, A. Lago, and R. Wallenstein, J. Opt. Soc. Am. B **4**, 1753 (1987).

72. R. Hilbig and R. Wallenstein, Appl. Opt. **21**, 913 (1982).

73. R. Hilbig and R. Wallenstein, Opt. Comm. **44**, 283 (1983).

74. R. Hilbig, A. Lago, and R. Wallenstein, Opt. Comm. **49**, 297 (1984).

75. E. Cromwell, T. Trickl, Y. T. Lee, and A. H. Kung, Rev. Sci. Instrum. **60**, 2888 (1989).

76. R. T. Hodgson, P. P. Sorokin, and J. J. Wynne, Phys. Rev. Lett. **32**, 343 (1974).

77. W. Jamroz, P. E. LaRocque, and B. P. Stoicheff, Opt. Lett. **7**, 148 (1982).

78. R. Mahon and F. S. Tompkins, IEEE J. Quantum Electron. **18**, 913 (1982).

79. P. R. Herman and B. P. Stoicheff, Opt. Lett. **10**, 502 (1985).

80. R. Hilbig and R. Wallenstein, IEEE J. Quantum Electron. **17**, 1566 (1981).

81. C. E. M. Strauss and D. J. Funk, Opt. Lett. **16**, 1192 (1991).

82. D. Cotter, Opt. Comm. **31**, 397 (1979).

83. J. Reintjes, C. Y. She, and R. C. Eckardt, IEEE J. Quantum Electron. **14**, 581 (1978).

84. K. S. E. Eikema, W. Ubachs, W. Vassen, and W. Hogervorst, Phys. Rev. Lett. **71**, 1690 (1993).

85. W. R. Hunter, in *Spectrometric Techniques*, edited by G. Vanasse (Academic Press, New York, 1985), Vol. IV, p. 63.

86. R. L. Johnson, in *Handbook of Synchrotron Radiation*, edited by E. E. Koch (1983), Vol. 1, p. 173.

87. J. A. R. Samson, in *Methods of Experimental Physics*, edited by D. Williams (Academic Press, New York, 1976), Vol. 13, p. 204.

88. E. W. Palmer, M. C. Hutley, A. Franks, J. F. Verrill, and B. Gale, Rep. Prog. Phys. **38**, 975 (1975).

89. M. C. Hutley, *Diffraction Gratings* (Academic Press, New York, 1982).

90. G. Schmahl and D. Rudolf, in *Progress in Optics*, edited by E. Wolf (North-Holland, Amsterdam, 1976), Vol. XIV, p. 197.

91. K. Ito, T. Namioka, Y. Morioka, T. Sasaki, H. Noda, K. Goto, T. Katayama, and M. Koike, Appl. Opt. **25**, 837 (1986).

92. K. Ito and T. Namioka, Rev. Sci. Instr. **60**, 1573 (1989).

93. K. Yoshino, D. E. Freeman, and W. H. Parkinson, Appl. Opt. **19**, 66 (1980).

94. J. Nordgren, H. Ågren, L. Pettersson, L. Selander, S. Griep, C. Nordling, and K. Siegbahn, Phys. Scr. **20**, 623 (1979).

95. J. Reader and N. Acquista, J. Opt. Soc. Am. **69**, 1285 (1979).

96. W. C. Cash, Appl. Opt. **21**, 710 (1982).

97. M. Neviere, D. Maystre, and W. R. Hunter, J. Opt. Soc. Am. **68**, 1106 (1978).

98. M. C. Hettrick, Appl. Opt. **24**, 1251 (1985).

99. W. C. Cash, Appl. Opt. **22**, 3971 (1983).

100. M. C. Hettrick and S. Bowyer, Appl. Opt. **22**, 3921 (1983).

101. A. P. Thorne, C. J. Harris, I. Wynne-Jones, R. C. M. Learner, and G. Cox, J. Phys. E **20**, 54 (1987).

102. A. P. Thorne, G. Cox, P. L. Smith, and W. H. Parkinson, Proc. SPIE **2282**, 58 (1994).

103. S. Chakrabarti, D. M. Cotton, J. S. Vickers, and B. C. Bush, Appl. Opt. **33**, 2596 (1994).

104. J. Harlander, R. J. Reynolds, and F. L. Roesler, Astro-

phys. J. **396**, 730 (1992).

105. R. A. Kruger, L. W. Anderson, and F. L. Roesler, J. Opt. Soc. Am. **62**, 938 (1972).

106. C. B. Johnson, Proc. SPIE **1243**, 2 (1990).

107. O. H. W. Siegmund, M. A. Gummin, J. Stock, and D. Marsh in *Photon Detectors for Space Instrumentation*, edited by T. D. Guyenne and J. Hunt, ESA SP-356 (European Space Agency, Paris, 1992).

108. J. G. Timothy and R. P. Madden, in *Handbook on Synchrotron Radiation*, edited by E.-E. Koch (North-Holland, Amsterdam, 1983), Vol. 1, p. 315.

109. J. G. Timothy, in *Photoelectronic Imaging Devices*, edited by B. L. Morgan, (I. O. P., Bristol, 1991), p. 85.

110. M. L. Furst, R. M. Graves, L. R. Canfield, and R. E. Vest, Rev. Sci. Instrum. **66**, 2257 (1995).

111. J. H. Moore, C. C. Davis, and M. A. Coplan, *Building Scientific Apparatus: A Practical Guide to Design and Construction* (Addison-Wesley, Reading, 1989), 2nd. ed.

112. R. W. Engstrom, *Photomultiplier Handbook* (RCA Electro Optics and Devices, Lancaster, 1980).

113. J. B. Birks, in *The Physics and Chemistry of the Organic Solid State*, edited by D. Fox, M. M. Labes, and A. Weissberger (Interscience, New York, 1965), pp. 434–509.

114. J. A. R. Samson and G. N. Haddad, J. Opt. Soc. Am. **64**, 1346 (1974).

115. E. B. Saloman, Nucl. Instrum. Meth. 172, 79 (1980).

116. L. R. Canfield and N. Swanson, J. Res. Nat. Bur. Stand. **92**, 97 (1987).

117. J. L. Wiza, Nucl. Instrum. Meth. **162**, 587 (1979).

118. J. V. Vallerga and M. Lampton, Proc. SPIE **868**, 25 (1987).

119. O. H. W. Siegmund, E. Everman, J. V. Vallerga, J. Sokolowski, and M. Lampton, Appl. Opt. **26**, 3607 (1987).

120. O. H. W. Siegmund, E. Everman, J. V. Vallerga, S. Labov, J. Bixler, and M. Lampton, Proc. SPIE **687**, 117 (1987).

121. J. G. Timothy and R. L. Bybee, Rev. Sci. Instrum. **48**, 292 (1977).

122. J. G. Timothy, Rev. Sci. Instrum. **52**, 1131 (1981).

123. C. L. Cromer, J. M. Bridges, J. R. Roberts, and T. B. Lucatorto, Appl. Opt. **24**, 2996 (1985).

124. C. Martin, P. Jelinsky, M. Lampton, and R. F. Malina, Rev. Sci. Instrum. **52**, 1067 (1981).

125. J. G. Timothy, G. H. Mount, and R. L. Bybee, SPIE Space Opt. **183**, 169 (1979).

126. J. G. Timothy, R. L. Bybee, Proc. SPIE **687**, 109 (1986).

127. W. E. McClintock, C. A. Barth, R. E. Steele, G. M. Lawrence, and J. G. Timothy, Appl. Opt. **21**, 3071 (1982).

128. L. R. Canfield, J. Kerner, and R. Korde, Appl. Opt. **28**, 3940 (1989).

129. L. R. Canfield, J. Kerner, and R. Korde, Proc. SPIE **1344**, 372 (1990).

130. R. Korde, L. R. Canfield, and B. Wallis, Proc. SPIE **932**, 153 (1988).

131. R. Korde, J. S. Cable, and L. R. Canfield, IEEE Trans. Nucl. Sci. **40**, 1655 (1993).

132. L. R. Canfield, R. Vest, T. N. Woods, and R. Korde, Proc. SPIE **2282**, 31 (1994).

133. P. F. Morrisey, S. R. McCandliss, P. D. Feldman, and S. D. Friedman, Appl. Opt. **33**, 2534 (1994).

134. J. Janesick, D. Campbell, T. Elliot, T. Daud, Opt. Eng. **26**, 853 (1987).

135. R. A. Stern, R. C. Catura, R. Kimble, A. F. Davidsen, M. Winzenread, M. M. Blouke, R. Hayes, D. M. Walton, and J. L. Culhane, Opt. Eng. **26**, 875 (1987).

136. J. A. R. Samson, J. Opt. Soc. Am. **54**, 6 (1964).

137. E. D. Palik, *Handbook of Optical Constants of Solids* (Academic Press, New York) (1985).

138. T. B. Lucatorto, T. J. McIlrath, and J. R. Roberts, Appl. Opt. **18**, 2505 (1979).

139. F. R. Powell, P. W. Vedder, J. F. Lindblom, and S. F. Powell, Opt. Eng. **26**, 614 (1990).

140. W. R. Hunter, in *Physics of Thin Films*, edited by G. Hass, M. H. Francombe, and R. W. Hoffman (Academic Press, New York, 1973), Vol. 7, pp. 43–114.

141. P. W. Vedder, J. V. Vallerga, O. H. W. Siegmund, J. Gibson, and J. Hull, SPIE **1159**, 392 (1989).

142. B. K. Flint, Adv. Space. Res. **2**, 135 (1983).

143. W. R. Hunter, in *Physics of Thin Films*, edited by G. Hass and M. H. Francombe (Academic Press, New York, 1980), Vol. 11, pp. 1–34.

144. M. C. Hettrick, S. A. Flint, and J. Edelstein, Appl. Opt. **24**, 3682 (1985).

145. M. Zukic, D. G. Torr, J. F. Spann, and M. R. Torr, Appl. Opt. **29**, 4284 (1990).

146. M. Zukic, D. G. Torr, J. F. Spann, and M. R. Torr, Appl. Opt. **29**, 4293 (1990).

147. M. Zukic and D. G. Torr, in *Thin Films*, edited by K. H. Guenther (Springer-Verlag, Berlin, 1991), Ch. 7.

148. M. Zukic and D. G. Torr, Appl. Opt. **31**, 1588 (1992).

149. M. Zukic, D. G. Torr, J. Kim, J. F. Spann, and M. R. Torr, Opt. Eng. **32**, 3069 (1993).

150. T. W. Barbee, S. Mrowka, and M. C. Hettrick, Appl. Opt. **24**, 883 (1985).

151. J. V. Bixler, T. W. Barbee, and D. D. Dietrich, Proc. Soc. Photo-Opt. Instrum. Eng. **1160**, 648 (1989).

152. E. Spiller, Appl. Opt. **15**, 2333 (1976).

153. T. W. Barbee, Phys. Scr. T**31**, 147 (1990).

154. R. J. Thomas, R. A. M. Keski-Kuha, W. M. Neuport, C. E. Condor, and J. S. Gunn, Appl. Opt. **30**, 2245 (1991).

155. J. F. Seely and C. M. Brown, Appl. Opt. **32**, 6288 (1993).

156. J. M. Slaughter, D. W. Schulze, C. R. Hills, A. Mirone, R. Stalia, R. N. Watts, C. Tarrio, T. B. Lucatorto, M. Krumrey, T. B. Mueller, and C. M. Falco, J. Appl. Phys. **76**, 2144 (1994).

157. D. G. Stearns, R. S. Rosen, and S. P. Vernon, Appl. Opt. **32**, 6952 (1993).

158. S. Chwirot and J. Slevin, Meas. Sci. Tech. **4**, 1305 (1994).

159. P. Zetner, A. Pradhan, W. B. Westerveld, and J. W. McConkey, Appl. Opt. **22**, 2210 (1983).

160. P. Zetner, K. Becker, W. B. Westerveld, and J. W. McConkey, Appl. Opt. **23**, 3184 (1984).

161. R. Hippler, M. Faust, R. Wolf, H. Kleinpoppen, and H. O. Lutz, Phys. Rev. A **31**, 1399 (1985).

162. H. Winter and H. W. Ortjohann, Rev. Sci. Instrum. **58**, 359 (1987).

163. W. B. Westerveld, K. Becker, P. W. Zetner, J. J. Orr, and J. W. McConkey, Appl. Opt. **24**, 2256 (1985).

43

Elastic Scattering: Classical, Quantal, and Semiclassical

M. R. Flannery

School of Physics, Georgia Institute of Technology, Atlanta, Georgia 30332-0430

43.1 CLASSICAL SCATTERING FORMULAE

Central Field. The total energy $E > 0$ and orbital angular momentum L of relative planar motion are conserved. For a particle of mass M with coordinates (R, ψ), a symmetric potential $V(R)$, and asymptotic speed v,

$$E = \frac{p^2(R)}{2M} + V(R) + \frac{L^2}{2MR^2} = \frac{1}{2}Mv^2 = \text{constant}, \tag{43.1}$$

$$L^2 = (2ME)b^2 = M^2 v^2 b^2 = \left[MR^2(t)\dot\psi(t) \right]^2 = \text{constant}. \tag{43.2}$$

Equation (43.2) implies constant areal velocity.
Radial momentum $p(R)$:

$$p(R; E, b) = (2ME)^{1/2} \left[1 - \frac{V(R)}{E} - \frac{b^2}{R^2} \right]^{1/2}. \tag{43.3}$$

Effective potential:

$$V_{\text{eff}}(R) = V(R) + \frac{L^2}{2MR^2} = V(R) + \frac{b^2}{R^2}E. \tag{43.4}$$

The turning points $R_i(E, b)$ are the roots of $E = V_{\text{eff}}(R)$. The smallest R_i is the distance of closest approach $R_c(E, b)$ at the pericenter. The maximum impact parameter b_X and angular momentum L_X for approach to within a distance R_X are

$$b_X^2 = R_X^2 \left[1 - V(R_X)/E \right], \tag{43.5a}$$
$$L_X^2 = 2MR_X^2 \left[E - V(R_X) \right]. \tag{43.5b}$$

The trajectory is defined by $R = R(t; E, b)$, $\psi = \psi(t; E, b)$, where R is the distance from the scattering center O and ψ is measured with respect to the apse line OA joining O to the pericenter R_c. Taking $t(R_c) = 0$, Eq. (43.1) implies

$$t(R) = \left[\frac{M}{2E} \right]^{1/2} \int_{R_c}^{R} \left[1 - \frac{V(R)}{E} - \frac{b^2}{R^2} \right]^{1/2} dR, \tag{43.6}$$

which implicitly provides $R = R(t; E, b)$.

$$\psi(t; E, b) = \frac{L}{M} \int_0^t R^{-2}(t)\, dt$$
$$= \frac{(2MEb^2)^{1/2}}{M} \int_0^t R^{-2}(t)\, dt. \tag{43.7}$$

Orbit integral ($0 \le \psi \le \pi$): ψ is symmetrical about and measured from the apse line joining O and R_c.

$$\psi(R; E, b) = b \int_{R_c}^{R} \frac{dR/R^2}{[1 - V(R)/E - b^2/R^2]^{1/2}} \tag{43.8}$$

$$= -\frac{\partial}{\partial b} \int_{R_c}^{R} \left[1 - \frac{V(R)}{E} - \frac{b^2}{R^2} \right]^{1/2} dR. \tag{43.9}$$

For large b and/or small $V(R)/E \ll 1$, (43.9) reduces to

$$\psi(R; E, b) = \frac{\pi}{2} - \sin^{-1}\frac{b}{R} + \frac{1}{2E}\frac{\partial}{\partial b} \int_b^R \frac{V(R)\, dR}{[1 - b^2/R^2]^{1/2}}, \tag{43.10}$$

$$\psi(R \to \infty; E, b) = \frac{\pi}{2} + \frac{b}{2E} \int_b^\infty \left(\frac{dV}{dR} \right) \frac{dR}{(R^2 - b^2)^{1/2}}. \tag{43.11a}$$

For a straight-line path, $R^2 = b^2 + Z^2$, where Z is the distance along the scattering axis, and

$$\psi(R \to \infty; E, b) = \frac{\pi}{2} + \frac{1}{4E}\frac{\partial}{\partial b} \int_{-\infty}^\infty V(b, Z)\, dZ. \tag{43.11b}$$

43.1.1 Deflection Functions

The deflection function $\chi(E, b)$, $(-\infty \le \chi < \pi)$, is defined to be $\chi(E, b) = \pi - 2\psi(R \to \infty; E, b)$. Then

$$\chi(E, b) = \pi - 2b \int_{R_c}^\infty \frac{dR/R^2}{[1 - V(R)/E - b^2/R^2]^{1/2}}, \tag{43.12}$$

$$= \pi - 2 \int_0^1 \left[\left\{ \frac{1 - V(R_c/x)/E}{1 - V(R_c)/E} \right\} - x^2 \right]^{-1/2} dx. \tag{43.13}$$

An expression which avoids spurious divergences is

$$\chi(E, b) = \pi + 2\frac{\partial}{\partial b} \int_{R_c}^\infty \left[1 - V(R)/E - b^2/R^2 \right]^{1/2} dR. \tag{43.14}$$

Small-angle scattering, $V(R_c)/E \ll 1$, $b \sim R_c$:

$$\chi(E, b) = \left(\frac{R_c}{E} \right) \int_{R_c}^\infty \frac{[V(R_c) - V(R)]\, R\, dR}{(R^2 - R_c^2)^{3/2}} \tag{43.15a}$$

$$= \frac{1}{E} \int_0^1 \frac{[V(R_c) - V(R_c/x)]\, dx}{(1 - x^2)^{3/2}}, \tag{43.15b}$$

where $x = R_c/R$. From (43.10)-(43.11b), other forms are

$$\chi(E, b) = -\frac{1}{E}\frac{\partial}{\partial b} \int_b^\infty \frac{V(R)\, dR}{(1 - b^2/R^2)^{1/2}} \tag{43.16a}$$

$$= -\frac{b}{E} \int_b^\infty \left(\frac{dV}{dR} \right) \frac{dR}{(R^2 - b^2)^{1/2}} \tag{43.16b}$$

$$= -\frac{1}{2E}\frac{\partial}{\partial b} \int_{-\infty}^\infty V(b, Z)\, dZ. \tag{43.16c}$$

For straight-line paths $R^2 = b^2 + v^2 t^2$, Eq. (43.12) yields the impulse-momentum result

$$\chi(E, b) = (Mv)^{-1} \int_{-\infty}^{\infty} F_\perp(t)\, dt = \Delta p_\perp / p_\infty, \quad (43.17)$$

where $\Delta p_\perp$ is the momentum transferred perpendicular to the incident direction and $F_\perp = -(\partial V / \partial R)(b/R)$ is the impulsive force causing scattering. Special cases are

Head-on collisions ($b = 0$): $\quad \chi(E, 0) = \pi$.
Overall repulsion: $\qquad\qquad 0 < \chi \leq \pi$.
Overall Attraction: $\qquad\quad -\infty \leq \chi \leq 0$.
Forward Glory: $\qquad\qquad\;\; \chi = -2n\pi$.
Backward Glory: $\qquad\qquad \chi = -(2n - 1)\pi$,
$\qquad\qquad\qquad\qquad\qquad\; n = 0, 1, 2, \dots .$
Rainbow Scattering: $\qquad (d\chi/db) = 0$ at $\chi_r < 0$.
Deflection Range: $\qquad\quad\; \chi_r \leq \chi \leq \pi$.
Orbiting Collisions: $\qquad\; cf.$ Eq. (43.34).
Diffraction Scattering: $\quad\; \chi \to 0$ as $b \to \infty$.

The scattered particle may wind or spiral many times around ($\chi \to -\infty$) the scattering center. The experimentally observed quantity is the scattering angle θ ($0 \leq \theta \leq \pi$) which is associated with various deflections

$$\chi_i = +\theta, -\theta, -2\pi \pm \theta, -4\pi \pm \theta, \dots (i = 1, 2, \dots n)$$

resulting from n different impact parameters b_i.

Gauss-Mehler Quadrature Evaluation of the Deflection Function.

$$\chi(E, b) = \pi \left[1 - \left(\frac{b}{R_c} \right) \frac{1}{n} \sum_{j=1}^{n} a_k g(a_j) \right], \quad (43.18)$$

where

$$a_k = \cos\left(\frac{2j - 1}{4n} \pi \right), \quad k = n + 1 - j, \text{ and}$$

$$g(x) = \left[1 - V(R_c/x)/E - b^2 x^2 / R_c^2 \right]^{-1/2}, \quad 0 \leq x \leq 1.$$

43.1.2 Elastic Scattering Cross Section

Differential cross section:

$$\frac{d\sigma(\theta, E)}{d\Omega} \equiv I(\theta; E) \equiv \sigma(\theta; E),$$

$$\sigma(\theta, E) = \sum_{i=1}^{n} \left| \frac{b_i\, db_i}{d(\cos \chi_i)} \right| = \sum_{i=1}^{n} I_i(\theta). \quad (43.19)$$

Integral cross section for scattering by angles $\theta \geq \theta_0$:

$$\sigma_0(E) = 2\pi \int_{\theta_0}^{\pi} I(\theta; E)\, d(\cos \theta) = 2\pi \int_0^{b_0(\theta_0)} b\, db, \quad (43.20)$$

where θ_0 results from one $b_0 = b(\theta_0)$. When θ_0 results from three impact parameters b_1, b_2, b_3, for example, then

$$\sigma_0(E) = 2\pi \int_0^{b_1} b\, db + 2\pi \int_{b_2}^{b_3} b\, db = \pi \left[b_1^2 + b_3^2 - b_2^2 \right].$$

$$(43.21)$$

Diffusion (momentum-transfer) cross section:

$$\sigma_d(E) = 2\pi \int_0^{\pi} \left[1 - \cos \theta(E, b) \right] I(\theta)\, d(\cos \theta), \quad (43.22a)$$

$$= 2\pi \int_0^{\infty} \left[1 - \cos \theta(E, b) \right] b\, db, \quad (43.22b)$$

$$= 4\pi \int_0^{\infty} \sin^2 \left[\tfrac{1}{2}\theta(E, b) \right] b\, db. \quad (43.22c)$$

Viscosity cross section:

$$\sigma_v(E) = 2\pi \int_0^{\pi} \left[1 - \cos^2 \theta(E, b) \right] I(\theta)\, d(\cos \theta), \quad (43.23a)$$

$$= 2\pi \int_0^{\infty} \left[1 - \cos^2 \theta(E, b) \right] b\, db, \quad (43.23b)$$

$$= 2\pi \int_0^{\infty} \left[\sin^2 \theta(E, b) \right] b\, db. \quad (43.23c)$$

Small-Angle Diffraction Scattering. The small-angle scattering regime is defined by the conditions $V(R_c)/E \ll 1$, $b \geq R_c$, where $\theta = |\chi|$. The main contribution to $d\sigma/d\Omega$ for small-angle scattering arises from the asymptotic branch of the deflection function χ at large impact parameters b, and is primarily determined by the long-range (attractive) part of the potential $V(R)$ (see Sect. 43.3.6).

Large-Angle Scattering. The main contribution to $d\sigma/d\Omega$ for large-angle scattering arises from the positive branch of χ at small b and is mainly determined by the repulsive part of the potential.

43.1.3 Center-of-Mass to Laboratory Coordinate Conversion

Let ψ_1, ψ_2 be the angles for scattering and recoil, respectively, of the projectile by a target initially at rest in the lab frame. Then

$$\sigma_1(\psi_1) d\Omega_1 = \sigma_2(\psi_2) d\Omega_2 = \sigma_{cm}^{(1,2)}(\theta) d\Omega_{cm}, \quad 0 \leq \theta \leq \pi;$$
$$(43.24)$$
$$\sigma_{cm}^{(2)}(\theta, \phi) = \sigma_{cm}^{(1)}(\pi - \theta, \phi + \pi). \quad (43.25)$$

(A) Two-body elastic scattering process without conversion of translational kinetic energy into internal energy: $(1) + (2) \to (1) + (2)$.

$$\sigma_1(\psi_1) = \sigma_{cm}(\theta) \frac{(1 + 2x \cos \theta + x^2)^{3/2}}{|1 + x \cos \theta|}; \quad (43.26)$$

$$\sigma_2(\psi_2) = \sigma_{cm}(\theta) \left| 4 \sin \tfrac{1}{2}\theta \right|;$$
$$\psi_2 = \tfrac{1}{2}(\pi - \theta), \quad 0 \leq \psi_2 \leq \tfrac{1}{2}\pi; \quad (43.27)$$

$$\tan \psi_1 = \frac{\sin \theta}{(x + \cos \theta)}, \quad x = M_1/M_2. \quad (43.28)$$

$M_1 > M_2$: As $0 \le \theta \le \theta_c = \cos^{-1}(-M_2/M_1)$,
$0 \le \psi_1 \to \psi_1^{\max} = \sin^{-1}(M_2/M_1) < \frac{1}{2}\pi$.
As $\theta_c \le \theta \to \pi$, $\psi_1^{\max} \le \psi_1 \to 0$.
θ is a double-valued function of ψ.

$M_1 = M_2$: $\sigma_1(\psi_1) = (4\cos\psi_1)\sigma_{\rm cm}(\theta = 2\psi_1)$,
$0 \le \psi_1 \le \frac{1}{2}\pi$, $\psi_1 + \psi_2 = \frac{1}{2}\pi$;
no backscattering.

$M_1 \ll M_2$: $\sigma_1(\psi_1) = \sigma_{\rm cm}(\theta = \psi_1)$
lab and c.m. frames identical.

(B) Two-body elastic scattering process with conversion of translational kinetic energy into internal energy: $(1) + (2) \to (3) + (4)$. For conversion of internal energy ε_i so that kinetic energy of relative motion (in the c.m. frame) increases from E_i to $E_f = E_i + \varepsilon_i$. For $j = 3, 4$,

$$\sigma_j(\psi_j) = \sigma_{\rm cm}(\theta)\frac{\left[1 + 2x_j\cos\theta + x_j^2\right]^{3/2}}{|1 + x_j\cos\theta|}, \quad (43.29)$$

$$x_3 = \left[\frac{M_1 M_3 E_i}{M_2 M_4 E_f}\right]^{1/2}, \quad x_4 = -\left[\frac{M_1 M_4 E_i}{M_2 M_3 E_f}\right]^{1/2},$$

$$\tan\psi_3 = \frac{\sin\theta}{(x_3 + \cos\theta)}, \quad \tan\psi_4 = \frac{\sin\theta}{(|x_4| - \cos\theta)}.$$

43.1.4 Glory and Rainbow Scattering

Glory. The deflection function χ passes through $-2n\pi$ (forward glory) or $-(2n+1)\pi$ (backward glory) at finite impact parameters $b_{\rm g}$. Then $\sin\theta \to 0$ as $\theta \to \theta_g$ so that classical cross section diverges as

$$\sigma(E, \theta) = \left(\frac{2b_{\rm g}}{\sin\theta}\right)\left|\frac{db}{d\chi}\right|_{\rm g} \quad \text{as } \theta \to \theta_{\rm g}. \quad (43.30)$$

Rainbow. The deflection function χ passes through a negative minimum at $b = b_{\rm r}$; $(d\chi/db)_{\rm r} \to 0$ so that

$$\chi(b) = \chi(b_{\rm r}) + \omega_{\rm r}(b - b_{\rm r})^2, \quad (43.31)$$

$$\omega_{\rm r} = \frac{1}{2}\left(\frac{d^2\chi}{db^2}\right)_{b_{\rm r}} > 0. \quad (43.32)$$

The classical cross section diverges as

$$\sigma(E, \theta) = \frac{b_{\rm r}}{2\sin\theta}\left[\omega_{\rm r}(\theta_{\rm r} - \theta)\right]^{-1/2}, \quad \theta < \theta_{\rm r}, \quad (43.33)$$

and is augmented by the contribution from the positive branch of $\chi(b)$.

43.1.5 Orbiting and Spiraling Collisions

Attractive interactions $V(R) = -C/R^n$ ($n \ge 2$) can support quasibound states with positive energy within the angular momentum barrier. Particles with $b < b_0$ spiral towards the scattering center. Those with $b = b_0$ are in unstable circular orbits of radius R_0. The radius R_0 is determined from the two conditions

$$\left(\frac{dV_{\rm eff}}{dR}\right)_{R_0} = 0, \quad E = V_{\rm eff}(R_0), \quad (43.34)$$

which, when combined, yields

$$E = V_{\rm eff}(R_0) = V(R_0) + \frac{1}{2}R_0\left(\frac{dV}{dR}\right)_{R_0}. \quad (43.35)$$

The angular momentum L_0 of the circular orbit is

$$L_0^2 = (2ME)b_0^2 = 2MR_0^2\left[E - V(R_0)\right]. \quad (43.36)$$

Thus $b_0^2 = R_0^2 F$, where

$$F = 1 - \frac{V(R_0)}{E} = \frac{1}{2}\left(\frac{R_0}{E}\right)\left(\frac{dV}{dR}\right)_{R_0} \quad (43.37)$$

is the *focusing factor*. The orbiting and spiraling cross section is then

$$\sigma_{\rm orb}(E) = \pi b_0^2 = \pi R_0^2 F. \quad (43.38)$$

43.1.6 Quantities Derived from Classical Scattering

The semiclassical phase $\eta(E, b) \equiv \eta(E, \lambda)$, with $\lambda = (\ell + \frac{1}{2}) = kb$ is a function of b or λ. The quantities $p(R) \equiv p(R; E, L)$ and $p_0(R) \equiv p_0(R; E, L)$ are radial momenta in the presence and absence of the potential $V(R)$, respectively.

$$\eta^{\rm SC}(E, b) = \frac{1}{\hbar}\left[\int_{R_c}^{\infty} p(R)\,dR - \int_b^{\infty} p_0(R)\,dR\right] \quad (43.39)$$

$$= k\int_{R_c}^{\infty}\left[1 - V/E - b^2/R^2\right]^{1/2}dR$$

$$- k\int_b^{\infty}\left[1 - b^2/R^2\right]^{1/2}dR. \quad (43.40)$$

Asymptotic speed v: $E = \frac{1}{2}Mv^2 = \hbar^2 k^2/2M$, $k/2E = 1/\hbar v$.

Jeffrey-Born phase function:
For small V/E and $b \sim R_c$,

$$\eta_{\rm JB}(E, b) = -\frac{k}{2E}\int_b^{\infty}\frac{V(R)\,dR}{(1 - b^2/R^2)^{1/2}}. \quad (43.41)$$

Eikonal phase function:
For small V/E and a linear trajectory $R^2 = b^2 + Z^2$,

$$\eta_{\rm E}(E, \mathbf{b}) = -\frac{k}{4E}\int_{-\infty}^{\infty} V(\mathbf{b}, Z)\,dZ. \quad (43.42)$$

Semiclassical cross sections:

$$\sigma(E) = 8\pi\int_0^{\infty}\left[\sin^2\eta(E, b)\right]b\,db, \quad (43.43)$$

$$= (8\pi/k^2)\int_0^{\infty}\sin^2\eta(E, \lambda)\lambda\,d\lambda. \quad (43.44)$$

Landau-Lifshitz cross section:

$$\sigma_{\mathrm{LL}}(E) = 8\pi \int_0^\infty \left[\sin^2 \eta_{\mathrm{JB}}(E,b)\right] b\,db. \qquad (43.45)$$

Massey-Mohr cross section:

$$\eta(E,b_0) = \tfrac{1}{2}, \quad \left\langle \sin^2 \eta(E, b < b_0)\right\rangle = \tfrac{1}{2}, \qquad (43.46)$$

$$\sigma_{\mathrm{MM}}(E) = 2\pi b_0^2 + 8\pi \int_{b_0}^\infty \eta_{\mathrm{JB}}^2(E,b) b\,db. \qquad (43.47)$$

Schiff cross section:

$$\sigma_c(E) = 4 \int_{-\infty}^\infty dX \int_{-\infty}^\infty dY \left[\sin^2 \eta_{\mathrm{E}}(X,Y)\right] \qquad (43.48)$$

This reduces to (43.45) for spherical $V(\mathbf{R})$; $b \equiv (x, y)$.

Random-phase approximation (RPA):
For angle α,

$$4\pi \int_0^\infty P(b)\sin^2\alpha(b) b\,db = 2\pi \int_0^{b_c} P(b) b\,db, \qquad (43.49)$$

where $\alpha(b_c) = 1/\pi$.

Collision delay time function:

$$\tau(E,\lambda) = 2\hbar \frac{\partial \eta(E,\lambda)}{\partial E} = \frac{2}{v}\frac{\partial \eta_\ell(k)}{\partial k}. \qquad (43.50)$$

Deflection-angle phase function relation:

$$\chi(E,\lambda) = \frac{2}{k}\frac{\partial \eta(E,b)}{\partial b} = 2\frac{\partial \eta(E,\lambda)}{\partial \lambda}. \qquad (43.51)$$

43.1.7 Collision Action

The classical collision action along a classical path with deflection $\chi = \chi(E,L)$, measured relative to the action along the path of the undeflected particle, is

$$S^{\mathrm{C}}(E,L;\chi) = S_{\mathrm{R}}(E,L) - L\chi(E,L) \qquad (43.52a)$$
$$= 2\eta^{\mathrm{SC}}(E,L)\hbar - L\chi. \qquad (43.52b)$$

Radial component of collision action S^{C}:

$$S_{\mathrm{R}}(E,L) = 2\int_{R_c(E,L)}^\infty p(R)dR - 2\int_{b(E,L)}^\infty p_0(R)\,dR \qquad (43.53a)$$
$$= 2\eta^{\mathrm{SC}}(E,L)\hbar \qquad (43.53b)$$

Collision delay time function:

$$\tau(E,L) = 2M\left[\int_{R_c}^\infty \frac{dR}{p(R)} - \int_b^\infty \frac{dR}{p_0(R)}\right] \qquad (43.54a)$$
$$= \left(\frac{\partial S_{\mathrm{R}}}{\partial E}\right)_L. \qquad (43.54b)$$

Deflection angle function:

$$\chi(E,L) = \pi - 2L \int_{R_c}^\infty \frac{dR/R^2}{p(R)} \qquad (43.55a)$$
$$= \left(\frac{\partial S_{\mathrm{R}}}{\partial L}\right)_E. \qquad (43.55b)$$

Radial collision action change:

$$dS_{\mathrm{R}} = \tau(E,L)dE + \chi(E,L)dL = 2\hbar d\eta^{\mathrm{SC}} \qquad (43.56)$$

43.2 QUANTAL SCATTERING FORMULAE

The basic quantity in *quantal* elastic scattering is the complex scattering amplitude $f(E,\theta)$, expressed in terms of the phase shifts $\eta_\ell(E)$ associated with scattering of the ℓth partial wave. Derived quantities are the diagonal elements of the scattering matrix S, transition matrix T and reactance matrix K.

Reduced Energy: $k^2 = (2M/\hbar^2)E$.
Reduced Potential: $U(R) = (2M/\hbar^2)V(R)$.

43.2.1 Basic Formulae

Wave function: As $R \to \infty$,

$$\Psi(\mathbf{R}) \to \exp(ikZ) + \frac{1}{R}f(\theta)\exp(ikR) \qquad (43.57)$$

for symmetric interactions $V = V(R)$.

Elastic scattering differential cross sections (DCS):

$$\frac{d\sigma}{d\Omega} = I(\theta) = |f(\theta)|^2. \qquad (43.58)$$

Scattering, transition and reactance matrix elements in terms of η_ℓ:

$$S_\ell(k) = \exp(2i\eta_\ell), \qquad (43.59a)$$
$$T_\ell(k) = \sin\eta_\ell \exp(i\eta_\ell), \qquad (43.59b)$$
$$K_\ell(k) = \tan\eta_\ell. \qquad (43.59c)$$

Scattering amplitudes $f(\theta)$:

$$f(\theta) = \frac{1}{2ik}\sum_{\ell=0}^\infty (2\ell+1)\left[\exp(2i\eta_\ell) - 1\right]P_\ell(\cos\theta)$$
$$= \sum_{\ell=0}^\infty f_\ell(\theta), \qquad (43.60a)$$

$$f(\theta) = \frac{1}{2ik}\sum_{\ell=0}^\infty (2\ell+1)\left[S_\ell(k) - 1\right]P_\ell(\cos\theta), \qquad (43.60b)$$

$$f(\theta) = \frac{1}{k}\sum_{\ell=0}^\infty (2\ell+1)T_\ell(k)P_\ell(\cos\theta), \qquad (43.60c)$$

$$f(\theta) = \frac{1}{2ik}\sum_{\ell=0}^\infty S_\ell(k)P_\ell(\cos\theta), \quad \theta \neq 0, \qquad (43.60d)$$

$$= \frac{1}{k}\sum_{\ell=0}^\infty (2\ell+1)T_\ell(k), \quad \theta = 0. \qquad (43.60e)$$

Integral cross sections $\sigma(E)$:

$$\sigma(E) = 2\pi \int_0^\pi I(\theta) d(\cos\theta) \,, \qquad (43.61a)$$

$$= \frac{4\pi}{k^2} \sum_{\ell=0}^\infty (2\ell+1)\sin^2\eta_\ell \,. \qquad (43.61b)$$

Optical theorem:

$$\sigma(E) = (4\pi/k)\Im f(0) \,. \qquad (43.62)$$

Partial cross sections $\sigma_\ell(E)$:

$$\sigma(E) = \sum_{\ell=0}^\infty \sigma_\ell(E) \qquad (43.63)$$

$$\sigma_\ell(E) = \frac{4\pi}{k^2}(2\ell+1)\sin^2\eta_\ell \qquad (43.64a)$$

$$= \frac{4\pi}{k^2}(2\ell+1)|T_\ell|^2 = \frac{2\pi}{k}(2\ell+1)\left[1 - \Re S_\ell\right] \qquad (43.64b)$$

Upper limit:

$$\sigma_\ell(E) \le (4\pi/k^2)(2\ell+1) \,.$$

Unitarity, flux conservation, η_ℓ is real:

$$|S_\ell|^2 = 1, \quad |T_\ell|^2 = \Im T_\ell \,.$$

Differential cross sections (DCS):

$$\frac{d\sigma(E,\theta)}{d\Omega} = |f(\theta)|^2 = I(\theta) = A(\theta)^2 + B(\theta)^2 \,, \qquad (43.65)$$

$$A(\theta) = \Re f(\theta) = \frac{1}{2k}\sum_{\ell=0}^\infty (2\ell+1)\sin 2\eta_\ell P_\ell(\cos\theta) \,,$$

$$B(\theta) = \Im f(\theta) = \frac{1}{2k}\sum_{\ell=0}^\infty (2\ell+1)\left[1 - \cos 2\eta_\ell\right]P_\ell(\cos\theta) \,.$$

$$\int A(\theta)^2 d\Omega = \frac{4\pi}{k^2}\sum_{\ell=0}^\infty (2\ell+1)\sin^2\eta_\ell \cos^2\eta_\ell \qquad (43.66a)$$

$$\int B(\theta)^2 d\Omega = \frac{4\pi}{k^2}\sum_{\ell=0}^\infty (2\ell+1)\sin^4\eta_\ell \qquad (43.66b)$$

$$\frac{d\sigma(E,\theta)}{d\Omega} = \frac{1}{k^2}\sum_{L=0}^\infty a_L(E)P_L(\cos\theta) \qquad (43.67a)$$

$$a_L = \sum_{\ell=0}^\infty \sum_{\ell'=|\ell-L|}^{\ell+L} (2\ell+1)(2\ell'+1)(\ell\ell'00 \mid \ell\ell'L0)^2$$

$$\times \sin\eta_\ell \sin\eta_{\ell'}\cos(\eta_\ell - \eta_{\ell'}) \qquad (43.67b)$$

where $(\ell\ell'mm' \mid \ell\ell'LM)$ are the Clebsch-Gordan Coefficients.

Three-term expansion:

$$\frac{d\sigma(E,\theta)}{d\Omega} = \frac{1}{k^2}\left[(a_0 - \frac{1}{2}a_2) + a_1\cos\theta + \frac{3}{2}a_2\cos^2\theta\right] \qquad (43.68)$$

$$a_0(E) = \sum_{\ell=0}^\infty (2\ell+1)\sin^2\eta_\ell \,, \qquad (43.69a)$$

$$a_1(E) = 6\sum_{\ell=0}^\infty (\ell+1)\sin\eta_\ell \sin\eta_{\ell+1}$$
$$\times \cos(\eta_{\ell+1} - \eta_\ell) \,, \qquad (43.69b)$$

$$a_2(E) = 5\sum_{\ell=0}^\infty \left[b_\ell \sin^2\eta_\ell + c_\ell \sin\eta_\ell \sin\eta_{\ell+2}\right.$$
$$\left.\times \cos(\eta_{\ell+2} - \eta_\ell)\right] \,. \qquad (43.69c)$$

with coefficients

$$b_\ell = \frac{\ell(\ell+1)(2\ell+1)}{(2\ell-1)(2\ell+3)} \,, \qquad (43.70a)$$

$$c_\ell = \frac{3(\ell+1)(\ell+2)}{(2\ell+3)} \,. \qquad (43.70b)$$

S, P wave ($\ell = 0,1$) net contribution:

$$\frac{d\sigma}{d\Omega} = \frac{1}{k^2}\left[\sin^2\eta_0 + [6\sin\eta_0 \sin\eta_1 \cos(\eta_1 - \eta_0)]\cos\theta\right.$$
$$\left. + 9\sin^2\eta_1 \cos^2\theta\right] \,, \qquad (43.71)$$

$$\sigma(E) = \frac{4\pi}{k^2}\left[\sin^2\eta_0 + 3\sin^2\eta_1\right] \,. \qquad (43.72)$$

For pure S-wave scattering, the DCS is isotropic. For pure P-wave scattering, the DCS is symmetric about $\theta = \pi/2$, where it vanishes; the DCS rises to equal maxima at $\theta = 0,\pi$. For combined S- and P-wave scattering, the DCS is asymmetric with forward-backward asymmetry.

Transport cross sections ($n \ge 1$):

$$\sigma^{(n)}(E) = 2\pi \left[1 - \frac{1 + (-1)^n}{2(n+1)}\right]^{-1}$$
$$\times \int_0^\pi [1 - \cos^n\theta]I(\theta)d(\cos\theta) \,. \qquad (43.73)$$

The diffusion and viscosity cross sections (43.22a) and (43.23a) are given by the transport cross sections $\sigma^{(1)}$ and $\frac{2}{3}\sigma^{(2)}$, respectively.

$$\sigma^{(1)}(E) = \frac{4\pi}{k^2}\sum_{\ell=0}^\infty (\ell+1)\sin^2(\eta_\ell - \eta_{\ell+1}) \qquad (43.74a)$$

$$\sigma^{(2)}(E) = \frac{4\pi}{k^2}\left(\frac{3}{2}\right)\sum_{\ell=0}^\infty \frac{(\ell+1)(\ell+2)}{(2\ell+3)}\sin^2(\eta_\ell - \eta_{\ell+2}) \,, \qquad (43.74b)$$

$$\sigma^{(3)}(E) = \frac{4\pi}{k^2}\sum_{\ell=0}^\infty \frac{(\ell+1)}{(2\ell+5)}\left[\frac{(\ell+2)(\ell+3)}{(2\ell+3)}\sin^2(\eta_\ell - \eta_{\ell+3})\right.$$
$$\left. + \frac{3(\ell^2+2\ell-1)}{(2\ell-1)}\sin^2(\eta_\ell - \eta_{\ell+1})\right] \,, \qquad (43.74c)$$

$$\sigma^{(4)}(E) = \frac{4\pi}{k^2} \left(\frac{5}{4}\right) \sum_{\ell=0}^{\infty} \frac{(\ell+1)(\ell+2)}{(2\ell+3)(2\ell+7)}$$

$$\times \left[\frac{(\ell+3)(\ell+4)}{(2\ell+5)} \sin^2(\eta_\ell - \eta_{\ell+4}) \right.$$

$$\left. + \frac{2(2\ell^2+6\ell-3)}{(2\ell-1)} \sin^2(\eta_\ell - \eta_{\ell+2}) \right] . \quad (43.74d)$$

Collision integrals: Averages of $\sigma^{(n)}(E)$ over a Maxwellian distribution at temperature T are

$$\Omega^{(n,s)}(T) = \left[(s+1)!(kT)^{s+2} \right]^{-1} \int_0^\infty \sigma^{(n)}(E)$$

$$\times \exp(-E/kT)E^{s+1}dE . \quad (43.75)$$

Normalization factors are chosen so that the above expressions for $\sigma^{(n)}$ and $\Omega^{(n,s)}$ reduce to πd^2 for classical rigid spheres of diameter d.

Mobility: The mobility K of ions of charge e in a gas of density N is given by the Chapman-Enskog formula

$$K = \frac{3e}{8N} \left(\frac{\pi}{2MkT}\right)^{1/2} \left[\Omega^{(1,1)}(T) \right]^{-1} . \quad (43.76)$$

Phase shifts η_ℓ can be determined from the numerical solution of the radial Schrödinger equation (43.91), from an integral equation (43.177b), from solving a nonlinear first-order differential equation (43.177a), from Logarithmic Derviatives (see Sect. 43.2.5) or from variational techniques (see Sect. 45.2.4).

43.2.2 Identical Particles: Symmetry Oscillations

Colliding Particles, Each with Spin s, in a Total Spin S_t Resolved State in the Range $(0 \rightarrow 2s)$.
Particle interchange: $\Psi(\mathbf{R}) = (-1)^{S_t}\Psi(-\mathbf{R})$

$$I_{A,S}(\theta) = \frac{1}{2}|f(\theta) \mp f(\pi - \theta)|^2 , \quad (43.77)$$

$$\Psi_{A,S}(\mathbf{R}) \rightarrow [\exp(ikZ) \mp \exp(-ikZ)]$$

$$+ \frac{1}{R}[f(\theta) \mp f(\pi-\theta)]\exp(ikR) , \quad (43.78)$$

$$I_{A,S}(\theta) = \frac{1}{4k^2} \left| \sum_{\ell=0}^{\infty} \omega_\ell(2\ell+1)\left[\exp 2i\eta_\ell - 1\right]P_\ell(\cos\theta) \right|^2 , \quad (43.79)$$

$$\sigma_{A,S}(E) = \frac{4\pi}{k^2} \sum_{\ell=0}^{\infty} \omega_\ell(2\ell+1)\sin^2\eta_\ell , \quad (43.80)$$

where A and S denote antisymmetric and symmetric wave functions (with respect to particle interchange) for collisions of identical particles with odd and even total spin S_t.

A: S_t odd $\omega_\ell = 0$ (ℓ even); $\omega_\ell = 2$ (ℓ odd);
S: S_t even $\omega_\ell = 2$ (ℓ even); $\omega_\ell = 0$ (ℓ odd).

Spin-States S_t Unresolved. S/A combination:

$$I(\theta) = g_A I_A(\theta) + g_S I_S(\theta) , \quad (43.81)$$

$$\sigma(E) = g_A \sigma_A(E) + g_S \sigma_S(E) . \quad (43.82)$$

where g_A and g_S are the fractions of states with odd and even total spins $S_t = 0, 1, 2, \ldots, 2s$. For Fermions (F) with half integer spin s, and Bosons (B) with integral spin s

F: $g_A = (s+1)/(2s+1)$, $g_S = s/(2s+1)$,
B: $g_A = s/(2s+1)$, $g_S = (s+1)/(2s+1)$,

so that (43.81) and (43.82) have the alternative forms

$$I(F) = |f(\theta)|^2 + |f(\pi-\theta)|^2 - \mathcal{I} , \quad (43.83a)$$

$$\sigma(F) = \frac{1}{2}[\sigma_S + \sigma_A] - \frac{1}{2}[\sigma_S - \sigma_A]/(2s+1) , \quad (43.83b)$$

$$I(B) = |f(\theta)|^2 + |f(\pi-\theta)|^2 + \mathcal{I} , \quad (43.83c)$$

$$\sigma(B) = \frac{1}{2}[\sigma_S + \sigma_A] + \frac{1}{2}[\sigma_S - \sigma_A]/(2s+1) , \quad (43.83d)$$

where the interference term is

$$\mathcal{I} = \left(\frac{2}{2s+1}\right) \Re\left[f(\theta)f^*(\pi-\theta)\right] . \quad (43.84)$$

Example: For fermions with spin $1/2$,

$$\sigma(E) = \frac{2\pi}{k^2}\left[\sum_{\ell=\text{even}}^{\infty} (2\ell+1)\sin^2\eta_\ell \right.$$

$$\left. + 3\sum_{\ell=\text{odd}}^{\infty} (2\ell+1)\sin^2\eta_\ell \right] . \quad (43.85)$$

Symmetry oscillations originate from the interference between unscattered incident particles in the forward $(\theta = 0)$ direction and backward scattered particles $(\theta = \pi, \ell = 0)$. Symmetry oscillations are sensitive to the repulsive wall of the interaction.

Resonant Charge Transfer and Transport Cross Sections for A^+–A Collisions.

The phase shifts for elastic scattering by the gerade (g) and ungerade (u) potentials of A_2^+ are, respectively, η_ℓ^g and η_ℓ^u. The charge transfer (X) and transport cross sections are

$$\sigma_X(E) = \frac{\pi}{k^2} \sum_{\ell=0}^{\infty} (2\ell+1)\sin^2(\eta_\ell^g - \eta_\ell^u) \quad (43.86a)$$

$$\sigma_{A,S}^{(1)}(E) = \frac{4\pi}{k^2} \sum_{\ell=0}^{\infty} (\ell+1)\sin^2(\beta_\ell - \beta_{\ell+1}) \quad (43.86b)$$

$$\sigma_{A,S}^{(2)}(E) = \frac{4\pi}{k^2}\left(\frac{3}{2}\right) \sum_{\ell=0}^{\infty} \frac{(\ell+1)(\ell+2)}{(2\ell+3)} \sin^2(\beta_\ell - \beta_{\ell+2}) . \quad (43.86c)$$

A: $\beta_\ell = \eta_\ell^g$ (ℓ even),   or η_ℓ^u (ℓ odd),
S: $\beta_\ell = \eta_\ell^u$ (ℓ even),   or η_ℓ^g (ℓ odd).

$\sigma_{A,S}^{(1)}$ contains (g/u) interference; $\sigma_{A,S}^{(2)}$ does not. When nuclear spin degeneracy is acknowledged, the cross sections $\sigma_{A,S}$ are summed according to (43.83b) or (43.83d).

Since there is no coupling between molecular states of different electronic angular momentum, the scattering by the $^2\Sigma_{g,u}$ pair and the $^2\Pi_{g,u}$ pair of Ne_2^+ potentials (for example) is independent and

$$\sigma_X(E) = \frac{1}{3}\sigma_\Sigma(E) + \frac{2}{3}\sigma_\Pi(E) \qquad (43.87)$$

Singlet-Triplet Spin Flip Cross Section.

$$\sigma_{ST}(E) = \frac{\pi}{k^2}\sum_{\ell=0}^{\infty}(2\ell+1)\sin^2(\eta_\ell^s - \eta_\ell^t), \qquad (43.88)$$

where $\eta_\ell^{s,t}$ are the phase shifts for individual scattering by the singlet and triplet potentials, respectively.

43.2.3 Partial Wave Expansion

$$\Psi(\mathbf{R}) = \frac{1}{kR}\sum_{\ell=0}^{\infty}A_\ell v_\ell(kR)P_\ell(\cos\theta), \qquad (43.89)$$

$$A_\ell = i^\ell(2\ell+1)\exp(i\eta_\ell). \qquad (43.90)$$

Radial Schrödinger equation (RSE):

$$\frac{d^2 v_\ell}{dR^2} + \left[k^2 - U(R) - \frac{\ell(\ell+1)}{R^2}\right]v_\ell(R) = 0 \qquad (43.91)$$

where v_ℓ is normalized so that

$$v_\ell(R) \overset{R \geq R_0}{=} \cos\eta_\ell F_\ell(kR) + \sin\eta_\ell G_\ell(kR) \qquad (43.92)$$
$$\to \sin(kR - \tfrac{1}{2}\ell\pi + \eta_\ell) \text{ as } R \to \infty.$$

The regular (nonsingular) solution (zero at $R = 0$) of the field-free RSE (43.91) with $U(R) = 0$ is

$$F_\ell(kR) = (kR)j_\ell(kR) = \left(\tfrac{1}{2}\pi kR\right)^{1/2}J_{\ell+1/2}(kR) \qquad (43.93)$$
$$\to \begin{cases} (kR)^{\ell+1}/(2\ell+1)!!, & R \to 0 \\ \sin(kR - \tfrac{1}{2}\ell\pi), & R \to \infty, \end{cases} \qquad (43.94)$$

where j_ℓ is the spherical Bessel function. Equation (43.89) with $v_\ell = F_\ell$ is the partial-wave expansion for the incident plane-wave $\exp(ikZ)$.

The irregular solution (divergent at $R = 0$) of the field-free RSE is

$$G_\ell(kR) = -(kR)n_\ell(kR)$$
$$= \left(\tfrac{1}{2}\pi kR\right)^{1/2}J_{-(\ell+1/2)}(kR) \qquad (43.95)$$
$$\to \begin{cases} (2\ell-1)!!/(kR)^\ell, & R \to 0 \\ \cos(kR - \tfrac{1}{2}\ell\pi), & R \to \infty, \end{cases} \qquad (43.96)$$

where n_ℓ is the spherical Neumann function.

The full asymptotic scattering solution is the combination (43.92) of the regular and irregular solutions. The mixture depends upon:

Forms of Normalization for v_ℓ. In Eq. (43.89), possible choices of normalization are:

(a) $\quad A_\ell = i^\ell(2\ell+1)\exp i\eta_\ell,$ $\qquad$ (43.97a)

$\quad v_\ell(R) \sim \sin(kR - \tfrac{1}{2}\ell\pi + \eta_\ell);$ $\qquad$ (43.97b)

(b) $\quad A_\ell = i^\ell(2\ell+1)\cos\eta_\ell,$ $\qquad$ (43.98a)

$\quad v_\ell(R) \sim \sin(kR - \tfrac{1}{2}\ell\pi) + K_\ell\cos(kR - \tfrac{1}{2}\ell\pi);$ $\qquad$ (43.98b)

(c) $\quad A_\ell = i^\ell(2\ell+1),$ $\qquad$ (43.99a)

$\quad v_\ell(R) \sim \sin(kR - \tfrac{1}{2}\ell\pi) + T_\ell e^{i(kR-\ell\pi/2)};$ $\qquad$ (43.99b)

(d) $\quad A_\ell = \tfrac{1}{2}i^{\ell+1}(2\ell+1),$ $\qquad$ (43.100a)

$\quad v_\ell(R) \sim e^{-i(kR-\ell\pi/2)} - S_\ell e^{i(kR-\ell\pi/2)};$ $\qquad$ (43.100b)

$$S_\ell = 1 + 2iT_\ell; \qquad K_\ell = T_\ell/(1 + iT_\ell); \qquad (43.101)$$
$$T_\ell = \sin\eta_\ell e^{i\eta_\ell}; \qquad 1 + iT_\ell = \cos\eta_\ell e^{i\eta_\ell}. \qquad (43.102)$$

Significance of η_ℓ, K_ℓ, T_ℓ, and S_ℓ. The effect of scattering is therefore to: (1) introduce a phase shift η_ℓ in Eq. (43.97b) to the regular standing wave, (2) leave the regular standing wave alone and introduce either an irregular standing wave of real amplitude K_ℓ in Eq. (43.98b) or, a spherical outgoing wave of amplitude T_ℓ in Eq. (43.99b), and (3) to convert in Eq. (43.100b) an incoming spherical wave of unit amplitude to an outgoing spherical wave of amplitude S_ℓ.

Levinson's Theorem. A local potential $U(R)$ can support n_ℓ bound states of angular momentum ℓ and energy E_n such that

$$\lim_{k\to 0}\eta_0(k) = \begin{cases} n_0\pi, & E_n < 0 \\ (n_0 + \tfrac{1}{2})\pi, & E_{n+1} = 0, \end{cases} \qquad (43.103)$$
$$\lim_{k\to 0}\eta_\ell(k) = n_\ell\pi, \quad \ell > 0. \qquad (43.104)$$

43.2.4 Scattering Length and Effective Range

Blatt-Jackson Effective Range Formula. For short-range potentials,

$$k\cot\eta_0 = -\frac{1}{a_s} + \frac{1}{2}R_e k^2 + \mathcal{O}(k^4). \qquad (43.105)$$

Effective range:

$$R_e = 2\int_0^\infty\left[u_0^2(R) - v_0^2(R)\right]dR, \qquad (43.106)$$

where $u_0 = \sin(kR + \eta_0)/\sin\eta_0$ is the $k = 0$ limit of the potential-free $\ell = 0$ radial wave function and normalized so that $u_0(R)$ goes to unity as $k \to 0$. The potential distorted $\ell = 0$ radial function v_0 is normalized at large R

to $u_0(R)$. The effective range is a measure of the distance over which v_0 differs from u_0. The outside factor of 2 in Eq. (43.106) is chosen such that $R_e = a$ for a square well of range a.

Scattering length:

$$a_s = -\lim_{k \to 0} f(\theta). \quad (43.107)$$

Relation with $k \to 0$ Elastic Cross Section.

$$\sigma(k \to 0) = \frac{4\pi}{k^2} \sin^2 \eta_0$$

$$= 4\pi a_s^2 \left\{ \left[1 - \tfrac{1}{2}k^2 a_s R_e\right]^2 + k^2 a_s^2 \right\}^{-1} \quad (43.108)$$

$$\sim 4\pi a_s^2 \left[1 + a_s k^2 (R_e - a_s)\right] \quad (43.109)$$

Relation with Bound Levels. If a $\ell = 0$ bound level of energy $E_n = -\hbar^2 k_n^2/2M$ lies sufficiently near the dissociation limit the effective range and scattering lengths, R_e and a_s, respectively, are related by,

$$-\frac{1}{a_s} = -k_n + \frac{1}{2}R_e k_n^2 + \cdots \quad (43.110)$$

Wigner Causality Condition. If η_ℓ provides the dominant contribution to $f(\theta)$ then

$$\frac{\partial \eta_\ell(k)}{\partial k} \geq -a_s \quad (43.111)$$

where a_s is the scattering length ($\ell = 0$) and is a measure of the range of the interaction.

Effective Range Formulae. The Blatt-Jackson formula must be modified [1–5] for long-range interactions as follows.

(1) Modified Coulomb potential: $V(R) \sim Z_1 Z_2 e^2/R$

$$2(K/a_0) = -(1/a_s) + \tfrac{1}{2}R_e k^2 \quad (43.112)$$

$$K = \frac{\pi \cot \eta_0}{e^{2\pi\alpha} - 1} - \ln\beta - 0.5772$$

$$+ \beta^2 \sum_{n=1}^{\infty} \left[n(n^2 + \beta^2)\right]^{-1} \quad (43.113)$$

where $\beta = Z_1 Z_2 e^2/\hbar v = Z_1 Z_2/(ka_0)$.

(2) Polarization potential: $V(R) = -\alpha_d e^2/2R^4$

$$\tan \eta_0 = -a_s k - \frac{\pi}{3}C_4 k^2 - \frac{4}{3}C_4 a_s k^3 \ln(ka_0) + Dk^3 + Fk^4, \quad (43.114)$$

$$\tan \eta_1 = \frac{\pi}{15}C_4 k^2 - a_s^{(1)} k^3, \quad (43.115)$$

$$\tan \eta_\ell = \frac{\pi C_4 k^2}{(2\ell+3)(2\ell+1)(2\ell-1)} + \mathcal{O}(k^{2\ell+1}), \quad (43.116)$$

for $\ell > 1$, where

$$C_4 = \frac{2M}{\hbar^2}\left(\frac{\alpha_d e^2}{2}\right) = \left(\frac{\alpha_d}{a_0}\right)\left(\frac{M}{m_e}\right). \quad (43.117)$$

Example: e^-–Ar low energy collisions: The values
$a_s = -1.459a_0$; $\quad D = 68.93a_0^3$
$a_s^{(1)} = 8.69a_0^3$; $\quad F = -97a_0^4$
provide an accurate fit to recent measurements [6] of (43.74a) for the diffusion cross section σ_d.

(3) Van der Waals potential: $V(R) = -C/R^6$

$$k \cot \eta_0 = -\frac{1}{a_s} + \frac{1}{2}R_e k^2 - \frac{\pi}{15a_s^2}\left[\frac{2MC}{\hbar^2}\right]k^3$$

$$- \frac{4}{15a_s}\left[\frac{2MC}{\hbar^2}\right]k^4 \ln(ka_0) + \mathcal{O}(k^4) \quad (43.118)$$

e–Atom Collisions with Polarization Attraction. As $k \to 0$, the differential cross section is

$$\frac{d\sigma}{d\Omega} = a_s^2\left[1 + \frac{C_4}{a_s}k\sin\frac{\theta}{2} + \frac{8}{3}C_4 k^2 \ln(ka_0) + \cdots\right] \quad (43.119)$$

and the elastic and diffusion cross sections are

$$\sigma(k \to 0) = 4\pi a_s^2\left[1 + \frac{2\pi C_4 k}{3a_s} + \frac{8}{3}C_4 k^2 \ln(ka_0) + \cdots\right] \quad (43.120)$$

$$\sigma_d(k \to 0) = 4\pi a_s^2\left[1 + \frac{4\pi C_4 k}{5a_s} + \frac{8}{3}C_4 k^2 \ln(ka_0) + \cdots\right] \quad (43.121)$$

For e^-–noble gas collisions, the scattering lengths are

	He	Ne	Ar	Kr	Xe
a_s (a_0)	1.19	0.24	-1.459	-3.7	-6.5

For atoms with $a_s < 0$, a *Ramsauer-Townsend* minimum appears in both σ and σ_d at low energies, provided that scattering from higher partial waves can be neglected, because from Eq. (43.114), $\eta_0 \simeq 0$ at $k = -3a_s/\pi C_4$.

Semiclassical Scattering Lengths. For heavy particle collisions, $\eta_0^{SC}(E \to 0, b)$ tends to

$$\eta_0^{SC} = \left(\frac{2M}{\hbar^2}\right)^{1/2}\int_{R_0}^{\infty}|V(R)|^{1/2}dR. \quad (43.122)$$

(a) Hard-core + well:

$$V(R) = \begin{cases} \infty, & R < R_0 \\ -V_0, & R_0 \leq R < R_1 \\ 0, & R_1 < R \end{cases}$$

$$a_s = \left[1 - \tan\eta_0^{SC}/(kR_1)\right]R_1 \quad (43.123)$$

$$\eta_0^{SC} = k(R_1 + R_0), \quad k^2 = 2MV_0/\hbar^2. \quad (43.124)$$

The phase-averaged scattering length is $\langle a_s \rangle = R_1$.

(b) Hard-core + power-law ($n > 2$):

$$V(R) = \begin{cases} \infty, & R < R_0 \\ \pm C/R^n, & R > R_0 \end{cases} \quad (43.125)$$

Repulsion (+): with $\gamma^2 = 2MC/\hbar^2$,

$$a_{\rm s}^{(+)} = \left(\frac{\gamma}{n-2}\right)^{2/(n-2)} \Gamma\left(\frac{n-3}{n-2}\right) \Big/ \Gamma\left(\frac{n-1}{n-2}\right) . \tag{43.126}$$

Attraction ($-$): with $\theta_n = \pi/(n-2)$,

$$a_{\rm s}^{(-)} = a_{\rm s}^{(+)}\left[1 - \tan\theta_n \tan\left(\eta_0^{\rm SC} - \tfrac{1}{2}\theta_n\right)\right] \cos\theta_n , \tag{43.127}$$

$$\eta_0^{\rm SC} = \gamma \int_{R_0}^{\infty} R^{-n/2}\, dR = \frac{2\gamma R_0^{1-n/2}}{n-2} , \tag{43.128}$$

$$\left\langle a_{\rm s}^{(-)} \right\rangle = a_{\rm s}^{(+)} \cos\theta_n . \tag{43.129}$$

Number of bound states:

$$N_b = {\rm int}\left\{\frac{1}{\pi}\left[\eta_0^{\rm SC} - \frac{1}{2}(n-1)\theta_n\right]\right\} + 1 , \tag{43.130}$$

where ${\rm int}[x]$ denotes the largest integer of the real argument x. For integer x, $a_{\rm s}^{(-)}$ is infinite and a new bound state appears at zero energy.

43.2.5 Logarithmic Derivatives

Phase shift calculations can be based on the logarithmic derivative at $R = a$ separating internal and external regions. Two equivalent forms using sets (R_ℓ, j_ℓ, n_ℓ) or (v_ℓ, F_ℓ, G_ℓ) when $R_\ell = v_\ell/kR$ are

$$K_\ell(k) = \frac{kj'_\ell(ka) - \gamma_\ell(k)j_\ell(ka)}{kn'_\ell(ka) - \gamma_\ell(k)n_\ell(ka)} = \tan\eta_\ell , \tag{43.131a}$$

$$= -\frac{kF'_\ell(ka) - L_\ell(k)F_\ell(ka)}{kG'_\ell(ka) - L_\ell(k)G_\ell(ka)} , \tag{43.131b}$$

where the logarithmic derivative of the internal solution at $R = a$ appropriate to either set, is

$$\gamma_\ell = \left[R_\ell^{-1}\, dR_\ell/dR\right]_{R=a} , \tag{43.132}$$

or alternatively, $L_\ell = \left[v_\ell^{-1}\, dv_\ell/dR\right]_{R=a}$. The primes denote differentiation with respect to the argument, i.e.

$$B'_\ell(ka) = \left[\frac{dB_\ell(x)}{dx}\right]_{x=ka} = \frac{1}{k}\left[\frac{dB_\ell(kR)}{dR}\right]_{R=a} , \tag{43.133}$$

where B_ℓ denotes the functions F_ℓ, G_ℓ, j_ℓ, and n_ℓ.

Decomposition of the S-Matrix Element.

$$S_\ell(k) = e^{2i\eta_\ell} = \left[\frac{\gamma_\ell - (r_\ell - is_\ell)}{\gamma_\ell - (r_\ell + is_\ell)}\right] e^{2i\eta_\ell^H} , \tag{43.134}$$

where

$$\eta_\ell^H(k) = -\frac{j_\ell(ka) - in_\ell(ka)}{j_\ell(ka) + in_\ell(ka)} , \tag{43.135}$$

$$r_\ell + is_\ell = k\left[\frac{j'_\ell(ka) + in'_\ell(ka)}{j_\ell(ka) + in_\ell(ka)}\right] . \tag{43.136}$$

Decomposition of the Phase Shift.

$$\eta_\ell = \eta_\ell^H + \delta_\ell , \tag{43.137}$$

where η_ℓ^H is determined by (43.135), and where δ_ℓ is determined by

$$\tan\delta_\ell = \frac{s_\ell}{\gamma_\ell - r_\ell} , \tag{43.138}$$

which depends on the shape of U via the logarithmic derivative γ_ℓ of Eq. (43.132), and can vary rapidly with k, thereby giving rise to resonances.

Examples:

(1) Hard sphere: if $V(R) = \infty$ for $R < a$, and $V(R) = 0$ for $R > a$, then $\gamma_\ell = \infty$, and

$$K_\ell^{\rm (HS)} = \tan\eta_\ell^{\rm (HS)}(k) = \frac{j_\ell(ka)}{n_\ell(ka)} \tag{43.139}$$

$$\rightarrow \begin{cases} -(ka)^{2\ell+1}/\left[(2\ell+1)!!(2\ell-1)!!\right], & ka \ll 1 \\ -\tan(ka - \tfrac{1}{2}\ell\pi), & ka \gg 1 \end{cases} \tag{43.140}$$

$$S_\ell^{\rm (HS)} = \exp\left[2i\eta_\ell^{\rm (HS)}\right] = -\frac{j_\ell(ka) - in_\ell(ka)}{j_\ell(ka) + in_\ell(ka)}$$

$$= 1 + 2iT_\ell^{\rm (HS)} . \tag{43.141}$$

The phase shift η_ℓ^H in the decomposition (43.137) is therefore identified as $\eta_\ell^{\rm (HS)}$ for hard sphere scattering.

$$\sigma(E \to 0) = 4\pi a^2 . \tag{43.142}$$

Diffraction pattern: As $E \to \infty$,

$$\frac{d\sigma}{d\Omega} \to \tfrac{1}{4}a^2\left[1 + \cot^2(\tfrac{1}{2}\theta)\, J_1^2(ka\sin\theta)\right] , \tag{43.143}$$

$$\sigma(E) \to 2\pi a^2 . \tag{43.144}$$

Classical hard sphere scattering and diffraction about the sharp edge each contribute πa^2 to σ.

(2) Spherical Well: if $U(R) = -U_0$ for $R \le a$, and $U(R) = 0$ for $R > a$, then

$$\gamma_\ell(k) = \kappa\frac{j'_\ell(\kappa a)}{j_\ell(\kappa a)}; \qquad \kappa^2 = U_0 + k^2 \equiv k_0^2 + k^2 . \tag{43.145}$$

S-wave ($\ell = 0$) properties:

$$\eta_0(k) = -ka + \tan^{-1}\left[(k/\kappa)\tan\kappa a\right] . \tag{43.146}$$

As $k \to 0$, $\sigma_0(E) \to 4\pi a_{\rm s}^2$, where the scattering length is

$$a_{\rm s} = \left[1 - \tan(k_0 a)/(k_0 a)\right] a . \tag{43.147}$$

For a shallow well $k_0 a \ll 1$: $\sigma_0(E) = (4\pi/9)U_0^2 a^6$, which agrees with the Born result (43.169) as $k \to 0$.

The condition for $\ell = 0$ bound state with energy $E_n = -(\hbar^2 k_n^2/2M)$ is

$$k_n \tan \kappa' a = -\kappa', \quad \kappa'^2 = k_0^2 - k_n^2. \qquad (43.148)$$

As the well is further deepened, $\sigma_0(E)$ oscillates between zero, where $\tan k_0 a = k_0 a$, and ∞, where $k_0 a = n\pi/2$, the condition both for appearance of a new level n at energy E and for $a_0 \to \infty$. In the neighborhood of these infinite resonances,

$$\sigma_0(E) = \frac{4\pi}{k^2 + \kappa'^2}, \qquad (43.149)$$

where $\kappa' = \kappa/\tan \kappa a$.

43.2.6 Coulomb Scattering

Direct solution of RSE (43.91) yields

$$v_\ell \sim \sin(kR - \tfrac{1}{2}\ell\pi + \eta_\ell^{(C)} - \beta \ln 2kR), \qquad (43.150)$$

$$\beta = Z_1 Z_2 e^2/\hbar v = Z_1 Z_2/(ka_0). \qquad (43.151)$$

Coulomb phase shift:

$$\eta_\ell^{(C)} = \arg\Gamma(\ell + 1 + i\beta) = \Im \ln\Gamma(\ell + 1 + i\beta). \qquad (43.152)$$

Coulomb S-matrix element:

$$S_\ell^{(C)} = \exp\left[2i\eta_\ell^{(C)}\right] = \frac{\Gamma(\ell + 1 + i\beta)}{\Gamma(\ell + 1 - i\beta)}. \qquad (43.153)$$

Coulomb scattering amplitude:

$$f_C(\theta) = -\frac{\beta \exp\left[2i\eta_\ell^{(C)} - i\beta \ln\left(\sin^2 \tfrac{1}{2}\theta\right)\right]}{2k \sin^2 \tfrac{1}{2}\theta}. \qquad (43.154)$$

Coulomb differential cross section:

$$\frac{d\sigma}{d\Omega} = \frac{\beta^2}{4k^2 \sin^4 \tfrac{1}{2}\theta} = \frac{Z_1^2 Z_2^2 e^4}{16E^2} \csc^4 \tfrac{1}{2}\theta, \qquad (43.155)$$

which is the Rutherford scattering cross section.

Mott Formula. For the Coulomb scattering of two identical particles: From (43.83a) and (43.83c)
(a) spin-zero bosons (e.g. ^{4}He – ^{4}He):

$$\frac{d\sigma}{d\Omega} = \frac{\beta^2}{4k^2}\left[\csc^4 \tfrac{1}{2}\theta + \sec^4 \tfrac{1}{2}\theta + 2\csc^2 \tfrac{1}{2}\theta \sec^2 \tfrac{1}{2}\theta \cos\Gamma\right], \qquad (43.156)$$

(b) spin-$\tfrac{1}{2}$ fermions (e.g. H^+ – H^+, $e^\pm$ – $e^\pm$)

$$\frac{d\sigma}{d\Omega} = \frac{\beta^2}{4k^2}\left[\csc^4 \tfrac{1}{2}\theta + \sec^4 \tfrac{1}{2}\theta - \csc^2 \tfrac{1}{2}\theta \sec^2 \tfrac{1}{2}\theta \cos\Gamma\right], \qquad (43.157)$$

where $\Gamma = 2\beta \ln(\tan \tfrac{1}{2}\theta)$.

43.2.7 Resonance Scattering

Zero-Energy Broad Resonances. The spherical well example (43.145) serves to illustrate broad resonances. When the well depth U_0 is strong enough to accomodate the $(n_0 + 1)$th energy level at zero energy, the bound state condition (43.148) implies that $\eta_0(k \to 0) = (n_0 + 1)\pi$, illustrating Levinson's theorem (43.103). As k increases, η_0 generally decreases through either $(2n-1)\pi/2$, or $(n-1)\pi$, where σ_0 has, respectively, a maximum value $4\pi/k^2$ and a minimum value of zero. If the phase shifts η_ℓ for $\ell > 0$ are small, then a nonzero minimum value in $\sigma(E)$ is evident. This is the *Ramsauer-Townsend minimum* manifest when the potential is just strong enough to introduce one or more wavelengths into the well with no observable scattering. Since the rate of decrease of η_0 cannot be arbitrarily rapid, [*cf.* Eq. (43.103)], broad resonances will be exhibited in contrast to narrow (Breit-Wigner) resonances when η_ℓ increases *rapidly* through $(2n-1)\pi/2$ over a small energy range ΔE.

Narrow Resonances. The general decomposition (43.137) can be used to analyze narrow resonances. When γ_ℓ varies rapidly within an energy width Γ about a resonance energy E_r then δ_ℓ increases through odd multiples of $\pi/2$ and

$$\delta_\ell = \delta_\ell^r = \tan^{-1}\frac{\Gamma}{2(E_r - E)}, \qquad (43.158)$$

so that (43.60a) with (43.59a) and (43.59b) is

$$f_\ell = \frac{(2\ell + 1)}{k}\left[T_\ell^{(HS)} + S_\ell^{(HS)}\frac{\Gamma/2}{E_r - E - \tfrac{i}{2}\Gamma}\right]P_\ell(\cos\theta). \qquad (43.159)$$

Breit-Wigner Formula. For a pure resonance with no background phase shift, $S_\ell^{(HS)} = 1$ and the cross section has the Lorentz shape

$$\sigma(E) = \frac{4\pi(2\ell + 1)}{k^2}\left[\frac{\Gamma^2/4}{(E - E_r)^2 + \Gamma^2/4}\right]. \qquad (43.160)$$

Shape Resonances. (Also called quasibound state or tunneling resonances). At very low impact energies near or below the energy threshold for orbiting, sharp spikes superimposed on the glory oscillations may be evident in the E-dependance of $\sigma(E)$. These are due to quasibound states with positive energy $E_{n\ell}$ supported by the effective potential $V(r) + L^2/(2Mr^2)$. In heavy particle collisions, quasibound levels are the continuation of the bound rotational levels to positive energies $E_{n\ell} > 0$.

Systems in quasibound states $(n\ell)$, with nonresonant eigenenergy $E_{n\ell}$ and phase shift $\eta_\ell^{(0)}$, have

$$E = E_{n\ell} - \frac{i}{2}\Gamma_{n\ell}, \qquad (43.161\text{a})$$

$$\eta_\ell = \eta_\ell^{(0)} + \eta_\ell^{(\text{r})}, \qquad (43.161\text{b})$$

$$\eta_\ell = \eta_\ell^{(0)} + \tan^{-1}\frac{\Gamma_{n\ell}}{2(E_{n\ell} - E)} ; \qquad (43.161\text{c})$$

$$S_\ell(E) = e^{2i\eta_\ell} = \left[\frac{E - E_{n\ell} - \frac{i}{2}\Gamma_{n\ell}}{E - E_{n\ell} + \frac{i}{2}\Gamma_{n\ell}}\right] e^{2i\eta_\ell^{(0)}} . \quad (43.162)$$

The phase shift $\eta_\ell^{(\text{r})}$ increases by π as E increases through $E_{n\ell}$ at a rate determined by $\Gamma_{n\ell}$. The dominant amplitude shifts from the external to the quasibound internal regions at $E = E_{n\ell}$.

Partial-Wave Scattering Amplitude.

$$f_\ell(\theta) = \frac{1}{2ik}(2\ell + 1)\exp(2i\eta_\ell^0)P_\ell(\cos\theta)$$

$$\times \left[1 - \frac{i\Gamma_{n\ell}}{(E - E_{n\ell} + \frac{i}{2}\Gamma_{n\ell})}\right] ; \qquad (43.163)$$

$$= \text{background potential scattering amplitude}$$
$$+ \text{resonance scattering amplitude.}$$

The partial-wave cross section is composed of the following: potential resonance and interference contributions to $\sigma_\ell = |f_\ell(\theta)|^2$:

$$\sigma_\ell = \frac{4\pi}{k^2}(2\ell + 1)\left[\sin^2\eta_\ell^{(0)}\right.$$

$$\left. + \frac{\Gamma_{n\ell}^2\cos 2\eta_\ell^{(0)} + 2\Gamma_{n\ell}(E_{n\ell} - E)\sin 2\eta_\ell^{(0)}}{4(E - E_{n\ell})^2 + \Gamma_{n\ell}^2}\right]$$

$$\qquad (43.164\text{a})$$

$$= \frac{4\pi}{k^2}(2\ell + 1)\left\{\sin^2\eta_\ell^{(0)}\left[\frac{(E - E_{n\ell})^2}{(E - E_{n\ell})^2 + (\Gamma_{n\ell}/2)^2}\right]\right.$$

$$+ \cos^2\eta_\ell^{(0)}\left[\frac{(\Gamma_{n\ell}/2)^2}{(E - E_{n\ell})^2 + (\Gamma_{n\ell}/2)^2}\right]$$

$$\left. + \sin 2\eta_\ell^{(0)}\left[\frac{(\Gamma_{n\ell}/2)(E_{n\ell} - E)}{(E - E_{n\ell})^2 + (\Gamma_{n\ell}/2)^2}\right]\right\} \quad (43.164\text{b})$$

$$\sigma(E) = \sum_{\ell=0}^{\infty}\sigma_\ell = \sigma_0(E) + \sigma_{\text{res}}(E) \qquad (43.164\text{c})$$

Resonance Shapes. Resonance shapes depend on the value of the background phase shift $\eta_\ell^{(0)}$. The case $\eta_\ell^{(0)} = 0$ gives a Lorentz line shape through the Breit-Wigner formula

$$\sigma_\ell = \frac{4\pi(2\ell + 1)}{k^2}\frac{\Gamma_{n\ell}^2/4}{(E - E_{n\ell})^2 + \Gamma_{n\ell}^2/4} . \qquad (43.165)$$

The other cases from Eq. (43.164b) are

$\eta_\ell^{(0)} = n\pi$	large positive spike;
$\eta_\ell^{(0)} = (n + \frac{1}{2})\pi$	large negative spike;
$n\pi < \eta_\ell^{(0)} < (n + \frac{1}{2})\pi$	positive then negative;
$(n + \frac{1}{2})\pi < \eta_\ell^{(0)} < (n + 1)\pi$	negative then positive.

$$(43.166)$$

Time Delay.

$$\tau = 2\hbar\left(\frac{\partial\eta_\ell}{\partial E}\right)_\ell = 2\hbar\left(\frac{\partial\eta_\ell^{(0)}}{\partial E}\right)_\ell + \frac{4\hbar}{\Gamma_{n\ell}} . \qquad (43.167)$$

The time for capture into quasibound levels is $\tau_{\text{c}} = 4\hbar/\Gamma_{n\ell}$, and the capture frequency is $\nu_{\text{c}} = \Gamma_{n\ell}/4\hbar$.

43.2.8 Integral Equation for Phase Shift

$$\sin\eta_\ell = -\frac{1}{k}\int_0^\infty F_\ell(kR)U(R)v_\ell^{(A)}(kR)\,dR, \quad (43.168\text{a})$$

$$K_\ell = \tan\eta_\ell = -\frac{1}{k}\int_0^\infty F_\ell(kR)U(R)v_\ell^{(B)}(kR)\,dR,$$

$$(43.168\text{b})$$

$$T_\ell = e^{i\eta_\ell}\sin\eta_\ell = -\frac{1}{k}\int_0^\infty F_\ell(kR)U(R)v_\ell^{(C)}(kR)\,dR,$$

$$(43.168\text{c})$$

$$S_\ell - 1 = \exp(2i\eta_\ell) - 1$$

$$= -\frac{1}{k}\int_0^\infty F_\ell(kR)U(R)v_\ell^{(D)}(kR)\,dR, \quad (43.168\text{d})$$

where $v_\ell^{(A)}, v_\ell^{(B)}, v_\ell^{(C)}, v_\ell^{(D)}$ are so chosen to have asymptotes prescribed by Eqs. (43.97b)–(43.100b), respectively.

Born Approximation for Phase Shifts. Set $v_\ell = F_\ell$ in Eqs. (43.168a)–(43.168d) to obtain

$$\tan\eta_\ell^B(k) = -k\int_0^\infty U(R)\left[j_\ell(kR)\right]^2 R^2\,dR. \quad (43.169)$$

For $\lambda = (\ell + 1/2) \gg ka$, substitute

$$\left\langle k^2R^2\left[j_\ell(kR)\right]^2\right\rangle = \frac{1}{2}\left[1 - \lambda^2/k^2R^2\right]^{-1/2} . \quad (43.170)$$

For the Jeffrey-Born (JB) phase shift function, $\ell \gg ka$,

$$\tan\eta_{\text{JB}}(\lambda) = -\frac{k}{2E}\int_{\lambda/k}^\infty\frac{V(R)dR}{[1 - \lambda^2/(kR)^2]^{1/2}} , \quad (43.171)$$

which agrees with (43.41) since $bk = \ell + \frac{1}{2} = \lambda$. For linear trajectories $R^2 = b^2 + Z^2$, the eikonal phase (43.42) is recovered.

Born S-Wave Phase Shift.

$$\tan\eta_0^B(k) = -\frac{1}{k}\int_0^\infty U(R)\sin^2(kR)dR \qquad (43.172)$$

Examples : (i) $U = U_0\dfrac{e^{-\alpha R}}{R}$, (ii) $U = \dfrac{U_0}{(R^2 + R_0^2)^2}$;

(i) $\tan \eta_0^{\mathrm{B}} = -\dfrac{U_0}{4k} \ln \left[1 + 4k^2/\alpha^2\right]$, $\qquad$ (43.173)

(ii) $\tan \eta_0^{\mathrm{B}} = -\dfrac{\pi U_0}{4k R_0^3} \left[1 - (1 + 2kR_0)e^{-2kR_0}\right]$. $\quad$ (43.174)

Born Phase Shifts (Large ℓ). For $\ell \gg ka$,

$$\tan \eta_\ell^{\mathrm{B}} = -\frac{k^{2\ell+1}}{[(2\ell+1)!!]^2} \int_0^\infty U(R) R^{2\ell+2} dR, \quad (43.175)$$

valid only for finite range interactions $U(R > a) = 0$. Example: $U = -U_0$, $R \le a$ and $U = 0$, $R > a$.

$$\tan \eta_\ell^{\mathrm{B}}(\ell \gg ka) = U_0 a^2 \frac{(ka)^{2\ell+1}}{[(2\ell+1)!!]^2 (2\ell+3)}. \quad (43.176)$$

For $\ell \gg ka$, $\eta_{\ell+1}/\eta_\ell \sim (ka/2\ell)^2$.

43.2.9 Variable Phase Method

The phase function $\eta_\ell(R)$ is defined to be the scattering phase shift produced by the part of the potential $V(R)$ contained within a sphere of radius R. It satisfies the nonlinear differential equation for $\eta_\ell(R)$

$$\frac{d\eta_\ell}{dR} = -kR^2 U(R) \left[\cos \eta_\ell(R) j_\ell(kR) - \sin \eta_\ell(R) n_\ell(kR)\right]^2.$$
$$(43.177a)$$

The corresponding integral equation for $\eta_\ell(R)$ is

$$\eta_\ell(R) = -k \int_0^R \left[\cos \eta_\ell(R) j_\ell(kR)\right.$$
$$\left. - \sin \eta_\ell(R) n_\ell(kR)\right]^2 U(R) R^2 dR. \quad (43.177b)$$

The Born approximation (43.169) is recovered by substituting $\eta_\ell = 0$ on the RHS of Eq. (43.177b) as $R \to \infty$.

43.2.10 General Amplitudes

For a general potential $V(\mathbf{R})$, define the reduced potential $U(\mathbf{R}) = (2M/\hbar^2)V(\mathbf{R})$. The plane wave scattering states are

$$\phi_{\mathbf{k}}(\mathbf{R}) = \exp(i\mathbf{k}\cdot\mathbf{R}) = \phi_{-\mathbf{k}}^*(\mathbf{R}), \quad (43.178)$$

and the full scattering solutions have the form

$$\Psi_{\mathbf{k}}^{(\pm)}(\mathbf{R}) \sim \phi_{\mathbf{k}}(\mathbf{R}) + \frac{f(\mathbf{k}, \mathbf{k}')}{R} \exp(\pm ikR), \quad (43.179)$$

where the scattering amplitude is

$$f(\mathbf{k}, \mathbf{k}') = -\frac{1}{4\pi} \left\langle \phi_{\mathbf{k}'}(\mathbf{R}) \mid U(\mathbf{R}) \mid \Psi_{\mathbf{k}}^{(+)}(\mathbf{R}) \right\rangle \quad (43.180a)$$

$$= -\frac{1}{4\pi} \left\langle \Psi_{\mathbf{k}'}^{(-)}(\mathbf{R}) \mid U(\mathbf{R}) \mid \phi_{\mathbf{k}}(\mathbf{R}) \right\rangle \quad (43.180b)$$

$$\equiv -\frac{1}{4\pi} \left\langle \phi_{\mathbf{k}'}(\mathbf{R}) \mid T \mid \phi_{\mathbf{k}}(\mathbf{R}) \right\rangle. \quad (43.180c)$$

The last equation defines the T-matrix element.
First Born Approximation.
Set $\Psi_{\mathbf{k}}^+ = \phi_{\mathbf{k}}$ in (43.180a). Then

$$f_{\mathrm{B}}(K) = -\frac{1}{4\pi} \int U(\mathbf{R}) \exp(i\mathbf{K}\cdot\mathbf{R}) d\mathbf{R}, \quad (43.181)$$

where the momentum change is $\mathbf{K} = \mathbf{k} - \mathbf{k}'$, and $K = 2k \sin \frac{1}{2}\theta$. For a symmetric potential,

$$f_{\mathrm{B}}(K) = -\int \frac{\sin KR}{KR} U(R) R^2 dR. \quad (43.182)$$

Connection with partial wave analysis:

$$\frac{\sin KR}{KR} = \sum_{\ell=0}^\infty (2\ell+1) [j_\ell(kR)]^2 P_\ell(\cos\theta). \quad (43.183)$$

which is consistent with (43.169).

The static e^-–atom scattering potential and Born scattering amplitude are

$$V(R) = -\frac{Ze^2}{R} + e^2 \int \frac{|\psi(\mathbf{r})|^2 d\mathbf{r}}{|\mathbf{R} - \mathbf{r}|}, \quad (43.184)$$

$$f_{\mathrm{B}}(K) = \frac{2Me^2}{\hbar^2} \frac{[Z - F(K)]}{K^2}. \quad (43.185)$$

where the *elastic form factor* is

$$F(K) = \int |\psi(\mathbf{r})|^2 \exp(i\mathbf{K}\cdot\mathbf{r}) d\mathbf{r}.$$

For a pure Coulomb field, $F(K) = 0$ and $\sigma_{\mathrm{B}}(\theta, E) = |f_{\mathrm{B}}(K)|^2$ reduces to Eq. (43.155).
Two Potential Formulae. For scattering from the combined potential $U(\mathbf{R}) = U_0(\mathbf{R}) + U_1(\mathbf{R})$,

$$f(\mathbf{k}, \mathbf{k}') = -\frac{1}{4\pi} \left[\left\langle \phi_{\mathbf{k}'}(\mathbf{R}) \mid U_0(\mathbf{R}) \mid \chi_{\mathbf{k}}^{(+)}(\mathbf{R}) \right\rangle \right.$$
$$\left. + \left\langle \chi_{\mathbf{k}'}^{(-)}(\mathbf{R}) \mid U_1(\mathbf{R}) \mid \Psi_{\mathbf{k}}^{(+)}(\mathbf{R}) \right\rangle \right], \quad (43.186a)$$

where $\chi_{\mathbf{k}}^{(\pm)}(\mathbf{R})$ and $\Psi_{\mathbf{k}}^{(\pm)}(\mathbf{R})$ are full solutions for scattering by V_0 and $V_0 + V_1$, respectively. For symmetric interactions,

$$f(\theta) = \frac{1}{k} \sum_{\ell=0}^\infty (2\ell+1) \left[T_\ell^{(0)} + T_\ell^{(1)}\right] P_\ell(\cos\theta), \quad (43.187)$$

$$T_\ell^{(0)} = \exp\left[i\eta_\ell^{(0)}\right] \sin \eta_\ell^{(0)} \quad (43.188a)$$

$$= -\frac{1}{k} \int_0^\infty dR \left[F_\ell(R) U_0(R) u_\ell(R)\right], \quad (43.188b)$$

$$T_\ell^{(1)} = \exp\left[2i\eta_\ell^{(0)}\right] \exp\left[i\eta_\ell^{(1)}\right] \sin \eta_\ell^{(1)}, \quad (43.189a)$$

$$T_\ell^{(1)} = -\frac{1}{k} \int_0^\infty dR \left[u_\ell(R) U_1(R) v_\ell(R)\right], \quad (43.189b)$$

where u_ℓ and v_ℓ are the radial wave functions in (43.91), with phase-shifts $\eta_\ell^{(0)}$ and $\eta_\ell = \eta_\ell^{(0)} + \eta_\ell^{(1)}$, for scattering by V_0 and $V_0 + V_1$, respectively, normalized according to (43.99b).

Distorted-Wave Approximation.

$$\Psi_{\mathbf{k}}^{(+)}(\mathbf{R}) \sim \chi_{\mathbf{k}}^{(+)}(\mathbf{R})$$

$$f(\mathbf{k}, \mathbf{k}') = -\frac{1}{4\pi} \left[\left\langle \phi_{\mathbf{k}'}(\mathbf{R}) \mid U_0(\mathbf{R}) \mid \chi_{\mathbf{k}}^{(+)}(\mathbf{R}) \right\rangle \right.$$
$$\left. + \left\langle \chi_{\mathbf{k}'}^{(-)}(\mathbf{R}) \mid U_1(\mathbf{R}) \mid \chi_{\mathbf{k}}^{(+)}(\mathbf{R}) \right\rangle \right]. \quad (43.190)$$

43.3 SEMICLASSICAL SCATTERING FORMULAE

43.3.1 Scattering Amplitude: Exact Poisson Sum Representation

Poisson Sum Formula: With $\lambda = \ell + \frac{1}{2}$,

$$\sum_{\ell=0}^{\infty} F\left(\ell + \tfrac{1}{2}\right) = \sum_{m=-\infty}^{\infty} (-1)^m \int_0^\infty F(\lambda) e^{i2m\pi\lambda} d\lambda. \quad (43.191)$$

When applied to (43.60b),

$$f(\theta) = (ik)^{-1} \sum_{m=-\infty}^{\infty} (-1)^m \int_0^\infty \lambda \left[e^{2i\eta(\lambda)} - 1 \right]$$
$$\times P_{\lambda-\frac{1}{2}}(\cos\theta) e^{i2m\pi\lambda} \, d\lambda, \quad (43.192)$$

where $\eta(\lambda)$ and $P_{\lambda-1/2} \equiv P(\lambda, \theta)$ are now phase functions and Legendre functions of the *continuous* variable λ, being interpolated from discrete to continuous ℓ. This *infinite-sum-of-integrals* representation for $f(\theta)$ is in principle exact. It is the appropriate technique for conversion from a sum over (quantal) discrete values of a variable to a continuous integration over that variable which classically can assume any value. The particular merit here is that the index m labels the classical paths that have encircled the (attractive) scattering center m times, and that the terms with $m < 0$ have no regions of stationary phase (SP). For deflections χ in the range $-\pi < \chi < \pi$, the only SP contribution is the $m = 0$ term.

43.3.2 Semiclassical Procedure

Semiclassical analysis [7–9] involves reducing (43.192) by the three approximations represented by cases A to C below. Since the integrands can oscillate very rapidly over large regions of λ, the main contributions to the integrals arise from points λ_i of stationary phase of each integrand. The amplitude can then be evaluated by the *method of stationary phase*, the basis of semiclassical analysis.

A. Legendre Function Asymptotic Expansions

Main range: $\sin\theta > \lambda^{-1}$, θ not within λ^{-1} of zero or π.

$$P_\ell(\cos\theta) = [2/(\pi\lambda\sin\theta)]^{1/2} \cos[\lambda\theta - \pi/4]. \quad (43.193)$$

Forward formula: θ within λ^{-1} of zero.

$$P_\ell(\cos\theta) = [\theta/\sin\theta]^{1/2} J_0(\lambda\theta), \quad (43.194a)$$

$$J_0(\lambda\theta) = \frac{1}{\pi} \int_0^\pi e^{-i\lambda\theta\cos\phi} d\phi. \quad (43.194b)$$

Backward formula: θ within λ^{-1} of π.

$$P_\ell(\cos\theta) = \left[\frac{\pi - \theta}{\sin\theta} \right]^{1/2} J_0\left[\lambda(\pi - \theta) \right] e^{-i\pi(\lambda-1/2)}. \quad (43.195)$$

Equations (43.193)–(43.195) are useful for analysis of caustics (rainbows), diffraction and forward and backward glories, respectively. Also, a useful identity is

$$\sum_{\ell=0}^{\infty} (2\ell + 1) P_\ell(\cos\theta) = \begin{cases} 4\delta(1 - \cos\theta), & \theta > 0 \\ 0 & \theta = 0 \end{cases}$$

where $\delta(x)$ is the Dirac delta function.

B. JWKB Phase Shift Functions

$$\eta(\lambda) = \int_{R_c}^\infty k_\lambda(R) \, dR - \int_{\lambda/k}^\infty \left[k^2 - \frac{\lambda^2}{R^2} \right]^{1/2} dR \quad (43.196a)$$

$$= \lim_{R\to\infty} \left[\int_{R_c}^\infty k_\ell(R') \, dR' - kR \right] + \frac{1}{2}\lambda\pi. \quad (43.196b)$$

Local wave number:

$$k_\lambda^2(R) = k^2 - U(R) - \lambda^2/R^2. \quad (43.197)$$

The *Langer modification*:

$$b = \frac{\sqrt{\ell(\ell+1)}}{k} = \frac{\ell + 1/2}{k} = \frac{\lambda}{k}. \quad (43.198)$$

Useful identity: As $R \to \infty$,

$$\sin\left\{ \int_{\lambda/k}^R \left[k^2 - \lambda^2/R^2 \right]^{1/2} dR + \frac{\pi}{4} \right\}$$
$$\to \sin\left(kR - \tfrac{1}{2}\ell\pi \right) \quad (43.199)$$

JWKB phase functions are valid when variation of the potential over the local wavenumber $k_\lambda^{-1}(R)$ is a small fraction of the available kinetic energy $E - V(R)$. Many wavelengths can then be accomodated within a range ΔR for a characteristic potential change ΔV. The classical method is valid when $(1/k)(dV/dR) \ll (E - V)$.

Phase–Deflection Function Relation.

$$\chi(\lambda) = 2 \frac{\partial \eta(\lambda)}{\partial \lambda} \qquad (43.200)$$

C. Stationary Phase Approximations (SPA) to Generic Integrals

$$A^{\pm}(\theta) = \int_{-\infty}^{\infty} g(\theta; \lambda) \exp\left[\pm i\gamma(\theta; \lambda)\right] \, d\lambda \qquad (43.201)$$

for parametric θ. In cases where the phase function γ has two stationary points, a phase minimum γ_1 at λ_1 and a phase maximum γ_2 at λ_2, then $\gamma_i' = 0$, $\gamma_1'' > 0$, $\gamma_2'' < 0$ where $\gamma_i' = (d\gamma/d\lambda)$ at λ_i and $\gamma_i'' = (d^2\gamma/d\lambda^2)$ for $i = 1, 2$. Since g is real, $A^- = (A^+)^*$, $g_i(\theta) = g(\theta, \lambda_i)$.

Uniform Airy Result.

$$A^+(\theta) = a_1(\theta)e^{i(\gamma_1 + \pi/4)} F^*(\gamma_{21})$$
$$+ a_2(\theta)e^{i(\gamma_2 - \pi/4)} F(\gamma_{21}) \, , \qquad (43.202)$$

$$a_i(\theta) = [2\pi / |\gamma_i''|]^{1/2} g_i(\theta) \, , \quad i = 1, 2 \, , \qquad (43.203)$$

$$\gamma_{21}(\theta) = \gamma_2 - \gamma_1$$
$$\equiv \tfrac{4}{3} |z(\theta)|^{3/2} > 0 \, , \qquad (43.204)$$

$$F[\gamma_{21}(\theta)] = \left[z^{1/4} \text{Ai}(-z) + iz^{-1/4} \text{Ai}'(-z) \right] \sqrt{\pi}$$
$$\times e^{-i(\gamma_{21}/2 - \pi/4)} \, , \qquad (43.205)$$

where Ai and Ai′ are the Airy function and its derivative.

This result holds for all separations $(\lambda_2 - \lambda_1)$ in location of stationary phases including a caustic (or rainbow), which is a point of inflection in γ, i.e. $\gamma_1 = \gamma_2$, $\gamma_i' = 0 = \gamma_i''$. An equivalent expression is [9]

$$A^+(\theta) = \left[(a_1 + a_2) z^{1/4} \text{Ai}(-z) \right.$$
$$\left. - i(a_1 - a_2) z^{-1/4} \text{Ai}'(-z) \right] \sqrt{\pi} e^{i\bar{\gamma}} \, , \qquad (43.206)$$

where the mean phase is $\bar{\gamma} = \tfrac{1}{2}(\gamma_1 + \gamma_2)$. The first form (43.202) is useful for analysis of widely separated regions of stationary phase when $\gamma_{21} \gg 0$ and $F \to 1$. The equivalent second form (43.206) is valuable in the neighborhood of caustics or rainbows when the stationary phase regions coalesce as $a_1 \to a_2$.

Primitive Result. For widely separated regions λ_1 and λ_2, $F \to 1$ and

$$A^{\pm}(\epsilon) = \left[a_1(\epsilon) \mp i a_2(\epsilon)e^{\pm i\gamma_{21}} \right] e^{\pm i(\gamma_1 + \pi/4)} \, , \qquad (43.207a)$$

$$A^{\pm}(\epsilon) = a_1(\epsilon) \exp\left[\pm i\left(\gamma_1 + \frac{\pi}{4}\right) \right]$$
$$+ a_2(\epsilon) \exp\left[\pm i\left(\gamma_2 - \frac{\pi}{4}\right) \right] \, . \qquad (43.207b)$$

Note that the minimum phase γ_1 is increased by $\pi/4$ and the maximum phase γ_2 is reduced by $\pi/4$.

Transitional Airy Result. In the neighborhood of a caustic or rainbow where $\gamma'' = 0$, at the inflection point $\lambda_1 = \lambda_2 = \lambda_r$, then

$$A^{\pm}(\theta) = 2\pi \left| \frac{2}{\gamma'''(\lambda_r)} \right|^{1/3} g(\theta; \lambda_r) \text{Ai}(-z) e^{\pm i\gamma(\theta; \lambda_r)} \quad (43.208)$$

$$z = \left| \frac{2}{\gamma'''(\lambda_r)} \right|^{1/3} \gamma'(\theta; \lambda_r) \, . \qquad (43.209)$$

Only over a very small angular range does this result agree in practice with the uniform result (43.202), which uniformly connects Eqs. (43.207a) and (43.208). These stationary-phase formulae are not only applicable to integrals involving (λ, θ) but also to (t, E) and (R, p) combinations which occur in the Method of Variation of Constants and in Franck-Condon overlaps of vibrational wave functions, respectively.

43.3.3 Semiclassical Amplitudes: Integral Representation

A. Off-Axis Scattering: $\sin\theta > \lambda^{-1}$.

Except in the forward and backward directions, Eq. (43.192) with (43.193) reduces to

$$f(\theta) = -\frac{1}{k(2\pi \sin\theta)^{1/2}} \sum_{m=-\infty}^{\infty} (-1)^m \int_0^{\infty} d\lambda \, \lambda^{1/2}$$
$$\times \left[e^{i\Delta^+(\lambda, m)} - e^{i\Delta^-(\lambda; m)} \right] \, , \qquad (43.210)$$

$$\Delta^{\pm}(\lambda; m) = 2\eta(\lambda) + 2m\pi\lambda \pm \lambda\theta \pm \pi/4 \qquad (43.211)$$
$$\equiv S^C \pm \pi/4 \, , \qquad (43.212)$$

where S^C is the classical action (43.52b) divided by $\hbar$.

The stationary phase condition $d\Delta^{\pm}/d\lambda = 0$ yields the deflection function χ to scattering angle θ relation

$$\chi(\lambda_i) = \mp\theta - 2m\pi \, , \qquad (43.213)$$

where λ_i are points of stationary phase (SP). Since $\pi \geq \chi \geq -\infty$, integrals with $m < 0$ have no SP's and vanish under the SPA. For cases involving no orbiting (where $\chi \to -\infty$) and when $\pi > \chi > -\pi$, then integrals with $m > 0$ also vanish under SPA so that the only remaining contribution from $m = 0$ to (43.212) is

$$f(\theta) = -\frac{1}{k(2\pi \sin\theta)^{1/2}} \int_0^{\infty} \lambda^{1/2} \left[e^{i\Delta^+(\lambda)} - e^{i\Delta^-(\lambda)} \right] d\lambda \, , \qquad (43.214)$$

$$\Delta^{\pm}(\lambda) = 2\eta(\lambda) \pm \lambda\theta \pm \pi/4 \, . \qquad (43.215)$$

The attractive branch Δ^+ contributes only negative deflections and the repulsive branch Δ^- contributes only positive deflections and has one SP point at λ_1 where Δ^- is maximum.

Rainbow angle θ_r: $(\Delta^+)''_{\lambda_r} = 0$, so that $\chi'(\lambda_r) = 0$ where $\chi(\lambda_r) < 0$ has reached its most negative value.

$\theta < \theta_r$: Δ^+ has two SP points $\lambda_{2,3}$;
          a maximum at λ_2 and
          a minimum at λ_3.
$\theta = \theta_r$: $\lambda_2 = \lambda_3$: SP's coalesce.
$\theta > \theta_r$:   no classical attractive scattering.
          Δ^+ has no SP points.

B. Forward Amplitude: $\sin\theta \sim \theta < \lambda^{-1}$.

$$f(\theta) = \frac{1}{ik}\left(\frac{\theta}{\sin\theta}\right)^{1/2}\sum_{m=-\infty}^{\infty}e^{-im\pi}\int_0^\infty \lambda J_0(\lambda\theta)$$

$$\times\left[e^{2i\eta(\lambda)}-1\right]e^{2im\pi\lambda}d\lambda\,. \qquad (43.216)$$

Stationary phase points: $\gamma'(\lambda_m) = 0$.

$$\chi(\lambda_m) = 2\left(\frac{\partial\eta}{\partial\lambda}\right) = -2m\pi\,. \qquad (43.217)$$

Terms with $m < 0$ therefore make no SP contribution to $f(\theta)$ since $\chi \leq \pi$. The $m = 0$ term provides diffraction due to $\chi \to 0$, $\chi' \to 0$ at long range, and a forward glory due to $\chi \to 0$ at a finite λ_g and nonzero χ'_g.

C. Backward Amplitude: $\theta \sim \pi - \mathcal{O}(\lambda^{-1})$.

$$f(\theta) = \frac{1}{k}\left(\frac{\pi-\theta}{\sin\theta}\right)^{1/2}\sum_{m=-\infty}^{\infty}e^{im\pi}\int_0^\infty \lambda J_0\left[\lambda(\pi-\theta)\right]$$

$$\times\, e^{i[2\eta(\lambda)+(2m-1)\pi\lambda]}\,d\lambda\,. \qquad (43.218)$$

Stationary phase points:

$$\chi(\lambda_m) = 2\left(\frac{\partial\eta}{\partial\lambda}\right) = -(2m-1)\pi\,. \qquad (43.219)$$

There are no SP for $m < 0$. The $m = 0$ term provides a normal backward amplitude due to repulsive collisions ($\chi = \pi$), and $m > 0$ terms are due to attractive half-windings.

D. Eikonal Amplitude.

The $m = 0$ term of Eq. (43.216) gives

$$f_E(\theta) = \frac{1}{ik}\int_0^\infty \lambda\left[e^{2i\eta(\lambda)}-1\right]J_0(\lambda\theta)\,d\lambda \qquad (43.220a)$$

$$= -ik\int_0^\infty\left[e^{2i\eta(b)}-1\right]J_0(kb\theta)b\,db\,. \qquad (43.220b)$$

From the optical theorem,

$$\sigma_E(E) = 8\pi\int_0^\infty \sin^2\eta(b,E)b\,db \qquad (43.221)$$

For potentials with cylindrical symmetry, $kb\theta$ can be replaced by $2kb\sin\frac{1}{2}\theta = \mathbf{K}\cdot\mathbf{b}$, and

$$f_E(\theta) = -\frac{ik}{2\pi}\int\left[e^{2i\eta(b)}-1\right]J_0(\mathbf{K}\cdot\mathbf{b})\,d\mathbf{b}\,. \qquad (43.222)$$

43.3.4 Semiclassical Amplitudes and Cross Sections

Amplitude addition:

$$f(\theta) = \sum_{j=1}^{N}f_j(\theta)\,, \qquad (43.223)$$

where each classical path $b_j = b_j(\theta)$ or SP-point $\lambda_j = \lambda_j(\theta)$ contributes $f_j(\theta)$ to the amplitude.

Primitive amplitudes:

$$f_j(\theta) = -i\alpha_j\beta_j\sigma_j^{1/2}(\theta)\exp\left[iS_j^C(\theta)\right] \qquad (43.224)$$

$$\chi'_j = (d\chi/d\lambda)_j \qquad (43.225)$$

$$\alpha_j = e^{\pm i\pi/4};\quad (+),\ \chi'_j > 0;\quad (-),\ \chi'_j < 0 \qquad (43.226a)$$

$$\beta_j = e^{\pm i\pi/4};\quad (+),\ \chi_j > 0;\quad (-),\ \chi_j < 0 \qquad (43.226b)$$

Classical cross section:

$$\sigma_j(\theta) = \left|\frac{b\,db}{d(\cos\chi)}\right|_{\chi_j} = \frac{1}{k^2}\frac{\lambda_j}{\sin\theta\,|\chi'_j|}\,. \qquad (43.227)$$

N classical deflections χ_j provide the same θ:

$$\chi_j = \chi(\lambda_j) = \theta, -\theta, -2\pi\pm\theta, -4\pi\pm\theta,\ldots. \qquad (43.228)$$

Classical collision action $S^C(E,L;\chi)/\hbar$:

$$S_j^C = 2\eta(\lambda_j) - \lambda_j\chi(\lambda_j) \qquad (43.229a)$$

$$= 2\eta(\lambda_j) - \lambda_j\theta,\quad 0\leq\chi\leq\pi \qquad (43.229b)$$

$$= 2\eta(\lambda_j) + \lambda_j\theta - m\pi,\quad \chi < 0, \qquad (43.229c)$$

where $m = 0, 1, 2,\ldots$ is the number of times the ray has traversed the backward direction during its attractive windings about the scattering center.

A. Amplitude Addition: For Three Well Separated Regions of Stationary Phase $\lambda_1 < \lambda_2 < \lambda_3$

A scattering angle θ in the range $0 < \theta < \theta_r$ (rainbow angle) typically results from deflections χ_j at three impact parameters b (or λ): $\theta = \{\chi(b_1), -\chi(b_2), -\chi(b_3)\} \equiv \{\chi_j\}$. Scattering in the range $\theta_r \leq \theta < \pi$ results from one deflection at b_1. b_1 is in the positive branch (inner repulsion) and $b_{2,3}$ are in the negative branch (outer attraction) of the deflection function $\chi(b)$ such that $b_1 < b_2 < b_3$. $kb = (\ell + 1/2) = \lambda$. Thus

$$f(\theta) = \sum_{j=1}^{3} f_j(\theta) = \sum_{j=1}^{3} [\sigma_j(\theta)]^{1/2} \exp(iS_j), \quad (43.230)$$

where the overall phases of each f_j are

$$S_1 = 2\eta(\lambda_1) - \lambda_1\theta - \pi/2, \quad (43.231a)$$
$$S_2 = 2\eta(\lambda_2) + \lambda_2\theta - \pi, \quad (43.231b)$$
$$S_3 = 2\eta(\lambda_3) + \lambda_3\theta - \pi/2, \quad (43.231c)$$

which are appropriate, respectively, to deflections $\chi_1 = \theta$ at λ_1, $\chi_2 = -\theta$ at λ_2 and $\chi_3 = -\theta$ at λ_3, within the range $-\pi \leq \chi \leq \pi$.

The elastic differential cross section

$$\sigma(\theta) = \sum_{j=1}^{3} \sigma_j(\theta) + 2\sum_{i<j}^{3} [\sigma_i(\theta)\sigma_j(\theta)]^{1/2} \cos(S_i - S_j)$$
$$\equiv \sigma_c(\theta) + \Delta\sigma(\theta) \quad (43.232)$$

exhibits interference effects. The first term σ_c is the classical background DCS with no oscillations. The second term $\Delta\sigma$ provides the oscillatory structure which originates from interference between classical actions associated with the different trajectories resulting in a given θ. The part of $S_{ij} = S_i - S_j$ most rapidly varying with θ are the angular action changes $(\lambda_1 + \lambda_2)\theta$, $(\lambda_1 + \lambda_3)\theta$ and $(\lambda_2 - \lambda_3)\theta$. Interference oscillations between the action phases S_1 and S_2 or between S_1 and S_3 then have angular separations

$$\Delta\theta_{1;(2,3)} = \frac{2\pi n}{(\lambda_1 + \lambda_{2,3})}, \quad (43.233)$$

which are much smaller than the separation

$$\Delta\theta_{2;3} = \frac{2\pi n}{(\lambda_2 - \lambda_3)} \quad (43.234)$$

for interference between phases S_2 and S_3. The oscillatory structure in $\Delta\sigma(\theta)$ is composed therefore of *supernumerary rainbow* oscillations with large angular separations $\Delta\theta_{2;3}$ from S_2 and S_3 interference, with superimposed rapid oscillations with smaller separation $\Delta\theta_{1,(2,3)}$ from interference between S_1 and S_2 or S_1 and S_3.

For deflections $\chi_j = \theta, -\theta, -2\pi \mp \theta, -4\pi \mp \theta, \cdots$, then the Δ^+-branch of (43.210) provides additional contributions to (43.230) with phases

$$S_{2m}^{\pm} = 2\eta(\lambda_{2m}) \pm \lambda_{2m}\theta - 2m\pi - \pi, \quad (43.235a)$$
$$S_{3m}^{\pm} = 2\eta(\lambda_{3m}) \pm \lambda_{3m}\theta - 2m\pi - \pi/2, \quad (43.235b)$$

for $m = 1, 2, 3, \ldots$.

B. Uniform Airy Result: For Two Regions of Stationary Phase Which Can Coalesce

The combined contribution $f_{23}(\theta)$ from the λ_2 and λ_3 attractive regions in Δ^+ branch is

$$f_{23}(\theta) = \sigma_2^{1/2} e^{iS_2} F_{23} + \sigma_3^{1/2} e^{iS_3} F_{23}^*, \quad (43.236a)$$
$$F_{23} = [A + iB]\, e^{-i(S_{23}/2)}, \quad (43.236b)$$
$$S_{23} = S_2 - S_3$$
$$= 2(\eta_2 - \eta_3) + (\lambda_2 - \lambda_3)\theta - \tfrac{1}{2}\pi, \quad (43.236c)$$
$$A(z) = \pi^{1/2} z^{1/4} \mathrm{Ai}(-z), \quad (43.236d)$$
$$B(z) = \pi^{1/2} z^{-1/4} \mathrm{Ai}'(-z), \quad (43.236e)$$
$$\tfrac{4}{3}|z|^{3/2} = S_2^C - S_3^C = S_{23} + \tfrac{1}{2}\pi. \quad (43.236f)$$

The amplitude f_{23} tends to the primitive result $f_2(\theta) + f_3(\theta)$ in the limit of well-separated regions ($z \gg 1$) when $F_{23} \to 1$. An equivalent form of (43.236a) is

$$f_{23}(\theta) = \left[A(\sigma_2^{1/2} + \sigma_3^{1/2}) + iB(\sigma_2^{1/2} - \sigma_3^{1/2})\right] \exp(i\bar{S}), \quad (43.237)$$

where the mean phase $\bar{S} = \tfrac{1}{2}(S_2 + S_3)$. This form is useful for analysis of caustic regions at $\theta \sim \theta_r$ where $z \to 0$.

C. Transitional Result: Neighborhood of Caustic or Rainbow at $(\theta_r, b_r, \lambda_r)$

In the vicinity of rainbow angle $\theta \approx \theta_r$,

$$\chi' = \frac{\partial\chi}{\partial\lambda} = [2(\theta_r - \theta)\chi''(\lambda_r)]^{1/2} \quad (43.238)$$
$$z = (\theta_r - \theta)[2/\chi''(\lambda_r)]^{1/3} > 0 \quad (43.239)$$
$$S_r = \tfrac{1}{2}(S_1 + S_2) = 2\eta(\lambda_r) + \lambda_r\theta_r - \tfrac{3}{4}\pi. \quad (43.240)$$

The scattering amplitude

$$f_{23}(\theta_r) = \left[\frac{2\pi\lambda_r}{k^2 \sin\theta_r}\right]^{1/2} \left[\frac{2}{\chi''(\lambda_r)}\right]^{1/3} \mathrm{Ai}(-z) e^{iS_r}, \quad (43.241)$$

is finite at the rainbow angle θ_r. In Eq. (43.237), the $(\theta_r - \theta)^{-1/4}$ divergence in $|\chi'_\ell|^{1/2}$ of Eq. (43.238) arising in the constructive interference term $(\sigma_2^{1/2} + \sigma_3^{1/2})$ is exactly balanced by the $z^{1/4}$ term of $A(z)$. Also $(\sigma_2^{1/2} - \sigma_3^{1/2}) \to 0$ in (43.237) more rapidly than $z^{-1/4}$ in $B(z)$ so that (43.237) at θ_r is finite and reproduces (43.241).

The uniform semiclassical DCS

$$\frac{d\sigma}{d\Omega} = \left| \sigma_1^{1/2}(\theta) e^{iS_1} + \sigma_2^{1/2}(\theta) F_{23} e^{iS_2} + \sigma_3^{1/2} F_{23}^* e^{iS_3} \right|^2 \quad (43.242)$$

contains, in addition to the S_i/S_j interference oscillations in the primitive result (43.232), the θ-variation of the Airy Function $|\text{Ai}(z)|^2$, which has a principal finite (rainbow) maximum at $\theta \leq \theta_{\text{r}}$, the classical rainbow angle, and subsidiary maxima (supernumary rainbows) at smaller angles. The DCS decreases exponentially as θ increases past θ_{r} into the classical forbidden region and tends to $\sigma_1(\theta)$ at larger angles. For $\theta_{\text{r}} < \theta < \pi$,

$$f(\theta) = \sigma_1^{1/2}(\theta) \exp(iS_1). \qquad (43.243)$$

43.3.5 Diffraction and Glory Amplitudes

Diffraction. Diffraction arises from the outer (attractive) part of the potential. Many contributions arise from the attractive Δ^+ branch appropriate for negative χ at large b where η is small. Here χ, χ' both tend to zero.

Glory. The deflection function χ passes through zero at a finite λ_{g}. A confluence of the two maxima of each phase shift from the positive and negative branches of $\chi(b)$ occurs at $b_1 = b_2 = b_{\text{g}} = \lambda_{\text{g}}/k$. η_{g} is maximum for $\chi = 0$. In general $\chi(b_m) = -2m\pi$ (forward glory); $\chi(b_m) = -(2m-1)\pi$ (backward glory); $m = 0, 1, 2, \ldots$. There is only a forward glory at $\chi = 0$ when the deflection at the rainbow is $|\chi_{\text{r}}| < 2\pi$. In contrast to diffraction, the glory contribution can be calculated by the *stationary phase approximation*.

Transitional Results for Forward and Backward Glories

Forward Glories. Contributions arise from $\chi = \pm\theta$, $-2\pi \pm \theta, \cdots, -2m\pi \pm \theta$ as $\theta \to 0$. The stationary phase points λ_m are located at

$$\chi(\lambda_m) = \chi_m = -2m\pi; \quad m \geq 0. \qquad (43.244)$$

The phase function in the neighborhood of a glory is

$$\eta(\lambda) = \eta_m - m\pi(\lambda - \lambda_m) + \tfrac{1}{4}\chi'_m(\lambda - \lambda_m)^2. \qquad (43.245)$$

The $m = 0$ term provides zero deflection χ due to a net balance of attractive and repulsive scattering for a finite impact parameter b_{g} or λ_{g} where $\eta(\lambda)$ attains its maximum value η_m. The glories at θ are due to a confluence of the two contributions from the deflections $\chi_m = -2m\pi \pm \theta$ at the stationary phase points $\lambda_{mn} = \lambda_{m1}$ and λ_{m2}. SP integration of (43.216) with (43.245) yields the forward glory amplitude

$$f_{\text{FG}} = \frac{1}{k} \sum_{n=1}^{2} \sum_{m=0}^{\infty} \lambda_{mn} \left[\frac{2\pi}{|\chi'_m|}\right]^{1/2} J_0(\lambda_{mn}\theta)e^{iS_{mn}^{(g)}}, \qquad (43.246)$$

$$S_{mn}^{(g)} = 2\eta(\lambda_{mn}) + m\pi(\lambda_{mn} - 1) - \tfrac{3}{4}\pi. \qquad (43.247)$$

Backward Glories. Contributions arising from $\chi = -\pi \pm \alpha$, $-3\pi \pm \alpha, \cdots, -(2m-1)\pi \pm \alpha$ coalesce as $\alpha \equiv \pi - \theta \to 0$. The phase function near a backward glory is

$$\eta(\lambda) = \eta(\lambda_m) + \tfrac{1}{2}\chi_m(\lambda - \lambda_m) + \tfrac{1}{4}\chi'_m(\lambda - \lambda_m)^2. \qquad (43.248)$$

The $m = 0$ term provides the normal backward amplitude due to head-on ($b = 0$) repulsive collisions. $m > 0$ terms provide contributions from attractive collisions for which there are two points λ_{mn} of stationary phase for each m in $\chi_m = -(2m-1)\pi \pm \alpha$.

The backward glory amplitude at $\theta = \pi - \alpha$ is

$$f_{\text{BG}} = \frac{1}{k} \sum_{n=1}^{2} \sum_{m=0}^{\infty} \lambda_{mn} \left[\frac{2\pi}{|\chi'_{mn}|}\right]^{1/2} J_0(\lambda_{mn}\alpha)e^{iS_{mn}^{(g)}}, \qquad (43.249)$$

$$S_{mn}^{(g)} = 2\eta(\lambda_{mn}) + \pi(2m-1)(\lambda_{mn} - \tfrac{1}{2}) - \tfrac{3}{4}\pi. \qquad (43.250)$$

In contrast to the Bessel amplitudes (below), these transitional formulae do not uniformly connect with the primitive semiclassical results for $(f_1 + f_2)$ away from the critical glory angles.

Uniform Bessel Amplitude for Glory Scattering

The combined contributions from $\chi_1 = -N\pi + \theta$ and $\chi_2 = -N\pi - \theta$, where $N = 2m$, for forward and $N = 2m - 1$ for backward scattering, yield [9]

$$\begin{aligned} f_{\text{G}}(\theta) = \frac{\alpha_j}{2i}e^{-iN\pi/2} &\left[\pi S_{21}^{(C)}\right]^{1/2} \exp i\bar{S}^{(C)}(\theta) \\ &\times \left[(\sigma_1^{1/2} + \sigma_2^{1/2})J_0\left(\tfrac{1}{2}S_{21}^{(C)}\right) \right. \\ &\left. - i(\sigma_1^{1/2} - \sigma_2^{1/2})J_1\left(\tfrac{1}{2}S_{21}^{(C)}\right)\right], \qquad (43.251) \end{aligned}$$

where $S_{21}^{(C)}(\theta) = S_2^{(C)} - S_1^{(C)}$ is the difference of the collision actions (43.229a),

$$S_i^{(C)}(\theta) = 2\eta(\lambda_i) - \lambda_i\chi_i, \quad i = 1, 2, \qquad (43.252)$$

with mean

$$\bar{S}_{21}^{(C)}(\theta) = \tfrac{1}{2}\left[S_2^{(C)} + S_1^{(C)}\right], \qquad (43.253)$$

and phases

$$\alpha_j = e^{\pm i\pi/4}; \quad (+), \; \chi'_j > 0; \quad (-), \; \chi'_j < 0, \qquad (43.254)$$

and the ordinary Bessel functions $J_n(Z)$ satisfy the relationships $J_1(z) = -J'_0(z)$, $J_1(-z) = -J_1(z)$. This formula, valid for both forward ($\theta \sim 0$) and backward ($\theta \sim \pi$) glories, does uniformly connect the *primitive result* for $(f_1 + f_2)$, valid when $S_{21}^{(C)} \gg 1$ to the *transitional results* (43.246) and (43.249), valid only in the vicinity of the glories.

43.3.6 Small-Angle (Diffraction) Scattering

Diffraction originates from scattering in the forward direction by the long-range attractive tail of $V(R)$ where χ, χ' and $\eta \to 0$. The main contributions to (43.220a) arise from a large number of small $\eta(\lambda)$ at large λ. The Jeffrey-Born phase function (43.171) can therefore be used in (43.220b) for $f(\theta)$ and in (43.45) and (43.47), respectively, for $\sigma(E)$. A finite forward diffraction peak as $\theta \to 0$ is obtained for $f(\theta)$ in contrast to the classical infinite result.

Integral Cross Sections

For $V(R) = -C/R^n$, the Landau-Lifshitz (LL) and Massey-Mohr (MM) cross sections are [*cf.* Eq. (43.344)]

$$\sigma_{LL}(E) = \pi \left[\frac{2CF(n)}{(n-1)\hbar v} \right]^{2/(n-1)}$$
$$\times \pi \left[\sin\left(\frac{\pi}{n-1}\right) \Gamma\left(\frac{2}{n-1}\right) \right]^{-1} \quad (43.255)$$

$$\sigma_{MM}(E) = \pi \left[\frac{2CF(n)}{(n-1)\hbar v} \right]^{2/(n-1)} \left(\frac{2n-3}{n-2}\right) \quad (43.256)$$

where $F(n) = \sqrt{\pi}\, \Gamma(\frac{1}{2}n + \frac{1}{2})/\Gamma(\frac{1}{2}n)$ and v is the relative speed. For σ_{MM}, the phases are $\eta(\lambda) = \frac{1}{2}$ $(0 < \lambda < \lambda_0)$ and $\eta(\lambda) = \eta_{JB}$ $(\lambda > \lambda_0)$. For σ_{LL}, phases are η_{JB} for all λ. Both σ_{LL} and σ_{MM} have the general form

$$\sigma_D(E) = \gamma \left(\frac{C}{\hbar v}\right)^{2/(n-1)} . \quad (43.257)$$

Ion-Atom Collisions. For $n = 4$ attraction at low energy, $\sigma_D \sim v^{-2/3}$. $\gamma_{LL} = 11.373$, $\gamma_{MM} = 10.613$. For $n = 12$ repulsion at high energy, $\sigma_D \sim v^{-2/11}$. $\gamma_{LL} = 6.584$, $\gamma_{MM} = 6.296$.

Atom-Atom Collisions. For $n = 6$ (attraction), $\sigma_D \sim v^{-2/5}$, $\gamma_{LL} = 8.083$, $\gamma_{MM} = 7.547$ (see Fig. 43.1).

Exact numerical calculations favor σ_{LL} over σ_{MM} (see Ref. [10], pp. 1325 for details).

Differential Cross Section

$$\frac{d\sigma}{d\Omega} = f_i^2(\theta) + f_r^2(\theta) , \quad (43.258a)$$

$$f_i = \frac{2}{k} \int_0^\infty \lambda \sin^2 \eta(\lambda) \left[1 - \frac{1}{4}\lambda^2\theta^2\right] d\lambda \quad (43.258b)$$

$$= \frac{k\sigma_D(E)}{4\pi} \left[1 - \left(\frac{k^2\sigma_D}{16\pi}\right) g_1(n)\theta^2\right] , \quad (43.258c)$$

$$f_r = \frac{1}{k} \int_0^\infty \lambda \sin 2\eta(\lambda) \left[1 - \frac{1}{4}\lambda^2\theta^2\right] d\lambda \quad (43.258d)$$

$$= \frac{k\sigma_D(E)}{4\pi} \left[1 - \left(\frac{k^2\sigma_D}{16\pi}\right) g_2(n)\theta^2\right] \tan\left(\frac{\pi}{n-1}\right) , \quad (43.258e)$$

where σ_D is given by (43.257), and

$$g_j(n) = \pi^{-1} \tan\left[\frac{j\pi}{n-1}\right] \frac{\{\Gamma[2/(n-1)]\}^2}{\Gamma[4/(n-1)]} . \quad (43.259)$$

The optical theorem (43.62) is satisfied, and

$$f_D(\theta \sim 0) = \sigma_D^{1/2}(E) e^{iS_D(n)} , \quad (43.260)$$

where the (energy-independent) phase is

$$S_D(n) = \frac{\pi(n-3)}{2(n-1)} . \quad (43.261)$$

43.3.7 Small-Angle (Glory) Scattering

Amplitude and Cross Section. The other contribution to forward scattering is the forward glory, which originates from the combined null effect of attraction and repulsion at a specified glory impact parameter $b_g = \lambda_g/k$, where $\eta(\lambda)$ attains a maximum value of η_g. The $m = 0$ term of (43.246) yields

$$f_G(\theta) = \sigma_G^{1/2}(\theta) \exp\left[iS_G(E)\right] , \quad (43.262a)$$

$$\sigma_G(\theta) = \frac{\lambda_g^2}{k^2} \left(\frac{2\pi}{|\chi_g'|}\right) J_0^2(\lambda_g\theta) , \quad (43.262b)$$

$$S_G(E) = 2\eta_g(E) - \frac{3}{4}\pi , \quad (43.262c)$$

where $J_0^2(x) \sim 1 - \frac{1}{4}x^2 + \cdots$. The classical result (43.30) is recovered by averaging (43.262b) over several oscillations with $\langle J_0^2(x)\rangle = (\pi x)^{-1}$.

Diffraction-Glory Oscillations.

$$\sigma(E) = \frac{4\pi}{k}\Im\left[f_D(0) + f_G(0)\right] \quad (43.263a)$$

$$= \sigma_D(E) + \Delta\sigma_G(E) , \quad (43.263b)$$

where the diffraction cross section is (43.257), and where

$$\Delta\sigma_G(E) = \frac{4\pi}{k^2}\lambda_g \left[\frac{2\pi}{|\chi_g'|}\right]^{1/2} \sin\left(2\eta_g(E) - \frac{3}{4}\pi\right) \quad (43.264)$$

oscillates with E.

For sufficiently deep attractive wells, the phase shift η_g successively decreases with increasing E through a series of multiples of $\pi/2$. Writing $\eta_g(E) = \pi(N - \frac{3}{8})$, maxima appear at $N = 1, 2, \ldots$, and minima at $N = \frac{3}{2}, \frac{5}{2}, \frac{7}{2}, \ldots$. The glories are indexed by N in order of appearance, starting at high energies. $\eta_g(E \to 0)$ is related to the number n of bound states by Levinson's theorem: $\eta_0(E \to 0) = (n + \frac{1}{2})\pi$. Diffraction-glory oscillations also occur in the DCS at a frequency governed entirely by the energy variation of $\eta_g(E)$ and n of (43.261).

JWKB Formulae for Shape Resonances and Tunneling Predissociation

For the three classical turning points $R_1 < R_2 < R_3$ at energies E below the orbiting threshold $V_{\max}$ at R_X, the JWKB phase shift

$$\eta_\ell = \left[\eta_\ell^{(0)} - \tfrac{1}{2}\phi(\gamma_\ell)\right] + \eta_\ell^{(r)} \qquad (43.265)$$

is composed of (a) the phase shift

$$\eta_\ell^{(0)} = \lim_{R \to \infty}\left[\int_{R_3}^\infty k(R)\,dR - kR + \tfrac{1}{2}(\ell + \tfrac{1}{2})\pi\right] \quad (43.266)$$

appropriate to one turning point at R_3, (b) a contribution $\eta^{(r)}$ arising from the region between the two inner turning points R_2 and R_3 due to penetration of the centrifugal barrier and given by

$$\tan\eta_\ell^{(r)}(E) = \left[\frac{\{1 + \exp(-2\gamma_\ell)\}^{1/2} - 1}{\{1 + \exp(-2\gamma_\ell)\}^{1/2} + 1}\right]$$
$$\times \tan\left(\alpha_\ell - \tfrac{1}{2}\phi_\ell\right), \qquad (43.267)$$

and (c) a phase correction factor

$$\phi_\ell(\gamma_\ell) = \arg\Gamma(\tfrac{1}{2} + i\epsilon) - \epsilon\ln|\epsilon| + \epsilon, \qquad (43.268)$$

where $\epsilon = -\gamma_\ell/\pi$. The radial action $J_R(E)$ is $2\hbar\alpha_\ell(E)$. For motion within the potential well α_ℓ is

$$\alpha_\ell(E) = \int_{R_1}^{R_2} k(R)\,dR, \qquad (43.269)$$

and is

$$\gamma_\ell(E < E_{\max}) = \int_{R_2}^{R_3} |k(R)|\,dR \qquad (43.270)$$

in the classically forbidden region of the potential hump. The above expressions also hold for energies $E > V_{\max}$, except that (43.270) is replaced by

$$\gamma_\ell(E > E_{\max}) = -i\int_{R_-}^{R_+} k(R)\,dR, \qquad (43.271)$$

where $R_\pm$ are the complex roots of $k_\ell(R) = 0$. For the quadratic form

$$V(R) = V_{\max} - \tfrac{1}{2}M\omega_*^2(R - R_{\max})^2, \qquad (43.272)$$

appropriate in the vicinity of the potential hump, γ for both cases reduces to

$$\gamma = \pi(V_{\max} - E)/\hbar\omega_*. \qquad (43.273)$$

The deflection function $\chi_\ell = 2(\partial\eta_\ell/\partial\ell)$ no longer diverges at the orbiting angular momentum ℓ_0 or impact

parameter b_0. The singularities in η_ℓ of Eq. (43.51) are exactly canceled by $-\tfrac{1}{2}(\partial\phi/\partial\ell)$ in Eq. (43.268).

Limiting cases:

(a) $E \gg V_{\max}$. Then $\gamma_\ell \to -\infty$ and $\phi \to -(\pi/24\gamma_\ell) \to 0$, so that $\eta_\ell^{(r)} \to \alpha_\ell$ and η_ℓ reduces to the single turning point result (43.266) with $R_3 = R_1$.

(b) $E \ll V_{\max}$. Then $\gamma_\ell \gg 1$ and

$$\eta_\ell^{(r)}(E) = \tan^{-1}\left[\tfrac{1}{2}e^{-2\gamma_\ell}\tan\left(\alpha_\ell - \tfrac{1}{2}\phi_\ell\right)\right], \quad (43.274)$$

which remains negligible except for those energies E close to quasibound energy levels $E_{n\ell}$ determined via the *Bohr quantization condition*

$$\alpha_\ell(E) - \tfrac{1}{2}\phi_\ell(E) = (n + \tfrac{1}{2})\pi. \qquad (43.275)$$

As E increases past each $E_{n\ell}$, $\eta_\ell^{(r)}$ increases rapidly by π. Since $(\partial J/\partial E)_{n\ell} = \nu_{n\ell}^{-1} = 2\pi/\omega_{n\ell}$, the time period for radial oscillation within the potential barrier, the level spacing is $\hbar\omega_{n\ell} = h\nu_{n\ell} = \pi(\partial E/\partial\alpha)_{n\ell}$.

Shape Resonance. In the neighborhood of $E_{n\ell} \sim E$,

$$\alpha_\ell(E) = \alpha_{n\ell}(E_{n\ell}) + \left(\frac{\pi}{\hbar\omega_{n\ell}}\right)(E - E_{n\ell}), \quad (43.276)$$

and, under the assumption that the energy variation of ϕ_ℓ can be neglected, (valid for E not close to $V_{\max}$), then η_ℓ reduces to the Breit-Wigner form

$$\eta_\ell(E) = \eta_\ell^{(0)}(E) + \tan^{-1}\left[\frac{\Gamma_{n\ell}/2}{E_{n\ell} - E}\right], \qquad (43.277)$$

with resonance width

$$\Gamma_{n\ell} = 2\left(\frac{\hbar\omega_{n\ell}}{\pi}\right)\frac{[1 + \exp(-2\gamma_{n\ell})]^{1/2} - 1}{[1 + \exp(-2\gamma_{n\ell})]^{1/2} + 1}, \quad (43.278)$$

where $\gamma_{n\ell} = \gamma(E_{n\ell})$. The partial cross sections are then determined by (43.164a)–(43.164b) with $\eta_\ell^{(0)}$ replaced by $\eta_\ell^{(0)} - \tfrac{1}{2}\phi(\gamma)$ of (43.266).

Gamow's Result. For $E \ll V_{\max}$, $\gamma_{n\ell} \gg 1$, then

$$\Gamma_{n\ell} \xrightarrow{\gamma \gg 1} \left(\frac{\hbar\omega_{n\ell}}{2\pi}\right)\exp(-2\gamma_{n\ell}). \qquad (43.279)$$

The probabilities of transmission through and reflection from a barrier for unit incident flux from the left are

Transmission Probability:

$$T = [1 + e^{2\gamma}]^{-1} \xrightarrow{\gamma \gg 1} e^{-2\gamma}. \qquad (43.280)$$

Reflection Probability:

$$R = [1 + e^{-2\gamma}]^{-1} \xrightarrow{\gamma \gg 1} 1. \qquad (43.281)$$

Frequency of leakage:

$$\nu_T = \Gamma_{n\ell}/\hbar = \left(\frac{\omega_{n\ell}}{2\pi}\right)e^{-2\gamma}. \qquad (43.282)$$

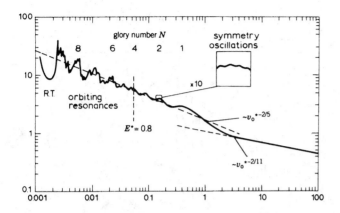

Figure 43.1. Illustration of all the various oscillatory effects for elastic scattering by a Lennard-Jones (12,6) potential of well depth ϵ and equilibrium distance R_e. Ordinate $\sigma^* = \sigma/(2\pi R_e^2)$, abscissa $v^* = \hbar v/(\epsilon R_e)$.

43.3.8 Oscillations in Elastic Scattering

Figure 43.1 is an illustration [11] of all the various oscillatory structure and effects — Ramsauer-Townsend minimum (see Sect. 43.2.4), orbiting resonances (43.338), diffraction-glory oscillations (43.263b) and symmetry oscillations (43.80) — for elastic scattering by a Lennard-Jones (12, 6) potential. Note the shift of velocity dependence from $v^{-2/5}$ at low v to $v^{-2/11}$ at high v. $\sigma = 2\pi R_e^2$ is the averaged cross section $2\pi b_0^2$ in (43.47) at $b_0 = R_e$. The region $\sigma^* > 1$ probes the attractive part of the potential at low speeds and $\sigma^* < 1$ probes the repulsive part at high speeds. The four distinct types of structure originate from nonrandom behavior of $\sin^2 \eta$ in (43.45). Orbiting trajectories exist for $E < 0.8\epsilon$ (see Sect. 43.5).

43.4 ELASTIC SCATTERING IN REACTIVE SYSTEMS

All nonelastic processes (e.g. inelastic scattering and rearrangement collisions/chemical reactions) can be viewed as a net absorption from the incident beam current vector $\mathbf{J}$ and modeled by a complex optical potential

$$V(R) = V_r(R) + iV_i(R) . \tag{43.283}$$

The continuity equation is then

$$\nabla \cdot \mathbf{J} = -\frac{2}{\hbar} V_i(R) \left| \Psi(\mathbf{R}) \right|^2 , \tag{43.284}$$

so that particle absorption implies $V_i > 0$ and particle creation $V_i < 0$. Since particle conservation implies $\left| S_\ell \right|^2 = 1$, the phase shift

$$\delta_\ell(k) = \eta_\ell(k) + i\gamma_\ell(k) \tag{43.285}$$

is also complex since then

$$S_\ell = A_\ell(k) \exp(2i\eta_\ell) , \tag{43.286}$$

where the absorption (inelasticity) factor is

$$A_\ell = \exp(-2\gamma_\ell) \leq 1 . \tag{43.287}$$

43.4.1 Quantal Elastic, Absorption, and Total Cross Sections

$$f_{el}(\theta) = \frac{1}{2ik} \sum_{\ell=0}^{\infty} (2\ell + 1) \left[A_\ell e^{2i\eta_\ell} - 1 \right] P_\ell(\cos\theta) , \tag{43.288a}$$

$$\sigma_{el}(k) = \frac{\pi}{k^2} \sum_{\ell=0}^{\infty} (2\ell + 1) \left| A_\ell e^{2i\eta_\ell} - 1 \right|^2 , \tag{43.288b}$$

$$\sigma_{abs}(k) = \frac{\pi}{k^2} \sum_{\ell=0}^{\infty} (2\ell + 1) \left[1 - A_\ell^2 \right] , \tag{43.288c}$$

$$\begin{aligned}
\sigma_{tot}(k) &= \sigma_{el}(k) + \sigma_{abs}(k) \\
&= \frac{2\pi}{k^2} \sum_{\ell=0}^{\infty} (2\ell + 1) \left[1 - A_\ell \cos 2\eta_\ell \right] .
\end{aligned} \tag{43.288d}$$

Upper limits to the partial cross sections are

$$\sigma_\ell^{el} \leq \frac{4\pi}{k^2} (2\ell + 1), \quad \sigma_\ell^{abs} \leq \frac{\pi}{k^2} (2\ell + 1), \tag{43.289a}$$

$$\sigma_\ell^{tot} \leq \frac{4\pi}{k^2} (2\ell + 1) = \frac{4\pi}{k} \Im f_\ell^{el}(\theta = 0) . \tag{43.289b}$$

For pure elastic scattering with no absorption, $A_\ell = 1$. All nonelastic processes ($0 \leq A_\ell < 1$) are always accompanied by elastic scattering, even in the ($A_\ell = 0$) limit of full absorption.

Eikonal Formulae for Forward Reactive Scattering.

$$f_{el}(\theta) = -ik \int_0^\infty \left[e^{2i\delta} - 1 \right] J_0(2kb \sin \tfrac{1}{2}\theta) b \, db, \tag{43.290a}$$

$$\sigma_{el}(k) = 2\pi \int_0^\infty \left| \left[1 - e^{-2\gamma} e^{2i\eta} \right] \right|^2 b \, db, \tag{43.290b}$$

$$\sigma_{abs}(k) = 2\pi \int_0^\infty \left[1 - e^{-4\gamma} \right] b \, db, \tag{43.290c}$$

$$\sigma_{tot}(k) = 4\pi \int_0^\infty \left[1 - e^{-2\gamma} \cos 2\eta \right] b \, db, \tag{43.290d}$$

where the phase shift function $\delta = \eta + i\gamma$ at impact parameter b can be either the Jeffrey-Born phase

$$\delta_{JB}(b) = -\frac{1}{2k} \int_b^\infty \frac{U(R) \, dR}{(1 - b^2/R^2)^{1/2}} , \tag{43.291}$$

where $kb = \lambda = (\ell + 1/2)$, or the eikonal phase

$$\delta_{\mathrm{E}}(b) = -\frac{1}{4k} \int_{-\infty}^{\infty} U(b, Z)\, dZ, \qquad (43.292)$$

where the reduced interaction is $U = (2m/\hbar^2)V$.

Fraunhofer Diffraction by a Black Sphere. For a complex spherical well U

$$U = \begin{cases} U_{\mathrm{r}} + iU_i, & R < a \\ 0, & R > a. \end{cases} \qquad (43.293)$$

The eikonal phase function (43.292) is

$$\delta(b) = \begin{cases} (U/2k)(a^2 - b^2)^{1/2}, & 0 \le b \le a \\ 0 & b > a. \end{cases} \qquad (43.294)$$

The absorption factor is

$$A(b)^2 \equiv e^{-4\gamma} = \exp\left[-2(a^2 - b^2)^{1/2}/\lambda\right], \qquad (43.295)$$

where $\lambda = k/U_i$ is the mean free path towards absorption. For strong absorption, $\lambda \ll a$, so that

$$f_{\mathrm{el}}(\theta) = ik \int_0^a J_0(2kb \sin \tfrac{1}{2}\theta)\, b\, db, \qquad (43.296)$$

$$\frac{d\sigma_{\mathrm{el}}}{d\Omega} = (ka)^2 \left[\frac{J_1(2ka \sin \tfrac{1}{2}\theta)}{2ka \sin \tfrac{1}{2}\theta}\right]^2 a^2, \qquad (43.297)$$

which has a diffraction shaped peak of width $\sim \theta \le (ka)^{-1}$ about the forward direction, and

$$\sigma_{\mathrm{tot}} = \frac{4\pi}{k} \Im f_{\mathrm{el}}(\theta = 0) = 2\pi a^2 \qquad (43.298)$$

is composed of πa^2 for classical absorption and πa^2 for edge diffraction or shadow (nonclassical) elastic scattering. This result also holds for the perfectly reflecting sphere (πa^2 for classical elastic and πa^2 for edge diffraction).

43.5 RESULTS FOR MODEL POTENTIALS

Exact results for various quantities in classical, quantal, and semiclassical elastic scattering are obtained for the model potentials (a)–(s) in Table 46.1.

Classical Deflection Functions for Model Potentials

(a) Hard Sphere.

$$\theta(b; E) = \chi = \begin{cases} \pi - 2\sin^{-1} b/a, & b \le a; \\ 0, & b > a. \end{cases} \qquad (43.299)$$

$$b(\theta) = a \cos \tfrac{1}{2}\theta, \qquad (43.300)$$

$$\sigma(\theta) = \frac{d\sigma}{d\Omega} = \frac{1}{4}a^2; \quad \text{isotropic}, \qquad (43.301)$$

$$\sigma = \pi a^2 = \text{geometric cross section}; \qquad (43.302)$$

Table 43.1. Model interaction potentials.

Potential		$V(R)$
(a)	Hard sphere	$\infty,\ R \le a;\ 0,\ R > a$
(b)	Barrier	$V_0,\ R \le a;\ 0,\ R > a$
(c)	Well	$-V_0,\ R \le a;\ 0,\ R > a$
(d)	Coulomb ($\pm$)	$\pm k/R$
(e)	Finite-range Coulomb	$-k/R + k/R_s\ R \le R_s;\ 0,\ R > R_s$
(f)	Pure dipole	$\pm \alpha/R^2$
(g)	Finite-range dipole	$\pm \alpha\left(\dfrac{1}{R^2} - \dfrac{1}{a^2}\right),\ R \le a;\ 0,\ R > a$
(h)	Dipole + hard sphere	$\pm \alpha/R^2,\ R \le a;\ 0,\ R > a$
(i)	Power law attractive	$-C/R^n,\ (n > 2)$
(j)	Fixed dipole + polarization	$-\dfrac{De \cos \theta_{\mathrm{d}}}{R^2} - \dfrac{\alpha_{\mathrm{d}} e^2}{2R^4}$
(k)	Fixed dipole + Coulomb	$-\dfrac{De \cos \theta_{\mathrm{d}}}{R^2} + \dfrac{e^2}{R}$
(l)	Lennard-Jones $(n, 6)$	$\dfrac{\epsilon n}{n - 6}\left[\dfrac{6}{n}\left(\dfrac{R_{\mathrm{e}}}{R}\right)^n - \left(\dfrac{R_{\mathrm{e}}}{R}\right)^6\right]$
(m)	Polarization $(n, 4)$	$\dfrac{\epsilon n}{n - 4}\left[\dfrac{4}{n}\left(\dfrac{R_{\mathrm{e}}}{R}\right)^n - \left(\dfrac{R_{\mathrm{e}}}{R}\right)^4\right]$
(n)	Multiple-term power law	$\dfrac{C_m}{R^m} - \dfrac{C_n}{R^n} = V_m(R) - V_n(R)$
(o)	Exponential	$V_0 \exp(-\alpha R)$
(p)	Screened Coulomb	$V_0 \exp(-\alpha R)/R$
(q)	Morse	$\epsilon\left[e^{2\beta(R_{\mathrm{e}} - R)} - 2e^{\beta(R_{\mathrm{e}} - R)}\right]$
(r)	Gaussian	$V_0 \exp(-\alpha^2 R^2)$
(s)	Polarization finite	$-V_0/(R^2 + R_0^2)^2$

θ, $\sigma(\theta)$ and σ are all independent of energy E.

(b) Potential Barrier.

For $E < V_0$, classical scattering is the same as for hard sphere reflection as given by Eqs. (43.299)–(43.302). For $E > V_0$ and $\theta = \chi$, define $n^2 = 1 - V_0/E$, $b_0 = na$. Then

$$\theta(b) = \begin{cases} 2\left[\sin^{-1}(b/na) - \sin^{-1}(b/a)\right] & 0 \le b \le b_0 \\ \pi - 2\sin^{-1}(b/a), & b_0 \le b \le a \end{cases}$$
$$\qquad (43.303)$$

and $\theta_{\max} = 2\cos^{-1} n$. For a given θ, the two impact parameters which contribute are

$$b_1(\theta) = \frac{an \sin \tfrac{1}{2}\theta}{[1 - 2n \cos \tfrac{1}{2}\theta + n^2]^{1/2}}, \quad 0 < b_1 \le b_0 \qquad (43.304)$$

$$b_2(\theta) = a \cos \tfrac{1}{2}\theta, \qquad b_0 < b_2 \le a \qquad (43.305)$$

For $0 \leq \theta \leq \theta_{\max}$,

$$\frac{d\sigma}{d\Omega} = \frac{1}{4}a^2 + \frac{a^2 n^2 (n \cos\frac{1}{2}\theta - 1)(n - \cos\frac{1}{2}\theta)}{4 \cos\frac{1}{2}\theta \left[1 + n^2 - 2\cos\frac{1}{2}\theta\right]}, \quad (43.306)$$

and $d\sigma/d\Omega = 0$ for $\theta_{\max} \leq \theta \leq \pi$. Finally,

$$\sigma = \int_{\theta=0}^{\theta_{\max}} \left(\frac{d\sigma}{d\Omega}\right) d\Omega = \pi a^2. \quad (43.307)$$

(c) Potential Well.

Results are similar to the potential barrier case above, except that there is only a single scattering trajectory with $\theta = -\chi$, and $n = (1 + V_0/E)^{1/2}$ is the effective index of refraction for the equivalent problem in geometrical optics. Refraction occurs on entering and exiting the well. Then

$$\theta(b) = -2\left[\sin^{-1}(b/na) - \sin^{-1}(b/a)\right], \quad (43.308)$$

$$\theta(b = a) = \theta_{\max} = 2\cos^{-1}(1/n), \quad (43.309)$$

$$b(\theta) = \frac{-an\sin\frac{1}{2}\theta}{[1 - 2n\cos\frac{1}{2}\theta + n^2]^{1/2}}, \quad (43.310)$$

$$\frac{d\sigma}{d\Omega} = \frac{a^2 n^2 \left[n \cos\frac{1}{2}\theta - 1\right]\left[n - \cos\frac{1}{2}\theta\right]}{4\cos\frac{1}{2}\theta \left[n^2 + 1 - 2n\cos\frac{1}{2}\theta\right]^2}, \quad (43.311)$$

$$\sigma = \pi a^2 \quad (43.312)$$

(d) Rutherford or Coulomb.

$$\theta(b, E) = |\chi| = 2\csc^{-1}\left[1 + (2bE/k)^2\right]^{1/2}, \quad (43.313)$$

$$b(\theta, E) = (k/2E)\cot\frac{1}{2}\theta, \quad (43.314)$$

$$\sigma(\theta) = \frac{d\sigma}{d\Omega} = \left(\frac{k}{4E}\right)^2 \csc^4\frac{1}{2}\theta. \quad (43.315)$$

(e) Finite Range Coulomb.

$$R_0(E) = \frac{k}{2E}, \quad \alpha(E) = R_0(E)/R_s,$$

$$\frac{d\sigma}{d\Omega} = \frac{R_0^2}{4}\left[\frac{1+\alpha}{\alpha^2 + (1+2\alpha)\sin^2\frac{1}{2}\theta}\right]^2. \quad (43.316)$$

(f) Pure Dipole. $R_0^2(E) = \alpha/E$.

Repulsion $(+)$: $\chi > 0$, $\chi = \theta$.

$$b^2(\chi) = R_0^2\left[\left(\frac{1}{\chi} + \frac{1}{2\pi - \chi}\right)\frac{\pi}{2} - 1\right], \quad (43.317)$$

$$\frac{d\sigma}{d\Omega} = \frac{\pi R_0^2}{4\sin\theta}\left|\frac{1}{\theta^2} - \frac{1}{(2\pi - \theta)^2}\right|. \quad (43.318)$$

Attraction $(-)$: $\chi < 0$.

$$b^2(\chi) = R_0^2\left[\left(\frac{1}{|\chi|} - \frac{1}{|\chi| + 2\pi}\right)\frac{\pi}{2} + 1\right]. \quad (43.319)$$

There is an infinite number of (negative) deflections $\chi = \chi_n^\pm$ associated with a given scattering angle θ:

$$|\chi_n^+| = 2\pi n + \theta, \quad n = 0, 1, 2, \ldots, \quad (43.320a)$$

$$|\chi_n^-| = 2\pi n - \theta, \quad n = 1, 2, 3, \ldots. \quad (43.320b)$$

The infinite sum over contributions from $b_n^\pm = b(\chi_n^\pm)$ for the attractive dipole yields

$$\frac{d\sigma}{d\Omega} = \frac{\pi R_0^2}{4\sin\theta}\left|\frac{1}{\theta^2} + \frac{1}{(2\pi - \theta)^2}\right|. \quad (43.321)$$

(g) Finite Range Dipole Scattering.

$R_0^2 = \alpha/E$, $(R_c^\pm)^2 = b^2 \pm R_0^2$, $(b_0^\pm)^2 = a^2 \pm R_0^2$.
Repulsion $(+)$: for $b \leq a$,

$$\chi(b) = \frac{\pi(R_c^+ - b)}{R_c^+} + \frac{2b}{R_c^+}\sin^{-1}\left(\frac{R_c^+}{b_0^+}\right) - 2\sin^{-1}\left(\frac{b}{a}\right), \quad (43.322)$$

$$\chi(0) = \pi, \quad \chi(b \geq a) = 0, \quad \sigma = \pi a^2.$$

Attraction $(-)$: for $b > R_0$,

$$\chi(b) = \frac{\pi(R_c^- - b)}{R_c^-} + \frac{2b}{R_c^-}\sin^{-1}\left(\frac{R_c^-}{b_0^-}\right) - 2\sin^{-1}\left(\frac{b}{a}\right), \quad (43.323)$$

$$\chi(R_0) \to \infty, \quad \chi(b \geq a) = 0, \quad \sigma = \pi a^2.$$

(h) Dipole Plus Hard Sphere Scattering.
$R_0^2 = \alpha/E$, $(R_c^\pm)^2 = b^2 \pm R_0^2$, $(b_0^\pm)^2 = a^2 \pm R_0^2$. Repulsion $(+)$: for $0 \leq b \leq b_0$,

$$\chi(b) = \frac{\pi(R_c^+ - b)}{R_c^+} + \frac{2b}{R_c^+}\sin^{-1}\left(\frac{R_c^+}{a}\right) - 2\sin^{-1}\left(\frac{b}{a}\right), \quad (43.324)$$

$$(43.325)$$

$$b_0 \leq b \leq a; \quad \chi(b) = \pi - 2\sin^{-1}(b/a), \quad (43.326)$$

$$\chi(0) = \pi, \quad \chi(b \geq a) = 0, \quad \sigma = \pi a^2.$$

Attraction $(-)$: for $b > R_0$,

$$\chi(b) = \frac{\pi(R_c^- - b)}{R_c^-} + \frac{2b}{R_c^-}\sin^{-1}\left(\frac{R_c^-}{a}\right) - 2\sin^{-1}\left(\frac{b}{a}\right), \quad (43.327)$$

$$\chi(b) = \chi_{\min} \text{ at } b = a,$$

$$\chi(0) = \pi, \quad \chi(b \geq a) = 0, \quad \sigma = \pi a^2.$$

Orbiting or Spiraling Collisions

From Sect. 43.1.7, the parameters are

Orbiting radius:	R_0.
Focusing factor:	$F = [1 - V(R_0)/E]$.
Orbiting cross section:	$\sigma_{\text{orb}} = \pi R_0^2 F$.

(i) Attractive Power Law Potentials.

$$V_{\text{eff}}(R_0) = (1 - \tfrac{1}{2}n)V(R_0), \quad n > 2,$$

$$R_0(E) = \left[\frac{(n-2)C}{2E}\right]^{1/n}, \quad F = \left[\frac{n}{(n-2)}\right], \quad (43.328)$$

$$\sigma_{\text{orb}}(E) = \pi\left[\frac{n}{(n-2)}\right]\left[\frac{(n-2)C}{2E}\right]^{2/n}. \quad (43.329)$$

For the case $n = 4$ with $V(R) = -\alpha_d e^2/2R^4$, this gives the Langevin cross section

$$\sigma_L(E) = 2\pi R_0^2 = 2\pi\left(\frac{\alpha_d e^2}{2E}\right)^{1/2} \quad (43.330)$$

for orbiting collisions, and the Langevin rate

$$k_L = v\sigma_L(E) = 2\pi(\alpha_d e^2/M)^{1/2}, \quad (43.331)$$

which is independent of E.

The case $n = 6$ with $V(R) = -C/R^6$ is the van der Waals potential for which

$$\sigma_{\text{orb}}(E) = \tfrac{3}{2}\pi R_0^2 = \tfrac{3}{2}\pi\,(2C/E)^{1/3}. \quad (43.332)$$

(j) Fixed Dipole Plus Polarization Potential.

$$R_0^2(E) = \left(\frac{\alpha_d e^2}{2E}\right)^{1/2}, \quad (43.333)$$

$$\sigma_{\text{orb}}(E) = 2\pi\left(\frac{\alpha_d e^2}{2E}\right)^{1/2} + \left(\frac{\pi De \cos\theta_d}{E}\right). \quad (43.334)$$

For a locked-in dipole, the orientation angle is $\theta_d = 0$, and $\sigma_{\text{orb}}(E) > 0$ for all θ_d when $E > E_c = (D^2/2\alpha_d)$.

On averaging over all θ_d from 0 to $\theta_{\max} = \left[\tfrac{1}{2}\pi + \sin^{-1}(2ER_0^2/De)\right]$, which satisfies $\sigma_{\text{orb}}(E) > 0$ for all E, then

$$\langle\sigma_{\text{orb}}(E)\rangle_{\theta_d} = \pi\left[\left(\frac{\alpha_d e^2}{2E}\right)^{1/2} + \left(\frac{\alpha_d e^2}{2E_c}\right)^{1/2}\right]$$

$$+ \frac{\pi De}{4E}\left[1 - \frac{E}{E_c}\right] \quad (43.335a)$$

$$\to \sigma_L(E) \text{ as } E \to E_c. \quad (43.335b)$$

(k) Fixed Dipole Plus Coulomb Repulsion.

$$R_0^2(E) = e^2/2E. \quad (43.336)$$

For all E and fixed rotations in the range $0 \leq \theta_d \leq \theta_{\max} = \cos^{-1}(e^2/2De)$,

$$\sigma_{\text{orb}}(E) = (\pi De \cos\theta_d)/E - \pi R_0^2(E). \quad (43.337)$$

(l) Lennard-Jones $(n,6)$. For the following two interactions, there are two roots of $E = V_{\text{eff}}(R_0) = V(R_0) + \tfrac{1}{2}R_0V'(R_0)$. They correspond to stable and unstable circular orbits [with different angular momenta associated with the minimum and maxima of the different $V_{\text{eff}}(R)$]. Analytical expressions can only be derived for

the orbiting cross section at the critical energy $E_{\max}$ above which no orbiting can occur.

For the Lennard-Jones $(n,6)$ potential, orbiting occurs for $E < E_{\max} = 2\epsilon\,[4/(n-2)]^{6/(n-6)}$. The orbiting radius at $E_{\max}$ is

$$R_0(E_{\max}) = R_e\,[(n-2)/4]^{1/(n-6)}.$$

The orbiting cross section at $E_{\max} = 2\epsilon(R_e/R_0)^6$ is

$$\sigma_{\text{orb}}(E_{\max}) = \pi b_0^2(E_{\max}) = \frac{3}{2}\pi R_0^2\left(\frac{n}{n-2}\right). \quad (43.338)$$

$n = 12$: $E_{\max} = 4\epsilon/5$, $R_0 = 1.165R_e$, $\sigma_{\text{orb}} = 2.4\pi R_e^2$.

(m) Polarization $(n,4)$. As discussed for case (l),

$$E_{\max} = \epsilon\left[\frac{2}{n-2}\right]^{4/(n-4)}, \quad (43.339)$$

$$R_0(E_{\max}) = R_e\left[\frac{n-2}{2}\right]^{1/(n-4)}, \quad (43.340)$$

$$\sigma_{\text{orb}}(E_{\max}) = 2\pi R_0^2\frac{n}{(n-2)}, \quad (43.341)$$

$n = 12$: $E_{\max} = \epsilon/\sqrt{5}$; $R_0 = 1.22R_e$; $\sigma_{\text{orb}} = 3.6\pi R_e^2$.

Small-Angle Scattering

For the power law potential $V(R) = -C/R^n$, Eq. (43.12) can be expanded in powers of $V(R)/E$ to obtain analytic expressions for χ and η_{JB}. The general form is

$$\chi(b) = \sum_{j=1}^{\infty}\left[\frac{V(b)}{E}\right]^j F_j(n) = \frac{2}{k}\frac{\partial\eta}{\partial b}, \quad (43.342)$$

$$F_j(n) = \frac{\pi^{1/2}\Gamma\left(\tfrac{1}{2}jn + \tfrac{1}{2}\right)}{\Gamma(j+1)\Gamma\left(\tfrac{1}{2}jn - j + 1\right)}. \quad (43.343)$$

The leading $j = 1$ terms equivalent to (43.16b) are $F_1(n) \equiv F(n)$, as defined following Eq. (43.255). Then to first order in V/E,

$$\eta_{\text{JB}} = -\left(\frac{k}{2E}\right)\left[\frac{CF(n)}{n-1}\right]b^{1-n}, \quad (43.344)$$

$$\frac{d\sigma}{d\Omega} = I_c(\theta) = \left[\frac{CF(n)}{E\theta}\right]^{2/n}\frac{1}{n\theta\sin\theta}. \quad (43.345)$$

From a log-log plot of $\sin\theta(d\sigma/d\Omega)$ versus E, C and n can both be determined.

The integral cross sections for scattering by $\theta \geq \theta_0$ is

$$\sigma(E) = 2\pi\int_{\theta_0}^{\pi} I_c(\theta)d(\cos\theta) = 2\pi\int_0^{b_{\max}} b\,db$$

$$= \pi\left[\frac{CF(n)}{E\theta_0}\right]^{2/n}, \quad (43.346)$$

where θ_0 is the smallest measured scattering angle corresponding to a trajectory with impact parameter, $b_{\max} = [CF(n)/E\theta_0]^{1/n}$. A plot of $\ln \sigma(E)$ versus $\ln E$ is a straight line with slope $(-2/n)$.

The Landau-Lifshitz cross section (43.255) and the Massey-Mohr cross section (43.256) follow from use of the JB phases (43.344).

The diffusion cross section in the Random Phase Approximation (43.49) is

$$\sigma_{\mathrm{d}}(E) = 4\pi \int_0^{b_c} \langle \sin^2 \theta/2 \rangle\, b\, db, \quad |\chi(b_c)| = \frac{2}{\pi}$$

$$= \pi(C/2E)^{2/n} [\pi F(n)]^{2/n}. \tag{43.347}$$

(n) Multiple-Term Power-Law Potentials.

$$\chi(E,b) = \frac{1}{E} [V_m(b)F(m) - V_n(b)F(n)]. \tag{43.348}$$

For example. a Lennard-Jones $(n,6)$ potential (see Table 43.5) has the following features:

Forward Glory: $\chi = 0$ when $b_g = \alpha_n^{1/(n-6)} R_e$,
where $\alpha_n = 6F(n)/[nF(6)]$.
Rainbow: $d\chi/db = 0$ at $b_r = (n\alpha_n/6)^{1/(n-6)} R_e$.

$$\chi_r = -F(n)(\epsilon/E)(R_e/b_r)^n, \tag{43.349}$$

$$\omega_r = \frac{1}{2} (d^2\chi_r/db^2)_r = \frac{3n}{b_r^2} |\chi(b_r)|. \tag{43.350}$$

(o) Exponential Potential.

$$\eta_{\mathrm{JB}}(E,b) = -\frac{1}{2}kb\frac{V_0}{E}K_1(\alpha b) \xrightarrow{\text{large } b} -\frac{1}{2}kb\frac{V(b)}{E}\left(\frac{\pi b}{2\alpha}\right)^{1/2}.$$

(p) Screened Coulomb Potential.

$$\chi(E,b) = \alpha (V_0/E) K_1(\alpha b)$$
$$\xrightarrow{\text{large } b} (\tfrac{1}{2}\pi\alpha b)^{1/2} V(b)/E, \tag{43.351}$$

$$\eta_{\mathrm{JB}}(E,b) = -\frac{k}{2E}V_0 K_0(\alpha b)$$
$$\xrightarrow{\text{large } b} -\frac{k}{2E}V(b)\left(\frac{\pi b}{2\alpha}\right)^{1/2}. \tag{43.352}$$

(q) Morse Potential.

$$\chi(E,b) = (2\beta b)\left(\frac{\epsilon}{E}\right)\left[e^{2\beta R_e}K_0(2\beta b) - e^{\beta R_e}K_0(\beta b)\right]$$
$$\xrightarrow{\text{large } b} (\pi\beta b)^{1/2}\left(\frac{\epsilon}{E}\right)\left[e^{2\beta(R_e-b)} - \sqrt{2}e^{\beta(R_e-b)}\right], \tag{43.353}$$

$$b_r = R_e + (2\beta)^{-1}\ln 2,$$
$$\chi_r = -(\pi\beta b_r)^{1/2}(\epsilon/2E),$$
$$\omega_r = \beta^2 |\chi_r| R_e^2.$$

Large-Angle Scattering

For power law potentials $V(R) = C/R^n$,

$$\chi(b) = \pi - \sum_{j=1}^n \left[\frac{E}{V(b)}\right]^{(2j-1)/n} G_j(n), \tag{43.354}$$

$$G_j(n) = \frac{(-1)^{j-1}}{\Gamma(j)\Gamma(k)}\left(\frac{2\pi^{1/2}}{n}\right)\Gamma\left(\frac{2j-1}{n}\right), \tag{43.355}$$

with $k = [(2j-1)/n] - j - \frac{1}{2}$. For the $j = 1$ term,

$$\chi(b) = \pi - \left[\frac{E}{C}\right]^{1/n} G_1(n)b, \tag{43.356}$$

$$I_c(\theta) = \frac{d\sigma}{d\Omega} = \left[\frac{C}{E}\right]^{2/n} G_1^{-2}(n), \tag{43.357}$$

which is isotropic. Including both $j = 1$ and 2 terms provides a good approximation to the entire repulsive branch of the deflection function χ. Series (43.354) for large angles and (43.342) for small angles eventually diverge for impact parameters $b < b_c$ and $b > b_c$, respectively, where

$$b_c = n^{1/2}\left(\frac{C}{2E}\right)^{1/n}\frac{|n-2|^{1/n}}{|n-2|^{1/2}}.$$

43.5.1 Born Amplitudes and Cross Sections for Model Potentials

$$k^2 = 2ME/\hbar^2, \quad K = 2k\sin\tfrac{1}{2}\theta,$$
$$U_0 = 2MV_0/\hbar^2, \quad U_0/k^2 = V_0/E.$$

(a) Exponential. $V(R) = V_0 \exp(-\alpha R)$.

$$f_{\mathrm{B}}(\theta) = -\frac{2\alpha U_0}{(\alpha^2 + K^2)^2}, \tag{43.358}$$

$$\sigma_{\mathrm{B}}(E) = \frac{16}{3}\pi U_0^2\left[\frac{3\alpha^4 + 12\alpha^2 k^2 + 16k^4}{\alpha^4(\alpha^2 + 4k^2)^3}\right]$$
$$\xrightarrow{E\to\infty} \frac{4}{3}\pi\left(\frac{V_0}{E}\right)\left(\frac{U_0}{\alpha^4}\right). \tag{43.359}$$

(b) Gaussian. $V(R) = V_0 \exp(-\alpha^2 R^2)$.

$$f_{\mathrm{B}}(\theta) = -\left(\frac{\pi^{1/2}U_0}{4\alpha^2}\right)\exp(-K^2/4\alpha^2), \tag{43.360}$$

$$\sigma_{\mathrm{B}}(E) = \left(\frac{\pi^2 U_0}{8\alpha^4}\right)\left(\frac{V_0}{E}\right)[1 - \exp(-2k^2/\alpha^2)]. \tag{43.361}$$

(c) Spherical Well. $V(R) = V_0$ for $R < a$, $V(R) = 0$ for $R > a$.

$$f_{\mathrm{B}}(\theta) = -\frac{U_0}{K^3}[\sin Ka - Ka\cos Ka], \tag{43.362}$$

$$\sigma_{\mathrm{B}}(E) = \frac{\pi}{2}\frac{V_0}{E}(U_0 a^4)\left[1 - (ka)^{-2} + (ka)^{-3}\sin 2ka\right.$$
$$\left. - (ka)^{-4}\sin^2 2ka\right]. \tag{43.363}$$

(d) Screened Coulomb. $V(R) = V_0 \exp(-\alpha R)/R$, $V_0 = Ze^2$, $U_0 = 2Z/a_0$.

$$f_B(\theta) = -\frac{U_0}{\alpha^2 + K^2}, \tag{43.364}$$

$$\sigma_B(E) = \frac{4\pi U_0^2}{\alpha^2(\alpha^2 + 4k^2)} \to \pi\left(\frac{V_0}{E}\right)\left(\frac{U_0}{\alpha^2}\right). \tag{43.365}$$

When $\alpha \to 0$, then $f_B(\theta) = -U_0/K^2$.

(e) e^--Atom.

$$V(R) = -Ne^2\left[Z/a_0 + 1/R\right]\exp(-2ZR/a_0), \tag{43.366}$$

H(1s): $N = 1$, $Z = 1$; He($1s^2$): $N = 2$; $Z = 27/16$.

$$f_B(\theta) = \frac{2N}{a_0}\left[\frac{2\alpha^2 + K^2}{(\alpha^2 + K^2)^2}\right], \quad \alpha = 2Z/a_0, \tag{43.367}$$

$$\sigma_B(E) = \frac{\pi a_0^2 N^2 \left[12Z^4 + 18Z^2k^2a_0^2 + 7k^4a_0^4\right]}{3Z^2(Z^2 + k^2a_0^2)^3}. \tag{43.368}$$

Also, f_B decomposes [*cf.* (43.185)] as

$$f_B(K) = f_B^{eZ}(K) + f_B^{ee}(K)F(K), \tag{43.369}$$

where f_B^{ij} are two-body Coulomb amplitudes for (i, j) scattering, and where

$$F(K) = \int |\Psi_0(\mathbf{R})|^2 \exp\left(i\mathbf{K}\cdot\mathbf{R}\right) d\mathbf{R} \tag{43.370}$$

is the elastic *form factor*.

(f) Dipole. $V(R) = V_0/R^2$.

$$f_B(\theta) = \pi U_0/2K. \tag{43.371}$$

(g) Polarization Potential. $V(R) = V_0/(R^2 + R_0^2)^2$.

$$f_B(\theta) = -\frac{1}{4}\pi\left(\frac{U_0}{R_0}\right)\exp(-KR_0), \tag{43.372}$$

$$\sigma_B(E) = \left(\frac{\pi^3 U_0}{32R_0^4}\right)\left(\frac{V_0}{E}\right)\left[1 - (1 + 4kR_0)\exp(-4kR_0)\right]. \tag{43.373}$$

Acknowledgments

This research is supported by the U. S. Air Force Office of Scientific Research under Grant No. F49620-94-1-0379. I wish to thank Dr. E. J. Mansky for numerous discussions on the content and form of this chapter and without whose expertise in computer typesetting this chapter would not have been possible.

GENERAL REFERENCES

(A) Textbooks on Scattering Theory.

N. F. Mott and H. S. W. Massey, *The Theory of Atomic Collisions* (Clarendon Press, Oxford, 1965), 3rd ed., Chaps. 2–5.

R. G. Newton, *Scattering Theory of Waves and Particles* (McGraw-Hill, New York, 1966).

L. S. Rodberg and R. M. Thaler, *Introduction to the Quantum Theory of Scattering* (Academic Press, New York, 1967).

Electronic and Ionic Impact Phenomena, edited by H. S. W. Massey, E. H. S. Burhop, and H. B. Gilbody (Clarendon Press, Oxford, 1969-1974), Vols. 1–5.

M. R. C. McDowell and J. P. Coleman, *Introduction to the Theory of Ion-Atom Collisions* (North-Holland, Amsterdam, 1970).

M. S. Child, *Molecular Collision Theory* (Academic Press, New York, 1974).

C. J. Joachain, *Quantum Collision Theory* (North-Holland, Amsterdam, 1975).

L. D. Landau and E. M. Lifshitz, *Course of Theoretical Physics* (Pergamon Press, Oxford, 1976), 3rd ed., Vol. 1, Chap. 4, p. 41.

H. Eyring, S. H. Lin, and S. M. Lin, *Basic Chemical Kinetics* (Wiley, New York, 1980), Chap. 3, p. 81.

B. H. Bransden, *Atomic Collision Theory* (Benjamin-Cummings, Menlo Park, 1983), 2nd ed.

B. H. Bransden and C. J. Joachain, *Physics of Atoms and Molecules* (Wiley, New York, 1983).

E. E. Nikitin and S. Ya. Umanskiĭ, *Theory of Slow Atomic Collisions* (Springer-Verlag, Berlin, 1984).

E. W. McDaniel, *Atomic Collisions: Electron and Photon Projectiles* (Wiley, New York, 1989), Chaps. 3 and 4, p. 71.

M. S. Child, *Semiclassical Mechanics with Molecular Applications* (Clarendon Press, Oxford, 1991).

A. G. Sitenko, *Scattering Theory* (Springer-Verlag, Berlin, 1991).

E. W. McDaniel, J. B. A. Mitchell, and M. E. Rudd, *Atomic Collisions: Heavy Particle Projectiles* (Wiley, New York, 1993), Chap. 1, p. 1.

(B) Chapters in Edited Books.

E. H. Burhop, *Theory of Collisions*, in *Quantum Theory I. Elements*, edited by D. R. Bates (Academic Press, New York, 1961), Chap. 9, p. 300.

E. A. Mason and J. T. Vanderslice, *High-Energy Elastic Scattering of Atoms, Molecules and Ions*, in *Atomic and Molecular Processes*, edited by D. R. Bates (Academic Press, New York, 1962), Chap. 17, p. 663.

B. L. Moiseiwitsch, *Elastic Scattering of Electrons*, in *Atomic and Molecular Processes*, edited by D. R. Bates (Academic Press, New York, 1962), Chap. 9, p. 281.

A. Dalgarno, *Theory of Ion-Molecule Collisions*, in *Ion-Molecule Reactions*, edited by E. W. McDaniel, V. Čermák, A. Dalgarno, E. E. Ferguson, and L. Friedman (Wiley, New York, 1970) Chap. 3, p. 159.

K. W. Ford and J. A. Wheeler, *Semiclassical Descrip-*

tion of Scattering, Ann. Phys. (NY) **7**, 259 (1959).

K. W. Ford and J. A. Wheeler, *Applications of Semiclassical Scattering Analysis, ibid.* 287 (1959).

M. V. Berry and K. E. Mount, *Semiclassical Approximations in Wave Mechanics*, Rep. Prog. Phys. **35**, 315 (1972).

W. H. Miller, *Classical-Limit Quantum Mechanics and the Theory of Molecular Collisions*, Adv. Chem. Phys. **25**, 69 (1974).

W. H. Miller, *The Classical S-Matrix in Molecular Collisions*, Adv. Chem. Phys. **30**, 77 (1975).

M. S. Child, *Semiclassical Methods in Molecular Collision Theory*, in *Dynamics of Molecular Collisions: Part B*, edited by W. H. Miller (Plenum Press, New York, 1976), Chap. 4, p. 171.

H. Pauly, *Elastic Scattering Cross Sections I: Spherical Potentials*, in *Atom-Molecule Collision Theory: A Guide for the Experimentalist*, edited by R. B. Bernstein (Plenum Press, New York, 1979), Chap. 4, p. 111.

S. Stolte and J. Reuss, *Elastic Scattering Cross Sections II: Noncentral Potentials* in *Atom-Molecule Collision Theory: A Guide for the Experime?ntalist*, edited by R. B. Bernstein (Plenum Press, New York, 1979), Chap. 5, p. 231.

J. N. L. Connor, *Semiclassical Theory of Elastic Scattering* in *Semiclassical Methods in Molecular Scattering and Spectroscopy*, edited by M. S. Child (D. Reidel, Boston, 1980), p. 45.

R. E. Johnson, *Atomic and Molecular Collisions* in *Encylopedia of Physical Science and Technology* (Academic Press, New York, 1987), Vol. 2, p. 224.

J. J. H. van Biesen, *Elastic Scattering I: Integral Cross Sections, ibid.* Vol. I, Chap. 19, p. 472.

U. Buck, *Elastic Scattering II: Differential Cross Sections* in *Atomic and Molecular Beam Methods*, edited by G. Scoles (Oxford University Press, New York, 1988), Vol. I, Chap. 20, p. 499.

REFERENCES

1. T. F. O'Malley, L. Spruch, and L. Rosenberg, J. Math. Phys. **2**, 491 (1961).
2. T. F. O'Malley, Phys. Rev. **130**, 1020 (1962).
3. L. M. Biberman and G. E. Norman, Soviet Phys. (JETP) **18**, 1353 (1964).
4. O. Hinckelmann and L. Spruch, Phys. Rev. A **3**, 642 (1971).
5. G. F. Gribakin and V. V. Flambaum, Phys. Rev. A **48**, 546 (1993).
6. Z. L. Petović , T. F. O'Malley, and R. W. Crompton, J. Phys. B: At. Mol. Opt. Phys. **28**, 3309 (1995).
7. K. W. Ford and J. A. Wheeler, Ann. Phys. (NY) **7**, 259 (1959).
8. K. W. Ford and J. A. Wheeler, *ibid.* 287 (1959).
9. M. V. Berry and K. E. Mount, Rep. Prog. Phys. **35**, 315 (1972).
10. *Electronic and Ionic Impact Phenomena*, edited by H. S. W. Massey, E. H. S. Burhop, and H. B. Gilbody (Clarendon Press, Oxford, 1969-1974), Vols. 1–5.
11. E. W. McDaniel, J. B. A. Mitchell, and M. E. Rudd, *Atomic Collisions: Heavy Particle Projectiles* (Wiley, New York, 1993), Chap. 1, p. 47.

44

Orientation and Alignment in Atomic and Molecular Collisions

Nils Andersen

Niels Bohr Institute, Ørsted Laboratory, Universitetsparken 5, DK-2100 Copenhagen, Denmark

44.1 INTRODUCTION

This chapter deals with the concepts of orientation and alignment in atomic and molecular physics. The terms refer to parameters related to the shape and dynamics of an excited atomic or molecular level, as it is manifested in a nonstatistical population of the magnetic sublevels. To take full advantages of the possibilities of this approach, one utilizes "third generation experiments," i.e., scattering experiments which exploit the planar scattering symmetry, contrary to an angular differential cross section determination (a "second generation experiment") having cylindrical symmetry, or a total cross section measurement (a "first generation experiment") integrating over all scattering angles. In this way one is often able to probe atomic collision theories

at a more fundamental level, and in favorable cases approach a "perfect scattering experiment" in which the complex quantal scattering amplitudes are completely determined. This term was coined by Bederson [1] and has since served as an ideal towards which scattering experiments attempt to strive.

The study of anisotropies has a long history in atomic physics, with light polarization, or Stokes parameter analysis, as a prominent example. A pioneering review by Fano and Macek [2] laid the mathematical and conceptual foundation for most of the later work. Advances in coincidence techniques, laser preparation methods, and development of efficient sources for polarized electrons [3] have boosted the field further. Parallel developments of powerful computational codes have enabled a matching theoretical effort. In this way, detailed insights into the

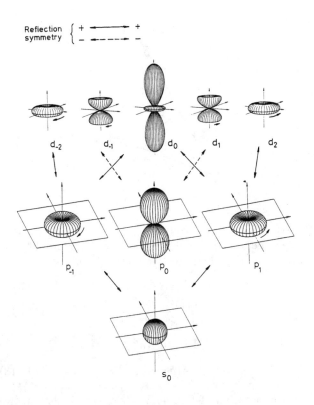

Figure 44.1. Illustration of the reflection symmetry of the simplest spherical harmonics, corresponding to S, P, and D states.

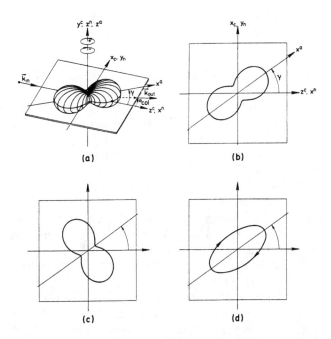

Figure 44.2. (a) shows the shape, or the angular part, of a P-state with positive reflection symmetry with respect to the scattering plane. Some relevant coordinate frames are also indicated. (b) shows a cut of this shape in the scattering plane. (c) is the angular distribution of photons emitted in the scattering plane. (d) is the polarization ellipse for light emitted in the direction perpendicular to the plane, as observed from above.

collision dynamics have been obtained, such as the locking radius model for low energy atomic collisions, propensity rules for orientation in fast electronic and atomic collisions, and spin effects in polarized electron-heavy atom scattering. Presentations of these developments are contained in Ref. [4] and Chaps. 36, 49, 48, 61, 62, and 64. A comprehensive and critical review of the literature and the mathematical formalism was initiated by NIST [5–7].

Space limitations allow only presentation of the formalism for the simplest case of excitation. Recent developments in the description of processes involving polarized beams are included. The presentation is restricted to atomic outer-shell excitation. Other reviews describe excitation of inner shells [8] and molecular levels [9]. Related material on density matrices is contained in Chap. 7 (see also Ref. [10]).

44.2 COLLISIONS INVOLVING UNPOLARIZED BEAMS

44.2.1 The Fully Coherent Case

Consider first the simplest nontrivial case of $S \to P$ excitation. A general property of an atomic collision is that the total reflection symmetry with respect to the

scattering plane of the total wave function describing the system is conserved. In simple cases, such as electron impact excitation of He singlet states, or atom excitation by fast, heavy particle impact, the projectile acts as a structureless particle, with only the target atom changing its quantum state. In this case, the reflection symmetry of the target wave function alone is conserved.

Figure 44.1 shows the angular parts of the simplest $|lm\rangle$ states with $l = 0, 1, 2$. The arrows indicate how the two families of states with positive or negative reflection symmetry may couple internally. Thus, in an $S \to P$ transition, an atom initially in an S-state may be excited to the (p_{+1}, p_{-1}) subspace, while the p_0 state is not accessible. A characteristic feature of the corresponding electron charge cloud is that it has "zero height", i.e., zero electron density in the direction perpendicular to the scattering plane. Furthermore, the expectation value of the orbital angular momentum has a nonvanishing component in this direction only.

The resulting electron charge cloud may thus have a shape as shown on Fig. 44.2(a). We shall now discuss the parametrization of the wave function of this state and analyze the connection between the wave function and the corresponding photon radiation pattern emitted when the state decays back to the initial S-state, using the properties of electric dipole radiation.

Basic Definitions and Coordinate Frames

The coordinate frames of particular use in describing the wave function are as follows. The *collision frame*, (x^c, y^c, z^c), is defined by $\hat{z}^c \parallel \mathbf{k}_{in}$ and $\hat{y}^c \parallel \mathbf{k}_{in} \times \mathbf{k}_{out}$. The *natural frame*, (x^n, y^n, z^n), is defined by $\hat{x}^n \parallel \mathbf{k}_{in}$ and $\hat{z}^n \parallel \mathbf{k}_{in} \times \mathbf{k}_{out}$. Finally, the *atomic frame* is identical to the natural frame, except that the frame is rotated by an angle γ around $z^n = z^a$ such that x^a is parallel to the major symmetry axis of the P-state charge cloud, as shown in Fig. 44.2. Most scattering calculations use the collision coordinate system as reference frame, while mathematical analysis is often most conveniently performed in the natural frame, where the algebra is particularly simple. Many expressions are even simpler in the atomic frame, but it has the disadvantage that the angle γ varies with collision parameters, such as impact velocity, impact parameter, etc.

The expansion of the P-state wave function in the three coordinate systems is

$$|\Psi\rangle = a^c_{+1}|p^c_{+1}\rangle + a^c_0|p^c_0\rangle + a^c_{-1}|p^c_{-1}\rangle \quad (44.1)$$
$$= a^n_{+1}|p^n_{+1}\rangle + a^n_{-1}|p^n_{-1}\rangle \quad (44.2)$$
$$= a^a_{+1}|p^a_{+1}\rangle + a^a_{-1}|p^a_{-1}\rangle . \quad (44.3)$$

Conservation of reflection symmetry in the scattering plane implies

$$a^c_{+1} = -a^c_{-1}, \quad (44.4)$$
$$a^n_0 = 0, \quad (44.5)$$
$$a^a_0 = 0. \quad (44.6)$$

The normalization condition implies

$$|a^c_0|^2 + 2|a^c_{+1}|^2 = |a^n_{+1}|^2 + |a^n_{-1}|^2 = |a^a_{+1}|^2 + |a^a_{-1}|^2 = 1. \quad (44.7)$$

Finally, the a_m coefficients are related to the scattering amplitudes f_m and the differential cross section $\sigma(\theta)$ by

$$a_m = f_m (k_{out}/k_{in})^{\frac{1}{2}} \sigma(\theta)^{-\frac{1}{2}} \quad (44.8)$$
$$\sigma(\theta) = (k_{out}/k_{in}) (|f^c_0|^2 + 2|f^c_{+1}|^2), \text{ etc.} \quad (44.9)$$

Except for an arbitrary phase factor, the pure state (44.1) may thus be characterized by two dimensionless parameters. Traditionally, they have been chosen as (λ, χ) defined as [11]

$$\lambda = \frac{|a^c_0|^2}{|a^c_0|^2 + 2|a^c_{+1}|^2}, \quad (44.10)$$
$$\chi = arg(a^c_{+1} a^{c*}_0). \quad (44.11)$$

An alternative parametrization $(L_\perp, \gamma)$ is given by [5]

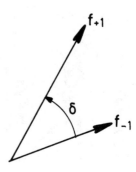

Figure 44.3. Fully coherent $S \rightarrow P$ excitation may be described by two independent scattering amplitudes, characterized by their relative size and phase.

$$L_\perp = \frac{|a^n_{+1}|^2 - |a^n_{-1}|^2}{|a^n_{+1}|^2 + |a^n_{-1}|^2}, \quad (44.12)$$
$$\gamma = \tfrac{1}{2} arg(a^n_{-1} a^{n*}_{+1}) \pm \frac{\pi}{2},$$
$$= -\tfrac{1}{2}(\delta \pm \pi), \quad (44.13)$$

with the notation of Fig. 44.3. Referring to the natural coordinate frame, the expectation value of the electronic orbital angular momentum is thus

$$\langle\Psi|\mathbf{L}|\Psi\rangle = (0, 0, L_\perp). \quad (44.14)$$

Coherence Analysis: Stokes Parameters

We now discuss the information obtainable from a polarization analysis of the emitted light. In the notation of classical optics (see, e.g., Born and Wolf [12]) the components of the Stokes vector (P_1, P_2, P_3) measured in the direction $+y^c$ $(+z^n)$ perpendicular to the scattering plane and defined by

$$IP_1 = I(0°) - I(90°), \quad (44.15)$$
$$IP_2 = I(45°) - I(135°), \quad (44.16)$$
$$IP_3 = I(RHC) - I(LHC), \quad (44.17)$$

with

$$I = I(0°) + I(90°)$$
$$= I(45°) + I(135°)$$
$$= I(RHC) + I(LHC), \quad (44.18)$$

are given by

$$P_1 = 2\lambda - 1, \quad (44.19)$$
$$P_2 = -2\sqrt{\lambda(1-\lambda)} \cos\chi, \quad (44.20)$$
$$P_3 = 2\sqrt{\lambda(1-\lambda)} \sin\chi, \quad (44.21)$$

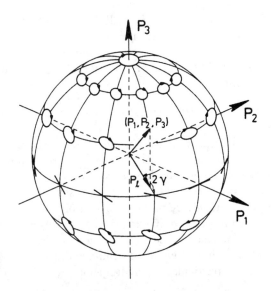

Figure 44.4. The Poincaré sphere. The Stokes vector for a pure state $|\Psi\rangle$ corresponds to a characteristic polarization ellipse, represented by a point on the unit sphere.

or, alternatively,

$$P_1 = P_l \cos 2\gamma, \tag{44.22}$$
$$P_2 = P_l \sin 2\gamma, \tag{44.23}$$
$$P_3 = -L_\perp, \tag{44.24}$$

with

$$P_l = \sqrt{P_1^2 + P_2^2}. \tag{44.25}$$

Here, $I(\theta)$ is the intensity transmitted through an ideal linear polarizer with transmission direction tilted by an angle θ with respect to the z^c or x^n direction. Similarly, RHC (LHC) refers to photons with negative (positive) helicity. Inspection of Eqs. (44.19) to (44.25) shows that determination of the Stokes vector in the direction perpendicular to the collision plane determines the wave function (44.1) completely. A determination of the Stokes vector thus constitutes a "perfect scattering experiment."

The Stokes vector (P_1, P_2, P_3) is a unit vector characterizing the state $|\Psi\rangle$, and it may be represented by a point on the Poincaré sphere, see Fig. 44.4.

Correlation Analysis

Another experimental approach is to map the angular distribution of the photons emitted in the subsequent $S \to P$ decay. For this purpose we first note that the angular part $\Upsilon(\theta, \phi)$ of the electron probability density

$\langle\Psi|\Psi\rangle$ corresponding to the wave function (44.1) may be written as

$$\Upsilon(\theta, \phi) = \tfrac{1}{2}[1 + P_l \cos 2(\phi - \gamma)] \sin^2 \theta. \tag{44.26}$$

Figure 44.2(b) shows a cut of this charge cloud in the scattering plane,

$$\Upsilon(\tfrac{1}{2}\pi, \phi) = \tfrac{1}{2}[1 + P_l \cos 2(\phi - \gamma)]. \tag{44.27}$$

The relative length, l, and width, w of the charge cloud are given by

$$l = \tfrac{1}{2}(1 + P_l), \tag{44.28}$$
$$w = \tfrac{1}{2}(1 - P_l). \tag{44.29}$$

The degree of linear polarization P_l is thus a measure of the shape of the charge cloud, since

$$P_l = \frac{l - w}{l + w}. \tag{44.30}$$

According to the properties of electric dipole radiation, the pattern $I(\phi)$ in the scattering plane is

$$I(\phi) = \tfrac{1}{2}[1 - P_l \cos 2(\phi - \gamma)]. \tag{44.31}$$

A mapping of the radiation pattern in the collision plane determines γ and P_l, and thereby the absolute value of the angular momentum by

$$|L_\perp| = \sqrt{1 - P_l^2}. \tag{44.32}$$

However, the sign of $L_\perp$ cannot be determined by correlation analysis, and this approach accordingly does not qualify for classification as "a perfect scattering experiment".

Density Matrix Representation

Recalling that $\rho_{mn} = a_m a_n^*$, the density matices in the various basis sets introduced above are

$$\rho^c = \frac{1}{\sqrt{8}} \begin{pmatrix} \frac{1}{\sqrt{2}}(1 - P_1) & -P_2 + iP_3 & -\frac{1}{\sqrt{2}}(1 - P_1) \\ -P_2 - iP_3 & \sqrt{2}(1 + P_1) & P_2 + iP_3 \\ -\frac{1}{\sqrt{2}}(1 - P_1) & P_2 - iP_3 & \frac{1}{\sqrt{2}}(1 - P_1) \end{pmatrix}, \tag{44.33}$$

$$\rho^n = \frac{1}{2} \begin{pmatrix} 1 + L_\perp & 0 & -P_l\, e^{-2i\gamma} \\ 0 & 0 & 0 \\ -P_l\, e^{2i\gamma} & 0 & 1 - L_\perp \end{pmatrix}, \tag{44.34}$$

$$\rho^a = \frac{1}{2} \begin{pmatrix} 1 + L_\perp & 0 & -P_l \\ 0 & 0 & 0 \\ -P_l & 0 & 1 - L_\perp \end{pmatrix}. \tag{44.35}$$

The density matrices illustrate the algebraic simplifications obtained using the natural coordinate frame. In the following, this frame will be used unless otherwise stated, and we therefore suppress the superscript n below.

Postcollisional Depolarization due to Fine Structure and Hyperfine Structure Effects

After the collision, the isolated atom evolves under the influence of internal forces, such as fine structure and hyperfine structure, until the optical decay. While typical collision times are of the order 10^{-15} s, the light emission occurs after a time interval of 10^{-9} s. This is long compared with the Larmor precession time of the electron spin ($\sim 10^{-12}$ s) and one can thus assume that the atom has completely relaxed into, e.g., its $^2P_{1/2}$ and $^2P_{3/2}$ states before the photon emission. For the simplest nontrivial case of electron spin $S = \frac{1}{2}$, the Stokes vector $\mathbf{P}(S)$ of the subsequent $^2P \to {}^2S$ transition is modified to become

$$P_{1,2}(\tfrac{1}{2}) = \tfrac{3}{7} P_{1,2}(0), \qquad (44.36)$$

$$P_3(\tfrac{1}{2}) = P_3(0), \qquad (44.37)$$

and the Stokes vector is thus no longer a unit vector. In general, the Stokes vector components are reduced by depolarization factors c_i (a table of c_i coefficients for the most common values of electron and nuclear spin is given in Appendix B of Ref. [5])

$$P_i(S) = c_i P_i(0). \qquad (44.38)$$

This suggests introduction of *reduced polarizations*

$$\overline{P}_i = P_i(S)/c_i. \qquad (44.39)$$

The *reduced* Stokes vector $(\overline{P}_1, \overline{P}_2, \overline{P}_3)$ is again a unit vector, and the formalism developed above for the spinless case can then be applied. We shall assume that this correction has been made, and the "bar" will be dropped.

44.2.2 The Incoherent Case with Conservation of Atomic Reflection Symmetry

The picture outlined above has to be modified in cases where the experiment sums over several, in principle distinguishable, channels, each of which, however, conserves reflection symmetry. Prototype examples are electron impact excitation of hydrogen or light alkali atoms. Here, the excitation process is described by singlet and triplet scattering amplitudes respectively, and we have the possibility of direct and exchange scattering. This doubles the number of scattering amplitudes from two to four. The experimental results are thus an incoherent sum over these channels (singlet and triplet), the unraveling of which would require application of spin-polarized beams, see Sect. 44.3. The atomic P-state can no longer be described as a single pure state, Eq. (44.1), but as a mixed state (see Chap. 7). The expressions (44.33) to (44.35) for the density matrix are unchanged, but the matrix elements are now sums of the contributions from the individual channels (We shall discuss its decomposition below). P_l and $L_\perp$ are now independent quantities, and the (reduced) Stokes vector $\mathbf{P}$ is generally no longer a unit vector. The degree of polarization P may thus be less than one, i.e.,

$$P^2 = P_1^2 + P_2^2 + P_3^2 = P_l^2 + L_\perp^2 \le 1. \qquad (44.40)$$

For electron impact excitation, deviation of the parameter P from unity may thus serve as a measure of the effect of electron exchange. We have accordingly three independent observables, e.g., $(L_\perp, \gamma, P_l)$. This set of variables is frame-independent. A photon correlation experiment in the scattering plane may extract the (γ, P_l) pair, while coherence analysis in the z direction provides the complete set. An alternative set of (frame-dependent) parameters used in particular for hydrogen is (λ, R, I), with

$$\lambda = \tfrac{1}{2}(1 + P_1), \qquad (44.41)$$

$$R = -\frac{1}{\sqrt{8}} P_2, \qquad (44.42)$$

$$I = \frac{1}{\sqrt{8}} P_3. \qquad (44.43)$$

44.2.3 The Incoherent Case without Conservation of Atomic Reflection Symmetry

In the general case, the assumption of positive reflection symmetry with respect to the scattering plane for the wave function of the excited atom cannot be maintained. For example, for electron impact excitation of a P-state of a heavy atom, spin-orbit effects may be so strong that they flip the electron spin, thereby allowing population of the $|p_0^n\rangle$ state. Thus, the total number of scattering amplitudes is now six. Typical examples are mercury, or the heavy noble gases. The total angular momentum of the excited state is $J = 1$, the fine structure being completely resolved in these cases. Strictly speaking, the density matrix elements no longer describe the electronic charge cloud, but rather the excited state $(J = 1)$ distribution. Similarly, $L_\perp$ should be replaced by $J_\perp$, but for simplicity we keep the notation. This state radiates as a set of classical oscillators, completely analogous to the $^1P_1$ state described above, so we shall maintain our previous notation, simply replacing the term charge cloud density by oscillator density. The main difference from the cases considered above is that the density now displays a *height*, see Fig. 44.5.

Blum, da Paixão and collaborators, Refs. [13, 14], were the first to formulate a parametrization of the general case. Here, we shall use as starting point the expression for the density matrix in the natural frame

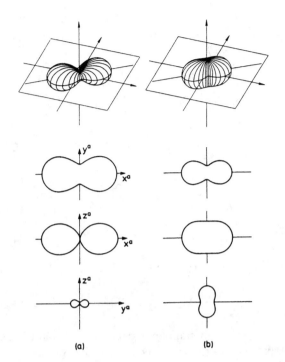

Figure 44.5. The top row shows classical oscillator densities for height parameters $h = 0$ and $h = 1/3$, respectively. The alignment angle is $\gamma = 35°$ and $P_l^+ = 0.6$ in both cases. Below are shown cuts along the symmetry axes.

and decompose it into the two terms with positive and negative reflection symmetry, respectively,

$$
\rho^n = (1 - h)\,\frac{1}{2}
\begin{pmatrix}
1 - P_3 & 0 & -P_l^+ e^{-2i\gamma} \\
0 & 0 & 0 \\
-P_l^+ e^{2i\gamma} & 0 & 1 + P_3
\end{pmatrix}
$$

$$
+ h
\begin{pmatrix}
0 & 0 & 0 \\
0 & 1 & 0 \\
0 & 0 & 0
\end{pmatrix},
\qquad (44.44)
$$

where the linear polarization P_l^+ is labeled with a "+" referring to the positive reflection symmetry. Similarly, we define

$$
L_\perp^+ \equiv -P_3\,, \qquad (44.45)
$$

$$
P^{+2} \equiv P_1^2 + P_2^2 + P_3^2 \le 1\,. \qquad (44.46)
$$

The shape of the density is now characterized by the three parameters

$$
l = (1 - h)\,\tfrac{1}{2}(1 + P_l)\,, \qquad (44.47)
$$

$$
w = (1 - h)\,\tfrac{1}{2}(1 - P_l)\,, \qquad (44.48)
$$

$$
h = \rho_{00}\,, \qquad (44.49)
$$

with $l + w + h = 1$. There are thus four independent observables, chosen as $(L_\perp^+, \gamma, P_l^+, h)$. Again, this set

Table 44.1. Summary of cases of increasing complexity, and the orientation and alignment parameters necessary for unpolarized beams. N_{p} is the number of independent parameters, and N_{d} is the number of observation directions required.

Variable	Sect. 44.2.1	Sect. 44.2.2	Sect. 44.2.3
Forces	Coulomb	+exchange	+spin-orbit
Representation	wave func.	ρ_{mn}	ρ_{mn}
Refl. symmetry	+	+	+, −
Ang. mom.	$L_\perp$	$L_\perp$	$L_\perp^+$
Align. angle	γ	γ	γ
Linear pol.	P_l	P_l	P_l^+
Degree of pol.	$P = 1$	$P \le 1$	$P^+ \le 1$
Height	$h = 0$	$h = 0$	$h \ge 0$
N_{p}	2	3	4
N_{d}	1	1	2

of variables is frame-independent. Determination of the height parameter evidently requires observation from a direction other than the z^n-direction. Traditionally, the Stokes parameter $(P_4, 0, 0)$ is measured from the y^n-direction to obtain

$$
h = \frac{(1 + P_1)(1 - P_4)}{4 - (1 - P_1)(1 - P_4)}\,. \qquad (44.50)
$$

Thus, all four parameters may be obtained from analysis of the light coherence, but two directions of observation are necessary. Similarly, photon correlation analysis in two planes are required to extract (γ, P_l^+, h), see, e.g., [5] for a discussion.

The various cases with unpolarized beams discussed in this Section are summarized in Table 44.1.

44.3 COLLISIONS INVOLVING SPIN-POLARIZED BEAMS

In this section we discuss the additional information that can be gained by application of a polarized beam compared with an unpolarized beam. Collisions with polarized electron beams are discussed in particular, but most of the ideas presented are easily generalized to beams of atoms in polarized, or otherwise prepared, states. We shall keep the order of increasing complexity introduced in the previous section.

44.3.1 The Fully Coherent Case

As seen in Sect. 44.2.1, a simple example of a fully coherent excitation process is He $1\,^1S \to n\,^1P$ excitation, for which complete information can be obtained from experiments using unpolarized beams. Consequently,

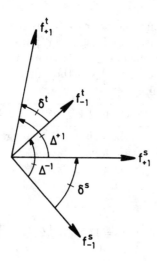

Figure 44.6. For $^2S \to {}^2P$ electron impact excitation of hydrogen or light alkali atoms, four scattering amplitudes come into play.

in this case application of polarized electrons will add nothing new. Also, the polarization of the scattered electron can be trivially predicted, since no change is possible in the scattering process.

44.3.2 The Incoherent Case with Conservation of Atomic Reflection Symmetry

The targets of interest here are hydrogen or light alkali atoms with an electron spin $S = \frac{1}{2}$. Since we now have the possibility of triplet (t) and singlet (s) scattering, this doubles the number of scattering amplitudes from two to four (recall that $f_0 = 0$), see Fig. 44.6.

The amplitudes of interest are

$$f^t_{+1} = \alpha_+ e^{i\phi_+} , \qquad (44.51)$$

$$f^t_{-1} = \alpha_- e^{i\phi_-} , \qquad (44.52)$$

$$f^s_{+1} = \beta_+ e^{i\psi_+} , \qquad (44.53)$$

$$f^s_{-1} = \beta_- e^{i\psi_-} . \qquad (44.54)$$

Neglecting an overall phase, seven independent parameters are needed to characterize the amplitudes completely. Traditionally, one is chosen as the differential cross section σ_u corresponding to unpolarized particles. Six additional dimensionless parameters may be defined: three to characterize the relative lengths of the four vectors, and three to define their relative phase angles.

The density matrix is parametrized, in analogy to the unpolarized beam case, according to [15]

$$\rho^t = \sigma^t \frac{1}{2} \begin{pmatrix} 1 + L^t_\perp & 0 & -P^t_l e^{-2i\gamma^t} \\ 0 & 0 & 0 \\ -P^t_l e^{2i\gamma^t} & 0 & 1 - L^t_\perp \end{pmatrix} , \qquad (44.55)$$

$$\rho^s = \sigma^s \frac{1}{2} \begin{pmatrix} 1 + L^s_\perp & 0 & -P^s_l e^{-2i\gamma^s} \\ 0 & 0 & 0 \\ -P^s_l e^{2i\gamma^s} & 0 & 1 - L^s_\perp \end{pmatrix} , \qquad (44.56)$$

where

$$\sigma^t = \alpha^2_+ + \alpha^2_- , \qquad (44.57)$$

$$\sigma^s = \beta^2_+ + \beta^2_- , \qquad (44.58)$$

$$L^t_\perp = \frac{1}{\sigma^t}(\alpha^2_+ - \alpha^2_-) , \qquad (44.59)$$

$$L^s_\perp = \frac{1}{\sigma^s}(\beta^2_+ - \beta^2_-) , \qquad (44.60)$$

$$P^t_l e^{2i\gamma^t} = P^t_1 + i P^t_2 = -\frac{2\alpha_+\alpha_-}{\sigma^t} e^{-i\delta^t} , \qquad (44.61)$$

$$P^s_l e^{2i\gamma^s} = P^s_1 + i P^s_2 = -\frac{2\beta_+\beta_-}{\sigma^s} e^{-i\delta^s} . \qquad (44.62)$$

In the case of an unpolarized beam, the total density matrix becomes the weighted sum of the two matrices ρ^s and ρ^t, i.e.,

$$\begin{aligned}
\rho_u &= \sigma_u \frac{1}{2} \begin{pmatrix} 1 + L_\perp & 0 & -P_l e^{-2i\gamma} \\ 0 & 0 & 0 \\ -P_l e^{2i\gamma} & 0 & 1 - L_\perp \end{pmatrix} \\
&= 3w^t \sigma_u \frac{1}{2} \begin{pmatrix} 1 + L^t_\perp & 0 & -P^t_l e^{-2i\gamma^t} \\ 0 & 0 & 0 \\ -P^t_l e^{2i\gamma^t} & 0 & 1 - L^t_\perp \end{pmatrix} \\
&+ w^s \sigma_u \frac{1}{2} \begin{pmatrix} 1 + L^s_\perp & 0 & -P^s_l e^{-2i\gamma^s} \\ 0 & 0 & 0 \\ -P^s_l e^{2i\gamma^s} & 0 & 1 - L^s_\perp \end{pmatrix}
\end{aligned} \qquad (44.63)$$

where

$$w^t = \frac{\sigma^t}{\sigma^s + 3\sigma^t} = \frac{\sigma^t}{4\sigma_u} , \qquad (44.64)$$

$$w^s = \frac{\sigma^s}{\sigma^s + 3\sigma^t} = \frac{\sigma^s}{4\sigma_u} = 1 - 3w^t , \qquad (44.65)$$

$$\sigma_u = (3w^t + w^s)\sigma_u = \tfrac{3}{4}\sigma^t + \tfrac{1}{4}\sigma^s . \qquad (44.66)$$

The six parameters $\sigma_u, w^t, L^t_\perp, L^s_\perp, \gamma^t,$ and γ^s have now been introduced, leaving one parameter still to be chosen. Inspection of Fig. 44.6 suggests, for example, the angle Δ^+. The fourth angle, Δ^-, is then fixed through the relation

$$\Delta^+ - \Delta^- = \delta^t - \delta^s = 2(\gamma^s - \gamma^t) . \qquad (44.67)$$

The following set of six dimensionless parameters is thus complete:

$$(w^t, L^t_\perp, L^s_\perp, \gamma^t, \gamma^s, \Delta^+) . \qquad (44.68)$$

Detailed recipes for extraction of the parameters from coherence experiments are somewhat complicated, and we refer to discussions in the literature [7, 15, 16]. Using spin-polarized electrons and spin-polarized targets, all parameters may be determined, except for information about singlet-triplet phase differences, such as Δ^+.

The reduced Stokes vector $\mathbf{P}$ of the unpolarized beam experiment is given by the singlet and triplet (unit) Stokes vectors $\mathbf{P}^{s,t}$ as

$$\mathbf{P} = 3 w^t \mathbf{P}^t + w^s \mathbf{P}^s, \qquad (44.69)$$

from which the set of parameters $(L_\perp, \gamma, P_l)$ for the unpolarized beam experiment may be evaluated from

$$L_\perp = 3 w^t L_\perp^t + w^s L_\perp^s, \qquad (44.70)$$
$$P_l\, e^{2i\gamma} = 3 w^t P_l^t\, e^{2i\gamma^t} + w^s P_l^s\, e^{2i\gamma^s}. \qquad (44.71)$$

Since in general, $L_\perp^t \neq L_\perp^s$ and $\gamma^t \neq \gamma^s$, this causes the (reduced) degree of polarization P to be smaller than unity.

To summarize this section, Stokes parameter analysis may provide five dimensionless parameters, the relative phase between the two f_{+1}^t and f_{-1}^t amplitudes and the relative phase between the two f_{+1}^s and f_{-1}^s amplitudes, as well as the relative sizes of all four amplitudes. However, none of the relative phases between any triplet and singlet amplitude can be determined, and coherence analysis alone is thus not able to provide a "perfect scattering experiment". The missing phase may be extracted from the *STU* parameters of the scattered electron, *cf.* Refs. [16, 17] and Chap. 7.

44.3.3 The Incoherent Case without Conservation of Atomic Reflection Symmetry

The results of coherence analysis of a $J = 0^e \rightarrow J = 1^o$ transition will now be discussed. We only analyze the photon polarization in the exit channel, not the electron spin parameters. There are six independent scattering amplitudes for a $J = 0 \rightarrow J = 1$ transition, see Fig. 44.7, thereby requiring the determination of one *absolute* differential cross section, five *relative* magnitudes, and another five *relative* phases of the scattering amplitudes.

This large number of independent parameters leads to considerable complications. Nevertheless, the natural coordinate system enables disentangling of the scattering amplitudes and generalization of the parametrization of the density matrix for the case of unpolarized beams in a straightforward way [7, 18]. The nonvanishing amplitudes $f^n(M_f, m_f, m_i)$ in the natural frame, see Fig. 44.7, for a $J = 0 \rightarrow J = 1$ transition are

$$f^n(\ 1,\ \tfrac{1}{2},\ \tfrac{1}{2}) \equiv f_{+1}^\uparrow = \alpha_+ e^{i\phi_+}, \qquad (44.72)$$
$$f^n(\ 1, -\tfrac{1}{2}, -\tfrac{1}{2}) \equiv f_{+1}^\downarrow = \beta_+ e^{i\psi_+}, \qquad (44.73)$$
$$f^n(-1,\ \tfrac{1}{2},\ \tfrac{1}{2}) \equiv f_{-1}^\uparrow = \alpha_{-e}^{i\phi_-}, \qquad (44.74)$$

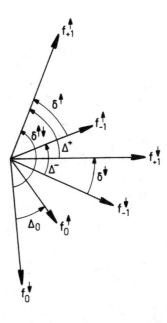

Figure 44.7. For $J = 0 \rightarrow J = 1$ electron impact excitation of heavy atoms six scattering amplitudes come into play since spin-flip may occur.

$$f^n(-1, -\tfrac{1}{2}, -\tfrac{1}{2}) \equiv f_{-1}^\downarrow = \beta_- e^{i\psi_-}, \qquad (44.75)$$
$$f^n(\ 0,\ \tfrac{1}{2}, -\tfrac{1}{2}) \equiv f_0^\downarrow = \beta_0 e^{i\psi_0}, \qquad (44.76)$$
$$f^n(\ 0, -\tfrac{1}{2},\ \tfrac{1}{2}) \equiv f_0^\uparrow = \alpha_{0e}^{i\phi_0}, \qquad (44.77)$$

where we have omitted $J_i = M_i = 0$ and $J_f = 1$. Equations (44.72)–(44.75) represent *noflip* amplitudes that leave the projectile spin unchanged while (44.76) and (44.77) describe the cases where the electron spin is *flipped*.

We first assume a polarization perpendicular to the scattering plane, i.e., along the z-direction. In Eq. (44.44), the density matrix for heavy atoms such as Xe or Hg was decomposed into a pair of matrices with one having positive reflection symmetry with respect to the scattering plane and the other one having negative reflection symmetry, respectively. The extension of this decomposition to the case of polarized electron beams is a pair of density matrices, one for spin-up electron impact excitation and one for spin-down excitation where "up" and "down" correspond to the initial spin component orientation with respect to the scattering plane. Hence,

$$
\begin{aligned}
\rho_u &= \sigma_u \left[(1-h)\frac{1}{2} \begin{pmatrix} 1+L_\perp^+ & 0 & -P_l^+ e^{-2i\gamma} \\ 0 & 0 & 0 \\ -P_l^+ e^{2i\gamma} & 0 & 1-L_\perp^+ \end{pmatrix} \right. \\
&\qquad \left. + h \begin{pmatrix} 0 & 0 & 0 \\ 0 & 1 & 0 \\ 0 & 0 & 0 \end{pmatrix} \right] \\
&= w^\uparrow \rho^\uparrow + w^\downarrow \rho^\downarrow
\end{aligned}
$$

$$= w^\uparrow \sigma_u \left[(1 - h^\uparrow) \frac{1}{2} \begin{pmatrix} 1 + L_\perp^{+\uparrow} & 0 & -P_l^{+\uparrow} e^{-2i\gamma^\uparrow} \\ 0 & 0 & 0 \\ -P_l^{+\uparrow} e^{2i\gamma^\uparrow} & 0 & 1 - L_\perp^{+\uparrow} \end{pmatrix} \right.$$

$$\left. + h^\uparrow \begin{pmatrix} 0 & 0 & 0 \\ 0 & 1 & 0 \\ 0 & 0 & 0 \end{pmatrix} \right]$$

$$+ w^\downarrow \sigma_u \left[(1 - h^\downarrow) \frac{1}{2} \begin{pmatrix} 1 + L_\perp^{+\downarrow} & 0 & -P_l^{+\downarrow} e^{-2i\gamma^\downarrow} \\ 0 & 0 & 0 \\ -P_l^{+\downarrow} e^{2i\gamma^\downarrow} & 0 & 1 - L_\perp^{+\downarrow} \end{pmatrix} \right.$$

$$\left. + h^\downarrow \begin{pmatrix} 0 & 0 & 0 \\ 0 & 1 & 0 \\ 0 & 0 & 0 \end{pmatrix} \right]. \tag{44.78}$$

Here we have defined

$$L_\perp^{+\uparrow} = \frac{\alpha_+^2 - \alpha_-^2}{\alpha_+^2 + \alpha_-^2} = -P_3^\uparrow, \tag{44.79}$$

$$L_\perp^{+\downarrow} = \frac{\beta_+^2 - \beta_-^2}{\beta_+^2 + \beta_-^2} = -P_3^\downarrow, \tag{44.80}$$

$$P_l^{+\uparrow} e^{2i\gamma^\uparrow} = P_1^\uparrow + iP_2^\uparrow = -\frac{2\alpha_+\alpha_- e^{i(\phi_- - \phi_+)}}{\alpha_+^2 + \alpha_-^2}, \tag{44.81}$$

$$P_l^{+\downarrow} e^{2i\gamma^\downarrow} = P_1^\downarrow + iP_2^\downarrow = -\frac{2\beta_+\beta_- e^{i(\psi_- - \psi_+)}}{\beta_+^2 + \beta_-^2}, \tag{44.82}$$

$$\sigma^\uparrow = \alpha_+^2 + \alpha_-^2 + \alpha_0^2, \tag{44.83}$$

$$\sigma^\downarrow = \beta_+^2 + \beta_-^2 + \beta_0^2, \tag{44.84}$$

$$\sigma_u = \frac{1}{2}\left(\alpha_+^2 + \alpha_-^2 + \alpha_0^2 + \beta_+^2 + \beta_-^2 + \beta_0^2\right)$$
$$= \frac{1}{2}\left(\sigma^\uparrow + \sigma^\downarrow\right), \tag{44.85}$$

$$h^\uparrow = \alpha_0^2/\sigma^\uparrow, \tag{44.86}$$

$$h^\downarrow = \beta_0^2/\sigma^\downarrow, \tag{44.87}$$

$$w^\uparrow = \sigma^\uparrow/(2\sigma_u), \tag{44.88}$$

$$w^\downarrow = \sigma^\downarrow/(2\sigma_u) = 1 - w^\uparrow. \tag{44.89}$$

From these definitions it follows that

$$(1 - h) L_\perp^+ = w^\uparrow (1 - h^\uparrow) L_\perp^{+\uparrow}$$
$$+ w^\downarrow (1 - h^\downarrow) L_\perp^{+\downarrow}, \tag{44.90}$$

$$(1 - h) P_l^+ e^{2i\gamma} = w^\uparrow (1 - h^\uparrow) P_l^{+\uparrow} e^{2i\gamma^\uparrow}$$
$$+ w^\downarrow (1 - h^\downarrow) P_l^{+\downarrow} e^{2i\gamma^\downarrow}, \tag{44.91}$$

$$h = w^\uparrow h^\uparrow + w^\downarrow h^\downarrow$$
$$= (\alpha_0^2 + \beta_0^2)/(2\sigma_u), \tag{44.92}$$

$$P_l^{+\uparrow 2} + L_\perp^{+\uparrow 2} = P_l^{+\uparrow 2} + P_3^{\uparrow 2} = 1, \tag{44.93}$$

$$P_l^{+\downarrow 2} + L_\perp^{+\downarrow 2} = P_l^{+\downarrow 2} + P_3^{\downarrow 2} = 1. \tag{44.94}$$

Extraction of these parameters is facilitated by introduction of "generalized Stoke parameters" [18]. In this way, $L_\perp^{+\uparrow}, L_\perp^{+\downarrow}, \gamma^\uparrow, \gamma^\downarrow$ may be determined. If, in addition, h is known, e.g., by polarization analysis in the y-direction, the following set of seven dimensionless independent parameters can be derived from the generalized Stokes parameters in the z-direction:

Table 44.2. Summary of cases of increasing complexity for spin-polarized beams. The number of independent dimensionless parameters N_p is listed, along with N_{OA}, the number determined from orientation and alignment only. N_d is the number of observation directions required.

Variable	Sect. 44.3.1	Sect. 44.3.2	Sect. 44.3.3
Forces	Coulomb	+exchange	+spin-orbit
Representation	wave func.	$\rho_{mn}^{t,s}$	$\rho_{mn}^{\uparrow,\downarrow}$
Refl. symmetry	+	+	+, −
N_p	2	6	10
N_{OA}	2	5	9
N_d	1	1	2

$$L_\perp^{+\uparrow}, L_\perp^{+\downarrow}, h^\uparrow, h^\downarrow, w^\uparrow; \gamma^\uparrow, \gamma^\downarrow. \tag{44.95}$$

This leaves three relative phases unknown. In the notation of Fig. 44.7, we see from inspection that

$$\Delta^+ - \Delta^- = \delta^\uparrow - \delta^\downarrow = 2\left(\gamma^\downarrow - \gamma^\uparrow\right), \tag{44.96}$$

in analogy to Eq. (44.67). A convenient choice for the remaining phase angles is $(\Delta^+, \Delta^0, \delta^{\uparrow\downarrow})$, with

$$\delta^{\uparrow\downarrow} \equiv \phi_+ - \psi_0. \tag{44.97}$$

A complete set of dimensionless independent parameters is then given by

$$(w^\uparrow, L_\perp^{+\uparrow}, L_\perp^{+\downarrow}, h^\uparrow, h^\downarrow, \gamma^\uparrow, \gamma^\downarrow, \Delta^+, \Delta^0, \delta^{\uparrow\downarrow}). \tag{44.98}$$

Information about the remaining three phase angles may be obtained in experiments with *in-plane* spin polarization. Further analysis shows that the generalized Stokes parameters in the y (or x) direction with in-plane spin polarization P_y or P_x provides two additional phases. None of the relative phases Δ^+ between $f_{+1}^\uparrow$ and $f_{+1}^\downarrow$, etc. enter. Determination of the final remaining angle requires determination of generalized STU parameters, describing the electron spin in the exit channel.

Table 44.2 summarizes the various cases with polarized beams discussed in this Section.

44.4 EXAMPLE

44.4.1 The First Born Approximation

As a simple, illustrative example, consider the predictions of the first Born approximation (FBA). Here, $S \to P$ excitation by electron impact is described as creation of a pure p-orbital along the direction of the linear momentum transfer $\Delta \mathbf{k} = \mathbf{k}_{in} - \mathbf{k}_{out}$, along which there is axial symmetry. Evidently

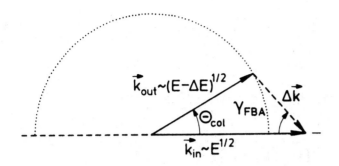

Figure 44.8. Diagram for evaluation of the alignment angle γ in the first Born approximation.

$$L_\perp^{\text{FBA}} = 0 \,, \qquad (44.99)$$

and consequently

$$P_l^{\text{FBA}} = 1 \,. \qquad (44.100)$$

The alignment angle is found from simple geometrical considerations, see Fig. 44.8. Denoting the incident energy by E and the energy loss by ΔE, the relation between the projectile scattering angle Θ_{col} and the alingment angle γ is directly read from the figure,

$$\tan \gamma^{\text{FBA}} = \frac{\sin \Theta_{\text{col}}}{\cos \Theta_{\text{col}} - x} \,, \qquad (44.101)$$

where $x = [E/(E - \Delta E)]^{1/2}$. For $\Delta E > 0$, γ^{FBA} is always negative, with its minimum value when $\mathbf{\Delta k} \perp \mathbf{k}_{\text{out}}$. Any theoretical effort beyond the FBA involves serious computations.

44.5 SUMMARY

A selection of fundamental formulas describing orientation and alignment in atomic collisions is given, with emphasis on the simplest case, $S \to P$ excitation. More details, and generalization to the case of higher angular momentum [19], may be found in the references below.

REFERENCES

1. B. Bederson, Comments At. Mol. Phys. **1**, 41 and 65 (1969).
2. U. Fano and J. H. Macek, Rev. Mod. Phys. **45**, 553 (1973).
3. J. Kessler, *Polarized Electrons* (Springer, New York, 1985).
4. I. V. Hertel, H. Schmidt, A. Bähring, and E. Meyer, Rep. Prog. Phys. **48**, 375 (1985).
5. N. Andersen, J. W. Gallagher, and I. V. Hertel, Phys. Rep. **165**, 1 (1988).
6. N. Andersen, J. T. Broad, E. E. B. Campbell, J. W. Gallagher, and I. V. Hertel, Phys. Rep. (to be published).
7. N. Andersen, K. Bartschat, J. T. Broad, and I. V. Hertel, Phys. Rep. (to be published).
8. U. Wille and R. Hippler, Phys. Rep. **132**, 129 (1986).
9. C. H. Greene and R. N. Zare, Ann. Rev. Phys. Chem. **33**, 119 (1982).
10. K. Blum, *Density Matrix Theory and Applications* (Plenum, New York, 1981).
11. J. Macek and D. H. Jaecks, Phys. Rev. A **4**, 2288 (1971).
12. M. Born and E. Wolf, *Principles of Optics* (Pergamon, New York, 1970).
13. K. Blum, F. T. da Paixão, and G. Csanak, J. Phys. B **13**, L257 (1980).
14. F. T. da Paixão, *et al.*, Phys. Rev. Lett. **45**, 1164 (1980).
15. I. V. Hertel, M. H. Kelley, and J. J. McClelland, Z. Phys. D **6**, 163 (1987).
16. N. Andersen and K. Bartschat, Comments At. Mol. Phys. **29**, 157 (1993).
17. K. Bartschat, Phys. Rep. **180**, 1 (1989).
18. N. Andersen and K. Bartschat, J. Phys. B **27**, 3189 (1994).
19. E. E. B. Campbell, J. Phys. Chem. **95**, 8177 (1991).

45

Electron-Atom, Electron-Ion, and Electron-Molecule Collisions

P. G. Burke

Department of Applied Mathematics and Theoretical Physics, The Queen's University of Belfast, Belfast BT7 1NN, Northern Ireland

45.1 INTRODUCTION

This chapter reviews the theory of electron collisions with atoms, ions and molecules. Section 45.2 discusses elastic, inelastic and ionizing collisions with atoms and atomic ions from close to threshold to high energies where the Born series becomes applicable. Section 45.3 extends the theory to treat electron collisions with molecules. Finally in Sect. 45.4 the theory of electron atom collisions in intense laser fields is discussed. This chapter will not present detailed comparisons of theoretical predictions with experiment. Such comparisons are given in recent review articles [1–3] and in Chap. 61.

45.2 ELECTRON-ATOM AND ELECTRON-ION COLLISIONS

45.2.1 Low-Energy Elastic Scattering and Excitation

In this section we consider the process

$$ e^- + A_i \rightarrow e^- + A_j \,, \tag{45.1} $$

where A_i and A_j are bound states of the target atom or ion and where the velocity of the incident or scattered electron is of the same order or less than that of the target electrons actively involved in the collision.

Assume initially that all relativistic effects can be neglected, which restricts the treatment to low-Z atoms and ions. The Schrödinger equation describing the scattering of an electron by a target atom or ion containing N electrons and nuclear charge Z is then

$$H_{N+1}\Psi = E\Psi, \qquad (45.2)$$

where E is the total energy of the system. The $(N+1)$-electron nonrelativistic Hamiltonian H_{N+1} is given in atomic units by

$$H_{N+1} = \sum_{i=1}^{N+1}\left(-\frac{1}{2}\nabla_i^2 - \frac{Z}{r_i}\right) + \sum_{i>j=1}^{N+1}\frac{1}{r_{ij}}, \qquad (45.3)$$

where $r_{ij} = |\mathbf{r}_i - \mathbf{r}_j|$, and $\mathbf{r}_i$ and $\mathbf{r}_j$ are the vector coordinates of electrons i and j relative to the origin of coordinates taken to be the target nucleus, which is assumed to have infinite mass.

The target eigenstates Φ_i and the corresponding eigenenergies w_i satisfy the equation

$$\langle\Phi_i|H_N|\Phi_j\rangle = w_i\delta_{ij}, \qquad (45.4)$$

where H_N is defined by (45.3) with $N+1$ replaced by N. The calculation of accurate target states is discussed in Chap. 21. The solution of (45.2), corresponding to the process (45.1), then has the asymptotic form

$$\Psi_i \underset{r\to\infty}{\sim} \Phi_i\chi_{\frac{1}{2}m_i}e^{ik_iz} + \sum_j \Phi_j\chi_{\frac{1}{2}m_j}f_{ji}(\theta,\phi)e^{ik_jr}.$$
$$\qquad (45.5)$$

In (45.5), $\chi_{\frac{1}{2}m_i}$ and $\chi_{\frac{1}{2}m_j}$ are the spin eigenfunctions of the incident and scattered electrons, where the direction of spin quantization is usually taken to be the incident beam direction, and $f_{ji}(\theta,\phi)$ is the scattering amplitude, the spherical polar coordinates of the scattered electron being denoted by r, θ and ϕ. Also the wave numbers k_i and k_j are related to the total energy of the system by

$$E = w_i + \frac{1}{2}k_i^2 = w_j + \frac{1}{2}k_j^2. \qquad (45.6)$$

The outgoing wave term in (45.5) contains contributions from all target states that are energetically allowed; i.e., for which $k_j^2 \geq 0$. If the energy is above the ionization threshold, this includes target continuum states. For an atomic ion, a logarithmic phase factor is also needed as discussed below.

The differential cross section for a transition from an initial state $|i\rangle = |\mathbf{k}_i, \Phi_i, \chi_{\frac{1}{2}m_i}\rangle$ to a final state $|j\rangle = |\mathbf{k}_j, \Phi_j, \chi_{\frac{1}{2}m_j}\rangle$ is given by

$$\frac{d\sigma_{ji}}{d\Omega} = \frac{k_j}{k_i}|f_{ji}(\theta,\phi)|^2, \qquad (45.7)$$

and the total cross section is obtained by averaging over initial spin states, summing over final spin states and integrating over all scattering angles.

In order to solve the Schrödinger equation to obtain the scattering amplitude and cross section at low energies, we make a partial wave expansion of the total wave function

$$\Psi_j^\Gamma(\mathbf{X}_{N+1})$$
$$= \mathcal{A}\sum_{i=1}^n \overline{\Phi}_i^\Gamma(\mathbf{x}_1,\ldots,\mathbf{x}_N;\hat{\mathbf{r}}_{N+1}\sigma_{N+1})r_{N+1}^{-1}F_{ij}^\Gamma(r_{N+1})$$
$$+ \sum_{i=1}^m \chi_i^\Gamma(\mathbf{x}_1,\ldots,\mathbf{x}_{N+1})b_{ij}^\Gamma, \qquad (45.8)$$

where $\mathbf{X}_{N+1} \equiv \mathbf{x}_1,\mathbf{x}_2\cdots\mathbf{x}_{N+1}$ represents the space and spin coordinates of all $N+1$ electrons, $\mathbf{x}_i \equiv \mathbf{r}_i\sigma_i$ represents the space and spin coordinates of the ith electron and $\mathcal{A}$ is the operator that antisymmetrizes the first summation with respect to exchange of all pairs of electrons in accordance with the Pauli exclusion principle. The channel functions $\overline{\Phi}_i^\Gamma$, assumed to be n in number, are obtained by coupling the orbital and spin angular momenta of the target states Φ_i with those of the scattered electron to form eigenstates of the total orbital and spin angular momenta, their z components and the parity π, where

$$\Gamma \equiv LM_LSM_S\pi \qquad (45.9)$$

is conserved in the collision. The square integrable correlation functions χ_i^Γ allow for additional correlation effects not included in the first expansion in (45.8) that goes over a limited number of target eigenstates, and possibly pseudostates.

By substituting (45.8) into the Schrödinger equation (45.2), projecting onto the channel functions $\overline{\Phi}_i^\Gamma$ and onto the square integrable functions χ_i^Γ, and eliminating the coefficients b_{ij}^Γ, we obtain n coupled integrodifferential equations satisfied by the reduced radial functions F_{ij}^Γ representing the motion of the scattered electron of the form

$$\left(\frac{d^2}{dr^2} - \frac{\ell_i(\ell_i+1)}{r^2} + \frac{2(Z-N)}{r} + k_i^2\right)F_{ij}^\Gamma(r)$$
$$= 2\sum_\ell \{V_{i\ell}^\Gamma(r)F_{\ell j}^\Gamma(r)$$
$$+ \int_0^\infty [K_{i\ell}^\Gamma(r,r') + X_{i\ell}^\Gamma(r,r')]F_{\ell j}^\Gamma(r')\,dr'\}. \quad (45.10)$$

Here ℓ_i is the orbital angular momentum of the scattered electron, and $V_{i\ell}^\Gamma$, $W_{i\ell}^\Gamma$ and $X_{i\ell}^\Gamma$ are the local direct, nonlocal exchange and nonlocal correlation potentials respectively. If the correlation potential which arises from the χ_i^Γ terms in (45.8) is not included, then (45.10) are called the close coupling equations.

The direct potential can be written as

$$V_{ij}^{\Gamma}(r_{N+1})$$

$$= \left\langle \overline{\Phi}_i^{\Gamma}(\mathbf{x}_1, \ldots, \mathbf{x}_N; \hat{\mathbf{r}}_{N+1}\sigma_{N+1}) \left| \sum_{i=1}^{N} \frac{1}{r_{iN+1}} - \frac{N}{r_{N+1}} \right| \right.$$

$$\left. \times \overline{\Phi}_j^{\Gamma}(\mathbf{x}_1, \ldots, \mathbf{x}_N; \hat{\mathbf{r}}_{N+1}\sigma_{N+1}) \right\rangle, \qquad (45.11)$$

where the integral is taken over all electron space and spin coordinates, except the radial coordinate of the $(N+1)$th electron. This potential has the asymptotic form

$$V_{ij}^{\Gamma}(r) = \sum_{\lambda=1}^{\lambda\,\mathrm{max}} a_{ij}^{\lambda} r^{-\lambda-1}, \; r \geq a \qquad (45.12)$$

where a is the range beyond which the orbitals in the target states Φ_i included in the first expansion in (45.8), are negligible. The $\lambda = 1$ term in (45.12) gives rise, in second-order, to the long-range attractive polarization potential

$$V(r) \underset{r \to \infty}{\to} -\frac{1}{2}\frac{\alpha}{r^4} \qquad (45.13)$$

seen by an electron incident on an atom. For an s-state atom in a state Φ_0 the dipole polarizability α is given by

$$\alpha = 2 \sum_j \frac{|\langle \Phi_0|(\frac{4\pi}{3})^{1/2}\sum_{i=1}^{N} r_i Y_{10}(\hat{\mathbf{r}}_i)|\Phi_j\rangle|^2}{w_j - w_0}. \qquad (45.14)$$

These long-range potentials have a profound influence on low-energy scattering.

The exchange and correlation potentials, unlike the direct potential, are both nonlocal and the exchange potential vanishes exponentially for large r. Explicit expressions for these potentials are too complicated to write down, except in the case of e^--H scattering where the direct and exchange potentials were first given by Percival and Seaton [4]. Instead they are determined by general computer programs.

The scattering amplitude and cross section can be obtained by solving (45.10) for all relevant conserved quantum numbers Γ subject to the following K-matrix asymptotic boundary conditions

$$F_{ij}^{\Gamma} \underset{r \to \infty}{\sim} k_i^{-\frac{1}{2}}(\sin \theta_i \delta_{ij} + \cos \theta_i K_{ij}^{\Gamma}),$$
$$\text{open channels } k_i^2 \geq 0$$

$$F_{ij}^{\Gamma} \underset{r \to \infty}{\sim} 0 \, , \text{ closed channels } k_i^2 < 0 \qquad (45.15)$$

where

$$\theta_i = k_i r - \frac{1}{2}\ell_i \pi + \frac{z}{k_i}\ln(2k_i r) + \sigma_i \qquad (45.16)$$

with $z = Z - N$, and $\sigma_i = \arg\Gamma(\ell_i + 1 - iz/k_i)$. The S-matrix and T-matrix are related to the K-matrix defined by (45.15) by the matrix equations

$$\mathbf{S}^{\Gamma} = \frac{\mathbf{I} + i\mathbf{K}^{\Gamma}}{\mathbf{I} - i\mathbf{K}^{\Gamma}}, \qquad \mathbf{T}^{\Gamma} = \mathbf{S}^{\Gamma} - \mathbf{I} = \frac{2i\mathbf{K}^{\Gamma}}{\mathbf{I} - i\mathbf{K}^{\Gamma}}, \qquad (45.17)$$

where the dimensions of the matrices in these equations are $n_a \times n_a$, where n_a is the number of open channels at the energy under consideration for the given Γ. The Hermiticity and time reversal invariance of the Hamiltonian ensures that $\mathbf{K}^{\Gamma}$ is real and symmetric, and $\mathbf{S}^{\Gamma}$ is unitary and symmetric.

The scattering amplitude defined by (45.5) can be expressed in terms of the T-matrix elements. For a neutral target,

$$f_{ji}(\theta, \phi)$$

$$= i\left(\frac{\pi}{k_i k_j}\right)^{\frac{1}{2}} \sum_{\substack{LS\pi \\ \ell_i \ell_j}} i^{\ell_i - \ell_j}(2\ell_i + 1)^{\frac{1}{2}}(L_i M_{L_i}\ell_i 0|L_i\ell_i L M_L)$$

$$\times (S_i M_{S_i}\tfrac{1}{2}m_i|S_i\tfrac{1}{2}S M_S)(L_j M_{L_j}\ell_j m_{\ell_j}|L_j\ell_j L M_L)$$

$$\times (S_j M_{S_j}\tfrac{1}{2}m_j|S_j\tfrac{1}{2}S M_S)T_{ji}^{\Gamma}Y_{\ell_j m_{\ell_j}}(\theta, \phi) \, , \qquad (45.18)$$

which describes a transition from an initial state $\alpha_i L_i S_i M_{L_i} M_{S_i} m_i$ to a final state $\alpha_j L_j S_j M_{L_j} M_{S_j} m_j$, where α_i and α_j represent any additional quantum numbers required to completely define the initial and final states. The corresponding total cross section, obtained by averaging over the initial magnetic quantum numbers, summing over the final magnetic quantum numbers, and integrating over all scattering angles, is

$$\sigma_{\mathrm{tot}}(i \to j) = \frac{\pi}{k_i^2}\sum_{\substack{LS\pi \\ \ell_i\ell_j}}\frac{(2L+1)(2S+1)}{2(2L_i+1)(2S_i+1)}|T_{ji}^{\Gamma}|^2,$$

$$(45.19)$$

which describes a transition from an initial target state $\alpha_i L_i S_i$ to a final target state $\alpha_j L_j S_j$. In applications, it is also useful to define a collision strength by

$$\Omega(i, j) = k_i^2 (2L_i + 1)(2S_i + 1)\sigma_{\mathrm{tot}}(i \to j), \qquad (45.20)$$

which is dimensionless and symmetric with respect to interchange of the initial and final states denoted by i and j. For scattering by an ion, the above expression for $f_{ji}(\theta, \phi)$ is modified by the inclusion of the Coulomb scattering amplitude when the initial and final states are identical.

For incident electron energies insufficient to excite the atom or ion, only elastic scattering is possible and the above expressions simplify. Consider low energy elastic electron scattering by a neutral atom is a ^1S ground state. Then the expression for the scattering amplitude (45.18) reduces to

$$f(\theta) = \frac{1}{2ik}\sum_{\ell=0}^{\infty}(2\ell+1)\left(e^{2i\delta_\ell} - 1\right)P_\ell(\cos\theta), \qquad (45.21)$$

where $\ell\,(=L=\ell_i=\ell_j)$ is the angular momentum of the scattered electron, k is its wave number, and the phase shift δ_ℓ can be expressed in terms of the K-matrix, which now has only one element since $n_a=1$, by

$$\tan\delta_\ell = K_{11}^\Gamma. \qquad (45.22)$$

The corresponding expression for the total cross section is then

$$\sigma_{\text{tot}} = \frac{4\pi}{k^2}\sum_{\ell=0}^\infty (2\ell+1)\sin^2\delta_\ell. \qquad (45.23)$$

A diffusion, or momentum transfer cross section, can also be defined as

$$\sigma_D = 2\pi\int_0^\pi |f(\theta)|^2 (1-\cos\theta)\sin\theta\,d\theta$$
$$= \frac{4\pi}{k^2}\sum_{\ell=0}^\infty (\ell+1)\sin^2(\delta_{\ell+1}-\delta_\ell), \qquad (45.24)$$

which is important when considering the diffusion of electrons through gases.

At low incident electron energies, the behavior of the phase shift for an atom in an s-state is dominated by the long-range polarization potential (45.13). O'Malley *et al.* [5] showed that for s-wave scattering $k\cot\delta_s$ satisfies the effective range expansion

$$k\cot\delta_0 = -\frac{1}{a_s} + \frac{\pi\alpha}{3a_s^2}k + \frac{2\alpha}{3a_s}k^2\ln\left(\frac{\alpha k^2}{16}\right) + O\left(k^2\right), \qquad (45.25)$$

where a_s is the scattering length, while for $\ell\geq 1$

$$k^2\cot\delta_\ell = \frac{8\left(\ell+\frac{3}{2}\right)\left(\ell+\frac{1}{2}\right)\left(\ell-\frac{1}{2}\right)}{\pi\alpha} + \cdots. \qquad (45.26)$$

It follows that close to threshold, the total elastic cross section has the form

$$\sigma_{\text{tot}} = 4\pi a_s^2 + \frac{8}{3}\pi^2\alpha a_s k + \cdots. \qquad (45.27)$$

When an electron is elastically scattered by a positive or negative ion, then these formulae for the low-energy behavior of the phase shift are modified. For scattering by a positive ion, Seaton [6] has shown that

$$\frac{\cot\delta_\ell(k)}{1-e^{2\pi\eta}} = \cot\left[\pi\mu(k^2)\right], \qquad (45.28)$$

where $\eta = -z/k$, and where $\mu\left(k^2\right)$ is the analytic continuation of the quantum defects of the electron-ion bound states to positive energies. This quantum defect theory enables spectroscopic observations of bound state energies to be extrapolated to positive energies to yield electron-ion scattering phase shifts. For a negative ion, where the Coulomb potential is repulsive, the phase shift behaves as

$$\delta_\ell \underset{k\to 0}{\longrightarrow} \exp(2\pi z/k), \qquad (45.29)$$

which vanishes rapidly as k tends to zero since z is now negative.

45.2.2 Relativistic Effects for Heavy Atoms and Ions

As the nuclear charge Z of the target increases, relativistic effects become important even for low energy scattering. There are two ways in which relativistic effects play a role. First, there is a direct effect corresponding to the relativistic distortion of the wave function describing the scattered electron by the strong nuclear Coulomb potential. Second, there is an indirect effect caused by the change in the charge distribution of the target due to the use of relativistic wave functions discussed in Chap. 22. We will concentrate on the direct effect in this section.

For atoms and ions with small Z, the K-matrices can first be calculated in LS coupling, neglecting relativistic effects. The K-matrices are then recoupled to yield transitions between fine-structure levels. We introduce the pair-coupling scheme

$$\mathbf{L}_i + \mathbf{S}_i = \mathbf{J}_i, \quad \mathbf{J}_i + \boldsymbol{\ell}_i = \mathbf{K}_i, \quad \mathbf{K}_i + \mathbf{s} = \mathbf{J}, \qquad (45.30)$$

where $\mathbf{J}_i$ is the total angular momentum of the target, $\boldsymbol{\ell}_i$ is the orbital angular momentum of the scattered electron, $\mathbf{s}$ is its spin, and $\mathbf{J}$ is the total angular momentum, which with the parity π is conserved in the collision. The transition from LS coupling involves the recoupling coefficient

$$\langle[(L_iS_i)J_i,\ell_i]K_i,\tfrac{1}{2};JM_J|(L_i\ell_i)L,(S_i\tfrac{1}{2})S;JM_J\rangle$$
$$= [(2J_i+1)(2L+1)(2K_i+1)(2S+1)]^{\frac{1}{2}}$$
$$\times W(L\ell_iS_iJ_i;L_iK_i)W(LJS_i\tfrac{1}{2};SK_i), \qquad (45.31)$$

and the corresponding K-matrix transforms as

$$K_{ij}^{J\pi}$$
$$= \sum_{LS}\langle((L_iS_i)J_i,\ell_i)K_i,\tfrac{1}{2};JM_J|(L_i\ell_i)L,(S_i\tfrac{1}{2})S;JM_J\rangle$$
$$\times K_{ij}^\Gamma$$
$$\times \langle(L_j\ell_j)L,(S_j\tfrac{1}{2})S;JM_J|((L_jS_j)J_j,\ell_j)K_j,\tfrac{1}{2};JM_J\rangle. \qquad (45.32)$$

This transformation has been implemented in a computer program by Saraph [7].

For intermediate-Z atoms and ions, relativistic effects can be included by adding terms from the Breit-Pauli Hamiltonian to the nonrelativistic Hamiltonian (Jones [8], Scott and Burke [9]). We write

$$H_{N+1}^{\text{BP}} = H_{N+1}^{\text{nr}} + H_{N+1}^{\text{rel}} \qquad (45.33)$$

where H_{N+1}^{nr} is defined by (45.3) and H_{N+1}^{rel} consists of one- and two-body relativistic terms. The one-body terms are (*cf.* Sect. 21.1)

$$H_{N+1}^{\text{mass}} = -\frac{1}{8}\alpha^2 \sum_{i=1}^{N+1} \nabla_i^4 \qquad \text{mass-correction term},$$

$$H_{N+1}^{\text{D}_1} = -\frac{1}{8}\alpha^2 Z \sum_{i=1}^{N+1} \nabla_i^2 \left(\frac{1}{r_i}\right) \quad \text{one-body Darwin term},$$

$$H_{N+1}^{\text{so}} = \frac{1}{2}\alpha^2 \sum_{i=1}^{N+1} \frac{1}{r_i}\frac{\partial V}{\partial r_i}\left(\boldsymbol{\ell}_i\cdot\mathbf{s}_i\right) \qquad \text{spin-orbit term}.$$

The two-body terms are less important and have not yet been included in collision calculations.

The modified Schrödinger equation defined by (45.2), with H_{N+1} replaced by H_{N+1}^{BP}, is solved by adopting an expansion similar in form to (45.8), but now using the pair coupling scheme in the definition of the channel functions and quadratically integrable functions. We then obtain coupled integrodifferential equations similar in form to (45.10), from which the K-matrix, S-matrix and T-matrix can be obtained. The corresponding total cross section in the pair-coupling scheme analogous to (45.19) is

$$\sigma_{\text{tot}}\left(i \to j\right) = \frac{\pi}{2k_i^2\left(2J_i+1\right)} \sum_{\substack{J\pi \\ K_i K_j \ell_i \ell_j}} \left(2J+1\right)\left|T_{ji}^{J\pi}\right|^2,$$

$$(45.34)$$

which describes a transition from an initial target state $\alpha_i J_i$ to a final target state $\alpha_j J_j$. The corresponding collision strength is

$$\Omega\left(i,j\right) = k_i^2\left(2J_i+1\right)\sigma_{\text{tot}}\left(i \to j\right). \qquad (45.35)$$

For high-Z atoms and ions, the Dirac Hamiltonian [10, 11] (*cf.* Sect. 22.2)

$$H_{N+1}^{\text{D}} = \sum_{i=1}^{N+1}\left(c\boldsymbol{\alpha}\cdot\mathbf{p}_i + \beta'c^2 - \frac{Z}{r_i}\right) + \sum_{i>j=1}^{N+1}\frac{1}{r_{ij}} \quad (45.36)$$

must be used instead of (45.3), where $\beta' = \beta - 1$ and $\boldsymbol{\alpha}$ and β are the usual Dirac matrices. The expansion of the total wave functions for a particular $JM_J\pi$ takes the general form of (45.8). However, now both the bound orbitals in the target, and correlation functions and the orbitals representing the scattered electron are represented by Dirac orbitals. These are defined in terms of large and small components $P(r)$ and $Q(r)$ by

$$\phi\left(\mathbf{r},\sigma\right) = \frac{1}{r}\begin{pmatrix} P_a\left(r\right)\chi_{\kappa m}\left(\hat{\mathbf{r}},\sigma\right) \\ Q_a\left(r\right)\chi_{-\kappa m}\left(\hat{\mathbf{r}},\sigma\right) \end{pmatrix} \qquad (45.37)$$

for the bound orbitals, and

$$F\left(\mathbf{r},\sigma\right) = \frac{1}{r}\begin{pmatrix} P_c\left(r\right)\chi_{\kappa m}\left(\hat{\mathbf{r}},\sigma\right) \\ Q_c\left(r\right)\chi_{-\kappa m}\left(\hat{\mathbf{r}},\sigma\right) \end{pmatrix} \qquad (45.38)$$

for the continuum orbitals, where $a = n\kappa m$, $c = k\kappa m$ and the spherical spinor

$$\chi_{\kappa m}\left(\hat{\mathbf{r}},\sigma\right) = \sum_{m_\ell m_i}\left(\ell m_\ell \tfrac{1}{2}m_i | \ell \tfrac{1}{2} jm\right)Y_{\ell m_\ell}\left(\theta,\phi\right)\chi_{\frac{1}{2}m_i}\left(\sigma\right),$$

$$(45.39)$$

where $\kappa = j + \frac{1}{2}$ when $\ell = j + \frac{1}{2}$, and $\kappa = -j - \frac{1}{2}$ when $\ell = j - \frac{1}{2}$. We can now derive coupled integrodifferential equations for the functions $P_c\left(r\right)$ and $Q_c\left(r\right)$ in a similar way to the derivation of (45.10), except that these are now coupled first-order equations instead of coupled second-order equations. The K-matrix, and hence the S-matrix and T-matrix, can be obtained from the asymptotic form of these equations. The total cross section in the j-j coupling scheme, is then given by (45.34), and the corresponding collision strength is given by (45.35).

45.2.3 Multichannel Resonance Theory

General resonance theories have been developed by Fano [12], [13], Feshbach [14], [15], and Brenig and Haag [16]. They are also discussed in Chap. 25. Here we will limit our discussion to the effect that resonances have on electron collision cross sections.

Following Feshbach, we introduce the projection operators P and Q, where P projects onto a finite set of low energy channels in (45.8) and Q projects onto the orthogonal space, where we restrict our consideration to the space corresponding to a particular set of conserved quantum numbers Γ. In this space we have

$$P^2 = P, \quad Q^2 = Q, \quad P + Q = 1. \qquad (45.40)$$

The Schrödinger equation (45.2) can then be written as

$$P\left(H - E\right)\left(P + Q\right)\Psi = 0 \qquad (45.41)$$

and

$$Q\left(H - E\right)\left(P + Q\right)\Psi = 0 \qquad (45.42)$$

where we have omitted the subscript $N+1$ on H and the superscript Γ on Ψ. After solving (45.42) for $Q\Psi$ and substituting into (45.41), we find that

$$P\left(H - PHQ\frac{1}{Q\left(H - E\right)Q}QHP - E\right)P\Psi = 0,$$

$$(45.43)$$

where the term

$$V_{\text{op}} = -PHQ\frac{1}{Q\left(H - E\right)Q}QHP, \qquad (45.44)$$

called the *optical potential*, allows for propagation in the Q-space channels.

We now introduce the eigenfunctions ϕ_i and eigenvalues ε_i of the operator QHQ by

$$QHQ\phi_i = \varepsilon_i\phi_i. \qquad (45.45)$$

It follows that the discrete eigenvalues ε_i each give rise to poles in V_{op} at ε_i. If the energy E is in the neighborhood of an isolated pole or bound state ε_i, we can rewrite (45.43) as

$$\left(PHP - \sum_{j \neq i} PHQ \frac{|\phi_j\rangle\langle\phi_j|}{\varepsilon_j - E} QHP - E \right) P\Psi$$
$$= PHQ \frac{|\phi_i\rangle\langle\phi_i|}{\varepsilon_i - E} QHP\Psi, \qquad (45.46)$$

where the rapidly varying part of the optical potential has been separated and put on the right-hand-side of (45.46). This equation can be solved by introducing the Green's function G_0 and the solutions ψ_{0j} of the operator on the left-hand side of (45.46). We find that the pole term on the right-hand side of this equation gives rise to a Feshbach resonance whose position is

$$E_i = \varepsilon_i + \triangle_i - \tfrac{1}{2}i\Gamma_i = E_{i,r} - \tfrac{1}{2}i\Gamma_i, \qquad (45.47)$$

where the resonance shift is

$$\triangle_i = \langle\phi_i|QHP\, G_0\, PHQ|\phi_i\rangle, \qquad (45.48)$$

and the resonance width is

$$\Gamma_i = 2\pi \sum_j |\langle\phi_i \mid QHP \mid \psi_{0j}\rangle|^2, \qquad (45.49)$$

where the summation in this equation is taken over all continuum states corresponding to the operator on the left-hand-side of (45.46) and these states are normalized to a delta function in energy.

In the neighborhood of the resonance energy $E_{i,r}$, the S-matrix is rapidly varying with the form

$$\mathbf{S} = \mathbf{S}_0^{\frac{1}{2}} \left(\mathbf{I} - i\Gamma \frac{\boldsymbol{\gamma}_i \times \boldsymbol{\gamma}_i}{E - E_{i,r} + \tfrac{1}{2}i\Gamma_i} \right) \mathbf{S}_0^{\frac{1}{2}}, \qquad (45.50)$$

where $\mathbf{S}_0$ is the slowly varying nonresonant or background S-matrix corresponding to $\boldsymbol{\psi}_0$, and the partial widths $\boldsymbol{\gamma}_i$ are defined by

$$\langle\phi_i \mid QHP \mid \boldsymbol{\psi}_0\rangle = \Gamma_i^{\frac{1}{2}} \boldsymbol{\gamma}_i \cdot \mathbf{S}_0^{\frac{1}{2}}, \qquad (45.51)$$

where $\boldsymbol{\gamma}_i \cdot \boldsymbol{\gamma}_i = 1$. A corresponding resonant expression can be derived for the K-matrix (see, e.g., Burke [17]).

Let us now diagonalize the S-matrix as follows:

$$\mathbf{S} = \mathbf{A} \exp\left(2i\boldsymbol{\Delta}\right) \mathbf{A}^{\mathrm{T}}, \qquad (45.52)$$

where $\mathbf{A}$ is an orthogonal matrix and $\boldsymbol{\Delta}$ is a diagonal matrix whose diagonal elements, δ_i, $i = 1, \ldots, n_a$, are called the eigenphases. If we define the eigenphase sum δ_{sum} by

$$\delta_{\mathrm{sum}} = \sum_{i=1}^{n_a} \delta_i, \qquad (45.53)$$

then we can show from (45.50) that in the neighborhood of the resonance

$$\delta_{\mathrm{sum}}(E) = \delta_{0,\mathrm{sum}}(E) + \tan^{-1}\left[\frac{\tfrac{1}{2}\Gamma_i}{E_{i,r} - E} \right], \qquad (45.54)$$

where $\delta_{0,\mathrm{sum}}$ is the slowly varying background eigenphase sum obtained by replacing $\mathbf{S}$ by $\mathbf{S}_0$ in (45.52) and (45.53). It follows from (45.54) that the eigenphase sum increases by π radians in the neighborhood of the resonance energy. If there are m resonances, that may be overlapping, (45.54) generalizes to

$$\delta_{\mathrm{sum}}(E) = \delta_{0,\mathrm{sum}}(E) + \sum_{i=1}^{m} \tan^{-1}\left[\frac{\tfrac{1}{2}\Gamma_i}{E_{i,r} - E} \right]. \qquad (45.55)$$

This result has proved to be very useful in analyzing closely spaced resonances to obtain the individual resonance positions and widths.

45.2.4 Multichannel Quantum Defect Theory

When an electron scatters from a positive ion, the operator QHQ in 45.45 has an infinite series of bound states, supported by the long-range attractive Coulomb potential, converging to the lowest energy threshold of QHQ. Multichannel quantum defect theory (MQDT), developed by Seaton [6, 18, 19], relates the S-matrices and K-matrices above and below thresholds using general analytic properties of the Coulomb wave function. In this way the cross sections above and below threshold can be related and whole Rydberg series of resonances rather than individual resonances can be predicted.

When all the channels are open, the S-matrix is related to the K-matrix by (45.17), rewritten here as

$$\mathbf{S} = (i\mathbf{I} - \mathbf{K})(i\mathbf{I} + \mathbf{K})^{-1}. \qquad (45.56)$$

In the energy region where some channels are closed, Seaton showed that the S-matrix can be written as

$$\mathbf{S} = (i\mathbf{I} - \mathbf{K})(\mathbf{t}\mathbf{I} + \mathbf{K})^{-1}, \qquad (45.57)$$

where $\mathbf{t}$ is a diagonal matrix with nonzero matrix elements given by

$$\begin{aligned} t_{ii} &= i, & \text{open channels } k_i^2 \geq 0 \\ t_{ii} &= \tan\pi\nu_i, & \text{closed channels } k_i^2 < 0 \end{aligned} \qquad (45.58)$$

where ν_i is defined in the closed channels by

$$k_i^2 = -z^2/\nu_i^2. \qquad (45.59)$$

When some channels are closed, Rydberg series of resonances occur because of the terms $\tan\pi\nu_i$. We now introduce the matrix

$$\chi = (i\mathbf{I} - \mathbf{K})(i\mathbf{I} + \mathbf{K})^{-1}, \qquad (45.60)$$

which is analytic through the thresholds. When all the channels are open then clearly $\mathbf{S} = \chi$. Let us partition the S-matrix and the χ-matrix into open (subscript o) and closed (subscript c) channel matrices as follows:

$$\mathbf{S} = \begin{pmatrix} \mathbf{S}_{oo} & \mathbf{S}_{oc} \\ \mathbf{S}_{co} & \mathbf{S}_{cc} \end{pmatrix}, \quad \chi = \begin{pmatrix} \chi_{oo} & \chi_{oc} \\ \chi_{co} & \chi_{cc} \end{pmatrix}. \quad (45.61)$$

Then eliminating $\mathbf{K}$ between (45.57) and (45.60) and using (45.61), the open-open submatrix of $\mathbf{S}$ is

$$\mathbf{S}_{oo} = \chi_{oo} - \chi_{oc} \left(\chi_{cc} - e^{-2\pi i \, \boldsymbol{\nu}_c} \mathbf{I}\right)^{-1} \chi_{co}. \quad (45.62)$$

This is an expression for the open channel S-matrix in terms of quantities that can be analytically continued through the thresholds.

A similar expression can be obtained for the open channel, or contracted, K-matrix $\mathbf{K}_{op}$ defined in analogy with (45.56) by

$$\mathbf{S}_{oo} = (i\mathbf{I} - \mathbf{K}_{op})(i\mathbf{I} + \mathbf{K}_{op})^{-1}. \quad (45.63)$$

Partitioning the K-matrix into open and closed channels as

$$\mathbf{K} = \begin{pmatrix} \mathbf{K}_{oo} & \mathbf{K}_{oc} \\ \mathbf{K}_{co} & \mathbf{K}_{cc} \end{pmatrix}, \qquad (45.64)$$

and substituting into (45.57) yields the expression

$$\mathbf{K}_{op} = \mathbf{K}_{oo} - \mathbf{K}_{oc}(\mathbf{K}_{cc} + \tan\pi\boldsymbol{\nu}_c)^{-1}\mathbf{K}_{co}. \quad (45.65)$$

This equation enables cross sections and the resonance parameters in an energy region where some channels are closed to be predicted by calculating $\mathbf{K}$ at a few energies where all channels are open and then extrapolating it to the region where some channels are closed.

A simple one channel example has already been discussed following (45.28). In this case, it follows from (45.22) that $K = \tan\delta_\ell$. Also, from (45.57), a pole in the S-matrix occurs below threshold when

$$\tan\pi\nu + K = 0. \qquad (45.66)$$

Since from (45.59), the bound state energies correspond to $\nu = n - \mu$, where n is an integer and μ is the quantum defect, then (45.66) shows that δ_ℓ extrapolates below threshold to give $\pi\mu$. This result agrees with (45.28) since $1 - \exp(2\pi\eta) \approx 1$ close to threshold.

The above theory has been extended by Gailitis [20] who showed that the collision strengths defined by (45.20) can also be extrapolated through thresholds. Provided that the separation between the resonances is large compared with their widths, then the collision strength averaged over resonances below a threshold denoted by $\overline{\Omega}(i, f)$, can be written as

$$\overline{\Omega}(i, f) = \Omega^>(i, f) + \sum_j \frac{\Omega^>(i, j)\Omega^>(j, f)}{\sum_k \Omega^>(j, k)}, \quad (45.67)$$

where $\Omega^>$ are the collision strengths determined above the threshold and extrapolated to energies below this threshold. Also in (45.67), j is summed over the degenerate closed channels of the new threshold and k is summed over all open channels. This result is particularly useful in applications to plasmas where it is often only the collision strength averaged over a Maxwellian distribution of electron energies that is required.

Multichannel quantum defect theory has been extended to molecules by Fano [21], where it has proved to be important in the analysis of resonances occuring in partial cross sections involving rovibrational and dissociating channels as well as electronic channels [22].

45.2.5 Solution of the Coupled Integrodifferential Equations

This section considers methods that have been developed for solving the coupled integrodifferential equations (45.10). These equations arise in low-energy electron-atom and electron-ion collisions. Section 45.3.2 shows that similar equations also arise in low energy electron-molecule collisions in the fixed-nuclei approximation. Thus the methods discussed in this section are also applicable for electron-molecule collisions.

R-Matrix Method

This method was first introduced in nuclear physics [23, 24] in a study of resonance reactions. It has now been applied to wide range of atomic, molecular and optical processes, as reviewed in Ref. [25].

This method starts by partitioning configuration space into two regions by a sphere of radius a, chosen so that the direct potential has achieved its asymptotic form given by (45.12) and the exchange and correlation potentials are negligible for $r \geq a$. The objective is then to calculate the R-matrix $R_{ij}^\Gamma(E)$, which is defined by

$$F_{ij}^\Gamma(a) = \sum_{\ell=1}^n R_{i\ell}^\Gamma(E) \left(a\frac{dF_{\ell j}^\Gamma}{dr} - b_\ell F_{\ell j}^\Gamma\right)_{r=a}, \quad (45.68)$$

by solving (45.10) in the internal region.

The collision problem is solved in the internal region by expanding the wave function, in analogy with (45.8), in the form

$$\Psi_k^\Gamma(\mathbf{X}_{N+1})$$
$$= \mathcal{A}\sum_{ij} \overline{\Phi}_i^\Gamma(\mathbf{x}_1, \dots, \mathbf{x}_N; \hat{\mathbf{r}}_{N+1}\sigma_{N+1}) r_{N+1}^{-1} u_j(r_{N+1}) a_{ijk}^\Gamma$$

$$+ \sum_i \chi_i^\Gamma \left(\mathbf{x}_1, \ldots \mathbf{x}_{N+1} \right) b_{ik}^\Gamma, \qquad (45.69)$$

where the u_j are radial basis functions defined over the range $0 \leq r \leq a$. For radial basis functions u_j satisfying arbitrary boundary conditions at $r = a$, the Hamiltonian H_{N+1} defined by (45.3) is not Hermitian in the internal region due to the kinetic energy operators. It can however be made Hermitian by adding the Bloch operator [26]

$$L_b = \sum_{i=1}^n | \overline{\Phi}_i^\Gamma \rangle \frac{1}{2} \delta \left(r - a \right) \left(\frac{d}{dr} - \frac{b_i - 1}{r} \right) \langle \overline{\Phi}_i^\Gamma | \quad (45.70)$$

to H_{N+1}, where b_i is an arbitrary parameter. The Schrödinger equation (45.2) then becomes

$$\left(H_{N+1} + L_b - E \right) \Psi^\Gamma = L_b \Psi^\Gamma, \qquad (45.71)$$

which can be formally solved, giving

$$\Psi^\Gamma = \left(H_{N+1} + L_b - E \right)^{-1} L_b \Psi^\Gamma. \qquad (45.72)$$

Next, expand the Green's function $\left(H_{N+1} + L_b - E \right)^{-1}$ in terms of the basis Ψ_k^Γ, where the coefficients a_{ijk}^Γ and b_{ik}^Γ in (45.69) are chosen to diagonalize $H_{N+1} + L_b$, giving

$$\langle \Psi_k^\Gamma | H_{N+1} + L_b | \Psi_{k'}^\Gamma \rangle = E_k^\Gamma \delta_{kk'}. \qquad (45.73)$$

Equation (45.72) can then be written as

$$| \Psi^\Gamma \rangle = \sum_k \frac{| \Psi_k^\Gamma \rangle \langle \Psi_k^\Gamma |}{E_k^\Gamma - E} L_b | \Psi^\Gamma \rangle. \qquad (45.74)$$

Finally, we project this equation onto the channel functions $\overline{\Phi}_i^\Gamma$, and evaluate it at $r = a$. Assuming Ψ^Γ is given by (45.8), we retrieve (45.68), where the R-matrix can be calculated from the expansion

$$R_{ij}^\Gamma (E) = \frac{1}{2a} \sum_k \frac{w_{ik}^\Gamma w_{jk}^\Gamma}{E_k^\Gamma - E}, \qquad (45.75)$$

and where we have introduced the surface amplitudes

$$w_{ik}^\Gamma = \sum_j u_j (a) \, a_{ijk}^\Gamma. \qquad (45.76)$$

The main part of the calculation involves setting up and diagonalizing the matrix given by (45.73). This has to be carried out once to determine the R-matrix for all energies E.

In the external region $r \geq a$, (45.10) reduces to coupled differential equations, coupled by the potential $V_{ij}^\Gamma (r)$ which has achieved its asymptotic form (45.12). These equations can be integrated outwards from $r = a$ for each energy of interest, subject to the boundary conditions (45.68), to yield the K-matrix given by (45.15), and hence the S-matrix and collision cross sections.

Kohn Variational Method

The application of variational methods in electron atom collision theory has been reviewed by Nesbet [27]. An S-matrix or complex Kohn version of this method has been developed [28], which has eliminated singularities present in earlier K-matrix versions of the theory. This new method has been particularly important in electron-molecule collision calculations [29].

This approach starts from the basic expansion given by (45.8), where the reduced radial functions are chosen to satisfy the T-matrix asymptotic boundary conditions

$$F_{ij}^\Gamma \underset{r \to \infty}{\sim} k_i^{-\frac{1}{2}} \left(\sin \theta_i \delta_{ij} + (2i)^{-1} e^{i\theta_i} T_{ij}^\Gamma \right), \qquad (45.77)$$

which follows by taking linear combinations of the n_a solutions defined by (45.15) and using (45.17). We then define the integral

$$\mathbf{I}^\Gamma = -\frac{1}{2} \int_0^\infty \left(\mathbf{F}^\Gamma \right)^T \mathbf{L} \, \mathbf{F}^\Gamma \, dr, \qquad (45.78)$$

where $\mathbf{L}$ is the integrodifferential operator given by (45.10) with all terms taken onto the left-hand side of the equation.

Now consider variations of the integral $\mathbf{I}^\Gamma$ resulting from arbitrary variations $\delta \mathbf{F}^\Gamma$ of the functions $\mathbf{F}^\Gamma$ about the exact solution of (45.10), where these solutions satisfy the boundary conditions (45.77), and the variations satisfy the boundary condition

$$\delta \mathbf{F}^\Gamma \underset{r \to \infty}{\sim} (2i)^{-1} \mathbf{k}^{-\frac{1}{2}} e^{i\boldsymbol{\theta}} \delta \mathbf{T}^\Gamma. \qquad (45.79)$$

The corresponding variation in $\mathbf{I}^\Gamma$ to first-order is

$$\delta \mathbf{I}^\Gamma = -\frac{1}{2} \int_0^\infty \left[\left(\delta \mathbf{F}^\Gamma \right)^T \mathbf{L} \, \mathbf{F}^\Gamma + \left(\mathbf{F}^\Gamma \right)^T \mathbf{L} \delta \mathbf{F}^\Gamma \right] dr, \qquad (45.80)$$

which after some manipulation yields

$$\delta \mathbf{I}^\Gamma = (4i)^{-1} \delta \mathbf{T}^\Gamma. \qquad (45.81)$$

It follows that the functional

$$[\mathbf{T}^\Gamma] = \mathbf{T}^\Gamma - 4i \, \mathbf{I}^\Gamma \qquad (45.82)$$

is stationary for small variations about the exact solution. This is the complex Kohn variational principle.

This variational principle can be used to solve (45.10) by representing the reduced radial functions F_{ij}^Γ by the expansion

$$F_{ij}^\Gamma (r) = w_{1i}^\Gamma (r) \delta_{ij} + (2i)^{-1} w_{2i}^\Gamma (r) T_{ij}^\Gamma + \sum_k \phi_k^\Gamma (r) c_{ijk}^\Gamma, \qquad (45.83)$$

where w_{1i}^{Γ} and w_{2i}^{Γ} are zero at the origin and have the asymptotic forms

$$w_{1i}^{\Gamma}(r) \underset{r \to \infty}{\sim} k_i^{-\frac{1}{2}} \sin \theta_i , \qquad (45.84)$$

$$w_{2i}^{\Gamma}(r) \underset{r \to \infty}{\sim} k_i^{-\frac{1}{2}} \exp i\theta_i , \qquad (45.85)$$

while ϕ_k^{Γ} are square integrable basis functions. The coefficients c_{ijk} and the T-matrix elements T_{ij}^{Γ} can then be determined as variational parameters in the variational principle (45.82).

Schwinger Variational Method

This method has been used to calculate electron-molecule collision cross sections by McKoy and co-workers [30]. The method starts from the Lippmann-Schwinger integral equation, which corresponds to the integrodifferential equations (45.10), together with the K-matrix or T-matrix boundary condition defined by (45.15) or (45.77) respectively. The T-matrix form of the Lippmann-Schwinger integral equation is

$$F_{ij}^{\Gamma}(r) = w_{1i}^{\Gamma}(r) \delta_{ij} + \sum_{k\ell} \int_0^{\infty} \int_0^{\infty} G_{ik}^{\Gamma(+)}(r,r')$$
$$\times U_{k\ell}^{\Gamma}(r',r'') F_{\ell j}^{\Gamma}(r'') dr' dr'' \qquad (45.86)$$

where $U_{k\ell}^{\Gamma}$ represents the sum of the potential terms $2\left(V_{k\ell}^{\Gamma} + K_{k\ell}^{\Gamma} + X_{k\ell}^{\Gamma}\right)$ in (45.10), and the multichannel outgoing wave Green's function $G_{ik}^{\Gamma(+)}$ is defined assuming all channels are open by

$$G_{ij}^{\Gamma(+)}(r,r') = \begin{cases} -w_{1i}^{\Gamma}(r) w_{2i}^{\Gamma}(r') \delta_{ij}, & r < r' \\ -w_{2i}^{\Gamma}(r) w_{1i}^{\Gamma}(r') \delta_{ij}, & r \geq r' \end{cases} \qquad (45.87)$$

where w_{1i}^{Γ} and w_{2i}^{Γ} are solutions of the Coulomb equation

$$\left(\frac{d^2}{dr^2} - \frac{\ell_i(\ell_i+1)}{r^2} + \frac{2(Z-N)}{r} + k_i^2\right) w_i(r) = 0 \qquad (45.88)$$

that satisfy the asymptotic boundary conditions (45.84) and (45.85), respectively.

An integral expression for the T-matrix can be obtained by comparing (45.77) and (45.86). This gives

$$T_{ij}^{\Gamma} = -2i \sum_k \int_0^{\infty} \int_0^{\infty} w_{1i}^{\Gamma}(r) U_{ik}^{\Gamma}(r,r') F_{kj}^{\Gamma}(r') dr dr' \qquad (45.89)$$

or, rewriting this equation using Dirac bracket notation,

$$\mathbf{T} = -2i\langle \mathbf{w}_1 \mid \mathbf{U} \mid \mathbf{F}^{(+)}\rangle , \qquad (45.90)$$

where the plus sign indicates that $\mathbf{F}^{(+)}$ satisfies the outgoing wave boundary condition (45.77), and where for convenience we have suppressed the superscript Γ. In a similar way, the Lippmann-Schwinger equation corresponding to the ingoing wave boundary condition

$$F_{ij}^{\Gamma} \underset{r \to \infty}{\sim} k_i^{-\frac{1}{2}} \left(\sin \theta_i \delta_{ij} - (2i)^{-1} e^{-i\theta_i} T_{ij}^{\Gamma *}\right) , \qquad (45.91)$$

can be introduced, from which follows the integral expression

$$\mathbf{T} = -2i\langle \mathbf{F}^{(-)} \mid \mathbf{U} \mid \mathbf{w}_1\rangle , \qquad (45.92)$$

where $\mathbf{F}^{(-)}$ satisfies the ingoing wave boundary condition (45.91). A further integral expression for the T-matrix is obtained by substituting for $\mathbf{w}_1$ in (45.92) from (45.86) giving

$$\mathbf{T} = -2i\langle \mathbf{F}^{(-)} \mid \mathbf{U} - \mathbf{U}\mathbf{G}^{(+)}\mathbf{U} \mid \mathbf{F}^{(+)}\rangle . \qquad (45.93)$$

Hence, a combination of (45.90), (45.92) and (45.93) yields the functional

$$[\mathbf{T}] = -2i\langle \mathbf{w}_1 \mid \mathbf{U} \mid \mathbf{F}^{(+)}\rangle$$
$$\times \left[\langle \mathbf{F}^{(-)} \mid \mathbf{U} - \mathbf{U}\mathbf{G}^{(+)}\mathbf{U} \mid \mathbf{F}^{(+)}\rangle\right]^{-1}$$
$$\times \langle \mathbf{F}^{(-)} \mid \mathbf{U} \mid \mathbf{w}_1\rangle . \qquad (45.94)$$

This functional is stationary for small variations of $\mathbf{F}^{(+)}$ and $\mathbf{F}^{(-)}$ about the exact solution of (45.10) satisfying the boundary conditions (45.77) and (45.91) respectively, and forms the basis of numerical calculations.

Linear Algebraic Equations Method

In this method, the integrodifferential equations (45.10) are reduced directly to a set of linear algebraic (LA) equations [31], or alternatively, equations (45.10) are first converted to integral form, given by (45.86), which are then reduced to a set of linear algebraic equations [32].

A direct approach [31] has been widely used for electron-atom and electron-ion collisions. As in the R-matrix method, configuration space is first divided into two regions. A mesh of N points is then used to span the internal region $r \leq a$ where

$$r_1 = 0; \quad r_{k-1} < r_k, \quad k = 1, \ldots, N; \quad r_N = a. \qquad (45.95)$$

In addition, two further mesh points r_{N+1} and r_{N+2} are introduced to enable the solution in the internal region to be matched to the solution in the external region $r \geq a$. In the external region, (45.10) reduces to coupled differential equations that can be solved by one of the same methods as adopted in the R-matrix approach.

The n functions $F_{ij}^{\Gamma}(r)$, $i = 1, \ldots, n$, in (45.10) are represented by their values at the mesh points r_k. Using a finite difference representation of the differential and integral operators, (45.10) then reduces to a set of linear algebraic equations for the unknown values $F_{ij}^{\Gamma}(r_k)$ in the internal region. These equations can be solved using standard methods for each linearly independent solution $j = 1, \ldots, n_a$ defined by the asymptotic boundary conditions (45.15).

45.2.6 Intermediate and High Energy Elastic Scattering and Excitation

For electron-atom and electron-ion collisions at electron impact energies greater than the ionization threshold of the target, an infinite number of channels is open so that they cannot all be included explicitly in the expansion of the total wave function. Several approaches have been developed to treat collisions at these energies, such as extensions of low energy methods based on expansion (45.8) to intermediate energies, the development of optical potentials that take account of loss of flux into the infinity of open channels in some average way, and extensions of the Born approximation to lower energies by including higher-order terms in the Born Series.

Modified Low-Energy Methods

In this approach, expansion (45.8) is extended by including a set of pseudostates $\bar{\Phi}_i$ as well as the target eigenstates Φ_i of interest. These pseudostates are chosen to represent in some average way the high lying Rydberg states and continuum states of the target. They can be chosen to satisfy an equation analagous to (45.4), viz.

$$\langle \bar{\Phi}_i \mid H_N \mid \bar{\Phi}_j \rangle = \bar{w}_i \delta_{ij} \, . \qquad (45.96)$$

However, the pseudoenergies $\bar{w}_i$ now span the continuum spectrum of the target, rather than representing only the target bound state eigenenergies w_i as in (45.4). Thus, ionization cross sections can also be obtained.

These pseudostates can be chosen to represent both long-range polarization and short-range electron correlation effects (see, e.g., [33–36]). However their choice is not unique. For example, the R-matrix method has been extended to treat intermediate energy collisions [37, 38], and recently an approach where the pseudostates are represented by a Sturmian basis, first suggested by Rotenberg [39], has been shown to be particularly effective for e^-–H like collisions over a wide energy range [40–42].

Optical Potential Methods

An approach which is also capable in principle of including the effect of all excited and continuum states is the optical potential method [43]. Adopting Feshbach projection operators P and Q as in (45.40), an optical potential V_{op} is defined by (45.44), where now P projects onto those low-energy channels that can be treated exactly, for example by solving equations (45.10), and Q allows for the remaining infinity of coupled channels, including the continuum.

At sufficiently high energies, it is appropriate to make a perturbation expansion of V_{op}, where the second-order term is given by

$$V^{(2)} = PVQ \frac{1}{E - \mathcal{T}_e - H_N + i\epsilon} QVP \, , \qquad (45.97)$$

V being the electron-atom (ion) interaction potential, $\mathcal{T}_e$ the kinetic energy operator of the scattered electron, and H_N the target Hamiltonian. Byron and Joachain [44] converted the lowest-order terms of perturbation theory into an *ab initio* local complex potential for the elastic scattering of electrons and positrons from a number of atoms.

The optical potential calculated in second-order has also been used by Bransden *et al.* [45, 46] to describe e^-–H collisions, while McCarthy [47] has studied an optical-potential approximation that goes beyond second-order, and also makes allowance for exchange. This method is called the coupled channels optical (CCO) model. Finally, Callaway and Oza [48] have constructed an optical potential using a set of pseudostates to evaluate the sum over intermediate states and have obtained encouraging results for elastic scattering and excitation of the $n = 2$ states in e^-–H collisions.

Born Series Methods

In the high energy domain, which can usually be assumed to extend from several times the ionization threshold of the target upwards, methods based on the Born series give reliable results. Ignoring electron exchange for the moment, the Born series for the direct scattering amplitude can be written as [49, 50]

$$f = \sum_{n=1}^{\infty} f_{n,\text{B}} \, , \qquad (45.98)$$

where the nth Born term $f_{n,\text{B}}$ contains the interaction V between the scattered electron and the target atom or ion n times, and the Green's function

$$G_0^{(+)} = (H_N + \mathcal{T}_e - E - i\epsilon)^{-1} \qquad (45.99)$$

$n-1$ times, where $\mathcal{T}_e$ and H_N are defined following (45.97).

It is important to retain consistently all terms in the Born series with similar energy and momentum-transfer $\Delta = |\mathbf{k}_i - \mathbf{k}_f|$ dependencies. For elastic scattering, the scattering amplitude converges to the first Born approximation for all Δ, but at lower energies it is necessary to include $\Re(f_{3,\mathrm{B}})$, as well as the second Born terms, to obtain the cross section correct to k^{-2}. In the forward direction, convergence to the first Born approximation is slow because of the contribution from $\Im(f_{2,\mathrm{B}})$ which, from the optical theorem, corresponds to loss of flux into all open channels.

For inelastic scattering, the first Born approximation does not give the correct high energy limit for large Δ. Instead this comes from $\Im(f_{2,\mathrm{B}})$. Physically, this can be understood by noting that inelastic scattering at large angles involves two collisions: a collision of the incident electron with the nucleus to give the large scattering angle, followed or preceded by an inelastic collision with the bound electrons to give excitation. Again, in the forward direction, it is necessary to retain $\Re(f_{3,\mathrm{B}})$ to obtain the cross section correct to k^{-2}.

Since the third Born term is difficult to calculate, Byron and Joachain [51] suggested that to third-order, the scattering amplitude should be calculated using the eikonal Born series (EBS) approximation

$$f_{\mathrm{EBS}} = f_{1,\mathrm{B}} + f_{2,\mathrm{B}} + f_{3,\mathrm{G}} + g_{\mathrm{och}}, \quad (45.100)$$

where $f_{3,\mathrm{G}}$ is the third-order term in the expansion of the Glauber amplitude [52] in powers of V, which can be more easily calculated than $f_{3,\mathrm{B}}$, and g_{och} is the Ochkur electron exchange amplitude [53]. The EBS method has been very successful when perturbation theory converges rapidly; namely, at high energies, at small and intermediate scattering angles, and for light atoms.

Distorted Wave Methods

Distorted wave methods are characterized by a separation of the interaction into two parts, one which is treated exactly and the other which is treated in first-order. The usual approach is based on the integral expressions for the T-matrix given by (45.90) or (45.92). The distorted wave approximation is obtained by replacing the exact solution of (45.10), denoted by $\mathbf{F}^{(+)}$ and $\mathbf{F}^{(-)}$ in (45.90) and (45.92) respectively, by approximate solutions, usually obtained by omitting all channels except the final or initial channels of interest. In addition, the potential interaction $\mathbf{U}$ is often approximated by just the direct term in (45.10).

This method becomes more accurate at intermediate energies as $Z - N$ increases, and as the angular momentum of the scattered electron becomes large. However, it gives poor results at low energies where the coupling between the channels in (45.10) is strong, and where resonances are often important. It has proved to be a very useful way of calculating electron-ion total and differential cross sections at intermediate and high energies well removed from threshold.

45.2.7 Ionization

This section discusses processes where electrons are ejected from the target during the collision, giving rise to ionization. The main focus is single ionization, or (e,2e) processes given by

$$e^- (E_i) + A_i \to A_j^+ + e^- (E_A) + e^- (E_B), \quad (45.101)$$

where E_i, E_A and E_B are the incident, scattered and ejected electron energies respectively. They are related through the energy conservation relation

$$E_i = E_A + E_B + \varepsilon, \quad (45.102)$$

where ε is the binding energy of the ejected atomic electron. The threshold behavior of the ionization cross section leading to the Wannier threshold law is treated in Chap. 50.

The scattering amplitude for the direct ionization process is [54]

$$f_i(\mathbf{k}_A, \mathbf{k}_B) = -(2\pi)^{\frac{1}{2}} e^{i\chi(\mathbf{k}_A,\mathbf{k}_B)} \langle (H_{N+1} - E)\Phi \mid \Psi_i^{(+)} \rangle \quad (45.103)$$

where $\Psi_i^{(+)}$ is an exact solution of (45.2) satisfying plane wave plus outgoing wave boundary conditions given by (45.5), and where Φ is a solution of the equation

$$(H_{N+1} - V - E)\Phi = 0 \quad (45.104)$$

satisfying ingoing wave boundary conditions. Also, $\mathbf{k}_A$ and $\mathbf{k}_B$ are the momenta of the two outgoing electrons defined in terms of E_A and E_B by

$$\tfrac{1}{2}k_A^2 = E_A, \qquad \tfrac{1}{2}k_B^2 = E_B. \quad (45.105)$$

In practice the potential V is chosen so that Φ has a simple form. Thus, for electron scattering by H-like ions with nuclear charge Z, V is often chosen as

$$V = -\frac{Z - Z_A}{r_1} - \frac{Z - Z_B}{r_2} + \frac{1}{r_{12}}, \quad (45.106)$$

so that

$$\Phi(\mathbf{r}_1, \mathbf{r}_2) = \Psi_{\mathrm{C}}^{(-)}(Z_A, \mathbf{k}_A, \mathbf{r}_1)\Psi_{\mathrm{C}}^{(-)}(Z_B, \mathbf{k}_B, \mathbf{r}_2), \quad (45.107)$$

where the $\Psi_{\mathrm{C}}^{(-)}$ are Coulomb wave functions satisfying ingoing-wave boundary conditions, and Z_A and Z_B are the effective charges seen by the two electrons. In order that the scattering amplitude does not contain a divergent phase factor, Z_A and Z_B must satisfy

$$\frac{Z_A}{k_A} + \frac{Z_B}{k_B} = \frac{Z}{k_A} + \frac{Z}{k_B} - \frac{1}{|\mathbf{k}_A - \mathbf{k}_B|}, \quad (45.108)$$

and the phase factor in (45.103) is then given by

$$\chi\left(\mathbf{k}_A, \mathbf{k}_B\right) = \frac{Z_A}{k_A} \ln \frac{k_A^2}{k_A^2 + k_B^2} + \frac{Z_B}{k_B} \ln \frac{k_B^2}{k_A^2 + k_B^2} \,.$$

$$(45.109)$$

The exchange ionization amplitude $g_i\left(\mathbf{k}_A, \mathbf{k}_B\right)$ can be obtained from the Peterkop theorem [55]

$$g_i\left(\mathbf{k}_A, \mathbf{k}_B\right) = f_i\left(\mathbf{k}_B, \mathbf{k}_A\right)\,. \qquad (45.110)$$

This result follows from the fact that the amplitudes $f_i\left(\mathbf{k}_B, \mathbf{k}_A\right)$ and $g_i\left(\mathbf{k}_A, \mathbf{k}_B\right)$ describe the same physical process where the electron at $\mathbf{r}_{N+1}$ has momentum $\mathbf{k}_A$, and the electron at $\mathbf{r}_N$ has momentum $\mathbf{k}_B$. In practice, the Peterkop theorem is not valid if approximate wave functions are used to calculate the ionization amplitude. For this reason, a relative phase $\tau_i\left(\mathbf{k}_A, \mathbf{k}_B\right)$ between these two amplitudes is sometimes introduced by

$$g_i\left(\mathbf{k}_A, \mathbf{k}_B\right) = e^{i\tau(\mathbf{k}_A, \mathbf{k}_B)} f_i\left(\mathbf{k}_B, \mathbf{k}_A\right)\,. \qquad (45.111)$$

This adds an element of arbitrariness into the calculation since different choices of this phase leads to different cross sections [56].

The ionization cross sections can be obtained directly from the scattering amplitudes. For random electron spin orientations, the triple differential cross section (TDCS) for ionization of a target with one active electron from an initial state denoted by $|i\rangle$ is given by

$$\frac{d^3\sigma_i}{d\Omega_A d\Omega_B dE} = \frac{k_A k_B}{k_i} \left[\frac{1}{4}|f_i + g_i|^2 + \frac{3}{4}|f_i - g_i|^2\right]\,.$$

$$(45.112)$$

By integrating the TDCS with respect to $d\Omega_A, d\Omega_B$ or dE, we can form three different double differential cross sections, and three different single differential cross sections. The total ionization cross section obtained by integrating (45.112) over all outgoing electron scattering angles and energies is

$$\sigma_i = \frac{1}{k_i} \int_0^{E/2} dE\, k_A k_B \int d\Omega_A$$
$$\times \int d\Omega_B \left[\frac{1}{4}|f_i + g_i|^2 + \frac{3}{4}|f_i - g_i|^2\right]\,, \qquad (45.113)$$

where the upper limit of integration over the energy variable is $E/2$ because the two outgoing electrons are indistinguishable.

The integral expression (45.103) for the direct ionization amplitude provides a starting point for the calculation of the cross section when some approximation is made for $\Psi_i^{(+)}$. The simplest approximation is to represent $\Psi_i^{(+)}$ by the incident plane wave term in (45.5). This gives a variety of first-Born approximations, depending on the choice of phase in (45.111). However, this approximation can only be expected to be reliable at high energies for the total cross section, and second-order Born

terms must be included, as in the case of excitation, to obtain accurate results at all momentum transfers.

In order to obtain cross sections that are reliable at lower energies, more accurate approximations must be made for $\Psi_i^{(+)}$. For example, $\Psi_i^{(+)}$ in (45.103) has been represented by coupled eigenstate or pseudostate expansions of the form (45.8). Another approximation has proved useful for electron impact ionization of ions [57, 58]. Starting from (45.103), the N-electron target in the initial state and the $(N-1)$-electron residual ion plus ejected electron in the final state are both expanded as in (45.8), while the ionizing electron is represented by distorted waves or Coulomb waves in both the initial and final states. This approximation is appropriate for situations where the ejected electron moves more slowly than the ionizing electron. It is particularly useful when the ejected electron can be captured into an autoionizing state before being emitted, giving rise to the important 'excitation-autoionization' processes [59]. This approximation can be regarded as a member of a class of distorted-wave Born approximations (DWBA) that have been widely used to describe electron-atom ionization processes over a wide energy range [60].

Finally, an important recent development is the realization that accurate ionization cross sections can be extracted directly from expansion (45.8) if sufficient pseudostates are retained in it to accurately discretize the continuum. Using the Sturmian basis discussed in Sect. 45.2.6, Bray and Stelbovics [41] have obtained accurate e$^-$–H ionization cross sections in this way.

45.3 ELECTRON-MOLECULE COLLISIONS

45.3.1 Laboratory Frame Representation

The processes that occur in electron collisions with molecules are more varied than those that occur in electron collisions with atoms and atomic ions because of the possibility of exciting degrees of freedom associated with the motion of the nuclei. In addition, the multicenter and nonspherical nature of the electron molecule interaction considerably complicates the solution of the collision problem by reducing its symmetry and by introducing multicenter integrals that are more difficult to calculate than those occurring for atoms and ions.

We first consider the derivation of the equations describing the collision in the laboratory frame of reference discussed by Arthurs and Dalgarno [61]. The Schrödinger equation describing the electron-molecule system is

$$\left(H_{\mathrm{m}} + \mathcal{T}_{\mathrm{e}} + V\right)\Psi = E\Psi\,, \qquad (45.114)$$

where H_{m} is the molecular Hamiltonian, $\mathcal{T}_{\mathrm{e}}$ is the kinetic energy operator of the scattered electron and V is the electron-molecule interaction potential

$$V\left(\mathbf{R},\mathbf{r}_{\mathrm{m}},\mathbf{r}\right)=\sum_i\frac{1}{|\mathbf{r}-\mathbf{r}_i|}-\sum_i\frac{Z_i}{|\mathbf{r}-\mathbf{R}_i|}\,. \quad (45.115)$$

Here, $\mathbf{R}$ represents the coordinates $\mathbf{R}_i$ of all the nuclei, $\mathbf{r}_{\mathrm{m}}$ represents the coordinates $\mathbf{r}_i$ of the electrons in the target molecule, and $\mathbf{r}$ represents the coordinates of the scattered electron. The total energy E in (45.114) refers to the frame of reference where the c.m. of the whole system is at rest.

As in the case of electron-atom and electron-ion collisions, introduce target eigenstates, and possibly pseudostates Φ_i, by the equation

$$\langle\Phi_i\,|H_{\mathrm{m}}|\,\Phi_j\rangle=w_i\delta_{ij}\,, \quad (45.116)$$

and then expand the total wave function Ψ, in analogy with (45.8), in the form

$$\Psi_j=\mathcal{A}\sum_i\Phi_i\left(\mathbf{R},\mathbf{r}_{\mathrm{m}}\right)\mathcal{F}_{ij}\left(\mathbf{r}\right)+\sum_i\chi_i\left(\mathbf{R},\mathbf{r}_{\mathrm{m}},\mathbf{r}\right)b_{ij}\,,$$

$$(45.117)$$

where the spin variables have been suppressed for notation simplicity, and we have not carried out a partial-wave decomposition of the wave function $\mathcal{F}_{ij}$ representing the scattered electron. The subscripts i and j now represent the rotational and vibrational states of the molecule as well as its electronic states.

Coupled equations for the functions $\mathcal{F}_{ij}$ can be obtained by substituting expansion (45.117) into (45.114), and projecting onto the target states Φ_i and onto the square integrable functions χ_j. After eliminating the coefficients b_{ij}, the coupled integrodifferential equations

$$\left(\nabla^2+k_i^2\right)F_{ij}\left(r\right)=2\sum_\ell\left(V_{i\ell}+W_{i\ell}+X_{i\ell}\right)F_{\ell j}\left(r\right)$$

$$(45.118)$$

are obtained, where $k_i^2=2\left(E-w_i\right)$ and where $V_{i\ell}$, $W_{i\ell}$, and $X_{i\ell}$ are the direct, nonlocal exchange, and nonlocal correlation potentials. By expanding F_{ij} in partial waves, a set of coupled radial integrodifferential equations result, analogous to (45.10) for atoms and ions.

The scattering amplitude and cross section for a transition from an initial state $|\,i\,\rangle=|\,\mathbf{k}_i,\Phi_i,\chi_{\frac{1}{2}m_i}\,\rangle$ to a final state $|\,j\,\rangle=|\,\mathbf{k}_j,\Phi_j,\chi_{\frac{1}{2}m_j}\,\rangle$ is then given by (45.7), where now the subscripts i and j refer collectively to the ro-vibrational and electronic states of the molecule.

45.3.2   Molecular Frame Representation

The theory described in the previous section is completely general, and has been the basis of a number of early calculations for simple diatomic molecules such as H_2. However, major computational difficulties arise because of the very large number of rovibrational channels that need to be retained in expansion (45.117) for all but the simplest low-energy calculations.

This difficulty can be overcome by making a Born-Oppenheimer separation of the electronic and nuclear motion. The electronic motion is first determined with the nuclei held fixed. This is referred to as the fixed-nuclei approximation. The molecular rotational and vibrational motion is then included in a second step of the calculation. This procedure owes its validity to the large ratio of the nuclear mass to the electronic mass, and can be adopted when the collision time is much shorter than the periods of molecular rotation and vibration. Thus it is expected to be valid when the scattered electron energy is not close to a threshold, or when the energy does not coincide with that of a narrow resonance. In these cases, further developments described below are needed to obtain reliable cross sections.

In order to formulate the collision process in this representation, adopt a frame of reference that is rigidly attached to the molecule. The fixed-nuclei approximation then starts from the Schrödinger equation

$$\left(H_{\mathrm{el}}+\mathcal{T}_{\mathrm{e}}+V\right)\psi=E\psi\,, \quad (45.119)$$

where H_{el} is the electronic part of the target Hamiltonian obtained by assuming that the target nuclei have fixed coordinates denoted collectively by $\mathbf{R}$. It follows that H_{el} is related to H_{m} in (45.114) by

$$H_{\mathrm{m}}=H_{\mathrm{el}}+\mathcal{T}_{\mathrm{R}}\,, \quad (45.120)$$

where $\mathcal{T}_{\mathrm{R}}$ is the kinetic energy operator for the rotational and vibrational motion of the nuclei. The remaining quantities $\mathcal{T}_{\mathrm{e}}$ and V are the same as in (45.114).

The solution of (45.119) proceeds in an analogous way to the solution of (45.2) for electron collisions with atoms and ions. We adopt an expansion similar to (45.8), where we now expand the function representing the motion of the scattered electron in terms of symmetry-adapted angular functions that transform as an appropriate irreducable representation (IRR) of the molecular point group (Burke *et al.* [62]). Substituting this expansion into (45.119), and projecting onto the corresponding channel functions and onto the square integrable functions, yields a set of coupled integrodifferential equations with the form given by (45.10), where now the channel indices i, j and ℓ represent the component of the IRR, as well as the electronic state of the target, and where Γ represents the conserved quantum numbers that now include the IRR and the total spin.

The final step is to solve these coupled integrodifferential equations for each set of nuclear coordinates $\mathbf{R}$ of importance in the collision, using one of the methods discussed in Sect. 45.2.5. This yields the K-matrices, S-matrices and cross sections for fixed $\mathbf{R}$. For scattering calculations where only the ground electronic state has been included in the expansion, a number of approaches have been developed that replace the nonlocal exchange and correlation potentials by local potentials (see, e.g., Ref. [2]). These approaches have proved particularly important in describing electronically elastic collisions of electrons with polyatomic molecules.

45.3.3 Inclusion of the Nuclear Motion

This section discusses how observables involving the nuclear motion, such as rotational and vibrational excitation cross sections, and dissociative attachment cross sections, can be obtained from the solutions of the fixed-nuclei equations.

The most widely used approach is the adiabatic-nuclei approximation [63, 64]. In the case of diatomic molecules in a $^1\Sigma$ state, the scattering amplitude for a transition between electronic, vibrational and rotational states represented by $ivjm_j$ and $i'v'j'm'_j$ is given by

$$f_{i'v'j'm'_j, ivjm_j}\left(\hat{\mathbf{k}}\cdot\hat{\mathbf{r}}\right)$$
$$= \langle \chi_{i'v'}(R) Y_{j'm'_j}(\hat{\mathbf{R}}) \mid f_{i'i}(\hat{\mathbf{k}}\cdot\hat{\mathbf{r}}; \mathbf{R}) \mid \chi_{iv}(R) Y_{jm_j}(\hat{\mathbf{R}}) \rangle, \tag{45.121}$$

where $f_{i'i}\left(\hat{\mathbf{k}}\cdot\hat{\mathbf{r}}; \mathbf{R}\right)$ is the fixed-nuclei scattering amplitude, which depends parametrically on the nuclear coordinates $\mathbf{R}$, and χ_{iv} and Y_{jm_j} are the molecular vibrational and rotational eigenfunctions respectively. This approximation is valid provided that the collision time is short compared with the vibration and/or rotation times, and is widely used in such situations.

The cross section is usually averaged over the degenerate sublevels m_j and summed over m'_j, giving the cross section for the transition ivj to $i'v'j'$. This leads to the relation

$$\frac{d\sigma_{i'v'j', ivj}}{d\Omega} = \sum_{j_t=|j-j'|}^{j+j'} [(j0j_t0 \mid jj_tj'0)]^2 \frac{d\sigma_{i'v'j_t, iv0}}{d\Omega}, \tag{45.122}$$

provided that the small differences in the wave numbers for the different rotational channels can be neglected. A similar relation holds for symmetric top molecules such as NH_3. For spherical top molecules such as CH_4, the equivalent relation is [65]

$$\frac{d\sigma_{i'v'j', ivj}}{d\Omega} = \frac{2j'+1}{2j+1} \sum_{j_t=|j-j'|}^{j+j'} \frac{1}{2j_t+1} \frac{d\sigma_{i'v'j_t, iv0}}{d\Omega}. \tag{45.123}$$

The sum in (45.122) or (45.123) over the final rotational state j' is independent of the initial state j. Also, if the

cross section is multiplied by the transition energy and then sum over j', the result, which is in the mean energy loss by the incident electron, is still independent of j.

The adiabatic-nuclei approximation breaks down close to threshold or in the neighborhood of narrow resonances [66]. A straightforward way of including nonadiabatic effects that arise in vibrational excitation is to retain the vibrational terms in the Hamiltonian, but still to treat the rotational motion adiabatically. Hence, instead of (45.119), the equation

$$(H_{\text{el}} + \mathcal{T}_{\text{vib}} + \mathcal{T}_e + V)\tilde{\psi} = E\tilde{\psi} \tag{45.124}$$

is solved, where $\mathcal{T}_{\text{vib}}$ is the kinetic energy operator for the nuclear vibrational motion, and where the other quantities have the same meaning as in (45.119). Adopting a frame of reference in which the molecule has fixed spacial orientation, and separating out the angular variables of the scattered electron, coupled integrodifferential equations coupling the target vibrational states as well the electronic states can be obtained. This approach has been adopted with success [67], but is computationally demanding since the number of coupled channels can become very large.

Vibrational excitation and dissociative attachment are particularly important in resonance regions when the scattered electron spends an appreciable time in the neighborhood of the molecule. As a result, a number of approaches has been developed describing these processes based on electron molecule resonance theories (e.g., [68–71]). The basic idea is that a series of fixed-nuclei resonance states $\psi_n^{(r)}$ are introduced for a range of values of R, either by imposing Siegert outgoing wave boundary conditions [72], or by introducing Feshbach projection operators [14, 15]. The amplitude for a transition from an initial electronic-vibrational state iv to a final state $i'v'$ is then given by

$$T_{i'v', iv} = \sum_n \langle \chi_{i'v'}(R') \zeta_{ni'}(R') \mid G_n^{(r)}(R', R)$$
$$\times \mid \zeta_{ni}(R) \chi_{iv}(R) \rangle, \tag{45.125}$$

where χ_{iv} are the vibrational eigenfunctions, ζ_{ni} are the "entry amplitudes" from the initial or final electronic states into the resonance states $\psi_n^{(r)}$, and $G_n^{(r)}$ are the Green's functions that describe the propagation in the intermediate resonance states $\psi_n^{(r)}$. Dissociative attachment can be described by a straightforward extension of this theory.

The fixed-nuclei R-matrix method has also been extended to treat vibrational excitation and dissociative attachment [73]. A generalized R-matrix is introduced by an equation which, in analogy with (45.125), can be written as

$$R_{i'v', iv} = \frac{1}{2a} \sum_k \langle \chi_{i'v'}(R') w_{i'k}^\Gamma(R') \mid G_k^{RM}(R', R)$$

$$\times \mid w_{ik}^{\Gamma}(R)\,\chi_{iv}(R)\rangle\,, \qquad (45.126)$$

where the surface amplitudes $w_{ik}^{\Gamma}(R)$ are defined by (45.76), and the Green's function G_k^{RM} now describes the propagation in the intermediate R-matrix states defined by (45.73). Once the generalized R-matrix has been determined, the final step in the calculation is to solve the collision problem in the external region which, for diatomic molecules, is defined by the condition that the scattered electron coordinate r is greater than some given radius a, and the internuclear coordinate R is greater than some given radius A.

Another approach that includes nonadiabatic effects is the energy-modified adiabatic approximation introduced by Nesbet [74]. In this approach, the S-matrix elements connecting the vibrational states are defined by

$$S_{i'v',iv} = \langle \chi_{i'v'} \mid S_{i'i}(E - H_i; R) \mid \chi_{iv}\rangle\,, \qquad (45.127)$$

where $S_{i'i}(E - H_i; R)$ is the S-matrix calculated in the fixed nuclei approximation at the internuclear separation R at an energy defined by the operator $H_i = E_i(R) + \mathcal{T}_{\mathrm{vib}}$. This has the effect of including the internal energy of the target into the S-matrix elements, giving the correct threshold energies.

Finally, we mention an off-shell T-matrix approach for including nonadiabatic effects discussed first by Shugard and Hazi [75]. This approach can also extend the range of validity of the adiabatic-nuclei approximation, while retaining much of its inherent simplicity. This has recently been applied with success to low-energy vibrational excitation of H_2 and CH_4 [76].

45.4 ELECTRON-ATOM COLLISIONS IN A LASER FIELD

Electron-atom collisions in the presence of an intense laser field have recently attracted considerable attention because of the importance of these processes in applications such as laser plasma interactions, and also because of their fundamental interest in atomic collision theory. This section summarizes the basic theory, commencing with the scattering of electrons by a potential in the presence of a laser field. The discussion is then generalized to the scattering of electrons by complex atoms and ions, where 'dressing' of the atomic eigenstates by the laser field must be considered, and where simultaneous electron photon excitation (SEPE) processes, defined by

$$nh\nu + e^- + A_i \rightarrow e^- + A_j\,, \qquad (45.128)$$

can occur. Recent reviews of aspects of this subject have been given by Mason [3], Newell [77] and Mittleman [78].

45.4.1 Potential Scattering

We adopt a semiclassical description of the collision process in which the electrons and the target atom are described by the nonrelativistic Schrödinger equation, and the laser field is described classically. This is valid for most high intensity fields of current interest, where a typical coherence volume of the field contains a very large number of photons [78].

The time-dependent Schrödinger equation describing an electron scattered by a potential $V(\mathbf{r})$ in the presence of an external electromagnetic (laser) field is then

$$i\frac{\partial}{\partial t}\Psi(\mathbf{r},t) = \left(-\frac{1}{2}\nabla^2 - \frac{i}{c}\mathbf{A}(\mathbf{r},t)\cdot\boldsymbol{\nabla}\right.$$
$$\left.+\frac{1}{2c^2}\mathbf{A}^2(\mathbf{r},t) + V(\mathbf{r})\right)\Psi(\mathbf{r},t)\,, \quad (45.129)$$

in the Coulomb gauge such that the vector potential satisfies $\boldsymbol{\nabla}\cdot\mathbf{A} = 0$. We also assume that the laser field is monochromatic, monomode, linearly polarized, and spacially homogeneous (i.e., its wavelength is large compared with the range of the potential, or more generally with the size of the atom). Hence we can write

$$\mathbf{A}(\mathbf{r},t) = \mathbf{A}(t) = \hat{\boldsymbol{\epsilon}}A_0\cos\omega t\,, \qquad (45.130)$$

where $\hat{\boldsymbol{\epsilon}}$ is a unit vector along the field polarization direction and ω is the angular frequency. The $\mathbf{A}^2$ term in (45.129) can be removed by the unitary transformation

$$\Psi(\mathbf{r},t) = \exp\left(-\frac{i}{2c^2}\int^t \mathbf{A}^2(t')\,dt'\right)\Psi_V(\mathbf{r},t)\,,$$
$$(45.131)$$

where Ψ_V satisfies the Schrödinger equation in the velocity gauge given by

$$i\frac{\partial}{\partial t}\Psi_V(\mathbf{r},t) = \left(-\frac{1}{2}\nabla^2 - \frac{i}{c}\mathbf{A}(t)\cdot\boldsymbol{\nabla} + V(\mathbf{r})\right)\Psi_V(\mathbf{r},t)\,.$$
$$(45.132)$$

The corresponding equation for the free electron with $V = 0$ is readily solved to give the Volkov wave function [79]

$$\chi_{\mathbf{k}}(\mathbf{r},t) = (2\pi)^{-\frac{3}{2}}\exp\left[i(\mathbf{k}\cdot\mathbf{r} - \mathbf{k}\cdot\boldsymbol{\alpha}_0\sin\omega t - Et)\right]\,, \qquad (45.133)$$

where $\mathbf{k}$ is the wave vector and $E = \frac{1}{2}k^2$ the kinetic energy. The quantity $\boldsymbol{\alpha}(t)$ is defined by

$$\boldsymbol{\alpha}(t) = \frac{1}{c}\int^t \mathbf{A}(t')\,dt' = \boldsymbol{\alpha}_0\sin\omega t\,, \qquad (45.134)$$

where $\boldsymbol{\alpha}_0 = \boldsymbol{\mathcal{E}}_0/\omega^2$, $\boldsymbol{\mathcal{E}}_0$ being the electric field strength.

To solve (45.132), introduce the causal Green's function $G_0^{(+)}(\mathbf{r}, t; \mathbf{r}', t')$ satisfying the equation

$$\left(i\frac{\partial}{\partial t} + \frac{1}{2}\nabla^2 + \frac{i}{c}\mathbf{A}(t)\cdot\boldsymbol{\nabla}\right) G_0^{(+)}(\mathbf{r}, t; \mathbf{r}', t')$$
$$= \delta(\mathbf{r} - \mathbf{r}')\,\delta(t - t')\,. \qquad (45.135)$$

This Green's function is given by

$$G_0^{(+)}(\mathbf{r}, \mathbf{t}; \mathbf{r}', t') = -i\theta(t - t')\int \chi_{\mathbf{k}}(\mathbf{r}, t)\,\chi_{\mathbf{k}}^*(\mathbf{r}', t')\,d\mathbf{k}\,, \qquad (45.136)$$

where $\theta(x) = 1$ for $x > 0$ and $\theta(x) = 0$ for $x < 0$. The corresponding causal outgoing wave solution of (45.132) is

$$\Psi_{\mathbf{k}}^{(+)}(\mathbf{r}, t) = \chi_{\mathbf{k}}(\mathbf{r}, t) + \int_{-\infty}^{t} dt'\int d\mathbf{r}'\,G_0^{(+)}(\mathbf{r}, t; \mathbf{r}', t')$$
$$\times V(\mathbf{r}')\,\Psi_{\mathbf{k}}^{(+)}(\mathbf{r}', t)\,, \qquad (45.137)$$

and the S-matrix element for a transition $\mathbf{k}_i \to \mathbf{k}_f$ in the presence of the laser field is given by

$$S_{\mathbf{k}_f, \mathbf{k}_i} = -i\left\langle \chi_{\mathbf{k}_f}\left|V\right|\Psi_{\mathbf{k}_i}^{(+)}\right\rangle\,, \qquad (45.138)$$

where an integration is carried out over all space and time in this matrix element. The time integration in (45.138) can be performed using the relation

$$\exp(ix\sin u) = \sum_{n=-\infty}^{\infty} J_n(x)\exp(inu)\,, \qquad (45.139)$$

where $J_n(x)$ is an ordinary Bessel function of order n. Then

$$S_{\mathbf{k}_f, \mathbf{k}_i} = -2\pi i\sum_{n=-\infty}^{\infty}\delta\left(E_{k_f} - E_{k_i} - n\omega\right)T_{\mathbf{k}_f, \mathbf{k}_i}^n \qquad (45.140)$$

where the delta function ensures energy conservation, and n is the number of photons absorbed or emitted. The differential cross section for the scattering process $\mathbf{k}_i \to \mathbf{k}_f$ with exchange of n photons can then be defined in terms of the T-matrix elements $T_{\mathbf{k}_f, \mathbf{k}_i}^n$ by

$$\frac{d\sigma_n}{d\Omega} = (2\pi)^4\frac{k_f}{k_i}\mid T_{\mathbf{k}_f, \mathbf{k}_i}^n\mid^2\,. \qquad (45.141)$$

There are two limiting cases in which considerable simplification in this expression occurs. First, at high energies or for weak potentials, the first Born approximation can be used to describe scattering by the potential $V(\mathbf{r})$. In this case (45.141) reduces to

$$\frac{d\sigma_n^{B1}}{d\Omega} = (2\pi)^4\frac{k_f}{k_i}J_n^2(\boldsymbol{\Delta}\cdot\boldsymbol{\alpha}_0)\mid V(\boldsymbol{\Delta})\mid^2\,, \qquad (45.142)$$

where $\boldsymbol{\Delta} = \mathbf{k}_i - \mathbf{k}_f$ is the momentum transfer vector and

$$\mathcal{V}(\boldsymbol{\Delta}) = (2\pi)^{-3}\int \exp(i\boldsymbol{\Delta}\cdot\mathbf{r})\,V(\mathbf{r})\,d\mathbf{r}\,. \qquad (45.143)$$

It follows immediately that

$$\frac{d\sigma_n^{1,\mathrm{B}}}{d\Omega} = J_n^2(\boldsymbol{\Delta}\cdot\boldsymbol{\alpha}_0)\frac{d\sigma^{1,\mathrm{B}}}{d\Omega}\,, \qquad (45.144)$$

where $d\sigma^{1,\mathrm{B}}/d\Omega$ is the field-free first Born differential cross section. Using the sum rule

$$\sum_{n=-\infty}^{\infty} J_n^2(x) = 1\,, \qquad (45.145)$$

Eq. (45.144) immediately yields

$$\sum_{n=-\infty}^{\infty}\frac{d\sigma_n^{1,\mathrm{B}}}{d\Omega} = \frac{d\sigma^{1,\mathrm{B}}}{d\Omega}\,. \qquad (45.146)$$

The second limiting case is the low frequency (soft photon) limit, where the laser photon energy ω is small compared with the electron energy E_{k_i}. In this limit, the T-matrix element is given by [80]

$$T_{\mathbf{k}_f, \mathbf{k}_i}^n = J_n(\boldsymbol{\Delta}\cdot\boldsymbol{\alpha}_0)\langle\chi_{\mathbf{k}_f'}\mid\mathcal{T}(E_{k_i'})\mid\chi_{\mathbf{k}_i'}\rangle + O(w^2)\,, \qquad (45.147)$$

where $\mathbf{k}_i'$ and $\mathbf{k}_f'$ are the shifted wave vectors

$$\mathbf{k}_i' = \mathbf{k}_i + \frac{n\omega}{\boldsymbol{\Delta}\cdot\boldsymbol{\alpha}_0}\boldsymbol{\alpha}_0\,, \quad \mathbf{k}_f' = \mathbf{k}_f + \frac{n\omega}{\boldsymbol{\Delta}\cdot\boldsymbol{\alpha}_0}\boldsymbol{\alpha}_0 \qquad (45.148)$$

and $\mathcal{T}(E_{k_i'})$ is the T-operator in the absence of the laser corresponding to the energy $E_{k_i'} = k_i'^2/2$. The differential cross section for the transfer of n photons is then

$$\frac{d\sigma_n}{d\Omega} = \frac{k_f}{k_i}J_n^2(\boldsymbol{\Delta}\cdot\boldsymbol{\alpha}_0)\frac{d\sigma}{d\Omega}(\mathbf{k}_f', \mathbf{k}_i') + O(\omega^2)\,, \qquad (45.149)$$

where $d\sigma(\mathbf{k}_f', \mathbf{k}_i')/d\Omega$ refers to the transition $\mathbf{k}_i' \to \mathbf{k}_f'$ in the absence of the laser. If the frequency is small enough to neglect the n dependence of $\mathbf{k}_i'$ and $\mathbf{k}_f'$, then using the sum rule (45.145), (45.149) becomes

$$\sum_{n=-\infty}^{\infty}\frac{d\sigma_n}{d\Omega} = \frac{d\sigma}{d\Omega}\,, \qquad (45.150)$$

where $d\sigma/d\Omega$ is the field-free differential cross section. The Kroll-Watson result (45.149) has been found to be surprisingly accurate, even for cases where $\omega/E_{k_i} \approx 0.5$ [81]. A nonrigorous extension of the Kroll-Watson result to inelastic processes has been considered by a number of authors, and has been found to give qualitative agreement with experiments on helium [82].

45.4.2 Scattering by Complex Atoms and Ions

The time-dependent Schrödinger equation describing an electron scattered by an N-electron atom or ion in the presence of a laser field can be written in analogy with (45.129) as

$$i\frac{\partial}{\partial t}\Psi\left(\mathbf{X}_{N+1},t\right) = \left(H_{N+1} - \frac{i}{c}\sum_{i=1}^{N+1}\mathbf{A}\left(\mathbf{r}_i,t\right)\cdot\boldsymbol{\nabla}_i\right.$$
$$\left. + \frac{1}{2c^2}\sum_{i=1}^{N+1}\mathbf{A}^2\left(\mathbf{r}_i,t\right)\right)\Psi\left(\mathbf{X}_{N+1},t\right),$$
$$(45.151)$$

where H_{N+1} is the $(N+1)$-electron Hamiltonian defined by (45.3) and $\mathbf{X}_{N+1}$ represents the space and spin coordinates of all $N+1$ electrons defined as in (45.8).

With the same assumptions made in the reduction of (45.129) to (45.132), (45.145) can be rewritten in the velocity gauge form

$$i\frac{\partial}{\partial t}\Psi_V\left(\mathbf{X}_{N+1},t\right)$$
$$= \left(H_{N+1} + \frac{1}{c}\mathbf{A}\left(t\right)\cdot\mathbf{P}_{N+1}\right)\Psi_V\left(\mathbf{X}_{N+1},t\right), \quad (45.152)$$

where

$$\mathbf{P}_{N+1} = \sum_{i=1}^{N+1} -i\boldsymbol{\nabla}_i \qquad (45.153)$$

is the total momentum operator.

The solution of (45.152) for e^-–H scattering in a strong laser field at high electron impact energies has been considered by Francken and Joachain [83]. They discussed the use of the Born series and the eikonal Born series (EBS) approximations, given by (45.100), to describe the electron-atom collision, and included the dressed wave function of the atomic hydrogen target to first-order in the field strength $\mathcal{E}_0$. Important effects related to the dressing of the target are the appearance in the cross sections of asymmetries between the absorption and the emission of a given number of laser photons, as well as the appearance of new resonance structures in the cross sections.

So far, detailed studies of low-energy electron-atom and electron-ion collisions in laser fields have been very limited. Pioneering work on e^-–H^+ collisions by Dimou and Faisal [84], and by Collins and Csanak [85] have shown important resonance effects caused by the field coupling bound states to the continuum. In addition, multichannel quantum defect theory has been applied by Zoller and co-workers [86,87] to study the behavior of Rydberg states in laser fields. We conclude this section by briefly describing the R-matrix-Floquet method [88]

for treating electron collisions with complex atoms and ions in a laser field, based on the R-matrix method discussed in Sect. 45.2.5.

In this approach, configuration space is divided into internal and external regions, as in the field-free case. In the internal region, a further gauge transformation of the field is made to the length gauge, so that the time-dependent Schrödinger equation (45.152) now becomes

$$i\frac{\partial}{\partial t}\Psi_L\left(\mathbf{X}_{N+1},t\right) = [H_{N+1} + \boldsymbol{\mathcal{E}}\left(t\right)\cdot\mathbf{R}_{N+1}]\Psi_L\left(\mathbf{X}_{N+1},t\right),$$
$$(45.154)$$

where $\mathbf{R}_{N+1} = \sum_{i=1}^{N+1}\mathbf{r}_i$, and we assume that the electric field $\boldsymbol{\mathcal{E}}\left(t\right)$ is given by

$$\boldsymbol{\mathcal{E}}\left(t\right) = -\frac{1}{c}\frac{d}{dt}\mathbf{A}\left(t\right) = \hat{\boldsymbol{\epsilon}}\mathcal{E}_0\cos\omega t. \qquad (45.155)$$

In order to solve (45.154), we introduce the Floquet-Fourier expansion [89,90]

$$\Psi_L\left(\mathbf{X}_{N+1},t\right) = e^{-iEt}\sum_{n=-\infty}^{\infty}e^{-in\omega t}\Psi_n^L\left(\mathbf{X}_{N+1}\right). \quad (45.156)$$

Substituting this equation into (45.154), using (45.155) and equating the coefficients of $\exp[-i\left(E+n\omega\right)t]$ to zero gives

$$\left(H_{N+1} - E - n\omega\right)\Psi_n^L + D_{N+1}\left(\Psi_{n-1}^L + \Psi_{n+1}^L\right) = 0,$$
$$(45.157)$$

where we have introduced the operator

$$D_{N+1} = \frac{1}{2}\mathcal{E}_0\hat{\boldsymbol{\epsilon}}\cdot\mathbf{R}_{N+1}. \qquad (45.158)$$

The functions Ψ_n^L can be regarded as the components of a vector Ψ^L in photon space. Equation (45.157) can then be written in this space as

$$\left(\mathbf{H}_F - E\mathbf{I}\right)\Psi^L = 0, \qquad (45.159)$$

where the Floquet Hamiltonian $\mathbf{H}_F$ is an infinite tridiagonal matrix.

In order to solve (45.159) in the internal region, the components Ψ_n^L are expanded in a basis which, in analogy with (45.69), has the form

$$\Psi_{kn}^L\left(\mathbf{X}_{N+1}\right) = \mathcal{A}\sum_{\Gamma ij}\overline{\Phi}_i^{\Gamma}\left(\mathbf{x}_1,\ldots,\mathbf{x}_N;\hat{\mathbf{r}}_{N+1}\sigma_{N+1}\right)r_{N+1}^{-1}$$
$$\times u_j\left(r_{N+1}\right)a_{ijkn}^{\Gamma}$$
$$+ \sum_{\Gamma i}\chi_i^{\Gamma}\left(\mathbf{x}_1,\ldots,\mathbf{x}_{N+1}\right)b_{ikn}^{\Gamma}, \quad (45.160)$$

where the summation over Γ is required since the total orbital angular momentum L and the total parity π in (45.9) are no longer conserved. The coefficients a_{ijkn}^{Γ}

and b_{ikn}^{Γ} are then determined by diagonalizing $\mathbf{H}_F + \mathbf{L}_b$, where $\mathbf{L}_b$ is an appropriate Bloch operator.

In the external region, the wave function describing the scattered electron is transformed to the velocity gauge, and the corresponding coupled equations integrated outwards from the internal region boundary for each energy of interest [91]. After a further transformation to the acceleration frame (or Kramers-Henneberger frame [92]) the wave function can be fitted to an asymptotic form to yield the K-matrix, S-matrix and collision cross sections. This approach is now being used to calculate multiphoton ionization and electron collisions by a number of complex atoms and ions [93, 94].

REFERENCES

1. P. G. Burke and A. Crowe, *Encyclopedia of Applied Physics* (1993), Vol. 5, p. 499.
2. F. A. Gianturco and A. Jain, Phys. Rep. **143**, 347 (1986).
3. N. J. Mason, Rep. Prog. Phys. **56**, 1275 (1993).
4. I. C. Percival and M. J. Seaton, Proc. Camb. Phil. Soc. **53**, 654 (1957).
5. T. F. O'Malley, L. Spruch, and L. Rosenberg, J. Math. Phys. **2**, 491 (1961).
6. M. J. Seaton, Rep. Prog. Phys. **46**, 167 (1983).
7. H. E. Saraph, Comp. Phys. Commun. **3**, 256 (1972); *ibid.* **15**, 247 (1978).
8. M. Jones, Phil. Trans. R. Soc. London A **277**, 587 (1975).
9. N. S. Scott and P. G. Burke, J. Phys. B **13**, 4299 (1980).
10. J. J. Chang, J. Phys. B **8**, 2327 (1975).
11. P. H. Norrington and I. P. Grant, J. Phys. B **14**, L261 (1981).
12. U. Fano, Phys. Rev. **124**, 1866 (1961).
13. U. Fano, Rep. Prog. Phys. **46**, 97 (1983).
14. H. Feshbach, Ann. Phys. (N. Y.) **5**, 357 (1958).
15. H. Feshbach, Ann. Phys. (N. Y.) **19**, 287 (1962).
16. W. Brenig and R. Haag, Fortschr. Phys. **7**, 183 (1959).
17. P. G. Burke, Adv. At. Mol. Phys. **4**, 173 (1968).
18. M. J. Seaton, Proc. Phys. Soc. **88**, 801, 815 (1966).
19. M. J. Seaton, J. Phys. B **2**, 5 (1969).
20. M. Gailitis, Sov. Phys. JETP **17**, 1328 (1963).
21. U. Fano, Phys. Rev. A **2**, 353 (1970).
22. U. Fano, Comm. At. Mol. Phys. **10**, 223 (1981); *ibid.* **13**, 157 (1983).
23. E. P. Wigner, Phys. Rev. **70**, 15, 606 (1946).
24. E. P. Wigner and L. Eisenbud, Phys. Rev. **72**, 29 (1947).
25. P. G. Burke and K. A. Berrington, *Atomic and Molecular Processes: An R-matrix Approach* (Institute of Physics, Bristol and Philadelphia 1993).
26. C. Bloch, Nucl. Phys. **4**, 503 (1957).
27. R. K. Nesbet, *Variational Methods in Electron-Atom Scattering Theory* (Plenum Press, New York, 1980).
28. W. H. Miller, Comm. At. Mol. Phys. **22**, 115 (1988).
29. B. I. Schneider and T. N. Rescigno, Phys. Rev. A **37**, 3749 (1988).
30. K. Takatsuka and V. McKoy, Phys. Rev. A **24**, 2743 (1981); *ibid.* **30**, 1734 (1984).
31. M. J. Seaton, J. Phys. B **7**, 1817 (1974).

32. B. I. Schneider and L. A. Collins, Comp. Phys. Rep. **10**, 49 (1989).
33. P. G. Burke and H. M. Schey, Phys. Rev. **126**, 147 (1962).
34. R. Damburg and E. Karule, Proc. Phys. Soc. **90**, 637 (1967).
35. P. G. Burke and T. G. Webb, J. Phys. B **3**, L131 (1970).
36. J. Callaway and J. W. Wooten, Phys. Lett. A **45**, 85 (1973).
37. P. G. Burke, C. J. Noble, and M. P. Scott, Proc. Roy. Soc. A **410**, 289 (1987).
38. T. T. Scholz, H. R. J. Walters, P. G. Burke, and M. P. Scott, J. Phys. B **24**, 2097 (1991).
39. M. Rotenberg, Ann. Phys. (N. Y.) **19**, 262 (1962).
40. I. Bray and A. T. Stelbovics, Phys. Rev. A **46**, 6995 (1992).
41. I. Bray and A. T. Stelbovics, Phys. Rev. Lett. **69**, 53 (1992); *ibid.* **70**, 746 (1993).
42. I. Bray, Phys. Rev. A **49**, 1066 (1994).
43. M. H. Mittleman and K. M. Watson, Phys. Rev. **113**, 198 (1959).
44. F. W. Byron Jr. and C. J. Joachain, J. Phys. B **14**, 2429 (1981).
45. B. H. Bransden and J. P. Coleman, J. Phys. B **5**, 537 (1972).
46. B. H. Bransden, T. Scott, R. Shingal, and R. K. Raychoudhurry, J. Phys. B **15**, 4605 (1982).
47. I. E. McCarthy, Comm. At. Mol. Phys. **24**, 343 (1990).
48. J. Callaway and D. H. Oza, Phys. Rev. A **32**, 2628 (1985).
49. H. R. J. Walters, Phys. Rep. **116**, 1 (1984).
50. C. J. Joachain, in *Collision Theory for Atoms and Molecules*, edited by F. A. Gianturco (Plenum Press, New York, 1989), p. 59.
51. F. W. Byron Jr. and C. J. Joachain, Phys. Lett. C **44**, 233 (1977).
52. R. J. Glauber, in *Lectures in Theoretical Physics*, edited by W. E. Brittin (Interscience, New York, 1959), Vol. 1, p. 315.
53. V. I. Ochkur, Sov. Phys. JETP **18**, 503 (1964).
54. M. R. H. Rudge and M. J. Seaton, Proc. Roy. Soc. A **283**, 262 (1965).
55. R. K. Peterkop, Proc. Phys. Soc. **77**, 1220 (1961).
56. M. R. H. Rudge, Rev. Mod. Phys. **40**, 564 (1968).
57. H. Jackubowicz and D. L. Moores, J. Phys. B **14**, 3733 (1981).
58. K. Bartschat and P. G. Burke, J. Phys. B **20**, 3191 (1987).
59. G. Müller, in *The Physics of Electronic and Atomic Collisions*, edited by A. Dalgarno, R. S. Freund, P. M. Koch, M. S. Lubell, and T. B. Lucatarto (AIP Conf. Series 250) (American Institute of Physics, New York, 1990), p. 418.
60. T. Rösel, J. Röder, L. Frost, K. Jung, H. Ehrhardt, S. Jones, and D. H. Madison, Phys. Rev. A **46**, 2539 (1992).
61. A. M. Arthurs and A. Dalgarno, Proc. Roy. Soc. A **256**, 540 (1960).
62. P. G. Burke, N. Chandra, and F. A. Gianturco, J. Phys. B **5**, 2212 (1972).
63. S. I. Drozdov, Sov. Phys. JETP **1**, 591 (1955); *ibid.* **3**, 759 (1956).
64. D. M. Chase, Phys. Rev. **104**, 838 (1956).
65. I. Shimamura, Chem. Phys. Lett. **73**, 328 (1980); J. Phys. B **15**, 93 (1982).

66. M. A. Morrison, Adv. At. Mol. Phys. **24**, 51 (1987).

67. N. Chandra and A. Temkin, Phys. Rev. A **13**, 188 (1976); *ibid.* **14**, 507 (1976).

68. A. Herzenberg and F. Mandl, Proc. Roy. Soc. A **270**, 48 (1962).

69. A. Herzenberg, in *Electron Molecule Collisions*, edited by I. Shimamura and K. Takayanagi (Plenum, New York, 1984), p. 351.

70. J. N. Bardsley, J. Phys. B **1**, 349 (1968); *ibid.* **1**, 365 (1968).

71. W. Domcke, Phys. Rep. **208**, 97 (1991).

72. A. F. J. Siegert, Phys. Rev. **56**, 750 (1939).

73. B. I. Schneider, M. Le Dourneuf, and P. G. Burke, J. Phys. B **22**, L365 (1979).

74. R. K. Nesbet, Phys. Rev. A **19**, 551 (1979).

75. M. Shugard and A. U. Hazi, Phys. Rev. A **12**, 1895 (1975).

76. T. N. Rescigno, in *Electron Collisions with Molecules, Clusters and Surfaces*, edited by H. Ehrhardt and L. A. Morgan (Plenum Press, New York, 1994), p. 1.

77. W. R. Newell, Comm. At. Mol. Phys. **28**, 59 (1992).

78. M. H. Mittleman, *Introduction to the Theory of Laser-Atom Interactions* (Plenum Press, New York, 1993), 2nd ed.

79. C. J. Joachain, in *Fundamentals of Laser Interactions*, edited by F. Ehlotzky (Springer Verlag, Berlin, 1985).

80. N. M. Kroll and K. M. Watson, Phys. Rev. **8**, 804 (1973).

81. R. Shakeshaft, Phys. Rev. A **28**, 667 (1983); *ibid.* **29**, 383 (1984).

82. S. Geltman and A. Maquet, J. Phys. B **22**, L419 (1989).

83. P. Francken and C. J. Joachain, J. Opt. Soc. Am. **7**, 554 (1990).

84. L. Dimou and F. H. M. Faisal, Phys. Rev. Lett. **59**, 872 (1987).

85. L. A. Collins and G. Csanak, Phys. Rev. **44**, R5343 (1991).

86. A. Giusti-Suzor and P. Zoller, Phys. Rev. A **36**, 5178 (1987).

87. P. Marte and P. Zoller, Phys. Rev. A **43**, 1512 (1991).

88. P. G. Burke, P. Francken, and C. J. Joachain, J. Phys. B **24**, 761 (1991).

89. J. H. Shirley, Phys. Rev. **138**, B979 (1965).

90. R. M. Potvliege and R. Shakeshaft, in *Atoms in Intense Laser Fields*, edited by M. Gavrila (Academic Press, 1993), p. 373.

91. M. Dörr, M. Terao-Dunseath, J. Purvis, C. J. Noble, P. G. Burke, and C. J. Joachain, J. Phys. B **25**, 2809 (1992).

92. W. C. Henneberger, Phys. Rev. Lett. **21**, 838 (1968).

93. J. Purvis, M. Dörr, M. Terao-Dunseath, C. J. Joachain, P. G. Burke, and C. J. Noble, Phys. Rev. Lett. **71**, 3943 (1993).

94. M. Dörr, M. Terao-Dunseath, P. G. Burke, C. J. Joachain, C. J. Noble, and J. Purvis, J. Phys. B (to be published).

46

Positron Collisions

R. P. McEachran and A. D. Stauffer

Department of Physics and Astronomy, York University, Toronto, Ontario M3J 1P3, Canada

46.1 INTRODUCTION

The positron is the antiparticle of the electron, having the same mass but opposite charge. Since beams of low-energy positrons are difficult to obtain, there is considerably less experimental data available for positrons than for electrons. This is particularly true for quantites which require large incident positron fluxes, such as differential scattering cross sections and coincidence parameters.

Positrons undergo collisions with atomic and molecular systems in much the same way as electrons do. Thus the standard scattering theory for electrons (see Chap. 45) can also be applied to positron scattering. However, there are several important differences from electron scattering which we outline below.

Since the positron is distinguishable from atomic electrons, it cannot undergo exchange processes with the bound electrons during a collision, as is possible with electrons. Thus the nonlocal exchange terms which arise in the description of electron scattering are not present for positrons. This leads to a simplification of the scattering equations. However, there are scattering channels available with positron scattering which do not exist with electrons. These are dealt with in Sect. 46.2.

Throughout this chapter we will employ atomic units unless otherwise noted.

46.2 SCATTERING CHANNELS

Positrons colliding with atomic and molecular systems have the same scattering channels available as for electrons, *viz.* elastic, excitation, ionization, and for molecules, dissociation. However, two channels exist for positrons which do not exist for electrons: positronium formation and annihilation.

46.2.1 Positronium Formation

Positronium, a bound state of an electron-positron pair (see Chap. 27), can be formed during the collision of a positron with an atomic or molecular target. The positronium 'atom' can escape to infinity, leaving the target in a singly ionized state. Thus, this process is difficult to distinguish experimentally from true ionization, where both the incident positron and the ionized electron are asymptotically free particles. The positronium atom can exist in its ground state or in any one of an infinite number of excited states after the collision. The level structure of positronium is, to order α^2, identical to that of hydrogen, but with each level having half the energy of the corresponding hydrogenic state.

Positronium formation is a rearrangement channel, and thus is a two-centre problem. Because positronium is a light particle, having a reduced mass one-half of that of an electron, the semiclassical type of approximation used in ion-atom collisions (see Chap. 49) is not applicable here. We will discuss various theoretical approaches to this process in Sect. 46.3.

Positronium formation in the ground state has a threshold which is $\frac{1}{2}$Ry (6.80285 eV) below the ionization threshold of the target. This means that it is normally the lowest inelastic channel in positron scattering from neutral atoms. For atoms with a small ionization potential, such as the alkalis, this channel is always open. The energy range between the positronium threshold and the first excited state of the atom is known as the Ore gap. In this range, positronium formation is the only possible inelastic process.

46.2.2 Annihilation

Annihilation is a process in which an electron-positron pair is converted into two or more photons. It can occur either directly with a bound atomic electron or after positronium formation has taken place. The direct annihilation cross section for a positron of momentum **k** colliding with an atomic or molecular target can be written as [1]

$$\sigma_{\mathrm{a}} = \frac{\alpha^3 Z_{\mathrm{eff}}}{k} \left(\pi a_0^2 \right), \qquad (46.1)$$

where Z_{eff} can be thought of as the effective number of electrons in the target with which the positron can annihilate. If $\Psi(\mathbf{r}_1, \mathbf{r}_2, \ldots, \mathbf{r}_N, \mathbf{x})$ is the wave function for the system of a positron, with coordinate **x**, colliding with an N-electron target, then

$$Z_{\mathrm{eff}} = \sum_{i=1}^{N} \int d\mathbf{r}_1 \, d\mathbf{r}_2 \cdots d\mathbf{r}_N \left| \Psi(\mathbf{r}_1, \mathbf{r}_2, \ldots, \mathbf{r}_N; \mathbf{r}_i) \right|^2. \qquad (46.2)$$

While this formula can be naively derived by assuming that the positron can only annihilate with an electron if it is at the identical location, it actually follows from a quantum electrodynamical treatment of the process [2, 3]. If the wave function Ψ is approximated by the product of the undistorted target wave function times a positron scattering function $F(\mathbf{x})$, then

$$Z_{\mathrm{eff}} = \int d\mathbf{r} \, \rho(\mathbf{r}) \left| F(\mathbf{r}) \right|^2, \qquad (46.3)$$

where ρ is the electron number density of the target. Thus, in the Born approximation, where F is taken as a plane wave, Z_{eff} simply becomes the total number of electrons Z in the target. When a positron annihilates with an atomic electron, two 511 keV photons are the most likely result if the positron-electron pair are in a singlet spin state (parapositronium). In the centre-of-mass frame of the pair, the photons are emitted in opposite directions to conserve momentum. However, in the laboratory frame, the bound electron has a momentum distribution which is reflected in the photon directions not being exactly 180 degrees apart. This slight angular deviation, called the angular correlation, can be measured and gives information about the momentum distribution of the bound electrons. This quantity is given by [2]

$$S(\mathbf{q}) = \sum_{i=1}^{N} \int d\mathbf{r}_1 \cdots d\mathbf{r}_{i-1} \, d\mathbf{r}_{i+1} \cdots d\mathbf{r}_N$$
$$\times \int d\mathbf{r}_i \, d\mathbf{x} \, e^{i\mathbf{q} \cdot \mathbf{x}} \left| \Psi(\mathbf{r}_1, \mathbf{r}_2, \ldots, \mathbf{r}_N; \mathbf{x}) \, \delta(\mathbf{r}_i - \mathbf{x}) \right|^2, \qquad (46.4)$$

where **q** is the resultant momentum of the annihilating pair. In evaluating this quantity, the positron is assumed to be thermalized in the gas before undergoing annihilation. Experimentally, only one component of **q** is measured so that $S(\mathbf{q})$ is integrated over the other two components of the momentum to obtain the measured quantity. The spin triplet component of an electron-positron pair (orthopositronium) can only decay with the emission of three or more photons which do not have well defined energies. This is a much less probable process than the two-photon decay from the singlet component.

46.3 THEORETICAL METHODS

The basic theoretical approaches to the calculation of positron scattering from atoms and molecules were originally developed for electron scattering, and later applied to the positron case. Thus, we emphasize here only the differences that arise between the electron and positron cases, both in the theoretical formulations and, in later sections, in the nature of the results.

The lowest-order interaction between a free positron and an atomic or molecular target is the repulsive static potential of the target

$$V_{\mathrm{s}} = \langle \psi_0 | V | \psi_0 \rangle, \qquad (46.5)$$

where ψ_0 is the unperturbed target wave function and V is the electrostatic interaction potential between the positron and the target. Since this interaction has the opposite sign from that for electron scattering, the static

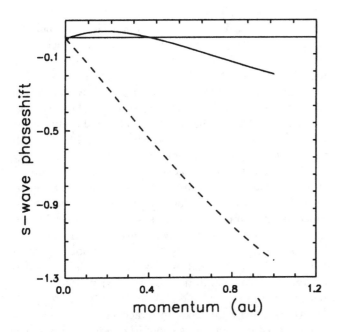

Figure 46.1. Variational *s*-wave phase shifts for electron [4] (– – –) and positron [5] (——) scattering from helium atoms.

Figure 46.2. Variational *p*-wave phase shifts for electron [4] (– – –) and positron [7] (——) scattering from helium atoms.

potential also has the opposite sign in these two cases. On the other hand, the next higher-order of interaction is polarization, which arises from the distortion of the atom by the incident particle. If we represent this distortion of the target to first-order by the wave function ψ_1, as, for example, in the polarized-orbital approximation [6], then the polarization interaction can be represented by the potential

$$V_{\mathrm{p}} = \langle \psi_0 | V | \psi_1 \rangle . \tag{46.6}$$

This potential is attractive for both positron and electron scattering and has an asymptotic form with leading term $-\alpha_{\mathrm{d}}/2r^4$, where α_{d} is the static dipole polarizability (see Sect. 23.3.3) of the target. Thus, the static and polarization potentials for positron scattering from ground state systems are of opposite sign and tend to cancel one another. This leads to very different behavior from the electron case where they are of the same sign. In particular, the elastic scattering cross sections for positron scattering from an atom are much smaller than for electron scattering, and the phase shifts (see Sect. 45.2.1) have very different magnitudes and dependence on energy. We illustrate this for the case of scattering from helium in Figs. 46.1–46.3, where we show the results of highly accurate variational calculations for scattering by electrons and positrons. Note the difference in sign between the electron and positron *s*-wave phase shifts for very small values of the incident momentum. The fact that the positron phase shift goes through zero leads to the Ramsauer minimum in the positron total cross sec-

tion, as shown in Fig. 46.3. The large difference in magnitudes between the electron and positron *s*- and *p*-wave phase shifts leads to the large difference in the total elastic cross sections as shown.

Higher-order terms in the interaction potential may also give important contributions to scattering cross sections. For a detailed discussion, see Drachman and Temkin [8].

A simple potential scattering calculation using the sum of the static and polarization potentials and without the exchange terms that are present for the electron case can be applied to elastic scattering calculations for closed shell systems (Sect. 46.4.2).

The potentials defined above can also be used in a distorted wave approximation (see Chap. 45) which can be applied to excitation and ionization by positron impact. Once again, the complicated exchange terms which arise in electron scattering are absent here. Sienkiewicz and Baylis [9] have included such potentials in a Dirac-Fock formulation of positron scattering which treats the positron as an electron with negative energy.

For positrons with high enough incident energies ($\sim$ 1keV) the first Born approximation will become valid (see Chap. 45). Since the first Born approximation is independent of the sign of the charge of the incident particles, this indicates that as the incident energy increases, the corresponding cross sections for electron and positron scattering eventually merge. From flux conservation arguments this means that the positronium formation cross section rapidly decreases as the incident

Figure 46.3. Total elastic cross sections for electron (– – –) and positron (——) scattering from helium atoms calculated from the phase shifts shown in Figs. 46.1 and 46.2. The higher-order phase shifts were calculated from effective range theory (see Sect. 45.2.1).

energy increases. In fact, from experimental measurements, the total cross sections (summed over all possible channels) for electron and positron scattering appear to merge at a much lower energy than the cross sections for individual channels [10]. More elaborate calculations for high-energy scattering have been carried out using the eikonal-Born series [11] (see Chap. 45). These approximations allow for both polarization and absorption (i.e., inelastic processes) and yield good agreement with elastic experimental measurements of differential cross sections at energies above 100 eV. A detailed analysis [11, 12] of the various contributions to the scattering indicates that absorption effects, due to the various open inelastic channels, play a much more important role here than for electron scattering.

A more elaborate treatment of positron scattering is based on the close-coupling approximation (see Chap. 45), where the wave function for the total system of positron plus target is expanded using a basis set comprised of the wave functions of the target. Once again, there are no exchange terms involving the positron and, in principle, a complete expansion including the continuum states of the target would include the possibility of positronium formation. However, such an expansion is not practicable if one wants to calculate explicit cross sections for positronium formation. Even in cases where such cross sections are not required, the considerable effect that the positronium formation channels can have on the other scattering cross sections is best included through a close-coupling expansion that includes terms representing positronium states plus the residual target ion. There is a potential for double counting of states in such an expansion, but in practice this does not appear to be a problem if the number of states in the expansion is not large. Also, in many cases, additional pseudostates have to be included in the expansion in order to correctly represent the long-range polarization interaction.

A close-coupling expansion including positronium states is a two-centre problem, i.e., it includes the centres of mass of both the target and the positronium states. Since positronium is a light system, the semiclassical approach often used to treat rearrangement collisions between heavy systems (see Chap. 48) is not applicable here. This means that one is faced with a problem of considerable computational complexity which is only recently being treated successfully [13–17].

Another way to take into account the effects of open inelastic channels without the complications of a full close-coupling approach is to use optical potentials. These are often based on a close-coupling formalism [12, 18] and lead to a complex potential, the real part of which represents distortions of the target (such as polarization) while the complex part allows for absorption (i.e., flux into open channels not explicitly represented).

Recently, Bray and Stelbovics [19] have applied the convergent close-coupling method to the scattering of positrons from atoms. This method includes contributions from the continuum states of the target and a sufficient number terms in the expansion to ensure numerical convergence. At the present time, however, positronium states have not been explicitly included.

Finally, there is the variational method (see Chap. 45) which uses an analytic form of trial wave function to represent the total system. The parameters of this analytic function are determined as part of the method. Given a trial wave function with sufficient flexibility and a large enough number of parameters, essentially exact results can be obtained in the elastic energy range and the Ore gap. Because the complexity of the trial function increases as the square of the number of electrons in the target, only positron scattering from hydrogen, helium and lithium and the hydrogen molecule have been treated by this method to date [5, 20].

In the case of ionization, there appears to be quite a distinct threshold behavior of the cross sections for electron and positron collisions. For electrons, the Wannier threshold law (see Sect. 50.3.1) has an exponent of 1.127, while a similar analysis for positrons [21] yields an exponent of 2.651. However, the existence of the positronium formation channel leaves a question as to whether this analysis will give the dominant term at threshold.

There has been a recent investigation [22] of the

behavior of the elastic cross sections at the positronium formation threshold which predicts the occurrence of a Wigner cusp for the heavier noble gases.

46.4 PARTICULAR APPLICATIONS

46.4.1 Atomic Hydrogen

Because of the difficulty of making measurements in atomic hydrogen, the available experimental data are restricted, so far, to total cross sections, as well as to total ionization and positronium formation cross sections. Essentially exact variational calculations have been carried out in the elastic energy regime and the Ore gap [5]. Ionization cross sections have been calculated in a number of approximations. Relatively simple Born and distorted-wave treatments of the total ionization cross section have been carried out [23]. However, disagreements between the experimental measurements mean that there is at present no reliable way of assessing the various approximations used. More elaborate calculations with asymptotically correct wave functions have been used to determine triple differential cross sections for ionization [24, 25]. However, the task of integrating these to produce total cross sections is a formidable one.

In order to determine reliable positronium formation cross sections, the explicit positronium states have to be included. Several close-coupling calculations have been carried out [14–17]. These indicate the necessity of explicitly including positronium formation channels in the expansion of the total wave function in order to obtain accurate results, even for elastic scattering.

As is the case for electron scattering, positron cross sections exhibit resonances (see Sect. 45.2.3). These have been extensively studied by Ho [26] using variational and complex rotation methods.

46.4.2 Noble Gases

Because the noble gases are convenient experimental targets, a good deal of effort has gone into calculations for these targets, particularly for elastic scattering. In the purely elastic energy range, i.e., for energies below the positronium formation threshold, the simple potential scattering approach using the static and polarization potentials defined above yields quite good results. Since the long-range behavior of the sum of the potentials is attractive, the scattering phase shifts for positrons must be positive for sufficiently low energies. However, as its incident energy increases, the positron probes the repulsive inner part of the potential and the phase shifts become negative. This behavior leads to the Ramsauer minimum in the integral elastic cross sections (see Chap. 45) for the lighter noble gases (helium, neon

and possibly argon) [10]. This differs from electron scattering where some low-energy phase shifts can be negative (modulo π) because of the existence of bound orbitals of the same symmetry.

Another difference between positron and electron scattering is exhibited by the differential cross sections (see Chap. 45). For electrons, the shape of the cross section is determined by a few dominant phase shifts, whereas for positrons, many phase shifts contribute to the final shape [27]. Because of this behavior, in the inelastic scattering regime, the existence of open channels has a much more marked effect on the shape of the differential cross sections for positron scattering than for electrons [28]. Thus the simple potential scattering approximation is no longer sufficient, and the inelastic channels have to be taken into account via, for example, the use of an optical potential [11, 12].

There are little experimental data for inelastic scattering and the theory is limited to some simple distorted wave calculations (see Chap. 45).

Total ionization cross sections have been measured for all of the noble gases. For these cross sections in helium, distorted wave methods seem to yield good results [29].

There exist some measurements of Z_{eff} and angular correlation parameters for these gases. Calculations have also been made for them in the polarized-orbital approximation [30].

46.4.3 Alkali Atoms

In the case of positron scattering from alkali atoms, the positronium formation channel is always open and the simple potential scattering approach does not yield reliable results. Close-coupling methods are the most suitable methods here; and several such calculations have been carried out [31, 32]. As for scattering from hydrogen, inclusion of positronium formation channels is important. Substantial resonance features have also been found in the cross sections [33].

46.4.4 Molecules

The hydrogen molecule is sufficiently simple that it can be treated by variational methods in the elastic regime and Ore gap [5]. Potential scattering calculations for positron scattering from the diatomic molecules N_2 have been carried out and more recently there have been some close-coupling calculations [34, 35]. Measurements of total cross sections and Z_{eff} on these and other more complicated molecules can be carried out quite readily [36, 37].

46.5 REVIEWS

For a number of years a Positron Workshop has been held as a satellite of the International Conference on the Physics of Electronic and Atomic Collisions. Their proceedings [38–44] give an excellent summary of the state of positron scattering research, both experimental and theoretical.

There are several review articles on positron scattering in gases. The most comprehensive of these, Refs. [1, 45], give a summary of the earlier results in the field.

REFERENCES

1. P. A. Fraser, Adv. At. Mol. Phys. **4**, 63 (1968).
2. R. Ferrell, Rev. Mod. Phys. **28**, 308 (1956).
3. P. R. Wallace, Solid State Phys. **10**, 1 (1960).
4. R. K. Nesbet, Phys. Rev. A **20**, 58 (1979).
5. E. A. G. Armour and J. W. Humberston, Phys. Rep. **204**, 165 (1991).
6. R. P. McEachran and A. D. Stauffer, J. Phys. B **10**, 663 (1977).
7. J. W. Humberston and R. I. Campeanu, J. Phys. B **13**, 4907 (1980).
8. R. J. Drachman and A. Temkin, in *Case Studies in Atomic Collision Physics II*, edited by E. W. McDaniel and M. R. C. McDowell (North-Holland, Amsterdam, 1972), Chap. 6.
9. J. E. Sienkiewicz and W. E. Baylis, Phys. Rev. A **43**, 1331 (1991).
10. W. E. Kauppila and T. S. Stein, Can. J. Phys. **60**, 471 (1981).
11. C. J. Joachain and R. M. Potvliege, Phys. Rev. A **35**, 4873 (1987).
12. K. Bartschat, R. P. McEachran, and A. D. Stauffer, J. Phys. B **21**, 2789 (1988).
13. R. N. Hewitt, C. J. Noble, and B. H. Bransden, J. Phys. B **26**, 3661 (1993).
14. K. Higgins and P. G. Burke, J. Phys. B **26**, 4269 (1993).
15. J. Mitroy, J. Phys. B **26**, 4861 (1993).
16. N. K. Sarkar and A. S. Ghosh, J. Phys. B **27**, 759 (1994).
17. G. Liu and T. T. Gien, Phys. Rev. A **46**, 3918 (1992).
18. I. E. McCarthy, K. Ratnavelu, and Y. Zhou, J. Phys. B **26**, 2733 (1993).
19. I. Bray and A. Stelbovics, Phys. Rev. A **49**, R2224 (1994).
20. J. W. Humberston, J. Phys. B **25**, L491 (1992).
21. H. Klar, J. Phys. B **14**, 4165 (1981).
22. G. Laricchia, J. Moxom, M. Charlton, A. Kover, and W. E. Meyerhoff, Hyperfine Interactions **89**, 209 (1994).
23. P. Acacia, R. I. Campeanu, M. Horbatsch, R. P. McEachran, and A. D. Stauffer, Phys. Lett. A **179**, 205 (1993).
24. M. Brauner, J. S. Briggs, and H. Klar, J. Phys. B **22**, 2265 (1989).
25. S. Jetzke and F. H. M. Faisal, J. Phys. B **25**, 1543 (1992).
26. Y. K. Ho, Hyperfine Interactions **73**, 109 (1992).
27. R. P. McEachran and A. D. Stauffer, in *Positron (Electron)-Gas Scattering*, edited by W. E. Kauppila, T. S. Stein, and J. M. Wadehra (World Scientific, Singapore, 1986), pp. 122–130.
28. W. E. Kauppila and T. S. Stein, Hyperfine Interactions, **40**, 87 (1990).
29. R. I. Campeanu, R. P. McEachran, and A. D. Stauffer, J. Phys. B **21**, 1411 (1988).
30. D. M. Schrader and R. E. Svetic, Can. J. Phys. **60**, 517 (1982).
31. J. Ward, M. Horbatsch, R. P. McEachran, and A. D. Stauffer, J. Phys. B **22**, 1845 (1989).
32. M. T. McAlinden, A. A. Kernoghan, and H. R. J. Walters, Hyperfine Interactions **89**, 161 (1994).
33. S. J. Ward, M. Horbatsch, R. P. McEachran, and A. D. Stauffer, J. Phys. B **22**, 3763 (1989).
34. A. S. Ghosh, T. Mukherjee, and J. W. Darewych, Hyperfine Interactions **89**, 319 (1994).
35. G. Danby and J. Tennyson, J. Phys. B **24**, 3517 (1991).
36. W. E. Kauppila and T. S. Stein, Adv. At. Mol. Opt. Phys. **26**, 1 (1990).
37. K. Iwata, R. G. Greaves, and C. M. Surko, Hyperfine Interactions **89**, 271 (1994).
38. Can. J. Phys. **40**, 461 (1981).
39. *Positron Scattering in Gases*, edited by J. W. Humberston and M. R. C. McDowell (Plenum, New York, 1984).
40. *Positron (Electron)-Gas Scattering*, edited by W. E. Kauppila, T. S. Stein, and J. M. Wadehra (World Scientific, Singapore, 1986).
41. *Atomic Physics with Positrons*, edited by J. W. Humberston and E. A. G. Armour (Plenum, New York, 1987).
42. *Annihilation in Gases and Galaxies*, edited by R. J. Drachman (NASA Conference Publication 3058, Washington, 1990).
43. *Positron Interactions with Gases*, edited by L. A. Parcell, Hyperfine Interactions **73**, 1 (1992).
44. *Positron Interactions with Atoms, Molecules and Clusters*, edited by W. Raith and R. P. McEachran, Hyperfine Interactions **89**, 1 (1994).
45. B. H. Bransden, *Case Studies in Atomic Collision Physics I*, edited by E. W. McDaniel and M. R. C. McDowell (Wiley, New York, 1969), Chap. 4.

47

Adiabatic and Diabatic Collision Processes at Low Energies

E. E. Nikitin

Department of Chemistry, Technion, Israel Institute of Technology, Haifa 32000, Israel

47.1 BASIC DEFINITIONS

47.1.1 Slow Quasiclassical Collisions

Slow collisions of atoms or molecules (neutral or charged) are defined as collisions for which the velocity of the relative motion of colliding particles v is substantially lower than the velocity of valence electrons v_e:

$$v/v_e \ll 1. \tag{47.1}$$

If v_e is estimated as $v_e \approx 1$ au $\approx 10^8$ cm/s, then Eq. (47.1) is fulfilled for medium mass nuclei (~ 10 amu) up to several keV.

Quasiclassical collisions are those for which the de Broglie wavelength λ_{dB} for the relative motion is substantially smaller than the range parameter a of the interaction potential:

$$\lambda_{\mathrm{dB}} \ll a. \qquad (47.2)$$

The two conditions (47.1) and (47.2) define the energy range within which collisions are slow and quasiclassical. For medium mass nuclei, this energy range covers collision energies above room temperature and below hundreds of eV. The paramater a should not be confused with another important parameter L_0 which characterizes the extent of the interaction region. For instance, for the exchange interaction between two atoms, L_0 corresponds to the distance of closest approach of the colliding particles, while a is the range of the exponential decrease of the interaction. Typically, L_0 noticeably exceeds a.

47.1.2   Adiabatic and Diabatic Electronic States

Let r refer to a set of electronic coordinates in a body-fixed frame related to the nuclear framework of a colliding system, and let $\mathcal{R}$ refer to a set of nuclear coordinates determining the relative position of nuclei in this system. A configuration of electrons and nuclei in a frame fixed in space is completely determined by r, $\mathcal{R}$, and the set of Euler angles Ω, which relate the body-fixed frame to the space-fixed frame. If the total Hamiltonian of the system is $\mathcal{H}(r, \mathcal{R}, \Omega)$, the stationary state wave function satisfies the equation

$$\mathcal{H}(r, \mathcal{R}, \Omega)\Psi_E(r, \mathcal{R}, \Omega) = E\Psi_E(r, \mathcal{R}, \Omega). \qquad (47.3)$$

The electronic adiabatic Hamiltonian $H(r; \mathcal{R})$ is defined to be the part of $\mathcal{H}(r, \mathcal{R}, \Omega)$ in which the kinetic energy is ignored. The adiabatic electronic functions $\psi_n(r; \mathcal{R})$ are defined as eigenfunctions of $H(r; \mathcal{R})$ at a fixed nuclear configuration $\mathcal{R}$:

$$H(r; \mathcal{R})\psi_n(r; \mathcal{R}) = U_n(\mathcal{R})\psi_n(r; \mathcal{R}). \qquad (47.4)$$

The eigenvalues $U_n(\mathcal{R})$ are called adiabatic potential energy surfaces (adiabatic PES). In the case of a diatom, the set $\mathcal{R}$ collapses into a single coordinate, the internuclear distance R, and the PES become potential energy curves, $U_n(R)$. The functions $\psi_n(r; \mathcal{R})$ depend explicitly on $\mathcal{R}$ and implicitly on the Euler angles Ω. The significance of the adiabatic PES is related to the fact that in the limit of very low velocities, a system of nuclei will move across a single PES. In this approximation, called the adiabatic approximation, the function $U_n(\mathcal{R})$ plays the part of the potential energy which drives the motion of the nuclei.

An electronic diabatic Hamiltonian is defined formally as a part of H, i.e., $H_0 = H + \Delta H$. The partitioning of H into H_0 and ΔH is dictated by the requirement that the eigenfunctions of H_0, called diabatic electronic functions ϕ_n, depend weakly on the configuration $\mathcal{R}$. The physical meaning of this weak dependence is different for different problems. A perfect diabatic basis set $\phi_n(r)$ is $\mathcal{R}$-independent; for practical purposes one can use a diabatic set which is considered as $\mathcal{R}$-independent within a certain region of the configuration space $\mathcal{R}$.

Two basis sets ψ_n and ϕ_n generate the matrices

$$\begin{aligned}
\langle \phi_m | H | \phi_n \rangle &= H_{mn}, \\
\langle \phi_m | \mathcal{H} | \phi_n \rangle &= H_{mn} + D_{mn}, \qquad (47.5)\\
\langle \psi_m | \mathcal{H} | \psi_n \rangle &= U_n(\mathcal{R})\delta_{mn} + \mathcal{D}_{mn}.
\end{aligned}$$

The eigenvalues of the matrix H_{mn} are U_n. D_{mn} is the matrix of dynamic coupling in the diabatic basis, and $\mathcal{D}_{mn}$ is the matrix of dynamic coupling in the adiabatic basis; the former matrix vanishes for a perfect diabatic basis. All the above matrices are, in principle, of infinite order. For low-energy collisions, the use of finite matrices of moderate dimension, will usually suffice.

Diabatic PES are defined as the diagonal elements H_{nn}. The significance of the diabatic PES is that for velocities which are high [but still satisfy Eq. (47.1)] the system moves preferentially across diabatic PES, provided that the additional conditions discussed in Sect. 47.3 are fulfilled.

For a given finite adiabatic basis $\psi_n(r; \mathcal{R})$, a perfect diabatic basis $\phi_{n'}(r)$ can be constructed by diagonalizing the matrix $\mathcal{D}_{nm}(\mathcal{R})$. The two basis sets are related by a unitary transformation

$$\psi_n(r; \mathcal{R}) = \sum_{n'} C_{nn'}(\mathcal{R})\phi_{n'}(r). \qquad (47.6)$$

47.1.3   Nonadiabatic Transitions: The Massey Parameter

Deviations from the adiabatic approximation manifest themselves in transitions between different PES which are induced by the dynamic coupling matrix $\mathcal{D}$. At low energies, the transitions usually occur in localized regions of nonadiabatic coupling (NAR). In these regions, the motion of nuclei in different electronic states is coupled, and in general it cannot be interpreted as being driven by a single potential.

An important simplifying feature of slow adiabatic collisions is that typically the distance between different NAR is substantially larger than the extents of each NAR. This makes it possible to formulate simple models for the coupling in isolated NAR, and subsequently to incorporate the solution for nonadiabatic coupling into the overall dynamics of the system.

For a system of s nuclear degrees of freedom, there are the following possibilities for the behavior of PES within NAR:

(i) If two s-dimensional PES correspond to electronic states of different symmetry, they can cross along an

$(s-1)$-dimensional line. For a system of two atoms, $s = 1$, and so two potential curves of different symmetry can cross at a point.

(ii) If two s-dimensional PES correspond to electronic states of the same symmetry, they can cross along an $(s-2)$-dimensional line. For a system of two atoms, $s = 1$, and so two potential curves of different symmetry cannot cross. If they have a tendency to cross, they will exhibit a pattern which is called an avoided crossing or a pseudocrossing.

(iii) If two s-dimensional PES correspond to electronic states of the same symmetry in the presence of spin-orbit coupling, they can cross along an $(s-3)$-dimensional line.

Statement (ii) applied to a two-atom system is known as the Wigner-Witmer noncrossing rule.

The efficiency of the nonadiabatic coupling between two adiabatic electronic states is determined, according to the adiabatic principle of mechanics (both classical and quantum), by the value of the Massey parameter ζ, which represents the product of the electronic transition frequency ω_{el} and the time τ_{nuc} that characterizes the rate of change of electronic function due to nuclear motion. Putting $\omega_{el} \approx \Delta U(\mathcal{R})/\hbar$, ($\Delta U$ is the spacing between any two adiabatic PES), and $\tau_{nuc} = \Delta L/v(\mathcal{R})$, ($\Delta L$ is a certain range which depends on the type of coupling), we get

$$\zeta(\mathcal{R}) = \omega_{el}\tau_{nuc} = \Delta U(\mathcal{R})\Delta L/\hbar v(\mathcal{R}). \quad (47.7)$$

The nonadiabatic coupling is inefficient at those configurations $\mathcal{R}$ where $\zeta(\mathcal{R}) \gg 1$. If $\zeta(\mathcal{R})$ is less than or of the order of unity, the nonadiabatic coupling is efficient, and a change in adiabatic dynamics of nuclear motion is very substantial.

The following relations usually hold for the parameters $\Delta L, a, L_0$ for slow collisions:

$$\Delta L \ll a \ll L_0. \quad (47.8)$$

When the nonadiabatic coupling is taken into account, the total (electronic and nuclear) wave function Ψ_E can be represented as a series expansion in ψ_n or ϕ_n:

$$\Psi_E(r;\mathcal{R}) = \sum_n \psi_n(r;\mathcal{R})\chi_{nE}(\mathcal{R}) = \sum_n \phi_n(r)\kappa_{nE}(\mathcal{R}). \quad (47.9)$$

Here $\chi_{nE}(\mathcal{R})$ and $\kappa_{nE}(\mathcal{R})$ are the functions which have to be found as solutions to the coupled equations formulated in the adiabatic or diabatic electronic basis, repectively [1,2]. In general, different contributions to the first sum in Eq. (47.9) can be associated with nonadiabatic transition probabilities between different electronic states.

A practical means of calculating functions $\chi_{nE}(\mathcal{R})$ [or $\kappa_{nE}(\mathcal{R})$] consists of expanding them over certain basis functions $\Xi_{n\nu}(\mathcal{R}')$, where $\mathcal{R}'$ denotes all coordinates $\mathcal{R}$ except for the interparticle distance R. Writing

$$\chi_{nE}(\mathcal{R}) = \sum_\nu \Xi_{n\nu}(\mathcal{R}')\xi_{n\nu E}(R), \quad (47.10)$$

one arrives at a set of coupled second-order equations for the unknown functions $\xi_{n\nu E}(R)$ (the scattering equations) [1]. In the semiclassical approximation, these equations become a set of first-order equations for the amplitudes of the WKB counterparts of $\xi_{n\nu}(R)$ which depend on time t; the latter variable is related to R via the classical trajectory $R = R(t)$ [2]. In the adiabatic approximation, the total wave function is represented by a single term in the first sum of Eq. (47.9):

$$\Psi_E(r;\mathcal{R}) = \psi_n(r;\mathcal{R})\chi_{nE}(\mathcal{R}). \quad (47.11)$$

47.2 TWO-STATE APPROXIMATION

47.2.1 Relation Between Adiabatic and Diabatic Basis Functions

In the two-state approximation, the basis of electronic functions consists of two states. In this case, the elements of the matrix C in Eq. (47.6) are expressed through a single parameter only, a mixing or rotation angle θ:

$$\psi_1(r;\mathcal{R}) = \cos\theta(\mathcal{R})\phi_1(r) + \sin\theta(\mathcal{R})\phi_2(r),$$
$$\quad (47.12)$$
$$\psi_2(r;\mathcal{R}) = -\sin\theta(\mathcal{R})\phi_1(r) + \cos\theta(\mathcal{R})\phi_2(r).$$

The rotation angle, $\theta(\mathcal{R})$, is expressed via the diagonal and off-diagonal matrix elements of the adiabatic Hamiltonian H in the diabatic basis ϕ_1, ϕ_2:

$$\tan 2\theta(\mathcal{R}) = \frac{2H_{12}(\mathcal{R})}{H_{11}(\mathcal{R}) - H_{22}(\mathcal{R})}. \quad (47.13)$$

The eigenvalues of H in terms of H_{ik} are

$$U_{1,2}(\mathcal{R}) = [H_{11}(\mathcal{R}) + H_{22}(\mathcal{R})]/2 \pm \Delta U(\mathcal{R})/2, \quad (47.14)$$

where

$$\Delta U(\mathcal{R}) = \{[H_{11}(\mathcal{R}) - H_{22}(\mathcal{R})]^2 + 4H_{12}^2(\mathcal{R})\}^{1/2}. \quad (47.15)$$

The matrix elements H_{ik} are expressed via the adiabatic potentials and the rotation angle by

$$H_{11}(\mathcal{R}) + H_{22}(\mathcal{R}) = U_1(\mathcal{R}) + U_2(\mathcal{R}),$$
$$H_{11}(\mathcal{R}) - H_{22}(\mathcal{R}) = \Delta U(\mathcal{R})\cos 2\theta(\mathcal{R}), \quad (47.16)$$
$$H_{12}(\mathcal{R}) = (1/2)\Delta U(\mathcal{R})\sin 2\theta(\mathcal{R}).$$

47.2.2 Coupled Equations and Transition Probabilities

A two-state nonadiabatic wave function $\Psi(r, \mathcal{R})$ can be written as an expansion into either adiabatic or diabatic electronic wave functions:

$$\begin{aligned} \Psi(r; \mathcal{R}) &= \psi_1(r; \mathcal{R})\alpha_1(\mathcal{R}) + \psi_2(r; \mathcal{R})\alpha_2(\mathcal{R}), \\ \Psi(r; \mathcal{R}) &= \phi_1(r)\beta_1(\mathcal{R}) + \phi_2(r)\beta_2(\mathcal{R}), \end{aligned} \quad (47.17)$$

in which the nuclear wave functions satisfy two coupled s-dimensional Schrödinger equations [1].

In the semiclassical approximation, the motion of the nuclei is described by the classical trajectory, i.e., by a one-dimensional manifold $\mathcal{Q}(t)$ embedded in the s-dimensional manifold $\mathcal{R}$.

A semiclassical counterpart of Eq. (47.17) is

$$\begin{aligned} \Psi(r, t) &= \psi_1[r; \mathcal{Q}(t)]a_1(t) + \psi_2[r; \mathcal{Q}(t)]a_2(t), \\ \Psi(r; t) &= \phi_1(r)b_1(t) + \phi_2(r)b_2(t). \end{aligned} \quad (47.18)$$

The adiabatic expansion coefficients $a_k(t)$ satisfy the set of equations

$$\begin{aligned} i\hbar\frac{da_1}{dt} &= U_1(\mathcal{Q})a_1 + i\dot{\mathcal{Q}}g(\mathcal{Q})a_2, \\ i\hbar\frac{da_2}{dt} &= -i\dot{\mathcal{Q}}g(\mathcal{Q})a_1 + U_2(\mathcal{Q})a_2, \end{aligned} \quad (47.19)$$

where $g(\mathcal{Q}) = \langle\psi_1|\partial/\partial\mathcal{Q}|\psi_2\rangle = d\theta/d\mathcal{Q}$, and $\mathcal{Q} = \mathcal{Q}(t)$. The diabatic expansion coefficients $b_k(t)$ satisfy the set of equations

$$\begin{aligned} i\hbar\frac{db_1}{dt} &= H_{11}(\mathcal{Q})b_1 + H_{12}(\mathcal{Q})b_2, \\ i\hbar\frac{db_2}{dt} &= H_{21}(\mathcal{Q})b_1 + H_{22}(\mathcal{Q})b_2. \end{aligned} \quad (47.20)$$

Clearly, for a system of two atoms, $\mathcal{Q} \equiv R$.

Solutions to Eqs. (47.19) and (47.20) are equivalent, provided that the initial conditions are matched, and the transition probability is properly defined.

For a given trajectory, it is customary to identify the center of the NAR with a value of $\mathcal{Q} = \mathcal{Q}_p$ which corresponds to the real part of the complex-valued coordinate $\mathcal{Q}_c$ at which two adiabatic PES cross. The crossing conditions in the adiabatic and diabatic representations are

$$U_1(\mathcal{Q}_c) - U_2(\mathcal{Q}_c) = 0, \quad (47.21)$$

or

$$[H_{11}(\mathcal{Q}_c) - H_{22}(\mathcal{Q}_c)]^2 + 4H_{12}^2(\mathcal{Q}_c) = 0. \quad (47.22)$$

Since $\mathcal{Q}$ represents a one-dimensional manifold, the crossing condition (47.22) is satisfied for a complex value

of $\mathcal{Q} = \mathcal{Q}_s$ unless $H_{12} = 0$. Then, by definition, the location of the NAR center is identified with $\mathcal{Q}_p$ through

$$\mathcal{Q}_p = \Re(\mathcal{Q}_c), \quad (47.23)$$

where $\mathcal{Q}_c$ is that value of $\mathcal{Q}_s$ which possesses the smallest imaginary part, and $\Re$ denotes the real part.

For the case when the regions of nonadiabatic coupling are well localized, the function $g(\mathcal{Q})$ possesses a pronounced maximum at (or close to) $\mathcal{Q}_p$, and the two Eqs. (47.19) decouple on both sides of this maximum. A solution of the equations for the nonadiabatic coupling across an isolated maximum of $g(\mathcal{Q})$ yields the so-called single-passage (or one-way) transition amplitude and transition probability. For this problem, the time $t = 0$ can be assigned to the maximum point of $g[\mathcal{Q}(t)]$. Assuming that away from $t = 0$ the decoupling occurs rapidly enough, the nonadiabatic transition probability

$$P_{12} = |a_2(\infty)|^2, \quad (47.24)$$

provided that a solution to Eq. (47.19) corresponds to the initial conditions,

$$a_1(-\infty) = 1, \quad a_2(-\infty) = 0. \quad (47.25)$$

In the limit of almost adiabatic conditions where P_{12} is very small, the following equation holds [3]:

$$P_{12} = \exp\left(-\frac{2}{\hbar}\left|\Im\int_{t_r}^{t_c}\{U_1[\mathcal{Q}(t)] - U_2[\mathcal{Q}(t)]\}dt\right|\right), \quad (47.26)$$

where t_c is a root of

$$\mathcal{Q}(t_c) = \mathcal{Q}_c. \quad (47.27)$$

Here, $\Im$ denotes the imaginary part, and t_r is any real-valued time. Equation (47.26) is valid when the exponent is large, so that P_{12} is exponentially small.

The property of the function $g(\mathcal{Q})$ to pass through a single maximum ensures that the rotation angle away from the maximum tends to constant values, and adiabatic functions in these regions are expressed by certain linear combinations of diabatic functions with constant mixing coefficients. These linear combinations should serve as the initial condition on one side of the coupling region, and as the proper final state on the other, when the problem of a nonadiabatic transition between adiabatic states is treated in the diabatic representation.

47.2.3 Selection Rules for Nonadiabatic Coupling

In the general case, the coupling between adiabatic states or diabatic states is controlled by certain selection rules. The most detailed selection rules exist for a

Table 47.1. Selection rules for the coupling between diabatic and adiabatic states of a diatomic quasimolecule ($w = g, u; \sigma = +, -$).

Interaction	$^{2S+1}\Lambda_w^{(\sigma)}$ nomenclature	$\Omega_w^{(\sigma)}$ nomenclature
Configuration interaction (electrostatic)	$\Delta\Lambda = 0, \Delta S = 0$ $g \neq u, + \neq -$	$\Delta\Omega = 0$ $g \neq u, + \neq -$
Spin-orbit interaction	$\Delta\Lambda = 0, \pm 1, \Delta S = 0, \pm 1$ $g \neq u, + \rightleftharpoons -$	$\Delta\Omega = 0$ $g \neq u, + \neq -$
Radial motion	$\Delta\Lambda = 0, \Delta S = 0$ $g \neq u, + \neq -$	$\Delta\Omega = 0$ $g \neq u, + \neq -$
Rotational motion	$\Delta\Lambda = \pm 1, \Delta S = 0$ $g \neq u, + \neq -$	$\Delta\Omega = \pm 1$ $g \neq u, + \neq -$
Hyperfine interaction	$\Delta\Lambda = 0, \pm 1, \Delta S = 0, \pm 1$ $g \rightleftharpoons u, + \rightleftharpoons -$	$\Delta\Omega = 0, \pm 1$ $g \rightleftharpoons u, + \rightleftharpoons -$

system of two colliding atoms, since this system possesses a high symmetry ($C_{\infty v}$ or, for identical atoms, $D_{\infty h}$ point symmetry in the adiabatic approximation). In the adiabatic representation, the coupling is due to the elements of the matrix $\mathcal{D}$. They fall into two different categories: those proportional to the radial nuclear velocity (coupling by radial motion or radial coupling), and those proportional to the angular velocity of rotation of the molecular axis (coupling by rotational motion or Coriolis coupling).

In a diabatic representation, provided that the effect of the D matrix is neglected, the coupling is due to the parts of the interaction potential neglected in the definition of the diabatic Hamiltonian H_0. In typical cases, these parts are the electrostatic interaction between different electronic states constructed as certain electronic configurations (H_0 corresponds to a self-consistent field Hamiltonian); spin-orbit interaction (H_0 corresponds to a nonrelativistic Hamiltonian); hyperfine interaction (H_0 ignores the magnetic interaction of electronic and nuclear spins as well as the electrostatic interaction between electrons and nuclear quadrupole moments). The selection rules for the above interactions in the case of two atoms

are listed in Table 47.1 for two conventional nomenclatures for molecular states: Hund's case (a), $^{2S+1}\Lambda_w^{(\sigma)}$ and Hund's case (c), $\Omega_w^{(\sigma)}$ [3].

For molecular systems with more than two nuclei, the selection rules cannot be put in a detailed form since, in general, the symmetry of the system is quite low. For the important case of three atoms, a general configuration is planar (C_s symmetry); particular configurations correspond to an isosceles triangle if two atoms are identical (C_{2v} symmetry), to an equilateral triangle for three identical atoms (D_{3h} symmetry) or to a linear configuration. For the last case, the selection rules are the same as for a system of two atoms. The selection rules for the dynamic coupling between adiabatic states classified according to the irreducible representations of the C_s and C_{2v} groups are listed in Table 47.2. In this table, z and y refer to two modes of the relative nuclear motion in the system plane, R_z and R_y refer to two rotations about principal axes of inertia lying in the system plane, and R_x refers to a rotation about the principal axis of inertia perpendicular to the system plane.

47.3 SINGLE-PASSAGE TRANSITION PROBABILITIES: ANALYTICAL MODELS

47.3.1 Crossing and Narrow Pseudocrossing of Potential Energy Curves: The Landau-Zener Model

A case of crossing and narrow pseudocrossing of adiabatic PES can be considered within a unified model since a narrow pseudocrossing of adiabatic PES corresponds

Table 47.2. Selection rules for dynamic coupling between adiabatic states of a system of three atoms.

C_s		A'		A''	
	C_{2v}	A_1	B_1	A_2	B_2
A'	A_1	z	y, R_x	R_z	R_y
	B_1	y, R_x	z	R_y	R_z
A''	A_2	R_z	R_y	z	y, R_x
	B_2	R_y	R_z	y, R_x	z

to a crossing of diabatic PES. Therefore, in both cases, one considers crossing of zero order PES (adiabatic or diabatic) and the interaction between them (dynamic or static). However, the definition of transition probability is different for pseudocrossing or crossing adiabatic PES.

In the framework of the Landau-Zener model [4, 5], the two zero-order PES which cross at a point Q_p along a trajectory $Q(t)$ are approximated by functions linear in $\Delta Q = Q - Q_\mathrm{p}$, and the off-diagonal matrix element is assumed to be a constant.

For the pseudocrossing adiabatic PES (crossing diabatic PES), the matrix H_{jk} within a NAR is approximated as

$$
\begin{aligned}
H_{11}(Q) &= E_0 - F_1(Q - Q_\mathrm{p})\,, \\
H_{22}(Q) &= E_0 - F_2(Q - Q_\mathrm{p})\,, \\
H_{12}(Q) &= V = \text{constant}\,,
\end{aligned}
\qquad (47.28)
$$

from which the spacing between adiabatic PES is

$$
\Delta U = [(\Delta F)^2(Q - Q_\mathrm{p})^2 + 4|V|^2]^{1/2}\,, \qquad (47.29)
$$

with $\Delta F = |F_1 - F_2|$.

The classical trajectory is assumed to be a linear function of t,

$$
Q = Q_\mathrm{p} + v_\mathrm{p}t\,, \qquad (47.30)
$$

where v_p is the velocity of Q-motion at point Q_p.

For this model, adiabatic wave functions on both sides of the nonadiabaticity region (in the limits $-\infty < t < +\infty$) coincide with diabatic functions, but their ordering is reversed. Explicitly,

$$
\begin{aligned}
\psi_1 &= \phi_1; \quad \psi_2 = \phi_2 \ \text{for } t \to -\infty\,, \\
\psi_1 &= \phi_2; \quad \psi_2 = -\phi_1 \ \text{for } t \to +\infty\,.
\end{aligned}
\qquad (47.31)
$$

The transition probability between pseudocrossing adiabatic curves for the Hamiltonian (47.28) and the trajectory (47.30) is given by the Landau-Zener formula

$$
P_{12}^{\mathrm{psc}} = P_{12}^{\mathrm{LZ}} = \exp(-2\pi\zeta^{\mathrm{LZ}}); \quad \zeta^{\mathrm{LZ}} = V^2/(\hbar\Delta F v_\mathrm{p})\,,
\qquad (47.32)
$$

where ζ^{LZ} is the appropriate Massey parameter at the pseudocrossing point. Note that for the LZ model, the single-passage transition probability depends on one dimensionless parameter ζ^{LZ}. A remarkable property of the Landau-Zener model is that the probability P_{12}^{LZ} is given by Eq. (47.26) for an arbitrary value of the exponent, and not only for large ones when the probability is very low.

With a change of the velocity from very low to very high values, the transition probability varies from zero to unity. It is the latter property of the LZ transition

probability that allows one to interpret diabatic energies H_{11} and H_{22} as the potentials which drive the nuclear motion at high velocities.

The region of applicability of the Landau-Zener formula is determined by the condition that the extension of the region of nonadiabatic interaction ΔQ should be small compared with the range a over which potential curves deviate substantially from linear functions. The condition $\Delta Q \ll a$ actually implies the two conditions [2]

$$
2V/\Delta F \ll a\,, \qquad (47.33)
$$

and

$$
2\left(\hbar v_\mathrm{p}/\Delta F\right)^{1/2} \ll a\,. \qquad (47.34)
$$

Clearly, the range parameter a does not enter the LZ formula since it controls the behavior of adiabatic curves away from the crossing point.

The constant velocity approximation (47.30) imposes yet another condition:

$$
V \ll \mu v_\mathrm{p}^2/2\,. \qquad (47.35)
$$

The application of Eq. (47.32) to other situations requires the specification of V and v_p for each particular trajectory $Q(t)$. For the case of avoided crossing between two potential curves of a diatom, V does not depend on the trajectory and represents, according to Table 47.1, the matrix element of the electrostatic interaction, spin-orbit interaction or hyperfine interaction.

For crossing adiabatic PES, the Landau-Zener model assumes the following approximation for adiabatic potentials and the dynamic coupling:

$$
\begin{aligned}
U_1(Q) &= E_0 - F_1(Q - Q_\mathrm{c})\,, \\
U_1(Q) &= E_0 - F_2(Q - Q_\mathrm{c})\,, \\
\mathcal{D}_{12}(Q) &= D = \text{constant}\,,
\end{aligned}
\qquad (47.36)
$$

with the trajectory parametrization given by Eq. (47.30) where v_p is replaced by v_c. Since the ordering of adiabatic PES for crossing and pseudocrossing is reversed on one side of a NAR, the following relation exists between transition probabilities for the crossing case P_{12}^{c}, and the survival probability for the pseudocrossing case $1 - P_{12}^{\mathrm{psc}}$:

$$
P_{12}^{\mathrm{c}} = 1 - P_{12}^{\mathrm{psc}}\,, \qquad (47.37)
$$

provided that D and v_c in the crossing situation are replaced by V and v_p in the pseudocrossing situation. Conditions (47.33) and (47.34) applied to the case of the dynamic coupling often imply that this coupling is weak [2]. Therefore, Eq. (47.37) yields

$$
P_{12}^{\mathrm{c}} = \frac{2\pi D^2}{\hbar\Delta F v_\mathrm{c}}\,. \qquad (47.38)
$$

Usually, the matrix element D is related to the Coriolis coupling, and it is proportional to the angular velocity of rotation of the molecular frame at the crossing point Q_c.

47.3.2 Arbitrary Pseudocrossing and Diverging Potential Energy Curves: The Nikitin Model

The restrictions (47.33) and (47.34) for the LZ model are relaxed in a more general model suggested by Nikitin [6].

In a diabatic basis, the model is formulated with the Hamiltonian

$$
\begin{aligned}
H_{11}(\mathcal{Q}) &= U_0(\mathcal{Q}) - \Delta E/2 + (A/2)\cos 2\vartheta \exp(-\alpha\mathcal{Q}), \\
H_{22}(\mathcal{Q}) &= U_0(\mathcal{Q}) + \Delta E/2 - (A/2)\cos 2\vartheta \exp(-\alpha\mathcal{Q}), \\
H_{12}(\mathcal{Q}) &= (A/2)\sin 2\vartheta \exp(-\alpha\mathcal{Q}).
\end{aligned}
\tag{47.39}
$$

The spacing between adiabatic PES is

$$
\begin{aligned}
\Delta U = \Delta E \{ 1 &- 2\cos 2\vartheta \exp[-\alpha(\mathcal{Q}-\mathcal{Q}_p)] \\
&+ \exp[-2\alpha(\mathcal{Q}-\mathcal{Q}_p)] \}^{1/2},
\end{aligned}
\tag{47.40}
$$

where $\mathcal{Q}_p$ is introduced instead of A via Eqs. (47.21) and (47.23). At the center of NAR, where $\mathcal{Q} = \mathcal{Q}_p$, the spacing between adiabatic energy levels $\Delta U_p = \Delta U(\mathcal{Q}_p)$ is

$$
\Delta U_p = 2\Delta E \sin \vartheta.
\tag{47.41}
$$

The classical trajectory within the NAR is taken in a form identical to Eq. (47.30) in which v_p is now the velocity of $\mathcal{Q}$ motion at the center of the coupling region $\mathcal{Q}_p$.

For this model, adiabatic wave functions coincide with diabatic functions before entering the coupling region (in the limit $\alpha(\mathcal{Q}-\mathcal{Q}_p) \gg 1$), but after exiting the coupling region (in the limit $\alpha(\mathcal{Q}-\mathcal{Q}_p) \ll -1$) they are linear combinations of the diabatic functions

$$
\begin{aligned}
\psi_1(\mathbf{r};\mathcal{Q}) &= \phi_1(\mathbf{r}), \quad \psi_2(\mathbf{r};\mathcal{Q}) = \phi_2(\mathbf{r}), \\
&\text{for } \alpha(\mathcal{Q}-\mathcal{Q}_p) \gg 1; \\
\psi_1 &= \phi_1 \cos\vartheta + \phi_2 \sin\vartheta, \\
\psi_2 &= -\phi_1 \sin\vartheta + \phi_2 \cos\vartheta, \\
&\text{for } \alpha(\mathcal{Q}-\mathcal{Q}_p) \ll -1.
\end{aligned}
\tag{47.42}
$$

The latter equation identifies the parameter ϑ that enters into the definition of the diabatic Hamiltonian in Eq. (47.39) with the asymptotic value of the mixing angle θ [see Eq. (47.12)] for $\alpha(\mathcal{Q}-\mathcal{Q}_p) \ll -1$.

The transition probability P_{12} between adiabatic PES for the Hamiltonian (47.39) and the trajectory (47.30) is

$$
P_{12}^{N} = \exp(-\pi\zeta_p)\frac{\sinh(\pi\zeta - \pi\zeta_p)}{\sinh(\pi\zeta)},
\tag{47.43}
$$

where $\zeta = \Delta E/(\hbar\alpha v_p)$ and $\zeta_p = \zeta\sin^2\vartheta$. With the change in velocity from very low to very high values, the transition probability varies from zero to $\cos^2\vartheta$. As ϑ changes from very small values to π, the pattern of

adiabatic potential curves changes from narrow to wide pseudocrossing and ultimately to strong divergence.

Since the single-passage transition probability (47.43) depends on two parameters, the Nikitin model is more versatile than the Landau-Zener one.

In two limiting cases, $\vartheta \ll 1$, $\zeta \gg 1$ and $\vartheta = \pi/2$, for arbitrary ζ, Eq. (47.43) may be simplified. In the first case, the diabatic Hamiltonian (47.39) becomes the Landau-Zener Hamiltonian (47.28). Also, Eq. (47.43) reduces to a single exponential which gives the LZ transition probability,

$$
P_{12}^{N} = P_{12}^{LZ} = \exp(-2\pi\zeta_p),
\tag{47.44}
$$

with ζ_p identical to ζ^{LZ}. The two conditions $\vartheta \ll 1$ and $\zeta \gg 1$ are equivalent to the two conditions (47.33) and (47.34). In the second case ($\vartheta = \pi/2$), the diabatic Hamitonian reads

$$
\begin{aligned}
H_{11}(R) &= E_0 - \Delta E/2, \\
H_{22}(R) &= E_0 + \Delta E/2, \\
H_{12}(R) &= (A/2)\exp(-\alpha\mathcal{Q}).
\end{aligned}
\tag{47.45}
$$

The transition probability in this case is given by the Rosen-Zener-Demkov formula [7,8]

$$
P_{12}^{RZD} = \frac{\exp(-\pi\zeta)}{1 + \exp(-\pi\zeta)}, \quad \zeta = \Delta E/(\hbar v_p \alpha).
\tag{47.46}
$$

With the change of the velocity from very low to very high values, the Rosen-Zener-Demkov transition probability varies from zero to one-half. For this model, diabatic energies do not represent potentials for the nuclear motion at high energies.

47.4 DOUBLE-PASSAGE TRANSITION PROBABILITIES AND CROSS SECTIONS

47.4.1 Mean Transition Probability and the Stückelberg Phase

In the case of an atomic collision, the set $\mathcal{R}$ shrinks into a single coordinate R. If there is only one NAR over the whole range of R, the colliding system traverses it twice, as the atoms approach and then recede. In this case, there are two different paths between the center of the NAR, R_p, and the turning points R_{t1} and R_{t2} on the adiabatic potential curves $U_1(R)$ and $U_2(R)$. The double-passage transition probability $\mathcal{P}_{12}$ is expressed via the single-passage transition probability P_{12}, the single-passage survival probability $1 - P_{12}$, and the Stückelberg interference term $\cos\Delta\Phi_{12}$ [9],

$$
\begin{aligned}
\mathcal{P}_{12} &= 2P_{12}(1 - P_{12})(1 - \cos\Delta\Phi_{12}) \\
&= 4P_{12}(1 - P_{12})\sin^2(\Delta\Phi_{12}/2).
\end{aligned}
\tag{47.47}
$$

The Stückelberg phase $\Delta\Phi_{12}/2$ is expressed as the phase difference $\Delta\Phi_{12}^{(0)}/2$ which is accumulated during the motion of a diatom from the center of the NAR to the turning points, together with an additional phase ϕ_{12} by

$$\Delta\Phi_{12}/2 = \Delta\Phi_{12}^{(0)}/2 + \phi_{12}. \tag{47.48}$$

Generally, Eq. (47.48) is valid provided $\Delta\Phi_{12}/2 \gg 1$. For transitions between electronic states of the same axial symmetry, the relative angular momentum ℓ of the colliding atoms is conserved, and we have

$$
\Delta\Phi_{12}^{(0)}/2
$$
$$
= \int_{R_p}^{R_{t1}} [2\mu E - \hbar^2\ell(\ell+1)/R^2 - 2\mu U_1(r)]^{1/2} dR/\hbar
$$
$$
- \int_{R_p}^{R_{t2}} [2\mu E - \hbar^2\ell(\ell+1)/R^2 - 2\mu U_2(r)]^{1/2} dR/\hbar, \tag{47.49}
$$

where R_{t1} and R_{t2} are the turning points for adiabatic motion on potential curves U_1 and U_2, E is the total energy, and μ is the reduced mass of the colliding atoms. Once R_p is chosen, $\Delta\Phi_{12}^{(0)}/2$ is well-defined and is independent of the dynamic details of a nonadiabatic transition. On the other hand, P_{12} and ϕ_{12} do depend on these details. In many applications, one can use the mean transition probability $\langle\mathcal{P}_{12}\rangle$, which is obtained from $\mathcal{P}_{12}$ by averaging over several oscillations:

$$\langle\mathcal{P}_{12}\rangle = 2P_{12}(1 - P_{12}). \tag{47.50}$$

The important limiting cases of the double-passage Nikitin model are:

(i) Double-passage Landau-Zener-Stückelberg equation,

$$\mathcal{P}_{12}^{LZS} = 4\exp(-2\pi\zeta^{LZ})[1 - \exp(-2\pi\zeta^{LZ})]$$
$$\times \sin^2(\Delta\Phi_{12}^{(0)}/2 + \phi_{12}^{LZS}), \tag{47.51}$$

where ζ^{LZ} is given by Eq. (47.32) and $\Delta\Phi_{12}^{(0)}/2$ by Eq. (47.49). The dynamic phase is

$$\phi_{12}^{LZS} = \pi/4 + \zeta^{LZ}\ln\zeta^{LZ} - \zeta^{LZ} - \arg\Gamma(1 - i\zeta^{LZ}), \tag{47.52}$$

where Γ is the gamma function. When ζ^{LZ} changes from zero to infinity, $\langle\mathcal{P}_{12}^{LZS}\rangle$ passes through a maximum, $\langle\mathcal{P}_{12}\rangle_{max} = 1/2$, and ϕ_{12}^{LZS} decreases from $\pi/4$ to zero.

(ii) Double-passage Rosen-Zener-Demkov equation:

$$\mathcal{P}_{12}^{RZD} = \frac{\sin^2(\Delta\Phi_{12}^{(0)}/2 + \phi_{12}^{RZD})}{\cosh^2(\pi\zeta/2)}. \tag{47.53}$$

where ζ is given by Eq. (47.46). Under certain conditions [2,8], the Stückelberg phase in Eq. (47.53) can be

identified with the phase accumulated during the motion of a diatom from infinitely large distance to the turning points.

(iii) Double-passage equation for a resonance process ($\Delta E = 0$):

$$\mathcal{P}_{12}^{res} = \sin^2(2\vartheta)\sin^2(\Delta\Phi_{12}/2). \tag{47.54}$$

The general resonance case (zero energy change, $\Delta E = 0$) is also called an *accidental* resonance. For the accidental resonance, the diagonal diabatic matrix elements are not equal to each other. A particular case of an accidental resonance is a *symmetric* resonance, for which the diabatic matrix elements are the same. For the Nikitin model, symmetric resonance corresponds to $\vartheta = \pi/4$, and Eq. (47.54) reads

$$\mathcal{P}_{12}^{symm} = \sin^2(\Delta\Phi_{12}/2). \tag{47.55}$$

Equation (47.55) also follows from Eq. (47.53) in the limit $\zeta \to 0$. Actually, Eq. (47.55) is valid for any symmetric resonance case [not necessarily for the model Hamiltonians (47.39) and (47.45)] and for the arbitrary values of the phase $\Delta\Phi_{12}/2$. This phase can be identified with $\Delta\Phi_{12}^{(0)}/2$ from Eq. (47.49) provided R_p is taken to be infinitely large.

The restrictions on the semiclassical approximation, and the constant velocity approximation Eq. (47.35) in particular, can be lifted for the Landau-Zener Hamiltonian (see Refs. [2] and [10]) and the Rosen-Zener Hamiltonian, Eq. (47.45) [11].

47.4.2 Approximate Formulae for the Transition Probabilities

Several approximate formulae are available for $\langle\mathcal{P}_{12}\rangle$ in the case where H_{12} depends on time in a bell-shaped manner, and $\Delta H = H_{11} - H_{22}$ can be represented as $\hbar\omega + \Delta V$, with ΔV also having a bell-shaped form. Define

$$v = \left(\frac{1}{\hbar}\right)\int_{-\infty}^{+\infty} H_{12}(t)\,dt;$$
$$w_0 = \left(\frac{1}{\hbar}\right)\int_{-\infty}^{+\infty} H_{12}(t)\exp(i\omega t)\,dt,$$
$$w = \left(\frac{1}{\hbar}\right)\int_{-\infty}^{+\infty} H_{12}(t)\exp\left[(i/\hbar)\int_0^t \Delta H(t)dt\right]dt,$$
$$u = \left(\frac{1}{2\hbar}\right)\int_{-\infty}^{+\infty} \Delta V_{12}(t)\,dt, \tag{47.56}$$

and

$$S(t) = (1/\hbar)\int_0^t [\Delta H^2(t) + 4H_{12}^2(t)]^{1/2}\,dt. \tag{47.57}$$

Then the various approximate formulae, as suggested by different authors [2], read

$$\mathcal{P}_{12} \cong (w_0/v)^2 \sin^2 v \,, \tag{47.58}$$

$$\mathcal{P}_{12} \cong \sin^2 w \,, \tag{47.59}$$

$$\mathcal{P}_{12} \cong \frac{w_0^2}{u_0^2 + w_0^2} \sin^2 \sqrt{(u_0^2 + w_0^2)} \,, \tag{47.60}$$

$$\mathcal{P}_{12} \cong \left| \int_{-\infty}^{+\infty} H_{12}(t) \exp[iS(t)]dt/\hbar \right|^2 \,, \tag{47.61}$$

$$\mathcal{P}_{12} \cong \frac{\sin^2 S_\mathrm{c}'}{\cosh^2 S_\mathrm{c}''} \,. \tag{47.62}$$

In Eq. (47.62), S_c' and S_c'' are the real and imaginary parts of the complex quantity $S_\mathrm{c} = S(t_\mathrm{c})$ from Eq. (47.57). The complex-valued time t_c is found from

$$\Delta H^2(t_\mathrm{c}) + 4H_{12}^2(t_\mathrm{c}) = 0 \,, \tag{47.63}$$

under the condition that t_c possesses the smallest imaginary part of all roots of this equation.

47.4.3 Integral Cross Sections for a Double-Passage Transition Probability

In the semiclassical approximation, the classical trajectory of relative motion is labeled by the initial collision velocity v and initial impact paramter b. These two parameters determine all the quantities which enter into the double-passage transition probability $\mathcal{P}_{12}$. The inelastic integral cross section σ_{12} for the transition between states 1 and 2 is related to $\mathcal{P}_{12}$ by

$$\sigma_{12} = 2\pi \int_0^\infty \mathcal{P}_{12} b \, db \,. \tag{47.64}$$

In calculating σ_{12}, one usually neglects the Stückelberg oscillating term and sets the upper limit in the integral in Eq. (47.64) to a value $b = b^*$ beyond which the integrand begins to fall off quickly. Yet another simplification is possible in the framework of the impact parameter approximation, when the relative motion of atoms is described by a rectilinear trajectory with constant velocity:

$$R(t) = [b^2 + v^2 t^2]^{1/2} \,. \tag{47.65}$$

For instance, for the Landau-Zener model $b^* = R_\mathrm{p}$, and the cross section depends on the one-dimensional parameter $\gamma = 2\pi V^2/(\Delta F \hbar v)$ according to

$$\sigma_{12}(\gamma) = 2\pi R_\mathrm{p}^2 \int_0^1 \exp(-\gamma/\sqrt{x}\,)[1 - \exp(-\gamma/\sqrt{x}\,)]dx$$
$$= 4\pi R_\mathrm{p}^2 [E_3(\gamma) - E_3(2\gamma)] \,, \tag{47.66}$$

where $E_\nu(z)$ is the exponential integral.

For the symmetric resonance process, b^* is found from the Firsov criterion $\Delta \Phi_{12}(b^*, v) = 2/\pi$ [2], and the cross section is

$$\sigma_{12}(v) = (\pi/2)[b^*(v)]^2 \,. \tag{47.67}$$

This cross section slowly decreases with increasing collision velocity.

47.5 MULTIPLE-PASSAGE TRANSITION PROBABILITIES

47.5.1 Multiple Passage in Atomic Collisions

In the case of atomic collisions, there is only one nuclear coordinate R. If there exist several NAR on the R-axis, those which are classically accessible (for given total energy E and total angular momentum J) can be traversed several times. In the semiclassical approximation, the multiple-passage transition amplitude A_{if} for a given transition between inital state i and final state f can be calculated as a sum of transition amplitudes $A_{if}^{\mathcal{L}}$, over all possible classical ways $\mathcal{L}$ which connect these states, and which run along a one-dimensional manifold R:

$$A_{if} = \sum_{\mathcal{L}} A_{if}^{\mathcal{L}} \,, \tag{47.68}$$

where each $A_{if}^{\mathcal{L}}$ can be expressed through the probability $\mathcal{P}_{if}^{\mathcal{L}}$ and the phase $\Phi_{if}^{\mathcal{L}}$ by [12]

$$A_{if}^{\mathcal{L}} = [\mathcal{P}_{if}^{\mathcal{L}}]^{1/2} \exp(i\Phi_{if}^{\mathcal{L}}) \,. \tag{47.69}$$

The net transition probability is then

$$\mathcal{P}_{if} = |A_{if}|^2$$
$$= \sum_{\mathcal{L}} \mathcal{P}_{if}^{\mathcal{L}} + \sum_{\mathcal{L},\mathcal{L}'}' [\mathcal{P}_{if}^{\mathcal{L}} \mathcal{P}_{if}^{\mathcal{L}'}]^{1/2} \cos[\Phi_{if}^{\mathcal{L}} - \Phi_{if}^{\mathcal{L}'}] \,. \tag{47.70}$$

The first sum runs over all different paths, and the second (primed) over all different *pairs* of paths. The primed sum usually yields a contribution to the transition probability which oscillates rapidly with a change of the parameters entering into $\mathcal{P}_{if}$ (i.e., E and J) and represents a multiple-passage counterpart to the Stückelberg oscillations.

If the Stückelberg oscillations are neglected, $\mathcal{P}_{if}$ is equivalent to a mean transition probability $\langle \mathcal{P}_{if} \rangle$:

$$\langle \mathcal{P}_{if} \rangle = \sum_{\mathcal{L}} \mathcal{P}_{if}^{\mathcal{L}} \,. \tag{47.71}$$

For one NAR, there are two equivalent paths, and $\mathcal{P}_{if}^{(i)} = \mathcal{P}_{if}^{(ii)} = P_{if}(1 - P_{if})$. Then Eq. (47.71) yields Eq. (47.50).

47.5.2 Multiple Passage in Molecular Collisions

For molecular collsions, Eqs. (47.68) and (47.70) apply as well. However, the manifold of $\mathcal{R}$ to which a trajectory $\mathcal{Q}(t)$ belongs now comprises $3N - 5$ (for a linear arrangement of nuclei) or $3N - 6$ (for a nonlinear arrangement) degrees of freedom, where N is the number of atoms in the system. The approximation (47.71) is called, in the context of inelastic molecular collisions, the surface-hopping approximation [13–15]. Each time a trajectory reaches a NAR, it bifurcates, and the system makes a hop from one PES to another with a certain probability. Keeping track of all the bifurcations and associated probabilities, one calculates $\mathcal{P}_{if}^{\mathcal{L}}$ along a path $\mathcal{L}$ made up of different portions of trajectories running across different PES. Because of the complicated sequence of nonadiabatic events leading from the initial state to the final state, each $\mathcal{P}_{if}^{\mathcal{L}}$ is a complicated function of different single passage transition probabilities P_{nm}, and survival probabilities $1 - P_{nm}$. Even if all P_{nm} are known in analytical form, the calculation of $\langle \mathcal{P}_{if} \rangle$ requires numerical computations to keep track of individual nonadiabatic events [14].

The manifold $\mathcal{R}$ can be reduced in size if one treats other degrees of freedom, besides electronic ones, on the same footing. In this way one introduces adiabatic vibronic (vibrational + electronic) states and adiabatic vibronic PES, and considers nonadiabatic transitions between them [16,17]. In the vibronic representation, the formal theory remains the same; however its implementation is more difficult since there are many more possibilities for trajectory branching. Finally, under certain conditions, one can use a fully adiabatic description of all degrees of freedom save one – the intermolecular distance R. This approach provides a basis for the statistical adiabatic channel model (SACM) of unimolecular reactions [18] where the receding fragments are scattered adiabatically in the exit channels after leaving the region of a statistical complex. In the exit channels, away from the boundary of the statistical complex, the total wave function is

$$\Psi_E(r; \mathcal{R}) = \psi_n(r; \mathcal{R})\Xi_{n\nu}(\mathcal{R}')\xi_{n\nu E}(\mathcal{R}), \qquad (47.72)$$

which is a particular form of the electronically adiabatic function in Eq. (47.10).

Acknowledgment

This work was supported in part by Deutsche Forschungsgemeinschaft Sonderforschungsbereich 216 *Polarisation and Korrelation in Atomaren Stosskomplexen*. The author is grateful to Prof. J. Hinze for his hospitality at the Department of Chemistry, University of Bielefeld.

REFERENCES

1. *Atom-Molecule Collision Theory: A Guide for the Experimentalist*, edited by R. B. Bernstein (Plenum, New York, 1979).
2. E. E. Nikitin and S. Ya. Unamskii, *Theory of Slow Atomic Collisions* (Springer, Heidelberg, 1984).
3. L. D. Landau and E. M. Lifshitz, *Quantum Mechanics* (Pergamon, Oxford, 1977).
4. L. D. Landau, Phys. Z. Sowjetunion **2**, 46 (1932).
5. C. Zener, Proc. Roy. Soc. A**137**, 396 (1932).
6. E. E. Nikitin, Discuss. Faraday Soc. **33**, 14 (1962).
7. N. Rosen and C. Zener, Phys. Rev. **40**, 502 (1932).
8. Yu. N. Demkov, Sov. Phys. JETP **18**, 138 (1964).
9. E. C. G. Stückelberg, Helv. Phys. Acta **5**, 369 (1932).
10. C. Zhu, H. Nakamura, N. Re, and V. Aquilanti, J. Chem. Phys. **97**, 1892 (1992); *ibid.* **97**, 8497 (1992).
11. V. I. Osherov and A. I. Voronin, Phys. Rev. A **49**, 265 (1994).
12. W. H. Miller, Adv. Chem. Phys. **30**, 77 (1975).
13. A. Bjerre and E. E. Nikitin, Chem. Phys. Lett. **1**, 179 (1967).
14. J. C. Tully, Adv. Chem. Phys. **42**, 63 (1980).
15. S. Chapman, Adv. Chem. Phys. **82**, 423 (1992).
16. V. Sidis, Adv. At. Opt. Phys. **26**, 161 (1990).
17. H. Nakamura, Adv. Chem. Phys. **82**, 243 (1992).
18. J. Troe, Adv. Chem. Phys. **82**, 485 (1992).

48

Ion-Atom and Atom-Atom Collisions

A. L. Ford and J. F. Reading

Department of Physics, Texas A&M University, College Station, Texas 77843

48.1 INTRODUCTION

This chapter summarizes the principal features of theoretical treatments of ion-atom and atom-atom collisions. This is a broad topic and the goal here   is a general overview that introduces the main concepts, terminology, and methods in the field. Attention will focus on intermediate and high collision velocities, for which the relative velocity between the projectile and target is on the order of, or larger than, the orbital speed of the electrons active in the transition. Low energy collisions are treated in Chap. 47. Charge transfer reactions are treated in Chap. 49. We will emphasize therefore excitation and ionization transitions, and discuss charge transfer only as it is interrelated with these transitions. The description of these heavy particle collisions has many features in common with electron and positron collisions with atoms and ions (Chaps. 45 and 46) and this Chapter should be studied in parallel with those. There are also chapters dealing with special phenomena in ion-atom collisions: excitation at high collision energies and the Thomas peak (Chap. 55), electron emission in high energy ion-atom collisions (Chap. 51), and alignment and

orientation (Chap. 44). Other chapters deal with certain specific theoretical methods: continuum distorted wave (CDW) approximations (Chap. 50), the binary encounter approximation (Chap. 54), and classical trajectory Monte Carlo (CTMC) techniques (Chap. 56). The emphasis of the present chapter is coupled-states calculations of excitation and ionization. There are several recent review articles and monographs [1, 2].

The collisions considered here involve a projectile ion or atom and a target atom. The collision kinematics can be described in the lab frame, where the target atom is assumed to be initially at rest and the collision energy is the kinetic energy of the projectile when it is far from the target prior to the collision, or in the center of mass frame. The primary quantities of interest are the cross sections for producing various final states of the system for given initial states of the target and projectile. The total cross sections depend on the initial and final quantum state of the target and projectile and on the collision energy. Let A denote the projectile atom or ion with ionic charge q, and B denote the target atom. Let A^* and B^* denote excited states. Some examples of processes for which the cross section is of interest are

$A^{q+} + B \rightarrow$

$$\begin{cases} (A^{q+})^* + B\,, & \text{projectile excitation} \\ A^{q+} + B^*\,, & \text{target excitation} \\ A^{q+} + B^+ + \text{e}^-\,, & \text{target single ionization} \\ A^{(q+1)+} + B + \text{e}^-\,, & \text{projectile single ionization} \\ A^{(q-1)+} + B^+\,, & \text{single e}^- \text{ charge transfer.} \end{cases}$$

For a multi-electron collision system, combinations of the above quantities are possible. A few representative examples are

$A^{q+} + B \rightarrow$

$$\begin{cases} A^{q+} + B^{++} + 2\text{e}^-\,, & \text{target double ionization} \\ A^{q+} + (B^+)^* + \text{e}^-\,, & \text{target excitation-ionization} \\ A^{(q-1)+} + (B^+)^*\,, & \text{transfer-excitation} \\ A^{(q-1)+} + B^{++} + \text{e}^-\,, & \text{transfer-ionization}. \end{cases}$$

For each of these processes, the projectile and/or target can be initially in excited states. A class of ion-atom collisions that has received much theoretical attention because of the relative simplicity is the one where the projectile ion is initially bare, with charge $q = Z_p$, the projectile nuclear charge. One can also consider various differential cross sections: differential in the projectile scattering angle, in the energy and angle of emission of ionized electrons, in recoil momentum of the target, etc.

Theoretical calculations of these cross sections can be classified according to the approximations and/or methods used. We will discuss four such classifications: treatment of the heavy particle motion, independent particle model (IPM) versus inclusion of correlation in multi-electron systems, analytical approximations (PWBA, SCA, Glauber, etc.) versus numerical methods (such as coupled-states) for obtaining cross sections, and treatments of the ionization continuum.

48.2   TREATMENT OF HEAVY PARTICLE MOTION

Only at very low energies must the motion of the nuclei be described quantum mechanically. At the intermediate and high collision velocities considered here the semiclassical approximation, in which the motion of the nuclei can be described classically and only the electrons need be described by quantum mechanical wave functions, is accurate [3]. The projectile nucleus then moves on a predetermined classical path. At high collision energies this path can often be taken to be a straight line path with constant speed. At somewhat lower energies the deflection and change in speed of the projectile due to the projectile-target interaction is often incorporated [4]. The Coulomb trajectory due to the nucleus-nucleus interaction can be used, and the screening effects of the projectile and target electrons

can be included. For a bare positive ion projectile, the Coulomb trajectory effects increase the distance of closest approach for a given impact parameter and reduce the projectile speed in the interaction region. Projectile trajectory effects are particularly strong for projectiles less massive than protons (positive or negative muons [5], for example). The semiclassical approach with trajectory effects has even been used for electron impact excitation and ionization [5]. Coulomb trajectories are also important when small impact parameter collisions are considered. The recoil of the target nucleus can also be treated classically and this recoil can affect the cross sections [6].

In the semiclassical approximation, the vector $\mathbf{R}(b,t)$ that locates the projectile relative to the target nucleus is a function of the impact parameter b and time t. The specific functional dependence is determined by the trajectory being used. For a straight line, constant velocity $\mathbf{v}$ path, $\mathbf{R} = \mathbf{b} + \mathbf{v}t$. For the case of a bare projectile ion and a one-electron target, the time-dependent projectile-target interaction is given by

$$V(b,t) = \frac{-Z_P e^2}{|\mathbf{r} - \mathbf{R}(b,t)|} + \frac{Z_P Z_T e^2}{R(b,t)}, \qquad (48.1)$$

where Z_P and Z_T are the projectile and target nuclear charges and $\mathbf{r}$ is the position vector of the electron relative to the target nucleus. If the collision calculation starting from Eq. (48.1) is done exactly the nuclear repulsion term $Z_P Z_T e^2 / R(b,t)$ does not involve the electronic coordinates and makes no contribution to total cross sections for anything other than elastic scattering. This term makes a nonzero contribution in an approximate calculation, such as a first-order perturbation theory calculation with nonorthogonal initial and final states, and this is a defect of such calculations.

A deficiency of the semiclassical approximation as described above is that, since a predetermined classical path is used, there is no coupling between the energy and momentum given to the target electron and that lost by the projectile. A simple improvement, in cases where the energy lost by the projectile when the target transition occurs is an appreciable fraction of its total energy, is to use some average of the projectile's initial and final speeds as the asymptotic projectile speed. This results in a projectile trajectory that depends on the cross section being calculated. This lack of coupling between the projectile motion and the states of the electrons can be a particular deficiency when cross sections differential in the scattering angle Θ of the projectile are computed. If the collision energy is large and the projectile is scattered primarily by the static potential of the target, a classical treatment of the scattering can be used to relate b to Θ. Even when straight line, constant speed projectile paths are used to calculate the transition probabilities, differential cross sections can be extracted by relating b to Θ. At lower energies, where the de Broglie wavelength of the projectile is not small compared with the range of the interaction, eikonal methods can be used to convert impact parameter dependent probabilities to differential cross sections [7]. A particularly striking example where there is strong coupling between the projectile motion and the final quantum state of the electron is proton impact single and double ionization of helium differential in the projectile scattering angle [8]. The projectile can be deflected by interaction with the target nucleus. In this case the projectile scattering angle can be related to the impact parameter by the equations of classical mechanics, and the projectile can be scattered by up to 180° as b is decreased to zero. But the projectile can also be scattered by a close interaction with a target electron, and the kinematics then can be quite different. Such an interaction has a low probability for scattering the projectile to a large angle. Classically, from energy and momentum considerations, the maximum angle through which a proton can be scattered by an electron initially at rest is about 0.5 mrads; and if the proton is scattered through 0.5 mrads, the electron acquires large energy and momentum. The combined effects of both projectile scattering mechanisms are difficult to treat in a semiclassical model [9].

48.3 INDEPENDENT-PARTICLE MODELS VERSUS MANY-ELECTRON TREATMENTS

In the semiclassical impact parameter method, as described in Sect. 48.2, for a single electron collision system one has to calculate a single particle wave function and from it single-electron transition amplitudes. A multi-electron collision problem can be treated as an effective single-electron problem if only one projectile-electron interaction is considered and the other electrons merely provide an effective single-particle potential in which the active electron moves.

In an independent particle model (IPM), the electron-electron interactions are replaced by effective single-electron potentials. Since the projectile-target interaction is a sum of single-electron interactions, the many-electron collision problem reduces to an uncoupled set of single electron problems. Their solution at each impact parameter gives single electron transition amplitudes a_{ij} and transition probabilities $\rho_{ij} = |a_{ij}|^2$. Cross sections for multi-electron transitions can still be calculated by combining the single-electron amplitudes in an appropriate way [10]. In doing this, it is important to distinguish between inclusive and exclusive processes. For an inclusive cross section, one final orbital occupancy, or a few final orbital occupancies, is specified, but the final states of the remaining electrons are not specified and all possibilities are summed over. For a totally exclusive process,

all final orbital occupancies are specified. For example, consider a p + Ne collision. For the inclusive cross section for K-shell vacancy production, at least one K-shell vacancy in the final state is specified. The other electrons could remain in their original orbitals, there could be two K-shell vacancies, the K-shell vacancy could be accompanied by any number of L-shell vacancies, etc., and all these possible final states are summed over. An example of an exclusive cross section is the cross section for producing a single K-shell vacancy with the specification that the final state of the target have no additional vacancies.

In calculating inclusive cross sections from single-electron transition amplitudes in an IPM it is important to take proper account of time-ordering and Pauli exclusion effects, as well as all multi-electron processes that lead to the specified final state. Again using the p + Ne collision example, a two-electron process that leads to a K-shell vacancy is for the projectile to first ionize an L-shell electron, and then in the same collision to excite a K-shell electron into the L-shell hole just produced. But the K-shell electron must have the same spin component as the L-shell electron that was removed, and the K to L excitation can occur only after the L-hole has been made. This can lead to correlations among the single-electron amplitudes that have been called Pauli correlations or Pauli blocking effects.

For a totally inclusive process where only one final occupancy is specified, all the multi-electron contributions cancel in any IPM and the probability for producing, for example, a hole in the initially occupied orbital labeled 1, without any other specification of final state orbital occupancies, is given by

$$\rho_1 = \sum_{k < N} |a_{k1}|^2 , \qquad (48.2)$$

where k runs over the unoccupied bound orbitals of the target, the bound orbitals of the projectile, and the continuum orbitals. This expression says that the inclusive probability of producing a hole in the initially occupied orbital 1 is given by the sum of the probabilities of exciting an electron from this orbital to any initially unoccupied orbitals of the projectile-target system. The cross section is then exactly the same as if only a single target electron were interacting with the projectile. However, for an inclusive process in which two vacancies in orbitals of the same spin are specified in the final state, say orbitals 1 and 2, then the probability of producing this final state is not the product of the independent probabilities of producing a vacancy in orbital 1 and in orbital 2 [10]. The correlation produced by antisymmetry destroys the independence of probabilities for this process. Specific expressions, and also a discussion of how to derive the correct expressions to use for any particular inclusive process, can be found in the literature [10, 11]. If these Pauli correlations are neglected, and products of independent single-electron transition probabilities are used, then the distribution of multiple vacancies within a shell, such as the L-shell, are given in terms of a binomial distribution [12].

The IPM treatment is inadequate if electron correlation plays an important role in the collision. One notable set of examples is two-electron transitions in which the projectile-target interaction is weak, such as at high collision energies. It is useful to characterize multi-electron transitions as being either externally or internally induced [13]. In an externally induced two-electron process, the projectile interacts with both electrons and induces the transition of each. In the internal process, energy is transferred to just one electron; this energy is then shared with the other electron through the internal action of the correlating electron-electron force.

Consider double ionization. At sufficiently high energies, the external reaction has an amplitude proportional to Z_P^2. To lowest order, the reaction proceeds through two consecutive first-order Born collisions, perhaps more correctly described as a two-particle second-order Born collision. This is called a two step mechanism, TS2. The internal reaction amplitude, again at sufficiently high energies, is proportional to Z_P; one first-order Born reaction initiates the process. The sharing of energy with the second electron can be thought of in two ways. If the first electron leaves without interacting with the second, there is a change in the screening that can be thought of as being responsible for shaking the second electron off. The main role of the electron-electron force here is to introduce correlation into the initial ground state wave function. A second way for both electrons to be ionized is for the first electron to strike the second on the way out. This has been variously called interception or TS1 [14,15]. The role of the electron-electron force here is to introduce correlation into the final two-electron wave function. At asymptotic energies the first Born term will eventually dominate, and the internal correlated mechanism is solely responsible for double ionization. At lower energies, the external and internal amplitudes are of the same order, and thus they interfere. The interference produces terms in the transition probability that are proportional to odd powers of Z_P, and therefore produces a cross section that changes when the sign of the projectile charge changes. One example is the double ionization of helium in the MeV/amu collision energy region, where antiprotons ($Z_P = -1$) produce a double ionization cross section a factor of two larger than that for protons ($Z_P = +1$) around 1 MeV/amu [15]. It is possible to account for electron correlation directly by putting the electron-electron interaction into the Hamiltonian and solving for the many-electron correlated wave function of the collision system. In coupled-states calculations (Sect. 48.4), this approach is computationally intractable if more than a few coupled channels are important, which is usually the case at intermediate collision energies; but some full electron calculations have been carried out [16]. In low-energy collisions where molecular basis expansions are appropriate, only a few channels can be important, i.e., those involved in potential curve crossings or near crossings, and full electron calculations are more feasible. Some alternatives to full electron quantum calculations have been developed, such as the independent event model [17], classical dynamics calculations (Chap. 56), time-dependent Hartree-Fock (TDHF) [18], and the forced impulse method (FIM) [15]. The FIM divides the collision time into short sequential segments such that an impulse approximation is forced to be valid. The system is allowed to collapse back into a fully correlated eigenstate at the end of each segment, but the electrons propagate independently during each segment.

48.4 ANALYTICAL APPROXIMATIONS VERSUS NUMERICAL CALCULATIONS

There is a wide variety of techniques used for computing cross sections for ion-atom and atom-atom collisions. One class uses approximate analytical methods. The expressions and the wave functions that enter into them may be sufficiently complicated that they must be evaluated on a computer, but they can be written down as closed form expressions. One important example is the first Born approximation, in which the projectile-target interaction is treated in first-order. Hence, the first Born approximation is most accurate for weak interactions, for example for small projectile charge and large collision energy. The first Born approximation is commonly expressed in either of two forms: the plane wave Born approximation (PWBA) [19], where the projectile nuclear motion wave function is taken to be a plane wave, or the semiclassical approximation (SCA) [3] which is an impact parameter approach that uses a classical path for the projectile motion. At the high collision velocities where the first Born approximation is accurate, and for projectiles that are at least as massive as a proton, the PWBA and SCA give identical results. Since it is based on first-order perturbation theory, the first Born approximation only requires computation of the matrix element of the projectile interaction between initial and final state wave functions. For example, the first Born transition amplitude for a bare ion (charge Z_P, mass M) incident on a one electron target is

$$f_{ji} = \frac{2MZ_Pe^2}{\hbar^2 K^2} \int \phi_j^*(\mathbf{r}) e^{i\mathbf{K}\cdot\mathbf{r}} \phi_i(\mathbf{r}) \, d^3\mathbf{r}, \qquad (48.3)$$

where $\hbar\mathbf{K}$ is the momentum transfer vector and ϕ_i, ϕ_j are the initial and final state wave functions. For a hydrogenic system, analytical expressions for the PWBA amplitudes for bound-bound transitions can be derived

[20] and for bound-continuum transitions (ionization) the cross section can be expressed in terms of a two-dimensional integral over energy and momentum transfer, where the integrand is given by a simple analytical expression [21]. First Born calculations are also carried out with relativistic wave functions; relativistic target wave function effects are important for the inner shells of heavy atoms [22].

There is a number of other approximate methods that lead to simple analytic expressions, such as the Glauber approximation [23], continuum distorted wave (CDW) and eikonal methods (Chap. 50), the binary encounter approximation (BEA, Chap. 54), and second Born calculations [24]. A difficulty with all these methods is that they do not admit to a sequence of successive improvements; there is no procedure for systematically driving them to convergence.

One class of approximation methods very widely used for inner shell processes, such as inner shell vacancy production, is based on perturbed stationary state (PSS) methods. One variation is the *energy loss and Coulomb deflection effects perturbed stationary state approximation with relativistic correction* (ECPSSR) theory, developed originally by Brandt and co-workers and since extended and applied by others [25]. This approximation involves procedures for correcting the first Born approximation. It is quite successful in fitting K-, L-, and M-shell vacancy production cross sections.

Another approach is to calculate the electronic wave function for the collision system by expanding it in some basis set [26]. This procedure is called the coupled-channels, coupled-states, or close-coupling method. It is convenient to diagonalize the target and projectile systems in the chosen basis to produce matrix eigenfunctions for the target and projectile. Putting this expansion into the time-dependent Schrodinger equation leads to a set of coupled first-order equations for the time-dependent expansion coefficients. This set of equations is solved subject to the boundary condition that long before the collision, the expansion coefficient for the initial state wave function is unity and all other coefficients are zero. Then the expansion coefficients a long time after the collision give the transition amplitudes for transitions into those states.

Several types of expansions are used. The notation appropriate to a single-electron problem will be used for simplicity.

48.4.1 Single-Centered Expansion

In the single-centered expansion [27], the wave function is expanded in a set of target centered basis functions. Diagonalization of the target Hamiltonian in this basis yields a set of target eigenfunctions $\phi_j^{\mathrm{T}}(\mathbf{r})$ and energies E_j. The expansion of the time-dependent wave function for the collision system is then

$$\psi_i(\mathbf{r}, b, t) = \sum_j a_{ji}(b, t) e^{-i(E_j - E_i)t/\hbar} \phi_j^{\mathrm{T}}(\mathbf{r}). \quad (48.4)$$

If the electron is initially in the target state i, the initial boundary condition is $a_{ji}(b, -\infty) = \delta_{ji}$. The expansion coefficients at times long after the collision, $a_{ji}(b, +\infty)$, give the transition amplitudes for transitions from target state i to j. Ionization probabilities and the description of the ionization continuum will be described in the next section. It will suffice to say here that some of the $\phi_j(\mathbf{r})$ represent ionized states and account for transitions that leave the electron no longer in a target bound state. This single centered expansion does not directly allow for calculation of charge transfer amplitudes. However, it has been shown that if the number of angular momenta included in the set of $\phi_j(\mathbf{r})$ is large enough, the basis can accurately describe electron loss from the target bound states even in cases where the charge transfer probability is large [27]. If the charge transfer amplitude is small, it can be extracted from $\psi_i(\mathbf{r}, b, t)$ by use of a first-order T-matrix expression of the form

$$a_{ki}^{\mathrm{P}}(b, +\infty) = \langle \phi_k^{\mathrm{P}}(\mathbf{r} - \mathbf{R}) | U(\mathbf{r}, \mathbf{R}) | \psi_i(\mathbf{r}, b, t) \rangle, \quad (48.5)$$

where ϕ_k^{P} is a projectile state eigenfunction and $U(\mathbf{r}, \mathbf{R})$ is the projectile-target interaction [28].

48.4.2 Two-Centered Expansion

In a two-centered expansion [26, 29], the time-dependent wave function for the collision system is expanded in two sets of functions: a set of eigenfunctions centered on the target $\phi_j^{\mathrm{T}}(\mathbf{r})$ and a set of projectile centered eigenfunctions $\phi_k^{\mathrm{P}}(\mathbf{r} - \mathbf{R})$,

$$\psi_i(\mathbf{r}, b, t) = \sum_j a_{ji}^{\mathrm{T}}(b, t) \exp\left[-i(E_j - E_i)t/\hbar\right] \phi_j^{\mathrm{T}}(\mathbf{r})$$
$$+ \sum_k a_{ki}^{\mathrm{P}}(b, t) \exp\left[-i(E_k - E_i)t/\hbar\right]$$
$$\times \exp\left[i\mathbf{p}\cdot\mathbf{r}/\hbar\right] \phi_k^{\mathrm{P}}(\mathbf{r} - \mathbf{R}). \quad (48.6)$$

Here $\exp[i\mathbf{p}\cdot\mathbf{r}/\hbar]$ is an electron translation factor that accounts for the momentum $\mathbf{p}$ that the electron has relative to the target when it moves with the projectile. The energies E_k also contain the kinetic energy in the target frame that the electron has when it moves with the projectile. The problem of electron translation factors when a two centered molecular basis is used instead of the atomic basis has been extensively addressed [26].

The coefficients $a_{ji}^{\mathrm{T}}(b, +\infty)$ and $a_{ki}^{\mathrm{P}}(b, +\infty)$ are the transition amplitudes for transitions into target and projectile states. The two-centered expansion allows for charge transfer amplitudes to be calculated directly and is capable of an accurate description of the electron flux loss from the target due to charge transfer, but the computational difficulty is significantly greater. It is difficult in practice to include enough basis functions for convergence. Also, computational linear dependence can arise if the projectile and target are close because the projectile-centred basis set can become nearly identical to the target-centred basis set.

Calculations have also been done with a triple-centered expansion [30]. Bound atomic states are centered on each nucleus and on a third center (the center of charge) in order to simulate the molecular character of slow collisions. The orbitals on this third center represent the united-atom character of the wave function.

48.4.3 One-and-a-Half Centered Expansion

To account for flux loss from the target region due to charge transfer, while retaining the computational efficiency of the single-centered expansion, a hybrid method called the one-and-a-half centered expansion (OHCE) was developed [31]. It is of the same form as the two-centered expansion except that the coefficients $a_{ki}^{P}(b,t)$ of the projectile-centered states have predetermined rather than variable time dependence. The time dependence of the $a_{ki}^{P}(b,t)$ must be such that the a_{ki}^{P} are all zero as $t \to -\infty$, and as $t \to +\infty$ they become constants that are identified as the charge transfer amplitudes. This ansatz greatly reduces the computational cost of propagating the coupled equations forward in time, but the projectile-centered functions are still present to allow explicitly for charge transfer channels at large times. At intermediate times, when the projectile and target separation is within the range of the target-centered basis, the target-centered part of the expansion, if enough basis functions are used, is sufficient to describe the time dependence of $\psi_i(\mathbf{r},b,t)$ and the projectile-centered part of the expansion is redundant. The OHCE has been applied to several collision systems with good success [32].

48.5 DESCRIPTION OF THE IONIZATION CONTINUUM

In ion-atom and atom-atom collision calculations description of the ionization continuum presents particular difficulties. For hydrogenic systems, the exact continuum wave functions are known analytically, and these may be used in analytic approximations such as the PWBA. For nonhydrogenic systems, screened hydrogenic wave functions or numerical wave functions computed from some local effective potential, such as the Hartree-Fock-Slater approximation, can be used [33]. But for coupled states calculations, that involve expansion of the system wave function in terms of a discrete set of functions, some discretization of the continuum must be performed.

A simple procedure for generating a discrete representation of the continuum is to diagonalize the target (or projectile) Hamiltonian in a finite basis of square integrable basis functions. Some commonly used basis functions are Sturmian [26], Gaussian, or Slater-type orbitals. This diagonalization produces a set of discrete matrix eigenvalues and eigenvectors, called pseudostates.

For a one-electron or effective one-electron system, the exact energy spectrum of the Hamiltonian consists of an infinite set of discrete bound states and an ionization continuum and the interpretation of the pseudostates is straightforward. The N matrix eigenvalues below the first ionization threshold represent the bound states. By the Hylleraas-Undheim Theorem, they are upper bounds to the first N exact energies (see Sect. 11.3.1). The matrix eigenvectors with positive eigenvalues provide a discrete representation of the ionization continuum [34]. Each pseudostate is accurately proportional to the exact continuum wave function at the pseudostate energy for values of r out to the range of the basis. The proportionality constant gives the energy width of the pseudostate — the energy region of the continuum represented by that pseudostate [28]. These widths are approximately equal to the energy spacing between pseudostates. The pseudostates represent the continuum in the sense that the total ionization cross section is equal to the sum of the continuum pseudostate cross sections. Coupled-states calculations have also been carried out with wave packets constructed at discrete energies from the exact continuum wave functions [35]. What all these discretization methods do is to provide a quadrature rule for evaluating the integration over the energy of the ionized electron. Differential cross sections, differential in the ionized electron's energy, have also been extracted from discretized continuum calculations [36].

The situation is more complex for two-electron or multi-electron systems. The energy spectrum of a two-electron atom consists of an infinite set of bound states, an infinite set of overlapping single ionization continua that correspond to different residual ion states, and overlapping the higher energy part of these single ionization continua, a double ionization continuum where both electrons are unbound. The exact wave function for two electrons in the continuum of an ion is not known, but there are a number of approximate forms [37]. The interpretation of two electron pseudostates is also complicated. For energies below the second single ionization threshold the energy of a pseudostate clearly identifies whether it represents a bound state or an ionized state. But in the energy region where there are overlapping continua there are ambiguities. For example, a pseudostate with energy above the double ionization threshold can represent a doubly ionized state, a singly ionized state of the same total energy, or some admixture of the two. A method has been developed to solve this pseudostate interpretation problem [15].

REFERENCES

1. B. H. Bransden, *Atomic Collision Theory* (Benjamin-Cummings, Reading, 1983).

2. J. S. Briggs and J. H. Macek, Adv. At. Mol. Opt. Phys. **28**, 1 (1991).

3. J. M. Hansteen, Phys. Scr. **42**, 299 (1990).

4. G. L. Swafford, J. F. Reading, A. L. Ford, and E. Fitchard, Phys. Rev. A **16**, 1329 (1977); A. Jakob, F. Rösel, D. Trautmann, and G. Bauer, Z. Phys. A **309**, 13 (1982); W. Fritsch, J. Phys. B **15**, L389 (1982).

5. M. H. Martir, A. L. Ford, J. F. Reading, and R. L. Becker, J. Phys. B **15**, 1729 (1982).

6. P. A. Amudsen, J. Phys. B **11**, 3179 (1978); M. Kleber and K. Unterseer, Zeit. Phys. A **292**, 311 (1979); F. Rösel, D. Trautmann, and G. Bauer, Nucl. Instrum. Methods **192**, 42 (1983).

7. B. H. Bransden and C. J. Noble, Phys. Lett. **70A**, 404 (1979); J. M. Wadehra and R. Shakeshaft, Phys. Rev. A **26**, 1771 (1982); R. Gayet and A. Salin, Nucl. Instrum. Methods **B56/57**, 82 (1991).

8. A. Salin, J. Phys. B **24**, 3211 (1991); L. Meng, R. E. Olson, R. Dörner, J. Ullrich, and H. Schmidt-Böcking, J. Phys. B **26**, 3387 (1991).

9. X. Fang and J. F. Reading, Nucl. Instrum. Meth. **B53**, 453 (1991).

10. J. Reinhardt, B. Muller, W. Greiner, and G. Soff, Phys. Rev. Lett. **43**, 1307 (1979); J. F. Reading and A. L. Ford, Phys. Rev. **21**, 124 (1980).

11. R. L. Becker, A. L. Ford, and J. F. Reading, Phys. Rev. A **29**, 3111 (1984).

12. J. H. McGuire and P. Richard, Phys. Rev. A **8**, 1874 (1973); J. H. McGuire, Adv. At. Mol. Opt. Phys. **29**, 217 (1991); Yu S. Sayasov, J. Phys. B **26**, 1197 (1993).

13. J. F. Reading and A. L. Ford, Comments At. Mol. Phys. **23**, 301 (1990).

14. J. H. McGuire, Phys. Rev. Lett. **49**, 1153 (1982); Adv. At. Mol. Opt. Phys. **29**, 217 (1991).

15. J. F. Reading and A. L. Ford, J. Phys. B **20**, 3747 (1987).

16. W. Fritsch and C. D. Lin, Phys. Rev. A **41**, 4776 (1990); L. F. Errea, L. Mendez, and A. Riera, Phys. Rev. A **27**, 3357 (1983); F. Martin, A. Riera, and M. Yanez, Phys. Rev. A **34**, 4675 (1986); H. A. Slim, B. H. Bransden, and D. R. Flower, J. Phys. B **26**, L159 (1993); K. Moribayashi, K. Hino, M. Matsuzawa, and M. Kimura, Phys. Rev. A **44**, 7234 (1991).

17. D. P. Marshall, C. LeSech, and D. S. F. Crothers, J. Phys. B **26**, L219 (1993).

18. J. D. Garcia, Nucl. Instrum. Methods **A240**, 552 (1985); W. Stich, J. J. Lüdde, and R. M. Driezler, J. Phys. B **18**, 1195 (1985).

19. E. Merzbacher and H. W. Lewis, *Encyclopedia of Physics* (Springer-Verlag, Berlin, 1958), Vol. 34, p. 166.

20. D. R. Bates and G. Griffing, Proc. Phys. Soc. A **66**, 961 (1953).

21. O. Benka and A. Kropf, At. Data Nucl. Data Tables **22**, 219 (1978).

22. M. H. Chen and B. Crasemann, At. Data Nucl. Data Tables **33**, 217 (1985).

23. R. Glauber in *Lectures in Theoretical Physics* (Interscience, New York, 1958), Vol. 2; V. Franco and B. K. Thomas, Phys. Rev. A **4**, 945 (1971); J. Binstock and J. F. Reading, Phys. Rev. A **11**, 1205 (1975).

24. B. H. Bransden and D. P. Dewangen, J. Phys. B **12**, 1371 (1979).

25. G. Basbas, W. Brandt, and R. H. Ritchie, Phys. Rev. A **7**, 1971 (1973); W. Brandt and G. Lapicki, Phys. Rev. A **23**, 1717 (1981).

26. W. Fritsch and C. D. Lin, Phys. Rep. **202**, 1 (1991).

27. C. O. Reinhold, R. E. Olson, and W. Fritsch, Phys. Rev. A **41**, 4837 (1990); A. L. Ford, J. F. Reading, and K. A. Hall, J. Phys. B **26**, 4537 (1993); **26**, 4553 (1993).

28. J. F. Reading, A. L. Ford, G. L. Swafford, and A. Fitchard, Phys. Rev. A **20**, 130 (1979); R. L. Becker, A. L. Ford, and J. F. Reading, J. Phys. B **13**, 4059 (1980).

29. R. Shakeshaft, Phys. Rev. A **18**, 1930 (1978); W. Fritsch and C. D. Lin, Phys. Rev. A **27**, 3361 (1983); A. M. Ermolaev, J. Phys. B **23**, L45 (1990); H. A. Slim, J. Phys. B **26**, L743 (1993).

30. M. J. Antal, D. G. M. Anderson, and M. B. McElroy, J. Phys. B **8**, 1513 (1975); T. G. Winter and C. D. Lin, Phys. Rev. A **29**, 567 (1984).

31. J. F. Reading, A. L. Ford, and R. L. Becker, J. Phys. B **14**, 1995 (1981).

32. J. F. Reading, A. L. Ford, and R. L. Becker, J. Phys. B **15**, 625 (1982); **15**, 3257 (1982).

33. B. H. Choi, Phys. Rev. A **11**, 2004 (1975).

34. A. L. Ford, E. Fitchard, and J. F. Reading, Phys. Rev. A **16**, 133 (1977).

35. G. Mehler, W. Greiner, and G. Soff, J. Phys. B **20**, 2787 (1987).

36. T. Mukoyama, C. D. Lin, and W. Fritsch, Phys. Rev. A **32**, 2490 (1985).

37. M. Brauner, J. S. Briggs, and H. Klar, J. Phys. B **22**, 2265 (1989); J. S. Briggs, Phys. Rev. A **41**, 539 (1990); C. Dal Cappello and H. Le Rouzo, Phys. Rev. A **43**, 1395 (1991).

49

Ion-Atom Charge Transfer Reactions at Low Energies

M. Gargaud* and R. McCarroll[†]

*Observatoire de Bordeaux (INSU/CNRS:URA 352), B.P. 89, 33270-Floirac, France
[†]Laboratoire de Dynamique Moléculaire et Atomique (CNRS:URA 774) Université P. et M. Curie, 4 Place Jussieu, 75252-Paris Cedex 05, France

49.1 INTRODUCTION

Ion-atom charge exchange reactions contribute significantly to the ionization balance of complex ions in many natural environments (astrophysical photoionized plasmas such as planetary nebulae, nova shells, etc) [1]. For electron temperatures lower than 10^6 K, the abundance of neutral atomic H or He can be sufficient for the charge exchange rate to exceed the radiative (direct or dielectronic) recombination rate for ions with charge $q \geq 2$ [2]. In the case of singly charged ions, accidental resonance conditions are required for the charge exchange rate to be large at thermal energies. In this chapter, we shall limit discussion to the case of $q \geq 2$. We may distinguish two types of reaction: type I, for which electron capture takes place with no change in the configuration of the core electrons, and type II, for which electron capture is accompanied by a rearrangement of the core electron configuration.

We shall concentrate on the low energy range (less than a few hundred eV), where an adiabatic representation of the collision complex can successfully describe the collision dynamics [3]. Nonadiabatic transitions tend to occur in the vicinity of avoided potential energy crossings. For the transition probability to be appreciable, the energy separation Δ_X at the crossing radius R_X must be neither too large nor too small ($0.1 < \Delta_X < 3$ eV). Since the size of Δ_X is determined by electron exchange, whose effect varies exponentially with internuclear distance, only crossings in the range $3 \leq R_X \leq 15a_0$ induce nonadiabatic transitions. Because of this constraint, the number of effective curve crossings is quite small and the electron capture process is very state selective [3]. In general, electron capture is of type I for $q \geq 3$ and of type II for $q = 2$. In a few cases, type II processes can also occur for $q = 3$.

Although there are a few direct experimental measurements at eV energies [4], most measurements have been carried out at collision energies exceeding a few 100 eV [5]. Unfortunately, the processes which dominate in the higher energy range become negligible at eV energies (and *vice versa*). Most estimates of the cross sections in the thermal–eV range are based on theoretical models. Since at thermal energies, orbiting and tunneling effects may be considerable, a quantum mechanical description of the collision is preferable.

The dynamic coupling matrix elements between different molecular states depend on the origin of the electronic coordinates, and the corresponding scattering equations are not invariant with respect to a Galilean transformation. In a semiclassical formulation of the dynamics, this defect can be removed by the introduction of translation factors [6–12]. However, in a quantum mechanical formulation, it is more appropriate to introduce reaction coordinates of a more flexible form than those used traditionally in the Born-Oppenheimer approximation. The most convenient is the Eckart coordinate system [9], which leads to a simple modification of the nonadiabatic coupling matrix elements, and which is well adapted to cases where long range avoided crossings are dominant [10]. If a semiclassical approximation is introduced, the scattering equations become almost identical to those obtained directly using translation factors of the CTF (common translation factor) type [11, 12].

Three representative systems are chosen to illustrate the basic features of the problem: Al^{3+}/H and B^{3+}/He which are typical of type I, and O^{2+}/H which is typical of type II. In the Al^{3+}/H system, charge transfer takes place via the reaction [13]

$$Al^{3+}(3s^2)^2S + H \rightarrow Al^{2+}(3s^23p)^2P + H^+ \qquad (49.1)$$

involving a network of two Σ and one Π adiabatic states. In the B^{3+}/He system, charge exchange occurs via a network of three Σ and one Π states involving two avoided crossings [14]. The two possible capture channels are

$$B^{3+}(1s^2) + He \rightarrow B^{2+}(1s^2nl) + He^+ , \quad nl = 2s, 2p .$$
$$(49.2)$$

The O^{2+}/H system is a good example of a type II reaction which plays an important role in astrophysical plasmas [15, 16]. The dominant reaction at low energies

$$O^{2+}(2s^22p^2)^3P + H \rightarrow O^+(2s2p^4)^4P + H^+ \qquad (49.3)$$

involves only quartet states.

49.2 MOLECULAR STRUCTURE CALCULATIONS

The construction of the network of adiabatic states of the molecular ion complex constitutes the first step

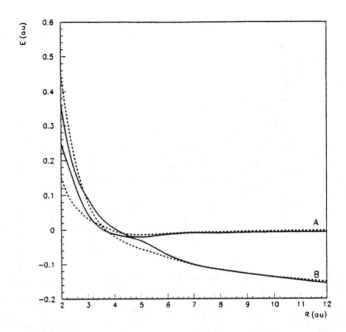

Figure 49.1. Adiabatic potential energy curves of the $^4\Sigma^-$ and $^4\Pi$ states of the OH^{2+} molecular ion. The solid curves designate $^4\Pi$ states, the dashed curves $^4\Sigma^-$ states. The dissociation limits A and B correspond respectively to $[O^{2+}(^3P) + H(^2S)]$ and $[O^+(^4P) + H^+]$.

in analyzing the dynamics of the ion-atom system. In principle, standard techniques of quantum chemistry can be employed. For doubly charged ions, where type II processes dominate and the effective avoided crossings take place at relatively short range, *ab initio* methods are required. On the other hand, for trebly and more highly charged ions, where type I processes dominate, the effective avoided crossings occur at long range where *ab initio* methods can be quite inaccurate. Model potential methods are then often more satisfactory.

49.2.1 *Ab Initio* Methods

Both molecular orbital and valence-bond methods have been extensively used and (provided a sufficient number of configurations are included) should be satisfactory in principle. A simple test is the location of the long range crossings, which is very sensitive to the energy difference of the initial and final channels. But this test is insufficient. For example, in the C^{3+}/H system, for which both the molecular orbital and valence bond methods yield comparable results in regions well away from avoided crossings, there are significant discrepancies in the important region of the avoided acrossings [17, 18]. In the present state of the art, the absolute accuracy of theoretical calculations in the crossing region is difficult

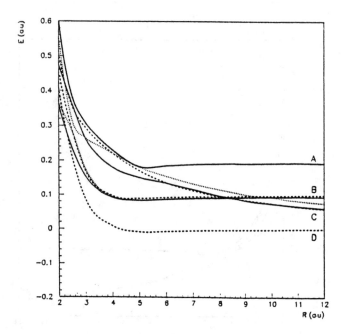

Figure 49.2. Adiabatic potential energy curves of the $^2\Sigma^+$, $^2\Pi$ and $^2\Delta$ states of the OH^{2+} molecular ion. The solid curves designate $^2\Sigma^+$ states, the dashed curves $^2\Pi$ states and the dotted curve a $^2\Delta$ state. The dissociation limits A, B, C and D correspond respectively to [O^{2+}(^1S) + H(^2S)], [O^{2+}(^1D) + H(^2S)], [O$^+$(^2D) + H$^+$] and [O^{2+}(^3P) + H(^2S)].

to ascertain. Of course, since the computed electron capture cross sections depend sensitively on the minimum energy separation, experimental data at low collision energies makes some tests possible. Unfortunately, there is not yet much reliable data at low eV energies.

Figures 49.1 and 49.2 present the adiabatic network of the $^4\Sigma^-$, $^4\Pi$ and $^2\Sigma^+$, $^2\Pi$, $^2\Delta$ states of the O^{2+}/H system. These have been obtained by means of a configuration interaction molecular orbital method with several hundred configurations [15]. It is clear from these calculations that the only effective avoided crossings are those involving the $^4\Sigma^-$ and $^4\Pi$ at 3.7 and 4.8 a_0, for which the energy separation is $\sim$0.1 eV. The avoided crossings involving the doublet states at 8.0, 8.5 and 10 a_0, for which the energy separation is $\sim$0.02 eV, are diabatic. The avoided crossings for type II transitions occur at much shorter distances than for type I transitions.

49.2.2 Model Potential Methods

Model potential methods offer an attractive alternative to *ab initio* methods in treating type I reactions. To illustrate the method [19, 20], we consider an effective one-electron system composed of a spherically symmetric

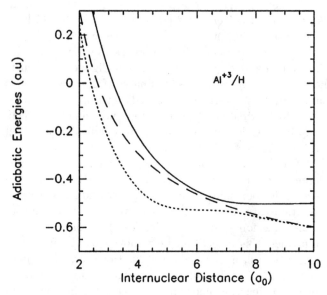

Figure 49.3. Adiabatic potential energies of AlH^{3+}. The full curve designates the $^2\Sigma$ state correlated to the [Al^{3+}(3p) + H(1s)] entry channel. The dotted and dashed curves designate respectively the $^2\Sigma$ and $^2\Pi$ states correlated to the [Al^{2+}(3p) + H$^+$] electron capture channel.

ion X^{q+} and a hydrogen atom. The model Hamiltonian of the molecular ion XH^{q+} is written as

$$H = T + V_X(r_b) - \frac{1}{r_a} + V_X(R), \qquad (49.4)$$

where T is the electronic kinetic energy, $\mathbf{r}_a$ and $\mathbf{r}_b$ are respectively the position vectors of the Rydberg electron with respect to nuclei A and B, and $V_X(r)$ is the effective potential of the ion core. The latter is usually expressed in the parametric form

$$V_X(r) = -[q + (Z - q)(1 + \alpha r + \beta r^2 + \gamma r^3)e^{-\delta r}], \qquad (49.5)$$

where Z is the ionic nuclear charge and the parameters $\alpha, \beta, \gamma, \delta, \ldots$ are optimized to the spectroscopic data so that the asymptotic energies of those Rydberg states of $X^{(q-1)+}$ which govern the charge transfer process are essentially exact.

The eigenvalues of H for a given internuclear distance R are determined by standard variational techniques, using a basis set of Slater-type orbitals in prolate spheroidal coordinates [3]. As an example, we present in Fig. 49.3 the potential energy curves of the Al^{3+} system which presents one $\Sigma - \Sigma$ avoided crossing around $R_X = 7.2 \ a_0$. Model potential methods can also be used to treat multielectron systems such as $(X\text{He})^{q+}$. In long range collisions, the transition probabilities are primarily deter-

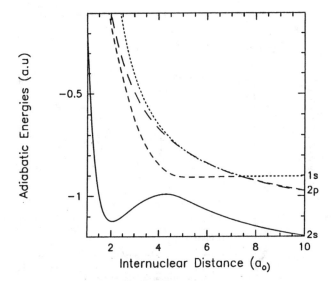

Figure 49.4. Adiabatic potential energies of BHe^{3+}. The dotted curve designates the $^1\Sigma$ state correlated to the $[\text{B}^{3+} + \text{He}(1s^2)]$ entry channel. The short and long dashed curves designate respectively the $^1\Sigma$ and $^1\Pi$ states correlated to the $[\text{B}^{2+}(2p) + \text{He}^+(1s)]$ electron capture channel. The full curve designates the $^1\Sigma$ state correlated to the $[\text{B}^{2+}(2s) + \text{He}^+(1s)]$ entry channel.

mined by the asymptotic form of the electron wave functions far from the nuclei. Thus even when multielectron targets are involved, the effect of dynamic correlation can be small. Since the asymptotic form of the wave functions may be easily generated by model potential techniques, the effect of static correlation can be taken into account in a rather simple way. The method proposed here is quite similar to that used by Grice and Herschbach [21] to treat the long range configuration interaction of ionic and covalent states in neutral atom-atom collisions.

For example, in the system X^{q+} He, where as previously X^{q+} is a closed shell ion, the energy separation at a long range avoided crossing is given by [22]

$$\Delta\varepsilon(R_c) \equiv 2\sqrt{2}\langle b_{1\sigma}|b'_{1s}\rangle\langle a_{nl\lambda}|V_X + q/R|b\rangle, \qquad (49.6)$$

where the molecular orbitals $a_{nl\lambda}$ (dissociating to $X_{nl}^{(q-1)+}$) and $b_{1\sigma}$ (dissociating to a $1s$ orbital of He) are generated from a model Hamiltonian of the form

$$H = T + V_X(r_b) - V_{\text{He}}(r_a) + V_{X,\text{He}}(R), \qquad (49.7)$$

where V_{He} is an effective potential describing the $1s$ orbital of He. The orbital b'_{1s} is the $1s$ orbital of He$^+$. This expression is the optimal form of $\Delta\varepsilon(R_c)$ which can be achieved without the use of explicitly correlated wave functions. Since both the $a_{nl\lambda}$ and $b_{1\sigma}$ orbitals are generated from the same model Hamiltonian, the calculation of the interaction matrix element in Eq. (49.6)

is simple. As an example, Fig. 49.4 presents the adiabatic energies of the B^{3+}/He system which has two avoided $\Sigma - \Sigma$ crossings, one at 4.6 a_0, the other at 7.4 a_0.

A similar extension to treat the case of capture by an ion with one electron in a p shell can be made along the same lines. See [23] for an application to the reaction

$$\text{O}^{3+}(2p)^2 P + \text{H} \rightarrow \text{O}^{2+}(2p3p)^3 P + \text{H}^+. \qquad (49.8)$$

A combination of model potential methods to represent the core electrons and *ab initio* molecular orbital methods for the valence electrons has been successfully developed [24] to treat some complex systems such as C^{3+}/H where both type I and II transitions take place. The flexibility of the model potential parameters makes it possible to choose them so that the energies of the initial and final states are accurately reproduced.

49.2.3 Empirical Estimates

Electron capture cross sections at low energies are largely controlled by two parameters: the crossing radius R_X and the minimum energy separation Δ_X. These are the basic parameters required for the Landau-Zener model, and it is useful to have a simple way of estimating them without having recourse to a complex molecular structure calculation. Many empirical estimates [25–27] have been proposed. However, it is generally necessary to take into account the strong ℓ-dependence of the electron capture channel states. Defining $\alpha = \sqrt{I_t/13.6}$, where I_t is the ionization potential of the target (in eV), then Taulbjerg's formula is [27]

$$\Delta_X(R_c) = \frac{18.26}{\sqrt{q}} f_{nl} \exp(-1.324\alpha R_c\sqrt{q}), \qquad (49.9)$$

where

$$f_{nl} = \frac{(-1)^{n+l-1}\sqrt{2l+1}\,\Gamma(n)}{[\Gamma(n+l+1)\Gamma(n-l)]^{1/2}} \qquad (49.10)$$

For type I systems, Eq. (49.9) is generally quite satisfactory when the energy levels of the capture states are well separated [28]. In that case, it can be used to predict the main electron capture reaction windows and obtain a reliable first estimate of the cross sections (in the energy range for which the Landau-Zener model is valid). On the other hand, in cases where there is a near degeneracy of the l states, serious errors can occur [28] and the predictions are less satisfactory.

For type II reactions, Eq. (49.9) must be modified by a corrective factor to take account of the simultaneous excitation of an electron from a $2s$ to $2p$ orbital. One such modification, proposed by Butler and Dalgarno [29], may give some useful idea of the main electron capture processes, but its precision is uncertain.

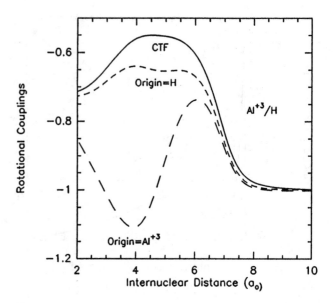

Figure 49.5. Radial coupling matrix elements between the two $^2\Sigma$ states of AlH^{3+}. The full curve designates the matrix element of the CTF (or Eckart) type. The short and long dashed curves designate the radial derivative with the origin of electron coordinates respectively on the Al and H nuclei.

49.3 DYNAMICS OF THE COLLISION

The first step is to introduce a suitable set of scattering coordinates which can automatically describe both the excitation and rearrangement channels. The particular choice of coordinate system is conditioned by practical considerations. We have found Eckart coordinates to be convenient [10]. Their application is straightforward since it involves nonadiabatic matrix elements which can be calculated by the conventional techniques of quantum chemistry. The practical implementation of Eckart coordinates leads to the introduction of an adiabatic variable ξ defined as

$$\xi = \sqrt{\mu}\left[\mathbf{R} + \tfrac{1}{\mu}\mathbf{s}\right], \quad \text{where } \mathbf{s} = \frac{\mathbf{r}\cdot\mathbf{R}}{R^2}\left[\mathbf{r} - \frac{\mathbf{r}\cdot\mathbf{R}}{2R^2}\mathbf{R}\right],$$

(49.11)

μ is the reduced mass of the colliding system, $\mathbf{r}$ the coordinate of the active electron with respect to the c.m. of the colliding system and $\mathbf{R}$ the relative position vector of the nuclei.

The adiabatic Eckart and Born-Oppenheimer equations differ only by terms of the order of $1/\mu$. It may be assumed that the Eckart states are given to sufficient accuracy by the Born-Oppenheimer adiabatic states designated by $\chi_i(\mathbf{r}; R)$. We expand the total wave function of the system in the form

Figure 49.6. Rotational coupling matrix elements between the $^2\Sigma$ and $^2\Pi$ states correlated to the electron capture channel. The designation of the curves is the same as for Fig. 49.5.

$$\Psi(\mathbf{r}, \xi) = \sum \chi_i(\mathbf{r}; \xi) F_i(\xi)\,.$$

(49.12)

(For a many-electron system, $\mathbf{r}$ represents the ensemble of electron coordinates.) Decomposing $F_i(\xi)$ on a basis set of symmetric top functions according to

$$F_i(\xi) = \sum_{K,M}(-1)^K\left[\frac{2K+1}{4\pi}\right]^{1/2}D^K_{\Lambda,M}(\theta,\phi)\frac{g_i^{(k)}(\xi)}{\xi}\,,$$

(49.13)

where $\{\theta,\phi\}$ are the spherical polar coordinates of ξ, then the radial functions $g_i^{(k)}(\xi)$ are solutions of the equation

$$\frac{d^2}{d\xi^2}\mathbf{g}^{(K)} + 2\mathbf{A}\frac{d}{d\xi}\mathbf{g}^{(K)} + \mathbf{W}\mathbf{g}^{(K)} = 0\,,$$

(49.14)

where

$$A_{mn}(\xi) = \left\langle\chi_m\left|\frac{\partial}{\partial\xi} + \frac{z}{\xi}\frac{\partial}{\partial z}\right|\chi_n\right\rangle\delta(\Lambda_m,\Lambda_n)\,,$$

(49.15)

$$B_{mn} = \left\langle\chi_m\left|\frac{\partial^2}{\partial\xi^2}\right|\chi_n\right\rangle\,,$$

(49.16)

$$W_{mn} = \left[2\mu(E-\varepsilon_n) - \frac{[K(K+1)-\Lambda_n^2]}{R^2}\right]\delta_{mn}$$
$$+ B_{mn} + \frac{2}{R^2}\sqrt{K(K+1)}\,L_{mn}\,,$$

(49.17)

$$L_{mn} = \mp\left\langle\chi_m\left|-2x\frac{\partial}{\partial z}\right|\chi_n\right\rangle\delta(\Lambda_m,\Lambda_m\pm1)\,,$$

(49.18)

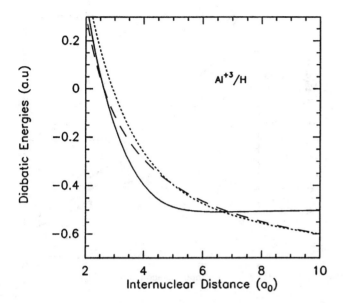

Figure 49.7. Diagonal elements of the diabatic matrix of ALH^{3+}.

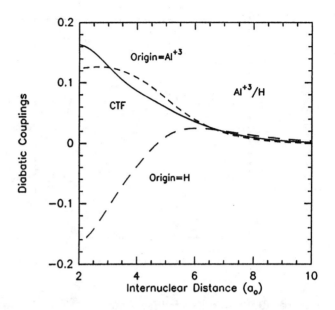

Figure 49.8. Off-diagonal diabatic matrix element between the two $^2\Sigma$ states of AlH^{3+}.

z, x are the components of $\mathbf{r}$ parallel to and perpendicular to the direction of ξ in the classical collision plane, respectively.

Since the ratio $1/\mu$ is small, it is legitimate to replace the matrix element $\langle \chi_m | \partial/\partial \xi | \chi_n \rangle$ by $\langle \chi_m | \partial/\partial R | \chi_n \rangle$ in Eqs. (49.15) and (49.16). Furthermore, all the matrix elements A_{mn} and L_{mn} vanish asymptotically to first order in $1/\mu$. The modified radial and rotational matrix elements are identical to those obtained in the semiclassical formalism with common translation factors.

The simplest way to solve equation (49.14) is to eliminate the first-order derivative by transforming to a diabatic representation in which the radial matrix elements vanish [30]

$$\mathbf{g}^{(K)} = \mathbf{C}\mathbf{h}^{(K)}, \qquad (49.19)$$

where

$$\frac{d}{d\xi}\mathbf{C} + \mathbf{A}\mathbf{C} = 0, \qquad \mathbf{C}(\infty) = \mathbf{I}. \qquad (49.20)$$

Equation (49.14) then reduces to

$$\frac{d^2}{d\xi^2}\mathbf{h}^{(K)} - 2\mu\mathbf{V}^d\mathbf{h}^{(K)} + \left[2\mu E - \frac{K(K+1)}{\xi^2}\right]\mathbf{h}^{(K)} = 0 \qquad (49.21)$$

with

$$\mathbf{V}^d = \mathbf{V}_E^d - \frac{\sqrt{K(K+1)}}{\mu R^2}\mathbf{V}_R^D, \qquad (49.22)$$

$$\mathbf{V}_E^d = \mathbf{C}^{-1}\varepsilon\mathbf{C}, \qquad \mathbf{V}_R^D = \mathbf{C}^{-1}\mathbf{L}\mathbf{C}. \qquad (49.23)$$

The solution of (49.21) and the subsequent extraction of the scattering matrix S may be carried out using an extension of the log derivative method [31], adapted to the case of a repulsive Coulomb potential in one or more of the scattering channels [32]. This method is particularly stable and advantageous to use at low energies for problems of the type considered here. Of course, at energies exceeding a few hundred eV, rapid oscillations of the radial function render the method rather time consuming. But the method is usable up to keV energies. However, for higher energies exceeding 1 keV, semiclassical methods are preferable.

49.4 RADIAL AND ROTATIONAL COUPLING MATRIX ELEMENTS

Two independent methods are employed for the determination of radial coupling matrix elements A_{mn}: the first based on a direct numerical differentiation of the expansion coefficients of the wave function, the second on a variant of the Hellman-Feynman (HF) theorem. In principle, the HF method is three times faster than the direct numerical method, since the eigenvectors need to be calculated only at the value of R concerned. However, the real gain in computing is not always appreciable, the HF theorem being more sensitive to errors in the wave function than the direct method. To achieve comparable accuracy, a larger basis set would be required in the molecular calculations, and this offsets much of the theoretical gain in computing time. The rotational

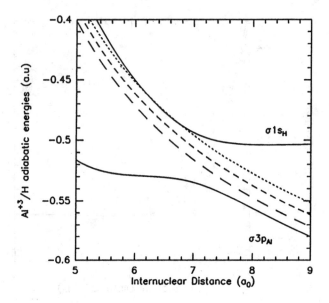

Figure 49.9. Results of three state (two $^2\Sigma$ and one $^2\Pi$) calculations for electron capture in the Al^{3+}/H system. The cross sections are plotted as a function of the parameter Δ (see text). The extreme sensitivity to Δ illustrates the necessity of knowing the $^2\Sigma - {}^2\Pi$ energy separation to high accuracy. The numerical values on the curves designate the collision energies in units of eV/amu.

Figure 49.10. Schematic diagram of the nondiagonal matrix elements of AlH^{3+} in the vicinity of the curve crossing. The three broken curves correspond to different $^2\Pi$ state potentials shifted from their calculated value by small amounts.

coupling matrix elements L_{mn} are calculated numerically for each internuclear distance.

Typical results for the radial and rotational matrix elements are presented in Figs. 49.5 and 49.6 for the Al^{3+}/H system, illustrating the influence of translation effects. The origin dependence of the matrix element (without inclusion of translation) is weak in the vicinity of an avoided crossing. Away from the crossing, the origin dependence can be considerable. The corresponding diabatic energies and couplings are given in Figs. 49.7 and 49.8. For internuclear distances on the inward side of the crossing, the diabatic states correspond to a mixing of adiabatic states: it is clear that the mathematical definition (49.20) of the diabatic representation does not correspond to the empirical definition of the Landau-Zener model (see Sect. 49.6).

49.5 TOTAL ELECTRON CAPTURE CROSS SECTIONS

Although the selection rules for electron capture at low energies are primarily governed by the avoided crossings between adiabatic states of the same symmetry, the combined effects of radial and rotational coupling can be quite complex. In general, rotational coupling effects are weak when capture occurs via an S state, but they can be strong for capture to P, D and higher L states.

In the case of Al^{3+}/H, where capture to the $(3p)\,^2P$

state of Al^{2+} takes place via a three-state network of two Σ and one Π states, rotational coupling between the Σ and Π states is strong in the vicinity of the avoided crossing and enhances the capture cross section considerably. This phenomenon is illustrated in Fig. 49.9, which shows the calculated total cross section at different energies as a function of Δ, the energy separation of the quasidegenerate diabatic $^2\Sigma$ and $^2\Pi$ exit channels at the $^2\Sigma$-$^2\Sigma$ crossing radius. The total cross section has a maximum for $\Delta = 0.08$ eV. The cause can be easily seen from the corresponding adiabatic potential energies (Fig. 49.10). For $\Delta = 0.8$ eV, the potential energies of the adiabatic $^2\Sigma$ entry channel and the $^2\Pi$ exit channel become tangential to one another, thereby inducing a resonant effect.

In the case of the B^{3+}/He system, four molecular states are implicated in the collision process: three Σ states (Σ_1 correlated to the ^2S exit channel, Σ_2 correlated to the ^2P exit channel and Σ_3 correlated to the entry channel) and one Π state (correlated to the ^2P exit channel). In order to understand better the rotational coupling mechanism, Figs. 49.11 and 49.12 show the ^2S and ^2P electron capture transition amplitudes as a function of angular momentum (impact parameter) for an energy of 6 keV, where rotational coupling is of major importance [14, 33]. The ^2P electron capture transition amplitude exhibits 2 maxima, one for an impact parameter of 1.6 a_0, the other for an impact parameter of 5.7 a_0, corresponding respectively

Figure 49.11. Transition amplitudes for ^2S and ^2P electron capture as a function of impact parameter in B^{3+}/He collisions. The solid curve refers to the ^2P contribution, the dotted curve to the ^2S contribution.

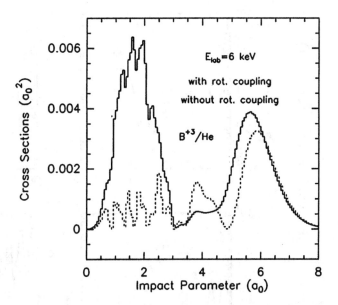

Figure 49.12. Influence of rotational coupling on the ^2P electron capture cross section in B^{3+}/He collisions. The solid curve refers to the complete calculation, the dotted curve to the calculation with only radial coupling included.

to capture via $\Sigma_2 - \Pi$ crossing around $R = 2\ a_0$ and the outer $\Sigma_2 - \Sigma_3$ crossing around $R = 7.4\ a_0$. It is clear from Fig. 49.12 that at very short internuclear distances, rotational coupling is much more important than radial coupling. At the outer crossing (which is nearly diabatic), radial coupling is dominant. This result may be generalized. When the avoided crossing has a largely diabatic character, as for $3p$ capture in Si^{2+}/H [34] or $3d$ capture in C^{4+}/H [10], the inclusion of rotational coupling is fairly weak, affecting principally the population of the sub-m levels and less appreciably the total capture cross section into a given l state. In this case, translation effects are of more importance than rotational coupling.

In the case of O^{2+}/H (a particularly important system in astrophysical plasmas), the existence of several adiabatic states correlated to the entry channel leads to many interesting features, typical of open p shell ions. The favored reaction channel via the $^4\Sigma^-$ and $^4\Pi$ states involves simultaneous electron capture into a $2p$ orbital and excitation of a $2s$ orbital. The avoided crossing is due to electron correlation. Rotational mixing of the $^4\Sigma^-$ and $^4\Pi$ states leads to a large enhancement of the cross section at energies exceeding a few tens of eV [16]. On the other hand, the avoided crossings involving the $^2\Sigma^+$ and $^2\Pi$ states around $8\ a_0$ are too diabatic to contribute to an electronic transition. As a consequence, the metastable ^1D ions can only react via a curve crossing at small distances ($2.5\ a_0$). The cross section is much smaller

than for ground ^3P state capture.

49.6 LANDAU-ZENER APPROXIMATION

If a rapid estimation of the cross sections is all that is needed, the Landau-Zener method can be used with advantage, provided that the molecular structure parameters are known accurately. This method, based on an approximate solution of the dynamical equations in a semiclassical formalism (see Chap. 47), is satisfactory for the dominant channels. Aside from the transition being assumed to be localized at the crossing point, the method can easily take account of trajectory effects at low energies. On the other hand, it is unreliable for the weaker channels and it makes allowance neither for translation effects nor for rotational coupling.

The cross section for capture into a state n via a curve crossing located at R_X is given by

$$Q_n^{\text{LZ}} = 2\pi \int_0^{\rho_{\max}} 2p(1-p)\rho\,d\rho, \qquad (49.24)$$

where p, the probability for a single passage trajectory (impact parameter) through the crossing, is given by

$$p = \exp\left[-2\pi\left(\frac{\Delta E_{nl}}{2}\right)^2 \frac{1}{v\Delta F}\right], \qquad (49.25)$$

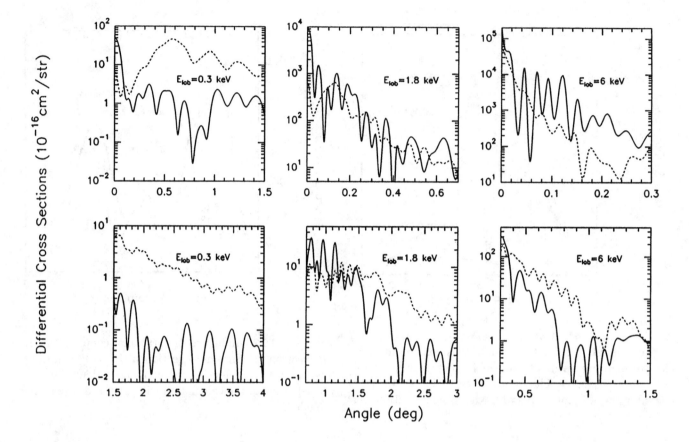

Figure 49.13. Differential cross sections (10^{-16} cm^{2}sr^{-1}) for electron capture in B^{3+}/He collisions as a function of scattering angle for ion energies of 0.3, 1.8 and 6 keV. The solid curves refer to ^2S capture, the dotted curves to ^2P capture.

ΔF being the difference in slope of the covalent and ionic diabatic curves at R_X, Δ_X the energy separation of the diabatic curves at R_X and v the radial velocity at R_X. It should be recalled that there is no very rigorous definition of the diabatic curves (which are required to obtain DF). For long distance crossings, the simplest (and probably the most satisfactory) estimate is that based on the asymptotic forms of the diabatic states

$$\Delta F = \frac{\partial}{\partial R}\left[V_1(R) - \frac{q-1}{R}\right]_{R=R_X} , \quad V_1(R) = -\frac{q^2\alpha_d}{2R^4},$$

(49.26)

where α_d is the polarizability of the target. The radial velocity is given by

$$v = \left\{\frac{2E}{\mu}\left[1 - \frac{\rho^2}{R_X^2} - \frac{V_1(R_X)}{E}\right]\right\}^{1/2}.$$

(49.27)

The inclusion of the attractive polarization potential V_1 considerably increases the cross section at low energies, since trajectories with impact parameters much greater than R_X contribute to the cross section. This effect can introduce a negative energy dependence of the

cross section in the limit of low energies, of the same kind as the Langevin model for ion-molecule reactions.

49.7 DIFFERENTIAL CROSS SECTIONS

Differential cross sections at large scattering angles (corresponding to small impact parameters) enable one to probe details of the collision dynamics not readily obtainable from total cross sections, which tend to be dominated by the contribution from small angle scattering (large impact parameters).

Their determination is staightforward once the S matrix elements have been extracted from the asymptotic solution of the coupled equations (49.21). The scattering amplitude $f_{fi}(\vartheta)$ for scattering through the c.m. angle ϑ is given by

$$f_{fi}(\vartheta) = \frac{1}{2i(k_ik_f)^{1/2}}\sum_{l=0}^{\infty}(2l+1)[S_{fi}^l - \delta_{fi}]$$
$$\times P_l(\cos\vartheta)\exp[i\alpha_f(l)],$$

(49.28)

where k_i and k_f are the wave numbers of the initial and

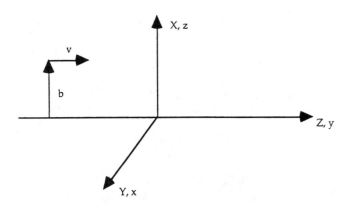

Figure 49.14. Coordinate systems for scattering.

final channels, S^l_{fi} the S-matrix element for the i to f transition, and $\alpha_f(l)$ the partial wave Coulomb phase shift for the final channel:

$$\alpha_f(l) = \arg \Gamma(l + 1 + i\gamma_f), \quad \gamma_f = \frac{\mu q_1 q_2}{k_f}. \quad (49.29)$$

The charges of the two collision partners in the final state are designated q_1 and q_2, and μ is the nuclear reduced mass. The differential cross section is evaluated as

$$\frac{d\sigma_{fi}}{d\Omega}(\vartheta) = \frac{k_f}{k_i}|f_{fi}(\vartheta)|^2. \quad (49.30)$$

Figure 49.13 shows some typical differential cross sections for capture to the $(2s)^2S$ and $(2p)^2P$ states of B^{2+} in the B^{3+}/He system. The oscillations observed in the differential cross sections are of Stückelberg type. A knowledge of differential cross sections is essential for estimating acceptance angles in laboratory experiments.

49.8 ORIENTATION EFFECTS

Recent experiments on the B^{3+}/He system [35, 36] show that there is a strong tendency for electron capture to produce strongly oriented states at small scattering angles. This propensity for orientation is a direct measure of rotational coupling between the Σ and Π molecular states converging to the ^2P asymptotic state [37].

The orientation and alignment parameters, which characterize the polarization of the emitted photons, can be simply expressed in terms of the scattering amplitudes for electron capture to the magnetic sublevels. These scattering amplitudes can be directly obtained from the scattering matrix. However, care must be exercised in the definition of the scattering amplitudes. In most applications where an adiabatic representation of the collision dynamics is used, the quantization axis is taken to be in

Figure 49.15. Right-hand circular polarization (full curve) and left-hand circular polarization (dashed curve) for ^2P electron capture in B^{3+}/He collisions as a function of scattering angle for $E = 1.5$ keV.

the direction of the internuclear axis. For most polarization measurements, it is more convenient to define the quantization axis with respect to an axis perpendicular to the collison plane. But it is straightforward to express the orientation and alignment parameters in terms of the scattering amplitudes obtained with respect to the molecular frame.

In accordance with customary conventions, in Fig. 49.14, the scattering plane contains the X and Z axes. The Y axis, perpendicular to the scattering plane, is taken to be the quantization axis. Let XYZ be the laboratory frame and xyz the body-fixed frame defined as above. The scattering amplitudes calculated in Sect. 49.4 are expressed with respect to the body-fixed frame xyz. The scattering amplitudes $f_{M_Y = \pm 1}$ in the laboratory frame are related to the amplitudes $f_{\Sigma z}$, $f_{\Pi_z^+}$ in the body-fixed frame by

$$f_{M_Y = \pm 1} = \frac{1}{\sqrt{2}}\left[f_{\Sigma z} \mp f_{\Pi_z^+}\right]. \quad (49.31)$$

The right-hand and left-hand circular polarizations (RHC and LHC respectively) are then defined as

$$RHC = |f_{M_Y = -1}|^2$$
$$= \frac{1}{2}\left[|f_{\Sigma z}|^2 + |f_{\Pi_z^+}|^2\right] + \Im\left[f_{\Sigma z}f^*_{\Pi_z^+}\right], \quad (49.32)$$
$$LHC = |f_{M_Y = +1}|^2$$
$$= \frac{1}{2}\left[|f_{\Sigma z}|^2 + |f_{\Pi_z^+}|^2\right] - \Im\left[f_{\Sigma z}f^*_{\Pi_z^+}\right], \quad (49.33)$$

and the circular polarization as

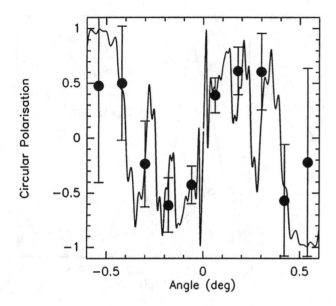

Figure 49.16. Circular polarization for ^2P electron capture in B^{3+}/He collisions as a function of scattering angle for $E = 1.5$ keV. The solid circles (with the error bars) are taken from Ref. [35].

$$L = \frac{RHC - LHC}{RHC + LHC} = \frac{2\Im\left[f_{\Sigma_z} f^*_{\Pi_z^+}\right]}{|f_{\Sigma_z}|^2 + |f_{\Pi_z^+}|^2}. \quad (49.34)$$

Figures 49.15 and 49.16 show the RHC, LHC and L quantities for an incident ion energy of 1.5 keV, where comparison with experiments [35,36] can be made. The strong propensity for orientation of the ^2P state is clearly exhibited for small scattering angle ($\vartheta < 0.2°$). At larger angles (smaller impact parameters), the propensity decreases and even reverses ($\vartheta > 0.4°$). We have also plotted on Fig. 49.15 the experimental data of Roncin *et al.* [35]. The agreement with experiment is very satisfactory.

REFERENCES

1. A. Dalgarno, S. E. Butler, Comments At. Mol. Phys. **7**, 129 (1978).
2. D. Pequignot, S. M. V. Aldrovandi, G. Stasinska, Astron. Astrophys. **63**, 313 (1978).
3. M. Gargaud, J. Hanssen, R. McCarroll, and P. Valiron, J. Phys. B **14**, 2259 (1981).
4. V. H. S. Kwong and Z. Fang, Phys. Rev. Lett. **71**, 4127 (1993).
5. T. K. McLaughlin, S. M. Wilson, R. W. McCullough, and H. B. Gilbody, J. Phys. B**23**, 737 (1990).
6. D. R. Bates and R. McCarroll, Proc. Roy. Soc. London A **245**, 175 (1958).
7. S. B. Schneiderman and A. Russek, Phys. Rev. A **181**, 311 (1969).
8. W. R. Thorson and J. B. Delos, Phys. Rev. A **18**, 117 (1978).
9. R. McCarroll and D. S. F. Crothers, Adv. At. Mol. Opt. Phys. **32**, 253 (1994).
10. M. Gargaud, R. McCarroll, and P. Valiron, J. Phys. B **20**, 1555 (1987).
11. L. F. Errea, L. Mendez, and A. Riera, J. Phys. B **15**, 2255 (1982).
12. L. F. Errea, C. Harel, H. Jouin, L. Mendez, B. Pons, and A. Riera, J. Phys. B **27**, 3603 (1994).
13. M. Gargaud, R. McCarroll, and L. Opradolce, J. Phys. B **21**, 521 (1988).
14. M. Gargaud, F. Fraija, M. C. Bacchus-Montabonel, and R. McCarroll, J. Phys. B **29**, 179 (1994).
15. P. Honvault, M. C. Bacchus-Montabonel, and R. McCarroll, J. Phys. B **27**, 3115 (1994).
16. P. Honvault, M. Gargaud, and R. McCarroll, Astron. Astrophys. (to be published).
17. D. L. Cooper, M. J. Ford, J. Gerratt, and M. Raimondi, Phys. Rev. A **34**, 1752 (1986).
18. S. Bienstock, T. G. Heil, and A. Dalgarno, Phys. Rev. A **25**, 2850 (1982).
19. C. Bottcher and A. Dalgarno, Proc. Roy. Soc. A **340**, 187 (1974).
20. P. Valiron, R. Gayet, R. McCarroll, F. Masnou-Seeuws, and M. Philippe, J. Phys. B **12**, 53 (1979).
21. R. Grice and D. R. Herschbach, Mol. Phys. **27**, 159 (1974).
22. M. Gargaud and R. McCarroll, Phys. Scr. **51**, 752 (1995).
23. M. Gargaud, R. McCarroll, and L. Opradolce, Astron. Astrophys. **208**, 251 (1989).
24. L. F. Errea, B. Herrero, L. Méndez, O. Mó, and A. Riera, J. Phys. B **24**, 4049 (1991).
25. R. E. Olson and A. Salop, Phys. Rev. A **14**, 579 (1976).
26. M. Kimura *et al.*, J. Phys. Soc. (Japan) **53**, 2224 (1984).
27. K. Taulbjerg, J. Phys. B **19**, L367 (1986).
28. M. Gargaud, Thèse d'État, Université de Bordeaux (France) (1987).
29. S. E. Butler and A. Dalgarno, Astrophys. J. **241**, 838 (1980).
30. F. T. Smith, Phys. Rev. **179**, 111 (1969).
31. B. R. Johnson, J. Comput. Phys. **13**, 445 (1973).
32. P. Valiron, Thèse de 3ème cycle, Université de Bordeaux (France) (1976).
33. F. Fraija, M. C. Bacchus-Montabonel, and M. Gargaud, Z. Phys. D **29**, 179 (1994).
34. M. Gargaud, R. McCarroll, and P. Valiron, Astron. Astrophys. **106**, 197 (1982).
35. P. Roncin, C. Adjouri, N. Andersen, M. Barat, A. Dubois, M. N. Gaboriaud, J. P. Hansen, S. E. Nielsen, and S. Z. Szilagyi, J. Phys. B **27**, 3079 (1994).
36. P. Roncin, C. Adjouri, M. N. Gaboriaud, L. Guillemot, M. Barat, and N. Andersen, Phys. Rev. Lett. **65**, 3261 (1990).
37. M. Gargaud, M. C. Bacchus-Montabonel, T. Grozdanov, and R. McCarroll, J. Phys. B **27**, 4675 (1994).

50

Continuum Distorted Wave and Wannier Methods

D. S. F. Crothers, F. B. M. Copeland, and J. T. Glass

The Queen's University of Belfast, Belfast BT7 1NN, Northern Ireland

50.1 INTRODUCTION

The continuum distorted wave model has been extensively applied to charge transfer and ionization processes. We present both the perturbative and variational capture theories as well as highlighting the suitability of this model in describing the continuum final states in both heavy and light particle ionization. We then develop the Wannier theory for threshold ionization, and further theoretical work which led to the modern semiclassical approximation. This very successful theory has provided the first absolute cross sections which are in good agreement with experiment.

50.2 CONTINUUM DISTORTED WAVE METHOD

50.2.1 Perturbation Theory

Continuum distorted wave theory (CDW) is one of the most advanced and complete perturbative theories of heavy particle collisions which has been formulated to date. It was originally introduced by Cheshire [1] to model the process of charge transfer during the collision of an atom/ion with an ion [specifically the resonant process of $p + H(1s) \rightarrow H(1s) + p$]. These types of three-body collisions are made amenable to the perturbative approach when the ratio of the projectile impact velocity v to the electron initial bound state mean velocity v_b satisfies

$$\frac{v}{v_b} \gtrsim 3 \,. \tag{50.1}$$

The criterion for nonrelativistic collisions, in which electron capture is a dominant process, is that both v and v_b are small compared with the speed of light. The theoretical description of collisions which involve the disturbance of a bound electron of mass m_e by the field of a fast moving heavy particle of mass M can be greatly simplified by exploiting the fact that since the ratio m_e/M is so small, the heavy particle follows a straight-line trajectory throughout the collision.

This allows the parametrization of the internuclear vector $\mathbf{R}$ in terms of an impact parameter $\mathbf{b}$, such that

$$\mathbf{R} = \mathbf{b} + \mathbf{v}t \,. \tag{50.2}$$

This impact parameter picture (IPP) of the collision is equivalent to the full quantal or wave treatment when the eikonal criterion for small angle scattering is satisfied [2].

It has become standard to work in a generalized nonorthogonal coordinate system in which the vectors $\mathbf{r}_T$ ($\mathbf{r}_P$) from the target (projectile) to the electron are treated, along with $\mathbf{R}$, as independent variables [3]. Working in the frame centered on the target nucleus and using atomic units, the Lagrangian is given by

$$H - i\frac{d}{dt_{\mathbf{r}_T}} = -\frac{1}{2}\nabla^2_{\mathbf{r}_T} + V_T(\mathbf{r}_T) - i\frac{\partial}{\partial t} - \frac{1}{2}\nabla^2_{\mathbf{r}_P} + V_P(\mathbf{r}_P)$$
$$+ i\mathbf{v}\cdot\boldsymbol{\nabla}_{\mathbf{r}_P} + V_{TP}(\mathbf{R}) - \boldsymbol{\nabla}_{\mathbf{r}_P}\cdot\boldsymbol{\nabla}_{\mathbf{r}_T}\,, \quad (50.3)$$

where $d/dt_{\mathbf{r}_T}$ refers to differentiation with respect to t, keeping $\mathbf{r}_T$ fixed, and where

$$V_T = \frac{-Z_T}{r_T}, \quad V_P = \frac{-Z_P}{r_P}, \quad V_{TP} = \frac{Z_T Z_P}{R}\,. \quad (50.4)$$

Since these potentials are pure Coulomb potentials, they continue to affect the relevant wave functions even at infinity. These long range Coulomb boundary conditions are defined in Eq. (32.10). The $+i\mathbf{v}\cdot\boldsymbol{\nabla}_{\mathbf{r}_P}$ term gives rise to the Bates-McCarroll electron translation factors which are required to satisfy Galilean invariance.

The Lagrangian above has been written in such a way as to highlight the three two-body decompositions exploited in CDW, with the $-\boldsymbol{\nabla}_{\mathbf{r}_P}\cdot\boldsymbol{\nabla}_{\mathbf{r}_T}$ term, the so-called nonorthogonal kinetic energy, coupling the systems. The essence of CDW is to treat the bound electron as simultaneously being in the continuum of the other heavy particle.

The initial wave function can be written as

$$\xi_i^{\pm} = D_{-\mathbf{v}}^{\pm}(\mathbf{r}_P)\Phi_i(\mathbf{r}_T,t)C(\mathbf{R},t)\,, \quad (50.5)$$

where $D_{-\mathbf{v}}^{\pm}$ is the distortion from the projectile, and C is due to the internuclear potential, V_{TP}. The bound state $\Phi_i = \phi_i(\mathbf{r}_T)\exp[-i\epsilon_i t]$, where $\phi_i(\mathbf{r}_T)$ is the initial eigenstate, and ϵ_i is the initial eigenenergy.

In this form, the action of the Lagrangian can be split into three separate differential equations plus a residual interaction. This gives the following solutions: for the distortion D,

$$D_{-\mathbf{v}}^{+} = N(\zeta_P){}_1F_1(i\zeta_P;1;i\mathbf{v}\cdot\mathbf{r}_P + ivr_P)\,,$$
$$D_{\mathbf{v}}^{-} = (D_{-\mathbf{v}}^{+})^*\,, \quad (50.6)$$

with

$$N(\zeta) = \exp(\pi\zeta/2)\Gamma(1-i\zeta), \quad \zeta_{T,P} = Z_{T,P}/v; \quad (50.7)$$

and for the internuclear function C,

$$C(\mathbf{R},t) = \exp\left[i\frac{Z_P Z_T}{v}\ln(vR - v^2 t)\right]\,. \quad (50.8)$$

Similarly it can be shown that

$$\xi_f^{\pm} = D_{\mathbf{v}}^{\pm}(\mathbf{r}_T)\Phi_f(\mathbf{r}_P,t)C^*(\mathbf{R},-t)\exp\left(i\mathbf{v}\cdot\mathbf{r}_T - i\frac{v^2}{2}t\right)\,, \quad (50.9)$$

where $i\mathbf{v}\cdot\mathbf{r}_T - i\frac{v^2}{2}t$ results from the Galilean transformation to the target frame. The superscripts plus and minus refer to outgoing and incoming Coulomb boundary conditions respectively. These are determined by the asymptotic form of the wave functions

$$\lim_{t\to-\infty}\xi_i^+ \sim \Phi_i(\mathbf{r}_T,t)\exp\left[i\frac{Z_P(Z_T-1)}{v}\ln(vR-v^2t)\right]\,, \quad (50.10)$$

and

$$\lim_{t\to+\infty}\xi_f^- \sim \Phi_f(\mathbf{r}_P,t)\exp\left(i\mathbf{v}\cdot\mathbf{r}_T - i\frac{v^2}{2}t\right)$$
$$\times \exp\left[-i\frac{Z_T(Z_P-1)}{v}\ln(vR+v^2t)\right]\,. \quad (50.11)$$

Of course, ξ_i^+ and ξ_f^- are not exact solutions of the three-body Schrödinger equation; in fact,

$$\left(H - i\frac{d}{dt_{\mathbf{r}_T}}\right)\xi_i^+ = W_i\xi_i^+ = -\boldsymbol{\nabla}_{\mathbf{r}_P}D_{-\mathbf{v}}^+\cdot\boldsymbol{\nabla}_{\mathbf{r}_T}\Phi_i\,, \quad (50.12)$$

and

$$\left(H - i\frac{d}{dt_{\mathbf{r}_T}}\right)\xi_f^- = W_f\xi_f^-$$
$$= -e^{i\mathbf{v}\cdot\mathbf{r}_T - i\frac{v^2}{2}t}\,\boldsymbol{\nabla}_{\mathbf{r}_T}D_{\mathbf{v}}^-\cdot\boldsymbol{\nabla}_{\mathbf{r}_P}\phi_f\,. \quad (50.13)$$

The CDW transition amplitude is written as

$$A_{if} = -i\int_{-\infty}^{+\infty}dt\langle\xi_f^- \mid T_{CDW} \mid \xi_i^+\rangle\,. \quad (50.14)$$

A perturbative expansion via the distorted wave Lippmann-Schwinger equation can be made for T_{CDW}, either in the post form

$$T_{CDW}^+ = W_f^\dagger(1 + G_V W_i) + T_{CDW}^+ G_i V G_V W_i\,, \quad (50.15)$$

or in the prior form

$$T_{CDW}^- = (1 + W_f^\dagger G_V)W_i + W_f^\dagger G_V V G_f T_{CDW}^-\,, \quad (50.16)$$

where the Green functions are given by

$$G_{i,f} = \left(i\frac{d}{dt_{\mathbf{r}_T}} - H + W_{i,f} + i\epsilon\right)^{-1}\,, \quad (50.17)$$

$$G_V = \left(i\frac{d}{dt_{\mathbf{r}_T}} - H + V + i\epsilon\right)^{-1}\,, \quad (50.18)$$

and V is any potential which ensures that the kernels of the integral equations for T_{CDW} are continuous [4].

By taking the first term in the expansions (50.15) and (50.16), we get the post and prior forms of the CDW1 amplitude as used by Cheshire. When calculating these amplitudes, the separable nature of the CDW wave function is best exploited by using Fourier transforms to move to the time-independent wave picture. A similar transformation is not suitable in the coupled channel approach discussed in Sect. 50.2.2.

Crothers [4], working in the wave treatment, has calculated the second order CDW2 amplitude using various approximations for the Green functions, and has shown that the CDW perturbation series has converged very well to first-order in most parts of the differential cross section. This is in contrast to the standard Born or Brinkman-Kramers approximations which do not start to converge until expanded to second-order. CDW1 is the only first-order perturbation theory, apart from asymmetric hybrid models derived from it, which produces a Thomas peak. Unfortunately, due to the accidental cancellation of the leading order terms, CDW1 has an extreme dip at the Thomas angle, a defect removed in CDW2 [5]. This is illustrated in Fig. 50.1 which also includes both folded and unfolded versions of the asymmetric Target CDW (TCDW) theory discussed below, as well as experimental data [6].

Anomalously large cross sections are obtained at low energies if the CDW wave function is not normalized at all times throughout the collision [3]. This is best demonstrated by the presence of the $N(\zeta)$ terms in

$$\lim_{t \to +\infty} \xi_i^+ = N(\zeta_{\mathrm{P}}) \Phi_i(\mathbf{r}_{\mathrm{T}}, t) \exp\left[i \frac{Z_{\mathrm{P}} Z_{\mathrm{T}}}{v} \ln(vR - v^2 t) \right]$$

(50.19)

and

$$\lim_{t \to -\infty} \xi_f^- = N^*(\zeta_{\mathrm{T}}) \Phi_f(\mathbf{r}_{\mathrm{P}}, t) \exp\left(i \mathbf{v} \cdot \mathbf{r}_{\mathrm{T}} - i \frac{v^2 t}{2} \right)$$
$$\times \exp\left[-i \frac{Z_{\mathrm{T}} Z_{\mathrm{P}}}{v} \ln(vR + v^2 t) \right] . \quad (50.20)$$

Using

$$\lim_{v \to 0} | N(Z/v) |^2 \sim \frac{2\pi Z}{v}$$

(50.21)

it is clear that the problem gets worse as v decreases. It can be corrected by defining

$$\hat{\xi}_{i,f}^{\pm} = \xi_{i,f}^{\pm} \langle \xi_{i,f}^{\pm} | \xi_{i,f}^{\pm} \rangle^{-1/2} .$$

(50.22)

Simpler distorted wave models can be generated through further approximations. Two of the most popular are Target CDW($D_{-\mathbf{V}}(\mathbf{r}_{\mathrm{P}}) \to 1$) and Projectile CDW($D_{\mathbf{V}}(\mathbf{r}_{\mathrm{T}}) \to 1$). These approximations are justified

when $Z_{\mathrm{T}} > Z_{\mathrm{P}}$ and $Z_{\mathrm{P}} > Z_{\mathrm{T}}$, respectively, and are particularly simple to calculate when the simple Born-like residual interaction is used rather than the full CDW form.

The asymptotic forms of the CDW wave functions can be used throughout the collision, ensuring normalization, and this leads to the eikonal or symmetric eikonal models.

50.2.2 Variational CDW

As the ratio of v/v_{b} decreases, perturbation theory starts to fail. This is due to the effective interaction time between the projectile and target atoms being long enough for strong three-body coupling. In this environment variational methods have proved successful. This procedure ensures both gauge invariance and unitarity – two fundamental attributes perturbation theory usually cannot guarantee. Continuing in the IPP, we use the Sil variational principle, which gives

$$\delta \int_{-\infty}^{+\infty} dt \langle \Psi | H - i \frac{d}{dt}{}_{\mathbf{r}_{\mathrm{T}}} | \Psi \rangle = 0 .$$

(50.23)

In the two-state CDW approximation we may assume

$$\Psi_{\mathrm{CDW}} = c_0(t) \xi_i^+ + c_1(t) \xi_f^-$$

(50.24)

subject to the boundary conditions $c_0(-\infty) = 1$ and $c_1(-\infty) = 0$. Variation of c_0^* and c_1^* gives the standard coupled equations

$$i N_{00} \dot{c}_0 + i N_{01} \dot{c}_1 = H_{00} c_0 + H_{01} c_1 , \quad (50.25)$$
$$i N_{10} \dot{c}_0 + i N_{11} \dot{c}_1 = H_{10} c_0 + H_{11} c_1 , \quad (50.26)$$

where

$$N_{00} = \langle \xi_i^+ | \xi_i^+ \rangle ,$$
$$N_{11} = \langle \xi_f^- | \xi_f^- \rangle ,$$
$$N_{01} = \langle \xi_i^+ | \xi_f^- \rangle = N_{10}^* ,$$

and

$$H_{00} = \langle \xi_i^+ | H - i \frac{d}{dt}{}_{\mathbf{r}_{\mathrm{T}}} | \xi_i^+ \rangle ,$$
$$H_{01} = \langle \xi_i^+ | H - i \frac{d}{dt}{}_{\mathbf{r}_{\mathrm{T}}} | \xi_f^- \rangle ,$$
$$H_{10} = \langle \xi_f^- | H - i \frac{d}{dt}{}_{\mathbf{r}_{\mathrm{T}}} | \xi_i^+ \rangle ,$$
$$H_{11} = \langle \xi_f^- | H - i \frac{d}{dt}{}_{\mathbf{r}_{\mathrm{T}}} | \xi_f^- \rangle .$$

Equations (50.25) and (50.26) can clearly be written as a matrix equation,

$$i \mathbf{N} \dot{\mathbf{c}} = \mathbf{H} \mathbf{c} ,$$

(50.27)

which is then easily generalized for larger expansions of Ψ. By using an orthogonalized basis set of normalized

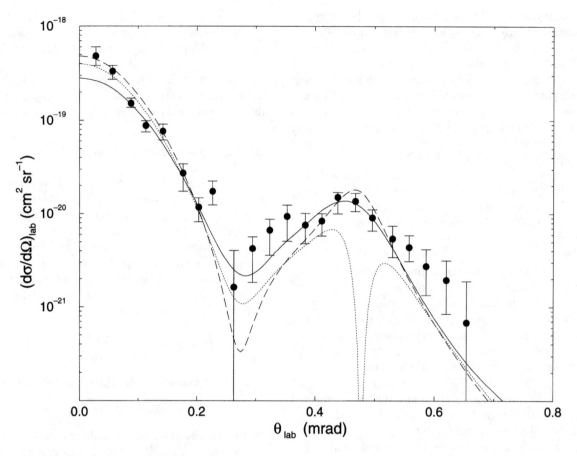

Figure 50.1. Differential cross sections for electron capture in the collision $H^+ + H(1s) \rightarrow H(1s) + H^+$ as a function of laboratory scattering angle (θ_{lab}) for impact energy of 5 MeV: —— TCDW, folded over the experimental resolution of Vogt *et al.* [6]. Unfolded theoretical results: - - - TCDW; · · · · CDW1. Experimental data; •, Vogt *et al.* [6].

functions in the manner of Löwdin [7, 8], the **N** matrix reduces to the unit matrix. This will be understood when considering expansions for Ψ from now on.

Another interesting, but potentially ruinous, result of the asymptotic forms (50.19) and (50.20) is their failure to obey the correct long-range Coulomb boundary conditions; compare this with their expressions at the opposite time extreme in Eqs. (50.10) and (50.11). This has no consequence until second-order VCDW is calculated in which divergent integrals arise as a direct result of this feature of the wave functions. These terms are analogous to the well-known intermediate elastic divergences which occur in Born-type expansions which do not have the correct Coulomb phases.

A novel way to avoid this problem [9, 10] is to split the time plane into two parts, allowing the well-behaved set $\{\xi^+\}$ to be used exclusively for $t \leq 0$, while the set $\{\xi^-\}$ forms the basis for $t \geq 0$. This phase integral halfway house VCDW is based on the factorization of the scattering matrix **S** into a product of two Møller matrices,

$$\mathbf{S} = \mathbf{\Omega}_-^\dagger \, \mathbf{\Omega}_+ , \qquad (50.28)$$

where $\mathbf{\Omega}_-^\dagger$ represents the propagation of the initial state from $t = -\infty$ to $t = 0_-$, while $\mathbf{\Omega}_+$ represents the propagation of the final state from $t = +\infty$ to $t = 0_+$.

The total wave function is similarly split into two expansions over an orthogonal basis ψ, with

$$\Psi^- = \mathbf{c}^- \psi^+ , \qquad t < 0 , \qquad (50.29)$$
$$\Psi^+ = \mathbf{c}^+ \psi^- , \qquad t > 0 , \qquad (50.30)$$

where the superscripts on the Ψ correspond to the respective heavy-particle motion. This in turn divides the coupled equations into two sets

$$i\dot{\mathbf{c}}^- = \mathbf{H}^{++} \mathbf{c}^- , \qquad t < 0 , \qquad (50.31)$$
$$i\dot{\mathbf{c}}^+ = \mathbf{H}^{--} \mathbf{c}^+ , \qquad t > 0 , \qquad (50.32)$$

where

$$\mathbf{H}^{\pm\pm} = \langle \Psi^\pm \mid H - i\frac{d}{dt}_{\mathbf{r}_\mathrm{T}} \mid \Psi^\pm \rangle . \qquad (50.33)$$

The coefficents $c^\pm$ then have to be matched over a local discontinuity in the total wave function at $t = 0$, such that $\mathbf{c}^+(0) = \mathbf{c}^-(0)$. Halfway house VCDW has all the appealing attribues of a variational theory but, by explicitly satisfying the long-range Coulomb boundary conditions, it is divergence free.

50.2.3 Ionization

CDW, by treating the Coulomb interactions to such a high degree, has obvious attractions for modeling the ionization process. Single ionization of an electron from an atom by a high-energy projectile is a perturbative process and the Born approximation will match experimental total cross sections pretty well. However CDW-like representations of the initial and final states generate better results at lower energies, as well as producing features in the differential cross sections which are beyond the reach of the first Born approximation.

In full CDW ionization theory, the initial state is given by the usual charge transfer wave function ξ_i [Eq. (50.5)], while the final state takes the form

$$
\begin{aligned}
\xi_f^- = (2\pi)^{-2/3} &\exp\left(i\mathbf{k}\cdot\mathbf{r}_\mathrm{T} - i\frac{k^2}{2}t\right) \\
&\times \exp\left[-i\frac{Z_\mathrm{T}Z_\mathrm{P}}{v}\ln(vR + \mathbf{v}\cdot\mathbf{R})\right] \\
&\times N^*(Z_\mathrm{T}/k)_1F_1(-iZ_\mathrm{T}/k; 1; -i\mathbf{k}\cdot\mathbf{r}_\mathrm{T} - ikr_\mathrm{T}) \\
&\times N^*(Z_\mathrm{P}/p)_1F_1(-iZ_\mathrm{P}/p; 1; -i\mathbf{p}\cdot\mathbf{r}_\mathrm{P} - ipr_\mathrm{P}),
\end{aligned}
\tag{50.34}
$$

where $\mathbf{k}$ ($\mathbf{p} = \mathbf{k} - \mathbf{v}$) is the momentum of the electron relative to the target (projectile) nucleus.

CDW ionization theories presented prior to 1982 produced spuriously large results due to the unnormalized initial state. Since the matrix element $\langle \xi_i^+ \mid \xi_i^+ \rangle$ is computationlly expensive to calculate as a function of $\mathbf{b}$ and t, a very successful alternative is to take the initial state as an eikonal distorted state [11], thus ensuring normalization. The initial state in this CDW-EIS theory is taken to be

$$
\xi_i^{\mathrm{EIS}} = \hat{D}_{-\mathbf{v}}^+(\mathbf{r}_\mathrm{P})\Phi_i(\mathbf{r}_\mathrm{T}, t)C(\mathbf{R}, t),
\tag{50.35}
$$

where now

$$
\hat{D}_{-\mathbf{v}}^+(\mathbf{r}_\mathrm{P}) = \exp\left[-i\frac{Z_\mathrm{P}}{v}\ln(vr_\mathrm{P} + \mathbf{v}\cdot\mathbf{r}_\mathrm{P})\right].
\tag{50.36}
$$

The final state remains as in Eq. (50.34).

The CDW final state is most effective when differential cross sections are studied. The most interesting features are in the forward $\theta = 0$ ejection angle; i.e., the soft collision peak ($k \simeq 0$), the capture to the continuum peak ECC ($k \simeq v$) and the binary encounter peak ($k \simeq 2v$). CDW theories are especially suited to the description of the ECC peak which results theoretically from the presence of the $N(Z_\mathrm{P}/p)$ factor in the wave function. This peak can be analyzed in detail via a multipole

expansion and much theoretical work has centered on the dipole parameter β. A negative β is strongly suggested by both experiment and physical intuition. CDW, CDW-EIS and halfway house VCDW all predict different values for β with the last theory being the only one which gives a high energy limit which remains negative [12].

Double ionization in general remains an extremely difficult area for theoretical models based on perturbative expansions – even those with explicit distortions built in. In this process, the explicit correlation between the electrons in the target atom are vitally important. However, one example of where CDW theory can overcome these problems is in the bound state wave function of Pluvinage [13]. This very successful treatment, which also includes a variationally determined parameter, is just the CDW analogy in bound state theory, although this appeared well before Cheshire's paper on scattering. The Pluvinage wave function for the ground state of a two-electron atom is given by

$$
\begin{aligned}
\phi = c(\kappa)&\left(\frac{Z_\mathrm{T}^3}{\pi}\right)\exp\left[-Z_\mathrm{T}(r_1 + r_2) + i\kappa r_{12}\right] \\
&\times {}_1F_1\left(1 + \frac{1}{2i\kappa}; 2; 2i\kappa r_{12}\right),
\end{aligned}
\tag{50.37}
$$

where $r_{1,2}$ are the distances of the two electrons from the target nucleus, and r_{12} is the interelectron separation. The constant κ is variationally determined to be 0.41, giving the normalization constant $c = (0.36405)^{1/2}$.

An analogous wave function for two electrons in the target continuum was derived [3] and implemented later in a successful CDW treatment of ionization by electrons and positrons [14]. This BBK theory, so named in deference to the authors, demonstrates the high resolution CDW final states obtained, right down to triply differential cross sections. However, this model suffers from the low-energy normalization problems associated with CDW, and is also unable to describe the threshold effects which are in the domain of Wannier theory.

50.3 WANNIER METHOD

50.3.1 The Wannier Threshold Law

In 1953, Wannier [15] deduced the relationship between the cross section at the threshold of a reaction, and the excess-of-threshold energy of the incident particle for a three-body ionization problem, where the final state consists of a residual unit positive charge and two electrons, with each body moving in the continuum of the other two. This extended to three bodies the earlier two-body threshold law derived by Wigner [16] (see Sect. 58.4.5).

For final states with $L=0$, Wannier employed hyperspherical coordinates $(\rho, \alpha, \theta_{12})$, where

$$\rho^2 = r_1^2 + r_2^2, \quad \alpha = \tan^{-1}\left(\frac{r_2}{r_1}\right), \quad \theta_{12} = \cos^{-1}(\hat{\mathbf{r}}_1 \cdot \hat{\mathbf{r}}_2).$$

$$(50.38)$$

Here we assume that the residual ion is infinitely massive and at rest with respect to the two electrons. By converting the two-electron problem to the case of motion of a single point in six-dimensional space, we can take the hyperradius ρ as the 'size' of the hypersphere, α as the radial correlation of the electrons and θ_{12} as their angular correlation. This allows the Schrödinger equation for the final state to be written as (in a.u.)

$$\left[h_0 - \frac{l^2(\hat{\mathbf{r}}_1)}{\rho^2 \cos^2 \alpha} - \frac{l^2(\hat{\mathbf{r}}_2)}{\rho^2 \sin^2 \alpha} + 2E \right.$$
$$\left. + \frac{2Z(\alpha, \theta_{12})}{\rho}\right] \Psi(\mathbf{r}_1, \mathbf{r}_2) = 0,$$
$$(50.39)$$

where

$$h_0 = \frac{1}{\rho^5}\frac{\partial}{\partial \rho}\left(\rho^5 \frac{\partial}{\partial \rho}\right)$$
$$+ \frac{1}{\rho^2 \sin^2 \alpha \cos^2 \alpha}\frac{\partial}{\partial \alpha}\left(\sin^2 \alpha \cos^2 \alpha \frac{\partial}{\partial \alpha}\right) \quad (50.40)$$

and

$$Z(\alpha, \theta_{12}) = \frac{1}{\cos \alpha} + \frac{1}{\sin \alpha} - \frac{1}{(1 - \cos \theta_{12} \sin 2\alpha)^{\frac{1}{2}}},$$
$$(50.41)$$

is the potential surface on which the particle is moving [17].

The most likely configuration of the electrons leading to double escape at threshold corresponds to the region $\mathbf{r}_1 = -\mathbf{r}_2$, i.e., the two electrons escape in opposite directions from the reaction zone, corresponding to the saddle point of the potential surface $Z(\alpha, \theta_{12})$, and defined as the Wannier ridge. Also, dynamic screening between the two electrons means there would be equal partitioning of the available energy for the two particles, and so they would have equal but opposite velocities on escape. In hyperspherical coordinates, the most important region for double escape is, therefore, $\alpha = \pi/4$, and $\theta_{12} = \pi$.

The main conclusion of Wannier's theory is that the total cross section for electron impact single ionization scales as

$$\sigma = kE^{m_{12}},$$
$$(50.42)$$

with

$$m_{12} = -\frac{1}{4} + \frac{1}{4}\left(\frac{100Z - 9}{4Z - 1}\right)^{\frac{1}{2}},$$
$$(50.43)$$

and where E is the excess-of-threshold energy; and m_{12} is 1.127 for unit residual charge, with $m_{12} \rightarrow 1$ as $Z \rightarrow \infty$.

50.3.2 Peterkop's Semiclassical Theory

The Wannier threshold law has been verified both semiclassically [18] and quantum mechanically [19]. Peterkop [18] adopted a semiclassical JWKB approach to the problem, by using the three-dimensional WKB ansatz

$$\Psi_0 = P^{\frac{1}{2}} \exp\left(\frac{iS}{\hbar}\right),$$
$$(50.44)$$

for the final-state wave function, where S and P are, respectively, the solutions of the Hamilton-Jacobi equation,

$$(\nabla_1 S)^2 + (\nabla_2 S)^2 = 2(E - V),$$
$$(50.45)$$

and the continuity equation,

$$\nabla_1(P\nabla_1 S) + \nabla_2(P\nabla_2 S) = 0.$$
$$(50.46)$$

In hyperspherical coordinates, these equations become

$$\left(\frac{\partial S}{\partial \rho}\right)^2 + \frac{1}{\rho^2}\left(\frac{\partial S}{\partial \alpha}\right)^2 + \frac{4}{\rho^2 \sin^2 2\alpha}\left(\frac{\partial S}{\partial \theta_{12}}\right)^2 = 2E + \frac{2Z}{\rho}$$
$$(50.47)$$

and

$$D_0\left(P\frac{\partial S}{\partial \rho}\right) + \frac{1}{\rho^2}\left(D_1\left(P\frac{\partial S}{\partial \alpha}\right) + D_2\left(\frac{\partial S}{\partial \theta_{12}}\right)\right) = 0,$$
$$(50.48)$$

where

$$D_0 f = \frac{1}{\rho^5}\frac{\partial}{\partial \rho}(\rho^5 f),$$
$$(50.49)$$

$$D_1 f = \frac{1}{\sin^2 2\alpha}\frac{\partial}{\partial \alpha}(f \sin^2 2\alpha),$$
$$(50.50)$$

$$D_2 f = \frac{4}{\sin^2 2\alpha}\frac{1}{\sin \theta_{12}}\frac{\partial}{\partial \theta_{12}}(f \sin \theta_{12}). \quad (50.51)$$

Following Wannier's hypothesis, solutions of these equations are found in the region $\alpha = \pi/4, \theta_{12} = \pi$. Taking the Taylor expansion for $Z(\alpha, \theta_{12})$ as

$$Z(\alpha, \theta_{12}) = Z_0 + \frac{1}{2}Z_1(\Delta\alpha)^2 + \frac{1}{8}Z_2(\Delta\theta_{12})^2 + \cdots,$$
$$(50.52)$$

where $\Delta\alpha = \alpha - \pi/4$ and $\Delta\theta_{12} = \theta_{12} - \pi$, it follows from (50.41) that

$$Z_0 = \frac{3}{\sqrt{2}}, \quad Z_1 = \frac{11}{\sqrt{2}}, \quad Z_2 = -\frac{1}{\sqrt{2}}.$$
$$(50.53)$$

Similarly, taking the solution of (50.47) in the form

$$S = S_0(\rho) + \frac{1}{2}S_1(\rho)(\Delta\alpha)^2 + \frac{1}{8}S_2(\rho)(\Delta\theta_{12})^2 + \cdots$$
$$(50.54)$$

gives

$$\frac{dS_0}{d\rho} = \omega\,, \qquad (50.55)$$

$$\omega \frac{dS_i}{d\rho} + \frac{S_i^2}{\rho^2} = \frac{Z_i}{\rho}\,, \quad i = 1, 2\,, \qquad (50.56)$$

where $\omega = (2E + 2Z_0/\rho)^{\frac{1}{2}}$. The solutions are

$$S_0 = \rho\omega + \frac{Z_0}{\chi} \ln \frac{\rho(\chi + \omega)^2}{2Z_0}\,, \qquad (50.57)$$

$$S_i = \rho^2 \omega \frac{1}{u_i} \frac{du_i}{d\rho}\,, \quad i = 1, 2\,, \qquad (50.58)$$

where $\chi = (2E)^{\frac{1}{2}}$ and

$$u_i = C_{i1} u_{i1} + C_{i2} u_{i2}\,, \qquad (50.59)$$

$$u_{ij} = \rho^{m_{ij}}\,_2F_1\left(m_{ij}, m_{ij} + 1; 2m_{ij} + \frac{3}{2}; \frac{-E\rho}{Z_0}\right)\,, \qquad (50.60)$$

$$m_{i1} = -\frac{1}{4} - \frac{1}{2}\mu_i\,, \quad m_{i2} = -\frac{1}{4} + \frac{1}{2}\mu_i\,, \qquad (50.61)$$

$$\mu_i = \frac{1}{2}\left(1 + \frac{8Z_i}{Z_0}\right)^{\frac{1}{2}}\,. \qquad (50.62)$$

where, for $i = 2$, the principal branch is understood.

Expanding P in the same form as S, and restricting the solution to finding P_0, gives

$$P_0 = \frac{C}{\rho^5 \omega u_1 u_2^2}\,, \qquad (50.63)$$

where $C \sim C_{12} E^{1-m_{12}}$. By solving these equations, Peterkop extracted the Wannier cross section behavior by matching the exact wave function with an approximate one for which the energy dependence is known at some arbitrarily finite value r_0 of ρ, giving the total cross section as

$$\sigma_{\text{tot}} \sim E^{1.127}\,, \qquad (50.64)$$

as required. However, neither this method of Peterkop nor the quantum mechanical approach of [19] were able to deduce the constant of proportionality.

50.3.3   The Semiclassical Approximation

As can be seen from the form of Eq. (50.63) for P_0, u_2 vanishes in the double limit $\rho \to +\infty, E \to 0$, and so, the semiclassical theory breaks down at the very configuration of importance. To avoid this problem, Crothers [20] adopted a change of the dependent variable to obtain a uniform JWKB approximation. Taking $\alpha = \pi/4$, (i.e., $\Delta\alpha = 0$) as the natural barrier, he set the final-state wave function as

$$\Psi^{-*} = \frac{x|\sin(\alpha - \pi/4)|^{1/2}}{\rho^{5/2}\sin\alpha\cos\alpha(\sin\theta_{12})^{1/2}}\,, \qquad (50.65)$$

so that

$$\left[\frac{\partial^2}{\partial\rho^2} + \frac{1}{\rho^2 \sin|\Delta\alpha|}\frac{\partial}{\partial\alpha}\left(\sin|\Delta\alpha|\frac{\partial}{\partial\alpha}\right)\right.$$
$$+ \frac{1}{\rho^2 \sin^2\alpha\cos^2\alpha}\frac{\partial^2}{\partial\theta_{12}^2} + 2E + \frac{2Z}{\rho}$$
$$\left.+ \frac{(\frac{1}{4} + \frac{1}{4}\csc^2\Delta\theta_{12})}{\rho^2 \sin^2\alpha\cos^2\alpha} - \frac{\csc^2|\Delta\alpha|}{4\rho^2}\right] x = 0\,, \quad (50.66)$$

where $|\Delta\alpha|$ and θ_{12} are, respectively, the polar and azimuthal angles. Near $\theta_{12} = \pi$ and $\alpha = \pi/4$, (i.e., at $\Delta\theta_{12} = \Delta\alpha = 0$), the term $(4\rho^2)^{-1}$ is negligible compared with $\csc^2\theta_{12}/(4\rho^2)$. Also the θ_{12} pseudopotential is clearly attractive while the α potential is repulsive, and both potentials are large just at the region of importance, i.e., at $\Delta\alpha = 0 = \Delta\theta_{12}$.

Again, following the method of Peterkop, the final state wave function is written in the form (50.44), (50.47) and (50.48), where now the action perturbation expression S must be generalized to

$$S = s_0 \ln|\Delta\alpha| + s_1 \ln(\Delta\theta_{12}) + S_0(\rho)$$
$$+ \frac{1}{2}S_1(\rho)(\Delta\alpha)^2 + \frac{1}{8}S_2(\rho)(\Delta\theta_{12})^2 + \dots\,, \qquad (50.67)$$

where the extra logarithmic phases indicate long-range Coulomb potentials. By applying the Kohn variational principle perturbatively, and invoking the Jeffreys' [21] connection formula on the Wannier ridge with $\rho = 0$ as the classical turning point, the final state wave function is [20]

$$\Psi_f^{-*} = \frac{c^{1/2}E^{m_{12}/2}\rho^{m_{12}/2+\frac{1}{4}}(2Z_0)^{1/4}(-\chi/2\pi)^{1/2}}{(2Z_0/\rho)^{1/4}\rho^{5/2}\sin\alpha\cos\alpha}$$
$$\times \delta(\hat{\mathbf{k}}_1 - \hat{\mathbf{r}}_1)\delta(\hat{\mathbf{k}}_2 - \hat{\mathbf{r}}_2)\exp\left[4i(8Z_0\rho)^{-\frac{1}{2}}(\Delta\theta_{12})^{-2}\right]$$
$$\times \left\{\exp\left[-i(8Z_0\rho)^{1/2} - \frac{1}{2}i(\Delta\alpha)^2(2Z_0\rho)^{1/2}m_{12}\right.\right.$$
$$\left.\left.- \frac{1}{8}i(\Delta\theta_{12})^2(2Z_0\rho)^{1/2}m_{21} - \frac{1}{4}i\pi\right] - \text{c.c}\right\}\,, \qquad (50.68)$$

where $\chi = 2\pi\Im(m_{21})$.

Taking the total cross section for distinguishable particles as

$$\sigma = \frac{\pi^2 a_0^2}{k_0}\iint d\hat{\mathbf{k}}_1 d\hat{\mathbf{k}}_2 \frac{\pi Z_2 \tanh\chi}{(2E)^{1/2}}|f(\hat{\mathbf{k}}_1, \hat{\mathbf{k}}_2)|^2$$
$$\times \exp\left(\frac{-Z_2}{4(2E)^{1/2}}(\Delta\Theta_{12})^2\pi\tanh\chi\right)\,, \qquad (50.69)$$

where f is the scattering amplitude, then the corresponding triple-differential cross section is

$$\frac{d^3\sigma}{d\hat{\mathbf{k}}_1 d\hat{\mathbf{k}}_2 d(\frac{1}{2}k_1^2)} = \frac{2\pi^2 a_0^2}{k_0}\frac{d}{dE}\frac{\pi Z_2 \tanh\chi}{(2E)^{\frac{1}{2}}}$$
$$\times \exp\left[\frac{-Z_2}{4(2E)^{1/2}}(\Theta_{12} - \pi)^2\pi\tanh\chi\right]$$
$$\times |f(\hat{\mathbf{k}}_1, \hat{\mathbf{k}}_2)|^2\,. \qquad (50.70)$$

Assuming the contribution from triplet states is negligible, $|f|^2$ can be written as

$$\frac{1}{4}\left| f(\hat{\mathbf{k}}_1, \hat{\mathbf{k}}_2) + f(\hat{\mathbf{k}}_2, \hat{\mathbf{k}}_1) \right|^2, \qquad (50.71)$$

where $f(\hat{\mathbf{k}}_1, \hat{\mathbf{k}}_2)$ [and by permutation $f(\hat{\mathbf{k}}_2, \hat{\mathbf{k}}_1)$, hereafter referred to as f and g respectively] is given by

$$f \simeq \frac{2i}{\pi} \int d\mathbf{r}_1 d\mathbf{r}_2 d\mathbf{r}_3 \Psi_f^{-*} \phi_f(2, \mathbf{r}_3)$$
$$\times (H - E) \exp(i\mathbf{k}_0 \cdot \mathbf{r}_1) \psi_i(\mathbf{r}_2, \mathbf{r}_3). \qquad (50.72)$$

As a test of the above formulation for the process $e^- + \text{He} \rightarrow \text{He}^+ + 2e^-$ near the ionization threshold, Crothers [20] used an independent-electron open-shell wave function for the initial bound state helium target, written as

$$\psi_i(\mathbf{r}_2, \mathbf{r}_3) = \frac{\phi(\mathbf{r}_2, z_o)\phi(\mathbf{r}_3, \beta) + \phi(\mathbf{r}_3, z_o)\phi(\mathbf{r}_2, \beta)}{[2(1+S)]^{1/2}},$$
$$(50.73)$$

where

$$\phi(\mathbf{r}, z_0) = z_0^{3/2} \pi^{-1/2} \exp(-z_0 r), \qquad (50.74)$$

$$S = \left[\int \phi(\mathbf{r}, z_0)\phi(\mathbf{r}, \beta) d\mathbf{r} \right]^2$$
$$= \left(\frac{4z_0 \beta}{(z_0 + \beta)^2} \right)^3, \qquad (50.75)$$

and z_0 and β take the physical values $z_0 = 1.8072^{1/2}$ and $\beta = 2$. The total singlet cross section was found to be (in atomic units)

$$\sigma = 2.37 E^{m_{12}} a_0^2, \qquad (50.76)$$

in line with Wannier's threshold law, and with experiment [24], while the corresponding absolute triple differential cross sections (TDCS) were expressed as

$$\frac{d^3\sigma}{d\hat{\mathbf{k}}_1 d\hat{\mathbf{k}}_2 d(\frac{1}{2}k_1^2)} = \frac{70cz_0^2 2^{1/2} \chi \tanh \chi}{\pi R_\infty Z_0^{1/2} m_{12}} |f + g|^2$$
$$\times \left\{ \frac{d}{dE} E^{m_{12}-1/2} \exp\left[\frac{-Z_2(\Theta_{12} - \pi)^2 \pi \tanh \chi}{4(2E)^{1/2}} \right] \right\}$$
$$(50.77)$$

in units of $10^{-19}\,\text{cm}^2\,\text{sr}^{-2}\,\text{eV}^{-1}$, where

$$c = \frac{\Gamma(m_{12} + 3/2)\Gamma(m_{12} + 1)}{2\pi Z_0^{m_{12}} \Gamma(2m_{12} + 3/2)}, \qquad (50.78)$$

and where f (and similarly g for Θ_2) is given by

$$f = \int_0^\infty d\rho\, \rho^{3/2 + m_{21}/2 + 1/4} \sum_{L=0}^{L_{max}} i^L (2L+1) j_L \left(\frac{\rho z_0}{2^{1/2}} \right)$$
$$\times P_L(\cos \Theta_1) \exp[\tfrac{1}{8} \Im m_{21}(\Theta_{12} - \pi)^2 (2Z_0 \rho)^{1/2}]$$
$$\times 2\cos[(8Z_0 \rho)^{1/2} + \tfrac{1}{8} \Re m_{21}(\Theta_{12} - \pi)^2 (2Z_0 \rho)^{1/2}]$$
$$\times r(\rho, \Theta_{12}) \qquad (50.79)$$

with

$$(1+S)^{1/2} r(\rho, \Theta_{12})$$
$$= \exp\left(-\frac{\rho z_0}{2^{1/2}}\right)\left(2^{1/2} z_0 - \frac{1}{(1 - \cos\Theta_{12})^{1/2}} \right)$$
$$+ \exp(-2^{1/2}\rho)\left[\frac{64(2)^{1/2}(z_0 - 1)}{(z_0 + 2)^3} + \frac{32}{(z_0 + 2)^3} \right.$$
$$\left. \times [2(2)^{1/2} + (z_0 + 2)\rho] \exp\left(\frac{-\rho(z_0 + 2)}{2^{1/2}} \right) \right]. \qquad (50.80)$$

These results have been found to compare favorably with both the relative experimental results of [25] and the absolute experimental results of [23]. The Crothers semiclassical approximation has been successfully applied to other threshold (e, 2e) and (photon, 2e) collisions, namely two-electron photodetachment from H^- [26], $\text{He}^-(^4P_{5/2}^0)$ [27] and K^- [28]. Further investigations of the TDCS for helium at threshold have since been carried out. The $^3P^0$ triplet contribution to the TDCS was studied in [29], where small but notable improvements in the comparison with experimental results [25] and [23] were achieved for most configurations of the angles θ_1 and θ_2. The inclusion of contributions from $^3D^{e,0}$ or $^3F^0$ to the absolute TDCS were found to be negligible in comparison with the effect of the $^3P^0$ [30], although the admittedly non-Wannier effective-charge investigation of $\theta_{12} = \pi$ by Pan and Starace [31] suggests that $^3F^0$ may be important at $\theta_1 = \pi/6$ or $5\pi/6$, in line with the experiment of Rösel et al. [23]. Another aspect which is thought to contribute to the TDCS is explicit correlation in the initial bound state wave function for the helium target, in which the interelectronic distance r_{23} is explicitly contained. Absolute singlet triple-differential cross sections have been obtained [22], using a helium ground state wave function developed by Le Sech [32]. Again, excellent agreement with the singlet results of [20] has been achieved, and, in most configurations, notable improvements with the corresponding experimental data are obtained, as shown in Fig. 50.2 for scattering angle $\theta_1 = 60^o$ (a) and 90° (b), indicating that electron correlation should also be considered if a full picture of threshold ionization is to be achieved.

REFERENCES

1. I. M. Cheshire, Proc. Phys. Soc. **84**, 89 (1964).
2. B. H. Bransden and M. R. C. McDowell, *Charge Exchange and the Theory of Ion-Atom Collisions* (Clarendon Press, Oxford, 1992).
3. D. S. F. Crothers, J. Phys. B **15**, 2061 (1982).
4. D. S. F. Crothers, J. Phys. B **18**, 2893 (1985).
5. D. S. F. Crothers and J. F. McCann, J. Phys. B **17**, L177 (1984).
6. H. Vogt, R. Schuch, E. Justiniano, M. Schultz, and W. Schwab, Phys. Rev. Lett. **57**, 2256 (1986).
7. P. O. Löwdin, Ark. Nat. Astron. Phys. **35A**, 918 (1947).

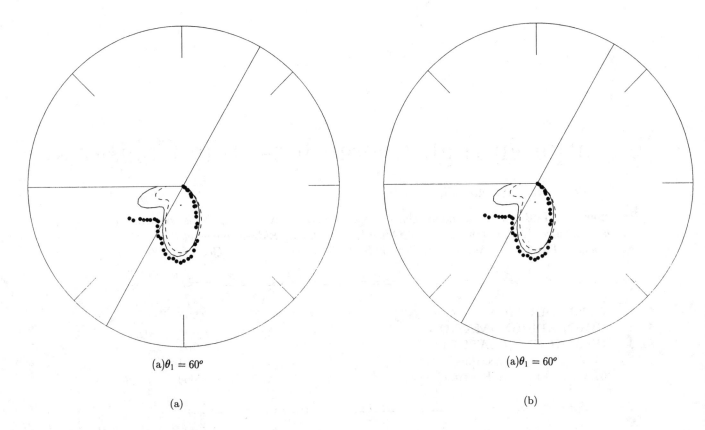

(a)$\theta_1 = 60^o$ (a)$\theta_1 = 60^o$

(a)   (b)

Figure 50.2. Helium triply differential ionization cross section for coplanar geometry, $E_1 = E_2 = 1$ eV and $E = 2$ eV, calculated (full curve) TDCS of Copeland and Crothers [22] in polar coordinates, with polar coordinates as θ_2, for scattering angles $\theta_1 = 60°$ (a) and $\theta_1 = 90°$ (b), in comparison with absolute experimental (•) data of Rösel *et al.* [23], and theoretical results (broken curve) of Crothers [20]. The radius of each circle is 1.0×10^{-19} cm^2 sr^{-2} eV^{-1}.

8. P. O. Löwdin, J. Chem. Phys. **18**, 365 (1950).
9. D. S. F. Crothers, Nucl. Instrum. Meth. B **27**, 555 (1987).
10. D. S. F. Crothers and L. J. Dube, Adv. At. Mol. Phys. **30**, 287 (1993).
11. D. S. F. Crothers and J. F. McCann, J. Phys. B **16**, 3229 (1993).
12. D. S. F. Crothers and S. F. C. O'Rourke, J. Phys. B **25**, 2351 (1992).
13. P. Pluvinage, Ann. Phys. N. Y. **5**, 145 (1950).
14. M. Brauner, J. S. Briggs, and H. Klar, J. Phys. B **22**, 2265 (1989).
15. G. H. Wannier, Phys. Rev. **90**, 817 (1953).
16. E. P. Wigner, Phys. Rev. **73**, 1002 (1948).
17. M. R. H. Rudge and M. J. Seaton, Proc. R. Soc. A **283**, 262 (1965).
18. R. K. Peterkop, J. Phys. B **4**, 513 (1971).
19. A. R. P. Rau, Phys. Rev. A **4**, 207 (1971).
20. D. S. F. Crothers, J. Phys. B **19**, 463 (1986).
21. H. Jeffreys, Proc. Lond. Math. Soc. **23**, 428 (1923).
22. F. B. M. Copeland and D. S. F. Crothers, J. Phys. B **27**, 2039 (1994).
23. T. Rösel, J. Röder, L. Frost, K. Jung, and H. Ehrhardt, J. Phys. B **25**, 3859 (1992).
24. S. Cvejanović and F. H. Read, J. Phys. B **7**, 1841 (1974).
25. P. Selles, A. Huetz, and J. Mazeau, J. Phys. B **20**, 5195 (1987).
26. J. F. McCann and D. S. F. Crothers, J. Phys. B **19**, L399 (1986).
27. D. S. F. Crothers and D. J. Lennon, J. Phys. B **21**, L409 (1988).
28. D. R. J. Carruthers and D. S. F. Crothers, J. Phys. B **24**, L199 (1991).
29. D. R. J. Carruthers and D. S. F. Crothers, Z. Phys. D **23**, 365 (1992).
30. D. R. J. Carruthers, Ph. D. thesis, Queen's University Belfast, (1993).
31. C. Pan and A. F. Starace, Phys. Rev. A **45**, 4588 (1992).
32. L. D. A. Siebbeles, D. P. Marshall, and C. Le Sech, J. Phys. B **26**, L321 (1993).

51

Ionization in High Energy Ion-Atom Collisions

J. H. Macek * and S. T. Manson †

* Department of Physics and Astronomy, University of Tennessee, Knoxville, Tennessee 37996-1501
  and Oak Ridge National Laboratory, Post Office Box 2009, Oak Ridge, Tennessee 37831
† Department of Physics and Astronomy, Georgia State University, Atlanta, Georgia 30303

51.1 INTRODUCTION

Atomic species moving at high velocities form an important component of ionizing radiation. When these species interact with matter, they effect chemical and biological changes which originate with primary collision processes, usually the ejection of electrons. This chapter gives an overview of this primary process that has emerged from studies of the energy and angular distribution of electrons ejected by the impact of atomic or ionic projectiles on atomic targets. It seeks to highlight those features which are most ubiquitous [1, 2]. Atomic units [3] are used, although e, m, and $\hbar$ are sometimes exhibited explicitly to show the connection with standard treatments of the first Born approximation [1].

When high-velocity projectiles strike atomic targets, electrons are ejected from the target atoms, and, if the projectiles have some electrons, from the projectile ions also. For partially ionized projectiles, there are two groups of electrons, one from the target and one from the projectile. In principle, these two groups of electrons cannot be separately identified, but in practice each group often predominates in separate energy and angular regions, and can be identified with the help of computed distributions. These distributions are expressed in terms of doubly differential cross sections (DDCS). Two somewhat different cross sections are used. The Galilean in-variant cross sections are differential in the wave vectors $\mathbf{k}$ of the Schrödinger waves for the ejected electrons and are denoted by $d^3\sigma/dk^3$. The wave vectors $\mathbf{k}$ refer to the laboratory or target frame. The Galilean invariant cross sections take the same form in any reference frame, including the projectile frame. In this frame the electron wave vectors are denoted by primes, $\mathbf{k}'$. The DDCS is given in terms of an alternative expression in Sect. 51.2, Eq. (51.9).

51.2 BORN APPROXIMATION

The first Born approximation, often referred to as FBA or B1, provides an excellent framework to understand qualitatively most of the prominent features observed in fast ion-atom collisions. For a bare ion projectile of charge Z_P impinging upon a neutral atom target of nuclear charge Z_T, the Hamiltonian of the system is written as

$$H = H_0 + H_1 \,, \qquad (51.1)$$

where

$$H_0 = \frac{P_P^2}{2\mu} + \sum_{j=1}^{Z_T} \left(\frac{p_j^2}{2m} - \frac{Z_T e^2}{r_j} + \sum_{i>j}^{Z_T} \frac{e^2}{|\mathbf{r}_i - \mathbf{r}_j|} \right) \,,$$

$$H_1 = \frac{Z_\mathrm{P} Z_\mathrm{T} e^2}{R} - \sum_{j=1}^{Z_\mathrm{T}} \frac{Z_\mathrm{P}^2}{|\mathbf{R} - \mathbf{r}_j|}, \qquad (51.2)$$

and where μ is the projectile reduced mass, m the electron mass, $\mathbf{R}$ the position vector of the projectile with respect to the target nucleus, $\mathbf{P}_\mathrm{P}$ the corresponding momentum operator, $\mathbf{r}_j$ the position vector of the ith target electron and $\mathbf{p}_j$ the corresponding momentum operator. The first Born approximation consists of breaking H as in Eq. (51.1): the wave functions employed are antisymmetric products of the target eigenfunctions of H_0 and a plane wave of relative motion for the target and projectile. Transitions are induced by the interaction term H_1. For an inelastic collision, the target wave function is orthogonal to the initial state wave function and the matrix element of the first term in H_1, $Z_\mathrm{P} Z_\mathrm{T} e^2 / R$, vanishes. Using Bethe's integral

$$\int \frac{e^{i\mathbf{q}\cdot\mathbf{R}}}{|\mathbf{R} - \mathbf{r}|} d^3 r = \frac{4\pi}{q^2} e^{i\mathbf{q}\cdot\mathbf{R}}, \qquad (51.3)$$

the inelastic cross section becomes [4]

$$\sigma_\mathrm{if} = \frac{4\mu^2 e^4}{\hbar^4} \frac{K_\mathrm{f}}{K_\mathrm{i}} \int d\Omega_{\mathbf{K}_\mathrm{f}} \frac{Z_\mathrm{P}^2}{q^4} |\langle \Phi_\mathrm{f}| \sum_j \exp(i\mathbf{q}\cdot\mathbf{r}_j) |\Phi_\mathrm{i}\rangle|^2, \qquad (51.4)$$

where $\hbar\mathbf{K}_\mathrm{i}$ and $\hbar\mathbf{K}_\mathrm{f}$ are the initial and final momenta of the projectile, $\hbar\mathbf{q} = \hbar(\mathbf{K}_\mathrm{f} - \mathbf{K}_\mathrm{i})$ is the momentum transferred from the projectile to the target, $\Phi_{(\mathrm{f},\mathrm{i})}$ represents the wave functions for the final (f) and initial (i) states of the target electrons, $\Omega_{\mathbf{K}_\mathrm{f}}$ is the solid angle of the scattered projectile, j sums over all target electron coordinates $\mathbf{r}_j$, and an average over initial and sum over final states is assumed [4]. The integral over $d\Omega_{\mathbf{K}_\mathrm{f}}$ can be performed by making the change of variables

$$d\Omega_{\mathbf{K}_\mathrm{f}} \equiv \sin\theta_{\mathbf{K}_\mathrm{f}} d\theta_{\mathbf{K}_\mathrm{f}} d\phi_{\mathbf{K}_\mathrm{f}} = \frac{1}{K_\mathrm{i} K_\mathrm{f}} q \, dq \, d\phi_{\mathbf{K}_\mathrm{f}} \qquad (51.5)$$

and substituting $\hbar K_\mathrm{i} = \mu v_\mathrm{i}$ to obtain

$$\sigma_\mathrm{if} = \frac{8\pi a_0^2 Z_\mathrm{P}^2}{(v_\mathrm{i}/v_\mathrm{B})^2} \int_{K_-}^{K_+} \frac{dq}{q} \frac{|\langle \Phi_\mathrm{f}| \sum_{j=1}^{Z_\mathrm{T}} \exp(i\mathbf{q}\cdot\mathbf{r}_j) |\Phi_\mathrm{i}\rangle|^2}{(q a_0)^2}, \qquad (51.6)$$

where v_i is the initial relative velocity and $v_\mathrm{B} = \alpha c$ [1] is the atomic unit of velocity. From the definition of $\mathbf{q}$, the limits of integration are $K_\pm = |K_\mathrm{i} \pm K_\mathrm{f}|$. The cross section is independent of the mass of the incident projectile.

The dimensionless generalized oscillator strength (GOS) is defined by [5]

$$f_\mathrm{if}(q) \equiv \frac{\Delta E}{R_\infty} \frac{|\langle \Phi_\mathrm{f}| \sum_j \exp(i\mathbf{q}\cdot\mathbf{r}_j) |\Phi_\mathrm{i}\rangle|^2}{(q a_0)^2}, \qquad (51.7)$$

---

[1] $(v_\mathrm{i}/v_\mathrm{B})^2 = T/R$ in the notation of Ref. [4].

where ΔE is the energy lost by the projectile in the collsion; again, an average over initial and sum over final magnetic substates is assumed. In terms of the GOS, the cross section becomes

$$\sigma_\mathrm{if} = \frac{8\pi a_0^2 Z_\mathrm{P}^2}{(v_\mathrm{i}/v_\mathrm{B})^2} \int_{K_-}^{K_+} \frac{f_\mathrm{if}(q)}{\Delta E/R_\infty} \frac{dq}{q}. \qquad (51.8)$$

The DDCS, as defined, is related to the Galilean invariant cross section $d^3\sigma/dk^3$ by [6]

$$\frac{d^3\sigma_\mathrm{if}}{dE_\mathbf{k} d\Omega_\mathbf{k}} = \left(\frac{m}{\hbar^2}\right) k \frac{d^3\sigma_\mathrm{if}}{d^3 k}, \qquad (51.9)$$

but it is more useful for describing the low energy ejected electrons since it connects smoothly with cross sections for discrete excitations of the target. The first Born approximation for both excitation and ionization by a bare projectile scale as Z_P^2. The advantage of the GOS is that, in the limit $q \to 0$, it approaches the optical oscillator strength — a relationship that connects ionization in fast ion-atom collisions with photoionization [7].

For an ionization process where an electron of momentum $\hbar k$ is ejected from the target, the DDCS, differential in ejected electron energy $E_\mathbf{k}$ and angle $\Omega_\mathbf{k}$, denoted by $d^3\sigma_\mathrm{if}/dE_\mathbf{k} d\Omega_\mathbf{k}$, is also given by Eq. (51.4) with the proviso that the Φ_f becomes Φ_f^-, representing a target ion plus an ejected (continuum) electron with incoming wave boundary conditions [5], normalized on the energy scale [3], i.e.,

$$\langle \Phi_\mathrm{f}^-(E)|\Phi_\mathrm{f}^-(E')\rangle = \delta(E - E'). \qquad (51.10)$$

If the energy is in Rydbergs, the asymptotic form of the radial part, $r \to \infty$, of the continuum wave function of angular momentum ℓ is given by [7]

$$\frac{1}{r}\left(\frac{1}{\pi k}\right)^{\frac{1}{2}} \sin(kr - \ell\pi/2 + k^{-1}\ln 2kr + \sigma_\ell + \delta_\ell), \qquad (51.11)$$

where $\sigma_\ell \equiv \arg\Gamma(\ell + 1 - i/k)$ is the Coulomb phase shift, and δ_ℓ is the non-Coulomb phase shift due to the short range part of the potential. For normalization on the energy scale, in atomic units (1 au = 2 R_∞), this asymptotic form is multiplied by $\sqrt{2}$, and for normalization on the k-scale, Eq. (51.11) is multiplied by $\sqrt{2k}$ [3].

One can also obtain the single differential cross section (SDS), $d\sigma_\mathrm{if}/dE_\mathbf{k}$, by integrating over $\Omega_\mathbf{k}$. This cross section can be expressed in terms of the differential GOS density in the continuum [5],

$$\frac{df_\mathrm{if}(q)}{dE_\mathbf{k}} = \frac{\Delta E}{R_\infty} \frac{1}{(q a_0)^2}$$

$$\times \int |\langle \Phi_\mathrm{f}^-| \sum_{j=1}^{Z_\mathrm{T}} \exp(i\mathbf{q}\cdot\mathbf{r}_j) |\Phi_\mathrm{i}\rangle|^2 \, d\Omega_\mathbf{k}, \qquad (51.12)$$

as

$$\frac{d\sigma_{\text{if}}}{dE_{\mathbf{k}}} = \frac{8\pi a_0^2}{(v_{\text{i}}/v_{\text{B}})^2} \int \frac{df_{\text{if}}(q)/dE_{\mathbf{k}}}{\Delta E/R} \frac{dq}{q}, \quad (51.13)$$

with $\Delta E = E_{\mathbf{k}} + I_{\text{T}}$, and I_{T} is the ionization energy of the target. Finally, the total ionization cross section is obtained by integration of $d\sigma_{\text{if}}/dE_{\mathbf{k}}$ over $E_{\mathbf{k}}$,

$$\sigma_{\text{if}} = \frac{8\pi a_0^2 Z_{\text{P}}^2}{(v_{\text{i}}/v_{\text{B}})^2} \int Z_{\text{P}}^2 \frac{df_{\text{if}}(q)/dE_{\mathbf{k}}}{\Delta E/R} \frac{dq}{q} dE_{\mathbf{k}}. \quad (51.14)$$

Let us now consider a collision in which the projectile brings in its own N_{P} electrons. The Hamiltonian for the system becomes

$$H = H_0 + H_1$$

$$H_0 = \frac{p_{\text{P}}^2}{2\mu} + \sum_{j=1}^{Z_{\text{T}}} \left(\frac{p_j^2}{2m} - \frac{Z_{\text{T}} e^2}{r_j} + \sum_{i>j}^{Z_{\text{T}}} \frac{e^2}{|\mathbf{r}_i - \mathbf{r}_j|} \right)$$

$$+ \sum_{k=1}^{N_{\text{P}}} \left(\frac{p_k^2}{2m} - \frac{Z_{\text{P}} e^2}{r_k} + \sum_{i>k}^{N_{\text{P}}} \frac{e^2}{|\mathbf{r}_i - \mathbf{r}_k|} \right) \quad (51.15)$$

$$H_1 = \frac{Z_{\text{P}} Z_{\text{T}} e^2}{R} - \sum_{j=1}^{Z_{\text{T}}} \frac{Z_{\text{P}} e^2}{|\mathbf{R} - \mathbf{r}_j|} - \sum_{k=1}^{N_{\text{P}}} \frac{Z_{\text{T}} e^2}{|\mathbf{R} + \mathbf{r}_k|}$$

$$+ \sum_{j=1}^{Z_{\text{T}}} \sum_{k=1}^{N_{\text{P}}} \frac{e^2}{|\mathbf{R} + \mathbf{r}_k - \mathbf{r}_j|}$$

where $\mathbf{p}_k$ and $\mathbf{r}_k$ refer to projectile electrons. Under these conditions, the solutions of H_0 include the wave function of the projectile electrons in the antisymmetric product. For an inelastic collision, the matrix element of the first term of H_1 vanishes as before. However, there are two extra terms in H_1: the interaction of the projectile electrons with the target nucleus, and the interaction of the target electrons with projectile electrons. These projectile electrons open physically distinct, alternative possibilities for the ionization process [2]: (a) target ionization, projectile remains in initial state; (b) target ionization, projectile excited (including ionization); (c) projectile ionization, target remains in initial state; (d) projectile ionization, target excited (including ionization).

For process (a), evaluation of the first Born cross section gives Eq. (51.4) with the projectile charge Z_{P}^2 replaced by $|Z_{\text{P}} - F_{\text{ii}}(q)|^2$, where the elastic form factor $F_{\text{ii}}(q)$ is given by

$$F_{\text{ii}}(q) = \langle \Phi_{\text{i}}^{\text{P}} | \sum_{k=1}^{N_{\text{P}}} \exp(i\mathbf{q} \cdot \mathbf{r}_k) | \Phi_{\text{i}}^{\text{P}} \rangle, \quad (51.16)$$

and $\Phi_{(\text{f,i})}^{\text{P}}$ represents the wave function for the final and initial electron state of the projectile. Thus, the effect of the projectile electrons on process (a) is to screen the projectile nucleus, and the dynamical screening depends

Figure 51.1. Theoretical double differential cross sections (DDCS) for the ionization of He by equal velocity H^+, He^{++} and He^+ (target ionization with no projectile excitation only) as a function of ejected electron energy in Ry for an ejection angle of 60°. The incident velocity corresponds to 0.5 MeV H^+ and 2.0 MeV He^+ and He^{++} which all have the same velocity as a 20 Ry (272 eV) electron.

upon the momentum transfer q. Clearly from Eq. (51.16), $F_{\text{ii}}(0) = N_{\text{P}}$ and $F_{\text{ii}}(\infty) = 0$. Thus, for small energy transfer, which implies small q (large impact parameter), $Z_{\text{P}} \rightarrow Z_{\text{P}} - N_{\text{P}}$, i.e., full screening of the projectile by its electrons. For large energy transfer, which implies q (small impact parameter), Z_{P} remains unmodified, i.e., no screening.

To illustrate this effect, a calculation of process (a) is presented in Fig. 51.1 for the DDCS for 2 MeV He^+ + He collisions, along with a DDCS calculation for equal velocity H^+ and He^{++} projectiles [8]. From these results it is seen that for small energy transfer, the He^+ projectile behaves almost like a heavy proton. With increasing ejected electron energy, however, it is seen to approach He^{++}-like behavior. Note that this result is process (a) alone. Looking at process (b), the situation is rather different. Now two electrons change their state so that transition matrix elements of all terms except the electron-electron term in H_1 vanish. In this case, Z_{P}^2 in Eq. (51.4) is replaced by the square of the inelastic projectile form factor $|F_{\text{fi}}|^2$, where

$$F_{\text{fi}}(q) = \langle \Phi_{\text{f}}^{\text{P}} | \sum_{k=1}^{N_{\text{P}}} \exp(i\mathbf{q} \cdot \mathbf{r}_k) | \Phi_{\text{i}}^{\text{P}} \rangle. \quad (51.17)$$

Owing to the orthogonality of initial and final projectile states, this form factor vanishes in both the limits, $q \rightarrow 0$

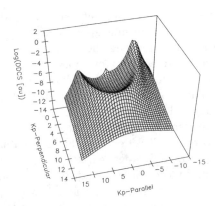

Figure 51.2. Plot of the Galilean invariant DDCS for electrons ejected from atomic hydrogen by 1.5 keV proton impact computed in the distorted wave strong potential Born approximation (DSPB).

and $q \to \infty$. Thus, the total target DDCS is process (a) plus the sum over all of the possible projectile excitations (including projectile ionization) of process (b). Since there is an infinite number of projectile excitations, it is useful to have a method for summing over all of them. Using the closure relation, the sum over all projectile excitations at fixed q is given by [2]

$$\sum_{f \neq i} |F_{fi}(q)|^2 = N_P - |F_{ii}(q)|^2$$

$$+ \langle \Phi_i^P | \sum_{j, j', j \neq j'}^{N_P} \exp[i\mathbf{q} \cdot (\mathbf{r}_j - \mathbf{r}_{j'})] | \Phi_i^P \rangle, \quad (51.18)$$

where only the initial state wave function of the projectile electrons appears. While this sum rule is exact, it cannot be substituted exactly into Eq. (51.4) because of the transformation between $\cos \theta_{\mathbf{K}_f}$ and q, and hence the limits of integration over q vary with the excitation energy of the final state of the projectile. Various ways of choosing approximate integration limits have been suggested [9].

Projectile ionization alone, and with target excitation, process (c) and (d) above, are handled exactly as (a) and (b), but in the projectile reference frame. Thus, ionization of the projectile by the target is calculated using the methodology detailed above, after which the results are transformed into the laboratory frame using the invariance of $d^3\sigma_{if}/d^3k$ and $\mathbf{k}_L = \mathbf{k}_P + \mathbf{K}_i/\mu$. The subscripts L and P refer to the laboratory and projectile frame, respectively.

51.3 PROMINENT FEATURES

A plot of $d^3\sigma/dk^3$ superimposes electrons from the target and electrons from the projectile. To sort out

the main features of the two groups of electrons, first consider impact of bare ions on neutral targets, where the DDCS exhibits only electrons ejected from the target. Figure 51.2 shows a computed cross section for protons on H. The wave vectors are resolved into two components: a component $\mathbf{k}_\parallel$, which is parallel to the wave vector $\mathbf{K}_i$ of the projectile in the laboratory frame, and a component $\mathbf{k}_\perp$ perpendicular to $\mathbf{K}_i$. In most cases, integration is over the direction of $\mathbf{K}_f$ so that $\mathbf{K}_i$ is an axis of symmetry. Then the DDCS is independent of the direction of $\mathbf{k}_\perp$. Because Fig. 51.2 was computed using an approximate theory described in Sect. 51.3.1, the details are not necessarily accurate, but these computations are qualitatively reliable over an electron energy and angular range which excludes very low electron energies, i.e., the region around $k = 0$.

51.3.1 Target Electrons

The Bethe Ridge

Fig. 51.2 provides a useful starting point to examine electron energy and angular distributions. The most prominent feature is the semicircular ridge, called the Bethe ridge, surrounding a valley. The center of the semicircle is at $\mathbf{k} = \mathbf{v}$, where $\mathbf{v} = \mathbf{K}_i/M_P$ is the velocity of the projectile in the lab frame. In the projectile frame, where $\mathbf{k}' = \mathbf{k} - \mathbf{v}$, the center of the circle is at $\mathbf{k}' = 0$ and the ridge is at $k' = v$. This region corresponds to ionization events where the momementum $\mathbf{J} = \mathbf{q} - \mathbf{k}$ of the recoiling target ion vanishes. Momentum conservation implies that

$$\mathbf{K}_i = \mathbf{K}_f + \mathbf{k} + \mathbf{J}, \quad (51.19)$$

while energy conservation is expressed as

$$\frac{K_i^2}{2M_P} + \varepsilon_i = \frac{K_f^2}{2M_P} + \frac{k^2}{2} + \frac{J^2}{2M_T}, \quad (51.20)$$

where M_T is the mass of the ionized target. At the Bethe ridge, $\mathbf{J} \approx 0$ so that Eqs. (51.19) and (51.20) combine to determine the value of $\mathbf{k}$ at this point, called $\mathbf{k}_B$;

$$[(1 + m/M_P)\mathbf{k}_B - \mathbf{v}]^2 = v^2 + (1 + m/M_P)\varepsilon_i \quad (51.21)$$

where ε_i is the initial electron eigenenergy. In the approximation that m/M_P is set equal to zero and the inital binding energy is small compared with $\frac{1}{2}v^2$, Eq. (51.21) shows that the prominent ridge seen in Fig. 51.2 at $k' = v$ corresponds to collisions where all of the momentum lost by the projectile is transferred to the ejected electron. This ridge extends from $E_\mathbf{k} = 0$ up to electron energies of the order of $2v^2$. Generally, ejected electrons move in the combined field of both the target and the projectile. Electrons with momenta $k_\parallel$ in the region $0 < k_\parallel < vZ_P/(Z_P + Z_P)$ will be referred to as low

Figure 51.3. Plot of the angular distribution of 13.6 eV electrons ejected from He by 5 keV proton impact. From Ref. [10]. The solid curve is the Born approximation and the points are the measured values.

energy electrons since they move in regions where the target potential is stronger than that of the projectile. For very fast projectiles, many more electrons are in the low energy region than at higher energies, thus this region is of special interest. Since the target potential influences the motion significantly, the ionization process in this region represents a continuation of excitation across the ionization threshold. To analyze this region, cross sections differential in $E_{\mathbf{k}}$ and angle $\Omega_{\mathbf{k}}$ in the target frame are preferable since they connect smoothly with cross sections for exciting target states [see Eq. (51.4)]. Because the DDCS is independent of the azimuthal angle $\varphi_{\mathbf{k}}$, the DDCS integrated over $\varphi_{\mathbf{k}}$ is often employed.

As the variables $E_{\mathbf{k}}$, $\theta_{\mathbf{k}}$ suggest, the main features of the low energy electrons emerge in plots of energy and angular distributions. For fixed energy, the Bethe ridge appears at an angle θ_B given by Eq. (51.21) and can be written as a relation between $\cos\theta_B$ and $E_{\mathbf{k}}$ as

$$\cos\theta_B = \left(\frac{E_{\mathbf{k}} + \varepsilon_i}{\sqrt{2E_{\mathbf{k}}}}\right)\frac{1}{v}. \qquad (51.22)$$

For fixed $E_{\mathbf{k}} \neq 0$, the DDCS maximizes at an angle that approaches 90° as $v \to \infty$; in particular this angle is obtained when $E_{\mathbf{k}} = -\varepsilon_i$. Figure 51.3 shows the angular distribution of 13.6 eV electrons ejected from He by 5 MeV proton impact [10]. The cross section maximizes at, and is nearly symmetric about, 90°. This feature is understandable from the Born approximation. For small deflection angles, Eq. (51.20) gives

$$q_{\parallel} \approx K_i - K_f \approx \frac{E_{\mathbf{k}} + \varepsilon_i}{v}. \qquad (51.23)$$

For electron energies $E_{\mathbf{k}}$ of the order of the initial binding energy $|\varepsilon_i|$ and large v, the parallel component of $\mathbf{q}$ is much smaller than the magnitude given by Eq. (51.19), so that $\mathbf{q}$ is predominantly perpendicular to the beam direction. From Eq. (51.4), electrons ejected from isotropic

inital states are distributed symmetrically about this axis, i.e., about 90°, in first approximation. Because $\mathbf{q}$ is averaged over in forming the cross section, all directions contribute to some extent, and the distribution is not completely symmetric. The solid curve in Fig. 51.3 represents a distribution computed in the Bethe-Born approximation, and it is seen to peak at 90° but is somewhat asymmeteric about this angle, reflecting effects of averaging over the direction of $\mathbf{q}$. Such distributions are well described by only a few partial waves for the outgoing electron. Indeed, the theoretical distribution is well described by a combination of s and p waves. The calculations fit the experimental data except in the forward direction where the experimental distribution increases sharply. Such sharp increases must reflect the presence of many partial waves not described by the Born approximation. Since the main disagreement between theory and experiment is confined to a small angular region near 0° which contributes only a small part to the total cross section, the Bethe-Born approximation represents a well-founded theoretical framework for interpreting features that appear in the low-electron-energy portion of the DDCS.

While the main contribution to total ionization cross sections comes from the low energy region, electrons in the high energy region ($v/2 < k$) carry more energy per electron, and therefore play a prominent role in processes initiated by fast electrons. The fast electrons on the Bethe ridge correspond to collisions where the momentum $\mathbf{q}$ lost by the incident projectile is transferred mainly to the ejected electron. Because the momentum $\mathbf{k}$ of the ejected electrons is much larger than the mean value of the electron momentum in the initial state, given by $\sqrt{2\varepsilon_i}$, this portion of the spectrum can be interpreted in terms of a binary collision between the incident projectile and the target electron treated as quasifree. By quasifree we understand that the electron has a nonzero binding energy ε_i, but is otherwise regarded as a free electron with an initial momentum $\mathbf{s}$, distributed over a range of values centered around the mean value. In the projectile reference frame, the electron has the momentum $\mathbf{s} - \mathbf{v}$. The electron scatters elastically from the projectile and emerges with a final momentum $\mathbf{k}' = \mathbf{k} - \mathbf{v}$. This simple picture is known as the *elastic scattering model*, and is often employed for processes involving weakly bound electrons [11]. There are several quantal versions of this picture, indeed any theory of ionization must reduce to the electron scattering model in the limit that the initial binding energy ε_i vanishes.

When the projectile P is a bare ion so that the binary scattering process is just Rutherford scattering, the Bethe-Born approximation is in accord with this picture. It works because the electrons are fast so that effects of the target potential in the final state are unimportant, and because a first-order computation of Rutherford scattering gives the same scattering cross section as the exact Rutherford amplitude. For that reason the domain of applicability of the first Born approximation includes the Bethe ridge, even in the high energy region.

The binary encounter peak shows new features when the projectile carries electrons. The target electrons scatter from a partially screened ion and the Bethe ridge reflects properties of the elastic scattering cross section for the complex projectile species. In contrast to the Rutherford cross section, elastic scattering cross sections for complex species may have maxima and minima as functions of the electron energy and angle. Such features have been identified in collisions of highly charged ions with neutral atomic targets [12].

The Continuum Electron Capture Cusp

Fig. 51.2 shows a sharp cusp-like structure when the momentum of the ejected electron in the projectile frame k' vanishes. Then the ejected electron moves with a velocity exactly equal to the projectile velocity. For charged projectiles, it is virtually impossible to determine whether such electron states are Rydberg states of high principal quantum number n', or continuum states with $E_{\mathbf{k}'} = \frac{1}{2}k'^2$ much less than ε_i. Indeed, the physical similarity of such atomic states implies that the cross section for electron capture to states of high n' differs from the ionization cross section near $k' = 0$ only by a density of states factor. This factor is just $dE_{\mathbf{k}'}$ for ionization processes and $dE_{n'} = Z_P^2 n'^{-3}$ for capture. It follows that the cross section $d^3\sigma/dE_{\mathbf{k}'}d\Omega_{\mathbf{k}'}$ is a smooth function of $E_{\mathbf{k}'}$ which is nonzero and finite at $E_{\mathbf{k}'} = 0$. The Galilean invariant cross section of Fig. 51.2, however, is not smooth; rather it is given by

$$\frac{d^3\sigma}{dk^3} = \frac{d^3\sigma}{dk'^3} = \frac{1}{k'}\frac{d^3\sigma}{dE_{\mathbf{k}'}d\Omega_{\mathbf{k}'}}. \quad (51.24)$$

Since $d^3\sigma/dE_{\mathbf{k}'}d\Omega_{\mathbf{k}'}$ is nonzero at $k' = 0$, it follows that the DDCS has a k'^{-1} singularity at $k' = 0$. Experiments measure cross sections averaged over this singular factor. The averaging results in the cusp-shaped feature at $k' = 0$ seen in Fig. 51.2 [13–15].

The theoretical description of the cusp feature requires that the electron in the final state move mainly in the field of the projectile rather than that of the target. While the appropriate final state wave function can be introduced into Eq. (51.4), the result is quantitatively inaccurate. A more accurate amplitude emerges when the exact amplitude is expanded in powers the interaction potential V_T of the electron with the final state target ion [16], since V_T has a much smaller effect than the interaction V_P. Approximate evaluation of the ionization amplitude in an independent particle approximation gives [17]

Figure 51.4. Free-free transition picture of ionization. (a) Schematic representation of a target electron with momentum $\mathbf{s} - \mathbf{v}$ in the projectile frame. (b) A free-free transition from the projectile eigenstate $\psi_{\mathbf{s}-\mathbf{v}}^-$ to the final eigenstate $\psi_{\mathbf{k}'}^+$ occurs with the target interaction as the transition operator.

$$T_{\mathrm{fi}} = \int d^3s\, \tilde{V}_T(\mathbf{J} + \mathbf{s})\tilde{\varphi}_i(\mathbf{s})$$
$$\times \langle \psi_{\mathbf{k}'}^-(\mathbf{r})| \exp[i(\mathbf{J} + \mathbf{s})]|\psi_{\mathbf{s}-\mathbf{v}}(\mathbf{r})\rangle, \quad (51.25)$$

and

$$\frac{d^3\sigma}{dE_{\mathbf{k}}d\Omega_{\mathbf{k}}} = (2\pi)^4\mu^2 \int |T_{\mathrm{fi}}|^2 d\Omega_{\mathbf{K}_{\mathrm{f}}} \quad (51.26)$$

where μ is the reduced mass given by

$$\mu = M_P(M_P + 1)/(M_P + M_T + 1), \quad (51.27)$$

and $\tilde{\varphi}_i(\mathbf{s})$ is the momentum space independent-particle orbital wave function for the active electron in the intial state, $\tilde{V}_T$ is the corresponding effective interaction potential, and the $\psi_{\mathbf{k}'}^\pm$ are normalized on the momentum scale. Fig. 51.4 gives a pictorial interpretation of this amplitude. The initial momentum distribtution of the electron is given by $\tilde{\varphi}_i(\mathbf{s})$. In the projectile frame, this electron has momentum $\mathbf{s} - \mathbf{v}$ and moves in a projectile continuum state represented by $\psi_{\mathbf{s}-\mathbf{v}}^+(\mathbf{r})$. Owing to the interaction with the target represented by $\exp[i(\mathbf{J} + \mathbf{s})] \tilde{V}_T(\mathbf{J} + \mathbf{s})$, a transition to the final projectile state $\psi_{\mathbf{k}'}^-$ occurs as illustrated in Fig. 51.4 (b). The final state $\psi_{\mathbf{k}'}^-$ has a $1/\sqrt{k'}$ normalization that gives rise to the cusp at $\mathbf{k} = \mathbf{v}$ seen in Fig. 51.2. This picture describes ionization in terms of the free-free transition, $\psi_{\mathbf{s}-\mathbf{v}}^- \to \psi_{\mathbf{k}'}^+$.

The amplitude in Eq. (51.25), known as the distorted wave strong potential Born amplitude (DSPB) [18], or the projectile impulse approximation amplitude [19], also describes binary encounter electrons. Thus the regions of applicability of Eqs. (51.4) and (51.25) overlap considerably. In the binary encounter region, where $k_{\parallel} > v/2$ and $\mathbf{J} \approx 0$, the cross section given by Eqs. (51.9) and (51.25) reduces to the elastic scattering model (ESM) [20], where the incident electron with projectile frame momentum $-\mathbf{J} - \mathbf{v}$ quasi-elastically scatters from the projectile into a final state with projectile frame momentum $\mathbf{k}'$. This quasi-elastic cross section is averaged over the initial momentum distribution of the electron $|\tilde{\varphi}_i(-\mathbf{J})|^2 d^3 J$ so that

$$d^5\sigma^{\mathrm{ESM}} = \left[(2\pi)^4 \frac{k'}{v} |T^{\mathrm{elas}}(\mathbf{k}', -\mathbf{J} - \mathbf{v})|^2 |d\Omega_{\mathbf{k}'} \right]$$
$$\times |\tilde{\varphi}_i(-\mathbf{J})|^2 \, d^3 J \, . \qquad (51.28)$$

The projectile frame cross section is recovered upon using the relation $d^3 J = \mu_f \mu_i v \, d\Omega_{\mathbf{K}_f} k'^2 \, dk'$ in Eq. (51.25).

The two approximations (51.4) and (51.25) describe all of the major features that emerge in the ionization of one-electron targets by fully ionized projectiles. Both theories also incorporate some multi-electron effects that appear when either the target or projectile (or both) have several electrons. Equation (51.28) is useful for describing effects of target ionization by multi-electron projectiles when the projectile electrons remain in their ground state since T^{elas} incorporates, in principle, the exact amplitude for electrons to scatter from the projectile in its ground state. Of course, the multi-electron projectile can also become excited or ionized owing to its interaction with the target electrons. These processes are not incorporated in the ESM cross section Eq. (51.28); rather the Born amplitude of Eq. (51.4) represents a more tractable theory to analyze these features.

51.3.2   Projectile Electrons

When the incident atomic species carries some attached electrons, these electrons may be removed in the collision with a target. Essentially all of the features that appear for the target electrons also appear for the projectile electrons; however the DDCS is shifted by the Galilean transformation to the projectile frame. The transformation from the laboratory frame to the projectile frame is an essential step for interpreting features of the projectile electrons. This transformation has the effect of spreading out the electron energy distribution since at $\theta_{\mathbf{k}} = 0°$ a small energy interval $\Delta E_{\mathbf{k}}$ in the lab frame relates to a small energy interval $\Delta E_{\mathbf{k}'}$ in the projectile frame according to

$$\Delta E_{\mathbf{k}} = \Delta E_{\mathbf{k}'} \left(1 + \frac{v}{\sqrt{2vE_{\mathbf{k}'}}} \right) . \qquad (51.29)$$

For example, when the projectile energy is 1.5 MeV/au, a projectile frame electron energy interval $0 < E_{\mathbf{k}'} < 0.1$ eV maps into an interval of ± 8.3 eV centered at $E_{\mathbf{k}} = 817$ eV. This amplification of both the electron energy and the electron energy interval proves useful for measuring features of projectile species with high resolution [21].

REFERENCES

1. U. Fano, Ann. Rev. Nucl. Sci. **13** (1963).
2. J. S. Briggs and K. Taulbjerg in *Structure and Collisions in Ions and Atoms*, edited by I. A. Sellin (Springer-Verlag, Berlin, 1978), pp. 105–153.
3. H. A. Bethe and E. E. Salpeter, *Quantum Mechanics of One-and Two-Electron Atoms* (Plenum, New York, 1977).
4. M. Inokuti, Rev. Mod. Phys. **43**, 297 (1971).
5. S. Geltman, *Topics in Atomic Collision Theory* (Academic Press, NY, 1969), pp. 41–44.
6. F. Drepper and J. S. Briggs, J. Phys. B **9**, 2063 (1976).
7. A. F. Starace in *Handbuch der Physik*, edited by W. Mehlhorn (Springer-Verlag, Berlin, 1982), Vol. 31, pp. 14ff.
8. S. T. Manson and L. H. Toburen, Phys. Rev. Lett. **46**, 529 (1981).
9. H. M. Hartley and H. R. J. Walters, J. Phys. B **20**, 1983 (1987).
10. S. T. Manson, L. Toburen, D. Madison, and N. Stolterfoht, Phys. Rev. A **12**, 60 (1975).
11. M. M. Duncan and M. G. Menendez, Phys. Rev. A **16**, 1799 (1977).
12. D. H. Lee, P. Richard, T. J. M. Zouros, J. M. Sanders, J. L. Shinpaugh, and H. Hidmi, Phys. Rev. A **41**, 4816 (1990).
13. M. E. Rudd and J. H. Macek, Case Studies in Atomic Physics **3**, 47 (1972).
14. G. G. Crooks and M. E. Rudd, Phys. Rev. Lett. **25**, 1599 (1970).
15. K. G. Harrison and M. W. Lucas, Phys. Lett. **33 A**, 142 (1970).
16. J. S. Briggs, J. Phys. B **10**, 3075 (1977).
17. M. Brauner and J. H. Macek, Phys. Rev. A **46**, 2519 (1992).
18. K. Taulbjerg, R. Barrachina, and J. H. Macek, Phys. Rev. A **41**, 207 (1990).
19. D. Jakubassa-Amundsen, J. Phys. B **16**, 1767 (1983).
20. J. Macek and K. Taulbjerg, J. Phys. B **26**, 1353 (1993).
21. N. Stolterfoht, Phys. Rep. **146**, 315 (1987).

Electron-Ion and Ion-Ion Recombination

M. R. Flannery

School of Physics, Georgia Institute of Technology, Atlanta, Georgia 30332-0430

52.1 RECOMBINATION PROCESSES

52.1.1 Electron-Ion Recombination

This proceeds via the following four processes:
(a) radiative recombination (RR)

$$e^- + A^+(i) \to A(n\ell) + h\nu, \qquad (52.1)$$

(b) three-body collisional-radiative recombination

$$e^- + A^+ + e^- \to A + e^-, \qquad (52.2a)$$
$$e^- + A^+ + M \to A + M, \qquad (52.2b)$$

where the third body can be an electron or a neutral gas.
(c) dielectronic recombination (DLR)

$$e^- + A^{Z+}(i) \rightleftharpoons \left[A^{Z+}(k) - e^-\right]_{n\ell} \to A^{(Z-1)+}_{n'\ell'}(f) + h\nu, \qquad (52.3)$$

(d) dissociative recombination (DR)

$$e^- + AB^+ \to A + B^*. \qquad (52.4)$$

Electron recombination with bare ions can proceed only via (a) and (b), while (c) and (d) provide additional pathways for ions with at least one electron initially or for molecular ions AB^+. *Electron radiative capture* denotes the combined effect of RR and DLR.

52.1.2 Positive-Ion Negative-Ion Recombination

This proceeds via the following three processes:
(e) mutual neutralization

$$A^+ + B^- \to A + B^*, \qquad (52.5)$$

(f) three-body (termolecular) recombination

$$A^+ + B^- + M \to AB + M, \qquad (52.6)$$

(g) tidal recombination

$$AB^+ + C^- + M \to AC + B + M \qquad (52.7a)$$
$$\to BC + A + M, \qquad (52.7b)$$

where M is some third species (atomic, molecular or ionic). Although (e) always occurs when no gas M is present, it is greatly enhanced by coupling to (f). The dependence of the rate $\hat{\alpha}$ on density N of background gas M is different for all three cases, (e)–(g).

Processes (a), (c), (d) and (e) are elementary processes in that microscopic detailed balance (proper balance) exists with their true inverses, i.e., with photoionization (both with and without autoionization) as in (c) and (a), associative ionization and ion-pair formation as in (d) and (e), respectively. Processes (b), (f) and (g) in general involve a complex sequence of elementary energy-changing mechanisms as collisional and radiative cascade and their overall rates are determined by an input-output continuity equation involving microscopic continuum-bound and bound-bound collisional and radiative rates.

52.1.3 Balances

Proper balances are detailed microscopic balances between forward and reverse mechanisms that are direct inverses of one another, as in

(a) Maxwellian: $e^-(v_1) + e^-(v_2) \rightleftharpoons e^-(v_1') + e^-(v_2')$, $$\qquad (52.8)$$

where the kinetic energy of the particles is redistributed;

(b) Saha: $e^- + H(n\ell) \rightleftharpoons e^- + H^+ + e^-$ $\qquad (52.9)$

between direct ionization from and direct recombination into a given level $n\ell$;

(c) Boltzmann: $e^- + H(n\ell) \rightleftharpoons e^- + H(n', \ell')$ $\quad (52.10)$

between excitation and de-excitation among bound levels;

(d) Planck: $e^- + H^+ \rightleftharpoons H(n\ell) + h\nu$, $\qquad (52.11)$

which involves interaction between radiation and atoms in photoionization/recombination to a given level $n\ell$.

Improper balances maintain constant densities via production and destruction mechanisms that are *not* pure inverses of each other. They are associated with flux activity on a macroscopic level as in the transport of particles into the system for recombination and net production and transport of particles (i.e. e⁻, A^+) for ionization. Improper balances can then exist between dissimilar elementary production-depletion processes as in (a) *coronal balance* between electron-excitation into and radiative decay out of level n. (b) *radiative balance* between radiative capture into and radiative cascade out of level n. (c) *excitation saturation balance* between upward collisional excitations $n-1 \to n \to n+1$ between adjacent levels. (d) *de-excitation saturation balance* between downward collisional de-excitations $n+1 \to n \to n-1$ into and out of level n.

52.2 COLLISIONAL-RADIATIVE RECOMBINATION

Radiative Recombination. Process (52.1) involves a free-bound electronic transition with radiation spread over the recombination continuum. It is the inverse of photoionization without autoionization and favors high energy gaps with transitions to low $n \sim 1, 2, 3$ and low angular momentum states $\ell \sim 0, 1, 2$ at higher electron energies.

Three-Body Electron-Ion Recombination. Processes (52.2a,b) favor free-bound collisional transitions to high levels n, within a few k_BT of the ionization limit of $A(n)$ and collisional transitions across small energy gaps. Recombination becomes stabilized by *collisional-radiative cascade* through the lower bound levels of A. Collisions of the e⁻$-A^+$ pair with third bodies becomes more important for higher levels n and radiative emission is important down to and among the lower levels n. In optically thin plasmas this radiation is lost, while in optically thick plasmas it may be re-absorbed. At low electron densities radiative recombination dominates with predominant transitions taking place to the ground level. For process (52.2a) at high electron densities, three-body collisions into high Rydberg levels dominate, followed by cascade which is collision dominated at low electron temperatures T_e and radiation dominated at high T_e. For process (52.2b) at low gas densities N, the recombination is collisionally-radiatively controlled while, at high N, it eventually becomes controlled by the rate of diffusional drift (52.61) through the gas M.

Collisional-Radiative Recombination. Here the cascade collisions and radiation are coupled via the continuity equation. The population n_i of an individual excited level i of energy E_i is determined by the rate equations

$$\frac{dn_i}{dt} = \frac{\partial n_i}{\partial t} + \nabla \cdot (n_i \mathbf{v}_i) \qquad (52.12)$$

$$= \sum_{i \neq f} [n_f \nu_{fi} - n_i \nu_{if}] = P_i - n_i D_i, \qquad (52.13)$$

which involve temporal and spatial relaxation in (52.12) and collisional-radiative production rates P_i and destruction frequencies D_i of the elementary processes included in (52.13). The total collisional and radiative transition frequency between levels i and f is ν_{if} and the f-sum is taken over all discrete and continuous (c) states of the recombining species. The transition frequency ν_{if} includes all contributing elementary processes that directly link states i and f, e.g., collisional excitation and de-excitation, ionization ($i \to c$) and recombination ($c \to i$) by electrons and heavy particles, radiative recombination ($c \to i$), radiative decay ($i \to f$), possibly radiative absorption for optically thick plasmas, autoionization and dielectronic recombination.

Production Rates and Processes. The production rate for a level i is

$$P_i = \sum_{f \neq i} n_e n_f K_{fi}^c + n_e^2 N^+ k_{ci}^R + \sum_{f > i} n_f \left[A_{fi} + B_{fi} \rho_\nu \right]$$
$$+ n_e N^+ \left[\hat{\alpha}_i^{RR} + \beta_i \rho_\nu \right], \qquad (52.14)$$

where the terms in the above order represent (1) collisional excitation and de-excitation by e⁻$-A(f)$ collisions, (2) three-body e⁻$-A^+$ collisional recombination into level i, (3) spontaneous and stimulated radiative cascade, and (4) spontaneous and stimulated radiative recombination.

Destruction Rates and Processes. The destruction rate for a level i is

$$n_i D_i = n_e n_i \sum_{f \neq i} K_{if}^c + n_e n_i S_i + n_i \sum_{f < i} [A_{if} + B_{if} \rho_\nu]$$
$$+ n_i \sum_{f > i} B_{if} \rho_\nu + n_i B_{ic} \rho_\nu, \qquad (52.15)$$

where the terms in the above order represent (1) collisional destruction, (2) collisional ionization, (3) spontaneous and stimulated emission, (4) photo-excitation, and (5) photoionization.

52.2.1 Saha and Boltzmann Distributions

Collisions of $A(n)$ with third bodies such as e⁻ and M are more rapid than radiative decay above a certain excited level n^*. Since each collision process is accompanied by its exact inverse the principle of detailed balance determines the population of levels $i > n^*$.

Saha Distribution. This connects equilibrium densities $\tilde{n}_i$, $\tilde{n}_e$ and $\tilde{N}^+$ of bound levels i, of free electrons at temperature T_e and of ions by

$$\frac{\tilde{n}_i}{\tilde{n}_e \tilde{N}^+} = \left[\frac{g(i)}{g_e g_A^+} \right] \frac{h^3}{(2\pi m_e k T_e)^{3/2}} \exp(I_i / k_B T_e), \quad (52.16)$$

where the electronic statistical weights of the free electron, the ion of charge $Z+1$ and the recombined $e^- - A^+$ species of net charge Z and ionization potential I_i are $g_e = 2$, g_A^+ and $g(i)$, respectively. Since $n_i \leq \tilde{n}_i$ for all i, then the Saha-Boltzmann distributions imply that $n_1 \gg n_i$ and $n_e \gg n_i$ for $i \neq 1, 2$, where $i = 1$ is the ground state.

Boltzmann Distribution. This connects the equilibrium populations of bound levels i of energy E_i by

$$\tilde{n}_i/\tilde{n}_j = [g(i)/g(j)] \exp\left[-(E_i - E_j)/k_B T_e\right] . \quad (52.17)$$

52.2.2 Quasi-Steady State Distributions

The reciprocal lifetime of level i is the sum of radiative and collisional components and is therefore shorter than the pure radiative lifetime $\tau_R \sim 10^{-7} Z^{-4}$ s. The lifetime τ_1 for the ground level is collisionally controlled, is dependent upon n_e, and generally is within the range of 10^2 and 10^4 s for most laboratory plasmas and the solar atmosphere. The excited level lifetimes τ_i are then much shorter than τ_1. The (spatial) diffusion or plasma decay (recombination) time is then much longer than τ_i and the total number of recombined species is much smaller than the ground-state population n_1. The recombination proceeds on a timescale much longer than the time for population/destruction of the excited levels. The condition for quasisteady state, or QSS-condition, $dn_i/dt = 0$ for the bound levels $i \neq 1$, therefore holds. The QSS distributions n_i therefore satisfy $P_i = n_i D_i$.

52.2.3 Ionization and Recombination Coefficients

Under QSS, the continuity equation (52.13) then reduces to a finite set of simultaneous equations $P_i = n_i D_i$. This gives a matrix equation which is solved numerically for $n_i (i \neq 1) \leq \tilde{n}_i$ in terms of n_1 and n_e. The net ground-state population frequency per unit volume $(\text{cm}^{-3}\,\text{s}^{-1})$ can then be expressed as

$$\frac{dn_1}{dt} = n_e N^+ \hat{\alpha}_{CR} - n_e n_1 S_{CR} , \quad (52.18)$$

where $\hat{\alpha}_{CR}$ and S_{CR}, respectively, are the overall rate coefficients for recombination and ionization via the collisional-radiative sequence. The determined $\hat{\alpha}_{CR}$ equals the direct $(c \rightarrow 1)$ recombination to the ground level supplemented by the net collisional-radiative cascade from that portion of bound-state population which originated from the continuum. The determined S_{CR} equals direct depletion (excitation and ionization) of the ground state reduced by the de-excitation collisional radiative cascade from that portion of the bound levels accessed originally from the ground level. At low n_e, $\hat{\alpha}_{CR}$

and S_{CR} reduce, respectively, to the radiative recombination coefficient summed over all levels and to the collisional ionization coefficient for the ground level.

$\mathcal{C}$, $\mathcal{E}$ and $\mathcal{S}$ Blocks of Energy Levels

For the recombination processes (52.2a), (52.2b) and (52.6) which involve a sequence of elementary reactions, the $e^- - A^+$ or $A^+ - B^-$ continuum levels and the ground $A(n = 1)$ or the lowest vibrational levels of AB are therefore treated as two large particle reservoirs of reactants and products. These two reservoirs act as reactant and as sink blocks $\mathcal{C}$ and $\mathcal{S}$ which are, respectively, drained and filled at the same rate via a conduit of highly excited levels which comprise an intermediate block of levels $\mathcal{E}$. This $\mathcal{C}$ draining and $\mathcal{S}$ filling proceeds, via block $\mathcal{E}$, on a timescale large compared with the short time for a small amount from the reservoirs to be re-distributed within block $\mathcal{E}$. This forms the basis of QSS.

52.2.4 Working Rate Formulae

For electron–atomic-ion collisional-radiative recombination (52.2a), detailed QSS calculations can be fitted by the rate [1]

$$\hat{\alpha}_{CR} = \left[3.8 \times 10^{-9} T_e^{-4.5} n_e + 1.55 \times 10^{-10} T_e^{-0.63} \right.$$
$$\left. + 6 \times 10^{-9} T_e^{-2.18} n_e^{0.37} \right] \text{cm}^3\,\text{s}^{-1} \quad (52.19)$$

agrees with experiment for a Lyman α optically thick plasma with n_e and T_e in the range $10^9\,\text{cm}^{-3} \leq n_e \leq 10^{13}\,\text{cm}^{-3}$ and $2.5\,\text{K} \leq T_e \leq 4000\,\text{K}$. The first term is the pure collisional rate (52.49), the second term is the radiative cascade contribution, and the third term arises from collisional-radiative coupling.

For $(e^- - He_2^+)$ recombination in a high $(5-100\,\text{Torr})$ pressure helium afterglow the rate for (52.2b) is [2]

$$\hat{\alpha}_{CR} = \left[(4 \pm 0.5) \times 10^{-20} n_e \right] (T_e/293)^{-(4\pm 0.5)}$$
$$+ \left\{ (5 \pm 1) \times 10^{-27} n(He) + (2.5 \pm 2.5) \times 10^{-10} \right\}$$
$$\times (T_e/293)^{-(1\pm 1)} \text{cm}^3/\text{s}. \quad (52.20)$$

The first two terms are in accord with the purely collisional rates (52.49) and (52.52b), respectively.

52.3 MACROSCOPIC METHODS

52.3.1 Resonant Capture-Stabilization Model: Dissociative and Dielectronic Recombination

The electron is captured dielectronically (*cf.* (52.41)) into an energy-resonant long-lived intermediate collision complex of super-excited states d which can autoionize or be stabilized irreversibly into the final product channel f either by molecular fragmentation

$$e^- + AB^+(i) \underset{\nu_a}{\overset{k_c}{\rightleftharpoons}} AB^{**} \overset{\nu_s}{\to} A + B^*, \qquad (52.21)$$

as in direct dissociative recombination (DR), or by emission of radiation as in dielectronic recombination (DLR)

$$e^- + A^{Z+}(i) \underset{\nu_a}{\overset{k_c}{\rightleftharpoons}} \left[A^{Z+}(k) - e^-\right]_{n\ell}$$

$$\overset{\nu_s}{\to} A^{(Z-1)+}_{n'\ell'}(f) + h\nu. \qquad (52.22)$$

Production Rate of Super-Excited States d.

$$\frac{dn_d^*}{dt} = n_e N^+ k_c(d) - n_d^* \left[\nu_A(d) + \nu_S(d)\right]; \qquad (52.23)$$

$$\nu_A(d) = \sum_{i'} \nu_a(d \to i'), \qquad (52.24a)$$

$$\nu_S(d) = \sum_{f'} \nu_s(d \to f'). \qquad (52.24b)$$

Steady-State Distribution. For a steady-state distribution, the *capture volume* is

$$\frac{n_d^*}{n_e N^+} = \frac{k_c(d)}{\nu_A(d) + \nu_S(d)}. \qquad (52.25)$$

Recombination Rate and Stabilization Probability. The recombination rate to channel f is

$$\hat{\alpha}_f = \sum_d \left[\frac{k_c(d)\nu_s(d \to f)}{\nu_A(d) + \nu_S(d)}\right], \qquad (52.26a)$$

and the rate to all product channels is

$$\hat{\alpha} = \sum_d \frac{k_c(d)\nu_S(d)}{\nu_A(d) + \nu_S(d)}. \qquad (52.26b)$$

In the above, the quantities

$$P_f^S(d) = \nu_s(d \to f)/\left[\nu_A(d) + \nu_S(d)\right], \qquad (52.27)$$

$$P^S(d) = \nu_S(d)/\left[\nu_A(d) + \nu_S(d)\right], \qquad (52.28)$$

represent the corresponding stabilization probabilities.

Macroscopic Detailed Balance and Saha Distribution.

$$K_{di}(T) = \frac{\tilde{n}_d^*}{\tilde{n}_e \tilde{N}^+} = \frac{k_c(d)}{\nu_a(d \to i)} = k_c(d)\tau_a(d \to i) \qquad (52.29a)$$

$$= \frac{h^3}{(2\pi m_e k_B T)^{3/2}} \left[\frac{\omega(d)}{2\omega^+}\right] \exp\left[-E_{di}^*/k_B T\right], \qquad (52.29b)$$

where E_{di}^* is the energy of super-excited neutral levels AB^{**} above that for ion level $AB^+(i)$, and ω are the corresponding statistical weights.

Alternative Rate Formula.

$$\hat{\alpha}_f = \sum_d K_{di} \left[\frac{\nu_a(d \to i)\nu_s(d \to f)}{\nu_A(d) + \nu_S(d)}\right]. \qquad (52.30)$$

Normalized Excited State Distributions.

$$\rho_d = n_d^*/\tilde{n}_d^* = \frac{\nu_a(d \to i)}{\left[\nu_A(d) + \nu_S(d)\right]}, \qquad (52.31)$$

$$\hat{\alpha} = \sum_d k_c(d) P^S(d) = \sum_d K_{di}\rho_d \nu_S(d) \qquad (52.32a)$$

$$= \sum_d k_c(d) \left[\rho_d \nu_S(d)\tau_a(d \to i)\right]. \qquad (52.32b)$$

Although equivalent, Eqs. (52.26a) and (52.30) are normally invoked for (52.21) and (52.22), respectively, since $P^S \le 1$ for DR so that $\hat{\alpha}_{DR} \to k_c$; and $\nu_A \gg \nu_S$ for DLR with $n \ll 50$ so that $\hat{\alpha} \to K_{di}\nu_S$. For $n \gg 50$, $\nu_S \gg \nu_A$ and $\hat{\alpha} \to k_c$. The above results (52.26a) and (52.30) can also be derived from microscopic Breit-Wigner scattering theory for isolated (nonoverlapping) resonances.

52.3.2 Reactive Sphere Model: Three-Body Electron-Ion and Ion-Ion Recombination

Since the Coulomb attraction cannot support quasibound levels, three body electron-ion and ion-ion recombination do not in general proceed via time-delayed resonances, but rather by reactive (energy-reducing) collisions with the third body M. This is particularly effective for A–B separations $R \le R_0$, as in the sequence

$$A + B \underset{\nu_d}{\overset{k_c}{\rightleftharpoons}} AB^*(R \le R_0), \qquad (52.33a)$$

$$AB^*(R \le R_0) + M \underset{\nu_{-s}}{\overset{\nu_s}{\rightleftharpoons}} AB + M. \qquad (52.33b)$$

In contrast to (52.21) and (52.22) where the stabilization is irreversible, the forward step in (52.33b) is reversible. The sequence (52.33a) and (52.33b) represents a closed system where thermodynamic equilibrium is eventually established.

Steady State Distribution of AB^* Complex.

$$n^* = \left(\frac{k_c}{\nu_s + \nu_d}\right) n_A(t) n_B(t) + \left(\frac{\nu_{-s}}{\nu_s + \nu_d}\right) n_s(t). \qquad (52.34)$$

Saha and Boltzmann balances:

$$\begin{aligned}
\text{Saha:} \quad & \tilde{n}_A \tilde{n}_B k_c = \tilde{n}^* \nu_d, \\
\text{Boltzmann:} \quad & \tilde{n}_s \nu_{-s} = \tilde{n}^* \nu_s.
\end{aligned} \qquad (52.35)$$

$\tilde{n}^*$ is in Saha balance with reactant block $\mathcal{C}$ and in Boltzmann balance with product block $\mathcal{S}$.

Normalized Distributions.

$$\rho^* = \frac{n^*}{\tilde{n}^*} = P^{\mathrm{D}}\gamma_{\mathrm{c}}(t) + P^{\mathrm{S}}\gamma_{\mathrm{s}}(t)\,, \qquad (52.36\mathrm{a})$$

$$\gamma_{\mathrm{c}}(t) = \frac{n_A(t)n_B(t)}{\tilde{n}_A\tilde{n}_B}\,, \quad \gamma_{\mathrm{s}}(t) = \frac{n_{\mathrm{s}}(t)}{\tilde{n}_{\mathrm{s}}}\,. \qquad (52.36\mathrm{b})$$

Stabilization and Dissociation Probabilities.

$$P^{\mathrm{S}} = \frac{\nu_{\mathrm{s}}}{(\nu_{\mathrm{s}} + \nu_{\mathrm{d}})}\,, \quad P^{\mathrm{D}} = \frac{\nu_{\mathrm{d}}}{(\nu_{\mathrm{s}} + \nu_{\mathrm{d}})}\,. \qquad (52.37)$$

Time Dependent Equations.

$$\frac{dn_{\mathrm{c}}}{dt} = -k_{\mathrm{c}}P^{\mathrm{S}}\tilde{n}_A\tilde{n}_B\left[\gamma_{\mathrm{c}}(t) - \gamma_{\mathrm{s}}(t)\right]\,, \qquad (52.38\mathrm{a})$$

$$\frac{dn_{\mathrm{s}}}{dt} = -\nu_{-\mathrm{s}}P^{\mathrm{D}}\tilde{n}_{\mathrm{s}}\left[\gamma_{\mathrm{s}}(t) - \gamma_{\mathrm{c}}(t)\right]\,, \qquad (52.38\mathrm{b})$$

$$\frac{dn_{\mathrm{c}}}{dt} = -\hat{\alpha}_3 n_A(t)n_B(t) + k_{\mathrm{d}}n_{\mathrm{s}}(t)\,. \qquad (52.39)$$

where the recombination rate coefficient (cm^3/s) and dissociation frequency are, respectively,

$$\hat{\alpha}_3 = k_{\mathrm{c}}P^{\mathrm{S}} = \frac{k_{\mathrm{c}}\nu_{\mathrm{s}}}{(\nu_{\mathrm{s}} + \nu_{\mathrm{d}})}\,, \qquad (52.40)$$

$$k_{\mathrm{d}} = \nu_{-\mathrm{s}}P^{\mathrm{D}} = \frac{\nu_{-\mathrm{s}}\nu_{\mathrm{d}}}{(\nu_{\mathrm{s}} + \nu_{\mathrm{d}})}\,, \qquad (52.41)$$

which also satisfy the *macroscopic detailed balance* relation

$$\hat{\alpha}_3\tilde{n}_A\tilde{n}_B = k_{\mathrm{d}}\tilde{n}_{\mathrm{s}}\,. \qquad (52.42)$$

Time Independent Treatment. The rate $\hat{\alpha}_3$ given by the time dependent treatment can also be deduced by viewing the recombination process as a source block $\mathcal{C}$ kept fully filled with dissociated species A and B maintained at equilibrium concentrations $\tilde{n}_A$, $\tilde{n}_B$ (i.e. $\gamma_{\mathrm{c}} = 1$) and draining at the rate $\hat{\alpha}_3\tilde{n}_A\tilde{n}_B$ through a steady-state intermediate block $\mathcal{E}$ of excited levels into a fully absorbing sink block $\mathcal{S}$ of fully associated species AB kept fully depleted with $\gamma_{\mathrm{s}} = 0$ so that there is no backward re-dissociation from block $\mathcal{S}$. The frequency k_{d} is deduced as if the reverse scenario, $\gamma_{\mathrm{s}} = 1$ and $\gamma_{\mathrm{c}} = 0$, holds. This picture uncouples $\hat{\alpha}$ and k_{d}, and allows each coefficient to be calculated independently. Both dissociation (or ionization) and association (recombination) occur within block $\mathcal{E}$.

If $\gamma_{\mathrm{c}} = 1$ and $\gamma_{\mathrm{s}} = 0$, then

$$\rho^* = n^*/\tilde{n}^* = \nu_{\mathrm{d}}/(\nu_{\mathrm{s}} + \nu_{\mathrm{d}})\,, \qquad (52.43\mathrm{a})$$

$$K = \tilde{n}^*/\tilde{n}_A\tilde{n}_B = k_{\mathrm{c}}/\nu_{\mathrm{d}} = k_{\mathrm{c}}\tau_{\mathrm{d}}\,, \qquad (52.43\mathrm{b})$$

$$P^{\mathrm{S}} = \nu_{\mathrm{s}}/(\nu_{\mathrm{s}} + \nu_{\mathrm{d}}) = \rho^*\nu_{\mathrm{s}}\tau_{\mathrm{s}}\,, \qquad (52.43\mathrm{c})$$

and the recombination coefficient is

$$\hat{\alpha} = k_{\mathrm{c}}P^{\mathrm{S}} = k_{\mathrm{c}}\left[\rho^*\nu_{\mathrm{s}}\tau_{\mathrm{d}}\right] = K\rho^*\nu_{\mathrm{s}}\,. \qquad (52.44)$$

Microscopic Generalization. From (52.167), the microscopic generalizations of rate (52.40) and probability (52.43c) are, respectively,

$$\hat{\alpha} = \bar{v}\int_0^\infty \varepsilon e^{-\varepsilon}d\varepsilon \int_0^{b_0} 2\pi b\, db\, P^{\mathrm{S}}(\varepsilon, b; R_0)\,, \qquad (52.45\mathrm{a})$$

$$P^{\mathrm{S}}(\varepsilon, b; R_0) = \oint_{R_i}^{R_0} \rho_i(R)\nu_i^b(R)\, dt \equiv \langle\rho\nu_{\mathrm{s}}\rangle\,\tau_{\mathrm{d}}\,, \qquad (52.45\mathrm{b})$$

where $\rho_i(R) = n(\varepsilon, b; R)/\tilde{n}(\varepsilon, b; R)$; $\nu_i^{(b)}$ is the frequency (52.164a) of $(A{-}B){-}M$ continuum-bound collisional transitions at fixed $A{-}B$ separation R, R_i is the pericenter of the orbit, $|\,i\,\rangle \equiv |\,\varepsilon, b\,\rangle$, and

$$b_0^2 = R_0^2\left[1 - V(R)/E\right]\,, \quad \varepsilon = E/k_{\mathrm{B}}T\,, \qquad (52.45\mathrm{c})$$

$$\hat{\alpha} \equiv k_{\mathrm{c}}\langle P^{\mathrm{S}}\rangle_{\varepsilon, b}\,, \quad \bar{v} = (8k_{\mathrm{B}}T/\pi M_{AB})^{1/2}\,, \qquad (52.45\mathrm{d})$$

$$k_{\mathrm{c}} = \left\{\pi R_0^2\left[1 - V(R_0)/k_{\mathrm{B}}T\right]\bar{v}\right\}\,. \qquad (52.45\mathrm{e})$$

where M_{AB} is the reduced mass of A and B.

Low Gas Densities. Here $\rho_i(R) = 1$ for $E > 0$,

$$P^{\mathrm{S}}(\varepsilon, b; R_0) = \oint_{R_i}^{R_0} \nu(t)\, dt = \oint_{R_i}^{R_0} ds/\lambda_i\,. \qquad (52.46)$$

$\lambda_i = (N\sigma)^{-1}$ is the microscopic path length towards the $(A{-}B){-}M$ reactive collision with frequency $\nu = N v\sigma$. For λ_i constant, the rate (52.45a) reduces at low N to

$$\hat{\alpha} = (v\sigma_0 N)\int_0^{R_0}\left[1 - \frac{V(R)}{k_{\mathrm{B}}T}\right]4\pi R^2\, dR \qquad (52.47)$$

which is linear in the gas density N.

52.3.3 Working Formulae for Three-Body Collisional Recombination at Low Density

For three-body ion-ion collisional recombination of the form $A^+ + B^- + M$ in a gas at low density N, set $V(R) = -e^2/R$. Then (52.47) yields

$$\hat{\alpha}^{\mathrm{c}}(T) = \left(\frac{8k_{\mathrm{B}}T}{\pi M_{AB}}\right)^{1/2}\frac{4}{3}\pi R_0^3\left[1 + \frac{3}{2}\frac{R_{\mathrm{e}}}{R_0}\right](\sigma_0 N)\,, \qquad (52.48)$$

where $R_{\mathrm{e}} = e^2/k_{\mathrm{B}}T$, and the trapping radius R_0, determined by the classical variational method, is $0.41R_{\mathrm{e}}$, in agreement with detailed calculation. The special cases are:

(a) $\mathrm{e}^- + A^+ + \mathrm{e}^-$. Here, $\sigma_0 = \frac{1}{9}\pi R_{\mathrm{e}}^2$ for $(\mathrm{e}^- - \mathrm{e}^-)$ collisions for scattering angles $\theta \geq \pi/2$ so that

$$\hat{\alpha}_{ee}^{\mathrm{c}}(T) = 2.7 \times 10^{-20}\left(\frac{300}{T}\right)^{4.5}n_{\mathrm{e}}\ \mathrm{cm}^3\,\mathrm{s}^{-1} \qquad (52.49)$$

in agreement with Mansbach and Keck [3].

(b) $A^+ + B^- + M$. Here, $\sigma_0 \bar{v} \sim 10^{-9} \mathrm{cm}^3\,\mathrm{s}^{-1}$, which is independent of T for polarization attraction. Then

$$\hat{\alpha}_3(T) = 2 \times 10^{-25} \left(\frac{300}{T}\right)^{2.5} N \ \mathrm{cm}^3\,\mathrm{s}^{-1}. \quad (52.50)$$

(c) $e^- + A^+ + M$. Only a small fraction $\delta = 2m/M$ of the electron's energy is lost upon $(e^- - M)$ collision so that (52.45a) for constant λ is modified to

$$\hat{\alpha}_{eM} = \sigma_0 N \int_0^{R_0} 4\pi R^2\, dR \int_0^{E_m} \tilde{n}(R,E) v\, dE \quad (52.51a)$$

$$= \bar{v}_e \sigma_0 N \int_0^{R_0} 4\pi R^2\, dR \int_0^{\varepsilon_m} \left[1 - \frac{V(R)}{E}\right] \varepsilon e^{-\varepsilon}\, d\varepsilon \quad (52.51b)$$

where $\varepsilon = E/k_\mathrm{B}T$, and $E_m = \delta e^2/R = \varepsilon_m k_\mathrm{B}T$ is the maximum energy for collisional trapping. Hence,

$$\hat{\alpha}_{eM}(T_e) = 4\pi\delta \left(\frac{8k_\mathrm{B}T_e}{\pi m_e}\right)^{1/2} R_e^2 R_0 \left[\sigma_0 N\right] \quad (52.52a)$$

$$\sim \frac{10^{-26}}{M} \left(\frac{300}{T}\right)^{2.5} N \ \mathrm{cm}^3\,\mathrm{s}^{-1}, \quad (52.52b)$$

where the mass M of the gas atom is now in u. This result agrees with the energy diffusion result of Pitaevskiĭ [4] when R_0 is taken as the Thomson radius $R_\mathrm{T} = \frac{2}{3} R_e$.

52.3.4 Recombination Influenced by Diffusional Drift at High Gas Densities

Diffusional-Drift Current. The drift current of A^+ towards B^- in a gas under an A^+–B^- attractive potential $V(R)$ is

$$\mathbf{J}(R) = -D\nabla n(R) - \left[\frac{K}{e}\nabla V(R)\right] n(R) \quad (52.53a)$$

$$= -\left[D\tilde{N}_A\tilde{N}_B e^{-V(R)/k_\mathrm{B}T} \frac{\partial\rho}{\partial R}\right]\hat{\mathbf{R}}. \quad (52.53b)$$

Relative Diffusion and Mobility Coefficients.

$$D = D_A + D_B, \\ K = K_A + K_B, \quad De = K(k_\mathrm{B}T), \quad (52.54)$$

where the D_i and K_i are, respectively, the diffusion and mobility coefficients of species i in gas M.

Normalized Ion-Pair R-Distribution.

$$\rho(R) = \frac{n(R)}{\tilde{N}_A\tilde{N}_B \exp\left[-V(R)/k_\mathrm{B}T\right]}. \quad (52.55)$$

Continuity Equations for Currents and Rates.

$$\frac{\partial n}{\partial t} + \nabla\cdot\mathbf{J} = 0, \quad R \geq R_0 \quad (52.56a)$$

$$\hat{\alpha}_{\mathrm{RN}}(R_0)\rho(R_0) = \hat{\alpha}\rho(\infty) \quad (52.56b)$$

The rate of reaction for ion-pairs with separations $R \leq R_0$ is $\alpha_{\mathrm{RN}}(R_0)$. This is the recombination rate that would be obtained for a thermodynamic equilibrium distribution of ion pairs with $R \geq R_0$, i.e. for $\rho(R \geq R_0) = 1$.

Steady-State Rate of Recombination.

$$\hat{\alpha}\tilde{N}_A\tilde{N}_B = \int_{R_0}^{\infty} \left(\frac{\partial n}{\partial t}\right) d\mathbf{R} = -4\pi R_0^2 J(R_0). \quad (52.57)$$

Steady-State Solution.

$$\rho(R) = \rho(\infty)\left[1 - \frac{\hat{\alpha}}{\hat{\alpha}_{\mathrm{TR}}(R)}\right], \quad R \geq R_0 \quad (52.58a)$$

$$\rho(R_0) = \rho(\infty)\left[\hat{\alpha}/\hat{\alpha}_{\mathrm{RN}}(R_0)\right]. \quad (52.58b)$$

Recombination Rate.

$$\hat{\alpha} = \frac{\hat{\alpha}_{\mathrm{RN}}(R_0)\hat{\alpha}_{\mathrm{TR}}(R_0)}{\hat{\alpha}_{\mathrm{RN}}(R_0) + \hat{\alpha}_{\mathrm{TR}}(R_0)} \quad (52.59a)$$

$$\rightarrow \begin{cases} \hat{\alpha}_{\mathrm{RN}}, & N \rightarrow 0 \\ \hat{\alpha}_{\mathrm{TR}}, & N \rightarrow \infty. \end{cases} \quad (52.59b)$$

Diffusional-Drift Transport Rate.

$$\hat{\alpha}_{\mathrm{TR}}(R_0) = 4\pi D\left[\int_{R_0}^{\infty} \frac{e^{V(R)/k_\mathrm{B}T}}{R^2}\, dR\right]^{-1}. \quad (52.60)$$

With $V(R) = -e^2/R$,

$$\hat{\alpha}_{\mathrm{TR}}(R_0) = 4\pi K e\left[1 - \exp(-R_e/R_0)\right]^{-1}, \quad (52.61)$$

where $R_e = e^2/k_\mathrm{B}T$ provides a natural unit of length.

Langevin Rate. For $R_0 \ll R_e$, the transport rate

$$\hat{\alpha}_{\mathrm{TR}} \rightarrow \hat{\alpha}_L = 4\pi K e, \quad (52.62)$$

tends to the Langevin rate which varies as N^{-1}.

Reaction Rate. When R_0 is large enough that R_0–pairs are in (E, L^2) equilibrium (*cf.* (52.167)),

$$\hat{\alpha}_{\mathrm{RN}}(R_0) = \bar{v}\int_0^{\infty} \varepsilon e^{-\varepsilon}\, d\varepsilon \int_0^{b_0} 2\pi b\, db\, P^S(\varepsilon, b; R_0)$$

$$\equiv \bar{v}\int_0^{\infty} \varepsilon e^{-\varepsilon}\, d\varepsilon \left[\pi b_0^2 P^S(\varepsilon; R_0)\right] \quad (52.63a)$$

$$\equiv \bar{v}\pi b_{\max}^2 P^S(R_0), \quad (52.63b)$$

where

$$b_0^2 = R_0^2\left[1 - V(R_0)/E\right], \quad \varepsilon = E/k_\mathrm{B}T, \quad (52.64a)$$

$$\bar{v} = (8kT/\pi M_{AB})^{1/2}, \quad (52.64b)$$

$$b_{\max}^2 = R_0^2\left[1 - \frac{V(R_0)}{k_\mathrm{B}T}\right]. \quad (52.64c)$$

The probability P^S and its averages over b and (b, E) for reaction between pairs with $R \leq R_0$ is determined in (52.63a)–(52.63b) from solutions of coupled master equations. P^S increases linearly with N initially and tends to unity at high N. The recombination rate (52.59a) with (52.63a) and (52.61) therefore increases linearly with N initially, reaches a maximum when $\hat{\alpha}_{TR} \sim \hat{\alpha}_{RN}$ and then decreases eventually as N^{-1}, in accord with (52.3.4).

Reaction Probability. The classical absorption solution of (52.157) is

$$P^S(E, b; R_0) = 1 - \exp\left[-\oint_{R_i}^{R_0} \frac{ds_i}{\lambda_i}\right] . \quad (52.65)$$

With the binary decomposition $\lambda_i^{-1} = \lambda_{iA}^{-1} + \lambda_{iB}^{-1}$,

$$P^S = P_A + P_B - P_A P_B . \quad (52.66)$$

Exact b^2–Averaged Probability. With $V_c = -e^2/R$ for the A^+–B^- interaction in (52.63a), and at low gas densities N,

$$P_{A,B}(E, R_0) = \frac{\frac{4R_0}{3\lambda_{A,B}}\left[1 - \frac{3V_c(R_0)}{2E_i}\right]}{[1 - V_c(R_0)/E_i]} \quad (52.67)$$

appropriate for constant mean free path λ_i.

(E, b^2)–Averaged Probability. $P^S(R_0)$ in (52.63b) at low gas density is

$$P_{A,B}(R_0) = P_{A,B}(E = k_B T, R_0) . \quad (52.68)$$

Thomson Trapping Distance. When the kinetic energy gained from Coulomb attraction is assumed lost upon collision with third bodies, then bound $A - B$ pairs are formed with $R \leq R_T$. Since $E = \frac{3}{2}k_B T - e^2/R$, then

$$R_T = \frac{2}{3}\left(\frac{e^2}{k_B T}\right) = \frac{2}{3}R_e . \quad (52.69)$$

Thomson Straight-Line Probability. The $E \to \infty$ limit of (52.65) is

$$P_{A,B}^T(b; R_T) = 1 - \exp\left[-2(R_T^2 - b^2)/\lambda_{A,B}\right] . \quad (52.70)$$

The b^2–average is the Thomson probability

$$P_{A,B}^T(R_T) = 1 - \frac{1}{2X^2}\left[1 - e^{-2X}(1 + 2X)\right] \quad (52.71a)$$

for reaction of $(A - B)$ pairs with $R \leq R_T$. As $N \to 0$

$$P_{A,B}^T(R_T) \to \frac{4}{3}X\left[1 - \frac{3}{4}X + \frac{2}{5}X^2 - \frac{1}{6}X^3 + \cdots\right] \quad (52.71b)$$

and tends to unity at high N. $X = R_T/\lambda_{A,B} = N(\sigma_0 R_T)$. These probabilites have been generalized [13] to include hyperbolic and general trajectories.

Thomson Reaction Rate.

$$\hat{\alpha}_T = \pi R_T^2 \overline{v}\left[P_A^T + P_B^T - P_A^T P_B^T\right] \quad (52.72)$$

$$\to \begin{cases} \frac{4}{3}\pi R_T^3(\lambda_A^{-1} + \lambda_B^{-1}), & N \to 0 \\ \pi R_T^2 \overline{v}, & N \to \infty . \end{cases}$$

52.4 DISSOCIATIVE RECOMBINATION

52.4.1 Curve-Crossing Mechanisms

Direct Process. Dissociative recombination (DR) for diatomic ions can occur via a crossing at R_X between the bound and repulsive potential energy curves $V^+(R)$ and $V_d(R)$ for AB^+ and AB^{**}, respectively. Here, DR involves the two-stage sequence

$$e^- + AB^+(v_i) \underset{\nu_a}{\overset{k_c}{\rightleftharpoons}} (AB^{**})_R \overset{\nu_d}{\longrightarrow} A + B^* . \quad (52.73)$$

The first stage is dielectronic capture whereby the free electron of energy $\varepsilon = V_d(R) - V^+(R)$ excites an electron of the diatomic ion AB^+ with internal separation R and is then resonantly captured by the ion, at rate k_c, to form a repulsive state d of the doubly excited molecule AB^{**}, which in turn can either autoionize at probability frequency ν_a, or else in the second stage predissociate into various channels at probability frequency ν_d. This competition continues until the (electronically excited) neutral fragments accelerate past the crossing at R_X. Beyond R_X the increasing energy of relative separation reduces the total electronic energy to such an extent that autoionization is essentially precluded and the neutralization is then rendered permanent past the stabilization point R_X. This interpretation [6] has remained intact and robust in the current light of *ab initio* quantum chemistry and quantal scattering calculations for the simple diatomics (O_2^+, N_2^+, Ne_2^+, etc.). Mechanism (52.73) is termed the direct process which, in terms of the macroscopic frequencies in (52.73), proceeds at the rate

$$\hat{\alpha} = k_c P_S = k_c\left[\nu_d/(\nu_a + \nu_d)\right] , \quad (52.74)$$

where P_S is probability for $A - B^*$ survival against autoionization from the initial capture at R_c to the crossing point R_X. Configuration mixing theories of this direct process are available in the quantal [7] and semiclassical-classical path formulations [8].

Indirect Process. In the three-stage sequence

$$e^- + AB^+(v_i^+) \to [AB^+(v_f) - e^-]_n \to (AB^{**})_d$$
$$\to A + B^* \quad (52.75)$$

the so-called *indirect* process [7] might contribute. Here the accelerating electron loses energy by vibrational excitation $(v_i^+ \to v_f)$ of the ion and is then resonantly captured into a Rydberg orbital of the bound molecule AB^* in vibrational level v_f, which then interacts one way (via configuration mixing) with the doubly excited repulsive molecule AB^{**}. The capture initially proceeds via a small effect — vibronic coupling (the matrix element of the nuclear kinetic energy) induced by the breakdown of the Born-Oppenheimer approximation — at certain resonance energies $\varepsilon_n = E(v_f) - E(v_i^+)$ and, in the absence of the direct channel (52.73), would therefore be manifest by a series of characteristic very narrow Lorentz profiles in the cross section. Uncoupled from (52.73) the indirect process would augment the rate. Vibronic capture proceeds more easily when $v_f = v_i^+ + 1$ so that Rydberg states with $n \approx 7 - 9$ would be involved [for $H_2^+(v_i^+ = 0)$] so that the resulting longer periods of the Rydberg electron would permit changes in nuclear motion to compete with the electronic dissociation. Recombination then proceeds as in the second stage of (52.73), i.e., by electronic coupling to the dissociative state d at the crossing point. A multichannel quantum defect theory [9] has combined the direct and indirect mechanisms

Interrupted Recombination. The process

$$\mathrm{e}^- + AB^+(v_i) \overset{k_c}{\underset{\nu_a}{\rightleftharpoons}} (AB^{**})_d \overset{\nu_d}{\to} A + B^*$$
$$\nu_{nd} \updownarrow \nu_{dn}$$
$$[AB^+(v) - \mathrm{e}^-]_n \tag{52.76}$$

proceeds via the first (dielectronic capture) stage of (52.73) followed by a two-way electronic transition with frequency ν_{dn} and ν_{nd} between the d and n states. All (n, v) Rydberg states can be populated, particularly those in low n and high v since the electronic $d - n$ interaction varies as $n^{-1.5}$ with broad structure. Although the dissociation process proceeds here via a second order effect $(\nu_{dn}$ and $\nu_{nd})$, the electronic coupling may dominate the indirect vibronic capture and interrupt the recombination, in contrast to (52.75) which, as written in the one-way direction, feeds the recombination. Both dip and spike structure has been observed [10].

52.4.2 Quantal Cross Section

The cross section for direct dissociative recombination

$$\mathrm{e}^- + AB^+(v_i^+) \rightleftharpoons (AB^{**})_r \longrightarrow A + B^* \tag{52.77}$$

of electrons of energy ε, wavenumber k_e and spin statistical weight 2, for a molecular ion $AB^+(v_i^+)$ of electronic statistical weight ω_{AB}^+ in vibrational level v_i^+ is

$$\sigma_{\mathrm{DR}}(\varepsilon) = \frac{\pi}{k_\mathrm{e}^2}\left(\frac{\omega_{AB}^*}{2\omega^+}\right)|a_Q|^2 = \left(\frac{h^2}{8\pi m_\mathrm{e}\varepsilon}\right)\left(\frac{\omega_{AB}^*}{2\omega^+}\right)|a_Q|^2 . \tag{52.78}$$

Here ω_{AB}^* is the electronic statistical weight of the dissociative neutral state of AB^* whose potential energy curve V_d crosses the corresponding potential energy curve V^+ of the ionic state. The transition T-matrix element for autoionization of AB^* embedded in the (moving) electronic continuum of $AB^+ + \mathrm{e}^-$ is the quantal probability amplitude

$$a_Q(v) = 2\pi \int_0^\infty V_{d\varepsilon}^*(R) \left[\psi_v^{+*}(R)\psi_d(R)\right] dR \tag{52.79}$$

for autoionization. Here ψ_v^+ and ψ_d are the nuclear bound and continuum vibrational wave functions for AB^+ and AB^*, respectively, while

$$V_{d\varepsilon}(R) = \langle \phi_d \mid \mathcal{H}_\mathrm{el}(\mathbf{r}, R(t)) \mid \phi_\varepsilon(\mathbf{r}, \mathbf{R}) \rangle_{\mathbf{r}, \hat{\varepsilon}}$$
$$= V_{\varepsilon d}^*(R) \tag{52.80}$$

are the bound-continuum electronic matrix elements coupling the diabatic electronic bound state wave functions $\psi_d(\mathbf{r}, \mathbf{R})$ for AB^* with the electronic continuum state wave functions $\phi_\varepsilon(\mathbf{r}, \mathbf{R})$ for $AB^+ + \mathrm{e}^-$. The matrix element is an average over electronic coordinates $\mathbf{r}$ and all directions $\hat{\varepsilon}$ of the continuum electron. Both continuum electronic and vibrational wave functions are energy normalized (see Sect. 52.8.3), and

$$\Gamma(R) = 2\pi |V_{d\varepsilon}^*(R)|^2 \tag{52.81}$$

is the energy width for autoionization at a given nuclear separation R. Given $\Gamma(R)$ from quantum chemistry codes, the problem reduces to evaluation of continuum vibrational wave functions in the presence of autoionization. The rate associated with a Maxwellian distribution of electrons at temperature T is

$$\hat{\alpha} = \overline{v}_\mathrm{e} \int \varepsilon \, \sigma_{\mathrm{DR}}(\varepsilon) e^{-\varepsilon/k_\mathrm{B}T} \, d\varepsilon /(k_\mathrm{B}T)^2 \tag{52.82}$$

where $\overline{v}_e$ is the mean speed (see Sect. 52.9).

Maximum Cross Section and Rate. Since the probability for recombination must remain less than unity, $|a_Q|^2 \leq 1$ so that the maximum cross section and rates are

$$\sigma_{\mathrm{DR}}^{\max}(\varepsilon) = \frac{\pi}{k_\mathrm{e}^2}\left(\frac{\omega_{AB}^*}{2\omega^+}\right) = \left(\frac{h^2}{8\pi m_\mathrm{e}\varepsilon}\right)(2\ell + 1), \tag{52.83}$$

where ω_{AB}^* has been replaced by $2(2\ell + 1)\omega^+$ under the assumption that the captured electron is bound in a high level Rydberg state of angular momentum ℓ, and

$$\hat{\alpha}_{\max}(T) = \overline{v}_\mathrm{e} \, \sigma_{\mathrm{DR}}^{\max}(\varepsilon = k_\mathrm{B}T) \tag{52.84a}$$

$$\approx 5 \times 10^{-7} \left(\frac{300}{T}\right)^{1/2}(2\ell + 1) \ \mathrm{cm}^3/\mathrm{s}. \tag{52.84b}$$

Cross section maxima of $5(2\ell + 1)(300/T) \times 10^{-14} \ \mathrm{cm}^2$ are therefore possible, being consistent with the rate (52.84b).

First-Order Quantal Approximation. When the effect of autoionization on the continuum vibrational wave function $\psi_d(R)$ for AB^* is ignored, then a first-order undistorted approximation to the quantal amplitude (52.79) is

$$T_B(v^+) = 2\pi \int_0^\infty V_{d\varepsilon}^*(R) \left[\psi_v^{+*}(R) \psi_d^{(0)}(R) \right] dR \quad (52.85)$$

where $\psi_d^{(0)}$ is ψ_d in the absence of the back reaction of autoionization. Under this assumption, (52.78) reduces to

$$\sigma_c(\varepsilon, v^+) = \frac{\pi}{k_e^2} \left(\frac{\omega_{AB}^*}{2\omega^+} \right) |T_B(v^+)|^2 , \quad (52.86)$$

which is then the cross section for initial electron capture since autoionization has been precluded. Although the Born T-matrix (52.85) violates unitarity, the capture cross section (52.86) must remain less then the maximum value

$$\sigma_c^{\max} = \frac{\pi}{k_e^2} \left(\frac{\omega_{AB}^*}{2\omega^+} \right) = \left(\frac{h^2}{8\pi m_e \varepsilon} \right) \left(\frac{\omega_{AB}^*}{2\omega^+} \right), \quad (52.87)$$

since $|a_Q|^2 \leq 1$. So as to acknowledge after the fact the effect of autoionization, assumed small, and neglected by (52.85), the DR cross section can be approximated as

$$\sigma_{DR}(\varepsilon, v^+) = \sigma_c(\varepsilon, v^+) P_S , \quad (52.88)$$

where P_S is the probability of survival against autoionization on the V_d curve until stabilization takes place at some crossing point R_X.

Approximate Capture Cross Section. With the energy-normalized Winans-Stückelberg vibrational wave function

$$\psi_d^{(0)}(R) = |V_d'(R)|^{-1/2} \delta(R - R_c), \quad (52.89)$$

where R_c is the classical turning point for $(A - B^*)$ relative motion, (52.86) reduces to

$$\sigma_c(\varepsilon, v^+) = \frac{\pi}{k_e^2} \left(\frac{\omega_{AB}^*}{2\omega^+} \right) [2\pi\Gamma(R_c)] \left\{ \frac{|\psi_v^+(R_c)|^2}{|V_d'(R_c)|} \right\} \quad (52.90)$$

where the term inside the braces in (52.90) is the effective Franck-Condon factor.

Six Approximate Stabilization Probabilities.
(1) A unitarized T−matrix is

$$T = \frac{T_B}{1 + \left| \frac{1}{2} T_B \right|^2} , \quad (52.91)$$

so that $P_S = |T|^2 / |T_B|^2$ to give

$P_S(\text{low } \varepsilon)$

$$= \left[1 + \frac{1}{4} |T_B|^2 \right]^{-2}$$

$$= \left\{ 1 + \pi^2 \left| \int_0^\infty V_{d\varepsilon}^*(R) \left[\psi_v^{+*}(R) \psi_d^{(0)}(R) \right] dr \right|^2 \right\}^{-2} \quad (52.92a)$$

which is valid at low ε when only one vibrational level v^+, i.e., the initial level of the ion is repopulated by autoionization.

(2) At higher ε, when population of many other ionic levels v_f^+ occurs, then

$$P_S(\varepsilon) = \left[1 + \frac{1}{4} \sum_f |T_B(v_f^+)|^2 \right]^{-2} , \quad (52.92b)$$

where the summation is over all the open vibrational levels v_f^+ of the ion. When no intermediate Rydberg $AB^*(v)$ states are energy resonant with the initial $e^- + AB^+(v^+)$ state, i.e., coupling with the indirect mechanism is neglected, then (52.88) with (52.92b) is the direct DR cross section normally calculated.

(3) In the high-ε limit when an infinite number of v_f^+ levels are populated following autoionization, the survival probability, with the aid of closure, is then

$$P_S = \left[1 + \pi^2 \int_{R_c}^{R_X} |V_{d\varepsilon}^*(R)|^2 \left| \psi_d^{(0)}(R) \right|^2 dR \right]^{-2}. \quad (52.93)$$

(4) On adopting in (52.93) the JWKB semiclassical wave function for $\psi_d^{(0)}$, then

$$P_S(\text{high } \varepsilon) = \left[1 + \frac{1}{2\hbar} \int_{R_c}^{R_X} \frac{\Gamma(R)}{v(R)} dR \right]^{-2}$$

$$= \left[1 + \frac{1}{2} \int_{t_c}^{t_X} \nu_a(t) dt \right]^{-2} , \quad (52.94)$$

where $v(R)$ is the local radial speed of $A - B$ relative motion, and where the frequency $\nu_a(t)$ of autoionization is $\Gamma/\hbar$.

(5) A classical path local approximation for P_S yields

$$P_S = \exp\left(-\int_{t_c}^{t_X} \nu_a(t) dt \right) , \quad (52.95)$$

which agrees to first-order for small ν with the expansion of (52.94).

(6) A partitioning of (52.73) yields

$$P_S = \nu_d/(\nu_a + \nu_d) = (1 + \nu_a \tau_d)^{-1} , \quad (52.96)$$

on adopting macroscopic averaged frequencies ν_i and associated lifetimes $\tau_i = \nu_i^{-1}$. The six survival probabilities in (52.92a), (52.92b), (52.93)–(52.96) are all suitable for use in the DR cross section (52.88).

52.4.3 Noncrossing Mechanism

The dissociative recombination (DR) processes

$$e^- + H_3^+ \rightarrow H_2 + H$$
$$\rightarrow H + H + H \qquad (52.97)$$

at low electron energy ε, and

$$e^- + HeH^+ \rightarrow He + H(n=2) \qquad (52.98)$$

have spurred renewed theoretical interest because they both proceed at respective rates of $(2 \times 10^{-7}$ to $2 \times 10^{-8})$ cm^3 s^{-1} and 10^{-8} cm^3 s^{-1} at 300 K. Such rates are generally associated with the direct DR, which involves favorable curve crossings between the potential energy surfaces, $V^+(R)$ and $V_d(R)$ for the ion AB^+ and neutral dissociative AB^{**} states. The difficulty with (52.97) and (52.98) is that there are no such curve crossings, except at $\varepsilon \geq 8$ eV for (52.97). In this instance, the previous standard theories would support only extremely small rates when electronic resonant conditions do not prevail at thermal energies. Theories [11, 12] are currently being developed for application to processes such as (52.97).

52.5 MUTUAL NEUTRALIZATION

$$A^+ + B^- \rightarrow A + B. \qquad (52.99)$$

Diabatic Potentials. $V_i^{(0)}(R)$ and $V_f^{(0)}(R)$ for initial (ionic) and final (covalent) states are diagonal elements of

$$V_{if}(R) = \langle \Psi_i(\mathbf{r}, \mathbf{R}) \mid \mathcal{H}_{el}(\mathbf{r}, \mathbf{R}) \mid \Psi_f(\mathbf{r}, \mathbf{R}) \rangle_{\mathbf{r}}, \qquad (52.100)$$

where $\Psi_{i,f}$ are diabatic states and $\mathcal{H}_{el}$ is the electronic Hamiltonian at fixed internuclear distance R.

Adiabatic Potentials for a Two-State System.

$$V^{\pm}(R) = V_0(R) \pm \left[\Delta^2(R) + |V_{if}(R)|^2\right]^{1/2} \qquad (52.101a)$$

$$V_0(R) = \frac{1}{2}\left[V_i^{(0)}(R) + V_f^{(0)}(R)\right] \qquad (52.101b)$$

$$\Delta(R) = \left[V_i^{(0)}(R) - V_f^{(0)}(R)\right]. \qquad (52.101c)$$

For a single crossing of diabatic potentials at R_X then $V_i^{(0)}(R_X) = V_f^{(0)}(R_X)$ and the adiabatic potentials at R_X are,

$$V^{\pm}(R_X) = V_i^{(0)}(R_X) \pm V_{if}(R_X) \qquad (52.102)$$

with energy separation $2V_{if}(R_X)$.

52.5.1 Landau-Zener Probability for Single Crossing at R_X

On assuming $\Delta(R) = (R - R_X)\Delta'(R_X)$, where $\Delta'(R) = d\Delta(R)/dR$, the probability for single crossing is

$$P_{if}(R_X) = \exp\left[\eta(R_X)/v_X(b)\right] \qquad (52.103a)$$

$$\eta(R_X) = \left(\frac{2\pi}{\hbar}\right)\frac{|V_{if}(R_X)|^2}{\Delta'(R_X)} \qquad (52.103b)$$

$$v_X(b) = \left[1 - V_i^{(0)}(R_X)/E - b^2/R_X^2\right]^{1/2}. \qquad (52.103c)$$

Overall Charge-Transfer Probability. From the incoming and outgoing legs of the trajectory,

$$P^X(E) = 2P_{if}(1 - P_{if}). \qquad (52.104)$$

52.5.2 Cross Section and Rate Coefficient for Mutual Neutralization

$$\sigma_M(E) = 4\pi \int_0^{b_X} P_{if}(1 - P_{if})b\,db = \pi b_X^2 P_M, \quad (52.105a)$$

$$\pi b_X^2 = \pi\left[1 - \frac{V_i^{(0)}(R_X)}{E}\right]R_X^2$$

$$= \pi\left[1 + \frac{14.4}{R_X(\text{Å})E(\text{eV})}\right]R_X^2. \qquad (52.105b)$$

P_M is the b^2-averaged probability (52.104) for charge-transfer reaction within a sphere of radius R_X.

The rate is

$$\hat{\alpha}_M = (8k_BT/\pi M_{AB})^{1/2}\int_0^\infty \epsilon\sigma_M(\epsilon)e^{-\epsilon}d\epsilon \qquad (52.106)$$

where $\epsilon = E/k_BT$.

52.6 ONE-WAY MICROSCOPIC EQUILIBRIUM CURRENT, FLUX, AND PAIR-DISTRIBUTIONS

Notation:

M	reduced mass $M_A M_B/(M_A + M_B)$
R	internal separation of $A - B$
E	orbital energy $\frac{1}{2}Mv^2 + V(R)$
L	orbital angular momentum
L^2	$2MEb^2$ for $E > 0$
v_R	radial speed $|\dot{R}|$
$\bar{v}$	mean relative speed $(8kT/\pi M_{AB})^{1/2}$
ε	normalized energy E/k_BT
n_i	pair distribution function $n_i^+ + n_i^-$
$n_i^{\pm}$	component of n_i with $\dot{R} > 0$ (+) and $\dot{R} < 0$ (−).

All quantities on the RHS in the expressions (a)–(e) below are to be multiplied by $\tilde{N}_A \tilde{N}_B [\omega_{AB}/\omega_A \omega_B]$ where the ω_i denote the statistical weights of species i which are *not* included by the density of states associated with the E, L^2 orbital degrees of freedom.

Case (a) $|i\rangle \equiv |\mathbf{R}, E, L^2\rangle$.

Current: $j_i^\pm(R) = n^\pm(\mathbf{R}, E, L^2) v_R \equiv n_i^\pm v_R$

Flux: $4\pi R^2 j_i^\pm(R)\, dE\, dL^2 = \dfrac{4\pi^2 e^{-E/k_B T}}{(2\pi M k_B T)^{3/2}}\, dE\, dL^2$. (52.107)

This flux is independent of R. For dissociated pairs $E > 0$,

$$4\pi R^2 j_i^\pm(R)\, dE\, dL^2 = \left[\overline{v}\varepsilon e^{-\varepsilon} d\varepsilon\right] [2\pi b\, db] . \quad (52.108)$$

$(\mathbf{R}, E, L^2)$ – Distribution: $n(\mathbf{R}, E, L^2)\, d\mathbf{R}\, dE\, dL^2$

$$= \dfrac{(8\pi^2/v_R) e^{-E/k_B T}}{(2\pi M k_B T)^{3/2}} \left(\dfrac{d\mathbf{R}}{4\pi R^2}\right) dE\, dL^2 . \quad (52.109)$$

Case (b) $|i\rangle \equiv |\mathbf{R}, E\rangle$; L^2–integrated quantities.

Current: $j_i^\pm(R) = \frac{1}{2} v n^\pm(\mathbf{R}, E) \equiv \frac{1}{2} v n_i^\pm$, (52.110)

Flux: $4\pi R^2 j_i^\pm(R)\, dE = \left[\overline{v}\varepsilon e^{-\varepsilon} d\varepsilon\right] \pi b_0^2$, (52.111a)

$\pi b_0^2 = \pi R^2 \left[1 - V(R)/E\right]$, (52.111b)

$(\mathbf{R}, E)$–Distribution:

$$n(\mathbf{R}, E)\, d\mathbf{R}\, dE = \dfrac{2}{\sqrt{\pi}} \left[\dfrac{E - V(R)}{k_B T}\right]^{1/2} e^{-\varepsilon} d\varepsilon\, d\mathbf{R}$$

$$\equiv G_{MB}(E, R)\, d\mathbf{R} , \quad (52.112)$$

which defines the *Maxwell-Boltzmann velocity vistribution* G_{MB} in the presence of the field $V(R)$.

Case (c) (E, L^2)–integrated quantities.

Current: $j^\pm(R) = \frac{1}{4} \overline{v} e^{-V(R)/k_B T}$, (52.113)

Flux: $4\pi R^2 j^\pm(R) = \pi R^2 \overline{v} e^{-V(R)/k_B T}$, (52.114)

Distribution: $n(R) = e^{-V(R)/k_B T}$. (52.115)

When E–integration is only over dissociated states $(E > 0)$, the above quantities are

$j_d^\pm(R) = \frac{1}{4} \overline{v} \left[1 - V(R)/k_B T\right]$, (52.116)

$4\pi R^2 j_d^\pm(R) = \pi R^2 \left[1 - \dfrac{V(R)}{k_B T}\right] \overline{v} \equiv \pi b_{max}^2 \overline{v}$, (52.117)

$n(R) = \left[1 - V(R)/k_B T\right]$. (52.118)

Case (d) (E, L^2)–Distribution. For Bound Levels

$$n(E, L^2)\, dE\, dL^2 = \dfrac{4\pi^2 \tau_R(E, L)}{(2\pi M k_B T)^{3/2}} e^{-E/k_B T}\, dE\, dL^2 ,$$

(52.119)

where $\tau_R = \oint dt = (\partial J_R/\partial E)$ is the period for bounded radial motion of energy E and radial action $J_R(E, L) = M \oint v_R\, dR$.

Case (e) E–Distribution. For bound levels

$$n(E)\, dE = \dfrac{2e^{-\varepsilon}}{\sqrt{\pi}}\, d\varepsilon \int_0^{R_A} \left(\dfrac{E - V}{k_B T}\right)^{1/2} d\mathbf{R} , \quad (52.120)$$

where R_A is the turning point $E = V(R_A)$.

Example: For electron-ion bounded motion, $V(R) = -Ze^2/R$, $R_A = Ze^2/|E|$, $R_e = Ze^2/k_B T$, $\varepsilon = E/k_B T$. Then $\tau_R = 2\pi(m/Ze^2)^{1/2}(R_A/2)^{3/2}$,

$$\int_0^{R_A} \left[\dfrac{R_e}{R} - |\varepsilon|\right]^{1/2} d\mathbf{R} = \dfrac{\pi^2}{4} R_A^{5/2} R_e^{1/2} , \quad (52.121)$$

and

$$n^s(E)\, dE = \left[\dfrac{2e^{-\varepsilon}}{\sqrt{\pi}}\, d\varepsilon\right] \dfrac{\pi^2}{4} R_A^{5/2} R_e^{1/2} \quad (52.122)$$

$$= \left[\dfrac{2e^{-\varepsilon}}{\sqrt{\pi}}\, d\varepsilon\right] \left(\dfrac{\pi^2 R_e^3}{4 |\varepsilon|^{5/2}}\right) . \quad (52.123)$$

For closely spaced levels in a hydrogenic $e^- - A^{Z+}$ system,

$$n^s(p, \ell) = n(E, L^2) \left(\dfrac{dE}{dp}\right) \left(\dfrac{dL^2}{d\ell}\right) \quad (52.124a)$$

$$n^s(p) = n(E) \left(\dfrac{dE}{dp}\right) . \quad (52.124b)$$

Using $E = -(2p^2)^{-1}(Z^2 e^2/a_0)$ and $L^2 = (\ell + 1/2)^2 \hbar^2$ for level (p, ℓ) then

$$\tau_R(E, L) \dfrac{dE}{dp} \left[\dfrac{dL^2}{d\ell}\right] = \left[\dfrac{dJ_R}{dp}\right]\left[\dfrac{dL^2}{d\ell}\right] \quad (52.125)$$

$$= h \left[(2\ell + 1)\hbar^2\right] \quad (52.126)$$

$$\dfrac{n^s(p, \ell)}{n_e N^+} = \dfrac{2(2\ell + 1)}{2\omega_A^+} \dfrac{h^3}{(2\pi m_e k_B T)^{3/2}} e^{I_p/k_B T} , \quad (52.127a)$$

$$\dfrac{n^s(p)}{n_e N^+} = \dfrac{2p^2}{2\omega_A^+} \dfrac{h^3}{(2\pi m_e k_B T)^{3/2}} e^{I_p/k_B T} , \quad (52.127b)$$

in agreement with the Saha ionization formula (52.16) where N^+ is the equilibrium concentration of A^{Z+} ions in their ground electronic states. The spin statistical weights are $\omega_{eA} = \omega_e = 2$.

52.7 MICROSCOPIC METHODS FOR TERMOLECULAR ION-ION RECOMBINATION

At low gas density, the basic process

$$A^+ + B^- + M \rightarrow AB + M \qquad (52.128)$$

is characterized by nonequilibrium with respect to E. Dissociated and bound A^+–B^- ion pairs are in equilibrium with respect to their separation R, but bound pairs are *not* in E–equilibrium with each other. L^2–equilibrium can be assumed for ion-ion recombination but *not* for ion-atom association reactions.

At higher gas densities N, there is nonequilibrium in the ion-pair distributions with respect to R, E and L^2. In the limit of high N, there is only nonequilibrium with respect to R. See Ref. [5] and the appropriate reference list for full details of theory.

52.7.1 Time Dependent Method: Low Gas Density

Energy levels E_i of A^+–B^- pairs are so close that they form a quasicontinuum with a nonequilibrium distribution over E_i determined by the *master equation*

$$\frac{dn_i(t)}{dt} = \int_{-D}^{\infty} (n_i \nu_{if} - n_f \nu_{fi})\, dE_f, \qquad (52.129)$$

where $n_i\, dE_i$ is the number density of pairs in the interval dE_i about E_i, and $\nu_{if}\, dE_f$ is the frequency of i-pair collisions with M that change the i-pair orbital energy from E_i to between E_f and $E_f + dE_f$. The greatest binding energy of the A^+–B^- pair is D.

Association Rate.

$$R^A(t) = \int_{-D}^{\infty} P_i^{\mathrm{S}} \left(\frac{dn_i}{dt}\right) dE_i \qquad (52.130a)$$

$$= \hat{\alpha} N_A(t) N_B(t) - k n_{\mathrm{s}}(t), \qquad (52.130b)$$

where P_i^{S} is the probability for collisional stabilization (recombination) of i-pairs via a sequence of energy changing collisions with M. The coefficients for $\mathcal{C} \rightarrow \mathcal{S}$ recombination out of the $\mathcal{C}$–block with ion concentrations $N_A(t)$, $N_B(t)$ (in cm^{-3}) into the $\mathcal{S}$ block of total ion-pair concentrations $n_{\mathrm{s}}(t)$ and for $\mathcal{S} \rightarrow \mathcal{C}$ dissociation are $\hat{\alpha}$ (cm^3 s^{-1}) and $k(s^{-1})$, respectively.

One-Way Equilibrium Collisional Rate and Detailed Balance.

$$C_{if} = \tilde{n}_i \nu_{if} = \tilde{n}_f \nu_{fi} = C_{fi}, \qquad (52.131)$$

where the tilde denotes equilibrium (Saha) distributions.

Normalized Distribution Functions.

$$\gamma_i(t) = n_i(t)/\tilde{n}_i^{\mathrm{S}}, \quad \gamma_{\mathrm{s}}(t) = n_{\mathrm{s}}(t)/\tilde{n}_{\mathrm{s}}^B(t), \qquad (52.132)$$

$$\gamma_{\mathrm{c}}(t) = N_A(t) N_B(t)/\tilde{N}_A \tilde{N}_B, \qquad (52.133)$$

where $\tilde{n}_i^{\mathrm{S}}$ and $\tilde{n}^B$ are the Saha and Boltzmann distributions.

Master Equation for $\gamma_i(t)$.

$$\frac{d\gamma_i(t)}{dt} = -\int_{-D}^{\infty} [\gamma_i(t) - \gamma_f(t)]\, \nu_{if}\, dE_f. \qquad (52.134)$$

Quasi-Steady State (QSS) Reduction. Set

$$\gamma_i(t) = P_i^{\mathrm{D}} \gamma_{\mathrm{c}}(t) + P_i^{\mathrm{S}} \gamma_{\mathrm{s}}(t) \overset{t \rightarrow \infty}{\longrightarrow} 1 \qquad (52.135)$$

where P_i^{D} and P_i^{S} are the respective time-independent portions of the normalized distribution γ_i which originate, respectively, from blocks $\mathcal{C}$ and $\mathcal{S}$. The energy separation between the $\mathcal{C}$ and $\mathcal{S}$ blocks is so large that $P_i^{\mathrm{S}} = 0$ ($E_i \geq 0$, $\mathcal{C}$ block), $P_i^{\mathrm{S}} \leq 1$ ($0 > E_i \geq -S$, $\mathcal{E}$ block), $P_i^{\mathrm{S}} = 1$ ($-S \geq E_i \geq -D$, $\mathcal{S}$ block). Since $P_i^{\mathrm{S}} + P_i^{\mathrm{D}} = 1$, then

$$\frac{d\gamma_i(t)}{dt} = -[\gamma_{\mathrm{c}}(t) - \gamma_{\mathrm{s}}(t)] \int_{-D}^{\infty} \left(P_i^{\mathrm{D}} - P_f^{\mathrm{D}}\right) C_{if}\, dE_f. \qquad (52.136)$$

Recombination and Dissociation Coefficients.
Equation (52.135) in (52.130a) enables the recombination rate in (52.130b) to be written as

$$\hat{\alpha} \tilde{N}_A \tilde{N}_B = \int_{-D}^{\infty} P_i^{\mathrm{D}}\, dE_i \int_{-D}^{\infty} \left(P_i^{\mathrm{D}} - P_f^{\mathrm{D}}\right) C_{if}\, dE_f. \qquad (52.137)$$

The QSS condition ($dn_i/dt = 0$ in block $\mathcal{E}$) is then

$$P_i^{\mathrm{D}} \int_{-D}^{\infty} \nu_{if}\, dE_f = \int_{-D}^{E} \nu_{if} P_f^{\mathrm{D}}\, dE_f, \qquad (52.138)$$

which involves only *time independent* quantities. Under QSS, (52.137) reduces to the net downward current across bound level $-E$,

$$\hat{\alpha} \tilde{N}_A \tilde{N}_B = \int_{-E}^{\infty} dE_i \int_{-D}^{-E} \left(P_i^{\mathrm{D}} - P_f^{\mathrm{D}}\right) C_{if}\, dE_f, \qquad (52.139)$$

which is independent of the energy level ($-E$) in the range $0 \geq -E \geq -S$ of block $\mathcal{E}$.

The dissociation frequency k in (52.130b) is

$$k \tilde{n}_{\mathrm{s}} = \int_{-D}^{-E} dE_i \int_{-E}^{\infty} \left(P_i^{\mathrm{S}} - P_f^{\mathrm{S}}\right) C_{if}\, dE_f, \qquad (52.140)$$

and macroscopic detailed balance $\hat{\alpha} \tilde{N}_A \tilde{N}_B = k \tilde{n}_{\mathrm{s}}$ is automatically satisfied. $\hat{\alpha}$ is the direct ($\mathcal{C} \rightarrow \mathcal{S}$) collisional contribution (small) plus the (much larger) net collisional cascade downward contribution from that fraction of bound levels which originated in the continuum $\mathcal{C}$. k_d is the direct dissociation frequency (small) plus the net collisional cascade upward contribution from that fraction of bound levels which originated in block $\mathcal{S}$.

52.7.2 Time Independent Methods: Low Gas Density

QSS-Rate. Since recombination and dissociation (ionization) involve only that fraction of the bound state population which originated from the $\mathcal{C}$ and $\mathcal{S}$ blocks, respectively, recombination can be viewed as time independent with

$$N_A N_B = \tilde{N}_A \tilde{N}_B, \quad n_s(t) = 0, \tag{52.141a}$$

$$\rho_i = n_i/\tilde{n}_i \equiv P_i^{\mathrm{D}} \tag{52.141b}$$

$$\hat{\alpha}\tilde{N}_A\tilde{N}_B = \int_{-E}^{\infty} dE_i \int_{-D}^{-E} (\rho_i - \rho_f)\, C_{if}\, dE_f\,. \tag{52.141c}$$

QSS Integral Equation.

$$\rho_i \int_{-D}^{\infty} \nu_{if}\, dE_f = \int_{-S}^{\infty} \rho_f \nu_{if}\, dE_f \tag{52.142}$$

is solved subject to the boundary condition

$$\rho_i = 1(E_i \geq 0), \quad \rho_i = 0(-S \geq E_i \geq -D). \tag{52.143}$$

Collisional Energy-Change Moments.

$$D^{(m)}(E_i) = \frac{1}{m!} \int_{-D}^{\infty} (E_f - E_i)^m C_{if}\, dE_f\,, \tag{52.144}$$

$$D_i^{(m)} = \frac{1}{m!}\frac{d}{dt}\langle (\Delta E)^m \rangle\,. \tag{52.145}$$

Averaged Energy-Change Frequency. For an equilibrium distribution $\tilde{n}_i$ of E_i−pairs per unit interval dE_i per second,

$$D_i^{(1)} = \frac{d}{dt}\langle \Delta E \rangle\,.$$

Averaged Energy-Change per Collision.

$$\langle \Delta E \rangle = D_i^{(1)}/D_i^{(0)}\,.$$

Time Independent Dissociation. The time independent picture corresponds to

$$n_s(t) = \tilde{n}_s, \quad \gamma_c(t) = 0, \quad \rho_i = n_i/\tilde{n}_i \equiv P_i^{\mathrm{S}}, \tag{52.146}$$

in analogy to the macroscopic reduction of (52.38ab).

Variational Principle

The QSS-condition (52.135) implies that the fraction P_i^{D} of bound levels i with precursor $\mathcal{C}$ are so distributed over i that (52.137) for $\hat{\alpha}$ is a minimum. Hence P_i^{D} or ρ_i are obtained either from the solution of (52.142) or from minimizing the *variational functional*

$$\hat{\alpha}\tilde{N}_A\tilde{N}_B = \int_{-D}^{\infty} n_i\, dE_i \int_{-D}^{\infty} (\rho_i - \rho_f)\, \nu_{if}\, dE_f \tag{52.147a}$$

$$= \frac{1}{2}\int_{-D}^{\infty} dE_i \int_{-D}^{\infty} (\rho_i - \rho_f)^2\, C_{if}\, dE_f \tag{52.147b}$$

with respect to variational parameters contained in a trial analytic expression for ρ_i. Minimization of the quadratic functional (52.147b) has an analogy with the *principle of least dissipation* in the theory of electrical networks.

Diffusion-in-Energy-Space Method

Integral Equation (52.142) can be expanded in terms of energy-change moments, via a Fokker-Planck analysis to yield the differential equation

$$\frac{\partial}{\partial E_i}\left[D_i^{(2)} \frac{\partial \rho_i}{\partial E_i} \right] = 0\,, \tag{52.148}$$

with the QSS analytical solution

$$\rho_i(E_i) = \left[\int_{E_i}^{0} \frac{dE}{D^{(2)}(E)} \right] \left[\int_{-S}^{0} \frac{dE}{D^{(2)}(E)} \right]^{-1} \tag{52.149}$$

of Pitaevskiĭ [4] for $(\mathrm{e}^- + A^+ + M)$ recombination where collisional energy changes are small. This distribution does not satisfy the exact QSS condition (52.142). When inserted in the exact non-QSS rate (52.147b), highly accurate $\hat{\alpha}$ for heavy-particle recombination are obtained.

Bottleneck Method

The one-way equilibrium rate $(\mathrm{cm}^{-3}\,\mathrm{s}^{-1})$ across $-E$, i.e., Eq. (52.141c) with $\rho_i = 1$ and $\rho_f = 0$, is

$$\hat{\alpha}(-E)\tilde{N}_A\tilde{N}_B = \int_{-E}^{\infty} dE_i \int_{-D}^{-E} C_{if}\, dE_f\,. \tag{52.150}$$

This is an upper limit to (52.141c) and exhibits a minimum at $-E^*$, the *bottleneck location*. The least upper limit to $\hat{\alpha}$ is then $\hat{\alpha}(-E^*)$.

Trapping Radius Method

Assume that pairs with internal separation $R \leq R_{\mathrm{T}}$ recombine with unit probability so that the one-way equilibrium rate across the dissociation limit at $E = 0$ for these pairs is

$$\hat{\alpha}(R_{\mathrm{T}})\tilde{N}_A\tilde{N}_B = \int_{0}^{R_{\mathrm{T}}} d\mathbf{R} \int_{V(R)}^{0} C_{if}(R)\, dE_f\,, \tag{52.151}$$

where $V(R) = -e^2/R$, and $C_{if}(R) = \tilde{n}_i(R)\nu_{if}(R)$ is the rate per unit interval $(d\mathbf{R}\, dE_i)\, dE_f$ for the $E_i \to E_f$ collisional transitions at fixed R in

$$(A^+ - B^-)_{E_i, R} + M \to (A^+ - B^-)_{E_f, R} + M\,. \tag{52.152}$$

The concentration (cm^{-3}) of pairs with internal separation R and orbital energy E_i in the interval $d\mathbf{R}\, dE_i$

about $(\mathbf{R}, E_i)$ is $\tilde{n}_i(R)d\mathbf{R}\,dE_i$. Agreement with the exact treatment [5] is found by assigning $R_T = (0.48 - 0.55)(e^2/k_{\mathrm{B}}T)$ for the recombination of equal mass ions in an equal mass gas for various ion-neutral interactions. For further details on the above methods, see the appropriate references on termolecular recombination in Sect. 52.9.

52.7.3 Recombination at Higher Gas Densities

As the density N of the gas M is raised, the recombination rate $\hat{\alpha}$ increases initially as N to such an extent that there are increasingly more pairs $n_i^-(R, E)$ in a state of contraction in R than there are those $n_i^+(R, E)$ in a state of expansion; i.e., the ion-pair distribution densities $n_i^\pm(R, E)$ per unit interval $dE\,dR$ are not in equilibrium with respect to R in blocks $\mathcal{C}$ and $\mathcal{E}$. Those in the highly excited block $\mathcal{E}$ in addition are not in equilibrium with respect to energy E. Basic sets of coupled master equations have been developed [5] for the microscopic nonequilibrium distributions $n^\pm(R, E, L^2)$ and $n^\pm(R, E)$ of expanding $(+)$ and contracting $(-)$ pairs with respect to A–B separation R, orbital energy E and orbital angular momentum L^2. With $n(\mathbf{R}, E_i, L_i^2) \equiv n_i(R)$, and using the notation defined at the beginning of Sect. 52.6, the distinct regimes for the master equations discussed in Sect. 52.7.4 are:

Low N	Equilibrium in R, but not in E, L^2 $\rightarrow$ master equation for $n(E, L^2)$.
Pure Coulomb attraction	Equilibrium in L^2 $\rightarrow$ master equation for $n(E)$.
High N	Nonequilibrium in R, E, L^2 $\rightarrow$ master equation for $n_i^\pm(R)$.
Highest N	Equilibrium in (E, L^2) but not in R $\rightarrow$ macroscopic transport equation (52.56a) in $n(R)$.

Normalized Distributions.
For a state $|\,i\,\rangle \equiv |\,E, L^2\,\rangle$,

$$\rho_i(R) = \frac{n_i(R)}{\tilde{n}_i(R)}, \quad \rho_i^\pm(R) = \frac{n_i^\pm(R)}{\tilde{n}_i^\pm(R)},$$

$$\rho_i(R) = \frac{1}{2}(\rho_i^+ + \rho_i^-). \tag{52.153}$$

Orbital Energy and Angular Momentum.

$$E_i = \frac{1}{2}M_{AB}v^2 + V(R) \tag{52.154a}$$

$$E_i = \frac{1}{2}M_{AB}v_R^2 + V_i(R) \tag{52.154b}$$

$$V_i(R) = V(R) + \frac{L_i^2}{2M_{AB}R^2} \tag{52.154c}$$

$$L_i = |\mathbf{R} \times M_{AB}\mathbf{v}|, \quad L_i^2 = (2M_{AB}E_i)b^2, \; E_i > 0.$$

$$\tag{52.154d}$$

Maximum Orbital Angular Momenta.
(1) A specified separation R can be accessed by all orbits of energy E_i with L_i^2 between 0 and

$$L_{\mathrm{im}}^2(E_i, R) = 2M_{AB}R^2\left[E_i - V(R)\right]. \tag{52.155a}$$

(2) Bounded orbits of energy $E_i < 0$ can have L_i^2 between 0 and

$$L_{\mathrm{ic}}^2(E_i) = 2M_{AB}R_c^2\left[E_i - V(R_c)\right], \tag{52.155b}$$

where R_c is the radius of the circular orbit determined by $\partial V_i/\partial R = 0$, i.e., by $E_i = V(R_c) + \frac{1}{2}R_c(\partial V/\partial R)_{R_c}$.

52.7.4 Master Equations

Master Equation for $n_i^\pm(R) \equiv n^\pm(\mathbf{R}, E_i, L_i^2)$. [5]

$$\pm\frac{1}{R^2}\frac{\partial}{\partial R}\left[R^2 n_i^\pm(R)\,|v_R|\right]_{E_i, L_i^2}$$
$$= -\int_{V(R)}^\infty dE_f \int_0^{L_{fm}^2} dL_f^2 \left[n_i^\pm(R)\nu_{if}(R)\right.$$
$$\left. - n_f^\pm(R)\nu_{fi}(R)\right]. \tag{52.156}$$

The set of master equations [5] for n_i^+ is coupled to the n_i^- set by the boundary conditions $n_i^-(R_i^\mp) = n_i^+(R_i^\mp)$ at the pericenter R_i^- for all E_i and apocenter R_i^+ for $E_i < 0$ of the E_i, L_i^2-orbit.

Master Equations for Normalized Distributions. [5]

$$\pm|v_R|\frac{\partial\rho_i^\pm}{\partial R} = -\int_{V(R)}^\infty dE_f \int_0^{L_{fm}^2} dL_f^2$$
$$\times \left[\rho_i^\pm(R) - \rho_f^\pm(R)\right]\nu_{if}(R). \tag{52.157}$$

Corresponding Master equations for the L^2 integrated distributions $n^\pm(R, E)$ and $\rho^\pm(R, E)$ have been derived [5].

Continuity Equations.

$$J_i = \left[n_i^+(R) - n_i^-(R)\right]|v_R| = \left[\rho_i^+ - \rho_i^-\right]\tilde{j}_i^\pm \tag{52.158}$$

$$\frac{1}{R^2}\frac{\partial}{\partial R}(R^2 J_i) = -\int_{V(R)}^\infty dE_f \int_0^{L_{fm}^2} dL_f^2$$
$$\times \left[n_i(R)\nu_{if}(R) - n_f(R)\nu_{fi}(R)\right], \tag{52.159}$$

$$\frac{1}{2}|v_R|\frac{\partial\left[\rho_i^+(R) - \rho_i^-(R)\right]}{\partial R} = -\int_{V(R)}^\infty dE_f \int_0^{L_{fm}^2} dL_f^2$$
$$\times \left[\rho_i(R) - \rho_f(R)\right]\nu_{if}(R). \tag{52.160}$$

52.7.5   Recombination Rate

Flux Representation.

The $R_0 \to \infty$ limit of

$$\hat{\alpha}\tilde{N}_A\tilde{N}_B = -4\pi R_0^2 J(R_0) \qquad (52.161)$$

has the microscopic generalization

$$\hat{\alpha}\tilde{N}_A\tilde{N}_B = \int_{V(R_0)}^{\infty} dE_i \int_0^{L_{ic}^2} dL_i^2 \left[4\pi R_0^2 \tilde{j}_i^{\pm}(R_0)\right]$$
$$\times \left[\rho_i^-(R_0) - \rho_i^+(R_0)\right], \qquad (52.162)$$

where L_{ic}^2 is given by (52.155b) with $R_c = R_0$ for bound states and is infinite for dissociated states, and where

$$\rho_i^-(R_0) - \rho_i^+(R_0) = \oint_{R_i}^{R_0} \rho_i(R) \left[\nu_i^b(R) + \nu_i^c(R)\right] dt,$$
$$(52.163)$$

with

$$\rho_i(R)\nu_i^b(R) = \int_{V(R)}^{V(R_0)} dE_f \int_0^{L_{fm}^2} dL_f^2 \left[\rho_i(R) - \rho_f(R)\right]$$
$$\times \nu_{if}(R), \qquad (52.164a)$$

$$\rho_i(R)\nu_i^c(R) = \int_{V(R_0)}^{\infty} dE_f \int_0^{L_{fm}^2} dL_f^2 \left[\rho_i(R) - \rho_f(R)\right]$$
$$\times \nu_{if}(R). \qquad (52.164b)$$

Collisional Representation.

$$\hat{\alpha}\tilde{N}_A\tilde{N}_B = \int_{V(R_0)}^{\infty} dE_i \int_0^{L_{ic}^2} dL_i^2 \int_{R_i}^{R_0} \tilde{n}_i(R) \, d\mathbf{R}$$
$$\times \left[\rho_i(R)\nu_i^b(R)\right], \qquad (52.165)$$

which is the microscopic generalization of the macroscopic result $\hat{\alpha} = K\rho^*\nu_s = \alpha_{RN}(R_0)\rho(R_0)$.

The flux for dissociated pairs $E_i > 0$ is

$$4\pi R^2 |v_R| \tilde{n}_i^{\pm}(R) \, dE \, dL^2 = \left[\overline{v}\varepsilon e^{-\varepsilon} \, d\varepsilon\right] [2\pi b \, db] \, \tilde{N}_A\tilde{N}_B,$$
$$(52.166)$$

so the rate (52.165) as $R_0 \to \infty$ is

$$\hat{\alpha} = \overline{v} \int_0^{\infty} \varepsilon e^{-\varepsilon} \, d\varepsilon \int_0^{b_0} 2\pi b \, db \oint_{R_i}^{R_0} \rho_i(R)\nu_i^b(R) \, dt,$$
$$(52.167)$$

which is the microscopic generalization (52.45b) of the macroscopic result $\hat{\alpha} = k_c P^S$ of (52.44).

Reaction Rate $\alpha_{RN}(R_0)$. On solving (52.157) subject to $\rho(R_0) = 1$, then according to (52.56b), $\hat{\alpha}$ determined by (52.162) is the rate $\hat{\alpha}_{RN}$ of recombination within the $(A - B)$ sphere of radius R_0. The overall rate of

recombination $\hat{\alpha}$ is then given by the full diffusional-drift reaction rate (52.59b) where the rate of transport to R_0 is determined uniquely by (52.60).

For development of theory [5] and computer simulations, see the reference list on Termolecular Ion-Ion Recombination: Theory, and Simulations, respectively.

52.8   RADIATIVE RECOMBINATION

In the radiative recombination (RR) process

$$\mathrm{e}^-(E, \ell') + A^{Z+}(c) \to A^{(Z-1)+}(c, n\ell) + h\nu, \quad (52.168)$$

the accelerating electron e^- with energy and angular momentum (E, ℓ') is captured, via coupling with the weak quantum electrodynamical interaction $(e/m_e c)\mathbf{A}\cdot\mathbf{p}$ associated with the electromagnetic field of the moving ion, into an excited state $n\ell$ with binding energy $I_{n\ell}$ about the parent ion A^{Z+} (initially in an electronic state c). The simultaneously emitted photon carries away the excess energy $h\nu = E + I_{n\ell}$ and angular momentum difference between the initial and final electronic states. The cross section $\sigma_R^{n\ell}(E)$ for RR is calculated (a) from the Einstein A coefficient for free-bound transitions or (b) from the cross section $\sigma_I^{n\ell}(h\nu)$ for photoionization (PI) via the detailed balance (DB) relationship appropriate to (52.168).

The rates $\langle v_e\sigma_R\rangle$ and averaged cross sections $\langle\sigma_R\rangle$ for a Maxwellian distribution of electron speeds v_e are then determined from either

$$\hat{\alpha}_R^{n\ell}(T_e) = \overline{v}_e \int_0^{\infty} \varepsilon\sigma_R^{n\ell}(\varepsilon) \exp(-\varepsilon) \, d\varepsilon = \overline{v}_e \left\langle\sigma_R^{n\ell}(T_e)\right\rangle,$$
$$(52.169)$$

where $\varepsilon = E/k_B T_e$, or from the Milne DB relation (52.243) between the forward and reverse macroscopic rates of (52.168). Using the hydrogenic semiclassical σ_I^n of Kramers [13], together with an asymptotic expansion [14] for the g−factor of Gaunt [15], the quantal/semiclassical cross section ratio in (52.249), Seaton [16] calculated $\hat{\alpha}_R^{n\ell}$.

The rate of electron energy loss in RR is

$$\left\langle\frac{dE}{dt}\right\rangle_{n\ell} = n_e\overline{v}_e(k_B T_e) \int_0^{\infty} \varepsilon^2\sigma_R^{n\ell}(\varepsilon)e^{-\varepsilon} \, d\varepsilon, \quad (52.170)$$

and the radiated power produced in RR is

$$\left\langle\frac{d(h\nu)}{dt}\right\rangle_{n\ell} = n_e\overline{v}_e \int_0^{\infty} \varepsilon h\nu\sigma_R^{n\ell}(\varepsilon)e^{-\varepsilon} \, d\varepsilon. \quad (52.171)$$

Standard Conversions

$$E = p_e^2/2m_e = \hbar^2 k_e^2/2m_e = k_e^2 a_0^2 (e^2/2a_0) \quad (52.172a)$$

$$= \kappa^2(Z^2 e^2/2a_0) = \varepsilon(Z^2 e^2/2a_0), \quad (52.172b)$$

$$E_\nu = h\nu = \hbar\omega = \hbar k_\nu c = (I_n + E) \quad (52.172c)$$

$$\equiv \left[1 + n^2\varepsilon\right](Z^2 e^2/2n^2 a_0), \quad (52.172d)$$

$$h\nu/I_n = 1 + n^2\varepsilon, \quad k_e^2 a_0^2 = 2E/(e^2/a_0), \quad (52.172e)$$

$$k_\nu a_0 = (h\nu)\alpha/(e^2/a_0), \quad (52.172f)$$

$$k_\nu^2/k_e^2 = (h\nu)^2/(2Em_e c^2) \quad (52.172g)$$

$$= \alpha^2(h\nu)^2/\left[2E(e^2/a_0)\right], \quad (52.172h)$$

$$I_H = e^2/2a_0, \quad \alpha = e^2/\hbar c = 1/137.035\,9895,$$

$$\alpha^{-2} = m_e c^2/(e^2/a_0), \quad I_n = (Z^2/n^2)I_H. \quad (52.172i)$$

The electron and photon wavenumbers are k_e and k_ν, respectively.

52.8.1 Detailed Balance and Recombination-Ionization Cross Sections

Cross sections $\sigma_R^{n\ell}(E)$ and $\sigma_I^{n\ell}(h\nu)$ for radiative recombination (RR) into and photoionization (PI) out of level $n\ell$ of atom A are interrelated by the *detailed balance* relation

$$g_e g_A^+ k_e^2 \sigma_R^{n\ell}(E) = g_\nu g_A k_\nu^2 \sigma_I^{n\ell}(h\nu), \quad (52.173)$$

where $g_e = g_\nu = 2$. Electronic statistical weights of A and A^+ are g_A and g_A^+, respectively. Thus, using Eq. (52.172g) for k_ν^2/k_e^2,

$$\sigma_R^{n\ell}(E) = \left(\frac{g_A}{2g_A^+}\right)\left[\frac{(h\nu)^2}{Em_e c^2}\right]\sigma_I^{n\ell}(h\nu). \quad (52.174)$$

The statistical factors are:

   (a) For $(A^+ + e^-)$ state $c[S_c, L_c; \varepsilon, \ell', m']$:
     $g_A^+ = (2S_c + 1)(2L_c + 1)$.

   (b) For $A(n\ell)$ state $b[S_c, L_c; n, \ell]SL$:
     $g_A = (2S + 1)(2L + 1)$.

   (c) For $n\ell$ electron outside a closed shell:
     $g_A^+ = 1, g_A = 2(2\ell + 1)$.

Cross sections are averaged over initial and summed over final degenerate states. For case (c),

$$\sigma_I^n = \frac{1}{n^2}\sum_{\ell=0}^{n-1}(2\ell+1)\sigma_I^{n\ell}; \quad (52.175a)$$

$$\sigma_R^n = \sum_{\ell=0}^{n-1}2(2\ell+1)\sigma_R^{n\ell}. \quad (52.175b)$$

52.8.2 Kramers Cross Sections, Rates, Electron Energy-Loss Rates and Radiated Power for Hydrogenic Systems

These are all calculated from application of detailed balance (52.173) to the original $\sigma_I^n(h\nu)$ of Kramers [13].

Semiclassical (Kramers) Cross Sections

For hydrogenic systems,

$$I_n = \frac{Z^2 e^2}{2n^2 a_0}, \quad h\nu = I_n + E. \quad (52.176)$$

The results below are expressed in terms of the quantities

$$b_n = \frac{I_n}{k_B T_e}, \quad (52.177)$$

$$\sigma_{I0}^n = \frac{64\pi a_0^2 \alpha}{3\sqrt{3}}\left(\frac{n}{Z^2}\right)$$

$$= 7.907071 \times 10^{-18}\left(n/Z^2\right)\ \text{cm}^2, \quad (52.178)$$

$$\sigma_{R0}(E) = \left(\frac{8\pi a_0^2 \alpha^3}{3\sqrt{3}}\right)\frac{(Z^2 e^2/a_0)}{E}, \quad (52.179)$$

$$\hat{\alpha}_0(T_e) = \bar{v}_e\left(\frac{8\pi a_0^2 \alpha^3}{3\sqrt{3}}\right)\frac{(Z^2 e^2/a_0)}{k_B T_e} \quad (52.180)$$

PI and RR Cross Sections for Level n. In the Kramer (K) semiclassical approximation,

$$_K\sigma_I^n(h\nu) = \left(\frac{I_n}{h\nu}\right)^3 \sigma_{I0}^n = {}_K\sigma_I^{n\ell}(h\nu) \quad (52.181)$$

$$_K\sigma_R^n(E) = \sigma_{R0}(E)\left(\frac{2}{n}\right)\left(\frac{I_n}{I_n + E}\right) \quad (52.182)$$

$$= 3.897 \times 10^{-20}\left[n\varepsilon(13.606 + n^2\varepsilon^2)\right]^{-1}\ \text{cm}^2,$$

where ε is in units of eV and is given by

$$\varepsilon = E/Z^2 \equiv (2.585 \times 10^{-2}/Z^2)(T_e/300). \quad (52.183)$$

Equation (52.182) illustrates that RR into low n at low E is favored.

Cross Section for RR into Level $n\ell$.

$$_K\sigma_R^{n\ell} = \left[(2\ell+1)/n^2\right]{}_K\sigma_R^n. \quad (52.184)$$

Rate for RR into Level n.

$$\hat{\alpha}_R^n(T_e) = \hat{\alpha}_0(T_e)(2/n)b_n e^{b_n}E_1(b_n), \quad (52.185a)$$

which tends for large b_n (i.e., $k_B T_e \ll I_n$) to

$$\hat{\alpha}_R^n(T_e \to 0) = \hat{\alpha}_0(T_e)(2/n)$$
$$\times\left[1 - b_n^{-1} + 2b_n^{-2} - 6b_n^{-3} + \cdots\right]. \quad (52.185b)$$

The Kramers cross section for photoionization at threshold is σ_{I0}^n and

$$\sigma_{R0}^n = 2\sigma_{R0}/n; \quad \hat{\alpha}_0^n = 2\hat{\alpha}_0/n \qquad (52.186)$$

provide the corresponding Kramers cross section and rate for recombination as $E \to 0$ and $T_e \to 0$, respectively.

RR Cross Sections and Rates into All Levels $n \geq n_f$.

$$\sigma_R^T(E) = \int_{n_f}^{\infty} \sigma_R^n(E)\, dn$$
$$= \sigma_{R0}(E)\ln(1 + I_f/E), \qquad (52.187a)$$
$$\hat{\alpha}_R^T(T_e) = \hat{\alpha}_0(T_e)\left[\gamma + \ln b_f + e^{b_f}E_1(b_f)\right] \qquad (52.187b)$$

Useful Integrals.

$$\int_0^{\infty} e^{-x}\ln x\, dx = \gamma \qquad (52.188a)$$

$$\int_b^{\infty} x^{-1}e^{-x}\, dx = E_1(b) \qquad (52.188b)$$

$$\int_0^b e^x E_1(x)\, dx = \gamma + \ln b + e^b E_1(b) \qquad (52.188c)$$

$$\int_0^b \left[1 - xe^x E_1(x)\right] dx = \gamma + \ln b + e^b(1-b)E_1(b) \qquad (52.188d)$$

where $\gamma = 0.577\,2157$ is Euler's constant, and $E_1(b)$ is the first exponential integral such that

$$be^b E_1(b) \xrightarrow{b \gg 1} 1 - b^{-1} + 2b^{-2} - 6b^{-3} + 24b^{-4} + \cdots .$$

Electron Energy Loss Rate

Energy Loss Rate for RR into Level n.

$$\left\langle \frac{dE}{dt} \right\rangle_n = n_e\hat{\alpha}_R^n(T_e)k_B T_e \left[\frac{1 - b_n e^{b_n}E_1(b_n)}{e^{b_n}E_1(b_n)}\right], \qquad (52.189a)$$

which for large b_n (i.e. $(k_B T_e) \ll I_n$) tends to

$$n_e\hat{\alpha}_R^n(T_e)k_B T_e \left[1 - b_n^{-1} + 3b_n^{-2} - 13b_n^{-3} + \cdots\right] \qquad (52.189b)$$

with (52.185a) for $\hat{\alpha}_R^n$.

Energy Loss Rate for RR into All Levels $n \geq n_f$.

$$\left\langle \frac{dE}{dt} \right\rangle = n_e k_B T_e \hat{\alpha}_0(T_e)\left[\gamma + \ln b_f + e^{b_f}E_1(b_f)(1-b_f)\right] \qquad (52.190a)$$

$$= n_e(k_B T_e)\left[\hat{\alpha}_R^T(T_e) - \hat{\alpha}_0(T_e)b_f e^{b_f}E_1(b_f)\right] \qquad (52.190b)$$

with (52.187b) and (52.180) for $\hat{\alpha}_R^T$ and $\hat{\alpha}_0$.

Radiated Power

Radiated Power for RR into Level n.

$$\left\langle \frac{d(h\nu)}{dt} \right\rangle_n = n_e\hat{\alpha}_R^n(T_e)I_n \left[b_n e^{b_n}E_1(b_n)\right]^{-1}, \qquad (52.191a)$$

which for large b_n (i.e. $(k_B T_e) \ll I_n$) tends to

$$n_e\hat{\alpha}_R^n(T_e)I_n \left[1 + b_n^{-1} - b_n^{-2} + 3b_n^{-3} + \cdots\right]. \qquad (52.191b)$$

Radiated Power for RR into All Levels $n \geq n_f$.

$$\left\langle \frac{d(h\nu)}{dt} \right\rangle = n_e\hat{\alpha}_0(T_e)I_f . \qquad (52.192)$$

To allow n-summation, rather than integration as in (52.187a), to each of the above expressions is added $\frac{1}{2}\sigma_R^{n_f}$, $\frac{1}{2}\hat{\alpha}_R^{n_f}$, $\frac{1}{2}\left\langle\frac{dE}{dt}\right\rangle_{n_f}$ and $\frac{1}{2}\left\langle\frac{d(h\nu)}{dt}\right\rangle_{n_f}$, respectively. The expressions valid for bare nuclei of charge Z are also fairly accurate for recombination to a core of charge Z_c and atomic number Z_A, provided that Z is identified as $\frac{1}{2}(Z_A + Z_c)$.

Differential Cross Sections for Coulomb Elastic Scattering:

$$\sigma_c(E,\theta) = \frac{b_0^2}{4\sin^4\frac{1}{2}\theta}, \quad b_0^2 = (Ze^2/2E)^2 . \qquad (52.193)$$

The integral cross section for Coulomb scattering by $\theta \geq \pi/2$ at energy $E = (3/2)k_B T$ is

$$\sigma_c(E) = \pi b_0^2 = \tfrac{1}{9}\pi R_e^2, \quad R_e = e^2/k_B T . \qquad (52.194)$$

Photon Emission Probability.

$$P_\nu = \sigma_R^n(E)/\sigma_c(E) . \qquad (52.195a)$$

This is small and increases with decreasing n as

$$P_\nu(E) = \left(\frac{8\alpha^3}{3\sqrt{3}}\right)\frac{8}{n}\frac{E}{(e^2/a_0)}\left(\frac{I_n}{h\nu}\right) . \qquad (52.195b)$$

52.8.3 Basic Formulae for Quantal Cross Sections

Radiative Recombination and Photoionization Cross Sections

The cross section $\sigma_R^{n\ell}$ for recombination follows from the continuum-bound transition probability P_{if} per unit time. It is also provided by the detailed balance relation (52.173) in terms of $\sigma_I^{n\ell}$ which follows from P_{fi}. The number of radiative transitions per second is

$$\left[g_e g_A^+ \rho(E)\, dE\, d\hat{\mathbf{k}}_e\right] P_{if} \left[\rho(E_\nu)\, dE_\nu\, d\hat{\mathbf{k}}_\nu\right]$$

$$= g_e g_A^+ v_e \frac{d\mathbf{p}_e}{(2\pi\hbar)^3} \sigma_R(\mathbf{k}_e)$$

$$= g_\nu g_A c \frac{d\mathbf{k}_\nu}{(2\pi)^3} \sigma_I(\mathbf{k}_\nu)\,, \qquad (52.196)$$

where the electron current $(\text{cm}^{-2}\,\text{s}^{-1})$ is

$$\frac{v_e d\mathbf{p}_e}{(2\pi\hbar)^3} = \left(\frac{2mE}{h^3}\right) dE\, d\hat{\mathbf{k}}_e\,, \qquad (52.197)$$

and the photon current $(\text{cm}^{-2}\,\text{s}^{-1})$ is

$$c\frac{d\mathbf{k}_\nu}{(2\pi)^3} = c\frac{(h\nu)^2}{(2\pi\hbar c)^3} dE_\nu\, d\hat{\mathbf{k}}_\nu\,. \qquad (52.198)$$

Time Dependent Quantum Electrodynamical Interaction.

$$V(\mathbf{r}, t) = \frac{e}{mc}\mathbf{A}\cdot\mathbf{p} = ie\left(\frac{2\pi h\nu}{\mathcal{V}}\right)^{1/2} (\hat{\mathbf{e}}\cdot\mathbf{r})e^{-i(\mathbf{k}_\nu\cdot\mathbf{r}-\omega t)}$$

$$\equiv V(\mathbf{r})e^{(i\omega t)}\,. \qquad (52.199)$$

In the dipole approximation, $e^{-i\mathbf{k}_\nu\cdot\mathbf{r}} \sim 1$.

Continuum-Bound State-to-State Probability.

$$P_{if} = \frac{2\pi}{\hbar}|V_{fi}|^2 \delta\left[E_\nu - (E + I_n)\right]$$

$$V_{fi} = \langle\Psi_{n\ell m}(\mathbf{r})\,|\,V(\mathbf{r})\,|\,\Psi_i(\mathbf{r},\mathbf{k}_e)\rangle\,. \qquad (52.200)$$

Number of Photon States in Volume $\mathcal{V}$.

$$\rho(E_\nu, \hat{k}_\nu)\, dE_\nu\, d\hat{\mathbf{k}}_\nu = \mathcal{V}(h\nu)^2/(2\pi\hbar c)^3\, dE_\nu\, d\hat{\mathbf{k}}_\nu \quad (52.201\text{a})$$

$$= \mathcal{V}\left[\omega^2/(2\pi c)^3\right] d\omega\, d\hat{\mathbf{k}}_\nu\,. \quad (52.201\text{b})$$

Continuum-Bound Transition Rate.

On summing over the two directions $(g_\nu = 2)$ of polarization, the rate for transitions into all final photon states is

$$A_{n\ell m}(E, \hat{\mathbf{k}}_e) = \int P_{if}\rho(E_\nu)\, dE_\nu\, d\hat{\mathbf{k}}_\nu$$

$$= \frac{4e^2}{3\hbar}\frac{(h\nu)^3}{(\hbar c)^3}|\langle\Psi_{n\ell m}\,|\,\mathbf{r}\,|\,\Psi_i(\mathbf{k}_e)\rangle|^2 \quad (52.202)$$

Transition Frequency: Alternative Formula.

$$A_{n\ell m}(E, \hat{\mathbf{k}}_e) = (2\pi/\hbar)|D_{fi}|^2\,, \qquad (52.203)$$

where the dipole atom-radiation interaction coupling is

$$D_{fi}(\mathbf{k}_e) = \left(\frac{2\omega^3}{3\pi c^3}\right)^{1/2}\langle\Psi_{n\ell m}\,|\,e\,\mathbf{r}\,|\,\Psi_i(\mathbf{k}_e)\rangle\,. \quad (52.204)$$

RR Cross Section into Level (n, ℓ, m).

$$\sigma_R^{n\ell m}(E) = \frac{1}{4\pi}\int \sigma_R^{n\ell m}(\mathbf{k}_e)\, d\hat{\mathbf{k}}_e$$

$$= \frac{h^3\rho(E)}{8\pi m_e E}\int A_{n\ell m}(E, \hat{\mathbf{k}}_e)\, d\hat{\mathbf{k}}_e\,. \qquad (52.205)$$

RR Cross Section into Level (n, ℓ).

$$\sigma_R^{n\ell}(E) = \frac{8\pi^2}{3}\left[\frac{(\alpha h\nu)^3}{2(e^2/a_0)E}\right]\rho(E)R_I^{n\ell}(E)$$

$$R_I^{n\ell}(E) = \int d\hat{\mathbf{k}}_e \sum_m |\langle\Psi_{n\ell m}\,|\,\mathbf{r}\,|\,\Psi_i(\mathbf{k}_e)\rangle|^2\,. \qquad (52.206)$$

Transition T-Matrix for RR.

$$\sigma_R^{n\ell}(E) = \frac{\pi a_0^2}{(ka_0)^2}|T_R|^2 \rho(E)\,, \qquad (52.207)$$

$$|T_R|^2 = 4\pi^2 \int \sum_m |D_{fi}|^2\, d\hat{\mathbf{k}}_e\,. \qquad (52.208)$$

Photoionization Cross Section. From detailed balance in Eq. (52.196), $\sigma_I^{n\ell}$ is

$$\sigma_I^{n\ell}(h\nu) = \left(\frac{8\pi^2}{3}\right)\alpha h\nu\left(\frac{g_A^+}{g_A}\right)\rho(E)R_I^{n\ell}(E)\,. \quad (52.209)$$

Continuum Wave Function Expansion.

$$\Psi_i(\mathbf{k}_e, \mathbf{r}) = \sum_{\ell'm'} i^{\ell'} e^{i\eta_{\ell'}} R_{E\ell'}(r)Y_{\ell'm'}(\hat{\mathbf{k}}_e)Y_{\ell'm'}(\hat{\mathbf{r}})\,.$$

$$(52.210)$$

Energy Normalization. With $\rho(E) = 1$,

$$\int \Psi_i(\mathbf{k}_e; \mathbf{r})\Psi_i^*(\mathbf{k}_e'; \mathbf{r})\, d\mathbf{r} = \delta(E - E')\delta(\hat{\mathbf{k}}_e - \hat{\mathbf{k}}_e')\,.$$

$$(52.211)$$

Plane Wave Expansion.

$$e^{i\mathbf{k}\cdot\mathbf{r}} = 4\pi\sum_{\ell=0}^{\infty} i^\ell j_\ell(kr)Y_{\ell m}^*(\hat{\mathbf{k}})Y_{\ell m}(\hat{\mathbf{r}}) \qquad (52.212)$$

$$j_\ell(kr) \sim \sin(kr - \tfrac{1}{2}\ell\pi)/(kr)\,. \qquad (52.213)$$

For bound states,

$$\Psi_{n\ell m}(\mathbf{r}) = R_{n\ell}(r)Y_{\ell m}(\hat{\mathbf{r}})\,. \qquad (52.214)$$

RR and PI Cross Sections and Radial Integrals.

$$\sigma_R^{n\ell}(E) = \frac{8\pi^2}{3}\left[\frac{(\alpha h\nu)^3}{2(e^2/a_0)E}\right]\rho(E)R_I(E; n\ell)\,. \quad (52.215)$$

For an electron outside a closed core,

$$g_A^+ = 1\,, \quad g_A = 2(2\ell + 1)$$

$$\sigma_I^{n\ell}(h\nu) = \frac{4\pi^2\alpha h\nu\rho(E)}{3(2\ell+1)}R_I(E; n\ell)\,, \qquad (52.216\text{a})$$

$$R_{n\ell}^{\varepsilon,\ell'} = \int_0^\infty (R_{\varepsilon\ell'}\, r\, R_{n\ell})\ r^2\, dr\,, \qquad (52.216\text{b})$$

$$R_I(E; n\ell) = \ell\left|R_{n\ell}^{\varepsilon,\ell-1}\right|^2 + (\ell+1)\left|R_{n\ell}^{\varepsilon,\ell+1}\right|^2\,. \quad (52.216\text{c})$$

For an electron outside an unfilled core (c) in the process $(A^+ + e^-) \rightarrow A(n\ell)$, the weights are

State i: $[S_c, L_c; \varepsilon]$, $\quad g_A^+ = (2S_c + 1)(2L_c + 1)$

State f: $[(S_c, L_c; n\ell)S, L]$, $\quad g_A = (2S + 1)(2L + 1)$.

$$R_{\text{I}}(E; n\ell) = \frac{(2L+1)}{(2L_c+1)} \sum_{\ell'=\ell\pm 1} \sum_{L'} (2L'+1) \left\{ \begin{array}{ccc} \ell & L & L_c \\ L' & \ell' & 1 \end{array} \right\}^2$$
$$\times \ell_{\max} \left| \int_0^\infty (R_{\varepsilon\ell'} \, r \, R_{n\ell}) \, r^2 \, dr \right|^2 . \quad (52.217)$$

This reduces to (52.216c) when the radial functions $R_{i,f}$ do not depend on (S_c, L_c, S, L).

Cross Section for Dielectronic Recombination

$$\sigma_{\text{DLR}}^{n\ell}(E) = \frac{\pi a_0^2}{(ka_0)^2} |T_{\text{DLR}}(E)|^2 \rho(E), \quad (52.218)$$

$$|T_{\text{DLR}}(E)|^2 = 4\pi^2 \int d\hat{\mathbf{k}}_e$$
$$\times \sum_j \left| \frac{\langle \Psi_f \mid D \mid \Psi_j \rangle \langle \Psi_j \mid V \mid \Psi_i(\mathbf{k}_e) \rangle}{(E - \varepsilon_j + i\Gamma_j/2)} \right|^2, \quad (52.219)$$

which is the generalization of the T-matrix (52.208) to include the effect of intermediate doubly-excited autoionizing states $| \Psi_j \rangle$ in energy resonance to within width Γ_j of the initial continuum state Ψ_i. The electrostatic interaction $V = e^2 \sum_{i=1}^N (\mathbf{r}_i - \mathbf{r}_{N+1})^{-1}$ initially produces dielectronic capture by coupling the initial state i with the resonant states j which become stabilized by coupling via the dipole radiation field interaction $\mathbf{D} = (2\omega^3/3\pi c^3)^{1/2} \sum_{i=1}^{N+1} (e\mathbf{r}_i)$ to the final stabilized state f. The above cross section for (52.3) is valid for isolated, nonoverlapping resonances.

Continuum Wave Normalization and Density of States

The basic formulae (52.206) for $\sigma_R^{n\ell}$ depends on the density of states $\rho(E)$ which in turn varies according to the particular normalization constant N adopted for the continuum radial wave,

$$R_{E\ell}(r) \sim N \sin(kr - \tfrac{1}{2}\ell\pi + \eta_\ell)/r, \quad (52.220)$$

in (52.210) where the phase is

$$\eta_\ell = \arg \Gamma(\ell + 1 + i\beta) - \beta \ln 2kr + \delta_\ell. \quad (52.221)$$

The phase corresponding to the Hartree-Fock short-range interaction is δ_ℓ. The Coulomb phase shift for electron motion under $(-Ze^2/r)$ is $(\eta_\ell - \delta_\ell)$ with $\beta = Z/(ka_0)$.

For a plane wave $\phi_{\mathbf{k}}(\mathbf{r}) = N' \exp(i\mathbf{k}\cdot\mathbf{r})$,

$$\langle \phi_{\mathbf{k}}(\mathbf{r}) \mid \phi_{\vec{k}'}(\mathbf{r}) \rangle \, d\mathbf{k} = (2\pi)^3 |N'|^2 \rho(\mathbf{k}) \, d\mathbf{k} \, \delta(\mathbf{k} - \mathbf{k}')$$
$$\equiv \left(\frac{h^3}{mp} \right) |N'|^2 \rho(E, \hat{\mathbf{k}}) \, dE \, d\hat{\mathbf{k}} \, \delta(E - E')\delta(\hat{\mathbf{k}} - \hat{\mathbf{k}}'). \quad (52.222)$$

On integrating (52.222) over all E and $\hat{\mathbf{k}}$ for a single particle distributed over all $| E, \hat{\mathbf{k}} \rangle$ states, N' and ρ are then interrelated by

$$|N'|^2 \rho(E, \hat{\mathbf{k}}) = mp/h^3. \quad (52.223)$$

The incident current is

$$j \, dE \, d\hat{\mathbf{k}}_e = v \, |N'|^2 \rho(E, \hat{\mathbf{k}}) \, dE \, d\hat{\mathbf{k}}_e \quad (52.224a)$$
$$= (2mE/h^3) \, dE \, d\hat{\mathbf{k}}_e = v \, d\mathbf{p}_e/h^3. \quad (52.224b)$$

Radial Wave Connection. From Eqs. (52.210) and (52.212), $N = (4\pi N'/k)$, so that the connection between N of (52.220) and $\rho(E)$ is

$$|N|^2 \rho(E, \hat{\mathbf{k}}) = \frac{(2m/\hbar^2)}{\pi k} = \frac{(2/\pi)}{(ka_0)e^2}. \quad (52.225)$$

RR Cross Sections for Common Normalization Factors of Continuum Radial Functions

(a) $N = 1$; $\quad \rho(E) = \frac{(2m/\hbar^2)}{\pi k} = \frac{(2/\pi)}{(ka_0)e^2}$, $\quad (52.226)$

$$\sigma_R^{n\ell}(E) = \frac{8\pi^2 a_0^2}{(ka_0)^3} \int \sum_m |D_{fi}|^2 \, d\hat{\mathbf{k}}_e, \quad (52.227)$$

where D_{fi} of (52.204) is dimensionless.

(b) $N = k^{-1}$; $\quad \rho(E) = (2m/\hbar^2)(k/\pi)$, $\quad (52.228)$

$$\sigma_R^{n\ell}(E) = \frac{16\pi a_0^2}{3\sqrt{2}} \left(\frac{\alpha h\nu}{e^2/a_0} \right)^3 \sqrt{\frac{(e^2/a_0)}{E}} \left(\frac{R_{\text{I}}}{a_0^5} \right), \quad (52.229)$$

where (52.216b) and (52.216c) for R_{I} has dimension $[L^5]$.

(c) $N = k^{-1/2}$; $\quad \rho(E) = \frac{(2m/\hbar^2)}{\pi}$, $\quad (52.230)$

$$\sigma_R^{n\ell}(E) = \frac{8\pi a_0^2}{3} \left[\frac{\alpha^3 (h\nu)^3}{(e^2/a_0)^2 E} \right] \left(\frac{R_{\text{I}}}{a_0^4} \right), \quad (52.231)$$

where R_{I} has dimensions of $[L^4]$.

(d) $N = (2m/\hbar^2\pi^2 E)^{1/4}$; $\quad \rho(E) = 1$, $\quad (52.232)$

$$\sigma_R^{n\ell}(E) = \frac{4(\pi a_0)^2}{3} \left[\frac{\alpha^3 (h\nu)^3}{(e^2/a_0)^2 E} \right] \left(\frac{R_{\text{I}}}{e^2 a_0} \right), \quad (52.233)$$

where R_{I} has dimensions of $[L^2 E^{-1}]$.

52.8.4 Bound-Free Oscillator Strengths

For a transition $n\ell \to E$ to $E + dE$,

$$\frac{df_{n\ell}}{dE} = \frac{2}{3}\frac{(h\nu)}{(e^2/a_0)}\frac{1}{(2\ell+1)}\sum_m\sum_{\ell'm'}\left|\mathbf{r}_{n\ell m}^{\varepsilon\ell'm'}\right|^2, \quad (52.234)$$

$$R_{\mathrm{I}}(\varepsilon;n\ell) = \int d\hat{\mathbf{k}}_e \sum_m \left|\left\langle \Psi_{n\ell m}\mid \vec{r}\mid \Psi_i(E;\hat{\mathbf{k}}_e;\ell'm')\right\rangle\right|^2$$

$$= \sum_{m,\ell',m'}\left|\mathbf{r}_{n\ell m}^{\varepsilon\ell'm'}\right|^2, \quad (52.235)$$

$$\sigma_{\mathrm{R}}^{n\ell}(E) = 2\pi^2\alpha a_0^2 g_A\left(\frac{k_\nu^2}{k_e^2}\right)\left(\frac{e^2}{a_0}\right)\frac{df_{n\ell}}{dE}, \quad (52.236a)$$

$$\sigma_{\mathrm{I}}^{n\ell}(h\nu) = 2\pi^2\alpha a_0^2 g_A^+\left(\frac{e^2}{a_0}\right)\frac{df_{n\ell}}{dE}. \quad (52.236b)$$

Semiclassical Hydrogenic Systems

$$g_A = g_{n\ell} = 2(2\ell+1), \quad g_A^+ = 1,$$

$$\sigma_{\mathrm{R}}^n(E) = \sum_{\ell=0}^{n-1}\sigma_{\mathrm{R}}^{n\ell}(E) = 2\pi^2\alpha a_0^2\left(\frac{k_\nu^2}{k_e^2}\right)\frac{dF_n}{dE}, \quad (52.237)$$

$$\frac{dF_n}{dE} = \sum_{\ell=0}^{n-1} g_{n\ell}\frac{df_{n\ell}}{dE} = 2\sum_{\ell,m}\frac{df_{n\ell m}}{dE}. \quad (52.238)$$

Bound-Bound Absorption Oscillator Strength.
For a transition $n \to n'$,

$$F_{nn'} = 2\sum_{\ell m}\sum_{\ell'm'} f_{n\ell m}^{n'\ell'm'} \quad (52.239a)$$

$$= \frac{2^6}{3\sqrt{3}\pi}\left[\left(\frac{1}{n^2}-\frac{1}{n'^2}\right)^{-3}\right]\frac{1}{n^3}\frac{1}{n'^3}, \quad (52.239b)$$

$$\frac{dF_n}{dE} = \frac{2^5}{3\sqrt{3}\pi}n\frac{I_n^2}{(h\nu)^3} = 2n^2\frac{df_n}{dE}, \quad (52.239c)$$

$$\sigma_{\mathrm{R}}^n(E) = \frac{2^5\alpha^3}{3\sqrt{3}}\left[\frac{nI_n^2}{E(h\nu)}\right]\pi a_0^2, \quad (52.239d)$$

$$\sigma_{\mathrm{I}}^{n\ell}(h\nu) = \frac{2^6\alpha}{3\sqrt{3}}\frac{n}{Z^2}\left(\frac{I_n}{h\nu}\right)^3\pi a_0^2, \quad (52.239e)$$

$$= 7.907071\left(\frac{n}{Z^2}\right)\left(\frac{I_n}{h\nu}\right)^3\ \mathrm{Mb}. \quad (52.239f)$$

This semiclassical analysis yields exactly Kramers PI and associated RR cross sections in Sect. 52.8.2.

52.8.5 Radiative Recombination Rate

$$\hat{\alpha}_{\mathrm{R}}^{n\ell}(T_e) = \bar{v}_e\int_0^\infty \varepsilon\,\sigma_{\mathrm{R}}^{n\ell}(\varepsilon)e^{-\varepsilon}\,d\varepsilon \quad (52.240a)$$

$$\equiv \bar{v}_e\left\langle\sigma_{\mathrm{R}}^{n\ell}(T_e)\right\rangle, \quad (52.240b)$$

where $\varepsilon = E/k_BT$ and $\left\langle\sigma_{\mathrm{R}}^{n\ell}(T_e)\right\rangle$ is the Maxwellian-averaged cross section for radiative recombination.

In terms of the continuum-bound $A_{n\ell}(E)$,

$$\hat{\alpha}_{\mathrm{R}}^{n\ell}(T_e) = \frac{h^3}{(2\pi m_e k_B T)^{3/2}}\int_0^\infty\left(\frac{dA_{n\ell}}{d\varepsilon}\right)e^{-\varepsilon}\,d\varepsilon, \quad (52.241)$$

$$\frac{dA_{n\ell}}{dE} = \rho(E)\sum_m\int A_{n\ell m}(E,\hat{\mathbf{k}}_e)\,d\hat{\mathbf{k}}_e. \quad (52.242)$$

Milne Detailed Balance Relation.
In terms of $\sigma_{\mathrm{I}}^{n\ell}(h\nu)$,

$$\hat{\alpha}_{\mathrm{R}}^{n\ell}(T_e) = \bar{v}_e\left(\frac{g_A}{2g_A^+}\right)\left(\frac{k_B T_e}{mc^2}\right)\left(\frac{I_n}{k_B T_e}\right)^2\left\langle\sigma_{\mathrm{I}}^{n\ell}(T_e)\right\rangle, \quad (52.243)$$

where, in reduced units $\omega = h\nu/I_n$, $T = k_B T_e/I_n = b_n^{-1}$, the averaged PI cross section corresponding to (52.174) is

$$\left\langle\sigma_{\mathrm{I}}^{n\ell}(T)\right\rangle = \frac{e^{1/T}}{T}\int_1^\infty \omega^2\sigma_{\mathrm{I}}^{n\ell}(\omega)e^{-\omega/T}\,d\omega. \quad (52.244)$$

When $\sigma_{\mathrm{I}}^{n\ell}(\omega)$ is expressed in Mb (10^{-18} cm^2),

$$\hat{\alpha}_{\mathrm{R}}^{n\ell}(T_e) = 1.508\times10^{-13}\left(\frac{300}{T_e}\right)^{1/2}\left(\frac{I_n}{I_{\mathrm{H}}}\right)^2\left(\frac{g_A}{2g_A^+}\right)$$

$$\times\left\langle\sigma_{\mathrm{I}}^{n\ell}(T)\right\rangle\ \mathrm{cm}^3\,\mathrm{s}^{-1}. \quad (52.245)$$

When σ_{I} can be expressed in terms of the threshold cross section σ_0^n [Eq. (52.178)] as

$$\sigma_{\mathrm{I}}^{n\ell}(h\nu) = (I_n/h\nu)^p\sigma_0(n); \quad (p = 0,1,2,3), \quad (52.246)$$

then $\left\langle\sigma_{\mathrm{I}}^{n\ell}(T)\right\rangle = S_p(T)\sigma_0(n)$, where

$$S_0(T) = 1 + 2T + 2T^2, \quad S_1(T) = 1 + T, \quad (52.247a)$$

$$S_2(T) = 1, \quad (52.247b)$$

$$S_3(T) = \left[e^{1/T}/T\right]E_1(1/T) \quad (52.247c)$$

$$\overset{T\ll1}{\sim} 1 - T + 2T^2 - 6T^3. \quad (52.247d)$$

The case $p = 3$ corresponds to Kramers PI cross section (52.181) so that

$$_{\mathrm{K}}\hat{\alpha}_{\mathrm{R}}^{n\ell}(T_e) = \frac{(2\ell+1)}{n^2}\frac{2}{n}\hat{\alpha}_0(T_e)S_3(T) \quad (52.248a)$$

$$\equiv {}_{\mathrm{K}}\hat{\alpha}_{\mathrm{R}}^{n\ell}(T_e \to 0)S_3(T), \quad (52.248b)$$

such that $_{\mathrm{K}}\hat{\alpha}_{\mathrm{R}}^{n\ell} \sim Z^2/(n^3 T_e^{1/2})$ as $T = (k_B T_e/I_n) \to 0$.

52.8.6 Gaunt Factor, Cross Sections, and Rates for Hydrogenic Systems

The Gaunt factor $G_{n\ell}$ is the ratio of the quantal to Kramers (K) semiclassical PI cross section such that

$$\sigma_{\mathrm{I}}^{n\ell}(h\nu) = {}_{\mathrm{K}}\sigma_{\mathrm{I}}^{n}(h\nu)G_{n\ell}(\omega); \qquad (52.249)$$

$$\omega = h\nu/I_n = 1 + E/I_n.$$

(a) Radiative Recombination Cross Section.

$$\sigma_{\mathrm{R}}^{n\ell}(E) = \left(\frac{g_A}{g_A^+}\right)\left[\frac{\alpha^2(h\nu)^2}{2E(e^2/a_0)}\right]G_{n\ell}(\omega){}_{\mathrm{K}}\sigma_{\mathrm{I}}^{n\ell}(h\nu) \quad (52.250a)$$

$$= G_{n\ell}(\omega){}_{\mathrm{K}}\sigma_{\mathrm{R}}^{n\ell}(E) \qquad (52.250b)$$

$$= \left[\frac{(2\ell+1)}{n^2}G_{n\ell}(\omega)\right]{}_{\mathrm{K}}\sigma_{\mathrm{R}}^{n}(E), \qquad (52.250c)$$

$$\sigma_{\mathrm{R}}^{n}(E) = G_n(\omega){}_{\mathrm{K}}\sigma_{\mathrm{R}}^{n}(E) \qquad (52.250d)$$

where the quantum mechanical correction, or Gaunt Factor, to the semiclassical cross sections

$$G_{n\ell}(\omega) \rightarrow \begin{cases} 1, & \omega \rightarrow 1 \\ \omega^{-(\ell+1/2)}, & \omega \rightarrow \infty \end{cases} \qquad (52.251)$$

favors low $n\ell$ states. The ℓ-averaged Gaunt factor is

$$G_n(\omega) = (1/n^2)\sum_{\ell=0}^{n-1}(2\ell+1)G_{n\ell}(\omega). \qquad (52.252)$$

Approximations for G_n: as ε increases from zero,

$$G_n(\varepsilon) = \left[1 + \frac{4}{3}(a_n + b_n) + \frac{28}{18}a_n^2\right]^{-3/4} \qquad (52.253a)$$

$$\simeq 1 - (a_n + b_n) + \frac{7}{3}a_n b_n + \frac{7}{6}b_n^2 \qquad (52.253b)$$

where $E = \varepsilon(Z^2e^2/2a_0)$, $\omega = 1 + n^2\varepsilon$, and

$$a_n(\varepsilon) = 0.172825(1 - n^2\varepsilon)c_n(\varepsilon), \qquad (52.254a)$$

$$b_n(\varepsilon) = 0.04959\left[1 + \frac{4}{3}n^2\varepsilon + n^4\varepsilon^2\right]c_n^2(\varepsilon), \qquad (52.254b)$$

$$c_n(\varepsilon) = n^{-2/3}(1 + n^2\varepsilon)^{-2/3}. \qquad (52.254c)$$

(b) Radiative Recombination Rate.

$$\hat{\alpha}_{\mathrm{R}}^{n\ell}(T_e) = {}_{\mathrm{K}}\hat{\alpha}_{\mathrm{R}}^{n\ell}(T_e \rightarrow 0)F_{n\ell}(T), \qquad (52.255)$$

$$\hat{\alpha}_{\mathrm{R}}^{n\ell}(T_e \rightarrow 0) = \frac{(2\ell+1)}{n^2}\left(\frac{2}{n}\right)\hat{\alpha}_0(T_e), \qquad (52.256)$$

in accordance with (52.185b).

$$F_{n\ell}(T) = \frac{e^{1/T}}{T}\int_1^\infty \frac{G_{n\ell}(\omega)}{\omega}e^{-\omega/T}\,d\omega. \qquad (52.257)$$

The multiplicative factors F and G convert the semiclassical (Kramers) $T_e \rightarrow 0$ rate and cross section to their quantal values. Departures from the scaling rule $(Z^2/n^3T_e^{1/2})$ for RR rates is measured by $F_{n\ell}(T)$.

52.8.7 Exact Universal Rate Scaling Law and Results for Hydrogenic Systems

$$\hat{\alpha}_{\mathrm{R}}^{n\ell}(Z, T_e) = Z\hat{\alpha}_{\mathrm{R}}^{n\ell}(1, T_e/Z^2) \qquad (52.258)$$

as exhibited by (52.243) with (52.239e) and (52.244).

Recombination rates are greatest into low n levels and the $\omega^{-\ell-1/2}$ variation of $G_{n\ell}$ preferentially populates states with low $\ell \sim 2$–5. Highly accurate analytical fits for $G_{n\ell}(\omega)$ have been obtained for $n \leq 20$ so that (52.249) can be expressed in terms of known functions of fit parameters [17]. This procedure (which does not violate the S_2 sum rule) has been extended to nonhydrogenic systems of neon-like Fe XVII, where $\sigma_{\mathrm{I}}^{n\ell}(\omega)$ is a monotonically decreasing function of ω.

The variation of the ℓ-averaged values

$$n^{-2}\sum_{\ell=0}^{n-1}(2\ell+1)F_{n\ell}(T)$$

is close in both shape and magnitude to the corresponding semiclassical function $S_3(T)$, given by (52.257) with $G_{n\ell}(\omega) = 1$. Hence the ℓ-averaged recombination rate is

$$\hat{\alpha}_{\mathrm{R}}^{n}(Z, T) = (300/T)^{1/2}(Z^2/n)F_n(T)$$
$$\times 1.1932 \times 10^{-12}\,\mathrm{cm}^3\,\mathrm{s}^{-1},$$

where F_n can be calculated directly from (52.257) or be approximated as $G_n(1)S(T)$. A computer program based on a three-term expansion of G_n is also available [18]. From a three-term expansion for G, the rate of radiative recombination into all levels of a hydrogenic system is

$$\hat{\alpha}(Z, T) = 5.2 \times 10^{-14}Z\lambda^{1/2}\left[0.43 + \frac{1}{2}\ln\lambda + 0.47/\lambda^{1/2}\right], \qquad (52.259)$$

where $\lambda = 1.58 \times 10^5 Z^2/T$ and $[\hat{\alpha}] = \mathrm{cm}^3/\mathrm{s}$. Tables [19] exist for the effective rate

$$\hat{\alpha}_E^{n\ell}(T) = \sum_{n'=n}^\infty \sum_{\ell'=0}^{n'-1}\hat{\alpha}_{\mathrm{R}}^{n'\ell'}C_{n'\ell',n\ell} \qquad (52.260)$$

of populating a given level $n\ell$ of H via radiative recombination into all levels $n' \geq n$ with subsequent radiative cascade $(i \rightarrow f)$ with probability $C_{i,f}$ via all possible intermediate paths. Tables [19] also exist for the full rate

$$\hat{\alpha}_F^N(T) = \sum_{n=N}^\infty \sum_{\ell=0}^{n-1}\hat{\alpha}_{\mathrm{R}}^{n\ell} \qquad (52.261)$$

of recombination, into all levels above $N = 1, 2, 3, 4$, of hydrogen. They are useful in deducing time scales of radiative recombination and rates for complex ions.

52.9   USEFUL QUANTITIES

(a) Mean Speed.

$$\bar{v}_e = \left(\frac{8k_BT}{\pi m_e}\right)^{1/2} = 1.076\,042 \times 10^7 \left[\frac{T}{300}\right]^{1/2} \text{ cm/s}$$

$$= 6.692\,38 \times 10^7 T_{eV}^{1/2} \text{ cm/s}$$

$$\bar{v}_i = 2.511\,16 \times 10^5 \left[\frac{T}{300}\right]^{1/2} (m_p/m_i)^{1/2} \text{ cm/s}$$

where $(m_p/m_e)^{1/2} = 42.850\,352$, and $T = 11\,604.45\,T_{eV}$ relates the temperature in K and in eV.

(b) Natural Radius: $|V(R_e)| = e^2/R_e = k_BT$.

$$R_e = \frac{e^2}{k_BT} = 557\left(\frac{300}{T}\right)\text{\AA} = \left(\frac{14.4}{T_{eV}}\right)\text{\AA}.$$

(c) Boltzmann Average Momentum.

$$\langle p \rangle = \int_{-\infty}^{\infty} e^{-p^2/2mk_BT}\,dp = (2\pi m_e k_BT)^{1/2}.$$

(d) De Broglie Wavelength.

$$\lambda_{dB} = \frac{h}{\langle p \rangle} = \frac{h}{(2\pi m_e k_BT)^{1/2}} = \frac{7.453\,818 \times 10^{-6}}{T_e^{1/2}} \text{ cm}$$

$$= 43.035\left(\frac{300}{T_e}\right)^{1/2}\text{\AA} = \frac{6.9194}{T_{eV}^{1/2}}\text{\AA}.$$

Acknowledgments

This research is supported by the U. S. Air Force Office of Scientific Research under Grant No. F49620-94-1-0379. I wish to thank Dr. E. J. Mansky for numerous discussions on the content and form of this chapter and without whose expertise in computer typesetting this chapter would not have been possible.

GENERAL REFERENCES

General Recombination

M. R. Flannery, *Electron-Ion and Ion-Ion Recombination Processes*, Adv. At. Mol. Phys. **32**, 117 (1994).

M. R. Flannery, *Recombination Processes in Molecular Processes in Space*, edited by T. Watanabe, I. Shimamura, M. Shimiza, and Y. Itikawa (Plenum Press, New York, 1990), Ch. 7, p. 145.

D. R. Bates, *Recombination* in *Electronic and Atomic Collisions*, edited by H. B. Gilbody, W. R. Newell, F. H. Read and A. C. H. Smith (North-Holland, Amsterdam, 1988), p. 3.

D. R. Bates, *Recombination* in *Case Studies in Atomic Physics*, edited by E. W. McDaniel and M. R. C. McDowell, (North-Holland, Amsterdam, 1974), Vol. 4, p. 57.

D. R. Bates and A. Dalgarno, *Electronic Recombination* in *Atomic and Molecular Processes*, edited by D. R. Bates (Academic Press, New York, 1962)

D. R. Bates, *Aspects of Recombination*, Adv. At. Mol. Phys. **15**, 235 (1979).

Recombination of Atomic Ions, edited by W. G. Graham (Plenum Press, New York, 1992).

J. N. Bardsley, *Recombination Processes in Atomic and Molecular Physics* in *Atomic and Molecular Collision Theory*, edited by F. A. Gianturco (Plenum Press, New York, 1980), p. 123.

J. Dubau and S. Volonte, *Dielectronic Recombination and its Applications in Astronomy*, Rep. Prog. Phys. **43**, 199 (1980).

Y. Hahn, *Theory of Dielectronic Recombination*, Adv. At. Mol. Phys. **21** 123 (1985).

Y. Hahn and K. J. LaGattuta, *Dielectronic Recombination and Related Processes*, Phys. Rep. **116** 195 (1988).

E. W. McDaniel and E. J. Mansky, *Guide to Bibliographies, Books, Reviews and Compendia of Data on Atomic Collisions*, Adv. At. Mol. Opt. Phys. **33** 389 (1994).

Three-Body Electron-Ion Collisional-Radiative Recombination: Theory

D. R. Bates, A. E. Kingston, and R. W. P. McWhirter, *Recombination between Electrons and Atomic Ions I. Optically Thin Plasmas, II. Optically Thick Plasmas*, Proc. R. Soc. (London) Ser. A **267**, 297 (1962); **270**, 155 (1962).

D. R. Bates and S. P. Khare, *Recombination of Positive Ions and Electrons in a Dense Neutral Gas*, Proc. Phys. Soc. **85**, 231 (1965).

D. R. Bates, V. Malaviya, and N. A. Young, *Electron-Ion Recombination in a Dense Molecular Gas*, Proc. R. Soc. (London) Ser. A **320**, 437 (1971).

A. Burgess and H. P. Summers, *The Recombination and Level Populations of Ions I. Hydrogen and Hydrogenic Ions*, Mon. Not. R. Astron. Soc. **174**, 345 (1976).

H. P. Summers, *The Recombination and Level Populations of Ions II. Resolution of Angular Momentum States*, Mon. Not. R. Astron. Soc. **178**, 101 (1977).

N. N. Ljepojevic, R. J. Hutcheon, and J. Payne, *COLRAD – A Program to Calculate Population Densitites of the Excited Atomic Levels of Hydrogen-like Ions in a Plasma*, Comp. Phys. Commun. **44**, 157 (1987).

Electron-Ion Recombination: Molecular Dynamics Simulations

W. L. Morgan, *Electron-ion Recombination in Water Vapor*, J. Chem. Phys. **80**, 4564 (1984).

W. L. Morgan, *Molecular Dynamics Simulation of Electron-ion Recombination in a Nonequilibrium Weakly Ionized Plasma*, Phys. Rev. A **30**, 979 (1984).

W. L. Morgan and J. N. Bardsley, *Monte-Carlo Simulation of Electron-ion Recombination at High Pressure*, Chem. Phys. Lett. **96**, 93 (1983).

W. L. Morgan, in *Recent Studies in Atomic and Molecular Processes*, edited by A. E. Kingston, (Plenum Press, New York, 1987), p. 149.

Ion-Ion Recombination: Review Articles

M. R. Flannery, *Three-Body Recombination between Positive and Negative Ions* in *Case Studies in Atomic Collision Physics*, edited by E. W. McDaniel and M. R. C. McDowell, (North-Holland, Amsterdam, 1972), Vol. 2, Ch. 1, p. 1.

M. R. Flannery, *Ionic Recombination* in *Atomic Processes and Applications*, edited by P. G. Burke and B. L. Moiseiwitsch, (North-Holland, Amsterdam, 1976), Ch. 12, p. 407.

M. R. Flannery, *Ion-Ion Recombination in High Pressure Plasmas* in *Applied Atomic Collision Physics*, edited by E. W. McDaniel and W. L. Nighan (Academic Press, New York, 1983), Vol. 3 *Gas Lasers*, Ch. 5, p. 393.

M. R. Flannery, *Microscopic and Macroscopic Perspectives of Termolecular Association of Atomic Reactants in a Gas* in *Recent Studies in Atomic and Molecular Processes*, edited by A. E. Kingston (Plenum Press, London, 1987), p. 167.

D. R. Bates, *Ion-ion Recombination in an Ambient Gas*, Adv. At. Mol. Phys. **20**, 1 (1985).

B. H. Mahan, *Recombination of Gaseous Ions*, Adv. Chem. Phys. **23**, 1 (1973).

J. T. Moseley, R. E. Olson, and J. R. Peterson, in *Case Studies in Atomic Collision Physics*, edited by M. R. C. McDowell and E. W. McDaniel (North-Holland, Amsterdam, 1972), p. 1.

D. Smith and N. G. Adams, *Studies of Ion-Ion Recombination using Flowing Afterglow Plasmas* in *Physics of Ion-Ion and Electron-Ion Collisions*, edited by F. Brouillard and J. W. McGowan, (Plenum Press, New York, 1982), p. 501.

Termolecular Ion-Ion Recombination: Theory

(A) Low Gas Densities: Linear Region

M. R. Flannery and E. J. Mansky, *Termolecular Recombination at Low Gas Density: Strong Collision, Bottleneck, and Exact Treatments*, J. Chem. Phys. **88**, 4228 (1988).

M. R. Flannery and E. J. Mansky, *Termolecular Recombination: Coupled Nearest-Neighbor Limit and Uncoupled Intermediate Levels Limit*, J. Chem. Phys. **89**, 4086 (1988).

M. R. Flannery, *Variational Principle for Termolecular Recombination in a Gas*, J. Chem. Phys. **89**, 214 (1988).

M. R. Flannery, *Diffusional Theory of Termolecular Recombination and Association of Atomic Species in a Gas*, J. Chem. Phys. **87**, 6947 (1987).

M. R. Flannery, *Ion-Ion Recombination in $(X^+ + Y^- + Z)$ Systems at Low Gas Densities I. Symmetrical Resonance Charge-Transfer Contribution*, J. Phys. B **13**, 3649 (1980).

M. R. Flannery, *Ion-Ion Recombination in $(X^+ + Y^- + Z)$ Systems at Low Gas Densities II. Elastic Ion-Neutral Collisions*, J. Phys. B **14**, 915 (1981).

D. R. Bates and I. Mendaš, *Termolecular Ionic Recombination at Low Ambient Gas Density for the Case of a Polarization Interaction*, J. Phys. B **15**, 1949 (1982).

D. R. Bates, P. B. Hays, and D. Sprevak, *Kinetic Energy of Ion Pairs in Three-Body Ionic Recombination*, J. Phys. B **4**, 962 (1971).

D. R. Bates and M. R. Flannery, *Recombination of Positive and Negative Ions II. General Third Body*, Proc. R. Soc. (London) Ser. A **302**, 367 (1968).

D. R. Bates and R. J. Moffett, *Recombination of Positive and Negative Ions I. Ions Recombining in Their Parent Gas*, Proc. R. Soc. (London) Ser. A **291**, 1 (1966).

(B) All Gas Densities: Non-Linear Region

M. R. Flannery, *Microscopic and Macroscopic Theories of Termolecular Recombination between Atomic Ions*, in *Dissociative Recombination*, edited by B. R. Rowe, J. B. A. Mitchell, and A. Canosa (Plenum Press, New York, 1993), p. 205.

M. R. Flannery, *Transport-Collisional Master Equations for Termolecular Recombination as a Function of Gas Density*, J. Chem. Phys. **95**, 8205 (1991).

M. R. Flannery, *Theory of Ion-Ion Recombination*, Phil. Trans. R. Soc. (London) Ser. A **304**, 447 (1982).

D. R. Bates and I. Mendaš, *Ionic Recombination in an Ambient Gas I. Extension of Quasi-Equilibrium Statistical Method into Nonlinear Region*, Proc. R. Soc. (London) Ser. A **359**, 275 (1978).

J. J. Thomson, *Recombination of Gaseous Ions, the Chemical Combination of Gases and Monomolecular Reactions*, Phil. Mag. **47**, 337 (1924).

Ion-Ion Recombination: Monte-Carlo Simulations

P. J. Feibelman, *Numerical Study of the Three-Body Ion-Ion Recombination*, J. Chem. Phys. **42**, 2462 (1965).

A. Jones and J. L. J. Rosenfeld, *Monte-Carlo simulation of Hydrogen-Atom Recombination*, Proc. R. Soc. (London) Ser. A **333**, 419 (1973).

D. R. Bates and I. Mendaš, *Ionic Recombination in an Ambient Gas II. Computer Experiment with Specific Allowance for Binary Recombination*, Proc. R. Soc. (London) Ser. A **359**, 287 (1978).

D. R. Bates, *Universal Curve for Termolecular Ionic Recombination Coefficients*, Chem. Phys. Lett. **75**, 409 (1980).

J. N. Bardsley and J. M. Wadehra, *Monte-Carlo Simulation of Three Body Ion-Ion Recombination*, Chem.

Phys. Lett. **72**, 477 (1980).

D. R. Bates, *Mutual Neutralization Coefficient in an Ambient Gas*, J. Phys. B **14**, 4207 (1981).

D. R. Bates, *Effect of Inelastic Collisions on the Rate of Termolecular Ionic Recombination*, J. Phys. B **14**, 2853 (1981).

D. R. Bates and I. Mendaš, *Rate Coefficients for Termolecular Ionic Recombination*, Chem. Phys. Lett. **88**, 528 (1982).

W. L. Morgan, J. N. Bardsley, J. Lin, and B. L. Whitten, *Theory of Ion-Ion Recombination in Plasmas*, Phys. Rev. A **26**, 1696 (1982).

B. L. Whitten, W. L. Morgan, and J. N. Bardsley, *Monte-Carlo Calculations of Two- and Three-Body Ionic Recombination*, J. Phys. B **15**, 319 (1982).

Ion-Ion Tidal Recombination: Molecular Dynamics Simulations

D. R. Bates and W. L. Morgan, *New Recombination Mechanism: Tidal Termolecular Ionic Recombination*, Phys. Rev. Lett. **64**, 2258 (1990).

W. L. Morgan and D. R. Bates, *Tidal Termolecular Ionic Recombination*, J. Phys. B **25**, 5421 (1992).

Radiative Recombination: Theory

M. J. Seaton, *Radiative Recombination of Hydrogenic Ions*, Mon. Not. R. Astron. Soc. **119**, 81 (1959).

D. R. Flower and M. J. Seaton, *A Program to Calculate Radiative Recombination Coefficients of Hydrogenic Ions*, Comp. Phys. Commun. **1**, 31 (1969).

A. Burgess and H. P. Summers, *Radiative Gaunt Factors*, Mon. Not. R. Astron. Soc. **226**, 257 (1987).

Y. S. Kim and R. H. Pratt, *Direct Radiative Recombination of Electrons with Atomic Ions: Cross Sections and Rate Coefficients*, Phys. Rev. A **27**, 2913 (1983).

D. J. McLaughlin and Y. Hahn, *Scaling Behavior of Radiative Recombination Cross Sections and Rate Coefficients*, Phys. Rev. A **43**, 1313 (1991).

F. D. Aaron, A. Costescu, and C. Dinu, *Direct Radiative Recombination and Photoeffect Cross Sections for Arbitrary Atomic Coulomb Bound States*, J. Phys. II (Paris) **3** 1227 (1993).

Dissociative Recombination: Theory and Experiment

Dissociative Recombination: Theory, Experiment and Applications III, edited by D. Zajfman, J. B. A. Mitchell, B. R. Rowe, and D. Schwalin (World Scientific, Singapore, 1996).

D. R. Bates, *Dissociative Recombination: Crossing and Tunneling Modes*, Adv. At. Mol. Phys. **34**, 427 (1994).

Dissociative Recombination: Theory, Experiment and Applications II, edited by B. R. Rowe, J. B. A. Mitchell and A. Canosa (Plenum Press, New York, 1993).

Dissociative Recombination: Theory, Experiment and Applications I, edited by J. B. A. Mitchell and S. L. Guberman (World Scientific, Singapore, 1989).

J. B. A. Mitchell, *The Dissociative Recombination of Molecular Ions*, Phys. Rep. **186**, 215 (1990).

A. Giusti-Suzor, *Recent Developments in the Theory of Dissociative Recombination and Related Processes* in *Atomic Processes in Electron-Ion and Ion-Ion Collisions*, edited by F. Brouillard, (Plenum Press, New York, 1986).

J. N. Bardsley and M. A. Biondi, *Dissociative Recombination*, Adv. At. Mol. Phys. **6**, 1 (1970).

REFERENCES

1. J. Stevefelt, J. Boulmer, and J-F. Delpech, Phys. Rev. A **12** 1246 (1975).
2. R. Deloche, P. Monchicourt, M. Cheret, and F. Lambert, Phys. Rev. A **13**, 1140 (1976).
3. P. Mansbach and J. Keck, Phys. Rev. **181**, 275 (1965).
4. L. P. Pitaevskiĭ, Sov. Phys. JETP **15**, 919 (1962).
5. M. R. Flannery, J. Chem. Phys. **95**, 8205 (1991).
6. D. R. Bates, Phys. Rev. **78**, 492 (1950).
7. J. N. Bardsley, *The Theory of Dissociative Recombination*, J. Phys. A: Proc. Phys. Soc. [2] **1**, 365 (1968).
8. M. R. Flannery, in *Atomic Collisions: A Symposium in Honor of Christopher Bottcher*, edited by D. R. Schultz, M. R. Strayer, and J. H. Macek (American Institute of Physics, New York, 1995), p. 53.
9. A. Giusti, J. Phys. B **13**, 3867 (1980).
10. P. van der Donk, F. B. Yousif, J. B. A. Mitchell, and A. P. Hickman, Phys. Rev. Lett. **68**, 2252 (1992).
11. S. L. Guberman, Phys. Rev. A **49**, R4277 (1994).
12. M. R. Flannery, Int. J. Mass Spectrom. Ion Processes, **149/150**, 597 (1995).
13. H. A. Kramers, Phil. Mag. **46**, 836 (1923).
14. A. Burgess, Mon. Not. R. Astron. Soc. **118**, 477 (1958).
15. J. A. Gaunt, Phil. Trans. Roy. Soc. A **229**, 163 (1930).
16. M. J. Seaton, Mon. Not. R. Astron. Soc. **119**, 81 (1959).
17. B. F. Rozsnyai and V. L. Jacobs, Astrophys. J. **327**, 485 (1988).
18. D. R. Flower and M. J. Seaton, Comp. Phys. Commun. **1**, 31 (1969).
19. P. G. Martin, Astrophys. J. Supp. Ser. **66**, 125 (1988).

53

Dielectronic Recombination

M. S. Pindzola,* D. C. Griffin,† and N. R. Badnell‡

* Department of Physics, Auburn University, Auburn, Alabama 36849
† Department of Physics, Rollins College, Winter Park, Florida 32789
‡Department of Physics, University of Strathclyde, Glasgow G40NG, United Kingdom

53.1 INTRODUCTION

Electron-ion recombination into a particular final recombined state may be schematically represented as

$$e^- + A_i^{q+} \rightarrow A_f^{(q-1)+} + \hbar\omega \,, \qquad (53.1)$$

and

$$e^- + A_i^{q+} \rightarrow \left[A_j^{(q-1)+} \right] \rightarrow A_f^{(q-1)+} + \hbar\omega, \qquad (53.2)$$

where q is the charge on the atomic ion A, ω is the frequency of the emitted light, and the brackets in Eq. (53.2) indicate a doubly-excited resonance state. The first process is called radiative recombination (RR), while the second is called dielectronic recombination (DR). Both recombination mechanisms are the inverse of photoionization. At sufficiently high electron density, three-body recombination becomes possible. The three-body mechanism is the inverse of electron impact ionization.

The review article by Seaton and Storey [1] includes an interesting history of the theoretical work on dielectronic recombination. The process was first referred to as dielectronic recombination by Massey and Bates [2],

after a suggestion of its possible importance in the ionosphere by Sayers in 1939. However, estimates of the rate coefficient for this process indicated that DR is not an important process in the ionosphere, where the temperatures are too low to excite anything but the lower energy resonance states.

In 1961, Unsold, in a letter to Seaton, suggested that DR might account for a well-known temperature discrepancy in the solar corona. Seaton initially concluded that DR would not significantly increase recombination in the solar corona. However, he had only included the lower energy resonance states in his analysis; Burgess [3] showed that when one includes the high members of the Rydberg series of resonance states that are populated at coronal temperatures, DR can indeed explain this discrepancy.

DR has since received much theoretical attention due, in part, to its importance in modeling high temperature plasmas. Various approaches to the theory are discussed in a review by Hahn [4] and in Ref. [5]

Interest in dielectronic recombination has increased dramatically in the last decade. Mitchell *et al.* [6] published the DR cross section for C$^+$ using a merged electron-ion beams apparatus, and Belic *et al.* [7] reported on a crossed beams measurement of the DR cross section for Mg$^+$. Also, Dittner *et al.* [8] published merged beams measurements of the DR cross section for the multiply charged ions B^{2+} and C^{3+}. Since that time, atomic physics experiments carried out using heavy-ion traps, accelerators, and storage-cooler rings have produced high-resolution mappings of the resonance structures associated with electron-ion recombination. The experiments have been carried out using a wide range of facilities and technologies, such as the test storage ring (TSR) at Heidelberg, the experimental storage ring (ESR) at Darmstadt, the accelerator-cooler ring facility at Aarhus, the electron beam ion trap (EBIT) at Livermore, and the electron beam ion source (EBIS) at Kansas State. A good review of the dramatic experimental progress in DR measurements is again found in the NATO proceedings [5].

53.2 THEORETICAL FORMULATION

In the independent-processes approximation, the two paths for recombination are summed incoherently. The radiative recombination cross section for Eq. (53.1), in lowest-order of perturbation theory, is given by

$$\sigma_{\rm RR} = \frac{8\pi^2}{k^3} \sum_{\ell j} \sum_{JM} \sum_{M_0} \frac{1}{2g_1} |\langle \alpha_0 J_0 M_0 | D | \alpha_1 J_1 \ell j JM \rangle|^2 .$$

(53.3)

The set $(\alpha_1 J_1)$ represents the quantum numbers for the N-electron target ion state, $(\alpha_0 J_0 M_0)$ represents the quantum numbers for the $(N+1)$-electron recombined ion state, $(k\ell j)$ represents the quantum numbers for the continuum electron state, (JM) represents the quantum numbers for the $(N+1)$-electron system of target plus free electron state, and g_1 is the statistical weight of a J_1 level. The dipole radiation field operator is given by

$$D = \sqrt{\frac{2\omega^3}{3\pi c^3}} \sum_{s=1}^{N+1} \vec{r}_s .$$

(53.4)

Continuum normalization is chosen as one times a sine function, and atomic units ($e = \hbar = m = 1$) are used. In the isolated-resonance approximation, the dielectronic recombination cross section for Eq. (53.2), in lowest-order perturbation theory, is given by

$$\sigma_{\rm DR} = \frac{8\pi^2}{k^3} \sum_{\ell j} \sum_{JM} \sum_{M_0} \frac{1}{2g_1}$$

$$\times \sum_{\alpha_i J_i M_i} \left| \frac{\langle \alpha_0 J_0 M_0 | D | \alpha_i J_i M_i \rangle \langle \alpha_i J_i M_i | V | \alpha_1 J_1 \ell j JM \rangle}{E_0 - E_i + i\Gamma_i/2} \right|^2 ,$$

(53.5)

where the set $(\alpha_i J_i M_i)$ represents the quantum numbers for a resonance state with energy E_i and total width

Γ_i, and the electrostatic interaction between electrons is given by

$$V = \sum_{s=1}^{N} |\vec{r}_s - \vec{r}_{N+1}|^{-1} .$$

(53.6)

By the principle of detailed balance, $\sigma_{\rm RR}$ of Eq. (53.3) is proportional to the photoionization cross section from the bound state, while the energy-averaged $\sigma_{\rm DR}$ may be written as

$$\langle \sigma_{\rm DR} \rangle = \frac{2\pi^2}{\Delta \epsilon k^2} \sum_{\alpha_i J_i M_i} \frac{1}{2g_1} \frac{A_{\rm a} A_{\rm r}}{\Gamma_i} ,$$

(53.7)

where the autoionization decay rate $A_{\rm a}$ is given by

$$A_{\rm a} = \frac{4}{k} \sum_{\ell j} \sum_{JM} |\langle \alpha_1 J_1 \ell j JM | V | \alpha_i J_i M_i \rangle|^2 ,$$

(53.8)

the radiative decay rate $A_{\rm r}$ is given by

$$A_{\rm r} = 2\pi \sum_{M_0} |\langle \alpha_0 J_0 M_0 | D | \alpha_i J_i M_i \rangle|^2 ,$$

(53.9)

and $\Delta \epsilon$ is the energy bin width. Each resonance level in Eq. (53.7) makes a contribution at a fixed continuum energy $k^2/2$; thus $\langle \sigma_{\rm DR} \rangle$ plotted as a function of energy is a histogram.

53.3 COMPARISONS WITH EXPERIMENT

53.3.1 Low-Z Ions

For the most part, the agreement between the recent high-resolution measurements and theoretical calculations based on the independent-processes and isolated-resonance (IPIR) approximations is quite good [9–25]. We illustrate the agreement with experiment obtained by calculations employing the IPIR approximations with three examples from the Li isoelectronic sequence. The pathways for dielectronic capture, in terms of specific levels, are given by

$$e^- + A^{q+}(2{\rm s}_{1/2}) \rightarrow \left[A^{(q-1)+}(2{\rm p}_{1/2}nlj) \right]$$
$$\searrow \left[A^{(q-1)+}(2{\rm p}_{3/2}nlj) \right] ,$$

(53.10)

where a $1s^2$ core is assumed to be present. Both Rydberg series autoionize by the reverse of the paths in Eq. (53.10), while for sufficiently high n, the $2{\rm p}_{3/2}nlj$ levels may autoionize to the $2{\rm p}_{1/2}$ continuum. Both series radiatively stabilize by either a $2p \rightarrow 2s$ core orbital transition, or by a $nlj \rightarrow n'l'j'$ valence orbital transition, where $2p_j n'l'j'$ with $j = \frac{1}{2}, \frac{3}{2}$ is a bound level.

Figure 53.1. Dielectronic Recombination for O^{5+}. Calculations were performed for fields of 0 (dotted curve), 3 V/cm (dashed curve), 5 V/cm (chain curve), and 7 V/cm (solid curve).

Dielectronic recombination cross section calculations [17] in the IPIR approximation for O^{5+} are compared with experiment in Fig. 53.1. Fine-structure splitting of the two series of Eq. (53.10) is minimal for this light ion, so there appears only one Rydberg series. The $2p\,6\ell$ resonances are located at 2.5 eV, the $2p\,7\ell$ at 5.0 eV, and so on; accumulating at the series limit around 11.3 eV. Electric field effects on the high-n resonances are strong in O^{5+}, so that calculations were done for fields of 0, 3, 5, and 7 V/cm. Since the precise electric field strength in the experiment is not known, the accuracy of an electric field dependent theory, in this case, has yet to be determined. However, the effects of state mixing by extrinsic fields in the collision region for the dielectronic recombination of Mg^{+} have been investigated both experimentally [26] and theoretically [27].

53.3.2 High-Z Ions and Relativistic Effects

Dielectronic recombination cross section calculations [20] in the IPIR approximation for Cu^{26+} are compared with experiment in Fig. 53.2. The two fine-structure Rydberg series are now clearly resolved; the fine-structure splitting is about 27 eV for Cu^{26+}. The $2p_{1/2}\,13\ell$ resonances are just above threshold, while the $2p_{3/2}\,11\ell$ resonances are found around 5.0 eV. Electric fields in the range 0–50 V/cm have little effect on the Cu^{26+} spectrum. Overall, the agreement between theory and experiment is excellent.

Electron-ion recombination cross section calculations [21] in the IPIR approximation for low-lying resonances in Au^{76+} are compared with experiment in Fig. 53.3. The $2p_{1/2}\,nlj$ series limit is at 217 eV, while the $2p_{3/2}\,nlj$ series limit is at 2.24 keV; yielding a fine structure splitting

Figure 53.2. Dielectronic Recombination for Cu^{26+}.

of 2.03 keV. QED effects alone shift the $2p_{3/2}\,nlj$ series limit by 22.0 eV. Thus accurate atomic structure calculations must be made to locate the $2p_{3/2}\,6lj$ resonances in the 0–50 eV energy range of the experiment. The figure shows that the perturbative relativistic, semirelativistic, and fully relativistic calculations for the dielectronic recombination cross section ride on top of a strong radiative recombination background. In principle, the fully relativistic theory contains the most physics, and thus it is comforting that on the whole it is in good agreement with the experiment. It is instructive, however, to see how well the computationally simpler perturbative relativistic and semirelativistic theories do for such a highly charged ion.

53.4 RADIATIVE-DIELECTRONIC RECOMBINATION INTERFERENCE

There has been a great deal of effort in recent years to develop a more general theory of electron-ion recombination which would go beyond the IPIR approximation to include radiative–dielectronic recombination interference and overlapping (and interacting) resonance structures [28–35]. In almost all cases, the interference between a dielectronic recombination resonance and the radiative recombination background is quite small and difficult to observe. The best possibility for observation of RR-DR interference appears to be in highly charged atomic ions. In the cases studied to date, the combination of electron and photon continuum coupling selection rules and the requirement of near energy degeneracy make the overlapping (and interacting) resonance effects small and diffi-

Figure 53.3. Dielectronic Recombination for Au^{76+}. The curves show (a) perturbative relativistic, (b) semirelativistic, and (c) fully relativistic calculations for the dielectronic recombination cross section.

cult to observe. Heavy ions in relatively low stages of ionization are the best place to look, since there are resonance series attached to the large numbers of LS terms or fine structure levels.

The distorted wave approximation (see Chap. 50) has been so successful in describing dielectronic recombination cross sections for most atomic ions because, for low charged ions, the DR cross section is proportional to the radiative rate, while for highly charged ions the DR cross section is proportional to the autoionization rate. Thus the weakness of the distorted wave method in calculating accurate autoionization rates for low charged ions is masked by a DR cross section that is highly dependent on radiative atomic structure. As one moves to more highly charged ions, the DR cross section becomes more sensitive to the autoionization rates, but at the same time, the distorted-wave method becomes increasingly more accurate.

53.5 DIELECTRONIC RECOMBINATION IN PLASMAS

Dielectronic recombination is an important atomic process that is included in the theoretical modeling of the ionization state and emission of radiating ions, which is fundamental to the interpretation of spectral emission from both fusion and astrophysical plasmas (see Chap. 78, 82, and 83). The dielectronic recombination rate coefficient, into a particular final recombined state, is given by

$$\alpha_{DR} = \sqrt{\frac{2}{\pi T^3}} k^2 \Delta \epsilon \left\langle \sigma_{DR} (i \to f) \right\rangle \exp \left(\frac{-k^2}{2T} \right),$$
(53.11)

where T is the electron temperature. Dielectronic recombination rate coefficients, from the ground and metastable states of a target ion into fully resolved low-lying states and bundled high-lying states of a recombined ion, are required for a generalized collisional radiative treatment [36, 37] of highly populated metastable states, the influence of finite plasma density on excited state populations, and of ionization in dynamic plasmas.

REFERENCES

1. M. J. Seaton and P. J. Storey, in *Atomic processes and applications*, edited by P. G. Burke and B. L. Moiseiwitsch (North-Holland, 1976), p. 133.
2. H. S. W. Massey and D. R. Bates, Rep. Prog. Phys. **9**, 62 (1942).
3. A. Burgess, Astrophys. J. **139**, 776 (1964).
4. Y. Hahn, Adv. At. Mol. Phys. **21**, 123 (1985).
5. *Recombination of Atomic Ions*, edited by W. G. Graham, W. Fritsch, Y. Hahn, and J. A. Tanis, NATO ASI Series B **296** (Plenum, New York, 1992).
6. J. B. A. Mitchell, C. T. Ng, J. L. Forand, D. P. Levac, R. E. Mitchell, A. Sen, D. B. Miko, and J. Wm. McGowan, Phys. Rev. Lett. **50**, 335 (1983).
7. D. S. Belic, G. H. Dunn, T. J. Morgan, D. W. Mueller, and C. Timmer, Phys. Rev. Lett. **50**, 339 (1983).
8. P. F. Dittner, S. Datz, P. D. Miller, C. D. Moak, P. H. Stelson, C. Bottcher, W. B. Dress, G. D. Alton, and N. Neskovic, Phys. Rev. Lett. **51**, 31 (1983).
9. G. Kilgus, J. Berger, P. Blatt, M. Grieser, D. Habs, B. Hochadel, E. Jaeschke, D. Kramer, R. Neumann, G. Neureither, W. Ott, D. Schwalm, M. Steck, R. Stokstad, E. Szmola, A. Wolf, R. Schuch, A. Müller, and M. Wagner, Phys. Rev. Lett. **64**, 737 (1990).
10. M. S. Pindzola, N. R. Badnell, and D. C. Griffin, Phys. Rev. A **42**, 282 (1990).
11. D. R. DeWitt, D. Schneider, M. W. Clark, M. H. Chen, and D. Church, Phys. Rev. A **44**, 7185 (1991).
12. L. H. Andersen, P. Hvelplund, H. Knudsen, and P. Kvistgaard, Phys. Rev. Lett. **62**, 2656 (1989).

13. N. R. Badnell, M. S. Pindzola, and D. C. Griffin, Phys. Rev. A **41**, 2422 (1990).
14. L. H. Andersen, G. Y. Pan, H. T. Schmidt, N. R. Badnell, and M. S. Pindzola, Phys. Rev. A **45**, 7868 (1992).
15. D. A. Knapp, R. E. Marrs, M. B. Schneider, M. H. Chen, M. A. Levine, and P. Lee, Phys. Rev. A **47**, 2039 (1993).
16. R. Ali, C. P. Bhalla, C. L. Cocke, and M. Stockli, Phys. Rev. Lett. **64**, 633 (1990).
17. D. C. Griffin, M. S. Pindzola, and P. Krylstedt, Phys. Rev. A **40**, 6699 (1989).
18. L. H. Andersen, J. Bolko, and P. Kvistgaard, Phys. Rev. A **41**, 1293 (1990).
19. L. H. Andersen, G. Y. Pan, H. T. Schmidt, M. S. Pindzola, and N. R. Badnell, Phys. Rev. A **45**, 6332 (1992).
20. G. Kilgus, D. Habs, D. Schwalm, A. Wolf, N. R. Badnell, and A. Müller, Phys. Rev. A **46**, 5730 (1992).
21. W. Spies, A. Müller, J. Linkemann, A. Frank, M. Wagner, C. Kozhuharov, B. Franzke, K. Beckert, F. Bosch, H. Eickhoff, M. Jung, O. Klepper, W. König, P. H. Mokler, R. Moshammer, F. Nolden, U. Schaaf, P. Spädtke, M. Steck, P. Zimmerer, N. Grün, W. Scheid, M. S. Pindzola, and N. R. Badnell, Phys. Rev. Lett. **69**, 2768 (1992).
22. N. R. Badnell, M. S. Pindzola, L. H. Andersen, J. Bolko, and H. T. Schmidt, J. Phys. B **24**, 4441 (1991).
23. A. Lampert, D. Habs, G. Kilgus, D. Schwalm, A. Wolf, N. R. Badnell, and M. S. Pindzola, in *AIP Conference Proceedings # 274, The VIth International Conference on the Physics of Highly-Charged Ions, 1992* (American Institute of Physics, 1992), p. 537.
24. D. R. DeWitt, D. Schneider, M. H. Chen, M. W. Clark, J. W. McDonald, and M. B. Schneider, Phys. Rev. Lett. **68**, 1694 (1992).
25. M. B. Schneider, D. A. Knapp, M. H. Chen, J. H. Scofield, P. Beiersdorfer, C. L. Bennett, J. R. Henderson, R. E. Marrs, and M. A. Levine, Phys. Rev. A **45**, R1291 (1992).
26. A. Muller, D. S. Belic, B. D. DePaola, N. Djuric, G. H. Dunn, D. W. Mueller, and C. Timmer, Phys. Rev. Lett. **56**, 127 (1986).
27. C. Bottcher, D. C. Griffin, and M. S. Pindzola, Phys. Rev. A **34**, 860 (1986).
28. P. C. W. Davies and M. J. Seaton, J. Phys. B **2**, 757 (1969).
29. E. Trefftz, J. Phys. B **3**, 763 (1970).
30. R. H. Bell and M. J. Seaton, J. Phys. B **18**, 1589 (1985).
31. G. Alber, J. Cooper, and A. R. P. Rau, Phys. Rev. A **30**, 2845 (1984).
32. K. J. LaGattuta, Phys. Rev. A **40**, 558 (1989).
33. S. L. Haan and V. L. Jacobs, Phys. Rev. A **40**, 80 (1989).
34. N. R. Badnell and M. S. Pindzola, Phys. Rev. A **45**, 2820 (1992).
35. M. S. Pindzola, N. R. Badnell, and D. C. Griffin, Phys. Rev. A **46**, 5725 (1992).
36. H. P. Summers and M. B. Hooper, Plasma Phys. **25**, 1311 (1983).
37. N. R. Badnell, M. S. Pindzola, W. J. Dickson, H. P. Summers, and D. C. Griffin, Astrophys. J. **407**, L91 (1993).

54

Rydberg Collisions: Binary Encounter, Born and Impulse Approximations

E. J. Mansky

School of Physics, Georgia Institute of Technology, Atlanta, Georgia 30332-0430
*Controlled Fusion Atomic Data Center, Oak Ridge National Laboratory, Oak Ridge, Tennessee 37831-6372

54.1   INTRODUCTION

Rydberg collisions are collisions of electrons, ions and neutral particles with atomic or molecular targets which are in highly excited Rydberg states characterized by large principal quantum numbers ($n \gg 1$). Rydberg collisions of atoms and molecules with neutral and charged particles includes the study of collision-induced transitions both to and from Rydberg states and transitions among Rydberg levels. The basic quantum mechanical structural properties of Rydberg states are given in Chap. 14. This Chapter collects together many of the equations used to study theoretically the collisional properties of both charged and neutral particles with atoms and molecules in Rydberg states or orbitals. The primary theoretical scattering approximations enumerated in this Chapter are the *impulse approximation*, *binary encounter approximation* and the *Born approximation*. The theoretical techniques used to study Rydberg collisions complement and supplement the *eigenfunction expansion* approximations used for collisions with target atoms and molecules in their ground ($n = 1$) or first few excited states ($n > 1$), as discussed in Chap. 45.

Direct application of eigenfunction expansion techniques to Rydberg collisions, wherein the target particle can be in a Rydberg orbital with principal quantum number in the range $n \geq 100$, is prohibitively difficult due to the need to compute numerically and store wave functions with n^3 nodes. For $n = 100$ this amounts to $\sim 10^6$ nodes for each of the wave functions represented in the eigenfunction expansion. Therefore, a variety of approximate scattering theories have been developed to deal specifically with the pecularities of Rydberg collisions.

54.1.1   Rydberg Collision Processes

(a) State-Changing Collisions.
Quasi-elastic ℓ-mixing collisions:

$$A^*(n\ell) + B \rightarrow A^*(n\ell') + B. \tag{54.1}$$

Quasi-elastic J-mixing collisions: Fine structure transitions with $J = |\ell \pm 1/2| \rightarrow J' = |\ell \pm 1/2|$ are

$$A^*(n\ell J) + B \rightarrow A^*(n\ell J') + B. \tag{54.2}$$

Energy transfer n-changing collisions:

---

*present address

$$A^*(n) + B(\beta) \rightarrow A^*(n') + B(\beta'), \tag{54.3}$$

where, if B is a molecule, the transition $\beta \rightarrow \beta'$ represents an inelastic energy transfer to the rotational-vibrational degrees of freedom of the molecule B from the Rydberg atom A^*.

Elastic scattering:

$$A^*(\gamma) + B \rightarrow A^*(\gamma) + B, \tag{54.4}$$

where the label γ denotes the set of quantum numbers n, ℓ or n, ℓ, J used.

Depolarization collisions:

$$A^*(n\ell m) + B \rightarrow A^*(n\ell m') + B, \tag{54.5a}$$
$$A^*(n\ell JM) + B \rightarrow A^*(n\ell JM') + B. \tag{54.5b}$$

(b) Ionizing Collisions.
Direct and associative ionization:

$$A^*(\gamma) + B(\beta) \rightarrow \left\{ \begin{array}{l} A^+ + B(\beta') + e^- \\ BA^+ + e^-. \end{array} \right. \tag{54.6}$$

Penning ionization:

$$A^*(\gamma) + B \rightarrow A + B^+ + e^-. \tag{54.7}$$

Ion pair formation:

$$A^*(\gamma) + B \rightarrow A^+ + B^-. \tag{54.8}$$

Dissociative attachment:

$$A^*(\gamma) + BC \rightarrow A^+B^- + C. \tag{54.9}$$

54.2   GENERAL PROPERTIES OF RYDBERG STATES

Table 54.1 displays the general n-dependence of a number of key properties of Rydberg States and some specific representative values for hydrogen.

54.2.1   Dipole Moments

Definition. $\mathbf{D}_{i \rightarrow f} = -e\mathbf{X}_{i \rightarrow f}$ where

$$\mathbf{X}_{i \rightarrow f} = \langle \phi_f | \sum_j e^{i\mathbf{k} \cdot \mathbf{r}_j} \mathbf{r}_j | \phi_i \rangle. \tag{54.10}$$

Table 54.1. General n-dependence of characteristic properties of Rydberg states. Adapted from Ref. [1].

Property	n-Dependence	$n = 10$	$n = 100$	$n = 500$	$n = 1\,000$
Radius (cm)	$n^2 a_0/Z$	5.3×10^{-7}	5.3×10^{-5}	1.3×10^{-3}	5.3×10^{-3}
Velocity (cm/s)	$v_B Z/n$	2.18×10^7	2.18×10^6	4.4×10^5	2.18×10^5
Area (cm^2)	$\pi a_0^2 n^4/Z^2$	8.8×10^{-13}	8.8×10^{-9}	5.5×10^{-6}	8.8×10^{-5}
Ionization potential (eV)	$Z^2 R_\infty/n^2$	1.36×10^{-1}	1.36×10^{-3}	5.44×10^{-5}	1.36×10^{-6}
Radiative lifetime (s)[a]	$n^5(3\ln n - \frac{1}{4})/(A_0 Z^4)$	8.4×10^{-5}	17	7.3×10^4	7.22 hours
Period of classical motion (s)	$2\pi/\omega_{n,n\pm1} = hn^3/(2Z^2 R_\infty)$	1.5×10^{-13}	1.5×10^{-10}	1.9×10^{-8}	1.5×10^{-7}
Transition frequency (s^{-1})	$\omega_{n,n\pm1} = 2Z^2 R_\infty/(\hbar n^3)$	4.1×10^{13}	4.1×10^{10}	3.3×10^8	4.1×10^7
Wavelength (cm)	$\lambda_{n,n\pm1} = 2\pi c/\omega_{n,n\pm1}$	4.6×10^{-3}	4.6	570	4.5609×10^3

$^a A_0 = \left[8\alpha^3/(3\sqrt{3}\pi) \right] (v_B/a_0)$

Hydrogenic Dipole Moments. See Bethe and Salpeter [2] and the references by Khandelwal and co-workers [3–6] for details and tables.

Exact Expressions. In the limit $|\mathbf{k}| \to 0$, the dipole allowed transitions summed over final states are

$$|X_{1s \to n}|^2 = \frac{2^8}{3} n^7 \frac{(n-1)^{2n-5}}{(n+1)^{2n+5}}, \tag{54.11a}$$

$$|X_{2s \to n}|^2 = \frac{2^5}{3n^3} \frac{\left(\frac{1}{2} - \frac{1}{n}\right)^{2n-7}}{\left(\frac{1}{2} + \frac{1}{n}\right)^{2n+7}} \left(\frac{1}{4} - \frac{1}{n^2}\right) \left(1 - \frac{1}{n^2}\right), \tag{54.11b}$$

$$|X_{2p \to n}|^2 = \frac{2^5}{144} \frac{1}{n^3} \frac{\left(\frac{1}{2} - \frac{1}{n}\right)^{2n-7}}{\left(\frac{1}{2} + \frac{1}{n}\right)^{2n+7}} \left(11 - \frac{12}{n^2}\right). \tag{54.11c}$$

Asymptotic Expressions. For $n \gg 1$,

$$n^3 |X_{1s \to n}|^2 \sim 1.563 + \frac{5.731}{n^2} + \frac{13.163}{n^4} + \frac{24.295}{n^6} + \frac{39.426}{n^8} + \frac{58.808}{n^{10}}, \tag{54.12a}$$

$$n^3 |X_{2s \to n}|^2 \sim 14.658 + \frac{180.785}{n^2} + \frac{1435.854}{n^4} + \frac{9341.634}{n^6} + \frac{54208.306}{n^8} + \frac{292202.232}{n^{10}}, \tag{54.12b}$$

$$n^3 |X_{2p \to n}|^2 \sim 13.437 + \frac{218.245}{n^2} + \frac{2172.891}{n^4} + \frac{17118.786}{n^6} + \frac{117251.682}{n^8} + \frac{731427.003}{n^{10}}. \tag{54.12c}$$

54.2.2 Radial Integrals

Definition.

$$R_{n\ell}^{n'\ell'} \equiv \int_0^\infty R_{n\ell}(r) \, r \, R_{n'\ell'}(r) \, r^2 \, dr, \tag{54.13}$$

where $R_{n\ell}(r)$ are solutions to the radial Schrödinger equation. See Chap. 9 for specific representations of $R_{n\ell}$ for hydrogen.

Exact Results for Hydrogen. For $\ell' = \ell - 1$ and $n \neq n'$ [7],

$$R_{n\ell}^{n'\ell-1} = \frac{a_0}{Z} \frac{(-1)^{n'-\ell}(4nn')^{\ell+1}(n-n')^{n+n'-2\ell-2}}{4(2\ell-1)!(n+n')^{n+n'}}$$

$$\times \left[\frac{(n+\ell)!(n'+\ell-1)!}{(n'+\ell)!(n-\ell-1)!} \right]^{1/2}$$

$$\times \left\{ {}_2F_1(-n+\ell+1, -n'+\ell; 2\ell; Y) \right.$$

$$\left. - \left(\frac{n-n'}{n+n'}\right)^2 {}_2F_1(-n+\ell-1, -n'+\ell; 2\ell; Y) \right\}, \tag{54.14}$$

where $Y = -4nn'/(n-n')^2$. For $n = n'$,

$$R_{n\ell}^{n\ell-1} = (a_0/Z)\frac{3}{2}n\sqrt{n^2 - \ell^2}. \tag{54.15}$$

Semiclassical Quantum Defect Representation (see Ref. [8]).

$$\left| R_{n\ell}^{n'\ell'} \right|^2 = \left(\frac{a_0}{Z}\right)^2 \left| \frac{n_c^2}{2\Delta} \left[\left(1 - \frac{\Delta\ell\,\ell_>}{n_c}\right) \mathbf{J}_{\Delta-1}(-x) \right. \right.$$

$$- \left(1 + \frac{\Delta\ell\,\ell_>}{n_c}\right) \mathbf{J}_{\Delta+1}(-x)$$

$$\left. \left. + \frac{2}{\pi}\sin(\pi\Delta)(1-e) \right] \right|^2, \tag{54.16}$$

where

$$n_c = 2n^* n^{*\prime}/(n^* + n^{*\prime}), \tag{54.17a}$$

$$\Delta = n^{*\prime} - n^*, \tag{54.17b}$$

$$\Delta\ell = \ell' - \ell, \quad \ell_> = \max(\ell, \ell'), \tag{54.17c}$$

$$x = e\Delta, \quad e = \sqrt{1 - (\ell_>/n_c)^2}, \tag{54.17d}$$

and $\mathbf{J}_n(y)$ is the Anger function.

The energies of the states $n\ell$ and $n'\ell'$ are given in terms of the quantum defects by

$$E_{n\ell} = -Z^2 R_\infty/n^{*2}, \quad n^* = n - \delta_\ell, \tag{54.18a}$$

$$E_{n'\ell'} = -Z^2 R_\infty/n^{*'2}, \quad n^{*'} = n' - \delta_{\ell'}. \tag{54.18b}$$

Sum Rule. For hydrogen

$$\sum_{n'} \left| R_{n\ell}^{n'\ell-1} \right|^2 = \sum_{n'} \left| R_{n\ell}^{n'\ell+1} \right|^2 \tag{54.19a}$$

$$= \frac{n^2 a_0^2}{2 Z^2} \left[5n^2 + 1 - 3\ell(\ell+1) \right]. \tag{54.19b}$$

See §61 of Ref. [2] for additional sum rules.

54.2.3 Line Strengths

Definition.

$$S(n'\ell', n\ell) = e^2(2\ell + 1) \left| \mathbf{r}_{n'\ell', n\ell} \right|^2 \tag{54.20a}$$

$$= e^2 \max(\ell, \ell') \left| R_{n\ell}^{n'\ell'} \right|^2, \tag{54.20b}$$

where $\ell' = \ell \pm 1$. For hydrogen

$$S(n', n) = 32 \left(\frac{ea_0}{Z} \right)^2 (nn')^6 \frac{(n - n')^{2(n+n')-3}}{(n + n')^{2(n+n')+4}}$$

$$\times \left\{ \left[{}_2F_1(-n', -n + 1; 1; Y) \right]^2 \right.$$

$$\left. - \left[{}_2F_1(-n' + 1, -n; 1; Y) \right]^2 \right\}, \tag{54.21}$$

where $Y = -4nn'/(n - n')^2$.

Semiclassical Representation [9].

$$S(n', n) = \frac{32}{\pi\sqrt{3}} \left(\frac{ea_0}{Z} \right)^2 \frac{(\varepsilon\varepsilon')^{3/2}}{(\varepsilon - \varepsilon')^4} G(\Delta n), \tag{54.22}$$

where $\varepsilon = 1/n^2$, $\varepsilon' = 1/n'^2$. and the *Gaunt factor* $G(\Delta n)$ is given by

$$G(\Delta n) = \pi\sqrt{3} |\Delta n| \, \mathbf{J}_{\Delta n}(\Delta n)\mathbf{J}'_{\Delta n}(\Delta n), \tag{54.23}$$

where the prime on the Anger function denotes differentiation with respect to the argument Δn. Equation (54.23) can be approximated to within 2% by the expression

$$1 - \frac{1}{4|\Delta n|}. \tag{54.24}$$

Relation to Oscillator Strength.

$$S(n', n) = \sum_{\ell, \ell'} S(n'\ell', n\ell)$$

$$= 3e^2 a_0^2 \frac{R_\infty}{\hbar\omega} \sum_{\ell, \ell'} f_{n'\ell', n\ell}, \tag{54.25}$$

Connection with Radial Integral.

$$- f_{n'\ell', n\ell} = \frac{\hbar\omega}{3R_\infty} \frac{\max(\ell, \ell')}{(2\ell + 1)} \left| R_{n\ell}^{n'\ell'} \right|^2. \tag{54.26}$$

Density of Line Strengths. For bound-free $n\ell \rightarrow E\ell'$ transitions in a Coulomb field, the semiclassical representation [1] is

$$\frac{d}{dE} S(n\ell, E) = 2n(2\ell + 1) \left(\frac{R_\infty}{\hbar\omega} \right)^2 \left[\mathbf{J}'_\Delta(e\Delta)^2 \right.$$

$$\left. + \left(1 - \frac{1}{e^2} \right) \mathbf{J}_\Delta(e\Delta)^2 \right] \frac{e^2 a_0^2}{R_\infty}, \tag{54.27}$$

where $\Delta = \hbar\omega n^3/2R_\infty$ and $e = \sqrt{1 - (\ell + \frac{1}{2})^2/n^2}$. Asymptotic expression for $\Delta \gg 1$:

$$\frac{d}{dE} S(n\ell, E) = \frac{2(2\ell + 1)}{3\pi^2} \left(\frac{R_\infty}{\hbar\omega} \right)^2 \frac{(\ell + \frac{1}{2})^4}{n^3}$$

$$\times \left[K_{1/3}^2(\eta) + K_{2/3}^2(\eta) \right] \frac{e^2 a_0^2}{R_\infty}, \tag{54.28}$$

where $\eta = (E/R_\infty)(\ell + 1/2)^3/6$ and the $K_\nu(x)$ are Bessel functions of the third kind.

Line Strength of Line n.

$$S_n \equiv S(n) = \sum_{k \neq 0} S(n + k, n) \frac{1}{k^3}. \tag{54.29}$$

Born Approximation to Line Strength S_n [1].

$$S_n^B = \frac{Z^2 R_\infty}{E} \left\{ \frac{1}{2} \ln(1 + \varepsilon_e/\varepsilon) \sum_{k \neq 0} \left(1 - \frac{1}{4k} \right) \frac{1}{k^4} \right.$$

$$\left. + \frac{4}{3} \frac{\varepsilon_e}{\varepsilon + \varepsilon_e} \sum_{k \neq 0} \left(1 - \frac{0.60}{k} \right) \frac{1}{k^3} \right\}$$

$$= \frac{Z^2 R_\infty}{E} \left[0.82 \ln \left(1 + \frac{\varepsilon_e}{\varepsilon} \right) + \frac{1.47\varepsilon_e}{\varepsilon + \varepsilon_e} \right], \tag{54.30}$$

where $\varepsilon = |E_n| Z^2/R_\infty$ and $\varepsilon_e = \varepsilon/Z^2 R_\infty$.

54.2.4 Form Factors

$$F_{n'n}(Q) = \sum_{\ell, m} \sum_{\ell', m'} \left| \langle n\ell m | e^{i\mathbf{Q}\cdot\mathbf{r}} | n'\ell'm' \rangle \right|^2. \tag{54.31}$$

Connection with Generalized Oscillator Strengths.

$$f_{n'n}(Q) = \frac{Z^2 \Delta E}{n^2 Q^2 a_0^2} F_{n'n}(Q). \tag{54.32}$$

Semiclassical Limit.

$$\lim_{Q \to 0} f_{n'n}(Q) = \frac{32}{3n^2} \left[\frac{nn'}{\Delta n(n + n')} \right]^3 \Delta n J_{\Delta n}(\Delta n) J'_{\Delta n}(\Delta n) \tag{54.33}$$

where $J_m(y)$ denotes the Bessel function.

Representation as Microcanonical Distribution.

$$F_{n',n\ell}(Q) = (2\ell+1)\frac{2Z^2R_\infty}{n'^3}\int d\mathbf{p}\, |g_{n\ell}(p)|^2$$
$$\times\, \delta\left(\frac{(\mathbf{p}-\hbar\mathbf{Q})^2}{2m} - \frac{p^2}{2m} - E_{n'} - E_{n\ell}\right),$$
$$(54.34)$$

$$F_{n',n}(Q) = \frac{4Z^2R_\infty{}^2}{(nn')^3}\int \frac{d\mathbf{p}\,d\mathbf{r}}{(2\pi\hbar)^3}\delta\left(\frac{p^2}{2m} - \frac{Ze^2}{r} - E_n\right)$$
$$\times\, \delta\left[\frac{(\mathbf{p}-\hbar\mathbf{Q})^2}{2m} - \frac{Ze^2}{r} - E_{n'}\right], \quad (54.35)$$

$$= \frac{2^9}{3\pi(nn')^3}\frac{\kappa^5}{(\kappa^2+\kappa_+^2)^3(\kappa^2+\kappa_-^2)^3}, \quad (54.36)$$

where $\kappa = Qa_0/Z$ and $\kappa_\pm = |1/n \pm 1/n'|$.

54.2.5 Impact Broadening

The total broadening cross section of a level n is

$$\sigma_n = (\pi a_0^2/Z^4)n^4 S_n. \quad (54.37)$$

The width of a line $n \to n+k$ is [10]

$$\gamma_{n,n+k} = n_e\left[\langle v\sigma_n\rangle + \langle v\sigma_{n+k}\rangle\right], \quad (54.38)$$

where n_e is the number density of electrons, and

$$\langle v\sigma_n\rangle = \sum_{k\neq 0}\langle v\sigma_{n+k,n}\rangle = \frac{n^4}{Z^3}K_n \quad (54.39a)$$

$$= \frac{n^4\pi a_0^2 v_B}{Z^3\theta^{3/2}}\int_0^\infty e^{-E/k_BT}S_n\frac{E\,dE}{(Z^2R_\infty)^2}, \quad (54.39b)$$

where $\theta = k_BT/Z^2R_\infty$. See Chap. 57 for collisional line broadening.

54.3 CORRESPONDENCE PRINCIPLES

These are used to connect quantum mechanical observables with the corresponding classical quantities in the limit of large n. See Ref. [11] for details on the equations in this section.

54.3.1 Bohr-Sommerfeld Quantization

$$A_i = J_i\Delta w_i \oint p_i dq_i = 2\pi\hbar(n_i + \alpha_i), \quad (54.40)$$

where $n_i = 0, 1, 2, \ldots$ and $\alpha_i = 0$ if the generalized coordinate q_i represents rotation, and $\alpha_i = 1/2$ if q_i represents a libration.

54.3.2 Bohr Correspondence Principle

$$E_{n+s} - E_n = h\nu_{n+s,n} \sim s\hbar\omega_n, \quad s = 1, 2, \ldots \ll n,$$
$$(54.41)$$

where $\nu_{n+s,n}$ is the line emission frequency and ω_n is the angular frequency of classical orbital motion. The number of states with quantum numbers in the range Δn is

$$\Delta N = \prod_{i=1}^{D}\Delta n = \prod_{i=1}^{D}(\Delta J_i\Delta w_i)/(2\pi\hbar)^D$$
$$= \prod_{i=1}^{D}(\Delta p_i\Delta q_i)/(2\pi\hbar)^D, \quad (54.42)$$

for systems with D degrees of freedom, and the mean value $\bar{F}$ of a physical quantity $F(q)$ in the quantum state Ψ is

$$\bar{F} = \langle\Psi|\,F(q)\,|\Psi\rangle = \sum_{n,m}a_m^*a_nF_{mn}^{(q)}e^{i\omega_{mn}t}, \quad (54.43)$$

where the $F_{mn}^{(q)}$ are the quantal matrix elements between time independent states.

The first order S-matrix is

$$S_{fi} = -\frac{i\omega}{2\pi\hbar}\int_{-\infty}^{\infty}dt\int_0^{2\pi/\omega}V[\mathbf{R}(t), r(t_1)]e^{is\omega(t_1-t)}dt_1,$$
$$(54.44)$$

where $\mathbf{R}$ denotes the classical path of the projectile and r the orbital of the Rydberg electron.

54.3.3 Heisenberg Correspondence Principle

For one degree of freedom [11],

$$F_{mn}^{(q)}(\mathbf{R}) = \int_0^\infty \phi_m^*(r)F(r,\mathbf{R})\phi_n(r)\,dr \quad (54.45)$$

$$= \frac{\omega}{2\pi}\int_0^{2\pi/\omega}F^{(c)}[r(t)]e^{is\omega t}dt. \quad (54.46)$$

The three-dimensional generalization is [11]

$$F_{\mathbf{n},\mathbf{n'}}^{(q)} \sim F_\mathbf{s}^{(c)}(\mathbf{J}) = \frac{1}{8\pi^3}\int F^c[\mathbf{r}(\mathbf{J},\mathbf{w})]e^{i\mathbf{s}\cdot\mathbf{w}}d\mathbf{w}, \quad (54.47)$$

where $\mathbf{n}$, $\mathbf{n'}$ denotes the triple of quantum numbers (n, ℓ, m), (n', ℓ', m'), respectively, and $\mathbf{s} = \mathbf{n} - \mathbf{n'}$.

The correspondence between the three dimensional quantal and classical matrix elements in (54.47) follows from the general Fourier expansion for any classical function $F^{(c)}(\mathbf{r})$ periodic in $\mathbf{r}$,

$$F^{(c)}[\mathbf{r}(t)] = \sum_{\mathbf{s}} F_{\mathbf{s}}^{(c)}(\mathbf{J}) \exp(-i\mathbf{s}\cdot\mathbf{w}), \qquad (54.48)$$

where $\mathbf{J}, \mathbf{w}$ denotes the action-angle conjugate variables for the motion. For the three dimensional Coulomb problem, the action-angle variables are

$$
\begin{aligned}
J_n &= n\hbar, & w_n &= \left(\frac{\partial E}{\partial J_n}\right)t + \delta, \\
J_\ell &= (\ell + \tfrac{1}{2})\hbar, & w_\ell &= \psi_{\mathrm{E}}, \\
J_m &= m\hbar, & w_m &= \phi_{\mathrm{E}},
\end{aligned}
\qquad (54.49)
$$

where ψ_{E} is the Euler angle between the line of nodes and a direction in the plane of the orbit (usually taken to be the direction of the perihelion or perigee), and is constant for a Coulomb potential. The Euler angle ϕ_{E} is the angle between the line of nodes and the fixed x-axis. See Ref. [11] for details.

The first order S-matrix is

$$S_{fi} = -\frac{i\omega}{2\pi\hbar} \int_0^{2\pi/\omega} dt_{\mathrm{e}} \int_{-\infty}^{\infty} dt\, V[\mathbf{R}(t), r(t + t_{\mathrm{e}})]e^{is\omega t_{\mathrm{e}}}, \qquad (54.50)$$

with $s = i - f$, $\mathbf{R}$ is the classical path of projectile, and $r(t_{\mathrm{e}})$ is the classical internal motion of the Rydberg electron.

54.3.4 Strong Coupling Correspondence Principle

The S-matrix is

$$
\begin{aligned}
S_{fi} = \frac{\omega}{2\pi} \int_0^{2\pi/\omega} dt_{\mathrm{e}} \exp\Big\{ &i(s\omega t_{\mathrm{e}}) \\
&- \frac{i}{\hbar} \int_{-\infty}^{\infty} V[\mathbf{R}(t), r(t + t_{\mathrm{e}})]\, dt \Big\}.
\end{aligned}
\qquad (54.51)
$$

See Refs. [11–14] for additional details.

54.3.5 Equivalent Oscillator Theorem

$$\sum_n a_n(t) V_{fn}(t) e^{i\omega_{fn}t} = \sum_{d=-f}^{} a_{d+f}(t) V_d(t) e^{-id\omega t}. \qquad (54.52)$$

The S-matrix is

$$
\begin{aligned}
S_{\mathbf{n}',\mathbf{n}} &= a_{\mathbf{n}'}(t \to \infty) & (54.53) \\
&= \int_0^{2\pi} \frac{d\mathbf{w}}{8\pi^3} \exp\left[i\mathbf{s}\cdot\mathbf{w} - \frac{i}{\hbar} \int_{-\infty}^{\infty} V(\mathbf{w} + \omega t, t) dt\right].
\end{aligned}
$$

54.4 DISTRIBUTION FUNCTIONS

The function $W_\alpha(x)dx$ characterizes the probability (distribution) of finding an electron in a Rydberg orbital α within a volume dx centered at the point x in phase space. Integration of the distribution function W_α over all phase space volumes dx yields, depending upon the normalization chosen, either unity or the density of states appropriate to the orbital α.

54.4.1 Spatial Distributions

Distribution over n, ℓ, m [1]:

$$
\begin{aligned}
&W_{n\ell m}(r,\theta)r^2 \sin\theta\, dr\, d\theta \\
&= \frac{r^2 \sin\theta\, dr\, d\theta}{\pi^2 a^2 r\{[\mathrm{e}^2 - (1 - r/a)^2]\left[\sin^2\theta - (m/\ell)^2\right]\}^{1/2}},
\end{aligned}
$$
$$(54.54)$$

where $a = Ze^2/2\,|E| = n^2\hbar^2/mZe^2\hbar^2$ is the semimajor axis, and $\mathrm{e}^2 = 1 - (\ell/n)^2$ is the eccentricity.

Distribution over n, ℓ:

$$
\begin{aligned}
&W_{n\ell}(r,\theta)r^2 \sin\theta\, dr\, d\theta \\
&= g(n\ell)\frac{r^2 \sin\theta\, dr\, d\theta}{2\pi a^2 r\left[\mathrm{e}^2 - (1 - r/a)^2\right]^{1/2}},
\end{aligned}
\qquad (54.55)
$$

where $g(n\ell) = 2\ell$.

Distribution over n:

$$W_n(r)r^2 dr = g(n)\frac{2}{\pi}\left[1 - \left(1 - \frac{r}{a}\right)^2\right]^{1/2}\frac{r\, dr}{a^2}, \qquad (54.56)$$

with $g(n) = n^2$.

54.4.2 Momentum Distributions

Distribution over n, ℓ [1]:

$$W_{n\ell}(p)p^2 dp = g(n\ell)\frac{4}{\pi}\frac{dx}{(1 + x^2)^2}, \qquad (54.57)$$

where $x = p/p_n$ and $p_n^2 = 2m\,|E|$.

Distribution over n:

$$W_n(p)p^2 dp = g(n)\frac{32}{\pi}\frac{x^2 dx}{(1 + x^2)^4}. \qquad (54.58)$$

Sum Rules.

$$\frac{1}{n^2}\sum_{\ell,m}|G_{n\ell m}(\mathbf{k})|^2 = \left(\frac{na_0}{Z}\right)^3\frac{8}{\pi^2(x^2 + 1)^4}, \qquad (54.59a)$$

$$\frac{1}{n^2}\sum_{\ell=0}^{n-1}(2\ell + 1)\,|g_{n\ell}(k)|^2\, k^2 = \frac{32na_0 x^2}{\pi Z(x^2 + 1)^4}, \qquad (54.59b)$$

where $x = nka_0/Z$, and

$$G_{n\ell m}(\mathbf{k}) = g_{n\ell}(k)Y_{\ell m}(\hat{\mathbf{k}}), \tag{54.60a}$$

$$g_{n\ell}(k) = \left(\frac{2}{\pi}\frac{(n-\ell-1)!}{(n+\ell)!}\right)^{1/2}\left(\frac{a_0}{Z}\right)^{3/2}2^{2(\ell+1)}n^2\ell!$$

$$\times \frac{(-ix)^\ell}{(x^2+1)^{\ell+2}}C_{n-\ell-1}^{(\ell+1)}\left(\frac{x^2-1}{x^2+1}\right), \tag{54.60b}$$

where $C_i^{(j)}(y)$ is the associated Gegenbauer polynomial. See Chap. 9 for additional details on hydrogenic wave functions.

Quantum Defect Representation [15].

$$g_{n\ell}(k) = -\left[\frac{2}{\pi}\frac{\Gamma(n^*-1)}{\Gamma(n^*+\ell+1)}\right]^{1/2}n^*(a_0/Z)^{3/2}2^{2(\ell+1)}$$

$$\times \frac{(\ell+1)!(-ix)^\ell}{(x^2+1)^{\ell+2}}\mathcal{J}(n^*,\ell+1;X), \tag{54.61}$$

where $n^* = n - \delta$, δ being the quantum defect, and $x = n^*ka_0/Z$. The function $\mathcal{J}$ is given by the recurrence relation

$$\mathcal{J}(n^*,\ell+1;X) = -\frac{1}{2(2\ell+2)}\frac{\partial}{\partial X}\mathcal{J}(n^*,\ell;X), \tag{54.62}$$

$$\mathcal{J}(n^*,0;X) = -\frac{n^*\sin[n^*(\beta-\pi)]}{\sin(\beta-\pi)}$$

$$-\frac{\sin n^*\pi}{\pi}\int_0^1\frac{(1-s^2)s^{n^*}}{(1-2Xs+s^2)}ds, \tag{54.63}$$

where $X = (x^2-1)/(x^2+1)$, and $\beta = \cos^{-1}X$. In the limit $\ell \ll n^*$, Eq. (54.61) becomes

$$|g_{n\ell}(k)|^2 = 4\left(\frac{n^*a_0}{Z}\right)^3\frac{1-(-1)^\ell\cos[2n^*(\beta-\pi)]}{\pi x^2(x^2+1)^2}. \tag{54.64}$$

Classical Density of States.

$$\rho(E) = \int\delta[E-H(\mathbf{p},\mathbf{r})]\frac{d\mathbf{p}\,d\mathbf{r}}{(2\pi\hbar)^3} = \frac{n^5\hbar^2}{mZ^2e^4}. \tag{54.65}$$

54.5 CLASSICAL THEORY

The classical cross section for energy transfer ΔE between two particles, with arbitrary masses m_1, m_2 and charges Z_1, Z_2, is given by [16]

$$\sigma_{\Delta E}(\mathbf{v}_1,\mathbf{v}_2) = \frac{2\pi(Z_1Z_2e^2V)^2}{v^2\,|\Delta E|^3}\left[1+\cos^2\bar\theta+\frac{\Delta E}{\mu vV}\cos\bar\theta\right] \tag{54.66}$$

valid for $-1 \le \cos\bar\theta - \Delta E/(\mu vV) \le 1$, and $\sigma_{\Delta E}(\mathbf{v}_1,\mathbf{v}_2) = 0$ otherwise, where

$$\mathbf{v} = \mathbf{v}_1 - \mathbf{v}_2, \tag{54.67a}$$

$$\mathbf{V} = (m_1\mathbf{v}_1 + m_2\mathbf{v}_2)/M, \tag{54.67b}$$

$$\cos\bar\theta = \frac{1}{vV}\mathbf{v}\cdot\mathbf{V}, \tag{54.67c}$$

and $\mu = m_1m_2/M$, $M = m_1 + m_2$. If particle 2 has an isotropic velocity distribution in the lab frame, the effective cross section averaged over the direction $\hat{\mathbf{n}}_2$ of $\mathbf{v}_2$ is

$$v_1\sigma_{\Delta E}^{(\text{eff})}(\mathbf{v}_1,\mathbf{v}_2) = \frac{1}{4\pi}\int d\hat{\mathbf{n}}_2\,|\mathbf{v}_1-v_2\hat{\mathbf{n}}_2|\,\sigma_{\Delta E}(\mathbf{v}_1,\mathbf{v}_2). \tag{54.68}$$

If $\mathbf{v}_1$ is also isotropic, then the average of (54.68), together with (54.66), gives for the special case of a Coulomb potential

$$\sigma_{\Delta E}^{(\text{eff})}(\mathbf{v}_1,\mathbf{v}_2)$$

$$= \frac{\pi(Z_1Z_2e^2)^2}{4\,|\Delta E|^3\,v_1^2v_2}\Big[(v_1^2-v_2^2)(v_2'^2-v_1'^2)(v_l^{-1}-v_u^{-1})$$

$$+(v_1^2+v_2^2+v_1'^2+v_2'^2)(v_u-v_l)-\tfrac{1}{3}(v_u^3-v_l^3)\Big], \tag{54.69}$$

where

$$v_1' = \left(v_1^2-2\Delta E/m_1\right)^{1/2}, \tag{54.70}$$

$$v_2' = \left(v_2^2+2\Delta E/m_2\right)^{1/2}, \tag{54.71}$$

and v_u, v_l are defined below for cases (1)–(4). With the definitions

$$\Delta\varepsilon_{12} = 4m_1m_2(E_1-E_2)/M^2, \quad \Delta m_{12} = |m_1-m_2|,$$

$$\Delta\tilde\varepsilon_{12} = \frac{4m_1m_2}{M^2}\left(E_1\frac{v_2}{v_1}-E_2\frac{v_1}{v_2}\right),$$

the four cases are

(1) $\Delta E \ge \Delta\varepsilon_{12} + |\Delta\tilde\varepsilon_{12}| \ge 0$, and $2m_2v_2 \ge \Delta m_{12}v_1$:

$$v_l = v_2' - v_1', \quad v_u = v_1' + v_2', \quad \Delta E \ge 0; \tag{54.72a}$$

$$v_l = v_2 - v_1, \quad v_u = v_1 + v_2, \quad \Delta E \le 0. \tag{54.72b}$$

If $2m_2v_2 < \Delta m_{12}v_1$, then $\sigma_{\Delta E}^{\text{eff}}(v_1,v_2) = 0$,

(2) $\Delta\varepsilon_{12} - \Delta\tilde\varepsilon_{12} \le \Delta E \le \Delta\varepsilon_{12} + \Delta\tilde\varepsilon_{12}$, and $m_1 > m_2$:

$$v_l = v_2' - v_1', \quad v_u = v_1 + v_2, \quad \Delta E \ge 0; \tag{54.72c}$$

$$v_l = v_2 - v_1,, \quad v_u = v_1' + v_2', \quad \Delta E \le 0 \tag{54.72d}$$

(3) $\Delta E \le \Delta\varepsilon_{12} - |\Delta\tilde\varepsilon_{12}| \le 0$, and $2m_1v_1 \ge \Delta m_{12}v_2$:

$$v_l = v_1 - v_2, \quad v_u = v_1 + v_2, \quad \Delta E \ge 0; \tag{54.72e}$$

$$v_l = v_1' - v_2', \quad v_u = v_1' + v_2', \quad \Delta E \le 0. \tag{54.72f}$$

If $2m_1v_1 < \Delta m_{12}v_2$ then $\sigma_{\Delta E}^{\text{eff}}(v_1,v_2) = 0$,

(4) $\Delta\varepsilon_{12} + \Delta\tilde{\varepsilon}_{12} \le \Delta E \le \Delta\varepsilon_{12} - \Delta\tilde{\varepsilon}_{12}$, and $m_1 < m_2$:

$$v_l = v_1 - v_2, \quad v_u = v_1' + v_2', \quad \Delta E \ge 0; \quad (54.72\text{g})$$
$$v_l = v_1' - v_2', \quad v_u = v_1 + v_2, \quad \Delta E \le 0. \quad (54.72\text{h})$$

If $2m_1 v_1 < \Delta m_{12} v_2$ then $\sigma_{\Delta E}^{\text{eff}}(v_1, v_2) = 0$.

Since v_1' and v_2', given by (54.70) and (54.71) respectively, must be real, $\sigma_{\Delta E}(\mathbf{v}_1, \mathbf{v}_2) = 0$ for ΔE outside the range

$$-\tfrac{1}{2} m_2 v_2^2 \le \Delta E \le \tfrac{1}{2} m_1 v_1^2, \quad (54.73)$$

which simply expresses the fact that the particle losing energy in the collision cannot lose more than its initial kinetic energy.

The cross section (54.69) must be integrated over the classically allowed range of energy transfer ΔE and averaged over a prescribed speed distribution $W(v_2)$ before comparison with experiment can be made. See Refs. [16, 17] for details.

Classical Removal Cross Section [18]. The cross section for removal of an electron from a shell is given by

$$\sigma_{\text{r}}(V) = \int_0^\infty f(v) \sigma_{\Delta E}(\mathbf{v}_1, \mathbf{v}_2) dv. \quad (54.74)$$

Total Removal Cross Section [18]. In an independent electron model,

$$\sigma_{\text{r}}^{\text{total}}(V) = N_{\text{shell}} \sigma_{\text{r}}(V), \quad (54.75)$$

where N_{shell} is the number of equivalent electrons in a shell. In a shielding model,

$$\sigma_{\text{r}}^{\text{total}}(V) = \left[1 - \frac{(N_{\text{shell}} - 1)}{4\pi\bar{r}^2} \sigma_{\text{r}}(V) \right] N_{\text{shell}} \sigma_{\text{r}}(V), \quad (54.76)$$

where $\bar{r}^2$ is the root mean square distance between electrons within a shell. Experiment [19] favors (54.76) over (54.75). See Fig. 4a–e of Ref. [18] for details.

Classical trajectory and Monte-Carlo methods are covered in Chap. 56.

54.6 WORKING FORMULAE FOR RYDBERG COLLISIONS

54.6.1 Inelastic n, ℓ-Changing Transitions

$$A^*(n\ell) + B \to A^*(n') + B + \Delta E_{n',n\ell}, \quad (54.77)$$

where $\Delta E_{n',n\ell} = E_{n'} - E_{n\ell}$ is the energy defect. The cross section for (54.77) in the *quasifree electron model* [20] is

$$\sigma_{n',n\ell}(V) = \frac{2\pi a_{\text{s}}^2}{(V/v_{\text{B}})^2 n'^3} f_{n',n\ell}(\lambda), \quad \ell \ll n, \quad (54.78)$$

where a_{s} is the scattering length for $e^- + B$ scattering, $\lambda = n^* a_0 \omega_{n',n\ell}/V$, $\omega_{n',n\ell} = |\Delta E_{n',n\ell}|/\hbar$, $E_{n'} = -R_\infty/n'^2$, and $E_{n\ell} = -R_\infty/n^{*2}$, with $n^* = n - \delta_\ell$. Also, v_{B} is the atomic unit of velocity, and

$$f_{n',n\ell}(\lambda) = \frac{2}{\pi} \left[\tan^{-1}\left(\frac{2}{\lambda}\right) - \frac{\lambda}{2} \ln\left(1 + \frac{4}{\lambda^2}\right) \right]. \quad (54.79)$$

Limiting cases: $f_{n',n\ell}(\lambda) \to 1$ as $\lambda \to 0$, and $f_{n',n\ell}(\lambda) \sim 8/(3\pi\lambda^3)$ for $\lambda \gg 1$. Then

$$\sigma_{n',n\ell} \sim \begin{cases} \dfrac{2\pi a_{\text{s}}^2}{(V/v_{\text{B}})^2 n'^3}, & \lambda \to 0, \\[3mm] \dfrac{16 a_{\text{s}}^2 V n^3}{3 v_{\text{B}} |\delta_\ell + \Delta n|^3}, & \lambda \gg 1. \end{cases} \quad (54.80)$$

Rate Coefficients.

$$\langle \sigma_{n',n\ell}(V) \rangle \equiv \langle V \sigma_{n',n\ell}(V) \rangle / \langle V \rangle \quad (54.81\text{a})$$
$$= \frac{2\pi a_{\text{s}}^2}{(V_T/v_{\text{B}})^2 n'^3} \varphi_{n',n\ell}(\lambda_T), \quad (54.81\text{b})$$

where $V_T = \sqrt{2k_{\text{B}}T/\mu}$, $\lambda_T = n^* a_0 \omega_{n',n\ell}/V_T$, $\Delta n = n' - n$, and μ is the reduced mass of A–B. The function $\varphi_{n',n\ell}(\lambda_T)$ in (54.81b) is given by

$$\varphi_{n',n\ell}(\lambda_T) = e^{\lambda_T^2/4} \text{erfc}\left(\tfrac{1}{2}\lambda_T\right)$$
$$- \frac{\lambda_T}{\pi} \int_0^\infty \frac{du}{\sqrt{u}} e^{-u} \ln\left(1 + \frac{4}{\lambda_T^2}\right) \quad (54.82\text{a})$$
$$= \begin{cases} 1 - \dfrac{\lambda_T}{\sqrt{\pi}} \ln(1/\lambda_T^2), & \lambda_T \to 0 \\[3mm] 2/(\sqrt{\pi}\lambda_T^3), & \lambda_T \gg 1 \end{cases} \quad (54.82\text{b})$$

and $\text{erfc}(x)$ is the complementary error function.

54.6.2 Inelastic $n \to n'$ Transitions

$$A^*(n) + B \to A^*(n') + B + \Delta E_{n'n}. \quad (54.83)$$

(a) Cross Sections.

$$\sigma_{n'n} = \sum_{\ell\ell'} \frac{(2\ell + 1)}{n^2} \sigma_{n'\ell',n\ell}, \quad (54.84)$$

$$\sigma_{n',n}(V) = \frac{2\pi a_{\text{s}}^2}{(V/v_{\text{B}})^2 n'^3} F_{n'n}(\lambda), \quad (54.85)$$

where $\lambda = n a_0 \omega_{n'n}/V = |\Delta n| v_{\text{B}}/(n^2 V)$, and

$$F_{n'n}(\lambda) = \frac{2}{\pi} \left[\tan^{-1}\left(\frac{2}{\lambda}\right) - \frac{2\lambda(3\lambda^2 + 20)}{3(4 + \lambda^2)^2} \right]. \quad (54.86)$$

Limiting cases:

$$\sigma_{n'n} \sim \begin{cases} \dfrac{2\pi a_{\text{s}}^2}{(V/v_{\text{B}}^2 n'^3)}, & \lambda \ll 1 \\[3mm] \dfrac{256 \sigma_{e^--B}^{\text{elastic}}(V/v_{\text{B}})^3 n^7}{15\pi |\Delta n|^5}, & \lambda \gg 1 \end{cases} \quad (54.87)$$

where $\sigma_{e-B}^{\text{elastic}}$ is the elastic cross section for $e^- + B$ scattering.

(b) Rate Coefficients.

$$K_{n'\ell',n\ell}(T) = \langle V \sigma_{n'\ell',n\ell} \rangle , \qquad (54.88\text{a})$$

$$K_{n'n}(T) = \sum_{\ell,\ell'} \frac{(2\ell+1)}{n^2} K_{n'\ell',n\ell} , \qquad (54.88\text{b})$$

$$K_{n'n}(T) = \frac{v_B \sigma_{e-B}^{\text{elastic}}}{\sqrt{\pi} n^3 (V_T/v_B)} \Phi_{n'n}(\lambda_T) , \qquad (54.88\text{c})$$

where

$$\Phi_{n'n}(\lambda_T) = e^{\lambda_T^2/8} \left\{ e^{\lambda_T^2/8} \text{erfc}(\tfrac{1}{2}\lambda_T) - \frac{\lambda_T^2}{\sqrt{2\pi}} D_{-3}\left(\frac{\lambda_T}{2}\right) \right.$$
$$\left. - \frac{5\lambda_T}{\sqrt{\pi}} D_{-4}\left(\frac{\lambda_T}{\sqrt{2}}\right) \right\} \qquad (54.89\text{a})$$

$$\sim \begin{cases} 1 - 8\lambda_T/3\sqrt{\pi}, & \lambda_T \ll 1 \\ 2^6/(\sqrt{\pi}\lambda_T^5), & \lambda_T \gg 1 \end{cases} , \qquad (54.89\text{b})$$

where $D_{-\nu}(y)$ denotes the parabolic cylinder function.

Limiting cases:

$$K_{n'n}(T) \sim \begin{cases} \left(\dfrac{\mu R_\infty}{\pi m_e k_B T}\right)^{1/2} \dfrac{v_B \sigma_{e-B}^{\text{elastic}}}{n^3}, & \lambda_T \to 0 \\[2ex] \dfrac{2^6 v_B \sigma_{e-B}^{\text{elastic}} n^7}{\pi |\Delta n|^5} \left(\dfrac{2k_B T}{\mu v_B^2}\right)^2, & \lambda_T \gg 1 . \end{cases}$$
$$(54.90)$$

Born Results

$$\sigma_{n'n} = \frac{8\pi}{k^2} \frac{1}{n^2} \int_{|k-k'|}^{k+k'} F_{n'n}(Q) \frac{d(Qa_0)}{(Qa_0)^3} . \qquad (54.91)$$

(a) Electron-Rydberg Atom Collision.

$$\sigma_{n'n} = \frac{8\pi a_0^2 R_\infty}{Z^2 E n^2} \left\{ \left[1 - \frac{1}{4\Delta n}\right] \frac{(\varepsilon\varepsilon')^{3/2}}{(\Delta\varepsilon)^4} \ln(1+\varepsilon_e/\varepsilon) \right.$$
$$\left. + \left[1 - \frac{0.6}{\Delta n}\right] \frac{\varepsilon_e}{\varepsilon + \varepsilon_e} \frac{(\varepsilon')^{3/2}}{(\Delta\varepsilon)^2} \left[\frac{4}{3\Delta n} + \frac{1}{\varepsilon}\right] \right\} \qquad (54.92)$$

for $n' > n$, where $\varepsilon_e = E/(Z^2 R_\infty)$, $\varepsilon = 1/n^2$, $\varepsilon' = 1/n'^2$, and $\Delta\varepsilon = \varepsilon - \varepsilon'$.

(b) Heavy Particle-Rydberg Atom Collision.

$$\sigma_{n'n} = \frac{8\pi a_0^2 \mathcal{Z}^2}{Z^4 n^2 \varepsilon_e} \left\{ \left[1 - \frac{1}{4\Delta n}\right] \frac{(\varepsilon\varepsilon')^{3/2}}{(\Delta\varepsilon)^4} \ln(1+\varepsilon_e/\varepsilon) \right.$$
$$\left. + \left[1 - \frac{0.6}{\Delta n}\right] \frac{\varepsilon_e}{\varepsilon + \varepsilon_e} \frac{(\varepsilon')^{3/2}}{(\Delta\varepsilon)^2} \left[\frac{4}{3\Delta n} + \frac{1}{\varepsilon}\right] \right\} , \qquad (54.93)$$

where $\varepsilon_e = m\varepsilon/MZ^2 R_\infty$ with heavy particle mass and charge denoted above by M and $\mathcal{Z}$, respectively, and all other terms retain their meaning as in Eq. (54.92).

54.6.3 Quasi-elastic ℓ-Mixing Transitions

$$\sigma_{n\ell}^{(\ell\text{-mixing})} \equiv \sum_{\ell' \neq \ell} \sigma_{n'\ell',n\ell} \qquad (54.94\text{a})$$

$$\sim \begin{cases} \sigma_{\text{geo}} = 4\pi a_0^2 n^4, & n \ll n_{\text{max}} \\ 2\pi a_s^2 v_B^2/V^2 n^3, & n \gg n_{\text{max}} . \end{cases} \qquad (54.94\text{b})$$

The two limits correspond to strong (close) coupling for $n \ll n_{\text{max}}$, and weak coupling for $n \gg n_{\text{max}}$, and expressions (54.94b) are valid when the quantum defect δ_ℓ of the initial Rydberg orbital $n\ell$ is small. n_{max} is the principal quantum number, where the ℓ-mixing cross section reaches a maximum [21],

$$n_{\text{max}} \sim \left(\frac{v_B |a_s|}{V a_0}\right)^{2/7} . \qquad (54.95)$$

For Rydberg atom–noble gas atom scattering, $n_{\text{max}} = 8$ to 20, while for Rydberg atom–alkali atom scattering $n_{\text{max}} = 15$ to 30.

54.6.4 Elastic $n\ell \to n\ell'$ Transitions

$$A^*(n\ell) + B \to A^*(n\ell') + B . \qquad (54.96)$$

(a) Cross Sections.

$$\sigma_{ns}^{\text{elastic}}(V) = \frac{2\pi C_{ss} a_s^2}{(V/v_B)^2 n^{*4}} , \qquad (54.97)$$

valid for $n^* \gg [v_B |a_s|/(4V a_0)]^{1/4}$ with

$$C_{ss} = \frac{8}{\pi^2} \int_0^{1/\sqrt{2}} [K(k)]^2 \, dk , \qquad (54.98)$$

where $K(k)$ denotes the complete elliptic integral of the first kind.

(b) Rate Coefficients [22] (three cases).

With the definitions

$$\nu_B = v_B/v_{\text{rms}}, \quad v_{\text{rms}} = \sqrt{(8k_B T)/\mu\pi} , \qquad (54.99)$$
$$f(y) = y^{-1/2}(1 - (1-y)e^{-y}) + y^{3/2}\text{Ei}(y) , \qquad (54.100)$$
$$y = (\nu_B a_s)^2/(4\pi a_0^2 n^{*8}) , \qquad (54.101)$$
$$n_1 = (|a_s|\nu_B/4a_0)^{1/4} , \qquad (54.102)$$
$$n_2 = 0.7 \left[|a_s|\nu_B^{5/6}/(\alpha_d a_0^3)^{1/6}\right]^{1/3} , \qquad (54.103)$$

where α_d is the dipole polarizability of A^*, then

$$\langle \sigma_{ns}^{\text{el}} \rangle \sim \begin{cases} 8\pi a_0^2 n^{*4}, & n^* \leq n_1 \\[1ex] 4\pi^{1/2} a_0 |a_s| \nu_B f(y), & n_1 \leq n^* \leq n_2 \\[1ex] 7(\alpha_d \nu_B)^{2/3} + \dfrac{4a_s^2 \nu_B^2}{n^{*4}} \\ \quad - \dfrac{2.7 a_s^2 \nu_B^2 (\alpha_d \nu_B)^{1/3}}{a_0 n^{*6}}, & n^* \geq n_2 . \end{cases}$$
$$(54.104)$$

54.6.5 Fine Structure $n\ell J \to n\ell J'$ Transitions

$$A^*(n\ell J) + B \to A^*(n\ell J') + B + \Delta E_{J'J} . \qquad (54.105)$$

(a) Cross Sections (two cases).

$$\sigma_{n\ell J}^{n\ell J'}(V) = \frac{2J'+1}{2(2\ell+1)} c_{\text{norm}} 4\pi a_0^2 n^{*4} , \qquad (54.106)$$

valid for $n^* \leq n_0(V)$, and

$$\sigma_{n\ell J}^{n\ell J'}(V) = \frac{2\pi C_{J'J}^{(\ell)} a_s^2 v_{\text{B}}^2}{V^2 n^{*4}} \varphi_{J'J}^{(\ell)}(\nu_{J'J}) \left[1 - \frac{n_0^8(V)}{2n^{*8}} \right] , \qquad (54.107)$$

valid for $n^* \geq n_0(V)$, where the quantity $n_0(V)$ is the effective principal quantum number such that the impact parameter ρ_0 of B (moving with relative velocity V) equals the radius $2n^{*2}a_0$ of the Rydberg atom A^*. $n_0(V)$ is given by the solution to the following transcendental equation

$$n_0^8(V) = \frac{(2\ell+1)C_{J'J}^{(\ell)}}{2(2J'+1)c_{\text{norm}}} \left(\frac{v_{\text{B}} a_s}{V a_0} \right)^2 \varphi_{J'J}^{(\ell)}(\nu_{J'J} [n_0(V)]) . \qquad (54.108)$$

The constant c_{norm} in (54.106) is equal to 5/8 if $\sigma_{\text{geo}} = \pi \langle r^2 \rangle_{n\ell}$, or 1 if $\sigma_{\text{geo}} = 4\pi a_0^2 n^{*4}$. The function $\varphi_{J'J}^{(\ell)}(\nu_{J'J})$ in (54.107) is given in general by [23]

$$\varphi_{J'J}^{(\ell)}(\nu_{J'J}) = \xi_{J'J}^{(\ell)}(\nu_{J'J}) / \xi_{J'J}^{(\ell)}(0) , \qquad (54.109a)$$

$$\xi_{J'J}^{(\ell)}(\nu_{J'J}) = \sum_{s=0}^{\ell} A_{\ell J',\ell J}^{(2s)} \int_{\nu_{J'J}}^{\infty} j_s^2(z) \mathbf{J}_s^2(z) z \, dz , \qquad (54.109b)$$

$$\nu_{J'J} = |\delta_{\ell J'} - \delta_{\ell J}| \frac{v_{\text{B}}}{V n^*} , \qquad (54.109c)$$

where $j_s(z)$ is the spherical Bessel function and the coefficients $C_{J'J}^{(\ell)}$ and $A_{\ell J',\ell J}^{(2s)}$ in (54.107) and (54.109b), respectively, are given in table 5.1 of Beigman and Lebedev [1]. The quantum defect of Rydberg state $n\ell J$ is $\delta_{\ell J}$. For elastic scattering, $\nu_{JJ} = 0$, and $\varphi_{JJ}^{(\ell)}(0) = 1$.

Symmetry relation:

$$\xi_{JJ'}^{(\ell)}(\nu_{J'J}) = \frac{2J+1}{2J'+1} \xi_{J'J}^{(\ell)}(\nu_{J'J}) . \qquad (54.110)$$

(b) Rate Coefficients.

$$\left\langle \sigma_{n\ell J}^{n\ell J'} \right\rangle = \left[\frac{c_{\text{norm}}(2J'+1)C_{J'J}^{(\ell)}}{2(2\ell+1)} \right]^{1/2}$$
$$\times \pi a_0^2 F(\zeta) \left(\frac{v_{\text{B}} |a_s|}{V_T a_0} \right) , \qquad (54.111)$$

where $\zeta = n_0^8(V_T)/n^{*8}$, and

$$F(\zeta) \equiv \sqrt{\zeta} \left[\text{E}_2(\zeta) + \frac{1}{\zeta}(1 - e^{-\zeta}) \right] , \qquad (54.112)$$

where $\text{E}_2(x)$ is an exponential integral.

Limiting cases:

$$\left\langle \sigma_{n\ell J}^{n\ell J'} \right\rangle = \begin{cases} \dfrac{2J'+1}{2(2\ell+1)} c_{\text{norm}} 4\pi a_0^2 n^{*4} , & n^* \ll n_{\text{max}}^* \\[2ex] \dfrac{2\pi C_{J'J}^{(\ell)} a_s^2 v_{\text{B}}^2}{V_T^2 n^{*4}} , & n^* \gg n_{\text{max}}^* \end{cases}$$
$$(54.113)$$

where $n_{\text{max}}^* = (3/2)^{1/8} n_0(V)$ if $\nu_{J'J} \ll 1$.

54.7 IMPULSE APPROXIMATION

54.7.1 Quantal Impulse Approximation

Basic Formulation [24]

Consider a Rydberg collision between a projectile (1) of charge Z_1 and a target with a valence electron (3) in orbital ψ_i bound to a core (2). The full three-body wave function for the system of projectile + target is denoted by Ψ_i. The relative distance between 1 and the center-of-mass of $2-3$ is denoted by $\boldsymbol{\sigma}$, while the separation of 2 from the center-of-mass of $1-3$ is $\boldsymbol{\rho}$.

Formal Scattering Theory.

$$\Psi_i^{(+)} = \Omega^{(+)} \psi_i , \qquad (54.114)$$

where the Möller scattering operator $\Omega^{(+)} = 1 + G^+ V_i$, and $V_i = V_{12} + V_{13}$.

Let χ_m be a complete set of free-particle wave functions satisfying

$$(H_0 - E_m)\chi_m = 0 , \qquad (54.115)$$

and define operators $\omega_{ij}^+(m)$ by

$$\omega_{ij}^+(m)\chi_m = \left(1 + \frac{1}{E_m - H_0 - V_{ij} + i\epsilon} V_{ij} \right) \chi_m , \qquad (54.116)$$

where V_{ij} denotes the pairwise interaction potential between particles i and j ($i,j = 1,2,3$). Then the action of the full Green's function G^+ on the two-body potential V_{ij} is

$$G^+ V_{ij} = \left[\omega_{ij}^+(m) - 1 \right]$$
$$+ G^+ \left\{ (E_m - E) + V_{12} + V_{13} + V_{23} - V_{ij} \right\}$$
$$\times \left[\omega_{ij}^+(m) - 1 \right] . \qquad (54.117)$$

Projection Operators.

$$b_{ij}^+(m) = \omega_{ij}^+(m) - 1, \qquad (54.118a)$$

$$b_{ij}^+ = \sum_m b_{ij}^+(m) |\chi_m\rangle \langle\chi_m| , \qquad \omega_{ij}^+ = b_{ij}^+ + 1 . \qquad (54.118b)$$

$$G^+ V_{ij} |\psi_i\rangle = \sum_m G^+ V_{ij} |\chi_m\rangle \langle\chi_m | \psi_i\rangle \qquad (54.119a)$$

$$= \{b_{ij}^+ + G^+ [V_{23}, b_{ij}^+]$$
$$+ G^+ [V_{12} + V_{13} - V_{ij}] b_{ij}^+(m)\} |\psi_i\rangle . \qquad (54.119b)$$

Möller Scattering Operator.

$$\Omega^+ = (\omega_{13}^+ + \omega_{12}^+ - 1) + G^+ [V_{23}, (b_{13}^+ + b_{12}^+)]$$
$$+ G^+(V_{13}b_{12}^+ + V_{12}b_{13}^+) . \qquad (54.120)$$

Exact T-matrix.

$$T_{if} = \langle\psi_f| V_f |(\omega_{13}^+ + \omega_{12}^+ - 1)\psi_i\rangle$$
$$+ \langle\psi_f| V_f |G^+ [V_{23}, (b_{13}^+ + b_{12}^+)] \psi_i\rangle$$
$$+ \langle\psi_f| V_f |G^+(V_{13}b_{12}^+ + V_{12}b_{13}^+)\psi_i\rangle . \qquad (54.121)$$

The impulse approximation to the exact T-matrix element (54.121) is obtained by ignoring the second term involving the commutator of V_{23}.

$$\Psi_i \longrightarrow \Psi_i^{\text{imp}} = (\omega_{13}^+ + \omega_{12}^+ - 1)\psi_i . \qquad (54.122)$$

Impulse Approximation: Post Form.

$$T_{if}^{\text{imp}} = \langle\psi_f| V_f |(\omega_{13}^+ + \omega_{12}^+ - 1)\psi_i\rangle . \qquad (54.123)$$

The impulse approximation can also be expressed using incoming-wave boundary conditions by making use of the *prior operators*

$$\omega_{ij}^-(m)\chi_m = \left(1 + \frac{1}{E_m - H_0 - V_{ij} - i\epsilon}V_{ij}\right)\chi_m , \qquad (54.124a)$$

$$\omega_{ij}^- = \sum_m \omega_{ij}^-(m) |\chi_m\rangle \langle\chi_m| . \qquad (54.124b)$$

The impulse approximation (54.123) is exact if $V_{23} = $ const. since the commutator of V_{23} vanishes in that case.

Applications [24]

(1) Electron Capture. $X^+ + H(i) \rightarrow X(f) + H^+$.

$$T_{if}^{\text{imp}} = \langle\psi_f | V_{12} + V_{23}(\omega_{12}^+ + \omega_{13}^+ - 1)\psi_i\rangle . \qquad (54.125)$$

Wave functions: $\psi_i = e^{i\mathbf{k}_i \cdot \boldsymbol{\sigma}}\varphi_i(\mathbf{r})$, $\psi_f = e^{i\mathbf{k}_f \cdot \boldsymbol{\rho}}\varphi_f(\mathbf{x})$, $\chi_m = (2\pi)^{-3}\exp[i(\mathbf{K}\cdot\mathbf{x} + \mathbf{k}\cdot\boldsymbol{\rho})]$, where the φ_n are hydrogenic wave functions.

If X above is a heavy particle, the V_{12} term in (54.125) may be omitted due to the difference in mass between the projectile 1 and the bound Rydberg electron 3. See [24] and references therein for details. With the definitions

$$z = \frac{4\alpha\delta^2}{(T - 2\delta)(T - 2\alpha\delta)} , \qquad T = \beta^2 + P^2$$

$$\delta = i\beta K - \mathbf{p}\cdot\mathbf{K}, \qquad \nu = aZ_1/K$$

$$\mathbf{t}_1 = \mathbf{K}/a + \mathbf{v}, \qquad N(\nu) = e^{\pi\nu/2}\Gamma(1 - i\nu)$$

$$a = \frac{M_1}{M_1 + m_e}, \qquad b = \frac{M_2}{M_2 + m_e}$$

$$\mathbf{k} = a\mathbf{k}_2 - (1-a)\mathbf{k}_f, \qquad \mathbf{K} = a\mathbf{k}_1 - (1-a)\mathbf{k}_i,$$

$$\mathbf{t} = (\mathbf{K} - \mathbf{p})/a, \qquad \mathbf{p} = a\mathbf{k}_f - \mathbf{k}_i,$$

$$\beta = aZ_1/n,$$

and n is the principal quantum number, the impulse approximation to the T-matrix becomes, in this case,

$$T_{if}^{\text{imp}} \sim \langle\psi_f| V_{23} |\omega_{13}^+\psi_i\rangle \qquad (54.126)$$

$$= \frac{-1}{2\pi^2 a^3} \int \frac{d\mathbf{K}}{t^2} N(\nu)g_i(\mathbf{t}_1)\mathcal{F}(f, \mathbf{K}, \mathbf{p}) , \qquad (54.127)$$

where, for a general final s-state,

$$\mathcal{F}(f, \mathbf{K}, \mathbf{p}) = \int \varphi_f^*(\mathbf{x})e^{i\mathbf{P}\cdot\mathbf{x}} {}_1F_1[i\nu, 1; i(Kx - \mathbf{K}\cdot\mathbf{x})] \, d\mathbf{x} , \qquad (54.128)$$

and $g_i(\mathbf{t}_1)$ denotes the Fourier transform of the initial state. The normalization of the Fourier transform is chosen such that momentum and coordinate space hydrogenic wave functions are related by $\varphi_n(\mathbf{r}) = (2\pi)^{-3}\int\exp(i\mathbf{t}_1\cdot\mathbf{r})g_n(\mathbf{t}_1)\,d\mathbf{t}_1$. For the case $f = 1s$,

$$\mathcal{F}(1s, \mathbf{K}, \mathbf{p}) = -\frac{\beta^{3/2}}{\sqrt{\pi}}\frac{\partial}{\partial\beta}\mathcal{I}(\nu, 0, \beta, -\mathbf{K}, \mathbf{p})$$

$$= 8\sqrt{\pi}\beta^{3/2}\left[\frac{(1 - i\nu)\beta}{T^2} \right.$$

$$\left. + \frac{i\nu(\beta - iK)}{T(T - 2\delta)}\left(\frac{T}{T - 2\delta}\right)^{i\nu}\right] \qquad (54.129)$$

evaluated at $\beta = aZ_1$. For the case $f = 2s$,

$$\mathcal{F}(2s, \mathbf{K}, \mathbf{p}) = -\frac{\beta^{3/2}}{\sqrt{\pi}}\left[\left(\frac{\partial}{\partial\beta} + \beta\frac{\partial^2}{\partial\beta^2}\right)\mathcal{I}(\nu, 0, \beta, -\mathbf{K}, \mathbf{p})\right] \qquad (54.130)$$

evaluated at $\beta = aZ_1/2$. For a general final ns−state f,

$$\mathcal{I}(\nu, \alpha, \beta, \mathbf{K}, \mathbf{p}) = \frac{4\sqrt{\pi}}{T}\left(\frac{T - 2\alpha\delta}{T}\right)^{i\nu}$$
$$\times (U\cosh\pi\nu \pm iV\sinh\pi\nu) , \qquad (54.131)$$

where the complex quantity $U + iV$ is

$$U + iV = (4z)^{i\nu}\frac{\Gamma(\frac{1}{2} + i\nu)}{\Gamma(1 + i\nu)}{}_2F_1(-i\nu, -i\nu; 1 - 2i\nu; 1/z) . \qquad (54.132)$$

646 / *Atomic, Molecular, & Optical Physics Handbook*

(2) Electron Impact Excitation.

$$e^- + H(i) \rightarrow e^- + H(f). \quad (54.133)$$

Neglecting V_{12} and exchange yields the approximate T-matrix element

$$T_{if}^{\rm imp} \sim \langle \psi_f | V_{13} | \omega_{13}^+ \psi_i \rangle \quad (54.134)$$

$$= \frac{-Z_1}{(2\pi a)^3} \int d\mathbf{x} \int d\mathbf{r}\, e^{i\mathbf{q}\cdot\boldsymbol{\sigma}} \varphi_f^*(\mathbf{r}) \frac{1}{x}$$

$$\times \int d\mathbf{K}\, N(\nu) g_i(\mathbf{t}_1) e^{i\mathbf{t}_1\cdot\mathbf{r}}\,_1F_1(i\nu,1; iKx - i\mathbf{K}\cdot\mathbf{x})$$

$$= \frac{-Z_1}{(2\pi a)^3} \int d\mathbf{K}\, N(\nu) g_i(\mathbf{t}_1) g_f^*(\mathbf{t}_2) \mathcal{I}(\nu,0,0,-\mathbf{K},-\mathbf{q}), \quad (54.135)$$

where

$$\mathcal{I}(\nu,0,0,-\mathbf{K},-\mathbf{q})$$

$$= \lim_{\beta \to 0} \frac{4\pi}{\beta^2 + q^2} \left[\frac{\beta^2 + q^2}{\beta^2 + q^2 + 2\mathbf{q}\cdot\mathbf{K} - 2i\beta K} \right]^{i\nu} \quad (54.136)$$

$$= \frac{4\pi}{q^2} \left| 1 + \frac{2K}{q}\cos\theta \right|^{-i\nu} A(\cos\theta), \quad (54.137)$$

with

$$A(\cos\theta) = \begin{cases} 1, & \cos\theta > -q/2K \\ e^{-\pi\nu}, & \cos\theta < -q/2K \end{cases} \quad (54.138)$$

and $\cos\theta = \hat{\mathbf{K}}\cdot\hat{\mathbf{q}}$, $\mathbf{t}_2 = \mathbf{t}_1 + b\mathbf{q}$ and $\mathbf{q} = \mathbf{k}_i - \mathbf{k}_f$.

(3) Heavy Particle Excitation [25].

$$H^+ + H(1s) \rightarrow H^+ + H(2s). \quad (54.139)$$

$$T_{if}^{\rm imp} = -\frac{Z_1 2^{15/2} b^5}{\pi a^3 q^2} \int_0^\infty dK\, N(K) K^2 \int_{-1}^1 d(\cos\theta)$$

$$\times \left| 1 + \frac{2K}{q}\cos\theta \right|^{-i\nu} \tilde{A}(\cos\theta), \quad (54.140)$$

where

$$\tilde{A}(\cos\theta) = \frac{2\pi}{D^4} \left[\frac{\alpha D(D - 2b^2)}{(\alpha^2 - \beta^2)^{3/2}} + \frac{8(3b^2 - D)}{(\alpha^2 - \beta^2)^{1/2}} \right.$$

$$- \frac{48\gamma^2 D^2 b^2}{(\gamma^2 - \delta^2)^{5/2}} + \frac{16D\left[\gamma D - (3\gamma + 4\alpha)b^2\right]}{(\gamma^2 - \delta^2)^{3/2}}$$

$$\left. + \frac{32(D - 3b^2)}{(\gamma^2 - \delta^2)^{1/2}} \right] A(\cos\theta), \quad (54.141)$$

with $A(\cos\theta)$ given by (54.138), and

$$\alpha = b^2 + v^2 + \frac{K^2}{a^2} + \frac{K}{aq}\left(\frac{q^2}{\mu} + \Delta E\right)\cos\theta, \quad (54.142a)$$

$$\beta = \frac{K}{aq}\sin\theta \left[4v^2 q^2 - \left(\frac{q^2}{\mu} + \Delta E\right)^2 \right]^{1/2}, \quad (54.142b)$$

$$\delta = 4\beta, \quad \gamma = 4\alpha + D, \quad (54.142c)$$

$$D = \frac{4bq}{a}(q + 2K\cos\theta), \quad (54.142d)$$

while ν and $N(\nu)$ retain their meaning from (54.127).

(4) Ionization: $e^- + H(i) \rightarrow e^- + H^+ + e^-$.

$$T_{if}^{\rm imp} \sim -\frac{4\pi}{q^2} N(\nu) g_i(\mathbf{k} - b\mathbf{q}) \left(\frac{q^2}{q^2 - 2\mathbf{q}\cdot\mathbf{K}}\right)^{i\nu}, \quad (54.143)$$

where $\mathbf{K} = a(\mathbf{k} - b\mathbf{q} - \mathbf{v})$ and $\mathbf{q} = \mathbf{k}_i - \mathbf{k}_f$ and exchange and V_{12} are neglected.

(5) Rydberg Atom Collisions: [11, 26]

$$A + B(n) \rightarrow A + B(n') \quad (54.144)$$

$$\rightarrow A + B^+ + e^-. \quad (54.145)$$

Consider a Rydberg collision between a projectile (3) and a target with an electron (1) bound in a Rydberg orbital to a core (2) (see [11, 26] for details).

Full T-matrix element:

$$T_{fi}(\mathbf{k}_3, \mathbf{k}_3')$$

$$= \left\langle \phi_f(\mathbf{r}_1) e^{i\mathbf{k}_3'\cdot\mathbf{r}_3} \middle| V(\mathbf{r}_1, \mathbf{r}_3) \middle| \Psi_i^{(+)}(\mathbf{r}_1, \mathbf{r}_3; \mathbf{k}_3) \right\rangle, \quad (54.146)$$

with primes denoting quantities *after* the collision, and where the potential V is

$$V(\mathbf{r}_1, \mathbf{r}_3) = V_{13}(\mathbf{r}) + V_{3C}(\mathbf{r}_3 + a\mathbf{r}_1), \quad \mathbf{r} = \mathbf{r}_1 - \mathbf{r}_3, \quad (54.147)$$

with $a = M_1/(M_1 + M_2)$, while the subscript C labels the core. The impulse approximation to the full, outgoing wave function $\Psi_i^{(+)}$ is written

$$\Psi_i^{\rm imp} = (2\pi)^{3/2} \int g_i(\mathbf{k}_1) \Phi(\mathbf{k}_1, \mathbf{k}_3; \mathbf{r}_1, \mathbf{r}_3) d\mathbf{k}_1, \quad (54.148)$$

$$g_i(\mathbf{k}_1) = \frac{1}{(2\pi)^{3/2}} \int \phi_i(\mathbf{r}_1) e^{-i\mathbf{k}_1\cdot\mathbf{r}_1} d\mathbf{r}_1. \quad (54.149)$$

Impulse approximation:

$$T_{fi}^{\rm imp}(\mathbf{k}_3, \mathbf{k}_3') = \int d\mathbf{k}_1 \int d\mathbf{k}_1'\, g_f^*(\mathbf{k}_1') g_i(\mathbf{k}_1) T_{13}(\mathbf{k}, \mathbf{k}')$$

$$\times \delta\left[\mathbf{P} - (\mathbf{k}_1' - \mathbf{k}_1)\right], \quad (54.150)$$

where T_{13} is the exact off-shell T-matrix for 1–3 scattering,

$$T_{13}(\mathbf{k}, \mathbf{k}') = \langle \exp(i\mathbf{k}'\cdot\mathbf{r}) | V_{13}(\mathbf{r}) | \psi(\mathbf{k}, \mathbf{r}) \rangle. \quad (54.151)$$

The delta function in (54.150) ensures linear momentum, $\mathbf{K} = \mathbf{k}_1 + \mathbf{k}_3 = \mathbf{k}_1' + \mathbf{k}_3'$, is conserved in 1–3 collisions, with

$$\mathbf{k}_1' = \mathbf{k}_1 + (\mathbf{k}_3 - \mathbf{k}_3') \equiv \mathbf{k}_1 + \mathbf{P}, \quad (54.152a)$$

$$\mathbf{k}' = \frac{M_3}{M_1 + M_3}(\mathbf{k}_1 + \mathbf{k}_3) - \mathbf{k}_3' = \mathbf{k} + \mathbf{P}. \quad (54.152b)$$

Elastic scattering:

$$T_{ii}(\mathbf{k}_3, \mathbf{k}_3) = \int g_f^*(\mathbf{k}_1) g_i(\mathbf{k}_1) T_{13}(\mathbf{k}, \mathbf{k}) d\mathbf{k}_1 \,, \quad (54.153)$$

where $\mathbf{k} = (M_3/M)\mathbf{k}_1 + (M_1/M)\mathbf{k}_3$ and $M = M_1 + M_3$.

Integral cross section: for 3–(1,2) scattering,

$$\sigma_{if}(k_3) = \left(\frac{M_{AB}}{M_{13}} \right)^2 \frac{k_3'}{k_3} \int |\langle g_f(\mathbf{k}_1 + \mathbf{P})|$$

$$\times f_{13}(\mathbf{k}, \mathbf{k}') |g_i(\mathbf{k}_1)\rangle|^2 \, d\hat{\mathbf{k}}_3' \,, \quad (54.154)$$

where M_{AB} is the reduced mass of the 3–(1,2) system, M_{13} the reduced mass of 1–3. The 1–3 scattering amplitude f_{13} is given by

$$f_{13}(\mathbf{k}, \mathbf{k}') = -\frac{1}{4\pi} \left(\frac{2M_{13}}{\hbar^2} \right) T_{13}(\mathbf{k}, \mathbf{k}') \,. \quad (54.155)$$

Six Approximations to Eq. (54.150)

(1) Optical Theorem.

$$\sigma_{\text{tot}}(k_3) = \frac{1}{k_3} \frac{2M_{AB}}{\hbar^2} T_{ii}(\mathbf{k}_3, \mathbf{k}_3')$$

$$= \frac{1}{v_3} \int |g_i(\mathbf{k}_1)|^2 \left[v_{13} \sigma_{13}^{\mathrm{T}}(v_{13}) \right] d\mathbf{k}_1 \,, \quad (54.156)$$

where σ_{13}^{T} is the total cross section for 1–3 scattering at relative speed v_{13}. The resultant cross section (54.156) is an upper limit and contains no interference terms.

(2) Plane-Wave Final State.

$$\phi_f(\mathbf{r}_1) = (2\pi)^{-3/2} \exp(i\boldsymbol{\kappa}_1' \cdot \mathbf{r}_1) \,, \quad (54.157)$$

$$g_f(\mathbf{k}_1') = \delta(\mathbf{k}_1' - \boldsymbol{\kappa}_1') \,, \quad (54.158)$$

$$T_{fi}(\mathbf{k}_3, \mathbf{k}_3') = g_i(\mathbf{k}_1) T_{13}(\mathbf{k}, \mathbf{k}') \,, \quad \mathbf{k}_1 = \boldsymbol{\kappa}_1 - \mathbf{P} \,, \quad (54.159)$$

$$\frac{d\sigma_{if}}{d\hat{\mathbf{k}}_3' d\mathbf{k}_1'} = \left(\frac{M_{AB}}{M_{13}} \right)^2 \frac{k_3'}{k_3} |g_i(\mathbf{k}_1)|^2 |f_{13}(\mathbf{k}, \mathbf{k}')|^2 \,. \quad (54.160)$$

(3) Closure.

$$\sum_f g_f(\mathbf{k}_1') g_f^*(\mathbf{k}_1'') = \delta(\mathbf{k}_1' - \mathbf{k}_1'') \,, \quad (54.161)$$

$$\frac{d\sigma_i^T}{d\hat{\mathbf{k}}_3'} = \frac{\bar{k}_3'}{k_3} \left(\frac{M_{AB}}{M_{13}} \right)^2 \int |g_i(\mathbf{k}_1)|^2 |f_{13}(\mathbf{k}, \mathbf{k}')|^2 \, d\mathbf{k}_1 \,, \quad (54.162)$$

where $\mathbf{k}_1' = (M_3/M)(\mathbf{k}_1 + \mathbf{k}_3) - \mathbf{k}_3'$.

Conditions for validity of (54.162): (a) k_3 is high enough to excite all atomic bound and continuum states, and (b) $k_3'^2 = (k_3^2 - 2\varepsilon_{fi}/M_{AB})$ can be approximated by k_3, or by an averaged wavenumber $\bar{k}_3' = (k_3^2 - 2\bar{\varepsilon}_{fi}/M_{AB})^{1/2}$, where the averaged excitation energy is

$$\bar{\varepsilon}_{fi} = \ln \langle \varepsilon_{fi} \rangle = \sum_j f_{ij} \ln \varepsilon_{ij} \Big(\sum_j f_{ij} \Big)^{-1} \,, \quad (54.163)$$

and the f_{ij} are the oscillator strengths.

(4) Peaking Approximation.

$$T_{fi}^{\text{peak}}(\mathbf{k}_3, \mathbf{k}_3') = F_{fi}(\mathbf{P}) T_{13}(\mathbf{k}, \mathbf{k}') \,, \quad (54.164)$$

where F_{fi} is the *inelastic form factor*

$$F_{fi}(\mathbf{P}) = \int g_f^*(\mathbf{k}_1 + \mathbf{P}) g_i(\mathbf{k}_1) d\mathbf{k}_1 \quad (54.165)$$

$$= \langle \psi_f(\mathbf{r}) | \exp(i\mathbf{P} \cdot \mathbf{r}) | \psi_i(\mathbf{r}) \rangle \,. \quad (54.166)$$

(5) $T_{13} = T_{13}(\mathbf{P})$:

$$T_{fi}(\mathbf{k}_3, \mathbf{k}_3') = T_{13}(\mathbf{P}) F_{fi}(\mathbf{P}) \,. \quad (54.167)$$

(6) $T_{13} = $ constant:

$f_{13} \equiv a_{\mathrm{s}} = $ constant scattering length.

$$\sigma_{if}(k_3) = \frac{2\pi a_{\mathrm{s}}^2}{k_3^2} \left(\frac{M_{AB}}{M_{13}} \right)^2 \int_{k_3 - k_3'}^{k_3 + k_3'} |F_{fi}(\mathbf{P})|^2 \, P \, dP \quad (54.168)$$

$$\sigma_{\text{tot}}(k_3) = \begin{cases} 4\pi a_{\mathrm{s}}^2 \,, & v_3 \gg v_1 \\ \langle v_1 \rangle 4\pi a_{\mathrm{s}}^2 / v_3 \,, & v_3 \ll v_1 \,. \end{cases} \quad (54.169)$$

Validity Criteria

(a) Intuitive Formulation [26].

(i) Particle 3 scatters separately from 1 and 2, i.e., $r_{12} \gg A_{1,2}$; the relative separation of (1,2) $\gg$ the scattering lengths of 1 and 2.

(ii) $\lambda_{13} \ll r_{12}$, i.e., the reduced wavelength for 1–3 relative motion $\ll r_{12}$. Interference effects of 1 and 2 can be ignored and 1, 2 treated as *independent* scattering centers.

(iii) 2–3 collisions do not contribute to inelastic 1–3 scattering.

(iv) Momentum impulsively transferred to 1 during collision (time τ_{coll}) with 3 $\gg$ Momentum transferred to 1 due to V_{12}, i.e.,

$$P \gg \langle \psi_{n\ell} | - \nabla V_{12} | \psi_{n\ell} \rangle \tau_{\text{coll}} \,. \quad (54.170)$$

For a precise formulation of validity criteria based upon the two-potential formula see the Appendix of Ref. [26].

Two classes of interaction in A–$B(n)$ Rydberg collisions justify use of the impulse approximation for the T-matrix for 1–3 collisions: (i) quasiclassical binding with $V_{\text{core}} = $ const., and (ii) weak binding with

$$E_3 \gg \Delta E_c \sim \langle \psi_n(r) | V_{1C}(\mathbf{r}) | \psi_n(\mathbf{r}) \rangle \quad (54.171a)$$

$$\langle \psi_n(\mathbf{r}_1) | - \frac{\hbar^2}{2M_{12}} \nabla_1^2 | \psi_n(\mathbf{r}_1) \rangle \sim |\varepsilon_n| \,, \quad (54.171b)$$

where E_3 is the kinetic energy of relative motion of 3, and ΔE_c is the energy shift in the core. The fractional error is [27]

$$\frac{f_{13}}{\lambda}\frac{\Delta E_c}{\hbar}\left(\frac{\hbar}{E_3}+\tau_{\text{delay}}\right)\ll 1\,,\qquad (54.172)$$

where $\lambda\sim k_3^{-1}$ is the reduced wavelength of 3, f_{13} is the scattering amplitude for 1–3 collisions and τ_{delay} is the time delay associated with 1–3 collisions.

Special Case. for nonresonant scattering with $\tau_{\text{delay}}=0$

$$\frac{f_{13}}{\lambda}\frac{|\varepsilon_n|}{E_3}\ll 1\,,\qquad (54.173)$$

which follows from (54.172) upon identifying the shift in the core energy ΔE_c with the binding energy $|\varepsilon|$.

Condition (54.173) is less restrictive than (54.171a) or (54.171b) since f_{13} can be either less than or greater than λ.

54.7.2 Classical Impulse Approximation

(a) Ionization. For electron impact on heavy particles [28], the cross section for ionization of a particle moving with velocity t by a projectile with velocity s is

$$Q(s,t)=\frac{1}{u^2}\frac{2s}{m_2}\int_1^{s^2}\frac{dz}{z^2}\left[\frac{A(z)}{z}+B(z)\right]\,,\qquad (54.174)$$

where

$$A(z)=\frac{1}{2st^3}\left[\tfrac{1}{3}(x_{02}^{3/2}-x_{01}^{3/2})-2(s^2+t^2)(x_{02}^{1/2}-x_{01}^{1/2})\right]$$
$$-\frac{(s^2-t^2)^2}{2ts^3}(x_{02}^{-1/2}-x_{01}^{-1/2})\qquad (54.175a)$$

$$B(z)=\frac{1}{2m_2st^3}\left[(m_1+m_2)(s^2-t^2)(x_{02}^{-1/2}-x_{01}^{-1/2})\right.$$
$$\left.-(m_2-m_1)(x_{02}^{1/2}-x_{01}^{1/2})\right]\,.\qquad (54.175b)$$

For electron impact, (54.175b) is evaluated at $m_1=1$. The remaining terms above are given by

$$s^2=v_2^2/v_0^2\,,\quad t^2=v_1^2/v_0^2=E_1/u\,,\quad E_2=m_2v_2^2\,,$$
$$u=v_0^2=\text{Ionization potential of target}\,,$$
$$x_{0i}=(s^2+t^2-2st\cos\theta_i)\,,\quad i=1,2\,,$$
$$\cos\theta_i=\begin{cases}\kappa_0\pm\kappa_1\,,&|\kappa_0\pm\kappa_1|\le 1\\ 1\,,&\kappa_0\pm\kappa_1>1\\ -1\,,&\kappa_0\pm\kappa_1<-1\end{cases}\,,$$
$$\kappa_0\pm\kappa_1=-\frac{1}{2}\Big(1-\frac{m_1}{m_2}\Big)\frac{z}{st}\pm\sqrt{\Big(1+\frac{z}{t^2}\Big)\Big(1-\frac{z}{s^2}\Big)}\,.$$

Equal Mass Case. $(m_1=m_2)$

$$Q(s,t)=\begin{cases}\dfrac{4}{3s^2}\dfrac{1}{u^2}\dfrac{2(s^2-1)^{3/2}}{t}\,,&1\le s^2\le t^2+1\\[3mm]\dfrac{4}{3s^2}\dfrac{1}{u^2}\left[2t^2+3-\dfrac{3}{s^2-t^2}\right]\,,&s^2\ge t^2+1\,.\end{cases}$$
$$(54.176)$$

Integrating over the speed distribution (see Sect. 54.4),

$$Q(s)=\frac{32}{\pi}\frac{1}{u^2}\int_0^\infty\frac{Q(s,t)t^2dt}{(t^2+1)^4}\,,\qquad (54.177)$$

which is then numerically evaluated. For electrons, the integral can be done analytically with the result

$$Q(y)=\frac{8}{3\pi y^2(y+1)^4}$$
$$\times\left[(5y^4+15y^3-3y^2-7y+6)(y-1)^{1/2}\right.$$
$$+(5y^5+17y^4+15y^3-25y^2+20y)\times\tan^{-1}(y-1)^{1/2}$$
$$\left.-24y^{3/2}\ln\left|\frac{\sqrt{y}+\sqrt{y-1}}{\sqrt{y}-\sqrt{y-1}}\right|\right]\,,\qquad (54.178)$$

with $y=s^2$.

Thomson's Result.

$$Q_T(y)=\frac{4}{y}\frac{1}{u^2}\left(1-\frac{1}{y}\right)\,.\qquad (54.179)$$

(b) Electron Loss Cross Section [18].

$$A(V)+B(u)\to A^++e^-+B(f)\,,\qquad (54.180)$$

where $B(f)$ denotes that the target B is left in any state (either bound or free) after the collision with the projectile A. The initial velocity of the projectile is V while the velocity of the Rydberg electron relative to the core is u, and the ionization potential of the target B is I.

$$\sigma_{\text{loss}}=\frac{1}{3\pi\nu^2}\int_{\tau/4\nu}^\infty dx\,\sigma_T(x\bar u)\left[\frac{8\nu x-1-(\nu-x)^2-2\tau}{[1+(\nu-x)^2]^3}\right.$$
$$\left.+\frac{1}{[1+(\nu+x)^2-\tau]^2}\right]\,,\qquad (54.181)$$

where $\nu=V/\bar u$, $\tau=I/\frac{1}{2}m_e\bar u^2$, $\bar u=\sqrt{2I/m_e}$, and σ_T is the total electron scattering cross section at speed $x\bar u$. The cross section (54.181) is valid only for particles being stripped (or lost from the projectile) which are not strongly bound. See Refs. [18, 29, 30] for details and numerous results.

(c) Capture Cross Section from Shell i [18].

$$\sigma_{\text{capture}}^i(V)=\frac{2^{5/2}N_i\pi}{3V^7}\int_0^{r_i}dr\int_{\mathcal{C}(-1)}^{\mathcal{C}(+1)}d(\cos\eta')\,[P_i(r)]^2$$
$$\times\frac{\sqrt{1+y^2}\,[4\varepsilon^2-(\varepsilon^2-y^2)(1+\varepsilon^2+a^2-y^2)]}{r^{3/2}\varepsilon^{9/2}(1+a^2)^3\sqrt{y^2-a^2}}\,,$$
$$(54.182)$$

where $\mathcal{C}$ denotes that the integration range $[-1, +1]$ is restricted such that the integrand is real and positive and that $|1 - \varepsilon| < \sqrt{y^2 - a^2}$. The dimensionless variables a and y above are defined,

$$y^2 = \frac{2}{m_e}|V(R)|V^2, \quad a^2 = \frac{2}{m_e}I_i V^2, \quad (54.183)$$

and with $P_i(r)/r$ representing the Hartree-Fock-Slater radial wave function for shell i, with normalization

$$\int_0^{r_i} [P_i(r)]^2\, dr = 1. \quad (54.184)$$

The ionization potential and number of electrons in shell i are denoted above, respectively, by I_i and N_i.

(d) Total Capture Cross Section [18].

$$\sigma_{\text{capture}}^{\text{total}}(V) = \sum_i \sigma_{\text{capture}}^i. \quad (54.185)$$

(e) Universal Capture Cross Section [18].

A universal curve independent of projectile mass M and charge Z is obtained from the above expressions for the capture cross section by plotting the scaled cross section

$$\widetilde{\sigma}_{\text{capture}}^{\text{total}} = \frac{E^{11/4}}{M^{11/4} Z^{7/2} \lambda^{3/4}} \sigma_{\text{capture}}^{\text{total}} \quad (54.186)$$

versus the scaled energy

$$\widetilde{E} = \frac{m_e}{M}\frac{E}{I}, \quad (54.187)$$

where m_e is the mass of the particle captured, which is usually taken to be a single electron, and I is the ionization potential of the target. The term λ in (54.186) is the coupling constant in the target potential, $V(R) = m_e\lambda/R^2$, which the electron being captured experiences during the collision. See Fig. 11 of [18] for details.

54.7.3 Semiquantal Impulse Approximation

Basic Expression [26, 31].

$$\frac{d\sigma}{d\varepsilon\, dP\, dk_1\, dk\, d\phi_1} = \frac{k_1'^2}{J_{55}}\frac{k_3'}{k_3}\left(\frac{M_3}{M_{13}}\right)^2 |g_i(\mathbf{k}_1)|^2 |f_{13}(\mathbf{k},\mathbf{k}')|^2. \quad (54.188)$$

J_{55} is the 5-dimensional Jacobian of the transformation

$$(P, \varepsilon, k_1, k, \phi_1) \to (\hat{\mathbf{k}}_3', \mathbf{k}_1'), \quad (54.189a)$$

$$J_{55} = \frac{\partial(P, \varepsilon, k_1, k, \phi_1)}{\partial(\cos\theta_3'\phi_3', k_1', \cos\theta_1', \phi_1')}. \quad (54.189b)$$

Expression for Elemental Cross Section [26].
In the $(P, \varepsilon, k_1, k, \phi_1)$ representation,

$$d\sigma = \frac{d\varepsilon\, dP}{M_{13}^2 v_3^2}\left[\frac{|g_i(\mathbf{k}_1)|^2\, k_1^2\, dk_1\, d\phi_1}{v_1}\right]\frac{|f_{13}(\mathbf{k},\mathbf{k}')|^2\, dg^2}{\sqrt{(g_+^2 - g^2)(g^2 - g_-^2)}} \quad (54.190)$$

where $g_\pm^2 = \frac{1}{2}B \pm \sqrt{\frac{1}{4}B^2 - C}$, and

$$\begin{aligned}B &= B(\varepsilon, P, v_1; v_3)\\ &= \frac{a}{(1+a)^2}\frac{P^2}{M_{13}^2} + \left(v_1^2 + v_1'^2 + v_3^2 + v_3'^2 + \frac{2\Delta_3}{M_{13}}\right)\\ &\quad - \frac{4\varepsilon(\varepsilon + \Delta_3)}{P^2},\end{aligned}$$

$$\begin{aligned}C &= C(\varepsilon, P, v_1; v_3)\\ &= \frac{v_1^2 + av_3^2}{1+a}\frac{P^2}{M_{13}^2} + (v_1^2 - v_3^2)(v_1'^2 - v_3'^2)\\ &\quad + \frac{2\Delta_3}{M_{13}}(v_1^2 + v_3^2) + \frac{4\Delta_3}{P^2}\left[v_1^2(\varepsilon + \Delta_3) - \varepsilon v_3^2\right],\end{aligned}$$

$$a = \frac{M_2 M_3}{M_1(M_1 + M_2 + M_3)},$$

$$\widetilde{M}_1 = M_1(1 + M_1/M_2),$$

$$v_1'^2 = v_1^2 + \frac{2\varepsilon}{\widetilde{M}_1}, \quad v_3'^2 = v_3^2 - \frac{2(\varepsilon + \Delta_3)}{M_{AB}},$$

and Δ_3 is the change in internal energy of particle 3, while ε denotes the energy change in the target $1-2$.

Hydrogenic Systems. $g_i(\mathbf{k}_1) = g_{n\ell}(k_1)Y_{\ell m}(\theta_1, \phi_1)$.

The $g_{n\ell}$ are the hydrogenic wave functions in momentum space. See Chap. 9 for details on hydrogenic wave functions.

Elemental Cross Sections. (m-averaged and ϕ_1-integrated)

$$d\sigma = \frac{d\varepsilon\, dP}{M_{13}^2 v_3^2}\frac{W_{n\ell}(v_1)\, dv_1}{2v_1}\frac{|f_{13}(P, g)|^2\, dg^2}{\sqrt{(g_+^2 - g^2)(g^2 - g_-^2)}}, \quad (54.191)$$

where the speed distribution $W_{n\ell}$ is given by (54.57).

Two Representations for 1–3 Scattering Amplitude [26].
(i) $f_{13} = f_{13}(P, g)$ is a function of momentum transferred and relative speed. Then

$$\begin{aligned}\sigma(v_3) = \frac{1}{M_{13}^2 v_3^2}\int_{\varepsilon_1}^{\varepsilon_2} d\varepsilon \int_{v_{10}}^{\infty}\frac{W_{n\ell}(v_1)\, dv_1}{v_1}\int_{P^-}^{P^+} dP\\ \times \int_{g_-}^{g_+}\frac{|f_{13}(P, g)|^2\, dg^2}{\sqrt{(g_+^2 - g^2)(g^2 - g_-^2)}},\end{aligned} \quad (54.192)$$

where $v_{10}^2(\varepsilon) = \max[0, (2\varepsilon/M)]$, and the limits to the P integral are

$$\begin{aligned}P^+ &= P^+(\varepsilon, v_1; v_3)\\ &= \min[M(v_1' + v_1), M_{AB}(v_3' + v_3)], \quad (54.193a)\end{aligned}$$

$$\begin{aligned}P^- &= P^-(\varepsilon, v_1; v_3)\\ &= \max[M|v_1' - v_1|, M_{AB}|v_3' - v_3|], \quad (54.193b)\end{aligned}$$

and unless $P^+ > P^-$, the P integral is zero.

(ii) $f_{13} = f_{13}(g, \psi)$ is a function of relative speed and scattering angle. Then

$$\sigma(v_3) = \frac{1}{v_3^2} \int_{\varepsilon_1}^{\varepsilon_2} d\varepsilon \int_{v_{10}}^{\infty} \frac{W_{n\ell}(v_1)\, dv_1}{v_1} \int_{g_-}^{g_+} \frac{g\, dg}{S(v_1, g; v_3)}$$
$$\times \int_{\psi^-}^{\psi^+} \frac{|f_{13}(g, \psi)|^2\, d(\cos\psi)}{\sqrt{(\cos\psi^+ - \cos\psi)(\cos\psi - \cos\psi^-)}},$$
$$\text{(54.194)}$$

where

$$S(v_1, g; v_3) = \frac{M_{13}}{1+a}\left[(1+a)(v_1^2 + av_3^2) - ag^2\right]^{1/2}.$$

Scattering angle ψ-limits,

$$\cos\psi^{\pm} = \cos\psi^{\pm}(\varepsilon, v_1, g; v_3) \qquad \text{(54.195)}$$
$$= \frac{g}{g'}\frac{1}{\alpha^2 + \beta^2}\left\{\alpha(\alpha + \tilde{\varepsilon}) \pm \beta\left[\omega^2(\alpha^2 + \beta^2)\right.\right.$$
$$\left.\left. -(\alpha + \tilde{\varepsilon})^2\right]^{1/2}\right\}, \qquad \text{(54.196)}$$

where

$$\alpha = \alpha(v_1, g; v_3) = \tfrac{1}{2}M_{13}\left[v_1^2 - v_3^2 + \left(\frac{1-a}{1+a}\right)g^2\right],$$
$$\beta = \beta(v_1, g; v_3) = \tfrac{1}{2}M_{13}\left[(2v_1^2 + 2v_3^2 - g^2)g^2\right.$$
$$\left. -(v_1^2 - v_3^2)^2\right]^{1/2},$$
$$\omega = g'/g, \quad \tilde{\varepsilon} = \varepsilon + \frac{a}{1+a}\Delta_3.$$

Special Case. $f_{13} = f_{13}(P)$.

$$\sigma(v_3) = \frac{\pi}{M_{13}^2 v_3^2}\int_{\varepsilon_1}^{\varepsilon_2} d\varepsilon \int_{v_{10}}^{\infty} \frac{W_{n\ell}(v_1) dv_1}{v_1}$$
$$\times \int_{P^-}^{P^+} |f_{13}(P)|^2\, dP. \qquad \text{(54.197)}$$

54.8 BINARY ENCOUNTER APPROXIMATION

The basic assumption of the binary encounter approximation is that an excitation or ionization process is caused solely by the interaction of the incoming charged or neutral projectile with the Rydberg electron bound to its parent ion. If, for example, the cross section depends only on the momentum transfer P to the Rydberg electron (as in the Born approximation), then the total cross section is obtained by integrating σ_P over the momentum distribution of the Rydberg electron. The basic cross sections required are given in the following section. For further details see [32] and references therein.

54.8.1 Differential Cross Sections

Cross Section per Unit Momentum Transfer

Let the masses, velocities and charges of the particles be $(m_1, \mathbf{v}_1, Z_1, e)$ and $(m_2, \mathbf{v}_2, Z_2, e)$, with $\mathbf{v} = |\mathbf{v}_1 - \mathbf{v}_2|$ denoting the relative velocity and quantities *after* the collision are denoted by primes. Then for distinguishable particles,

$$\sigma_P = \frac{8\pi Z_1^2 Z_2^2 e^4 P}{v^2}\left|\frac{\exp(i\eta_P)}{P^2}\right|^2, \qquad \text{(54.198)}$$

where the phase shift η_P is

$$\eta_P = -2\gamma\ln(P/2\mu v) + 2\eta_0 + \pi, \qquad \text{(54.199)}$$

and with

$$\mu = \frac{m_1 m_2}{m_1 + m_2}, \quad \gamma = \frac{Z_1 Z_2 e^2}{\hbar v}, \quad e^{2i\eta_0} = \frac{\Gamma(1 + i\gamma)}{\Gamma(1 - i\gamma)}.$$
$$\text{(54.200)}$$

For identical particles,

$$\sigma_P^{\pm} = \frac{8\pi Z_1^2 Z_2^2 e^4 P}{v^2}\left|\frac{e^{i\eta_P}}{P^2} \pm \frac{e^{i\eta_S}}{S^2}\right|^2, \qquad \text{(54.201)}$$

where η_P is given by (54.199) and η_S is

$$\eta_S = -2\gamma\ln(S/2\mu v) + 2\eta_0 + \pi, \qquad \text{(54.202)}$$

while η_0 is given by (54.200). The momenta $\mathbf{P}$ and $\mathbf{S}$ transferred by direct and exchange collisions, respectively, are given by

$$\mathbf{P} = m_1(\mathbf{v}_1 - \mathbf{v}_1') = m_2(\mathbf{v}_2 - \mathbf{v}_2'), \qquad \text{(54.203a)}$$
$$\mathbf{S} = m_1(\mathbf{v}_1 - \mathbf{v}_2') = m_2(\mathbf{v}_1' - \mathbf{v}_2). \qquad \text{(54.203b)}$$

The collision rates (in cm^3/s) are

$$\hat{\alpha}_P = v_1 \sigma_P, \quad \hat{\alpha}_P^{\pm} = v_1 \sigma_P^{\pm}. \qquad \text{(54.204)}$$

Cross Section per Unit Momentum Transferred per Unit Sterradian

Differential relationships:

$$\alpha_{E,P} = \frac{d^2\alpha}{dP\, dE} = \frac{d^2\alpha}{dP\, d\varphi}\frac{d\varphi}{dE} = \alpha_{P,\varphi}\frac{d\varphi}{dE}. \qquad \text{(54.205)}$$

For distinguishable particles,

$$\alpha_P = 2\pi v_1 \sigma_{P,\varphi} = 2\pi\alpha_{P,\varphi}, \qquad \text{(54.206a)}$$
$$\alpha_{P,\varphi} = \frac{4Z_1^2 Z_2^2 e^4 P}{v}\left|\frac{e^{i\eta_P}}{P^2}\right|^2, \qquad \text{(54.206b)}$$
$$\alpha_{E,P} = v_1 \sigma_{E,P} = \frac{8Z_1^2 Z_2^2 e^4}{v_1 v_2 \sqrt{X}}\left|\frac{e^{i\eta_P}}{P^2}\right|^2. \qquad \text{(54.206c)}$$

For identical particles,

$$\alpha_{P,\varphi}^{\pm} = \frac{4Z_1^2 Z_2^2 e^4 P}{v}\left|\frac{e^{i\eta_P}}{P^2}\pm\frac{e^{i\eta_S}}{S^2}\right|^2, \quad (54.207a)$$

$$\alpha_{E,P}^{\pm} = v_1\sigma_{E,P}^{\pm} = \frac{8Z_1^2 Z_2^2 e^4}{v_1 v_2\sqrt{X}}\left|\frac{e^{i\eta_P}}{P^2}\pm\frac{e^{i\eta_S}}{S^2}\right|^2, \quad (54.207b)$$

where

$$X = -\cos^2\phi + 2(\hat{\mathbf{v}}_1\cdot\hat{\mathbf{P}})(\hat{\mathbf{v}}_2\cdot\hat{\mathbf{P}})\cos\phi + 1$$
$$- (\hat{\mathbf{v}}_1\cdot\hat{\mathbf{P}})^2 - (\hat{\mathbf{v}}_2\cdot\hat{\mathbf{P}})^2 \quad (54.208a)$$
$$= (\cos\phi_{\min}-\cos\phi)(\cos\phi-\cos\phi_{\max}) \quad (54.208b)$$
$$= \left(\frac{v}{v_1 v_2 P}\right)^2 (E_{\max}-E)(E-E_{\min}), \quad (54.208c)$$

with ϕ being the angle between the velocity vectors $\mathbf{v}_1$ and $\mathbf{v}_2$.

For the special case of electron impact, $M_2 = m_e$, $Z_2 = -1$, and

$$\sigma_{E,P}(\phi) = \frac{8Z_1^2 e^4}{v_1^2 v_2 P^4\sqrt{X}}, \quad (54.209)$$

$$\sigma_{E,P}^{\pm}(\phi) = \frac{8e^4}{v_1^2 v_2\sqrt{X}}\left[\frac{1}{P^4}+\frac{1}{S^4}\pm\frac{2\cos(\eta_P-\eta_S)}{P^2 S^2}\right], \quad (54.210)$$

where $\eta_P - \eta_S = -2\gamma\ln(P/S) = (2e^2/\hbar v)\ln(S/P)$, and X is given by (54.208b).

Integrated Cross Sections

For incident heavy particles:

$$\sigma_{E,P} = \int_0^\pi \sigma_{E,P}(\phi)\frac{1}{2}\sin\phi\,d\phi = \frac{4\pi Z_1^2 e^4}{v_1^2 v_2 P^4}. \quad (54.211)$$

For incident electrons:

$$\sigma_{E,P}^{\pm} = \int_0^\pi \sigma_{E,P}^{\pm}(\phi)\frac{1}{2}\sin\phi\,d\phi \quad (54.212a)$$
$$= \frac{4\pi e^4}{v_1^2 v_2}\left[\frac{1}{P^4}+\frac{v_1^2+v_2^2-P^2/2m_e^2-2E^2/P^2}{m_e^4|v_1^2-v_2^2-2E/m_e|^3}\right.$$
$$\left.\pm\frac{2\Phi}{m_e^2 P^2|v_1^2-v_2^2-2E/m_e|}\right], \quad (54.212b)$$

where Φ can be approximated [33] by

$$\Phi \sim \cos\left(\left|\frac{R_\infty}{E_3-E_2}\right|^{1/2}\ln\left|\frac{E}{E_3-E_2-E}\right|\right). \quad (54.213)$$

and E_3 is defined in Ref. [33].

Cross Sections per Unit Energy

For incident heavy particles (three cases):

$$\sigma_E = \frac{2\pi Z_1^2 e^4}{m_e v_1^2}\left(\frac{1}{E^2}+\frac{2m_e v_2^2}{3E^2}\right), \quad (54.214)$$

which is valid for $2v_1 \geq v_2+v_2'$, $E \leq 2m_e v_1(v_1-v_2)$, or

$$\sigma_E = \frac{\pi Z_1^2 e^4}{3v_1^2 v_2 E^3}\left[4v_1^3-\frac{1}{2}(v_2'-v_2)^2\right], \quad (54.215)$$

which is valid for $v_2'-v_2 \leq 2v_1 \leq v_2'+v_2$, $2m_e v_1(v_1-v_2) \leq E \leq 2m_e v_1(v_1+v_2)$, or otherwise, $\sigma_E = 0$ for $E \geq m_e v_1(v_2'+v_2)$.

For incident electrons (two cases):

$$\sigma_E^{\pm} = \frac{2\pi e^4}{m_e v_1^2}\left[\frac{1}{E^2}+\frac{2m_e v_2^2}{3E^3}+\frac{1}{D^2}+\frac{2m_e v_2^2}{3D^3}\pm\frac{2\Phi}{ED}\right], \quad (54.216)$$

which is valid for $m_e(v_2-v_2') \leq m_e(v_1'-v_1)$, $m_e(v_2'+v_2) \leq m_e(v_1+v_1')$, $D \geq 0$, or

$$\sigma_E = \frac{2\pi e^4}{m_e v_1^2}\left[\frac{1}{E^2}+\frac{2m_e v_1'^2}{3E^3}+\frac{1}{D^2}+\frac{2m_e v_1'^2}{3|D|^3}\right.$$
$$\left.\pm\frac{2\Phi}{E|D|}\right]\frac{v_1'}{v_1}, \quad (54.217)$$

which is valid for $m_e(v_2'-v_2) \leq m(v_1-v_1')$, $m_e(v_1+v_1') \leq m_e(v_2'+v_2)$, $D \leq 0$. In the expressions above, the exchange energy D transferred during the collision is

$$D = \tfrac{1}{2}m_e v_1^2-\tfrac{1}{2}m_e v_2'^2 = \tfrac{1}{2}m_e v_1^2-\tfrac{1}{2}m_e v_2^2-E. \quad (54.218)$$

54.8.2 Integral Cross Sections

$$e^-(T)+A(E_2)\rightarrow e^-(E)+A^++e^-, \quad (54.219)$$

where T is the initial kinetic energy of the projectile electron, while the Rydberg electron, initially bound in potential U_i to the core A^+, has kinetic energy E_2. The cross section per unit energy E is denoted below by σ_E. See the review by Vriens [32] for details.

For electron impact, there are two collision models: the *unsymmetrical collision model* of Thomson and Gryzinski assumes that the incident electron has zero potential energy, and the *symmetrical collision model* of Thomas and Burgess assumes that the incident electron is accelerated initially by the target (and thereby gains kinetic energy) while losing an equal amount of potential energy.

Unsymmetrical model (two cases):

$$\sigma_E = \frac{\pi e^4}{T}\left[\frac{1}{E^2}+\frac{4E_2}{3E^3}+\frac{1}{D^2}+\frac{4E_2}{3D^3}-\frac{\Phi}{ED}\right], \quad (54.220)$$

which is valid for $D = T - E_2 - E \geq 0$ or,

$$\sigma_E = \frac{\pi e^4}{T} \left[\frac{1}{E^2} + \frac{4T'}{3E^3} + \frac{1}{D^2} + \frac{4T'}{3|D|^3} \right.$$
$$\left. - \frac{\Phi}{E|D|} \right] \left(\frac{T'}{E_2} \right)^{1/2} \qquad (54.221)$$

which is valid for $D \leq 0$ and $T \geq E$; and where $T' \equiv T - E$.

Symmetrical model (two cases):

$$\sigma_E = \frac{\pi e^4}{T_i} \left[\frac{1}{E^2} + \frac{4E_2}{3E^3} + \frac{1}{X_i^2} + \frac{4E_2}{3X_i^3} - \frac{\Phi}{EX_i} \right], \quad (54.222)$$

which is valid for $X_i \equiv T + U_i - E \geq 0$, with $T_i \equiv T + U_i + E_2$, and

$$\sigma_E = \frac{\pi e^4}{T_i} \left[\frac{1}{E^2} + \frac{4T_i'}{3E^3} + \frac{1}{X_i^2} + \frac{4T_i'}{3|X_i|^3} \right.$$
$$\left. - \frac{\Phi}{E|X_i|} \right] \left(\frac{T_i'}{E_2} \right)^{1/2} \qquad (54.223)$$

which is valid for $0 \leq T_i' \leq E_2$, $T \geq 0$, with $T_i' \equiv T_i - E$, and where Φ is given by (54.213).

For incident heavy particles, the unsymmetrical model (54.220) should be used.

Single Particle Ionization

The total ionization cross section per atomic electron for incident heavy particles is

$$Q_i = \frac{2\pi Z_1^2 e^4}{m_e v_1^2} \left[\frac{1}{U_i} + \frac{m_e v_2^2}{3U_i^2} - \frac{1}{2m_e(v_1^2 - v_2^2)} \right], \quad (54.224)$$

which is valid for $U_i \leq 2m_e v_1(v_1 - v_2)$, or

$$Q_i = \frac{\pi Z_1^2 e^4}{m_e v_1^2} \left\{ \frac{1}{2m_e v_2(v_1 + v_2)} + \frac{1}{U_i} + \frac{m_e}{3v_2 U_i^2} \left[2v_1^3 + v_2^3 \right. \right.$$
$$\left. \left. - \left(2U_i/m_e + v_2^2 \right)^{3/2} \right] \right\}, \qquad (54.225)$$

which is valid for $2m_e v_1(v_1 - v_2) \leq U_i \leq 2m_e v_1(v_1 + v_2)$, or otherwise $Q_i = 0$ for $U_i \geq 2m_e v_1(v_1 + v_2)$.

For electron impact,

$$Q_i = \frac{1}{2} \left[Q_i^{\text{dir}} + Q_i^{\text{ex}} + Q_i^{\text{int}} \right]. \qquad (54.226)$$

In the unsymmetrical model, Q_i^{ex} diverges, hence the exchange and interference terms above are omitted in the unsymmetrical model for electrons to obtain

$$Q_i^{\text{dir}} = \frac{\pi e^4}{T} \left[\frac{1}{U_i} + \frac{2E_2}{3U_i^2} - \frac{1}{T - E_2} \right], \qquad (54.227)$$

which is valid for $T \geq E_2 + U_i$, or

$$Q_i^{\text{dir}} = \frac{2\pi e^4}{3T} \frac{(T - U_i)^{3/2}}{U_i^2 \sqrt{E_2}}, \qquad (54.228)$$

which is valid for $U_i \leq T \leq E_2 + U_i$.

In the symmetrical model,

$$Q_i^{\text{dir}} = Q_i^{\text{ex}}$$
$$= \frac{\pi e^4}{T_i} \left[\frac{1}{U_i} - \frac{1}{T} + \frac{2}{3} \left(\frac{E_2}{U_i^2} - \frac{E_2}{T^2} \right) \right], \qquad (54.229)$$

$$Q_i^{\text{int}} = -\frac{\pi e^4}{T_i} \left[\frac{2\Phi'}{T + U_i} \ln \frac{T}{U_i} \right], \qquad (54.230)$$

where Φ' can be approximated by [33]

$$\Phi' = \cos \left[\left(\frac{R_\infty}{E_1 + U_i} \right)^{1/2} \ln \frac{E_1}{U_i} \right]. \qquad (54.231)$$

and E_1 is defined in [33].

The sum of (54.229) and (54.230) yields

$$Q_i = \frac{\pi e^4}{T_i} \left[\frac{1}{U_i} - \frac{1}{T} + \frac{2}{3} \left(\frac{E_2}{U_i} - \frac{E_2}{T^2} \right) - \frac{\Phi'}{T + U_i} \ln \frac{T}{U_i} \right], \qquad (54.232)$$

which is also obtained by integrating the expression (54.223) for σ_E,

$$Q_i = \int_{U_i}^{\frac{1}{2}(T + U_i)} \sigma_E \, dE. \qquad (54.233)$$

Ionization Rate Coefficients. For heavy particle impact [28],

$$\langle Q \rangle = \frac{a_0^2}{\kappa^2} \left\{ \frac{128}{9} \left(\kappa^3 b^3 - b^{3/2} \right) + \frac{1}{3} \lambda b \left(35 - \frac{58}{3} b - \frac{8}{3} b^2 \right) \right.$$
$$+ \frac{2}{3} \kappa a b \left[(5 - 4\kappa^2)(3a^2 + \frac{3}{2} ab + b^2) \right.$$
$$\left. - \lambda \kappa (\frac{15}{2} + 9a + 5b) \right] - 16\kappa a^4 \ln(4\kappa^2 + 1)$$
$$\left. + \theta \left[\frac{35}{6} - \kappa^2 a (\frac{5}{2} + 3a + 4a^2 + 8a^3) \right] \right\}, \qquad (54.234)$$

where

$$\kappa = v_1/v_0, \quad \lambda = \kappa - (4\kappa)^{-1}, \quad \theta = \pi + 2\tan^{-1}\lambda,$$
$$a = (1 + \kappa^2)^{-1}, \quad b = (1 + \lambda^2)^{-1}.$$

$$\langle \sigma_{E,P} \, dP \, dE \rangle = \frac{64 e^4 v_0^5}{3 v_1^2 P^4} \left(\left| \frac{E}{P} - \frac{P}{2m_e} \right|^2 + v_0^2 \right)^{-3} dP \, dE, \qquad (54.235)$$

where $\frac{1}{2} m_e v_0^2$ is the ionization energy of H(1s).

Scaling Laws. Given the binary encounter cross section for ionization by protons of energy E_1 of an atom with binding energy u_a, the cross section for ionization of an atom with different binding energy u_b and scaled proton energy E_1' can be determined to be [18]

$$\sigma_{\text{ion}}(E_1', u_b) = \left(\frac{u_a^2}{u_b^2}\right) \sigma_{\text{ion}}(E_1, u_a), \qquad (54.236)$$

$$E_1' = (u_b/u_a)E_1, \qquad (54.237)$$

where $\sigma_{\text{ion}}(E, u)$ is the ionization cross section for removal of a single electrom from an atom with binding energy u by impact with a proton with initial energy E.

Double Ionization. See Ref. [34] for binary encounter cross section formulae for the direct double ionization of two-electron atoms by electron impact.

Excitation

Excitation is generally less violent than ionization and hence binary encounter theory is less applicable. Binary encounter theory can be applied to *exchange excitation* transitions, e.g. $e^- + \text{He}(n^1 L) \rightarrow e^- + \text{He}(n'^3 L)$, with the restriction of large incident electron velocities. The cross section is

$$
\begin{aligned}
Q_e^{\text{ex}} &= \int_{U_n}^{U_{n+1}} \sigma_{E,\text{ex}} dE \\
&= \frac{\pi e^4}{T_i} \left\{ \frac{1}{\mathcal{T}_{n+1}} - \frac{1}{\mathcal{T}_n} + \frac{2}{3}\left[\frac{E_2}{\mathcal{T}_{n+1}^2} - \frac{E_2}{\mathcal{T}_n^2} \right] \right\},
\end{aligned}
$$
$$(54.238)$$

valid for $T \geq U_{n+1}$, with $\mathcal{T}_n \equiv T + U_i - U_n$ and $\mathcal{T}_{n+1} \equiv T + U_i + U_{n+1}$, or

$$
\begin{aligned}
Q_e^{\text{ex}} &= \int_{U_n}^{T} \sigma_{E,\text{ex}} dE \\
&= \frac{\pi e^4}{T_i} \left\{ \frac{1}{U_i} - \frac{1}{\mathcal{T}_n} + \frac{2}{3}\left[\frac{E_2}{U_i^2} - \frac{E_2}{\mathcal{T}_n^2} \right] \right\},
\end{aligned}
$$
$$(54.239)$$

valid for $U_n \leq T \leq U_{n+1}$. U_n and U_{n+1} denote the excitation energies for levels n and $n + 1$, respectively.

54.8.3 Classical Ionization Cross Section

Applying the classical energy-change cross section result (54.69) of Gerjuoy [16] to the case of electron-impact ionization yields the four cases [17]

$$
\begin{aligned}
\sigma_{\text{ion}}(v_1, v_2) &\sim \int_{\Delta E_\ell}^{\Delta E_u} \sigma_{\Delta E}^{\text{eff}}(v_1, v_2; m_1/m_2)\, d(\Delta E) \\
&= \frac{\pi (Z_1 Z_2 e^2)^2}{3 v_1^2 v_2}\left[\frac{-2v_2^3}{(\Delta E)^2} - \frac{6v_2}{m_2 \Delta E} \right],
\end{aligned}
$$
$$(54.240)$$

which is valid for $0 < \Delta E < b$, or

$$
\sigma_{\text{ion}}(v_1, v_2)
$$
$$
= \frac{\pi (Z_1 Z_2 e^2)^2}{3 v_1^2 v_2}\left[\frac{4(v_1 - 2v_1')}{m_1^2(v_1 - v_1')^2} + \frac{4(v_2 - 2v_2')}{m_2^2(v_2 - v_2')^2} \right], (54.241)
$$

which is valid for $b < \Delta E < a$, or

$$
\sigma_{\text{ion}}(v_1, v_2) = \frac{\pi (Z_1 Z_2 e^2)^2}{3 v_1^2 v_2}\left[\frac{-2v_1'^3}{(\Delta E)^2} \right], \qquad (54.242)
$$

which is valid for $\Delta E > a$, $2m_2 v_2 > |m_1 - m_2|v_1$, or otherwise is zero for $\Delta E > a$, $2m_2 v_2 < |m_1 - m_2|v_1$.

The limits $\Delta E_{\ell,u}$ to the ΔE integration in each of the four cases is indicated in the appropriate validity conditions. The constants a and b above are given by

$$
a = \frac{4m_1 m_2}{(m_1 + m_2)^2}\left[E_1 - E_2 + \tfrac{1}{2}v_1 v_2 (m_1 - m_2) \right],
$$

$$
b = \frac{4m_1 m_2}{(m_1 + m_2)^2}\left[E_1 - E_2 - \tfrac{1}{2}v_1 v_2 (m_1 - m_2) \right].
$$

The expressions above for $\sigma_{\text{ion}}(v_1, v_2)$ must be averaged over the speed distribution of v_2 before comparison with experiment. See [17] for explicit formulae for the case of a delta function speed distribution.

54.8.4 Classical Charge Transfer Cross Section

Applying the classical energy-change cross section result (54.69) of Gerjuoy [16] to the case of charge-transfer yields the four cases [17]

$$
\begin{aligned}
\sigma_{\text{CX}}(v_1, v_2) &\sim \int_{\Delta E_\ell}^{\Delta E_u} \sigma_{\Delta E}^{(\text{eff})}(\mathbf{v}_1, \mathbf{v}_2)\, d\Delta E \\
&= \frac{\pi e^4}{3 v_1^2 v_2}\left[-\frac{2v_2^3}{(\Delta E)^2} - \frac{6v_2/m_2}{\Delta E} \right] (54.243)
\end{aligned}
$$

which is valid for $0 < \Delta E < b$, or

$$
\sigma_{\text{CX}}(v_1, v_2) = \frac{\pi e^4}{3 v_1^2 v_2}\left[3\frac{v_1/m_1 - v_2/m_2}{\Delta E} \right.
$$
$$
\left. + \frac{(v_2'^3 - v_2^3) - (v_1^3 + v_1'^3)}{(\Delta E)^2} \right], \qquad (54.244)
$$

which is valid for $b < \Delta E < a$, or

$$
\sigma_{\text{CX}}(v_1, v_2) = \frac{\pi e^4}{3 v_1^2 v_2}\left(-\frac{2v_1'^2}{(\Delta E)^2} \right), \qquad (54.245)
$$

which is valid for $\Delta E > a$, $m_e v_2 > (m_1 - m_e)v_1$, or otherwise $\sigma_{\text{CX}} = 0$ when $\Delta E > a$, $m_e v_2 < (m_1 - m_e)v_1$. The above expressions for $\sigma_{\text{CX}}(v_1, v_2)$ must be averaged over the speed distribution $W(v_2)$ before comparison with experiment. See [17] for details. The constants a and b above are as defined in Sect. 54.8.3, and the limits $\Delta E_{\ell,u}$ are given by

$$
\Delta E_\ell = \tfrac{1}{2}m_e v_1^2 + U_A - U_B, \qquad (54.246a)
$$
$$
\Delta E_u = \tfrac{1}{2}m_e v_1^2 + U_A + U_B, \qquad (54.246b)
$$
$$
v_2 = \sqrt{2U_A/m_e}, \qquad (54.246c)
$$

where $U_{A,B}$ are the binding energies of atoms A and B. The expressions above for σ_{CX} diverges for some $v_1 > 0$ if $U_A < U_B$. If $U_A = U_B$ then σ_{CX} diverges at $v_1 = 0$. To avoid the divergence, employ *Gerjuoy's modification*, $\Delta E_\ell = \tfrac{1}{2}m_e v_1^2 + U_A$.

54.9 BORN APPROXIMATION

See reviews [35, 36], as well as any standard textbook on scattering theory, for background details on the Born approximation, and [37–40] for extensive tables of Born cross sections.

54.9.1 Form Factors

The basic formulation of the first Born approximation for high energy heavy particle scattering is discussed in Sect. 51.2. For the general atom-atom or ion-atom scattering process

$$A(i) + B(i') \to A(f) + B(f'), \qquad (54.247)$$

with nuclear charges Z_A and Z_B respectively, let $\hbar\mathbf{K}_i$ and $\hbar\mathbf{K}_f$ be the initial and final momenta of the projectile A, and $\hbar\mathbf{q} = \hbar\mathbf{K}_f - \hbar\mathbf{K}_i$ be the momentum transferred to the target. Then Eq. (51.6) can be written in the generalized form

$$\sigma_{if}^{i'f'} = \frac{8\pi a_0^2}{s^2} \int_{t_-}^{t_+} \frac{dt}{t^3} \left| Z_A \delta_{if} - F_{if}^A(t) \right|^2$$
$$\times \left| Z_B \delta_{i'f'} - F_{i'f'}^B(t) \right|^2, \qquad (54.248)$$

where the momentum transfer is $t = qa_0$, and $s = v/v_B$ is the initial relative velocity in units of v_B. The form factors are

$$F_{if}^A(t) = \left\langle \Phi_f^A \left| \sum_{k=1}^{N_A} \exp(it \cdot \mathbf{r}_a/a_0) \right| \Phi_i^A \right\rangle, \qquad (54.249)$$

where N_A is the number of electrons associated with atom A, and similarly for $F_{if}^B(t)$. The limits of integration are $t_\pm = |K_f \pm K_i| a_0$. For heavy particle collisions, $t_+ \sim \infty$ and

$$t_- = (K_i - K_f) a_0 \simeq \frac{\Delta E_{if}}{2s} \left(1 + \frac{m_e \Delta E_{if}}{4Ms^2} \right), \qquad (54.250)$$

where $M = M_A M_B/(M_A + M_B)$.

Limiting Cases. As discussed in Sect. 51.2, for the case $i = f$, $F_{if}^A(t) \to N_A$ as $t \to 0$, so that $Z_A - F_{if}^A(t) \to Z_A - N_A$. For the case $i \neq f$, $F_{if}^A(t) \to 0$ as $t \to 0$ and $t \to \infty$.

54.9.2 Hydrogenic Form Factors

Bound-Bound Transitions. In terms of $\tau = t/Z$,

$$|F_{1s,1s}| = \frac{16}{(4 + \tau^2)^2} \qquad (54.251a)$$

$$|F_{1s,2s}| = 2^{17/2} \frac{\tau^2}{(4\tau^2 + 9)^3} \qquad (54.251b)$$

$$|F_{1s,2p}| = 2^{15/2} \frac{3\tau}{(4\tau^2 + 9)^3} \qquad (54.251c)$$

$$|F_{1s,3s}| = 2^4 3^{7/2} \frac{(27\tau^2 + 16)\tau^2}{(9\tau^2 + 16)^4} \qquad (54.251d)$$

$$|F_{1s,3p}| = 2^{11/2} 3^3 \frac{(27\tau^2 + 16)\tau}{(9\tau^2 + 16)^4} \qquad (54.251e)$$

$$|F_{1s,3d}| = 2^{17/2} 3^{7/2} \frac{\tau^2}{(9\tau^2 + 16)^4}. \qquad (54.251f)$$

Bound-Continuum Transitions. In terms of the scaled wave vector $\kappa = ka_0/Z$ for the ejected electron,

$$|F_{1s,\kappa}|^2 = \frac{2^8 \kappa \tau^2 (1 + 3\tau^2 + \kappa^2) \exp(-2\theta/\kappa)}{3 \left[1 + (\tau - \kappa)^2 \right]^3 \left[1 + (\tau + \kappa)^2 \right]^3 \left(1 - e^{-2\pi/\kappa} \right)} \qquad (54.252)$$

where $\theta = \tan^{-1}[2\kappa/(1 + \tau^2 - \kappa^2)]$. Expressions for the bound-continuum Form Factors for the L-shell ($2\ell \to \kappa$) and M-shell ($3\ell \to \kappa$) transitions can be found in Refs. [41] and [42], respectively. See also §4 of [43] for further details.

General Expressions and Trends:

For final ns states

$$|F_{1s,ns}|^2 = \frac{2^4 n \left[(n-1)^2 + n^2\tau^2 \right]^{n-1}}{\tau^2 \left[(n+1)^2 + n^2\tau^2 \right]^{n+1}}$$
$$\times \sin^2(n \tan^{-1} x + \tan^{-1} y), \qquad (54.253)$$

where

$$x = \frac{2\tau}{n(\tau^2 + 1 - n^{-2})}, \quad y = \frac{2\tau}{\tau^2 - 1 + n^{-2}}. \qquad (54.254)$$

For final $n\ell$ states [44]

$$F_{1s,n\ell}(\tau)$$
$$= (i\tau)^\ell 2^{3(\ell+1)} \sqrt{2\ell+1} (\ell+1)! \left(\frac{(n-\ell-1)!}{(n+\ell)!} \right)^{1/2}$$
$$\times n^{\ell+1} \frac{\left[(n-1)^2 + n^2\tau^2 \right]^{(n-\ell-3)/2}}{\left[(n+1)^2 + n^2\tau^2 \right]^{(n+\ell+3)/2}} \left\{ a_{n\ell} C_{n-\ell-1}^{(\ell+2)}(x) \right.$$
$$\left. - b_{n\ell} C_{n-\ell-2}^{(\ell+2)}(x) + c_{n\ell} C_{n-\ell-3}^{(\ell+2)}(x) \right\}, \qquad (54.255)$$

with coefficients $a_{n\ell}$, $b_{n\ell}$ and $c_{n\ell}$ given by

$$a_{n\ell} = (n+1) \left[(n-1)^2 + n^2\tau^2 \right],$$
$$b_{n\ell} = 2n \sqrt{\left[(n-1)^2 + n^2\tau^2 \right] \left[(n+1)^2 + n^2\tau^2 \right]},$$
$$c_{n\ell} = (n-1) \left[(n+1)^2 + n^2\tau^2 \right],$$

and argument

$$x = \frac{n^2 - 1 + n^2\tau^2}{\sqrt{\left[(n+1)^2 + n^2\tau^2 \right] \left[(n-1)^2 + n^2\tau^2 \right]}}.$$

Summation over final ℓ states:

$$|F_{1s,n}|^2 = \sum_\ell |F_{1s,n\ell}|^2$$

$$= 2^8 n^7 \tau^2 \left[\tfrac{1}{3}(n^2 - 1) + n^2\tau^2 \right]$$

$$\times \frac{\left[(n-1)^2 + n^2\tau^2\right]^{n-3}}{\left[(n+1)^2 + n^2\tau^2\right]^{n+3}} \, . \qquad (54.257)$$

which becomes for large n,

$$|F_{1s,n}|^2 \sim \frac{2^8 \tau^2 (3\tau^2 + 1)}{3n^3 (\tau^2 + 1)^6} \exp\left(\frac{-4}{(\tau^2 + 1)}\right) \, . \qquad (54.258)$$

For initial $2s$ and $2p$ states,

$$|F_{2s,n}|^2 = 2^4 n^7 \tau^2 \left[-\tfrac{1}{3} + \tfrac{1}{2}n^2 - \tfrac{3}{16}n^4 + \tfrac{1}{48}n^6 \right.$$

$$+ n^2\tau^2 (\tfrac{1}{3} - \tfrac{2}{3}n^2 + \tfrac{19}{48}n^4) + n^4\tau^4 (\tfrac{5}{3} - \tfrac{7}{6}n^2)$$

$$\left. + n^6\tau^6 \right] \frac{\left[(\tfrac{1}{2}n - 1)^2 + n^2\tau^2\right]^{n-4}}{\left[(\tfrac{1}{2}n + 1)^2 + n^2\tau^2\right]^{n+4}} \, , \qquad (54.259)$$

$$|F_{2p,n}|^2 = \frac{2^4 n^9 \tau^2}{3} \left[\tfrac{1}{4} - \tfrac{7}{24}n^2 + \tfrac{11}{192}n^4 \right.$$

$$- n^2\tau^2 (\tfrac{5}{6} - \tfrac{23}{24}n^2) + \tfrac{1}{4}n^4\tau^4 \right]$$

$$\times \frac{\left[(\tfrac{1}{2}n - 1)^2 + n^2\tau^2\right]^{n-4}}{\left[(\tfrac{1}{2}n + 1)^2 + n^2\tau^2\right]^{n+4}} \, . \qquad (54.260)$$

Power Series Expansion. $\tau^2 \ll 1$ [4]

$$|F_{1s,ns}(\tau)|^2 = A(n)\tau^4 + B(n)\tau^6 + C(n)\tau^8 + \cdots , \qquad (54.261)$$

where

$$A(n) = \frac{2^8 n^9 (n-1)^{2n-6}}{3^2 (n+1)^{2n+6}} \, ,$$

$$B(n) = -\frac{2^9 n^{11} (n^2 + 11)(n-1)^{2n-8}}{3^2 5 (n+1)^{2n+8}} \, ,$$

$$C(n) = -\frac{2^8 n^{13} (313n^4 - 1654n^2 - 2067)(n-1)^{2n-10}}{3^2 5^2 7 (n+1)^{2n+10}} \, .$$

For analytical expressions for $A(n)$, $B(n)$ and $C(n)$ for final np and nd states see Refs. [45, 46].

General Trends in Hydrogenic Form Factors [47].

The inelastic form factor $|F_{n\ell \to n'\ell'}|$ oscillates with ℓ' on an increasing background until the value

$$\ell'_{max} = \min\left\{ (n'-1), n\left(\frac{2(n+3)}{(n+1)}\right)^{1/2} - \frac{1}{2} \right\} \qquad (54.262)$$

is reached, after which a rapid decline for $\ell > \ell'_{max}$ occurs. See [47] for illustrative graphs.

54.9.3 Excitation Cross Sections

Atom-Atom Collisions [48]

Single Excitation. For the process

$$A(i) + B \to A(f) + B \, , \qquad (54.263)$$

Eq. (54.248) reduces to

$$\sigma_{if} = \frac{8\pi a_0^2}{s^2} \int_{t_-}^{\infty} \frac{dt}{t^3} \left|F_{if}^A\right|^2 \left|Z_B - F_{i'i'}^B\right|^2 \, . \qquad (54.264)$$

Double Excitation. For the process

$$\mathrm{H}(1s) + \mathrm{H}(1s) \to \mathrm{H}(n\ell) + \mathrm{H}(n'\ell') \, , \qquad (54.265)$$

$$\sigma_{1s,n\ell}^{1s,n'l'} = \frac{8\pi a_0^2}{s^2} \int_{t_-}^{\infty} \frac{dt}{t^3} |F_{1s,n\ell}|^2 |F_{1s,n'\ell'}|^2 \, . \qquad (54.266)$$

Special cases are [49]

$$\sigma_{1s,2s}^{1s,2s} = \frac{2^{30}\pi a_0^2 (880t_-^4 + 396t_-^2 + 81)}{495 s^2 (4t_-^2 + 9)^{11}} \, , \qquad (54.267a)$$

$$\sigma_{1s,2p}^{1s,2p} = \frac{2^{30} 3^4 \pi a_0^2}{11 s^2 (4t_-^2 + 9)^{11}} \, , \qquad (54.267b)$$

$$\sigma_{1s,2s}^{1s,2p} = \frac{2^{29} 3^2 (44t_-^2 + 9)}{55 s^2 (4t_-^2 + 9)^{11}} \, , \qquad (54.267c)$$

with $t_-^2 = [9/(16s^2)][1 + 3m_e/(4Ms^2) + \cdots]$.

Ion-Atom Collisions

For the proton impact process

$$\mathrm{H}^+ + \mathrm{H}(1s) \to \mathrm{H}^+ + \mathrm{H}(n\ell) \, , \qquad (54.268)$$

Eq. (54.248) reduces to

$$\sigma_{1s,n\ell} = \frac{8\pi a_0^2}{s^2} \int_{t_-}^{\infty} \frac{dt}{t^3} |F_{1s,n\ell}(t)|^2 \qquad (54.269)$$

with $t_- = (1 - n^{-2})/(2s)$.

Asymptotic Expansions.

$$\sigma_{1s,ns} = \frac{4\pi a_0^2 (n^2-1)|X_{1s \to ns}|^2}{24 s^2 n^2} \left[C_s(n) - \frac{1}{s^2} + \frac{n^2+11}{20n^2 s^4} \right.$$

$$\left. + \frac{313n^4 - 1654n^2 - 2067}{8400 n^4 s^6} \right] \, , \qquad (54.270)$$

$$\sigma_{1s,np} = \frac{2^{10}\pi a_0^2 n^7 (n-1)^{2n-5}}{3 s^2 (n+1)^{2n+5}} \left[C_p(n) + \ln s^2 + \frac{n^2+11}{10n^2 s^2} \right.$$

$$\left. + \frac{313n^4 - 1654n^2 - 2067}{5600 n^4 s^4} \right] \, , \qquad (54.271)$$

$$\sigma_{1s,nd} = \frac{2^{11}\pi a_0^2 (n^2-4)n^5 (n^2-1)^2 (n-1)^{2n-7}}{3^2 5 s^2 (n+1)^{2n+7}}$$

$$\times \left[C_d(n) - \frac{1}{s^2} + \frac{11n^2 + 13}{28 n^2 s^4} \right] \, , \qquad (54.272)$$

where $C_s(2) = 16/5$, $C_s(3) = 117/32$, $C_s(4) \approx 3.386$, and

$$\gamma_n C_p(n) = \frac{1.3026}{n^3} + \frac{1.7433}{n^5} + \frac{16.918}{n^7}, \quad (54.273)$$

$$\gamma_n C_d(n) = \frac{2.0502}{n^3} + \frac{7.6125}{n^5}, \quad (54.274)$$

with

$$\gamma_n \equiv \frac{2^8 n^7 (n-1)^{(2n-5)}}{3(n+1)^{(2n+5)}}. \quad (54.275)$$

Further asymptotic expansion results can be found in Refs. [3–6].

A general expression for Born excitation and ionization cross sections for hydrogenic systems in terms of a parabolic coordinate representation (see Chap. 9) is given in Ref. [50].

Number of Independent Transitions $\mathcal{N}_i$ between levels n and n' [50].

$$\mathcal{N}_i = n^2 \left[n' - \left(\frac{n}{3} \right) \right] + \left(\frac{n}{3} \right) \quad (54.276)$$

Validity Criterion. The Born approximation is valid provided that [51]

$$nE \gg \ln \left(\frac{4E}{J_n} \right) \quad (54.277)$$

for transitions $n \to n'$ when n, $n' \gg 1$ and $|n - n'| \sim 1$. The constant J_n is undetermined (see [51] for details) but is generally taken to be the ionization potential of level n.

54.9.4 Ionization Cross Sections

$$e^-(k) + H \to e^-(k') + H^+ + e^-(\kappa) \quad (54.278)$$

The general expression for the Born differential ionization cross section can be evaluated in closed form using screened hydrogenic wave functions. The differential cross section per incident electron scattered into solid angle $d\Omega_{k'}$, integrated over directions κ for the ejected electron (treated as distinguishable) is [52, 53]

$$I(\theta, \phi) \, d\Omega_{k'} \, d\kappa' = \frac{4k'}{kq^4 a_0 \tilde{Z}_B^4} |F_{n\ell, \kappa'}(q)|^2 \, d\Omega_{k'} \, d\kappa', \quad (54.279)$$

where the form factor is given by Eq. (54.253) for the case $n\ell = 1s$, with the ejected electron wavenumber κ and momentum transferred q in the collision, $\kappa' = \kappa a_0 / \tilde{Z}_B$, $\mathbf{q} = (\mathbf{k}' - \mathbf{k}) a_0 / \tilde{Z}_B$, being scaled by the screened nuclear charge $\tilde{Z}_B$ appropriate to the $n\ell$-shell from which the electron is ejected. The total Born ionization cross section per electron is

$$\sigma_{\text{ion}}^B = \int_0^{\kappa_{\max}} d\kappa' \int_{k-k'}^{k+k'} I(q, \kappa') \, dq, \quad (54.280)$$

which is generally evaluated numerically.

54.9.5 Capture Cross Sections

Electron Capture.

$$A^+ + B(n\ell) \to A(n'\ell') + B^+. \quad (54.281)$$

In the Oppenheimer-Brinkman-Kramers (OBK) approximation [54],

$$\sigma_{n\ell, n'\ell'} = \frac{M^2}{2\pi\hbar^3} \frac{v_f}{v_i} \int_{-1}^1 d(\cos\theta) \, |F_{n\ell \to n'\ell'}|^2, \quad (54.282)$$

where $\mathbf{v}_i = v_i \hat{\mathbf{n}}_i$, $\mathbf{v}_f = v_f \hat{\mathbf{n}}_f$, θ is the angle between $\hat{\mathbf{n}}_i$ and $\hat{\mathbf{n}}_f$, $M = M_A M_B / (M_A + M_B)$, and

$$|F_{n\ell, n'\ell'}| = \int\int d\mathbf{r} \, d\mathbf{s} \, \varphi_i(\mathbf{r}) \varphi_f^*(\mathbf{s}) \left(\frac{Z_A e^2}{r} \right) e^{i(\boldsymbol{\alpha} \cdot \mathbf{r} + \boldsymbol{\beta} \cdot \mathbf{s})}, \quad (54.283)$$

with

$$\boldsymbol{\alpha} = k_f \hat{\mathbf{n}}_f + k_i \hat{\mathbf{n}}_i \frac{M_A}{M_A + m_e},$$

$$\boldsymbol{\beta} = -k_i \hat{\mathbf{n}}_i - k_f \hat{\mathbf{n}}_f \frac{M_B}{M_B + m_e},$$

$$k_i = \frac{v_f}{\hbar} \frac{M_B (M_A + m_e)}{(M_A + M_B + m_e)},$$

$$k_f = \frac{v_f}{\hbar} \frac{(M_B + m_e) M_A}{(M_A + M_B + m_e)}.$$

The Jackson-Schiff correction factor [55] is

$$\gamma_{\text{JS}} = \frac{1}{192} \left(127 + \frac{56}{p^2} + \frac{32}{p^4} \right)$$
$$- \frac{\tan^{-1} \frac{1}{2} p}{48p} \left(83 + \frac{60}{p^2} + \frac{32}{p^4} \right)$$
$$+ \frac{(\tan^{-1} \frac{1}{2} p)^2}{24p^2} \left(31 + \frac{32}{p^2} + \frac{16}{p^4} \right), \quad (54.284)$$

and the capture cross section is

$$\sigma(n_i \ell_i, n_f \ell_f) = \frac{\gamma_{\text{JS}} \pi a_0^2}{p^2} C(n_i \ell_i, n_f \ell_f)$$
$$\times \int_x^\infty F(n_i \ell_i, n_f \ell_f; x) \, dx, \quad (54.285)$$

with

$$p = \frac{m v_i}{\hbar}, \quad a = \frac{Z_A}{n_i}, \quad b = \frac{Z_B}{n_f},$$
$$x = \left[p^2 + (a+b)^2 \right] \left[p^2 + (a-b)^2 \right] / 4p^2.$$

The coefficients C in Eq. (54.285) are given in Table 54.2, while the functions F are given in Table 54.3 [54]. In Table 54.3, the appropriate value of a and b is indicated by the quantum numbers n_i, ℓ_i and n_f, ℓ_f.

Table 54.2. Coefficients $C(n_i\ell_i \to n_f\ell_f)$ in the Born capture cross section formula (54.285).

$n_f\ell_f$	$C(1s \to n_f\ell_f)$
$1s$	$2^8 Z_A^5 Z_B^5$
$2s$	$2^5 Z_A^5 Z_B^5$
$2p$	$2^5 Z_A^5 Z_B^7$
$3s$	$2^8 Z_A^5 Z_B^5 / 3^3$
$3p$	$2^{13} Z_A^5 Z_B^7 / 3^6$
$3d$	$2^{15} Z_A^5 Z_B^9 / 3^9$
$4s$	$2^2 Z_A^5 Z_B^5$
$4p$	$5 Z_A^5 Z_B^7$
$4d$	$Z_A^5 Z_B^9$
$4f$	$Z_A^5 Z_B^{11} / 20$

$$C(2s \to n_f\ell_f) = C(1s \to n_f\ell_f)/8$$
$$C(2p \to n_f\ell_f) = C(1s \to n_f\ell_f)/24$$

Table 54.3. Functions $F(n_i\ell_i \to n_f\ell_f; x)$ in the Born capture cross section formula (54.285).

$n_f\ell_f$	$F(n_i\ell_i, n_f\ell_f; x)$
$1s$	x^{-6}
$2s$	$(x - 2b^2)^2 x^{-8}$
$2p$	$(x - b^2) x^{-8}$
$3s$	$(x^2 - \frac{16}{3}b^2 x + \frac{16}{3}b^4)^2 x^{-10}$
$3p$	$(x - b^2)(x - 2b^2)^2 x^{-10}$
$3d$	$(x - b^2)^2 x^{-10}$
$4s$	$(x - 2b^2)^2 (x^2 - 8b^2 x + 8b^4)^2 x^{-12}$
$4p$	$(x - b^2)(x^2 - \frac{24}{5}b^2 x + \frac{24}{5}b^4)^2 x^{-12}$
$4d$	$(x - b^2)^2 (x - 2b^2)^2 x^{-12}$
$4f$	$(x - b^2)^3 x^{-12}$

$$F(2s, n_f\ell_f; x) = (x - 2a^2)^2 x^{-2} F(1s, n_f\ell_f; x)$$
$$F(2p, n_f\ell_f; x) = (x - a^2) x^{-2} F(1s, n_f\ell_f; x)$$

Acknowledgments

The author thanks Prof. M. R. Flannery for many helpful discussions and Prof. E. W. McDaniel for access to his collection of reprints. The author would also like to thank Dr. D. R. Schultz of ORNL for the time necessary to complete this work.

This work was begun at the Georgia Institute of Technology (GIT) and completed at the Controlled Fusion Atomic Data Center at Oak Ridge National Laboratory (ORNL). The work at GIT was supported by US AFOSR grant No. F49620-94-1-0379. The work at ORNL was supported by the Office of Fusion Energy, US DOE contract No. DE-AC05-84OR21400 with Lockheed-Martin Energy Systems, Inc. and by ORNL Research Associates Program administered jointly by ORNL and ORISE.

REFERENCES

1. I. L. Beigman and V. S. Lebedev, Phys. Rep. **250**, 95 (1995).
2. H. A. Bethe and E. E. Salpeter, *Quantum Mechanics of One- and Two-Electron Atoms* (Springer-Verlag, Berlin, 1957).
3. G. S. Khandelwal and B. H. Choi, J. Phys. B **1**, 1220 (1968).
4. G. S. Khandelwal and E. E. Fitchard, J. Phys. B **2**, 1118 (1969).
5. G. S. Khandelwal and J. E. Shelton, J. Phys. B **4**, L109 (1971).
6. G. S. Khandelwal and B. H. Choi, J. Phys. B **2**, 308 (1969).
7. W. Gordon, Ann. Phys. (Leipzig) **2**, 1031 (1929).
8. V. A. Davidkin and B. A. Zon, Opt. Spectrosk. **51**, 25 (1981).
9. L. A. Bureeva, Astron. Zh. **45**, 1215 (1968).
10. H. Griem, Astrophys. J. **148**, 547 (1967).
11. M. R. Flannery, in *Rydberg States of Atoms and Molecules*, edited by R. F. Stebbings and F. B. Dunning (Cambridge University Press, 1983), Ch. 11.
12. A. Burgess and I. C. Percival, Adv. At. Mol. Phys. **4**, 109 (1968).
13. I. C. Percival, in *Atoms and Molecules in Astrophysics*, edited by T. R. Carson and M. J. Roberts (Academic Press, New York, 1972), p. 65.
14. I. C. Percival and D. Richards, Adv. At. Mol. Phys. **11**, 1 (1975).
15. M. Matsuzawa, J. Phys. B **8**, 2114 (1975).
16. E. Gerjuoy, Phys. Rev. **148**, 54 (1966).
17. J. D. Garcia, E. Gerjuoy, and J. E. Welker, Phys. Rev. **165**, 66 (1968).
18. D. R. Bates and A. E. Kingston, Adv. At. Mol. Phys. **6**, 269 (1970).
19. J. M. Khan and D. L. Potter, Phys. Rev. **133**, A890 (1964).
20. V. S. Lebedev and V. S. Marchenko, Sov. Phys. JETP **61**, 443 (1985).
21. A. Omont, J. de Phys. **38**, 1343 (1977).
22. B. Kaulakys, J. Phys. B **17**, 4485 (1984).
23. V. S. Lebedev, J. Phys. B: At. Mol. Opt. Phys. **25**, L131 (1992); Soviet Phys. (JETP) **76**, 27 (1993).
24. J. P. Coleman, in *Case Studies in Atomic Collision Physics I*, edited by E. W. McDaniel and M. R. C. McDowell (North Holland, Amsterdam, 1969), Ch. 3.
25. J. P. Coleman, J. Phys. B **1**, 567 (1968).
26. M. R. Flannery, Phys. Rev. A **22**, 2408 (1980).
27. M. L. Goldberger and K. M. Watson, *Collision Theory* (Wiley, New York, 1964), ch. 11.
28. M. R. C. McDowell, Proc. Phys. Soc. **89**, 23 (1966).
29. D. R. Bates and J. C. G. Walker, Planetary Spac. Sci. **14**, 1367 (1966).
30. D. R. Bates and J. C. G. Walker, Proc. Phys. Soc. **90**, 333 (1967).
31. M. R. Flannery, Ann. Phys. (NY) **61**, 465 (1970), *ibid*. **79**, 480 (1973).
32. L. Vriens, in *Case Studies in Atomic Collision Physics I*, edited by E. W. McDaniel and M. R. C. McDowell (North Holland, Amsterdam, 1969), Ch. 6.

33. L. Vriens, Proc. Phys. Soc. **89**, 13 (1966).

34. B. N. Roy and D. K. Rai, J. Phys. B **5**, 816 (1973).

35. A. R. Holt and B. Moiseiwitsch, Adv. At. Mol. Phys. **4**, 143 (1968).

36. K. L. Bell and A. E. Kingston, Adv. At. Mol. Phys. **10**, 53 (1974).

37. L. C. Green, P. P. Rush, and C. D. Chandler, Astrophys. J. Suppl. Ser. **3**, 37 (1957).

38. W. B. Sommerville, Proc. Phys. Soc. **82**, 446 (1963).

39. A. Burgess, D. G. Hummer, and J. A. Tully, Phil. Trans. Roy. Soc. A **266**, 255 (1970).

40. C. T. Whelan, J. Phys. B **19**, 2343 and 2355 (1986).

41. M. C. Walske, Phys. Rev. **101**, 940 (1956).

42. G. S. Khandelwal and E. Merzbacher, Phys. Rev. **144**, 349 (1966).

43. E. Merzbacher and H. W. Lewis, *Handbuch der Physik* **34/2** *X-ray Production by Heavy Charged Particles*, pp. 166, Springer-Verlag (1958).

44. H. Bethe, *Quantenmechanik der Ein- und Zwei-Elektronenprobleme*, in *Handbuch der Physik* **24/1** (Springer-Verlag, Berlin, 1933), p. 502.

45. M. Inokuti, Argonne Nat. Lab. Radio Phys. Div. A Report No. ANL-7220 (1965), pp. 1–10.

46. M. Inokuti, Rev. Mod. Phys. **43**, 297 (1971).

47. M. R. Flannery and K. J. McCann, Astrophys. J. **236**, 300 (1980).

48. D. R. Bates and G. Griffing, Proc. Phys. Soc. **66A**, 961 (1953), *ibid.* **67A**, 663 (1954).

49. D. R. Bates and A. Dalgarno, Proc. Phys. Soc. **65A**, 919 (1952).

50. K. Omidvar, Phys. Rev. **140**, A26 and A38 (1965).

51. A. N. Starostin, Sov. Phys. JETP **25**, 80 (1967).

52. E. H. S. Burhop, Proc. Camb. Phil. Soc. **36**, 43 (1940); J. Phys. B **5**, L241 (1972).

53. N. F. Mott and H. S. W. Massey, *The Theory of Atomic Collisions* (Clarendon Press, Oxford, 1965), pp. 489–490.

54. D. R. Bates and A. Dalgarno, Proc. Phys. Soc. **66A**, 972 (1953).

55. J. D. Jackson and H. Schiff, Phys. Rev. **89**, 359 (1953).

<u>55</u>

Mass Transfer at High Energies: Thomas Peak

J. H. McGuire,* Jack C. Straton,† and T. Ishihara‡

*Department of Physics, Tulane University, New Orleans, Louisiana 70118
†Department of Physics, Portland State University, Portland, Oregon 97207-0751
‡Institute of Applied Physics, University of Tsukuba, Tsukuba, Ibaraki 305, Japan

55.1 INTRODUCTION

Transfer of mass is a quasiforbidden process. Simple transfer of a stationary mass M_2 to a moving mass M_1 is forbidden by conservation of energy and momentum. If $M_1 < M_2$ then M_1 rebounds, if $M_1 = M_2$ then M_1 stops and M_2 continues on, and if $M_1 > M_2$ then M_2 leaves faster than M_1. In none of these cases do M_1 and M_2 leave together. L. H. Thomas [1] understood this in 1927 and further realized that transfer of mass occurs only when a third mass is present and all three masses interact. The simplest allowable process is a two-step process now called a Thomas process [2, 3].

Quantum mechanically [4], the second Born term at high energies is the largest Born term and corresponds to the simplest allowed classical process, namely the Thomas process. While the classically forbidden first Born term is not zero (saved by the uncertainty principle), the first Born cross section varies at high v as v^{-12}, in contrast to the second Born cross section which varies as v^{-11}, thus dominating. Higher Born terms correspond

to multistep processes that are unlikely in fast collisions where there is not enough time for complicated processes. The higher Born terms ($n > 2$) are also smaller than the second Born term.

55.2 THE CLASSICAL THOMAS PROCESS

The basic Thomas process is shown in Fig. 55.1. Here, the entire collision is coplanar since particles 1 and 2 go off together (that is what is meant by mass transfer). We assume that all the masses and the incident velocities $\mathbf{v}$ are known. Thus, there are six unknowns, $\mathbf{v}'$, $\mathbf{v}_f$ and $\mathbf{v}_3$, each with two components. Conservation of momentum gives two equations of constraint for each collision. Conservation of overall energy gives a fifth constraint, and conservation of energy in the intermediate state (which only holds classically) gives a sixth constraint. With six equations of constraint, all six unknowns may be completely determined. The allowed values of $\mathbf{v}'$, $\mathbf{v}_f$ and

$$| + (2,3) \longrightarrow (1,2) + 3$$

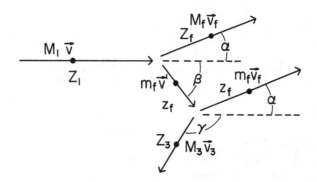

Figure 55.1. Diagram for mass transfer via a Thomas two-step process.

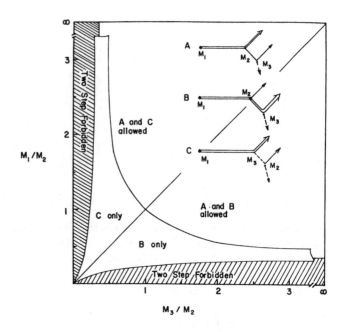

Figure 55.2. Lieber diagram for mass transfer.

$\mathbf{v}_3$ depend on the masses M_1, M_2 and M_3. For example, in the case of the transfer of an electron from atomic hydrogen to a proton, i.e., $p^+ = H \rightarrow H + p^+$, it is easily verified that the angles are $\alpha = (M_2/M_1)\sin 60°$, $\beta = 60°$ and $\gamma = 120°$, where $m' = M_2 = m_e$ and $M_1 = M_3 = M_p = 1836 m_e$. For the case $e^+ + H \rightarrow Ps + p^+$, it may be verified that $\alpha = 45°$, $\beta = 45°$ and $\gamma = 90°$.

In general, the intermediate mass m' may be equal to M_1, M_2 or M_3. We shall regard these as different Thomas processes, and label them B, A, and C respectively. The standard Thomas process (the one actually considered by Thomas in 1927) is case A, and corresponds to the first example given above. Lieber diagrams for the Thomas processes A, B, and C [5] are illustrated in Fig. 55.2. In the Lieber diagram, mass regions in which solutions exist for processes A, B, and C are shown. (An equivalent diagram was given earlier by Detmann and Liebfried [3].) There are some regions in which two-step processes are forbidden. In these regions the theory of mass transfer is not fully understood at present.

55.3 QUANTUM DESCRIPTION

55.3.1 Uncertainty Effects

In quantum mechanics, energy conservation in the intermediate states may be violated within the limits of the uncertainty principle, $\Delta E \geq \hbar/\Delta t$, where Δt is the uncertainty in time of mass transfer. It is not possible to determine if mass transfer actually occurs at the beginning, in the middle, or at the end of the collision. Thus, we choose $\Delta t = \bar{r}/\bar{v}$, where $\bar{r}$ is the size of the collision region and $\bar{v}$ is the mean collision velocity. Taking $\bar{r} \approx a_0/Z_{\text{target}}$ and $\bar{v} \approx v$, the projectile velocity,

we have

$$\Delta E \approx \frac{\hbar}{\Delta t} = \frac{\hbar \bar{v}}{\bar{r}} = \frac{\hbar v}{a_0/Z_{\text{target}}} = \frac{\hbar Z_{\text{target}}}{a_0}. \quad (55.1)$$

Here, a_0 is the Bohr radius and Z_{target} is the nuclear charge of the target in units of the electron charge. Within this range of energy ΔE, the constraint of energy conservation in the intermediate state does not apply.

55.3.2 Conservation of Overall Energy and Momentum

Conservation of overall energy and momentum then gives three equations of constraint on the four unknowns $\mathbf{v}_f$ and $\mathbf{v}_3$ [6], namely,

$$M_1\mathbf{v} = (M_f + \tilde{M}_f)\mathbf{v}_f + M_3\mathbf{v}_3, \quad (55.2)$$

$$M_1 v^2 = (M_f + \tilde{M}_f)v_f^2 + M_3 v_3^2, \quad (55.3)$$

where M_f ($\tilde{M}_f$) is the mass of the upper (lower) particle in the final state of the bound system shown in Fig. 55.1, in which m' is the mass of the intermediate particle M_1, M_2 or M_3. From Eqs. (55.2) and (55.3) it may be shown that the velocity of the recoil particle is constrained by the condition

$$2\hat{\mathbf{v}}_3 \cdot \hat{\mathbf{v}} = 2\cos\gamma = \frac{M_1 + M_2 + M_3}{M_1}\frac{v_3}{v} - \frac{M_2}{M_3}\frac{v}{v_3}. \quad (55.4)$$

Thus, the magnitude and the direction of $\hat{\mathbf{v}}_3$ are not independent. Specifying either v_3 or $\hat{\mathbf{v}}_3$ is sufficient,

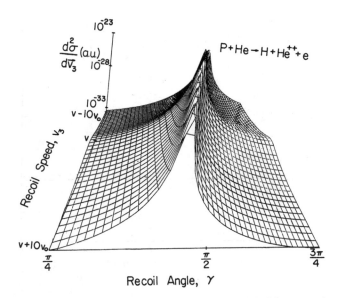

Figure 55.3. Counting rate (or cross section) on the vertical axis versus recoil speed v_3 and recoil angle γ for a Thomas process in which a proton (projectile) picks up an electron from helium (target). The captured electron bounces off the second target electron.

together with the equations of constraint, to determine the energies and directions of all particles in the final state. Similarly, one may express the equations of constraint in terms of v_f and $\hat{\mathbf{v}}_f$.

55.3.3 Conservation of Intermediate Energy

In a classical two-step process, the projectile hits a particle in the target, and the intermediate mass m' then propagates and subsequently undergoes a second collision. Quantum mechanically, this corresponds to a second-Born term represented by $V_1 G_0 V_2$, where V represents an interaction and G_0 is the propagator of the intermediate state, namely,

$$
\begin{aligned}
G_0 &= (E - H_0 + i\epsilon)^{-1} \\
&= -i\pi\delta(E - H_0) + \wp\,\frac{1}{E - H_0}\,,
\end{aligned} \tag{55.5}
$$

where $\wp$ is the Cauchy principal value of G_0 which excludes the singularity at $E = H_0$. This singularity corresponds to conservation of energy in the intermediate state. It is this singularity which gives rise to the weaker secondary ridge in Fig. 55.3 at $v_3 = v$. The width of the secondary ridge is given approximately by $\Delta E = \hbar/\Delta t$, as discussed above. The intersection of the ridges is the Thomas peak. At very high collision velocities, the Thomas peak dominates the total cross section for mass transfer.

The constraint imposed by conservation of intermediate energy may be expressed by replacing the speed of the recoil particle v_3 by the scaled variable $K = M_3 v_3 / m'v$. Then it may be shown that the conservation of intermediate energy may be expressed in the form [6]:

$$
\left(\frac{M_3 v_3}{m'v}\right)^2 \equiv K^2 = 1\,. \tag{55.6}
$$

The constraints of conservation of overall energy and momentum, i.e., Eq. (55.4), may be easily written in terms of K as

$$
2\cos\gamma = r\,\frac{m'}{M_2}\,K - \frac{M_2}{m'}\,\frac{1}{K}\,, \tag{55.7}
$$

where $r = (M_1 + M_2 + M_3)M_2/(M_1 M_3)$.

55.3.4 Example: Proton-Helium Scattering

The effect of the constraints of conservation of overall energy and momentum may be seen in Fig. 55.3, where a sharp ridge is evident in the reaction

$$
\mathrm{p}^+ + \mathrm{He} \rightarrow \mathrm{H} + \mathrm{He}^{++} + \mathrm{e}^-\,, \tag{55.8}
$$

and $M_1 = M_\mathrm{p}$, $M_2 = M_3 = m_\mathrm{e}$. Here v_3 is the speed of the recoiling ionized target electron, and the target nucleus is not directly involved in the reaction. The width of the sharp ridge is due to the momentum spread of the electrons in helium and may be regarded as being caused by the uncertainty principle since this momentum (or velocity) spread corresponds to $\Delta p = \hbar/\Delta r$ where Δr is taken as the radius of the helium atom.

The locus of the sharp ridge in Fig. 55.3, corresponding to conservation of overall energy and momentum, is given by Eq. (55.8). The locus of the weaker ridge, corresponding to the conservation of energy in the intermediate state, is given by Eq. (55.7). The intersection of these two loci gives the unique classical result suggested by Thomas. The width of these ridges may be estimated from the uncertainty principle as described above.

Experimental evidence for the double ridge structure has been reported by Palinkas *et al.* [7] corresponding to the calculations given in Fig. 55.3, but at a collision energy of 1 MeV, as shown below.

The data in Fig. 55.4 corresponds to a slice across the sharp ridge of Fig. 55.3 at $v = v_3$. The solid line is a second Born calculation [8] at 1 MeV. The bump of data above a smooth backgroud is the indication of the ridge structure.

55.4 OFF-ENERGY-SHELL EFFECTS

In Eq. (55.6), the Green's function G_0 contains an energy-conserving term $i\pi\delta(E - H_0)$ which is imaginary,

Figure 55.4. Observation of a slice of the Thomas ridge structure in $p^+ + He \rightarrow H + He^{++} + e^-$ at 1 MeV by Palinkas *et al.* [7].

Figure 55.5. Energy-conserving (on-shell) and energy-nonconserving contributions to the second Born scattering amplitude.

and a real energy-nonconserving term $\wp[1/(E - H_0)]$. The latter does not occur classically; it is permitted by the uncertainty principle and represents the contribution of virtual (off-the-energy-shell or energy-nonconserving) states within $\pm \Delta E = \hbar/\Delta t$ about the classical value $E = H_0$. This quantum term also represents the effect of time-ordering in the second Born amplitude [9]. In plane wave second-Born calculations, the off-energy-shell term gives the real part of the scattering amplitude f_2, while the on-shell (energy conserving) term gives the imaginary part of f_2. These two contributions are shown in Fig. 55.5.

Half of the Thomas peak comes from energy-nonconserving contributions which are not included in a classical description. Also, the energy-nonconserving contribution plays a significant role in determining the shape of the standard Thomas peak, which has been observed [10].

55.5 DISPERSION RELATIONS

Because of the form of the Green's function of Eq. (55.6), the second Born contribution f_2 to the scattering amplitude has a single pole in the lower half of the complex plane. Consequently it obeys the dispersion relation

$$\Re f_2(\lambda) = -\frac{1}{\pi} \wp \int_{-\infty}^{+\infty} \frac{\Im f_2(\lambda')}{\lambda - \lambda'} d\lambda',$$

$$\Im f_2(\lambda) = \frac{1}{\pi} \wp \int_{-\infty}^{+\infty} \frac{\Re f_2(\lambda')}{\lambda - \lambda'} d\lambda'. \tag{55.9}$$

where $\Re f_2$ and $\Im f_2$ denote the real and imaginary parts of f_2. Thus the energy-nonconserving part of f_2 is related to an integral over the energy-conserving part and vice versa. In the case of the dielectric constant it is well known that the real and imaginary parts of ϵ are also related by a dispersion relation, namely the Kramers-Kronig relation [11].

Resonances are usually a function of energy E. The width of a resonance gives the lifetime τ of the resonance. Classically, τ is how long the projectile orbits the target before it leaves, corresponding to a delay or shift in time of the projectile during the interaction. If the width of the resonance is ΔE, then the lifetime is $\tau = \hbar/\Delta E$. E and τ are conjugate variables. The Thomas peak is an overdamped resonance in scattering angle, corresponding to a shift in the impact parameter of the scattering event [12]. However, unlike energy resonances, our Thomas resonance in the scattering angle seems to have no classical analog [13].

55.6 DESTRUCTIVE INTERFERENCE OF AMPLITUDES

It was noted in the Introduction that the location of the Thomas peaks depends on the mass of the collision partners. For the process

$$p^+ + \text{Atom} \rightarrow \text{Atom} + \text{H}, \tag{55.10}$$

there are two separate Thomas peaks [2, 14] corresponding to cases A and B in the Lieber diagram (Fig. 55.2). Experimental evidence exists for both peaks. The standard Thomas peak occurs at small forward angles [10],

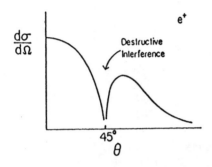

Figure 55.6. Change of position and nature of the Thomas peaks with decreasing projectile mass.

while the second peak [15] occurs at about 60°. If the mass of the projectile is reduced, the positions of these Thomas peaks move toward one another [16] as illustrated in Fig. 55.6.

When $M_1 = M_2$, then both Thomas peaks occur at 45°. This occurs in positronium formation where $M_1 = M_2 = m_e$, i.e.,

$$e^+ + He \rightarrow Ps + He^{++} + e^- . \qquad (55.11)$$

In cases A and B of Fig. 55.2, the two $V_1 G_0 V_2$ second Born terms are of opposite sign because V_2 is of opposite sign in diagrams A and B. This leads to destructive interference for $1s - 1s$ electron capture (which is dominant at high velocities) as was first discussed by Shakeshaft and Wadehra [16]. Consequently, the observed Thomas peak structure is expected [17, 18] to be quite different for e^+ impact than for impact of p^+ or other projectiles heavier than an electron. The double ridge structure for transfer ionization of helium by e^+ is expected to differ significantly from the structure shown in Fig. 55.3. Understanding such destructive interference between resonant amplitudes may give deeper insight into the physical

nature of the intermediate states in this special few-body collision system.

55.7 SUMMARY

Thomas peaks correspond to singular second-order quantum effects whose location may be determined by classical two step kinematics. The widths of these peaks (or ridges) may be estimated using the uncertainty principle. A second-order quantum calculation is required to obtain the magnitude of these peaks. Thomas peaks and ridges have been observed in various reactions in atomic and molecular collisions involving mass transfer and also ionization.

REFERENCES

1. L. H. Thomas, Proc. Roy. Soc. A **114**, 561 (1927).
2. R. Shakeshaft and L. Spruch, Rev. Mod. Phys. **51**, 369 (1979).
3. K. Detmann and G. Liebfried, Z. Phys. **218**, 1 (1968).
4. P. R. Simony, Ph. D. thesis, Kansas State University, (1981).
5. M. Lieber, private communication (1987).
6. J. H. McGuire, J. C. Straton, and T. Ishihara, in *The Application of Many-Body Theory to Atomic Physics*, edited by M. S. Pindzola and J. J. Boyle, (Cambridge University Press, Cambridge, 1994).
7. J. Palinkas, R. Schuch, H. Cederquist, and O. Gustafsson, Phys. Rev. Lett. **22**, 2464 (1989).
8. J. H. McGuire, J. C. Straton, W. C. Axmann, T. Ishihara, and E. Horsdal, Phys. Rev. Lett. **62**, 2933 (1989); J. S. Briggs and K. Taulbjerg, J. Phys. B **12**, 2565 (1979).
9. J. H. McGuire, Adv. At. Mol. and Opt. Phys. **29**, 217 (1991).
10. E. Horsdal-Pederson, C. L. Cocke, and M. Stöckli, Phys. Rev. Lett. **57**, 2256 (1986).
11. J. D. Jackson, *Classical Electrodynamics* (John Wiley, New York, 1975), p. 286.
12. O. L. Weaver and J. H. McGuire, Phys. Rev. A **32**, 1435 (1985).
13. J. H. McGuire and O. L. Weaver, J. Phys. **17**, L583 (1984).
14. J. H. McGuire, Indian J. Phys. **62B**, 261 (1988).
15. E. Horsdal-Pederson, P. Loftager, and J. L. Rasmussen, J. Phys. B **15**, 7461 (1982).
16. R. Shakeshaft and J. Wadehra, Phys. Rev. A **22**, 968 (1980).
17. A. Igarashi and N. Toshima, Phys. Rev. A **46**, R1159 (1992).
18. J. H. McGuire, N. C. Sil, and N. C. Deb, Phys. Rev. A **34**, 685 (1986).

56

Classical Trajectory and Monte Carlo Techniques

R. E. Olson

Physics Department, University of Missouri, Rolla, Missouri 65401

56.1 INTRODUCTION

The classical trajectory Monte Carlo (CTMC) method originated with Hirschfelder, who studied the H + D$_2$ exchange reaction using a mechanical calculator [1]. With the availability of computers, the CTMC method was actively applied to a large number of chemical systems to determine reaction rates, and final state vibrational and rotational populations (see, e.g., Karplus *et al.* [2]). For atomic physics problems, a major step was introduced by Abrines and Percival [3] who employed Kepler's equations and the Bohr-Sommerfield model for atomic hydrogen to investigate electron capture and ionization for intermediate velocity collisions of H$^+$ + H. An excellent description is given by Percival and Richards [4]. The CTMC method has a wide range of applicability to strongly-coupled systems, such as collisions by multiply-charged ions [5]. In such systems, perturbation methods fail, and basis set limitations of coupled-channel molecular- and atomic-orbital techniques have difficulty in representing the multitude of active excita-

tion, electron capture, and ionization channels. Vector- and parallel-processors now allow increasingly detailed study of the dynamics of the heavy projectile and target, along with the active electrons.

56.2 THEORETICAL BACKGROUND

56.2.1 Hydrogenic Targets

For a simple three-body collision system comprised of a fully-stripped projectile (a), a bare target nucleus (b), and an active electron (c), one begins with the classical Hamiltonian for the system,

$$H = p_a^2/2m_a + p_b^2/2m_b + p_c^2/2m_c$$
$$+ Z_a Z_b/r_{ab} + Z_a Z_c/r_{ac} + Z_b Z_c/r_{bc}, \quad (56.1)$$

where p_i are the momenta and $Z_i Z_f/r_{if}$ are the Coulomb potentials between the individual particles. From Eq. (56.1), one obtains a set of 18 coupled, first-order differential equations arising from the necessity to determine the time evolution of the Cartesian coordinates of each particle,

$$dq_i/dt = \partial H/\partial p_i,\qquad(56.2)$$

and their corresponding momenta,

$$dp_i/dt = -\partial H/\partial q_i.\qquad(56.3)$$

Five random numbers, constrained by Kepler's equation, are then used to initialize the plane and eccentricity of the electron's orbit, and another is used to determine the impact parameter within the range of interaction [4,5]. A fourth-order Runge-Kutta integration method is suitable because of its ease of use and its ability to vary the time step-size. This latter requirement is essential since it is not uncommon for the time step to vary by three orders of magnitude during a single trajectory.

In essence, the CTMC method is a computer experiment. Total cross sections for a particular process are determined by

$$\sigma_R = (N_R/N)\pi b_{\max}^2,\qquad(56.4)$$

where N is the total number of trajectories run within a given maximum impact parameter $b_{\max}$, and N_R is the number of positive tests for a reaction, such as electron capture or ionization. Angle and energy differential cross sections are easily generalized from the above. As in an experiment, the cross section given by Eq. (56.4) has a standard deviation of

$$\Delta\sigma_R = \sigma[(N-N_R)/NN_R]^{1/2},\qquad(56.5)$$

which for large N is proportional to $1/N_R^{1/2}$. Here lies one of the major difficulties associated with the CTMC method: it takes considerable computation time to determine minor or highly differential cross sections. Present day desktop workstations can provide a partial remedy of this statistics problem. To decrease the statistical error of a cross section by a factor of two, four times as many trajectories must be evaluated.

56.2.2 Nonhydrogenic One-Electron Models

For many-electron target atoms, it is sometimes adequate to treat the problem within a one-electron model and employ the independent electron approximation to approximate atomic shell structure [6]. For an accurate calculation, it is necessary to use an interaction potential that simulates the screening of the target nucleus by the electrons. One can simply apply a Coulomb potential with an effective charge Z_{eff} obtained from, for example, Slater's rules. Then, the computational procedure is the same as for the hydrogenic case. However, the boundary

conditions for the long- and short-range interactions are poorly satisfied.

To improve the electronic representation of the target, potentials derived from quantum mechanical calculations are now routinely used. Here, the simple solution of Kepler's equation cannot be applied. However, Peach *et al.* [7] and Reinhold and Falcón [8] have provided the appropriate methods that yield a target representation that is correct under the microcanonical distribution. The method of Reinhold and Falcón is popular because of its ease of use and generalizability. For the effective interaction potential, Garvey *et al.* [9] have performed a large set of Hartree-Fock calculations, and have parametrized their results in the form

$$V(R) = -[Z - NS(R)]/R,\qquad(56.6)$$

with the screening of the core given by

$$S(R) = 1 - \{(\eta/\xi)[\exp(\xi R)-1]+1\}^{-1/2},\qquad(56.7)$$

where Z and N denote the nuclear charge and number of nonactive electrons in the target core, and η and ξ are screening parameters. Screening parameters are given in Ref. [9] for all ions and atoms with $Z \leq 54$. This potential can also be used for the representation of partially-stripped projectile ions.

56.2.3 Multiply-Charged Projectiles and Many-Electron Targets

Multiple ionization and electron capture mechanisms in energetic collisions between multiply-charged ions and many-electron atoms is poorly understood because major approximations must be made to solve a many-electron problem associated with transitions between two centers. For a representative collision system such as

$$A^{q+} + B \rightarrow A^{(q-j)+} + B^{i+} + (i-j)e^-,\qquad(56.8)$$

it is essential that the theoretical method be able to predict simultaneously the various charge states of the projectile and recoil ions, and also the energy and angular spectra of the ejected electrons. Theoretical methods based on the independent electron model fail because the varying ionization energies of the electrons are not well represented by a constant value, especially for outer shells. To present, only the nCTMC method, which is a direct extension of the hydrogenic CTMC method to an n-electron system, has been able to make reasonable predictions of the cross sections and scattering dynamics of such strongly-coupled systems [10]. As such, the number of coupled equations rises to $6n + 12$, where n is the number of electrons included in the calculation. However, computing time does not increase linearly, since modern vector-processors become very efficient with coupled equations for multiples of 64.

In the nCTMC technique, all interactions of the projectile and target nuclei with each other and the electrons are implicitly included in the calculations. The inclusion of all the particles then allows a direct determination of their angular scattering, along with an estimate of the energy deposition to electrons and heavy particles. Post collision interactions are included between projectile and recoil ions with the electrons; however, electron-electron interactions are introduced only in the bound initial state via a screening factor in a central-field approximation. This theoretical model has been very successful in predicting the single and double differential cross sections for the ionized electron spectra. Moreover, since a fixed target nucleus approximation is not used, this method has been the only one available to help understand and predict the results for the new field of recoil-ion momentum spectroscopy [11].

Only a modest amount of work has been completed on molecular targets. At present, only an H_2 target has been formulated for application to the CTMC method [12]. For H_2, a fixed internuclear axis is assumed which is then randomly orientated for each trajectory. The electrons are initialized in terms of two one-electron microcanonical distributions constructed from the quantum mechanical wave function for the ground state of H_2. Total cross sections for a variety of projectiles are in reasonable accord with experiment. The effect of the orientation of the molecular axis on the cross sections was also investigated, along with tests as to the validity of assuming that the cross sections for H_2 are simply the product of twice the H values.

56.3   REGION OF VALIDITY

The CTMC method has a demonstrated region of applicability for ion-atom collisions in the intermediate velocity regime, particularly in the elucidation of both heavy-particle and electron collision dynamics. The method can be termed a semiclassical method in that the initial conditions for the electron orbits are determined by quantum mechanically determined interaction potentials with the parent nucleus. Since the method is most applicable to strongly-coupled systems, it has been applied successfully to a variety of intermediate energy multiply-charged ion collisions. Figure 56.1 describes pictorially the regions of validity of theoretical models. Both the atomic orbital (AO) and molecular orbital (MO) basis set expansion methods (see Chap. 48) work well until ionization strongly mediates the collision, since the theoretical description of the ionization continuum is not well-founded and relies on pseudostates to span all ejected electron energies and angles. We have arbitrarily limited these methods to a projectile charge to target charge ratio of, Z_P/Z_T, $\simeq 8$, since above this value the number of terms in the basis set becomes prohibitively

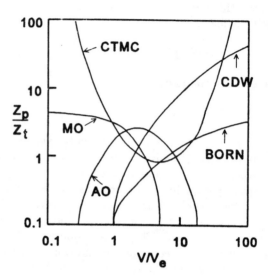

Figure 56.1. Approximate regions of validity of various theoretical methods. Z_P/Z_T is the ratio of the projectile charge to the target charge, and v/v_e is the ratio of the collision velocity to the velocity of the active target electron. Theoretical methods: molecular orbital (MO), atomic orbital (AO), classical trajectory Monte Carlo (CTMC), continuum distorted wave (CDW), and first-order perturbation theory (BORN).

large. The CTMC method does not include molecular effects, and thus it is restricted from low velocities, except in the case of high-charge-state projectiles that capture electrons into high-lying Rydberg states which are well-described classically. Likewise, at high velocities the method is inapplicable in the perturbation regime where quantum tunneling is important, and thus is restricted to strongly coupled systems. The continuum distorted wave (CDW) method (see Chap. 50) greatly extends the region of applicability of first-order perturbation methods and has demonstrated validity in high-charge state ionization collisions.

56.4   APPLICATIONS

56.4.1   Hydrogenic Atom Targets

The original application of the CTMC method to atomic physics collisions were done on the $H^+ + H$ system [3]. Here, the electron capture and ionization total cross sections were found to be in very good accord with experiment. The Abrines and Percival procedure casts the coupled equations into the c.m. coordinate system to reduce the three-body problem to 12 coupled equations. However, this reduction complicates extensions of the code to laser processes and collisions in electric fields or with many electrons.

An ideal application for the CTMC method is for collisions involving excited targets. Such processes are well-described classically, and basis set expansion methods show limited applicability due to computer memory constraints. Considerable early work has been done on Rydberg atom collisions which includes state-selective electron capture, ionization, and electric fields [13–15]. Presently, there is a resurgence of work on Rydberg atom collisions because new crossed-field experimental techniques allow the production of these atoms with specific spatial orientations and eccentricities [16].

For hydrogenic ion-ion collision processes, one must be careful to apply the CTMC method only for projectile charges $Z_P \geq Z_T$ because after the initialization of the active electron's orbit and energy, there is no classical constraint on the orbital energy of a captured electron. For a low-charge-state ion colliding with a ground state highly charged ion, one will obtain unphysical results because a captured electron will tend to preserve its original binding energy. Thus, excess probability will be calculated for electron orbits that lead to unrealistic deeply bound states of the projectile.

56.4.2 Pseudo One-Electron Targets

Collisions involving alkali atoms are of interest because of their relevance to applied programs, such as plasma diagnostics in tokamak nuclear fusion reactors. They are also a testing ground for theoretical methods since experimental benchmarks are difficult to realize with hydrogenic targets, but are amenable for such cases as alkali atoms. In such collisions, it is essential that a theoretical formalism be used that correctly simulates the screening of the nucleus by the core electrons (see Sect. 56.2.2), since a simple $-Z_{\text{eff}}/R$ Coulomb potential is inadequate for both large and small R.

One can also apply the methods of Sect. 56.2.2 to partially or completely filled atomic shells. This works reasonably well for collisions with a low charge state ion such as a proton, but fails for strong collisions involving multiply-charged ions. The reason is that the independent electron approximation [6] must be applied to the calculated transition probabilities in order to simulate the shell structure. This latter method can only maintain its validity if the transition probability is low. Otherwise, it will greatly overestimate the multiple electron removal processes since the first ionization potential is inherently assumed for each subsequent electron that is removed from the shell, leading to an underestimate of the energy deposition.

56.4.3 State-Selective Electron Capture

One of the powers of the CTMC method is that it can be applied to electron capture and excitation of high-lying states that are not accessible with basis set expansion techniques [17]. For the C^{4+} + Li system, where AO calculations and experimental data exist, the CTMC method agrees quite favorably with both [18].

The procedure is first to define a classical number n_c related to the calculated binding energy E of the active electron to either the projectile (electron capture) or target nucleus (excitation) as

$$E = -Z^2/(2n_c^2). \tag{56.9}$$

Then, n_c is related to the principal quantum number n of the final state by the condition [19]

$$[n(n-\tfrac{1}{2})(n-1)]^{1/3} < n_c \leq [n(n+\tfrac{1}{2})(n+1)]^{1/3}. \tag{56.10}$$

From the electron's normalized classical angular momentum $l_c = (n/n_c)(\mathbf{r} \times \mathbf{k})$, l_c is related to the orbital quantum number l of the final state by

$$l < l_c \leq l+1. \tag{56.11}$$

The magnetic quantum number can be sorted in a similar manner. In principle, it is also possible to analyze the final-state distributions from the effective quantum number

$$n^* = n - \delta_l, \tag{56.12}$$

where δ_l is the quantum defect. In this latter case, it is necessary to sort the angular momentum quantum numbers first, then sort the principal quantum numbers.

56.4.4 Exotic Projectiles

The study of collisions involving antimatter projectiles, such as positrons and antiprotons, is a rapidly growing field which is being spurred on by recent experimental advances. Such scattering processes are of basic interest, and they also contribute to a better understanding of normal matter-atom collisions. Antimatter-atom studies highlight the underlying differences in the dynamics of the collision, as well as on the partitioning of the overall scattering. In the Born approximation, ionization cross sections depend on the square of the projectile's charge and are independent of its mass. Thus, the comparison of the cross sections for electron, positron, proton and antiproton scattering from a specific target gives a direct indication of higher-order corrections to scattering theories.

Early work using the CTMC method concentrated on the spectra of ionized electrons for antimatter projectiles

[20]. Later work focused on the angular scattering of the projectiles during electron removal collisions, such as positronium formation, on ratios of the electron removal cross sections, and on ejected electron 'cusp' and 'anticusp' formation. A recent review that compares various theoretical results and available experiments is given in Ref. [21].

56.4.5 Heavy Particle Dynamics

A major attribute of the CTMC method is that it inherently includes the motion of the heavy particles after the collision. A straight-line trajectory for the projectile is not assumed, nor is the target nucleus constrained to be fixed. This allows one to compute easily the differential cross sections for projectile scattering or the recoil momenta of the target nucleus. As a computational note, the angular scattering of the projectile should be computed from the momentum components, not the position coordinates after the collision, since faster convergence of the cross sections using the projectile momenta is obtained. For recoil ion momentum transfer studies, one must initialize the target atom such that the c.m. of the nucleus plus its electrons has zero momentum so that there is no initial momentum associated with the target. A common error is to initialize only the target nucleus momenta to zero. Then the target atom after a collision has an artificial residual momentum that is associated with the Compton profile of the electrons because target-electron interactions are included in the calculations. Examples of recoil and projectile scattering cross sections are given in Refs. [10, 11].

The new field of recoil ion momentum spectroscopy [22] is rapidly expanding, and so far, only the CTMC theoretical method has had an impact on the interpretation and understanding of the experimental results. For the systems studied, it has been conclusively shown that a central potential is unable to reproduce the experimental results that also give information about the out-of-plane scattering in ionization collisions [23]. It is necessary that a theoretical method be able to follow all ejected electrons and the heavy particle after a collision. By so doing, it is possible to observe the backward recoil of the target in electron capture reactions, due to the active electron being transferred from the target to the projectile frame of reference [24]. Very recently, the direct signature of the electron-electron and the electron-nuclear interactions could be observed in the recoil ion's perpendicular and longitudinal momentum distributions [25].

56.5 CONCLUSIONS

In many ways it is surprising that a classical model can be successful in a quantum mechanical world, especially since the classical radial distribution for the hydrogen atom is described so poorly. However, hydrogen's classical momentum distribution is exactly equivalent to the quantum one, and since collision processes are primarily determined by velocity matching between projectile and electron, reasonable results can be expected. Moreover, the CTMC method preserves conservation of flux, energy, and momentum; and Coulomb scattering is the same in both quantal and classical frameworks.

REFERENCES

1. J. Hirschfelder, H. Eyring, and B. Topley, J. Chem. Phys. **4**, 170 (1936).
2. M. Karplus, R. N. Porter, and R. D. Sharma, J. Chem. Phys. **43**, 3259 (1965).
3. R. Abrines and I. C. Percival, Proc. Phys. Soc. **88**, 861 (1966).
4. I. C. Percival and D. Richards, Adv. At. Mol. Phys. **11**, 1 (1975).
5. R. E. Olson and A. Salop, Phys. Rev. A **16**, 531 (1977).
6. J. H. McGuire and L. Weaver, Phys. Rev. A **16**, 41 (1977).
7. G. Peach, S. L. Willis, and M. R. C. McDowell, J. Phys. B **18**, 3921 (1985).
8. C. O. Reinhold and Falcón, Phys. Rev. A **33**, 3859 (1986).
9. R. H. Garvey, C. H. Jackman, and A. E. S. Green, Phys. Rev. A **12**, 1144 (1975).
10. R. E. Olson, J. Ullrich, and H. Schmidt-Böcking, Phys. Rev. A **39**, 5572 (1989).
11. C. L. Cocke and R. E. Olson, Phys. Rep. **205**, 153 (1991).
12. L. Meng, C. O. Reinhold, and R. E. Olson, Phys. Rev. A **40**, 3637 (1989).
13. R. E. Olson, J. Phys. B **13**, 483 (1980).
14. R. E. Olson, Phys. Rev. Lett. **43**, 126 (1979).
15. R. E. Olson and A. D. MacKellar, Phys. Rev. Lett. **46**, 1451 (1981).
16. D. Delande and J. C. Gay, Europhys. Lett. **5**, 303 (1988).
17. R. E. Olson, Phys. Rev. A **24**, 1726 (1981).
18. R. Hoekstra *et al.*, J. Phys. B **26**, 2029 (1993).
19. R. C. Becker and A. D. MacKellar, J. Phys. B **17**, 3923 (1984).
20. R. E. Olson and T. J. Gay, Phys. Rev. Lett. **61**, 302 (1988).
21. D. R. Schultz, R. E. Olson, and C. O. Reinhold, J. Phys. B **24**, 521 (1991).
22. W. Wu *et al.*, Phys. Rev. A **48**, 3617 (1993).
23. R. Dörner, J. Ullrich, H. Schmidt-Böcking, and R. E. Olson, Phys. Rev. Lett. **63**, 147 (1989).
24. V. Frohne *et al.*, Phys. Rev. Lett. **71**, 696 (1993).
25. R. Dörner *et al.*, Phys. Rev. Lett. **72**, 3166 (1994).

57

Collisional Broadening of Spectral Lines

G. Peach

Department of Physics and Astronomy, University College London, London WC1E 6BT, United Kingdom

57.1 SCOPE AND TERMINOLOGY

One-photon processes only are discussed and aspects of line broadening directly related to collisions between the emitting (or absorbing) atom and one perturber are considered. Molecular lines and bands are not considered here. Pointers to other aspects are included and a comprehensive bibliography of work on atomic line shapes, widths, and shifts already exists [1–5]. The perturber may be an electron, a neutral atom or an atomic ion and can interact weakly or strongly with the emitter. The emitter is either a hydrogenic or nonhydrogenic atom that is either neutral or ionized. In general, transitions in nonhydrogenic atoms can be treated as isolated, that is

the separation between neighboring lines is much greater than the width of an individual line. When the emitter is hydrogen or a hydrogenic ion, the additional degeneracy of the energy levels with respect to orbital angular momentum quantum number means that lines overlap and are coupled.

Pressure broadening is a general term that describes any broadening and shift of a spectral line produced by fields generated by a background gas or plasma. The term Stark broadening implies that the perturbers are atomic ions and/or electrons, and collisional broadening implies that the 'collision' model is appropriate; this term is often used to describe an isolated line perturbed by electrons. Neutral atom broadening indicates neutral atomic perturbers; this implies short-range emitter-perturber interactions which in turn influence approximations made.

General reviews of the theory of pressure broadening have been given [6–8], and Chaps. 2, 10, 14, 19, 45, 43, and 82 discuss topics relevant to the theory of collisional broadening of spectral lines. The International Conference on Spectral Line Shapes (ICSLS), is devoted exclusively to this subject.

57.2 IMPACT APPROXIMATION

If the perturbers are rapidly moving, the broadening and shift of the line arises from a series of binary collisions between the atom and one of the perturbers. The theory assumes that although weak collisions may occur simultaneously, strong collisions are relatively rare and only occur one at a time. The impact approximation is valid if

$$w\tau \ll 1, \quad \bar{V}\tau/\hbar \ll 1, \qquad (57.1)$$

where w is the half width at half maximum, τ is the average time of collision and $\bar{V}$ is the average emitter–perturber interaction. It is not only widely applicable to electron and neutral atom broadening, but also, for certain plasma conditions, to broadening by atomic ions. The power radiated per unit time and per unit interval in circular frequency ω, in terms of the line profile $I(\omega)$, is

$$P(\omega) = \frac{4}{3}\frac{\omega^4}{c^3} I(\omega). \qquad (57.2)$$

For an isolated line produced by a transition from an upper energy level i to a lower level j, the line profile is Lorentzian with a shift d:

$$I(\omega) = \frac{1}{\pi}\langle\!\langle if^* |\Delta| if^*\rangle\!\rangle \frac{w}{(\omega - \omega_{if} - d)^2 + w^2}, \qquad (57.3)$$

and if the profile is for a transition between an upper set of levels i, i' and a lower set f, f' with quantum numbers $J_i M_i$, $J_{i'} M_{i'}$, $J_f M_f$, and $J_{f'} M_{f'}$, then

$$I(\omega) = \frac{1}{\pi}\Re \sum_{ii'ff'} \langle\!\langle if^* |\Delta| i'f'^*\rangle\!\rangle$$
$$\times \left\langle\!\!\left\langle i'f'^* \left| [\mathbf{w} + i\mathbf{d} - i(\omega - \omega_{if})\mathbf{I}]^{-1} \right| if^* \right\rangle\!\!\right\rangle, \quad (57.4)$$

where $\mathbf{I}$ is the unit operator and $\mathbf{w}$ and $\mathbf{d}$ are width and shift operators. In (57.4) Δ is an operator corresponding to the dipole line strength defined by

$$\langle\!\langle if^* |\Delta| i'f'^*\rangle\!\rangle \equiv q^2 \langle i|\mathbf{r}|f\rangle \cdot \langle i'|\mathbf{r}|f'\rangle, \qquad (57.5)$$

where $\mathbf{r}$ represents the internal emitter coordinates, $q^2 = e^2/(4\pi\epsilon_0)$ is the square of the electronic charge, and ϵ_0 is

the permittivity of vacuum in SI units of J m. On taking the average over all degenerate magnetic sublevels,

$$I(\omega) = \frac{1}{\pi}\Re \sum_{ii'ff'} \langle\!\langle if^* \|\Delta\| i'f'^*\rangle\!\rangle$$
$$\times \left\langle\!\!\left\langle i'f'^* \left\| [\mathbf{w} + i\mathbf{d} - i(\omega - \omega_{if})\mathbf{I}]^{-1} \right\| if^* \right\rangle\!\!\right\rangle \quad (57.6)$$

in terms of reduced matrix elements that are independent of magnetic quantum numbers. They are defined by

$$\langle\!\langle if^* |\Delta| i'f'^*\rangle\!\rangle = D_{ifi'f'} \langle\!\langle if^* \|\Delta\| i'f'^*\rangle\!\rangle, \qquad (57.7)$$

where

$$D_{ifi'f'} = \sum_\mu (-1)^{J_i + J_{i'} - M_i - M_{i'}}$$
$$\times \begin{pmatrix} J_i & 1 & J_f \\ -M_i & \mu & M_f \end{pmatrix} \begin{pmatrix} J_{i'} & 1 & J_{f'} \\ -M_{i'} & \mu & M_{f'} \end{pmatrix}, \qquad (57.8)$$

$$\langle\!\langle i'f'^* \| [\mathbf{w} + i\mathbf{d} - i(\omega - \omega_{if})\mathbf{I}]^s \| if^*\rangle\!\rangle$$
$$= \sum_{M_i, M_{i'}, M_f, M_{f'}} D_{ifi'f'} \langle\!\langle i'f'^* | [\mathbf{w} + i\mathbf{d} - i(\omega - \omega_{if})\mathbf{I}]^s | if^*\rangle\!\rangle$$
$$(57.9)$$

with $s = -1, 1$. For the line profile (57.3), the width and shift matrices each have a single element:

$$\gamma = 2w = \langle\!\langle if^* \|2\mathbf{w}\| if^*\rangle\!\rangle, \quad d = \langle\!\langle if^* \|\mathbf{d}\| if^*\rangle\!\rangle, \qquad (57.10)$$

where γ is the full width at half maximum.

57.3 ISOLATED LINES

57.3.1 Semiclassical Theory

The motion of the perturber relative to the emitter is treated classically and is assumed to be independent of the internal states of the emitter and perturber. This common trajectory is specified by a emitter–perturber separation

$$\mathbf{R} \equiv \mathbf{R}(\mathbf{b}, \mathbf{v}, t), \quad \mathbf{b}\cdot\mathbf{v} = 0, \qquad (57.11)$$

where $\mathbf{v}$ is the relative velocity and $\mathbf{b}$ is the impact parameter. The time-dependent wave equation for the emitter-perturber system is

$$i\hbar \frac{d\Psi}{dt} = H\Psi \qquad (57.12)$$

and the eigenfunctions ψ_i for the unperturbed emitter obey

$$H_0 \psi_i = E_i \psi_i, \quad i = 0, 1, 2, \ldots . \qquad (57.13)$$

If $\Psi(\mathbf{r}, \mathbf{R})$ is expanded in the form

$$\Psi(\mathbf{r}, \mathbf{R}) = \sum_j a_{ji}(t) \psi_j(\mathbf{r}) \exp(-iE_j t/\hbar), \quad (57.14)$$

where initially at time $t = -\infty$

$$a_{ji}(-\infty) = \delta_{ji}, \qquad (57.15)$$

and (57.12)–(57.14) give

$$i\hbar \frac{da_{ji}}{dt} = \sum_k a_{ki} V_{jk} \exp(i\omega_{jk}t), \quad i, j, k = 0, 1, 2, \ldots,$$
$$(57.16)$$

where

$$V_{jk}(\mathbf{R}) = \int \psi_j^*(\mathbf{r}) V(\mathbf{r}, \mathbf{R}) \psi_k(\mathbf{r}) \, d\mathbf{r} \qquad (57.17)$$

in which $V(\mathbf{r}, \mathbf{R})$ is the emitter-perturber interaction and

$$\hbar\omega_{jk} = E_j - E_k. \qquad (57.18)$$

Integration of equations (57.16) for $-\infty \le t \le \infty$ gives the unitary scattering matrix $\mathbf{S}$, with elements

$$S_{ji}(\mathbf{b}, \mathbf{v}) \equiv a_{ji}(\infty), \quad i, j = 0, 1, 2, \ldots . \quad (57.19)$$

Then

$$w + id = 2\pi N \int_0^\infty v f(v) \, dv$$
$$\times \int_0^\infty \left[\delta_{i'i} \delta_{f'f} - S_{i'i}(\mathbf{b}, \mathbf{v}) S_{f'f}^*(\mathbf{b}, \mathbf{v}) \right]_{av} b \, db,$$
$$(57.20)$$

where $J_{i'} = J_i$ and $J_{f'} = J_f$, N is the perturber density and $[\cdots]_{av}$ denotes an average over all orientations of the collision and over the magnetic sublevels [see (57.9)]. In (57.20), $f(v)$ is the Maxwell velocity distribution at temperature T for an emitter-perturber system of reduced mass μ:

$$f(v) = 4\pi v^2 \left(\frac{\mu}{2\pi k_B T} \right)^{3/2} \exp\left(-\frac{\mu v^2}{2 k_B T} \right),$$
$$\int_0^\infty f(v) \, dv = 1. \qquad (57.21)$$

57.3.2 Simple Formulae

These are useful for making quick estimates, but in individual cases may give results in error by a factor of two or more. If it is assumed in (57.17) that

$$V_{ij}(\mathbf{R}) = V_{jj}(\mathbf{R}) \delta_{ij}, \qquad (57.22)$$

where $V_{jj}(\mathbf{R})$ is a simple central potential

$$V_{jj}(\mathbf{R}) = \hbar C_j R^{-p}, \quad j = i, f, \qquad (57.23)$$

and C_j depends only on the state j of the emitter, and if the relative motion is along the straight line

$$\mathbf{R} = \mathbf{b} + \mathbf{v}t, \qquad (57.24)$$

then

$$\left[S_{i'i}(\mathbf{b}, \mathbf{v}) S_{f'f}^*(\mathbf{b}, \mathbf{v}) \right]_{av} = \exp\left[2i(\eta_i - \eta_f) \right], \quad (57.25)$$

where the phase shifts are

$$\eta_j(b, v) = -\frac{1}{2\hbar} \int_{-\infty}^{\infty} V_{jj}(R) \, dt, \quad j = i, f. \quad (57.26)$$

Equations (57.20)–(57.26) give

$$w + id = \pi N \bar{v} \left[\frac{\beta_p |C_p|}{\bar{v}} \right]^{2/(p-1)}$$
$$\times \Gamma\left(\frac{p-3}{p-1} \right) \exp\left(\pm \frac{i\pi}{p-1} \right) \alpha_p, \qquad (57.27)$$

where

$$\alpha_p = \Gamma\left(\frac{2p-3}{p-1} \right) \left(\frac{\pi}{4} \right)^{-1/(p-1)},$$
$$\beta_p = \sqrt{\pi} \frac{\Gamma\left[\frac{1}{2}(p-1) \right]}{\Gamma\left(\frac{1}{2}p \right)},$$
$$C_p = C_i - C_f,$$
$$\bar{v} = \int_0^\infty v f(v) \, dv = \left(\frac{8 k_B T}{\pi\mu} \right)^{1/2}.$$

In (57.27) the $\pm$ sign indicates the sign of C_p, $\alpha_p \simeq 1$ for $p \ge 3$, and $\Gamma(\cdots)$ is the gamma function.

The cases $p = 3, 4$, and 6 correspond to resonance, quadratic Stark and van der Waals broadening, respectively. This approximation is invalid for the dipole case ($p = 2$) for which (57.27) is not finite. The dipole-dipole interaction ($p = 3$) occurs when emitter and perturber are identical atoms (apart from isotopic differences). If the level i is connected to the ground state by a strong allowed transition with absorption oscillator strength f_{gi}, and the perturbation of the level f can be neglected by comparison, then

$$C_3 = c_d \frac{q^2 f_{gi}}{2 m_e |\omega_{gi}|},$$
$$c_d = 1 + \frac{1}{2\sqrt{3}} \ln\left(2 + \sqrt{3} \right) = 1.380\,173. \qquad (57.28)$$

The constant c_d may be replaced by an empirical value related to the ratio of statistical weights for the levels

$$c_d = \frac{4}{\pi} \left(\frac{g_g}{g_i} \right)^{1/2} , \qquad (57.29)$$

and this gives a width correct to about 10% [9]. Equation (57.27) does not predict a finite shift. Quadratic Stark broadening occurs when a nonhydrogenic emitter is polarized by electron perturbers. Then

$$C_4 = -\frac{q^2}{2\hbar} (\alpha_i - \alpha_f) , \qquad (57.30)$$

where α_i and α_f are the dipole polarizabilities of states i and f, respectively. Van der Waals broadening occurs when the emitter and perturber are nonidentical neutral atoms. If energy level separations of importance in the perturbing atom are much greater than those of the emitter (e.g. alkali spectra broadened by noble gases), C_6 is given by

$$C_6 = -\frac{q^2}{\hbar} \alpha_d \left(\overline{r_i^2} - \overline{r_f^2} \right) , \qquad (57.31)$$

where α_d is the dipole polarizability of the perturber. The mean square radii can be calculated from the normalized radial wave functions $\frac{1}{r} P_{n_j^* l_j} (r)$ or estimated from

$$\overline{r_j^2} = \int_0^\infty P_{n_j^* l_j}^2 (r) r^2 \, dr \simeq \frac{n_j^{*2} a_0^2}{2z^2} \left[5n_j^{*2} + 1 - 3l_j (l_j + 1) \right] ,$$
$$j = i, f \qquad (57.32)$$

in which the effective principal quantum numbers n_j^* are given by

$$E_j \equiv -\frac{z^2}{n_j^{*2}} I_h , \quad z = Z_e + 1 , \qquad (57.33)$$

where $I_h = R_\infty hc$ is the Rydberg energy, Z_e is the charge on the emitter and $z = 1$ in this case.

57.3.3 Perturbation Theory

An approximate solution of (57.16) is given by [6–8, 10]

$$S_{ji} (\mathbf{b}, \mathbf{v}) = \delta_{ji} - \frac{i}{\hbar} \int_{-\infty}^\infty V_{ji} (t) \exp (i\omega_{ji} t) \, dt$$
$$- \frac{1}{\hbar^2} \sum_k \left[\int_{-\infty}^\infty V_{jk} (t) \exp (i\omega_{jk} t) \, dt \right.$$
$$\left. \times \int_{-\infty}^t V_{ki} (t') \exp (i\omega_{ki} t') \, dt' \right] . \quad (57.34)$$

This gives a cross section for the collisional transition $i \to j$:

$$\sigma_{ij} (v) = 2\pi \int_0^\infty [P_{ij} (\mathbf{b}, \mathbf{v})]_{av} \, b \, db , \quad i, j = 0, 1, 2, \dots ,$$
$$(57.35)$$

where

$$P_{ij} (\mathbf{b}, \mathbf{v}) = |\delta_{ji} - S_{ji} (\mathbf{b}, \mathbf{v})|^2 , \quad (57.36)$$
$$2\Re [1 - S_{ii} (\mathbf{b}, \mathbf{v})] = \sum_j P_{ij} (\mathbf{b}, \mathbf{v}) , \qquad (57.37)$$

correct to second-order in $V(\mathbf{r}, \mathbf{R})$ on both sides. Using (57.10) (57.20) and (57.35)–(57.37), the full width is

$$\gamma = N \int_0^\infty v f(v) \, dv \left[\sum_{j \neq i} \sigma_{ij} (v) + \sum_{j \neq f} \sigma_{fj} (v) + \tilde{\sigma}_{if} (v) \right] ,$$
$$(57.38)$$

where the sums are taken over all energy-changing transitions, the tilde indicates an interference term, and

$$\tilde{\sigma}_{if} (v) = 2\pi \int_0^\infty \left[\tilde{P}_{if} (\mathbf{b}, \mathbf{v}) \right]_{av} b \, db , \qquad (57.39)$$

in which

$$\tilde{P}_{if} (\mathbf{b}, \mathbf{v}) = \left| \frac{1}{\hbar} \int_{-\infty}^\infty [V_{ii} (t) - V_{ff} (t)] \, dt \right|^2 . \quad (57.40)$$

57.3.4 Broadening by Charged Particles

The total emitter-perturber interaction is

$$\frac{z Z_p q^2}{R} - \frac{Z_p q^2}{|\mathbf{R} - \mathbf{r}|} \simeq V_0 (R) + V (\mathbf{r}, \mathbf{R}) , \qquad (57.41)$$

where Z_p is the charge on the perturber and

$$V_0 (R) = \frac{Z_e Z_p q^2}{R} , \quad V (\mathbf{r}, \mathbf{R}) = -Z_p q^2 \frac{\mathbf{r} \cdot \mathbf{R}}{R^3} . \quad (57.42)$$

If $Z_e = 0$, the relative motion is described by (57.24), but if $Z_e \neq 0$, the trajectory is hyperbolic and is given by

$$\mu \frac{d^2 \mathbf{R}}{dt^2} = -\nabla V_0 = \frac{Z_e Z_p q^2}{R^3} \mathbf{R} , \qquad (57.43)$$

with the resulting hyperbola characterized by a semimajor axis a and an eccentricity ϵ, where

$$b^2 = a^2 (\epsilon^2 - 1) , \quad a = \frac{|Z_e Z_p| q^2}{\mu v^2} . \qquad (57.44)$$

On using (57.17), (57.20), (57.35)–(57.40) and (57.42),

$$V_{ii} (t) = 0 , \quad \tilde{P}_{if} (\mathbf{b}, \mathbf{v}) = 0 , \quad \tilde{\sigma}_{if} (v) = 0 , \quad (57.45)$$

and

$$w + id = 2\pi N \int_0^\infty v f(v) \, dv$$
$$\times \left[\int_0^\infty \sum_{j \neq i} Q_{ij} (b, v) + \sum_{j \neq f} Q_{fj} (b, v) \right] b \, db ,$$
$$(57.46)$$

where

$$Q_{ij}(v) = \frac{4Z_p^2 I_h^2 a_0^2}{\hbar m_e |\omega_{ij}|} \frac{f_{ij}}{b^2 v^2} [A(\beta, \xi) + iB(\beta, \xi)],$$

$$2\Re Q_{ij}(v) = [P_{ij}(\mathbf{b}, \mathbf{v})]_{av}. \tag{57.47}$$

If

$$\xi \equiv \frac{a|\omega_{ij}|}{v}, \quad \beta \equiv \xi\epsilon, \quad \delta \equiv \frac{\epsilon^2 - 1}{\epsilon^2}, \tag{57.48}$$

the functions $A(\beta, \xi)$ and $B(\beta, \xi)$ in (57.47) are given by

$$A(\beta, \xi) = \delta \exp(\mp\pi\xi)\,\beta^2 \left[\left| K'_{i\xi}(\beta) \right|^2 + \delta \left| K_{i\xi}(\beta) \right|^2 \right] \tag{57.49}$$

$$B(\beta, \xi) = \frac{2\beta}{\pi}\wp \int_0^\infty \frac{A(\beta', \xi)\,d\beta'}{(\beta^2 - \beta'^2)}. \tag{57.50}$$

In (57.49), the $\mp$ sign corresponds to $Z_e Z_p = \pm|Z_e Z_p|$, and in (57.50), $\wp$ indicates the Cauchy principal value. If $Z_e = 0$, then

$$\xi = 0, \quad \beta = b\frac{|\omega_{ij}|}{v}, \quad \delta = 1 \tag{57.51}$$

in (57.49) and (57.50).

Approximation (57.47) breaks down at small values of b because of assumption (57.42) and the lack of unitarity of $\mathbf{S}$ as given by (57.34). This problem is discussed elsewhere [6, 10], and all methods used involve choosing a cutoff at $b = b_0$, where $b_0^2 \gtrsim r_f^2$, and using (57.47) only for $b > b_0$. For $b \le b_0$, an effective constant probability is introduced and the method works well as long as the contribution from $b \le b_0$ is small. For $b > b_0$ (or $\beta > \beta_0$), the contribution to $\sigma_{ij}(v)$ in (57.38) is evaluated using (57.46)–(57.49) and (57.51), where

$$\int_{b_0}^\infty A(\beta, \xi)\frac{db}{b} = -e^{\mp\pi\xi}\beta_0 K'_{i\xi}(\beta_0) K_{i\xi}(\beta_0). \tag{57.52}$$

A similar treatment exists for the quadrupole contribution to $V(\mathbf{r}, \mathbf{R})$ in (57.42) [6, 10].

57.3.5 Empirical Formulae

An empirical formula based on the theory of Sect. 37.3.4 for the width of an atomic line Stark broadened by electrons has been developed [11]. The full half-width is given by (57.38), where $\tilde{\sigma}_{if} = 0$ and

$$\int_0^\infty v f_e(v) \sum_{k \ne j} \sigma_{jk}(v)\,dv$$

$$= \frac{8\pi^2}{3\sqrt{3}}\left(\frac{\hbar}{m_e}\right)^2 \overline{v^{-1}} \left[T_j g(x_j) + \sum_{l_{j'} = l_j \pm 1} \tilde{T}_{jj'}\tilde{g}(x_{jj'}) \right],$$

$$j = i, f, \tag{57.53}$$

where

$$\overline{v^{-1}} = \int_0^\infty v^{-1} f_e(v)\,dv = \left(\frac{2m_e}{\pi k_B T}\right)^{1/2}, \tag{57.54}$$

and $f_e(v) = f(v)$ with $\mu = m_e$. In (57.53),

$$T_j = \left(\frac{3n_j^*}{2z}\right)^2 \frac{1}{9}\left[n_j^{*2} + 3l_j(l_j + 1) + 11\right] \tag{57.55}$$

$$\tilde{T}_{jj'} = \frac{l_>}{(2l_j + 1)}R_{jj'}^2\left(n_{l_>}^*, n_{l_<}^*, l_>\right), \tag{57.56}$$

$$l_< = \min(l_j, l_{j'}), \quad l_> = \max(l_j, l_{j'}),$$

with

$$R_{jj'}\left(n_{l_>}^*, n_{l_<}^*, l_>\right) \equiv a_0^{-1}\int_0^\infty P_{n_{l_>}^* l_>}(r)\,r P_{n_{l_<}^* l_<}(r)\,dr. \tag{57.57}$$

The radial matrix element (57.57) can be written as

$$R_{jj'}\left(n_{l_>}^*, n_{l_<}^*, l_>\right) = \tilde{R}_{jj'}\left(n_{l_>}^*, l_>\right)\phi\left(n_{l_>}^*, n_{l_<}^*, l_>\right)$$

$$\equiv \frac{3n_{l_>}^*}{2z}\left(n_{l_>}^{*2} - l_>^2\right)^{1/2}\phi\left(n_{l_>}^*, n_{l_<}^*, l_>\right) \tag{57.58}$$

and $\phi\left(n_{l_>}^*, n_{l_<}^*, l_>\right)$ is tabulated elsewhere [12]. The effective principal quantum numbers $n_{l_>}^*$ and $n_{l_<}^*$ of the states j and j' in (57.56)–(57.58) both correspond to principal quantum number n_j and $\phi \simeq 1$ for $\left(n_{l_>}^* - n_{l_<}^*\right) \ll 1$. The effective Gaunt factors $g(x)$ and $\tilde{g}(x)$ are given by

$$\tilde{g}(x) = 0.7 - 1.1/z + g(x), \quad x = \frac{3k_B T}{2\Delta E}, \tag{57.59}$$

where

x	≤ 2	3	5	10	30	100
$g(x)$	0.20	0.24	0.33	0.56	0.98	1.33

is used for $x < 50$, and for $x > 50$

$$\tilde{g}(x) = g(x) = \frac{\sqrt{3}}{\pi}\left[\frac{1}{2} + \ln\left(\frac{4}{3z}\frac{|E|}{I_h}x\right)\right], \tag{57.60}$$

with (57.59) and (57.60) joined smoothly near $x = 50$. The energy $E = E_j$ is given by (57.33) and x_j and $x_{jj'}$ in (57.53) are evaluated using

$$\Delta E_j = \frac{2z^2}{n_j^{*3}}I_h, \quad \Delta E_{jj'} = |E_j - E_{j'}|. \tag{57.61}$$

For $Z_e = 2$ and 3, (57.53) is generally accurate to within $\pm 30\%$ and $\pm 50\%$. For $Z_e \ge 4$, (57.53) is less accurate, as relativistic effects and resonances become more important. Accuracy increases for transitions to higher Rydberg levels as long as the line remains isolated.

Tables in appendices IV and V of [6] give widths for atoms with $Z_e = 0, 1$ and other semi-empirical formulas based on detailed calculations have been developed by Seaton [13] for use in the Opacity Project where simple estimates of many thousands of line widths are required.

57.4 OVERLAPPING LINES

57.4.1 Transitions in Hydrogen and Hydrogenic Ions

The most important case is that of lines of hydrogenic systems emitted by a plasma with overall electrical neutrality, broadened by perturbing electrons and atomic ions. The line profile is given by (57.4)–(57.9) in which (57.20) is generalized to give

$$
\begin{aligned}
&\langle\!\langle i'f'^{*}\,\|\mathbf{w}+i\mathbf{d}\|\,if^{*}\rangle\!\rangle \\
&= 2\pi N \int_{0}^{\infty} vf(v)\,dv \\
&\quad \times \int_{0}^{\infty} \big[\delta_{i'i}\delta_{f'f} - S_{i'i}(\mathbf{b},\mathbf{v})\,S_{f'f}^{*}(\mathbf{b},\mathbf{v})\big]_{\mathrm{av}}\, b\,db \quad (57.62)
\end{aligned}
$$

The superscripts and suffices e and i will be used to denote electron and ion quantities, and em indicates that averaging over magnetic quantum numbers has not been carried out. In the impact approximation, electron and ion contributions evaluated using (57.62) are additive and the matrix to be inverted is of order $n_i n_f$. However, under typical conditions in a laboratory plasma, e.g., a hydrogen plasma with $N_e = N_i = 10^{22}$ m^{-3}, perturbing atomic ions cannot be treated using the impact approximation. The ions collectively generate a static field at the emitter which produces first-order Stark splitting of the upper and lower levels. The ions are randomly distributed around the emitter and the field distribution $W(\mathbf{F})$ used assumes that each ion is Debye screened by electrons; allowance is made for these heavy composite perturbers interacting with each other as well as with the emitter. If the ion field has a slow time variation, ion dynamic effects on the line are produced [6,7,14].

The shift produced by electron perturbers is very small and the usual model adopted is to assume that the ions split the line into its Stark components, and that each component is broadened by electron impact. In both cases, only the dipole interactions in (57.42) are included and the profile is symmetric. Then (57.4) takes the form

$$
\begin{aligned}
I(\omega) = \frac{1}{\pi}\Re \int W(\mathbf{F})\,d\mathbf{F} \sum_{ii'ff'} &\langle\!\langle if^{*}\,|\boldsymbol{\Delta}|\,i'f'^{*}\rangle\!\rangle \\
&\times \big\langle\!\big\langle i'f'^{*}\,\big|[\mathbf{w}+i\mathbf{d}-i(\omega-\omega_{if})\mathbf{I}]^{-1}\big|\,if^{*}\big\rangle\!\big\rangle,
\end{aligned}
$$
$$(57.63)$$

and the destruction of the degeneracy by the ion field means that the matrix to be inverted in (57.63) is of order $(n_i n_f)^2$. Inclusion of higher multipoles in $V(\mathbf{r},\mathbf{R})$ in (57.42) introduces small asymmetries. The Stark representation for the hydrogenic wave functions is often used because it diagonalizes the shift matrix in (57.63).

The transformation is given by (see Sects. 9.2.2 and 13.4.2)

$$
\begin{aligned}
|n_j K_j m_j\rangle = &\sum_{l_j=|m_j|}^{n_j-1} (-1)^{K}\,(2l_j+1)^{1/2} \\
&\times \begin{pmatrix} N & N & l_j \\ M_1 & M_2 & -m_j \end{pmatrix} |n_j l_j m_j\rangle, \\
&\qquad j = i, i', f, f', \qquad (57.64)
\end{aligned}
$$

where quantum number K_j replaces l_j and

$$
\begin{aligned}
n_j &= K_j + K_j' + |m_j| + 1, & N &= \tfrac{1}{2}(n_j-1), \\
K &= \tfrac{1}{2}\big(2K_j' + |m_j| + m_j\big) + 1, & 0 &\le K_j \le (n_j-1), \\
M_1 &= \tfrac{1}{2}\big(m_j + K_j' - K_j\big), & M_2 &= \tfrac{1}{2}\big(m_j + K_j - K_j'\big).
\end{aligned}
$$

For the electron impact broadening, it is convenient to separate the energy-changing and the zero energy-change transitions, so that (57.38) is generalized to give

$$
\gamma_{em} = \gamma_{em}^{0} + \tilde{\gamma}_{em} \equiv \langle\!\langle i'f'^{*}\,|2\mathbf{w}_e|\,if^{*}\rangle\!\rangle, \quad (57.65)
$$

where

$$
\begin{aligned}
\gamma_{em}^{0} &\equiv \langle\!\langle i'f'^{*}\,|2\mathbf{w}_e^{0}|\,if^{*}\rangle\!\rangle = N_e \int_{0}^{\infty} vf_e(v)\,dv \\
&\quad \times \left[\sum_{j\neq i}\sigma_{ij}^{em}(v) + \sum_{j\neq f}\sigma_{fj}^{em}(v)\right]\delta_{ii'}\delta_{ff'}, \quad (57.66)
\end{aligned}
$$

$$
\tilde{\gamma}_{em} \equiv \langle\!\langle i'f'^{*}\,|2\tilde{\mathbf{w}}_e|\,if^{*}\rangle\!\rangle = N_e \int_{0}^{\infty} vf_e(v)\,dv\,\tilde{\sigma}_{i'f'if}^{em}(v).
$$
$$(57.67)$$

In (57.66), $|n_i - n_j| \neq 0$ and $|n_f - n_j| \neq 0$ in the first and second terms, respectively, and in (57.67) $(n_{i'} - n_i) = (n_{f'} - n_f) = 0$. The matrix element (57.66) can be evaluated using (57.35), (57.47), (57.49), and (57.51) as before. In (57.66),

$$
\sigma_{ij}^{em}(v) = 2\pi \int_{0}^{\infty} \big[P_{ij}^{e}(\mathbf{b},\mathbf{v})\big]_{\mathrm{av}_0}\, b\,db, \quad i,j = 0,1,2,\ldots,
$$
$$(57.68)$$

and in (57.67)

$$
\tilde{\sigma}_{i'f'if}^{em}(v) = 2\pi \int_{0}^{\infty} \big[\tilde{P}_{i'f'if}^{e}(\mathbf{b},\mathbf{v})\big]_{\mathrm{av}_0}\, b\,db \quad (57.69)
$$

by analogy with (57.35) and (57.39), where av_0 indicates an average over all orientations of the collision only. On using (57.47)–(57.49), (57.58) and (57.8) with $j_i = l_i$, $j_{i'} = l_{i'}$, $j_f = l_f$, and $j_{f'} = l_{f'}$,

$$
\big[\tilde{P}_{i'f'if}^{e}(\mathbf{b},\mathbf{v})\big]_{\mathrm{av}_0} = \frac{8I_h a_0^2}{3m_e b^2 v^2} D_{ifi'f'}\tilde{R}_{i'f'if}A(0,0),
$$
$$(57.70)$$

where

$$\tilde{R}_{i'f'if}$$

$$\equiv \left[\sum_{l_j=l_i\pm 1} \tilde{R}_{ij}^2 (n_i, l_{i>}) + \sum_{l_j=l_f\pm 1} \tilde{R}_{fj}^2 (n_f, l_{f>}) \right] \delta_{l_{i'}l_i}\delta_{l_{f'}l_f}$$

$$- 2\tilde{R}_{i'i} (n_i, l_{i>}) \tilde{R}_{f'f} (n_f, l_{f>}) \delta_{l_{i'}l_i\pm 1}\delta_{l_{f'}l_f\pm 1} . \quad (57.71)$$

From (57.44),

$$A(0,0) = \delta ,$$

$$\int_{b_0}^{b_1} A(0,0) \frac{db}{b} = \begin{cases} \ln (\epsilon_0/\epsilon_1) , & Z_e \neq 0 \\ \ln (b_1/b_0) , & Z_e = 0, \end{cases} \quad (57.72)$$

[*cf.* Eq. (57.52)]. The impact approximation neglects electron-electron correlations and the finite duration of collisions, so the long-range dipole interaction leads to a logarithmic divergence at large impact parameters in (57.72). Therefore, a second cutoff parameter is introduced which is chosen to be the smaller of the Debye length b_D and $v\tau$:

$$b_1 = \min \left[b_D \equiv \left(\frac{k_B T}{4\pi q^2 N_e} \right)^{1/2} , v\tau \right] , \quad (57.73)$$

but estimating τ in this case is not straightforward; it depends on the splitting of the Stark components [6].

57.4.2 Infrared and Radio Lines

If the density N_i is low enough, the impact approximation becomes valid for the perturbing atomic ions, and since impact shifts are unimportant, (57.6) gives

$$I(\omega) = \frac{1}{\pi} \Re \sum_{ii'ff'} \langle\!\langle if^* \|\mathbf{\Delta}\| i'f'^* \rangle\!\rangle$$

$$\times \langle\!\langle i'f'^* \| [\mathbf{w}_e + \mathbf{w}_i - i(\omega - \omega_{if})\mathbf{I}]^{-1} \| if^* \rangle\!\rangle$$
$$(57.74)$$

where in (57.74)

$$\gamma_{e,i} = \gamma_{e,i}^0 + \tilde{\gamma}_{e,i} \equiv \langle\!\langle i'f'^* \|2\mathbf{w}_{e,i}\| if^* \rangle\!\rangle , \quad (57.75)$$

$$\gamma_{e,i}^0 \equiv \langle\!\langle i'f'^* \|2\mathbf{w}_{e,i}^0\| if^* \rangle\!\rangle = N_{e,i} \int_0^\infty v f_{e,i}(v) \, dv$$

$$\times \left[\sum_{j\neq i} \sigma_{ij}^{e,i}(v) + \sum_{j\neq f} \sigma_{fj}^{e,i}(v) \right] \delta_{i'i}\delta_{f'f} , \quad (57.76)$$

$$\tilde{\gamma}_{e,i} \equiv \langle\!\langle i'f'^* \|2\tilde{\mathbf{w}}_{e,i}\| if^* \rangle\!\rangle$$

$$= N_{e,i} \int_0^\infty v f_{e,i}(v) \, dv \, \tilde{\sigma}_{i'f'if}^{e,i}(v) . \quad (57.77)$$

In general, cross sections for electron and heavy-particle impact are roughly comparable for the same velocity and hence different impact energies. Therefore, using (57.76) and (57.77),

$$\gamma_e^0 \gg \gamma_i^0 , \quad \tilde{\gamma}_e \ll \tilde{\gamma}_i , \quad (57.78)$$

and this result is consistent with approximation (57.63) for high density plasmas. If $(n_i - n_f) = 1, 2$, say, as n_f increases, the relative contributions from (57.76) and (57.77) decrease because there is increasing coherence, and hence cancellation in $\tilde{\sigma}_{i'f'if}^e(v)$ and $\tilde{\sigma}_{i'f'if}^i(v)$ between the effects of levels i, i' and f, f'.

Radio lines of hydrogen are observed in galactic HII regions where principal quantum numbers are of the order of $n_f \simeq 100$, temperatures are $T_e = T_i \simeq 10^4$ K, and densities are $N_e = N_i \simeq 10^9$ m^{-3}. If γ is the full-half width and $\tilde{\gamma}$ is the full-half width when only contributions (57.77) are retained, the effect of cancellation is illustrated by

$n_i - n_f = 1$	electrons	protons + electrons
n_f	$\tilde{\gamma}/\gamma$	$\tilde{\gamma}/\gamma$
5	0.81	0.99
10	0.44	0.95
15	0.21	0.87
20	0.11	0.75
25	0.06	0.61
50	0.00	0.32
100	0.00	0.16

57.5 QUANTUM-MECHANICAL THEORY

57.5.1 Impact Approximation

The scattering amplitude for a collisional transition $i \to j$ is given in terms of elements of the transition matrix $\mathbf{T} = \mathbf{1} - \mathbf{S}$ by

$$f(\mathbf{k}_j, \mathbf{k}_i) \equiv f(\chi_j M_j \mathbf{k}_j, \chi_i M_i \mathbf{k}_i)$$

$$= \frac{2\pi i}{(k_i k_j)^{1/2}} \sum_{lml'm'} i^{l-l'} Y_{lm}^* \left(\hat{\mathbf{k}}_i\right) Y_{l'm'} \left(\hat{\mathbf{k}}_j\right)$$

$$\times T(\chi_j M_j l'm'; \chi_i M_i lm) , \quad (57.79)$$

where the quantities $\mathbf{k}_i lm$ and $\mathbf{k}_j l'm'$ refer to the motion of the perturber relative to the emitter before and after the collision, and χ_i and χ_j represent all nonmagnetic quantum numbers associated with the unperturbed states i and j of the emitter. The total energy of the emitter-perturber system is given by

$$E_J = E_j + \varepsilon_j , \quad \varepsilon_j = \frac{\hbar^2}{2\mu} k_j^2 ,$$

$$(J,j) = (I,i), (F,f) , \quad (57.80)$$

and for an isolated line, γ is given by (57.38), where

$$\sigma_{ij}(v) = \frac{k_j}{k_i} \frac{1}{4\pi} \frac{1}{g_i} \sum_{M_i M_j} \int |f(\chi_j M_j \mathbf{k}_j, \chi_i M_i \mathbf{k}_i)|^2 \, d\hat{\mathbf{k}}_i d\hat{\mathbf{k}}_j ,$$

$$k_i = \mu v/\hbar , \quad (57.81)$$

and the interference term $\tilde{\sigma}_{if}(v) \equiv \tilde{\sigma}_{ifif}(v)$ is given by

$$\tilde{\sigma}_{if}(v) = \frac{1}{4\pi} \sum_{M_i M_{i'} M_f M_{f'}} D_{ifi'f'} \int |f(\chi_i M_{i'} \mathbf{k}', \chi_i M_i \mathbf{k})$$

$$-f(\chi_f M_{f'} \mathbf{k}', \chi_f M_f \mathbf{k})|^2 \, d\hat{\mathbf{k}} \, d\hat{\mathbf{k}}', \qquad (57.82)$$

with $k = k' = \mu v / \hbar$. From (57.8) and (57.20),

$$w + id = \pi \left(\frac{\hbar}{\mu}\right)^2 N \int_0^\infty \frac{1}{v} f(v) \, dv$$

$$\times \sum_{l=0}^\infty (2l+1) \left[1 - S_{ii}(l,v) S_{ff}^*(l,v)\right], \qquad (57.83)$$

where

$$(\mu v b)^2 \Longrightarrow \hbar^2 l(l+1) \qquad (57.84)$$

and the integral over b has been replaced by a summation over l. In (57.83), $S_{ii}(l,v) S_{ff}^*(l,v)$ is given by

$$S_{ii}(l,v) S_{ff}^*(l,v)$$

$$= \frac{1}{(2l+1)} \sum_{\substack{M_i M_{i'} M_f M_{f'} \\ mm'}} D_{ifi'f'} S_I(\chi_{i'} M_{i'} lm'; \chi_i M_i lm)$$

$$\times S_F^*(\chi_{f'} M_{f'} lm'; \chi_f M_f lm) \qquad (57.85)$$

and subscripts I and F are introduced to emphasize that the **S**-matrix elements correspond to different total energies E_I and E_F defined by (57.80). If scattering by the emitter in a state j is treated using a central potential, the amplitude for elastic scattering is

$$f(k',k) = \frac{i}{2k} \sum_{l=0}^\infty (2l+1) T_{jj}(l,v) P_l\left(\hat{\mathbf{k}}' \cdot \hat{\mathbf{k}}\right), \qquad (57.86)$$

where

$$1 - T_{jj}(l,v) = S_{jj}(l,v) = \exp\left[2i\eta_j(l,k)\right],$$
$$j = i, f, \qquad (57.87)$$

[*cf.* Eqs. (57.23), (57.25), and (57.26)]. For the case of overlapping lines, (57.62) becomes

$$\langle\langle i'f'^* \|\mathbf{w} + i\mathbf{d}\| if^* \rangle\rangle$$

$$= \pi \left(\frac{\hbar}{\mu}\right)^2 N \int_0^\infty \frac{1}{v} f(v) \, dv$$

$$\times \sum_{l=0}^\infty (2l+1) \left[\delta_{i'i}\delta_{f'f} - S_{i'i}(l,v) S_{f'f}^*(l,v)\right] \qquad (57.88)$$

on generalizing (57.83) and using (57.85). Formulae (57.83) and (57.88) have been obtained by assuming that a collision produces no change in the angular momentum of the relative emitter-perturber motion. This corresponds to the assumption of a common trajectory in semiclassical theory, and means that the total angular momentum of the emitter-perturber system is not conserved. This assumption is removed in the derivation of the more general expressions given in the following sections.

57.5.2 Broadening by Electrons

Different coupling schemes can be used to describe the emitter-perturber collision. For LS coupling,

$$\chi_j M_j \Rightarrow \chi_j^0 L_j M_j S M_S, \quad j = i, i', f, f', \qquad (57.89)$$

in (57.79), where χ_j^0 denotes all other quantum numbers required to describe state j that do not change during the collision. Then

$$\left| L_j M_j S M_S l m \tfrac{1}{2} m_s \right\rangle = \sum_{L_j^T M_j^T S^T M_S} C_{M_j m M_S^T}^{L_j l L_j^T} C_{M_S m_s M_S^T}^{S \frac{1}{2} S^T}$$

$$\times \left| L_j S l \tfrac{1}{2} L_j^T M_j^T S^T M_S^T \right\rangle, \qquad (57.90)$$

where $C_{m_1 m_2 m_3}^{j_1 j_2 j_3}$ is a vector coupling coefficient, the superscript T denotes quantum numbers of the emitter-perturber system, and $\tfrac{1}{2}$, m_s are the spin quantum numbers of the scattered electron. On using (57.90), (57.88) is replaced by

$$\langle\langle i'f'^* \|\mathbf{w} + i\mathbf{d}\| if^* \rangle\rangle$$
$$= \pi (\hbar/m_e)^2 N$$

$$\times \sum_{L_i^T L_f^T S^T ll'} (-1)^{L_i + L_{i'} + l + l'} (2L_i^T + 1)(2L_f^T + 1)$$

$$\times \frac{(2S^T + 1)}{2(2S+1)} \left\{\begin{matrix} L_f^T & L_i^T & 1 \\ L_i & L_f & l \end{matrix}\right\} \left\{\begin{matrix} L_f^T & L_i^T & 1 \\ L_{i'} & L_{f'} & l' \end{matrix}\right\}$$

$$\times \int_0^\infty \frac{1}{v} f_e(v) \, dv \left[\delta_{l'l}\delta_{L_{i'}L_i}\delta_{L_{f'}L_f}\right.$$

$$- S_I(L_{i'} S l' \tfrac{1}{2} L_i^T S^T; L_i S l \tfrac{1}{2} L_i^T S^T)$$

$$\left. \times S_F^*(L_{f'} S l' \tfrac{1}{2} L_f^T S^T; L_f S l \tfrac{1}{2} L_f^T S^T)\right], \qquad (57.91)$$

where, for an isolated line, the width and shift are given by (57.91) with $L_{i'} = L_i$ and $L_{f'} = L_f$. For hydrogenic systems, where states i, i' and f, f' with different angular momenta are degenerate, a logarithmic divergence occurs for large values of l and l' [*cf.* Eq. (57.72)], and must be removed by using (57.73) and (57.84).

If a jj coupling scheme is used $\chi_j M_j \Rightarrow \chi_j^0 J_j M_j$, $j = i, i', f, f'$ in (57.79), and

$$\left| J_j M_j l m \tfrac{1}{2} m_s \right\rangle$$

$$= \sum_{J_j^T M_j^T j m'} C_{M_j m' M_j^T}^{J_j j J_j^T} C_{m m_s m'}^{l \frac{1}{2} j} \left| J_j l j J_j^T M_j^T \right\rangle, \qquad (57.92)$$

then (57.91) becomes

$$\langle\!\langle i'f'^* \,\|\mathbf{w} + i\mathbf{d}\|\, if^*\rangle\!\rangle = \pi\,(\hbar/m_{\rm e})^2\,N$$
$$\times \sum_{J_i^T J_f^T jj'll'} (-1)^{J_i+J_{i'}+2J_f^T+j+j'}\,\tfrac{1}{2}(2J_i^T+1)(2J_f^T+1)$$
$$\times \left\{ \begin{matrix} J_f^T & J_i^T & 1 \\ J_i & J_f & j \end{matrix} \right\}\left\{ \begin{matrix} J_f^T & J_i^T & 1 \\ J_{i'} & J_{f'} & j' \end{matrix} \right\}$$
$$\times \int_0^\infty \frac{1}{v}f_{\rm e}(v)\,dv\Big[\delta_{l'l}\delta_{j'j}\delta_{J_{i'}J_i}\delta_{J_{f'}J_f}$$
$$- S_I(J_{i'}l'j'J_i^T;\,J_i l j J_i^T)\,S_F^*(J_{f'}l'j'J_f^T;\,J_f l j J_f^T)\Big], \tag{57.93}$$

where $J_{i'} = J_i$ and $J_{f'} = J_f$ for an isolated line. If the spectrum of the emitter is classified using LS coupling, it is often sufficient to use energies defined by

$$E_{L_jS} = \sum_{J_j} \frac{(2J_j+1)}{(2L_j+1)(2S_j+1)}E_{L_jSJ_j}, \tag{57.94}$$

and obtain the **S**-matrix elements in an LS coupling scheme. They are then transformed to jj coupling by using the algebraic transformation

$$S\left(J_{j'}l'j'J_j^T;\,J_j l j J_j^T\right)$$
$$= [(2J_j+1)(2J_{j'}+1)(2j+1)(2j'+1)]^{1/2}$$
$$\times \sum_{L_j^T S^T}(2L_j^T+1)(2S^T+1)$$
$$\times \left\{ \begin{matrix} L_j & l & L_j^T \\ S & \tfrac{1}{2} & S^T \\ J_j & j & J_j^T \end{matrix} \right\}\left\{ \begin{matrix} L_{j'} & l' & L_j^T \\ S & \tfrac{1}{2} & S^T \\ J_{j'} & j' & J_j^T \end{matrix} \right\}$$
$$\times S\left(L_{j'}Sl'\tfrac{1}{2}L_j^TS^T;\,L_jSl\tfrac{1}{2}L_j^TS^T\right), \tag{57.95}$$

and introducing the splitting of the fine structure components in (57.4) or (57.6). If in LS coupling the line is isolated, but nevertheless the broadened fine structure components overlap significantly, then the interference terms in (57.6) must be included.

57.5.3 Broadening by Atoms

The formal result is very similar to (57.93), but in this case, the relative motion only gives rise to orbital angular momentum. Thus

$$\langle\!\langle i'f'^* \,\|\mathbf{w} + i\mathbf{d}\|\, if^*\rangle\!\rangle = \pi\,(\hbar/\mu)^2\,N$$
$$\times \sum_{J_i^T J_f^T ll'} (-1)^{J_i+J_{i'}+2J_f^T+l+l'}(2J_i^T+1)(2J_f^T+1)$$
$$\times \left\{ \begin{matrix} J_f^T & J_i^T & 1 \\ J_i & J_f & l \end{matrix} \right\}\left\{ \begin{matrix} J_f^T & J_i^T & 1 \\ J_{i'} & J_{f'} & l' \end{matrix} \right\}$$
$$\times \int_0^\infty \frac{1}{v}f(v)\,dv\big[\delta_{l'l}\delta_{J_{i'}J_i}\delta_{J_{f'}J_f}$$
$$- S_I\left(J_{i'}l'J_i^T;\,J_i l J_i^T\right)S_F^*\left(J_{f'}l'J_f^T;\,J_f l J_f^T\right)\big]. \tag{57.96}$$

For many cases of practical interest, transitions of type $J_i \to J_f$ are isolated and so have line profiles given by (57.3), (57.10), and (57.96), where $J_i = J_{i'}$ and $J_f = J_{f'}$ (*cf.* Ref. [9]). In order to obtain the **S**-matrix elements in (57.96), it is usually sufficient to use adiabatic potentials for the emitter-perturber system that have been calculated neglecting fine structure. Since T is typically a few hundred degrees, only coupling between adiabatic states that tend to the appropriate separated-atom limit are retained in the scattering problem. The coupled scattering equations are then solved with fine structure introduced by applying an algebraic transformation to the adiabatic potentials, and using the observed splittings of the energy levels. The Born-Oppenheimer approximation is valid, and details are given in [15].

57.6 ONE-PERTURBER APPROXIMATION

57.6.1 General Approach and Utility

If only one perturber is effective in producing broadening, $I(\omega)$ can be obtained by considering a dipole transition between initial and final states I and F of the emitter-perturber system. Then $P(\omega)$ is given by (57.2), where

$$I(\omega) = [\delta(\omega - \omega_{IF})\,\langle\!\langle IF^*\,|\mathbf{\Delta}|\,IF^*\rangle\!\rangle]_{\rm av},$$
$$\hbar\omega_{IF} = E_I - E_F, \tag{57.97}$$

and av denotes an average over states I and a sum over states F [7]. Wave functions Ψ_J are given by

$$\Psi_J(\mathbf{r},\mathbf{R}) = \mathcal{O}\sum_j \psi_j(\mathbf{r})\,\phi(\mathbf{k}_j,\mathbf{k}_{j_0};\mathbf{R}), \tag{57.98}$$

where $J = I, F$, and $\mathcal{O}$ is an operator that takes account of any symmetry properties of the emitter-perturber system. The energies E_I and E_F are given by (57.80). The perturber wave functions for initial state j_0 and final state j are expanded in the form

$$\phi(\mathbf{k}_j,\mathbf{k}_{j_0};\mathbf{R}) = 2\pi i \sum_{l_{j_0}m_{j_0}l_j m_j} i^{l_{j_0}} k_{j_0}^{-1/2} Y_{l_{j_0}m_{j_0}}^*\left(\hat{\mathbf{k}}_{j_0}\right)$$
$$\times Y_{l_j m_j}\left(\hat{\mathbf{R}}\right)\frac{1}{R}F(\Gamma_j,\Gamma_{j_0};R) \tag{57.99}$$

where Γ_j denotes a channel characterized by

$$\Gamma_j = \chi_j M_j l_j m_j, \quad j = 0,1,2,\ldots, \tag{57.100}$$

[see Eq. (57.79)]. In (57.99), the radial perturber wave function has the limiting forms

$$F(\Gamma_j,\Gamma_{j_0};R) \overset{R\to 0}{\sim} R^{l_j+1}$$
$$\overset{R\to\infty}{\sim} k_j^{-1/2}\big[\delta_{\Gamma_j\Gamma_{j_0}}\exp(-i\theta_j)$$
$$- S_J(\Gamma_j;\Gamma_{j_0})\exp(i\theta_j)\big], \tag{57.101}$$

with

$$\theta_j = k_j R - \frac{1}{2}l_j\pi - \frac{z}{k_i}\ln(2k_j R) + \arg\Gamma\left(l_j + 1 + i\frac{z}{k_j}\right),$$

$$z = \frac{\mu q^2}{\hbar^2}Z_{\rm e}Z_{\rm p}. \tag{57.102}$$

The coupled equations obtained by using (57.12), (57.13), (57.98), and (57.99), where

$$\Psi(\mathbf{r},\mathbf{R},t) = \Psi_J(\mathbf{r},\mathbf{R})\exp(-iE_J t/\hbar), \tag{57.103}$$

are integrated to give functions $F(\Gamma_j,\Gamma_{j_0};R)$. Using (57.98) and (57.99), (57.97) becomes

$$I(\omega) = \frac{1}{2}N\sum_{\substack{\Gamma_{i_0}\Gamma_i\Gamma_{i'} \\ \Gamma_{f_0}\Gamma_f\Gamma_{f'}}} u_{i_0}\langle\!\langle\Gamma_i\Gamma_f{}^*|\mathbf{\Delta}|\Gamma_{i'}\Gamma_{f'}{}^*\rangle\!\rangle$$

$$\times\int_0^\infty \frac{1}{v}f(v)\,dv\mathcal{F}(\Gamma,v), \tag{57.104}$$

where

$$\mathcal{F}(\Gamma,v) = \int_0^\infty F^*(\Gamma_i,\Gamma_{i_0};R)F(\Gamma_f,\Gamma_{f_0};R)\,dR$$

$$\times\int_0^\infty F(\Gamma_{i'},\Gamma_{i_0};R)F^*(\Gamma_{f'},\Gamma_{f_0};R)\,dR, \tag{57.105}$$

$$u_{i_0} = g_{i_0}\Big/\sum_{i'_0}g_{i'_0}, \quad v = \frac{\hbar k_{i_0}}{\mu}. \tag{57.106}$$

The one-perturber approximation is valid when

$$\Delta\omega \equiv |\omega - \omega_{if}| \gg w; \quad V \gg \bar{V}, \tag{57.107}$$

where V is the effective interaction potential required to produce a shift $\Delta\omega$. In the center of the line, many-body effects are always important and the one-perturber approximation diverges as $\Delta\omega \to 0$. In many cases, there is a region of overlap where criteria (57.1) and (57.107) are all valid, but when $\Delta\omega\tau \gg 1$, (57.97) is a static approximation, since the average time between collisions is $\Delta\omega^{-1}$.

57.6.2 Broadening by Electrons

If LS coupling is used, definition of channel Γ_j in (57.100) is replaced by

$$\Gamma_j = L_j S l_j \tfrac{1}{2}L_j^T S^T, \quad j = i_0, i, i', f_0, f, f', \tag{57.108}$$

[*cf.* Eqs. (57.89) and (57.90)]. Then assuming that the weights u_{i_0} of all the levels i_0 that effectively contribute to the line are the same, (57.104) becomes

$$I(\omega)$$

$$= \frac{1}{2}N_{\rm e}\sum_{\substack{\Gamma_{i_0}\Gamma_i\Gamma_{i'} \\ \Gamma_{f_0}\Gamma_f\Gamma_{f'}}}\langle\!\langle L_iS(L_fS)^*\|\mathbf{\Delta}\|L_{i'}S(L_{f'}S)^*\rangle\!\rangle$$

$$\times\,\delta_{l_i l_f}\delta_{l_{i'}l_{f'}}\delta_{L_{i_0}^T L_i^T}\delta_{L_{f_0}^T L_f^T}\delta_{L_{i_0}^T L_{i'}^T}\delta_{L_{f_0}^T L_{f'}^T}$$

$$\times\,(-1)^{L_i+L_{i'}+l_i+l_{i'}}\frac{(2S^T+1)}{2(2S+1)}(2L_i^T+1)(2L_f^T+1)$$

$$\times\left\{\begin{array}{ccc}L_f^T & L_i^T & 1 \\ L_i & L_f & l\end{array}\right\}\left\{\begin{array}{ccc}L_f^T & L_i^T & 1 \\ L_{i'} & L_{f'} & l_{i'}\end{array}\right\}$$

$$\times\int_0^\infty\frac{1}{v}f_{\rm e}(v)\,dv\mathcal{F}(\Gamma,v), \tag{57.109}$$

where $\mathcal{F}(\Gamma,v)$ is defined by (57.105) and (57.108). If the functions $F(\Gamma_{j'},\Gamma_{j_0};R)$ in (57.105) are replaced by their asymptotic forms (57.101), then

$$\mathcal{F}(\Gamma,v) \simeq \Delta\omega^{-2}(\hbar/m_{\rm e})^2$$

$$\times\left[\delta_{\Gamma_{i_0}\Gamma_i}\delta_{\Gamma_{f_0}\Gamma_f} - S_I(\Gamma_i;\Gamma_{i_0})S_F^*(\Gamma_f;\Gamma_{f_0})\right]$$

$$\times\left[\delta_{\Gamma_{i_0}\Gamma_{i'}}\delta_{\Gamma_{f_0}\Gamma_{f'}} - S_I^*(\Gamma_{i'};\Gamma_{i_0})S_F(\Gamma_{f'};\Gamma_{f_0})\right]. \tag{57.110}$$

On substituting (57.110) into (57.109), summing over Γ_{i_0} and Γ_{f_0} and using the unitary property of the $\mathbf{S}$-matrix,

$$I(\omega) = \frac{1}{\pi\Delta\omega^2}\sum_{L_i^T L_f^T S^T l l'}\langle\!\langle L_iS(L_fS)^*\|\mathbf{\Delta}\|L_{i'}S(L_{f'}S)^*\rangle\!\rangle$$

$$\times\langle\!\langle i'f'^*\|\mathbf{w}\|if^*\rangle\!\rangle, \tag{57.111}$$

where $\langle\!\langle i'f'^*\|\mathbf{w}\|if^*\rangle\!\rangle$ is given by (57.91). Line shape (57.111) is identical to that obtained from (57.6) when $\Delta\omega \gg w$. If the jj coupling scheme specified by (57.92) is used, and channel Γ_j is defined by

$$\Gamma_j = J_j l_j j_j J_j^T, \quad j = i_0, i, i', f_0, f, f', \tag{57.112}$$

(57.104) becomes, [*cf.* (57.93)]

$$I(\omega) = \frac{1}{2}N_{\rm e}\sum_{\substack{\Gamma_{i_0}\Gamma_i\Gamma_{i'} \\ \Gamma_{f_0}\Gamma_f\Gamma_{f'}}}\langle\!\langle J_i(J_f)^*\|\mathbf{\Delta}\|J_{i'}(J_{f'})^*\rangle\!\rangle$$

$$\times\,\delta_{l_i l_f}\delta_{l_{i'}l_{f'}}\delta_{j_i j_f}\delta_{j_{i'}j_{f'}}\delta_{J_{i_0}^T J_i^T}\delta_{J_{f_0}^T J_f^T}\delta_{J_{i_0}^T J_{i'}^T}\delta_{J_{f_0}^T J_{f'}^T}$$

$$\times\,(-1)^{J_i+J_{i'}+2J_f^T+j_i+j_{i'}}\frac{1}{2}(2J_i^T+1)(2J_f^T+1)$$

$$\times\left\{\begin{array}{ccc}J_f^T & J_i^T & 1 \\ J_i & J_f & j_i\end{array}\right\}\left\{\begin{array}{ccc}J_f^T & J_i^T & 1 \\ J_{i'} & J_{f'} & j_{i'}\end{array}\right\}$$

$$\times\int_0^\infty\frac{1}{v}f_{\rm e}(v)\,dv\mathcal{F}(\Gamma,v), \tag{57.113}$$

where $\mathcal{F}(\Gamma,v)$ is given by (57.105) and (57.112).

57.6.3 Broadening by Atoms

In the wings of a line where $\hbar|\Delta\omega| \simeq |E_{J_i} - E_{J_{i'}}|$, $j = i_0, i, i', f_0, f, f'$, coupling between the fine structure levels is important. If channel Γ_j is defined by

$$\Gamma_j = J_j l_j J_j^T, \quad j = i_0, i, i', f_0, f, f', \quad (57.114)$$

(57.104) becomes [*cf.* Eq. (57.95)]

$$
\begin{aligned}
I(\omega) = \frac{1}{2} N \sum_{\substack{\Gamma_{i_0}\Gamma_i\Gamma_{i'} \\ \Gamma_{f_0}\Gamma_f\Gamma_{f'}}} & \langle\langle J_i (J_f)^* \| \mathbf{\Delta} \| J_{i'} (J_{f'})^* \rangle\rangle \\
& \times \delta_{l_i l_f} \delta_{l_{i'} l_{f'}} \delta_{J_{i_0}^T J_i^T} \delta_{J_{f_0}^T J_f^T} \delta_{J_{i_0}^T J_{i'}^T} \delta_{J_{f_0}^T J_{f'}^T} \\
& \times (-1)^{J_i + J_{i'} + 2J_f^T + l_i + l_{i'}} (2J_i^T + 1)(2J_f^T + 1) \\
& \times \left\{ \begin{matrix} J_f^T & J_i^T & 1 \\ J_i & J_f & l_i \end{matrix} \right\} \left\{ \begin{matrix} J_f^T & J_i^T & 1 \\ J_{i'} & J_{f'} & l_{i'} \end{matrix} \right\} \\
& \times \int_0^\infty \frac{1}{v} f(v) \, dv \, \mathcal{F}(\Gamma, v), \quad (57.115)
\end{aligned}
$$

where $\mathcal{F}(\Gamma, v)$ is given by (57.105) and (57.114). In the far wings, where $\hbar|\Delta\omega| \gg |E_{J_i} - E_{J_{i'}}|$, $j = i_0, i, i', f_0, f, f'$, an adiabatic approximation is valid. Adiabatic states of the diatomic molecule formed by the emitter-perturber system are considered in which the total spin is assumed to be decoupled from the total orbital angular momentum of the electrons. The coupling between rotational and electronic angular momentum can also be neglected, because typically, contributions to the line profile come from $0 \le l_j \lesssim 400$, whereas $\Lambda_j \lesssim 2$. Therefore, transitions take place between channels defined by

$$\Gamma_j = \Lambda_j L_j S L_j, \quad j = i, f, \quad (57.116)$$

where the unperturbed emitter in state j has quantum numbers $L_j S$, the quantum number Λ_j represents the projection of the orbital angular momentum on the internuclear axis, and (57.98) is replaced by

$$\Psi_J(\mathbf{r}, \mathbf{R}) = \mathcal{O} \sum_j \psi_j(\mathbf{r}; \mathbf{R}) \phi(\mathbf{k}_j, \mathbf{k}_{j_0}; \mathbf{R}), \quad (57.117)$$

where $k_j = k_{j_0}$ and $\psi_j(\mathbf{r}; \mathbf{R})$ is the wave function for molecular state Λ_j. In (57.117), the only molecular states retained are those that correlate with emitter states i and f. The scattering is described by

$$\left[\frac{d^2}{dR^2} - \frac{l_j(l_j+1)}{R^2} - \frac{2z}{R} - \frac{2\mu}{\hbar^2} V_{\Lambda_j}(R) + k_j^2 \right] F_j(R) = 0,$$

where $k_j = k_{j_0}$ and

$$F_j(R) = P_{v_j l_j}(R) \text{ or}$$
$$F_j(R) \equiv F(\Gamma_j, \Gamma_j; R) = F_{k_i l_i}(R)$$

for vibrational or free states, respectively, and $V_{\Lambda_j}(R)$ is the potential energy of state Λ_j. Free-free transitions always contribute to the line profile, but bound-free and free-bound transitions only contribute on the red and blue wings, respectively. On using (57.80), (57.97), and (57.107), $\hbar\Delta\omega = \varepsilon_i - \varepsilon_f$, and ε_j becomes the energy of bound state j with vibrational quantum number v_j when $\varepsilon_j < 0$. If

$$\mathcal{G}(\Gamma, \varepsilon_i, \varepsilon_f) \equiv \left| \int_0^\infty F_i^*(R) \bar{\Delta}(R) F_f(R) \, dR \right|^2 \quad (57.118)$$

[*cf.* Eq. (57.105)], where

$$\bar{\Delta}(R) = -q \int \psi_i^*(\mathbf{r}; \mathbf{R}) \mathbf{r} \psi_f(\mathbf{r}; \mathbf{R}) \, d\mathbf{r} \quad (57.119)$$

is the dipole moment, then using (57.117), the free-free contribution is given by

$$I_0(\omega) = \frac{N}{2} \sum_{\Gamma_i\Gamma_f} u_{\Lambda_i} \delta_{l_i l_f} (2l_i + 1) \int_0^\infty \frac{f(v)}{v} \, dv \, \mathcal{G}(\Gamma, \varepsilon_i, \varepsilon_f),$$

where $\varepsilon_i = \frac{1}{2}\mu v^2$, and u_{Λ_i} is the relative weight of state Λ_i [*cf.* Eqs. (57.104)–(57.106)]. The bound-free contribution is

$$I_1(\omega) = \frac{N}{2} \sum_{\Gamma_i\Gamma_f} u_{\Lambda_i} \delta_{l_i l_f} (2l_i + 1) \sum_i g(\varepsilon_i) \mathcal{G}(\Gamma, \varepsilon_i, \varepsilon_f)$$

and the free-bound contribution is

$$
\begin{aligned}
I_2(\omega) = \frac{1}{2} N \sum_{\Gamma_i\Gamma_f} u_{\Lambda_i} \delta_{l_i l_f} (2l_i + 1) \exp\left(-\frac{\hbar\Delta\omega}{k_B T} \right) \\
\times \sum_f g(\varepsilon_f) \mathcal{G}(\Gamma, \varepsilon_i, \varepsilon_f),
\end{aligned}
$$

where

$$
\begin{aligned}
g(\varepsilon) &= 2\pi \left(\frac{\hbar}{\mu} \right)^2 \frac{f(v)}{v^2} \\
&= 8\pi^2 \left(\frac{\hbar}{\mu} \right)^2 \left(\frac{\mu}{2\pi k_B T} \right)^{3/2} \exp\left(-\frac{\varepsilon}{k_B T} \right),
\end{aligned}
$$

on using (57.21) with $\varepsilon = \frac{1}{2}\mu v^2$. The full line profile is then given by

$$I(\omega) = \sum_{j=0}^2 I_j(\omega). \quad (57.120)$$

The satellite features that are often seen in line wings, arise because turning points in the difference potential $[V_{\Lambda_i}(R) - V_{\Lambda_f}(R)]$ produce a phenomenon analogous to the formation of *rainbows* in scattering theory. The JWKB approximation is often used for the functions $F_{k_j l_j}(R)$, and can be shown to lead to the correct static limit in which transitions take place at fixed values of R called 'Condon points', i.e., the Franck-Condon principle is valid. Further details are given in [7, 8, 16].

57.7 UNIFIED THEORIES AND CONCLUSIONS

The pressure broadening of spectral lines is in general a time-dependent many-body problem and as such cannot be solved exactly. After all, even the problem of two free electrons scattered by a proton is still a subject of active research. There is no practical theory that leads to the full static profile in the limit of high density (or low temperature) and to the full impact profile in the limit of low density (or high temperature). As with so many problems in physics, it is the intermediate problem that is intractable because no particular feature can be singled out as providing a weak perturbation on a known physical situation. However, much progress has been made over the last thirty years in developing theories that take into account many of the key features of the intermediate problem and they are often successful in predicting line profiles for practical applications [6–8, 16]. More recently, time-dependent many-body problems have been tackled using computer-oriented approaches that invoke Monte Carlo and other simulation methods to study line broadening in dense, high-temperature plasmas, see for example [17]. In this chapter, the emphasis has been on aspects of the subject that relate directly to electron-atom and low-energy atom-atom scattering. Many experts in the fields of electron-atom and atom-atom collisions are still not exploiting the direct applicability of their work to line broadening. It is hoped that this contribution will encourage more research workers to study these fascinating problems that not only provide links with plasma physics and in particular with the physics of fusion plasmas, but also with a quite distinct body of laboratory-based experimental data.

REFERENCES

1. J. R. Fuhr, W. L. Wiese, and L. J. Roszman, *Bibliography on Atomic Line Shapes and Shifts*, National Bureau of Standards, Special Publication 366 (1972).
2. J. R. Fuhr, L. J. Roszman, and W. L. Wiese, *Bibliography on Atomic Line Shapes and Shifts*, National Bureau of Standards, Special Publication 366, Supplement 1 (1974).
3. J. R. Fuhr, G. A. Martin, and B. J. Specht, *Bibliography on Atomic Line Shapes and Shifts*, National Bureau of Standards, Special Publication 366, Supplement 2 (1974).
4. J. R. Fuhr, B. M. Miller, and G. A. Martin, *Bibliography on Atomic Line Shapes and Shifts*, National Bureau of Standards, Special Publication 366, Supplement 3 (1978).
5. J. R. Fuhr and A. Lesage, *Bibliography on Atomic Line Shapes and Shifts*, National Institute of Standards and Technology, Special Publication 366, Supplement 4 (1992).
6. H. R. Griem, *Spectral Line Broadening in Plasmas* (Academic Press, New York, 1974).
7. G. Peach, Adv. Phys. **30**, 367 (1981).
8. N. Allard and J. R. Kielkopf, Rev. Mod. Phys. **54**, 1103 (1982).
9. E. L. Lewis, Phys. Rep. **58**, 1 (1980).
10. S. Sahal-Bréchot, Astron. & Astrophys. **1**, 91 (1969); *ibid.* **2**, 322 (1969).
11. M. S. Dimitrijević and N. Konjević, J. Quant. Spectrosc. Radiat. Transfer **24**, 451 (1980).
12. G. K. Oertel and L. P. Shomo, Astrophys. J. Suppl. **16**, 175 (1969).
13. M. J. Seaton, J. Phys. B **21**, 3033 (1988); *ibid.* **22**, 3603 (1989).
14. V. I. Kogan, V. S. Lisitsa, and G. V. Sholin, Rev. Plasma Phys. **13**, 261 (1987).
15. P. J. Leo, G. Peach, and I. B. Whittingham, J. Phys. B **28**, 591 (1995).
16. J. Szudy and W. E. Baylis, J. Quant. Spectrosc. Radiat. Transfer **15**, 641 (1975); **17**, 269 (1977); *ibid.* Phys. Rep. (to be published).
17. A. Calisti, L. Godbert, R. Stamm, and B. Talin, J. Quant. Spectrosc. Radiat. Transfer **51**, 59 (1994).

58

Photodetachment

David J. Pegg

Department of Physics, University of Tennessee, Knoxville, Tennessee 33796 and Physics Division, Oak Ridge National Laboratory, Oak Ridge, Tennessee 37831

58.1 INTRODUCTION

Interest in photon-ion interactions has grown rapidly over the past few decades due primarily to the increased availability of laser and synchrotron light sources. Below a photon energy of about 1 keV, the dominant radiative process is the electric-dipole induced photoelectric effect (see Sect. 59.1). In the gaseous phase, the photoelectric effect is referred to as either photoionization (atoms and positive ions) or photodetachment (negative ions). This chapter briefly reviews the current status of the field of photodetachment. The emphasis is on experimental studies of atomic negative ions.

Photodetachment is a fundamental ionization mechanism in which electrons are detached from negative ions following absorption of a photon of frequency ω. Typical quantities measured are the electron affinity A_e, the asymmetry parameter β, the photoabsorption cross section $\sigma_\gamma(\omega, \theta)$ and parameters associated with autodetaching resonances. Most work has concentrated on the sim-

plest process in which the absorption of a single photon results in the ejection of a single valence shell electron. Inner shell electron detachment in negative ions has not yet been studied. Higher-order processes involving the absorption of more than one photon and the detachment of more than one electron have been made possible by the use of intense and energetic photon sources (see Sect. 50.3.3). The monographs of Massey [1] and Smirnov [2] offer good introductions to the subject of negative ions and photodetachment. Reviews by Lineberger [3], Miller [4] and Esaulov [5] are also valuable sources of information. Other reviews have emphasized the subject of the photodetachment of molecular negative ions [6].

58.2 THEORETICAL CONCEPTS

Photodetachment involves a bound-free radiative transition of the general type

$$A^- + \hbar\omega \longrightarrow A_n + e^-, \qquad (58.1)$$

where A_n denotes the residual neutral atom or molecule, possibly in excited state n. Since the recoil energy of A_n can be neglected, conservation of energy gives

$$\hbar\omega = A_e + \Delta E_n + E_f, \qquad (58.2)$$

where ΔE_n is the excitation energy of A_n, and E_f is the excess energy carried off by the free electron. The threshold energy for channel n is thus $A_e + \Delta E_n$.

For electric-dipole induced photodetachment, the selection rules for electric dipole transitions apply (see Sect. 23.3.1). In the independent particle approximation, this restricts the angular momentum l_f of the outgoing electron partial waves to $l_f = l_i \pm 1$, where l_i is the initial angular momentum of the ejected electron.

Electron-electron correlation effects (see Sect. 23.3.1) play an important role in the stabilization of a negative ion A^-, and the photodetachment process because there is no long-range Coulomb force to bind the attached electron. Instead, it moves in the short-range polarization forces of A, both before and after detachment. Correlation is even more important in double photoexcitation and detachment. A doubly excited state of A^- consists of a pair of strongly correlated electrons moving in the field of the A^+ core. In double photodetachment, the incident photon energy $\hbar\omega$ exceeds $A_e + I_P$ where I_P is the ionization potential of A. Just above threshold, the pair of strongly correlated electrons slowly recedes from the A^+ core. This threshold region is of long-standing theoretical interest (see Sect. 50.3.1).

58.3 EXPERIMENTAL TECHNIQUES

58.3.1 Sources of Ions

Photodetachment experiments are limited by the difficulty of producing a source of negative ions with a density sufficiently high for spectroscopic measurements. Negative ions are weakly bound systems that are easily destroyed in inter-particle collisions. The production of a practical source is a challenge since processes involved in ion creation must compete with more probable destruction processes. Sources are designed to concentrate negative ions into a region of space that can be efficiently overlapped with a spatially small but intense beam of photons from a laser or other light source. For example, negative ions have been stored in an electromagnetic trap [7]. Here, photodetachment can be investigated by monitoring the depletion of the number of ions in the trap in the presence of laser radiation. However, the most commonly used and versatile spectroscopic source of negative ions is a tenuous beam produced by an accelerator.

In an accelerator-based apparatus, a beam of negative ions of essentially any chosen energy can be produced either directly in the ion source of the accelerator or indirectly by doubly charge changing a beam of positive

Figure 58.1. Schematic of a crossed beam apparatus used in photodetachment measurements.

ions in flight. After production, the fragile negative ions travel through a high vacuum environment to the photon-ion interaction region. Accelerator beams have the advantages that they are essentially unidirectional and monoenergetic, and it is relatively easy to manipulate the ions of a beam, in flight, by applying external electric or magnetic fields. The beams can, for example, be made elementally and isotopically pure by high resolution mass selection in a magnetic field.

58.3.2 Interacting Beams

The well-defined spatial dimensions of an ion beam (limited radially by apertures) readily permit an efficient overlap to be made with a spatially small but intense beam of monochromatic photons from a laser. The two interacting beams are usually mated in either a crossed beam geometry ($\theta = 90°$) or a collinear beam geometry ($\theta = 0°$ or $180°$). Figure 58.1 is a schematic diagram of the crossed beam apparatus used by Pegg *et al.* [8]. The choice of a particular interacting beam arrangement depends on the type of particles to be detected, the detection geometry and the level of sensitivity and resolution required in the experiment. Both continuous wave (CW) and pulsed lasers have been used. The time structure of pulsed lasers makes them a popular choice since they are naturally suited to synchronous detection schemes that discriminate against background contributions by gating the detector on and off in coincidence with the presence or absence of the laser radiation. This procedure partially compensates for the low duty cycle of a pulsed laser. In addition, the higher peak power available from pulsed lasers is indispensable in some experiments. Continuous wave lasers are preferred in high resolution measurements. In the case of CW lasers, the power in the interaction region can be enhanced by performing experiments in which the ion beam passes through the laser cavity [9].

The energy resolution obtainable in interacting beam experiments is generally determined by either instrumental broadening associated with an electron energy analyzer, or kinematic broadening associated with the motion of the ions. The linewidth and divergence of the laser beam are, in comparison, typically negligible. In the crossed beam geometry, kinematic broadening is primarily associated with the angular divergence of the ion beam. Much higher resolution can be obtained by use of the collinear beam geometry. In this case, the broadening associated with ion beam divergence is small. Instead the dominant source of broadening is the finite spread in the velocities of the ions in the direction of the laser beam. However, this spread is kinematically compressed in the process of accelerating the beam ions [10] and very narrow linewidths can be obtained if the acceleration voltage is well stabilized. In addition, the larger overlap of the two beams in the collinear geometry reduces broadening due to the finite interaction time, and simultaneously enhances sensitivity. Thus the collinear beam arrangement has advantages over the crossed-beam arrangement in both sensitivity and resolution. For practical reasons, however, essentially all photoelectron channel-specific measurements are performed in the crossed-beam geometry since it is difficult to collect electrons efficiently over an extended interaction region. There are, of course, exceptions. Mead *et al.* [11] have reported on a successful collinear beam experiment that was designed to collect threshold electrons over a 30 cm interaction length. The trajectories of these low energy electrons were constrained by a longitudinal magnetic field to the forward direction where they were efficiently collected. One important advantage of the crossed beam arrangement is that it is well suited to photoelectron angular distribution measurements using linearly polarized radiation.

58.3.3 Detection Schemes

In an interacting beam apparatus, the photodetachment process is monitored by detecting one of the product particles, i.e., photoelectrons or neutral atoms (molecules). In general, neutral atoms can be collected with a much higher efficiency than photoelectrons since they move in essentially a single direction – the direction

of motion of the ion beam. In this case, the sensitivity can also be enhanced by the choice of a collinear beam arrangement. However, measurements employing neutral particle detection are not channel-specific and are therefore limited to determining the sum of the partial cross sections for all open channels, unless some method is employed to monitor the excitation state of the neutral. State selective detection is most efficiently achieved by the process of resonance ionization [12].

The detection of photoelectrons is preferable to neutral atom detection if more than one channel is open (i.e., for energies above the first excited state threshold). In this case, energy analysis of the electrons permits one to isolate a particular channel and determine channel-specific quantities such as partial cross sections, branching ratios and asymmetry parameters. The crossed-beam geometry is favored in photoelectron spectroscopy measurements since the detached electrons are more easily collected from the relatively small interaction volume. The electron detector is usually placed either perpendicular or parallel to the ion beam direction. However, the fraction of electrons collected is generally quite low since electron emission is characterized by a broad angular distribution. Consequently, only a small fraction of the detached electrons can be collected by a detector set at a particular angle. There is an exception however. In the laboratory frame, kinematics associated with the motion of the ions amplifies the ion-frame energies of the detached electrons and also causes them to be swept into the direction of motion of the ion beam. Thus, the detection of low energy electrons, detached from a fast ion beam, can be made quite efficient if the electrons are collected in the forward direction. In addition, since the electrons are emitted in essentially a single direction, the laser and ion beams can, in principle, be superposed collinearly to enhance the sensitivity and resolution of the measurement. The major drawback in collecting electrons in the direction of the ion beam is that the background from collisionally-generated electrons is large. However, these electrons arrive randomly in time at the detector, in contrast to the photoelectrons that arrive in pulses synchronized to the time structure of the laser. It is therefore a simple matter to discriminate against them using a gated detection scheme.

58.3.4 Kinematic Modifications

As a consequence of the fast and unidirectional nature of the beam source, the kinetic energies, yields, and energy and angular distributions of photoelectrons ejected in flight from the moving ions are modified by kinematics. Kinematic peak shifting and doubling can be exploited. Kinematic effects are most pronounced for high energy ion beams and forward-directed electron collection. Details of how these effects can be exploited in photodetachment experiments using high energy beams have been reported by Pegg *et al.* [13].

58.4 RESULTS

58.4.1 Perspectives

Pioneering experiments to measure absolute photodetachment cross sections began in the late 1950s with the development of a crossed beam apparatus by Branscomb and Fite [14], as reviewed by Branscomb [15]. These challenging pre-laser experiments involved the use of a tenuous beam of negative ions from a small accelerator that was crossed perpendicularly with a beam of photons from an arc lamp. A second generation of experiments began in the late 1960s, when the conventional light source was replaced by a laser. This modification greatly enhanced the photon fluxes and resolution, and made it possible to perform energy- and angle-resolved spectroscopic measurements on photoelectrons detached from beams of both atomic and molecular negative ions [9]. Most of the experiments of this period were designed either to measure electron affinities and asymmetry parameters, or to determine relative changes in cross sections, in the vicinity of thresholds and resonances, as reviewed by Lineberger *et al.* [16]. The advent of the laser also made it possible in 1965 to investigate multiphoton detachment processes [17]. The first double-photodetachment measurement was performed in 1982 [18].

58.4.2 Electron Affinities

The electron affinity A_e of an atom or molecule is numerically equal to the binding energy of the attached electron and so it is also an indication of the stability of the ion. The A_e of atoms, for example, range from about 0.02 eV for Ca to 3.6 eV for F. The first detachment thresholds of most negative ions can therefore be accessed by the use of infrared or visible light sources. The A_e's are at least an order of magnitude smaller than the corresponding I_P's of atoms and positive ions. Hotop and Lineberger [19] have published a critical compilation of electron affinities of atoms, and corresponding values for molecules have also been tabulated [20].

Two experimental methods have been developed to determine A_e. The most direct method, and potentially the most accurate, is called the laser photodetachment threshold (LPT) method. Here, the onset of production of one of the photodetachment products, i.e., photoelectrons or neutral atoms (molecules) is monitored as a laser is tuned across the threshold for the opening of a detachment channel. By fitting the data just above threshold to the appropriate Wigner law (see Sect. 58.4.5) one can

accurately extrapolate to determine the photon energy at threshold. The sharper onset associated with s-wave detachment makes this task easier than for p-wave detachment. The accuracy of the method is, in principle, limited only by kinematic broadening, the linewidth of the laser, and calibration of the wavelength scale. The most accurate measurements involve the use of collinear beams [11]. In these experiments, the first-order Doppler shift is eliminated by separately determining the threshold energy using both co-propagating and counter-propagating beams. Narrow bandwidth CW lasers are usually used to enhance the resolution, but Hanstorp *et al.* [21] have demonstrated that accurate LPT measurements can also be made using pulsed lasers.

A second method of determining A_e is called laser photodetached electron spectrometry (LPES). Here, a fixed frequency laser can be used. If the photon energy is greater than the threshold value, the excess energy E_f appears as kinetic energy of the detached electron [see Eq. (58.2)]. Energy analysis of the photoelectrons serves to determine the binding energy of the electron and hence A_e. Kinematic corrections are applied to the spectral data in order to determine the ion-frame energies of the detached electrons. The accuracy of the LPES method is often limited by the energy resolution of the electron energy analyzer, although broadening associated with the motion of the beam ions may become important in high resolution measurements. Breyer *et al.* [22], for example, have reported on an LPES study of O^- and S^- with an energy resolution of 5 meV. This was sufficiently high to resolve fine structure components.

58.4.3 Photoelectron Angular Distributions

Several theoretical treatments of the subject of angular distributions (see Sect. 59.1.1) have been published [23]. The angular distribution of photoelectrons ejected in a single-photon detachment process takes the form $\sigma_\gamma(\omega, \theta) = \sigma_\gamma^{tot}(\omega) \, f(\omega, \theta)$, where

$$f(\omega, \theta) = \frac{1 + \beta(\omega)P_2(\cos\theta)}{4\pi} \qquad (58.3)$$

when a randomly oriented target of negative ions is illuminated with linearly polarized electric-dipole radiation. Here, θ is the angle between the polarization vector of the laser beam and the emission direction of the ejected photoelectron and β is the asymmetry parameter, a quantity that completely characterizes the shape of the emission pattern. β is a measure of the deviation of the distribution from isotropy ($\beta = 0$), and the range is restricted to $-1 \le \beta \le 2$. The value $\beta = 2$ corresponds to the classical distribution of $\cos^2\theta$ in which the electrons are preferentially ejected in the direction of the polarization vector. In general, the two partial waves $l_f = l_i \pm 1$ (see Sect.

58.2) interfere to determine values of $\beta \neq 2$. Cooper and Zare [24] have derived an analytical expression (59.8) for β based on the central-field approximation.

A crossed beam geometry facilitates the measurement of $f(\theta)$, and has been used in most experiments to date (an exception is the collinear beam measurement of Hanstorp *et al.* [25]). In the crossed-beam arrangement, the polarization vector of the laser beam lies in a plane containing the ion beam and it can easily be rotated in this plane by the use of a $\lambda/2$ phase retarder such as a double Fresnel rhomb. The electron collection direction is fixed either perpendicular or parallel to the ion beam direction. In this manner the photoelectron emission pattern is effectively swept past the detector along with the rotating polarization vector. The angular scale of the polarization rotator can be calibrated by use of a known $\cos^2 \theta$ distribution, such as in the case of Li$^-$ photodetachment. Once the calibration is made it is sufficient to measure the photoelectron yield at just two angles, e.g., $\theta = 0°$ and $90°$. Corrections are necessary for the finite solid angle of collection, deviations of the linear polarization fraction from unity, and kinematic transformations of angles. Blondel *et al.* [26] recently reported the results of the first angular distribution measurements on photoelectrons ejected in multiphoton detachment events.

58.4.4 Photodetachment Cross Sections

The $\sigma_\gamma(\omega)$ as a function of ω begin at zero at threshold, rise to a maximum at an energy equal to a few times the threshold energy, and then decrease monotonically at higher energies. Peak cross sections for atomic negative ions typically lie in the range 1–100 Mb (1 Mb = 10^{-18} cm^2). Partial cross sections refer to a particular final state channel while total cross sections are the sum of partial cross sections for all open channels. Figure 58.2 shows the total cross section for photodetachment of H$^-$ as calculated by Broad and Reinhardt [27].

In the case of interacting beams, σ_γ is determined by measuring the yield $Y(\omega, \theta)$ of product particles per unit volume for known values of the ion beam density ρ and photon flux $I(\omega)$ according to

$$\sigma_\gamma(\omega, \theta) = \frac{GY(\omega, \theta)}{\rho I(\omega)}. \qquad (58.4)$$

Here, G incorporates the efficiency for collection and detection of the product particles as well as the spatial and temporal overlap of the laser and ion beams. In the case of photoelectron detection, measurements at two angles are needed to determine $f(\theta)$, and hence σ_γ^{tot} (see Eq. 58.3). The first absolute measurements of σ_γ were made by Smith and Branscomb [28], who detected photoelectrons perpendicular to the plane of

Figure 58.2. Theoretical single-photodetachment and double-photodetachment cross sections for H$^-$ [27].

crossed ion and photon beams. Small electric and magnetic fields were used to ensure that all the electrons produced in the photodetachment event were collected. The signal-to-background ratio was enhanced by the use of a synchronous detection scheme. The result of this impressive pre-laser experiment was suprisingly accurate (10%) for an absolute measurement. Since then, nearly all absolute measurements have involved the detection of neutral atoms using a collinear beam arrangement. In the Li$^-$ cross section measurement by Bae and Peterson [29], for example, the two beams were merged over a path length of 30 cm and the residual atoms were monitored by detecting the secondary electrons produced when the atoms impinged on a glass plate. The quoted accuracy of this type of absolute measurement is typically 30%. In all absolute cross section experiments, the accuracy is ultimately limited by the difficulty of measuring the quantity G. This problem can be circumvented by determining, under the same experimental conditions, the cross section of the species of interest relative to that of a reference species (*cf.* Sect. 58.3.4). The ratio is normalized to the known cross section of the reference species. For example, in a crossed beam experiment to determine σ_γ for Li$^-$, Dellwo *et al.* [30] used energy- and angle-resolved spectroscopy to measure the yields and angular distributions of photoelectrons produced by Li$^-$ ions and D$^-$ ions. Tight collimation was necessary to keep the overlap of the interacting beams essentially unaltered during the relative measurement. Under this condition, the yield ratio was assumed to be independent of the beam overlap. Corrections were made to account for quantities that differ for the two species such as ion beam intensities and velocities, photoelectron angular distributions and kinematically-modified emission solid angles. The dependence of the ratio on the electron collection and detection efficiency was eliminated by choosing the energies of the two ion beams such that the kinematics shifted the energies of the two groups of electrons to the same value in the laboratory frame. The ratio $\sigma_\gamma(\text{Li}^-)/\sigma_\gamma(\text{D}^-)$ was measured with an estimated accuracy of about 7%. The ratio was normalized to a theoretical $\sigma_\gamma(\text{H}^-)$, known to be better than 3%. The estimated accuracy of the measured $\sigma_\gamma(\text{Li}^-)$ is ~10%.

Saturation effects become important if the probability p of photodetachment per ion approaches unity, and the ratio $Y(\omega, \theta)/I(\omega)$ in Eq. (58.3) is no longer independent of $I(\omega)$. In this case, a saturation curve in which Y is measured as a function of I can be used to determine σ_γ. For example, Dellwo *et al.* [30] in the above measurements for Li$^-$ and D$^-$ used the functional form $Y(I) \propto 1 - e^{-p(I)}$ to determine $p(I)$ for each case. An apparent cross section ratio was determined from the two fitted curves for each value of the photon flux. The change in the ratio with photon flux demonstrated the effects of saturation. The ratio was extrapolated to zero photon flux and found to be in good agreement with the value obtained from data taken solely in the linear (unsaturated) regime.

Single-photon detachment cross sections have been measured over a limited range of photon energies for less than two dozen atomic species, and for even fewer molecular species. Theoretical information on photodetachment cross sections is even more sparse. Calculations of photodetachment cross sections involve evaluating the dipole matrix element connecting the initial and final states of the bound-free radiative transition. The process of photodetachment is more sensitive to electron correlation both in the initial and final states than the process of photoionization. As a consequence, it is important to incorporate correlation directly into the wave functions representing both the bound and continuum states. Accurate calculations have been limited primarily to the pure two-electron ion H$^-$ [31] and effective two-electron systems, such as He$^-$ [32], and Li$^-$ [33] and the alkali metals negative ions [34]. Recently, there has been

renewed interest in multiphoton detachment (MPD) following the pioneering measurement in 1965 of the two-photon detachment generalized cross section for the I$^-$ ion [17]. The current status of this field has been recently reviewed [35]. Blondel and co-workers [36], for example, have measured the cross sections for three-photon detachment of F$^-$, Br$^-$, and I$^-$, and four-photon detachment of Cl$^-$ using a crossed beam apparatus. Kwon *et al.* [37] used a Penning trap source for their measurements of the cross section for two- and three-photon detachment of F$^-$. Cross sections for the multiphoton detachment of atomic negative ions are small since the process is necessarily nonresonant. The lack of bound excited states in negative ions preclude resonant enhancement of MPD cross sections. Recently, observations of excess photon detachment have been reported [38]. Here, intense laser fields cause an electron to absorb more photons than are necessary for detachment. The process is analogous to above-threshold ionization (see Sect. 72.2) previously observed in atoms except that the simple structure of a negative ion can make interpretation less complicated.

58.4.5 Threshold Behavior

Several valuable reviews have been published on this subject [39]. The threshold behavior for photodetachment differs fundamentally from that for photoionization (see Sects. 50.3.1 and 59.1.2) due to the difference in the nature of the final state interaction in the two cases. In single-electron detachment processes, an electron moves in the field of the residual core, which is neutral in the case of photodetachment and positively charged in the case of photionization. Wigner [40] showed that for a two-particle final state, the behavior of the reaction cross section in the close vicinity of a channel opening depends solely on the asymptotic form of the final state interaction. The dynamics of escape are determined by the longest range interaction in the final state. In photoionization, the interaction is Coulombic and the cross section starts with a finite value at threshold. In contrast, in photodetachment the longest range interaction is usually the repulsive centrifugal barrier force. As a consequence of this relatively short range final state interaction, the threshold behavior in photodetachment exhibits a dependence on l_f. The Wigner threshold law for photodetachment can be written as

$$\sigma \propto E_f^{l_f + 1/2} .$$

The validity of the Wigner law has been verified in the case of s-wave and p-wave single-photon detachment [41]. The Wigner threshold law is really only the first term of a power series expansion in the electron momentum, and deviations appear when higher order terms become important. O'Malley [42] showed that the

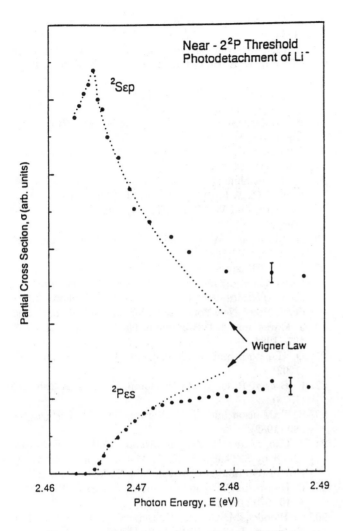

Figure 58.3. Anomalous behavior in the vicinity of the first excited state threshold in the photodetachment of Li⁻ [43].

range of validity of the law will be affected by short-range interactions such as induced polarization, static multipole moments of the residual neutral atom and the presence of nearby resonances. The Li⁻ excited-state threshold data of Dellwo *et al.* [43] appears to deviate from the expected $E^{1/2}$ threshold dependence about 7 meV above threshold. In contrast, in the analogous transition in Cs⁻, the deviation was observed [44] to occur only 25 μeV above threshold. This reduction in the range of validity for Cs⁻ is caused by the enhanced polarization associated with the heavier alkali-metal atom and the presence of a Feshbach resonance.

The Li⁻ cross section studied by Dellwo *et al.* [30] exhibits anomalous behavior in the $2\,^2S\epsilon p$ channel at the threshold for the opening of the $2\,^2P\epsilon s$ channel. Figure 58.3 shows the *Wigner cusp*, which arises as a result of the two photodetachment channels being strongly coupled via configuration interaction. The partial cross section

for *p*-wave detachment is seen to drop sharply as a result of the *s*-wave detachment channel opening with an infinite slope at threshold.

Single-photon detachment threshold behavior is expected to change in the presence of an external electric or magnetic field [45]. Oscillations are predicted in the near-threshold cross section in both cases. In the magnetic field case, these oscillations have been observed [46] using trapped ions. The observation of a threshold shift due to the presence of an intense laser field has also been reported by Trainham *et al.* [47]. In this two-color experiment, an intense beam of infrared radiation was used to detach the Cl⁻ ion via two-photon absorption, and a weaker beam of ultraviolet radiation was used to probe the resulting threshold shift. These investigators also demonstrated that the Wigner law holds for multiphoton as well as for single-photon detachment thresholds. Later, Tang *et al.* [48] confirmed the validity of the Wigner law in multiphoton detachment involving two- to five-photon absorption by the H⁻ ion.

The process of double photodetachment, and its strongly correlated dynamics, are discussed in Sect. 58.2 and in Chap. 50. The theoretical double photodetachment cross section for H^- is shown by the dotted line in Fig. 58.2. The Wannier theory [49–51] (see Sect. 50.3.1) predicts a threshold behavior of $\sigma_\gamma \propto E_f^{1.127}$. Recently, Temkin [52] suggested a different escape mechanism based on an asymmetric model in which one of the electrons remains closer to the core than the other during escape. The threshold behavior in this case should exhibit an oscillatory energy dependence. Three separate double-photo-detachment experiments [53] have been performed to date, but in all cases, due to the presence of large backgrounds, it was not possible to determine conclusively whether the threshold data was best fit by the Wannier or Temkin threshold laws.

58.4.6 Resonances

Resonance structure is often superposed on photodetachment cross section curves over narrow energy bands in the vicinity of detachment thresholds to form excited states A_n of the residual neutral atom. These resonances are called either Feshbach or shape resonances depending on whether they lie below or above the threshold, respectively. Both types of resonances can be seen at an energy of $\hbar\omega \simeq 10$ eV in the theoretical cross section curve for the photodetachment of H⁻ shown in Fig. 58.2. Resonance structure arises from an interference between the amplitudes for direct single-photon detachment of the ion and the transient photoexcitation and subsequent decay of a doubly-excited state of the negative ion by autodetachment (see Sect. 25.1.2). The energy and width of the resonance state can be found by fitting the experimental data to the Fano function [54]. Doubly excited states are particularly interesting since they owe their very existence to strong correlation. The electrons drastically affect each other's behavior and the independent electron model of atomic structure begins to break down.

Feshbach resonances in photodetachment cross sections were first observed in the heavier alkali metal negative ions. Patterson, *et al.* [55] reported very narrow window resonances appearing just below the first excited state thresholds. A number of autodetaching resonances in the vicinity of the $n = 2 - 7$ thresholds of H have been reported by Bryant and his co-workers [56]. There exists good agreement between theory and experiment for this simple pure two-electron system [57]. Analogous Feshbach resonances have also been recently observed by Berzinsh *et al.* [58] in the effective two-electron ion, Li$^-$. In the metastable He$^-$ ion a shape resonance was predicted [59] to lie approximately 11 meV above the He $(2\ ^3P)$ threshold. This large resonance (the peak cross section is about 10^4 Mb) has been experimentally observed by Peterson and Bae [60].

REFERENCES

1. H. S. W. Massey, *Negative Ions* (Cambridge University Press, London, 1976).
2. B. M. Smirnov, *Negative Ions* (McGraw-Hill, New York, 1972).
3. T. M. Miller, Adv. Electron. Electron Phys. **55**, 119 (1981).
4. W. C. Lineberger, Appl. At. Coll. Phys. **5**, 239 (1982).
5. A. V. Esaulov, Ann. Phys. Fr. **11**, 493 (1986).
6. R. R. Cordermann and W. C. Lineberger, Ann. Rev. Phys. Chem. **30**, 347 (1979); R. D. Mead, A. E. Stevens, and W. C. Lineberger, Gas Phase Ion Chem. **3**, 213 (1984); B. K. Janousek and J. I. Braumann, Gas Phase Ion Chem. **2**, 53 (1979); P. S. Drzaic, J. Marks, and J. I. Braumann, Gas Phase Ion Chem. **3**, 167 (1984).
7. D. J. Larson and R. Stoneman, Phys. Rev. A **31**, 2210 (1985); B. K. Janousek and J. I. Braumann, Gas Phase Ion Chem. **2**, 53 (1979).
8. D. J. Pegg, J. S. Thompson, R. Dellwo, R. N. Compton, and G. D. Alton, Phys. Rev. Lett. **64**, 278 (1990).
9. B. Brehm, M. A. Gusinow, and J. L. Hall, Phys. Rev. Lett. **19**, 737 (1967); M. W. Siegel, R. J. Celotta, J. L. Hall, J. Levine, and R. A. Bennett, Phys. Rev. A **6**, 607 (1972); R. J. Celotta, R. A. Bennett, J. L. Hall, M. W. Siegel, and J. Levine, Phys. Rev. A **6**, 631 (1972).
10. S. L. Kauffman, Opt. Comm. **17**, 309 (1976).
11. R. D. Mead, K. R. Lykke, W. C. Lineberger, J. Marks, and J. I. Braumann, J. Chem. Phys. **81**, 4883 (1984).
12. P. Balling, P. Kristensen, H. Staplefeldt, T. Anderson, and H. K. Haugen, J. Phys. B. **26**, 3531 (1993).
13. D. J. Pegg, J. S. Thompson, R. N. Compton, and G. D. Alton, Nucl. Instrum. Methods B **40/41**, 221 (1989); D. J. Pegg, in *The Physics of Electronic and Atomic Collisions*, edited by A. Dalgarno, R. Freund, P. Koch, M. Lubell, and T. Lucatorto (American Institute of Physics, New York, 1990), p. 233.

14. L. M. Branscomb and W. L. Fite, Phys. Rev. **93**, 615A (1954).
15. L. M. Branscomb in *Atomic and Molecular Processes*, edited by D. R. Bates (Academic Press, New York, 1962). p. 100.
16. W. C. Lineberger in *Laser Spectroscopy*, edited by R. Brewer and A. Mooradian (Plenum, New York, 1974); R. R. Corderman and W. C. Lineberger, Ann. Rev. Phys. Chem. **30**, 347 (1979).
17. J. L. Hall, E. J. Robinson, and L. M. Branscomb, Phys. Rev. Lett. **14**, 1013 (1965).
18. J. B. Donahue, P. A. M. Gram, M. V. Hynes, R. W. Hamm, C. A. Frost, H. C. Bryant, K. B. Butterfield, D. A. Clark, and W. W. Smith, Phys. Rev. Lett. **48**, 1538 (1982).
19. H. Hotop and W. C. Lineberger, J. Phys. Chem. Ref. Data **14**, 731 (1985).
20. A. A. Christodoulides, D. L. McCorkle, and L. G. Christophorou, in *Electron-Molecule Interactions and their Applications*, edited by L. G. Christophorou (Academic Press, New York, 1984), Vol. 2, p. 423; P. S. Drzaic, J. Marks, and J. I. Braumann, Gas Phase Ion Chem. **3**, 167 (1984).
21. D. Hanstorp and M. Gustafsson, J. Phys. B **25**, 1773 (1992).
22. F. Breyer, P. Frey, and H. Hotop, Z. Phys. A **286**, 133 (1978).
23. S. T. Manson and A. F. Starace, Rev. Mod. Phys. **54**, 389 (1982).
24. J. Cooper and R. Zare in *Atomic Collision Processes*, edited by S. Geltman, K. T. Mahanthappa, and W. E. Brittin (Gordon and Breach, New York, 1969). p. 317.
25. D. Hanstorp, C. Bengtsson, and D. J. Larson, Phys. Rev. A **40**, 670 (1989).
26. C. Blondel, M. Crance, C. Delsart, A. Girand, and R. Trainham, J. Phys. B **23**, L685 (1990).
27. J. T. Broad and W. P. Reinhardt, Phys. Rev. A **14**, 2159 (1976).
28. L. M. Branscomb and S. J. Smith, Phys. Rev. **98**, 1028 (1955).
29. Y. K. Bae and J. R. Peterson, Phys. Rev. A **32**, 1917 (1985).
30. J. Dellwo, Y. Liu, C. Y. Tang, D. J. Pegg, and G. D. Alton, Phys. Rev. A **46**, 392 (1992).
31. For example, J. T. Broad, and W. P. Reinhardt, Phys. Rev. A **14**, 2159 (1976); H. R. Sadeghpour and C. H. Greene, Phys. Rev. Lett. **65**, 313 (1990); C. D. Lin, Adv. At. Mol. Opt. Phys. **22**, 77 (1986); Y. K. Ho, Phys. Rev. A bf 45, 148 (1992).
32. For example, A. U. Hazi, and K. Reed, Phys. Rev. A **24**, 2269 (1981); S. Watanabe, Phys. Rev. A **25**, 2074 (1982); H. P. Saha and R. N. Compton, Phys. Rev. Lett. bf 64, 1510 (1990).
33. C. Pan, A. F. Starace, and C. H. Greene, J. Phys. B **27**, L137 (1994); E. Lindroth, Phys. Rev. A **52**, 2737 (1995).
34. D. L. Moores and D. W. Norcross, Phys. Rev. A **10**, 1646 (1974).
35. M. Crance, Comments At. Mol. Phys. **24**, 95 (1990); M. D. Davidson, H. G. Muller, and H. B. van den Heuvell, Comments At. Mol. Phys. **29**, 65 (1993).
36. C. Blondel, R.-J. Champeau, M. Crance, A. Crubellier,

C. Delsart, and D. Marinesai, J. Phys. B **22**, 1335 (1989).

37. N. Kwon, P. S. Armstrong, T. Olsson, R. Trainham, and D. J. Larson, Phys. Rev. A **40**, 676 (1989).

38. H. Stapelfeldt, P. Balling, C. Brink, H. K Hangen, Phys. Rev. Lett. **67**, 1731 (1991); M. D. Davidson, H. G. Muller, H. B. van den Heuvell, Phys. Rev. Lett. **67**, 1712 (1991); C. Blondel, M. Crance, C. Delsart, and A. Giraud, J. Phys. B **24**, 3575 (1991).

39. A. R. P. Rau, Comments At. Mol. Phys. **14**, 285 (1984).

40. E. Wigner, Phys. Rev. **73**, 1002 (1948).

41. See for example, W. C. Lineburger, and B. W. Woodward, Phys. Rev. Lett. **25**, 424 (1970) (S-wave); H. Hotop and W. C. Lineburger, J. Chem. Phys. **58**, 2379 (1973) (P-wave).

42. T. F. O'Malley, Phys. Rev. **137**, A1668 (1965).

43. J. Dellwo, Y. Liu, D. J. Pegg, and G. D. Alton, Phys. Rev. A **45**, 1544 (1992).

44. R. D. Mead, K. R. Lykke, and W. C. Lineberger, in *Electronic and Atomic Collisions*, edited by J. Eichler, I. Hertel, and N. Stolterfoht (North Holland, Amsterdam, 1984), p. 721.

45. C. H. Greene in *Atomic Spectra and Collisions in External Fields*, edited by K. T. Taylor, M. H. Neyfeh, and C. W. Clark (Plenum, New York, 1988), p. 233; H. C. Bryant *et al.* Phys. Rev. Lett. **58**, 2412 (1987); See also Q. Wang and A. Starace, Phys. Rev. A **48**, R1741 (1993).

46. W. A. M. Blumberg, R. M. Jopson, and D. J. Larson, Phys. Rev. Lett. **40**, 1320 (1978).

47. R. Trainham, G. D. Fletcher, and D. J. Larson, J. Phys. B **20**, L777 (1987).

48. C. Y. Tang, P. G. Harris, A. H. Mohagheghi, H. C. Bryant, C. R. Quick, J. B. Donahue, R. A. Reeder, S. Cohen, W. W. Smith, and J. E. Stewart, Phys. Rev. A **39**, 6068 (1989).

49. G. H. Wannier, Phys. Rev. **90**, 817 (1953).

50. R. Peterkop, J. Phys. B **4**, 513 (1971).

51. A. R. P. Rau, J. Phys. B **9**, L283 (1976).

52. A. Temkin, Phys. Rev. Lett. **49**, 365 (1982).

53. Y. K. Bae, M. J. Coggiola, and J. R. Peterson, Phys. Rev. A **28**, 3378 (1983); Y. K. Bae and J. R. Peterson, Phys. Rev. A **37**, 3254 (1988), and Ref. 19.

54. U. Fano, Phys. Rev. **124**, 1866 (1961).

55. T. A. Patterson, H. Hotop, A. Kasdan, D. W. Norcross, and W. C. Lineberger, Phys. Rev. Lett. **32**, 189 (1974).

56. H. C. Bryant, K. B. Butterfield, D. A. Clark, C. A. Frost, J. B. Donahue, P. A. M. Gram, M. E. Hamm, R. W. Hamm, and W. W. Smith in *Atomic Physics*, edited by D. Kleppner and F. M. Pipkin (Plenum, New York, 1981), Vol. 7, p. 29; P. G. Harris, H. C. Bryant, A. H. Mohagheghi, R. A. Reeder, H. Sharifian, C. Y. Tang, H. Tootoochi, J. B. Donahue, C. R. Quick, D. C. Rislove, W. W. Smith, and J. E. Stewart, Phys. Rev. Lett. **65**, 309 (1990).

57. For example, H. Sadeghpour, C. H. Greene, and M. Cavagnero, Phys. Rev. A **45**, 1587 (1992).

58. U. Berzinsh, G. Haeffler, D. Hanstorp, A. Klinkmüller, E. Lindroth, U. Ljungblad, and D. J. Pegg, Phys. Rev. Lett. **74**, 4795 (1995).

59. A. M. Hazi and K. Reed, Phys. Rev. A **24**, 2269 (1981); H. Saha and R. N. Compton, Phys. Rev. Lett. **64**, 1510 (1990).

60. J. R. Peterson, M. J. Coggiola, and Y. K. Bae, Phys. Rev. Lett. **50**, 664 (1983).

59

Photon-Atom Interactions: Low Energy

C. D. Caldwell* and M. O. Krause[†]

*Department of Physics, University of Central Florida, Orlando, Florida 32816-2385
[†]Oak Ridge National Laboratory, Oak Ridge, Tennessee 37831-6201

59.1 THEORETICAL CONCEPTS

Scattering of low-energy photons proceeds predominantly through the photoelectric effect. In the dipole limit, a photon γ of energy $h\nu$, angular momentum $j_\gamma = 1$, and parity $\pi_\gamma = -1$ interacts with a free atom or molecule A, having total energy E_i, angular momentum J_i, and parity π_i to produce an electron of energy ε, spin $s = 1/2$, orbital angular momentum ℓ, total angular momentum j, and parity $\pi_e = (-1)^\ell$ and an ion A^+ with final total energy E_f, angular momentum J_f, and parity π_f. This process can be written as the reaction

$$\gamma(h\nu, j_\gamma = 1, \pi_\gamma = -1) + A(E_i, J_i, \pi_i)$$
$$\rightarrow A^+(E_f, J_f, \pi_f) + e^-[\varepsilon, \ell sj, \pi_e = (-1)^\ell]. \quad (59.1)$$

Conservation laws require that

$$h\nu + E_i = \varepsilon + E_f,$$
$$\mathbf{J}_i + \mathbf{j}_\gamma = \mathbf{J}_f + \mathbf{s} + \boldsymbol{\ell}, \quad (59.2)$$
$$\pi_i \cdot \pi_\gamma = \pi_f \cdot \pi_e = (-1)^\ell \cdot \pi_f.$$

Since $E_f - E_i$ becomes quite large for inner shells or deep core levels, scattering of low-energy photons involves the removal of an electron from a valence or shallow core

level. In the low-energy regime, from the first ionization threshold to $h\nu \approx 1$ keV, the photoelectric effect accounts for more than 99.6 % of the photon interactions in the elements, with elastic scattering contributing the remainder [1, 2]. Ionization by inelastic scattering, the Compton effect, assumes increasing importance with the higher photon energies and the lower Z elements. Above the first ionization potential, the total photoabsorption cross section and the photoionization cross section are essentially equivalent at the lower photon energies.

The cross section σ_{if} for producing a given final ionic state in the photoionization process is given by

$$\sigma_{if} = \frac{4\pi^2 \alpha^2}{k} \sum |\langle \Psi_f \,|\, \hat{T} \,|\, \Psi_i \rangle|^2, \quad (59.3)$$

where k is the photon momentum, $\hat{T}$ is the transition operator, Ψ_i and Ψ_f are the wave functions of the initial and final states, and the summation includes an average over all initial states and a summation over all the unobserved variables in the final state. A detailed derivation of this expression, including the different forms for $\hat{T}$, is given in the articles by Fano and Cooper [3] and Starace [4] and in Chap. 24. The total cross section is given by the sum of all these different partial cross sections, σ_{if}.

59.1.1 Differential Analysis

Detailed information about the photoionization process can be obtained most directly in emission measurements, especially those involving the photoelectron. The resulting *photoelectron spectrum* yields the energy and intensity for a given interaction. Further differentiation is obtained by varying the angle of observation and by a spin analysis of the photoelectron. Hence, electron emission analysis can reveal all energetically allowed photoprocesses connecting an initial atomic state i to a final ionic state f and yield their dynamic properties. When averaged over the spin, the differential cross section $d\sigma_{if}/d\Omega$ is given in terms of the partial cross section σ_{if} and an expression involving an expansion in Legendre polynomials of order n with the coefficients B_n:

$$\frac{d\sigma_{if}}{d\Omega} = \left(\frac{\sigma_{if}}{4\pi}\right) \sum_n B_n P_n(\cos\theta)\,, \qquad (59.4)$$

where the angle θ is measured between the direction of the emitted electron and the unpolarized incoming photon beam. In the dipole approximation, only the terms containing P_0 and P_2 contribute. Then Eq. (59.4) reduces, for a photon beam with linear polarization p, to

$$\frac{d\sigma_{if}}{d\Omega} = \left(\frac{\sigma_{if}}{4\pi}\right) \left[1 + \frac{\beta_{if}}{4}(1 + 3p\cos 2\theta)\right]\,, \qquad (59.5)$$

where the angle θ lies in the plane perpendicular to the direction of propagation and is measured with respect to the major axis of the polarization ellipse [5]. Then, the differential cross section, or photoelectron angular distribution, is characterized by the single angular distribution or anisotropy parameter β_{if} for a particular process $i \to f$. For observation at the so-called pseudomagic angle θ_{m}, defined as

$$\theta_{\mathrm{m}} = \tfrac{1}{2}\cos^{-1}\left(-\tfrac{1}{3}p\right)\,, \qquad (59.6)$$

the differential cross section $d\sigma_{if}/d\Omega$ becomes proportional to the angle-integrated, or partial, cross section σ_{if}.

In the absence of correlation effects, the partial cross section σ_{if} for the production of an individual final state and the corresponding anisotropy parameter β_{if} are given by simple expressions derived from a single-particle model [1, 4]. For the central field potential,

$$\sigma_{if} = \frac{4\pi^2\alpha}{3}a_o^2 N_{n\ell}h\nu\left[\left(\frac{\ell}{2\ell+1}\right)R_-^2 + \left(\frac{\ell+1}{2\ell+1}\right)R_+^2\right]\,, \qquad (59.7)$$

where $N_{n\ell}$ is the occupation number of the subshell, and

$$\beta_{if} =$$

$$\frac{\ell(\ell-1)R_-^2 + (\ell+1)(\ell+2)R_+^2 - 6\ell(\ell+1)R_+R_-\cos\Delta}{(2\ell+1)[\ell R_-^2 + (\ell+1)R_+^2]}\,. \qquad (59.8)$$

The subscripts $+$ and $-$ refer to the $(\ell+1)$ and $(\ell-1)$ channels respectively, and $\Delta = \delta_+ - \delta_-$ is the difference in phase shift between these two allowed outgoing waves. The parameter $R_\pm$ is the radial dipole matrix element connecting the electron in the bound orbital with orbital angular momentum ℓ with the outgoing wave having orbital angular momentum $\ell \pm 1$.

Effects of the electron correlation on the direct photoionization process can result in values for β which are not reproduced by the Cooper-Zare expression (59.8) (See Ref. [4]). The contribution of the different partial waves to the outgoing wave function can, however, be ascertained through the angular momentum transfer formalism developed by Fano and Dill [6]. In this approach, one defines the angular momentum transferred from the photon to the unobserved variables j_t as

$$\mathbf{j}_t = \mathbf{j}_\gamma - \boldsymbol{\ell} = \mathbf{J}_f + \mathbf{s} - \mathbf{J}_i\,, \qquad (59.9)$$

where the second portion of the equality results from the conservation of angular momentum. For each allowed value of j_t the associated transfer can be defined as either parity favored or parity unfavored according to whether the product $\pi_i\pi_f$ is equal to $+(-1)^{j_t}$ or $-(-1)^{j_t}$ respectively. All symbols have the same definition as in Eq. (59.1). Calculation of the partial cross section for the production of a given final state characterized by the values $\mathbf{J}_f$ and $\mathbf{s}$ is then determined from the cross section corresponding to each angular momentum transfer according to

$$\sigma_{if} = \sum_{j_t} \sigma(j_t)\,. \qquad (59.10)$$

The associated anisotropy parameter β_{if} is derived from a similar sum:

$$\sigma_{if}\beta_{if} = \sum_{j_t=\mathrm{fav}} \sigma(j_t)_{\mathrm{fav}}\,\beta(j_t)_{\mathrm{fav}} - \sum_{j_t=\mathrm{unfav}} \sigma(j_t)_{\mathrm{unfav}}\,. \qquad (59.11)$$

The second equation derives from the fact that $\beta(j_t)$ for each parity-favored value must be calculated separately, whereas for the parity-unfavored case, $\beta(j_t) = -1$ always.

The physical effect described by the angular momentum transfer approach is the interaction between an electron and the anisotropic distribution of the other electrons in the atom. Thus, it becomes most useful in the case of ionization from an open-shell atom having an extra electron or a hole in a shell with $\ell \neq 0$. An illustrative example is the $3s$ ionization of chlorine. Here $\beta_{if} = 2$ identically in Eq. (59.8) because only the single value $\ell = 1$ is allowed in the single-particle model. However, the three possible values, $j_t = 0, 1, 2$ are allowed, of which only the first corresponds to the Cooper-Zare or single-particle, central field result [4].

It is generally the case for ionization of elements which are found naturally in the atomic state that there is an equal population in all the fine-structure components of the initial state. This is because of the relatively small energies associated with the fine-structure splitting. (This does not necessarily apply to atomic species generated through a process of molecular dissociation or high-temperature metal vaporization).

Thus, the determination of all cross sections and angular distributions involves an average over these fine-structure components. However, it is possible to generate atoms in which one of the fine-structure components is preferentially populated. In this case there can also be a preferential ionization to a particular J-component of the final ionic state even in the limit in which the electron correlation is neglected, i.e., the geometrical limit. The partial intensities for the production of a given ionic state characterized by the angular momenta $L_f S_f J_f$ by removal of an electron from an orbital ℓ of a state characterized by $L_i S_i J_i$ are given by

$$R_\ell^{L_f S_f J_f} = \frac{[J_f][L_f][S_f]}{[1/2][\ell]} g(\ell, L_i, S_i, L_f, S_f)$$
$$\times \sum_{j=\ell-1/2}^{j=\ell+1/2} [j] \left\{ \begin{matrix} \ell & 1/2 & j \\ L_i & S_i & J_i \\ L_f & S_f & J_f \end{matrix} \right\}^2 \quad (59.12)$$

Here the term in curly brackets is a 9-j symbol, and the notation $[J] = 2J + 1$ is used. The quantities $g(\ell, L_i, S_i, L_f, S_f)$ are weighting factors determined solely by the initial-state wave function. For the case in which ℓ represents a closed shell, these factors are equal to unity [7].

In situations in which the target atoms possess an initial orientation, i.e., have an average value $\langle J_z \rangle \neq 0$, or if the ionization is performed with circularly polarized radiation, the electrons which are produced have a net spin [8]. It is also possible that unpolarized atoms which are ionized by unpolarized photons can have a net spin, provided that the detection is carried out at a specific angle, and the ionization is from a given fine-structure component of the initial state to a given fine-structure component of the final state. In the latter case, the transverse spin polarization is given by

$$P = -\frac{2\xi \sin\theta \cos\theta}{1 + \beta P_2(\cos\theta)}, \quad (59.13)$$

for linearly polarized radiation, and by

$$P = \frac{2\xi \sin\theta \cos\theta}{2 - \beta P_2(\cos\theta)}, \quad (59.14)$$

for unpolarized radiation [9]. The angle θ is the same as in the angular distribution measurement; the parameter ξ is the spin parameter analogous to β; and $P_2(\cos\theta)$ is the Legendre polynomial of order 2.

Yet another parameter which describes the differentiation inherent in the photoionization process is the alignment A, which reflects an anisotropy in the quadrupole distribution of the angular momentum J_f of the ion [10]. For the cylindrically symmetric coordinate system appropriate to dipole photon excitation, only one moment A_0 of the distribution is nonzero. This is defined by

$$A_0 = \frac{\sum_{m_j} [3m_j^2 - J_f(J_f+1)] \sigma(m_j)}{J_f(J_f+1) \sum_{m_j} \sigma(m_j)}, \quad (59.15)$$

where $\sigma(m_j)$ is the partial cross section for production of a given m_j component of J_f. A very useful approach to the interpretation of the alignment can be obtained through the angular-momentum-transfer formalism [11]. In this approach the angular momentum transfer j_t is defined as

$$\mathbf{j}_t = \mathbf{j}_\gamma - \mathbf{J}_f. \quad (59.16)$$

In contrast to the case of the electron angular distribution, it is possible to derive an alignment for each value of j_t as a function of the angular momentum J_f of the ion. The net alignment is then the incoherent sum of the contributions corresponding to each:

$$A_0 = \sum_{j_t} A_0(j_t) \sigma(j_t) \Big/ \sum_{j_t} \sigma(j_t) \quad (59.17)$$

If the photoionization produces an ion in an excited state which decays by photoemission, the parameter A_0 is reflected either in the angular distribution of the fluorescence photons $I(\theta)$ or the linear polarization P measured at one angle, typically 90°, according to

$$I(\theta) = I_0 \left[1 - \tfrac{1}{2} h^{(2)} A_0 P_2(\cos\theta) \right.$$
$$\left. + \tfrac{3}{4} h^{(2)} A_0 \sin^2\theta \cos(2\chi) \cos(2\eta) \right] \quad (59.18)$$

or, for $\theta = \pi/2$ and $\chi = 0$,

$$P = \frac{I(\eta = 0) - I(\eta = \pi/2)}{I(\eta = 0) + I(\eta = \pi/2)} = \frac{3h^{(2)} A_0}{4 + h^{(2)} A_0}, \quad (59.19)$$

respectively. The angle θ is the angle at which the fluorescence is determined, and the angle χ is measured between the axis of the polarization selected by the detector and the quantization axis. The polarization of the fluorescence is given by $\zeta = (\cos\eta, i\sin\eta, 0)$. The quantity $h^{(2)}$ is a ratio of 6-j symbols depending on the angular momenta J_f of the intermediate ion and the final state J_f':

$$h^{(2)} = (-1)^{J_f - J_f'} \left\{ \begin{matrix} J_f & J_f & 2 \\ 1 & 1 & J_f' \end{matrix} \right\} \Big/ \left\{ \begin{matrix} J_f & J_f & 2 \\ 1 & 1 & J_f \end{matrix} \right\}. \quad (59.20)$$

When it is energetically allowed, an inner-shell hole will undergo Auger decay, emitting an electron with an energy ε_A determined by the energy difference between the energy E_f of the ion and E'_f of the state of the doubly-charged ion to which the decay occurs. Angular analysis of the Auger electrons reflects the alignment of the intermediate ionic state, which is different from, and does not bear a one-to-one relationship to, the angular distribution parameter β of the photoelectrons. Normally, Auger decay is regarded as a two-step process in which the first step is the production of the hole and the release of the primary photoelectron, followed by the decay and the release of the second electron. Within this approximation [12], the angular distribution of the Auger electrons takes on the simple form

$$I(\theta) = \left(\frac{I_0}{4\pi}\right)[1 + \alpha_2 A_0 P_2(\cos\theta)]. \qquad (59.21)$$

Here $P_2(\cos\theta)$ is the second-order Legendre polynomial, and α_2 is the matrix element corresponding to the Auger decay. For the specific case in which the Auger decay is to a final ionic state of $^1S_0$ symmetry, α_2 is purely geometric, and a measurement of the angular distribution leads directly to a determination of the alignment. Correspondingly, if the alignment of a specific state can be determined through such a decay, then analysis of the angular distribution of the decay to other states provides a value for α_2.

59.1.2 Electron Correlation Effects

The primary focus of advanced studies in photoionization is to determine the role played by electron correlation in the structure and dynamics of electron motion above the lowest ionization threshold. Because the form of the interaction potential for the Coulomb interaction is very well known, theory can focus on the many-body aspects of the process (see Chap. 23). Electron correlation manifests itself in many ways. Most prominent are the appearance of autoionization structure due to the excitation of one or two electrons, the production of correlation satellites due primarily to the ionization of one electron accompanied by the excitation of another, and the creation of two continuum electrons in a double ionization process.

Autoionization resonances are perhaps the oldest known features associated with electron correlation (see Chap. 25). These features arise when the absorption of a photon creates a localized state which lies in energy above at least one ionization limit. This state is then degenerate in energy with a state of an electron in the continuum, and the interaction between these states results in the decay of the quasi-localized state into the continuum. Such resonance states appear in an absorption spectrum in the form of strong, localized variations over an energy range characteristic of the width Γ of the feature, which is in turn related to the lifetime τ of the state by

$$\Gamma = \hbar/\tau. \qquad (59.22)$$

In contrast to absorption features between bound states, autoionization resonances are characterized by having an asymmetric line shape. When only one localized state and one continuum are involved, these line shapes can be derived analytically, as first shown by Fano [13] and later by Shore [14], resulting in simple parametrized forms which are suitable for numerical calculation of overlap integrals for determining widths. For the Fano profile

$$\sigma(\epsilon) = \sigma_a \frac{(\epsilon + q)^2}{\epsilon^2 + 1} + \sigma_b, \qquad (59.23)$$

with

$$\epsilon = \frac{E - E_r}{(\Gamma/2)}, \qquad (59.24)$$

the parameter q describes the asymmetry of the line, E_r is the resonance position, and Γ is the width of the line. The parameters σ_a and σ_b reflect the relative contributions to states in the continuum which do and do not interact with the autoionizing state respectively. The energy E_r does not correspond to the peak energy E_m of the resonance feature but is related to the maximum through

$$E_m = E_r + \frac{\Gamma}{2q}. \qquad (59.25)$$

The Shore profile,

$$\sigma(\epsilon) = C(\epsilon) + \frac{A\epsilon + B}{\epsilon^2 + 1}, \qquad (59.26)$$

describes the same phenomenon except that the interpretation of the parameters A and B is different. In this case, they represent products of dipole and Coulomb matrix elements. $C(\epsilon)$ is the continuum contribution.

From an experimental point of view, the parametrized forms for the Fano and Shore profiles are very useful as a basis for fitting autoionization spectra. However, they both have the limitation that they only describe the interaction of an isolated state with the continuum. While they can be extended to include several continuua [4], they do not allow for an interaction among two or more localized states [15, 16]. Nevertheless, it is possible to use these functions to achieve fits of states which do interact with each other as these functions are mathematical representations of localized resonances in a continuous spectral distribution. If this is done, however, the parameters no longer have the physical meaning which they have for the noninteracting case.

The process of autoionization is discussed in more detail in Chap. 25. In Fig. 59.1 an example is shown of the set of $2s^2 2p^3(^4S) \rightarrow 2s2p^3np$, $n \geq 3$ autoionization resonances which decay into the ^3P ground state of the N$^+$ ion [17]. The energies E_n of these resonances are related to the ionization limit E_∞ of the series by the Rydberg formula

ORNL-DWG 94M-11916
162c12nr.gra

Figure 59.1. Autoionization resonances $2s^2 2p^3(^4S) \rightarrow 2s2p^3np$ in atomic nitrogen.

$$E_n = E_\infty - R_\infty/(n - \mu_s)^2 \qquad (59.27)$$

where n is the principal quantum number and μ_s is the quantum defect characteristic of a given series and reflecting the short-range electrostatic interactions of the electron with the ion core. Values of μ_s for s,p,d, and f electrons have been calculated for atoms and ions up to $Z = 50$ [18]. For high precision work, the reduced Rydberg constant

$$R_M = \frac{R_\infty}{1 + 5.485\,799 \times 10^{-4}/(M_A - m_e)} \qquad (59.28)$$

should be used instead of the value R_∞ for infinite nuclear mass (see Chap. 1). The atomic mass M_A and the electron mass m_e are in a.u.

A process closely related to the autoionization phenomenon is resonant Auger decay. This process differs from the ordinary Auger process [19] in that an electron from an inner shell is not ionized but *excited* to either a partially filled or an empty subshell. It may be viewed either as an Auger process or as autoionization. Such an inner-shell excited state of a neutral atom (molecule) lies above one or more of the ionization limits of the singly ionized species and consequently must decay by electron emission unless the decay is forbidden by selection rules. As a result, resonance structure will appear superimposed on the continua of direct photoionization from the various subshells. From a most general point of view, the resonant Auger process can be considered as resonances in the continua of single photoionization, while the ordinary Auger process can be regarded as resonances in the continua of double photoionization. If excitation proceeds to a partially filled subshell *within* a principal shell, as, for example, Mn $3p \rightarrow 3d$ [20], interference between the direct photoionization channels and the indirect resonance channel may be strong, and the lineshapes are given by Eq. (59.23) with arbitrary q values and σ_a/σ_b

ratios. If, however, the excitation proceeds to an empty shell, as, for example, Mg $2p \rightarrow ns$ or nd, $n \geq 4$, interference with the direct channels is likely to be negligible, and the resulting resonances are distinguished by essentially Lorentzian line shapes (as for normal Auger lines) with $q \gg 1$ and $\sigma_a/\sigma_b \gg 1$ in Eq. (59.23). For a given excitation state a number of resonance peaks may arise because more than one ionization channel is usually available and, in addition, the excited electron can change its orbital from n to $n' = n\pm1,2,\dots$ in a shakeup or shakedown process [21].

As a consequence of the electron-electron interactions which occur simultaneously with the electron-photon interaction, ions are produced in states which do not correspond to those which would be expected based on an interpretation using an independent particle model, which allows for only a single-electron transition (see Chap. 24). Evidence for these states appears as correlation satellites in the photoelectron spectrum, the Auger electron spectrum or the x-ray spectrum [22]. Figure 59.2 [23] presents as an example the photoelectron spectrum of argon produced by photons with h$\nu = 60.6$ eV. In addition to the $3s$ main line of single electron photoionization (and the $3p$ main lines not shown) numerous satellite lines are seen as the manifestation of two-electron transitions involving ionization-with-excitation correlations. It is convenient to categorize the satellites in a photoelectron spectrum according to various electron correlations, as, for example, initial state interactions which mix different configurations into the initial state, and final state interactions, which include core relaxation and electron-electron interactions in the final ionic state, electron-continuum, and continuum-continuum interactions. While initial-state correlations are essentially independent of the photon energy, final-state correlations depend on the energy of the photon through the interactions with the continuum channels. However, the heuristic value of placing correlation effects into a strict classification scheme is limited by the fact that their relative strengths depend on the basis set used in a particular theoretical model and its expansion into a "fully correlated" system within a given gauge [3, 4, 24, 25] (see also Chap. 24).

Another manifestation of double-electron processes is the simultaneous *excitation* of two electrons to bound states. These states may decay by electron or photon emission and are seen as resonance structures above the thresholds of inner-shell ionization or near autoionizing members of Rydberg series. As single or double ionization continua are usually strong in the spectral range of the double excitations, interference occurs, and the lineshapes can display dispersion forms. Typically, the cross section for the sum of all correlation processes is between 10 and 30% of that for single photoionization, but may exceed this range considerably in special cases.

Photon scattering near thresholds is complex because of the possibility of strong interactions between the

Figure 59.2. Photoelectron spectrum (PES) of $3s, 3p$ satellites in argon at a photon energy of 60.6 eV. Note the reduced intensity of the satellites compared to the $3s$ main line.

various particles created and the different modes of deexcitation (see Chap. 60). In the case of ionization-with-excitation processes, the threshold cross section is finite, as it is for single electron photoionization, in accord with Wigner's theorem [26] (see Sect. 58.4.5). In the case of double photoionization, the cross section is zero at threshold and then rises according to Wannier's law [27] (see Chap. 50). For the motion of two electrons with essentially zero kinetic energies in the field of the ionic core,

$$\sigma \propto E^{(2\mu-1)/4}, \qquad (59.29)$$

where μ depends on the value of the nuclear charge Z through

$$\mu = \frac{1}{2}\left[\frac{(100Z-9)}{(4Z-1)}\right]^{\frac{1}{2}}. \qquad (59.30)$$

For $Z = 1$ the Wannier exponent has the value 1.127.

In the case of Auger decay following ionization at threshold, interaction between the two electrons results in a shift in the energy of the Auger electron and a corresponding shift in that of the photoelectron (to conserve energy), as well as an asymmetry in the shape of the Auger electron peak and a corresponding asymmetry in the photoelectron peak shape. In this so-called post-collision interaction (see also Chap. 60), the lineshape, averaged over angles, has the form [28]

$$K(\varepsilon) = \frac{(\Gamma/2\pi)}{(\varepsilon - \varepsilon_A)^2 + (\Gamma/2)^2} f(\varepsilon) \qquad (59.31)$$

with

$$f(\varepsilon) = \frac{\pi\psi}{\sinh(\pi\psi)} \exp\left[2\psi \tan^{-1}\left(\frac{\varepsilon - \varepsilon_A}{(\Gamma/2)}\right)\right]. \qquad (59.32)$$

In the above equations, ε is the energy of the Auger line, ε_A is the nominal Auger energy, Γ is the initial hole-state width, and the parameter $\psi = 1/\sqrt{2\varepsilon_e} - 1/\sqrt{2\varepsilon_A}$, with ε_e being the energy of the photoelectron, and $\varepsilon_A \geq \varepsilon_e$.

59.2 EXPERIMENTAL METHODS

An overview of the experimental approaches to the study of photon interactions at low energies is given in Fig. 59.3. The sketch emphasizes the interaction of a polarized photon beam with a small static or particle-beam target of atoms, ions, molecules, or clusters, and the detection of the reaction products at various angles in a plane perpendicular to the direction of propagation of the photon beam, where the general Eq. (59.5) is valid. Emission products, such as electrons, ions, or photons, may be studied by way of the total yields, which can be related to the total photoionization cross section, or by differential analysis in a spectrometer according to energy, intensity, emission angle, and polarization. The various particles may be measured independently, simultaneously, or in coincidence. The photon monitor provides the information for normalization of the data with regard to flux and polarization. The photon monitor can also be used for a measurement of the total photoionization cross section, equivalent to the photoabsorption or photoattenuation coefficient at low photon energies. For this purpose, the size of the target source is advantageously increased in the direction of the photon beam. While experimental apparatus differs, sometimes drastically, for the photon sources as well as for the spectrometry of electrons, ions, and fluorescence photons, many features are common, and the relationships of the measured quantities to basic properties of the atoms and the photon-atom interaction are similar. Thus, the following will place emphasis only on the roles of the synchrotron radiation source and photoelectron spectrometry, whereas specific references to other methods can be found elsewhere [5, 29–33].

59.2.1 Synchrotron Radiation Source

The primary source of photons for experiments in the vuv and soft x-ray region of the spectrum is the synchrotron radiation source [34, 35]. In a synchrotron or electron storage ring, radiation is produced as the electrons are bent to maintain the closed orbit. Such bending magnet radiation is emitted in a broad continuous spectrum which begins in the ir and ends sharply at a critical photon energy given by $h\nu_c = \kappa E_e^2$, where E_e is the energy of the electrons in the ring and κ is a constant characteristic of the ring. Synchrotron radiation can also be generated by introducing additional magnetic field structures into the ring, such as undulators or wigglers, which produce a deviation of the electron

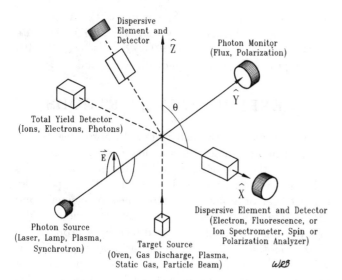

Figure 59.3. Generic arrangement for detection of particles in an emission measurement. The incoming radiation is assumed to be linearly polarized along the z-axis.

motion from a straight path in a well-defined manner. Wiggler radiation has the same spectrum as a bending magnet, except that the critical energy is generally much higher because the effective magnetic fields can be larger than those of the bending magnet. Undulator radiation is very different in that it consists of a sharp spiked profile of about a 1% bandwidth at energies determined by the magnetic field within the undulator and by the electron beam energy in the storage ring.

Synchrotron radiation, no matter what the magnetic field structure of the source, requires monochromatization before it can be used for experiments. For the wavelengths of interest in low-energy photon scattering, this can be achieved by using grating instruments with a metallic coating on the grating surface. The highest resolution possible is obtained through the use of a normal incidence monochromator (NIM) with a plane grating set at normal incidence. However, because the reflectivity of the metallic coating at normal beam incidence decreases drastically as the photon energy increases, use of a NIM has an upper limit of about 40 eV. At energies above this, up to about 1 keV, gratings can still be used but must be mounted at grazing incidence. There is a number of functional designs for these grazing-incidence instruments which vary in the shape of the grating — spherical, toroidal, or plane surfaces (SGM, TGM, PGM) — and the associated optics. Above 1 keV, gratings are no longer suitable, and crystal diffraction must be used. While the radiation emerging from a beamline which couples the monochromator to a bending magnet, wiggler, or undulator has a high degree of linear polarization, varying from 80 to 99% in the plane of the electron orbit, a

useful flux of circularly polarized radiation can be derived from out-of-plane radiation [29] or by the use of multiple reflection optics.

59.2.2 Photoelectron Spectrometry

The primary particle emitted in photoionization is the photoelectron. Hence, a photoelectron spectrum provides a detailed view of the photon interaction by (a) specifying the individual processes from an initial state i to a final state f by way of the electron energy, (b) determining their differential and partial cross sections by recording the number of electrons as a function of emission angle, and (c) measuring the polarization of the electrons by a spin analysis (spin polarimetry). The experimental approach is governed largely by the relations (59.3), (59.5), and (59.6). The number of electrons $N_{if}(e)$ detected per unit time at an angle θ within an energy interval $d\varepsilon$ and within a solid angle $d\Omega$ is given by

$$N_{if}(e) = GN(h\nu)N(A)f(h\nu)f(\varepsilon)\frac{d\sigma_{if}}{d\Omega}d\varepsilon \quad (59.33)$$

where G is a geometry factor, which includes the source dimensions, $N(h\nu)$ the number of photons, $N(A)$ the number of atoms in the source, $f(h\nu)$ and $f(\varepsilon)$ efficiency factors depending, respectively, on photon and electron kinetic energies, and $d\sigma_{if}/d\Omega$ is the differential cross section for a particular transition $i \rightarrow f$. Equation (59.33) assumes that $d\Omega$ and $d\varepsilon$ are sufficiently small that integration over the pertinent parameters is not needed. Since $N_{if} \propto d\sigma_{if}/d\Omega$, a measurement at two angles, e.g., $\theta = 0°$ and $90°$, yields the electron angular distribution parameter β_{if} according to Eq. (59.5), and a measurement at θ_m [Eq. (59.6)] yields the partial cross section σ_{if}. In the case of closed-shell atoms, $n\ell j$ notation is sufficient to designate single ionization to an $\varepsilon\ell$ continuum, e.g., $3p_{1/2,3/2} \rightarrow \varepsilon s$ or εd in argon, but for open-shell atoms LSJ notation is required, e.g., $3p^5(^2P^o_{3/2}) \rightarrow 3p^4(^3P^e_{2,1,0},{}^1D^e_2,{}^1S^e_0)\varepsilon\ell(^2D^e,{}^2P^e,{}^2S^e)$ in chlorine. Similarly, for ionization-with-excitation transitions, the final state requires an open-shell designation.

The sum of the partial cross sections is equal to the total photoionization or absorption cross section

$$\sigma_{\text{tot}} = \sum_{i,f} \sigma_{if} \quad (59.34)$$

where the σ_{if} encompass (a) single ionization events in all energetically accessible subshells $n\ell j$, or the LSJ multiplet components, (b) ionization-with-excitation events (shakeup or shakedown), and (c) double ionization events (shakeoff). All σ_{if} can be determined from a photoelectron spectrum from its discrete peaks (*cf.* Fig. 59.2) and from the continuum distribution of multiple ionization. However, the latter process is measured more readily by observing the multiply charged ions in a mass spectrometer.

The differentiation afforded by measuring the various partial cross sections, and the associated β parameters, can be augmented by differentiating the continuum channels according to the spin using a spin polarimeter [29]. In closed-shell atoms, this allows for an experimental determination of the relevant matrix elements and phase shifts, and hence for a direct comparison with theory at the most basic level [30, 31]. In a more global measurement, the cross section σ_{tot} is obtained by ion or mass spectrometry from

$$\sigma_{\text{tot}} = \sigma(A^+) + \sigma(A^{2+}) + \sigma(A^{3+}) + \cdots . \quad (59.35)$$

Generally, the charge states can be correlated with the various initial photoionization processes if allowance is made for Auger transitions, and the fluorescence yield upon exceeding the binding energies of core levels.

If the charge states are not distinguished, as in a total ion yield measurement, σ_{tot} is obtained directly. Similarly, a direct measurement of the global quantity σ_{tot} is obtained by the total electron yield, although care must be exercised to avoid discrimination by angular distribution effects. At photon energies below about 1 keV, ionization, absorption, and attenuation are equivalent, and σ_{tot} can also be determined in an ion chamber setup [32] or in a photoabsorption measurement in which the number of photons ΔN absorbed in a source of length d and having an atom density n is given by

$$\Delta N = N(\text{ph})[1 - \exp(-\sigma_{\text{tot}} n d)] \quad (59.36)$$

with $N(\text{ph})$ being the flux of incident photons. As a rule, in all experiments employing the relation (59.5), the total, partial, or differential cross sections are determined on a relative rather than an absolute scale because it is very difficult to know accurately such factors as the geometry of the source volume and the number density. However, once a single absolute value of σ_{tot} or any σ_{if} is available, all relative values of the other quantities can be converted to absolute values.

An electron spectrometric experiment can be carried out in three different operational modes as defined in Fig. 59.4. In the most conventional mode, PES, the photon energy is fixed, and a scan of the electron kinetic energy reveals all the electron-emission processes possible and yields their properties. The CIS (constant ionic state) mode is especially suited to follow continuously a selected process as a function of photon energy by locking onto a given state E_f - E_i (denoted by E_B) which requires a strict synchronization of the photon energy ($h\nu$) and electron kinetic energy (ε) during a scan. This mode is particularly advantageous to elucidate resonance features, such as autoionization resonances. Finally, a CKE (constant kinetic energy) scan allows one to access various processes sequentially or, most

Figure 59.4. Energy relationship among the three different operational modes of the technique of photoelectron spectrometry. E_B is the binding energy of the level.

importantly, follow a process of fixed energy, such as an Auger transition, as a function of photon energy. This description also includes the technique of zero-kinetic-energy measurements. Most frequently, the PES and CIS modes are employed, and Fig. 59.5 gives a self-explanatory example of an actual experiment directed at the characterization of the argon $3s \rightarrow np$ autoionizing resonances. It should be stressed that the cross section σ_{tot} can be partitioned into its components by CIS scans that differentiate between the $3p_{1/2}$ and $3p_{3/2}$ doublet states.

Energy analysis can be performed either by electrostatic energy analyzers or by time-of-flight techniques. The latter is well-suited to those electrons which have very low kinetic energy, including threshold electrons with $\varepsilon \approx 0$ eV. Of the electrostatic energy analyzers now in use, two designs are prevalent, the cylindrical mirror analyzer (CMA) and the hemispherical analyzer, where the latter readily lends itself to the application of multichannel detectors.

59.2.3 Resolution and Natural Width

The details that can be gleaned from an experiment using photons depend on the resolution achievable with the particular photon source and spectrometry used, the particular excitation or analysis modes and the target conditions chosen, and, ultimately, on the natural width of either the levels or transitions examined as well as any fine structure present. Generally, the instrumental and operational resolution should approach, but need not exceed by much, the natural width inherent in the photoprocess under scrutiny. The demands are most severe for processes involving outer levels because of their typically very narrow widths, and are relatively mild for processes involving inner levels [5, 31]. It is desirable that in the former case the resolving power (the inverse of the resolution) of the instrument exceed 10^5, while in the latter case 10^4 may suffice. If the target atoms move randomly, a resolution limit is set by the thermal motion, namely

$$\Delta\varepsilon = 0.723\,(\varepsilon T/M)^{1/2}\ (\text{meV})\,, \qquad (59.37)$$

where, in an experiment involving photoelectron spectrometry, ε is the kinetic energy of the photoelectron in eV, T the temperature in K, and M the mass in a.u. of the target atom. This contribution can be limited in first order by employing a suitably directed atomic beam.

The experimental peak-width and shape generally contain the natural width; the extent to which instrumental factors enter depends on the specific experiment. In a measurement of the total or partial photoionization cross section in which either the fluorescence photons, the electrons, or the ions are monitored, the resolution of the photon source (often called the bandpass) is the only instrumental contributor. This applies specifically to the CIS mode of electron spectrometry, in which features are scanned in photon energy and the electron serves solely as a monitor. However, in such a CIS study, or a corresponding fluorescence study, the resolution of the electron or fluorescence spectrometer must be adequate to be able to *distinguish* adjacent processes. For the example of Fig. 59.5, the $3p_{3/2}$ and $3p_{1/2}$ levels of Ar need to be separated if the partial photoionization cross sections are to be determined across the resonances. In such a case of more than one open ionization channel, the natural widths of the features will be identical in all channels [4], but the shapes may be different. In emission processes subsequent to initial photoionization, namely electron (Auger) decay or photon (x-ray) emission, the resolution of only the spectrometer performing the detection counts on the instrumental side. In the PES mode (Fig. 59.4) the observed lines contain contributions from all sources, the photon source or photon monochromator, the electron analyzer, thermal broadening, and the natural level width. In the special case of photoprocesses near inner thresholds, the post-collision interaction influences the position and shape of photoelectron and Auger lines (see Chap. 60).

Excluding threshold regions and the resonant Raman effect, the line profile observed in the various experiments is given by the Voigt function

$$V(\omega,\omega_0) = \int L(\omega'-\omega)G(\omega'-\omega_0)d\omega'\,. \qquad (59.38)$$

In this the Lorentzian function L represents the natural level or transition profile, and the Gaussian function G is

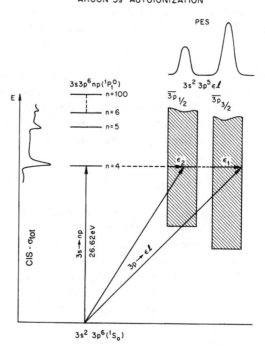

ORNL-DWG 84-12236R

ARGON 3s AUTOIONIZATION

Figure 59.5. Connection between the PES and CIS techniques as illustrated by the $3s \to np$ autoionization resonances in argon.

representative of the window functions of the dispersive apparatus. Although the integral representing the Voigt profile has no analytic form, it can be represented for practical purposes by the analytic Pearson-7 function [36]

$$P_7(\varepsilon) = A\left[1 + \frac{(\varepsilon-\varepsilon_0)^2}{B^2 C}\right]^{-C}\,, \qquad (59.39)$$

where A is the peak height, ε_0 the peak position, B the nominal half-width-half-maximum of the peak, and C the shape of the peak. In the limit in which $C = 1$, this function is identically a Lorentzian; in the limit $C \to \infty$, the function is essentially Gaussian. Use of this function allows one to fit the resulting photoelectron spectrum using standard numerical techniques. If the width of the feature is the only quantity of interest, the simple approximate expression

$$\frac{\Gamma_L}{\Gamma_V} = 1 - \left(\frac{\Gamma_G}{\Gamma_V}\right)^2 \qquad (59.40)$$

which relates the Voigt width Γ_V with the Lorentzian width Γ_L and the Gaussian width Γ_G can be used to determine either Γ_L or Γ_G from the measured Γ_V [5]. Often the observed feature exhibits a dispersive shape given by the Fano or Shore profile. In this instance the instrumental function must be convoluted with the resonance profiles given by Eq. (59.23) or Eq. (59.26) in order to fit the data and to extract the parameters [37].

For the special case of resonant Auger decay in which the bandpass of the exciting radiation is very narrow compared with the natural width of the excited state, the experimental linewidth is governed by the width of the exciting radiation, and will be more narrow than the natural width of the line. The resulting lineshape in this resonant Raman effect is then the simple product [19]

$$L'(\omega, \omega_0) = L(\omega)G(\omega - \omega_0), \qquad (59.41)$$

where $L(\omega)$ is the line profile as determined by the natural width, typically Lorentzian for resonant Auger decay, and $G(\omega - \omega_0)$ is the, usually, Gaussian function representing the bandpass of the exciting radiation.

59.3   SPECIAL TOPICS

Although low-energy photon interactions are well described nonrelativistically in the dipole approximation, relativistic and higher multipole effects which are important at higher energies become discernible even below 1 keV. Then the use of the full relativistic version of Eq. (59.3) is indicated [38]. Spin-orbit effects [39] and relativistic effects [40] are of special significance even at low energies in Cooper minima, where one of the transition matrix elements becomes zero. Level energies of heavy elements require also a relativistic treatment [41] (see Chap. 22), and it is natural to employ relativistic formulations for calculating the spin parameters appearing in photoionization [42]. Photoelectron angular distributions are influenced because in Eq. (59.4) the coefficients B_n which are calculated relativistically [43] can assume finite values for $n \geq 2$, even at photon energies of a few hundred eV. As a result, Eq. (59.5) needs to be expanded, and a convenient pseudomagic angle Eq. (59.6) can no longer be defined. Experiments and theory have shown higher multipole effects to be present, but small, at $h\nu < 1$ keV [43, 44]. Measurements that take the dipole formulations as a basis [Eqs. (59.5) and (59.6)] will incur a small error, usually less than 5% below 1 keV in both the differential $d\sigma_{if}/d\Omega$ and partial cross sections σ_{if}. A more accurate determination of $d\sigma_{if}/d\Omega$ can be made [45] on the basis of the equation

$$\frac{d\sigma_{if}}{d\Omega} = \left(\frac{\sigma_{if}}{4\pi} \right) [1 + \beta P_2(\cos\theta) \\ + (\delta + \gamma \cos\theta)\, \sin\theta \cos\phi]\,, \qquad (59.42)$$

using linearly polarized radiation, $P_2(\cos\theta)$ is the second-order Legendre polynomial, θ is the angle between electron emission direction and the electric vector, ϕ the angle between electron and photon directions, β is the angular distribution parameter related to B_2, and the parameters δ and γ are related to B_1 and B_3 in Eq. (59.4).

The parameters β, δ, and γ have been calculated [45] for most subshells of the noble gases for photoelectron energies between 100 eV and 5 keV.

Acknowledgment

This work has been supported in part by the National Science Foundation under grant PHY-9207634 and in part by the US Department of Energy, Division of Chemical Sciences, under contract with Martin Marietta Energy Systems, Inc., DE-AC0584OR21400.

REFERENCES

1. W. J. Veigele, At. Data Nucl. Data Tables **5**, 51 (1973).
2. J. H. Hubbell and M. J. Berger, in *Engineering Compendium on Radiation Shielding*, edited by R. G. Jaeger (Springer, Berlin, 1968), Vol. 1, p. 167.
3. U. Fano and J. W. Cooper, Rev. Mod. Phys. **40**, 441 (1963).
4. A. F. Starace, in *Handbuch der Physik*, edited by S. Flügge and W. Mehlhorn (Springer, Berlin, 1982), Vol. XXI, p. 1.
5. M. O. Krause in *Synchrotron Radiation Research*, edited by H. Winick and S. Doniach (Plenum, New York, 1980), Chap. 5.
6. U. Fano and D. Dill, Phys. Rev. A **6**, 185 (1972).
7. J. Schirmer, L. S. Cederbaum, and J. Kiessling, Phys. Rev. A **22**, 2696 (1990).
8. V. W. Hughes, R. L. Long, Jr., M. S. Lubell, M. Posner, and W. Raith, Phys. Rev. A **5**, 195 (1972); U. Fano, Phys. Rev. **178**, 131 (1969).
9. N. A. Cherepkov, Adv. At. Mol. Phys. **19**, 395 (1983).
10. U. Fano and J. H. Macek, Rev. Mod. Phys. **45**, 553 (1973).
11. C. H. Greene and R. N. Zare, Phys. Rev. A **25**, 2031 (1982).
12. W. Mehlhorn, in *Atomic Inner-Shell Physics*, edited by B. Crasemann (Plenum, New York, 1985), Chap. 4.
13. U. Fano, Phys. Rev. **124**, 1866 (1961).
14. B. Shore, J. Opt. Soc. Am. **57**, 881 (1967).
15. F. H. Mies, Phys. Rev. **175**, 164 (1968).
16. J. P. Connerade and A. M. Lane, Rep. Prog. Phys. **51**, 1439 (1988).
17. S. J. Schaphorst, S. B. Whitfield, H. P. Saha, C. D. Caldwell, and Y. Azuma, Phys. Rev. A **47**, 3007 (1993).
18. C. E. Theodosiou, M. Inokuti, and S. T. Manson, At. Data Nucl. Data Tables **35**, 473 (1986).
19. T. Åberg and G. Howat, in *Handbuch der Physik*, edited by S. Flügge and W. Mehlhorn (Springer, Berlin, 1982), Vol. XXI, p. 469.
20. S. B. Whitfield, M. O, Krause, P. van der Meulen, and C. D. Caldwell, Phys. Rev. A **50**, 1269 (1994).
21. S. B. Whitfield, J. Tulkki, and T. Åberg, Phys. Rev. A **44**, R6985 (1994).
22. M. O. Krause, J. Phys. (Paris) **32**, C4-67 (1971).

23. M. O. Krause, S. B. Whitfield, C. D. Caldwell, J.-Z. Wu, P. van der Meulen, C. A. de Lange, and R. W. C. Hansen, J. Elec. Spectros. Rel. Phen. **58**, 79 (1992).

24. R. L. Martin and D. A. Shirley, Phys. Rev. A **13**, 1475 (1976).

25. T. Åberg, Nucl. Instrum. Methods B **87**, 5 (1994).

26. E. P. Wigner, Phys. Rev. **73**, 1002 (1948).

27. G. H. Wannier, Phys. Rev. **90**, 817 (1953).

28. M. Y. Kuchiev and S. A. Sheinermann, Sov. Phys. JETP **63**, 986 (1986).

29. U. Heinzmann, J. Phys. B **13**, 4353 (1980).

30. V. Schmidt, Rep. Prog. Phys. **55**, 1483 (1992).

31. *VUV and Soft X-ray Radiation Studies*, edited by U. Becker and D. A. Shirley (Plenum, New York, 1996).

32. J. A. R. Samson, in *Handbuch der Physik*, edited by S. Flügge and W. Mehlhorn (Springer, Berlin, 1982), Vol. XXI, p. 123.

33. B. Sonntag and P. Zimmermann, Rep. Prog. Phys. **55**, 911 (1992).

34. H. Winick and G. Brown, Nucl. Instrum. Methods **195**, 1 (1987).

35. E. E. Koch, *Handbook of Synchrotron Radiation*, (North-Holland, Amsterdam, 1983).

36. S. B. Whitfield, C. D. Caldwell, D. X. Huang, and M. O. Krause, J. Phys. B **25**, 4755 (1992).

37. J. Jiménez-Mier, J. Quant. Spectros. Radiat. Transfer **51**, 741 (1994).

38. W. R. Johnson and K. T. Cheng, Phys. Rev. A **20**, 978 (1979).

39. U. Fano, Phys. Rev. **178**, 131 (1969).

40. S. T. Manson, Z. J. Lee, R. H. Pratt, I. B. Goldberg, B. R. Tambe, and A. Ron, Phys. Rev. A **28**, 2885 (1983).

41. I. P. Grant, Aust. J. Phys. **39**, 649 (1986).

42. K. N. Huang, W. R. Johnson, and K. T. Cheng, At. Data Nucl. Data Tables **26**, 33 (1981).

43. H. K. Tseng, R. H. Pratt, S. Yu, and A. Ron, Phys. Rev. A **17**, 1060 (1978).

44. M. O. Krause, Phys. Rev. **177**, 151 (1969).

45. J. W. Cooper, Phys. Rev. A **47**, 1841 (1993).

60

Photon-Atom Interactions: 1 keV – 1 MeV

Bernd Crasemann

Department of Physics, University of Oregon, Eugene, Oregon 97403

60.1 OVERVIEW

60.1.1 Scope of Chapter

The main photon-atom interaction processes in the energy range from ∼1 keV to ∼1 MeV are outlined. The atomic response to inelastic photon scattering is discussed; essential aspects of radiative and radiation-less transitions are described, including hole-state widths and fluorescence yields, in the two-step approximation which applies well above threshold where atomic de-excitation is nearly independent of the excitation process. Multi-electron photoexcitation that transcends the independent-electron model is briefly considered.

Threshold phenomena that arise when the two-step model breaks down, so that excitation and de-excitation must be treated as a single second-order quantum process are described. The time-independent resonant scattering approach that leads to a unified description of photoexcitation, ionization, and Auger electron or x-ray emission is outlined. In terms of this theory, Raman processes and post-collision interaction are described.

60.1.2 Photon-Atom Processes

It is convenient to divide photon-atom interactions into elastic and inelastic processes, depending on whether or not the photon energy changes in the c.m. frame. This separation, while useful, is somewhat arbitrary [1]: the radiative corrections of quantum electrodynamics (Chap. 27) and the possibility of emitting very soft photons, as well as target recoil, make all processes in fact inelastic, while experimental comparison of incident and scattered photon energies is limited by source bandwidth and detector resolution.

Elastic photon-atom interactions (Chap. 43) include scattering from bound electrons (Rayleigh scattering) with transition amplitude A^{R} and nuclear Thomson scattering, Delbrück scattering, and nuclear resonance scattering, with amplitudes A^{NT}, A^{D}, and A^{NR}, respectively. These amplitudes are coherent, i.e., not physically distinguishable: they add, retaining relative phase information, in the total atom elastic scattering amplitude A [1]. In the commonly used approximation, that is however "neither unique nor exact" [2], we have

$$A = A^{\mathrm{R}} + A^{\mathrm{NT}} + A^{\mathrm{D}} + A^{\mathrm{NR}}. \qquad (60.1)$$

The total atom elastic cross section is proportional to A^2.

Inelastic photon-atom interactions are photoexcitation (including ionization) (Chap. 24), Compton scattering, and pair production. The pair production threshold at $2m_e c^2$ lies above the energy region considered here (Chap. 27). Inelastic scattering processes are generally incoherent, e.g., ejected Compton electrons originating from different orbitals can be distinguished [1]. However, coherence can enter even between photoionization and Compton scattering in the infrared-divergence limit [3–5].

The cross sections for the main elastic and inelastic photon-atom processes comprising the photoeffect are illustrated in Fig. 60.1 for Cu [6] in which, however, resonant pieces of the inelastic (Compton) cross section are omitted. Elastic exceeds inelastic scattering up to well above the K edge; this is true for all atomic numbers. The total cross section is dominated by the photoeffect up to photon energies that increase with atomic number, from $\sim$6 keV for He [5] to $\sim$25 keV for C and $\sim$700 keV for U [6, 7]. (See also Sects. 24.3.2 and 88.3.2).

60.2 ELASTIC PHOTON-ATOM SCATTERING

60.2.1 Rayleigh Scattering

The term *Rayleigh scattering* is applied to elastic photon scattering from bound atomic electrons (see also Sect. 66.7.1). The Rayleigh scattering amplitude for neutral atoms dominates the total atom elastic scattering amplitude, at all angles, for photon energies below $\sim$1 MeV and dominates forward scattering at higher energies [1]. At angles other than 0°, Rayleigh scattering occurs increasingly from inner atomic shells as the photon energy is raised. Depending on the energy, the process can thus sensitively probe atomic structure in distinct regions [1].

Rayleigh scattering has been reviewed by Henke [8], Gavrila [9], Kissel and Pratt [1], and Kane *et al.* [10].

Figure 60.1. Contributions of the photoeffect, elastic, and inelastic scattering to the total photon interaction cross section of copper. (From [6], by permission, University of California, Lawrence Livermore National Laboratory; work performed under the auspices of the U.S. Department of Energy).

The process has commonly been estimated in a simple *form-factor approximation*, which represents scattering from a free charge distribution; this underlies comprehensive tabulations of elastic scattering cross sections and attenuation coefficients [7]. The form factor relates cross sections to classical theory, and has been derived from both nonrelativistic and relativistic quantum mechanics. It is attractive because of calculational simplicity [2, 11].

The form factor $f(q)$ for a spherically symmetric charge number density $\rho(r)$ is the Fourier transform of the charge density:

$$f(q) \equiv \int \rho(\mathbf{r}) e^{i\mathbf{q}\cdot\mathbf{r}} d\mathbf{r} = \int_0^\infty \rho(r) \frac{\sin(qr)}{qr} r^2 dr, \quad (60.2)$$

where

$$\hbar q = (2\hbar\omega/c)\sin(\theta/2) \qquad (60.3)$$

is the momentum transfer, and θ the scattering angle. In elastic photon scattering, $f(q)$ corrects the Thomson point charge scattering cross section for scattering from an extended charge distribution. In the form factor approximation, the differential Rayleigh scattering cross section (neglecting other coherent processes) for unpolarized photons, averaged over final polarization, is

$$\frac{d\sigma}{d\Omega} = \tfrac{1}{2} r_e^2 (1 + \cos^2\theta)|f(q)|^2, \qquad (60.4)$$

where $r_e = e^2/(m_e c^2)$ is the classical electron radius and the terms multiplying $|f(q)|^2$ are the Thomson cross section.

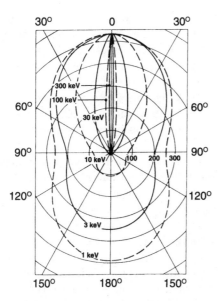

Figure 60.2. Differential elastic photon-atom scattering cross section (in b/atom) for neutral lead in form factor approximation. The contribution of nuclear Thomson scattering (see Sect. 60.2.2) to the amplitude is included. (From [1]).

Even with modifications and corrections (see [1, 2]), the range of validity for the form factor approximation for elastic photon scattering is limited [11]. Predictions based on this approach are often wrong in quantitative detail, although they lend insight into qualitative features of the Rayleigh amplitude [2] (see, e.g., Fig. 60.2).

The shortcomings of the form factor approximation arise from the absence of binding effects in the Thomson cross section, which pertains to photon scattering from a free charged particle. Specifically, accurate Rayleigh amplitudes must include the interaction of the electron with the atomic potential in the initial, intermediate, and final states [1]. The form factor approximation fails, especially for photons of low energy compared with the electron binding energy, and for large scattering angles (high momentum transfer), but $\mathcal{O}(Z\alpha^2)$ discrepancies remain at all energies and angles [2].

Substantial progress in the understanding of elastic photon-atom scattering was made with the advent of numerical calculations of the amplitudes based on the second-order S-matrix of quantum electrodynamics and relativistic wave functions [2, 12]. These numerical partial wave calculations include binding effects in the intermediate state exactly to all orders. At present they constitute the state-of-the-art approach that leads to amplitudes with errors of $\mathcal{O}(1\%)$. However, some limitations do exist: (1) great computer time requirements make it necessary to employ simpler (e.g., modified form factor) methods for outer shell contributions, (2) only lowest-order terms of the S-matrix expansion in powers of e

are included, and (3) most electron-electron correlation effects are not taken into account [1].

Measurements of Rayleigh scattering in the energy range here considered are rather scarce [10]. Gamma rays from radioactive sources have been used in most experiments, but discrimination from inelastic contributions has been a problem [13]. Comparison of measurements on medium- and high-Z targets with theory, albeit incomplete, has however been encouraging [1, 10, 13].

60.2.2 Nuclear Scattering

The nucleus contributes to the total atom elastic scattering amplitude through nuclear Thomson, Delbrück, and nuclear resonance scattering. Only the first two of these processes are important in the energy range considered here.

Nuclear Thomson Scattering

Photon scattering by the nuclear charge can be approximated by treating the nucleus as a rigid, spinless sphere of charge Ze, mass M, and radius r_N. In the simplest point-charge form factor approximation, nuclear Thomson scattering is included by replacing $|f(q)|$ in Eq. (60.4) by $|f(q) + Z^2(m_e/M)|$. If the finite nuclear radius r_N is taken into account, the nuclear form factor is approximately [1]

$$f_N \sim Z^2 \left(1 - \frac{\omega^2}{3c^2} \langle r_N^2 \rangle \right). \qquad (60.5)$$

The nuclear Thomson scattering cross section is

$$\frac{d\sigma^{\text{NT}}}{d\Omega} = \frac{r_e^2}{2} \left| \frac{m_e}{M} f_N \right|^2 (1 + \cos^2 \theta) \qquad (60.6)$$

if other coherent processes can be neglected. For some energies, scattering angles, and target atoms, the nuclear Thomson scattering contribution to elastic scattering can dominate. A few measurements exist that confirm the preceding predictions with $f_N \cong Z^2$ [1].

Delbrück Scattering

Scattering of photons by virtual electron-positron pairs created in the screened nuclear Coulomb potential is called *Delbrück scattering* [1, 14]. The imaginary part of the amplitude describes photon absorption due to pair production; it vanishes for $\hbar\omega \leq 2m_ec^2$. The real part of the amplitude, which adds coherently to the Rayleigh and nuclear Thomson amplitudes, arises from vacuum polarization. Together with photon-photon scattering, photon splitting and coalescence in an external field, Delbrück scattering belongs to the nonlinear effects of quantum electrodynamics that have no classical analog (see Sect. 27.3) [15]. In the photon energy range 0.1–1 MeV, the Delbrück amplitude can be significant at intermediate scattering angles for high-Z targets [1]. Theory and experiment have been reviewed by Papatzacos and Mork [14] and more recently by Kissel and Pratt [1] and by Milstein and Schumacher [16]. Lowest-order Born approximation calculations of the Delbrück amplitude scale as $\alpha(Z\alpha^2)$. Experiments to verify the process have been performed since 1933, but only recently has the necessary accuracy been attained [16]; a major obstacle in the interpretation was the lack of accurate predictions of the Rayleigh amplitude, a problem that is now being solved to a considerable extent (see Sect. 60.2.1). Recent measurements agree with the Born approximation predictions to a few percent below $Z = 60$; but for heavier targets, *Coulomb corections* appear to play a significant role that have only recently been calculated [17].

60.3 INELASTIC PHOTON-ATOM INTERACTIONS

60.3.1 Photoionization

Photoexcitation and ionization are important processes for the investigation of atomic structure and dynamics (see also Chap. 59). Early reviews of the theory of the atomic photoelectric effect are by Fano and Cooper [18] (low photon energy) and by Pratt *et al.* [19] ($\hbar\omega > 10$ keV). See also recent comprehensive monographs by Starace [20] and Amusia [21]. In the present volume, the theory of atomic photoionization for photon energies below 1 keV is summarized in Chap. 24. Experimental aspects have been reviewed, e.g., by Samson [22] and by Siegbahn and Karlsson [23].

An extensive compilation of measured x-ray attenuation coefficients or total absorption cross sections, as well as calculated photoelectric cross sections (computed with relativistic Hartree-Slater wave functions), has been published by Saloman *et al.* [24]. Based on the National Bureau of Standards (now the National Institute of Standards and Technology) data base, this work covers the photon energy range from 0.1–100 keV and includes all elements with atomic numbers $1 \le Z \le 92$; an extensive bibliography is included. Very useful tables of theoretical *subshell* photoionization cross sections covering all elements and the photon energy range 1–1500 keV have been computed by Scofield [25]; the atomic electrons are treated relativistically as moving in a Hartree-Slater central potential, and all relevant multipoles as well as retardation effects are included. Accuracy of the results is borne out by systematic comparisons with experiment and detailed tests [26]. A useful energy range for applications in electron spectrometry is covered by a tabula-

tion of subshell cross sections calculated by Yeh and Lindau [27] in the dipole approximation with nonrelativistic Hartree-Fock-Slater wave functions. Because they were obtained in a frozen-core model, these cross sections like those of Scofield [25] automatically include all multiple excitations [28]. Such multi-electron processes (see Sect. 60.4.4) can make substantial contributions, e.g., $\sim 40\%$ of the $1s$ photoionization in the threshold region of Kr [29].

The usefulness of photoexcitation and ionization for experimental studies of atomic structure and processes has been greatly enhanced by the advent of *synchrotron radiation sources* beginning in the 1960's [30, 31] . At first, experiments were performed mostly 'parasitically'. Later, they were done with dedicated sources, taking advantage of the high brightness, narrow bandwidth and wide tunability of the radiation (with suitable monochromators, from the visible to the hard x-ray regime), its high degree of polarization and sharp time structure [32]. In the 1990's, "third-generation" sources are being commissioned in which radiation is primarily derived from *insertion devices*, wigglers and undulators [33], entailing further increase of the power of this experimental tool. There is a growing literature on research applications of synchrotron radiation in atomic and molecular physics, including handbooks [34–36], proceedings of the series of conferences on x-ray and atomic inner-shell physics [37], and monographs [38–41]. Much of the physics described in the remainder of this chapter is being explored with synchrotron radiation.

60.3.2 Compton Scattering

Compton scattering denotes the scattering of photons from free electrons. The term is also used for inelastic photon scattering from bound electrons, which approaches the free electron case when the photon energy greatly exceeds the electron binding energy [1]. The theory of the process is discussed in standard textbooks [15, 42, 43]. Photon scattering from a free electron at rest is expressed, within lowest-order relativistic quantum electrodynamics, by the Klein-Nishina formula. A low energy approximation to the Klein-Nishina cross section is

$$\sigma^{\mathrm{KN}} = \sigma^{\mathrm{T}} \left(1 + \frac{2\hbar\omega}{m_{\mathrm{e}}c^2} + \cdots \right), \quad \hbar\omega \ll m_{\mathrm{e}}c^2, \quad (60.7)$$

where σ^{T} is the classical total Thomson cross section,

$$\sigma^{\mathrm{T}} = (8\pi/3)r_{\mathrm{e}}^2. \quad (60.8)$$

The energy $\hbar\omega'$ of a photon scattered through an angle θ is related to the incident photon energy $\hbar\omega$ by the Compton relation

$$\hbar\omega' = \frac{\hbar\omega}{1 + (\hbar\omega/m_{\mathrm{e}}c^2)(1 - \cos\theta)}. \quad (60.9)$$

The differential cross section for Compton scattering from free electrons, averaged over initial and summed over final photon and electron polarizations, is

$$\frac{d\sigma^{\text{KN}}}{d\Omega} = \frac{r_e^2}{2}\left(\frac{\omega'}{\omega}\right)^2\left(\frac{\omega}{\omega'} + \frac{\omega'}{\omega} - \sin^2\theta\right). \quad (60.10)$$

In the limit $\omega' \to \omega$, this reduces to the Thomson differential cross section,

$$\frac{d\sigma^{\text{T}}}{d\Omega} = \frac{r_e^2}{2}(1 + \cos^2\theta) \quad (60.11)$$

contained in Eq. (60.4).

Compton scattering from bound electrons has extensive applications in the study of electron momentum distributions in atoms and solids [44]. An exact second-order S-matrix code for the relativistic numerical calculation of cross sections for Compton scattering of photons by bound electrons within the independent particle approximation has now been developed [45].

A systematic treatment of elastic and inelastic photon scattering, starting from many-body formalism, has recently been described by Pratt, Kissel, and Bergstrom [46].

60.4 ATOMIC RESPONSE TO INELASTIC PHOTON-ATOM INTERACTIONS

60.4.1 Auger Transitions

Atomic inner shell vacancy states tend to decay predominantly through *radiationless* or *Auger* transitions, which are autoionization processes (see also Chaps. 25 and 59) that arise from the Coulomb interaction between electrons [47–52]. Radiationless transitions dominate over radiative ones, except for $1s$ vacancies in atoms with atomic numbers $Z < 30$, primarily because of the magnitude of the matrix elements and because far fewer channels are allowed by electric dipole x-ray emission selection rules (*cf.* Sects. 60.4.2 and 60.4.3) than by the selection rules for Auger transitions, which in pure LS coupling are

$$\Delta L = \Delta S = \Delta M_L = \Delta M_S = 0,$$
$$\Delta J = \Delta M = 0, \quad \Pi_i = \Pi_f. \quad (60.12)$$

Thus, for example, 2784 electron-electron interaction matrix elements can contribute to the radiationless decay of a $2p_{3/2}$ hole state in Hg.

In the simplest approach, ionization and subsequent Auger decay are treated as two distinct steps. This approximation is valid if the electron which is ejected in the ionization process is sufficiently energetic so that it does not interact significantly with the Auger electron,

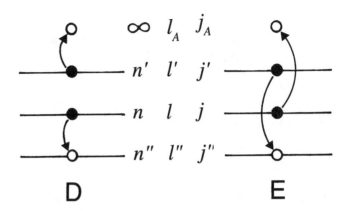

Figure 60.3. Energy levels involved in the direct (D) and exchange (E) Auger processes, and notation for the principal, orbital angular momentum, and total angular momentum quantum numbers that characterize the pertinent electron states.

and the interaction of the core hole state with the Auger continuum is weak. Then the hole state can be considered to be quasi-stationary and the decay rate can be expressed according to *Wentzel's ansatz*, formulated in 1927 and later known as *Fermi's Golden Rule No. 2* of time dependent perturbation theory [47]. In the independent-electron central-field approximation, this leads to the following nonrelativistic matrix element for a direct Auger transition:

$$D = \iint \psi^*_{n''\ell''f''}(1)\psi^*_{\infty\ell_A j_A}(2)\left|\frac{e^2}{\mathbf{r}_1 - \mathbf{r}_2}\right|$$
$$\times \psi_{n\ell j}(1)\psi_{n'\ell'j'}(2)d\tau_1 d\tau_2, \quad (60.13)$$

where the quantum numbers n, ℓ, j characterize electrons that are identified schematically in Fig. 60.3. The state of the continuum (Auger) electron is labeled by $\infty\ell_A j_A$. In the physically indistinguishable exchange process, described by a matrix element E, the roles of electrons $n\ell j$ and $n'\ell'j'$ are interchanged (Fig. 60.3). The radiationless transition probability per unit time is

$$w_{fi} = \frac{2\pi}{\hbar}|D - E|^2 \rho(E_f), \quad (60.14)$$

where $\rho(E_f)$ is the density of final states for the energy E_f that satisfies energy conservation. With the continuum wave function normalized to one electron ejected per unit time, we have $\rho(E_f) = (2\pi\hbar)^{-1}$ (see [50], Appendix E).

The matrix elements D and E in Eq. (60.14) can be separated into radial and angular factors. Evaluation of the angular factors depends upon the choice of an appropriate angular momentum coupling scheme [49, 51], ranging from the $(LSJM)$ representation of Russell-Saunders coupling for the lightest atoms in which the spin-orbit

Table 60.1. Nomenclature for vacancy states. The subscripts in the column headings are the values of $j = l \pm s$, and n is the principal quantum number.

n	$s_{1/2}$	$p_{1/2}$	$p_{3/2}$	$d_{3/2}$	$d_{5/2}$	$f_{5/2}$	$f_{7/2}$
1	K						
2	L_1	L_2	L_3				
3	M_1	M_2	M_3	M_4	M_5		
4	N_1	N_2	N_3	N_4	N_5	N_6	N_7

interaction can be neglected, through intermediate coupling, to j-j coupling in the high-Z limit. The coupling scheme particularly affects intensity ratios among multiplet components of Auger spectra, which can vary by more than an order of magnitude depending upon its choice. Furthermore, realistic calculations of Auger energies and rates call for the use of relativistic wave functions and inclusion of correlation and relaxation effects [52]. It is this extreme sensitivity of Auger transitions to the details of the atomic model that makes them an exceedingly useful probe of atomic structure (see also Chaps. 23 and 59).

Classification of Auger transitions in the central field model is conventionally based on the nomenclature summarized in Table 60.1. The spectroscopic notation s, p, d, f, ... is employed for orbital angular momentum quantum numbers $\ell = 0, 1, 2, 3, \ldots$, and shells with principal quantum numbers $n = 1, 2, 3, \ldots$ are denoted by $K, L, M, \ldots$. Let an inner shell vacancy be created initially in a subshell X_ν and consider a radiationless transition in which an electron from subshell Y_μ fills that vacancy and a Z_ξ-electron is ejected, in the direct process. This transition is denoted by X_ν-$Y_\mu Z_\xi$, which stands for $(n''\ell''j'')$-$(n\ell j)(n'\ell'j')$ in terms of the vacancies indicated in Fig. 60.3

Coster-Kronig transitions are a subclass of Auger transitions in which a vacancy "bubbles up" among subshells of the same shell, i.e., X_ν-$X_\mu Y_\xi$. These are exceptionally fast because the low energy of the ejected electron tends to prevent cancellations of its wave function overlap with the other factors in the matrix element (60.13) and the overlap of the bound state wave functions can be large owing to their similarity. Coster-Kronig transitions can therefore lead to hole-state widths in excess of 10 eV (lifetimes $< 10^{-16}$ s) for the $L_{1,2}$ subshells of the heaviest elements, for example. McGuire has coined the name *super-Coster-Kronig transitions* for the (even faster) type X_ν-$X_\mu X_\xi$ [48]. Here the initial state can no longer be considered quasi-stationary, the width becomes energy-dependent and the two-step model breaks down.

The *Auger transition energy* or kinetic energy of the ejected electron is, within the central field model,

$$E_A = E_{n''\ell''j''} - E_{n\ell j, n'\ell'j'} , \qquad (60.15)$$

where, in the notation of Fig. 60.3, $E_{n''\ell''j''}$ is the absolute value of the energy of the atom with an $n''\ell''j''$ vacancy, with reference to the neutral atom energy, and $E_{n\ell j, n'\ell'j'}$ is the energy of the atom with *two* vacancies in the states indicated by the subscripts. This latter energy can be approximated by a mean of measured single-ionization energies of the atom under consideration and that with the next higher atomic number (the "$Z+1$ rule" [47]); with present-day computers it is readily calculated through one of the nonrelativistic [53] or relativistic [54] self-consistent-field codes.

Photoionization and, even more so, ionization by charged particle impact, often produces more than one vacancy in the target atom (see Sects. 59.1.2 and 60.4.4). The ensuing Auger spectrum can then exhibit *satellite lines* shifted in energy, due to the presence of *spectator vacancies and electrons*, with respect to the *diagram lines* that arise from the decay of a singly ionized atom [48, 51].

60.4.2 X-Ray Emission

The emission of x-rays by atoms ionized in inner shells was studied long before the discovery of the much more probable radiationless transitions, and has played a crucial role in the investigation of atomic structure and interactions with the environment [55]. At present, x-ray spectrometry is providing important insights into the structure of highly ionized species produced, e.g., in electron beam ion traps [56] or plasmas [57], and for the very precise experimental evaluation of inner-vacancy level energies [58].

A classical treatment of multipole radiation by atomic and nuclear systems is given by Jackson [42], who shows by a simple argument that for atoms the electric dipole transition rate is $\sim (137/Z_{\text{eff}})^2$ times more intense than electric quadrupole or magnetic dipole emission, where Z_{eff} is the effective (screened) nuclear charge. The quantum theory of atomic radiation is presented in the treatises by Sakurai [59], Mizushima [60], Berestetskii *et al.* [15], and Sobel'man [61], for example. X-ray emission has been reviewed by Scofield [62] and, with emphasis on relativistic effects, by Chen [52, 63]. In the present volume, radiative transitions are treated in Chap. 32 to which the reader is referred for details.

60.4.3 Widths and Fluorescence Yields

The energy profile of an atomic hole state that decays exponentially with time is found to be the same from quantum as from classical theory [64]; it has the resonance or *Lorentz* shape

$$I(\omega)d\omega = \frac{I_0(\Gamma/2\pi)d(\hbar\omega)}{(\hbar\omega - \hbar\omega_0)^2 + \frac{1}{4}\Gamma^2} . \qquad (60.16)$$

The full width Γ at half maximum of this energy distribution is proportional to the total decay rate τ^{-1} of the state, in accordance with Heisenberg's uncertainty principle $\Gamma\tau = \hbar$.

The decay rate $\tau^{-1} = \Gamma/\hbar$ of an atomic hole state is commonly given in units of eV/$\hbar$ or in atomic units (a.u.) of inverse time, with the corresponding level width Γ in eV or a.u. Thus, for the decay rate,

$$\frac{1}{\tau} = 1 \text{ a.u.} = 4.134\,14 \times 10^{16} \text{ s}^{-1} = 27.2116 \, \frac{\text{eV}}{\hbar}\,. \quad (60.17)$$

K-level (1s hole-state) widths increase monotonically with atomic number, from 0.24 eV for Ne to 96 eV for U, closely following the approximate relation [47]

$$\Gamma(1s) = 1.73 \, Z^{3.93} \times 10^{-6} \text{ eV}\,. \quad (60.18)$$

L-subshell widths, as functions of atomic number, exhibit sharp discontinuities that correspond to energetic cutoffs and onsets of intense Coster-Kronig channels (Sect. 60.4.1) [65].

If there are several decay channels, their partial widths Γ_i add:

$$\Gamma = \sum_i \Gamma_i\,. \quad (60.19)$$

The main decay channels for inner shell hole states are radiative, with a partial width Γ_R, and radiationless or Auger, with a partial width Γ_A.

The *fluorescence yield* ω_i of a hole state i is defined as the relative probability that the state decays radiatively:

$$\omega_i \equiv \Gamma_R(i)/\Gamma(i)\,. \quad (60.20)$$

Macroscopically, for example, the K-shell fluorescence yield ω_K is the number of characteristic K x-ray photons emitted from a sample divided by the number of primary K-shell vacancies created in the sample. This ratio rises from 0.04 for Al to 0.97 for U [47]. The definition of fluorescence yields of shells above the K-shell is more complicated because the higher shells consist of several subshells and Coster-Kronig transitions (Sect. 60.4.1) can shift primary vacancies to higher subshells before they are filled radiatively.

The concept of fluorescence yields is useful in many applications, particularly where the transport of radiant energy through matter is an issue, as in space sciences, medical radiology, and some fields of engineering as well as in physics. Tables of fluorescence yields can be found in the review by Bambynek *et al.* [47] and in the compilations by Krause [66] and Hubbell *et al.* [67]. Considerable uncertainty persists in the fluorescence yields of higher shells.

60.4.4   Multi-Electron Excitations

In atomic inner-shell photoionization, more than one electron can be excited with significant probability. Final states are thus produced that can be described approximately by removal of a core electron and excitation of additional electrons to higher bound states (shakeup) or to the continuum (shakeoff). These multiple excitation processes exhibit themselves through satellites in photoelectron spectra (Sect. 59.1.2) and in the Auger and x-ray spectra emitted when the excited states decay, as well as in some cases by features in absorption spectra [68]. As the photon-electron interaction is described by a one-electron operator, the frozen-core, central field model does not predict changes of state under photon impact by more than one electron. Direct multiple excitation processes are thus a result of electron-electron correlation (see Chap. 23 and Sect. 59.1.2).

Cross sections for photoexcitation and ionization of two inner shell electrons have been evaluated [68] by the multichannel multiconfiguration Dirac-Fock (MMCDF) method [69]. Within this model, the correlation effects that cause direct two-electron processes are (i) relaxation or core rearrangement, (ii) initial state configuration interaction, (iii) final ionic state configuration interaction, and (iv) final continuum state configuration, or final state channel interaction.

Mechanisms that contribute to two-electron photoionization of the outermost shells of noble gas atoms have been separated within many-body perturbation theory [70]. In first order of the combined perturbations by the photon field and electron correlations, the important contributions are core rearrangement, initial state correlations, virtual Auger transitions, and "direct collisions" by the photoelectron with another orbital electron. If treated nonperturbatively, these mechanisms belong within the MMCDF scheme to categories (i)–(iv) described above. The different mechanisms have been found to be of varying importance, depending on the photon energy relative to the double ionization threshold energy, the orbitals which are ionized, and the relative final state energies of the active electrons. Interest in the theory [71, 72] has been enhanced by experiments with synchrotron radiation on the double ionization of He [73], a process that epitomizes the problem. Above ~5 keV photon energy, the double ionization of He has been shown to be dominated by Compton scattering, not photoeffect [74].

60.5   THRESHOLD PHENOMENA

During the creation of a core vacancy by photoexcitation or ionization, the escape of the photoelectron from an inner shell involves complex dynamics of electron excitations with multiple correlational aspects. The entire process of inner shell photoionization and subsequent de-excitation has drastically different characteristics near threshold from that in the high energy limit. In the latter case, discussed in the preceding sections, if an x-ray photon promotes an inner shell electron to an energetic continuum state, the atom first relaxes in its excited (hole) state. In a practically distinct second step, the hole is then filled—most often by a radiation-less transition, under emission of an Auger electron with *diagram* energy that can readily be calculated from the wave function of the stationary, real intermediate state [51].

On the other hand, in the vicinity of core-level energy thresholds, atomic photoexcitation and the ensuing x-ray or Auger electron emission occur in a single second-order quantum process, the *resonant Raman effect*. Here, the intermediate states are virtual and there is no relaxation phase. The resonant Raman regime, comprising the energy region just below threshold, is linked by the *post-collision interaction* regime (Sect. 60.5.2) above threshold to the (asymptotically) two-step regime.

The primary process of photoionization and Auger electron emission can be viewed as resonant double pho-toionization mediated by a complete set of intermediate states; these correspond to an intermediate virtual inner shell hole and an electron in excited bound or continuum states. A unified treatment of the process, as well as of resonant x-ray Raman scattering, has been developed within the framework of relativistic time independent res-onant scattering theory [75–77]. As the energy of the incoming photons is increased, sweeping through an in-ner shell threshold, continuous asymmetric electron dis-tributions associated with different two-hole multiplets evolve into Auger-electron lines with characteristic ener-gies. The shapes and energies of the characteristic lines continue to change until the energy of the photoelectron exceeds that of the Auger electron. The nonresonant two-electron emission amplitude is usually negligible in this region, compared with the resonant amplitude.

60.5.1 Raman Processes

Resonant Raman transitions to bound final states are the inelastic analog of resonance fluorescence [64], with which they share two characteristic features: (1) the emitted radiation can have a narrower bandwidth than the natural lifetime width of the corresponding diagram line, and (2) the energy of the emitted radiation exhibits linear dispersion with excitation energy. Radiative reso-nant Raman scattering was first observed by Sparks in 1974 [78] and subsequently explored with synchrotron ra-diation by Eisenberger *et al.* (see [77,79]). Radiationless resonant Raman scattering was identified by Brown *et al.* [80] and in subsequent studies [81] of the deforma-tion of Auger-electron lines near threshold (Fig. 60.4). The connection between resonant Raman and Compton scattering has been pointed out by Bergstrom *et al.* [45].

According to a generalization of time independent res-onant scattering theory [75,76], a transition matrix ele-

Figure 60.4. Energy of the Xe $L_3 - M_4M_5$ ($^1G_4$) Auger electron peak, as a function of exciting photon energy. Near threshold Auger satellites, caused by the spectator photoelectron in bound orbits, exhibit linear Raman dispersion. The post-collision interaction shift (right-hand scale) vanishes only in the asymptotic limit. (From Ref. [81]).

ment can be constructed that accounts for photoexcita-tion and either radiative or radiationless de-excitation, including resonance structure [77]. The wave functions represent both the atomic and photon fields and include all electrons throughout the process. The conventional two-step model of x-ray fluorescence and Auger electron emission is but a special case of the transition amplitude

$$T_{\beta\alpha} = \left\langle \psi_{\beta\varepsilon\varepsilon'}^{-} \left| V_\gamma + \sum_\nu \int d\tau \frac{(H-E)|\psi_{\tau\nu}\rangle\langle\psi_{\tau\nu}|V_\gamma}{E - \mathcal{E}_{\tau\nu}(E)} \right| \psi_\alpha \right\rangle$$

(60.21)

where H is the total Hamiltonian, V_γ the photon-electron interaction operator, and $E = \hbar\omega_1 + E_\alpha$ is the total en-ergy. The first term in $T_{\beta\alpha}$ represents direct nonresonant Raman scattering; the resonance behavior is embedded in the second amplitude. The complex eigenenergies $\mathcal{E}_{\tau\nu}(E)$ satisfy a complicated secular equation that involves diag-onal and nondiagonal matrix elements of the level shifts and widths [3,77]. The initial state wave function ψ_α is a direct product of the initial atomic state wave function and a one-photon state $|\mathbf{k}_1,\mathbf{e}_1\rangle$, where $\mathbf{k}_1$ is the wave vector and $\mathbf{e}_1$, the polarization vector. The final state scattering wave function $\psi_{\beta\varepsilon\varepsilon'}^{-}$ corresponds, in the radia-tive case, to a scattered photon in a definite state $|\mathbf{k}_2,\mathbf{e}_2\rangle$ and an electron plus ion in a given quantum state, which may involve the linear momentum $\hbar\mathbf{k}_e$ of the electron if the latter is in the continuum. In the radiationless case, there are an ion and two electrons, both of which may be in a continuum state $|\mathbf{k}_e, \mathbf{k}'_e\rangle$. The wave function $\psi^-_{\beta \varepsilon \varepsilon'}$ fulfills the ingoing-wave boundary condition with respect to the electron(s) and accounts for final state channel interaction effects.

Work on inner shell Raman processes is as yet in its infancy. Synchrotron radiation experiments, interpreted in light of the theory outlined above, promise to become a useful tool in atomic physics and materials science [82].

60.5.2 Post-Collision Interaction

In the radiationless decay of an atom that has been photoionized near an inner shell threshold, the Coulomb field of the receding photoelectron perturbs the Auger electron energy and line shape. This *post-collision interaction* provides continuity in the energy evolution of inner shell dynamics between the Raman and asymptotic two-step regimes. In Auger decay, following photoionization, the Auger electron initially screens the ionic Coulomb field seen by the receding photoelectron. This screening subsides when the (usually fast) Auger electron passes the slow photoelectron. Distortion of the Auger line shape results, and the Auger energy is raised at the expense of the photoelectron energy (Fig. 60.4).

A semiclassical potential curve model of post-collision interaction leads to intuitive insight and surprisingly accurate predictions [83]. In simple terms, the attractive potential well in which the slow photoelectron moves is deepened suddenly when the residual singly ionized atom undergoes Auger decay and its net charge changes from $+e$ to $+2e$. The photoelectron sinks down into this deeper well, there being no time for it to speed up, and the energy lost by the photoelectron is transferred to the Auger electron. The photoelectron slows down and may even be recaptured by the atom from which it was emitted [85].

Quantum theoretically, post-collision interaction can be treated from the point of view of resonant scattering theory (Sect. 60.5.1). A lowest-order line shape formula, corresponding to a *shakedown mechanism*, can be shown to emerge from approximations of the general multichannel transition matrix element [84]; the phenomenon thus arises as a consequence of a resonant rearrangement collision in which a photon and an atom in the initial channel turn into an ion and two electrons (one of which has nearly characteristic energy) in the final channel. In this lowest order, the results can be interpreted in terms of an analytic line shape formula based on asymptotic Coulomb wave functions [84]. The line shape depends only on the excess photon energy, the lifetime of the initial state of the Auger process, and the change of the ionic charge during Auger electron emission.

The interaction between the photoelectron and the Auger electron in the final state can be included by rein-terpreting the ionic charge seen by the photoelectron on the basis of asymptotic properties of the continuum wave function pertaining to the two outgoing electrons [86]. This results in a procedure which is consistent with semiclassical models that account for the time required for the Auger electron to overtake the photoelectron. Calculations based on this approach agree very well with measurements performed with synchrotron radiation [86].

Acknowledgments

The author gratefully acknowledges helpful suggestions from Teijo Åberg, M. Ya. Amusia, Mitio Inokuti, L. D. Kissel, and R. H. Pratt. This work was supported in part by the National Science Foundation through Grant No. PHY-9203779.

REFERENCES

1. L. Kissel and R. H. Pratt, in *Atomic Inner-Shell Physics*, edited by B. Crasemann (Plenum, New York, 1985), Chap. 11.
2. L. Kissel, R. H. Pratt, and S. C. Roy, Phys. Rev. A **22**, 1970 (1980).
3. T. Åberg and J. Tulkki, in Ref. [1], Chap. 10.
4. V. A. Bushuev and R. N. Kuz'min, Sov. Phys. Usp. **20**, 406 (1977).
5. K.-I. Hino, P. M. Bergstrom, Jr., and J. H. Macek, Phys. Rev. Lett. **72**, 1620 (1994).
6. D. E. Cullen, Lawrence Livermore National Laboratory Report No. UCRL-50400, 1989.
7. J. H. Hubbell, W. H. McMaster, N. Kerr-Del Grande, and J. H. Mallett, *International Tables for X-Ray Crystallography* (Kynoch Press, Birmingham, 1974), Vol. VI.
8. B. L. Henke, in *American Institute of Physics Conference Proceedings*, No. 15, edited by D. T. Atwood and B. L. Henke (American Institute of Physics, New York, 1981).
9. M. Gavrila, in *American Institute of Physics Conference Proceedings*, No. 94, edited by B. Crasemann (American Institute of Physics, New York, 1982).
10. P. P. Kane, L. Kissel, R. H. Pratt, and S. C. Roy, Phys. Rep. **140**, 75 (1986).
11. L. Kissel, B. Zhou, S. C. Roy, S. K. SenGupta, and R. H. Pratt, Acta Crystallogr. **A51**, No. 3 (1985).
12. W. R. Johnson and K.-T. Cheng, Phys. Rev. A **13**, 692 (1976).
13. W. Mückenheim and M. Schumacher, J. Phys. G **6**, 1237 (1980).
14. P. Papatzacos and K. Mork, Phys. Rep. **21**, 81 (1975).
15. V. B. Berestetskii, E. M. Lifshitz, and L. P. Pitaevskii, *Quantum Electrodynamics* (Pergamon, New York, 1982), 2nd edition.
16. A. I. Milstein and M. Schumacher, Phys. Rep. **243**, 183 (1994).
17. R. Solberg, K. Mork, and I. Øverbø, Phys. Rev. A **51**, 359 (1995).
18. U. Fano and J. W. Cooper, Rev. Mod. Phys. **40**, 441 (1968).

19. R. H. Pratt, A. Ron, and H. K. Tseng, Rev. Mod. Phys. **45**, 273 (1973).

20. A. F. Starace, in *Handbuch der Physik*, edited by S. Flügge and W. Mehlhorn (Springer, Berlin, 1982), Vol. XXI, p. 1.

21. M. Ya. Amusia, *Atomic Photoeffect* (Plenum, New York, 1990).

22. J. A. R. Samson, in Ref. [20], p. 123.

23. H. Siegbahn and L. Karlsson, in Ref. [20], p. 215.

24. E. B. Saloman, J. H. Hubbell, and J. H. Scofield, At. Data Nucl. Data Tables **38**, 1 (1988).

25. J. H. Scofield, Lawrence Livermore Laboratory Report No. UCRL-51326, 1973 (unpublished).

26. J. H. Hubbell and W. J. Veigele, National Bureau of Standards (U. S.) Technical Report No. 901 (U. S. GPO, Washington, DC, 1976).

27. J. J. Yeh and I. Lindau, At. Data Nucl. Data Tables **32**, 1 (1985).

28. T. Åberg, in *Photoionization and Other Probes of Many-Electron Interactions*, edited by F. J. Wuilleumier (Plenum, New York, 1976), p. 49.

29. D. L. Wark *et al.*, Phys. Rev. Lett. **67**, 2291 (1991).

30. B. Crasemann and F. Wuilleumier, in Ref. [1], Chap. 7.

31. B. Crasemann, in *Electronic and Atomic Collisions*, edited by W. R. MacGillivray, I. E. McCarthy, and M. C. Standage (Adam Hilger, Bristol, 1992), p. 69, and references therein.

32. *Synchrotron Radiation Research*, edited by H. Winick and S. Doniach (Plenum, New York, 1980).

33. Much information on these developments is contained in the proceedings of the series of conferences on synchrotron-radiation instrumentation, Nucl. Instrum. Methods A **246**, 1 (1986); **261**, 1 (1987); **266**, 1 (1988); **282**, 369 (1989); **291**, 1 (1990); **303**, 397 (1991); **319**, 1 (1992).

34. *Synchrotron Radiation–Techniques and Applications*, edited by C. Kunz (Springer, Berlin, 1979).

35. *Handbook on Synchrotron Radiation*, edited by E.-E. Koch (North-Holland, Amsterdam, 1983).

36. J. F. Moulder *et al.*, *Handbook of X-Ray Photoelectron Spectroscopy* (Perkin-Elmer, Eden Prairie, 1992).

37. Most recent in the series are *X-Ray and Inner-Shell Processes*, edited by P. Lagarde, F. J. Wuilleumier, and J. P. Briand, J. Phys. (Paris) C **9**, 48 (1987); *idem*, *American Institute of Physics Conference Proceedings*, No. 215, edited by T. A. Carlson, M. O. Krause, and S. T. Manson, (American Institute of Physics, New York, 1990).

38. T. A. Carlson, *Photoelectron and Auger Spectroscopy* (Plenum, New York, 1975).

39. T. A. Carlson, *X-Ray Photoelectron Spectroscopy* (Dowden, Hutchinson & Ross, Stroudsburg, 1978).

40. *Electron Spectroscopy: Theory, Techniques, and Applications*, edited by C. R. Brundle and A. D. Baker (Academic, New York, 1977), Vol. 1 & Vol. 2.

41. J. Berkowitz, *Photoabsorption, Photoionization, and Photoelectron Spectroscopy* (Academic, New York, 1979).

42. J. D. Jackson, *Classical Electrodynamics* (Wiley, New York, 1975), 2nd edition.

43. J. D. Bjorken and S. D. Drell, *Relativistic Quantum Mechanics* (McGraw-Hill, New York, 1964).

44. *Compton Scattering*, edited by B. Williams (McGraw-Hill, New York, 1977).

45. P. M. Bergstrom, Jr., T. Surić, K. Pisk, and R. H. Pratt, Phys. Rev. A **48**, 1134 (1993).

46. R. H. Pratt, L. Kissel, and P. M. Bergstrom, Jr., in *X-Ray Anomalous (Resonant) Scattering: Theory and Experiment*, edited by G. Materlik, C. J. Sparks, and K. Fischer, (North Holland, Amsterdam, 1994).

47. W. Bambynek *et al.*, Rev. Mod. Phys. **44**, 716 (1972).

48. E. J. McGuire, in *Atomic Inner-Shell Processes*, edited by B. Crasemann (Academic, New York, 1975), Vol. 1, Chap. 7.

49. D. Chattarji, *The Theory of Auger Transitions* (Academic, New York, 1976).

50. T. Åberg and G. Howat, in Ref. [20], p. 469.

51. W. Mehlhorn, in Ref. [1], Chap. 4.

52. M. H. Chen, in Ref. [1], Chap. 2.

53. C. Froese Fischer, Comput. Phys. Commun. **14**, 145 (1978); **64**, 369 (1991).

54. I. P. Grant *et al.*, Comput. Phys. Commun. **21**, 207 (1980).

55. A. Meisel, G. Leonhardt, and R. Szargan, *X-Ray Spectra and Chemical Binding* (Springer, Berlin, 1989).

56. P. Beiersdorfer, in *American Institute of Physics Conference Proceedongs*, No. 215, edited by T. A. Carlson, M. O. Krause, and S. T. Manson (American Institute of Physics, New York, 1990), p. 648.

57. E. Källne and J. Källne, Phys. Scr. **T17**, 152 (1987).

58. R. D. Deslattes and E. G. Kessler, Jr., in Ref. [1], Chap. 5.

59. J. J. Sakurai, *Advanced Quantum Mechanics* (Addison-Wesley, Reading, 1967).

60. M. Mizushima, *Quantum Mechanics of Atomic Spectra and Atomic Structure* (Benjamin, New York, 1970).

61. I. I. Sobel'man, *Introduction to the Theory of Atomic Spectra* (Pergamon, Oxford, 1972).

62. J. H. Scofield, in Ref. [48], Vol. 1, Chap. 6.

63. M. H. Chen, in Ref. [56], p. 391.

64. W. Heitler, *The Quantum Theory of Radiation* (Clarendon, Oxford, 1954), 3rd edition, Chap. V.

65. M. H. Chen *et al.*, At. Data Nucl. Data Tables **19**, 97 (1977).

66. M. O. Krause, J. Phys. Chem. Ref. Data **8**, 307 (1979).

67. J. H. Hubbell *et al.*, J. Phys. Chem. Ref. Data **23**, 339 (1994).

68. S. J. Schaphorst *et al.*, Phys. Rev. A **47**, 1953 (1993).

69. J. Tulkki *et al.*, Phys. Rev. A **46**, 1357 (1992).

70. T. N. Chang, T. Ishihara, and R. T. Poe, Phys. Rev. Lett. **27**, 838 (1971).

71. C. Pan and H. P. Kelly, Phys. Rev. A **41**, 3624 (1990).

72. K.-I. Hino *et al.*, Phys. Rev. A **48**, 1271 (1993).

73. J. C. Levin *et al.*, Phys. Rev. A **47**, R16 (1993).

74. J. A. R. Samson, C. H. Greene, and R. J. Bartlett, Phys. Rev. Let. **71**, 201 (1993).

75. T. Åberg, Phys. Scr. **21**, 495 (1980).

76. T. Åberg, Phys. Scr. **T41**, 71 (1992).

77. T. Åberg and B. Crasemann, in Ref. [46].

78. C. J. Sparks, Jr., Phys. Rev. Lett. **33**, 262 (1974).

79. P. L. Cowan, in Ref. [46].

80. G. S. Brown *et al.*, Phys. Rev. Lett. **45**, 1937 (1980).

81. G. B. Armen *et al.*, Phys. Rev. Lett. **54**, 1142 (1985).

82. H. Wang *et al.*, Phys. Rev. A **50**, 1359 (1994).
83. A. Niehaus, J. Phys. B **10**, 1845 (1977); A. Niehaus and C. J. Zwakhals, J. Phys. B **16**, L135 (1983).
84. W. Eberhardt *et al.*, Phys. Rev. A **38**, 3808 (1988).

85. J. Tulkki *et al.*, Z. Phys. D **5**, 241 (1987).
86. G. B. Armen, J. Tulkki, T. Åberg, and B. Crasemann, Phys. Rev. A **36**, 5606 (1987).

61

Electron-Atom and Electron-Molecule Collisions

S. Trajmar,* J. W. McConkey,† and I. Kanik*

*Jet Propulsion Laboratory, California Institute of Technology, Pasadena, California 91109
†Department of Physics, University of Windsor, Windsor, Ontario N9B 3P4, Canada

61.1 INTRODUCTION

Electron-atom and electron-molecule collision processes play a prominent role in a variety of systems ranging from discharge or electron-beam lasers and plasma processing devices to aurorae and solar plasmas. Early studies of these interactions contributed significantly to the understanding of the quantum nature of matter. Experimental activities in this field, initiated by Franck and Hertz [1], flourished in the 1930s and, after a dormant period of about a quarter of a century, have had a renaissance in recent years.

When electrons collide with atomic or molecular targets, a large variety of reactions can take place (see Sect. 61.2.1). We limit our discussion to electron collisions with gaseous targets, where single collision conditions prevail. Furthermore, we discuss only low-energy (threshold to few hundred eV) impact processes where the interaction between the valence-shell electrons

of the target and the free electron dominates.

Comprehensive discussions on electron-atom (molecule) collision physics can be found in the books of Massey et al. [2], McDaniel [3] and the volumes of Advances in Atomic and Molecular (since 1990 Atomic, Molecular and Optical) Physics. The latest developments are usually published in Physical Review A, Journal of Physics B and the Journal of Chemical Physics, and are presented at the biannual International Conference on the Physics of Electronic and Atomic Collisions.

61.2 BASIC CONCEPTS

61.2.1 Electron Impact Processes

Low energy electrons can very effectively interact with the valence-shell electrons of atoms and molecules, in part because they have similar speeds. In elastic scattering, the continuum electron changes direction and transfers momentum to the target. Inelastic collisions include also a transfer of kinetic energy to the target, such as excitation of valence electrons to discrete energy levels, to the ionization continuum, and, in the case of molecules, excitation of nuclear motion (rotational, vibrational) and excitation to states which dissociate into neutral or ionic fragments. Various combinations of these processes are also possible, e.g., dissociative attachment, excitation or ionization. Excitation of more than one valence electron at the same time, or excitation of electrons from intermediate and inner shells, may also occur but these processes are more likely at impact energies of a few keV. These excitations lie above the first ionization limit and lead, therefore, with high probability, to autoionization, except for heavy elements where x-ray emission is an important competing process.

61.2.2   Definition of Cross Sections

The parameters which characterize collision processes are the cross sections. Electron collision cross sections depend on impact energy E_0 and scattering polar angles θ and ϕ. The differential cross section, for a specific well-defined excitation process indicated by the index n is defined as

$$\frac{d\sigma_n(E_0, \Omega)}{d\Omega} = \frac{k_{\rm f}}{k_{\rm i}} |f_n(E_0, \Omega)|^2 , \qquad (61.1)$$

where Ω is the polar angle of detection, $k_{\rm i}$ and $k_{\rm f}$ are the initial and final electron momenta, and f_n is the complex scattering amplitude ($n = 0$ refers to elastic scattering). Integration over the energy loss profile is assumed. If the energy-loss spectrum is broad, differentiation with respect to energy loss also has to be included. For certain processes it may be necessary to define differential cross sections with respect to angle and energy for both primary and secondary particles.

Integration over all scattering angles yields the integral cross sections

$$\sigma_n(E_0) = \int_0^{2\pi} \int_0^\pi \frac{d\sigma_n(E_0, \Omega)}{d\Omega} \sin\theta \, d\theta \, d\phi . \qquad (61.2)$$

In the case of elastic scattering, the momentum transfer cross section is defined as

$$\sigma_0^{\rm M}(E_0) = \int_0^{2\pi} \int_0^\pi \frac{d\sigma_0(E_0, \Omega)}{d\Omega} (1 - \cos\theta) \sin\theta \, d\theta \, d\phi . $$

$$(61.3)$$

The total electron scattering cross section is obtained by summing all integral cross sections:

$$\sigma_{\rm tot}(E_0) = \sum_n \sigma_n(E_0) + \sum_m \sigma_m(E_0) , \qquad (61.4)$$

where σ_m are the cross sections for other possible channels. Experimental cross sections typically represent averages over indistinguishable processes (e.g., magnetic sublevels, hyperfine states, rotational states etc.). The cross section obtained this way corresponds to an average over initial and sum over final indistinguishable states with equal weight given to the initial states. (This may not always be true, as discussed later.) If the target molecules are randomly oriented, the cross section averaged over these orientations is independent of ϕ. In addition, there is an averaging over the finite energy and angular resolution of the apparatus. It is important, therefore, to specify clearly the nature of the measured cross section, otherwise their utilization and comparison with other experimental and theoretical cross sections become meaningless. We denote the conventionally measured differential and integral cross sections by $D_n(E_0, \theta)$ and $Q_n(E_0)$, with the various averagings implied. Similarly, $Q^{\rm M}(E_0)$ and $Q_{\rm tot}(E_0)$ are the corresponding momentum transfer and total scattering cross sections.

The collision strength for a process $i \to j$, which is the particle equivalent of the oscillator strength, is defined by

$$\Omega_{ij}(E_0) = q_i E_0 \sigma_{ij}(E_0) , \qquad (61.5)$$

where q_i is the statistical weight of the initial state $[q_i = (2L_i + 1)(2J_i + 1)]$, E_0 is in Rydbergs and σ_{ij} is in units of πa_0^2.

The rate for a specific collision process (e.g., excitation) for electrons of energy E_0 is given as

$$R_{ij}(E_0) = N I(E_0) \sigma_{ij}(E_0) , \qquad (61.6)$$

where N is the density of the target particles (m^{-3}), and $I(E_0)$ is the electron flux (m^{-2}s^{-1}); σ_{ij} is in m^2, yielding R_{ij} in units of m^{-3}s^{-1}. For nonmonoenergetic electron beams, Eq. (61.6) must be integrated over E_0 to get the average rate.

61.2.3   Scattering Measurements

Most scattering experiments, are carried out in a beam-beam arrangement (see Fig. 61.1). A beam of nearly monoenergetic electrons is formed by extracting electrons from a hot filament and selecting a narrow segment of the thermal energy distribution. For the formation and control of the electron beam, electrostatic lenses are used and the energy selection is achieved with electrostatic energy selectors. A magnetic field may also be applied to obtain a magnetically collimated electron beam. The target beam is formed by letting the sample gas effuse from an orifice, tube or capillary array with various degrees of collimation. Target species which are in the condensed phase at room temperature need to be placed in a crucible and evaporated by heating. Extensive discussion of this technique has been given

Figure 61.1. A schematic diagram for electron scattering measurements.

by Scoles [4]. The electron beam intersects the target beam at a 90° angle and electrons scattered into a specific direction, over a small solid angle ($\sim 10^{-3}$ sr), are detected. However the scattered electron is not necessarily the same as the incoming electron. Exchange with the target electrons may occur, and is required for spin-forbidden transitions in light elements. The detector system consists of electron lenses and energy analyzers similar to those used in the electron gun. The actual detector is an electron multiplier which generates a pulse for each electron. In the scattering process, secondary species (electrons, photons, ions, neutral fragments) may also be generated and can be detected individually or in various coincidence schemes. The experiments are carried out in a vacuum chamber and it is important to minimize stray electric and magnetic fields. More details about the apparatus and procedures can be found in a review by Trajmar and Register [5]. The primary information gained in these experiments is the energy and angular distribution of the scattered electrons.

There are several methods used to carry out scattering measurements. In the most commonly used energy-loss mode, the impact energy and scattering angle are fixed, and the scattering signal as a function of energy lost by the electron is measured by applying pulse counting and multichannel scaling techniques. The result of such an experiment is an energy-loss spectrum. A typical example is shown in Fig. 61.2. The elastic scattering feature appears at zero energy loss; the other features correspond to various excitation processes and to ionization. Energy-loss spectra can also be generated in the constant residual energy mode. In this case, the detector is set to detect only electrons with a specific residual energy $E_R = E_0 - \Delta E$ at a fixed scattering angle, and E_0

and ΔE are simultaneously varied. Each feature in the energy-loss spectrum is obtained then at the same energy above its own threshold.

The energy loss spectrum becomes equivalent to the photoabsorption spectrum in the limit of small momentum transfer K, where $\mathbf{K} = \mathbf{k}_f - \mathbf{k}_i$ (i.e., high impact energy, small scattering angle). The equivalence of electrons and photons in this limit follows from the Born approximation, and it can be used to obtain optical absorption and ionization cross sections. The correspondence is defined through the Limit Theorem

$$\lim_{K \to 0} f_n^{\mathrm{G}}(K) = f_n^{\mathrm{opt}}, \qquad (61.7)$$

$$f_n^{\mathrm{G}}(K) = \frac{\Delta E}{2} \frac{k_i}{k_f} K^2 \frac{d\sigma_n}{d\Omega}(k), \qquad (61.8)$$

where f_n^{G} is the generalized oscillator strength for excitation process n, and f_n^{opt} is the corresponding optical f-value. Equation (61.7) was originally derived by Bethe [6] from the Born cross section. It was extended later by Lassettre et al. [7] to cases where the Born approximation does not hold. In this case, $f_n^{\mathrm{G}}(K)$ is replaced by $f_n^{\mathrm{app}}(K)$, the apparent generalized oscillator strength. The Limit Theorem implies that, in the limit of small K, optical selection rules apply to electron impact excitation. The practical problem is that the limit is nonphysical and the extrapolation to zero K involves some arbitrariness. When K is significantly different from zero, optical-type selection rules do not apply to electron-impact excitation. As can be seen from Fig. 61.3, spin and/or symmetry forbidden transitions then readily occur and can be an efficient way of producing metastable species.

Selection rules for electron impact excitations can be derived from group theoretical arguments [8, 9]. For atoms, the selection rule $S_g \leftrightarrow S_u$ applies in general and scattering to 0° and 180° is forbidden if $(L_i + \Pi_i + L_f + \Pi_f)$ is odd. Here, L_i and L_f are the angular momenta and Π_i and Π_f are the parities. For molecules, selection rules can be derived under two special conditions: (a) rules concerning 0° and 180° scattering for arbitrary orientation of the molecule, and (b) rules concerning scattering to any angle but for specific orientation of the molecule. An important example of the first case is the $\Sigma^- \not\leftrightarrow \Sigma^+$ selection rule for linear molecules at 0° and 180° scattering angles.

The energy dependence of cross sections is obtained by fixing the energy-loss value (scattering channel) and studying the variation of scattering signal with impact energy at a given angle or integrated over all scattering angles. In general, cross sections associated with spin forbidden and optically allowed transitions peak near and at several times the threshold impact energy respectively, and they usually vary smoothly with impact energy. However, resonances may appear at certain specific impact energies. These sudden changes are associated

Figure 61.2. Energy-loss spectrum of He at an impact energy of 29.6 eV and scattering angle of 20°. The elastic peak is at zero energy loss (not shown). The inelastic features with the corresponding principal quantum numbers (and for $n = 2$ with the spectroscopic term symbols) are shown.

with temporary electron capture and are the result of quantum mechanical interference between two indistinguishable paths. An example is shown in Fig. 61.4 for He in the elastic channel at 19.37 eV impact energy.

Integral cross sections can be obtained from extrapolation of the measured $D_n(E_0, \theta)$ to 0° and 180° scattering angles and integration over all angles. In certain cases it is possible to measure integral cross sections directly by detecting secondary products such as photons and ions. These procedures and the resulting cross sections will be discussed in some detail in Sect. 61.3.

61.3 COLLISION PROCESSES

In addition to the basic elastic and inelastic processes defined in Sect. 61.2.2, we now also explicitly include dissociation (to neutral and charged fragments) cross sections $Q_D(E_0)$; and ionization cross sections $Q_I(E_0)$. Each of these is now considered separately.

61.3.1 Total Scattering Cross Sections

Total electron scattering cross sections represent the sum of all integral cross sections:

$$Q_{\text{tot}}(E_0) = \sum_n Q_n(E_0) + Q_I(E_0) + Q_D(E_0), \quad (61.9)$$

$Q_{\text{tot}}(E_0)$ are useful for checking the validity of scattering theories, and the consistency of available data, for normalization of integral and differential cross sections, and as input to the Boltzmann equation. At low impact energies, elastic scattering dominates, while at intermediate and high impact energies, electronic excitations and ionization become major contributors to Q_{tot}. Figure 61.5 shows the various cross sections for electron-helium collisions. The data are from the recommended values of Trajmar and Kanik [10].

Two methods are commonly used for measuring $Q_{\text{tot}}(E_0)$: the *transmission method* and the *target recoil method* (for details see Refs. [5, 11]). Total scattering

716 / *Atomic, Molecular, & Optical Physics Handbook*

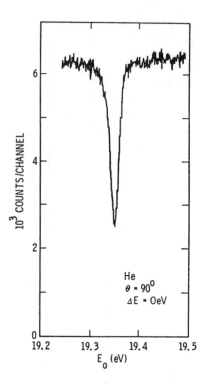

Figure 61.4. The 19.37 eV He resonance observed in the elastic channel at 90° scattering angle.

Figure 61.3. Variation of energy-loss spectra (and DCSs) with scattering angle for He at 40 eV impact energy. Spectra are shown at 5°, 50° and 125° scattering angles.

cross sections measured by these techniques are, in general, accurate to within a few percent.

61.3.2 Elastic Scattering Cross Sections

Elastic scattering cross sections $Q_0(E_0)$ are not as readily available as $Q_{tot}(E_0)$. They are obtained from differential scattering experiments over limited angular ranges by extrapolation and integration of the measured values. Typical error limits are 5 to 20%. For molecular species rotational excitation is usually not resolved but is included in the $D_0(E_0, \theta)$ and $Q_0(E_0)$ values. In order to obtain the absolute $D_0(E_0, \theta)$ directly from the scattering signal, one has to know the electron flux, the number of scattering species, the scattering geometry and the overall response function of the apparatus. A direct measurement of these parameters can be made at high

energies (>100 eV). However, at low electron energies, this approach is not feasible. A number of methods have been devised to derive relative $D_0(E_0, \theta)$ from the measured scattering intensities and then to normalize the $D_0(E_0, \theta)$ to the absolute scale. We briefly outline here only the most commonly used procedure.

The most practical and reliable method of obtaining the absolute $D_0(E_0, \theta)$ is the *relative flow technique* in which scattering signals for a known standard gas and an unknown test gas are compared at each energy and angle [5, 12–14]. The He elastic cross section is the natural choice of standard since it is known accurately over a wide energy and angular range, and He is experimentally easy to handle. Only the relative electron beam flux and molecular beam densities (and their distributions) need be known in the two measurements. The flow rate of the test gas is adjusted so that the flux and density distribution patterns of the two gases are identical, and all geometrical factors cancel in the scattering intensity ratios. The absolute $D_0(E_0, \theta)$ for the sample gas is obtained from the measured scattering intensity, target density, and electron beam intensity ratios and the standard $D_0(E_0, \theta)$ value. See Refs. [5, 14, 15] for a detailed discussion of this technique.

Figure 61.5. Cross sections for various processes in the electron-helium collision (see text for data sources).

61.3.3 Momentum Transfer Cross Sections

$Q^{M}(E_0)$ can be obtained both from the elastic DCSs and from swarm measurements. At low electron-impact energies (from 0.05 to a few eV), where only a few collision channels are open, the electron swarm technique is the most accurate ($\approx 3\%$) way to determine the momentum transfer cross sections. Beam-beam experiments are mandatory at higher energies. A detailed discussion of these techniques is given by Trajmar and Register [5].

61.3.4 Excitation Cross Sections

$D_n(E_0, \theta)$ and $Q_n(E_0)$ can be derived from energy-loss spectra obtained in beam-beam scattering experiments. The relative $D_n(E_0, \theta)$ is usually normalized to $D_0(E_0, \theta)$ which in turn is normalized to the helium $D_0(E_0, \theta)$ by the relative flow technique described in Sect. 61.3.2. There are, however, complications and uncertainties associated with this technique because of the sensitivity of the instrument response function to the residual energy of the scattered electrons. For more details, see Trajmar and McConkey [15]. Data obtained by this procedure are rather limited, partly due to experimental difficulties and partly due to the time required to carry out such measurements.

For cases where an excited state j is formed which can radiatively decay by means of a short-lived (dipole-allowed) transition to a lower lying state i, the intensity of the resultant radiation is directly related to the cross section for production of the excited state in the original collision process. An optical emission cross section, $Q_{ji}(E_0)$, is defined by

$$Q_{ji}(E_0) = \frac{N_j \Gamma_{ji}}{I n_0 \tau_j}, \qquad (61.10)$$

where N_j and n_0 are the densities in the excited and ground states, respectively, Γ_{ji} is the branching ratio for radiative decay from state j to state i, I is the electron beam flux, and τ_j is the natural radiative lifetime of state j. Since the excited state may be produced either by direct electron impact or by cascade from higher-lying states k, also formed in the collision process, we may define the direct excitation cross section $Q_j^d(E_0)$ by

$$Q_j^d(E_0) = \sum_i Q_{ji}(E_0) - \sum_k Q_{kj}(E_0). \qquad (61.11)$$

The last term subtracts the cascade contribution from higher lying states. The quantity $Q_j^a(E_0) = \sum_i Q_{ji}(E_0)$ is known as the *apparent* excitation cross section for level j. Clearly, to obtain $Q_j^d(E_0)$ from $Q_j^a(E_0)$, the cascade contribution must be known.

In Eq. (61.10), $N_j \Gamma_{ji}/\tau_j$ gives the steady state number of $j \rightarrow i$ photons per unit time per unit volume emitted from the interaction region. Since observation is made in a particular direction care must be taken to correct for any anisotropy in the radiation pattern. Alternatively, if observation is made at the so-called 'magic' angle (54° 44') to the electron beam direction, the emission intensity per unit solid angle is equal to the average intensity per unit solid angle irrespective of the polarization of the emitted radiation. However, even at this magic angle, care must be taken to avoid problems with polarization sensitivity of the detection equipment (see Refs. [16, 17]).

The phenomenon of *radiation trapping* is often a problem if the radiative decay channels of the excited state include a dipole allowed channel to the ground state. Repeated absorption and re-emission of the radiation can occur and can lead to a diffuse emitting region much larger than the original interaction region, and the polarization of the emitted light can also be altered. Often a study of the variation of the emitted intensity or polarization with the target gas pressure is sufficient to reveal the presence of radiation trapping or other secondary effects.

The emission cross sections of certain lines have been measured with great care and now serve as bench-marks for other work. Examples of these are the measurements of van Zyl *et al.* [18] on the $n\,^1S$ levels of He in the visible spectral region or the measurements of the cross section for production of Lyman α from H_2 in the vuv region (see Ref. [19] for a full discussion of this; including many references). Use of secondary standards is particularly important when crossed-beam measurements are being carried out because of the cancellation of geometrical and other effects which occur.

The Bethe-Born theory [20] provides a convenient calibration of the detection system for optically allowed transitions of known oscillator strength. At sufficiently high energies, the excitation cross section, Q_n, of level n is given by

$$Q_n = \frac{4\pi a_0^2}{E_0/R_\infty} \frac{f_n^{\text{opt}}}{\Delta E_n/R_\infty} \ln(4c_n E_0/R_\infty). \quad (61.12)$$

Here ΔE_n is the excitation energy, and c_n is a constant dependent on the transition. A plot of $Q_n E_0$ versus $\ln E_0$ is a straight line with a slope proportional to f_n^{opt} and the intercept with the $\ln E_0$ axis yields an experimental value for c_n independent of the normalization. For example, the He $n\,^1P$–$1\,^1S$ optical oscillator strengths are very accurately known, as are cascade contributions. Thus accurate normalization of the slope of the Bethe plot can be made, yielding accurate excitation cross sections.

As mentioned above, the excitation cross sections display characteristic shapes as a function of energy. For optically allowed transitions, the cross section rises relatively slowly from threshold to a broad maximum approximately five times the threshold energy. At higher energies the $(\ln E_0)/E_0$ dependence of the cross section predicted by Eq. (61.12) is observed. For exchange processes, e.g., a triplet excited state from a singlet ground state, the cross section peaks sharply close to threshold and falls off at high energy as E^{-3}. If the excitation is spin allowed but optically forbidden, e.g., He $n\,^1D$ from $1\,^1S$, then the Bethe theory predicts an E^{-1} dependence of the cross section at high energies.

When excitation occurs to a long-lived (metastable or Rydberg) state following electron impact, it is often possible to detect the excited particle directly. Time-of-flight (T.O.F.) techniques are used to distinguish the long lived species from other products, e.g., photons, produced in the collisions.

61.3.5 Dissociation Cross Sections

Dissociation of a molecular target can result in fragments which may be excited or ionized. Such processes may be studied using the techniques discussed in the previous section or in the following section, where charged particle detection is considered. Because a repulsive state of the molecule is accessed, the fragments can leave the interaction region with considerable kinetic energy (several eV). If the fragment is in a long-lived metastable or Rydberg state, T.O.F. techniques may be used to distinguish the long-lived species from other products such as photons, and also to measure the energies of the excited fragments, and thus provide information on the repulsive states responsible for the dissociation. For further discussion see the reviews by Compton and Bardsley [21], Freund [22], and Zipf [23]. If the detector can be made sensitive to a particular excited species, its excitation can be isolated and studied. An example is the work of LeClair and McConkey [24] on $O(^1S)$ production

from various molecules. The detection of unexcited neutral fragments is more challenging. One early method was to trap selectively the dissociation products using a getter and measure the resulting pressure decrease. In a more sophisticated approach, Cosby [25] produced a fast (≈ 1 keV) target molecular beam by resonant charge exchange and subjected it to electron impact dissociation. The fast dissociation products were detected by conventional particle detectors in a time correlated measurement. Laser techniques, such as laser-induced fluorescence or multiphoton ionization, have also been used recently to detect the dissociation products.

The Franck-Condon principle largely governs molecular dissociation. The principle states that if the excitation takes place on a time scale which is short compared with vibrational motion of the atomic nuclei the transition occurs vertically between potential energy curves. Since dissociation rapidly follows a vertical transition to the repulsive part of a potential energy curve, compared with the period of molecular rotation, the dissociation products tend to move in the direction of vibrational motion. Since the excitation probability depends on the relative orientation of the electron beam and the molecule, dissociation products often demonstrate pronounced anisotropic angular distributions. The angular distributions have been analyzed by Dunn [26] using symmetry considerations.

61.3.6 Ionization Cross Sections

Tate and Smith [27] some 60 years ago developed the basic techniques for measuring total ionization cross sections. These were later improved by Rapp and Englander-Golden [28]. Full details of the experimental methods are given in the reviews and books already cited. Märk and Dunn [29] reviewed the situation as it existed in the mid 1980s. In the basic 'parallel plate' method, the electron beam is directed through a beam or a static target gas between collector plates which detect the resultant ions. Unstable species can be studied by the 'fast neutral beam' technique [30], in which the neutral target species is formed by charge neutralization from a fast ion beam, and is subsequently ionized by a crossed electron beam. For the determination of partial ionization cross sections specific to a given ion species in a given ionization stage, mass spectroscopic (quadrupole mass spectrometer, electrostatic or magnetic charged particle analyzer or time of flight) methods are used. Absolute total ionization cross sections have been measured for a large number of species with an accuracy of better than 10%.

A large number of mechanisms can contribute to the ionization of atoms and molecules by electron impact. For targets with only a few atomic electrons, the dominant process is single ionization of the outer shell, with the resultant ion being left in its ground state. The process is direct and is characterized by large impact parameters b and small momentum transfers. The cross section varies with incident electron energy in a way very similar to the optically allowed excitation processes discussed in Sect. 61.3.4. Processes involving ionization of more than one outer shell electron become more important as the size of the target increases. These events are associated with small b and electron-electron correlations are usually strong. Autoionization increases in significance for heavier targets. Here also, collisions with small b dominate and electron-electron correlations are strong. For heavier targets, inner shell effects, such as Auger electron or x-ray emission, become progressively more important. For molecular targets, dissociative ionization (either directly or through a highly excited intermediate state) and ion pair formation also play a significant role.

In addition to measurement of gross ion production, it is also possible to study the ionization process by monitoring the electron(s) ejected or scattered inelastically. Conventional electron spectroscopic techniques are used for this purpose. The addition of coincidence techniques (e-2e measurements) in which the momenta of all the electrons involved are completely specified has allowed many of the fine details of the ionization process to be extracted [31].

61.4 COINCIDENCE AND SUPERELASTIC MEASUREMENTS

The cross section measurements described so far do not yield complete information on electron scattering processes. As mentioned in 61.2.2, these cross sections do not distinguish for magnetic sublevels, electron spin etc. and represent summation of cross sections over these experimentally indistinguishable processes (summation of the square moduli of the corresponding scattering amplitudes). The quantum mechanical description of a scattering process is given in terms of scattering amplitudes and under certain conditions requires summing up amplitudes and squaring the sum. This leads to interference terms which arise from the coherent nature of the scattering process. A complete characterization of a scattering process, therefore, requires knowledge of the complex scattering amplitudes.

Sophisticated experimental techniques have been developed in recent years, which go beyond the conventional scattering cross section measurements and yield information on magnetic sublevel specific scattering amplitudes and the polarization (alignment and orientation) of the excited atomic ensemble. The experimental techniques fall into two main categories: a) electron-photon coincidence measurements, and b) superelastic scattering measurements involving coherently excited species. (We still

Figure 61.6. Schematic illustration of a collisionally induced charge cloud in a p-state atom. The scattering plane is fixed by the direction of incoming $\mathbf{k}_{in}$ and outgoing $\mathbf{k}_{out}$ momentum vectors of the electrons. The atom is characterized by the relative length (l), width (w), and height (h) of the charge cloud, by its alignment angle γ, and by its inherent angular momentum $L_{\perp}$. The coordinate frame is the natural frame with the z-axis perpendicular to the scattering plane and with the x- and y-axes defined as shown in the figure relative to $\mathbf{k}_{in}$ and $\mathbf{k}_{out}$.

consider unpolarized electron beams in the description of these two techniques here and will address the question of spin polarization in the following section.)

In electron-photon coincidence measurements, the radiation pattern emitted by the excited atom is determined for a given direction of the scattered electron. A scattering plane is defined by $\mathbf{k}_i$ and $\mathbf{k}_f$, and hence the symmetry is lowered from cylindrical (around the incident beam direction) to planar, see Fig. 61.6. It is now possible to determine, at least in principle, both the atomic *alignment* (i.e., the shape of the excited state charge cloud and its alignment in space) and its *orientation* (i.e., the angular momentum transferred to the atom during the course of the collision). Complete sets of excitation amplitudes for the coherently excited atomic states and their relative phases have been measured in some cases. A comparison with theory can then be made at the most fundamental level. See Ref. [32] for full discussion and analysis. For more recent work see Refs. [33, 34].

The electron-photon coincidence measurements can

be carried out in two ways: (1) measuring polarization correlations, and (2) measuring angular correlations. In (1), polarization analysis of the emitted photon in a given direction occurs, while in (2), the angular distribution of the emitted photons is determined without polarization analysis. We will describe here only method (1) in some detail.

Method (1) has the advantage that it measures directly the angular momentum (perpendicular to the scattering plane) transferred in the collision. For P-state excitation from a $^1S_0$ ground state, four parameters plus a cross section are needed to describe fully the collisionally excited P-state. The natural parameters introduced by Andersen *et al.* [32] are defined as follows (see Fig. 61.6): γ is the alignment angle of the excited state charge cloud relative to the electron beam axis, P_ℓ^+ is the linear polarization in the scattering plane, $L_\perp^+$ is the orbital angular momentum perpendicular to the scattering plane that is transferred to the atom in the collision, and ρ_{00} is the relative height of the charge cloud perpendicular to the scattering plane at the point of origin. The $+$ superscript indicates positive reflection symmetry with respect to the scattering plane.

In polarization correlation experiments, one typically measures two linear (P_1, P_2) and one circular (P_3) polarization correlation parameters perpendicular to the scattering plane. One additional linear polarization correlation parameter P_4 is measured in the scattering plane. Each parameter is the result of two intensity measurements for different orientations of the polarization analyzer:

$$P_1 = \frac{I(0°) - I(90°)}{I(0°) + I(90°)}\beta^{-1},$$

$$P_2 = \frac{I(45°) - I(135°)}{I(45°) + I(135°)}\beta^{-1},$$

$$P_3 = \frac{I^R - I^L}{I^R + I^L}\beta^{-1},$$ (61.13)

$$P_4 = \frac{I(0°) - I(90°)}{I(0°) + I(90°)}\beta^{-1}.$$

Here $I(\alpha)$ denotes the photon intensity measured for a polarizer orientation α with respect to the electron beam axis, I^R and I^L refer to right- and left-handed circularly polarized light and β denotes the polarization sensitivity of the polarization analyzer. The relationships between the experimentally determined polarization correlation parameters and the natural parameters are given by

$$\gamma = \tfrac{1}{2}\tan^{-1}(P_2/P_1),$$
$$P_\ell^+ = [P_1^2 + P_2^2]^{1/2},$$
$$L_\perp^+ = -P_3,$$ (61.14)
$$\rho_{00} = \frac{(1 + P_1)(1 - P_4)}{4 - (1 - P_1)(1 - P_4)}.$$

The total polarization P_{tot}^+, which is defined as

$$P_{\text{tot}}^+ = [(P_\ell^+)^2 + (L_\perp^+)^2]^{1/2}$$
$$= [P_1^2 + P_2^2 + P_3^2]^{1/2},$$ (61.15)

is a measure for the degree of coherence in the excitation process. In the absence of atomic depolarizing effects due to, for example, fine and/or hyperfine interactions, a value of $P_{\text{tot}}^+ = +1$ for the emitted radiation indicates total coherence of the excitation process.

Much of the earlier work involved excitation of helium. Here the situation is simplified as L-S coupling applies strictly: $P_4 = 1$ and ρ_{00} is zero. Excitation of the $2\,^1P$ state, for example, is fully coherent and hence the excitation is completely specified by just two parameters, γ and $L_\perp$ (or P_ℓ since $P_\ell = (1 - L_\perp^2)^{1/2}$). More recently, the techniques have been applied to heavier targets and more complicated excitation processes [34, 35].

The superelastic scattering experiments could be looked at as time inverse electron-photon coincidence experiments (although this is not exactly the case). In these experiments, a laser beam is utilized to prepare a coherently excited, polarized ensemble of target atoms for the electron scattering measurement. The superelastic scattering intensity is then measured as a function of laser-beam polarization and/or angle with respect to a reference direction. Linearly polarized laser light produces an aligned target (uneven population in magnetic sublevels for quantum numbers of different $|M_J|$ value). Circularly polarized laser light produces oriented targets (uneven population in $M_J = +m$ and $M_J = -m$ magnetic sublevels). From these measurements the same electron impact coherence parameters can be deduced as from the coincidence experiments. This approach has been applied to atomic species (mainly metal atoms) which are conveniently excited with available lasers. Detailed descriptions of the experimental techniques, the underlying theoretical background, and the interpretation of the experimental data are given in Refs. [36–42].

It should be noted that electron scattering by coherently excited atoms can be utilized not only for obtaining electron impact coherence parameters for inelastic processes originating from ground state but for elastic, inelastic, and superelastic transitions involving excited states. These measurements yield information on creation, destruction, and transfer of alignment and orientation in electron collision processes which is needed, e.g., in the application of plasma polarization spectroscopy [43].

61.5 EXPERIMENTS WITH POLARIZED ELECTRONS

So far we have considered the utilization of unpolarized electron beams which yield spin averaged cross sections. Little information on spin dependent interactions is gained from these experiments. However, these interactions can be studied using polarized electron beam techniques. Developments on both the production and detection of spin-polarized electron beams have resulted in a wide range of elegant experiments probing these effects. The theory is also highly developed. For a detailed discussion see the works of Kessler [44, 45], Blum and Kleinpoppen [46], Hanne [47–49] and Bartschat [50] and the references therein. Some basic concepts are presented here.

The degree of polarization P of an electron beam is given by

$$P = \frac{N(\uparrow) - N(\downarrow)}{N(\uparrow) + N(\downarrow)}, \qquad (61.16)$$

where $N(\uparrow)$ and $N(\downarrow)$ are the numbers of electrons with spins respectively parallel and antiparallel to a particular quantization direction. Measurements of P both before and after the collision enable one to probe directly for specific spin dependent processes. For example, in elastic scattering from heavy spinless atoms any changes in the polarization of the electrons must be caused by spin-orbit interactions alone since, in this instance, it is not possible to alter the polarization of the electron beam by electron exchange. Measurements have been carried out for Hg and Xe and both direct (f) and spin-flip (g) scattering amplitudes, as well as their phase differences, have been determined [45].

The spin-orbit interaction for the continuum electron caused by the target nucleus leads to different scattering potentials and consequently to different cross sections for spin-up and spin-down electrons (called Mott scattering). Consequently, an initially unpolarized electron beam can become spin polarized after scattering by a specific angle according to

$$\mathbf{P}' = S_p(\theta)\hat{\mathbf{n}}, \qquad (61.17)$$

where $\hat{\mathbf{n}}$ is the unit vector normal to the scattering plane, $S_p(\theta)$ is the polarization function and $\mathbf{P}'$ is the polarization of the scattered beam. For the same reasons, when a spin-polarized electron beam is scattered by an angle θ to the left and to the right, an asymmetry is found in the scattering cross sections. Furthermore, an existing polarization $\mathbf{P}'$ can be detected through the left-right asymmetry A in the differential cross section, which is given by

$$A \equiv \frac{\sigma_l(\theta) - \sigma_r(\theta)}{\sigma_l(\theta) + \sigma_r(\theta)} = S_A(\theta)\mathbf{P}'\cdot\hat{\mathbf{n}}, \qquad (61.18)$$

where $\sigma_l(\theta)$ and $\sigma_r(\theta)$ are the differential cross sections for scattering at an angle θ relative to the incident beam axis to the left and to the right, respectively. For elastic scattering, the polarization function S_P, and the asymmetry function S_A are identical and are called the *Sherman function*.

When electron exchange is studied under conditions where other explicitly spin-dependent forces can be neglected, the cross sections for scattering of polarized electrons from polarized atoms depend on the relative orientation of the polarization vectors according to [45]

$$\sigma(\theta) = \sigma_u(\theta)[1 - A_{\text{ex}}(\theta)\,\mathbf{P}_e\cdot\mathbf{P}_A], \qquad (61.19)$$

where $\mathbf{P}_e$ and $\mathbf{P}_A$ are the electron and atom polarization vectors, and $\sigma_u(\theta)$ is the cross section for unpolarized electrons. Hence, an 'exchange asymmetry' $A_{\text{ex}}(\theta)$ can be defined by

$$A_{\text{ex}}(\theta)\,\mathbf{P}_e\cdot\mathbf{P}_A = \frac{\sigma_{\uparrow\downarrow}(\theta) - \sigma_{\uparrow\uparrow}(\theta)}{\sigma_{\uparrow\downarrow}(\theta) + \sigma_{\uparrow\uparrow}(\theta)}, \qquad (61.20)$$

where $\sigma_{\uparrow\uparrow}(\theta)$ and $\sigma_{\uparrow\downarrow}(\theta)$ denote the cross sections for parallel and antiparallel polarization vectors respectively. As Bartschat [50] points out, an asymmetry can occur even if the scattering angle is not defined. In this case the function $A_{\text{ex}}(\theta)$ is averaged over all angles. Differential and integral measurements of this kind have been performed for elastic scattering, excitation and ionization.

For heavy target systems it is necessary to consider a combination of effects together with a description of the target states in the intermediate or fully coupled scheme. Consequently, the number of independent parameters can become very large and the 'complete' experiments, which disentangle the various contributions to any observed asymmetry in the scattering, are rarely possible.

Even for very light target atoms, where conventional Mott scattering is negligible, Hanne [47] has shown that the 'fine structure' effect, in which electron scattering from individual fine structure levels of a multiplet occurs, can lead to polarization effects. In fact it can be a dominating effect for inelastic collision processes.

For full details of these various mechanisms and how density matrix theory and other theoretical techniques have been applied to scattering involving polarization effects, the reader is referred to the review articles cited.

In certain cases, experiments involving spin polarized electron beams coupled with coincidence (or superelastic) measurements allow one to extract the maximum possible information for a given process, and are termed as *complete* or *perfect* in the sense defined by Bederson [51].

61.6 ELECTRON COLLISIONS WITH EXCITED SPECIES

There are many plasma systems where electron collisions with excited atoms and molecules play a prominent role, e.g., electron beam and discharge pumped lasers, planetary and astrophysical plasmas. Especially important are electron collisions with metastable species because of the long lifetime, large cross section and large amount of excitation energy associated with them. Electron collision studies and cross section data in this area are scarce mainly due to experimental difficulties associated with the production of target beams with sufficiently high densities of excited species. With the application of lasers for the preparation of the excited species this problem can be overcome. However, this approach has not yet been extensively exploited. Recent reviews of this field are given by Lin and Anderson [52] and Trajmar and Nickel [53].

Since electron collisions with excited species necessarily involve a method of preparation, they are two step processes. Excitation and ionization in these cases are frequently referred to as stepwise excitation and ionization. The target preparation leads to mixed beams containing both ground and excited atoms or molecules. Preparation of excited atoms is achieved by electron impact or photoabsorption. Fast metastable beams can be produced by near-resonance charge exchange. For more details see Ref. [53]. Electron impact excitation is simple and effective but highly nonspecific, and characterization of the composition of the mixture is difficult. Laser excitation is more involved but very well defined. Specific fine and hyperfine levels of individual isotopes can be excited. When laser excitation is used in conjunction with superelastic electron scattering, an energy resolution of 10^{-8} eV is easily achieved, compared with the 10^{-2} eV resolution possible in conventional electron scattering.

Depending on the method of preparation, the population distribution in the magnetic sublevels of the target atoms may be uneven and some degree of polarization (alignment or orientation) may be present. The scattering will then be ϕ-dependent. For polarized target atoms the measured electron collision cross sections do not correspond to the conventional cross sections (which are summed over final and averaged over initial experimentally indistinguishable states, with equal populations in the initial states). One, therefore, has to characterize precisely the state of the target beam in order to be able to deduce a well defined, meaningful cross section. Polarization of atoms can be conveniently controlled, in the case of excitation with laser light, through the control of the laser light polarization (as discussed in Sect. 61.4). Since the atomic ensemble is coherently excited in this case, the scattering cross sections will depend on the azimuthal scattering angle ϕ. These considerations also come into play when one tries to relate measured inelastic and corresponding inverse superelastic cross sections by the principle of detailed balancing.

REFERENCES

1. J. Franck and G. Hertz, Verh. dtsch. Physik. Ges. **16**, 457 (1914).

2. H. S. N. Massey, E. H. S. Burhop, and H. B. Gilbody, *Electronic and Ionic Impact Phenomena* (Clarendon Press, Oxford, 1969), Vols. 1 & 2.
3. E. W. McDaniel, *Atomic Collisions, Electron and Photon Projectiles* (John Wiley and Sons, New York, 1989).
4. G. Scoles, *Atomic and Molecular Beam Methods* (Oxford University Press, New York, 1988), Vol. 1.
5. S. Trajmar and D. F. Register, in *Electron Molecule Collisions*, edited by I. Shimamura and K. Takayanagi (Plenum Press, New York, 1984).
6. H. A. Bethe, Ann. Phys. **5**, 325 (1930).
7. E. N. Lassettre, A. Skerbele, and M. A. Dillon, J. Chem. Phys. **50**, 1829 (1969).
8. D. C. Cartwright, S. Trajmar, W. Williams, and D. L. Huestis, Phys. Rev. Lett. **27**, 704 (1971).
9. W. A. Goddard, D. L. Huestis, D. C. Cartwright, and S. Trajmar, Chem. Phys. Lett. **11**, 329 (1971).
10. S. Trajmar and I. Kanik, in *Atomic and Molecular Processes in Fusion Edge Plasmas*, edited by R. K. Janev (Plenum Press, New York, 1995).
11. B. Bederson and L. J. Kieffer, Rev. Mod. Phys. **43**, 601 (1971).
12. S. Srivastava, A. Chutjian, and S. Trajmar, J. Chem. Phys. **63**, 2659 (1975).
13. J. C. Nickel, C. Mott, I. Kanik, and D. C. McCollum, J. Phys. B **21**, 1867 (1988).
14. J. C. Nickel, P. Zetner, G. Shen, and S. Trajmar, J. Phys. E **22**, 730 (1989).
15. S. Trajmar and J. W. McConkey, Adv. At. Mol. Opt. Phys. **33**, (1994).
16. P. N. Clout and D. W. O. Heddle, J. Opt. Soc. Am. **59**, 715 (1969).
17. F. G. Donaldson, M. A. Hender, and J. W. McConkey, J. Phys. B **5**, 1192 (1972).
18. B. van Zyl, G. H. Dunn, G. Chamberlain, and D. W. O. Heddle, Phys. Rev. A **22**, 1916, (1980).
19. A. R. Filippelli, C. C. Lin, L. W. Anderson, and J. W. McConkey, Adv. At. Mol. Opt. Phys. **33**, (1994).
20. M. Inokuti, Rev. Mod. Phys. **43**, 297 (1971).
21. R. N. Compton and J. N. Bardsley, in *Electron-Molecule Collisions*, edited by I. Shimamura and K. Takayanagi (Plenum, New York, 1984).
22. R. S. Freund, *Rydberg States of Atoms and Molecules*, edited by R. F. Stebbings and F. B. Dunning (Cambridge University Press, 1983).
23. E. C. Zipf, in *Electron-Molecule Interactions and Their Applications*, edited by L. G. Christophorou (Academic Press, New York, 1984), Vol. 1.
24. L. R. LeClair and J. W. McConkey, J. Chem. Phys. **99**, 4566 (1993).
25. P. C. Cosby, J. Chem. Phys. **98**, 7804, 9544, and 9560 (1993).
26. G. H. Dunn, Phys. Rev. Lett. **8**, 62 (1962).
27. J. T. Tate and P. T. Smith, Phys. Rev. **39**, 270 (1932).
28. D. Rapp and P. Englander-Golden, J. Chem. Phys. **43**, 1464 (1965).
29. T. D. Märk and G. H. Dunn, *Electron Impact Ionization* (Springer-Verlag, Vienna, New York, 1985).
30. A. J. Dixon, M. F. A. Harrison, and A. C. H. Smith, J. Phys. B **9**, 2617 (1986).
31. H. Ehrhardt, K. Jung, G. Knoth, and P. Schlemmer, Z.

Phys. D **1**, 3 (1986).

32. N. Andersen, J. W. Gallagher, and I. V. Hertel, Phys. Rep. **165**, 1 (1988).

33. J. A. Slevin and S. Chwirot, J. Phys. B **23**, 165 (1990).

34. K. Becker, A. Crowe, and J. W. McConkey, J. Phys. B **25**, 3885 (1992).

35. M. Sohn and G. F. Hanne, J. Phys. B **25**, 4627 (1992).

36. I. V. Hertel and W. Stoll, Adv. At. Mol. Phys. **13**, 113 (1977).

37. P. W. Zetner, S. Trajmar, and G. Csanak, Phys. Rev. A **41**, 5980 (1990).

38. U. Fano, Rev. Mod. Phys. **29**, 74 (1957).

39. K. Blum, *Density Matrix Theory and Applications*, (Plenum Press, New York, 1981).

40. U. Fano and J. H. Macek, Rev. Mod. Phys. **45**, 553 (1973).

41. J. H. Macek and I. V. Hertel, J. Phys. B **7**, 2173 (1974).

42. W. R. MacGillivray and M. C. Standage, Phys. Rep. **168**, 1 (1988).

43. S. A. Kazantsev and J.-C. Hénoux, *Polarization Spectroscopy of Ionized Gases*, (Kluwer Academic Publishers, Dordrecht, 1995).

44. J. Kessler, *Polarized Electrons* (Springer-Verlag, Berlin, 1985), 2nd ed.

45. J. Kessler, Adv. At. Mol. Opt. Phys. **27**, 81 (1991).

46. K. Blum and H. Kleinpoppen, Adv. At. Mol. Phys. **19**, 187 (1983).

47. G. F. Hanne, Phys. Rep. **95**, 95 (1983).

48. G. F. Hanne, *Coherence in Atomic Collision Physics*, edited by H. J. Beyer, K. Blum, and R. Hippler (Plenum, New York, 1988), p. 41.

49. G. F. Hanne, in *Proceedings of the 17th ICPEAC, Electronic and Atomic Collisions, 1992*, edited by W. R. MacGillivray, I. E. McCarthy, and M. C. Standage (Hilger, Bristol, 1992), p. 199.

50. K. Bartschat, Comments At. Mol. Phys. **27**, 239 (1992).

51. B. Bederson, Comments At. Mol. Phys. **1**, 41 (1969).

52. C. C. Lin and L. W. Anderson, Adv. At. Mol. Opt. Phys. **29**, 1 (1992).

53. S. Trajmar and J. C. Nickel, Adv. At. Mol. Opt. Phys. **30**, 45 (1993).

62

Ion-Atom Scattering Experiments: Low Energy

Ronald Phaneuf

Department of Physics, University of Nevada, Reno, Nevada 89557-0058

62.1 INTRODUCTION

This chapter outlines the physical principles and experimental methods used to investigate low energy ion-atom collisions. A low energy collision is here defined as one in which the initial ion-atom relative velocity is less than the mean orbital velocity $\langle v_e \rangle$ of the electrons affected by the collision. For outer or valence electrons, $\langle v_e \rangle \simeq v_B$, where $v_B = 2.1877 \times 10^8$ cm/s is the Bohr velocity. In terms of the energy of a projectile ion, v_B corresponds to 24.8 KeV/N, where N is the projectile nuclear number (e.g., 16 for O^+).

The theory and results of ion-atom scattering studies are further discussed in Chaps. 36, 37, and 45 to 49. The focus here is on the experimental techniques. Since several of these depend on the characteristics of a specific process, the following section presents a summary of the physics of low-energy ion-atom collisions. See also Chaps. 49 and 48 for more detailed information.

62.1.1 Low Energy Ion-Atom Collision Processes

The most important and widely studied inelastic ion-atom collision process in the low energy region is *electron capture* (also referred to as charge exchange, charge transfer or electron transfer) represented by

$$A^{+q} + B \rightarrow A^{+q-k} + B^{k+} + Q , \qquad (62.1)$$

where Q is the potential energy difference between the initial and final states. For an exoergic process, $Q > 0$ and this energy appears as excess kinetic energy of the products after the collision. For an endoergic process, $Q < 0$ and must be provided by the initial kinetic energy of the reactants, so that the corresponding cross section is

Figure 62.1. Typical cross section variation with collision energy for electron capture by a multiply charged ion from hydrogen. The low energy behavior depends on the structure of the quasimolecule formed during the collision.

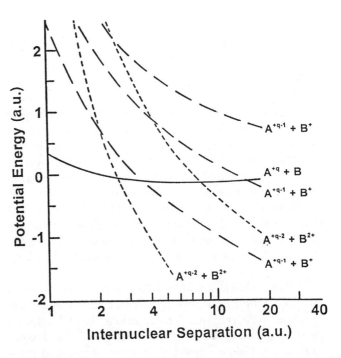

Figure 62.2. Schematic representation of potential energy curves for the collision of a multiply charged ion A^{q+} with a multielectron atom B.

usually small at low collision energies. Cross sections for electron capture are appreciable even at very low energies if Q is zero or very small (resonant or near-resonant process). Electron capture by multiply charged ions from atoms is predominantly an exoergic process, for which cross sections may also be large at low energies. In this case, electrons are preferentially captured into excited levels of A^{+q-k}. The typical cross section behavior for single electron capture ($k = 1$) by a multiply charged ion from atomic hydrogen is shown in Fig. 62.1. The initial ionic charge is the major determinant of the cross section at intermediate and high collision energies, whereas the cross section at low energies depends strongly on the structure of the transient quasimolecule formed during the collision.

Multiple electron capture ($k > 1$) from multielectron atoms occurs predominantly into multiply excited levels, which stabilize either radiatively, leading to stabilized or "true" double capture, or via autoionization. The latter process is usually referred to as *transfer ionization*,

$$A^{+q} + B \rightarrow A^{+q-k} + B^{m+} + (m-k)\mathrm{e}^- + Q. \quad (62.2)$$

At low energies, transfer ionization is particularly important in collisions of highly charged ions with multielectron atoms.

Diabatic potential energy curves representing the collision of a multiply charged ion with a neutral atom are presented in Fig. 62.2. For such collisions, the Coulomb repulsion in the final state produces avoided crossings of the initial and final state potential curves, the positions of which are determined by the binding energy of the atomic electron and electronic energy level structure of the product ion. Generally, reaction channels that are moderately exoergic produce curve crossings at internuclear separations where there is sufficient overlap of the electron clouds for electron capture to be a likely process. The Landau-Zener curve-crossing model [1] (see Chap. 47), the classical over-barrier model for single capture [2] and the extended classical over-barrier model for multiple capture [3], are useful in predicting the important final product states, and in providing a semiquantitative interpretation. In the case of single electron capture by a bare multiply charged ion (of charge $q = Z$) from a hydrogen atom, the principal quantum number n_p of the most probable final ionic state is given in this model [2] by

$$n_p = \sqrt{\frac{2Z^{1/2} + 1}{Z + 2Z^{1/2}}}. \quad (62.3)$$

The internuclear separation R_p at which the potential

curves cross in this case is given by

$$R_p = \frac{2(Z-1)}{Z^2/n_p^2 - 1}.$$ (62.4)

There has been much discussion of the role of electron correlation in the multiple electron capture process. At issue is the relative importance of the mechanism whereby several electrons are transferred (in a correlated manner) at a single curve crossing compared with that whereby single electrons are transferred successively at different curve crossings. Experimental evidence exists for both mechanisms, with the relative importance depending on the electronic structure of the transient quasi-molecule that is formed during the collision. Measurements of the distribution of final ion product electronic states provides the major insight into such collision mechanisms [4].

Other inelastic ion-atom collision processes, such as direct electronic excitation and ionization, are endothermic, with relatively small cross sections that fall off with decreasing energy below a few tens of keV/N. Exceptions are collisions involving Rydberg atoms and collisional excitation of fine structure transitions, for which the required energy transfer is relatively small. Relatively little experimental data are available for direct excitation and ionization processes at low collision energies.

62.2 EXPERIMENTAL METHODS FOR TOTAL CROSS SECTION MEASUREMENTS

In the present context, a total cross section measurement refers to an integration or summation over scattering angles, product kinetic energies and (frequently) electronic states. The total cross section is usually measured as a function of relative collision energy or velocity.

62.2.1 Gas Target Beam Attenuation Method

The attenuation of a collimated ion beam of incident intensity I_0 in a differentially pumped gas target cell or gas jet is related to the collision cross section σ by

$$I = I_0 e^{-\sigma NL},$$ (62.5)

where I is the itensity after traversing an effective length L of the target gas, and N is the number density of target atoms. For a gas target cell with entrance and exit apertures of diameter d_1 and d_2 which are much less than the physical length z of the gas cell, L is given to a good approximation by $z + (d_1 + d_2)/2$. This is valid under molecular flow conditions, for which the mean free path between collisions of target atoms is much larger than the dimensions of the gas cell.

In designing the gas cell for measurements of total cross sections, d_2 and the beam detector must be large enough that elastic scattering may be eliminated as a contributor to the measured attenuation. Usually d_2 is made larger than d_1. Measurement of the gas pressure in a target cell is usually made using a capacitance manometer connected to the cell via a tube whose conductance is much larger than that of the gas cell apertures so that, to a good approximation, the pressure will be the same in both the manometer and the gas cell. For gas jet targets, the effective target thickness NL is usually determined by *in situ* normalization to some well-known cross section.

The quantity σ in Eq. (62.5) refers to an *effective* cross section for removing projectile ions from the incident beam, which is the sum of cross sections for all such processes. In many cases, a single process (e.g., electron capture) is dominant, and σ primarily describes that process. Whether a collision process removes a projectile ion from the reactant beam or not depends on the configuration of the experiment. For example, the projectile particle may remain physically in the beam after passing through the target, but with a changed charge due to a collision. This would be registered as an attenuation of the primary ion beam if the beam is charge analyzed after the reaction.

62.2.2 Gas Target Product Growth Method

The product growth method is similar to the beam attenuation method; the major difference that the growth of reaction products is measured rather than the loss of reactant projectiles. The products may be derived from either the projectile beam or the gas target, or both. The main advantage of this method is its higher degree of selectivity of a specific collision process. In addition, the reactants and products can usually be registered simultaneously, or in some cases in coincidence, eliminating the sensitivity of the measurement to temporal variation of ion beam intensity.

An important criterion is that the target gas density be low enough that single collision conditions prevail (i.e. that the likelihood of an ion passing through the gas target and interacting with more than one target atom is negligibly small). This must in general be satisfied in order for Eq. (62.5) to relate correctly the measured attenuation to the collision cross section of interest, and is critical to the product growth method [5]. In this case, under single collision conditions, one may set the number of products $I_p = I_0 - I$, and Eq. (62.5) then may be written as

$$\frac{I_p}{I_0} = 1 - e^{-\sigma NL} \approx \sigma NL.$$ (62.6)

The approximate expression is useful for $\sigma NL \ll 1$, which is a requirement for single collision conditions. It is also important that the products not be lost in a subsequent collision in the gas cell, so the magnitudes of cross sections for loss of products in the target gas must also be considered. The products in such an experiment may be derived from either the projectile beam or the target gas (e.g., collection of slow product ions in a gas cell), or from both in coincidence to enhance the specificity of the method. The method may in principle be used to determine either total or differential cross sections, depending on the degree of selectivity of collision products.

62.2.3 Crossed Ion and Thermal Beams Method

Replacement of the gas target cell by an effusive thermal beam is advantageous for studying collisions of ions with reactive species such as atomic hydrogen, as well as for collecting slow ion products. The use of accelerating electrodes or grids for slow charged products allows coincident detection of fast and slow products, permitting measurement of ionization as well as electron-capture cross sections. Use of an effusive source or gas jet precludes accurate measurement of the effective target thickness, and *in situ* normalization to the cross section for some well known process is usually employed. A comprehensive discussion of such methods as applied to collisions of multiply charged ions with atomic hydrogen is given by Gilbody [6].

62.2.4 Fast Merged Beams Method

Cross sections for ion-atom collision processes at very low energies have been measured by merging fast beams of ions and neutral atoms [7], as in Fig. 62.3. In this case, σ is determined from experimental parameters by

$$\sigma = \frac{R}{\epsilon}\frac{e}{I_+ I_0}\frac{v_+ v_0}{|v_+ - v_0|}F, \qquad (62.7)$$

where R is the number of products detected per second, ϵ is the product detection efficiency, e is the electronic charge, I_+ is the ion current, I_0 is the flux of atoms, v_+ and v_0 are the laboratory velocities of the ion and atom beams, $v_{\rm rel}$ is their relative velocity, and F is the form factor that describes the spatial overlap of the two beams. If the z-axis is chosen to be the direction of propagation of the beams, the form factor has units of length and is given by

$$F = \frac{\iint I_+(x,y,z)\,dx\,dy \; \iint I_0(x,y,z)\,dx\,dy}{\iiint I_+(x,y,z)\,I_0(x,y,z)\,dx\,dy\,dz}. \qquad (62.8)$$

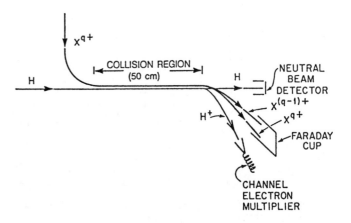

Figure 62.3. Schematic of the merged beams arrangement used by Havener *et al.* [7] to study low energy electron capture collisions of multicharged ions with H atoms.

The two-dimensional integrals in the numerator represent the total intensities of the two beams, which are independent of z, so that F is also independent of z.

The relative collision energy $E_{\rm rel}$ in eV/u is given by

$$\frac{E_{\rm rel}}{\mu} = \frac{E_+}{m_+} + \frac{E_0}{m_0} - 2\sqrt{\frac{E_+}{m_+}\frac{E_0}{m_0}}\cos\theta. \qquad (62.9)$$

where E_+, m_+ and E_0, m_0 are the energies and masses of the ion and atom, respectively, and μ is their reduced mass. For collinear merged beams, $\theta = 0$, and $E_{\rm rel}$ can be reduced to zero by making the two beam velocities the same. In practice, the finite divergences of the beams place a lower limit on the energy and the energy resolution. The fast neutral atom beam is created by neutralizing an accelerated ion beam by electron capture by a positive ion beam in a gas, by electron detachment of a negative ion beam either in a gas, or using a laser beam. In gas collisions, a small fraction of the neutral beam is produced in excited n-levels (with an n^{-3} distribution), which may influence the measurements.

With fast colliding beams, the maximum effective beam densities are invariably much smaller than the background gas density, even under ultrahigh vacuum conditions. For example, a typical 10 keV proton beam with a circular cross section of diameter 3 mm and a current $I = 10\ \mu$A has an average effective density $n = I/(eAv) = 1.6 \times 10^6$ cm^{-3} (A is the beam cross sectional area and v is its velocity). It is therefore necessary either to modulate the beams or to use coincidence techniques to separate signal events due to beam-beam collisions from background events produced by collisions of either beam with background gas. A typical two-beam modulation scheme is shown in Fig. 62.4. To eliminate the production of spurious signals, the detector gates are

Figure 62.4. Fast two-beam modulation scheme to separate the signal due to beam-beam collisions from events due to beam collisions with residual gas or surfaces [7].

delayed for a short time after the beams are switched, and the beam modulation period is made much shorter than the pressure time constant of the vacuum system.

Absolute electron-capture cross sections have been measured for O^{5+} + H collisions to energies below 1eV/N, where the attractive ion - induced-dipole (polarization) interaction is expected to play a role [7]. The inverse velocity dependence of the cross section in this region is suggestive of the classical *Langevin orbiting* model for ion-neutral collisions [8].

62.2.5   Trapped Ion Method

The trapped ion method is used to determine rate coefficients and effective cross sections for ion-atom collisions at near thermal energies. The technique involves storing ions in an electrostatic or electromagnetic trap, and measuring the rate of loss of ions from the trap after a small quantity of neutral gas is admitted [9]. Like the beam attenuation method, the trap technique cannot distinguish different processes that cause ions to be lost from the trap. The mean collision energy is estimated from an analysis of the ion dynamics in the trap.

62.2.6   Swarm Method

The swarm method, using the flowing afterglow, drift tube or selected-ion flow tube, has been used successfully to study ion-atom collisions at very low energies [10]. Ions are injected into a homogeneous electric field and drift through a suitable low-density buffer gas such as

helium. The ions move as a swarm whose mean energy depends on the applied electric field and on collisions with the buffer gas, and can be varied from the near-thermal region to tens of eV. The method involves measuring the additional attenuation of the directed ion swarm by a known quantity of added reactant gas, and is the major technique that has been used for the study of ion-atom collisions at near-thermal energies [11]. As with ion beam attenuation and ion trap methods, the drift tube is not selective of the process that leads to attenuation of the ion swarm.

62.3   METHODS FOR STATE AND ANGULAR SELECTIVE MEASUREMENTS

Three principal methods have been developed and applied to the measurement of partial cross sections for population of specific product electronic states in ion-atom collisions involving electron capture [12]. These are based on spectroscopic measurements of photon emission, translational energy spectroscopy and electron emission in collisions of ion beams with gas targets.

62.3.1   Photon Emission Spectroscopy

Since electron capture from atoms by multiply charged ions populates excited levels, photon emission spectroscopy may be employed to determine state-selective partial cross sections [13]. An ion beam is directed through a gas cell or jet, and a photomultiplier or suitable detector registers photons analyzed by an optical filter, a grating or a crystal spectrometer. The measured photon signal at a given angle depends on the detection solid angle, the polarization of the emitted radiation, the absolute efficiencies of the dispersive device and detector, and (for emission by fast ions) the lifetime of the radiating state. Depending on the spectral region, the photon detection system may be absolutely calibrated using a standard photon source or detector, or by using a reference ion or electron beam and well established cross section data for photon emission [14]. Cascading from higher populated levels must also be considered whenever measured spectral line emission intensities are used to infer cross sections for populating specific energy levels. Successful state-selective cross section measurements have been made for transitions in the visible, vuv and x-ray spectral regions.

62.3.2   Translational Energy Spectroscopy

Translational energy loss or energy gain spectroscopy provides a convenient method to determine the distribution of final states in low energy ion-atom collisions that are either endoergic or exoergic ($Q \neq 0$). For example, this method has been used extensively by the Belfast group [15] to study the predominantly exoergic process of electron capture by multiply charged ions from hydrogen atoms. An ion beam with a well defined energy is directed through a gas target, and the energy of the product ion beam is energy analyzed after the collision. Since the energy gain or loss to be measured is only a very small fraction of the initial kinetic energy, it is usually necessary to reduce the initial energy spread to a few tenths of an eV by decelerating the reactant ion beam prior to energy selection by an electrostatic analyzer. If the scattering angle of the ion is very small, its energy change is approximately equal to Q. Since the ion beam is attenuated by deceleration and energy analysis, cross sections for collisionally populating specific states are determined by normalizing the measured product-state distributions to total cross section data. The attainable state resolution is not as good as for photon emission spectroscopy.

62.3.3 Electron Emission Spectroscopy

As noted in Sect. 62.1.1, multiple electron capture by multiply charged ions from atoms at low energies occurs primarily into multiply excited states, which decay either radiatively or via autoionization (with a branching ratio) [4]. The latter decay pathway (transfer ionization) provides an experimental method to determine the product ionic states by ejected-electron spectroscopy. Analysis of electrons emitted into the forward (ion beam) direction (zero degree spectroscopy) offers significant advantages for analysis of low energy electrons with high resolution [16]. Since a gas jet is often employed and absolute electron collection and spectrometer efficiencies are difficult to determine, some normalization procedure is usually employed to determine state-selective cross sections by this method. Electrons and product projectile or recoil ions have also been detected in coincidence to increase the specificity of the method.

62.3.4 Angular Differential Measurements

The measurement of angular distributions of scattered ions in low energy ion-atom collisions has been facilitated by the availability of position-sensitive particle detectors consisting of a microchannel plate and a resistive or segmented anode [17]. The method for processes that have forward-peaked angular distributions involves directing a highly collimated ion beam through a gas target cell, and counting the scattered projectile ions on a position-sensitive particle detector. Product ions produced by electron capture can be selected by the use of electrostatic retarding grids mounted immediately in front of the detector, to reject the primary ion beam.

REFERENCES

1. F. W. Meyer, A. M. Howald, C. C. Havener, and R. A. Phaneuf, Phys. Rev. Lett. **54**, 2663 (1985).
2. H. Ryufuku, K. Sasaki, and T. Watanabe, Phys. Rev. A **23**, 745 (1980).
3. A. Niehaus, J. Phys. B **19**, 2925 (1986).
4. M. Barat and P. Roncin, J. Phys. B **25**, 2205 (1992).
5. H. B. Gilbody, Adv. At. Mol. Phys. **22**, 143 (1986).
6. F. W. Meyer, A. M. Howald, C. C. Havener, and R. A. Phaneuf, Phys. Rev. A **32**, 3310 (1985).
7. C. C. Havener, M. S. Huq, H. F. Krause, P. A. Schulz, and R. A. Phaneuf, Phys. Rev. A **39**, 1725 (1989).
8. G. Gioumoussis and D. P. Stevenson, J. Chem. Phys. **29**, 294 (1958).
9. D. A. Church, Phys. Rep. **228**, 254 (1993).
10. Y. Kaneko, Comm. At. Mol. Phys. **10**, 145 (1981).
11. W. Lindinger, Phys. Scrip. T**3**, 115 (1983).
12. R. K. Janev and H. Winter, Phys. Rep. **117**, 266-387 (1985).
13. D.Ćirić, A. Brazuk, D. Dijkkamp, F. J. de Heer, and H. Winter, J. Phys. B **18**, 3639 (1985).
14. B. Van Zyl, G. H. Dunn, G. Chamberlain, and D. W. O. Heddle, Phys. Rev. A **22**, 1916 (1980).
15. H. B. Gilbody, Adv. At. Mol. Opt. Phys. **32**, 149 (1994).
16. N. Stolterfoht, Phys. Rep. **146**, 316-424 (1987).
17. L. N. Tunnell, C. L. Cocke, J. P. Giese, E. Y. Kamber, S. L. Varghese, and W. Waggonner, Phys. Rev. A **35**, 3299 (1987).

63

Ion Atom Collisions – High Energy

C. L. Cocke

Department of Physics, Kansas State University, Manhattan, Kansas 66506

63.1 INTRODUCTION

This chapter deals with inelastic processes which occur in collisions between fast, often highly charged, ions and atoms. Fast collisions are here defined to be those for which $V/v_e \geq 1$, where V is the projectile velocity and v_e the orbital velocity of this electron. For processes involving outer shell target electrons, this implies $V \gtrsim 1$ a.u., or the projectile energy $\gtrsim 25$ keV/amu. For inner shell electrons, typically, $V \gtrsim Z_2/n$ a.u., where Z_2 is the target nuclear charge and n the principal quantum number of the active electron. A useful relationship is $V = 6.35\sqrt{E/M}$, where V is in a.u., E is in MeV, and M is in a.m.u. Fast collisions involving outer shell processes can be studied using relatively small accelerators, while those involving inner shell processes require larger

van de Graaffs, LINACs, etc. Inner shell processes, studied with the latter accelerators lie within the scope of this chapter, even when $V/v_e < 1$. Because the motion of the inner shell electrons is dominated by the nuclear Coulomb field of the target, and because transitions involving these electrons take place rather independently of what transpires with the outer shell electrons, it has proven somewhat easier to understand one electron processes involving inner shell electrons. Thus a great deal of the work on fast ion-atom collisions has concentrated on inner shell processes. While this field continues to be refined, present investigations go beyond the one-electron picture to include the influence of the electron-electron interaction. The present chapter outlines some of the developments in this area over a very active past few decades. The literature is vast, and only a small sampling of references is given. Emphasis is on experimental results (for the theory see Chaps. 55–43).

63.2 BASIC ONE-ELECTRON PROCESSES

63.2.1 Inner Shell and Perturbative Processes

Ionization

For ion-atom collisions involving projectile and target nuclear charges Z_1 and Z_2 respectively, the parameters $\eta_1 = [\hbar V/(Z_1 e^2)]^2$ and $\eta_2 = [\hbar v_e/(Z_2 e^2)]^2$ are useful in characterizing the strength of the interaction between Z_1, Z_2, and the target electron. If $\eta_2 \ll \eta_1$, (i.e., $Z_1/Z_2 \ll V/v_e$), the effect of the projectile on the target wave function can be treated perturbatively. Perturbation treatments of inner shell ionization by lighter projectiles have been extensively studied and reviewed [1–7]. Two well-known formulations have been used: the plane wave Born approximation (PWBA) [1–3], and the semiclassical approximation (SCA) [8,9]. The former represents the nuclear motion with plane waves, while the latter is formulated in terms of the impact parameter b with the nuclear motion treated classically. For straight line motion of the nuclei, the results are equivalent [10]. The differential cross section for ionizing the K-shell of a target of charge Z_2 by a projectile of charge Z_1 is given within the PWBA by

$$\sigma_i = (8\pi Z_1^2/Z_2^4 \eta_2)\, f(\theta_2, \eta_2)\, a_0^2, \qquad (63.1)$$

where $\theta_2 = 2U_2 n^2/Z_2^2$ and U_2 is the target binding energy. The function f rises rapidly for $V < v_e$, reaching a value near unity near $V = v_e$ and falling very slowly thereafter. Tables of f for K- and L-shell ionization are given in Ref. [3]. Figure 63.1 shows a comparison of experimental data for K vacancy production by protons with PWBA calculations, and with a classical binary encounter approximation [11] for a large range of proton data [5]. For larger Z_1, corrections to the PWBA and SCA must be made for the effective increase of U_2 due to the presence of the projectile during the ionization, for nuclear projectile deflection, for relativistic corrections, and for the polarization of the electron cloud, as reviewed in Refs. [12–16]. Total cross section measurements for inner shell vacancy production in the perturbative region are reviewed in [14,15].

In the SCA treatment, the heavy particle motion is taken to be classical, and the evolution of the electronic wave function under the influence of the projectile field is calculated by time-dependent perturbation theory. The assumption of classical motion is valid if the Bohr parameter $K = 2Z_1 Z_2 e^2/(\hbar V)$ is much larger than

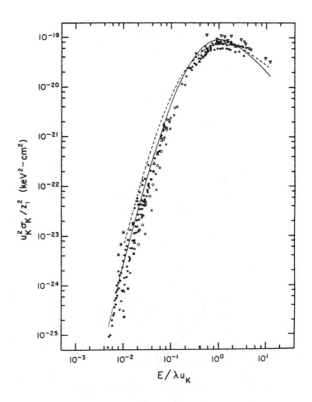

Figure 63.1. Comparison of experimental cross sections for K-shell vacancy production with PWBA (dashed) and binary encounter (solid) theories. U_k is the target binding energy in keV and λ the projectile/electron mass ratio [5].

unity [17]. If this condition is satisfied, the projectile scattering angle can be associated with a particular b through a classical deflection function. For K-shell ionization, the action occurs typically at sufficiently small b that a screened Coulomb potential is sufficient for calculating the deflection. In the absence of screening, $\theta = r_0/b$, where $r_0 = Z_1 Z_2 e^2/E$ with θ and E expressed in either the laboratory or c.m. system. Calculations for K- and L-shell ionization have been carried out [9]. The typical ionization probability $P(b)$ for $V \sim v_e$ and $b = 0$ is $P(0) \sim (Z_1/Z_2)^2$. For $V < v_e$, $P(b)$ decreases with increasing b with a characteristic scale length of $r_{ad} = V/\omega$, the *adiabatic radius*, where $\hbar\omega$ is the transition energy. For $V > v_e$, $P(b)$ cuts off near the K-shell radius of the target. A more sophisticated relativistic SCA program has been written [18], and is widely used for calculating $P(b)$, cross sections, and probabilities differential in final electron energy and angle. Experimentally, the probability $P(b)$ for inner shell ionization can be determined from

$$P(b) = \frac{1}{\epsilon\omega\Delta\Omega}\left(\frac{Y}{N[\theta(b)]}\right), \qquad (63.2)$$

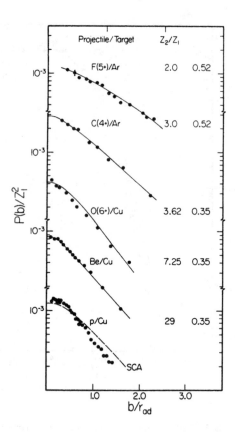

Figure 63.2. $P(b)$ for K-shell vacancy production versus b/r_{ad} for several systems (see text). The ratio V/v_e is designated as "V" in this figure. For protons, agreement with the SCA theory is found [9], while for higher Z_1/Z_2, $P(b)$ moves to larger impact parameters as one leaves the perturbative region [21].

where Y is the coincidence yield for the scattering of $N(\theta)$ ions into a well-defined angle $\theta(b)$ accompanied by x-ray (or Auger electron) emission with fluorescence yield ω into a detector of efficiency ϵ and solid angle $\Delta\Omega$ [19]. The necessary ω can be obtained from calculations for neutral targets [20] (see Chap. 60). However, they must be corrected for changes due to extensive outer shell ionization during the collision. Such corrections are particularly important for targets with low fluorescence yields, for Z_2 below 30, and for collisions in which the L-shell is nearly depleted in the collision [14, 15]. Values of $P(b)$ have been measured for many systems and generally show good agreement (better than 10%) with the SCA for fast light projectiles such as protons, with increasing deviation as higher Z_1 or slower V are used [21]. Examples of $P(b)$ for K vacancy production for several systems are shown in Fig. 63.2, showing the evolution away from the SCA as the collision becomes less perturbative.

Figure 63.3. Energy loss spectrum for 50 keV protons in atomic hydrogen, showing excitation to discrete states in H proceeding smoothly into ionization at the continuum limit [25].

Excitation

Inner shell excitation can be treated within the same perturbative framework, which leads to a cross section given in terms of the generalized oscillator strength for the transition [22–24]. For inner shell vacancy production by light projectiles, the excitation is generally much smaller than the ionization, since the strongest oscillator strengths are to low-lying occupied orbitals, as reviewed by Inokuti [23]. Excitation cross sections can be deduced from photon production cross sections and from inelastic energy loss experiments. An example of the cross section for excitation of the $n = 2$ level of H by protons, measured by the latter technique, is shown in Fig. 63.3 [25].

Capture

As Z_1/Z_2 is raised, the probability for direct transfer of inner shell electrons from projectile to target becomes competitive with, and can even exceed, that for ionization of the target electron into the continuum. The first-order perturbation treatment for electron capture, given by Oppenheimer [26] and by Brinkman and Kramers [27] (OBK) (Ref. [10], p. 379) results in a cross section per atom

$$\sigma_{\text{OBK}} = 2^9 \pi (Z_1 Z_2)^5 / 5 \, V^2 \nu^5 n^3 \beta^5 \, a_0^2 , \qquad (63.3)$$

from a filled shell ν to all final states n, where

$$\beta = \tfrac{1}{4}V^2 \left[V^4 + 2V^2(Z_2^2/\nu^2 + Z_1^2/n^2) \right.$$
$$\left. + (Z_2^2/\nu^2 - Z_1^2/n^2)^2 \right] . \qquad (63.4)$$

Both the PWBA/SCA and the OBK cross sections maximize near matching velocity, but the OBK falls off much more strongly with increasing V beyond this, eventually falling as V^{-12}, while the ionization cross section only falls as $V^{-2}\ln V$. The OBK amplitude for capture is simply the momentum space overlap of the initial wave function with the final state wave function, where the latter is simply a bound state on the projectile but moving at a velocity V relative to the initial bound state. The integral is done only over the transverse momentum, since the longitudinal momentum transfer is fixed by energy conservation [10]. This capture amplitude thus depends heavily on there being enough momentum present in the initial and/or final wave function to enable the transfer, and the loss of this match is what leads to the steep decrease in the OBK cross section above velocity matching. Cross sections for K-shell capture have been measured by detection of K Auger electrons and K x-rays in coincidence with charge capture by the projectiles [21, 28, 29]. Figure 63.4 shows a comparison of the data with the OBK results for proton projectiles. On the basis of these and many other data on electron capture, the OBK is a factor of approximately three too large [28–30]. This factor comes from a fundamental failure of first-order perturbation theory for electron capture. As pointed out already in 1927 by Thomas [31], who proposed a classical two-collision mechanism for capture, it is essential that the electron interact with both electrons during the collision in order to be captured (see Chap. 55). In quantum theories, this corresponds to the fundamental need to include second-order terms (and higher) in the capture amplitude. In the limit of large V, the second-order cross section decreases more slowly than the OBK term, as V^{-10}, and thus is asymptotically larger than the first-order term [32]. At large V, the coefficient of the V^{-12} term, the dominant one at most experimentally reachable V, is 0.29 times the OBK cross section when the theory is carried out to second-order in the projectile potential [32, 33]. Roughly speaking, this provides an explanation for the factor of three. Much more sophisticated treatments of high velocity capture are now available [34–42]. The underlying role of the second-order scattering process was confirmed experimentally by the detection of the Thomas peak in the angular distribution of protons capturing electrons from He and H [43] (see Chap. 55).

In spite of the basic importance of second-order amplitudes in perturbative capture, the OBK gives an excellent account of the relative contributions from and to different final shells over a large range of V above v_e, and is thus, when appropriately reduced, still useful as an

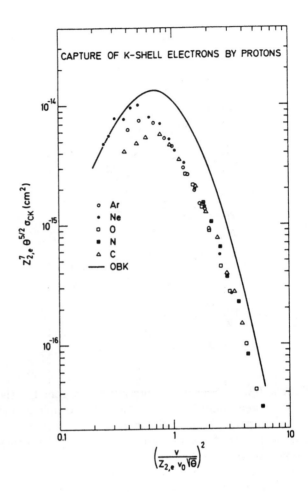

Figure 63.4. Cross sections for capture from the K-shell of various targets by protons. The shielding parameter $\theta = U_k/2Z_2 e^2$ is near unity (U_k is the target binding energy in a.u.), so that the abscissa is approximately V/v_e. The scaling of the ordinate is such that, in the OBK approximation, the data should fall on a universal curve. The OBK calculation is shown as the solid line [28].

estimate for perturbative capture cross sections between well defined ν and n for large V.

63.2.2 Nonperturbative Processes

Fano-Lichten Model

When the collision becomes very nonperturbative, either due to low V or large Z_1/Z_2, the perturbation treatment is replaced by a molecular orbital treatment. Fano and Lichten [44] pointed out that the ratio V/v_e can be small for inner orbitals even for V of several a.u., and thus an adiabatic picture of the collision holds. K vacancy production cross sections become much larger than

Figure 63.5. Schematic correlation diagram for the Cl-Ar system, indicating the rotational coupling and radial couplings important for K vacancy production and the $4f\sigma$ orbital whose promotion leads to L vacancy production [50].

the perturbation treatments above predict and extend to much larger b. In the molecular orbital picture, the collision system is described in terms of time-dependent molecular orbitals (MO) formed when the inner shells of the systems overlap. Vacancy production occurs due to rotational, radial, and potential coupling terms between these orbitals during the collision. The independent electron model is used, but the results in any specific collision are quite sensitive to the occupation numbers (or vacancies) in the initial orbitals. These are very difficult to control in ion-atom collisions in solids and even problematic in gases, since outer shell couplings can produce vacancies at large internuclear distances which then enable transfers at smaller distances. Numerous reviews of the subject are available, including Refs. [45–49]. The most famous MO ionization mechanism involves the promotion of the $4f\sigma$ orbital in a symmetric collision (see Fig. 63.5), which promotes both target and projectile L electrons to higher energies where they are easily lost to the continuum during the collision. There are now many treatments of inner shell vacancy production mechanisms based on MO expansions (see Chaps. 49, 48). For the case of K vacancy production in quasisymmetric collisions, an important MO mechanism is the transfer of L vacancies in the projectile to the K-shell of the target through the

rotational coupling between $2p\pi$ and $2p\sigma$ orbitals which correlate to the L- and K-shells respectively of the separated systems (see Fig. 63.5) [51, 52]. The process can be dynamically altered by the sharing at large b between L vacancies of target and projectile through a radial coupling mechanism [53, 54]. This sharing mechanism can also give rise to the direct transfer of K vacancies from projectile to target (KK sharing). In symmetric systems, the KK sharing results in an oscillation of the K vacancy back and forth between target and projectile during the collision, and leads to an oscillatory behavior of the transfer probability with V and b [55, 56]. Both of the above vacancy production mechanisms are electron transfer processes rather than direct ionization processes, in that no inner shell electron need be liberated into the continuum. Between the perturbation region and the full MO region the importance of transfer increases relative to ionization. While the MO correlation diagrams and mechanisms are qualitatively useful, actual close coupling calculations for both inner and outer shell processes are often carried out using atomic orbitals instead of molecular orbitals, as well as other basis sets (see Chaps. 49, 48).

Outer Shell Processes

For He and H targets, or for the outer shells of any neutral targets, excitation, ionization or capture by fast protons can be treated by perturbation theory. For $Z_1/V > 1$, the calculated probabilities can exceed unity, in which case they can be "unitarized" through the solution of population balance equations [57]. This procedure provides good estimates, but proceeding any further requires a more sophisticated approach, such as a close coupling calculation (see Chaps. 49, 48) or CTMC (Chap. 56).

63.3 MULTI-ELECTRON PROCESSES

In a single collision between multi-electron partners, two or more electrons may be simultaneously excited or ionized. The electric fields created during a violent ion-atom collision are so large that the probability of such multi-electron processes can be of order unity. While there are many similarities between ion-atom collisions and the interaction of atoms with photons (x-rays or short laser pulses) or electrons, the dominance of multi-electron processes is very much less common in the photon and electron cases. As an example, when a K-shell electron is removed from a target atom by the passage of a fast highly charged ion through its heart, the probability that L-shell electrons will be removed at the same time can be large. This gives rise to target x-ray and Auger-electron spectra which are dominated by satellite structure [19]. For example, the spectrum

Figure 63.6. K x-ray spectrum of Ti for various projectiles, showing dominance of multi-electron transitions when K vacancies are collisionally produced by heavily ionizing projectiles (^{16}O beam) [58].

of x-rays from Ti bombarded by 30 MeV oxygen shows that the production of the K vacancy is accompanied by multiple L vacancy production, and that the dominant K x-rays are those of systems which are missing several L electrons [58] (see Fig. 63.6). With the help of theoretical fluorescence yields from highly ionized systems, such spectra can be used to deduce the probability $P_L(b)$ for L-shell ionization for small b, where the requirement that a K hole be created means that $b \sim 0$. The distribution of satellite intensities is then given by binomial statistics in terms of $P_L(0)$ [59]. For lightly ionizing systems, P_L can be calculated from the SCA and binary encounter models; for heavily ionizing systems $P_L(0)$ is empirically found to be correlated with Z_1/V, but unitarization of these theories becomes necessary.

When a gas target is used, the recoil target ion is heavily ionized and/or excited electronically without receiving much translational kinetic energy. In the impulse approximation the transverse momentum $\Delta p_\perp$ received by the target from a projectile passing at impact parameter b is given in a.u. by $\Delta p_\perp = 2Z_1Z_2/bV$. This expression ignores the exchange of electronic translational momentum but gives a good estimate. The resulting recoil energies are typically quite small, ranging from thermal to a few eV. This subject has been reviewed

in Refs. [60, 61]. These slow moving recoils have been used to provide information about the primary collision dynamics, and as secondary highly charged ions from a fast-beam-pumped ion source. Such an ion source has, for moderately charged ions, a high brightness and has been used extensively for energy-gain measurements. The primary recoil production process is difficult to treat without the independent electron model, and even in this model the nonperturbative nature of the collision makes the theory difficult. The most successful treatments have been the CTMC (see Chap. 56) and a solution of the Vlasov equation [62].

For He, the isolation of mechanisms becomes more tractable. The ratio of double to single ionization in charged particle ionization and capture has been well studied, and reviewed in Ref. [63]. Important mechanisms include double interactions with the projectile, which dominates total cross sections at low V, and interactions of the outgoing primary electron with the spectator, which gives a common ratio of 0.26% at large V for all systems [64]. For processes which force the primary outgoing electron to be fast, such as hard projectile-electron or electron capture, the ratio is much higher, from 2 to 4%. Important mechanisms include shakeoff as well as initial and final state correlation effects. Many parallels between the mechanisms in play here and those which produce double ionization of He by photons can be drawn, and this issue has been at the heart of much work and controversy in recent years [63, 65].

63.4 ELECTRON SPECTRA IN ION-ATOM COLLISIONS

63.4.1 General Characteristics

An ionizing collision between a single ion and a neutral atom ejects electrons into the continuum via two major processes. Electrons ejected during the collision form broad features or continua, and are traditionally referred to as *delta rays*; electrons ejected after the collisions from the Auger decay of vacancies created during the collision form sharp lines in the spectra. The distributions of energy and angle of all electrons determine the electronic stopping power and characteristics of track formation of ions in matter (Chap. 87), and the study of these distributions in the binary encounter of one ion with one atom form the basis of any detailed understanding of these averaged quantities. Figure 63.7 shows a typical electron spectrum from the collision of a fast O ion with O_2 [66]. Electrons from the projectile can be identified in the cusp peak (Electron Loss, P) or ELC, and the O-K-Auger (P) peak. Electrons from the target include the soft (large b) collision electrons (T) which are ejected directly by Coulomb ionization by the projectile, the binary collision (or encounter) electrons coming from hard collisions

Figure 63.7. Electron spectrum from 30 MeV O on molecular oxygen. See text for explanation of features [66].

between projectile and quasifree target electrons, and the target O-K-Auger (T) electrons. The electron loss peak is widely called the *cusp* peak because the doubly differential cross section in the laboratory $d^2\sigma/dEd\Omega$ becomes infinite, in principle, if it is finite in the projectile frame. In general, this peak may also contain capture to the continuum. All of these features have been heavily studied; some reviews are [67, 68]. Capture to the continuum [69] is an extension of normal capture into the continuum of the projectile, and is not a weak process. Both it and ELC produce a heavy density of events in the electron momentum space centered on the projectile velocity vector, and thus appear strongly only at or near zero degrees in the laboratory and at $v_e \simeq V$. This peak is closely related to the "convoy" electron peak produced when fast ions pass through thin foils [70]. The binary encounter electrons at forward angles (backwards scattering in the projectile frame) have recently been found to display strong diffraction structure in their angular distributions caused by the non-Coulomb nature of the projectile potential in which they are scattered [71, 72], and is now being used to deduce information on the scattering of quasifree electrons off the potentials of highly

stripped ions. This is a difficult experiment do to with free electrons because of the low rates which attend the crossed beam experiments. The broader electron continua have been well studied for light projectiles [67, 68] and are describable in terms of perturbation theory for simple cases. For nonperturbative collisions, both data and understanding have large gaps (see Chap. 51).

63.4.2 High Resolution Measurements

The Auger electron spectra provide detailed information about inner shell vacancy production mechanisms. When coupled with fluorescence yields, Auger electron production probabilities and cross sections can be converted into the corresponding quantities for vacancy production [14, 68]. This is best done when sufficient resolution can be obtained to isolate individual Auger lines. The Auger spectra in ion-atom collisions are often completely different from those obtained from electron or photon bombardment because of the multiple outer shell ionization which attends the inner shell vacancy producing event. Figure 63.8 shows a succession of target K Auger spectra from Ne bombarded with increasingly more heavily ionizing beams showing that the diagram lines (decay from a single K-hole and filled L-shell, proton bombardment) become completely replaced by satellites, which reflects heavy L-shell ionization in the collisions [73]. Projectile Auger electron spectra suffer from kinematic broadening due to the finite solid angle of the spectrometer and velocity of the emitter, but at 0° to the beam this problem vanishes, and the resolution in the emitter frame is actually enhanced by the projectile motion, such that for electrons with eV energies in the projectile frame, resolutions in the meV region are possible [74, 75]. The highest resolution Auger lines from ion-atom collisions has been done on the projectiles. A sample spectrum is shown in Fig. 63.9. From such high resolution spectra, one-electron processes in which one electron is excited, captured or ionized can be distinguished from the configuration of the emitting state.

63.5 QUASI-FREE ELECTRON PROCESSES IN ION-ATOM COLLISIONS

At sufficiently low V, those electrons not actively involved in a transition play only a passive role in screening the Coulomb potential between the nuclei, and thereby create a coherent effective potential for their motion. However, at high V the colliding electrons begin behaving as incoherent quasifree particles capable of inducing transitions directly via the electron-electron interaction. Such processes signal their presence through their free-particle kinematics, as if the parent nucleus

Figure 63.9. High resolution Auger electron spectrum from Li-like B on H_2, showing resolved lines from excited projectiles (doublet and quartet), and from resonant recombination (singlet and triplet) lying on top of a continuum due to binary encounter electrons [76].

Figure 63.8. K Auger electron spectrum of Ne showing the developing dominance of satellite structure in the spectra as the ionizing power of the projectile is raised [68].

were not present. For example, a projectile ionization process requiring energy U has a threshold at $\frac{1}{2}m_e V^2 \simeq U$, in collisions with light targets where the quasifree picture is meaningful. The threshold is not sharp, due to the momentum distribution or Compton profile of the target electrons. Within the impulse approximation, the cross section for any free electron process can be related to the corresponding cross section for the ion-atom process by folding the free electron cross section into the Compton profile [77–79].

63.5.1 Radiative Electron Capture

The first quasifree electron process to be observed was radiative electron capture (REC), the radiative capture of a free electron by an ion. Conservation of energy and momentum is achieved by the emission of

a photon which carries away the binding energy. The cross section exceeds that for bound state capture at high V. Radiative electron capture was observed in ion-atom collisions long before the corresponding free electron process was seen [80], through the x-ray spectra from fast heavy projectiles for which the electrons of light targets appear to be 'quasi-free', and has recently been heavily studied in EBIT [81], cooler [82], and storage rings [83]. Total cross sections for REC have also been deduced from measured total capture cross sections at large V where REC dominates bound state capture [84]. At high velocities, the cross section for radiative capture to the K-shell of a bare projectile is given approximately by

$$\sigma_n = [n/(\kappa^{-2} + \kappa^{-4})] \times 2.1 \times 10^{-22} \text{ cm}^2, \qquad (63.5)$$

where [85] $\kappa = \sqrt{E_B/E_0}$, E_B is the binding energy of the captured electron, E_0 the energy of the initial electron in the ion frame and n the principal quantum number of the captured electron. The theory seems to be in good agreement with experiment for capture to all shells of fast bare projectiles, although a small unexplained discrepancy between theory and experiment exists for capture to the K-shell [83].

63.5.2 Resonant Transfer and Excitation

Dielectronic recombination in electron-ion collisions is the process whereby an incident electron excites one target electron and, having suffered a corresponding energy loss, drops into a bound state on the projectile (see Chap. 53). If the doubly excited state so populated decays radiatively, resonant radiative recombination is achieved (DR); if it Auger decays, resonant elastic scattering has occurred. The process has long been known to be important as a recombination process in hot plasmas [86], but was not observed in the laboratory until 1983 [87–89]. The corresponding ion-atom process, known as resonant transfer and excitation (RTE) was seen a bit earlier by Tanis *et al.* [90]. (See Refs. [91, 92] for reviews of both DR and RTE.)

63.5.3 Excitation and Ionization

Excitation and ionization of inner shells of fast projectile ions by the quasifree electrons of light targets (usually He or H_2) have been identified and studied. This process competes with excitation and ionization by the target nucleus [93], and special signatures must be sought to distinguish the processes. In the case of excitation, the e-e excitation populates states through the exchange part of the interaction which is excluded for the nuclear excitation, and this has been used to separate this mechanism [94]. For ionization, enhancements of the ionization cross section above the Born result for nuclear ionization [95], coincident charge exchange measurements [96], and projectile [97] and recoil ion momentum spectroscopy [98, 99] have been used to distinguish the two processes. Rapid development in the production of good sources of beams of highly charged ions (EBIS/T, ECR; see Ref. [100]) have made these studies possible. Continued study of this field in heavy ion storage rings is now achieving resolutions of meV and opening broad new opportunities for data of unprecedented high quality for electron-ion collisions [101].

63.6 RECOIL AND PROJECTILE ION MOMENTUM SPECTROSCOPY

The transverse momentum $\Delta p_\perp$ given to the projectile in a close ion-atom collision is usually dominated by the Coulomb repulsion between the nuclei and provides information on the b dependence of the process. Traditionally, $\Delta p_\perp$ is found by measuring the projectile scattering angle, which is typically quite small, down to ~ 100 μrad [43, 102]. This kind of transverse momentum spectroscopy has been widely applied to K vacancy production (see Sect. 63.2.1) to obtain angular dependences for electron capture [102, 103] and to distinguish between

mechanisms for capture and ionization [43, 103–105]. On the other hand, $\Delta p_\parallel$ provides information related to the internal energy transfer Q in the collision. The most common form of longitudinal momentum spectroscopy is the measurement of the energy change of the projectile, which for small angles is equal to Q. The subject is reviewed in Ref. [106].

Measurements of the recoil ion momentum can in many cases be superior to corresponding measurements on the projectile [107]. Because the recoil energies are so small, the resolution is limited by the thermal motion of the target; and thus recoil ion momentum spectroscopy (RIMS) had for years been largely considered inferior to projectile measurements [106]. The use of cooled targets [108] has changed this situation, to the extent that RIMS is now becoming a standard tool in ion-atom collisions. For the transverse momentum, $\Delta p_\perp$ is more closely correlated to b than is the projectile momentum transfer [109].

In the beam direction, $\Delta p_\parallel$ is related to the Q value by

$$\Delta p_\parallel = -Q/V - nmV/2 - P_{e\parallel}\,, \qquad (63.6)$$

where Q is the electronic energy release (positive if exoergic), n the number of electrons captured, m the electron mass, and $P_{e\parallel}$ the longitudinal momentum of any continuum electrons after the collision [10]. If no electrons are left in the continuum, which is the case for pure electron capture, $P_{e\parallel} = 0$, and a measurement of $\Delta p_\parallel$ is equivalent to a measurement of Q. For example, if a He target is used with a thermal motion of 0.1 a.u. in momentum (0.2 K) the corresponding energy resolution for a 10 MeV/amu beam is 54 eV. If the beam were Ne, to obtain such a resolution with translational energy spectroscopy would require an unrealistic resolution ($\Delta E/E$) of 5×10^{-7} on both incident and product projectiles. Only one high resolution (10^{-4}) experiment of this type has been reported [110]. A sample Q measurement with RIMS is shown in Fig. 63.10. A similar gain is found in the transverse direction. RIMS has the added advantage that it can be used in a storage ring geometry. When the final state has electrons in the continuum, the technique cannot be used to measure Q due to the $P_{e\parallel}$ term in Eq. (63.6), but it has been used to distinguish between mechanisms for projectile ionization [98, 99]. The potential of RIMS for use with synchrotron light sources is just being explored [108, 111].

63.7 SOME EXOTIC PROCESSES

63.7.1 Molecular Orbital X-Rays

A typical time duration for an ion-atom encounter is $\sim 10^{-17}$ s shorter than Auger and x-ray lifetimes, so that hard characteristic radiation is emitted by the products

Figure 63.10. Longitudinal momentum spectrum of recoil ions from 0.25 MeV He^{2+} capturing a single electron from a cold He target, showing clear resolution of capture to $n = 1$ from that which leaves target or projectile excited [108].

long after the collision. There remains, however, a small but finite probabiltiy that x-rays or Auger electrons can be emitted during the collision, in which case the radiation proceeds between the time-dependent molecular orbitals formed in the collision and reflects the time evolution of the energies and transition strengths between the orbitals. Such molecular orbital x-radiation (MOX) has now been observed in many collision systems [112] and is reviewed in Refs. [113–115]. MOX spectra have been studied in total cross sections as well as a function of impact parameter. In the latter case, interference structures in the MOX spectra are seen [114, 116], due to the interference between amplitudes for the emission of x-rays with the same energy on the incoming and outgoing parts of the trajectory. Few experiments have tried to extend this work to low velocities [117] where MOX spectra should be most easily seen. The corresponding MO Auger electrons have not been positively identified [118].

63.7.2 Positron Production from Atomic Processes

The formation of transient molecular orbitals in head-on collisions between highly charged ions provides opportunities for studying the electrodymamics of very highly charged systems [119]. For example, two uranium nuclei passing within one K-shell radius of each other form a transient molecule whose energy levels resemble those of an atom of charge 184. All perturbative treatments are off in such a situation. Reinhart *et al.* [120] pre-

dicted that, for such highly charged species, the binding energy of the united atom K-shell exceeds twice the rest mass energy of the electron, and that if a K vacancy is either brought into the collision or created during it, spontaneous electron-positron pair production occurs (the decay of the charged vacuum) with the electron filling the K-hole. However, futher analysis showed that the dominant mechanism for positron production (other than those resulting from the decay of nuclear excitations) likely results from the dynamic time dependence of the fields during the collision [121]. Experiments showed evidence for such positron production in collisions at 6 MeV/u [122], but in addition showed unexpected positrons of well defined energy [123–125]. The source of these has never been identified, although exotic theories abound. Attempts are in progress to substantiate or explain these results through new experiments.

Present theories for the production of lepton pairs in the close collision of two highly charged systems predict that the cross section grows rapidly with collision energy. Electrons produced in such a process may end up in bound states on either collision partner, and thus represent a new charge changing mechanism. At highly relativistic velocities, the cross section for this process exceeds that for any other charge changing process. In a heavy ion collider, such as RHIC, this process could limit the ultimate storage time for the counter-propagating beams, since charge-exchanged ions are lost. The cross section has been measured recently by Vane *et al.* [126], for 6.4 TeV S on several targets (the highest energy ion-atom collision experiment performed to date), and are in good agreement with theory [127]. The bound state capture has been measured at lower energy by Belkacem *et al.* [128], with similarly good agreement. The extension of ion-atom collisions to such extreme velocities has opened the field for the study of processes not even imagined a short time ago.

Acknowledgments

This work was supported by the Chemical Sciences Division, Basic Energy Sciences, Office of Energy Research, U.S. Department to Energy.

REFERENCES

1. E. Merzbacher and H. W. Lewis, *Handbuch der Physik* (Springer, 1958), Vol. 34, p. 166.

2. D. Madison and E. Merzbacher in, *Atomic Inner Shell Processes*, edited by B. Craseman (Academic Press, 1975).

3. B. S. Khandelwal, B. H. Choi, and E. Merzbacher, At. Data **1**, 103 (1969); **5**, 291 (1973).

4. *Atomic Inner Shell Processes*, edited by B. Crasemann (Academic Press, 1975).

5. J. D. Garcia, R. J. Fortner, and T. M. Kavanagh, Rev. Mod. Phys. **45**, 111 (1973).
6. P. Richard, Meth. Expt. Physics, *Atomic Physics: Accelerators* (Academic Press, 1980), Vol. 17.
7. D. Williams, Meth. Expt. Physics, *Spectroscopy* (Academic Press, 1976), Vol. 13.
8. J. Bang and J. M. Hansteen, Kgl. Dan. Viden. Mat.-Fys. Medd. **31**, 13 (1973).
9. J. M. Hansteen, O. M. Johnsen, and L. Kocbach, At. Data Nucl. Data Tables **15**, 305 (1975).
10. M. R. C. McDowell and J. P. Coleman, *Introduction to the Theory of Ion-Atom Collisions* (North Holland, 1970).
11. J. D. Garcia, Phys. Rev. A **4**, 955 (1971).
12. R. Anholt, Phys. Rev. A **17**, 983 (1978).
13. G. Basbas, W. Brandt, and R. Laubert, Phys. Rev. A **7**, 983 (1973).
14. T. J. Gray in Ref. [6], 193.
15. H. Paul and J. Muhr, Phys. Rep. **135**, 47 (1986).
16. W. Brandt and G. Lapicki, Phys. Rev. A **10**, 474 (1974).
17. N. Bohr, K. Dan. Vidensk. Selsk. Mat. Fys. Medd. **18**, No. 8 (1948).
18. M. Pauli, D. Trautmann, J. Phys. B **11**, 667 (1978); M. Pauli, F. Mösel, and D. Trautmann, J. Phys. B **11**, 2511 (1978).
19. R. L. Kauffman and P. Richard in Ref. [6], 148.
20. W. Bambynek *et al.*, Rev. Mod. Phys. **44**, 716 (1972).
21. C. L. Cocke, in [6], 303.
22. N. F. Mott and S. N. Massey, *The Theory of Atomic Collisions* (Oxford U. P., 1965).
23. M. Inokuti, Rev. Mod. Phys. **43**, 297 (1971).
24. H. Bethe, Ann. Phys. **5**, 325 (1930).
25. J. T. Park, J. E. Aldag, J. M. Geroge, and J. L. Peacher, Phys. Rev. A **14**, 608 (1976).
26. J. R. Oppenheimer, Phys. Rev. **31**, 349 (1928).
27. H. S. Brinkman and H. A. Kramers, Proc. Acad. Sci. Amsterdam **33**, 973 (1930).
28. M. Rødbro, E. Horsdal Pedersen, C. L. Cocke, and J. R. Macdonald, Phys. Rev. A **19**, 1936 (1979).
29. J. R. Macdonald, in Ref. [6], p. 279.
30. Nikolaev, Zh. Eksp. Teor. Fiz. **51**, 1263 (1966); Sov. Phys. JETP **24**, 847 (1967).
31. L. H. Thomas, Proc. R. Soc. London A **114**, 561 (1927).
32. R. Shakeshaft and L. Spruch, Rev. Mod. Phys. **51**, 369 (1979).
33. B. H. Bransden and I. M. Cheshire, Proc. Phys. Soc. **81**, 820 (1963).
34. P. R. Simony and J. H. McGuire, J. Phys. B **14**, L737 (1981).
35. J. Macek and S. Alston, Phys. Rev. A **26**, 250 (1982).
36. J. Macek and K. Taulbjerg, Phys. Rev. Lett. **46**, 170 (1981).
37. A. L. Ford, J. R. Reading, and R. L. Becker, Phys. Rev. A **23**, 510 (1981).
38. Dz. Belkic and A. Salin, J. Phys. B **11**, 3905 (1978).
39. J. Eichler and T. Chan, Phys. Rev. A **20**, 104 (1978).
40. D. P. Dewangan and J. Eichler, Comm. At. Mol. Phys. **27**, 317 (1992).
41. H. Marxer and J. S. Briggs, J. Phys. B **18**, 3823 (1992).
42. J. S. Briggs, J. H. Macek, and K. Taulbjerg, Comments At. Mol. Phys. **12**, 1 (1983).
43. E. Horsdal-Pedersen, C. L. Cocke, and M. Stöckli, Phys. Rev. Lett. **50**, 1910 (1983); H. Vogt *et al.*, Phys. Rev. Lett. **57**, 2256 (1986).
44. U. Fano and W. Lichten, Phys. Rev. Lett. **14**, 627 (1965).
45. U. Wille and R. Hippler, Phys. Rep. **132**, 131 (1986).
46. Q. Kessel and B. Fastrup, Case Stud. At. Phys. **3**, 137 (1973).
47. J. Briggs, Rep. Prog. Phys. **39**, 217 (1976).
48. W. E. Meyerhof and K. Taulbjerg, Ann. Rev. Nucl. Sci. **27**, 279 (1977).
49. I. A. Sellin, *Structure and Collisions of Ions and Atoms*, Topics in Current Physics 5 (Springer-Verlag, 1978).
50. C. L. Cocke, R. R. Randall, S. L. Varghese, and B. Curnutte Phys. Rev. A **14**, 2026 (1976).
51. K. Taulbjerg and J. S. Briggs, J. Phys. B **8**, 1895 (1975).
52. J. S. Briggs and J. Macek, J. Phys. B **5**, 579 (1972).
53. Y. Demkov, Sov. Phys. JETP **18**, 138 (1964).
54. W. E. Meyerhof, Phys. Rev. Lett. **31**, 1973 (1341).
55. G. T. Lockwood and E. Everhart, Phys. Rev. **125**, 567 (1962).
56. S. Hagmann *et al.*, Phys. Rev. A **25**, 1918 (1982).
57. V. A. Sidorovich, V. S. Nikolaev, and J. H. McGuire, Phys. Rev. A **31**, 2193 (1985).
58. C. F. Moore, M. Senglaub, B. Johnson, and P. Richard, Phys. Rev. Lett. **40**, 107 (1972).
59. J. H. McGuire and O. L. Weaver, Phys. Rev. A **16**, 41 (1977).
60. I. A. Sellin, in Ref. [49], Chap. 7.
61. C. L. Cocke and R. E. Olson, Phys. Rep. **205**, 153 (1991).
62. M. Horbatsch, Z. Phys. D **21**, 63 (1991).
63. J. H. McGuire, Adv. At. Mol. Opt. Phys. **29**, 217 (1992).
64. J. Ullrich *et al.*, Phys. Rev. Lett. **71**, 1697 (1993).
65. J. H. McGuire *et al.*, J. Phys. B (to be published).
66. N. Stolterfoht, D. Schneider, D. Burch, H. Wieman, and J. S. Risely, Phys. Rev. Lett. **33**, 59 (1974).
67. M. E. Rudd, Y.-K. Kim, D. H. Madison, and T. J. Gay, Rev. Mod. Phys. **64**, 441 (1992).
68. N. Stolterfoht, in Ref. [49], Chap. 5.
69. G. B. Crooks and M. E. Rudd, Phys. Rev. Lett. **25**, 1599 (1970).
70. I. A. Sellin, in *Physics of Electron and Atom Collisions*, edited by S. Datz (North-Holland, 1982), Vol. XII, p. 195.
71. C. O. Reinhold *et al.*, Phys. Rev. Lett. **66**, 1842 (1991).
72. C. Liao *et al.*, in *AIP Conference Proceedings No. 281, Highly Charged Ions, 1992*, edited by P. Richard *et al.* (American Institute of Physics, 1992).
73. N. Stolterfoht, D. Schneider, R. Mann, and F. Folkmann, J. Phys. B **10**, L281 (1977).
74. A. Itoh *et al.*, J. Phys. B **16**, 3965 (1983).
75. D. H. Lee *et al.*, Phys. Rev. A **41**, 4816 (1990).
76. P. Richard (private communication).
77. P. Richard, in *Proceedings of 15th International Conference on X-Ray and Inner Shell Processes, 1990*, edited by T. Carlson (American Institute of Physics, 1990).
78. C. L. Cocke in Ref. [129], 49.
79. I. A. Sellin, in Ref. [130], 4.
80. H. W. Schnopper *et al.*, Phys. Rev. Lett. **29**, 898 (1972).
81. R. W. Marrs, Phys. Rev. Lett. **60**, 1757 (1988).

82. L. H. Andersen and J. Bolko, J. Phys. B **23**, 3167 (1990).

83. Th. Stöhlker, *et al.*, Z. Phys. D **23**, 121 (1992).

84. H. Gould, *et al.*, Phys. Rev. Lett. **52**, 180 (1984).

85. H. A. Bethe and E. E. Salpeter, in *Encyclopidia of Physics*, edited by S. Fluegge (Springer, 1957), Vol. 35, p. 408.

86. A. Burgess, Astrophy. J. **139**, 776 (1964).

87. J. B. A. Mitchell *et al.*, Phys. Rev. Lett. **50**, 335 (1983).

88. D. S. Belic *et al.*, Phys. Rev. Lett. **50**, 339 (1983).

89. P. F. Dittner *et al.*, Phys. Rev. Lett. **51**, 31 (1983).

90. J. A. Tanis *et al.*, Phys. Rev. Lett. **47**, 828 (1981).

91. *Recombination of Atomic Ions*, edited by W. G. Graham *et al.*, NATO ASI Series B (Plenum, 1992), p. 296.

92. J. A. Tanis, in Ref. [91], 241.

93. D. R. Bates and G. Griffing, Proc. Phys. Soc. London A **66**, 961 (1955); **67**, 663 (1954); **68**, 90 (1955).

94. T. Zouros, D. H. Lee, and P. Richard in Ref. [131], 568.

95. W. E. Meyerhof *et al.*, Phys. Rev. A **43**, 5907 (1991).

96. E. C. Montenegro *et al.*, Phys. Rev. Lett. **69**, 3033 (1992).

97. E. C. Montenegro *et al.*, Phys. Rev. A **47**, 1045 (1993).

98. W. Wu *et al.*, Phys. Rev. Lett. **72**, 3170 (1994).

99. R. Doerner *et al.*, Phys. Rev. Lett. **72**, 3166 (1994).

100. C. L. Cocke, in Ref. [132], 138.

101. R. Schuch, in Ref. [132], 169.

102. F. J. Martin *et al.*, Phys. Rev. A **23**, 2858 (1981).

103. E. Kamber *et al.*, Phys. Rev. Lett. **60**, 20261 (1988).

104. H. Vogt *et al.*, Phys. Rev. Lett. **57**, 2256 (1986).

105. R. Dorner, J. Ullrich, H. Schmidt-Böcking, and R. E. Olson, Phys. Rev. Lett. **63**, 147 (1989).

106. B. Fastrup in Ref. [6], p. 149.

107. H. Schmidt-Böcking, in [130], p. 268.

108. V. Mergel *et al.*, Phys. Rev. Lett. **74**, 220 (1990).

109. J. Ullrich *et al.*, in *AIP Conference Proceedings No. 251, Highly Charged Ions, 1992*, edited by P. Richard *et al.* (American Institute of Physics, New York, 1992).

110. R. Schuch and H. Schöne *et al.*, Phys. Rev. Lett. **60**, 925 (1988).

111. J. A. R. Samson *et al.*, Phys. Rev. Lett. **72**, 3329 (1994).

112. F. W. Saris, W. F. van der Weg, H. Tawara, and R. Laubert, Phys. Rev. Lett. **28**, 717 (1972).

113. P. O. Mokler, in Ref. [60], p. 245.

114. R. Schuch *et al.*, Phys. Rev. A **37**, 3313 (1988).

115. R. Anholt, Rev. Mod. Phys. **57**, 995 (1985).

116. I. Tserruya *et al.*, Phys. Rev. Lett. **50**, 30 (1983).

117. M. H. Prior *et al.*, Phys. Rev. A **47**, 2964 (1993).

118. R. Koch, Ph. D. dissertation, University of Frankfurt, 1990.

119. W. Pieper and W. Geiner, Z. Phys. **218**, 126 (1969).

120. J. Reinhardt, U. Müller, B. Müller, and W. Greiner, Z. Phys. A **303**, 173 (1981).

121. G. Soff, J. Reinhard, B. Müller, and W. Greiner, Z. Phys. A **294**, 137 (1980).

122. U. Müller-Nehrer and G. Soff, Phys. Rep. **246** (1994).

123. P. Kienle, Ann. Rev. Nucl. Part. Sci. **36**, 605 (1986).

124. T. Cowan *et al.*, Phys. Rev. Lett. **56**, 444 (1986).

125. W. König *et al.*, Z. Phys. A **328**, 129 (1987).

126. C. R. Vane *et al.*, Phys. Rev. Lett. **69**, 1911 (1992).

127. C. Bottcher and M. Strayer, Phys. Rev. D**39**, 1330 (1989).

128. A. Belkacem *et al.*, Phys. Rev. Lett. **71**, 1514 (1993).

129. *Physics of Electron and Atom Collisions*, edited by W. R. MacGillivray, *et al.* (Adam Hilger, 1992).

130. Lecture Notes in Physics, 376, edited by D. Berenyi and D. Hock (Springer, 1990).

131. *AIP Conference Proceedings No. 205, Physics of Electron and Atom Collisions, 1989*, edited by A. Dalgarno *et al.* (American Institute of Physics, New York, 1989).

132. *Review of Fundemental Processes and Application of Ions and Atoms*, edited by C. D. Lin (World Scientific, 1993).

64

Reactive Scattering

Arthur G. Suits and Yuan T. Lee

Lawrence Berkeley Laboratory, University of California, Berkeley, California 94720

64.1 INTRODUCTION

This chapter presents a résumé of the methods commonly employed in scattering experiments involving neutral molecules at chemical energies, i.e., less than about 10 eV. These experiments include the study of intermolecular potentials, the transfer of energy in molecular collisions, and elementary chemical reaction dynamics. Closely related material is presented in Chaps. 34, 36, and 47 as well as in other chapters on quantum optics.

64.2 EXPERIMENTAL METHODS

64.2.1 Molecular Beam Sources

The development of molecular beam methods in the past two decades has transformed the study of chemical physics [1]. Supersonic molecular beam sources allow one to prepare reagents possessing a very narrow velocity distribution with very low internal energies, ideal for use in detailed studies of intermolecular interactions. Early experiments generally employed continuous beam sources, but in recent years intense pulsed beam sources have come into common usage [2]. The advantages of pulsed beams primarily arise from the lower gas loads associated with their use, hence reduced demands on the pumping system. If any component of the experiment is pulsed (pulsed laser detection, for example) then considerable advantage may be obtained by also pulsing the beam. Although the theoretical descriptions of pulsed and continuous expansions are essentially equivalent, in practice some care is required in employing pulsed beams because the temperature and velocity distributions may change dramatically through the course of the pulse. Free jet expansions are supersonic because the dramatic drop in the local temperature in the beam is associated with a drop in the local speed of sound. A detailed description of the supersonic expansion may be found in references [3–5]. In practice, many of the detailed features associated with a supersonic expansion may be ignored and one may assume an isentropic expansion into the vacuum. For an isentropic nozzle expansion of an ideal gas, the maximum terminal velocity is given by

$$v_{\max} = \sqrt{2\hat{C}_p T_0}\,, \qquad (64.1)$$

where, for an ideal gas, the heat capacity is

$$\hat{C}_p = \left(\frac{\gamma}{\gamma - 1}\right)\frac{R}{m}\,, \qquad (64.2)$$

R is the gas constant, m is the molar molecular mass, T_0 the temperature in the stagnation region, and γ the heat capacity ratio. For ideal gas mixtures, and assuming C_p independent of temperature for the range encountered in the expansion, one may use

$$\bar{C}_p = \sum_i X_i C_{p_i} = \sum_i X_i \left(\frac{\gamma_i}{\gamma_i - 1}\right) R\,, \qquad (64.3)$$

and the average molar mass

$$\bar{m} = \sum_i X_i m_i\,, \qquad (64.4)$$

where X_i is the mole fraction of component i, to obtain an estimate of the maximum velocity for a mixture:

$$v_{\max} = \sqrt{2\bar{C}_p T_0 / \bar{m}}\,. \qquad (64.5)$$

By seeding heavy species in light gases one may accelerate them to superthermal energies. Supersonic beams are characterized by the speed ratio, i.e., the mean velocity divided by the velocity spread:

$$S \equiv \frac{v}{\sqrt{2kT/m}}\,, \qquad (64.6)$$

where T is the local translational temperature, or by the Mach number

$$M \equiv \frac{v}{\sqrt{\gamma kT/m}}\,. \qquad (64.7)$$

Table 64.1. Collision numbers for coupling between different modes. V, R, T refer to vibrational, rotational, and translational energy, respectively. Each entry is the typical range of Z_{A-B}.

	V	R	T
V	$10^{.5-3}$	10^{3-4}	10^{5-6}
R		10^{0-1}	10^{2-3}

For the purpose of order-of-magnitude calculations, the number density on axis far from the nozzle may be estimated as

$$n \approx f n_0 \left(d/x\right)^2\,, \qquad (64.8)$$

where n_0 is the number density in the stagnation region, d is the nozzle diameter, x is the distance from the nozzle, and f is a scaling parameter equal to 0.15 for $\gamma = 1.67$, dropping to 0.059 for $\gamma = 1.29$ [6]. The number density vs. speed distribution of a nozzle beam is well described as a Gaussian characterized by the speed ratio S and a parameter $\alpha = v_0/S$, where v_0 is the most probable velocity:

$$n(v) = v^2 \exp\left[-\left(v/\alpha - S\right)^2\right]\,. \qquad (64.9)$$

Cooling efficiencies for the various internal degrees of freedom correlate with the efficiency of coupling of these modes with translation, hence they vary widely. Coupling of modes A and B is expressed by the collision number Z_{A-B}:

$$Z_{A-B} \equiv Z \tau_{A-B}\,, \qquad (64.10)$$

where τ_{A-B} is the bulk relaxation time, and Z the collision frequency. This represents the number of collisions between effective inelastic events. Typical values are summarized in Table 64.1. R-T coupling is relatively efficient, while V-T coupling is quite inefficient, so that vibrational excitation may not be effectively cooled in the expansion.

64.2.2 Reagent Preparation

Molecular beam methods may be used in conjunction with a variety of other techniques to prepare atoms or molecules in excited or polarized initial states (see Chap. 44), to generate unstable molecules or radicals [7,8] or to produce beams of refractory materials such as transition metals or carbon [9,10]. Some of the common techniques are outlined below. Optical pumping of atoms to excited electronic states is a useful means of reagent preparation, and this topic is presented in detail in Chap. 10. This technique further allows one, using polarized lasers, to explore the influence of angular momentum polarization in the reagents on the collision dynamics [11]. Most of these studies have been performed using alkali and alkaline earth metals since there exist strong electronic transitions and convenient narrow-band visible lasers suitable for use with these systems. Laser excitation may also be used to generate vibrationally excited molecules in their ground electronic states. The techniques employed include direct ir excitation using an HF chemical laser [12], population depletion methods [13] and various Raman techniques [14].

Metastable atoms may also be prepared by laser photolysis of a suitable precursor. $O(^1D)$ is readily prepared by photolysis of ozone or N_2O, for example [15]. Alternatively, rf or microwave discharges may be used to produce metastable species or reactive atoms or radicals [16]. These techniques may also be used to prepare ground state atoms; for example, hot H atom beams are frequently produced by photolysis of HI or H_2S [17]. Such atomic or molecular radical beams may also be generated by pyrolysis in the nozzle. In this case care must be taken to minimize recombination through careful choice of the temperature, nozzle geometry, and transit time through the heated region. Beams of refractory materials are now commonly generated using laser ablation sources [9, 10]. Typically these employ a rod or disk of the substrate of interest which is simultaneously rotated and translated to provide a fresh surface for ablation at each laser pulse. A laser beam is focused on the substrate and timed to fire just as a carrier gas pulse passes over. Laser power and wavelength must be optimized for a given substrate. Lasers operating in the ir, visible, and uv have all been employed. Aligned or oriented molecules have been prepared using multipole focussing [18, 19], and more recently using strong electric fields ("brute force") [20]. In the former case, specific quantum states are focused by the field. In the latter case, so called pendular states are prepared from the low rotational levels of molecules possessing large dipole moments and small rotational constants. The ability to orient these molecules can be estimated on the basis of the Stark parameter $\omega = \mu E/B$, where μ is the dipole moment, E the electric field strength, and B the rotational constant. Orientation is feasible for low rotational levels of molecules when the Stark parameter is on the order of 10 or higher [20].

64.2.3 Detection of Neutral Products

Broadly speaking, detection of neutral molecules is accomplished either by optical (spectroscopic) or nonoptical techniques. Nonoptical methods usually involve nonspecific ionization of neutral particles, most commonly by electron impact, followed by mass selection and ion counting. Thermal detectors such as cryogenic bolometers are also finding widespread application in molecular beam experiments owing to their remarkable sensitivity [21]. In general, optical methods may rely on resonant or nonresonant processes, hence they may or may not enjoy quantum state selectivity. Both photoionization and laser-induced fluorescence methods are now in common usage, usually in applications where quantum state resolved information is desired. The advantage of nonoptical methods is primarily one of generality: all neutral molecules may be detected, and branching into different channels readily measured. Quantum state resolution is more difficult to achieve using nonoptical detection methods, but both vibrationally- and rotationally-resolved measurements have been obtained by these means [22, 23].

The primary advantage of spectroscopic detection is the aforementioned possibility of quantum state specificity. Another unique opportunity afforded by spectroscopic probes is the measurement of product alignment and orientation. In addition, in some cases background interference may be reduced or eliminated using state-specific probes, thereby affording enhanced signal-to-noise ratios.

Nonoptical Techniques

Detectors based on nonspecific ionization remain the most commonly used in molecular beam experiments, owing to the ease of subsequent mass selection, and the convenience and sensitivity of ion detection. Surface ionization is a sensitive means of detecting alkali atoms and other species exhibiting low ionization potentials [24]. Surface ionization occurs when a neutral atom or molecule with a low ionization potential sticks on a surface with a high work function and is subsequently desorbed. Typically these detectors employ a hot platinum or oxidized tungsten wire or ribbon for formation and subsequent desorption of the ions, which is surrounded by an ion collector. They are very efficient for the detection of alkali atoms and molecules whose ionization potentials are $\lesssim 6$ eV.

All neutral gas molecules may be ionized by collision with energetic electrons, and electron beam ionizers may be produced that couple conveniently to quadrupole mass spectrometers [25]. Collision of a molecule with a 100–200 eV electron leads predominantly to formation of the positive ion and a secondary electron. Other processes also occur and can be very significant: doubly or triply charged ions may be formed and, importantly, molecules can fragment yielding many daughter ions in addition to the parent ion. These fragmentation patterns vary with different molecules, and may further show a strong dependence on molecular internal energy, so particular care must be taken to determine the role of these phenomena in each particular application. It is often necessary to record data for the parent ion and daughter ions for a given product channel and compare them to eliminate contributions arising from cracking of the parent molecule or other species [26]. Electron impact ionization probabilities for most species exhibit a similar dependence on electron energy, rising rapidly from the ionization potential to a peak at 80–100 eV, then falling more slowly with increasing collision energy. The ionization cross section for different species scales with molecular polarizability according to a well-established empirical relation [27]:

$$\sigma_{\mathrm{ion}} = 36\sqrt{\alpha} - 18, \qquad (64.11)$$

where σ_{ion} is in Å^2 and α, the molecular polarizability, is in Å^3. This relation can be used to estimate branching ratios in the absence of any other means of calibrating the relative contributions of two different channels. The ionization rate is given by

$$\frac{d[M^+]}{dt} = I_e \sigma [M], \qquad (64.12)$$

where I_e is the electron beam intensity, typically 10 mA/cm^2 or 6×10^{16} electrons/cm^{2}s, and $[M]$ is the number density of molecules M in the ionizer. If one assumes an ionization cross section σ_{ion} of 10^{-16} cm^2 for collision with 150 eV electrons (a typical value for a small molecule), the ionization probability for molecules residing in the ionizer is then

$$\frac{d[M^+]}{dt}\frac{1}{[M]} = I_e\sigma = 6 \times 10^{16} \times 10^{-16} = 6\ \mathrm{s}^{-1}. \quad (64.13)$$

However, product molecules arriving in the detector are not stationary. Typically, product velocities are on the order of 500 m/s. If the ionization region has a length of 1 cm, the residence time τ of a product molecule is on the order of 2×10^{-5} s. Consequently, the ionization probability of product molecules passing through the ionizer is

$$\frac{d[M^+]}{dt}\frac{\tau}{[M]} = 2 \times 10^{-5} \times 6 = 1.2 \times 10^{-4}. \quad (64.14)$$

Although this does not appear very efficient (indeed, it is 4 orders of magnitude less so than surface ionization), nevertheless, if the background count rate is sufficiently low, then good statistics may be obtained with signal levels as low as 1 Hz. Thus, for detection based on electron impact ionization, a key factor determining the sensitivity of the experiment is the background count rate at the masses of interest.

Spectroscopic Detection

Spectroscopic detection methods usually involve either laser-induced fluorescence (LIF) or resonant pho-toionization (REMPI) (see Chap. 42). Alternative techniques such as laser-induced grating methods and nonresonant vuv photoionization are also being applied to scattering experiments. Essential to the use of spectroscopic methods for reactive scattering studies is an understanding of the spectrum of the species of interest. This may be challenging for many reactive systems because the products may be produced in highly excited vibrational or electronic states that may not be well characterized. Additional spectroscopic data may be required. Franck-Condon factors are necessary to compare the intensities of different product vibrational states, while a calibration of the relative intensities of different electronic bands requires a measure of the electronic transition moments. In some cases, one must include the specific dependence of the electronic transition moment on the internuclear distance by integrating over the vibrational wave function. Populations corresponding to different rotational lines may be compared after the appropriate correction, which is represented by the Hönl-London factors, only for isotropic irradiation and detection. This is certainly not the case for most laser-based experiments. Generally, the detailed dependence of the excitation and detection on the relevant magnetic sublevels must be considered [28–30]. Caution is required in using any spectroscopic method involving a level that is predissociated. This may lead to a dramatic decrease in the associated fluorescence or photoionization yield if the predissociation rate approaches or exceeds the rate of fluorescence or subsequent photoionization. An important question in any experiment based on spectroscopic detection is whether product flux or number density is probed. This question is considered in detail in several articles [14,31]. It depends on the lifetime of the state that is probed, the relative time that the molecule is exposed to the probe laser field, and its residence time in the interaction region. Saturation phenomena are also important, yet not necessarily easily anticipated. Complete saturation does not readily occur because excitation in the wings of the laser beam profile becomes more significant as the region in the center of the beam becomes saturated [32].

LIF is currently the most widely used spectroscopic technique in inelastic and reactive scattering experiments [28,33,34]. It has been used to measure state-resolved total cross sections [35] and differential cross sections in electronic [36], vibrational and rotationally inelastic scattering [13] as well as reactive scattering [37].

With the development of high-power tunable lasers and the discovery of useful photoionization schemes, REMPI is becoming a more general   technique [38,39]. REMPI has the advantages associated with ion detection, namely considerable convenience in mass selection and efficient detection, in addition to the capability for quantum state selectivity. Disadvantages associated with REMPI arise primarily from higher laser power employed compared with LIF. Caution is required in attempting to extract quantitative information from REMPI spectra if one or several of the steps involved in the ionization process are saturated. This is of particular concern at the high laser powers necessary for multiple photon transitions. An alternative to direct photoionization involves excitation of products to metastable Rydberg states, followed by field ionization some distance from the interaction region. This technique has the advantage of very low background and is capable of extraordinary time-of-flight resolution. Remarkable results have recently been obtained for the reaction D+H_2 using this method [40]. Photoionization techniques are becoming more widely used in scattering experiments as the basis for product imaging detection schemes discussed below.

64.2.4 A Typical Signal Calculation

For a crossed-beam system in which a beam of atoms A collides with a beam of molecules B yielding products C and D, the rate of formation of C is given by

$$\frac{dN_C}{dt} = n_A n_B \sigma_r g \Delta V \,, \qquad (64.15)$$

where n_A and n_B are the number densities of the respective reagents at the interaction region, σ_r is the reaction cross section, g the magnitude of the relative velocity between the reactants, and ΔV the volume of intersection of the beams. For a typical experiment employing continuous supersonic beams, the number densities of the atomic and molecular reactants are $\sim 10^{11} - 10^{12}$ cm^{-3} and the scattering volume 10^{-2} cm^3. For $g = 10^5$ cm/s and $\sigma_r = 10^{-15}$ cm^2, the rate of product formation $dN_C/dt = 10^{11}$ molecules/s. The kinematics and energetics of the reaction then determines the range of laboratory angles into which the products scatter, and the magnitude of the scattered signal.

If the products scatter into 1 sr of solid angle, and the detector aperture is 3×10^{-3} sr (roughly 1 degree in both directions perpendicular to the detector axis), then the detector receives 3×10^7 product molecules/s. Given the detection probability obtained above, 3600 product ions/s are detected. This is adequate to obtain very good statistics in a short time as long as the background count rate is not considerably higher.

For a nonspecific detection technique, such as electron bombardment ionization coupled with mass filtering, it is necessary to use ultrahigh vacuum (10^{-10} torr) in the detector region to minimize interference from background gases. The residual gases are then primarily H_2 and CO, with number densities on the order of 10^6 cm^3. Differential pumping stages, each of which may reduce the background by 2 orders of magnitude, are generally used to lower the background from gases whose partial pressures are lower than the ultrahigh vacuum limit of the detector chamber. However, this differential pumping helps only for those molecules that do not follow a straight trajectory through the detector. The contribution from the latter is given by

$$n' = \frac{nA}{4\pi x^2} \,, \qquad (64.16)$$

where n is the number density of molecules effusing from an orifice of area A, and n' is their number density at a distance x on axis downstream. For a distance of 30 cm and a main chamber pressure of 3×10^{-7} torr, this corresponds to a steady state density of 10^5 molecules/cm^3 at the ionizer, a reduction of 6 orders of magnitude. Three stages of differential pumping are thus the maximum useful under these conditions, since the primary source of background is then molecules following a straight trajectory from the main chamber. A liquid helium cooled surface opposite the detector entrance may then be useful to minimize scattering of background molecules into the ionizer.

64.3 EXPERIMENTAL CONFIGURATIONS

64.3.1 Crossed-Beam Rotatable Detector

The configuration illustrated in Fig. 64.1 represents a standard now widely used [25], usually with two continuous beams fixed at 90 degrees. The molecular beam sources are differentially pumped and collimated to yield an angular divergence of about 2 degrees. The beams cross as close as possible to the nozzles, with a typical interaction volume of 3 mm^3. Scattered products pass through an aperture on the front of the detector, thence through several stages of differential pumping before reaching the ionizer. Ions formed by electron impact on the neutral products are then extracted into a quadrupole mass spectrometer with associated ion counter. A chopper wheel is generally used at the entrance to the detector to provide a time origin for recording time-of-flight spectra. Pseudorandom sequence chopper disks provide optimal counting statistics while maintaining a high duty cycle (50%) [41]. The detector may be rotated about the interaction region, typically through a range of 120° or so, allowing one to examine products scattered at a range of laboratory angles. In addition to time-of-flight detection, one of the beams may be gated on and off for background subtraction and the detector moved to record the integrated signal at each laboratory angle.

Two kinds of measurements are typically made in these experiments: time-of-flight spectra and angular distributions. Usually one is interested in obtaining the complete product-flux vs. velocity contour map, since this contains full details of the scattering process. This is

Figure 64.1. Experimental arrangement for $F+D_2 \rightarrow DF+F$ reactive scattering. Pressure (in Torr) indicated in each region. Components are (1) effusive F atom source; (2) velocity selector; (3) cold trap; (4) D_2 beam source; (5) heater; (6) liquid nitrogen feedline; (7) skimmer; (8) tuning fork chopper; (9) synchronous motor; (10) cross-correlation chopper; (11) ultrahigh vacuum differentially pumped mass spectrometer detector.

obtained by measuring a full angular distribution as well as time-of-flight data at many laboratory angles. The results are then simulated using a forward convolution fitting procedure to obtain the underlying contour map [42–44]. Because scattering of isotropic reagents exhibits cylindrical symmetry about the relative velocity vector, it is sufficient to measure products scattered in any plane containing this vector to determine the product distribution. This is not true for structured particles (e.g. involving atoms in P states); however, this azimuthal anisotropy has been used to explore the impact parameter dependence of the reaction dynamics [45].

In a typical reactive scattering experiment, $A+BC \rightarrow AB + C$, either of the two products may be detected. Conservation of linear momentum requires that the c.m. frame momenta of the two products must sum to zero. It is thus only necessary to obtain the contour map for one of the products. The choice of detected product is usually dictated by kinematic considerations, although one may choose to detect a product that is kinematically disfavored if its partner happens to have a mass with a large natural background in the detector. Kinematic considerations can be critical in assessing the suitability of a given system for study. It is very important that one of the products be scattered entirely within the viewing range of the detector in order to obtain a complete picture of the reaction dynamics.

The advantages of crossed-beams employed in conjunction with an electron impact ionizer-mass spectrometer detector derive primarily from the universality of the detector. No spectroscopic information is required and there are no invisible channels, such as may occur with spectroscopic detection methods. In addition, the resolution of these machines may be increased almost arbitrarily; indeed, even rotationally inelastic scattering has been studied [46]. The disadvantages are complementary to the advantages: the universal detector implies that quantum state resolution is not achieved directly, although in favorable cases the product vibrational states may be resolved in the translational energy distributions [22,47]. In addition, if the product of interest represents a mass that receives interference from one of the beam masses, background interference may be problematic. Kinematic considerations mentioned above may also preclude study of certain systems. However, the kinematic requirements for the Doppler and imaging approach discussed below are complementary to those of the rotatable-detector configuration.

64.3.2 Doppler Techniques

Spectroscopic detection methods in crossed-beam experiments allows the measurement of state-resolved differential cross sections, and thus the ultimate level of insight into the reaction dynamics. A method developed by Kinsey and others [48,49] determines differential cross sections by measurement of product Doppler profiles using LIF (called ADDS for Angular Distribution by Doppler Spectroscopy). For a laser directed parallel to the relative velocity vector, a particle scattered with a c.m. velocity of **u** perceives the photon as having the Doppler shifted frequency

$$\nu' = \nu [1 - (\mathbf{u} + \mathbf{V}_{cm}) \cdot \hat{\mathbf{n}}/c] , \qquad (64.17)$$

where ν is the laser frequency and $\mathbf{V}_{cm}$ is the velocity of the center of mass, both in the laboratory frame of refernce, and $\hat{\mathbf{n}}$ is the unit vector in the probe laser direction. For the case of a single possible recoil speed, one may obtain the full differential cross section directly in the c.m. frame by reconstruction of a single Doppler profile [48]. In this case, the angular resolution is a maximum for the sideways scattered products, and a minimum at the poles. An alternative approach is to measure the Doppler profile with a laser perpendicular to the relative velocity vector. This approach (PADDS for Perpendicular ADDS) affords complementary angular resolution, but folds the forward and backward scattered products into a single symmetric component [49]. For the case in which the detected product does not possess a known recoil speed (for example if the thermodynamics of the process is not known, or if one probes the atomic fragment in an $A + BC \rightarrow AB + C$ reaction), a single

Figure 64.2. Schematic view of crossed molecular beam apparatus with LIF-Doppler detection.

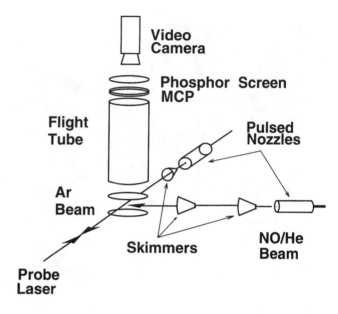

Figure 64.3. Schematic view of crossed molecular beam apparatus with product imaging detection.

Doppler profile is insufficient to reconstruct the double differential cross sections. Nevertheless, Kinsey's earliest experimental results were for one such example: the reaction $H + NO_2 \rightarrow OH + NO$ [50].

More recently Mestdagh *et al.* [51] have studied electronically inelastic collision processes using this approach by measuring the Doppler profiles over a range of probe laser angles, as illustrated in Fig. 64.2. A beam of barium atoms is crossed at 90 degrees by a beam of some molecular perturber. At the interaction region, the barium atoms are electronically excited using a narrow band dye laser. Scattered barium atoms that have undergone a specific electronic transition as a result of the collision are probed at the interaction region using a second dye laser, which is scanned across the Doppler profile. The product-flux vs. velocity contour map is then reconstructed by means of a forward convolution simulation procedure analogous to that described in the preceding section.

In addition to state-resolved detection, another difference between the Doppler methods and the traditional crossed-beam configuration is that the kinematic considerations favor detection of fast particles, and almost any system that is spectroscopically suitable may be considered. The primary disadvantage of Doppler methods is the limited angular and translational energy resolution possible. Often, however, modest angular resolution is sufficient to achieve a global picture of the reaction dynamics. Much current work involves the study of photoinitiated reactions in cells, relying on the short excited state lifetimes to guarantee single collision conditions,

and using iterative fitting procedures to probe product velocity distributions and angular momentum polarization [15]. Angular momentum polarization can have a profound effect on the measured distributions and can afford a powerful additional means of exploring the collision dynamics. Examples of 3- and 4-vector correlation experiments approach a "complete description" of the scattering process (see Chap. 44) [51–53].

64.3.3 Product Imaging

Another spectroscopic technique is based on direct imaging of the scattered product distribution. The technique was first used to record state-resolved angular distributions of methyl radicals from the photodissociation of methyl iodide [54]. The method has since been widely employed to study photodissociation, and more recently to record state-resolved inelastic scattering in a crossed-beam experiment [55]. Recently it has been applied to a crossed-beam reactive scattering system [56]. The crossed-beam configuration used by Houston and coworkers is shown schematically in Fig. 64.3. The two skimmed supersonic beams cross at right angles, and scattered products are state-selectively ionized on the axis of a time-of-flight mass spectrometer using resonant photoionization. The ion cloud thus formed continues to expand with its nascent recoil velocity as it drifts through the flight tube. The ions then strike a microchannel plate coupled to a phosphor screen. The latter is viewed by a video camera gated to record the signal at the mass of interest. The images are thus two-dimensional projections of the nascent three dimensional product distributions.

There now exist two alternatives for regenerating the three dimensional distribution from the projection. The first, a tomographic reconstruction using an inverse Abel transform, is widely used in photodissciation studies [57, 58]. It is a direct inversion procedure feasible for cases in which the image is the projection of a cylindrically symmetric object, with its axis of symmetry parallel to the image plane. This analysis yields a unique product contour map directly from the image, but it is difficult to incorporate apparatus functions, and is sensitive to noise in the data. The second alternative is a forward convolution fitting method. A Monte Carlo based simulation has the advantage that one may treat the averaging over experimental parameters quite rigorously.

The advantages of the imaging method again derive from its reliance on a spectroscopic probe, so that quantum state resolution is possible and background interference may be avoided. In addition, it possesses a multiplexing advantage since the velocity distribution is recorded for all angles simultaneously. Imaging relies exclusively on photoionization, unlike the Doppler methods which may use either photoionization or LIF. This is somewhat disadvantageous since the available photoionization schemes are limited and often high laser power is necessary to achieve adequate signal intensity. As a result, background ions can be a problem. In general, resonantly enhanced two-photon ionization, i.e. [1+1], detection schemes are thus preferable.

64.3.4 Laboratory to Center-of-Mass Transformation

Angular and velocity distributions measured in the laboratory frame must be transformed to the c.m. frame for theoretical interpretation. Accounts of this transformation and details concerning the material presented below may be found in Refs. [59–62], among others. The Newton diagram is useful to aid in visualizing the transformation, and in understanding the kinematics of a given collision system. For the scattering of F+D$_2$ for example, shown in Fig. 64.4, a beam of fluorine atoms with a velocity $\mathbf{v}_\mathrm{F}$ is crossed by a beam of D$_2$, velocity $\mathbf{v}_{\mathrm{D}_2}$, at 90 degrees. The relative velocity between the two reactants is $\mathbf{g} = \mathbf{v}_\mathrm{F} - \mathbf{v}_{\mathrm{D}_2}$, and the velocity of the c.m. of the entire system is

$$\mathbf{V}_\mathrm{cm} = \frac{M_\mathrm{F}\mathbf{v}_\mathrm{F} + M_{\mathrm{D}_2}\mathbf{v}_{\mathrm{D}_2}}{M_\mathrm{F} + M_{\mathrm{D}_2}}. \qquad (64.18)$$

$\mathbf{V}_\mathrm{cm}$ divides the $\mathbf{g}$ into two segments corresponding to the c.m. velocities of the two reagents. The magnitude of these vectors, u_F and u_{D_2} are inversely proportional to the respective masses. If scattered DF products are formed with a laboratory scattering angle Θ and a laboratory velocity $\mathbf{v}_\mathrm{DF}$ as shown in Fig. 64.4, this

Figure 64.4. Newton diagram for collision of F with H$_2$ with superimposed c.m. flux vs. velocity contour map.

corresponds to DF *backscattered* with respect to the incident F atom, in the c.m. system. It is common to refer the scattering frame direction to the atomic reagent in an $A + BC \rightarrow AB + C$ reaction, for example, to make clear the dynamics of the process. In this case the backscattered DF arises as a result of a direct rebound collision. Some useful kinematic quantities are summarized here. For beams A and BC intersecting at 90 degrees, the angle of the c.m. velocity vector with respect to A is given by

$$\Theta_\mathrm{cm} = \arctan \frac{M_{BC} v_{BC}}{M_A v_A}. \qquad (64.19)$$

For an arbitrary Newton diagram with angle α between the two beams the magnitude of the relative velocity is

$$g^2 = v_A^2 + v_{BC}^2 - 2 v_A v_{BC} \cos\alpha, \qquad (64.20)$$

the relative velocity vector is

$$\mathbf{g} = \mathbf{v}_A - \mathbf{v}_{BC}, \qquad (64.21)$$

and the collision energy is

$$E_\mathrm{coll} = \tfrac{1}{2}\mu_i g^2, \qquad (64.22)$$

where μ_i is the reduced mass of the initial collision system. The magnitude of the c.m. frame velocity of particle A before collision is

$$\mathbf{u}_A = \frac{m_{BC}}{m_A + m_{BC}}\mathbf{g}. \qquad (64.23)$$

The final relative velocity is

$$\mathbf{g}' = \mathbf{v}_{AB} - \mathbf{v}_C, \qquad (64.24)$$

with magnitude

$$g' = \sqrt{2E_\mathrm{avail}/\mu_\mathrm{F}}, \qquad (64.25)$$

where the available energy E_avail is

$$E_{\text{avail}} = E_{\text{coll}} + E_{\text{int,reac}} + E_{\text{exo}} - E_{\text{int,prod}}, \quad (64.26)$$

in which $E_{\text{int,reac}}$ is the internal energy of the reactants, E_{exo} is the exoergicity of the reaction, and $E_{\text{int,prod}}$ is the internal energy of the products.

One must transform the laboratory intensity $I(\Omega) \equiv d^2\sigma/d^2\Omega$ into $I(\omega)$ ($\equiv d^2\sigma/d^2\omega$), the corresponding c.m. quantity. For the crossed-beam configuration described in 64.3.1, the laboratory distributions are distorted by a transformation Jacobian that arises because the laboratory detector views different c.m. frame solid angles depending on the scattering angle and recoil velocity. For the spectroscopic experiments described in 64.3.2 and 64.3.3, the Jacobian is unity (the c.m. velocity represents a simple frequency offset of the Doppler profiles, for example); however, the transformation of the scattering distributions from the recorded quantities (2–dimensional projections or intensity vs. wavelength) to recoil velocity distributions may be complex. Two cases must be considered for the configuration discussed in 64.3.1: one in which discrete velocities result (such as elastic or state-resolved scattering experiments), and one in which continuous final velocities are measured. For the first case, the laboratory and c.m. differential cross sections are independent of the respective product velocities v and u and these quantities are related by

$$\frac{d^2\sigma}{d^2\Omega} = J\frac{d^2\sigma}{d^2\omega}, \quad (64.27)$$

so that the transformation Jacobian is given by

$$J = \frac{d^2\omega}{d^2\Omega}. \quad (64.28)$$

For discrete recoil velocities, the c.m. solid angle is

$$d^2\omega = \frac{dA}{u^2}, \quad (64.29)$$

where dA is a surface element of the product Newton sphere. The laboratory solid angle corresponding to this quantity is

$$d^2\Omega = \frac{\cos(\mathbf{u}, \mathbf{v})}{v^2}dA, \quad (64.30)$$

so that the Jacobian for the first case is given by

$$J = \frac{v^2}{u^2 \cos(\mathbf{u}, \mathbf{v})}. \quad (64.31)$$

For the case of continuous final velocities, the σ are velocity-dependent and are related by

$$\frac{d^3\sigma}{d^2\Omega\, dv} = J\frac{d^3\sigma}{d^2\omega\, du}, \quad (64.32)$$

so that here the Jacobian is given by

$$J = \frac{d^2\omega\, du}{d^2\Omega\, dv}. \quad (64.33)$$

In this case we consider a recoil volume element $d\tau$ (in velocity space), which must be the same in both coordinate frames:

$$d\tau_{\text{cm}} = u^2\, du\, d^2\omega = d\tau_{\text{lab}} = v^2\, dv\, d^2\Omega, \quad (64.34)$$

so that the Jacobian is

$$J = v^2/u^2. \quad (64.35)$$

The laboratory intensity is then related to that in the c.m. frame by

$$I_{\text{lab}}(v, \Theta) = (v^2/u^2)\, I_{\text{cm}}(\theta, u). \quad (64.36)$$

For a mass spectrometer detector with electron bombardment ionizer, one measures number density of particles rather than flux, so that the recorded signal is given by

$$N_{\text{lab}}(v, \Theta) = \frac{I_{\text{lab}}(v, \Theta)}{v} = \frac{v}{u^2}\, I_{\text{cm}}(u, \theta). \quad (64.37)$$

The usual flux vs. velocity contour map is a polar plot of the quantity $I_{\text{cm}}(u, \theta)$. The product velocity distributions are then

$$I(u) = \iint I(\theta, u)\, \sin\theta\, d\theta\, d\phi = 2\pi \int_0^\pi I(u, \theta)\, \sin\theta\, d\theta, \quad (64.38)$$

and the translational energy distributions are

$$I(E_T) = I(u)\left|\frac{du}{dE_T}\right|. \quad (64.39)$$

64.4 ELASTIC AND INELASTIC SCATTERING

When particles collide, they may exchange energy or recouple it into different modes, they may change their direction of motion, and they may even change their identity. The study of these processes reveals a great deal of information about the forces acting between the particles and their internal structure. It is useful to begin with a summary of the dominant features of elastic and inelastic scattering.

64.4.1 The Differential Cross Section

Figure 64.5 illustrates the relation between the deflection function χ and the impact parameter b for a realistic potential containing an attractive well and a repulsive core. For large b there is no interaction, hence no deflection. At smaller values of b, the attractive part of the potential is experienced and some positive deflection results. At a smaller value of b, b_r, the influence of the attractive component of the potential reaches a maximum,

$$E_T^{2/s}\theta^{2(1+1/s)}I(\theta) = \text{const.} \qquad (64.44)$$

For a potential exhibiting a minimum, the rainbow angle θ_r is proportional to the collision energy, and clearly resolved when the collision energy is 3 to 5 times the well depth. In addition, supernumery rainbows and quantum mechanical "fast osillations" occur in the $d\sigma$, and these provide a sensitive probe of the interaction. Accurate interatomic potentials are routinely obtained from elastic scattering experiments [61, 64].

64.4.2 Rotationally Inelastic Scattering

Classical scattering involving an anisotropic potential results in another rainbow phenomenon, distinct from that seen in pure elastic scattering and notable in that it does not require an attractive component in the potential. These rotational rainbows are equivalently seen in a plot of integral cross section against change in rotational angular momentum Δj, or in the differential cross section for a particular value of Δj. The rotational rainbow peaks arise from the range of possible orientation angles γ in a collision involving an anisotropic potential. When there is a minimum in $d\gamma/d\theta$ for a given Δj, the differential cross section reaches a maximum [65]. The rotational rainbow peak occurs at the most forward classically allowed value of the scattering angle, and $d\sigma$ drops rapidly at smaller angles. The rainbow moves to more backward angles with increasing Δj because the larger j-changing collisions require greater momentum transfer, hence must arise from lower impact parameter collisions. For heteronuclear molecules, two rainbow peaks may be observed, corresponding to scattering off either side of the molecule. One can relate the location of the rainbow peak to the shape of the potential using a classical hard ellipsoid model [66]:

$$A - B = \frac{j}{p_0}\left[2\sin\left(\frac{\theta_{\mathrm{r,cl}}}{2}\right)\right]^{-1}, \qquad (64.45)$$

where j is the rotational angular momentum, p_0 is the inital linear momentum, $\theta_{\mathrm{r,cl}}$ is the classical rainbow position, and A and B are the semimajor and semiminor axes of a hard ellipse potential. The classical rainbow positions occur somewhat behind the quantum mechanical and experimental rainbow positions, so the classical rainbow may be estimated as the point at which the peak has fallen to 44% of the experimental value. Real molecular potentials may be far from ellipsoids, however, so detailed quantitative insight into the potential requires a comparison of scattering data with trajectory calculations.

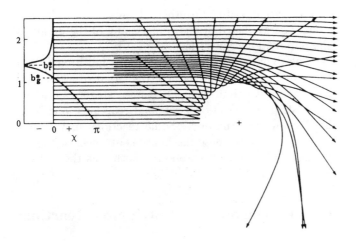

Figure 64.5. Schematic diagram showing the relation between impact parameter b and deflection function χ.

giving the greatest positive deflection: this is the *rainbow angle* by analogy with the optical phenomenon. There is another value of the b for which point the attractive and repulsive parts of the potential balance, yielding no net deflection. This is the *glory* impact parameter b_g. For yet smaller values of b, the interaction is dominated by the repulsive core and rebound scattering gives a negative deflection function.

The important expressions related to the differential cross section are summarized here [63]. For scattering involving an isotropic potential, the deflection angle is $\Theta = |\chi|$. The differential cross section $d\sigma$ gives the rate of all collisions leading to deflection angles in the solid angle element $d\omega$:

$$\frac{dN(\theta)}{dt} \propto I(\theta)\, d\omega = I(\theta)2\pi\sin(\theta)\, d\theta. \qquad (64.40)$$

The incremental cross section is $d\sigma = I(\theta)\, d\omega = 2\pi b\, db$, so

$$I(\theta) = \frac{b}{\sin\theta\, (d\theta/db)}. \qquad (64.41)$$

For classical particles, the relation between the deflection function and the potential is

$$\chi = \pi - 2b\int_{R_0}^{\infty}\frac{dR}{R^2}\left[1 - \frac{V(R)}{E_T} - \frac{b^2}{R^2}\right]^{-1/2}, \qquad (64.42)$$

where $V(R)$ is the potential as a function of interparticle distance R, R_0 is the turning point of the collision, and E_T the collision energy.

In the high energy limit, for large $b \approx R_0$,

$$\chi(b, E_T) \propto V(b)/E_T. \qquad (64.43)$$

For a long-range potential $V(R)$ proportional to R^{-s},

64.4.3 Vibrationally Inelastic Scattering

There has been no direct observation of the differential cross section of T-V or V-T energy transfer involving neutral molecules owing to the small cross sections for these processes. Integral cross section data are available, however. Above threshold, the latter has shown a linear dependence of σ on collision energy for $\Delta\nu = 1$, quadratic for $\Delta\nu = 2$ and cubic for $\Delta\nu = 3$ [67]. In addition, a great deal of information on vibrational relaxation processes has been obtained in cell experiments [68].

64.4.4 Electronically Inelastic Scattering

A wealth of information is available on electronically inelastic scattering systems, since these in general exhibit much larger cross sections than V-T processes [69, 70]. In addition, spectroscopic methods may be used to overcome some of the background problems that hamper the study of the latter. Often, quenching of electronically excited states involves curve crossing mechanisms, so that very effective coupling of electronic to vibrational energy may occur. Spin-orbit changing collisions of $Ba(^1P)$ with O_2 or NO, for example, occur by a near-resonant process and result in almost complete conversion of electronic energy to vibrational excitation of the product [71]. The analogous collisions with N_2 and H_2, however, reveal very repulsive energy release with little concomitant vibrational excitation. Both processes likely occur via curve crossings of the relevant electronic states, but the near-resonant mechanism occurs by way of an ionic intermediate.

64.5 REACTIVE SCATTERING

Reactive differential cross sections reveal several distinct aspects of the chemical encounter. The angular distributions themselves may be used to infer the lifetime of the collision intermediate: long-lived complexes exhibit forward-backward symmetry along the relative velocity vector. In this case "long-lived" means on the order of several rotational periods. The rotational period of the complex may thus be used as a clock to study the energy dependence of the intermediate's lifetime. The angular distributions further reveal the relation of initial and final orbital angular momentum. Sharply peaked angular distributions generally indicate strongly correlated initial and final orbital angular momentum vectors. Finally, the product translational energy release contains the details of the energy disposal, and reveals a wealth of information about the thermodynamics of the process, the existence of barriers, and sometimes even the geometry of the transition state. Together, the angular and translational energy distributions reveal many of the details of the potential energy surface.

The dynamics of reactive collisions fall broadly into three main categories characterized by distinct angular and energy distributions. The three categories are harpoon/stripping reactions, rebound reactions, and long-lived complex formation. Some reactions may exhibit more than one of these mechanisms at once, or the dynamics may change from one to another as the collision energy is varied.

64.5.1 Harpoon and Stripping Reactions

It was known in the 1930s that collisions of alkali atoms with halogen molecules exhibit very large cross sections and yield highly excited alkali halide products. These observations were accounted for by the *harpoon* mechanism proposed by M. Polanyi. Because alkali atoms have low ionization potentials and halogen molecules large electron affinities, as the alkali atom approaches the molecule, electron transfer may occur at long range. These processes are considered in detail in Chap. 47 [72, 73]. The harpooning distance R_c at which this curve crossing takes place may be estimated simply as the distance at which the Coulomb attraction of the ion pair is sufficient to compensate for the endoergicity of the electron transfer:

$$R_c = e^2/(I_P - A_e), \qquad (64.46)$$

where I_P and A_e represent the ionization potential and electron affinity of the electron donor and recipient, respectively. For R in Å and E in eV, this relation is

$$R_c = 14.4/(I_P - A_e). \qquad (64.47)$$

Owing to the large Coulombic attraction between the ion pair, reaction proceeds immediately following electron transfer. The crossing distance may then be used to estimate the effective reaction cross section. The vertical A_e is not necessarily the appropriate value to use in estimating these crossing distances; stretching of the halogen bond may occur during approach, so the effective A_e is generally somewhere between the vertical and adiabatic values. Often there exists some repulsion between the atoms in the resulting halogen molecular ion, so that electron transfer is accompanied by dissociation of the molecule in the strong field of the ion pair. The alkali ion, having sent out the electron as the "harpoon", then reels in the negative ion, leaving the neutral halogen atom nearly undisturbed as a spectator. Because these events occur at long range, there is no momentum transfer to the spectator atom, and it is a simple matter to estimate the anticipated angular and translational energy distributions in this *spectator stripping* limit. The product molecule is scattered forward (relative to the direction of the incident atom) and for the reaction $A+BC \rightarrow AB+C$, the final c.m. velocity for the product AB is given by

$$\mathbf{u}'_{AB} = -M_C \mathbf{u}_{BC}/M_{AB}, \qquad (64.48)$$

where $\mathbf{u}_{BC}$ is the initial c.m. velocity of the BC molecule. This spectator stripping mechanism may occur in systems other than harpoon reactions, and is useful to remember as a limiting case.

The likelihood of electron transfer at these crossings may be estimated using a simple Landau-Zener model [73] (see Chap. 47). For relative velocity g, impact parameter b and crossing distance R_c, the probability for undergoing a transition from one adiabatic curve to another (that is, the probability for remaining on the diabatic curve) is given by

$$p = 1 - e^{-\delta}, \qquad (64.49)$$

where

$$\delta = \frac{2\pi H_{12}^2 R_c^2}{g} \left[1 - \frac{b^2}{R_c^2}\right]^{-\frac{1}{2}}, \qquad (64.50)$$

and H_{12} is the coupling matrix element between the two curves. H_{12} may be estimated from an empirical relation which is accurate within a factor of three over a range of 10 orders of magnitude [74]. In atomic units

$$H_{12} = \sqrt{I_1 I_2} R_c^* e^{-0.86 R_c^*}, \qquad (64.51)$$

where

$$R_c^* = (\sqrt{I_1} + \sqrt{I_2}) \qquad (64.52)$$

is the reduced crossing distance, and I_1 and I_2 are the initial and final ionization potentials of the transferred electron. One finds electron transfer probabilities near unity for curve crossing distances below about 5 Å, dropping to zero for crossing at distances greater than about 8 Å. These estimates are based on electron transfer in atom–atom collisions, and it is important to remember that atom–molecule collisions occur on surfaces rather than curves, so the crossing seam may cover a broad range of internuclear distances.

64.5.2 Rebound Reactions

Another common direct reaction mechanism is the *rebound* reaction exemplified by $F+D_2 \rightarrow DF + D$ [22]. The c.m. product flux vs. velocity contour map obtained for this reaction is shown in Fig. 64.4. Owing to the favorable kinematics and energetics in this case, the FD product vibrational distribution is clearly resolved, and peaks at $v = 2$. The dominant $v = 2$ product peaks

at a c.m. angle of 180 degrees (referred to the direction of the incident F atom.) This rebound scattering is characteristic of reactions exhibiting a barrier in the entrance channel. Rebound scattering implies small b collisions, and this serves to couple the translational energy efficiently into overcoming the barrier. Small b collisions have necessarily smaller cross sections however, since cross section scales quadratically with b.

64.5.3 Long-Lived Complexes

A third important reaction mechanism involves the formation of an intermediate that persists for some time before dissociating to give products. If the collision complex survives for many rotational periods ($\sim 10^{-11}$ s), then the c.m. angular distribution exhibits a characteristic forward-backward symmetry, usually with peaking along the poles. The latter occurs because the initial and final orbital angular momenta tend to be parallel (and perpendicular to the initial relative velocity vector). When there exist dynamical constraints enforcing some other relation, as in the case $F+C_2H_4$, then sideways scattering may be observed, despite a lifetime of several rotational periods [75–77]. For some systems exhibiting this long-lived behavior, the rotational period may be used as a *molecular clock* to monitor the lifetime of the complex. By increasing the collision energy until the distribution begins to lose its forward-backward symmetry, one can investigate the internal energy of the system just when its lifetime is on the order of a rotational period.

Systems that have an inherent symmetry may exhibit this forward-backward symmetry in the scattering distributions despite lifetimes that are considerably shorter than a rotational period. This is the case for $O(^1D)$ reacting with H_2, for example [78]. This reaction involves insertion of the O atom into the H_2 bond resulting in an intermediate that accesses the deep H_2O well and contains considerable vibrational excitation. Trajectory calculations show that the complex dissociates after a few vibrational periods, but the distribution exhibits forward-backward symmetry because the O atom is equally likely to depart with either H atom.

REFERENCES

1. Y. T. Lee, Science, **236**, 793 (1987).
2. W. R. Gentry, *Atomic and Molecular Beam Methods*, edited by G. Scoles (Oxford University Press, New York, 1988) Chap. 3, pp. 54 ff.
3. J. B. Anderson, R. P. Andres, and J. B. Fenn, Adv. Chem. Phys. **10**, 275 (1966).
4. D. R. Miller, in *Atomic and Molecular Beam Methods*, Edited by G. Scoles, (Oxford University Press, New York, 1988) Chap. 2, pp. 14 ff.
5. R. Campargue, J. Phys. Chem. **88**, 275 (1984).

6. P. C. Engelking, Chem. Rev. **91**, 399 (1991).

7. H. F. Davis, B. Kim, H. S. Johnston, and Y. T. Lee, J. Phys. Chem. **97**, 2172 (1993).

8. P. Chen, in *Unimolecular and Bimolecular Reaction Dynamics*, edited by C. Y. Ng, T. Baer, and I Powis (John Wiley and Sons, New York, 1994) Chap. 8 pp. 371 ff.

9. R. E. Smalley, Laser Chem. **2**, 167 (1983).

10. M. D. Morse, J. B. Hopkins, P. R. R. Langridge-Smith, and R. E. Smalley, J. Chem. Phys. **79**, 5316 (1983).

11. I. V. Hertel and W. Stoll, Adv. At. Mol. Phys. **13**, 113 (1978).

12. J. G. Pruett and R. N Zare, J. Chem. Phys. **64**, 1774 (1976).

13. K. Bergmann, U. Hefter, and J. Witt, J. Chem. Phys. **72**, 4777 (1980).

14. K. Bergmann, in *Atomic and Molecular Beam Methods*, edited by G. Scoles (Oxford University Press, New York, 1988) Chap. 12.

15. M. Brouard, S. Duxon, P. A. Enriquez, and J. P. Simons, J. Chem. Soc. Faraday Trans. **89**, 1435 (1991).

16. S. J. Sibener, R. J. Buss, P. Cassavechia, T. Hirooka, and Y. T. Lee, J. Chem. Phys. **72**, 4341 (1980).

17. R. E. Continetti, B. A. Balko, and Y. T. Lee, J. Chem. Phys. **93**, 5719 (1990).

18. P. R. Brooks, Science **193**, 11 (1976).

19. S. Stolte, Ber. Bunsenges. Phys. Chem. **86**, 413 (1982).

20. B. Friedrich and D. R. Herschbach, Nature **353**, 412 (1991).

21. D. Bassi, A. Boschetti, M. Scotoni, and M. Zen, Appl. Phys. B **26**, 99 (1981).

22. D. M. Neumark, A. M. Wodtke, G. N. Robinson, C. C. Hayden, K. Shobatake, R. K. Sparks, T. P. Schaefer, and Y. T. Lee, J. Chem. Phys. **82**, 3067 (1985).

23. M. Faubel, Adv. At. Mol. Phys. **19**, 345 (1983).

24. T. R. Touw and J. W. Trischka, J. Appl. Phys. **34**, 3635 (1963).

25. Y. T. Lee, J. D. McDonald, P. R. LeBreton, and D. Herschbach, Rev. Sci. Intstrum. **40**, 1402 (1969).

26. Y. T. Lee, in *Atomic and Molecular Beam Methods*, edited by G. Scoles (Oxford University Press, New York, 1988) Chap. 22, p. 563.

27. R. E. Center and A. Mandl, J. Chem. Phys. **57**, 4104 (1972).

28. C. H. Greene and R. N. Zare, J. Chem. Phys. **78**, 6741 (1983).

29. D. A. Case, G. M. McLelland, and D. R. Herschbach, Mol. Phys. **35**, 541 (1978).

30. G. Hall and P. Houston, Ann. Rev. Phys. Chem. **40**, 375 (1989).

31. D. M. Sonnenfroh and K. Liu, Chem. Phys. Lett. **176**, 183 (1991).

32. N. Billy, B. Girard, G. Gouédard, and J. Vigué, Mol. Phys. **61**, 65 (1987).

33. R. N. Zare and P. J. Dagdigian, Science **185**, 739 (1974).

34. R. Altkorn and R. N. Zare, Ann. Rev. Phys. Chem. **35**, 265 (1984).

35. H. Joswig, P. Andresen, and R. Schinke, J. Chem. Phys. **85**, 1904 (1986).

36. J. M. Mestdagh, J. P. Visticot, and A. G. Suits, in *The Chemical Dynamics and Kinetics of Small Radicals*, edited by K. Liu and W. Wagner (World Scientific, 1994).

37. S. D. Jons, J. E. Shirley, M. T. Vonk, C. F. Giese, and W. R. Gentry, J. Chem. Phys. **92**, 7831 (1992).

38. D. L. Feldman, R. K. Lengel, and R. N. Zare, Chem. Phys. Lett. **52**, 413 (1977).

39. E. E. Marinero, C. T. Rettner, and R. N. Zare, J. Chem. Phys. **80**, 4142 (1984).

40. L. Schneider, K. Seekamp-Rahn, F. Liedeker, H. Stewe, and K. H. Welge, Farad. Discuss. Chem. Soc. **91**, 259 (1991).

41. G. Comsa, R. David, and B. J. Schumacher, Rev. Sci. Instrum. **52**, 789 (1981).

42. E. Entemann and D. R. Herschbach, J. Chem. Phys. **55**, 4872 (1971).

43. R. Buss, Ph. D. thesis, University of California, Berkeley, 1972.

44. R. T Pack, J. Chem. Phys. **81**, 1841 (1984).

45. A. G. Suits, H. Hou, H. F. Davis, and Y. T. Lee, J. Chem. Phys. **95**, 8207 (1991).

46. M. Faubel, K.-H. Kohl, J. P. Toennies, K. T. Tang, and Y. Y. Yung, Faraday Discuss. Chem. Soc. **73**, 205 (1982).

47. R. E. Continetti, B. A. Balko, and Y. T. Lee, J. Chem. Phys. **93**, 5719 (1990).

48. J. L. Kinsey, J. Chem. Phys. **66**, 2560 (1977).

49. J. A. Serri, J. L. Kinsey, and D. E. Pritchard, J. Chem. Phys. **75**, 663 (1981).

50. E. L. Murphy, J. H. Brophy, G. S. Arnol, W. L. Dimpfl, and J. L. Kinsey, J. Chem. Phys. **70**, 5910 (1979).

51. J. M. Mestdagh, J. P. Visticot, P. Meynadier, O. Sublemontier, and A. G. Suits, J. Chem. Soc. Faraday Trans. **89**, 1413 (1993).

52. C. J. Smith, E. M. Spain, M. J. Dalberth, S. R. Leone, and J. P. J. Driessen, J. Chem. Soc. Faraday Trans. **89**, 1401 (1993).

53. H. A. J. Meijer, T. J. C. Pelgrim, H. G. M. Heideman, R. Morgenstern, and N. Andersen, J. Chem. Phys. **90**, 738 (1989).

54. D. W. Chandler and P. L. Houston, J. Chem. Phys. **87**, 1445 (1987).

55. L. S. Bontuyan, A. G. Suits, P. L. Houston, and B. J. Whitaker, J. Phys. Chem. **97**, 6342 (1993).

56. T. A. Kitsopoulos, D. P. Baldwin, M. A. Buntine, R. N. Zare, and D. W. Chandler, Science **260**, 1605 (1993).

57. R. N. Strickland and D. W. Chandler, Appl. Opt. **30**, 1811 (1990).

58. K. R. Castleman, *Digital Image Processing* (Prentice Hall, Englewood Cliffs, 1979).

59. T. T. Warnock and R. B. Bernstein, J. Chem. Phys. **49**, 1878 (1968).

60. R. K. B. Helbing, J. Chem. Phys. **48**, 472 (1968).

61. J. P. Toennies, *in Physical Chemistry, an Advanced Treatise*, edited by W. Jost, (Academic Press, New York, 1974), Vol. VIA, Chap. 5.

62. M. Faubel and J. P. Toennies, Adv. At. Mol. Phys. **13**, 229 (1978).

63. R. D. Levine and R. B. Bernstein, *Molecular Reaction Dynamics and Chemical Reactivity* (Oxford, New York, 1987) Chap. 3.

64. J. M. Farrar, T. P. Schafer, and Y. T. Lee, in *AIP Conference Proceedings No. 11, Transport Phenomena*, edited by J. Kestin (American Institute of Physics, New

York, 1973).

65. R. Schinke and J. Bowman, in *Molecular Collision Dynamics*, edited by J. Bowman (Springer-Verlag, Berlin, 1983).
66. S. Bosanac, Phys. Rev. A **22**, 2617 (1980).
67. G. Hall, K. Liu, M. J. McAuliffe, C. F. Giese, and W. R. Gentry, J. Chem. Phys. **81**, 5577 (1984).
68. X. Yang and A. Wodtke, Int. Rev. Phys. Chem. **12**, 123 (1993).
69. I. V. Hertel, Adv. Chem. Phys. **50**, 475 (1982).
70. W. H. Breckenridge and H. Umemoto, Adv. Chem. Phys. **50**, 325 (1982).
71. A. G. Suits, P. de Pujo, O. Sublemontier, J. P. Visticot, J. Berlande J. Cuvellier, T. Gustavsson, J. M. Mestdagh, P. Meynadier, and Y. T. Lee, J. Chem. Phys. **97**, 4094 (1992).

72. J. Los and A. W. Kleyn, in *Alkali Halide Vapors*, edited by P. Davidovits and D. L. McFadden (Academic Press, New York, 1979) Chap. 8.
73. E. A. Gislason, in *Alkali Halide Vapors*, edited by P. Davidovits and D. L. McFadden (Academic Press, New York, 1979) Chap. 13.
74. R. E. Olson, F. T. Smith, and E. Bauer, Appl. Optics **10**, 1848 (1971).
75. W. B. Miller, S. A. Safron, and D. R. Herschbach, Discuss. Faraday Soc. **44**, 108 (1967).
76. W. B. Miller, Ph.D. thesis, Harvard University, 1969.
77. J. M. Farrar and Y. T. Lee, J. Chem. Phys. **65**, 1414 (1976).
78. R. J. Buss, P. Casavecchia, T. Hirooka, and Y. T. Lee, Chem. Phys. Lett. **82**, 386 (1981).

65

Ion-Molecule Reactions

James M. Farrar

Department of Chemistry, University of Rochester, Rochester, New York 14627

65.1 INTRODUCTION

The observation of ion-molecule reactions has a history that goes back to the beginning of the twentieth century, when J. J. Thomson discovered that operating his positive ray parabola apparatus in a hydrogen atmosphere produced signals at a mass to charge ratio of 3, which he correctly attributed to the species H_3 [1]. Later studies showed that this species was produced by a reaction between the primary ionization product H_2^+ and molecular hydrogen. Most ion-molecule reactions proceed without an activation barrier and their cross sections are governed by the long range attractive potential of the approaching reactants (see Sect. 62.2.4). Reaction rates based on long range potential capture models [2] predict rates in excess of 10^{-9} cm^{3}molecule^{-1}s^{-1}, corresponding to thermal energy cross sections (see Sect. 45.2.7) of 10^{-16}–10^{-15} cm^2. The importance of ion-molecule reactions in such widely diverse areas as planetary atmospheres, (see Sect. 80.1), electrical discharges and plasmas (see Sect. 83.2.4), particularly in semiconductor processing, in the formation of molecules in interstellar space (see Chap. 78), and in flames and combustion systems

(see Sect. 84.1), has borne out that prediction. This chapter discusses applications of single-collision scattering methods to the study of reactive collision dynamics of ionic species with neutral partners.

A number of different physical processes can be categorized as ion-molecule reactions, with examples such as

$$A^{\pm} + BC \rightarrow A + BC^{\pm}, \qquad \text{charge transfer}$$
$$A^{\pm} + BC \rightarrow A + (B+C)^{\pm}, \quad \text{dissociative charge transfer}$$
$$A^{\pm} + BC \rightarrow AB^{\pm} + C. \qquad \text{particle transfer}$$

The $\pm$ superscript indicates charge appropriate to anions and cations. The parentheses indicate that the charge can reside on either the B or C fragment. Particle transfer reactions often involve the transfer of a hydrogen atom or a proton, but heavy particle transfer processes are often important ones as well. An interesting example in which new carbon-carbon bonds are formed, termed a condensation reaction, is the following:

$$C^+ + CH_4 \rightarrow C_2H_3^+ + H.$$

An additional process unique to anionic systems is detachment, occurring when the intermediate collision complex is internally excited above its autodetachment threshold:

$$A^- + BC \rightarrow [ABC^-]^* \rightarrow A + BC + e^-, \quad \text{detachment}$$
$$A^- + BC \rightarrow [ABC^-]^* \rightarrow ABC + e^-. \quad \text{associative detachment}$$

For exothermic reactions at low collision energies where the long range attraction dominates the interaction potential, cross sections are consistent with Langevin orbiting, generally having $E^{-1/2}$ energy dependence. At higher energies, cross sections drop below this limit, as surface crossings and short-range repulsive features become important. Endothermic reactions exhibit cross section thresholds, as illustrated in Sect. 65.4.6 on the $N^+ + D_2$ system.

Scattering measurements probe the potential energy surface, or surfaces, governing the collision dynamics with techniques that measure the fluxes from specific reactant quantum states into product quantum states, scattering angles, and product translational energies. Scattering experiments define more precisely than in a bulb the initial and final conditions in a collision [3]. A scattering experiment measures a cross section (see Sect. 87.2.1) rather than a rate constant. The measured cross section represents an average over initial conditions and a sum over final states inherent in the technique used.

Cross sections of various forms can be defined with respect to the rate of formation of product states under single-collision conditions. The total rate of product formation $dN_\pm/dt$ for a beam of ions with number density n_1 intersecting a gas of number density n_2 within a scattering volume ΔV, defined by the overlap of the ion beam with the target gas, is given by

$$dN_\pm/dt = \sigma v_{\text{rel}} n_1 n_2 \Delta V, \quad (65.1)$$

where σ is the total reaction cross section (see Sect. 45.2.7), and v_{rel} is the relative speed of the collision partners. Using the experimental ion beam current $I_\pm$, cross section σ, neutral number density n_2, and attenuation length L, this expression converts to the particularly useful form

$$dN_\pm/dt = 6.25 \times 10^{-7} \sigma I_\pm n_2 L \quad (65.2)$$

for computing signal levels, where σ is expressed in Å^2, $I_\pm$ in nA, n_2 in cm^{-3}, and L in cm. An ion beam of current 1 nA, intersecting a target of length 1 cm at a pressure of 10^{-3} torr, corresponding to a number density of 3.5×10^{13} cm^{-3} at STP, and reacting with a cross section of 1 Å^2 yields a total rate of product formation of 2×10^7 s^{-1}.

The detector observes only a fraction of this total rate. If reaction products are scattered isotropically over 4π steradians in the laboratory, and the detector entrance slit of area dS is located a distance r from the collision center, subtending a solid angle dS/r^2, the

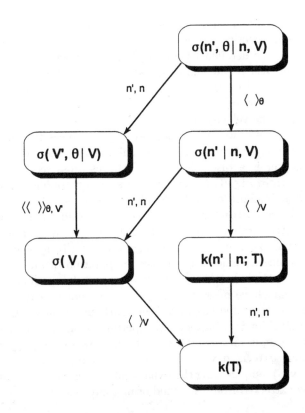

Figure 65.1. Relationships among differential and total cross sections, and rate constants. Brackets denote averages over indicated variables, or averages over initial states and summations over final states.

fraction of the total signal scattered into the slit is $dS/(4\pi r^2)$ [4]. The fraction of the collision volume, or of the product state distribution accessible to a particular experimental method, determines the tradeoffs between resolution and signal level, and therefore the feasibility of a given experiment.

In an "ideal" experiment, one collides reactants, with well specified quantum numbers collectively denoted $\mathbf{n}$, at a precisely defined relative velocity $\mathbf{v}_{\text{rel}}$, resolving products in quantum states $\mathbf{n}'$ scattered through center-of-mass scattering angle θ. The resulting detailed differential cross section (DCS) (see Sect. 45.2.1) is denoted by $\sigma(\mathbf{n}', \theta | \mathbf{n}, \mathbf{v}_{\text{rel}})$. However, most experiments involve at least partial averages over initial states and/or summations over final states. Figure 65.1 shows the result of averaging $\sigma(\mathbf{n}', \theta | \mathbf{n}, \mathbf{v}_{\text{rel}})$ over θ to yield the *state-to-state cross section* at fixed $\mathbf{v}_{\text{rel}}$, denoted by $\sigma(\mathbf{n}' | \mathbf{n}, \mathbf{v}_{\text{rel}})$.

Averaging this cross section over a Maxwell-Boltzmann distribution of molecular speeds at a specified temperature T yields the detailed state-to-state rate constant $k(\mathbf{n}'|\mathbf{n}, T)$, while summation over the final states $\mathbf{n}'$ and averaging over the initial states $\mathbf{n}$ yields the thermal rate constant $k(T)$. These latter two quantities are thermally averaged, multiple collision properties and not the subject of this chapter, although they play an important role in practical applications. Another pathway for averaging the detailed DCS $\sigma(\mathbf{n}',\theta|\mathbf{n},\mathbf{v}_{\text{rel}})$ arises from averaging over $\mathbf{n}$ and summing over $\mathbf{n}'$ to yield cross sections differential in product velocity $\mathbf{v}'_{\text{rel}}$ and θ at fixed $\mathbf{v}_{\text{rel}}$, denoted by $\sigma(\mathbf{v}'_{\text{rel}},\theta|\mathbf{v}_{\text{rel}})$. An average over θ and $\mathbf{v}'_{\text{rel}}$ yields the velocity dependent total cross section $\sigma(\mathbf{v}_{\text{rel}})$, and its Maxwell-Boltzmann average produces the thermal rate constant $k(T)$ once again. The subject of this chapter is a discussion of the various cross sections σ shown in Fig. 65.1. As one moves from more highly averaged quantities to the detailed cross sections, more sophisticated reactant preparation and product detection schemes are required to extract the desired information, at the expense of decreased signal levels. Technological advances, particularly in laser preparation of quantum state-selected reactants and in state-specific product detection, have made the "ideal" experiment a near reality in favorable circumstances, particularly in neutral-neutral interactions, e.g., in experiments on $H + H_2$ and its isotopic variants. This chapter discusses the cross sections $\sigma(\mathbf{n}',\theta|\ \mathbf{n},\mathbf{v}_{\text{rel}})$, $\sigma(\mathbf{v}'_{\text{rel}},\theta|\mathbf{v}_{\text{rel}})$, $\sigma(\mathbf{n}'|\ \mathbf{n},\mathbf{v}_{\text{rel}})$, and $\sigma(\mathbf{v}_{\text{rel}})$, emphasizing the dynamical information that can be extracted from each kind of measurement.

65.2 INSTRUMENTATION

Instrumentation for studying ion-molecule reactions is quite diverse, and numerous literature sources are available for further discussion [5]. A typical instrument has an ionization source, a primary mass selector, a collision region, and detector, consisting of a mass spectrometer or employing a spectroscopic technique allowing molecular identification.

65.3 KINEMATIC ANALYSIS

The transformation of laboratory measured speeds, angles and intensities to their center of mass (c.m.) counterparts can be accomplished with appropriate geometric constructions [6]. The geometric relationships can be understood by considering a kinematic Newton diagram for the collision process A + BC, as shown in Fig. 65.2. The diagram is constructed for the special case in which the reactant beams intersect at 90°. The laboratory scattering angle and velocity are Θ and v respectively, while the corresponding center of mass quantities are θ and u. The beam velocity vectors $\mathbf{v}_A$ and $\mathbf{v}_{BC}$ define the initial conditions, and the relative velocity vector $\mathbf{v}_{\text{rel}}$ is defined by their vector difference. The velocity $\mathbf{c}$ of the center of mass, or centroid, of the collision system is determined by conservation of linear momentum; the vector $\mathbf{c}$ divides the $\mathbf{v}_{\text{rel}}$ in inverse proportion to the masses of A and BC:

$$\mathbf{c} = \frac{m_A \mathbf{v}_A + m_{BC}\mathbf{v}_{BC}}{M}, \tag{65.3}$$

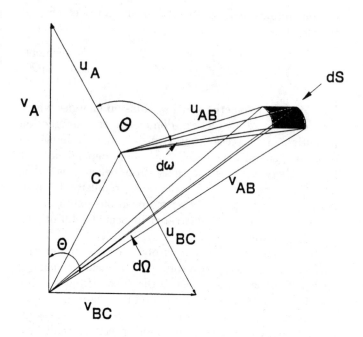

Figure 65.2. Kinematic Newton diagram showing laboratory and center of mass velocities and scattering angles.

$$\mathbf{u}_A = -\frac{m_{BC}}{M}\mathbf{v}_{\text{rel}}, \tag{65.4}$$

$$\mathbf{u}_{BC} = \frac{m_A}{M}\mathbf{v}_{\text{rel}}, \tag{65.5}$$

where M is the total mass of the reactants with masses m_A and m_{BC}. An observer moving away from the laboratory origin in the direction of the center of mass vector $\mathbf{c}$ would see the reactants A and BC approach along the direction of $\mathbf{v}_{\text{rel}}$, with the products retreating along the direction of $\mathbf{v}'_{\text{rel}}$. The angle θ between these vectors defines the center of mass scattering angle. For a single Newton diagram, detection of products at recoil speed u and scattering angle θ requires that measurements be made at laboratory coordinates (v,Θ). For monoenergetic incident beams, the measurement of both laboratory scattering angle and speed results in a unique lab to c.m. coordinate transformation. Linear momentum conservation also relates the center of mass speeds of products AB and C to the relative velocity of the separating products according to

$$\mathbf{u}'_{AB} = \frac{m_C}{M}\mathbf{v}'_{\text{rel}}, \tag{65.6}$$

$$\mathbf{u}'_C = -\frac{m_{AB}}{M}\mathbf{v}'_{\text{rel}}. \tag{65.7}$$

The final relative kinetic energy of the separating products is

$$\mathcal{T}'_{\text{rel}} = \tfrac{1}{2}\mu'\mathbf{v}'_{\text{rel}}\cdot\mathbf{v}'_{\text{rel}}, \tag{65.8}$$

where μ' is the reduced mass of the products.

The total energy of a collision system is computed from the energies of the incident reactants and the energy accessible to the products:

$$E_{\text{tot}} = \mathcal{T}_{\text{rel}} + E_{\text{int}} - \Delta D_0^o = \mathcal{T}'_{\text{rel}} + E'_{\text{int}} ; \qquad (65.9)$$

$\mathcal{T}_{\text{rel}}$ and E_{int} refer to the incident kinetic and internal excitation energies of the reactants and the primed quantities correspond to the products. ΔD_0^o is the zero point energy difference of reactants and products. Specification of the vibrational and rotational energy of a product determines the relative velocity $\mathbf{v}'_{\text{rel}}$ with which the products separate, according to

$$\mathbf{v}'_{\text{rel}} = \left[\left(\frac{2}{\mu'} \right) \left(\mathcal{T}_{\text{rel}} + E_{\text{int}} - \Delta D_0^o - E'_{\text{int}} \right) \right]^{1/2} . \quad (65.10)$$

The quantization of internal energy of the product AB leads to a series of concentric circles about the centroid that describe the loci of final translational speeds for AB produced in specific internal states. An example of the kinematic resolution of product vibrational states in the reaction $O^- + HF \rightarrow F^- + OH$ is shown in Fig. 65.3 [7]. At a collision energy of 0.47 eV (45.0 kJ mol^{-1}), the laboratory flux distribution shows structure that is attributable to the formation of OH in $v' = 0$, 1, and 2 states. The kinematic Newton diagram at the bottom of the figure shows the concentric circles corresponding to the formation of F^- in concert with OH in specific vibrational states having quantum numbers $v' = 0$, 1, and 2. For laboratory scattering angles in the range $0° \leq \Theta \leq 11°$ and $80° \leq \Theta \leq 90°$ energy scans only intersect the kinematic circles corresponding to OH in $v' = 0$ and 1 states, while data in the intermediate range of angles $18° \leq \Theta \leq 70°$ show contributions from all three states. The laboratory speeds at which these vibrational states appear, for a given Θ, are marked in the figure, showing clear correspondence with the structure in the experimental data. Rotational excitation in the products will broaden the contributions from individual vibrational states in experiments lacking rotational resolution. The best resolution achieved in electrostatic energy analyzers is approximately 5 meV (40 cm^{-1}); and consequently, scattering measurements of product fluxes based on kinetic energy analysis generally do not have rotational state resolution. However, photoelectron spectroscopy measurements on H_2 [8] have yielded spectra of H_2^+ energy levels with rotational resolution.

The transformation of intensities and cross sections between laboratory and c.m. coordinate systems relies on conservation of flux in a transformation between two coordinate systems moving with respect to one another. Figure 65.2 shows the nature of this conservation: in laboratory coordinates, the flux into solid angle $d\Omega$ is given by $I_{\text{lab}}(\Omega)d\Omega$, where the laboratory intensity I_{lab} is the flux per unit solid angle, while the corresponding flux

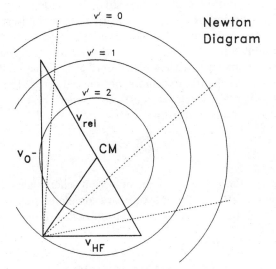

Figure 65.3. F^- reactive fluxes in the $O^- + HF$ system at a collision energy of 45 kJ mol^{-1}. Reprinted by permission from Ref. [7].

in the c.m. frame is $I_{\text{cm}}(\omega)d\omega$, where I_{cm} is the c.m. flux per unit solid angle. The solid angles subtended by the detector in the laboratory and c.m. frames expressed in velocity space may be computed from the surface element dS subtended by the detector:

$$d\omega = \frac{dS}{u^2}, \quad d\Omega = \frac{dS}{v^2} . \qquad (65.11)$$

The flux equality $I_{\text{lab}}(\Omega)d\Omega = I_{\text{cm}}(\omega)d\omega$ leads to the intensity transformation

$$I_{\text{lab}}(v, \Theta) = \frac{v^2}{u^2} I_{\text{cm}}(u, \theta) . \qquad (65.12)$$

The widths of the beam velocity distributions must be accounted for in extracting accurate c.m. cross sections from laboratory data. In the case where the laboratory flux at a given Θ is comprised of contributions from individual quantum states of the products, the laboratory flux is recovered from the relation

$$I_{\text{lab}}(v, \Theta) = v^2 \int_0^\infty dv_2 f_2(v_2) \int_0^\infty dv_1 f_1(v_1) \frac{\mathbf{v}_{\text{rel}}}{u^2}$$
$$\times \left[\sum_{n'} \sigma(\mathbf{n}', \theta | \mathbf{n}, \mathbf{v}_{\text{rel}}) \delta(u - u_{\mathbf{n}'}) \right] . \quad (65.13)$$

In this expression, the velocity distributions for the primary ion beam and the secondary neutral beam are given by $f_1(v_1)$ and $f_2(v_2)$ respectively. The final product internal states are labeled by the index $\mathbf{n}'$; the c.m. cross section for producing collective quantum state $\mathbf{n}'$ is given by $\sigma(\mathbf{n}', \theta | \mathbf{n}, \mathbf{v}_{\text{rel}})$ and the c.m. speed corresponding to the formation of this state is given by $u_{\mathbf{n}'}$. The solution of this equation for $\sigma(\mathbf{n}', \theta | \mathbf{n}, \mathbf{v}_{\text{rel}})$ can be accomplished both by forward convolution integration fitting procedures [9] and by iterative unfolding [10]. Forward convolution methods assume parametric forms for $\sigma(\mathbf{n}', \theta | \mathbf{n}, \mathbf{v}_{\text{rel}})$ that are substituted into Eq. (65.13). The parameters are then varied until the calculated fluxes agree with the data within experimental error. Iterative deconvolution methods generally extract the c.m. cross section summed over final states,

$$\sum_{\mathbf{n}'} \sigma(\mathbf{n}', \theta | \mathbf{n}, \mathbf{v}_{\text{rel}}), \quad (65.14)$$

but the finite energy of the detection scheme also determines whether quantum states are completely resolved or if the data represent a summation over product energy levels.

65.4 SCATTERING CROSS SECTIONS

Measurements of the single-collision cross sections shown schematically in Fig. 65.1 require a variety of sophisticated techniques. Each of these are discussed in the context of measurement techniques and information content, with illustrative examples.

65.4.1 State-to-State Differential Cross Sections

Information most diagnostic of the potential energy surface for chemical reaction comes from $\sigma(\mathbf{n}', \theta | \mathbf{n}, \mathbf{v}_{\text{rel}})$. The experimental data of Fig. 65.3 on the $O^- + HF$ system provide an example of a case in which $\mathbf{n}'$ refers to product OH vibrations in the ground electronic state, but without resolution of product rotations. Iterative deconvolution of the laboratory fluxes results in a flux distribution that is a sum of cross sections, as described in Eq. (65.14). Figure 65.4 shows this distribution in c.m. velocity space as a function of u and θ. In this representation, the relative velocity vector lies along the $0° - 180°$ line, and the symmetric peaks of the data near $0°$ (forward scattering) and $180°$ (backward scattering)

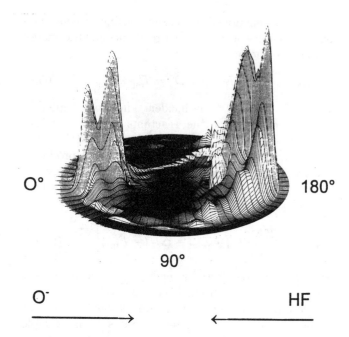

Figure 65.4. Axonometric plot of F^- fluxes in velocity space in the reaction $O^- + HF \rightarrow OH + F^-$ at 45 kJ mol^{-1}. Data from Ref. [7].

indicate that the reaction proceeds through a transient collision complex living for at least several rotational periods. The symmetry of the flux distribution with respect to $90°$ is the most important diagnostic for the participation of a long-lived transient complex. Whether the distribution is forward-backward peaked, as in the present example, or more isotropic reflects the geometry of the complex (i.e., oblate or prolate symmetric top) and the manner in which angular momentum is partitioned in the products [11]. In Fig. 65.4, the forward-backward scattering is indicative of the fact that the orbital angular momentum $\mathbf{L}$ of the approaching reactants is partitioned preferentially into orbital angular momentum of the products, $\mathbf{L}'$, through a near-linear $[O \cdots H \cdots F]^-$ intermediate. The various peaks in the flux distribution of Fig. 65.4 can be assigned to OH vibrations $v' = 0$, 1, and 2. Integration of the cross sections over the appropriate angular and velocity ranges yields the result $P(v' = 0) : P(v' = 1) : P(v' = 2) = 0.38 : 0.43 : 0.18$.

Crossed-beam experiments (see Sect. 58.3.2) with sufficient angular and kinetic energy resolution yield angular distributions for individual vibrational states. Similarly, laser-induced fluorescence experiments [12] yield rotational distributions. Recent work on the $Ar^+ + N_2 \rightarrow Ar + N_2^+(v' = 0, J')$ system illustrates this capability [13] (see also Sects. 62.1.1 and 37.4.1). However, these experiments probe N_2^+ in the collision volume, thereby averaging over all possible scattering angles, and are more properly examples of product-state resolved cross sections, discussed in Sect. 65.4.4. Experiments that provide complete product vibrational-rotational state specification and product scattering angles have yet to be carried out.

In the above examples the neutral beam is produced by supersonic expansion, producing reactants in their ground vibrational states, with rotational temperatures of only a few kelvins. Experiments with reactants prepared in excited states, either by laser absorption [14] or photoionization, and with full final state selection and angular resolution, remain a major goal of ion-molecule chemistry.

The production of reagent ions in selected vibrational and vibrational-rotational states by resonance enhanced multiphoton ionization (REMPI) (see Sect. 72.2.2) has been accomplished in a number of systems [14,15]. The use of photoelectron spectroscopy to assess the purity of the state-selection process is essential [16].

65.4.2 Velocity-Angle Differential Cross Sections

The velocity-angle DCS $\sigma(\mathbf{v}'_{rel}, \theta | \mathbf{v}_{rel})$ obtained by measuring kinetic energy distributions as a function of Θ can be plotted as a distribution in u and θ. Figure 65.5 shows $\sigma(\mathbf{v}'_{rel}, \theta | \mathbf{v}_{rel})$ for the reaction $C^+ + H_2O \rightarrow [COH]^+ + H$ at a collision energy of 2.14 eV (206.5 kJ mol^{-1}) [17]. The scattering is predominately backward, and the data indicate a reaction taking place in part through a transient complex, but principally through low impact parameter direct collisions leading to backward scattered products. Collisions leading to such backward scattered products are called rebound collisions and are dominated by the repulsive part of the potential surface. Although the reaction of C^+ with H_2O is exothermic, high kinetic energy release in the rebound component suggests that a potential energy barrier is in the exit channel, i.e., it acts as the products separate.

At lower collision energies, the forward peak in the flux distribution becomes more pronounced for the $C^+ + H_2O$ reaction, suggesting that the reaction is mediated by a collision complex living for a fraction of a rotational period. Under these conditions, it is possible to extract collision complex lifetimes from the angular distribution asymmetry through the osculating model for chemical reactions [18]. In this model, the lifetime is a parametric function of the complex's rotational period, evidenced by the forward-backward asymmetry

$$I(180°)/I(0°) = 1/\cosh(\tau_R/2\tau). \quad (65.15)$$

In this expression, τ is the lifetime of the complex and τ_R is the rotational period, estimated from the moment of inertia I and the maximum orbital angular momentum L_{max} of the complex. The total reaction cross section σ can be used to estimate this latter quantity from

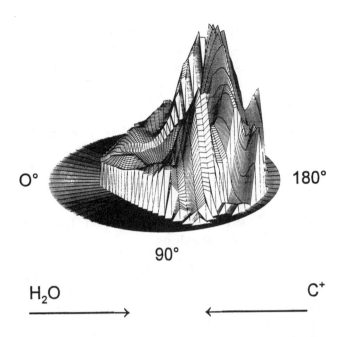

Figure 65.5. Axonometric plot of $[COH]^+$ fluxes in velocity space in the reaction $C^+ + H_2O \rightarrow [COH]^+ + H$ at 2.14 eV (206.5 kJ mol^{-1}). Data from Ref. [17]

$$\sigma = \left[\frac{\pi\hbar^2}{2\mu\mathcal{T}_{rel}}\right](L_{max} + 1)^2. \quad (65.16)$$

In the present case, the lifetime estimate for the $[CHOH]^+$ complex is less than 1×10^{-13} s (see Chap.34).

At high energies, rebound collisions dominate the dynamics of the $C^+ + H_2O$ reaction. Another limit often encountered in high energy ion-molecule reactions is the stripping process in which the incoming ionic projectile removes a particle from its molecular collision partner, and the new ionic product travels in the same direction as the incident ionic reactant. The spectator stripping limit [19] for a reaction of the type $A + BC \rightarrow AB + C$ corresponds to the case in which the c.m. speed of product C is unchanged from the c.m. speed of reactant BC, i.e., C "spectates" as the reaction occurs.

65.4.3 Total Cross Sections with State-Selected Reactants

State-selected cross sections $\sigma(\mathbf{n}'|\mathbf{n}, \mathbf{v}_{rel})$ were first measured by Chupka and collaborators [20] in the endothermic reaction of $H_2^+ + He \rightarrow HeH^+ + H$, with H_2^+ prepared in states $v \geq 2$ by photoionization, and HeH^+ detected without product state analysis. Thus, $\mathbf{n}'$ represents a summation over all accessible product states in state-selected reactant experiments. At a given vuv ionization wavelength, H_2^+ is prepared, subject to Franck-Condon factor limitations (see Sect. 32.6.1), in all possible vibrational states up to the maximum allowed by the photon energy. Therefore, extraction of cross sections for individual vibrational states requires vuv wavelength dependent studies in which the number of reactant vibrational states is increased smoothly from the ground state up to the limit allowed by instrumental resolution or photon sources. A number of reactive and charge transfer systems have been studied with photoionization techniques at the reactant state-selected level with unresolved product states. Data are typically in the form of $P(v')$ vs. v', as a function of collision energy. A comprehensive review by Ng is available [21].

65.4.4 Product-State Resolved Total Cross Sections

The resolution of product states in cross section measurements can be accomplished in a variety of ways. We will first discuss the case in which the reactants have not been state-selected. In principle, spectroscopic methods such as laser-induced fluorescence can be used for complete product quantum state specification. Earlier reference to N_2^+ vibration-rotation state-resolved charge transfer experiments [13] provides an excellent example of the state of the art of such methods, and their extension to chemical reactions represents an important future development. A beautiful example is mass spectrometric detection of N_2^+ in the $Ar^+ + N_2$ charge transfer system [22]. Guided ion-beam production and detection methods [23] in conjunction with a supersonic beam of N_2 allow 4π detection of the charge transfer products as a function of collision energy, yielding accurate total cross sections. The results, plotted in Fig. 65.6, show thresholds for the production of excited vibrational states of N_2^+ ($X\,^2\Sigma_g, v' = 1, 2,$ and 4) as well as the formation of the excited $B\,^2\Sigma_g$ state. The detection of chemiluminescence from electronically excited states of ion-molecule reactions (see Sect. 10.21.2) is also a powerful method for determining such cross sections; methods and results are reviewed in Ref. [24].

65.4.5 State-to-State Total Cross Sections

The most detailed probe of an ion-molecule reaction is the state-to-state cross section. The measurement of angularly-resolved state-to-state cross sections has only been achieved for reactants in their ground states, but true state-to-state total cross sections have now been measured in a few favorable cases. A particularly novel method for product-state determination with reactant-state selection has been developed [21]. In the *differential reactivity method*, product vibrational states are distinguished by their differing cross sections for charge transfer with selected molecules. Although the method is

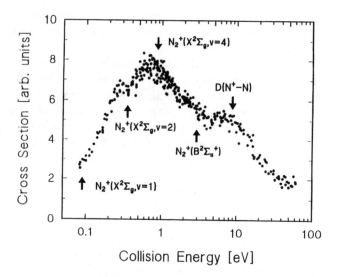

Figure 65.6. Energy dependent cross section for production of N_2^+ in states as indicated in the $Ar^+ + N_2$ charge transfer system. Reprinted by permission from Ref. [22].

limited at present because state-selected charge transfer cross sections have been measured in only a few systems, the technique has great potential. The charge transfer reaction $H_2^+(v) + H_2 \rightarrow H_2 + H_2^+(v')$, in which H_2^+ is prepared in $v = 0$ and 1 by vuv photoionization, provides a particularly good example of its power. The H_2^+ charge transfer products, formed in states $v' = 0 - 3$, are first mass-analyzed, accelerated to 10 eV, and then passed through a collision cell containing N_2, Ar, or CO, where charge transfer occurs once again. The N_2^+, Ar^+, or CO^+ products are mass analyzed and the cross sections σ_m for forming these ions by charge transfer from the H_2^+ reaction products are measured. The charge transfer cross sections for H_2^+ with these gases have different dependences on v', as shown in Fig. 65.7, and therefore, the gases can be used to probe the product states of H_2^+ formed in the symmetric charge exchange reaction. Letting $X_{v'}$ denote the fraction of H_2^+ formed in the vibrational state v', n denote the number density of the neutral collision gas, and l denote the attenuation length, the following set of simultaneous equations can be solved for $X_{v'}$, since the cross sections $\sigma_{v'}$ are known and the σ_m are measured:

$$X_0 + X_1 + X_2 + X_3 = 1 \qquad (65.17)$$

and, for a particular ion (e.g., N_2^+, Ar^+, CO^+),

$$X_0 n l \sigma_0 + X_1 n l \sigma_1 + X_2 n l \sigma_2 + X_3 n l \sigma_3 = n l \sigma_m. \qquad (65.18)$$

(An equation like (65.18) will exist for each collision gas.) An extensive set of data at collision energies from 2 eV up to 16 eV has been obtained [25]. When

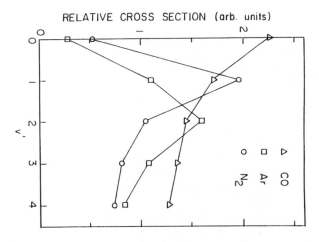

RELATIVE CROSS SECTION (arb. units)

Figure 65.7. Cross sections for charge transfer of $H_2^+(v')$ vs. v' with N_2, Ar, and CO at a collision energy of 10 eV. Reprinted from Ref. [21] by permission of John Wiley & Sons, Inc.

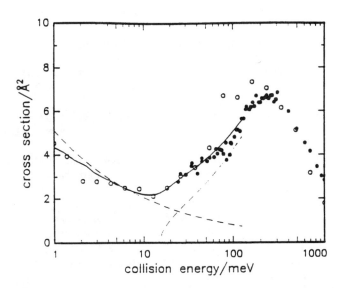

Figure 65.8. Total cross section for ND^+ formation in the $N^+ + D_2$ system, showing a low energy threshold at 15 meV. Reprinted by permission from Ref. [26].

the reactant ions are in the vibrational ground state, at low kinetic energies the charge transfer product is also in the ground state, indicating that resonant charge transfer is the dominant process. At increasing collision energies, X_1 increases from 0.0 to 0.17. For vibrationally excited H_2^+ reactants, inelastic relaxation to form H_2^+ in $v' = 0$ is important at all collision energies, increasing in magnitude with increasing collision energy. Of particular interest is the fact that inelastic relaxation forming $v' = 0$ is substantially more important than inelastic excitation producing $v' = 1$ or 2. This trend is predicted by theory, but underestimated at lower collision energies.

65.4.6 Energy Dependent Total Cross Sections

Although non-state-selected cross sections, denoted $\sigma(v)$, lack information about product energy disposal, important features of the potential energy surface can be obtained from their measurement. Guided ion beam methods have been especially important in the determination of accurate cross sections, particularly those that employ a supersonic beam rather than a gas as a neutral target. Energy-dependent cross sections provide crucial information on ion and neutral thermochemistry through accurate measurements of reaction thresholds, and help to elucidate important potential surface features such as thresholds, barriers, and crossings [27]. A particularly illustrative example of threshold formation concerns the reaction $N^+ (^3P) + H_2 \rightarrow NH^+ + H$ and its isotopic variants. This reaction is important at low (< 0.1 eV) collision energies as the first step in the chain reaction that leads to formation of ammonia in dense interstel-

lar clouds, but the bond energy of NH^+ is uncertain enough to prevent knowing whether the above reaction is endothermic or exothermic. Recent total cross section measurements on the isotopic variant $N^+ (^3P) + D_2 \rightarrow ND^+ + D$ at very low collision energies now appear to answer this question [26]. Figure 65.8 shows experimental results performed on two different instruments, one a guided ion beam-crossed neutral beam machine and the other a guided merged-beam apparatus. The data show a very clear threshold at 15 meV, demonstrating that the reaction is endothermic and allowing for a more accurate estimate of the ND^+ bond energy. This experiment shows that the high energy resolution afforded in total cross section measurements employing crossed and merged beams, rather than thermal collision cells, can answer thermochemical questions of importance in a wide variety of applications, including astrophysics, combustion, electrical discharges, and atmospheric processes.

65.5 CONCLUDING REMARKS

Progress in the study of ion-molecule dynamics is strongly dependent on technology. As ion preparation methods continue to develop, yielding higher fluxes of state-selected ions, and as detection methods become ever more sensitive, progress toward the "ideal" experiment will continue, and the applicablity of these methods to additional systems of importance will continue as well.

REFERENCES

1. J. J. Thomson, *Rays of Positive Electricity* (Longmans Green, New York, 1913).
2. T. Su and M. T. Bowers, *Gas Phase Ion Chemistry*, edited by M. T. Bowers (Academic Press, New York, 1979), Vol. 1, p. 84.
3. R. D. Levine and R. B. Bernstein, *Molecular Reaction Dynamics and Chemical Reactivity* (Oxford, New York, 1987).
4. Y. T. Lee, in *Atomic and Molecular Beam Methods*, edited by G. Scoles and U. Buck (Oxford, New York, 1988), p. 553.
5. See, for example, *Techniques for the Study of Ion-Molecule Reactions*, edited by J. M. Farrar and W. H. Saunders, Jr. (Wiley-Interscience, New York, 1988).
6. T. T. Warnock and R. B. Bernstein, J. Chem. Phys. **49**, 1878 (1968); G. L. Catchen, J. Husain, and R. N. Zare, *ibid.* **69**, 1737 (1978).
7. D. J. Levandier, D. F. Varley, M. A. Carpenter, and J. M. Farrar, J. Chem. Phys. **99**, 148 (1993); M. A. Carpenter, M. T. Zanni, D. J. Levandier, D. F. Varley, and J. M. Farrar, Can. J. Chem. **72**, 828 (1994).
8. J. E. Pollard, D. J. Trevor, J. E. Reutt, Y. T. Lee, and D. A. Shirley, J. Chem. Phys. **77**, 34 (1982).
9. P. E. Siska, Ph. D. thesis, Harvard University, 1969. Available from University Microfilms, Ann Arbor, MI.
10. P. E. Siska, J. Chem. Phys. **59**, 6052 (1973).
11. S. A. Safron, W. B. Miller, and D. R. Herschbach, Discuss. Faraday Soc. **44**, 108 (1967).
12. J. L. Kinsey, Ann. Rev. Phys. Chem. **28**, 349 (1977).
13. L. Hüwel, D. R. Guyer, G.-H. Lin, and S. R. Leone, J. Chem. Phys. **81**, 3520 (1984).
14. S. L. Anderson, Adv. Chem. Phys. **82**, 177 (1992).
15. L. A. Posey, R. D. Guettler, N. J. Kirchner, and R. N. Zare, J. Chem. Phys. **101**, 3772 (1994).
16. R. N. Compton and J. C. Miller, in *Laser Applications in Physical Chemistry*, edited by D. K. Evans (Marcel Dekker, New York, 1989), p. 221.
17. D. M. Sonnenfroh, R. A. Curtis, and J. M. Farrar, J. Chem. Phys. **83**, 3958 (1985).
18. M. K. Bullitt, C. H. Fisher, and J. L. Kinsey, J. Chem. Phys. **64**, 1914 (1974).
19. A. Henglein, K. Lacmann, and G. Jacobs, Ber. Bunsenges. Phys. Chem. **69**, 279 (1965).
20. W. A. Chupka and M. E. Russell, J. Chem. Phys. **48**, 1527 (1968); *ibid.* **49**, 5426 (1968).
21. C. Y. Ng, Adv. Chem. Phys. **82**, 401 (1992).
22. P. Tosi, O. Dmitrijev, and D. Bassi, Chem. Phys. Lett. **200**, 483 (1992).
23. D. Gerlich, Adv. Chem. Phys. **82**, 1 (1992).
24. M. Tsuji, in *Techniques for the Study of Ion-Molecule Reactions*, edited by J. M. Farrar and W. H. Saunders, Jr. (Wiley-Interscience, New York, 1988), p. 489.
25. C.-L. Liao and C. Y. Ng, J. Chem. Phys. **84**, 197 (1986).
26. P. Tosi, O. Dmitriev, D. Bassi, O. Wick, and D. Gerlich, J. Chem. Phys. **100**, 4300 (1994).
27. P. B. Armentrout, Ann. Rev. Phys. Chem. **41**, 313 (1990).

66

Light-Matter Interaction

Pierre Meystre

Optical Sciences Center, University of Arizona, Tucson, Arizona 85721

66.1 INTRODUCTION

Optical physics is concerned with the dynamical interactions of atoms and molecules with electromagnetic fields. Semiclassical theories, which study the interac-tion of atoms with classical fields, are often said to comprise *optical physics*, while *quantum optics* treats the interaction of atoms or molecules with quantized electromagnetic fields. A significant part of optical physics and quantum optics is the study of near-resonant atom-field interactions, and concentrates on nonperturbative dynamics, where the effects of the optical fields have to be kept to all orders. The atomic properties themselves are assumed to be known.

The vast majority of problems in light-matter interactions can be treated quite accurately within semiclassical theories. However, an important class of problems where this is not the case are presented in Chap.75. While most of optical physics and quantum optics ignores the effects of the electromagnetic fields on the center-of-mass motion of the atoms, important topics such as atomic trapping and cooling (Chap. 73) and de Broglie optics (Chap. 74) rely in an essential way on such *mechanical effects* of light. The present chapter deals with more "traditional" aspects of optical physics, where these effects are ignored.

66.2 MULTIPOLE EXPANSION

Consider a test charge q of mass m localized within an atom and acted upon by an external electromagnetic field with electric field $\mathbf{E}(\mathbf{r}, t)$ and magnetic field $\mathbf{B}(\mathbf{r}, t)$. In the *multipole expansion* formalism [1] [1, 2], the electric and magnetic interaction energies between the charge and the electromagnetic field are

$$V_e = V_{E0}(t) - q \int_0^{\mathbf{r}} d\mathbf{s}\cdot\mathbf{E}(\mathbf{R} + \mathbf{s}, t)\,, \qquad (66.1)$$

$$V_m = -q \int_0^{\mathbf{r}} d\mathbf{s}\cdot\mathbf{v} \times \mathbf{B}(\mathbf{R} + \mathbf{s}, t)\,, \qquad (66.2)$$

respectively. These energies correspond to the work done by the electric and magnetic components of the Lorentz force in first moving the charge to a stationary origin of coordinates at a point $\mathbf{R}$ and then to a location $\mathbf{r}$ relative to $\mathbf{R}$. Here, $V_{E0}(t)$ represents the energy of the charge when located at the reference point $\mathbf{R}$. It may be expressed in terms of the electrostatic potential $\phi(\mathbf{R}, t)$ as $V_{E0}(t) = +q\phi(\mathbf{R}, t)$.

A Taylor series expansion of $V_e(t)$ and $V_m(t)$ about $\mathbf{r} = 0$ yields

$$V_e(t) = V_{E0}(t) - q \sum_{n=1}^{\infty} \frac{1}{n!} \left(\mathbf{r}\cdot\frac{\partial}{\partial\mathbf{R}} \right)^{n-1} \mathbf{r}\cdot\mathbf{E}(\mathbf{R}, t)\,,$$

$$(66.3)$$

$$V_m(t) = -\frac{q\hbar}{m} \sum_{n=1}^{\infty} \frac{1}{(n+1)!} \left(\mathbf{r}\cdot\frac{\partial}{\partial\mathbf{R}} \right)^{n-1} \boldsymbol{\ell}\cdot\mathbf{B}(\mathbf{R}, t)\,,$$

$$(66.4)$$

where $\hbar\boldsymbol{\ell} = \mathbf{r} \times \mathbf{p}$ is the angular momentum of the test charge relative to the coordinate origin $\mathbf{R}$. Here, use of the mechanical momentum $\mathbf{p} = m\dot{\mathbf{r}}$, instead of

the canonical momentum, neglects the electromagnetic component of the momentum responsible for diamagnetic effects.

In addition to the electromagnetic interaction, electrons and nuclei are characterized by a spin magnetic moment $\mathbf{m}_s = (q\hbar/2m)g_s\mathbf{s}$, where $\mathbf{s}$ is the spin of the test charge and g_s its gyromagnetic factor, equal to $2.002\ldots$ for electrons. The factor $q\hbar/2m$ is the particle's magneton. The spin magnetic moment yields an additional term to the magnetic energy V_m, which becomes

$$V_m(t) = -\frac{e\hbar}{2m} \sum_{n=1}^{\infty} \frac{1}{n!} \left(\mathbf{r}\cdot\frac{\partial}{\partial\mathbf{R}} \right)^{n-1}$$

$$\times \left(\frac{2}{n+1} g_\ell \boldsymbol{\ell} + g_s \mathbf{s} \right) \cdot \mathbf{B}(\mathbf{R}, t)\,, \qquad (66.5)$$

where the orbital g-factor is $g_\ell = q/e$ ($g_\ell = -1$ for an electron). For an ensemble $\{\alpha\}$ of charged particles q_α in an atom, these expressions are to be summed over all particles. Thus, the electric energy becomes

$$V_e(t) = \sum_{\alpha} q_\alpha \phi(\mathbf{R}, t) - \sum_{i=1}^{3} \sum_{\alpha} q_\alpha r_i(\alpha) E_i(\mathbf{R}, t)$$

$$-\frac{1}{2} \sum_{i,j=1}^{3} \left[\sum_{\alpha} q_\alpha r_i(\alpha) r_j(\alpha) \right] \frac{\partial}{\partial R_j} E_i(\mathbf{R}, t) + \cdots$$

$$\equiv V_{E0}(t) + V_{E1}(t) + V_{E2}(t) + \cdots\,, \qquad (66.6)$$

and the magnetic energy becomes

$$V_m(t) = \sum_{i=1}^{3} B_i(\mathbf{R}, t) \sum_{\alpha} \frac{e\hbar}{2m_\alpha} [g_\ell(\alpha)\ell_i(\alpha) + g_s(\alpha)s_i(\alpha)]$$

$$-\sum_{i,j=1}^{3} \frac{\partial B_i(\mathbf{R}, t)}{\partial R_j} \sum_{\alpha} \frac{e\hbar}{2m_\alpha} \left[\tfrac{2}{3} g_\ell(\alpha)\ell_i(\alpha) r_j(\alpha) \right.$$

$$\left. + g_s(\alpha)s_i(\alpha) \right] + \cdots$$

$$\equiv V_{M1}(t) + V_{M2}(t) + \cdots\,. \qquad (66.7)$$

66.2.1 Electric Dipole (E1) Interaction

For optical fields whose wavelength is large compared with the interacting atom, only the first few terms in the Taylor expansions of $V_e(t)$ and $V_m(t)$ need be retained. The first term, $V_{E0}(t)$, of $V_e(t)$ is the net charge of the atom, which vanishes for neutral atoms. The second term, $V_{E1}(t)$, is the *electric dipole interaction* energy. Introducing the *electric dipole* (E1) *moment*,

$$\mathbf{d} = \sum_{\alpha} q_\alpha \mathbf{r}(\alpha)\,, \qquad (66.8)$$

or

$$\mathbf{d} = \int d^3 r \rho(\mathbf{r})\mathbf{r}\,, \qquad (66.9)$$

---

[1]The material of this chapter is discussed in detail in a number of texts and review articles. We cite such references rather than the original sources whenever possible.

for a charge distribution, this contribution to the inter-action energy may be re-expressed as

$$V_{E1}(t) = -\mathbf{d}\cdot\mathbf{E}(\mathbf{R}, t)\,. \tag{66.10}$$

The E1 term dominates most optical phenomena.

66.2.2 Electric Quadrupole (E2) Interaction

The $V_{E2}(t)$ contribution to $v_e(t)$, describes *electric quadrupole* (E2) interactions. In terms of the quadrupole tensor

$$Q = 3\int d^3 r \rho(\mathbf{r}) r_i r_j\,, \tag{66.11}$$

$V_{E2}(t)$ becomes

$$V_{E2}(t) = -\frac{1}{6}\sum_{i,j=1}^{3} Q_{ij}\frac{\partial}{\partial R_i} E_j(\mathbf{R}, t)\,. \tag{66.12}$$

Alternatively, E2 interactions can be expressed in terms of the traceless quadrupole tensor $Q_{ij}^{(2)} = \int d^3 r \rho(\mathbf{r}) \times [3r_i r_j - \delta_{ij} r^2]$. E2 interactions are typically weaker than E1 interactions by a factor a_0/λ, where a_0 is the Bohr radius and λ is the wavelength of the transition. Since a_0/λ is very small for optical transitions, E2 interactions are typically neglected in quantum optics.

66.2.3 Magnetic Dipole (M1) Interaction

The first term in the multipole expansion the magnetic interaction (66.7) is the *magnetic dipole M1 interaction*,

$$V_{M1} = -\mathbf{m}\cdot\mathbf{B}(\mathbf{R}, t)\,, \tag{66.13}$$

of a magnetic moment $\mathbf{m}$ in a magnetic field, where

$$\mathbf{m} = \sum_{\alpha}\left(\frac{q_\alpha \hbar}{2m_\alpha}\right)[g_\ell(\alpha)\boldsymbol{\ell}(\alpha) + g_s(\alpha)\mathbf{s}(\alpha)]$$
$$= -\mu_B[\mathbf{L} + 2\mathbf{S}]\,, \tag{66.14}$$

and we have used the fact that for electrons $g_\ell = -1$ and $g_s \simeq -2$. The *Bohr magneton* μ_B is

$$\mu_B = \frac{e\hbar}{2mc} = \frac{\alpha e a_0}{2}\,, \tag{66.15}$$

where α is the fine structure constant. Thus M1 interactions tend to be smaller than E1 interactions by a factor of order $\alpha/2$. The connection between $\mathbf{m}$ and angular momentum $\mathbf{J}$ is $\mathbf{m} = \gamma\mathbf{J}$, where γ is the gyromagnetic ratio.

66.3 LORENTZ ATOM

The Lorentz atom consists of a classical electron harmonically bound to a proton. It provides a framework for understanding a number of elementary aspects of the electric dipole interaction between a single atom and light [3–7]. Assuming that the c.m. motion of the atom is unaffected by the field, and neglecting magnetic effects, the equation of motion of the electron is

$$\left(\frac{d^2}{dt^2} + \frac{\Gamma_0}{2}\frac{d}{dt} + \omega_0^2\right)\mathbf{r} = -\frac{e}{m}\mathbf{E}(\mathbf{R}, t)\,, \tag{66.16}$$

where ω_0 is the electron's natural oscillation frequency, and Γ_0 represents a frictional decay rate that accounts for the effects of radiative damping. For the classical Lorentz atom

$$\Gamma_0 = 2\omega_0^2 r_0/3c\,, \tag{66.17}$$

where $r_0 = e^2/4\pi\epsilon_0 mc^2$ is the classical electron radius. This damping arises physically from the radiation reaction of the field radiated by the atom on itself. In the E1 approximation, the electric field is evaluated at the location $\mathbf{R}$ of the atomic c.m.

66.3.1 Complex Notation

The study of light-matter interactions is simplified by the introduction of complex variables [8, 9]. For example, an electric field

$$\mathbf{E}(\mathbf{R}, t) = \sum_{n,\mu}\boldsymbol{\epsilon}_\mu \mathcal{E}_n \cos(\omega_n t)\,, \tag{66.18}$$

where $\boldsymbol{\epsilon}_\mu$ is the polarization vector of the field Fourier component at frequency ω_n is expressed as

$$\mathbf{E}(\mathbf{R}, t) = \mathbf{E}^+(\mathbf{R}, t) + \mathbf{E}^-(\mathbf{R}, t)\,, \tag{66.19}$$

where the *positive frequency part* of the field is

$$\mathbf{E}^+(\mathbf{R}, t) = \frac{1}{2}\sum_{n,\mu}\boldsymbol{\epsilon}_\mu \mathcal{E}_n \exp[i(\mathbf{k}_n\cdot\mathbf{R} - \omega_n t)]\,. \tag{66.20}$$

Due to the linearity of Eq. (66.16), it is sufficient to study the response of the Lorentz atom to a plane monochromatic electric field of frequency ω, complex amplitude $\mathcal{E}$ and polarization $\boldsymbol{\epsilon}$. Introducing the *complex dipole moment*

$$\mathbf{d} = -e\mathbf{r} = \boldsymbol{\epsilon}\mathcal{P}\exp[i(\mathbf{k}\cdot\mathbf{R} - \omega t)] + \text{c.c.}\,, \tag{66.21}$$

where $\mathcal{P}$ is in general complex for $\mathcal{E}$ real, and the complex polarizability $\alpha(\omega)$ via $\mathcal{P} = \alpha(\omega)\mathcal{E}$, then

$$\alpha(\omega) = \frac{e^2/m}{\omega_0^2 - \omega^2 - i\gamma\omega}\,. \tag{66.22}$$

66.3.2 Index of Refraction

From the Maxwell wave equation

$$\left[\nabla^2 - \frac{1}{c^2}\frac{\partial^2}{\partial t^2}\right]\mathbf{E}(\mathbf{R},t) = \frac{1}{\epsilon_0 c^2}\frac{\partial^2 \mathbf{P}(\mathbf{R},t)}{\partial t^2}, \qquad (66.23)$$

where $\mathbf{P}(\mathbf{R},t)$ is the electric polarization, given by the electric dipole density of the medium $\mathbf{P} = N\mathbf{d}$, N being the atomic density, the plane wave dispersion relation is

$$k^2 = \frac{\omega^2}{c^2}n^2(\omega), \qquad (66.24)$$

where the *index of refraction* $n(\omega)$ is

$$n(\omega) = \sqrt{1 + \frac{N\alpha(\omega)}{\epsilon_0}}. \qquad (66.25)$$

66.3.3 Beer's Law

Since the polarizability $\alpha(\omega)$ is normally complex, so is the index of refraction. Its real part leads to dispersive effects, while its imaginary part leads to absorption. Specifically, $\Re[n(\omega)] - 1$ has the form of a standard dispersion curve, positive for $\omega - \omega_0 < 0$ and negative for $\omega - \omega_0 > 0$, while $\Im[n(\omega)]$ is a Lorentzian curve peaked at $\omega = \omega_0$. The intensity absorption coefficient $a(\omega)$ is

$$\begin{aligned} a(\omega) &= 2\Im\left[n(\omega)\right]\omega/c \\ &= \frac{2\omega}{c}\Im\left[1 + \left(\frac{Ne^2}{m\epsilon_0}\right)\frac{i\gamma\omega + (\omega_0^2 - \omega^2)}{(\omega_0^2 - \omega^2)^2 + \gamma^2\omega^2}\right]^{1/2}. \end{aligned}$$

$$(66.26)$$

For atomic vapors, the corrections to the vacuum index of refraction are normally small, so that the square root in Eq. (66.26) can be expanded to first order, giving

$$a(\omega) = \left(\frac{Ne^2}{\epsilon_0 mc}\right)\frac{\gamma\omega^2}{(\omega_0^2 - \omega^2)^2 + \gamma^2\omega^2}. \qquad (66.27)$$

The intensity of a monochromatic field propagating along the z-direction through a gas of Lorentz atoms is therefore attenuated according to *Beer's law* given by

$$I(\omega,z) = I(\omega,0)e^{-a(\omega)z}. \qquad (66.28)$$

If the index of refraction at a given frequency becomes purely imaginary, no electromagnetic wave can propagate inside the medium. This is the case for field frequencies smaller than the *plasma frequency*

$$\omega_{\rm p} = \sqrt{\frac{Ne^2}{m\epsilon_0}}. \qquad (66.29)$$

While the Lorentz atom model gives an adequate description of absorption and dispersion in a weakly excited absorbing medium, it fails to predict the occurence of important phenomena such as saturation and light amplification. This is because, in this model, the phase of the induced atomic dipoles with respect to the incident field is always such that the polarization field adds destructively to the incident field. The description of light amplification requires a quantum treatment of the medium, which gives a greater flexibility to the possible relative phases between the incident and polarization fields.

66.3.4 Slowly-Varying Envelope Approximation

Light-matter interactions often involve *quasi-monochromatic fields* for which the electric field (taken to propagate along the z-axis) can be expressed in the form

$$\mathbf{E}(\mathbf{R},t) = \frac{1}{2}\boldsymbol{\epsilon}\mathcal{E}^+(\mathbf{R},t)e^{i(kz-\omega t)} + {\rm c.c.}, \qquad (66.30)$$

such that

$$\left|\frac{\partial \mathcal{E}^+}{\partial t}\right| \ll \omega|\mathcal{E}^+|, \quad \left|\frac{\partial \mathcal{E}^+}{\partial z}\right| \ll k|\mathcal{E}^+|. \qquad (66.31)$$

It is further consistent within this approximation to assume that the polarization takes the form

$$\mathbf{P}(\mathbf{R},t) = \boldsymbol{\epsilon}\mathcal{P}^+(\mathbf{R},t)e^{i(kz-\omega t)} + {\rm c.c.}, \qquad (66.32)$$

with

$$\left|\frac{\partial \mathcal{P}^+}{\partial t}\right| \ll \omega|\mathcal{P}^+|. \qquad (66.33)$$

Under these conditions, known as the *slowly-varying envelope approximation* (see, e.g., Ref. [5]), Maxwell's wave equation reduces to

$$\left(\frac{\partial}{\partial z} + \frac{1}{c}\frac{\partial}{\partial t}\right)\mathcal{E}^+(z,t) = -\frac{k}{2\epsilon_0}\mathcal{P}^+(z,t). \qquad (66.34)$$

Hence, in the slowly-varying envelopes approximation we ignore the backward propagation of the field [10]. The *slowly-varying amplitude and phase approximation* is essentially the same, except that it expresses the electric field envelope in terms of a real amplitude and phase.

66.4 TWO-LEVEL ATOMS

A large number of optical phenomena can be understood by considering the interaction between a quasi-monochromatic field of central frequency ω and a *two-level atom*, which simulates a (dipole-allowed) atomic transition [5, 7, 11–13]. This approximation is well justified for near-resonant interactions; i.e., $\omega \simeq \omega_0$. The next three sections discuss the model Hamiltonian for this system in the *semiclassical approximation* where the electromagnetic field can be described classically. The formal results are then extended to the case of a quantized field, where the electric field is treated as an operator.

66.4.1 Hamiltonian

In the absence of dissipation mechanisms, the dipole interaction between a quasi-monochromatic classical field and a two-level atom is

$$H = \hbar\omega_e |e\rangle\langle e| + \hbar\omega_g |g\rangle\langle g| - \mathbf{d}\cdot\mathbf{E}(\mathbf{R}, t), \qquad (66.35)$$

where $|e\rangle$ and $|g\rangle$ label the upper and lower atomic levels, of frequencies ω_e and ω_g, respectively, with $\omega_e - \omega_g = \omega_0$, and $\mathbf{R}$ is the location of the center of mass of the atom. The electric dipole operator (see Eq. 66.8), couples the excited and ground levels, and may be expressed as

$$\mathbf{d} = \epsilon_d d \left(|e\rangle\langle g| + |g\rangle\langle e| \right), \qquad (66.36)$$

where ϵ_d is a unit vector in the direction of the dipole and d the matrix element of the electric dipole operator between the ground and excited state, which we take to be real for simplicity. We also neglect the vector character of $\mathbf{d}$ and $\mathbf{E}(\mathbf{R}, t)$ in the following, assuming, for example, that both ϵ_d and ϵ are parallel to x-axis. The Hamiltonian (66.35) may then be expressed as

$$\begin{aligned} H = \hbar\omega_e |e\rangle\langle e| + \hbar\omega_g |g\rangle\langle g| - d \left(|e\rangle\langle g| + |g\rangle\langle e| \right) \\ \times \left[E^+(\mathbf{R}, t) + E^-(\mathbf{R}, t) \right], \end{aligned} \qquad (66.37)$$

where we have generalized the notation of Eq. (66.20) in an obvious way. One can introduce the *pseudo-spin operators*

$$\begin{aligned} s_z &= \left(|e\rangle\langle e| - |g\rangle\langle g| \right)/2, \\ s_+ &= s_-^\dagger = |e\rangle\langle g|, \end{aligned} \qquad (66.38)$$

and redefine the zero of atomic energy to introduce the commonly used form

$$H = \hbar\omega_0 s_z - d(s_+ + s_-)\left[E^+(\mathbf{R}, t) + E^-(\mathbf{R}, t) \right]. \qquad (66.39)$$

66.4.2 Rotating Wave Approximation

Under the influence of a monochromatic electromagnetic field of frequency ω, atoms undergo transitions between their lower and upper states by interacting with either the positive or the negative frequency part of the field. The corresponding contributions to the atomic dynamics oscillate at frequencies $\omega_0 - \omega$ and $\omega_0 + \omega$, respectively, and their contributions to the probability amplitudes involve denominators containing this same frequency dependence. For near-resonant atom-field interactions, the rapidly oscillating contributions lead to small

corrections, the first-order one being the *Bloch-Siegert shift*, whose value near resonance, is $\omega \simeq \omega_0$ (see, e.g., Ref. [13]).

$$\delta\omega_{eg} = -\frac{(d|E^+|/\hbar)^2}{4\omega} \qquad (66.40)$$

to lowest-order in $d\mathcal{E}/\hbar\omega$. The neglect of these terms is the *Rotating Wave Approximation* (RWA). Note that it is normally inconsistent to regard an atom as a two-level system and not to perform the RWA. In the RWA, the atomic system is described by the Hamiltonian

$$H = \hbar\omega_0 s_z - d\left[s_+ E^+(\mathbf{R}, t) + s_- E^-(\mathbf{R}, t) \right], \qquad (66.41)$$

or, in a frame rotating at the frequency ω of the field,

$$H = \hbar\Delta s_z - \tfrac{1}{2}d\left(s_+ \mathcal{E} e^{i\mathbf{k}\cdot\mathbf{R}} + \text{h.c.} \right), \qquad (66.42)$$

where $\Delta = \omega_0 - \omega$ is the atom-light detuning. (Note that the alternate definition $\delta = \omega - \omega_0$ is frequently used in the literature.) In the rest of this Chapter, we consider atoms placed at $\mathbf{R} = 0$.

66.4.3 Rabi Frequency

The dynamics of the two-level atom is conveniently expressed in terms of its density operator ρ, whose evolution is given by the Schrödinger equation

$$\frac{d\rho}{dt} = -\frac{i}{\hbar}[H, \rho], \qquad (66.43)$$

where $\rho_{ee} = \langle e|\rho|e\rangle$ and $\rho_{gg} = \langle g|\rho|g\rangle$ are the upper and lower state populations P_e and P_g, respectively, while the off-diagonal matrix elements $\rho_{eg} = \langle e|\rho|g\rangle = \rho_{ge}^\star$ are called the *atomic coherences*, or simply *coherences*, between levels $|e\rangle$ and $|g\rangle$. These coherences play an essential role in optical physics and quantum optics, since they are proportional to the expectation value of the electric dipole operator.

The evolution of $P_g(t)$ and $P_e(t) = 1 - P_g(t)$, is characterized by oscillations at the *generalized Rabi frequency*

$$\Omega = \left(\Omega_1^2 + \Delta^2 \right)^{1/2}, \qquad (66.44)$$

where the *Rabi frequency* Ω_1 is $\Omega_1 = d\mathcal{E}/\hbar$, (or $\Omega_1 = d\mathcal{E}(\epsilon_d\cdot\epsilon)/\hbar$ when the vector character of the electric field and dipole moment are included). Specifically, assuming that the atom is initially in its ground state $|g\rangle$, the probability that it is in the excited state $|e\rangle$ at a subsequent time t is given by *Rabi's formula*

$$P_e(t) = (\Omega_1/\Omega)^2 \sin^2(\Omega t/2). \qquad (66.45)$$

At resonance ($\Delta = 0$), the generalized Rabi frequency Ω reduces to the Rabi frequency Ω_1. (In addition to the texts on quantum optics already cited, see also Ref. [14].)

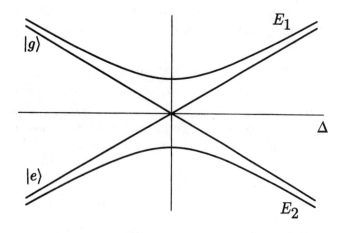

Figure 66.1. Dressed levels of a two-level atom driven by a classical monochromatic field as a function of the detuning $\Delta = \omega_0 - \omega$.

66.4.4 Dressed States

Semiclassical Case

The atomic dynamics can alternatively be described in terms of a *dressed states* basis instead of the *bare states* $|e\rangle$ and $|g\rangle$ (see especially Ref. [13]). The dressed states $|1\rangle$ and $|2\rangle$ are eigenstates of the Hamiltonian (66.42), and, by convention, the state $|1\rangle$ is the one with the greatest energy. They are conveniently expressed in terms of the bare states via the *Stückelberg angle* $\theta/2$ as

$$\begin{aligned} |1\rangle &= \sin\theta|g\rangle + \cos\theta|e\rangle\,, \\ |2\rangle &= \cos\theta|g\rangle - \sin\theta|e\rangle\,, \end{aligned} \qquad (66.46)$$

where $\sin(2\theta) = -\Omega_1/\Omega$, $\cos(2\theta) = \Delta/\Omega$. The corresponding eigenenergies are

$$\begin{aligned} E_1 &= +\tfrac{1}{2}\hbar\Omega\,, \\ E_2 &= -\tfrac{1}{2}\hbar\Omega\,. \end{aligned} \qquad (66.47)$$

These energies are illustrated in Fig. 66.1 as a function of the field frequency ω. The dressed levels repel each other and form an anticrossing at resonance $\omega = \omega_0$. As the detuning Δ varies from positive to negative values, state $|1\rangle$ passes continuously from the excited state $|e\rangle$ to the bare ground state $|g\rangle$, with both bare states having equal weights at resonance. The distances between the perturbed levels and their asymptotes for $|\Delta| \gg \Omega_1$ represent the *ac Stark shifts*, or *light shifts*, of the atomic states when coupled to the laser. From Fig. 66.1, the ac Stark shift of $|g\rangle$ is positive for $\Delta < 0$ and negative for $\Delta > 0$, while the $|e\rangle$ state shift is negative for $\Delta < 0$ and positive for $\Delta > 0$.

Quantized Field

The concept of dressed states can readily be generalized to a two-level atoms interacting with a single-mode quantized field in the dipole and rotating wave approximations. The atom and its dipole interaction with the field are still described by the Hamiltonian (66.41), except that the positive and negative frequency components of the field are now operators, and the free field Hamiltonian must be included. The Hamiltonian of the total atom-field system becomes

$$H = \hbar\omega_0 s_z + \hbar\omega(a^\dagger a + \tfrac{1}{2}) + \hbar g(s_+ a + a^\dagger s_-)\,, \qquad (66.48)$$

where the creation and annihilation operators $a^\dagger$ and a obey the boson commutation relation $[a, a^\dagger] = 1$ (see Chap. 6), and the coupling constant

$$g = d\sqrt{\frac{\omega}{2\epsilon_0 \hbar V}} \qquad (66.49)$$

is the *vacuum Rabi frequency.*, with V being a photon normalization volume. This Hamiltonian defines the *Jaynes-Cummings model*, (see, e.g., Refs. [7, 15]) which is discussed in more detail in Chap. 75.

The dressed states of the atom-field system are the eigenstates of the Jaynes-Cummings model. Since, in the RWA, the dipole interaction only couples states of same "excitation number", e.g., $|e, n\rangle$ and $|g, n+1\rangle$, where $|n\rangle$ is an eigenstate of the photon number operator, $a^\dagger a|n\rangle = n|n\rangle$, with n an integer, the diagonalization of the Jaynes-Cummings model reduces to that of the semiclassical driven two-level atom in each of these manifolds. Hence, the dressed states are

$$\begin{aligned} |1, n\rangle &= \sin\theta_n|g, n+1\rangle + \cos\theta_n|e, n\rangle, \\ |2, n\rangle &= \cos\theta_n|g, n+1\rangle - \sin\theta_n|e, n\rangle, \end{aligned} \qquad (66.50)$$

with

$$\tan(2\theta_n) = -2g\sqrt{n+1}/\Delta\,. \qquad (66.51)$$

(The factor of 2 difference between this and the semiclassical case is due to the use of a running waves quantization scheme, while the semiclassical discussion was for standing waves.) The corresponding eigenenergies are

$$\begin{aligned} E_{1n} &= \hbar(n + \tfrac{1}{2})\omega - \hbar R_n\,, \\ E_{2n} &= \hbar(n + \tfrac{1}{2})\omega + \hbar R_n\,, \end{aligned} \qquad (66.52)$$

where

$$R_n = \tfrac{1}{2}\sqrt{\Delta^2 + 4g^2(n+1)}\,. \qquad (66.53)$$

Chap. 75 shows that by including the effects of spontaneous emission, this picture yields a straightforward interpretation of a number of effects, including the *Burshtein-Mollow resonance fluorescence spectrum*. Dressed states also help to elucidate the interaction between two-level atoms and quantized single-mode fields, as occur for example in cavity QED (see, e.g., Ref. [15] and Chap. 75). Their generalization to the case of moving atoms offers simple physical interpretations of several aspects of laser cooling, see Chap. 73 and Ref. [16].

66.4.5   Optical Bloch Equations

Introducing the density operator matrix elements $\rho_{ab} = \langle a|\rho|b\rangle$, where a, b can be either e or g, as well as the real quantities

$$
\begin{aligned}
U &= \rho_{eg}e^{i\omega t} + \text{c.c.}\,, \\
V &= i\rho_{eg}e^{i\omega t} + \text{c.c.}\,, \\
W &= \rho_{ee} - \rho_{gg}\,,
\end{aligned}
\tag{66.54}
$$

the equations of motion for the density matrix elements $\rho_{ij} = \langle i|\rho|j\rangle$ may be expressed, with Eq. (66.43), as

$$
\begin{aligned}
\frac{dU}{dt} &= -\Delta V\,, \\
\frac{dV}{dt} &= \Delta U + \Omega_1 W\,, \\
\frac{dW}{dt} &= -\Omega_1 V\,.
\end{aligned}
\tag{66.55}
$$

These are the *optical Bloch equations*, as discussed extensively in Refs. [5, 13]. Physically, U describes the component of the *atomic coherence* in phase with the driving field, V the component in quadrature with the field, and W the atomic inversion. The optical Bloch equations have a simple geometrical interpretation offered by thinking of U, V and W as the three components of a vector called the *Bloch vector* $\mathbf{U}$, whose equation of motion is

$$
\frac{d\mathbf{U}}{dt} = \boldsymbol{\Omega} \times \mathbf{U}\,,
\tag{66.56}
$$

where $\boldsymbol{\Omega} = (-\Omega_1, 0, \Delta)$. Thus $\mathbf{U}$ precesses about $\boldsymbol{\Omega}$, of length Ω, while conserving its length. The evolution of a two-level atom driven by a monochromatic field is thus mathematically equivalent to that of a spin-$\frac{1}{2}$ system in two magnetic fields $\mathbf{B}_0$ and $2\mathbf{B}_1\cos\omega t$ which are parallel to the z- and x-axis, respectively, and whose amplitudes are such that the Larmor spin precession frequencies around them are ω and $2\Omega_1\cos\omega t$, respectively. In optics, this vectorial picture is often referred to as the *Feynman-Vernon-Hellwarth picture* [17]. It is very useful in discussing the coherent transient phenomena discussed in Chap. 71.

66.5   RELAXATION MECHANISMS

In addition to their coherent interaction with light fields, atoms suffer incoherent relaxation mechanisms, whose origin can be as diverse as elastic and inelastic collisions and spontaneous emission. Collisional broadening is discussed in Chap. 57, while a QED microscopic discussion of spontaneous emission is described in Chap. 75 in terms of reservoir theory. One advantage of describing the atomic state in terms of a density operator ρ is that the physical interpretation of its elements allows us phenomenologically to add various relaxation terms directly to its elements.

66.5.1   Relaxation Toward Unobserved Levels

If the relaxation mechanisms transfer populations or atomic coherences toward uninteresting or unobserved levels, their description can normally be given in terms of a Schrödinger equation, but with a complex Hamiltonian. In contrast, if all levels involved in the relaxation mechanism are observed, a more careful description, e.g. in terms of a master equation, is required. Specifically, in the case of relaxation to unobserved levels, the evolution of the atomic density operator, restricted to the levels of interest, is of the general form [12]

$$
\frac{d\rho}{dt} = -\frac{i}{\hbar}[H_{\text{eff}}\rho - \rho H_{\text{eff}}^{\dagger}]\,,
\tag{66.57}
$$

where

$$
H_{\text{eff}} = H + \hat{\Gamma}\,,
\tag{66.58}
$$

H being the atom-field Hamiltonian and $\hat{\Gamma}$ the non-Hermitian relaxation operator, defined by its matrix elements

$$
\langle n|\hat{\Gamma}|m\rangle = \frac{\hbar}{2i}\gamma_n\delta_{nm}\,.
\tag{66.59}
$$

Both *inelastic collisions* and spontaneous emission to unobserved levels can be described by this form of evolution. In the framework of this Chapter, inelastic, or strong, collisions are defined as collisions that can induce atomic transitions into other energy levels.

66.5.2   Relaxation Toward Levels of Interest

A master equation description is necessary when all involved levels are observed (see, e.g., Refs. [7, 13]). This master equation can rapidly take a complicated form if more than two levels are involved. We give results only for the case of a two-level atom and upper to lower-level spontaneous decay and *elastic* or soft collisions; i.e., collisions that change the separation of energy levels during the collision, but leave the level populations unchanged. In that case, the atomic master equation takes the form

$$\frac{d\rho}{dt} = -\frac{i}{\hbar}[H, \rho] - \frac{\Gamma}{2}[s_+ s_- \rho + \rho s_+ s_- - 2s_- \rho s_+]$$
$$- \frac{1}{2}\gamma_{ph}\rho + 2\gamma_{ph}s_z \rho s_z , \qquad (66.60)$$

where the free-space spontaneous decay rate Γ is found from QED to be

$$\Gamma = \frac{1}{4\pi\epsilon_0} \frac{4d^2\omega_0^3}{3\hbar c^3} , \qquad (66.61)$$

and γ_{ph} is the decay rate due to elastic collisions.

It is possible to express the classical decay rate (66.17) in terms of the quantum spontaneous emission rate (66.61) as

$$\Gamma = \Gamma_{cl} f_{ge} , \qquad (66.62)$$

where f_{ge} is the *oscillator strength* of the transition. The various oscillator strengths characterizing the dipole-allowed transitions from a ground state $|e\rangle$ to excited levels $|e\rangle$ obey the Thomas-Reiche-Kuhn sum rule

$$\sum_e f_{ge} = 1 , \qquad (66.63)$$

where the sum is on all levels dipole-coupled to $|g\rangle$. Assuming that $\mathbf{d}$ and the polarization of the field are both parallel to the x-axis, this gives

$$f_{ge} = \frac{2m\omega_0}{\hbar}|\langle g|x|e\rangle|^2 . \qquad (66.64)$$

66.5.3 Optical Bloch Equations with Decay

In general, the optical Bloch equations cannot be generalized to cases where relaxation mechanisms are present. There are, however, two notable exceptions corresponding to situations where

1. the upper level spontaneously decays to the lower level only, while the atom undergoes only elastic collisions;

2. spontaneous emission between the upper and lower levels can be ignored in comparison with decay to unobserved levels, which occur at equal rates $\gamma_e = \gamma_g = 1/T_1$.

Under these conditions, Eq. (66.55) generalizes to

$$\frac{dU}{dt} = -U/T_2 - \Delta V ,$$
$$\frac{dV}{dt} = -V/T_2 + \Delta U + \Omega_1 W , \qquad (66.65)$$
$$\frac{dW}{dt} = -(W - W_{eq})/T_1 - \Omega_1 V ,$$

where we have introduced the *longitudinal* and *transverse relaxation times* T_1 and T_2, with $T_1 = 1/\Gamma$ and $T_2 = (1/2T_1 + \gamma_{ph})^{-1}$ in the first case, and $T_2 = (1/T_1 + \gamma_{ph})^{-1}$ in the second case. The equilibrium inversion W_{eq} is equal to zero in the second case since the decay is to unobserved levels.

66.5.4 Density Matrix Equations

In the general case, it is necessary to consider the density operator equation (66.43) instead of the optical Bloch equations. The equations of motion for the components of ρ become, for the general case of complex Ω_1,

$$\frac{d\rho_{ee}}{dt} = -\gamma_e \rho_{ee} - \frac{1}{2}(i\Omega_1^* \tilde{\rho}_{eg} + \text{c.c.}) ,$$
$$\frac{d\rho_{gg}}{dt} = -\gamma_g \rho_{gg} + \frac{1}{2}(i\Omega_1^* \tilde{\rho}_{eg} + \text{c.c.}) , \qquad (66.66)$$
$$\frac{d\tilde{\rho}_{eg}}{dt} = -(\gamma + i\Delta)\tilde{\rho}_{eg} - i\frac{\Omega_1}{2}(\rho_{ee} - \rho_{gg}) ,$$

where $\gamma = (\gamma_e + \gamma_g)/2 + \gamma_{ph}$, and $\tilde{\rho}_{eg} = \rho_{eg} e^{i\omega t}$. In the case of spontaneous decay from the upper to the lower level, these equations become

$$\frac{d\rho_{ee}}{dt} = -\Gamma \rho_{ee} - \frac{1}{2}(i\Omega_1^* \tilde{\rho}_{eg} + \text{c.c.}) ,$$
$$\frac{d\rho_{gg}}{dt} = +\Gamma \rho_{ee} + \frac{1}{2}(i\Omega_1^* \tilde{\rho}_{eg} + \text{c.c.}) , \qquad (66.67)$$
$$\frac{d\tilde{\rho}_{eg}}{dt} = -(\gamma + i\Delta)\tilde{\rho}_{eg} - i\frac{\Omega_1}{2}(\rho_{ee} - \rho_{gg}) ,$$

where $\gamma = \Gamma/2 + \gamma_{ph}$. Equations (66.67) are completely equivalent to the optical Bloch equations (66.65).

66.6 RATE EQUATION APPROXIMATION

If the coherence decay rate γ is dominated by elastic collisions, and hence is much larger than the population decay rates γ_e and γ_g, $\tilde{\rho}_{eg}$ can be adiabatically eliminated from the equations of motion (66.66) and (66.67) to obtain the *rate equations* (see also Sect. 67.3)

$$\frac{d\rho_{ee}}{dt} = -\gamma_e \rho_{ee} - R(\rho_{ee} - \rho_{gg}) ,$$
$$\frac{d\rho_{gg}}{dt} = -\gamma_g \rho_{gg} + R(\rho_{ee} - \rho_{gg}) , \qquad (66.68)$$

and

$$\frac{d\rho_{ee}}{dt} = -\Gamma \rho_{ee} - R(\rho_{ee} - \rho_{gg}) ,$$
$$\frac{d\rho_{gg}}{dt} = +\Gamma \rho_{gg} + R(\rho_{ee} - \rho_{gg}) , \qquad (66.69)$$

respectively, where the transition rate is

$$R = |\Omega_1|^2 \mathcal{L}(\Delta)/(2\gamma) , \qquad (66.70)$$

and we have introduced the dimensionless Lorentzian

$$\mathcal{L}(\Delta) = \gamma^2/(\gamma^2 + \Delta^2) . \qquad (66.71)$$

The transitions between the upper and lower state are thus described in terms of simple rate equations.

Adding phenomenological pumping rates Λ_e and Λ_g on the right-hand side of Eqs. (66.68) provides a description of the excitation of the upper and lower levels from some distant levels, as would be the case in a laser. The equations then form the basis of conventional, single-mode laser theory. In the absence of such mechanisms, the atomic populations eventually decay away.

66.6.1 Steady State

In the case of upper to lower-level decay, the state populations reach a steady state with inversion [5, 13]

$$W_{\mathrm{st}} = -\frac{\Gamma}{\Gamma + 2R} = -\frac{1}{1+s}, \qquad (66.72)$$

where s is the *saturation parameter*. In the case of *pure radiative decay*, $\gamma_{ph} = 0$, s is given by

$$s = \frac{\Omega_1^2/2}{\Gamma^2/4 + \Delta^2}. \qquad (66.73)$$

In steady state, the other two components of the Bloch vector $\mathbf{U}$ are given by

$$U_{\mathrm{st}} = -\frac{2\Delta}{\Omega_1}\left(\frac{s}{1+s}\right) \qquad (66.74)$$

and

$$V_{\mathrm{st}} = \frac{\Gamma}{\Omega_1}\left(\frac{s}{1+s}\right). \qquad (66.75)$$

U_{st} varies as a dispersion curve as a function of the detuning Δ, while V_{st} is a Lorentzian of *power-broadened* half-width at half maximum $(\Gamma^2/4 + \Omega_1^2)^{1/2}$.

66.6.2 Saturation

As the intensity of the driving field, or Ω_1^2, increases, U_{st} and V_{st} first increase linearly with Ω_1, reach a maximum, and finally tend to zero as $\Omega_1 \to \infty$. The inversion W_{st}, which equals -1 for $\Omega_1 = 0$, first increases quadratically, and asymptotically approaches $W_{\mathrm{st}} = 0$ as $\Omega_1 \to \infty$. At this point, where the upper and lower state populations are equal, the transition is said to be *saturated*, and the medium becomes effectively transparent, or bleached. (This should not be confused with self-induced transparency discussed in Chap. 71.) The inversion is always negative, which means in particular that no steady-sate light amplification can be achieved in this system. This is one reason why external pump mechanisms are required in lasers.

66.6.3 Einstein A and B Coefficients

When atoms interact with broadband radiation instead of the monochromatic fields considered so far, Eqs. (66.69) still apply, but the rate R becomes

$$R \to B_{eg}\varrho(\omega), \qquad (66.76)$$

where $\varrho(\omega)$ is the spectral energy density of the inducing radiation. Einstein's A and B coefficients apply to an atom in thermal equilibrium with the field, which is described by Planck's black-body radiation law

$$\varrho(\omega) = \frac{\hbar\omega^3}{\pi^2 c^3}\frac{1}{e^{\hbar\omega/k_{\mathrm{B}}T} - 1}, \qquad (66.77)$$

where T is the temperature of the source and k_{B} is Boltzmann's constant. Invoking the *principle of detailed balance*, which states that at thermal equilibrium, the average number of transitions between arbitrary states $|i\rangle$ and $|k\rangle$ must be equal to the number of transitions between $|k\rangle$ and $|i\rangle$, one finds

$$\frac{A_{ki}}{B_{ki}} = \frac{\hbar\omega^3}{\pi^2 c^3}, \qquad (66.78)$$

where A_{ki} is the rate of spontaneous emission from $|k\rangle$ to $|i\rangle$, and $\Gamma_k = \sum_i A_{ki}$ is the level width.

66.7 LIGHT SCATTERING

Far from resonance, the approximation of a two- or few-level atom is no longer adequate. Two limiting cases, which are always far from resonance, are Rayleigh scattering for low frequencies, and Thomson scattering for high frequencies.

66.7.1 Rayleigh Scattering

Rayleigh scattering is the elastic scattering of a monochromatic electromagnetic field of frequency ω, wave vector $\mathbf{k}$ and polarization $\boldsymbol{\epsilon}$ by an atomic system in the limit where ω is very small compared with its excitation energies [4, 13]. To second-order in perturbation theory, the Rayleigh scattering differential cross section into the solid angle Ω' about the wave vector $\mathbf{k}'$ with $k' = k$, and polarization $\boldsymbol{\epsilon}'$ is

$$\frac{d\sigma}{d\Omega'} = r_0^2 \omega^4 (\boldsymbol{\epsilon}\cdot\boldsymbol{\epsilon}')^2 \left(\sum_e \frac{f_{ge}}{\omega_{eg}^2}\right)^2, \qquad (66.79)$$

where r_0 is the classical electron radius, the sum is over all states $|e\rangle$, f_{ge} is the E1 oscillator strength (66.64) and ω_{ge} is the transition frequency. The corresponding total cross section is

$$\sigma = \frac{8\pi r_0^2 \omega^4}{3} \left(\sum_e \frac{f_{ge}}{\omega_{eg}^2}\right)^2. \qquad (66.80)$$

66.7.2 Thomson Scattering

Thomson scattering is the corresponding elastic photon scattering by an atom in the limit where ω is very large compared with the atomic ionization energy, yet small enough compared to $\alpha mc^2/\hbar$, that the dipole approximation can be applied. The differential cross section for this process is [4, 13]

$$\frac{d\sigma}{d\Omega'} = r_0^2 (\epsilon \cdot \epsilon')^2 \,, \qquad (66.81)$$

and the total cross section is

$$\sigma = \tfrac{8}{3} \pi r_0^2 \,. \qquad (66.82)$$

This is a completely classical result, which exhibits no frequency dependence.

66.7.3 Resonant Scattering

We finally consider elastic scattering in the limit where ω is close to the transition frequency ω_0 between $|g\rangle$ and $|e\rangle$. Provided that no other level is near-resonant with the ground state, the resonant scattering differential cross section is [6, 13]

$$\frac{d\sigma}{d\Omega'} = \frac{9}{16\pi^2} \lambda_0^2 (\epsilon \cdot \epsilon')^2 \frac{(\Gamma/2)^2}{\Delta^2 + (\Gamma/2)^2} \,, \qquad (66.83)$$

where $\lambda_0 = 2\pi c/\omega_0$ is the wavelength of the transition and Γ is the spontaneous decay rate (66.61). The total elastic scattering cross section is

$$\sigma = \frac{3}{2\pi} \lambda_0^2 \,. \qquad (66.84)$$

In contrast to the nonresonant Rayleigh and Thomson scattering cross sections, which scale as the square of the classical electron radius, the resonant scattering cross section scales as the square of the wavelength. For optical frequencies, $\lambda_0/r_0 \simeq 10^4$, giving a *resonant enhancement*

of about eight orders of magnitude. This illustrates why near resonant phenomena, which form the bulk of the following Chapters, are so important in optical physics and quantum optics.

REFERENCES

1. B. W. Shore, *The Theory of Coherent Atomic Excitation* (Wiley-Interscience, New York, 1990).
2. C. Cohen-Tannoudji, J. Dupont-Roc, and G. Grynberg, *Photons and Atoms: Introduction to Quantum Electrodynamics* (Wiley-Interscience, New York, 1989).
3. A. Sommerfeld, *Optics* (Academic Press, New York, 1967).
4. J. D. Jackson, *Classical Electrodynamics* (Wiley, New York, 1975).
5. L. Allen and J. H. Eberly, *Optical Resonance and Two-Level Atoms* (Dover, New York, 1987).
6. P. W. Milonni and J. H. Eberly, *Lasers* (Wiley-Interscience, New York, 1988).
7. P. Meystre and M. Sargent III, *Elements of Quantum Optics* (Springer, Berlin, 1991), 2nd ed.
8. M. Born and E. Wolf, *Principles of Optics* (Pergamon Press, Oxford, 1970), 4th ed.
9. R. J. Glauber, in *Quantum Optics and Electronics*, edited by C. DeWitt, A. Blandin, and C. Cohen-Tannoudji (Gordon and Breach, New York, 1995).
10. Y. R. Shen, *The Principles of Nonlinear Optics* (Wiley, New York, 1984).
11. M. Sargent III, M. O. Scully, Jr., and W. E. Lamb, *Laser Physics* (Wiley, Reading, 1977).
12. S. Stenholm, *Foundations of Laser Spectroscopy* (Wiley-Interscience, New York, 1984).
13. C. Cohen-Tannoudji, J. Dupont-Roc, and G. Grynberg, *Atom-Photon Interactions: Basic Processes and Applications* (Wiley-Interscience, New York, 1992).
14. P. L. Knight and P. W. Milonni, Phys. Rep. **66**, 21 (1980).
15. *Cavity QED*, edited by P. P. Berman (Academic Press, Boston, 1994).
16. C. Cohen-Tannoudji, in *Fundamental Systems in Quantum Optics*, edited by J. Dalibard, J. M. Raimond, and J. Zinn-Justin (North-Holland, Amsterdam, 1992).
17. R. P. Feynman, F. L. Vernon, and R. W. Hellwarth, J. Appl. Phys. **28**, 49 (1957).

67

Absorption and Gain Spectra

Stig Stenholm

Research Institute for Theoretical Physics, P. O. Box 9, FIN-00014 University of Helsinki, Finland

67.1 INTRODUCTION

This chapter develops theoretical techniques to describe absorption and emission spectra, using concepts introduced in Chap. 66, and density matrix methods from Chap. 7. The simplest cases are treated, compatible with the physics involved, and more realistic applications are referred to in other chapters. Vector notation is not used, but it can be inserted as required. Only steady-state spectroscopy is covered; for time-resolved transient techniques see Chap. 71.

Laser technology has greatly expanded the potential of atomic and molecular spectroscopy, but the same techniques for describing the interaction of light with matter also apply to the traditional arc lamps and flash discharges, and the more recent synchrotron radiation sources.

In many cases, departures from the *thin sample limit*, such as beam attenuation, light scattering and radiation trapping (see Sect. 67.3) may be important. However, the properties of laser devices themselves depend in an essential way on these effects, making a self-consistent treatment of their properties necessary.

At the other extreme, the spectroscopy of dilute gases is well characterized by ensemble averages over the properties of the individual particles, interrupted by occasional brief collisions. However, ensemble averages may no longer apply to recent experiments probing a single atomic particle in a trap, as discussed in Chap. 73.

67.2 INDEX OF REFRACTION

As discussed in Sect. 66.3.2, the complex index of refraction for a medium containing harmonically bound charges (electrons) [1] with natural frequency ω_0 is

$$
\begin{aligned}
n(\omega) &= \sqrt{1 + \frac{N\alpha(\omega)}{\varepsilon_0}} \approx 1 + \frac{N\alpha(\omega)}{2\varepsilon_0} \\
&= 1 + \frac{Ne^2}{2m\varepsilon_0}\left(\frac{i\gamma\omega + (\omega_0^2 - \omega^2)}{(\omega_0^2 - \omega^2)^2 + \gamma^2\omega^2}\right) \\
&\equiv n' + i\,n''.
\end{aligned} \tag{67.1}
$$

The expansion is valid when the density of atoms N is low.

A plane wave can be written in the form

$$
\mathcal{E} \propto e^{ikz} = e^{i\omega n z/c} = e^{i\omega n' z/c}e^{-\omega n'' z/c}. \tag{67.2}
$$

The absorption of light through the medium then shows a resonant behavior near $\omega \approx \omega_0$ determined by

$$n'' = \frac{Ne^2}{2m\varepsilon_0}\left(\frac{\gamma\omega}{(\omega_0^2-\omega^2)^2+\gamma^2\omega^2}\right)$$

$$\approx \frac{\pi Ne^2}{4m\varepsilon_0\omega_0}\left(\frac{\gamma/2\pi}{(\omega-\omega_0)^2+\gamma^2/4}\right). \quad (67.3)$$

This is called an *absorptive lineshape*. When the single electron is harmonically bound, all its interaction with radiation is found in this response. For a real atom, the response of the electron is divided among the various transitions to other states. The fraction assigned to one single transition is characterized by the *oscillator strength* f_n as discussed in Sect. 66.3.2.

In ordinary *linear spectroscopy*, the laser is tuned through the resonance $\omega \approx \omega_0$, and the value of ω_0 is determined from the lineshape (67.3). Several closely spaced resonances can be *resolved* if their spacing is larger than their widths

$$|\omega_0^{(1)}-\omega_0^{(2)}| > \gamma. \quad (67.4)$$

This defines the *spectral resolution* (see Chap. 10).

The velocity of light in the medium is seen to be given by the expression

$$c_{\text{eff}} = \frac{c}{n'} \approx c\left[1-\frac{Ne^2}{2m\varepsilon_0}\left(\frac{\omega_0^2-\omega^2}{(\omega_0^2-\omega^2)^2+\gamma^2\omega^2}\right)\right]. \quad (67.5)$$

This expression shows a *dispersive behavior* around the position $\omega = \omega_0$, where the modification of the velocity disappears. Below resonance $\omega < \omega_0$, the velocity of light is lower than in vacuum. This derives from the fact that the polarization is in phase with the driving field. Thus, by storing the incoming energy, the driving field retards the propagation of the radiation.

For a harmonically bound charge, the refractive index (67.1) always stays absorptive and it is independent of the intensity of the laser radiation. This no longer holds for atomic discrete level systems. In order to see this, we consider the *two-level atom* in Sect. 67.3 (see also Sect. 66.4).

67.3 DENSITY MATRIX TREATMENT OF THE TWO-LEVEL ATOM

The response of atoms to light is conveniently expressed in terms of the density matrix ρ. In addition to the direct physical meaning of the density matrix elements discussed in Sect. 66.4.3, the density matrix formalism is advantageous because the various relaxation mechanisms affecting the atomic resonances can be introduced phenomenologically into its equations of motion (see Sect. 66.5), and theoretical derivations often provide *master equations* for the density matrix (see Chap. 7).

The two-level Hamiltonian [Eq. (66.35)] can be written

$$\mathbf{H} = \begin{bmatrix} \hbar\omega_0/2 & -d\mathcal{E}(R,t) \\ -d\mathcal{E}(R,t) & -\hbar\omega_0/2 \end{bmatrix}. \quad (67.6)$$

The equation of motion for the density matrix (66.36) is then

$$\frac{d}{dt}\rho_{ee} = -\Gamma\rho_{ee} + \frac{id\mathcal{E}}{\hbar}\cos\omega t\,(\rho_{ge}-\rho_{eg}),$$

$$\frac{d}{dt}\rho_{gg} = \Gamma\rho_{ee} - \frac{id\mathcal{E}}{\hbar}\cos\omega t\,(\rho_{ge}-\rho_{eg}), \quad (67.7)$$

$$\frac{d}{dt}\rho_{eg} = -(\gamma+i\omega_0)\rho_{eg} + \frac{id\mathcal{E}}{\hbar}\cos\omega t\,(\rho_{gg}-\rho_{ee}).$$

Here Γ is the spontaneous decay rate given by Eq. (66.61) and

$$\gamma = \Gamma/2 + \gamma_{\text{ph}}, \quad (67.8)$$

where γ_{ph} derives from all processes that tend to randomize the phase between the quantum states $|e\rangle$ and $|g\rangle$, such as collisions (Chap. 7 and 19), noise in the laser fields and thermal excitation of the environment in solid state spectroscopy. The Greek letters Γ and γ correspond to the *longitudinal relaxation rate* (T_1-process) and the *transverse relaxation rate* (T_2-process) respectively (see Sect. 66.5.3).

In the rotating wave approximation (RWA) discussed in Sect. 66.4.2, the density matrix equations (67.7) become identical to Eqs. (66.67) with

$$\rho_{eg} = \tilde{\rho}_{eg}e^{-i\omega t}. \quad (67.9)$$

Using the conservation of probability condition

$$\rho_{ee} + \rho_{gg} = 1, \quad (67.10)$$

the steady state solutions to (66.67) are [2]

$$\rho_{ee} = \frac{\Omega_1^2\gamma}{2\Gamma}\left(\frac{1}{\Delta^2+\gamma^2+\Omega_1^2\gamma/\Gamma}\right), \quad (67.11)$$

$$\tilde{\rho}_{eg} = \frac{i\Omega_1}{2}\frac{(\rho_{gg}-\rho_{ee})}{\gamma+i\Delta}$$

$$= \frac{i\Omega_1}{2}\left(\frac{\gamma-i\Delta}{\Delta^2+\gamma^2+\Omega_1^2\gamma/\Gamma}\right), \quad (67.12)$$

where Ω_1 is the Rabi frequency from Eq. (66.44), and $\Delta = \omega_0 - \omega$ is the detuning. The induced polarization is then

$$\mathcal{P} = N\,\text{Tr}(\hat{d}\,\rho) = N\,\text{Tr}\left[\begin{pmatrix} 0 & d \\ d & 0 \end{pmatrix}\begin{pmatrix} \rho_{ee} & \rho_{eg} \\ \rho_{ge} & \rho_{gg} \end{pmatrix}\right]$$

$$= Nd\left(e^{-i\omega t}\tilde{\rho}_{eg}+e^{i\omega t}\tilde{\rho}_{ge}\right)$$

$$= N\left[\alpha\mathcal{E}^{(+)}+\alpha^*\mathcal{E}^{(-)}\right], \quad (67.13)$$

where N is the density of active two-level atoms. Setting

$$\mathcal{E}^{(+)} = \frac{1}{2}\mathcal{E}e^{-i\omega t}, \qquad (67.14)$$

the complex polarization is

$$\alpha(\omega) = \frac{d^2}{\hbar}\left(\frac{i\gamma + \Delta}{\Delta^2 + \gamma^2 + \Omega_1^2\gamma/\Gamma}\right), \qquad (67.15)$$

and from Eq. (67.1), the complex index of refraction is

$$\begin{aligned}n(\omega) &= 1 + \frac{N\alpha(\omega)}{2\varepsilon_0}\\ &= 1 + \frac{\pi Ne^2}{4\varepsilon_0 m\omega_0}\left(\frac{f_0}{\pi}\right)\frac{i\gamma + \Delta}{\Delta^2 + \gamma^2 + \Omega_1^2\gamma/\Gamma}, \quad (67.16)\end{aligned}$$

where $f_0 = 2d^2 m\omega_0/\hbar e^2$ is the oscillator strength. Summing over all possible transitions yields the *f-sum rule* (66.63) (see also Chap. 21).

The imaginary part of Eq. (67.16) shows exactly the same absorptive behavior as in the harmonic oscillator model of Sect. 67.1 [see, e.g., Eq. (67.3)]. However, the additional factor of $(\Omega_1^2\gamma/\Gamma)$ in the denominator makes the line appear broader than in the harmonic case; the line is *power broadened*. Physically, this derives from a *saturation* of the two-level system in which the population of the upper level becomes an appreciable fraction of that of the lower level. In the limit $\Omega_1 \to \infty$, Eq. (67.16) shows that $n(\omega) \to 1$ and the atom-field interaction effectively vanishes. In this limit, $\rho_{ee} \to \frac{1}{2}$ [see Eq. (67.11)], and the field induces as many upward transitions as downward transitions.

When radiation at the frequency ω propagates in a medium of two-level atoms, the energy density is

$$I(z) \propto \mathcal{E}^*\mathcal{E} \propto \exp\left[-\left(\frac{Nd^2\omega}{\hbar\varepsilon_0 c}\right)\frac{\gamma z}{(\omega-\omega_0)^2 + \gamma^2}\right]. \qquad (67.17)$$

Far from resonance ($|\omega - \omega_0| \gg \gamma$), the medium is transparent; but near resonance, damping is observed. The impinging radiation energy is deposited in the medium and propagation is impeded. This is called *radiation trapping*. In spectroscopy, the phenomenon is seen as a prolongation of the radiative decay time; the spontaneously emitted energy is seen to emerge from the sample more slowly than the single atom lifetime implies.

67.4 LINE BROADENING

The effective width of a spectral line from Eqs. (67.16) is

$$\gamma_{\text{eff}} = \sqrt{\gamma^2 + \Omega_1^2\frac{\gamma}{\Gamma}} \simeq \frac{1}{2}\Gamma + \gamma_{\text{ph}} + \frac{\Omega_1^2}{2\Gamma} + O\left(\Omega_1^4\right). \qquad (67.18)$$

The various contributions are as follows [3]. The term γ contains all the transverse relaxation mechanisms. If decay to additional levels occurs, these must be included (see Sect. 67.5). The term γ_{ph} contains all perturbing effects affecting each single atom. For low enough pressures, collisional perturbations are proportional to the density of perturbing atoms so that

$$\gamma_{\text{ph}} = \eta_{\text{coll}}\, p, \qquad (67.19)$$

where p is the pressure of the perturbing gas and η_{coll} is a constant of proportionality. This is called *collision* or *pressure broadening* (see Chap. 57) whose order of magnitude can be estimated to be the inverse of the average free time between collisions. For large pressure (usually of the order of torrs), the linearity in Eq. (67.19) breaks down. When identical atoms collide, *resonant exchange* of energy may also take place. The third term in Eq. (67.18) is the *power broadening* term. It derives from the effect of the laser field on each individual atom. All such relaxation processes that are active on each and every individual atom separately are called *homogeneous broadening* processes. For a detailed discussion of *line broadening*, consult Chap. 19.

In contrast to homogeneous broadening, Doppler broadening is characterized by an atomic velocity parameter v which varies over the observed assembly. In a thermal assembly at temperature T (e.g., a gas cell), the velocity distribution is

$$P(v) = \frac{1}{\sqrt{2\pi u^2}}\exp\left(-\frac{v^2}{2u^2}\right), \qquad (67.20)$$

where $u^2 = k_B T/M$. A particular atom with velocity v in the direction of the optical beam with wave vector k then experiences the Doppler-shifted frequency $\omega - kv$ relative to a stationary atom, and the effective detuning becomes

$$\Delta' = \Delta + kv, \qquad (67.21)$$

replacing Δ. The population in the lower level from Eq. (67.11) is then

$$\rho_{gg} = 1 - \frac{\Omega_1^2\gamma}{2\Gamma}\left(\frac{1}{(\Delta+kv)^2 + \gamma^2 + \Omega_1^2\gamma/\Gamma}\right). \qquad (67.22)$$

The atoms in the lower level, originally distributed according to (67.20), are now depleted of the velocity group around

$$v = (\omega - \omega_0)/k. \qquad (67.23)$$

The width of the depleted region is given by γ_{eff} of Eq. (67.18). This region is called a *Bennett hole*. When the laser frequency ω is tuned, the hole sweeps over the velocity distribution of the atoms. The atomic response is saturated at the velocity group of the hole, indicating that *spectral hole burning* has occurred.

The observed spectrum is obtained by averaging the single atom response (67.15) over the velocity distribution. From the imaginary part, the absorption response is

$$\alpha''(\omega) = \frac{d^2}{\hbar\sqrt{2\pi u^2}} \int_{-\infty}^{+\infty} \frac{\gamma\, e^{-v^2/2u^2}}{(\Delta + kv)^2 + \gamma^2 + \Omega_1^2\gamma/\Gamma}\, dv. \tag{67.24}$$

In the limits $\Omega_1 \to 0$ (no saturation), and $\gamma \ll ku$ (the Doppler limit), the Lorentzian line shape sweeps over the entire velocity profile, finding a resonant velocity group according to (67.23) as long as $v \le u$. Thus the linear spectroscopy sees a *Doppler broadened line* of width ku. This is called *inhomogeneous broadening*.

In the unsaturated regime, the atomic response function (67.24) is proportional to the imaginary part of the function

$$V(z) = \frac{1}{\sqrt{2\pi\sigma^2}} \int_{-\infty}^{+\infty} \frac{\exp\left(-x^2/2\sigma^2\right)}{z - x}\, dx, \tag{67.25}$$

at $z = -\Delta - i\gamma$ and $\sigma = ku$. This is the Hilbert transform of the Gaussian, and its shape is called a *Voigt profile* [4]. For $\gamma \ll ku$, it traces over the Gaussian, but for large detunings it always goes to zero as slowly as the Lorentzian; i.e., as Δ^{-1}. The profile has been widely used to interpret the data of *linear spectroscopy*. The function is tabulated [5] and its expansion is

$$V(z) = \sqrt{-\frac{\pi}{2\sigma^2}} \sum_{n=0}^{\infty} \frac{(iz/\sqrt{2}\sigma)^n}{\Gamma(1 + n/2)}, \tag{67.26}$$

and it has the continued fraction representation,

$$V(z) = \cfrac{1}{z - \cfrac{\sigma^2}{z - \cfrac{2\sigma^2}{z - \cfrac{3\sigma^2}{z - \cfrac{4\sigma^2}{z - \cdots}}}}}. \tag{67.27}$$

In addition to velocity, any other parameter shifting the individual atomic resonance frequencies ω_0 by different amounts for the different individuals leads to *inhomogeneous broadening*. The detuning Δ is then different for different members of the observed assembly, and a line shape similar to Eq. (67.24) applies. The distribution function must be replaced by the one relevant for the problem. In practice a Gaussian is almost always assumed.

An example of inhomogeneous broadening is the influence of the lattice environment on impurity spectroscopy in solids. The resonant light excites selectively atoms at those particular positions which make the atoms resonant. Thus only these spatial locations are saturated, and the phenomenon of *spatial hole burning* occurs. This has been investigated as a method for storing information, signal processing and volume holography.

It is, however, possible that the effects of collisions can counteract the inhomogeneous broadening. In order to see this we observe that the induced atomic dipole is proportional to the induced density matrix element, from Eq. (67.9),

$$\rho_{eg} = \tilde{\rho}_{eg} e^{i(kz(t)-\omega t)} \propto e^{ikvt}. \tag{67.28}$$

If the atoms now experience collisions characterized by an average free time of flight τ, the phase kvt cannot build up coherently for times longer than this duration, the atomic velocity is quenched on the average, and the full Doppler profile cannot be observed. To see how this comes about, consider a time $t \gg \tau$. During this period the atom experiences on the average $\bar{n} = t/\tau$ collisions. Assuming a Poisson distribution, the probability of n collisions in time t is

$$p_n = \frac{e^{-\bar{n}}}{n!}\, \bar{n}^n. \tag{67.29}$$

Taking the average of (67.28) over the time $t = n\tau$ with the distribution p_n yields

$$\begin{aligned} \bar{\rho}_{eg} &\propto \sum_{n=0}^{\infty} e^{ikv\tau n} \left(\frac{e^{-\bar{n}}}{n!}\, \bar{n}^n\right) \\ &= e^{-\bar{n}} \exp(\bar{n}\, e^{ikv\tau}) \\ &\approx e^{ikvt} \exp(-\tfrac{1}{2}tk^2v^2\tau). \end{aligned} \tag{67.30}$$

This heuristic derivation suggests that for long enough interaction times ($t \gg \tau$), the large velocity components are suppressed. This tends to prevent the tails of the velocity distribution from contributing to the observed spectral profile. The effect is called *collisional narrowing* or *Dicke narrowing*. It is an observable effect, but the narrowing cannot be very large. To overcome the Doppler broadening one has to turn to nonlinear laser methods (see Sect. 67.6).

67.5 THE RATE EQUATION LIMIT

Consider now a generalized theory for the case of several incoming electromagnetic fields of the form

$$\mathcal{E}(R,t) = \sum_i \frac{1}{2}\mathcal{E}_i(R)\, e^{-i\omega_i t + i\varphi_i} + \text{c.c.}. \tag{67.31}$$

The index i may range over several laser sources, the output of a multimode laser or the multitude of components of a flashlight or a thermal source. Each component carries its own amplitude $\mathcal{E}_i$.

In steady state, the generalization of Eq. (67.12) becomes

$$\rho_{eg} = \frac{i}{2} \sum_i \frac{d\mathcal{E}_i}{\hbar} \frac{(\rho_{gg} - \rho_{ee})}{\gamma + i\Delta_i}\, e^{-i\omega_i t + i\varphi_i}, \tag{67.32}$$

where the detuning is

$$\Delta_i = \omega_0 - \omega_i. \qquad (67.33)$$

The response of the atom now separates into individual contributions oscillating at the various frequencies ω_i according to

$$\rho_{eg} = \sum_i \rho_{eg}^{(i)} e^{-i\omega_i t + i\varphi_i}. \qquad (67.34)$$

This resolution is of key importance to the theory.

With the multimode field, Eqs. (66.67) for the level occupation probabilities become

$$\frac{d}{dt}\rho_{ee} = -\frac{d}{dt}\rho_{gg}$$

$$= -\Gamma\rho_{ee} + \frac{i}{2}\sum_i \left(\frac{d\mathcal{E}_i}{\hbar} e^{-i\omega_i t + i\varphi_i} \rho_{ge} - \text{c.c.} \right). \qquad (67.35)$$

Insertion of the steady state result (67.32) into this equation yields a closed set of equations for the level occupation probabilities. These are called *rate equations.*

To justify the above steps, consider the single frequency case again. The off-diagonal time derivatives can be neglected when

$$\left| \frac{d}{dt}\rho_{eg} \right| \ll |\gamma + i\Delta| |\rho_{eg}|. \qquad (67.36)$$

This can be surmised to hold when the phase relaxation contributions to γ are large [*cf.* Eq. (67.8)] or the detuning $|\Delta|$ is large. Insertion of (67.32) into (67.35) for the single mode case gives

$$\frac{d}{dt}(\rho_{ee} - \rho_{gg}) = -\Gamma\rho_{ee} - W(\rho_{ee} - \rho_{gg}), \qquad (67.37)$$

where the rate coefficient is given by

$$W = 2\pi \left(\frac{d\mathcal{E}}{2\hbar} \right)^2 \frac{\gamma/\pi}{\Delta^2 + \gamma^2}. \qquad (67.38)$$

Multiplying Eq. (67.37) by the density of active atoms N then produces the conventional rate equations for the populations $N_{ee} = N\rho_{ee}$ and $N_{gg} = N\rho_{gg}$.

Two physical effects can be discerned in Eq. (67.37): *induced* and *spontaneous emission.* The term with Γ gives the spontaneous emission which forces all the population to the lower level. The terms proportional to W describe induced emission, with upward transitions proportional to $W\rho_{gg}$ and downward transitions proportional to $W\rho_{ee}$. In the absence of spontaneous emission, they strive to equalize the population on the two levels. Using Eq. (67.10), the steady-state solution is

$$\rho_{ee} = \frac{W}{\Gamma + 2W}$$

$$= \frac{1}{2}\left(\frac{d\mathcal{E}}{\hbar} \right)^2 \frac{\gamma}{\Gamma} \frac{1}{\Delta^2 + \gamma^2 + (d\mathcal{E}/\hbar)^2\gamma/\Gamma}. \qquad (67.39)$$

This is clearly seen to agree with the solution (67.11) as is expected in steady state.

Although the rate equations were derived in the limit (67.36), the rate coefficient (67.38) has a special significance in the limit $\gamma \to 0$. In this limit, the factor

$$\frac{1}{\hbar} \lim_{\gamma \to 0} \left(\frac{\gamma/\pi}{\Delta^2 + \gamma^2} \right) = \frac{1}{\hbar}\delta(\Delta)$$

$$= \delta(\hbar\omega - \hbar\omega_0), \qquad (67.40)$$

enforces energy conservation in the transition. Using the field in Eq. (67.14), and the interaction from Eq. (67.6), the off-diagonal matrix element is

$$|\langle e | H | g \rangle| = \tfrac{1}{2}d\mathcal{E}. \qquad (67.41)$$

With these results, Eq. (67.38) can be written in the form of *Fermi's Golden Rule*

$$W = \frac{2\pi}{\hbar} |\langle e | H | g \rangle|^2 \delta(\hbar\omega - \hbar\omega_0), \qquad (67.42)$$

usually derived from time dependent perturbation theory.

Returning to the multimode rate equations, an incoherent *broad band light source* has many components that contribute to the sum over field frequencies. In the case of flash pulses, thermal light sources or free-running multimode lasers the spectral components are uncorrelated. Inserting (67.32) into (67.35) yields

$$\frac{d}{dt}\rho_{ee} = -\frac{d}{dt}\rho_{gg}$$

$$= -\Gamma\rho_{ee} + \frac{d^2}{2\hbar^2}(\rho_{gg} - \rho_{ee})$$

$$\times \sum_{i,j} \mathcal{E}_i \mathcal{E}_j e^{i(\omega_j - \omega_i)t} e^{i(\varphi_i - \varphi_j)} \frac{\gamma}{\Delta_j^2 + \gamma^2}. \qquad (67.43)$$

The contributions from the different terms $i \neq j$ average to zero either by beating at the frequencies $|\omega_i - \omega_j|$ or by incoherent effects from the random phases φ_j. Thus only the coherent sum survives to give

$$\frac{d}{dt}\rho_{ee} = -\frac{d}{dt}\rho_{gg}$$

$$= -\Gamma\rho_{ee} - \sum_i W^{(i)}(\rho_{gg} - \rho_{ee}), \qquad (67.44)$$

where the rate coefficients $W^{(i)}$ are given by (67.38) with the appropriate detunings $\Delta_j = \omega_0 - \omega_j$. This is a *rate equation* in the limit of many uncorrelated components of light; i.e., for a *broad band light source.* In this case the incoherence between the different components justifies the use of a rate approach, and no assumption like (67.36) is needed. Thus the limit $\gamma \to 0$ [see Eq. (67.40)] is legitimate, and the $W^{(i)}$ can be calculated in time dependent perturbation theory from Fermi's golden rule.

In the limit of an incoherent broad band light source, the sum in Eq. (67.44) can be replaced by an integral. In particular, this is allowed for incandescent light sources as used in *optical pumping* experiments [6]. Pumping of lasers by strong lamps or flashes are also describable by the same rate equations.

In amplifiers and lasers, the atoms must be brought into states far from equilibrium by incoherent optical excitation or resonant transfer of excitation energy in collisions (see Chap. 68). In the two-level description, the atomic levels are constantly replenished. The normalization condition (67.10) is then no longer appropriate; often the density matrix is normalized so that $\text{Tr}(\rho)$ directly gives the density of active atoms.

With pumping into the levels, one must allow for decay out of the two-level system in order to prevent the atomic density from growing in an unlimited way. This decay takes the atom to unobserved levels. In the rate equation approximation, the pumping and decay processes can be described by terms added to the equations of the form [*cf.* Eq. (66.68)]

$$\frac{d}{dt}\rho_{ee} = \lambda_e - \gamma_e \rho_{ee},$$
$$\frac{d}{dt}\rho_{gg} = \lambda_g - \gamma_g \rho_{gg}. \tag{67.45}$$

In a laser, the level $|g\rangle$ is usually not the ground state of the system.

From Eq. (67.45), the steady state population is

$$\rho_{gg}^{(0)} - \rho_{ee}^{(0)} = \lambda_g/\gamma_g - \lambda_e/\gamma_e. \tag{67.46}$$

A *population inversion* exists when this is negative. The population difference (67.46) is modified when the effects of spontaneous and induced processes are added, as in Eq. (67.37). Then the transitions saturate because of the *induced processes*.

Using (67.46) in (67.12), the calculated polarizability without saturation is

$$\alpha(\omega) = \left(\frac{d^2}{\hbar}\right)\left(\frac{i\gamma + \Delta}{\gamma^2 + \Delta^2}\right)\left(\rho_{gg}^{(0)} - \rho_{ee}^{(0)}\right). \tag{67.47}$$

According to (67.1), the index of refraction is

$$n'(\omega) = 1 + \frac{N\alpha'(\omega)}{2\varepsilon_0}$$
$$= 1 - \left(\frac{Nd^2}{2\varepsilon_0\hbar}\right)\frac{(\omega - \omega_0)}{(\omega - \omega_0)^2 + \gamma^2}\left(\rho_{gg}^{(0)} - \rho_{ee}^{(0)}\right). \tag{67.48}$$

For weakly excited atoms, $\rho_{gg} \approx 1$, and Eq. (67.47) agrees with the unsaturated limit of (67.16). The dispersion of a light signal behaving according to (67.48) is called *normal*, i.e., according to the harmonic model in Sect. 67.1. Below resonance ($\omega < \omega_0$), n' is larger than unity, implying a reduction of the velocity of light.

As a function of ω, the curve [Eq. (67.48)] starts above unity, and passes below unity for $\omega > \omega_0$. This is *normal dispersion*.

However, for an *inverted medium* ($\rho_{gg}^{(0)} < \rho_{ee}^{(0)}$), n' is less than unity for low frequencies and goes through unity with a positive slope. This is called *anomalous dispersion* and signifies the presence of a *gain profile*. In such a medium, $\alpha'' = \Im\alpha(\omega)$ is of opposite sign, as seen from Eq. (67.47); in the inverted medium α'' becomes negative near $\Delta = 0$. From Eq. (67.2), this indicates a growing electromagnetic field; i.e., an *amplifying medium*. The amplitude grows, and the assumption of a small signal becomes invalid. Then saturation has to be included, either at the rate equation level or by performing a full density matrix calculation. This regime describes a laser with *saturated gain*. In steady state, the two levels become nearly equally populated ($\rho_{ee} \approx \rho_{gg}$), and the operation is stable. The theory of the laser is discussed in detail in Chap. 68.

67.6 TWO-LEVEL DOPPLER-FREE SPECTROSCOPY

The *linear absorption* of a scanned laser signal defines *linear spectroscopy* and gives information characterizing the sample. However, Sect. 67.4 shows that inhomogeneous broadening masks the desired information by dominating the line shape. The availability of laser sources has made it possible to overcome this limitation, and use the *saturation* properties of the medium to perform *nonlinear spectroscopy*. This Section discusses how Doppler broadening can be eliminated to achieve *Doppler-free spectroscopy*. Similar techniques may be used to overcome other types of *inhomogeneous line broadening*; a general name is then *hole-burning spectroscopy*; see the discussion in Sect. 67.4. Other aspects of nonlinear matter-light interaction are found in Chap. 70.

Equation (67.24) shows that a single laser cannot resolve beyond the Doppler width. However, if a strong laser is used to pump the transition, a weak *probe signal* can see the hole burned into the spectral profile by the pump. This technique is called *pump-probe spectroscopy*. Because the probe is taken to be weak, perturbation theory may be used to calculate the induced polarization to lowest order in the probe amplitude only.

In the field expansion (67.31), define the strong pump amplitude to be $\mathcal{E}_1$ and the weak probe $\mathcal{E}_2$ at frequency ω_2. From the resolution (67.34), the component $\rho_{eg}^{(2)}$ carries the information about the linear response at frequency ω_2. If the field $\mathcal{E}_2$ propagates in a direction opposite to that of $\mathcal{E}_1$, its detuning is

$$\Delta_2' = \Delta_2 - kv \tag{67.49}$$

(as compared with $\Delta_1' = \Delta_1 + kv$ for $\mathcal{E}_1$). Since $\omega_1 \simeq \omega_2$, the two k-vectors are nearly equal in magnitude.

The linear response now becomes

$$\rho_{eg}^{(2)} = \left(\frac{id\mathcal{E}_2}{2\hbar}\right)\frac{1}{\gamma + i(\Delta_2 - kv)}(\rho_{gg} - \rho_{ee}).\quad(67.50)$$

With only the signal $\mathcal{E}_1$ present, the population difference follows directly from Eq. (67.22). The linear response at frequency ω_2 is then

$$\rho_{eg}^{(2)} = \left(\frac{d\mathcal{E}_2}{2\hbar}\right)\frac{i\gamma + (\Delta_2 - kv)}{\gamma^2 + (\Delta_2 - kv)^2},$$
$$\times\left[1 - \frac{\Omega_1^2\gamma}{\Gamma}\left(\frac{1}{(\Delta_1 + kv)^2 + \gamma^2 + \Omega_1^2\gamma/\Gamma}\right)\right](67.51)$$

This is the linear response of atoms moving with velocity v. To obtain the polarization of the whole sample, we must average over the velocity distribution using the Gaussian weight (67.20). The first term in (67.51) gives the linear response in the form of a *Voigt profile* as discussed in Sect. 67.4. This part of the response carries no Doppler-free information. The second terms contain the nonlinear response. This shows the details of the homogeneous features under the Doppler line shape. For simplicity we assume the Doppler limit, $\gamma \ll ku$, and neglect the variation of the Gaussian over the atomic line shape. We also neglect the power broadening due to the field $\mathcal{E}_1$ and obtain, using Eqs. (67.13), (67.20), and (67.51),

$$\alpha''(\omega) = -\left(\frac{d^2}{\hbar}\right)\frac{\Omega_1^2\gamma^2}{\sqrt{2\pi}\Gamma u}$$
$$\times\int_{-\infty}^{+\infty}\frac{dv}{\left((\Delta_2 - kv)^2 + \gamma^2\right)\left((\Delta_1 + kv)^2 + \gamma^2\right)}$$
$$= -\left(\frac{d^2}{\hbar}\right)\frac{\sqrt{2\pi}\Omega_1^2}{4\Gamma ku}\frac{\gamma}{(\omega - \omega_0)^2 + \gamma^2}.\quad(67.52)$$

This denotes the energy absorbed from the field $\mathcal{E}_1$, as induced nonlinearly by the intensity $\mathcal{E}_1^2$. The resonance is still at $\omega = \omega_0$, but with a homogeneous atomic line shape. In the Doppler limit, the Doppler broadening is only seen in the prefactor

$$\frac{\Omega_1^2}{\Gamma ku} = \left(\frac{\Omega_1^2}{\Gamma\gamma}\right)\frac{\gamma}{ku},\quad(67.53)$$

which shows that only the fraction (γ/ku) of all atoms can contribute to the resonant response. The first factor on the right-hand side of Eq. (67.53) is the dimensionless saturation factor.

The Doppler-free character of this spectroscopy derives from the fact that the two fields burn their two separate Bennett holes at the velocities

$$kv_1 = -\Delta_1, \quad kv_2 = \Delta_2.\quad(67.54)$$

When these two groups coincide $v_1 = v_2$, the probe $\mathcal{E}_2$ sees the absorption saturated by the pump field $\mathcal{E}_1$ and

a decreased absorption is observed. For $\omega_1 = \omega_2$, the two holes meet in the middle at zero velocity. With two different frequencies, one can make the holes meet at a nonzero velocity to one side of the Doppler profile. The decreased absorption seen in these experiments is called an *inverted Lamb dip* (see also Chap. 68).

The results derived here are based on a simplified view of the pump-probe response based on the rate equation approach. Certain coherent effects are neglected, which considerably complicate the treatment [2]. Section 67.7 discusses these effects in the three-level system where they are more important.

In addition to measuring the probe absorption induced by the pump field, it is also possible to observe the dispersion of the probe signal caused by the saturation induced by the pump. Assuming that the pump introduces the velocity dependent population difference

$$\rho_{ee} - \rho_{gg} = \Delta\rho(v),\quad(67.55)$$

then from Eq. (67.50), $\Re(n)$ is

$$n' = 1 + \frac{Nd^2}{2\hbar\varepsilon_0}\int\frac{\Delta_2 - kv}{(\Delta_2 - kv)^2 + \gamma^2}\Delta\rho(v)\,dv.\quad(67.56)$$

From Eq. (67.2), the phase of the electromagnetic signal feels the value of n' through the factor

$$\mathcal{E} \propto \exp(i\omega n'z/c).\quad(67.57)$$

Thus, by modifying $\Delta\rho(v)$, a pump laser can control the phase acquired by light traversing the sample, and thereby control the optical length of the sample.

A real atom has magnetic sublevels, which are coupled to light in accordance with *dipole selection rules* (see the discussion in Chap. 32). If a pump laser is used to affect populations in the various sub-levels differently, the optical paths experienced by the differently circularly polarized components of a linearly polarized probe signal are different. Thus its plane of polarization will turn, corresponding to the *Faraday effect*. By tuning the pump and the probe over the spectral lines of the sample, the turning of the probe polarization provides a signal to investigate the atomic level structure. This method of *polarization spectroscopy* can be used both with two-level and three-level systems [7].

67.7 THREE-LEVEL SPECTROSCOPY

New nonlinear phenomena appear when one of the levels in the two-state configuration, $|e\rangle$ say, is coupled to a final level $|f\rangle$ by a weak probe. This may be above the level $|e\rangle$ (the cascade configuration denoted by Ξ) or below $|e\rangle$ (the lambda configuration Λ) (if the third level were coupled to $|g\rangle$ we would talk about the inverted lambda or V configuration) [8]. For simplicity, only the Ξ configuration is discussed here.

Assume that the level pair $|g\rangle \leftrightarrow |e\rangle$ is pumped by the field $\mathcal{E}_1$ and its effect probed by the field $\mathcal{E}_2$ coupling $|e\rangle \leftrightarrow |f\rangle$. The dipole matrix element is $d_2 = \langle f|H|e\rangle$. The RWA is now achieved by introducing slowly varying quantities through the definitions

$$\rho_{fe} = e^{i(k_2 z - \omega_2 t)} \tilde{\rho}_{fe},$$
$$\rho_{fg} = e^{i[(k_1 + k_2)z - (\omega_1 + \omega_2)t]} \tilde{\rho}_{fg}, \qquad (67.58)$$

and omitting all components oscillating at multiples of the optical frequencies. From the equation of motion for the density matrix, the steady state equations are [2]

$$(\Delta_2 + k_2 v - i\gamma_{fe}) \tilde{\rho}_{fe} = \frac{d_2 \mathcal{E}_2}{2\hbar} \rho_{ee} - \frac{d_1 \mathcal{E}_1}{2\hbar} \tilde{\rho}_{fg},$$
$$[(\Delta_1 + \Delta_2) + (k_1 + k_2)v - i\gamma_{fg}] \tilde{\rho}_{fg} \qquad (67.59)$$
$$= \frac{d_2 \mathcal{E}_2}{2\hbar} \tilde{\rho}_{eg} - \frac{d_1 \mathcal{E}_1}{2\hbar} \tilde{\rho}_{fe},$$

where the second step detuning is now

$$\Delta_2 = \omega_{fe} - \omega_2. \qquad (67.60)$$

These equations contain the lowest order response proportional to $\mathcal{E}_2$, including some coherence effects. The coherence effects remain to lowest order, even when the last term in Eq. (67.59) is neglected to obtain the solution

$$\tilde{\rho}_{fe} = \left(\frac{d_2 \mathcal{E}_2}{2\hbar}\right) \frac{\rho_{ee}}{\Delta_2 + k_2 v - i\gamma_{fe}} - \left(\frac{d_1 \mathcal{E}_1}{2\hbar}\right) \left(\frac{d_2 \mathcal{E}_2}{2\hbar}\right)$$
$$\times \frac{\tilde{\rho}_{eg}}{(\Delta_2 + k_2 v - i\gamma_{fe})((\Delta_1 + \Delta_2) + (k_1 + k_2)v - i\gamma_{fg})}. \qquad (67.61)$$

The result to lowest order in $\mathcal{E}_2$ follows by replacing the density matrix elements for the two-level system $|e\rangle \leftrightarrow |g\rangle$ by their results calculated without this field. This shows that the polarization induced at the frequency ω_2 consists of two parts: one induced by the population excited to level $|e\rangle$ by the field $\mathcal{E}_1$ proportional to ρ_{ee}, and the other one is induced by the coherence $\tilde{\rho}_{eg}$ created by the pump. Retaining only the former produces a rate equation approximation, called a *two-step process*. This misses important physical features which are included in the second term called a *two-photon* or *coherent* process.

In order to see the effects of the two terms most clearly, consider the two-level matrix elements from Eqs. (67.11 and 67.12) in lowest perturbative order with respect to the pump field $\mathcal{E}_1$; i.e.,

$$\rho_{ee} = \left(\frac{d_1 \mathcal{E}_1}{\hbar}\right)^2 \frac{\gamma_{eg}}{2\gamma_{ee}} \frac{1}{(\Delta_1 + k_1 v)^2 + \gamma_{eg}^2},$$
$$\tilde{\rho}_{eg} = \frac{d_1 \mathcal{E}_1}{2\hbar} \frac{1}{\Delta_1 + k_1 v - i\gamma_{eg}}. \qquad (67.62)$$

From Eq. (67.61), these matrix elements give

$$\tilde{\rho}_{fe} = -\left(\frac{A}{\Delta_1 + k_1 v - i\gamma_{eg}}\right)$$
$$\times \left[\left(\frac{2\gamma_{eg}}{\gamma_{ee}}\right) \frac{1}{(\Delta_1 + k_1 v + i\gamma_{eg})(\Delta_2 + k_2 v - i\gamma_{fe})}\right.$$
$$\left. - \frac{1}{(\Delta_2 + k_2 v - i\gamma_{fe})((\Delta_1 + \Delta_2) + (k_1 + k_2)v - i\gamma_{fg})}\right], \qquad (67.63)$$

where

$$A = -\left(\frac{d_2 \mathcal{E}_2}{2\hbar}\right)\left(\frac{d_1 \mathcal{E}_1}{2\hbar}\right)^2. \qquad (67.64)$$

The imaginary part of this yields the absorptive part of the polarization at the frequency ω_2. The first term becomes the product of two Lorentzians, and is an incoherent rate contribution. It dominates when the induced population ρ_{ee} decays much more slowly than the induced coherence $\tilde{\rho}_{eg}$; i.e., when $\gamma_{ee} \ll \gamma_{eg}$.

The significance of the second term in Eq. (67.63) is evident in the limit when no phase perturbing processes intervene, so that

$$\gamma_{ee} = \Gamma, \qquad \gamma_{eg} = \tfrac{1}{2}\Gamma,$$
$$\gamma_{fg} = \tfrac{1}{2}\gamma_{ff}, \qquad \gamma_{fe} = \tfrac{1}{2}(\gamma_{ff} + \Gamma). \qquad (67.65)$$

For $v = 0$,, $\tilde{\rho}_{fe}$ then becomes

$$\tilde{\rho}_{fe} = -\frac{A}{\Delta_1^2 + (\Gamma/2)^2}\left(\frac{1}{\Delta_1 + \Delta_2 - i\tfrac{1}{2}\gamma_{ff}}\right). \qquad (67.66)$$

Neglecting the decay of the final level $|f\rangle$, $\gamma_{ff} \to 0$, the absorption becomes proportional to

$$\Im \tilde{\rho}_{fe} = \frac{\pi|A|}{\Delta_1^2 + (\Gamma/2)^2} \delta(\Delta_1 + \Delta_2). \qquad (67.67)$$

In this limit, strict energy conservation between the ground state and the final state must prevail; the final state must be reached by the absorption of exactly two quanta. The delta function in Eq. (67.67) indicates precisely this:

$$\Delta_1 + \Delta_2 = \omega_{fe} + \omega_{eg} - (\omega_1 + \omega_2)$$
$$= \omega_{fg} - \omega_1 - \omega_2$$
$$= 0. \qquad (67.68)$$

The detuning and the width of the intermediate state affect the total transition rate, but not the condition of energy conservation. The presence of the second term in (67.63) makes the resonance contributions at

$$\Delta_2 = \omega_{fg} - \omega_2 = 0 \qquad (67.69)$$

cancel exactly. Only a *two-photon transition* remains; in the Λ configuration this would be a *Raman process* (see Chaps. 60 and 70).

If the velocity dependence in Eq. (67.63) is retained, the nonlinear response of an atomic sample must be averaged over the velocity distribution given by the Gaussian (67.20). The computations become involved, but the results show more or less well resolved resonances around the two positions (67.68) and (67.69); i.e., the *coherent two-photon process* and the single step rate process $|e\rangle \to |f\rangle$ appearing due to a previous single step process $|g\rangle \to |e\rangle$; this is the *two-step process*.

A special situation arises when the intermediate step is so much detuned that no velocity group is in resonance; i.e., $|\Delta_1| \approx |\Delta_2| \gg kv$ for all velocities contributing significantly to the spectrum. Then the second coherent term of Eq. (67.63) can be written in the form

$$\tilde{\rho}_{fe} = -\frac{A}{\Delta_1^2} \left(\frac{1}{(\Delta_1 + \Delta_2) + (k_1 + k_2)\,v - i\gamma_{fg}} \right).$$

(67.70)

If the two fields $\mathcal{E}_1$ and $\mathcal{E}_2$ have the same frequency but are counterpropagating, then $k_1 + k_2 = 0$, and no velocity dependence occurs in (67.70) for the two-photon resonance. All atoms in the sample contribute to the strength of the resonance, and then polarization is obtained directly by multiplication with the total atomic density. Equation (67.13) then gives

$$\alpha_2'' = \frac{d_2^2}{\hbar} \left(\frac{d_1 \mathcal{E}_1}{2\hbar\Delta_1} \right)^2 \frac{\gamma_{fg}}{(2\omega - \omega_{fg})^2 + \gamma_{fg}^2}.$$

(67.71)

This is a sharp *Doppler-free resonance* on the two-photon transition ω_{fg}. The advantage is that all atoms contribute with the sharp line width γ_{fg}, which is not easily affected by phase perturbations because of the two-photon nature of the transition. The disadvantage is the weakness of the transition, which is caused by the large detuning, making $d_1\mathcal{E}_1/|\Delta_1| \ll 1$ in most cases. However, with tunable lasers this Doppler-free spectroscopy method has been used successfully in many cases.

67.8 SUMMARY OF THE LITERATURE

Much of the material needed to formulate the basic theory of interaction between light and matter can be found in the text book [2]. A comparison between the harmonic model and the two-level model is given by Feld [1]. The density matrix formulation is presented in detail in Ref. [2]. The influence of various line broadening mechanisms on laser spectroscopy is discussed in the book [3]. The Voigt profile is related to the error function, which is treated in the compilation [5]. The numerical evaluation of the Voigt profile is discussed in Ref. [4]. Rate equations are commonly used in laser theory and they are derived for optical pumping and laser-induced processes in the lectures [6]. The Doppler-free spectroscopy was developed in the 1960s and 1970s by many authors following the initial discovery of the Benett hole by Bill R. Bennett Jr. and the Lamb dip by Willis E. Lamb Jr. Much of the pioneering work can be found in the book [3]. The three-level work has been reviewed by Chebotaev [8]. Various applications of lasers in spectroscopy are treated by Levenson and Kano [7]. For references to other topics, we refer to the specialized chapters of the present book.

REFERENCES

1. M. S. Feld, in *Frontiers in Laser Spectroscopy*, edited by R. Balian, S. Haroche, and S. Liberman (North-Holland, Amsterdam, 1977), p. 203.
2. S. Stenholm, *Foundations of Laser Spectroscopy* (John Wiley & Sons, New York, 1984).
3. V. S. Letokhov and V. P. Chebotaev, *Nonlinear Laser Spectroscopy* (Springer-Verlag, Heidelberg, 1977).
4. W. J. Thompson, Computers in Physics, **7**, 627 (1993).
5. M. Abramowitz and I. E. Stegun, *Handbook of Mathematical Functions* (Dover Publ. New York, 1970).
6. C. Cohen-Tanoudji, in *Frontiers in Laser Spectroscopy*, edited by R. Balian, S. Haroche, and S. Liberman (North-Holland, Amsterdam, 1977), p. 1.
7. M. D. Levenson and S. S. Kano, *Introduction to Nonlinear Laser Spectroscopy* (Academic Press, New York, 1988).
8. V. P. Chebotaev, in *High-Resolution Spectroscopy*, edited by K. Shimoda (Springer-Verlag, Heidelberg, 1976).

68

Laser Principles

P. W. Milonni

Theoretical Division (T-4), Los Alamos National Laboratory, Los Alamos, New Mexico 87545

68.1 INTRODUCTION

Despite their great variety and range of power, wavelength, and temporal characteristics, all lasers involve certain basic concepts, such as gain, threshold, and electromagnetic modes of oscillation [1–3]. In addition to these universal characteristics are features, such as Gaussian beam modes, that are important to such a wide class of devices that they must be included in any reasonable compendium of important laser concepts and formulas. We have therefore included here both generally applicable results as well as some more specific but widely applicable ones.

68.2 GAIN, THRESHOLD, AND MATTER–FIELD COUPLING

All lasers involve some medium that amplifies an electromagnetic field within some band of frequencies. At the simplest level of description, the amplifying medium changes the intensity I of a field according to the equation

$$\frac{dI}{dz} = gI \,, \tag{68.1}$$

where z is the coordinate along the direction of propagation and g is the gain coefficient, typically expressed in

cm^{-1}. Amplification occurs as a consequence of stimulated emission of radiation from the upper state (or band of states) of a transition for which a population inversion exists; i.e., for which an upper state has greater likelihood of occupation than a lower state. Different types of lasers may be classified by the pump mechanisms used to achieve population inversion (see Chap. 69). In the case that the amplifying transition involves two discrete energy levels, E_1 and $E_2 > E_1$, the gain coefficient at the frequency ν is given by

$$g(\nu) = \frac{\lambda^2 A}{8\pi n^2} \left(N_2 - \frac{g_2}{g_1} N_1 \right) S(\nu) \,. \tag{68.2}$$

Here $\lambda = c/\nu$ is the transition wavelength, A (s^{-1}) is the Einstein A coefficient for spontaneous emission for the transition, and g_2, g_1 are the degeneracies of the upper and lower energy levels. These quantities in nearly every case are fixed characteristics of the medium, independent of the laser intensity or the pump mechanism. N_2 and N_1 are the population densities (cm^{-3}) of the upper and lower levels, respectively, and $S(\nu)$ is the normalized transition lineshape function (see Chaps. 19 and 67). n is the refractive index at frequency ν of the "background" host medium and in general has contributions from all nonlasing transitions. Equation (68.2) describes either amplification or absorption, depending on whether $N_2 - (g_2/g_1)N_1$ is positive (amplification) or negative (absorption).

By far the most common configuration is that in which the gain medium is contained in a cavity bounded on two sides by reflecting surfaces. The mirrors allow feedback; i.e., the redirection of the field back into the gain medium for multipass amplification and sustained laser action. The two mirrors allow the field to build up along the directions parallel to the "optical axis" and to form a pencil-like beam of light. In order to sustain laser action, the gain in intensity due to stimulated emission must equal or exceed the loss due to imperfect mirror reflectivities, scattering, absorption in the host medium, and diffraction.

Typically the imperfect mirror reflectivities dominate the other sources of loss. If the mirror reflectivities are r_1 and r_2, then the intensity I is reduced by the factor $r_1 r_2$ in a round trip pass through the cavity, while according to Eq. (68.1) the gain medium causes the intensity to increase by a factor $\exp(2g\ell)$ in the two passes through the gain cell of length ℓ. Equating of the gain and loss factors leads to the *threshold condition* for laser oscillation: $g \geq g_t$, where the threshold gain is

$$g_t = -\frac{1}{2\ell}\ln(r_1 r_2) + \alpha\,, \qquad (68.3)$$

α being an attenuation coefficient associated with any loss mechanisms that may exist in addition to reflection losses at the mirrors.

Suppose, for example, that a laser has a 50 cm gain cell and mirrors with reflectivities 0.99 and 0.97, and that absorption within the host medium of the gain cell is negligible. Then the threshold gain is $g_t = 4\times10^{-4}$ cm^{-1}. If the lasing transition is the 6328 Å Ne transition of the He–Ne laser, we have $A \cong 1.4 \times 10^6$ s^{-1}, $n \cong 1.0$ and, assuming a pure Doppler lineshape,

$$S(\nu) = \left(\frac{4\ln 2}{\pi}\right)^{1/2}\frac{1}{\delta\nu_D} \qquad (68.4)$$

at line center, where $\delta\nu_D$ is the width (FWHM) of the Doppler lineshape (see Sect. 67.4). For $T = 400$ K and the Ne atomic weight, $\delta\nu_D \cong 1500$ MHz and $S(\nu) \cong 6.3\times10^{-10}$ s. Then the threshold population difference required for laser oscillation is

$$\left(N_2 - \frac{g_2}{g_1}N_1\right)_t = \frac{8\pi n^2 g_t}{\lambda^2 A S(\nu)} \cong 2.8 \times 10^9 \text{ cm}^{-3}\,. \quad (68.5)$$

This is a typical result: the population inversion required for laser oscillation is small compared with the total number of active atoms.

Calculations of population inversions and other properties of the gain medium are based on rate equations, or more generally, the density matrix ρ. In many instances, the medium is fairly well described in terms of two energy eigenstates, other states appearing only indirectly

through pumping and decay channels. In this case, ρ is a 2×2 matrix whose elements satisfy [3] [*cf.* Eqs. (66.66) and (66.67)]

$$\dot\rho_{22} = -(\Gamma_2 + \Gamma)\rho_{22} - \tfrac{1}{2}i(\Omega_1^*\tilde\rho_{21} - \Omega_1\tilde\rho_{12})\,,$$
$$\dot\rho_{11} = -\Gamma_1\rho_{11} + \Gamma\rho_{22} + \tfrac{1}{2}i(\Omega_1^*\tilde\rho_{21} - \Omega_1\tilde\rho_{12})\,, \qquad (68.6)$$
$$\dot{\tilde\rho}_{21} = -(\gamma + i\Delta)\tilde\rho_{21} - \tfrac{1}{2}i\Omega_1(\rho_{22} - \rho_{11})\,,$$

with $\tilde\rho_{12} = \tilde\rho_{21}^*$. Here, Γ_2 and Γ_1 are, respectively, the rates of decay of the upper and lower states due to all processes other than the spontaneous decay from state 2 to state 1 described by the rate $\Gamma = A$. γ, which is 2π times the homogeneous linewidth (HWHM) of the transition, is the rate of decay of off-diagonal coherence due to both elastic and inelastic processes; in general, $\gamma \geq (\Gamma_1+\Gamma_2+\Gamma)/2$. $\Omega_1 = \mathbf{d}_{21}\cdot\boldsymbol{\mathcal{E}}/\hbar$ is the Rabi frequency[1] (see Sect. 66.4.3 and Chap. 71), with $\boldsymbol{\mathcal{E}}$ the complex amplitude of the electric field; i.e., the electric field is

$$\mathbf{E}(\mathbf{r},t) = \Re[\boldsymbol{\mathcal{E}}(\mathbf{r},t)e^{i(\mathbf{k}\cdot\mathbf{r}-\omega t)}] \cong \Re[\boldsymbol{\epsilon}\mathcal{E}(\mathbf{r},t)e^{ikz}e^{-i\omega t}]\,.$$
$$(68.7)$$

It is assumed that $\mathcal{E}$ is slowly varying in time compared with the oscillations at frequency ω ($= 2\pi\nu$), and that the wave vector $\mathbf{k}$ is approximately $k\hat{\mathbf{z}} = (n\omega/c)\hat{\mathbf{z}}$, where $\hat{\mathbf{z}}$ is a unit vector pointing in the direction of propagation. Finally, $\Delta = \omega_0 - \omega$ in Eq. (68.6) is the detuning of ω from the central transition frequency $\omega_0 = (E_2 - E_1)/\hbar$ of the lasing transition. Rapidly oscillating terms involving $\omega_0 + \omega$ are ignored in the *rotating wave approximation* that pervades nearly all of laser theory (see Sect. 66.4.2).

In most lasers, γ is so large compared with the diagonal decay rates that the off-diagonal elements of ρ may be assumed to relax quickly to the quasisteady values obtained by setting $\dot\rho_{12} = 0$ in Eq. (68.6). Then the diagonal density matrix elements satisfy the *rate equations* [*cf.* Eq. (67.37)]

$$\dot\rho_{22} = -(\Gamma_2 + \Gamma)\rho_{22} - \frac{|\Omega_1|^2\gamma/2}{\Delta^2 + \gamma^2}(\rho_{22} - \rho_{11})\,,$$
$$\dot\rho_{11} = -\Gamma_1\rho_{11} + \Gamma\rho_{22} + \frac{|\Omega_1|^2\gamma/2}{\Delta^2 + \gamma^2}(\rho_{22} - \rho_{11})\,. \qquad (68.8)$$

Such rate equations, usually expressed equivalently in terms of population densities N_2, N_1 rather than occupation probabilities ρ_{22}, ρ_{11}, are the basis of most practical computer models of laser oscillation. These equations, or, more generally, the density matrix equations, must also include terms accounting for the pump mechanism. In the simplest model of pumping, one adds a constant pump rate Λ_2 to the right-hand side of the equation for ρ_{22} to obtain [*cf.* Eq. (67.45)]

---

[1] The Rabi frequency is often defined as $2\mathbf{d}_{21}\cdot\boldsymbol{\mathcal{E}}/\hbar$.

$$\dot{\rho}_{22} = \Lambda_2 - (\Gamma_2 + A)\rho_{22} - \frac{|\Omega_1|^2 \gamma/2}{\Delta^2 + \gamma^2}(\rho_{22} - \rho_{11}). \quad (68.9)$$

In the case of an inhomogeneously broadened laser transition (see Sect. 67.4), equations of the type (68.6) and (68.8) apply separately to each detuning Δ arising from the distribution of atomic or molecular transition frequencies. In writing these equations, we have assumed a nondegenerate electric dipole transition. The generalization to magnetic or multiphoton transitions, or to a case where the amplification is due, for instance, to a Raman process, is straightforward but of less general interest.

A more realistic treatment of the electromagnetic field than that based on Eq. (68.1) proceeds from the Maxwell equations, which, for a homogeneous and nonmagnetic medium, lead to the equation

$$\frac{1}{2ik}\nabla_T^2 \mathcal{E} + \left(\frac{\partial}{\partial z} + \frac{1}{c}\frac{\partial}{\partial t}\right)\mathcal{E} = \frac{4\pi i\omega}{nc}N\mu^*\rho_{21}. \quad (68.10)$$

Here N is the density of active atoms, $\mu \equiv (\mathbf{d}_{12} \cdot \boldsymbol{\epsilon})^*$, and $\nabla_T^2 \equiv \partial^2/\partial x^2 + \partial^2/\partial y^2$. The result (68.10) assumes the validity of the rotating wave approximation as well as the assumption that $\mathcal{E}$ is slowly varying compared with $\exp(ikz)$ and $\exp(-i\omega t)$. In the plane wave approximation, Eq. (68.10) becomes [2]

$$\left(\frac{\partial}{\partial z} + \frac{1}{c}\frac{\partial}{\partial t}\right)\mathcal{E} = \frac{4\pi i\omega}{nc}N\mu^*\rho_{21}. \quad (68.11)$$

Equations (68.6) or (68.8) and (68.10) or (68.11) are coupled matter–field equations whose self consistent solutions determine the operating characteristics of the laser. The density matrix or rate equations must be modified to include pumping, as in Eq. (68.9), and the field equations must be supplemented by boundary conditions and loss terms. With these modifications, the equations are the basis of *semiclassical laser theory*, wherein the particles constituting the gain medium are treated quantum mechanically whereas the field is treated according to classical electromagnetic theory [4]. Aside from fundamental linewidth considerations and photon statistics (see Sects. 68.3 and 68.5), very few aspects of lasers require the quantum theory of radiation.

68.3 CONTINUOUS WAVE, SINGLE-MODE OPERATION

In the case of steady state, continuous wave (cw) operation, the appropriate matter–field equations are those obtained by setting all time derivatives equal to zero. Equation (68.11), for instance, becomes

---

[2]More generally, the velocity c on the left sides of Eqs. (68.10) and (68.11) should be replaced by the group velocity v_g associated with nonresonant transitions. If there is substantial group velocity dispersion, it is sometimes necessary to include a term involving the second derivative of $\mathcal{E}$ with respect to t.

$$\frac{d\mathcal{E}}{dz} = \frac{2\pi\omega N|d|^2}{3n\hbar c}\frac{1}{\gamma + i\Delta}(\rho_{22} - \rho_{11})\mathcal{E}, \quad (68.12)$$

or, in terms of the intensity I,

$$\begin{aligned}
\frac{dI}{dz} &= \frac{4\pi\omega N|d|^2}{3n\hbar c}\frac{\gamma}{\Delta^2 + \gamma^2}(\rho_{22} - \rho_{11})I \\
&= \frac{\lambda^2 A}{8\pi n^2}(N_2 - N_1)S(\nu)I = g(\nu)I \quad (68.13)
\end{aligned}$$

for the nondegenerate case under consideration. Here, $|d|^2 = 3|\mathbf{d}_{12}\cdot\boldsymbol{\epsilon}|^2$, $N_j = N\rho_{jj}$, $S(\nu) = \gamma/(\Delta^2 + \gamma^2)$ is the Lorentzian lineshape function for homogeneous broadening, and $A = 4\omega^3|d|^2 n/3\hbar c^3$ is the spontaneous emission rate in the host medium of (real) refractive index n. Local (Lorentz–Lorenz) field corrections will in general modify these results, but such corrections are ignored here [5].

The steady state solution of the density matrix or rate equations gives, similarly,

$$g(\nu) = \frac{g_0(\nu)}{1 + I/I_{\text{sat}}}, \quad (68.14)$$

where the saturation intensity I_{sat}, like the small signal gain coefficient $g_0(\nu)$, depends on decay rates and other characteristics of the lasing species. Thus, in the plane wave approximation, the growth of intensity in a homogeneously broadened laser medium is typically described by the equation

$$\frac{dI}{dz} = \frac{g_0(\nu)I}{1 + I/I_{\text{sat}}}. \quad (68.15)$$

This equation, supplemented by boundary conditions at the mirrors, and possibly other terms on the right side to account for any distributed losses within the medium, determines the intensity in cw, single-mode operation.

The simplest model for calculating output intensity assumes that the intensity is uniform throughout the laser cavity. In steady state, the gain exactly compensates for the loss; i.e., $g(\nu) = g_t$, the *gain clamping condition* for cw lasing. Eq. (68.14) then implies that the steady state intracavity intensity is

$$I = I_{\text{sat}}\left[g_0(\nu)/g_t - 1\right]. \quad (68.16)$$

If I is assumed to be the sum of the intensities of waves propagating in the $+z$ and $-z$ directions, i.e. $I = I_+ + I_-$, then the output intensity from the laser is

$$I_{\text{out}} = t_2 I_+ + t_1 I_-, \quad (68.17)$$

where t_2, t_1 are the mirror transmissivities at the right and left mirrors, respectively. The uniform intensity approximation implies that $I_+ = I_- = I/2$ and

$$I_{\text{out}} = \tfrac{1}{2}(t_2 + t_1)I_{\text{sat}}\left[g_0(\nu)/g_t - 1\right]. \quad (68.18)$$

Suppose one of the mirrors is perfectly transmitting, so that $t_1 = 0$ and $t_2 = t > 0$. Furthermore, if the reflectivity r of the transmitting mirror is close to unity, then $g_t \cong (1/2\ell)(1 - r) = (1/2\ell)(t + s)$, where s is the fraction of the incident beam power that is scattered or absorbed at the output mirror. Then

$$I_{\text{out}} \cong \frac{1}{2} I_{\text{sat}} t \left(\frac{2g_0(\nu)\ell}{t + s} - 1 \right) , \qquad (68.19)$$

and it follows that the optimal output coupling, i.e., the transmissivity that maximizes the output intensity, is

$$t_{\text{opt}} = \sqrt{2g_0(\nu)\ell s} - s . \qquad (68.20)$$

This output coupling gives the output intensity

$$I_{\text{out}}^{\max} = I_{\text{sat}} \left[\sqrt{g_0(\nu)\ell} - \sqrt{s/2} \right]^2 . \qquad (68.21)$$

$g_0(\nu)I_{\text{sat}}$ is the largest possible power per unit volume extractable as output laser radiation at the frequency ν.

More generally, when mirror reflectivities are not necessarily close to unity, $I_+ \neq I_-$ and both I_+ and I_- vary with the axial coordinate z. In this more general case, Eqs. (68.14) and (68.15) are replaced by

$$g(\nu, z) = \frac{g_0(\nu)}{1 + [I_+(z) + I_-(z)]/I_{\text{sat}}} \qquad (68.22)$$

and

$$\frac{dI_+}{dz} = g(\nu, z)I_+ , \quad \frac{dI_-}{dz} = -g(\nu, z)I_- , \qquad (68.23)$$

where it is assumed that $g_0(\nu)$ is independent of z, and that all cavity loss processes occur at the mirrors. The solution of these equations with the boundary conditions $I_-(L) = r_2 I_+(L)$, $I_+(0) = r_1 I_-(0)$ for mirrors at $z = 0$ and $z = L$ gives, for $I_{\text{out}} = t_1 I_-(0) + t_2 I_+(L)$, the formula [3,6]

$$I_{\text{out}} = I_{\text{sat}} \left(t_2 + \sqrt{\frac{r_2}{r_1}} \, t_1 \right) \frac{\sqrt{r_1}}{(\sqrt{r_1} + \sqrt{r_2})(1 - \sqrt{r_1 r_2})}$$
$$\times \, [g_0(\nu)\ell + \ln \sqrt{r_1 r_2}] . \qquad (68.24)$$

Analysis of this result gives an optimal output coupling that reduces to Eq. (68.19) in the limit $t + s \ll 1$. Curves for optimal output coupling and I_{out} as a function of g_0 and s are given by Rigrod [6]. These results are based on several assumptions and approximations: the gain medium is assumed to be homogeneously broadened and to saturate according to the formula (68.14); g_0 and I_{sat} are taken to be constant throughout the medium; the field is approximated as a plane wave; field loss processes occur only at the mirrors; and interference between the left- and right-going waves is ignored.

Interference of the counterpropagating waves in a standing wave, single-mode laser modifies the gain saturation formula (68.14) as follows:

$$g(\nu, z) = \frac{g_0(\nu)}{1 + (2I_+/I_{\text{sat}}) \sin^2 kz} \qquad (68.25)$$

in the case of small output coupling, where $I_+ \cong I_-$ as assumed in Eq. (68.18). (The general case of arbitrary output coupling with spatial interference of counterpropagating waves is somewhat complicated and is not considered here.) The $\sin^2 kz$ term is responsible for *spatial hole burning*: "holes" are "burned" in the curve of $g(\nu, z)$ versus z at points where $\sin^2 kz$ is largest. This spatially dependent saturation acts to reduce the output intensity, typically by as much as about 30%, compared with the case where interference of counterpropagating waves is absent or ignored. Spatial hole burning tends to be washed out by atomic motion in gas lasers, and is absent entirely in pure traveling wave ring lasers. If complete spatial hole burning based on Eq. (68.25) is assumed, the output intensity is

$$I_{\text{out}} = \frac{t}{2} I_{\text{sat}} \left(\frac{g_0(\nu)}{g_t} - \frac{1}{4} - \sqrt{\frac{g_0(\nu)}{2g_t} + \frac{1}{16}} \right) \qquad (68.26)$$

in the case where one mirror is perfectly reflecting.

In inhomogeneously broadened media, the gain coefficient is obtained by integrating the contributions from all possible values of ν_0. The different contributions saturate differently, depending on the detuning of ν_0 from the cavity mode frequency ν. If spatial hole burning and power broadening (see Sect. 67.4) are ignored, then, to a good approximation, the gain saturates as

$$g(\nu) = \frac{g_0(\nu)}{\sqrt{1 + I/I_{\text{sat}}}} \qquad (68.27)$$

in typical inhomogeneously broadened media.

Oscillation on a single longitudinal mode (see Sect. 68.4) may be realized simply by making the cavity length L small enough that the mode spacing $c/2nL$ (see Eq. 68.28) exceeds the spectral width of the gain curve. This is possible in many gas lasers where the spectral width is small, and in semiconductor lasers, where L is very small. More generally the gain clamping condition $g(\nu) = g_t$ implies that the cavity mode frequency having the largest small-signal gain $g_0(\nu)$ saturates the gain $g(\nu)$ down to the threshold value g_t, while the gain at all other mode frequencies then lies below g_t. In other words, the gain clamping condition implies single-mode oscillation. However, this conclusion assumes homogeneous broadening and also that spatial hole burning is unimportant, so that the gain saturates uniformly throughout the cavity. High pressure gas lasers, where the line broadening is due primarily to collisions and therefore is homogeneous, tend to oscillate on a single mode because spatial hole burning is largely washed out by atomic motion. On the other hand, homogeneously broadened solid state lasers can be multi-mode as a consequence of spatial hole burning.

Single-mode oscillation in inhomogeneously broadened media is generally more difficult to achieve because of *spectral* hole burning (see Sect. 67.4), which makes the simple gain clamping argument inapplicable. However, single-mode oscillation can be enforced in any case by introducing, in effect, an additional loss mechanism for all mode frequencies except one. This is commonly done with a Fabry–Perot etalon having a free spectral range that is large compared with the spectral width of the gain curve. By choosing the tilt angle appropriately, a particular resonant frequency of the etalon can be brought close to the center of the gain curve while all other resonance frequencies lie outside the gain bandwidth.

Laser oscillation at a fixed polarization can likewise be achieved by discriminating against the orthogonal polarization, as is done when Brewster windows are employed.

Laser oscillation, in general, does not occur precisely at one of the allowed cavity mode frequencies. Associated with the $\sin kz$ dependence of the intracavity field is the condition $kL = N\pi$, or

$$\nu = N\frac{c}{2nL} \quad (N \text{ an integer}), \tag{68.28}$$

for the cavity mode frequencies in the plane wave approximation. If the gain medium of refractive index n does not fill the entire length L between the mirrors, then Eq. (68.28) must be replaced by

$$\nu = \frac{Nc/2}{n\ell + (L - \ell)}, \tag{68.29}$$

or

$$\frac{\ell}{L}[n(\nu) - 1]\nu = \nu_N - \nu, \tag{68.30}$$

where $\ell\ (\leq L)$ is the length of the gain medium and $\nu_N = Nc/2L$ is an empty cavity mode frequency. The laser oscillation frequency will therefore be different, in general, from any of the allowed empty-cavity mode frequencies. If the refractive index $n(\nu)$ is attributable primarily to the lasing transition, as opposed to the host material, or other nonlasing transitions, then the following relation between $n(\nu)$ and $g(\nu)$ may be used in the case of homogeneous broadening [3]:

$$n(\nu) - 1 = -\frac{\lambda_o}{4\pi}\frac{\nu_o - \nu}{\delta\nu_o}g(\nu), \tag{68.31}$$

where λ_0, ν_0, and $\delta\nu_0$ are the wavelength, frequency, and homogeneous linewidth (HWHM), respectively, of the lasing transition. This implies that

$$\nu = \frac{\nu_0\delta\nu_c + \nu_N\delta\nu_0}{\delta\nu_c + \delta\nu_0} \tag{68.32}$$

for the laser oscillation frequency ν, where

$$\delta\nu_c \equiv \frac{cg(\nu)\ell}{4\pi L} \tag{68.33}$$

is the *cavity bandwidth*. Thus, the actual lasing frequency is not simply one of the allowed empty-cavity frequencies ν_N, but rather is "pulled" away from ν_N toward the center of the gain profile. This is called *frequency pulling*.

If spatial hole burning is absent or ignored, then $g(\nu) = g_t$ in steady state oscillation, and the cavity bandwidth $\delta\nu_c = cg_t\ell/4\pi L$ is largest for lossy cavities. Most lasers fall into the "good cavity" category, that is $\delta\nu_c \ll \delta\nu_0$, so that (68.32) may be approximated by

$$\nu \cong \nu_N + (\nu_0 - \nu_N)\delta\nu_c/\delta\nu_0. \tag{68.34}$$

Similar results apply to inhomogeneously broadened lasers. For a Doppler broadened medium, for instance, the frequency pulling formula is

$$\nu \cong \nu_N + (\nu_0 - \nu_N)(\delta\nu_c/\delta\nu_D)\sqrt{4\ln 2/\pi} \tag{68.35}$$

for good cavities. These results show that frequency pulling is most pronounced in lasers with large peak gain coefficients and narrow gain profiles, as observed experimentally.

Spectral hole burning leads to especially interesting consequences in Doppler broadened gas lasers. Since two traveling waves propagating in opposite ($\pm z$) directions will strongly saturate spectral packets of atoms with oppositely Doppler shifted frequencies, a standing wave field will burn two holes on opposite sides of the peak of the Doppler profile. When the mode frequency is exactly at the center of the Doppler profile, however, the two holes merge, the field now being able to saturate strongly only those atoms with zero velocity along the z-direction. In this case, since the field "feeds" off a single spectral packet of more strongly saturated atoms, there is a dip in the output power compared with the case when the mode frequency is detuned from line center. This dip in the output power at line center is called the *Lamb dip*. It can be used to determine whether a gas laser is Doppler broadened, and more importantly, to stabilize the laser frequency to the center of the dip. Lamb dip frequency stabilization employs a feedback circuit to control the bias voltage across a piezoelectric element used to vary the cavity length, and thereby sweep the laser frequency.

The above semiclassical laser theory suggests that cw laser radiation should be perfectly monochromatic, since the amplitude and phase of the field given by Eq. (68.7) are time independent. However, when quantum electrodynamical considerations are built into laser theory, it is found that spontaneous emission, which adds to the number of photons put in the lasing mode by stimulated emission, causes a phase diffusion that results in a Lorentzian linewidth (HWHM)

$$\Delta\nu = \frac{N_2}{N_2 - N_1}\frac{8\pi h\nu}{P_{\text{out}}}(\delta\nu_c)^2, \tag{68.36}$$

where P_{out} is the output laser power. This is the *Schawlow-Townes linewidth*, which implies a fundamentally quantum mechanical, finite linewidth that persists no matter how small various sources of "technical noise," such as mirror jitter, are made. Although the Schawlow-Townes linewidth has been observed in highly stabilized gas lasers, it is negligible compared with technical noise in conventional lasers. But in semiconductor lasers, L is very small and consequently $\delta\nu_c$ is large, and quantum noise associated with the Schawlow-Townes formula can be the dominant contribution to the laser linewidth.

However, the 10–100 MHz linewidths typically observed in semiconductor lasers are too large to be explained by the Schawlow-Townes formula [see Eq. (68.36)], and two modifications to this formula are necessary, each of them involving a multiplication of the Schawlow-Townes linewidth $\Delta\nu$ by a certain factor:

$$\Delta\nu \to \Delta\nu' = (1 + \alpha^2)K\Delta\nu, \qquad (68.37)$$

where α is called the "Henry α parameter" and is associated with a coupling between phase and intensity fluctuations above the laser threshold [7]. Values of α between about 4 and 6 are typical in semiconductor injection lasers, and consequently, the correction to the Schawlow-Townes linewidth due to the Henry α parameter is substantial. The K factor [8] arises as a consequence of the deviation from the spatially uniform intracavity intensity assumed in the derivation of the Schawlow-Townes formula [9]. (Intracavity intensities along the optical axis are approximately uniform only in the case of low output couplings.) The fundamental quantum mechanical linewidth under consideration can be associated with vacuum field fluctuations, which, according to general fluctuation–dissipation ideas, will increase as the cavity loss increases. This explains why the "Petermann K factor" deviates increasingly from unity as the output coupling (cavity loss) increases. Values of K between 1 and 2 appear to be typical for lossy, stable resonators [10], but much larger values are possible for unstable resonators [11] (see also Siegman [2]).

68.4 LASER RESONATORS

The assumption that the complex field amplitude $\mathcal{E}(\mathbf{r}, t)$ is slowly varying in z compared with $\exp(ikz)$ leads to the *paraxial wave equation* [*cf.* Eq. (68.10)]

$$\nabla_T^2 \mathcal{E} + 2ik\frac{\partial \mathcal{E}}{\partial z} = 0 \qquad (68.38)$$

for a monochromatic field in vacuum. If $\mathcal{E}(x, y, z)$ satisfies the paraxial wave equation and is specified in the plane $(x, y, z = 0)$, it follows that

$$\mathcal{E}(x, y, z) = -\frac{i}{\lambda z} \iint dx'dy' \mathcal{E}(x', y', 0)$$
$$\times e^{ik[(x-x')^2 + (y-y')^2]/2z}. \qquad (68.39)$$

Thus, in the case of a laser resonator, the field $\mathcal{E}(x, y, L)$ at the mirror at $z = L$ is related to the field $\mathcal{E}(x, y, 0)$ at the mirror at $z = 0$ by

$$\mathcal{E}(x, y, L) = -\frac{i}{\lambda L} \iint dx'dy' \mathcal{E}(x', y', 0)$$
$$\times e^{ik[(x-x')^2 + (y-y')^2]/2L}$$
$$\equiv \iint dx'dy' K(x, y; x', y')\mathcal{E}(x', y', 0). \qquad (68.40)$$

Similarly, the field at $z = 0$ after one round trip pass through the resonator is

$$\mathcal{E}(x, y, 0) = \iint dx'dy' K(x, y; x', y')\mathcal{E}(x', y', L)$$
$$= \iint dx'dy' K(x, y; x', y') \iint dx''dy''$$
$$\times K(x', y'; x'', y'')\mathcal{E}(x'', y'', 0)$$
$$\equiv \iint dx''dy'' \tilde{K}(x, y; x'', y'')\mathcal{E}(x'', y'', 0). \qquad (68.41)$$

By definition, a *mode* of the resonator is a field distribution that does not change on successive round-trip passes through the resonator. More precisely, since an empty cavity is assumed, a mode will be such that the field spatial pattern remains the same except for a constant decrease in amplitude per pass. A *longitudinal* mode is defined by the value of k in $\exp(ikz)$, whereas a *transverse* mode is defined by the corresponding (x, y) dependence and satisfies the integral equation

$$\gamma\mathcal{E}(x, y, z) = \int dx'dy' \tilde{K}(x, y; x', y')\mathcal{E}(x', y', z), \quad (68.42)$$

where z defines any plane between the mirrors and $|\gamma| < 1$. Iterative numerical solutions of this equation for laser resonator modes were first discussed by Fox and Li [12].

Laser resonators may be classified as *stable* or *unstable* according to whether a paraxial ray traced back and forth through the resonator remains confined in the resonator or escapes. This leads to the condition

$$0 \leq g_1 g_2 \leq 1 \qquad (68.43)$$

for stability, where the g parameters are defined in terms of the (spherical) mirror curvatures R_i and the mirror separation L by $g_i \equiv 1 - L/R_i$. R_i is defined as positive or negative depending on whether the mirror i is concave or convex, respectively, with respect to the interior of the resonator. Plane–parallel mirrors ($R_i \to \infty$) can be

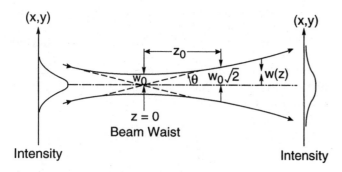

Figure 68.1. Variation of the spot size $w(z)$ of a Gaussian beam with the propagation distance z from the beam waist.

used, but they are difficult to keep aligned and have much larger diffractive losses.

Among stable configurations, the *symmetric confocal resonator* with $R_1 = R_2 = L$ has the smallest mode spot sizes at the mirrors, while the *concentric resonator* with $R_1 = R_2$ slightly greater than $L/2$, has the smallest beam waist (see Fig. 68.1). The widely used *hemispherical resonator* ($R_1 = \infty$, R_2 slightly greater than L) is relatively easy to keep aligned and allows the spot size at mirror 2 to be adjusted by slight changes in L.

The fundamental Gaussian beam modes of stable resonators may be constructed from the free space solutions of the paraxial wave equation. The most important (lowest-order) solution for this purpose is

$$\mathcal{E}(x,y,z) = \frac{Ae^{-i\phi(z)}e^{ik(x^2+y^2)/2R(z)}e^{-(x^2+y^2)/w^2(z)}}{\sqrt{1+z^2/z_0^2}},$$

(68.44)

where A is a constant, $\phi(z) = \tan^{-1}(z/z_0)$, and $R(z)$, $w(z)$, and z_0 are the radius of curvature of surfaces of constant phase, the spot size, and the *Rayleigh range*, respectively. The *confocal parameter*, $2z_0$, is also used to characterize Gaussian beams. Here $z_0 \equiv \pi w_0^2/\lambda$, where w_0 is the spot size at the beam waist at $z = 0$ (see Fig. 68.1), and $R(z)$ and $w(z)$ vary with the distance z as follows:

$$R(z) = z + z_0^2/z, \quad w(z) = w_0\sqrt{1+z^2/z_0^2}. \quad (68.45)$$

The divergence angle of a Gaussian beam (Fig. 68.1) is given by $\theta = \lambda/\pi w_0$.

The ABCD law for Gaussian beams allows the effects of various optical elements on Gaussian beam propagation to be calculated in a relatively simple fashion [1–3]. For instance, a Gaussian beam incident on a lens of focal length f at its waist is focused to a new waist at a distance $d = f/(1+f^2/z_0^2)$ behind the lens, and the spot size at the new waist is

$$w_0' = \frac{f\lambda}{\pi w_0} \frac{1}{\sqrt{1+f^2/z_0^2}}, \quad (68.46)$$

which is approximately $f\lambda/\pi w_0 = f\theta$ for tight focusing. This shows that a Gaussian beam can be focused to a very small spot. On the other hand, beam expanders consisting of two appropriately spaced lenses may be used to increase the spot size by the ratio of the focal lengths.

Gaussian beam modes of laser resonators have radii of curvature that match in magnitude those of the mirrors. The spot sizes at the mirrors and the location of the beam waist with respect to the mirrors may be expressed in terms of λ, L, and g_1g_2. The empty-cavity mode frequencies are given by

$$\nu_N = \frac{c}{2L}\left(N + \frac{1}{\pi}\cos^{-1}\sqrt{g_1g_2}\right), \quad (68.47)$$

where N is an integer. For a host medium of refractive index n, $c/2L$ is replaced by $c/2nL$, and Eq. (68.47) then generalizes the plane wave result (68.28) to account for both longitudinal and transverse effects in the determination of the cavity mode frequencies.

The assumption of Gaussian modes presupposes that the resonator mirrors are large enough to intercept the entire beam without any "spillover"; i.e., that $a \gg w_1, w_2$, where w_1, w_2 are the spot sizes at the mirrors and a is an effective mirror cross sectional radius. This implies that the *Fresnel number* $N_F \equiv a^2/\lambda L \gg 1$. Diffraction losses generally increase with decreasing Fresnel numbers.

Higher-order Gaussian modes, where the Gaussian functions of x and y in Eq. (68.44) are replaced by higher-order Hermite polynomials, are often more difficult to realize than the fundamental lowest-order mode because their larger spot sizes imply higher diffractive losses, and beyond a certain mode order the spot sizes are too large to satisfy the no-spillover condition for a pure Gaussian mode.

It is not possible in general to write closed form expressions for laser modes. There are at least two reasons for this, the first being that the gain medium cannot in general be regarded as a simple amplifying element that preserves the basic empty-cavity mode structure. In low power gas lasers, the spatial variations of the gain and refractive index are sufficiently mild that the lasing modes can be accurately described as Gaussians, but more generally, there can be strong gain and index variations that themselves play an important role in determining the modes of the laser, as in "index-guided" and "gain-guided" semiconductor lasers.

Secondly, the resonator structure itself may introduce complications that preclude closed form solutions even for the empty cavity. This is generally true, for instance, of unstable resonators, where iterative numerical solutions of the integral equation (68.42) are necessary for accurate predictions of modes. In such computer simulations, the mode structure must be determined self consistently with the numerical solutions of density matrix or, more commonly, rate equations for the gain medium.

Unstable resonators, though inherently lossy, offer some important advantages for high-gain lasers. Thus, whereas the Gaussian modes of stable resonators typically involve very small beam spot sizes, the distinctly non-Gaussian modes of unstable resonators can have large mode volumes and make more efficient use of the available gain volume. Unstable resonators also tend to yield higher output powers when they oscillate on the lowest loss transverse mode, whereas in stable resonators higher output powers are generally associated with multitransverse mode oscillation. For very high power lasers, the fact that unstable resonators involve all-reflective optics, as opposed to transmissive output coupling, can be important in avoiding optical damage.

68.5   PHOTON STATISTICS

Optical fields may be characterized and distinguished by their photon statistical properties (see Chap. 75 and References [13, 14]). In a photon counting experiment, the number of photons registered at a photodetector during a time interval T is measured and used to infer the probability $P_n(T)$ that n photons are counted in a time interval T. If the probability of counting a photon at the time t in the interval dt is denoted $\alpha I(t) dt$, where $I(t)$ is the intensity of the field and α is a factor depending on the microscopic details of the photoelectric process and on the phototube geometry, then it may be shown from largely classical considerations that [3, 13, 14]

$$P_n(T) = \left\langle \frac{1}{n!} \left[\alpha \int_0^T dt'\, I(t') \right]^n e^{-\alpha \int_0^T dt'\, I(t')} \right\rangle , \quad (68.48)$$

where the average $\langle \cdots \rangle$ is over the intensity variations during the counting interval. This is *Mandel's formula*. In the simplest case of constant intensity, P_n follows the Poisson distribution:

$$P_n = \bar{n}^n e^{-\bar{n}} / n! , \quad (68.49)$$

where $\bar{n} = \alpha I T$. Since a single-mode laser field can be thought of approximately as a "classical stable wave" [14], it is not surprising that its photocount distribution is found both theoretically and experimentally to satisfy Eq. (68.49), which is characteristic of a *coherent state* of the field (see Chap. 75). A thermal light source, by contrast, follows the Bose-Einstein distribution,

$$P_n = \bar{n}^n / (1 + \bar{n})^{n+1} , \quad (68.50)$$

if the time interval T is short compared with the coherence time of the light; i.e., if $T\Delta\nu << 1$, where $\Delta\nu$ is the bandwidth. If $T\Delta\nu >> 1$, P_n is again Poissonian.

Thus, if a quasi-monochromatic beam of light is made from a natural source by spatial and spectral filtering, it has measurably different photon counting statistics from a single-mode laser beam of exactly the same bandwidth and average intensity. Laser radiation approaches the ideal coherent state that, of all possible quantum mechanical states of the field, most closely resembles the "classical stable wave." These differences are exhibited in other experiments, such as the measurement of intensity correlations of the Brown–Twiss type [3, 13, 14]. In such experiments, thermal photons have a statistical tendency to arrive in pairs (*photon bunching*), whereas the photons from a laser arrive independently.

68.6   MULTI-MODE AND PULSED OPERATION

Multi-mode laser theory is generally much more complicated than single-mode theory, particularly in the case of inhomogeneous broadening with both spectral and spatial hole burning. In certain situations, however, considerable simplification is possible. For instance, when the cavity-mode frequency spacing is small compared with the homogeneous linewidth $\delta\nu_0$, the gain tends to saturate homogeneously, and the total output power on all modes is well described by the Rigrod analysis outlined in Sect. 68.3.

Pulsed laser operation adds the further complication of temporal variations to the cw theory outlined in the preceding sections. It is possible, nevertheless, to understand some of the most important types of multi-mode and pulsed operation using relatively simple models.

One method of obtaining short, high power laser pulses is *Q-switching*. In a very lossy cavity, the gain can be pumped to large values before the threshold condition is met and gain saturation occurs. If the cavity loss is suddenly decreased, there will be a rapid build-up of intensity because the small-signal gain is far above the (now reduced) threshold value. The switching of the cavity loss is called *Q-switching*, the "quality factor" Q being defined as $\nu/2\delta\nu_c$. This type of Q-switching produces intense pulses of duration typically in the range 10–100 ns, as dictated by the fact that a light pulse must make several passes through the gain cell for amplification. Q-switching requires that the gain medium be capable of retaining a population inversion over a time much larger than the Q-switched pulse duration, and in particular that the spontaneous emission lifetime be relatively long. The pulse duration can be reduced to approximately a single cavity round trip time by Q-switching from low reflectivity mirrors to 100% reflectivities, and then switching the reflectivity of the outcoupling mirror from 100% to 0% at the peak of the amplified pulse. In addition to this *pulsed*

(a)

(b)

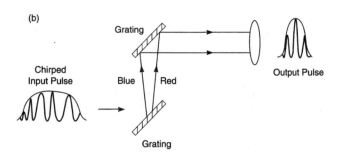

Figure 68.2. Nonlinear pulse compression by frequency chirping. In (a), the nonlinear refractive index of a glass fiber results in a time dependent frequency of the transmitted pulse, and in (b) a pair of diffraction gratings is used to produce frequency dependent path delays such as to temporally compress the pulse.

transmission mode is *cavity dumping*, where both mirrors have nominally 100% reflectivity and the intracavity power is "dumped" by an acousto-optic or electro-optic intracavity element that deflects the light out of the cavity. The pulse duration achieved in this way is again roughly a round trip time. Cavity dumping can be employed with cw lasers and does not require the long energy storage times necessary for ordinary Q-switching.

Shorter pulses can be realized by *mode-locking*, where the phases of N longitudinal cavity modes are locked together. In the simplest model, assuming equal amplitudes and phases of the individual modes, the net field amplitude is proportional to

$$X(t) = \sum_{n=-(N-1)/2}^{(N-1)/2} X_0 \sin[(\omega_0 + n\Delta)t + \phi_0]$$

$$= X_0 \sin(\omega_0 t + \phi_0) \frac{\sin(N\Delta t/2)}{\sin(\Delta t/2)} . \qquad (68.51)$$

The temporal variation described by this function for large N is a train of "spikes" of amplitude NX_0 at times $t_m = 2\pi m/\Delta$, $m = 0, \pm 1, \pm 2, \cdots$, the width of each spike being $2\pi/(N\Delta)$. In the case of a mode-locked laser, $\Delta = 2\pi(c/2L)$, and the output field consists of a train of pulses separated in time by $T = 2\pi/\Delta = 2L/c$. The peak amplitude of the spikes is proportional to N, and the duration of each spike is approximately $2L/cN$.

The maximum number $N_{\max}$ of modes that can actually be phase-locked is limited by the spectral width $\Delta\nu_g$ of the gain curve:

$$N_{\max} = \frac{\Delta\nu_g}{c/2L} = \frac{2L}{c}\Delta\nu_g . \qquad (68.52)$$

Similarly, the shortest pulse duration is

$$\tau_{\min} = \frac{2L}{cN_{\max}} = \frac{1}{\Delta\nu_g} . \qquad (68.53)$$

Mode-locking thus requires a gain bandwidth large compared with the cavity mode spacing, and the shortest and most intense mode-locked pulse trains are obtained in gain media having the largest gain bandwidths. Trains of picosecond pulses are routinely obtained with liquid dye and solid gain media having gain bandwidths $\Delta\nu_g \sim 10^{12}$ s^{-1} or more.

Various techniques, employing acoustic or electro-optic modulation or saturable absorbers, are used to achieve mode-locking [1–3]. The different methods all rely basically on the fact that a modulation of the gain or loss at the mode separation frequency $c/2L$ tends to cause the different modes to oscillate in phase. Such a modulation is achieved "passively" when a saturable absorber is placed in the laser cavity: the multi-mode intensity oscillates with a beat frequency that is impressed on the saturated loss coefficient. Dye lasers, with their large gains across a broad range of optical frequencies, are often employed to generate mode-locked picosecond pulses.

Ultrashort pulse generation is possible with additional nonlinear or frequency chirping techniques [15]. The *colliding-pulse laser* [16] is a three-mirror ring laser in which two mode-locked pulse trains propagate in opposite senses and overlap in a very thin ($\sim 10~\mu$m) saturable absorber placed in the ring in addition to the gain cell. The cavity loss is least when the two pulses synchronize to produce the highest intensity, and therefore the lowest loss coefficient in the saturable absorber. The short length of the absorbing cell forces the pulses to overlap within a very small distance and therefore to produce very short pulses (10 μm/c$\sim$ 30 fs pulse duration with $c/L \sim 100$ MHz repetition rate).

Another method of ultrashort pulse generation relies on *frequency chirping*; i.e., a time dependent shift of the frequency of an optical pulse [17]. In a medium (e.g., a glass fiber) with linear and nonlinear refractive index coefficients n_0 and n_2, the refractive index is

$$n = n_0 + n_2 I , \qquad (68.54)$$

so that there is an instantaneous phase shift $\phi(t)$ that depends on the instantaneous intensity. Therefore, as $I(t)$ increases toward the peak intensity of the pulse, $\phi(t)$ increases (assuming $n_2 > 0$), whereas $\phi(t)$ decreases as

$I(t)$ decreases from its peak value. The frequency shift $\dot{\phi}(t)$ is such that the instantaneous frequency of the pulse is smaller at the leading edge and larger at the trailing edge of the pulse (see Fig. 68.2), resulting in a stretching of the pulse bandwidth.

Following this spectral broadening by the nonlinear medium, the pulse can be compressed in time by means of a frequency dependent delay line such that the smaller frequencies, say, are delayed more than the higher frequencies. The trailing edge of the pulse can therefore "catch up" to the leading edge, resulting in a shorter pulse whose duration is given by the inverse of the chirp bandwidth. The delay line can be realized with a pair of diffraction gratings (see Fig. 68.2). Using this pulse compression technique, 40 fs amplified pulses from a colliding pulse laser have been compressed to 8 fs, corresponding to about four optical cycles [18].

In *chirped-pulsed-amplification* (CPA) *lasers* [19] a laser pulse is chirped, temporally stretched, and then passed through an amplifier. The lengthening in time of the pulse prior to amplification allows greater energy extraction from the amplifier. After amplification, pulse compression is performed with a grating pair. The Ti:sapphire (Ti:Al$_2$O$_3$) amplifier is particularly attractive for femtosecond CPA because of its very large spectral width and high saturation fluence.

The wavelength dependence of the linear refractive index n_0 in Eq. (68.54) results in a group velocity

$$v_g = c[n_0 - \lambda dn_0/d\lambda]^{-1} \qquad (68.55)$$

that, if $dn_0/d\lambda < 0$, is such that higher frequencies propagate more rapidly than lower frequencies. Assuming $n_2 > 0$, on the other hand, the nonlinear part of the index causes a delay of higher frequencies with respect to lower ones, as discussed above. This leads to soliton solutions of the wave equation, such that the opposing effects of the linear and nonlinear dispersion are balanced and the pulse propagates without distortion. *Soliton lasers*, with pulse durations ranging from picoseconds down to ~ 100 fs, depending on the fiber length, have been made with solid state lasers and intracavity optical fibers [20].

68.7 INSTABILITIES AND CHAOS

Mode-locked pulses and solitons exemplify an ordered dynamics, as opposed to the erratic and seemingly random intensity fluctuations that are sometimes observed in the output of a laser. In fact, it is possible, under certain circumstances, for laser oscillation to exhibit deterministic chaos; i.e., an effectively random behavior that can nevertheless be described by purely deterministic equations of motion [21–23].

Lasers are nonlinear and dissipative systems, and as such exhibit essentially all the modes of behavior characteristic of such systems. It was shown by Haken [24] that the equations (68.6) and (68.11), with $\Delta = 0$ and with pumping and field loss terms included, can be put into the form of the Lorenz model for chaos. For a single-mode, homogeneously broadened ring laser, the Lorenz model instability requires a "bad cavity" in the sense that the field loss rate is larger than the sum of the homogeneous linewidth of the lasing transition and the population decay rate. It also requires the gain medium to be pumped at least nine times above the threshold gain value, a condition sometimes referred to as a "second laser threshold."

When the single-mode laser equations corresponding to the Lorenz model are generalized to the case of inhomogeneous broadening, an instability occurs at small–signal gain values much less than nine times threshold. In particular, the cw output of the laser gives way to an oscillatory intensity, even though the pumping and loss terms in the equations are assumed to be time independent. As the pumping and loss parameters are varied, this *self-pulsing instability* can give way to chaotic behavior, and numerical studies of the set of coupled matter–field equations reveal period doubling, two-frequency, and intermittency routes to chaos in different regimes [21]. This unstable behavior of single-mode inhomogeneously broadened lasers was first discovered experimentally and analyzed by Casperson [25] for low pressure, 3.51 μm He–Xe lasers. Arecchi *et al.* [26] reported the first observation and characterization of chaotic behavior in a laser system; by modulating the cavity loss of a CO$_2$ laser they observed a period doubling route to chaos as the modulation frequency was varied.

Extensive experimental and theoretical work on unstable and chaotic behavior in a wide variety of other laser systems has been reported [21–23, 27], including work aimed at the control of chaotic laser oscillation by the so-called occasional proportional feedback technique [28]. Instabilities of single *transverse mode* dynamics have also been studied, especially in connection with spontaneous spatial pattern formation [27].

REFERENCES

1. A. Yariv, *Quantum Electronics* (Wiley, New York, 1989), 2nd ed.
2. A. E. Siegman, *Lasers* (University Science Books, Mill Valley, 1986).
3. P. W. Milonni and J. H. Eberly, *Lasers* (Wiley, New York, 1988).
4. P. W. Milonni, Phys. Rep. **25**, 1 (1976).
5. B. DiBartolo, *Optical Interactions in Solids* (Wiley, N. Y., 1968), p. 405.
6. W. W. Rigrod, J. Appl. Phys. **36**, 27 (1965).
7. C. H. Henry, IEEE J. Quantum Electron. **18**, 259 (1982).
8. K. Petermann, IEEE J. Quantum Electron. **19**, 1391 (1979); A. E. Siegman, Phys. Rev. A **39**, 1253 (1989); **39**, 1264 (1989).

9. P. Goldberg, P. W. Milonni, and B. Sundaram, Phys. Rev. A **44**, 1969 (1991); **44**, 4556 (1991).

10. W. A. Hamel and J. P. Woerdman, Phys. Rev. Lett. **64**, 1506 (1990).

11. K.-J. Cheng, P. Mussche, and A. E. Siegman, IEEE J. Quantum Electron. **30**, 1498 (1994).

12. A. G. Fox and T. Li, Bell System Tech. J. **40**, 453 (1961).

13. R. J. Glauber, Phys. Rev. **130**, 2529 (1963); **131**, 2766 (1963); L. Mandel and E. Wolf, Rev. Mod. Phys. **37**, 231 (1965).

14. R. Loudon, *The Quantum Theory of Light* (Clarendon Press, Oxford, 1983), 2nd ed.; P. Meystre and M. Sargent III, *Elements of Quantum Optics* (Springer–Verlag, Berlin, 1991), 2nd ed.

15. E. B. Treacy, IEEE J. Quantum Electron. **5**, 454 (1969).

16. R. L. Fork, B. I. Greene, and C. V. Shank, Appl. Phys. Lett. **38**, 671 (1981).

17. D. Grischkowsky and A. C. Balant, Appl. Phys. Lett. **41**, 1 (1982); L. F. Mollenauer, R. H. Stolen, J. P. Gordon, and W. J. Tomlinson, Opt. Lett. **8**, 289 (1983); J. G. Fujimoto, A. M. Weiner, and E. P. Ippen, Appl. Phys. Lett. **44**, 832 (1984).

18. C. V. Shank, Science **233**, 1276 (1986), and references therein.

19. P. Maine, D. Strickland, P. Bado, M. Pessot, and G. Mourou, IEEE J. Quantum Electron. **24**, 398 (1988).

20. L. F. Mollenauer and R. H. Stolen, Opt. Lett. **9**, 13 (1984).

21. P. W. Milonni, M.-L. Shih, and J. R. Ackerhalt, *Chaos in Laser–Matter Interactions* (World Scientific, Singapore, 1987).

22. Special issues of J. Opt. Soc. Am. B **2**, 1 (1985); **5**, 5 (1988); **7**, 6 (1990); **7**, 7 (1990).

23. *Instabilities and Chaos in Quantum Optics II*, edited by N. B. Abraham, F. T. Arecchi, and L. A. Lugiato (Plenum, New York, 1988).

24. H. Haken, Phys. Lett. A **53**, 77 (1975).

25. L. W. Casperson, IEEE J. Quantum Electron. **14**, 756 (1978); see also L. W. Casperson, in *Laser Physics*, edited by J. D. Harvey and D. F. Walls (Springer–Verlag, Berlin, 1983).

26. F. T. Arecchi, R. Meucci, G. Puccione, and J. Tredicce, Phys. Rev. Lett. **49**, 1217 (1982).

27. L. A. Lugiato, W. Kaige, and N. B. Abraham, Phys. Rev. A **49**, 2049 (1994), and references therein.

28. R. Roy, T. W. Murphy, Jr., T. D. Maier, and Z. Gills, Phys. Rev. Lett. **68**, 1259 (1992); Z. Gills, C. Iwata, R. Roy, I. B. Schwartz, and I. Triandaf, Phys. Rev. Lett. **69**, 3169 (1992); see also Optics and Photonics News (May, 1994).

69

Types of Lasers

Richard C. Powell

Optical Sciences Center, University of Arizona, Tuscon, Arizona 85715

69.1   INTRODUCTION

The availability of coherent light sources (i.e., lasers) has revolutionized atomic, molecular, and optical science. Since its invention in 1960, the laser has become the basic tool for atomic and molecular spectroscopy and for elucidating fundamental properties of optics and optical interactions with matter. The unique properties of laser light have spawned new types of spectroscopy, as discussed in Chaps. 70 to 77.

There are now literally hundreds of different types of lasers. However, only a few of these are commercially available, and lasers tailor made with operational properties optimized for specific applications are often needed. Chapter 68 describes the principles of laser operation leading to specific output characteristics. This chapter summarizes the current status of the development of different types of lasers, emphasizing those that are commercially available. There are several ways to categorize types of lasers; for example, in terms of spectral range,

Table 69.1. Categories of Lasers

Gas Lasers	Special Lasers	Solid State Lasers	Liquid Lasers	Miscellaneous
Atomic	Metal vapor	Transition metal ion	Organic dye	X-ray
Ionic	Chemical	Rare earth ion	Rare earth chelate	Nuclear pumped
Molecular		Color center	Inorganic solvents with	Free electron
Excimer		Semiconductor	rare earth ions	

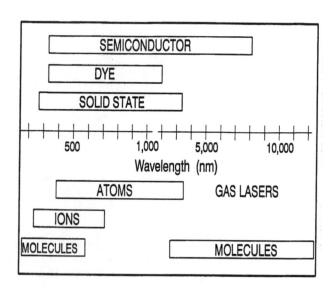

Figure 69.1. Spectral range of laser emission.

temporal characteristics, pumping mechanism, or lasing media. For this chapter, the types of lasers are categorized in terms of the lasing media, as shown in Table 69.1.

The variety of different laser types offers a wide range of beam parameters. The spectral output of lasers covers the x-ray to far ir regions as shown in Fig. 69.1. Spectral linewidths can be as narrow as 20 Hz and diffraction limited beam quality with coherence lengths up to 10 m can be obtained. Temporal pulse widths of a few femtoseconds have been generated. Some lasers can produce peak powers of over 10^{13} W and average powers of 10^5 W with pulse energies greater than 10^4 J. The important operational characteristics, such as frequency range and output power, are given for each of the types of lasers described in the following sections. Extensive tables of laser properties have been published in several different handbooks [1–5] and all of these details for every laser are not repeated here due to space limitations. The data that are quoted come from these reference books and from recent proceedings of conferences such as CLEO (Conference on Lasers and Electro-Optics) and ASSL (Advanced Solid State Laser Conference). The reader is referred to these sources for further details of specific laser operation parameters.

69.2 GAS LASERS

Gas lasers can be separated into subclasses based on the lasing media: neutral atoms; ions; and molecules. In addition, molecular lasers contain the special classes of excimer lasers and chemical lasers. Except for chemical lasers, the pumping mechanism is generally an electrical discharge in a gas filled tube. This discharge causes the acceleration of electrons that transfer their kinetic energy to the lasing species through collisions, leaving them in a variety of excited states. These relax back to the ground state with different rates, resulting in the possibility of a population inversion for some transitions. These can be electronic, vibrational, or rotational transitions with wavelengths ranging from the near ultraviolet (uv) through the far infrared (ir) spectral regions. Systems lasing at short wavelengths generally operate only in the pulsed mode because of the short radiative lifetimes of the transitions involved. Because of their low gain, gas lasers usually have relatively long linear cavity designs. Gas lasers generally have narrow spectral emission lines with the possibility of lasing at several different wavelengths. They are broadened both by collisions (homogeneous broadening) and the Doppler effect of the motion of the atoms or molecules (inhomogeneous broadening). To date, 51 elements in the periodic table have shown either ionic or neutral ion gas laser emission.

69.2.1 Neutral Atom Lasers

These lasers generally emit in the visible and near ir spectral range. The first and most common laser of this class is the He-Ne laser. Its most prominent emission line is at 632.8 nm. It is usually operated in the continuous wave (cw) configuration with typical power outputs between 0.5 to 50 mW, although powers of over 100 mW have been achieved. Excitation occurs through electrical discharge which pumps both the He and Ne atoms to excited states. The more abundant He atoms transfer their energy to several excited states of Ne atoms. Several radiative transitions of Ne are available for laser transitions. These provide emission at 543.36, 632.8, 1152.27, and 3391.32 nm. Final relaxation back to the ground state occurs through collisions with the walls of the gas tube. The desired laser line can be selected by adjusting the reflectivity of the cavity mirrors to discriminate against unwanted transitions. In addition, it is possible to adjust the discharge current, gas ratio, and pressure to optimize a specific emission transition. Using an external magnetic field to produce a Zeeman effect can also be helpful in tuning the laser emission. In addition to the normal red He-Ne lasers, green lasers are now available. Typical He-Ne lasers operate with a coherence length of 0.1–0.3 m, a beam divergence of 0.5–2 mrad, and a stability of 5%/hr. The gain at the red line is 0.5 dB/m. This line can be operated as a single mode with a linewidth of 0.0019 nm and a coherence length between 20 and 30 cm.

69.2.2   Ion Lasers

The most important ion lasers are based on noble gas ions such as Ar, Kr, Ne, or Xe in various states of ionization. These operate in either a pulsed or cw mode, and their emission covers the wavelength range from the near uv through the visible part of the spectrum. In the electrical-discharge excitation process, electrons collide with neutral atoms in their ground states, transferring enough energy to ionize them and leave the ions in several possible excited states. For example, low discharge currents produce Ar^+ giving rise to visible emission lines, while high discharge currents produce Ar^{2+} giving rise to uv emission lines. Radiative emission then occurs to lower excited levels of the ions, followed by subsequent spontaneous emission to the ground state of the ion, and then radiationless relaxation back to the neutral atom ground state. This transition scheme limits the wall plug efficiency to about 0.1% for visible and 0.01% for uv operation. Heat management is accomplished through either water cooling or air cooling.

In the visible region of the spectrum, argon lasers have blue and green emission lines with the strongest ones at 488 and 515 nm, respectively. Krypton lasers have several strong emission lines in the green and red, with the most prominent ones at 521, 568, and 647 nm. Mixed gas lasers can produce all of these lines. In the near uv, argon has a strong laser line at 351 nm as well as several other emission lines down to 275 nm. The same methods for selecting specific laser lines on neutral atom lasers, discussed in Sect. 69.2.1, are used for ion lasers. These lasers can operate at powers of over 20 W of cw emission in the visible and at powers of several watts cw in the uv. The Doppler-limited linewidth of noble gas ion lasers is generally $\sim 5-10$ GHz. By using special techniques for stabilization of the cavity, linewidths of ~ 500 MHz with drifts of 100 MHz/hr can be obtained. By mode-locking argon or krypton lasers with 10 GHz linewidths, it is possible to produce trains of pulses with pulse lengths of 100 ps and peak powers of 1 kW, at a pulse repetition frequency of 150 MHz. Cavity-dumping

produces narrow pulses at pulse repetition frequencies of ~ 1 MHz with peak powers over 100 times the cw power. Lower power cw lasers of this type typically have beam divergences of 1.5 mrad and a stability of 5%, while high power lasers have beam divergences of ~ 0.4 mrad with a stability of 0.5%.

69.2.3   Metal Vapor Lasers

These lasers can operate with either neutral atoms or ions. Their excitation process begins with vaporizing a solid or liquid to produce the gas for lasing, followed by normal electrical discharge pumping. Either cw or pulsed operation can be obtained, with laser emission lines in the near uv and visible spectral regions.

One important ion laser of this class is the Helium-Cadmium laser. For the excitation processes, metallic Cd is evaporated and mixed with He. Then a d.c. electric discharge excites the He ions and ionizes the Cd. The excited He atoms transfer their energy to the Cd atoms and the laser transitions take place between electronic levels of the Cd atom. The main emission line of a He-Cd laser is the blue line at 441.6 nm. This typically has a cw output from 130 mW for single-mode operation up to 150 mW for multimode operation. The laser linewidth can be as narrow as 0.003 nm. This system also has an important laser emission at 325.029 nm, which typically has cw powers between 5 and 10 mW single-mode and 100 mW for multimode emission. The wall plug efficiency is between 0.002% and 0.02%.

The most important neutral atom metal vapor laser of this type is the copper vapor laser. This has important emission lines in the green at 510.55 nm and in the yellow at 578.21 nm. These lasers operate in the pulsed mode with temporal pulse widths between 10 and 20 ns at pulse repetition rates of up to 20 000 pps. Typical pulse energies are ~ 1 mJ, yielding average powers of 20 W. It is possible to increase the repetition rate significantly to achieve average powers of 120 W or even higher. However, this laser is self-terminating since the lower levels of the laser transitions are metastable. This restricts the pulse sequencing of the laser and requires fast discharge risetimes. Copper vapor lasers have high gain (10% to 30%/cm), and very high wall plug efficiency (~ 1.0%). Gold vapor lasers have similar properties but operate in the red at 624 nm at several watts of power.

69.2.4   Molecular Lasers

There are several types of molecular lasers that can be classified with respect to their spectral emission range, their mode of excitation, or the energy levels involved in the lasing transition. In the far ir, molecular lasers operate on transitions between rotational energy levels. These include water vapor lasers that emit between

17 and 200 μm, cyanide lasers at 337 μm, methyl fluoride lasers emitting between 450 and 550 μm, and ammonia lasers that operate at 81 μm. These are generally excited through optical pumping by a CO_2 laser. They are built with a metal or dielectric wave guide cavity. The former design results in lower thresholds but gives multimode, mixed polarization output, while the latter design results in propagation losses, giving higher thresholds but linearly polarized outputs.

CO_2 lasers are some of the most widely used, with a variety of medical and industrial applications. They emit in the mid-ir range at 10.6 and 9.6 μm. The CO_2 molecules are excited by electrical discharge, and it is common to mix CO_2 with other gases, such as nitrogen, to enhance the efficiency of the excitation process through energy transfer, or with helium, to keep the average electron energy high and to depopulate the lower levels of the laser transition. Both the initial and final states of the lasing transition are vibrational levels that have many rotational sublevels. This allows discrete tuning of the output within the 9.4 and 10.4 μm bands. Using high pressures of the gas broadens the laser line into a continuum so that the emission can be continuously tuned over several microns near 10 μm. Without any frequency selective element in the cavity, the system oscillates on the transition with highest gain, which is near 10.6 μm. Under pulsed operating conditions, CO_2 lasers can also emit at bands near 4.3 μm and at several bands between 11 and 18 μm.

These laser systems typically operate at a few milliwatts of power to over 100 kW cw. Waveguide and slab cavity designs have been developed for heat removal. In the pulsed or Q-switched mode of operation, pulse widths are between a few microseconds and a few milliseconds, with energies as high as 10 000 J/pulse. This leads to peak powers more than 100 times higher than cw powers. It is also possible to mode-lock these systems to get a train of nanosecond pulses. In the transverse-excitation-atmospheric-pressure (TEA) configuration, CO_2 lasers are pumped very rapidly compared with the lifetime of the metastable state, resulting in a large population inversion in the gain medium when the electromagnetic field builds up in the cavity. This results in an intense "gain-switched pulse" of between 100 and 200 ns in duration followed by a lower intensity emission due to continued pumping of the upper state. These systems provide up to 3 J/pulse at pulse repetition frequencies of up to 50 Hz. At much lower pulse repetition frequencies, energies as high as 1000 J/pulse have been obtained. The typical beam divergence of these lasers is less than 3 mrad.

Carbon monoxide lasers operate on the vibrational levels of CO. They can be excited either through electrical discharge or chemical reaction (as discussed in Sect. 69.2.7). CO lasers have a tunable emission between 5 and 7 μm, operating with powers as high as 1 kW cw. In the pulsed mode, the typical energy emission is 10 mJ/pulse with 1 μs pulses at a pulse repetition frequency of 10 Hz. N_2O lasers extend the molecular laser wavelength range to beyond 10 μm.

Nitrogen lasers operate in the near uv at 337.1 nm. These are based on transitions between electronic energy levels of the N_2 molecule excited by electrical discharge. The typical emission from an N_2 laser is a single pulse of 10 ns in duration. The peak pulse power can be as high as 1 MW with 10 mJ/pulse at a pulse repetition frequency of less than 100 Hz. It is also possible to design these systems to obtain picosecond pulses. H_2 operates in a similar way in the 120 to 160 nm region of the uv.

69.2.5 Excimer Lasers

The most important lasers in the near uv to vuv for industrial and medical applications are based on rare-gas halide excimers such as XeF at 351 nm, XeCl at 308 nm, KrF at 248 nm, ArF at 193 nm, and F_2 at 153 nm. The major problem with these systems is the corrosive nature of the gases. They can be pumped by electric discharges, electron or proton beams, or optical excitation. Using electrical discharge excitation, the electrons ionize the noble gas molecules, and these react by pulling an atom off the halide molecule to create an excited state dimer molecule (excimer) that radiates to an unstable lower state where dissociation occurs. The short radiative lifetimes of excimers result in laser pulses of 10 to 50 ns duration. Systems can be configured to have pulse durations ranging from picoseconds to microseconds. These lasers typically operate at high pulse repetition frequencies between 200 and 1000 Hz. The energy per pulse of discharge-pumped excimer lasers ranges from several mJ to 0.1 J, with typical average powers up to 200 W. Single shot, electron-beam-pumped excimer lasers can have as high as 10^4 J/pulse for an average power of almost 1 kW.

Typical emission bands for excimer lasers are approximately 100 cm^{-1} wide due to vibrational sublevels. With the use of etalons and gratings, the laser line can be narrowed to 0.3 cm^{-1}. These frequency-selective elements can be used to tune the emission over several nanometers. Special configurations of excimer lasers have been developed to obtain specific operational parameters. For example, pulses as short as 45 fs have been obtained by mode-locking or by using excimer amplifiers to amplify frequency shifted dye laser pulses [3, 4]. Stimulated Brillouin scattering has been used as a phase conjugate mirror to minimize phase-front distortion in excimer lasers. Both master oscillator power amplifiers and injection locked resonator techniques have been used to achieve low spatial divergence with narrow bandwidths [3, 4]. Input pulses for the latter system can be from either frequency shifted Nd:YAG or dye lasers.

One interesting recent development has been the

demonstration of laser operation of solid state excimer systems [6]. This is discussed in Sect. 69.3.4.

69.2.6 Nonlinear Mixing

Another way to produce tunable vuv laser transitions is to use nonlinear four-wave sum mixing in atomic vapors and molecular gases. This requires optical pumping with two sources and can be achieved with any of a variety of laser combinations such as excimer lasers with dye lasers, or Nd:YAG lasers with dye lasers. This technique has been used to obtain laser emission in the 57 to 195 nm spectral range with visible to vuv conversion efficiencies as high as 10^{-3}.

69.2.7 Chemical Lasers

It is possible for some chemical reactions between molecules to leave the final molecule in an excited state. This type of "pumping" can result in a population inversion with respect to one of the lower states and laser transitions can occur between vibrational or rotational states of the molecule. Using mixtures of different types of molecules, pumping can be enhanced through energy transfer. Both pulsed and cw laser operation have been obtained with chemical lasers. Typical emission is of the order of a few hundred watts in the ir with a tunable output wavelength. Visible emission from chemical lasers has been demonstrated [7]. This type of laser provides the possibility of a system with a self contained chemical power supply for use in remote environments. The problems associated with handling hazardous chemicals have restricted the applications of chemical lasers.

The HF chemical laser is associated with an exothermic chain reaction between H_2 and F_2 molecules to yield vibrationally excited HF [4]. The fluoride atoms are generated in an electrical discharge tube from the dissociation of SF_6. These are injected into the optical resonator along with the H_2 or D_2 gas which flows perpendicular to the lasing direction. Controlling the gas flow for mixing the reactants is critical to the laser design. The chemical reaction for direct excitation is $F + H_2 \rightarrow HF^*$ + H. For this reaction, $\Delta H = -32\,\text{kcal/mole}$, resulting in laser emission energies of between 100 and 400 kJ/kg, at wavelengths between 2.5 and 3.7 μm. This corresponds to vibrational-rotational transitions in the HF molecule. It is possible to select a single line for the laser output and then tune the laser output wavelength by selecting different lines. Single line output for cw operation can produce up to 100 W of power. If D_2 replaces H_2, the emission shifts to between 3.6 and 4.2 μm, and the single line cw output power drops to about 50 W. The output power for a cw laser of this type, operating in the multiple line mode, can be as high as 2.2 MW, while in the

pulsed mode of operation pulse energies of 5 kJ can be obtained with multiple line emission.

An example of energy transfer pumping of chemical lasers is the $DFCO_2$ system. Pumping of vibrational-rotational transitions of DF occurs through multiple chemical reactions of fluorine and deuterium, followed by energy transfer to excited states of the CO_2 molecules. This exhibits laser emission with kilowatts of power at 10.6 μm, as described above. Another important laser system using energy transfer pumping is the chemical-oxygen-iodine laser (COIL). This is based on transitions between electronic levels in which singlet oxygen is excited and transfers its energy to a metastable state of iodine. Emission occurs at 1.3 μm and cw powers of up to 25 kW have been obtained.

One example of chemical laser action in the visible region is excited GeO transferring its energy to atomic Tl, which lases at 535 nm. Only a few systems of this type have been demonstrated, and none are developed to the level of commercial availability [7].

69.3 SOLID STATE LASERS

Solid state lasers are based on luminescence centers randomly distributed in a crystalline or glass host material. These can be classified in terms of the type of their laser active centers: transition metal ion lasers; rare earth ion lasers; and color center lasers. The use of dye molecules in plastic host media is a new type of solid state laser that is discussed in Sect. 69.5.1. The active ions are substitutionally "doped" into the host during crystal growth or glass melting, whereas the lattice defects that produce color centers are generally produced by post-growth radiation or heat treatments. The excitation mechanism is through optical pumping by either another laser or lamps. The spectral range covered by solid state lasers spans the visible and near ir. The variety of combinations of active centers and hosts provides the ability for both pulsed and cw operation with either narrow band or broad band emission. The latter type can provide frequency tunable output with the appropriate frequency-selective element in the cavity. Temperature tuning of narrow emission lines is also possible over a limited range. The spectral lines are broadened by internal strains in the host lattice (inhomogeneous broadening) and by radiationless relaxation and scattering processes involving thermal vibrations of the host (homogeneous broadening). Standard Q-switching and mode-locking techniques can be used with most solid state lasers.

69.3.1 Transition Metal Ion Lasers

The laser ions of this type have optically active electrons in unfilled $3d^n$ electron configurations. They include the positively charged ions Cr^{3+}, Cr^{4+}, Co^{2+}, V^{2+},

Ni^{2+}, and Ti^{3+}. The transitions involved in pumping and lasing are associated with the optically active electrons. Because these are outer shell electrons, they are sensitive to their local crystal field environment. Typical host materials are ionic crystals formed from oxides such as sapphire, emerald, chrysoberyl, forsterite, and garnets, or fluorides such as MgF_2, $KMgF_3$, $LiCaAlF_6$, and $LiSrAlF_6$. The host determines the bulk optical, mechanical, and thermal properties of the laser material, and it influences the spectroscopic properties of the active ions.

The most successful transition metal laser ion is Cr^{3+}. It has been made to lase in many different types of host crystals with both strong and weak crystal field environments. The most common strong field host is sapphire, and Al_2O_3:Cr^{3+}, commonly known as *ruby*, which was the first laser invented. The typical characteristic of a strong field laser material is a sharp laser line associated with a spin-flip electronic transition between states of the same crystal field configuration. In ruby this occurs at 699.7 nm. Because of its strong, broad absorption bands, ruby can be efficiently pumped by lamps and operates in either a pulsed or cw mode. Typical cw power output is a few watts with an efficiency of $\sim 0.1\%$. Ruby can be Q-switched to produce 10 ns pulses with several joules of energy per pulse, and mode-locked to produce pulses that are 5 ps in duration.

The typical characteristic of a weak field laser material is a broad gain curve associated with vibronic transitions between states of different crystal field configurations. Although chrysoberyl is a host with intermediate crystal field, $BeAl_2O_4$:Cr^{3+}, commonly known as *alexandrite*, is sufficiently close to a weak field case to operate as a laser in this regime. Using frequency selective elements in the cavity, alexandrite lasers can be tuned in the 700 to 820 nm range. The gain of alexandrite increases with temperature, with the cross section at the peak of the gain curve near 10^{-20} cm^2. Typical laser outputs are 4.5 J/pulse at 20 Hz pulse repetition rate and 90 W average power with 2% overall efficiency. In the Q-switched mode, 40 ns pulses with 2 J/pulse are obtained and the pulse width can be stretched to much longer values. Alexandrite lasers can also be mode-locked to obtain 28 ps pulses with 0.5 mJ/pulse.

Fluoride crystals, such as $LiCaAlF_6$ and $LiSrAlF_6$, are also weak field hosts for tunable Cr^{3+} lasers in the near ir. The latter is termed Cr:LiSAF and has a peak stimulated emission cross section of 0.4 x 10^{-19} cm^2 with a tuning range from 780 to 1020 nm. This system can be either flashlamp pumped or diode laser pumped. Single pulses with energies of 75 J have been generated by these lasers. Kerr lens mode-locking has produced pulses shorter than 100 fs.

One of the most interesting tunable solid state laser systems is Ti-sapphire (Al_2O_3:Ti^{3+}) because it has the broadest tuning range of any ion, extending from about 660 nm to about 1180 nm. Pumping can be provided by either lasers or flashlamps, resulting in either cw or pulsed operation. This results in a versatile source of excitation in this spectral region and sub-picosecond pulses can be obtained through mode-locking. The single $3d$ electron of Ti^{3+} gives a simplified energy level scheme which minimizes losses due to excited state absorption, which can be a problem in Cr^{3+} lasers. However, the metastable state lifetime of Ti^{3+} is significantly shorter than that of Cr^{3+}, and therefore Cr^{3+} lasers have much greater energy storage capability. Thus, Ti-sapphire lasers are difficult to pump by flashlamp, and are therefore generally pumped by argon lasers or frequency-doubled Nd-YAG lasers. Ti-sapphire has a very high peak gain cross section of about 4×10^{-19} cm^2. On the other hand, it is difficult to grow Ti-sapphire crystals with high concentrations of Ti^{3+} ions because of valance state stability.

69.3.2 Rare Earth Ion Lasers

All of the trivalent lanthanide ions (Ce^{3+}, Pr^{3+}, Nd^{3+}, Pm^{3+}, Sm^{3+}, Eu^{3+}, Gd^{3+}, Tb^{3+}, Dy^{3+}, Ho^{3+}, Er^{3+}, Tm^{3+}, Yb^{3+}) and the divalent ions Sm^{2+}, Dy^{2+}, Tm^{2+} have been used as active ions in solid state lasers. These ions are characterized by unfilled $4f^n$ electron configurations and the most common source of laser emission comes from electronic transitions among their energy levels. Because the inner shell $4f$ electrons are shielded by outer shell electrons, the energy levels are not strongly affected by the environment of the local host material. This leads to sharp lines in both absorption and emission. One exception to this is vibronic emission from Ho^{3+} which can produce tunable laser emission in the near ir. In some cases, transitions between $5d$ and $4f$ levels are involved in the laser emission. Since the $5d$ levels are broadened by the environment, broadly tunable laser emission can be obtained. Examples are Ce^{3+} in the uv and Sm^{2+} in the near ir The only actinide ion that has been made into a laser is U^{3+}.

Both crystals and glasses can be used as host materials for rare earth ion lasers. Common oxide crystal hosts include the garnets such as $Y_3Al_5O_{12}$ (commonly referred to as YAG) and a typical fluoride crystal host is $YLiF_3$. A wide variety of glass hosts has been used, including silicates, phosphates, heavy metal fluorides, and mixtures of these. The major difference between crystal and glass hosts is that crystals provide similar crystal field sites for every dopant ion, leading to a minimum of inhomogeneous broadening, while the disorder associated with glass structure gives many different types of local crystal field sites for the dopant ions and thus significant inhomogeneous broadening.

Because of the abundance of their energy levels, many trivalent rare earth ions have more than one metastable state, and laser emission is possible from several transitions. This results in over 100 possible laser emission lines ranging from the near uv through the visible and near ir. Both pulsed and cw operation can be obtained. The standard configuration for a rare earth solid state laser is a rod of laser material pumped by a lamp. Other configurations are used for special situations, such as a slab of laser material for high power glass lasers where heat management is a problem, and microchip lasers for photonics applications. Glass *fiber lasers* and amplifiers are becoming important configurations for some applications.

A major problem is the inefficiency of coupling the excitation energy of a lamp source with a broad spectral output into the spectrally sharp absorption bands of the trivalent rare earth ions. This can be overcome by using a laser as a pump source. One of the major advances in solid state laser technology has been the development of bars of high power diode laser arrays as pump sources. This has significantly increased the efficiency and decreased the thermal problems in these lasers. Currently available diode laser pump sources cover a limited range of wavelengths and thus can only be used to excite a limited set of metastable states. Several schemes have been adapted to excite other metastable states. One of these is *up-conversion pumping* [8] in which an ion is excited to a low-lying metastable state by a photon from the pump source and then re-excited to a higher metastable state either by another photon from the pump source or by energy transfer interaction with a neighboring ion that has also been excited to the low-lying metastable state. *Avalanche pumping* [9] relies on thermal fluctuations to populate a low-lying energy level of an ion which can then be re-excited by a pump photon to a high energy metastable state. Another method involves the addition of a second dopant ion to the host. This "sensitizer" ion is one with broad pump bands (such as Cr^{3+}) that can efficiently absorb the energy from the pump lamp. The excited sensitizer then interacts with the "activator" (lasing) ion through a radiationless, resonant energy transfer process. This deactivates the sensitizer and excites the activator ion.

Nd^{3+} has been made to lase in a greater number of host materials than any other active ion. $Y_3Al_5O_{12}:Nd^{3+}$, commonly referred to as Nd-YAG, is one of the most successful commercially available lasers. Although it is possible for Nd-YAG lasers to operate at several different wavelengths around 1 μm, the standard lasers emit at 1.06 μm. Continuous wave powers of 250 W are available and pulsed performance of several megawatts at 10 Hz and 1 J/pulse can be obtained. To obtain visible (532 nm) and near uv (354 and 266 nm) emission, nonlinear optical crystals are used to modify the near ir output through second, third, and fourth harmonic generation. The optimum concentration of Nd^{3+} in YAG is a few percent, and above this amount, concen-

tration quenching of the emission occurs through energy transfer and cross-relaxation processes. Attempts to co-dope Nd-YAG with Cr^{3+} to enhance pumping through energy transfer have not been successful [10]. However, in other garnet crystal hosts such as $Gd_3Sc_2Ga_3O_{12}$ (GSGG), the Cr-Nd energy transfer is strong enough to enhance pumping efficiency. Diode laser pumping of Nd-YAG has significantly increased pumping efficiency.

Other crystal hosts such as the pentaphosphates do not exhibit concentration quenching, so materials with 100% Nd^{3+} can be used as "stoichiometric laser materials". These can be important for some mini-laser applications. In glass hosts, Nd^{3+} ions can produce pulses with energies of ~ 100 kJ/pulse, but these systems must operate at very low pulse repetition rates of one pulse every few minutes to allow for heat dissipation. New athermal glass compositions decrease the problems with thermal lensing for high power laser operation. Nd:YVO lasers have been operated in a microchip configuration pumped by diode lasers.

A major area of recent development involves rare earth doped crystal lasers operating at specific wavelengths in the near ir [11]. This includes both the "eye safe" region between 1.35 and 2.2 μm for applications involving atmospheric transmission, and the 2 to 3 μm range to match water overtone absorption bands for medical applications. The ions of most interest are Er^{3+}, Tm^{3+}, and Ho^{3+}. Er^{3+} can lase on 13 transitions at wavelengths ranging from 0.56 to 4.8 μm. Both Tm^{3+} and Ho^{3+} exhibit laser transitions near 2 μm, but they are difficult to pump efficiently. One successful laser in this spectral region is triply doped Cr;Tm;Ho:YAG. The flashlamp energy is absorbed by the Cr^{3+} ions and transferred to Tm^{3+} ions. Cross-relaxation between two thulium ions doubles the quantum efficiency by leaving two Tm^{3+} ions in the excited state. Energy transfer among the thulium ions with transfer to Ho^{3+} then occurs and the lasing transition occurs on the holmium ions.

Fiber lasers have been developed that consist of trivalent rare earth ions doped in either oxide or fluoride glass fibers and pumped by other lasers [12]. A useful application of active fibers is Er^{3+} and Pr^{3+} optical amplifiers for fiber communication systems. The extended length of the gain media and the nonlinear dispersion effects in fiber transmission allow precise tailoring of laser emission properties. Heavy metal fluoride fibers give improved ir transmission in the 2 to 3 μm spectral region. An important recent development involves writing laser-induced gratings in fibers to produce distributed feedback lasers with stable, single-mode operation [13]. Nd-doped fiber lasers pumped by diode lasers have produced 5 W of power. Mode-locked fiber ring lasers have produced solitons. Efficient up-conversion laser operation has been achieved in fibers. In addition, the efficient nonlinear optical properties of fibers has led to fiber Raman lasers.

69.3.3 Color Center Lasers

In color center lasers, the optically active center is a point defect in the lattice. For example, in alkali halide host crystals, such as NaCl, a typical color center consists of an electron trapped at a halide ion vacancy. Similar color centers occur in oxide host crystals such as diamond and sapphire. Color centers can be produced by thermal treatment or exposure to radiation. Many times, these centers are stable only at low temperatures due to ion and electron mobility. In some cases, impurity ions act to stabilize the defect center. A neutral Tl atom at a cation site next to a anion vacancy in KCl is an example of this. A major recent advance in color center lasers involves the development of room temperature stable pulsed laser systems. Systems based on the vibrational transitions of molecular defects such as CN^- have been demonstrated to operate as lasers in the 5 μm spectral region.

Color center absorption generally occurs in the visible, and they are optically pumped by Ar, Kr, or Nd:YAG lasers. Typical color center emission occurs as a broad band in the near ir between 0.8 and 4.0 μm. The emission is based on allowed transitions with high oscillator strengths leading to high gain cross sections.

The homogeneously broadened emission band of color centers allows for efficient, tunable laser emission and single mode operation. Optically pumped cw output powers of ~ 2 W have been obtained, and mode-locked pulses of less than 100 fs and 1 MW peak power at repetition rates of 100 MHz and hundreds of milliwatts average power have been generated. Laser linewidths of less than 4 kHz have been obtained.

One problem with a high gain medium such as a color center crystal is a tendency for multimode laser operation. In a linear standing wave cavity, a primary oscillating mode reaches gain saturation and burns spatially periodic holes in the population inversion of the gain medium. The high gain allows secondary modes to oscillate with peaks at the nodes of the primary mode. A grating/etalon combination can be used to select and tune the laser output frequency [4]. The etalon selects one cavity mode and the grating selects one order of the etalon. This results in single mode tunable output. The single mode power output is 70% of the multimode laser power due to energy loss in the hole burning mode. A ring laser configuration in a traveling wave operation can be used to give uniform saturation of the gain medium, and thus no hole burning. A ring laser cavity needs additional optics such as a Faraday rotator and an optically active plate to force oscillation in only one direction. Active frequency stabilization circuits are necessary to obtain linewidths of less than 4 kHz.

Synchronously pumping a color center laser with a mode-locked Nd:YAG laser gives mode-locked output with pulses typically between 5 and 15 ps. If passive mode-locking is obtained through use of a saturable absorber, pulses of ~ 200 fs can be obtained. Additive pulse mode-locking consists of two coupled cavities, one with a color center gain medium and the other with a single mode optical fiber. Self phase modulation in the fiber gives a broader frequency spectrum to the pulses and thus shorter time widths. This scheme has generated pulses of about 75 fs [14]. Soliton lasers are also obtained by coupling a color center laser with a fiber laser.

69.3.4 New Types of Solid State Laser Systems

One type of new solid state laser system being studied involves the fourth and fifth row transition metal ions. So far, stimulated emission and gain have been reported for Rh^{3+} [15] under strong pumping conditions, but no laser operation has been achieved.

Solid state dye lasers [16] consist of organic dye molecules doped in crystal or glass host materials. The first systems of this type used the same class of organic dyes as in liquid lasers (such as rhodamine 6G) and host materials such as sol-gels or polymethylmethacrylate. These systems have traditionally had a problem with photo-degradation of the material after a limited number of shots. However, recent combinations of new dyes and new host materials have produced outputs of tens of millijoules per pulse, over 50% slope efficiency, and a degradation to 60% of the initial output after 30 000 pulses. One example of these new materials is pyrromethene-BF_2 complex dye doped in an acrylic plastic host [16]. Another material system that has exhibited good laser performance involves xerogel hosts doped with perylene or pyrromethene dyes [17]. The performance of some of these new systems has reached the point that they may be useful for tunable laser applications in the visible spectral region.

Another new type of system can be described as a *solid state excimer laser* [6]. These materials consist of noble gas solids as host crystals, such as Ar and Ne, doped with excimer molecules, such as XeF. The emission lines occur in the uv and visible, with major lines at 286, 411, and 540 nm. These systems have very high stimulated emission cross sections and large gain coefficients.

69.3.5 Frequency Shifters

Since solid state lasers are commercially available at only a limited number of wavelengths, it is sometimes easier to use nonlinear optical techniques to shift the frequency of an available laser than to develop a new primary laser system. The techniques for this include harmonic frequency generation, frequency mixing, optical parametric oscillators and amplifiers, and Raman shifting. Significant advances have been made recently [18] in developing new types of materials for these applications. For frequency mixing, harmonic generation and OPOs in the visible and near uv, important new crystals include $KTiOPO_4$ (KTP), BaB_2O_4 (BBO), and LiB_3O_5 (LBO). In the 3 to 5 μm region, new materials for frequency mixing and OPOs include KTA (the arsinate analog of KTP) and $ZnGeP_2$. Gas phase Raman cells have been commercially available for solid state laser systems for many years. Recently, it has been demonstrated that crystals such as $Ba(NO_3)_2$ can be used as efficient solid state Raman shifters. New waveguide configurations with periodic poling for quasiphase matching have greatly enhanced the efficiency of harmonic generation [19]. Optical damage threshold is still the limiting parameter for nonlinear optical materials.

69.4 SEMICONDUCTOR LASERS

The light emission from semiconductor diode lasers is generally associated with the radiative recombination of electrons and holes. This occurs at the junction of an n-type material with excess electrons and a p-type material with excess holes. The excitation is provided by an external electric field applied across the p-n junction that causes the two types of charges to come together. The most common semiconductor laser emission lines occur in the near or mid-ir. These are generally made of III-V compounds such as gallium arsenide in the red and near ir, and lead salts in the mid-ir region. Wide bandgap II-VI materials are currently being explored for use in the green and blue spectral regions. The spectral lines are generally narrow with broadening due to lattice defects (inhomogeneous broadening) and radiationless relaxation and scattering processes associated with the thermal vibration of the host (homogeneous broadening). Temperature tuning can be used to change the output wavelength over a narrow spectral range. Direct modulation of the laser output can be achieved by modulating the external current. This is an important feature of electric current pumped semiconductor lasers, and leads to applications where high frequency modulation is required. Modulation bandwidths in excess of 11 GHz have been obtained.

The ability to design and grow specialized structures one atomic layer at a time using techniques such as molecular beam epitaxy (MBE) has led to the design of quantum well lasers with enhanced properties, as well as more esoteric designs of quantum wires and quantum boxes [20]. Quantum well microlasers can be as small as 100 μm and operate with milliamps of current at a few volts. These are generally based on material systems such as GaAs/GaAlAs or InP/InGaAsP. Heterostructure lasers have layers of aluminum, indium, and phosphorus on the sides of the junction to confine the electronic current to the junction region to minimize the amount of current required, and thus minimize heat dissipation compared with homostructures of the same materials. Special device structures can be fabricated to produce gain-guiding and index-guiding to enhance the operating characteristics of the lasers. Grating structures can be fabricated to give distributed feedback lasers that narrow the laser linewidths to ~ 1 MHz. External cavity lasers with gratings have achieved linewidths as low as 1 kHz. CW operating powers of up to 10 mW have been achieved from a single p-n junction, while phased arrays have reached combined powers of well over 10 W.

Gallium arsenide (GaAs) was the first compound semiconductor diode laser. It can produce laser emission at wavelengths between 750 and 870 nm. The development of strained-layer technology has allowed the use of mixed compounds of gallium aluminum arsenide ($Ga_{1x}Al_xAs$) to fabricate different types of laser structures, and the concentration of aluminum determines the laser emission wavelength. Wavelengths from 620 nm to 905 nm have been obtained. The most common diode laser structures are simple double-heterostructure lasers, and monolithic arrays of laser stripes can be fabricated for higher power. In a typical GaAlAs laser, the active layer is sandwiched between two layers having larger bandgaps and lower refractive indices. The former characteristic produces electrical confinement and the latter produces optical confinement. This improves the efficiency and allows cw laser operation. In the conventional horizontal-cavity structure, cleaved end facets of the chip produce the optical feedback required for laser oscillation. Fabricating quantum well structures in the active layer produces improved confinement, and thus higher efficiency operation. For low power lasers, high beam quality is achieved through an index-guiding structure that concentrates the optical beam in the laser stripe. In high power lasers, the current is concentrated in the laser stripe to achieve gain-guiding. Laser arrays can generate cw powers of the order 20 W and in a quasicw mode they can produce peak powers of 100 W. Stacking diode bars in planar arrays can generate kilowatts of power. Another approach to obtaining high powers is a master oscillator power amplifier (MOPA) configuration. This has the advantage of maintaining high beam quality, and gallium arsenide MOPAs have produced single frequency operation at 1 W of power.

Long wavelength ir diode lasers are made of IV-VI compounds such as PbS. These lasers operate at cryogenic temperatures and provide tunable emission from 4 to 32 μm. The tunability is achieved by changing temperature or current.

One of the most important areas of research in semiconducting lasers is the development of new device configurations such as *vertical cavity surface emitting lasers* (VCSELs) [21]. These have lower round trip gain but significantly reduced divergence of the output beam. This configuration allows for the fabrication of two dimensional arrays of independently modulated lasers.

Another major research area is generating new laser wavelengths. Using strained layer technology, a variety of different combinations of direct bandgap materials can be made into semiconductor lasers [22]. The range of available bandgaps can conceivably result in lasers with emission wavelengths spanning the visible and near ir spectral regions. There is currently special emphasis on the development of lasers in the blue and green spectral regions using wide bandgap II-VI materials such as ZnSe [23].

69.5 LIQUID LASERS

There are three classes of liquid lasers. The most widely used are based on organic solvents with organic dye molecules as the active laser species. The other two types are based on rare earth ions for the lasing entity. In one case, the lasing system involves rare earth chelates in organic solvents, while in the other, the rare earth ions are in inorganic solvents. These systems are optically pumped with either flashlamps or other lasers.

69.5.1 Organic Dye Lasers

Dye lasers are generally based on fluorescent dyes in liquid solvents, using optical pumping by either flash-lamps or other lasers as the mechanism for excitation. They can operate in either a pulsed or cw mode at wave-lengths as short as 310 nm out to about 1.5 μm. Dye lasers provide the versatility of varying wavelength, band-width, and pulse length as desired.

The fluorescence emission of dye molecules appears as broad spectral bands due to coupling of the electronic energy levels with molecular vibrations. This gives a broadband gain curve for lasing, and thus with a dispersive element such as a grating, prism, filter, or etalon in the cavity, dye laser outputs can be tuned over a range of several hundred angstroms (30–60 nm). There are now over 200 organic laser dyes. One of the most successful dyes is rhodamine 6G which covers the spectral range 570–630 nm. Alcohol is a typical solvent. The various types of cavity designs include folded cavities and ring cavities. Flowing dye configurations are useful for heat management. Oscillator/amplifier configurations are used to suppress amplified spontaneous emission. Dispersive elements plus frequency stabilization have been used for spectral line narrowing to single mode cw operation. Frequencies as narrow as 10 GHz or less have been obtained. For cw operation, output powers of a few watts can be obtained, while in the pulsed mode the energy per pulse can be up to 100 mJ in 10 ns pulses. With mode-locking, trains of femtosecond pulses with intervals of 20 ns can be produced.

Mode-locking takes advantage of the broad emission spectrum of the dye molecules to get short pulses. The standard techniques of synchronous pumping, active, and passive mode-locking have been used. In addition, colliding pulse mode-locking has been used in a ring configuration. In this case, pump beams going the opposite direction in the cavity collide in an absorber jet dye to produce interference fringes and saturation. The gain dye is located half way around the ring from the absorber dye. Fiber compression techniques have also been used with dye laser systems. The shortest pulses obtained so far are 6 fs [4]. Hybrid mode-locking utilizing synchronous pumping plus a saturable absorber and prism dispersion compensators has been employed to achieve powers of 350 mW and greater tunability than colliding pulse systems [4].

Several new technological advances have increased the spectral coverage of dye lasers. These include bet-ter pump sources, such as the increased power of argon pump lasers and the availability of uv pump lines, and the use of Ti-sapphire pump lasers. Combining these pump sources with new dyes has provided extended dye laser output in both the blue and near ir [4]. Also, improved nonlinear crystals have allowed coverage of the near uv from 260 to 960 nm through harmonic frequency gen-eration and frequency mixing the dye laser output with pump laser wavelengths. Another developing technology is *solid state dye lasers.* As mentioned in Sect. 69.3.4, significant progress has been made recently in decreas-ing the photodegradation problems associated with dye molecules doped in solid host materials [16, 17].

69.5.2 Rare Earth Chelate Lasers

In these systems, the active lasing center is a rare earth complex with organic molecules in an organic sol-vent. The chelate ligands are organic phosphates, car-boxylate ions, or β-diketonate. The optical pump energy is absorbed by the ligand and efficiently transferred to the rare earth ion. Energy transfer quenching from the rare earth ion to the organic molecule vibrational levels decreases the efficiency of these lasers. This is especially true for Nd^{3+}. The three ions that have been most effec-tive in these systems are Eu^{3+}, Tb^{3+}, and Nd^{3+}.

69.5.3 Inorganic Rare Earth Liquid Lasers

In these systems, the active lasing center is a rare earth ion inorganic complex of heavy metal halides or oxyhalides (the optical pumping is directly into the rare earth ion). Nd^{3+} lasers can produce several hundred joules of energy per pulse in the long pulse mode and peak powers of 180 MW in a Q-switched mode. These lasers have also been mode-locked to obtain 3 ps long pulses having 1 GW of peak power. Self mode-locking and self *Q*-switching is also observed in these systems.

69.6 OTHER TYPES OF LASERS

Several more complex laser systems have been developed that have significant interest for scientific studies, but so far have had limited applications outside the laboratory. These include x-ray lasers, particle-beam-pumped lasers, and free electron lasers.

69.6.1 X-Ray and Extreme uv Lasers

These systems are based on highly ionized ions produced by powerful laser sources. Examples are krypton at 93 nm, molybdenum at 13 nm, and carbon at 18 nm. There is significant interest in developing x-ray lasers for applications in lithography and medical imaging, but so far the lack of reliable x-ray optics for the laser cavities has limited the technology.

69.6.2 Nuclear Pumped Lasers

These are gas lasers excited by high energy charged particles or gamma rays resulting from nuclear reactions. Either nuclear reactors or nuclear explosives are used as pump sources. These can operate in either a pulsed or cw mode and produce emission covering the spectral range from the uv through the ir [4]. Typical gases range from Xe_2^* at 170 nm to CO at 5.4 μm. Other gases that have been used include argon and nitrogen. Typical pulsed outputs produce 10 ns pulses with energies from 2×10^{-7} J to 3 J. There are significant problems with radiation damage of the laser components in these systems.

69.6.3 Free Electron Lasers

These lasers are based on a high-energy beam of electrons in a spatially varying magnetic field. The varying field causes the electrons to oscillate, and thus to emit radiation at the oscillation frequency. The stimulated emission produced under these conditions provides the laser output. Since the electrons are making transitions between continuum states instead of discrete states, these systems can give high power output over the entire spectral range from the vuv to the far ir [24]. Powers as high as 1 GW and efficiencies as high as 35% have been obtained. Beam spread is controlled by the use of tapered instead of uniform wigglers. Both cw and picosecond pulsed emission can be obtained.

There are three types of free electron laser configurations [24]. The first is a master oscillator power amplifier (MOPA), in which an electron beam is injected into a wiggler in synchronism with the signal to be amplified. The external radiation source to drive the amplifier is a master oscillator such as a conventional laser system. This is a single pass, high gain system. The second configuration is an oscillator. This is designed with reflection at the ends of the wiggler so that the signal makes multiple passes in the cavity. This can operate with low gain. Since it amplifies spontaneous noise, no injected signal is necessary. The third configuration is a super-radiant amplifier. In this configuration, shot noise is amplified over a single pass through the wiggler. These systems require high current accelerators to drive them. Because their operation is based on broad band shot noise, super-radiant amplifier radiation has a broader band than radiation from a master oscillator power amplifier.

Different accelerator configurations can be used to produce the electron beam, storage rings, induction linacs, pulse line accelerators, etc. [24]. These give different beam properties such as quality, current, and energy. They also each give a limited range of wavelengths and temporal structure of the output.

Advances have been made recently in designing smaller and less complex free electron lasers [24]. As this trend continues, these systems will find important applications in medicine and industry.

REFERENCES

1. *Handbook of Laser Science and Technology*, edited by M. J. Weber (CRC Press, Boca Raton, 1982), Vol. 1.
2. *Handbook of Laser Science and Technology*, edited by M. J. Weber (CRC Press, Boca Raton, 1991), Supp. 1.
3. *Laser Handbook*, edited by M. Bass and M. L. Stitch (North Holland, Amsterdam, 1985), Vol. 5.
4. *Encyclopedia of Lasers and Optical Technology*, edited by R. A. Meyers (Academic Press, San Diego, 1991).
5. *Handbook of Solid State Lasers*, edited by P. K. Cheo (Marcel Dekker, New York, 1989).
6. N. Schwentner and V. A. Apkarian, Chem. Phys. Lett. **154**, 413 (1989); G. Zerza, G. Sliwinski, and N. Schwentner, Appl. Phys. **B55**, 331 (1992); *ibid.* **A56**, 156 (1993).
7. W. H. Crumly, J. L. Gole, and D. A. Dixon, J. Chem. Phys. **76**, 6439 (1982); S. H. Cobb, J. R. Woodward, and J. L. Gole, Chem. Phys. Lett. **143**, 205 (1988); ibid. **157**, 197 (1989).
8. R. M Macfarlane, F. Tong, A. J. Silversmith, and W. Lenth, Appl. Phys. Lett. **52**, 1300 (1988); R. A. Macfarlane, Appl. Phys. Lett. **54**, 2301, (1989); T. Hebert, R. Wannemacher, W. Lenth, and R. M. Macfarlane, Appl. Phys. Lett. **57**, 1727 (1990); R. A. Macfarlane, Opt. Lett. **16**, 1397 (1991).
9. M. E. Koch, A. W. Kueny, and W. E. Case, J. Appl. Phys. **56**, 1083, (1990).
10. N. Karayianis, D. E. Wortman, and C. A. Morrison, Solid State Comm. **18**, 1299 (1976); W. F. Krupke, M. D. Shinn, J. E. Marion, J. A. Caird, and S. E. Stokowski, J. Opt. Soc. Am. **B3**, 102, (1986).

11. M. J. Weber, M. Bass, and G. A. deMars, J. Appl. Phys. **42**, 301 (1971); G. J. Quarles, A. Rosenbaum, C. L. Marquardt, and L. Esterowitz, Opt. Lett. **15**, 42 (1990); L. Esterowitz, Opt. Eng. **29**, 676 (1990).

12. P. Urquhart, IEE Proc. **J135**, 385 (1988).

13. I. M. Jauncey, L. Reekie, R. J. Mears, D. N. Payne, C. J. Rowe, D. C. J. Reid, I. Bennion, and C. Edge, Electron. Lett. **22**, 987 (1986).

14. L. F. Mollenhauer and R. H. Stolen, Opt. Lett. **9**, 12 (1984).

15. R. C. Powell, G. J. Quarles, J. J. Martin, C. A. Hunt, and W. A. Sibley, Opt. Lett. **10**, 212 (1985).

16. R. E. Hermes, T. H. Allik, S. Chandra, and J. A. Hutchinson, Appl. Phys. Lett. **63**, 877 (1993).

17. B. Dunn, F. Nishida, R. Toda, J. I. Zink, T. H. Allik, S. Chandra, and J. A. Hutchinson, Mat. Res. Soc. Symposium, Proc. **329**, 267 (1994).

18. V. G. Dmitriev, G. G. Girzodyan, and D. N. Nikogosyan, *Handbook of Nonlinear Optical Crystals* (Springer-Verlag, Berlin 1991).

19. E. J. Lim, M. M. Fejer, R. L. Byer, and W. J. Kozlovsky, Electron. Lett. **25**, 731 (1989).

20. Y. Arakawa, K. Vahala, and A. Yariv, Surf. Sci. **174**, 155 (1986); K. Vahala, IEEE J. Quantum Electron. **24**, 523 (1988).

21. R. E. Slusher, Optics, and Photonics News **4**, 8 (1993), and references therein.

22. W. W. Chow, S. W. Koch, and M. Sargent III, *Semiconductor Laser Physics* (Springer-Verlag, Berlin 1994).

23. M. A. Hasse, J. Qui, J. M. DePuydt, and H. Cheng, Appl. Phys. Lett. **58**, 1272 (1991).

24. C. A. Brau, *Free Electron Lasers* (Academic Press, San Diego 1990).

70

Nonlinear Optics

Alexander L. Gaeta* and Robert W. Boyd†

*School of Applied and Engineering Physics, Cornell University, Ithaca, New York 14853
†Institute of Optics, University of Rochester, Rochester, New York 14627

70.1 INTRODUCTION

Nonlinear optics is concerned with the propagation of intense beams of light through a material system. The optical properties of the medium can be modified by the intense light beam, leading to new processes that would not occur in a material that responded linearly to an applied optical field. These processes can lead to the modification of the spectral, spatial, or polarization properties of the light beam, or the creation of new frequency components. More complete accounts of nonlinear optics including the origin of optical nonlinearities can be found in references [1–4].

Both the Gaussian and MKS system of units are commonly used in nonlinear optics. Thus, we have chosen to express the equations in this chapter in both the Gaussian and MKS systems. Each equation can be interpreted in the MKS system as written or in the Gaussian system by omitting the prefactors (e.g., $1/4\pi\varepsilon_0$) that appear in square brackets at the beginning of the expression on the right-hand-side of the equation.

70.2 NONLINEAR SUSCEPTIBILITY

In linear optics it is customary to describe the response of a material in terms of a macroscopic polarization $\tilde{\mathbf{P}}$ (i.e., dipole moment per unit volume) which is linearly related to the applied electric field $\tilde{\mathbf{E}}$ through the linear susceptibility $\chi^{(1)}$. In order to extend the relationship between $\tilde{\mathbf{P}}$ and $\tilde{\mathbf{E}}$ into the nonlinear regime, the polarization is expanded in a power series of the electric field strength. We express this relationship mathematically by first decomposing the field and the polarization into their frequency components such that

$$\tilde{\mathbf{E}}(\mathbf{r},t) = \sum_l \mathbf{E}(\mathbf{r},\omega_l)e^{-i\omega_l t},$$
$$\tilde{\mathbf{P}}(\mathbf{r},t) = \sum_l \mathbf{P}(\mathbf{r},\omega_l)e^{-i\omega_l t}, \tag{70.1}$$

where the summations are performed over both positive and negative frequencies. The reality of $\tilde{\mathbf{E}}$ and $\tilde{\mathbf{P}}$ is then assured by requiring that $\mathbf{E}(\mathbf{r},\omega_l) = \mathbf{E}^*(\mathbf{r},-\omega_l)$ and $\mathbf{P}(\mathbf{r},\omega_l) = \mathbf{P}^*(\mathbf{r},-\omega_l)$. In this case the general expression for the Cartesian component i of the polarization at frequency ω_σ is given by

$$P_i(\omega_\sigma) = [\varepsilon_0] \Big[\sum_j \chi_{ij}^{(1)}(\omega_\sigma)E_j(\omega_\sigma)$$
$$+ \sum_{jk} \sum_{(mn)} \chi_{ijk}^{(2)}(\omega_\sigma;\omega_m,\omega_n)E_j(\omega_m)E_k(\omega_n)$$
$$+ \sum_{jkl} \sum_{(mno)} \chi_{ijkl}^{(3)}(\omega_\sigma;\omega_m,\omega_n,\omega_o)E_j(\omega_m)E_k(\omega_n)E_l(\omega_o)$$
$$+ \cdots \Big], \tag{70.2}$$

where $ijkl$ refer to field components, and the notation (mn), for example, indicates that the summation over n and m should be performed such that $\omega_\sigma = \omega_m + \omega_n$ is held constant. Inspection of Eq. (70.2) shows that the $\chi^{(n)}$ can be required to satisfy intrinsic permutation symmetry, i.e., the Cartesian components and the corresponding frequency components [e.g., (j,ω_j) but not (i,ω_σ)] associated with the applied fields may be permuted without changing the value of the susceptibility. For example, for the second-order susceptibility,

$$\chi_{ijk}^{(2)}(\omega_\sigma;\omega_m,\omega_n) = \chi_{ikj}^{(2)}(\omega_\sigma;\omega_n,\omega_m). \tag{70.3}$$

If the medium is lossless at all the field frequencies taking part in the nonlinear interaction, then the condition of full permutation symmetry is necessarily valid. This condition states that the pair of indices associated with the Cartesian component and the frequency of the nonlinear polarization [i.e., (i,ω_σ)] may be permuted along with the pairs associated with the applied field components. For example, for the second-order susceptibility, this condition implies that

$$\chi_{ijk}^{(2)}(\omega_\sigma;\omega_m,\omega_n) = \chi_{kji}^{(2)}(-\omega_n;\omega_m,-\omega_\sigma). \tag{70.4}$$

If full permutation symmetry holds, and in addition all the frequencies of interest are well below any of the transition frequencies of the medium, the $\chi^{(n)}$ are invariant upon free permutation of all the Cartesian indices. This condition is known as the *Kleinman symmetry condition*.

70.2.1 Tensor Properties

The spatial symmetry properties of a material can be used to predict the tensor nature of the nonlinear susceptibility. For example, for a material that possesses inversion symmetry, all the elements of the even-ordered susceptibilities must vanish (i.e., $\chi^{(n)} = 0$ for n even). The number of independent elements of the nonlinear susceptibility for many materials can be substantially fewer than than the total number of elements. For example, in general $\chi^{(3)}$ consists of 81 elements, but for the case of isotropic media such as gases, liquids, and glasses, only 21 elements are nonvanishing and only three of these are independent. The nonvanishing elements consist of the following types: $\chi_{iijj}^{(3)}$, $\chi_{ijij}^{(3)}$, and $\chi_{ijji}^{(3)}$, where $i \neq j$. In addition, it can be shown that

$$\chi_{iiii}^{(3)} = \chi_{iijj}^{(3)} + \chi_{ijij}^{(3)} + \chi_{ijji}^{(3)}. \tag{70.5}$$

70.2.2 Nonlinear Refractive Index

For many materials, the refractive index n is intensity-dependent such that

$$n = n_0 + n_2 I\,, \qquad (70.6)$$

where n_0 is the linear refractive index, n_2 is the nonlinear refractive index coefficient, and $I = [4\pi\varepsilon_0]n_0 c|E|^2/2\pi$ is the intensity of the optical field. For the case of a single, linearly polarized light beam traveling in an isotropic medium or along a crystal axis of a cubic material, n_2 is related to $\chi^{(3)}$ by

$$n_2 = \left[\frac{1}{16\pi^2\varepsilon_0}\right]\frac{12\pi^2}{n_0^2 c}\chi_{iiii}^{(3)}(\omega;\omega,\omega,\omega,-\omega)\,. \qquad (70.7)$$

For the common situation in which n_2 is measured in units of cm^2/W and $\chi^{(3)}$ is measured in Gaussian units, the relation becomes

$$n_2\left(\frac{\mathrm{cm}^2}{\mathrm{W}}\right) = \frac{12\pi^2 \times 10^7}{n_0^2 c}\chi_{iiii}^{(3)}(\omega;\omega,\omega,\omega,-\omega)\,. \qquad (70.8)$$

There are various physical mechanisms that can give rise to a nonlinear refractive index. For the case of induced molecular orientation in CS_2, $n_2 = 3\times 10^{-14}\ \mathrm{cm}^2/\mathrm{W}$. If the contribution to the nonlinear refractive index is electronic in nature (e.g., glass), then $n_2 \sim 2\times 10^{-16}\ \mathrm{cm}^2/\mathrm{W}$.

70.2.3 Quantum Mechanical Expression for $\chi^{(n)}$

The general quantum mechanical perturbation expression for the $\chi^{(n)}$ in the nonresonant limit is[1]

$$\chi_{i_0\cdots i_n}^{(n)}(\omega_\sigma;\omega_1,\ldots,\omega_n) = \left[\frac{1}{\varepsilon_0}\right]\frac{N}{\hbar^n}\mathcal{P}_F\sum_{ga_1\cdots a_n}\rho_0(g)$$

$$\times \frac{\mu_{ga_1}^{i_0}\mu_{a_1 a_2}^{i_1}\cdots\mu_{a_{n-1}a_n}^{i_{n-1}}\mu_{a_n g}^{i_n}}{(\omega_{a_1 g}-\omega_1-\cdots-\omega_n)(\omega_{a_2 g}-\omega_2-\cdots-\omega_n)\cdots(\omega_{a_n g}-\omega_n)}\,, \qquad (70.9)$$

where $\omega_\sigma = \omega_1 + \cdots + \omega_n$, N is the density of atoms or molecules that compose the material, $\rho_0(g)$ is the probability that the atomic or molecular population is initially in the state g in thermal equilibrium, $\mu_{a_1 a_2}^{i_1}$ is the i_1th Cartesian component of the $(a_1 a_2)$ dipole matrix element, $\omega_{a_1 g}$ is the transition frequency between the states a_1 and g, and $\mathcal{P}_F$ is the full permutation operator which is defined such that the expression that follows it is to be summed over all permutations of the pairs (i_0,ω_σ), $(i_1,\omega_1)\cdots(i_n,\omega_n)$ and divided by the number of permutations of the input frequencies. Thus the full expression for $\chi^{(2)}$ consists of six terms and that for $\chi^{(3)}$ consists of 24 terms.

---

[1] Under conditions of resonant excitation, relaxation phenomena must be included in the treatment, and the density matrix formalism must be used [4]. The resulting equation for the nonlinear susceptibility is then more complicated.

In the limit in which the frequencies of all the fields are much smaller than any resonance frequency of the medium, the value of $\chi^{(n)}$ can be estimated to be

$$\chi^{(n)} \simeq \left[\frac{1}{\varepsilon_0}\right]\left(\frac{2\mu}{\hbar\omega_0}\right)^n N\mu\,, \qquad (70.10)$$

where μ is a typical value for the dipole moment and ω_0 is a typical value of the transition frequency between the ground state and the lowest-lying excited state. For the case of $\chi^{(3)}$ in Gaussian units, the predicted value is $\chi^{(3)} = 3 \times 10^{-14}$, which is consistent with the measured values of many materials (e.g., glass) in which the nonresonant electronic nonlinearity is the dominant contribution.

70.2.4 The Hyperpolarizability

The nonlinear susceptibility relates the macroscopic polarization $\mathbf{P}$ to the electric field strength $\mathbf{E}$. A related microscopic quantity is the hyperpolarizability, which relates the dipole moment $\mathbf{p}$ induced in a given atom or molecule to the electric field $\mathbf{E}^{\mathrm{loc}}$ (the *Lorentz local field*) that acts on that atom or molecule. The relationship between $\mathbf{p}$ and $\mathbf{E}^{\mathrm{loc}}$ is

$$p_i(\omega_\sigma) = [\varepsilon_0]\Bigg[\sum_j \alpha_{ij}(\omega_\sigma)E_j^{\mathrm{loc}}(\omega_\sigma)$$

$$+ \sum_{jk}\sum_{(mn)}\beta_{ijk}(\omega_\sigma;\omega_m,\omega_n)E_j^{\mathrm{loc}}(\omega_m)E_k^{\mathrm{loc}}(\omega_n)$$

$$+ \sum_{jkl}\sum_{(mno)}\gamma_{ijkl}(\omega_\sigma;\omega_m,\omega_n,\omega_o)$$

$$\times E_j^{\mathrm{loc}}(\omega_m)E_k^{\mathrm{loc}}(\omega_n)E_l^{\mathrm{loc}}(\omega_o) + \cdots\Bigg]\,, \qquad (70.11)$$

where α_{ij} is the linear polarizability, β_{ijk} is the first hyperpolarizability, and γ_{ijkl} is the second hyperpolarizability. The nonlinear susceptibilities and hyperpolarizabilities are related by the number density of molecules N and by local-field factors, which account for the fact that the field $\mathbf{E}^{\mathrm{loc}}$ that acts on a typical molecule is not in general equal to the macroscopic field $\mathbf{E}$. Under many circumstances, it is adequate to relate $\mathbf{E}^{\mathrm{loc}}$ to $\mathbf{E}$ through use of the Lorentz approximation

$$\mathbf{E}^{\mathrm{loc}}(\omega) = \mathbf{E}(\omega) + \left[\frac{1}{4\pi\varepsilon_0}\right]\frac{4\pi}{3}\mathbf{P}(\omega)\,. \qquad (70.12)$$

To a good approximation, one often needs to include only the linear contribution to $\mathbf{P}(\omega)$, and thus the local electric field becomes

$$\mathbf{E}^{\mathrm{loc}}(\omega) = \mathcal{L}(\omega)\mathbf{E}(\omega)\,, \qquad (70.13)$$

where $\mathcal{L}(\omega) = \{[\varepsilon_0^{-1}]\varepsilon(\omega) + 2\}/3$ is a the local field correction factor and $\varepsilon(\omega)$ is the linear dielectric constant. Since $\mathbf{P}(\omega) = N\mathbf{p}(\omega)$, Eq. (70.2) and Eqs. (70.11) through (70.13) relate the $\chi^{(n)}$ to the hyperpolarizabilities through

$$\chi_{ij}^{(1)}(\omega_\sigma) = \mathcal{L}(\omega_\sigma) N \alpha_{ij}(\omega_\sigma) \,, \qquad (70.14)$$

$$\chi_{ijk}^{(2)}(\omega_\sigma; \omega_m, \omega_n) = \mathcal{L}(\omega_\sigma)\mathcal{L}(\omega_m)\mathcal{L}(\omega_n) \\ \times N\beta_{ijk}(\omega_\sigma; \omega_m, \omega_n) \,, \qquad (70.15)$$

$$\chi_{ijk}^{(3)}(\omega_\sigma; \omega_m, \omega_n, \omega_o) = \mathcal{L}(\omega_\sigma)\mathcal{L}(\omega_m)\mathcal{L}(\omega_n)\mathcal{L}(\omega_o) \\ \times N\gamma_{ijkl}(\omega_\sigma; \omega_m, \omega_n, \omega_o) \,. \quad (70.16)$$

For simplicity, the analysis above ignores the vector character of the interacting fields in calculating $\mathcal{L}(\omega)$. A generalization that does include these effects is given in Ref. [5].

70.3 WAVE EQUATION IN NONLINEAR OPTICS

70.3.1 Coupled-Amplitude Equations

The propagation of light waves through a nonlinear medium is described by the wave equation

$$\nabla^2 \tilde{\mathbf{E}} - \frac{1}{c^2}\frac{\partial^2}{\partial t^2}\tilde{\mathbf{E}} = \left[\frac{1}{4\pi\varepsilon_0}\right]\frac{4\pi}{c^2}\frac{\partial^2}{\partial t^2}\tilde{\mathbf{P}} \,. \qquad (70.17)$$

For the case in which $\tilde{\mathbf{E}}$ and $\tilde{\mathbf{P}}$ are given by Eq. (70.1), the field amplitudes associated with each frequency component can be decomposed into their plane wave components such that

$$\mathbf{E}(\mathbf{r}, \omega_l) = \sum_l \mathbf{A}_n(\mathbf{r}, \omega_l)e^{i\mathbf{k}_n \cdot \mathbf{r}} \,, \\ \mathbf{P}(\mathbf{r}, \omega_l) = \sum_l \mathbf{P}_n(\mathbf{r}, \omega_l)e^{i\mathbf{k}_n \cdot \mathbf{r}} \,, \qquad (70.18)$$

where $k_n = n(\omega_l)\omega_l/c$ is the magnitude of the wavevector $\mathbf{k}_n$. The amplitudes $\mathbf{A}_n$ and $\mathbf{P}_n$ are next decomposed into vector components whose linear optical properties are such that the polarization associated with them does not change as the field propagates through the material. For example, for a uniaxial crystal these *eigenpolarizations* could correspond to the ordinary and extraordinary components. In order to describe the propagation and the nonlinear coupling of these eigenpolarizations, the vector field amplitudes are expressed as

$$\mathbf{A}_n(\mathbf{r}, \omega_l) = \hat{\mathbf{u}}_{ln} A_n(\mathbf{r}, \omega_l) \,, \\ \mathbf{P}_n(\mathbf{r}, \omega_l) = \hat{\mathbf{u}}_{ln} \mathcal{P}_n(\mathbf{r}, \omega_l) \,, \qquad (70.19)$$

where $\hat{\mathbf{u}}_{ln}$ is the unit vector associated with the eigenpolarization of the spatial mode n at frequency ω_l. If the fields are assumed to travel along the z-direction, and the slowly-varying amplitude approximation $\partial^2 A_n/\partial z^2 \ll 2k_n\partial A_n/\partial z$ is made, the change in the

amplitude of the field as it propagates through the nonlinear medium with no linear absorption is described by the differential equation

$$\frac{dA_n(\omega_l)}{dz} = \pm\left[\frac{1}{4\pi\varepsilon_0}\right]\frac{i2\pi\omega_l}{n(\omega_l)c}\mathcal{P}_n^{\mathrm{NL}}(\omega_l) \,, \qquad (70.20)$$

where $\mathcal{P}_n^{\mathrm{NL}}$ is the nonlinear contribution to the polarization amplitude $\mathcal{P}_n$, $n(\omega_l)$ is the linear refractive index at frequency ω_l, and the plus (minus) sign indicates propagation in the positive (negative) z-direction. Sections 70.4 and 70.5 give expressions for the $\mathcal{P}_n^{\mathrm{NL}}$ for various second- and third-order nonlinear optical processes. Equation (70.20) is used to determine the set of coupled-amplitude equations describing a particular nonlinear process. For example, for the case of sum-frequency generation, the two fields of frequency ω_1 and ω_2 are combined through second-order nonlinear interaction to create a third wave at frequency $\omega_3 = \omega_1 + \omega_2$. Assuming full permutation symmetry, the amplitudes of the nonlinear polarization for each of the waves are

$$\mathcal{P}^{\mathrm{NL}}(z, \omega_1) = [\varepsilon_0]2\chi_{\mathrm{eff}}^{(2)}A(z, \omega_3)A^*(z, \omega_2)e^{-i\Delta kz} \,, \quad (70.21)$$

$$\mathcal{P}^{\mathrm{NL}}(z, \omega_2) = [\varepsilon_0]2\chi_{\mathrm{eff}}^{(2)}A(z, \omega_3)A^*(z, \omega_1)e^{-i\Delta kz} \,, \quad (70.22)$$

$$\mathcal{P}^{\mathrm{NL}}(z, \omega_3) = [\varepsilon_0]2\chi_{\mathrm{eff}}^{(2)}A(z, \omega_1)A(z, \omega_2)e^{i\Delta kz} \,, \qquad (70.23)$$

where $\Delta k = k_1 + k_2 - k_3$ is the wavevector mismatch (see Sect. 70.3.2) and $\chi_{\mathrm{eff}}^{(2)}$ is given by

$$\chi_{\mathrm{eff}}^{(2)} = \sum_{ijk}\chi_{ijk}^{(2)}(\hat{\mathbf{u}}_1^*)_i(\hat{\mathbf{u}}_2)_j(\hat{\mathbf{u}}_3)_k \,, \qquad (70.24)$$

where $(\hat{\mathbf{u}}_l)_i = \hat{\mathbf{u}}_l \cdot \hat{\mathbf{i}}$. For simplicity, the subscripts on each of the field amplitudes have been dropped, since only one spatial mode at each frequency contributed. The resulting coupled amplitude equations are

$$\frac{dA(\omega_1)}{dz} = \left[\frac{1}{4\pi}\right]\frac{i4\pi\omega_1\chi_{\mathrm{eff}}^{(2)}}{n(\omega_1)c}A(\omega_3)A^*(\omega_2)e^{-i\Delta kz} \,, \quad (70.25)$$

$$\frac{dA(\omega_2)}{dz} = \left[\frac{1}{4\pi}\right]\frac{i4\pi\omega_2\chi_{\mathrm{eff}}^{(2)}}{n(\omega_2)c}A(\omega_3)A^*(\omega_1)e^{-i\Delta kz} \,, \quad (70.26)$$

$$\frac{dA(\omega_3)}{dz} = \left[\frac{1}{4\pi}\right]\frac{i4\pi\omega_3\chi_{\mathrm{eff}}^{(2)}}{n(\omega_3)c}A(\omega_1)A(\omega_2)e^{i\Delta kz} \,. \quad (70.27)$$

70.3.2 Phase Matching

For many nonlinear optical processes (e.g., harmonic generation) it is important to minimize the wave vector mismatch in order to maximize the efficiency. For example, if the field amplitudes $A(\omega_1)$ and $A(\omega_2)$ are constant, the solution to Eq. (70.27) yields for the output intensity

$$I(L,\omega_3) = \left[\frac{1}{64\pi^3\varepsilon_0}\right]\frac{32\pi^3\left[\chi_{\text{eff}}^{(2)}\right]^2\omega_3^2 I(\omega_1)I(\omega_2)L^2}{n(\omega_1)n(\omega_2)n(\omega_3)c^3}$$
$$\times \text{sinc}^2(\Delta k L/2), \tag{70.28}$$

in terms of $\text{sinc}\,x = (\sin x)/x$, where $I(L,\omega_3) = [4\pi\varepsilon_0]n(\omega_3)c|A(L,\omega_3)|^2/2\pi$, and $I(\omega_1)$ and $I(\omega_2)$ are the corresponding input intensities. Clearly, the effect of the wavevector mismatch is to reduce the efficiency of the generation of the sum frequency wave. The maximum propagation distance over which efficient nonlinear coupling can occur is given by the coherence length

$$L_c = \frac{2}{\Delta k}. \tag{70.29}$$

As a result of the dispersion in the linear refractive index that occurs in all materials, achieving phase matching over typical interaction lengths (e.g., 5 mm) is nontrivial. For the case in which the nonlinear material is birefringent, it is sometimes possible to achieve phase matching by insuring that the interacting waves possess some suitable combination of ordinary and extraordinary polarization. Other techniques for achieving phase matching include quasiphase matching [5] and the use of the mode dispersion in waveguides [6].

However, the phase matching condition is automatically satisfied for certain nonlinear optical processes, such as two-photon absorption (see Sect. 70.5.6) and Stokes amplification in stimulated Raman scattering (see Sect. 70.6.1). One can tell when the phase matching condition is automatically satisfied by examining the frequencies that appear in the expression for the nonlinear susceptibility. For a nonlinear susceptibility of the sort $\chi^{(3)}(\omega_1;\omega_2,\omega_3,\omega_4)$ the wave vector mismatch is given in general by $\Delta\mathbf{k} = \mathbf{k}_2 + \mathbf{k}_3 + \mathbf{k}_4 - \mathbf{k}_1$. Thus, for the example of Stokes amplification in stimulated Raman scattering, the nonlinear susceptibility is given by $\chi^{(3)}(\omega_1;\omega_1,\omega_0,-\omega_0)$ where $\omega_0(\omega_1)$ is the frequency of the pump (Stokes) wave, and consequently the wave vector mismatch vanishes identically.

70.3.3 Manley-Rowe Relations

Under conditions of full permutation symmetry, there is no flow of power from the electromagnetic fields to the medium, and thus the total power flow of the fields is conserved. The flow of energy among the fields can be described by the Manley-Rowe relations. For example, for the case of sum-frequency generation, one can deduce from Eqs. (70.25)–(70.27) that

$$\frac{d}{dz}\left[\frac{I(\omega_1)}{\omega_1}\right] = \frac{d}{dz}\left[\frac{I(\omega_2)}{\omega_2}\right] = -\frac{d}{dz}\left[\frac{I(\omega_3)}{\omega_3}\right]. \tag{70.30}$$

The expressions in square brackets are proportional to the flux of photons per unit area per unit time, and

imply that the creation of a photon at ω_3 must be accompanied by the annihilation of photons at both ω_1 and ω_2. Similar relations can be formulated for other nonlinear optical processes that are governed by a nonlinear susceptibility that satisfies full permutation symmetry. Since this behavior occurs at the photon level, nonlinear optical processes can lead to the generation of light fields that have esoteric quantum statistical properties (see Chaps. 75 and 77).

A nonlinear optical process that satisfies the Manley-Rowe relations is called a *parametric process*. Conversely, a process for which field energy is not conserved, and thus Manley-Rowe relations cannot be formulated, is said to be *nonparametric*. Thus, parametric processes are described by purely real $\chi^{(n)}$, whereas nonparametric processes are described by complex $\chi^{(n)}$.

70.3.4 Pulse Propagation

If the optical field consists of ultrashort (<100 ps) pulses, it is more convenient to work with the temporally varying amplitude, rather than with the individual frequency components. Thus, for a linearly polarized plane wave pulse propagating along the z-axis, the field is decomposed into the product of a slowly varying amplitude $A(z,t)$ and a rapidly varying oscillatory term such that

$$\tilde{E}(\mathbf{r},t) = A(z,t)e^{i(k_0 z - \omega_0 t)} + c.c., \tag{70.31}$$

where $k_0 = n_0\omega_0/c$. For a pulse propagating in a material with an intensity-dependent refractive index, the propagation can be described by the nonlinear Schrödinger equation

$$\frac{\partial A}{\partial z} + \frac{i\beta_2}{2}\frac{\partial^2 A}{\partial\tau^2} = i\gamma|A|^2 A, \tag{70.32}$$

where $\beta_2 = (d^2k/d\omega^2)|_{\omega=\omega_0}$ is the group velocity dispersion parameter, $\tau = t - z/v_g$ is the local time for the pulse, $v_g = [(dk/d\omega)|_{\omega=\omega_0}]^{-1}$ is the group velocity, and $\gamma = [4\pi\varepsilon_0]n_2 n_0\omega_0/2\pi$ is the nonlinear refractive index parameter.

70.4 SECOND-ORDER PROCESSES

Second-order nonlinear optical processes occur as a consequence of the second term in expression (70.2), i.e., processes whose strength is described by $\chi^{(2)}(\omega_\sigma;\omega_m,\omega_n)$. These processes entail the generation of a field at frequency $\omega_\sigma = \omega_m + \omega_n$ in response to applied fields at (positive and/or negative) frequencies ω_m and ω_n. Several examples of such processes are described in this Section.

70.4.1 Sum Frequency Generation

Sum frequency generation produces an output field at frequency $\omega_3 = \omega_1 + \omega_2$ for ω_1 and ω_2 both positive. It is useful, for example, for the generation of tunable radiation in the uv if ω_1 and/or ω_2 are obtained from tunable lasers in the visible range. Sum frequency generation is described in detail in Sects. 70.3.1 – 70.3.3.

70.4.2 Second Harmonic Generation

Second harmonic generation is routinely used to convert the output of a laser to a higher frequency. It is described by $\chi^{(2)}(2\omega; \omega, \omega)$. Let η be the power conversion efficiency from frequency ω to 2ω. Assuming that phase matching is perfect, and the pump wave at frequency ω is undepleted by the interaction, a derivation analogous to that for Eq. (70.28) yields

$$\eta = \tanh^2(z/l)\,, \qquad (70.33)$$

where the characteristic conversion length l is given by

$$l = [4\pi] \frac{c\sqrt{n(\omega)n(2\omega)}}{4\pi\omega\chi^{(2)}|A_1(0)|}\,. \qquad (70.34)$$

Note that the conversion efficiency asymptotically approaches unity. In practice, conversion efficiencies exceeding 80% can be achieved.

70.4.3 Difference Frequency Generation

Difference frequency generation can be used to create light in the infrared and far infrared by generating the difference frequency $\omega_2 = \omega_3 - \omega_1$ (where ω_3 and ω_1 are positive and $\omega_3 > \omega_1$) of two incident lasers. Consider the case in which a strong (undepleted) pump wave at frequency ω_3 and a weak (signal) wave at ω_1 are incident on a nonlinear medium described by $\chi^{(2)}(\omega_2; \omega_3, -\omega_1) = \chi^{(2)}(\omega_1; \omega_3, -\omega_2)$. The amplitude $A(\omega_3)$ of the strong wave can be taken as a constant, and thus the interaction can be described by finding simultaneous solutions to Eqs. (70.25) and (70.26) for $A(\omega_1)$ and $A(\omega_2)$. In the limit of perfect phase matching (i.e., $\Delta k = 0$), the solutions are

$$A(z, \omega_1) = A(0, \omega_1)\cosh\kappa z\,, \qquad (70.35)$$

$$A(z, \omega_2) = i\sqrt{\frac{n_1\omega_2}{n_2\omega_1}} \frac{A(\omega_3)}{|A(\omega_3)|} A^*(0, \omega_1)\sinh\kappa z\,, \quad (70.36)$$

where

$$\kappa^2 = \left[\frac{1}{16\pi^2}\right] \frac{16\pi^2[\chi^{(2)}]^2\omega_1^2\omega_2^2}{k_1 k_2 c^4}|A(\omega_3)|^2\,. \qquad (70.37)$$

Equation (70.36) describes the spatial growth of the difference frequency signal.

70.4.4 Parametric Amplification and Oscillation

For the foregoing case of a strong wave at frequency ω_3 and a weak wave with $\omega_1 < \omega_3$ incident on a second-order nonlinear optical material, the lower frequency input wave is amplified by the nonlinear interaction; this process is known as *parametric amplification*. Difference frequency generation is a consequence of the Manley-Rowe relations, as described above in Sect. 70.3.3. Since $\omega_3 = \omega_1 + \omega_2$, the annihilation of an ω_3 photon must be accompanied by the simultaneous creation of photons ω_1 and ω_2.

An *optical parametric oscillator* can be constructed by placing the nonlinear optical material inside an optical resonator that provides feedback at ω_1 and/or ω_2. When such a device is excited by a wave at ω_3, it can produce output frequencies ω_1 and ω_2 that satisfy $\omega_1 + \omega_2 = \omega_3$. Optical parametric oscillators are of considerable interest as sources of broadly tunable radiation [7].

70.4.5 Focused Beams

For conceptual clarity, much of the discussion so far has assumed that the interacting beams are plane waves. In practice, the incident laser beams are often focused into the nonlinear material to increase the field strength within the interaction region and consequently to increase the nonlinear response. However, it is undesirable to focus too tightly, because doing so leads to a decrease in the effective length of the interaction region. In particular, if w_0 is the radius of the laser beam at the beam waist, the beam remains focused only over a distance of the order $b = 2\pi w_0^2/\lambda$ where λ is the laser wavelength measured in the nonlinear material. For many types of nonlinear optical processes, the optimal nonlinear response occurs if the degree of focusing is adjusted so that b is several times smaller than the length L of the nonlinear optical material.

70.5 THIRD-ORDER PROCESSES

A wide variety of nonlinear optical processes are possible as a result of the nonlinear contributions to the polarization that are third-order in the applied field. These processes are described by $\chi^{(3)}(\omega_\sigma; \omega_m, \omega_n, \omega_o)$ [see Eq. (70.2)] and can lead not only to the generation of new field components (e.g., third-harmonic generation) but can also result in a field affecting itself as it propagates (e.g., self-phase modulation). Several examples are described in this section.

70.5.1 Third-Harmonic Generation

Assuming full-permutation symmetry, the nonlinear polarization amplitudes for the fundamental and third-harmonic beams are

$$\mathcal{P}^{\mathrm{NL}}(z,\omega) = [\varepsilon_0]3\chi_{\mathrm{eff}}^{(3)}A(z,3\omega)[A^*(z,\omega)]^2 e^{-i\Delta kz},$$

$$\mathcal{P}^{\mathrm{NL}}(z,3\omega) = [\varepsilon_0]\chi_{\mathrm{eff}}^{(3)}[A(z,\omega)]^3 e^{i\Delta kz}, \qquad (70.38)$$

where $\Delta k = 3k(\omega) - k(3\omega)$ and $\chi_{\mathrm{eff}}^{(3)}$ is the effective third-order susceptibility for third-harmonic generation and is defined in a manner analogous to the $\chi_{\mathrm{eff}}^{(2)}$ in Eq. (70.24). If the intensity of the fundamental wave is not depleted by the nonlinear interaction, the solution for the output intensity $I(L, 3\omega)$ of the third-harmonic field for a crystal of length L is

$$I(L,3\omega) = \left[\frac{1}{256\pi^4\varepsilon_0^2}\right]\frac{48\pi^2\omega^2\left[\chi_{\mathrm{eff}}^{(3)}\right]^2}{n(3\omega)n(\omega)^3c^4}$$
$$\times\, I(\omega)^3 L^2 \sinh^2(\Delta kL/2), \qquad (70.39)$$

where $I(\omega)$ is the input intensity of the fundamental field. As a result of the typically small value of $\chi_{\mathrm{eff}}^{(3)}$ in crystals, it is generally more efficient to generate the third harmonic by using two $\chi^{(2)}$ crystals in which the first crystal produces second harmonic light and the second crystal combines the second harmonic and the fundamental beams via sum-frequency generation. It is also possible to use resonant enhancement of $|\chi^{(3)}|$ in gases to increase the efficiency of third-harmonic generation [8].

70.5.2 Self-Phase and Cross-Phase Modulation

The nonlinear refractive index leads to an intensity-dependent change in the phase of the beam as it propagates through the material. If the medium is lossless, the amplitude of a single beam at frequency ω propagating in the positive z-direction can be expressed as

$$A(z,\omega) = A(0,\omega)e^{i\phi^{\mathrm{NL}}(z)}, \qquad (70.40)$$

where the nonlinear phase shift $\phi^{\mathrm{NL}}(z)$ is given by

$$\phi^{\mathrm{NL}}(z) = \frac{\omega}{c}n_2 Iz, \qquad (70.41)$$

and $I = [4\pi\varepsilon_0]n_0 c|A(0,\omega)|^2/2\pi$ is the intensity of the laser beam. If two fields at different frequencies ω_1 and ω_2 are traveling along the z-axis, the two fields can affect each other's phase; this effect is known as *cross-phase modulation*. The nonlinear phase shift $\phi_{1,2}^{\mathrm{NL}}(z)$ for each of the waves is given by

$$\phi_{1,2}^{\mathrm{NL}}(z) = \frac{\omega_{1,2}}{c}n_2(I_{1,2} + 2I_{2,1})z. \qquad (70.42)$$

For the case of a light pulse, the change in the phase of the pulse inside the medium becomes a function of time. In this case the solution to Eq. (70.32) for the time-varying amplitude $A(z,\tau)$ shows that in the absence of group-velocity dispersion (GVD) (i.e., $\beta_2 = 0$) that the solution for $A(z,\tau)$ is of the form of Eq. (70.40), except that the temporal intensity profile $I(\tau)$ replaces the steady-state intensity I in Eq. (70.41). As the pulse propagates through the medium, its frequency becomes time dependent, and the instantaneous frequency shift from the central frequency ω_0 is given by

$$\delta\omega(\tau) = -\frac{\partial\phi^{\mathrm{NL}}(\tau)}{\partial\tau} = -\frac{\omega n_2 z}{c}\frac{\partial I}{\partial t}. \qquad (70.43)$$

This time-dependent self-phase modulation leads to a broadening of the pulse spectrum and to a frequency chirp across the pulse.

If the group velocity dispersion parameter β_2 and the nonlinear refractive index coefficient n_2 are of opposite sign, the nonlinear frequency chirp can be compensated by the chirp due to group velocity dispersion, and Eq. (70.32) admits soliton solutions. For example, the fundamental soliton solution is

$$A(z,t) = \sqrt{\frac{1}{L_{\mathrm{D}}}}\mathrm{sech}\left(\frac{\tau}{\tau_p}\right)e^{iz/2L_{\mathrm{D}}}, \qquad (70.44)$$

where τ_p is the pulse duration and $L_{\mathrm{D}} = \tau_p^2/|\beta_2|$ is the dispersion length. As a result of their ability to propagate in dispersive media without changing shape, optical solitons show a great deal of promise in applications such as optical communications and optical switching. For further discussion of optical solitons see Ref. [9].

70.5.3 Four-Wave Mixing

Various types of four-wave mixing processes can occur among different beams. One of the most common geometries is backward four-wave mixing used in nonlinear spectroscopy and optical phase conjugation. In this interaction, two strong counterpropagating pump waves with amplitudes A_1 and A_2 and with equal frequencies $\omega_{1,2} = \omega$ are injected into a nonlinear medium. A weak wave, termed the probe wave, (with frequency ω_3 and amplitude A_3) is also incident on the medium. As a result of the nonlinear interaction among the three waves, a fourth wave with an amplitude A_4 is generated which is counterpropagating with respect to the probe wave and with frequency $\omega_4 = 2\omega - \omega_3$. For this case, the third-order nonlinear susceptibilities for the probe and conjugate waves are given by $\chi^{(3)}(\omega_{3,4};\omega,\omega,-\omega_{4,3})$. For constant pump wave intensities and full permutation symmetry, the amplitudes of the nonlinear polarization for the probe and conjugate waves are given by

$$\mathcal{P}^{\mathrm{NL}}(z,\omega_{3,4}) = \pm[\varepsilon_0]6\chi^{(3)}\big[(|A_1|^2 + |A_2|^2)A_{3,4}$$
$$+ A_1 A_2 A_{4,3}^* e^{i\Delta kz}\big], \qquad (70.45)$$

where $\Delta k = k_1 + k_2 - k_3 - k_4$ is the phase mismatch, which is nonvanishing when $\omega_3 \neq \omega_4$. For the case of optical phase conjugation by degenerate four-wave mixing (i.e., $\omega_3 = \omega_4 = \omega$ and $A_4(L) = 0$), the phase conjugate reflectivity R_{PC} is

$$R_{\mathrm{PC}} = \frac{|A_4(0)|^2}{|A_3(0)|^2} = \tan^2(\kappa L), \qquad (70.46)$$

where $\kappa = [1/16\pi^2\varepsilon_0][24\pi^2\omega\chi^{(3)}/(n_0c)^2]\sqrt{I_1 I_2}$ and $I_{1,2}$ are the intensities of the pump waves. Phase-conjugate reflectivities greater than unity can be routinely achieved by performing four-wave mixing in atomic vapors or photorefractive media.

70.5.4 Self-Focusing and Self-Trapping

Typically a laser beam has a transverse intensity profile that is approximately Gaussian. In a medium with an intensity-dependent refractive index, the index change at the center of the beam is different from the index change at the edges of the beam. The gradient in the refractive index created by the beam can allow it to self-focus for $n_2 > 0$. For this condition to be met, the total input power of the beam must exceed the critical power P_{cr} for self-focusing which is given by

$$P_{\mathrm{cr}} = \frac{\pi(0.61\lambda)^2}{8n_0 n_2}, \qquad (70.47)$$

where λ is the vacuum wavelength of the beam. For powers much greater than the critical power, the beam can break up into various filaments, each with a power approximately equal to the critical power. For a more extensive discussion of self-focusing and self-trapping see Refs. [10, 11].

70.5.5 Saturable Absorption

When the frequency ω of an applied laser field is sufficiently close to a resonance frequency ω_0 of the medium, an appreciable fraction of the atomic population can be placed in the excited state. This loss of population from the ground state leads to an intensity-dependent saturation of the absorption and the refractive index of the medium (see Sect. 67.3 for more detailed discussion) [4]. The third-order susceptibility as a result of this saturation is given by

$$\chi^{(3)} = \left[\frac{1}{\varepsilon_0}\right] \frac{|\mu|^2 T_1 T_2 \alpha_0 c}{3\pi\omega_0\hbar^2} \frac{\delta T_2 - i}{[1 + (\delta T_2)^2]^2}, \qquad (70.48)$$

where μ is the transition dipole moment, T_1 and T_2 are the longitudinal and transverse relaxation times, respectively (see Sect. 66.5.3), α_0 is the line-center weak-field intensity absorption coefficient, and $\delta = \omega - \omega_0$ is

the detuning. For the $3s \leftrightarrow 3p$ transition in atomic sodium vapor at $300\,^\circ\mathrm{C}$, the nonlinear refractive index $n_2 \sim 10^{-7}\mathrm{cm}^2/\mathrm{W}$ for a detuning $\delta T_2 = 300$.

70.5.6 Two-Photon Absorption

When the frequency ω of a laser field is such that 2ω is close to a transition frequency of the material, it is possible for two-photon absorption (TPA) to occur. This process leads to a contribution to the imaginary part of $\chi^{(3)}(\omega;\omega,\omega,-\omega)$. In the presence of TPA, the intensity $I(z)$ of a single, linearly polarized beam as a function of propagation distance is

$$I(z) = \frac{I(0)}{1 + \beta I(0)z}, \qquad (70.49)$$

where $\beta = [1/16\pi^2\varepsilon_0]24\pi^2\omega\Im[\chi^{(3)}]/(n_0c)^2$ is the TPA coefficient. For wide-gap semiconductors such as ZnSe at 800 nm, $\beta \sim 10^{-8}\mathrm{cm}/\mathrm{W}$.

70.5.7 Nonlinear Ellipse Rotation

The polarization ellipse of an elliptically polarized laser beam rotates but retains its ellipticity as the beam propagates through an isotropic nonlinear medium. Ellipse rotation occurs as a result of the difference in the nonlinear index changes experienced by the left- and right-circular components of the beam, and the angle θ of rotation is

$$
\begin{aligned}
\theta &= \tfrac{1}{2}\Delta n \omega z/c \\
&= \left[\frac{1}{16\pi^2\varepsilon_0}\right] \frac{12\pi^2}{n_0^2 c} \chi^{(3)}_{xyyx}(\omega;\omega,\omega,-\omega)(I_+ - I_-)z,
\end{aligned}
$$
$$(70.50)$$

where $I_\pm$ are the intensities of the circularly polarized components of the beam with unit vectors $\hat{\sigma}_\pm = (\hat{\mathbf{x}} \pm i\hat{\mathbf{y}})/\sqrt{2}$. Nonlinear ellipse rotation is a sensitive technique for determining the nonlinear susceptibility element $\chi^{(3)}_{xyyx}$ for isotropic media and can be used in applications such as optical switching.

70.6 STIMULATED LIGHT SCATTERING

Stimulated light scattering occurs as a result of changes in the optical properties of the material that are induced by the optical field. The resulting nonlinear coupling between different field components is mediated by some excitation (e.g., acoustic phonon) of the material that results in changes in its optical properties. The nonlinearity can be described by a complex susceptibility and a nonlinear polarization that is of third order in the interacting fields. Various types of stimulated scattering can occur. Discussed below are the two processes that are most commonly observed.

70.6.1 Stimulated Raman Scattering

In stimulated Raman scattering (SRS), the light field interacts with a vibrational mode of a molecule. The coupling between the two optical waves can become strong if the frequency difference between them is close to the frequency ω_v of the molecular vibrational mode. If the pump field at ω_0 and another field component at ω_1 are propagating in the same direction along the z-axis, the steady-state nonlinear polarization amplitudes for the two field components are given by

$$\mathcal{P}^{\mathrm{NL}}(z,\omega_{0,1}) = [\varepsilon_0]6\chi_{\mathrm{R}}(\omega_{0,1})|A(z,\omega_{1,0})|^2 A(z,\omega_{0,1}),$$
(70.51)

where $\chi_{\mathrm{R}}(\omega_{0,1}) \equiv \chi^{(3)}(\omega_{0,1};\omega_{0,1},\omega_{1,0},-\omega_{1,0})$, the *Raman susceptibility*, actually depends only on the frequency difference $\Omega = \omega_0 - \omega_1$ and is given by

$$\chi_{\mathrm{R}}(\omega_{0,1}) = \left[\frac{1}{\varepsilon_0}\right]\frac{N(\partial\alpha/\partial q)_0^2}{6\mu_M}\frac{1}{\omega_v^2 - \Omega^2 \mp 2i\gamma\Omega}, \quad (70.52)$$

where the minus (plus) sign is taken for the ω_0 (ω_1) susceptibility, μ_M is the reduced nuclear mass, and $(\partial\alpha/\partial q)_0$ is a measure of the change of the polarizability of the molecule with respect to a change in the intermolecular distance q at equilibrium. If the intensity of the pump field is undepleted by the interaction with the ω_1 field and is assumed to be constant, the solution for the intensity of the ω_1 field at $z = L$ is given by

$$I(L,\omega_1) = I(0,\omega_1)e^{G_{\mathrm{R}}}, \quad (70.53)$$

where the SRS gain parameter G_{R} is

$$\begin{aligned} G_{\mathrm{R}} &= \left[\frac{1}{16\pi^2\varepsilon_0}\right]48\pi^2\frac{\omega_1}{(n_1 c)^2}\Im[\chi_{\mathrm{R}}(\omega_1)]I_0 L \\ &= g_{\mathrm{R}}I_0 L, \end{aligned} \quad (70.54)$$

g_{R} is the SRS gain factor, and I_0 is the input intensity of the pump field. For $\omega_1 < \omega_0$ ($\omega_1 > \omega_0$), the ω_1 field is termed the Stokes (anti-Stokes) field, and it experiences exponential amplification (attenuation). For sufficiently large gains (typically $G_{\mathrm{R}} \gtrsim 25$), the Stokes wave can be seeded by spontaneous Raman scattering and can grow to an appreciable fraction of the pump field. For a complete discussion of the spontaneous initiation of SRS see Ref. [12]. For the case of CS_2, $g_{\mathrm{R}} = 0.024$ cm/MW.

Four-wave mixing processes that couple a Stokes wave having $\omega_1 < \omega_0$ and an anti-Stokes wave having $\omega_2 > \omega_0$, where $\omega_1 + \omega_2 = 2\omega_0$, can also occur [4]. In this case, additional contributions to the nonlinear polarization are present and are characterized by a Raman susceptibility of the form $\chi^{(3)}(\omega_{1,2};\omega_0,\omega_0,-\omega_{2,1})$. The technique of coherent anti-Stokes Raman spectroscopy is based on this four-wave mixing process [13].

70.6.2 Stimulated Brillouin Scattering

In stimulated Brillouin scattering (SBS), the light field induces and interacts with an acoustic wave inside the medium. The resulting interaction can lead to extremely high amplification for certain field components (i.e., Stokes wave). For many optical media, SBS is the dominant nonlinear optical proccess for laser pulses of duration > 1 ns. The primary applications for SBS are self-pumped phase conjugation and pulse compression of high-energy laser pulses.

If an incident light wave with wave vector $\mathbf{k}_0$ and frequency ω_0 is scattered from an acoustic wave with wave vector $\mathbf{q}$ and frequency Ω, the wave vector and frequency of the scattered wave are determined by conservation of momentum and energy to be $\mathbf{k}_1 = \mathbf{k}_0 \pm \mathbf{q}$ and $\omega_1 = \omega_0 \pm \Omega$, where the $(+)$ sign applies if $\mathbf{k}_0 \cdot \mathbf{q} > 0$ and the $(-)$ applies if $\mathbf{k}_0 \cdot \mathbf{q} < 0$. Here, Ω and $\mathbf{q}$ are related by the dispersion relation $\Omega = v|\mathbf{q}|$ where v is the velocity of sound in the material. These Bragg scattering conditions lead to the result that the Brillouin frequency shift $\Omega_{\mathrm{B}} = \omega_1 - \omega_0$ is zero for scattering in the forward direction (i.e., in the $\mathbf{k}_0$ direction) and reaches its maximum for scattering in the backward direction given by

$$\Omega_{\mathrm{B}} = 2\omega_0 v n_0/c, \quad (70.55)$$

where n_0 is the refractive index of the material.

The interaction between the incident wave and the scattered wave in the Brillouin-active medium can become nonlinear if the interference between the two optical fields can coherently drive an acoustic wave, either through electrostriction or through local density fluctuations resulting from the absorption of light and consequent temperature changes. The following discussion treats the more common electrostriction mechanism.

Typically, SBS occurs in the backward direction (i.e., $\mathbf{k}_0 = k_0\hat{\mathbf{z}}$ and $\mathbf{k}_1 = -k_1\hat{\mathbf{z}}$), since the spatial overlap between the Stokes beam and the laser beam is maximized under these conditions and, as mentioned above, no SBS occurs in the forward direction. The steady-state nonlinear polarization amplitudes for backward SBS are

$$\mathcal{P}^{\mathrm{NL}}(z,\omega_{0,1}) = [\varepsilon_0]6\chi_{\mathrm{B}}(\omega_{0,1})|A(z,\omega_{1,0})|^2 A(z,\omega_{0,1}),$$
(70.56)

where $\chi_{\mathrm{B}}(\omega_{0,1}) \equiv \chi^{(3)}(\omega_{0,1};\omega_{0,1},\omega_{1,0},-\omega_{1,0})$, the *Brillouin susceptibility*, depends only on $\Omega = \omega_0 - \omega_1$ and is given by

$$\chi_{\mathrm{B}}(\omega_{0,1}) = \left[\frac{1}{\varepsilon_0}\right]\frac{\omega_0^2\gamma_{\mathrm{e}}^2}{24\pi^2 c^2\rho_0}\frac{1}{\Omega_{\mathrm{B}}^2 - \Omega^2 \mp i\Gamma_{\mathrm{B}}\Omega}, \quad (70.57)$$

where the minus (plus) sign is taken for the ω_0 (ω_1) susceptibility, γ_e is the electrostrictive constant, ρ_0 is the mean density of the material, and Γ_B is the Brillouin linewidth given by the inverse of the phonon lifetime. If the pump field is undepleted by the interaction with the ω_1 field and is assumed to be constant, the solution for the output intensity of the ω_1 field at $z = 0$ is given by

$$I(0, \omega_1) = I(L, \omega_1)e^{G_B}, \qquad (70.58)$$

where the Brillouin gain coefficient G_B is given by

$$
\begin{aligned}
G_B &= \left[\frac{1}{16\pi^2\varepsilon_0}\right] 48\pi^2 \frac{\omega_1}{(n_0 c)^2} \mathrm{Im}[\chi_B(\omega_1)]I_0 L, \\
&= g_0 \frac{\Omega\Omega_B\Gamma_B^2}{[\Omega_B^2 - \Omega^2]^2 + (\Omega\Gamma_B)^2} I_0 L \\
&= g_B I_0 L, \qquad (70.59)
\end{aligned}
$$

g_B is the SBS gain factor, I_0 is the input intensity of the pump field, and

$$g_0 = \left[\frac{1}{\varepsilon_0^2}\right] \frac{\omega_0^2\gamma_e^2}{n_0 c^3 \rho_0 v \Gamma_B} \qquad (70.60)$$

is the line-center (i.e., $\Omega = \pm\Omega_B$) SBS gain factor. For $\Omega > 0$ ($\Omega < 0$), the ω_1 field is termed the Stokes (anti-Stokes) field, and it experiences exponential amplification (attenuation). For sufficiently large gains (typically $G_B \gtrsim 25$), the Stokes wave can be seeded by spontaneous Brillouin scattering and can grow to an appreciable fraction of the pump field. For a complete discussion of the spontaneous initiation of SBS see Ref. [14]. For CS_2, $g_0 = 0.15$ cm/MW.

70.7 OTHER NONLINEAR OPTICAL PROCESSES

70.7.1 High-Order Harmonic Generation

If full permutation symmetry applies and the fundamental field ω is not depleted by nonlinear interactions, then the intensity of the qth harmonic is given by

$$
\begin{aligned}
I(z, q\omega) = &\left[\frac{1}{4\pi(4\pi\varepsilon_0)^{(q-1)/2}}\right] \frac{2\pi q^2\omega^2}{n^2(q\omega)c} \left[\frac{2\pi I(\omega)}{n(\omega)c}\right]^q \\
&\times |\chi^{(q)}(q\omega; \omega, \ldots, \omega)J_q(\Delta k, z_0, z)|^2, \quad (70.61)
\end{aligned}
$$

where $\Delta k = [n(\omega) - n(q\omega)]\omega/c$,

$$J_q(\Delta k, z_0, z) = \int_{z_0}^{z} \frac{e^{i\Delta k z'}dz'}{(1 + 2iz'/b)^{q-1}}, \qquad (70.62)$$

$z = z_0$ at the input face of the nonlinear medium, and b is the confocal parameter (see Sect. 70.4.5) of the fundamental beam. Defining $L = z - z_0$, the integral J_q

can be easily evaluated in the limits $L \ll b$ and $L \gg b$. The limit $L \ll b$ corresponds to the plane-wave limit in which case

$$|J_q(\Delta k, z_0, z)|^2 = L^2 \mathrm{sinc}^2\left(\frac{\Delta k L}{2}\right). \qquad (70.63)$$

The limit $L \gg b$ corresponds to the tight-focusing configuration in which case

$$J_q(\Delta k, z_0, z) = \begin{cases} 0, & \Delta k \le 0, \\ \dfrac{\pi b}{(q-2)!}\left(\dfrac{b\Delta k}{2}\right)^{q-2} e^{-b\Delta k/2}, & \Delta k > 0. \end{cases} \qquad (70.64)$$

Note that in this limit, the qth harmonic light is only generated for positive phase mismatch. Reintjes *et al.* [15] observed both the fifth and seventh harmonics in helium gas which exhibited a dependence on $I(\omega)$ which is consistent with the $I^q(\omega)$ dependence predicted by Eq. (70.61). However, more recent experiments in gas jets have demonstrated the generation of extremely high-order harmonics which do not depend on the intensity in this simple manner (see Chap. 72 for further discussion of this nonperturbative behavior).

70.7.2 Electro-Optic Effect

The electro-optic effect corresponds to the limit in which the frequency of one of the applied fields approaches zero. The linear electro-optic effect (or Pockels effect) can be described by a second-order susceptibility of the form $\chi^{(2)}(\omega; \omega, 0)$. This effect produces a change in the refractive index for light of certain polarizations which depends linearly on the strength of the applied low-frequency field. More generally, the linear electro-optic effect induces a change in the amount of birefringence present in an optical material. This electrically controllable change in birefringence can be used to construct amplitude modulators, frequency shifters, optical shutters, and other optoelectronic devices. Materials commonly used in such devices include KDP and lithium niobate [16]. If the laser beam is propagating along the optic axis (i.e., z-axis) of the material of length L and the low-frequency field E_z is also applied along the optic axis, the nonlinear index change $\Delta n = n_y - n_x$ between the components of the electric field polarized along the principal axes of the crystal is given by

$$\Delta n = \left[\frac{1}{4\pi}\right] n_0^3 r_{63} E_z \qquad (70.65)$$

where r_{63} is one of the electro-optic coefficients.

The quadratic electro-optic effect produces a change in refractive index that scales quadratically with the applied dc electric field. This effect can be described by a third-order susceptibility of the form $\chi^{(3)}(\omega; \omega, 0, 0)$.

70.7.3 Photorefractive Effect

The photorefractive effect leads to an optically induced change in the refractive index of a material. In certain ways this effect mimics that of the nonlinear refractive index described in Sect. 70.2.2, but it differs from the nonlinear refractive index in that the change in refractive index is independent of the overall intensity of the incident light field, and depends only on the degree of spatial modulation of the light field within the nonlinear material. In addition, the photorefractive effect can occur only in materials that exhibit a linear electro-optic effect, and contain an appreciable density of trapped electrons and/or holes that can be liberated by the application of a light field. Typical photorefractive materials include lithium niobate, barium titanate, and strontium barium niobate.

A typical photorefractive configuration might be as follows: two beams interfere within a photorefractive crystal to produce a spatially modulated intensity distribution. Bound charges are ionized with greater probability at the maxima than at the minima of the distribution and, as a result of the diffusion process, carriers tend to migrate away from regions of large light intensity. The resulting modulation of the charge distribution leads to the creation of a spatially modulated electric field that produces a spatially modulated change in refractive index as a consequence of the linear electro-optic effect. For a more extensive discussion see Ref. [17].

REFERENCES

1. N. Bloembergen, *Nonlinear Optics* (Benjamin, New York, 1964).

2. Y. R. Shen, *Nonlinear Optics* (Wiley, New York, 1984).

3. P. N. Butcher and D. Cotter, *The Elements of Nonlinear Optics* (Cambridge University Press, 1990).

4. R. W. Boyd, *Nonlinear Optics* (Academic, Boston, 1992).

5. J. A. Armstrong, N. Bloembergen, J. Ducuing, and P. S. Pershan, Phys. Rev. **127**, 1918 (1962).

6. See, for example, G. I. Stegeman, in *Contemporary Nonlinear Optics*, edited by G. P. Agrawal and R. W. Boyd (Academic, Boston, 1992), Chap. 1.

7. See, for example, the Special Issue on *Optical Parametric Oscillation and Amplification* in J. Opt. Soc. Am. B **10**, No. 11 (1993).

8. R. B. Miles and S. E. Harris, IEEE J. Quant. Electron. **9**, 470 (1973).

9. G. P. Agrawal, *Nonlinear Fiber Optics* (Academic, Boston, 1989).

10. S. A. Akhmanov, R. V. Khokhlov, and A. P. Sukhorukov, in *Laser Handbook*, edited by F. T. Arecchi and E. O. Schulz-Dubois, (North-Holland, 1972).

11. J. H. Marburger, Prog. Quant. Electr. **4**, 35 (1975).

12. M. G. Raymer and I. A. Walmsley, in *Progress in Optics*, edited by E. Wolf (North-Holland, Amsterdam, 1990), Vol. 28.

13. M. D. Levenson and S. Kano, *Introduction to Nonlinear Spectroscopy* (Academic, Boston, 1988).

14. R. W. Boyd, K. Rzazewski, and P. Narum, Phys. Rev. A **42**, 5514 (1990).

15. J. Reintjes, C. Y. She, R. C. Eckardt, N. E. Karangelen, R. C. Elton, and R. A. Andrews, Phys. Rev. Lett. **37**, 1540 (1976); Appl. Phys. Lett. **30**, 480 (1977).

16. I. P. Kaminow, *An Introduction to Electro-optic Devices* (Academic, New York, 1974).

17. *Photorefractive Materials and Their Applications*, edited by P. Günter and J.-P. Huignard (Springer-Verlag, Berlin, Part I, 1988; Part II, 1989).

71

Coherent Transients

J. H. Eberly and C. R. Stroud, Jr.

Department of Physics and Astronomy, and Institute of Optics, University of Rochester, Rochester, New York 14627

71.1 INTRODUCTION

Coherent optical transients are excited in atomic and molecular systems when a stable phase relation persists between an exciting light field and the system's electronic response. The extreme sensitivity of phase-dependent effects is responsible for the many applications of optical transient techniques in atomic and molecular physics [1–4].

The theory of coherent transients distinguishes carefully between two types of relaxation: homogeneous and inhomogeneous. Relaxation occurs whenever the environment of a physical system fluctuates randomly. By random environment one means the combination of all interactions that are too complex to be treated fundamentally, and that can be seen to lead to degradation of the degree of coherence of a particular interaction of main interest. The time scale of environmental fluctuations then provides the division between the two types.

When environmental fluctuations are sufficiently rapid that all dynamical systems in a macroscopic sample experience the whole range of fluctuations in a time short compared with the time of an experiment, the resultant relaxation is called homogeneous. If environmental fluctuations exist randomly over a macroscopic sample, but change relatively slowly in time, then the relaxation is called inhomogeneous. For example, weak distant collisions are experienced constantly by all atoms at thermal equilibrium in a vapor cell, and give rise to homogeneous relaxation. If the vapor is sufficiently dilute, the same atoms may nevertheless retain for long times their own individual velocities. These velocities are relatively fixed in time, but they are random over the Maxwellian distribution of velocities and so give rise to inhomogeneous relaxation. Fundamentally, the distinction between homogeneous and inhomogeneous relaxation is artificial, depending on a separation of time scales that may not always exist. Nevertheless, when it exists, the distinction provides an extremely useful way to classify coherent transients. It is one of the foundations of the subject.

71.2 OPTICAL BLOCH EQUATIONS

A very weakly excited dipole transition in an atom responds linearly to an applied time-dependent electric field. This is the basis of classical Lorentzian dielectric theory, but because any transition can be inverted, an atom is more than a classical linear oscillator [5]. The three atomic variables that describe the primary coherent optical transients in a dipole-allowed transition include the intrinsically quantum inversion variable, as well as the components of the expectation value of the atomic dipole moment that are in-phase and in-quadrature with the field.

We write the time-dependent atomic dipole moment of a transition excited by light near exact resonance in the form

$$-\langle ex(t)\rangle = -\langle \Psi(t)|ex|\Psi(t)\rangle = \Re\{d(U - iV)e^{-i\omega t}\},$$
(71.1)

where d is the transition dipole matrix element, ω is the frequency of the impressed optical field, and U and V are the time-dependent amplitudes of the in-phase and in-quadrature dipole components. The impressed field is taken in the quasi-monochromatic form

$$E(t) = \tfrac{1}{2}(\mathcal{E}e^{-i\omega t} + \text{c.c.})$$
(71.2)

Both dipole moment and field will be taken to be real scalars to emphasize that complications of vector notation add little to a first discussion of the principles of coherent transients.

Section 66.4.5 shows that U and V are dynamically coupled to each other and to the inversion W through the optical Bloch equations (OBE). When relaxation terms are included, the OBEs are given by Eqs. (66.55). These are

$$\frac{dU}{dt} = -\Delta V - U/T_2,$$
$$\frac{dV}{dt} = \Delta U + \Omega_1 W - V/T_2,$$
(71.3)
$$\frac{dW}{dt} = -\Omega_1 V - (W - W_{eq})/T_1,$$

the key equations of the theory of optical transients [1, 2].

As in Chap. 66, $\Delta = (E_e - E_g)/\hbar - \omega$ is the detuning and $\Omega_1 = d\mathcal{E}/\hbar$ is the Rabi frequency. It is the dipole interaction energy in frequency units, but has a significance beyond this, as discussed in Sect. 71.4.1.

71.3 NUMERICAL ESTIMATES OF PARAMETERS

The nature of the coherent interaction between an atom or molecule and an optical field is controlled by the relative size of a number of frequencies or rates. In the case of single photon transitions they include: Δ and Ω_1, the detuning and Rabi frequency defined above, $1/T_2$ the transverse, and $1/T_1$ the longitudinal damping rates, $1/T^*$ the inhomogenenous linewidth, and $2\pi/\tau_p$ the transform bandwidth of the optical pulse. All of these frequencies with the exception of the last are defined in Chap. 66. In the case of multiphoton transitions and simultaneous excitation by a number of resonant laser fields, the appropriately generalized versions of these same parameters apply.

A laser pulse with $\tau_p \geq 1$ ns and with an intensity less than about 1 GW/cm² can be tuned to an isolated atomic resonance and the interaction can be described in terms of a simple two-level theory. Laser pulses as short as 7 fs in duration, or with intensities as high as 10^{19} W/cm², have been produced and such extreme pulses quasi-resonantly excite more than one upper level. A 1 ps pulse has a bandwidth of approximately 20 cm⁻¹, while a 1 fs pulse has a bandwidth of about 20 000 cm⁻¹. If a 1 ps pulse were tuned so that it resonantly excited the $n = 95$ Rydberg state of an atom, it would simultaneously and coherently excite all the levels from $n = 67$ to the continuum limit, while a 10 fs pulse could excite all the levels from $n = 4$ to the continuum.

Similarly, when laser pulses are intense enough so that the electric field amplitude approaches that of the Coulomb field holding the atom together — in hydrogen this occurs at an intensity of 3.6×10^{16} W/cm² — a Rabi frequency on the order of 200 000 cm⁻¹ is generated, again much more than enough to excite a coherent superposition of all atomic bound states.

71.4 HOMOGENEOUS RELAXATION

Homogeneous relaxation is dominant in well-collimated atomic and molecular beams as well as in high-pressure vapor cells. In the absence of a laser field ($\Omega_1 = 0$) the solutions of the OBEs are

$$(U - iV) = (U - iV)_0 e^{-(1/T_2 + i\Delta)t},$$
$$W = -1 + (W_0 + 1)e^{-t/T_1},$$
(71.4)

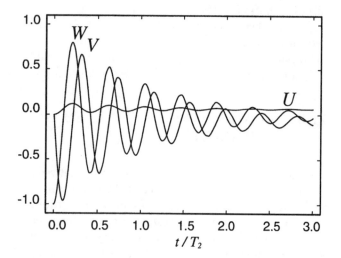

Figure 71.1. Damped Rabi oscillations of the atomic variables after sudden turn-on of the field. In this example $T_1 = T_2$, $\Delta T_2 = 1$, and $\Omega_1 T_2 = 15$.

where the subscript denotes values at $t = 0$. The roles of T_1 and T_2 as relaxation times are clear. They are homogeneous because they apply to each atom individually.

71.4.1 Rabi Oscillations

The OBEs predict coherent damped oscillations of the inversion with the angular Rabi frequency Ω_1 if Ω_1 is large enough, such that $\Omega_1 T_1 \gg 1$ and $\Omega_1 T_2 \gg 1$. These oscillations were originally called *optical nutations* following the terminology of nuclear magnetic resonance, however they are now usually called *Rabi oscillations*. Fig. 71.1 shows the behavior of the atomic variables undergoing Rabi oscillations in a representative case.

71.4.2 Bloch Vector and Bloch Sphere

Coherent dynamical behavior is simplest for times much shorter than the relaxation times T_1 and T_2. In this case, the damping terms can be dropped from the OBEs and the resulting equations written in the form (see Sect. 66.4.5)

$$\frac{d\mathbf{U}}{dt} = \mathbf{\Omega} \times \mathbf{U}, \qquad (71.5)$$

where $\mathbf{U} = (U, V, W)$ is the Bloch vector, and $\mathbf{\Omega} = (-\Omega_1, 0, \Delta)$ acts as a torque vector defining the axis and rate of precession. By conservation of probability, $\mathbf{U} \cdot \mathbf{U} = 1$.

All possible quantum states of the two-level atom are mapped onto a unit sphere in U-V-W space. Conventionally, W defines the polar axis with the atomic ground

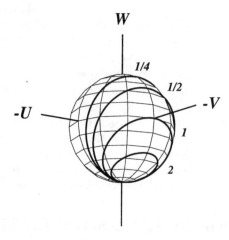

Figure 71.2. Orbits of the Bloch vector on the unit sphere for various ratios of the detuning Δ to the Rabi frequency Ω_1

state the south pole, and the excited state the north pole. Points on the sphere between the poles are coherent superpositions of the two states. The azimuthal angle ϕ represents the phase between the expectation value of the dipole moment and the optical field. In Fig. 71.2 the solutions to Eqs. (71.5) are shown for the case of a square pulse applied to an atom in its ground state at $t = 0$. The solutions in this case are

$$U(t, \Delta) = \frac{\Omega_1}{\Omega} \sin \Omega t,$$

$$V(t, \Delta) = -\frac{\Delta \Omega_1}{\Omega^2}(1 - \cos \Omega t), \qquad (71.6)$$

$$W(t, \Delta) = -1 + \frac{\Omega_1^2}{\Omega^2}(1 - \cos \Omega t).$$

For any Δ the solution orbit is a circle on the surface of the sphere with the orbit passing through the south pole. The rate at which the system precesses about the circle is given by the *generalized Rabi frequency* $\Omega \equiv \sqrt{\Delta^2 + \Omega_1^2}$.

71.4.3 Pi Pulses and Pulse Area

The exactly resonant ($\Delta = 0$) undamped OBEs can be solved analytically even for arbitrarily time dependent laser pulse envelopes. The solutions are

$$U(t, 0) = 0,$$
$$V(t, 0) = -\sin \theta(t), \qquad (71.7)$$
$$W(t, 0) = -\cos \theta(t),$$

a rotation of the Bloch vector in the V-W plane. The angle θ is called the *pulse area*, defined by

$$\theta(t) \equiv \int_{-\infty}^{t} dt' \Omega_1(t') = \frac{d}{\hbar} \int_{-\infty}^{t} dt' \mathcal{E}(t'). \qquad (71.8)$$

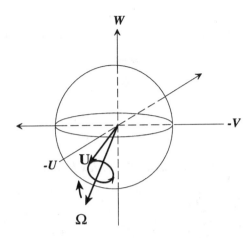

Figure 71.3. In adiabatic inversion the Bloch vector of each atom precesses in a small cone about the torque vector as the torque vector goes from straight down to straight up.

The area under the envelope of the Rabi frequency is thus the same as the angle through which an exactly-resonant Bloch vector turns due to the pulse. If $\theta = \pi$, the atom is driven from the ground state exactly to the excited state. This is "π-pulse" inversion. A "2π-pulse" takes the atom from the ground state through the excited state and back to the ground state. For $\Delta = 0$, the Bloch vector rotation angle does not depend upon the shape of the field pulse, only on the area of the pulse.

71.4.4   Adiabatic Following

The Bloch vector picture is used for semiquantitative predictions. These are reliable even if Δ and Ω_1 are time-dependent, if the parameters change slowly (adiabatically). For example, if $\mathbf{\Omega}$ is moved slowly the Bloch vector follows closely. It is possible to achieve complete inversion smoothly in this way. If the field is initially tuned far below resonance so $\Delta \ll \Omega_1$ then $\mathbf{\Omega}$ points approximately toward the south pole of the Bloch sphere. As shown in Fig. 71.3, the Bloch vector then precesses in a very small circle about the torque vector. If the field frequency is now slowly changed (chirped) so that Δ goes from a large negative value to a large positive value then every atomic Bloch vector will continue to precess rapidly around the torque vector, and follow it as it proceeds from pointing straight down to pointing straight up. In this way the population is transferred between the two levels.

71.5   INHOMOGENEOUS RELAXATION

The fact that the various atoms in a sample may have different resonance frequencies produces a number of novel phenomena. Given a distribution $g(\Delta)$ of detunings in a dilute gas of density N, the macroscopic polarization can be written

$$P(t) = -N\langle ex(t)\rangle$$
$$= Nd \int g(\Delta)\Re\left[(U - iV)e^{-i\omega t}\right] d\Delta, \quad (71.9)$$

where $U - iV$ generally depends on both t and Δ.

71.5.1   Free Induction Decay

Free induction refers to evolution of the polarization in the absence of a laser field. For $\Omega_1 = 0$, the Bloch vector of an atom with $\Delta < 0$ precesses counterclockwise in the U-V plane. In a macroscopic sample, there are many values of Δ and about as many are positive as negative. Thus an oriented collection of Bloch vectors, all pointing in the V direction at $t = 0$, will rapidly fan out in the U-V plane due to differing precession rates, and after a short time the net V value will be zero, as will the net U value. This is free induction decay (FID) of polarization.

More precisely, if all atoms are first exposed to a θ_0 pulse, so that at $t = 0$

$$\begin{aligned} U(0,\Delta) &= 0\,, \\ V(0,\Delta) &= -\sin\theta_0\,, \quad (71.10) \\ W(0,\Delta) &= -\cos\theta_0\,, \end{aligned}$$

then if $\mathcal{E} = 0$ for $t > 0$, $U - iV$ for an individual atom with detuning Δ evolves according to (71.4):

$$U - iV = i\sin\theta_0 e^{-(1/T_2 - i\Delta)t}\,. \quad (71.11)$$

The macroscopic polarization is found by summing the individual $(U - iV)$ values over the detuning distribution $g(\Delta)$.

For simplicity, in this section we will assume Lorentzian (instead of the usual Maxwellian) lineshape

$$g(\Delta) = \left(\frac{1}{\pi T^*}\right)\frac{1}{\Delta^2 + (1/T^*)^2}\,, \quad (71.12)$$

and the result is

$$P(t) = Nd\sin\theta_0\sin\omega t \exp\left[-(T_2^{-1} + T^{*-1})t\right]\,. \quad (71.13)$$

The macroscopic decay of polarization occurs at the rate $1/T^*$, the width of the distribution of detuning frequencies. The detuning "inhomogeneity" in the sample leads to dephasing of the collective dipole moment. This is illustrated in Fig. 71.4. For a typical room temperature gas a visible transition has a width given by $1/(2\pi T^*) \approx 1.5$ GHz so that $T^* \approx 10^{-10}$ s.

71.5.2 Photon Echoes

A photon echo is generated by recovery of a nonzero $P(t)$ *after* $P(t) \to 0$ due to free induction decay (FID). This analog of the spin echo effect is possible because each atom retains its own detuning for a relatively long time, usually up to an average collision time T_2.

During FID, the U-V projection of every atom's Bloch vector precesses steadily clockwise or counterclockwise depending on the sign of its Δ. Thus the Bloch vectors could be rephased if they could all be forced at the same moment to reverse their relative sense of precession. The prototypical echo scenario has FID beginning at $t = 0$, with $P(t) \to 0$ for $t \gg T^*$, followed by a π-pulse at the time t', where $t' \gg T^*$ but $t' \ll T_1, T_2$. The effect of the π-pulse is to reverse the sign of V and W [recall Eq. (71.10)], in effect flipping the equatorial plane of the Bloch sphere upside down. Thus for $t \gg t'$ we have Bloch vectors fanning back together. The macroscopic polarization obeys

$$P(t) = Nd \sin \theta_0 \sin \omega t \exp \left[- \frac{|t - 2t'|}{T^*} \right] e^{-t/T_2} . \quad (71.14)$$

The result is illustrated in Fig. 71.4. An "echo" of the initial excitation at $t = 0$ appears at the time $t = 2t'$. After this, FID occurs again, and this second decay can also be reversed by applying another π-pulse, and so on, until $t \approx T_2$, at which time the inevitable and irreversible homogeneous relaxation cannot be avoided. The scenario of π-pulse reversal is only the most ideal, leading to the most complete echo, and other pulse areas will also lead to echos; a more important factor is that the reversing pulse must be short enough that negligible dephasing takes place during its application.

71.6 RESONANT PULSE PROPAGATION

71.6.1 Maxwell-Bloch Equations

Time-dependent atomic dipole moments created by applied fields are themselves a source of fields, another form of coherent transient. We limit discussion to plane-wave propagation in the z-direction. Note that we use z rather than Z for convenience, although in the dipole approximation the coordinate entering our equations is the coordinate of the center of mass of the atom rather than the internal electron coordinate. The field is generalized from (71.2) to

$$E(t, z) = \tfrac{1}{2} [\mathcal{E}(t, z) e^{i(Kz - \omega t)} + \text{c.c.}], \quad (71.15)$$

and the macroscopic polarization is the correspondingly generalized form of (71.9):

Free Induction Decay - Dipoles Dephase

$t = 0$ $t = t' - \epsilon$

After π Pulse - Dipoles Rephase

$t = t' + \epsilon$ $t = 2t'$

Figure 71.4. The ensemble of dipole moments spreads due to the distribution of resonance frequencies. The distribution of Bloch vectors in the U-V plane is shown at various times after the initial short pulse excitation. By the time t', the dipoles have spread uniformly around the unit circle. A π-pulse then flips the relative orientation of the dipoles so that they subsequently rephase.

$$P(t, z) = Nd \int g(\Delta) \Re \left\{ (U - iV) e^{i(Kz - \omega t)} \right\} d\Delta .$$

$$(71.16)$$

The difference between K and $k = \omega/c$ indicates that the refractive index is nonzero. The traveling-pulse rotating frame is also obtained by replacing ωt by $\omega t - Kz$.

If the field envelope $\mathcal{E}$ is slowly varying, its second derivatives can be dropped when $E(t, z)$ is substituted into Maxwell's wave equation. The resulting dispersive and absorptive *reduced wave equations* are

$$\left[K^2 - k^2 \right] \mathcal{E} = 4\pi k^2 Nd \int g(\Delta) U(t, z, \Delta) d\Delta ,$$

$$(71.17)$$

$$\left[K \frac{\partial}{\partial z} + k \frac{\partial}{\partial (ct)} \right] \mathcal{E} = 2\pi k^2 Nd \int g(\Delta) V(t, z, \Delta) d\Delta ,$$

The Bloch equations along with these reduced wave equations form the *self-consistent Maxwell-Bloch equations* that are used to treat most resonant propagation problems in quantum optics and laser theory [1, 2, 5].

71.6.2 Index of Refraction and Beers Law

If a weak pulse of duration τ propagates in a medium of ground-state atoms ($W \approx -1$), the Bloch equations have simple quasisteady-state solutions ($\tau \gg T_2$)

$$U = \frac{\Omega_1 \Delta}{\Delta^2 + 1/T_2^2},$$
$$V = \frac{-\Omega_1/T_2}{\Delta^2 + 1/T_2^2}. \qquad (71.18)$$

When U and V are substituted back into the reduced wave equations (71.17), the dispersive equation gives the index of refraction $n = K/k$ due to the ground-state atoms:

$$n^2 - 1 = \frac{4\pi N d^2}{\hbar} \int \frac{\Delta g(\Delta)}{\Delta^2 + 1/T_2^2} d\Delta, \qquad (71.19)$$

and the absorptive equation predicts steady state field attenuation during propagation:

$$\frac{\partial}{\partial z} \mathcal{E} = -\frac{1}{2}\alpha_B \mathcal{E}. \qquad (71.20)$$

The constant α_B given by

$$\alpha_B = \frac{4\pi N d^2 \omega}{\hbar c T_2} \int \frac{g(\Delta)}{\Delta^2 + 1/T_2^2} d\Delta \qquad (71.21)$$

is called the *extinction coefficient*, or the *reciprocal Beers length*.

Since field intensity I is proportional to $\mathcal{E}^2$, the solution to the absorptive equation is

$$I(z,t) = I(0,t)e^{-\alpha_B z}. \qquad (71.22)$$

This relation is called Beers Law. Both the dispersive and absorptive results are familiar from classical physics [6], with the important distinction that here the $\hbar$-dependent oscillator strength enters naturally rather than as the empirical parameter it is in Lorentzian dielectric theory [5].

71.6.3 The Area Theorem and Self-Induced Transparency

A form of pulse propagation with no classical analog arises in the short-pulse limit ($\tau \ll T_2, T_1$, but $\tau \gg T_2^*$). By integration over the entire pulse, the absorptive Maxwell equation becomes an equation for $\partial\theta/\partial z$, where θ is the pulse area defined in Eq. (71.8). In the short-pulse limit, the relaxation terms in the OBEs can be ignored and when substituting from them we obtain the *McCall-Hahn Area Theorem* [1]:

$$\frac{\partial}{\partial z}\theta(z) = -\frac{1}{2}\alpha_B \sin\theta(z). \qquad (71.23)$$

This predicts the same exponential attenuation as Eq. (71.20) in the case of a small area pulse, $\theta(z) \ll \pi$, but in the case of larger area pulses the behavior is quite different. In general, the area decreases during propagation for areas in the range $0 < \theta(z) < \pi$, but it *increases*

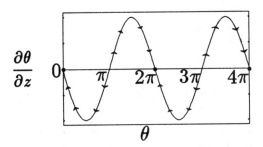

Figure 71.5. McCall-Hahn area theorem for an absorbing medium. On propagation the area of the pulse will follow the arrows toward one of the stable values, $0, 2\pi, 4\pi, \ldots$.

for areas $\pi < \theta(z) < 2\pi$. As seen from Fig. 71.5, this change of area with propagation causes the pulse area to evolve to one of the stable values $0, 2\pi, 4\pi, \ldots$.

There is one special pulse, a *soliton* solution with area exactly 2π, which propagates without shape change in the short pulse limit, given by

$$\mathcal{E}(t,z) = \frac{2\hbar}{\tau d}\operatorname{sech}\left(\frac{t - z/v}{\tau}\right), \qquad (71.24)$$

where τ is the pulse duration, which is arbitrary but must satisfy the short-pulse inequality $\tau \ll T_1, T_2$. The soliton's group velocity is determined by the corresponding soliton solutions to the OBE's and the dispersive wave equation to be

$$v = \frac{c}{1 + \frac{1}{2}\alpha_B c\tau}, \qquad (71.25)$$

where α_B is to be taken in the limit $T_2^* \ll T_2$. The group velocity can be slower than the speed of light by orders of magnitude if $\alpha_B c\tau \gg 1$.

71.7 MULTI-LEVEL GENERALIZATIONS

71.7.1 Rydberg Packets and Intrinsic Relaxation

A short laser pulse can populate a band of excited states whose probability amplitudes will exhibit mutual coherence. This single-atom coherence is transient, even without collisions or other external perturbations to disrupt it, and its decay can be called *intrinsic relaxation*. The decay is basically a dephasing. The dipole moments associated with the excited band interfere due to the wide variety of resonance frequencies of the states in the superposition. Because of the discreteness of the energy levels of any bounded quantum system, this relaxation has its own unique characteristics, including similarities with both homogeneous and inhomogeneous decay.

The wave function for a coherently excited atom can be expressed in the interaction picture in the form

$$\Psi(r,t) = a(t)\psi_g(r) + \sum_n b_n(t)e^{-i\omega_n t}\psi_n(r), \quad (71.26)$$

where $\psi_g(r)$ is the ground state wave function, and n labels the states in a band with excitation frequencies $\omega_n \approx \omega$. If $|\omega_n - \omega| \ll \omega_n$, the transition frequency ω_n can be expanded about the principal quantum number $\overline{n}$ of the resonant excited state $E_{\overline{n}} = \hbar\omega$ to obtain

$$\omega_n = \omega_{\overline{n}} + (n - \overline{n})\frac{\partial \omega_n}{\partial n} + \frac{1}{2!}(n - \overline{n})^2\frac{\partial^2 \omega_n}{\partial n^2} + \cdots$$

$$= \omega_{\overline{n}} + (n - \overline{n})\frac{2\pi}{T_K} + (n - \overline{n})^2\frac{2\pi}{T_R} + \cdots. \quad (71.27)$$

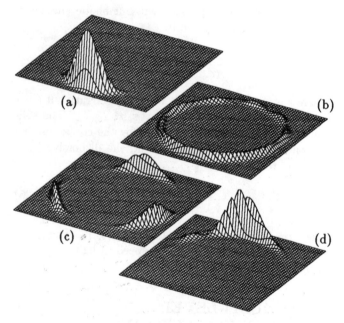

(a)

(b)

(c)

(d)

Figure 71.6. The free evolution of a Rydberg wave packet made up of a superposition of circular-orbit states centered about $n = 360$. (a) Initially the wave packet is to good approximation a minimum uncertainty wave packet in all three dimensions, but after 12 orbits (b) the packet has spread all the way around the orbit so that the head and tail of the wave packet overlap, producing interference fringes. (c) After 40 orbits, $t = T_R/3$, the fringes have produced the one-third fractional revival in which three miniature replicas of the original wave packet are equally spaced around the orbit. (d) After 120 orbits, $t = T_R$, the complete wave packet revival occurs.

Thus $2\pi/T_K$ is the mean frequency separating neighboring levels, i.e., $T_K/2\pi = \hbar\rho(E)$, where $\rho(E)$ is the density of excited states, and $2\pi/T_R$ is the mean change in this frequency separation. T_K is the same as the *Kepler period* of a classical orbit, and T_R is the *revival* time. Substituting the Bohr frequencies into the definitions for T_K and T_R yields $T_R = \overline{n}T_K/3$.

For times $t \leq T_R$ the expansion can be truncated after the third term. Then the wave function is given approximately by

$$\Psi(r,t) \approx a(t)\psi_g(r) + e^{-i\omega_{\overline{n}}t}\sum_m b_{\overline{n}+m}(t)e^{-i2\pi mt/T_K}$$

$$\times e^{-i2\pi m^2 t/T_R}\psi_{\overline{n}+m}(r), \quad (71.28)$$

where $m = n - \overline{n}$. For high Rydberg states, $\overline{n} \gg 1$, the time scales associated with the two exponentials inside the sum are quite different. For times $t \ll T_K$ the individual levels are not resolved, thus the laser excites what is effectively a continuum with a density of states $\rho(E)$. In that case the ground state population simply decays exponentially at the rate given by first-order perturbation theory, $\Gamma = (2\pi/\hbar)d^2\mathcal{E}^2\rho(E)$. At longer times $t \approx T_K$, the first exponential contributes, but the second does not, giving a simple Fourier series time dependence. In this regime the evolution of the wave function is just periodic motion of a wave packet around a Kepler orbit, as is illustrated in Fig. 71.6(a).

The coherent quantum wave packet behaves like a classical particle for many Kepler periods, gradually spreading out as the second exponential in the sum (71.28) begins to contribute. This spreading of the wave packet produces the intrinsic relaxation of the collective dipole moments from the various transitions. However, because the levels are discrete, this decay is not permanent, but is reversed and leads to a spontaneous "revival" of the original wave packet, without the need for a π-pulse to produce an "echo."

In its evolution toward the revival, the wave packet passes through a number of *fractional revivals* in which miniature replicas of the original wave packet are equally spaced around the orbit, each traveling at the velocity of a particle traveling in a corresponding classical Kepler orbit. This complex time evolution arises from the spreading of the wave packet all the way around the orbit so that the head and tail of the packet interfere with each other, producing interference fringes. The further evolution of this fringe pattern produces the various revival phenomena shown in Fig. 71.6.

71.7.2 Multiphoton Resonance and Two-Photon Bloch Equations

Multiphoton transitions [see Fig. 71.7(a)] introduce new coherent transient phenomena. If levels $|g\rangle$ and

$|e\rangle$ have the same parity, two photons from the same laser field are sufficient to excite level $|e\rangle$. For simplicity we regard $|e\rangle$ as a single state, but any number of intermediate levels $|j\rangle$ of opposite parity may be present.

Substituting the state vector

$$|\Psi(t)\rangle = a_g(t)|g\rangle + \sum_j b_j(t)e^{-i\omega t}|j\rangle + a_e(t)e^{-i2\omega t}|e\rangle \, ,$$

(71.29)

into the Schrödinger equation yields

$$i\frac{da_g}{dt} = -\frac{1}{2}\sum_j \Omega_{gj}b_j \, ,$$

(71.30)

$$i\frac{db_j}{dt} = \Delta_{jg}b_j - \frac{1}{2}[\Omega_{jg}a_g + \Omega_{je}a_e] \, ,$$

(71.31)

$$i\frac{da_e}{dt} = \Delta_{eg}a_e - \frac{1}{2}\sum_j \Omega_{ej}b_j \, ,$$

(71.32)

where the Δ's are the detunings and the Ω's are the Rabi frequencies for the dipole-allowed transitions. For example, $\Omega_{gj} = d_{gj}\mathcal{E}/\hbar$, and $\Delta_{jg} = (E_j - E_g)/\hbar - \omega$ and $\Delta_{eg} = (E_e - E_g)/\hbar - 2\omega$.

If the states $|j\rangle$ are not too close to resonance, the b_j oscillate rapidly and to a first approximation average to zero. A better approximation is to retain the small nonzero solution for b_j obtained by setting $db_j/dt = 0$ in Eq. (71.31) to obtain

$$b_j = -\frac{[\Omega_{jg}a_g + \Omega_{je}a_e]}{2\Delta_{jg}} \, ,$$

(71.33)

which can be used to eliminate b_j from the equations for a_g and a_e. This is called *adiabatic elimination* of dipole coherence. In this approximation, levels $|g\rangle$ and $|e\rangle$ are directly coupled to each other and *two-photon coherence* arises. The coupling of levels $|g\rangle$ and $|e\rangle$ is similar to the two-level coupling described in Sects. 71.2 – 71.5 and *two-photon Bloch equations* analogous to Eqs. (71.5) are the result:

$$\frac{dU^{(2)}}{dt} = -\Delta^{(2)}V^{(2)} \, ,$$

$$\frac{dV^{(2)}}{dt} = \Delta^{(2)}U^{(2)} + \Omega^{(2)}W^{(2)} \, ,$$

(71.34)

$$\frac{dW^{(2)}}{dt} = -\Omega^{(2)}V^{(2)} \, ,$$

Here the superscript (2) indicates that the variables are identified with the two-photon $|g\rangle \to |e\rangle$ transition.

The various coefficients are similarly generalized [7]. For example, the two-photon Rabi frequency is given by

$$\Omega^{(2)} \equiv \tfrac{1}{2}\sum_j \frac{d_{gj}d_{je}}{\hbar^2\Delta_{jg}}\mathcal{E}^2 \, .$$

(71.35)

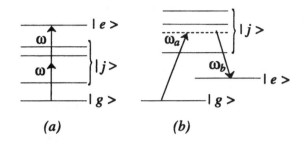

Figure 71.7. Model two-photon resonances. Two photons couple the ground state $|g\rangle$ with an excited state $|e\rangle$. Many intermediate nonresonant levels $|j\rangle$ are present. In (a) we have the cascade system, and in (b) the **Λ** pump-probe system.

and the *two-photon detuning* $\Delta^{(2)}$ incorporates the laser-induced level shifts

$$\Delta^{(2)} \equiv \Delta_{ej} + \Delta_{jg}$$
$$+ \tfrac{1}{4}\sum_j \frac{|d_{ej}|^2\mathcal{E}^2(t)}{\hbar^2\Delta_{ej}} + \tfrac{1}{4}\sum_j \frac{|d_{jg}|^2\mathcal{E}^2(t)}{\hbar^2\Delta_{jg}} \, .$$

(71.36)

The last two terms give the difference in the *ac Stark shifts* of the upper and lower levels produced by the laser field.

$W^{(2)}$ is the inversion as before, but $U^{(2)}$ and $V^{(2)}$ are somewhat different. They cannot be directly tied to the expectation value of a dipole moment because levels $|g\rangle$ and $|e\rangle$ have the same parity. Thus, while the quantity $U^{(2)} - iV^{(2)}$ is the two-photon analog of $U - iV$ in the original OBE's, it cannot serve as a source term in the Maxwell equation.

In the case of cw applied fields, the solutions to the two-photon OBEs are formally identical to those for a two-level atom. In the case of pulsed fields, however, the detuning $\Delta^{(2)}(t)$ is automatically "chirped" in frequency by the Stark shifts. This chirping may significantly modify the dynamics. Multiphoton generalizations of the Bloch equations can be made for other arrangements and numbers of levels. The two-photon version applies as well to three-level **Λ** and **V** configurations as to the *cascade* system shown in Fig. 71.7(a), for which they were derived.

71.7.3 Pump-Probe Resonance and Dark States

Dark states or *trapping states* occur whenever a field-dependent linear combination of active levels is dynamically disconnected from the other levels. This occurs, e.g., in a pump-probe interaction, which fits the scenario of Fig. 71.7(b) if two lasers instead of one are used to excite level $|e\rangle$ from the ground level via two-photon resonance. A strong steady laser a is applied for the $|g\rangle \rightarrow |j\rangle$ transitions and a weak tunable "probe" laser b for the $|j\rangle \rightarrow |e\rangle$ transitions. In the simplest format, $\Delta_{eg} = 0$, and all the $|j\rangle$ levels can be combined into a single level labeled $|2\rangle$.

The three-level state vector can be written in terms of field-free states

$$|\Psi(t)\rangle = a_g|g\rangle + b_2|2\rangle + a_e|e\rangle \,, \qquad (71.37)$$

or, in terms of field-dependent dressed states

$$|\Psi(t)\rangle = A_T|T\rangle + b_2|2\rangle + A_S|S\rangle \,, \qquad (71.38)$$

where $\Omega|T\rangle \equiv \Omega_a|e\rangle - \Omega_b|g\rangle$, $\Omega|S\rangle \equiv \Omega_a|g\rangle + \Omega_b|e\rangle$ and

$$\begin{aligned} A_S(t) &\equiv \Omega^{-1}[\Omega_a a_g + \Omega_b a_e] \,, \\ A_T(t) &\equiv \Omega^{-1}[\Omega_a a_e - \Omega_b a_g] \,. \end{aligned} \qquad (71.39)$$

The normalizing factor is $\Omega \equiv \sqrt{\Omega_a{}^2 + \Omega_b{}^2}$.

The state $|T\rangle$ is an eigenvector of the three-level RWA Hamiltonian, with eigenvalue zero, and the amplitude $A_T(t)$ is a constant of motion. Thus $|T\rangle$ is termed a *trapping state*, and population in $|T\rangle$ is inaccessible to the (possibly very strong) laser fields. At two-photon resonance this conclusion is robust, not depending strongly on the idealized conditions assumed here. In fact, A_T, the trapping state amplitude, is an adiabatic invariant, remaining constant to first order even under slow changes in Ω_a and Ω_b. In a pump-probe experiment, this trapping state is observed as an abrupt drop in probe absorption as the probe frequency is tuned through two-photon resonance. Since only two-photon resonance is required (both transitions can be equally detuned) this coherent transient effect has no analog in two-level physics.

The ideal method for exciting the trapping state from the ground state uses another coherent transient process called *counter-intuitive pulse sequencing* in which pulse b is turned on first. The trapping state $|T\rangle$ is essentially the ground state $|g\rangle$ if $\Omega_a = 0$. Thus if Ω_b is turned on first, and then Ω_a turned on later, the ground state adiabatically becomes the trapping state and all initial probability flows smoothly with it.

An essential point is the ease with which pump-probe adiabaticity is maintained, particularly for strong fields on resonance, in contrast to one-photon adiabaticity, which is never achieved at strong-field resonance. In the pump-probe case one must only satisfy the inequality

$$\left| \frac{d\Omega_a}{dt}\Omega_b - \frac{d\Omega_b}{dt}\Omega_a \right| \ll \left(\Omega_a^2 + \Omega_b^2 \right)^{3/2} \,, \qquad (71.40)$$

which is automatically accomplished by counter-intuitive pulse sequencing. The inequality allows one to tolerate rapid change of Ω_b while $\Omega_a = 0$. Then after Ω_b has

reached a high value, Ω_a can also be turned on very rapidly because the right side of (71.40) is already very large. This is "counter-intuitive" excitation because if the population is in level $|g\rangle$ it is "natural" to turn on pulse Ω_a first, not Ω_b. It can be dramatically beneficial to use counter-intuitive excitation when it is important to avoid relaxation associated with level $|2\rangle$.

71.7.4 Three-Level Transparency

The foregoing results for three-level excitation can be extended to resonant pulse propagation in three-level media. The equations governing simultaneous two-pulse evolution in the local-time coordinates $cT \equiv ct - z$ and $\zeta \equiv z$ are

$$\begin{aligned} \frac{\partial \mathcal{E}_a}{\partial \zeta} &= i\frac{\hbar}{d_a}\mu_a a_g^* b_2 \,, \\ \frac{\partial \mathcal{E}_b}{\partial \zeta} &= i\frac{\hbar}{d_b}\mu_b a_e^* b_2 \,, \end{aligned} \qquad (71.41)$$

where

$$\mu_a = \frac{4\pi d_a^2 N \omega_a}{\hbar c} \,, \quad \text{etc.} \qquad (71.42)$$

Note that the bilinear combination $2a_g^* b_2$ corresponds to $U - iV$ in Eq. (71.16).

Soliton-like pulses can propagate in three-level media. Both pulses must compete for interaction with level $|2\rangle$. They depend only on a single variable $Z \equiv \zeta - uT$ where u is the pulse's constant velocity in the moving frame. Soliton solutions are given (for $\mu_a = \mu_b$) for Λ media by

$$\mathcal{E}_a = \hbar/d_a = A \operatorname{sech} KZ \,, \quad \mathcal{E}_b = \hbar/d_b = B \tanh KZ \,, \qquad (71.43)$$

and

$$\begin{aligned} a_g &= -\tanh KZ \,, \\ b_2 &= \frac{-2iKu}{A} \operatorname{sech} KZ \,, \\ a_e &= \frac{B}{A} \operatorname{sech} KZ \,, \end{aligned} \qquad (71.44)$$

where the parameters A, B, Ku are nonlinearly related to the pulse length τ:

$$Ku \equiv 1/\tau \quad \text{and} \quad (2/\tau)^2 = A^2 - B^2 \,. \qquad (71.45)$$

The moving frame velocity is given by $1/u = 2\mu/A^2$ and the expression for the lab frame velocity V is $1/V = 1/c + 2\mu/A^2$.

If $B \to 0$, then $\mathcal{E}_a \to (2\hbar/\tau d_a)\mathrm{sech}KZ$, which is the exact McCall–Hahn formula for the two-level one-pulse soliton amplitude [1]. No adiabatic condition was invoked in obtaining the soliton solutions. The physical measure of adiabaticity comes from the pulse duration τ. If τ is short, an appreciable population appears transiently in level $|2\rangle$, but if τ is long (an adiabatic pulse), the population skips level $|2\rangle$ and goes directly from $|g\rangle$ to $|e\rangle$ and back again during the pulse.

Note that the sech and tanh functions are ideally counter-intuitive, with pulse b starting infinitely far ahead of pulse a. In practice, the infinite leading edge of the tanh function plays no role and can be truncated to several times τ without appreciable change in the character of the pulse pair.

REFERENCES

1. L. Allen and J. H. Eberly, *Optical Resonance and Two-Level Atoms* (Dover, 1987).
2. B. W. Shore, *Theory of Coherent Atomic Excitation* (Wiley, 1990), Vols. 1 & 2.
3. P. Meystre and M. Sargent III, *Elements of Quantum Optics* (Springer-Verlag, 1990).
4. C. Cohen-Tannoudji, J. Dupont-Roc, and G. Grynberg, *Atom-Photon Interactions* (Wiley, 1992).
5. P. W. Milonni and J. H. Eberly, *Lasers* (Wiley 1988).
6. J. D. Jackson, *Classical Electrodynamics* (Wiley 1975), 2nd ed.
7. F. H. M. Faisal, *Theory of Multiphoton Processes* (Plenum, 1987).

72

Multiphoton and Strong-Field Processes

K. C. Kulander*[†] and M. Lewenstein**[††]

*Physics Division, Lawrence Livermore National Laboratory, Livermore, California 94551
[†]Joint Institute for Laboratory Astrophysics, University of Colorado, Boulder, Colorado 80309
**Service des Photons, Atomes et Molécules, Centre d'Etudes de Saclay, Gif-sur-Yvette 91191, France
[††]Centre for Theoretical Physics, Polish Academy of Sciences, 02-668 Warsaw, Poland

72.1 INTRODUCTION

The excitation of atoms by intense laser pulses can be divided into two broad regimes determined by the characteristics of the laser pulse relative to the atomic response. The first regime involves relatively weak optical laser fields of long duration (>1 ns), and the second involves strong fields of short duration (<10 ps). These will be referred to as the weak-long (WL) and strong-short (SS) cases respectively.

In the case of atomic excitation by WL pulses, the intensity is presumed to be high enough for multiphoton transitions to occur. The resulting spectroscopy of absorption to excited states is potentially much richer than single-photon excitation because it is not limited by the single-photon selection rules for radiative transitions. However, the intensity is still low enough for a theoretical description based on perturbations of field-free atomic states to be valid, and the time dependence of the field amplitude does not play an essential role.

The SS case is fundamentally different in that the atomic electrons are strongly driven by fields too large to be treated by perturbation theory, and the time dependence of the pulse as it switches on and off must be taken into account. Atoms may absorb hundreds of photons, leading to the emission of one or more electrons, as well as photons of both lower and higher energy. Because the flux of incident photons is high, a classical description of the laser field is adequate, but the time dependent Schrödinger equation (TDSE) must be solved directly to obtain an accurate representation of the atom-field interaction.

For SS pulses of optical wavelength, it is sufficient in most cases to consider only the electric dipole (E1) interaction term defined in Chap. 66. The atom-field interaction can then be expressed in either the length gauge or the velocity gauge [1] (see Chap. 21). In the length gauge, the TDSE is

$$i\hbar\frac{\partial\Psi(\mathbf{r},t)}{\partial t} = [H_0 + e\mathbf{r}\cdot\mathbf{E}(t)]\,\Psi(\mathbf{r},t)\,, \tag{72.1}$$

where H_0 is the field-free atomic Hamiltonian, $\mathbf{r}$ the collective coordinate of the electrons, and $\mathbf{E}(t)$ the electric field of the laser given by

$$\mathbf{E}(t) = \mathcal{E}(t)[\hat{\mathbf{x}}\cos(\omega t + \varphi) + \epsilon\hat{\mathbf{y}}\sin(\omega t + \varphi)]\,. \tag{72.2}$$

Here φ is the phase and ϵ defines the polarization: linear if $\epsilon = 0$ and circular if $|\epsilon| = 1$. In the velocity gauge, the TDSE is

$$i\hbar\frac{\partial\psi(\mathbf{r},t)}{\partial t} = \left[H_0 - \frac{ie\hbar}{mc}\mathbf{A}(t)\cdot\nabla + \frac{e^2}{2mc^2}A^2(t)\right]\psi(\mathbf{r},t)\,. \tag{72.3}$$

Here $\mathbf{A}(t) = -c\int^t \mathbf{E}(t')dt'$ is the vector potential of the laser field. The solutions of Eqs. (72.1) and (72.3) are related by the phase transformation

$$\Psi(\mathbf{r},t) = \exp\left[\frac{ie}{\hbar c}\mathbf{r}\cdot\mathbf{A}(t)\right]\psi(\mathbf{r},t)\,. \tag{72.4}$$

Since lasers usually must be focussed to reach the strong field regime, measured electron and ion yields include contributions from a distribution of field *strengths*. The photoemission spectrum, on the other hand, contains a coherent component due to the macroscopic polarization of all the atoms and therefore is sensitive also to the laser *phase* variations within the focal volume. In this chapter methods for solving Eqs. (72.1) and (72.3)

are discussed along with details of the atomic emission processes.

References [1–3] are three recent books which provide excellent introductions to this subject. Further developments are well described in the proceedings of the International Conferences on Multiphoton Physics [4–6] and the NATO workshop on Super-Intense Laser-Atom Physics [7].

72.2 WEAK FIELD MULTIPHOTON PROCESSES

72.2.1 Perturbation Theory

Since atomic ionization energies are generally $\gtrsim 10$ eV, while optical photons have energies of only a few eV, several photons must be absorbed to produce ionization, or even electronic excitation in the case of the noble gases. For WL pulses, the electronic states are only weakly perturbed by the electromagnetic field. The rate of an n-photon transition can then be calculated using nth order perturbation theory for the atom-field interaction. For an incident photon number flux ϕ of frequency ω, the rate is

$$W_{i\to f}^{(n)} = 2\pi\left(\frac{2\pi\alpha\phi\omega}{e^2}\right)^n |T_{i\to f}^{(n)}|^2\delta(\omega_i + n\omega - \omega_f)\,, \tag{72.5}$$

where

$$T_{i\to f}^{(n)} = \langle f\,|\,dG[\omega_i + (n-1)\omega]\,dG[\omega_i + (n-2)\omega] \\ \cdots dG[\omega_i + \omega]\,d\,|\,i\rangle\,, \tag{72.6}$$

$|i\rangle$ is the ith eigenstate of the field-free atomic Hamiltonian, $d = e\hat{\boldsymbol{\epsilon}}\cdot\mathbf{r}$, with $\hat{\boldsymbol{\epsilon}}$ the polarization direction and

$$G(\omega) = \sum_j \frac{|j\rangle\langle j|}{(\omega - \omega_j + i\Gamma_j/2)}\,. \tag{72.7}$$

The sum over j includes an integration over the continuum for all sequences of E1 transitions allowed by angular momentum and parity selection rules. Methods for calculating cross sections and rates in the weak-field regime are described in Refs. [1, 8] and in Chap. 24.

72.2.2 Resonant Enhanced Multiphoton Ionization

For multiphoton ionization, ω can be continuously varied because the final state in Eq. (72.5) lies in the continuum. If ω is tuned so that $\omega_i + m\omega \simeq \omega_j$ for some contributing intermediate state $|j\rangle$ in (72.7), then that state lies an integer m photons above the initial state, and the corresponding denominator vanishes (to within the level width Γ_j), producing a strongly peaked resonance. Since it takes $k = n - m$ additional photons for ionization, the process is called m, k *resonant enhanced multiphoton ionization* (REMPI). Measurements of the photoelectron angular distribution are useful in characterizing the resonant intermediate state.

Calculations using the semi-empirical multichannel quantum defect theory to provide the needed matrix elements have been very successful in describing experimental results. This technique is discussed in more detail in Chap. 24.

The perturbation equation 72.5 indicates that the rate for nonresonant multiphoton ionization scales as ϕ^n for an n-photon process [9]. However, this is not the case for REMPI since the resonant transition saturates and Eq. (72.5) no longer applies. Then the rate can be controlled either by the m-photon resonant excitation step, or by the number of photons k needed for the ionization step.

72.2.3 Multi-Electron Effects

Multiply excited states can play a role in multiphoton excitation dynamics. These states are particularly important if their energies are below or not too far above the first ionization potential. Configuration expansions including these states have been used successfully in studies, for example, of the alkaline earth atoms, which have many low-lying doubly excited states. The presence of these states also can enhance the direct double ionization of an atom [10].

72.2.4 Autoionization

The configuration interaction between a bound state and an adjacent continuum leads to an absorption profile in the single photon ionization spectrum with a Fano lineshape. The actual lineshape reflects the interference between the two pathways to the continuum. Autoionizing states can also be probed via multiphoton excitation [11]. Because, in the strong field regime, coupling strengths and phases change with intensity, the lineshapes can be strongly distorted by changing the incident intensity. At particular intensities, the phases of the excited levels can be manipulated to prevent autoionization completely. Then a trapped population with energy above the ionization limit can be created [12].

72.2.5 Coherence and Statistics

Real laser fields exhibit various kinds of fluctuations, and so are never perfectly coherent. The effects of such fluctuations on the complex electric field amplitude

$$E(t) = \mathcal{E}(t) \exp[-i\omega t + \varphi(t)] \qquad (72.8)$$

can be modeled by a variety of stochastic processes [13], depending on the conditions [8, 14–16], as follows.

For cw lasers, a *phase diffusion model* (PDM) is often used for which $\mathcal{E}(t) = \mathcal{E} = $ const. and

$$\dot{\varphi}(t) = \sqrt{2b}F(t), \qquad (72.9)$$

where $F(t)$ describes *white noise* by a real Gaussian function [13] characterized by the averaged values $\langle F(t) \rangle = 0$, $\langle F(t)F(t') \rangle = 2b\delta(t - t')$. The stochastic electric field then has an exponential autocorrelation function

$$\langle E(t)E^*(t') \rangle = \mathcal{E}^2 \exp\left[-b|t - t'| - i\omega(t - t')\right], \quad (72.10)$$

and a Lorentzian spectrum of width b. Far off resonance, such a Lorentzian spectrum often gives unrealistic results, and the model (72.9) is then replaced by an *Ornstein-Uhlenbeck* process,

$$\ddot{\varphi}(t) = -\beta\dot{\varphi}(t) + \sqrt{2b\beta}F(t), \qquad (72.11)$$

where the parameter β for $\beta \ll b$ plays the role of a cut-off of the Lorentzian spectrum.

A multimode laser with a large number M of independent modes has a field of the form $E(t) = \sum_{j=1}^{M} \mathcal{E}_j \exp[-i\omega_j t + i\varphi_j(t)]$, and according to the central limit theorem [13], can be described for large M as a complex Gaussian process defined to be a *chaotic field*,

$$\dot{E}(t) = -(b + i\omega)E(t) + \sqrt{2b\langle|E(t)|^2\rangle}F(t), \qquad (72.12)$$

where $F(t)$ is now a complex white noise, and ω is the central frequency of the field. The field, (72.12) has an exponential autocorrelation function, and a Lorentzian spectrum of width b.

Various other stochastic models have been discussed in the literature. These include Gaussian fluctuations of the real amplitude of the field $\mathcal{E}(t)$; Gaussian chaotic fields with non-Lorentzian spectra; non-Gaussian, non-linear diffusion processes (that describe for instance a laser close to threshold [13]); multiplicative stochastic processes (that describe a laser with pump fluctuations [15]) and jump-like Markov processes [17, 18]. Statistical properties of laser fields can sometimes be controlled experimentally to a great extent [16].

72.2.6 Effects of Field Fluctuations

Since the response of systems undergoing multiphoton processes is in general a nonlinear function of the field intensity (and, in particular, of the field amplitude), it depends in a complex manner on the statistics of the field. The enhancement of the nonresonant multiphoton ionization rate illustrates the point. According to the perturbation equation (72.5), the rate of an n-photon process is proportional to ϕ^n; i.e., to I^n, where I is the field intensity. For fluctuating fields, the average response is thus

$$W_{i \to f}^{(n)} \propto \langle I^n \rangle \propto \langle |E(t)|^{2n} \rangle . \qquad (72.13)$$

Phase fluctuations (as described by PDM) do not affect the average. On the other hand, for complex chaotic fields, the average is

$$\langle I^n \rangle \simeq n! \langle I \rangle^n , \qquad (72.14)$$

i.e., significant enhancement of the rate for $n > 1$.

Field fluctuations lead to more complex effects in resonant processes. Two well-studied examples are the enhancement of the ac Stark shift in resonant multiphoton ionization [19], and the spectrum of double optical resonance — a process in which the ac Stark splitting of the resonant line is probed by a slightly detuned fluctuating laser field [15]. Double optical resonance is very sensitive not only to the bandwidth of the probing field, but also to the shape of its frequency spectrum.

72.2.7 Excitation with Multiple Laser Fields

The simultaneous application of more than one laser field produces interesting and novel effects. If a laser and its second (2ω) or third (3ω) harmonic are combined and the relative phase between the fields controlled, product state distributions and yields can be altered dramatically. The effects include reducing the excitation or ionization rates in the $\omega - 3\omega$ case [20] or altering the photoelectron angular distributions and the harmonic emission parity selection rules using $\omega - 2\omega$ [21].

A laser field can *dress* or strongly mix the field-free excited states, including the continuum, of an atom. This can produce a number of effects depending on how the dressed system is probed. By coupling a bound, excited state with the continuum, ionization strengths and dynamics are altered, resulting in new resonance-like structures where none existed before. This effect is called *laser-induced continuum structure*, or LICS [12,22]. This general idea has been exploited to design schemes for *lasers without inversion* [23] in which the dressed atom can have an inverted population, allowing gain even though in terms of the undressed states the lower level has the largest population. A laser can produce dramatic changes in the index of refraction of an atomic medium [24], creating, at specific frequencies, *laser-induced transparency* for a second, probe laser field.

Multistep ionization, where each step is driven by a laser at its resonant frequency, has resulted in two useful applications. These are: efficient atomic *isotope separation* [18]; and the detection of small numbers of atoms in a sample, called *single-atom detection*. This technique is extremely sensitive because the use of exact resonance for each step yields very large cross sections for ionization, and the efficiency of collecting ions is high [25].

72.3 STRONG-FIELD MULTIPHOTON PROCESSES

Recently developed laser systems can produce very short pulses, some as short as a few to tens of femtoseconds, while at the same time maintaining the pulse energy so that the peak power becomes very high. Focused intensities up to 10^{19} W/cm^2 have been achieved. Because the pulses are short, atoms survive to much higher intensities before ionizing, making possible studies of laser-atom interactions in an entirely new regime. A discussion of the status of short pulse laser development is given in Chap. 69 and in Ref. [26].

With increasing intensity, higher-order corrections to Eq. (72.6) contribute to the transition rate. The next order correction comes from transitions involving two additional photons, one absorbed and one emitted, leading to the same final state. One effect of these terms is to shift the energies of the excited states in response to the oscillating field. This is called the *dynamic* or *ac Stark shift*. The ac Stark shift of the ground state tends to be small because of the large detuning from the excited states for long wavelength photons. On the other hand, in strong fields the shift of the higher states and the continuum can become appreciable. Electrons in highly excited states respond to the oscillating field in the same manner as a free electron. Their energies shift with the continuum by the amount

$$U_{\mathrm{p}} = \frac{(1 + \epsilon)e^2 \mathcal{E}^2}{4m\omega^2} , \qquad (72.15)$$

where U_{p} is the cycle-averaged kinetic energy of a free electron in the field and ϵ defines the polarization of the field in Eq. (72.2). U_{p} is called the *ponderomotive* or *quiver* energy of the electron. For strong laser fields, U_{p} can be several eV or more, meaning that during a pulse, many states shift through resonance as their energies change by an amount larger than the incident photon energy. The resulting intensity-induced resonances can dominate the ionization dynamics.

Electrons promoted into the continuum acquire the ponderomotive energy, oscillating in phase with the field. In a linearly polarized field, the amplitude of the quiver motion of a free electron, given by $e\mathcal{E}/4m\omega^2$, can become many times larger than the bound state orbitals. If the initial velocity of an electron is small after ionization, it can be accelerated by the field back into the ion core. The subsequent rescattering changes the photoelectron energy and angular distributions, and allows the emission of high energy photons [27,28]. This simple dynamical picture forms the basis of the current understanding of many strong-field multiphoton processes.

72.3.1 Nonperturbative Multiphoton Ionization

The breakdown of perturbation theory for nth-order multiphoton processes occurs when the higher-order correction terms become comparable to the nth-order term. Assuming that the dipole strength is $\sim ea_0$, where a_0 is the Bohr radius, and the detuning is $\delta \sim \omega$, the ratio of an $(n+2)$-order contribution to the nth-order term from Eq. (72.6) is [1]

$$R_{n+2,n} \simeq \left(\frac{2\pi\alpha\phi\omega}{e^2}\right)\left(\frac{ea_0}{\omega}\right)^2 = \left(\frac{I}{I_\gamma}\right)\left(\frac{\omega_a}{2\omega}\right)^2 , \quad (72.16)$$

where $\hbar\omega_a \simeq 27.2114$ eV is the atomic unit of energy e^2/a_0, and $I_\gamma \simeq 3.50945 \times 10^{16}$ W/cm^2 is the intensity corresponding to an atomic unit of field strength, given by $E_a = \alpha c(m/a_0^3)^{1/2} \simeq 5.1422 \times 10^9$ V/cm. The atomic unit of intensity itself is defined by

$$I_a = \phi_a\hbar\omega_a = \alpha c E_a^2 , \quad (72.17)$$

which is $6.436414(4) \times 10^{15}$ W/cm^2. Thus $I_\gamma = I_a/(8\pi\alpha)$. For photon energies of 1 eV, $R_{n+2,n}$ becomes unity for $I \sim 10^{14}$ W/cm^2. Because of the large number of $(n+2)$-order terms, perturbation theory actually fails for $I > 10^{13}$ W/cm^2. Above this *critical intensity*, nonresonant n-photon ionization ceases to scale with the ϕ^n dependence predicted by perturbation theory.

72.3.2 Tunneling Ionization

At sufficiently high intensity and low frequency, a tunneling mechanism changes the character of the ionization process. For lasers in the ir or optical range, a strongly bound electron can respond to the instantaneous laser field since the oscillating electric field varies slowly on the time scale of the electron. The Coulomb attraction of the ion core combines with the laser electric field to form an oscillating barrier through which the electron can escape by tunneling, if the amplitude of the laser field is large enough. The dc rate for this process is $e\mathcal{E}/\sqrt{2mI_P}$, where I_P is the ionization potential of the electron. When this rate is comparable to the laser frequency, tunneling becomes the most probable ionization mechanism [29]. The ratio of the incident laser frequency to the tunneling rate is called the *Keldysh parameter*, and is given by

$$\gamma = \sqrt{I_P/2U_P} , \quad (72.18)$$

which is less than unity when tunneling dominates and larger than unity when multiphoton ionization dominates.

72.3.3 Multiple Ionization

Excitation and ionization dynamics are dominated by single electron transitions in the strong field regime. Although atoms can lose several electrons during a single pulse, the electrons are released sequentially. There is no convincing evidence of significant collective excitation in atoms in strong fields, even though it has been extensively sought. Once one electron is excited in an atom, the remaining electrons have much higher binding energies. As a result, the laser field is unable to affect them significantly until it reaches much higher intensity. By that time the first electron has been emitted.

Simultaneous ejection of two electrons occurs as a minor channel (<1 %) in strong field multiple ionization. Although it is possible that doubly excited states of atoms could assist in the double ionization, in the helium and neon cases studied, these states are unlikely to be contributors [30].

72.3.4 Above Threshold Ionization

In strong optical and ir laser fields, electrons can gain more than the minimum amount of energy required for ionization. Rather than forming a single peak, the emitted electron energy spectrum contains a series of peaks separated by the photon energy. This is called *above threshold ionization*, or ATI [31, 32]. The peaks appear at the energies

$$E_s = (n + s)\hbar\omega - I_P , \quad (72.19)$$

where n is the minimum number of photons needed to exceed I_P, and $s = 0, 1, \ldots$ is called the number of excess photons or above threshold photons carried by the electron. Calculations in the perturbative regime for ATI are given for hydrogen in Ref. [9].

Peak Shifting. As the intensity approaches the nonperturbative regime, the ac Stark shift of the atomic states begins to play a significant role in the structure of the ATI spectrum. The first effect is a shift of the ionization potential, given roughly by the ponderomotive energy U_P. Additional photons may then be required in order to free the electron from the atom; i.e., enough to exceed $I_P + U_P$. If the emitted electron escapes from the focal volume while the laser is still on, it is accelerated by the gradient of the field. The quiver motion is converted into radial motion, increasing the kinetic energy by U_P and exactly canceling the shift of the continuum. The electron energies are still given by Eq. (72.19). However, when U_P exceeds the photon energy, the lowest ATI peaks disappear from the spectrum. In this long pulse limit, no electron is observed with energy less than U_P. This is called *peak shifting* in that the dominant peak in the ATI spectrum moves to higher order as the intensity increases.

ATI Resonance Substructure. If the laser pulse is short enough (<1 ps for the typical laser focus), the field turns off before the electron can escape from the focal volume. Then the quiver energy is returned to the field and the ATI spectrum becomes much more complicated. The observed electron energy corresponds to the energy

$$E_{\mathrm{s}}(\text{shortpulse}) = (n+s)\hbar\omega - (I_{\mathrm{P}} + U_{\mathrm{p}}). \quad (72.20)$$

relative to the *shifted* ionization potential. Electrons from different regions of the focal volume are thus emitted with different ponderomotive shifts, introducing substructure in the spectrum which can be directly associated with ac Stark-shifted resonances [31].

ATI in Circular Polarization. The above discussion is appropriate for the case of linear polarization where the excited states of the atom can play a significant role in the excitation. In a circularly polarized field, the orbital angular momentum L must increase one unit with each photon absorbed so that multiphoton ionization is allowed only to states which have high L, and hence a large centrifugal barrier. The lower energy scattering states then cannot penetrate into the vicinity of the initial state. Thus the ATI spectrum in circular polarization peaks at high energy and is very small near threshold.

72.3.5 High Harmonic Generation

High-order harmonic generation (HG) in noble gases is a rapidly developing field of laser physics [33]. When an SS pulse interacts with an atomic gas, the atoms respond in a nonlinear way, emitting coherent radiation at frequencies that are integral multiples of the laser frequency. Due to the inversion symmetry of the atom, only odd harmonics of the fundamental are emitted. In the high intensity ($>10^{13}$ W/cm^2), low frequency regime, the harmonic strengths fall off for the first few orders, followed by a broad plateau of nearly constant conversion efficiency, and then a rather sharp cut-off [33]. The plateaux extend to well beyond the hundredth order of the 800–1000 nm incident wavelengths, using the light noble gases as the active medium. There has also been experimental evidence of HG from ions. Harmonic generation provides a source of very bright, short-pulse, coherent xuv radiation which can have several advantages over the other known sources, such as the synchrotron.

Plateau and Cut-off. A recently developed two-step model [27, 28], which combines quantum and classical aspects of laser-atom physics, accounts for many strong field phenomena. In this model, the electron first tunnels [34] from the ground state of the atom through the barrier formed by the Coulomb potential and the laser field. Its subsequent motion can be treated classically, and primarily consists of oscillatory motion in

phase with the laser field. If the electron returns to the vicinity of the nucleus with kinetic energy $\mathcal{T}$, it may recombine into the ground state with the emission of a photon of energy $(2n+1)\hbar\omega \leq \mathcal{T} + I_{\mathrm{P}}$, where n is an integer. The maximum kinetic energy of the returning electron turns out to be $\mathcal{T} \simeq 3.2 U_{\mathrm{p}}$, resulting in a cut-off in the harmonic spectrum at the harmonic of order

$$N_{\max} \simeq (I_{\mathrm{P}} + 3.2 U_{\mathrm{p}})/\hbar\omega. \quad (72.21)$$

Theoretical Methods. Calculation of harmonic strengths requires the evaluation of the time-dependent dipole moment of the atom,

$$\mathbf{d}(t) = \langle \Psi(t)|e\mathbf{r}|\Psi(t)\rangle. \quad (72.22)$$

The strength of harmonics emitted by a single atom are then related to Fourier components of $\mathbf{d}(t)$, or more precisely, its second time derivative, $\ddot{\mathbf{d}}(t)$.

The induced dipole moment $\mathbf{d}(t)$ can be directly evaluated from the numerical [35] or Floquet [21] solutions of the TDSE. Good agreement with numerical and experimental data is also obtained using a strong field approximation discussed below and a *Landau-Dyhne formula*. This approach can be considered to be a quantum mechanical implementation of the two-step model [36].

Propagation and Phase Matching Effects. A single atom response is not sufficient to determine the macroscopic response of the atomic medium. Because different atoms interact with different parts of the focused laser beam, they feel different peak field intensities and phases (which actually undergo a rapid π shift close to the focus). The total harmonic signal results from coherently adding contributions from single atoms, accounting for propagation and interference effects. The latter effects can wipe out the signal completely unless a constructive *phase matching* takes place.

The propagation and phase matching effects in the strong field regime [37] can be studied by solving the Maxwell's equations for a given harmonic component of the electric field $\mathcal{E}_M(\mathbf{r})$ [*cf.* Eq. (66.23)],

$$\nabla^2 \mathcal{E}_M(\mathbf{r}) + n_M(\mathbf{r})\mathcal{E}_M(\mathbf{r}) = -\frac{1}{\epsilon_0}\left(\frac{M\omega}{c}\right)^2 \mathcal{P}_M(\mathbf{r}), \quad (72.23)$$

where $n_M(\mathbf{r})$ is the refractive index of the medium (which depends on atomic, electronic and ionic dipole polarizabilities), while $\mathcal{P}_M(\mathbf{r})$ is the polarization induced by the fundamental field only. It can be expressed as

$$\mathcal{P}_M(\mathbf{r}) \propto N(\mathbf{r})d_M(\mathbf{r})\exp(-iM\Delta(\mathbf{r})), \quad (72.24)$$

where $N(\mathbf{r})$ is the atomic density, $d_M(\mathbf{r})$ is the Mth Fourier component of the induced dipole moment, and $\Delta(\mathbf{r})$ is a phase shift coming from the phase dependence of the fundamental beam due to focusing. All of these quantities may have a slow time dependence, reflecting the temporal envelope of the laser pulse.

Phase matching is most efficient in the forward direction. In general, the strength and spatial properties of an harmonic depend in a very complex way on the focal parameters, the medium length and the coherence length of a given harmonic. Propagation and phase matching effects can lead to a shift of the location of the cut-off in the harmonic spectrum [38].

Harmonic Generation by Elliptically Polarized Fields. The two-step model implies that for harmonic emission it is necessary that the tunneling electrons return to the nucleus and recombine into their initial state. According to classical mechanics, there are many trajectories in a linearly polarized field that involve one or more returns to the origin. However, there are practically no such trajectories in elliptically polarized fields. As a result, the two-step model predicts a strong decrease of the harmonic strengths as a function of the laser ellipticity. This prediction has been confirmed experimentally [39].

72.3.6 Stabilization of Atoms in Intense Laser Fields

It has been argued [40] that in very intense laser fields of high frequency, atoms undergo dynamical stabilization and do not ionize. The stabilization effect can be explained by gauge transforming the TDSE (72.1) to the Kramers-Henneberger (K-H) frame; i.e., a noninertial oscillating frame which follows the motion of the free electron in the laser field. The K-H transformation consists of replacing $\mathbf{r} \rightarrow \mathbf{r} + \boldsymbol{\alpha}(t)$, where, for the linearly polarized monochromatic laser field, $\boldsymbol{\alpha}(t) = \hat{\mathbf{x}} \alpha_0 \cos(\omega t - \varphi)$; $\alpha_0 = e\mathcal{E}/(m\omega^2)$ is the *excursion amplitude* of a free electron, and $\hat{\mathbf{x}}$ is the polarization direction. The TDSE in the K-H frame is

$$i\hbar \frac{\partial \Psi(\mathbf{r}, t)}{\partial t} = \left[-\frac{\hbar^2 \nabla^2}{2m} + V(\mathbf{r} + \boldsymbol{\alpha}(t)) \right] \Psi(\mathbf{r}, t), \quad (72.25)$$

i.e., it describes the motion of the electron in an oscillatory potential. In the high frequency limit, this potential may be replaced by its time average

$$V_{\text{K-H}}(\mathbf{r}) = \frac{\omega}{2\pi} \int_0^{2\pi/\omega} dt\, V(\mathbf{r} + \boldsymbol{\alpha}(t)), \quad (72.26)$$

and the remaining Fourier components of $V(\mathbf{r} + \boldsymbol{\alpha}(t))$ treated as a perturbation. When α_0 is large, the effective potential (72.26) has two minima close to $\mathbf{r} = \pm\hat{\mathbf{x}}\alpha_0$. The corresponding wave functions of the bound states are centered near these minima, thus exhibiting a *dichotomy*. The ionization rates from the K-H bound states are induced by the higher Fourier components of $V(\mathbf{r}+\boldsymbol{\alpha}(t))$. For large enough α_0, the rates decrease if either the laser intensity increases or the frequency decreases.

Numerical solutions of the TDSE [41] show that stabilization indeed occurs for laser field strengths and frequencies of the order of one atomic unit. More importantly, stabilization is possible even when the laser excitation is not monochromatic, but rather, is produced by a short laser pulse. Physically, free electrons in a monochromatic laser field cannot absorb photons due to the constraints imposed by energy and momentum conservation. Absorption is possible only in the vicinity of a potential, such as the Coulombic attraction of the nucleus. In the case of strong excitation, i.e., when α_0 is much larger than the Bohr radius, the electron spends most of the time very far from the nucleus, and therefore does not absorb energy from the laser beam. Thus stabilization, as viewed from the K-H frame, has a classical analog. Other mechanisms of stabilization based on the quantum mechanical effects of destructive interference between various ionization paths have also been proposed [42].

Due to classical scaling (see Sect. 72.3.8) stabilization is predicted to occur for much lower laser frequencies if the atoms are initially prepared in highly excited states. If additionally, the initial state has a large orbital angular momentum corresponding to classical trajectories that do not approach the nucleus, stabilization is even more easily accomplished. The stabilization of a *5g* Rydberg state of neon has recently been reported [43].

72.3.7 Molecules in Intense Laser Fields

Molecular systems are more complex than atoms because of the additional degrees of freedom resulting from nuclear motion. Even in the presence of a laser field, the electron and nuclear degrees of freedom can be separated by the *Born-Oppenheimer approximation*, and the dynamics of the system can be described in terms of motions on potential energy surfaces. In strong fields, the Born-Oppenheimer states become dressed, or mixed by the field, creating new molecular potentials. Because of avoided crossings between the dressed molecular states, the field induces new potential wells in which the molecules become trapped. These states, known as *laser-induced bound states*, are stable against dissociation, but exist only while the laser field is present [44]. Their existence affects the spectra of photoelectrons, photons, and the fragmentation dynamics. If the field is strong enough, many electrons can be ejected from a molecule before dissociation, producing highly charged, energetic fragments [45]. Such experiments are similar to beam-foil Coulomb explosion studies of molecular structure. However, because of changes from the field-free equilibrium geometries in laser dissociation, the energies of the fragments lie systematically below the corresponding values from Coulomb explosion studies.

72.3.8 Microwave Ionization of Rydberg Atoms

Similar phenomena appear in the ionization of highly excited hydrogen-like (Rydberg) atoms by microwave fields [46], but the dynamical range of the parameters involved is different from the case of tightly bound electrons. Recent developments have greatly extended techniques for the preparation and detection of Rydberg states. Since, according to the *equivalence principle*, highly excited Rydberg states exhibit many classical properties, a classical perspective of ionization yields useful insights (see Sect. 72.4.5).

Classical Scaling. The classical equations of motion for an electron in both a Coulomb field and a monochromatic laser field polarized along the z-axis are invariant with respect to the following scaling transformations:

$$\mathbf{p} \propto n_0^{-1}\tilde{\mathbf{p}}, \qquad \mathbf{r} \propto n_0^2 \tilde{\mathbf{r}},$$
$$t \propto n_0^3 \tilde{t}, \qquad \varphi \propto \tilde{\varphi},$$
$$\omega \propto n_0^{-3}\tilde{\omega}, \qquad \mathcal{E} \propto n_0^{-4}\tilde{\mathcal{E}}. \qquad (72.27)$$

In the scaled units, the Hamiltonian $\tilde{H} = n_0^{-2} H$ of the system becomes

$$\tilde{H} = \frac{\tilde{p}^2}{2m} - \frac{1}{\tilde{r}} + \tilde{z} e\tilde{\mathcal{E}} \cos(\tilde{\omega}\tilde{t} + \tilde{\varphi}), \qquad (72.28)$$

i.e., it depends only on $\tilde{\omega}$ and $\tilde{\mathcal{E}}$. In experiments, the principal quantum number n_0 of the prepared initial state typically ranges from 1 to 100.

Classical scaling extends to the fields of other polarization and to pulsed excitation, provided that the number of cycles in the rise, top and fall of the pulse is kept fixed. This scaling does not hold for a quantum Hamiltonian, unless one also rescales Planck's constant, $\tilde{\hbar} = \hbar/n_0$. In practice, increasing n_0, keeping $\tilde{\mathcal{E}}$ and $\tilde{\omega}$ constant, corresponds to a decrease in the effective $\hbar$ toward the classical limit. In view of this classical scaling, experimental and theoretical results are usually analyzed in terms of the scaled variables. Since the classical dynamics generated by the Hamiltonian (72.28) exhibits *chaotic behavior* in some regimes, the dynamics of the corresponding quantum system is frequently referred to as an example of *quantum chaos* [47].

Regimes of Response. By varying the initial n_0, several regimes of the scaled parameters can be covered. The experimentally measured response of Rydberg atoms in microwave fields can be divided into six categories:

The tunneling regime. For $\tilde{\omega} \leq 0.07$, the response of the system is accurately represented as tunneling through the slowly oscillating potential barrier composed of the Coulomb and microwave potentials.

The low frequency regime. For $0.05 \leq \tilde{\omega} \leq 0.3$, ionization probability exhibits structure (bumps, steps,

changes of slope) as a function of the field strength. The quantum probability curves might be lower or higher than the corresponding classical counterparts calculated with the aid of the phase averaging method (see Sect. 72.4.5).

The semiclassical regime. For $0.1 \leq \tilde{\omega} \leq 1.2$, the ionization probabilities agree well for most frequencies with the results obtained from the classical theory. In particular, the onset of ionization and appearance intensities (i.e., the intensities at which a given degree of ionization is achieved) coincide with the onset of chaos in the classical dynamics. Resonances in the ionization probabilities appear that correspond to the *classical trapping resonances* [47, 48].

The transition region. For $1 \leq \tilde{\omega} \leq 2$, the differences between the quantum and classical results are visible. Quantum ionization probabilities are frequently lower and appearance intensities higher than their classical counterparts.

The high frequency regime. For $\tilde{\omega} \geq 2$, quantum results for ionization probabilities are systematically lower and appearance intensities higher than their classical counterparts. This apparent stability of the quantum system has been attributed to three kinds of effects: quantum localization [48], quantum scars [49], and perhaps to the stabilization of atoms in intense laser fields (see Sect. 72.3.6).

The photoeffect regime. When the scaled frequency becomes greater than the single photon ionization threshold, the system undergoes single photon ionization (the *photoeffect*).

Quantum Localization. The classical dynamics changes as the field increases. Chaotic trajectories start to fill phase space and, as the KAM tori (describing periodic orbits) [47] break down, the motion becomes stochastic, resembling a random walk. This process, in which the mean energy grows linearly in time, is termed *diffusive ionization*. In the quantum theory, diffusion corresponds to a random walk over a ladder of suitably chosen quantum levels. However, both diagonal and off-diagonal elements of the evolution operator, which describe quantum mechanical amplitudes for transitions between the levels, depend in a quasiperiodic manner on the quantum numbers of the levels involved. Such quasiperiodic behavior is quite analogous to a random one. Electronic wave packets that initially spread in accordance with the classical laws tend to remain localized for longer times due to destructive quantum interference effects. *Quantum localization* is an analog of the Anderson localization of electronic wave functions propagating in random media [48].

Quantum Scars. Even in the fully chaotic regime, classical phase space contains periodic, though unstable, orbits. Nevertheless, quantum mechanical wave function amplitudes can become localized around these unstable orbits, resulting in what are called *quantum scars*. The increased stability of the hydrogen atom at $\tilde{\omega} \simeq 1.3$ has been in fact attributed [49] to the effects of quantum scars. These effects are very sensitive to fluctuations in the driving laser field. Control of the laser noise therefore provides a powerful spectroscopic tool to study such quantal phenomena [50]. Using this tool, it has recently become possible to demonstrate the effects of quantum scars in the intermediate regime of the scaled frequencies (less than but close to 1).

72.4 STRONG-FIELD CALCULATIONAL TECHNIQUES

The SS pulse regime requires a nonperturbative solution of the TDSE. Two approaches have been developed: the explicit numerical solution of the TDSE and the Floquet expansion technique. In addition to these, several approximate methods have been proposed.

72.4.1 Floquet Theory

The excitation and ionization dynamics of an atom in a strong laser field can be determined by turning the problem into a time-independent eigenvalue problem [21,51]. From *Floquet's theorem*, the eigenfunctions for a perfectly periodic Hamiltonian of the form

$$H = H_0 + \sum_{N \neq 0} H_N e^{-iN\omega t} \qquad (72.29)$$

can be expressed in the form

$$\Psi(t) = e^{-iXt/\hbar} \sum_N e^{-iN\omega t} \psi_N . \qquad (72.30)$$

Putting this into the time-dependent Schrödinger equation results in an infinite set of coupled *Floquet equations* for the harmonic components ψ_N. In the velocity gauge, the Floquet equations are

$$(X + N\hbar\omega - H_0)\psi_N = V_+\psi_{N-1} + V_-\psi_{N+1} , \qquad (72.31)$$

where, for a vector potential of amplitude $\mathcal{A}$,

$$V_+ = -\frac{e}{2mc}\mathcal{A}\cdot\mathbf{p} , \qquad (72.32)$$

and $V_- = V_+^\dagger$. The equations (72.31) have been solved, after truncation to a manageable number of terms, using many techniques to provide what are called the quasi-energy states of the laser-atom system. The eigenvalues X of these equations are complex, with $\Im(X)$ giving the decay or ionization rate for the system. The generated rates are found to be very accurate as long as the pulse length of the laser field is not too short, at least hundreds of cycles. The eigenfunctions provide the amplitudes

for the photoelectron energy spectra, and the time-dependent dipole of the state can be related to the photo-emission spectrum of the system. Yields for slowly varying pulses can be constructed by combining the results from the individual, fixed-intensity calculations [21].

The Floquet method can be applied for any periodic Hamiltonian. In strong enough fields of high frequency, the Floquet equations can be truncated to a very small set in the K-H frame [40].

72.4.2 Direct Integration of the TDSE

Methods for the direct solution of the time-dependent Schrödinger equation are described in general in Chap. 8 and in [52,53] for multiphoton processes. The wave functions are defined on spatial grids or in terms of an expansion in basis functions. The time evolution is obtained by either explicit or implicit time propagators. All these methods are capable of generating numerically exact results for an atom with a single electron in a short pulsed field for a wide range of pulse shapes, wavelengths and intensities. The solutions are time-dependent wave functions for the electrons which can be analyzed to obtain excitation and ionization rates, photoelectron energies, angular distributions, and photoemission yields. The ability to generate an explicit solution of the TDSE allows the study of arbitrary pulse shapes and provides insight into the excitation dynamics.

For multi-electron atoms, one generally has to limit the calculations to that for a single electron in effective potentials which represent, as well as possible, the influences of the remaining atomic electrons. This approach is called the *single active electron approximation*, and it gives generally accurate results for systems with no low-lying doubly excited states, for example, the noble gases [52]. In these cases, the excitation dynamics are dominated by the sequential promotion of a single electron at a time.

72.4.3 Volkov States

A laser interacting with free a electron superimposes an oscillatory motion on its drift motion in response to the field. The wave function for an electron with drift velocity $\mathbf{v} = \hbar\mathbf{k}/m$ is given by

$$\Psi_V(\mathbf{r}, t) = \exp\left(-\frac{i}{2m\hbar}\int^t \left[\hbar\mathbf{k} - \frac{e}{c}\mathbf{A}(t')\right]^2 dt'\right)$$
$$\times e^{i[\mathbf{k} - e\mathbf{A}(t)/\hbar c]\cdot\mathbf{r}} , \qquad (72.33)$$

where $\mathbf{A}(t) = -c\int^t \mathbf{E}(t')dt'$ is the vector potential of the field. Ψ_V is called a *Volkov state*. In a linearly polarized field, the electron oscillates along the direction of polarization with an amplitude $\alpha_0 = e\hbar\mathcal{A}/(mc\omega)$. In the strong field regime, this amplitude can greatly exceed the size of a bound state orbital. Volkov states provide a useful tool that can be applied in various strong field approximations discussed in the next Section.

72.4.4 Strong Field Approximations

There have been several attempts to solve the TDSE in the strong field limit using approximate, but analytic methods. Such *strong field approximations* (SFA) typically neglect all the bound states of the atom except for the initial state. In the tunneling regime ($\gamma < 1$), and a quasistatic limit ($\omega \to 0$), one can use a theory [34] in which the ionization occurs due to the tunneling through the Coulomb barrier distorted by the electric field of the laser. The wave function is constructed as a combination of a bare initial wave function of the electron (close to the nucleus) and a wave function describing a motion of the electron in a quasistatic electric field (far from the nucleus). In an second approach [29], the elements of the scattering matrix $\hat{S}$ are calculated assuming that initially the electronic wave function corresponds to a bare bound state. On the other hand, the final, continuum states of the electron are described by dressed wave functions that account for the free motion of the electron in the laser field. In the simplest case, such dressed states are Volkov states (72.33). Alternatively, the time-reversed $\hat{S}$-Matrix is obtained by dressing the initial state and using field-free scattering states for the final state.

Yet another method consists of expanding the electronic continuum-continuum dipole matrix elements in terms corresponding to matrix elements for free electrons plus corrections due to the potential [36]. In the latter version of SFA, the amplitude of the electronic wave function $b(\mathbf{p})$ corresponding to outgoing momentum $\mathbf{p}$ is given by

$$b(\mathbf{p}) = i \int_0^{t_F} dt \, \mathbf{d}(\mathbf{p} - e\mathbf{A}(t)/c) \cdot \mathbf{E}(t) \exp[-iS(t_F, t)/\hbar].$$

(72.34)

Here $\mathbf{d}(\mathbf{p} - e\mathbf{A}(t)/c)$ denotes the dipole matrix element for the transition from the initial bound state to the continuum state in which the electron has the kinetic momentum $\mathbf{p} - e\mathbf{A}(t)/c$, t_F is the switch-off time of the laser pulse, and

$$S(t_F, t) = \int_t^{t_F} dt' \left[\frac{(\mathbf{p} - e\mathbf{A}(t')/c)^2}{2m} + I_{\mathrm{P}} \right]$$

(72.35)

is a quasiclassical action for an electron which is born in the continuum at t and propagates freely in the laser field. The form of the expression (72.34) is generic to the SFA.

72.4.5 Phase Space Averaging Method

The methods of classical mechanics are particularly useful in describing the microwave excitation of highly excited (Rydberg) atoms [46] (see Sect. 72.3.8), but have also been applied to describe high harmonic generation, stabilization of atoms in super intense fields and two electron ionization [54].

The classical *phase space averaging method* [55] solves Newton's equations of motion

$$\dot{\mathbf{r}} = \mathbf{p}/m,$$ (72.36)
$$\dot{\mathbf{p}} = -\nabla V(\mathbf{r}) - e\mathbf{E}(t),$$ (72.37)

for the electron interacting with the ion core and the laser field. A distribution of initial conditions in phase space is chosen to mimic the initial quantum mechanical state of the system, and a sample of classical trajectories generated. Quantum mechanical averages of physical observables are then identified with ensemble averages of those observables over the initial distribution. Since the dynamics of multiphoton processes is very complex, the neglected phases in this approach generally cause negligible errors and the results can be in quite good agreement with quantum calculations. Additionally, an examination of the trajectories provides details of the excitation dynamics which are often difficult to extract from a complex time-dependent wave function.

Acknowledgments

This work was supported in part by the U.S. DOE under contract number W-7405-ENG-48.

REFERENCES

1. F. H. M. Faisal, *Theory of Multiphoton Processes* (Plenum, New York, 1987).
2. *Atoms in Intense Laser Fields*, edited by M. Gavrila (Academic Press, San Diego, 1992).
3. M. H. Mittleman, *Theory of Laser–Atom Interactions* (Plenum, New York, 1993), 2nd ed.
4. *Proceedings of the International Conference of Multiphoton Processes III, 1984*, edited by P. Lambropoulos and S. J. Smith (Springer-Verlag, Berlin, 1984), Vol. 2.
5. *Proceedings of the International Conference on Multiphoton Processes IV, 1988*, edited by S. J. Smith and P. L. Knight (Cambridge University Press, Cambridge, 1988), Vol. 8.
6. *Proceedings of the International Conference on Multiphoton Processes V, 1991*, edited by G. Mainfray and P. Agostini (Centre d'Etudes de Saclay, Saclay, 1991).
7. *Super-Intense Laser–Atom Physics*, edited by B. Piraux, A. L'Huillier, and K. Rzc@ażewski (Plenum Press, New York, 1993), Vol. 316.
8. P. Lambropoulos, Adv. At. Mol. Phys. **12**, 87-158 (1976).
9. Y. Gontier and M. Trahin, J. Phys. B **13**, 4383 (1980).

10. X. Tang and P. Lambropoulos, Phys. Rev. Lett. **58**, 108 (1987).

11. K. Rzc@ażewski and J. H. Eberly, Phys. Rev. Lett. **47**, 408 (1981); P. Lambropoulos and P. Zoller, Phys. Rev. A **24**, 379 (1981).

12. P. L. Knight, M. A. Lauder, and B. J. Dalton, Phys. Rep. **190**, 1 (1990).

13. H. Risken, in *The Fokker–Planck Equation: Methods of Solution and Applications*, edited by H. Haken (Springer-Verlag, Berlin, 1984).

14. J. H. Eberly, in *Laser Spectroscopy*, edited by H. Walther and K. W. Rothe (Springer-Verlag, Berlin, 1979), p. 80.

15. P. Zoller in Ref. [4] pp. 68–75.

16. D. S. Elliot in Ref. [4] pp. 76–81; D. S. Elliot *et al.*, Phys. Rev. A **32**, 887 (1985).

17. A. I. Burshtein *et al.*, Sov. Phys. JETP **21**, 597 (1965); *ibid.* **22**, 939 (1996).

18. B. W. Shore, *The Theory of Coherent Atomic Excitation* (J. Wiley and Sons, New York, 1990).

19. L. A. Lompré, G. Mainfray, C. Manus, and J. P. Marinier, J. Phys. B **14**, 4307 (1981).

20. C. Chen and D. S. Elliot, Phys. Rev. Lett. **65**, 1737 (1990).

21. R. M. Potvliege and R. Shakeshaft in Ref. [2] pp. 373–434.

22. O. Faucher *et al.*, Phys. Rev. Lett. **70**, 3004 (1993).

23. A. Imamoğlu, J. E. Field, and S. E. Harris, Phys. Rev. Lett. **66**, 1154 (1991).

24. S. E. Harris, Phys. Rev. Lett. **70**, 552 (1993).

25. G. S. Hurst, M. G. Payne, S. D. Kramer, and J. P. Young, Rev. Mod. Phys. **52**, 767 (1979).

26. M. D. Perry and G. Mourou, Science **264**, 917 (1994).

27. K. C. Kulander, K. J. Schafer, and J. L. Krause in Ref. [7] pp. 95–110.

28. P. B. Corkum, Phys. Rev. Lett. **73**, 1995 (1993).

29. L. V. Keldysh, Sov. Phys. JETP **20**, 1307 (1965); H. R. Reiss, Phys. Rev. A **22**, 1786 (1980); F. Faisal, J. Phys. B **6**, L312 (1973).

30. D. Fittinghoff, P. R. Bolton, B. Chang, and K. C. Kulander, Phys. Rev. A **49**, 2174 (1994).

31. H. G. Muller, P. Agostini, and G. Petite, in Ref. [2] pp. 1–42; R. R. Freeman *et al.*, in Ref. [2] pp. 43–65.

32. J. H. Eberly, J. Javanainen, and K. Rzc@ażewski, Phys. Rep. **204**, 331 (1991).

33. A. L'Huillier, L.-A. Lompré, G. Mainfray, and C. Manus in Ref. [2], pp. 139–205; for recent references see for instance Y. Liang, M. V. Ammosov, S. L. Chin, J. Phys. B **27**, 1269 (1994).

34. M. V. Ammosov, N. B. Delone, and V. P. Kraïnov Sov.

Phys. JETP **64**, 1191 (1986).

35. J. L. Krause, K. J. Schafer, and K. C. Kulander, Phys. Rev. A **45**, 4998 (1992).

36. M. Lewenstein, Ph. Balcou, M. Yu. Ivanov, A. L'Huillier, and P. Corkum, Phys. Rev. A **49**, 2117 (1994).

37. A. L'Huillier, K. J. Schafer, and K. C. Kulander, J. Phys. B **24**, 3315 (1991).

38. A. L'Huillier, M. Lewenstein, P. Salières, Ph. Balcou, M. Yu. Ivanov, J. Larsson, and C. G. Wahlström, Phys. Rev. A **48**, 4091 (1993).

39. K. S. Budil, P. Salières, A. L'Huillier, T. Ditmire, and M. D. Perry, Phys. Rev. A **48**, R3437 (1993).

40. M. Gavrila in Ref. [2] pp. 435–510 and references therein.

41. Q. Su, J. H. Eberly, and J. Javanainen, Phys. Rev. Lett. **64**, 862 (1990); K. C. Kulander, K. J. Schafer, and J. L. Krause, Phys. Rev. Lett. **66**, 2601 (1991).

42. M. V. Fedorov in Ref. [7] pp. 245–259.

43. M. P. de Boer, J. H. Hoogenraad, R. B. Vrijen, L. D. Noordam, and H. Muller, Phys. Rev. Lett. **71**, 3263 (1993).

44. *Molecules in Laser Fields*, edited by A. D. Bandrauk (M. Dekker, New York, 1994).

45. D. Normand and C. Cornaggia in Ref. [7] pp. 351–362.

46. P. M. Koch in Ref. [6], pp. 305–316; P. M. Koch, in *Proceedings of the Eigth South African Summer School in Physics, 1993* (Springer-Verlag, Berlin, 1993).

47. M. C. Gutzwiller, *Chaos in Classical and Quantum Mechanics* (Springer-Verlag, Berlin, 1990); F. Haake, *Quantum Signatures of Chaos* (Springer-Verlag, Berlin, 1991); G. Casati, B. Chirikov, D. L. Shepelyansky, and I. Guarnieri, Phys. Rep. **154**, 77 (1987).

48. G. Casati, I. Guarneri, and D. L. Shepelyansky, Physica A **163**, 205 (1990) and Refs. therein.

49. R. V. Jensen, M. M. Sanders, M. Saraceno, and B. Sundaram, Phys. Rev. Lett. **63**, 2771 (1989).

50. L. Sirko, M. R. W. Bellermann, A. Haffmans, P. M. Koch, and D. Richards, Phys. Rev. Lett. **71**, 2895 (1993).

51. S. I. Chu, Adv. Chem. Phys. **73**, 739 (1989).

52. K. C. Kulander, K. J. Schafer, and J. L. Krause in Ref. [2] pp. 247–300.

53. K. Burnett, V. C. Reed, and P. L. Knight, J. Phys. B **26**, 561 (1993).

54. M. Lewenstein, K. Rzc@ażewski, and P. Salières in Ref. [7] pp. 425–434; P. B. Lerner, K. LaGattuta, and J. S. Cohen, *ibid.* pp. 413–424; V. Vèniard, A. Maquet, and T. Mènis, *ibid.* pp. 225–232.

55. J. G. Leopold and I. C. Percival, J. Phys. B **12**, 709 (1979).

73

Cooling and Trapping

Juha Javanainen

Department of Physics, University of Connecticut, Storrs, Connecticut 06269-3046

73.1 INTRODUCTION

Interactions of light with an atomic particle are accompanied by exchange of momentum between the electromagnetic field and the atom. Narrow-band resonance radiation from tunable lasers enhances the ensuing mechanical effects of light to the extent that it is now literally possible to stop atoms emanating from a thermal gas, and to trap atoms with light.

References [1] and [2] are two widely accessible sources on laser cooling and trapping of the traditional two-state model atom. A number of articles on cooling and trapping of atoms with the inclusion of angular momentum degeneracy are contained in [3]. While the development based on the atom-field dressed states is followed sparingly in the present Chapter, an authoritative survey of this approach is given in [4].

Reviews on traps for charged particles include [5] and [6]. These articles, as well as [2], also discuss cooling of trapped particles. Ref. [7] contains entries to more recent

literature. In particular, the introductory paper [8] is a useful summary of trapped-ion physics.

Additional references are given for the newest developments that cannot be traced from the sources cited so far. The brief discussions of applications in Sect. 73.5 are augmented with further references, often including one of the most recent articles (at this writing) on each topic. All references are intended solely as entries to the literature. No assignment of credit or priority is implied.

73.1.1 Notation

In this Chapter, the lower and upper states of an optical transition are denoted by the respective labels g and e, for "ground" and "excited". The notation $J_g \rightarrow J_e$ stands for a transition in which the lower and upper levels have the angular momentum degeneracies $2J_g + 1$ and $2J_e + 1$. The resonance frequency of the transition is ω_0.

The detuning of the driving monochromatic light of frequency ω from the atomic resonance is $\Delta = \omega_0 - \omega$.

Γ is the spontaneous decay rate. Spontaneous emission is taken to be the sole mechanism of line broadening, so that the HWHM linewidth of the transition is $\gamma = \Gamma/2$. The Rabi frequency is $\Omega = \mathcal{D}\mathcal{E}/\hbar$, where $\mathcal{D}$ is the reduced dipole moment matrix element that would apply to a transition with unit Clebsch-Gordan coefficient, and $\mathcal{E}$ is the electric field amplitude of the laser. The corresponding intensity scale is the saturation intensity

$$I_{\rm s} = 4\pi^2 \hbar c \Gamma/(3\lambda^3)\,, \qquad (73.1)$$

defined in such a way that the light intensity I satisfies

$$\Omega = \Gamma \quad \Rightarrow \quad I = I_{\rm s}\,. \qquad (73.2)$$

If multiple laser beams are explicitly mentioned, laser intensity and Rabi frequency are quoted for each of the equally intense beams.

The momentum of a photon with the wave vector $\mathbf{k}$ is $\hbar\mathbf{k}$. The *recoil velocity*

$$v_{\rm r} = \hbar k/m \qquad (73.3)$$

equals the change of the velocity of an atom of mass m when it absorbs a photon with wave number $k = 2\pi/\lambda$. The kinetic energy of an atom with velocity $v_{\rm r}$ and the corresponding frequency,

$$R = \tfrac{1}{2}mv_{\rm r}^2\,, \qquad \varepsilon_{\rm r} = R/\hbar \qquad (73.4)$$

are referred to as *recoil energy* and *recoil frequency*. Two temperatures, the *Doppler limit* $T_{\rm D}$ and the *recoil limit* $T_{\rm r}$ are often cited in laser cooling. They are

$$T_{\rm D} = \hbar\gamma/k_{\rm B}\,, \qquad T_{\rm r} = R/k_{\rm B}\,, \qquad (73.5)$$

where $k_{\rm B}$ is the Boltzmann constant.

Table 73.1 lists numerical values of pertinent parameters for laser cooling and trapping using the D_2 line of the alkalis and hydrogen.

73.2 CONTROL OF ATOMIC MOTION BY LIGHT

73.2.1 General Theory

Hamiltonian. The mechanical effects of light may be derived from the Hamiltonian

$$\hat{H} = \hat{H}_{\rm A} + \hat{H}_{\rm cm} + \hat{H}_{\rm F} - \hat{\mathbf{d}}\cdot\hat{\mathbf{E}}(\hat{\mathbf{r}})\,, \qquad (73.6)$$

where $\hat{H}_{\rm A}$, $\hat{H}_{\rm cm}$ and $\hat{H}_{\rm F}$ are the Hamiltonians for the internal degrees of freedom of the atom, center-of-mass (c.m.) motion of the atom, and free electromagnetic field. The quantized electric field is $\hat{\mathbf{E}}(\hat{\mathbf{r}})$, where $\hat{\mathbf{r}}$ is the c.m. position operator. The dipole operator $\hat{\mathbf{d}}$ acts on the internal degrees of freedom of the atom. Since the $\hat{\mathbf{d}}\cdot\hat{\mathbf{E}}(\hat{\mathbf{r}})$ term couples all degrees of freedom, the possibility of

influencing c.m. motion by light immediately follows. The inclusion of the quantized c.m. motion is the essential ingredient not contained in traditional theories of light-matter interactions. For an atom with mass m trapped in a possibly anisotropic harmonic oscillator potential with frequencies ν_i ($i = x, y, z$), the c.m. Hamiltonian is

$$H_{\rm cm} = \frac{\hat{\mathbf{p}}^2}{2m} + \sum_{i=x,y,z} \frac{m\nu_i^2\hat{\mathbf{r}}_i^2}{2}\,, \qquad (73.7)$$

where $\hat{\mathbf{p}}$ is the c.m. momentum operator. For a free atom, $\nu_i = 0$.

Master Equation. With the aid of Markov and Born approximations, the vacuum modes of the electromagnetic field may be eliminated as described in Sect. 75.8. This gives a master equation for the reduced density operator $\hat{\rho}$ that contains the internal and c.m. degrees of freedom of the atom. Relaxation terms proportional to Γ and γ are all that is left of the quantized fields. The term *generalized Bloch equations* is also used for the master equation, as the inclusion of the c.m. motion is a generalization beyond the ordinary optical Bloch equations (Sect. 66.4.5).

Consider as an example a two-state atom in a traveling wave of light with the electric field strength

$$E(\mathbf{r}, t) = \tfrac{1}{2}\mathcal{E}e^{i(\mathbf{k}\cdot\mathbf{r}-\omega t)} + \text{c.c.}\,. \qquad (73.8)$$

Master equations are conveniently written using *Wigner functions* to represent the c.m. motion. Given the internal-state labels i and $j = g$ or e, and the three-dimensional variables $\mathbf{r}$, $\mathbf{p}$, the Wigner functions are defined as

$$\rho_{ij}(\mathbf{r}, \mathbf{p}) = \frac{1}{(2\pi\hbar)^3} \int d^3u\, e^{i\mathbf{u}\cdot\mathbf{p}/\hbar} \langle \mathbf{r} - \tfrac{1}{2}\mathbf{u}|\langle i|\hat{\rho}|j\rangle|\mathbf{r} + \tfrac{1}{2}\mathbf{u}\rangle\,. \qquad (73.9)$$

The Wigner function is one of the quantum mechanical quasiprobability distributions (Sect. 75.6), with the special property that the marginal distribution of $\mathbf{r}$ obtained by integrating over $\mathbf{p}$ coincides with the correct quantum probability distribution for position, and *vice versa* with $\mathbf{r}$ and $\mathbf{p}$ interchanged. In the rotating wave approximation (Sect. 66.4.2), the master equations are

$$\frac{d}{dt}\rho_{ee}(\mathbf{p}) = -\Gamma\rho_{ee}(\mathbf{p}) + i\frac{\Omega}{2}\big[e^{i\mathbf{k}\cdot\mathbf{r}}\hat{\rho}_{ge}(\mathbf{p}-\tfrac{1}{2}\hbar\mathbf{k})$$
$$- e^{-i\mathbf{k}\cdot\mathbf{r}}\hat{\rho}_{eg}(\mathbf{p}-\tfrac{1}{2}\hbar\mathbf{k})\big]\,, \qquad (73.10)$$

$$\frac{d}{dt}\rho_{gg}(\mathbf{p}) = \Gamma\int d^2n\, W(\hat{\mathbf{n}})\,\rho_{ee}(\mathbf{p}+\hbar k\hat{\mathbf{n}}) - \frac{i\Omega}{2}\big[e^{i\mathbf{k}\cdot\mathbf{r}}\hat{\rho}_{ge}$$
$$\times(\mathbf{p}+\tfrac{1}{2}\hbar\mathbf{k}) - e^{-i\mathbf{k}\cdot\mathbf{r}}\hat{\rho}_{eg}(\mathbf{p}+\tfrac{1}{2}\hbar\mathbf{k})\big]\,, \qquad (73.11)$$

$$\frac{d}{dt}\hat{\rho}_{ge}(\mathbf{p}) = -(\gamma - i\Delta)\hat{\rho}_{ge}(\mathbf{p}) - \frac{i\Omega}{2}\big[e^{-i\mathbf{k}\cdot\mathbf{r}}\rho_{gg}(\mathbf{p}-\tfrac{1}{2}\hbar\mathbf{k})$$
$$- e^{i\mathbf{k}\cdot\mathbf{r}}\rho_{ee}(\mathbf{p}+\tfrac{1}{2}\hbar\mathbf{k})\big]\,, \qquad (73.12)$$

$$\frac{d}{dt}\hat{\rho}_{eg}(\mathbf{p}) = -(\gamma + i\Delta)\rho_{eg}(\mathbf{p}) + \frac{i\Omega}{2}\big[e^{i\mathbf{k}\cdot\mathbf{r}}\rho_{gg}(\mathbf{p}-\tfrac{1}{2}\hbar\mathbf{k})$$
$$- e^{-i\mathbf{k}\cdot\mathbf{r}}\rho_{ee}(\mathbf{p}+\tfrac{1}{2}\hbar\mathbf{k})\big]\,. \qquad (73.13)$$

Table 73.1. Laser cooling parameters for the lowest $S_{1/2} - P_{3/2}$ transition of hydrogen and most alkalis (the D_2 line). Also shown are the nuclear spin I and the ground state hyperfine splitting $\Delta\nu_{\text{hfs}}$. Γ is typically known to within a few per cent, so these values of Γ, T_D and I_s may not all be accurate to the full displayed precision.

Parameter	^{1}H	^{6}Li	^{7}Li	^{23}Na	^{39}K	^{85}Rb	^{87}Rb	^{133}Cs	Units
m	1.67	9.99	11.7	38.2	64.7	141	144	221	10^{-27} kg
λ	121.6	670.8	670.8	589.0	766.5	780.0	780.0	852.1	nm
v_r	326	9.89	8.48	2.95	1.34	0.602	0.589	0.352	cm/s
Γ	98.9	5.92	5.92	9.90	6.16	5.89	5.89	5.22	2π MHz
T_D	2390	142	142	238	148	141	141	125	μK
ε_r	13396	73.7	63.2	25.0	8.72	3.86	3.77	2.07	2π kHz
T_r	643	3.54	3.03	1.20	0.418	0.185	0.181	0.0992	μK
I_s	14509	5.13	5.13	12.7	3.58	3.24	3.24	2.21	mW/cm^2
I	1/2	1	3/2	3/2	3/2	5/2	3/2	7/2	
$\Delta\nu_{\text{hfs}}$	1420	228.2	803.5	1772	461.7	3036	6835	9193	2π MHz

Here the convective derivative that describes the motion of the atom in the absence of light is

$$\frac{d}{dt} = \frac{\partial}{\partial t} + \frac{\mathbf{p}}{m} \cdot \frac{\partial}{\partial \mathbf{r}} - \sum_i m\nu_i r_i \frac{\partial}{\partial p_i} ; \qquad (73.14)$$

cf. H_{cm} in Eq. (73.7). $W(\hat{\mathbf{n}})$ is the angular distribution of spontaneous photons, and the integral runs over the unit sphere. Representative expressions for $W(\hat{\mathbf{n}})$ are

$$W(\hat{\mathbf{n}}) = \frac{1}{4\pi}, \quad \frac{3}{8\pi}\left[1 - (\hat{\mathbf{e}}\cdot\hat{\mathbf{n}})^2\right], \quad \frac{3}{16\pi}\left[1 + (\hat{\mathbf{e}}\cdot\hat{\mathbf{n}})^2\right]. \qquad (73.15)$$

These apply, respectively, for isotropic spontaneous emission, for spontaneous emission in a $\Delta m = 0$ transition, and in $\Delta m = \pm 1$ transitions; $\hat{\mathbf{e}}$ stands for the unit vector in the direction of the quantization axis for angular momentum. Only the $\mathbf{p}$ dependence has been denoted explicitly in the Wigner functions, as the recoil effects displayed on the right-hand side of Eqs. (73.10)–(73.13) take place at a fixed position $\mathbf{r}$.

Semiclassical Theory. Suppose that $v_{\text{cm}} \gg v_r$ and $\tau \ll \tau_{\text{cm}}$, where τ and τ_{cm} are the time scales for light-driven changes of the internal state and c.m. motion of the atom. Then the internal degrees of freedom may be eliminated adiabatically from the master equations in favor of the position-momentum distribution for the c.m. motion,

$$f(\mathbf{r}, \mathbf{p}, t) = \sum_i \rho_{ii}(\mathbf{r}, \mathbf{p}, t), \qquad (73.16)$$

where the sum runs over the internal states of the atom. Technically, the recoil velocity v_r is treated as an asymptotically small expansion parameter. The result is the *Fokker-Planck equation* for the c.m. motion

$$\frac{d}{dt} f = -\frac{\partial}{\partial \mathbf{p}} \cdot (\mathbf{F}f) + \sum_{i,j} \frac{\partial^2}{\partial p_i \partial p_j} (D_{ij}f). \qquad (73.17)$$

In this *semiclassical* theory the c.m. motion of the atom is regarded as classical. The atom moves under the optical *force* $\mathbf{F}(\mathbf{r}, \mathbf{p}, t)$, which models the coarse-grained flow of momentum between the electromagnetic field and the atom. $D_{ij}(\mathbf{r}, \mathbf{p}, t)$, with $i, j = x, y, z$, is the *diffusion tensor*. Diffusion is an attempt to model quantum mechanics with a classical stochastic process, including discreteness of recoil kicks, random directions of spontaneous photons, and random timing of optical absorption and emission processes.

A general prescription exists for calculating the force and the diffusion tensor for an arbitrary atomic level scheme and light field [9]. However, the analysis of diffusion amounts to an involved analysis of photon statistics of the scattered light, and here only the force is considered explicitly. Let $\hat{V}(\mathbf{r})$ be the dipole interaction operator coupling the driving field and the internal state for an atom at position $\mathbf{r}$. By assumption, $\hat{V}(\mathbf{r})$ has been rendered slowly varying in time with the aid of a suitable rotating wave approximation. To compute the force, one takes an atom that travels along a hypothetical trajectory unperturbed by light in such a way that at time t it arrives at the phase space point $(\mathbf{r}, \mathbf{p})$, whereupon the density operator of the internal degrees of freedom is $\hat{\varrho}$. The force is then

$$F_i(\mathbf{r}, \mathbf{p}, t) = -\text{Tr}\left[\hat{\varrho} \frac{\partial \hat{V}}{\partial r_i}\right]. \qquad (73.18)$$

Quantum Theory. When either $v_{\text{cm}} \lesssim v_r$ or $\tau \gtrsim \tau_{\text{cm}}$, the full quantum theory of cooling and trapping is needed. Master equations such as (73.10)–(73.13) must then be solved without the assumption that v_r is small. Most practical calculations have been numerical case studies [10]. A truncated basis, e.g. of plane waves, is used to expand the c.m. state. Density matrix equations are solved numerically either directly, or by resorting to quantum Monte-Carlo simulations (Sect. 75.12).

Qualitative Origin of Laser Cooling. Velocity dependent dissipative forces are needed for cooling. They arise because the evolution of the internal state of a moving atom has a finite response time τ. The atom conveys the memory of the field it has sampled over the length $\ell = v\tau$ on its past trajectory. If $\ell \ll \lambda$, a nonequilibrium component proportional to ℓ is present in the density operator of the internal state of the atom. Further interactions with light convert this component into a velocity dependent force of the form

$$F = -m\beta v, \qquad \beta \propto I\tau. \qquad (73.19)$$

If the *damping constant* β is positive, Eq. (73.19) describes exponential damping of the velocity on the time scale β^{-1}. In the contrary case $\ell \gg \lambda$, when the atom travels many wavelengths during the memory time, linear dependence of force on velocity breaks down. The watershed is the *critical velocity* or *velocity capture range*

$$v_c \approx \lambda/\tau. \qquad (73.20)$$

One-Dimensional Considerations. Most specific results cited here are one-dimensional. By default, the propagation direction of light and the direction of vector quantities other than light polarization is $\hat{\mathbf{e}}_x$. The relevant components of position, velocity and momentum are denoted by x, v, and p.

The general one-dimensional Fokker-Planck equation for a particle trapped in a harmonic oscillator potential with a c.m. oscillation frequency ν is

$$\left(\frac{\partial}{\partial t} + \frac{p}{m}\frac{\partial}{\partial x} - m\nu^2 x\frac{\partial}{\partial p} \right) f = -\frac{\partial}{\partial p}(Ff) + \frac{\partial^2}{\partial p^2}(Df).$$

$$(73.21)$$

For the force (73.19) with constant $\beta = \beta_0$ and $D(z, p) = D_0$, the steady state of the Fokker-Planck equation is a thermal distribution of the form

$$f(x, p) = K \exp\left[-\frac{\beta_0 m}{D_0}\left(\frac{p^2}{2m} + \frac{m\nu^2 x^2}{2} \right) \right], \quad (73.22)$$

where K is a normalization coefficient.

Since Wigner functions give correct quantum mechanical marginal distributions for $\mathbf{r}$ and $\mathbf{p}$, expectation values of kinetic and potential energy may be calculated from the distribution function (73.22) as if it were a classical phase space density. For a *free* atom with $\nu = 0$, the temperature is directly proportional to the kinetic energy,

$$T = \frac{D_0}{\beta_0 m k_B}. \qquad (73.23)$$

However, for a *trapped* particle with $\nu \neq 0$ the Fokker-Planck equation may be valid all the way to the quantum mechanical zero-point energy. Then temperature and energy are no longer directly proportional to one another.

For a trapped particle, the safe interpretation of (73.22) is that the total c.m. energy of the particle is

$$E = \frac{D_0}{\beta_0 m}. \qquad (73.24)$$

73.2.2 Two-State Atoms

A *two-state* or *two-level* atom, discussed in detail in Sect. 66.4, stands for a closed (recycling) transition with one lower state and one excited state. In practice, a two-state system is often realized by driving a $J \to J+1$ transition with circularly polarized light. This leads to optical pumping to the states with maximal (or minimal) component of angular momentum along the quantization axis, say, to the transition $m = J \to J+1$.

Two types of force are generally distinguished: *light pressure*, or *scattering force*, or *spontaneous force*, and *dipole*, or *gradient*, or *induced force*. However, the distinction is neither exclusive, nor exhaustive. Here the two types of force are approached by way of examples.

Traveling Waves

Light Pressure. Consider a cycle of absorption and spontaneous emission. In an absorption, the atom receives a photon recoil kick in the direction of the laser beam, while in spontaneous emission the recoil kick has a random direction and zero average. The atom is on the average left with a velocity change equal to v_r. The corresponding force is along $\mathbf{k}$, and is given by

$$F = F_m \frac{\Omega^2/2}{\gamma^2 + \Delta^2(v) + \Omega^2/2}. \qquad (73.25)$$

Here the maximum of light pressure force, a convenient scale for optical forces, is

$$F_m = \tfrac{1}{2} M v_r \Gamma, \qquad (73.26)$$

and

$$\Delta(v) = \Delta + kv \qquad (73.27)$$

is the effective detuning, which includes the Doppler shift experienced by the moving atom.

Diffusion. For a traveling wave, the diffusion coefficient accompanying light pressure is

$$\frac{D}{\hbar^2 k^2 \Gamma} = \frac{(1+\alpha)\Omega^2}{4[\Delta^2(v) + \gamma^2 + \Omega^2/2]} - \frac{[\Delta^2(v) - 3\gamma^2]\Omega^4}{4[\Delta^2(v) + \gamma^2 + \Omega^2/2]^3},$$

$$(73.28)$$

where

$$\alpha = \int d^2 n\, W(\hat{\mathbf{n}})\, (\hat{\mathbf{e}}_x \cdot \hat{\mathbf{n}})^2 \qquad (73.29)$$

depends on $W(\hat{\mathbf{n}})$, see (73.15). Representative values are $\alpha = 1/3$ for isotropic spontaneous emission, and $\alpha = 2/5$ ($\alpha = 3/10$) for spontaneous emission with $\Delta m = 0$ ($\Delta m = \pm 1$) with respect to a quantization axis that is perpendicular (parallel) to the direction $\hat{\mathbf{e}}_x$.

Spontaneously emitted photons cover all of the 4π solid angle, and so do the directions of photon recoil kicks on the atom. Absorption from a light wave traveling in a particular direction leads to *transverse diffusion* also in the orthogonal directions, which is not accounted for by the one-dimensional Eq. (73.28).

Phenomenology in Multimode Fields

Doppler Cooling in Standing Waves. Next take an atom in two counterpropagating plane waves of light. At low intensity, $\Omega \ll \Gamma$, forces of the form (73.25) for the two beams may be added when averaged over a wavelength. For velocities well below the critical velocity

$$v_{\mathrm{c,D}} = \Gamma/k \,, \qquad (73.30)$$

the wavelength average force is of the form of Eq. (73.19),

$$F = -m\bar{\beta}v \,, \qquad \bar{\beta} = \frac{4\Omega^2\gamma\Delta}{(\Delta^2 + \gamma^2)^2}\, \varepsilon_{\mathrm{r}} \,. \qquad (73.31)$$

When light is tuned *below* the atomic resonance ("red detuning" with $\Delta > 0$), exponential damping of the atomic velocity with the time constant $\bar{\beta}^{-1}$ ensues. No matter which way the atom moves, it is always Doppler tuned toward resonance with the light wave that propagates opposite to its velocity, and away from resonance with the light wave that propagates along its velocity. Net momentum transfer therefore opposes the motion of the atom. This is known as *Doppler cooling*.

Optical Molasses. For three pairs of counterpropagating waves in three orthogonal directions, Eq. (73.31) is valid in all coordinate directions, and hence as a vector equation between the force $\mathbf{F}$ and velocity $\mathbf{v}$. For $\Delta > 0$ an atom experiences an isotropic viscous damping force, as if it were moving in a thick liquid. Such a field configuration is dubbed *optical molasses*. Two counterpropagating beams make a one-dimensional optical molasses.

Limit of Doppler Cooling. Under the conditions of Eq. (73.31), the diffusion coefficients for the two counterpropagating beams averaged over a wavelength may be added, and the $v = 0$ form suffices for slow atoms. This yields

$$\frac{\bar{D}(v=0)}{\hbar^2 k^2 \Gamma} = \frac{(1+\alpha)\,\Omega^2}{2(\Delta^2 + \gamma^2)} \,. \qquad (73.32)$$

The random diffusive motion of the atom corresponds to *diffusive heating* that competes with Doppler cooling. In equilibrium, the temperature is

$$T = \frac{\bar{D}(v=0)}{m\bar{\beta}k_{\mathrm{B}}} = \frac{\hbar\gamma}{4k_{\mathrm{B}}}(1+\alpha)\left(\frac{|\Delta|}{\gamma} + \frac{\gamma}{|\Delta|}\right) \,. \qquad (73.33)$$

Equation (73.33) also applies to three-dimensional Doppler cooled molasses, provided one uses $\alpha = 1$ corresponding to added transverse diffusion. The minimum temperature is reached at

$$\Delta = \gamma = \Gamma/2 \,. \qquad (73.34)$$

For three-dimensional molasses, the Doppler limit T_{D} of Eq. (73.5) is obtained.

For $\Omega > \Gamma$, the performance of Doppler cooling deteriorates. Qualitatively, power broadening increases the effective linewidth γ.

Dipole Forces. Dipole forces are the resonant analog of ponderomotive forces discussed in Sect. 72.3. They arise from successions of absorption and induced emission driven by photons with different momenta. Such processes occur only if there is more than one wave vector present in the field, i.e., if there is an intensity gradient. For a zero-velocity two-state atom, the gradient force in a monochromatic field with the local total intensity $I(\mathbf{r})$ is

$$\mathbf{F}_{\mathrm{g}}(\mathbf{r}) = \frac{4\hbar\Delta\gamma^2}{\Delta^2 + \gamma^2[1 + 2I(\mathbf{r})/I_{\mathrm{s}}]}\, \frac{\nabla I(\mathbf{r})}{I_{\mathrm{s}}} \,. \qquad (73.35)$$

The dipole force may be derived from the potential energy

$$V_{\mathrm{g}}(\mathbf{r}) = -2\hbar\Delta \ln\left[1 + \frac{2\gamma^2 I(\mathbf{r})/I_{\mathrm{s}}}{\Delta^2 + \gamma^2}\right] \,. \qquad (73.36)$$

The atoms are strong field seekers for $\Delta > 0$, and weak field seekers when $\Delta < 0$.

Dipole Force Trap. Gradient forces are utilized in the *dipole force trap* of atoms. A common configuration consists of a focused laser beam tuned below resonance. The intensity is highest at the focus, but light pressure pushes the trap center slightly downstream.

Light fields used for dipole force traps cause light shifts of the atomic states. A dipole force trap therefore complicates Doppler cooling. This problem is often avoided by alternating the light fields between cooling and trapping configurations at a suitable frequency, so that the atoms have time to cool during the cooling field period, but not enough time to fly away from the trap.

Induced Diffusion. Random motion of atoms in velocity space owing to absorptions and induced emissions of photons with different momenta leads to *induced diffusion*. Contrary to diffusion in a traveling wave as in Eq. (73.28), induced diffusion does not saturate at high intensity. Instead, the diffusion coefficient continues to grow linearly with I. Induced diffusion is another reason why the lowest Doppler cooling temperatures are generally reached at low ($I < I_{\mathrm{s}}$) light intensities.

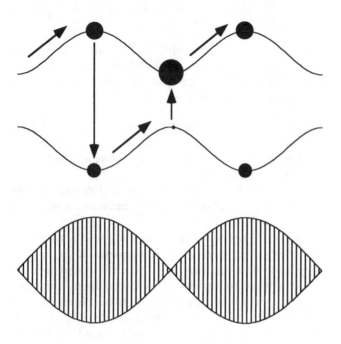

Figure 73.1. Qualitative origin of Sisyphus effect. The hatched pattern represents a standing light wave. The energies of the two dressed states are drawn as a function of position, along with a few filled circles representing the admixture of the gound state in each dressed state at selected field positions. Larger circles correspond to larger ground state admixtures, and hence, larger equilibrium populations of the dressed state. This figure applies for tuning of the laser above the atomic resonance ("blue detuning").

Sisyphus Effect. In a standing wave at high intensity and large detuning, another kind of optical force becomes important that cannot be categorized either as light pressure or gradient force.

As explained in Sect. 66.4.4 one may diagonalize the Hamiltonian to obtain the dressed atom-field states. Because the light field depends on position, so do the energies of the dressed states and their decompositions into plain atomic states. In Fig. 73.1 drawn for $\Delta < 0$, the dressed state with a minimum (maximum) at the field nodes coincides with the bare ground state (excited state) at the nodes. At the antinodes the admixtures of ground and excited states are evened out to some extent.

The energy of a dressed state acts as potential energy for the c.m. motion of an atom residing in that particular state. In fact, the gradient force is the force derived from these potential energies, averaged over the occupation probabilities of the dressed states. The occupation probability is larger for the dressed state with a larger ground state admixture. From Fig. 73.1 one therefore sees that the atom predominantly resides in the dressed

state that has a minimum of energy at the nodes. The atom is a weak-field seeker, as it should for $\Delta < 0$.

Spontaneous emission remains to be considered. It gives rise to transitions between the dressed states. These transitions may go both ways between the dressed states, because the states are in general superpositions of the bare ground state and the excited state. Rate of spontaneous transitions from one dressed state to another increases (decreases) with the excited (ground) state admixture of the initial state.

In reference to Fig. 73.1, suppose that the atom is coming from the left in the upper dressed state. The probability that the atom makes a transition to the lower dressed state, as marked by the downward vertical arrow, is largest at the node. If this transition takes place, near the next antinode the most probable transition is as shown by the upward vertical arrow. In this manner the atom spends most of its time at an uphill climb against the potential, and is therefore slowed down. In reference to Greek mythology, this is called the *Sisyphus effect*. In the two-state model atom, cooling takes place when the laser frequency is higher than the atomic resonance frequency.

Exact Results for Standing Waves

Force. The force on a two-state atom in a one-dimensional standing light wave may be expanded analytically to first order in velocity. With the field phase chosen so that the antinode is at $x = 0$, the force is

$$F(x, v) = F_g(x) - m\beta(x)v, \qquad (73.37)$$

where the gradient force is

$$F_g(x) = -\frac{\hbar k \, \Delta \, \Omega^2 \sin 2kx}{d}, \qquad (73.38)$$

and the damping coefficient is

$$\begin{aligned}\beta(x) = 8\Delta \, \Omega^2 \varepsilon_r \gamma^{-1} d^{-3} (1 - \cos^2 kx) \\ \times (\Delta^2\gamma^2 + \gamma^4 - 2\gamma^2\Omega^2\cos^2 kx - 2\Omega^4\cos^4 kx),\end{aligned} \qquad (73.39)$$

with $d = \Delta^2 + \gamma^2 + 2\Omega^2 \cos^2 kx$.

Diffusion. For a standing wave, the $v = 0$ form of the diffusion coefficient is

$$\begin{aligned}\frac{D(v=0)}{\hbar^2 k^2 \Gamma} &= \frac{\Omega^2 (\alpha\gamma^2\cos^2 kx + \gamma^2\sin^2 kx + 2\Omega^2\sin^2 kx \cos^2 kx)}{2\gamma^2 d} \\ &\quad - \frac{\Delta^2\Omega^4 \sin^2 kx \cos^2 kx \, (\Delta^2 + 5\gamma^2 + 4\Omega^2\cos^2 kx)}{\gamma^2 d^3}.\end{aligned}$$

$$(73.40)$$

Semiclassical Versus Quantum Theory

When $\gamma \gg \varepsilon_{\rm r}$, the r.m.s. velocity of a cooled two-state atom is always $\gg v_{\rm r}$, and semiclassical theory is valid. Under the same condition $\gamma \gg \varepsilon_{\rm r}$, the Doppler-limit r.m.s. velocity also is less than the critical velocity $v_{\rm c,D}$ from Eq. (73.30). Velocity expansions such as in Eqs. (73.37) and (73.40) are then justified.

In the contrary case, $\gamma \lesssim \varepsilon_{\rm r}$, full quantum theory of trapping and cooling must be employed. The cooled velocity distribution is not thermal, and temperature is ill-defined. The lowest expectation value of kinetic energy for a two-state atom in a linearly polarized standing wave occurs at low I for $\Delta = 4.4\varepsilon_{\rm r}$, and is equal to $0.53R$.

73.2.3 Multistate Atoms

Energy levels of atomic systems usually have angular momentum degeneracy. In addition, the polarization of light in general depends on position. A combination of these aspects leads to phenomena beyond the two-state atomic model.

Polarization Gradient Cooling

As explained in connection with Eq. (73.19), a finite memory time of the internal atomic state may lead to damping of the c.m. motion. For a two-state atom, internal equilibration arises from spontaneous emission. The time scale is $\tau_{\rm D} \sim \Gamma^{-1}$, and Doppler cooling ensues. However, an atom whose ground state has angular momentum degeneracy is also subject to optical pumping. If the polarization of light varies as a function of position, optical pumping is needed to reach local equilibrium. The pumping time scale $\tau_{\rm p} \propto I^{-1}$ then becomes relevant for a moving atom. The associated cooling is known as *polarization gradient cooling*. Its hallmark is that, for low I, the damping coefficient $\beta \propto I\tau_{\rm p}$ is independent of intensity.

Two detailed mechanisms of polarization gradient cooling have been described [11], although in three-dimensional light fields they are intertwined. The *Sisyphus effect* works like the Sisyphus effect for a two-state atom, except that it relies on light shifts and optical pumping within the ground state manifold. *Induced orientation cooling* is analogous to Doppler cooling. Velocity dependence of optical pumping in counterpropagating waves leads to pumping to a state for which the force due to the wave propagating opposite to the atom exceeds the force due to the wave propagating along with the atom.

Lin $\perp$ Lin Molasses. One-dimensional lin $\perp$ lin molasses consists of two counterpropagating waves with orthogonal linear polarizations. The net polarization varies from linear – via elliptical – to circular over a distance of $\lambda/8$. With the phases chosen such that light is linearly polarized at $x = 0$, in the limit of nonsaturating intensity, and at low velocity, the semiclassical force on an atom with a $J = 1/2 \to 3/2$ transition is

$$F = -\hbar k \Delta [\tfrac{1}{3} s \sin 4kx - \tfrac{2}{3} qv\gamma^{-1}(1 + \cos 4kx)] . \quad (73.41)$$

Here the saturation parameter s is

$$s = \frac{\Omega^2/2}{\Delta^2 + \gamma^2} . \quad (73.42)$$

In this configuration, only the Sisyphus effect contributes to cooling. Cooling takes place for $\Delta > 0$, and the resulting temperature obtained from the position-averaged quantities is

$$T = \frac{135\Delta^2 + 296\gamma^2}{1080(\Delta^2 + \gamma^2)} \left[\frac{2I\Gamma}{I_{\rm s}|\Delta|} \right] T_{\rm D} . \quad (73.43)$$

σ^+–σ^- Molasses. One-dimensional σ^+–σ^- molasses consists of two counterpropagating waves with opposite circular polarizations. The net polarization is linear everywhere, but the direction of polarization rotates as the point of observation is displaced along the propagation axis; hence, the alternative name, *corkscrew molasses*. At low intensity and low velocity, the force on an atom with a $J = 1 \to 2$ transition is

$$F = -\frac{60}{17} \frac{\Delta\gamma}{5\gamma^2 + \Delta^2} \hbar k^2 v . \quad (73.44)$$

In this configuration only induced orientation cooling contributes. Cooling again takes place for $\Delta > 0$, and the resulting temperature is

$$T = \frac{29\Delta^2 + 1045\gamma^2}{300(\Delta^2 + \gamma^2)} \left[\frac{2I\Gamma}{I_{\rm s}|\Delta|} \right] T_{\rm D} . \quad (73.45)$$

Lin $\parallel$ Lin Molasses. The designation lin $\parallel$ lin denotes a standing wave with the same linear polarization for both counterpropagating beams. In one dimension there is no polarization gradient, but three lin $\parallel$ lin pairs in orthogonal directions (often with mutually orthogonal polarizations) make a three-dimensional optical molasses with potential polarization gradient cooling.

Experimental Molasses. For $I\Gamma/I_{\rm s}|\Delta| < 1$ and $|\Delta| > \Gamma$, the temperatures (73.43) and (73.45) both reduce to the form

$$T = C \frac{\hbar\Omega^2}{k_{\rm B}|\Delta|} = C \frac{2I\Gamma}{I_{\rm s}|\Delta|} T_{\rm D} . \quad (73.46)$$

Under these conditions the same scaling is approximately observed also in three-dimensional six-beam optical molasses operating with atoms that have a degenerate ground state. The constant C depends on the degeneracy of the transitions and on the polarizations of the molasses beams. Measured values are mostly in the range $0.25 < C < 0.5$ [12].

846 / *Atomic, Molecular, & Optical Physics Handbook*

Limit of Cooling. While the expressions (73.43), (73.45) and (73.46) suggest that T goes all the way to zero as $I \to 0$ or $\Delta \to \infty$, there is a lower limit of T reached in polarization gradient cooling. T eventually starts to rise abruptly when $|\Delta|$ is increased or I is decreased. The empirical rule of thumb is that $T \sim 10\,T_\mathrm{r}$ is the lowest temperature one can expect.

Semiclassical Versus Quantum Theory. According to the semiclassical theory, $T \propto I$, so the r.m.s. velocity of cooled atoms is proportional to $I^{1/2}$. Now, the critical velocity of polarization gradient cooling, estimated roughly as

$$v_{\mathrm{c,p}} = \frac{\lambda}{\tau_\mathrm{p}} \approx \frac{\lambda \gamma \Omega^2}{\Delta^2 + \gamma^2}, \qquad (73.47)$$

is proportional to I. At low enough I, $v_{\mathrm{c,p}}$ is therefore smaller than the velocity width of the cooled atoms. Expansions of force and diffusion in velocity are no longer useful, and temperature predictions of the type (73.46) fail. This occurs, at the latest, when the r.m.s. velocity equals a few recoil velocities.

Semiclassical theory does not lead to predictions that grossly violate its key premise that the ensuing velocity distribution is much broader than v_r. However, reliable theoretical limits of temperature for polarization gradient cooling may only be obtained from the full quantum treatment.

Cooling Below Recoil Limit

Velocity-Dependent Coherent Population Trapping. Multistate atoms may exhibit nonabsorbing states, i.e. superpositions of ground states that do not couple to light. A nonabsorbing superposition of joint internal and c.m. states is also possible. An atom then experiences recoil kicks in absorptions and spontaneous emissions, until it by chance ends up in the nonabsorbing state and remains trapped thereafter. This scheme is known as *velocity-dependent coherent population trapping* [13]. An example is provided by a $J = 1 \to 1$ atom in one-dimensional σ^+–σ^- molasses, for which the entangled state

$$|\psi\rangle = \tfrac{1}{\sqrt{2}} \left(|+1\rangle | + v_\mathrm{r}\rangle + e^{i\varphi} |-1\rangle | - v_\mathrm{r}\rangle \right) \qquad (73.48)$$

of the ground-state atom with magnetic quantum numbers $m = \pm 1$ and c.m. states with velocities $\pm v_\mathrm{r}$ is nonabsorbing. The proper phase φ depends on the phases of the laser beams.

In practice, a trapped state is only in the neighborhood of the correct $|\psi\rangle$. After a random delay, the atom manages to absorb a photon, and a burst of absorptions and spontaneous emissions ensues. The activity ceases for a while when the atom lands in a quiescent state near $|\psi\rangle$, only to start all over again. The term *Lévy flights* is

used for this kind of a stochastic process. On the average, the atom is in a mixed state. The measured velocity distribution of the atom shows well-resolved peaks at $\pm v_\mathrm{r}$, which is interpreted as a signature of sub-recoil temperature.

Raman Cooling. A narrow resonance of the atomic system is associated with the nonabsorbing state. The two-photon resonance in a three-level Λ configuration is an example. Doppler shifts in induced Raman transitions, driven by two counterpropagating beams between the ground states of the Λ, enable a sharp discrimination between atomic velocities. *Raman cooling* based on this velocity selectivity, using judiciously crafted pulse sequences, has been employed to produce atomic temperatures below the recoil limit [14].

Magneto-Optical Trap

Since spontaneous forces are generally stronger than dipole forces, use of light pressure to trap neutral atoms appears desirable. However, the *optical Earnshaw theorem* states that (in the limit of low I) the spontaneous force on a two-state atom is sourceless. While confinement may be possible in some directions, escape routes for atoms remain open in others. Three-dimensional trapping of a two-state atom with light pressure is not possible.

Magneto-optical traps defeat the Earnshaw theorem by relying on angular momentum degeneracy. Consider an atom with a $J = 0 \to 1$ transition in a magnetic field $\mathbf{B}$ that depends linearly on position around the zero at $x = 0$. Suppose that the gradient of $\mathbf{B}$ is chosen in such a way that the $m = 1$ ($m = -1$) magnetic substate of the excited state has the higher (lower) energy for $x > 0$, and that the atom is illuminated by $\sigma^{\pm}$ polarized beams propagating in the $\pm x$-directions, tuned below resonance. When the atom is displaced from $x = 0$ in either direction, it is closer to resonance with the beam that pushes it back toward $x = 0$. This makes the restoring force responsible for trapping.

A magneto-optical trap can be set up also in three dimensions. A quadrupole magnetic field of the form

$$\mathbf{B}(\mathbf{r}) \simeq \left. \frac{\partial B_z}{\partial z} \right|_{\mathbf{r}=0} \left(z\,\hat{\mathbf{e}}_z - \tfrac{1}{2}x\,\hat{\mathbf{e}}_x - \tfrac{1}{2}y\,\hat{\mathbf{e}}_y \right) \qquad (73.49)$$

is produced by reversing the direction of current in one of the two Helmholtz coils. Three orthogonal pairs of light beams, each in the σ^+–σ^- configuration, complete the trap. The magnetic field is sourceless. To compensate for the ensuing signs of the field gradients, one of the σ^+–σ^- corkscrews has the opposite handedness from the other two.

The mechanism of the magneto-optical trap for the $J = 0 \to 1$ configuration is the same as the mechanism for Doppler cooling, except that position dependent level shifts of the excited states take the place of velocity dependent Doppler shifts. The restoring force and the damping coefficient of Doppler cooling are closely related. For the coordinate directions $u = x, y, z$ the relation is

$$F_u = -\kappa_u u, \quad \kappa_u = \beta \frac{m g_e \mu_B}{\hbar k} \left| \frac{\partial B_u}{\partial u} \right|, \qquad (73.50)$$

where g_e is the Landé factor for the excited state.

A magneto-optical trap may similarly be based on the induced orientation mechanism of polarization gradient cooling. In that case, the Landé factor of the ground state, g_g, should probably be used in Eq. (73.50). This may be the true mechanism of most magneto-optical traps, but insufficient quantitative understanding precludes firm conclusions.

Atoms in a well-aligned magneto-optical trap reside near the zero of **B**, which therefore has little effect on polarization gradient cooling. Trapping and cooling are achieved simultaneously.

73.3 MAGNETIC TRAP FOR ATOMS

The magnitude $B(\mathbf{r})$ of a magnetic field may have a minimum in free space, as in (73.49). A particle with a magnetic dipole moment $\boldsymbol{\mu}$ then experiences a trapping potential $U(\mathbf{r}) = \mu B(\mathbf{r})$ if $\boldsymbol{\mu}$ and **B** are antiparallel. $\boldsymbol{\mu}$ remains locked antiparallel to **B** if the field seen by the moving dipole satisfies the adiabatic condition

$$\frac{1}{B} \left| \frac{dB}{dt} \right| \ll \frac{\mu B}{\hbar} \qquad (73.51)$$

(see Sect. 71.4.4). However, if $B(\mathbf{r}) = 0$ at the minimum, the adiabatic condition is violated and the dipole may flip (*Majorana transition*). The particle may end up in a repulsive potential, and get expelled from the trap. This becomes a problem at low temperatures, when the particles accumulate near the minimum of the potential. Trap configurations are therefore designed in which $B(\mathbf{r}) \neq 0$ at the minimum. Alternatively, the particles may be kept away from the zero of B, for instance, with the aid of gradient forces of light.

Evaporative Cooling. A magnetic trap is often combined with *evaporative cooling*. The most energetic atoms from the tail of the thermal distribution escape from the trap, whereupon the average energy of the remaining atoms decreases. Successful operation of evaporative cooling requires a high enough rate of elastic collisions so that the atoms thermalize in a time short compared with the lifetime of the sample. In order to sustain the rate of evaporation, the effective depth of the trap is lowered as the atoms cool [15].

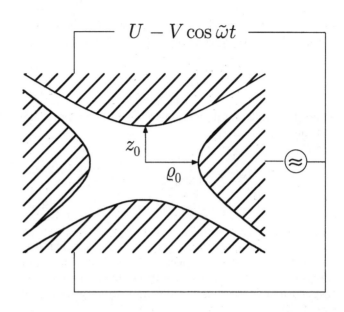

Figure 73.2. Electrode configuration and voltages of an ideal hyperboloid Paul trap.

73.4 TRAPPING AND COOLING OF CHARGED PARTICLES

Since the potential $\Phi(\mathbf{r})$ of a static electric field satisfies Laplace's equation, $\Phi(\mathbf{r})$ cannot have an extremum in free space. A static electric field therefore cannot serve as an ion trap (Earnshaw's theorem). *Paul* and *Penning* traps circumvent this limitation by making use of an alternating electric field and a magnetic field, respectively. Cooling is often employed to assist trapping.

73.4.1 Paul Trap

Trapping

Configuration. Consider an ideal trap whose surfaces are hyperboloids of revolution; see Fig. 73.2. The two "endcaps" and the intervening "ring" are equipotential surfaces of the quasistatic electric potential

$$\Phi(x, y, z) = \frac{\Phi_0(t)(2z^2 - x^2 - y^2)}{2\varrho_0^2}, \qquad (73.52)$$

where ϱ_0 is the distance from the center to the ring, $z_0 = \varrho_0/\sqrt{2}$ is the distance to the endcaps, and $\Phi_0(t)$ is a voltage applied between the endcaps and the ring,

$$\Phi_0(t) = U - V \cos \tilde{\omega} t. \qquad (73.53)$$

Motion of an Ion. In the ideal three-dimensional Paul trap, Newton's equations of motion for the coordinates $u = x$, y or z may be recast as *Mathieu's equations*,

$$\frac{d^2u}{d\tau^2} + (a_u - 2q_u \cos 2\tau)u = 0, \qquad (73.54)$$

where $\tau = \tilde{\omega}t/2$ is a dimensionless quantity proportional to time, the parameters are

$$a_z = -2a_{x,y} = -\frac{8qU}{m\tilde{\omega}^2 \varrho_0^2}, \qquad (73.55)$$

$$q_z = -2q_{x,y} = -\frac{4qV}{m\tilde{\omega}^2 \varrho_0^2}, \qquad (73.56)$$

and m and q are the mass and charge of the particle. Stable trapping ensues when a_u and q_u are such that the motion of the ion is stable in all directions. A Paul trap normally operates in the first stability region of Eqs. (73.54).

Stable motion may be qualitatively divided into forced *micromotion* at the frequency $\tilde{\omega}$ of the external drive, and into slower *secular motion* of the center of the micromotion. If $U = 0$, the secular motion takes place in an effective ponderomotive potential U_P (Sect. 72.3) equal to the cycle-averaged kinetic energy in the micromotion. Explicitly,

$$U_P(\mathbf{r}) = \frac{q^2\mathcal{E}^2(\mathbf{r})}{4m\tilde{\omega}^2} = \frac{q^2V^2(x^2 + y^2 + 4z^2)}{4m\tilde{\omega}^2\varrho_0^4}, \qquad (73.57)$$

where $\mathcal{E}(\mathbf{r})$ is the ac field amplitude. This is an anisotropic harmonic oscillator potential characterized by the oscillation frequencies

$$\nu_z = 2\nu_{x,y} = \frac{\sqrt{2}qV}{m\tilde{\omega}\varrho_0^2}. \qquad (73.58)$$

Quantization of C.M. Motion. The separation of micromotion and secular motion is excellent, and the trap is stable, when $\nu_{x,y,z} \ll \tilde{\omega}$. Ignoring the micromotion, the c.m. motion of the ions in the potential $U_P(\mathbf{r})$ may be quantized readily. The energy of a state with $n_{x,y,z}$ quanta in the coordinate directions x, y, z is

$$E = \sum_{i=x,y,z} \hbar\nu_i \left(n_i + \tfrac{1}{2}\right). \qquad (73.59)$$

Variations of Paul Trap. Little practical advantage usually arises from a realization of the ideal shape. Even a single electrode with an applied ac voltage may work as a Paul trap. A *linear trap* is basically a two-dimensional Paul trap with an added static longitudinal potential to prevent escape of the ions from the ends of the trap. A closed *race track* Paul trap is obtained by bending a linear trap into a ring.

Cooling

Laser Cooling in One Dimension. The secular motion of an ion may be cooled using lasers. Consider the motion of the ion in one of the principal-axis directions x, y, z, with ν denoting the corresponding c.m. frequency. In the common case where $\gamma \gg \nu$, *Doppler cooling* works basically as with a free atom. In the contrary case, $\nu \gg \gamma$, cooling may be achieved by tuning the laser to $\omega = \omega_0 - \nu$. Resonant photoabsorption starting with n c.m. quanta decreases the quantum number from n to $n - 1$, and subsequent spontaneous emission on the average leaves the c.m. energy untouched. The net effect is reduction of the c.m. energy by $\hbar\nu$ in such a Raman process. Since the oscillating ion sees a frequency-modulated laser with sidebands, this method of ion cooling is called *sideband cooling*.

For one-dimensional motion of a two-state ion in a traveling light wave at low I, the velocity damping rate is

$$\beta = \frac{2\Omega^2\gamma\Delta}{[(\Delta+\nu)^2+\gamma^2][(\Delta-\nu)^2+\gamma^2]}\varepsilon_r, \qquad (73.60)$$

and the expectation value of the c.m. energy is

$$E = \frac{\hbar}{4\Delta}\left\{\Delta^2 + \gamma^2 + \nu^2 \right.$$
$$\left. + \alpha\frac{[(\Delta-\nu)^2+\gamma^2][(\Delta+\nu)^2+\gamma^2]}{\Delta^2+\gamma^2}\right\}, \qquad (73.61)$$

where α characterizes the angular distribution of spontaneous emission, see Eq. (73.29). The result (73.61) is useful when either $\varepsilon_r \ll \gamma$ or $\varepsilon_r \ll \nu$. The limit $\nu \ll \gamma$ is for Doppler cooling; the temperature from Eq. (73.61) coincides with Eq. (73.33) for a free atom. The case with both $\nu \gg \gamma$ and $\varepsilon_r \ll \nu$ corresponds to sideband cooling in the *Lamb-Dicke regime*, in which the cooled ion is confined to a region much smaller than λ.

In connection with sideband cooling, it is convenient to cite the expectation number of harmonic oscillator quanta $\langle n \rangle$ instead of energy or temperature; the latter are

$$E = \hbar\nu(\langle n\rangle + \tfrac{1}{2}), \quad T = \frac{\hbar\nu}{k_B}\left[\ln\left(1 + \langle n\rangle^{-1}\right)\right]^{-1}. \quad (73.62)$$

For optimal sideband cooling, $\nu \gg \gamma$ and $\Delta = \nu$, the result is

$$\langle n\rangle = \tfrac{1}{4}(1 + 4\alpha)(\gamma/\nu)^2 + O[(\gamma/\nu)^4]. \qquad (73.63)$$

In principle, by increasing the trapping frequency ν, the ion may be put arbitrarily close to the ground state of the c.m. harmonic oscillator.

Laser Cooling in Three Dimensions. Either by design or chance, no two of the ν_i are precisely degenerate. If the damping rate β and the trap frequencies ν_i satisfy

$$\beta \lesssim |\nu_i - \nu_j|, \quad i \neq j, \tag{73.64}$$

the motion of the ion in each principal axis direction of the trap is cooled independently of the other directions. For $\gamma \gg \nu_{x,y,z}$, a single laser beam propagating approximately in the direction $\frac{1}{\sqrt{3}}(\hat{\mathbf{e}}_x + \hat{\mathbf{e}}_y + \hat{\mathbf{e}}_z)$ suffices to cool all components of the secular motion.

Energy in Micromotion. Possibly, with the aid of compensating static electric fields, *one* cooled ion may be confined near the zero of the trapping ac electric fields. Then the energy in the micromotion is comparable to the energy in the secular motion.

73.4.2 Penning Trap

Trapping

Configuration. In the Penning trap, a dc voltage U is applied between the endcaps and the ring, and a constant magnetic field $\mathbf{B}$ in the direction of the trap axis z is added. The magnetic field forces an ion escaping toward the ring to turn back.

Motion of an Ion. The motion of an ion is a superposition of three periodic components. For the same ideal hyperboloid shape that was discussed with the Paul trap, Eq. (73.52) and Fig. 73.2, the three components are completely decoupled. Firstly, in the axial direction, the ion executes oscillations at the *axial frequency*

$$\nu_z = \left(\frac{2qU}{m\varrho_0^2}\right)^{1/2}. \tag{73.65}$$

Secondly, the ion undergoes cyclotron motion in the plane perpendicular to the trap axis. As a result of the electric field, the frequency of the cyclotron motion

$$\nu_c' = \tfrac{1}{2}\nu_c + \left(\tfrac{1}{4}\nu_c^2 - \tfrac{1}{2}\nu_z^2\right)^{1/2} \tag{73.66}$$

is displaced from the *cyclotron frequency* $\nu_c = qB/m$ of a free ion. Thirdly, the guiding center of cyclotron motion rotates about the trap axis at the *magnetron frequency*

$$\nu_m = \tfrac{1}{2}\nu_c - \left(\tfrac{1}{4}\nu_c^2 - \tfrac{1}{2}\nu_z^2\right)^{1/2}. \tag{73.67}$$

The frequencies typically satisfy

$$\nu_m \ll \nu_z \ll \nu_c'. \tag{73.68}$$

Magnetron motion has unusual properties. It takes up the majority of the electrostatic energy in the transverse directions, which in the absence of the magnetic field would lead to expulsion of the ion. Relative to a stationary ion at the trap center, the energy of the magnetron motion is bounded from above by zero. The radius, as well as velocity and kinetic energy of magnetron motion, *decreases* with increasing total energy. The energy for a state with n_c, n_z and n_m quanta in the cyclotron, axial and magnetron motions is therefore

$$E = \hbar\nu_c'(n_c + \tfrac{1}{2}) + \hbar\nu_z(n_z + \tfrac{1}{2}) - \hbar\nu_m(n_m + \tfrac{1}{2}). \tag{73.69}$$

Cooling

Laser Cooling. For ions in a practical Penning trap, the frequencies ν_c', ν_z and ν_m are $< \gamma$. If $\mathbf{k}$ is not orthogonal to either the cyclotron motion or the axial motion, Doppler cooling proceeds essentially as for a free atom. However, energy should be added to the magnetron motion in order to reduce the magnetron radius and velocity. The solution is to aim a finite-size laser beam off the center of the trap in such a way that an ion experiences a higher (lower) intensity over the part of its magnetron orbit in which it travels in the direction of (opposite to) the laser beam. With a proper choice of the parameters, the ensuing addition of energy overcomes Doppler cooling of the magnetron motion.

Other Cooling Methods. Precision measurements are carried out in Penning traps with objects that do not have an internal level structure suitable for laser cooling; and thus, other cooling methods are used.

For light particles such as electrons, characteristic times of *radiative damping* of the cyclotron motion are in the subsecond regime, and hence, so are the equilibration times with blackbody radiation. Cooling is accomplished by enclosing the trap in a low-temperature (e.g. liquid helium) environment. For protons and heavier particles, the equilibration times of the cyclotron motion with the environment are impractically long, and the same applies to the axial and magnetron motions even for electrons.

A workable cooling scheme for the axial motion is based on the charges that the oscillating particle induces on the endcaps. The charges generate currents in an external circuit connecting the endcaps. The endcaps are thus coupled to a cooled resonant circuit tuned to the axial frequency, and axial motion relaxes to thermal equilibrium with the resonant circuit. A variant of this *resistive cooling*, in which the ring is split into electrically insulated segments, is used to cool the cyclotron motion of protons and heavier ions.

Magnetron motion of an electron or proton is cooled by *sideband cooling*. An electric field with components in both the z-direction and xy-plane, and tuned to $\omega = \nu_z + \nu_m$, drives transitions which may either increase or decrease the number of quanta in each mode. However, the matrix elements favor transitions with $\Delta n_z = 1$ and $\Delta n_m = -1$. Pumping of the axial motion is canceled by axial cooling, while an equilibrium with low kinetic energy ensues for the magnetron motion. Ideally, the ratio of kinetic energies becomes

$$\mathcal{T}_{\text{kin,m}}/\mathcal{T}_{\text{kin,z}} = \nu_m/\nu_z. \tag{73.70}$$

73.4.3 Collective Effects in Ion Clouds

As soon as there is more than one ion in the trap, Coulomb interactions between the ions profoundly shape the physics [16].

Ion Crystal. In the standard Paul trap *radio frequency heating* due (presumably) to transfer of energy from micromotion to secular motion limits the number of ions that can be cooled efficiently by a laser to about ten. At a low temperature, the ions settle to equilibrium positions corresponding to a minimum of the joint trapping and Coulomb potentials, and form a small "crystal". Depending on the trap parameters, the ions may also execute quasiperiodic or chaotic collective motion, or move nearly independently of one another. Changes between crystalline and liquid forms of the ion cloud resembling phase transitions are observed.

Strongly Coupled Plasma. Cooling of a large number of ions is possible in a Penning trap. However, magnetron motion becomes uniform rotation of the entire cloud, and Coulomb interactions set a lower limit on the attainable radius of the cloud. This leads to a lower limit on the kinetic energy and second-order Doppler shift.

In a co-rotating frame, the ions behave like a one-component plasma on a neutralizing background. The characteristic parameter for a one-component plasma with the charge per particle q and density n is

$$\Gamma_P = \left(\frac{4\pi n}{3}\right)^{1/3} \frac{q^2}{4\pi\epsilon_0 k_B T}, \qquad (73.71)$$

essentially the ratio of the Coulomb energy between two nearest-neighbor ions divided by the thermal kinetic energy. $\Gamma_P > 1$ indicates a strongly coupled plasma; for $\Gamma_P > 2$ and $\Gamma_P > 170$ solid and liquid phases are expected in an infinite plasma. Experiments with a Penning trap have produced $\Gamma_P \sim 340$, at which the plasma is loosely arranged in concentric ion shells.

Sympathetic Cooling. In a trap that holds two or more species of charged particles, cooling of the motion of one species is transferred by Coulomb interactions to the other species. This *sympathetic cooling* broadens the scope of ion cooling methods.

73.5 APPLICATIONS OF COOLING AND TRAPPING

Trapping and cooling offer increased interaction times between the atoms/ions and the light. This leads to reduced transit time broadening, and indeed to macroscopic (> 1 s) interaction times. Laser cooling in a magneto-optical trap routinely gives temperatures so low that the Doppler width is below the natural linewidth of the cooling transition. A homogeneously broadened atomic sample is thus prepared. Cooling also enables reduction of the second-order Doppler effect.

Various frequency measurements are the prime beneficiary of cooling and trapping. Potential applications range from detection of the change of natural constants

in time to such feats of technology as the Global Positioning System.

73.5.1 Neutral Atoms

Experimental Considerations. Currently experiments start with longitudinal deceleration and cooling of an atomic beam by a counterpropagating laser beam. To compensate for the change of the Doppler shift, the laser frequency is chirped downward as the atoms slow down. Alternatively, the position dependent magnetic field of a tapered solenoid shifts the transition frequency of the atom to keep them near resonance while they move down the solenoid. A magneto-optical trap then scoops some atoms, cools them, and captures them. A magneto-optical trap may also be loaded in a gas cell by letting it collect atoms from the low-velocity wing of the Maxwell distribution. Depths of neutral-atom traps are below 1 K. Storage times are of the order of 1 s, limited at high densities by exothermal binary collisions and at low densities by collisions with the atoms in the background gas.

The temperature of cooled atoms may be measured by the *time-of-flight* method. All cooling and trapping beams are turned off, whereupon the atoms fall freely under gravity. The distribution of arrival times of atoms at a probe laser beam underneath the initial molasses is compared with a numerical model for the disintegration of the molasses. A fit gives the temperature. In a recent derivative of the method, video of the disintegrating molasses is analyzed.

The current focus is on Rb and Cs, to a large extent because the required laser frequencies can be generated using inexpensive diode lasers. Hyperfine structure of the ground state of the alkalis complicates experiments because the atoms may end up in an inert hyperfine level outside the active cooling/trapping transitions. To counteract this, a second appropriately tuned *repump* laser is added to return such atoms to circulation. An increasing number of experiments is being made with metastable atoms of noble gases, some isotopes of which do not have hyperfine structure.

Cold Collisions. In the molecular picture of a collision involving a laser, optical excitation takes place at the interatomic distance for which the difference between the potential curves of the incoming and excited states equals the energy of a laser photon [17]. The end products of an inelastic collision (fine or hyperfine structure changing collision, associative ionization, radiative escape,...) are normally determined at shorter interatomic distances, when the potential curve of the excited state has an anticrossing with the potential curve of the product channel. The novel feature of ultracold collisions is the long duration due to the low velocity of the collision partners. Spontaneous emissions and other phenomena irrelevant at room temperature may take place *during* the collision.

On the scale of typical resonance widths, at very low temperatures the collision partners are in effect in a single continuum state with zero energy. This facilitates *photoassociation spectroscopy*. Laser-induced transitions from the initial continuum state to bound vibrational states of the molecule are observed.

Collisions are of practical importance in that they limit the achievable atom density in a trap: an inelastic collision may release more kinetic energy than the trap can contain, which results in a loss of an atom (or two atoms) from the trap. Collision rates are actually measured by monitoring the loss rate of atoms as a function of density [18]. Precise energies of the molecular vibrational states from photoassociation spectroscopy are used to study the *s*-wave scattering length [19], possibly a relevant parameter for Bose condensation.

Frequency Standards. *Atomic fountain* starts with an optical molasses or a magneto-optical trap. The laser beams are then manipulated to give an upward push to the atoms. The atoms fly up against gravity for a few tens of centimeters, then turn back and, because of the initial transverse velocities, fan out to a "fountain". A fountain of Cs might erupt through a microwave cavity driving the same hyperfine transition that is used in current atomic clocks [20]. The atoms in a fountain offer an interrogation time of the order of one second and velocities of the order of one meter per second.

Spin-Polarized Hydrogen. Spin-polarized hydrogen (Sect. 7.3) is a long-standing subject of trapping and cooling. Spin polarization prevents two-body recombination to H_2, and also permits trapping to a magnetic trap. Evaporative cooling is currently the cooling method of choice [21]. One of the aims is to realize Bose condensation of a weakly interacting Bose gas.

Alkalis in Magnetic Traps. A major multi-laboratory effort is invested in realizing Bose condensation of various alkali atoms in a magnetic trap using evaporative cooling. At this writing, the first report of a successful experiment has appeared [22].

73.5.2 Trapped Particles

Experimental Considerations. Both Paul and Penning traps behave like a conservative potential, and scatter rather than confine a particle coming from the outside. One method to load a trap is to generate the ions *in situ*, e.g., by letting a beam of atoms and electrons collide inside the trap. Time dependent electric potentials are another loading method. The trapped species is injected thorough a hole in the endcap, and the opposing endcap is raised to an electric potential that makes the entering particles stop. The potential is then lowered before it ejects the particles. A single electron, proton, or ion may be loaded. Typical depths of ion traps are ~ 1 eV or $\sim 10^4$ K. With the aid of cooling,

the storage time may be made infinite for all practical purposes.

Trap frequencies are measured by observing the resonance excited by added ac fields. For instance, an electric field near the axial c.m. resonance frequency may be coupled between the ring and one endcap. A resonance circuit coupled between the ring and the other endcap is used to detect the resonance. Alternatively, ejection of the driven ions is monitored.

For an electron in a Penning trap, the cyclotron frequency is in the extreme microwave region. Detection of the resonance is achieved indirectly. The uniform magnetic field is perturbed with a piece of a ferromagnet to make a magnetic bottle. The axial motion and the cyclotron motion are then coupled. A resonant microwave drive adds energy to the cyclotron motion, which detectably alters the axial frequency.

The three trap frequencies satisfy

$$\nu_c^2 = \nu_c'^2 + \nu_m^2 + \nu_z^2 . \tag{73.72}$$

This relation remains valid even if the magnetic field is misaligned with respect to the trap axis, and is also insensitive to small imperfections in the cylindrical symmetry of the electrodes. The bare cyclotron frequency may therefore be deduced accurately.

For an ion with a dipole-allowed resonance transition, fluorescence of a single ion is readily detected. Even absorption of a single ion may be measurable. Various methods of finding the temperature have been devised. At temperatures of 1 K and higher, Doppler broadening of a dipole-allowed optical transition is observable. The size of the single-ion cloud is a measure of temperature. Finally, motional sidebands in the absorption of a narrow transition ($\gamma \ll \nu$), not necessarily the same transition as the one used for cooling, may be measured to find $\langle n \rangle$. In the Lamb-Dicke regime only the carrier absorption at $\Delta = 0$ and sidebands at $\Delta = \pm\nu$ are significant, and the ratios of the peak absorptions are

$$\alpha_- : \alpha_0 : \alpha_+ = \langle n \rangle \frac{\varepsilon_r}{\nu} : 1 : (1 + \langle n \rangle) \frac{\varepsilon_r}{\nu} . \tag{73.73}$$

Quantum Jumps. Studies of the fundamentals of quantum mechanics are a prominent line of applications of ion trapping and cooling. *Quantum jumps* are a case in point. Suppose that, in addition to an optically driven two-level system, a single ion has a third *shelving state*. The ion infrequently makes a transition to the shelving state, stays there for a long time compared with the time scale of spontaneous emission of the active system, and then returns to the two-level system. When the ion makes a transition to the shelving state, fluorescence from the two-level system suddenly ceases; and the fluorescence, equally abruptly, reappears when the ion returns to the two-level system. The jumps in light scattering are the quantum jumps [23]. They are a method to detect a weak transition with an enormous amplification: a single transition to or from the shelving state may mean the difference between the presence or absence of billions of fluorescence photons.

Frequency Standards. A 303 MHz hyperfine transition in sympathetically cooled Be$^+$ in a Penning trap is probably the most thoroughly studied microwave frequency standard based on ion trapping [24]. An optical frequency standard might be built around a single ion, e.g. In$^+$, in a Paul trap. The ion may be cooled to the Doppler limit of an optical transition, placed accurately at the zero of all fields, and is unperturbed by other ions. A linear trap with several ions crystallized to a string along the trap axis [25] might offer similar advantages, plus improved signal-to-noise ratio.

g-2 Measurements. For an electron, the cyclotron frequency ν_c and the spin-flip frequency ν_s are related by

$$\nu_s = \tfrac{1}{2} g \nu_c . \tag{73.74}$$

Due to quantum electrodynamic corrections $g \neq 2$, and so the anomaly frequency $\nu_a = \nu_c - \nu_s \sim 10^{-3} \nu_c$ is nonzero. A magnetic field at the anomaly frequency causes a simultaneous flip of the spin and a loss or gain of one quantum of energy in the cyclotron motion. In a magnetic bottle, the change in the cyclotron motion causes a change in the axial resonance frequency, which is detected. The anomaly frequency can thus be measured accurately. Together with a measurement of the cyclotron frequency, this yields a measurement of the g- factor of the electron, or positron [26].

Measurements of Mass Ratios. As the cyclotron frequency is inversely proportional to the mass of the ion (or electron, positron, proton, antiproton, ...), an accurate measurement of the cyclotron frequencies of two species in the same Penning trap amounts to an accurate measurement of the ratio of the masses [27]. A sufficient resolution to weigh molecular bonds is conceivable.

REFERENCES

1. *The Mechanical Effects of Light*, special issue, J. Opt. Soc. Am. B **2**, Num. 11 (1985).
2. S. Stenholm, Rev. Mod. Phys. **58**, 699 (1986).
3. *Laser Cooling and Trapping of Atoms*, special issue, J. Opt. Soc. Am. B **6**, Num. 11 (1989).
4. C. Cohen-Tannoudji, in *Fundamental Systems in Quantum Optics*, edited by J. Dalibard, J. M. Raymond, and J. Zinn-Justin (Elsevier, Amsterdam, 1991), p. 1.
5. D. J. Wineland, W. M. Itano, and R. S. Van Dyck, Jr., Adv. At. Mol. Phys. **19**, 135 (1984).
6. L. S. Brown and G. Gabrielse, Rev. Mod. Phys. **58**, 233 (1986).
7. *Physics of trapped ions*, special issue, J. Mod. Opt., **39**, Num. 2 (1992).
8. R. Blatt, P. Gill, and R. C. Thompson, J. Mod. Opt. **39**, 193 (1992).
9. J. Javanainen, Phys. Rev. A **46**, 5819 (1992).
10. Y. Castin and K. Mølmer, Phys. Rev. Lett. **74**, 3772 (1995); M. R. Doery, E. J. D. Vredenbregt, and T. Bergeman, Phys. Rev. A **51**, 4881 (1995).
11. J. Dalibard and C. Cohen-Tannoudji, J. Opt. Soc. Am. B **6**, 2023 (1989).
12. C. Gerz, T. W. Hodapp, P. Jessen, K. M. Jones, W. D. Phillips, C. I. Westbrook, and K. Mølmer, Europhys. Lett. **21**, 661 (1993).
13. J. Lawall, F. Bardou, B. Saubamea, K. Shimizu, M. Leduc, A. Aspect, and C. Cohen-Tannoudji, Phys. Rev. Lett. **73**, 1915 (1994).
14. N. Davidson, H.-J. Lee, M. Kasevich, and S. Chu, Phys. Rev. Lett. **72**, 3158 (1994).
15. K. B. Davis, M.-O. Mewes, M. A. Joffe, M. R. Andrews, and W. Ketterle, Phys. Rev. Lett. **74**, 5202 (1995).
16. H. Walther, Adv. At. Mol. Opt. Phys. **31**, 137 (1993); J. J. Bollinger, D. J. Wineland, and D. H. E. Dubin, Phys. Plasmas **1**, 1403 (1994).
17. P. S. Julienne, A. M. Smith, and K. Burnett, Adv. At. Mol. Opt. Phys. **30**, 141 (1993).
18. T. Walker and P. Feng, Adv. At. Mol. Opt. Phys. **34**, 125 (1994).
19. E. R. I. Abraham, W. I. McAlexander, C. A. Sackett, and R. G. Hulet, Phys. Rev. Lett. **74**, 1315 (1995).
20. K. Gibble and S. Chu, Phys. Rev. Lett. **70**, 1771 (1993).
21. J. M. Doyle, J. C. Sandberg, I. A. Yu, C. L. Cesar, D. Kleppner, and T. J. Greytak, Phys. Rev. Lett. **67**, 603 (1991).
22. M. H. Anderson, J. R. Ensher, M. R. Matthews, C. E. Wieman, and E. A. Cornell, Science **269**, 198 (1995).
23. R. Blatt and P. Zoller, Eur. J. Phys. **9**, 250 (1988).
24. J. J. Bollinger, D. J. Heinzen, W. M. Itano, S. L. Gilbert, and D. J. Wineland, IEEE Trans. Instrum. Meas. **40**, 126 (1991).
25. M. G. Raizen, J. M. Gilligan, J. C. Bergquist, W. M. Itano, and D. J. Wineland, Phys. Rev. A **45**, 6493 (1992).
26. R. S. Van Dyck Jr., P. B. Schwinberg, and H. G. Dehmelt, Phys. Rev. Lett. **59**, 26 (1987).
27. F. DiFilippo, V. Natarajan, K. R. Boyce, and D. E. Pritchard, Phys. Rev. Lett. **73**, 1481 (1994).

74

De Broglie Optics

Martin Wilkens

Fakultät für Physik, Universität Konstanz, 78434 Konstanz, Germany

74.1 INTRODUCTION

Any collection of particles of mass M and momentum p has a de Broglie wavelength $\lambda_{dB} = 2\pi\hbar/p$. Originally meant to explain the orbits of a single Coulomb-bound electron, the ultimate wave character of matter has been confirmed for all fundamental particles, and also for composite particles such as ions, atoms and molecules.

In terms of the Bohr radius a_0, the fine-structure constant α, and the electron mass m_e

$$\lambda_{dB} = \frac{2\pi\alpha a_0}{(v/c)(M/m_e)} = 2\pi\left(\frac{E_h}{2T}\frac{m_e}{M}\right)^{1/2} a_0, \quad (74.1)$$

853

where $v = p/M$ is the velocity, $\mathcal{T} = \frac{1}{2}Mv^2$ is the kinetic energy, and $E_h = 27.211\,396$ eV is the atomic unit of energy. For electrons, $\lambda_{dB} \approx 1.226\,426 \times 10^{-9}/\mathcal{T}^{1/2}$ m, with $\mathcal{T}$ in eV. The *thermal de Broglie wavelength* is defined in Eq. (74.78).

De Broglie optics concerns the propagation of matter waves, their reflection, refraction, diffraction and interference. The main subfields of de Broglie optics are electron optics [1], neutron optics [2], and atom optics. Well-established applications are found in electron diffraction and microscopy [3], electron holography [4], neutron diffraction and interferometry [5]. The subject of atom optics is relatively new, and applications are currently being developed in precision spectroscopy, precision measurement, atom lithography, atom microscopy, and atom interferometry [6]. This chapter concentrates on the *principles* of de Broglie optics. Illustrations of these principles will be presented mainly in the framework of atom optics.

A typical de Broglie optical experiment involves a source, a beam of particles produced by that source, an array of optical elements, possibly a probe, other optical elements placed behind the probe, and finally a detector. Optical elements are collimators, apertures, lenses, mirrors, and beam-splitters. A probe could be a crystal, a biological sample, or just another unknown optical element whose properties are to be investigated.

To avoid proliferation of notation, we refer to any object placed in the beam path simply as an "optical element".

74.1.1 Theoretical Overview

Regardless of the particle species, the theory of de Broglie optics divides into two distinct parts: the *theory of dispersion* and the *theory of the optical phenomena*. In the theory of dispersion an effective Hamiltonian is derived which describes the interaction of the particles with the optical elements. In the theory of optical phenomena, the ensuing Schrödinger equation is solved and put into an optical context (e.g., by relating distributions of intensity on the detector screen to properties of a given optical element).

Dispersion of de Broglie Waves

The theory of dispersion is highly particle specific, since it depends on any *internal degrees of freedom* which may give rise to permanent magnetic moments or to induced electromagnetic moments either in static or in optical light fields. Interaction with the latter fields also may give rise to *spontaneous emission*. Spontaneous emission has important consequences for the *coherence properties* of atomic matter waves because of the *random recoil* associated with it (see Chap. 73).

Dispersion of matter waves differs from that of light waves in a number of regards. First, the dispersion relation of free particles is quadratic in the wavenumber giving rise to spatial *spreading* of wavepackets even in one-dimensional configurations. Second, particles may be brought to rest, which is impossible for photons. And finally, particles in beams may show *self-interaction* giving rise to *nonlinear* optical phenomena even for freely traveling particle waves. Electrons and ions, for example, experience a density dependent Coulomb broadening in the focus of a lens (*Boersch effect* [4]).

In atom optics, nonlinear phenomena are possible due to atom-atom interactions in the beam, the nature of which may be significantly influenced by laser light. In the presence of laser light, atom-atom interactions mainly result from photon exchange which, in most cases, leads to a repulsive interaction. Details of the microscopic basis of atom-atom interactions, in particular, those concerning *cold collisions*, are presented in Sect. 73.5.1. A theory of *nonlinear atom optics* is currently under development; see e.g., Ref. [7].

Optics of de Broglie Waves

In contrast to the theory of dispersion, the theory of optical phenomena of matter waves is quite universal and bears strong resemblance to ordinary light optics. This resemblance is closest if the particles can be described by a *scalar wave function* of a structureless point particle. However, in the presence of resonant laser fields, electronic levels, their Zeeman sublevels and possible hyperfine structure may play an important role, in which case a multicomponent *spinor wave function* must be used to describe the atom optical phenomena properly.

74.2 HAMILTONIAN OF DE BROGLIE OPTICS

A large class of phenomena of particle optics is well-described by an effectively one-particle Hamiltonian model in which the particles are not assumed to mutually interact. In such a model the ensemble of particles is described by a wave function $\psi(\mathbf{x}, t) = \langle \mathbf{x}|\psi(t)\rangle$ whose time evolution is governed by the Schrödinger equation

$$i\hbar \frac{\partial}{\partial t}|\psi(t)\rangle = H(t)|\psi(t)\rangle \qquad (74.2)$$

with a one-particle Hamiltonian of the generic form

$$H(t) = \frac{[\hat{\mathbf{p}} - \boldsymbol{\mathcal{A}}(\hat{\mathbf{x}}, t)]^2}{2M} + \mathcal{U}(\hat{\mathbf{x}}, t). \qquad (74.3)$$

Here, $\hat{\mathbf{p}}$ and $\hat{\mathbf{x}}$ denote the canonically conjugate operators of the c.m. momentum and position, respectively. The cartesian components of $\hat{\mathbf{p}}$ and $\hat{\mathbf{x}}$ obey the fundamental commutation relation

$$[\hat{x}_i, \hat{p}_j] = i\hbar\delta_{ij}\,. \qquad (74.4)$$

The vector potential $\boldsymbol{\mathcal{A}}(\mathbf{x}, t)$ and scalar potential $\mathcal{U}(\mathbf{x}, t)$ account for the interaction of the particle with all the optical elements, probes, and samples which are placed between the source and the detector, including the effects of gravitation and rotation.

74.2.1 Gravitation and Rotation

All particles are subject to the influence of *gravitation* and *rotation*, which both may be viewed as being special cases of an *accelerated* frame of reference. Effects of uniform gravitation are described by

$$\mathcal{U}(\mathbf{x}) = -M\mathbf{g}\cdot\mathbf{x}, \quad \boldsymbol{\mathcal{A}}(\mathbf{x}) = 0\,, \qquad (74.5)$$

where $\mathbf{g}$ describes the direction and magnitude of gravitational acceleration. Effects of uniform rotation are described by

$$\mathcal{U}(\mathbf{x}) = -\frac{M}{2}\left(\boldsymbol{\Omega}\times\mathbf{x}\right)^2, \quad \boldsymbol{\mathcal{A}}(\mathbf{x}) = M\left(\boldsymbol{\Omega}\times\mathbf{x}\right)\,, \quad (74.6)$$

where the direction and magnitude of $\boldsymbol{\Omega}$ refer to the orientation of the axis of rotation and the angular velocity, respectively. Here, $\mathcal{U}(\mathbf{x})$ is the potential of the *centrifugal force* while $\boldsymbol{\mathcal{A}}(\mathbf{x})$ is the potential of the *Coriolis force*.

74.2.2 Charged Particles

The interaction of charged particles (electrons, ions) with the electromagnetic field is described by

$$\mathcal{U}(\mathbf{x}, t) = q\Phi(\mathbf{x}, t), \quad \boldsymbol{\mathcal{A}}(\mathbf{x}, t) = (q/c)\mathbf{A}(\mathbf{x}, t)\,, \quad (74.7)$$

where q is the particle charge ($q = -e$ for electrons), and $\mathbf{A}$ and Φ denote the gauge potentials of the electromagnetic fields $\mathbf{E}$ and $\mathbf{B}$, respectively. The fields are given by

$$\mathbf{E}(\mathbf{x}, t) = -\frac{1}{c}\frac{\partial}{\partial t}\mathbf{A}(\mathbf{x}, t) - \nabla\Phi(\mathbf{x}, t)\,, \qquad (74.8)$$

$$\mathbf{B}(\mathbf{x}, t) = \nabla\times\mathbf{A}(\mathbf{x}, t)\,. \qquad (74.9)$$

74.2.3 Neutrons

For neutrons interacting with a spatially homogeneous gas, liquid, or amorphous solid contained in a volume V, a common model is

$$\mathcal{U}(\mathbf{x}) = \begin{cases} \mathcal{U}_0 & \text{inside } V\,, \\ 0 & \text{outside } V\,, \end{cases} \qquad (74.10)$$

where $\mathcal{U}_0 = 2\pi\hbar^2\varrho b/M$, ϱ being the number density of scatterers in the volume, and b the *bound coherent scattering length*.

For neutrons interacting with perfect crystals, $\mathcal{U}(\mathbf{x})$ is a smooth periodic function with the same periodicity as the lattice and an average value $\mathcal{U}_0$ within a single unit cell (see Ref. [2] for details).

74.2.4 Spins

The interaction of the spin related magnetic moment $\boldsymbol{\mu}$ with the electromagnetic field is described by

$$\hat{\mathcal{U}}(\mathbf{x}, t) = -\hat{\boldsymbol{\mu}}\cdot\mathbf{B}(\mathbf{x}, t), \quad \hat{\boldsymbol{\mathcal{A}}}(\mathbf{x}, t) = [\hat{\boldsymbol{\mu}}\times\mathbf{E}(\mathbf{x}, t)]/c\,. \qquad (74.11)$$

Here, the vector potential is due to the *motional correction* of the magnetic dipole interaction. Usually, it is neglected. However, it does play an important role for the *Aharonov-Casher effect* (see Sect. 74.6).

For fundamental particles, one has

$$\hat{\boldsymbol{\mu}} = g\frac{\hbar e}{4Mc}\boldsymbol{\sigma}\,, \qquad (74.12)$$

where $\boldsymbol{\sigma} \equiv 2\mathbf{s}$ is the vector of Pauli matrices, and g is the *g-factor* of the particle (see Sect. 73.5.2).

Here and in what follows, a hat on $\mathcal{U}$ and $\boldsymbol{\mathcal{A}}$ indicates the matrix character of the hatted quantity, the matrix indices referring to the internal degrees of freedom, like spin. Similarly, $\Psi(\mathbf{x}, t)$ denotes a spinor-valued wave function. The wave function of a spin-$\frac{1}{2}$ particle, for example, is displayed in the form

$$\Psi(\mathbf{x}, t) = \begin{pmatrix} \psi_\uparrow(\mathbf{x}, t) \\ \psi_\downarrow(\mathbf{x}, t) \end{pmatrix}\,, \qquad (74.13)$$

where $\psi_\uparrow$ ($\psi_\downarrow$) is the scalar wave function of the $\sigma_3 = +1(-1)$ component of the state $|\Psi(t)\rangle$.

74.2.5 Atoms

Many optical elements in atom optics are based on the *mechanical effects of the radiation interaction*. In the electric dipole approximation, the interaction of a single atom with the electromagnetic field is described by

$$\hat{\mathcal{U}}(\mathbf{x}, t) = -\hat{\mathbf{d}}\cdot\mathbf{E}(\mathbf{x}, t), \quad \hat{\boldsymbol{\mathcal{A}}}(\mathbf{x}, t) = \mathbf{B}(\mathbf{x}, t)\times\hat{\mathbf{d}}/c\,, \qquad (74.14)$$

where $\hat{\mathbf{d}}$ is the operator of electric dipole transition. The vector potential $\hat{\boldsymbol{\mathcal{A}}}$ is due to the *motional correction* of the electric dipole interaction. Usually it is neglected; see however, the paragraph *Electric Dipole Phase* in Sect. 74.6.

For a single mode of frequency ω

$$\mathbf{E}(\mathbf{x}, t) = \mathbf{E}^{(+)}(\mathbf{x})e^{-i\omega t} + \text{h.c.}, \qquad (74.15)$$

where $\mathbf{E}^{(+)}(\mathbf{x})$ defines both polarization and spatial characteristics of the mode. A standing wave laser field with optical axis in the x-direction and linear polarization ϵ, for example, is described by

$$\mathbf{E}^{(+)}(\mathbf{x}) = \epsilon \,\mathcal{E}_0 f(x, y, z)\cos(kx), \qquad (74.16)$$

where $\mathcal{E}_0$ is the electric field amplitude, $k = \omega/c$ is the wavenumber, and the slowly varying function $f(x, y, z)$ accounts for the transverse *profile* of the laser field.

The electric dipole operator $\hat{\mathbf{d}}$ acts in the Hilbert space of electronic states of the atom. In the particular case that the polarization of the laser field is spatially uniform and that spontaneous emission does not play a role, two states are generally sufficient and the atom may be modeled by a two-level atom with electronic levels $|e\rangle$ and $|g\rangle$ of energy E_e and E_g, respectively. With the spinor representation

$$|e\rangle = \begin{pmatrix} 1 \\ 0 \end{pmatrix}, \quad |g\rangle = \begin{pmatrix} 0 \\ 1 \end{pmatrix}, \qquad (74.17)$$

the electric dipole operator assumes the form

$$\hat{\mathbf{d}} = d\,\epsilon_d \begin{pmatrix} 0 & 1 \\ 1 & 0 \end{pmatrix}, \qquad (74.18)$$

where d is the matrix element of the dipole transition and ϵ_d denotes its polarization.

The laser field is assumed to be near resonant with the $e \leftrightarrow g$ transition at $\omega_0 = (E_e - E_g)/\hbar$, and we denote by $\Delta \equiv \omega_0 - \omega$ the atom-laser detuning. Using the rotating wave approximation (see Sect. 66.4.2) in an interaction picture with respect to the laser frequency, the Hamiltonian describing the atomic dynamics — both internal and center-of-mass — is given by

$$H = \frac{\hat{\mathbf{p}}^2}{2M} - \frac{\hbar}{2}\begin{pmatrix} -\Delta & \Omega_1(\hat{\mathbf{x}}) \\ \Omega_1^{\dagger}(\hat{\mathbf{x}}) & \Delta \end{pmatrix}, \qquad (74.19)$$

where

$$\Omega_1(\mathbf{x}) = 2d\,\epsilon_d\cdot\mathbf{E}^{(+)}(\mathbf{x})/\hbar \qquad (74.20)$$

is the spatially dependent *bare Rabi frequency*. In the field (74.16), $\Omega_1(\mathbf{x})$ is cosinusoidal with peak value $\Omega_0 = 2d\mathcal{E}_0/\hbar$.

Atom Optical Stern-Gerlach Effect

The Hamiltonian (74.19) may be written in the form

$$H = \frac{\hat{\mathbf{p}}^2}{2M} + \frac{\hbar}{2}\sigma\cdot\mathbf{B}_{\text{eff}}(\hat{\mathbf{x}}), \qquad (74.21)$$

where

$$\mathbf{B}_{\text{eff}}(\mathbf{x}) = (-\Re\{\Omega_1(\mathbf{x})\}, \Im\{\Omega_1(\mathbf{x})\}, \Delta). \qquad (74.22)$$

As it stands, the Hamiltonian (74.21) describes the precession and c.m. motion of a fictitious spin in an external "magnetic field" $\mathbf{B}_{\text{eff}}(\mathbf{x})$. Spatial variations of this field give rise to the *Stern-Gerlach effect*, i.e., the splitting of the atomic c.m. wave function (see also Sect. 74.5.3).

Adiabatic Approximation

In the position representation, the Hamiltonian (74.19) acts on state vectors of the form of a bispinor

$$\Psi(\mathbf{x}, t) = \begin{pmatrix} \psi_e(\mathbf{x}, t) \\ \psi_g(\mathbf{x}, t) \end{pmatrix}. \qquad (74.23)$$

Alternatively, this state vector may be expanded in terms of the *local eigenvectors* $\alpha_{\pm}(\mathbf{x})$, $\beta_{\pm}(\mathbf{x})$ of the interaction matrix, also called *dressed states*,

$$\Psi(\mathbf{x}, t) = \psi_+(\mathbf{x}, t)\begin{pmatrix} \alpha_+(\mathbf{x}) \\ \beta_+(\mathbf{x}) \end{pmatrix} + \psi_-(\mathbf{x}, t)\begin{pmatrix} \alpha_-(\mathbf{x}) \\ \beta_-(\mathbf{x}) \end{pmatrix}, \qquad (74.24)$$

where

$$\hat{\mathcal{U}}(\mathbf{x})\begin{pmatrix} \alpha_{\pm}(\mathbf{x}) \\ \beta_{\pm}(\mathbf{x}) \end{pmatrix} = \pm\frac{\hbar\Omega(\mathbf{x})}{2}\begin{pmatrix} \alpha_{\pm}(\mathbf{x}) \\ \beta_{\pm}(\mathbf{x}) \end{pmatrix}, \qquad (74.25)$$

with eigenvalues determined by the *dressed Rabi frequency*

$$\Omega(\mathbf{x}) = \sqrt{|\Omega_1(\mathbf{x})|^2 + \Delta^2}, \qquad (74.26)$$

and coefficients (we suppress the $\mathbf{x}$-dependence for notational clarity)

$$\begin{pmatrix} \alpha_+ & \alpha_- \\ \beta_+ & \beta_- \end{pmatrix} = \begin{pmatrix} \cos\frac{\vartheta}{2} & -e^{-i\varphi}\sin\frac{\vartheta}{2} \\ e^{i\varphi}\sin\frac{\vartheta}{2} & \cos\frac{\vartheta}{2} \end{pmatrix} \equiv \hat{S}. \qquad (74.27)$$

Here, the *Stückelberg angle* $\vartheta \equiv \vartheta(\mathbf{x})$ and phase angle $\varphi \equiv \varphi(\mathbf{x})$ are defined in terms of the polar representation of the effective magnetic field

$$\mathbf{B}_{\text{eff}} = \Omega\,(\cos\varphi\sin\vartheta, \sin\varphi\sin\vartheta, \cos\vartheta). \qquad (74.28)$$

In the dressed state basis, the transformed Hamiltonian assumes the form

$$\tilde{H} \equiv \hat{S}^{\dagger}\hat{H}\hat{S} = \frac{\left[\hat{\mathbf{p}} - \hat{\boldsymbol{\mathcal{A}}}(\hat{\mathbf{x}})\right]^2}{2M} + \frac{\hbar}{2}\Omega(\hat{\mathbf{x}})\sigma_3, \qquad (74.29)$$

with matrix-valued vector potential

$$\hat{\boldsymbol{\mathcal{A}}}(\mathbf{x}) = i\hbar S^{\dagger}(\mathbf{x})\left[\nabla S(\mathbf{x})\right]. \qquad (74.30)$$

This matrix is not diagonal; its off-diagonal elements describe *nonadiabatic transitions* between the dressed states. If the detuning is sufficiently large, and the atom moves sufficiently slowly, these nonadiabatic transitions may be neglected in a first approximation. In such an approximation, which is akin to the *Born-Oppenheimer approximation* in molecular physics, the dynamics of the atom is described by two decoupled Hamiltonians of the generic form (74.3), with scalar potentials $\mathcal{U}$ given by

$$\mathcal{U}(\mathbf{x}) = \pm \frac{\hbar}{2}\Omega(\mathbf{x}) \qquad (74.31)$$

and vector potentials $\boldsymbol{\mathcal{A}}(\mathbf{x})$ given by the diagonal elements of (74.30). The vector potential is usually neglected. If included, it describes the *Berry phase* of the mechanical effects of the radiation interaction of a two-level atom.

The idea behind the adiabatic approximation is that the internal state of the atom, which is described by any of the locally varying dressed states, has enough time to adjust smoothly to the motion of the atom. For the important case of strong detuning $|\Delta| \gg |\Omega_1(\mathbf{x})|$, this assumption is usually well justified. In this case, an atom in the dressed ground state experiences a potential which is approximately given by

$$\mathcal{U}(\mathbf{x}) = -\frac{\hbar|\Omega_1(\mathbf{x})|^2}{4\Delta}. \qquad (74.32)$$

For red detuning we have $\Delta > 0$, in which case the atom is attracted towards regions of high intensity (*high field seeker*). For blue detuning, the atom is repelled from such regions (*low field seeker*).

74.3 PRINCIPLES OF DE BROGLIE OPTICS

Since the Schrödinger equation is a linear partial differential equation, de Broglie optics shares most of its principles with principles of other wave phenomena, and in particular with the optical principles of electromagnetic waves.

74.3.1 Light Optics Analogy

The analogy of de Broglie optics and light optics becomes particularly transparent for monoenergetic beams of scalar particles. Such beams are described by a time harmonic wave function $\psi(\mathbf{x}, t) = e^{-iEt/\hbar}\psi_E(\mathbf{x})$, where $\psi_E(\mathbf{x})$ obeys the *stationary Schrödinger equation*

$$H\psi_E = E\psi_E. \qquad (74.33)$$

Setting $\boldsymbol{\mathcal{A}}(\mathbf{x}) = 0$ in Eq. (74.3) for simplicity, this equation assumes the form

$$\left\{\nabla^2 + \tilde{k}_0^2\left[1 - \frac{\mathcal{U}(\mathbf{x})}{E}\right]\right\}\psi_E(\mathbf{x}) = 0, \qquad (74.34)$$

where the wavenumber $\tilde{k}_0$ is related to the energy E via the *dispersion relation*

$$E \equiv \frac{\hbar^2 \tilde{k}_0^2}{2M}. \qquad (74.35)$$

If $\mathcal{U} = 0$ at the entrance to the interaction region, E is the kinetic energy of the freely traveling de Broglie wave and $\tilde{k}_0$ is the related wavenumber.

Comparing (74.34) with the *scalar Helmholtz equation* of electromagnetic theory, and identifying

$$\tilde{n}_E(\mathbf{x}) \equiv [1 - \mathcal{U}(\mathbf{x})/E]^{1/2} \qquad (74.36)$$

as an *index of refraction* for matter waves, one observes the complete analogy of scalar optics of stationary matter waves and monochromatic light waves. This analogy can be generalized for spinor valued wave functions which would correspond to vector wave optics in nonisotropic index media. However, in contrast to light, spinor-valued wave functions must not obey a transversality condition.

In Eqs. (74.34)–(74.36), the parameter E describes the kinetic energy of the incoming beam. Thus, E is positive, and therefore $\tilde{n}_E < 1$ for positive values of the potential, while $\tilde{n}_E > 1$ for negative values of the potential. For neutrons, one generally has $\tilde{n}_E < 1$. This contrasts to the index of refraction for light waves which is generally larger than one. For electrons, ions, and atoms both $\tilde{n}_E < 1$ and $\tilde{n}_E > 1$ may be realized.

74.3.2 WKB Approximation

Waves are described by amplitude and phase. Particles are described by position and momentum. The link between these concepts is provided by *Hamilton's ray optics*. For scalar matter waves, a ray follows a *classical trajectory*. The optical signature of the ray is the phase associated with it. The quantum mechanical version of Hamilton's ray optics is obtained in the *WKB approximation* of the stationary Schrödinger equation (74.33).

Any solution of (74.33) may be written in the form

$$\psi_E(\mathbf{x}) = A(\mathbf{x})e^{iW(\mathbf{x})} \qquad (74.37)$$

with real-valued $A(\mathbf{x})$ and $W(\mathbf{x})$. In the WKB approximation $A(\mathbf{x}) \approx 1$, and

$$W(\mathbf{x}) = \tilde{k}_0 \int_{P_i}^{P} \tilde{n}_E(s)ds + \hbar^{-1}\int_{\mathbf{x}_i}^{\mathbf{x}} \boldsymbol{\mathcal{A}}(\mathbf{x}')\cdot d\mathbf{x}', \qquad (74.38)$$

which is called *eikonal* in Hamilton's ray-optics. In this expression, $\tilde{n}_E(s) \equiv \tilde{n}_E(\mathbf{x}(s))$, where $\mathbf{x}(s)$ denotes the classical trajectory of energy E connecting the point P_i with the point P, $\mathbf{x}_i$ and $\mathbf{x}$ are the coordinates of P_i

and P, respectively, and $ds \equiv |d\mathbf{x}|$ is an infinitesimal arc length measured along the classical trajectory. Note that the second contribution in Eq. (74.38) is generally gauge-dependent. However, for closed loops which are frequently encountered in interferometry, the gauge-dependence disappears by virtue of Stokes theorem which transforms the path integral into an area integral over the rotor of $\boldsymbol{\mathcal{A}}$.

The eikonal (74.38) may also be written in the form of a *reduced action*

$$W(\mathbf{x}) = \frac{1}{\hbar} \int_{\mathbf{x}_i}^{\mathbf{x}} \mathbf{p}(\mathbf{x}') \cdot d\mathbf{x}' = \int_{\mathbf{x}_i}^{\mathbf{x}} \tilde{\mathbf{k}}(\mathbf{x}') \cdot d\mathbf{x}', \quad (74.39)$$

where $\mathbf{p}(\mathbf{x})$ is the local value of the *canonical momentum* of the particle, $\tilde{\mathbf{k}}(\mathbf{x}) \equiv \mathbf{p}(\mathbf{x})/\hbar$ is the corresponding wave vector, and the integral is evaluated along the classical trajectory of the particle. Note that in the presence of a vector potential, $\mathbf{p}(\mathbf{x})$ and $d\mathbf{x}$ are no longer parallel as a result of the difference between canonical momentum $\mathbf{p}$ and *kinetic momentum* $M d/dt \, \mathbf{x} \equiv \mathbf{p} - \boldsymbol{\mathcal{A}}$.

The WKB approximation becomes invalid in the vicinity of *caustics* where neighboring rays intersect. There, *connection formulae* are used to find the proper phase factors picked up by the ray in traversing the caustics. For further details see Ref. [8].

74.3.3 Phase and Group Velocity

The velocity of a particle which traverses a region of negative potential increases so that $p(\mathbf{x}) > p_0$, but the phase becomes retarded: $\delta W = \int [\mathbf{p}_0 - \mathbf{p}(\mathbf{x})] \cdot d\mathbf{x} < 0$. In quantum mechanics, the classical velocity corresponds to the *group velocity*, while the evolution of the phase is determined by the *phase velocity*. The phase and group velocities of de Broglie waves are given by

$$v_{\mathrm{p}}(\mathbf{x}) \equiv \frac{E}{p(\mathbf{x})} = \frac{1}{\tilde{n}_E(\mathbf{x})} \sqrt{\frac{E}{2M}}, \quad (74.40)$$

$$v_{\mathrm{g}}(\mathbf{x}) \equiv \frac{\partial E}{\partial p(\mathbf{x})} = \tilde{n}_E(\mathbf{x}) \sqrt{\frac{2E}{M}}, \quad (74.41)$$

respectively. Note that the product $v_{\mathrm{p}} v_{\mathrm{g}} = E/M$ is independent of $\tilde{n}_E(\mathbf{x})$.

74.3.4 Paraxial Approximation

The paraxial approximation is useful in describing the evolution of wave-like properties and/or distortion of wavefronts in the immediate neighborhood of an optical ray.

Let the z-axis be the central *optical axis* of symmetry along which the optical elements are aligned. Using the

Ansatz $\psi_E(\mathbf{x}) = e^{i\tilde{k}_0 z} \phi(x, y; z)$, and dropping $\partial^2/\partial z^2 \phi$ in a *slowly varying envelope approximation*, one obtains

$$i\hbar v_0 \frac{\partial}{\partial z} \phi(x, y; z) = \left\{ -\frac{\hbar^2}{2M} \nabla_\perp^2 + \mathcal{U}(x, y; z) \right\} \phi(x, y; z), \quad (74.42)$$

where $\nabla_\perp^2 = \partial^2/\partial x^2 + \partial^2/\partial y^2$, and $v_0 = \hbar \tilde{k}_0/M$ is the *longitudinal velocity* of incoming particles. This equation has exactly the form of a time-dependent Schrödinger equation in two dimensions, with z/v_0 playing the role of a fictitious time t. With this interpretation, the spatial evolution of phase fronts along z can be analyzed in dynamical terms of particles moving in the xy-plane.

74.3.5 Raman-Nath Approximation

In the *Raman-Nath approximation* (RNA) (also called the short-time, thin-hologram, or thin-lens approximation), the $\nabla_\perp^2$ term in Eq. (74.42) is neglected. The potential $\mathcal{U}(x, y; z)$ then acts as a pure phase structure, and the solution of Eq. (74.42) becomes

$$\phi(x, y; z) = \exp\left\{ -\frac{i}{\hbar v_0} \int_{z_i}^{z} dz' \mathcal{U}(x, y; z') \right\} \phi(x, y; z_i). \quad (74.43)$$

In terms of a classical particle moving under the influence of $\mathcal{U}$, the approximation loses validity for $\frac{1}{2} M v_\perp^2 \gtrsim \mathcal{U}$, which is just a quarter cycle for a harmonic oscillator.

74.4 REFRACTION AND REFLECTION

Consider a particle beam of energy E incident on a medium with constant index of refraction $\tilde{n}_E$. The boundary plane at $z = 0$ in Fig. 74.1 divides the vacuum from the medium. At the boundary, the beam is partially reflected and partially transmitted, with the angles determined by *Snell's law of refraction*

$$\sin \alpha = \tilde{n}_E \sin \beta, \quad (74.44)$$

and the *law of reflection*

$$\alpha = \alpha'. \quad (74.45)$$

The coefficients of *reflectivity* $\tilde{R}$, and *transmittivity* $\tilde{T} = 1 - \tilde{R}$ are given by the *Fresnel formula*

$$\tilde{R} = \left| \frac{\cos \alpha - \tilde{n}_E \cos \beta}{\cos \alpha + \tilde{n}_E \cos \beta} \right|^2. \quad (74.46)$$

For $\tilde{n}_E > 1$, the interface is "attractive" and $\tilde{R} \ll 1$, with $\tilde{R} \to 1$ only for *glancing incidence* $\alpha \to \pi/2$. For $\tilde{n}_E < 1$, the interface is "repulsive" and *total reflection*

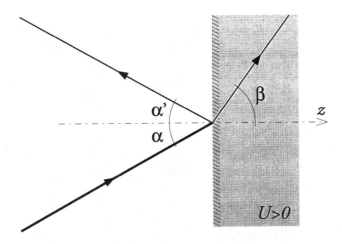

Figure 74.1. Reflection geometry.

($\tilde{R} = 1$) occurs for $\alpha \geq \tilde{\alpha}_c$, where $\tilde{\alpha}_c = \sin^{-1} \tilde{n}_E$ is the *critical angle*. For $\alpha > \tilde{\alpha}_c$, the de Broglie wave becomes *evanescent*, with $\psi_E(z > 0) \sim e^{-\tilde{\kappa}z}$ inside the medium ($z > 0$), where

$$\tilde{\kappa} = \tilde{k}_0 \left[\sin^2 \alpha - \sin^2 \tilde{\alpha}_c \right]^{1/2} . \qquad (74.47)$$

For thermal neutrons, $\pi/2 - \tilde{\alpha}_c \sim 3 \times 10^{-3}$ radians.

If $E < \mathcal{U}$, then $\tilde{n}_E$ is imaginary and total reflection occurs for all α. In neutron optics, this *total mirror reflection* requires *ultracold neutrons* ($T \sim 0.5$ mK). It has important applications for storage of ultracold neutrons in material cavities, and neutron microscopy using spherical mirrors. For details see Ref. [2].

74.4.1 Atomic Mirrors

Inelastic processes such as diffuse scattering absorption inhibit coherent reflection of atoms from bare surfaces. The surface must therefore be coated either with material of low adsorptivity (noble gas, see Ref. [9]) or light (evanescent light, see below).

Reflection of Atoms by Evanescent Laser Light

Evanescent light fields are produced by total internal reflection of a light beam at a dielectric-vacuum interface. In the vacuum, the field decays exponentially away from the interface on a characteristic length κ^{-1} where

$$\kappa = k \left[n^2 \sin^2 \theta_i - 1 \right]^{-1/2} . \qquad (74.48)$$

Here, n is the light index of refraction of the dielectric, k is the wave number of the light beam *in vacuo*, and θ_i is its angle of incidence.

If the light is blue-detuned from the atomic resonance, an incident beam of ground state atoms experiences the repulsive potential

$$\mathcal{U}(\mathbf{x}) = \frac{\hbar}{2} \left[\left(\Omega_0^2 e^{-2\tilde{\kappa}|z|} + \Delta^2 \right)^{1/2} - |\Delta| \right] . \qquad (74.49)$$

For $\alpha > \tilde{\alpha}_c$, the evanescent field acts as a nearly perfect mirror, the imperfections being due to nonadiabatic transitions into the "wrong" dressed state, and possible spontaneous emission. Reflection of atoms by evanescent laser light was demonstrated by Balykin *et al.* [10].

74.4.2 Atomic Cavities

Atomic reflections are used in the two kinds of cavities proposed so far: the trampoline cavity and the Fabry-Perot resonator.

The *trampoline cavity*, also called the *gravito-optical cavity*, consists of a single evanescent mirror facing upwards, the second mirror being provided by gravitation. A stable cavity is realized with the evanescent laser field of a parabolically shaped dielectric-vacuum interface [11].

In the *atomic Fabry-Perot resonator* both mirrors are realized by laser light [12].

74.4.3 Atomic Lenses

De Broglie waves may be focused by *refraction* from a parabolic potential or by *diffraction*, e.g., by a Fresnel zone plate (see Sect. 74.5.2). Consider focusing by the parabolic potential

$$\mathcal{U}(x, y; z) = \begin{cases} \frac{1}{2} M \omega_f^2 (x^2 + y^2) & , \quad -w \leq z \leq 0 \\ 0 & , \quad \text{otherwise} . \end{cases} \qquad (74.50)$$

For ground state atoms, such a potential is realized in the vicinity of the node of a blue detuned standing wave laser field of transverse width w. In this case

$$\omega_f = \Omega_0 \left(\omega_{\text{rec}}/|\Delta| \right)^{1/2} , \qquad (74.51)$$

where $\omega_{\text{rec}} = \hbar k^2 / 2M$ is the recoil frequency.

Comparison with the Raman-Nath approximation (74.43) at $z = 0$, with the phase fronts of a spherical wave converging towards a point $\mathbf{x}_f = (0, 0, f)$, shows that $\mathcal{U}$ describes a lens of focal length

$$f = \frac{v_0^2}{w \omega_f^2} . \qquad (74.52)$$

The Raman-Nath approximation is only valid for a *thin lens* $w \ll f$, and breaks down for $w > w_{\text{RN}} = \pi v_0 / 2\omega_f$. In the latter case, oscillations of the particles in the harmonic potential become relevant, a phenomenon sometimes called *channeling*. Channeling may be used to realize *thick lenses* with focal length $f = w_{\mathrm{RN}}$ corresponding to a quarter oscillation period.

Focusing of a metastable helium beam using the antinode of a *large period standing wave laser field* has been demonstrated [13]. Such a field is produced by reflecting a laser beam from a mirror under glancing incidence. The standing wave forms normal to the mirror surface.

74.5 DIFFRACTION

The diffraction of matter waves is described by the solution of the Schrödinger equation (74.33) subject to the boundary conditions imposed by the diffracting object. For a plane screen Σ made of opaque portions and apertures, the solution in the source-free region behind the screen is given by the Rayleigh-Sommerfeld formulation of the *Huygens principle*

$$\psi_E(\mathbf{x}) = \frac{\tilde{k}_0}{i} \int_\Sigma \frac{d\xi d\eta}{2\pi} \frac{e^{i\tilde{k}_0 R}}{R} \left(1 + \frac{i}{\tilde{k}_0 R}\right) \frac{\mathbf{n}\cdot\mathbf{R}}{R} \psi(\boldsymbol{\xi}) , \tag{74.53}$$

where $\boldsymbol{\xi} = (\xi, \eta, \zeta)$ denotes coordinates of points on Σ, $\mathbf{n}$ is an inwardly directed normal to Σ at a point $\boldsymbol{\xi}$, and $\mathbf{R} = \mathbf{x} - \boldsymbol{\xi}$.

A diffraction pattern only becomes manifest in the *diffraction limit* $r \gg d$, where r is the distance to the observation point, and d is the length scale of the diffracting system. The two diffraction regimes are then the *Fraunhofer limit* $r \gg d^2/\lambda_{\mathrm{dB}}$ and the *Fresnel regime* $r \approx d^2/\lambda_{\mathrm{dB}}$, also called *near-field optics*.

74.5.1 Fraunhofer Diffraction

In the Fraunhofer limit, the field at position (x, y) on a screen at a distance L downstream from the diffracting object is given by

$$\psi(x, y; L) = \tilde{k}_0 \frac{e^{ik_0 L}}{iL} \int \frac{d\xi d\eta}{2\pi} e^{-i(\tilde{k}_x \xi + \tilde{k}_y \eta)} \psi(\xi, \eta; 0) , \tag{74.54}$$

where $\tilde{k}_x = \tilde{k}_0 x/L$, $\tilde{k}_y = \tilde{k}_0 y/L$. The field at the observation screen is thus given by the Fourier transform of the field in the object plane; i.e. the *momentum representation* of the diffracted state. Since most diffraction experiments in atom optics are performed in the Fraunhofer limit, most calculations are done in the momentum representation.

Atomic diffractions from microfabricated transmission gratings [14] and double slits [15] have been observed.

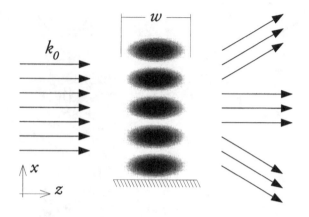

Figure 74.2. Geometry of the Kapitza-Dirac effect.

74.5.2 Fresnel Diffraction

Typical applications of Fresnel diffraction are Fresnel zone plates and the effects of Talbot and Lau. *Fresnel zone plates* are microfabricated concentric amplitude structures which act like lenses. They are frequently employed in optics of α-particles and neutrons. In atom optics, focusing with a Fresnel zone plate was first demonstrated by Carnal *et al.* [16].

The *Talbot effect* and the related *Lau effect* refer to the *self-imaging* of a grating of period d, which appears downstream at distances that are integral multiples of the *Talbot length* $L = 2d^2/\lambda_{\mathrm{dB}}$. For a discrete set of smaller distances than the Talbot length, images of the grating appear with higher periods d/n, $n = 2, 3 \cdots$. For applications in matter wave interferometry, see Ref. [17]; for applications in atom lithography see Ref. [18].

74.5.3 Near-Resonant Kapitza-Dirac Effect

The *near resonant Kapitza-Dirac effect* refers to the diffraction of two-level atoms from a standing wave laser field with a spatially uniform polarization. The dynamics of the effect is described by the Hamiltonian (74.19) with the mode function of the laser field given by (74.16). Consider atoms traveling predominantly in the z-direction, i.e., orthogonal to the axis of the laser field, with energy $\frac{1}{2}Mv_0^2 \gg \hbar\Omega_0$ (see Fig. 74.2). Kapitza-Dirac diffraction is then observed in transmission, which in the theory of diffraction is called *Laue geometry*. In the paraxial approximation (see Sect. 74.3.4) for motion in the z-direction, and assuming that the laser profile is homogeneous in the y-direction, the description reduces to an effectively one-dimensional model for the quantum mechanical motion along the x-axis of the laser field.

Due to the periodicity of the standing wave light field, the transverse momenta of the transmitted atom waves differ by multiples of $\hbar k$ from the transverse momentum of the undiffracted wave. For the important case of strong detuning, and assuming that the incoming atoms are in their electronic ground state moving with transverse momentum $p_x = 0$, the p_x-distribution of the outgoing wave is given in the RNA (Sect. 74.3.5) by

$$\text{Prob}(p_x = 2n\hbar k) \propto \left| J_n\left(\frac{\Omega_0^2}{8|\Delta|}\tau\right)\right|^2 , \qquad (74.55)$$

where J_n is a Bessel function of order n, $\tau = w/v_0$ is an effective interaction time, w being the width of the laser field, and v_0 the longitudinal velocity of the atoms. The distribution (74.55) was observed by Gould *et al.* [19].

For $\tau \gtrsim \tau_{\text{RN}}$, where $\tau_{\text{RN}} = (\omega_{\text{rec}}\Omega_0^2/|\Delta|)^{-1/2}$, the Raman-Nath approximation becomes invalid. As a result of Doppler related phase-mismatch, the momentum spread saturates and shows a sequence of collapse and revival as a function of τ [20].

If the detuning is too small to allow for a scalar description, the two-level character of the atoms must be taken into account. For the particular case of $\Delta = 0$, the ground state evolves into an equal superposition of the two *diabatic states* $\frac{1}{\sqrt{2}}(|e\rangle \pm |g\rangle)$ while entering the interaction region. Inside the interaction region, these states experience potentials which differ only by their sign. For atomic beams with a small spatial spread $\delta x \ll 2\pi/k$, the diabatic states experience opposite forces, leading to a splitting of the atomic beam called the *atom-optical Stern-Gerlach effect* [21].

In the general case of arbitrary Δ, the Kapitza-Dirac Hamiltonian (74.21) is most conveniently analyzed using band theoretical methods of solid state theory [22].

74.5.4 Atom Beam Splitters

Beam splitters are optical devices which divide an incoming beam into two outgoing beams traveling in different directions. For thermal neutrons, beam splitters may be realized by diffraction from perfect crystals in Laue geometry. For atoms, they can be realized using diffraction from crystalline surfaces, microfabricated structures (see Sect. 74.5.1), or by using diffraction from an optical standing wave.

The Kapitza-Dirac effect, for example, may be exploited to split an atomic beam coherently using *Bragg reflection* at the "lattice planes" provided by the periodic intensity variations of a standing wave laser field [23]. This process is resonant for an incoming atomic beam traveling with transverse momentum $p_x = \hbar k$ because it is energetically degenerate with the diffracted beam traveling with transverse momentum $\bar{p}_x = -\hbar k$. This level degeneracy is lifted while the atoms enter the interaction region. In the *Bragg regime*, the lifting happens slowly enough that only the momentum states $|\pm\hbar k\rangle$ participate

in the diffraction (*two-beam resonance*), and their populations show *Pendellösung* type oscillations as a function of the transit time. The frequency of the oscillations is given by $\delta E/\hbar$, where δE is the energy splitting of the two beams inside the interaction region. For a transit time given by a quarter period of the *Pendellösung*, a 50% beam splitting is observed [23].

In principle, Bragg resonances may also be realized for higher diffraction order $p_x = n\hbar k \leftrightarrow \bar{p}_x = -n\hbar k$. However, in this case, intermediate momentum states become populated (*multibeam resonance*), which makes the higher-order Bragg resonances less suitable for beam splitting purposes.

More promising for the realization of an atomic beam splitter is the *magneto-optical diffraction* which refers to the diffraction of three-level atoms from a laser field with a periodic polarization gradient (lin ⊥ lin configuration) (Chap. 73), and a magnetic field aligned parallel to the optical axis of the laser field. This configuration realizes an interaction potential in the form of a *blazed grating*, i.e. , a phase grating with an approximately triangular variation of phase. In an experiment by Pfau *et al.* [24], transverse splitting of a beam of metastable helium by $42\hbar k$ was observed (see also Ref. [20]).

Diffraction from an evanescent standing wave involves Bragg reflection of atoms under glancing incidence from the periodic grating of a blue detuned evanescent standing wave laser field (see Ref. [25]).

74.6 INTERFERENCE

While the overall phase of a wave function ψ is not observable, interferometry makes detectable the *relative phases* of two components ψ_1, ψ_2 in a superposition $\psi = \psi_1 + \psi_2$. Two types of interferometers are most common: the *Young double slit* as a paradigm for interferometers based on division of wavefront, and the *Mach-Zehnder interferometer* as a paradigm for interferometers based on division of amplitude. In de Broglie optics, the latter type is realized in the form of a three-grating interferometer, since division of amplitude is achieved by diffraction at gratings rather than by semitransparent mirrors.

From a fundamental point of view, any interferometer is a ring. At the entrance port of a three-grating interferometer displayed in Fig. 74.3, for example, the wave function is split into two coherent parts which spatially evolve along different paths and subsequently come together at the exit port where they are superimposed to produce two outgoing waves

$$\psi_\pm = \frac{1}{\sqrt{2}}\left(\psi_1 \pm \psi_2\right) , \qquad (74.56)$$

where the components from path 1 and 2 are given by

$$\psi_1 = A_1 \exp(iW_1) , \quad \psi_2 = A_2 \exp(iW_2) . \qquad (74.57)$$

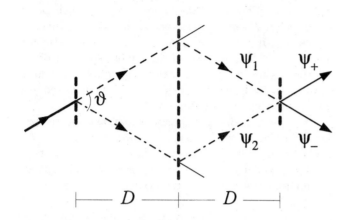

Figure 74.3. Geometry of the three-grating interferometer.

For simplicity, assume $A_1 = A_2 = A_0/\sqrt{2}$. The relative flux of the outgoing waves is then

$$I_\pm = \tfrac{1}{2}\left(1 \pm \cos\chi\right), \qquad (74.58)$$

where $\chi = W_1 - W_2$ is the relative phase of the two components. The sinusoidal variations of $I_\pm$ with varying χ are called *interference fringes*, and $\chi/2\pi$ is called the *fringe order number*.

In the WKB approximation, the phases W_1 and W_2 are given by

$$W_i = W_0 + \int_i \tilde{\mathbf{k}}(\mathbf{x})\cdot d\mathbf{x}, \quad i = 1, 2, \qquad (74.59)$$

where the integrals extend over the classical paths 1 and 2, respectively, and $d\mathbf{x}$ is an element of displacement along the paths. Using Eq. (74.59), the relative phase is

$$\chi = \int_C \tilde{\mathbf{k}}(\mathbf{x})\cdot d\mathbf{x}, \qquad (74.60)$$

where C is the closed *interferometer loop*. Note, that on path 2, the path element $d\mathbf{x}$ and $\tilde{\mathbf{k}}$ are antiparallel.

Usually, the absolute value of χ is not measured, but only variations, called *phase shifts*, which result from displacements of the diffraction gratings or placement of optical elements into the beam path. Phase shifts come in two categories: *dispersive* and *geometric*. If a phase shift χ depends on v_0, it is called dispersive, otherwise it is called geometric. Geometric phases depend only on the geometry of the interferometer loop. The *Sagnac effect* (see below), for example, may be geometric. A phase which depends neither on v_0 nor on the geometry of the interferometer loop is called *topological*. The *Aharonov-Bohm effect* (see below) is topological.

Using Eq. (74.38), χ becomes

$$\chi = \chi_0 + \chi\{\mathcal{U}\} + \chi\{\boldsymbol{\mathcal{A}}\}, \qquad (74.61)$$

where for weak potentials $\mathcal{U}$ and $\boldsymbol{\mathcal{A}}$,

$$\chi_0 = \frac{Mv_0}{\hbar}\left(L_1 - L_2\right), \qquad (74.62)$$

$$\chi\{\mathcal{U}\} = -\frac{1}{\hbar v_0}\oint \mathcal{U}(\mathbf{x}(s))\,ds, \qquad (74.63)$$

$$\chi\{\boldsymbol{\mathcal{A}}\} = \frac{1}{\hbar}\oint \boldsymbol{\mathcal{A}}(\mathbf{x})\cdot d\mathbf{x}, \qquad (74.64)$$

and L_i is the geometric length of the path i. For a constant potential $\mathcal{U}_0$ intersecting the interferometer on a length w, $\chi\{\mathcal{U}\}$ is given by

$$\chi\{\mathcal{U}\} = -\frac{\mathcal{U}_0 w}{\hbar v_0}. \qquad (74.65)$$

Using Stokes theorem, $\chi\{\boldsymbol{\mathcal{A}}\}$ may be written in the manifestly gauge-invariant form

$$\chi\{\boldsymbol{\mathcal{A}}\} = \frac{1}{\hbar}\int (\nabla \times \boldsymbol{\mathcal{A}}(\mathbf{x}))\cdot d\mathbf{a}, \qquad (74.66)$$

where the integral extends over the area enclosed by the interferometer, and $d\mathbf{a}$ is an infinitesimal area element.

Sagnac Effect. The Sagnac effect refers to $\chi\{\boldsymbol{\mathcal{A}}\}$ in a rotating interferometer. Inserting Eq. (74.6) into Eq. (74.66), and assuming that the axis of rotation is oriented perpendicular to the plane of the interferometer, the Sagnac phase shift is given by

$$\chi_{\text{Sa}} = 4M\Omega A/\hbar, \qquad (74.67)$$

where A is the geometric area enclosed by the interferometer loop. χ_{Sa} may be dispersive or geometric depending on the type of interferometer. In a Young double slit, A is independent of energy and χ_{Sa} is geometric. In a three-grating interferometer, the area is $A \approx \vartheta D^2$, where ϑ is the splitting angle; see Fig. 74.3. In this case, χ_{Sa} is dispersive because of the velocity dependence of $\vartheta \approx 2\hbar k/Mv_0$. The Sagnac effect for de Broglie waves was first observed by Werner *et al.* [26] using a neutron interferometer.

Aharonov-Bohm Effect. The Aharonov-Bohm effect refers to the $\chi\{\boldsymbol{\mathcal{A}}\}$ of charged particles encircling a magnetic flux line [27]. Inserting $\boldsymbol{\mathcal{A}}$ from Eq. (74.7) into (74.66), and assuming particles of charge q encircling a line of flux Φ once, one finds

$$\chi_{\text{AB}} = q\Phi/\hbar. \qquad (74.68)$$

A characteristic feature of the Aharonov-Bohm effect is that the particles actually never "see" the magnetic field of the flux line which is confined to some region inaccessible to the particles. χ_{AB} is strictly topological, and only depends on the linking number of the interferometer loop and the flux line. Its appearance is characteristic for all gauge theories. For further details and a summary on its experimental verification see Ref. [4].

Aharonov-Casher Effect. The Aharonov-Casher effect refers to $\chi\{\mathcal{A}\}$ of a magnetic spin encircling an electric line charge [28]. Inserting $\mathcal{A}$ from Eq. (74.11) into (74.66), one obtains for proper alignment of $\boldsymbol{\mu}$ and $\mathbf{E}$

$$\chi_{\rm AC} = 2\pi \frac{|\boldsymbol{\mu}|}{\mu_{\rm B}} \frac{r_0}{\xi}, \qquad (74.69)$$

where r_0 is the classical electron radius, $\mu_{\rm B}$ is the Bohr-magneton, and $\xi = e/\rho_{\rm el}$, $\rho_{\rm el}$ being the electric line charge density. $\chi_{\rm AC}$ is topological only if the spin is aligned parallel to the electric line charge and both are oriented perpendicular to the plane of the interferometer. $\chi_{\rm AC}$ for atoms has been observed by Sangster *et al.* [29].

Electric Dipole Phase. Electric dipole phase refers to the $\chi\{\mathcal{A}\}$ of an *electric* dipole moment encircling a *magnetic* line charge [30]. Inserting $\mathcal{A} = \mathbf{B} \times \mathbf{d}$ into Eq. (74.66), one obtains for proper alignment of $\mathbf{d}$ and $\mathbf{B}$

$$\chi_{\rm dE} = 2\pi \frac{|\mathbf{d}|}{ea_0} \frac{a_0}{\xi}, \qquad (74.70)$$

where a_0 is the Bohr radius, and $\xi = \Phi_0/\varrho_{\rm mg}$, Φ_0 being the magnetic flux unit, and $\varrho_{\rm mg}$ being the magnetic line charge density. In analogy to the Aharonov-Casher effect, $\chi_{\rm dE}$ is topological, provided that $\mathbf{d}$ is aligned parallel to the magnetic line charge, and both are oriented perpendicular to the interferometer plane.

74.6.1 Internal State Interferometry

Manipulation of the internal state of atoms by means of electromagnetic fields makes it possible to realize interferometric setups which involve separation of paths in internal space rather than in real space. Examples of such interferometers are the *Optical Ramsey interferometer* [31], the *stimulated Raman interferometer* [32], and the interferometers using static electric and magnetic fields [33, 34].

74.6.2 Manipulation of Cavity Fields by Atom Interferometry

The entanglement of atomic states and quantized field states opens novel possibilities for manipulating and/or measuring nonclassical field states in a cavity. In the adiabatic limit, for example, and assuming sufficient detuning between the atom and the cavity field, the interaction and c.m. motion of an atom traversing a cavity is well described by the potential [see also Eq. (74.32)]

$$\mathcal{U}(\mathbf{x}) = -\hbar \frac{g^2}{\Delta} f(\mathbf{x})^2 a^{\dagger} a, \qquad (74.71)$$

where g is the *vacuum Rabi frequency*, $f(\mathbf{x})$ is a cavity mode function, and a, $a^{\dagger}$ denote cavity photon annihilation and creation operators, respectively.

Because of the presence of the photon-number operator $a^{\dagger}a$ in Eq. (74.71), the deflection and phase shift of an atom traversing the cavity is quantized, displaying essentially the photon number statistics in the cavity. The quantized deflection is sometimes called the *inverse Stern-Gerlach effect*.

Due to the entanglement of atom and cavity states, and the position dependence of the interaction strength, the phase shift induced by $\mathcal{U}(\mathbf{x})$ in a standing wave cavity may be used to measure either the atomic position via homodyne detection of the cavity field [35, 36], or the photon statistics via atom interferometry [37, 38]. In a ring cavity, the entanglement of c.m. motion and cavity field may be used to measure the atomic momentum [39] via homodyne detection of the cavity field. For further details see Ref. [40].

74.7 COHERENCE OF SCALAR MATTER WAVES

The general solution of the free Schrödinger equation (74.2) may be written in the form

$$\psi(\mathbf{x}, t) = \int d^3\tilde{k} \, a(\tilde{\mathbf{k}}) e^{i(\tilde{\mathbf{k}} \cdot \mathbf{x} - \omega(\tilde{\mathbf{k}})t)}, \qquad (74.72)$$

where $\omega(\tilde{\mathbf{k}}) \equiv E/\hbar = \hbar \tilde{k}^2/2M$. If the coefficients $a(\tilde{\mathbf{k}})$ are known, the state represented by $\psi(\mathbf{x}, t)$ is called a *pure state*. Otherwise it is called a *mixed state*, and physical quantities are obtained by an *ensemble average* over the possible realizations of $a(\tilde{\mathbf{k}})$.

The *degree of coherence* of matter waves is described by the autocorrelation function of $\Psi(\mathbf{x}, t)$:

$$\Gamma(\mathbf{x}, t; \mathbf{x}', t') \equiv \overline{\Psi(\mathbf{x}', t')^* \Psi(\mathbf{x}, t)}, \qquad (74.73)$$

where the overline $\overline{(\cdots)}$ denotes the ensemble average over the possible realizations of $a(\tilde{\mathbf{k}})$.

In light optics, $\Gamma(\mathbf{x}, t; \mathbf{x}', t')$ is called the *mutual coherence function*. In particular, for equal times, $\Gamma(\mathbf{x}, t; \mathbf{x}', t)$ describes the *spatial coherence*, and for equal positions, $\Gamma(\mathbf{x}, t; \mathbf{x}, t')$ describes the *temporal coherence*.

For a beam of particles coherence may be either *longitudinal* (measured along the beam) or *transverse* (measured across the beam). In contrast to light optics, there is no simple relation between longitudinal coherence and temporal coherence because the dispersion relation of matter waves is quadratic in the wavenumber.

The spatial coherence function is intimately related to the quantum mechanical density operator of the particles (see also Chap. 7)

$$\rho(t) = \overline{|\Psi(t)\rangle \langle \Psi(t)|}. \qquad (74.74)$$

In the position representation, one has

$$\langle \mathbf{x}|\rho(t)|\mathbf{x}'\rangle \equiv \rho(\mathbf{x}, \mathbf{x}'; t) = \Gamma(\mathbf{x}, t; \mathbf{x}', t). \qquad (74.75)$$

Longitudinal and temporal coherence of a particle beam is determined mainly by the source of the beam. The thermal fission reactors used in neutron optics and the ovens used in atom optics are analogous to blackbody sources in light optics. In contrast, the transverse coherence is mainly determined by the way the particles are extracted from the oven to form a beam.

74.7.1 Atomic Sources

To describe thermal sources, consider a single particle in an oven of temperature T and volume V, assuming that $a(\tilde{\mathbf{k}})$ and $a(\tilde{\mathbf{k}}')$ are statistically independent:

$$\overline{a(\tilde{\mathbf{k}})^* a(\tilde{\mathbf{k}}')} = \rho(\tilde{k})\delta(\tilde{\mathbf{k}} - \tilde{\mathbf{k}}'), \qquad (74.76)$$

where

$$\rho(\tilde{k}) = \frac{\lambda_{\text{th}}^3}{V} \exp\left\{ -\frac{\hbar^2 \tilde{k}^2}{2Mk_{\text{B}}T} \right\} \qquad (74.77)$$

accounts for the thermal distribution of wavenumbers, and

$$\lambda_{\text{th}} = \left(\frac{2\pi\hbar^2}{Mk_{\text{B}}T} \right)^{1/2} \qquad (74.78)$$

denotes the *thermal de Broglie wavelength*.

Using (74.76)–(74.78) in Eq. (74.73), the mutual coherence function becomes

$$\Gamma(\mathbf{x}, t; \mathbf{x}', t') = \frac{1}{V} \frac{1}{\{1 + [i(t - t')/\tau_{\text{th}}]\}^{\frac{3}{2}}}$$
$$\times \exp\left\{ -\pi \frac{(\mathbf{x} - \mathbf{x}')^2}{\lambda_{\text{th}}^2 (1 + [i(t - t')/\tau_{\text{th}}])} \right\},$$

$$(74.79)$$

where

$$\tau_{\text{th}} = \frac{\hbar}{k_{\text{B}}T}, \qquad (74.80)$$

is the *thermal coherence time*.

According to (74.79), the spatial coherence of a thermal state falls off in a Gaussian manner on a scale given by λ_{th}. The temporal coherence, in contrast, falls off algebraically on a time scale given by τ_{th}. Expressed in physical units, one has

$$\lambda_{\text{th}} = \frac{1.74(5) \times 10^{-9}\text{m}}{\sqrt{(M/\text{u})(T/\text{K})}}, \quad \tau_{\text{th}} = \frac{7.63 \times 10^{-12}\text{s}}{(T/\text{K})}, \quad (74.81)$$

where u is the atomic mass unit.

Atomic Beams

Effusive Beams. Effusive beams are produced from thermal sources by a suitable set of collimators placed in front of the opening of the oven. This produces a Maxwell-Boltzmann distribution of atomic velocities in the longitudinal direction. The coherence properties in the transverse direction are described by the *van Cittert-Zernike theorem*; for details see any textbook on classical optics (see also Ref. [41]).

Supersonic Beams. Supersonic beams are produced by supersonic expansion of a high pressure gas which is forced through an appropriately designed nozzle. The expansion produces a velocity distribution in the longitudinal direction which is approximately Gaussian with a *velocity ratio* $v/\delta v \approx 10\text{–}20$.

Pulsed Beams. Pulsed beams are produced by chopping any of the beams described above. Important applications for pulsed beams are the resolution of temporal coherence and the mapping of the relative phases of the $a(\tilde{\mathbf{k}})$ in matter wave interferometry.

Laser-Like Source of Atoms. These are proposed sources where many atoms with integral spin (Bosons) occupy one and the same quantum state of motion [42]. Their operational principle is rooted in the quantum statistical effects of indistinguishability. It may be viewed in close analogy to the operational principle of an ordinary laser (see Chap. 68) and the mechanism underlying Bose-Einstein Condensation (see Sect. 73.5.1).

REFERENCES

1. P. W. Hawkes and E. Kasper, *Principles of Electron Optics* (Academic Press, London, 1989).
2. V. F. Sears, *Neutron Optics*, Oxford Series on Neutron Scattering in Condensed Matter (Oxford University Press, Oxford, 1989).
3. L. Reimer, *Transmission Electron Microscopy* in *Springer Series in Optical Sciences* (Springer, Berlin, 1989), 2nd ed., Vol. 36.
4. A. Tonomura, *Electron Holography* in *Springer Series in Optical Sciences*, edited by K. Shimoda (Springer, Berlin, 1993), Vol. 70.
5. D. M. Greenberger and A. W. Overhauser, Rev. Mod. Phys. **51**, 43 (1979).
6. C. S. Adams, M. Sigel, and J. Mlynek, Phys. Rep. **240**, 143 (1994).
7. G. Lenz, P. Meystre, and E. M. Wright, Phys. Rev. Lett. **71**, 3271 (1993).
8. A. B. Migdal and V. Krainov, in *Approximation Methods in Quantum Mechanics, Frontiers in Physics Series*, edited by D. Pines (W. A. Benjamin, New York, 1969).
9. J. J. Berkhout *et al.*, Phys. Rev. Lett. **63**, 1689 (1989).
10. V. I. Balykin, V. S. Letokhov, Y. B. Ovchinnikov, and A. I. Sidorov, Phys. Rev. Lett. **60**, 2137 (1988).
11. H. Wallis, J. Dalibard, and C. Cohen-Tannoudji, Appl. Phys. **B54**, 407 (1992).

12. M. Wilkens, E. Goldstein, B. Taylor, and P. Meystre, Phys. Rev. **A47**, 2366 (1993).
13. T. Sleator, T. Pfau, V. Balykin, and J. Mlynek, Appl. Phys. **B54**, 375 (1992).
14. D. W. Keith, M. L. Schattenburg, H. I. Smith, and D. E. Pritchard, Phys. Rev. Lett. **61**, 1580 (1988).
15. O. Carnal and J. Mlynek, Phys. Rev. Lett. **66**, 2689 (1991).
16. O. Carnal *et al.*, Phys. Rev. Lett. **67**, 3231 (1991).
17. J. F. Clauser and M. Reinsch, Appl. Phys. **B54**, 380 (1992).
18. U. Janicke and M. Wilkens, Journ. Physique II **4**, 1975 (1994).
19. P. L. Gould, G. A. Ruff, and D. E. Pritchard, Phys. Rev. Lett. **56**, 827 (1986).
20. U. Janicke and M. Wilkens, Phys. Rev. **A50**, 3265 (1994).
21. T. Sleator *et al.*, Phys. Rev. Lett. **68**, 1996 (1992).
22. M. Wilkens, E. Schumacher, and P. Meystre, Phys. Rev. **A44**, 3130 (1991).
23. P. J. Martin, B. G. Oldaker, A. H. Miklich, and D. E. Pritchard, Phys. Rev. Lett. **60**, 515 (1988).
24. T. Pfau *et al.*, Phys. Rev. Lett. **71**, 3427 (1993).
25. R. Deutschmann, W. Ertmer, and H. Wallis, Phys. Rev. **A47**, 2169 (1993).
26. S. A. Werner, J. Staudenmann, and R. Colella, Phys. Rev. Lett. **42**, 1102 (1979).
27. Y. Aharonov and D. Bohm, Phys. Rev. **115**, 485 (1959).
28. Y. Aharonov and A. Casher, Phys. Rev. Lett. **53**, 319 (1984).
29. K. Sangster, E. A. Hinds, S. M. Barnett, and E. Riis, Phys. Rev. Lett. **71**, 3641 (1993).
30. M. Wilkens, Phys. Rev. Lett. **72**, 5 (1994).
31. C. J. Bordé, Phys. Lett. **A140**, 10 (1989).
32. M. Kasevich and S. Chu, Phys. Rev. Lett. **67**, 181 (1991).
33. Y. L. Sokolov and V. P. Yakovlov, Sov. Phys. JETP **56**, 7 (1982).
34. J. Robert *et al.*, Europhys. Lett. **16**, 29 (1991).
35. M. A. M. Marte and P. Zoller, Appl. Phys. **B54**, 477 (1992).
36. P. Storey, M. Collett, and D. Walls, Phys. Rev. Lett. **68**, 472 (1992).
37. P. Meystre, E. Schumacher, and S. Stenholm, Opt. Commun. **73**, 443 (1989).
38. A. M. Herkommer, V. M. Akulin, and W. P. Schleich, Phys. Rev. Lett. **69**, 3298 (1992).
39. T. Sleator and M. Wilkens, Phys. Rev. **A48**, 3286 (1993).
40. S. Haroche and J. M. Raimond, in *Cavity Quantum Electrodynamics, Advances in Atomic, Molecular, and Optical Physics*, edited by P. R. Berman (Academic Press, Boston, 1994), pp. 123–170, supp. 2.
41. B. Taylor, K. J. Schernthanner, G. Lenz, and P. Meystre, Opt. Commun. **110**, 569 (1994).
42. R. J. C. Spreeuw, T. Pfau, U. Janicke, and M. Wilkens, Europhys. Lett. **32**, 469 (1995).

75

Quantized Field Effects

M. Freyberger,* K. Vogel,* W. Schleich,* and R. F. O'Connell†

*Universität Ulm, Abteilung für Quantenphysik, 89069 Ulm, Federal Republic of Germany
†Department of Physics and Astronomy, Louisiana State University, Baton Rouge, Louisiana 70803

75.1   INTRODUCTION

The electromagnetic field appears almost everywhere in physics. Following the introduction of Maxwell's equations in 1864, Max Planck initiated quantum theory when he discovered $h = 2\pi\hbar$ in the laws of black-body radiation. In 1905 Albert Einstein explained the photoelectric effect on the hypothesis of a corpuscular nature of radiation and in 1917 this paradigm led to a description of the interaction between atoms and electromagnetic radiation.

The study of quantized field effects requires an understanding of the quantization of the field which leads to the concept of a quantum of radiation, the *photon*. Specific nonclassical features arise when the field is prepared in particular quantum states, such as squeezed states. When the radiation field interacts with an atom, there is an important difference between a classical field and a quantized field. A classical field can have zero amplitude, in which case it does not interact with the atom. On the other hand a quantized field *always* interacts with the atom, even if all the field modes are in their ground states, due to *vacuum fluctuations*. These lead to various effects such as *spontaneous emission* and the *Lamb shift*.

The interaction of an atom with the many modes of the radiation field can conveniently be described in an approximate manner by a master equation where the radiation field is treated as a reservoir. Such a treatment gives a microscopic and quantum mechanically consistent description of damping.

75.2   FIELD QUANTIZATION

This section provides the basis for the quantized field effects discussed in this Chap. [1]. We expand the field in a complete set of normal modes which reduces the problem of field quantization to the quantization of a one dimensional harmonic oscillator corresponding to each normal mode.

The classical free electromagnetic field, i.e., the field in a region without charge and current densities, obeys the Maxwell equations

$$\nabla \cdot \mathbf{B} = 0 \,, \tag{75.1}$$

$$\nabla \cdot \mathbf{D} = 0 \,, \tag{75.2}$$

$$\nabla \times \mathbf{E} + \frac{\partial \mathbf{B}}{\partial t} = 0 \,, \tag{75.3}$$

$$\nabla \times \mathbf{H} - \frac{\partial \mathbf{D}}{\partial t} = 0 \,, \tag{75.4}$$

where $\mathbf{B} = \mu_0 \mathbf{H}$, $\mathbf{D} = \epsilon_0 \mathbf{E}$. The magnetic permeability μ_0 connects the magnetic induction $\mathbf{B}$ with the magnetic field $\mathbf{H}$ and the electric permittivity ϵ_0 of free space connects the displacement $\mathbf{D}$ with the electric field $\mathbf{E}$. In the case of a free field, $\mathbf{E}$ and $\mathbf{B}$ may be obtained from

$$\mathbf{B} = \nabla \times \mathbf{A} \,, \tag{75.5}$$

$$\mathbf{E} = -\frac{\partial \mathbf{A}}{\partial t} \,, \tag{75.6}$$

where the vector potential $\mathbf{A}$ obeys the Coulomb gauge condition $\nabla \cdot \mathbf{A} = 0$ and satisfies a wave equation. In order to solve this wave equation we expand the vector potential

$$\mathbf{A}(\mathbf{x}, t) = \sum_{\mathbf{k}} \sum_{\sigma=1}^{2} \left(\frac{\hbar}{2\omega_{\mathbf{k}}\epsilon_0 V} \right)^{1/2}$$
$$\times \left[\alpha_{\mathbf{k}\sigma} \boldsymbol{\epsilon}_{\mathbf{k}\sigma} e^{i(\mathbf{k}\cdot\mathbf{x}-\omega_{\mathbf{k}}t)} + \text{c.c.} \right] \tag{75.7}$$

in a set of normal modes $V^{-1/2}\exp(i\mathbf{k}\cdot\mathbf{x})\boldsymbol{\epsilon}_{\mathbf{k}\sigma}$ which are orthonormal in the volume V. Due to the gauge condition $\nabla \cdot \mathbf{A} = 0$, we obtain two orthogonal polarization vectors $\boldsymbol{\epsilon}_{\mathbf{k}1}$ and $\boldsymbol{\epsilon}_{\mathbf{k}2}$ with $\boldsymbol{\epsilon}_{\mathbf{k}\sigma}\cdot\mathbf{k} = 0$ for each wave vector $\mathbf{k}$. The dispersion relation is $\omega_{\mathbf{k}} = c|\mathbf{k}|$. The Fourier amplitudes $\alpha_{\mathbf{k}\sigma}$ are complex numbers in the classical theory.

The field is quantized by replacing the classical amplitude $\alpha_{\mathbf{k}\sigma}$ by the mode annihilation operator $a_{\mathbf{k}\sigma}$. The complex conjugate $\alpha_{\mathbf{k}\sigma}^{*}$ is replaced by the mode creation operator $a_{\mathbf{k}\sigma}^{\dagger}$. They obey the commutation relation

$$[a_{\mathbf{k}\sigma}, a_{\mathbf{k}'\sigma'}^{\dagger}] = \delta_{\mathbf{k}\mathbf{k}'}\delta_{\sigma\sigma'} \,. \tag{75.8}$$

The representation of the electric field operator

$$\mathbf{E}(\mathbf{x}, t) = i \sum_{\mathbf{k},\sigma} \left(\frac{\hbar\omega_{\mathbf{k}}}{2\epsilon_0 V} \right)^{1/2} [a_{\mathbf{k}\sigma} \boldsymbol{\epsilon}_{\mathbf{k}\sigma} e^{i(\mathbf{k}\cdot\mathbf{x}-\omega_{\mathbf{k}}t)} - \text{h.c.}]$$
$$\equiv \mathbf{E}^{+}(\mathbf{x}, t) + \mathbf{E}^{-}(\mathbf{x}, t) \tag{75.9}$$

in terms of these operators follows from Eq. (75.6) and the operator for the vector potential. Note also the often used decomposition of the electric field operator into the positive and negative frequency parts $\mathbf{E}^{+}$ and $\mathbf{E}^{-}$ respectively. A similar relation holds for the operator describing the magnetic induction $\mathbf{B}$.

Using the operators for the electric and magnetic field, one can transform the field energy

$$H = \frac{1}{2} \int dV \left(\epsilon_0 \mathbf{E}^2 + \mathbf{B}^2 / \mu_0 \right) \tag{75.10}$$

into the form

$$H = \sum_{\mathbf{k},\sigma} \hbar \omega_{\mathbf{k}} (a_{\mathbf{k}\sigma}^\dagger a_{\mathbf{k}\sigma} + 1/2), \tag{75.11}$$

which is a sum of independent harmonic oscillator Hamiltonians corresponding to each mode $(\mathbf{k}, \sigma)$. The number operator $N_{\mathbf{k}\sigma} = a_{\mathbf{k}\sigma}^\dagger a_{\mathbf{k}\sigma}$ represents the number of photons in the mode $(\mathbf{k}, \sigma)$, while $\hbar\omega_{\mathbf{k}}/2$ is the energy of the vacuum fluctuations.

Hence each mode of the electromagnetic field is equivalent to a harmonic oscillator. In the next section we discuss specific states of a single mode. The general quantum state of the electromagnetic field consisting of many modes is given by a superposition of product states that are composed out of these single mode states.

75.3 FIELD STATES

This section summarizes the properties of several important states of the electromagnetic field. From the independence of the normal modes, the discussion may be restricted to a single normal mode. With the mode index $(\mathbf{k}, \sigma)$ suppressed, a single mode Hamiltonian is

$$H = \hbar\omega(a^\dagger a + 1/2)$$
$$= \frac{\hbar\omega}{2}p^2 + \frac{\hbar\omega}{2}x^2. \tag{75.12}$$

In the second step, the quadrature operators

$$x = \frac{1}{\sqrt{2}}(a + a^\dagger), \tag{75.13}$$

$$p = \frac{1}{i\sqrt{2}}(a - a^\dagger) \tag{75.14}$$

are introduced, which are equivalent to scaled position and momentum operators of a massive particle in a harmonic potential. The quadratures of a quantized field are measurable with the help of homodyne detection as discussed in Sect. 75.5.1.

We shall now describe several states of this quantized field mode: number states, coherent states, squeezed states, Schrödinger cats, and phase states. A quantized field in a coherent state shows the most classical behavior. A superposition of two coherent states, which is a Schrödinger cat, already shows nonclassical features. Number states and squeezed states are further typical examples of nonclassical states.

75.3.1 Number States

The eigenstates of the Hamiltonian (75.12) are the eigenstates of the number operator $N = a^\dagger a$,

$$N|n\rangle = n|n\rangle, \tag{75.15}$$

where $n = 0, 1, 2, \ldots$ denotes the excitations or the number of photons in the mode. The vacuum state of the mode $|0\rangle$, is defined by

$$a|0\rangle = 0. \tag{75.16}$$

The ladder of excitations can be climbed up and down via the application of creation and annihilation operators

$$a^\dagger|n\rangle = \sqrt{n+1}|n+1\rangle, \tag{75.17}$$
$$a|n\rangle = \sqrt{n}|n-1\rangle, \tag{75.18}$$

on a Fock state $|n\rangle$. These number or Fock states form a complete and orthonormal set of states so that

$$\sum_{n=0}^{\infty} |n\rangle\langle n| = 1, \quad \langle n|k\rangle = \delta_{nk}. \tag{75.19}$$

Their quadrature representations are

$$\langle x|n\rangle = (\sqrt{\pi}2^n n!)^{-1/2} H_n(x) e^{-x^2/2}, \tag{75.20}$$
$$\langle p|n\rangle = (\sqrt{\pi}2^n n!)^{-1/2} (-i)^n H_n(p) e^{-p^2/2}. \tag{75.21}$$

The states $|x\rangle$ and $|p\rangle$ are eigenstates of the quadrature operators x and p, Eqs. (75.13) and (75.14).

Number states provide a frequently used representation of a pure quantum state

$$|\psi\rangle = \sum_{n=0}^{\infty} c_n|n\rangle, \tag{75.22}$$

or a mixed quantum state given by the density operator

$$\rho = \sum_{n,k} \rho_{nk}|n\rangle\langle k| \tag{75.23}$$

(see Chap. 7).

75.3.2 Coherent States

The coherent state is a specific superposition of number states. In contrast to a number state, a coherent state does not possess a definite number of photons: the photon distribution is Poissonian. For a large average photon number, the electric and magnetic fields have rather well defined amplitudes and phases with vanishing relative quantum fluctuations. Hence the Poissonian photon distribution frequently serves as a borderline between classical and nonclassical field states. Nonclassical states show a sub-Poissonian behavior. An extreme example is a field prepared in a number state. A parameter which quantifies the deviations from Poissonian behavior is the Q parameter introduced by Mandel [2].

We define the coherent state $|\alpha\rangle$ as an eigenstate of the annihilation operator

$$a|\alpha\rangle = \alpha|\alpha\rangle \qquad (75.24)$$

with the complex amplitude $\alpha = |\alpha|e^{i\theta}$. The coherent state can be represented by

$$|\alpha\rangle = e^{\alpha a^\dagger - \alpha^* a}|0\rangle = D(\alpha)|0\rangle \,, \qquad (75.25)$$

that is, by the action of the displacement operator $D(\alpha)$ on the vacuum.

The number state representation of $|\alpha\rangle$ reads

$$|\alpha\rangle = e^{-|\alpha|^2/2} \sum_{n=0}^{\infty} \frac{\alpha^n}{\sqrt{n!}}|n\rangle \,. \qquad (75.26)$$

A coherent state $|\alpha_0\rangle$ that evolves in time according to the free field Hamiltonian Eq. (75.12) stays coherent, i.e.,

$$|\psi(t)\rangle = \exp[-iHt/\hbar]|\alpha_0\rangle = e^{-i\omega t/2}|\alpha(t)\rangle \,, \qquad (75.27)$$

with amplitude $\alpha(t) = \alpha_0 \exp[-i\omega t]$.

Another important representation of a coherent state is the x representation

$$\langle x|\alpha\rangle = \pi^{-1/4} \exp\{-[\Re(\alpha)]^2\} \exp(-x^2/2 + \sqrt{2}\alpha x) \,, \qquad (75.28)$$

where $|x\rangle$ denotes again the x quadrature eigenstate.

The photon distribution in a coherent state

$$|\langle n|\alpha\rangle|^2 = \frac{|\alpha|^{2n}e^{-|\alpha|^2}}{n!} \qquad (75.29)$$

is a Poisson distribution with average photon number $\langle N\rangle = |\alpha|^2$ and variance $(\Delta N)^2 = \langle N^2\rangle - \langle N\rangle^2 = |\alpha|^2$. Hence the relative fluctuations $(\Delta N)/\langle N\rangle = \langle N\rangle^{-1/2}$ vanish for a large average photon number.

The Mandel Q parameter

$$Q \equiv \frac{(\Delta N)^2 - \langle N\rangle}{\langle N\rangle} \qquad (75.30)$$

vanishes for a field in a coherent state. A nonclassical field may show sub-Poissonian behavior with $Q < 0$. As an example, the Schrödinger cat state is a *macroscopic* superposition of two coherent states

$$|\text{cat}\rangle = \left[2 + 2\cos(\alpha^2 \sin\phi)e^{-2\alpha^2 \sin^2(\phi/2)}\right]^{-1/2}$$
$$\times \left[|\alpha e^{i\phi/2}\rangle + |\alpha e^{-i\phi/2}\rangle\right] \,, \qquad (75.31)$$

where α is assumed to be real. The Q parameter for this superposition state, shown in Fig. 75.1, takes on negative values for specific angles ϕ. The nonclassical behavior

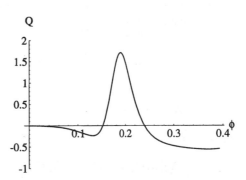

Figure 75.1. The Q parameter for a Schrödinger cat state, Eq. (75.31), with amplitude $\alpha = 4$.

of such a $|\text{cat}\rangle$-state can be explained [3] as a result of quantum interference between the two coherent states present in Eq. (75.31). The incoherent superposition described by the density operator

$$\rho = \frac{1}{2}\left[|\alpha e^{i\phi/2}\rangle\langle\alpha e^{i\phi/2}| + |\alpha e^{-i\phi/2}\rangle\langle\alpha e^{-i\phi/2}|\right] \qquad (75.32)$$

does not have this nonclassical character: its Q parameter vanishes. Coherent states have a direct physical significance: the quantum state of a stabilized laser operating well above threshold can be approximated by a coherent state.

75.3.3 Squeezed States

Squeezed states [4] minimize the uncertainty product of the quadrature components of the electromagnetic field. The quadrature components x and p of the single mode field are defined in Eqs. (75.13) and (75.14). They obey the commutation relation $[x, p] = i$. Their uncertainties $(\Delta x)^2 \equiv \langle x^2\rangle - \langle x\rangle^2$ and $(\Delta p)^2 \equiv \langle p^2\rangle - \langle p\rangle^2$ fulfill the Heisenberg inequality

$$\Delta x \Delta p \geq 1/2 \,. \qquad (75.33)$$

The coherent state is a special minimum uncertainty state with equal uncertainties $\Delta x = \Delta p = 1/\sqrt{2}$. Squeezed states comprise a more general class of minimum uncertainty states with reduced uncertainty in one quadrature at the expense of increased uncertainty in the other. These states $|\alpha, \epsilon\rangle$ are obtained by applying the displacement operator $D(\alpha)$ and the unitary squeeze operator

$$S(\epsilon) = e^{\frac{1}{2}\epsilon^* a^2 - \frac{1}{2}\epsilon a^{\dagger 2}} \qquad (75.34)$$

to the vacuum

$$|\alpha, \epsilon\rangle = D(\alpha)S(\epsilon)|0\rangle \,. \qquad (75.35)$$

The squeeze operator $S(\epsilon)$ transforms a and $a^\dagger$ according to

$$S^\dagger(\epsilon)aS(\epsilon) = a\cosh r - a^\dagger e^{-2i\phi}\sinh r\,, \qquad (75.36)$$

$$S^\dagger(\epsilon)a^\dagger S(\epsilon) = a^\dagger\cosh r - a e^{2i\phi}\sinh r\,, \qquad (75.37)$$

where $\epsilon = re^{-2i\phi}$. The rotated quadratures

$$X_1 = x\cos\phi - p\sin\phi\,, \qquad (75.38)$$

$$X_2 = p\cos\phi + x\sin\phi\,, \qquad (75.39)$$

transform according to

$$S^\dagger(\epsilon)(X_1 + iX_2)S(\epsilon) = X_1 e^{-r} + iX_2 e^{r}\,, \qquad (75.40)$$

which yields the uncertainties

$$\Delta X_1 = e^{-r}/\sqrt{2}\,, \qquad (75.41)$$

$$\Delta X_2 = e^{r}/\sqrt{2}\,. \qquad (75.42)$$

In particular, for $\phi = 0$ the squeezed state $|\alpha, r\rangle$ is a minimum uncertainty state for the quadratures x and p with $\Delta x = e^{-r}/\sqrt{2}$ and $\Delta p = e^{r}/\sqrt{2}$. The degree of squeezing in the quadrature x is determined by the squeeze factor r.

The average photon number

$$\langle N \rangle = |\alpha|^2 + \sinh^2 r \qquad (75.43)$$

of a squeezed state and its photon number variance

$$(\Delta N)^2 = \left| \alpha\cosh r - \alpha^* e^{-2i\phi}\sinh r \right|^2 + 2\cosh^2 r\sinh^2 r \qquad (75.44)$$

contain the coherent contribution α as well as squeezing contributions expressed by r and ϕ. In particular, for $\phi = 0$, the Q parameter becomes negative for a large enough amplitude α and $r > 0$. The photon number distribution $W_n = |\langle \alpha, r|n\rangle|^2$ becomes narrower than the one for the corresponding coherent state with the same α. This sub-Poissonian behavior is one of the nonclassical features of a squeezed state. Furthermore, W_n shows oscillations [5] for larger squeezing. The two regimes with sub-Poissonian and oscillating photon statistics W_n are shown in Fig. 75.2.

A second representation of squeezed states has been introduced by Yuen [6]. In his notation, a squeezed state is an eigenstate of the operator

$$b = \mu a + \nu a^\dagger\,, \qquad (75.45)$$

with $|\mu|^2 - |\nu|^2 = 1$ and eigenvalue β. This eigenstate can be written in the form

$$|\epsilon, \beta\rangle = S(\epsilon)D(\beta)|0\rangle\,, \qquad (75.46)$$

which connects the squeezing operator $S(re^{-2i\phi})$ with the parameters $\mu = \cosh r$ and $\nu = e^{-2i\phi}\sinh r$. In contrast to the definition (75.35), the displacement operator $D(\beta)$ and the squeezing operator $S(\epsilon)$ are applied now in reversed order. Nevertheless, the two equations

Figure 75.2. The photon number distribution W_n of a squeezed state $|\alpha, r\rangle$ with the coherent amplitude $\alpha = 7$. For a squeezing parameter $r = 0$, the Poisson distribution of a coherent state $|\alpha = 7\rangle$ is just visible. When r increases the photon distribution first becomes sub-Poissonian and then oscillatory.

(75.35) and (75.46) define the same state if the relation $\alpha = \beta\mu + \beta^*\nu$ is fulfilled.

Several experiments have demonstrated the generation of squeezed light. Slusher *et al.* [7] obtained squeezing in the sidemodes of a four-wave mixing process. An optical parametric oscillator below threshold has been used by Wu *et al.* [8] in order to generate squeezed light. Nonclassical features can also be found in a down conversion process. This second-order process creates so-called signal and idler photons from one pump photon. Signal and idler beam are distinguished by frequency or polarization. Heidmann *et al.* [9] have shown that the difference intensity of these twin beams may exhibit reduced quantum fluctuations. Pulsed twin beams also contain reduced noise in the difference of their intensities.

75.3.4 Phase States

The problem of a correct quantum mechanical description of phase has a long history in quantum mechanics [10]. First attempts to define a quantum phase are due to London and Dirac. The London phase state is

$$|\phi\rangle = \frac{1}{\sqrt{2\pi}}\sum_{n=0}^{\infty} e^{in\phi}|n\rangle\,, \qquad (75.47)$$

which is an eigenstate of the exponential phase operator

$$\widehat{e^{i\phi}} = \sum_{n=0}^{\infty} |n\rangle\langle n+1|\,. \qquad (75.48)$$

Since this operator is not unitary, it does not define a Hermitian operator $\hat{\phi}$ for the phase. Nevertheless many treatments of the phase of a quantum state $|\psi\rangle = \sum c_n|n\rangle$ are based on the *London phase distribution*

$$Pr(\phi) = |\langle\phi|\psi\rangle|^2 = \frac{1}{2\pi}\left|\sum_n c_n e^{-in\phi}\right|^2 . \qquad (75.49)$$

Later treatments [11] rely on the Hermitian operators

$$\widehat{\sin\phi} = \frac{1}{2i}\left(\widehat{e^{i\phi}} - \widehat{e^{i\phi}}^\dagger\right), \qquad (75.50)$$

$$\widehat{\cos\phi} = \frac{1}{2}\left(\widehat{e^{i\phi}} + \widehat{e^{i\phi}}^\dagger\right), \qquad (75.51)$$

for the sine and cosine function of the phase.

Recently [12] a Hermitian phase operator was constructed starting from the phase state Eq. (75.47), restricted to a finite Hilbert space. An operational phase description has been proposed [13] in which a classical phase measurement is translated to the quantum realm by using an eight-port homodyne detector.

75.4 QUANTUM COHERENCE THEORY

This section introduces the correlation functions of the electromagnetic field. Ideal photon correlation measurements can bring out the phenomenon of photon bunching and antibunching.

75.4.1 Correlation Functions

Correlation functions were originally introduced to describe an ideal photodetection process. Glauber [14] has presented a treatment based on an absorption mechanism in the detector which is sensitive to the positive frequency part E^+ of the electric field evaluated at the detector's space-time position $x \equiv (\mathbf{x}, t)$. This leads to an average field intensity

$$I(x) = \text{Tr}(\rho E^-(x)E^+(x)) \qquad (75.52)$$

at point x. Here the density operator ρ describes the state of the field. The ordering of the operators, i.e., $E^-E^+ \sim a^\dagger a$, is known as normal ordering with all annihilation operators to the right of all creation operators.

The expression (75.52) now immediately generalizes to the correlation function of first order

$$G^{(1)}(x_1, x_2) = \text{Tr}[\rho E^-(x_1)E^+(x_2)], \qquad (75.53)$$

with $x_1 = (\mathbf{x}_1, t_1)$ and $x_2 = (\mathbf{x}_2, t_2)$. The classical interference experiments, such as Young's double slit experiment, can be described in terms of $G^{(1)}$. Furthermore, the correlation function of first order is connected to the power spectrum $S(\omega)$ of a quantized field via the Wiener-Khintchine theorem [15]. Under the assumption of a stationary process, i.e., when the autocorrelation

function $\langle E^-(t)E^+(t')\rangle$ depends only on the time difference $\tau = t - t'$, then

$$S(\omega) = \frac{1}{2\pi}\int_0^\infty d\tau\langle E^-(\tau)E^+(0)\rangle e^{-i\omega\tau} + \text{c.c.} \qquad (75.54)$$

This relation between the spectrum and the first-order correlation function is known as Wiener-Khintchine theorem.

In order to analyze the Hanbury-Brown and Twiss experiment [16] it is necessary to define higher order correlation functions. The general nth order correlation function is defined by

$$G^{(n)}(x_1, \ldots, x_n, x_{n+1}, \ldots, x_{2n})$$
$$= \text{Tr}[\rho E^-(x_1)\cdots E^-(x_n)E^+(x_{n+1})\cdots E^+(x_{2n})], \qquad (75.55)$$

where the field operators are again normal ordered.

These correlation functions fulfill a generalized Schwartz inequality

$$G^{(1)}(x_1, x_1)G^{(1)}(x_2, x_2) \geq |G^{(1)}(x_1, x_2)|^2, \qquad (75.56)$$

which becomes, for the nth order functions,

$$G^{(n)}(x_1, .., x_n, x_n, .., x_1)G^{(n)}(x_{n+1}, .., x_{2n}, x_{2n}, .., x_{n+1})$$
$$\geq |G^{(n)}(x_1, .., x_n, x_{n+1}, .., x_{2n})|^2 . \qquad (75.57)$$

A field is said to be first-order coherent when its normalized correlation function

$$g^{(1)}(x_1, x_2) = \frac{G^{(1)}(x_1, x_2)}{\left[G^{(1)}(x_1, x_1)G^{(1)}(x_2, x_2)\right]^{1/2}} \qquad (75.58)$$

satisfies $|g^{(1)}(x_1, x_2)| = 1$. In a Young type experiment, this case gives maximum fringe visibility. A more general definition of first order coherence is the condition that $G^{(1)}(x_1, x_2)$ factorizes

$$G^{(1)}(x_1, x_2) = \mathcal{G}^*(x_1)\mathcal{G}(x_2), \qquad (75.59)$$

where $\mathcal{G}$ denotes some complex function. This definition can be readily generalized to the nth order case. The nth order coherence applies when the relation

$$G^{(n)}(x_1, \ldots, x_{2n}) = \mathcal{G}^*(x_1)\cdots\mathcal{G}^*(x_n)\mathcal{G}(x_{n+1})\cdots\mathcal{G}(x_{2n}) \qquad (75.60)$$

holds. A field in a coherent state possesses nth order coherence.

75.4.2 Photon Correlations

The Young experiment demonstrates the appearance of first-order correlations. However, experiments that can distinguish between the classical and quantum domains have to be based on measurements of second-order correlations. These experiments are of the Hanbury-Brown and Twiss type, and determine the arrival of a photon at detector position $\mathbf{x}$ and time t and another photon at time $t + \tau$. Following the theory of Glauber, the second-order correlation function

$$G^{(2)}(\tau) = \langle E^-(t)E^-(t+\tau)E^+(t+\tau)E^+(t)\rangle \quad (75.61)$$

is measured. In this formula we have omitted the variable for the position $\mathbf{x}$. Usually the normalized correlation function

$$g^{(2)}(\tau) = \frac{G^{(2)}(\tau)}{[G^{(1)}(0)]^2} \quad (75.62)$$

is introduced. The function $g^{(2)}$ is always positive, which is true for classical as well as for quantum fields; but there exists a purely quantum domain given by

$$0 \le g^{(2)}(0) < 1. \quad (75.63)$$

For example, for a number state $|n\rangle$,

$$g^{(2)}_{|n\rangle}(0) = 1 - 1/n, \quad (75.64)$$

with $n \ge 1$. In contrast, a coherent state $|\alpha\rangle$ yields $g^{(2)}_{|\alpha\rangle}(0) = 1$.

75.4.3 Photon Bunching and Antibunching

In a realistic theory (but not in an oversimplified one-mode model), the correlation function $G^{(2)}(\tau)$ always factorizes on a sufficiently long time scale, and $g^{(2)}(\tau) \to 1$. The photons are then no longer correlated, and they arrive randomly as in the case of coherent light; see for example Eq. (75.193).

If $g^{(2)}(0) > 1$, the photons show a tendency to arrive in bunches, an effect known as *photon bunching*. This effect has been observed for chaotic light. The opposite situation with $0 \le g^{(2)}(0) < 1$ demonstrates the reverse effect, namely photon antibunching. As seen from Eq. (75.63), this is a regime only accessible to nonclassical light. An example is given by the resonance fluorescence of a two-level atom, treated in Sect. 75.14. Note that we can rewrite $g^{(2)}(0)$ with the help of Mandel's Q parameter

$$g^{(2)}(0) = \frac{\langle a^\dagger a^\dagger aa\rangle}{\langle a^\dagger a\rangle^2} = 1 + Q. \quad (75.65)$$

Hence, a field state with $Q < 0$ shows the effect of photon antibunching.

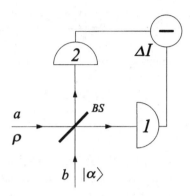

Figure 75.3. The principle of optical homodyning.

75.5 PHOTODETECTION THEORY

So far we have used a very simple theory of photodetection: any absorbed photon leads to a photoelectric emission which can be observed. But in any real experiment, these photons are counted over some time interval T and the observed photoelectric emissions are dominated by two statistics: (i) the statistics of photoelectric emission which is also present for a classical field and (ii) the specific quantum statistics of a quantized field. A detailed discussion of the quantum theory of photoelectric detection has been given by Kelley and Kleiner [17]. A central result is the formula

$$p(n, t, T) = \left\langle : \frac{I^n}{n!}\exp(-I) : \right\rangle \quad (75.66)$$

for the probability of counting n photoelectrons in the time interval from t to $t + T$. This photocounting distribution contains the integrated intensity operator

$$I = \eta \int_t^{t+T} dt'\, E^-(t')E^+(t') \quad (75.67)$$

containing the quantum efficiency η of the detector. The notation $\langle : \cdots : \rangle$ indicates a quantum average where the operators have to be normally ordered and time ordered. This operator ordering reflects the process on which a photodetector is based. It annihilates or absorbs photons, one after the other. A good treatment of photoelectric detection can be found in Ref. [18].

75.5.1 Homodyne and Heterodyne Detection

These detection methods allow the extraction of specific quantum features of a single mode quantum field, the *signal field*. Figure 75.3 summarizes the principle of optical homodyning. Two quantum fields described by the annihilation operators a and b are mixed at a 50/50 beam splitter *BS*. Both fields have the same frequency. The mode *a* represents the signal mode whose quantum state is given by the density operator ρ. Mode *b* serves as a reference field, the *local oscillator*. The coherent state $|\alpha\rangle = ||\alpha|e^{i\theta}\rangle$ determines the quantum state of the local oscillator. Two ideal photodetectors 1 and 2 measure the number of photons in the output modes of the beam splitter. For a highly excited coherent state, i.e., a classical local oscillator, the statistics of the photocurrent difference ΔI can be described by the moments of the signal mode operator

$$X_\theta = \frac{1}{\sqrt{2}}(ae^{-i\theta} + a^\dagger e^{i\theta}). \qquad (75.68)$$

For example, the photocurrent difference

$$\langle \Delta I \rangle \sim \langle X_\theta \rangle = \text{Tr}(\rho X_\theta) \qquad (75.69)$$

is proportional to the expectation value of X_θ. In particular, for $\theta = 0$ and $\theta = \frac{\pi}{2}$ one is able to measure all the moments of the two quadratures x and p [Eqs. (75.13) and (75.14)] of the signal mode. In general, the statistics of the photocurrent ΔI reveal the probability distribution

$$Pr(X_\theta) = \langle X_\theta | \rho | X_\theta \rangle \qquad (75.70)$$

of the observable X_θ when the signal mode is in the state ρ. The states

$$|X_\theta\rangle = \pi^{-1/4}\exp[-X_\theta^2/2]\sum_{n=0}^{\infty}\frac{1}{\sqrt{2^n n!}}H_n(X_\theta)e^{in\theta}|n\rangle \qquad (75.71)$$

are eigenstates of the operator X_θ, and are known as *rotated quadrature states*.

The heterodyne technique [19] relies on a similar mixing of a signal field with a local oscillator at a beam splitter, but this time the local oscillator frequency is offset by the intermediate frequency $\Delta\omega$ with respect to the frequency ω_0 of the signal mode. Filters select the beat frequency components in the photocurrent of the detectors. This photocurrent contains the quantum statistics of the two quadratures of the signal field [19].

75.6 QUASI-PROBABILITY DISTRIBUTIONS

Quasi-probability distributions play an important role in quantum optics for three reasons. First, they are a complete representation of the density operator of a quantum field. Second, they allow one to calculate expectation values in the spirit of classical statistical physics. Third, they offer the possibility of converting a master equation for the density operator into an equivalent *c*-number partial differential equation. In this section, we relate a specific quasi-probability function to a specific operator ordering.

75.6.1 *s*-Ordered Operators

A normally ordered product of *a* and $a^\dagger$ is a product of the form $(a^\dagger)^m a^n$: the annihilation operators *a* stand to the right of the creation operators $a^\dagger$. In an antinormally ordered product like $a^n(a^\dagger)^m$, the order of *a* and $a^\dagger$ has changed. A generalized *s*-ordered product can be defined as

$$\{a^m(a^\dagger)^n\}_s \equiv \left(\frac{\partial}{\partial\xi}\right)^n\left(-\frac{\partial}{\partial\xi^*}\right)^m D(\xi,\xi^*,s)\Big|_{\xi=\xi^*=0} \qquad (75.72)$$

with the generalized displacement operator

$$D(\xi,\xi^*,s) \equiv \exp[\xi a^\dagger - \xi^* a + \tfrac{1}{2}s\xi\xi^*]. \qquad (75.73)$$

For $s = 1$ we find again normal ordering. The values $s = 0$ and $s = -1$ produce symmetric and antinormal ordered products. As an example we note $\{a^\dagger a\}_s = a^\dagger a - (s-1)/2$.

Expectation values of those *s*-ordered products are easily derived from the characteristic function

$$\chi(\xi,\xi^*,s) = \text{Tr}\big[\rho D(\xi,\xi^*,s)\big] \qquad (75.74)$$

via differentiation

$$\left(\frac{\partial}{\partial\xi}\right)^n\left(-\frac{\partial}{\partial\xi^*}\right)^m\chi(\xi,\xi^*,s)\Big|_{\xi=\xi^*=0} = \langle\{(a^\dagger)^n a^m\}_s\rangle. \qquad (75.75)$$

The Fourier transform of χ yields the quasi-probability distribution of Cahill and Glauber [20]

$$W(\alpha,s) = \frac{1}{\pi^2}\int d^2\xi\,\chi(\xi,\xi^*,s)\,e^{\alpha\xi^* - \alpha^*\xi}, \qquad (75.76)$$

where $d^2\xi = d\Re(\xi)d\Im(\xi)$. With this distribution one is able to calculate the expectation value of any *s*-ordered operator product

$$\langle\{(a^\dagger)^n a^m\}_s\rangle = \int (\alpha^*)^n \alpha^m W(\alpha,s)\,d^2\alpha. \qquad (75.77)$$

We concentrate now on three important quasi-probability distributions, namely the cases $s = 1$, $s = 0$, and $s = -1$, corresponding to the Glauber-Sudarshan distribution, the Wigner function, and the Q function respectively. A detailed discussion of these three functions can be found in Ref. [21].

75.6.2 The *P* Function

The quasi-probability function $P(\alpha)$ was introduced [22, 23] as a diagonal representation of the density operator

$$\rho = \int P(\alpha)|\alpha\rangle\langle\alpha|\, d^2\alpha \qquad (75.78)$$

in terms of coherent states $|\alpha\rangle$, Eq. (75.26). It is related to the Cahill-Glauber function $W(\alpha, s)$ via

$$P(\alpha) = W(\alpha, s = 1). \qquad (75.79)$$

The expectation value of any normally ordered operator product

$$\langle (a^\dagger)^m a^n \rangle = \int (\alpha^*)^m \alpha^n P(\alpha)\, d^2\alpha \qquad (75.80)$$

has a particularly simple form in terms of $P(\alpha)$. From Eq. (75.78) the P function of a coherent state $|\alpha_0\rangle$ becomes

$$P(\alpha) = \delta^{(2)}(\alpha - \alpha_0) \qquad (75.81)$$
$$= \delta\big(\Re(\alpha) - \Re(\alpha_0)\big)\delta\big(\Im(\alpha) - \Im(\alpha_0)\big).$$

Quantized fields for which the P function is positive do not show nonclassical effects such as squeezing and antibunching. For nonclassical states, such as number states or squeezed states, the P function only exists in terms of generalized functions, such as delta functions and their derivatives, which have a highly singular character.

The positive P-representation $P(\alpha, \beta)$ for a nondiagonal decomposition of a density operator is [24, 25]

$$\rho = \int\int d^2\alpha\, d^2\beta\, \frac{|\alpha\rangle\langle\beta^*|}{\langle\beta^*|\alpha\rangle} P(\alpha, \beta). \qquad (75.82)$$

The function $P(\alpha, \beta)$ is a direct generalization of the Glauber-Sudarshan function $P(\alpha)$, Eq. (75.78). $P(\alpha, \beta)$ exists for any physical density operator ρ [24] and is given by

$$P(\alpha, \beta) = \frac{1}{4\pi^2} \exp\left[-\frac{1}{4}|\alpha - \beta^*|^2\right]$$
$$\times \left\langle \frac{1}{2}(\alpha + \beta^*)\Big|\rho\Big|\frac{1}{2}(\alpha + \beta^*)\right\rangle. \qquad (75.83)$$

Here the state $|1/2(\alpha+\beta^*)\rangle$ denotes a coherent state with the complex amplitude $1/2(\alpha + \beta^*)$. Note that $P(\alpha, \beta)$ is always positive.

75.6.3 The Wigner Function

This quasi-probability was first introduced by Wigner [26] and may be defined as the distribution function for a symmetrically ordered operator product which is obtained in the case $s = 0$. The Wigner function plays an important role in other branches of physics, such as quantum chaology, and in particular in any semiclassical phenomenon when one considers the transition from quantum mechanics to classical mechanics.

Consider the Wigner function of a quantum mechanical particle of position

$$x = \sqrt{\frac{\hbar}{2m\omega}}(a + a^\dagger), \qquad (75.84)$$

and momentum

$$p = i\sqrt{\frac{m\hbar\omega}{2}}(a^\dagger - a). \qquad (75.85)$$

The Wigner function may be written in terms of position and momentum variables

$$W(x, p) = \frac{1}{\pi\hbar}\int_{-\infty}^{\infty} dy\, e^{-2ipy/\hbar}\langle x + y|\rho|x - y\rangle, \qquad (75.86)$$

where $|x \pm y\rangle$ denotes position eigenstates. The $s = 0$ Cahill-Glauber definition and the above definition of the Wigner function are related by

$$W(x, p) = \frac{1}{2}W\left(\alpha = \frac{m\omega x + ip}{\sqrt{2m\hbar\omega}}, s = 0\right). \qquad (75.87)$$

The position and momentum distributions of a particle, or equivalently the quadruture distributions in the case of a quantized field mode, are

$$Pr(x) = \int_{-\infty}^{\infty} W(x, p)\, dp, \qquad (75.88)$$

$$Pr(p) = \int_{-\infty}^{\infty} W(x, p)\, dx. \qquad (75.89)$$

Furthermore, the scalar product

$$|\langle\psi_1|\psi_2\rangle|^2 = 2\pi\hbar\int\int dx\, dp\, W_{|\psi_1\rangle}(x, p)W_{|\psi_2\rangle}(x, p) \qquad (75.90)$$

of two quantum states is expressed by the phase space overlap of the two corresponding Wigner functions. Consequently, any Wigner function $W(x, p)$ has to obey the necessary condition

$$\int\int dx\, dp\, W(x, p)W_{|\psi\rangle}(x, p) \geq 0 \qquad (75.91)$$

for all $W_{|\psi\rangle}$ representing a pure state. For a normalized state $|\psi\rangle$,

$$\int\int dx\, dp\, \Big[W_{|\psi\rangle}(x, p)\Big]^2 = \frac{1}{2\pi\hbar}. \qquad (75.92)$$

Instead of solving the Schrödinger equation for the dynamics of a massive particle in a potential $V(x)$, we can try to solve the equation

$$\frac{\partial W}{\partial t} = -\frac{p}{m}\frac{\partial W}{\partial x} + \sum_{r=1,3,..} \frac{1}{r!}\left(\frac{i\hbar}{2}\right)^{r-1}\frac{\partial^r V}{\partial x^r}\frac{\partial^r W}{\partial p^r}$$

$$(75.93)$$

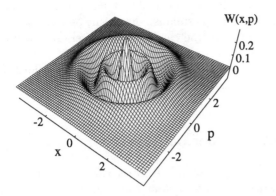

Figure 75.4. The Wigner function, Eq. (75.96), of a Fock state $|n = 4\rangle$. The negative parts can be seen clearly.

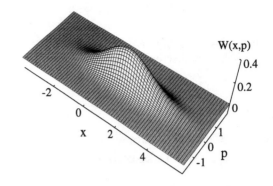

Figure 75.5. The Wigner function of a squeezed state $|\alpha, r\rangle$ with coherent amplitude $\alpha = 1$ and squeezing $e^{2r} = 0.25$. For these values the phase space ellipse is oriented along the x-axis and squeezed in the p-direction. Note that this function is positive everywhere.

for its Wigner function $W(x, p, t)$. Note that here only the odd derivatives of the potential V enter. This equation is the quantum analogue of the classical Liouville equation, to which it reduces in the limit of $\hbar \to 0$. However, the initial distribution $W(x, p, t = 0)$ has to be a Wigner function in the sense of Eq. (75.86).

Furthermore the Wigner function of an energy eigenfunction in the potential $V(x)$ may be obtained from the equations

$$\left[\frac{p^2}{2m} + V(x) - \frac{\hbar^2}{8m}\frac{\partial^2}{\partial x^2} - \frac{\hbar^2}{8}\frac{\partial^2 V}{\partial x^2}\frac{\partial^2}{\partial p^2} \right] W(x, p)$$
$$+ \sum_{r=4,6,..} \frac{1}{r!}\left(\frac{i\hbar}{2}\right)^r \frac{\partial^r V}{\partial x^r}\frac{\partial^r W}{\partial p^r} = EW(x, p), \quad (75.94)$$

and

$$\left[-\frac{p}{m}\frac{\partial}{\partial x} + \frac{\partial V}{\partial x}\frac{\partial}{\partial p} \right] W(x, p)$$
$$+ \sum_{r=3,5,..} \frac{1}{r!}\left(\frac{i\hbar}{2}\right)^{r-1} \frac{\partial^r V}{\partial x^r}\frac{\partial^r W}{\partial p^r} = 0. \quad (75.95)$$

The Wigner function has negative parts for most quantum states. For example, the Wigner function of a Fock state $|n\rangle$,

$$W_{|n\rangle}(\bar{x}, \bar{p}) = \frac{(-1)^n}{\pi} e^{-\bar{x}^2 - \bar{p}^2} L_n(2\bar{x}^2 + 2\bar{p}^2), \quad (75.96)$$

clearly becomes negative due to the oscillating Laguerre polynomial L_n as shown in Fig. 75.4. Note that we have introduced the dimensionless position $\bar{x} = \sqrt{m\omega/\hbar}\,x$ and momentum $\bar{p} = 1/\sqrt{m\omega\hbar}\,p$. On the other hand, the Wigner function

$$W_{|\alpha, r\rangle}(\bar{x}, \bar{p}) = \frac{1}{\pi} \exp\{ -e^{2r}[\bar{x} - \sqrt{2}\,\Re(\alpha)]^2$$
$$- e^{-2r}[\bar{p} - \sqrt{2}\,\Im(\alpha)]^2 \} \quad (75.97)$$

of a squeezed state Eq. (75.35) is always positive as shown in Fig. 75.5. It is a long thin ellipse in phase space (i.e. a Gaussian cigar). Concerning the negative parts of the Wigner function, the Hudson theorem [27] states that a necessary and sufficient condition for the Wigner function of a pure state $|\psi\rangle$ to be nonnegative is that it can be described by a wave function of the form

$$\langle x|\psi\rangle = \exp\left[-\tfrac{1}{2}(ax^2 + bx + c) \right]. \quad (75.98)$$

Here a, b and c denote some constants with $\Re(a) > 0$.

Finally, the Kirkwood distribution function

$$K(x, p) = \frac{1}{\pi\hbar}\int_{-\infty}^{\infty} dy\, e^{-2ipy/\hbar}\langle x|\rho|x - 2y\rangle \quad (75.99)$$

is a phase space function that resembles the Wigner function. In the case of a pure state $\rho = |\psi\rangle\langle\psi|$ this function reduces to

$$K(x, p) = \psi(x)\tilde{\psi}(p)e^{-ixp/\hbar}, \quad (75.100)$$

where $\tilde{\psi}(p)$ denotes the Fourier transform of $\psi(x)$.

75.6.4 The Q Function

The Q function is defined by the diagonal matrix elements

$$Q(\alpha) = \langle\alpha|\rho|\alpha\rangle/\pi \quad (75.101)$$

of the density operator ρ, where $|\alpha\rangle$ denotes a coherent state. The $Q(\alpha)$ function is always a positive and bounded function, which exists for any density operator ρ. The Q function is also known as Husimi's function. It allows one to calculate expectation values of antinormally ordered operator products of the form

$$\langle a^n (a^\dagger)^m \rangle = \int d^2\alpha \, \alpha^n (\alpha^*)^m Q(\alpha) . \qquad (75.102)$$

Moreover, since the Q function corresponds to the case $s = -1$ of the Cahill-Glauber distribution,

$$Q(\alpha) = W(\alpha, -1) . \qquad (75.103)$$

75.6.5 Relations Between Quasi-Probabilities

In general, the relation

$$W(\alpha, s) = \frac{2}{\pi(s'-s)} \int d^2\beta \exp\left[-\frac{2|\alpha - \beta|^2}{s'-s}\right] W(\alpha, s') \qquad (75.104)$$

holds between two Cahill-Glauber distributions with the parameters $s' > s$. In particular, the non-negative Q function

$$Q(\alpha) = \frac{2}{\pi} \int d^2\beta \exp[-2|\alpha - \beta|^2] W(\beta, s = 0) \qquad (75.105)$$

turns out to be a smoothed Wigner function $W(\beta, s = 0)$. It is this smoothing process that washes out possible negative parts in the Wigner function.

75.7 RESERVOIR THEORY

Reservoir theory treats the interaction of one system with a few degrees of freedom, called the system, with another system with many degrees of freedom, called the reservoir. A typical application of reservoir theory is a microscopic theory of damping: the system interacts with a reservoir, called the heat bath. The system dissipates energy into the heat bath whereas the heat bath introduces additional fluctuations to the system. Since the present chapter focuses on quantized field effects, the reservoir consists of the many modes of the radiation field in free space. Such a reservoir is modeled by a large number of independent harmonic oscillators

$$H_r = \sum_i \hbar\omega_i \left(b_i^\dagger b_i + \frac{1}{2}\right) , \qquad (75.106)$$

where b_i and $b_i^\dagger$ are the annihilation and creation operators for the ith harmonic oscillator of the reservoir. For convenience the interaction with the system is frequently approximated by a Hamiltonian of the form

$$H_{\text{int}} = \hbar \sum_i \left(g_i A b_i^\dagger + g_i^* A^\dagger b_i\right) , \qquad (75.107)$$

where A is an operator of the small system and g_i is the coupling strength of this system to the ith oscillator of the reservoir. For example, A may be an annihilation operator if the system is a harmonic oscillator or a Pauli spin matrix in the case of a two-level atom coupled to the free space radiation field.

Reservoir theory has important applications, and a detailed discussion can be found in various books, for example Refs. [15, 18, 24, 25, 28–30].

75.7.1 Thermal Reservoir

The most commonly used reservoir is the *thermal reservoir* or *thermal heat bath*. Its characteristic properties are

$$\langle b_i \rangle = \langle b_i^\dagger \rangle = \langle b_i b_j \rangle = \langle b_i^\dagger b_j^\dagger \rangle = 0 , \qquad (75.108)$$
$$\langle b_i^\dagger b_j \rangle = \bar{n}_i \delta_{ij} .$$

Here

$$\bar{n}_i = \frac{1}{\exp(\hbar\omega_i/k_B T) - 1} \qquad (75.109)$$

is the average number of photons at frequency ω_i, T is the temperature of the reservoir, and k_B denotes the Boltzmann constant.

75.7.2 Squeezed Reservoir

Another example of a reservoir is a *squeezed vacuum* or *squeezed reservoir*. If, for example, multiwave mixing is used to squeeze the radiation field, conjugate pairs of the reservoir operators b are correlated. Therefore, the expectation values $\langle b_i b_j \rangle$ and $\langle b_i^\dagger b_j^\dagger \rangle$ may be nonvanishing. Apart from the average number $\bar{n}_i$ of photons at frequency ω_i, which take into account nonvanishing expectation values $\langle b_i^\dagger b_i \rangle$, additional complex squeezing parameters are needed to describe the reservoir [25, 30, 31]. The characterization of a squeezed reservoir based on noise operators is discussed in Sect. 75.11.

75.8 MASTER EQUATION

In quantum mechanics, density operators are used to describe mixed states, and are discussed in Chap. 7. Here we introduce the concept of the reduced density operator

$$\rho_s = \text{Tr}_r (\rho_{sr}) , \qquad (75.110)$$

which is the density operator ρ_{sr} of the complete system traced over the degrees of freedom of the reservoir. The equation of motion for ρ_s in the Schrödinger picture is

$$\dot{\rho}_s(t) = -\frac{i}{\hbar} \text{Tr}_r \{[H_{sr}, \rho_{sr}(t)]\} . \qquad (75.111)$$

In the Born-Markov approximation the trace over the reservoir can be evaluated and leads to an equation of motion for ρ_s which no longer contains reservoir operators. This equation of motion is usually called the master equation. The Born-Markov approximation consists of two different parts:

1. Born approximation: The coupling to the reservoir is assumed to be sufficiently weak to allow a perturbative treatment of the interaction between the reservoir and the system.

2. Markov approximation: The correlations of the reservoir are assumed to decay very rapidly on a typical time scale of the system, or equivalently, the reservoir has a very broad spectrum. This approximation involves the assumption that the modes of the reservoir are spaced closely together, so that the frequency ω_i is a smooth function of i.

Since a general treatment is rather technical, we consider two typical examples. A more general discussion can be found in Refs. [15, 18, 24, 25, 28–30]

75.8.1 Damped Harmonic Oscillator

The universally accepted Hamiltonian in nonrelativistic QED for a harmonic oscillator of frequency ω coupled to a reservoir consisting of a large number of harmonic oscillators is given by the total Hamiltonian [32]

$$H_{\text{sr}} = \hbar\,\omega\left(a^\dagger a + \tfrac{1}{2}\right) + \sum_i \hbar\omega_i \left(b_i^\dagger b_i + \tfrac{1}{2}\right)$$
$$+ H_{\text{lc}} + H_{\text{si}}\,, \qquad (75.112)$$

with the linear coupling term

$$H_{\text{lc}} = \hbar \sum_i g_i \left(a + a^\dagger\right)\left(b_i + b_i^\dagger\right)\,, \qquad (75.113)$$

and the self-interaction term

$$H_{\text{si}} = \sum_i \frac{\hbar g_i^2}{\omega_i}\left(a + a^\dagger\right)^2\,. \qquad (75.114)$$

The approach used in quantum optics is to drop the term H_{si} and to make the rotating-wave approximation, that is, to drop the terms ab_i and $a^\dagger b_i^\dagger$, see also Chap. 66. Then the approximate total Hamiltonian reads

$$H_{\text{sr}} = \hbar\left(a^\dagger a + \tfrac{1}{2}\right) + \sum_i \hbar\omega_i\left(b_i^\dagger b_i + \tfrac{1}{2}\right)$$
$$+ \hbar \sum_i g_i\left(ab_i^\dagger + a^\dagger b_i\right)\,. \qquad (75.115)$$

Despite the problems with this approximate Hamiltonian (see Sect. V.D of Ref. [32] for a discussion) we adopt it in the present context because it leads to the widely used master equation for the damped harmonic oscillator. We consider two reservoirs: a thermal bath and a squeezed bath.

Harmonic Oscillator in a Thermal Bath

Within the Born-Markov approximation the master equation is

$$\dot{\rho} = \tfrac{1}{2}\gamma(\bar{n}+1)\left(2a\rho a^\dagger - a^\dagger a\rho - \rho a^\dagger a\right)$$
$$+ \tfrac{1}{2}\gamma\bar{n}\left(2a^\dagger\rho a - aa^\dagger\rho - \rho aa^\dagger\right)\,, \qquad (75.116)$$

where

$$\rho(t) = e^{i\omega a^\dagger a(t-t_0)}\rho_s(t)e^{-i\omega a^\dagger a(t-t_0)} \qquad (75.117)$$

is the reduced density operator in the interaction picture. The damping constant γ is given by

$$\gamma = 2\pi\mathcal{D}(\omega)|g(\omega)|^2\,, \qquad (75.118)$$

where $g(\omega)$ denotes the coupling strength at frequency ω. The number of thermal photons at frequency ω is

$$\bar{n} = \frac{1}{\exp\left(\hbar\omega/k_{\text{B}}T\right)-1}\,. \qquad (75.119)$$

Thus the Born-Markov approximation replaces the discrete reservoir modes by a continuum of modes with a density $\mathcal{D}(\omega)$.

Harmonic Oscillator in a Squeezed Bath

Within the Born-Markov approximation, the reduced density operator (75.117) in the interaction picture satisfies the master equation

$$\dot{\rho} = \tfrac{1}{2}\gamma(\bar{n}+1)\left(2a\rho a^\dagger - a^\dagger a\rho - \rho a^\dagger a\right)$$
$$+ \tfrac{1}{2}\gamma\bar{n}\left(2a^\dagger\rho a - aa^\dagger\rho - \rho aa^\dagger\right)$$
$$- \tfrac{1}{2}\gamma\bar{m}\left(2a^\dagger\rho a^\dagger - a^\dagger a^\dagger\rho - \rho a^\dagger a^\dagger\right)$$
$$- \tfrac{1}{2}\gamma\bar{m}^*\left(2a\rho a - aa\rho - \rho aa\right)\,. \qquad (75.120)$$

Here γ is again given by Eq. (75.118). The squeezed reservoir is characterized by a real number $\bar{n}$ and a complex number $\bar{m}$. Physically $\bar{n}$ is the number of photons at frequency ω, i.e., similar to the thermal reservoir, it measures the average energy at frequency ω. The complex number $\bar{m}$ determines the amount of squeezing. In general, the positivity of the density operator requires

$$|\bar{m}|^2 \le \bar{n}(\bar{n}+1)\,. \qquad (75.121)$$

A more quantitatively definition of $\bar{n}$ and $\bar{m}$ in terms of noise operators is given in Sect. 75.11.

75.8.2 Damped Two-Level Atom

The interaction of a two-level atom with a classical electromagnetic field is already discussed in Chap. 66.

For a quantum mechanical treatment of the field we only have to replace the classical field by its quantum mechanical counterpart Eq. (75.9). We then find in the rotating-wave approximation (see Chap. 66), that the dynamics of a two-level atom with a transition frequency ω_0 coupled to a reservoir consisting of a large number of harmonic oscillators is approximately described by the total Hamiltonian

$$H_{\mathrm{sr}} = \tfrac{1}{2}\hbar\omega_0\sigma_z + \sum_i \hbar\omega_i \left(b_i^\dagger b_i + \tfrac{1}{2} \right)$$
$$+ \hbar \sum_i \left(g_i\sigma_- b_i^\dagger + g_i^*\sigma_+ b_i \right) , \quad (75.122)$$

where

$$\sigma_+ = \begin{pmatrix} 0 & 1 \\ 0 & 0 \end{pmatrix}, \quad \sigma_- = \begin{pmatrix} 0 & 0 \\ 1 & 0 \end{pmatrix},$$
$$\sigma_z = \begin{pmatrix} 1 & 0 \\ 0 & -1 \end{pmatrix} . \quad (75.123)$$

Again, two reservoirs are considered: a thermal bath and a squeezed bath.

Two-Level Atom in a Thermal Bath

Within the Born-Markov approximation, the master equation is

$$\dot{\rho} = \tfrac{1}{2}\gamma\left(\bar{n} + 1 \right)\left(2\sigma_-\rho\sigma_+ - \sigma_+\sigma_-\rho - \rho\sigma_+\sigma_- \right)$$
$$+ \tfrac{1}{2}\gamma\bar{n}\left(2\sigma_+\rho\sigma_- - \sigma_-\sigma_+\rho - \rho\sigma_-\sigma_+ \right) , \quad (75.124)$$

where

$$\rho(t) = e^{i\omega_0\sigma_z(t-t_0)/2}\rho_{\mathrm{s}}(t)e^{-i\omega_0\sigma_z(t-t_0)/2} \quad (75.125)$$

is the reduced density operator in the interaction picture, and γ and $\bar{n}$ are given by Eqs. (75.118) and (75.119).

Two-Level Atom in a Squeezed Bath

Within the Born-Markov approximation, the reduced density operator in the interaction picture, Eq. (75.125), satisfies the master equation

$$\dot{\rho} = \tfrac{1}{2}\gamma\left(\bar{n} + 1 \right)\left(2\sigma_-\rho\sigma_+ - \sigma_+\sigma_-\rho - \rho\sigma_+\sigma_- \right)$$
$$+ \tfrac{1}{2}\gamma\bar{n}\left(2\sigma_+\rho\sigma_- - \sigma_-\sigma_+\rho - \rho\sigma_-\sigma_+ \right)$$
$$- \gamma\bar{m}\sigma_+\rho\sigma_+ - \gamma\bar{m}^*\sigma_-\rho\sigma_- , \quad (75.126)$$

where γ, $\bar{n}$ and $\bar{m}$ have the same meaning as in Eq. (75.120).

75.9 SOLUTION OF THE MASTER EQUATION

75.9.1 Damped Harmonic Oscillator

We consider only a thermal reservoir and present the solution of the master equation (75.116). For $\bar{n} = 0$ it can be solved in terms of coherent states, see Eq. (75.26). For $\bar{n} \neq 0$ we give solutions in terms of quasi-probability distributions.

Coherent States

For $\bar{n} = 0$, which is a good approximation for optical frequencies, if the system is initially in a coherent state $|\alpha_0\rangle$ with a density operator

$$\rho(t_0) = |\alpha_0\rangle\langle\alpha_0| , \quad (75.127)$$

then there exists a simple analytical solution of the master equation (75.116)

$$\rho(t) = |\alpha_0 e^{-\gamma(t-t_0)/2}\rangle\langle\alpha_0 e^{-\gamma(t-t_0)/2}| . \quad (75.128)$$

A coherent state thus remains a coherent state with an exponentially decaying amplitude $\alpha_0 e^{-\gamma(t-t_0)/2}$. According to Eq. (75.78) a general solution

$$\rho(t) = \int d^2\alpha_0\, P(\alpha_0)|\alpha_0 e^{-\gamma(t-t_0)/2}\rangle\langle\alpha_0 e^{-\gamma(t-t_0)/2}| \quad (75.129)$$

can be constructed for an initial density operator

$$\rho(t_0) = \int d^2\alpha_0\, P(\alpha_0)|\alpha_0\rangle\langle\alpha_0| . \quad (75.130)$$

If the system is initially in a superposition

$$|\psi(t_0)\rangle = \sum_i c_i|\alpha_i\rangle \quad (75.131)$$

of coherent states, the time evolution is given by

$$\rho(t) = \sum_{i,k} c_i c_k^* \exp\left[-\frac{1}{2}\left(1 - e^{-\gamma(t-t_0)} \right)|\alpha_i - \alpha_k|^2 \right]$$
$$\times \exp\left[i\left(1 - e^{-\gamma(t-t_0)} \right)\Im\left(\alpha_i\alpha_k^* \right) \right]$$
$$\times |\alpha_i e^{-\gamma(t-t_0)/2}\rangle\langle\alpha_k e^{-\gamma(t-t_0)/2}| . \quad (75.132)$$

For $\gamma(t - t_0) \ll 1$, the interference terms $|\alpha_i\rangle\langle\alpha_k|, i \neq k$ decay with an effective decay constant $\gamma|\alpha_i - \alpha_k|^2/2$. Thus the damping constant is modified by the separation of the two coherent states in phase space.

Fokker-Planck Equation

A widely used procedure for solving the master equation for a damped harmonic oscillator, Eq. (75.116), or for similar problems, is to derive an equation of motion for the quasi-probability distributions $W(\alpha, \alpha^*; s)$ defined in Eq. (75.76) from the master equation. The operators a and $a^\dagger$ are replaced by appropriate differential operators. The substitution rules can be derived from Eqs. (75.73), (75.74) and (75.76) and are

$$\left(a^\dagger\right)^k a^\ell \rho \to \left(\alpha^* - \frac{s+1}{2}\frac{\partial}{\partial\alpha}\right)^k \left(\alpha - \frac{s-1}{2}\frac{\partial}{\partial\alpha^*}\right)^\ell W,$$

$$\rho\left(a^\dagger\right)^k a^\ell \to \left(\alpha - \frac{s+1}{2}\frac{\partial}{\partial\alpha^*}\right)^\ell \left(\alpha^* - \frac{s-1}{2}\frac{\partial}{\partial\alpha}\right)^k W,$$

$$a\rho a^\dagger \to \left(\alpha - \frac{s-1}{2}\frac{\partial}{\partial\alpha^*}\right)\left(\alpha^* - \frac{s-1}{2}\frac{\partial}{\partial\alpha}\right) W,$$

$$a^\dagger\rho a \to \left(\alpha^* - \frac{s+1}{2}\frac{\partial}{\partial\alpha}\right)\left(\alpha - \frac{s+1}{2}\frac{\partial}{\partial\alpha^*}\right) W,$$

$$a^\dagger\rho a^\dagger \to \left(\alpha^* - \frac{s+1}{2}\frac{\partial}{\partial\alpha}\right)\left(\alpha^* - \frac{s-1}{2}\frac{\partial}{\partial\alpha}\right) W,$$

$$a\rho a \to \left(\alpha - \frac{s-1}{2}\frac{\partial}{\partial\alpha^*}\right)\left(\alpha - \frac{s+1}{2}\frac{\partial}{\partial\alpha^*}\right) W. \tag{75.133}$$

In general, this procedure leads to equations of motion which involve higher derivatives of W as exemplified by the quantum mechanical Liouville equation (75.93) for the Wigner function. For simple Hamiltonians, however, this equation has the form of a Fokker-Planck equation which is well known in classical stochastic problems [24, 33], (see also Sect. 75.11.2). In particular, for a damped harmonic oscillator described by the master equation (75.116), one obtains

$$\frac{\partial W}{\partial t} = \frac{\gamma}{2}\left[\frac{\partial}{\partial\alpha}(\alpha W) + \frac{\partial}{\partial\alpha^*}(\alpha^* W)\right] + \gamma\bar{n}_s \frac{\partial^2 W}{\partial\alpha\partial\alpha^*}, \tag{75.134}$$

where

$$\bar{n}_s = \bar{n} + \frac{1-s}{2} = \frac{1}{\exp(\hbar\omega/k_B T)-1} + \frac{1-s}{2}. \tag{75.135}$$

The time-dependent solution of this Fokker-Planck equation has the form

$$W(\alpha, \alpha^*, t; s) = \int G(\alpha, \alpha^*, t|\alpha', \alpha'^*, t'; s)$$
$$\times W(\alpha', \alpha'^*, t'; s)\, d^2\alpha', \tag{75.136}$$

where

$$G(\alpha, \alpha^*, t|\alpha', \alpha'^*, t'; s) = \frac{\exp\left[-\frac{|\alpha-\alpha' e^{-\gamma(t-t')/2}|^2}{\bar{n}_s\left[1-e^{-\gamma(t-t')}\right]}\right]}{\pi\bar{n}_s\left[1-e^{-\gamma(t-t')}\right]}$$

$$\tag{75.137}$$

is the Green's function of the Fokker-Planck equation (75.134). The steady-state solution is

$$W(\alpha, \alpha^*, t\to\infty; s) = \frac{1}{\pi\bar{n}_s} e^{-|\alpha|^2/\bar{n}_s}, \tag{75.138}$$

which is the distribution function of a harmonic oscillator in thermal equilibrium with a reservoir of temperature T.

75.9.2 Damped Two-Level Atom

The density operator

$$\rho = \begin{pmatrix} \rho_{ee} & \rho_{eg} \\ \rho_{ge} & \rho_{gg} \end{pmatrix} \tag{75.139}$$

for a two-level atom can be written as

$$\rho = \begin{pmatrix} \frac{1}{2}[1+\langle\sigma_z\rangle] & \langle\sigma_-\rangle \\ \langle\sigma_+\rangle & \frac{1}{2}[1-\langle\sigma_z\rangle] \end{pmatrix}. \tag{75.140}$$

Thus, a two-level atom is completely described by the expectation values

$$\langle\sigma_z\rangle = \rho_{ee} - \rho_{gg},$$
$$\langle\sigma_+\rangle = \rho_{ge},$$
$$\langle\sigma_-\rangle = \rho_{eg}. \tag{75.141}$$

Hence the master equation (75.126) can be cast into the equations of motions for these expectation values

$$\frac{d}{dt}\langle\sigma_+\rangle = -\gamma\left(\bar{n}+\tfrac{1}{2}\right)\langle\sigma_+\rangle - \gamma\bar{m}^*\langle\sigma_-\rangle,$$
$$\frac{d}{dt}\langle\sigma_-\rangle = -\gamma\left(\bar{n}+\tfrac{1}{2}\right)\langle\sigma_-\rangle - \gamma\bar{m}\langle\sigma_+\rangle,$$
$$\frac{d}{dt}\langle\sigma_z\rangle = -2\gamma\left(\bar{n}+\tfrac{1}{2}\right)\langle\sigma_z\rangle - \gamma, \tag{75.142}$$

which can easily be solved for arbitrary initial conditions. In contrast to a thermal reservoir ($\bar{m}=0$), a squeezed reservoir results in two different transverse decay constants $\gamma\left(\bar{n}+\tfrac{1}{2}+|\bar{m}|\right)$ and $\gamma\left(\bar{n}+\tfrac{1}{2}-|\bar{m}|\right)$ [31].

75.10 QUANTUM REGRESSION HYPOTHESIS

In the Schrödinger picture, time-dependent expectation values for system operators A_j can be calculated from the reduced density operator $\rho_s(t)$ via

$$\langle A_j\rangle = \mathrm{Tr}_s\{A_j\rho_s(t)\}. \tag{75.143}$$

The reduced density operator, however, is not sufficient to calculate two-time correlation functions such as $\langle A_j(t+\tau)A_k(t)\rangle$. For a definition of two-time correlation functions, the Heisenberg picture is more appropriate. Here, expectation values follow from

$$\langle A_j \rangle = \text{Tr}_{\text{sr}} \left\{ U_{\text{sr}}^{\dagger}(t, t_0) A_j(t_0) U_{\text{sr}}(t, t_0) \rho_{\text{sr}}(t_0) \right\} , \quad (75.144)$$

where $U_{\text{sr}}(t, t_0)$ describes the unitary time evolution of the complete system and $\rho_{\text{sr}}(t_0)$ is the density operator in the Heisenberg picture. Similarly, two-time correlation functions such as $\langle A_j(t+\tau) A_k(t) \rangle$ can be defined as

$$\begin{aligned} &\langle A_j(t+\tau) A_k(t) \rangle \\ &= \text{Tr}_{\text{sr}} \left\{ U_{\text{sr}}^{\dagger}(t+\tau, t_0) A_j(t_0) U_{\text{sr}}(t+\tau, t_0) \right. \\ &\quad \left. \times U_{\text{sr}}^{\dagger}(t, t_0) A_k(t_0) U_{\text{sr}}(t, t_0) \rho_{\text{sr}}(t_0) \right\} . \end{aligned} \quad (75.145)$$

The quantum regression hypothesis avoids the calculation of $U_{\text{sr}}(t, t_0)$. Two equivalent formulations exist, one based on the master equation for ρ_{s} and another based on the equation of motion for the expectation values $\langle A_j \rangle$, see for example Refs. [18, 24, 25, 28, 30].

75.10.1 Two-Time Correlation Functions and Master Equation

It follows from their definition (75.145) in the Heisenberg picture that two-time correlation functions $\langle A_j(t+\tau) A_k(t) \rangle$ for system operators A_j and A_k can be calculated with the help of the operator

$$R_{\text{s}}(t+\tau, t) = \text{Tr}_{\text{r}} \left\{ U_{\text{sr}}(t+\tau, t) A_k \rho_{\text{sr}}(t) U_{\text{sr}}^{\dagger}(t+\tau, t) \right\} , \quad (75.146)$$

where $U_{\text{sr}}(t+\tau, t)$ describes the unitary time evolution of the complete system between t and $t+\tau$. We find

$$\langle A_j(t+\tau) A_k(t) \rangle = \text{Tr}_{\text{s}} \left\{ A_j R_{\text{s}}(t+\tau, t) \right\} . \quad (75.147)$$

Note, that in Eqs. (75.146) and (75.147) we interpret A_j and A_k as operators in the Schrödinger picture and have omitted the argument t_0. Because the reduced density operator

$$\rho_{\text{s}}(t) = \text{Tr}_{\text{r}} \left\{ U_{\text{sr}}(t, t_0) \rho_{\text{sr}}(t_0) U_{\text{sr}}^{\dagger}(t, t_0) \right\} \quad (75.148)$$

satisfies the master equation, it is plausible to assume, that when the time derivative is taken with respect to τ, the operator $R_{\text{s}}(t+\tau, t)$ also satisfies the master equation for ρ_{s}, subject to the initial condition $R_{\text{s}}(t, t) = A_k \rho_{\text{s}}(t)$. However, this requires the additional assumption that the approximations made in the derivation of the master equation for $\rho_{\text{s}}(t)$ are also valid for $R_{\text{s}}(t+\tau, t)$.

75.10.2 Two-Time Correlation Functions and Expectation Values

A second formulation of the quantum regression hypothesis asserts that two-time correlation functions $\langle A_j(t+\tau) A_k(t) \rangle$ obey

$$\frac{\partial}{\partial \tau} \langle A_j(t+\tau) A_k(t) \rangle = \sum_{\ell} G_{j\ell}(\tau) \langle A_{\ell}(t+\tau) A_k(t) \rangle , \quad (75.149)$$

provided that the expectation values of a set of system operators A_j satisfy

$$\frac{\partial}{\partial t} \langle A_j(t) \rangle = \sum_{\ell} G_{j\ell}(t) \langle A_{\ell}(t) \rangle . \quad (75.150)$$

This is the form of the quantum regression hypothesis that was first formulated by Lax [34].

The equivalence of the two formulations follows from the interpretation of $R_{\text{s}}(t+\tau, t)$ on the right side of Eq. (75.147) as a "density operator". Then $\text{Tr}_{\text{s}} \left\{ A_j R_{\text{s}}(t+\tau, t) \right\}$ is an "expectation value" for which we assume that Eq. (75.150) is valid; i.e.,

$$\begin{aligned} &\frac{\partial}{\partial \tau} \text{Tr}_{\text{s}} \left\{ A_j R_{\text{s}}(t+\tau, t) \right\} \\ &= \sum_{\ell} G_{j\ell}(\tau) \text{Tr}_{\text{s}} \left\{ A_{\ell} R_{\text{s}}(t+\tau, t) \right\} . \end{aligned} \quad (75.151)$$

According to Eq. (75.147), this is identical to Eq. (75.149).

75.11 QUANTUM NOISE OPERATORS

The master equation is based on the Schrödinger picture in quantum mechanics: the state of the system described by a density operator is time-dependent, whereas operators corresponding to observables are time independent. If we use the Heisenberg picture instead and make similar approximations as in the derivation of the master equation, we arrive at equations of motion for the Heisenberg operators, see for example Refs. [15, 25, 28, 29]. Due to the interaction with a reservoir these equations have additional *noise terms* and *damping terms*.

75.11.1 Quantum Langevin Equations

Again consider a damped harmonic oscillator. The equation of motion for the annihilation operator

$$\tilde{a}(t) = e^{i\omega(t-t_0)} a(t) \quad (75.152)$$

in the interaction picture follows from the Heisenberg equations for the operators $a, a^{\dagger}, b_i$ and $b_i^{\dagger}$ and reads

$$\begin{aligned} \frac{d\tilde{a}}{dt} = &- \sum_i |g_i|^2 \int_{t_0}^{t} e^{-i(\omega_i - \omega)(t-t')} \tilde{a}(t') \, dt' \\ &- i \sum_i g_i^* e^{-i(\omega_i - \omega)(t-t_0)} b_i(t_0) . \end{aligned} \quad (75.153)$$

In general, the noise operator

$$F(t) = -i \sum_i g_i^* e^{-i(\omega_i - \omega)(t - t_0)} b_i(t_0) \qquad (75.154)$$

is not delta-correlated, and there are also memory effects in Eq. (75.153). The noise operator $F(t)$ can be used to classify the reservoir: if it is delta-correlated, that is, if the reservoir has a very broad spectrum, one speaks of white noise, see below. If the correlation time is finite so that there are memory effects, one speaks of colored noise.

If the spectrum of the noise is very broad (as in the derivation of the master equation for the reduced density operator), the operator $\tilde{a}(t)$ satisfies the quantum Langevin equation

$$\frac{d\tilde{a}}{dt} = -\frac{\gamma}{2}\tilde{a}(t) + F(t), \qquad (75.155)$$

with a damping term $-\gamma\tilde{a}(t)/2$ and a noise term $F(t)$. Note that a simple damping equation such as

$$\dot{\tilde{a}}(t) = -\frac{\gamma}{2}\,\tilde{a}(t) \qquad (75.156)$$

is unphysical since it does not preserve the commutation relation $[\tilde{a}, \tilde{a}^\dagger] = 1$. It is the noise term which saves the commutation relation.

For a thermal reservoir with a sufficiently small correlation time, the standard derivations [29] give

$$\langle F(t)\rangle = \langle F^\dagger(t)\rangle = \langle F(t)F(t')\rangle = \langle F^\dagger(t)F^\dagger(t')\rangle = 0\,,$$
$$\langle F^\dagger(t)F(t')\rangle = \gamma\overline{n}\delta(t - t')\,,$$
$$\langle F(t)F^\dagger(t')\rangle = \gamma(\overline{n} + 1)\delta(t - t')\,, \qquad (75.157)$$

where the averages are taken over the reservoir. The damping constant γ and the number of thermal photons $\overline{n}$ are given in Eqs. (75.118) and (75.119). For more general relations, see Ref. [32], where it is shown explicitly that correlation functions involving the fluctuation force do not in fact depend on the oscillator frequency. The condition of a sufficiently small reservoir correlation time requires that $\tau_c \approx \hbar/(k_B T)$ is small compared with the time scales of the systems. The only time scale in Eq. (75.155) is γ^{-1}. The relevant condition is therefore $\tau_c \ll \gamma^{-1}$. For typical applications in quantum optics, a is the annihilation operator and $a^\dagger$ is the creation operator of a single-mode cavity field. Here one can have quality factors of the cavity on the order of $Q = \omega/\gamma \approx 10^6$. In terms of the quality factor, the condition of sufficiently small reservoir correlation times requires $\hbar\omega/(k_B T) \ll Q$. For optical frequencies ($\omega \approx 3 \times 10^{15}$ Hz) and $T \approx 300$ K one has $\hbar\omega/(k_B T) \approx 75$. In the microwave regime ($\omega \approx 30$ GHz) one can have temperatures as low as $T \approx 3$ mK and still have $\hbar\omega/(k_B T) \approx 75$. Therefore the assumption of delta-correlated noise is a good approximation for typical applications in quantum optics.

Similarly, for a squeezed reservoir one has

$$\langle F(t)\rangle = \langle F^\dagger(t)\rangle = 0\,,$$
$$\langle F(t)F(t')\rangle = \gamma\overline{m}\delta(t - t')\,,$$
$$\langle F^\dagger(t)F^\dagger(t')\rangle = \gamma\overline{m}^*\delta(t - t')\,,$$
$$\langle F^\dagger(t)F(t')\rangle = \gamma\overline{n}\delta(t - t')\,,$$
$$\langle F(t)F^\dagger(t')\rangle = \gamma(\overline{n} + 1)\delta(t - t')\,, \qquad (75.158)$$

which gives a quantitative definition of the parameters $\overline{n}$ and $\overline{m}$ in the master equations (75.120) and (75.126). Again, a detailed discussion in Ref. [32] shows that correlation functions involving the fluctuation forces do not depend on the oscillator frequency.

The Langevin equation (75.155) is based on the use of the approximate Hamiltonian given in (75.115), i.e., it is based on the rotating-wave approximation and the neglect of self-interaction terms. The corresponding Langevin equation for $x = \sqrt{\hbar/(2m\omega)}\left(a + a^\dagger\right)$ may be calculated and, not unexpectedly, it disagrees with the Abraham-Lorentz equation which Ford and O'Connell [35] showed could be derived systematically using the exact Hamiltonian (75.112). In fact, Ford and O'Connell showed that an improved equation for the radiating electron (improved in the sense that it is second-order and is not subject to the analyticity problems and the problems with runaway solutions associated with the Abraham-Lorentz equation) may be obtained by generalizing the Hamiltonian (75.112) to include electron structure. The implications following from these different equations are presently under study.

75.11.2 Stochastic Differential Equations

In Sect. 75.11.1 we discussed one of the simplest quantum systems with dissipation, the damped harmonic oscillator. For more complicated systems the noise term can also contain system operators. In such cases there are two different ways to interpret the Langevin equation. In order to give a feeling for the two possible interpretations, we discuss the one dimensional classical Langevin equation

$$\frac{dx}{dt} = g(x, t) + h(x, t)F(t) \qquad (75.159)$$

for the stochastic variable $x(t)$ with delta-correlated noise $\langle F(t)F(t')\rangle = \delta(t - t')$. Due to the singular nature of delta-correlated noise, such a Langevin equation does not exist from a strictly mathematical point of view. A mathematically more rigorous treatment is based on stochastic differential equations [24, 25, 33]. The variable $x(t)$ is said to obey a stochastic differential equation

$$dx(t) = g(x, t)\,dt + h(x, t)F(t)\,dt$$
$$= g(x, t)\,dt + h(x, t)\,dW(t), \qquad (75.160)$$

if, for all times t and t_0, $x(t)$ is given by

$$x(t) = x(t_0) + \int_{t_0}^t g(x(t'), t')\, dt' + \int_{t_0}^t h[x(t'), t']\, dW(t').$$

(75.161)

Here the last term is a Riemann-Stieltjes integral defined by

$$\int_{t_0}^t h[x(t'), t']\, dW(t')$$

$$= \lim_{n\to\infty} \sum_{i=0}^{n-1} h[x(\tau_i), \tau_i]\, [W(t_{i+1}) - W(t_i)],$$

(75.162)

where τ_i is in the interval (t_i, t_{i+1}).

There are two different approaches to such problems: the Ito approach and the Stratonovich approach. They differ in the definition of stochastic integrals.

In the Stratonovich approach, one evaluates $h[x(\tau_i), \tau_i]$ at $\tau_i = (t_i + t_{i+1})/2$, whereas in the Ito approach one evaluates $h[(x(\tau_i), \tau_i)]$ at $\tau_i = t_i$. This slightly different definition of τ_i leads to different results because, as a consequence of the delta-correlated noise term, $x(t)$ is not a continuous path. However, there is a relation between the solution of a Stratonovich stochastic differential equation and an Ito stochastic differential equation. Suppose $x(t)$ is a solution of the Stratonovich stochastic differential equation

$$dx(t) = g(x, t)\, dt + h(x, t)\, dW(t).$$

(75.163)

Then $x(t)$ satisfies the Ito stochastic differential equation

$$dx(t) = \left[g(x, t) + \frac{1}{2} h(x, t) \frac{\partial h(x, t)}{\partial x} \right] dt + h(x, t)\, dW(t).$$

(75.164)

Instead of dealing with stochastic differential equations, one can derive a Fokker-Planck equation for the conditional probability $P(x, t | x_0, t_0)$. For the Stratonovich stochastic differential equation (75.163), the Fokker-Planck equation is

$$\frac{\partial P}{\partial t} = -\frac{\partial}{\partial x} \left[g(x, t) + \frac{1}{2} h(x, t) \frac{\partial h(x, t)}{\partial x} \right] P$$

$$+ \frac{1}{2} \frac{\partial^2}{\partial x^2}\, h^2(x, t) P,$$

(75.165)

which takes the form

$$\frac{\partial P}{\partial t} = -\frac{\partial}{\partial x}\, g(x, t) P + \frac{1}{2} \frac{\partial^2}{\partial x^2}\, h^2(x, t) P,$$

(75.166)

if Eq. (75.163) is interpreted as a stochastic differential equation in the Ito sense.

The two approaches have the following properties: (i) in most of the models used in physics the Stratonovich definition of a stochastic integral is needed to give correct results, (ii) rules from ordinary calculus are applicable only in the Stratonovich approach, and (iii) for Langevin equations with $h(x, t) = $ const, as in Eq. (75.155), the Stratonovich interpretation and the Ito interpretation of stochastic integrals are equivalent.

75.12   QUANTUM MONTE CARLO FORMALISM

The quantum Monte Carlo formalism was developed to solve numerically master equations of the Lindblad type [18, 36, 37]

$$\dot{\rho}_s = -\frac{i}{\hbar} [H_s, \rho_s]$$

$$+ \frac{1}{2} \sum_j \left(2C_j \rho_s C_j^\dagger - C_j^\dagger C_j \rho_s - \rho_s C_j^\dagger C_j \right).$$

(75.167)

Here C_j are arbitrary system operators.

As an illustrative example, consider

$$\dot{\rho}_s = -\frac{i}{\hbar} [H_s, \rho_s] + \frac{1}{2} \left(2C \rho_s C^\dagger - C^\dagger C \rho_s - \rho_s C^\dagger C \right),$$

(75.168)

where C is an arbitrary system operator. Instead of solving the master equation, one defines quantum trajectories or stochastic wave functions as follows. Starting from $|\psi(t)\rangle$, there are two possibilities for the time evolution during the interval dt:

1. The system evolves according to the non-Hermitian Hamiltonian $H_s - \frac{i\hbar}{2} C^\dagger C$; i.e.

$$|\psi(t + dt)\rangle = \frac{\left[1 - i\, dt \left(H_s - \frac{i\hbar}{2} C^\dagger C \right) / \hbar \right] |\psi(t)\rangle}{\sqrt{1 - \langle \psi(t) | C^\dagger C | \psi(t)\rangle\, dt}}.$$

(75.169)

2. The system makes a jump; i.e.

$$|\psi(t + dt)\rangle = \frac{C |\psi(t)\rangle}{\sqrt{\langle \psi(t) | C^\dagger C | \psi(t)\rangle}}.$$

(75.170)

Since both possibilities describe a nonunitary time evolution, $|\psi\rangle$ must be normalized after each step. For each time interval dt one of these two possibilities is randomly chosen according to the probability

$$P(t)\, dt = \langle \psi(t) | C^\dagger C | \psi(t)\rangle\, dt$$

(75.171)

to make a jump between t and $t + dt$. We can now define a density operator

$$\rho_s(t) = |\psi(t)\rangle \langle \psi(t)|$$

(75.172)

for a specific quantum trajectory $|\psi(t)\rangle$. The density operator

$$\overline{\rho_{\rm s}(t)} = \overline{|\psi(t)\rangle\langle\psi(t)|} \qquad (75.173)$$

averaged over all trajectories (indicated by the bar) is then a solution of the master equation (75.168).

This method can easily be generalized to master equations of the form (75.167).

75.13 SPONTANEOUS EMISSION IN FREE SPACE

Consider an atom which is initially in one of its excited states and which interacts with the quantized electromagnetic field of free space. Even if none of the modes of the electromagnetic field is excited, there are still the vacuum fluctuations which "interact" with the atom and give rise to important effects:

1. Spontaneous emission: the atom spontaneously emits a photon and decays from the excited state.

2. Natural linewidth: due to the finite lifetime of the atomic levels, the radiation from an atomic transition has a finite linewidth, called the natural linewidth.

3. Lamb shift: the energy levels of the atom are shifted.

The standard theory of spontaneous emission is the Wigner-Weisskopf theory [15, 29]. Here an initially excited atomic state $|\ell\rangle$ decays exponentially according to

$$|c_\ell(t)|^2 = e^{-\Gamma_\ell t}, \qquad (75.174)$$

where the decay constant Γ_ℓ is given by

$$\Gamma_\ell = \sum_i \frac{\omega_{\ell i}^3 |\mathbf{d}_{\ell i}|^2}{3\pi\epsilon_0 \hbar c^3}, \qquad (75.175)$$

and the sum is over all atomic states with an energy E_i lower than the energy E_ℓ of the state $|\ell\rangle$. $\omega_{\ell i} = (E_\ell - E_i)/\hbar$ is the transition frequency for the transition $|\ell\rangle \to |i\rangle$, and $\mathbf{d}_{\ell i} = e\langle\ell|\mathbf{r}|i\rangle$ is the corresponding dipole moment.

The same decay constant Γ_ℓ is also observed as a linewidth in the spectrum of the radiation scattered by an atom when the incoming photon excites the atom to the level $|\ell\rangle$.

The energy level shift is more troublesome and needs the concept of mass renormalization, a standard problem in quantum electrodynamics. The theory and results are discussed in Chaps. 27 and 28.

Recent calculations of Pachucki [38] based on fully relativistic quantum electrodynamics and including two-loop corrections predict $1\,057\,838(6)$ kHz for the energy difference between the $2s_{1/2}$-state and the $2p_{1/2}$-state which is in excellent agreement with the experimental result of $1\,057\,839(12)$ kHz [39].

75.14 RESONANCE FLUORESCENCE

Consider a two-level atom driven by a continuous monochromatic wave which is treated classically. The excited state of the atom can decay by spontaneous emission into vacuum modes of the electromagnetic field. This emission is called resonance fluorescence. Of particular interest are the properties of the emitted light. For a detailed discussion of resonance fluorescence, see for example Refs. [18, 28–30].

The far field at position $\mathbf{R}$ emitted by an atom at the origin is proportional to its dipole moment and can be expressed in terms of the dipole operators σ_+ and σ_- according to the relation [18]

$$\mathbf{E}^+(\mathbf{R}, t) = -\frac{\omega_0^2 \, (\mathbf{d} \times \mathbf{R}) \times \mathbf{R}}{4\pi\epsilon_0 c^2 R^3} \, \sigma_-(t - r/c),$$

$$\mathbf{E}^-(\mathbf{R}, t) = -\frac{\omega_0^2 \, (\mathbf{d}^* \times \mathbf{R}) \times \mathbf{R}}{4\pi\epsilon_0 c^2 R^3} \, \sigma_+(t - r/c), \qquad (75.176)$$

where

$$\mathbf{d} = e\langle g|\mathbf{r}|e\rangle \qquad (75.177)$$

is the atomic dipole matrix element and the field operators $\mathbf{E}^+(\mathbf{R}, t)$ and $\mathbf{E}^-(\mathbf{R}, t)$ as well as the dipole operators $\sigma_+(t)$ and $\sigma_-(t)$ are in the Heisenberg picture. Knowledge of the operators $\sigma_+(t)$ and $\sigma_-(t)$ is therefore sufficient to study the properties of the emitted light in the far field.

75.14.1 Equations of Motion

The total Hamiltonian for the system reads

$$H_{\rm sr} = \tfrac{1}{2}\hbar\omega_0\sigma_z + \sum_i \hbar\omega_i \left(b_i^\dagger b_i + \frac{1}{2}\right)$$

$$+ \, \hbar \sum_i \left(g_i \sigma_- b_i^\dagger + g_i^* \sigma_+ b_i\right)$$

$$- \, \tfrac{1}{2}\left(d\,\mathcal{E}^* \sigma_- e^{i\omega_0 t} + d^* \mathcal{E}\,\sigma_+ e^{-i\omega_0 t}\right), \qquad (75.178)$$

where a resonant driving term has been added to the Hamiltonian (75.122). Here d is the projection of the dipole matrix element $e\langle g|\mathbf{r}|e\rangle$ onto the polarization vector of the driving field with an amplitude $\mathcal{E}$. The corresponding master equation in the interaction picture is

$$\dot{\rho} = -i\tfrac{1}{2}\Omega_1\left[\sigma_+ + \sigma_-, \rho\right]$$

$$+ \tfrac{1}{2}\gamma\left(2\sigma_-\rho\sigma_+ - \sigma_+\sigma_-\rho - \rho\sigma_+\sigma_-\right), \qquad (75.179)$$

where $\Omega_1 = -\mathcal{E}\,d^*/\hbar$ is the Rabi frequency associated with the driving field. The vacuum modes of the field are described by a thermal reservoir at zero temperature.

The equations of motion for the expectation values $\langle\sigma_+\rangle$, $\langle\sigma_-\rangle$, and $\langle\sigma_z\rangle$ are the optical Bloch equations with radiative damping (see also Chap. 66) and are

$$\frac{d}{dt}\langle\sigma_+\rangle = -\frac{\gamma}{2}\langle\sigma_+\rangle - i\frac{\Omega_1}{2}\langle\sigma_z\rangle,$$

$$\frac{d}{dt}\langle\sigma_-\rangle = -\frac{\gamma}{2}\langle\sigma_-\rangle + i\frac{\Omega_1}{2}\langle\sigma_z\rangle,$$

$$\frac{d}{dt}\langle\sigma_z\rangle = -\gamma(\langle\sigma_z\rangle + 1) - i\Omega_1(\langle\sigma_+\rangle - \langle\sigma_-\rangle). \quad (75.180)$$

These expectation values determine the density operator (75.140) of the two-level atom. Because Eqs. (75.180) are a system of linear differential equations for $\langle\sigma_+\rangle$, $\langle\sigma_-\rangle$, and $\langle\sigma_z\rangle$, they can be solved analytically. Furthermore, the quantum regression hypothesis allows one to calculate two-time correlation functions as shown in Sect. 75.10.

75.14.2   Intensity of Emitted Light

According to Eqs. (75.52) and (75.176), the intensity of the fluorescence light at position $\mathbf{R}$ is given by

$$I = \langle E^-(\mathbf{R},t)E^+(\mathbf{R},t)\rangle \propto \langle\sigma_+\sigma_-\rangle, \quad (75.181)$$

and can be decomposed into two parts: the coherent intensity

$$I_{\mathrm{coh}} \propto \langle\sigma_+\rangle\langle\sigma_-\rangle \quad (75.182)$$

originating from the mean motion of the dipole, and the incoherent intensity

$$I_{\mathrm{inc}} \propto \langle\sigma_+\sigma_-\rangle - \langle\sigma_+\rangle\langle\sigma_-\rangle, \quad (75.183)$$

which is due to fluctuations of the dipole motion around its average value. The steady state intensities are

$$I_{\mathrm{coh}} \propto \frac{\Omega_1^2\gamma^2}{(\gamma^2 + 2\Omega_1^2)^2}, \quad (75.184)$$

and

$$I_{\mathrm{inc}} \propto \frac{2\Omega_1^4}{(\gamma^2 + 2\Omega_1^2)^2}. \quad (75.185)$$

For weak laser intensities (Ω_1 small) the intensity of the fluorescence light is dominated by the coherent part whereas for high intensities (Ω_1 large) it is dominated by the incoherent part.

75.14.3   Spectrum of the Fluorescence Light

The Wiener-Khintchine theorem (75.54) allows one to express the steady state spectrum of the fluorescence light as the Fourier transform of the correlation function $\langle\sigma_+(\tau)\sigma_-(0)\rangle_{\mathrm{ss}}$ in the form

$$S(\omega) = \frac{1}{2\pi}\int_0^\infty e^{-i\omega\tau}\langle E^-(\tau)E^+(0)\rangle_{\mathrm{ss}}\,d\tau + \mathrm{c.c}$$

$$\propto \frac{1}{2\pi}\int_0^\infty e^{-i(\omega-\omega_0)\tau}\langle\sigma_+(\tau)\sigma_-(0)\rangle_{\mathrm{ss}}\,d\tau + \mathrm{c.c.}$$

$$(75.186)$$

Again it consists of two contributions: a coherent part $S_{\mathrm{coh}}(\omega)$, and an incoherent part $S_{\mathrm{inc}}(\omega)$. The coherent part is

$$S_{\mathrm{coh}}(\omega) \propto \frac{\Omega_1^2\gamma^2}{(\gamma^2 + 2\Omega_1^2)^2}\,\delta(\omega - \omega_0). \quad (75.187)$$

The incoherent part of the fluorescence light has two qualitatively different spectra. For $\Omega_1 < \gamma/4$, it has a single peak at ω_0, whereas it consists of three peaks for $\Omega_1 > \gamma/4$. For $\Omega_1 \gg \gamma/4$ it is given by

$$S_{\mathrm{inc}}(\omega) \propto \frac{1}{2\pi\gamma}\left[\frac{(\gamma/2)^2}{(\omega-\omega_0)^2 + (\gamma/2)^2}\right.$$

$$+ \frac{1}{3}\frac{(3\gamma/4)^2}{(\omega-\omega_0+\Omega_1)^2 + (3\gamma/4)^2}$$

$$\left. + \frac{1}{3}\frac{(3\gamma/4)^2}{(\omega-\omega_0-\Omega_1)^2 + (3\gamma/4)^2}\right]. \quad (75.188)$$

The central peak at $\omega = \omega_0$ has a width of $\gamma/2$ whereas the width of the two side peaks at $\omega = \omega_0 \pm \Omega_1$ is $3\gamma/4$. Their heights are one third of the height of the central peak. This spectrum was predicted by Mollow [40] and experimentally confirmed by Wu *et al.* [41] and Hartig *et al.* [42].

This triplet can be explained in terms of the dressed states $|1,n\rangle$ and $|2,n\rangle$ introduced in Chap. 66, see Eq. (66.50). If the driving field is resonant with the atomic transition, these states have the energies

$$E_{1,n} = \hbar\left(n + \tfrac{1}{2}\right)\omega_0 - \hbar R_n,$$

$$E_{2,n} = \hbar\left(n + \tfrac{1}{2}\right)\omega_0 - \hbar R_n, \quad (75.189)$$

see Eqs. (66.52). The energy differences between the allowed transitions are

$$E_{2,n} - E_{2,n-1} = \hbar\omega_0 + \hbar R_n - \hbar R_{n-1} \approx \hbar\omega_0,$$

$$E_{2,n} - E_{1,n-1} = \hbar\omega_0 + \hbar R_n + \hbar R_{n-1} \approx \hbar\omega_0 + \hbar\Omega_1,$$

$$E_{1,n} - E_{2,n-1} = \hbar\omega_0 - \hbar R_n - \hbar R_{n-1} \approx \hbar\omega_0 - \hbar\Omega_1,$$

$$E_{1,n} - E_{1,n-1} = \hbar\omega_0 + \hbar R_n - \hbar R_{n-1} \approx \hbar\omega_0, \quad (75.190)$$

where we have made the approximations

$$R_n - R_{n-1} \approx 0,$$

$$R_n + R_{n+1} \approx 2g\sqrt{\bar{n}+1} \approx \Omega_1. \quad (75.191)$$

This is a good approximation for an intense driving field which can approximated by a highly excited coherent

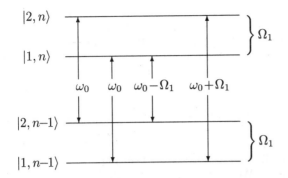

Figure 75.6. Energy level diagram of dressed states. The transition frequencies are ω_0, $\omega_0 - \Omega_1$ and $\omega_0 + \Omega_1$.

state with an average photon number $\bar{n}$. Figure 75.6 shows these energy levels and the allowed transition. Obviously, the transitions correspond to frequencies ω_0, $\omega_0 - \Omega_1$ and $\omega_0 + \Omega_1$. The dressed state picture also explains the 2:1 ratio for the *integrated* intensities of the central peak and the side peak in Eq. (75.188).

75.14.4   Photon Correlations

In addition to the spectrum which is based on the correlation function $\langle E^-(\tau)E^+(0)\rangle_{\text{ss}}$ in Sect. 75.14.3, the second-order correlation function

$$G_{\text{ss}}^{(2)}(\tau) = \langle E^-(0)E^-(\tau)E^+(\tau)E^+(0)\rangle_{\text{ss}}$$
$$\propto \langle \sigma_+(0)\sigma_+(\tau)\sigma_-(\tau)\sigma_-(0)\rangle_{\text{ss}} \quad (75.192)$$

can be measured to gain more insight into the fluorescence light, see also Sect. 75.4.2. Experimentally this is done by measuring the joint probability for detecting a photon at time $t = 0$ and a subsequent photon at time $t = \tau$. Again the result can be obtained from the quantum regression hypothesis and reads

$$g^{(2)}(\tau) = \frac{G_{\text{ss}}^{(2)}(\tau)}{|G_{\text{ss}}^{(1)}(0)|^2}$$
$$= 1 - e^{-3\gamma\tau/4}\left(\cos\delta\tau + \frac{3\gamma}{4\delta}\sin\delta\tau\right), \quad (75.193)$$

where δ is given by

$$\delta = \sqrt{\Omega_1^2 - \gamma^2/4}. \quad (75.194)$$

For $\tau = 0$, $g^{(2)}(0) = 0$, indicating a tendency of photons to be separated. This tendency is known as photon antibunching and was first predicted by Carmichael and Walls [43] and experimentally verified by Kimble *et al.* [44]. Photon antibunching of radiation emitted from a two-level atom has a simple explanation: After the atom has emitted a photon it is in the ground state and must first be excited again before it can emit another photon.

Acknowledgment

We want to thank Prof. I. Bialynicki-Birula for stimulating discussions and a critical reading of the manuscript. The work of RFOC was supported in part by the U.S. Army Research Office under grant No. DAAH04-94-G-0333.

REFERENCES

1. C. Cohen-Tannoudji, J. Dupont-Roc, and G. Grynberg, *Photons, and Atoms. An Introduction to Quantum Electrodynamics* (Wiley, New York, 1989).
2. L. Mandel, Opt. Lett. **4**, 205 (1979).
3. W. Schleich, M. Pernigo, and Fam Le Kien, Phys. Rev. A **44**, 2172 (1991).
4. See the special issues on Squeezed States, J. Opt. Soc. Am. B **4**, No. 10 (1987); J. Mod. Opt. **34**, No. 6 (1987); Appl. Phys. B **55**, No. 3 (1992).
5. W. Schleich and J. A. Wheeler, J. Opt. Soc. Am. B **4**, 1715 (1987).
6. H. P. Yuen, Phys. Rev. A **13**, 2226 (1976).
7. R. E. Slusher, L. W. Hollberg, B. Yurke, J. C. Mertz, and J. F. Valley, Phys. Rev. Lett. **55**, 2409 (1985).
8. L. A. Wu, M. Xiao, and H. J. Kimble, J. Opt. Soc. Am. B **4**, 1465 (1987).
9. A. Heidmann, R. J. Horowicz, S. Reynaud, E. Giacobino, and C. Fabre, Phys. Rev. Lett. **59**, 2555 (1987).
10. See the special issue on *Quantum Phase and Phase Dependent Measurements*, Phys. Scr. **48T** (1993).
11. P. Carruthers and M. M. Nieto, Rev. Mod. Phys. **40**, 411 (1968).
12. D. T. Pegg and S. M. Barnett, Phys. Rev. A **39**, 1665 (1989).
13. J. W. Noh, A. Fougères, and L. Mandel, Phys. Rev. A **45**, 424 (1992).
14. R. J. Glauber, Phys. Rev. **130**, 2529 (1963).
15. W. H. Louisell, *Quantum Statistical Properties of Radiation* (Wiley, New York, 1973).
16. R. Hanbury Brown and R. Q. Twiss, Nature **177**, 27 (1956).
17. P. L. Kelley and W. H. Kleiner, Phys. Rev. **136**, 316 (1964).
18. H. Carmichael, *An Open Systems Approach to Quantum Optics* (Springer, Berlin, 1993).
19. H. Yuen and H. P. Shapiro, IEEE Trans. Inf. Theory **26**, 78 (1980); J. H. Shapiro and S. S. Wagner, IEEE J. Quantum Electron. **20**, 803 (1984).
20. K. E. Cahill and R. J. Glauber, Phys. Rev. A **177**, 1882 (1969).
21. M. Hillery, R. F. O'Connell, M. O. Scully, and E. P. Wigner, Phys. Rep. **106**, 121 (1984).
22. R. J. Glauber, Phys. Rev. **131**, 2766 (1963).
23. E. C. G. Sudarshan, Phys. Rev. Lett. **10**, 277 (1963).
24. C. W. Gardiner, *Handbook of Stochastic Methods* (Springer, Berlin, 1985).
25. C. W. Gardiner, *Quantum Noise* (Springer, Berlin, 1991).
26. E. Wigner, Phys. Rev. **40**, 749 (1932).
27. R. L. Hudson, Rep. Math. Phys. **6**, 249 (1974).

28. P. Meystre and M. Sargent III, *Elements of Quantum Optics* (Springer, Berlin 1991).

29. C. Cohen-Tannoudji, J. Dupont-Roc, and G. Grynberg, *Atom-Photon Interactions* (Wiley, New York, 1992).

30. D. F. Walls and G. J. Milburn, *Quantum Optics* (Springer, Berlin 1994).

31. C. W. Gardiner, Phys. Rev. Lett. **56**, 1917 (1986).

32. G. W. Ford, J. T. Lewis, and R. F. O'Connell, Phys. Rev. A **37**, 4419 (1988); R. F. O'Connell, in *Proceedings of the Second International Workshop on Squeezed States and Uncertainty Relations*, edited by D. Han, Y.-S. Kim, and V. I. Man'ko (NASA Conference Publication 3219, 1993).

33. H. Risken, *The Fokker-Planck Equation* (Springer, Berlin 1989).

34. M. Lax, in *Brandeis University Summer Institute Lectures*, edited by M. Chretin, E. P. Gross, and S. Deser (Gordon and Breach, New York, 1966) Vol. 2.

35. G. W. Ford and R. F. O'Connell, Phys. Lett. A **157**, 217 (1991).

36. J. Dalibard, Y. Castin, and K. Mølmer, Phys. Rev. Lett. **68**, 580 (1992); K. Mølmer, Y. Castin, and J. Dalibard, J. Opt. Soc. Am. B **10**, 524 (1993).

37. R. Dum, P. Zoller, and H. Ritsch, Phys. Rev. A **45**, 4879 (1992); C. W. Gardiner, A. S. Parkins, and P. Zoller, Phys. Rev. A **46**, 4363 (1992); R. Dum, A. S. Parkins, P. Zoller, and C. W. Gardiner, Phys. Rev. A **46**, 4382 (1992).

38. K. Pachucki, Phys. Rev. Lett. **72**, 3154 (1994).

39. E. W. Hagley and F. M. Pipkin, Phys. Rev. Lett. **72**, 1172 (1994).

40. B. R. Mollow, Phys. Rev. **188**, 1969 (1969).

41. F. Y. Wu, R. E. Grove, and S. Ezekiel, Phys. Rev. Lett. **35**, 1426 (1975).

42. W. Hartig, W. Rasmussen, R. Schieder, and H. Walther, Z. Phys. A **278**, 205 (1976).

43. H. J. Carmichael and D. F. Walls, J. Phys. B **9**, L43 (1976); J. Phys. B **9**, 1199 (1976).

44. H. J. Kimble, M. Dagenais, and L. Mandel, Phys. Rev. Lett. **39**, 691 (1977); Phys. Rev. A **18**, 201 (1978).

76

Entangled Atoms and Fields: Cavity QED

Dieter Meschede* and Axel Schenzle[†]

*Institut für Angewandte Physik der Rheinische Friedrich-Wilhelms-Universität Bonn, Wegelerstraße 8, D-53125 Bonn, Germany
[†]Sektion Physik der Ludwig-Maximilians-Universität München, Theresienstraße 37, D-80333 München, Germany

76.1 INTRODUCTION

Although the concept of a "free atom" is of use as a first approximation, a full quantum description of the interaction of atoms with an omnipresent electromagnetic radiation field is necessary for a proper account of spontaneous emission and radiative level shifts such as the Lamb shift (see Chap. 27). This chapter is concerned with the changes in the atom-field interaction that take place when the radiation field is modified by the presence of a cavity. An atom in the vicinity of a plane perfect mirror serves as an example of *cavity quantum electrodynamics* [1, 2].

The primary focus in this chapter is the two extreme cases of weak coupling and strong coupling, as exemplified by spontaneous emission. In the weak coupling

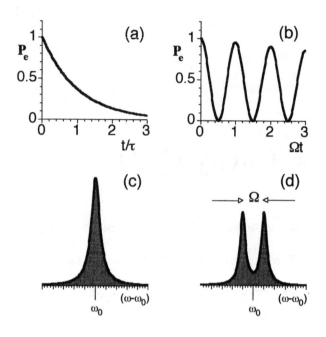

Figure 76.1. Upper row: Excitation probability of an excited atom. (a) Exponential decay in free space or bad cavities in the weak coupling limit. (b) Oscillatory evolution in good cavities or in the strong coupling case. Lower row: The spectral signature of exponential decay is a Lorentzian line shape (c) while the so called vacuum Rabi splitting (d) is observed in the strong coupling case.

regime, the coupling of an excited atom to a broad continuum of radiation modes leads to exponential decay [Fig. 76.1(a)], as first described by Weisskopf and Wigner [3]. Spontaneous emission may be enhanced or suppressed in structures such as waveguides or "bad" cavities. Cavities also introduce van der Waals forces and the subtle *Casimir* level shifts [4].

In the strong coupling regime, the excited atom is strongly coupled to an isolated resonant cavity mode. In the absence of damping, an oscillatory exchange of energy between the atom and the field replaces exponential decay [Fig. 76.1(b)] with a coherent evolution in time. Experimental realization of these effects has recently become possible [5] with the development of suitable resonators and techniques for producing atoms with long lived excited states and strong dipole transition moments.

76.2 ATOMS AND FIELDS

76.2.1 Atoms

The essential features of cavity QED are elucidated by the two-level model atom discussed in Chaps. 66, 67,

68 and (see also Ref. [6]). A ground state $|g\rangle$ and an excited state $|e\rangle$ are coupled to the radiation field by a dipole interaction. Using the formal equivalence to a spin-1/2 system, the Pauli spin operators are

$$
\begin{aligned}
\sigma_x &= \sigma^\dagger + \sigma, \\
\sigma_y &= -i\left(\sigma^\dagger - \sigma\right), \\
\sigma_z &= \sigma^\dagger\sigma - \sigma\sigma^\dagger = [\sigma^\dagger, \sigma],
\end{aligned} \tag{76.1}
$$

with $\sigma^\dagger = |e\rangle\langle g|$ and $\sigma = |g\rangle\langle e|$. The quadratures (out of phase components) of the atomic polarization are given by σ_x and σ_y, while σ_z is the occupation number difference. The free atom Hamiltonian is

$$
\mathcal{H}_{\text{atom}} = \tfrac{1}{2}\hbar\omega_0\sigma_z, \tag{76.2}
$$

where $\hbar\omega_0 = E_e - E_g$ is the transition energy.

76.2.2 Electromagnetic Fields

Classical Fields

Classical electromagnetic fields have longitudinal and transverse components

$$
\mathbf{E}(\mathbf{r}, t) = \mathbf{E}^l(\mathbf{r}, t) + \mathbf{E}^t(\mathbf{r}, t). \tag{76.3}
$$

In the Coulomb gauge, the longitudinal part is the instantaneous electric field. The transverse part is the radiation field which obeys the wave equation

$$
\left(\nabla^2 - \frac{1}{c^2}\frac{\partial^2}{\partial t^2}\right)\mathbf{E}^t(\mathbf{r}, t) = \frac{1}{\epsilon_0 c^2}\frac{\partial}{\partial t}\mathbf{j}(\mathbf{r}, t). \tag{76.4}
$$

In empty space the driving current density $\mathbf{j}(\mathbf{r}, t)$ vanishes, and the field may be expanded in a set of orthogonal modes,

$$
\mathbf{E}^t(\mathbf{r}, t) = \sum_\mu E_\mu(t)e^{-i\omega_\mu r}\mathbf{u}_\mu(\mathbf{r}) + \text{c.c.}, \tag{76.5}
$$

with slowly varying amplitudes $E_\mu(t)$. Spatial distributions $\mathbf{u}_\mu(\mathbf{r})$ obey the vector Helmholtz equation

$$
\left[\nabla^2 + \left(\frac{\omega_\mu}{c}\right)^2\right]\mathbf{u}_\mu(\mathbf{r}, t) = 0, \tag{76.6}
$$

depending on geometric boundary conditions as imposed by conductive or dielectric mirrors, waveguides and resonators. In free space, plane wave solutions $\mathbf{u}_\mu(\mathbf{r}, t) = \mathbf{u}_\epsilon e^{i\mathbf{k}\mathbf{r}}$ have a continuous index $\mu = (\mathbf{k}, \epsilon)$ with wave vector $\mathbf{k}$ and an index ϵ for the two independent polarizations. The orthogonality relation

$$
\frac{1}{V}\int_V \mathbf{u}_\mu \cdot \mathbf{u}_\nu^* d^3\mathbf{r} = \delta_{\mu\nu} \tag{76.7}
$$

applies. For a closed cavity, V is the resonator volume. In waveguides and free space an artificial boundary is introduced and taken to infinity at the end of a calculation, such that the final results do not depend on the V.

Quantum Fields

The quantum analog of the classical transverse field in Eq. (76.4) is obtained through a quantization of its harmonic modes leading to a number state expansion. Field operators obey standard commutation relations $[a_\mu, a_\nu^\dagger] = \delta_{\mu,\nu}$, and for a single mode with index μ, the amplitude E_μ in Eq. (76.5) is replaced by the corresponding operator

$$E_\mu(t) = \mathcal{E}_\mu a_\mu e^{-i\omega t}, \quad E_\mu^\dagger(t) = \mathcal{E}_\mu^* a_\mu^\dagger e^{i\omega t}. \quad (76.8)$$

The normalization factor $\mathcal{E}_\mu$ is chosen such that the energy difference between number states $|n\rangle_\mu$ and $|n{+}1\rangle_\mu$ in the volume V is $\hbar\omega_\mu$, giving

$$\mathcal{E}_\mu \mathcal{E}_\mu^\dagger = \frac{\hbar\omega_\mu}{2\epsilon_0 V}. \quad (76.9)$$

The Hamiltonian of the free field is

$$\mathcal{H}_{\text{Field}} = \sum_\mu \hbar\omega_\mu (a_\mu^\dagger a_\mu + \tfrac{1}{2}). \quad (76.10)$$

In the Coulomb gauge the vector potential $\mathbf{A}(\mathbf{r})$ is related to the electric field $\mathbf{E} = -\partial\mathbf{A}/\partial t$ by

$$\mathbf{A}_\mu(\mathbf{r}, t) = -\frac{\mathcal{E}_\mu}{\omega_\mu} \left(a_\mu e^{-i\omega t} + a_\mu^\dagger e^{i\omega t}\right) \mathbf{u}_\mu(\mathbf{r}). \quad (76.11)$$

The ground state $|0\rangle_\mu$ is called the *vacuum* state. While the expectation value $\langle n|\mathbf{E}|n\rangle = 0$ for a number state, the variance is not zero, since $\langle n|\mathbf{E}\mathbf{E}^*|n\rangle > 0$, giving rise to nonvanishing "fluctuations" of the free electromagnetic field.

Dipole Coupling of Fields and Atoms

The combined system of atoms and fields can be described in product quantum states $|a, n\rangle$ of atoms ($|a\rangle$) and fields ($|n\rangle$). The interaction Hamiltonian $\mathcal{H}_I$ of the atom and the radiation field is given by [1]

$$\mathcal{H}_I = -\frac{q}{m}\mathbf{p}\cdot\mathbf{A}(\mathbf{r}) + \frac{q^2}{2m}\mathbf{A}^2(\mathbf{r}). \quad (76.12)$$

This interaction causes the atom to exchange energy with the radiation field. In the dipole approximation the coupling strength is proportional to the component of the atomic dipole moment $\mathbf{d}_{eg} = q\langle e|\mathbf{r}_{eg}|g\rangle$ along the electric field, with coupling constant

$$g_\mu(\mathbf{r}) = |\mathbf{d}_{eg}\cdot\mathbf{u}_\mu(\mathbf{r})\mathcal{E}_\mu|/\hbar. \quad (76.13)$$

In the rotating wave approximation (RWA) (see Chaps. 66, 67, 68),

$$\mathcal{H}_{\text{RWA}} = \sum_\mu \hbar(g_\mu \sigma^\dagger a_\mu + g_\mu^* a_\mu^\dagger \sigma), \quad (76.14)$$

where we have used the the atomic operators of Eq. (76.1).

In a continuous electromagnetic spectrum, the atom interacts with a large number of modes having quantum numbers μ, yielding exponential decay of an excited atomic level at the rate [3]

$$\Gamma_{eg} = \frac{2\pi}{\hbar^2} \sum_{\tilde{\mu}} \sum_{\mathbf{k}} |g_\mu|^2 \delta(\omega_\mu - \omega_0). \quad (76.15)$$

Here we have separated the discrete ($\tilde{\mu}$) and the continuous part (wave vector $\mathbf{k}$) of the mode index μ. If g_μ [Eq. (76.13)] does not vary much across a narrow resonance, then

$$\Gamma_{eg} \simeq 2\pi |g_\mu(\omega_0)|^2 \sum_{\tilde{\mu}} \rho_{\tilde{\mu}}(\omega_0). \quad (76.16)$$

The density of states coresponding to the continuous mode index $\mathbf{k}$ of dimension ν can be evaluated on an ν-dimensional fictitious volume $V^{(\nu)}$ as

$$\rho_{\tilde{\mu}} = \sum_{\mathbf{k}} \delta(\omega_{\tilde{\mu}\mathbf{k}} - \omega) \rightarrow \frac{V^{(\nu)}}{(2\pi)^\nu} \int_0^\infty d^\nu k \, \delta(\omega_{\tilde{\mu}\mathbf{k}} - \omega), \quad (76.17)$$

provided $\omega(\mathbf{k})$ is known, and by converting the sum[2] over plane wave vectors $\mathbf{k}$ into an integral.

The Rate of Spontaneous Emission

In free space $[\omega(\mathbf{k})^2 = (c\mathbf{k})^2]$, the sum in Eq. (76.16) contributes a factor of two, due to polarization, to the total density of states in free space, $\rho_{\text{free}}(\omega) = V\omega^2/\pi^2 c^3$. When the vector coupling of atom and field [Eq. (76.13)] is replaced by its $1/3$ average in isotropic free space, the result

$$\Gamma_{eg} = A_{eg} = \frac{e^2 r_{eg}^2 \omega^3}{3\pi\epsilon_0 \hbar c^3} \quad (76.18)$$

is obtained for the decay rate A_{eg} as measured by the *natural linewidth* Γ_{eg}.

76.3 WEAK COUPLING IN CAVITY QED

The regime of weak Cavity QED generally applies when an atom is coupled to a continuum of radiation

---

[1] The $\mathbf{A}^2$-term plays an important role in energy shifts and can only be neglected when radiative processes involving energy exchange are considered.

[2] This is formally accomplished by taking the limit of $\Delta k = 2\pi/l$ for large l, where l is a linear dimension of an artificial resonator, and the resonator volume is $V = l^3$. If the relation between mode spacing Δk and geometric dimension is nonlinear in a more complex geometry, this analysis can be very complicated.

modes. This is always the case with mirrors, waveguides or bad cavities. The signatures of weak cavity QED are modifications of the rate of spontaneous emission, as well as the existence of van der Waals and Casimir forces. Formally this regime is well described by perturbation theory.

76.3.1 Radiating Atoms in Waveguides

Within the continuous spectrum of a waveguide, radiative decay of an excited atomic level remains exponential, and Γ_{eg} may be determined as in the preceding section. Consider the modifications of spontaneous decay in a parallel plate waveguide. According to Eq. (76.16), the theoretical problem is reduced to a geometric evaluation of mode densities. Between a pair of mirrors it is convenient to distinguish TE_{nk} and TM_{nk} modes, where n is the number of half waves across the gap of width d. The dispersion relation $\omega(\mathbf{k})$ reflects the discrete standing wave part $(n\pi/d)$ and a running wave part as in free space,

$$\omega_{n,\mathbf{k}}^2 = c^2(|\mathbf{k}|^2 + n\pi/d)^2 \quad \begin{matrix} n = 0, 1, 2, \ldots & \text{TM} \\ n = 1, 2, \ldots & \text{TE}. \end{matrix}$$

$$(76.19)$$

The average mode density [$\mathbf{du} = 1$, Eq. (76.13)] is evaluated [Eq. (76.17), $\nu = 2$] with an appropriate quantization volume containing the area of the plates, $V = Ad$, giving

$$\rho^{TE}(\omega) = \frac{\omega_c[\omega]}{2\omega_c^2}\rho_{\text{free}}(\omega_c),$$

$$\rho^{TM}(\omega) = \frac{\omega_c[\omega + 1]}{2\omega_c^2}\rho_{\text{free}}(\omega_c),$$

$$(76.20)$$

where $[x]$ is the largest integer in x, and $\omega_c = \pi c/d$ gives the waveguide cutoff frequency. Below ω_c, the TE-mode density clearly vanishes and, with the pictorial notion of *turning off the vacuum* introduced by Kleppner [7], inhibition of radiative decay is obvious. Figure 76.2 shows the calculated mode density for a parallel plate waveguide. The decay rate can be calculated from Eq. (76.16), with the spatial variation of g_μ included. This configuration was used for experiments in both the microwave and the near optical frequency domain [8].

Boundary conditions imposed by conductive surfaces may also be simulated by appropriately positioned image charges. Inspired by classical electrodynamics, this *image charge model* is successfully used to determine modifications of radiative properties in confined space.

Figure 76.2. Modification of average vacuum spectral density $(\rho^{TE} + \rho^{TM})$ in a parallel plate cavity (thick line) compared with free space (thin line).

76.3.2 Radiating Atoms in Resonators

Resonators

In a resonator, the electromagnetic spectrum is no longer continuous and the discrete mode structure can be resolved also experimentally. While a resonator is only weakly coupled to external electromagnetic fields, it still interacts with a large thermal reservoir through currents induced in its walls. The total damping rate is due to resistive losses in the walls (κ_{wall}) and also due to transmission at the radiation ports, $1/\tau_\mu = \kappa_\mu = \kappa_{\text{wall}} + \kappa_{\text{out}}$. An empty resonator stores energy for times

$$\tau_\mu = Q/\omega_\mu, \qquad (76.21)$$

and the power transmission spectrum is a Lorentzian with width $\Delta\omega_\mu = \omega_\mu/Q_\mu$. The index μ for instance represents the TE_{lm} and TM_{lm} modes of a "pillbox" microwave cavity, or the TEM_{klm} modes of a Fabry-Perot interferometer (see Fig. 76.3).

When cavity damping remains strong, $\Gamma_\mu \gg \Gamma_{eg}$, the atomic radiation field is "immediately" absorbed and Weisskopf-Wigner perturbation theory remains valid. In this so called *bad cavity limit*, resonator damping can be accounted for by an effective mode density of Lorentzian width $\Delta\omega_\mu$ for a single isolated mode,

$$\rho_\mu(\omega) = \frac{1}{\pi} \frac{\omega_\mu/2Q_\mu}{(\omega - \omega_\mu)^2 + (\omega_\mu/2Q_\mu)^2}. \qquad (76.22)$$

Bad and Good Cavities

The modification of spontaneous decay is again calculated from Eq. (76.16). For an atomic dipole aligned

(a) open resonator

(b) closed resonator

Figure 76.3. Two frequently used resonator types for cavity QED: (a) Open Fabry-Perot optical cavity. (b) Closed "pillbox" microwave cavity.

parallel to the mode polarization, and right at resonance, $\omega_\mu = \omega_0$, the enhancement of spontaneous emission is found to be proportional to the Q-value of a selected resonator mode

$$\frac{\Gamma_{eg}^{\text{cav}}}{\Gamma_{eg}^{\text{free}}} = \frac{\rho_\mu |\mathbf{u}(\mathbf{r})|}{\rho_{\text{free}}} = \frac{3Q\lambda^3}{4\pi^2 V}|\mathbf{u}(r)|^2 = \frac{3Q\lambda^3}{4\pi^2 V_{\text{eff}}}, \quad (76.23)$$

where the *effective mode volume* is $V_{\text{eff}} = V/|\mathbf{u}(\mathbf{r})|^2$. The lowest possible value $V_{\text{eff}} \simeq \lambda^3$ is obtained for ground modes of a closed resonator. For an atom located at the waist of an open Fabry-Perot cavity with length L, it is much larger. Special limiting cases for concentric and confocal cavities are $V_{\text{eff}}^{\text{conc}} = \lambda^2 L(R/D)$ and $V_{\text{eff}}^{\text{conf}} = \lambda L^2/2\pi$, respectively, where (R/D) gives the ratio of mirror radius to cavity diameter.

At resonance, the atomic decay rate Γ_μ grows with Q_μ, whereas the resonator damping time constant κ_μ is reduced. Eventually, the energy of the atomic radiation field is stored for such a long time that reabsorption becomes possible. Perturbative Weisskopf-Wigner theory is no longer valid in this *good cavity limit*, which is separated from the regime of *bad cavities* by the more formal condition

$$\Gamma_{eg}^{\text{cav}} > \kappa_\mu. \quad (76.24)$$

The strong coupling case is explicited in Sect. 76.4.

Antenna Patterns

Since the reflected radiation field of an atomic radiator is perfectly coherent with the source field, the combined radiation pattern modifies the usual dipole distribution of a radiating atom. The new radiation pattern can be understood in terms of antenna arrays [9]. For a single atomic dipole in front of a reflecting mirror for example, one finds a quadrupole type pattern due to the

superposition of a second, coherent image antenna. In some of the earliest experimental investigations [10] on radiating molecules in cavities, modifications of the radiation pattern were observed.

76.3.3 Radiative Shifts and Forces

When the radiation field of an atom is reflected back onto its source, an energy or *radiative shift* is caused by the corresponding self polarization energy. An atom in the vicinity of a plane mirror (see Fig. 76.4) again makes a simple model system. Since the energy shift depends on the atom wall separation z, it is equivalent to a dipole force $\mathbf{F}_{\text{dip}}$ whose details depend on the role of retardation. Spontaneous forces may also be present if radiation energy is exchanged between the atom and the field.

The Unretarded Limit: van der Waals Forces

When the radiative round trip time $t_{\text{r}} = 2z/c$ is short compared with the characteristic atomic revolution period $2\pi/\omega_{eg}$, retardation is not important. In this quasistatic limit, van der Waals energy shifts for decaying atomic dipoles vary as z^{-3} with the atom-wall separation. Such a shift is also present for a nonradiating atom in its ground state. In perturbation theory, the van der Waals energy shift of an atomic level $|a\rangle$ is

$$\Delta_{\text{vdW}} = -\frac{\langle a|q^2\left((\mathbf{d}^2 \cdot \hat{\mathbf{x}}_t)^2 + 2(\mathbf{d}^2 \cdot \hat{\mathbf{z}})^2\right)|a\rangle}{64\pi\epsilon_0 z^3}. \quad (76.25)$$

Since the van der Waals force is anisotropic for electronic components parallel ($\hat{\mathbf{z}}$) and perpendicular ($\hat{\mathbf{x}}_t$) to the mirror normal, the degeneracy of magnetic sublevels in an atom is lifted near a surface. The total energy shift is of ~ 1 kHz for a ground state atom at 1 μm separation, and very difficult to detect. However, the energy shifts grow as n^4 since the transition dipole moment scales as n^2. With Rydberg atoms, van der Waals energy shifts have been successfully observed in spectroscopic experiments [11].

The Retarded Limit: Casimir Forces

At large separation, retardation becomes relevant, since the contributions of individual atomic oscilllation frequencies in Eq. (76.25) cancel by dephasing, thus reducing the Δ_{vdW}. A residual Casimir-Polder [12] shift may be interpreted as the polarization energy of a slowly fluctuating field with squared amplitude $\langle \mathcal{E}^2 \rangle = 3\hbar c/64\epsilon_0 z^4$ originating from the vacuum field noise

$$\Delta_{\text{CP}} = -\frac{1}{4\pi\epsilon_0}\frac{3\hbar c \alpha_{\text{st}}}{8\pi z^4}, \quad (76.26)$$

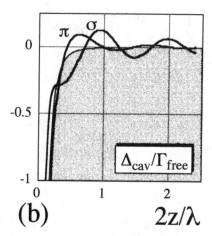

Figure 76.4. (a) Normalized rate of modified spontaneous emission in the vicinity of a perfectly reflecting wall for σ and π orientation of the radiating dipole. (b) Corresponding energy shift of the resonance frequency. Shaded area indicates contribution of static van der Waals interaction.

where α_{st} is the static electric polarizability. The vacuum field noise Δ_{CP} replaces Δ_{vdW} at distances larger than characteristic wavelengths, and is even smaller. Only indirect observations have been possible to date, relying on a deflection of polarizable atoms by this force [13].

Radiative Forces in a Cavity

The Casimir-Polder force can also be regarded as an ultimate, cavity induced consequence of the mechanical action of light on atoms [14]. It is an example of the conservative and dispersive dipole force which is even capable of binding a polarizable atom to a cavity [15]. Spontaneous emission of atoms in the vicinity of a reflecting wall is an example of cavity induced modification of the dissipative type of light forces, or radiation pressure. The photon emitted by the atom can reach a detector directly, or following a reflection off the wall. Since the two paths are indistinguishable, the atom is left in a superposition of two recoil momentum states.

76.3.4 Experiments on Weak Coupling

Perhaps the most dramatic experiment is the suppression of spontaneous emission in Rydberg atoms, and for a low-lying near infrared atomic transition. It is essential to prepare atoms in a single decay channel. In addition, the atoms must be oriented in such a way that they are only coupled to a single decay mode (see the model waveguide in Fig. 76.3). This may be interpreted as an anisotropy of the electromagnetic vacuum, or as a specific antenna pattern.

The main problem in detecting radiative shifts arises from their inhomogeneity due to the dependence on

atom–wall separation. This difficulty has been overcome by controlling the atom–wall separation at microscopic distances through light forces [11], and by spectroscopic techniques with sensitivity to a thin layer of surface atoms only [16]. In this way, van der Waals shifts have been clearly resolved.

Conceptually most attractive and experimentally most difficult to detect is the elusive Casimir interaction. Only for atomic ground states is this effect observable free from other much larger shifts. The influence of the corresponding Casimir force on atomic motion has been observed in a variant of a scattering experiment, confirming the existence of this force in neutral atoms [13].

The success of this experiment shows that spectroscopic techniques involving exchange of photons are not suitable for the Casimir problem. A notable exception could be Raman spectroscopy of the magnetic substructure in the vicinity of a surface. In general, scattering or atomic interferometry experiments are more promising methods. The recent experiment by Brune *et al.* [17] may be interpreted in this way.

76.3.5 Cavity QED in Dielectrics

Cavities with dimensions comparable to the wavelength promise the most dramatic modification of radiative atomic properties. Micrometer sized cavities for optical frequencies with highly reflecting walls are difficult to manufacture. However, dielectric materials with a periodic modulation of the index of refraction have *photonic bandgaps* analogous to the electronic bandgaps in a periodic crystal [18]. The electronic phenomena of solid state physics then apply to photons. If an excited atom such as a crystal dopant radiates into a photonic bandap, the radiation cannot propagate. The bandgap behaves like an empty resonator, thereby opening the regime of strong coupling to photonic structures.

76.4 STRONG COUPLING IN CAVITY QED

Strong coupling of atoms and fields is realized in a good cavity when $\Gamma_\mu < \Gamma_{eg}$ [see Eq. (76.24)]. The Hilbert space of the combined system is then the product space of a single two level atom and the countable set of Fock-states of the field,

$$\mathcal{H} = \mathcal{H}_{\text{atom}} \otimes \mathcal{H}_{\text{field}}, \qquad (76.27)$$

which is spanned by the states

$$|n; a\rangle = |n\rangle |a\rangle. \qquad (76.28)$$

The interaction of a single cavity mode with an isolated atomic resonance is now characterized by the Rabi nutation frequency, which gives the exchange frequency of the energy between atom and field. For an amplitude $\mathcal{E}$ corresponding to n photons

$$\Omega(n) = g_\mu \sqrt{n+1}. \qquad (76.29)$$

This is the simplest possible situation of a strongly coupled atom–field system. The new energy eigenvectors are conveniently expressed in the dressed atom model [19]:

$$\begin{aligned} |+, n\rangle &= \cos\theta |g, n+1\rangle + \sin\theta |e, n\rangle, \\ |-, n\rangle &= -\sin\theta |g, n+1\rangle + \cos\theta |e, n\rangle, \end{aligned} \qquad (76.30)$$

with $\tan 2\theta = 2g_\mu \sqrt{n+1}/(\omega_0 - \omega_\mu)$. The separate energy structures of free atom and empty resonator are now replaced by the combined system of Fig. 76.5. At resonance, the new eigenstates are separated by $2\hbar\Omega_R$, where $\Omega_R = g_\mu$ is the vacuum Rabi frequency.

76.5 STRONG COUPLING IN EXPERIMENTS

In order to achieve strong coupling experimentally, it is necessary to use a high Q resonator in combination with a small effective mode volume. This condition was first realized for ground modes of a closed microwave cavity [5], and later also for open cavity optical resonators (see Fig. 76.3) [20].

76.5.1 Rydberg Atoms and Microwave Cavities

At microwave frequencies, very low loss superconducting niobium cavities are available with $Q \approx 10^{10}$.

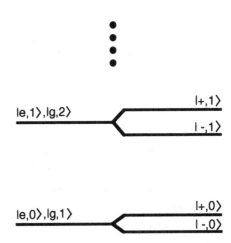

Figure 76.5. Level diagram for combined states of noninteracting atoms and fields (a) which are degenerate at resonance. Degeneracy is lifted by strong coupling of atoms and fields (b) yielding new "dressed" eigenstates.

Resonator frequencies are typically of order several tens of GHz and can be matched by atomic dipole transitions between two highly excited Rydberg states. By selective field ionization, the excitation level of Rydberg atoms can be detected, and hence it is possible to measure whether a transition between the levels involved has occured. The efficiency of this method appoaches unity, so that experiments can be performed at the single atom level. The interaction or transit time T is usually much shorter than the lifetime τ_{Ryd} of the Rydberg states involved. For this reason, circular Rydberg states with quantum numbers $l = m = n - 1$ are particularly suitable.

Rydberg atoms [21] are prepared in an atomic beam, selectively excited to an upper level, and and then sent through a microwave cavity where the upper and lower levels are coupled by the electromagnetic field. If the atom is detected in the lower of the coupled levels as it leaves the resonator, the excitation energy has been stored in the resonator field. Thus the evolution of the resonator field is recorded as a function of the atomic interaction.

A microwave cavity in interaction with a single or few Rydberg atoms is called a *micromaser* (formerly a *one atom maser*) [5]. The experimental conditions may be summarized as

$$g_\mu > 1/T > 1/\tau_{\text{Ryd}} > \kappa_\mu. \qquad (76.31)$$

76.5.2 Strong Coupling in Open Optical Cavities

At optical wavelengths, a cavity with small V_{eff} in Eq. (76.23) is clearly more difficult to construct than at centimeter wavelengths. However, dielectric coatings are now available which allow very low damping rates ω_μ/Q_μ for optical cavities. Very high *finesse* $\mathcal{F} \simeq 10^7$ (which is a more convenient measure for the damping rate of an optical Fabry Perot interferometer) has been achieved. By reducing the volume of such a high Q cavity mode, strong coupling of atoms and fields at optical frequencies has been demonstrated [20]. In open structures, the atoms can also decay into the continuum states.

76.6 MICROSCOPIC MASERS AND LASERS

In a microscopic laser, simple atoms are strongly coupled to a single mode of a resonant or near resonant radiation field. Collecting atomic and field operators from Eqs. (76.2), (76.10), and (76.14), this situation is described by the *Jaynes Cumming Model*, [22]

$$\mathcal{H}_{JC} = \mathcal{H}_{atom} + \mathcal{H}_{field} + \mathcal{H}_{RWA}$$
$$= \tfrac{1}{2}\hbar\omega_0\sigma_z + \hbar\omega_\mu(a_\mu^\dagger a_\mu + \tfrac{1}{2}) + \hbar g_\mu(\sigma^\dagger a_\mu + a_\mu^\dagger\sigma).$$
(76.32)

76.6.1 The Jaynes Cummings Model

The Jaynes-Cummings model [Eq. (76.32)] represents the most basic and at the same time the most informative model of strong coupling in quantum optics. It consists of a single two-level atom interacting with a single mode of the quantized cavity field. The time evolution of the system is determined by

$$i\hbar\frac{\partial\psi}{\partial t} = H\psi.$$
(76.33)

This model can be solved exactly due to the existence of the additional constant of motion

$$N = a^\dagger a + \sigma_z + 1,$$
(76.34)

i.e., conservation of the "number of excitations." Its eigenvalues are the integers N which are twofold degenerate except for $N = 0$. The simultaneous eigenstates of H and N are the pairs of dressed states defined in (76.30) which are not degenerate with respect to the energy H. The initial state problem corresponding to Eq. (76.33) is solved by elementary methods in terms of the expansion:

$$|\Psi(t)\rangle = \sum_{n=0}^{\infty}\sum_{j=1}^{2}C_n^j(t)|n,j\rangle,$$
(76.35)

where the expansion coefficients are

$$C_n^1(t) = \left\{C_n^1(0)\left[\cos[\Omega(n)t] - i\frac{\delta}{2\Omega(n)}\sin[\Omega(n)t]\right]\right.$$
$$\left. - i\frac{\sqrt{n}g_\mu}{\Omega(n)}C_{n-1}^2(0)\sin[\Omega(n)t]\right\}$$
$$\times \exp\left[-i\omega_\mu\left(n - \tfrac{1}{2}\right)t\right]$$
(76.36)

and

$$C_n^2(t) = \left\{C_n^2(0)\left[\cos[\Omega(n+1)t]\right.\right.$$
$$\left. + i\frac{\delta}{2\Omega(n+1)}\sin[\Omega(n+1)t]\right]$$
$$\left. - i\frac{g_\mu\sqrt{n+1}}{\Omega(n+1)}C_{n+1}^1(0)\sin[\Omega(n+1)t]\right\}$$
$$\times \exp\left[-i\omega_\mu\left(n + \tfrac{1}{2}\right)t\right],$$
(76.37)

with $\delta = \omega_\mu - \omega_0$ the detuning between atom and cavity and $\Omega(n) = \tfrac{1}{2}\sqrt{\delta^2 + 4g_\mu^2 n}$ is the generalized Rabi frequency. The coefficients $C_n^j(0)$ are determined by the initial preparation of atom and cavity mode. The result simplifies considerably for $\delta = 0$ to

$$|\Psi(t)\rangle = \sum_{m=0}^{\infty}\{C_m^1(0)e^{-i\omega_\mu(m-1/2)t}$$
$$\times [\cos(g_\mu\sqrt{m}\,t)|m;1\rangle - i\sin(g_\mu\sqrt{m}\,t)|m-1;2\rangle]$$
$$+ C_m^2(0)e^{-i\omega_\mu u(m+1/2)t}[\cos(g_\mu\sqrt{m+1}\,t)|m;2\rangle$$
$$- i\sin(g_\mu\sqrt{m+1}\,t)|m+1;1\rangle]\}.$$
(76.38)

The coefficients $C_n^j(0)$ represent any initial state of the system from uncorrelated product states to entangled states of atom and field. There exist numerous generalizations of this model which include more atomic levels and several coherent fields.

76.6.2 Fock States, Coherent States, and Thermal States

We now illustrate the properties of the Jaynes-Cummings model by specifying the initial state. Assume that atom and field are brought into contact at time $t = 0$ and suppress all correlations that might exist due to previous interactions.

Rabi Oscillations

If the atom is in initially the excited state and the field contains precisely m quanta, then

$$C_n^j(t=0) = \delta_{n,m}\delta_{j,2}.$$
(76.39)

The solution of Eq. (76.33) assumes the form

$$|\Psi(t)\rangle = e^{-i\omega_\mu(m+1/2)t}\left[\cos(g_\mu\sqrt{m+1}\,t)|m;2\rangle\right.$$
$$\left. - i\sin(g_\mu\sqrt{m+1}\,t)|m+1;1\rangle\right].$$
(76.40)

The occupation probabilities of the atomic states evolve in time according to

$$n_2(t) = \langle \Psi(t)|2\rangle\langle 2|\Psi(t)\rangle = \cos^2(g_\mu\sqrt{m+1}\,t)\,, \quad (76.41)$$

$$n_1(t) = \langle \Psi(t)|1\rangle\langle 1|\Psi(t)\rangle = \sin^2(g_\mu\sqrt{m+1}\,t)\,. \quad (76.42)$$

The photon number and its variance are

$$\langle n(t)\rangle = \langle \Psi(t)a^\dagger a\Psi(t)\rangle = m + \sin^2(g_\mu\sqrt{m+1}\,t)\,, \quad (76.43)$$

$$\langle \Delta^2 n\rangle = \langle \Psi(t)(a^\dagger a - \langle a^\dagger a\rangle)^2\Psi(t)\rangle = \frac{\sin^2(2g_\mu\sqrt{m+1}\,t)}{4}\,. \quad (76.44)$$

In the limit of large m, $g_\mu\sqrt{m+1}$ is proportional to the field amplitude and the classical Rabi oscillations in a resonant field are recovered. Nonclassical features of the state are characterized by Mandel's parameter

$$Q_M = \frac{\langle \Delta^2 n\rangle - \langle n\rangle}{\langle n\rangle} \geq -1\,. \quad (76.45)$$

For the present example

$$Q_M = -1 + \frac{1}{4}\frac{\sin^2(2g_\mu\sqrt{m+1}\,t)}{m + \sin^2(g_\mu\sqrt{m+1}\,t)}\,. \quad (76.46)$$

$Q_M \geq 0$ indicates the classical regime, while $Q \leq 0$ can only be reached by a quantum process.

The Coherent State

Consider the case where the field is initially prepared in a coherent state

$$|\alpha\rangle = \exp(\alpha a^\dagger - \alpha^* a)|0\rangle = e^{-|\alpha|^2/2}\sum_{n=0}^{\infty}\frac{\alpha^n}{\sqrt{n!}}|n\rangle\,, \quad (76.47)$$

while the atom starts from the excited state

$$C_n^j(0) = e^{-|\alpha|^2/2}\frac{|\alpha|^n}{\sqrt{n!}}\delta_{j,2}\,. \quad (76.48)$$

In this case the general solution specializes to

$$|\Psi(t)\rangle = \sum_{n=0}^{\infty}\frac{\alpha^n}{\sqrt{n!}}e^{-i\omega(n+1/2)t}e^{-|\alpha|^2/2}$$
$$\times\left[\cos(g_\mu\sqrt{n+1}\,t)|n;2\rangle - i\sin(g_\mu\sqrt{n+1}\,t)|n+1;1\rangle\right]\,, \quad (76.49)$$

and the occupation probability of the excited state is

$$n_2(t) = \frac{1}{2}\left(1 + \sum_{n=0}^{\infty}\frac{|\alpha|^{2n}}{n!}e^{-|\alpha|^2}\cos(2g_\mu\sqrt{n+1}\,t)\right)\,. \quad (76.50)$$

From here, detailed quantitative results can only be obtained by numerical methods [23]. However, if the coherent state contains a large number of photons $|\alpha|^2 \gg 1$, the essential dynamics can be determined by elementary methods. Initially, the population oscillates with the Rabi frequency $\Omega_1 \sim g_\mu|\alpha|$ proportional to the average amplitude of the field, as expected from its classical counterpart. With increasing time, the coherent oscillations tend to cancel due to the destructive interference of the different Rabi frequencies in the sum

$$n_2(t) = \frac{1}{2}\left[1 + \cos(2g_\mu|\alpha|\,t)e^{-(gt)^2/2}\right]\,. \quad (76.51)$$

However, strictly aperiodic relaxation of $n_2(t)$ is impossible since the exact expressions (76.35 and 76.36) represent a *quasiperiodic function* which, given enough time, approaches its initial value with arbitrary accuracy.

For short times, the oscillating terms in the sum cancel each other due to the slow evolution of their frequency with n. However, consecutive terms interfere constructively for larger times t_r such that the phases satisfy

$$\phi_{n+1}(t_r) - \phi_n(t_r) = 2\pi\,. \quad (76.52)$$

For $|\alpha|^2 \gg 1$ the increment of the arguments is

$$\phi_{n+1} - \phi_n = g_\mu t_r/|\alpha|\,, \quad (76.53)$$

and therefore the first revival of the Rabi oscillations occurs approximately at $t_r = \pi|\alpha|/g_\mu$. A clear distinction of Rabi oscillation, collapses and revivals requires a clear separation of the three time scales

$$t_1 \ll t_2 \ll t_3\,, \quad (76.54)$$

where $t_1 \sim (g_\mu|\alpha|)^{-1}$ for Rabi oscillation, $t_2 \sim g_\mu^{-1}$ for collapse, and $t_3 \sim |\alpha|/g_\mu$ for revival.

The typical features of the transient evolution starting from a coherent state are shown in Fig. 76.6. With time increasing even further, revivals of higher order occur which spread in time, and finally can no longer be separated order by order.

Thermal State

Consider a microwave resonator brought into thermal contact with a reservoir, inducing loss on a time scale κ^{-1}, and thermal excitation. The dissipative time evolution is described by the master equation

$$\dot\rho = (L_0 + L)\rho$$
$$\equiv i[H,\rho]/\hbar + \kappa(n_{th} + 1)\{[a,\rho a^\dagger] + [a\rho, a^\dagger]\}$$
$$+ \kappa n_{th}\{[a^\dagger,\rho a] + [a^\dagger\rho, a]\}\,, \quad (76.55)$$

where $n_{th} = [\exp(\beta\hbar\omega) - 1]^{-1}$ at $T = k_B\beta^{-1}$ is the equilibrium population of the cavity mode, L_0 symbolizes the unitary evolution according to the Jaynes-Cummings dynamics and L is a dissipation term.

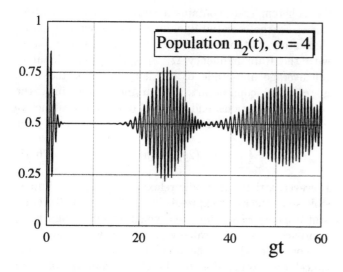

Figure 76.6. Rabi oscillations, dephasing, and quantum revival.

The solution of this model can be expressed in terms of an eigenoperator expansion of the equation

$$L\rho = -\lambda\rho. \qquad (76.56)$$

The eigenvalues λ that determine the relaxation rates, as well as the eigenoperators, are known in closed form for the case of vanishing temperature [24]. Since energy is exchanged between the nondecaying atom and the decaying cavity mode, cavity damping is modified in a characteristic way due to the presence of the atom. The technical details can be found in Ref. [25].

76.6.3 Vacuum Splitting

In the classical case, the eigenvalues of the interaction free Hamiltonian are degenerate at resonance. The atom–field interaction splits the eigenvalues and determines the Rabi frequency of oscillation between the two states. A consequence is the side bands of the resonance fluorescence spectrum [26]. In the quantum case, the field itself is treated as a quantized dynamical variable determined from a self-consistent solution for the complete system of atom plus field. The vacuum Rabi frequency $\Omega_{\rm vac} = g_\mu$ remains finite, and accounts for the spontaneous emission of an excited atom placed in a vacuum. In the limiting case of a single atom interacting with the quantized field, the photon number n can only change by ± 1, and the population oscillates with the frequency $\Omega(n)$ given by (76.29). For an ensemble of N atoms, n can in principle change by up to $\pm N$. However, if the field and atoms are only weakly excited, the collective frequency of the ensemble is determined by the linearized Maxwell-Bloch equations. The eigenfrequencies are given by

$$\lambda^\pm = \tfrac{1}{2}\left[i(\gamma_\perp + \kappa) \pm \sqrt{4g_\mu^2 N - (\gamma_\perp - \kappa)^2}\right], \qquad (76.57)$$

where $\gamma_\perp^{-1}$ is the phase relaxation time of the atom and κ^{-1} the decay time of the resonator. This is the polariton dispersion relation in the neighborhood of the polariton gap. The spectral transmission

$$T(\omega) = T_0\left|\frac{\kappa[\gamma + i(\omega_0 - \omega)]}{(\omega - \lambda^+)(\omega - \lambda^-)}\right|^2 \qquad (76.58)$$

of an optical cavity containing a resonant atomic ensemble of N atoms reveals the internal dynamics of the coupled system and a splitting of the resonance line occurs. T_0 is the peak transmission of the empty cavity. The splitting increases either with the number of photons, approaching $\sqrt{n+1}$ in the presence of a single atom, or with the number of atoms, approaching $\sqrt{N}$ in the resonator when the field is weak. The latter case is demonstrated in Fig. 76.7 [20] for an optical resonator with 1–10 atoms interacting with a field that on average contains much less than a single photon.

76.7 MICROMASERS

Sustained oscillations of a cavity mode in a microwave resonator can be achieved by a weak beam of Rydberg atoms excited to the upper level of a resonant transition. For a cavity with a $Q \sim 10^{10}$, much less than a single atom at a time, on average, suffices to balance the cavity losses. Operation of a single atom maser has been demonstrated [5]. The atoms enter the cavity at random times according to the Poisson statistics of a thermal beam, and interact with the field only for a limited time. In order to restrict the fluctuations of the atomic transit time, the velocity spread reduced. This is achieved either by Fizeau chopping techniques, or by making use of Doppler velocity selection in the initial laser excitation process. Since most of the time no atom is present, it is natural to separate the dynamics into two parts [27]:

1. For the short time while an atom is present, the state evolves according to the Jaynes-Cummings dynamics, where H is defined in Eq. (76.32)

$$\dot\rho(t) = i[H, \rho]/\hbar, \qquad (76.59)$$

and damping can safely be neglected. The formal solution is abbreviated by $\rho(t) = F(t - t_0)\rho(t_0)$.

2. In the time interval between successive atoms, the cavity field relaxes freely toward the thermal equilibrium according to Eq. (76.55) with $L_0 = 0$:

$$\dot\rho(t) = L\rho, \qquad (76.60)$$

with the formal solution $\rho(t) = \exp(L(t - t_0))\rho(t_0)$.

$$\langle n|\mathrm{Tr}_{\mathrm{atom}}\rho|m\rangle = P_n\delta_{n,m}\,. \tag{76.63}$$

Due to continuous injection of atoms, the field never becomes time independent, but may relax toward a stroboscopic state defined by

$$\langle\rho(t_{i+1})\rangle = \langle\rho(t_i)\rangle\,. \tag{76.64}$$

The state of the cavity field can be determined in closed form by iteration:

$$P_n = N\prod_{k=1}^{n}\frac{n_{\mathrm{th}}\kappa + A_k}{(n_{\mathrm{th}}+1)\kappa}\,, \tag{76.65}$$

where N guarantees normalization of the trace and $A_k = (R/n)\sin^2(g_\mu\tau\sqrt{n})$, and exact resonance between cavity mode and atom is assumed. Since all off-diagonal elements vanish in steady state, Eq. (76.65) provides a complete description for the photon statistics of the field.

76.7.1 Maser Threshold

The steady state distribution determines the mean photon number of the resonator as a function of the operating conditions:

$$\langle n\rangle = \sum_{n=0}^{\infty} nP_n\,. \tag{76.66}$$

A suitable dimensionless control parameter is

$$\Theta = \tfrac{1}{2}g_\mu\tau\sqrt{R/\kappa}\,, \tag{76.67}$$

For $\Theta \ll 1$, the energy input is insufficient to counterbalance the loss of the cavity, effectively resulting in a negligible photon number. With increasing pump rate R, a threshold is reached at $\Theta \simeq 1$ where $\langle n\rangle$ increases rapidly with R. In contrast to the behavior of the usual laser, the single atom maser displays multiple thresholds with a sequence of minima and maxima of $\langle n\rangle$ as a function of Θ [28]. This can be related to the rotation of the atomic Bloch vector. When the atom undergoes a rotation of about π during the transit time τ, maximum energy is transfered to the cavity and $\langle n\rangle$ is maximized. The converse applies if the average rotation is a multiple of 2π. This behavior is shown in Fig. 76.8. The minima in $\langle n\rangle$ are at $\Theta \simeq 2n\pi$.

76.7.2 Nonclassical Features of the Field

Fluctuations can be of classical or of quantum origin. The variance of the photon number

$$\sigma^2 = (\langle n^2\rangle - \langle n\rangle^2) \tag{76.68}$$

is a measure of the randomness of the field intensity. Classical Poisson statistics require that $\sigma^2 \geq \langle n\rangle$. A

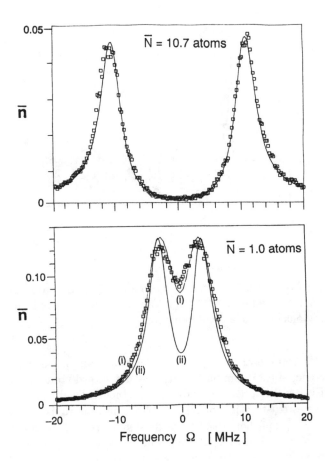

Figure 76.7. Intracavity photon number (measured from a transmission experiment, Ref. [20]) as a function of probe frequency detuning, and for two values of N, the average number of atoms in the mode. Thin lines give theoretical fits to the data, including atomic number and position fluctuations. Curve (ii) in the lower graph is for a single intracavity atom with optimal coupling g_μ.

The time development of the micromaser therefore consists of an alternating sequence of unitary $F(t)$ and dissipative $\exp(Lt)$ evolutions. Atoms enter the cavity one by one at random times t_i. The evolution t_i until the next atom enters at time t_{i+1} is given by

$$\rho(t_{i+1}) = \exp(Lt_p)F(\tau)\rho(t_i)\,. \tag{76.61}$$

where $t_p = t_{i+1} - t_i - \tau$, and τ is the transit time. If $\tau \ll t_{i+1} - t_i$ on average, then $t_p \approx t_{i+1} - t_i$. After averaging Eq. (76.61) over the Poisson distribution $P(t) = R\exp(-Rt_p)$ for t_p, where R is the injection rate, the mean propagator from atom to atom is

$$\langle\rho(t_{i+1})\rangle = \frac{R}{R-L}F(\tau)\langle\rho(t_i)\rangle\,. \tag{76.62}$$

After excitation, the reduced density matrix of the field alone becomes diagonal after several relaxation times κ^{-1}:

Figure 76.8. Average photon number as a function of the normalized transit time defined by Eq. (76.67)

.

Figure 76.9. Variance normalized on the average photon number $\sigma^{2<n>}/\langle\sigma\rangle$. Values below unity indicate regions of nonclassical behavior.

value below unity indicates quantum behavior which has no classical analog. In Fig. 76.9 the variance is plotted as a function of Θ. Regions of enhanced fluctuations $\sigma^2 > \langle n \rangle$ alternate with regions with sub-Poissonian character $\sigma^2 < \langle n \rangle$ [29]. When $\langle n \rangle$ approaches a local maximum it is accompanied by large fluctuations, while at points of minimum field strength the fluctuations are reduced below the classical limit. This feature is repeated with a period of $\Theta \simeq 2\pi$, but finally washes out at large values of Θ.

The large variance of n is caused by a splitting of the photon distribution P_n into two peaks which gives rise to bistability in the transient response [30]. The sub-Poisson behavior of the field is reflected in an increased regularity of the atoms leaving the cavity in the ground state.

76.7.3 Trapping States

If cavity losses are neglected, operating conditions exist which lead directly to nonclassical, i.e., Fock states. If the cavity contains precisely n_q photons, an atom that enters the resonator in the excited state leaves it again in the same state provided the condition [31]

$$g_\mu \tau \sqrt{n_q + 1} = 2q\pi \qquad (76.69)$$

is satisfied, i.e., the Bloch vector of the atom undergoes q complete rotations. If the maser happens to reach such a *trapping state* $|n_q\rangle$, the photon number n_q can no longer increase irrespective of the flux of pump atoms. With the inclusion of cavity damping at zero temperature, n_q

still represents an upper barrier that cannot be overcome, since damping only causes downward transitions. Even in the presence of dissipation, generalized trapping states exist with a photon distribution that vanishes for $n > n_q$ and has a tail towards smaller photon numbers $n \leq n_q$. However, thermal fluctuations at finite temperatures destabilize the trapping states since they can momentarily increase the photon number and allow the distribution to jump over the barrier $n = n_q$. Nevertheless, even for $n_{th} < 10^{-7}$, remnants of the trapping behavior persist, and can be seen in the transient response of the micromaser (see Sect. 76.7.4).

76.7.4 Atom Counting Statistics

Direct measurements of the field in a single atom maser resonator are not possible, because detector absorption would drastically degrade its quality. However, the field can be deduced from the statistical signature of atoms leaving the resonator.

The probability $P(n)$ of finding n atoms in a beam during an observation interval t is given by the classical Poisson distribution

$$P(n) = (Rt)^n e^{-Rt}/n! . \qquad (76.70)$$

Information on the field inside is then revealed by the conditional probability $W(n, |g\rangle, m, |e\rangle, T)$ of finding n atoms in the ground state and m atoms in the excited state during a time t. Since there are only two states, it is sufficient to determine the probability

$$W(n, |g\rangle, t) = \sum_{m=0}^{\infty} W(n, |g\rangle, m, |e\rangle, t) \qquad (76.71)$$

for being in the ground state [32]. For $n = 0$, the probability for observing no atom in the ground state during the period t is

$$W(0, |g\rangle, t) = \mathrm{Tr}(\rho_{stst}) \exp\{L + R[O_{|g\rangle} + (1 - \eta)O_{|e\rangle} - 1]t\}, \quad (76.72)$$

where $O_{|j\rangle} = \langle j|F(\tau)|j\rangle$ [*cf.* Eq. (76.59)] and ρ_{stst} is the steady state of the maser field. This probability is closely related to the waiting time statistic $P_2(0, |g\rangle, t)$ between two successive ground state atoms, a property which is easily determined in a start-stop experiment. For an atom detector with finite quantum efficiency η for state selective detection, the waiting time probability is

$$P_2(0, |g\rangle, t) = \big[\mathrm{Tr}(\rho_{stst})O_{|g\rangle} \exp\{L + R[O_{|g\rangle} + (1 - \eta)O_{|e\rangle} - 1]T\}O_{|g\rangle}\big] / \big[\mathrm{Tr}(\rho_{stst})O_{|g\rangle}\big]^2 \quad (76.73)$$

How a specific field state is reflected in the atom counting statistics will be illustrated for two situations: the region of sub-Poisson statistics and the region where the trapping condition is satisfied. Increased regularity of the cavity field $Q_M \leq 0$ manifests itself in increased regularity of ground state atoms in the beam. The statistical behavior exhibits "anti-bunching", i.e., $P_2(0, |g\rangle, t)$ has a maximum at finite t indicating "repulsion" between successive atoms in comparison with a Poissonian beam. If the transit time τ is chosen in such a way that $g\tau \simeq 2\pi$, the chance of observing an initially excited atom in the ground state is negligible. However, at some point an unlikely thermal fluctuation occurs, adding a photon. The rotation angle of the Bloch vector suddenly increases to $2\pi\sqrt{2} \simeq 3\pi$, and the atoms tend to leave the cavity in the ground state. After a typical cavity life time, the field decays and the trapping condition is restored again. Under this operation condition, the statistics of ground state atoms is governed by two time constants:

1. a short interval, in which successive atoms leave the cavity in the ground state after a thermal fluctuation;

2. a long time interval, in which the trapping condition is maintained and all atoms leave the resonator in their excited state until the next fluctuation occurs.

The probability $P_2(0, |g\rangle, t)$ is plotted in Fig. 76.10. The plot clearly shows the two time regimes that govern the imperfect trapping situation.

76.8 QUANTUM THEORY OF MEASUREMENT

When the object of interest consists of only a few atoms and a few photons, the puzzling consequences of quantum mechanical measurement become visible. In

Figure 76.10. Waiting time probability for atoms in the ground state while cavity is operated at vacuum trapping-state condition.

the case of the micromaser, the information on the state of the field is imprinted in a subtle way on the atomic beam. While photon counting is normally a destructive operation, the dispersive part of the photon-atom interaction may be used to determine the photon number inside a resonator without altering it on average. Dispersive effects shift the phase of an oscillating atomic dipole without changing its state.

The phase shift due to the field in the resonator can be measured in a Ramsey-type experiment [33]. Consider an atom with two transitions $|g\rangle \rightarrow |e\rangle$ and $|e\rangle \rightarrow |i\rangle$. The first is far from resonance with the cavity and the second is close to resonance, but with a detuning $\delta_{ie} = \omega - \omega_{ie}$ large enough so as not to change the cavity photon number as the atom passes through. The dynamic Stark effect of the $|g\rangle \rightarrow |e\rangle$ transition frequency due to state $|i\rangle$ is then

$$\Delta\omega_{eg} = [g_{ie}\sqrt{n+1}]^2/\delta_{ie}. \quad (76.74)$$

If the resonator is now placed between the two Ramsey cavities tuned to $\omega_R \approx \omega_{eg}$, such that the polarization of the $|e\rangle \rightarrow |g\rangle$ transition is rotated by $\sim \pi/2$, then the additional phase shift $\Delta\omega_{eg}\tau$, where τ is the transit time through the optical resonator, can be measured, and hence the photon number n. Since Rydberg states have a large coupling constant g_μ, the phase shift due to a single atom is detectable [33].

A complete measurement of n requires a sequence of N atoms, because a single Ramsey measurement only determines whether the atom is in state $|e\rangle$ or $|g\rangle$, and hence $\Delta\omega_{eg}\tau$ to within $\pm\pi/2$. Since each measurement provides one binary bit of information, a sequence of N measurements can in principle distinguish 2^N possible

Fock states for the photon field. However, with a monoenergetic beam, integral multiples of 2π remain undetermined. A distribution of velocities, and hence transit times, is therefore desirable. An *entropy reduction* strategy for selecting an optimal velocity distribution, based on the outcome of previous measurements, is described in Ref. [34].

As a consequence of the uncertainty principle, a measurement of the photon number destroys all information about the phase of the field. In the present case, the noise in the conjugate variable (the phase) is prevented from coupling back on the measured one, and hence the measurement is called a *quantum nondemolition experiment*. Many other aspects of phase diffusion, entangled states, and quantum measurements in the micromaser are discussed in Ref. [35].

REFERENCES

1. Cavity QED is in detail reviewed in: S. Haroche, in *Fundamental Systems in Quantum Optics*, edited by J. Dalibard *et al.* (Elsevier, Amsterdam, 1992); D. Meschede, Phys. Rep. **211**, 201-250 (5) (1992).

2. An introduction into the more general framework of low energy Quantum Electrodynamics may be found in the recent textbook *The Quantum Vacuum*, edited by P. Milonni (Academic, Boston, 1994).

3. V. Weisskopf and E. Wigner, Z. Phys. **63**, 54 (1930).

4. *Long-Range Casimir Forces*, edited by F. Levin and D. Micha (Plenum Press, New York, 1993).

5. D. Meschede, H. Walther, and G. Müller, Phys. Rev. Lett. **54**, 551 (1985).

6. L. Allen and J. Eberly, *Optical Resonance and Two-Level-Atoms* (Dover, New York, 1987, reprint of the original 1975 edition).

7. D. Kleppner, Phys. Rev. Lett. **47**, 233 (1981).

8. R. G. Hulet, E. Hilfer, and D. Kleppner, Phys. Rev. Lett. **55**, 2137 (1985); W. Jhe, A. Anderson, E. Hinds, D. Meschede, L. Moi, and S. Haroche, Phys. Rev. Lett. **58**, 666 (1987).

9. J. Dowling, in *Quantum Measurements in Optics*, edited by P. Tombesi and D. Walls (Plenum Press, New York, 1992).

10. K. H. Drexhage, in *Progress in Optics*, edited by E. Wolf (North-Holland, Amsterdam, 1974), Vol. 12.

11. V. Sandoghar, C. Sukenik, E. Hinds, and S. Haroche, Phys. Rev. Lett. **68**, 3432 (1992).

12. H. Casimir and D. Polder, Phys. Rev. **73**, 360 (1948).

13. C. Sukenik, M. Boshier, D. Cho, V. Sandoghar, and E. Hinds, Phys. Rev. Lett. **70**, 560 (1993); A. Shih, D. Raskin, and P. Kusch, Phys. Rev. A **9**, 652 (1974).

14. C. Cohen-Tannoudji, in *Fundamental Systems in Quantum Optics*, edited by J. Dalibard *et al.* (Elsevier, Amsterdam, 1992).

15. S. Haroche, M. Brune, and J. M. Raimond, Europhys. Lett. **14**, 19 (1991).

16. M. Chevrollier, M. Fichet, M. Oria, G. Rahmat, D. Bloch, and M. Ducloy, J. Phys. (Paris) **2**, 631 (1992).

17. M. Brune, P. Nussenzveig, F. Schmidt-Kaler, R. Bernadot, A. Maali, J. M. Raimond, and S. Haroche, Phys. Rev. Lett. **72**, 3339 (1994).

18. E. Yablonovitch, Phys. Rev. Lett. **58**, 2059 (1987); a compilation of current work can be found in the feature issue on *Photonic Bandgaps* of J. Opt. Soc. Am. B **10** (1993).

19. C. Cohen-Tannoudji in *Frontiers in Laser Spectroscopy*, edited by R. Balian *et al.* (North-Holland, Amsterdam, 1977).

20. R. Thompson, G. Rempe, and H. Kimble, Phys. Rev. Lett. **68**, 1132 (1992).

21. *Rydberg States of Atoms and Molecules*, edited by R. F. Stebbings and F. B. Dunning (Cambridge University Press, Cambridge, 1983).

22. E. T. Jaynes and F. W. Cummings, Proc. IEEE **51**, 89 (1963); H. Paul, Ann. Phys. **11**, 411 (1963).

23. J. H. Eberly, N. B. Narozhny, and J. J. Sanchez-Mondragon, Phys. Rev. Lett. **44**, 1323 (1980).

24. H. J. Briegel and B. G. Englert, Phys. Rev. A **47**, 3311 (1993).

25. C. Ginzel, H. J. Briegel, U. Martini, B. G. Englert, and A. Schenzle, Phys. Rev. A **48**, 732 (1993).

26. B. R. Mollow, Phys. Rev. **188**, 1969 (1969).

27. P. Filipowicz, J. Javanainen, and P. Meystre, Phys. Rev. A **34**, 3077 (1986).

28. G. Rempe and H. Walther, Phys. Rev. A **42**, 1650 (1990).

29. G. Rempe, F. Schmidt-Kaler, and H. Walther, Phys. Rev. Lett. **64**, 2783 (1990).

30. U. Benson, G. Raithel, and H. Walther, Phys. Rev. Lett. **72**, 3506 (1994).

31. P. Meystre, G. Rempe, and H. Walther, Opt. Lett. **13**, 1078 (1988).

32. C. Wagner, A. Schenzle, and H. Walther, Opt. Commun. **107**, 318 (1994).

33. M. Brune, S. Haroche, V. Lefevre, J. M. Raimond, and N. Zagury, Phys. Rev. Lett. **65**, 976 (1990).

34. R. Schack, A. Breitenbach, and A. Schenzle, Phys. Rev. A **45**, 3260 (1992).

35. C. Wagner, A. Schenzle, and H. Walther, Phys. Rev. A**47**, 5068 (1993).

77

Quantum Optical Tests of the Foundations of Physics

A. M. Steinberg,* P. G. Kwiat,† and R. Y. Chiao

Department of Physics, University of California, Berkeley, California 94720
* Currently at National Institute of Standards and Technology, PHYS A167, Gaithersburg, Maryland 20899
† Currently at Los Alamos National Laboratory, Physics Div. P-23, Los Alamos, New Mexico 87545

77.1 INTRODUCTION: THE PHOTON HYPOTHESIS

Quantum mechanics began with the solution of the problem of blackbody radiation by Planck's quantum hypothesis: in the interaction of light with matter, energy can only be exchanged between the light in a cavity and the atoms in the walls of the cavity by the discrete amount $E = h\nu$, where h is Planck's constant and ν is the frequency of the light. Einstein, in his treatment of the photoelectric effect, reinterpreted this equation to mean that a beam of light consists of particles ("light quanta") with energy $h\nu$. The Compton effect supported this particle viewpoint of light by demonstrating that photons carried momentum, as well as energy. In this way, the wave-particle duality of quanta made its first appearance in connection with the properties of light.

It might seem that the introduction of the concept of the photon as a particle would necessarily also introduce the concept of locality into the quantum world. However, in view of observed violations of Bell's inequalities, exactly the opposite seems to be true. Here we review some recent results in quantum optics which elucidate nonlocality and other fundamental issues in physics.

In spite of the successes of quantum electrodynamics, and of the standard model in particle physics, there is still considerable resistance to the concept of the photon as a particle. Many papers have been written trying to explain all optical phenomena semiclassically, i.e., with the light viewed as a classical wave, and the atoms treated quantum mechanically [1–4]. We first present some quantum optics phenomena which exclude this semiclassical viewpoint.

In an early experiment, Taylor reduced the intensity of a thermal light source in Young's two-slit experiment, until, on the average, there was only a single photon passing through the two slits at a time. He then observed a two-slit interference pattern which was identical to that for a more intense classical beam of light. In Dirac's words, the apparent conclusion is that "each photon then interferes only with itself" [5]. However, a coherent state, no matter how strongly attenuated, always remains a coherent state (see Sect. 75.4); since a thermal light source can be modeled as a statistical ensemble of coherent states, a stochastic classical wave model yields complete agreement with Taylor's observations. The one-by-one darkening of grains of film can be explained by treating the matter alone quantum mechanically [2]; consequently, the concept of the photon need not be invoked, and the claim that this experiment demonstrates quantum interference of individual photons is unwarranted [6].

This weakness in Taylor's experiment can be removed by the use of nonclassical light sources; as discussed by Glauber [7], classical predictions diverge from quantum ones only when one considers counting statistics, or photon *correlations*. In particular, two-photon light sources, combined with coincidence detection, allow the production of single-photon ($n = 1$ Fock) states with near certainty. In the first such experiment [8], two photons, produced in an atomic cascade within nanoseconds of each other, impinged on two beam splitters, and were then detected in coincidence by means of four photomultipliers placed at all possible exit ports.

In a simplified version of this experiment [9], one of the beam splitters and its two detectors are replaced with a single detector D_1 (see Fig. 77.1). We define the anticorrelation parameter

$$\alpha \equiv N_{123}N_1/N_{12}N_{13}, \qquad (77.1)$$

where N_{123} is the rate of triple-coincidences between detectors D_1, D_2 and D_3; N_1 is the singles rate at D_1; and N_{12} and N_{13} are double-coincidence rates. Then from Schwarz's inequality [9, 6, 10], $\alpha \geq 1$ for any classical wave. In essence, since the wave divides smoothly, the coincidence rate between D_2 and D_3 is never smaller than the "accidental" coincidence rate, even when measurements are conditioned on an event at D_1. (The Hanbury-Brown and Twiss experiment [11] *can* be explained classically, because the thermal fluctuations lead to "bunching," or a mean coincidence rate which is *greater* than the mean accidental rate; *cf.* Sect.75.4.3.) By contrast, the indivisibility of the photon leads to strong anticorrelations between D_2 and D_3, making α arbitrarily small. In agreement with this quantum mechanical picture, Grangier *et al.* observed a 13-standard-deviation violation of the inequality [9], corroborating the notion of the "collapse of the wave packet" as proposed by Heisenberg [12] (see also Sect. 73.5.2).

77.2 QUANTUM PROPERTIES OF LIGHT

77.2.1 Vacuum Fluctuations: Cavity QED

The above considerations necessitate the quantization of the electromagnetic field, which in turn leads to the concept of vacuum fluctuations [4] (see Sect. 75.2). Difficulties with this idea, such as the implied infinite zero-point energy of the universe, have led some researchers to attempt to dispense with this concept altogether, along with that of the photon, in every explanation of electromagnetic interactions with matter. Of course, it is impossible to explain all phenomena, such as spontaneous emission and the Lamb shift, without some kind of fluctuating electromagnetic fields (see Chap. 75 and Sect. 76.3.3), but one can go a long way with an *ad hoc* ambient classical electromagnetic noise-field filling all of space, in conjunction with the radiation reaction [1, 4].

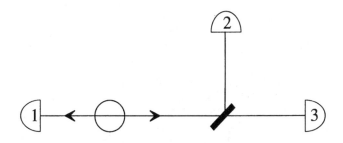

Figure 77.1. Triple-coincidence setup of Grangier *et al.* [9].

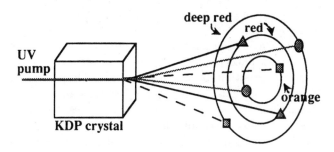

Figure 77.2. Conical emissions of down-conversion from a nonlinear crystal. Photon energy depends on the cone opening angle, and conjugate photons lie on opposite sides of the axis, e.g., the inner "circle" orange photon is conjugate to the outer "circle" deep-red photon, etc.

In particular, even the Casimir attraction between two conducting plates (see Sect. 76.1) can be explained semiclassically in terms of dipole forces between electrons in each plate as they undergo zero-point motion and induce image charges in the other plate (see Sect. 76.3.3). Nevertheless, the effects of cavity QED (Chap. 76) [13–15], including the influence of cavity-induced boundary conditions on energy levels and spontaneous emission rates, are most easily unified via quantization of the electromagnetic field.

77.2.2 The Down-Conversion Two-Photon Light Source

The quantum aspects of electromagnetism are made more striking with a two-photon light source, in which two highly correlated photons are produced in spontaneous parametric down-conversion, or parametric fluorescence [16–18]. In this process, an ultraviolet "pump" photon produced in a laser spontaneously decays inside a crystal with a $\chi^{(2)}$ nonlinearity into two highly correlated red photons, conventionally called the "signal" and the "idler" (see Sect. 70.4.4). As shown in Fig. 77.2, a rainbow of colored cones is produced around an axis defined by the direction of the uv beam (for the case of type-I phase-matching), with the correlated down-conversion photons always emitted on opposite sides of the uv beam (see Sect. 70.3.2). Their emission times are within femtoseconds of each other, so that detection of one photon implies with near certainty that there is exactly one quantum present in the conjugate mode.

Energy and momentum are conserved in this process:

$$\hbar\omega_0 = \hbar\omega_1 + \hbar\omega_2, \tag{77.2}$$

$$\hbar\mathbf{k}_0 \approx \hbar\mathbf{k}_1 + \hbar\mathbf{k}_2, \tag{77.3}$$

where $\hbar\omega_0$ ($\hbar\mathbf{k}_0$) is the energy (momentum) of the parent photon, and $\hbar\omega_1$ ($\hbar\mathbf{k}_1$) and $\hbar\omega_2$ ($\hbar\mathbf{k}_2$) are the energies (momenta) of the daughter photons; $\mathbf{k}_1$ and $\mathbf{k}_2$ sum to $\mathbf{k}_0$ to within an uncertainty given by the reciprocal of the crystal length [19]. Since there are many ways of

partitioning the parent photon's energy, each daughter photon may have a broad spectrum, and hence a wave packet narrow in time. However, $\omega_1 + \omega_2 = \omega_0$ is extremely well-defined, so that the *difference* in the daughter photons' arrival times, and the *sum* of their energies can be simultaneously known to high precision.

In type-I phase matching the correlated photons share the same polarization, while in type-II phase matching they have orthogonal polarizations. This latter possibility has recently permitted the production of *unpolarized* down-conversion photons that nevertheless display near-perfect polarization correlations [20]; see Sect. 77.5.2.

77.2.3 Squeezed States of Light

The creation of correlated photon pairs is closely related to the process of quadrature-squeezed light production (see Sect. 75.3, and the review in Ref. [21]). For example, when the gain arising from parametric amplification in a down-conversion crystal becomes large, there is a transition from spontaneous to stimulated emission of pairs. This gain is dependent on the phase of amplified light relative to the phase of the pump light. As a result, the vacuum fluctuations are reduced ("squeezed") below the standard quantum limit in one quadrature, but increased in the other, in such a way as to preserve the minimum uncertainty-principle product [22]. This periodicity of the fluctuations at 2ω is a direct consequence of the fact that the light is a superposition of states differing in energy by $2\hbar\omega$ — the quadrature-squeezed vacuum state

$$|\xi\rangle = \exp\left(\tfrac{1}{2}\xi^* aa - \tfrac{1}{2}\xi a^\dagger a^\dagger\right)|0\rangle \tag{77.4}$$

represents a vacuum state transformed by the creation ($a^\dagger a^\dagger$) and destruction (aa) of photons two at a time.

Essentially any optical processes operating on photon pairs (e.g., four-wave mixing [23, 24]) can also produce such squeezing.

Amplitude squeezing involves preparation of states with well-defined photon number, i.e., states lacking the Poisson fluctuations of the coherent state. The possibility of producing such states (e.g., via a constant-current-driven semiconductor laser [25]) demonstrates that "shot noise" in photodetection should not be thought of as merely the result of the probabilistic (à la Fermi's Golden Rule, *cf.* Sect. 67.5) excitation of quantum mechanical atoms in a classical field, but as representing real properties of the electromagnetic field, accessible to experimental control.

77.2.4 Quantum Nondemolition

The uncertainty principle between the number of quanta N and phase ϕ of a beam of light,

$$\Delta N \Delta \phi \geq 1/2, \qquad (77.5)$$

implies that to know the number of photons exactly, one must give up all knowledge of the phase of the wave. In theory, a *quantum nondemolition* (QND) process is possible [26]: without annihilating any of the light quanta, one can count them. It might seem that this would make possible successive measurements on noncommuting observables of a single photon, in violation of the uncertainty principle; it is the unavoidable introduction of phase uncertainty by any number measurement which prevents this.

The most common QND schemes [27] use the intensity-dependent index of refraction arising from the optical Kerr effect (see Sect. 70.5.3) — the change in the index due to the intensity of the "signal" beam changes the optical phase shift on a "probe" beam [28–30]. Still other QND proposals use Rydberg atoms to give indirect information concerning the number of photons in a microwave cavity through which they have passed [31, 32] (see Sect. 76.5.1), or the Aharonov-Bohm effect to sense photons via the phase shift their fields induce in passing electrons [33, 34].

77.3 NONCLASSICAL INTERFERENCE

77.3.1 Single-Photon Interference and Energy-Time Uncertainty

The simplest one-photon interference experiment [9] uses the cascade source discussed in Sect. 77.1. One of the photons is directed to a "trigger" detector, while the other, thus prepared in an $n = 1$ Fock state, is sent through a Mach-Zehnder interferometer. The output

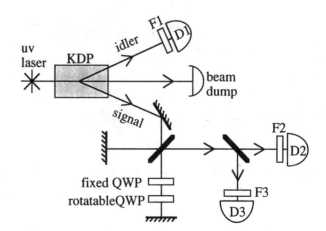

Figure 77.3. The energy-time uncertainty relation and wave function collapse were studied by investigating the effect of various filters before the detectors in a single-photon interference experiment [9, 35].

photon, detected in coincidence with the trigger photon, showed fringes with a visibility > 98%. Dirac's statement that a single photon interferes with itself is thus verified.

The energy-time uncertainty principle,

$$\Delta E \Delta t \geq \hbar/2, \qquad (77.6)$$

has recently been tested in a related interference experiment. While the canonical commutator $[x, p_x] = i\hbar$ gives rise to $\Delta x \Delta p_x \geq \hbar/2$, the corresponding commutator $[t, \mathcal{H}] = i\hbar$ does not exist. If the t operator existed, it would be the generator of translations of either sign in energy (just as $\mathcal{H}$ is the generator of translations in x), and could thus generate arbitrarily large negative energies — but unlike momentum, energy is bounded from below[1]. Thus, in the time-dependent Schrödinger equation, time is a c-number parameter [36] (as is position as well in relativistic theory), rather than an observable.

Due to energy conservation in the down-conversion process, the daughter photons of a parent photon of sharp energy E_0 are in an energy-"entangled" state, a nonfactorizable sum of product states [37]:

$$|\Psi\rangle = \int_0^{E_o} dE\, A(E)|E\rangle |E_0 - E\rangle, \qquad (77.7)$$

where $A(E)$ is the probability amplitude for the production of two photons of energies E and $E_0 - E$. A measurement of the energy of one of the photons to be E_1 causes an instantaneous "collapse" of the system to the state $|E_1\rangle |E_0 - E_1\rangle$, implying an instantaneous increase of the width of the other photon's wave packet.

---

[1]As can be seen from the fact that $\mathcal{H} = (N + 1/2)\hbar\omega$ and $\phi = \omega t$, these problems are closely related to the number-phase uncertainty relation and the difficulties of defining a phase operator (Sect. 75.3.4).

In a modification [35] of the previous experiment (see Fig. 77.3), an interference filter F1 is used to sharply measure the frequency of the trigger photon 1, while a Michelson interferometer is used to measure the coherence length of the conjugate photon 2. If the trigger photon passes through the filter of narrow width ΔE and is detected, then photon 2 occupies a broad wave packet of duration $\Delta t \approx \hbar/\Delta E$, whereas the absence of fringes without triggering implies a much shorter wave packet. This is a nonlocal effect in that the photons can be arbitrarily far away from each other when the collapse occurs. (The same apparatus was also used to demonstrate that Berry's phase in optics has a quantum origin [38].)

77.3.2 Two-Photon Interference

In the above experiments, interference occurs between two paths taken by a *single* photon. An early experiment to demonstrate *two*-photon interference using the down-conversion light source was performed by Ghosh and Mandel [39]. They looked at the counting rate of a detector illuminated by both of the twin beams. No interference was observed at the detector, because although the sum of the phases of the two beams emitted in parametric fluorescence is well-defined (by the phase of the pump), their *difference* is not, due to the number-phase uncertainty principle (see Sect. 75.4). However, when Ghosh and Mandel looked at the rate of *coincidence* detections between two such detectors whose separation was varied, they observed high-visibility interference fringes. Whereas in the standard two-slit experiment, interference occurs between the two paths a single photon could have taken to reach a given point on a screen, in this case it occurs between the possibility that the signal photon reached detector 1 and the idler photon detector 2, and the possibility that the reverse happened. This experiment provides a manifestation of quantum nonlocality; interference occurs between alternate global histories of a system, not between local fields. At a null of the coincidence fringes, the detection of one photon at detector 1 excludes the possibility of finding the conjugate photon at detector 2.

Such interference becomes clearer in the related interferometer of Hong, Ou, and Mandel [40] (see Fig. 77.4).

The identically-polarized conjugate photons from a down-conversion crystal are directed to opposite sides of a 50-50 beam splitter, such that the transmitted and reflected modes overlap. If the difference in the path lengths ΔL prior to the beam splitter is larger than the two-photon correlation length (of the order of the coherence length of the down-converted light), the photons behave independently at the beam splitter, and coincidence counts between detectors in the two output ports are observed half of the time — the other half of the

Figure 77.4. Simplified setup for a Hong-Ou-Mandel (HOM) interferometer [40]. Coincidences may result from both photons being reflected, or both being transmitted. When the path lengths to the beam splitter are equal, these processes destructively interfere, causing a null in the coincidence rate. In a modified scheme, a half waveplate in one arm of the interferometer (at "A") serves to distinguish these otherwise interfering processes, so that no null in coincidences is observed. Using polarizers before the detectors, one can "erase" the distinguishability, thereby restoring interference [41]. (See Sects. 77.4.1 and 77.5.2.)

time both photons travel to the same detector. However, when $\Delta L \approx 0$, such that the photon wave packets overlap at the beam splitter, the probability of coincidences is reduced, in principle to zero if $\Delta L = 0$. One can explain the coincidence null at zero path-length difference using the Feynman rules for calculating probabilities: add the probability amplitudes of *indistinguishable* processes which lead to the same final outcome, and then take the absolute square. The two indistinguishable processes here are both photons being reflected at the beam splitter (with Feynman amplitude r·r) and both photons being transmitted (with Feynman amplitude t·t). The probability of a coincidence detection is then

$$P_c = |\mathrm{r}\cdot\mathrm{r} + \mathrm{t}\cdot\mathrm{t}|^2 = \left| \frac{i}{\sqrt{2}}\cdot\frac{i}{\sqrt{2}} + \frac{1}{\sqrt{2}}\cdot\frac{1}{\sqrt{2}} \right|^2 = 0 \ ,$$

(77.8)

assuming a real transmission amplitude, and where the factors of i come from the phase shift upon reflection at a beam splitter [42, 43].

The possibility of a perfect null at the center of the dip is indicative of a nonclassical effect. Indeed, classical field predictions allow a maximum coincidence-fringe visibility of only 50% [44]. The tendency of the photons to travel off together at the beam splitter can be thought of as a manifestation of the Bose-Einstein statistics for the photons [45]. In practice, the bandwidth of the photons, and hence the width of the null, is determined by filters and/or irises before the detectors [19]. Widths as small as 5 μm have been observed, corresponding to time delays of only 15 fs [46]. Consequently, one application is the determination of single-photon propagation times with extremely high time resolution (see Sect. 77.7).

77.4   COMPLEMENTARITY

77.4.1   Quantum Eraser

The complementary nature of wave-like and particle-like behavior is frequently interpreted as follows: due to the uncertainty principle, any attempt to measure the position (particle aspect) of a quantum leads to an uncontrollable, irreversible disturbance in its momentum, thereby washing out any interference pattern (wave aspect) [47, 48]. This picture is incomplete though; no "state reduction," or "collapse," is necessary to destroy interference, and measurements which do not involve reduction can be reversible. One must view the loss of coherence as arising from an *entanglement* of the system wave function with that of the measuring apparatus (MA) [49]. Previously interfering paths can thereby become distinguishable, such that no interference is observed. Interference may be regained, however, if one manages to "erase" the distinguishing information. This is the physical content of quantum erasure [50, 51]. The primary lesson is that one must consider the total physical state, including any MA with which the interfering quantum has become entangled. If the coherence of the MA is maintained, then interference may be recovered.

One demonstration of a quantum eraser is based on the interferometer in Fig. 77.4 [41]. A half waveplate inserted into one of the paths before the beam splitter serves to rotate the polarization of light in that path. In the extreme case, the polarization is made orthogonal to that in the other arm, and the r·r and t·t processes become distinguishable; hence, the destructive interference which led to a coincidence null does not occur. The distinguishability can be erased, however, by using polarizers just before the detectors. In particular, if the initial polarization of the photons is horizontal, and the waveplate rotates one of the photon polarizations to vertical, then polarizers at 45° before both detectors restore the original interference dip. If one polarizer is at 45° and the other at −45°, interference is once again seen, but now in the form of a peak instead of a dip (see Fig. 77.5). There are four basic measurements possible on the MA (here the polarization) — two of which yield which-path information, one of which recovers the initial interference fringes (here the coincidence dip), and one of which yields interference anti-fringes (the peak instead of the dip). In some implementations, the decision to measure wave-like or particle-like behavior may even be delayed until *after detection* of the original quantum, an irreversible process[2] [55, 56]. But in all cases, one must correlate the

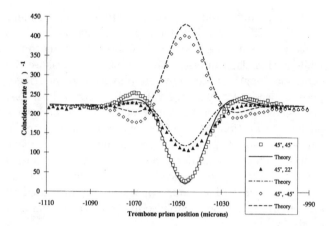

Figure 77.5. Experimental data and scaled theoretical curves (adjusted to fit observed visibility of 91%) with polarizer 1 at 45° and polarizer 2 at various angles. Far from the dip, there is no interference and the angle is irrelevant [41].

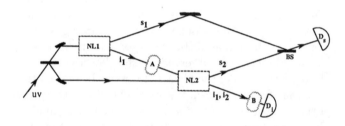

Figure 77.6. Schematic of setup used in [58]. The idler photons from the two crystals are indistinguishable; consequently, interference fringes may be observed in the signal singles rate at detector D_s. Additional elements at A and B can be used to make a quantum eraser.

results of measurements on the MA with the detection of the originally interfering system.

77.4.2   Vacuum-Induced Coherence

A somewhat different demonstration [57, 58] of complementarity involves two down-conversion crystals, NL1 and NL2, aligned such that the trajectories of the idler photons from each crystal overlap (see Fig. 77.6). A beam splitter acts to mix the signal modes. If the path lengths are adjusted correctly, and the idler beams overlap precisely, there is no way to tell, even in principle, from which crystal a photon detected at D_s originated. Interference appears in the signal *singles* rate at D_s, as any of the path lengths is varied. If the idler beam from crystal NL1 is prevented from entering crystal NL2, then the interference vanishes, because the presence or absence of

---

[2]This is an extension of the original delayed-choice discussion by Wheeler [52], and the experiments by Hellmuth *et al.* and Alley *et al.* [53, 54], in which the decision to display wave-like or particle-

like aspects in a light beam may be delayed until after the beam has been split by the appropriate optics.

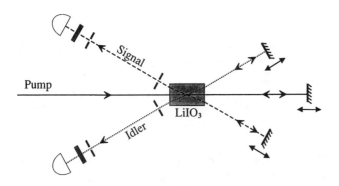

Figure 77.7. Schematic of the experiment to demonstrate enhancement and suppression of spontaneous down-conversion [59].

an idler photon at D_i then "labels" the parent crystal. Thus one explanation for the effect of blocking this path is that coherence is established by the idler-mode *vacuum* field seen by both crystals.

77.4.3 Suppression of Spontaneous Parametric Down-Conversion

A modification [59] of this two-crystal experiment uses only a single nonlinear crystal (see Fig. 77.7). A given pump photon may down-convert in its initial rightward passage through the crystal, or in its left-going return trip (or not at all, the most likely outcome). As in the previous experiment, the idler modes from these two processes are made to overlap; moreover, the signal modes are also aligned to overlap. Thus, the left-going and right-going production processes are indistinguishable and interfere. The result is that fringes are observed in all of the counting rates (i.e., the coincidence rate and both singles rates) as any of the mirrors is translated. One interpretation is as a change in the spontaneous emission of the down-converted photons. In contrast to the cavity QED demonstrations discussed in Sects. 77.2.1 and 76.1, here the distances to the mirrors are much longer than the coherence lengths of the spontaneously emitted photons.

With the inclusion of waveplates to label the photons' paths, and polarizers to erase this information, an improved quantum eraser experiment was recently completed [56]. The arrangement has also shown that in a sense, there are always photons between the down-conversion crystal and the mirrors, even in the case of complete suppression of the spontaneous emission process [60].

77.5 THE EPR PARADOX AND BELL'S INEQUALITIES

77.5.1 Generalities

Nowhere is the nonlocal character of the quantum mechanical entangled state as evident as in the "paradox" of Einstein, Podolsky, and Rosen (EPR) [61], the version of Bohm [62], and the related inequalities by Bell [63, 64]. Consider two photons traveling off back-to-back, described by the entangled singlet-like state

$$|\psi\rangle = \left(|H_1, V_2\rangle - |V_1, H_2\rangle\right)/\sqrt{2}, \qquad (77.9)$$

where the letters denote horizontal (H) or vertical (V) polarization, and the subscripts denote photon propagation direction. This state is isotropic — it has the same form regardless of what basis is used to describe it. Measurement of any polarization component for one of the particles will yield a count with 50% probability; individually, each particle is unpolarized. Nevertheless, if one measures the polarization component of particle 1 in any basis, one can predict with certainty the polarization of particle 2 in the same basis, seemingly without disturbing it, since it may be arbitrarily remote. Therefore, according to EPR, to avoid any nonlocal influences one should ascribe an "element of reality" to every component of polarization. A quantum mechanical state cannot specify that much information, and is consequently an incomplete description, according to the EPR argument. The intuitive explanation implied by EPR is that the particles leave the source with definite, correlated properties, determined by some local "hidden variables" not present in quantum mechanics (QM).

For two entangled particles, a local hidden variable (LHV) theory can be made which correctly describes perfect correlations or anti-correlations (i.e., measurements made in the same polarization basis). The choice of an LHV theory versus QM is then a philosophical decision, not a physical one. In 1964 John Bell [63] discovered that QM gives different statistical predictions than does *any* LHV theory, for situations of nonperfect correlations (i.e., analyzers at intermediate angles). His inequality constraining various joint probabilities given by any local realistic theory was later generalized to include any model incorporating locality [65, 66], and also extended to apply to real experimental situations [67, 68]; the Clauser-Horne (CH) form relates the directly observable singles rates S_1 and S_2 and the coincidence rate C_{12}, rather than "inferred" probabilities, by

$$C_{12}(a, b) + C_{12}(a, b') + C_{12}(a', b) - C_{12}(a', b')$$
$$\leq S_1(a) + S_2(b), \qquad (77.10)$$

where a and a' are any pair of analyzer (e.g., polarizer) settings at detector 1, and b and b' a pair of settings at detector 2. For certain choices of a, a', b, and b', QM predicts a violation of this inequality.

In fact, inequality (77.10) has not yet been violated in any experiment because the minimum detector efficiency required is 67% [69]. Experiments thus far have employed an additional assumption, roughly equivalent to the "fair-sampling" assumption that the fraction of pairs detected is representative of the entire ensemble (see [70] for a fuller discussion). Moreover, no experiment has fulfilled the additional requirement that the analyzer settings be switched randomly and rapidly. Two efforts are now underway to attempt a direct violation of Eq. (77.10) with no auxiliary assumptions. One utilizes down-converted photons [71] and newly developed high-efficiency ($> 85\%$) single-photon detectors [72, 73]; the other uses dissociated mercury dimers as the correlated particles, the advantage being detection efficiencies of 95% by photoionizing the atoms and detecting the photoelectrons [74].

77.5.2 Polarization-Based Tests

With the caveat of supplementary assumptions, Bell's inequalities have now been tested many times, and the vast majority of experiments have violated these inequalities, in support of QM. The first such test was performed with pairs of photons produced via an atomic cascade, and a later version incorporated rapid (albeit periodic) switching of the analyzers [75, 76]. Unfortunately, the angular correlation of the cascade photons is not very strong. In contrast, the strong correlations of the down-converted photons make them ideal for such tests, the first of which were performed using setups essentially identical to that already discussed in connection with quantum erasure [77, 78], (see Fig. 77.4). With the condition of equal path lengths, and with the half wave-plate oriented to rotate the polarization from horizontal to vertical, the state of light after the 50-50 beam splitter is

$$|\psi\rangle \approx \tfrac{1}{2} \left[|H_1 V_2\rangle - |V_1 H_2\rangle + i|V_1 H_1\rangle + i|V_2 H_2\rangle \right],$$
(77.11)

which reduces to Eq. (77.9) if one considers only the terms which can lead to coincidence detections (*viz.*, the first two). The probability of coincidence detection depends nonlocally on the difference in the orientation angles of the two polarizers: $P_c \propto \sin^2(\theta_2 - \theta_1)$. A more recent experiment [79] using the orthogonally-polarized photons from collinear type-II down-conversion (see Sect. 70.3.2) has produced a violation of a Bell's inequality by 22 standard deviations. Yet more recently, a *non*-collinear type-II down-conversion experiment has produced true polarization-entangled states (i.e., without the need to discard half of the terms) and produced a 100 standard deviation violation of a Bell's inequality.

Figure 77.8. Outline of apparatus used to demonstrate a violation of a Bell's inequality based on momentum entanglement [80].

77.5.3 Nonpolarization Tests

The advent of parametric down-conversion has also led to the appearance of several nonpolarization-based tests of Bell's inequalities, using, for example, an entanglement of the photon momenta (see Fig. 77.8) [80]. By use of small irises (labeled 'A' in the figure), only four down-conversion modes are examined: 1s, 1i, 2s, and 2i. Beams 1s and 1i correspond to one pair of conjugate photons; beams 2s and 2i correspond to a different pair. Photons in beams 1s and 2s have the same wavelength, as do photons in beams 1i and 2i. With proper alignment, after the beam splitters there is no way to tell whether a pair of photons came from the 1s-1i or the 2s-2i paths. Consequently, the coincidence rates display interference, although the singles rates at the four detectors indicated in Fig. 77.8 remain constant. This interference depends on the difference of phase shifts induced by rotatable glass plates P_i and P_s in paths 1i and 2s, respectively, and is formally equivalent to the polarization case considered above, in which it is the difference of polarization-analyzer angles that is relevant. By measuring the coincidence rates for two values for each of the phase shifters — a total of four combinations — the experimenters were able to violate an appropriate Bell's inequality. One interpretation is that the paths taken by a given pair of photons are not elements of reality.

Several groups [81, 82] have implemented a different Bell's inequality proposal [83], based on energy-time entanglement of the photons. One member of each down-converted pair is directed into an unbalanced Mach-Zehnder-like interferometer, allowing both a short and long path to the final beam splitter; the other photon is directed into a separate but similar interferometer. There arises interference between the indistinguishable processes ("short-short" and "long-long") which could lead to coincidence detection. Using fast detectors to select out only these processes, the reduced state [*cf.* Eq. (77.9)] is

$$|\psi\rangle = \frac{1}{2} \left(|S_1, S_2\rangle - e^{i\phi}|L_1, L_2\rangle \right),$$
(77.12)

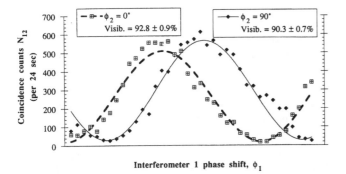

Figure **77.9**. High-visibility coincidence fringes in a Franson dual-interferometer experiment [83] for two values of the phase in interferometer 2 as the phase in interferometer 1 is slowly varied. The curves are sinusoidal fits.

where the letters indicate the short or long path, and the phase is the sum of the relative phases in each interferometer. Although no fringes are seen in any of the singles count rates, the high-visibility coincidence fringes (Fig. 77.9) lead to a violation of an appropriate Bell's inequality. One conclusion is that it is incorrect to ascribe to the photons a definite time of emission from the crystal, or even a definite energy, unless these observables are explicitly measured. A more general interpretation, applicable to all violations of Bell's inequalities, is that the predictions of QM cannot be reproduced by any completely local theory. It must be that the results of measurements on one of the photons depend on the results for the other, and these correlations are not merely due to a common cause at their creation [84, 85].

77.6   RELATED ISSUES

77.6.1   Nonlocality Without Inequalities

In the above experiments for testing nonlocality, the disagreement between quantum predictions and Bell's constraints on local realistic theories are only statistical. Recently, Greenberger, Horne, and Zeilinger (GHZ) pointed out that in some systems involving *three* or more entangled particles [86, 87], a contradiction could arise even at the level of perfect correlations. A schematic of one version of the GHZ *Gedankenexperiment* is shown in Fig. 77.10. The source at the center is posited to emit trios of correlated particles. Just as the Rarity-Tapster experiment selected two pairs of photons (*cf.* Fig. 77.8), the GHZ source selects two trios of photons; these are denoted by abc and $a'b'c'$. Hence, the state coming from the source may be written

$$|\psi\rangle = (|abc\rangle + |a'b'c'\rangle) / \sqrt{2}. \qquad (77.13)$$

Figure **77.10**. A three-particle gedanken experiment to demonstrate the inconsistency of quantum mechanics and any local realistic theory. All beam splitters are 50-50 [57].

After passing through a variable phase shifter (e.g., ϕ_a), each primed beam is recombined with the corresponding unprimed beam at a 50-50 beam splitter. Detectors (denoted by Greek letters) at the output ports signal the occurrence of triple coincidences. The following simplified argument conveys the spirit of the GHZ result.

Given the state (77.13), one can calculate from standard QM the probability of a triple coincidence as a function of the three phase shifts:

$$P(\phi_1, \phi_2, \phi_3) = \tfrac{1}{8} \left[1 \pm \sin(\phi_a + \phi_b + \phi_c) \right], \qquad (77.14)$$

where the plus sign applies for coincidences between all unprimed detectors, and the minus sign for coincidences between all primed detectors. For the case in which all phases are 0, it will occasionally happen (1/8th of the time) that there will be a triple coincidence of all primed detectors. Using a "contrafactual" approach, we ask what would have happened if ϕ_a had been $\pi/2$ instead. By the locality assumption, this would not change the state from the source, nor the fact that detectors β' and γ' went off. But from (77.14) the probability of a triple coincidence for primed detectors is zero in this case; therefore, we can conclude that detector α would have "clicked" if ϕ_a had been $\pi/2$. Similarly, if ϕ_b or ϕ_c had been $\pi/2$, then detectors β or γ would have clicked. Consequently, if all the phases had been equal to $\pi/2$, we would have seen a triple coincidence between unprimed detectors. But according to Eq. (77.14) this is impossible: the probability of triple coincidences between unprimed detectors when all three phases are equal to $\pi/2$ is strictly zero! Hence, if one believes the quantum mechanical predictions for these cases of perfect correlations, it is not possible to have a consistent local realistic model. By similar arguments, one can even demonstrate nonlocality without inequalities, at least in a *gedanken* experiment, using just two particles [88, 89]. When the arguments are suitably modified to deal with real experiments [90], inequalities once again result. Nevertheless, these new schemes underscore the vast disagreement between quantum theory and locality.

77.6.2 Information Content of a Quantum

The inherent nonlocality of particles in an entangled state cannot be used to transmit superluminal messages. If A and B receive an EPR pair of photons, which A then collapses in a certain basis by performing a polarization measurement, B can only extract one bit of information from a measurement on his photon — this bit corresponds not to A's choice of basis, but to the (random) *outcome* of A's measurement. However, instantaneous communication *would* be possible if one could make copies ("clones") of a single photon in an unknown polarization state: by performing measurements on n copies of his photon, B could determine its polarization to a resolution of n bits, thereby accurately determining A's choice of basis. Taking into account quantum fluctuations [91], one finds that no amplifier can make a sufficiently faithful copy for such a scheme to work — it is impossible to clone photons [92].

This makes the recent work by Bennett *et al.* on "quantum teleportation" and related effects all the more remarkable. For example, a single photon can be used to transfer two bits of information, when it is part of an entangled EPR pair [93]. Again consider A and B, each possessing one photon of such a pair. By manipulating only her photon (via a polarization rotator and a phase shifter), A can convert the joint state into any of the four two-particle "Bell states" $[|\psi_{AB}^{\pm}\rangle = (|H_A, V_B\rangle \pm |V_A, H_B\rangle)/\sqrt{2}, \ |\phi_{AB}^{\pm}\rangle = (|H_A, H_B\rangle \pm |V_A, V_B\rangle)/\sqrt{2}]$, (such states are analogous to measuring J and m for a pair of spin-1/2 particles, rather than the individual spin projections) and then send her photon to B. By making a suitable measurement on both photons, B can then in principle determine which of the four states A produced [94]: A's single photon carried two bits.

In the even more striking *quantum teleportation* effect [95], an unknown polarization state f (with its in-principle infinite amount of information) can be "teleported" from B to A, if each already possesses one photon of an EPR pair (in the singlet-state ψ_{AB}^{-}). First, B jointly measures his EPR photon and the photon F (whose state f is to be teleported) in the basis defined

by the Bell states of these two photons. Via a mere two bits of classical information, B then informs A which of the four (equally probable) Bell states he actually measured. With this information, A can transform the state of her EPR particle into f.

If B found the singlet-state ψ_{BF}^{-}, then the polarization of his EPR photon must have been orthogonal to that of F (because the polarizations of particles in a singlet-state are always perfectly anticorrelated, regardless of the quantization basis). But because the two EPR photons were initially also in a singlet-state, *their* polarizations must also be orthogonal, so A's EPR photon is already in state f. If instead B found ϕ_{BF}^{-}, for example, then A simply makes the same transformation (with a polarization rotator) that would have changed ψ_{AB}^{-} into ϕ_{AB}^{-}, again leaving her photon in the state f. Thus, although one may only extract one bit of (normally useless) information from an EPR particle, the perfect correlations may be used to transfer an infinite amount of information, i.e., precise specification of a point in the state space of the particle (the Poincaré sphere for a photon or the Bloch sphere for an electron). The "no cloning" theorem is not violated, since B irrevocably alters the state of F by the measurement he performs, leaving only one particle in f.

77.6.3 Cryptography

Although EPR schemes cannot send signals superluminally, they have other potential applications in cryptography. In the "one-time pad" of classical cryptography [96], two collaborators share a secret "key" (a random string of binary digits) in order to encode and decode a message. Such a key may provide an absolutely unbreakable code, provided that it is unknown to an eavesdropper. The problem arises in key distribution: any classical distribution scheme is subject to noninvasive eavesdropping, e.g., using a fiber-coupler to tap the line, without disturbing the transmitted classical signal. In quantum cryptography proposals, security is guaranteed by using single-photon states [97, 98], some of the schemes employing particles prepared in an EPR-entangled state. Each collaborator receives one member of each correlated pair, and measures the polarization in a random basis. After repeating the process many times, the two then discuss publicly which bases were used for each measurement, but not the actual measurement results. The cases where different bases were chosen are not used for conveying the key, and may be discarded, along with instances where one party detected no photon. In cases where the same bases were used, however, the participants will now have correlated information, from which a random, shared key can be generated. As long as single photons are used, any attempt at eavesdropping, even one relying on QND, will necessarily introduce errors due to the uncertainty principle. If the eavesdropper uses the wrong basis to study a photon before sending it on to the real recipient, the very act of measuring will disturb the original state.

77.6.4 Issues in Causality

Outstanding causal paradoxes in optics include the paradox of Barton and Scharnhorst [99], closely related to the Casimir effect (recently given indirect confirmation [14]; see Sect. 76.3.3). In this case, the amplitude for light-by-light scattering is modified by the presence of closely-spaced parallel plates. It appears that this can lead to propagation of light in vacuum (albeit a vacuum "colored" by the presence of the Casimir plates) faster than *c*. Similar paradoxes have been pointed out in connection with localization of any particle in a quantum field theory [100] and with the Glauber theory of photodetection (see Sect. 75.5) [101]. It may be that the high momentum/energy components introduced by localizing a detector to within a wavelength lead to spurious "dark" detections [102], which can be mistaken for causality-violating events [103].

77.7 THE SINGLE-PHOTON TUNNELING TIME

77.7.1 An Application of EPR Correlations to Time Measurements

In this section we discuss experiments involving the quantum propagation of light in matter. Due to the sharp time correlations of the paired photons from spontaneous down-conversion, one can use the HOM interferometer (see Sect. 77.3.2, and Fig. 77.4) to measure very short relative propagation delay times for the signal and idler photons. One early application was therefore to confirm that single photons in glass travel at the group velocity [46]. At least until recently, the only quantum theory of light in dispersive media was an *ad hoc* one [104]. The shift of the interference dip resulting from a medium introduced into one of the interferometer arms can be accurately measured by determining how much the path lengths must be changed to compensate the shift and recover the dip. This result suggests that when looking for a microscopic description of dielectrics, it is unnecessary to consider the medium as being polarized by an essentially classical electric field due to the collective action of all photons present, and reradiating accordingly. Linear dielectric response is not a collective effect in this sense — each photon interferes only with itself (as per Dirac's dictum) as it is partially

scattered from the atoms in the medium. The single-photon group velocity thus demonstrates "wave-particle unity."

The standard limitation for measurements of short-time phenomena is that to have high time-resolution, one needs short pulses (or at least short coherence lengths), but these in turn require broad bandwidths and are therefore very susceptible to dispersive broadening. It is a remarkable consequence of the EPR energy correlations of the down-conversion photons (Sect. 77.3.1) that time measurements made with the HOM interferometer are essentially immune to such broadening [105, 106, 46]. In effect, the measurement is sensitive to the *difference* in emission times while the broadening is sensitive to the *sum* of the frequencies. While frequency and time cannot both be specified for a given pulse, the crucial feature of EPR correlations such as those exhibited by down-converted photons is that this difference and this sum correspond to commuting observables, and both may be arbitrarily well defined. The photon which reaches detector 1 could either have traversed the dispersive medium and been transmitted, or traversed the empty path and been reflected, leaving its twin to traverse the medium. The medium thus samples *both* of the (anticorrelated) frequencies, leading to an automatic cancellation of any first-order (and in fact, all odd-order) dispersive broadening. Measurements can be more than 5 orders of magnitude more precise than would be possible via electronic timing of direct detection events, and in principle better than those performed with nonlinear autocorrelators (which rely on the same fundamental physics as down-conversion, but do not benefit from a cancellation of dispersive broadening).

77.7.2 Superluminal Tunneling Times

Another well-known problem in the theory of quantum propagation is the delay experienced by a particle as it tunnels. There are difficulties associated with calculating the "duration" of the tunneling process, since evanescent waves do not accumulate any phase [107–109]. First, the kinetic energy in the barrier region is negative, so the momentum is imaginary. Second, the transit time of a wavepacket peak through the barrier, defined in the stationary phase approximation by

$$\tau^{(\phi)} \equiv \partial[\arg t(\omega)e^{ikd}]/\partial\omega \ , \qquad (77.15)$$

tends to a constant as the barrier thickness diverges, in seeming violation of relativistic causality[3]. For example,

---

[3]Actually, it is shown in [43] that such saturation of the delay time is a natural consequence of time-reversal symmetry, and in [110] that one can deduce from the principle of causality itself that *every* system possesses a superluminal group delay, at least at the frequency where its transmission is a minimum.

the transmission function for a rectangular barrier,

$$t(k,\kappa) = \frac{e^{-ikd}}{\cosh\kappa d + i\frac{\kappa^2 - k^2}{2k\kappa}\sinh\kappa d}, \qquad (77.16)$$

leads in the opaque limit ($\kappa d \gg 1$) to a traversal time of $2m/\hbar k\kappa$, independent of the barrier width d. The same result applies to photons undergoing frustrated total internal reflection [43], when the mass m is replaced by $n^2\hbar\omega/c^2$, and similar results apply to other forms of tunneling.

Some researchers have therefore searched for some more meaningful "interaction time" for tunneling, which might accord better with relativistic intuitions and perhaps have implications for the ultimate speed of devices relying on tunneling [111]. The "semiclassical time" corresponds to treating the *magnitude* of the (imaginary) momentum as a real momentum. This time is of interest mainly because it also arises in Büttiker and Landauer's calculation of the critical timescale in problems involving oscillating barriers, which they take to imply that it is a better measure of the duration of the interaction than is the group delay [107]. The Larmor time [112] is one of the early efforts to attach a "clock" to a tunneling particle, in the form of a spin aligned perpendicular to a small magnetic field confined to the barrier region. The basic idea is that the amount of Larmor precession experienced by a transmitted particle is a measure of the time spent by that particle in the barrier. This clock turns out to contain components corresponding both to the distance-independent "dwell time" and the linear-in-distance semiclassical time. Curiously, the most common theories for tunneling times become superluminal in certain cases anyway, whether or not they deal with the motion of wave packets.

Here, we will restrict ourselves to discussing the time of appearance of a peak of a single-photon wave packet. While other tunneling-time experiments have been performed in the past [113], optical tests offer certain unique advantages [114], including the ease of construction of a barrier with no dissipation, very little energy-dependence, and a superluminal group delay. The transmitted wave packets suffer little distortion, and are essentially indistinguishable from the incident wave packets. At a theoretical level, the fact that photons are described by Maxwell's (fully relativistic) equations is an important argument against interpreting superluminal tunneling predictions as a mere artifact of the nonrelativistic Schrödinger equation. Also, one is denied the recourse suggested by some workers [115] of interpreting the superluminal appearance of transmitted peaks to mean that only the high-energy components (which, for matter waves, traveled faster even before reaching the barrier [116]) were transmitted.

Figure 77.11. Transmission probability for the tunnel barrier used in [117] (light solid curve); heavy, dashed, and dotted curves show group delay, Larmor time, and semiclassical time. $d/c = 3.6$ fs is shown for comparison.

77.7.3 Tunneling Delay in a Multilayer Dielectric Mirror

A suitable optical tunnel barrier can be a standard multilayer dielectic mirror. The alternating layers of low and high index material, each one quarter-wave thick at the design frequency of the mirror, lead to a photonic bandgap [118] analogous to that in the Kronig-Penney model of solid state physics (see Sect. 76.3.5). The gap represents a forbidden range of energies, in which the multiple reflections will interfere constructively so as to exponentially damp any incident wave. The analogy with tunneling in nonrelativistic quantum mechanics[4] arises because of the exponential decay of the field *envelope* within the periodic structure, i.e., the imaginary value of the quasimomentum. The same qualitative features arise for the transmission time: for thick barriers, it should saturate at a constant value, as was verified in a recent experiment employing short *classical* pulses [125].

The phenomenon was investigated at the single-photon level by using the high time-resolution techniques discussed in Sect. 77.7.1 to measure the relative delay experienced by down-conversion photons [117] when such a tunnel barrier (consisting of 11 layers) was introduced

---

[4]A more direct analogy, that of waveguides beyond cutoff, yielded similar results in a classical microwave experiment [119], while another paper has reported superluminal effects related to the penetration of diffracted or "leaky" microwaves into a shadow region [120]. All these experiments involve very small detection probabilities, just as in Chu and Wong's pioneering experiment on propagation within an absorption line [121] (see Sect. i67.5). However, it has been predicted that superluminal propagation could occur *without* high loss or reflection [122–124], by operating outside the resonance line of an *inverted* medium (see Sect. 68.2). One can understand the effect as off-resonance "virtual amplification" of the leading edge of a pulse.

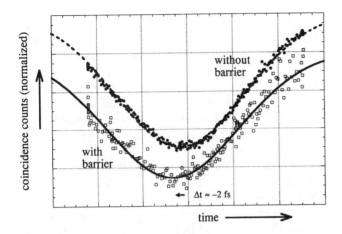

Figure 77.12. Coincidence profiles with and without the tunnel barrier map out the single-photon wave packets. The lower profile shows the coincidences with the barrier; this profile is shifted by ≈ 2 fs to *negative* times relative to the one with no barrier (upper curve): the average particle which tunnels arrives *earlier* than the one which travels the same distance in air.

into one arm of a HOM interferometer. The transmissivity of the barrier was relatively flat throughout the bandgap (extending from 600 nm to 800 nm; see Fig. 77.11), with a value of 1% at the gap center (700 nm), where the experiment was performed. The HOM coincidence dip was measured both with the barrier (and its substrate) and with the substrate alone (see Fig. 77.12). Each dip was subsequently fitted to a Gaussian, and the difference between their centers was calculated. When several such runs were combined, it was found that the tunneling peak arrived 1.47 ± 0.21 fs *earlier* than the one traveling through air, in reasonable agreement with the theoretical prediction of approximately 1.9 fs. Taking into account the 1.1-μm thickness of the barrier, this implies an effective photon tunneling velocity of $1.7c$. The results exclude the "semiclassical" time but are consistent with the group delay. A recent investigation of the energy-dependence of the tunneling time has been performed by angling the dielectric mirror, thus shifting its bandgap [126]. The data confirm the group delay in this limit as well, and rule out identification of Büttiker's Larmor time with a peak propagation time.

77.7.4 Interpretation of the Tunneling Time

Even though a wave packet peak may appear on the far side of a barrier sooner than it would under allowed propagation, it is important to stress that no information is transmitted faster than c, nor on average is any energy. These effects occur in the limit of low transmission, where

the transmitted wavepacket can be considered as a "reshaped" version of the leading edge of the incident pulse [127, 121, 128]. At a physical level, the reflection from a multilayer dielectric is due to destructive interference among coherent multiple reflections between the different layers. At times before the field inside the structure reaches its steady-state value, there is little interference, and a non-negligible fraction of the wave is transmitted. This preferential treatment of the leading edge engenders a sort of "optical illusion," shifting the transmitted peak earlier in time. A signal, such as a sharp onset, relies on high-frequency components, which would not benefit from this illusion, but instead travel slower than c. Even for a smooth wave packet, no energy travels faster than light; most is simply reflected by the barrier. Only if one considers the Copenhagen interpretation of quantum mechanics, with its instantaneous collapse, does one find superluminal propagation of *those particles* which happen to be transmitted. This leads to the question of whether it is possible to ask which part of a wave packet a given particle comes from.

One paper argued that transmitted particles do in fact stem only from the leading edge of the wave packet [129]. While it is true that the transmission only depends on causally connected portions of the incident wave packet, further analysis revealed that *simultaneous* discussion of such particle-like questions and the wave nature of tunneling ran afoul of the complementarity principle [130]. In essence, labeling the initial positions of a tunneling particle destroys the careful interference by which the reshaping occurs (as in the quantum eraser). However, one picture in which the transmitted particles really *do* originate earlier is the Bohm–de Broglie model of quantum mechanics [131, 132]. This theory considers Ψ to be a real field (residing however in configuration space, thus incorporating nonlocality) which guides pointlike particles in a deterministic manner. It reproduces all the predictions of quantum mechanics without incorporating any randomness; the probabilistic predictions of QM arise from a range of initial conditions. Bohm's equation of motion has the form of a fluid-flow equation, $\mathbf{v}(\mathbf{x}) = \hbar \nabla \arg \Psi(\mathbf{x})/m$, implying that particle trajectories may never cross, as velocity is a single-valued function of position. Consequently, all transmitted particles originate earlier in the ensemble than all reflected particles [133, 130]. This approach yields trajectories with well-defined (and generally subluminal) dwell times in the barrier region. However, the fact that the mean tunneling delay of Bohm particles diverges as the incident bandwidth becomes small, along with other interpretational issues [134, 135], leaves open the question of whether time scales as defined by the Bohm model have any physical meaning.

The "weak measurement" approach of Aharonov *et al.* [136, 137] or equivalently, complex conditional probabilities obeying Bayes's theorem [138, 139], can be used to address the question of tunneling interaction times in an experimentally unambiguous way. The real part of the resulting complex times determine the effect a tunneling particle would have on a "clock" to which it coupled, while the imaginary part indicates the clock's back-action on the particle. They unify various approaches such as the Larmor and Büttiker-Landauer times, as well as Feynman-path methods. In addition, they allow one to discuss separately the *histories* of particles which have been transmitted or reflected by a barrier, rather than discussing only the wave function as a whole. Interestingly, these calculations do *not* support the assertion that transmitted particles originate in the leading edge of a wave packet.

77.8 THE REALITY OF THE WAVE FUNCTION

77.8.1 Status of the Wave Function in Measurement

The odd behavior of tunneling wave packets pushes us to reexamine our interpretation of wave functions. We customarily use the wave function only as a calculational tool, but we have also learned that it is in some sense physical, and should not be regarded merely as some distribution from classical statistics. A recent proposal [140, 141] suggests that the wave function of a single particle should be regarded as a real entity. When a state is "protected" from change, e.g. by an energy gap, and measurements are performed sufficiently "gently", one should be able to determine not just the expectation value of position, but the wave function at many different positions, without altering the state of the particle[5]. No violation of the uncertainty principle or of the no-cloning theorem (see Sect. 77.6.2) arises from this, as the ability to "protect" a state relies on some preexisting knowledge about the state; but it assigns a deeper significance to the wave function, one Aharonov terms "ontological," as opposed to merely epistemological (but see also [143]).

77.8.2 "Interaction-Free" Measurements

The real (but undetectable) field of the Bohm model, reminiscent of the "vacuum fluctuations" of quantum field theory, can produce observable effects. This can be seen particularly well in considering a recently proposed *gedanken* experiment [144], although it can be described perfectly well by standard QM. Consider a simple single-particle interferometer, with particles injected one at a time. The path lengths are adjusted so that all the particles leave a given output port (A), and never the other (B). Now suppose that a nontransmitting object is inserted into one of the interferometer's two arms — to emphasize the result, we consider an infinitely sensitive "bomb", such that interaction with even a single photon will cause it to explode. By classical intuition, any attempt to check for the presence of the bomb involves interacting with it in some way, and by hypothesis this will inevitably set it off.

Quantum mechanics, however, allows one to be certain some fraction of the time that the bomb is in place, *without* setting it off. After the first beam splitter of the interferometer, a photon has a 50% chance of heading towards the bomb, and thus exploding it. On the other hand, if the photon takes the path without the bomb, there is no more interference, since the *non*explosion of the bomb provides *welcher Weg* ("which way") information (see Sect. 77.4). Thus the photon reaches the final beam splitter and chooses randomly between the two exit ports. Some of the time (25%), it leaves by output port B, something which never happened in the absence of the bomb. This immediately implies that the bomb (or some object) is in place — even though (since the bomb is unexploded) it has not interacted with any photon. It is the possibility that the bomb *could* have interacted with a photon which destroys interference. An experimental implementation of these ideas [145] used down-conversion to prepare the single photon states (see Sect. 77.1), and a single-photon detector as the "bomb".

How can the final state depend on what is in an arm where no particle has passed? According to the Bohm theory the real Ψ-field *does* interact with the bomb, but in the absence of the particle is incapable of exploding it. At the final detector, Ψ, which now depends on whether or not the bomb is present, "pilots" the particle accordingly. Thus the "empty" Bohm field here plays a role very similar to that of the vacuum fluctuations in the two-crystal *welcher Weg* experiment described in Sect. 77.4.2. In such a picture for this experiment, it is the vacuum (incapable of triggering the bomb) which interferes with the fields at the beam splitter [146, 147], influencing which port a photon exits. The placement of a bomb in the path of the field destroys the coherence on which interference relies[6]. The choice the photons make at the final beam splitter is thus affected not by direct action of the bomb on any photons, but by its influence on the fluctuating fields permeating the vacuum.

An improved method has recently been discovered with which one can make the fraction of interaction-free measurements arbitrarily close to 1 [145]. For example, consider a photon initially in cavity #1 of two identical cavities coupled by a lossless beam splitter

---

[5]This idea of measuring the entire wave function of a single particle is often erroneously conflated with Raymer *et al.*'s fascinating work on the reconstruction of the quantum state of a light field by repeated sampling of a large ensemble; see [142] and Sect. 75.5.1.

---

[6]Even if the bomb's trigger is of the QND sort, the number-phase uncertainty relation guarantees this.

whose reflectivity $R = cos^2(\pi/2N)$. If the photon's coherence length is shorter than the cavity length, after N cycles the photon will with certainty be located in cavity #2, due to an interference effect. However, if cavity #2 instead contains an absorbing object (e.g., the ultra-sensitive bomb), at each cycle there is only a small chance ($= 1 - R$) that the photon will be absorbed; otherwise, the *non*absorption projects the photon wave packet entirely back onto cavity #1. After all N cycles, the total probability for the photon to be absorbed by the object is $1 - R^N$, which goes to 0 as N becomes large. The photon effectively becomes trapped in cavity #1, thus indicating unambiguously the presence of the object in cavity #2.

77.9 GRAVITATIONAL ANTENNAS

According to general relativity, gravitational radiation can be produced and detected by moving mass distributions [148–150]. However, gravitational radiation is coupled only to time-varying mass *quadrupole* moments in lowest order, since the mass dipole moment is $\sum_j m_j \mathbf{r}_j = M\mathbf{R}_{cm}$ and $\mathbf{R}_{cm}$ for a closed system can only exhibit uniform rectilinear motion. Current efforts focus on detecting gravitational waves (typically at 100 Hz to 1 kHz) from astrophysical sources, such as supernovae or collapsing binary stellar systems. For example, it is expected that in the nearby Virgo cluster of galaxies, several such events should occur per year, each yielding a fractional strain ($\Delta L/L$) of 10^{-21} on Earth. However, there are large uncertainties in this estimate.

There are at present two main efforts toward gravitational radiation detection[7]. A detector of the first type is known as a resonant-mass detector, or "Weber bar" after its inventor. It utilizes a cylindrical mass, whose fundamental mode of acoustical oscillation is excited by time-varying tidal forces produced by the passage of a gravitational wave. The induced motions are typically detected by piezoelectric crystals, or by SQUIDs (superconducting quantum interference devices) [152], yielding strain sensitivities better than $10^{-18}/\sqrt{\text{Hz}}$. Such detectors have been in operation since before 1977 [150], but to date no incontrovertible detections have been reported. Attempts to improve the signal-to-noise ratio here led to the consideration of back-action-evading sensors [a special case of QND measurements; see (77.2.4)] to circumvent the standard quantum noise limit [153].

The second kind of detector utilizes a Michelson interferometer. A passing gravitational wave alters the relative path length in the interferometer arms, thereby slightly shifting the output fringes. Although the gravitational mass of the light is much smaller than that of the Weber bar, very long interferometer arms (planned as 4 km, with a Fabry-Perot cavity in each arm to increase the effective length) more than make up for this disadvantage. The US initiative, called LIGO (Laser Interferometer Gravitational-Wave Observatory) and the European version, VIRGO, are expected to have initial sensitivities of $10^{-21}/\sqrt{\text{Hz}}$ with an ultimate goal of $10^{-23}/\sqrt{\text{Hz}}$. The signal-to-noise ratio for the detection of a fringe shift depends on the power of the light; an additional external mirror will be used to recirculate the unmeasured light, thus increasing the stored light power up to 10 kW. Because the standard quantum noise limit of these detectors is ultimately determined by the vacuum fluctuations incident on the unused input ports of the interferometers, it is in principle possible to achieve reduced noise levels by using *squeezed* vacuum instead [146, 150] (Sect. 77.2.3). This has been demonstrated experimentally [154]. At present, however, serious problems involving seismic and thermal isolation, absorption and heating, intrinsic thermal noise and optical quality are being addressed. Coincidence operation between two LIGO interferometers located respectively in Hanford, Washington and Livingston, Louisiana is expected around the year 2000 [155].

Acknowledgments

This work was supported by ONR grant number N00014-90-J-1259. We would like to thank Grant McKinney for his assistance in preparing this manuscript.

REFERENCES

1. E. T. Jaynes, Phys. Rev. A **2**, 260 (1970).
2. J. F. Clauser, Phys. Rev. A **6**, 49 (1972).
3. A. O. Barut and J. P. Dowling, Phys. Rev. A **41**, 2284 (1990).
4. Peter W. Milonni, *The quantum vacuum: an introduction to quantum electrodynamics* (Academic Press, Boston, 1994).
5. P. A. M. Dirac, *The Principles of Quantum Mechanics* (Oxford University Press, London, 1958), 4th ed.
6. A. Aspect, P. Grangier, and G. Roger, J. Opt. (Paris) **20**, 119 (1989).
7. R. J. Glauber, in *Quantum Optics and Electronics*, edited by C. DeWitt *et al.* (Les Houches, Haute-Savoie, France, 1964).
8. J. F. Clauser, Phys. Rev. D **9**, 853 (1974).
9. P. Grangier, G. Roger, and A. Aspect, Europhys. Lett. **1**, 173 (1986).

---

[7]Like radio antennas, gravitational antennas must obey the principle of reciprocity, so they could in principle also be used as sources, with equal efficiencies. Currently this is infeasible, due to the small value of the coupling constant between mass and field (Newton's constant G), and the resulting miniscule efficiencies of the detection systems considered here. If a practical system overcoming these limitations were found [151], it would allow one to transmit and receive gravitational radiation, and perform a Hertz-type experiment.

10. R. Y. Chiao, P. G. Kwiat, I. H. Deutsch, and A. M. Steinberg, in *Recent Developments in Quantum Optics*, edited by R. Inguva (Plenum Press, New York, 1993), p. 145.

11. R. Hanbury-Brown and R. Q. Twiss, Proc. Roy. Soc. London **A248**, 199 (1958).

12. W. Heisenberg, *The Physical Principles of the Quantum Theory* (Dover, New York, 1930), p. 39.

13. P. R. Berman, ed., *Cavity Quantum Electrodynamics* (Adv. At. Mol. Opt. Phys., supplement 2), Academic Press, 1994.

14. C. I. Sukenik, M. G. Boshier, D. Cho, V. Sandoghdar, and E. A. Hinds, Phys. Rev. Lett. **70**, 560 (1993).

15. G. Rempe, R. J. Thompson, and H. J. Kimble, Phys. Scr. **T51**, 67 (1994).

16. D. C. Burnham and D. L. Weinberg, Phys. Rev. Lett. **25**, 84 (1970).

17. S. E. Harris, M. K. Oshman, and R. L. Byer, Phys. Rev. Lett. **18**, 732 (1967).

18. D. N. Klyshko, Pis'ma Zh. Eksp. Teor. Fiz. Sov. Phys. JETP **6**, 490 (1967); JETP Lett. **6**, 23 (1967).

19. A. Joobeur, B. Saleh, and M. Teich, Phys. Rev. A **50**, 3349 (1994).

20. P. G. Kwiat, K. Mattle, H. Weinfurter, A. Zeilinger, A. V. Sergienko, and Y. H. Shih, Phys. Rev. Lett. **75**, 4337 (1995).

21. H. J. Kimble and D. F. Walls, J. Opt. Soc. Am. B **4**, 1450 (1987).

22. J. G. Rarity, P. R. Tapster, J. A. Levenson, J. C. Garreau, I. Abram, J. Mertz, T. Debuisschert, A. Heidmann, C. Fabre, and E. Giacobino, Appl. Phys. B **55**, 250 (1992).

23. R. E. Slusher, L. W. Hollberg, B. Yurke, J. C. Mertz, and J. F. Valley, Phys. Rev. Lett. **55**, 2409 (1985).

24. A. Lambrecht, T. Coudreau, A. Steinberg, and E. Giacobino, to appear in *Coherence and Quantum Optics VII*, edited by J. Eberly, L. Mandel, and E. Wolf (Plenum Pub. Corp., 1995).

25. S. Machida, Y. Yamamoto, and Y. Itaya, Phys. Rev. Lett. **58**, 1000 (1987).

26. V. B. Braginsky, Y. I. Vorontsov, and K. S. Thorne, Science, **209**, 547 (1980).

27. J.-F. Roch, G. Roger, P. Grangier, J. Courty, and S. Reynaud, Appl. Phys. B **55**, 291 (1992).

28. N. Imoto, H. A. Haus, and Y. Yamamoto, Phys. Rev. A **32**, 2287 (1985).

29. M. Kitagawa and Y. Yamamoto, Phys. Rev. A **34**, 3974 (1986).

30. P. Grangier, J.-P. Poizat, and J.-F. Roch, Phys. Scr. **T51**, 51 (1994).

31. S. Haroche, M. Brune, and J. M. Raimond, in *Atomic Physics 13*, edited by H. Walther, T. W. Hänsch, and D. Niezart (American Institute of Physics, New York, 1993), p. 261.

32. H. Walther, in *Atomic Physics 13*, edited by H. Walther, T. W. Hänsch, and D. Niezart (American Institute of Physics, New York, 1993), p. 287.

33. R. Y. Chiao, Phys. Lett. A **33**, 177 (1970).

34. B. Lee, E. Yin, T. K. Gustafson, and R. Y. Chiao, Phys. Rev. A **45**, 4319 (1992).

35. R. Y. Chiao, P. G. Kwiat, and A. M. Steinberg, in *Workshop on Squeezed States and Uncertainty Relations, 1991*, edited by D. Han, Y. S. Kim, and W. W. Zachary (NASA Conference Publication 3135, 1991), Vol. 3135, p. 61.

36. Y. Aharonov and D. Bohm, Phys. Rev. **122**, 1649 (1961).

37. E. Schrödinger, Proc. Am. Phil. Soc. **124**, 323 (1980).

38. P. G. Kwiat and R. Y. Chiao, Phys. Rev. Lett. **66**, 588 (1991).

39. R. Ghosh and L. Mandel, Phys. Rev. Lett. **59**, 1903 (1987).

40. C. K. Hong, Z. Y. Ou, and L. Mandel, Phys. Rev. Lett. **59**, 2044 (1987).

41. P. G. Kwiat, A. M. Steinberg, and R. Y. Chiao, Phys. Rev. **A45**, 7729 (1992).

42. Z. Y. Ou and L. Mandel, Phys. Rev. Lett. **57**, 66 (1989).

43. A. M. Steinberg and R. Y. Chiao, Phys. Rev. A **49**, 3283 (1994).

44. L. Mandel, Phys. Rev. A **28**, 929 (1983).

45. H. Fearn, in *New Frontiers in Quantum Electronics and Quantum Optics*, edited by A. O. Barut (Plenum Press, New York, 1990).

46. A. M. Steinberg, P. G. Kwiat, and R. Y. Chiao, Phys. Rev. Lett. **68**, 2421 (1992).

47. N. Bohr, in *Quantum Theory and Measurement*, edited by J. A. Wheeler and W. H. Zurek (Princeton University Press, Princeton, 1983).

48. R. P. Feynman, R. B. Leighton, and M. Sands, *The Feynman Lectures on Physics* (Addison-Wesley, Reading, 1965).

49. J. von Neumann, *Mathematical Foundations of Quantum Mechanics* (Princeton University Press, 1955).

50. M. Hillery and M. O. Scully in *Quantum Optics, Experimental Gravitation, and Measurement Theory*, edited by P. Meystre and M. O. Scully (Plenum Press, New York, 1983).

51. M. O. Scully, B.-G. Englert, and H. Walther, Nature **351**, 111 (1991).

52. J. A. Wheeler, in *Problems in the Formulation of Physics*, edited by G. T. diFrancia (North-Holland, Amsterdam, 1979).

53. T. Hellmuth, H. Walther, A. Zajonc, and W. Schleich, Phys. Rev. A **35**, 2532 (1987).

54. C. O. Alley, O. G. Jakubowicz, and W. C. Wickes, in *Proceedings of the Second International Symposium on the Foundations of Quantum Mechanics, 1986*, edited by M. Namiki *et al.* (Physical Society of Japan, Tokyo, 1986), p. 36.

55. P. G. Kwiat, A. M. Steinberg, and R. Y. Chiao, Phys. Rev. A **49**, 61 (1994).

56. T. Herzog, P. G. Kwiat, H. Weinfurter, and A. Zeilinger, Phys. Rev. Lett. **75**, 3034 (1995),

57. D. M. Greenberger, M. A. Horne, and A. Zeilinger, Phys. Today **22** (1993).

58. X. Y. Zou, L. J. Wang, and L. Mandel, Phys. Rev. Lett. **67**, 318 (1991).

59. T. J. Herzog, J. G. Rarity, H. Weinfurter, and A. Zeilinger, Phys. Rev. Lett. **72**, 629 (1994).

60. H. Weinfurter, T. Herzog, P. G. Kwiat, J. G. Rarity, A. Zeilinger, and M. Zukowski, in *Fundamental Problems on Quantum Theory*, edited by D. M. Greenberger and

A. Zeilinger, Ann. N. Y. Acad. Sci., Vol. 755 (1995).

61. A. Einstein, B. Podolsky, and N. Rosen, Phys. Rev. **47**, 777 (1935).
62. D. Bohm, *Quantum Theory and Measurement* edited by J. A. Wheeler and W. H. Zurek (Princeton Univ. Press, Princeton, 1983).
63. J. S. Bell, Physics **1**, 195 (1964).
64. J. S. Bell, *Speakable and Unspeakable in Quantum Mechanics* (Cambridge Univ. Press, Cambridge, 1987).
65. H. P. Stapp, Phys. Rev. D **3**, 1303 (1971).
66. P. H. Eberhard, Nuovo Cimento B **38**, 75 (1977).
67. J. F. Clauser, M. A. Horne, A. Shimony, and R. A. Holt, Phys. Rev. Lett. **23**, 880 (1969).
68. J. F. Clauser and M. A. Horne, Phys. Rev. D **10**, 526 (1974).
69. P. H. Eberhard, Phys. Rev. A **47**, R747 (1993).
70. J. F. Clauser and A. Shimony, Rep. Prog. Phys. **41**, 1881 (1978).
71. P. G. Kwiat, P. H. Eberhard, A. M. Steinberg, and R. Y. Chiao, Phys. Rev. A **49**, 3209 (1994).
72. P. G. Kwiat, A. M. Steinberg, R. Y. Chiao, P. H. Eberhard, and M. D. Petroff, Appl. Opt. **33**, 1844 (1994).
73. P. H. Eberhard, P. G. Kwiat, M. D. Petroff, M. G. Stapelbroek, and H. H. Hogue, in *Proceedings of the IEEE International Conference on Applications of Photonics Technology (ICAPT), 1994*, edited by C. P. Wang (STS Press, McLean, Virginia, 1995).
74. E. S. Fry, T. Walther, and S. Li, Phys. Rev. A **52**, 4381 (1995).
75. S. J. Freedman and J. F. Clauser, Phys. Rev. Lett. **28**, 938 (1972).
76. A. Aspect, J. Dalibard, and G. Roger, Phys. Rev. Lett. **49**, 1804 (1982).
77. Y. H. Shih and C. O. Alley, Phys. Rev. Lett. **61**, 2921 (1988).
78. Z. Y. Ou and L. Mandel, Phys. Rev. Lett. **61**, 50 (1988).
79. T. E. Kiess, Y. H. Shih, A. V. Sergienko, and C. O. Alley, Phys. Rev. Lett. **71**, 3893 (1993).
80. J. G. Rarity and P. R. Tapster, Phys. Rev. Lett. **64**, 2495 (1990).
81. J. Brendel, E. Mohler, and W. Martienssen, Europhys. Lett. **20**, 575 (1992).
82. P. G. Kwiat, A. M. Steinberg, and R. Y. Chiao, Phys. Rev. A **47**, R2472 (1993).
83. J. D. Franson, Phys. Rev. Lett. **62**, 2205 (1989).
84. J. Jarrett, Nôus **18**, 569 (1984).
85. A. Shimony, *An Exposition of Bell's Theorem* (Plenum Press, New York, 1990).
86. N. D. Mermin, Phys. Rev. Lett. **65**, 1838 (1990).
87. D. M. Greenberger, M. A. Horne, A. Shimony, and A. Zeilinger, Am. J. Phys. **58**, 1131 (1990).
88. L. Hardy, Phys. Rev. Lett. **68**, 2981 (1992).
89. L. Hardy, Phys. Lett. A **167**, 17 (1992).
90. J. Torgerson, D. Branning, and L. Mandel, Appl. Phys. B **60**, 267 (1995).
91. R. Glauber, Ann. N. Y. Acad. Sci. **480**, 336 (1986).
92. W. K. Wootters and W. H. Zurek, Nature **299**, 802 (1982).
93. C. Bennett and S. J. Wiesner, Phys. Rev. Lett. **69**, 2881 (1992).

94. S. L. Braunstein, A. Mann, and M. Revzen, Phys. Rev. Lett. **68**, 3259 (1992).
95. C. H. Bennett, G. Brassard, C. Crepeau, R. Jozsa, A. Peres, and W. K. Wooters, Phys. Rev. Lett. **70**, 1895 (1993).
96. *Modern Cryptology: A Tutorial*, edited by G. Brassard (Springer-Verlag, New York, 1988).
97. A. K. Ekert, Phys. Rev. Lett. **67**, 661 (1991).
98. C. H. Bennett, Phys. Rev. Lett. **68**, 3121 (1992).
99. G. Barton and K. Scharnhorst, J. Phys. A **26**, 2037 (1993).
100. G. C. Hegerfeldt, Phys. Rev. Lett. **54**, 2395 (1985).
101. V. P. Bykov and V. I. Tatarskii, Phys. Lett. A **136**, 77 (1989).
102. S.-T. Ho, *OSA Annual Meeting Technical Digest, 1993* (Optical Sociaty of America, Washington D. C., 1993), Vol. 16, p. 258.
103. P. H. Eberhard and R. R. Ross, Found. Phys. Lett. **2**, 127 (1989).
104. R. J. Glauber and M. Lewenstein, Phys. Rev. A **43**, 467 (1991).
105. J. D. Franson, Phys. Rev. A **45**, 3126 (1992).
106. A. M. Steinberg, P. G. Kwiat, and R. Y. Chiao, Phys. Rev. A **45**, 6659 (1992).
107. M. Büttiker and R. Landauer, Phys. Rev. Lett. **49**, 1739 (1982).
108. E. H. Hauge and J. A. Støvneng, Rev. Mod. Phys. **61**, 917 (1989).
109. E. P. Wigner, Phys. Rev. **98**, 145 (1955).
110. E. L. Bolda, R. Y. Chiao, and J. C. Garrison, Phys. Rev. A **48**, 3890 (1993).
111. M. Büttiker and R. Landauer, Phys. Scr. **32**, 429 (1985).
112. M. Büttiker, Phys. Rev. B **27**, 6178 (1983).
113. R. Landauer, Nature **341**, 567 (1989).
114. T. Martin and R. Landauer, Phys. Rev. A **45**, 2611 (1992).
115. R. S. Dumont and T. L. Marchioro II, Phys. Rev. A **47**, 85 (1993).
116. E. H. Hauge, J. P. Falck, and T. A. Fjeldly, Phys. Rev. B **36**, 4203 (1987).
117. A. M. Steinberg, P. G. Kwiat, and R. Y. Chiao, Phys. Rev. Lett. **71**, 708 (1993).
118. E. Yablonovitch, J. Opt. Soc. Am. B **10** (2), 283 (1993), and other articles in this special issue on photonic bandgaps.
119. A. Enders and G. Nimtz, J. Phys. (Paris) **3**, 1089 (1993).
120. A. Ranfagni, P. Fabeni, G. P. Pazzi, and D. Mugnai, Phys. Rev. E **48**, 1453 (1993).
121. S. Chu and S. Wong, Phys. Rev. Lett. **48**, 738 (1982).
122. R. Y. Chiao, Phys. Rev. A **48**, R34 (1993).
123. A. M. Steinberg and R. Y. Chiao, Phys. Rev. A **49**, 2071 (1994).
124. E. L. Bolda, J. C. Garrison, and R. Y. Chiao, Phys. Rev. A **49**, 2938 (1994).
125. Ch. Spielmann, R. Szipöcs, A. Stingl, and F. Krausz, Phys. Rev. Lett. **73**, 2308 (1994).
126. A. M. Steinberg and R. Y. Chiao, Phys. Rev. A **51**, 3525 (1995).
127. C. G. B. Garrett and D. E. McCumber, Phys. Rev. A **1**, 305 (1970).
128. A. E. Siegman, *Lasers* (University Science Books, Mill

Valley, 1986).

129. J. M. Deutch and F. E. Low, Ann. Phys. **228**, 184 (1993).

130. A. M. Steinberg, P. G. Kwiat, and R. Y. Chiao, Found. Phys. Lett. **7**, 223 (1994).

131. D. Bohm and B. J. Hiley, *The Undivided Universe: An Ontological Interpretation of Quantum Mechanics* (Routledge, London, 1993).

132. P. R. Holland, *The Quantum Theory of Motion* (Cambridge Univ. Press, Cambridge, 1993).

133. C. R. Leavens and G. C. Aers, in *Scanning Tunneling Microscopy III*, edited by R. Wiesendanger and H.-J. Güntherodt (Springer-Verlag, Berlin, 1993).

134. B.-G. Englert, M. O. Scully, G. Süssmann, and H. Walther, Z. Naturforsch. A **47**, 1175 (1992).

135. C. Dewdney, L. Hardy, and E. J. Squires, Phys. Lett. A **184**, 6 (1993).

136. Y. Aharonov and L. Vaidman, Phys. Rev. A **41**, 11 (1990).

137. Y. Aharonov, D. Z. Albert, and L. Vaidman, Phys. Rev. Lett. **60**, 1351 (1988).

138. A. M. Steinberg, Phys. Rev. A **52**, 32 (1995).

139. A. M. Steinberg, Phys. Rev. Lett. **74**, 2405 (1995).

140. Y. Aharonov and J. Anandan, Phys. Rev. A **47**, 4616 (1993).

141. Y. Aharonov and L. Vaidman, Phys. Lett. A **178**, 38 (1993).

142. D. T. Smithey, M. Beck, M. G. Raymer, and A. Faridani, Phys. Rev. Lett. **70**, 1244 (1993).

143. W. G. Unruh, Phys. Rev. A **50**, 882 (1994).

144. A. C. Elitzur and L. Vaidman, Found. Phys. **23**, 987 (1993).

145. P. G. Kwiat, H. Weinfurter, T. Herzog, A. Zeilinger, and M. A. Kasevich, Phys. Rev. Lett. **74**, 4763 (1995).

146. C. M. Caves, Phys. Rev. D **23**, 1693 (1981).

147. C. M. Caves, Opt. Lett. **12**, 971 (1987).

148. S. Weinberg, *Gravitation and Cosmology* (Wiley, New York, 1972).

149. C. W. Misner, K. S. Thorne, and J. A. Wheeler *Gravitation* (Freeman, San Francisco, 1987).

150. K. S. Thorne, in *Three Hundred Years of Gravitation*, edited by S. W. Hawking and W. Israel (Cambridge Univ. Press, Cambridge, 1987), p. 400.

151. R. Y. Chiao, J. Boyce, and J. C. Garrison, in *Proceedings of the Conference "Fundamental Problems in Quantum Theory" in Honor of J. A. Wheeler*, edited by D. M. Greenberger and A. Zeilinger, Ann. N. Y. Acad. of Sci. **755**, 400 (1995).

152. P. F. Michelson, J. C. Price, and R. C. Taber, Science **237**, 150 (1987).

153. C. M. Caves, K. S. Thorne, W. P. Drever, V. D. Sandberg, and M. Zimmerman, Rev. Mod. Phys. **52**, 341 (1980).

154. M. Xiao, L.-A. Wu, and H. J. Kimble, Phys. Rev. Lett. **59**, 278 (1987).

155. R. E. Spero and S. E. Whitcomb, Opt. and Photon. News **6**, 35 (1995).

78

Applications of Atomic and Molecular Physics to Astrophysics

A. Dalgarno* and S. Lepp[†]

*Harvard-Smithsonian Center for Astrophysics, 60 Garden Street, Cambridge, Massachusetts 02138
[†]Department of Physics, University of Nevada, Las Vegas, 4505 Maryland Pkwy., Las Vegas, Nevada 89154

78.1 INTRODUCTION

Almost all our information about the Universe reaches us in the form of photons. Observational astronomy is based on measurements of the distribution in frequency and intensity of the photons that are emitted by astronomical objects and detected by instrumentation on ground-based and space-borne telescopes. Information about the earliest stages in the evolution of the Universe before galaxies and stars had formed is carried to us by blackbody background photons that attended the beginning of the Universe. The photons that are the signatures of astronomical phenomena are the result of many processes of nuclear physics, plasma physics and atomic, molecular and optical physics. The processes that modify the photons on their journey from distant origins through intergalactic and interstellar space to the Earth belong mostly to the domain of atomic, molecular, and optical physics, as do the instruments that detect and measure the arriving photons and their spectral distribution. The spectra are used to classify galaxies and stars and to identify the astronomical entities and phenomena such as quasars, active galactic nuclei, gravitational lensing, jets and outflows, pulsars, supernovae, novae, supernova remnants, nebulae, masers, protostars, shocks, molecular clouds, circumstellar shells, accretion disks and black holes.

Quantitative analyses of the spectra of astronomical sources of photons and of the atomic, molecular, and optical processes that populate the atomic and molecular energy levels and give rise to the observed absorption and emission require accurate data on transition frequencies and wavelengths, oscillator strengths, cross sections for electron impact, rate coefficients for radiative, dielectronic and dissociative recombination, and cross sections for heavy particle collisions involving charge transfer, excitation, ionization, dissociation, fine structure, and hyperfine structure transitions, collision-induced absorption and line broadening. Data on radiative association and ion-molecule and neutral particle reaction rate coefficients are central to the interpretation of measurements of chemical composition in molecular clouds, circumstellar shells and supernova ejecta.

The range of physical conditions of density, temperature, and radiation fields encountered in astrophysical environments is extreme and can rarely be reproduced in a laboratory setting. It is not only reliable data on known processes that are needed but also a deep understanding so that the relevant processes can be identified and the influence of the conditions in which they occur fully taken into account.

We present here a summary of the processes that take place in photoionized gas, collisionally ionized gas, the diffuse interstellar medium, molecular clouds, circumstellar shells, supernova ejecta, shocked regions and the early Universe.

78.2 PHOTOIONIZED GAS

The Universe contains copious sources of energetic photons most often in the form of hot stars, and much of the material of the Universe exists as photoionized gas. Photoionized gas produces the visible emission from emission nebulae, planetary nebulae, nova shells, starburst galaxies and probably active galactic nuclei [2].

Emission nebulae are extended regions of luminosity in the sky. They arise from the absorption of stellar radiation by the gas surrounding one or more hot stars.The gas is ionized by the photons and excited and heated by the electrons released in the photoionizing events. A succession of ionization zones is created in which highly ionized regions give way to less ionized gas with increasing distance from the central star as the photon flux is diminished by geometrical dilution and by absorption. The outer edge of a nebula is a front of ionization pushing out into the neutral interstellar gas. The densities are typically between 100 and 10 000 cm^{-3} and the temperatures between 5000 and 15 000 K. Nebulae are also called H$\textsc{ii}$ regions. At low densities, the luminosity is low, but the ionized regions can still be detected by radio observations.

Planetary nebulae are smaller in extent and more dense. They have a passing similarity in appearance to planets. Planetary nebulae are produced by the photoionization of shells of gas that have been ejected from the parent star as it evolved to its final white dwarf stage. Because the core of the parent star is very hot the irradiated gas is more highly ionized than are emisssion nebulae and has a distinctive spectrum.

Photoionized gas is also found around novae. Novae are stars that have undergone spasmodic outbursts and they are surrounded by faint shells of ejected gas, photoionized and excited by the stellar radiation. Some supernova remnants, which are what remains after a massive star has exploded, have spectra that also appear

to be emanating from photoionized gas. The source of ionization may be synchrotron radiation. Figure 78.1 shows the x-ray emission spectrum of a supernova remnant.

The nuclei of starburst galaxies have spectra like those of emission nebulae. They result from gas photoionized by radiation from hot stars created in a period of rapid star formation. Active galactic nuclei, such as quasars, have a different spectrum characterized by broad lines indicating a large range of velocities. Photoionized gas is the most likely interpretation.The ionizing source may be an accretion disk around a compact object such as a black hole.

The ionization structure in a photoionized gas is determined by a balance of photoionization

$$X^{(m-1)+} + h\nu \rightarrow X^{m+} + e^- \tag{78.1}$$

and radiative

$$X^{m+} + e^- \rightarrow X^{(m-1)+} + h\nu \tag{78.2}$$

and dielectronic

$$X^{m+} + e^- \rightarrow (X^{(m-1)+})^* \rightarrow X^{(m-1)+} + h\nu \tag{78.3}$$

recombination, and in plasmas with a significant population of neutral hydrogen and helium, by charge transfer recombination

$$X^{m+} + \text{H} \rightarrow X^{(m-1)+} + \text{H}^+ \tag{78.4}$$

$$X^{m+} + \text{He} \rightarrow X^{(m-1)+} + \text{He}^+ . \tag{78.5}$$

Many detailed calculations of the ionization structure of photoionized regions have been carried out [2].

The ionizing source spectra of hot stars can be obtained from calculations of stellar atmospheres (Sect. 78.6). Approximate values of cross sections for photoionization for a wide range of atomic and ionic systems in many stages of ionization are available [3–6]. Calculations of higher precision and reliability that incorporate the contributions from autoionizing resonance structures exist for specific systems [7]. They are undergoing continual improvements as increasingly powerful computational techniques are brought to bear on the calculations.

The cross sections for radiative recombination are obtained by summing the cross sections for capture into the ground and excited states of the recombining system. Because of the contribution from highly excited states which are nearly hydrogenic, the rate coefficients are similar for different ions of the same excess nuclear charge. They vary slowly with temperature. In contrast, dielectronic recombination is a specific process whose efficiency depends on the energy level positions of the resonant states. For nebular temperatures, the rate coefficients vary exponentially with temperature. Explicit calculations have been carried out for many ionic systems [8–10]. Because

Figure 78.1. X-ray spectrum of the supernova remnant Puppis A as observed by the Einstein satellite. Note the high level of ionization with hydrogen-like ions of oxygen and neon, suggesting a high temperature. This figure is from Ref. [1].

the photoionization cross sections of the major cosmic gases hydrogen and helium diminish rapidly at high frequencies, multiply charged ions and neutral gas coexist in cosmic plasmas produced by energetic photons and charge transfer recombination may control the ionization structure. For multiply charged ions with excess charge greater than two, charge transfer is rapid. For doubly charged and singly charged ions, the cross sections are sensitive to the details of the potential energy curves of the quasimolecule formed in the approach of the ion and the neutral particle. Few reliable data exist. Some recent calculations may be found in the papers [11–13].

Photoionized gas is heated by collisions of the energetic photoelectrons and cooled by electron impact excitation of metastable levels, principally of O^+ and O^{++}, N^+ and N^{++}, and S^+ and S^{++}, followed by emission of photons which escape from the nebula. Considerable attention has been given to the determination of the rate coefficients [14]. The resulting cooling rates increase exponentially with temperature and keep the temperature of the gas between narrow limits. Some contribution to cooling occurs from recombination and from free-free emission by electrons moving in the field of the positive ions.

The luminosity of the photoionized gas comes from the photons emitted in the cooling processes and from radiative and dielectronic recombination. The radiative recombination spectrum of hydrogen extends from the Ly α line at 121.6 nm to radio lines at meter wavelengths.

The recombination spectrum can be predicted to high accuracy, and calculations for a wide range of temperature, density, and radiation environments have been carried out for diagnostic purposes [15–18]. Electron impact and proton impact induced transitions are important in determining the energy level populations and the resulting spectrum. Stimulated emission often affects the intensities of the radio lines, especially those from extragalactic sources. Comparisons of the predicted intensities in the visible and infrared with theoretical predictions yield information on interstellar extinction in the nebula and along the line of sight.

The relative intensities of the lines emitted by different metastable levels depend exponentially on the temperature. The relative intensities of the lines at 500.7 nm and 436.3 nm originating in the $^1D_2$ and $^1S_0$ levels of O^{++} vary as $\exp(33170/T)$, and are commonly used to derive the temperature T.

The electron density can be inferred from the lines emitted from neighboring levels with different radiative lifetimes for which there occurs a competition between spontaneous emission and quenching by electron impact. There are many possible combinations of lines. The lines at 372.89 nm and 372.62 nm emitted by the $^2D_{3/2}$ and $^2D_{5/2}$ levels of N^+ are readily observable and their relative intensity yields the electron density.

Radiative and dielectronic recombination lines are often seen in the spectra, as are a few lines due to charge transfer recombination. Fluorescence of starlight and resonance fluorescence of lines emitted in the nebula (called Bowen fluorescence by astronomers) also contribute to the spectra of photoionized gases. Many data are needed to adequately interpret the observations.

78.3   COLLISIONALLY IONIZED GAS

Hot gas is found in the coronae of stars and particularly the Sun, and in young supernova remnants, in the hot phase of the interstellar medium, and in intergalactic space. In a hot gas the ionization is produced by the impact ionization of the fast thermal electrons and recombination is radiative and dielectronic [20]. The rate coefficients for electron impact ionization and for recombination for any given ionization stage are functions only of temperature, and hence so is the resulting ionization distribution. When ionization and recombination balance, coronal equilibrium is attained in which the ionization structure is specified by the temperature.

Recombination at high temperatures is dominated by dielectronic recombination. At high temperatures, dielectronic recombination is stabilized by transitions in which the core electrons are the active electrons. The associated emission lines lie close in frequency to that of the resonant transition of the parent ion. They are called satellite lines. Together with lines generated by electron impact excitation, they provide a powerful diagnostic probe of density and temperature. In many circumstances such as in supernova remnants, coronal equilibrium does not hold, and the ionization and recombination must be followed as functions of time. The temperature also evolves as the hot plasma is cooled by electron impact excitation and ionization.

The recombining gas produces x-rays and extreme uv radiation which modify the ionization structure. There is a particular need for more reliable data on high energy photoionization cross sections, on collision cross sections for electrons and positive ions, and on the energy levels and transition probabilities of highly stripped complex ions. Figure 78.2 shows the emissivity of coronal gas.

78.4   DIFFUSE MOLECULAR CLOUDS

Diffuse molecular clouds are intermediate between the hot phase of the galaxy and the giant molecular clouds where much of the gas resides. They are called diffuse because they have optical depths of order unity, so photons can penetrate from outside the cloud and affect the chemical composition. The atoms and molecules are observed in absorption against background stars. Translucent clouds with optical depths between about 2 and 5 are intermediate between diffuse and dark clouds where photons from the outside still affect the chemistry.

They can be observed both in absorption against a background source or in emission in the radio.

The temperature is 100–200 K at the edges of a diffuse cloud with a density of about 100 cm^{-3}. In a typical diffuse cloud the temperature decreases to about 30 K at the center while the density increases to about 300–800 cm^{-3}. The chemistry is driven by ionization from interstellar uv photons and from cosmic rays.

Interstellar uv photons ionize species which have ionization potentials less then that of atomic hydrogen. Atomic hydrogen is so pervasive in the galaxy that uv photons with energies higher than 13.6 eV are absorbed near the source. The uv flux is a very important parameter in determining the composition of a diffuse cloud. Photodissociation provides destruction which limits the buildup of more complex species and so diffuse clouds are dominated by simpler diatomic species.

Species with ionization potentials greater then hydrogen are mainly ionized by cosmic rays. Cosmic rays are high energy nuclei which stream through the galaxy. The cosmic ray ionization rate, the number of cosmic ray ionizations per second per particle, is an important parameter in interstellar chemistry. A lower limit to the cosmic ray ionization rate may be set by measured high energy cosmic rays reaching earth, giving an ionization rate of $\approx 10^{-17}$ s^{-1}. More realistic estimates of the cosmic ray ionization rate from looking at recombination lines suggest values of a few $\times 10^{-17}$ s^{-1}.

The hydroxyl radical OH is produced in a manner similar to that discussed below in Sect. 78.5, and removed by photodissociation. Thus in diffuse clouds, OH may be used to measure the cosmic ray ionization rate, subject to uncertainties in the OH photodissociation rate and the H_3^+ recombination rate. The OH abundances give rates of several $\times 10^{-17}$ s^{-1} for many diffuse clouds.

The carbon chemistry begins with the ionization of C by uv photons:

$$C + h\nu \rightarrow C^+ + e^- . \qquad (78.6)$$

The carbon ion cannot react directly with H_2 by

$$C^+ + H_2 \rightarrow CH^+ + H , \qquad (78.7)$$

as this reaction is exothermic by 0.4 eV. Instead, the chemistry proceeds by the slow radiative association process

$$C^+ + H_2 \rightarrow CH_2^+ + h\nu . \qquad (78.8)$$

The CH_2^+ ion may either dissociatively recombine

$$CH_2^+ + e^- \rightarrow CH + H , \qquad (78.9)$$

or react with molecular hydrogen

$$CH_2^+ + H_2 \rightarrow CH_3^+ + H . \qquad (78.10)$$

The CH_3^+ then undergoes dissociative recombination

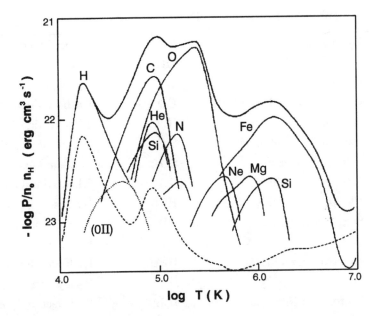

Figure 78.2. Total emissivity and emissivity by element as a function of temperature in coronal equilibrium. The heavy solid curve is total emissivity and the lighter lines are contributions from individual elements. This figure is from Ref. [19].

$$CH_3^+ + e \rightarrow C + H_2 + H \qquad (78.11)$$
$$\rightarrow CH + H + H \qquad (78.12)$$
$$\rightarrow CH_2 + H \qquad (78.13)$$
$$\rightarrow CH + H_2 , \qquad (78.14)$$

where the products are listed in order of decreasing likelihood. The CH is removed by photodissociation

$$CH + h\nu \rightarrow C + H \qquad (78.15)$$

and by photoionization

$$CH + h\nu \rightarrow CH^+ + e^- . \qquad (78.16)$$

CH may also be removed by reactions with oxygen or nitrogen atoms to form CO and CN respectively.

One of the outstanding problems in diffuse clouds is to understand the large abundance of CH$^+$ relative to CH. The problem is producing the CH$^+$ without producing additional CH. Since most reaction paths go through reaction (78.7), this is the most likely candidate. What is needed is some extra energy to overcome the endothermicity. This energy must come from either hot C$^+$ or from hot or vibrationally excited H$_2$. The most popular model is gas heated by a shock, possibly a magnetic shock in which ions stream relative to the neutrals, giving a high effective energy. Unfortunately, though these shock models can reproduce the CH$^+$ abundances, they also predict relative velocities between the CH$^+$ and CH which are not often observed. Recently there have been suggestions that turbulence in the cloud could account for the CH$^+$ abundance.

The most comprehensive models of diffuse clouds are by van Dishoeck and Black [21]. A collection of photodissociation rates and photoionization rates is given in Roberge *et al.* [22].

The uv flux is predominantly from stars and may be as much as 10^5 times larger near an HII region (Sect. 78.2) than it is in the general interstellar medium. Regions in which the chemistry is dominated by photons are referred to as *photon dominated regions* or *photodissociation regions* (PDR's). In the presence of high uv flux the cloud is much warmer than in a typical diffuse cloud. Temperatures may reach 1000 K near the edge of the cloud and 100 K far into the cloud. The chemistry differs from traditional diffuse cloud chemistry in that the high temperatures allow endothermic reactions to proceed. Sternberg and Dalgarno [23] have published a comprehensive model of photodominated regions.

78.5 DARK MOLECULAR CLOUDS

Much of the mass of the galaxy is in the form of dark molecular clouds. The molecular clouds are sites of forming new stars. They are composed primarily of hydrogen, with about 10% helium and trace amounts of heavier elements. They have densities of approximately 10^3 or 10^4 cm^{-3} and temperatures between 10 and 20 K, and often contain denser clumps. The clouds are optically thick and so photons from the outside are absorbed on the surface of the clouds. The interiors are heated and ionized by cosmic rays which penetrate deep

Table 78.1. Molecules observed in interstellar clouds.

H_2	Hydrogen	CH	Methylidyne
CH^+	Methylidyne Ion	OH	Hydroxyl
C_2	Carbon	CN	Cyanogen
CO	Carbon Monoxide	CO^+	Carbon Monoxide Ion
NH	Amidogen	NO	Nitric Oxide
CS	Carbon Monosulphide	SiO	Silicon Monoxide
SO	Sulphur Monoxide	SO^+	Sulphur Monoxide Ion
NS	Nitrogen Sulphide	SiS	Silicon Sulphide
PN	Phosphorus Nitride	HCl	Hydrogen Chloride
SiN	Silicon Nitride	NH_2	Amino Radical
H_2O	Water	C_2H	Ethynyl
HCN	Hydrogen Cyanide	HNC	Hydrogen Isocyanide
HCO	Formyl	HCO^+	Formyl Ion
N_2H^+	Protonated Nitrogen	H_2S	Hydrogen Sulphide
HNO	Nitroxyl	OCS	Carbonyl Sulphide
SO_2	Sulphur Dioxide	HCS^+	Thioformyl Ion
C_2O	Carbon Suboxide	C_2S	Dicarbon Sulphide
N_2O	Nitrous Oxide	H_2CN	Methylene Amidogen
H_2CO	Formaldehyde	H_2CS	Thioformaldehyde
NH_3	Ammonia	HCNS	Isothiocyanic Acid
HNCO	Isocyanic Acid	$HOCO^+$	Protonated Carbon Dioxide
C_3H	Propynylidyne	C_3N	Cyanoethynyl
C_3S	Tricarbon Sulphide	C_3O	Tricarbon Monoxide
C_2H_2	Acetylene	H_3O^+	Hydronium Ion
$HCNH^+$	Protonated Hydrogen Cyanide	C_3H_2	Cyclopropenylidene
CH_4	Methane	H_2CCC	Propadienylidene
HCOOH	Formic Acid	CH_2CO	Ketene
HC_3N	Cyanoacetylene	HNCCC	Cyanoacetylene Isomer
HCCNC	Ethynyl Isocyanide	C_4H	Butadinyl
NH_2CH	Cyanamide	CH_2CN	Cyanomethyl Radical
CH_2NH	Methanimine	CH_3CH	Methyl Cyanide
H_2CCCC	Butatrienylidene	CH_3SH	Methyl Mercaptan
C_5H	Pentynylidyne	HCC_2HO	Propynal
CH_3OH	Methyl Alcohol	HC_3NH^+	Protonated Cyanoacetylene
NH_2CHO	Formamide	CH_3C_2H	Methyl Acetylene
CH_2CHCN	Vinyl Cyanide	HC_5N	Cyanodiacetylene
C_6H	Hexatrinyl	CH_3NH_2	Methylamine
CH_3CHO	Acetaldehyde	$HCOOCH_3$	Methyl Formate
CH_3C_3N	Methyl Cyanoacetylene	CH_3C_4H	Methyl Diacetylene
$(CH_3)_2O$	Dimethyl Ether	CH_3CH_2CN	Ethyl Cyanide
HC_7N	Cyanohexatriyne	CH_3CH_2OH	Ethyl Alcohol
HC_9N	Cyano-octatetra-yne	$HC_{11}N$	Cyano-decapenta-yne

into the cloud.

The temperatures are too low to sustain much neutral chemical activity in the clouds, and cosmic ray ionization is important in driving the chemistry. In dense clouds, the cosmic rays both initiate the chemistry and limit it through the production of He^+ and through cosmic ray induced photons.

Table 78.1 is a list of molecules that have been observed in the interstellar medium, many of which have also been found in other galaxies. It is likely that all but H_2 are formed in the gas phase by ion-molecule reaction sequences initiated by cosmic ray ionization. The fact that isomers such as HCN and HNC are seen in approximately equal abundances suggests a low density gas phase environment. Reactions on surfaces and the formation of grains are not well understood, but are surely important.

The chemistry of molecular clouds is dominated by ion-molecule reactions driven by cosmic ray ionization. The cosmic rays primarily ionize H_2:

$$H_2 \rightarrow H_2^+ + e^- , \qquad (78.17)$$

producing both H_2^+ and fast electrons. The fast electrons produce additional ionizations. The H_2^+ quickly reacts with H_2 to form H_3^+

$$H_2^+ + H_2 \rightarrow H_3^+ + H . \qquad (78.18)$$

The H_3^+ reacts with other species by proton transfer, which then drives much of the interstellar chemistry.

As an example of the production of more complex molecules in interstellar chemistry, we examine the reaction networks leading to the production of water H_2O and the hydroxl radical OH. The H_3^+ ions formed by cosmic ray ionization react with atomic oxygen to form OH^+

$$O + H_3^+ \rightarrow OH^+ + H_2 , \qquad (78.19)$$

which quickly reacts with H_2 to form H_3O^+ in an abstraction sequence

$$OH^+ + H_2 \rightarrow H_2O^+ + H , \qquad (78.20)$$
$$H_2O^+ + H_2 \rightarrow H_3O^+ + H . \qquad (78.21)$$

The H_3O^+ then undergoes dissociative recombination to form water and OH

$$H_3O^+ + e^- \rightarrow H_2O + H , \qquad (78.22)$$
$$\rightarrow OH + H_2 . \qquad (78.23)$$

The water is removed by reactions with neutral or ionized carbon, which eventually lead to the production of CO. OH is primarily removed by reactions with atomic oxygen leading to O_2. The CO and O_2 are removed by reactions with He^+. The He^+, generated by cosmic ray ionization of helium, does not react with H_2 and so is available to remove species by reactions such as

$$CO + He^+ \rightarrow C^+ + O + He . \qquad (78.24)$$

Water and OH are also removed by uv photons generated within the cloud. The clouds are too thick for external uv photons to penetrate, but cosmic rays excite H_2 into electronically excited states which decay through emission of uv photons. These internally generated photons play an important role in determining the composition of the cloud. Gredel *et al.* [24] have compiled a list of the photodissociation and photoionization rates for cosmic ray induced photons.

Modern chemical networks for molecular clouds include several hundred species and several thousand reactions. A standard set of reaction rates is provided by the UMIST (University of Manchester Institute of Science and Technology) dataset [25, 26]. The dataset may be obtained from the UMIST Astrophysics Group homepage (http://saturn.ma.umist.ac.uk:8000/).

The clouds also contain dust particles as evidenced by the extinction curves for clouds and the observed depletions of heavier elements. The importance of surface chemistry on these dust particles to interstellar clouds is still uncertain. Dust particles are the best candidate to be the site of formation of molecular hydrogen, because known gas phase reactions fail to produce H_2 in the quantities observed.

78.6 CIRCUMSTELLAR SHELLS AND STELLAR ATMOSPHERES

The continuum emission from a star is very nearly that of a blackbody. This emission is then absorbed and redistributed by the atmosphere of the star. The spectrum of the star is thus determined by its atmosphere. In the hottest stars, most of the material is ionized and the absorption lines are predominantly those of ions, while in the coldest stars, molecular lines are prominent. Kurucz has calculated models with continuum spectra and the inclusion of a large number of absorption lines [27]. There are two major projects for calculating the required atomic data. In 1984 an international collaboration named the *Opacity Project* was set up to calculate accurate atomic data needed for opacity calculations [28, 29]. The other earlier project is called OPAL. The two sources of data are compared in Refs. [30, 31].

Low and intermediate mass stars eject circumstellar envelopes in their red giant phase near the end of their evolution. Circumstellar envelopes are an important part of astronomy and they are a likely location for dust formation. They provide an interesting environment for studying molecules because they represent a transition between very high density stellar atmosphere environments to low density interstellar environments. These objects evolve to become planetary nebulae (see Sect. 78.2).

We are fortunate in having one example, IRC 10216, which is very close to the Sun. The brightest 10 μm source beyond the solar system, IRC 10216 is a carbon-rich star surrounded by dust and gas it ejected in a strong stellar wind. The central star is so shielded that it is almost undetectable at optical wavelengths, and was not discovered until the 2 μm survey. IRC 10216 is where most of the circumstellar molecules are detected, and has greatly increased our understanding of circumstellar envelopes.

The envelopes are ejected by the red giant in its final phase of evolution. The mass loss rates increase to $\sim 10^{-4}$ solar masses per year and temperatures in the envelope are of order 1000 K. Close to the star, the density is high and the chemistry is characteristic of thermal equilibrium. The situation is quite unlike any interstellar environments. For example in IRC 10216, the HNC is over one hundred times less abundant than HCN, whereas in molecular clouds, they have about the same abundance.

If the star is oxygen-rich, large amounts of H_2O are formed and if it is carbon-rich, C_2H_2. This high temperature environment forms both molecules and grains. The high obscuration of the central source indicates that grains are formed in these envelopes. Polycyclic aromatic hydrocarbons (PAH's) or some similar species are observed in carbon rich planetary nebulae. These large molecules must have been produced when the object was a carbon rich circumstellar envelope.

As the material flows out from the star the density and temperature decrease. As the density becomes lower, three body reactions become less important and at some point the products of these reactions are frozen out in a similar manner to the evolution of molecules in the early universe (see Sect. 78.9). In the outermost portions of the circumstellar envelope, molecules are dissociated by interstellar uv photons. The penetrating uv radiation is shielded by dust, H_2, and CO. The relative abundances can vary rapidly with radius, and observations provide abundance and radial distribution, a wealth of data for modelers. Circumstellar chemistry is reviewed by Omont [32] and recent chemical models are given in Refs. [33–35].

78.7 SUPERNOVA EJECTA

A supernova, the explosion of a massive star following core collapse, is one of the most spectacular displays in the Universe. The explosion occurs when the iron core of a massive star collapses to form a neutron star and the rebound shock and neutrino flux eject the outer portion of the star. The ejected portions of the star are rich in heavy elements produced in the interior of the progenitor star.

We are fortunate to have had in our lifetime a supernova which was close and in an unobscured line of sight. Supernova 1987A, the first supernova observed in 1987, went off in the Large Magellanic Cloud, a small satellite galaxy to our own. It was the first supernova visible to the naked eye in nearly 400 years (since the Kepler supernova in 1604). Using the full range of modern astronomical instruments has allowed us to get detailed spectra of the evolving ejecta which has greatly increased our knowledge of supernovae. We will use SN1987A as an example of supernova ejecta.

Initially the temperature of the ejecta of SN1987A was high, $\approx 10^6$ K, but it quickly cooled through adiabatic expansion and radiation from the photosphere. The temperature leveled off at several thousand degrees because of heating by radioactive nuclei, first ^{56}Ni and then ^{56}Co, formed in the explosion. The dynamics is homologous free expansion: the velocity scales linearly with the radius $r(t) = vt$ where v the velocity and t the time since the explosion.

The ejecta at first were optically thick and the spectrum resembled that of a hot star continuum with absorption lines from the surface. After a few days, the temperature dropped, but the ejecta remained optically thick and continued to show strong continuum emission. As the ejecta expand, the temperature drops, the ejecta become optically thin, and the spectrum is dominated by strong emission lines, superficially resembling an emission nebula (see Sect. 78.2). The emission is dominated by neutral atoms and singly ionized species.

The gas is heated and ionized by the gamma rays from radioactive decay. The gamma rays Compton scatter, producing x-rays and fast electrons. The x-rays further ionize the gas and produce multiply charged ions through the Auger process. These multiply charged ions recombine through charge transfer with neutral atoms. Further charge transfer determines the relative ionization of different species, with the lowest ionization potential species more ionized than the higher ionization potential species. The development of the infrared and optical spectrum of Supernova 1987A has been recently reviewed by McCray [36].

One of the great surprises in the spectrum of Supernova 1987A was the discovery of molecules in the infrared region. CO, SiO and possibly H_3^+ have been identified. In the absence of grains, molecules must be formed through either three-body or radiative processes. In the supernova ejecta, the densities are too low for three-body processes to be effective and molecules are formed through radiative association reactions. The molecules are removed by reactions with He^+ and the molecular abundances put a constraint on how much helium can be mixed back into the region with carbon and oxygen [37].

78.8   SHOCKED GAS

Shock waves occur in compressible fluids when the pressure gradients are large enough to generate supersonic motion, or when a disturbance is propagating through the fluid at supersonic velocities. Because information about the disturbance cannot propagate upstream in the fluid faster than the speed of sound, the fluid cannot respond dynamically until the shock arrives. The shock then compresses, heats, and accelerates the fluid. The boundary separating the hot compressed gas and the upstream gas is the shock front in which the energy of directed motion of the shock is converted to random thermal energy.

Shocks are ubiquitous in the interstellar medium where they are driven by the ionization fronts of expanding HII regions or nebulae, by outflowing gas accompanying stellar birth and evolution, and by supernova explosions. If the shock velocity is above 50 km s^{-1}, the shock gas is excited, dissociated, and ionized. The subsequent recombination and cooling radiation produces photons that may ionize and dissociate the gas components ahead of and behind the shock. This precursor radiation modifies the effects of the shock and influences its dynamical and thermal evolution. Fast shocks destroy all molecules by dissociating H_2 by collisions with H, H_2 and He and with electrons. Exchange reactions with H atoms destroy the other molecular species. At low densities, radiative stabilization occurs and dissociation is less efficient. Molecules reform in the cooling postshock gas. Slower shocks do not cause ionization or dissociation, but the chemical composition and the ion composition are modified by reactions taking place in the warm gas. The response of the interstellar gas to slow shocks is significantly affected by the presence of a magnetic field. In some ionization conditions, a magnetic precursor may occur in which a magnetosonic wave carries information about the shock, and the ionized and neutral components of the gas react differently to the shock. Many different kinds of shock have been identified [38].

A very fast shock with a velocity of hundreds of km s^{-1} such as are driven by supernova explosions, creates a hot dilute cavity in the interstellar medium with a temperature of millions of degrees. The density is low and the gas cools and recombines slowly. Overlapping supernova-induced cavities may be responsible for the hot gas that occupies a considerable volume of the interstellar medium in the Galaxy and in some external galaxies. The conditions are far from coronal equilibrium as the gas cools more rapidly than it recombines. The cooling radiation appears as soft x-rays and uv emission lines with a characteristic spectrum.

As the gas cools below 10 000 K, molecular formation occurs. Molecular hydrogen is formed on the surfaces of grains as in molecular clouds and by the negative ion sequence that is effective in the early Universe (Sect.

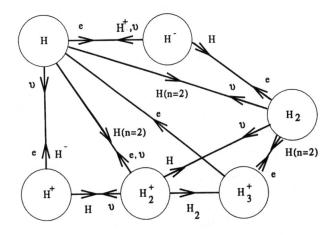

Figure 78.3. Diagram showing the important reactions in the production of hydrogen molecules in the early Universe.

78.9). With the formation of H_2 in a still warm gas, the chemistry is driven by exothermic and endothermic reactions with H_2. Thus OH is produced by the reaction of O atoms, and H_2O by the further reaction of OH with H_2. Enhanced abundances of other neutral and ionic molecules are the products of subsequent reactions with OH. The reactions of S^+ and S with OH lead to SO^+ and SO, and their simultaneous presence may be an indicator of a dissociative shock. There are in addition physical indicators of shocks, such as asymmetric line profiles indicating high velocities.

In a nondissociative shock in a molecular gas, reactions with warm H_2 dominate the chemistry as it does in the cooling zone of a dissociative shock. The composition is controlled by the post shock temperature and the H/H_2 ratio. The warm H_2 changes the ionic composition by converting C^+ into CH^+. Evidence for a nondissociative shock is the infrared emission from H_2. The thermal emission from collisionally excited vibrational levels in shock-heated gas is readily distinguished from that discussed in Sect. 78.4 arising from uv pumping in a PDR. Emission from H_2 has been detected in numerous objects in the Galaxy and in many distant external galaxies. In external galaxies, x-rays may contribute to the H_2 infrared sprectrum through heating the gas and through excitation by photoelectron pumping to excited states followed by a downward cascade [39, 40].

78.9   THE EARLY UNIVERSE

Molecules appeared first in the Universe after the adiabatic expansion had reduced the matter and radiation temperature to a few thousand degrees and recombination occurred, creating a nearly neutral Universe. The small fractional ionization that remained was essential to the formation of molecules. Molecular hydrogen formed through the sequences

$$H^+ + H \rightarrow H_2^+ + h\nu \qquad (78.25)$$
$$H_2^+ + H \rightarrow H_2 + H^+ \qquad (78.26)$$

and

$$H + e^- \rightarrow H^- + h\nu \qquad (78.27)$$
$$H^- + H \rightarrow H_2 + e^- , \qquad (78.28)$$

the protons and electrons acting as catalysts. Many other atomic and molecular processes occurred (see Fig. 78.3), some involving excited hydrogen atoms. Thus

$$H^* + H_2 \rightarrow H_3^+ + e^- \qquad (78.29)$$

was a source of H_3^+.

The Universe contained trace amounts of deuterium and ^7Li nuclei with which heteronuclear molecules could be made. Molecules with dipole moments may leave an imprint on the cosmic blackbody background radiation that occupies the Universe. The deuterated molecules HD form from

$$D^+ + H_2 \rightarrow HD + H^+ , \qquad (78.30)$$

and H_2D^+ from

$$D + H_3^+ \rightarrow H_2D^+ + H , \qquad (78.31)$$
$$HD^+ + H_2 \rightarrow H_2D^+ + H , \qquad (78.32)$$

and

$$HD + H_2^+ \rightarrow H_2D^+ + H . \qquad (78.33)$$

Lithium hydride is formed through

$$Li + H \rightarrow LiH + h\nu \qquad (78.34)$$
$$Li + H^- \rightarrow LiH + e^- \qquad (78.35)$$
$$Li^- + H \rightarrow LiH + e^- . \qquad (78.36)$$

There are many destruction processes, of which

$$LiH + H \rightarrow Li + H_2 \qquad (78.37)$$

may be the most severe, though its rate coefficient is uncertain. The chemistry of the early Universe is summarized in Ref. [41].

The formation of molecules was a crucial step in the fragmentation of the first gravitationally collapsing objects which separated out of the cosmic flow. Three-body recombination

$$H + H + H \rightarrow H_2 + H \qquad (78.38)$$
$$Li + H + H \rightarrow LiH + H \qquad (78.39)$$

may be a major source of molecules as the density increases.

78.10   OTHER READING

Astronomy is one of the oldest sciences and one of the fastest evolving. Advances in technology are rapidly increasing the sensitivity and resolution of our instruments and so new observations and more sophisticated models lead to an ever greater understanding of the Universe. This means that books will often be somewhat dated when they appear. However, the series *Annual Review of Astronomy and Astrophysics* is a good source of recent review articles.

In addition, good introductions or overviews of a particular field are given in Refs. [2, 42–47]. Many sources of atomic and molecular data are listed and discussed in Ref. [7]. For details on atomic spectroscopy, see Refs. [48, 49]. For details on molecular spectroscopy, see Refs. [50, 51].

REFERENCES

1. P. F. Winkler, G. W. Clark, T. H. Markert, K. Kalata, H. W. Schnopper, and C. R. Canizares, Astrophys. J. Lett. **246**, 27L (1981).
2. D. E. Osterbrock, *Astrophysics of Gaseous Nebulae and Active Galactic Nuclei* (University Science Books, Mill Valley, CA, 1989).
3. R. F. Reilman and S. T. Manson, Astrophys. J. Suppl. Ser. **40**, 815 (1979).
4. B. L. Henke, P. Lee, T. J. Tanaka, R. L. Shimabukoro, and B. K. Fujikawa, Atom. Nucl. Data Tables **27**, 1 (1982).
5. M. Balucinska-Church and D. McCammon, Astrophys. J. **400**, 699 (1992).
6. D. A. Verner, D. G. Yakovlav, J. M. Band, and A. B. Trzhaskavskaya, Atom. Nucl. Data Tables **55**, 233 (1993).
7. Special issue of Revista Mexicana de Astronomia y Astrofisica (March 23 1992).
8. H. Nussbaumer and P. J. Storey, Astron. Astrophys. **178**, 324 (1978).
9. H. R. Ramadan and Y. Hahn, Phys. Rev. **A39**, 3350 (1989).
10. N. R. Badnell, Phys. Scripta **T28**, 33 (1989).
11. M. C. Bacchus-Montabonel and K. Amezian, Z. Phys. **D25**, 323 (1993).
12. P. Honvault, M. C. Bacchus-Montabonel, and R. McCarroll, J. Phys. B **27**, 3115 (1994).
13. B. Herrero, I. L. Cooper, A. S. Dickinson, and D. R. Flower, J. Phys. B **28**, 711 (1995).
14. C. Mendoza, in *Planetary Nebulae IAU Symp. 103*, edited by D. R. Flower (Reidel, Dordrecht, 1983), p. 143.
15. M. Brocklehurst, and M. Salem, Comp. Phys. Comm. **13**, 39 (1977).
16. M. Salem and M. Brocklehurst, Astrophys. J. Suppl. Ser. **39**, 633 (1979).
17. P. G. Martin, Astrophys. J. Suppl. Ser. **66**, 125 (1988).
18. P. J. Storey and D. G. Hummer, Mon. Not. R. Astron. Soc. **272**, 41 (1995).

19. T. J. Gaetz and E. E. Salpeter, Astrophys. J. Suppl. Ser. **52**, 155 (1983).

20. M. Arnaud and R. Rothenflug, Astron. Astophys. Suppl. **60**, 425 (1985).

21. E. van Dishoeck and J. Black, Astrophys. J. Suppl. Ser. **62**, 109 (1987).

22. W. G. Roberge, D. Jones, S. Lepp, and A. Dalgarno, Astrophys. J. Suppl. Ser. **77**, 287 (1991).

23. A. Sternberg and A. Dalgarno, Astrophys. J. Suppl. Ser. **99**, 565 (1995).

24. R. Gredel, S. Lepp, A. Dalgarno, and E. Herbst, Astrophys. J. **347**, 289 (1989).

25. T. J. Millar, A. Bennett, J. M. C. Rawlings, P. D. Brown, and S. B. Charnley, Astron. Astophys. Suppl. **87**, 585 (1991).

26. P. R. A. Farquhar and T. J. Millar, CCP7 Newsletter No. 18, 6 (1993).

27. R. L. Kurucz, Astrophys. J. Suppl. Ser. **40**, 1 (1979).

28. M. J. Seaton, J. Phys. B **20**, 6363 (1987).

29. A. E. Lynas-Gray, M. J. Seaton and P. J. Storey, J. Phys. B **28**, 2817 (1995).

30. M. J. Seaton, Y. Yan, B. Mihalas, and A. K. Pradhan, Mon. Not. R. Astron. Soc. **266**, 805 (1994).

31. C. A. Iglesias and F. J. Rogers, Astrophys. J. **443**, 460 (1995).

32. A. Omont, *Circumstellar Chemistry* in *Chemistry in Space*, edited by J. M. Greenberg and V. Pirronello (Kluwer Academic Publishers, Dordrecht, 1991), p. 171.

33. G. A. Mamon, A. E. Glassgold, and A. Omont, Astrophys. J. **323**, 306 (1987).

34. L. A. M. Nejad and T. J. Millar, Astron. Astrophys. **183**, 279 (1987).

35. T. J. Millar and E. Herbst, Astron. Astrophys. **288**, 561 (1994).

36. R. McCray, Ann. Rev. Astron. Astrophys. **31**, 175 (1993).

37. W. Liu, S. Lepp, and A. Dalgarno, Astrophys. J. **396**, 679 (1992).

38. B. T. Draine and C. F. McKee, Ann. Rev. Astron. Astrophys. **31**, 373 (1993).

39. S. Lepp and R. McCray, Astrophys. J. **269**, 560 (1983).

40. R. Gredel and A. Dalgarno, Astrophys. J. **446**, 852 (1995).

41. A. Dalgarno and J. Fox, *Ion Chemistry in Atmospheric and Astrophysical Plasmas* in *Unimolecular and Bimolecular Reaction Dynamics*, edited by C.-Y. Ng, T. Baer and I. Powis (Wiley, New York, 1994).

42. C. W. Allen *Astrophysical Quantities* (Athlone Press, London, 1973).

43. K. R. Lang, *Astrophysical Formula* (Springer-Verlag, Berlin, 1980).

44. W. W. Duley and D. A. Williams, *Interstellar Chemistry* (Academic Press, London, 1984).

45. L. Spitzer *Physical Processes in the Interstellar Medium* (Wiley, New York, 1978).

46. A. Dalgarno and D. R. Layzer, *Spectroscopy of Astrophysical Plasmas* (Cambridge University Press, Cambridge, 1987).

47. T. Hartquist, *Molecular Astrophysics* (Cambridge University Press, Cambridge, 1990).

48. R. D. Cowan, *The Theory of Atomic Structure and Spectra* (University of California Press, Berkeley, 1981).

49. I. I. Sobelman, *Atomic Spectra and Radiative Transitions* (Springer-Verlag, Berlin, 1979).

50. G. Herzberg, *Molecular Spectra and Molecular Structure* (Prentice-Hall, New York, 1939).

51. P.F. Bernath *Spectra of Atoms and Molecules* (Oxford University Press, Oxford, 1995).

79

Comets

Paul D. Feldman

Department of Physics and Astronomy, The Johns Hopkins University, Baltimore, Maryland 21218

79.1 INTRODUCTION

Comets are small bodies of the solar system believed to be remnants of the primordial solar disk. Formed near the orbits of Uranus and Neptune and subsequently ejected into an "Oort cloud" of some 40 000 AU in extent, these objects likely preserve a record of the volatile composition of the early outer solar system, and so are of great interest for the physical and chemical modeling of solar system formation. The comets arrive in the inner solar system as a result of galactic perturbations. The cometary volatiles are vaporized as their orbits bring them closer to the sun and it is solar radiation that initiates all of the processes that lead to the extended *coma*. Gas vaporization also leads to the release of dust into the coma, and the scattering of sunlight by dust is the major source of the visible coma and *dust tail* of comets. Somewhat fainter, and much more extended, is the *ion tail*, resulting from photoionization by solar

extreme uv radiation of the neutral volatiles and their subsequent interaction with the solar wind.

With the exception of the *in situ* measurements made by the *Giotto* and *Vega* spacecraft at comet P/Halley (the P/ signifies a periodic comet) during March 1986, all determinations of the volatile composition of the coma are derived from spectroscopic analyses. Detailed modeling is then used to infer the volatile composition of the cometary *nucleus*. This chapter focuses on the principal atomic and molecular processes that lead to the observed spectrum as well as the needs for basic atomic and molecular data in the interpretation of these spectra. The largely collisionless and low density coma, with no gravity or magnetic field, is a unique spectroscopic laboratory, as evidenced by the discovery of C_3 before its identification in terrestrial laboratories [1]. Many key discrepancies remain to be resolved concerning the basic molecular composition and the elemental abundances of both the volatile and refractory components of the cometary nucleus, as well as the comet-to-comet variation (particularly between "new" and evolved periodic comets) of these quantities. These issues (and many others) are discussed in the recent analytical review of Festou, Rickman and West [2] or in the compendia of P/Halley results [3]. The former also contains a comprehensive bibliography. Other sources concentrating largely on the physics and chemistry of comets include the volumes edited by Wilkening [4] and Huebner [5] and the pre-Halley review of Mendis, Houpis and Marconi [6].

79.2 OBSERVATIONS

In a review in 1965, Arpigny [7] summarized the known molecular and atomic emissions detected in the visible region of the spectrum (here defined as 3000 to 11 000 Å) as follows:

radicals: OH, NH, CN, CH, C_3, C_2, and NH_2; ions: OH^+, CH^+, CO_2^+, CO^+, and N_2^+; metals: Na, Fe.

The only known atomic feature was the O I forbidden red doublet at 6300 and 6364 Å. From the radicals and ions one could infer the presence of their progenitor "parent" molecules such as H_2O, NH_3, HCN, CO and CO_2, directly vaporizing from the comet's nucleus. The metals, seen only in comets passing close to the sun, were assumed to come from the vaporization of refractory grains. The inventory of metals was soon expanded to include K, Ca^+, Ca, V, Cr, Mn, Ni and Cu, from observations of the sun-grazing comet Ikeya-Seki (1965 VIII) [8] and H_2O^+ was identified in comet Kohoutek (1973 XII). This latter comet was also the first to be extensively studied at wavelengths both shortward and longward of the visible spectral range.

The first parent molecule to be directly identified was CO which fluoresces in the Fourth Positive system ($A\,^1\Pi_u - X\,^1\Sigma^+$) in the vuv [9], although the *in situ* neutral mass spectrometer measurements made of Halley disclosed the presence of an extended, dominant source of CO [10] whose origin is still being debated. Ideally, the molecular species should be detectable through their radio and sub-mm rotational transitions or through the detection of vibrational bands or individual ro-vibrational lines in the near ir. Water was first directly detected through ro-vibrational lines near 2.7 μm in comet Halley and again in comet Wilson (1987 VII) [11]. However, due to the low column densities of the other expected species, typically $\sim$1% or less than that of H_2O, the direct detection of species such as H_2CO, H_2S and CH_3OH has only recently been made possible by the development of more sensitive instrumental techniques [12]. Tentative identifications of other species, such as CH_4, remain to be confirmed.

The ultimate result of solar photolysis (and to a lesser degree, the interaction with the solar wind) is the reduction of all of the cometary volatiles to their atomic constituents. The atomic inventory is somewhat easier to derive as the resonance transitions of the cosmically abundant elements H, O, C, N and S all lie in the vuv and, in principle, the total content of these species in the coma can be determined by an instrument with a suitably large field of view. Of course, a fraction of the atomic species of each element will be produced directly in ionic form, and will not be counted using this approach. In addition, another fraction exists in the coma in the solid grains, and this component will also not be included, except for a small amount volatilized by evaporation or sputtering by energetic particles. The composition of the grains, though not the absolute abundance, has been determined from *in situ* measurements made by the Halley encounter spacecraft [13], and can be inferred, though not unambiguously, from reflection spectroscopy of cometary dust in the 3–5 μm range.

The advent of space-borne platforms for observations in the vuv has produced a wealth of new information about the volatile constituents of the coma. The $A\,^2\Sigma^+ - X\,^2\Pi$ (0,0) band of the OH radical at $\sim$3085 Å was well known from ground-based spectroscopic observations, but as this wavelength lies very close to the edge of the atmospheric transparency window, the strength of this feature (relative to that of other species) was not appreciated until 1970 when comet Bennett (1970 II) was observed from space by the *Orbiting Astronomical Observatory* (OAO-2). The OAO-2 spectrum also showed a very strong, broadened H I Ly-α emission from H, the other principal dissociation product of H_2O. The broad shape of Ly-α seen in the OAO-2 spectrum is due to the large spatial extent of the atomic H envelope, the result of a high velocity acquired in the photodissociation process and a long lifetime against ionization. Later, at the apparition of comet Kohoutek (1973 XII), atomic O and C were identified in the spectra and direct uv images of the H coma, as well as of the O I and C I emissions, were obtained from sounding rocket experiments. These experiments were repeated for comet West (1976 VI) and led to the first detection of CO [9].

Since 1978, over 50 comets have been observed spectroscopically over the wavelength range 1200–3400 Å by the International Ultraviolet Explorer (IUE) satellite observatory [15]. Most of the spectra were obtained at moderate resolution ($\Delta\lambda = 6$–10 Å), although high dispersion echelle spectra ($\Delta\lambda = 0.2$–0.3 Å) are useful for some studies, particularly those of fluorescence equilibrium (see Sect. 79.3.2). For Halley alone, over 200 uv spectra were obtained from September 1985 to July 1986. The launch of the *Hubble Space Telescope* (HST) in 1990, whose aberrated optics were recently corrected, marks another advance in sensitivity as well as the ability to observe in a small field-of-view very close to the nucleus. Already this has yielded the first detection of CO Cameron band emis-

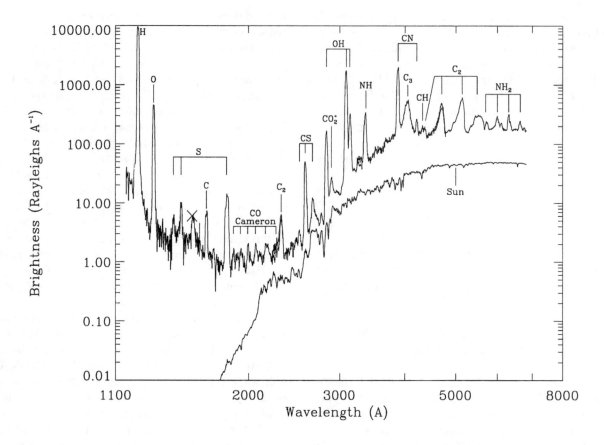

Figure 79.1. Composite FOS spectrum of comet P/Hartley 2. From [14].

sion, a direct measure of CO_2 being vaporized from the nucleus [16]. For an overview of a cometary spectrum, a composite spectrum of P/Hartley 2 (1991 XV) spanning the region from H I Ly-α to 7000 Å taken with the Faint Object Spectrograph of HST, using five separate gratings, is shown in Fig. 79.1.

An unusual comet was the focus of world-wide attention in July 1994 when approximately 20 fragments of the tidally disrupted comet P/Shoemaker-Levy 9 (1993e) plunged into the Jovian atmosphere. The appearance of the fragment train (discovered in March 1993) spread out along the orbit of the original body's center-of-mass, was that of a "string of pearls", each with its own coma. Orbit analysis showed that the original body had broken up at the time of perijove in July 1992 and would collide with Jupiter at its next perijove. Early observations set out to determine the diameters of the largest fragments so as to predict the kinetic energy input into the Jovian atmosphere at the relative impact velocity of $\sim$60 km s^{-1}. Current best estimates place the size of the largest of the fragments at no more than a few kilometers. An attempt was also made, both with HST [17] and from ground-based observatories [18], to spectroscopically detect cometary volatiles such as OH, CN or CO$^+$ to determine if the fragments were in fact comets. These

yielded only upper limits which were consistent with the expected vaporization rate of kilometer-size icy nuclei at the heliocentric distance of Jupiter, $\sim$5.4 AU. However, spectra taken of the plumes or impact sites immediately following entry of a fragment into the Jovian atmosphere revealed the presence of metallic line emission, both in the uv (Mg, Mg$^+$, Si, Fe, and Fe$^+$) [19] and visible (Na, K, Li, Ca and Fe) [20], material clearly of cometary origin.

79.3 EXCITATION MECHANISMS

79.3.1 Basic Phenomenology

Coma abundances may be derived from spectrophotometric measurements of either the total flux or the surface brightness in a given spectral feature. The uncertainty in the derived abundances includes not only the measurement uncertainty, but also uncertainties in the atomic and molecular parameters and, in the case of surface brightness measurements, uncertainties in the model parameters used. Thus relative abundances, derived from observations of different comets with the same instrument and under similar geometrical conditions, are often more reliable.

Atoms and ions in the cometary coma emit radiation primarily by means of resonance re-radiation of solar photons. For the cosmically abundant elements H, C, N, O, and S, their strongest resonance transitions are in the vuv. The few exceptions are noted below. Assume that the coma is optically thin in these transitions. The total number of species i in the coma is

$$M_i = Q_i \tau_i(r), \qquad (79.1)$$

where Q_i is the production rate (atoms or molecules s^{-1}) of species i and $\tau_i(r)$ is its lifetime at heliocentric distance r, $\tau_i(r) = \tau_i(1\,\text{AU})r^2$. The r dependence arises from photolytic destruction processes induced by solar uv radiation, and to a lesser degree by the solar wind, as described in Sect. 79.4.1.

The luminosity, in photons s^{-1}, in a given transition at wavelength λ, is then

$$L_{i\lambda} = M_i g_{i\lambda}(r), \qquad (79.2)$$

where the fluorescence efficiency, or "g-factor", $g_{i\lambda}(r)$ $= g_{i\lambda}(1\,\text{AU})r^{-2}$, is

$$g_{i\lambda}(1\,\text{AU}) = \frac{\pi e^2}{mc^2}\lambda^2 f_\lambda \pi F_\odot \tilde{\omega} \quad \text{photons s}^{-1}\text{ atom}^{-1},$$
$$(79.3)$$

where f_λ is the absorption oscillator strength, $\pi F_\odot$ is the solar flux per unit wavelength interval at 1 AU and $\tilde{\omega}$ is the albedo for single scattering, defined for a line in an atomic multiplet as

$$\tilde{\omega} = \frac{A_j}{\sum_j A_j}, \qquad (79.4)$$

where A_j is the decay rate. If a given multiplet is not resolved, then $\tilde{\omega} = 1$. For diatomic molecules, fluorescence to other vibrational levels becomes important and the evaluation of $\tilde{\omega}$ depends on the physical conditions in the coma, as discussed in Sect. 79.3.2. Thus, for a comet at a geocentric distance Δ, the total flux from the coma for the transition is

$$F_{i\lambda} = \frac{L_{i\lambda}}{4\pi\Delta^2} = \frac{Q_i g_{i\lambda}(r)\tau_i(r)}{4\pi\Delta^2}, \qquad (79.5)$$

and the product $g_{i\lambda}(r)\tau_i(r)$ is independent of r.

Unfortunately, the scale lengths (the product of lifetime and outflow velocity) of almost all of the species of interest in the uv are $\sim 10^5$–10^6 km at 1 AU. Thus, total flux measurements require fields of view ranging from several arc-minutes to a few degrees. This has been done only in the case of a few isolated sounding rocket experiments. Most information about the uv spectra comes from observations made by orbiting satellite observatories whose spectrographs have small apertures (e.g., $10'' \times 20''$ for IUE) and thus sample only a very small

part of the total coma. In this case, again assuming an optically thin coma, the measured flux $F'_{i\lambda}$ in the aperture can be converted to an average surface brightness $B_{i\lambda}$ (in units of Rayleighs):

$$B_{i\lambda} = 4\pi 10^{-6} F'_{i\lambda}\Omega^{-1}, \qquad (79.6)$$

where Ω is the solid angle subtended by the aperture. The brightness, in turn, is related to $\overline{N}_i$, the average column density of species i within the field of view by

$$B_{i\lambda} = 10^{-6} g_{i\lambda}(r)\overline{N}_i. \qquad (79.7)$$

The evaluation of Q_i from $\overline{N}_i$ requires the use of a model of the density distribution of the species i (Sect. 79.4.2).

79.3.2 Fluorescence Equilibrium

In the case of low resolution spectroscopy, where an atomic multiplet or molecular band is unresolved, the evaluation of the g-factor (79.3) does not require knowledge of the population of either atomic fine structure levels or molecular rotational levels in the ground state of the transition. Furthermore, the assumption that the solar flux does not vary over the multiplet or band allows us to use the total transition oscillator strength. This assumption is more often than not invalid because of the Fraunhofer structure of the solar spectrum in the near uv and visible region and the emission line nature of the spectrum below 2000 Å. For high resolution spectra, the g-factors for each individual line must be calculated separately and the relative populations of the ground state levels must be included. There are three cases to be considered:

(1) The g-factor, or probability of absorption of a solar photon, is less than the probability that the species will be dissociated or ionized, i.e., $g_{i\lambda} < (\tau_i)^{-1}$. In this case, the ground state population is not affected by fluorescence and a Boltzmann distribution at a suitable temperature (typically $\sim$200 K at 1 AU) corresponding to the production of the species may be used. This is often the case in the far uv, where the solar flux is low, such as for the Fourth Positive band system of CO. For atomic transitions from triplet ground states, such as is found with O, C and S, downward fine structure transitions are fast enough to effectively depopulate all but the lowest fine structure level.

(2) The species undergoes many photon absorption and emission cycles in its lifetime, and the ground state population is determined (usually after 5 or 6 cycles) by the fluorescence branching ratios. This is the condition of *fluorescence equilibrium*, which applies for almost all radicals observed in the visible and near uv regions. The general procedure is to solve a set of coupled equations of the form

$$\frac{dn_a}{dt} = -n_a \sum_{b=1}^{N} p_{ab} + \sum_{b=1}^{N} n_b p_{ba}, \qquad (79.8)$$

where n_a is the relative population of level a and p_{ab} and p_{ba} are transition rates out of and into this level, respectively. The n_a are normalized to unity. The steady state, obtained after many cycles, is given by $dn_a/dt = 0$. Since the downward transition rates are determined only by quantum mechanics, while the absorption rates depend on the magnitude of the solar flux, the steady state population varies with distance from the sun with higher rotational levels (as for the case of CN [7]) populated closer to the sun. In some cases, where only a few cycles occur, the equations are integrated numerically. As the g-factor varies with time, it also effectively varies with the position of a species in the coma. Care must also be taken when spectra taken with small apertures are analyzed as the transit time for an atom or molecule to cross the aperture may be $\lesssim g_{i\lambda}(r)^{-1}$. In practice, these considerations are often not important.

(3) The same as **(2)** except that the density is sufficiently high that collisional transitions must be included in addition to the radiative transitions between levels. However, as the collisional rates are poorly known, in practice a "collision sphere" is defined such that a molecule traveling radially outward from this sphere suffers only one collision with other molecules or atoms. Outside this sphere, fluorescence equilibrium is assumed to hold, while inside a thermal distribution of ground state levels is used. A rough estimate of the radius R_c of the collision sphere, based on the radial outflow model of Sect. 79.4.2 is given by [2]

$$R_c = 10^3 \frac{Q}{10^{29}} \text{ km}, \qquad (79.9)$$

where Q is the total production rate in molecules s^{-1}.

79.3.3 Swings and Greenstein Effects

Swings [22] first pointed out that because of the Fraunhofer absorption lines in the visible region of the solar spectrum, the absorption of solar photons in a molecular band would vary with the comet's heliocentric velocity $\dot{r}$, leading to differences in the structure of a band at different values of $\dot{r}$ when observed at high resolution. In Eq. (79.8), this corresponds to evaluating the $p_{ba} = p_{ba}(\dot{r})$. For typical comets observed near 1 AU, $\dot{r}$ can range from −30 to +30 km s^{-1}, while in certain cases of comets with small perihelia the range can be twice as large. This effect of the Doppler shift between the sun and the comet is commonly referred to as the *Swings effect*. Even for observations at low resolution, the Swings effect must be taken into account in the calculation of the total band g-factor, and this has

Figure 79.2. OH (0,0) band g-factor as a function of heliocentric velocity. From Ref. [21].

Figure 79.3. Solar flux and fluorescence efficiency for O I $\lambda 1302$ as a function of heliocentric velocity. From Ref. [23].

been done for a number of important species such as OH, CN and NH. A particularly important case, that of the OH $A\,^2\Sigma^+ - X\,^2\Pi$ (0,0) band at $\sim$3085 Å, which is often used to derive the water production rate of a comet, is illustrated in Fig. 79.2, which also shows the dependence of fluorescence equilibrium on heliocentric distance [21].

While this effect was first recognized in the spectra of radicals in the visible range, a similar phenomenon occurs in the excitation of atomic multiplets below 2000 Å, where the solar spectrum makes a transition to an emission line spectrum. For example, the three lines of O I $\lambda 1302$ have widths of $\sim$0.1 Å, corresponding to a velocity of $\sim$25 km s^{-1}, so that knowledge of exact solar line shapes is essential for a reliable evaluation of the g-factor for this transition [23], as illustrated in Fig. 79.3.

A differential Swings effect occurs in the coma since atoms and molecules on the sunward side of the coma, flowing outward towards the sun, have a net velocity that is different from those on the tailward side, and so, if the absorption of solar photons takes place on the edge of a line, the g-factors will be different in the two directions. Differences of this type appear in long-slit spectra in which the slit is placed along the sun-comet line (the *Greenstein effect* [24]). Again, an analogue in the far uv has been observed in the case of O I $\lambda1302$ [25], as can be seen in Fig. 79.3. Although it is also possible to explain the observation by nonuniform outgassing, this was considered unlikely as all of the other observed emissions had symmetric spatial distributions. The measurement of the Greenstein effect leads immediately to a determination of the mean outflow velocity of the given species.

79.3.4 Bowen Fluorescence

Figure 79.3 also demonstrates that for heliocentric velocities >30 km s^{-1}, the Doppler shift reduces the solar flux at the center of the absorption line to a very small value, so that the O I $\lambda1302$ line is expected to appear weakly, if at all, in the observed spectrum. Thus, it was a surprise that this line appeared fairly strongly in two comets, Kohoutek (1973 XII) and West (1976 VI), whose values of $\dot{r}$ were both >45 km s^{-1} at the times of observation. The explanation invoked the accidental coincidence of the solar H I Ly-β line at 1025.72 Å with the O I $^3D - {}^3P$ transition at 1025.76 Å, cascading through the intermediate ^3P state as shown in the simplified energy level diagram of Fig. 79.4 [23]. This mechanism, well known in the study of planetary nebulae, is referred to as *Bowen fluorescence* [26]. The g-factor due to Ly-β pumping is an order of magnitude smaller than that for resonance scattering, as shown in Fig. 79.3, but sufficient to explain the observations and to confirm that H$_2$O is the dominant source of the observed oxygen in the coma. However, to this date the cascade O I $\lambda8447$ ($^3D - {}^3P$) line has not been reported in any cometary spectrum. Ly-β is also coincident with the P1 line of the (6,0) band of the H$_2$ Lyman system ($B\,{}^1\Sigma_u^+ - X\,{}^1\Sigma_g^+$) leading to fluorescence in the same line of several $(6,v'')$ bands, the strongest being that of the (6,13) band at 1608 Å [27]. While H$_2$ fluorescence has recently been observed in the atmosphere of Jupiter, it has not been detected in any comet spectrum including the recent spectra taken with HST.

79.3.5 Electron Impact Excitation

The photoionization of the parent molecules and their dissociation products leads to the formation of a cometary ionosphere whose characteristics are only poorly known. Planetary ionospheres serve only as a poor model since the cometary atmosphere is gravitationally

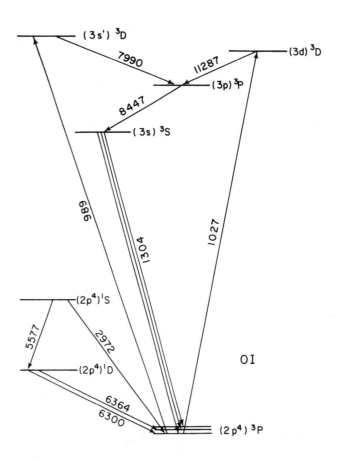

Figure 79.4. Simplified O I energy level diagram showing transitions of interest in cometary spectra.

unbound and there is no constraining magnetic field. The *in situ* measurements of Halley provided order of magnitude confirmation of the theoretical modeling. In principle, electron impact excitation, which is often the dominant source of airglow in the atmospheres of the terrestrial planets (see Chap. 80), also contributes to the observed emissions, particularly in the uv, and so must be accounted for in deriving column densities from the observed emission brightnesses. However, one can use a very simple argument, based on the known energy distribution of solar uv photons, to demonstrate that electron impact excitation is only a minor source for the principal emissions. Since the photoionization rate of water (and of the important minor species such as CO and CO$_2$) is $\sim10^{-6}$ s^{-1} at 1 AU, and the efficiency for converting the excess electron energy into excitation of a single emission is of the order of a few percent, the effective excitation rate for any emission will be $\sim10^{-8}$ s^{-1} or less at 1 AU [28]. Since the efficiencies for resonance scattering or fluorescence for almost all the known cometary emissions are much larger, electron impact may be safely neglected except in a few specific cases.

The cases of interest are those of forbidden transitions, where the oscillator strength, and consequently the g-factor, is very small. Examples include the O I $^5S_2$ – $^3P_{2,1}$ doublet at 1356 Å, which has been observed in comets West and Halley by rocket-borne spectrographs, the O I ^1D – ^3P red lines at 6300 and 6364 Å, observed in many comets, and the recently discovered CO Cameron bands (Sect. 79.2). However, the excitation of these latter two is dominated by prompt emission in the inner coma, the same region of the coma where electron excitation is important, as described in the next section. The Lyman-Birge-Hopfield system of N_2 has a similar excitation cross section to that of O I $\lambda 1356$, and also appears in the far uv, but to date has not been detected. The derived upper limit on the relative abundance of N_2 in Halley is consistent with a more sensitive limit ($< 10^{-3}$ of H_2O) derived from the weakness of the N_2^+ fluorescence at 3914 Å.

79.3.6 Prompt Emission

In cases where the dissociation or ionization of a molecule leaves the product atom or molecule in an excited state, the decay of this state with the prompt emission of a photon provides a useful means for tracing the spatial distribution of the parent molecule in the inner coma. The products of interest for the water molecule are described in Sect. 79.4.1. Prompt emission includes both allowed radiative decays (such as from the $A\,^2\Sigma^+$ state of OH) as well as those from metastable states such as O(^1D), since the latter will move ~150 km (for a comet at a geocentric distance of 1 AU, 1 arc-second corresponds to a projected distance of 725 km) in its lifetime. The O I ^1D – ^3P transition at 6300 and 6364 Å (see Fig. 79.4) has been used extensively as a ground-based monitor of the water production rate with the caveat that other species such as OH, CO and CO_2 may also contribute to the observed red line emission. In addition, when the density of H_2O is sufficient to produce observable red line emission, it is also sufficient to produce collisional quenching of the ^1D state, and this must also be considered in the interpretation of the observations. The analogous ^1D – ^3P transitions in carbon occur at 9823 and 9849 Å and should provide similar information about the production rate of CO. Carbon atoms in the ^1D state, whose lifetime is ~4000 s, are known to be present from the observation of the resonantly scattered $^1P^o$ – ^1D transition at 1931 Å [9], but so far have not been detected via their infrared decay photons.

The OH $A\,^2\Sigma^+$ – $X\,^2\Pi$ prompt emission competes with that produced by the resonance fluorescence of OH and is difficult to detect, except close to the nucleus (inside 100 km) where the density of water molecules exceeds that of OH by a few orders of magnitude [29]. Again, the reason is that at the wavelengths

below the threshold for simultaneous dissociation and excitation, the sun has much less flux than at the resonance wavelength. On the other hand, only a few rotational lines are excited in fluorescence equilibrium [21], while the prompt emission is characterized by a very "hot" rotational distribution, so in principle the two components may be separated. Even so, observations at very high spatial resolution are required.

79.3.7 OH Level Inversion

An important consequence of fluorescence equilibrium in the OH radical is the uv pumping of the hyperfine and Λ-doublet levels of the $X\,^2\Pi_{3/2}$ ($J = 3/2$) ground state, which results in a deviation of the population from statistical equilibrium [21]. Depending on the heliocentric velocity, this departure may be either "inverted" or "anti-inverted" giving rise to either stimulated emission or absorption against the galactic background at 18 cm wavelength. This technique has been used extensively since 1974 to monitor the OH production rate in comets, even of those that appear close to the sun [30]. The resulting radio emissions are easily quenched by collisions with molecules and ions, the latter giving rise to a fairly large R_c that must be accounted for in interpreting the derived OH column density. Nevertheless, the radio and uv measurements give reasonably consistent results [30].

79.4 COMETARY MODELS

79.4.1 Photolytic Processes

As an example of the photolytic destruction processes occurring in the coma, consider the dominant molecular species, water. Water vapor is assumed to leave the surface of the nucleus with some initial velocity v_0 and flow radially outward, expanding into the vacuum, and increasing its velocity according to thermodynamics [6]. Even though collisions are important at distances typically up to 10^4 km (depending on the density and consequently, on the total gas production rate), the net flow of H_2O molecules is radially outward, such that the density varies as R^{-2} near the nucleus, where R is the cometocentric distance. This is the basis for spherically symmetric coma models (the number of particles flowing through a spherical surface is conserved), which assume isotropic gas production, but appears to hold equally well for the case of Halley, which clearly was not outgassing uniformly over its surface [3].

The photolysis of H_2O can proceed by:

a	$H_2O + h\nu$	$\rightarrow$	OH + H	2424.6 Å
a'		$\rightarrow$	OH($A\,^2\Sigma^+$) + H	1357.1 Å
b		$\rightarrow$	$H_2 + O(^1D)$	1770 Å
b'		$\rightarrow$	$H_2 + O(^1S)$	1450 Å

b''		$\rightarrow$	$H + H + O(^3P)$	1304	Å
c		$\rightarrow$	$H_2O^+ + e^-$	984	Å
d		$\rightarrow$	$H + OH^+ + e^-$	684.4	Å
e		$\rightarrow$	$H_2 + O^+ + e^-$	664.4	Å
f		$\rightarrow$	$OH + H^+ + e^-$	662.3	Å

The right-hand column gives the energy threshold for each reaction, in wavelength units. The products are subsequently removed by

g	$OH + h\nu$	$\rightarrow$	$O + H$	2823.0	Å
h		$\rightarrow$	$OH^+ + e^-$	928	Å
i	$H_2 + h\nu$	$\rightarrow$	$H + H$	844.79	Å
j		$\rightarrow$	$H_2^+ + e^-$	803.67	Å
k		$\rightarrow$	$H + H^+ + e^-$	685.8	Å
l	$O + h\nu$	$\rightarrow$	$O^+ + e^-$	910.44	Å
m	$H + h\nu$	$\rightarrow$	$H^+ + e^-$	911.75	Å

Reactions l and m can also occur by resonant charge exchange with solar wind protons. Reactions a', b and b' correspond to the production of prompt emission, as discussed in Sect. 79.3.6. The determination of column densities of H, O and OH simultaneously was convincing evidence that the dominant volatile species in the cometary nucleus was H_2O, long before the direct infrared detection of this species in the coma.

Detailed cross sections for the absorption of uv photons by each of the reactants, including proper identification of the final product states, is necessary for the evaluation of the photodestruction rate in the solar radiation field of each of the above reactions. These rates are evaluated at 1 AU using whole disk measurements of the solar flux by integrating the cross section

$$J_d = \int_0^{\lambda_{th}} \pi F_\odot \sigma_d \, d\lambda \, . \qquad (79.10)$$

Another quantity of interest in coma modeling is the excess velocity (or energy) of the dissociation or ionization products, and this requires knowledge of the partitioning of energy between internal and translational modes for each reaction [31].

Qualitatively, the photodissociation and photoionization rates can be estimated from the threshold energies given in the table above, since the solar flux is decreasing very rapidly to shorter wavelengths, as can be seen in Fig. 79.1. It is customary to specify the lifetime against photodestruction, τ_i, of species i, which is just

$$(\tau_i)^{-1} = \sum_j (J_d)_j \, , \qquad (79.11)$$

where the sum is over all possible reaction channels, as well as the lifetimes into specific channels. Processes with thresholds near 3000 Å have lifetimes $\sim 10^4$ s, those with thresholds near 2000 Å an order of magnitude longer, while those with thresholds below Ly-α, such as most photoionization channels, have lifetimes $\sim 10^6$ s, all at

1 AU. In addition to uncertainties in the details of the absorption cross sections, further uncertainty is introduced into the calculation of J_d by the lack of knowledge of the solar flux at the time of a given observation due to the variability of the solar radiation below 2000 Å, and most importantly, below Ly-α, where there have not been continuous space observations for more than a decade. The solar uv flux is known to vary considerably both with the 27-day solar rotation period and with the 11-year solar activity cycle. Also, at any given point in its orbit, a comet sees a different hemisphere of the sun than what is seen from Earth. Huebner, Keady and Lyon [32] have compiled an extensive list of useful photodestruction rates using mean solar fluxes to represent the extreme conditions of solar minimum and solar maximum. They also include the excess energies of the dissociation products. A detailed analysis of the rates for H_2O and OH, using surrogate solar indices such as the 10.7 cm solar radio flux, or the equivalent width of the He I line at 1.083 μm, is in good agreement with observations [33]. Similar analyses still remain to be carried out for other important species, such as CO and NH_3.

79.4.2 Density Models

For parent molecules produced directly by sublimation from the surface of the comet, a spherically symmetric radial outflow model is often adopted. Such a model assumes a steady-state gas production rate Q_i and a constant outflow velocity v, and gives rise to a density distribution as a function of cometocentric distance R given by

$$n_i(R) = \frac{Q_i}{4\pi v R^2} e^{-R/\beta_i} \, , \qquad (79.12)$$

where $\beta_i = v\tau_i$ is the scale length of species i. The basic validity of this model was demonstrated by the *Giotto* neutral mass spectrometer measurements of H_2O and CO_2 [3], although detailed analysis revealed that the velocity of the water molecules increased from 0.8 km s^{-1} at about 1000 km from the nucleus to 1.1 km s^{-1} at a radial distance of 10 000 km. The dependence of outflow velocity on heliocentric distance remains uncertain, although Delsemme [34] has suggested an $r^{-1/2}$ dependence based on thermodynamic arguments.

For the dissociation products, the modeling is more complex. The simplest model assumes continued radial outflow, although at a different velocity, such as to maintain a constant flux of the initial particle across an arbitrary spherical surface surrounding the nucleus [35]. This model, which is valid only at distances equal to a few β_i, is widely used as the densities can be easily expressed in analytical form. However, as surface brightness measurements are often made with small fields of view close to the nucleus, this model can lead to a factor of two errors from the neglect of the dissociation kinematics. Since the solar photodissociation often leaves the product fragments with 1–2 eV of kinetic energy [31, 32], the resultant motion (which is assumed to be isotropic in the parent molecule's rest frame), will contain a large nonradial component. Several approaches have been developed to account properly for the kinematics; notably, the vectorial model of Festou [36], and the average random walk model (a Monte Carlo method) of Combi and Delsemme [37]. The latter model has an advantage in the calculation of "granddaughter" densities, such as that of oxygen produced by the consecutive dissociations of H_2O and OH.

In addition to the photodestruction chains, chemical reactions, particularly ion-molecule reactions, can alter the composition within the collision zone defined in Sect. 79.3.2. While such reactions may produce numerous minor species, they do not erase the signatures of the original parent molecules. In fact, detailed chemical models have clearly demonstrated the need for complex molecules to serve as the parents of the observed C_2 and C_3 radicals in the coma, strengthening the connection between comet formation and molecular cloud abundances. Thus, the photochemical chains provide a valid means of relating the coma composition to that of the nucleus.

79.4.3 Radiative Transfer Effects

The results of a model calculation for the density of a species must be integrated over the line of sight to obtain the column density at a given projected distance from the nucleus, and then integrated over the instrumental field of view for comparison with the observed average surface brightness or derived average column density $\overline{N_i}$. This assumes that the coma is *optically thin*, and that all atoms or molecules have an equal probability of absorbing a solar photon. In practice, this is true for all molecular emissions except perhaps within 1000 km of the nucleus (i.e., for observations made at better than 1″ resolution). Since the cross sections at line center can be very large for an atomic resonance transition, the optical depth for the abundant species can exceed unity and radiative transfer along both the line of sight to the sun and that to the Earth must be considered. This is not a trivial problem as the velocity distribution of the atoms, particularly the component due to the excess energy of the photodestruction process, must be well known, as must be the shape of the exciting solar line. The only case studied to date is that of H I Ly-α, whose angular extent, in direct images, can exceed several degrees on the sky [38].

An interesting case arises for resonance transitions between an excited $^3S_1$ state and the ground $^3P_{2,1,0}$ state, as for O and S, particularly the latter as its concentration near the nucleus can be quite large due to the rapid decay

of one of its parents, CS_2. For S I the three lines at 1807, 1820 and 1826 Å are not observed to have their statistical intensity ratio of 5:3:1, except at large distances from the nucleus. This is explained by noting that fine structure transitions will lead to all of the S atoms reaching the $J = 2$ ground state in a time short compared with that for absorbing a solar photon, and that the emitted 1807 Å photons will be re-absorbed and can then branch into the other two lines. The detailed solution to this problem is complicated by the Greenstein effect. Two other S I multiplets are observed in moderately bright comets at small heliocentric velocity (there is a very strong Swings effect) at 1425 and 1474 Å. The relative intensities of the three multiplets, however, differ strongly from the expectations of the optically thin g-factors, casting doubt on the reliability of the derived column densities. To date, satisfactory observations resolving the multiplet structure of the strong O I $\lambda1302$ and C I $\lambda1657$ multiplets have not been made.

79.5 SUMMARY

Two classes of problems remain: those related to the actual determination of relative abundances from the available data, and those concerned with accommodating the abundances within the framework of cosmogonic models. Examples of the former include resolving the discrepant derivations of NH_3 abundance from observations of NH or NH_2, or the determination of the S column density from the different S I transitions, discussed in Sect. 79.4.3. Of the latter, we cite the problem of CO/N_2 abundance in Halley or the unique observation of S_2 in a single comet, IRAS-Araki-Alcock (1983 VII).

More "parent" molecules should be detected directly in the infrared and sub-mm, while the far- and extreme-uv should yield detections of noble gases, the O^+ ion, and H_2. The Hubble Space Telescope is capable of an accurate measurement of the important D/H ratio (or alternatively, the OD/OH ratio), as well as the possibility of studying photodissociation dynamics near the nucleus.

REFERENCES

1. G. Herzberg, *The Spectra and Structures of Simple Free Radicals: An Introduction to Molecular Spectroscopy.* (Dover Publications, New York, 1971).
2. M. C. Festou, H. Rickman, and R. M. West, Astron. Astrophys. Rev. **4**, 363 (1993); **5**, 37 (1993).
3. D. Krankowsky, in *Comets in the Post-Halley Era*, edited by R. L. Newburn, Jr., M. Neugebauer, and J. Rahe (Kluwer, Dordrecht, 1989), p. 855.
4. *Comets*, edited by L. L. Wilkening (University of Arizona Press, Tucson, 1982).
5. M. F. A'Hearn and M. C. Festou, in *Physics and Chemistry of Comets*, edited by W. F. Huebner (Springer-Verlag, New York, 1990) p. 69.

6. D. A. Mendis, H. L. F. Houpis, and M. L. Marconi, Fund. Cosmic Phys. **10**, 1 (1985).

7. C. Arpigny, Ann. Rev. Astron. Astrophys. **3**, 351 (1965).

8. G. W. Preston, Astrophys. J. **147**, 718 (1967); C. D. Slaughter, Astron. J. **74**, 929 (1969).

9. P. D. Feldman and W. H. Brune, Astrophys. J. Lett. **209**, L45 (1976).

10. P. Eberhardt, D. Krankowsky, W. Schulte, U. Dolder, P. Lämmerzahl, J. J. Berthelier, J. Woweries, U. Stubbemann, R. R. Hodges, J. H. Hoffman, and J. M. Illiano, Astron. Astrophys. **187**, 484 (1987).

11. M. J. Mumma, H. A. Weaver, H. P. Larson, D. S. Davis, and M. Williams, Science **232**, 1523 (1986); H. P. Larson, H. A. Weaver, M. J. Mumma, and S. Drapatz, Astrophys. J. **338**, 1106 (1989).

12. J. Crovisier, in *Asteroids, Comets, Meteors*, edited by A. Milani, M. DiMartino, and A. Cellino (Kluwer, Dordrecht, 1994), p. 313.

13. E. K. Jessberger and J. Kissel, in *Comets in the Post-Halley Era*, edited by R. L. Newburn, Jr., M. Neugebauer, and J. Rahe (Kluwer, Dordrecht, 1989), p. 1075.

14. H. A. Weaver and P. D. Feldman, in *Science with the Hubble Space Telescope*, edited by P. Benvenuti and E. Schreier (ESO, Garching, 1992), p. 475.

15. M. C. Festou and P. D. Feldman, in *Exploring the Universe with the IUE Satellite*, edited by Y. Kondo (Reidel, Dordrecht, 1987) p. 101; M. C. Festou, *International Ultraviolet Explorer—Uniform Low Dispersion Data Archive Access Guide: Comets*, ESA SP-1134 (ESA, Noordwijk, 1990).

16. H. A. Weaver, P. D. Feldman, J. B. McPhate, M. F. A'Hearn, C. Arpigny, and T. E. Smith, Astrophys. J. **422**, 374 (1994).

17. H. A. Weaver, M. F. A'Hearn, C. Arpigny, D. C. Boice, P. D. Feldman, S. M. Larson, Ph. Lamy, D. H. Levy, B. G. Marsden, K. J. Meech, K. S. Noll, J. V. Scotti, Z. Sekanina, C. S. Shoemaker, E. M. Shoemaker, T. E. Smith, S. A. Stern, A. D. Storrs, J. T. Trauger, D. K. Yeomans, and B. Zellner, Science **267**, 1282 (1995).

18. A. L. Cochran, A. L. Whipple, P. J. MacQueen, P. J. Shelus, R. W. Whited, and C. F. Claver, Icarus **112**, 528 (1994).

19. K. S. Noll, M. A. McGrath, L. M. Trafton, S. K. Atreya, J. J. Caldwell, H. A. Weaver, R. V. Yelle, C. Barnet, and S. Edgington, Science **267**, 1307 (1995).

20. M. Roos-Serote, A. Barucci, J. Crovisier, P. Drossart, M. Fulchignoni, J. Lecacheux, and F. Roques, Geophys. Res. Lett. **22**, 1621 (1995).

21. D. G. Schleicher and M. F. A'Hearn, Astrophys. J. **331**, 1058 (1988).

22. P. Swings, Lick. Obs. Bull. XIX, 131 (1941).

23. P. D. Feldman, C. B. Opal, R. R. Meier, and K. R. Nicolas, in *The Study of Comets*, edited by B. Donn *et al.*, NASA SP-393 (NASA, Washington, 1976), p. 773.

24. J. L. Greenstein, Astrophys. J. **128**, 106 (1958).

25. K. F. Dymond, P. D. Feldman, and T. N. Woods, Astrophys. J. **338**, 1115 (1989).

26. I. S. Bowen, Publ. Astron. Soc. Pacific **59**, 196 (1947).

27. P. D. Feldman and W. G. Fastie, Astrophys. J. Lett. **185**, L101 (1973).

28. T. E. Cravens and A. E. S. Green, Icarus **33**, 612 (1978).

29. J.-L. Bertaux, Astron. Astrophys. **160**, L7 (1986).

30. E. Gérard, Astron. Astrophys. **230**, 489 (1990).

31. H. Okabe, *The Photochemistry of Small Molecules*, (Wiley, New York, 1978).

32. W. F. Huebner, J. J. Keady, and S. P. Lyon, *Solar Photo Rates for Planetary Atmospheres and Atmospheric Pollutants*, (Springer-Verlag, New York, 1992).

33. S. A. Budzien, M. C. Festou, and P. D. Feldman, Icarus **107**, 164 (1994).

34. A. H. Delsemme, in *Comets*, edited by L. L. Wilkening (University of Arizona Press, Tucson, 1982) p. 85.

35. L. Haser, Bull. Acad. R. Sci. Liège **43**, 740 (1957).

36. M. C. Festou, Astron. Astrophys. **95**, 69 (1981).

37. M. R. Combi and A. H. Delsemme, Astrophys. J. **237**, 633 (1980).

38. H. U. Keller, Space Sci. Rev. **18**, 641 (1976).

80

Aeronomy

J. L. Fox

Institute for Terrestrial and Planetary Atmospheres and Marine Sciences Research Center,
The University at Stony Brook, Stony Brook, New York 11794-5000

80.1 BASIC STRUCTURE OF ATMOSPHERES

80.1.1 Introduction

In a stationary atmosphere, the force of gravity is balanced by the plasma pressure gradient force in the vertical direction, and the variation of pressure $P(z)$ with altitude above the surface z is governed by the hydrostatic relation

$$\frac{dP(z)}{dz} = -\rho(z)g(z), \qquad (80.1)$$

where $\rho(z) = n(z)m(z)$ is the mass density, $n(z)$ is the number density, and $m(z)$ is the average mass of the atmospheric constituents. In general, variables such as P, ρ, g, n and even m are functions of altitude, although it will often not be shown explicitly in the equations that follow for the sake of compactness. The acceleration of gravity g is usually taken to be the vector sum of the gravitational attraction per unit mass and the centrifugal acceleration due to the rotation of the planet:

$$g(r) = GM/r^2 - \omega^2 r \cos^2 \phi, \qquad (80.2)$$

where $r = r_0 + z$ is the distance from the center of the planet, r_0 is the planetary radius, M is the planetary mass, $G = 6.670 \times 10^{-8}$ dyne cm^2 g^{-2} is the gravitational constant, ϕ is the latitude, and ω is the angular velocity of the planet.

When the hydrostatic relation (80.1) is combined with the ideal gas law in the form

$$P = nk_B T, \qquad (80.3)$$

where k_B is Boltzmann's constant and T is the temperature, and integrated, the the barometric formula

$$P(z) = P_0 \exp\left(-\int_{z_0}^{z} \frac{1}{H(z')} dz'\right), \qquad (80.4)$$

for the pressure $P(z)$ above a reference level (denoted by the subscript 0) as a function of altitude results. The pressure scale height $H(z)$ is defined as

$$H(z) = \frac{k_B T}{mg}. \qquad (80.5)$$

In the lower and middle atmosphere, the mass m in Eq. (80.5) is the weighted average mass of the atmospheric constituents.

When the ideal gas law (80.3) is substituted into the barometric formula (80.4), the altitude distribution

$$n(z) = n_0 \frac{T_0}{T(z)} \exp\left(-\int_{z_0}^{z} \frac{1}{H(z')} dz'\right). \qquad (80.6)$$

for the number density $n(z)$ above a reference altitude is obtained. Integration of (80.1) or (80.6) from z to infinity shows that the column density above that altitude is approximately $N(z) = n(z)H(z)$. Thus the scale height can be thought of as the effective thickness of the atmosphere.

In its lower and middle regions, the homosphere, the atmosphere is well-mixed by convection and/or turbulence. The upper boundary of this region is called the homopause (or turbopause), and above this level, the major transport process is diffusion. The homopause is defined as the level at which the time constants for mixing and diffusion are equal, and usually occurs at $n(z) \sim 10^{11} - 10^{13}$ cm^{-3}, depending on the strength of vertical mixing for a given planet. Since molecular diffusion coefficients vary from one species to another, the exact altitude of the homopause is species-dependent, with smaller species having lower homopause altitudes. In the terrestrial atmosphere, the homopause is near 100 km at $n(z) \sim 10^{13}$ cm^{-3}. Below the homopause, the mixing ratios (or fractions by number) of the constituent gases, apart from those minor or trace species whose density profiles are determined by photochemistry or physical loss processes, are fairly constant with altitude.

Throughout the atmosphere, gravity exerts a force on each particle that is proportional to its mass. Below the homopause, however, the tendency of the species

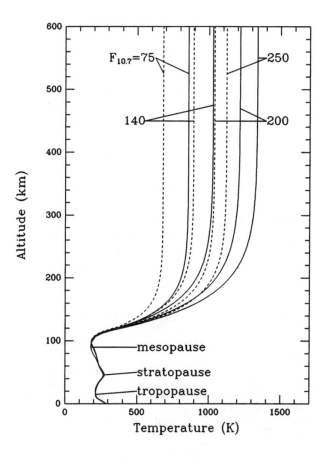

Figure 80.1. Vertical distribution of temperature in the terrestrial atmosphere. The altitudes of the tropopause, stratopause and mesopause are indicated. The thermospheric temperatures depend on solar activity and profiles are shown for four values of the $F_{10.7}$ index, from 75 (low solar activity) to 250 (high solar activity). The solid and dashed curves are for noon and midnight, respectively. From the MSIS model of Hedin [15].

to separate out under the force of gravity is overpowered by large scale mixing processes, such as convection and turbulence. Above the homopause, each species is distributed according to its own scale height. Characteristics of the homopauses of the planets are presented in Table 80.1.

80.1.2 Atmospheric Regions

The division of atmospheres into regions is based on the temperature structure of the terrestrial atmosphere, which is shown in Fig. 80.1. In the troposphere of a planet, above the boundary layer, T decreases at close to the adiabatic lapse rate (Γ) for the constituent gases from the surface to the tropopause. For an atmosphere that is a mixture of ideal gases, $\Gamma = g/c_p$, where c_p is the specific heat of the gas mixture at constant pressure.

Table 80.1. Homopause characteristics of planets and satellites.

Planet	Altitude (km)	K cm^2 s^{-1}	T (K)	n_t (cm^{-3})	P (μbar)	Composition (Fraction by number)
Venus[b]	135	4(7)[a]	199	1.4(11)	4.7(-3)	CO_2(76%), N_2(7.6%), O(9.3%), CO(6.7%), N(0.16%), C(0.01%)
Earth[c]	100	1(6)	185	1.3(13)	0.3	N_2(77%), O_2(18%), O(3–4%), Ar(0.7%), He(9.5ppm), H(1.3 ppm)
Mars[d]	120–125	5(7)	154	1.3(11)	2.8(-3)	CO_2(95%), N_2(2.5%), Ar(1.5%), O(0.9%), CO(0.42%), O_2(0.12%), NO(0.007%)
Jupiter[e]	440[f]	1.4(6)	200	1.4(13)	0.4	H_2(95%), He(4.1%), H(0.055%), CH_4(40 ppb), C_2H_2(1.2 ppb), C_2H_4(2.5 ppb), C_2H_6(0.12 ppb)
Saturn[g]	1100[f]	1.3(8)	200	1.2(11)	3(-3)	H_2(94%), He(6%), CH_4(178 ppm)
Uranus[h]	300[f]	1(4)	130	1(15)	20	H_2(85%), He(15%), CH_4(20 ppm), C_2H_2(10 ppb), C_2H_6(0.1 ppb), C_4H_2(0.05 ppt)
Neptune[i]	750[f]	2(7)	280	1.2(12)	3.9(-2)	H_2(83%), He(16%), CH_4(3 ppm)
Titan[j]	1050	1(9)	175	2.5(9)	1(-4)	N_2(93%), CH_4(6%), C_2H_2(0.42%), C_2H_4(0.194%), C_2H_6(404 ppm), C_4H_2(105 ppm)
Triton[k]	30	3(3)	40	3.3(14)	1.8	N_2(99.9%), H_2(190 ppm), CH_4(37 ppm), H(3.1 ppm), N(4.1 ppb)

[a] Read as 4×10^7.
[b] K from von Zahn *et al.* [1] and model atmosphere from Hedin *et al.* [3], for 1500 hrs, 15° N latitude, $F_{10.7}$ =150.
[c] From The US Standard Atmosphere [6].
[d] Viking model from Ref. [8].
[e] From Ref. [2] and photochemical model of Kim [12].

[f] Altitude above the 1 bar level.
[g] From Ref. [2].
[h] From Refs. [4, 5].
[i] ¿From Ref.[7].
[j] From Refs. [9–11].
[k] From Refs. [13, 14].

The presence of a condensible constituent, such as water vapor in the terrestrial troposphere, and ammonia or methane in the atmospheres of the outer planets and satellites, decreases Γ because upward motion leads to cooling and condensation, which releases latent heat. On Earth, the dry adiabatic lapse rate is about 10 K km^{-1} and the moist adiabatic lapse rate is about 4–6 K km^{-1} in the lower to middle troposphere. The average lapse rate is about 6.5 K km^{-1}, and the altitude of the tropopause varies from about 9 to 16 km from the poles to the equator. The composition of the lower atmosphere of the Earth is given in Table 80.2.

Above the terrestrial tropopause lies the stratosphere, a region of increasing T that is terminated at the stratopause, near 50 km. This increase in T is caused by absorption of solar near uv radiation by ozone in the Hartley bands and continuum (200–310 nm). In the terrestrial mesosphere, which lies above the stratosphere, T decreases again to an absolute minimum at the mesopause, where $T \sim 180$ K and $n(z) \sim 10^{14}$ cm^{-3}. Above the mesopause, in the thermosphere, T increases rapidly to a constant value, the exospheric temperature, T_∞. The value of T_∞ in the terrestrial atmosphere depends on solar activity and is usually between about 700 and 1500 K. Fig. 80.1 also shows altitude profiles of the noon and midnight thermospheric temperature for four values of the $F_{10.7}$ index, (the 2800 MHz flux in units of

10^{-22} W m^{-2} Hz^{-1} at 1 AU), which represent different levels of solar activity.

The exosphere is a nearly collisionless region of the thermosphere that is bounded from below by the exobase. A particle traveling upward at or above the exobase will, with high probability, escape from the gravitational field of the planet. The exobase on Earth is located at about 450–500 km, depending upon solar activity.

The surface P and T on Mars are about 6 mbar and 230 K, respectively. Due to the effect of dust storms, the extent of the Martian troposphere is highly variable, with a lapse rate that is 2–3 K km^{-1} compared with the adiabatic lapse rate of 4.5 K km^{-1} and a variable thickness of 20–50 km [17]. The atmosphere of Mars, like many planetary atmospheres, does not have a stratosphere. A roughly isothermal mesosphere extends from the tropopause to the base of the thermosphere at about 90 km. The thermospheric T is sensitive to solar activity and, since Mars has a very eccentric orbit, to heliocentric distance; T_∞ varies from about 180 to 350 K.

Near the surface of Venus, $T \gtrsim 700$ K and $P \sim 95$ bar. T decreases with a mean lapse rate of 7.7 K km^{-1}, compared with the adiabatic lapse rate of 8.9 K km^{-1}, from the surface to about 50 km. The region from 50 to 60 km contains the major cloud layer, and the tropopause is usually considered to be at about 60 km. In the mesosphere, between about 60 and 85 km, T decreases

Table 80.2. Molecular weights and fractional composition of dry air in the terrestrial atmosphere.[a]

Species	Molecular weight (g mole^{-1})	Fraction by volume
N_2	28.0134	0.78084
O_2	31.9988	0.209476
Ar	39.948	0.00934
CO_2	44.00995	0.00035624[b]
Ne	20.183	0.00001818
He	4.0026	0.00000524
Kr	83.80	0.00000114
Xe	131.30	0.000000087
CH_4	16.04303	0.000002
H_2	2.01594	0.0000005

[a] Taken from The US Standard Atmosphere [6], except as noted.
[b] 1992 annual average value. The CO_2 mixing ratio is increasing at an annual rate of 0.2–0.4%. Value is from Ref. [16].

Table 80.3. Composition of the lower atmospheres of Mars and Venus.[a]

Species	Mixing ratio Mars	Venus
CO_2	0.953	0.96
N_2	0.027	0.04
^{40}Ar	0.016	50–120 ppm[b]
O_2	0.0013	20–40 ppm
CO	0.0008	20–30 ppm
H_2O	0.0003[c]	30 ppm[d]
He	1.1 ppm[e]	10 ppm
Ne	2.5 ppm	5–13 ppm
Kr	0.3 ppm	<0.02–0.4 ppm
Xe	0.08 ppm	—
SO_2	—	150 ppm
H_2S	—	1–3 ppm

[a] From Ref. [18], except as noted.
[b] Includes all isotopes of Ar.
[c] Variable.
[d] From Ref. [19].
[e] From Krasnopolsky *et al.* [20].

Table 80.4. Composition of the lower atmospheres of Jupiter and Saturn.[a]

Species	Mixing ratio Jupiter	Saturn
H_2	0.89	0.94
He	0.11	0.06
CH_4	0.00175	0.0045
NH_3	180 ppm	
H_2O	1–30 ppm	
C_2H_6	5 ppm	4.8 ppm
PH_3	0.6 ppm	2 ppm
C_2H_2	0.02 ppm	0.11 ppm

[a] After Strobel [21].

slowly from about 250 K to 180 K, and is nearly constant from 85 km to the mesopause at 100 km. The daytime exospheric temperature is only weakly dependent on solar activity, varying from about 230 to 300 K from low to high solar activity. The slow retrograde rotation of the planet, which results in a period of darkness that lasts 58 days, leads to the relative isolation of the nightside thermosphere, where T is found to decrease above the mesopause to an exospheric $T \sim 100$ K. Because of this, the nightside Venus thermosphere has been called the "cryosphere". The compositions of the lower atmospheres of Mars and Venus are given in Table 80.3.

The giant planets, Jupiter, Saturn, Uranus and Neptune do not have solid surfaces, so their atmospheric regions are defined either in terms of pressure, altitude above the 1 bar level, or altitude above the cloud tops. The temperature structures of all but Uranus are influenced by internal heat sources that the terrestrial planets do not possess. The temperature structures near the tropopause can be determined from ir observations and radio occultation data, and at thermospheric altitudes from ultraviolet solar and stellar occultations performed by the Voyager spacecraft. In between these regions, there is a substantial gap in which only average temperatures can be inferred. Thus the location and temperature of the mesopauses are largely unknown.

Below 300 mbar on Jupiter, the lapse rate is close to adiabatic (1.9 K km^{-1}). T at 1 bar is about 165 K, and the tropopause occurs near 140 mbar, where $T \sim 110$ K. At 1 mbar, T again reaches 160–170 K. Temperature inversions have been reported in the stratosphere, and are probably due to absorption of solar radiation by dust or aerosols. Temperatures derived from the Voyager uv stellar and solar occultations show that T increases from about 200 K near 1 μbar to an exospheric value of about 1100 K [2].

For $P > 500$ mbar on Saturn, the lapse rate approaches the adiabatic value of 0.9 K km^{-1}, and the tropopause, near the 100 mbar level, is characterized by $T \sim 80$ K. Above the tropopause the temperature increases to about 140 K near a $P \sim 1$ mbar, and above that altitude there are no measurements of T up to a pressure of about 10^{-8} bars, about 1000 km above the 1 bar level, where T is again about 140 K. Application of the hydrostatic equation to the altitude range 300–1000 km yields an average temperature near 140 K for the region. Above 1000 km, T increases to a $T_\infty \sim 800$ K [2]. The mixing ratios of the species in the lower atmospheres of Jupiter and Saturn are given in Table 80.4.

The tropopauses on both Uranus and Neptune occur

Table 80.5. Composition of the lower atmospheres of Uranus and Neptune.[a]

Species	Mixing ratio	
	Uranus	Neptune
H_2	~0.825	~0.80
He	~0.152	~0.19
CH_4	~0.023	~0.01–0.02
HD	~148 ppm	~192 ppm
CH_3D	~8.3 ppm	~12 ppm
C_2H_6	~1–20 ppb	~1.5 ppm
C_2H_2	~10 ppb	~60 ppb

[a] After Fegley [22].

Table 80.6. Composition of the lower atmosphere of Titan.[a]

Species	Mixing ratio
N_2	0.90–0.98
CH_4	0.026–0.034[b]
H_2	2.0×10^{-3}
CO	60–150 ppm
C_2H_6	20 ppm
C_3H_8	4 ppm
C_2H_2	2 ppm
C_2H_4	0.4 ppm
HCN	0.2 ppm

Titan's atmosphere may also contain up to 14% Ar [9].
[a] From Ref. [23], except as noted.
[b] Tropopause value. From Ref. [9].

Table 80.7. Composition of the atmosphere of Triton.[a]

Species	Mixing ratio	Comments
N_2	0.99±0.01	Below ~ 200 km
CO	0.0001–0.01	Uncertain
CH_4	113 ppm	
H_2	75 ppm	
N	3.8×10^{-5} ppm	
N	290 ppm	100 km (near peak)
H	0.092 ppm	
H	1 ppm	30 km (near peak)
C_2H_4	3.9×10^{-4} ppm	
C_2H_4	2.9×10^{-2} ppm	26 km (near peak)

[a] From Ref. [13]. Values are at 10 km, except as noted.

Table 80.8. Number densities of species at the surface of Mercury.[a]

Species	Number density (cm^{-3})
H	230
hot H	23
He	6×10^3
O	4.4×10^4
Na	$(1.7 - 3.8) \times 10^4$
K	5×10^2

[a] From Hunten *et al.* [25].

near 100 mbar, where $T \sim 50$ K. The lapse rates in the troposphere are 0.7 and 0.85 K km^{-1} for Uranus and Neptune, respectively. The temperatures in the Uranus thermosphere range from 500 K near 10^{-7} bar (about 1000 km above the 1 bar level) to an exospheric value of about 800 K. At 300 km on Neptune, T attains a nearly constant value in the range 150 to 180 K. At 600 km where $P \sim 1$ μbar, T increases again to a value that is probably about 600 K. Because the Voyager data have not been fully analyzed, the value of T_∞ is uncertain [11]. The compositions of the lower atmospheres of Uranus and Neptune are given in Table 80.5.

Titan, which is a satellite of Saturn, has an N_2/CH_4 atmosphere of intermediate oxidation state. The mixing ratios of components of the lower atmosphere are given in Table 80.6. The surface P and T are 1.496 bar and 94 K, respectively. T decreases above the surface to about 71 K at the tropopause, which occurs at an altitude of 42 km and a pressure of 128 mbar. There are no measurements of T in the mesosphere, but a model has predicted the existence of a mesopause at about 0.1 μbar (600 km), with $T \sim 130$–140 K [24]. Above 600 km, T rises to a

T_∞ of 175–185 K.

Triton is a satellite of Neptune. It also has an N_2 atmosphere with small amounts of methane, CO, H_2, and other species. The mixing ratios at 10 km are given in Table 80.7. The surface P is about 14 μbar. Methane in the troposphere is in equilibrium with a surface methane frost at about 38 K. The tropopause temperature is about 36 K, and occurs in the 8 to 12 km region. The base of the thermosphere is near the tropopause, and in the thermosphere, T rises to an exospheric value of about 100 K.

Mercury does not have a troposphere, mesosphere, or stratosphere; the surface of the planet is the exobase. Nonetheless, several atomic species have been identified in fluorescence. They are listed along with the inferred densities in Table 80.8.

The atmosphere of Pluto is pure N_2, and the surface P and T are about 1 μbar and 80 K, respectively. Although the surface P is about the same as that at the base of the thermosphere on most planets, the structure of the atmosphere is influenced by the large thermal escape flux at the top of the atmosphere and by adiabatic cooling. T maximizes near two Pluto radii at ~ 130 K due to absorption of solar uv radiation. Above that altitude, T decreases asymptotically to a value of about 100 K [11].

80.2 DENSITY DISTRIBUTIONS OF NEUTRAL SPECIES

80.2.1 The Continuity Equation

The density distribution of a minor neutral species j in an atmosphere is determined by the continuity equation:

$$\frac{\partial n_j}{\partial t} + \nabla \cdot \mathbf{\Phi}_j = P_j - L_j , \qquad (80.7)$$

where $\mathbf{\Phi}_j$ is the flux of species j, and P_j and L_j are the chemical production and loss rates, respectively. If only the vertical direction is considered, the divergence of the flux becomes $\partial \Phi_j / \partial z$, and $\Phi_j = n_j w_j$ where w_j is the vertical velocity of the species and n_j is its number density. In one-dimensional models, transport due to turbulence and other macroscopic motions of air masses is often parametrized like molecular diffusion, using an eddy diffusion coefficient K in place of the molecular diffusion coefficient D_j. The total transport velocity w_j is then the sum of the diffusion velocity w_j^{D} and the eddy diffusion velocity w_j^{K}:

$$w_j = w_j^{\mathrm{D}} + w_j^{\mathrm{K}} . \qquad (80.8)$$

If there are no net flows of major constituents, w_j^{D} and w_j^{K} satisfy the equations

$$w_j^{\mathrm{D}} = -D_j \left[\frac{1}{n_j} \frac{dn_j}{dz} + \frac{1}{H_j} + \frac{(1+\alpha_j^{\mathrm{T}})}{T} \frac{dT}{dz} \right] , \qquad (80.9)$$

$$w_j^{\mathrm{K}} = -K \left[\frac{1}{n_j} \frac{dn_j}{dz} + \frac{1}{H_{\mathrm{avg}}} + \frac{1}{T} \frac{dT}{dz} \right] . \qquad (80.10)$$

In these expressions, α_j^{T} is the thermal diffusion factor (the ratio of the thermal diffusion coefficient to the molecular diffusion coefficient), and the pressure scale height H_{avg} for a mixed atmosphere is given by Eq. (80.5) with $m = m_{\mathrm{avg}}$, the average molecular mass.

For a stationary atmosphere, if molecular diffusion greatly exceeds eddy diffusion and if photochemistry can be neglected, then $w_j^{\mathrm{D}} = 0$. The resulting number density distribution is called diffusive equilibrium, and is given by

$$n_j(z) = n_j(z_0) \left(\frac{T_0}{T} \right)^{(1+\alpha_j^{\mathrm{T}})} \exp\left(-\int_{z_0}^{z} \frac{dz'}{H_i} \right) . \qquad (80.11)$$

When mixing processes dominate and $w_j^{\mathrm{K}} = 0$, the distribution is given by Eq. (80.6), with $H = H_{\mathrm{avg}}$.

80.2.2 Diffusion Coefficients

In the thermosphere of a planet, above the homopause, the major transport mechanism is diffusion, or transport by random molecular motions. The characteristic time τ_{D} for molecular diffusion is approximately H_j^2 / D_j. The diffusion coefficient for a species j in a multicomponent mixture is usually taken as a weighted mean of binary diffusion coefficients D_{jk}:

$$\frac{1}{D_j} = \sum_{k \neq j} \frac{f_k}{D_{jk}} , \qquad (80.12)$$

where f_k is the mixing ratio of species k. The binary diffusion coefficient can be expressed as

$$D_{jk} = \frac{3k_{\mathrm{B}}T}{16 n_t \mu_{jk} \Omega_{jk}} , \qquad (80.13)$$

where μ_{jk} is the reduced mass:

$$\mu_{jk} = \frac{m_j m_k}{m_j + m_k} \qquad (80.14)$$

and $n_t = n_j + n_k$ is the total number density. The collision integral Ω_{jk} is given by

$$\Omega_{jk} = \frac{1}{2\pi^{1/2}} \left(\frac{\mu}{2k_{\mathrm{B}}T} \right)^{5/2}$$
$$\times \int_0^\infty Q^{\mathrm{D}}(v) v^5 \exp(-\mu v^2 / 2k_{\mathrm{B}}T) dv \qquad (80.15)$$

where v is the relative velocity of the particles, $Q^{\mathrm{D}}(v)$ is the diffusion or momentum transfer cross section

$$Q^{\mathrm{D}}(v) = 2\pi \int_0^\pi \sigma_{jk}^{el}(\theta, v)(1 - \cos\theta) \sin\theta \, d\theta , \qquad (80.16)$$

and $\sigma_{jk}^{\mathrm{el}}(\theta, v)$ is the differential cross section for elastic scattering of species j and k through angle θ.

In practice, D_{jk} is often expressed as b_{jk}/n_t where n_t is the total number density and b_{jk} is the binary collision parameter, which is usually given in tabulations in the semi-empirical form $b = AT^s$. Here A and s ($0.5 \leq s \leq 1.0$) are parameters that are fit to the data. The binary collision parameter appears, for example, in the expression for the diffusion limited flux of a light species to the exobase of a planet (see Sect. 80.7).

80.3 INTERACTION OF SOLAR RADIATION WITH THE ATMOSPHERE

80.3.1 Introduction

The source for all atmospheric processes is ultimately the interaction of solar radiation, either photons or particles, with atmospheric gases. Since visible photons arise from the photosphere of the sun, which is characterized by $T \sim 6000$ K, the solar spectrum in the visible and ir is similar to that of a black body at 6000 K. At longer

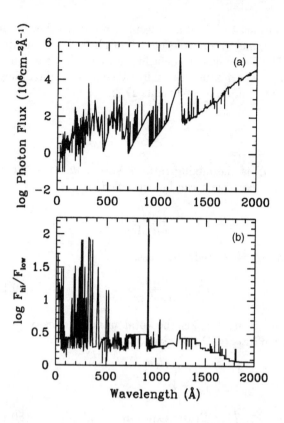

Figure 80.2. (a) Solar spectrum at 1 AU for 18–2000 Å. (b) Ratio of the flux at high solar activity to low solar activity. Plotted with data from Hinteregger [26].

Figure 80.3. Downward electron flux as a function of energy measured by electron spectrometers on board a rocket traversing an auroral arc near Poker Flat, Alaska. Reprinted from Ref. [27] with kind permission from Elsevier Science Ltd., The Boulevard, Langford Lane, Kidlington, OX5 1GB, UK.

(radio) and shorter (uv and x-ray) wavelengths, the photons arise from parts of the chromosphere and corona where the temperatures are higher (10^4 to 10^6 K). Thus the photon fluxes differ substantially from those which would be predicted for a 6000 K black body. Photons in the extreme and far uv regions of the spectrum are absorbed in the terrestrial thermosphere and x-rays in the lower thermosphere and mesosphere. The solar Lyman α line at 1216 Å penetrates through a window in the O_2 absorption cross sections to about 75 km. Near uv photons are absorbed by ozone in the stratosphere, and visible radiation is not appreciably attenuated by the atmosphere.

The wavelength ranges that are most important for aeronomy are the uv and x-ray regions. A solar spectrum in the uv and soft x-ray regions at low solar activity is presented in Fig. 80.2a, and the ratio of a high solar activity photon fluxes to those at low solar activity is shown in Fig. 80.2b. The ratio is near unity at wavelengths longward of 2000 Å, but increases to factors that range between 2 and 3 over much of the extreme uv. At wavelengths between about 100 and 550 Å, the ratio of high to low solar activity fluxes reaches values as high as 100. The fluxes at x-ray wavelengths arise principally from solar flares and can increase by orders of magnitude

from low to high solar activity.

The sun also emits a stream of charged particles, the solar wind, which flows radially outward in all directions, and consists mostly of protons, electrons, and alpha particles. The average number density of solar wind protons is about 5 cm^{-3}, and the average speed is about 400–450 km s^{-1} at Earth orbit (1 AU). The interaction of these particles with the magnetic field (either induced or intrinsic) of a planet, and ultimately with the atmosphere, is the source of auroral activity. Terrestrial auroras arise mostly from precipitation of electrons with energies in the kilovolt range, although measured spectra vary widely. An example of a primary electron auroral spectrum is shown in Fig. 80.3. Terrestrial auroral emissions maximize in the midnight sector, but dayside cusp auroras are produced by lower energy electrons, and diffuse proton auroras are also observed.

Since charged particles are constrained to move along magnetic field lines, for planets with intrinsic magnetic fields, auroras usually occur in an oval near the magnetic poles, where the dipole field lines enter the atmosphere. For Venus, which has no intrinsic magnetic field, auroras are seen as diffuse and variable emissions on the nightside of the planet. On Earth, low latitude auroras, which arise from heavy particle precipitation, have also been observed. The primary particles that are responsible for Jovian aurora may be heavy ions originating from its satellite Io, protons, or electrons. Due to charge transfer, heavy particles spend part of their lifetime as neutral species, and their paths may then diverge from magnetic field lines. In any case, a large fraction of the effects of auroral precipitation is due to secondary electrons, regardless of the identity of the primary particles. In addition to producing emissions of atmospheric species in the visible, uv and ir portions of the spectrum, auroral particles ionize and dissociate atmospheric species and contribute to heating the neutrals, ions and electrons.

80.3.2 Interaction of Solar Photons with Atmospheric Gases

The number flux of solar photons in a small wavelength interval around λ at an altitude z can, for the most part, be computed from the Beer-Lambert absorption law

$$F_\lambda(z) = F_\lambda^\infty \exp[-\tau(\lambda, z)], \qquad (80.17)$$

where F_λ^∞ is the solar photon flux outside the atmosphere, and $\tau(\lambda, z)$ is the optical depth which, in the plane parallel approximation, is given by

$$\tau(\lambda, z) = \sum_j \int_z^\infty n_j(z')\sigma_j^a(\lambda) \sec\chi \, dz'. \qquad (80.18)$$

Here, $\sigma_j^a(\lambda)$ is the absorption cross section of species j at wavelength λ, and the solar zenith angle χ is the angle of the sun with respect to the local vertical.

For χ greater than about $75°$, the variation of the solar zenith angle along the path of the radiation cannot be neglected; the optical depth must be computed by numerical integration along this path in spherical geometry. For $\chi \leq 90°$ the optical depth is

$$\tau(\lambda, z) = \sum_j \int_z^\infty n_j(z')\sigma_j^a(\lambda)$$
$$\times \left[1 - \left(\frac{r_o + z}{r_o + z'}\right)^2 \sin^2\chi\right]^{-0.5} dz'. \qquad (80.19)$$

For χ larger than $90°$, the optical depth is given by

$$\tau(\lambda, z)$$
$$= \sum_j \left\{ 2\int_{z_s}^\infty n_j(z')\sigma_j^a(\lambda)\left[1 - \left(\frac{r_o + z_s}{r_o + z'}\right)^2\right]^{-0.5} dz' \right.$$
$$\left. - \int_z^\infty n_j(z')\sigma_j^a(\lambda)\left[1 - \left(\frac{r_o + z}{r_o + z'}\right)^2 \sin^2\chi\right]^{-0.5} dz' \right\}, \qquad (80.20)$$

where z_s is the tangent altitude, the point at which the solar zenith angle is $90°$ for the path of solar radiation through the atmosphere.

In a one-species atmosphere, the rate of absorption of solar photons of wavelength λ is

$$q^a(\lambda) = F_\lambda \sigma^a(\lambda) n. \qquad (80.21)$$

For an isothermal atmosphere in which $H(z) \approx$ const., the absorption maximizes where $\tau(\lambda, z) = 1$. This is a fairly good approximation even for regions of the atmosphere where the $H(z)$ is not constant. The altitude of unit optical depth is shown for wavelengths from x-rays to the near uv for overhead sun in the terrestrial atmosphere in Fig. 80.4a. Similar plots for Venus, Mars and Jupiter are shown in Fig. 80.4b, Fig. 80.4c, and Fig. 80.4d, respectively. N_2 does not absorb longward of about 100 nm, so in the terrestrial atmosphere, O_2 and O_3 are the primary absorbers between 100 and 220 nm, while ozone dominates the absorption for wavelengths in the range 220–320 nm. On Venus and Mars, CO_2 is the main absorber of fuv and euv radiation, although at wavelengths less than about 100 nm, N_2, CO, and O also contribute. On Titan, methane is the primary absorber of uv radiation between 1400 Å and the absorption threshold of N_2, near 1000 Å.

The interaction of uv photons with atmospheric gases produces ions and photoelectrons through photoionization, which may be represented as

$$\mathrm{X} + h\nu \rightarrow \mathrm{X}^+ + \mathrm{e}^-, \qquad (80.22)$$

and photodissociative ionization

$$\mathrm{AB} + h\nu \rightarrow \mathrm{A}^+ + \mathrm{B} + \mathrm{e}^-. \qquad (80.23)$$

In these equations, X represents any atmospheric species; A is either an atom or a molecular fragment and AB a molecule. The energy of the photoelectron in reaction (80.22) is given by

$$E_{\mathrm{pe}} = h\nu - I_{\mathrm{X}} - E_{\mathrm{ex}} \qquad (80.24)$$

and in reaction (80.23) is

$$E_{\mathrm{pe}} = h\nu - E_{\mathrm{d}} - I_{\mathrm{A}} - E_{\mathrm{ex}}, \qquad (80.25)$$

where I_j is the ionization potential of species j, E_{d} is the dissociation energy of molecule AB, and E_{ex} is the internal excitation energy of the products. Neutral fragments, which may be reactive radicals, are also produced in photodissociation

$$\mathrm{AB} + h\nu \rightarrow \mathrm{A} + \mathrm{B}. \qquad (80.26)$$

The rate of ionization of a species j by a photon of wavelength λ at an altitude z is given by

$$q_j^i(\lambda, z) = F_\lambda(z)\sigma_j^i(\lambda)n_j(z), \qquad (80.27)$$

where $\sigma_j^i(\lambda)$ is the photoionization cross section. The rate for photodissociation is given by a similar expression,

Figure 80.4. The altitude where $\tau = 1$ versus wavelength (a) Earth [28] (b) Venus (c) Mars [29] (d) Jupiter (Y. H. Kim, unpublished).

with the photoionization cross section replaced by the photodissociation cross section. The expression above must be integrated over the solar spectrum to give the total rate. In addition, it is often necessary to take into account ionization and/or dissociation to different final internal states of the products, so the partial cross sections or yields are needed.

In the atmospheres of magnetic planets, photoelectrons may travel upward along the magnetic field lines to the conjugate point, where the field line re-enters the atmosphere. In order to model this effect, the differential (with respect to angle) cross sections for photoionization $\sigma_j^i(\lambda, \theta)$ are necessary. The differential cross section is sometimes expressed as

$$\sigma_j^i(\lambda, \theta) = \frac{\sigma_j^i(\lambda)}{4\pi}[1 - \tfrac{1}{2}\beta(\lambda)P_2(\cos\theta)], \qquad (80.28)$$

where θ is the angle between the incident photon beam and the ejected electron, P_2 is a Legendre polynomial, and β is an asymmetry parameter.

80.3.3 Interaction of Energetic Electrons with Atmospheric Gases

Suprathermal electrons, which are denoted here e^{*-}, and include both photoelectrons and auroral primary electrons, can also ionize species through electron-impact ionization

$$X + e^{*-} \rightarrow X^+ + e^- + e'^- \qquad (80.29)$$

and electron-impact dissociative ionization

$$AB + e^{*-} \rightarrow A^+ + B + e^- + e'^-. \qquad (80.30)$$

In these reactions, e^- represents the energy degraded photoelectron or primary electron, and e'^- the secondary electron. The energy of the secondary electron $E_{e'}$ in an electron-impact ionization process (80.29) is given by

$$E_{e'} = E_{e*} - I_X - E_{ex} - E_e, \qquad (80.31)$$

where E_{e*} is the energy of the primary or photoelectron, E_e is the energy of the degraded primary or photoelectron, and E_{ex} is the internal excitation energy of the product ions and/or neutral fragments. For the dissociative ionization process (80.30) the dissociation energy of the molecule must also be subtracted as well.

Energetic electrons can also dissociate atmospheric species. In this process

$$AB + e^{*-} \rightarrow A + B + e^- \qquad (80.32)$$

the energy of the degraded electron is

$$E_e = E_{e*} - D_{AB} - E_{ex} \qquad (80.33)$$

where D_{AB} is the dissociation energy of molecule AB. Collisions with suprathermal electrons can also promote species to excited electronic, vibrational or rotational states:

$$AB + e^{*-} \rightarrow AB^{\dagger} + e^- , \qquad (80.34)$$

where the dagger denotes internal excitation. The energy lost by the electron is thus the excitation energy of the species.

In determining the rate of ionization, dissociation and excitation by photoelectrons, the local energy loss approximation, that is, the assumption that the electrons lose their energy at the same altitude where they are produced, is fairly good near the altitude of peak photoelectron production. The mean free path of an electron near 150 km is about 30 m. Substantially above the altitude of peak production of photoelectrons, transport of electrons from below is important, and use of the local energy loss assumption causes the excitation, ionization, and dissociation rates to be underestimated. For keV auroral electrons, the computation of the energy deposition of the electrons must consider their transport through the atmosphere. Thus the elastic total and differential cross sections for electrons colliding with neutral species must be employed, as well as the inelastic cross sections, and the angles through which the electrons scatter must be taken into account.

In general, the excitation rate $q_j^k(z)$ of a species j to an excited level k with a threshold energy E_k at an altitude z by electron impact is given by:

$$q_j^k(z) = n_j(z) \int_{E_k}^{\infty} \sigma_j^k(E) \frac{dF(z,E)}{dE} dE , \qquad (80.35)$$

where $\sigma_j^k(E)$ is the excitation cross section at electron energy E, and $dF(z,E)/dE$ is the differential flux of electrons (between energies E and $E + dE$). The ionization rate $q_j^i(z)$ of a species with ionization potential I_j due to electron impact is given by

$$q_j^i(z)$$
$$= n_j(z) \int_{I_j}^{\infty} \int_0^{(E-I_j)/2} \frac{d\sigma_j^i(E)}{dW_s} \frac{dF(z,E)}{dE} dW_s dE ,$$
$$\qquad (80.36)$$

where $d\sigma_j^i(E)/dW_s$ is the differential cross section for production of a secondary electron with energy W_s by a primary electron with energy E. The integral over secondary energies W_s terminates at $(E - I_j)/2$ because the secondary electron is by convention considered to be the one with the smaller energy. Since the average energy of photoelectrons is less than 20 eV, the error incurred in cutting off the integrals in equations (80.35) and (80.36) at 200 eV or so, rather than $(E - I_j)/2$ is not serious, although for high energy auroral electrons a larger upper limit may be required.

An estimate of the number of ionizations in a gas produced by a primary electron with energy E_p is E_p/W_{ip}, where W_{ip} is the energy loss per ion pair produced, which approaches a constant value as the energy of the electron increases. Empirical values are available for W_{ip} for many gases, and usually fall in the range 30–40 eV [30].

The total loss function or stopping cross section for an electron with incident energy E in a gas j is given by the expression

$$L_j(E) = \sum_k \sigma_j^k(E) W_j^k$$
$$+ \int_0^{(E-I_j)/2} (I_j + W_s) \frac{d\sigma_j^i(E)}{dW_s} dW_s , \quad (80.37)$$

where W_j^k is the energy loss associated with excitation of species j to excited state k. The differential cross section is usually adopted from an empirical formula that is normalized so that

$$\sigma_j^i(E) = \int_0^{(E-I_j)/2} \frac{d\sigma_j^i(E)}{dW_s} dW_s , \qquad (80.38)$$

where $\sigma_j^i(E)$ is the total ionization cross section at primary electron energy E. One formula in common use is that employed by Opal, Peterson and Beaty [31] to fit to their data:

$$\frac{d\sigma_j^i(E)}{dW_s} = \frac{A(E)}{1 + \left(W_s/\overline{W}\right)^{2.1}} , \qquad (80.39)$$

where $A(E)$ is a normalization factor and $\overline{W}$ is an empirically determined constant, which has been found to be equal to within a factor of about 50% to the ionization potential for a number of species.

For energy loss due to elastic scattering by thermal electrons, an analytic form of the loss function such as that proposed by Swartz *et al.* [32] may be used:

$$L_e(E) = \frac{3.37 \times 10^{-12}}{E^{0.94} n_e^{0.03}} \left(\frac{E - k_B T_e}{E - 0.53 k_B T_e}\right)^{2.36} , \quad (80.40)$$

where T_e is the electron temperature and n_e is the number density of ambient thermal electrons.

For high energy auroral electrons, the rate of energy loss per electron per unit distance over the path s of the electrons in the atmosphere can be estimated using the continuous slowing down approximation (CSDA) as

$$-\frac{dE}{ds} = \sum_j n_j(z) L_j(E) \sec\theta + n_e(z) L_e(E) \sec\theta ,$$
$$\qquad (80.41)$$

where θ is the angle between the path of the primary electron s and the local vertical. In the CSDA, all the electrons of a given energy are assumed to lose their energy continuously and at the same rate. The rate of energy loss $(-dE/ds)$ is integrated numerically over the path of the electron, which degrades in energy until it is thermalized. In this approximation, inelastic processes are assumed always to scatter the electrons forward, so cross sections that are differential in angle are not required. Because electrons actually lose energy at different rates, however, and because elastic and inelastic scattering processes do change the direction of the electrons, the CSDA gives an estimate for the rates of electron energy loss processes that is increasingly inaccurate as the energy of the electron decreases.

In practice, discrete energy loss of electrons can be easily treated numerically if the local energy loss approximation is valid. The spectrum of electrons is divided into energy bins that are smaller than the energy losses for the processes, and the integrals in equations (80.35)–(80.36) are replaced by sums over energy bins. Since elastic scattering of electrons by neutrals changes mostly the direction of the incident electron, and not its energy, only inelastic processes need be considered. In order to compute excitation and dissociation rates, only integral cross sections are required; the scattering angle is unimportant. For ionization, of course, the energy distribution of the secondary electrons must be considered, but not the scattering angles of either the primary or secondary electron. Below the lowest thresholds for excitations, energetic electrons lose their energy in elastic collisions with thermal electrons. The process of energy loss to thermal electrons is often approximated as continuous, rather than discrete.

The collision frequency ν_j^k for a discrete electron-impact excitation process k of a species j is given by

$$\nu_j^k(E) = n_j(z)v_e(E)\sigma_j^k(E).\tag{80.42}$$

For energy loss due to elastic scattering from thermal electrons, a pseudo-collision frequency ν_e may be defined as

$$\nu_e(E) = \frac{1}{\Delta E}\left(-\frac{dE}{dt}\right),\tag{80.43}$$

where ΔE is the grid spacing in the calculation, and the energy loss rate is

$$-\frac{dE}{dt} = v_e(E)n_eL_e(E),\tag{80.44}$$

where L_e is taken from Eq. (80.40).

Since the energy bins should be smaller than the typical energy loss in order to obtain accurate rates for the excitation processes, it is often convenient to treat rotational excitation also as a continuous process, with a pseudo-collision frequency similar to that for elastic scattering from ambient electrons [Eq. (80.43)] with

$$-\frac{dE}{dt} = v_e(E)n_jL_j^{rot}(E),\tag{80.45}$$

where the loss function for rotational excitation is given by

$$L_j^{rot}(E) = \sum_J \eta_j^J \sum_{J'} \sigma_j^{J,J'}(E)W_j^{J,J'}.\tag{80.46}$$

In this expression, η_j^J is the fraction of molecules j in rotational state J, $\sigma_j^{J,J'}(E)$ is the measured or computed cross section for electron-impact excitation of species j from rotational state J to rotational state J', and $W_j^{J,J'}$ is the associated energy loss.

The slowing down of high energy auroral primary electrons or photoelectrons arises from both elastic and inelastic scattering processes, and cannot be treated using the local energy loss approximation. In solving the equations for electron transport, the angle through which the primary electron is scattered, as well as the change in energy of the primary electron and the production of any secondaries, must be taken into account. Thus differential cross sections for the elastic and inelastic scattering of electrons by neutral species are required. The detailed equations for electron transport have been presented by, for example, Rees [33].

Several methods for approximating the energy deposition of auroral electrons are currently in use. The CSDA has already been discussed, but it provides only a rough approximation to the depth of penetration of the electrons, and the rates of excitation, ion production, and other energy loss processes. In the *two-stream approximation*, the electrons are assumed to be scattered in either the forward or backward direction [34]. Implementation of this method requires only the backscattering probabilities, rather than complete differential cross sections. The method has been generalized to multi-stream models, in which the solid angle range of the electrons is divided into 20 or more intervals, so more or less complete differential cross sections are required (e.g., [35, 36]). Monte Carlo methods have also been used to model auroral precipitation (e.g., [37]).

80.4 IONOSPHERES

80.4.1 Ionospheric Regions

The division of the ionosphere into regions is based on the structure of the terrestrial ionosphere, which consists of overlapping layers of ions. These layers are the result of changes both in the composition of the thermosphere and in the sources of the ionization, and are shown schematically in Fig. 80.5. The major molecular ion layer is the F_1 layer, which is produced by absorption of euv (100–1000 Å) photons by the major thermospheric species, and occurs where the ion production maximizes. The E layer is below the F_1 layer and is produced by shorter and longer wavelength photons that are absorbed

Figure 80.5. Ionospheric regions and primary ionization sources. From Bauer [38].

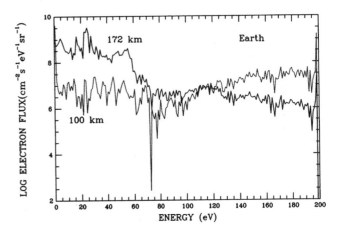

Figure 80.6. Primary photoelectron spectrum for the terrestrial atmosphere at 172 km (near the F_1 peak) and at 100 km. The spectrum at 100 km is significantly harder than that at 172 km.

deeper in the atmosphere: soft x-rays and Lyman β, which can ionize O_2 and NO (see Fig. 80.4a). In the D region, the densities of negative ions become appreciable and large densities of positive cluster ions appear. These ions are produced by harder x-rays, with $\lambda \lesssim 10$ Å, and Lyman α, which penetrates to about 75 km, where it ionizes NO. The highest altitude peak in the terrestrial ionosphere is the F_2 peak, which occurs near or slightly below 300 km, where the major ion is O^+. The peak density occurs where the chemical lifetime of the ion is equal to the characteristic time for transport by diffusion ($\sim H^2/D$).

80.4.2 Sources of Ionization

As discussed in Sect. 80.3, ionization can be produced either by solar photons and photoelectrons during the daytime or by energetic particles and secondary electrons during auroral events. Photoelectrons have sufficient energy to carry out further ionization if they are produced by photons with $\lambda \lesssim 500$ Å. These photons penetrate further and exhibit larger solar activity variations than longer-wavelength ionizing photons. Thus the ionization rate due to photoelectrons peaks below the main photoionization peak. Primary flux spectra of photoelectrons produced near the F_1 peak (172 km) and below the ion peak (100 km) are shown in Fig. 80.6. The primary spectrum at the ion peak consists mostly of low energy electrons, whereas at 100 km, the low energy primaries

are depleted, and there are relatively larger fluxes of electrons with $E \gtrsim 50$ eV. Fig. 80.7 shows the primary and steady-state photoelectron spectra near the ion peaks on Venus and Titan.

The major ions produced in the ionospheres of the earth and planets are usually those from the major thermospheric species: N_2^+, O_2^+, and O^+ on Earth; CO_2^+, O^+, N_2^+, and CO^+ on Venus and Mars; and H_2^+, H^+, and He^+ on the outer planets; N_2^+, N^+, and CH_4^+ in the ionosphere of Titan, and N_2^+, N^+, and C^+ in the ionosphere of Triton. In the presence of sufficient neutral densities, however, ion-molecule reactions transform ions whose parent neutrals have high ionization potential to ions whose parent neutrals have low ionization potentials. This is a rigorous rule only for charge transfer reactions, but it applies more often than not in other ion-molecule reactions as well.

Because of transformations by ion-molecule reactions, the major ions in the F_1 regions of the ionospheres of Earth, Venus and Mars are O_2^+ and NO^+, in spite of the large differences in composition between the thermosphere of the earth and the thermospheres of Venus and Mars. A diagram illustrating the ion chemistry in the terrestrial ionosphere is shown in Fig. 80.8. The vertical positions of the ions in this figure represent the relative ionization potentials of the parent neutrals. In regions where there are sufficient neutral densities the ionization flows downward.

Table 80.9 shows ionization potentials (I_P) for several major and minor species present in planetary thermospheres. Major atmospheric species generally have $I_P \gtrsim 12$–13 eV ($\lambda < 900$–1000 Å). Only a few species can be ionized by the strong solar Lyman alpha line (1216 Å, 10.2 eV), including NO, and a few small hydrocarbons and radicals, such as CH_3 and C_2H_5. Metal

Figure 80.8. Diagram illustrating the ion chemistry in the terrestrial ionosphere. The numbers under the ions indicate the ionization potential of the parent neutral. In the presence of sufficient neutral densities, the ionization flows downward.

Figure 80.7. Computed primary and steady-state spectra for photoelectrons near the F_1 peak on Venus at 1 eV resolution (top), and Titan at 0.5 eV resolution (bottom). The steady-state spectra are averaged over three intervals in both plots.

Table 80.9. Ionization potentials (I_P) of common atmospheric species[a]. Units are eV.

High I_P		Medium I_P		Ionized by Ly α	
Species	I_P	Species	I_P	Species	I_P
He	24.59	CH_4	12.51	CH_3	9.84
Ar	15.76	O_2	12.07	C_3H_6	9.73
N_2	15.58	C_2H_6	11.52	NO	9.26
H_2	15.43	C_2H_2	11.40	C_2H_5	8.13
N	14.53	C	11.26	C_3H_7	8.09
CO	14.01	C_3H_8	10.95	Mg	7.65
CO_2	13.77	CH	10.64	trans-HCNH	7.0[b]
O	13.62	C_2H_4	10.51	cis-HCNH	6.8[b]
H	13.60	CH_2	10.40	Ca	6.11
HCN	13.60	S	10.35	Na	5.14

[a] Computed with data taken from Ref. [39], except as noted.
[b] From Ref. [40].

atoms, which are produced in the lower thermospheres and mesospheres of planets from ablation of meteors, have very low ionization potentials, and some can be ionized by photons with wavelengths longer than 2000 Å. In ionospheres where hydrogen is abundant and sufficient neutral densities are present, ionization flows to species formed by protonation of neutrals that have large proton affinities. There are no *in situ* measurements of the ion composition of the outer planets, but models predict that H_3^+ and hydrocarbon ions dominate the lower ionospheres. In regions where meteor ablation occurs, metal ions may also be found.

Many ion-molecule reactions proceed at or near gas kinetic (or collision) rates. The interaction of an ion with a nonpolar molecule is dominated by the ion-induced-dipole interaction, for which the interaction potential is $-\frac{1}{2}\alpha_d q^2/r^4$, where α_d is the polarizability of the neutral, q is the charge on the ion, and r is the distance between the particles (see Sect. 11.5.2). The Langevin rate coefficient is then given by

$$k_L = 2\pi q(\alpha_d/\mu)^{1/2}, \qquad (80.47)$$

where μ is the reduced mass of the two species. For a singly charged ion, with α_d in Å^3 and μ in atomic mass units, this formula reduces to $2.34 \times 10^{-9}(\alpha_d/\mu)^{1/2}$. The rate coefficient for an ion with a polar molecule is

$$k_d = \frac{2\pi q}{\mu^{1/2}}\left[\alpha_d^{1/2} + c\mu_d\left(\frac{2}{\pi k_B T}\right)^{1/2}\right], \qquad (80.48)$$

where μ_d is the dipole moment and c is a constant that is unity in the locked dipole approximation and is about 0.1 in the average dipole orientation (ADO)

Figure 80.9. Model thermospheres for the Earth and planets. The curves are number density profiles and are labeled by the species they represent. (a) Earth, based on the MSIS model of Ref. [15] for a latitude of 45°, a local time of noon for low (top) and high (bottom) solar activities; (b) High solar activity model of Venus, based on the model of Ref. [3] for 15° N latitude 15 hrs local time; (c) Low solar activity model of Mars, based on Viking 1 measurements [8]. Adapted from Ref. [41]; (d) Jupiter. From Refs. [12,2]. (e) Saturn, from Ref. [2].

Figure 80.9. Continued. (f) Uranus, from Ref. [42]; (g) Neptune, [43]; (h) Pluto, [11]. The abscissa is given in units of Pluto radii, and the ordinate in the top plot is number density in cm^{-3}; the bottom is temperature in K. (i) Titan, from Ref. [9, 11]; and (j) Triton model, adapted from Ref. [13]. The short dashed curves are for solar ionization only, and the solid and long-dashed curves are for different assumptions about the interaction of Triton's thermosphere with electrons from Neptune's magnetosphere. The solid curves are the recommended model.

Figure 80.10. Ionospheric density profiles. (a) Measured profiles for Earth [44]; Computed profiles for (b) Venus, for high solar activity and a solar zenith angle of 45°; (c) Mars, for a low solar activity model for Viking 1 conditions and a solar zenith angle of 45 °; the dashed curves are measured profiles from Viking [45]; (d) Jupiter, from Ref. [46]; (e) Saturn, adapted from Ref. [47]; (f) Neptune, [43]. See also Ref. [48].

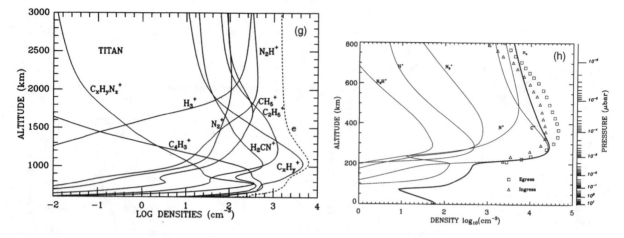

Figure 80.10. Continued. (g) Titan, from Fox and Yelle (unpublished, 1995); and (h) Triton [14].

theory. Theories for ion-quadrupole interactions have also been developed, and the resulting formulas can be found in, for example, the review by Su and Bowers [49]. Measured rate coefficients for ion-molecule reactions have been compiled by Anicich *et al.* [50] and Ikezoe *et al.* [51].

Loss of ionization in planetary atmospheres proceeds mainly by dissociative recombination of molecular ions, which may be represented by

$$AB^+ + e^- \rightarrow A + B. \qquad (80.49)$$

Dissociative recombination coefficients are characteristically large, about 10^{-7} cm^3 s^{-1}, at the electron temperatures T_e typical of planetary ionospheres, which are usually within a factor of two or so of the neutral $T \sim 200$–2000 K near the molecular ion density peak. Daytime peak electron densities are usually in the range 10^4–10^6 cm^{-3}, and fractional ionizations are small, about 10^{-5}, near the F_1 peak.

The relative importance of ion-molecule reactions and dissociative recombination in the destruction of a particular ion is determined by the relative densities of electrons and neutrals with which the ions can react. In general, molecular ions whose parent neutrals have high I_P's are transformed by ion-molecule reactions preferentially to loss by dissociative recombination, and their peak densities occur higher in the atmosphere. Ions for which dissociative recombination is an important loss mechanism near the ion peaks of the terrestrial planets include NO$^+$ and O$_2^+$, and in the atmospheres of the outer planets, H$_3^+$ and hydrocarbon ions. For ions with very high I_P's, such as N$_2^+$ and H$_2^+$, dissociative recombination is rarely important as a loss process, except at very high altitudes. It may, however, be important as a source of vibrationally or electronically excited fragments or hot atoms.

Atomic ions may be destroyed by radiative recombination:

$$X^+ + e^- \rightarrow X + h\nu, \qquad (80.50)$$

but the rate coefficients are small, about 10^{-12} cm^3 s^{-1} at the typical T_e's of planetary ionospheres [52]. Atomic ions may dominate at high altitudes, where neutral densities are low, but in such regions, loss by downward diffusion is more important than chemical recombination. The ions diffuse downward to altitudes where the neutral densities are higher and are then destroyed in ion-molecule reactions. The major ions in the topside ionospheres of the planets tend to be atomic ions: O$^+$ in the ionospheres of Earth and Venus, and H$^+$ in the ionospheres of the outer planets. On Mars, however, the O$^+$ peak density does not exceed that of O$_2^+$ even at high altitudes.

Model thermospheres for Earth and selected planets and satellites are shown in Figs. 80.9a–j. Measured or computed ion density profiles for the Earth and selected planets and satellites are shown in Fig. 80.10a–h.

80.4.3 Nightside Ionospheres

Nightside ionospheres can result from several sources including remnant ionization from dayside, like O$^+$ in the terrestrial ionosphere. While the lower molecular ion layers recombine, the F_2 peak persists through the night, although it rises and the peak density is reduced by a factor of 10. Electron density profiles for day and night at high and low solar activities are shown in Fig. 80.11. In the auroral regions of the Earth, the precipitating electrons may also produce significant ionization, which maximizes in the midnight sector of the auroral oval.

The nightside ionosphere of Venus is highly variable, but has been shown to contain the same ions as the dayside ionosphere. The densities are, however, lower by factors of 10 or more than those of the dayside ionosphere, and the average peak in the electron density profile is about $(1–2) \times 10^4$ cm^{-3}. It is produced by a combination of precipitation of suprathermal electrons that have been observed at high altitudes in the umbra, and transport of atomic ions (mostly O^+) at high altitudes from the dayside. For Mars, only a narrow range of solar zenith angles near the terminator at low solar activity has been measured by the radio occultation experiments on the Viking spacecraft [53]. The electron densities are apparently low, and no composition information is available. At this time, there is no information available about the nightside ionospheres of the other planets.

80.4.4 Ionospheric Density Profiles

Density profiles of molecular ions can often be approximated as idealized Chapman layers. A Chapman layer of ions is one in which the ions are produced by photoionization and lost locally by dissociative recombination. The ionization rate q^i in a one-species Chapman layer for monochromatic radiation is given by

$$q^i = F\sigma^i n, \tag{80.51}$$

where σ^i is the ionization cross section, and $F = F^\infty \exp[-\tau(z)]$ is the local solar flux. For an isothermal atmosphere, the scale height is approximately constant and therefore $n = n_0 \exp(-z/H)$. Sometimes an ionization efficiency η^i is defined such that

$$\sigma^i = \eta^i \sigma^a. \tag{80.52}$$

Near threshold, the ionization efficiency for molecules is usually about 0.3–0.7 but it increases rapidly to 1.0 at shorter wavelengths.

Since the maximum ionization rate in an isothermal atmosphere occurs where the optical depth ($\tau = nH\sigma^a \sec\chi$) is unity and therefore $n = 1/(\sigma^a H \sec\chi)$, the maximum ionization rate in a Chapman layer is

$$q^i_{\mathrm{max},\chi} = \frac{F^\infty}{e}\frac{\sigma^i}{\sigma^a H \sec\chi} = \frac{q^i_{\mathrm{max},0}}{\sec\chi}. \tag{80.53}$$

If the altitude of maximum ionization for overhead sun is defined as $z = 0$, then $n_0 = (\sigma^a H)^{-1}$, and, expressing F^∞ in terms of $q^i_{\mathrm{max},0}$, the ionization rate is

$$q^i(z) = q^i_{\mathrm{max},0} \exp\left[1 - \frac{z}{H} - \sec\chi\, e^{-z/H}\right]. \tag{80.54}$$

It is apparent that at high altitudes ($z \to \infty$) the ionization profile follows that of the neutral density, and below the peak ($z \to -\infty$), the ionization rate rapidly approaches zero. As the solar zenith angle increases, the peak rises and the magnitude of the density maximum decreases. Fig. 80.12 shows a production profile for an idealized Chapman layer on both linear and semilog plots. The asymmetry with respect to the maximum is more obvious for the semilog plot.

Figure 80.11. Typical midlatitude ionospheric electron density profiles for sunspot maximum and minimum, day and night. From Richmond [54].

If photochemical equilibrium prevails, the production rate of the major ion is equal to the loss rate due to dissociative recombination:

$$q^i(z) = \alpha_{\mathrm{dr}} n_{\mathrm{i}} n_{\mathrm{e}} = \alpha_{\mathrm{dr}} n_{\mathrm{i}}^2, \tag{80.55}$$

where α_{dr} is the dissociative recombination coefficient, n_{i} is the ion density, and n_{e} is the electron density. Therefore the density of an ion in a Chapman layer (in the photochemical equilibrium region) is given by

$$\begin{aligned}
n_{\mathrm{i}}(z) &= \left[\frac{q^i(z)}{\alpha_{\mathrm{dr}}}\right]^{1/2} \\
&= \left[\frac{q^i_{\mathrm{max},0}}{\alpha_{\mathrm{dr}}}\right]^{1/2} \exp\left[\frac{1}{2} - \frac{z}{2H} - \frac{1}{2}\sec\chi\, e^{-z/H}\right].
\end{aligned}$$
$$\tag{80.56}$$

Actual ionization profiles differ from the idealized Chapman profile for several reasons. First, ionization is produced by photons over a range of wavelengths, which do not all reach unit optical depth at the same altitude. Second, thermospheres are often not isothermal near the altitude of peak ion production. Third, photoionization is supplemented by photoelectron-impact ionization,

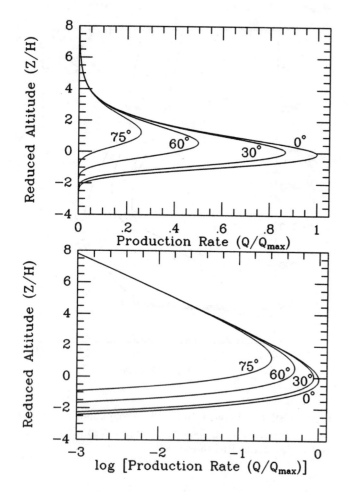

Figure 80.12. Chapman production profile as a function of altitude on linear (top) and semilog (bottom) plots. The production rate is divided by the maximum production rate and the altitude by the scale height H, which is assumed to be constant. The origin on the altitude scale corresponds to the point of maximum absorption for overhead sun.

which peaks lower in the atmosphere; and finally, the major ion produced is often transformed by ion-molecule reactions before it can recombine dissociatively. Nonetheless, the idealized concept of the Chapman profile is useful in understanding the general shape of ion profiles and their behavior as the solar zenith angle changes. In addition, ion layers produced by auroral precipitation may take on a similar appearance to a Chapman-type layer, although energetic electrons are not always extinguished, as are photons, in ion production.

80.4.5 Ion Diffusion

Above the photochemical equilibrium layer of the ionosphere, upward and downward transport of ions

must be considered. The motions of ions, neutrals, and electrons are coupled, and the momentum equation, which determines the fluxes or velocities of the ions must take into account these interactions. The interaction of an ion, denoted by a subscript i, with a neutral species denoted by a subscript n, is through the ion-induced-dipole attraction or, for the diffusion of an ion through its parent neutral, by resonant charge transfer. For the former process, the ion-neutral diffusion coefficient is given by

$$D_{in} = \frac{k_B T_i}{m_i \overline{\nu}_{\text{in}}}, \quad (80.57)$$

where μ_{in} is the reduced mass of the ion-neutral pair, and the ion-neutral momentum transfer collision frequency is

$$\overline{\nu}_{\text{in}} = 2.21\pi \frac{n_n m_n}{m_i + m_n} \left(\frac{\alpha_n e^2}{\mu_{\text{in}}} \right)^{1/2}, \quad (80.58)$$

where α_n is the polarizability of the neutral species (Dalgarno *et al.* [55]; Schunk and Nagy [56]). For the resonant charge transfer interaction, the diffusion coefficient is

$$D_{in}^{\text{ct}} = \frac{3(\pi/2)^{1/2}}{8 n_n \overline{Q}_{in}^{\text{ct}}} \left(\frac{k_B T_i}{m_i} \right)^{1/2} \frac{1}{\left(1 + T_n/T_i\right)^{1/2}}, \quad (80.59)$$

where $\overline{Q}_{in}^{\text{ct}}$ is the average charge transfer cross section [57].

The momentum transfer collision frequency for Coulomb interactions between an ion i and another ion or electron denoted by the subscript s is

$$\overline{\nu}_{is} = \frac{16\pi^{1/2}}{3} \frac{n_s m_s}{m_i + m_s} \left(\frac{\mu_{is}}{2 k_B T_{is}} \right)^{3/2} \frac{e_i^2 e_s^2}{\mu_{is}^2} \ln \Lambda, \quad (80.60)$$

where e_s is the charge on species s, $\ln \Lambda$ is related to the Debye shielding length, and $T_{is} = (m_s T_i + m_i T_s)/(m_i + m_s)$ is a reduced temperature. Numerically, $\ln \Lambda$ is about 15, and the collision frequency is approximately [56]

$$\overline{\nu}_{is} = 1.27 \frac{Z_i^2 Z_s^2 \mu_{is}^{1/2} n_s}{m_i T_{is}^{3/2}} \text{ s}^{-1}, \quad (80.61)$$

where Z is the species charge number, μ and m are in amu, and the number density is in units of cm^{-3}.

The ion densities can be computed by solving the ion continuity equation, which is similar to Eq. (80.7) for neutral species, and in one dimension is

$$\frac{\partial n_i}{\partial t} + \frac{\partial \Phi_i}{\partial z} = P_i - L_i, \quad (80.62)$$

where the ion flux is given by $\Phi_i = n_i w_i$. In general it is impossible to solve the momentum equation for the ion diffusion velocity w_i in closed form, except for the special cases of a single major ion and of a minor ion moving through a dominant ion species. If motion of the ions only parallel to magnetic field lines is considered, the vertical velocity of a dominant ion (for which $n_i \approx n_e$) moving through a stationary neutral atmosphere is

$$w_i = -D_a \sin^2 I \left[\frac{1}{n_i} \frac{dn_i}{dz} + \frac{m_i g}{k(T_e + T_i)} + \frac{1}{T_e + T_i} \frac{d(T_e + T_i)}{dz} \right] \quad (80.63)$$

where I is the magnetic dip angle and the ambipolar diffusion coefficient defined as

$$D_a = \frac{k_B(T_e + T_i)}{m_i \bar{\nu}_{in}}. \quad (80.64)$$

For a minor ion i diffusing through a major ion species j, its velocity is given by

$$w_i = -\frac{k_B T_i / m_i}{\bar{\nu}_{ij} + \bar{\nu}_{in}} \left[\frac{1}{n_i} \frac{dn_i}{ds} + \frac{T_e/T_i}{n_e} \frac{dn_e}{ds} + \frac{m_i g}{k_B T_i} + \frac{1}{T_i} \frac{d(T_e + T_i)}{ds} \right] - \frac{\bar{\nu}_{ij} w_j}{\bar{\nu}_{ij} + \bar{\nu}_{in}}. \quad (80.65)$$

For regions in which there are large gradients in the ion or electron temperatures, thermal diffusion may also be important in determining the ion density profiles, especially those of light ions such as H^+ and He^+. Equations for ion distributions in which thermal diffusion is included have been presented by, for example, Schunk and Nagy [56] and references therein.

80.5 NEUTRAL, ION, AND ELECTRON TEMPERATURES

The temperature distribution in planetary thermospheres/ionospheres can be modeled by solving the equation for conservation of energy, which, in simplified form in the vertical direction is

$$n_m \frac{\mathcal{N}}{2} k_B \frac{\partial T_m}{\partial t} - \frac{\partial}{\partial z} \left(\kappa_m \frac{\partial T_m}{\partial z} \right) = Q_m - L_m, \quad (80.66)$$

where $\mathcal{N}$ is the number of degrees of freedom (3 for an atom, and 5 for a diatomic molecule), the subscript m refers to the neutrals, ions or electrons, κ_m is the thermal conductivity, Q_m is the volume heating rate, and L_m is the volume cooling rate. If horizontal variations are considered, the model becomes multidimensional and advective terms must be added to the equations. The T_n are also affected by compression or expansion due to subsidence or upwelling, respectively. Viscous heating may be a factor where there are local regions of intense energy input, such as in auroral arcs. These terms are not shown in the energy equation above, but may be found in standard aeronomy texts, such as Banks and Kockarts [57] or Rees [33]. For planets with intrinsic

magnetic fields, the electrons and ions are constrained to move along magnetic field lines, and the second term on the left-hand side of Eq. (80.66) must be multiplied by a factor $\sin^2 I$.

The neutral thermospheres of planets are mostly heated by absorption of solar radiation in the 10 to 2000 Å range, although on planets with powerful auroras, electron precipitation may be an important source of heat. Absorption of euv radiation (100–1000 Å) largely results in ionization of the major thermospheric species, in which most of the excess energy is carried away by the photoelectron. The photoelectron, however, may produce further dissociation or excitation of neutral species along the path to thermalization, and these processes may result in neutral heating. Photons near and longward of ionization thresholds in the fuv may lose their energy in photodissociation, in which the excess energy of the photon appears as kinetic or internal energy of the fragments.

Chemical reactions that follow ionization or dissociation release much of the absorbed solar energy as heat. Although the partitioning of kinetic energy released between the product species can be determined easily by conservation of energy and momentum, the fraction of energy that appears as internal or kinetic energy must be determined by measurements or theoretical calculations. If vibrationally or electronically excited states are produced in these interactions, however, the energy may be radiated to space, thus producing cooling. This may occur promptly if the radiative lifetime is short, or subsequent to an energy transfer process from a long-lived metastable species to a species for which radiation to a lower state is allowed. If the metastable species is quenched, however, its energy can also appear as heat. Thus the energy partitioning in chemical reactions and in the interactions of photons and photoelectrons with atmospheric species is important in understanding the temperature structure of thermospheres.

A heating efficiency ϵ is often defined as the fraction of energy absorbed at a given altitude that appears locally as heat. The heating efficiencies are in the range 30–40% in the terrestrial lower thermosphere. Above 200 km, the heating efficiency decreases because the energy of the important metastable species $O(^1D)$ is lost as radiation rather than by quenching [58]. The heating efficiencies in the thermospheres of Venus and Mars are about 20% from 100 to 200 km [59,60], and on Titan, they range from 20 to 30% from 800 to 2000 km. A column averaged heating efficiency for the Jovian thermosphere has been computed as 53% [61].

On Venus and Mars, CO_2 is the major absorber of far uv radiation, whereas on the Earth, O_2 plays that role. In the F_1 regions of the ionospheres of the terrestrial planets, dissociative recombination of molecular ions tends to be the major source of heating. Below the F_1 peak, photodissociation and neutral-neutral reactions, includ-

Figure 80.13. Heating rates for the thermospheres of (a) Earth and (b) Mars. In (a), the curve labeled Q_n is the total heating rate; e-i is the heating rate due to collisions between the neutrals and electron and ions; iC and nc are the heating rates due to exothermic ion-neutral and neutral-neutral chemical reactions, respectively; J is that due to Joule heating for a superimposed electric field of 3.6 mV m^{-1}; A is that from auroral particle precipitation; $O(^1D)$ is the heating due to quenching of $O(^1D)$; SRC and SRB are the heating rates due to absorption in the Schumann Runge continuum and bands, respectively; O is the heating from recombination of atomic oxygen; O_3 is the heating rate due to absorption of photons by O_3 in the Hartley bands. Taken from Ref. [58]. In (b) the curve labeled "O_2^+ DR" is the heating rate due to dissociative recombination of O_2^+; that labeled "photodissociation" is heating due to the production of energetic neutrals in photodissociation; the curve labeled "quenching" is that due to quenching of metastable species, such as $O(^1D)$; and "chemical reactions" denotes heating due to exothermic chemical reactions other than quenching of metastable species. Taken from Ref. [60].

ing quenching of metastable species, dominate. Since CH_4 is a very strong absorber, the major heating mechanisms in the thermosphere of Titan are photodissociation and neutral-neutral reactions, both above and below the F_1 peak. The few data that exist suggest that electron-impact dissociation is unimportant as a source of neutral heating, although further measurements would certainly be of benefit. Profiles of the heat sources in the terrestrial thermosphere and that of Mars are shown in Fig. 80.13.

Important cooling processes in planetary thermospheres include downward transport of heat by molecular and eddy conduction and infrared cooling from rotational and vibrational excitation of ir active species such as NO, CO and CO_2. Excitation of the fine structure levels of atomic oxygen and subsequent emission at 63 and 147 μm also plays a role in cooling the neutral species in the thermospheres of the terrestrial planets. In the outer planets and their satellites, hydrocarbon molecules such as CH_4 and C_2H_2 are the primary thermospheric ir radiators. The global circulation may play a role in redistributing the heat that is deposited in the dayside or auroral thermosphere (*cf.* [62, 63]).

In order to model heating rates, cross sections for processes in which solar photons or photoelectrons interact with neutral species, and rate coefficients and product yields for chemical reactions of ions and neutral atmo-

spheric species are necessary. In addition, it is necessary to know, for example, how much of the energy released appears as internal energy of the products in chemical reactions and how much appears as kinetic energy of the products. Knowledge of energy transfer processes, including vibration-vibration ($V - V$) and translation-vibration ($T - V$) transfer between atmospheric species is also important. For example, a particularly important cooling process for the thermospheres of the terrestrial planets is excitation of the CO_2 15 μm bending mode in collisions with energetic O, and subsequent radiation [64]. The de-excitation rate is several percent of gas kinetic, which is anomalously large for a $V - T$ process [65].

In the lower ionosphere, the electrons and ions are in thermal equilibrium with the neutral species, but at higher altitudes the plasma temperatures deviate from the neutral temperatures. Near the F_1 peak, T_e is usually larger than T_i, but T_i begin to diverge from T_e at slightly higher altitudes. The energy source for the electrons on the dayside is largely photoionization, which, as discussed above, produces electrons with average energies in the 15 to 20 eV range. In slowing down, these energetic electrons lose their energy in inelastic processes with neutrals until $E \sim 1$–2 eV. At this point, elastic scattering by the thermal electron population becomes the dominant energy loss process for the suprathermal electrons and the major source of heat for the thermal electrons. Other electron heating mechanisms include deactivation of electronically or vibrationally excited species, and, for the terrestrial planets, quenching of the fine structure levels of O.

As for the neutrals, heat in the electron gas is redistributed by conduction at a rate that depends on the electron thermal conductivity. This quantity is inversely proportional to the sum of the momentum transfer collision frequencies of electrons with ions, neutrals, and ambient electrons. Cooling mechanisms for thermal electrons include Coulomb collisions with ions, rotational excitation of molecules, and, for the terrestrial planets, excitation of the fine structure levels of O. Because of the large mass difference, elastic collisions between neutrals and electrons are not effective in transferring kinetic energy.

T_i in the ionosphere is elevated above T_n at high altitudes principally because of Coulomb collisions with energetic electrons. Another potentially important source of heat input to the ions near the ion peak is exothermic ion-neutral reactions, including quenching of metastable ions, such as $O^+(^2D)$, by neutrals. In the presence of electric fields, joule heating may be important and can cause T_i to exceed T_e.

The ions cool in elastic collisions and resonant charge transfer with neutral species, which are characterized by lower temperatures than the ions. The cooling rate for elastic collisions is

$$L_{in} = -2n_i \frac{m_i}{m_i + m_n} \bar{\nu}_{in} \frac{3}{2} k_B (T_i - T_n). \quad (80.67)$$

Collisions between ions and neutrals (other than their parents) are dominated by the ion-induced-dipole interaction. The momentum transfer collision frequency is thus given by Eq. (80.58). Resonant charge transfer between an ion and its parent neutral, such as

$$O^{+*} + O \rightarrow O^* + O^+, \quad (80.68)$$

leads to very effective ion cooling, which dominates at sufficiently high temperatures.

Examples of T_i and T_e profiles are shown in Figs. 80.14a–c for Earth, Venus and Titan. The International Reference Ionosphere (IRI) temperatures profiles for the terrestrial thermosphere show the close coupling between the electrons, ions and neutrals at low altitudes and the ions and electrons at high altitudes. T_i increases with increasing solar activity at low altitudes and approaches T_e at high altitudes. The values of T_e and T_i are about 3000 K at 1000 km [66]. Electron and ion temperatures in the terrestrial ionosphere are discussed in Rees [33], Banks and Kockarts [57], and Whitten and Popoff [69].

T_e and T_i in the Venus atmosphere were measured by instruments on the Pioneer Venus spacecraft. T_i, which approaches values of 2000–2500 K at high altitudes, has been found to be insensitive to solar zenith angle, except near the antisolar point, where it increases to values of 5000-6000 K [67]. T_e also does not vary appreciably with solar zenith angle; the high altitude values are in the range 4000–6000 K [70].

The retarding potential analyzer (RPA) on the Viking spacecraft found that T_i on Mars decouples from the T_n near 180 km, and approach values of about 3000 K at high altitudes [45]. T_e is predicted to diverge from the T_n in the lower ionosphere, and to approach values of 3000–4000 K at high altitudes [71].

There are no measurements of plasma temperatures in the ionospheres of the outer planets. The plasma temperatures on Titan have been predicted by a model [68]. The computed T_i are greater than T_n near the n_e maximum near 1000 km, but approach values of about 300 K at high altitude. For the solar source only, T_e increases rapidly to a constant value of about 800 K near 1200 km. Electrons from Saturn's magnetosphere may interact with the Titan ionosphere during the part of its orbit that is within the magnetosphere, and in this case T_e's up to about 5000 K near 2000 km are predicted.

80.6 LUMINOSITY

The luminosity that originates in the atmospheres of the planets is generally classified as dayglow, nightglow, or aurora. Dayglow is the luminosity of the dayside atmosphere that occurs as a more or less direct result of the interaction of solar radiation with atmospheric gases. Among the sources of dayglow are photodissociative excitation and simultaneous photoionization and excitation. Dayglow may also include scattering of solar radiation by processes that are selective, such as resonance scattering by atoms and fluorescent scattering by molecules, but the term generally excludes nonselective scattering processes, such as Rayleigh scattering.

In resonance scattering, the absorption of a photon by an atom in the ground state, causes a (usually dipole allowed) transition to a higher electronic state:

$$A + h\nu \rightarrow A^*, \quad (80.69)$$

followed by the emission of a photon as the state decays back to the ground state:

$$A^* \rightarrow A + h\nu. \quad (80.70)$$

The wavelength of the emitted radiation is very nearly the same as the wavelength of the radiation absorbed. The cross section for absorption of a line in the solar spectrum is

$$\sigma_{12}^a(\nu) = \frac{\varpi_2}{\varpi_1} \frac{c^2}{8\pi\nu^2} A_{21} \phi(\nu), \quad (80.71)$$

where the subscript 1 indicates the lower state and 2 the upper state, ϖ is the statistical weight of the state and

Figure 80.14. Ion and electron temperature profiles. (a) Neutral (dot-dashed curve), ion (dashed curve) and electron (solid curve) temperatures for the terrestrial ionosphere from the International Reference Ionosphere for equinox, noon and low and high solar activities. The electron temperature is found not to vary substantially with solar activity. Reprinted from Bilitza and Hoegy [66] with kind permission from Elsevier Science Ltd., The Boulevard, Langford Lane, Kidlington 0X5 1GB, UK. (b) Smoothed median ion (dashed curve) and electron (solid curve) temperatures in the Venus ionosphere as measured by the PV retarding potential analyzer and the Langmuir probe, respectively. The electron temperature profile is essentially constant with solar zenith angle. The ion temperature profile applies to solar zenith angles between 0 and 90°. After Ref. [67]. (c) Computed electron and ion temperatures for the ionosphere of Titan, including only solar photoionization as the source of electron heating. From Ref. [68].

$\nu = c/\lambda$ is the frequency of the transition. A_{21} is the Einstein A coefficient for the transition, and $\phi(\nu)$ is the lineshape function, which in this equation is normalized so that the integral over all frequencies is unity.

If the linewidth is determined by the spread of velocities of the species, the lineshape $\phi(\nu)$ is a Doppler (Gaussian) profile

$$\phi_D(\nu) = \frac{c}{u\nu_0\sqrt{\pi}} \exp\left[-\left(\frac{\nu-\nu_0}{\nu_0}\right)^2 \frac{c^2}{u^2}\right], \quad (80.72)$$

where ν_0 is the frequency at line center. The variable $u = (2k_B T/m)^{1/2}$ is the modal velocity of a gas in thermal equilibrium at temperature T. The width of the line at half maximum, $\Delta\nu_D$ is

$$\Delta\nu_D = \frac{2\nu_0 u}{c}(\ln 2)^{1/2}. \quad (80.73)$$

If the linewidth is determined by the natural lifetime, the profile is a Lorentzian

$$\phi_L(\nu) = \frac{\Delta\nu_L/2\pi}{(\nu-\nu_0)^2 + (\Delta\nu_L/2)^2}, \quad (80.74)$$

where $\Delta\nu_L = \Gamma_R/2\pi$ is the line width at half maximum, and

$$\Gamma_R = \Gamma_2 + \Gamma_1, \quad (80.75)$$

where Γ_2 and Γ_1 are the inverse radiative lifetimes of the levels 2 and 1, respectively. Collisional broadening also results in a Lorentzian lineshape. If both Doppler and natural broadening mechanisms are important, the lineshape is a convolution of the two profiles, called a Voigt profile:

$$\phi_V(\nu) = \int_{-\infty}^{\infty} \phi_D(\nu')\phi_L(\nu-\nu')d\nu'. \quad (80.76)$$

The absorption cross section σ_{12}^a integrated over all frequencies is proportional to the absorption oscillator strength f_{12}:

$$\int_0^{\infty} \sigma_{12}^a(\nu)d\nu = \frac{\pi e^2}{m_e c}f_{12}, \quad (80.77)$$

where m_e is the mass of the electron. A_{21} is related to the oscillator strength through:

$$A_{21} = \frac{\varpi_1}{\varpi_2}\frac{8\pi^2 e^2 \nu^2}{m_e c^3}f_{12} = \frac{\varpi_1}{\varpi_2}\frac{8\pi^2 e^2}{m_e c\lambda^2}f_{12}. \quad (80.78)$$

The excitation rate q_2 of an upper level 2 by resonance scattering is given by

$$q_2 = F(\nu)\frac{\pi e^2}{m_e c}f_{12} = F(\lambda)\frac{\varpi_2}{\varpi_1}\frac{\lambda^4}{8\pi c}A_{21}, \quad (80.79)$$

where $F(\nu)$ is the solar flux in units of photons cm^{-2} s^{-1} Hz^{-1} and $F(\lambda)$ is the flux in units of photons cm^{-2} s^{-1} per unit wavelength interval. It should be noted that for radiative transfer purposes the photon flux that we have called $F(\nu)$ is sometimes denoted $\pi F(\nu)$. It is customary in aeronomy to define a "g-factor," which is the probability per atom that a photon will be resonantly scattered in a particular transition:

$$g_{21} = q_2 A_{21}/\sum_i A_{2i}, \quad (80.80)$$

where the sum in the denominator is over all the lower states i that are accessible from the upper state 2. The *g*-factor for unattenuated solar radiation is often quoted at the mean sun-earth distance or at a particular planet. The volume emission rate $\varepsilon_{21}(z)$ for resonance scattering of a solar photon is then given by

$$\varepsilon_{21}(z) = g_{21} n_1(z), \tag{80.81}$$

where $n_1(z)$ is the number density of atoms in level 1.

In fluorescent scattering, a photon is absorbed by a molecule in a vibrational state v producing an excited electronic state with a vibrational quantum number v'

$$AB(v) + h\nu \rightarrow AB^*(v'). \tag{80.82}$$

This is followed by emission, at wavelengths that are usually the same as or longer than that of the absorbed photon, to a range of vibrational levels v'' of a lower state

$$AB^*(v') \rightarrow AB(v'') + h\nu'. \tag{80.83}$$

The volume emission rate of a transition from a level v' of the upper electronic state to a vibrational level v'' of a lower electronic state at an altitude z is given by

$$\varepsilon_{v'v''}(z) = n(z) g_{v'v''} = n(z) q_{v'} \frac{A_{v'v''}}{\sum_{v''} A_{v'v''}}, \tag{80.84}$$

where $A_{v'v''}$ is the transition probability, $n(z)$ is the number density of the molecular species at altitude z, and $q_{v'}$ is the excitation rate of vibrational level v' of the upper electronic state from a range of lower states v. The latter quantity is

$$q_{v'} = \sum_v \eta_v F(\lambda) \frac{\pi e^2}{m_e c^2} \lambda^2 f_{vv'}, \tag{80.85}$$

where η_v is the fraction of molecules in the v vibrational level.

Dayglow also includes emissions that are the result of the interaction of atmospheric species with the photoelectrons produced in solar photoionization, either by direct excitation

$$A + e^{*-} \rightarrow A^* + e^-, \tag{80.86}$$

or by simultaneous dissociation and excitation

$$AB + e^{*-} \rightarrow A^* + B + e^-, \tag{80.87}$$

or ionization and excitation

$$X + e^{*-} \rightarrow X^{+*} + e^-. \tag{80.88}$$

Electron impact processes are particularly important in producing excited states that are connected to the ground state by dipole forbidden transitions, whereas resonance and fluorescent scattering are largely limited to transitions that are dipole allowed. Dayglow emissions may also result from *prompt* chemiluminescent reactions,

which occur when fragments or ions produced by dissociation or ionization recombine with the emission of a photon.

As an example, the dayglow spectrum of the earth from 1200 to 9000 Å is shown in Fig. 80.15. An ultraviolet spectrum of Mars as measured by the Mariner 9 spectrometer is shown in Fig. 80.16a, and the uv dayglow of Saturn and Uranus, which were measured by the Voyager spacecraft, are compared in Fig. 80.16b.

Nightglow arises from chemiluminescent reactions of species whose origin can be traced to species produced during the daytime or which have been transported from the dayside. For example, on Venus, O and N produced on the dayside are transported by the subsolar to antisolar circulation to the nightside, where they subside and radiatively associate:

$$N + O \rightarrow NO + h\nu, \tag{80.89}$$

producing emission in the δ and γ bands of NO (e.g., Stewart *et al.* [73]). The Venus uv nightglow spectrum is shown in Fig. 80.17.

Atomic oxygen emission in the red and green lines occurs in the nightglows of the terrestrial planets in part from the production of $O(^1D)$ and $O(^1S)$ in the dissociative recombination of O_2^+:

$$O_2^+ + e \rightarrow O^* + O^*. \tag{80.90}$$

Near the dusk terminator, the O_2^+ could be remnant O_2^+ from the dayside, but near the anti-solar point on Venus, it must be produced in situ; this can occur via the reaction of CO_2 with O^+ transported from the dayside:

$$O^+ + CO_2 \rightarrow O_2^+ + CO. \tag{80.91}$$

Auroral emissions are defined here as those produced by impact of particles other than photoelectrons. Although aurorae are usually thought of as confined to the polar regions of the Earth and the outer planets, Venus, which does not have an intrinsic magnetic field, exhibits uv emissions on the nightside that are highly variable and cannot be explained as nightglow. It has been proposed that the emissions are produced by precipitation of soft electrons into the nightside thermosphere. It is unknown whether Mars exhibits such phenomena.

The intensities of airglow and aurora are usually measured in units of brightness called Rayleighs. One Rayleigh is an apparent column emission rate at the source of 10^6 photons cm^{-2} s^{-1} integrated over all angles, or $10^6/4\pi$ photons cm^{-2} s^{-1} sr^{-1}.

A discussion of terrestrial airglow and auroral emissions can be found in Rees [33]. Meier [76] has reviewed spectroscopy and remote sensing of the terrestrial ultraviolet emissions. The airglows of Mars and Venus have been discussed by Barth [17], by Fox [77] and by Paxton and Anderson [29]. Airglow and auroral emissions on the outer planets have been reviewed by Atreya *et al.* [78], Atreya [2], and Strobel [4].

Figure 80.15 Terrestrial dayglow spectrum measured in a single 32-second exposure by the Arizona Imager/Spectrograph on board the Space Shuttle [72].

spectrum recorded by Mariner 9 at 15 Å resolution. From Ref.

Figure 80.16. Dayglow spectra of selected planets. (a) Martian airglow [74]. (b) Comparison of dayglow spectra from Uranus (heavy line) and Saturn (thin line) recorded by Voyager 2 [4].

80.7 PLANETARY ESCAPE

Escape of species from atmospheres can occur by thermal and nonthermal mechanisms. Thermal processes include Jeans escape and hydrodynamic escape. Jeans escape is essentially evaporation of the energetic tail of the Maxwell-Boltzmann distribution, while hydrodynamic escape is a large-scale "blow-off" of the atmosphere that occurs when the average molecular velocity is near or above the escape velocity. Although Jeans escape still occurs for light species in the thermospheres of small plan-

Figure 80.17. Far ultraviolet nightglow spectrum of Venus obtained with the Pioneer Venus orbiter ultraviolet spectrometer. The predicted responses for three different band systems are also shown. Reprinted with permission from Ref. [75], copyright 1979 American Association for the Advancement of Science.

ets, hydrodynamic escape is thought to have occurred only in the early history of the terrestrial planets when the solar flux in the uv was higher. Nonthermal escape mechanisms, which dominate for heavy species on smaller bodies such as Mars and Titan, and for all species on Venus and Earth, include both photochemical and mechanical processes. A pedagogical discussion of escape processes can be found in Chamberlain and Hunten [79]

Because of the exponential rate of change of density with altitude in atmospheres, escape is sometimes assumed to occur only at and above the exobase. The exobase is mathematically defined as the altitude where the mean free path $l = (n\sigma^c)^{-1}$ (where σ^c is the collision cross section) is equal to the atmospheric scale height. The probability that a particle, moving upward from the exobase with sufficient velocity will actually escape without suffering another collision is $1/e$. The condition $l = H$ therefore reduces to $nH\sigma = 1$ or, equivalently, to $N = (\sigma^c)^{-1}$, where N is the column density. Since a typical collision cross section is about 3×10^{-15} cm^2, the exobase is located near the altitude above which the column density is about 3.3×10^{14} cm^{-3}, although the collision cross section and thus the location of the exobase is different for different escaping species.

Whether the trajectory of a particle moving upward at the exobase is ballistic (bound) or escaping (free) is determined by its total energy E, which is the sum of its kinetic and potential energies:

Table 80.10. Exobase properties of the planets.

Planet	r_c (km)	g_c^a (cm s^{-2})	$v_{esc,c}^a$ (km s^{-1})	$H_{avg}^{a,b}$ (km)
Mercury[c]	2 439	378	4.29	–
Venus[d]	6 250	831	10.2	17
Earth[e]	6 878	842	10.8	71
Mars[f]	3 593	333	4.89	17
Jupiter[g]	73 000	2236	57.2	250
Saturn[h]	67 000	731	31.3	910
Uranus[i]	31 800	561	18.9	66
Neptune[j]	27 300	919	22.4	250
Titan[k]	4 175	51.4	2.07	116
Triton[l]	2 222	28.9	1.13	140

[a] Values given are those at the equatorial exobase, and assume that the thermosphere co-rotates with the planet.

[b] Average value computed from $H_{avg} = k_B T / m_{avg} g$, but does not represent the local pressure scale height, except for cases where there is one major constituent.

[c] Exobase is at the surface.

[d] Model from Hedin *et al.* [3] for $F_{10.7}$=150, 45°solar zenith angle.

[e] MSIS model for $F_{10.7}$=150, equator, 45 °Solar Zenith Angle [15].

[f] Model from Nier and McElroy [8], and pertains to low solar activity conditions.

[g] Model from Refs. [2,12].

[h] Model from Atreya [2].

[i] Model from Ref. [42].

[j] Model from [43].

[k] Model from Strobel *et al.* [9].

[l] Model from Krasnopolsky *et al.* [13]

$$E = \frac{1}{2}mv_c^2 + \int_{\infty}^{r_c} \frac{mGM}{r^2} dr , \qquad (80.92)$$

where the symbols have the same meaning as in equation (80.2), and the subscript c refers to the critical level or exobase. If $E < 0$, the particle is bound. Expression (80.92) reduces to

$$E = \frac{1}{2}mv_c^2 - mg_c r_c \qquad (80.93)$$

where g_c is the gravitational acceleration at the exobase. The escape velocity at the exobase, v_{esc}, is then defined by the condition $v_{esc} = (2g_c r_c)^{1/2}$. Particles with velocities greater than the escape velocity are assumed to escape if their velocity vector is oriented in the upward hemisphere and if they undergo no further collisions. The radius, gravitational acceleration, escape velocities and scale height at the equatorial exobases of the planets are given in Table 80.10.

In the Jeans process, escape occurs when particles in the high energy tail of the Maxwellian distribution attain the escape velocity. The escape flux, Φ_J is given by

$$\Phi_J = \frac{n_c u}{2\sqrt{\pi}} (1 + \lambda_c) \exp(-\lambda_c) , \qquad (80.94)$$

where $u = (2k_B T/m)^{1/2}$ is the modal velocity and λ is the gravitational potential energy in units of $k_B T$:

$$\lambda = \frac{GMm}{r k_B T} = \frac{mgr}{k_B T} = \frac{r}{H} , \qquad (80.95)$$

Sometimes a correction factor is applied to the expression for the escape flux to account for the suppression of the tail of the distribution due to the escape of the energetic particles (Chamberlain [80]; see also Hunten [81]).

Photochemical processes that produce energetic fragments include photodissociation and photodissociative ionization, photoelectron impact dissociation and dissociative ionization, as well as exothermic chemical reactions. The most important example of the latter are dissociative recombination reactions, which are very exothermic and tend to produce neutral fragments with large kinetic energies. Charge transfer processes such as

$$H^{+*} + O \rightarrow O^+ + H^* \qquad (80.96)$$

can produce fast neutrals if the ion temperature is larger than the neutral temperature, as is usually the case near the exobase of a planet. In modeling these processes, the kinetic energy distribution of the product species is important, as well as the cross sections or reaction rates.

Physical or collisional escape mechanisms include sputtering and "knock-on." Sputtering can occur when a heavy ion picked up by the solar wind collides with an atmospheric species near or above the exobase, and in the process produces a "back-splash" in which the accelerated neutral may be ejected from the atmosphere. In knock-on, hot atmospheric neutral species, such as O atoms produced in exothermic chemical reactions near the exobase can collide with a lighter species, such as H, imparting sufficient kinetic energy to allow it to escape. Modeling these processes requires knowledge of the ion-neutral or neutral-neutral collision cross sections.

The escape rate of a light species from a planetary atmosphere may be controlled by diffusion of the species from the lower atmosphere to the exobase, rather than by the escape process itself [82]. The limiting upward flux, ϕ_l of a species i with mixing ratio f_i can be estimated as

$$\phi_l \approx b_i f_i / H_a , \qquad (80.97)$$

where H_a is the average scale height of the atmosphere and b_i is the binary collision parameter introduced in Sect. 80.2.2. Equation (80.97) above is usually evaluated at the homopause, with the mixing ratio taken from a suitable altitude in the middle atmosphere, but above the cold trap (where the species condenses), if one exists. The limiting flux obtains if and only if the mixing ratio is constant with altitude. The effect of photochemistry can be accounted for if all chemical forms of the species are considered in the calculation of f_i.

Even if energetic particles released at the exobase of a planet do not have enough energy to escape, they may travel to great heights along ballistic orbits before falling back to the atmosphere. These particles are said to form a hot atom "corona". Hot H and O coronas have been found to surround the Earth and Venus, and have been predicted for Mars. Reviews of the H and O coronas of Venus have been presented by Fox and Bougher [63] and by Nagy *et al.* [83]. See Chamberlain and Hunten [79] for a detailed discussion of planetary coronal population processes.

REFERENCES

1. U. von Zahn, S. Kumar, J. Niemann, and R. Prinn, in *Venus* (University of Arizona Press, Tucson, 1983).
2. S. K. Atreya, *Atmospheres and Ionospheres of the Outer Planets and Their Satellites* (Springer Verlag, New York, 1986).
3. A. E. Hedin, H. B. Neimann, W. T. Kasprzak, and A. Seiff, J. Geophys. Res. **88**, 73 (1983).
4. D. F. Strobel, R. V. Yelle, D. E. Shemansky, and S. K. Atreya, in *Uranus*, edited by J. T. Bergstralh, E. D. Miner, and M. S. Matthews (University of Arizona Press, Tucson, 1991).
5. S. K. Atreya, B. R. Sandel, and P. N. Romani, in *Uranus*, edited by J. T. Bergstralh, E. D. Miner, and M. S. Matthews (University of Arizona Press, Tucson, 1991).
6. *The U. S. Standard Atmosphere* (National Oceanic and Atmospheric Administration, U. S. Government Printing Office, 1976).
7. R. V. Yelle, F. Herbert, B. R. Sandel, R. J. Vervack, and T. M Wentzel, Icarus **104**, 38 (1993).
8. A. O. Nier and M. B. McElroy, J. Geophys. Res. **82**, 4341 (1977).
9. D. F. Strobel, M. E. Summers, and X. Zhu, Icarus, **100**, 512 (1992).
10. Y. L. Yung, M. Allen, and J. P. Pinto, Astrophys. J. Suppl. **55**, 465 (1984).
11. R. V. Yelle, private communication.
12. Y. H. Kim, *The Jovian Ionosphere*, Ph. D. Thesis, State University of New York at Stony Brook (1991).
13. V. A. Krasnopolsky, B. R. Sandel, F. Herbert, and R. J. Vervack, J. Geophys. Res. **98**, 3065 (1993).
14. D. F. Strobel and M. E. Summers, in *Neptune*, edited by D. Cruikshank and M. S. Matthews (University of Arizona Press, Tucson, 1996).
15. A. E. Hedin, J. Geophys. Res. **96**, 1159–1172 (1991).
16. C. D. Keeling and T. P. Whorf, in *Trends '93: A Compendium of Data on Global Change*, edited by T. A. Boden, D. P. Kaiser, R. J. Sepanski, and F. W. Stoss ORNL/CDIAC-65 (Carbon Dioxide Information Analysis Center, Oak Ridge National Laboratory, Oak Ridge, 1993).
17. C. A. Barth, A. I. F. Stewart, S. W. Bougher, D. M. Hunten, S. J. Bauer, and A. F. Nagy, in *Mars*, edited by H. Kiefer, B. M. Jakosky, C. W. Snyder, and M. S. Matthews (University of Arizona Press, Tucson, 1992).
18. J. S. Lewis and R. G. Prinn, *Planets and their Atmospheres: Origin and Evolution* (Academic Press, Orlando, 1984).
19. T. M. Donahue and R. R. Hodges, Jr., Geophys. Res. Lett. **20**, 591 (1993).
20. V. A. Krasnopolsky, S. Bowyer, S. Chakrabarti, G. R. Gladstone, and J. S. McDonald, *Icarus, 109*, 337 (1994).
21. D. F. Strobel, in *The Photochemistry of Atmospheres: Earth, the Other Planets, and Comets*, edited by J. S. Levine (Academic Press, New York, 1985).
22. B. Fegley, Jr., in *Global Earth Physics: A Handbook of Physical Constants*, edited by T. Ahrens (American Geophysical Union, Washington, D. C., 1995), pp. 320–345.
23. D. M. Hunten, M. G. Tomasko, F. M. Flasar, JR. E. Samuelson, D. F. Strobel, and D. J. Stevenson, in *Saturn*, edited by T. Gehrels and M. S. Matthews (University of Arizona Press, 1984).
24. R. V. Yelle, Astrophys. J. **383**, 380–400 (1991).
25. D. M. Hunten, T. H. Morgan, and D. E. Shemansky, in *Mercury*, edited by F. Vilas, C. R. Chapman, and M. S. Matthews (University of Arizona Press, Tucson, 1988).
26. H. E. Hinteregger, private communication.
27. D. Lummerzheim, M. H. Rees, and H. R. Anderson, Planet. Space Sci. **37**, 109 (1989).
28. L. Herzberg, in *Physics of the Earth's upper atmosphere*, edited by C. Hines, I. Paghis, T. R. Hartz, and J. A. Fejer (Prentice-Hall, Englewood Cliffs, NJ, 1965).
29. L. J. Paxton and D. E. Anderson, in *Venus and Mars: Atmospheres, Ionospheres and Solar Wind Interactions*, Geophysical Monograph 66 (American Geophysical Union, Washington, D. C., 1992), pp. 113–190.
30. J. M. Valentine and S. C. Curran, Rep. Prog. Phys. **21**, 1 (1958).
31. C. B. Opal, W. K. Peterson, and E. C. Beaty, J. Chem. Phys. **55**, 4100 (1971).
32. W. E. Swartz, J. S. Nisbet, and A. E. S. Green, J. Geophys. Res. **74**, 6415 (1971).
33. M. H. Rees, *Physics and Chemistry of the Upper Atmosphere* (Cambridge University Press, Cambridge, 1989).
34. A. F. Nagy and P. M. Banks, J. Geophys. Res. **75**, 6260 (1970).
35. H. S. Porter, F. Varosi, and H. G. Mayr, J. Geophys. Res. **92**, 5933 (1987).
36. D. J. Strickland, R. E. Daniell, B. Basu, and J. R. Jasperse, J. Geophys. Res. **98**, 21533 (1993).
37. S. C. Solomon, Geophys. Res. Lett. 20, 185 (1993).
38. S. J. Bauer, *Physics of Planetary Ionospheres* (Springer-Verlag, New York, 1973).
39. S. G. Lias, J. E. Bartmess, J. F. Liebman, J. L. Holmes, R. D. Levin, and W. G. Mallard, Gas-phase ion, and neutral thermochemistry, J. Phys. Chem. Ref. Data **17**, Suppl. 1 (1988).
40. F. L. Nesbitt, G. Marston, L. J. Stief, M. A. Wickramaaratchi, W. Tao, and R. B. Klemm, J. Phys. Chem. **95**, 7613 (1991).
41. J. L. Fox and A. Dalgarno, J. Geophys. Res. **84**, 7315 (1979).
42. F. Herbert, B. R. Sandel, R. V. Yelle, J. B. Holberg, A. L. Broadfoot, and D. E. Shemansky, J. Geophys. Res. **92**, 15093 (1987).
43. J. R. Lyons, private communication.
44. C. Y. Johnson, J. Geophys. Res. **71**, 330 (1966).
45. W. B. Hanson, S. Sanatani, and D. R. Zuccaro, J. Geophys. Res. **82**, 4351 (1977).
46. Y. H. Kim and J. L. Fox, Icarus **112**, 310 (1994).

47. T. Majeed and J. C. McConnell, Planet. Space Sci. **39**, 1715 (1991).

48. J. R. Lyons, Nature **267**, 648 (1995).

49. T. Su and M. T. Bowers, in *Gas Phase Ion Chemistry*, edited by M. T. Bowers (Academic Press, New York, 1979), Vol. 1.

50. V. G. Anicich, Astrophys. J. Suppl. Ser. **62**, 553–672 (1993).

51. Y. Ikezoe, S. Matsuoka, M. Takabe, and A. Viggiano, *Gas Phase Ion-Molecule Rate Constants Through 1986* (Maruzen Co., Tokyo, 1986).

52. D. R. Bates and A. Dalgarno, in *Atomic and Molecular Processes*, edited by D. R. Bates (Academic Press, New York, 1962), pp. 245–271.

53. M. H. G. Zhang, J. G. Luhmann, and A. J. Kliore, J. Geophys. Res. **95**, 17095 (1990).

54. A. D. Richmond, The ionosphere, in *The Solar Wind and the Earth*, edited by S.-I. Akasofu and Y. Kamide (Terra Scientific Publishing Co., Tokyo, 1987).

55. A. Dalgarno, M. R. C. McDowell, and A. Williams, Phil. Trans. Roy. Soc. London, Ser. A **250**, 411 (1958).

56. R. W. Schunk and A. F. Nagy, Rev. Geophys. Space Phys. **18**, 813 (1980).

57. P. M. Banks and G. Kockarts, *Aeronomy* (Academic Press, New York, 1973).

58. R. G. Roble, E. C. Ridley, and R. E. Dickinson, On the global mean temperature of the thermosphere, J.Geophys. Res. **92**, 8745-8758 (1987).

59. J. L. Fox, Planet. Space Sci., **37**, 36 (1988).

60. J. L. Fox, P. Zhou, and S. W. Bougher, Adv. Space Res. **17**, (11)203-(11)218, 1995.

61. J. H. Waite, T. E. Cravens, J. Kozyra, A. F. Nagy, S. K. Atreya, and R. H. Chen, J. Geophys. Res. **88**, 6143 (1983).

62. R. G. Roble, Rev. Geophys. **21**, 217–233 (1983).

63. J. L. Fox and S. W. Bougher, Space Sci. Rev. **55**, 357 (1991).

64. S. W. Bougher, D. M. Hunten, and R. G. Roble, J. Geophys. Res. **99**, 14609 (1994).

65. R. P. Wintersteiner, R. H Picard, R. D. Sharma, J. R. Winick, and R. A. Joseph, J. Geophys. Res. **97**, 18083 (1992).

66. D. Bilitza and W. R. Hoegy, Adv. Space Res. **10**, (8)81–(8)90 (1990).

67. K. L. Miller, W. C. Knudsen, K. Spenner, R. C. Whitten, and V. Novak, J. Geophys. Res. **85**, 7759 (1980).

68. A. Roboz and A. F. Nagy, J. Geophys. Res. **99**, 2087 (1994).

69. R. C. Whitten and I. G. Poppoff, *Fundamentals of Aeronomy*, (Wiley, New York, 1971).

70. R. F. Theis, L. H. Brace, and H. G. Mayr, J. Geophys. Res. **85**, 7787 (1980).

71. W. B. Hanson and G. P. Mantas, J. Geophys. Res. **93**, 7538 (1988).

72. A. L. Broadfoot, private communication.

73. A. I. F. Stewart, J.-C. Gerard, D. W. Rusch, and S. W. Bougher, J. Geophys. Res. **85**, 7861 (1980).

74. C. A. Barth, A. I. Stewart, C. W. Hord, and A. L. Lane, Icarus **17**, 457 (1972).

75. A. I. F. Stewart and C. A. Barth, Science **205**, 59 (1979).

76. R. R. Meier, R. R., Space Sci. Rev. **58**, 1–187 (1991).

77. J. L. Fox, in *Venus and Mars: Atmospheres, Ionospheres and Solar Wind Interactions*, Geophysical Monograph 66 (American Geophysical Union, Washington, D. C., 1992), pp. 191–222.

78. S. K. Atreya, J. H. Waite Jr., T. M. Donahue, A. F. Nagy, and J. C. McConnell, in *Saturn*, edited by T. Gehrels and M. S. Matthews (University of Arizona Press, Tucson, 1984).

79. J. W. Chamberlain and D. M. Hunten, *Theory of Planetary Atmospheres* (Academic Press, New York, 1987).

80. J. W. Chamberlain, Planet. Space Sci. **11**, 901 (1963).

81. D. M. Hunten, Planet. Space Sci. **30**, 773 (1982).

82. D. M. Hunten, J. Atmos. Sci. **30**, 1481 (1973).

83. A. F. Nagy, J. Kim, and T. E. Cravens, Ann. Geophys. **8**, 251 (1990).

81

Applications of Atomic and Molecular Physics to Global Change

Kate P. Kirby

Harvard-Smithsonian Center for Astrophysics, 60 Garden Street, Cambridge, Massachusetts 02138

81.1 INTRODUCTION

81.1.1 Global Change Issues

Over the last several decades there has been increasing concern about the global environment and the effect of human perturbations on it. This whole area has become known as *Global Change,* involving a wide range of scientific disciplines. Knowledge of processes taking place in the atmosphere, oceans, land masses, and plant and animal populations, as well as the interactions between these various earth-system components is essential to an overall understanding of global change — both natural and human-induced.

The processes of atomic and molecular physics find greatest application in the area of atmospheric global change. The two major issues which have received significant attention in both the media and the scientific literature are: (1) global warming, due to the buildup of infrared-active gases; and (2) stratospheric ozone depletion due to an enhancement of destructive catalytic cycles. Although both of these problems are thought to be caused by atmospheric pollutants due to industrialized human society, the general problem of air pollution and its direct effects on plant and animal populations will not be addressed here.

While there has been a general understanding and appreciation of the science involved in both global warming and stratospheric ozone depletion by atmospheric scientists for some time, detailed understanding and rigorous proof has often been lacking. Over the last ten years there have been many advances made in filling in the details, and there will continue to be rapid advances in the future. This means that any article or book discussing this topic becomes out of date as soon as it is written. Nevertheless several recent books on these topics are recommended [1].

Atomic and molecular structure and spectroscopy, as well as collision processes involving atoms, molecules, ions, and electrons, are important to the study of all planetary atmospheres. For additional information on this topic, see Chap. 80 in this volume.

81.1.2 Structure of the Earth's Atmosphere

The vertical temperature structure of the earth's atmosphere shown in Fig. 81.1 provides an important nomenclature that is widely used [2]. The atmosphere is divided into regions called "-spheres", in which the sign of the temperature gradient with respect to altitude, dT/dz, is constant. The regions in which the temperature gradient changes sign are called "-pauses". The precise altitude pertaining to each of these regions can vary depending upon latitude and the time of year.

In the troposphere, covering the range from 0 to ~15 km above the earth, the temperature steadily decreases with altitude. This is the most complex region of the atmosphere, as it interacts directly with plant and animal life, land masses, and the oceans. It is the region in which weather occurs.

The change in sign of the temperature gradient at the tropopause to a positive dT/dz in the stratosphere is due to heating by absorption of solar ultraviolet radiation which photodissociates O_2 and O_3. The stratosphere, extending from ~15 to 50 km above the earth, contains the *ozone layer* which shields the earth's surface from harmful ultraviolet radiation in the range of 280–320 nm.

At the stratopause, the heating processes have become too weak to compete with the cooling processes, and throughout the mesosphere, approximately 50–85 km, the temperature again decreases with increasing altitude. Cooling processes, which will be discussed in Sect. 81.2.2, involve collisional excitation of molecular vibrational modes which decay by radiating to space. The coldest temperatures in the atmosphere are found at the mesopause, where the temperature gradient once again becomes positive. From approximately 70 km upward, a very diffuse plasma called the ionosphere exists due to photoionization of atoms and molecules by short wavelength (uv and euv) solar radiation.

Throughout the thermosphere, which extends from approximately 90 km upward, heating occurs because

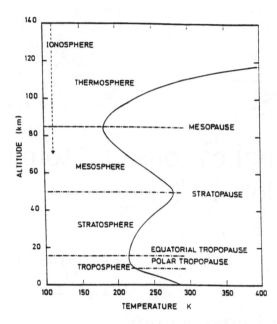

Figure 81.1. Vertical temperature profile of the atmosphere (courtesy of R. Roble).

the atmosphere has become so thin that there are very few collisions and thus inefficient equilibration of the highly translationally excited atoms and ions with the molecular species which can radiate in the infrared. This "bottleneck" for energy loss causes increased heating. In the thermosphere and ionosphere, the thermal inertia is very small and there are huge temperature variations, both diurnally, and with respect to solar activity.

The densities are low enough in the thermosphere, ionosphere, and mesosphere that the primary processes determining the chemical and physical characteristics of these regions are two-body processes and "half-collision events" discussed elsewhere in this volume: dissociative recombination, photoionization, photodissociation, charge transfer, and collisional excitation of molecular rotation and vibration. As the altitude decreases, the density increases. Then three-body interactions, interactions on surfaces (of aerosols and ices), and complex chemical cycles together with dynamical effects such as winds determine the chemical and physical characteristics of the stratosphere and troposphere.

81.2 ATMOSPHERIC MODELS AND DATA NEEDS

While models are absolutely essential to the study of any system as complex as the earth's atmosphere, they play a particularly fundamental role in exploring global change issues. Models not only provide predictions of future changes, but also allow exploration of sensitivities to particular parameters. Comparing the results of a model with observations ultimately tests and challenges scientific understanding. Of critical importance is the atomic and molecular data which goes into the models.

Generally, atmospheric models become increasingly complex as altitude decreases. Huge computer codes, known as General Circulation Models (GCMs), incorporating thousands of chemical reactions, global wind patterns, and abundances of large numbers of trace species, require fast supercomputers in order to model aspects of the troposphere and stratosphere.

81.2.1 Modeling the Thermosphere and Ionosphere

The data necessary for modeling of the thermosphere and ionosphere are described here. The primary components of such a model include: a solar spectrum of photon fluxes as a function of wavelength, concentrations of neutral species, photoabsorption, photodissociation and photoionization cross sections as a function of wavelength. The computer code brings all these elements together, calculating opacity as the solar radiation propagates downward through the atmosphere, keeping track of ion production and electron production. Additional steps are required to calculate the abundances of trace species such as NO^+, necessitating the inclusion of all relevant ion-neutral reactions. Electron energy degradation can be tracked by including inelastic collisions of electrons with ions, atoms, and molecules.

The primary neutral species are N_2, O_2, O, and He. Below about 100 km, the atmosphere is fully mixed by turbulence in the ratio 78% N_2, 21% O_2 and 1% trace species such as O_3 and CO_2. Above 100 km, turbulence dies out and the atmospheric species are in diffusive equilibrium, distributed by their molecular weight, with atomic oxygen dominating above $\sim$150 km and He dominating much higher. The major ions are N_2^+, O_2^+, O^+, and NO^+.

81.2.2 Heating and Cooling Processes

In the upper atmosphere, heating occurs through absorption of short wavelength solar radiation to produce ionization and dissociation, and is mediated by collisions between electrons, ions, and neutrals. Ions and electrons are created during the daytime and to a great extent disappear during the night with the absence of solar radiation. Processes such as dissociative recombination, the primary electron loss mechanism, heat the gas:

$$
\begin{aligned}
e + O_2^+ &\to O(^3P) + O(^3P) + 7 \text{ eV}\,, \\
e + N_2^+ &\to N(^4S) + N(^4S) + 6 \text{ eV}\,.
\end{aligned}
\tag{81.1}
$$

Cooling takes place when the kinetic energy of the gas is transformed through collisions into internal energy which can then be radiated away. The primary coolant above $\sim$200 km is the fine structure transition of atomic oxygen, $O(^3P_1) \overset{h\nu}{\to} O(^3P_2)$, which is excited by thermal collisions and radiates at 63 μm. From approximately 120 km to 200 km, the fundamental band of NO, $v = 1 \to v = 0$, which is excited by collisions with atomic oxygen and radiates at 5.3 μm, dominates the cooling. Below 120 km and throughout the mesosphere and stratosphere, the primary coolant is the ν_2 band of CO_2 radiating at 15 μm. This transition is excited by collisions of CO_2 and atomic oxygen. Cooling throughout most of the atmosphere is accomplished through trace species since the major molecular species N_2 and O_2 are not infrared active.

81.2.3 Atomic and Molecular Data Needs

Knowledge of rate coefficients for ion-neutral and neutral-neutral reactions as a function of vibrational and rotational excitation of the reactants is becoming increasingly important, as there is recent evidence of more internal excitation of molecular species than had previously been thought [3]. Accurate photoabsorption, photodissociation, and photoionization cross sections as a function of wavelength for all the relevant species are important parameters determining the reliability and ultimate accuracy of an atmospheric model. Compilations of data, such as that by Conway [4] and Kirby *et al.* [5] are very useful, but can become rapidly outdated. A group at Smithsonian Astrophysical Observatory [6] maintains a molecular line database for analysis of atmospheric spectra in the infrared, from 10 to 800 cm^{-1}. Another database, used extensively worldwide for atmospheric applications is HITRAN, maintained by L. Rothman [7]. Discussions of the needs for atomic and molecular data in the context of space astronomy, but including applications to atmospheric physics can be found in a book edited by Smith and Wiese [8].

81.3 TROPOSPHERIC WARMING/UPPER ATMOSPHERE COOLING

81.3.1 Incoming and Outgoing Energy Fluxes

The overall temperature of a planet is determined by a balance between incoming and outgoing energy fluxes. In a steady state, the planet must radiate as much energy as it absorbs from the sun. The Earth, radiating as a blackbody at an effective temperature T_E, obeys the

Stefan-Boltzmann law in which the energy emitted is expressed as $\sigma T_E^4 4\pi R_E^2$, where σ is the Stefan-Boltzmann constant, and R_E the radius of the earth. An equation expressing the equality of energy absorbed and energy emitted can be written [9] as

$$F_s \pi R_E^2 (1 - A) = \sigma T_E^4 4\pi R_E^2, \qquad (81.2)$$

where A is the albedo of the earth (the fraction of solar radiation reflected from, rather than absorbed by, the Earth), F_s is the solar flux at the edge of the earth's atmosphere, and πR_E^2 is the Earth's area normal to the solar flux. Solving this equation for T_E, one obtains $T_E \simeq 255$ K ($-18°$C).

The sun, which has a surface temperature of approximately 6000 K, emits most of its radiation in the 0.2–4.0 μm region of the spectrum (2 000–40 000 Å). The upper atmosphere of the Earth (thermosphere, ionosphere, mesosphere, and stratosphere) absorbs all the solar radiation shortward of 320 nm. The atmosphere of the earth absorbs only weakly in the visible region of the spectrum where the solar flux peaks.

The Earth, with an effective radiating temperature of 255 K, emits mainly long-wavelength radiation in the 4–100 μm region. Molecules naturally present in the atmosphere in trace amounts, such as carbon dioxide, water and methane, absorb strongly in this wavelength region [10]. Radiation coming from the Earth is thus absorbed, reradiated back to the surface, and thermalized through collisions with the ambient gas. This trapping of the radiation produces an additional warming of 33 K. Thus the mean surface temperature of the Earth is 288 K, not 255 K as found for T_E above. This effect of the Earth's atmosphere is known as the *greenhouse effect*. The greenhouse effect is what makes Earth habitable for life as we know it. Gases, both natural and man-made, which absorb strongly in the 4–100 μm region, are known collectively as greenhouse gases.

81.3.2   Tropospheric "Global" Warming

From air bubbles trapped at different depths in polar ice, it is possible to determine carbon dioxide and methane concentrations several thousand years ago. Over the last two hundred years, CO_2 levels have increased by 20%, from 280 to 330 ppm. Over the next century the total amount of CO_2 in the atmosphere since 1900 is expected to double to as much as 600 ppm [11]. This increase is due primarily to the burning of fossil fuels.

Although methane is present at levels several orders of magnitude less than CO_2, it is increasing much more rapidly. Methane concentrations have more than doubled over the last two hundred years due to industrial processes, fuels, and agriculture [11].

The man-made chlorofluorocarbons (CFCs), which have been widely used as refrigerants and in industry, have been increasing in the atmosphere at a rate of over 5% per year since the 1970s. Only recently has there been an indication that this trend is slowing down [12].

Ozone, which is a primary component of chemical smog, is a pollutant when it occurs in the troposphere and an effective greenhouse gas. It has been increasing worldwide also.

This buildup of CO_2, CH_4, CFCs and tropospheric O_3 causes a problem. In much of the spectral region from 5–100 μm, there is 100% absorption of radiation by the atmosphere — due mainly to naturally occurring water vapor. There is, however, a region of rather weak absorption, from $\sim$7–15 μm, known as the "atmospheric window". Increased concentrations of the greenhouse gases strengthen the absorption in this region, tending to "close" this window, thus increasing the infrared opacity of the atmosphere. The increased opacity causes an immediate decrease in the thermal radiation from the planet-atmosphere system, forcing the temperature to rise until the energy balance is restored [13].

There are some indications that this heating may already be occurring. Over the last century the mean global temperature has increased by approximately 0.5°C [11]. It is difficult, however, to prove that the buildup of greenhouse gases is the cause of this temperature rise. Other possible causes include slight changes in solar activity and irradiance, and changes in ocean currents, which may have a profound effect on global temperature and climate. These are areas of active research.

Given the increase in concentrations of greenhouse gases that has occurred and is predicted to continue, the change in radiative heating of the troposphere can be calculated. Models generally predict an increase in tropospheric temperatures ranging from 1.5 to 4.5°C, upon doubling the CO_2 concentrations over the next century. The 3°C range in predicted temperature increase is due to the ways that different models incorporate climate feedbacks. Climate feedbacks include water vapor, snow and sea ice, and clouds. Rising temperatures increase the concentration of water vapor, which is itself a greenhouse gas, producing further warming. Rising temperatures reduce the extent of reflective snow and ice, thus reducing the Earth's albedo. This leads to increased absorption of solar radiation, further increasing temperatures. Clouds both contribute to the albedo, thereby reducing the solar flux reaching the Earth, and absorb infrared radiation causing temperatures to rise. The modeling of clouds and their radiative properties is very difficult, and is one of the largest sources of uncertainty in the climate models. Understanding the role that the ocean, with its giant heat capacity, plays in global warming, and identifying and quantifying the various interactions occurring at the ocean-atmosphere interface, are vital areas of research which will affect the size of the predicted temperature increase. At present, there are few obvious opportunities for traditional atomic and molecular physics to play a significant part in global-warming research.

81.3.3   Upper Atmosphere Cooling

The buildup of CO_2 has an even greater effect on the temperature in the upper atmosphere than on that in the troposphere [14]. As discussed in Sect. 81.2.2, CO_2 is a coolant in the stratosphere, mesosphere, and thermosphere, but as a greenhouse gas is involved in heating the troposphere. The explanation for this revolves around the collision physics issue of quenching versus radiating.

In the troposphere, CO_2 absorbs infrared radiation coming from the Earth, exciting the ν_2 vibrational bending mode at 15 μm. The excited molecule can either reradiate or collisionally de-excite. In the lower atmosphere where densities are large, the lifetime against collisions is very short and the excited molecule is rapidly quenched. This transfer of energy from radiation through collisions into the kinetic energies of the colliding partners results in a net heating.

In the stratosphere and above, atomic oxygen collisions with CO_2 excite this same bending mode. But at these higher altitudes, densities are lower and quenching is greatly reduced. The excited molecule radiates and the radiation escapes to space. A net cooling results because the opacity is low at these altitudes.

Roble and co-workers [15, 16] have investigated the doubling of CO_2 and CH_4 concentrations (as predicted for the next century) in the mesosphere and at the lower boundary of the thermosphere. Using sophisticated atmospheric general circulation models, they predict that the stratosphere, mesosphere, and thermosphere will show significant cooling — the largest cooling of 40–50°C occurring in the thermosphere.

The extent of this cooling depends very much on the rate coefficient for the O + CO_2 excitation of the ν_2 bending mode. Rishbeth and Roble [15] assumed a value for this rate coefficient of 1×10^{-12} cm^3 s^{-1}, intermediate between the value of Sharma and Wintersteiner [17] (6×10^{-12} cm^3 s^{-1}) and an earlier value of 2×10^{-13} cm^3 s^{-1} used by Dickenson [18]. The Sharma and Wintersteiner value, based on observations of 15 μm emission in the atmosphere around 100–150 km, was recently confirmed by Rodgers *et al.* [19], but Pollock *et al.* [20] obtain a value of 1.2×10^{-12} cm^3 s^{-1} in laboratory experiments. Using the larger rate coefficient would result in even greater cooling [21].

The overall consequences of such a large temperature decrease in the upper atmosphere have not been fully explored — particularly, the question as to how the dynamics of the atmosphere will be affected. Since many chemical reactions depend on temperature, there may be considerable readjustments in the vertical distribution of minor species in the atmosphere. Cooler temperatures cause the atmosphere to contract, reducing densities and, consequently, satellite drag. Cooler temperatures may also increase the occurrence of polar stratospheric clouds, thereby affecting ozone depletion (see Sect. 81.4).

Most significantly, tropospheric warming and upper atmosphere cooling both result from a buildup of CO_2. The size of the predicted cooling is greater by an order of magnitude than the amount of the predicted heating. Thus it may be possible to monitor the global warming trend by observing the predicted cooling in the upper atmosphere.

There is evidence in the mesosphere that this cooling has already begun. Temperatures appear to have decreased by 3–4°C over the last decade [22, 23]. Gadsden [24] has also found that the frequency of occurrence of noctilucent clouds, the highest-lying clouds in the atmosphere, has more than doubled over the last twenty-five years. He has calculated that this change could result from a decrease of 6.4°C in the mean temperature at the mesopause during this time period. However, increased concentrations of water produced by oxidation of increased amounts of methane may be responsible for the more frequent appearance of the clouds. This is an ongoing area of research.

81.4   STRATOSPHERIC OZONE

81.4.1   Production and Destruction

Ozone production takes place continually in the stratosphere during daylight hours, as molecular oxygen is photodissociated and the resulting oxygen atoms undergo three-body recombination with O_2:

$$O_2 + h\nu \rightarrow 2O \,, \tag{81.3}$$

$$O + O_2 + M \rightarrow O_3 + M \,. \tag{81.4}$$

Ozone can be destroyed through photodissociation,

$$O_3 + h\nu \rightarrow O_2 + O \,, \tag{81.5}$$

but because an oxygen atom is produced which immediately recombines with another O_2 to form O_3, no net loss of O_3 results. The photodissociation of O_2 and O_3 are important heating processes in the stratosphere.

The amount of ozone in the stratosphere is quite variable, changing significantly with the seasons and with latitude. In the lower stratosphere, much of the ozone is created over the equatorial regions and then transported toward the poles.

Besides being photodissociated, O_3 is destroyed by reactions with radicals that are involved in catalytic cycles. The short-hand notation for the major cycles, NO_x, HO_x, ClO_x (BrO_x) refers to the catalytically active forms involved in the cycles. Our knowledge about the relative importance of these catalytic cycles in ozone destruction has increased dramatically over the last decade. A number of these cycles are given below, with the ozone-destroying step listed first, and the rate-limiting step closing the catalytic cycle and regenerating the ozone-destroying radical, listed last. The net effect in each of these cases is to convert ozone and atomic oxygen (otherwise known as odd-oxygen) into molecular oxygen:

$$NO + O_3 \rightarrow NO_2 + O_2$$
$$NO_2 + O \rightarrow NO + O_2$$
$$\text{NET: } O_3 + O \rightarrow 2O_2 \,, \tag{81.6}$$

and

$$NO + O_3 \rightarrow NO_2 + O_2$$
$$NO_2 + O_3 \rightarrow NO_3 + O_2$$
$$NO_3 + h\nu \rightarrow NO + O_2$$
$$\text{NET: } 2O_3 \rightarrow 3O_2 \,; \tag{81.7}$$

$$OH + O_3 \rightarrow HO_2 + O_2$$
$$HO_2 + O_3 \rightarrow OH + 2O_2$$
$$\text{NET: } 2O_3 \rightarrow 3O_2 \,, \tag{81.8}$$

$$Z + O_3 \rightarrow ZO + O_2$$
$$ZO + O \rightarrow Z + O_2$$
$$\text{NET: } O_3 + O \rightarrow 2O_2 \,. \tag{81.9}$$

The following series of reactions couples the HO_x and halogen cycles:

$$HO_2 + ZO \rightarrow HOZ + O_2$$
$$HOZ + h\nu \rightarrow OH + Z$$
$$Z + O_3 \rightarrow ZO + O_2$$
$$OH + O_3 \rightarrow HO_2 + O_2$$
$$\text{NET: } 2O_3 \rightarrow 3O_2 \,. \tag{81.10}$$

In the latter two sets of reactions, $Z = $ Cl or Br. Finally the reaction set

$$BrO + ClO \rightarrow Br + Cl + O_2$$
$$Br + O_3 \rightarrow BrO + O_2$$
$$Cl + O_3 \rightarrow ClO + O_2$$
$$\text{NET: } 2O_3 \rightarrow 3O_2 \tag{81.11}$$

is also important in the halogen destruction cycle. The coupling between these different cycles by reactions such as

$$HO_2 + NO \rightarrow OH + NO_2 \tag{81.12}$$

turns out to be very important in understanding the details of ozone destruction, such as how much each

mechanism contributes to the destruction as a function of altitude and in the presence of aerosols. Wennberg *et al.* [25] have recently shown that catalytic destruction by NO_2, which for two decades was considered to be the predominant loss process, accounted for less than 20% of the O_3 removal in the lower stratosphere during May 1993. They further show that the cycle involving the hydroxyl radical accounted for nearly 50% of the total O_3 removal and the halogen-radical chemistry was responsible for the remaining 33%.

The NO_x and HO_x cycles are naturally occurring, whereas the ClO_x and BrO_x cycles are due mainly to man-made chemicals — the CFCs and halons. The amplification that takes place through a catalytic cycle is the reason that these chemicals, which are only present at the level of parts per trillon, can have such a destructive effect.

It is useful to think in terms of a total chemical budget for a radical such as Cl which enters into a cycle. Chlorine is put into the stratosphere when chemicals such as CF_2Cl_2 are released into the atmosphere. Such compounds are chemically inert and insoluble in water, and therefore are not easily cleansed out of the lower atmosphere. In the stratosphere, however, the CF_2Cl_2 is subjected to solar uv radiation and is photodissociated, producing the Cl radical. Chlorine exists in the upper atmosphere in catalytically active forms, Cl and ClO, as well as in stable, reservoir species, HCl and $ClONO_2$. The total chlorine budget consists of both the catalytically active plus reservoir species. Reactions which reduce the formation of reservoir species, or convert reservoir species to catalytically active forms, contribute to the ozone destruction. Photolysis of stable reservoir species, such as $ClONO_2$, can produce catalytically active forms. Bromine has an identical cycle to that of chlorine, but is 50 to 100 times more destructive than Cl because it does not react readily to go into its reservoir form, HBr. A knowledge of the photodestruction rates of all such species is important to an understanding of the overall ozone photochemical depletion problem.

Studies of the Antarctic ozone hole show that gas phase photochemical cycles, as given above, are not the whole story with respect to ozone depletion. Heterogeneous chemistries taking place on the surfaces of ice crystals and sulfate aerosols play an important role also. These are discussed briefly in Sects. 81.4.2 and 81.4.4.

81.4.2 The Antarctic Ozone Hole

The ozone depletion problem was largely theoretical until the discovery of the ozone hole over Antarctica. Following the 1985 announcement by Farman *et al.* [26] of ground-based observations of significant decline in O_3 concentrations during springtime in the Southern Hemisphere, it was possible to map this event using archived satellite data beginning in 1979. The data depict a worsening event throughout the early 1980s. In 1987, 70% of the total O_3 column over Antarctica was lost during the month of September and early October, and the areal extent of the hole was $\sim$10% of the Southern Hemisphere. Recent data show that the hole continues to grow both in depth and width each year [27].

The causal link between the release and buildup of man-made CFCs and the ozone hole over Antarctica has been quite convincingly established by Anderson *et al.* [28] through *in situ* observations from high altitude aircraft flights into the polar vortex during the end of polar night and the beginning of Antarctic spring in 1987.

The polar vortex is a stream of air circling Antarctica in the winter, creating an isolated region which becomes very cold during the polar night. Flights into the vortex were able to document a heightened, increasing level of ClO and a monotonically decreasing O_3 concentration over a 3–4 week time period during late September and early October.

The mechanism which appears to be repartitioning the chlorine from its reservoir form into its catalytically active form is a heterogeneous process occurring on the surfaces of polar stratospheric clouds. At the cold temperatures during the polar night, polar stratospheric clouds form, consisting of ice and nitric acid trihydrate. Gaseous $ClONO_2$ collides with HCl that has been adsorbed onto the surface of the cloud crystals. Chlorine gas is liberated and the nitric acid formed in the reaction remains in the ice [28]:

$$HCl + ClONO_2 \rightarrow Cl_2(g) + HNO_3 . \qquad (81.13)$$

As solar radiation starts to penetrate the region at the beginning of spring, the Cl_2 molecules are rapidly photodissociated, producing Cl atoms which initiate the catalytic destruction of O_3.

As there are no oxygen atoms around to complete the catalytic cycles, several mechanisms for regenerating the Cl and Br radicals have been proposed which involve only the ClO and BrO molecules themselves.

Mechanism I [29]

$$ClO + ClO + M \rightarrow (ClO)_2 + M$$
$$(ClO)_2 + h\nu \rightarrow Cl + ClOO$$
$$ClOO + M \rightarrow Cl + O_2 + M$$
$$2 \times (Cl + O_3 \rightarrow ClO + O_2)$$
$$\text{NET: } 2O_3 \rightarrow 3O_2 ; \qquad (81.14)$$

Mechanism II [30]

$$ClO + BrO \rightarrow Cl + Br + O_2$$
$$Cl + O_3 \rightarrow ClO + O_2$$
$$Br + O_3 \rightarrow BrO + O_2$$
$$\text{NET: } 2O_3 \rightarrow 3O_2 . \qquad (81.15)$$

While Mechanism I accounts for 75% of the observed ozone loss, the sum of I and II yields a destruction rate in harmony with the observed O_3 loss rates [28].

81.4.3 Arctic Ozone Loss

The region around the North Pole does not appear to exhibit an ozone hole as severe as that found in the Antarctic. Several factors lessen the probability of a significant ozone hole developing in the Arctic. First, a stable polar vortex does not get well established due to increased atmospheric turbulence from the greater land surface area in the Northern Hemisphere. Second, temperatures during the Arctic winter do not get as cold as during the Antarctic winter, so that polar stratospheric clouds (PSCs) do not form as easily. As seen in the preceding section, the surfaces of PSCs play an essential role in the O_3 destruction mechanisms in the Antarctic.

However, ozone levels are showing 20–25% reductions during February and March [31] over a much larger area around the North Pole than in the South. Thus ozone destruction is taking place during the transition from polar winter to spring in the Arctic but the phenomenon is more widespread, diffuse, and not as well-contained as in the Antarctic.

81.4.4 Global Ozone Depletion

Over the last fifteen years, ozone levels have been steadily decreasing globally, especially at mid- to high-latitudes. Trends have been deduced from a statistical fit to the data [32] obtained by the Total Ozone Monitoring Spectrometer (TOMS) aboard the NIMBUS-7 satellite. This instrument obtained total ozone column measurements from October 1978 to May 1993. At 50° N, the annually averaged trend is -0.5%/year, and the trend in the Southern Hemisphere is even larger, consistent with dilution effects from the Antarctic ozone hole. Heterogeneous reactions on aerosol surfaces, as well as the homogeneous gas phase chemical cycles mentioned earlier, must be invoked to explain these large trends.

A particularly important reaction appears to be the hydrolysis of N_2O_5 on sulfate aerosols. This occurs very rapidly, converting reactive nitrogen, NO_2, into its reservoir species HNO_3:

$$N_2O_5 + H_2O \xrightarrow{\text{sulfate aerosol}} 2HNO_3 . \qquad (81.16)$$

The N_2O_5 is formed at night by reaction of NO_2 and NO_3. Following the hydrolysis of N_2O_5, there is less reactive NO_2 around to convert ClO into its reservoir species, $ClONO_2$, and less NO_2 around to convert OH into the reservoir species, HNO_3. A heightened sensitivity of the ozone to increasing levels of CFCs develops [33]. It has been shown that certain regions of significantly depleted ozone also show high concentrations of sulfate aerosols. In addition, measurements of the ratio of catalytically active nitrogen to total nitrogen can be reproduced using the above heterogeneous reaction, and not by using gas phase processes alone. Study of further mechanisms at varying altitudes and latitudes is an active area of research.

Record low global ozone measurements, 2% to 3% lower than any previous year, were reported beginning in 1992 [34] and continuing well into 1993. The increase in naturally occuring aerosols due to the eruption of Mount Pinatubo in June 1991 appears to explain this decline. During the winter of 1993–1994, total ozone levels returned to levels slightly above normal [35], presumably because the excess aerosols had been removed from the stratosphere by natural sedimentation processes.

The continuing buildup of CO_2 is predicted to contribute to increased cooling of the stratosphere. Declining temperatures in the stratosphere may increase the frequency of formation of polar stratospheric clouds which drive the destructive heterogeneous chemistry creating the Antarctic ozone hole. An increased occurrence of these clouds outside of the polar regions could affect ozone levels globally. There are also indications that certain ozone depletion chemistries taking place on the surface of sulfate aerosols may also be enhanced by lower temperatures [33].

Ozone itself is the dominant heat source in the lower stratosphere. Decreasing the amount of ozone will drive temperatures still lower. It is unfortunate that the two most significant atmospheric global change effects — the buildup of CO_2 and the enhanced ozone destruction due to man-made CFCs — both cause decreasing temperatures in the stratosphere which may further heighten the destructiveness of the ozone photochemical cycles.

81.5 ATMOSPHERIC MEASUREMENTS

Ground-based observations, as well as measurements made by instruments carried aloft in satellites, balloons, and high-flying aircraft, allow one to explore the atmosphere.

Measurements may be made either *in situ* or by remote-sensing techniques. The region of the atmosphere from ~60 km to 120 km, encompassing the mesosphere and lower thermosphere and ionosphere, cannot be studied *in situ* as it is too high for balloons and aircraft, and too low (i.e. too much drag) for satellites. For this region, remote sensing experiments are essential and a comprehensive book on the subject is recommended [36].

Most of the instruments used to make atmospheric measurements have been developed in molecular physics and spectroscopy laboratories. Even experiments utiliz-

ing sophisticated techniques, such as laser induced fluorescence, and instruments, such as Fourier transform spectrometers, are being flown on payloads. An excellent compendium of ozone-measuring instruments for stratospheric research has been assembled by Grant [37].

A combination of good laboratory experiments, theoretical calculations, and ingenuity are necessary to extract accurate information from measurements made in the atmosphere. For instance, in order to understand the complicated interactions of the different photochemical cycles involved in ozone chemistry, spectroscopic emissions of the many trace species are used to measure concentration profiles [38]. An accurate knowledge of the emission spectroscopy of species such as OH, HO_2, H_2O_2, H_2O, O_3, HNO_3, NO_2, N_2O, HCl, HOCl, and ClO is essential. Such measurements provide a rigorous test of atmospheric models.

In order to analyze the data and deconvolve some of the line profiles to give information on concentrations as a function of altitude, molecular data such as line strengths and pressure broadening coefficients are needed.

Until recently, it has been impossible for remote-sensing experiments to distinguish between ozone occurring in the stratosphere (where it is formed naturally) and ozone occurring in the troposphere (where it is a pollutant). Future instruments such as the ESA Global Ozone Monitoring Experiment (GOME) will have a broad enough spectral coverage and high enough resolution that the temperature dependence of the ozone absorption features from 300–340 nm, known as the Huggins bands, can be used to separate out the ozone concentrations in the middle and lower atmospheres [39].

REFERENCES

1. T. E. Graedel and P. J. Crutzen, *Atmospheric Change: An Earth System Perspective* (W. H. Freeman, New York, 1993); J. Houghton, *Global Warming: The Complete Briefing* (Lion Publishing, Oxford, 1994).
2. R. G. Roble, in *Encyclopedia of Applied Physics* (VCH Publishers, 1991), Vol. 2, pp. 201–224.
3. P. S. Armstrong, J. J. Lipson, J. A. Dodd, J. R. Lowell, W. A. M. Blumberg, and R. M. Nadile, Geophys. Res. Lett. **21**, 2425 (1994).
4. R. R. Conway, NRL Memorandum Report, 1988.
5. K. Kirby, E. R. Constantinides, S. Babeu, M. Oppenheimer, and G. A. Victor, At. Data Nucl. Data Tables **23**, 63 (1979).
6. K. Chance, K. W. Jucks, D. G. Johnson, and W. A. Traub, J. Quant. Spectrosc. Radiat. Transfer **52**, 447 (1994).
7. L. Rothman, R. R. Gamache, R. H. Tipping, C. P. Rinsland, M. A. H. Smith, D. C. Benner, V. M. Devi, J.-M. Flaud, C. Camy-Peyret, A. Perrin, A. Goldman, S. T. Massie, L. R. Brown, and R. A. Toth, J. Quant. Spectrosc. Radiat. Transfer **48**, 469 (1992).

8. P. L. Smith and W. L. Wiese, *Atomic and Molecular Data for Space Astronomy* (Springer-Verlag, Berlin, 1992).

9. R. P. Wayne, *Chemistry of Atmospheres* (Oxford University Press, New York, 1991), p. 41.

10. J. F. B. Mitchell, Rev. Geophys. **27**, 115 (1989).

11. J. Firor, *The Changing Atmosphere* (Yale University, New Haven, 1990).

12. J. N. Elkins, T. M. Thompson, T. H. Swanson, J. H. Butler, B. D. Hall, S. O. Cummings, A. Fisher, and A. G. Raffo, Nature **364**, 780 (1993).

13. J. Hansen, D. Johnson, A. Lacis, S. Lebedeff, P. Lee, D. Rind, and G. Russell, Science **213**, 957 (1981).

14. R. J. Cicerone, Nature **344**, 104 (1990).

15. H. Rishbeth and R. G. Roble, Planet. Space Sci. **40**, 1011 (1992).

16. R. G. Roble and R. E. Dickinson, Geophys. Res. Lett. **16**, 1441 (1989).

17. R. D. Sharma and P. P. Wintersteiner, Geophys. Res. Lett. **17**, 2201 (1990).

18. R. E. Dickenson, J. Atmos. Terr. Phys. **46**, 995 (1984).

19. C. D. Rodgers, F. W. Taylor, A. H. Muggeridge, M. Lopez-Puertas, and M. A. Lopez-Valverde, Geophys. Res. Lett. **19**, 589 (1992).

20. D. S. Pollock, G. B. I. Scott, and L. F. Phillips, Geophys. Res. Lett. **20**, 727 (1993).

21. S. W. Bougher, D. M. Hunten, and R. G. Roble, J. Geophys. Res. **99**, No. E7, 14609 (1994).

22. A. C. Aiken, M. L. Charin, J. Nash, and D. J. Kendig, Geophys. Res. Lett. **18**, 416 (1991).

23. A. Hauchecorne, M. L. Charin, and R. Keckhut, J. Geophys. Res. **96**, No. D8, 15297 (1991).

24. M. Gadsden, J. Atmos. Terr. Phys. **52**, 247 (1990).

25. P. O. Wennberg *et al.*, Science **266**, 398 (1994).

26. J. C. Farman, B. G. Gardiner, and J. D. Shankin, Nature **315**, 207 (1985).

27. R. A. Kerr, Science **266**, 217 (1994).

28. J. G. Anderson, D. H. Toohey, and W. H. Brune, Science **251**, 39 (1991).

29. L. T. Molina and M. J. Molina, J. Phys. Chem. **91**, 433 (1987).

30. M. B. McElroy, R. J. Salawitch, S. C. Wofsy, and J. A. Logan, Nature **321**, 759 (1986).

31. G. L. Manney, L. Froidevaux, J. W. Waters, R. W. Zurek, W. G. Read, L. S. Elson, J. B. Kumer, J. L. Mergenthaler, A. E. Roche, A. O'Neil, R. S. Harwood, I. MacKenzie, and R. Swinbank, Nature **370**, 429 (1994).

32. R. S. Stolarski, P. Bloomfield, R. D. McPeters, and J. R. Herman, Geophys. Res. Lett. **18**, 1015 (1991).

33. D. W. Fahey *et al.*, Nature **363**, 509 (1993).

34. J. F. Gleason *et al.*, Science **260**, 523 (1993).

35. D. J. Hofmann *et al.*, Geophys. Res. Lett. **21**, 1779 (1994).

36. J. T. Houghton, F. W. Taylor, and C. D. Rodgers, *Remote Sounding of Atmospheres* (Cambridge University Press, Cambridge, 1984).

37. *Ozone Measuring Instruments for the Stratosphere*, edited by W. B. Grant (Optical Society of America, Washington DC, 1989).

38. K. Chance, W. Traub, D. G. Johnson, K. W. Jucks, P. Ciarpallini, R. A. Stachnik, R. J. Salawitch, and H. A. Michelsen, J. Geophys. Res. (to be published).

39. K. Chance (private communication).

82

Atoms in Plasmas

Jon Weisheit

Department of Space Physics and Astronomy, Rice University, Houston, Texas 77251-1892

82.1 INTRODUCTION

Ionized gases, or *plasmas*, are the predominant form of matter throughout the universe. As Figure 82.1 shows, however, physical conditions in cosmic and laboratory plasmas vary greatly. No single experimental methodology or theoretical construct suffices to explore all aspects of this "fourth state" of matter.

Systematic study of plasmas began early in this century, but until recently the physics of atoms in plasmas has been largely synonymous with the physics of *isolated* ions. This perspective suffices as long as the interparticle spacing is very much larger than the relevant atomic dimensions, typically a few to a few tens of Bohr radii. For example, ions are isolated in this sense in interstellar space where the electron density $n_e \approx 1$ cm^{-3}, or even in a tokamak [1], where $n_e \approx 10^{14}$ cm^{-3}. For neutral and moderately charged atoms, data such as energy levels, oscillator strengths, and collision cross sections, have long been obtained from traditional kinds of experiments and quantal calculations discussed elsewhere in this book. Electron beam ion traps (EBIT)

[2] now make possible accurate measurements involving highly stripped atoms.

The focus of the present chapter is partially ionized plasmas in which important atomic phenomena are influenced by the plasma environment. Figure 82.1 reveals that the densities in many laboratory and astrophysical plasmas are high enough to invalidate the presumption of isolated systems. There are several periodic meetings devoted to this aspect of plasma science, and especially relevant ones include: *Atomic Processes in Plasmas*; *Radiative Properties of Hot, Dense Matter*; *Spectral Line Shapes*; and *Strongly Coupled Plasma Physics*. Printed proceedings of these conferences are an excellent guide to recent developments in the topics discussed here, as well as numerous other related ones.

82.1.1 Plasma Parameters

Most plasmas are charge neutral, so the mean number densities N_i and n_e of constituent ions (charge $Z_i e$, mass m_i) and electrons (charge $-e$, mass m_e) satisfy

Figure 82.1. Characteristic temperatures and densities of plasmas mentioned in text. Plasmas below the bold dashed line are strongly coupled.

$$\sum_a Z_a N_a = 0\,, \tag{82.1}$$

where the sum ranges over all particle species. Moreover, plasma conditions usually change slowly enough that each of the species is able to establish a thermal distribution of velocities, fixed by its temperature $\Theta_a = k_B T_a$ (in energy units). These distributions are Maxwellian as long as $\Theta_a \gg E_F$, where $E_F = \hbar^2(3\pi^2 N_a)^{2/3}/2m_a$ is the Fermi energy for species a, and N_a is a generic number density, with $N_a = n_e$ for the case $a = e$.

A few key quantities characterize such plasmas:

1. The plasma frequency

$$\omega_p = \left[\sum_a 4\pi N_a (Z_a e)^2/m_a\right]^{1/2} = \left[\sum_a \omega_a^2\right]^{1/2} \tag{82.2}$$

defines a timescale ($\sim \omega_e^{-1}$) for the damping of charge-density fluctuations.

2. Coulomb coupling parameters

$$\Gamma_a = (Z_a e)^2/R_a \Theta_a\,, \tag{82.3}$$

with $R_a = (3/4\pi N_a)^{1/3}$, give the average ratio of potential to kinetic energies for each species a. When Γ_a is greater (less) than 1, that species is said to be strongly (weakly) coupled.

3. The Debye length

$$\lambda_D = \left[\sum_a 4\pi N_a(Z_a e)^2/\Theta_a\right]^{-1/2}$$
$$= \left[\sum_a D_a^{-2}\right]^{-1/2} \tag{82.4}$$

is the distance beyond which plasma particles effectively screen any localized charge imbalance. The number of particles a in a sphere of radius D_a (a Debye sphere) is $[4\pi N_a D_a^3/3] = 1/(3\Gamma_a)^{3/2}$.

Derivations, and full discussions of the roles of these and other, auxiliary quantities can be found in standard plasma physics texts (see Refs. [3, 4]).

82.1.2 Electromagnetic Fields in Plasmas

A simple, yet useful description of the plasma environment is given by the *one-component plasma* (OCP) model [5], in which particles of a single kind (Z_a, m_a) move against a smooth background of matter having the opposite charge density, $\rho(\mathbf{r}) = -Z_a e N_a$. Most plasma phenomena can be described within the context of either this or the two-component model (electrons and their parent ions, with $N_i = n_e/Z_i$). When there are many particles in each Debye sphere, the electrostatic potential $\Phi(r)$ near a test charge q placed in an otherwise uniform, neutral plasma is exponentially reduced from the Coulomb expression, *viz.*

$$\Phi_D(r) = (q/r)\exp(-r/\lambda_D)\,. \tag{82.5}$$

The Debye screening obtains only in weakly coupled plasmas and applies only to a test charge at rest. The faster charge q moves, the less effective is the plasma at screening it, since only plasma particles with higher velocities can form the shielding cloud [3, 4].

In the opposite limit of large Γ-values, screening is better described by the ion-sphere (IS) picture [5], in which each stationary ion of charge $Z_i e$ is surrounded by Z_i electrons, uniformly distributed throughout a sphere of radius R_i, to produce the potential

$$\Phi_{IS}(r) = Z_i e[1/r - (1/2R_i)(3 - r^2/R_i^2)] \tag{82.6}$$

inside the sphere, and zero potential outside.

Consideration of a plasma's electric microfield illustrates several of these concepts. Moreover, microfields are a key ingredient in calculations of spectral line broadening in plasmas — an important subject recently reviewed elsewhere [6].

Local fluctuations in the density of any species about its mean value N_a create a microscopic electric field $\tilde{\mathbf{E}}_a(\mathbf{r}, t)$. There is a probability distribution $P(\tilde{E}_a)$ that characterizes the strengths of these microfields, which are quasistatic within time intervals short compared with the fluctuation timescale, $1/\omega_a$. Holtsmark first calculated this distribution at an arbitrary position in an infinite, isotropic OCP of noninteracting particles, and Chandrasekhar [7] gives a careful account of this famous stochastic problem. Holtsmark's formula is

$$P_\mathrm{H}(\varepsilon) = (2\varepsilon/\pi) \int_0^\infty x \sin(\varepsilon x) \exp[-x^{3/2}]\, dx\,, \quad (82.7)$$

where $\varepsilon = \tilde{E}_a/[(8\pi/25)^{1/3} E_a]$ is the scaled field, and $E_a = |Z_a|e/R_a^2$. The mean Holtsmark field is $\langle \varepsilon \rangle \simeq 2.99$, and for $\varepsilon \gg 1$, $P_\mathrm{H}(\varepsilon)$ is well approximated by the distribution of fields due to a single nearest-neighbor in an OCP [8],

$$P_\mathrm{nn}(\varepsilon) \simeq 3/(2\varepsilon^{5/2})\,. \quad (82.8)$$

Both of these distributions ignore the interactions among charged particles that become increasingly important as Γ_a grows, because particle positions then tend to be correlated. Further, in more realistic two-component, electron-ion plasmas, electron screening of ions is established on times $\omega_e^{-1} \ll \omega_i^{-1}$. Quasistatic ion microfields therefore become weaker, on average, than in the OCP case. For studies of microfields in plasmas with moderate Γ-values see Refs. [9, 10].

Besides its own microfield, which is always present, a plasma also has within it electromagnetic waves and oscillations resulting from collective effects, plus radiation produced in atomic transitions. Additional atom-plasma couplings arise when these or other, large-scale electromagnetic fields are important.

The interaction of intense laser radiation and solid matter produces rapidly evolving, hot, and dense plasmas (see Refs. [11, 12]) that mimic some of the most extreme conditions in nature, including the thermonuclear environment of stellar interiors; laser plasmas also can be transient gain media for amplified spontaneous emission at x-ray wavelengths [13, 14]. In the first area, x-ray line emission from highly stripped ions can be used to diagnose conditions in inertial confinement fusion (ICF) plasmas [15, 16]; in the second, the complicated interplay of collisional and radiative processes produces ionic population inversions (see Ref. [17]). Additional impetus for the study of atoms in plasmas now comes from experiments involving irradiation of solids by ultra-short (sub-picosecond) laser pulses [18, 19] (see Sect. 69.3.1). The moderate temperatures ($\sim$10s of eV) but high (near-solid) densities typical of these plasmas combine to produce severely perturbed bound ionic states in the strongly coupled plasma regime.

82.2 ATOMIC STRUCTURE

The short lifetimes of dense plasmas created by intense laser irradiation or by explosive electromagnetic

pinch devices [20] often preclude the establishment of a thermal distribution of atomic level populations; in some instances there is not even enough time to establish a Maxwellian distribution of particle velocities. Populations in these nonequilibrium plasmas must be found by solving rate equations [21, 22]. In contrast, dense cosmic plasmas — specifically, the interiors of stars and giant planets — are very large and have very long lifetimes. Their thermodynamic variables change only slowly with position or time and, hence, macroscopic regions can be considered as statistical systems evolving through a succession of states in local thermodynamic equilibrium (LTE).

For LTE plasmas, key quantum phenomena are the equation of state (EOS), and the radiation and heat transport coefficients, *viz.* the opacity and thermal conductivity. Review articles by More [23, 24] and by Weisheit [25] survey many of the high-density consequences for atomic structure and for coefficients of radiation and heat transport. Here, discussion focuses on the Thomas-Fermi model for hot, dense matter, plus comments on two other challenging topics — continuum lowering, and quantum effects in strongly magnetized plasmas.

82.2.1 Statistical Models of Ions

The Thomas-Fermi model, discussed in Chap. 20, describes atomic charge densities by treating bound electrons as a fully degenerate ($\Theta_e = 0$) Fermi gas, subject to a spherical electrostatic potential $\Phi_\mathrm{TF}(r)$ resulting from the nuclear charge $Z_n e$ and the electrons themselves. By introducing certain dimensionless variables, the Thomas-Fermi (TF) equation takes a universal form for all atoms [26, 27] (see Chap. 20 for further discussion). This equation predicts incorrectly that $\Phi_\mathrm{TF}(r)$ for isolated, neutral atoms falls asymptotically only as r^{-3}, but more realistically that isolated ions have a finite radius at which the electron density goes to zero. More [23] published a convenient table for finding ionization potentials and total energies, as predicted by the TF model, of ions with net charge $Z_i e$ between $0.1 Z_n e$ and $0.9 Z_n e$. Given the simplicity of the TF model, agreement with experiment is surprisingly good, usually well within a factor of 2 for the thousands of ions in the periodic table.

The TF model has been embellished with exchange (TF-Dirac) and finite temperature corrections, and used to attack problems as diverse as the stability of bulk matter and the ground state energy of white dwarf stars; these and numerous other aspects of TF theory are discussed in Spruch's engaging review [28]. Feynman *et al.* [29] first used the TF model to determine the EOS of dense matter, treating exchange and finite temperature effects separately as perturbations; later work (see Refs. [27, 30]) has led to the inclusion of both refinements simultaneously. The basic idea in all cases is that each nucleus is at the center of a spherical cell, and is surrounded by a neutralizing cloud of Z_n electrons. Cell radii R are fixed by the mean density N_n of nuclei, and are equal to ion sphere radii for fully stripped atoms. By symmetry, the electrostatic potential has a minimum at R. Hence, there is no net force on electrons at R, and the electron pressure at the cell boundary is directly related to the TF density of electrons there, $n_e(R)$. This connection yields an implicit PV relation for a specified Θ, i.e., an equation of state. In the TF model there is no distinction between bound and free electrons, but an average degree of ionization can be obtained from the relation $\langle Z_i \rangle = n_e(R)/N_n$, if one considers $n_e(R)$, which is the minimum density, to be that of the background of free electrons.

More elaborate statistical schemes for atoms in dense plasmas, based on solutions of self-consistent field equations for one-electron orbitals confined to spherical cells [24, 31, 32], yield the eigenvalue spectrum of an artificial, "average" atom — the average being over all charge states of the element and over all perturbing configurations of plasma particles. Such average atom models have been used extensively to construct opacity and EOS tables for hot, dense matter. More recent calculations have also employed screened potentials to represent the environment, but have taken account of the many different configurations of bound electrons [33].

82.2.2 Continuum Lowering

Because plasma screening attenuates the Coulomb interaction at long range, atoms and ions no longer have an infinite number of bound (Rydberg) states, and atomic partition functions are truncated in a natural way. The naive picture is that the onset of continuum energies has been "lowered" by some amount ΔI (relative to the atom's ground state), but the true situation seems far more complicated [34]. A variety of arguments has been put forth to quantify continuum lowering, including in particular:

1. determine ΔI from the last distinct spectral line near a series limit (the Inglis-Teller formula [35]);

2. determine ΔI from the atom's dipolar interaction with the plasma's microfield [36];

3. determine ΔI from the binding energy of the ground state in a specified, screened Coulomb potential [37–39];

4. determine ΔI from a rigorous, statistical mechanical treatment of the atomic partition functions [40].

When ΔI is a fixed quantity, continuum lowering eliminates bound states whose (unscreened) ionization potentials were less than ΔI, and moves all remaining

states closer to the continuum by this same amount. Thus, levels get shifted but spectral lines do not. The more sophisticated schemes that use an effective, single-particle potential to determine a spectrum of modified eigenstates produce distinct plasma shifts for different levels and, hence, spectral line shifts. However, the shifts are so small that some earlier measurements even differ in sign from more recent work. References [41, 42] review earlier work, and Refs. [43, 44] discuss more recent results.

82.2.3 Strong Magnetic Field Effects

For most magnetic fields encountered in laboratory and cosmic plasmas, the effects on atomic phenomena can be treated as perturbations because the Bohr magneton is so small ($\mu_B = e\hbar/2m_e c \approx 5.78 \times 10^{-9}$ eV/Gauss). However, neutron stars and a few white dwarf stars have huge fields, $B \sim 10^{12}$ G and $B \sim 10^8$ G, respectively [45], and also, in explosively driven pinch plasmas [20] and certain laser plasmas [46, 47], multi-megagauss fields exist temporarily.

Ions in the neutron star environment should have long, thin electron charge distributions. Motion parallel to **B** is controlled by relatively weak Coulomb forces, while transverse motion is well characterized by the Landau orbitals of unbound electrons in a magnetic field [48, 49] (see Sect. 13.1). The eigenfrequencies representative of these motions are very unequal, thus permitting an adiabatic approximation and separation of variables. This is the basis of self-consistent-field methods developed recently [50]. For lower field strengths, say $B \sim 10^9$ G, the adiabatic approximation is poor, and the problem of atomic structure is much more difficult. For the hydrogen atom, transition energies and oscillator strengths are now available for any B-value [51]. See Refs. [52, 53] for recent work on two-electron systems in very strong B-fields.

The TF model has been extended to treat many-electron atoms in magnetic fields that are intermediate to the Landau (intense field) and the Zeeman (weak field) regimes [54–56]. A statistical description cannot provide spectroscopic information, but it can form the basis of EOS and resistivity calculations for strongly magnetized plasmas [57]. The results quantify the dramatic effects that strong static magnetic fields have on atomic properties, and emphasize the need for further astronomical studies of strongly magnetized white dwarf stars, coupled with more refined diagnostics of laboratory plasmas having strong fields, to improve our general understanding of matter under these strange conditions.

82.3 ELEMENTARY PROCESSES

Atomic interactions are of little interest in an equilibrium plasma because they do not modify the plasma's physical state. However, the evolution of nonequilibrium plasmas is regulated by the time rate of change of quantities such as Θ and n_e, and these in turn can be analyzed using transport coefficients such as the radiative opacity, the electrical resistivity, and the thermal conductivity.

In a plasma, atomic transitions are induced by radiative and collisional interactions. Indeed, so many processes *can* occur that a major task is the identification of those which are most important in a particular (non-LTE) situation. The plasma environment may also alter rates applicable to isolated atoms, through the perturbation of the atomic states involved and/or the screening of long-range Coulomb forces. Further complicating the two-body collision picture is the close proximity of many scattering centers in a dense plasma. Within the Born approximation, multicenter effects can be couched in terms of dynamic structure factors, quantities which characterize density fluctuations at different space-time points [4, 58]. Presently, analysis of any of these many-body problems requires considerable simplification.

82.3.1 Radiative Transitions

For a radiative transition between states α and β, the absorption and emission rates are proportional to quantities of the form $(\Delta E)^n |\langle \alpha | \mathbf{d} | \beta \rangle|^2$ summed over degenerate substates, where $\mathbf{d}$ is the elctric dipole operator (see Chap. 10), and $n = 1$ or 3 for the Einstein B and A coefficients, respectively. Changes in these radiative quantities are due primarily to changes in the atomic wave functions. Theory predicts that plasma screening reduces line strengths, and the reduction factor increases toward the series limit [59, 60]. Also, the cross section for photoejection of an electron bound by any screened Coulomb potential must vanish at threshold — in marked contrast to the nonzero photoionization cross sections of isolated atoms and ions. Since the oscillator strength sum rule still holds, any diminution of the total bound-bound oscillator strength must be just matched by an increase in the bound-free contribution; continuum lowering partly accounts for this latter increase. Spectroscopic observations of the reduction of decay rates by plasma screening have been reported [61], but the plasma densities were evidently too low for this effect to manifest itself. Alternative explanations have since been given [62, 63]. What happens just below threshold also is a matter of dispute: there may or may not exist a *transparency window*, that is, a spectral region with negligible absorption strength per unit energy interval [60, 64]. In addition, there is the more fundamental question of whether *static* screening models are even appropriate for the description of radiative processes. This issue arises again in connection with inelastic collision processes in dense plasmas.

82.3.2 Momentum Transfer

Diffusion of charged particles in a plasma results from a large number of scatterings through small angles θ, and momentum transfer in the diffusion process is described by an angular integration of the differential cross section for elastic scattering $\sigma(\theta)$ with the weighting factor $(1 - \cos\theta)$. Because the pure Coulomb interaction has infinite range, its elastic and momentum transfer cross sections both diverge at small θ-values, and somehow a smallest relevant scattering angle must be identified.

Spitzer [66] used classical formulae to relate a minimum angle $\theta_{\min}$ to a maximum effective impact parameter $b = \lambda_D$. In this simple picture, Coulomb interactions at small (large) impact parameters are unscreened (completely screened). Rutherford's differential cross section $\sigma_C(\theta)$ for the scattering of charges $Z_a e$ and $Z_b e$ at relative energy $\mathcal{T}$ then yields

$$\sigma_m(\mathcal{T}) = \int_{\theta_{\min}} d\Omega \, (1 - \cos\theta) \, \sigma_C(\theta)$$
$$= \pi (Z_a Z_b e^2 / \mathcal{T})^2 \ln(2\lambda_D \mathcal{T} / |Z_a Z_b| e^2) \quad (82.9)$$

for the momentum transfer cross section. If, further, the kinetic energy $\mathcal{T}$ in the logarithm is replaced by its mean thermal value, this cross section takes the simple form

$$\sigma_m(\mathcal{T}) \approx \pi (Z e^2 / \mathcal{T})^2 \ln(\Lambda), \quad (82.10)$$

where $\ln(\Lambda)$ is known as the Coulomb logarithm. Equation (82.10) is used to estimate relaxation times in weakly coupled plasmas, where $\ln(\Lambda) \approx 10$; corrections to various transport coefficients in the intermediate-coupling regime $[\ln(\Lambda) \approx 1]$ are available [67].

One can argue that in a plasma, each collision is regulated by a screened interaction and that instead of a sharp cutoff at $\theta_{\min}$, one should adopt an appropriate screened potential to obtain various transport quantities [25]. For instance, if the plasma is weakly coupled, the Debye potential Φ_D may be used. For collision wavevectors k large enough that $k\lambda_D > 1$, the Born approximation for elastic scattering is reasonably accurate, and the corresponding σ_m is the same as Eq. (82.9) except that the argument of the logarithm is $(2k\lambda_D)$. The ion sphere potential describes the limiting case of a uniform distribution of screening electrons about each target ion. A more realistic description is given by the TF model, and the corresponding cross sections can be written in terms of an atomic form factor, i.e., the Fourier transform of $\Phi_{\mathrm{TF}}(r)$. Mott and Massey [68] tabulated the universal TF form factor for isolated, neutral atoms, but no analogous results are available for ions in plasmas. (Note that

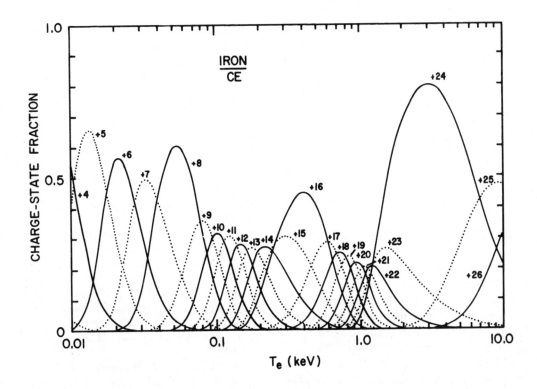

Figure 82.2. Fractional abundances of different charge states of an iron plasma in coronal equilibrium. The strongest peaks correspond to Fe ions with closed shell configurations: $Fe^{+24}(1s^2)$, $Fe^{+16}(2p^6)$, $Fe^{+8}(3p^6)$. [Copyright 1983 by the American Nuclear Society, La Grange Park, Illinois.]

the formulae in Ref. [25] actually describe the scattering of one unscreened charge by another charge that is screened, so they are most relevant to the scattering of fast electrons by slow ions.)

82.3.3 Collisional Excitation

The various screened plasma interactions can also be used to study plasma effects on inelastic scattering. Refs. [69,70] and citations therein use either Debye or ion-sphere potentials, and Born, distorted-wave, or close coupling approximations, to investigate excitation processes in plasmas; however, bound states are left unperturbed. More elaborate static potentials and perturbed bound states were treated by Davis and Blaha [71,72], but they did not self-consistently screen the interaction between projectile and target.

The use of static screening models is invalid when $\Delta E \gg \hbar\omega_e$, because the collision duration is too short for any average description of the plasma's screening to apply. In such cases, ionization being one example, one must consider the response of the target to electrodynamic disturbances [70].

Bremsstrahlung is another important plasma collision process for which static screening models have been extensively used [73–75]. Unfortunately, most

bremsstrahlung radiation emanating from a plasma represents free–free transitions in which $\Delta E \gg \hbar\omega_e$ (lower frequency emission being attenuated), and thus static screening models are invalid. On the other hand, the formation of laser plasmas occurs mainly through inverse bremsstrahlung (free–free absorption) under conditions such that $\Delta E = \hbar\omega_{\text{laser}} \approx \hbar\omega_e$ [11], and there exist still other experimental situations where absorption of low frequency radiation, $\omega < \omega_e$, is of interest, making static screening models relevant (see Ref. [64] and additional references therein).

82.4 IONIZATION BALANCE

Consider a nondegenerate plasma in thermal equilibrium at a temperature Θ (for instance, some region of a star's interior). The time independent ionization balance for each element is given by the Saha-Boltzmann formula [76]

$$\frac{N_{Z+1}}{N_Z} = \left(\frac{2G_{Z+1}}{G_Z n_e}\right)\left(\frac{m_e\Theta}{2\pi\hbar^2}\right)^{3/2} \exp(-I_Z/\Theta), \quad (82.11)$$

for the density ratio of successive charge states, where G_Z and I_Z are, respectively, the partition function and ionization potential for the Z-times ionized atom. Even in equilibrium plasmas, continuum lowering must be considered if the density is high, because it modifies partition functions as well as ionization potentials.

When plasma conditions change sufficiently slowly, the plasma evolves through a succession of "steady states" in which the relative abundances of different ion stages are determined by a balance of ionization and recombination rates. Then, in order to answer the straightforward question of what are the relative abundances of atomic ionization stages, one needs a vast data base of atomic energy levels, plus collisional and radiative rates, including high-density effects and involving excited states. A set of rate equations must then be solved to determine the populations $N_Z(\alpha)$ for each quantum state α of each ion stage Z. Each equation involves transitions to and from all other states. Continuum lowering determines the states which actually exist in the plasma and are to be included [77].

The balance of photoionizations and dielectronic plus radiative recombinations in low-density, steady-state plasmas is termed *nebular* equilibrium, because these are the conditions appropriate to astrophysical nebulae — regions of ionized gas surrounding hot stars. The balance of collisional ionizations and dielectronic plus radiative recombinations in low-density, steady-state plasmas is termed *coronal* equilibrium, because these are the conditions appropriate to the solar corona. Most tokamak plasmas also are in coronal equilibrium. Under coronal conditions, essentially all ions are in their respective ground states and the ionization balance is a function only of the plasma temperature. Figure 82.2 (from Ref. [78]) shows the coronal ionization balance for the element iron. As the plasma density increases, three-body collisions become important, and the resulting steady-state ionization balance, which depends on density as well as temperature, is termed *collisional-radiative* equilibrium [21, 22]. This situation exists in most ICF experiments.

Finally, conditions in sub-picosecond laser-plasma experiments can change so rapidly that none of the above simplifications apply. Ionization is strongly time-dependent, and involves multiphoton processes.

Acknowledgments

Support of the National Science Foundation, through its grant PHY-9321329 to Rice University, is gratefully acknowledged.

REFERENCES

1. J. Wesson, *Tokamaks* (Oxford University Press, Oxford, 1987).
2. R. E. Marrs, Comments At. Mol. Phys. **27**, 57 (1991).
3. N. A. Krall and A. W. Trivelpiece, *Principles of Plasma Physics* (McGraw-Hill, New York, 1973).
4. S. Ichimaru, *Statistical Plasma Physics* (Addison-Wesley, Redwood City, 1992), Vol. 1.
5. S. Ichimaru, Rev. Mod. Phys. **54**, 1017 (1982).
6. R. W. Lee, in *AIP Conference Proceedings #257, Atomic Processes in Plasmas* (American Institute of Physics, New York, 1992).
7. S. Chandrasekhar, Rev. Mod. Phys. **15**, 1 (1943).
8. H. Margenau and M. Lewis, Rev. Mod. Phys. **31**, 569 (1959).
9. C. F. Hooper, Jr., Phys. Rev. **165**, 215 (1968).
10. C. A. Iglesias, H. E. DeWitt, J. L. Lebowitz, D. MacGowan, and W. B. Hubbard, Phys. Rev. A**31**, 1698 (1985).
11. C. E. Max, in *Laser-Plasma Interaction* (North-Holland, Amsterdam, 1982).
12. C. Yamanaka, in *Physics of Laser Plasmas* (North-Holland, Amsterdam, 1991).
13. M. H. Key, Nature **316**, 314 (1985).
14. R. C. Elton, E. A. McLean, J. A. Stamper, C. K. Manka, and B. H. Ripin, in *AIP Conference Proceedings #206, Atomic Processes in Plasmas* (American Institute of Physics, New York, 1990).
15. C. De Michelis and M. Mattioli, Nucl. Fusion **21**, 677 (1981).
16. C J. Keane, B. A. Hammel, D. R. Kania, J. D. Kilkenny, R. W. Lee, A. L. Osterheld, R. C. Mancini, C. F. Hooper, Jr., and N. D. Delameter, Phys. Fluids B**5**, 3328 (1993).
17. I. I. Sobelman and A. V. Vinogradov, Adv. At. Mol. Phys. **20**, ?? (1985).
18. G. T. Schappert, D. E. Casperson, J. A. Cobble, J. C. Comly, L. A. Jones, G. A. Kyrala, K. J. LaGatuta, P. H. Y. Lee, G. L. Olson, and A. J. Taylor, in *AIP Conference Proceedings #206, Atomic Processes in Plasmas* (American Institute of Physics, New York, 1990).
19. P. Audebert, J. P. Geindre, J. C. Gauthier, R. Benattat, J. P. Chambaret, A. Mysyrowicz, and A. Antonetti, in *AIP Conference Proceedings #257, Atomic Processes in Plasmas* (American Institute of Physics, New York, 1992).
20. N. Miura and F. Herlach, in *Strong and Ultrastrong Magnetic Fields and Their Applications* (Springer-Verlag, Berlin, 1985).
21. D. R. Bates, A. E. Kingston, and R. W. P. McWhirter, Proc. Roy. Soc. (London) **A267**, 297 (1962).
22. G. J. Pert, J. Phys. B**23**, 619 (1990).
23. R. M. More, in *Atomic and Molecular Physics of Controlled Thermonuclear Fusion* (Plenum, New York, 1983).
24. R. M. More, in *Physics of Laser Plasmas* (North-Holland, Amsterdam, 1991).
25. J. C. Weisheit, in *Applied Atomic Collision Physics* (Academic Press, Orlando, 1984), Vol. 2.
26. H. A. Bethe and R. W. Jackiw, *Intermediate Quantum Mechanics* (W. A. Benjamin, Reading, 1986) 3rd ed.
27. S. Eliezer, A. Ghatak, and H. Hora, *An Introduction to Equations of State: Theory and Applications* (Cambridge University Press, Cambridge, 1986).
28. L. Spruch, Rev. Mod. Phys. **63**, 151 (1991).
29. R. P. Feynman, N. Metropolis, and E. Teller, Phys. Rev. **73**, 1561 (1949).
30. G. A. Baker, Jr. and J. D. Johnson, Phys. Rev. A**44**, 2271 (1991).

31. B. F. Rozsnyai, Phys. Rev. A**5**, 1137 (1972).
32. R. Cauble, M. Blaha, and J. Davis, Phys. Rev. A**29**, 3280 (1984).
33. F. J. Rogers and C. A. Iglesias, Science **263**, 50 (1994).
34. R. M. More, Adv. Am. Mol. Phys. **21**, 305 (1985).
35. D. R. Inglis and E. Teller, Astrophys. J. **90**, 439 (1939).
36. D. G. Hummer and D. Mihalas, Astrophys. J. **331**, 794 (1988).
37. F. J. Rogers, H. C. Graboske, and D. J. Harwood, Phys. Rev. A**1**, 1577 (1970).
38. J. C. Stewart and K. D. Pyatt, Astrophys. J. **144**, 1203 (1966).
39. M. C. W. Dharma-Wardana and F. Perrot, Phys. Rev. A**45**, 5883 (1992).
40. F. J. Rogers, Astrophys. J. **310**, 723 (1986).
41. S. Volonte, J. Phys. D **11**, 1615 (1978).
42. J. Cooper, D. E. Kelleher, and R. W. Lee, in *Proceedings of 2nd Intl. Conference on Radiative Properties of Hot Dense Matter* (World Scientific, Singapore, 1983).
43. H. Nguyen, M. Koenig, D. Benredjem, M. Caby, and G. Coulaud, Phys. Rev. A **33**, 1279 (1986).
44. H. R. Griem, C. A. Iglesias, and D. B. Boercker, Phys. Rev. A**44**, 5318 (1991).
45. G. Chanmugam, Ann. Rev. Astron. Astrophys. **30**, 143 (1992).
46. J. A. Stamper and B. H. Ripin, Phys. Rev. Lett. **34**, 138 (1975).
47. S. C. Wilks, W. L. Kruer, M. Tabak, and A. B. Langdon, Phys. Rev. Lett. **69**, 1383 (1992).
48. R. H. Garstang, Rep. Prog. Phys. **40**, 105 (1977).
49. P. Meszaros, *High-Energy Radiation from Magnetized Neutron Stars* (University of Chicago Press, Chicago, 1992).
50. M. C. Miller and D. Neuhauser, Mon. Not. Roy. Astron. Soc. **253**, 107 (1991).
51. H. Ruder, G. Wunner, H. Herold, and F. Geyer, *Atoms in Strong Magnetic Fields* (Springer-Verlag, Berlin, 1994).
52. C.-H. Park and A. F. Starace, Phys. Rev. A**29**, 442 (1984).
53. M. Demeur, P.-H. Heenen, and M. Godefroid, Phys. Rev. A**49**, 176 (1994).
54. B. B. Kadomtsev, Sov. Phys. JETP **31**, 945 (1970).
55. I. Fushiki, E. H. Gudmundsson, C. J. Pethic, and J. Yngvason, Ann. Phys. **216**, 29 (1992).
56. E. P. Lief and J. C. Weisheit, Contrib. Plasma Phys. **33**, 471 (1993).
57. A. M. Abrahams and S. L. Shapiro, Astrophys. J. **374**, 652 (1991).
58. L. Van Hove, Phys. Rev. **95**, 249 (1954).
59. J. C. Weisheit and B. W. Shore, Astrophys. J. **194**, 519 (1974).
60. L. G. D'yachkov, G. A. Kobzev, and P. M. Pankratov, J. Quant. Spectrosc. Radiat. Transfer **44**, 123 (1990).
61. Y. Chung, H. Hirose, and S. Suckewer, Phys. Rev. A**40**, 7142 (1989).
62. Y.-C. Chen and J. L. Libowitz, Phys. Rev. A**41**, 2127 (1990).
63. F. Aumayr, W. Lee, C. H. Skinner, and S. Suckewer, J. Phys. B**24**, 4489 (1991).
64. M. A. Berkovsky, D. Kelleher, Yu. K. Kurilenkov, and M. Skowronek, J. Phys. B**26**, 2475 (1993).
65. M. S. Murillo and J. C. Weisheit, J. Quant. Spectrosc. Radiat. Transfer **54**, 271 (1995).
66. L. Spitzer, Jr., *Physics of Fully Ionized Gases* (Wiley Interscience, New York, 1962), 2nd ed.
67. C.-K. Li and R. D. Petrasso, Phys. Rev. Lett. **70**, 3063 (1993).
68. N. F. Mott and H. S. W. Massey, *The Theory of Atomic Collisions* (Oxford Univ. Press, Oxford, 1965), 3rd ed.
69. B. L. Whitten, N. F. Lane, and J. C. Weisheit, Phys. Rev. A**29**, 945 (1984).
70. J. C. Weisheit, Adv. At. Mol. Phys. **25**, 101 (1988).
71. J. Davis and M. Blaha, J. Quant. Spectrosc. Radiat. Transfer **27**, 307 (1982).
72. J. Davis and M. Blaha, in *Physics of Electronic and Atomic Collisions* (North-Holland, Amsterdam, 1982).
73. I. P. Grant, Mon. Not. Roy. Astron. Soc. **118**, 241 (1958).
74. B. F. Rozsnyai, J. Quant. Spectrosc. Radiat. Transfer **22**, 337 (1979).
75. L. Kim, R. H. Pratt, and H. K. Tseng, Phys. Rev. A**32**, 1693 (1985).
76. Ya. B. Zeldovich and Yu. P. Raizer, *Physics of Shock Waves and High-Temperature Hydrodynamic Phenomena* (Academic Press, New York, 1966), Vol. I.
77. S. R. Stone and J. C. Weisheit, J. Quant. Spectrosc. Radiat. Transfer **35**, 67 (1986).
78. R. A. Hulse, Nucl. Tech. Fusion **3**, 259 (1983).

83

Conduction of Electricity in Gases

Alan Garscadden

Wright Laboratory, Wright Patterson Air Force Base, Ohio 45433

83.1 INTRODUCTION

The conduction of electricity through gases has played ubiquitous roles in science and technology. It was responsible for many of the fundamental discoveries in atomic and molecular physics; gas discharge lighting is essential to everynight operations; gas discharge lasers are still important in research and manufacturing; and all of advanced microelectronics depends on plasma enhanced processing. To a large extent, the efficiencies of the above cited applications of gaseous electronics depend on the maintenance of the distinct nonequilibrium between the electrons and the gas or vapor. This nonequilibrium can be achieved by operating at low pressures or under pulsed excitation, where the duration of the energy input is less than the energy equilibration time between the electrons and the heavy particles. The term *gas discharge* originally described a transient or spark condition, but has been extended to mean the continuous conduction of electricity through gases.

The physics and chemistry of the conduction of electricity in ionized gases involves the interactions of the electrons and ions in the gas among themselves, with ground-state and excited-state gas atoms or molecules, with any surfaces that may be present, and with any electric or magnetic fields that are externally applied or generated by movement of the charged species. The electron collision-induced excitation and dissociation can result in new compounds being formed at rates which are orders of magnitude larger than those without a plasma present. For a gas pressure of 1 torr, the electron mean free path λ_{mfp} between collisions is ~ 1 mm ($\sim 10^9$ collisions/second). The assumption that λ_{mfp} is small relative to the plasma physical dimensions is usually valid, however it is often comparable to, or larger than the space charge sheaths that form near electrodes and surfaces.

Section 83.2 treats the electron-velocity distribution and its effect on various measurements involving a swarm or distribution of electron velocities. In this section, low fractional ionization ($<10^{-6}$) is assumed; electron-ion collisions are negligible relative to electron-atom (and electron-molecule) collisions in so far as they affect electron mobility, diffusion or energy loss. Since the electron-atom interactions, in most cases of interest, have ranges shorter than the average gas-atom separation, they may be treated as binary collisions and the effects of neighboring gas atoms on a given electron-atom encounter can be neglected. Quantities involving such interactions in a partially ionized gas, where the electrons are distributed both in velocity and position, can be directly related to the corresponding quantities in beam experiments, where the collisions are binary and the electrons have a defined velocity. The measurements and calculations of the electron velocity distributions, or of the electron energy distribution functions (EEDF), are therefore of fundamental importance in the description of partially ionized gases.

Section 83.3 introduces the glow discharge and considers the cold cathode and hot cathode discharge phenomena. Section 83.4 discusses ionization by electron collision, electron attachment, ion mobility, ion-ion and electron-ion recombination, and other important processes that affect the conduction of electricity in gases. Section 83.5 illustrates the importance of gaseous electronics with several important phenomena and technical applications.

83.2 ELECTRON SCATTERING AND TRANSPORT PHENOMENA

83.2.1 Electron Scattering Experiments

In 1903, Lenard [1] determined the attenuation of a beam of monoenergetic electrons by several gases. He measured the fraction of an electron beam that was transmitted without scattering through a field-free region containing the gas. Let an electron beam of density n_e electrons/cm^3, traveling with velocity v, pass a distance dx through the gas. Let N be the number of atoms/cm^3. The loss of electrons by scattering per unit time due to collisions with atoms is then given by

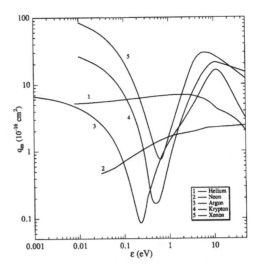

Figure 83.1. Momentum transfer cross section for noble gases.

$$\frac{dn_e}{dt} = -Nv\sigma n_e\,, \qquad (83.1)$$

where σ is the collision cross section for electrons of velocity v. Writing $v\,dt = dx$, we obtain

$$n_e = n_{e0}e^{-N\sigma x}\,. \qquad (83.2)$$

In relation to the conduction of electricity in gases, a more useful concept than σ is the momentum transfer cross section σ_{m}. In general, σ_{m} is not equal to σ. Experimental determinations of σ_{m}, generally from electron swarm experiments, are summarized in Ref. [2]. Results for the noble gases are presented in Fig. 83.1 and for representative diatomic gases in Fig. 83.2. Above energies of about 10 eV, σ_{m} decreases as the electron energy increases. Towards higher electron energies in the monatomic gases, σ_{m} is inversely proportional to the ionization potential and directly proportional to the gas polarizability. The fact that σ_{m} has extremely low values in certain gases at low energies is known as the *Ramsauer-Townsend* (RT) effect. This can be explained by partial wave scattering theory as the point where the radial wave function has exactly one more oscillation inside the interaction potential well than does the corresponding free function. Outside the well the two functions are indistinguishable and there is no $\ell = 0$ (s-wave) scattering [3] (see also Sect. 43.2.7). For the scattering potentials of the heavier noble gases, this occurs at low energies where the partial cross sections for the higher angular momenta ($\ell = 1, 2, \dots$) are still negligible. Methane and silane (Fig. 83.3, [4, 5]) and possibly other nonpolar polyatomic gases, also display Ramsauer-Townsend (RT) minima. There is great similarity between the σ_{m} curves for atoms and molecules with similar external electron arrangements.

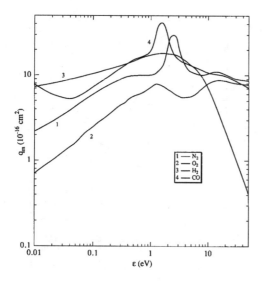

Figure 83.2. Momentum transfer cross section for some diatomic gases.

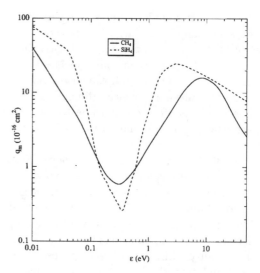

Figure 83.3. Momentum transfer cross section for CH_4 and SiH_4.

The σ_m curves for H_2, O_2, N_2 and CO (Fig. 83.2) show that these cross sections are large over the range 0.1 to 10 eV. Around 1.5 eV, N_2 and CO display resonances which are associated with the formation of temporary negative ions followed by efficient vibrational excitation of these molecules [6].

83.2.2   Electron Transport Phenomena

Electron transport phenomena in an ionized gas, such as diffusion under the influence of a density gradient and drift under the influence of an electric field, are directly related to the electron current density Γ_e which is caused by these influences. In order to calculate Γ_e, it is necessary to know the electron spatial and velocity density distribution $f(\mathbf{v}, \mathbf{r}, t)$. The electron density n_e and current density Γ_e are given by

$$n_e(\mathbf{r}, t) = \int f(\mathbf{v}, \mathbf{r}, t) d^3 v, \qquad (83.3)$$

$$\Gamma_e(\mathbf{r}, t) = \int v f(\mathbf{v}, \mathbf{r}, t) d^3 v. \qquad (83.4)$$

The electron drift velocity w_d is related to Γ_e and n_e by

$$w_d = \Gamma_e / n_e. \qquad (83.5)$$

The electron drift velocity is usually small compared with the electron random velocity. The other limiting case is important in connection with ionic mobilities. It can also occur in electron transport in Ramsauer gases or gas mixtures that are used in fast particle counters [7]. The theoretical analysis of electron motion in gases uses primarily the Boltzmann transport equation approach [8]. Mean free path methods [9] and Monte-Carlo methods [10] have also been applied.

83.2.3   The Boltzmann Equation

The electron density distribution $f(\mathbf{v}, \mathbf{r}, t)$ of a given kind of particle in phase space (configuration and velocity space) is determined by the combined effect of all the interactions to which the particle is subjected. If there are no sources or sinks, the number of particles in a volume element of the six-dimensional space which moves with the particles does not change with time. In configuration space, the continuity equation for $f(\mathbf{v}, \mathbf{r}, t)$ is

$$\frac{\partial f}{\partial t} + \nabla_r \cdot \mathbf{v} f = 0; \qquad (83.6)$$

in phase space, it has the form

$$\frac{\partial f}{\partial t} + \nabla_r \cdot \mathbf{v} f + \nabla_v \cdot \mathbf{a} f = 0, \qquad (83.7)$$

where the subscripts on the divergence operators denote the independent variable, $\mathbf{v}$ is the particle flow velocity in configuration space, and $\mathbf{a}$ is the particle acceleration (i.e., the flow velocity *in velocity space*). This may be written as $q(\mathbf{E} + \mathbf{v} \times \mathbf{B})/m$ for a particle of charge q and mass m in the presence of electric and magnetic fields, $\mathbf{E}$ and $\mathbf{B}$ respectively. If there are sources or sinks of particles, then terms representing these appear on the right-hand side of Eq. (83.7).

In principle, the effect of elastic and inelastic collisions may be included in the expressions for $\mathbf{E}$ and $\mathbf{B}$, but they are usually treated as source terms which transport the colliding particles *instantaneously* from one volume element of velocity space to another. This is a good approximation, for example, in the case of electron-atom collisions, but it is not valid for the longer range electron-ion and electron-electron interactions in the presence of high-frequency electric fields. Other collisions, such as ionizing collisions, act to change the total n_e as well as transporting electrons from one element of velocity space to another.

The electron-electron collisions will produce a Maxwellian velocity distribution when the fractional ionization is sufficiently large. For molecular gases, this generally requires $n_e/N_0 > 10^{-5}$. Since the electron-electron Coulomb collision frequency varies as v^{-3}, these collisions are especially effective in the low energy elastic regime where there are no inelastic energy loss processes, and they may have an influence for n_e/N_0 as low as 10^{-7} for the noble gases [11]. Spitzer [12], Dreicer [13], and Butler and Buckingham [14] treat cases where the electron-electron term is included explicitly. A full treatment of this subject is given by Shkarofsky [15].

Using $\nabla_v \cdot \mathbf{a} = 0$, and including collisions as a source term, Eq. (83.7) becomes

$$\frac{\partial f}{\partial t} + \mathbf{v} \cdot \nabla_r f + \mathbf{a} \cdot \nabla_v f = \left(\frac{\partial f}{\partial t} \right)_{\text{collisions}}, \qquad (83.8)$$

where the subscript 'collisions' identifies the source term due to all types of collisions.

83.2.4 Electron-Atom Elastic Collisions

For electrons in a weakly ionized gas, where only electron-atom elastic collisions need be explicitly considered, the collision term is given by the Boltzmann collision integral [16]. This integral describes the rate at which electrons are brought into and removed from an element of volume in velocity space in terms of the differential elastic scattering cross section.

The Boltzmann equation may be solved by a perturbation method [8] in which $\mathbf{E}$, $\mathbf{B}$, and the density gradients are the perturbations, starting from

$$f(\mathbf{v}, \mathbf{r}, t) \approx f_0(v, \mathbf{r}, t) + \frac{\mathbf{v}}{v} \cdot \mathbf{f}_1(v, \mathbf{r}, t), \qquad (83.9)$$

where f_0 is the unperturbed isotropic term and $\mathbf{f}_1$ is the anisotropy due to $\mathbf{E}$ and $\mathbf{B}$. Substituting into Eq. (83.8) and equating terms of the same angular type then yields

$$\frac{\partial f_0}{\partial t} + \frac{v}{3} \nabla_r \cdot \mathbf{f}_1 - \frac{e}{3mv^2} \frac{\partial}{\partial v} \left(v^2 \mathbf{E} \cdot \mathbf{f}_1 \right)$$
$$= \frac{g}{2v^2} \frac{\partial}{\partial v} \left[v^3 \nu_m \left(f_0 + \frac{k_B T_g}{mv} \frac{\partial f_0}{\partial v} \right) \right] \qquad (83.10)$$

and

$$\frac{\partial \mathbf{f}_1}{\partial t} + v \nabla_r f_0 - \frac{e\mathbf{E}}{m} \frac{\partial f_0}{\partial v} + \boldsymbol{\omega}_B \times \mathbf{f}_1 = -\nu_m \mathbf{f}_1, \quad (83.11)$$

where $\nu_m = e\mathbf{E}/(w_d m)$ is the momentum transfer collision frequency from the Boltzmann collisions integral, $\boldsymbol{\omega}_B = e\mathbf{B}/m$ is a vector whose magnitude is the *cyclotron*

frequency, and $g = 2m/(M + m)$ controls the partition of kinetic energy in an elastic collision between particles with masses M and m.

The right-hand term of Eq. (83.10) is the first-order correction due to elastic kinetic energy exchange between electrons and atoms. The $\mathbf{E} \cdot \mathbf{f}_1$ term arises from work done by $\mathbf{E}$ on the electrons, and the $\nabla_r \cdot \mathbf{f}_1$ term accounts for particle loss from the volume element in phase space due to electric field drift and density gradients. Similarly, in Eq. (83.11), the $\mathbf{E}$ term accounts for loss due to electric field acceleration, and the $v\nabla_r f_0$ term accounts for changes due to electron drift across a density gradient. The $-\nu_m \mathbf{f}_1$ term results from randomization of the electron velocity direction in electron-atom collisions.

The neglect of higher order terms in Eq. (83.9) is justified if $\delta T_e / T_e \ll 1$, where T_e is the gain in electron kinetic energy between collisions, and $\delta f_0 / f_0 \ll 1$, where δf_0 is the change in f_0 over λ_{mfp}. For the case of static $\mathbf{E}$ and $\mathbf{B}$ with $\mathbf{E} \perp \mathbf{B}$, the orbital time $1/\omega_B$ replaces the time between collisions if it is shorter, making the above conditions less stringent. However, for an ac field with the resonant frequency ω_B, the original conditions apply. The off-resonant case in between. See Refs. [8, 17] for a further discussion of validity.

83.2.5 The Electron Drift Current

Transport properties of a partially ionized gas are calculated from the moments of $f(\mathbf{v}, \mathbf{r}, t)$. To obtain $\boldsymbol{\Gamma}_e$, Eq. (83.11) must be solved for $\mathbf{f}_1$. Assume that $\omega_B = 0$ and that $\mathbf{E}$ is constant in time. Then, since f_0 usually changes slowly compared with ν_m, $\partial \mathbf{f}_1 / \partial t$ may be neglected to a first approximation, and

$$\mathbf{f}_1 \simeq -\frac{1}{\nu_m} \left[v \nabla_r (n_e f_v^0) - \frac{e\mathbf{E}}{m} n_e \frac{\partial f_v^0}{\partial v} \right], \qquad (83.12)$$

where $n_e(\mathbf{r}, t) f_v^0(v, \mathbf{r}, t) = f_0(v, \mathbf{r}, t)$. Use of Eq. (83.4) gives

$$\boldsymbol{\Gamma}_e = -\nabla_r (D_e n_e) - n_e \mu_e \mathbf{E}, \qquad (83.13)$$

which defines the electron free diffusion coefficient

$$D_e = \int \frac{v^2}{3\nu_m} f_v^0 4\pi v^2 dv = \frac{\langle v^2 \rangle}{3\nu_m}, \qquad (83.14)$$

and the static electron mobility

$$\mu_e = -\frac{e}{m} \int \frac{v}{3\nu_m} \frac{\partial f_v^0}{\partial v} 4\pi v^2 dv. \qquad (83.15)$$

In Eq. (83.13), ∇_r should be interpreted as acting only on that part of D_e whose spatial dependence enters through an energy change, such as a variation of T_e with position (as in the theory of striations [18], but not on variations due to pressure gradients which would affect ν_m [8]).

Electron Free Diffusion

When the ionization density is low enough, the electrons diffuse freely to the walls, being unaffected by the space charge field caused by the heavier ions, which diffuse more slowly than the electrons. Setting $\mathbf{E} = 0$ in Eq. (83.13), the electron current due to diffusion is

$$\mathbf{\Gamma}_e = -\nabla_r(D_e n_e). \qquad (83.16)$$

Since the electron heating and elastic recoil terms in (83.10) do not contribute to a net gain or loss of electrons, integration of (83.10) over $\mathbf{v}$ gives

$$\frac{\partial n_e}{\partial t} + \nabla_r \cdot \mathbf{\Gamma}_e = \frac{\partial n_e}{\partial t} - \nabla_r^2(D_e n_e) = 0, \qquad (83.17)$$

for a decaying electron density. This equation is identical in form to the thermal conductivity equation, which can be solved analytically for certain simple geometries. For the case of a long cylindrical container of radius R, the equation becomes

$$\frac{\partial n_e}{\partial t} = D_e \frac{1}{r} \frac{\partial}{\partial r}\left(r \frac{\partial n_e}{\partial r}\right), \qquad (83.18)$$

if D_0 is constant and n_e depends only on r. The general solution for n_e satisfying $n_e(R, t) = 0$ is

$$n_e = \sum_{m=0} b_m \exp\left(-\frac{t}{t_m}\right) J_0\left(\frac{a_m r}{R}\right), \qquad (83.19)$$

where $t_m = [D_e(a_m/R)^2]^{-1}$, and J_0 is the zeroth-order Bessel function with roots $J_0(a_m) = 0$, $a_{m+1} > a_m$. The b_m are determined by the radial dependence of n_e at $t = 0$. (If the coaxial wall of radius $r_0 < R$ confines the plasma to an annular region $r_0 < r < R$, the general solution must also include the zeroth-order Bessel function of the second kind, Y_0, to satisfy the boundary conditions). The time constants t_m decrease rapidly with increasing m so that for sufficiently long t, the density distribution relaxes (e.g., after the microwave or static excitation is switched off) to

$$n_e \approx b_0 \exp\left(-\frac{t}{t_0}\right) J_0\left(\frac{a_0 r}{R}\right). \qquad (83.20)$$

This is called the *fundamental diffusion mode*. The above treatment gives a good approximation to $n_e(r)$ provided that $\lambda_{mfp} \ll R$.

When ν_m is velocity-independent and f_0 is the Maxwellian of temperature T_e, Eq. (83.14) becomes

$$D_e = \frac{k_B T_e}{m \nu_m}. \qquad (83.21)$$

For the case where a magnetic field is applied to the ionized gas, parallel to the cylinder axis so as to impede

the electron diffusion to the walls, the free-diffusion coefficient in equation (83.18) is replaced by

$$D_{eB} = \int \frac{\nu_m v^2 f_v^0}{3(\nu_m^2 + w_B^2)} 4\pi v^2 dv = \left\langle \frac{\nu_m v^2}{3(\nu_m^2 + w_B^2)} \right\rangle. \qquad (83.22)$$

In this case, the electrons cannot diffuse to the walls unless they make collisions that interrupt their spiraling motion.

Electron DC Mobility in Static Fields

From Eq. (83.13), the electron particle current induced by an electric field $\mathbf{E}$ is $\mathbf{\Gamma}_e = -n_e \mu_e E$, and the electron drift velocity is

$$w_d = \frac{\mathbf{\Gamma}_e}{n_e} = -\mu_e E. \qquad (83.23)$$

Measurement of w_d by a time-of-flight measurement can in principle, through the use of Eq. (83.14), provide some information about ν_m, provided that f_v^0 is known. Since a static electric field may perturb considerably the distribution function, and since n_e is low in such measurements, f_v^0 should be determined using Eq. (83.10) after having eliminated $\mathbf{f}_1$ using Eq. (83.11). In the case where ν_m is independent of v, Eq. (83.15) after integration by parts, reduces to

$$\mu_e = \frac{e}{m\nu_m} \int f_v^0 4\pi v^2 dv = \frac{e}{m\nu_m}, \qquad (83.24)$$

which is independent of f_v^0.

In general, w_d exhibits a complex dependence on E/N, as shown for some example gases in Fig. 83.4 [19]. The high mobility of electrons in methane is of technical interest because of potential applications in plasma switches used for current interruption [20]. The local maximum of w_d in methane for $E/N = 3$ Td (1 Townsend $= 10^{-17}$ Vcm^2) and its decrease with further increase of E/N arise because of the onset of inelastic collisions due to vibrational excitation at an energy close to the energy of the RT minimum [21]. As shown in Fig. 83.4, the drift velocities are generally in the range of $10^6 - 10^7$ cm/s.

A detailed summary of electron drift velocities measured in many pure gases and gas mixtures of technical interest is given in Refs. [22, 23], and the techniques are elaborated in Refs. [24, 25].

The Ratio of D_e/μ_e for Electrons

From Eqs. (83.14) and (83.15),

$$\frac{D_e}{\mu_e} = -\frac{m}{e} \frac{\int_0^\infty (v^4/\nu_m) f_v^0 \, dv}{\int_0^\infty (v^3/\nu_m) \partial f_v^0/\partial v \, dv}. \qquad (83.25)$$

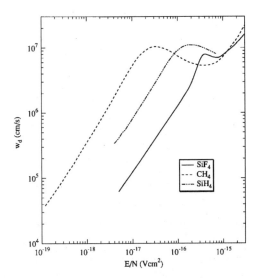

Figure 83.4. Electron drift velocities in polyatomic gases.

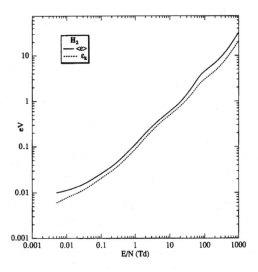

Figure 83.5. Calculated u_{av} and D_e/μ_e coefficients in H_2.

The above ratio of the free-diffusion coefficient to the mobility is termed the *characteristic energy* and it is a measure of the average electron energy in certain cases. When f_v^0 is Maxwellian, then

$$D_e/\mu_e = k_B T_e/e \qquad (83.26)$$

becomes a direct measure of the electron energy, independent of ν_m. Since the average electron energy $u_{av} = \frac{3}{2}k_B T_e$, Eq. (83.26) becomes

$$D_e/\mu_e = \tfrac{2}{3}u_{av}/e. \qquad (83.27)$$

In this form it is known as the *Einstein relation*. Furthermore, if ν_m is independent of velocity, Eq. (83.27) applies *regardless of the velocity distribution*.

Values of D_e/μ_e can be determined from experiments where a static electric field E_z is applied to electrons drifting under steady state conditions of diffusion. If D_e is taken to be independent of position, then the generalization of Eq. (83.17) is

$$\nabla^2 n_e = -\frac{\mu_e}{D_e} E_z \frac{\partial n_e}{\partial z}. \qquad (83.28)$$

Solutions to this equation for n_e were for many years used to deduce D_e/μ_e from the measurements [26]. However, Wagner *et al.* [27] showed that there are two effective diffusion coefficients D_L and D_T in the longitudinal and transverse directions with $D_L \ll D_T$. For H_2, N_2 and He, $D_L \approx D_T/2$, while for Ar, $D_L \approx D_T/8$. See Refs. [28, 29] for a quantitative discussion. The different coefficients arise because n_e and w_d are functions of $\partial n_e/\partial z$.

For H_2, f_v^0 is fairly well approximated by a Maxwellian distribution. A comparison of the average and characteristic energies u_{av} and D_e/μ_e calculated from the Boltzmann transport equation for H_2 is shown in Fig. 83.5.

Ambipolar Diffusion

In the case of electron free diffusion discussed above, diffusion of the electrons (and ions) is unaffected by space-charge fields caused by an imbalance of positive and negative charges. When the charge density is high, this no longer holds. The resulting electric field caused by space-charge separation retards the electron diffusion and enhances the positive ion motion. In the steady state the flows of positive ions and of electrons are equal, i.e., $\mathbf{\Gamma}_e = \mathbf{\Gamma}_+$. For electrons in the absence of temperature gradients,

$$\mathbf{\Gamma}_e = -D_e \nabla n_e - n_e \mu_e \mathbf{E}, \qquad (83.29)$$

where D_e and μ_e are given by Eqs. (83.14) and (83.15) respectively. The analogous equation for the positive ions is

$$\mathbf{\Gamma}_+ = -D_+ \nabla n_+ - n_+ \mu_+ \mathbf{E}. \qquad (83.30)$$

Eliminating $\mathbf{E}$ between the two equations and setting $n_e \approx n_+ = n$, and $\mathbf{\Gamma}_e \approx \mathbf{\Gamma}_+ = \mathbf{\Gamma}$ gives

$$\mathbf{\Gamma} = -\left(\frac{D_+ \mu_e + D_e \mu_+}{\mu_e + \mu_+}\right)\nabla n. \qquad (83.31)$$

The quantity in parentheses is a diffusion coefficient that is applicable both to the electrons and to the ions. Since they are interacting, they diffuse together. This quantity is termed the *ambipolar diffusion coefficient*

$$D_a = \frac{D_+ \mu_e + D_e \mu_+}{\mu_e + \mu_+}. \qquad (83.32)$$

In the case where the electrons have a temperature T_e and the ions a temperature T_+,

$$\frac{D_+}{\mu_+} = \frac{k_B T_+}{e}, \qquad \frac{D_e}{\mu_e} = \frac{k_B T_e}{e}, \qquad (83.33)$$

and since $\mu_e \gg \mu_+$,

$$D_a \approx D_+ \left(1 + \frac{T_e}{T_+}\right). \qquad (83.34)$$

If $T_e \gg T_+$, then $D_a \approx \mu - k_B T_e/e$, i.e., the ambipolar diffusion is determined by the ion mobility and the electron energy. Allis and Rose [30] have studied the transition from free to ambipolar diffusion as the electron density is increased.

For the case in which the electrons and ions are diffusing together in the radial direction across a static B_z, and the electron-atom and ion collision rates may be assumed to be independent of T_e, the magneto-ambipolar diffusion coefficient is given by

$$D_{aB} = \frac{D_a}{1 + \mu_+ \mu_e B^2}. \qquad (83.35)$$

Refs. [31, 32] review theoretical and experimental work on diffusion in a magnetic field.

Debye Shielding

Consider an ensemble of positive ions and electrons in thermal equilibrium. Each ion repels other ions in its neighborhood, but attracts electrons. The space charge around each ion (assumed stationary) is $e(n_+ - n_e)$, which by Poisson's equation creates a potential ϕ satisfying

$$\nabla^2 \phi = -\frac{e}{\epsilon_0}(n_+ - n_e), \qquad (83.36)$$

where n_+ is the average ion density. For time independent fields in the absence of collisions, Eq. (83.8) becomes

$$\mathbf{v} \cdot \nabla_r f + e\mathbf{E} \cdot \nabla_v f = \mathbf{v} \cdot \nabla_r f - e \nabla_r \phi \cdot \nabla_v f = 0. \quad (83.37)$$

Assuming a solution of the form

$$f = f_0 \exp\left(-\frac{e\phi}{k_B T_e}\right), \qquad (83.38)$$

where f_0 is Maxwellian, and using $\int f_0 d^3 v = n_e$ gives

$$\left(-\frac{e}{k_B T_e} \mathbf{v} \cdot \nabla_r \phi + e \nabla_r \phi \cdot \frac{\mathbf{v}}{k_B T}\right) f \equiv 0. \qquad (83.39)$$

Hence,

$$\int f d^3 v = n_e \exp\left(-\frac{e\phi}{k_B T_e}\right). \qquad (83.40)$$

Since the assembly is macroscopically neutral, $n_e = n_+$, and

$$\nabla^2 \phi = -\frac{e}{\epsilon_0} n_+ \left[1 - \exp\left(-\frac{e\phi}{k_B T_e}\right)\right]. \qquad (83.41)$$

If $e\phi \ll k_B T_e$, then

$$\nabla^2 \phi \approx \frac{e^2}{\epsilon_0} n_+ \frac{\phi}{k_B T_e}. \qquad (83.42)$$

If spherical symmetry is assumed,

$$\phi = \frac{e}{4\pi\epsilon_0 r} \exp\left(-\frac{r}{\lambda_D}\right), \qquad (83.43)$$

where

$$\lambda_D = \left(\frac{\epsilon_0 k_B T_e}{n_+ e^2}\right)^{1/2} \qquad (83.44)$$

is the Debye shielding length for the Coulomb potential. λ_D arises from the screening effect of the electron cloud about each ion. The dimensions of a plasma must be much larger than λ_D for the electrons to act collectively, i.e., to undergo plasma oscillations.

83.2.6 Cross Sections Derived from Swarm Data

The use of transport theory to unfold low energy electron collision cross sections from experimentally measured electron transport data was originally introduced by Townsend [26]. Beginning with trial input values for the cross sections as a function of energy, the electron transport coefficients are calculated and compared with experiments (such as w_d, D_e, ϵ_k, and ionization/attachment coefficients, measured as a function of E/N). The input cross sections are then adjusted in the appropriate energy range, and the comparison procedure is iterated until theory and experiment come into agreement. The above *trial and error* procedure was greatly advanced by the use of computers to solve the Boltzmann equation [33, 34]. A large body of fairly complete sets of low energy cross sections has been obtained in this way [23].

Assuming that the electron kinetics are accurately described by the solution of the Boltzmann equation, the above unfolding procedure is severely limited by the lack of uniqueness in the derived cross sections (especially in the molecular gases where many inelastic processes dominate in determining the electron energy distribution). For the noble gases, the method works well for He and Ne at low energies, where only elastic scattering occurs [24]. For Ar, Kr, and Xe, the method becomes questionable again for electron energies near the RT minimum. The slow redistribution of electron energies leads to lack of sensitivity of the calculated transport coefficients to the σ_m.

Recently it has been shown that the swarm analyses of He-Ramsauer noble gas mixtures lead to unique σ_m of the Ramsauer noble gases (Ar, Kr, Xe) [35]. The measured electron drift velocity data in He-Xe mixtures shown in

Figure 83.6. Measured electron drift velocities in He-Xe mixtures.

Fig. 83.6 satisfy the two important requirements (1) high sensitivity of the drift velocity data in gas mixtures over that in either pure He and Xe, and (2) the σ_m of He is very accurately known, to warrant the uniqueness of the derived σ_m of Xe.

Once the cross sections of the Ramsauer noble gases have been accurately defined by the above approach, the RT minimum can be used to advantage in solving the uniqueness problem for molecular gases. The electron transport data in molecular-gas–rare-mixtures exhibit large sensitivity due to the low energy inelastic collisions in the molecular gas (which increase the $\mathbf{f_1}$ component of the electron energy distribution) [36, 21]. Since the cross sections of the buffer noble gas are now accurately defined, the demanding fits to the mixture transport data can be conveniently exploited to enhance the accuracy of the cross sections of the molecular gases.

83.3 GLOW DISCHARGE PHENOMENA

83.3.1 Cold Cathode Discharges

The cold cathode discharge operates because of feedback by the electrons that are generated by ion bombardment of the cathode (secondary electron emission) in turn generating sufficient ions that flow to the cathode to keep the process going. There can be contributions to secondary electron emission from photons and metastable atoms which are created by electron collisions with the gas. Since the secondary emission coefficients are functions of the cathode material, ion species, and ion energy, the magnitude of the needed accelerating potential between the plasma and the cathode varies. This poten-

tial creates a space charge sheath, and various excitation features occur near the cold cathode. The potential required is typically 200–600 volts. The electrons from the cathode (called *primaries*) are accelerated through the sheath and acquire sufficient energy to excite the gas efficiently, resulting in a region termed the *negative glow*. As one moves further from the cathode, the light from the *negative glow* tapers off due to the degradation of the *primaries* in creating slow electrons (termed *secondaries* and *ultimates*). A sufficiently high density of electrons is usually created to satisfy the external circuit continuity at very low fields (or even slight field reversal) [37]. Since the low energy electrons do not create visible excitation, the region to the anode side of the negative glow forms the *Faraday dark space*. At even greater distance from the cathode, where the slow electron density has decayed by drift, diffusion, recombination, and in some gases by attachment, the electric field increases and the discharge develops long diffuse bright regions which occupy the remainder of the interelectrode space. This is termed the *positive column* [38]. If one increases the interelectrode spacing, the cathode regions remain essentially constant and the *positive column* extends with average uniform properties. Temporal variations called *moving striations* are common in all gases, and spatial variations called *standing striations*, or *ionization waves* are frequently observed in molecular gases. The different cathodic regions occur because the abrupt boundary conditions create local discharge conditions of excitation and ionization which are not in equilibrium with the local electric field. Even in the so-called uniform positive column, the ionization waves correspond to nonlocal equilibrium solutions of the discharge equations that permit more efficient discharge ionization and conductivity than would be given by local equilibrium with the electric field. The ionization waves can be absolutely or convectively stable, and have dispersion properties of forward or backward waves (the latter corresponding to opposite directions of phase and group velocities) [39].

83.3.2 Hot Cathode Discharges

Low pressure, high current discharges are important in thyratrons and other high current switches. The hot cathode permits much higher current densities than are usually obtained from a cold cathode. The voltage drop of the cathode sheath is much less, and a discharge will operate if the applied potential exceeds I_P for the gas. Hot cathode discharges with low P and Γ_e show three groups of electrons: the primaries, which are beam electrons that essentially have retained the energy acquired in the cathode sheath; secondaries which are randomized electrons with T_e approximately proportional to the energy of the primaries (~ 5 eV for 30 eV primaries) and approximately independent of the gas; and the ultimates

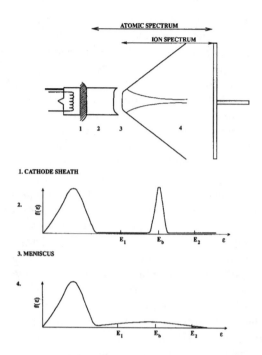

Figure 83.7. Illustration of the appearance of a hot cathode discharge; E_b is the sheath potential, E_1 and E_2 are the atomic and ionic excitation thresholds.

which are the bulk of the electrons with $T - E < 1$ eV. The primaries degrade slowly with distance from the cathode, provided that $\Gamma_e \leq 40$ mA for argon at 10 μm (or more generally when the density of the beam electrons is still less than n_e). Above this critical current (a function of the gas and its pressure), the discharge adopts a definite new structure. The beam is no longer quasi-homogeneous but displays a meniscus about 7 mm from the cathode (Fig. 83.7). Langmuir probe measurements show that the primary electrons are abruptly scattered in energy at the meniscus. Spectroscopic measurements in argon, for example, with a cathode fall of 26 V, show the Ar spectrum throughout the discharge, whereas the Ar^+ lines appear only to the anode side of the meniscus. The Ar^+ lines appear even though the minimum energy required to excite them is 34.8 to 39.9 eV and the total voltage across the discharge can be 5 to 10 V lower. Coincident with onset of the meniscus, distinct GHz and MHz frequency oscillations are detected. These frequencies have been interpreted as electron- and ion-plasma oscillations respectively [40].

83.4 ATOMIC AND MOLECULAR PROCESSES

If the cross section for a given process is σ_j, then the rate of excitation per electron for that process is

$$Z_j = N \int_0^\infty \sigma_j(v) v f_v(v, \mathbf{r}, t) 4\pi v^2 dv. \qquad (83.45)$$

For inelastic processes, $\sigma_i(v) = 0$ for velocities below the threshold energy. The ionization threshold is the ionization potential I_P. The ionization cross sections typically increase monotonically with energy from threshold to about 100 eV. An exception to this is the alkali metal group for which the ionization cross section maximum occurs below 15 eV.

Assuming that the discharge excitation conditions are spatially uniform, the electron conservation equation is

$$\frac{dn_e}{dt} = (Z_i - Z_a - Z_d)n_e - Z_r n_e^2, \qquad (83.46)$$

where $Z_i n_e$ is the total rate of ionization, $Z_a n_e$ is the rate of attachment, $Z_d n_e$ is the net diffusion rate out of the region considered, and $Z_r n_e^2$ is the rate of recombination, all per unit volume. The coefficients are usually functions of both P and E, either directly or indirectly. We consider these processes separately in Sects. 83.4.1 to 83.4.3.

83.4.1 Ionization

The ionization frequency (in s^{-1}) for a Maxwellian distribution of electron velocities is given by [41]

$$Z_i \approx 9 \times 10^7 a P \exp\left(-\frac{I_P}{\Theta}\right) \Theta^{1/2} I_P, \qquad (83.47)$$

where $\Theta = (k_B T_e/e)$ and a is the initial slope of the ionization efficiency curve in electron energy (ion pairs/torr/Volt/electron). (The ionization efficiency is defined as the number of ion pairs produced per electron per cm of path at 1 torr and 0°C. The ionization efficiency is proportional to the ionization cross section.) For a Maxwellian distribution, the quantity

$$\frac{dZ_i}{d\Theta} = Z_i \frac{(V_i + \Theta/2)}{\Theta^2}, \qquad (83.48)$$

together with Z_i, is useful in estimating the electron density and electric field for steady state discharges, or the onset of ionization instabilities [18].

If more accuracy is desired in estimating the ionization rate, a numerical solution of the Boltzmann equation is needed to calculate f_v^0. An energy balance calculation can then be performed by integrating the electron kinetic equation over all electron energies. The electron energy balance and the average electron energy are independent of n_e in a weakly ionized gas. Therefore the energy transfer into different excitations can be treated on a single electron basis. The total power density in the discharge as a function of E/N is then given by

$$\mathbf{J} \cdot \mathbf{E} = e w_d \frac{E}{N} N n_e. \qquad (83.49)$$

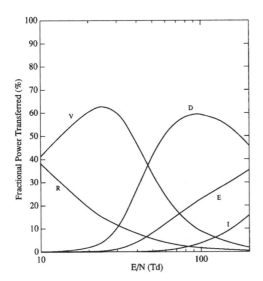

Figure 83.8. Fractional power deposited into the different inelastic modes of hydrogen as a function of E/N. The labels V, D, E, R, and I correspond to vibrational, dissociation, electronic, rotational, and ionization processes respectively. The remainder is transferred into elastic collisions.

The relative contributions of the various electron loss processes are then obtained. Figure 83.8 shows the fractional input power deposition into the principal electron loss channels of hydrogen [42] as a function of the normalized electric field. Over a wide range of E/N, a large fraction of the discharge energy goes into vibrational excitation. Significant energy deposition into dissociation occurs only above 40 Townsends. This explains why it is necessary to use a capillary tube discharge (or *Wood's tube*) to obtain the atomic spectrum, or to provide a source of atomic hydrogen. The higher diffusion losses of the narrow bore cause the discharge to run at high E/N in order to maintain the electron density. The high E/N also provides large dissociation rates.

To obtain the total rate of excitation at high Γ_e, it is necessary that an assessment of the excited state densities also be made. This is most important for excited states that are known to be metastable so that the effective lifetimes are determined by collisional quenching or diffusion. The metastables can be very important in the excitation of higher states and in ionization. First, the threshold to ionize a metastable is usually much less than the I_P of the ground state. Second, the cross section for ionization of an excited state is often much larger than that of the ground state. Evaluations [43] of the low pressure helium discharge above 1 torr show that at current densities above a few mA cm^{-2} the two-step ionization through metastable states (proportional to n_e^2) exceeds the single step ionization (proportional to n_e) from the ground state. A multi-temperature

approximation introduced by Vriens [44] has proven to be very useful in analyzing noble gas-alkali discharges and also noble gas discharges. The decreasing electric field required at higher discharge currents (higher n_e) is thus due to a combination of more efficient ionization through the metastables and increased gas heating.

Also, molecular dissociation changes the ionization rate through a combination of changes in f_v^0. At high energy density depositions ($\sim$0.1 eV/molecule), the ionization rate and the dissociation rate are influenced by the degree of vibrational excitation, and by the changed Franck-Condon probabilities. However, the dissociation rate is changed primarily by two processes: collisions of the second kind [41]

$$e_{(slow)} + N_2(v) \rightarrow e_{(fast)} + N_2(v = 0), \qquad (83.50)$$

which increase the number of fast electrons, and by increased anharmonic pumping of vibrational states to the dissociation limit [45]. Anharmonic pumping [46] occurs when large populations of vibrationally excited states are created in a molecular gas (usually diatomic) with a small amount of anharmonicity between adjacent levels, and the translational energy of the molecules is low. Because of the anharmonicity, the vibrational energy exchange favors the transfer from low vibrational levels to higher vibrational levels. This process is the basis of infrared laser action in carbon monoxide [46].

For discharges in gas mixtures, the component with the larger inelastic cross section *and* lower threshold tends to control the mean energy of the electron energy distribution. This situation is of common occurrence in discharge lasers. Numerical solutions of the Boltzmann equation have been obtained for a large number of gas mixtures of technical interest [47, 48]. Usually the two-term approximation in Eq. (83.9) is adequate, although there are circumstances where higher order terms are needed [49].

83.4.2 Electron Attachment

The phenomenon of electron attachment to a neutral atom or molecule to form a negative ion is a common occurrence for gases whose outer electronic shells are nearly filled. The energy of formation of a negative ion is called the *electron affinity* (E_A). This varies from about 3.5 eV (for the halogens) to nearly zero among the gases that exhibit electron attachment. Atoms having closed electronic shells do not form negative ions. These atoms, which have $^1S_0$ ground states, include the noble gases. Molecules in the $^1\Sigma$ ground state also do not form permanent negative ions.

A general experimental technique for measurement of attachment coefficients involves passing electrons through a gas target and measuring the attenuation of the electrons due to attachment, and usually also the negative ions produced [6]. Mass spectrometers are often included to identify the ions. A comprehensive survey of negative ions is available in the revised edition of Massey's text [50], and an update with emphasis on discharge lasers has been given by Chantry [51].

83.4.3 Recombination

One of the most common loss mechanisms for ions is the recombination of negative ions and electrons with positive ions. The loss of ions due to recombination is proportional to the product of their concentrations:

$$\frac{dn_+}{dt} = \frac{dn_-}{dt} = -\alpha n_- n_+ \,. \tag{83.51}$$

Here α is called the *recombination coefficient* and n_- is the negative ion or electron density. The value of α is quite different for negative ions and electrons. If there is only one negative ion species, and the positive and negative ion species have equal concentrations, then

$$\frac{dn}{dt} = -\alpha n^2 \,, \tag{83.52}$$

which on integration gives

$$n^{-1} = n_0^{-1} + \alpha t \,, \tag{83.53}$$

where n_0 is the initial ion concentration at $t = 0$. Recombination phenomena therefore often exhibit a linear relation between $1/n$ and t after switch-off of a discharge.

The three primary processes for positive ion–negative ion recombination are:
three body recombination

$$A^+ + B^- + M \to AB + M \,; \tag{83.54}$$

radiative recombination

$$A^+ + B^- \to AB + \text{photon} \,; \tag{83.55}$$

mutual neutralization

$$A^+ + B^- \to A^* + B^* \,. \tag{83.56}$$

There are two additional processes for positive ion-electron recombination:
dissociative recombination

$$e + AB^+ \to [AB]^* \to A + B \,; \tag{83.57}$$

dielectronic recombination

$$e + A^+ \to A^{**} \to A + \text{photon} \,. \tag{83.58}$$

83.5 ELECTRICAL DISCHARGE IN GASES: APPLICATIONS

83.5.1 High Frequency Breakdown

In the theory of high frequency breakdown, the electron energy distribution function is calculated as a function of the applied ac electric field **E**, using Eqs. (83.12) and (83.13). From this the ionization frequency is derived, so that the rate of ion production can be expressed in terms of **E**. At breakdown, the ionization rate equals the sum of the losses due to diffusion, attachment and recombination. The μ_e, ν_e and attachment coefficients, as well as the various rate constants producing new species, depend on **E**. The electron continuity equation is solved to obtain the breakdown field. [1]

Rewriting Eq. (83.13) for a constant ν_m, and assuming a Maxwellian velocity distribution, the electron current density becomes

$$\mathbf{J} = -e\Gamma_e = \frac{n_e e^2 E_0 e^{-i\omega t}}{m(\nu_m - i\omega)} \,, \tag{83.59}$$

where the time dependence of **E** is written explicity. The power P gained per unit volume is the time average

$$P = \langle \mathbf{J} \cdot \mathbf{E} \rangle = \frac{n_e e^2 \nu_m}{m(\nu_m^2 + \omega^2)} \frac{E_0^2}{2} \,. \tag{83.60}$$

For a given **E**, P has a maximum at $\nu_m = \omega$. Equation (83.60) can also be written as

$$P = n_e \frac{e^2 E_{eff}^2}{m\nu_m} \,, \tag{83.61}$$

where the *effective* electric field is defined by

$$E_{eff}^2 = \frac{E_0^2}{2}\left(\frac{\nu_m^2}{\nu_m^2 + \omega^2}\right) \,. \tag{83.62}$$

E_{eff} is useful for comparing the relative heating effect of alternating and static fields. These considerations are carried further by MacDonald [52], and are important in determining microwave antenna breakdown fields at high altitudes.

83.5.2 Parallel Plate Reactors and RF Discharges

A situation of interest for plasma deposition and etching is the discharge between two parallel plates

---

[1]In the absence of appreciable preionization, the electron loss is governed by free diffusion. When the electron loss is controlled by ambipolar diffusion, the *operating field* of the discharge is obtained. This effect, in addition to cumulative ionization and gas heating, gives an operating field lower than the breakdown field.

driven by a rf power supply. The industrial standard excitation frequency is 13.56 MHz. There have been several quite different approaches. One of the first models due to Bell [53] treats the discharge as a circuit element, and applies the boundary condition that the sum of the conduction and displacement currents remain constant:

$$\mathbf{J} = \sigma \mathbf{E} + \epsilon_0 \frac{\partial \mathbf{E}}{\partial t} . \qquad (83.63)$$

Substituting for σ, $\mathbf{E}$ and $\partial \mathbf{E}/t$ in Eq. (83.63) for the case where electrons are the major current carriers, and assuming a velocity-independent ν_m, the amplitude of the electric field as a function of position becomes

$$E = \frac{J}{\epsilon_0 \omega} \left[\frac{(1+q^2)}{(\delta-1)^2 + q^2} \right]^{\frac{1}{2}} , \qquad (83.64)$$

where $\delta = \omega_p^2(x)/\omega^2$ [see Eq. (82.2)] and $q = \nu_m/\omega$. At high electron densities or low frequency fields, $\delta \gg 1$ and

$$E = E_0 \left[\frac{(\delta_0 - 1)^2 + q^2}{(\delta - 1)^2 + q^2} \right]^{\frac{1}{2}} , \qquad (83.65)$$

where E_0, δ_0 denote values at the centre of the discharge. At low densities or high frequency fields, $\delta \ll 1$ and $E = J/(\epsilon_0 \omega)$, i.e., the system acts like a capacitor. However, it is also necessary to consider the self-bias that the electrodes acquire under the influence of an applied electric field. The bias arises because of the requirement that, averaged over time, no net charge can collect at the electrode. Since the anode current-voltage characteristic is very nonlinear, application of a sinusoidal voltage to the electrode creates at each electrode a negative dc bias approximately equal to the peak amplitude of the rf voltage. The system therefore acts like a hollow cathode discharge with modulated sheath voltages. The ions from the plasma are accelerated across these sheaths and, over most of the rf cycle, acquire sufficient energy to cause secondary electron emission at the electrodes.

The ion flux and energy distribution hitting the substrate depend on the plasma conditions: the values of the dc and rf bias voltages, and the ratio of the transit time of the ions across the sheath to the period of the rf excitation frequency. The anisotropy and selectivity of the etching processes are determined by the ion energy and its directionality, the etching radical, and the substrate materials. Above energies of several hundred volts, ion bombardment of the substrate causes sputtering, which is relatively unselective. Pure chemical etching by radicals alone may be selective, but it is relatively isotropic. The aim in discharge enhanced processing is to obtain plasma conditions that generate selected radicals with high efficiency, concurrent with bias voltages that provide highly directional ions to prepare the substrate for etching. The combination of these processes gives etch rates that are typically

an order of magnitude higher than the sum of the individual processes. The microstructures that can be fabricated using processing are critical to all of high density integrated circuit manufacturing. The achievements are illustrated by 0.2 micron wide trenches in silicon that are 4 microns deep with vertical sides [54].

On the other hand, the electrons form an energy-modulated fast electron beam that is injected into the plasma region. These fast electrons maintain much of the ionization in a symmetrically excited, equal electrode area reactor. In principle, it is possible to develop a complete self-consistent model of the rf discharge, consisting of the Boltzmann equation for the energy distribution function of the electrons, positive ions, and negative ions coupled to the Poisson's equation for the electric field. Simpler models based on moments of the Boltzmann equation have been developed for higher pressures [55] (where a fluid approximation is more appropriate) and particle-in-cell Monte-Carlo simulations for low pressures [56]. The above studies have led to the identification of four principal mechanisms affecting energy deposition. These involve:

1. the impedance of the bulk plasma;

2. the energy deposition by fast electrons created by secondary emission due to ion bombardment of the electrodes;

3. collisionless absorption due to the asymmetrical sheath boundary-plasma electron interaction; and,

4. wave-riding, or collisional sheath interactions causing electron heating in the sheath modulated electric field. The electron is regarded as surfing on the expanding sheath field.

Processes (3) and (4) are related, the difference being that in (4) electron collisions occur during the sheath expansion. There is also an additional interaction due to the changes in the complex impedance (capacitive sheaths and resistive bulk plasma) of the discharge and the consequent changes in the power transferred from the rf generator. Other less general plasma modes often occur due to resonance when the electric field and period are such that one electron transit requires one half period of the rf cycle. Under these conditions at very low pressures, when λ_{mfp} is larger than the gap spacing, and secondary electron emission can occur, one has the multipactor discharge mode. Otherwise when λ_{mfp} corresponds approximately to small integral fractions of the gap spacing, plasmoid modes are excited.

83.6   CONCLUSIONS

The study of the conduction of electricity in gases has had several renaissances because of application incentives. Discharge lighting technology science and development are described by Waymouth [57]. Atmospheric plasma physics and cosmic plasma physics are covered by Rees [58] and Alfvén [59] respectively. The role of plasma physics in gas discharge laser development was recently reviewed [60], and the fundamental interactions in laser generated plasmas are detailed by Hughes [61]. The scaling of discharge volumes to many liters is again of interest because of applications to large area surface treatments and thin films. The discharge enhanced chemistry of complex gas mixtures is of particular relevance to etch processing used most microelectronics fabrication. Work on discharge-generated and -trapped particulates in microelectronics processing plasmas has re-established links with cluster physics and space plasmas. The physics of many of the effects was formulated by Emeleus and Breslin [62] much earlier. Future prospects for further exploitation of gas discharge physics include waste/toxic hazard decontamination, fabrication of selected complex molecules and selected cluster morphologies, and the development of plasma flat panel displays.

Acknowledgment

This article is based on notes developed in cooperation with Dr. J. C. Ingraham. I wish also to thank Dr. R. Nagpal for assistance with the revisions.

REFERENCES

1. P. Lenard, Ann. Physik **12**, 714 (1903).
2. M. Hayashi, Technical Report No. 1PPJ-AM-19, Institute of Plasma Physics (Nagoya, Japan) Report, 1981 (unpublished).
3. H. S. W. Massey, *Handbuch der Physik* (Springer Verlag, Berlin, 1956), Vol. 36, p. 232.
4. W. J. Pollock, Trans. Farad. Soc. **64**, 2919 (1968).
5. R. Nagpal and A. Garscadden, J. Appl. Phys. **75**, 703 (1994).
6. G. J. Schulz, Rev. Mod. Phys. **45**, 423 (1973).
7. P. Kleban and H. T. Davis, Phys. Rev. Lett. **39**, 456 (1977).
8. W. P. Allis, *Handbuch der Physik* (Springer Verlag, Berlin, 1956), Vol. 21, p. 383.
9. G. Cavaleri and G. Sesta, Phys. Rev. **170**, 286 (1968).
10. G. L. Braglia, Physica **92C**, 91 (1979).
11. W. L. Nighan, Phys. Fluids **12**, 162 (1969).
12. L. Spitzer, *Physics of Fully Ionized Gases* (Wiley Interscience, New York, 1956).
13. H. Dreicer, Phys. Rev. **117**, 343 (1960).
14. S. T. Butler and M. J. Buckingham, Phys. Rev. **126**, 1 (1962).
15. I. P. Shkarofsky, T. W. Johnston, and M. P. Bachynski, *The Particle Kinetics of Plasmas* (Addison-Wesley, Reading-Massachusetts, 1966).
16. A. M. Weinberg and E. P. Wigner, *The Physical Theory of Neutron Chain Reactors* (University of Chicago Press, Chicago, 1958).

17. V. L. Ginzburg, *The Propagation of Electromagnetic Waves in Plasmas* (Addison-Wesley, Reading-Massachusetts, 1964).
18. L. Pekarek, Uspeki Fiz. Nauka **94**, 463 (1968).
19. T. L. Cottrell and I. C. Walker, Trans. Farad. Soc. **61**, 1585 (1965).
20. M. R. Hallada, P. Bletzinger, and W. F. Bailey, IEEE Trans. Plasma Sc. **PS-10**, 218 (1982).
21. W. H. Long Jr., W. F. Bailey, and A. Garscadden, Phys. Rev. A **13**, 471 (1976).
22. J. Dutton, J. Phys. Chem. Ref. Data **4**, 577 (1975).
23. S. R. Hunter and L. G. Christophorou, in *Electron-Molecule Interactions and Their Applications*, edited by L. G. Christophorou (Academic Press, New York, 1984), Vol. II, p. 89.
24. L. G. H. Huxley and R. W. Crompton, *The Diffusion and Drift of Electrons in Gases* (Wiley Interscience, New York, 1974).
25. A. L. Gilardini, *Low Energy Electron Collisions in Gases* (Wiley Interscience, New York, 1972).
26. J. S. E. Townsend, *Electricity in Gases* (Clarendon Press, Oxford, 1915).
27. E. B. Wagner, F. J. Davis, and G. S. Hurst, J. Chem. Phys. **47**, 3138 (1967).
28. J. H. Parker and J. J. Lowke, Phys. Rev. **181**, 290 (1969).
29. H. R. Skullerud, J. Phys. B: At. Mol. Opt. Phys. **2**, 696 (1969).
30. W. P. Allis and D. J. Rose, Phys. Rev. **93**, 84 (1954).
31. F. C. Hoh, Rev. Mod. Phys. **34**, 267 (1962).
32. F. F. Chen, *Introduction to Plasma Physics* (Plenum Press, New York, 1974).
33. L. S. Frost and A. V. Phelps, Phys. Rev. **127**, 1621 (1962).
34. A. G. Englehardt, A. V. Phelps, and C. G. Risk, Phys. Rev. **135**, 1566 (1964).
35. R. Nagpal and A. Garscadden, Phys. Rev. Lett. **73**, 1598 (1994).
36. M. Kurachi and Y. Nakamura, J. Phys. D: Appl. Phys. **21**, 602 (1988).
37. V. I. Kolobov and V. A. Godyak, IEEE Trans. Plasma Science, **23**, 503 (1995).
38. Y. P. Raizer, *Gas Discharge Physics* (Springer-Verlag, Berlin, 1991), Chap. 8.
39. A. Garscadden, *Ionization Waves*, in *Gaseous Electronics: Electrical Discharge*, edited by M. N. Hirsh and H. J. Oskam (Academic Press, New York, 1978), Chap. 2.
40. T. K. Allen, R. A. Bailey, and K. G. Emeleus, Brit. J. Appl. Phys. **6**, 320 (1955).
41. A. von Engel, *Ionized Gases* (Clarendon Press, Oxford, 1965).
42. A. Garscadden and W. F. Bailey, in *Progress in Astronautics and Aeronautics*, edited by S. S. Fisher (AIAA, New York, 1981), Vol. 74.
43. B. E. Cherrington, IEEE Trans. Elec. Devices **ED26**, 148 (1979).
44. L. Vriens, J. Appl. Phys. **45**, 1191 (1974).
45. M. Cacciatore, M. Capitelli, and C. Gorse, J. Phys. D: Appl. Phys. **13**, 575 (1980).
46. C. E. Treanor, J. W. Rich, and R. G. Rehm, J. Chem. Phys. **48**, 1798 (1968).
47. E. W. McDaniel and W. L. Nighan, *Gas Lasers* (Aca-

demic Press, New York, 1982), Vol. 3 Applied Atomic Collision Physics.

48. A. Garscadden and R. Nagpal, Plasma Sources Sci. Technol. **4**, 268 (1995).

49. L. C. Pitchford, S. V. ONeil, and J. R. Rumble Jr., Phys. Rev. A **23**, 294 (1981).

50. H. S. W. Massey, *Negative Ions*, 3 ed. (Cambridge University Press, Cambridge, 1976).

51. P. J. Chantry, in *Gas Lasers* (Academic Press, New York, 1982), Vol. 3 Applied Atomic Collision Physics. See Ref. [47].

52. A. D. MacDonald, *Microwave Breakdown in Gases* (Wiley Interscience, New York, 1969).

53. A. T. Bell, in *Engineering, Chemistry and Use of Plasma Reactors*, edited by J. E. Flinn (American Institute of Chemical Engineering, New York, 1971).

54. M. A. Lieberman and A. J. Lichtenberg, *Principles of Plasma Discharges and Materials Processing* (Wiley Interscience, New York, 1994).

55. F. F. Young and C. J. Wu, IEEE Trans. Plasma Sci. **21**, 312 (1993).

56. R. Krimke, H. M. Urbassek, D. Korzec, and J. Engemann, J. Phys. D: Appl. Phys. **27**, 1653 (1994).

57. J. F. Waymouth, *Electric Discharge Lamps* (MIT Press, Cambridge, 1971).

58. M. H. Rees, *Physics and Chemistry of the Upper Atmosphere* (Cambridge University Press, Cambridge, 1989).

59. H. Alfvén, *Cosmical Electrodynamics* (Clarendon Press, Oxford, 1950).

60. A. Garscadden, M. J. Kushner, and J. G. Eden, IEEE Trans. Plasma Science **19**, 1013 (1991).

61. T. P. Hughes, *Plasmas and Laser Light* (Wiley Interscience, New York, 1975).

62. K. G. Emeleus and A. C. Breslin, Int. J. Elect. **29**, 1 (1970).

84

Applications to Combustion

David R. Crosley

SRI International, Menlo Park, California 94025

84.1 INTRODUCTION

Combustion processes are vital to the operation of present-day society. Although there exist alternative energy sources, combustion of a variety of fuels is worldwide a major mode of energy production, and will remain so for many years to come. Despite a well-established technology, combustion is not without need of improvement. The pollutant emissions associated with burning pose major barriers to more widespread use. Increasingly stringent environmental regulations make this a current focus of combustion research and development.

Both scientific inquiry and empirical solutions will be important in advancing combustion in the future. Most of the progress in combustion technology has been made without benefit of an understanding of the science involved; rather, it has been accomplished through an approach of trial and error, which has led to many ingenious, innovative solutions to problems. In fact, as witnessed by the early development of thermodynamics, the technology has often driven the science rather than the other way around. Nonetheless, future advances in combustion are expected to rely more and more on a planned implementation of fundamental knowledge of physics and chemistry.

Combustion may be thought of as self-sustaining reactive flow, in which chemical energy is converted into ex-

tractable, useful heat. This is usually accompanied by an abrupt change in properties of the system in space and/or time, particularly the chemical composition and the temperature. The description of combustion processes involves the subdisciplines of thermochemistry, chemical kinetics, fluid mechanics, and transport. The major challenge is to apply the known principles from these to a description and understanding of the entire combustion process. This endeavor involves three approaches: experiment, computation, and theory, the latter two differentiated via the characteristic of numerical versus analytical solutions. Each part now plays an important role in combustion science.

The efficiency of combustion processes, be they steady state such as a burner flame or rocket motor, or transitory such as an explosion or internal combustion engine, is generally governed by the overall thermochemistry, fluid dynamics, and heat transfer characteristics of the system. Details of the chemical reactions and the role of trace intermediates are usually unimportant for this question. The latter do play a key part, however, when considering problems of pollutant formation (emission of NO_x, SO_x, soot, toxic organic compounds), and flame ignition and inhibition. The current challenges in combustion science are solutions to these problems of efficiency and pollution, with advancement likely to occur through two major thrusts: a better understanding of turbulent reactive flow, and the incorporation of detailed chemical kinetics into combustion models (as contrasted to simplified versions such as one- or two-step reaction schemes). Additionally, new areas of science involve combustion under unusual conditions (e.g., microgravity combustion aboard the space shuttle) and new applications (e.g., flame synthesis of exotic materials such as diamond films).

From the viewpoint of the basic science, most of the general, qualitative understanding of behavior at the molecular scale needed to describe combustion is at hand. In many engineering applications, an empirical approach is taken in which a quantitative description of that behavior is condensed and parametrized, using previously acquired experimental data acquired directly on the system of interest. As models become more detailed (often through computational advances), there is a need for further quantitative supporting molecular data, such as improved values of transport properties of molecules.

However, there have been two major advances in combustion research which contain new ideas and directions in detailed chemistry and physics. The first is the use of powerful computers for the numerical solutions to combustion problems, so that a predictive description can be built beginning with fundamental physical principles. The second is the use of laser diagnostic techniques for the determination of the detailed properties of the combustion system, especially temperature, velocity, and composition, including both major components and trace chemical intermediates. These have not only provided tests of the computational models, but have also furnished new insights and approaches to an understanding of combustion. This has occurred by measuring properties of the system not previously available, or via improved scales of spatial and temporal resolution.

This chapter concentrates on these two areas of physical models and laser diagnostics, and the outstanding physical questions that remain. Each is greatly influencing the development of the two important combustion science issues: describing turbulent flows and incorporating realistic chemistry.

84.2 COMBUSTION CHEMISTRY

The addition of detailed chemical kinetics to flame models will greatly advance our understanding of combustion, at least for nonturbulent flows. Detailed chemistry is becoming a part of codes describing systems of greater and greater complexity. Numerical integration packages, including full chemistry for a one-dimensional flow, are routinely available [1], and progress is being made for the two-dimensional case. A sensitivity analysis enables the modeler to examine the value predicted for some variable (e.g., concentration of a transient species) at a given point in the flame as a function of the rate

coefficients and other input variables. This facilitates interpreting the output in terms of physically meaningful variables, and extrapolating the model predictions to other applications.

The quality of the predictions of a combustion model is a strong function of the quality of the reaction rate coefficients for important elementary reaction steps, and the thermochemical properties of reactive species, especially free radicals. A flame chemical mechanism can be quite complex: a current methane/air model [2] containing reactions involving only hydrogen, oxygen and carbon (i.e., no nitrogen chemistry) contains 30 species and 177 reactions (plus their reverses). Although it is necessary to include most of these steps in the reaction scheme, a much smaller number play dominant roles in determining any given parameter such as flame speed. The determination of sufficiently accurate rate coefficients for those important reactions is made difficult by the complexity of the reactions, and the difficult conditions (high temperature) and reactants (usually free radicals) that are involved.

There are many unimolecular reactions and their reverses that play important roles in combustion chemistry, including the recombination of fuel-derived radicals such as $CH_3 + H + M \rightarrow CH_4 + M$, and low temperature decomposition/recombination reactions important in ignition, e.g., $H + O_2 + M \rightarrow HO_2 + M$, compared with chain-branching steps. Under most combustion conditions, the important reactions in this class are either in the low pressure limit, where the rate constant is proportional to total pressure P of collider M, or in the fall-off region between that limit and the P-independent high pressure limit. Knowledge of the collisional energy transfer characteristics for highly excited vibrational levels of polyatomic molecules is apropos to a fundamental understanding of these reactions. For purposes of combustion models, these rate constants can be formulated in a modified Arrhenius mode, requiring as many as nine parameters to describe correctly the combined P, T dependence needed for a realistic range of combustion conditions.

A further complication is posed by the multiple potential energy surfaces that can be accessed by many reactions. For example, two methyl radicals can combine to form not only C_2H_6, but also $C_2H_5 + H$. Measurements are often made of the overall rate coefficient (in this case, disappearance of CH_3) but seldom are determinations made of branching ratios into these multiple paths. Especially, the temperature dependence of that branching is rarely known accurately enough. Since reaction rate coefficients can change orders of magnitude over the range of temperatures encountered in combustion, simple extrapolations are seldom sufficient.

Theoretical chemistry has begun addressing these problems. Electronic structure calculations provide thermochemical properties of transient, reactive species. Using theories which include semi-empirical adjustments, enthalpies of formation of free radicals and potential barriers on reactive surfaces can be calculated with accuracies of about 2 kcal/mole, sufficient for modeling purposes. Multidimensional potential surfaces provide starting points for conventional transition state theory calculation of rate coefficients, as well as variational transition state approaches. Theoretical calculations may be useful for understanding the energy transfer questions of unimolecular and recombination rate theory. Also useful are recent laser experiments which examine the foundations of rate theory. These include laser photolysis investigations of the reaction rate coefficient as a function of energy just above threshold [4], and the formation of molecules from dissociating complexes on subpicosecond scales [3], as well as stimulated emission pumping experiments which probe the density of states and energy transfer in high vibrational levels of diatomic and polyatomic molecules [5].

We conclude this section with brief comments on some individual combustion systems. Flames of H_2 and CO burning in (moist) O_2 are relatively simple, and their mechanisms are important subsets of reaction networks for more complex systems. The chemistry of these flames is relatively well understood on a fundamental basis. So, generally, is that of methane, although there remain crucial questions, for example, the rates for CH_3 + OH, a multisurface reaction producing singlet and triplet CH_2 + H_2O, as well as CH_2OH. For hydrocarbon fuels more complex than methane, the complexity of the mechanism rapidly escalates, due to the many possible ways to combine radicals containing multiple carbon-carbon bonds. For a fuel such as isooctane (where few rate coefficients are known) it is necessary to use a computer algorithm simply to write down the mechanism without making mistakes or omissions [6].

Many pollutant emissions are produced by reactions of trace intermediate species. Nitric oxide has received much recent attention due to increasingly stringent environmental standards. NO is formed in combustion processes in three ways: (i) so-called thermal (or Zeldovich) NO, formed by reaction of $O + N_2$, $N + O_2$, and $OH + N_2$, where the atoms and OH are usually present at nonequilibrium, kinetically controlled concentrations; (ii) prompt NO, which begins with the $CH + N_2$ reaction breaking the strong N-N bond to form N atoms which react with O_2 and HCN, which is oxidized to NO in a series of steps; (iii) and fuel-nitrogen NO, which is formed from the nitrogen present in coal and coal-derived fuels, usually entering the combustion process as HCN. NO can be dealt with through catalytic reduction, staged combustion (burning at a series of mixing ratios), reburning at a later stage with additional fuel, and injection of compounds such as NH_3 or cyanuric acid, $(HOCN)_3$. Progress has been made in the determination of the rate coefficients of many pertinent elementary reactions for each of these processes, but significant gaps remain. Examples are the full T, P dependence of the complex-forming reaction $CH + N_2$, and the complicated reaction pathways in the $NH_2 + NO$ reaction important in de-NO_x through ammonia injection.

The mechanism of soot formation is a matter of dispute; unsaturated chains containing 4 or 5 carbons, aromatic rings, ions, and C_3H_3 radicals have all been suggested as precursors. Ignition (and engine knock) chemistry occurs at low temperatures and probably involves hydroperoxy (HO_2) and alkylperoxy (RO_2) radicals. The mechanism of production of toxic organic compounds such as aldehydes is poorly known, and few relevant rate coefficients are at hand.

84.3   LASER COMBUSTION DIAGNOSTICS

The advent of the laser has altered much of experimental combustion science. Laser Doppler velocimetry, which employs particle scattering to measure directed flow velocity, is available in the form of commercial systems. Coherent anti-Stokes Raman scattering (CARS) is widely used in engineering laboratories to study practical combustors, including jet engines, industrial furnaces, internal combustion engines, and burning propellants. Laser-induced fluorescence (LIF) has come into prominence for the study of combustion systems and related chemical kinetics.

The spectroscopically based techniques of most interest in atomic, molecular, and optical physics are CARS, LIF, and the related recently emerging method of degenerate four wave mixing (DFWM). Each may be used to determine individual molecular species or temperature. Each is highly selective, due to its spectroscopic nature; they provide good spatial resolution, often on the order of a mm^3 or less; the use of pulsed lasers yields excellent time resolution, ~ 10 ns. When used properly, they are nonintrusive, perturbing neither the flow dynamics nor flame chemistry. They can be used in very hostile environments (corrosive atmospheres, high temperatures) where physical probes would not survive.

LIF and CARS also have separate, complementary features. CARS is suitable for the determination of major species (fuel, O_2, CO_2, H_2O, and, in air-breathing combustion, N_2). It may be used for accurate temperature measurements, including measurements on a single laser shot when operated in a broadband mode. This makes CARS especially useful for problems in which fluid dynamics and flow patterns are most important. Because CARS produces a coherent, laser-like beam, spatial filtering can be used to discriminate against background radiation such as the intense thermal emission from hot combustor walls. LIF is highly sensitive and is useful for measurements on reaction intermediates present at low concentrations. It may be used in an imaging mode to obtain instantaneous two-dimensional patterns of radical species in rapidly time-varying flows.

Despite their utility, LIF and CARS are not user-friendly technologies. At present, considerable nurturing of the equipment, knowledge of the theory, and care in the data analysis are needed for accurate measurements, and nearly all such measurements are made by scientists trained in physics, chemistry, or physics-oriented engineering departments. Both methods (and DFWM) are examined here with an emphasis on the physics questions in their continued development and application.

84.3.1 Coherent Anti-Stokes Raman Scattering

CARS depends on the nonlinear polarization induced in molecules by the intense electric fields of lasers, exploiting resonant effects in the third order nonlinear susceptibility $\chi^{(3)}$. $\chi^{(3)}$ mixes the fields from a pump laser at frequency ω_1 and a Stokes beam at frequency ω_2 to form a third beam at $\omega_3 = 2\omega_1 - \omega_2$. $\chi^{(3)}$ is strongly enhanced when the difference $\omega_1 - \omega_2$ is tuned to a vibrationally or rotationally resonant frequency in the ground state of a molecule. This is usually accomplished using the intense green beam of a frequency doubled Nd:YAG laser for ω_1, and a dye laser Stokes shifted by the vibrational frequency of the molecule of interest for ω_2. In general, the signal in the visible region of the spectrum contains a combination of a nonresonant $\chi^{(3)}$ due to the electronic polarizability of the medium (mostly N_2 for air-breathing combustion) and the resonant part. For high concentrations c_A (e.g., ~70% N_2 in air based flames), the signal is a Lorentzian-shaped resonance on top of the flat nonresonant term, proportional to c_A^2/Γ where Γ is the linewidth. For lower concentrations (e.g., a few percent CO_2 in exhaust gases) the resonant and nonresonant amplitudes interfere, so that the signal is dispersion shaped with a magnitude $\propto c_A/\Gamma$.

A very useful and popular application of CARS operates the Stokes laser in a broadband mode. This generates the entire coherent anti-Stokes spectrum at once, which is then dispersed through a spectrometer and detected with an array. T may be determined from the rotational structure and the appearance of vibrational hot bands. These measurements are generally performed on the N_2, almost always present in large quantity in any combustion system of practical interest. This is the most reliable, accurate method for determination of pointwise temperatures with a single laser shot.

The nonlinear, coherent nature of CARS, in which interfering amplitudes are added, means that spectral details can have a significant influence on the observed spectrum. These include the concentration of all contributing species (including nonresonant background) and the widths of individual Raman lines. The linewidths must

be dealt with correctly in order to deduce temperatures accurately from the spectral shapes. It has only been in the past five years that linewidths as a function of rotational level have been available for N_2, at the accuracy needed for application of the spectral fitting codes used for T determination. As P is increased, the lines broaden enough to begin to overlap; at high enough P (several atm for N_2) collisions occur so rapidly that the spectrum collapses to a single broad line. To make T measurements at high pressure, one must deal with both these P broadening and collisional narrowing effects; this requires an understanding of the underlying collision dynamics and energy transfer, generally on a rotational level state specific basis. Both energy transfer and dephasing collisions contribute to the linewidths. For molecules other than N_2, there is scant information on Raman linewidths even at atmospheric pressure, at accuracies needed for modeling CARS spectra. Molecules of particular interest are O_2, CO_2, H_2O, CO, and, in the future, hydrocarbon fuels. In some cases, especially H_2O, further spectral information for high rotational lines (populated at high temperature) is also needed.

Most of the treatments of CARS spectra assume a monochromatic laser. Finite bandwidth effects of real lasers must be taken into account for accurate theoretical treatments (and thus accurate T and concentration determinations). There are several nonequivalent theoretical treatments to convolute the frequency dependence of $\chi^{(3)}$ with the bandwidths of the pump and Stokes lasers.

84.3.2 Laser-Induced Fluorescence

LIF is much more sensitive than CARS, relying on resonant excitation to a real, emitting electronically excited state. The use of a tunable laser to scan the molecular excitation spectrum makes LIF also highly selective for small species that have characteristic, identifiable line spectra. Thus LIF is suitable for measurement of reactive intermediates of importance in understanding flame chemistry. Concentrations determined by LIF range from parts per billion to parts per thousand. Detectable species are diatomics such as OH and CH, triatomics including HCO, NCO, and NH_2, and a few larger molecules including CH_3O. Most of the pertinent molecular electronic transitions are in the uv or blue region of the spectrum. Because vuv radiation cannot penetrate flame gases, atoms (such as H and O) must be detected via two-photon absorption to a high-lying electronic state of the same symmetry as the ground state, which then radiates in the visible or near ir to another excited state.

A list of all molecules through teratomics composed of the five main atoms naturally occurring in combustion (H, C, O, N, and S), and detected to date by LIF, is given in Table 84.1. This forms a large fraction of the small reactive intermediates important in combustion

Table 84.1. Combustion chemistry intermediates detectable by laser-induced fluorescence.

Molecule	Excitation Wavelength (nm)	Molecule	Excitation Wavelength (nm)
H*	205	NCO*	440
C*	280	HCO*	245
O*	226	HCN	189
N*	211	HNO	640
S	311	NH_2*	598
OH*	309	C_3	405
CH*	413	C_2O	665
NH*	336	S_2O	340
SH*	324	SO_2*	320
CN*	388	NO_2*	590
CO*	280	HSO	585
CS	258	CS_2	320
NO*	226	N_3	272
NS*	231	NCN	329
SO*	267	CCN	470
O_2*	217	NH_3*	305
S_2*	308	NO_3	570
C_2*	516	C_2H_2*	220
$^1CH_2$*	537	CH_2O*	320

An asterisk denotes that LIF detection has been performed in a flame.

chemistry, so that LIF has the ability to characterize fairly completely a combustion reaction mechanism. It would be particularly desirable to add certain other species; noteworthy are HO_2, C_2H, CH_3, and triplet CH_2, although suitable fluorescing electronic transitions for these molecules are not currently known. In addition to the species listed in Table 84.1, there are many other combustion-related compounds that can be seen with LIF. These include metal atoms and their compounds, species present in specialized combustion situations such as boron or chlorine containing radicals, and some polyatomic, partially oxidized hydrocarbon molecules.

LIF may be employed in several ways to study flames. With a pulsed laser, very rapid time resolution is achievable. Pointwise, single shot measurements (usually of OH) in turbulent flames may be interpreted on a statistical basis for comparison with flame models incorporating simplified chemistry. Two dimensional planar imaging (usually of OH, CH, or NO) is accomplished by focusing the laser into a sheet of radiation which passes through the flame. Imaging the illuminated region at right angles onto a two-dimensional array produces an instantaneous snapshot of the distribution of the radical throughout the flame. This may be used to understand flow patterns under conditions of rapid time variation, such as turbulence or flame spread following ignition.

On the other hand, turbulent flows are far too complex for an investigation of fine details of the combustion chemistry. This is best accomplished via experiments in laminar flames, where LIF is used to obtain spatial profiles (one- or two- dimensional) of reactive species, made as a function of height above a burner surface. Absolute or relative profiles provide highly constraining tests of detailed models of flame chemistry. Operation at reduced P spreads out the active flame front region, furnishing excellent spatial resolution for these profiles. The T profile, used as input to the model, is measured using rotational excitation scans. The OH radical, present throughout much of any given flame containing hydrogen and oxygen, is an ideal LIF thermometer. LIF may also be used in a semiquantitative way, so that simply the detection of some species at an approximate concentration can reveal new information about the chemical mechanism [7]. An example is LIF detection of NS in flames containing minor amounts of N and S compounds, indicating the importance of this radical in linking the chemistry of NO_x and SO_x formation.

LIF measurements involve quantitative knowledge of molecular spectroscopic characteristics and collisional behavior. Identification of the absorbing molecule requires assignment of the vibrational and rotational structure of the electronic transition in question. For weak signals, particularly in the presence of a strong absorber/fluorescor (e.g., CH_3O in the presence of OH near 310 nm), it may be necessary to scan many lines to ensure identification. The fluorescence spectrum is also valuable since selective detection of fluorescence may be used to discriminate between two molecules absorbing at the same wavelength. In many cases, the detailed spectroscopy of molecules of interest is well established in the literature; in other cases, it may be necessary to make studies via flow cells. A recent example is an LIF study of the B–X ultraviolet system of the HCO molecule [8], a radical whose reactions are important in controlling the hydrogen atom concentration in hydrocarbon flames. In some cases (e.g., NCO) LIF spectroscopic studies in flames themselves have been useful, in that high rotational and vibrational levels are populated and thus readily accessed. In addition to spectral line and band identification, rotational line strengths and vibrational band transition probabilities (including possible effects of an electronic transition moment that varies with internuclear distance) provide the Einstein A and B coefficients needed for the data analysis. Accurate values are especially important in the determination of T via excitation scans, since systematic errors in oscillator strengths can lead to errors of hundreds of degrees, but are not discernible through statistical goodness-of-fit criteria. Such large errors in T render meaningless any attempts to compare measurements with predictions that include detailed chemical kinetics.

Of considerable consequence for LIF measurements

is an understanding of the collisional effects governing the fluorescence quantum yield. For example, in a flame at atmospheric pressure, approximately three of each thousand OH molecules excited by the laser will fluoresce; the remainder is removed nonradiatively by collisions with the ambient flame gases. Furthermore, to reduce interference or background, fluorescence detection is often accomplished using a filter or spectrometer with a bandpass encompassing only a portion of the total fluorescence. Vibrational and rotational energy transfer collisions of the excited radical with the surrounding gases may determine the fraction of emission into that particular bandpass.

Because of the extreme importance of the OH radical to combustion chemistry (and also in the atmosphere), its collisional behavior has been investigated extensively, although not yet completely. The results show definite quantum state and translational energy dependence of energy transfer and quenching cross sections. These are not only crucial to quantitative measurements of this radical, but they are also important in understanding molecular collision dynamics. OH is small enough that its interactions are amenable to quality *ab initio* calculations, at least with simple colliders like noble gases and H_2. Quantum scattering calculations may then be compared with experimental results; the state-to-state detail provided by this open shell radical furnishes valuable tests of those calculations. Furthermore, they may be compared with spectroscopic characteristics and half-collision dynamics of van der Waals complexes of OH with noble gases (and, in the future, molecular partners).

Quenching of the $A\ ^2\Sigma^+$ excited state of OH is the major factor determining the fluorescence quantum yield. Measurements have been made over a wide range of temperatures, from 200 to 2500 K, using cooled and room temperature flow cells, laser flash heated cells, flames, and shock tubes [9]. For nearly all collision partners the quenching cross section decreases with increasing T, i.e., increasing collision velocity. This shows that attractive forces are involved in the collision, and the form of the variation for most colliders can be explained in terms of a collision complex formation mechanism. The $A\ ^2\Sigma^+$ state of NO behaves similarly, but quenching of the $A\ ^2\Delta$ state of CH increases with T, showing that a barrier exists on the potential surfaces; the $A\ ^3\Pi$ state of NH exhibits more varied behavior.

Quenching also varies with rotational level for OH and NH, as does vibrational transfer ($v = 1 \rightarrow v' = 0$) for OH. This appears to be a dynamic, not energetic, effect, and has been ascribed to a rotational averaging of the effects of the highly anisotropic surface on which these polar hydrides interact with colliders. Vibrational transfer in OH has been studied in both the excited and ground electronic states. It is found to be much faster in the upper $A\ ^2\Sigma^+$ state for all colliders, by factors of 100 to 1000. This is perhaps related to the fact that,

according to *ab initio* calculations on OH-Ar, there is a much deeper well for interaction of the excited radical than in the ground state. Rotational transfer in OH has been shown to have some definite propensities, including parity conservation for some collisions with He that is reproduced in quantum scattering calculations.

Such a wealth of detail is not yet available for other free radical combustion intermediates, although they are beginning to be investigated. Of particular current interest is NO and the major precursor to its formation, CH. The details of quenching by a variety of colliders are probably known well enough that quantum yields for OH, NO, and CH may be calculated reasonably accurately, given the local temperature and major species flame composition at the point of measurement. There are enough data to compute quenching rates to within about 30 to 50% for many flames; this is partly because H_2O, when present, accounts for the majority of the quenching for all three radicals. When the local environment (especially T) is not known, the situation is worse. This would arise for an instantaneous two-dimensional image of a radical in a flame. Even so, the net quenching rate for OH rarely varies more than threefold through the entire flame, so useful semiquantitative information can still be obtained.

If absolute LIF measurements are desired, some means of calibration is necessary. Stable compounds (NO, CO) can simply be introduced into the flames or corresponding cold flows. Free radicals pose a greater problem. OH is often present in sufficient quantity that absorption measurements in a stable, laminar one-dimensional flame can be performed, but this is generally not possible for other species. The recently introduced technique of cavity ring down spectroscopy may alter this situation [10]. Burnt gases, in thermal equilibrium at a measured T, furnish a reliable source of known concentrations of OH, H, and O but generally not other radicals (which are consumed earlier in the flame).

Two-photon excitation is used for detection of atomic species and the CO molecule. The high laser flux needed for two-photon absorption, together with the uv wavelengths required for the pertinent transitions, often leads to problems with photochemical interferences. Sometimes these produce the same species that is being measured, e.g., dissociation of vibrationally excited O_2 at the wavelength of atomic oxygen detection, with subsequent excitation of the spurious O atom. In other cases, especially in complex hydrocarbon fuels, a parent compound or partially oxidized fragment may be excited directly or photolyzed to yield an emitting product. Excitation scans and measurements as a function of laser power are needed to discern and avoid these interferences.

84.3.3 Degenerate Four-Wave Mixing

DFWM is a nonlinear process like CARS; but like LIF, it operates on real transitions. It combines some of the attributes of both methods: production of a coherent signal beam that can be spatially filtered to discriminate against background, and high sensitivity so that trace radical species may be detected [11]. It depends on $\chi^{(3)}$, as does CARS, and can be used in a broadband mode. Like LIF, it can produce two-dimensional images, with the proper laser sheet arrangement. DFWM adds some of its own advantages, especially a Doppler-free nature enhancing the spectral resolution and thus the molecular selectivity. Three laser beams are used to generate the DFWM signal, all the same frequency, and tuned to an absorption transition of the species of interest. Two pump beams create an interference pattern in the medium, and a probe beam scatters off the resulting grating to form the signal. The interference pattern may be in the ground/excited state populations or may exploit polarization phenomena. Rapid quenching of the excited state may produce a thermal grating in the medium; the probe beam can then scatter from the resulting variation in the index of refraction. DFWM depends only on absorption and may be used for sensitive detection in a nonfluorescing or poorly fluorescing case (e.g., a predissociative state or ir vibrational transition).

The interpretation of DFWM signals to obtain molecular concentrations or temperatures requires knowledge of the underlying physical phenomena. Because they are governed by $\chi^{(3)}$, the DFWM amplitudes interfere, and an understanding of molecular collisions, motion, and the effects of the laser power are needed for accurate modeling of DFWM spectra. Collisional effects are of particular interest due to the Doppler-free character of the technique. Signal intensities can be dramatically influenced by line broadening due to quenching, energy transfer, and dephasing collisions. The magnitude of the influence depends sensitively on the degree of optical saturation. A cohesive picture of these effects is currently being sought through a combination of measurement and theory. Furthermore, nearly all current theoretical treatments assume single-mode lasers; however, real lasers contain intensity and frequency fluctuations within each laser pulse. The effects of these fluctuations on the nonlinear wave-mixing process must be accounted for in a proper description of DFWM.

A likely important application will be for nonfluorescing transitions; furthermore, because collisions affect the signal only during the laser pulse, rapid (ps) measurements may provide excellent sensitivity under very high P conditions.

In the future, we will see further expansion of applications of CARS and LIF to combustors of practical interest, and LIF will be extended to a larger number of molecules of chemical importance. DFWM will find increased use on both laboratory and engineering scales; detection of nonfluorescing molecules in the ir is likely to become an important application of this method. The development of tunable solid state lasers and the extension of their available wavelengths into the uv through frequency conversion will enable the construction of rugged, more easily operated diagnostic systems, yet furthering the range of applications of laser combustion diagnostics.

The articles and books [7, 9, 11–18] are of a review or overview nature. In addition, archival articles illustrating the state of the combustion field can be found in the biennial International Symposia volumes published by the Combustion Institute, Pittsburgh.

REFERENCES

1. R. J. Kee, F. M. Rupley, and J. A. Miller, Sandia National Laboratory Report SAND89-8009, 1989.
2. W. Gardiner *et al.*, in *Proceedings of the 1995 International Gas Research Conference, Gas Research Institute, Chicago, 1995*.
3. H. Reisler, M. Noble, and C. Wittig, in *Molecular Photodissociation Dynamics*, edited by M. N. R. Ashfold and J. E. Baggot (Royal Society of Chemistry, London, 1987), p. 139.
4. A. H. Zewail, Science **242**, 1645 (1988).
5. X. Yang *et al.*, J. Phys. Chem. **97**, 3944 (1993).
6. J. Warnatz, in *Twenty-Fourth Symposium (International) on Combustion* (The Combustion Institute, Pittsburgh, 1992), p. 553.
7. D. R. Crosley, Comb. Flame **78**, 153 (1989).
8. A. D. Sappey and D. R. Crosley, J. Chem. Phys. **93**, 7601 (1990).
9. D. R. Crosley, J. Phys. Chem. **93**, 6273 (1989).
10. G. Meijer *et al.*, Chem. Phys. Lett. **217**, 112 (1994).
11. R. L. Farrow and D. J. Rakestraw, Science **257**, 1894 (1992).
12. D. R. Crosley, in *Turbulence and Molecular Processes in Combustion*, edited by T. Takeno (Elsevier, New York, 1993), p. 221.
13. A. C. Eckbreth, *Laser Diagnostics for Combustion Temperature and Species* (Abacus Press, Cambridge, 1988).
14. *Combustion Chemistry*, edited by W. C. Gardiner (Springer-Verlag, New York, 1984).
15. L. P. Goss, in *Instrumentation for Flows with Combustion*, edited by A. M. K. P. Taylor (Academic Press, New York, 1993), p. 251.
16. K. Kohse-Höinghaus, Prog. Energy Comb. Sci. **20**, 203 (1994).
17. A. Linan and F. A. Williams, *Fundamental Aspects of Combustion* (Oxford, New York, 1993).
18. J. A. Miller, R. J. Kee, and C. K. Westbrook, Ann. Rev. Phys. Chem. **41**, 345 (1990).

85

Surface Physics

E. T. Jensen

Department of Physics, University of Windsor, Windsor, Ontario N9B 3P4, Canada.

85.1 INTRODUCTION

This chapter describes various applications of atomic and molecular physics to phenomena that occur at surfaces. Particular attention is placed on the application of electron- and photon-atom scattering processes to obtain surface specific structural and spectroscopic information.

The study of surfaces and interfaces touches on many fields of pure and applied science. In particular there are applications in the fields of semiconductor processing, thin film growth, catalysis, corrosion and fundamental physics in two-dimensions. A number of recent texts cover surface physics in general [1–4] as well as specific areas such as experimental techniques [5], surface electron spectroscopies [6,7], and the application of synchrotron radiation to surface science [8,9]. Also a number of book series that deal with areas of particular interest are published at regular intervals, such as surface chemistry [10], surface vibrations [11] and stimulated desorption processes [12].

85.2 LOW ENERGY ELECTRONS AND SURFACE SCIENCE

To obtain information specific to the first few layers of atoms at the surface of a solid, techniques must be devised that discriminate between signals from the surface and from the bulk of the material. If a solid has dimensions of ~ 1 cm^3, then only approximately one atom in 10^7 is located at the surface. This makes the application of bulk techniques to the study of surfaces (e.g. x-ray diffraction [13]) problematic.

There are several experimental approaches to achieving surface sensitivity. If the probe used is an atom or molecule that can be scattered from or desorbed from the surface, then these species can be analyzed by techniques such as mass spectrometry or resonant ionization. A more recent innovation has been the use of scanning probe microscopies [14] (e.g. the scanning tunneling microscope and its variants) which exploit a surface sensitivity such as the surface valence electron density. How-

Figure 85.1. The variation of the inelastic mean free path with electron kinetic energy ('universal curve'). Based on Briggs and Seah [15].

ever the most common surface analytical techniques use the intrinsic surface sensitivity of low energy (10–2000 eV) electrons.

The surface specificity of low energy electrons arises from the very short inelastic mean free path (MFP) for these electrons in a solid. This property can be seen in Fig. 85.1 which plots measured values of the inelastic MFP for a number of materials and electron energies between 1 and 2000 eV. The curve is called 'universal' because the same general trend of short inelastic MFP is observed for nearly all materials [15]. The dominant energy loss mechanisms are valence band excitations (plasmons and electronic excitations), and since most materials have similar valence electron densities, the resultant inelastic scattering MFP is to a good approximation, material independent.

The very short inelastic MFPs for low energy electrons has the result that any that escape from the solid without having undergone inelastic scattering can only have originated very close to the surface, usually within a few atomic layer spacings. Many of the surface analysis techniques described in this chapter use this surface sensitivity to obtain surface structural and spectroscopic information. The same techniques are easily applicable to conducting and semiconducting materials, and can also be applied to insulating materials, if the charging effects are dealt with in some manner.

85.3 ELECTRON-ATOM INTERACTIONS

85.3.1 Elastic Scattering: Low Energy Electron Diffraction (LEED)

The elastic scattering of low energy (20–500 eV) electrons at surfaces is historically important in physics as

the experiments of Davisson and Germer provided early experimental evidence of the wave nature of electrons. After these initial experiments, the technique was largely unused until the advent of cleaner ultra-high vacuum systems and surface preparation techniques in the 1960's. LEED is one of the most important and widely used surface characterization techniques due to its surface sensitivity and wide utility [5].

Diffraction from a two-dimensional net of scatterers results in a two-dimensional array of reciprocal lattice 'rods' oriented normal to the surface. All kinematically allowed rods intersect the Ewald sphere at all energies— unlike the case in bulk x-ray diffraction. The short inelastic MFP for scattering low energy electrons yields the surface sensitivity of LEED for a crystal surface. The kinematic theory used in x-ray diffraction is not directly applicable to LEED since the low energy electrons can undergo several elastic collisions in the surface region. The elastic MFPs for low energy electrons are comparable in magnitude to the inelastic mean free paths shown in Fig. 85.1. This 'multiple scattering' does not affect the positions of the diffraction beams, but does alter their intensities due to interference effects.

Kinematic analysis is sufficient to determine the diffraction beam positions for a proposed structure, and this is the most common use for LEED. A diffraction pattern is often sufficient to determine the surface periodicity and unit mesh size. However determining the surface crystal basis from the intensities of the diffracted LEED beams requires moderately sophisticated calculations to be performed for proposed structures.

Quantitative LEED generally compares measured 'I(V) curves' (the intensity I of a particular diffraction beam as a function of electron energy measured in volts) with a calculated I(V) profile for a proposed surface crystal structure. These calculations determine the propagating electron wave functions Ψ_{LEED} that take into account electron-ion core scattering cross sections and phase shifts, available multiple scattering pathways, inelastic scattering cross sections and Debye-Waller effects. The structural parameters are varied systematically until agreement can be reached with the experimental I(V) curves [16, 17]. These calculations can yield surface atom positions to better than 0.1 Å vertically and 0.2 Å horizontally. Over 1000 surface structures have been determined using LEED and associated techniques [18].

Clean surfaces often have a different crystallography than simple termination of the bulk crystal structure. Due to the absence of neighbors above the surface, the surface atoms can undergo both relaxation (change in interlayer spacing) and reconstruction (changes in periodicity and bonding) [2]. While relaxation and reconstruction do occur on all types of surfaces, the reconstructions found on semiconductors are most striking. The strong covalent bonds that are broken when a semiconductor is cleaved give rise to high energy 'dangling bonds'. The surface energy is minimized by having the surface reconstruct to reduce the total number of these bonds. The resultant structure is formed through a balance between eliminating as many dangling bonds as possible and the resultant stress caused in other bonds due to the displacements of the surface atoms. One example of this is the clean Si(111) surface, in which 49 surface atoms form a new periodic arrangement to make up the reconstructed Si(111)–(7×7) unit cell [2]. The number of dangling bonds is thereby reduced from 49 to 19.

85.3.2 Inelastic Scattering: Electron Energy Loss Spectroscopy

Low energy surface excitations such as phonons, plasmons and electron-hole pair excitations can be studied using inelastic electron scattering. High resolution electron energy loss spectroscopy (HREELS) [19] uses incident electron beam energies of 1–300 eV with energy resolutions of 1–10 meV. Inelastic scattering phenomenon are divided into three types: dipole, impact and resonant scattering. Surface vibrations are also studied using infra-red absorption spectroscopy and inelastic atomic beam scattering techniques [20].

Dipole Scattering

In dipole scattering HREELS, the incident electron ($1 < E_i < 20$ eV) undergoes a long range Coulomb interaction with dipole fields associated with a dynamic dipole moment. The main characteristic of dipole scattering is that the inelastically scattered electrons have an angular distribution that is strongly peaked (width $\Delta\theta \approx \hbar\omega/E_i$) close to the specular direction ($\theta_i = \theta_f$). In this limit there is no momentum exchange with the surface vibrational mode (i.e. $\vec{k}_\parallel = 0$ in the bandstructure), so dipole scattering HREELS is similar to infrared spectroscopy in that only ir active vibrational modes at the zone center can be studied. At the surface of a conductor, dielectric screening guarantees that only ir active modes having a component perpendicular to the surface are visible—modes having dipole moments parallel to the surface are screened by the surface and so cannot be excited by dipole scattering. This 'quasi-selection rule' allows the site symmetry for some systems to be decided [19]. Dipole scattering is most commonly used in the study of vibrations of molecules on surfaces, which often have dipole-active vibrational modes that are both intrinsic and caused by adsorption (i.e. frustrated translations and rotations).

Impact Scattering

In impact scattering, the incident electron samples the short-range interatomic potential, and is not restricted to dipole-active vibrational modes [19]. Higher electron energies are used ($30 < E_i < 300$ eV) since the cross section for impact scattering generally increases with energy. The inelastically scattered electrons are distributed at all angles, and so there can be exchange of parallel momentum ($\vec{k}_\parallel$) with the surface, allowing the mapping of surface excitation bandstructure $\hbar\omega(\vec{k}_\parallel)$. The cross sections for impact scattering HREELS can be calculated using an approach similar to LEED, but now considering the normal displacements of atoms from equilibrium positions. The scattering potential $V(\vec{r}, [\vec{R}])$ (where $[\vec{R}]$ is the set of position vectors of the N atoms in the surface region) can be expanded in terms of the normal displacements $\vec{\mu}_i$ from equilibrium according to

$$V(\vec{r}, [\vec{R}]) = V(\vec{r}, [\vec{R}_0]) + \sum_{i=1}^{N} (\nabla_{\mu_i} V(\vec{r}, [\vec{R}]))|_{[\vec{R}_0]} \cdot \vec{\mu}_i + \cdots . \quad (85.1)$$

The first term is responsible for elastic scattering (LEED) and the second term is the dominant term for impact scattering. The inelastic cross section for mode i is then [21]

$$\frac{d\sigma}{d\Omega} = |\langle \Psi_{\text{LEED}}^*(\vec{k}_f)|\frac{\partial V}{\partial \mu_i}|\Psi_{\text{LEED}}(\vec{k}_i)\rangle|^2 \quad (85.2)$$

using electron wave functions calculated using the same formalism as LEED. The symmetry properties of this matrix element can be used to determine the polarization direction of surface vibrational modes by searching for systematic absences in the inelastic intensity in particular high symmetry direction of the surface [21]. Impact scattering studies are most commonly made in the study of surface phonon bandstructure and the vibrational modes of adsorbed molecules that are not dipole-active.

Resonant Scattering

Resonant electron scattering at surfaces [22, 23] is usually applied to the study of adsorbed molecules, and has much in common with resonant scattering from gas-phase atoms and molecules (see Chap. 61). In resonant scattering, the incident electron combines with a target molecule to form a short-lived molecular ion, which subsequently decays and can leave the molecule vibrationally or electronically excited. The cross sections for this process have a typical profile. For example, in a *shape resonance* the formation of the temporary negative ion intermediate corresponds to adding the incident electron to a particular unoccupied orbital of the target molecule. The study of the angular dependence of resonance scattering cross sections [23] can yield information on the orientation of molecules on surfaces, since the electron capture and emission cross sections are fixed by the molecular orientation and the resonance symmetry.

85.3.3 Auger Electron Spectroscopy

Auger electron spectroscopy (AES) is one of the most widely used surface science techniques due to its chemical and surface sensitivity [5, 15, 24]. Core holes are created in near surface atoms using a high energy electron beam (2–5 keV, 1–100 μA) or less commonly, an x-ray source. For these low binding energy core holes, the Auger decay mode is highly probable (see Sect. 59.1.2). Although the Auger energy and lineshape can give spectroscopic information, in surface science this is seldom used—rather it is the chemical fingerprint of the atoms which is of interest.

Auger electron spectroscopy is a surface sensitive technique by virtue of the low kinetic energy (50–1000 eV) of the emitted Auger electrons. Auger electrons from atoms more than a few Ångstroms below the surface are inelastically scattered and so not detected by the energy selective detector. The kinetic energy of the Auger electrons is

$$\mathcal{T}_\mathrm{e} = E_A - E_B - E_C - U \tag{85.3}$$

where E_A is the binding energy of the initial core electron and E_B, E_C are the binding energies of the other electrons (one or both are valence levels) involved in the Auger process. Energy shifts and relaxation are accounted in the term U which includes hole-hole interactions and atomic and solid state (dielectric) screening of the holes, and hence can be sensitive to the local chemical environment.

Auger spectra are typically obtained in derivative mode [i.e. $N'dN(E)/dE$] to separate the Auger transitions from the secondary electron background, and comparison is made to reference spectra [25]. The raw chemical sensitivity of AES is very high, and surface concentrations of ~1% of many common chemical species can be detected. Semiquantitative measurements may be made by comparison to these reference spectra, but such comparisons only give atomic concentrations within a factor of 2 (or worse) since the measured AES signal can be modified by a number of factors. More precise quantitative measurements can be made by calibration of the AES intensity for the atomic constituents at a surface [5]. The AES sensitivity to atoms A on a clean substrate (atoms B) can be determined if the absolute quantity of A can be established by some other means.

The chemical sensitivity of AES can also be exploited as a form of chemical microscopy (scanning Auger microscopy) since the exciting electron beam that is used may be focused to a very small size. The Auger signal from the small target volume can be analyzed for specific chemical components. By rastering the incident electron beam, a chemical map of the surface can be made.

85.4 PHOTON-ATOM INTERACTIONS

85.4.1 Ultraviolet Photoelectron Spectroscopy (UPS)

Photoelectron spectroscopy (PES) (see Sect. 59.1) has been historically divided into uv photoelectron spectroscopy for low photon energies (generally the study of valence electron states) and x-ray photoelectron spectroscopy (study of core electron levels). The use of low energy photons ($5 < h\nu < 50$ eV) for PES of solids has the advantage that the photons have a negligible momentum ($\vec{k} \approx 0$). This allows straightforward band mapping since the transitions are vertical in momentum space:

$$h\nu = E_f(\mathbf{k} + \mathbf{G}) - E_i(\mathbf{k}) \tag{85.4}$$

where $\mathbf{k}$ is the electron state wavevector and $\mathbf{G}$ is a reciprocal lattice vector.

Most UPS studies are done using angle-resolved photoelectron detection (also called angle resolved photoelectron spectroscopy or ARPES). The kinetic energy $\mathcal{T}_\mathrm{e}$ and emission angle θ of the photoelectrons are measured, allowing the initial state binding energy and momentum parallel to the surface ($\mathbf{k}_\parallel$) to be determined from

$$k_\parallel = \sqrt{\frac{2m\mathcal{T}_\mathrm{e}}{\hbar^2}} \sin\theta \,. \tag{85.5}$$

The mean free path of the photoelectrons from valence levels allows both surface and bulk electron states to be studied. Separation of the surface from bulk bands can be accomplished in several ways [5]. Since a crystal surface has two-dimensional symmetry, only the $\mathbf{k}_\parallel$ component of momentum is a good quantum number for the surface states. Also, in transporting the surface or bulk state photoelectron of momentum $\mathbf{k}$ through the surface to the detector, only the $\mathbf{k}_\parallel$ component is conserved. A bulk state disperses in three-dimensions, requiring knowledge of $\mathbf{k} = \mathbf{k}_\parallel + \mathbf{k}_\perp$. The magnitude of $\mathbf{k}_\perp$ cannot be determined unless the inner potential (change in potential normal to the surface) is known by some other means. This property allows bulk and surface states to be distinguished since surface states remain fixed in energy for a constant $\mathbf{k}_\parallel$, while bulk states disperse if $\mathbf{k}_\perp$ is changed while $\mathbf{k}_\parallel$ is kept constant. This is done by measuring the photoelectron spectrum for a range of uv photon energies at a fixed value of $\mathbf{k}_\parallel$ (often at $\theta = 0$ so $\mathbf{k}_\parallel = 0$). Bulk states disperse since $\mathbf{k}_\perp$ varies as the photon energy is changed, but the surface states remain at a fixed binding energy in the photoelectron spectrum.

The photoelectron transition matrix element between initial and final states $|i\rangle$ and $|f\rangle$ due to the incident photon vector potential $\mathbf{A}$ is

$$I \propto |\langle f|\mathbf{A}\cdot\mathbf{p} + \mathbf{p}\cdot\mathbf{A}|i\rangle|^2 \approx |\langle f|2\mathbf{A}\cdot\mathbf{p}|i\rangle|^2 . \quad (85.6)$$

The spatial variation of $\mathbf{A}$ near the surface (the surface photoeffect) can be neglected, although this is not strictly valid at these low photon energies [26].

Due to the low photoelectron kinetic energies in most UPS work, precise calculation of photoelectron spectrum intensities is rather difficult due to multiple scattering and phase shifts sensitive to valence electrons. However, the initial electron state symmetries can be determined using the symmetry properties of the matrix element [Eq. (85.6)]. Selection rules allow the state symmetries to be determined by measuring the photoemission spectrum along high symmetry directions of the surface using polarized uv radiation [5].

The valence electron states of adsorbed molecules can also be studied using UPS. Peaks in the photoelectron spectrum due to valence levels of adsorbed molecules tend to have larger linewidths (≈ 1 eV typically) than in the gas phase due to solid state and instrumental effects, so vibrational structure is seldom resolved. The positions of the molecular valence states are shifted in energy due to the surface work function and solid state relaxation (dielectric screening) effects. In addition to these rigid shifts, chemical shifts are also observed due to bonding (chemisorption) interactions between the molecule and surface. The molecular character of the valence orbitals is often retained from the gas phase, so the shifted levels can be identified by using the symmetry properties of the photoemission matrix element. For example, photoemission from gas phase CO shows three valence states: 5σ, 1π and 4σ in order of increasing binding energy. A series of photoemission spectra [27] for gas phase CO, solid CO and CO chemisorbed on several transition metal surfaces (in order of increasing CO–surface binding energy) is shown in Fig. 85.2. For solid CO and weakly chemisorbed CO/Ag(111) all three valence orbitals of molecular CO are well resolved, though the adsorbed CO spectrum is shifted rigidly in energy by relaxation effects. Since CO is chemisorbed more strongly on Cu(111) and Pd(111), the 1π and 5σ valence states shift relative to one another. The strongly overlapping 1π and 5σ states observed for chemisorbed CO can be individually resolved by using the photoemission selection rules and linearly polarized uv radiation [27]. The origin of the chemical shift between the 5σ and 1π levels has been ascribed to bond formation involving the CO 5σ level and σ-symmetry d-electron states on the transition metal surfaces [2].

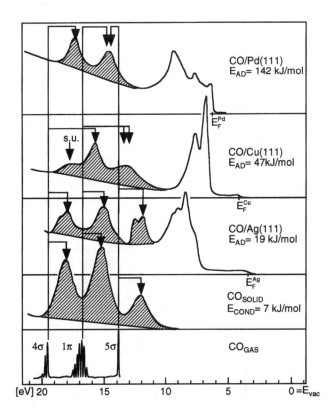

Figure 85.2. A set of UPS spectra for CO adsorbed on various surfaces, as well as gas phase and solid CO. The peak labeled 'su' is a shake-up satellite peak. From [27].

85.4.2 Inverse Photoemission Spectroscopy (IPES)

Inverse photoemission spectroscopy is properly classified as an incident electron technique but is included here as a natural companion to UPS. IPES utilizes low energy electrons (5–50 eV) incident on a surface. Transitions from a high-lying initial electron state in the continuum to a lower unoccupied state cause a uv photon to be emitted. By detecting the emitted photon intensity as a function of energy, the joint density of states is measured [7]. IPES is used to study the unoccupied portion of surface and bulk bandstructure, particularly the region between the Fermi level and the vacuum level, which is difficult to access by other means. This allows the study of unoccupied states of adsorbed molecules (e.g. antibonding molecular levels) as well as intrinsic surface states such as the Rydberg-like states of electrons trapped in the image potential at the surface [7].

85.4.3 X-Ray Photoelectron Spectroscopy (XPS)

X-ray photoelectron spectroscopy allows the study of the energy of atomic core levels via the Einstein photoelectric equation $\mathcal{T}_e = h\nu - E_b$, where $\mathcal{T}_e$ is the kinetic energy of the photoemitted electron and E_b is the binding energy of the core level. Atoms bound in different chemical environments (e.g. at particular sites on a surface) experience different chemical shifts, and so are measured at slightly different binding energies. This sensitivity to chemical environment allows XPS to characterize the different types of binding sites for similar atoms, and measure their abundance by measuring the relative intensities of XPS emission from different species.

Shifts in $\mathcal{T}_e$ arise from two sources: intra-atomic and inter-atomic relaxation shifts. The intra-atomic relaxation E_a is due to screening of the core hole by other electrons in the emitter atom. The inter-atomic relaxation E_r is important in solids (particularly for metals) and is due to screening of the core hole by the dielectric response of the surrounding medium. These relaxation shifts (on the order of a few eV) tend to increase the kinetic energy of the emitted photoelectron, so the kinetic energy in the adiabatic limit is

$$\mathcal{T}_e = h\nu - E_b + E_a + E_r. \tag{85.7}$$

Since photoemission is a rapid process, nonadiabatic processes lead to *shakeup* and *shakeoff* features in which other electrons are excited to higher energy levels, causing the photoelectron to have lower kinetic energy [5] (see Sect. 60.4.4). It is also possible to excite discrete excitations of the solid, such as plasmons and a continuum of low energy electron-hole excitations, causing the XPS peak to have an asymmetric lineshape. The overall XPS distributions of $\mathcal{T}_e$ thus contain contributions from the adiabatic channel, and lower energy photoelectrons in a series of discrete peaks or a continuum from nonadiabatic processes.

The chemical abundance of a species can be determined from the sum rule that the total XPS cross section is proportional to the sum of the adiabatic peak and all the shake-up and shake-off components. However it is usually only convenient to measure the intensity of the adiabatic peak. If comparisons are made between chemical species in different chemical environments, the shake-up and shake-off intensities may differ. Hence the adiabatic channel intensities might not reflect the true abundance. Very often this problem can be minimized by careful calibration, and chemical analyses can be made to an accuracy of a few percent.

The surface sensitivity of XPS is due to the short inelastic MFP for the photoelectrons, which can be made to have energies in the range 10–1000 eV by appropriate choice of the x-ray wavelength. The XPS signal is proportional to $\exp[-z/(\lambda \cos\theta)]$ where z is the depth of the emitter, λ is the inelastic mean free path of the photoelectron and θ is the angle measured from the surface normal.

The popularity of XPS as a surface analysis technique is due to the availability of convenient and sufficiently

Figure 85.3. XPS spectra of the $4f_{7/2}$ states of W(111) and Ta(111). Contributions from the bulk (b) and surface (S1,S2) atoms can be distinguished, and the surface atom peaks are shifted by the adsorption of hydrogen. From [28].

intense monochromatic x-rays from lab sources (usually Al and Mg K_α x-ray lines at 1486.6 and 1253.6 eV). The increasing availability of continuously tunable x-rays from synchrotron radiation sources allows improved measurements due to the ability to tune the x-rays to slightly above threshold, where the XPS cross-sections are maximized, the inelastic MFP is shorter (hence more surface specific) and electron monochromators can operate with higher resolution [5].

As an example of the chemical sensitivity of XPS, photoelectron spectra of the $4f_{7/2}$ core levels of W(111) and Ta(111) are shown in Fig. 85.3. The spectra show discrete peaks from both bulk and surface atoms. The binding energies of the surface atoms are affected by the adsorption of hydrogen.

XPS can also be used to give local structural information due to elastic scattering of the photoemitted electron in the region near the emitter [5]. One form of this is the use of forward scattering from a buried emitter. A high energy XPS photoelectron is focused in the direction of nearby atoms due to a lower effective potential close to the atomic core. Angular scanning of the XPS detector will then detect a more intense signal along the bond axis. This method has proved useful in studying the orientation of adsorbed molecules and the structure of hetero-epitaxial thin films [29].

85.4.4 X-Ray Absorption Methods

X-ray absorption methods [8] measure the decay of the core hole rather than the intensity of emitted photoelectrons, since that XPS process is complicated by a number of possible final state processes, as discussed above. To obtain surface sensitivity, the x-ray absorption is measured using a low energy electron emitting channel such as an Auger electron emission or the total electron yield, which are proportional to the overall x-ray absorption. These methods require an intense and tunable source of monochromatic x-radiation near the excitation edge for the core level of a particular atom, and so are usually performed using synchrotron radiation. Synchrotron sources have the additional benefit of linearly polarized light, which is crucial for NEXAFS and useful for SEXAFS discussed next.

SEXAFS: Measurement of Bond Lengths

The surface extended x-ray absorption fine structure (SEXAFS) technique is most commonly used to measure the bond lengths for atoms adsorbed on a surface. SEXAFS utilizes the elastic backscattering of the emitted XPS photoelectron from nearby atoms that surround the emitter. Elastically scattered waves arrive back at the emitter and add coherently (constructively and destructively) to the outgoing wave, thus modifying the matrix element for the transition to the final state [30]. Experimentally, the cross section $\sigma(h\nu)$ for x-ray absorption above the threshold photon energy is modified by an oscillatory structure. The 'atomic' contribution to the absorption cross section can be removed using

$$\chi = \frac{\sigma - \sigma_0}{\sigma_0}, \tag{85.8}$$

where σ and σ_0 are the measured surface and free atom x-ray absorption cross sections. If only single scattering events are involved in backscattering to the emitter then the fine structure function χ is

$$\chi(k) = \sum_{i=1}^{N} A_i(k) \sin[2kR_i + \phi_i(k)], \tag{85.9}$$

where k is the magnitude of the photoelectron wavevector and the sum is done for N 'shells' of atoms surrounding the emitter. The distance R_i is the radius of the ith shell and $2kR_i$ is the associated phase factor for the backscattered photoelectron, with an amplitude $A_i(k)$. The phase shift $\phi_i(k)$ is required due to the backscattering path through the potential surrounding the emitter and scattering from the atom at R_i. If the ϕ_i could be ignored, then a simple Fourier transform of $\chi(k)$ would reveal the radial distribution function and the bond lengths R_i [30]. In practice, the phase shifts cannot be neglected, but very

often these can be found by studying chemically similar systems in which the bond lengths are known. The phase shifts for photoelectrons well above the absorption edge (having kinetic energies greater than ~ 50 eV) are dominated by the atomic ion cores, and so are not sensitive to the valence electronic structure. The phase shifts can also be calculated in a straightforward way since this problem is essentially the same as done in LEED multiple scattering calculations.

The amplitudes $A_i(k)$ due to scattering from shell i at distance R_i from a point source emitter are [30]

$$A_i(k) = \frac{N_i^*}{kR_i^2} |f_i(\pi, k)| \exp(-2\langle u^2 \rangle_i k^2) \exp(-2R_i/\lambda),$$

$$\tag{85.10}$$

where N_i^* is an effective number of atoms and $\langle u^2 \rangle_i$ is the mean-square displacement of the atoms in shell i. The backscattering amplitude $|f_i(\pi, k)|$ has been separated from its phase factor $\phi_i(k)$ in Eq. (85.9). The inelastic mean free path for the photoelectrons reduces the contribution from successive shells by a factor $\exp(-2R_i/\lambda)$, and it is this term that allows the kinematic single-scattering approach to be used. Multiple scattering paths involve longer trajectories and so are more strongly attenuated by this exponential factor. If the near-edge energy region of the absorption cross section is not included in the analysis, then the scheme outlined in Eqs. (85.9) and (85.10) is reasonable. The near-edge region (within ~ 50 eV of the absorption edge) is troublesome, not only because of multiple scattering, but also because the phase shifts $\phi_i(k)$ for low energy photoelectrons are more sensitive to valence electron distributions, and so are more sensitive to the details of the local chemical environment.

The SEXAFS method is illustrated with the data of Fig. 85.4 for iodine adsorbed on Cu(111). Panel (a) shows the measured x-ray absorption $\sigma(h\nu)$ for both the surface system Cu(111)–I and bulk CuI. The extracted $\chi(k)$ and their Fourier transforms are shown in (b) and (c). The Cu–I nearest neighbor bondlength is clearly shown by the peak in (c), and is found to be 0.07 Å longer than in bulk CuI. SEXAFS is most commonly applied to atomic adsorption systems since molecular systems are difficult to analyze, as they can contain several similar bond lengths, and the shells containing different atoms are difficult to model.

NEXAFS: Molecular Orientation at Surfaces

In SEXAFS, the x-ray absorption close to the excitation edge is avoided due to the problems of multiple scattering, and phase shifts that are chemically sensitive to valence electrons. The study of near-edge x-ray absorption fine structure (NEXAFS) can avoid these difficulties by using only the symmetry properties of the transition

Figure 85.4. (a) X-ray absorption data taken near the iodine edge for bulk CuI and Cu(111)–I. (b) The extracted fine structure function $\chi(k)$, shown multiplied by k^2 to enhance the high k structure. (c) The Fourier transform of $\chi(k)$ showing the location of the first shell of Cu atoms from the I emitters. (d) Back-transformed $k^2\chi(k)$ after applying a filter to extract the nearest neighbor data. From [31].

Figure 85.5. Near-edge x-ray absorption data from the C K-edge from CO adsorbed on Ni(100) as function of incident photon angle θ. The molecular π (peak A) and σ (peak B) resonances are observed. The polarization dependence of the absorption allows the CO orientation to be determined. From [33].

matrix element without concern for the absolute amplitudes [32]. For isolated molecules, unoccupied molecular states having σ or π symmetry are very commonly found close to or just below the threshold for photoemission. These 'molecular resonances' often remain for adsorbed molecules, and the symmetry properties of the absorption intensity $I \propto |\langle f|\mathbf{A}\cdot\mathbf{p}|i\rangle|^2$ can be used to determine the molecular orientation.

For the overall matrix element to be nonzero, it must be totally symmetric. Using linearly polarized synchrotron radiation, the direction of polarization $\mathbf{A}$ can be varied by rotating the crystal. For example, the K-edge x-ray absorption of CO on the Ni(100) surface in Fig. 85.5 shows final state resonances A (π-symmetry) and B (σ-symmetry). The intensity of these features depends of the direction of polarization of the incident x-rays. X-rays with a polarization vector parallel to the surface ($\theta = 90°$) strongly excite the π-symmetry absorption while the σ-symmetry absorption is absent. For grazing incidence x-rays polarized normal to the surface, the σ-resonance is prominent. From dipole

selection rules, this polarization dependence of the x-ray absorption is evidence that the CO molecule is adsorbed with its bond perpendicular to the plane of the Ni(100) surface.

The σ-resonance final state in NEXAFS corresponds to multiple scattering of the photoelectron along the bond axis. The overall phase shift for this final state is approximately

$$\int \sqrt{E - V(r)}\,dr \approx \sqrt{E - V(r)}\,R = \text{const.} \quad (85.11)$$

where R is the bond length. This sensitivity of the final state phase shift to bond lengths allows the σ-resonance energy to be used as a measure the molecular bond length [34]. A plot of the σ-resonance energy vs. $1/R^2$ shows a linear relationship, as is found in the case of simple hydrocarbons in the gas phase and adsorbed on a Cu(100) surface [32].

85.5 ATOM-SURFACE INTERACTIONS

85.5.1 Physisorption

The binding interactions between atoms and surfaces can be classified as physisorption (long range attractive dispersion forces) and chemisorption, in which chemical bonds are formed. The long-range dispersion force between a polarizable atom and a conducting surface give rise to the leading term of the van der Waals potential $V(z) \propto -1/z^3$. At smaller atom-surface separations, the location of the reference image plane needs to be included, resulting in an attractive potential $V(z) = -C_v/|z - z_i|^3$ where z_i is the distance from the last atomic plane to the image plane, typically 2–3 Å. In principle, the constant C_v is calculable from the dielectric properties of the substrate and the atomic polarizability, but experiments have found values of C_v 40% less than expected [35]. The cause of this discrepancy has not been clarified, although contributions from surface roughness have been suggested.

At larger atom-surface separations, retardation effects must be taken into account in the dispersion interaction (the Casimir-Poldar force, Sect. 76.3.3), and here theory predicts a $1/z^4$ interaction potential. This form for the potential has been confirmed experimentally [36].

As the atom approaches the surface, wave function overlap eventually causes repulsion. In the absence of chemical bond formation, the repulsive potential is simply proportional to the surface electron density $n(z)$, resulting in the overall physisorption potential

$$V(z) = Kn(z) - \frac{C_v}{|z - z_i|^3}. \tag{85.12}$$

The surface charge density $n(z)$ decreases exponentially above the surface [2], and the constant K can be determined to reasonable accuracy by an effective medium theoretical approach [37]. This shallow physisorption potential well has been studied experimentally in atomic beam (often He) scattering experiments. Under certain scattering conditions, it is possible for an incident atom to be 'selectively adsorbed' on the surface by making transitions into and out of bound states of the potential well of Eq. (85.12). The incident atom with wavevector **k** is diffracted by a surface reciprocal lattice vector $\vec{G}_{hk}$ into a bound state E_n of the potential well [2]. The diffracted atomic beam intensities due to scattering via selective adsorption show strong variations which can be related to the quantum numbers (n, h, k), and give information on the shape of the physisorption potential well.

85.5.2 Chemisorption

Many atoms and molecules will chemisorb when brought sufficiently close to a surface, forming chemical bonds (covalent or ionic) that are much stronger than the physisorption bond [3]. In chemisorption, there is charge transfer between the adsorbate and the surface, modifying the electronic structure of both. Valence electronic levels of the adsorbate are shifted in energy and also broadened by resonant interactions with the delocalized valence electrons at the surface (e.g. free electron-like s-p states). For many transition metal surfaces, interactions with the more localized d-states are also important.

The adsorbate-surface bond energies are most commonly studied by thermal desorption spectroscopy (TDS) [38,39]. In this method, the adsorbate covered surface is heated using a linear temperature ramp, and the desorption rate of a particular species is measured using a mass spectrometer. By using a range of heating rates β, not only can the desorption energy be measured, but the kinetics governing the desorption process can be uncovered. For example, in the simplest case of a coverage-independent adsorption energy and first order kinetics, if the peak desorption rate occurs at a temperature T_0 then [40]

$$\frac{\nu}{\beta} = \frac{E_d}{k_B T_0^2} \exp\left(\frac{E_d}{k_B T_0}\right) \tag{85.13}$$

where E_d is the desorption energy and ν is the 'attempt frequency' for desorption. Since both ν and E_d are unknown, values for both can be found by measuring TDS spectra using two different heating rates β or by estimating ν (often $\nu \approx 10^{13}$ s^{-1}). More complex desorption kinetics are studied by utilizing the full desorption profile, and a range of heating rates and initial adsorbate coverages. These kinetic data are also applicable to adsorption if the adsorption and desorption are reversible processes. For irreversible adsorption systems, it is possible to measure the adsorption energies directly [41] by monitoring the ir radiation emitted from a surface as a submonolayer quantity of atoms or molecules is adsorbed.

REFERENCES

1. G. Ertl and J. Küppers, *Low Energy Electrons and Surface Chemistry* (VCH, Weinheim, FRG, 1985), 2nd ed.
2. A. Zangwill, *Physics at Surfaces* (Cambridge University Press, Cambridge, 1988).
3. *Interaction of Atoms and Molecules with Solid Surfaces*, edited by V. Bortolani, N. H. March, and M. P. Tosi (Plenum Press, New York, 1990).
4. H. Lüth, *Surfaces and Interfaces of Solids* (Springer-Verlag, Berlin, 1993).

5. D. P. Woodruff and T. A. Delchar, *Modern Techniques of Surface Science* (Cambridge University Press, Cambridge, 1994), 2nd ed.

6. *Angle-Resolved Photoemission: Theory and Current Applications*, edited by S. D. Kevan (Elsevier, Amsterdam, 1992).

7. J. C. Fuggle and J. E. Inglesfield, *Unoccupied Electronic States* (Springer-Verlag, Berlin, 1992).

8. *Synchrotron Radiation Research: Advances in Surface and Interface Science*, edited by R. Z. Bachrach (Plenum, New York, 1992), Vol. 1.

9. *Synchrotron Radiation Research: Advances in Surface and Interface Science*, edited by R. Z. Bachrach (Plenum, New York, 1992).

10. *Chemical Physics of Solid Surfaces and Heterogeneous Catalysis*, edited by D. A. King and D. P. Woodruff (Elsevier, New York, 1990).

11. *Vibrations at Surfaces*, edited by Y. J. Chabal, F. M. Hoffmann, and G. P. Williams (Elsevier, Amsterdam, 1990).

12. A. R. Burns, E. R. Stechel, and D. R. Jennison, *Desorption Induced by Electronic Transitions, DIET V* (Springer, Heidelberg, 1993).

13. P. H. Fuoss, K. S. Liang, and P. Eisenberger, in *Synchrotron Radiation Research: Advances in Surface and Interface Science*, edited by R. Z. Bachrach (Plenum, New York, 1992), Vol. 1.

14. R. Wiesendanger and H.-J. Güntherodt, *Scanning Tunneling Microscopy* (Springer, Berlin, 1993).

15. D. Briggs and M. P. Seah, *Practical Surface Analysis: Auger and X-Ray Photoelectron Spectroscopy* (Wiley, Chichester, 1990), Vol. 1.

16. P. J. Rous, Prog. Surface Science **39**, 3 (1992).

17. M. A. Van Hove, *Low-Energy Electron Diffraction* (Springer-Verlag, Berlin, 1986).

18. J. M. MacLaren, J. B. Pendry, P. J. Rous, D. K. Saldin, G. A. Somorjai, M. A. Van Hove, and D. Vvedensky, *Surface Crystallographic Information Service: A Handbook of Surface Structures* (Reidel, Dordrecht, 1987).

19. H. Ibach and D. L. Mills, *Electron Energy Loss Spectroscopy and Surface Vibrations* (Academic, 1982).

20. H. Ibach, in *Interaction of Atoms and Molecules with Solid Surfaces*, edited by V. Bortolani, N. H. March, and M. P. Tosi (Plenum, New York, 1990).

21. S. Y. Tong, C. H. Li and, D. L. Mills, Phys. Rev. B **24**, 806 (1981).

22. L. Sanche, J. Phys. B **23**, 1597 (1990).

23. R. E. Palmer and P. J. Rous, Rev. Mod. Phys. **64**, 383 (1992).

24. *Auger Electron Spectroscopy*, edited by C. L. Bryant and R. P. Messmer (Academic Press, 1988).

25. L. E. Davis, N. C. MacDonald, P. W. Palmberg, G. E. Riach, and R. E. Weber, *Handbook of Auger Spectroscopy* (Physical Electronics Div., Perkin Elmer, 1976).

26. K. L. Kliewer, in *Photoemission and the Electronic Properties of Surfaces*, edited by B. Feuerbacher, B. Fitton, and R. F. Willis (Wiley, Chichester, 1978).

27. H. J. Freund and M. Neumann, in *Angle-Resolved Photoemission*, edited by S. D. Kevan (Elsevier, Amsterdam, 1992), p. 319.

28. D. E. Eastman, F. J. Himpsel, and J. F. van der Veen, J. Vac. Sci. Technol. **20**, 609 (1982).

29. D. P. Woodruff, in *Angle-Resolved Photoemission*, edited by S. D. Kevan (Elsevier, Amsterdam, 1992).

30. J. E. Rowe, in *Synchrotron Radiation Research*, edited by R. Z. Bachrach (Plenum, New York, 1992), Vol. 1, p. 117.

31. P. H. Citrin, P. Eisenberger, and R. C. Hewitt, Phys. Rev. Lett. **45**, 1948 (1980).

32. A. Bianconi and A. Marcelli, in *Synchrotron Radiation Research*, edited by R. Z. Bachrach (Plenum, New York, 1992), Vol. 1, p. 63.

33. J. Stöhr, K. Baberschke, R. Jaeger, R. Treichler, and S. Brennan, Phys. Rev. Lett. **47**, 381 (1981).

34. J. Stöhr, F. Sette, and A. L. Johnson, Phys. Rev. Lett. **53**, 1684 (1984).

35. A. Shih and V. A. Parsegian, Phys. Rev. A **12**, 835 (1975).

36. C. I. Sukenik, M. G. Boshier, D. Cho, V. Sandoghdar, and E. A. Hinds, Phys. Rev. Lett. **70**, 560 (1993).

37. J. K. Norskov and N. D. Lang, Phys. Rev. B. **21**, 2136 (1980).

38. D. Menzel, in *Chemistry and Physics of Solid Surfaces IV*, edited by R. Vanselow and R. Howe (Springer, Berlin, 1982), p. 102.

39. H. J. Kreuzer and S. H. Payne, in *Dynamics of Gas-Surface Interactions*, edited by C. T. Rettner and M. N. R. Ashfold (Royal Society of Chemistry, Cambridge, 1991), p. 220.

40. P. A. Redhead, Vacuum **12**, 203 (1962).

41. J. T. Stuckless, N. Al-Sarraf, C. Wartnaby, and D. A. King, J. Chem. Phys. **99**, 2202 (1993).

Interface with Nuclear Physics

John D. Morgan III* and James S. Cohen†

*Department of Physics and Astronomy, University of Delaware, Newark, Delaware 19716
†Theoretical Division, Los Alamos National Laboratory, Los Alamos, New Mexico 87545

86.1   INTRODUCTION

That nuclei are not infinitesimally small, structureless particles causes small but perceptible shifts in the electronic structure of atoms and molecules. Even for nuclei with small nuclear charge Z, the effects are readily detectable through modern high precision spectroscopy, and their magnitude grows as $Z^{14/3}$. Conversely, the presence of electrons tightly bound to atomic nuclei can alter the ordering of nuclear energy levels or make them unstable to β decay. Atomic effects can also influence nuclear branching ratios into the product channels. Nominally small atomic effects have been shown to affect the complicated chain of nuclear reactions responsible for the generation of energy in the sun. Setting bounds to the rest mass of the neutrino from the endpoint of the β-decay spectrum of tritium requires a precise understanding of atomic and molecular structure and scattering processes.

The great disparity between nuclear scales of energy (several MeV) and distance (10^{-5} to 10^{-4} Å) and the corresponding atomic scales (several eV and 1 Å, respectively) usually allows the separate treatment of nuclear and atomic effects. However, since not absolute energies but energy differences determine the magnitudes of perturbative effects, near coincidences in energy differences can greatly enhance the interplay between the two regimes. Such comparable differences of energies account for the important role of nuclear structure in the Lamb shift splitting between the $2s_{1/2}$ and $2p_{1/2}$ states of hydrogen (see Chap. 27), and the influence of atomic structure on nuclear processes (see Sects. 86.2 and 86.3).

For the case of muonic atoms and molecules, the interplay is enhanced by the much larger mass of a muon relative to an electron ($m_\mu \approx 207 m_e$). This decreases the distance scale by a factor of m_e/m_μ and increases the energy scale by a factor of m_μ/m_e. Small corrections such as the vacuum polarization part of the Lamb shift are amplified even more ($\sim (m_\mu/m_e)^3$ for low Z). The shrinking of the internuclear separation in the muonic deuterium-tritium molecule increases the rate of *muon-catalyzed fusion* by 77 orders of magnitude, as discussed in Sect. 86.4.

86.2 NUCLEAR SIZE EFFECTS IN ATOMS

86.2.1 Nuclear Size Effects on Nonrelativistic Energies

Interest in the influence of a finite nuclear charge distribution on the energy levels of hydrogen stems back to the measurement of the Lamb shift [1], and even earlier indications that the fine structure of hydrogen did not quite agree with the predictions of the Dirac equation for a point nucleus [2]. The finite proton size does in fact raise the energy of the $2s_{1/2}$ state relative to $2p_{1/2}$, but the shift is only $\sim 0.012\%$ of the dominant electron self-energy contribution (see Chap. 27). It must nevertheless be taken into account in high precision tests of QED.

Early derivations were given by several authors [3–6] and generalized by Zemach [7] (see also Ref. [8]) to a form involving integrals over the nuclear electric and magnetic form factors. The basic result is illuminated by the following argument. Let $\rho(\mathbf{r})$ be the electron density, which may have no spatial symmetry properties in the particular case of a polyatomic molecule, and $\rho_n(\mathbf{r}_n)$ be the charge density of a nucleus, which obeys

$$\int d^3 r_n \, \rho_n(\mathbf{r}_n) = Z \,. \tag{86.1}$$

Assume that $\rho_n(\mathbf{r}_n)$ has no permanent electric dipole moment, so that

$$\int d^3 r_n \, \mathbf{r}_n \, \rho_n(\mathbf{r}_n) = \mathbf{0} \,. \tag{86.2}$$

By writing the Coulomb potential for a pointlike nucleus as

$$-\frac{Z}{r} = -\int d^3 r_n \, \frac{Z \delta^{(3)}(\mathbf{r}_n)}{|\mathbf{r} - \mathbf{r}_n|} \,, \tag{86.3}$$

the first-order shift of the electronic energy due to the replacement of the pointlike nucleus by an extended nucleus is

$$\Delta E_{\text{nuc}} = \int d^3 r \int d^3 r_n \, \frac{Z \delta^{(3)}(\mathbf{r}_n) - \rho_n(\mathbf{r}_n)}{|\mathbf{r} - \mathbf{r}_n|} \, \rho(\mathbf{r}) \,. \tag{86.4}$$

Since the Fourier transform, defined by

$$\hat{\rho}_n(\mathbf{k}) = \int d^3 r_n \, e^{-i\mathbf{k} \cdot \mathbf{r}_n} \, \rho_n(\mathbf{r}_n) \,, \tag{86.5}$$

preserves inner products within a factor of $(2\pi)^3$ and maps convolutions to simple products, the integral reduces to

$$(2\pi)^3 \int d^3 k \, [Z - \hat{\rho}_n(\mathbf{k})] \, \frac{4\pi}{k^2} \, \hat{\rho}(\mathbf{k}) \,, \tag{86.6}$$

where the hats denote the Fourier transforms of the densities and $4\pi/k^2$ is the Fourier transform of the Coulomb potential $1/r$. Since $Z = \int d^3 r_n \, \rho_n(\mathbf{r}_n)$, the energy shift can be reexpressed as

$$(2\pi)^3 \int d^3 k \int d^3 r_n \, (1 - e^{-i\mathbf{k} \cdot \mathbf{r}_n}) \, \rho_n(\mathbf{r}_n) \, \frac{4\pi}{k^2} \, \hat{\rho}(\mathbf{k}) \,, \tag{86.7}$$

which is still an exact first-order perturbation expression. Since typical nuclear length scales are much smaller than typical nonrelativistic atomic length scales, it is legitimate to expand the exponential in a Taylor series. The zeroth-order term, -1, is canceled by the $+1$. The linear term, $i\mathbf{k} \cdot \mathbf{r}_n$, contributes nothing by the hypothesis that the nuclear charge distribution has no permanent electric dipole moment. The first nonvanishing term is

$$(2\pi)^3 \int d^3 k \int d^3 r_n \, \frac{1}{2} (\mathbf{k} \cdot \mathbf{r}_n)^2 \, \rho_n(\mathbf{r}_n) \, \frac{4\pi}{k^2} \, \hat{\rho}(\mathbf{k}) \,. \tag{86.8}$$

If $\rho(\mathbf{r})$ is nonzero at the nucleus, then for large k the leading behavior of $\hat{\rho}(\mathbf{k})$ is that of a spherically symmetric s-wave with a radial dependence proportional to k^{-4}. The angular integration in the variable $\mathbf{k}$ leads to the replacement of $(\mathbf{k} \cdot \mathbf{r}_n)^2$ by its average value $\frac{1}{3} k^2 r_n^2$, so that expression (86.8) reduces to

$$\frac{2\pi}{3} (2\pi)^3 \int d^3 k \, \hat{\rho}(\mathbf{k}) \int d^3 r_n \, r_n^2 \, \rho_n(\mathbf{r}_n) \,, \tag{86.9}$$

which can be further simplified by observing that

$$(2\pi)^3 \int d^3 k \, \hat{\rho}(\mathbf{k}) = (2\pi)^3 \int d^3 \mathbf{k} \, e^{i 0 \cdot \mathbf{k}} \, \hat{\rho}(\mathbf{k}) = \rho(0) \,, \tag{86.10}$$

and by definition

$$\int d^3 r_n \, r_n^2 \, \rho_n(\mathbf{r}_n) = Z \langle r_n^2 \rangle \,, \qquad (86.11)$$

thus yielding the final expression

$$\Delta E_{\text{nuc}} = \frac{2\pi}{3} Z e^2 \rho(\mathbf{0}) \langle r_n^2 \rangle \,, \qquad (86.12)$$

with

$$\rho(0) = \left[\frac{\mu}{m_e}\right]^3 \frac{Z^3}{\pi n^3} a_0^{-3} \qquad (86.13)$$

for a hydrogenic ion with reduced mass μ. This derivation is independent of the specific nuclear model or the assumption of spherical symmetry of the electron density. Since $\langle r_n^2 \rangle$ scales as $Z^{2/3}$, ΔE_{nuc} then scales as $Z^{14/3}$. For a molecule with several nuclei, the contributions (86.12) from each nucleus should be summed.

In the case of helium, $\rho(0) = \langle \delta(\mathbf{r}_1) + \delta(\mathbf{r}_2) \rangle$ can be accurately calculated from high precision variational wave functions (see Chap. 11). For the $1s^2 \, {}^1S_0$ ground state, $\rho(0) \simeq (\mu/m_e)^3 [3.620\,8586 - 0.182\,37(\mu/M)] a_0^{-3}$ where M is the nuclear mass. Results for other states up to $n = 10$ are tabulated in Ref. [9]. Combined with high precision isotope shift measurements, the results can be used to extract differences in nuclear radii for pairs such as ^{3}He/^{4}He, ^{6}Li/^{7}Li, and H/D [10–12].

Expression (86.12) works well for atoms with small Z, since relativistic corrections to the electron density are small. However, it breaks down for heavier nuclei, for which relativistic wave functions are needed.

86.2.2 Nuclear Size Effects on Relativistic Energies

The preceding analysis breaks down for relativistic wave functions because they are singular at a point nucleus, making $\rho(\mathbf{0})$ infinite. In this case, the Dirac equations with Hamiltonians H_0 and H for the point nucleus and distributed nucleus cases, respectively, can be combined to obtain

$$(E - E_0)\Psi^\dagger \Psi_0 = \Psi^\dagger H \Psi_0 - \Psi^\dagger H_0 \Psi_0 \,. \qquad (86.14)$$

If a finite radius r_s is now chosen such that $H = H_0$ outside the sphere $r = r_s$, then this equation can be integrated from r_s outward to yield [13]

$$\Delta E_{\text{nuc}} = \frac{\hbar c \, (g f_0 - f g_0)_{r=r_s}}{\int_{r_s}^{\infty} (g g_0 + f f_0) \, dr} \qquad (86.15)$$

where f and g are the large and small radial components of Ψ (see Chap. 9), and the numerator is the surface term that remains after integrating by parts the $c\boldsymbol{\alpha}\cdot\mathbf{p}$ term in H. The units are $\hbar c/a_0 = \alpha m_e c^2$. The solutions can be further expanded in terms of Bessel functions, or the Dirac equation can simply be integrated numerically.

For hydrogenic ions up to moderately large Z, the results are reasonably well represented by [14, 15]

$$\Delta E_{\text{nuc}} = \frac{2}{3n^3}(Z\alpha)^2 m_e c^2 [\delta_{\ell,0} + C_2(Z\alpha)^2](Z^2 \langle r_n^2 \rangle / a_0^2)^\gamma$$

$$(86.16)$$

with $\gamma = [1 - (Z\alpha)^2]^{1/2}$, and $C_2 \simeq 0.50$, 1.38, and 0.1875 for the $1S_{1/2}$, $2S_{1/2}$, and $2P_{1/2}$ states, respectively. Extensions to higher-order terms are discussed in Ref. [16]. The above formula was used in the tabulations of Mohr [15] for $10 \leq Z \leq 40$, while Johnson and Soff [17] used the numerical integration method for Z up to 110. The nuclear electric and magnetization density distributions are tabulated in Ref. [18], nuclear moments in Ref. [19], and nuclear masses in Ref. [20]. In the absence of better data, the rms nuclear radius can be estimated from $\langle r_n^2 \rangle^{1/2} \approx 0.777 A^{1/3} + 0.778 \pm 0.06$ fm, where A is the atomic mass number.

86.2.3 Nuclear Size Effects on QED Corrections

Recent progress in the experimental study of transition energies in heavy ions stripped of most of their electrons [21] has inspired theoretical work on modifications of QED corrections due to an extended nuclear charge distribution. Calculations based on propagators expanded in terms of basis splines [22, 23] (*cf.* Sect. 8.2.1) have led to relatively rapid convergence with the number of angular functions.

86.3 ELECTRONIC STRUCTURE EFFECTS IN NUCLEAR PHYSICS

86.3.1 Electronic Effects on Closely Spaced Nuclear Energy Levels

The presence of a nearby cloud of electrons can significantly affect nuclear processes involving closely spaced nuclear energy levels. One of the most dramatic cases involves the β-decay process ^{187}Re $\to {}^{187}$Os $+ e^- + \bar{\nu}_e$, which is energetically forbidden by about 12 keV for bare nuclei, but becomes allowed for the neutral atoms when the difference in electronic binding energies is included. The nuclear charges are $Z = 75$ for ^{187}Re and $Z = 76$ for ^{187}Os. There is also the possibility of the electron being captured into a bound state of ^{187}Os, as opposed to the continuum β-decay process.

The electronic binding energies of heavy atoms can be estimated from the Thomas-Fermi result $E_{\text{TF}} \simeq -20.93 Z^{7/3}$ eV for a neutral atom of charge Z. The difference between the energies of two atoms with nuclear charges $Z + 1$ and Z, respectively, is then

$$E_{\mathrm{TF}}(Z+1) - E_{\mathrm{TF}}(Z) \simeq -48.83 Z^{4/3} \text{ eV}, \qquad (86.17)$$

which amounts to -15.4 keV at $Z = 75$. This is more than sufficient to overcome the 12 keV energy deficit in the otherwise energetically forbidden β-decay of ^{187}Re.

The general theory of bound state β-decay is discussed by Bahcall [24], who also calculated the ratio ρ of bound state β-decay to continuum β-decay for bare nuclei. In the case of ^{187}Re $\rightarrow^{187}$Os, ρ is of importance in estimating changes in the half-life for β-decay of ^{187}Re under various conditions of ionization, since the measured isotope ratios ^{187}Re/^{188}Re and ^{187}Os/^{188}Os from terrestrial rocks and meteorites can be used to determine not only the age of the solar system, but also the age of our galaxy [25]. Estimates based on a modified TF model [26] indicate that $\rho \simeq 0.01$, and further multiconfiguration Dirac-Fock calculations give $\rho = 0.005$ to 0.007 [27]. See Ref. [27] for further details and references.

86.3.2 Electronic Effects on Tritium Beta Decay

The mass of the neutrino, normally taken to be zero in the Standard Model, can be determined in principle from analysis of the β-decay process ^{3}T $\rightarrow^3$He$^+ + e^- + \bar{\nu}_e$. An early measurement based on this method [28] yielded a neutrino mass of ~ 25 eV. Several independent tests of this result were initiated soon thereafter. Since the experiments are performed not on bare tritons but on tritium gases and solids under various conditions, it is essential to understand quantitatively the atomic and molecular processes that affect the distribution of the highest-energy electrons produced from various initial states [29].

Martin and Cohen [30] used a Stieltjes imaging technique to calculate shake-up and shake-off probabilities for the β-decay of T_2 into ^{3}HeT$^+$. Simultaneously, extensive calculations were carried out [31] using potential energy curves for the reactant T_2 and TH molecules and the product ^{3}HeT$^+$ and ^{3}HeH$^+$ molecules and accounting for the production of electronically and rovibrationally ground and excited final states, as well as resonant states. Nuclear motion was found to have a small but detectable effect on the results, and solid-state effects for frozen T_2 were also investigated and found to be small. These calculations play a crucial role in the interpretation of the experiments [32], which so far indicate that the neutrino mass, if nonzero, is less than ~ 10 eV.

86.3.3 Electronic Screening of Low Energy Nuclear Reactions

The cross section $\sigma(E)$ for a nuclear reaction involving charged reactants drops very rapidly for collision energies E below the Coulomb barrier. A WKB treatment shows that for low collision energies, the dependence of $\sigma(E)$ can be conveniently expressed as

$$\sigma(E) = S(E) E^{-1} e^{-2\pi\eta}, \qquad (86.18)$$

where $S(E)$ is the astrophysical factor, and

$$\eta = Z_1 Z_2 \alpha c (\mu/2E)^{1/2} \qquad (86.19)$$

is the Sommerfeld parameter, which depends on the charge numbers Z_1 and Z_2 of the projectile and target nuclides, their reduced mass μ, and the c.m. energy E. For nuclear reactions involving light nuclei, it is found that $S(E)$ typically varies slowly with E except close to resonances. Thus the accurate determination of $S(E)$ at moderately low E can be used to extrapolate $\sigma(E)$ to much lower energies, which are beyond the reach of laboratory experiments but are of great relevance to the nuclear reactions that occur in stars.

However, electron screening effects can greatly enhance cross sections for nuclear reactions as measured in the laboratory at low energy [33], because at least the target nucleus is almost always surrounded by a cloud of electrons which screen the Coulomb repulsion between nuclei. The effect has been observed in various low-energy reactions such as ^{3}He$(d,p)^4$He, ^{6}Li$(p,\alpha)^3$He, ^{6}Li$(d,\alpha)^4$He, and ^{6}Li$(p,\alpha)^4$He [34, 35]. Since reactions in stars involve bare nuclei, the laboratory data must be carefully corrected for screening effects.

Analysis of the data for the ^{3}He$(d,p)^4$He reaction indicates that the effect of screening is always greater than that predicted in the adiabatic limit [36]. A more general theoretical treatment of the $d+^2$H and $d+^3$He reactions [37], using a time-dependent Hartree-Fock method for the electrons screening and classical motion for the nuclei found less enhancement than that observed. An improved treatment taking account of electron correlation and quantum-mechanical effects on the nuclear motion will likely be needed. This remains an important area of development for the future.

86.3.4 Atomic and Molecular Effects in Relativistic Ion-Atom Collisions

High-energy accelerators can now produce beams of atomic ions partly or completely stripped of their electrons, even for Z as high as 92. The collisions of such beams of highly charged ions with fixed targets involve a broad array of atomic and molecular processes, such as excitation, ionization, charge transfer, and, in the extreme relativistic case, pair production. A similarly broad array of theoretical techniques is required to study these topics. A thorough review of them, including comparisons with experimental data where available, is given in Refs. [38–40].

A topic of particular recent interest is the first experimental observation of the capture of electrons from electron-positron pair production in the extreme relativistic collision of a 0.96 GeV/nucleon U^{92+} beam with gold, silver, copper, and Mylar targets [41]. The energy and angular distributions of the positrons were also measured. For the gold target, the cross section for capture was nearly as large as that for pair production without capture, and it was found to vary with the nuclear charge Z_t of the target nucleus roughly as $Z_t^{2.8(0.25)}$. Neither the dependence on Z_t nor the relatively great probability for capture is in agreement with perturbation theory, which highlights the need for further exploration of this exotic system.

86.4 MUON-CATALYZED FUSION

Exotic muonic atoms and molecules are more suitable subjects than electronic atoms and molecules for probing some physical effects. The muon μ^- is a leptonic elementary particle like the electron except that it is 206.768 times more massive and has a finite lifetime ($\tau_0 = 1/\lambda_0$, where λ_0 is the rate of decay) of 2.197 μs. This lifetime is amply long for most experiments. In normal atoms, the fine-structure splitting (due to $\mathbf{L \cdot S}$ coupling) is much larger than the hyperfine splitting (due to $\mathbf{s}_{nuc} \cdot \mathbf{s}_e$ coupling); this relation is reversed in muonic atoms. Likewise, vacuum polarization,[1] relativistic, finite-nuclear-size, and nonadiabatic effects are enhanced. Remarkably, muonic molecules make nuclear fusion possible at room temperature. In the phenomenon of muon-catalyzed fusion (μCF), there are both indirect and direct interactions between the atomic and molecular physics and the nuclear physics. Indirectly, the atomic and molecular densities and transition rates control the nuclear fusion rates, and, in turn, the kinetic energies of the fusion products affect the atomic and molecular kinetics. Directly, the nuclear structure affects some molecular energy levels that determine important resonant rates and the boundary condition on the muonic wave functions used to calculate the muon "sticking" loss.

In comparison with μCF, hot fusion schemes are made difficult by the electrostatic (Coulomb) repulsion between nuclei. In the two conventional approaches to controlled fusion, magnetic and inertial confinement, this barrier is partially surmounted by energetic collisions.[2] On the other hand, in μCF the objective is to tunnel through the barrier without the benefit of kinetic energy. This feat is enabled by binding two hydrogenic nuclei (p, d, or t) in

an exotic molecule like H_2^+ with the electron replaced by a negative muon.

Since the molecular size is inversely proportional to the mass of the binding particle, the average distance between nuclei in $pp\mu$ is $\sim \frac{1}{200}$ Å (500 fm) instead of 1 Å as in ppe (i.e., H_2^+). This distance, which would be reached in a $d + d$ collision at $\sim$3 keV ($\approx 3 \times 10^7$ K), is still large compared with the separation of a few fm where the nuclear strong forces cause fusion, but fusion occurs rapidly because of the increased vibrational frequency and, more important, the increased probability of tunneling per vibration. The vibrational frequency is $(m_\mu/m_e)^{3/2} \approx 3 \times 10^3$ times faster than for the corresponding electronic molecule.[3] The effect on the tunneling probability depends on the nuclear masses; for $dt\mu$, which has the largest nuclear matrix element (astrophysical S factor, see Sect. 86.3.3), the increase is by a factor of $\sim 10^{77}$ compared with DT, and the consequential fusion rate is $\lambda_{dt\mu}^f \approx 10^{12}$ s^{-1}.

Just on the basis of the fusion rate, one would expect a yield of $\lambda_{dt\mu}^f/\lambda_0 \sim 10^6$ muon-catalyzed d-t fusions for the average muon. While this number indeed provides an upper limit, the actual average number of fusions, $\sim$150 for $dt\mu$, is much smaller and is determined by the *atomic and molecular physics* of the catalysis cycle (though the energy released in the nuclear fusion does play an important role here). Some of the atomic and molecular processes in the μCF cycle are quite ordinary, but others, like atomic capture and resonant molecular formation, have no counterpart with "normal" atoms.

Muon-catalyzed fusions of all pairs of hydrogen isotopes, except two protons, have been observed. Based on the experiments and theory, the reaction products are:

$$pd\mu \rightarrow \begin{cases} {}^3\text{He}(0.005) + \mu + \gamma(5.49) & (92\% - 89\%) \\ {}^3\text{He}(0.20) + \mu(5.29) & (\ 8\% - 11\%) \end{cases}$$

$$(86.20)$$

$$dd\mu \rightarrow \begin{cases} t(1.01) + p(3.02) + \mu & (41\% - 53\%) \\ {}^3\text{He}(0.82) + \mu + n(2.45) & (59\% - 47\%) \end{cases}$$

$$(86.21)$$

$$pt\mu \rightarrow \begin{cases} {}^4\text{He}(0.05) + \mu + \gamma(19.76) & (60\% - 76\%) \\ {}^4\text{He}(0.59) + \mu(19.22) & (23\% - 14\%) \\ {}^4\text{He} + \mu + e^+ + e^- \ (19.81 \text{ total}) & (17\% - 10\%)^4 \end{cases}$$

$$(86.22)$$

$$dt\mu \rightarrow {}^4\text{He}(3.54) + \mu + n(14.05) \qquad (86.23)$$

---

[1]The muonic Bohr radius $\hbar^2/m_\mu e^2 \approx \frac{1}{207} a_0$ is similar in size to the Compton wavelength $\hbar/m_e c \approx \frac{1}{137} a_0$.

[2]The particle densities N and confinement times τ in the hot plasmas ($T \gtrsim 10^8$ K) are typically more than ten orders of magnitude different for these two schemes, but the product of the two required for d-t fusion is $N\tau \gtrsim 10^{14}$ s/cm^3 in either case. For muon-catalyzed fusion, effectively $N\tau \approx 10^{25}$ s/cm^3, but this criterion doesn't tell the real story.

[3]For comparison, the muonic/electronic energy scales as m_μ/m_e and the rotational energy scales as $(m_\mu/m_e)^2$. These relations [42] are based on the Born-Oppenheimer approximation, which is not very accurate for muonic molecules.

[4]Theoretically predicted [43], but not yet observed.

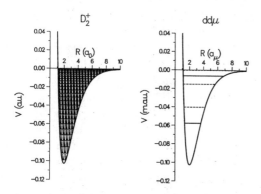

Figure 86.1. Rovibrational energy levels for D_2^+ and $dd\mu$. The $J = 0$ levels are shown as solid lines and the $J > 0$ levels are shown dashed. For D_2^+, all 28 vibrational levels are displayed, but the associated rotational levels are displayed only up to the next higher vibrational level. All levels of $dd\mu$ are displayed; the $(J=1, v=1)$ level is barely discernible below the $V = 0$ axis.

$$tt\mu \rightarrow {}^4He + \mu + 2n \ (11.33 \ total) . \tag{86.24}$$

Here the product particle kinetic energies (in MeV) are given in parentheses. A μ without an energy designated is a spectator, i.e. serves to bring the nuclei together but plays no significant role in the kinematics of the reaction — such a μ may actually be bound (stuck) to one of the product nuclei. As indicated, the branching fractions depend somewhat on the target parameters (isotopic composition, density, and temperature) [44–46].

Each reaction is of special interest in its own right: $pd\mu$ and $pt\mu$ for the contribution of μ conversion, and $tt\mu$ for the correlation of the two final state neutrons. Only the $dd\mu$ and $dt\mu$ molecular formations are resonant; i.e., their formation can occur in a one-body state because they, and only they, possess a loosely bound state such that the muonic binding energy can go into rovibrational energy of the electronic molecule. That the existence of such a state really is fortuitous can be seen in Fig. 86.1 where the bound rovibrational states of $dd\mu$ (1 m.a.u. = 5626.5 eV) are compared with the rovibrational states of D_2 (1 a.u. = 27.2 eV). Though both $dd\mu$ and $dt\mu$ can be formed resonantly, $dt\mu$ is unique in having a *rapid* (as compared with muon decay) formation rate *and* also in having a small sticking loss. The sticking loss is due to the possibility that the negatively charged muon may form a bound state with the positively charged fusion product. The relatively low branching fraction ($<1\%$) for $dt\mu \rightarrow {}^4He + n$ is due simply to the high speed of the outgoing ${}^4He$ (α particle).

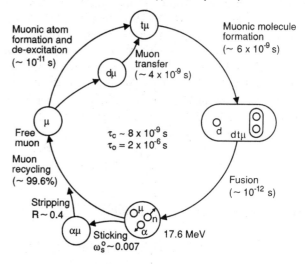

Figure 86.2. The simplified *d-t* muon-catalyzed fusion cycle. The times are for density $\phi = 1$ (liquid hydrogen density) and tritium fraction $c_t=0.4$. τ_c is the cycle time, and τ_0 is the muon-decay time.

86.4.1 The Catalysis Cycle

A diagram of the μCF cycle for a *d-t* mixture is shown in Fig. 86.2. The basic steps in the cycle are

1. Atomic capture to form $d\mu$ or $t\mu$ (initially in a highly excited state, $n \gtrsim 14$).
2. Transfer of the μ from d to t, if necessary.
3. Resonant molecular formation, shown schematically. Here the $dt\mu$ is so small (in reality) that is can be considered to be a pseudo-nucleus in the electronic molecule.
4. Nuclear fusion.
5. Sticking ($\alpha\mu$ formation) *or* recycling.

The reaction times shown are at liquid hydrogen density ($\phi = 1$ in the conventional LHD units) and a tritium fraction $c_t = 0.4$, which is close to the value that maximizes the number of cycles. The times for muonic atom formation and deexcitation, $\sim$ ps, are short compared with the times for muon transfer and molecular formation, $\sim$ ns, which in turn are short compared with the muon decay time, $\sim \mu$s.

Thus the time for a cycle is mainly given by the average time the μ spends as $d\mu$ waiting to transfer to t plus the average time it then spends as $t\mu$ waiting to form $dt\mu$,

$$\tau_c \approx \tau_{d\mu} + \tau_{t\mu} , \tag{86.25}$$

or, in terms of rates,

$$\frac{1}{\lambda_c} \approx \frac{q_{1s}c_d}{\lambda_{dt}c_t} + \frac{1}{\lambda_{dt\mu}c_d} , \tag{86.26}$$

where c_d and c_t are the fractions of deuterium and tritium ($c_d + c_t = 1$), λ_{dt} is the d-to-t transfer rate in the $1s$ state, q_{1s} is the fraction of $d\mu$ atoms reaching the $1s$ state (before transfer), and $\lambda_{dt\mu}$ is the molecular formation rate. The factor q_{1s} takes into account the fact that any transfer in excited states is rapid (of necessity, since it must compete with the rapid deexcitation).

The cycle rate λ_c along with the sticking fraction ω_s constitute the two basic parameters of the catalysis cycle. The average number of fusions per muon Y is given by[5]

$$\frac{1}{Y} \approx \frac{\lambda_0}{\lambda_c} + \omega_s \qquad (86.27)$$

where $\lambda_0 (\equiv 1/\tau_0)$ is the muon-decay rate. Coincidentally, the limits imposed on the yield by the cycling rate and by sticking are similar; for $\phi \approx 1$ and $T \approx 300$ K, $\lambda_c/\lambda_0 \approx 300$ and $1/\omega_s \approx 200$ for $dt\mu$. More than 100 muon-catalyzed d-t fusions per muon have been observed. Similar considerations apply to the $dd\mu$ cycle, and, by further coincidence, the two limits are similar there as well, $\lambda_c^{dd\mu}/\lambda_0 \approx 7$ and $1/\omega_s^{\text{eff}(dd\mu)} \approx 14$.[6] The four experimental "knobs" are the temperature (T), density (ϕ), isotopic fractions (c_t, c_d, and c_p), as well as the molecular fractions (c_{D_2}, c_{DT}, c_{T_2}, $\cdots$) in the case of a target not in chemical equilibrium.

Each stage of the cycle is discussed in the following sections. The reader is referred to reviews [42, 47–50] for details of the theoretical and experimental methods and extensive values of the relevant parameters.

86.4.2 Muon Atomic Capture

The μCF process starts with a free muon, injected into a mixture of hydrogen isotopes, being stopped to form a muonic atom (stopping power is discussed in Sect. 87.2.1). The slowing and capture occur primarily by ionization, e.g.,

$$\mu + D \rightarrow \begin{cases} \mu + d + e \\ d\mu(n) + e \,. \end{cases} \qquad (86.28)$$

It is now known that the muon is slowed to an energy of ~ 10 eV, near the target ionization potential, and then captured into an orbital with $n \gtrsim \sqrt{m_\mu/m_e} \approx 14$, which has about the same size and energy as that of the displaced electron.

Rigorous theory has been developed for μ^- capture by the H atom, but only rudimentary work has been done

Figure 86.3. Comparison of different capture cross sections for $\mu^- + H$ collisions: adiabatic ionization with straight-line trajectories(AI-slt), adiabatic ionization with curved trajectories (AI-ct), diabatic states with polarized orbital (DS), classical-trajectory Monte Carlo (CTMC), time-dependent Hartree-Fock (TDHF), and classical-quantal coupling (CQC).

so far on molecular effects. Methods for hydrogen and helium have been reviewed in [51]; the brief discussion here emphasizes the correct intuitive understanding. Until 1977 most calculations were done using the Born or Coulomb-Born approximation (see, e.g., [52]). These methods are not very accurate for μ^- at velocities below 1 a.u., but, more important, their implementation treated slowing down and capture inconsistently. The upshot was prediction of capture of muons typically with kinetic energies of hundreds of eV, whereas it turns out that most captures actually occur at energies below 20 eV.

The perturbative methods fail because of the great electron charge redistribution that occurs during the capture process. Six other approaches have led to accurate treatment:

1. Adiabatic ionization with straight-line trajectories (AI-slt) [53],
2. Adiabatic ionization with curved trajectories (AI-ct) [51,54],
3. Diabatic states (DS) [55],
4. Classical-trajectory Monte Carlo (CTMC) [54],
5. Time-dependent Hartree-Fock (TDHF) [56],
6. Classical-quantal coupling (CQC) [57].

The first three are models tailored for the muon capture problem (the first two of these specialized for the hydrogen-atom target). The results for all six methods are shown in Fig. 86.3.

The early study by Wightman [53] shed a great deal of light on the capture process. His method, known as adiabatic ionization (AI), followed on the observation of

---

[5]More precisely, W should appear in Eq. (86.27) in place of ω_s. W may include other losses, e.g., muon capture by impurities, but we will restrict the present discussion to ω_s, which is fundamental and normally dominant.

[6]The effective sticking probability (per cycle) in the $dd\mu$ cycle takes into account that in only 58% of the fusion reactions is a ^{3}He produced that can remove the μ by sticking. Sticking to t or p is possible but would facilitate rather than terminate the cycling.

Fermi and Teller [58] that there exists a critical strength of the dipole eR_c, formed by the negative muon and positive proton at distance $R_c = 0.639\,a_0$, for binding the electron. In collisions where the μ^- approaches closer than this distance, the electron is assumed to escape adiabatically, and, if the electron carries off more energy than the muon's initial kinetic energy, the $p\mu$ atom is formed. This cross section is thus

$$\sigma_{\text{AI-slt}} = \pi R_c^2, \tag{86.29}$$

and μ^- capture results if and only if $E < 0.5$ a.u., the target ionization energy.

The AI-slt model has three major shortcomings: (1) it does not take into account trajectory curvature, which is caused by the Coulomb attraction of μ^- toward the nucleus and can be large at the low trajectory velocities where capture usually occurs, (2) the adiabatically escaping electron takes off no kinetic energy, and (3) ionization occurs with unit probability if the approach is closer then R_c. The first failing is easy to remedy. The cross section with curved adiabatic trajectories (AI-ct) is just [54]

$$\sigma_{\text{AI-ct}} = \frac{\pi R_c^2}{E}\left(E + \frac{1}{R_c} - 0.5 \text{ a.u.}\right) \tag{86.30}$$

as long as the collision energy E in the center-of-mass system is greater than 0.03 a.u. (0.8 eV). Hence trajectory curvature increases the capture cross section by a factor $1 + \frac{1.06}{E}$ (for E in a.u.), which is over a factor of three even at the highest collision energy (0.5 a.u.) where adiabatic capture can occur. For $E < 0.03$ a.u., the centrifugal barrier in the effective potential,

$$V_{\text{a(eff)}}(R, b) = V_a(R) + \frac{b^2}{R^2}E, \tag{86.31}$$

restricts penetration and reduces the cross section below the value given by Eq. (86.30) [51].

Cures for the second and third failings are less trivial. These two assumptions can be avoided by using the diabatic-states (DS) model [55]. The adiabatic electronic energy no longer increases once it reaches the continuum ceiling; however, in view of the μ^- acceleration by the Coulomb attraction, the electron cloud actually does not have enough time to adjust adiabatically. In recognition of this situation the diabatic electronic energy crosses into the continuum at a distance larger than R_c and continues to rise smoothly. The concomitant probability of ionization is given by the ionization width, obtained by a Fermi-golden-rule-like formula. The first three approaches are somewhat specialized models, while the next three are general methods. The most economical in terms of computer time is the classical-trajectory Monte Carlo (CTMC) method [54] (see Chap. 56). This method treats the dynamics of all particles completely classically. The time-dependent Hartree-Fock (TDHF) method (see Sect. 48.3) is purely quantum mechanical [56], but

neglects correlation, which turns out to be important in the present problem. This deficiency is remedied by the classical-quantal coupling (CQC) method, which makes only the seemingly well-justified approximation of treating the muon classically while retaining the quantum treatment of the electron [57].

Real μ^- capture experiments are done with molecules (H_2, DT, etc.). The naive notion that the H_2 cross section is simply twice that of H is quite unrealistic for slow ($v \ll 1$ a.u.) collisions. A calculation of the total $\mu^- + H_2$ inelastic cross section (primarily capture for $E \lesssim 0.57$ a.u.) [59], using a generalization of the DS model, indicates that $\sigma(H_2)$ is actually smaller than $\sigma(H)$ for $E < 0.3$ a.u.;[7] at higher energies it is larger though not by nearly a factor of two.

86.4.3 Muonic Atom Deexcitation and Transfer

The muon is captured in a highly excited state but normally must reach the $1s$ configuration of the heavier isotope (in case of mixtures like D/T) before the muonic molecule is formed. In the $1s$ configuration there are two hyperfine levels — the ground state with the nuclear and μ^- spins antiparallel and an excited state with spins parallel. Resonant molecular formation rates in the two states can be quite different and also depend strongly on the atom's kinetic energy. Thus there are several types of muonic atom collisions that must be taken into account: (1) elastic scattering in the ground and excited states, (2) isotopic transfer in excited states, (3) deexciting transitions (which may also occur radiatively), (4) isotopic transfer in the $1s$ state, and (5) hyperfine transitions. Cross sections for most of these processes have been calculated. The bulk of the calculations have been done by expanding in adiabatic (or modified adiabatic) eigenfunctions, but there also exist some calculations using the coupled-rearrangement-channel, Faddeev, and hyperspherical approaches (see Ref. [42] for references).

The cascade of the initially formed muonic atom, especially in mixtures, is a complicated process not yet completely characterized. It constitutes a crucial part of the d-t μCF cycle in that it determines the parameter q_{1s} in Eq. (86.26). This parameter is essential to experimental analysis, but it was evident that early calculations yielded values of q_{1s} too small to be consistent with experiments. Recent calculations [60] suggest the explanation is that the excited muonic atoms are not thermalized. Epithermal atoms have three effects here:

---

[7]Adiabatic ionization cannot occur since the adiabatic limit as μ^- approaches H_2 is $p + H^-$, a bound negative ion. Although, in principle, the molecule could *dissociate* adiabatically, the adiabatic criterion is not satisfied for the nuclei with respect to the muon, so collisional dissociation is improbable. However, the molecule may be left in an excited state that subsequently predissociates.

(1) the normal transfer rates are smaller, (2) the transfer is reversible down to lower principal quantum numbers n where E still exceeds the threshold for excitation of the next-higher level,[8] and (3) excited-state [from $(t\mu)^*$] resonant formation of $(dt\mu)^*$ molecules that can predissociate back to $d\mu$ is enhanced [61]. The q_{1s} is determined by competition between transfer and deexcitation, which depend on the kinetic energies that result from further competition between superelastic deexcitation and thermalizing elastic collisions. It appears that the stage of the cascade most crucial to q_{1s} is $n \approx 4$ for normal muon transfer and $n = 2$ for the resonant sidepath.

For muons, the elastic cross sections are more difficult to calculate than the inelastic ones. The inelastic transitions occur at short range (a few a_μ) where the effects of *electronic* structure are negligible. However, electronic effects are not negligible for low energy (< 1 eV) elastic scattering where $\lambda_{dB} \sim 1 \, a_0$. They have been taken into account for ground state but not yet excited state scattering. In doing so, it is not necessary to solve the general many-body problem directly because of the following simplifications: (1) this energy is below the vibrational threshold so the molecular target can be taken as a rigid rotor and (2) the relative smallness of the muonic atoms makes the sudden approximation adequate.

If the $1s$ state is reached without muon transfer to the heavier isotope already having occurred, the transfer takes significant time and plays an important role in determining the tritium fraction c_t that optimizes the fusion yield. All of the $1s$ isotopic-exchange cross sections display the characteristic $\sim 1/v$ velocity dependence at thermal energies so that the corresponding rate $v\sigma$ is independent of temperature.

In muon-catalyzed d-d and d-t fusions, the resonant molecular formation rates in different hyperfine structure (hfs) states can differ by two or more orders of magnitude at low T due to their different energy levels. The hfs also has important effects on thermalization and diffusion via the different elastic cross sections. Under usual μCF experimental conditions, the hyperfine quenching (or "spin flip") is irreversible; the hfs splittings are 0.1820, 0.0485, and 0.2373 eV for $p\mu$, $d\mu$, and $t\mu$, respectively. Theoretically, it is expected that transitions between hfs levels mainly occur in symmetric collisions since muon exchange suffices in such collisions [62]; e.g.,

$$t\mu(\uparrow\uparrow) + t(\downarrow) \to t(\uparrow) + t\mu(\downarrow\uparrow) \qquad (86.32)$$

(the usual terminology here is "muon exchange" although it might seem more logical to refer to the reaction as "triton exchange" since it is the identity of the tritons that enables the reaction). As in the case of the isotopic exchange cross sections, the behavior is $\sim 1/v$ at thermal energies, so the rates are nearly independent

---

8 The isotopic energy splittings are $134.7/n^2$ for $d\mu$-$p\mu$, $182.8/n^2$ for $t\mu$-$p\mu$, and $48.0/n^2$ for $t\mu$-$d\mu$.

of temperature.

86.4.4 Muonic Molecule Formation

Until the prediction by Vesman [63] of a resonant formation for $dd\mu$, it was thought that all muonic molecules were formed by an Auger process of the type

$$d\mu + D_2 \to [(dd\mu)de]^+ + e^- . \qquad (86.33)$$

Unlike the resonant process for $dd\mu$ and $dt\mu$, the nonresonant process generally depends weakly on the temperature of the target, the hyperfine state of the muonic atom, and the "spectator" atom X in the molecule DX, where X can be H, D, or T. The nonresonant rate at low (liquid hydrogen) temperature for $dd\mu$ formation is about 3×10^4 s^{-1} and for $dt\mu$ is about 6×10^5 s^{-1}. These rates are competitive with the resonant rates at low T for $d\mu(\uparrow\downarrow)+D_2$ and $t\mu(\uparrow\uparrow)+D_2$, but are 2 to 3 orders of magnitude smaller than the resonant rates for $d\mu(\uparrow\uparrow)+D_2$ and $t\mu(\uparrow\downarrow) + D_2$, respectively. The $(\uparrow\downarrow)$ state is the ground state; thus hfs quenching plays an important role in low-temperature experiments, especially for $dd\mu$. At room temperature, resonant formation is dominant for both the ground and excited hyperfine states of $d\mu$ and $t\mu$.

In the Vesman mechanism the binding energy of the muonic molecule goes into rovibrational excitation of the electronic host molecule instead of into ionization of a molecular electron. The process is resonant since the collision energy must be tuned to match the energy of the final discrete state. For the compound molecule formed, two sets of rovibrational quantum numbers are needed, e.g.,

$$(t\mu)_F + [D_2]_{K_i \nu_i} \to [(dt\mu)_{Jv}^{FS} dee]_{K_f \nu_f} , \qquad (86.34)$$

where (K_i, ν_i), and (K_f, ν_f) are the initial and final quantum numbers of the electronic molecule, (J, v) are the quantum numbers of the muonic molecule, F is the spin of the muonic atom, and S is the total spin of the muonic molecule. The energetics of this process is shown in Fig. 86.4.

The resonant condition is achieved at the collision energy

$$\epsilon_{res}[t\mu + D_2] = \epsilon_{11}^{FS}[dt\mu] + E_{K_f \nu_f}[(dt\mu)dee] - E_{K_i 0}[D_2] , \qquad (86.35)$$

where it is explicitly recognized that $(J, v) = (1,1)$ is the only muonic level that can satisfy the resonant energy condition and that only $\nu_i = 0$ is populated at ordinary temperatures. Accurate calculations require values of ϵ_{res} to within about 0.1 meV. The rovibrational energies $E_{K\nu}$ of the electronic molecule, as well as the Coulomb contributions to the binding energy of the muonic molecule, ϵ_{11}^{FS}, are now known to this high accuracy. However, ϵ_{11}^{FS}

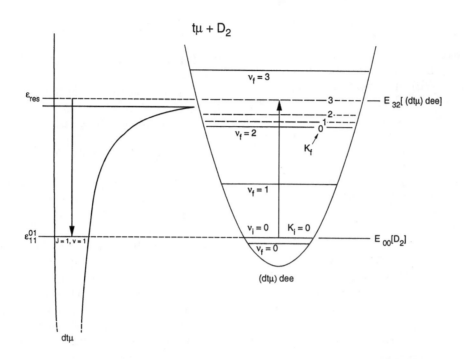

Figure 86.4. Diagram of energy levels for the resonant reaction $t\mu + [D_2]_{K_i=0,\nu_i=0} \rightarrow [(dt\mu)_{J=1,v=1}^{F=0,S=1} dee]_{K_f \nu_f}$. The rovibrational quantum numbers are designated by (J, v) for the muonic molecule and (K, ν) for the electronic molecules.

is subject to corrections due to relativity, vacuum polarization, nuclear charge distributions and polarizabilities, the hyperfine interaction, and the finite size and shape of the muonic molecule in the complex. The present overall accuracy is ~ 1 meV. Some of the resulting values of ϵ_{res} are given in Table 86.1. The calculated cross section for reaction (86.34) is sharply peaked at E_{res}, but must be averaged over a kinetic energy distribution (e.g. Maxwellian) to obtain the observable rate. Still, the rate will display a characteristic resonant dependence on T.

Because the E_{res} are different for each target molecule (D_2, DT, T_2, $\cdots$), the effective molecular formation rate in a mixture depends on the molecular composition in addition to the isotopic fractions (c_d, c_t, $\cdots$) if the target is not in chemical equilibrium.

The rate of $dd\mu$ and $dt\mu$ resonant molecular formation is calculated from

$$\lambda^{mf}(T) = N \sum_f \int d\epsilon \, 2\pi \, |\langle i|H'|f\rangle|^2 \, f(\epsilon, T) \, I(\epsilon - \epsilon_{if}, T),$$

(86.36)

where N is the target density, $\langle i|H'|f\rangle$ is a transition matrix element, $f(\epsilon, T)$ is the collisional energy distribution, ϵ_{if} is the energy of the unperturbed resonance, and $I(\Delta\epsilon, T)$ is the intensity at energy $\Delta\epsilon$ relative to the unperturbed energy. H' has usually been taken to be the dipole interaction ("post" form of the rearrangement-collision Hamiltonian using the $dt\mu$ bound state as the zeroth-order Hamiltonian [64, Appendix 1]),

$$H' = e^2 \mathbf{d} \cdot \mathbf{E},$$

(86.37)

where $\mathbf{d}$ is the dipole operator of the $dt\mu$ (or $dd\mu$) system and $\mathbf{E}$ is the electric field at the $dt\mu$ (or $dd\mu$) center of mass due to the "spectator" nucleus and electrons [65]. Conservation of angular momentum thus requires [66]

$$\mathbf{L} + \mathbf{K}_i = \mathbf{J} + \mathbf{K}_f$$

(86.38)

where $\mathbf{L}$ is the orbital angular momentum of relative motion for $t\mu + D_2$ in reaction (86.34). At low T, $L=0$ is predominant so that $K_f = K_i \pm 1$. This is simply the case for $d\mu + D_2 \rightarrow (dd\mu)dee$ where the most probable transition is $(K, \nu) = (0,0) \rightarrow (1,7)$. For $dt\mu$ the vibrational state of the electronic molecule changes by only $\Delta\nu_i = 2$ or 3 instead of 7, so the matrix element of (86.37) and the resulting rate are considerably larger than for $dd\mu$. However, it can be seen in Fig. 86.4 that if D_2 is in its ground state ($K_i=0$), the first level energetically accessible for $(dt\mu)dee$ has $K_f=3$. If, as proves to be adequate in the case of $dd\mu$, the intensity distribution I is taken to be a δ function, the lower levels are eliminated from Eq. (86.36). There are two possible solutions to this problem, whose importances have not been fully resolved: (1) the less likely $L > 0$ collisions contribute or (2) the levels with smaller K_f play a role even though they lie "below threshold".

The latter case is termed "quasiresonant". Theoretically the levels below threshold can contribute (1) *directly* if they are broadened so that they extend to positive energy [67–69] or (2) *indirectly* if configurations with

Table 86.1. Resonant (quasiresonant if negative) collision energies ϵ_{res} (in meV) calculated using Eq. (86.35).[a]

$d\mu + D_2$		$t\mu + D_2$				$t\mu + \text{DT}$	
$[(dd\mu)_{11}dee]_{\nu_f=7}$		$[(dt\mu)_{11}dee]_{\nu_f=2}$		$[(dt\mu)_{11}dee]_{\nu_f=3}$		$[(dt\mu)_{11}tee]_{\nu_f=3}$	
F, S, K_i, K_f	ϵ_{res}	F, S, K_i, K_f	ϵ_{res}	F, S, K_i, K_f	ϵ_{res}	F, S, K_i, K_f	ϵ_{res}
$\frac{1}{2}, \frac{1}{2}, 0, 1$	52.7	0, 1, 0, 1	−14.0	0, 1, 0, 1	277.1	0, 1, 0, 1	163.8
$\frac{1}{2}, \frac{3}{2}, 0, 1$	76.9	0, 1, 0, 2	−4.3	1, 0, 0, 1	223.5	1, 0, 0, 1	110.2
$\frac{3}{2}, \frac{1}{2}, 0, 1$	4.2	0, 1, 0, 3	10.3	1, 1, 0, 1	226.9	1, 1, 0, 1	113.6
$\frac{3}{2}, \frac{3}{2}, 0, 1$	28.4	0, 1, 1, 2	−11.7	1, 2, 0, 1	233.3	1, 2, 0, 1	120.0
$\frac{1}{2}, \frac{1}{2}, 1, 0$	40.9	0, 1, 1, 3	2.9				
$\frac{1}{2}, \frac{3}{2}, 1, 0$	65.1						
$\frac{1}{2}, \frac{1}{2}, 1, 2$	54.1						
$\frac{1}{2}, \frac{3}{2}, 1, 2$	78.3						

[a] Note: $kT = 1$ meV for $T = 11.6$ K.

different K_f are mixed [70]. Broadening can occur either inhomogeneously due to the finite lifetime (mainly with respect to Auger emission of an electron in the complex) or homogeneously due to collisions with neighboring molecules. Interactions with neighboring molecules also can mix the different K_f states, so the $K_f=3$ state may "borrow" some intensity from the lower K_f states. Three-body molecular formation facilitated by neighboring molecules leads to a density dependence of the formation rate (normalized to LHD) that has been observed in experiments.

86.4.5 Fusion

Usually the nuclear fusion rate in muonic molecules is calculated by a separable method; i.e., the united-atom limit of the molecular wave function is determined ignoring nuclear forces and then simply multiplied by a single number extracted from nuclear scattering experiments [71]. Fusion of d-t is strongly dominated by the $I^\pi = \frac{3}{2}^+$ resonance of ^{5}He. For $dt\mu$ we have

$$\lambda_{dt\mu}^f = A \lim_{r_{dt}\to 0} \int |\psi_{dt\mu}|^2 d^3 r_\mu \,, \qquad (86.39)$$

where A is simply related to the low-energy limit of the astrophysical S factor by

$$A = \frac{\hbar}{\pi e^2 M_r} \lim_{E\to 0} S(E) \,, \qquad (86.40)$$

where M_r is the reduced mass of the nuclei. $S(E)$ is usually obtained by fitting the $d + t$ fusion cross section observed in beam experiments to the form of Eq. (86.18).

The above formulation has yielded fusion rates in good agreement with another formulation — more accurate in principle — where a complex molecular wave function and energy are obtained directly incorporating the nuclear forces. In the latter approach,

$$\lambda_{dt\mu}^f = -2\,\text{Im}(E_{dt\mu})/\hbar \,, \qquad (86.41)$$

where $E_{dt\mu}$ is the complex eigenvalue (the imaginary part is $-\Gamma/2$, where Γ is the width). The nuclear effect in this formulation is taken into account by two different techniques: (1) using a complex optical potential [72, 73] (see Sect. 46.3), and (2) using the nuclear R-matrix as an interior boundary condition [74–76] (see Sect. 45.2.5). In the optical potential method, a short-range complex potential, determined by fitting experimental nuclear scattering data, is added to the three-body Coulomb potential; the real part describes elastic scattering and the imaginary (absorptive) part describes the fusion reaction. Then the eigenvalue problem is solved over all space with a regular boundary condition at the origin. In the R-matrix method, the same nuclear scattering data are used to determine a complex boundary condition at a distance characteristic of the nuclear forces, say $r_{dt} = a_{dt}$ where $a_{dt} \approx 5$ fm. The muonic eigenvalue problem is then formulated with the boundary condition at $r_{dt} = a_{dt}$ and solved over the space excluding $r_{dt} < a_{dt}$. The two methods can be used with similar basis sets to expand the wave function, which can also be used to calculate the sticking probability (see Sect. 86.4.6).

The relation of other μCF cross sections to normal beam experiments is somewhat more complicated. In the cases of $dd\mu$ and $tt\mu$, the fusion may occur in $J=1$ states, since the $J=1$ to $J=0$ transition is forbidden in molecules with identical nuclei. In this case the relevant information from beam experiments resides in the p-wave anisotropy, which is relatively small at low energies where σ is dominated by the s wave. It is then necessary to carry out the analysis in terms of the partial-wave transition amplitudes rather than the fit of the integrated σ via the S factor [77].

Fusions in $pd\mu$ and $pt\mu$ present a different complication [71]. There is a significant $E0$ contribution from muon conversion in addition to the $M1$ γ-ray contribution seen in $p + d$ and $p + t$ beam experiments (see Sect. 12.1 for discussion of multipole moments). The $E0$ contribution cannot be expressed through cross sections observed in the beam experiments, but has been determined using the bound-state nuclear wave functions of ^{3}He (or ^{4}He) and scattering wave functions of $p + d$ (or $p + t$). For $pd\mu$ there is the additional complication that two different p-d spin states contribute significantly. For $pt\mu$, theory [71] predicts that the probability of the fusion energy going into a e^+e^- pair is competitive with that of muon conversion, though the former has not yet been observed.

86.4.6 Sticking and Stripping

The fundamental mechanism of muon loss from the catalysis cycle, other than by particle decay, is via *sticking* to a helium nucleus, ^{4}He($\equiv \alpha$) or ^{3}He, produced in the fusion reaction. Especially in the case of $dt\mu$, where the charged particle is fast and the sticking probability is already small, subsequent collisions may *strip* the muon. Thus the sticking probability is determined by two steps

$$dt\mu \overset{1-\omega_s^0}{\underset{\omega_s^0}{\diagup\diagdown}} \begin{matrix} \alpha + \mu + n \\ \\ \alpha\mu\,(3.5\ \mathrm{MeV}) + n \end{matrix} \tag{86.42}$$

$$\quad\quad\quad\quad\quad \overset{R}{\underset{1-R}{\diagup\diagdown}} \begin{matrix} \alpha + \mu \\ \\ \alpha\mu\ (\mathrm{thermal})\,. \end{matrix}$$

The initial sticking probability ω_s^0 depends only on intramolecular dynamics, but the stripping conditional probability depends on collisions.[9] The net sticking is then

$$\omega_s = \omega_s^0 (1 - R)\,. \tag{86.43}$$

Since the nuclear reaction is very rapid compared with the atomic and molecular dynamics, the probability of sticking in a given state ν is given adequately by the *sudden approximation*,

$$P_\nu = |\langle \psi_\nu^{(f)} | \psi^{(i)} \rangle|^2\,, \tag{86.44}$$

where the initial wave function $\psi^{(i)}$ is the normalized molecular wave function in the limit $r_{dt} \to 0$ and the final wave function $\psi_\nu^{(f)}$ is given by

$$\psi_\nu^{(f)} = \phi_{n\ell m}(\mathbf{r})e^{i\mathbf{q}\cdot\mathbf{r}} \tag{86.45}$$

in which $\phi_{n\ell m}$ is an atomic wave function of $(^4$He$\mu)^+$ and the plane wave with momentum $\mathbf{q}$ represents its motion with respect to the initial molecule (recoil determined

---

[9]Note that ω_s^0 is *not* the sticking in the zero-density limit since R is still finite in this limit.

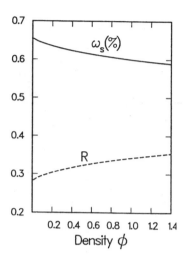

Figure 86.5. Theoretical sticking fraction (solid curve) and reactivation probability (dashed curve) for d-t μCF.

by conservation of energy and momentum). The total sticking is then

$$\omega_s^0 = \sum_\nu P_\nu\,. \tag{86.46}$$

The $(^4$He$\mu)^+$ wave function is known analytically since it is hydrogenic. Most of the labor goes into determination of the muonic molecule wave function. In the Born-Oppenheimer approximation this is simply "$(^5$He$\mu)^+$" and results in $(\omega_s^0)_{\mathrm{BO}} = 1.20\%$ for $dt\mu$ [78]. More accurate nonadiabatic calculations show that the muonic motion lags behind that of the nuclei and reduces ω_s^0 to 0.886% [79]. After inclusion of nuclear effects, the best current theoretical value of ω_s^0 is 0.912% [76, 80].

Since the ground state (Heμ)$^+$ ion is bound by 11 keV, it takes a quite energetic collision to strip off the muon. The reactivation fraction R is determined basically by competition between collisional processes that slow down the muonic ion and those that lead to stripping. Calculation of R requires a full kinetic treatment of the fast (Heμ)$^+$ ion, starting with its distribution among various states ($1s$, $2s$, $2p$, $\cdots$, $n \approx 10$). The most important processes are stopping power (due mainly to ionization of the medium) and muon ionization or transfer in collisions of the (Heμ)$^+$ ion with an isotope of H, but inelastic (excitation and deexcitation), Auger deexcitation, and ℓ-changing collisions as well as radiative deexcitation are also involved. The initial sticking occurs mostly in the $1s$ state (77% of the $\alpha\mu$'s from $dt\mu$ fusion), but the excited states have larger ionization cross sections. Most of the muons stripped from $\alpha\mu$ originally stuck in the $1s$ state, but a significant number are promoted to excited states before being ionized (so-called "ladder ionization"). The metastable $2s$ state is significant for its role in prolonging the excited

Table 86.2. Comparison of sticking values.[a]

Source	ϕ	$\omega_s(\%)$	$\omega_s^0(\%)$
		Theory	
[76, 80]	1.2	0.59	0.91
		Neutron Experiments	
LANL [81]	~ 1	$0.43\pm0.05\pm0.06$	0.66
PSI [82]	~ 1	$0.48\pm0.02\pm0.04$	0.74
KEK [83]	1.2	0.51 ± 0.004	0.78
		x-ray Experiments	
PSI [84]	1.2	0.39 ± 0.10	0.60
KEK [83]	1.2	0.34 ± 0.13	0.52
		$\alpha/\alpha\mu$ Experiments	
LANL [85]	0.001	–	$0.80\pm0.15\pm0.12$
PSI [86]	0.17	0.56 ± 0.04	0.80

[a]Experimental values of ω_s^0 without error bars were obtained assuming the theoretical stripping [87, 88]. In cases of two error estimates, the first is statistical and the second is systematic. The extraction of ω_s from the x-ray experiments requires theoretical scaling.

state populations. The resulting values of R and ω_s are shown as a function of density in Fig. 86.5.

Most experiments on *d-t* μCF have been done with neutron detection [89–91], where λ_c and the muon loss probability per cycle W can be deduced from the time structure of the neutron emissions. The analysis is indirect and requires a theoretical model. What is actually measured is the product $W\phi\lambda_c$; the extraction of ω_s requires corrections for other loss mechanisms and separate determination of λ_c. Thus it is desirable to have other experimental diagnostics. Two types of corroborating experiments detect either x-rays from the $\alpha\mu$ formed by sticking *or* detect the species $(\alpha)^{2+}$ and $(\alpha\mu)^+$ by the different effects of their double and single electrical charges.

The theoretical sticking is compared in Table 86.2 with that from all three types of experiments. For a more meaningful comparison of measurements at different densities ϕ, the theoretical R has been used to convert all values to ω_s^0. The theoretical values are slightly, but significantly, higher than the observations. This discrepancy has not yet been resolved.

One intriguing explanation for the lower-than-predicted value of ω_s may be that a significant fraction of the fusions might occur in *muonically* excited bound or resonant states for which the initial sticking is lower than in the ground state. This possibility has just begun to be investigated [92].

86.4.7 Prospectus

There is still a great deal to be learned from the less-studied μCF cycles, like those of *p-d*, *p-t*, and

t-t, but more quantitative work is also needed on some key processes in the *d-t* μCF cycle: in particular, three-body effects on molecular formation at high densities, the excited-state cross sections and kinetics that go into the determination of the cascade factor q_{1s}, and the remaining discrepancy in the sticking factor ω_s which might have a theoretical or experimental resolution. Experimentally it is of interest to push on to higher temperatures and densities to see if more surprises lurk there. There have been a few schemes proposed to enhance stripping of stuck muons artificially, but none has been subjected to experiments yet.

The currently observed yield of about 150 *d-t* fusions (releasing 17.6 MeV each) per muon produces an energy return 25 times the rest-mass energy of the muon, but is only about one-third of that required for breakeven in a pure-fusion reactor. This conclusion is based on the estimated energy cost of producing a muon, $\sim$8 GeV [93]. Other possible practical uses include a hybrid (fusion-fission) reactor [93] or an intense 14 MeV neutron source [94].

Apart from such technological applications, the study of μCF is fruitful for a number of reasons including (1) bridging the gap between atomic and nuclear physics, (2) enabling nuclear reactions (including *p*-waves) at room temperature, (3) allowing precise studies under unusual physical conditions, (4) observing a compound electronic-muonic molecular environment, and (5) exhibiting phenomena spanning nine orders of magnitude in distance and energy. The experimental possibilities are far from exhausted even though the holy grail of pure fusion energy now appears just beyond reach.

Acknowledgments

J.D.M. is grateful to the Institute for Nuclear Theory of the University of Washington for making it possible for him to spend a productive semester there in the spring of 1993. This work has also been supported by National Science Foundation grant PHY-9215442.

REFERENCES

1. W. E. Lamb and R. C. Retherford, Phys. Rev. **72**, 241; **79**, 549 (1950); **81**, 222 (1951); **86**, 1014 (1952).
2. W. V. Houston and Y. M. Hsieh, Phys. Rev. **45**, 263 (1934); W. V. Houston, Phys. Rev. **51**, 148 (1937); **51**, 446 (1937); R. C. Williams, Phys. Rev. **54**, 558 (1938); S. Pasternack, Phys. Rev. **54**, 1113 (1938).
3. E. C. Kemble and R. D. Present, Phys. Rev. **44**, 1031 (1933).
4. R. Karplus, A. Klein, and J. Schwinger, Phys. Rev. **86**, 301 (1951).
5. W. E. Lamb, Phys. Rev. **85**, 276 (1952).
6. E. E. Salpeter, Phys. Rev. **89**, 95 (1953).
7. A. C. Zemach, Phys. Rev. **104**, 1771 (1956).

8. S. Flügge, *Practical Quantum Mechanics* I (Springer, New York, 1971), pp. 191–192.

9. G. W. F. Drake and Z.-C. Yan, Phys. Rev. A **46**, 2378 (1992).

10. D. Shiner, R. Dixson, and V. Vedantham, Phys. Rev. Lett. **74**, 3553 (1995).

11. E. Riis *et al.*, Phys. Rev. A **49**, 207 (1994).

12. K. Pachucki, K. M. Weitz, and T. W. Hänsch, Phys. Rev. A **49**, 2255 (1994).

13. A. L. Schawlow and C. H. Townes, Phys. Rev. **100**, 1273 (1955).

14. A. R. Bodmer, Proc. Phys. Soc. A**66**, 1041 (1953).

15. P. J. Mohr, At. Data Nucl. Data Tables **33**, 456 (1983).

16. J. L. Friar, Ann. Phys. (N. Y.) **122**, 151 (1979).

17. W. R. Johnson and G. Soff, At. Data Nucl. Data Tables **33**, 407 (1985).

18. C. W. de Jager, H. de Vries, and C. de Vries, At. Data Nucl. Data Tables **14**, 479 (1974).

19. P. Raghaven, At. Data Nucl. Data Tables **42**, 189 (1989).

20. A. H. Wapstra and G. Audi, Nucl. Phys. A **432**, 1 (1985).

21. E.g., J. F. Seely *et al.*, Phys. Rev. Lett. **57**, 2924 (1986); T. E. Cowan *et al.*, Phys. Rev. Lett. **66**, 1150 (1991); J. Schweppe *et al.*, Phys. Rev. Lett. **66**, 1434 (1991).

22. S. A. Blundell and N. J. Snyderman, Phys. Rev. A **44**, R1427 (1991); S. A. Blundell, Phys. Rev. A **46**, 3762 (1992).

23. K. T. Cheng, W. R. Johnson, and J. Sapirstein, Phys. Rev. A **47**, 1817 (1993).

24. J. N. Bahcall, Phys. Rev. **124**, 495 (1961).

25. D. D. Clayton, Ap. J. **139**, 637 (1964); S. E. Woosley and W. A. Fowler, Ap. J. **233**, 411 (1979).

26. R. D. Williams, W. A. Fowler, and S. E. Koonin, Astrophys. J. **281**, 363 (1984).

27. Z. Chen, L. Rosenberg, and L. Spruch, Phys. Rev. A **35**, 1981 (1987); Adv. At. Mol. Opt. Phys. **26**, 297 (1989); Z. Chen and L. Spruch, in *AIP Conference Proceedings # 189, Relativistic, Quantum Electrodynamic, and Weak Interaction Effects in Atoms, 1989* (American Institute of Physics, New York, 1989), pp. 460–478.

28. V. A. Lubimov, E. G. Novikov, V. Z. Nozik, E. F. Tretyakov, and V. S. Kozik, Phys. Lett. B**94**, 266 (1980). See also S. Boris *et al.*, Phys. Rev. Lett. **58**, 2019 (1987); S. D. Boris *et al.*, Pis'ma Zh. Eksp. Teor. Fiz. **45**, 267 (1987) [Sov. Phys. JETP Lett. **45**, 333 (1987)].

29. K.-E. Bergkvist, Phys. Scr. **4**, 23 (1971); Nucl. Phys. B **39**, 317, 371 (1972).

30. R. L. Martin and J. S. Cohen, Phys. Lett. A **110**, 95 (1985).

31. W. Kolos, B. Jeziorski, J. Rychlewski, K. Szalewicz, H. J. Monkhorst, and O. Fackler, Phys. Rev. A **37**, 2297 (1988), and earlier references therein.

32. M. Fritschi *et al.*, Phys. Lett. B **173**, 485 (1986); J. F. Wilkerson *et al.*, Phys. Rev. Lett. **58**, 2023 (1987); R. G. H. Robinson *et al.*, Phys. Rev. Lett. **67**, 957 (1991); E. Holzschuh, M. Fritschi, and W. Kündig, Phys. Lett. B **287**, 381 (1992); C. Weinheimer *et al.*, Phys. Lett. B **300**, 210 (1993).

33. H. J. Assenbaum, K. Langanke, and C. Rolfs, Z. Phys. A **327**, 461 (1987).

34. S. Engstler, A. Krauss, K. Neldner, C. Rolfs, U. Schröder, and K. Langanke, Phys. Lett. B **202**, 179 (1988).

35. U. Schröder, S. Engstler, A. Krauss, K. Neldner, C. Rolfs, E. Somorjai, and K. Langanke, Nucl. Instrum. Meth. B **40/41**, 466 (1989); S. Engstler, G. Raimann, C. Angulo, U. Greife, C. Rolfs, U. Schröder, E. Somorjai, B. Kirch, and K. Langanke, Phys. Lett. B **279**, 20 (1992); Z. Phys. A **342**, 471 (1992).

36. G. Blüge, K. Langanke, H. G. Reusch, and C. Rolfs, Z. Phys. A **333**, 219 (1989); K. Langanke and D. Lukas, Ann. Phys. (Leipzig) **1**, 332 (1992); K. Langanke, Adv. Nucl. Phys. **21**, 179 (1994).

37. T. D. Shoppa, S. E. Koonin, K. Langanke, and R. Seki, Phys. Rev. C **48**, 837 (1993).

38. J. Eichler, Phys. Rep. **193**, 165 (1990).

39. R. Anholt and H. Gould, Adv. At. Mol. Phys. **22**, 315 (1986).

40. J. Stachel and P. Braun-Munsinger, Adv. Nucl. Phys. (to be published).

41. A. Belkacem, H. Gould, B. Feinberg, R. Bossingham, and W. E. Meyerhof, Phys. Rev. Lett. **71**, 1514 (1993).

42. J. S. Cohen, in *Review of Fundamental Processes and Applications of Atoms and Ions*, edited by C. D. Lin (World Scientific, Singapore, 1993), Chap. 2, pp. 61–110.

43. L. Bogdanova and V. Markushin, Nucl. Phys. A **508**, 29c (1990).

44. P. Ackerbauer *et al.*, Hyperfine Interact. **82**, 243 (1993).

45. D. V. Balin *et al.*, Muon Catal. Fusion **5/6**, 163 (1990/91).

46. P. Baumann *et al.*, Phys. Rev. Lett. **70**, 3720 (1993).

47. W. H. Breunlich, P. Kammel, J. S. Cohen, and M. Leon, Ann. Rev. Nucl. Part. Sci. **39**, 311 (1989).

48. S. S. Gershtein, Yu. V. Petrov, and L. I. Ponomarev, Usp. Fiz. Nauk **160**, 3 (1990) [Sov. Phys. Usp. **33**, 591 (1990)].

49. P. Froelich, Adv. Phys. **41**, 405 (1992).

50. C. Petitjean, Nucl. Phys. A **543**, 79c (1992).

51. J. S. Cohen, in *Electromagnetic Cascade and Chemistry of Exotic Atoms*, Proceedings of the International School of Physics of Exotic Atoms, 5th Course, Erice, Sicily, 1989, edited by L. M. Simons, D. Horvath, and G. Torelli (Plenum, New York, 1990), pp. 1–22.

52. P. K. Haff and T. A. Tombrello, Ann. Phys. (N. Y.) **86**, 178 (1974).

53. A. S. Wightman, Phys. Rev. **77**, 521 (1950).

54. J. S. Cohen, Phys. Rev. A **27**, 167 (1983).

55. J. S. Cohen, R. L. Martin, and W. R. Wadt, Phys. Rev. A **24**, 33 (1981); **27**, 1821 (1983).

56. J. D. Garcia, N. H. Kwong, and J. S. Cohen, Phys. Rev. A **35**, 4068 (1987).

57. N. H. Kwong, J. D. Garcia, and J. S. Cohen, J. Phys. B **22**, L633 (1989).

58. E. Fermi and E. Teller, Phys. Rev. **72**, 406 (1947).

59. G. Ya. Korenman and V. P. Popov, in *AIP Conference Proceedings # 181, Muon Catalyzed Fusion, 1989*, edited by S. E. Jones, J. Rafelski, and H. J. Monkhorst (American Institute of Physics, New York, 1989), p. 145.

60. W. Czaplinski *et al.*, Phys. Rev. A **50**, 525 (1994).

61. P. Froelich and J. Wallenius, Phys. Rev. Lett. **75**, 2108 (1995).

62. J. S. Cohen, Phys. Rev. A **43**, 4668 (1991).

63. E. A. Vesman, Pis'ma Zh. Eksp. Fiz. **5**, 113 (1967) [JETP Lett. **5**, 91 (1967)].

64. M. P. Faifman, L. I. Menshikov, and T. A. Strizh, Muon

Catal. Fusion **4**, 1 (1989).

65. J. S. Cohen and R. L. Martin, Phys. Rev. Lett. **53**, 738 (1984).

66. M. Leon, Phys. Rev. Lett. **52**, 605 (1984).

67. Yu. V. Petrov, Phys. Lett. B **163**, 28 (1985).

68. L. I. Menshikov and L. I. Ponomarev, Phys. Lett. B **167**, 141 (1986).

69. J. S. Cohen and M. Leon, Phys. Rev. A **39**, 946 (1989).

70. M. Leon, Phys. Rev. A **39**, 5554 (1989).

71. L. N. Bogdanova, Muon Catal. Fusion **3**, 359 (1988).

72. L. N. Bogdanova, V. E. Markushin, V. S. Melezhik, and L. I. Ponomarev, Yad. Phys. **34**, 1191 (1981) [Sov. J. Nucl. Phys. **34**, 662 (1981)].

73. M. Kamimura, in *AIP Conference Proceedings # 181, Muon-Catalyzed Fusion, 1989*, edited by S. E. Jones, J. Rafelski, and H. J. Monkhorst (AIP, New York, 1989), p. 330.

74. M. C. Struensee, G. M. Hale, R. T Pack, and J. S. Cohen, Phys. Rev. A **37**, 340 (1988).

75. K. Szalewicz *et al.*, Phys. Rev. A **42**, 3768 (1990).

76. C.-Y. Hu, G. M. Hale, and J. S. Cohen, Phys. Rev. A **49**, 4481 (1994).

77. G. M. Hale, Muon Catal. Fusion **5/6**, 227 (1990/91).

78. L.
Bracci and G. Fiorentini, Nucl. Phys. A **364**, 383 (1981).

79. S. E. Haywood *et al.*, Phys. Rev. A **37**, 3393 (1988); **43**, 5847 (1991).

80. B. Jeziorski *et al.*, Phys. Rev. A **43**, 1640 (1991).

81. S. E. Jones, S. F. Taylor, and A. N. Anderson, Hyperfine Interact. **82**, 303 (1993).

82. C. Petitjean *et al.*, Hyperfine Interact. **82**, 273 (1993).

83. K. Nagamine *et al.*, Hyperfine Interact. **82**, 343 (1993).

84. H. Bossy *et al.*, Phys. Rev. Lett. **59**, 2864 (1987).

85. M. A. Paciotti *et al.*, in *AIP Conference Proceedings # 181, Muon-Catalyzed Fusion, 1989*, edited by S. E. Jones, J. Rafelski, and H. J. Monkhorst (American Institute of Physics, New York, 1989), p. 38.

86. T. Case *et al.*, Hyperfine Interact. **82**, 295 (1993).

87. M. C. Struensee and J. S. Cohen, Phys. Rev. A **38**, 44 (1988).

88. C. D. Stodden, H. J. Monkhorst, K. Szalewicz, and T. G. Winter, Phys. Rev. A **41**, 1281 (1990).

89. S. E. Jones *et al.*, Phys. Rev. Lett. **56**, 588 (1986).

90. W. H. Breunlich *et al.*, Phys. Rev. Lett. **58**, 329 (1987).

91. K. Nagamine *et al.*, Muon Catal. Fusion **1**, 137 (1987).

92. P. Froelich and A. Flores-Riveros, Phys. Rev. Lett. **70**, 1595 (1993).

93. Yu. V. Petrov, Nature (London) **285**, 466 (1980); Muon Catal. Fusion **3**, 525 (1988).

94. T. Kase *et al.*, Muon Catal. Fusion **5/6**, 521 (1990/91).

87

Charged Particle-Matter Interactions

Hans Bichsel

1211 22nd Avenue East, Seattle, Washington 98112-3534

87.1 INTRODUCTION

In the description of the interaction of fast charged particles with matter, two aspects can be distinguished: the effects on the particle, usually energy losses and deflections, and the spatial distribution of the energy lost by the particle in the absorber.

This Chapter discusses principles needed in the operation of charged particle detectors. Specifically, the effects used for the instantaneous observation (i.e., within fractions of a millisecond) of the passage of a charged particle are described. Delayed effects, such as chemical reactions (e.g., biological effects, chemical dosimeters, photographic emulsions), metastable states, etc. are not discussed. It is assumed that particle speeds have been determined with, e.g., magnetic analyzers or by measurement of the time of flight. A measurement of particle ranges can also be used to determine the initial speed. The description is restricted to fast charged particles, defined by speeds $v > 6v_B$ (v_B is the Bohr speed), or $\beta = v/c > 0.04$. Interactions and cross sections at smaller speeds are discussed, e.g., by Rudd *et al.* [1].

The present Chapter primarily considers energy loss straggling, rather than the stopping powers discussed by Fano [2]. The methods described can be used to calculate reliable data for detector applications. No attempt is made to present a complete review. Anecdotal, qualitative examples of various effects are described. The following definitions are used:

1. The number of atoms or molecules per unit volume $N = N_A\rho/A$, with A the molecular weight (in g/mole) of the absorber (with Z_2 electrons per molecule), ρ its density, and $N_A = 6.022\,137 \times 10^{23}$ mol^{-1} is Avogadro's number;

2. The relativistic factors β and γ for particles with rest mass M_0, speed $v = \beta c$ and kinetic energy $\mathcal{T}$:
$$\beta^2 = (\mathcal{T}/M_0c^2)(2 + \mathcal{T}/M_0c^2)/(1 + \mathcal{T}/M_0c^2)^2,$$
$$\gamma = M/M_0 = 1 + \mathcal{T}/M_0c^2, \quad \gamma^2 = 1/(1 - \beta^2),$$
$$\beta^2\gamma^2 = \gamma^2 - 1;$$

3. The coefficient of the Rutherford equation:
$$k_{\mathrm{R}} = \frac{2\pi Z_1^2 e^4}{m_e c^2} = 2\pi r_0^2 m_e c^2 Z_1^2$$
$$= 2.54955 \times 10^{-19} \, Z_1^2 \text{ eV cm}^2 , \quad (87.1)$$

where m_e is the rest mass of an electron, $-e$ its charge, $Z_1 e$ the charge of the incident particle, and $r_0 = \alpha^2 a_0$ is the classical electron radius; and

4. The maximum energy loss of a heavy particle to an electron: $E_{\max} = 2mc^2\beta^2\gamma^2$.

87.2   EXPERIMENTAL ASPECTS

87.2.1   Energy Loss Experiments and Radiation Detectors

After passing through a thickness x of material, an initially monoenergetic beam of particles acquires a distribution of energies described by a probability density function
$$F(\Delta) = \frac{d\mathcal{N}(\Delta)}{d\Delta}, \quad (87.2)$$

where Δ is the energy per particle, and $d\mathcal{N}(\Delta)$ is the number of particles in the range Δ to $\Delta + d\Delta$. The *straggling function* $F(\Delta)$ represents the spectrum of energy losses Δ, such that $\int F(\Delta)\,d\Delta = \mathcal{N}$, the total number of particles observed. It is characterized by the quantities

1. the most probable energy loss Δ_{mp},

2. the full width at half maximum Γ,

3. the moments [6]
$$\mu_\nu = \frac{1}{\mathcal{N}} \int F(\Delta)\Delta^\nu\,d\Delta , \quad (87.3)$$

and the central moments
$$C_\nu = \frac{1}{\mathcal{N}} \int F(\Delta)(\Delta - \mu_1)^\nu\,d\Delta . \quad (87.4)$$

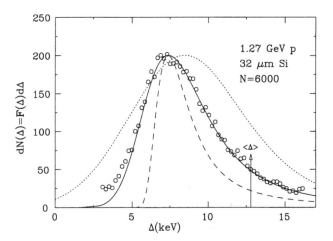

Figure 87.1. Energy loss straggling functions $F(\Delta)d\Delta$ for 1.27 GeV protons traversing a 32 μm silicon detector, with $d\Delta = 0.21$ keV. The experimental data are shown by circles. Three calculated functions normalized to the same peak height are shown for comparison [3]. The solid line was calculated with the Bethe cross section (see Fig. 87.3) with $\Delta_{\mathrm{mp}} = 7.4$ keV, $\Gamma = 5.2$ keV, and $\langle\Delta\rangle = 12.8$ keV. The dotted line is the Landau function [4] and the broken line is the Blunck–Leisegang modification [5].

Then $\mu_1 = \langle\Delta\rangle$ is the mean energy loss, $C_2 = \sigma^2$ is the variance and $\gamma_1 = C_3/C_2^{3/2}$ is the skewness. The *fluence spectrum* $\phi(\mathcal{T})$ is the complementary function describing the distribution of residual energies $\mathcal{T}$ of the particles.

As an example, Fig. 87.1 shows the $F(\Delta)\,d\Delta$ for 1.27 GeV protons passing through a 32 μm thick silicon wafer [3]. The measured quantity is the ionization J resulting from the creation of electron-hole pairs in the silicon, so that $\Delta = JW$ is the energy deposited, where W is the energy required to create an electron-hole pair. For $\Delta < 15$ keV, the energy deposited differs little from the energy lost by the protons. For larger Δ, some of the secondary electrons may escape from the silicon, making the apparent energy deposited less than the energy lost [7]. The spectrum has a long tail extending up to a maximum energy loss of 4.6 MeV. This accounts for the large value of $\langle\Delta\rangle = 12.8$ keV, and the even larger value $C_2^{1/2} = 43$ keV for the standard deviation, relative to $\Delta_{\mathrm{mp}} = 7.4$ keV and $\Gamma = 5.2$ keV.

In losing energy, the beam particles also suffer angular deflections, but for heavy particles, the deflections are usually small. For electrons, angular deflections are quite important, and are discussed in Chaps. 45, 62, and 63. Nuclear reactions cause large effects but are quite infrequent. Further examples of straggling spectra are given in Refs. [3, 8–10] for thin, extremely thin, moderately thick, and thick absorbers respectively.

The stopping power $S(T) = -dT/dx$ (i.e., the mean energy lost per unit thickness) provides a convenient measure of the energy loss process. For a beam with incident energy T_0 and $\langle\Delta\rangle < 0.1T_0$, $S(T)$ is often approximated by

$$S(T_0 - \langle\Delta\rangle/2) \simeq \langle\Delta\rangle/x, \qquad (87.5)$$

provided that $x \gg a_0$ (see Sect. 87.5). The stopping power $S(T)$ depends on the absorber properties, especially the electron density NZ_2. The radiation dose $D(x)$ at a distance x into the material is given by

$$D(x) = \int \phi(T,x)S(T)dT, \qquad (87.6)$$

where $\phi(T,x)$ is the fluence spectrum. $D(x)$ is used in radiation dosimetry and protection [11].

$S(T)$ ceases to be useful for very thin materials, such as microscopic biological specimens or thin ionization chambers, because $\Delta_{mp} \ll \langle\Delta\rangle$. In this case, $F(\Delta)$ can be used directly together with, for example, *Monte Carlo* simulations of the scattering events (see Sect. 87.7). If measurements of $F(\Delta)$ are not available for the thin sample, then it must be estimated theoretically.

A detailed theoretical description of the energy deposition process requires cross sections for the various scattering events in the target, the most important being collisions with electrons. The Rutherford cross section for the collision of a particle with a free electron is only useful for $\Delta \gtrsim 30U_Z$, where U_Z is the binding energy of a target electron ($\sim$500 eV for outer shell electrons). Various modifications to the Rutherford cross section for smaller Δ are discussed in Sect. 87.3.

87.2.2   Inelastic Scattering Events

If a beam of $\mathcal{N}$ monoenergetic particles with speed v passes through an absorber of infinitesimal thickness dx, the number of particles experiencing a collision with energy loss between E and $E + dE$ is given by

$$d\mathcal{N}(E) = \mathcal{N}N\sigma(E)\,dE\,dx, \qquad (87.7)$$

where $\sigma(E)$ is the collision cross section per molecule differential in E; it depends on β. Using this equation, $\sigma(E)$ can be obtained from measurements [12, 13] with extremely thin absorbers where particles on the average suffer less than one collision. If particles pass through a thicker absorber, they make several collisions with energy losses E_i; and the total energy loss is $\Delta = \sum E_i$.

Large differences exist between a gas and a solid of the same composition due to changes in the valence shell electrons. In isolated atoms [14], the smallest energy losses are to discrete excited electronic states. Ionizations occur for energy losses exceeding the binding energies U_l for each atomic shell l and the released electrons

Figure 87.2.   Comparison of the dipole oscillator strength spectra for atomic and solid silicon. Discrete atomic excitations are shown by vertical lines, continuum excitations by the dotted line. Data for the atom are based on calculations [15]. Data for the solid are from Ref. [3]. The broad peak at $\sim$17 eV represents the plasmon excitations. Uncertainties may exceed 10%. The discrete atomic excitations disappear in the solid. The photoionization cross section is $\sigma(E) = 109.8\,f(E)$ Mbarn [14].

are given a kinetic energy $\delta = E - U_l$. For ionization, energy losses are continuous (see Fig. 87.2). If atoms are brought closer together to form a liquid or a solid, their valence electrons come under the influence of the cores of surrounding atoms. A core is defined to consist of the nucleus and all the electrons inside the valence shell. For C, O, or H_2O, the core consists of two electrons in the K-shell, for Al or Si it contains two electrons in the K-shell and eight electrons in the L-shell. For metals, the valence electrons form a conduction band where they are nearly free. If a charged particle moves through the solid, the transfer of very small energies to the electron by Rutherford scattering is not observed. Instead, many thousands of them are excited at a time. For metals, this process is called *plasmon excitation*, for insulators, *collective excitation* (see Fig. 87.2). For most substances, the plasmon excitation energy E_p is much larger than the energy of the lowest excited state of the atom, E_1. For example, for Be, $E_1 = 3.6$ eV [15], while $E_p = 19$ eV [12]. Similarly, for silicon, $E_1 = 3.6$ eV [15] and $E_p = 16.7$ eV [3].

For molecules, measurements of electron energy losses [13] provide information about the difference in the structure of $\sigma(E)$ between gas and solid. For example, for benzene (C_6H_6) the vapor shows distinct structures for excitations to several discrete states, and a broad peak at about 16 eV which appears to be equivalent to a collective excitation. For the solid, the structures are broadened, and the major peak for the collective excitations shifts to 21 eV. For water, experimental data are incomplete.

For energy losses well above E_p, $\sigma(E)$ decreases smoothly until the ionization energy of the next electron shell is reached, as shown in Fig. 87.2. In a molecular substance such as $CaCO_3$, several broad peaks appear between 7 and 40 eV, followed by narrow peaks at 280 eV (carbon K-shell), 345 eV (Ca L-shells), 530 eV (oxygen K-shell) and 4 keV (Ca K-shell) [12].

While there are large changes in the structure of the excitation of the valence electrons as the atoms are coalesced into a solid (see Fig. 87.2), changes are less important for the inner shells because the binding energies of these shells are much larger than chemical energies. As an example, for an amorphous thin carbon film, the K-edge is at 284 eV [12], and K-shell excitation shows essentially just one peak at $\sim$296 eV. For the three biologically important molecules adenine, uracil and thymine, several peaks appear [16], all located between 284 and 300 eV. Thus there is at most a small change in the energy of the K-shell excitation, but the number of peaks as well as their locations changes considerably. For Al above the K-edge (1.56 keV), a structure with several peaks appears [17], with separations of about 40 eV. This structure, called extended x-ray absorption fine structure (EXAFS), is caused by the presence of nearby atomic cores which backscatter the photoelectrons and thus change their wave functions. The discrete excitations of the atom below the K-edge disappear completely. Thus solid state and chemical effects are very important for valence shell electrons, and less so for inner-shell electrons.

87.3 THEORY OF CROSS SECTIONS

The cross section for the collision of two free charged particles is given by the Rutherford expression. If charged particles collide with electrons bound in atoms, molecules or solids, the cross section can be written as a modified Rutherford cross section. A plausible way of describing these interactions is to consider the emission of virtual photons by the fast particle, which then are absorbed by the material in the *Fermi virtual photon method* (FVP) [18]. The collision cross section then is proportional to the photoabsorption cross section of the molecules. Bohr [19] described this as a "resonance" effect.

A variety of models has been used to obtain theoretical $\sigma(E)$ for bound electrons. Here, three of them will be described and compared with each other. Examples are shown in Fig. 87.3. Few analytic functions and methods are available to calculate cross sections (see Chap. 45). Usually, numerical calculations must be made to obtain reliable data for real absorbers.

Figure 87.3. Inelastic cross sections $\sigma(E)$ for single collisions in solid silicon, for incident protons with an energy $T = 100$ MeV, calculated using different theories. The horizontal line at 1.0 represents the Rutherford cross section from Eq. (87.8). The other curves are (—) Bethe theory; ($\cdots$) FVP approximation; (- - -) binary encounter approximation from Eq. (87.10) [26]. The curves extend to $E_{max} = 230$ keV.

87.3.1 Rutherford Cross Section

The cross section for close collisions of fast charged particles with loosely bound electrons is well approximated by the cross section for the collision of a charged particle with a free electron at rest. The nonrelativistic Rutherford cross section $\sigma_R(E)$ for an energy loss E in the collision of a charged particle with speed v in the laboratory frame is given by [21–23]

$$\sigma_R(E) = \frac{2\pi Z_1^2 e^4}{mv^2}\frac{1}{E^2} = \frac{k_R}{\beta^2}\frac{1}{E^2}. \qquad (87.8)$$

Since the secondary electron receives all the energy E lost by the incident particle, the momentum transfer is $q = \sqrt{2mE}$. The cross section $\sigma_R(E)$ does not depend on particle mass M. The leading relativistic correction[1] is

$$\sigma_R'(E) = \sigma_R(E)\left(1 - \beta^2\frac{E}{E_{max}}\right). \qquad (87.9)$$

87.3.2 Binary Encounter Approximation

A simple correction to $\sigma_R(E)$ can be achieved by taking into account the velocity of the bound electrons. With binding energy U_b, average kinetic energy T_e, and average speed u of the electrons, the expression is

---

[1]For $\gamma > M/m$, further terms must be included [24, 25].

$$\sigma_L(E) = \sigma_R(E)\left(1 + \frac{4}{3}\frac{\mathcal{T}_e}{E}\right), \quad E > U_b, \quad (87.10)$$

which is valid for $v \gg u$. The total cross section for a molecule includes contributions from each electron shell. Variants of this approach are described in Refs. [26, 27]. Fig. 87.3 shows an example.

87.3.3 Bethe Model of Cross Section

The Bethe model [28] derives from the first Born approximation for inelastic scattering, which becomes essentially exact at high energies. In terms of the *inelastic form factor* $|\mathcal{F}(E, \mathbf{K})|^2$, the energy loss cross section is

$$d\sigma(E, Q) = \sigma_R(Q)|\mathcal{F}(E, \mathbf{K})|^2\, dQ, \quad (87.11)$$

where $Q = q^2/2m_e$, and $\hbar\mathbf{K}$ is the momentum transfer vector. The *generalized oscillator strength* (GOS) is defined by

$$f(E, K) = E|\mathcal{F}(E, \mathbf{K})|^2/Q, \quad (87.12)$$

such that $f(E, K)$ reduces to the optical dipole oscillator strength $f(E, 0) \equiv f(E)$ in the limit $K \to 0$. Then

$$d\sigma(E, Q) = \sigma_R(E)Ef(E, K)\, d\ln Q. \quad (87.13)$$

For hydrogenic atoms, $f(E, K)$ is well known [28–31], and a model spectrum is shown in Fig. 10 of Ref. [23]. For more complicated atoms, many-electron effects introduce small corrections, as shown in Fig. 87.4 for Si using a Hartree-Fock approximation. Adding the contributions from all shells yields the Bethe result in Fig. 87.3 [3]. Similar results are given in Ref. [32] for Al. For the outermost electrons in metals, the electron gas model can be used to generate cross sections [33–35]. For semiconductors and insulators a model has been derived using the tight binding approximation for the ground state wave function and orthogonalized plane waves for excited states [34]. The dipole oscillator strength is derived from the data for optical absorption coefficients; for solids [36] and gases [37, 38] see Chap. 59.

87.3.4 Fermi Virtual Photon Method

In the Fermi virtual photon (FVP) method, $f(E, K)$ is approximated by $f(E, 0)$ for $Q < E$, with a delta function at $Q = E$ [39, 18, 40, 41] (see Fig. 87.4). Then $\sigma(E)$ is given by [39]

$$\sigma(E) = \sigma_R(E)\left[Ef(E, 0)\ln\left(\frac{2mv^2}{E}\right)\right.$$
$$\left. + \int_0^E f(E', 0)\, dE'\right], \quad (87.14)$$

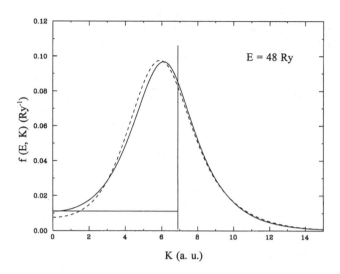

Figure 87.4. Generalized oscillator strength (GOS) $f(E, K)$ (solid line) for longitudinal excitations of the 2p-shell of Si atoms (with a binding energy $U_L = 8$ Ry), calculated with the Hartree-Fock-Slater potential. The energy transfer is $E = 650$ eV. The hydrogenic approximation is given by the broken line. $f(E, K)$ peaks at $(Ka_0)^2 \approx E - U_L$. In the FVP model, Eq. (87.14), the GOS is replaced by a δ-function at $Ka_0 = E^{1/2}$ and by $f(E, 0)$ for $0 < Ka_0 < E^{1/2}$ (straight lines).

and so only $f(E)$ need be known, or equivalently $\Im(-1/\epsilon)$, where ϵ is the complex dielectric constant of the absorber. Data can be extracted from a variety of optical measurements [37, 36, 42], and from electron energy loss measurements [43]. A cross section calculated with this model is given in Fig. 87.3. The $\sigma(E)$ differ by as much as 50% from the Bethe result, but the moments differ by at most 10% (see Tables 87.1 and 87.2). It is evidently a better approximation than that given by the binary encounter approximation of Eq. (87.10).

87.4 MOMENTS OF THE CROSS SECTION

Various moments of $\sigma(E)$ are defined by [45, 3]

$$M_\nu \equiv N \int E^\nu \sigma(E)\, dE. \quad (87.15)$$

where the range of integration covers all nonvanishing $\sigma(E)$, including a summation over discrete excited states. Then $M_0 = N\sigma_{tot} = 1/\lambda_{mfp}$, where σ_{tot} is the total collision cross section and λ_{mfp} the mean free path between collisions, and $M_1 = S$, the stopping power. M_2 and M_3 give the width and skewness of $F(\Delta)$.

Higher moments are not very useful [45] except for special applications [46, 47]. For incident electrons, a

Table 87.1. The coefficient $\tau(\beta) = M_0\beta^2/(NZk_R)$ for pions with $M_\pi = 139.567$ MeV/c^2, calculated in the FVP approximation. For comparison, the Si (GOS) results were calculated from the complete $F(E, K)$ generalized oscillator strengths. The estimated accuracies are ~1% for Si (GOS) and ~30% for the others. The values are the same for all heavy particles with the same β; they differ slightly for electrons. For Si ($\rho = 2.389$ g/cm^3), $NZk = 17.82$ eV/μm, for the gases (at STP), $NZk = 6.3828\,Z$ eV/cm. Units of $\tau(\beta)$ are eV^{-1}.

T(MeV)	β	Si(GOS)	Si	He	Ar	ethane	butane
1.	0.11907	0.1208	0.1333	0.1937	0.1081	0.2961	0.2855
3.	0.20406	0.1408	0.1528	0.2229	0.1242	0.3379	0.3259
10.	0.35950	0.1612	0.1738	0.2538	0.1412	0.3821	0.3686
30.	0.56791	0.1792	0.1917	0.2802	0.1557	0.4198	0.4050
100.	0.81276	0.1957	0.2082	0.3093	0.1717	0.4614	0.4451
300.	0.94825	0.2061	0.2186	0.3441	0.1908	0.5111	0.4931
1000.	0.99247	0.2127	0.2252	0.3958	0.2189	0.5843	0.5637
3000.	0.99901	0.2150	0.2275	0.4499	0.2468	0.6580	0.6333
10000.	0.99991	0.2155	0.2280	0.5057	0.2671	0.7034	0.6646
30000.	0.99999	0.2156	0.2281	0.5332	0.2737	0.7111	0.6697
100000.	1.00000	0.2156	0.2281	0.5407	0.2765	0.7128	0.6709

further averaging of energy loss over different paths must be performed because of multiple angular scattering.

87.4.1 Total Collision Cross Section M_0

A simple result for M_0 is obtained with the Rutherford cross section

$$_RM_0 = NZ_2\int \sigma_R(E)\,dE = NZ_2\frac{k_R}{\beta^2}\left[\frac{1}{E_{min}} - \frac{1}{E_{max}}\right].$$
(87.16)

Clearly, $_RM_0$ is very sensitive to the choice of E_{min}, and there is no simple prescription for choosing it, as is evident from Fig. 87.3. The same applies to the binary encounter approximation [26]. From the Bethe model, a good relativistic approximation to M_0 is given by [23, 48]

$$M_0 = NZ_2\frac{k_R}{\beta^2}g_0[\ln\beta^2\gamma^2 - \beta^2 + h_0 + 11.227]$$

$$= NZ_2\frac{k_R}{\beta^2}\tau(\beta).$$
(87.17)

Values for g_0 and h_0 may be found in Ref. [48].[2] For Si, He, Ar, ethane, and butane, values of τ_β, calculated by numerical integration of $\sigma(E)$ obtained with the FVP method, are given in Table 87.1. For Si, comparison values calculated with the Bethe model are also given. The quantity $\tau(\beta)$, which is closely related to the maximum $f(E)$ [37, 38], is more suitable for interpolation than M_0.

---

[2]The symbol g_0 is equivalent to M_{tot}^2/ZRy in Chap. 88.

87.4.2 Stopping Power

The stopping power is usually written in the form

$$M_1 = S = NZ_2\frac{k_R}{\beta^2}2B,$$
(87.18)

where B is called the stopping number. M_1 calculated with the Rutherford cross section is

$$_RM_1 = NZ_2\int_{E_{min}}^{E_{max}} E\sigma_R(E)\,dE = NZ_2\frac{k_R}{\beta^2}\ln\frac{E_{max}}{E_{min}}.$$
(87.19)

To obtain realistic values, it is evident from Fig. 87.3 that a small value must be chosen for E_{min} to compensate for the peaks in $\sigma(E)$.

The Bethe Model

In the Bethe model of stopping powers, $d\sigma(E, K)$ from Eq. (87.13) must be integrated over both E and $Q = \hbar^2K^2/2m$ [31]. Since $f(E, K)$ is nearly constant near $K = 0$ [49] (see Fig. 87.4), the integral may be broken into four parts according to

$$S = \frac{Nk_R}{\beta^2}\int dE\left\{\int_{Q_m}^{Q_1} f(E, 0)\frac{dQ}{Q} + \int_{Q_1}^{\infty} f(E, K)\frac{dQ}{Q}\right.$$

$$+ \int_0^{Q_1} [f(E, K) - f(E, 0)]\frac{dQ}{Q}$$

$$\left. - \int_0^{Q_m} [f(E, K) - f(E, 0)]\frac{dQ}{Q}\right\},$$
(87.20)

Table 87.2. Stopping number $B(\beta) = \beta^2 S/(2NZk_R)$ for pions in the FVP approximation. The Si(GOS) values are included for comparison (see Table 87.1). The values are the same for all particles with the same β, except for electrons, where $B(\beta)$ is much smaller [44].

T(MeV)	β	Si(GOS)	Si	He	Ar	ethane	butane
1.	0.11907	8.699	8.850	11.554	8.287	11.569	11.447
3.	0.20406	10.885	11.005	13.696	10.354	13.728	13.607
10.	0.35950	13.175	13.282	15.969	12.543	16.020	15.900
30.	0.56791	15.120	15.221	17.897	14.417	17.964	17.845
100.	0.81276	17.179	17.276	20.025	16.470	20.108	19.990
300.	0.94825	19.471	19.565	22.567	18.931	22.672	22.555
1000.	0.99247	22.557	22.650	26.329	22.568	26.441	26.316
3000.	0.99901	25.229	25.322	30.233	26.310	30.248	30.050
10000.	0.99991	27.687	27.780	34.389	30.187	33.716	33.314
30000.	0.99999	29.478	29.571	37.258	33.177	36.004	35.547
100000.	1.00000	31.030	31.123	39.238	35.756	37.897	37.388

where $Q_m \approx E^2/2mv^2$ [23], and Q_1 is chosen such that $f(E, K)$ differs little from $f(E, 0)$ in the interval $0 \leq K \leq \sqrt{2mQ_1}/\hbar$ [30, 50]. Collisions with $Q < Q_1$ are "distant" and those with $Q > Q_1$ are "close". The integrals simplify by interchanging the order of integration in the second and third terms, and by using the sum rule [23, 28]

$$\int_0^\infty f(E, K)\, dE = Z_2, \qquad (87.21)$$

for all K. The last term of (87.20) is then small and the second last vanishes exactly. The remaining two terms give

$$S = \frac{NZ_2k_R}{\beta^2} \left[\ln \frac{2mv^2 Q_1}{\mathcal{I}^2} + \ln \frac{Q_{max}}{Q_1} \right], \qquad (87.22)$$

where $\mathcal{I}$ is the logarithmic mean excitation energy defined by

$$Z_2 \ln \mathcal{I} = \int f(E) \ln E\, dE, \qquad (87.23)$$

and $Q_{max} = 2m_e c^2 \beta^2 \gamma^2$ is the maximum energy loss (for electrons, $Q_{max} = \frac{1}{4} m_e c^2 \beta^2 \gamma^2$ [2]). Including relativistic and other correction terms, the final result for $M_1 \gg m_e$ is

$$S = \frac{NZ_2k_R}{\beta^2} \left\{ B_0 - \frac{C(\beta)}{Z_2} + Z_1 L_1(\beta) + Z_1^2 L_2(\beta) \right. $$
$$\left. + \tfrac{1}{2} [G(M_1, \beta) - \delta(\beta)] \right\}, \qquad (87.24)$$

where

$$B_0 = \ln \frac{2mc^2 \beta^2 \gamma^2}{\mathcal{I}} - \beta^2 \qquad (87.25)$$

is the uncorrected stopping number. In the limit $\beta \to 0$, $2B_0$ reduces to the terms in brackets in Eq. (87.22). The other correction terms are as follows:

The Shell Correction. $C(\beta)$ accounts for the last term in Eq. (87.20), together with modifications of $d\sigma(E, Q)$ near $Q = E_{max}$ [2]. It can be estimated on a shell-by-shell basis [51–53] using nonrelativistic hydrogenic calculations for the K- and L-shells [31, 29], and rescaling methods for the outer shells which have not been calculated directly [54, 51]. The effects are important for small β, but simple formulas are not known.

The Barkas Term. $L_1(\beta)$ arises from polarization of the target electrons by the incident particle. Various approximating functions and fits to experimental data are described in Ref. [20].

The Bloch Term. $L_2(\beta)$ arises from corrections to the approximation that for close collisions, the electrons can be represented by plane waves [55]. Confinement of the electrons to the interior of a cylinder of atomic dimensions introduces transverse momentum components, resulting in the widely used correction

$$Z_1^2 L_2(\beta) = -q^2 \sum_{j=1}^\infty 1/[j(j^2 + y^2)] \qquad (87.26)$$

with $y = Z_1 \alpha/\beta$. For $y = 0$, the sum is $\zeta(3) \simeq 1.202\,057$. The need for the L_1 and L_2 terms was established experimentally in Ref. [56], and discussed in a more general context in Ref. [57].

The Mott Term. $G(M_1, \beta)$ is a kinematic recoil correction which becomes important for relativistic projectiles. In the limit of a point-like spinless nucleus, the correction is

$$G(M_1, \beta) = -\ln(1 + 2t) - t\beta^2/\gamma^2, \qquad (87.27)$$

where $t = m\gamma/M_1$ [24]. The correction is negligible for $\gamma < 100$ and $M_1 > m_p$.

The Density Correction. $\delta(\beta)$ arises from the dielectric response of a solid absorber as a whole to the electric field of the projectile, and the work done by the interaction [2, 24]. Sternheimer's algorithm [58] is usually used, as summarized in Ref. [24].

The remaining parameter to be discussed in Eq. (87.24) is $\mathcal{I}$. If the oscillator strength distribution $f(E)$ is known, then $\mathcal{I}$ can be calculated directly from Eq. (87.23). Results for many gases are given in Ref. [38]. The values give good agreement with experimental stopping powers [59]. For solids, the definition [2]

$$\ln \mathcal{I} = \frac{2}{\pi(\hbar\omega_{\mathrm{p}})^2} \int_0^\infty E \Im\left[\frac{-1}{\epsilon(E)}\right] \ln E \, dE \qquad (87.28)$$

may be used, where $\hbar\omega_{\mathrm{p}}$ is the plasma energy for *all* the electrons, defined by $(\hbar\omega_{\mathrm{p}})^2 = 830.4\,\rho Z/A$ with $\hbar\omega_p$ in eV. For metals, this is substantially larger than the plasmon energy associated with the conduction electrons. Only for Al [60] and partially for Si [3] have sufficiently good measurements of $\Im[-1/\epsilon(E)]$ been available to permit the use of this method. For other materials, $\mathcal{I}$ can be deduced from measurements of S, provided that the other corrections in Eq. (87.24) are known sufficiently well. A list of values for all elements and many compounds is given in Ref. [53]. See also Refs. [21, 44, 51, 52, 59, 61–65].

For rough estimates of $\mathcal{I}$, the approximation [66]

$$\mathcal{I} \approx \begin{cases} 11.7 + 11.2/Z_2 \text{ eV}, & Z_2 \le 13 \\ 9.5 \pm 1 \text{ eV}, & Z_2 > 13 \end{cases} \qquad (87.29)$$

is useful, together with the Bragg rule

$$n_{\mathrm{e}} \ln \mathcal{I} = \sum_i n_i \ln \mathcal{I}_i \qquad (87.30)$$

for compounds and composite materials, where n_i is the electron density associated with element i. However, chemical shifts may be as large as 10% [67].

The Fermi Virtual Photon Method

From Eq. (87.14), the nonrelativistic FVP approximation to the stopping number $B(\beta)$ is

$$B(\beta) = \frac{1}{2} \ln\left(\frac{2m_{\mathrm{e}}v^2}{\mathcal{I}}\right)$$
$$+ \frac{1}{2Z_2} \int_{E_{\min}}^{E_{\max}} \frac{dE}{E} \int_0^E f(E') \, dE'. \quad (87.31)$$

Although the integrals must now be calculated numerically, the full $f(E, K)$ is not required, and in the case of silicon, the results are in close agreement with the corresponding Bethe model. Stopping numbers for pions in various substances are listed in Table 87.2.

Stopping Power at Small Speeds

For small speeds, the various correction terms in the stopping number B, Eq. (87.24), become large compared with B_0 [51]. In particular, for $2mv^2 = I$, B_0 becomes zero. For example, for α-particles in U ($I = 840$ eV), $B_0 = 0$ at $T = 1.5$ MeV, and B then consists only of correction terms. For smaller energies, empirical approaches are used to describe S. Many of the tables referenced above give such data.

Mean Energy Loss per Collision

The quantity $\langle E \rangle = M_1/M_0$ is the mean energy loss per collision. For the substances listed in the tables, $\langle E \rangle \sim 50 - 100$ eV. It changes at most by a factor of 2 for all energies in the tables. Values calculated with the data in Tables 87.1 and 87.2 can be used to obtain preliminary estimates of numbers of collisions in given thicknesses of absorbers, and thus lead to a suitable choice of the method for calculating straggling functions.

87.4.3 Second Moment M_2

The relativistic result calculated with the Rutherford cross section is

$$\begin{aligned} {}_{\mathrm{R}}M_2 &= NZ_2 \frac{k_R}{\beta^2}(1 - \beta^2/2)E_{\max} = 2NZ_2 k_R mc^2 \frac{1 - \beta^2/2}{1 - \beta^2} \\ &= 0.156915 \frac{Z_1^2 Z_2}{A} \frac{1 - \beta^2/2}{1 - \beta^2} \text{ MeV}^2\,\mathrm{cm}^2/\mathrm{g}\,. \end{aligned}$$
$$(87.32)$$

For small β, ${}_{\mathrm{R}}M_2$ is practically independent of β. A better approximation can be achieved with the binary encounter method:

$$\begin{aligned} {}_{\mathrm{b}}M_2 = NZ_2 \frac{k_{\mathrm{R}}}{\beta^2}E_{\max}\Bigg[1 - \frac{\beta^2}{2} \\ + \sum_l \frac{4Z_l U_l}{3Z_2 E_{\max}}\left(\ln\frac{E_{\max}}{{}_l E_{\min}} - \beta^2\right)\Bigg], \end{aligned} \quad (87.33)$$

where the summation extends over atomic shells l. For ${}_l E_{\min}$ the ionization energy J_l for shell l can be used, while U_l represents the kinetic energy of the electrons in shell l. This approximation is only useful for relatively large x [3]. By using sum rules, Fano [2] [his Eq. (72)] achieved a better approximation, which corresponds to the Bethe approximation for the stopping power [68, 3]. Hydrogenic values of M_n are calculated in Ref. [45].

87.5 ENERGY LOSS STRAGGLING

87.5.1 Straggling Parameters

Parameters for Thick Absorbers

Straggling is due to the stochastic nature of the energy losses of the charged particles. Because the single collision spectrum is highly skewed (the most probable energy loss is ~20 eV, the mean value ~100 eV), straggling functions will also be skewed. Four parameters are useful in a preliminary study of straggling problems. The parameter

$$\xi = NZ_2 \frac{k_R}{\beta^2} x = 153.537 \frac{Z_1^2}{\beta^2} \frac{Z_2}{A} x \text{ keV cm}^2 \quad (87.34)$$

gives the energy loss [calculated with $\sigma_R(E)$] which, on the average, is exceeded once for each particle in its passage through an absorber of thickness x, i.e., for $I \ll \xi \ll E_{max}$,

$$xNZ_2 \int_{\xi}^{E_{max}} \sigma_R(E) \, dE = 1 \,. \quad (87.35)$$

The parameter $\kappa = \xi/E_{max}$ is related to the skewness of the straggling function: $\gamma_1^2 = 1/4\kappa$. The mean energy loss is

$$\langle \Delta \rangle = xM_1 = 2\xi 2B = 2\kappa E_{max}B \,, \quad (87.36)$$

and the standard deviation of $F(\Delta)$ is

$$\omega^2 = xM_2 \sim xNZ_2 \frac{k_R}{\beta^2} E_{max} = \xi E_{max} = \kappa E_{max}^2 \,. \quad (87.37)$$

For thick absorbers, the straggling function becomes Gaussian [19]. The requirement for this to occur is $\gamma_1 \to 0$, thus $\kappa \to \infty$, also $\langle \Delta \rangle \to \infty$, and $x \to \infty$.

Parameters for Thin Absorbers

The parameters described above are not suitable for describing $F(\Delta)$ for very thin absorbers. Instead, Δ_{mp} and Γ are used. Values for pions traversing Ar are given as a function of thickness x in Tables 87.3 and 87.4. However, before comparing with experimental data, values of Δ must be converted into ionization values, and the detector and amplifier noise must be added [3] (see Sect. 87.8).

Parameters for Extremely Thin Absorbers

If the number n_c of collisions in the absorber is less than about 16, $F(\Delta)$ still shows the details of $\sigma(E)$ (see Fig. 87.5). The most probable energy losses are much less than $\langle \Delta \rangle$. Functions of this type have been observed with electron microscopes [8]. For comparison, the Landau straggling function [4] is also shown. This result, derived from $\sigma_R(E)$, does not show the structure of a realistic spectrum.

Detector noise and energy loss to ionization conversions introduce important changes [69–72] to the energy loss spectrum. In particular, if $n_c < 4$, a large fraction of the particles do not lose any energy in the absorber.

Table 87.3. Calculated most probable energy loss Δ_{mp} of pions with $Z_1 = \pm 1$ and kinetic energy T passing through a distance x of argon gas (at 760 torr, 293 K, $\rho = 1.66$ g/dm^3). For heavy ions (α, C$^+$), the values scale as $Z_1^{1.3}$ (for a better approximation, see Table V in [3]). Units of Δ_{mp} are keV.

T (MeV)	x (cm)				
	0.5	1	2	4	8
1	29.373	63.341	130.662	265.062	533.811
3	9.820	21.598	47.000	101.329	216.013
10	3.104	6.842	14.950	32.558	69.966
30	1.153	2.687	5.927	12.883	28.023
100	0.376	1.274	2.928	6.412	13.884
300	0.288	0.973	2.313	5.099	10.993
1×10^3	0.303	1.042	2.445	5.356	11.494
3×10^3	0.344	1.217	2.783	6.044	12.926
1×10^4	0.384	1.390	3.145	6.784	14.495
3×10^4	0.653	1.515	3.405	7.322	15.663
1×10^5	0.667	1.605	3.595	7.725	16.575

Table 87.4. Calculated values of Γ (fwhm) of the straggling function $F(\Delta)$ (see Table 87.3). As a rough approximation, $\Gamma \sim x^{0.8}$ (Table VI in [3]).

T (MeV)	x (cm)				
	0.5	1	2	4	8
1	16.014	24.961	35.503	51.327	69.120
3	6.283	12.012	22.928	43.393	68.589
10	2.507	4.563	8.659	16.153	30.026
30	1.360	2.279	3.922	7.441	13.881
100	0.781	1.434	2.350	4.178	7.705
300	0.574	1.239	2.030	3.410	6.232
1×10^3	0.621	1.286	2.085	3.397	6.340
3×10^3	0.752	1.391	2.244	3.687	6.844
1×10^4	0.905	1.510	2.432	3.999	7.502
3×10^4	0.983	1.615	2.548	4.314	7.956
1×10^5	1.040	1.701	2.689	4.666	8.399

Figure 87.5. Calculated straggling functions $F(\Delta)$ for 1 GeV pions traversing four thicknesses x of silicon. The average number of collisions is n_c, with $\langle E \rangle = 106$ eV, the mean energy loss is $\langle \Delta \rangle = n_c \langle E \rangle$, shown by an arrow. The number of particles traversing the absorber without a collision is n_0. The peak heights do not follow a Poisson distribution because successive convolutions give a broader distribution. The Landau function [4] for $n_c = 4$ is shown by the broken line. For all, the most probable energy loss calculated according to Landau, Δ_L, is shown.

87.5.2 Analytic Methods for Calculating Energy Loss Straggling Functions

Three principal methods are convolutions, Laplace transforms and the use of moments [3]. They are practical for different thicknesses of absorbers, as indicated.

Convolutions

For this method, a complete collision cross section must be available [41], and lengthy numerical calculations are required. The probability density function for particles having suffered n collisions is given by the n-fold convolution of the single collision spectrum [3, 73]:

$$\sigma^{*n}(\Delta) = \int_0^{\Delta} \sigma(E) \sigma^{*(n-1)}(\Delta - E)\, dE \,, \qquad (87.38)$$

with

$$\sigma^{*0}(\Delta) = \delta(\Delta) \,, \quad \sigma^{*1}(\Delta) = \sigma(\Delta) \,. \qquad (87.39)$$

Assuming that successive collisions are statistically independent, the number of collisions is described by the Poisson distribution

$$P_n^{n_c} = \frac{n_c^n}{n!} e^{-n_c} \,, \qquad (87.40)$$

where $P_n^{n_c}$ gives the fraction of particles suffering n collisions, and $n_c = N x \sigma_{\text{tot}}$ is the average number of collisions for all particles. The complete straggling function is then

$$F(x, \Delta) = \sum_{n=0}^{\infty} P_n^{n_c} \sigma^{*n}(\Delta) \,. \qquad (87.41)$$

Usually, this method is only used for absorbers thin enough that the single collision cross section changes negligibly during the passage through the absorber.

This approach would become quite complex for large n_c. The *doubling method* [74] uses the following procedure: the function for an absorber of thickness $2x$ is calculated by convoluting $F(x, \Delta)$ with itself, *viz.*

$$F(2x, \Delta) = \int_0^{\Delta} F(x, \Delta - g) F(x, g)\, dg \,. \qquad (87.42)$$

An initial distribution is calculated for an extremely thin absorber of thickness dx from

$$F(dx, \Delta) = \delta(\Delta)(1 - \sigma_{\text{tot}} dx) + \sigma(\Delta) dx \,, \qquad (87.43)$$

where $\delta(\Delta) = F(0, \Delta)$ and $\sigma_{\text{tot}} dx \sim 0.001$.

For Si, the convolution method has been tested and compared with experiments for as many as 10^4 collisions [3]. The limitations of the algorithm are $\sim 10^5$ collisions. Since $P_0^{n_c} = e^{-n_c}$, there will be particles that pass through an absorber without making a collision.

Laplace Transforms

The Laplace transform method is mathematically equivalent to the convolution method. The method, as implemented by Landau [4], is based on $\sigma_R(E)$. The implicit number of collisions is therefore too large, and the Landau function is too narrow, as shown in Figs. 87.1 and 87.5. A refinement proposed in Ref. [5] gives straggling functions which are too wide (see Fig. 87.1).

Use of Moments

If the skewness γ_1 is not small, a Gaussian function is not a good approximation to a straggling function. This can be remedied by using a modified Gaussian of the form [47]

$$f(y) = \frac{\tilde{\delta}}{\sqrt{2\pi}} \frac{1}{\sqrt{y^2 + 1}} \exp\{-\tfrac{1}{2}[\tilde{\gamma} + \tilde{\delta} \ln(y + \sqrt{y^2 + 1})]^2\}, \tag{87.44}$$

where $y = (\Delta - \langle \Delta \rangle)/\sigma$ and $\tilde{\gamma}$ and $\tilde{\delta}$ are related to M_2 and M_3 [47] such that the second and third moments are reproduced exactly.

The moments can be calculated as a function of particle speed, and thus the method is useful for mean energy losses up to about $0.7T_0$. At these losses, some particles reach their full range and thus disappear from the beam. The moments are then distorted and other procedures are needed [9].

Thick Absorbers, Ranges and Range Straggling

In radiation therapy with charged particles, a broad beam is directed at the body. Usually, the energy T_0 of the particles is adjusted to penetrate a given distance into the body. A primary measure of this distance is the mean range R calculated with the continuous slowing down approximation (CSDA)

$$R(T_0) = \int_0^{T_0} \frac{dT}{S(T)}. \tag{87.45}$$

Because of energy-loss straggling, there also is straggling in range amounting to a few percent of R [62]. In the practical application, a *depth-dose curve* is needed. At a given depth, the dose D is given by Eq. (87.6). For the first 50% of the range, the spread in energy of the beam is relatively small, and the dose is approximated closely by the stopping power $S(T)$ for the mean energy of the beam. For larger thicknesses, a detailed transport calculation must be made in order to find $\phi(T, x)$ [9]. Range straggling functions can then also be obtained. Close agreement between measurement and calculation has been achieved.

Similarly, the mean energy loss T_1 in a thick absorber is given implicitly by

$$x = \int_{T_1}^{T_0} \frac{dT}{S(T)}. \tag{87.46}$$

Usually, values are obtained from a range energy table. Such tables may be found in Refs. [21, 44, 51, 52, 59, 61–65].

87.6 MULTIPLE SCATTERING AND NUCLEAR REACTIONS

Coulomb scattering of particles heavier than electrons by nuclei and electrons usually produces many, but very small angular deflections. The Molière theory [75] giving the distribution $f(\theta)$ of angular deflections θ agrees well with experimental data [76]. As a first approximation, a Gaussian distribution $f(\theta) = e^{-y^2}$, where $y = \theta/\theta_0$, can be used. The value θ_0 is given by $\theta_0^2 = \theta_1^2 B_M$, where

$$\theta_1^2 = 0.157 \frac{Z_2(Z_2 + 1)}{A} Z_1^2 \frac{x\gamma^2}{(Mc^2\beta^2)^2} \quad \text{radians} \tag{87.47}$$

and B_M has values of the order of 10. A more detailed description plus several tables is given in Ref. [62].

For a broad beam of particles, multiple scattering produces only a minor correction in the range. In a narrow "pencil" beam, it causes a broadening, e.g., a 320 MeV proton beam spreads to a width of about 2 cm after traveling through 40 cm of water [77].

It may be necessary to take into account the influence of nuclear interactions. As a first approximation, the total cross section is roughly [21]

$$\sigma_{\text{tot}} = \pi(1.3 \times 10^{-13} A^{1/3} + \lambda)^2 \text{ cm}^2, \tag{87.48}$$

where $\lambda = \lambda_{\text{dB}}/2\pi$.

87.7 MONTE CARLO CALCULATIONS

In Monte Carlo calculations, the interactions occurring during the passage of the particles through matter can be simulated one at a time, collision by collision, and including those for the δ-rays [78]. Particles travel random distances x_i between successive collisions, calculated by selecting a random number r_i and determining the distance to the next collision from

$$x_i = -\lambda_{\text{mfp}} \ln r_i. \tag{87.49}$$

The energy loss E_i is selected with a second random number from the integrated collision spectrum

$$Q(E) = \int_0^E \frac{\sigma(E)}{M_0} dE. \tag{87.50}$$

This process is repeated until $\sum x_i$ exceeds the absorber thickness. The total energy loss Δ of the particle is $\Delta = \sum_i E_i$.

87.8 DETECTOR CONVERSION FACTORS

For all detectors, a study must be made of the process in which energy lost by the particles is converted into the observable effect, see Chap. 88. The primary products of collisions are excited states, electron-ion and electron-hole pairs, and secondary electrons also called δ-rays. Secondary products are chemical species, fluorescent photons and Auger electrons. An important problem is the export of energy from the volume under observation by secondary radiations (δ-rays, mesons, x-rays, neutrons). Prominent among the effects [79] are ionization currents measured under applied electrical fields (in gaseous and solid state ionization chambers), photoeffects by fluorescent radiation (scintillators), measurements of chemical yields (ferrous sulphate dosimeters, photographic emulsions, etc.), and the release of light by stored crystalline defects in thermoluminescent measurements.

The effect J can be related to the energy deposited D by a conversion factor W as $J = D/W$, where W represents the amount of energy needed to produce a unit of observable effect (e.g., an electron-ion pair). W depends on the absorber, particle type, and speed. For ionization, typical values are $W \sim 25$ eV for gases, $W \sim 3.5$ eV for semiconductors [62,80].

For chemical effects [81], J is sometimes written as $J = YD$. Usually Y is given as the "yield per 100 eV". A typical value for a Fricke dosimeter is that, for fast electrons, 15 ferrous ions are converted into ferric ions per 100 eV of deposited energy. Usually, Y depends on time after the passage of the particles, and the full values are only reached after say 1 μs. For slow ions, Y is much smaller. Light emission (scintillations) has been observed from gases (typically, $W \sim 20$ eV), liquids and solids (typically, $W \sim 1000$ eV) [79].

The subsequent measurements of these radiation effects are done with current or charge amplifiers, proportional avalanche amplification, photo multipliers, channel plates, etc. [79].

REFERENCES

1. M. E. Rudd, Y.-K. Kim, D. H. Madison, and T. J. Gay, Rev. Mod. Phys. **64**, 441 (1992).
2. U. Fano, Ann. Rev. Nucl. Sci. **13**, 1 (1963).
3. H. Bichsel, Rev. Mod. Phys. **60**, 663 (1988).
4. L. Landau, J. Phys. (Moscow) **8**, 201 (1944).
5. O. Blunck and S. Leisegang, Z. Phys. **128**, 500 (1950).
6. H. Cramér, *The Elements of Probability Theory* (John Wiley and Sons, New York, 1955).
7. H. Bichsel, *Microsimetry*, Rad. Protection Dosim. **13**, 91 (1985).
8. J. Ph. Perez, J. Sevely, and B. Jouffrey, Phys. Rev. A **16**, 1061 (1977).
9. H. Bichsel and T. Hiraoka, Int. J. Quantum Chem. **23**, 565 (1989).
10. H. Bichsel, R. F. Mozley, and W. Aron, Phys. Rev. **105**, 1788 (1957).
11. AAPM report No. 16 *Protocol for heavy charged-particle therapy beam dosimetry*, published for the Amer. Assoc. of Physicists in Medicine by the A. I. P. (1986).
12. C. C. Ahn, O. L. Krivanek, R. P. Burgner, M. M. Disko, and P. R. Swann, *EELS Atlas*, HREM Facility (Arizona State University, Tempe, 1980).
13. U. Killat, Z. Phys. **263**, 83 (1973).
14. U. Fano and J. W. Cooper, Rev. Mod. Phys. **40**, 441 (1968).
15. J. L. Dehmer, M. Inokuti, and R. P. Saxon, Phys. Rev. A **12**, 102 (1975).
16. M. Isaacson, J. Chem. Phys. **56**, 1813 (1972).
17. P. Pianetta, T. W. Barbee, Jr., Nucl. Instrum. Meth. A **266**, 441 (1988).
18. E. Fermi, Z. Phys. **29**, 315 (1924).
19. N. Bohr, Dan. Mat. Fys. Medd. **18**, No. 8 (1953).
20. H. Bichsel, Phys. Rev. A **41**, 3642 (1990).
21. H. Bichsel, in *Radiation Dosimetry*, edited by F. H. Attix and Wm. C. Roesch (Academic Press, 1968), 2nd ed., Vol. 1, Chap. 4, p. 157.
22. R. D. Evans, *The Atomic Nucleus* (McGraw Hill, 1967), p. 838–851.
23. M. Inokuti, Rev. Mod. Phys. **43**, 297 (1971), **50**, 23 (1978).
24. S. P. Ahlen, Rev. Mod. Phys. **52**, 121 (1980).
25. E. A. Uehling, Ann. Rev. Nucl. Sci. **4**, 315 (1954).
26. H. Bichsel, *Scanning electron microscopy supplement*, 4, 147 (1990).
27. M. E. Rudd and J. H. Macek, Case Studies in Atomic Physics **3**, 47–136 (1972).
28. H. Bethe, Ann. Phys. **5**, 325 (1930).
29. H. Bichsel, Phys. Rev. A **28**, 1147 (1983). Calculations for L-shell electrons have not yet been published.
30. G. S. Khandelwal, Phys. Rev. A **26**, 2983 (1982).
31. M. C. Walske, Phys. Rev. **88**, 1283 (1952), and **101**, 940 (1956).
32. S. T. Manson, Phys. Rev. A **6**, 1013 (1972).
33. J. Lindhard and A. Winther, Mat. Fys. Medd. Dan. Vid. Selsk. **34**, No. 4 (1964).
34. C. J. Tung, R. H. Ritchie, J. C. Ashley, and V. E. Anderson, ORNL Report TM-5188, 1976.
35. R. E. Johnson and M. Inokuti, Comments At. Mol. Phys. **14**, 19 (1983).
36. E. D. Palik, *Handbook of optical constants of solids* (Academic Press, 1985).
37. J. W. Gallagher, C. E. Brion, J. A. R. Samson, and P. W. Langhoff, J. Phys. Chem. Ref. Data **17**, 9–153 (1988).
38. B. L. Jhanwar, Wm. J. Meath, and J. C. F. MacDonald, Rad. Res. **96**, 20 (1983).
39. W. W. M. Allison, J. H. Cobb, Ann. Rev. Nucl. Part. Sci. **30**, 253 (1980).
40. D. Liljequist, J. Phys. D **16**, 1567 (1983).
41. E. J. Williams, Proc. R. Soc. A **125**, 420 (1929).
42. J. H. Barkyoumb and D. Y. Smith, Phys. Rev. A **41**, 4863 (1990).
43. H. Raether, *Excitation of plasmons and interband transitions by electrons* (Springer Verlag, New York, 1980).
44. International Commission on Radiation Units and Measurements Report 37 (7910 Woodmont Ave, Bethesda, Maryland, 1984).
45. H. Bichsel, Phys. Rev. B **1**, 2854 (1970).
46. P. Sigmund and U. Haagerup, Phys. Rev. A **34**, 892 (1986).
47. C. Tschalär, Nucl. Instrum. Meth. **61**, 141, and **64**, 237–

243 (1968).

48. M. Inokuti, R. P. Saxon, and J. L. Dehmer, Int. J. Radiat. Phys. Chem. **7**, 109 (1975).

49. H. A. Bethe, L. M. Brown, M. C. Walske, Phys. Rev. **79**, 413 (1950).

50. J. M. Peek, Phys. Rev. A **27**, 2384 (1983).

51. H. Bichsel, Phys. Rev. A **46**, 5761 (1992); *Stopping power of charged particles for heavy elements*, Nat. Inst. Stand. Tech. Report, 4550 (1991).

52. J. F. Janni, At. Data Nucl. Data Tables **27**, 147 (1982).

53. International Commission on Radiation Units and Measurements Report No. 49 (7910 Woodmont Ave., Bethesda, Maryland, 1993).

54. J. O. Hirschfelder and J. L. Magee, Phys. Rev. **73**, 207 (1948).

55. F. Bloch, Ann. Phys. (5. Folge) **16**, 285 (1933).

56. H. H. Andersen, J. F. Bak, H. Knudsen, and B. R. Nielsen, Phys. Rev. A **16**, 1929 (1977).

57. G. Basbas, Nucl. Instrum. Meth. B **4**, 227 (1984).

58. R. M. Sternheimer, S. T. Seltzer, and M. J. Berger, Phys. Rev. B **26**, 6067 (1982); At. Data Nucl. Data Tables **30**, 261 (1984).

59. H. Bichsel and L. E. Porter, Phys. Rev. A **25**, 2499 (1982).

60. E. Shiles, T. Sasaki, M. Inokuti, and D. Y. Smith, Phys. Rev. A **22**, 1612 (1980).

61. H. H. Andersen and J. F. Ziegler, *HYDROGEN Stopping powers and ranges in all elements* (Pergamon, New York, 1977).

62. H. Bichsel, in *Am. Inst. Phys. Handbook*, edited by D. E. Gray (McGraw Hill, 1972).

63. F. Hubert, R. Bimbot, and H. Gauvin, At. Data Nucl. Data Tables **46**, 1 (1990).

64. *Coordinated Research Program on Atomic and Molecular Data for Radiotherapy and Related Research* Report IAEA-TECDOC-799, edited by M. Inokuti (International Atomic Energy Agency, Vienna, 1995). (Available upon request to the IAEA).

65. J. F. Ziegler, *HELIUM Stopping powers and ranges in all elements* (Pergamon, New York, 1977). Note that some of data in Table 1 of this reference are incorrect: apparently the "low energy electronic stopping" was used between 10 and 30 MeV.

66. P. Dalton and J. E. Turner, Health Phys. **15**, 257 (1968).

67. C. Tschalär and H. Bichsel, Phys. Rev. **175**, 476 (1968).

68. H. Bichsel, Phys. Rev. A **9**, 571 (1974).

69. J. Va'vra, P. Coyle, J. Kadyk, and J. Wise, Nucl. Instrum. Meth. A **324**, 113 (1993).

70. D. Srdoc, B. Obelic, and I. K. Bronic, J. Phys. B **20**, 4473 (1987).

71. G. W. Fraser and E. Mathieson, Nucl. Instrum. Meth. A **247**, 544 (1986).

72. B. Obelic, Nucl. Instrum. Meth. A **241**, 515 (1985).

73. J. R. Herring and E. Merzbacher, J. Elisha Mitchell Sci. Soc. **73**, 267 (1957).

74. A. M. Kellerer, *Mikrodosimetrie* (Strahlenbiologisches Institut der Universität München, G. S. F. Bericht B-1, 1968).

75. G. Molière, Z. Natf. **6**, 1013 (1955).

76. H. Bichsel, Phys. Rev. **112**, 182 (1958).

77. H. Bichsel, K. M. Hanson, and M. E. Schillaci, Phys. Med. Biol. **27**, 959 (1982).

78. H. Bichsel, Nucl. Instrum. Meth. B **52**, 136 (1990).

79. G. F. Knoll, *Radiation detection and measurement* (John Wiley and Sons, New York, 1989), 2nd. ed.

80. International Commission on Radiation Units and Measurements Report No. 31 (7910 Woodmont Ave, Bethesda Maryland, 1982).

81. Farhataziz and M. A. J. Rodgers, *Rad. Chem. Principles and Applications* (VCH Publishers, New York, 1987).

88

Radiation Physics

Mitio Inokuti

Argonne National Laboratory, Argonne, Illinois 60439

88.1 INTRODUCTION

Radiation physics entails studies of the interactions of ionizing radiation with matter. The term *ionizing radiation* refers to any energetic particles, either charged or uncharged, that can ionize atoms or molecules in matter. These particles include photons in the ultraviolet, x-ray, or γ-ray spectral region; electrons and positrons; mesons; protons and deuterons; α-particles; heavier ions including molecular ions; and neutrons. The matter under consideration includes substances in every phase of atomic aggregation (i.e., gas, liquid, solid, or plasma). Strictly speaking, the term "ionization" signifies an event in which at least one electron leaves an atom or molecule and eventually becomes free. This notion applies to a dilute gas, but not to condensed matter, for which it would

be more precise to use the term "electronic activation" encompassing all modes of excitation, as sketched in Sect. 88.3.5.

The motivation for studying the interactions of radiation with matter and their consequences is manifold. First, nuclear and particle physics experiments often require radiation detectors — devices that detect, score, or analyze energetic particles. The working principles of these detectors rest on knowledge of elementary processes of radiation interactions with matter [1]. Second, measurements of radiation fields, called dosimetry [2], are important to many purposes, including the protection of workers in industry and medicine, as well as of the general public; risk estimates for exposures to radiation from natural sources (such as cosmic rays and terrestrial radioactivity) or from artificial sources (such as accelerators, man-made radioisotopes, and nuclear-power facilities); and good performance of radiation diagnosis and therapy. The principles of dosimetry are largely based on radiation physics. Finally, ionizing radiation is useful for processing materials such as plastics and semiconductors, to endow desirable properties. To optimize methods of radiation processing, knowledge of radiation physics is highly valuable.

Atomic, molecular, and optical physics provides important underpinnings for radiation physics in two general contexts: instrumentation for the measurement of radiation and its effects on matter on the one hand, and elucidation of the atomic and molecular mechanisms leading to the effects on the other. The term *instrumentation* here means all devices that enable one to detect, identify, and quantify consequences of radiation interactions, such as particle kinetic energies and ionization, luminescence, and spectral analyses of atomic and molecular species produced. Conversely, needs in radiation physics have led to new atomic and molecular techniques such as resonance ionization spectroscopy [3], which was originally developed to determine the yield of excited atoms in gases exposed to radiation, but is now regarded as a branch of multiphoton spectroscopy valuable for studies of excited states of atoms and molecules (see Chap. 72). Thus, the instrumentation development constitutes a mutually beneficial interface between atomic, molecular, and optical physics and radiation physics.

The elucidation of microscopic mechanisms of radiation effects requires a grand synthesis of knowledge from various areas ranging from atomic, molecular, and optical physics to condensed matter physics, statistical physics, chemical kinetics, and molecular biology [4,5]. Yet, the role of atomic, molecular, and optical physics is most fundamental [6,7]. Interactions of radiation with matter lead to numerous collisions of energetic particles with atoms and molecules, and result in the production of various excited and ionized states, as well as electrons and other secondary particles. These have a wide range of kinetic energies, and in turn collide with atoms and molecules to generate further products. A detailed analysis of the chain of these early events demands knowledge in three topical areas: first, spectroscopic data (i.e., energy values and quantum numbers) of excited and ionized states; second, modes of decay and relaxation of these states, such as luminescence, molecular dissociation, and internal conversion and the branching ratios of each mode; and finally, the cross sections for all of the processes involved. All such knowledge is, in a broad sense, the fruit of work in atomic, molecular, and optical physics.

The following sections will concern key topics in basic radiation physics selected from a vast range of possibilities. The treatment will concentrate on basics and principles, and examples are meant to be illustrative rather than exhaustive; yet, the author hopes that the discussion will convey the charm and challenges of current research.

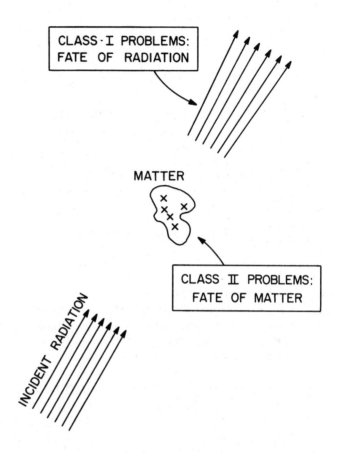

Figure 88.1. Schematic diagram showing the definition of Class I and Class II problems. Incident radiation comes from the lower left, interacts with matter, and travels toward the upper right, having been scattered. The crosses symbolize interaction events taking place within matter.

88.2 RADIATION ABSORPTION AND ITS CONSEQUENCES

88.2.1 Two Classes of Problems of Radiation Physics

It is useful to distinguish between two classes of problems in radiation physics [6], as illustrated in Fig. 88.1. Problems of Class I concern the fate of radiation after interactions with matter. Problems of Class II concern the fate of matter after interactions with radiation.

Problems of Class I are exemplified by the determination of the energy losses of particles penetrating matter, and the attenuation of a beam of photons in matter. Experimental studies on these problems are straightforward in principle; one only needs to analyze the kinetic energies of transmitted particles in the first problem and to count the number of transmitted photons in the second problem. Problems of Class I are also often simple to treat theoretically; for instance, in the Bethe theory of stopping power, discussed in Chap. 87, the use of sum rules enables one to bypass detailed knowledge about excited and ionized states. As a consequence, many of the problems of Class I have been solved. Furthermore, many of the applications of radiation physics are based on the established knowledge of solutions of problems of Class I. For instance, diagnostic uses of x-ray photons in medicine and industry rest on the firm knowledge of the attenuation of photons in matter (more precisely, on its dependence on the atomic number).

In sharp contrast, problems of Class II are fundamentally difficult; indeed, none of them has been solved completely. Reasons for the difficulty are manifold. First, any piece of irradiated matter is a new material that needs to be fully characterized. Second, irradiated matter is in general in a nonequilibrium state that changes physically and chemically with time. Finally, existing tools and techniques, both experimental and theoretical, for material characterization are limited.

To illustrate the nature of Class II problems, it is useful to consider a prototype problem that has been treated reasonably well, but certainly not completely. Suppose a single electron with energy 10 keV enters a dilute hydrogen gas of a sufficiently large volume. The incident electron will collide with a hydrogen molecule. Possible outcomes include excitation of molecular states, ionization leading to H_2^+ in various vibrational and rotational states, and ionization leading to dissociation into $H^+ + H$ (in various excited states or in the ground state). An electron resulting from an ionizing event may have enough kinetic energy to cause further excitation and ionization. Eventually, all the liberated electrons will lose enough energy to become *subexcitation* electrons unable to excite even the lowest electronic level of H_2 at 9.5 eV. Table 88.1 summarizes the various atomic and molecular species produced. To produce this answer, it was necessary to survey all the spectroscopic and electron collision data on H_2, and to carry out an analysis of electron transport. Problems of this kind will be further discussed in Sects. 88.3.2 and 88.3.3.

88.2.2 Photons

High-energy photons (i.e. x-rays, γ-rays, or a major part of synchrotron radiation), are important as the ionizing radiation most frequently used in applications. Their initial interactions with matter result in a spectrum of energetic electrons. Detailed discussion is given in Chaps. 59 and 60.

88.2.3 Charged Particles

Energetic charged particles penetrating matter cause ionization and electronic excitation of atoms and molecules as they lose kinetic energy (see Chap. 63). The

Table 88.1. The mean number N_j of initial species produced in molecular hydrogen upon complete degradation of an incident electron at 10 keV, and the energy absorbed $E_{abs.}$ (in percent).[a] The left column of this table indicates kinds of species tersely, For instance, "Lyman" here means the $B\,^1\Sigma_u^+$ state, which emits the Lyman band. The term "Werner" means the $C\,^1\Pi_u$ state, which emits the Werner band. The designation "H(2p)" means the production of hydrogen atoms in the 2p state. The designation "Slow H(2s)" means the production of hydrogen atoms in the 2s state with no appreciable kinetic energy. The designation "Fast H(2s)" means the production of hydrogen atoms with kinetic energies of several eV. The designation "Triplet" means all the triplet states combined. The designation "Subexcitation electrons" means all the electrons that are not energetic enough to cause further electronic excitation.

Initial species j	N_j	$E_{abs.}$
$H_2(B\,^1\Sigma_u^+)$ (Lyman)	96.8	11.7
$H_2(C\,^1\Pi_u)$ (Werner)	112.1	14.0
H(2p)	5.6	0.8
Slow H(2s)	29.6	4.4
Fast H(2s)	5.3	1.7
H ($n=3$)	3.2	0.5
Higher excited states	12.2	1.7
Ions	295.6	45.9
Triplets	102.6	10.0
Subexcitation electrons	295.6	9.4

[a] Adapted from Douthat [8].

most fundamental index to characterize the energy loss process is the total cross section for inelastic collisions, which is the mean number of inelastic collisions per unit path length in a material of unit molecular density. The next fundamental index is the stopping power, which is the mean energy loss per unit path length. For many purposes, a more detailed description of particle penetration is necessary. An example is energy loss straggling, fully discussed in Chap. 87.

Ionization is generally accompanied by the production of secondary electrons, which in general have a broad energy spectrum. This topic is reviewed by Rudd and coworkers [9–11].

88.2.4 Neutrons

Neutrons with kinetic energies in the MeV domain cause nuclear reactions, generally leading to the production of a number of charged particles. Neutrons of any kinetic energies collide with nuclei in matter elastically (viz., without causing nuclear reactions) and transfer substantial kinetic energy to nuclei. Consequently, neutrons may be regarded as a source of charged particles of various kinds, most importantly protons and light nuclei, all having broad energy spectra [12, 13].

88.3   ELECTRON TRANSPORT AND DEGRADATION

88.3.1   The Dominant Role of Electrons

In many Class II problems (i.e., radiation actions on molecular substances such as gases, molecular liquids or solids, and biological cells), energetic electrons generated by ionizing radiation play a central role because they are numerous, and deliver to atoms and molecules a large fraction of the total energy of radiation. What follows is a resume of our current understanding.

88.3.2   Degradation Spectra and Yields of Products

The prototype H_2 problem discussed in Sect. 88.2.1 may be generalized as follows. Consider the total number N_j of a particular molecular species j produced as a result of complete slowing down of electrons in a chemically pure medium consisting of n molecules per unit volume. The product species j could be ions or excited states specified by a set of quantum numbers. Let $\sigma_j(T)$ be the cross section for the production of j in a collision of an electron of kinetic energy T with a molecule. During its passage over an infinitesimal path length dx, an electron of kinetic energy T contributes the amount $dN_j = n\sigma_j(T)dx$ to N_j. Therefore, one may write

$$N_j = n \int \sigma_j(T)\,dx\,. \tag{88.1}$$

To carry out this integration, one must have a relation between T and x. To the extent that one regards x and T as related through an analytic function, one may write

$$dx = (dT/dx)^{-1}\,dT\,, \tag{88.2}$$

where dT/dx is the mean energy loss per unit path length traversed, or the stopping power as defined in Chap. 87. This treatment, called the continuous-slowing-down approximation (CSDA), is justified if T greatly exceeds the majority of energy losses upon individual collisions.

However, the CSDA is inadequate for electrons in general because the energy loss of an electron upon an individual collision is not always small compared with the current kinetic energy, and because the CSDA does not account for secondary electrons, which are abundantly produced. An appropriate formulation, introduced by Spencer and Fano [14], is obtained by writing

$$dx = y(T)\,dT \tag{88.3}$$

in place of Eq. (88.2) and determining the function $y(T)$ through full analysis of electron transport, production, and slowing down. This function, called the degradation spectrum, has been an object of extensive studies [6, 15]. Once $y(T)$ is determined, the yield N_j of product species j is evaluated as

$$N_j = n \int_{E_{1,j}}^{T_{\max}} \sigma_j(T)y(T)\,dT\,, \tag{88.4}$$

where $E_{1,j}$ is the threshold energy for the production of j, and $T_{\max}$ is the highest kinetic energy of electrons in the medium.

Three basic properties of $y(T)$ are as follows [15]. First, $y(T)$ is proportional to the electron energy distribution (as treated by the Boltzmann equation) multiplied by electron speed v, or $(2T/m)^{1/2}$ in the nonrelativistic case.

Second, $y(T)$ obeys an integral equation called the Spencer-Fano equation. This equation expresses the balance of the number of electrons arriving at kinetic energy T with the number of electrons departing from it. One may write the cross section for a collision process in which an electron of kinetic energy T collides with an atom or molecule and an electron of kinetic energy T' emerges as $\sigma(T \to T')$. Then, the expression $\int \sigma(T' \to T)y(T')\,dT' - \int \sigma(T \to T')\,dT'y(T)$ represents the change in the number of electrons at kinetic energy T due to all collisions in a medium of unit density. For convenience, one may write the above expression as $K_T y(T)$ using a linear operator K_T called the collision operator. In the simple case where a source steadily generates $u(T)dT$ electrons having kinetic energies between T and $T + dT$ in a medium consisting of n atoms or molecules of a single species, the Spencer-Fano equation takes the form

$$nK_T y(T) + u(T) = 0\,. \tag{88.5}$$

Third, in the domain $E_1 < T < T_{\max}$, where E_1 is the lowest electronic-excitation threshold energy of the medium, $y(T)$ is invariably bimodal, as exemplified by Fig. 88.2. The behavior at high T is largely understandable from the CSDA Eq. (88.2), which approximately holds at $T \gg E_1$; then, $y(T)$ is the reciprocal of the stopping power, which is given by the Bethe theory and corrections as explained in Chap. 87. For example, for electrons with T between 1 keV and 10 keV, the stopping power of molecular hydrogen is a monotonically decreasing function of T, and therefore $y(T)$ is a monotonically increasing function of T, as seen in Fig. 88.2. In contrast, the steep rise of $y(T)$ at low T occurs because most of the secondary electrons produced in an ionizing collision of any energetic charged particle (including an energetic electron) have low kinetic energies, corresponding to a few multiples of the ionization threshold energy, as fully documented by Rudd *et al.* [9–11].

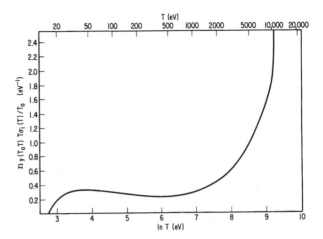

Figure 88.2. Electron degradation spectrum in molecular hydrogen. Source electrons of kinetic energy $T_{\max} = 10$ keV steadily enter the hydrogen gas at 0°C and 1 atmosphere at the rate of 1 electron $^{-1}$cm^{-3}. Data are taken from Douthat [8].

As T varies, the principal contribution to the yield expression (88.4) comes from ranges where both $\sigma_j(T)$ and $y(T)$ are large. A large $y(T)$ means a long path length of electrons at that T, and hence a modest energy loss per unit path length. From Eq. (88.2), the time interval dt during which electrons stay in the energy interval dT is

$$dt = (dt/dx)^{-1}dx = v^{-1}y(T)\,dT\,. \qquad (88.6)$$

The quantity $v^{-1}y(T)dT$ represents the sojourn time of electrons at energy T. Equation (88.4) expresses how N_j is determined by the competition between electron degradation, and the cross section $\sigma_j(T)$ for production of j upon individual collisions. The product $\sigma_j(T)y(T)$, called the yield spectrum for production of j, plays a central role in the theory.

Since T ranges from a few eV up to keV or MeV, it is convenient to rescale the yield spectrum to $\ln T$. In fact, $\ln T$ is more meaningful as a variable because it represents, in effect, the mean number of elastic collisions required to reduce T by a given amount. (In the neutron slowing down theory [16], $\ln T$ is called *lethargy* by Fermi.) Equation (88.4) then becomes

$$N_j = n \int T\sigma_j(T)y(T)\,d(\ln T)\,, \qquad (88.7)$$

with $T\sigma_j(T)y(T)$ regarded as a function of $\ln T$. The product $T\sigma_j(T)$ is proportional to the collision strength in the theory of atomic collisions, as treated in Chap. 45. It is also a key quantity in the Bethe theory [17].

Figure 88.3 is an example of a yield spectrum for ionization in molecular hydrogen. The area under the curve over any given interval of $\ln T$ represents the number of ions produced in collisions of electrons in that interval. One sees a sizable contribution from high

Figure 88.3. The yield spectrum for ionization in molecular hydrogen. Source electrons are the same as for Fig. 88.2. The vertical axis represents $ny(T_{\max}, T)T\sigma_i(T)/T_{\max}$, where n is the number density of molecules, $T_{\max} = 10$ keV is the source-electron energy, and $\sigma_i(T)$ is the ionization cross section of H_2 for electrons of energy T. Data are taken from Douthat [8].

energies (between 1 keV and 10 keV). Below 1 keV, where the spectrum is roughly constant, each decade of T contributes roughly the same amount to the total ionization yield. This observation applies to any product j that results from dipole-allowed transitions.

In contrast, the yield spectrum for a product j resulting from forbidden transitions is dominated by values at low T, where $\sigma_j(T)$ is appreciable, as Fig. 88.4 illustrates. An example is the production of two ground-state hydrogen atoms from a hydrogen molecule, which occurs solely from the lowest triplet repulsive state.

Finally, once electrons fall below E_1, they moderate much less rapidly than they do above E_1. These subexcitation electrons [18], lose energy through momentum transfer upon elastic collisions with molecules, rotational excitation, and vibrational excitation, as discussed in Chap. 45. The behavior of the subexcitation electrons is treated by an extension of the Spencer-Fano equation [15], until they reach kinetic energies comparable with the thermal energy of the medium molecules. Then, the transport, moderation, and eventual energy gain from medium molecules become important, as well as the energy loss to them. The three main energy domains of electronic excitation, subexcitation, and thermal [19] are illustrated in Fig. 88.5.

Because of the great disparity in the modes of interactions with the medium, electrons in each of the domains need different treatments.

Together with ions, molecular fragments, and other products of electron degradation, electrons in the thermal domain are important as precursors to subsequent

Figure 88.4. The yield spectra for three product species in molecular hydrgen. Source electrons of kinetic energy $\mathcal{T}_{max} = 1000$ eV steadily enter the hydrogen gas at 0°C and 1 atmosphere at the rate of 1 electron^{-1}cm^{-3}. The solid curve labeled "ION" represents ionization, the long-broken curve labeled "SINGLET" represents the production of the $B\,^1\Sigma_u^+$ state, and the short-broken curve labeled "TRIPLET" represents the production of the a $^3\Sigma_g^+$ state, which dissociates into two hydrogen atoms in the ground state (Courtesy: Mineo Kimura).

chemical reactions, and therefore have been the subject of extensive study [4, 20, 21].

88.3.3 Quantities Expressing the Yields of Products

The total ionization yield is often expressed in terms of the quantity W, defined to be the mean energy required to produce an ion pair, or equivalently,

$$W = \mathcal{T}_{abs.}/N_i, \qquad (88.8)$$

where $\mathcal{T}_{abs.}$ is the mean enery absorbed and N_i is the mean number of ions observed over many events. $\mathcal{T}_{abs.}$ is simply the initial injection energy if all the resulting electrons are degraded to kinetic energies below the ionization threshold. Since each event is stochastic, $\mathcal{T}_{abs.}$ and N_i in general fluctuate from one event to the next.

A simple index for the statistical fluctuations in N_I is the variance V of N_I from its mean. The ratio $F = V/N_i$, called the Fano factor [22].

Both measurements and theory indicate that W and F are approximately the same for different kinds of ionizing radiations of sufficiently high energy, and that they depend primarily on the material of the medium [15, 23, 24]. This fact is important as a basis for radiation

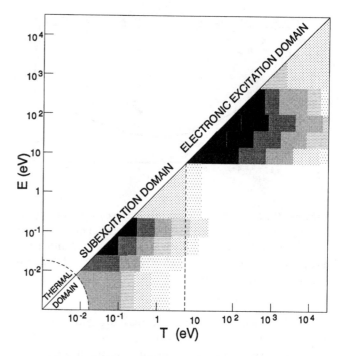

Figure 88.5. A schematic diagram for showing three distinct domains of electron transport in a molecular substance. The horizontal axis represents the kinetic energy $\mathcal{T}$ of an electron. The vertical axis represents the energy loss E upon a single collision with a molecule. The vertical broken line indicates the first electronic excitation threshold E_1. The shade and fade schematically represent the magnitudes of cross sections.

dosimetry through ionization measurements. The ratio W/I ranges from 1.7 to 3.2 depending on the nature of materials, where I is the ionization threshold energy. The ratio always exceeds unity because a fraction of radiation energy absorbed is expended to generate products other than ionization, such as discrete excited states, neutral molecular fragments, and subexcitation electrons.

The Fano factor F is restricted to $0 < F < 1$ on general theoretical grounds. The maximum value $F = 1$ would occur if the ionization process were characterized by the Poisson statistics. In reality, the ionization process is subject to the constraints that it must compete with nonionizing events, and that the available total energy is fixed; therefore the fluctuations in the ionization yield must be smaller than those given by the Poisson statistics, viz., $F < 1$. Both theory and experimental data indicate that F and W tend to vary together from one material to another; more precisely, Krajcar-Bronić [24] found an approximate empirical relation

$$F = a(W/I) + b, \qquad (88.9)$$

where a and b are constant. This relation is valid for electrons of sufficiently high energies.

For molecular products other than ions, it is customary to use G_j defined by

$$G_j = N_j/\mathcal{T}_{\text{abs.}}, \qquad (88.10)$$

where N_j is the mean number of molecular products of type j (see pp. 241–242 of Kimura *et al.* [15] for fuller discussion). Like W, G_j is often approximately the same for different kinds of ionizing radiations of sufficiently high energy, provided that the material, either liquid or solid, is homogeneous and that the absorbed energy is not extremely great. A counter example is the ferrous sulphate equeous solution, which is often used for dosimetry and in which the G value for the ferrous-ferric conversion depends appriciably on the kind of radiation. (See Chap. 87.) In the more general case of a heterogeneous material having complex molecular structure and aggregation, such as a biological cell, the yield of a product is not even approximately proportional to the absorbed energy, and the idea of a G value is useful only at the limit of vanishing absorbed energy.

In SI units, the G value has units of mol J^{-1}. This is related to the older unit, molecules (100 eV)$^{-1}$, by

$$1 \text{ mol J}^{-1} = 9.6485 \times 10^6 \text{ molecules } (100 \text{ eV})^{-1}. \qquad (88.11)$$

The meaning of the yield of a product depends on the process. The molecular species produced as a result of electron degradation (e.g., those given in Table 88.1) may be called initial products. Subsequently, they react with neighboring molecules or among themselves to form other species, which may be called secondary products, such as H_3^+, through the ion-molecule reaction

$$H_2^+ + H_2 \rightarrow H_3^+ + H. \qquad (88.12)$$

The yield of an initial product is seldom measured and is accessible by theory only. Most yield measurements concern a secondary product. To keep the distinction in mind, Platzman recommended [25] use of the symbol g for an initial product and the symbol G for a secondary product.

For data on G values, see Tabata *et al.* [4] and references cited therein.

88.3.4 Track Structures

The spatial distributions of initial products such as ions and excited states influence the kinetics of their subsequent chemical reactions, which lead to radiation effects. Spatial distribution is generally expected to have a role in a heterogeneous material; this role should be especially prominent when the scale characteristic of the distribution is comparable to the scale of the material inhomogeneity.

The simplest index of the distribution is the stopping power, which gives the linear density of energy lost from a particle along its path. From the point of view of radiation effects, or Class II problems, the linear density of energy imparted locally to matter is of greater interest. The distinction between the energy lost from a particle and the energy imparted locally to matter arises when one considers a volume having a linear scale less than the ranges of the majority of secondary electrons produced in the volume; sufficiently energetic secondary electrons will escape the volume and impart much of their kinetic energies elsewhere. This recognition led to the idea of the linear energy transfer (often abbreviated as LET), which is defined as the mean energy loss, excluding contributions from the production of secondary electrons having kinetic energies above a fixed value, such as 100 eV [26]. The LET, like the stopping power, depends on the particle charge and speed, as well as on the material, and is often used in radiation chemistry and biology to consider the role of the spatial distribution of initial products.

However, a given value of LET from particles of different charges and speeds does not necessarily lead to the same radiation effect. The aim of microdosimetry [27] is to provide a more detailed description of the energy imparted locally to a small volume of matter. For a fuller discussion, see Chap. 87.

The full representation of the spatial pattern of the initial products resulting from individual incident and secondary particles is called the *track structure*, because the pattern was first recognized in particle tracks visualized in cloud chambers, photographic emulsions, and other radiation detectors. Treatments of track structures and their consequences for the subsequent chemical reactions by Monte Carlo simulations are certainly informative and flexible for application to a wider range of problems, as exemplified by Refs. [28–33]. Such treatments are in principle sound, because they are independent of an assumption that the yield of a product is proportional or otherwise simply related to the energy lost or imparted.

The spatial distribution of the initial products and other consequences of radiation interactions with matter can be also studied by analytic methods using transport equations. An excellent exposition of the recent status of this approach is given in the NCRP Report No. 108 [34]. By solving relatively simple problems, the analytic methods are effective in elucidating principles and providing insights into the physics involved, while the Monte Carlo simulations are powerful in providing answers to complicated problems. In summary, the two approaches are complementary, rather than competitive.

Meaningful results from these treatments require a large volume of accurate cross section and other data as input. Various surveys of atomic and molecular cross section data are available to present recommended values and to identify needs for further studies, such as the report [35] sponsored by the International Atomic Energy Agency. A broader review of the current status of cross section determination, dissemination, and related topics is in Ref. [36].

88.3.5   Condensed Matter Effects

Beyond the interactions of ionizing particles with individual atoms and molecules in a gas or beam, the atomic or molecular aggregation in condensed matter has further important influences. A prelude to condensed matter effects is seen in studies of high-pressure gases [37] and in Chap. 38.

Many phenomena are related to the complex dipole-response function $\epsilon(E)$, which is the electric displacement generated in matter by a spatially uniform electric field of unit strength oscillating at angular frequency $\omega = E/\hbar$ (see, e.g., Landau and Lifshitz [38]). For light at this frequency, the real part $\epsilon_1(E)$ describes the dispersion, and the imaginary part $\epsilon_2(E)$ describes the absorption. The probability that a glancing collision of a fast charged particle transfers energy E to matter is proportional to

$$\Pi(E) = \Im[-1/\epsilon(E)] = \frac{\epsilon_2}{\epsilon_1^2(E) + \epsilon_2^2(E)} . \tag{88.13}$$

In a low-density material, such as a dilute gas, $\epsilon_1(E)$ is close to unity, and $\epsilon_2(E)$ is much smaller than unity for all E. Then, $\Pi(E)$ is practically the same as $\epsilon_2(E)$; in other words, the spectrum of energy transfer in a glancing collision of a fast charged particle is effectively the same as the spectrum of photoabsortion. In a high-density material, $\Pi(E)$ differs appreciably from $\epsilon_2(E)$ at some E. An extreme case occurs at E at which $\epsilon_1(E)$ changes its sign; there $\Pi(E)$ has a peak. This is the well-known plasma excitation in metals.

A general criterion for determining the effect of atomic or molecular aggregation on the dipole oscillator strength spectrum, and hence on $\Pi(E)$, was given by Fano [39]. One may qualitatively summarize his criterion as follows: If the density of the dipole oscillator strength of an atom or molecule in both space and energy (or frequency) is sufficiently high, the effects of atomic or molecular aggregation will be appreciable. In other words, the effect will be seen in a sufficiently dense material and in a generally limited spectral region where the oscillator strength spectrum of an isolated atom or molecule is especially intense. With the use of the Fano criterion, one understands the occurrence of collective excitation (electronic excitation involving more than one atom or molecule), such as excitons, plasmons, and other elementary modes of excitation in condensed matter. For a fuller and more recent discussion of this topic, see Fano [40].

Many manifestations of condensed matter effects on the absorption of radiation energy have been identified, as discussed, for instance, in Refs. [7] and [41]. Table 88.2 presents a summary of these effects.

88.4   CONNECTIONS WITH RELATED FIELDS OF RESEARCH

88.4.1   Astrophysics and Space Physics

Some of the ideas and methods sketched in Sects. 88.3.1 and 88.3.2 are readily applicable to studies on terrestrial and planetary auroras [42–46] and on effects of cosmic rays or x-rays on interstellar clouds and other astronomical objects [47–53]. These studies treat consequences of the interactions of energetic charged particles or high-energy photons on voluminous gases, at low densities and often partially ionized. Often the presence of electric and magnetic fields influence radiation degradation phenomena.

88.4.2   Material Science

As fully discussed earlier, radiation effects on gases, liquids, and molecular substances, including the biological cell, result mainly from electronic excitation. Radiation effects on insulators, including crystalline solids, are also initiated by electronic excitation [54, 55] and are therefore closely related to atomic, molecular, and optical physics.

In metals, electronic excitation *per se* is of little consequence, because most of the electronic excitation energy is rapidly converted into phonons (thermal energy). Generally, a small fraction of electronic excitation energy causes appreciable atomic displacement. Radiation effects on these materials occur predominantly via direct energy transfer to atoms from neutrons, protons, and heavier charged particles, causing atomic displacement from a regular crystalline site [56–58].

More generally, the role of atoms and ions having kinetic energies far exceeding the thermal energy is important in chemistry. This topic, known as hot-atom chemistry [59, 60], represents another application of the knowledge of atomic, molecular, and optical physics as discussed in Chaps. 62 and 63, 64, and 65.

Radiation effects on the surface of a solid are a subject of extensive study, because of their importance as means for material-structure probing and also for material processing (for example, the ion implantation in the manufacture of semiconductors and other devices). A major phenomenon of interest is sputtering, which is the ejection of atoms and molecules from the surface by the action of ionizing radiation [61–64].

Acknowledgment

This work was supported by the U.S. Department of Energy, Office of Energy Research, Office of Health

Table 88.2. Condensed matter effects.

Classification	Example	Remarks on Characteristics
Shift of oscillator strength toward higher energies	Water, hydrocarbons, and organics	Occurs over wide ranges of exitation energies.
Excitation of special modes of motion	Plasmons in metals	Occurs at specific energies and carries considerable strength.
	Excitons in molecular crystals and ionic crystals	Occurs at specific energies and at minor strengths.
Interaction of fast ejected electrons with other atoms and molecules	EXAFS (extended x-ray absorption fine structure)	Occurs at energies above 100 eV, slightly above the pertinent threshold, and at weak strengths, chiefly due to interference of electrons emerging from different atoms.
Resonances in electron interactions	N_2, C_2H_4, C_2H_2, and other unsaturated hydrocarbons	Occurs at energies slightly above the excitation threshold or even lower; leads to inelasticity.
Diffraction of the electron de Broglie wave	Materials with periodic structure	Leads to band structure effects, such as transmission without energy loss at certain electron kinetic energies.

and Environmental Research, under Contract W-31-109-ENG-38.

REFERENCES

1. G. S. Knoll, *Radiation Detection and Measurement* (John Wiley & Sons, New York, 1989), 2nd ed.
2. K. R. Kase, B. E. Bjarngard, and F. H. Attix, *The Dosimetry of Ionizing Radiation* (Academic Press, Orlando, 1985–1987).
3. G. S. Hurst, M. G. Payne, S. D. Kramer, and J. P. Young, Rev. Mod. Phys. **51**, 767 (1979).
4. *CRC Handbook of Radiation Chemistry*, edited by Y. Tabata, Y. Ito, and S. Tagawa (CRC Press, Boca Raton, 1991).
5. *Physical and Chemical Mechanisms in Molecular Radiation Biology*, edited by W. A. Glass and M. N. Varma (Plenum Press, New York, 1991).
6. M. Inokuti, in *Applied Atomic Collision Physics*, edited by S. Datz (Academic Press, Orlando, 1983), Vol. 4, p. 179.
7. M. Inokuti, in Ref. [5], p. 29.
8. D. A. Douthat, J. Phys. B **12**, 663 (1979).
9. M. E. Rudd, Y.-K. Kim, D. H. Madison, and J. W. Gallagher, Rev. Mod. Phys. **57**, 965 (1985).
10. M. E. Rudd, Nucl. Tracks Radiat. Meas. **16**, 213 (1989).
11. M. E. Rudd, Y.-K. Kim, D. H. Madison, and T. J. Gay, Rev. Mod. Phys. **64**, 441 (1992).
12. R. S. Caswell, Radiat. Res. **27**, 92 (1966).
13. R. S. Caswell, J. J. Coyne, H. M. Gerstenberg, and E. J. Axton, Radiat. Prot. Dosim. **23**, 41 (1988).
14. L. V. Spencer and U. Fano, Phys. Rev. **93**, 1172 (1954).
15. M. Kimura, M. Inokuti, and M. A. Dillon, in *Advances in Chemical Physics*, edited by I. Prigogine and S. A. Rice (John Wiley & Sons, New York, 1993), Vol. 84, p. 193.
16. A. M. Weinberg and E. P. Wigner, *The Physical Theory of Neutron Chain Reactors* (The University of Chicago Press, Chicago, 1958).
17. M. Inokuti, Rev. Mod. Phys. **43**, 297 (1971).
18. R. L. Platzman, Radiat. Res. **2**, 1 (1955).
19. M. Inokuti, Appl. Radiat. Isot. **42**, 979 (1991).
20. C. Ferradini and J.-P. Jay-Gerin, *Excess Electrons in Dielectric Media* (CRC Press, Boca Raton, 1991).
21. R. Holroyd, Ann. Rev. Phys. Chem. **40**, 439 (1989).
22. U. Fano, Phys. Rev. **72**, 26 (1947).
23. *Average Energy Required to Produce an Ion Pair*, ICRU Report 31 (International Commission on Radiation Units and Measurements, Bethesda, 1979).
24. I. Krajcar-Bronić, J. Phys. B **25**, L215 (1992).
25. R. L. Platzman, in *Proceedings of the Third International Congress of Radiation Research, 1966*, edited by G. Silini (North-Holland, Amsterdam, 1967), p. 20.
26. *Linear Energy Transfer*, ICRU Report 16 (International Commission on Radiation Units and Measurements, Bethesda, 1970).
27. *Microdosimetry*, ICRU Report 36 (International Commission on Radiation Units and Measurements, Bethesda, 1983).
28. H. G. Paretzke, in *Kinetics of Non-Homogenous Processes*, edited by G. R. Freeman (John Wiley & Sons, New York, 1987), p. 89.
29. *Monte Carlo Transport of Electrons and Photons*, edited by T. M. Jenkins, W. R. Nelson, and A. Rindi (Plenum Press, New York, 1988).
30. R. H. Ritchie, R. N. Hamm, J. E. Turner, H. A. Wright, and W. E. Bloch, in Ref. [5], p. 99.

31. M. Zaider, in Ref. [5], p. 137.

32. M. Terrissol and A. Beaudre, Radiat. Prot. Dosim. **31**, 175 (1990).

33. *Computational Approaches in Molecular Radiation Biology*, edited by M. N. Varma and A. Chatterjee (Plenum Press, New York, 1994).

34. *Conceptual Basis for Calculations of Absorbed-Dose Distributions*, NCRP Report No. 108 (National Council on Radiation Protection and Measurements, Bethesda, 1991).

35. *Coordinated Research Program on Atomic and Molecular Data for Radiotherapy and Related Research. Report IAEA-TECDOC-XXXX*, edited by M. Inokuti (International Atomic Energy Agency, Vienna, 1995). (Available free of charge upon request to the IAEA).

36. *Advances in Atomic, Molecular, and Optical Physics*: Special Volume on Cross Section Data, edited by M. Inokuti (Academic Press, San Diego, 1994), Vol. 33.

37. L. G. Christophorou, in [5], p. 183.

38. L. D. Landau and E. M. Lifshitz, *Electrodynamics of Continuous Media*, translated by J. B. Sykes and J. S. Bell (Pergamon Press London, 1960), Chap. IX.

39. U. Fano, Phys. Rev. **118**, 451 (1960).

40. U. Fano, Rev. Mod. Phys. **64**, 313 (1993).

41. M. Inokuti, Radiat. Effects Defects **117**, 143 (1991).

42. K. Takayanagi and Y. Itikawa, Space Sci. Rev. **11**, 380 (1970).

43. J. L. Fox and G. A. Victor, Planet. Space Sci. **36**, 329 (1988).

44. S. P. Slinker, R. D. Taylor, and A. W. Ali, J. Appl. Phys. **63**, 1 (1988).

45. S. P. Slinker, R. D. Taylor, and A. W. Ali, J. Appl. Phys. **64**, 982 (1988).

46. K. Onda, M. Hayash, and K. Takayanagi, The Institute of Space, and Astronautical Science Report **645** (1992).

47. T. E. Cravens, G. A. Victor, and A. Dalgarno, Planet. Space Sci. **23**, 1059 (1970).

48. A. Dalgarno and R. A. McCray, Ann. Rev. Astron. Astrophys. **10**, 375 (1972).

49. A. E. Glassgold and W. D. Langer, Astrophys. J. **186**, 859 (1973).

50. T. E. Cravens and A. Dalgarno, Astrophys. J. **219**, 750 (1978).

51. R. Gredel, S. Lepp, and A. Dalgarno, Astrophys. J. **323**, L137 (1987).

52. R. Gredel, S. Lepp, A. Dalgarno, and E. Herbst, Astrophys. J. **347**, 289 (1989).

53. G. M. Vogt, Astrophys. J. **377**, 158 (1991).

54. *Direct Processes Induced by Electronic Excitation in Insulators*, edited by N. Itoh (World Scientific, Singapore, 1989).

55. *Atomic Processes Induced by Electronic Excitation in Non-Metallic Solids*, edited by W. B. Fowler and N. Itoh (World Scientific, Singapore, 1990).

56. F. Seitz and J. S. Koehler, in *Solid State Physics. Advances in Research and Applications*, edited by F. Seitz and D. Turnbull (Academic Press, New York, 1956), p. 305.

57. M. W. Thompson, *Defects and Radiation Damage in Metals* (Cambridge University Press, London, 1969).

58. W. Schilling and H. Ullmaier, in *Materials Science and Technology. A Comprehensive Treatment*, edited by R. W. Cahn, P. Haasen, and E. J. Kramer (VCH Verlagsgesellschaft, Weinheim, 1994), Vol. 10B; *Nuclear Materials, Part 2*, edited by B. R. T. Frost (VCH Verlagsgesellschaft, Weinheim, 1994), p. 179.

59. T. Matsuura, *Hot Atom Chemistry. Recent Trends and Applications in the Physical and Life Sciences and Technology* (Kodansha, Tokyo, 1984).

60. *Handbook of Hot Atom Chemistry*, edited by J.-P. Adloff, P. P. Gasper, M. Imamura, A. G. Maddock, T. Matsuura, H. Sano, and K. Yoshihara (Kodansha, Tokyo, 1992).

61. *Sputtering by Particle Bombardment I. Physical Sputtering of Single-Element Solids*, edited by R. Behrisch (Springer-Verlag, Berlin, 1981).

62. *Sputtering by Particle Bombardment II. Sputtering of Alloys and Compounds, Electron and Neutron Sputtering, Surface Topography*, edited by R. Behrisch (Springer-Verlag, Berlin, 1983).

63. *Sputtering by Particle Bombardment III. Characteristics of Sputtered Particles, Technical Applications*, edited by R. Behrisch and K. Wittmaack (Springer-Verlag, Berlin, 1991).

64. *Fundamental Processes in Sputtering of Atoms and Molecules*, edited by P. Sigmund, K. Dan. Vidensk. Selsk. Mat. Fys. Medd. **43**, 1 (1993).

Subject Index

F

G